IMP	inosine-5′-monophosphate
IP_3	inositol-1,4,5-triphosphate
K_m	Michaelis constant
kb	kilobases
kDa	kilodalton
LDL	low-density lipoprotein
LHC	light harvesting complex
Man	mannose
NAA	nonessential amino acids
NAD^+	nicotinamide adenine dinucleotide (oxidized form)
NADH	nicotinamide adenine dinucleotide (reduced form)
$NADP^+$	nicotinamide adenine dinucleotide phosphate (oxidized form)
NADPH	nicotinamide adenine dinucleotide phosphate (reduced form)
ncRNA	noncoding RNA
NDP	nucleoside-5′-diphosphate
NMR	nuclear magnetic resonance
NO	nitric oxide
nt	nucleotides
NTP	nucleoside-5′-triphosphate
P_i	orthophosphate (inorganic phosphate)
PAPS	3′-phosphoadenosine-5′-phosphosulfate
PC	plastocyanin
PDGF	platelet-derived growth factor
PEP	phosphoenolpyruvate
PFK	phosphofructokinase
PIP_2	phosphatidylinositol-4,5-bisphosphate
PP_i	pyrophosphate
PRPP	phosphoribosylpyrophosphate
PS	photosystem
PQ(Q)	plastoquinone (oxidized)
PQH_2 (QH_2)	plastoquinone (reduced)
RER	rough endoplasmic reticulum
RF	releasing factor
RFLP	restriction-frament length polymorphism
RNA	ribonucleic acid
dsRNA	double-stranded RNA
mRNA	messenger RNA
rRNA	ribosomal RNA
siRNA	small interfering RNA
snRNA	small nuclear RNA
ssRNA	single-stranded RNA
tRNA	transfer RNA
snRNP	small ribonucleoprotein particles
RNase	ribonuclease
S	Svedberg unit
SAH	S-adenosylhomocysteine
SAM	S-adenosylmethionine
SDS	sodium dodecyl sulfate
SER	smooth endoplasmic reticulum
SRP	signal recognition particle
T	thymine
THF	tetrahydrofolate
TPP	thiamine pyrophosphate
U	uracil
UDP	uridine-5′-diphosphate
UMP	uridine-5′-monophosphate
UTP	uridine-5′-triphosphate
UQ	ubiquinone (coenzyme Q)(oxidized form)
UQH_2	ubiquinone (reduced form)
VLDL	very low density lipoprotein
XMP	xanthosine-5′ monophosphate

Biochemistry

THE MOLECULAR BASIS OF LIFE

Biochemistry

THE MOLECULAR BASIS OF LIFE

SIXTH EDITION

Trudy McKee
James R. McKee

New York Oxford
OXFORD UNIVERSITY PRESS

Oxford University Press is a department of the University of Oxford.
It furthers the University's objective of excellence in research,
scholarship, and education by publishing worldwide.

Oxford New York
Auckland Cape Town Dar es Salaam Hong Kong Karachi
Kuala Lumpur Madrid Melbourne Mexico City Nairobi
New Delhi Shanghai Taipei Toronto

With offices in
Argentina Austria Brazil Chile Czech Republic France Greece
Guatemala Hungary Italy Japan Poland Portugal Singapore
South Korea Switzerland Thailand Turkey Ukraine Vietnam

For titles covered by Section 112 of the US Higher Education Opportunity Act, please visit www.oup.com/us/he for the latest information about pricing and alternate formats.

Published by Oxford University Press
198 Madison Avenue, New York, New York 10016
http://www.oup.com

Library of Congress Cataloging-in-Publication Data
McKee, Trudy.
Biochemistry : the molecular basis of life / Trudy McKee, James R. McKee. -- Sixth edition.
pages cm
ISBN 978-0-19-020989-6
1. Biochemistry--Textbooks. I. McKee, James R. (James Robert), 1946- II. Title.
QD415.M36 2016
572--dc23

2015018555

Printing number: 9 8 7 6 5 4 3 2 1

Printed in the United States of America
on acid-free paper

Only the educated are free.
—Epictetus (Greek philosopher, 55–135 AD)

True freedom requires an education that begins with literacy and then evolves into compassion for all humans and respect for the natural world.

This book is dedicated to those individuals around the world who struggle to educate all children.

Brief Contents

Contents

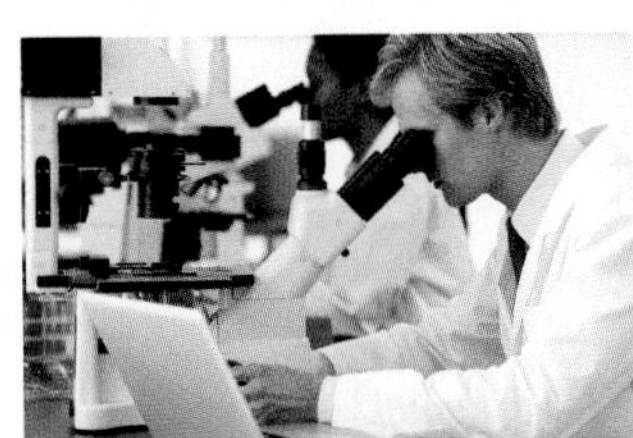

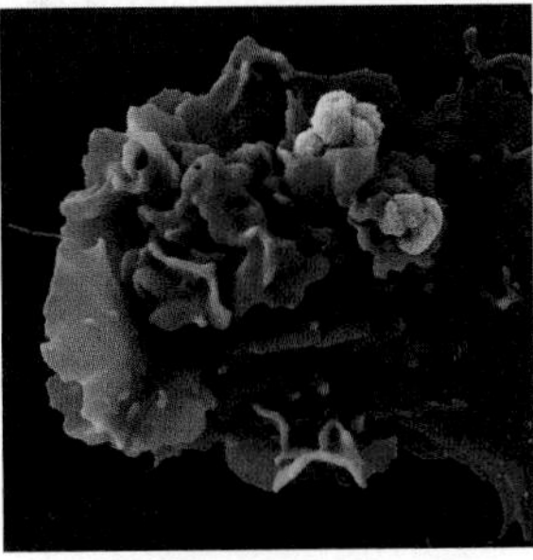

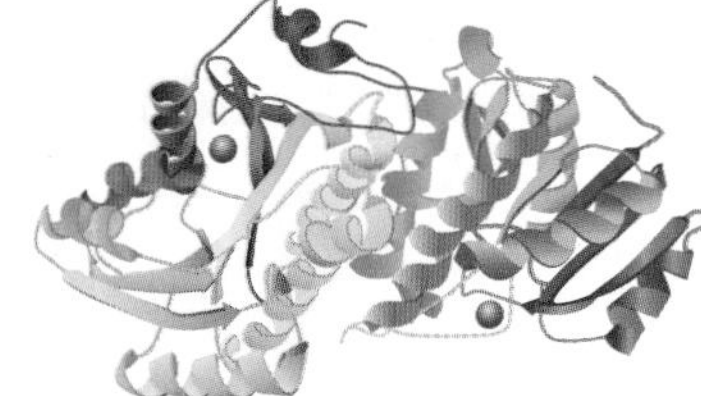

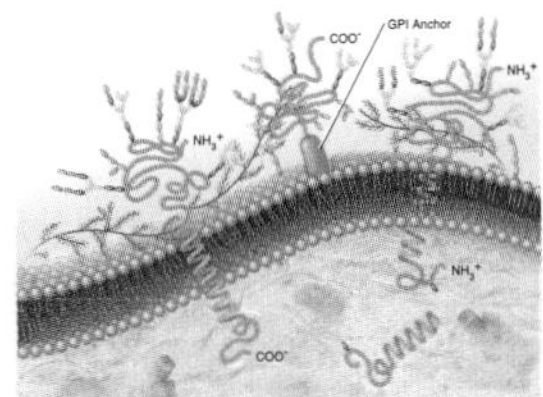
GPI Anchor

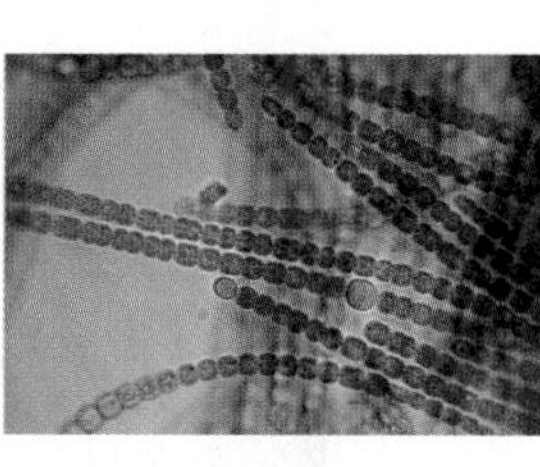

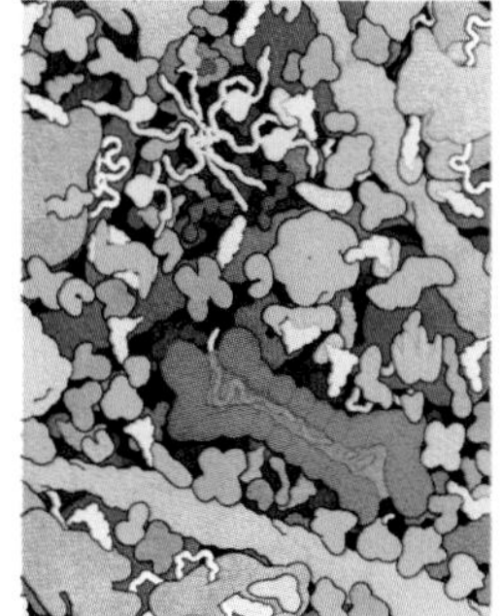

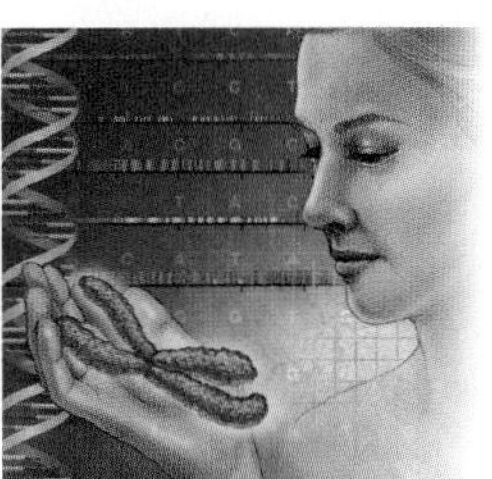

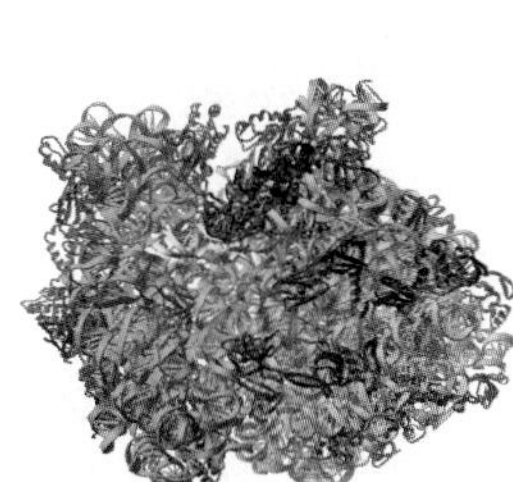

Chapter Opening Vignettes

Biochemistry in Perspective Essays

Biochemistry in Perspective Essays

Biochemistry in Perspective Essays

	Questions for Students	Key Points	Page No.
Chapter 17 Forensic Investigations	How is DNA analysis used in the investigation of violent crime?	Forensic scientists use PCR and other technologies to amplify crime scene DNA in order to generate the unique genetic profile that distinguishes one individual from all others.	649
Chapter 17 HIV Infection	How does HIV infect human cells?	HIV infection disrupts cell function. By suppressing some cellular genes and activating others, the HIV genome directs the host cell to produce new HIV particles that proceed to infect other cells.	655
Chapter 18 Carcinogenesis	What is cancer, and what are the biochemical processes that facilitate the transformation of normal cells into those with cancerous properties?	Carcinogenesis is the process whereby cells with a growth advantage over their neighbors are transformed by mutations in the genes that control cell division into cells that no longer respond to regulatory signals.	720
Chapter 19 Trapped Ribosomes: RNA to the Rescue!	How are ribosomes bound to damaged mRNAs retrieved so they can be recycled?	Living organisms, faced with the high metabolic cost of protein synthesis, have evolved the means of ensuring the efficiency of the process. Both prokaryotes and eukaryotes have methods for rescuing ribosomes trapped by association with damaged mRNAs.	746
Chapter 19 Context-Dependent Coding Reassignment	How are selenocysteine and pyrrolysine, two nonstandard amino acids, incorporated into polypeptides during protein synthesis?	In context-dependent codon reassignment, a specific tRNA, a tRNA synthetase, and other molecules are used to transform a stop codon into one that codes for the incorporation of a nonstandard amino acid.	752

Biochemistry in the Lab Boxes

Don't forget your favorite Biochemistry in Perspective essays from past editions, now available on the companion website at www.oup.com/us/mckee:

Organelles and Human Disease (Chapter 2)
Water, Abiotic Stress, and Compatible Solutes (Chapter 3)
The Extremophiles: Organisms That Make a Living in Hostile Environments (Chapter 4)
Protein Poisons (Chapter 5)
Protein Folding and Human Disease (Chapter 5)
Myosin: A Molecular Machine (Chapter 5)
Enzymes and Clinical Medicine (Chapter 6)
Quantum Tunneling and Catalysis (Chapter 6)
Scurvy and Ascorbic Acid (Chapter 7)
Sweet Medicine (Chapter 7)
Conversion of Fischer Structures into Haworth Structures (Chapter 7)
Fermentation: An Ancient Heritage (Chapter 8)
The Evolutionary History of the Citric Acid Cycle (Chapter 9)
Hans Krebs and the Citric Acid Cycle (Chapter 9)
Glucose-6-Phosphate Dehydrogenase Deficiency (Chapter 10)
Photosynthesis in the Deep (Chapter 13)
Starch and Sucrose Metabolism (Chapter 13)
The Amine Neurotransmitters (Chapter 14)
Parkinson's Disease and Dopamine (Chapter 14)
Heme and Chlorophyll Biosynthesis (Chapter 14)
The Essential Amino Acids (Chapter 14)
The Nucleotides: IMP Biosynthesis (Chapter 14)
Lead Poisoning (Chapter 14)
Hyperammonemia (Chapter 15)
Gout (Chapter 15)
Heme Biotransformation (Chapter 15)
Mammalian Hormones and the Hormone Cascade System (Chapter 16)
Metabolism in the Mammalian Body: Division of Labor (Chapter 16)
A Short History of DNA Research: The Early Years (Chapter 17)
The Meselson-Stahl Experiment (Chapter 18)
EF-Tu: A Motor Protein (Chapter 19)

Your favorite Biochemistry in the Lab boxes from past editions are also available on the companion website at www.oup.com/us/mckee:

Dialysis (Chapter 3)
Protein Sequence Analysis: The Edman Degradation (Chapter 5)
Glycomics (Chapter 7)
Photosynthetic Studies (Chapter 13)

Preface

Welcome to the sixth edition of *Biochemistry: The Molecular Basis of Life*. Although this textbook has been revised and updated to reflect the latest research in biochemistry, our original mission remains unchanged. We continue to believe that the cornerstone of an education in the life sciences is a coherent understanding of the basic principles of biochemistry. Once biochemical concepts have been mastered, students are then prepared to tackle the complexities of their chosen science fields. To that end, we have sought comprehensive coverage of biochemical systems, structures, and reactions, but within the context of the organism. We have thus sought a unique balance between chemistry, biology, and their applications to medicine and human health.

ORGANIZATION AND APPROACH

CHEMICAL AND BIOLOGICAL PRINCIPLES IN BALANCE. As with previous editions, the sixth edition is designed for both life science students and chemistry majors. We provide thorough coverage of biochemical principles, structures, and reactions, but within a biological context that emphasizes their relevance.

A REVIEW OF BASIC PRINCIPLES. Few assumptions have been made about a student's chemistry and biology background. To ensure that all students are sufficiently prepared for acquiring a meaningful understanding of biochemistry, the first four chapters review the principles of such topics as organic functional groups, noncovalent bonding, thermodynamics, and cell structure. These chapters can either be covered in class or assigned for self-study.

Several topics are introduced in these early chapters and then continued throughout the book. Examples include cell volume changes triggered by metabolic processes that alter osmotic balance across membranes, the self-assembly of biopolymers such as proteins into supermolecular structures, and the nature and function of molecular machines. New in this edition is coverage of proteostasis, the mechanism whereby cells protect their proteins. Other important concepts that are emphasized include the relationship between biomolecular structure and function and the dynamic, unceasing, and self-regulating nature of living processes. Students are also provided with overviews of the major physical and chemical techniques that biochemists have used to explore life at the molecular level.

REAL-WORLD RELEVANCE. Because students who take the one-semester biochemistry course come from a range of backgrounds and have diverse career goals, the sixth edition consistently demonstrates the fascinating connections between biochemical principles and the worlds of medicine, nutrition, agriculture, bioengineering, and forensics. Features such as the Biochemistry in Perspective essays and chapter opening vignettes, as well as dozens of examples integrated into the body of the text, help students see the relevance of biochemistry to their chosen fields of study.

SUPERIOR PROBLEM-SOLVING PROGRAM. Analytical thinking is at the core of the scientific enterprise, and mastery of biochemical principles requires consistent and sustained engagement with a wide range of problems. The sixth edition continues to present students with a complete problem-solving system. This includes the effective in-chapter worked problems, which illustrate how quantitative problems are solved, as well as dozens of practice questions throughout, when new concepts and high-interest topics are introduced. There is also a large set of end-of-chapter questions, 285 of which are new. These include review, fill-in-the-blank, short-answer, and thought questions, which will allow students to test their knowledge. Sapling Learning's Online Homework System provides a powerful and effective tool for teaching and learning.

SIMPLE, CLEAR ILLUSTRATIONS. Biochemical concepts often require a high degree of visualization, and we have crafted an art program that brings complex processes to life. More than 720 full-color figures fill the pages of the sixth edition, many newly enhanced for a more vivid presentation in three dimensions and consistent scale and color of chemical structures.

CURRENCY. The sixth edition has been updated to present recent developments in the field, while remaining focused on the "big-picture" principles that are the cornerstone of the one-term biochemistry course. These changes again reflect the goal of balanced and thorough coverage of chemistry within a biological context. A detailed list of updated material follows in the next section.

NEW IN THIS EDITION

As a result of the rapid pace of discovery in the life sciences and our commitment to provide students with the highest-quality learning system available in any biochemistry textbook, we have revised the sixth edition in the following ways:

- **Enhanced Supplements Program.** Adding to the already robust ancillary program, an all-new video and animation guide provides links to dozens of freely available and high-quality online resources. In addition, an expanded and enriched test-item file written completely by the textbook authors accompanies the sixth edition.
- **General and Organic Chemistry Review Primer.** This edition includes a helpful in-text review of foundational general chemistry and organic chemistry topics to help students refresh their memories and begin to apply their knowledge in new biochemical contexts.
- **New and Updated Applications.** The sixth edition includes two new Biochemistry in Perspective essays, "Carcinogenesis: The Warburg Effect and Metabolic Reprogramming" and "The Artificial Leaf: Biomimetic Photosynthesis," which will capture student interest. In addition, several essays have been updated to reflect recent research. These include "Myocardial Infarct: Ischemia and Reperfusion," "Atherosclerosis," "Diabetes Mellitus," and "HIV Infection."
- **Expanded Problem-Solving Program.** The sixth edition includes a total of 285 additional questions at the end of the chapters. The expanded problem sets span a range of difficulty, from basic practice problems to more challenging integrative exercises.
- **Brand-New Illustrations.** With 28 new figures, the sixth edition incorporates a superior and expanded art program designed to help students develop a strong visual grasp of biochemical processes. Many figures have been enhanced for vivid, clear, and consistent presentation in color and three dimensions.
- **Important Themes.** We have retained and strengthened the themes of macromolecular crowding and systems biology, introduced in the fourth edition, and introduced a new theme, referred to as proteostasis.

  - Macromolecular crowding, the dense packing of vast numbers of proteins and other molecules within cells, has a profound effect on a wide variety of living processes. The concept of macromolecular crowding provides students with a more realistic view of cell structure and function.
  - The relatively new field of systems biology is an approach to biochemical processes that is based on engineering principles. Developed in response to the overwhelmingly vast amounts of information now available to life scientists, systems biology is the computer-assisted investigation of the complex interactions among biomolecules. Our accessible introduction to systems biological principles provides students with newly revised and expanded insights into the basic patterns of biomolecular processes.
  - We offer new coverage of proteostasis, a diverse set of mechanisms whereby cells control the folding of the proteome, has a crucial impact on cellular processes such as gene expression and signaling pathways. It protects against the toxic effects of protein aggregation.
- **Increased Attention to Reaction Mechanisms.** Catalytic mechanisms provide students with an enhanced understanding of the means by which biochemical reactions occur. Examples include the rubisco and proline residue hydroxylation mechanisms. We have retained a description of the roles of amino acid side chains in the catalytic mechanisms of enzymes and the mechanisms of the nucleic acid polymerases and ribosome-catalyzed peptide bond formation. Here, too, we have sought to enhance the text's unique balance between chemistry and biology.
- **Current Topics.** The following is a list of some, but not all, of the updated content that has been introduced in the sixth edition:
  - **Chapter 1** contains a revision of the section devoted to systems biology. Explanations of the terms *network*, *module*, and *motif* facilitate students' efforts to understand biochemical principles. Building on the discussions of organic reaction mechanisms introduced in the Chemistry Primer, the revised description of biochemical reactions relates each reaction type to its basic organic mechanism.
  - **Chapter 2** contains an expanded basic themes section. The signal transduction section includes overviews of the role of calcium ions in signaling and the relationship between signaling, metabolism, and gene expression. A new basic theme, proteostasis, provides an introduction to the strategies that cells use to protect the integrity of their proteins. A newly revised discussion of the endomembrane system in eukaryotic cells provides insight into the relationship between cell architecture and biochemical functions such as signaling, protein processing, and gene expression. A revised section concerned with mitochondria includes descriptions of the effects of the ER-mitochondrial interactions on cell signaling and phospholipid synthesis. The Biochemistry in Perspective essay on primary cilia has been updated. The Biochemistry in the Lab reading now includes a brief description of live cell imaging.
  - In **Chapter 5** the section devoted to unstructured proteins has been expanded to include p53, a major tumor suppressor protein with an unstructured domain. The discussion of molecular chaperones, proteins that have a central role in proteostasis, has been expanded and updated. The description of hemoglobin allostery has been revised to improve clarity. The Biochemistry in the Lab reading on protein technology has been expanded to include the use of affinity chromatography in the purification of recombinant proteins and the use of top-down mass spectrometry and NMR spectroscopy in protein structure analysis.

- The discussion of transition state stabilization in **Chapter 6** has been revised using triose phosphate isomerase as an example.
- The section in **Chapter 8** devoted to glycolysis regulation has been expanded with a description of glucose-induced gene expression.
- **Chapter 9** has a new Biochemistry in Perspective reading that describes the relationship between aerobic glycolysis and carcinogenesis.
- A revised and updated Biochemistry in Perspective reading in **Chapter 10** describes the biochemical consequences of the reintroduction of oxygen to heart cells damaged by clot-induced inadequate blood flow.
- In **Chapter 11** the discussion of the plasma membrane protein CFTR has been updated to include a brief overview of the clinical use of molecules that improve folding and/or function of the receptor in cystic fibrosis patients.
- In **Chapter 12** the section devoted to lipolysis in adipocytes has been revised to reflect recent discoveries concerning the roles of perilipin A and several enzymes. The Biochemistry in Perspective reading on atherosclerosis now includes a description of the consequences of AGE formation in endothelial cells.
- In **Chapter 13** a new Biochemistry in Perspective reading, "The Artificial Leaf," describes the attempts of researchers to imitate natural photosynthesis for fuel synthesis.
- The discussion of nitrogen fixation in **Chapter 14** has been revised to reflect recent discoveries. The transamination reaction mechanism figure has been modified to improve clarity.
- In **Chapter 15** the section devoted to the ubiquitin proteasomal system has been revised to increase clarity.
- The section devoted to cell surface receptor function in **Chapter 16** has been revised and expanded. The Biochemistry in Perspective reading on diabetes mellitus has been updated.
- In **Chapter 17** the section devoted to histones and chromatin structure has been revised to improve clarity and expanded to include recent research. Descriptions of noncoding DNA in the human genome have been updated. In the Biochemistry in the Lab reading devoted to nucleic acid methods, descriptions of Illumina and ion torrent DNA sequencing have been added. The description of ncRNAs has been revised. The Biochemistry in Perspective on HIV infection has been updated.
- In the Biochemistry in the Lab on genomic methods in **Chapter 18**, the description of genome projects has been updated with an introduction to ENCODE (the Encyclopedia of DNA Elements). The discussion of eukaryotic RNA polymerase II now includes a brief description of transcription factories in the nucleus. The section devoted to eukaryotic transcription and posttranscriptional processing has been revised and expanded to reflect recent discoveries.
- A new section of **Chapter 19** describes the proteostasis network, which protects the integrity of proteins from their synthesis through folding, transport, and degradation when they are damaged or obsolete. The discussion includes brief descriptions of the heat shock response and the role of the proteostasis network in several human diseases.

LEARNING PACKAGE

We have created a comprehensive set of additional resources to accompany the sixth edition. These are designed to help students master the subject matter and to assist instructors in meeting this objective.

For Students

STUDENT STUDY GUIDE AND SOLUTIONS MANUAL. Written by the textbook authors, this manual provides the solutions to all of the exercises from the text that are not included in the book itself. Each solution has been independently checked for accuracy by a panel of expert reviewers.

NEW! ANIMATION AND VIDEO GUIDE. New to the sixth edition, the student companion website now includes a curated guide to biochemical animations. This set of more than 200 high-quality and freely available animations help students visualize complex biochemical processes.

WEB QUIZZES. At http://www.oup.com/us/mckee/, students seeking an online resource to test their knowledge of biochemistry can gain access to more than 600 questions written by Dan Sullivan (University of Nebraska at Omaha). Students receive a feedback summary with each graded quiz.

INTERACTIVE 3D MOLECULES. Todd Carlson (Grand Valley State University) has created more than 300 interactive 3D molecules in JMOL format to accompany this text. Students can manipulate and study individual molecules and their structures, take self-guided concept tutorials, and test their molecule-recognition abilities by working through the interactive self-quizzes at http://www.oup.com/us/mckee/.

For Instructors

SAPLING LEARNING ONLINE HOMEWORK SYSTEM. Sapling Learning's online homework system includes algorithmic questions, exercises and guided tutorials for molecule drawing, chemical equation entry, 2D and 3D atom selection, labeling diagrams, and graphing. Automatic homework grading, diagnostic feedback, and dedicated support from chemists provide instructors all the resources they need to finally assign homework their students will actually complete. To schedule a live demonstration of Sapling, go to http://www.oup.com/us/mckee/.

ANCILLARY RESOURCE CENTER. The Ancillary Resource Center (ARC) at **http://www.oup-arc.com** is a convenient, instructor-focused single destination for resources that accompany the text. Accessed online through individual user accounts, the ARC provides instructors with up-to-date ancillaries at any time while guaranteeing the security of grade-significant resources. The following instructor's resources are available on the McKee ARC:

- **All text images in electronic format.** Instructors who adopt the sixth edition gain access to every numbered illustration, photo, figure caption, and table from the text in high-resolution electronic format. Labels have been enlarged and multipart figures have been broken down into separate components for clearer projection in large lecture halls. Images are available on both the Instructor's Resource CD and the sixth edition website, http://www.oup.com/us/mckee/.
- **Computerized test-item file.** Written by the authors and completely revised for the sixth edition, the test-item file includes more than 1,400 questions provided

as editable Word files that can be easily customized. Using the test-authoring and -management tool Diploma, the computerized version of the test bank is designed for both novice and advanced users. Diploma enables instructors to create and edit questions, create randomized quizzes and texts with an easy-to-use drag-and-drop tool, publish quizzes and tests to online courses, and print quizzes and tests for paper-based assessments.

LECTURE NOTES SLIDES. This set of more than 1100 editable lecture notes slides makes preparing lectures faster and easier than ever. Available in PowerPoint format, the lecture notes for the sixth edition now include embedded links from the new curated animations and video guide.

COURSE MANAGEMENT SYSTEMS. All instructor and student resources, including the text images, the test bank files, and the online self-quiz questions, are compatible with a variety of management systems.

ACKNOWLEDGMENTS

We wish to express our appreciation for the efforts of the dedicated individuals who provided detailed content and accuracy reviews of the text and the supplemental materials of the sixth edition:

Erika L. Abel—Baylor University
Josephine Arogyasami—Southern Virginia University
Curt Ashendel—Purdue University
Sandra L. Barnes—Alcorn State University
Ruth Birch—Saint Louis University
Karl Bishop—Husson University
Albert Bobst—University of Cincinnati
Michael G. Borland—University of Cincinnati
John Brewer—University of Georgia
Weiguo Cao—Clemson University
Srikripa Chandrasekaran—Clemson University
Sulekha R. Coticone—Florida Gulf Coast University
Matthew R. Dintzner—Western New England University
Eric R. Gauthier—Laurentian University (Canada)
Joseph Hajdu—California State University, Northridge
Marlin Halim—California State University, East Bay
Angela Hoffman—University of Portland
Jeffrey Hoyt—Paradise Valley Community College
Christine A. Hrycyna—Purdue University
Holly Huffman—Arizona State University
Sajith Jayasinghe—California State University, San Marcos
Christa R. Koval—Colorado Christian University
Allison Lamanna—Boston University
Kristi L. McQuade—Bradley University
Kyle Murphy—Rutgers University
Michael Nosek—Fitchburg State University
Peter Oelkers—University of Michigan–Dearborn
Peter M. Palenchar—Villanova University
Dominic F. Qualley—Berry College
Niina J. Ronkainen—Benedictine University
Abbey Rosen—University of Minnesota, Morris
Vijay Singh—University of North Texas
Madhavan Soundararajan—University of Nebraska
Blair R. Szymczyna—Western Michigan University
Timothy Vail—Northern Arizona University
Terry Watt—Xavier University of Louisiana
Rosemary Whelan—University of New Haven
Ryan D. Wynne—St. Thomas Aquinas College

We would also like to thank the individuals who reviewed the first five editions of this text:

Gul Afshan—Milwaukee School of Engineering
Kevin Ahern—Oregon State University
Mark Annstron—Blackburn College
Donald R. Babin—Creighton University
Stephanie Baker—Erksine College
Bruce Banks—University of North Carolina at Greensboro
Thurston Banks—Tennessee Technological University
Ronald Bartzatt—University of Nebraska, Omaha
Deborah Bebout—The College of William and Mary
Werner Bergen—Auburn University
Steven Berry—University of Minnesota, Duluth
Allan Bieber—Arizona State University
Ruth E. Birch—Saint Louis University
Brenda Braaten—Framingham State College
John Brewer—University of Georgia
Martin Brock—Eastern Kentucky University
David W. Brown—Florida Gulf Coast University
Edward J. Carroll Jr. —California State University, Northridge
Jiann-Shin Chen—Virginia Tech
Alice Cheung—University of Massachusetts, Amherst
Oscar P. Chilson—Washington University

Randolph A. Coleman—The College of William and Mary
Sean Coleman—University of the Ozarks
Kim K. Colvert—Ferris State University
Sulekha Coticone—Florida Gulf Coast University
Elizabeth Critser—Columbia College
Michael Cusanovich—University of Arizona
Anjuli Datta—Pennsylvania State University
Bansidhar Datta—Kent State University
Danny J. Davis—University of Arkansas
Patricia DePra—Carlow University
Siegfried Detke—University of North Dakota
William Deutschman—State University of New York, Plattsburgh
Robert P. Dixon—Southern Illinois University–Edwardsville
Patricia Draves—University of Central Arkansas
Lawrence K. Duffy—University of Alaska, Fairbanks
Charles Englund—Bethany College
Paula L. Fischhaber—California State University, Northridge
Nick Flynn—Angelo State University
Clarence Fouche—Virginia Intermont College
Thomas Frielle—Shippensberg University
Matthew Gage—Northern Arizona University
Paul J. Gasser—Marquette University
Eric R. Gauthier—Laurentian College
Frederick S. Gimble—Purdue University
Mark Gomelsky—University of Wyoming
George R. Green—Mercer University
Gregory Grove—Pennsylvania State University
James Hawker—Florida State University
Terry Helser—State University of New York, Oneonta
Kristin Hendrickson—Arizona State University
Tamara Hendrickson—Wayne State University
Pui Shing Ho—Oregon State University
Charles Hosler—University of Wisconsin
Andrew J. Howard—Illinois Institute of Technology
Christine A. Hrycyna—Purdue University
Holly Huffman—Arizona State University
Larry L. Jackson—Montana State University
John R. Jefferson—Luther College
Craig R. Johnson—Carlow University
Gail Jones—Texas Christian University
Ivan Kaiser—University of Wyoming
Michael Kalafatis—Cleveland State University
Peter Kennelly—Virginia Tech University
Barry Kitto—University of Texas, Austin
Paul Kline—Middle Tennessee State University
James Knopp—North Carolina State University
Vijaya L. Korlipara—Saint John's University
Hugh Lawford—University of Toronto
C. Martin Lawrence—Montana State University
Carol Leslie—Union University
Duane LeTourneau—University of Idaho
Robley J. Light—Florida State University
Rich Lomneth—University of Nebraska at Omaha
Maria O. Longas—Purdue University, Calumet
Cran Lucas—Louisiana State University–Shreveport
Jerome Maas—Oakton Community College
Arnulfo Mar—University of Texas–Brownsville
Larry D. Martin—Morningside College
Carrie May—University of New Hampshire
Dougals D. McAbee—California State University, Long Beach
Martha McBride—Norwich University
Gary Means—Ohio State University
Alexander Melkozernov—Arizona State University
Joyce Miller—University of Wisconsin–Platteville
Robin Miskimins—University of South Dakota
David Moffet—Loyola Marymount University
Rakesh Mogul—California Polytechnic State University
Joyce Mohberg—Governors State University
Jamil Momand—California State University, Los Angeles
Bruce Morimoto—Purdue University
Alan Myers—Iowa State University
George Nemecz—Campbell University
Harvey Nikkei—Grand Valley State University
Treva Palmer—Jersey City State College
Ann Paterson—Williams Baptist College
Scott Pattison—Ball State University
Allen T. Phillips—Pennsylvania State University
Jerry L. Phillips—University of Colorado at Colorado Springs
Jennifer Powers—Kennesaw State University
Ramin Radfar—Wofford College
Rachel Roberts—Texas State University–San Marcos
Gordon Rule—Carnegie Mellon University
Tom Rutledge—Ursinus College
Ben Sandler—Ashford University
Richard Saylor—Shelton State Community College
Michael G. Sehorn—Clemson University
Steve Seibold—Michigan State University
Edward Senkbeil—Salisbury State University
Ralph Shaw—Southeastern Louisiana University
Andrew Shiemke—West Virginia University
Aaron Sholders—Colorado State University
Kevin R. Siebenlist—Marquette University
Ram P. Singhal—Wichita State University
Deana J. Small—University of New England
Maxim Sokolov—West Virginia University
Madhavan Soundararajan—University of Nebraska at Lincoln
Salvatore Sparace—Clemson University
David Speckhard—Loras College
Narasimha Sreerama—Colorado State University
Ralph Stephani—St. John's University
Dan M. Sullivan—University of Nebraska, Omaha

William Sweeney—Hunter College
Christine Tachibana—Pennsylvania State University
John M. Tomich—Kansas State University
Anthony Toste—Southwest Missouri State University
Toni Trumbo-Bell—Bloomsburg University of Pennsylvania
Craig Tuerk—Morehead State University
Sandra L. Turchi-Dooley—Millersville University
Shashi Unnithan—Front Range Community College
Harry van Keulan—Cleveland State University
Ales Vancura—Saint John's University
William Voige—James Madison University
Alexandre G. Volkov—Oakwood College
Justine Walhout—Rockford College
Linette M. Watkins—Southwest Texas State University
Athena Webster—California State University, East Bay
Lisa Wen—Western Illinois University
Kenneth O. Willeford—Mississippi State University
Alfred Winer—University of Kentucky
Beulah Woodfin—University of New Mexico
Kenneth Wunch—Tulane University
Les Wynston—California State University, Long Beach
Wu Xu—University of Louisiana at Lafayette
Laura S. Zapanta—University of Pittsburgh

We also wish to express our appreciation to Jason Noe, senior editor; Lauren Mine, developmental editor; Andrew Heaton, assistant editor; David Jurman, marketing manager; Frank Mortimer, director of marketing; Patrick Lynch, editorial director; and John Challice, vice president and publisher. The excellent efforts of the Oxford University Press production team are gratefully acknowledged. We are especially appreciative of the efforts of David Bradley, senior production editor; Lisa Grzan, production manager; and Michele Laseau, art director. We give a very special thank you to Susan Brown, whose consistent diligence on this project has ensured the accuracy of the text.

We wish to extend our deep appreciation to those individuals who have encouraged us and made this project possible: Ira and Jean Cantor and Josephine Rabinowitz. Finally we thank our son, James Adrian McKee, for his patience and encouragement.

Trudy McKee
James R. McKee

ABOUT THE AUTHORS

Trudy McKee is a biochemist who has taught Biochemistry at Thomas Jefferson University, Rosemont College, Immaculata College, and the University of the Sciences.

James R. McKee is Professor of Chemistry at the University of the Sciences.

Biochemistry

THE MOLECULAR BASIS OF LIFE

General and Organic Chemistry Review Primer

OUTLINE

BIOCHEMISTRY COURSES ARE ALWAYS FAST-PACED AND CHALLENGING.

It is for this reason that success is highly dependent on a student's background in general and organic chemistry. Although courses in these subjects are prerequisites, students often have trouble recalling the detailed chemical information that will help them understand the chemical processes in living organisms. This review is divided into two sections: general and organic chemistry. General chemistry topics include atomic structure, chemical bonding, acids and bases, and the chemical properties of the principal elements found in living organisms. Topics in the organic chemistry section include the structure and chemical properties of carbon-containing compounds, nucleophiles and electrophiles, functional group structure and chemical behavior, and organic reaction classes. Topics that are directly relevant to biochemistry (e.g., biomolecule classes, pH, buffers, kinetics, and thermodynamics) are described within the textbook.

GENERAL CHEMISTRY

Chemistry is the investigation of matter and the changes it can undergo. Matter, which can be described as physical substances that occupy space and have mass, is composed of various combinations of the chemical elements. Each chemical element is a pure substance that is composed of one type of atom. About 98 of the 118 known elements occur on earth and an even smaller number occur naturally in living organisms. These elements fall into three categories: *metals* (substances such as sodium and magnesium with high electrical and heat conductivity, metallic luster, and malleability), *nonmetals* (elements such as nitrogen, oxygen, and sulfur, which are defined as a group because of their lack of metallic properties), and *metalloids* (elements such as silicon and boron, which have properties intermediate between metals and nonmetals).

The review of general chemistry includes an overview of atomic structure, atomic electron configurations, the periodic table, chemical bonds, valence bond theory, chemical reaction types, reaction kinetics, and equilibrium constants.

Atomic Structure: The Basics

Atoms are the smallest units of an element that retain the property of that element. Atomic structure consists of a positively charged central nucleus surrounded by one or more negatively charged electrons. With the exception of the element hydrogen (H), the dense, positively charged nucleus contains positively charged protons and neutrons, which have no charge. (The hydrogen nucleus consists of a single proton.) Atoms are electrically neutral so the number of protons is equal to the number of electrons. When atoms gain or lose one or more electrons, they become charged particles called ions. Ions formed when atoms lose electrons, called *cations*, are positively charged because they have fewer electrons than protons. For example, when a sodium atom (Na) loses an electron, it becomes the positively charged ion Na^+. Ions formed by the gain of electrons, called *anions*, are negatively charged. Chlorine (Cl) gains an electron to form the chlorine ion Cl^-.

ATOMIC NUMBER AND MASS NUMBER Elements are identified by their atomic number and mass number. The *atomic number* of an element is the number of protons in its nucleus. The atomic number uniquely identifies an element. Carbon (C) has 6 protons in its nucleus, so its atomic number is 6. Any atom with 16 protons in its nucleus is an atom of sulfur (S).

The *mass number* of an element, measured in atomic mass units, is equal to the number of protons and neutrons. Calculating an element's mass number is complicated by the existence of *isotopes*, atoms of an element with the same number of protons but different numbers of neutrons.

Many naturally occurring elements exist as a mixture of isotopes. For example, carbon has three naturally occurring isotopes containing six, seven, and eight neutrons, called carbon-12, carbon-13, and carbon-14, respectively. Carbon-12, the most abundant carbon isotope, is used as a reference standard in the measurement of atomic mass. An atomic mass unit (μ) or dalton (Da), named after the chemist John Dalton, is defined as one twelfth of the mass of an atom of carbon-12. Because the isotopes of an element do not occur with equal frequency, the *average atomic mass unit* (the weighted average of the atomic masses of the naturally occurring isotopes) is used. For example, hydrogen has three isotopes: hydrogen-1, hydrogen-2 (deuterium), and hydrogen-3 (tritium), which contain zero, one, and two neutrons, respectively. The average atomic mass for hydrogen is 1.0078 μ. This number is very close to 1.0 because hydrogen-1 has an abundance of more than 99.98%.

RADIOACTIVITY Some isotopes are radioactive (i.e., they undergo *radioactive decay*, a spontaneous process in which an atomic nucleus undergoes a change that is accompanied by an energy emission). For example, relatively unstable carbon-14 undergoes a form of radioactive decay, referred to as β-decay. In β-decay one neutron in the atom's nucleus is converted into a proton and an electron. The new proton converts the carbon-14 atom to a stable nitrogen-14 atom. The newly created electron is emitted as a β-particle. Hydrogen-3 (tritium) also decays to form the more stable helium-3 (a rare isotope of helium with one neutron instead of two) by the emission of a β-particle. Essentially, the unstable tritium nucleus, which contains one proton and two neutrons, decays to form the helium-3 isotope (two protons and one neutron).

ATOMIC THEORY According to the Bohr model of atoms, electrons are in circular orbits with fixed energy levels that occur at specific distances from the nucleus. When an atom absorbs energy, an electron moves from its "ground state" to a higher energy level. The electron returns to its ground state when the atom releases the absorbed energy. As quantum theory revolutionized physics in the early twentieth century, it became apparent that the theory explained many properties of atoms that the Bohr model did not.

Quantum theory is based on the principle that both matter and energy have the properties of particles and waves. Using quantum theory, physicists and chemists eventually described an atomic model in which electrons are predicted to occur in complex orbitals that are essentially probability clouds. An orbital is a probability distribution (i.e., variations in an orbital's cloud density correlate with the probability of finding an electron). The different shapes and sizes of orbital clouds depend on the energy level of the electrons within them. Together, four quantum numbers describe the configuration of the electrons and the orbitals in an atom.

The *principal quantum number n* defines the average distance of an orbital from the nucleus where $n = 1, 2, 3$, etc. In other words, the quantum number n designates the principal energy shell. The higher its n value, the farther an electron is from the nucleus.

The *angular momentum quantum number l* (lower case L) determines the shape of an orbital. The l values of 0, 1, 2, 3, and 4 correspond to the s, p, d, and f subshells. Note that the value of n indicates the total number of subshells within the principal energy shell. So if $n = 3$, the atom's principal shell has three subshells with l values of 0, 1, and 2. In such an atom the principal energy shell would contain s, p, and d orbitals. Each subshell also has a specific shape. The s orbital is spherical with the nucleus at its center. Each p orbital is dumbbell-shaped and each d orbital is double dumbbell-shaped. The shape of f orbitals is extremely complex and is not discussed further.

The *magnetic quantum number m* describes an orbital's orientation in space. Values of m range from $-l$ to $+l$. With an s orbital, $l = 0$ so the value of m is 0. For p orbitals, the value of l is 1, so m is equal to -1, 0, or -1 (i.e., there are three orbitals designated p_x, p_y, and p_z (**Figure 1**). For d orbitals $l = 2$, so there are five possible orientations: -2, -1, 0, $+1$, or $+2$.

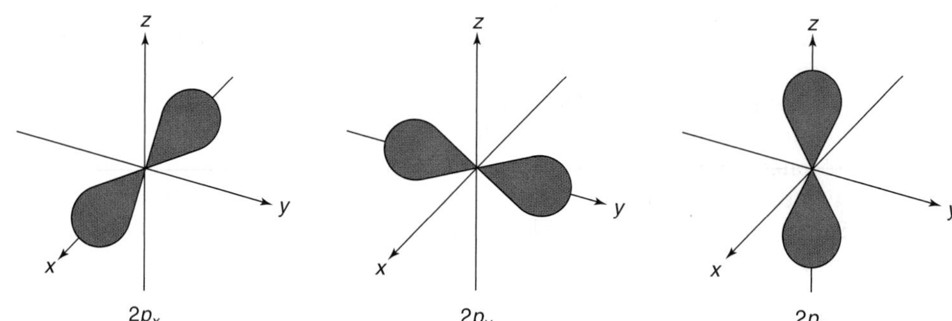

FIGURE 1
The 2p Orbitals
Three 2p orbitals are oriented at right angles to each other.

The fourth quantum number is the *spin quantum number* m_s, which describes the direction in which an electron is rotating: clockwise or counterclockwise. The values for m_s can be either $+1/2$ or $-1/2$. Because the *Pauli exclusion principle* states that each electron in an atom has a unique set of the four quantum numbers, it follows that when two electrons are in the same orbital, they must have opposite spins. Such spins are described as "paired." The spinning of an electron creates a magnetic field. *Diamagnetic* atoms such as nitrogen are not attracted to magnets because they have paired electrons (i.e., the magnetic fields of the paired electrons cancel out). Atoms that contain unpaired electrons (e.g., oxygen) are referred to as *paramagnetic* because they are attracted to magnets.

ELECTRON CONFIGURATION IN ATOMS Knowing how electrons are distributed in atoms is essential to any understanding of how chemical bonds are formed. There are several rules concerning electron distribution. The most basic rule is the *Aufbau principle*, which stipulates that electrons are put into orbitals, two at a time, in the order of increasing orbital energy (i.e., the inner orbitals are filled before the outer, higher-energy orbitals). Chemists use a shorthand method to illustrate how electrons are arranged around the nucleus of a ground-state atom (an atom at its lowest possible energy state). The electron configuration pattern useful for the elements relevant to living organisms is as follows: $1s^2 2s^2 2p^6 3s^2 3p^6 4s^2 3d^{10} 4p^6 5s^2 4d^{10} 5p^6 6s^2$.

The superscripts in the electron configuration pattern indicate the maximum number of electrons in each subshell. Note that because of orbital overlaps, the order of the orbitals being filled becomes more complicated as the filling pattern progresses. **Figure 2** is a diagram that will aid in recalling the order in which the subshells are filled.

FIGURE 2
Subshell-Filling Sequence
All of the subshells of a given value of n are on the same horizontal line. The filling sequence is determined by following the arrows starting at the lower left.

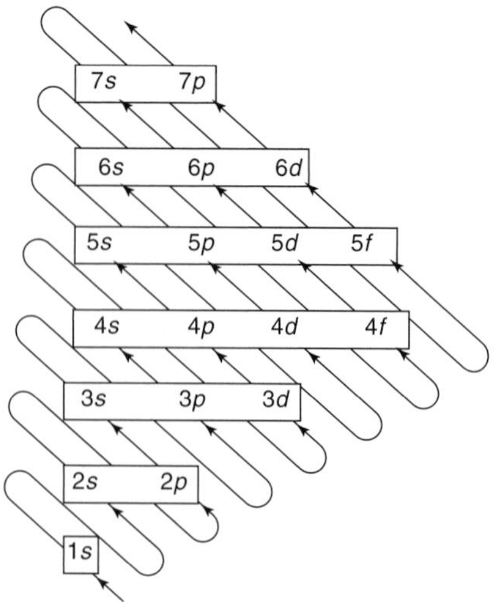

Determining an element's electron configuration requires knowing its atomic number (the number of protons), which is also equal to the number of electrons. Using the electron configuration pattern, the element's electrons are then used to fill in the orbitals beginning with the lowest energy level. For example, the electron configurations of hydrogen (1 electron) and helium (2 electrons) are $1s^1$ and $1s^2$, respectively. Similarly, the electron configurations for carbon (6 electrons) and chlorine (17 electrons) are $1s^2 2s^2 2p^2$ and $1s^2 2s^2 2p^6 3s^2 3p^5$, respectively. According to *Hund's rule*, when an energy subshell has more than one orbital (e.g., p and d orbitals) there is only 1 electron allowed in each orbital until all the orbitals have 1 electron. Such electrons have parallel spins. As additional electrons enter the orbitals, they will spin pair with the previously unpaired electrons. The orbital diagrams of nitrogen and oxygen illustrate this rule.

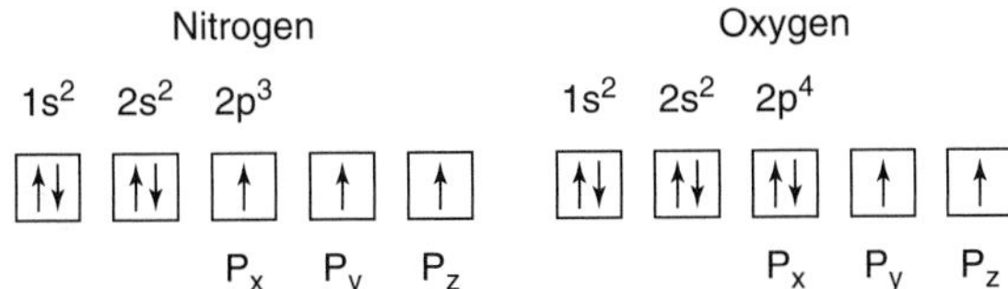

For many elements, an electron configuration also reveals how many valence electrons there are. *Valence electrons*, the electrons in the s and p orbitals of the outermost energy level, determine the element's chemistry (i.e., how it will react with other elements). For example, oxygen atoms with an electron configuration of $1s^22s^22p^4$ have six valence electrons (i.e., there are a total of six electrons in its 2s and 2p orbitals). Chlorine has seven valence electrons because there are seven electrons in its 3s and 3p orbitals. For many elements, atoms will react so that their outermost energy level or *valence shell* is filled, which is the most stable configuration they can have. The term *octet rule* is used to describe this phenomenon because the atoms of most elements react so that their valence shells contain eight electrons. Hydrogen and lithium are two obvious exceptions. Because a hydrogen atom only has one electron in its 1s orbital, it can gain one electron when it reacts to form a $1s^2$ orbital or it can give up an electron to form a proton (H^+). With three electrons, lithium (Li) has a $1s^22s^1$ configuration. By losing its one valence electron, lithium atoms gain stability by having a filled 1s shell (two electrons). In the reaction of lithium with chlorine forming lithium chloride (LiCl), lithium gives up one electron to become a lithium ion (Li^+). The lithium valence electron is donated to chlorine to form the chloride ion (Cl^-). Chlorine has thereby increased its valence shell from seven to eight electrons. Understanding the significance of electron configurations, valence, and other properties of the elements is enhanced by familiarity with the periodic table of the elements, which is discussed next. It should be noted that the term *oxidation state* is often used in reference to atoms that have gained or lost electrons. The lithium ion, for example, has a +1 oxidation state and the chloride ion has a −1 oxidation state.

THE PERIODIC TABLE The modern periodic table (**Figure 3**) is a chart based on the *periodic law*, which states that the electron configurations of the elements vary periodically with their atomic number. The properties of the elements that depend on their electronic configuration, therefore, also change with increasing atomic number in a periodic pattern. The periodic table is arranged in vertical rows called *groups* or *families* and horizontal rows called *periods*. Certain characteristics of the elements increase or decrease along the vertical or horizontal rows. These characteristics, which affect chemical reactivity, are atomic radius, ionization energy, electron affinity, and electronegativity. The *atomic radius* of a neutral atom is the distance from the nucleus to the outermost electron orbital. *Ionization energy* is defined as the amount of energy required to remove the highest energy electron from each atom in 1 mol of the atoms in the gaseous state (i.e., how strongly an atom holds on to its electrons). *Electron affinity* is the energy that is released when an electron is added to an atom. *Electronegativity* is the tendency of an atom to attract electrons to itself.

Each of the seven horizontal rows of the periodic table, called *periods*, begins with an element with a new shell with its first electron. For example, there is one electron in the 2s, 3s, and 4s subshells of lithium, sodium, and potassium (K), respectively. The atomic radii of the elements in groups 1, 2, and 13 to 18 decrease from left to right. As the number of positively charged protons at the center of the atom increases, the negatively charged electrons are attracted more strongly (i.e., the electrons are drawn closer to the nucleus). The same trend is not seen in

THE PERIODIC TABLE OF THE ELEMENTS

Periods	1	2	3	4	5	6	7	8	9	10	11	12	13	14	15	16	17	18	
Traditional groups (USA)	IA	IIA	IIIB	IVB	VB	VIB	VIIB	VIIIB			IB	IIB	IIIA	IVA	VA	VIA	VIIA	0	
1									1 H 1.008									2 He 4.003	1
2	3 Li 6.941	4 Be 9.012											5 B 10.81	6 C 12.01	7 N 14.01	8 O 16.00	9 F 19.00	10 Ne 20.18	2
3	11 Na 22.99	12 Mg 24.30											13 Al 26.98	14 Si 28.09	15 P 30.97	16 S 32.07	17 Cl 35.45	18 Ar 39.95	3
4	19 K 39.10	20 Ca 40.08	21 Sc 44.96	22 Ti 47.88	23 V 50.94	24 Cr 52.00	25 Mn 54.94	26 Fe 55.85	27 Co 58.93	28 Ni 58.69	29 Cu 63.55	30 Zn 65.39	31 Ga 69.72	32 Ge 72.61	33 As 74.92	34 Se 78.96	35 Br 79.90	36 Kr 83.80	4
5	37 Rb 85.47	38 Sr 87.62	39 Y 88.91	40 Zr 91.22	41 Nb 92.91	42 Mo 95.94	43 Tc [98.91]	44 Ru 101.1	45 Rh 102.9	46 Pd 106.4	47 Ag 107.9	48 Cd 112.4	49 In 114.8	50 Sn 118.7	51 Sb 121.8	52 Te 127.6	53 I 126.9	54 Xe 131.3	5
6	55 Cs 132.9	56 Ba 137.3	71 Lu 175.0	72 Hf 178.5	73 Ta 180.9	74 W 183.8	75 Re 186.2	76 Os 190.1	77 Ir 192.2	78 Pt 195.1	79 Au 197.0	80 Hg 200.6	81 Tl 204.4	82 Pb 207.7	83 Bi 209.0	84 Po [209]	85 At [210.0]	86 Rn [222.0]	6
7	87 Fr [223.0]	88 Ra [226.0]	103 Lr [260.1]	104 Unq [261.1]	105 Unp [262.1]	106 Unh [263.1]	107 Uns [262.1]	108 Uno [265.1]	109 Une [266.1]										7

57 La 138.9	58 Ce 140.1	59 Pr 140.9	60 Nd 144.2	61 Pm [144.9]	62 Sm 150.4	63 Eu 152.0	64 Gd 157.2	65 Tb 158.9	66 Dy 162.5	67 Ho 164.9	68 Er 167.3	69 Tm 168.9	70 Yb 173.0
89 Ac 227.0	90 Th 232.0	91 Pa 231.0	92 U 238.0	93 Np [237.0]	94 Pu [244.1]	95 Am [243.1]	96 Cm [247.1]	97 Bk [247.1]	98 Cf [251.1]	99 Es [252.1]	100 Fm [257.1]	101 Md [258.1]	102 No [259.1]

Key

Atomic number	27
Symbol	Co
Atomic mass	58.98

Values in brackets are masses of most stable isotopes.

FIGURE 3
The Periodic Table

In the modern periodic table elements are organized on the basis of their atomic numbers, electron configurations and recurring chemical properties. Note that the Lanthanides (elements 57–70) and the Actinides (elements 89–102) are not relevant to biochemistry and are not discussed.

the atomic radii of the elements in groups 3 to 12. There is little reduction in atomic size in these elements because of the repulsion between the 4s and 3d electrons.

The ionization energies of the elements in a period typically increase with increasing atomic number. As the atomic radii decrease across a period (i.e., the distance between the outer electron and the nucleus of the atoms decreases), more energy is required to remove the outer electron. For example, it is easier to remove an electron from lithium (atomic number = 3) than from nitrogen (N) (atomic number = 7).

Electron affinity, which can be thought of as the likelihood that a neutral atom will gain an electron, increases from left to right across periods. Metals such as sodium ($1s^22s^22p^63s^1$) have low electron affinities because they become more stable when they lose valence electrons. Elements on the right side of the periodic table have high electron affinities because of vacancies in their valence shells. Chlorine ($1s^22s^22p^63s^23p^5$) has a very high electron affinity because it releases a large amount of energy to become more stable as the chloride ion when it fills its valence shell by gaining one electron. The noble gases [e.g., helium (He), neon (Ne), and argon (Ar)] in group 18 do not conform to this trend because their valence shells are filled; hence, they are chemically unreactive.

Electronegativity, a measure of an atom's affinity for electrons in a chemical bond, increases across a period as atomic radii are decreasing. In the water molecule (H_2O), for example, oxygen is more electronegative than the hydrogen atoms because oxygen's larger nucleus strongly attracts its electrons. Hence, in water molecules the electrons in the bonds between each of the two hydrogen atoms and the oxygen atom are shared unequally.

The 18 vertical rows of the periodic table consist of elements with similar chemical and physical properties. The group 1 elements all have one electron in their outermost shell. With the exception of hydrogen, all of the group 1 elements [lithium (Li), sodium (Na), potassium (K), rubidium (Rb), cesium (Cs), and the rare radioactive francium (Fr)] are referred to as the *alkali metals* because they react vigorously with water to form hydroxides (e.g., NaOH). The alkali metals do so because they readily lose their single valence electron to form cations with a +1 charge. For example, sodium ($1s^22s^22p^63s^1$) reacts with water to form sodium hydroxide (NaOH) and hydrogen gas (H_2). NaOH then dissociates to form Na^+ and OH^-. Because the alkali metals donate their valence electron so readily, they are considered especially strong reducing agents. (*Reducing agents* are elements or compounds that donate electrons in chemical reactions.) Of all the alkali metals, only sodium and potassium have normal functions in living organisms. For example, the balance of sodium ions and potassium ions across the plasma membrane of neurons is critical to the transmission of nerve impulses.

The group 2 alkaline earth metals [beryllium (Be), magnesium (Mg), calcium (Ca), strontium (Sr), barium (Ba), and radium (Ra)] have two electrons in their outermost shell. The electron configurations for the biologically important group 2 elements magnesium (DNA structure and enzyme function) and calcium (bone structure and muscle contraction) are $1s^22s^22p^63s^2$ and $1s^22s^22p^63s^2\ 3p^64s^2$, respectively. With the exception of beryllium, the alkaline earth metals lose their two valence electrons to form cations with a +2 charge [e.g., they react with water to form metal hydroxides such as $Ca(OH)_2$]. Like the group 1 metals, the alkaline earth metals are strong reducing agents, although each element is somewhat less reactive than the alkali metal that precedes it.

Groups 3 to 12 are referred to as the d-block elements because electrons progressively fill the d orbitals. The majority of the d-block elements are the *transition elements*, which have incompletely filled d orbitals. Zinc (Zn, atomic number = 30) is not considered a transition metal because its 3d subshell has 10 electrons. Because electron configurations of elements with high atomic numbers are unwieldy, chemists use a simplification for an element's electron configuration that is an abbreviation for the electron configuration of the noble gas immediately preceding the element. For example, zinc's electron configuration can be described as $[Ar]3d^{10}4s^2$.

The transition elements are metals with special properties. Among these are the capacities to have more than one oxidation state and to form colored compounds. Iron is a transition metal found in a large number of proteins in all living organisms. It most notably occurs in hemoglobin, the oxygen-transport protein that gives blood its red color. Iron atoms ($[Ar]3d^64s^2$) can form a wide range of oxidation states from −2 to +6, but its most common oxidation states are +2 and +3. Neutral iron atoms can form the +2 ion because the 4s orbital and the 3d orbitals have very similar energies, so that the removal of two electrons to form Fe^{+2} ($[Ar]3d^6$) requires little energy. The loss of an additional electron to form Fe^{+3} {$[Ar]3d^5$} requires more energy.

In addition to iron, several other d-block elements are important in living organisms. Manganese (Mn) is found in numerous enzymes in all living organisms. Cobalt (Co) is an important component of vitamin B_{12} structure. Nickel (Ni) occurs in several enzymes in microorganisms and plants. Copper (Cu) is found in several energy generation proteins. Zinc (Zn) occurs in over 100 enzymes and has structural roles in numerous proteins. Molybdenum (Mo) has a vital role in nitrogen fixation.

The remaining elements in the standard periodic table, groups 13 to 18, are in the *p block,* so named because electrons progressively fill p orbitals. The p-block elements found in living organisms (carbon, nitrogen, phosphorus, sulfur, chlorine, and iodine) are nonmetals. Carbon ($1s^22s^22p^2$) in group 14 has four electrons available to form stable bonds both with other carbon atoms and with a variety of other elements (most notably hydrogen, oxygen, nitrogen, and sulfur). As a result, carbon can form an almost infinite number of compounds. Carbon is the crucial

WORKED PROBLEM 1

Consider the element potassium (atomic number = 19). What is its electronic configuration? How many electrons are in its valence shell? Is elemental potassium paramagnetic or diamagnetic?

SOLUTION

The number of electrons in the potassium atom is equal to its atomic number, 19. Using the Aufbau and Pauli exclusion principles and Hund's rule, the electronic configuration is $1s^2 2s^2 2p^6 3s^2 3p^6 4s^1$. Potassium has one valence electron because its outermost energy level (4s) has one electron. Elemental potassium is paramagnetic because it has one unpaired electron in its valence shell.

element in most biomolecules, with the exceptions of molecules such as water and ammonia and the electrolytes (e.g., Na^+, K^+, and Mg^{+2}). (An *electrolyte* is an ionic species that influences the distribution of electric charge and the flow of water across membranes.) Because nitrogen ($1s^2 2s^2 2p^3$) has five electrons in its outer shell, its valence is −3. In biomolecules nitrogen is found in amines ($R\text{-}NH_{2,}$ where R is a carbon-containing group) and amides (e.g., bonds between amino acids in proteins). Phosphorus ($1s^2 2s^2 2p^6 3s^2 3p^3$), in the nitrogen family, most commonly occurs in living organisms as phosphate ($PO_{4-}{}^{3}$); (e.g., in the nucleic acids DNA and RNA and as a structural component of bones and teeth). In living organisms, oxygen ($1s^2 2s^2 2p^4$) is found most abundantly in water molecules, where it has an oxidation state of −2. Oxygen atoms are also found in all the major classes of biomolecules (e.g., proteins, carbohydrates, fats, and nucleic acids). Sulfur ($[Ne]3s^2 3p^6$), the second member of the oxygen family, is found in proteins and small molecules such as the vitamin thiamine. It often occurs in biomolecules in the form of thiols (R-SH) and disulfides (R-S-S-R). Of all the members of the halogen family (group 17), only chlorine ($[Ne]3s^2 3p^5$) and iodine ($[Kr]5s^2 4d^{10} 5p^5$) routinely occur in living organisms. The functions of chlorine in the form of the chloride ion include the digestion of protein in the animal stomach [hydrochloric acid (HCl)] and its function as an electrolyte. Iodine is a component of the thyroid hormones, which regulate diverse metabolic processes in the animal body. It should be noted that phosphorus, sulfur, and chlorine are usually assigned valences of −3, −2, and −1, respectively. However, because they have vacant d orbitals, they can expand their valence shell and form different oxidation states. For example, phosphorus has a +5 valence in phosphoric acid (H_3PO_4), sulfur has a valence of +6 in the sulfate ion (SO_4^{2-}), and chlorine has a +1 valence in the hypochlorite ion (ClO^-).

Chemical Bonding

A chemical bond is a strong attractive force between the atoms in a chemical compound. Chemical bonds form because of interactions between atoms that involve the rearrangement of their outer shell electrons. According to the octet rule, atoms react so as to achieve the outer electron configuration of the noble gases. They do so because complete outer valence shells are stable because of a reduction in stored potential energy.

Two major types of chemical bonds are ionic and covalent. The bonds differ in how valence electrons are shared among the bonded atoms. *Ionic bonds* form when electrons are transferred from atoms with a tendency to release electrons (e.g., alkali or alkaline earth metals) to electronegative atoms that tend to gain electrons. The transfer process results in the formation of oppositely charged atoms called *ions*. The positively charged ion product is called a *cation*, and the negatively charged ion product is called an *anion*. For example, the transfer of sodium's single electron to chlorine (valence = 7) yields the cation Na^+ and the anion Cl^-. The ionic bond in NaCl is the electrostatic attraction between the positive and negative ions.

In *covalent bonds* electrons are shared between atoms with similar electronegativity values. A single covalent bond consists of two shared electrons. For example, there is one covalent bond in molecular hydrogen (H_2). The two hydrogen atoms, with one electron each, complete their valence shell by sharing their electrons. Elements such as carbon, nitrogen, and oxygen can form multiple covalent bonds. Carbon, for example, can form double and triple bonds. In the molecule ethylene a carbon-carbon double bond involves the sharing of two sets of valence electrons. The triple covalent bond in molecular nitrogen (N_2) is an example of the sharing of three sets of valence electrons.

Covalent bonds between atoms with moderate differences in electronegativity are referred to as *polar covalent bonds*. In such bonds the electrons are shared unequally, with the electron density shifted toward the atom with the greater electronegativity. The electrical asymmetry in such bonds causes one end of the molecule to possess a slightly negative charge and the other end a slightly positive charge. These partial charges are indicated by the lowercase Greek letter δ: δ^+ and δ^-. For example, in the water molecule H_2O the oxygen atom has a significantly larger electronegativity value than the hydrogen atoms. As a result, the electron pairs between oxygen and each of the hydrogen atoms are drawn closer to the oxygen atom. Each hydrogen atom has a partial positive charge (δ^+) and the oxygen has a partial negative charge (δ^-).

In a *coordinate covalent bond*, a shared pair of electrons in the bond comes from one atom. The reaction between ammonia (NH_3) and HCl provides a simple example. The product ammonium chloride (NH_4Cl) results when a covalent bond is formed between the nitrogen with its lone pair and the proton that has dissociated from HCl.

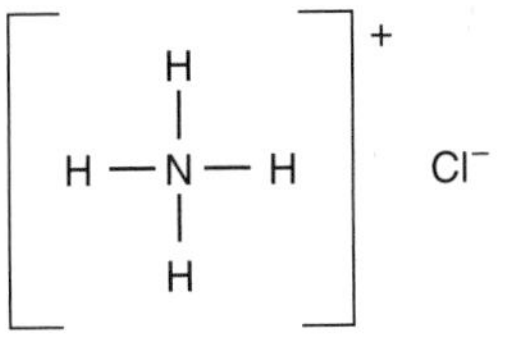

$$\left[\begin{array}{ccc} & H & \\ & | & \\ H - & N & - H \\ & | & \\ & H & \end{array}\right]^+ \quad Cl^-$$

LEWIS DOT NOTATION Chemists often describe chemical bonds using Lewis dot structures. Devised by the chemist G. N. Lewis, Lewis dot structures are a shorthand notation for explaining how the valence electrons of the atoms in various compounds combine to form covalent bonds. Molecular hydrogen (H_2) is a simple example. Because each hydrogen atom has one electron, the Lewis dot structure for the hydrogen atom is H·. The energy shell of H_2 has a maximum of two electrons, and the formation of H_2 is depicted as

$$H\cdot + H\cdot \rightarrow H{:}H$$

The following rules facilitate the drawing of Lewis structures for more complicated molecules:

(1) **Determine the number of valence electrons for each atom in the molecule.** For example, carbon dioxide (CO_2) has one carbon atom with four valence electrons and two oxygen atoms each with six valence electrons.

(2) **Determine the identity of the central atom in the Lewis structure.** This atom will often be the one with the lowest electronegativity. Recall that electronegativity decreases from right to left across and from top to bottom of the periodic table. In the case of CO_2, carbon is less electronegative than oxygen, so carbon is the central atom.

(3) **Arrange the electrons so that each atom donates one electron to a single bond between it and another atom and then count the electrons around each atom.** Are the octets complete? For CO_2, a first try would yield

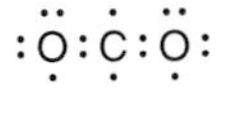

$$:\underset{\cdot}{\ddot{O}}:\underset{\cdot}{\dot{C}}:\underset{\cdot}{\ddot{O}}:$$

Note, however, that in this structure each carbon only has six electrons and each oxygen atom has seven electrons. Because the octets are incomplete, more electrons must be shared, an indication that there are double or triple bonds in the molecule. In the case of CO_2, rearranging the electrons results in the following Lewis dot structure in which there are two double bonds and all three atoms have a full octet of electrons.

O::C::O

In molecules such as ammonia (NH_3), it is obvious that although the nitrogen atom is more electronegative than the hydrogens, it is the only atom in the molecule that can form multiple bonds. Hence, the nitrogen atom is the central atom in ammonia molecules. Its Lewis dot structure is

H:N:H
H

For a large number of molecules, there is more than one valid Lewis dot structure. The nitrate ion (NO_3-) is a typical example. Considering that nitrogen has five valence electrons and each oxygen atom has six valence electrons, the following Lewis dot structure of the nitrate ion satisfies the octet rule:

[O :: :O:N:O:]$^-$

However, there is no reason why the double bond should appear where it does in this formula. It could easily appear in either of the two other locations around the nitrogen atom. Therefore, there are three valid Lewis dot structures for the nitrate ion.

[O :: :O:N:O:]$^-$ ⟷ [:O: O::N:O:]$^-$ ⟷ [:O: :O:N::O]$^-$

When this situation occurs, the ion or molecule is said to be a *resonance hybrid*. (The double-headed arrows are used in the representation of resonance structures.) In the case of the nitrate ion, it is considered to have a structure that is the average of these three states.

MOLECULAR STRUCTURE Molecules are three-dimensional arrangements of atoms. Understanding molecular structure, also referred to as molecular shape, is important because structure provides insight into the physical and chemical properties of molecules. Physical properties that are affected by molecular shape include boiling point, melting point, and water solubility. The shape of molecules also powerfully affects chemical reactivity.

According to the *valence shell electron pair repulsion* (VSEPR) *theory*, repulsive forces between valence shell bonding and nonbonding electrons (lone pairs) determine molecular geometry (molecular shape). In other words, the valence electron pairs on the central atom in a molecule orient themselves in space so that repulsion is minimized (i.e., their total energy is minimized). Lone pairs of electrons have a greater repulsive effect than bonding pairs. (A *lone pair* is a valence electron pair on a central atom that is not involved in bonding.) The term electron group is used in discussions of VSEPR theory. An *electron group* is defined as a set of valence electrons in a region around a central atom that exerts repulsion on other valence electrons. Electron groups include bonding and nonbonding electron pairs or the pairs of electrons in double or triple bonds.

WORKED PROBLEM 2

What is the Lewis electron dot formula for formaldehyde ($H_2C{=}O$)?

SOLUTION

The valence electrons for hydrogen, carbon, and oxygen are 2 (1 for each atom), 4, and 6, respectively, for a total of 12 electrons. Single bonds between the elements account for 6 electrons, leaving 6 electrons unaccounted for. Group the remaining 6 electrons around the most electronegative atom (oxygen) until a total of 8 electrons (bonding and nonbonding) is reached. Using one pair of these electrons to form a double bond between carbon and oxygen completes the carbon octet. The final Lewis structure is given below.

H
:C::O:
H

Ascertaining a molecule's three-dimensional shape begins with a correct Lewis dot structure. The molecule's geometry is then determined based on the number of bonding and nonbonding electrons on the central atom (**Figure 4**). If there are two electron pairs, the molecule has a linear shape. Carbon dioxide (CO_2), for example, is a *linear* molecule with two electron groups. Its bond angle is 180°. Formaldehyde ($H_2C{=}O$), with three electron groups, has *trigonal planar* geometry with bond angles of 120°. Molecules with a central atom with four pairs of electrons have a tetrahedral shape. Methane (CH_4), with its four carbon-hydrogen bonds, has bond angles of 109.5°. If one of the four electron groups in a tetrahedron is a lone pair, the molecular shape is *trigonal pyramidal.* Because of the strong repulsion of the lone pair, bond angles are less than 109.5°. For example, the lone pair in NH_3 forces the NH bonding electron pairs closer together with bond angles of 107.3°.

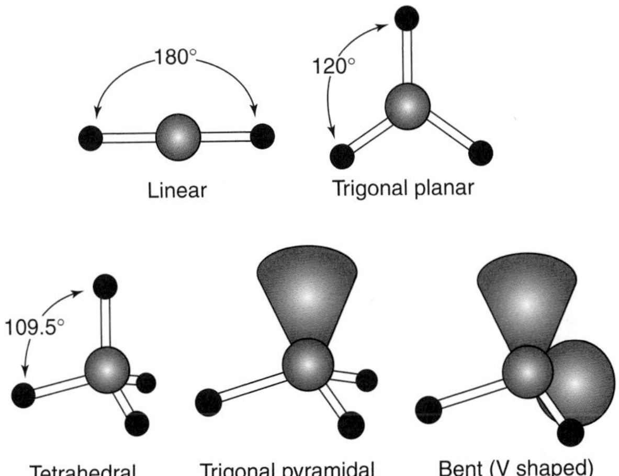

FIGURE 4
Common Molecular Geometrics

These structures illustrate the spatial orientations of electron groups. Note that unpaired electrons are indicated by an enlarged representation of an orbital.

Three-dimensional shape also affects molecular polarity. In polar covalent bonds there is an unequal sharing of electrons because the atoms have different electronegativities. This separation of charge is called a *dipole*. Although a polar molecule always contains polar bonds, some molecules with polar bonds are nonpolar. Molecular polarity requires an asymmetric distribution of polar bonds. For example, CO_2 contains two C—O dipole bonds. Carbon dioxide is a nonpolar molecule because of its linear shape (i.e., its bond dipoles are symmetrical and cancel each other out). Water, which also has two polar bonds (two O—H

WORKED PROBLEM 3

Dimethyl ether has the following formula: $CH_3—O—CH_3$. What is the shape of dimethyl ether? Does this molecule have a dipole moment?

SOLUTION

The oxygen in dimethyl ether has four electron pairs including two lone electron pairs. As a result, the oxygen has a tetrahedral shape with an overall bent shape for the dimethyl ether molecule. Because the molecule has a bent shape and the electron distribution is uneven, dimethyl ether has a dipole moment.

bonds), is a polar molecule because of its geometry. Water is tetrahedral because it contains four electron groups: two bonding pairs and two lone pairs. As a result of the greater repulsions from the lone pairs on the oxygen, however, the bond angle of a water molecule is 104.5°. Water's "bent" geometry (refer to **Figure 4**) makes it an asymmetric molecule and therefore polar.

VALENCE BOND THEORY AND ORBITAL HYBRIDIZATION Although the VSEPR theory accounts for molecular shape, it does not explain how the orbitals of the individual atoms interact to form the covalent bonds in molecules. The concept of *orbital hybridization*, the result of quantum mechanical calculations, explains how the mixing of atomic orbitals results in the formation of the more stable hybrid orbitals found in molecules. Each type of hybrid orbital corresponds to a type of electron group arrangement predicted by VSEPR theory. The three most common hybrid orbitals observed in biomolecules are sp^3, sp^2, and sp.

Carbon has an electron configuration of $1s^2 2s^2 2p^2$, which can also be represented as

↑↓	↑↓	↑	↑	
$1s^2$	$2s^2$	$2p_x$	$2p_y$	$2p_z$

It appears from this diagram that carbon only has two bonding electrons. The carbon atoms in molecules such as methane, however, are bonded to four hydrogen atoms in a tetrahedral arrangement. During methane formation, as a result of the attraction of each of the hydrogen nuclei (i.e., protons) for carbon's lower-energy valence electrons, the two 2s electrons move into 2p orbitals.

↑↓	↑	↑	↑	↑
$1s^2$	$2sp^3$	$2sp^3$	$2sp^3$	$2sp^3$

As they do so, they mix, forming four identical sp^3 orbitals (**Figure 5**).

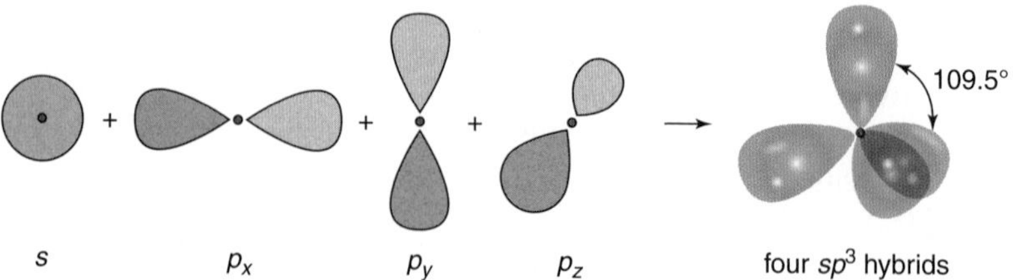

FIGURE 5
sp^3 Orbitals
Hybridization of an s orbital and all three p orbitals gives four identical sp^3 orbitals.

In the methane molecule (**Figure 6**), each of the four sp^3 hybrid orbitals overlaps with the 1s orbital of hydrogen to form a sigma bond. A *sigma bond* (σ), which is formed by the overlapping by the outermost orbitals of two atoms, is the strongest type of covalent bond.

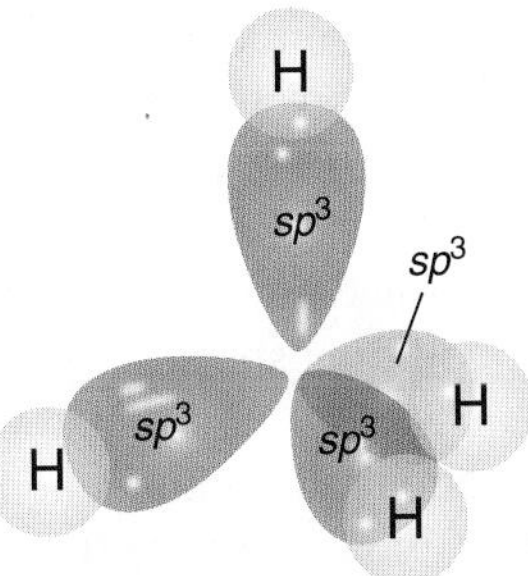

FIGURE 6

Structure of Methane

Methane (CH_4) has a tetrahedral geometry with four σ bonds formed by the overlap of four sp^3 orbitals of carbon with four 1s orbitals of hydrogen atoms.

Each of the two carbon atoms in the molecule ethene ($H_2C{=}CH_2$) is bonded to three atoms in trigonal planar geometry. Carbon's 2s orbital mixes with two of the three available 2p orbitals to form three sp^2 orbitals.

↑↓	↑	↑	↑	↑
$1s^2$	$2sp^2$	$2sp^2$	$2sp^2$	2p

Two of the three sp^2 orbitals of each carbon atom overlap the orbital of a hydrogen atom, forming a total of four σ bonds. The third sp^2 orbital of the two carbon atoms overlap to form a carbon-carbon σ bond. The p orbitals, one on each carbon, overlap to form a pi (π) bond (**Figure 7**). A *double bond* in molecules such as ethene consists of a σ bond and a π bond.

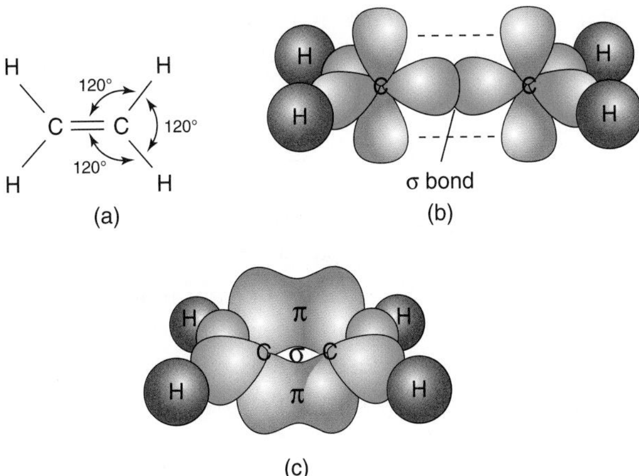

FIGURE 7

Ethene Structure

(a) Each carbon atom in ethane (also known as ethylene) has three sp^2 orbitals with bond angles of 120°, which have a trigonal planar geometry.
(b) Two of the sp^2 orbitals of each carbon (green) overlap with an s orbital of hydrogen (red) forming a total of 4 σ bonds. The remaining two sp^2 orbitals, one from each carbon, overlap to form a carbon-carbon σ bond.
(c) Two p orbitals (blue), one from each carbon atom, overlap to form a π bond.

Acetylene (C_2H_2) is a molecule with a triple bond with each carbon bonded to two other atoms in a linear geometry. Carbon's 2s orbital

↑↓	↑	↑	↑	↑
$1s^2$	sp	sp	2p	2p

mixes with one 2p orbital to form 2 sp hybrid orbitals. Each carbon also possesses two unhybridized 2p orbitals. Acetylene has a triple bond consisting of one σ bond and two π bonds. The carbon-carbon σ bond is formed by the overlap of an sp hybrid orbital from each carbon atom. Each π bond is formed by the overlap of two of carbon's 2p orbitals. Each of the two carbon-hydrogen σ bonds is formed by the overlap of carbon's other sp orbital with the 1s orbital of hydrogen.

Chemical Reactions

In chemical reactions the atoms in chemical substances are rearranged to form new substances as chemical bonds are broken and formed. According to the *collision theory*, the reaction rate in bimolecular reactions depends on the frequency of successful collisions between the chemical species. Successful collisions occur when there is sufficient energy at the moment of impact (called the *activation energy*) and the colliding species are oriented during the collision in a manner that favors the rearrangement of the atoms and electrons. *Catalysts* are substances that increase reaction rates without being affected by the reaction. They do so by lowering the activation energy of the reaction by providing an alternative pathway for the reaction. For example, the metal iron is used as a catalyst in the Haber process, the industrial method of converting nitrogen and hydrogen gases into ammonia (NH_3). As the N_2 and H_2 molecules become adsorbed onto the surface of the metal, where they are more likely to be in a favorable orientation for successful collisions, the bonds in both molecules are weakened. Once formed, the ammonia molecules desorb from the catalyst.

Chemical reactions are described with chemical equations. The substances undergoing the reaction, called *reactants*, appear on the left-hand side of the equation, and the *products* of the reaction are on the right-hand side. An arrow between the reactants and products symbolizes the chemical change that occurs as a result of the reaction. For example, the chemical equation for the reaction in which molecule A reacts with molecule B to form molecules C and D is

$$A + B \rightarrow C + D$$

Other symbols that may appear in a chemical equation provide information about the physical state of the reactants and products or a required energy source. The equation for the decomposition of calcium carbonate, for example, is

$$CaCO_3\ (s) \xrightarrow{\Delta} CaO\ (s) + CO_2\ (g)$$

In this equation the letter s indicates that the reactant $CaCO_3$ and the product CaO (calcium oxide) are solids. The letter g indicates that CO_2 is a gas. The uppercase Greek letter delta (Δ) above the arrow indicates that the reaction requires an input of energy in the form of heat. Reactions that require an input of energy in the form of heat are described as *endothermic*. If light energy is involved in a reaction, hν (ν is the lowercase Greek letter nu) is placed above the arrow.

All chemical equations must obey the *law of conservation of matter*, which states that during chemical reactions mass is neither created nor destroyed. In other words, the mass of the reactants must be equal to the mass of the products. For example, in the equation for the reaction in which methane (CH_4) reacts with molecular oxygen (O_2) to form carbon dioxide (CO_2) and water, the number of each type of atom on both sides of the arrow must be equal.

$$CH_4 + 2\,O_2 \rightarrow CO_2 + 2\,H_2O$$

In this balanced equation the same number of carbon, hydrogen, and oxygen atoms is on both sides of the arrow because the number 2 has been placed before the formulas for molecular oxygen and water.

WORKED PROBLEM 4

Consider the following reaction equation:

$$KClO + H_2S \rightarrow KCl + H_2SO_4$$

Balance the equation and identify the elements that are oxidized, reduced, or unchanged by the reaction. Use the generalizations that the oxidation state of hydrogen and group 1 metals is +1 and that of oxygen is −2.

SOLUTION

Balancing a chemical reaction equation requires that the number and types of atoms be the same. To satisfy this requirement, the number 4 is placed before the reactant KClO and the product KCl. The reaction becomes

$$4KClO + H_2S \rightarrow 4KCl + H_2SO_4$$

Using the oxidation state information given above, oxidation numbers are assigned to each element:

$$K(+1)Cl(+1)O(-2) + H_2(+1)S(-2) \rightarrow K(+1)Cl(-1) + H_2(+1)S(+6)O_4(-2)$$

Sulfur is the element that is oxidized (i.e., its oxidation number increases from −2 to +6). Chlorine is reduced because its oxidation number is lowered from +1 to −1. Elements whose oxidation numbers remain unchanged are hydrogen, potassium, and oxygen.

REACTION KINETICS As informative as a chemical reaction's equation is, it reveals nothing about several important properties of the reaction: (1) How fast does the reaction occur? (2) When the reaction ends, what will be the ratio of product to reactant molecules? (3) Does the reaction require or release energy? The science of chemical kinetics seeks to answer these and other questions about chemical reaction rates (i.e., the change in the number of product and reactant molecules as the reaction progresses).

Reaction rate is defined as the change in the concentration of reactant or product per unit time. For the general reaction

$$aA + bB \rightarrow cC + dD$$

the rate is equal to $k[A]^m[B]^n$, where k is the rate constant, and [A] and [B] are the concentrations of the reactants A and B, respectively. The exponents m and n are used to determine the reaction's order, a number that relates the rate at which a chemical reaction occurs to the concentrations of the reactants. For example, if m is equal to 1, the reaction rate doubles when the concentration of reactant A doubles. If m is equal to 2, the rate quadruples when the concentration of reactant A doubles. (Refer to pp. 198–200 in the textbook for a more detailed description of reaction order.) The rate constant and order of a reaction can only be determined by experiment. Experiments performed over the course of the past century have revealed that the following factors influence reaction rate:

(1) Reactant structure—the nature and the strength of chemical bonds affect reaction rates. For example, salt formation, the exchange of ions, is a fast process compared with the breaking and forming of covalent bonds.

(2) Reactant concentration—the number of molecules of a substance per unit volume affects the likelihood of collisions. Reaction rates increase as the reactant molecules become more crowded.

(3) Physical state—whether the reactants are in the same phase (solid, liquid, or gas) affects reaction rates because reactants must come into contact with each other. When reactants are in the aqueous phase, for example, thermal motion brings them into contact. When reactants are in different phases, contact only occurs at the interface between the phases. In such circumstances increasing interface surface area raises a reaction's rate. For example, when reactants are in the solid and liquid phases, grinding the solid into small pieces increases its surface area that is in contact with the liquid phase.

(4) Temperature—at higher temperatures molecules have more thermal energy and are, therefore, more likely to collide with each other.

(5) Catalysts—substances that accelerate a reaction's rate, but remain unchanged afterward. A catalyst provides a different pathway for the reaction, thereby lowering the activation energy.

CHEMICAL REACTIONS AND EQUILIBRIUM CONSTANTS Many chemical reactions are reversible (i.e., they occur in both forward and reverse directions). Reversible reactions are indicated in reaction equations with double arrows. When a reversible reaction begins (i.e., when the reactants are mixed together), reactants begin to be converted into product. At some point in time, which differs for each reaction, some product molecules are reconverted back into reactant molecules. Eventually, the reaction reaches a dynamic equilibrium state in which both the forward and the reverse reactions occur, but there is no net change in the ratio of reactant and product molecules. The extent to which the reaction proceeds to product is measured by an equilibrium constant (K_{eq}), which reflects the concentrations of reactants and products under specific conditions of temperature and pressure. For a reaction with the equation

$$aA + bB \rightleftharpoons cC + dD$$

K_{eq} is calculated as the ratio of the molar concentrations of product and reactant, each of which is raised to the power of its coefficient.

$$K_{eq} = [C]^c[D]^d/[A]^a[B]^b$$

Note that K_{eq} is also equal to k_f/k_r, the ratio of the forward and reverse rates of the reaction. A high K_{eq} value (significantly greater than 1) indicates that when a reaction reaches equilibrium, the concentration of the reactant is low (i.e., the reaction favors the production of product). If the K_{eq} value is lower than 1, then product concentration is lower than reactant concentration when equilibrium has been reached. When K_{eq} is greater than 1000, the reaction has gone to completion (i.e., almost all reactants have been converted to product).

In 1885 the French chemist Henri Louis Le Chatelier reported his discovery of a remarkable feature of systems at equilibrium. For a chemical reaction at equilibrium a change in the conditions of the reaction (e.g., temperature, pressure, or the concentrations of its components) triggers a shift in the equilibrium to counteract the change. Chemists and chemical engineers use *Le Chatelier's principle* to manipulate chemical reactions to maximize product synthesis. The Haber-Bosch process for making ammonia (NH_3) from N_2 and H_2 is a prominent example.

All living organisms require a source of useable nitrogen-containing molecules. As a result of the extraordinary difficulty in breaking the stable triple bond of N_2, *nitrogen fixation* (the conversion of N_2 to NH_3, a molecule that can be assimilated into organic molecules such as amino acids) is largely limited to a select group of microorganisms. Note that the synthesis of ammonia is an *exothermic reaction* (i.e., it releases heat energy):

$$N_2\,(g) \quad + \quad 3H_2\,(g) \quad \rightarrow \quad 2NH_3\,(g) \quad + \quad 98\ kJ$$

where a joule (J) is a unit of energy and a kilojoule (kJ) is 1000 joules.

The Haber-Bosch industrial process for synthesizing ammonia maximizes the reaction's yield in several ways:

(1) An iron-based catalyst (iron oxide with small amounts of other metal oxides), which increases the rate at which equilibrium is attained, converts a slow reaction to one that is fast enough to be commercially feasible.

(2) Ammonia, the product of the reaction, is removed from the reaction vessel. As a result, the system produces more NH_3 to reestablish equilibrium.

(3) An increase in the pressure within the reaction vessel (to 200 atm), obtained by decreasing volume, causes an increase in ammonia synthesis. Note that in this reaction 4 mol of reactant molecules are converted to 2 mol of product. The equilibrium shifts toward ammonia synthesis because there are fewer molecules of this gas.

(4) By lowering the temperature of the reaction (i.e., by removing heat from an exothermic reaction), the equilibrium is shifted toward more ammonia synthesis. There is a limit to how much the temperature can be lowered, however, because the catalyst requires heat to be efficient. As a result, the reaction vessel operates at 400°C, a temperature that is hot enough for the catalyst yet relatively cool for an industrial process.

ACID-BASE EQUILIBRIA AND pH When acids and bases dissolve in water, they dissociate, forming ions. Hydrochloric acid and acetic acid (CH_3COOH) are two well-known acids. HCl dissociates in water to yield chloride and hydrogen ions, and acetic acid dissociates to yield acetate (CH_3COO^-) and hydrogen ions. Sodium hydroxide (NaOH) and methylamine (CH_3NH_2) are examples of bases. In water NaOH dissociates to yield sodium and hydroxide ions and methylamine forms methylammonium ($CH_3NH_3^+$) and hydroxide ($-OH$) ions. The strength of an acid or a base is determined by the degree to which it dissociates. HCl is a strong acid because its dissociation in water is complete (i.e., 100% of HCl molecules dissociate into chloride and hydrogen ions). *Weak acids* and weak bases are so named because they dissociate only to a limited extent and establish a dynamic equilibrium with their ions. The general equation for the dissociation of a weak acid is

$$HA \rightleftharpoons A^- + H^+$$

where HA is the undissociated acid and A^- is the conjugate base of the acid. The degree to which a weak acid dissociates is expressed as an acid dissociation constant K_a, the quotient of the equilibrium concentrations of the ions A^- and H^+ and the undissociated acid (HA).

$$K_a = [A^-][H^+]/[HA]$$

The dissociation constants of weak acids and bases are usually expressed as the negative log of the equilibrium constant ($-\log K_a$ or $-\log K_b$), where the term $-\log$ is replaced by the letter p. The extent to which a weak acid dissociates is referred to as its pK_a value. The dissociation constant and pK_a for acetic acid at 25°C, for example, are 1.8×10^{-5} and 4.76, respectively. The behavior of weak acids and bases is especially important in biochemistry because many biomolecules possess carboxylate, amino, and other functional groups that can accept or donate hydrogen ions. For example, refer to pp. 138–140 for a description of the effect of hydrogen ion concentration on amino acids, the molecules used to construct proteins.

Water also has a slight capacity to dissociate into ions.

$$H_2O + H_2O \rightleftharpoons OH^- + H_3O^+$$

The hydrogen ion concentration of pure water at 25°C is 1.0×10^{-7} M. Because one hydroxide ion is produced for each hydrogen ion, the hydroxide ion concentration is also 1.0×10^{-7} M. The product of these two values (i.e., $[H^+][OH^-]$), referred to as the ion product for water, is 1×10^{-14}. The concentrations of hydrogen and hydroxide ions change depending on the substances that are dissolved in water, but their product is always 1×10^{-14}.

For weak acids and bases, hydrogen ion concentrations in aqueous solution can vary from 1 M to 1×10^{-14} M. For the sake of convenience, hydrogen ion concentrations are usually converted to pH values. The term pH simply means that the concentration of hydrogen ions in a solution has been converted to its negative log value (i.e., pH = $-\log [H^+]$. Refer to pp. 92–101 for a more detailed description of pH and the pH scale, a convenient means of expressing the acidity or alkalinity of substances.

REACTION TYPES There are several basic types of chemical reaction: synthesis reactions, decomposition reactions, displacement reactions, double displacement reactions, acid-base reactions, and redox reactions. Each is discussed briefly.

Synthetic reactions (also referred to as combination reactions) involve two or more substances that combine together to form a single new substance. The reaction of sulfur trioxide (SO_3) with water, for example, yields sulfuric acid (H_2SO_4).

$$SO_3 + H_2O \rightarrow H_2SO_4$$

WORKED PROBLEM 5

The K_a for acetic acid is 1.8×10^{-5}. Determine the hydrogen ion concentration of a 0.1 M solution of acetic acid in water. What is the pH of this solution?

SOLUTION

The equation for the dissociation of acetic acid is

$$K_a = [\text{acetate}][H^+]/[\text{acetic acid}]$$

Because acetic acid is a weak acid, it is assumed that the dissociation of acetic acid has no substantive effect on acetic acid concentration. The values of the concentrations of acetate and hydrogen ions are equal to each other and are set at x. The equation for determining the hydrogen ion concentration in a 0.1 M solution is

$$1.8 \times 10^{-5} = x^2/0.1, \text{ which becomes}$$
$$1.8 \times 10^{-6} = x^2$$

Solving for x yields 1.35×10^{-3}, the hydrogen ion concentration in the acetic acid solution.

The pH of the 0.1 M solution of acetic acid is calculated as follows:

$$\begin{aligned} pH &= -\log[H^+] \\ &= -\log(1.35 \times 10^{-3}) \\ &= 3 - 0.13 = 2.87 \end{aligned}$$

In *decomposition reactions*, a compound breaks down to form simpler products when the reactant absorbs enough energy so that one or more of its bonds breaks. For example, ammonium sulfate, $(NH_4)_2SO_4$, decomposes upon heating to yield ammonia (NH_3) and H_2SO_4.

$$(NH_4)_2SO_4 \xrightarrow{\Delta} 2\,NH_3 + H_2SO_4$$

In *displacement* or substitution reactions a more reactive element replaces a less active element. For example, if an iron nail is placed in an aqueous solution of copper (II) sulfate (i.e., copper with a +2 oxidation state), the color of the solution turns from blue to green because the iron displaces the copper from copper sulfate to yield iron sulfate.

$$Fe + CuSO_4 \xrightarrow{\Delta} FeSO_4 + Cu$$

The surface of the iron nail turns to reddish brown because of the deposition of metallic copper. Predicting whether a specific metal will displace another is accomplished by referring to the *activity series of metals*, a list of metals (found in general chemistry textbooks) that is arranged in order of strength of metal reactivity from highest to lowest.

In *double displacement reactions* two compounds exchange their ions to form two new compounds. For example, silver nitrate reacts with potassium bromide in aqueous solution to yield silver bromide and potassium nitrate.

$$AgNO_3 + KBr \rightarrow AgBr + KNO_3$$

The silver bromide product is insoluble in water and precipitates out of solution.

Acid-base reactions are a type of double displacement reaction. The *Bronsted-Lowry theory* defines acids and bases as proton donors and proton acceptors, respectively. For example, hydrogen chloride reacts with water to yield the hydronium ion H_3O^+ and chloride ion.

$$HCl + H_2O \rightarrow H_3O^+ + Cl^-$$

In this reaction the hydrogen chloride donates a proton (H^+) to H_2O (acting as a base because it accepts the proton) to form the H_3O^+ and chloride ion. In this reaction Cl^- is the *conjugate base* of the acid HCl. Together these two species constitute a *conjugate acid-base pair.* Similarly, H_3O^+ is the conjugate acid of H_2O. They also form a conjugate acid-base pair.

In another way of explaining acid-base reactions, referred to as the *Lewis acid and base theory*, acids and bases are defined in terms of atomic structure and bonding. A *Lewis acid* is a chemical species that accepts an electron pair and has a vacant low energy orbital. Examples of Lewis acids include cations such as Cu^{+2} and Fe^{+2} and molecules such as carbon monoxide (CO) with multiple bonds and atoms with different elctronegativities. A Lewis base is defined as a chemical species that donates an electron pair and possesses lone pair electrons. Examples include NH_3, OH^-, and cyanide ions (CN^-). The product of a Lewis acid-base reaction contains a new covalent bond.

$$A + :B \longrightarrow A\text{–}B$$

In the reaction of HCl with ammonia, HCl is polarized with the slightly positive hydrogen and the chloride slightly negative.

$$\overset{\delta^+}{H}-Cl^{\delta^-} \quad :\underset{H}{\overset{H}{N}}:H \longrightarrow \left[H:\underset{H}{\overset{H}{N}}:H \right]^+ Cl^-$$

Ammonia ($:NH_3$) acting as a Lewis base is attracted to the hydrogen atom. As the lone pair on the nitrogen approaches the HCl, the latter becomes more polarized (i.e., the hydrogen becomes more positive), eventually causing the formation of a coordinate covalent bond between the nitrogen and the hydrogen as the hydrogen-chloride bond breaks.

Redox reactions, also referred to as oxidation-reduction reactions, involve the exchange of electrons between chemical species. In the reaction of metallic zinc with molecular oxygen to form zinc oxide, for example, the zinc atoms are oxidized (i.e., they lose electrons) and oxygen atoms are reduced (i.e., they gain electrons).

$$2\ Zn(s) + O_2(g) \rightarrow 2\ ZnO(s)$$

Although oxidation and reduction occur simultaneously, for convenience they may be considered two separate half-reactions, one involving oxidation and the other reduction. The oxidation half-reaction is

$$2\ Zn \rightarrow 2\ Zn^{+2} + 4e^-$$

where the two zinc atoms lose two electrons each. In the reduction half-reaction

$$O_2 + 4\ e^- \rightarrow 2O^{2-}$$

the two atoms of oxygen gain a total of four electrons. In redox reactions, the species that gives up or "donates" electrons is referred to as the *reducing agent.* The species that accepts the electrons is referred to as the *oxidizing agent.* In the reaction of zinc with molecular oxygen, zinc is the reducing agent and molecular oxygen serves as the oxidizing agent. It should be noted that any type of reaction in which the oxidation state of the reactants changes could also be classified as a redox reaction. For example, in the Haber reaction

$$N_2\ (g) + H_2\ (g) \rightarrow 2\ NH_3\ (g)$$

in which molecular nitrogen reacts with molecular hydrogen to form ammonia, the oxidation number of nitrogen atoms changes from 0 to −3, and hydrogen atoms change from 0 to +1. The displacement reaction described on p. 19 in which iron displaces the copper ion in copper (II) sulfate, is also a redox reaction because the oxidation state of iron changes from 0 to +2 and that of copper changes from +2 to 0.

Combustion reactions are a type of redox reaction in which fuel molecules react with an oxidizing agent to release large amounts of energy, usually in the form of heat and light. Reactions that release energy are described as *exothermic.* The burning of the hydrocarbon methane (natural gas) is a typical combustion reaction.

$$CH_4\ (g) + O_2\ (g) \rightarrow CO_2\ (g) + 2\ H_2O\ (g)$$

The oxidation half-reaction is

$$CH_4 + O_2 \rightarrow CO_2 + 8e^- + 4\ H^+$$

The reduction half reaction is

$$O_2 + 4\ H^+ + 8e^- \rightarrow 2\ H_2O$$

Molecular oxygen is the oxidizing agent in the combustion of methane. The eight electrons removed from methane, the reducing agent, are used in combination with four protons to reduce the oxygen atoms to form two water molecules. It should be noted that cellular respiration, the biochemical mechanism whereby aerobic (oxygen-utilizing) living cells extract energy from fuel molecules such as the sugar glucose, is a slower kind of combustion reaction.

MEASURING CHEMICAL REACTIONS Chemists use the mole concept as a means of determining the amounts of the reactants and products in chemical reactions. A *mole* is defined as the amount of a substance that contains as many particles (e.g., atoms, molecules, or ions) as there are atoms in 12 g of carbon-12. This number, which is 6.022×10^{23} particles, is referred to as *Avogadro's number*. So there are 6.022×10^{23} molecules in 1 mol of H_2O and 6.022×10^{23} sodium ions in 1 mol of NaCl.

The molar mass of substances (mass per mole of particles) is used to determine the amounts of reactants and products in a reaction. For example, in the reaction of methane (CH_4) with O_2 to yield carbon dioxide and water, how much water is produced from the combustion of 8 g of methane? Solving this problem begins with a balanced equation:

$$CH_4 + 2O_2 \rightarrow CO_2 + 2\,H_2O$$

According to this reaction equation, the combustion of every mole of methane yields 2 mol of water. The number of moles of methane is calculated by dividing the mass of methane (8 g) by the molecular mass of methane, which is 16 g (the carbon atom has a mass of 12 g and each of the four hydrogens is 1 g). (Refer to the periodic table for atomic mass numbers.) By this calculation there are 0.5 mol of methane in the reaction. Because the ratio of methane to water is 1 to 2, the 0.5 mol of methane are multiplied by 2 to yield 1 mol of water. Because the molecular mass of water is 18 g, the combustion of 8 g of CH_4 produces 18 g of H_2O.

Moles are also used to express concentrations of substances in solution. *Molarity* is defined as the number of moles in 1 liter (l) of solution. For example, what is the molarity of a solution of 5 g of NaCl in 2 l of water? First, the number of moles of NaCl must be determined by dividing the mass of NaCl (5 g) by the formula weight of NaCl (58.5 g; i.e., 23 g for sodium and 35.5 g for chlorine). By this calculation (i.e., 5/58.5) there are 0.085 mol of NaCl in the 2 l of solution. Molarity of the solution is determined by dividing the number of moles by the number of liters. The molarity of the solution in this problem is 0.085 mol/2 l, which is equal to 0.0425 M (moles per liter, or molar). This number is rounded off to 0. 043 M because of the rule of significant figures. Refer to a general chemistry textbook for a discussion of significant figures.

WORKED PROBLEM 6

The empirical formula of the sugar glucose is $C_6H_{12}O_6$. (a) How many moles are there in 270 g of glucose? (b) Calculate the molarity of a solution of 324 g of glucose dissolved in 2.0 l of water.

SOLUTION

(a) The number of moles of glucose is calculated by dividing the molecular mass of glucose by its mass. First, the molecular mass of glucose must be determined by adding the sums of the masses of each atom in glucose.

Carbon: 12 g × 6 atoms = 72 g
Hydrogen: 1 g × 12 atoms = 12 g
Oxygen: 16 g × 6 atoms = 96 g

WORKED PROBLEM 6 CONT.

Adding these numbers, the molecular mass (m) of glucose is determined to be 180 g. The number of moles of glucose in 270 g is calculated by dividing the mass of glucose (270 g) by the molecular mass (180 g).

$$\text{Moles} = 270\text{ g}/\ 180\text{ g} = 1.5$$

There are 1.5 mol of glucose in 270 g of the substance.

(b) The molarity of the glucose solution is calculated by first determining the number of moles of glucose in 300 g.

$$\text{Moles} = 324/180 = 1.8\text{ mol}$$

The molarity of the glucose solution is then calculated by dividing the number of moles by the number of liters.

$$\text{Mol/l} = 1.8/2 = 0.9\text{ M.}$$

The molarity of the glucose solution is 0.9 M.

ORGANIC CHEMISTRY

Organic chemistry is the investigation of carbon-containing compounds. An entire field is devoted to molecules composed of carbon because of its astonishing versatility. In addition to its capacity to form stable covalent bonds with other carbon atoms to form long chains, branch chains, and rings, carbon also forms stable covalent bonds with a variety of other elements (e.g., hydrogen, oxygen, nitrogen, and sulfur). Carbon can also form carbon-carbon double and triple bonds. As a result of these properties, the possibilities for molecules with different arrangements of carbon and the other elements are virtually limitless. For students embarking on the study of biochemistry, a thorough understanding of the principles of organic chemistry is essential because, as stated previously, with the exceptions of inorganic molecules such as H_2O, O_2, NH_3, and CO_2 and several minerals (e.g., Na^+, Ca^{2+}, and Fe^{2+}), biomolecules are organic molecules. The structural and functional properties of proteins, nucleic acids (DNA and RNA), fats, and sugars can only be appreciated when students know how carbon-based molecules behave. This review will focus on the structures and the chemical properties of the major classes of organic molecules: the *hydrocarbons* (molecules only containing carbon and hydrogen) and *substituted hydrocarbons* (hydrocarbon molecules in which one or more hydrogens has been replaced with another atom or group of atoms).

Hydrocarbons

Because hydrocarbon molecules contain only carbon and hydrogen, they are nonpolar. They dissolve in nonpolar solvents such as hexane and chloroform, but not in water. Such molecules are described as *hydrophobic* ("water-hating"). The hydrocarbons are classified into four groups: (1) saturated hydrocarbons (molecules containing only single bonds), (2) unsaturated hydrocarbons (molecules with one or more carbon-carbon double or triple bonds), (3) cyclic hydrocarbons (molecules containing one or more carbon rings), and (4) aromatic hydrocarbons (molecules that contain one or more aromatic rings, which can be described as cyclic molecules with alternating double and single bonds).

The *saturated hydrocarbons*, referred to as the *alkanes*, are either normal (straight chains) or branched chains. These molecules are "saturated" because they will not react with hydrogen. The straight-chain alkanes belong to a homologous series of compounds that differ in the number of carbon atoms they contain. Their formula is C_nH_{2n+2}. The first six members of this series are methane (CH_4), ethane (C_2H_6), propane (C_3H_8), butane (C_4H_{10}), pentane (C_5H_{12}), and hexane (C_6H_{14}). Note that the prefix in each of these names indicates the number of carbon atoms (e.g., meth- = 1 carbon atom) and the suffix -ane indicates a saturated molecule. Hydrocarbon groups that are derived from alkanes are called *alkyl groups*. For example, a methyl group is a methane molecule with one hydrogen atom removed.

As their name suggests, the branched-chain hydrocarbons are carbon chains with branched structures. If one hydrogen atom is removed from carbon-2 of hexane, for example, and a methyl group is attached, the branched product is 2-methylhexane.

```
     H    H    H    H    H    H                H      H      H    H    H    H
     |    |    |    |    |    |                |      |      |    |    |    |
H — C1 — C2 — C3 — C4 — C5 — C6 — H        H — C ———— C ———— C —— C —— C —— C — H
     |    |    |    |    |    |                |      |      |    |    |    |
     H    H    H    H    H    H                H  H — C — H  H    H    H    H
                                                      |
                                                      H
            Hexane                                2-Methylhexane
```

Note that branched-chain molecules are named by first identifying the longest chain and that the number of the carbon that is bonded to the side-chain group is the lowest one possible.

One of the most remarkable features of the hydrocarbons is the capacity to form *isomers*, molecules with the same type and number of atoms that are arranged differently. For example, there are three molecules, each with its own set of properties, with the molecular formula (C_5H_{12}): pentane, 2-methylbutane, and 2,2-dimethylpropane.

```
                                                              CH3
                                                              |
CH3CH2CH2CH2CH3      CH3—CH—CH2CH3                      CH3—C—CH3
                          |                                   |
                          CH3                                 CH3

    Pentane            2-Methylbutane                  2,2-Dimethylpropane
```

The alkanes are unreactive except for combustion (p. 20) and halogenation reactions. In halogenation reactions alkane molecules react at elevated temperatures or in the presence of light, forming *free radicals* (atoms or molecules with an unpaired electron). For example, when methane reacts with chlorine gas (Cl_2) the molecule breaks down to form two chlorine radicals, which then initiate a chain reaction with methane molecules that yields several chlorinated products: CH_3Cl (methyl chloride), CH_2Cl_2 (methylene chloride), $CHCl_3$ (chloroform), and CCl_4 (carbon tetrachloride).

There are two types of unsaturated hydrocarbons: the *alkenes*, which contain one or more double bonds, and the *alkynes*, which contain one or more triple bonds. The double bond in alkenes is formed from the overlap of two carbon sp^2 orbitals (a σ bond) and the overlap of two unhybridized p orbitals (one from each carbon) to form a π *bond*. The homologous family of alkenes (formula = C_nH_{2n}) are named by taking the names of the alkane with the same number of carbons and substituting the suffix -ene for -ane. Ethene ($H_2C{=}CH_2$), also known by the older name ethylene, is the first member of the series. For alkenes with more than three carbons, the carbons are numbered in reference to the double bond so that the numbers are the lowest possible. For example, $CH_2{=}CH{—}CH_2{—}CH_2{—}CH_2{—}CH_3$ is named 1-hexene, not 5-hexene. Alkenes with four or more carbons have structural isomers in which the position of carbon-carbon double bond is different. For example, 1-butene and 2-butene are referred to as positional isomers. The rigidity of the carbon-carbon double bond prevents rotation, thereby producing another class of isomers: geometric isomers. *Geometric isomers* occur when each of the carbons in the double bond has two different groups on it. For example, there are two geometric isomers of 2-butene: *cis*-2-butane, in which the methyl groups are on the same side of the double bond, and *trans*-2-butene, in which the methyl groups are on opposite sides of the double bond.

$$\underset{Cis\text{-2-Butene}}{(H_3C)(H)C{=}C(CH_3)(H)} \qquad \underset{Trans\text{-2-Butene}}{(H_3C)(H)C{=}C(H)(CH_3)}$$

Note that 1-butene does not form geometric isomers because one of the double-bonded carbons does not have two different groups.

The alkynes such as ethyne (or acetylene)

$$H-C{\equiv}C-H$$

Acetylene

contain triple bonds composed of 1 σ bond and 2 π bonds. The carbon-carbon triple bond is rare in biomolecules and is not discussed further.

The principal reaction of alkenes is the *electrophilic addition reaction* in which an *electrophile* (an electron-deficient species) forms a bond by accepting an electron pair from a *nucleophile* (an electron-rich species).

$$H_2C{-}CH_2 + AB \rightleftharpoons H{-}CH(A){-}CH(B){-}H$$

Electrophiles have positive charges or they may have an incomplete octet. Examples include H^+, $CH_3{}^+$, and polarized neutral molecules such as HCl. Nucleophiles have negative charges (e.g., OH^-), contain atoms with lone pairs (e.g., H_2O and NH_3), or have π bonds.

Hydrogenation and hydration are two addition reactions that occur frequently in living organisms. In a laboratory or industrial hydrogenation reaction a metal catalyst (e.g., nickel or platinum) is required to promote the addition of H_2 to an alkene to yield an alkane.

$$H_2C{=}CH_2 + H_2 \rightleftharpoons H{-}CH_2{-}CH_2{-}H$$

Hydration reactions of alkenes are electrophilic addition reactions that yield alcohols. The reaction (**Figure 8**) requires a small amount of a strong acid catalyst such as sulfuric acid (H_2SO_4) because water is too weak an acid to initiate protonation of the alkene. In the presence of the sulfuric acid, the electron pair in the π cloud polarizes toward the hydronium ion (H_3O^+) and forms a new carbon-hydrogen ion. The newly formed *carbocation* (a molecule containing a positively charged carbon atom) is then attacked by the nucleophilic water molecule to yield an *oxonium ion* (a molecule containing an oxygen cation with three bonds). The alcohol product is formed as the oxonium ion transfers a proton to a water molecule. It is important to note that in hydration reactions of propene and larger alkenes, the most highly substituted carbocation will form (Markovnikov's rule). For example, the principal product of propene hydration is 2-propanol and not 1-propanol.

Carbocation

Oxonium ion

FIGURE 8

Acid Catalyzed Hydration of an Alkene

In the first step protonation of the double bond forms a carbocation. Nucleophilic attack by water gives a protonated alcohol. Deprotonation by a water molecule yields the alcohol. Note that the arrows in this and other reaction mechanisms indicate the movement of electrons.

CYCLIC HYDROCARBONS As their name suggests, the cyclic hydrocarbons are the cyclic counterparts of the alkanes and alkenes. The general formula of the cyclic alkanes, C_nH_{2n}, has two fewer hydrogens than the alkane formula C_nH_{2n+2}. As with the alkanes, the cycloalkanes undergo combustion and halogenation reactions.

Ring strain, observed in cycloalkane rings with three or four carbons, is caused by unfavorable bond angles that are the result of distortion of tetrahedral carbons. As a result, the carbon-carbon bonds in these molecules are weak and reactive. There is minimal or no ring strain in cycloalkane rings with five to seven carbons. Cyclohexane has no ring strain because it is puckered so that its bond angles are near the tetrahedral angles. The most stable puckered conformation is the chair form.

Because of greater ring strain, cyclopropene and cyclobutene are even less stable than cyclopropane and cyclobutane.

AROMATIC HYDROCARBONS Aromatic hydrocarbons are planar (flat) hydrocarbon rings with alternating single and double bonds. Heterocyclic aromatic compounds have two or more different elements in their rings. The simplest aromatic hydrocarbon is benzene. Cytosine, a pyrimidine base found in DNA and RNA, is an example of a heterocyclic aromatic molecule.

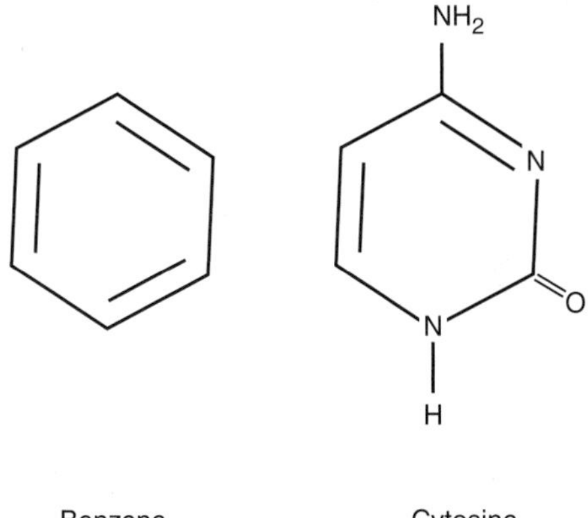

Benzene

Cytosine

Despite the presence of double bonds, benzene and the other aromatic molecules do not undergo reactions typical of the alkenes. In fact, aromatic compounds are remarkably stable. This stability is the result of the unique bonding arrangement of aromatic rings. Each carbon has three sp^2 orbitals that form three σ bonds with two other carbon atoms and with one hydrogen atom. The 2p orbital of each of the six carbon atoms overlaps side to side above and below the plane of the ring to form a continuous circular π bonding system. Instead of two alternate structures of benzene,

Alternate Structures of Benzene

benzene is instead a resonance hybrid, which is indicated by the fact that all of the carbon-carbon bonds in benzene are the same length. (In alkenes carbon-carbon double bonds are shorter than carbon-carbon single bonds.) Each carbon atom is joined to its neighbors by the equivalent of one and a half bonds. Resonance explains why aromatics do not undergo the addition reactions observed with alkenes: the delocalizing of the π electrons around an aromatic ring confers considerable stability to the molecule.

Not all cyclic compounds that contain double bonds are aromatic molecules. According to *Huckel's rule*, to be aromatic a ring molecule must be planar and have $4n + 2$ electrons in the π cloud, where n is a positive integer. For benzene, which has six π electrons, n is equal to 1 [$4(1) + 2 = 6$]. In addition, every atom in an aromatic ring has either a p orbital or an unshared pair of electrons. For example, cytosine is aromatic because the NH adjacent to the carbonyl group donates its lone pair of electrons to the ring's π electron cloud.

Despite their resistance to addition reactions, aromatic compounds are not inert. They can undergo substitution reactions. In *electrophilic aromatic substitution reactions*, an electrophile reacts with an aromatic ring and substitutes for one of the hydrogens. For example, benzene reacts with HNO_3 in the presence of H_2SO_4 to yield nitrobenzene and water (**Figure 9**). In the first step in the reaction a strong electrophile is generated. In this case the nitronium ion ($^+NO_2$) is created when the sulfuric acid protonates the nitric acid on the OH group and the resulting water molecule leaves. In the second step a pair of π electrons in the benzene attack the electrophile, resulting in the formation of a resonance-stabilized carbocation intermediate. In the final step the aromatic ring is regenerated when a water molecule abstracts a proton from the carbon atom bonded to the electrophile.

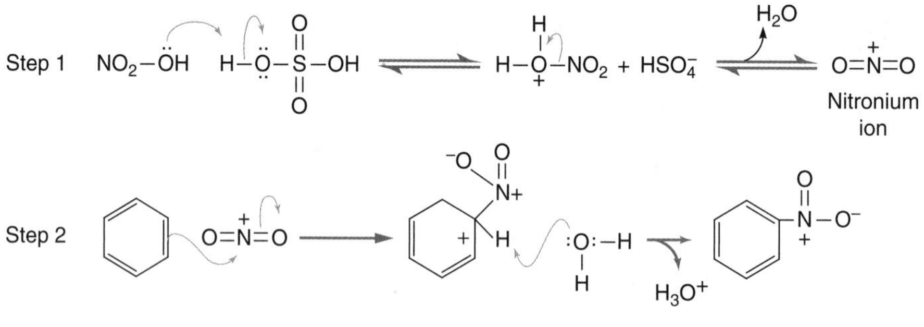

FIGURE 9

Nitration of Benzene

The nitronium ion, a powerful electrophile, is created by the protonation of HNO_3 by H_2SO_4. The product loses a water molecule to leave $0{=}\overset{+}{N}{=}0$. In the second step, the nitronium ion reacts with the nucleophilic benzene. The aromaticity of nitrobenzene is restored with the loss of a protein to a water molecule.

Substituted Hydrocarbons

Substituted hydrocarbons are produced by replacing one or more hydrogens on hydrocarbon molecules with functional groups. A *functional group* is a specific group of atoms within a molecule that is responsible for the molecule's chemical reactivity. Functional groups also separate the substituted hydrocarbons into families. For example, methanol (CH_3—OH), a member of the alcohol family of organic molecules, is the product when the functional group -OH is substituted for a hydrogen atom on methane (CH_4). There are three general classes of functional groups that are important in biomolecules: oxygen-containing, nitrogen-containing, and sulfur-containing molecules. The structural and chemical properties of each class are briefly discussed. Also refer to **Table 1.1** on p. 6 of the textbook for a brief overview of the functional groups. There are six major families of organic molecules that contain oxygen: alcohols, aldehydes, ketones, carboxylic acids, esters, and ethers. Amines and amides possess nitrogen-containing functional groups. The major type of sulfur-containing functional group is the sulfhydryl group, which occurs in thiols.

ALCOHOLS In alcohols the hydroxyl group (-OH) is bonded to an sp^3 hybridized carbon. The presence of the polar -OH group makes alcohol molecules polar, allowing them to form hydrogen bonds with each other and with other polar molecules. A *hydrogen bond* is an attractive force between a hydrogen atom attached to an electronegative atom (e.g., oxygen or nitrogen) of one molecule and an electronegative atom of a different molecule. For alcohols with up to four carbons (methanol, ethanol, propanol, and butanol), the polar OH group allows them to dissolve in water because hydrogen bonds form between the hydrogen of the OH group of the alcohol and the oxygen of a water molecule. Such molecules are described as *hydrophilic* ("water-loving"). Alcohols with five or more carbons are not water-soluble because the hydrophobic properties of hydrocarbon components of these molecules are dominant. Alcohols can be classified by the number of alkyl groups (designated as R groups) attached to the carbon adjacent to the -OH group. Ethanol (CH_3CH_2—OH) is a primary alcohol. In a secondary alcohol (RR′CH—OH) such as 2-propanol, the carbon atom bonded to the OH group is also attached to two alkyl groups. Tertiary alcohols (RR′R″C—OH) such as 2-methyl-2-propanol have three alkyl groups bonded to the carbon bearing the OH group. Alcohols are weak acids (i.e., a strong base can remove the proton from an alcohol's hydroxyl group to form the alkoxide ion $R—O^-$). Tertiary alcohols are less acidic than primary alcohols because the alkyl groups inhibit the solvation of the alkoxide ion. The increased electron density on the oxygen atom in these molecules also decreases proton removal.

Alcohols react with carboxylic acids to form esters. They can also be oxidized to give the carbonyl group-containing aldehydes, ketones, or carboxylic acids. The *carbonyl group* (C=O) in which a carbon is double-bonded to an oxygen atom is a structural feature of the aldehydes, ketones, carboxylic acids, and esters. (Amides, which contain both nitrogen atoms and carbonyl groups, are described on p. 30.) The carbonyl group is polar because of the difference in electronegativity between oxygen and carbon. The slightly positive carbon is electrophilic and therefore able to react with nucleophiles.

ALDEHYDES The functional group of the aldehydes is a carbonyl group bonded to a hydrogen atom [-(C=O)—H]. The simplest aldehyde is formaldehyde (also referred to as methanal) in which the aldehyde group is bonded to a hydrogen atom. In all other aldehydes the aldehyde group is bonded to an alkyl group. The general formula for aldehydes is abbreviated as R-CHO. Acetaldehyde (CH_3CHO) is the oxidation product of ethanol. The reaction of an aldehyde with an alcohol yields a hemiacetal (**Figure 10**).

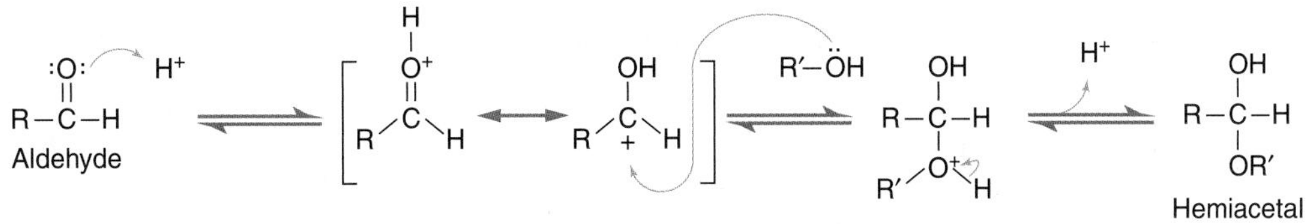

Aldehyde ⇌ [resonance-stabilized protonated carbonyl] ⇌ (R′–ÖH) intermediate ⇌ (H⁺) Hemiacetal

FIGURE 10

Hemiacetal Formation

The reaction begins with the acid catalyst protonating the carbonyl group. The alcohol, acting as a nucleophile, attacks the resonance stabilized carbocation. The hemiacetal product forms with the release of a proton from the positively charged intermediate.

Note that hemiacetals are unstable and their formation is readily reversible. The reaction of the aldehyde group of aldose sugars with an intramolecular OH group to form the more stable cyclic hemiacetals (see p. 243 in the textbook) is an important feature of the chemistry of carbohydrates (Chapter 7).

KETONES Ketones are molecules in which a carbonyl group is flanked by two R groups [R—(C=O)—R′]. The names of members of the ketone family end in -one. For example, dimethyl ketone is usually referred to by its original name, acetone. Ethylmethyl ketone is also referred to as 2-butanone.

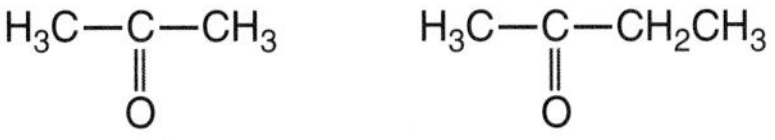

$H_3C—C(=O)—CH_3$ Acetone

$H_3C—C(=O)—CH_2CH_3$ Ethylmethylketone

In ketoses, sugars with a ketone group (most notably fructose), the carbonyl group reacts with an OH group on the sugar molecule to form a cyclic hemiketal.

CARBOXYLIC ACIDS Carboxylic acids (RCOOH) contain the carboxyl group, which is a carbonyl linked to an OH group. These molecules function as weak acids (i.e., they are proton donors) because the carboxylate group (COO^-), the conjugate base of a carboxylic acid, is resonance stabilized. Carboxylic acids react with bases to form carboxylate salts. For example, acetic acid reacts with sodium hydroxide to yield sodium acetate and water.

$$\underset{\text{Acetic Acid}}{CH_3\overset{\overset{O}{\|}}{C}{-}OH} + \underset{\text{Sodium Hydroxide}}{NaOH} \longrightarrow \underset{\text{Sodium Acetate}}{CH_3\overset{\overset{O}{\|}}{C}{-}O^-Na^+} + H_2O$$

The simplest carboxylic acid is formic acid (HCOOH), which is found in ant and bee stings. Carboxylic acids with more than two carbon atoms are often named using the hydrocarbon precursor name followed by the ending -oic acid. For example, the carboxylic acid derived from the four-carbon molecule butane is butanoic acid. In living organisms, the longer-chained carboxylic acids, called *fatty acids*, are important components of biological membranes and the triacylglycerols, a major energy-storage molecule.

ESTERS Found widely in nature, esters [R(C=O)—OR′] are responsible for the aromas of numerous fruits. An ester is the product of a *nucleophilic acyl substitution reaction* in which a carboxylic acid reacts with an alcohol. For example, isobutanol reacts with acetic acid to form isobutyl acetate, an ester found in cherries, raspberries, and strawberries.

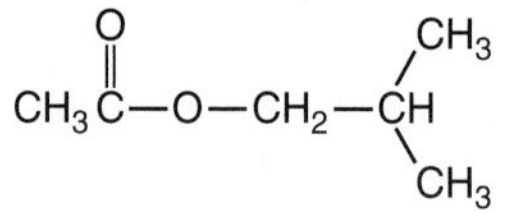

$$CH_3\overset{\overset{O}{\|}}{C}{-}O{-}CH_2{-}CH(CH_3)_2$$

Isobutyl acetate

The formation of the ester methyl acetate from acetic acid and methanol is illustrated in **Figure 11**.

Fats and vegetable oils, also called triacylglycerols (see p. 390 in the textbook), are triesters of the trialcohol molecule glycerol and three fatty acids.

ETHERS Ethers have the general formula R—O—R′. Diethyl ether (CH_3CH_2—O—CH_2CH_3), the best-known ether, was the first anesthetic used in surgery (late nineteenth century) and is still used as a solvent. Ethers are relatively inert chemically, but they do convert over time into explosive peroxides (e.g., diethyl ether hydroperoxide) when exposed to air.

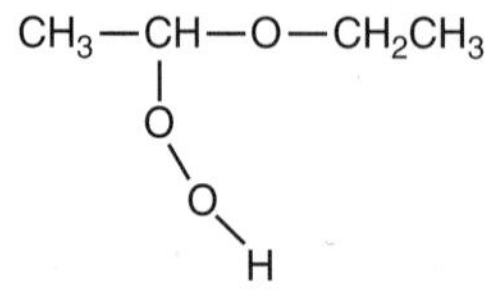

$$CH_3{-}\underset{\underset{O{-}O{-}H}{|}}{CH}{-}O{-}CH_2CH_3$$

Diethylether hydroperoxide

In living organisms the ether linkage occurs in biomolecules such as the carbohydrates.

AMINES Amines are organic molecules that can be considered derivatives of ammonia (NH_3). Primary amines (R—NH_2) are molecules in which only one of

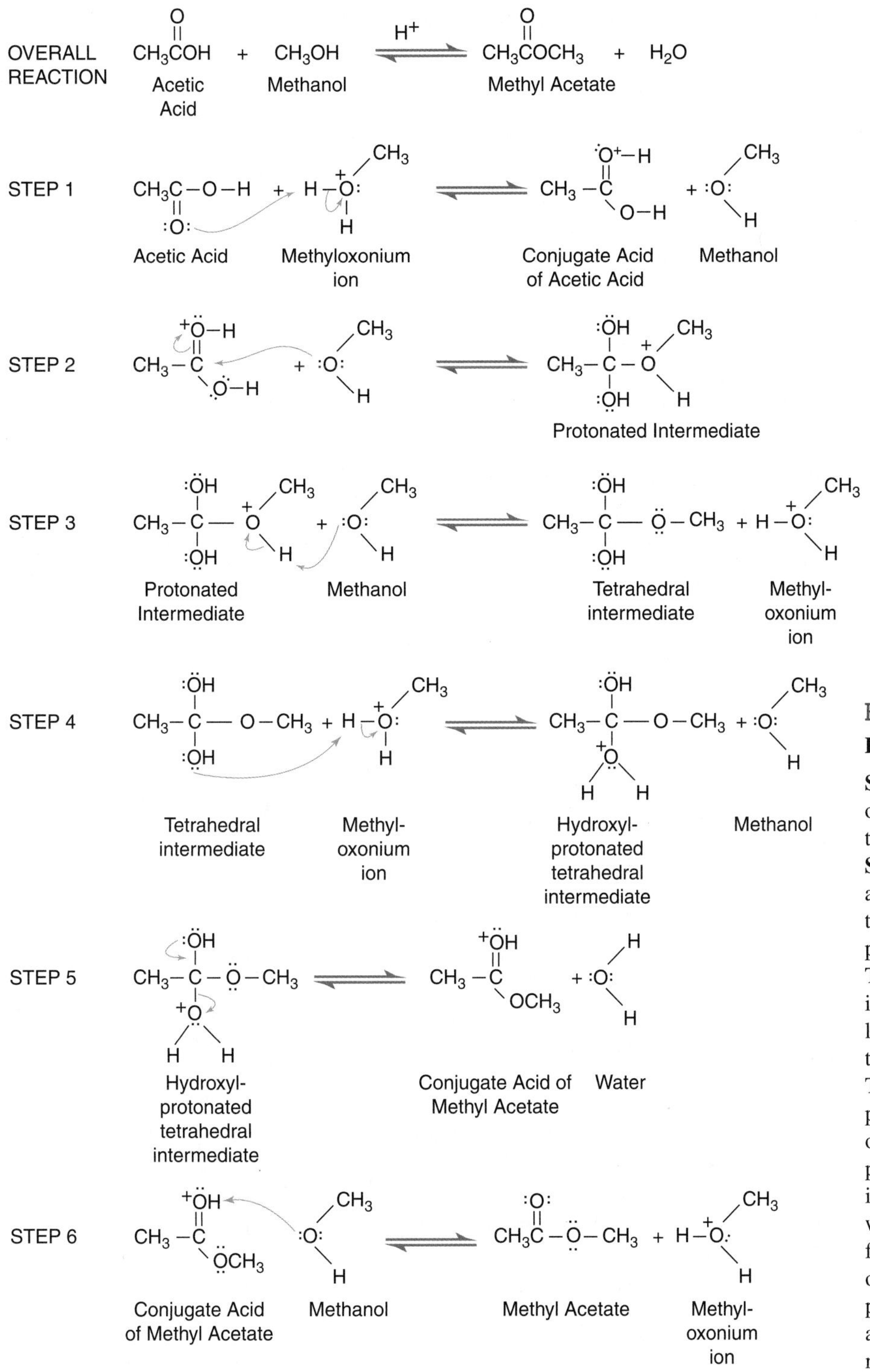

OVERALL REACTION

$$CH_3COOH + CH_3OH \underset{}{\overset{H^+}{\rightleftharpoons}} CH_3COOCH_3 + H_2O$$

Acetic Acid — Methanol — Methyl Acetate

STEP 1

$$CH_3C(=\ddot{O})\text{–}O\text{–}H + H\text{–}\overset{+}{O}(CH_3)\text{–}H \rightleftharpoons CH_3\text{–}C(=\overset{+}{O}\text{–}H)\text{–}O\text{–}H + :\ddot{O}(CH_3)\text{–}H$$

Acetic Acid — Methyloxonium ion — Conjugate Acid of Acetic Acid — Methanol

STEP 2

$$CH_3\text{–}C(=\overset{+}{\ddot{O}}\text{–}H)\text{–}\ddot{O}\text{–}H + :\ddot{O}(CH_3)\text{–}H \rightleftharpoons CH_3\text{–}C(:\ddot{O}H)(:\ddot{O}H)\text{–}\overset{+}{O}(CH_3)\text{–}H$$

Protonated Intermediate

STEP 3

$$CH_3\text{–}C(:\ddot{O}H)(:\ddot{O}H)\text{–}\overset{+}{O}(CH_3)\text{–}H + :\ddot{O}(CH_3)\text{–}H \rightleftharpoons CH_3\text{–}C(:\ddot{O}H)(:\ddot{O}H)\text{–}\ddot{O}\text{–}CH_3 + H\text{–}\overset{+}{O}(CH_3)\text{–}H$$

Protonated Intermediate — Methanol — Tetrahedral intermediate — Methyl-oxonium ion

STEP 4

$$CH_3\text{–}C(:\ddot{O}H)(:\ddot{O}H)\text{–}O\text{–}CH_3 + H\text{–}\overset{+}{O}(CH_3)\text{–}H \rightleftharpoons CH_3\text{–}C(:\ddot{O}H)(\overset{+}{O}H_2)\text{–}O\text{–}CH_3 + :\ddot{O}(CH_3)\text{–}H$$

Tetrahedral intermediate — Methyl-oxonium ion — Hydroxyl-protonated tetrahedral intermediate — Methanol

STEP 5

$$CH_3\text{–}C(:\ddot{O}H)(\overset{+}{O}H_2)\text{–}\ddot{O}\text{–}CH_3 \rightleftharpoons CH_3\text{–}C(=\overset{+}{\ddot{O}}H)\text{–}OCH_3 + :\ddot{O}H_2$$

Hydroxyl-protonated tetrahedral intermediate — Conjugate Acid of Methyl Acetate — Water

STEP 6

$$CH_3\text{–}C(=\overset{+}{\ddot{O}}H)\text{–}\ddot{O}CH_3 + :\ddot{O}(CH_3)\text{–}H \rightleftharpoons CH_3C(=\ddot{O})\text{–}\ddot{O}\text{–}CH_3 + H\text{–}\overset{+}{O}(CH_3)\text{–}H$$

Conjugate Acid of Methyl Acetate — Methanol — Methyl Acetate — Methyl-oxonium ion

FIGURE 11

Formation of Methyl Acetate

Step 1 Acetic acid is protonated on its carbonyl oxygen to form the conjugate acid of acetic acid. **Step 2** A molecule of methanol, acting as a nucleophile, attacks the electrophilic carbon of the protonated acetic acid. **Step 3** The oxonium ion (the protonated intermediate formed in step 2) loses a proton to form a neutral tetrahedral intermediate. **Step 4** The tetrahedral intermediate is protonated on one of its hydroxyl oxygens. **Step 5** The hydroxyl-protonated intermediate formed in step 4 loses a molecule of water to yield the protonated form of the ester. **Step 6** The loss of a proton from the protonated product of step 5 (the conjugate acid of methyl acetate) yields methyl acetate.

the hydrogen atoms of ammonia has been replaced by an organic group (e.g., alkyl or aromatic groups). Methylamine (CH_3NH_2) is an example of a primary amine. In secondary amines, such as dimethylamine (CH_3—NH—CH_3), two hydrogens have been replaced by organic groups. Tertiary amines such as triethylamine [$(CH_3CH_2)_3N$] are molecules in which all three hydrogens have been replaced with organic groups. Amines with small organic groups are water soluble, although the solubility of tertiary amines is limited because they do not have any hydrogen atoms bonded to the electronegative nitrogen atom. Like ammonia, amines are weak bases because of the lone pair of electrons on the nitrogen atom, which can accept a proton. Protonation of the nitrogen converts the amine into a cation.

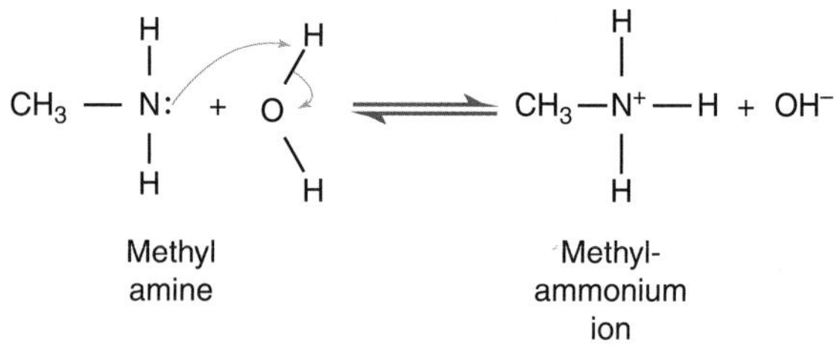

$$CH_3-NH_2 + H_2O \rightleftharpoons CH_3-N^+H_3 + OH^-$$

Methyl amine   Methylammonium ion

There are an enormous number of biomolecules that contain amine nitrogens. Examples include the amino acids (components of proteins), the nitrogenous bases of the nucleic acids, and the alkaloids (complex molecules produced by plants such as caffeine, morphine, and nicotine that have significant physiological effects on humans).

AMIDES Amides are amine derivatives of carboxylic acids with the general formula [R(C=O)—NR_2] where the R groups bonded to the nitrogen can be hydrogens or hydrocarbon groups. In contrast to the amines, amides are neutral molecules. The C—N bond is a resonance hybrid because of the attraction of the carbonyl group for the nitrogen's lone pair. As a result, amides are not weak bases (i.e., they have little capacity to accept protons).

$$R-C(=O)-NH_2 \longleftrightarrow R-C(-O^-)=N^+H_2$$

Resonance hybridization explains why the amide functional group is planar with the nitrogen's sp^2 orbital forming a π bond with the carbonyl carbon atom. In living organisms, the amide functional group is the linkage, referred to as the peptide bond, that connects amino acids in polypeptides. Amides are classified according to how many carbon atoms are bonded to the nitrogen atom. Molecules with the molecular formula R(C=O)—NH_2 are primary amides. The substitution of one of the nitrogen's hydrogens with an alkyl group yields a secondary amide [R(C=O)—NHR′]. Amides with two alkyl groups attached to the nitrogen are tertiary amides [R(C=O)—NR_2].

THIOLS A thiol is a molecule in which an sp^3 carbon is bonded to a sulfhydryl group (-SH). Although thiols are considered the sulfur analogs of alcohols, the low polarity of the SH bonds limits their capacity to form hydrogen bonds. As a result, thiols are not as soluble in water as their alcohol counterparts. Thiols are stronger acids than their alcohol equivalents, however, in part because of the weakness of the S—H bond. For the same reason, thiolates (R—S^-), the conjugate bases of the thiols, are weaker bases than the alkoxides (R—O^-). Thiolates are excellent nucleophiles because sulfur's 3p electrons are easily polarized. The sulfhydryl group of thiols is easily oxidized to form disulfides (RS—SR). For example, two molecules of the amino acid cysteine react to form cystine, which contains a disulfide bond.

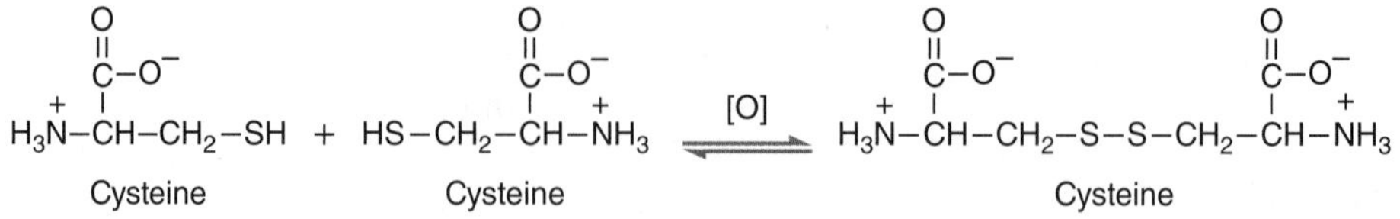

$$H_3N^+-CH(COO^-)-CH_2-SH + HS-CH_2-CH(COO^-)-N^+H_3 \overset{[O]}{\rightleftharpoons} H_3N^+-CH(COO^-)-CH_2-S-S-CH_2-CH(COO^-)-N^+H_3$$

Cysteine   Cysteine   Cysteine

WORKED PROBLEM 7

Cephalosporin c is one of a class of antibiotics, known as the cephalosporins, that kill bacteria by preventing the cross-linking of peptidoglycan, a key structural element in the cell wall. Its structure is shown below. Identify the indicated functional groups.

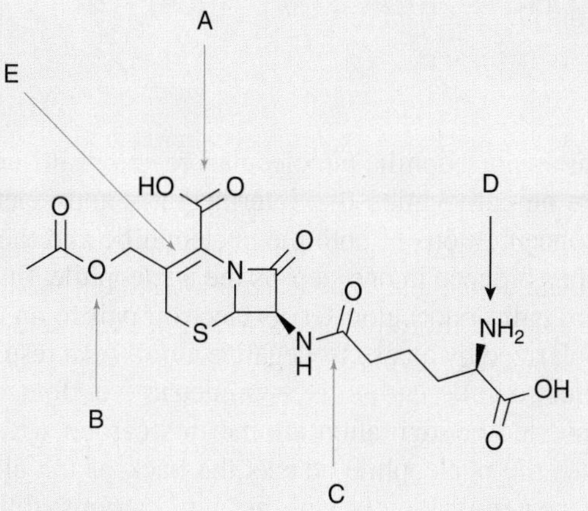

SOLUTION

A = carboxylic acid B = ester C = amide D = amine E = alkene

This reaction is especially important in numerous proteins that contain cysteine. The disulfide bond that forms when cysteine residues are linked is an important stabilizing feature of protein structure.

Organic Reactions: Substitutions and Eliminations

There are a very large number of organic reaction types. Electrophilic addition reactions (p. P-24), electrophilic aromatic substitution reactions (p. P-26), and nucleophilic acyl substitution (p. P-28) have already been described. There are two additional reaction classes, aliphatic substitution and elimination reactions, which students should be familiar with. (The term *aliphatic* refers to non-aromatic hydrocarbon compounds.)

SUBSTITUTION REACTIONS Aliphatic substitution reactions, which involve tetrahedral carbons, are designated as either S_N1 or S_N2. *S_N1* (*s*ubstitution *n*ucleophilic *uni*molecular) reactions, which often involve secondary or tertiary alkyl halides (molecules in which a halogen atom such as chloride has been substituted for a hydrogen in an alkane) or alcohols, proceed in two steps. In the first step of an S_N1 reaction, a planar carbocation forms as the leaving group (a stable ion or a neutral molecule) is displaced. S_N1 reactions are considered unimolecular because the rate of the reaction depends only on the rate of carbocation formation. In the second step the nucleophile attacks the electrophilic carbocation to form the product. Good reactants for S_N1 reactions are molecules with tertiary carbons that can form a stable carbocation when the leaving group is released. Examples of nucleophiles in S_N1 reactions include alcohols and water. Because carbocations are sp^2 hybridized and have an empty p orbital, the nucleophile can attack on either side of the ion. As a result, two isomeric products may be produced. A typical example of an S_N1 reaction is the reaction of *t*-butyl bromide with methanol (**Figure 12**).

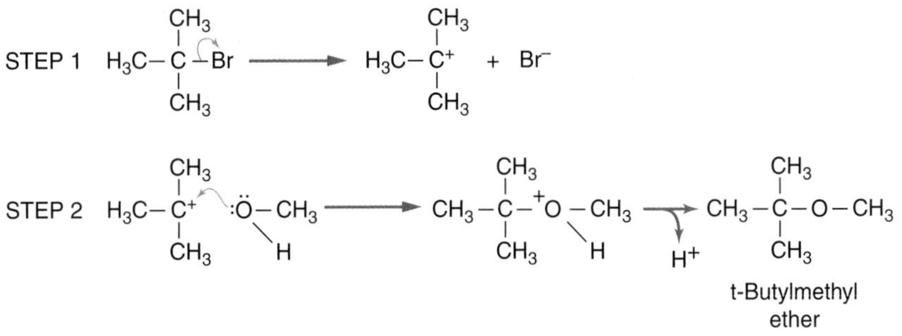

FIGURE 12

Example of S_N1 Reaction

In the first and slowest step the alkyl halide *t*-butyl bromide forms the *t*-butyl carbocation as the leaving group bromide is released. In step 2, the nucleophilic oxygen of methanol attacks the carbocation. The product methyl *t*-butyl ether is formed as a proton is released from the oxonium ion into the solvent.

S_N2 (*s*ubstitution *n*ucleophilic *bi*molecular) reactions differ from S_N1 reactions in that there are no carbocation-like intermediates and reaction rates are determined by the concentrations of both the nucleophilic and the electrophilic reactants. S_N2 reactions proceed in one step: as the nucleophile, functioning as a Lewis base (an electron pair donor), donates its electron pair to an electrophilic carbon that has been polarized by an electronegative atom. As a result, the leaving group leaves. If the electrophilic carbon is asymmetric (i.e., four different groups are attached), an inverted configuration around this carbon will be observed in the product. Because the nucleophile attacks the back of the electrophilic reactant, S_N2 reactions occur most rapidly with primary carbons, followed by secondary carbons. Molecules with tertiary carbons do not undergo S_N2 reactions because of *steric hindrance* (the reactive site on a molecule is blocked by adjacent groups).

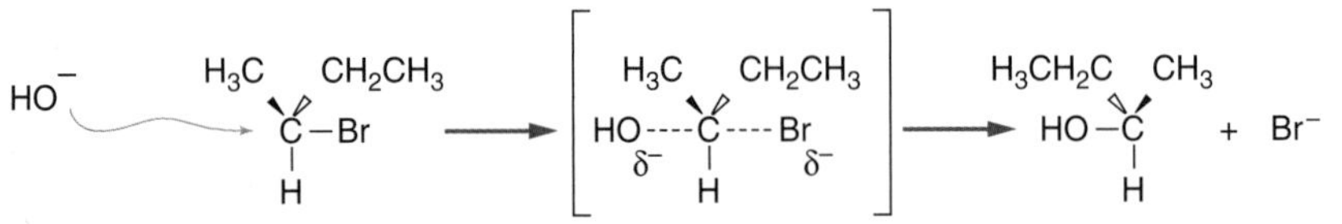

A class of enzymes referred to as the *S*-adenosylmethionine-dependent methyltransferases catalyzes some of the best known examples of S_N2 reactions in biochemistry. *S*-Adenosylmethionine (SAM) is widely used in methylation reactions as a methyl group donor. SAM's methyl group is readily donated because it is attached to an electron-withdrawing sulfur atom. The inactivation of the neurotransmitter epinephrine (adrenalin) catalyzed by catechol-*O*-methyltransferase (COMT) is illustrated in **Figure 13**.

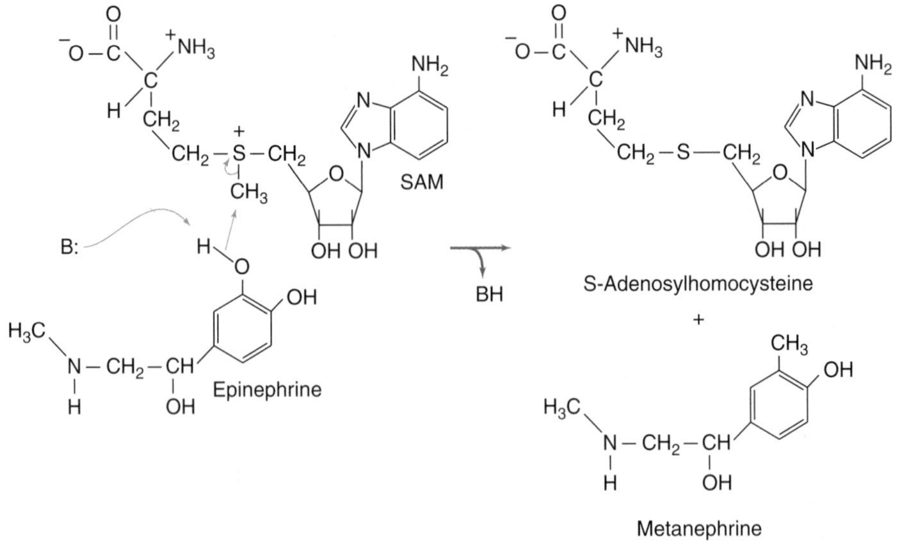

FIGURE 13

COMT-Catalyzed Methylation of Epinephrine: An S_N1 Mechanism

When epinephrine enters the active site of the enzyme, a basic amino acid R group deprotonates one of the hydroxyl groups of epinephrine, forming a nucleophilic alkoxide group. The alkoxide then attacks the methyl carbon atom of SAM. This carbon is electrophilic because it is bonded to an electron-withdrawing positively charged sulfur atom. As the alkoxide attacks, the sulfur-carbon bond begins to break to form the more stable sulfide leaving group. The products of the reactions are S-adenosylhomocysteine and epinephrine's inactive methylated derivative metanephrine.

ELIMINATION REACTIONS As their name suggests, elimination reactions involve the loss of two atoms or groups from a molecule. This loss is usually accompanied by the formation of a π bond. There are three types of elimination reactions: E1, E2, and E1cb.

In E1 (*e*limination *uni*molecular) reactions, the first and slowest step is carbocation formation. For this reason E1 reactions are similar to S_N1 reactions because the reaction rates of both depend on one molecule, the precursor of the carbocation. In the second and faster step a weak base (often a solvent molecule) abstracts a proton from a carbon atom next to the carbocation to yield a carbon-carbon double bond.

$CH_3-C(X)(CH_3)-C(H)(CH_3)-CH_3 \xrightarrow{-X^-} [CH_3-{}^+C(CH_3)-C(H)(CH_3)-CH_3 \;\; {-:B}] \xrightarrow{-BH} CH_3-C(CH_3)=C(CH_3)-CH_3$

Note that the carbon bonded to "X" in the reactant molecule is sp^3 hybridized, whereas the same carbon in the product is sp^2 hybridized. Molecules with tertiary carbons groups are good substrates for E1 reactions because reaction rates depend on the stability of the carbocation.

E2 (*e*limination *bi*molecular) reactions involve the simultaneous removal of a β-proton by a strong base and release of the leaving group. (A β-proton is bonded to the carbon adjacent to the carbon bearing the leaving group.) Molecules with tertiary carbons typically undergo E2 reactions.

$B: + H_3C-C(H)(CH_3)-C(H)(X)-H \longrightarrow H_2C=CH_2 + B-H + :X^-$

For an E2 reaction mechanism to be possible, the β-hydrogen and the leaving group must be anticoplanar (i.e., in a geometric arrangement in which they are 180° from each other on neighboring carbons).

Although all three forms of elimination reactions occur in living organisms, the E1cB is more commonly observed. E1cb (*e*limination *uni*molecular *c*onjugate *b*ase) reactions involve the formation of a *carbanion* (an organic ion bearing a negatively charged carbon). Carbanions are nucleophiles that are stabilized by adjacent electronegative atoms and resonance effects. The E1cb mechanism involves the removal of a proton from the reactant to form the carbanion, followed by the slower loss of the leaving group.

$:B + R-C(H)(R)-C(R)(X)-R \xrightarrow{-BH} [R-\ddot{C}^-(R)-C(R)(X)-R] \xrightarrow{-:X^-} R_2C=CR_2$

carbanion

The second phase, the loss of the leaving group, is the rate-limiting step in E1cb reactions. The presence of an electron-withdrawing atom or group (X in this illustration) makes the C—H acidic. Note that the carbanion form of the reactant is the conjugate base referred to in the term E1cb. The conversion of glycerate-2-phosphate, an intermediate in glucose degradation, to phosphoenolpyruvate provides an example of an E1cb mechanism (**Figure 14**). This reaction is catalyzed by the enzyme enolase.

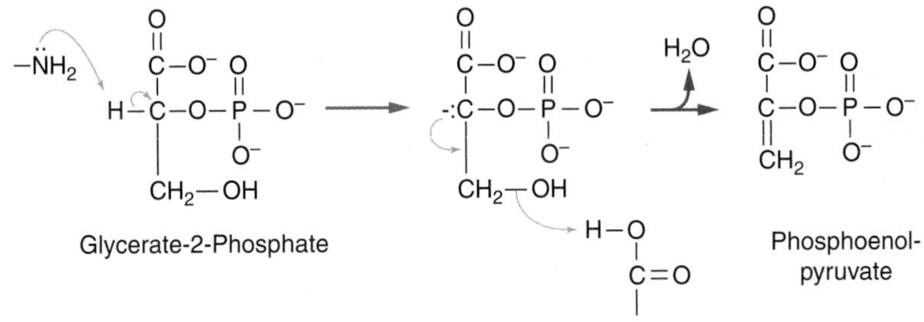

FIGURE 14

Enolase-Catalyzed Dehydration of Glycerate-2-Phosphate: An E1cb Mechanism

Within the active site of the enzyme enolase, the amino nitrogen of a precisely oriented side chain of an amino acid, acting as a base, abstracts the acidic hydrogen at carbon-2 to produce a carbanion. The carbanion electrons displace the hydroxyl (-OH) group from carbon-3, forming a carbon-carbon π bond. The actual leaving group is H_2O, the hydroxyl group having been protonated by a nearby carboxyl (-COOH) group.

CHAPTER

1 Biochemistry: An Introduction

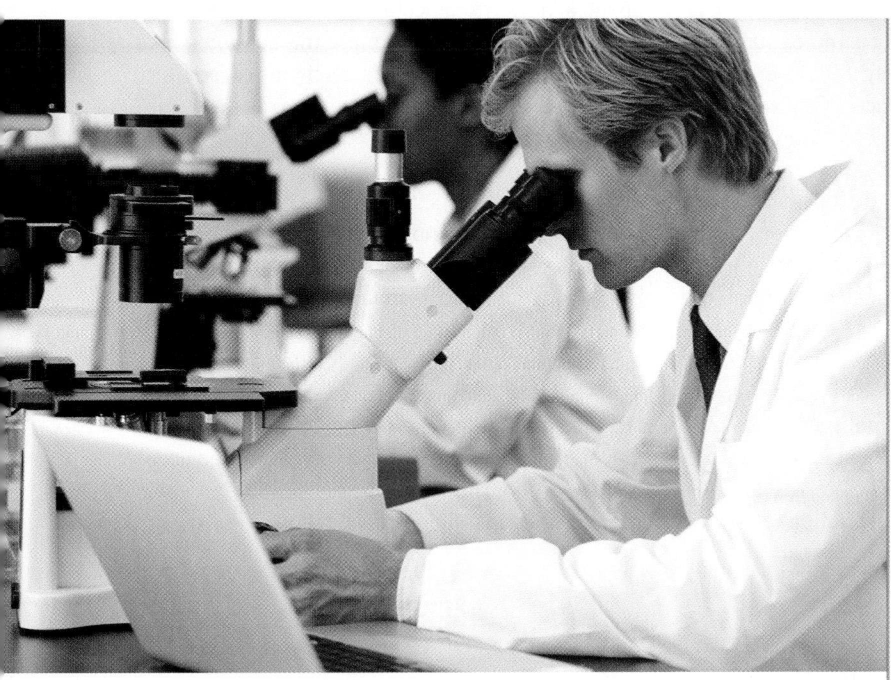

Life Scientists at Work
A thorough knowledge of biochemistry is a core requirement for life scientists.

OUTLINE

Why Study Biochemistry?

Why study biochemistry? For students embarking on careers in the life sciences, the answer should be obvious: biochemistry, the scientific discipline concerned with chemical processes within living organisms, is the bedrock upon which all of the modern life sciences are built. During the past two decades the influence of biochemistry and the allied field of molecular biology has increased exponentially. Life sciences as diverse as agronomy (the science of soil management and crop production), forensics, marine biology, plant biology, and ecology are now being explored with powerful biotechnological tools. As a result, there is now a vast array of career opportunities in federal or state government agencies and industry (e.g., pharmaceutical, biotechnology, and agribusiness companies) for recent graduates with life science degrees. Examples of such fields include biomedical and clinical research, forensic analysis, plant or animal genetics, environmental protection, and wildlife biology.

Economic conditions often dictate life science career choices. (The Occupational Outlook Handout on the U.S. Bureau of Labor Statistics website offers an unbiased assessment of future employment prospects.) No matter the economic conditions when students graduate, employment opportunities are always better for those who have undergraduate research experience. Developing a network of connections beginning with professors and expanding into the student's field or interests (e.g., by attending science career fairs or professional society conferences) also increases employment opportunities. Furthermore, writing, data analysis, problem solving, and communication are skills that employers always value highly. For students not interested in research careers, there are opportunities in science journalism, education, and software engineering. Other examples of alternate careers where a life science degree will be an asset include public policy (e.g., public health risk assessment and health product regulation), law (e.g., lawyers for pharmaceutical and biotech companies and environmental organizations), and marketing and sales (e.g., drugs and medical devices).

Overview

FROM MODEST BEGINNINGS IN THE LATE NINETEENTH CENTURY, THE SCIENCE OF BIOCHEMISTRY HAS PROVIDED INCREASINGLY MORE sophisticated intellectual and laboratory tools for the investigation of living processes. Today, in the early years of the twenty-first century, we find ourselves in the midst of a previously unimagined biotechnological revolution. Life sciences as diverse as medicine, agriculture, and forensics have generated immense amounts of information. The capacity to understand and appreciate the significance of this phenomenon begins with a thorough knowledge of biochemical principles. This chapter provides an overview of these principles. The chapters that follow focus on the structure and functions of the most important biomolecules and the major biochemical processes that sustain the living state.

This textbook is designed to provide an introduction to the basic principles of biochemistry. The opening chapter provides an overview of the major components of living organisms and the processes that sustain the living state. After a brief description of the nature of the living state, an introduction to the structures and functions of the major biomolecules is provided. This material is followed by an overview of the most important biochemical processes.

The chapter concludes with a brief discussion of the concepts of modern experimental biochemistry and an introduction to *systems biology*, an investigative strategy used to improve our understanding of living organisms as integrated systems rather than collections of isolated components and chemical reactions.

1.1 WHAT IS LIFE?

What is life? Despite the work of life scientists over several centuries, a definitive answer to this deceptively simple question has been elusive. Much of the difficulty in delineating the precise nature of living organisms lies in the overwhelming diversity of the living world and the apparent overlap in several properties of living and nonliving matter. Consequently, life has been viewed as an intangible property and is usually described in operational terms, such as movement, reproduction, adaptation, and responsiveness to external stimuli. The work of life scientists, made possible by the experimental approaches of biochemistry, has revealed that all organisms obey the same chemical and physical laws that rule the universe.

1. **Life is complex and dynamic**. All organisms are composed of the same set of chemical elements, primarily carbon, nitrogen, oxygen, hydrogen, sulfur, and phosphorus. **Biomolecules**, the molecules synthesized by living organisms, are organic (carbon based). Living processes, such as growth and development, involve thousands of chemical reactions in which vast quantities and varieties of vibrating and rotating molecules interact, collide, and rearrange into new molecules.
2. **Life is organized and self-sustaining**. Living organisms are hierarchically organized systems: they consist of patterns of organization from smallest (atom) to largest (organism) (**Figure 1.1**). In biological systems, the functional capacities of each level of organization are derived from the structural and chemical properties of the level below it. Biomolecules are composed of atoms, which in turn are formed from subatomic particles. Certain biomolecules become linked to form polymers called **macromolecules**. Examples include nucleic acids, proteins, and polysaccharides, which are formed from nucleotides, amino acids, and sugars, respectively. Cells are composed of a diversity of biomolecules and macromolecules that form into more complex supermolecular structures. At the molecular level, there are hundreds of biochemical reactions that together sustain the living state. Catalyzed by biomolecular catalysts called **enzymes**, these reactions are organized into pathways. (A *biochemical pathway* is a series of reactions in which a specific molecule is converted into a terminal product.) The sum total of all the reactions in a living organism is referred to as **metabolism**. The capacity of living organisms to regulate metabolic processes despite variability in their internal and external environments is called **homeostasis**. In multicellular organisms there are other levels of organization that include tissues, organs, and organ systems.
3. **Life is cellular**. Cells, the basic units of living organisms, differ widely in structure and function, but each is surrounded by a membrane that controls the transport of substances into and out of the cell. The membrane also mediates the response of the cell to the extracellular environment. If a cell is divided into its component parts, it will cease to function in a life-sustaining way. Cells arise only from the division of existing cells.
4. **Life is information-based**. Organization requires information. Living organisms can be considered information-processing systems because maintenance of their structural integrity and metabolic processes involves

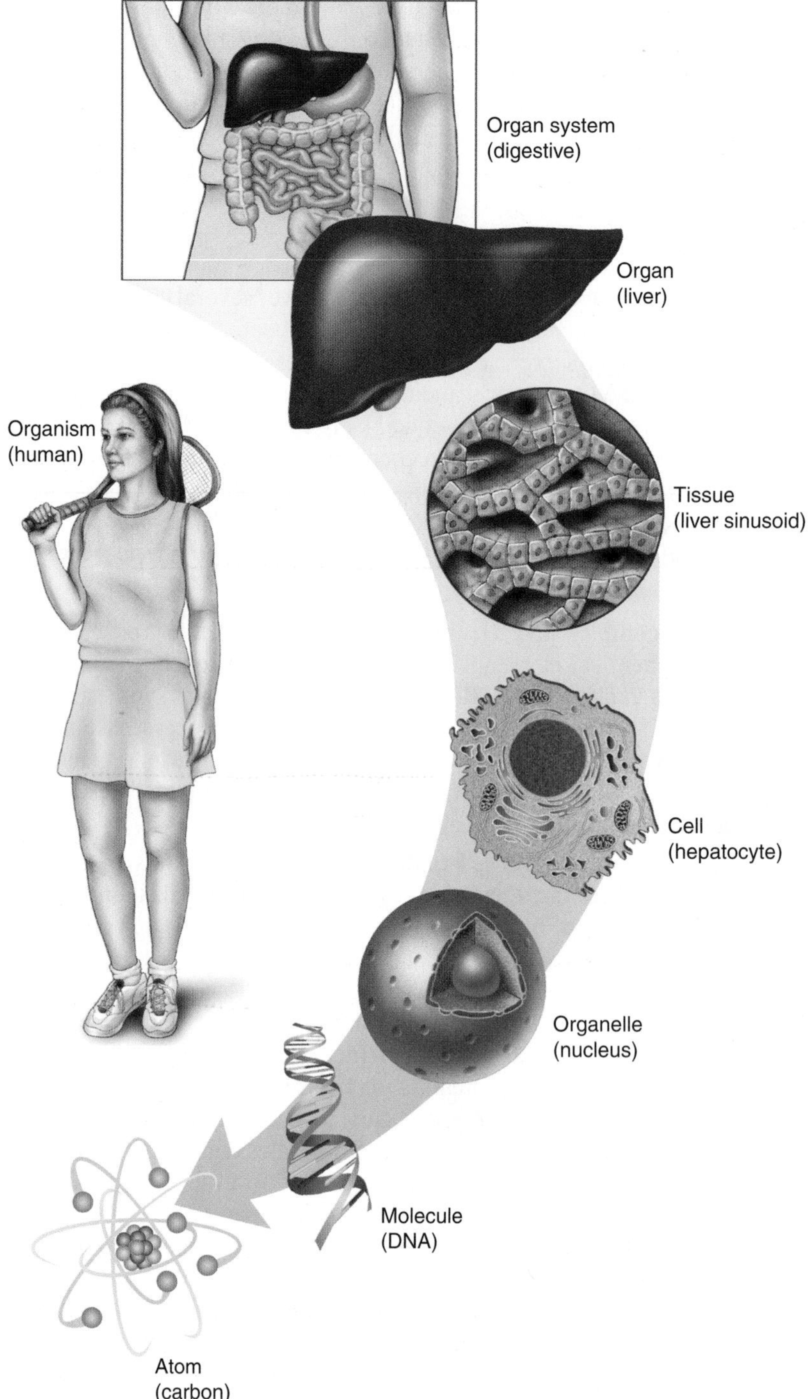

FIGURE 1.1
Hierarchical Organization of a Multicellular Organism: The Human Being

Multicellular organisms have several levels of organization: organ systems, organs, tissues, cells, organelles, molecules, and atoms. The digestive system and one of its component organs (the liver) are shown. The liver is a multifunctional organ that has several digestive functions. For example, it produces bile, which facilitates fat digestion, and it processes and distributes the food molecules absorbed in the small intestine to other parts of the body. DNA, one type of molecule found in cells, contains the genetic information that controls cell function.

interactions among a vast array of molecules within and between cells. Biological information is expressed in the form of coded messages that are inherent in the unique three-dimensional structure of biomolecules. Genetic information, which is stored in **genes**, the linear sequences of nucleotides in deoxyribonucleic acid (DNA), in turn specifies the linear sequence of amino acids in proteins and how and when those proteins are synthesized. Proteins perform their function by interacting with other molecules. The unique three-dimensional structure of each type of protein allows it to bind to, and interact with, a specific type of molecule that has a precise complementary shape. Information is transferred during the binding process. For example, the binding of the protein insulin to insulin receptor molecules on the surface of certain cells is a signal that initiates the uptake of the nutrient molecule glucose.

5. **Life adapts and evolves.** All life on earth has a common origin, with new forms arising from older forms. When an individual organism in a population reproduces itself, stress-induced DNA modifications and errors that occur when DNA molecules are copied can result in **mutations** or sequence changes. Most mutations are silent; that is, they either are repaired or have no effect on the functioning of the organism. Some, however, are harmful, serving to limit the reproductive success of the offspring. On rare occasions mutations may contribute to an increased ability of the organism to survive, to adapt to new circumstances, and to reproduce. A principal driving force in this process is the capacity to exploit energy sources. Individuals possessing traits that allow them to better exploit a specific energy source within their habitat may have a competitive advantage when resources are limited. Over many generations, the interplay of environmental change and genetic variation can lead to the accumulation of favorable traits and eventually to increasingly different forms of life.

KEY CONCEPTS

- All living organisms obey the chemical and physical laws.
- Life is complex, dynamic, organized, and self-sustaining.
- Life is cellular and information-based.
- Life adapts and evolves.

1.2 BIOMOLECULES

Living organisms are composed of thousands of different kinds of inorganic and organic molecules. Water, an inorganic molecule, may constitute 50 to 95% of a cell's content by weight, and ions such as sodium (Na^+), potassium (K^+), magnesium (Mg^{2+}), and calcium (Ca^{2+}) may account for another 1%. Almost all the other kinds of molecules in living organisms are organic. Organic molecules are principally composed of six elements: carbon, hydrogen, oxygen, nitrogen, phosphorus, and sulfur, and they contain trace amounts of certain metallic and other nonmetallic elements. The atoms of each of the most common elements found in living organisms can readily form stable covalent bonds, the kind that allow the formation of such important molecules as proteins.

The remarkable structural complexity and diversity of organic molecules are made possible by the capacity of carbon atoms to form four strong, single covalent bonds either to other carbon atoms or to atoms of other elements. Organic molecules with many carbon atoms can form complicated shapes such as long, straight structures or branched chains and rings.

Functional Groups of Organic Biomolecules

Most biomolecules can be considered derived from the simplest type of organic molecules, called the **hydrocarbons**. Hydrocarbons (**Figure 1.2**) are carbon- and hydrogen-containing molecules that are **hydrophobic**, or insoluble in water. All other organic molecules are formed by attaching other atoms or groups of atoms to the carbon backbone of the hydrocarbon. The chemical properties of these derivative molecules are determined by the specific arrangement of atoms called **functional groups** (**Table 1.1**). For example, alcohols result when hydrogen atoms are replaced by hydroxyl groups (—OH). Thus methane (CH_4), a component of natural gas, can be converted into methanol (CH_3OH), a toxic liquid that is used as a solvent in many industrial processes.

Most biomolecules contain more than one functional group. For example, many simple sugar molecules have several hydroxyl groups and an aldehyde group. Amino acids, the building-block molecules of proteins, have both an amino group and a carboxyl group. The distinct chemical properties of each functional group contribute to the behavior of any molecule that contains it.

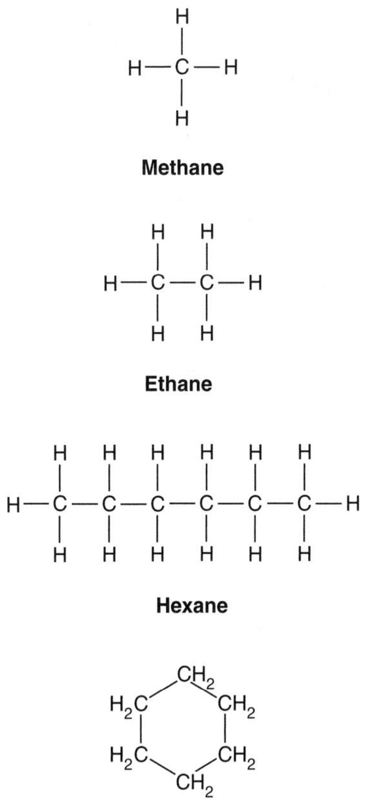

Methane

Ethane

Hexane

Cyclohexane

FIGURE 1.2
Structural Formulas of Several Hydrocarbons

TABLE 1.1 Important Functional Groups in Biomolecules

Family Name	Group Structure	Group Name	Significance
Alcohol	R—OH	Hydroxyl	Polar (and therefore water-soluble), forms hydrogen bonds
Aldehyde	R—C(=O)—H	Carbonyl	Polar, found in some sugars
Ketone	R—C(=O)—R′	Carbonyl	Polar, found in some sugars
Acids	R—C(=O)—OH	Carboxyl	Weakly acidic, bears a negative charge when it donates a proton
Amine	R—NH_2	Amino	Weakly basic, bears a positive charge when it accepts a proton
Amide	R—C(=O)—NH_2	Amido	Polar but does not bear a charge
Thiol	R—SH	Thiol	Easily oxidized; can form —S—S— (disulfide) bonds readily
Ester	R—C(=O)—O—R′	Ester	Found in certain lipid molecules
Alkene	RCH=CHR′	Double bond	Important structural component of many biomolecules (e.g., found in lipid molecules)

Major Classes of Small Biomolecules

Many of the organic compounds found in cells are relatively small, with molecular masses of less than 1000 daltons (Da). (One dalton, 1 atomic mass unit, is equal to 1/12 of the mass of one atom of ^{12}C.) Cells contain four families of small molecules: amino acids, sugars, fatty acids, and nucleotides (**Table 1.2**). Members of each group serve several functions. First, they are used in the synthesis of larger molecules, many of which are polymers. For example, proteins, certain carbohydrates, and nucleic acids are polymers composed of amino acids, sugars, and nucleotides, respectively. Fatty acids are components of lipid (water-insoluble) molecules of several types.

Second, some molecules have special biological functions. For example, the nucleotide adenosine triphosphate (ATP) serves as a cellular reservoir of

TABLE 1.2 Major Classes of Biomolecules

Small Molecule	Polymer	General Functions
Amino acids	Proteins	Catalysts and structural elements
Sugars	Carbohydrates	Energy sources and structural elements
Fatty acids	N.A.	Energy sources and structural elements of complex lipid molecules
Nucleotides	DNA	Genetic information
	RNA	Protein synthesis

chemical energy. Finally, many small organic molecules are involved in complex reaction pathways. Examples of each class of molecule are described next.

AMINO ACIDS AND PROTEINS There are hundreds of naturally occurring **amino acids**, each of which contains an amino group and a carboxyl group. Amino acids are classified α, β, or γ according to the location of the amino group in reference to the carboxyl group. In α-amino acids, the most common type, the amino group is attached to the carbon atom (the α-carbon) immediately adjacent to the carboxyl group (**Figure 1.3**). In β- and γ-amino acids, the amino group is attached to the second and third carbon, respectively, from the carboxyl group. Also attached to the α-carbon is another group, referred to as the side chain or R group. The chemical properties of each amino acid, once incorporated into protein, are determined largely by the properties of its side chain. For example, some side chains are hydrophobic (i.e., low solubility in water), whereas others are **hydrophilic** (i.e., dissolve easily in water). Several examples of α-amino acids are illustrated in **Figure 1.4**.

Twenty standard α-amino acids occur in proteins. Some standard amino acids have unique functions in living organisms. For example, glycine and glutamic acid function in animals as **neurotransmitters**, signal molecules released by nerve cells. Proteins also contain nonstandard amino acids that are modified versions of the standard amino acids. The structure and function of protein molecules are often altered by conversion of certain amino acid residues to derivatives via phosphorylation, hydroxylation, and other chemical modifications. (The term "residue" refers to a small biomolecule that is incorporated in a macromolecule, e.g., amino acid residues in a protein.) For example, many of the residues of proline are hydroxylated in collagen, the connective tissue protein. Many naturally occurring amino acids are not α-amino acids. Prominent examples include β-alanine, a precursor of the vitamin pantothenic acid, and γ-aminobutyric acid (GABA), a neurotransmitter found in the brain (**Figure 1.5**).

Amino acid molecules are used primarily in the synthesis of long, complex polymers known as **polypeptides**. Up to a length of about 50 amino acids, these molecules are called **peptides**. **Proteins** consist of one or more polypeptides. Polypeptides play a variety of roles in living organisms. Examples include transport proteins, structural proteins, and the enzymes (catalytic proteins).

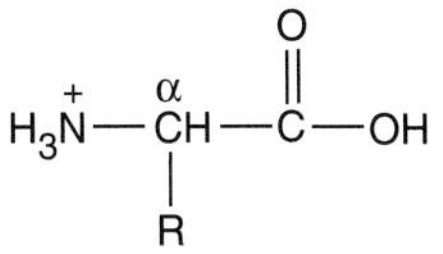

$H_3\overset{+}{N}-\overset{\alpha}{C}H(R)-C(=O)-OH$

FIGURE 1.3

General Formula for α-Amino Acids

For 19 of the 20 standard amino acids the α-carbon is bonded to a hydrogen atom, a carboxyl group, an amino group, and an R group.

Glutamine: $H_3\overset{+}{N}-CH[(CH_2)_2-C(=O)-NH_2]-C(=O)-OH$

Valine: $H_3\overset{+}{N}-CH[CH(CH_3)-CH_3]-C(=O)-OH$

Lysine: $H_3\overset{+}{N}-CH[(CH_2)_4-NH_2]-C(=O)-OH$

Glycine: $H_3\overset{+}{N}-CH(H)-C(=O)-OH$

Phenylalanine: $H_3\overset{+}{N}-CH(CH_2-C_6H_5)-C(=O)-OH$

Serine: $H_3\overset{+}{N}-CH(CH_2-OH)-C(=O)-OH$

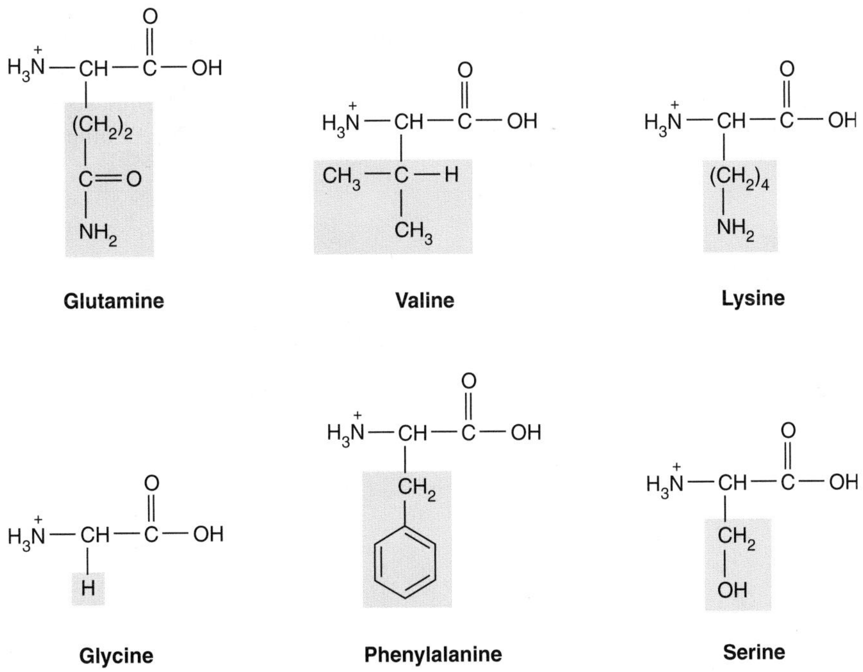

FIGURE 1.4

Structural Formulas for Several α-Amino Acids

An R group (highlighted) in an amino acid structure can be a hydrogen atom (e.g., in glycine), a hydrocarbon group (e.g., the isopropyl group in valine), or a hydrocarbon derivative (e.g., the hydroxymethyl group in serine).

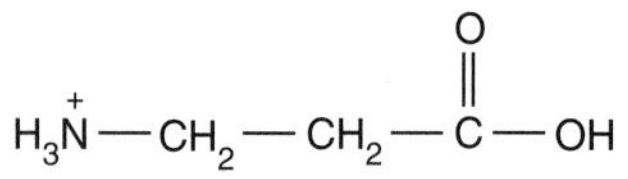

$H_3\overset{+}{N}-CH_2-CH_2-\overset{O}{\overset{||}{C}}-OH$

β-Alanine

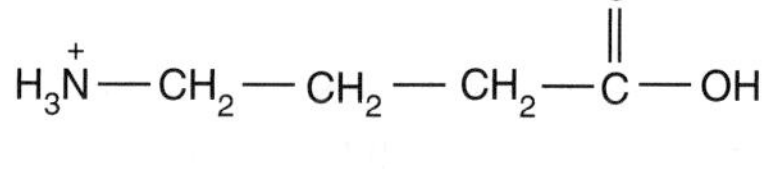

$H_3\overset{+}{N}-CH_2-CH_2-CH_2-\overset{O}{\overset{||}{C}}-OH$

GABA

FIGURE 1.5

Selected Examples of Naturally Occurring Amino Acids That Are Not α-Amino Acids: β-Alanine and γ-Aminobutyric Acid (GABA)

The individual amino acids are connected in peptides (**Figure 1.6**) and polypeptides by the peptide bond. **Peptide bonds** are amide linkages that form in a type of nucleophilic substitution reaction (p. 12) in which the amino group nitrogen of one amino acid attacks the carbonyl carbon in the carboxyl group of another. The final three-dimensional structure, and therefore biological function, of polypeptides results largely from interactions among the R groups (**Figure 1.7**).

WORKED PROBLEM 1.1

Living organisms generate a vast number of different biopolymers by linking monomers in different sequences. A set of tripeptides, each containing three amino acid residues, contains only two types of amino acids: A and B. How many possible tripeptides are in this set?

SOLUTION

The number of possible tripeptides is given by the formula $\mathbf{X}^n$, where
$\mathbf{X}$ = the number of types of constituent amino acid residues
and n = length of the peptide.
Substituting these values into the formula yields $2^3 = 8$. The eight tripeptides are as follows: AAA, AAB, ABA, BAA, ABB, BAB, BBA, and BBB. ■

SUGARS AND CARBOHYDRATES **Sugars**, the smallest carbohydrates, contain alcohol and carbonyl functional groups. They are described in terms of both carbon number and the type of carbonyl group they contain. Sugars that possess an aldehyde group are called *aldoses* and those that possess a ketone group are called *ketoses*. For example, the six-carbon sugar glucose (an important energy source in most living organisms) is an aldohexose; fructose (fruit sugar) is a ketohexose (Figure 1.8).

Sugars are the basic units of carbohydrates, the most abundant organic molecules found in nature. Carbohydrates range from the simple sugars, or **monosaccharides**, such as glucose and fructose, to the **polysaccharides**, polymers that contain thousands of sugar units. Examples of the latter include starch and cellulose in plants and glycogen in animals. Carbohydrates serve a variety of functions in living organisms. Certain sugars are important energy sources. Glucose is the principal carbohydrate energy source in animals and plants. Sucrose is used by many plants as an efficient means of transporting energy throughout their tissues. Some carbohydrates serve as structural materials. Cellulose is the major structural component of wood and certain plant fibers. Chitin, another type

FIGURE 1.6

Structure of Met-Enkephalin, a Pentapeptide

Met-enkephalin is one of a class of molecules that have opiate-like activity. Found in the brain, metenkephalin inhibits pain perception. (The peptide bonds are colored blue. The R groups are highlighted.)

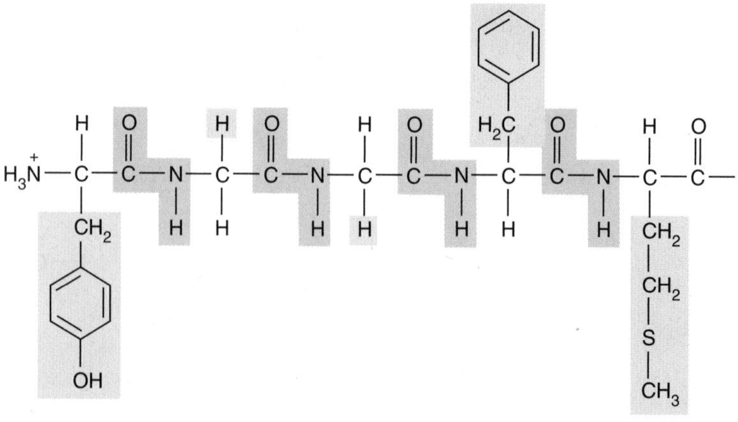

Unfolded

Folded

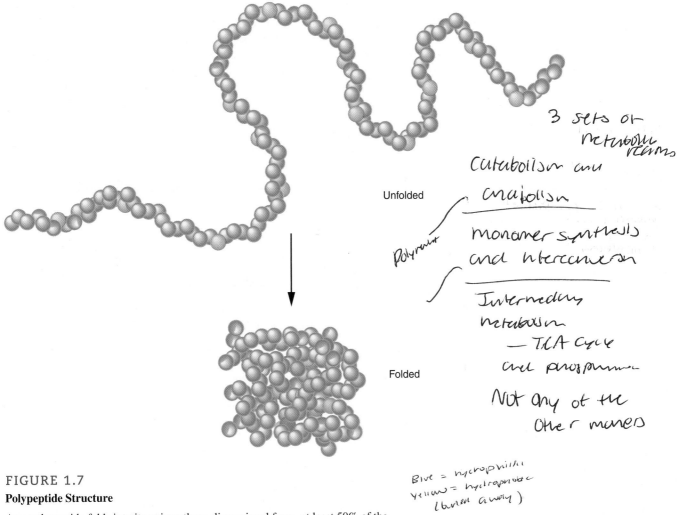

FIGURE 1.7

Polypeptide Structure

As a polypeptide folds into its unique three-dimensional form, at least 50% of the hydrophobic R groups (yellow spheres) become buried in the interior away from water. Hydrophilic groups usually occur on the surface.

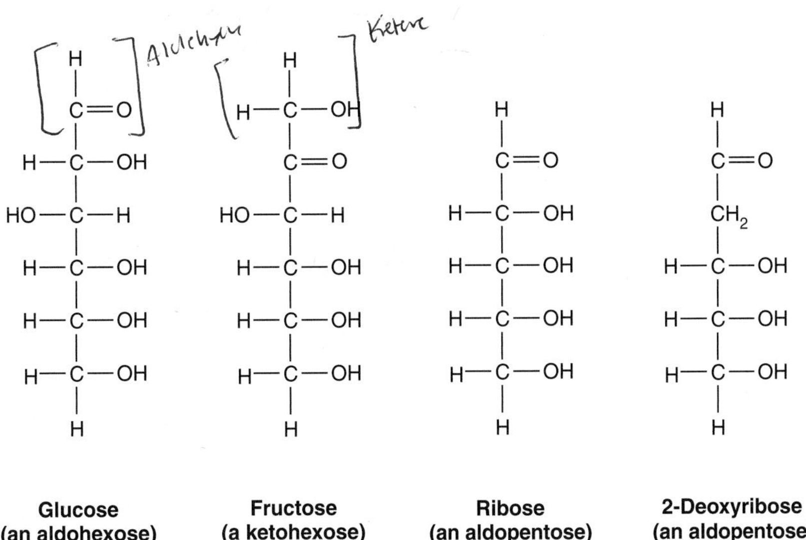

FIGURE 1.8

Some Biologically Important Monosaccharides

Glucose and fructose are important sources of energy in plants and animals. Ribose and deoxyribose are components of nucleic acids. These monosaccharides occur as ring structures in nature.

of polysaccharide, is found in the protective outer coverings of insects and crustaceans.

Some biomolecules contain carbohydrate components. Nucleotides, the building-block molecules of the nucleic acids, contain either of the pentoses ribose or deoxyribose. Certain proteins and lipids also contain carbohydrate. Glycoproteins and glycolipids occur on the external surface of cell membranes in multicellular organisms, where they play critical roles in the interactions between cells.

FATTY ACIDS **Fatty acids** are monocarboxylic acids that usually contain an even number of carbon atoms. Fatty acids are represented by the chemical formula R—COOH, in which R is an alkyl group that contains carbon and hydrogen atoms. There are two types of fatty acids: **saturated** fatty acids, which contain no carbon-carbon double bonds, and **unsaturated** fatty acids, which have one or more double bonds (**Figure 1.9**). Under physiological conditions the carboxyl group of fatty acids exists in the ionized state, $R—COO^-$. For example, the 16-carbon saturated fatty acid called palmitic acid usually exists as palmitate, $CH_3(CH_2)_{14}COO^-$. Although the charged carboxyl group has an affinity for water, the long nonpolar hydrocarbon chains render most fatty acids insoluble in water.

Fatty acids occur as independent (free) molecules in only trace amounts in living organisms. Most often they are components of several types of **lipid** molecules (**Figure 1.10**). Lipids are a diverse group of substances that are soluble in organic solvents such as chloroform or acetone, but are not soluble in water. For example, triacylglycerols (fats and oils), energy sources in some organisms, are esters containing glycerol (a three-carbon alcohol with three hydroxyl groups) and three fatty acids. Certain lipid molecules that resemble triacylglycerols, called phosphoglycerides, contain two fatty acids. In these molecules the third hydroxyl group of glycerol is coupled with phosphate, which is in turn attached to small polar compounds such as choline. Phosphoglycerides are an important structural component of cell membranes.

NUCLEOTIDES AND NUCLEIC ACIDS Each **nucleotide** contains three components: a five-carbon sugar (either ribose or deoxyribose), a nitrogenous base, and one or more phosphate groups (**Figure 1.11**). The bases in nucleotides

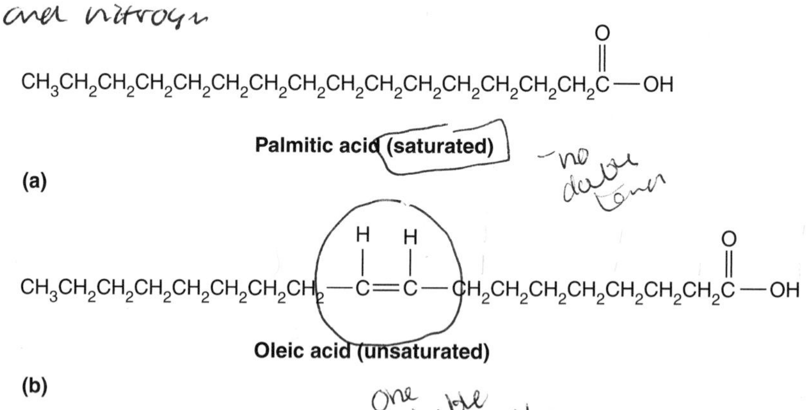

FIGURE 1.9

Fatty Acid Structure

(a) A saturated fatty acid. (b) An unsaturated fatty acid.

$$CH_3CH_2CH_2CH_2CH_2CH_2CH_2CH_2CH_2CH_2CH_2CH_2CH_2CH_2CH_2C(=O)—OH$$

Palmitic acid (saturated)

(a)

$$CH_3CH_2CH_2CH_2CH_2CH_2CH_2CH_2—CH{=}CH—CH_2CH_2CH_2CH_2CH_2CH_2CH_2C(=O)—OH$$

Oleic acid (unsaturated)

(b)

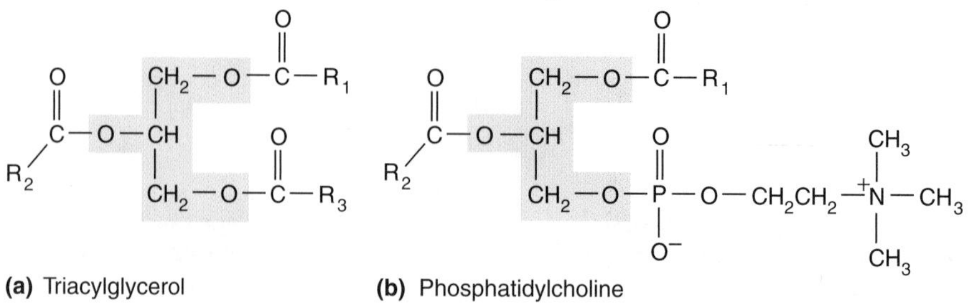

FIGURE 1.10

Lipid Molecules That Contain Fatty Acids

(a) Triacylglycerol. (b) Phosphatidylcholine, a type of phosphoglyceride.

(a) Triacylglycerol **(b)** Phosphatidylcholine

FIGURE 1.11
Nucleotide Structure
Each nucleotide contains a nitrogenous base (in this case, adenine), a pentose sugar (ribose), and one or more phosphates. This nucleotide is adenosine triphosphate.

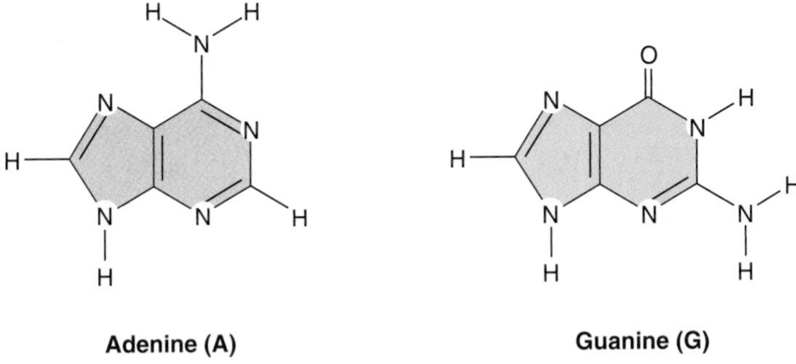

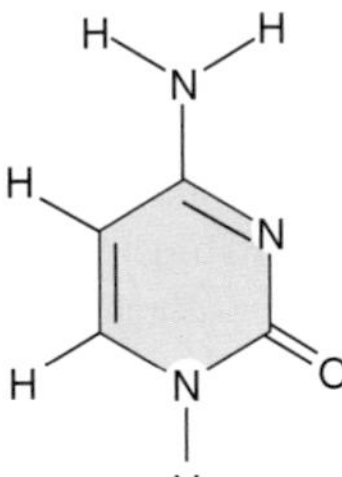
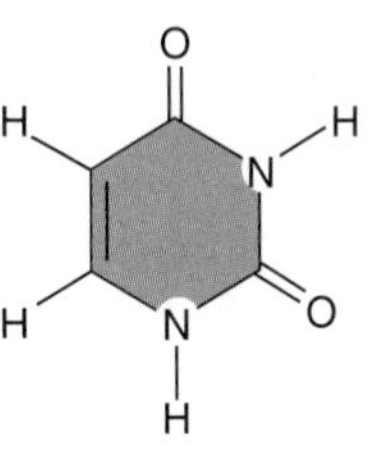

FIGURE 1.12
The Nitrogenous Bases
(a) The purines. (b) The pyrimidines.

are heterocyclic aromatic rings with a variety of substituents. There are two classes of base: the bicyclic purines and the monocyclic pyrimidines (Figure 1.12).

Nucleotides participate in a wide variety of biosynthetic and energy-generating reactions. For example, a substantial proportion of the energy obtained from food molecules is used to form the high-energy phosphate bonds of ATP. Energy is released when the phosphoanhydride bonds are hydrolyzed. Nucleotides also have an important role as the building-block molecules of the nucleic acids. In a **nucleic acid**, from hundreds to millions of nucleotides are linked by phosphodiester linkages to form long polynucleotide chains or strands. There are two types of nucleic acid: DNA and RNA.

DNA. DNA is the repository of genetic information. Its structure consists of two antiparallel polynucleotide strands wound around each other to form a right-handed double helix (**Figure 1.13**). In addition to the pentose sugar deoxyribose and phosphate, DNA contains bases of four types: the **purines** adenine and guanine and the **pyrimidines** thymine and cytosine; adenine pairs with thymine

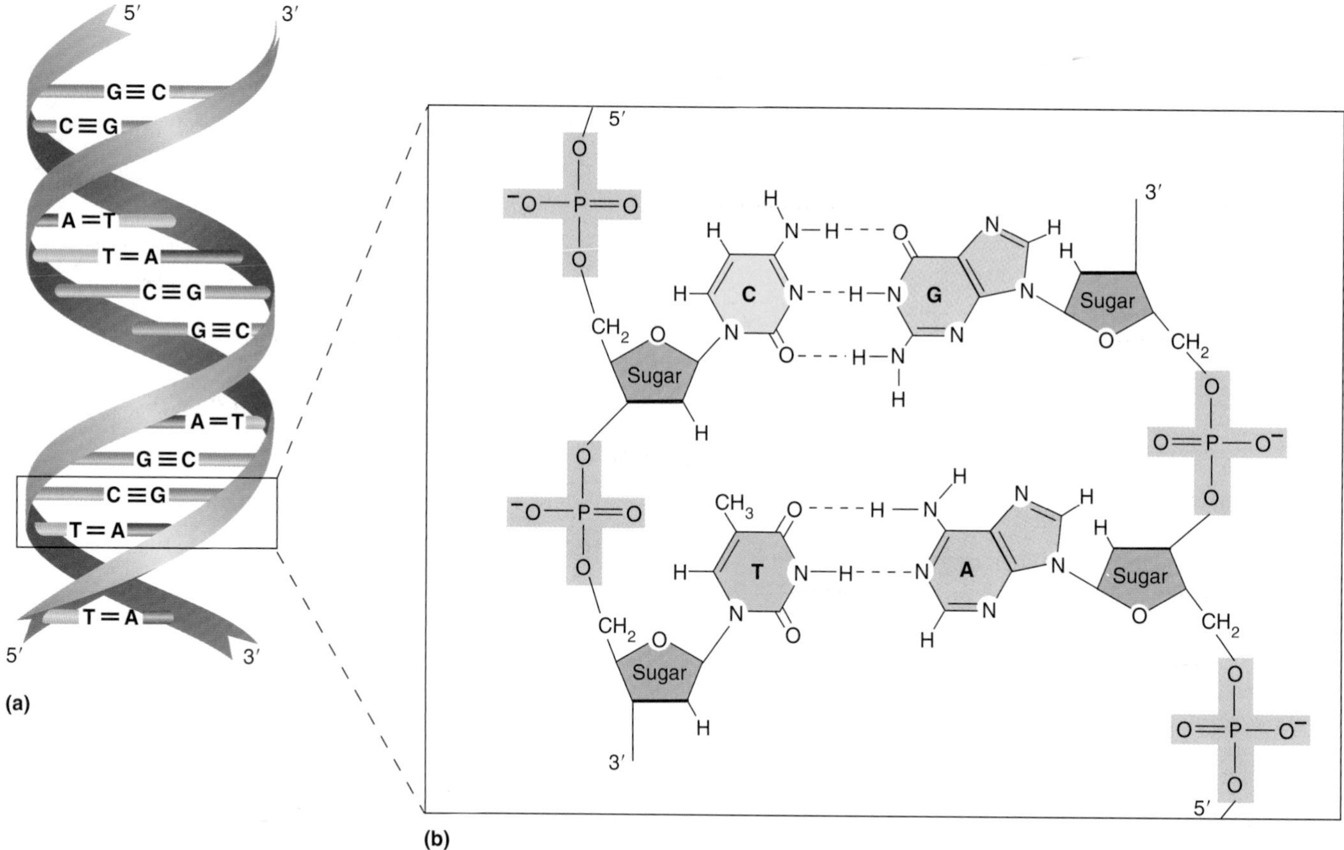

FIGURE 1.13
DNA

(a) A diagrammatic view of DNA. The sugar-phosphate backbones of the double helix are represented by colored ribbons. The bases attached to the sugar deoxyribose are on the inside of the helix. (b) An enlarged view of two base pairs. Note that the two DNA strands run in opposite directions defined by the 5′ and 3′ groups of deoxyribose. The bases on opposite strands form pairs because of hydrogen bonds. Cytosine always pairs with guanine; thymine always pairs with adenine.

and guanine pairs with cytosine. The double helix forms because of complementary pairing between the bases made possible by the formation of hydrogen bonds. A hydrogen bond is a force of attraction between a polarized hydrogen of one molecular group and the electronegative oxygen or nitrogen atoms of nearby aligned molecular groups.

An organism's entire set of DNA sequences is called its *genome*. DNA consists of both coding and noncoding sequences. Coding sequences, called genes, specify the structure of functional gene products such as polypeptides and RNA molecules. Some noncoding sequences have regulatory functions (e.g., controlling the synthesis of certain proteins), whereas the functions of others are as yet undetermined.

RNA. Ribonucleic acid (RNA) is a polynucleotide that differs from DNA in that it contains the sugar ribose instead of deoxyribose, and the base uracil instead of thymine. In RNA, as in DNA, the nucleotides are linked by phosphodiester linkages. In contrast to the double helix of DNA, RNA is single-stranded. RNA molecules fold into complex three-dimensional structures created by local regions of complementary base pairing. When the DNA double helix unwinds, one strand can serve as a template. During the process called **transcription**, RNA molecules are synthesized. Complementary base pairing specifies the nucleotide base sequence of the RNA molecule. There are three major types of RNA: messenger RNA (mRNA), ribosomal RNA (rRNA), and transfer RNA (tRNA). Each unique sequence or molecule of mRNA possesses the information that codes directly for the amino acid sequence in a specific polypeptide. Ribosomes, the large, complex, supramolecular structures composed of rRNA and protein molecules, convert the mRNA base sequence into the amino acid sequence of a polypeptide. Transfer RNA molecules transport amino acids to the ribosome during protein synthesis.

In recent years large numbers of RNA molecules have been discovered that are not directly involved in protein synthesis. These molecules, called *noncoding RNAs* (ncRNA), have roles in a great variety of cellular processes. Examples include short interfering RNAs (siRNA), micro RNAs (miRNA), small nuclear RNAs (snRNA), and small nucleolar RNAs (snoRNA). siRNAs are important components in *RNA interference*, an antiviral defense mechanism. miRNAs regulate the timing of mRNA synthesis, and snRNAs facilitate the process by which mRNA precursor molecules are transformed into functional mRNA. snoRNAs assist in the maturation of ribosomal RNA during ribosome formation.

KEY CONCEPTS

- Most molecules in living organisms are organic. The chemical properties of organic molecules are determined by specific arrangements of atoms called functional groups.
- Cells contain four families of small molecules: amino acids, sugars, fatty acids, and nucleotides.
- Proteins, polysaccharides, and the nucleic acids are biopolymers composed of amino acids, sugars, and nucleotides, respectively.

Gene Expression. **Gene expression** controls when or if the information encoded in a gene will be accessed. The process begins with transcription, the mechanism whereby the base sequence of a DNA segment is used to synthesize a gene product. A class of proteins called **transcription factors** regulates the expression of protein-coding genes when they bind to specific regulatory DNA sequences referred to as **response elements**. Transcription factors are synthesized and/or regulated in response to an information-processing mechanism initiated by a signal molecule (e.g., insulin, a protein that regulates several metabolic processes) or an abiotic factor such as light.

1.3 IS THE LIVING CELL A CHEMICAL FACTORY?

Even the simplest cells are so remarkable that cells have often been characterized as chemical factories. Like factories, living organisms acquire raw materials, energy, and information from their environment. Components are manufactured, and waste products and heat are discharged back into the environment. However, for this analogy to hold true, human-made factories would not only manufacture and repair all their structural and functional components, but also clone themselves, that is, manufacture new factories. The term **autopoiesis** has been created to describe the remarkable properties of living organisms. In this view, each living organism is considered an autonomous, self-organizing, and self-maintaining entity. Life emerges from a self-regulating network of thousands of biochemical reactions.

The constant flow of energy and nutrients through organisms and the functional properties of thousands of enzymes make possible the process of metabolism. The primary functions of metabolism are (1) acquisition and utilization of energy, (2) synthesis of molecules needed for cell structure and functioning (i.e., proteins, carbohydrates, lipids, and nucleic acids), (3) growth and development, and (4) removal of waste products. Metabolic processes require significant amounts of useful energy. This section begins with a review of the primary chemical reaction types and the essential features of energy-generating strategies observed in living organisms. A brief outline of metabolic processes and the means by which living organisms maintain ordered systems follows.

Biochemical Reactions

At first glance the thousands of reactions that occur in cells appear overwhelmingly complex. However, several characteristics of metabolism allow us to simplify this picture:

1. Although the number of reactions is large, the number of reaction types is relatively small.

2. Biochemical reactions have simple organic reaction mechanisms.
3. Reactions of central importance in biochemistry (i.e., those used in energy production and the synthesis and degradation of major cell components) are relatively few.

Among the most common reaction types encountered in biochemical processes are nucleophilic substitution, elimination, addition, isomerization, and oxidation-reduction.

NUCLEOPHILIC SUBSTITUTION REACTIONS In **nucleophilic substitution** reactions, as the name suggests, one atom or group is substituted for another:

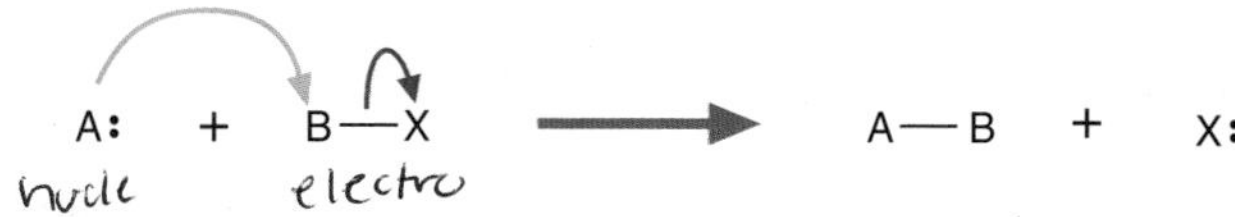

In the general reaction shown, the attacking species (A) is called a **nucleophile** ("nucleus lover"). Nucleophiles are anions (negatively charged atoms or groups) or neutral species possessing nonbonding electron pairs. **Electrophiles** ("electron lovers") are deficient in electron density and are therefore easily attacked by a nucleophile. As the new bond forms between A and B, the old one between B and X breaks. The outgoing nucleophile (in this case, X), called a **leaving group**, leaves with its electron pair. Several types of nucleophilic substitution reactions occur in living organisms. Examples include S_N2 reactions (e.g., the methylation of epinephrine, refer to p. P-32), acyl group transfers, and phosphoryl group transfers.

In nucleophilic substitution reactions involving acyl transfer, a nucleophile attacks the carbonyl carbon of a carboxylic acid derivative, forming a tetrahedral intermediate. The carbonyl group reforms as the tetrahedral intermediate collapses and the leaving group is ejected. Biologically important examples of carboxylic acid derivatives include carboxylates (deprotonated carboxylic acids), esters, amides, thioesters, and acyl phosphates. These derivatives vary in their reactivity with nucleophiles. Acyl phosphates are the most reactive, followed by thioesters, esters, amides, and finally carboxylates.

The biologically active form of fatty acids is the thioester of coenzyme A (p. 328). Carboxylates are not good substrates for nucleophilic substitution reactions because the carbonyl carbon is not sufficiently electrophilic. As a result, fatty acids must first be activated by the formation of an acyl adenosyl monophosphate derivative (**Figure 1.14**) at the expense of ATP bond energy. Once the activated fatty acyl-AMP is formed, its carbonyl carbon is easily attacked by the thiol sulfur of coenzyme A to yield the fatty acyl-SCoA product.

Hydrolysis reactions are nucleophilic acyl substitution reactions in which the oxygen of a water molecule serves as the nucleophile. The electrophile is usually the carbonyl carbon of an ester, amide, or anhydride. (An **anhydride** is a molecule containing two carbonyl groups linked through an oxygen atom.)

$$R-\underset{\displaystyle\overset{\|}{O}}{C}-O-R' + H_2O \longrightarrow R-\underset{\displaystyle\overset{\|}{O}}{C}-OH + R'OH$$

The digestion of many food molecules involves hydrolysis. For example, the amide linkages of proteins are hydrolyzed in the stomach in an acid-catalyzed reaction that yields amino acids.

The hydrolysis of ATP to yield ADP and inorganic phosphate (P_i) and the reaction of glucose with ATP provide two examples of nucleophilic substitution involving phosphoryl group transfer. The attack by the OH of water on the terminal phosphate of ATP (**Figure 1.15**) breaks the phosphoanhydride bond, thereby releasing energy that is used to drive many cellular processes.

The reaction of glucose with ATP, yielding glucose-6-phosphate and ADP, is the first step in the utilization of glucose as an energy source (**Figure 1.16**). The

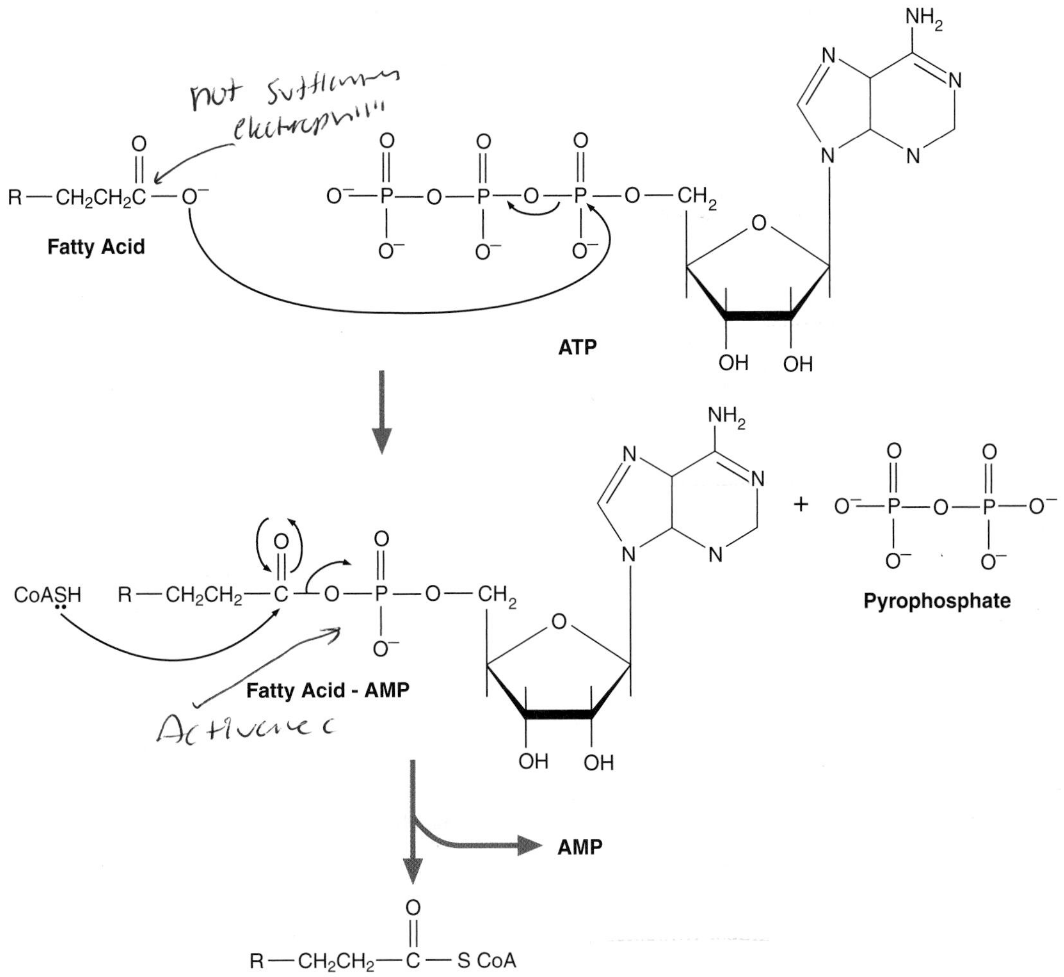

FIGURE 1.14
Activation of a Fatty Acid

Before a fatty acid can be degraded to yield energy or used in the synthesis of a triacylglycerol, it must first be activated. In the first step the carboxylate ion attacks a phosphate of ATP to form a fatty acyl-AMP intermediate and pyrophosphate (PP_i). In the second step the fatty acyl-AMP is attacked by the thiol group of coenzyme A (CoASH) to form the thioester fatty acyl-SCoA and AMP. The rapid hydrolysis of PP_i to form two phosphates (P_i) drives the reaction forward.

hydroxyl oxygen on carbon 6 of the sugar molecule is the nucleophile and phosphorus is the electrophile. Adenosine diphosphate is the leaving group.

ELIMINATION REACTIONS In **elimination reactions** a double bond is formed when atoms in a molecule are removed.

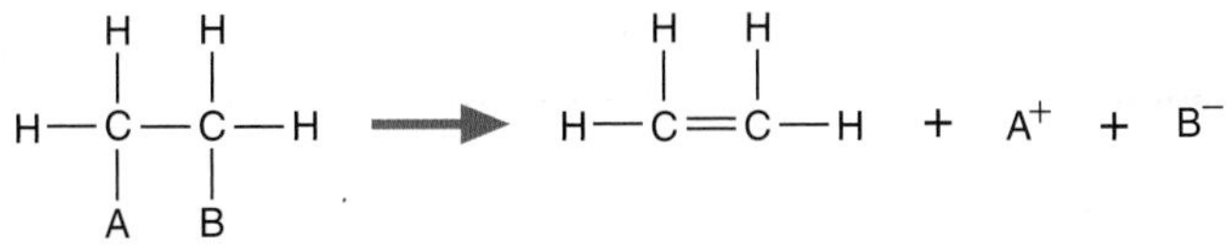

$$\mathrm{H{-}CH(A){-}CH(B){-}H} \longrightarrow \mathrm{H{-}CH{=}CH{-}H} + A^+ + B^-$$

The removal of H_2O from biomolecules containing alcohol functional groups is a commonly encountered reaction. A prominent example is the dehydration of 2-phosphoglycerate, a reaction in *glycolysis*, which is a biochemical pathway in carbohydrate metabolism (**Figure 1.17**). As illustrated on pp. P-33–P-34, this reaction occurs via an E1cB mechanism. Other products of elimination reactions include ammonia (NH_3), amines (RNH_2), and alcohols (ROH).

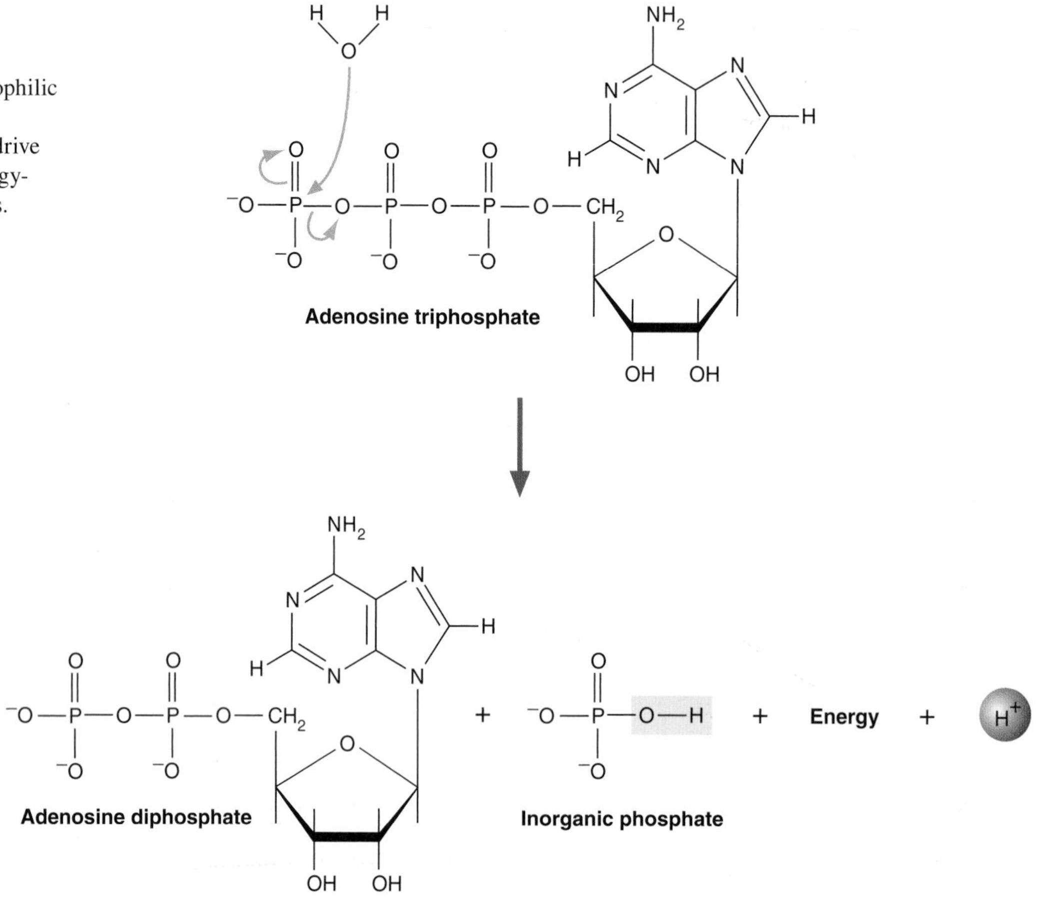

FIGURE 1.15

A Hydrolysis Reaction

The hydrolysis of ATP, a nucleophilic substitution reaction involving phosphoryl transfer, is used to drive an astonishing diversity of energy-requiring biochemical reactions.

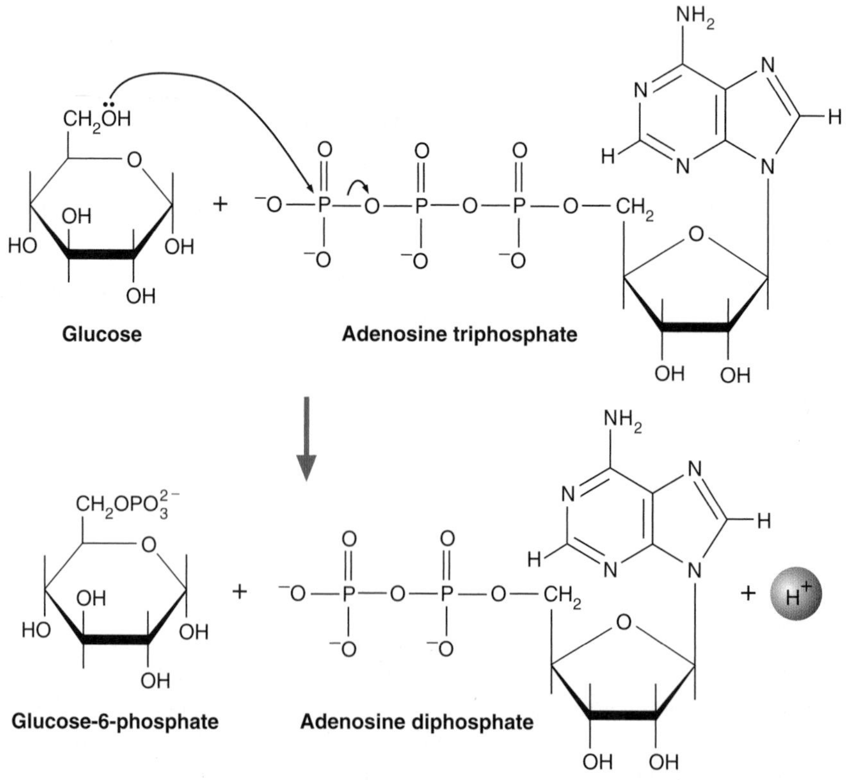

Adenosine triphosphate

Adenosine diphosphate + Inorganic phosphate + Energy + H^+

FIGURE 1.16

Example of Nucleophilic Substitution

In the reaction of glucose with ATP, the hydroxyl oxygen of glucose is the nucleophile. The phosphorus atom (the electrophile) is polarized by the oxygens bonded to it so that it bears a partial positive charge. As the reaction occurs, the unshared pair of electrons on the CH_2OH of the sugar attacks the phosphorus, resulting in the expulsion of ADP, the leaving group.

Glucose + Adenosine triphosphate

Glucose-6-phosphate + Adenosine diphosphate + H^+

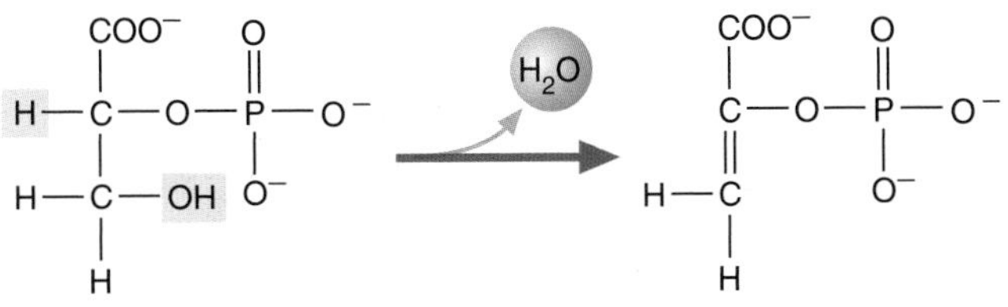

2-Phosphoglycerate → Phosphoenolpyruvate

FIGURE 1.17

An Elimination Reaction

When 2-phosphoglycerate is dehydrated, a double bond is formed. This reaction involves an E1cB mechanism, which is illustrated on p. P-34.

ADDITION REACTIONS In **addition reactions** two molecules combine to form a single product.

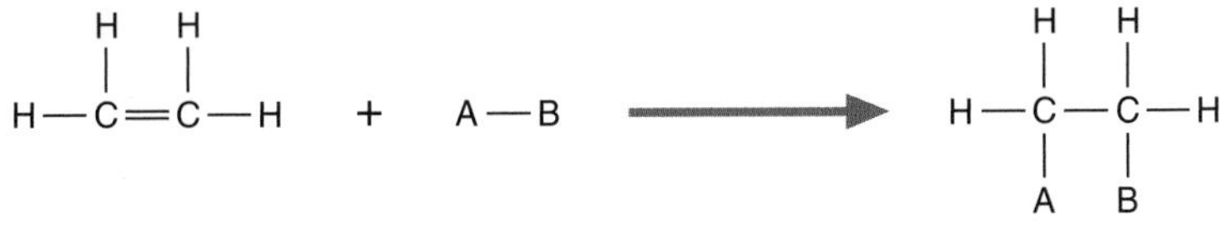

Hydration is one of the most common addition reactions. When water is added to an alkene, an alcohol results. The hydration of the metabolic intermediate fumarate to form malate is a typical example (**Figure 1.18**).

ISOMERIZATION REACTIONS In **isomerization** reactions, atoms or groups undergo intramolecular shifts. One of the most common biochemical isomerizations is the interconversion between aldose and ketose sugars (**Figure 1.19**). The isomerization of dihydroxyacetone phosphate to glyceraldehyde-3-phosphate (**Figure 1.19b**) is a reaction in glycolysis.

Fumarate + H_2O ⇌ Malate

(a)

Fumarate → Malate

(b)

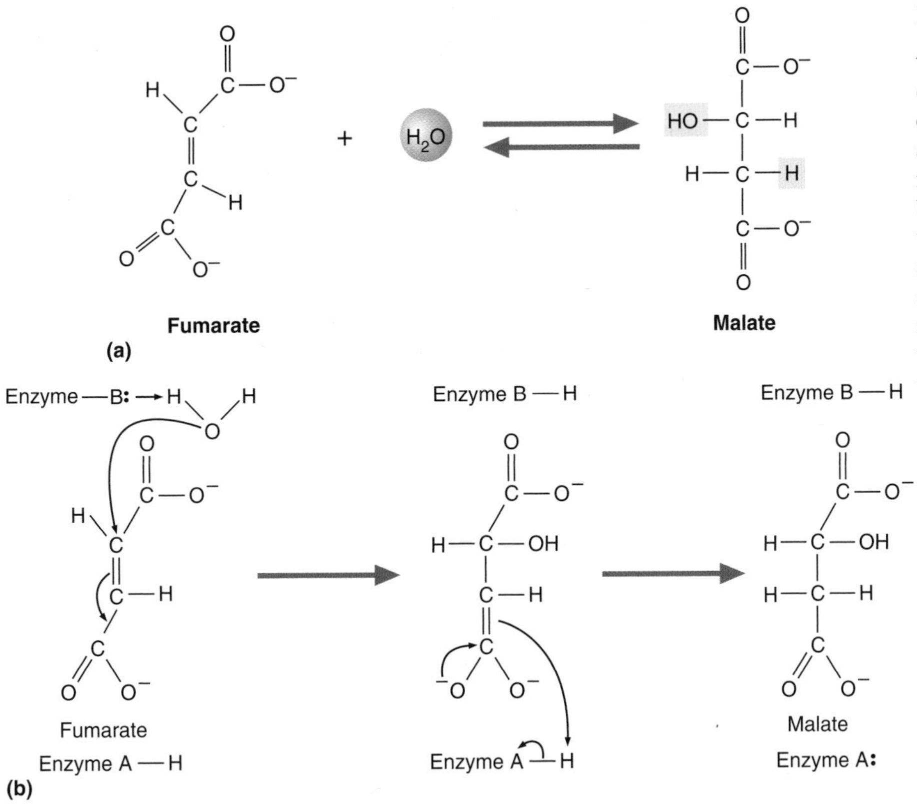

FIGURE 1.18

An Addition Reaction

(a) When water is added to a molecule that contains a double bond, such as fumarate, an alcohol results. (b) The hydration of fumarate, catalyzed by the enzyme fumarase, begins with the removal of a proton from a water molecule by an amino acid side chain acting as a base. The resulting nucleophile attacks the carbon-carbon double bond. The initial product, a resonance-stabilized ion, is then protonated by an acidic side chain of the enzyme to yield the product malate.

FIGURE 1.19
An Isomerization Reaction

(a) The reversible interconversion of aldose and ketose isomers is a commonly observed biochemical reaction type. (b) The isomerization of dihydroxacetone phosphate to form glyceraldehyde-3-phosphate begins when a basic side chain of the enzyme, triose phosphate isomerase, removes a proton from carbon 1 and an acidic side chain donates a proton to the carbonyl oxygen. The intermediate product is an *enediol* (a molecule in which a hydroxyl group is attached to each of the carbon atoms in a carbon-carbon double bond). In the second step the enediol is deprotonated by a basic side chain and an acidic side chain adds a proton to carbon 2, yielding the product glyceraldehyde-3-phosphate.

(a) Aldose ⇌ Ketose

(b) Dihydroxyacetone phosphate → enediol → Glyceraldehyde-3-phosphate

OXIDATION-REDUCTION REACTIONS **Oxidation-reduction (redox) reactions** occur when there is a transfer of electrons from a donor (called the **reducing agent**) to an electron acceptor (called the **oxidizing agent**). When reducing agents donate their electrons, they become **oxidized**. As oxidizing agents accept electrons, they become **reduced**. The two processes always occur simultaneously.

It is not always easy to determine whether biomolecules have gained or lost electrons. However, two simple rules may be used to ascertain whether a carbon atom in a molecule has been oxidized or reduced:

1. Oxidation has occurred if a carbon atom gains oxygen or loses hydrogen:

CH_3CH_2—OH → CH_3C(=O)—OH

Ethyl alcohol → Acetic acid

2. Reduction has occurred if a carbon atom loses oxygen or gains hydrogen:

In biological redox reactions, electrons are transferred to electron acceptors such as the nucleotide NAD^+/NADH (nicotinamide adenine dinucleotide in its oxidized/reduced form).

KEY CONCEPT

The most common reaction types encountered in biochemical processes are nucleophilic substitution, elimination, addition, isomerization, and oxidation-reduction.

Energy

Energy is defined as the capacity to do work, that is, to move matter. In contrast to human-made machines, which generate and use energy under harsh conditions such as high temperature, high pressure, and electrical currents, the relatively fragile molecular machines within living organisms must use more subtle mechanisms. Cells generate most of their energy by using redox reactions in which electrons are transferred from an oxidizable molecule to an electron-deficient molecule. In these reactions electrons are often removed or added as hydrogen atoms (H•) or hydride ions ($H:^-$). The more reduced a molecule is—that is, the more hydrogen atoms it possesses—the more energy it contains. For example, fatty acids contain proportionately more hydrogen atoms than sugars do and therefore yield more energy upon oxidation. When fatty acids and sugars are oxidized, their hydrogen atoms are removed by the redox coenzymes FAD (flavin adenine dinucleotide) or NAD^+, respectively. (Coenzymes are small molecules that function in association with enzymes by serving as carriers of small molecular groups, or in this case, electrons.) The reduced products of this process ($FADH_2$ or NADH, respectively) can then transfer the electrons to another electron acceptor.

Whenever an electron is transferred, energy is lost. Cells have complex mechanisms for exploiting this phenomenon in a way that permits some of the released energy to be captured for cellular work. The most prominent feature of energy generation in most cells is the electron transport pathway, a series of linked membrane-embedded electron carrier molecules. During a regulated process, energy is released as electrons are transferred from one electron carrier molecule to another. During several of these redox reactions, the energy released is sufficient to drive the synthesis of ATP, the energy carrier molecule that directly supplies the energy used to maintain highly organized cellular structures and functions.

Despite their many similarities, groups of living organisms differ in the precise strategies they use to acquire energy from their environment. **Autotrophs** are organisms that transform the energy of the sun (**photosynthesis**) or various chemicals (**chemosynthesis**) into chemical bond energy; they are called, respectively, **photoautotrophs** and **chemoautotrophs**. The **heterotrophs** obtain energy by degrading preformed food molecules obtained by consuming other organisms. **Chemoheterotrophs** use preformed food molecules as their sole source of energy. Some prokaryotes and a small number of plants (e.g., the pitcher plant, which digests captured insects) are **photoheterotrophs**; that is, they use both light and organic biomolecules as energy sources.

The ultimate source of the energy used by most life-forms on earth is the sun. Photosynthetic organisms such as plants, certain prokaryotes, and algae capture light energy and use it to transform carbon dioxide (CO_2) into sugar and other biomolecules. Chemotrophic species derive the energy required to incorporate CO_2 into organic biomolecules by oxidizing inorganic substances such as hydrogen sulfide (H_2S), nitrite (NO_2^-), or hydrogen gas (H_2). The biomass produced in both types of process is, in turn, consumed by heterotrophic organisms that use it as sources of energy and structural materials. At each step, as molecular bonds are rearranged, some energy is captured and used to maintain the organism's complex structures and activities. Eventually energy becomes disorganized and is released in the form of heat. The metabolic pathways by which energy is generated and used by living organisms are briefly outlined next, followed by the basic mechanisms by which cellular order is maintained.

KEY CONCEPT

In living organisms, energy, the capacity to move matter, is usually generated by redox reactions.

Overview of Metabolism

Metabolism is the sum of all the enzyme-catalyzed reactions in a living organism. These reactions, none of which occur in isolation, are organized into pathways (**Figure 1.20**) in which an initial reactant molecule is modified in a

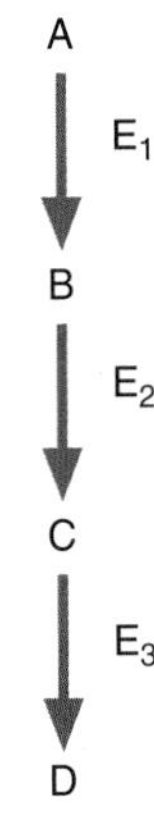

FIGURE 1.20
A Biochemical Pathway

In this three-step biochemical pathway biomolecule A is converted into biomolecule D in three sequential reactions. Each reaction is catalyzed by a specific enzyme (E).

step-by-step sequence into a product that can be used by the cell for a specific purpose. For example, glycolysis, the energy-generating pathway that degrades the six-carbon sugar glucose, is composed of 10 reactions. All of an individual organism's metabolic processes consist of a vast weblike pattern of interconnected biochemical reactions that are regulated such that resources are conserved and energy use is optimized. There are three classes of biochemical pathways: metabolic, energy transfer, and signal transduction.

METABOLIC PATHWAYS There are two types of metabolic pathway: anabolic and catabolic. In **anabolic** (biosynthetic) **pathways**, large complex molecules are synthesized from smaller precursors. Building-block molecules (e.g., amino acids, sugars, and fatty acids), either produced or acquired from the diet, are incorporated into larger, more complex molecules. Because biosynthesis increases order and complexity, anabolic pathways require an input of energy. Anabolic processes include the synthesis of polysaccharides and proteins from sugars and amino acids, respectively. During **catabolic pathways** large complex molecules are degraded into smaller, simpler products. Some catabolic pathways release energy. A fraction of this energy is captured and used to drive anabolic reactions.

The relationship between anabolic and catabolic processes is illustrated in **Figure 1.21**. As nutrient molecules are degraded, energy and reducing power (high-energy electrons) are conserved in ATP and NADH molecules, respectively. Biosynthetic processes use metabolites of catabolism, synthesized ATP and NADPH (reduced nicotinamide adenine dinucleotide phosphate, a source of reducing power), to create complex structure and function.

ENERGY TRANSFER PATHWAYS Energy transfer pathways capture energy and transform it into forms that organisms can use to drive biomolecular processes. The absorption of light energy by chlorophyll molecules and the energy-releasing redox reactions required for its conversion to chemical bond energy in a sugar molecule is a prominent example.

SIGNAL TRANSDUCTION **Signal transduction** pathways allow cells to receive and respond to signals from their surroundings. In the initial or reception phase,

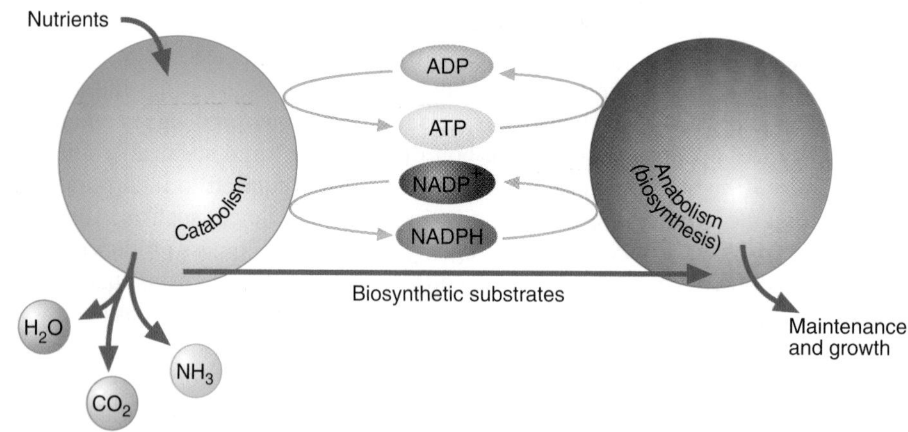

FIGURE 1.21

Anabolism and Catabolism

In organisms that use oxygen to generate energy, catabolic pathways convert nutrients to small-molecule starting materials. The energy (ATP) and reducing power (NADPH) that drive biosynthetic reactions are generated during catabolic processes as certain nutrient molecules are converted to waste products such as carbon dioxide, ammonia, and water.

a signal molecule such as a hormone or a nutrient molecule binds to a receptor protein. This binding event initiates the transduction phase, a cascade of intracellular reactions that triggers the cell's response to the original signal. For example, glucose binds to its receptor on pancreatic insulin-secreting cells, whereupon insulin is released into the blood. Most commonly, such responses are an increase or decrease in the activity of already existing enzymes or the synthesis of new enzyme molecules.

KEY CONCEPTS

- Metabolism is the sum of all the enzyme-catalyzed reactions in a living organism.
- There are three classes of biochemical pathway: metabolic (anabolic and catabolic), energy transfer, and signal transduction.

Biological Order

The coherent unity that is observed in all living organisms involves the functional integration of millions of molecules. In other words, life is highly organized complexity. Despite the rich diversity of living processes that contribute to generating and maintaining biological order, most can be classified into the following categories: (1) synthesis and degradation of biomolecules, (2) transport of ions and molecules across cell membranes, (3) production of force and movement, and (4) removal of metabolic waste products and other toxic substances.

SYNTHESIS OF BIOMOLECULES Cellular components are synthesized in a vast array of chemical reactions, many of which require energy, supplied directly or indirectly by ATP molecules. The molecules formed in biosynthetic reactions perform several functions. They can be assembled into supramolecular structures (e.g., the proteins and lipids that constitute membranes) or serve as informational molecules (e.g., DNA and RNA) or catalyze chemical reactions (i.e., the enzymes).

TRANSPORT ACROSS MEMBRANES Cell membranes regulate the passage of ions and molecules from one compartment to another. For example, the plasma membrane (the animal cell's outer membrane) is a selective barrier. It is responsible for the transport of certain substances such as nutrients from a relatively disorganized environment into the more orderly cellular interior. Similarly, ions and molecules are transported into and out of organelles during biochemical processes. For example, fatty acids are transported into organelles known as mitochondria so that they may be broken down to generate energy.

CELL MOVEMENT Organized movement is one of the most obvious characteristics of living organisms. The intricate and coordinated activities required to sustain life require the movement of cell components. Examples in eukaryotic cells include cell division and organelle movement, two processes that depend to a large extent on the structure and function of a complex network of protein filaments known as the *cytoskeleton*. The forms of cellular motion profoundly influence the ability of all organisms to grow, reproduce, and compete for limited resources. As examples, consider the movement of protists as they search for food in a pond or the migration of human white blood cells as they pursue infectious foreign cells. More subtle examples include the movement of specific enzymes along a DNA molecule during the chromosome replication that precedes cell division and the secretion of insulin by certain pancreatic cells.

WASTE REMOVAL All living cells produce waste products. For example, animal cells ultimately convert food molecules, such as sugars and amino acids, into CO_2, H_2O, and NH_3. These molecules, if not disposed of properly, can be toxic. Some substances are readily removed. In animals, for example, CO_2 diffuses out of cells and (after a brief and reversible conversion to bicarbonate by red blood cells) is quickly exhaled through the respiratory system. Excess H_2O is excreted through the kidneys. Other molecules, however, are so toxic that specific processes have evolved to provide for their disposal. The urea cycle (described in Chapter 15) provides a mechanism for converting free ammonia and excess amino nitrogen into urea, a less toxic molecule. The urea molecule is then removed from the body through the kidney as a major component of the urine.

Living cells also contain a wide variety of potentially toxic molecules that must be disposed of. Plant cells solve this problem by transporting such molecules into a vacuole, where they are either broken down or stored. Animals, however, must use disposal mechanisms that depend on water solubility (e.g., the formation of urine by the kidney). Hydrophobic substances such as steroid hormones, which cannot be broken down into simpler molecules, are converted during a series of reactions into water-soluble derivatives. This mechanism is also used to solubilize some exogenous organic molecules such as drugs and environmental contaminants.

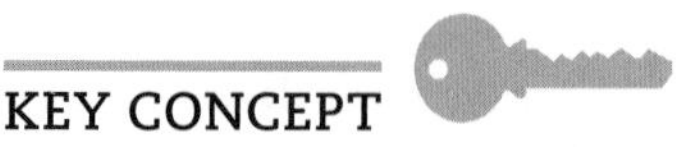

KEY CONCEPT

In living organisms, processes of highly ordered complexity are sustained by a constant input of energy.

1.4 SYSTEMS BIOLOGY

The discovery of the information in the overview of biochemical processes that you have just read about was made possible by a method of inquiry based on *reductionism.* In this powerful, mechanistic strategy a complex, living "whole" is studied by "reducing" it to its component parts. Each individual part is then further broken down so that the chemical and physical properties of its molecules and the connections between them can be determined. Most of the accomplishments of the modern life sciences would have been impossible without the reductionist philosophy. However, reductionism has its limitations because of the assumption that detailed knowledge of all the properties of the parts will of itself ultimately provide a complete understanding of the functioning of the whole. Despite intense efforts, a coherent understanding of dynamic living processes continues to elude investigators.

In recent decades a new approach called systems biology has been utilized to achieve a deeper understanding of living organisms. Based on the engineering principles originally developed to build jet aircraft, **systems biology** regards living organisms as integrated systems. Each system allows certain functions to be performed. One such system in animals is the digestive system, which comprises a group of organs that is tasked to break down food into molecules that can be absorbed by the body's cells.

Although human-engineered systems and living systems are remarkably similar in some respects, they are significantly different in others. The most important difference is the design issue. When engineers plan a complex mechanical or electrical system, each component is designed to fulfill a precise function, and there are no unnecessary or unforeseen interactions between network components. For example, the individual electrical wires in the cables that control aircraft are insulated to prevent damage caused by short circuits. In contrast, biological systems have evolved by trial and error over several billion years. Evolution, the adaptation of populations of living organisms in response to selection pressure, is made possible by the capacity to generate genetic diversity through various forms of mutation, gene duplications, or the acquisition of new genes from other organisms. The components of living organisms, unlike engineer-designed parts, have no fixed functions, and overlapping functions are permissible. Living systems have become increasingly more complex, in part because of the unavoidability of interactions among established system components and potentially useful new parts (e.g., derived from gene duplications followed by mutations).

The systems approach is especially useful because the human mind cannot analyze the hundreds of biochemical reactions that are taking place at once in a living organism. To tackle this problem, systems biologists have invented mathematical and computer models to derive from biochemical reaction pathways an understanding of how these processes operate over time and under varying conditions. The success of these models is reliant on huge data sets containing accurate information about cellular concentrations of biomolecules and the rates of biochemical reactions as they occur in living, functioning cells. Although these data sets are incomplete, this analytic method has produced some notable successes. The technology required to identify and quantify biomolecules of all types continues to be refined. System biologists have identified two core principles that underpin the complex and diverse biochemical pathways described in this textbook: emergence, and robustness. In addition, systems biologists organize the vast complexities of living cells with concepts such as systems, networks, modules, and motifs, which are also briefly described.

Emergence

As we have discovered, the behavior of complex systems cannot be understood simply by knowing the properties of constituent parts. At each level of organization of the system, new and unanticipated properties emerge from interactions among parts. For example, hemoglobin (the protein that transports oxygen in the blood to the body's cells) requires ferrous iron (Fe^{2+}) to be functional. Whereas iron easily oxidizes in the inanimate world, the iron in hemoglobin does not usually oxidize even though it is linked directly to oxygen during the transport process. The amino acid residues that line the iron-binding site protect Fe^{2+} from oxidation. The protection of ferrous iron in hemoglobin is an **emergent property**, that is, a property conferred by the complexity and dynamics of the system.

Robustness

Systems that remain stable despite diverse perturbations are described as *robust.* Autopilot systems in aircraft, for example, maintain a designated flight path despite expected fluctuations in conditions such as wind speed or the plane's mechanical functions. All robust systems are necessarily complex because failure prevention requires an integrated set of automatic fail-safe mechanisms. The robust (fail-safe) properties of human-made mechanical systems are created by *redundancy*, the use of duplicate parts (e.g., backup electric generators in an airplane). Although the design of living organisms does include some redundant parts, the robust properties of living systems are largely the result of **degeneracy**, the capacity of structurally different parts to perform the same or similar functions. The genetic code is a simple, well-recognized example. Of the 64 possible three-base sequences (called codons) on an mRNA molecule, 61 base triplets code for 20 amino acids during protein synthesis. Since most amino acids have more than one codon, degeneracy of the code provides a measure of protection against base substitution mutations.

Systems Biology Model Concepts

The research efforts of systems biologists have resulted in the development of simplifying models, which facilitate the efforts of life science researchers, as well as students, to understand the vast complexities of living organisms. Terminology used in systems biology includes system, network, module, and motif.

SYSTEM A *system* is defined as an interconnected and interacting assembly of biomolecules. Systems under investigations can be organisms, organs, cells, or organelles. For example, the mitochondrion is an organelle (a type of cell compartment in the cells of organisms such as animals and plants). It possesses structural features and biochemical pathways that convert the energy in food molecules into the chemical energy required to drive cell processes and synthesize numerous biomolecules, among other functions.

NETWORK Systems can be thought of as the dynamic interaction of networks, each of which is a group of interconnected molecules that performs one or more functions. Living organisms possess metabolic, signaling, and regulatory networks. A *metabolic network* consists of interconnected biochemical reaction pathways that synthesize and degrade biomolecules. Reactant and product molecules connect these pathways to each other. For example, glycolysis, the pathway that degrades the sugar glucose, is linked to energy capture pathways within mitochondria by pyruvate, the product of glycolysis. Pyruvate is transported into the mitochondrion, where biochemical reactions in another pathway begin the process of capturing the energy in its hydrogen atoms. Glycolysis is also linked to amino acid biosynthetic pathways because certain glycolytic intermediates serve as precursor molecules.

Living organisms must perceive and correctly respond to both their internal and external environments. Cells acquire and process information through vast, intricate *signaling networks* composed of receptor proteins that receive information and signaling pathways, whose components process it. For example, the binding of epinephrine (p. 309) to its receptor on the surface of muscle and liver cells initiates a signaling mechanism that results in the activation of enzymes that degrade glycogen.

Living organisms have elaborate, robust mechanisms that tightly control metabolic pathways. This control is accomplished by *regulatory networks* that switch on and off the genes that code for the synthesis of enzymes and all other biomolecules. For example, the binding of the hormone insulin to its receptor on the surface of its target cells sets in motion a signaling mechanism that alters the expression of numerous genes (e.g., enzymes in glycogen and triacylglycerol synthesis). However, gene regulatory networks in living organisms are inextricably integrated with other networks. For example, insulin receptor binding also triggers a signaling pathway that quickly modifies the activity of several biochemical pathways by stimulating the activity of certain enzymes while inhibiting others.

MODULE Complex systems are composed of *modules*, components or subsystems that perform specific functions. Living organisms utilize modules because they are easily assembled, rearranged, and repaired, as well as eliminated when necessary. Although modules (e.g., enzymes extracted from cells in the lab) can often be isolated with some or even most of their functional properties, their function is meaningful only within the context of the larger system. In living organisms, modularity occurs at all system levels. Examples within a cell include amino acids, proteins, and biochemical pathways. Modularity is especially important because it provides the capacity to limit damage to components that can be easily removed and replaced. For example, glycolysis can be considered a module. Functional relationships between modules in a system are managed by *protocols*, or sets of rules that specify how and whether modules will interact. The mechanism that facilitates pyruvate transport into a mitochondrion is an example of a protocol.

MOTIF Network motifs are recurring regulatory circuits that have many different uses. In living organisms the most common type is **feedback control** (Figure 1.22), a self-regulating mechanism in which the product of a process acts to modify the process, either negatively or positively. In *negative feedback*, the most common

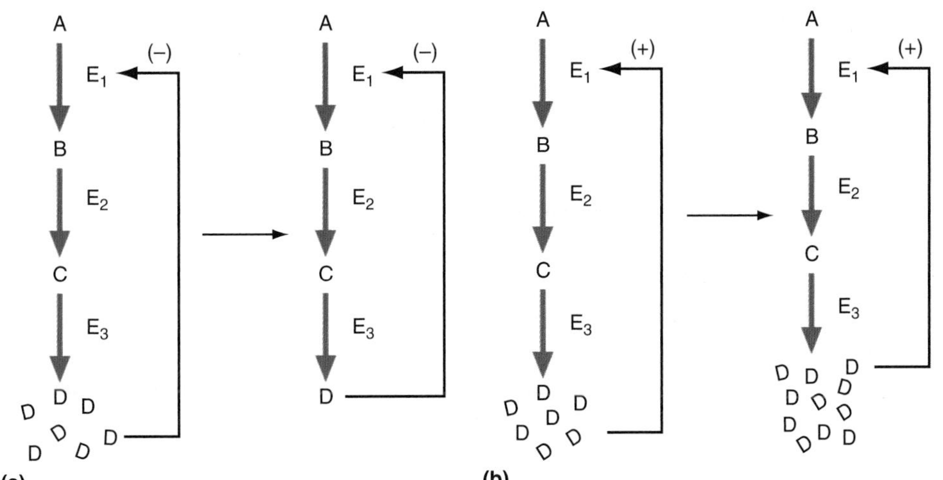

FIGURE 1.22
Feedback Mechanisms

(a) **Negative Feedback**. As a product molecule accumulates, it binds to and inhibits the activity of an enzyme in the pathway. The result is the decreased production of the product. (b) **Positive Feedback**. As product molecules accumulate, they stimulate an enzyme in the pathway, thereby causing an increased rate of product synthesis.

form, an accumulating product slows its own production. Many biochemical pathways are regulated by negative feedback. Typically a pathway product inhibits an enzyme near the beginning of the pathway. In *positive feedback* control, a product increases its own production. Positive feedback control is found less often in living organisms because the mechanism is potentially destabilizing. If not carefully controlled, the amplifying effect of a positive feedback loop can result in the collapse of the system. In blood clotting, for example, the platelet plug that seals a damaged blood vessel does not expand continuously because inhibitors are released by nearby undamaged blood vessel cells.

Living Organisms and Robustness

Fail-safe control mechanisms, whether in human-made systems or in living organisms, are expensive. Constraints such as energy availability require priorities in resource allocation. As a result, systems are generally protected from commonly encountered environmental changes, but they are vulnerable to unusual or rare events that cause damage. This vulnerability, referred to as *fragility*, is an inescapable feature of robust systems. Cancer, a group of diseases in which cell cycle control is disrupted, is an example of the "robust, yet fragile" nature of robust systems. Despite the elaborate controls on cell division in animal bodies, mutations in just a few of the genes that code for cell cycle regulatory proteins can result in the uncontrolled proliferation of the affected cell.

KEY CONCEPTS

- Systems biology is an attempt to reveal the functional properties of living organisms by developing mathematical models of interactions from available data sets.
- The systems approach has provided insights into the emergent properties, robustness, and modularity of living organisms.

Feedback control - product of a self regulating mechanism to modify the process either negatively or positively.

Biochemistry IN THE LAB

An Introduction

Biochemical technologies exploit the chemical and physical properties of biomolecules: chemical reactivity, size, solubility, net electrical charge, movement in an electric field, and absorption of electromagnetic radiation. As life science research has become increasingly more sophisticated, scientists have provided a progressively more coherent view of the living state. The Human Genome Project was a landmark event in this process. The goal of this international research effort, begun in the late 1980s, was to determine the nucleotide base sequence of human DNA. The subsequent development of automated DNA sequencing technology revolutionized life science research because it provided scientists with a "high-throughput" (i.e., rapid, high-volume, relatively inexpensive) means of investigating the information content of genomes, a field now referred to as **genomics**.

Genomics has been especially useful in medical research. A large number of human diseases have been linked to errors in one or more gene sequences or to faulty regulation of gene expression. Among the early benefits of this work are rapid, accurate tests for predisposition to pathological conditions such as cystic fibrosis, breast cancer, and some liver diseases. Several recently developed technologies have created additional opportunities to investigate the molecular basis of disease. For example, DNA microchips (thousands of DNA molecules arrayed on a solid surface) are now routinely used to monitor gene expression of cells. Proteins can also be rapidly analyzed with a combination of gel electrophoresis and mass spectrometry. Among the new fields created by high-throughput methods are **functional genomics** (the investigation of gene expression patterns) and **proteomics** (the investigation of protein synthesis patterns and protein-protein interactions). The science of **bioinformatics** is the computer-based field that facilitates the analysis of the massive amounts of protein and nucleic acid sequence data that are being generated. System biologists take advantage of all of these methods to decipher biological networks such as metabolic process control, gene regulation, and signal transduction (information processing) mechanisms.

In the past, biochemists and other scientists have often benefited from each other's work. For example, technologies created by physicists such as X-ray diffraction, electron microscopy, and radioisotope labeling made biomolecular structure investigations possible. In recent years, the life sciences have also benefited from the services provided by computer scientists, mathematicians, and engineers. As the biological knowledge base has continued to expand, it has become increasingly obvious that future advances in life science and medical research will require the efforts of multidisciplinary teams of scientists.

Chapter Summary

1. Biochemistry may be defined as the study of the molecular basis of life. Biochemists have contributed to the following insights into life: (1) life is complex and dynamic, (2) life is organized and self-sustaining, (3) life is cellular, (4) life is information-based, and (5) life adapts and evolves.
2. Animal and plant cells contain thousands of different types of molecules. Water constitutes 50 to 90% of a cell's content by weight, and ions such as Na^+, K^+, and Ca^{2+} may account for another 1%. Almost all the other kinds of biomolecules are organic.
3. Many of the biomolecules found in cells are relatively small, with molecular weights of less than 1000 daltons. Cells contain four families of small molecules: amino acids, sugars, fatty acids, and nucleotides.
4. DNA, consisting of two antiparallel polynucleotide strands, is the repository of genetic information in living organisms. DNA contains coding sequences, referred to as genes, and noncoding sequences, some of which have regulatory functions. RNA is a single-stranded polynucleotide that differs from DNA in that it contains the sugar ribose instead of deoxyribose and the base uracil instead of thymine. RNAs have numerous functions. Examples include protein synthesis and transcription regulation. Gene expression, the process that controls if or when a gene will be transcribed, involves the binding of transcription factors to specific regulatory DNA sequences called response elements.
5. All life processes consist of chemical reactions catalyzed by enzymes. Among the most common reaction types encountered in biochemical processes are nucleophilic substitution, elimination, addition, isomerization, and oxidation-reduction.
6. Living organisms require a constant flow of energy to prevent disorganization. The principal means by which cells obtain energy is oxidation of biomolecules or certain minerals.
7. Metabolism is the sum of all the reactions in a living organism. There are two types of metabolic pathway: anabolic and catabolic. Energy transfer pathways capture energy and transform it into forms that organisms can use to drive biomolecular processes. Signal transduction pathways, which

allow cells to receive and respond to signals from their environment, consist of three phases: reception, transduction, and response.

8. The complex structure of cells requires a high degree of internal order. This is accomplished by four primary means: synthesis of biomolecules, transport of ions and molecules across cell membranes, production of movement, and removal of metabolic waste products and other toxic substances.

9. Systems biology is a new field that attempts to provide understanding of the functional properties of living organisms by applying mathematical modeling strategies to amassed biological data. Among the early benefits of the systems approach are the insights associated with emergence, robustness, and modularity.

Take your learning further by visiting the **companion website** for Biochemistry at **www.oup.com/us/mckee** where you can complete a multiple-choice quiz on this introductory chapter to help you prepare for exams.

Suggested Readings

Campbell, N. A., and Reece, J. B., *Biology*, 9th ed., Benjamin Cummings, San Francisco, 2011.

Goodsell, D. S., *The Machinery of Life*, 2nd ed., Springer, New York, 2009.

Rothman, S., *Lessons from the Living Cell: The Limits of Reductionism*, McGraw-Hill, New York, 2002.

Tudge, C., *The Variety of Life: A Survey and a Celebration of All the Creatures That Have Ever Lived*, Oxford University Press, New York, 2000.

Key Words

addition reaction, *17*
amino acid, *7*
anabolic pathway, *19*
anhydride, *14*
autopoiesis, *13*
autotroph, *19*
bioinformatics, *26*
biomolecule, *3*
catabolic pathway, *20*
chemoautotroph, *19*
chemoheterotroph, *19*
chemosynthesis, *19*
degeneracy, *23*
electrophile, *14*
elimination reaction, *15*
emergent property, *23*
energy, *19*
enzyme, *3*
fatty acid, *10*
feedback control, *24*
functional genomics, *26*
functional group, *5*
gene, *4*
gene expression, *13*
genomics, *26*
heterotroph, *19*
homeostasis, *3*
hydration reaction, *17*
hydrocarbon, *5*
hydrolysis, *14*
hydrophilic, *7*
hydrophobic, *5*
isomerization, *17*
leaving group, *14*
lipid, *10*
macromolecule, *3*
metabolism, *3*
modules, *24*
monosaccharide, *8*
mutation, *5*
negative feedback, *25*
neurotransmitter, *7*
noncoding RNA, *13*
nucleic acid, *11*
nucleophile, *14*
nucleophilic substitution, *14*
nucleotide, *10*
oxidation-reduction (redox) reaction, *18*
oxidize, *18*
oxidizing agent, *18*
peptide, *7*
peptide bond, *8*
photoautotroph, *19*
photoheterotroph, *19*
photosynthesis, *19*
polypeptide, *7*
polysaccharide, *8*
positive feedback, *25*
protein, *7*
proteomics, *26*
purine, *11*
pyrimidine, *11*
reduce, *18*
reducing agent, *18*
reductionism, *21*
response element, *13*
robust, *23*
saturated, *10*
signal transduction, *20*
sugar, *8*
systems biology, *22*
transcription, *12*
transcription factor, *13*
unsaturated, *10*

Review Questions

These questions are designed to test your knowledge of the key concepts discussed in this chapter before moving on to the next chapter. You may like to compare your answers to the solutions provided in the back of the book and in the accompanying Study Guide.

1. Define the following terms:
   a. biomolecule
   b. macromolecule
   c. enzyme
   d. metabolism
   e. homeostasis
2. Define the following terms:
   a. gene
   b. mutation
   c. hydrocarbon
   d. hydrophobic
   e. hydrophilic
3. Define the following terms:
   a. functional group
   b. R group
   c. carboxyl group
   d. amino group
   e. hydroxyl group
4. Define the following terms:
   a. polypeptide
   b. peptide
   c. protein
   d. peptide bond
   e. standard amino acids
5. Define the following terms:
   a. sugar
   b. monosaccharide
   c. polysaccharide
   d. glucose
   e. cellulose
6. Define the following terms:
   a. fatty acid
   b. saturated fatty acid
   c. unsaturated fatty acid
   d. triacylglycerol
   e. phosphoglyceride
7. Define the following terms:
   a. nucleotide
   b. purine
   c. pyrimidine
   d. nucleic acid
   e. ribose
8. Define the following terms:
   a. DNA
   b. RNA
   c. double helix
   d. genome
   e. transcription
9. Define the following terms:
   a. mRNA
   b. tRNA
   c. rRNA
   d. siRNA
   e. miRNA
10. Define the following terms:
    a. transcription factor
    b. response element
    c. signal molecule
    d. RNA interference
    e. ribosome
11. Define the following terms:
    a. nucleophile
    b. electrophile
    c. leaving group
    d. adenosine triphosphate
    e. anhydride
12. Define the following terms:
    a. elimination reaction
    b. hydrolysis
    c. addition reaction
    d. dehydration reaction
    e. hydration reaction
13. Define the following terms:
    a. redox reaction
    b. oxidizing agent
    c. reducing agent
    d. NADH
    e. oxidized molecule
14. Define the following terms:
    a. FAD
    b. hydride ion
    c. energy
    d. electron transport pathway
    e. coenzyme
15. Define the following terms:
    a. autotroph
    b. chemoautotroph
    c. photoautotroph
    d. chemoheterotroph
    e. photoheterotroph
16. Define the following terms:
    a. metabolic pathway
    b. anabolic pathway
    c. catabolic pathway
    d. glycolysis
    e. signal transduction pathway

17. Define the following terms:
    a. system
    b. network
    c. emergent property
    d. degeneracy
    e. feedback control
18. List three Life Science fields that require a solid understanding of biochemical principles.
19. What are the six major elements present in living organisms?
20. Identify the functional groups in the following molecules.

(a) $CH_3C(=O)-H$

(b) $HO-C(=O)-CH_2CH_2CH(NH_2)-C(=O)-OH$

(c) $CH_3CH_2CH(SH)CH_3$

(d) $CH_3C(=O)-O-CH_3$

(e) $CH_3(H)C=C(H)CH_2CH_3$

(f) $CH_3C(=O)-N(H)-CH_2CH_3$

(g) $CH_3C(=O)CH_3$

(h) $CH_2OH-CH(OH)-CH_2OH$

21. Name four classes of small biomolecules. In what larger biomolecules are they found?
22. List two functions for each of the following biomolecules:
    a. fatty acids
    b. sugars
    c. nucleotides
    d. amino acids
23. What are the roles of DNA and RNA?
24. How do cells obtain energy from chemical bonds?
25. Identify the following reaction type.

$$C_6H_5C(=O)-NH_2 + H_2O \longrightarrow C_6H_5C(=O)-O^- + NH_4^+$$

26. Identify the oxidizing and reducing agents in the following reaction:

$$CH_3CH_2C(=O)-OH + NADH + H^+ \longrightarrow CH_3CH_2C(=O)H + NAD^+ + H_2O$$

27. What are the general characteristics of anabolic and catabolic processes?
28. Classify each of the following reactions as anabolic or catabolic:
    a. Glucose $\longrightarrow$ Cellulose
    b. Glucose + ADP + P_i $\xrightarrow{O_2}$ CO_2 + ATP + NADH
29. Name two common greenhouse gases.
30. List three types of biochemical reactions involving acyl transfer nucleophilic substitution.
31. What reaction is the first step in utilizing glucose as an energy source?
32. How do plants dispose of waste products?
33. Assign each of the following compounds to one of the major classes of biomolecule:

(a) indole ring (N–H) $-CH_2-CH(NH_2)-C(=O)-OH$

(b) pyranose ring with CH_2OH, HO, OH, OH, OH

(c) $CH_3-(CH_2)_9-CH_2-C(=O)-OH$

(d) $^-O-P(=O)(O^-)-O-CH_2$ ribose (OH, OH) attached to adenine (NH_2)

34. What are the primary functions of metabolism?
35. Give an example of each of the following reactions:
    a. nucleophilic substitution
    b. elimination
    c. oxidation-reduction
    d. addition
36. List several important ions that are found in living organisms.
37. Compare and contrast the features of an airplane autopilot system with a biological system.
38. Carbohydrates are widely recognized as sources of metabolic energy. What are two other critical roles that carbohydrates play in living organisms?
39. Describe several functions of polypeptides.
40. What are the largest biomolecules? What functions do they serve in living organisms?
41. Nucleotides have roles in addition to being components of DNA and RNA. Give an example.
42. Name several waste products that animal cells produce.
43. Provide several examples of emergent properties.
44. Compare the functions of mRNA, rRNA, and tRNA in protein synthesis.
45. Describe the significance of the phrase "robust yet fragile."
46. Compare and contrast the general features of human-designed complex systems and living systems.
47. Compare an autopoietic system with a factory that manufactures airplanes.

Fill in the Blank

48. The indicated molecule contains an ________ functional group.

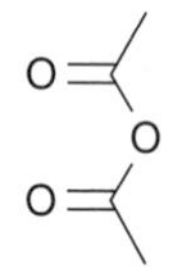

O O O

49. The biologically active form of a fatty acid is the thioester of ________________.
50. ________________ is the capacity of structurally different components of a system to perform the same or similar functions.
51 The sum of all reactions in a living organism is called its ________________.
52. The following is an example of a ________________ reaction.

$$CH_3CH_2OH \rightarrow CH_2 = CH_2 + H_2O$$

53. Sequence changes in DNA are called ____________.
54. The following is an example of which class of biomolecules?

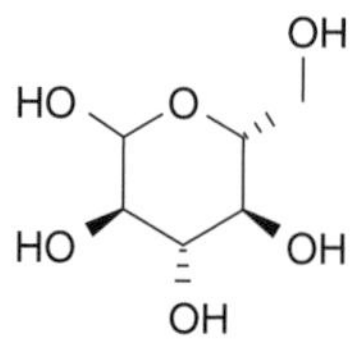

OH HO O HO OH OH

55. Each organism is considered an autonomous self-organizing, self-maintaining entity. This set of properties is referred to as ________________.
56. ________________ are organisms that transform the energy of the sun into chemical bond energy.
57. A property conferred by the complexity and dynamics of the system is called an ________________ property.

Short Answer

58. Distinguish between living organisms and human-made factories.
59. Which molecule contains more energy: carbon dioxide (CO_2) or ethanol (CH_3CH_2-OH)? Explain.
60. Heliobacteria convert light energy into chemical energy and use organic molecules as a carbon and energy source. How would you classify them?
61. Compare anabolic and catabolic pathways. What molecules link these two processes?
62. Why is cancer, which occurs as the result of disrupted cell cycle control, an example of the fragility of a robust system?

Thought Questions

These questions are designed to reinforce your understanding of all of the key concepts discussed in the book so far. They may not have one right answer! The authors have provided possible solutions to these questions in the back of the book and in the accompanying Study Guide for your reference.

63. Carboxylic acids that undergo nucleophilic acyl substitution reactions are often first converted to thioesters. For example, acetic acid forms thioester with a molecule called coenzyme A, which has a sulfhydryl group.

$$CH_3\overset{\overset{\displaystyle O}{\|}}{C}—S—\text{Coenzyme A}$$

What is the leaving group in these reactions?

64. Why are fatty acids the principal long-term energy reserve of the body?

65. When a substance such as sodium chloride is dissolved in water, the ions that form become completely surrounded by water molecules, which form structures called hydration spheres. When the sodium salt of a fatty acid is mixed with water, the carboxylate group of the molecule becomes hydrated but the hydrophobic hydrocarbon portion of the molecule is poorly hydrated, if at all. Using a circle to represent the carboxylate group and an attached squiggly line to represent the hydrocarbon chain of a fatty acid, draw a picture of how fatty acids interact in water.

66. The order of reactivity in nucleophilic substitution reactions is as follows: phosphate > thiols > esters > amides. Explain this order on the basis of their pK_a values: phosphoric acid (1×10^{-3}), hydrogen sulfide (1×10^{-7}), alcohols (1×10^{-16}), and ammonia (1×10^{-36}). (pK_a is an acid dissociation constant, which is a quantitative measure of the strength of an acid in solution, i.e., the tendency of an acid to lose a proton.)

67. Elements such as carbon, hydrogen, and oxygen that occur in biomolecules form stable covalent bonds. What would be the result if the bonds between these atoms were either slightly less or more stable than the naturally occurring bonds?

68. Humans synthesize most of the cholesterol required for cell membranes and for the synthesis of vitamin D and steroid hormones. What would you expect to happen if a person's diet is high in cholesterol? Provide a reason for your response.

69. Tay-Sachs disease is a devastating genetic neurological disorder caused by the lack of the enzyme that degrades a specific lipid molecule. When this molecule accumulates in brain cells, an otherwise healthy child undergoes motor and mental deterioration within months after birth and dies by the age of 3 years. In general terms, how would a systems biologist evaluate this phenomenon?

70 How does the statement "the whole is more than the sum of its parts" apply to living organisms? Give an example.

71. Unlike human-engineered systems, the components of biological systems often have multiple functions. Is this phenomenon a strength or a weakness of biological systems? Explain.

72. The cancerous cells in a tumor proliferate uncontrollably, and treatment often involves the use of toxic drugs in attempts to kill them. Often, however, after initial success (i.e., shrinkage of the tumor), the cancer returns because resistance to the drugs has developed. Biochemists have identified one of the major causes of this phenomenon, called multidrug resistance. One or more cells in the tumor have expressed the gene for P-glycoprotein, a membrane transport protein that pumps the drugs out of the cells. In the absence of the toxic drug molecules, these cells grow uncontrollably and eventually become the dominant cells in the tumor. What features of living organisms does this process illustrate?

73. Hundreds of thousands of proteins have been discovered in living organisms. Yet, as astonishing as this diversity is, these molecules constitute only a small fraction of those that are possible. Calculate the total number of possible decapeptides (molecules with 10 amino acid residues linked by peptide bonds) that could be synthesized from the 20 standard amino acids. If you were to spend 5 minutes writing out the molecular structure of each possible decapeptide, how long would the task take?

CHAPTER

2 Living Cells

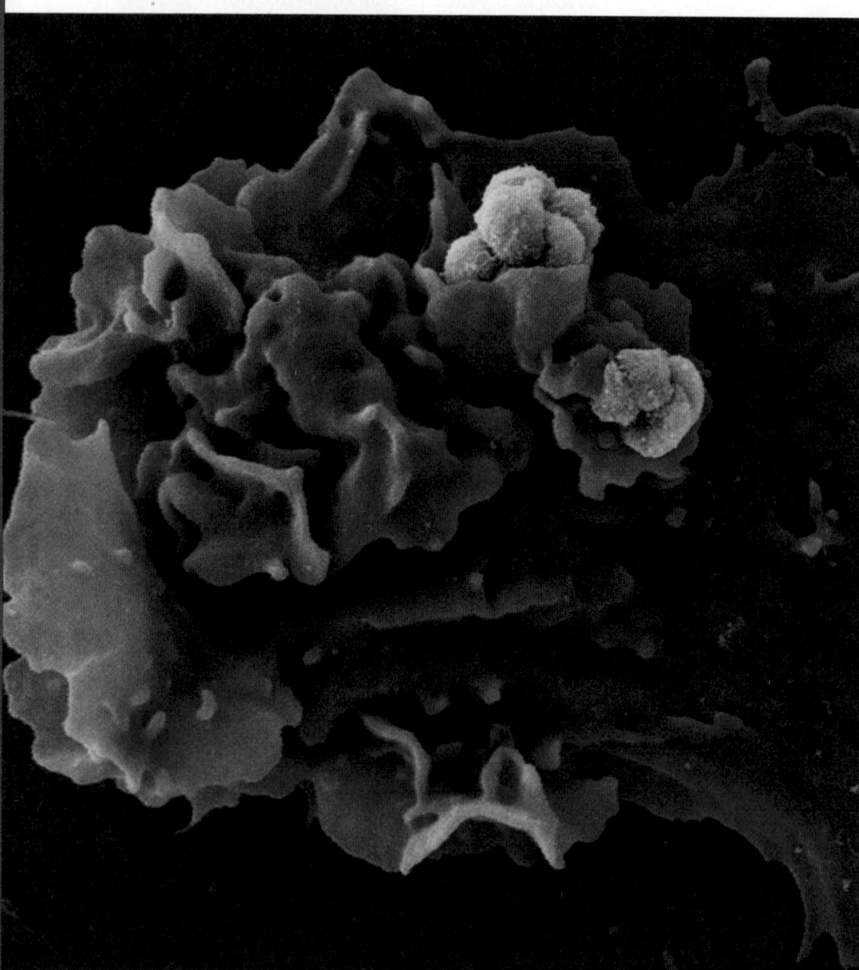

A Scavenger Cell Engulfing Bacterial Cells
In this pseudocolored scanning electron micrograph, a neutrophil (blue) is in the process of engulfing bacterial cells (yellow). Neutrophils, the most abundant type of white blood cell in mammals, are phagocytes, cells that are capable of ingesting microorganisms.

OUTLINE

Our Bodies, Our Selves

It would surprise most humans that of the estimated 100 trillion cells in our bodies only 10 trillion are actually ours. The other 90 trillion are microorganisms. Most of these organisms, referred to as an indigenous flora or **microbiota** (**Figure 2.1**), are bacteria with smaller numbers of *archaeans* (another type of prokaryotes) and fungi. Humans and their microbiota have evolved together into a dynamic, interdependent superorganism. This relationship is usually symbiotic (i.e., it is beneficial in some way) or commensal (nonharmful). However, a few species in the normal human microbiota are pathogens because they can cause disease if conditions permit (e.g., if the immune system is depressed).

Our bodies, sterile before birth, begin acquiring microbes as soon as the amniotic sac ruptures. As babies proceed down the birth canal, colonization begins as they are exposed to their mother's microbiota. Within a short time, a diverse array of microbes has taken up residence in all body surfaces that are exposed to the external environment: skin and certain parts of the respiratory, gastrointestinal (GI), and urogenital tracts. These ecosystems, each with its own set of environmental conditions (e.g., temperature, pH, and O_2 availability), eventually possess their own characteristic communities of microorganisms. For example, the bacterial flora in the intestine (between 500 and 1000 species by some estimates) provides a spectrum of beneficial services in exchange for a stable nutrient supply and favorable environmental conditions. Examples include improved digestion (e.g., about 5% of human energy requirements are normally generated by microbes that digest soluble dietary fiber molecules), vitamin synthesis (e.g., K and B_{12}), pathogen growth repression, and robust immune system development.

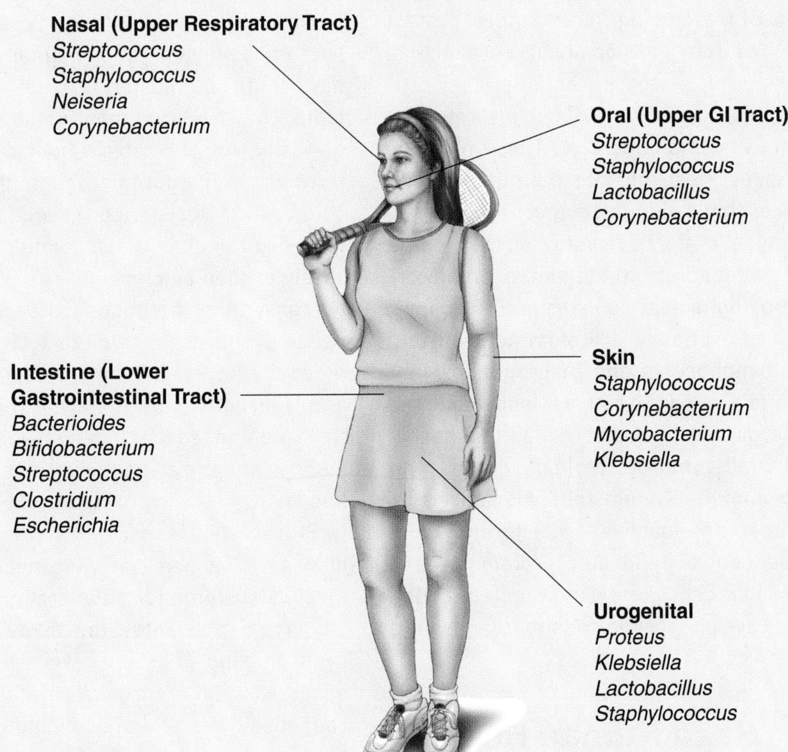

FIGURE 2.1
The Human Microbiota

Each healthy human is host to an exceptionally large and unique set of highly adapted microorganisms. Examples of the major groups of organisms observed in the normal flora are given for each ecosystem. Note that although some bacterial groups occur at different body locations, the species often differ.

Our Bodies, Our Selves cont.

Experiments with gnotobiotic (germ-free) mice have revealed that the absence of a microbiota has a profound effect on health. In addition to requiring 30% more calories to maintain body weight than conventional mice, germ-free mice were observed to have underdeveloped intestinal architecture and fat storage capacity, as well as smaller organs (heart, lungs, and liver). These mice are also highly susceptible to numerous infections caused by transient pathogens (e.g., respiratory or intestinal viruses) because their immune systems are immature.

Defense Mechanisms

Despite the many benefits of the human indigenous microbiota, the body must constantly protect itself from the microbiota's potentially unrestricted growth. Strategies used to prevent damage include impenetrable epithelial tissue barriers and immune system cells. Epithelial barriers in the GI tract lining, for example, protect internal organs from invasion of microorganisms. Epithelial cells also produce antimicrobial proteins. Examples include the α-defensins, released from cells of the small intestine. These peptides kill bacteria by inserting into membranes and forming pores that cause cell rupture.

The immune system normally strikes a fine balance of protection from pathogens and tolerance of nonpathogenic organisms and the body's own cells. Approximately 70% of immune system cells (e.g., lymphocytes and macrophages) are located in and around the GI tract (especially the small and large intestine), where maintaining this balance is vital. *Lymphocytes* (e.g., T cells and B cells) are immune system cells specialized for the recognition and inactivation of foreign cells and the production of antibodies (proteins used to identify foreign cells or virus-infected cells). *Macrophages* are phagocytic cells that ingest and digest foreign cells.

Gut Flora and Human Health

Each person has a unique microbiota that is the result of genetic inheritance, birth circumstances (e.g., vaginal delivery vs. Caesarean section and the mother's microbiota), diet, and environment (e.g., exposure to antibiotics or acid-suppressing medications). Although several eukaryotes such as yeast are present, bacteria dominate the human gut microbiota. In most humans, more than 90% of the microorganisms in the intestines belong to two phyla: the Firmicutes and Bacteroidetes. (A phylum is a primary division of a taxonomic domain, in this case the domain eubacteria.) A healthy gut flora is characterized as diverse, relatively stable, and resilient. The mechanism that maintains this condition, referred to as *colonization resistance*, protects the body from the challenge of exogenous and potentially pathogenic microbes. Colonization resistance is a constellation of tactics involving endogenous microbes (e.g., competition for space in the mucus layer lining the intestinal lumen and intermicrobial chemical warfare) and host defenses (e.g., peristalsis, which rapidly eliminates newly ingested microorganisms).

In recent years, antibiotic use and poor diets (e.g., those high in fat and sugar and low in fiber) have contributed to a rise in disrupted gut microbiota, characterized most notably by low species diversity. Antibiotic treatment causes major alterations of the gut microbiota, thereby reducing colonization resistance. Frequent antibiotic use can result in *dysbiosis*, a condition in which a severely eroded colonization resistance allows the overgrowth of pathogens. Poor diets can also cause dysbiosis since dietary factors play a significant role in shaping the microbiota. For example, in both humans and lab animals the switch from a balanced diet to a high-fat diet causes the loss of beneficial bacteria, some of which are known to protect cells in the gut lining. In addition to GI tract diseases such as inflammatory bowel disease, dysbiosis has also been linked to numerous systemic disease states, including obesity, type 2 diabetes (pp. 596–597), metabolic syndrome (pp. 607), and several autoimmune diseases. Low-level chronic systemic inflammation, caused by the leakage of bacterial molecules such as lipopolysaccharide (also called endotoxin) across a gut wall compromised by the loss of protective bacteria, is now believed to be a significant feature of most metabolic diseases.

Overview

CELLS ARE THE STRUCTURAL UNITS OF ALL LIVING ORGANISMS. ONE REMARKABLE FEATURE OF CELLS IS THEIR DIVERSITY: THE HUMAN BODY contains about 200 types of cells. This great variation reflects the variety of functions that cells can perform. However, no matter what their shape, size, or species, cells are also amazingly similar. They are all surrounded by a membrane that separates them from their environment. They are all composed of the same types of molecules.

The structural hierarchy of life on earth extends from the biosphere to biomolecules. Each level is inextricably linked to the levels above and below it. Cells, however, are considered the basic unit of life, since they are the smallest entities that are actually alive. Cells are complex, intricate molecular machines that can sense and respond to their environment, transform matter and energy, and reproduce themselves.

Living cells are classified as either prokaryotic or eukaryotic. Prokaryotes are single-celled organisms that lack a nucleus (*pro* = "before," *karyon* = "nucleus" or "kernel"). Analysis of the RNA in prokaryotes has revealed that there are two distinct types of prokaryotes: the Bacteria and the Archaea. Some bacterial species cause disease (e.g., cholera, tuberculosis, syphilis, and tetanus), whereas others are of practical interest for humans (e.g., those used to make foods such as yogurt, cheese, and sourdough bread). A prominent feature of the Archaea is their unsurpassed capacity to occupy and even thrive in very challenging habitats. The eukaryotes (*eu* = "true") are composed of relatively large cells that possess a nucleus, a membrane-bound compartment that contains the cell's DNA. Animals, plants, fungi, and single-celled protists are examples of the eukaryotes. Eukaryotes also differ from prokaryotes in size and complexity. The volume of a typical eukaryotic cell such as a hepatocyte (liver cell) is between 6000 and 10,000 μm^3. The volume of the bacterium *Escherichia coli* is significantly smaller at 2 to 4 μm^3. Although the structural complexity of prokaryotes is significant, that of eukaryotes is greater by several orders of magnitude, largely because of subcellular compartments called **organelles**. Each organelle is specialized to perform specific tasks. The compartmentalization afforded by organelles creates microenvironments in which biochemical processes can be efficiently regulated. In multicellular eukaryotes, complexity is increased by cellular specialization and intercellular communication mechanisms.

The common features of prokaryotic and eukaryotic cells include their similar chemical composition and the universal use of DNA as genetic material. This chapter provides an overview of cell structure. This review is a valuable exercise because biochemical reactions do not occur in isolation. Our understanding of living processes is incomplete without knowledge of their cellular context. After a brief discussion of some basic themes in cellular structure and function, the essential structural features of prokaryotic and eukaryotic cells are described in relation to their biochemical roles.

2.1 BASIC THEMES

Each living cell contains millions of densely packed biomolecules that perform at a frenetic pace the thousands of tasks that together constitute life. The application of biochemical techniques to investigations of living processes has provided

KEY CONCEPTS

- The chemical and physical properties of water make it an indispensable component of living organisms.
- Hydrophilic molecules interact with water. Hydrophobic molecules do not interact with water.

significant insights into the unique chemical and structural properties of biomolecules that make their functional properties possible. Understanding of the biological context of biochemical processes is enhanced by examining the following key concepts: water, biological membranes, self-assembly, molecular machines, macromolecular crowding, proteostasis, and signal transduction.

Water

Water dominates living processes. Its chemical and physical properties (described in Chapter 3) that result from its unique polar structure and its high concentration make it an indispensable component of living organisms. Among water's most important properties is its capacity to interact with a wide range of substances. In fact, the behavior of all other molecules in living organisms is defined by the nature of their interactions with water. **Hydrophilic** molecules, that is, those that possess positive or negative charges or contain relatively large numbers of electronegative oxygen or nitrogen atoms, interact easily with water. Examples of simple hydrophilic molecules include salts such as sodium chloride and sugars such as glucose. In contrast, **hydrophobic** molecules, such as the hydrocarbons, which possess few if any electronegative atoms, do not interact with water. Instead, when they are mixed with water, hydrophobic molecules spontaneously form clusters, minimizing contact between hydrocarbon chains and water molecules (**Figure 2.2**). Between the two extremes is an enormous group of both large and small biomolecules, each of which possesses its own unique pattern of hydrophilic and hydrophobic functional groups. Living organisms exploit the distinctive molecular structure of each of these biomolecules.

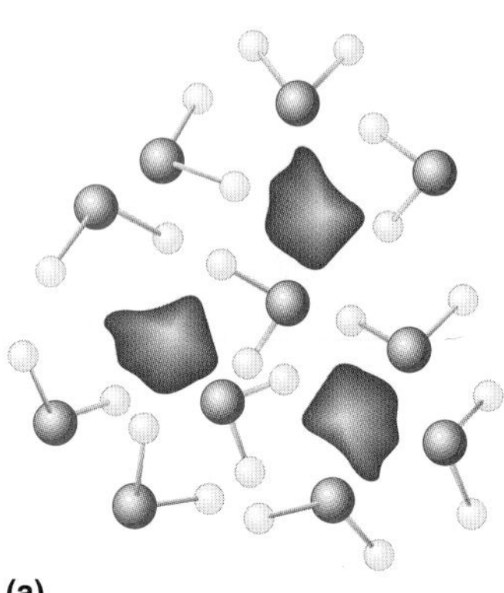

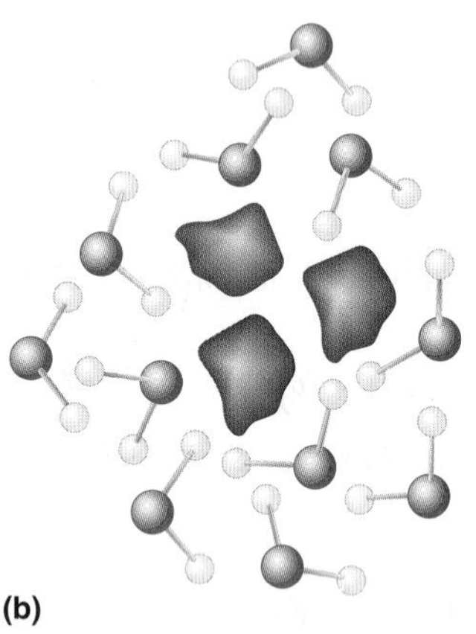

FIGURE 2.2

Hydrophobic Interactions between Water and a Nonpolar Substance

As soon as nonpolar substances (e.g., hydrocarbons) are mixed with water (a), they coalesce into droplets (b). Hydrophobic interactions between nonpolar molecules take effect only when the cohesiveness of water and other polar molecules forces nonpolar molecules or regions of molecules close together.

Biological Membranes

Biological membranes are thin, flexible, and relatively stable sheetlike structures that enclose all living cells and organelles. These membranes can be thought of as noncovalent two-dimensional supramolecular complexes (i.e., they are composed of molecules that are held together by noncovalent intermolecular forces; see pp. 79–81) that provide chemically reactive surfaces and exhibit unique transport functions between the extracellular and intracellular compartments. They are also versatile and dynamic cellular components that are intricately integrated into all living processes. Among the numerous crucial functions that have been assigned to membranes, the most basic is to serve as selective physical barriers. Membranes prevent the indiscriminate leakage of molecules and ions out of cells or organelles into their surroundings and allow the timely intake of nutrients and export of waste products. In addition, membranes have significant roles in information processing and energy generation.

Most biological membranes have the same basic structure: a lipid bilayer composed of phospholipids and other lipid molecules, into which various proteins are embedded or attached indirectly (**Figure 2.3**). Phospholipids have two features that make them ideally suited to their structural role: a hydrophilic charged or uncharged polar group (referred to as a "head group") and a hydrophobic group composed of two fatty acid chains (often called hydrocarbon "tails").

KEY CONCEPTS

- Each biological membrane is composed of a lipid bilayer into which proteins are inserted or attached indirectly.
- Biological membranes are inextricably integrated into all living processes.

There are two classes of membrane proteins: integral and peripheral. **Integral proteins** are embedded within the membrane because the amino acid residues in the membrane-spanning portions of these proteins are hydrophobic. **Peripheral proteins** are not embedded within the membrane. Rather, they are attached to it either by a covalent bond to a lipid molecule or by noncovalent interaction with a membrane protein or lipid. Membrane proteins perform a variety of functions. **Channel** and **carrier proteins** transport specific ions and molecules, respectively. **Receptors** are proteins with binding sites for extracellular ligands (signal molecules). The binding of a ligand to its cognate receptor triggers a cellular response.

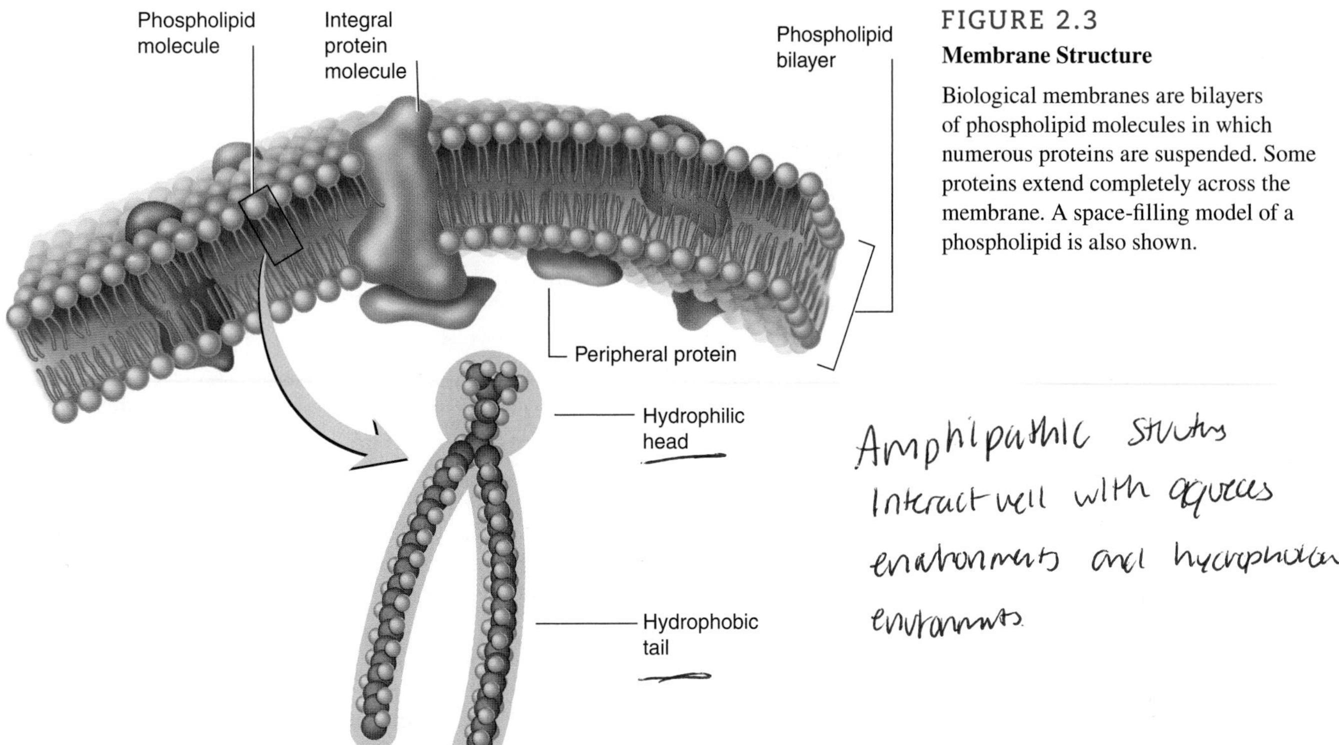

FIGURE 2.3
Membrane Structure

Biological membranes are bilayers of phospholipid molecules in which numerous proteins are suspended. Some proteins extend completely across the membrane. A space-filling model of a phospholipid is also shown.

Self-Assembly

Many of the working parts of living organisms are supramolecular structures. Prominent examples include *ribosomes* (the protein-synthesizing units that are formed from several different types of protein and RNA) and large protein complexes such as the sarcomeres in muscle cells and *proteosomes* (large protein complexes that degrade proteins). According to the principle of self-assembly, most molecules that interact to form stable and functional supramolecular complexes are able to do so spontaneously because they inherently possess the steric information required. They have or are predisposed to have intricately shaped surfaces with complementary structures, charge distributions, and/or hydrophobic regions that allow numerous relatively weak noncovalent interactions (**Figure 2.4**). Self-assembly of such structures involves a balance between the tendency of hydrophilic groups to interact with water and for water to be excluded by hydrophobic groups. In some cases self-assembly processes need assistance. For example, the folding of some proteins requires the aid of molecular chaperones, protein molecules that, among other functions, prevent inappropriate interactions during the folding process. The assembly of certain supramolecular structures (e.g., chromosomes and membranes) requires preexisting information; that is, a new structure must be created on a template of an existing structure.

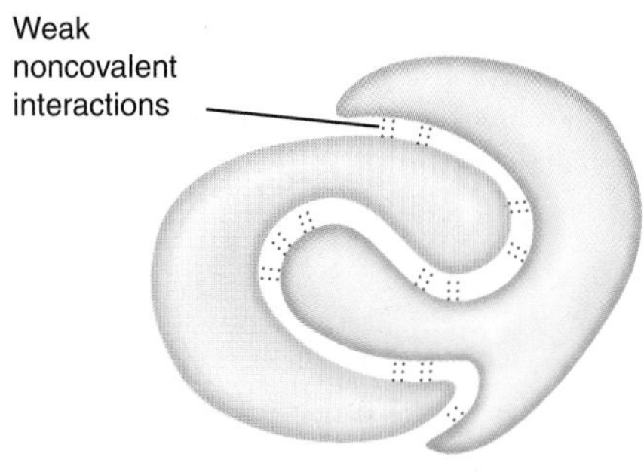

FIGURE 2.4
Self-Assembly

The information that permits the self-assembly of biomolecules consists of the complementary shapes and distributions of charges and hydrophobic groups in the interacting molecules. Large numbers of weak interactions are required for supramolecular structures to form. In this diagrammatic illustration, several weak noncovalent interactions stabilize the binding of two molecules that possess complementary shapes.

Molecular Machines

Researchers now recognize that many of the multisubunit complexes involved in cellular processes function as molecular machines: physical entities with moving parts that perform work, the product of force and distance. Like the mechanical devices used by humans, molecular machines ensure that precisely the correct amount of applied force results in the appropriate amount and direction of movement required for a specific task to be completed. Machines permit the accomplishment of tasks that often would be impossible without them.

Although biological machines are composed of relatively fragile molecules (primarily proteins) that cannot withstand the physical conditions (e.g., heat and

KEY CONCEPTS

- In living organisms the molecules in supramolecular structures assemble spontaneously.
- Biomolecules are able to self-assemble because of the steric information they contain.

FIGURE 2.5
Biological Machines

Proteins perform work when motor protein subunits bind and hydrolyze nucleotides such as ATP. The energy-induced change in the shape of a motor protein subunit causes an orderly change in the shapes of adjacent subunits. In this diagrammatic illustration, a motor protein complex moves attached cargo (e.g., a vesicle) as it "walks" along a cytoskeletal filament.

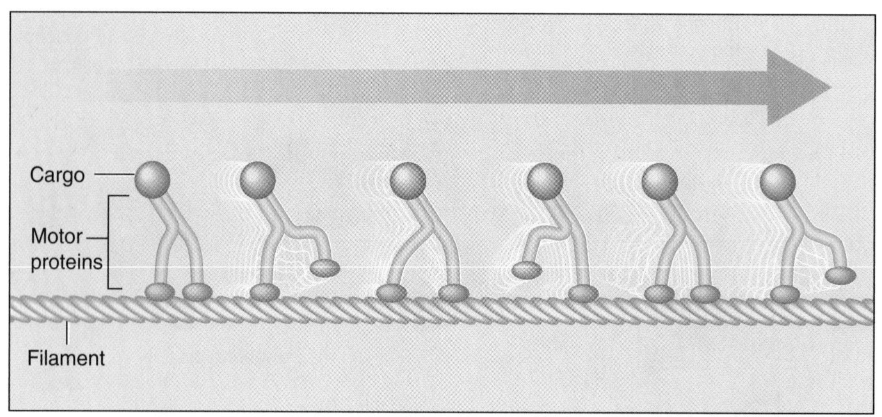

friction) associated with human-made machines, the two device classes do share important features. In addition to being composed of moving parts, both require energy-transducing mechanisms; that is, they both convert energy into directed motion. Despite the wide diversity of types of work performed by biological machines, they all share one key feature: energy-driven changes in the three-dimensional shapes of proteins. One or more components of biological machines bind nucleotide molecules such as ATP or GTP (guanosine triphosphate). The binding of nucleotide molecules to these protein subunits, referred to as **motor proteins**, and the release of energy that occurs when the nucleotide is hydrolyzed, result in a precisely targeted change in the subunit's shape (**Figure 2.5**). This wave of change is transmitted to nearby subunits in a process that resembles a series of dominoes falling. Biological machines are relatively efficient because the hydrolysis of nucleotides is essentially irreversible; therefore, the functional changes that occur in each machine occur in one direction only.

KEY CONCEPT

Many molecular complexes in living organisms function as molecular machines; that is, they are mechanical devices with moving parts that perform work.

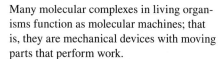

Macromolecular Crowding

The interior space within cells is dense and crowded. The concentrations of proteins, the dominant type of cellular macromolecules, may be as high as 200 to 400 mg/mL. The term "crowded" rather than "concentrated" is used because macromolecules of each type usually are present in low numbers. Estimates of the volume occupied by macromolecules, called the *excluded volume*, in individual cell types vary between 20 and 40%. As illustrated in **Figure 2.6**, nonspecific steric repulsion prevents the introduction of additional macromolecules under macromolecular crowding conditions. In contrast, the remaining 70% of the space is available to small molecules. The consequences of macromolecular crowding on living systems are significant. It is now believed to be an important factor in biochemical reaction rates, protein folding, protein-protein binding, chromosome structure, gene expression, and signal transduction.

KEY CONCEPT

Cells are densely crowded with macromolecules of diverse types. Macromolecular crowding is a significant factor in a wide variety of cellular processes.

Proteostasis

Each type of living cell has its own characteristic set of proteins, referred to as its **proteome**, which changes constantly in response to environmental conditions. Mammalian cells have an average of 10,000 types of protein, most of which are produced in multiple copies for an estimated total of 1 billion molecules per cell. Bacteria cells such as those of *E. coli* have about 2000 different types for a total of about 4 million molecules per cell. Following their synthesis on ribosomes, these enormous numbers of proteins must fold into their functional shapes, be transported to their proper destinations, and then be promptly degraded when they become damaged or obsolete. Adding to this complexity, cells must protect themselves from *proteotoxic stress*, a potentially lethal condition in which there is an accumulation of misfolded proteins caused by genetic variations or environmental insults such as oxidative stress (pp. 367–372), elevated temperatures, and exposure to toxins. Therefore, it is not surprising that all organisms have evolved stringent protein quality-control processes that prevent or correct protein misfolding and aggregation (the formation of usually toxic clumps of misfolded proteins) or, if necessary, destroy damaged proteins or even the cell itself.

Cells in which protein quality control is high are said to be in a state of protein homeostasis, or **proteostasis**. The processes that monitor and restore proteostasis are referred to as the **proteostasis network** (PN). The PN in mammalian cells consists of at least 2000 proteins. PN components include molecular chaperones (proteins that assist in protein folding or unfolding) (pp. 165–167), simple proteolytic enzymes, and elaborate pathways that degrade selected proteins or organelles. Examples of degradative processes include the *unfolded protein response* (p. 48), the *ubiquitin-proteasome system* (a mechanism in which the multiprotein proteasome complex destroys proteins that are covalently bound to ubiquitin) (p. 556), lysosomal degradation (p. 51), and *autophagy* (a mechanism that destroys unnecessary or dysfunctional cell components) (p. 558). Numerous signaling pathways detect unfolded proteins and the stressful conditions that threaten proteostasis. Significant research efforts have been devoted to proteostasis because PN deficiencies are an important feature of numerous human diseases. Examples include type 2 diabetes (p. 597), cardiovascular disease, lysosomal storage diseases (caused by deficiencies in degradative enzymes) (p. 396), and neurodegenerative diseases such as Alzheimer's, Parkinson's, and Huntington's.

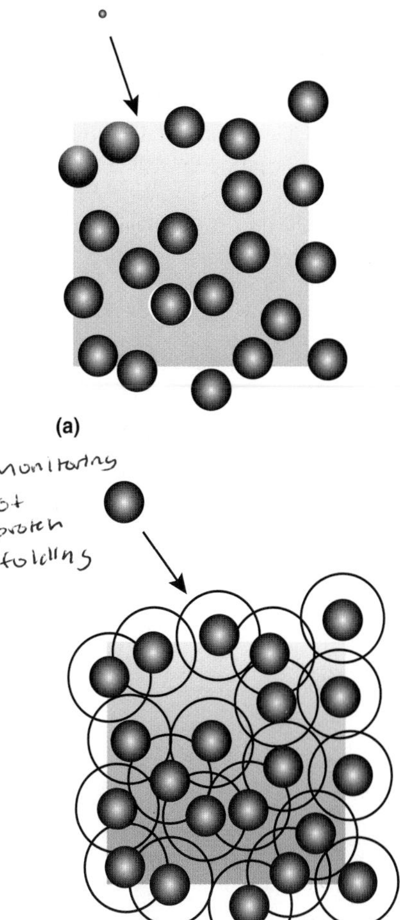

FIGURE 2.6
Volume Exclusion

Macromolecules and small molecules are depicted with large balls and small balls, respectively. Within each square, macromolecules occupy 30% of available space. (a) An introduced small molecule can penetrate into virtually all of the remaining 70% of the space. (b) Steric repulsion between macromolecules (open circles) limits these molecules' ability to approach each other. Although the macromolecules occupy only 30% of the volume, the introduced macromolecule is excluded.

Signal Transduction

If energy is the force that drives biochemical processes, then information is the power to specify what is done. Self-organizing living organisms are so complicated that they must have not only precise structural specifications for each type of biomolecule, but also specifications for how, when, and where each type is to be synthesized, utilized, and degraded. In other words, living organisms require both energy and information to create order. Survival requires that organisms process information from their environment. For example, bacterial cells track down food molecules, plants adapt to changing light levels, and animals seek to avoid predators. Information, or *signals*, comes in the form of molecules (e.g., nutrients) or physical stimuli (e.g., light). Although organisms are bombarded with signals, they can adapt to changing environmental conditions only if they can recognize, interpret, and respond to each type of message. The process that organisms use to receive and interpret information is referred to as **signal transduction**. Although both prokaryotes and eukaryotes process environmental information, most research efforts have been concerned with eukaryotic signal transduction. Consequently, the following discussion focuses on information processing in eukaryotes. Examples of eukaryotic signal molecules include **neurotransmitters** (products of neurons), **hormones** (products of glandular cells), and *cytokines* (products of white blood cells). All information-processing mechanisms can be divided into four phases:

1. **Reception**. A signal molecule, called a **ligand**, binds to and activates a receptor.
2. **Transduction**. Ligand binding triggers a change in the three-dimensional structure of the receptor that results in the conversion of a primary message or signal to a secondary message, often across a membrane barrier.
3. **Response**. Once initiated, the internal signal causes a **signaling cascade**, a series of reactions that involve covalent modifications (e.g., phosphorylation) of intracellular proteins. Results of this process include changes in enzyme activities and/or gene expression, cytoskeletal rearrangements, cell movement, or cell cycle progression (e.g., cell growth or division).

4. **Termination**. The efficiency and effectiveness of signal mechanisms requires that they be terminated in a timely manner. Living organisms use a variety of signal termination methods. For example, signaling molecules may be destroyed or removed (e.g., neurotransmitters such as acetylcholine and serotonin), activated proteins are inactivated by changes in covalent modification (e.g., removal of phosphate groups), and nonprotein signals are degraded by enzymes.

KEY CONCEPTS

- Living organisms receive, interpret, and respond to environmental information by means of the process of signal transduction.
- Signal transduction can be divided into four phases: reception, transduction, response, and termination.

The protein hormone insulin is a signal molecule. When released from the pancreas in response to high blood glucose levels, insulin binds to its receptor on a target cell. The insulin receptor is a member of a class of receptors called the *tyrosine kinase receptors*. They are so named because upon activation, these receptors initiate an intracellular response by catalyzing the transfer of phosphate groups to tyrosine (an amino acid residue that contains an OH group) in specific target proteins. Cellular responses triggered by insulin binding to its receptor include uptake of glucose into the cell and increased fat and glycogen synthesis.

CALCIUM IONS (Ca^{2+}): A UNIVERSAL SIGNALING DEVICE Cells respond to various external stimuli by increasing their cytoplasmic Ca^{2+} concentrations that are normally kept quite low (approximately 100 nM) (1 nM $= 1 \times 10^{-9}$ M) by ATP-driven pump complexes in the plasma membrane and in eukaryotes in the membrane of organelles such as the endoplasmic reticulum (p. 47). Calcium signal decoding is dependent on the amplitude and localization of brief spikes in the concentration of cytoplasmic calcium ions $[Ca^{2+}]_{cyt}$. Each type of stimulus triggers a specific signaling cascade composed of a set of Ca^{2+}-responsive proteins that change in both their shape and their functional properties when bound to the ion. Prevention of nonspecific activation of Ca^{2+}-dependent processes requires precise localization of Ca^{2+} release and then rapid clearance of the ion from the cytoplasm.

In animals, calcium ions are involved in an astonishingly diverse set of signaling processes that includes neurotransmitter release from nerve cells, hormone secretion, protein folding (assisted by calcium-dependent molecular chaperone proteins), and contraction of all muscle types. In insulin secretion, for example, the release of insulin from pancreatic β-cells is triggered by calcium ions. The detection of high blood glucose levels sets in motion an intracellular signal transduction process that causes cytoplasmic Ca^{2+} levels to rise near the β-cells plasma membrane. Insulin secretion occurs because the binding of Ca^{2+} to specific calcium-sensitive membrane proteins facilitates the fusion of the membrane of insulin-containing secretory granules with the plasma membrane in a process known as exocytosis (p. 50).

SIGNAL TRANSDUCTION AND METABOLISM Signal transduction mechanisms in living organisms are vital since their function is to detect relevant information in cell environments in which there is a profusion of stimuli, to integrate this information, and then to respond by executing an appropriate response. Such responses involve precise alterations in gene expression and the flow of metabolites in biochemical pathways. Research efforts over several decades have revealed that such signal transduction processes are hierarchical and immensely intricate. This textbook covers the most basic of signal transduction mechanisms, yet even these may be viewed as complicated. The most essential features of signal transduction processes and their effects on metabolic regulation (the effects of hormones and transcription factors on biochemical reactions) are introduced in Chapter 8 (Carbohydrate Metabolism). In later chapters devoted to metabolic networks (e.g., lipid and energy metabolism) other facets of signal transduction and metabolic regulation are introduced that will allow a more integrated understanding of cell and organ function. Finally, Chapter 16 provides an overview of a complex metabolic process (human digestion and the feeding-fasting cycle) and how it is regulated.

2.2 STRUCTURE OF PROKARYOTIC CELLS

The prokaryotes are an immense and heterogeneous group that are similar in external appearance: cylindrical or rodlike (bacillus), spheroidal (cocci), or helically coiled (spirilla). Prokaryotes are also characterized by their relatively small size (a typical rod-shaped bacterial cell has a diameter of 1 μm and a length of 2 μm), their capacity to move (i.e., whether they have flagella, whiplike appendages that propel them), and their retention of specific dyes. Most are identified on the basis of nutritional requirements, energy sources, chemical composition, and biochemical capacities. Despite their diversity, most prokaryotes possess the following common features: cell walls, a plasma membrane, circular DNA molecules, and no internal membrane-enclosed organelles. The anatomical features of a typical bacterial cell are illustrated in **Figure 2.7**.

Cell Wall

The prokaryotic cell wall is a complex semirigid structure that maintains the shape of the organism and protects it from mechanical injury. The cell wall's strength is largely caused by the presence of a polymeric network made up of *peptidoglycan*, a covalent complex of short peptide chains linking long carbohydrate chains. The thickness and chemical composition of the cell wall and its adjacent structures determine how avidly a cell wall takes up and/or retains specific dyes.

Most cells can be differentiated on the basis of whether they retain crystal violet stain during the Gram stain procedure. Those that retain the dye are called

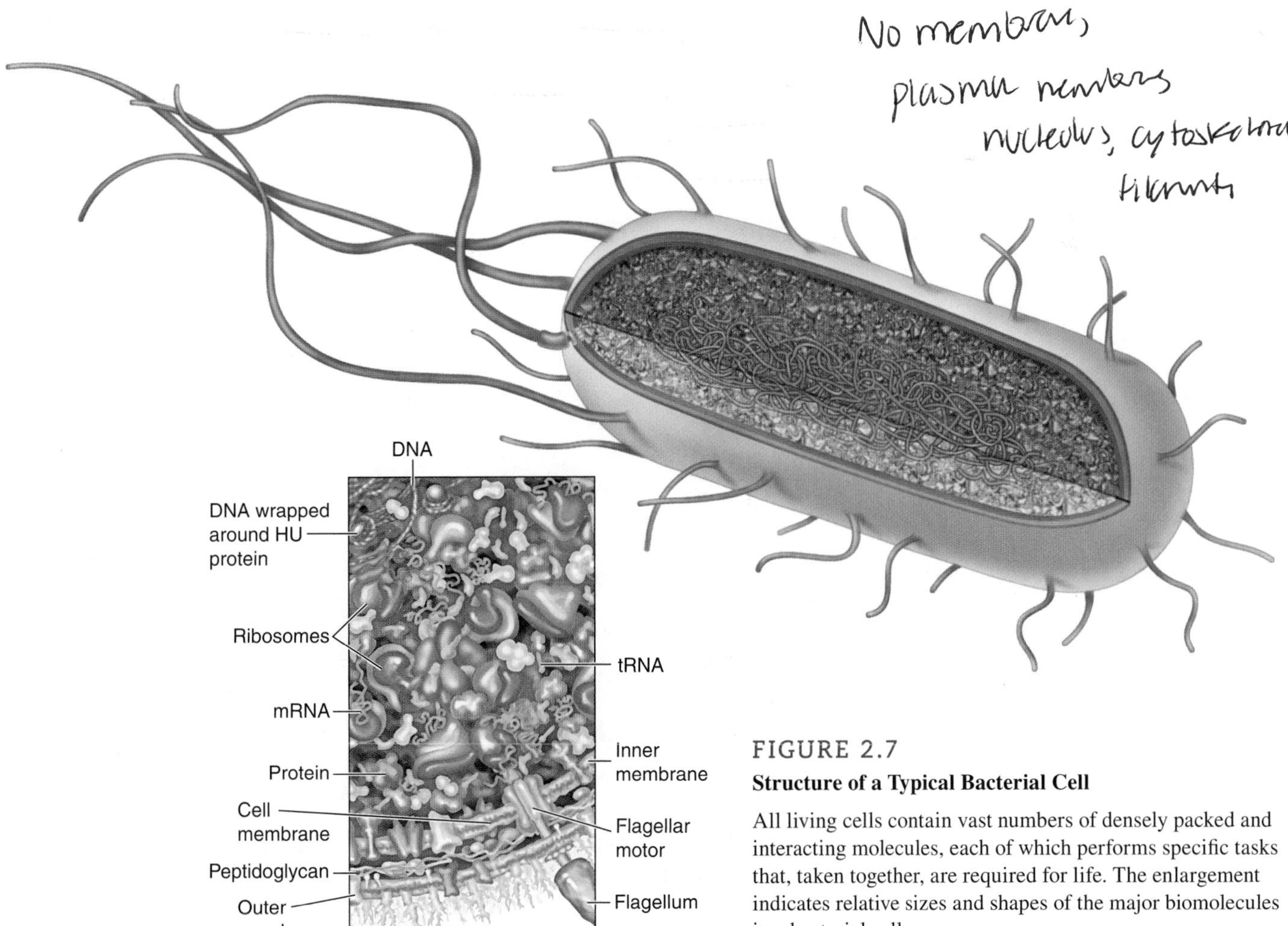

FIGURE 2.7

Structure of a Typical Bacterial Cell

All living cells contain vast numbers of densely packed and interacting molecules, each of which performs specific tasks that, taken together, are required for life. The enlargement indicates relative sizes and shapes of the major biomolecules in a bacterial cell.

Gram-positive; those that do not are called *Gram-negative*. The wall of a Gram-positive cell consists of a single, relatively thick peptidoglycan layer that lies outside the plasma membrane.

The cell walls of the Gram-negative bacterium illustrated in **Figure 2.7** is substantially more complex than that of Gram-positive cells. A thin peptidoglycan layer lies between the outer membrane and the plasma membrane and within the periplasmic space. The lipid component of the *outer membrane* is lipopolysaccharide instead of phospholipids. Lipopolysaccharide, which is composed of a membrane-bound lipid (lipid A) attached to a polysaccharide, acts as an endotoxin. So called because they are released when the cell disintegrates, *endotoxins* are responsible for symptoms such as fever and shock in animals infected by Gram-negative bacteria. The outer membrane is relatively permeable, and small molecules move across it through *porins*, transmembrane protein complexes that contain channels. The *periplasmic space*, the region between the outer membrane and the plasma membrane, is filled with a gelatinous fluid that contains, in addition to peptidoglycan, a variety of proteins. Many of these proteins participate in nutrient digestion, transport, or chemotaxis.

Some bacteria secrete substances such as polysaccharides and proteins, collectively known as the *glycocalyx*. Depending on the structure and composition of this material, which accumulates on the outside of the cell, the glycocalyx may also be referred to as a capsule or a slime layer. Some pathogenic (disease-causing) bacterial species possess capsules that allow them to avoid detection or damage by host immune systems and to attach to host cells to facilitate colonization. Slime layers, also referred to as *biofilms*, are disorganized accumulations of polysaccharides that form when microorganisms adhere to surfaces and grow. In time, as more cells and secreted material accumulate, biofilms become thicker. Biofilms provide microorganisms with a protective barrier and are a significant feature in a variety of medical conditions (e.g., tooth decay, cystic fibrosis, and tuberculosis). The bacteria in biofilms are very resistant to immune system attack and antibiotic therapy.

Plasma Membrane

Directly inside the cell wall of bacteria is the **plasma membrane** (**Figure 2.8**), also called the cytoplasmic membrane, a phospholipid bilayer that is reinforced

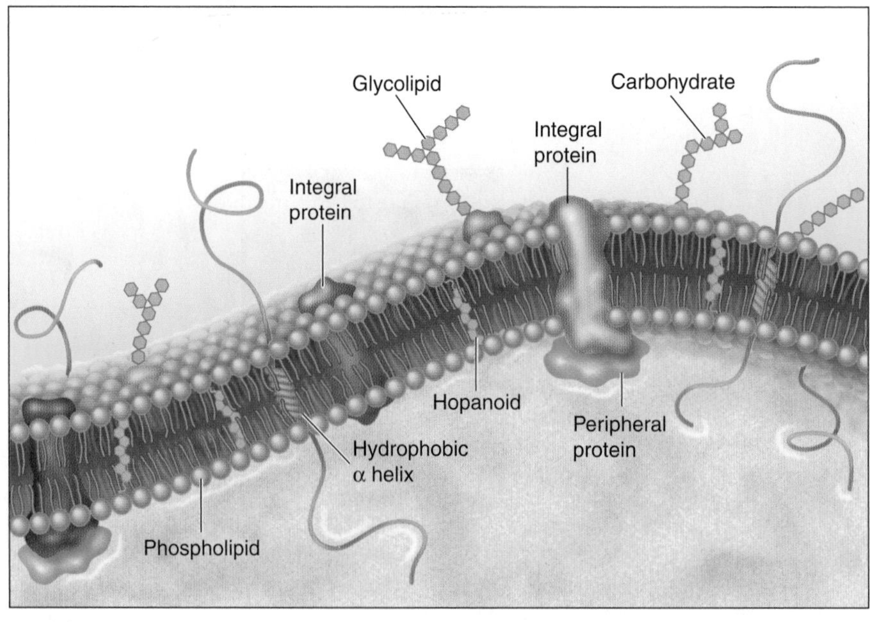

FIGURE 2.8
The Bacterial Plasma Membrane
Simplified view of the plasma membrane illustrating several classes of protein and lipid. Many of these proteins and certain lipids are covalently bound to carbohydrate molecules. (Glycolipids contain carbohydrate groups.) Hopanoids are complex lipid molecules that stabilize bacterial membranes.

with hopanoids, a group of relatively stiff molecules that resemble sterols (e.g., cholesterol) that stiffen membranes in eukaryotes. A diverse group of proteins are embedded in the lipid bilayer.

In addition to acting as a selective permeability barrier, the bacterial plasma membrane possesses receptor proteins that detect nutrients and toxins in their environment. Numerous types of transport proteins involved in nutrient uptake and waste product disposal also occur here. Depending on the species of organism, there may also be proteins involved in energy transduction processes such as **photosynthesis** (the conversion of light energy into chemical energy) and **respiration** (the oxidation of fuel molecules to generate energy).

Cytoplasm

Despite the absence of internal membranes, prokaryotic cells do appear to have functional compartments (**Figure 2.9a**). The most obvious of these is the **nucleoid** (**Figure 2.9b**), a spacious, irregularly shaped, centrally located region that contains a long, circular DNA molecule called a **chromosome**. The bacterial chromosome typically comprises numerous regions of highly coiled and uncoiled structures. Protein complexes involved in DNA synthesis and regulation of gene expression are also found within the nucleoid. Many bacteria also contain additional small circular DNA molecules called **plasmids** that exist outside the nucleoid and can replicate independently of the chromosome. Although they are not required for growth or cell division, plasmids usually provide the cell with a biochemical advantage over cells that lack plasmids. For example, DNA sequences that code for antibiotic resistance are often found on plasmids. In the presence of the antibiotic, resistant cells synthesize a protein that inactivates the antibiotic before it can damage the cell. Such cells continue to grow and reproduce, whereas susceptible cells die.

Under low magnification the cytoplasm of prokaryotes has a uniform, grainy appearance except for inclusion bodies, large granules that contain organic or inorganic substances. Some species use glycogen or poly-β-hydroxybutyric acid as carbon storage polymers. Polyphosphate inclusions are a source of phosphate for nucleic acid and phospholipid synthesis. Prokaryotes that derive energy by oxidizing reduced sulfur compounds form sulfur granules. The iron mineral magnetite (Fe_3O_4) forms inclusions, called magnetosomes, which allow some species of aquatic anaerobic prokaryotes to orient themselves with the earth's magnetic field. The remaining space in the cytoplasm is filled with **ribosomes** (molecular machines composed of RNA and proteins that synthesize polypeptides) and a diverse number of macromolecules and smaller metabolites.

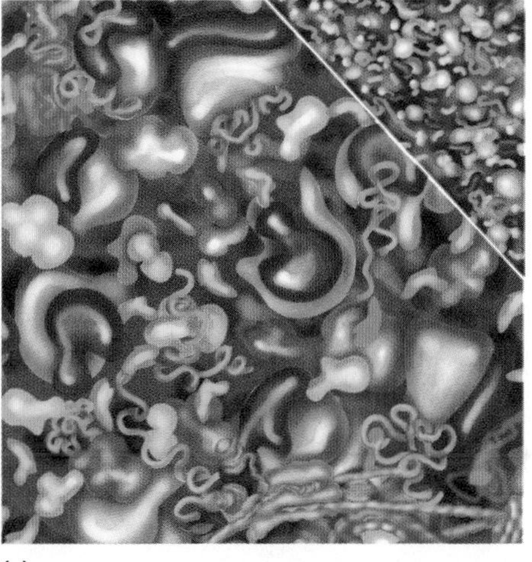

(a)

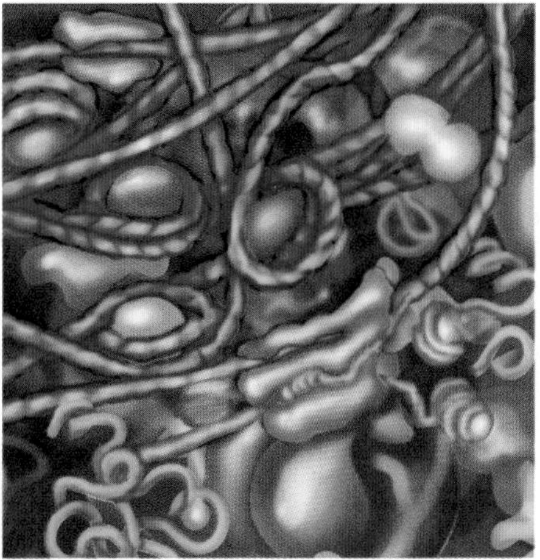

(b)

FIGURE 2.9

Bacterial Cytoplasm

(a) Cytoplasm is a complex mixture of proteins, nucleic acids, and an enormous variety of ions and small molecules. For clarity, the small molecules appear only in the upper right corner. (b) Close-up view of the nucleoid. Note that DNA is coiled and folded around protein molecules (brown).

Pili and Flagella

Many bacterial cells have external appendages. *Pili* (singular: pilus) are fine, hair-like structures that may allow cells to attach to food sources and host tissues. Sex pili are used by some bacteria to transfer genetic information from donor cells to recipients, a process called *conjugation*. In bacteria, the *flagellum* (plural: flagella) is a flexible corkscrew-shaped protein filament that is used for locomotion. Cells are pushed forward when flagella rotate in a counterclockwise direction, whereas clockwise rotation results in a stop-and-tumble motion, allowing the cell to reorient for a forward run. The filament of the flagellum is anchored into the cell by a protein complex (**Figure 2.7**). Motor proteins in this complex convert chemical energy into rotational motion.

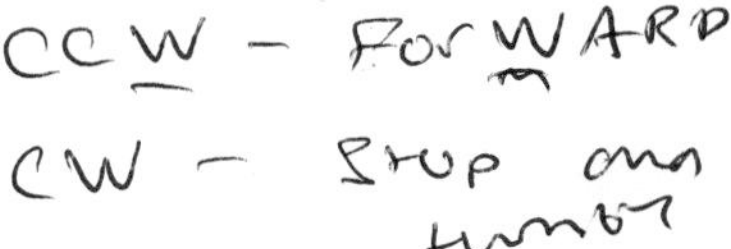

KEY CONCEPTS

- Prokaryotic cells are small and structurally simple. They are bounded by a cell wall and a plasma membrane. They lack a nucleus and other organelles.
- Their DNA molecules, which are circular, are located in an irregularly shaped region called the nucleoid.
- At low magnification ribosomes and inclusion bodies of several types appear to be present in an otherwise featureless cytoplasm.

QUESTION 2.1

A typical, roughly spheroidal, hepatocyte (liver cell) is a widely studied eukaryotic cell that has a diameter of about 20 μm. Calculate the volume of both a prokaryotic and a eukaryotic cell. To appreciate the magnitude of the size difference between the two cell types, estimate how many bacterial cells would fit inside the liver cell. [*Hint:* Use the expression $V = \pi r^2 h$ for the volume of a cylinder and $V = 4\pi r^3/3$ for the volume of a sphere.]

2.3 STRUCTURE OF EUKARYOTIC CELLS

The structural complexity of eukaryotic cells allows more sophisticated regulation of living processes than is possible in the prokaryotes. The most obvious features of eukaryotic cells are their large sizes (diameters of 10–100 μm) in comparison to prokaryotes. More importantly, the membrane surface area is greatly expanded by the presence of membrane-bound organelles. Each organelle within the cell contains a characteristic set of biomolecules and is specialized to perform specific functions. The biochemical processes within an organelle proceed efficiently because of locally high enzyme concentrations and because they can be individually regulated.

Most organelles are components of the **endomembrane system**, an extensive set of interconnecting internal membranes that divide the cell into functional compartments. The endomembrane system consists of the plasma membrane, endoplasmic reticulum, Golgi apparatus, lysosomes, and nucleus. Either through direct physical contact between compartments or by transport vesicles, the endomembrane system transports a vast array of molecules through cells, as well as to and from cell exteriors. **Vesicles** are membranous sacs that bud off from a donor membrane and then fuse with the membrane of a different compartment. Once formed, each vesicle acquires a "coat" of specific proteins that facilitates its transport or targets it to its destination. Other membrane-bound organelles are mitochondria and peroxisomes and the chloroplasts in plant cells.

In addition to membranous organelles, eukaryotic cells possess several components that are devoid of membranes. Included in this group are protein-synthesezing molecular machines called ribosomes and the cytoskeleton. The cytoskeleton is a complex, dynamic, and force-generating network of filaments that give eukaryotic cells shape, structural support, and the capacity for the directed movement of molecules and organelles.

Although most eukaryotic cells possess similar structural features, there is no "typical" eukaryotic cell. Each cell type has its own characteristic structural and functional properties. They are sufficiently similar, however, that a discussion of the basic components is useful. The generalized structures of cells from animals and plants, the major forms of multicellular eukaryotic organisms, are illustrated in Figures 2.10 and 2.11.

Plasma Membrane

The plasma membrane isolates the cell from the outside environment. It is composed of a lipid bilayer and an enormous number and variety of integral and peripheral proteins (**Figure 2.12**). Channels and carriers within the plasma membrane regulate the passage of various ions and molecules in and out of the cell. Immense numbers of receptors play key roles in signal transduction. The extracellular face of a eukaryotic cell is heavily "decorated" with carbohydrate; that is, much of the membrane protein and lipid contains covalently attached carbohydrate. This carbohydrate "coat" is referred to as the **glycocalyx** (Figure 2.13). The carbohydrate molecules play important roles in cell-cell recognition and adhesion, receptor specificity, and self-identity (an immune system

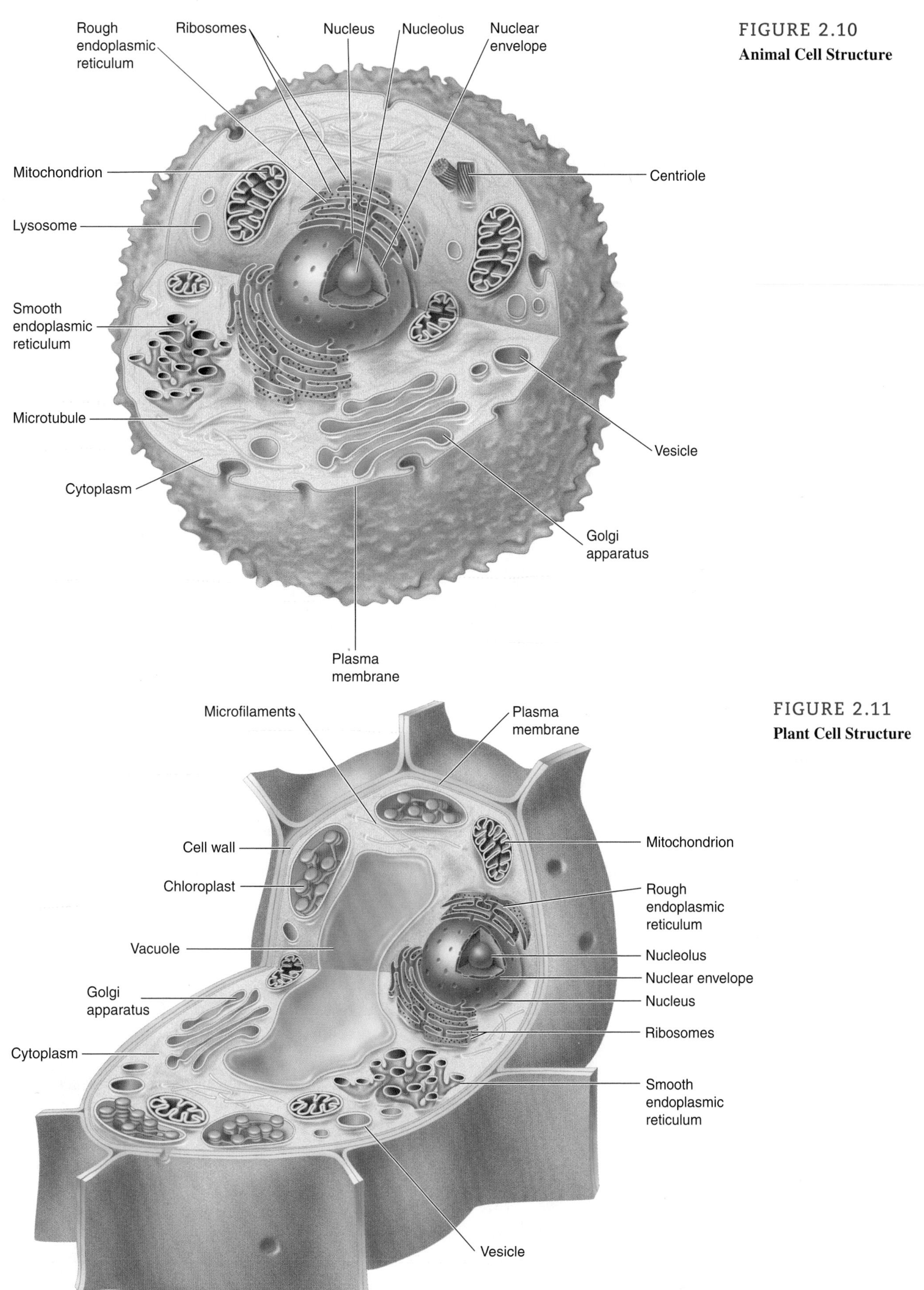

FIGURE 2.10
Animal Cell Structure

FIGURE 2.11
Plant Cell Structure

FIGURE 2.12

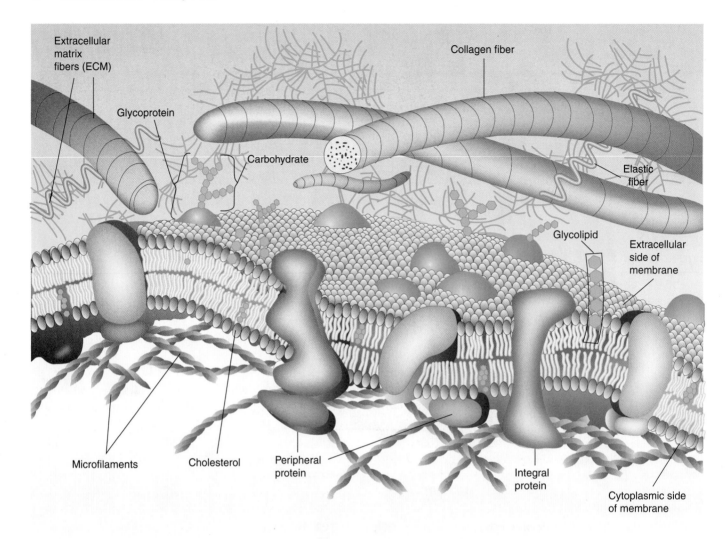

The Plasma Membrane of an Animal Cell

The plasma membrane (PM) is composed of a lipid bilayer in which a wide variety of integral proteins are embedded. Note that numerous integral proteins and lipid molecules are covalently attached to carbohydrate. Peripheral proteins are attached by noncovalent bonds to the cytoplasmic surface of the PM. Specialized cells of the connective tissue of higher animals called fibroblasts synthesize and secrete glycoproteins of the extracellular matrix (ECM). The inner surface of the PM is reinforced by the membrane skeleton, which is composed of a meshwork of actin microfilaments and other proteins linked to the cell's cytoskeleton.

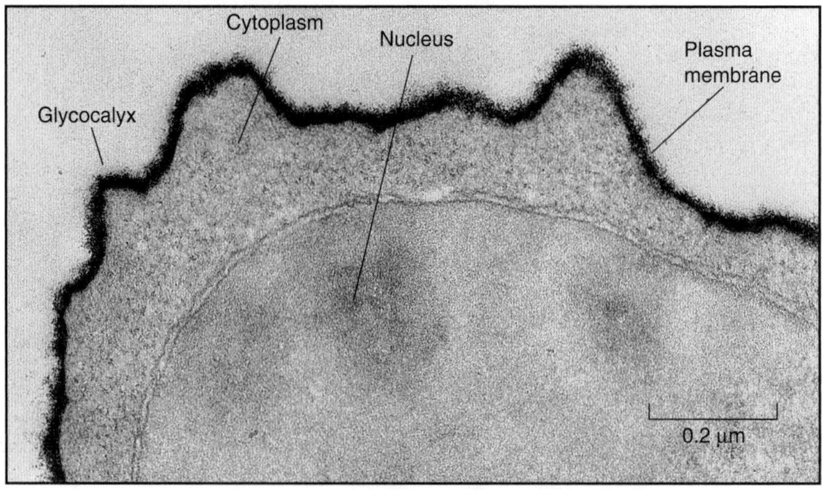

FIGURE 2.13

The Glycocalyx

Electron micrograph of the surface of a lymphocyte stained to reveal the glycocalyx (cell coat).

requirement). The basic blood group antigens are an example of this self-identity function.

In most eukaryotes the plasma membrane is protected with extracellular and intracellular structures (**Figure 2.12**). Within animal tissues, the specialized cells called fibroblasts synthesize and secrete structural proteins and complex carbohydrates that form the **extracellular matrix (ECM)**, a gelatinous material that binds cells together. In addition to its support and protective functions, the ECM plays roles in cell behavior regulation through the binding of some of its components to specific membrane receptors in chemical and mechanical signaling processes of various types. The inner surface of the eukaryotic plasma membrane is reinforced by a three-dimensional meshwork of proteins called the **membrane skeleton** that is attached to the membrane by extensive noncovalent bonding to peripheral proteins. In animal cells this protein network, composed of actin (p. 64), several types of actin-binding proteins, and spectrin (p. 408), provides mechanical strength to the plasma membrane and determines cell shape. Direct and indirect interactions among membrane skeleton components, plasma membrane integral proteins, and lipid molecules intermittently partition the membrane into compartments. The resulting temporary confinement of transmembrane proteins and membrane microdomains is believed to facilitate signal transduction processes.

KEY CONCEPTS

- Besides providing mechanical strength and shape to the cell, the plasma membrane is actively involved in selecting the molecules that can enter or exit the cell.
- Receptors on the plasma membrane's surface allow the cell to respond to external stimuli.

Endoplasmic Reticulum

The **endoplasmic reticulum (ER)** is a system of interconnected membranous tubules, vesicles, and large flattened sacs. A hint of its importance in cell function is that it often constitutes more than half a cell's total membrane. The repeatedly folded, continuous sheets of ER membrane enclose an internal space called the ER *lumen*. This compartment, which is often referred to as the *cisternal space*, is entirely separated from the cytoplasm by the ER membrane. The ER is responsible for several vital processes. Among these are the synthesis of several kinds of protein, a variety of membrane lipids and steroid molecules, and the storage of calcium ions.

ER comes in two interconnected forms: **rough ER (RER)** and **smooth ER (SER)** (**Figure 2.14**). The precise functional properties and relative sizes of both ER types vary with cell type and physiological conditions. RER is so named because of the numerous ribosomes that stud its cytoplasmic surface. Several protein classes are processed by the RER: membrane proteins, as well as water-soluble proteins destined for retention within the ER, transport to other organelles, or export out of the cell. Polypeptides enter the RER during ongoing protein synthesis as they are threaded, or translocated, through the membrane.

Transmembrane polypeptides (i.e., those that contain one or more hydrophobic sequence segments) remain embedded in the membrane because the translocation process is halted when hydrophobic segments enter the membrane. As water-soluble polypeptides emerge into the ER lumen, the folding process, facilitated by processing enzymes and *molecular chaperones* (proteins that facilitate protein folding), begins. Glycosylation reactions, the attachment of carbohydrate groups to specific amino acid residues, are a prominent example of ER-processing reactions. The binding of molecular chaperones to short hydrophobic segments in partially folded polypeptides facilitates efficient folding and prevents aggregation. The failure of polypeptides to fold within the RER, resulting in an accumulation of misfolded molecules, is a potential threat to the entire cell since overall cell function can be disrupted. This phenomenon, called **ER stress**, is caused by environmental factors such as metabolic stress (changes in metabolism triggered by injury, illness, or infection), oxidative stress (from oxygen radicals), and activated inflammatory signaling processes, as well as genetic factors. **ER-associated protein degradation** is a cellular mechanism that targets misfolded polypeptides and transports them into the cytoplasm, where they are degraded by proteasomes (p. 556). If stress is severe, the ER

FIGURE 2.14

The Endoplasmic Reticulum

There are two forms of endoplasmic reticulum (ER): RER, the rough endoplasmic reticulum, and SER, the smooth endoplasmic reticulum. Note that in living eukaryotic cells, RER and SER are interconnected.

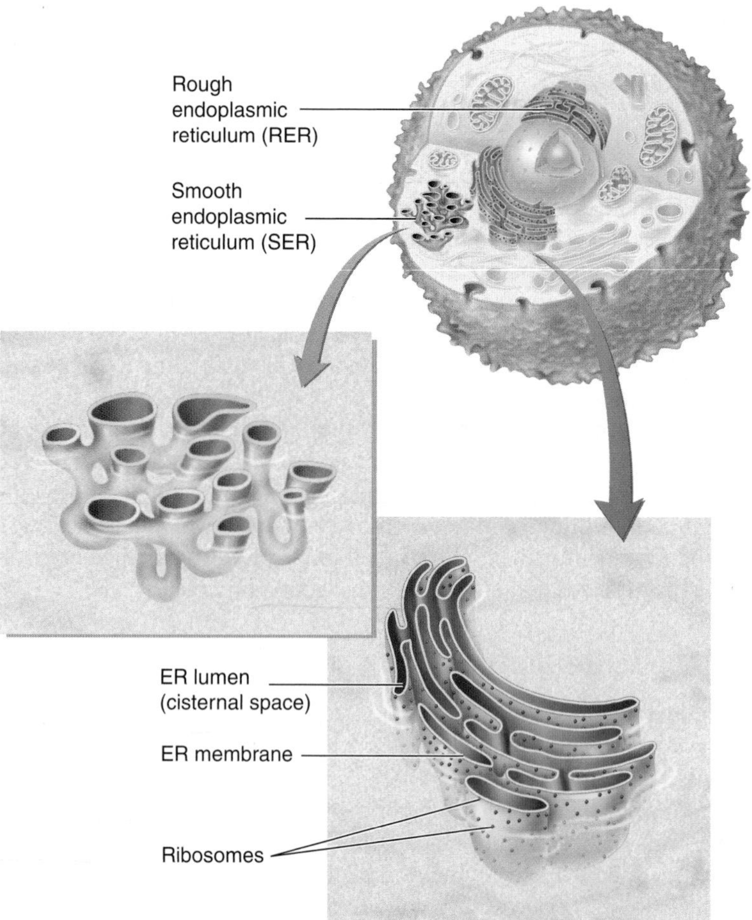

initiates the **unfolded protein response** in an attempt to restore proteostasis. Signals sent to the nucleus result in the inhibition of protein synthesis with the exception of additional molecular chaperones. In addition, ER membrane volume expands because of increased synthesis of membrane lipids. In addition to proteosomal protein destruction, *autophagy* (controlled digestion of damaged or unnecessary organelles or other cell components; see p. 558) can be utilized in an attempt to prevent cell death. If protein homeostasis cannot be achieved within a certain time period, *apoptosis*, a programmed cell death process, can be initiated.

Smooth ER lacks attached ribosomes, and its membranes are continuous with those of RER. The size and functional properties of SER vary considerably in different cell types from sparse to abundant. In most cells SER is involved in the synthesis of lipid molecules. The SER is especially noteworthy in hepatocytes and striated muscle cells. Hepatocyte SER performs a wide variety of functions, which include biotransformation and synthesis of the lipid components of very-low-density lipoproteins (water-soluble lipid transport complexes that deliver lipids to tissue cells). **Biotransformation reactions** convert an enormous variety of water-insoluble metabolites and xenobiotics (foreign and potentially toxic molecules) into more soluble products that can then be excreted. The SER in striated muscle is so highly specialized in both structure and function that it has a different name, the *sarcoplasmic reticulum* (SR). The SR membrane extends throughout the muscle cell and is in close proximity to all myofibrils, the organized arrays of contractile proteins. SR is a reservoir for calcium, the signal that triggers muscle contraction.

Newly synthesized protein and lipid molecules exit the ER in coated vesicles that bud off from exit sites in an ER subdomain referred to as the *transitional ER* (tER). The vesicular coat of *COPII* (*co*at *p*rotein complex II) and its adapter proteins

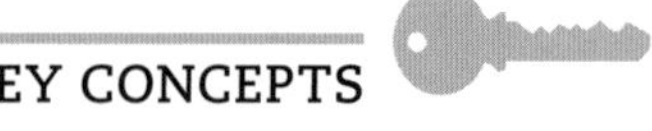

KEY CONCEPTS

- The RER is primarily involved in protein synthesis. The external surface of RER membrane is studded with ribosomes.
- SER lacks attached ribosomes and is involved in lipid synthesis, biotransformation, and Ca^{2+} storage.

ensure that the vesicle is directed to the correct target membrane. After exiting the tER, the vesicles are transported to the *ER-Golgi intermediate compartment* (ERGIC), a structure of membranous tubules and vesicles that facilitates the sorting of cargo molecules from resident ER proteins. Newly formed COPII-coated vesicles deliver the molecular cargo to the Golgi complex for further processing. The resident ER molecules, identified by retrieval signals in their structures, are recycled by returning to the ER via vesicles that have *COPI* (coat protein I) coats.

Golgi Apparatus

The **Golgi apparatus** (also known as the **Golgi complex**) is formed from relatively large, flattened, saclike membranous vesicles that resemble a stack of plates. The Golgi apparatus is involved in the processing, packaging, and distribution of cell products (e.g., glycoproteins) to internal and external compartments (**Figure 2.15**).

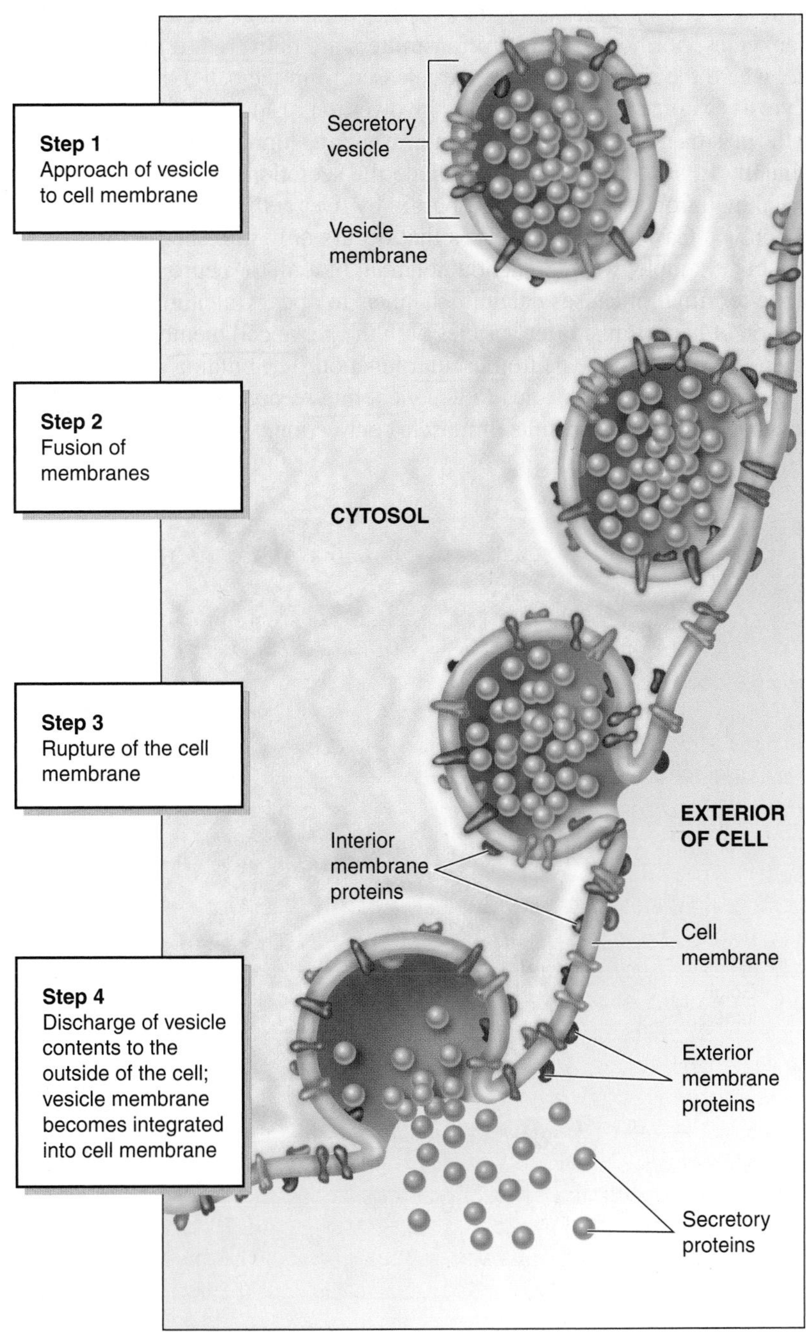

FIGURE 2.15
Exocytosis

Proteins destined to be secreted by a cell are produced in the ER and processed by the Golgi apparatus, where they are packaged into vesicles that migrate to the plasma membrane and merge with it.

The Golgi apparatus has two faces. The plate (or *cisterna*) positioned closest to the ER is on the forming (*cis*) face, whereas the one on the maturing (*trans*) face is typically close to the portion of the cell's plasma membrane that is engaged in secretion. Small membranous vesicles containing newly synthesized protein and lipid bud off from the ER and fuse with the *cis* Golgi membrane. Upon entering the Golgi apparatus, the vesicle's molecules are chemically modified (e.g., by the addition of sugar molecules, or sulfate or phosphate groups) as they are transported toward the *trans* face. A complex network of *trans*-Golgi membrane and vesicles, called the *trans-Golgi network* (TGN), sorts the processed molecules and packages them into vesicles coated with a protein called clathrin (p. 51). Clathrin adapter proteins link clathrin to membrane-bound receptors and target clathrin-coated vesicles to destinations such as endosomes (p. 51), lysosomes, and the plasma membrane in the case of the secretory process. Clathrin-coated vesicles are also used to transport vesicles from the plasma membrane to destinations such as endosomes and the TGN.

In the *secretory process*, secretory vesicles containing molecules such as digestive enzymes, hormones, or neurotransmitters are delivered to the plasma membrane, where the cargo molecules are released from the cell. Often referred to as **exocytosis** (**Figure 2.16**), this process involves the fusion of the vesicle membrane with the plasma membrane. *Constitutive exocytosis* (unregulated secretion) occurs continually in all cells. Examples include the secretion of the structural protein collagen by fibroblasts and serum albumin by liver cells. In regulated exocytosis, secretion is a Ca^{2+}-triggered process that occurs only in response to an external signal. For example, when an action potential of a motor neuron reaches the presynaptic terminal, it causes calcium channels to open. Calcium ions then trigger the fusion of neurotransmitter vesicles with the nerve cell membrane, thus releasing their contents into the neuromuscular junction. The binding of sufficient numbers of acetylcholine molecules to acetylcholine receptors on the surface of the postsynaptic muscle cell results in muscle contraction.

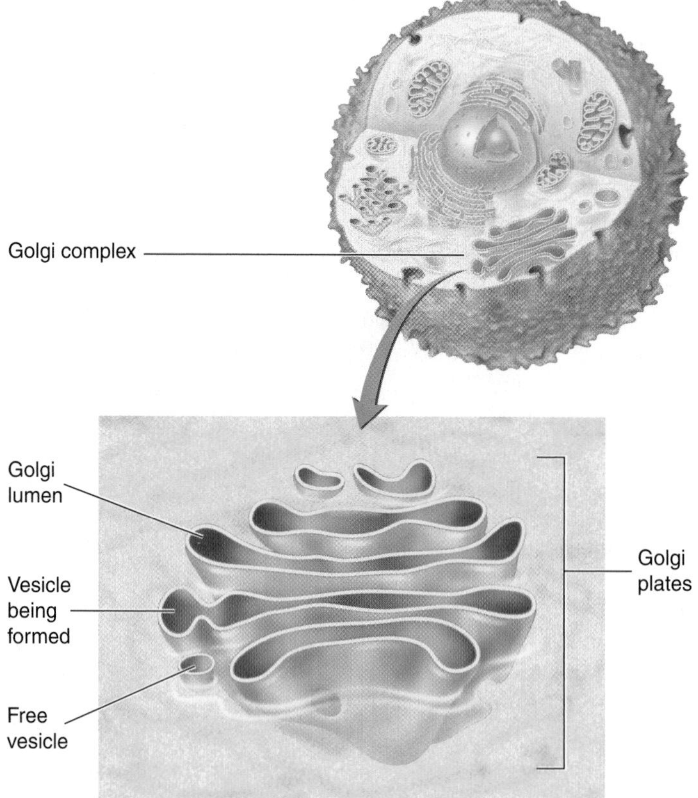

FIGURE 2.16
The Golgi Apparatus

The Golgi apparatus is essentially a factory that synthesizes and/or processes a diverse group of proteins and lipids. These biomolecules are then sorted for transport to their final destination.

Until recently it was believed that Golgi cisternae are relatively stationary and that protein and lipid vesicles were the mechanism of cargo transport from one Golgi sac to the next. In this view, as cargo molecules proceed through the Golgi system, Golgi enzymes process them further. It is now believed that Golgi cisternae move physically from the cis face position to the trans face as they simultaneously transport and modify their cargo. Transport vesicles recycle Golgi membrane and enzymes back to the newly formed cis Golgi cisterna. The term *anterograde transport* is used to describe the transport of newly synthesized molecules from the ER to the Golgi apparatus and then to other cell destinations or to the plasma membrane for secretion. Vesicles coated with COPII carry newly synthesized molecules from the ER to the ERGIC and then on to the cis face of the Golgi apparatus. Transport in the opposite direction, called *retrograde transport*, is a mechanism used to recycle lipid and protein molecules. Vesicles coated with COPI are used to return escaped ER-resident proteins back from the Golgi apparatus to their site of origin.

KEY CONCEPT

Formed from relatively large, flattened, saclike membranous vesicles, the Golgi apparatus functions in the packaging and secretion of cell products.

Vesicular Organelles and Lysosomes: The Endocytic Pathway

Endocytosis (**Figure 2.17**) is a cellular process in which plasma membrane protein receptors and lipids and exogenous substances are taken into cells, most notably involving the pinching off of regions of the plasma membrane. The newly made vesicles then enter the *endocytic pathway* when they fuse with a membrane-bound organelle called an early endosome. Usually located near the cell's periphery, *early endosomes* serve as the nexus or focal point of the endocytic pathway because it is here that the fate of internalized molecules is determined. An elaborate mechanism involving regulatory proteins ensures that internalized molecules are appropriately recycled back to the plasma membrane, delivered to the TGN for transport to locations throughout the cell, or degraded within organelles called lysosomes. **Lysosomes** are vesicles that contain granules consisting of digestive enzymes called *acid hydrolases*, which catalyze the attack of a water molecule on ester and amide linkages under acidic conditions. In addition to their role in endocytosis, lysosomes also contribute to the autophagic degradation of debris within cells (p. 558).

Early endosomes are tubular-vesicular networks that mature to form *late endosomes*, which are also called *multivesicular bodies* because they contain numerous closely packed vesicles. The maturation process is achieved in part by an increase in hydrogen ion concentration (i.e., a reduction in internal pH) through the activity of V-ATPase (an ATP-dependent proton pump) and the arrival of TGN vesicles containing lysosomal acid hydrolases and membrane proteins. Late endosomes are converted to fully functional lysosomes when the internal pH is less than 5, a circumstance that activates the acid hydrolases. Late endosomes may also fuse with existing lysosomes.

There are several forms of endocytosis. The best-researched examples are clathrin-dependent endocytosis and clathrin-independent caveolar endocytosis.

Clathrin-dependent endocytosis, also referred to as *receptor-mediated endocytosis*, is a versatile and widely used mechanism in which clathrin-coated vesicles containing cargo bound to membrane receptors are taken into cells. Examples of its uses include nutrient uptake [e.g., low-density lipoproteins (LDL), sources of lipids such as cholesterol (p. 400), and transferrin, an iron-binding protein], intercellular signal transduction, and membrane recycling. The process begins with the binding of a specific ligand to its cognate receptor on the external surface of the plasma membrane. Adapter proteins then bind to the cytoplasmic side of the receptor-ligand complex, after which clathrin is recruited. **Clathrin** is a soluble protein complex called a *triskelion* (three heavy chains and three light chains) because of its shape (**Figure 2.18**). As the clathrin triskelia bind to the adapter proteins, a basketlike latticework forms that forces the membrane into the

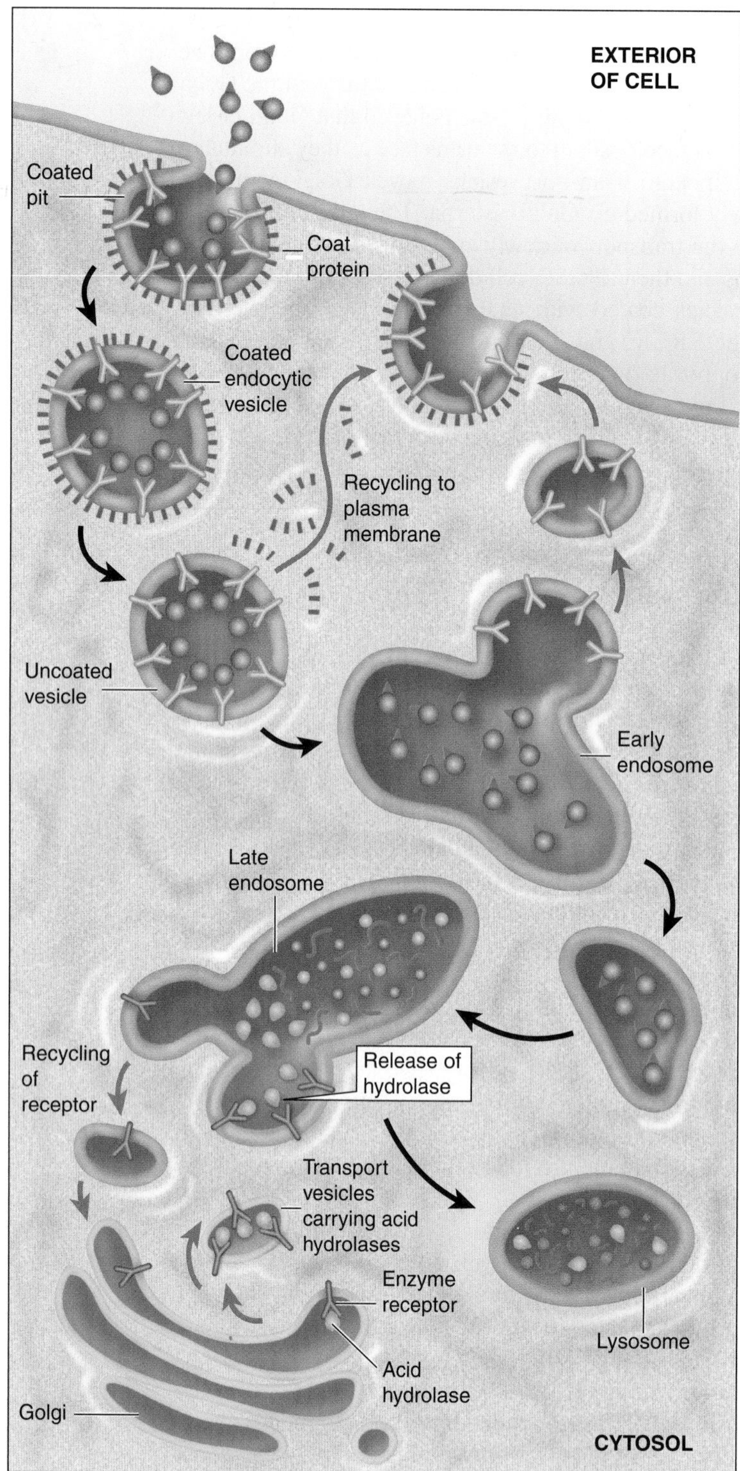

FIGURE 2.17

Receptor-Mediated Endocytosis

Extracellular substances may enter the cell during endocytosis, a process in which receptor molecules in the plasma membrane bind to the specific molecules or molecular complexes called ligands. Specialized regions of plasma membrane called coated pits (composed of clathrin triskelia, not shown) progressively invaginate to form closed vesicles. After the coat proteins are removed, the vesicle fuses with an early endosome, the precursor of lysosomes. The coat proteins are then recycled to the plasma membrane. During endosomal maturation, the proton concentration rises and the ligands are released from their receptors, which are subsequently also recycled back to the plasma membrane. As endosomal maturation continues, lysosomal hydrolases are delivered from the Golgi apparatus. Lysosomal formation is complete when all the hydrolases have been transferred to the late endosome and the Golgi membrane has been recycled back to the Golgi apparatus.

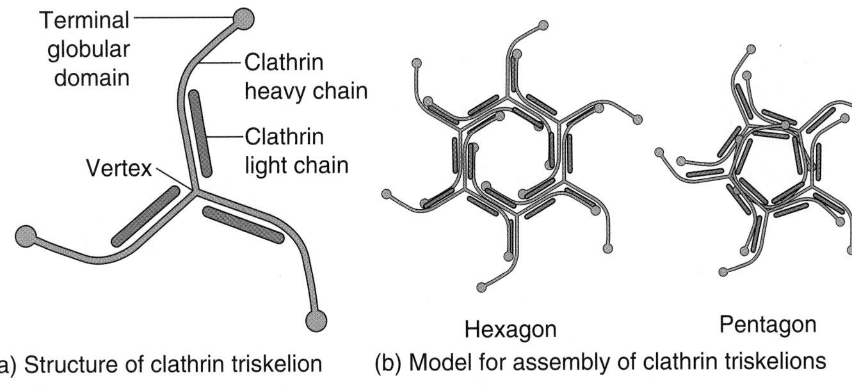

(a) Structure of clathrin triskelion (b) Model for assembly of clathrin triskelions

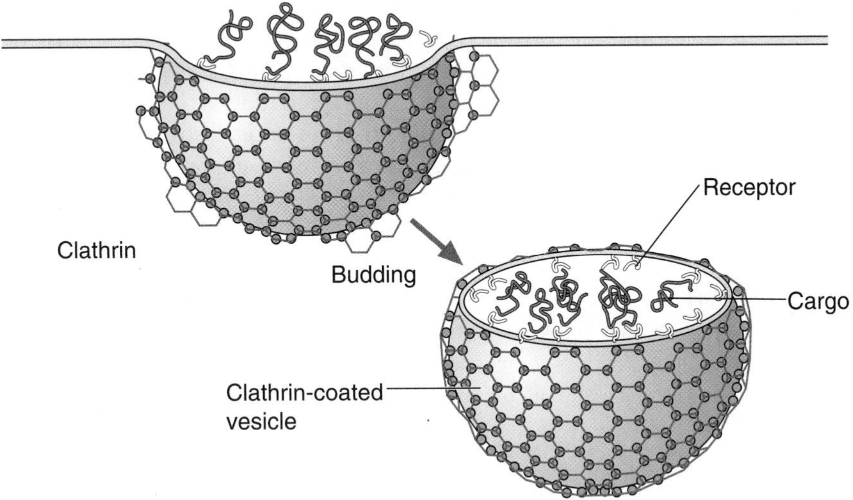

(c) Clathrin-coated vesicle formation

FIGURE 2.18
Clathrin-Dependent Endocytosis

(a) Each clathrin triskelion is formed by three heavy chains and three light chains. (b) Triskelia combine to form the hexagons and pentagons observed in the latticework of clathrin-coated vesicles. (c) Clathrin-coated vesicle formation is initiated by GTP-binding proteins (not shown) that recruit adaptor proteins that serve as binding sites for clathrin. The association of clathrin triskelia to form hexagons (in this figure) causes the membrane distortion of the vesicle-forming process.

shape of a bud. The clathrin-coated vesicle is then excised from the plasma membrane by dynamin, a guanosine triphosphate (GTP)-requiring protein that encircles and constricts the vesicle's neck until a fully formed coated vesicle is released from the plasma membrane. Vesicle fusion with the early endosome is preceded with removal of the clathrin coat. When vesicles, such as those containing LDLs, have fused with early endosomes, the reduction in pH releases the cargo from their receptors. The LDL receptors are recycled back to the plasma membrane, and LDL molecules (lipids and proteins) are degraded within lysosomes.

Some forms of endocytosis are clathrin-independent. The best known of these is **caveolar endocytosis. Caveolae** ("little caves") are relatively small invaginations formed from a specialized type of plasma membrane microdomain enriched in cholesterol, several types of membrane lipids, signaling molecules, and ion channel regulatory proteins. Caveolae formation requires *caveolin* (p. 409), an integral protein in the inner leaflet of the plasma membrane of numerous cells, but occurs most prominently in *endothelial cells* (cells that line the inner surface of blood and lymphatic vessels) and *adipocytes* (the principal cells in adipose tissue). Caveolar vesicle membrane curvature occurs when caveolin molecules associate to form an *oligomer* (a protein complex formed via noncovalent bonds). A group of cytoplasmic proteins called cavins assist in caveolae formation. The internalization of insulin receptors in adipocytes is an example of caveolar endocytosis.

Once believed to be a simple process, endocytosis is now recognized as being fully integrated into cellular signaling and regulation. Endocytosis in combination with exocytosis (p. 49), referred to as the **endocytic cycle**, plays a central role in cellular information processing. Until recently, the endocytic cycle has been regarded as a means of controlling a cell's response to a signal molecule by regulating the number of its cognate receptors in the plasma membrane. In this view endocytosis is a mechanism for downregulating receptors, thereby desensitizing cells to signaling molecules. Recent evidence indicates that endocytosis pathways contribute in other ways to signal transduction. For example, signaling from some receptors, such as insulin receptor and thyroid-stimulating hormone receptor, has been shown to continue after they enter endosomes. In addition, endosomes may serve as signaling platforms because they contain protein or lipid components not present in the plasma membrane, and endosomal pathways offer opportunities for signal diversification.

QUESTION 2.2

In many genetic disorders, a lysosomal enzyme required to degrade a specific molecule is missing or defective. These maladies, often referred to as *lysosomal storage diseases*, include Tay-Sachs disease. Afflicted individuals inherit from each parent a defective gene that codes for an enzyme that degrades a complex lipid molecule. Symptoms include severe mental retardation and death before the age of 5 years. What is the nature of the process that is destroying the patient's cells? [*Hint:* Synthesis of the lipid molecule continues at a normal rate.]

Nucleus

The **nucleus** (**Figure 2.19**) is the most prominent organelle in eukaryotic cells. It contains most of the cell's genetic information. Low-resolution micrographs reveal that nuclear structure consists of a seemingly amorphous nucleoplasm surrounded by membrane, the nuclear envelope. **Nucleoplasm** contains a network of

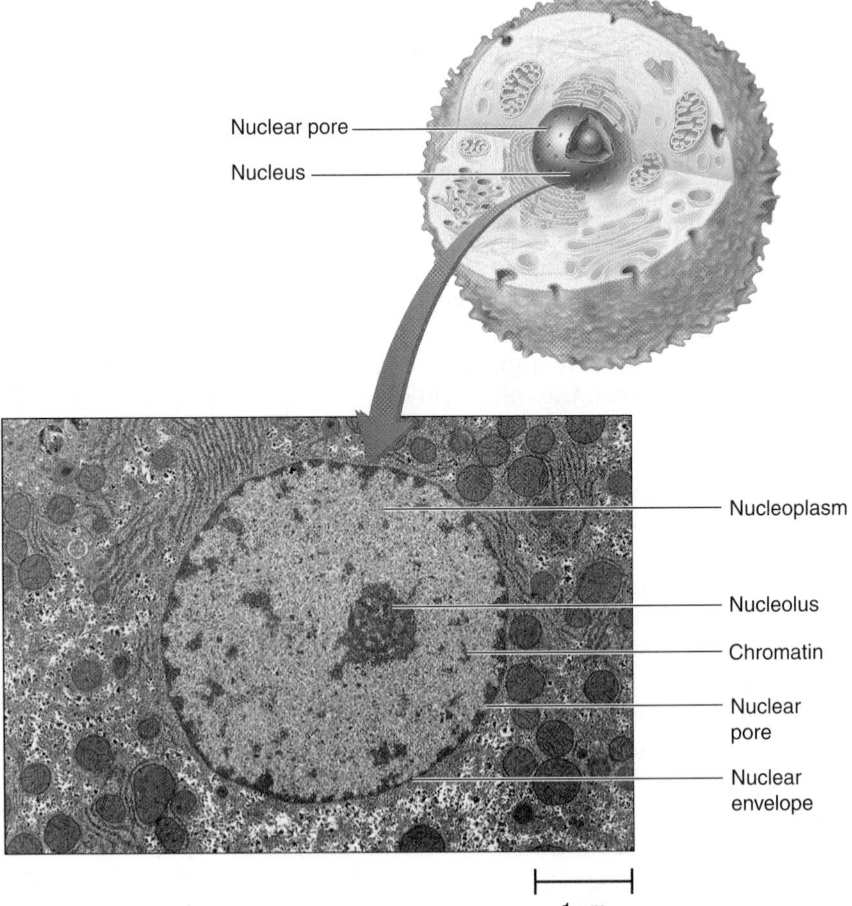

FIGURE 2.19
The Eukaryotic Nucleus
The nucleus is an organelle surrounded by a double membrane, the nuclear envelope. The nuclear envelope, a barrier that prevents the free passage of molecules between the nuclear compartment and the cytoplasm, plays a vital role in gene expression regulation.

chromatin fibers, which during the mitotic phase of the cell cycle are condensed to form the chromosomes that will be distributed to daughter cells. Chromatin is highly structured, consisting of DNA and DNA packaging proteins known as the *histones*. Chromatin has been traditionally classified according to how densely it is packaged. *Euchromatin*, the lightly packaged form, is usually gene rich and is easily accessed by transcription factors and transcription enzyme complexes. The other form, called *heterochromatin*, is tightly packed and inaccessible for transcription. *Constitutive heterochromatin*, containing the sequences for chromosomal structures such as centromeres and telomeres and remnants of ancient viruses, is highly condensed and permanently inactive. *Facultative heterochromatin* is capable of changing in response to specific signaling processes from a condensed inactive state into actively transcribed euchromatin. Each type of differentiated cell has its own specific set of facultative heterochromatin. Chromatin distribution within the nucleus is not random. Chromosomes occupy discrete locations called *chromosome territories*. In general, chromosome segments containing heterochromatin typically occur in the periphery of the nucleus. Actively transcribed, gene-dense euchromatin is located more centrally.

The **nuclear envelope** (NE) acts as a barrier that prevents the free passage of molecules between the nucleus and the cytoplasm. As a result, processes such as DNA replication and transcription are more easily regulated. The NE is composed of two concentric membranes. The **outer nuclear membrane** (ONM) is continuous with the rough ER, and ribosomes are attached to its cytoplasmic surface. Among the many ER proteins on the cytoplasmic side of the ONM, several (e.g., nesprins) bind to the filaments of the cytoskeleton. Unlike the outer membrane, the **inner nuclear membrane** (INM) contains integral proteins that are unique to the nucleus. In addition to stabilizing NE structure, the functions of these proteins include chromatin binding, chromatin remodeling protein recruitment, and various enzyme activities. The space between the two membranes, the **perinuclear space** (diameter 20–50 nM), is continuous with the lumen of the rough ER. The inner and outer membranes fuse at structures called *nuclear pores*, which are elaborate macromolecular structures that regulate molecular traffic between the cytoplasm and the nucleus. Nuclear pores, referred to as **nuclear pore complexes** (NPCs) (Figure 2.20), vary in number in vertebrates from 2000 to 4000 per nucleus. Each NPC is a 120-MDa structure (diameter = 120 nm) composed of 30 different proteins called the *nucleoporins*. The function of NPCs was once thought to be limited to nucleocytoplasmic transport. Recent research, however, has revealed that nucleoporins also have roles in chromatin organization and in DNA replication and repair.

The membrane-embedded ring-shape core of the NPC is attached to a basket-shape structure. Filaments that extend from the cytoplasmic and nucleoplasmic side of the NPC function as docking sites for large molecules that will subsequently be transported through the pore. A meshwork formed from flexible nucleoporins that line the central pore restricts transport through the NPC only to those macromolecules (e.g., RNAs and large proteins) bound to either import or export chaperone proteins. Small substances such as ions and small proteins (40 kDa or less) diffuse through the NPC, which has a functional diameter of about 9 nm. Driven by the hydrolysis of the nucleotide GTP, traffic through the NPC is brisk and efficient.

Proteins entering the nucleus must have a nuclear localization signal amino acid sequence, which is recognized by a cargo transport protein called an *importin*. In a process driven by the energy released by GTP hydrolysis, the protein-importin complex is then transported through the nuclear pore. The nuclear export process, also driven by GTP hydrolysis, is similar to the import process. Cargo leaving the nucleus, usually RNA molecules, binds to proteins with nuclear export signal sequences. The newly formed RNA-protein complexes then bind to *exportins*, the nuclear transport proteins that mediate cargo

KEY CONCEPTS

- The nucleus contains the cell's genetic information and the machinery for converting that information into a code for protein synthesis.
- The nucleolus plays an important role in the synthesis of ribosomal RNA.

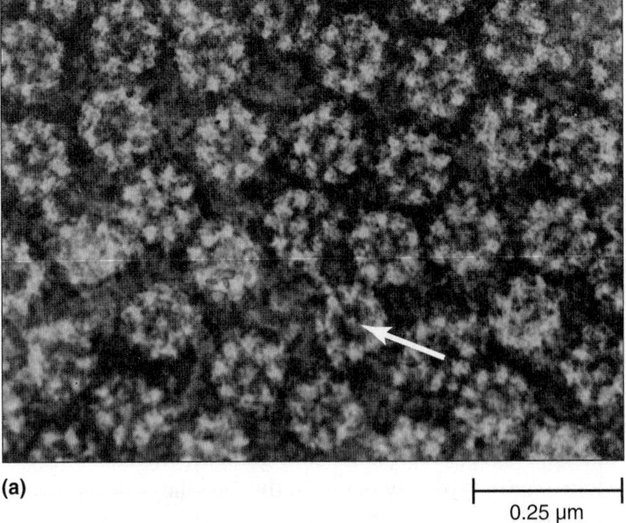

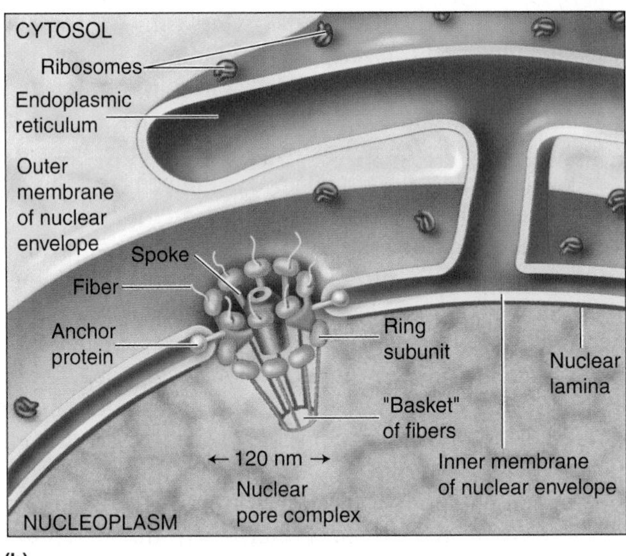

FIGURE 2.20
The Nuclear Pore Complex

(a) The nuclear envelope is studded with thousands of nuclear pore complex structures, one of which is indicated by the arrow. (b) The basic structure of NPCs is a donut-shaped scaffold seated on top of a basketlike structure, the center is a porelike opening. Unstructured polypeptides (not shown) in the center ensure the selective transport of cargo molecules bound to nuclear transport proteins.

transport through the NPC and into the cytoplasm. About 1000 macromolecules pass through each NPC per second.

The **nuclear lamina** (**Figure 2.21**) is a thin, dense protein meshwork attached to the inner surface of the inner nuclear membrane. Once thought to only provide shape and mechanical stability to the nuclear envelope, the nuclear lamina is now believed to have roles in numerous nuclear processes including DNA replication, transcription, and chromatin organization. The nuclear lamina consists largely of lamin filaments and lamin-associated proteins. As a result of linkages between lamin filaments, lamin-associated proteins, and the cell's cytoskeleton (pp. 62–66), mechanical force strong enough to deform the cytoskeleton can alter the shape of the NE and possibly change chromatin organization within the nucleus.

The lamins are intermediate filament proteins (see p. 65) that are classified as type A (lamins A and C) and type B (lamins B1 and B2). Type A and type B lamins each polymerize to form separate types of filament. Examples of lamin-associated proteins include the integral INM proteins emerin, LBR, and SUN-domain proteins. Emerin binds directly to lamin A filaments and to BAF, a DNA-bridging protein with roles in chromatin organization and gene expression. LBR (lamin B receptor) binds to both B lamin filaments and heterochromatin. It is noteworthy that the nuclear lamina and its associated heterochromatin do not

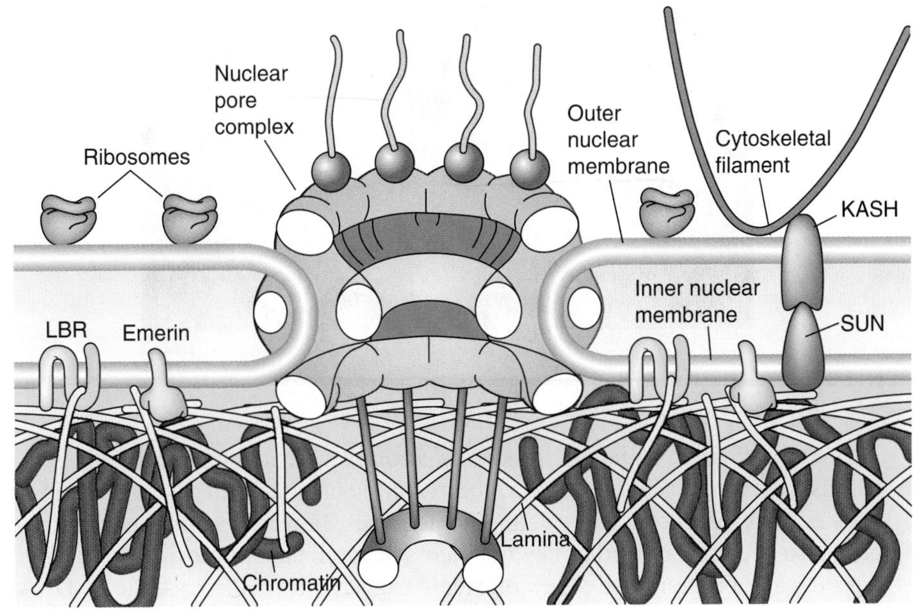

FIGURE 2.21
The Nuclear Lamina

A thin, dense network of lamin filaments is linked to the inner nuclear membrane via interactions with lamin-associated proteins such as emerin, LBR (lamin B receptor), and SUN-domain proteins. Note that emerin and LBR also bind to chromatin, although emerin does so indirectly via the protein BAF (not shown). SUN domain proteins, which are integral INM proteins, bind to the KASH domains of the ONM nesprins to form the LINC complex. The LINC complex thus connects the nuclear lamina to the cell's cytoskeleton.

extend to NPCs. INM SUN-domain proteins bind with one or more of the nesprins in the ONM to form LINC (*li*nks *n*ucleoskeleton and *c*ytoskeleton) complexes. Since the nesprins bind directly or indirectly to cytoskeletal filaments, LINC connects the nucleoplasm to the cytoskeleton.

Regions of the nucleus called *nuclear bodies* have been found to contain certain types of chromatin sequences and nuclear proteins. Examples include nucleoli, speckles, and Cajal bodies. The **nucleolus** is the largest of the nuclear bodies. Its best understood functions are the transcription of rRNA genes, rRNA processing reactions, and ribosomal subunit synthesis. After their transport into the cytoplasm, the ribosomal subunits bind to mRNAs to form ribosomes, the macromolecular complexes that synthesize proteins. The most recognizable feature of the nucleolus is the *nucleolar organizer region*, which is formed by the association of several chromosomal segments, each with multiple copies of rRNA genes. There are also discrete sites within the nucleoplasm, called *transcription factories*, in which active genes come together to be transcribed into polypeptides or small RNAs by multiple copies of transcription enzyme complexes. *Speckles*, as many as 50 per cell, are storage sites for transcription components. *Cajal bodies* (named after Ramon y Cajal, the Spanish histologist, 1852–1934) are sites for processing reactions of histone mRNAs and several ncRNAs.

The **nuclear matrix** (nucleoskeleton) is a scaffold within the nucleoplasm composed of a large number of proteins in which loops of chromatin are organized. It is believed to be analogous to the cell's cytoskeleton because various forms of cytoskeletal proteins (e.g., actins, actin binding proteins, and myosins; see p. 64) have been located within the nucleoplasm. Although numerous nuclear processes have been shown to require specific nucleoskeletal proteins, their structural properties remain unresolved.

Mitochondria

Mitochondria (singular: mitochondrion) are organelles that have long been recognized as the site of **aerobic metabolism**, the mechanism by which the chemical bond energy of food molecules is captured and used to drive the oxygen-dependent synthesis of adenosine triphosphate (ATP), the cell's energy storage molecule. They are often observed positioned near sites of high-energy demand. For example, most striated muscle cell mitochondria are positioned along the entire length of myofibrils. Mitochondria also play vital roles as central

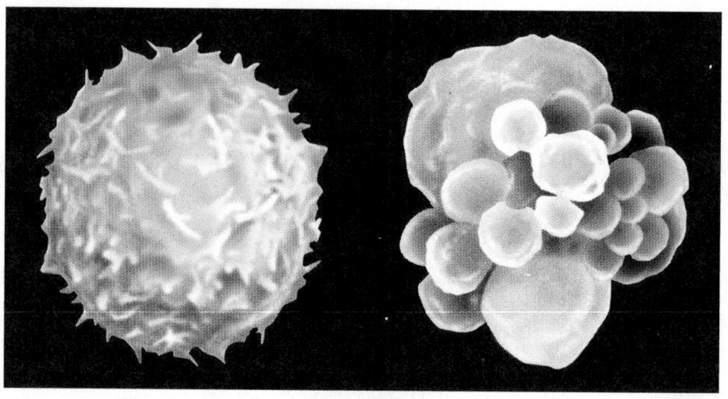

FIGURE 2.22
Apoptosis

White blood cells before (left) and during (right) apoptosis. In response to cell stress, mitochondria release a protein called cytochrome c into the cytoplasm, which facilitates the activation of enzymes that proceed to degrade cell components. The apoptotic cell shown is forming blebs that will eventually fragment into apoptotic bodies. Ultimately, the apoptotic bodies will be ingested by phagocytes (immune system cells that digest cell debris).

integrators of other metabolic processes. Prominent examples include the metabolism of amino acids and lipids, the synthesis of iron-sulfur clusters (p. 350) used in redox reactions, and calcium homeostasis. In recent years mitochondria have also been recognized as key regulators of **intrinsic apoptosis**, one form of a genetically programmed series of events triggered by cell stresses (e.g., DNA damage, hypoxia, or nutrient deprivation) that leads to cell death (**Figure 2.22**). They have traditionally been described as sausage-shaped structures with lengths ranging from 1 to 10 μm. This view has changed considerably as researchers have discovered that mitochondria have no fixed sizes. Mitochondria cannot be generated *de novo*, that is, new mitochondria are the result of *biogenesis*, the growth and division of pre-existing mitochondria. They are dynamic organelles that are continuously dividing (*fission*), branching, and merging (*fusion*) to form extended reticular networks.

In healthy cells there is a continuous remodeling of mitochondrial networks, most notably by continuous cycles of fission and fusion. **Mitochondrial fission** (**Figure 2.23**) allows biogenesis when cell energy requirements are high or when damaged or inactive portions of mitochondria are segregated prior to destruction. **Mitochondrial fusion** generates extended mitochondrial networks and facilitates the rescue of mitochondria with minor damage by allowing the mixing of their contents with healthy mitochondria. An autophagic process called mitophagy removes heavily damaged mitochondria.

Each **mitochondrion** is bounded by two membranes (**Figure 2.24**). The relatively porous smooth **outer membrane** is permeable to most molecules with masses less than 10,000 Da. The **inner membrane** is permeable to O_2, CO_2, and H_2O and impermeable to ions and a variety of organic molecules. It projects inward into folds that are called *cristae* (singular: crista). Embedded in this membrane are protein complexes and other molecules that comprise the *mitochondrial respiratory chain* (MRC). In a process referred to as *oxidative phosphorylation* (OXPHOS), the energy released by the oxidation of nutrient molecules (e.g., glucose, fatty acids, and amino acids) is transformed into the chemical bond energy of ATP. Also present are proteins that are responsible for the transport of specific molecules and ions.

Together, the inner and outer membranes create two separate compartments: (1) the *intermembrane space* and (2) the *matrix*. The intermembrane space contains several enzymes involved in nucleotide metabolism, whereas the gel-like matrix consists of high concentrations of enzymes and ions and a myriad of small organic molecules. The matrix also contains 2 to 10 circular DNA molecules.

Mitochondrial DNA (mt DNA) resembles bacterial DNA in that both molecules are "naked" (i.e., not packaged with histones) and are located within a nucleoid. The mitochondrial genome codes for 2rRNAs, 22tRNAs, and 13MRC protein components. About 95% of the genes that code for mitochondrial proteins are located on nuclear chromosomes.

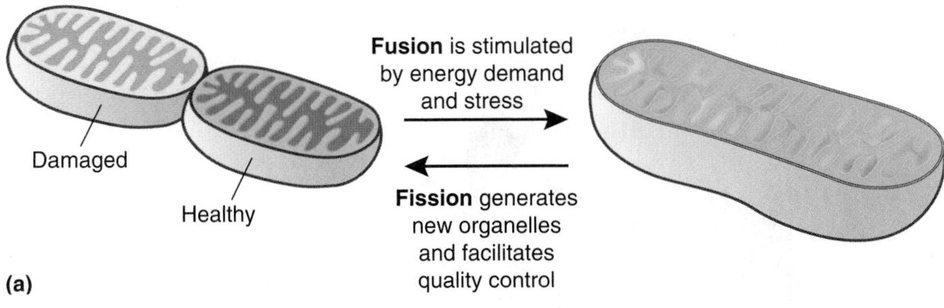

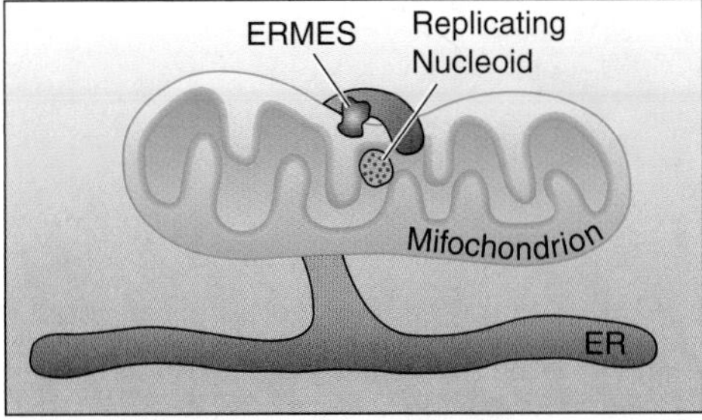

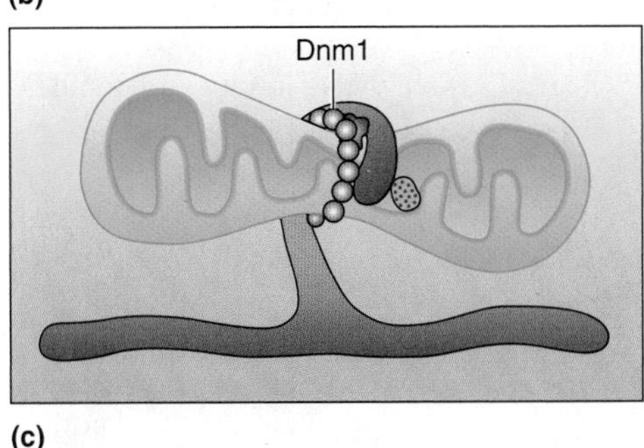

FIGURE 2.23

Mitochondrial Fission and Fusion

(a) Both fission and fusion help mitochondria remain functional when cells experience metabolic, oxidative, and other stresses. Fission creates new mitochondria and serves as a mechanism for removing damaged mitochondria. The damaged mitochondrial product of fission is then destroyed by mitophagy. Mildly damaged mitochondria can be rescued when they fuse with healthy mitochondria (i.e., components from a healthy mitochondrion compensate for those in the damaged organelle). (b) In yeast cells mitochondrial fission involves an interaction with an ER tubule, which is mediated in part by a protein tether called ERMES (ER-mitochondria encounter structure). (c) Fission is subsequently facilitated by the GTP-hydrolyzing protein dynamin-1 (Dnm1) (Drp1 in mammals), which then constricts and severs the membrane stalk between the two mitochondria.

The number of mitochondria a cell possesses is a function of its energy demands and physiological status. There is considerable variety across cell types. For example, human oocytes and hepatocytes may have as many as 200,000 and 2000 mitochondria, respectively. Most cell types have several hundred mitochondria. Erythrocytes (red blood cells) have none. Notably, the configuration of mitochondria changes with the physiological status of the cell. For example, the internal appearance of liver mitochondria has been observed to change dramatically during active respiration (**Figure 2.25**).

Mitochondria form stable contact sites with regions of the ER called *mitochondria-associated membranes* (MAMs). These contacts, which consist of closely apposed membrane segments tethered by proteins, have several functions that together regulate mitochondrial dynamics:

1. **Calcium signaling.** Calcium-dependent signaling cascades initiated by a hormone, growth factor, or other type of stimulus always involve the rapid transfer of calcium into mitochondria via MAMs. The subsequent activation of Ca^{2+}-responsive proteins in every phase of ATP synthesis ensures that sufficient energy is available for the processes that are initiated by the signaling event and quenching of the calcium signal by ATP-pump complexes in the ER and plasma membrane. Mitochondria have low-affinity calcium channels that also contribute to calcium signal quenching.
2. **Lipid exchange**. Although cellular phospholipid biosynthesis occurs predominantly in the ER, several reactions require mitochondrial enzymes. Bidirectional lipid transfer between mitochondria and the ER via MAMs, therefore, is required for maintenance of the unique lipid composition of both the inner and the outer mitochondrial membranes.

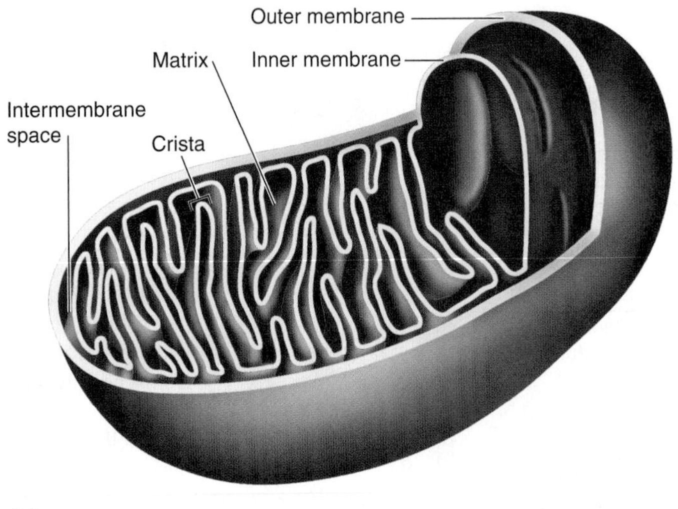

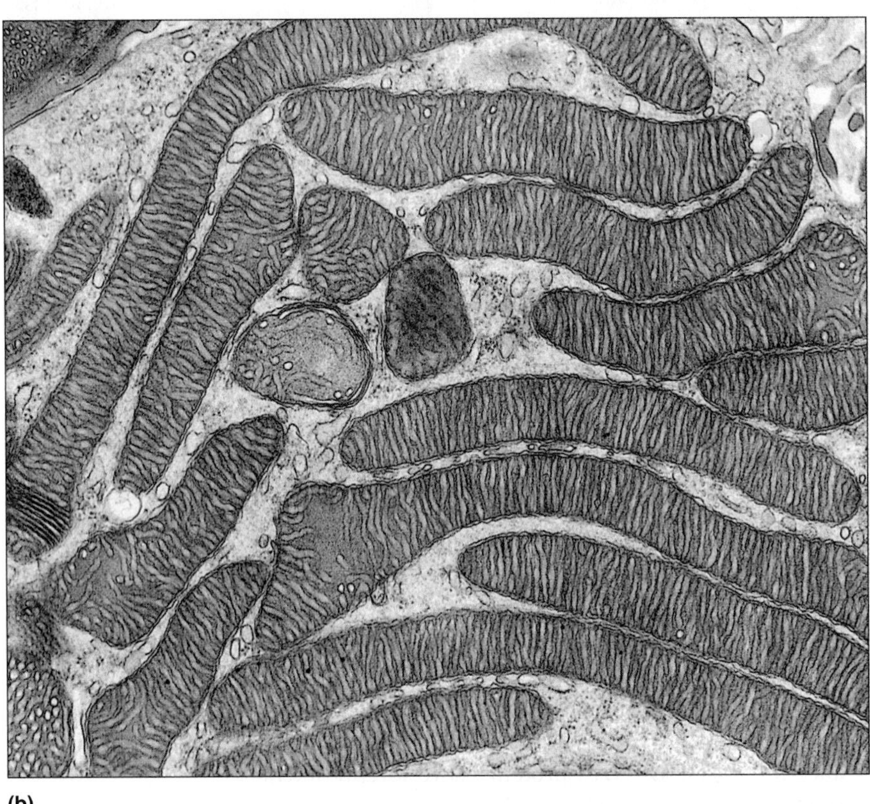

FIGURE 2.24

The Mitochondrion

(a) Membrane and crista. The internal structure depicted in this diagram is referred to as the baffle model because of the bellowlike shape of the crista. Electron tomography studies (a microscopic technique in which electron beams are used to create three-dimensional reconstructions of specimens) have revealed a more complex anatomy. Complex arrays of fusing and dividing inner membrane tubules have been observed in the mitochondria of some tissues. The functional significance of these structural features is unknown. (b) Mitochondria from adrenal cortex, the outer layer of cells of the adrenal glands located above the kidneys.

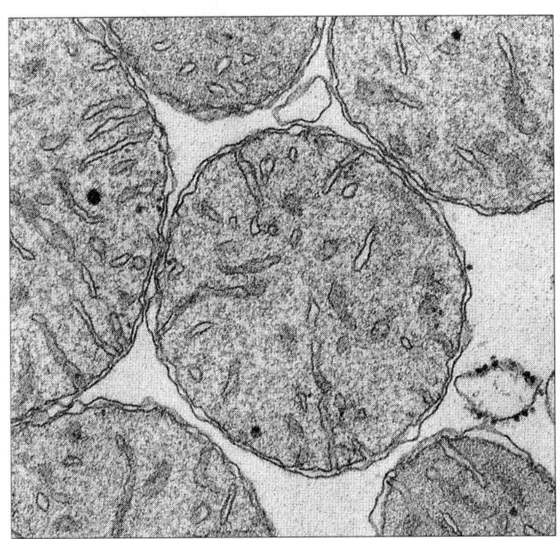
(a)

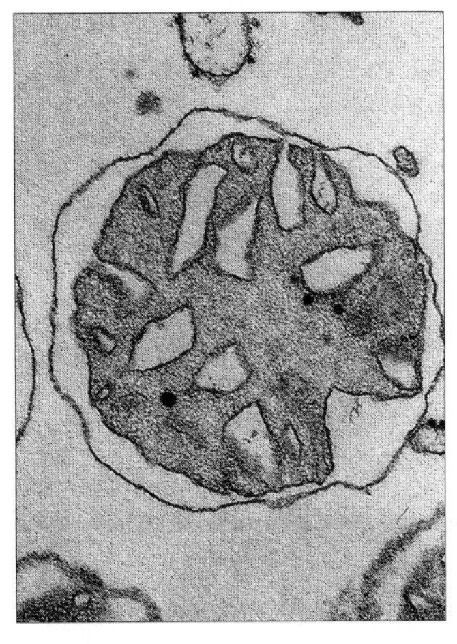
(b)

FIGURE 2.25
Rat Liver Mitochondria

(a) Low-energy (orthodox) and (b) high-energy (condensed) conformations.

3. **Mitochondrial fission regulation**. The wrapping of an ER tubule around a mitochondrion (refer to **Figure 2.23c**) is an early step in the fission process. It is believed that proteins at ER-mitochondria contact sites initiate fission by marking the division location on the mitochondrion and recruiting the fission proteins that cause constriction and division.

KEY CONCEPTS

- Aerobic respiration, the process that generates most of the energy required in eukaryotes, takes place in mitochondria.
- Embedded in the inner membrane of a mitochondrion are respiratory assemblies, where ATP is synthesized.

QUESTION 2.3

It has been estimated that mitochondria occupy 20% of the volume in the human body. For a 70-kg adult, the average number of mitochondria has been estimated to be 1×10^{16} (10,000 trillion). Using this information, provide a rough estimate of the average mass of a mitochondrion.

Peroxisomes

Peroxisomes are small, spherical organelles that consist of the single membrane that surrounds an enzyme-rich matrix. Found in all human cells except for erythrocytes (red blood cells), peroxisomes most notably occur in large numbers in the liver and kidney. Peroxisomal enzymes are involved in a variety of anabolic and catabolic pathways, including synthesis of certain membrane lipids, purine and pyrimidine bases (pp. 540–545), and bile acids (p. 461). Peroxisomes are also involved in the degradation of long-chain fatty acids and purine bases. As their name suggests, peroxisomes are most noted for their involvement in the generation and breakdown of toxic molecules known as peroxides. Hydrogen peroxide (H_2O_2) is generated when molecular oxygen (O_2) is used to remove hydrogen atoms from specific organic molecules.

$$RH_2 + O_2 \rightarrow R + H_2O_2$$

For example, the enzyme xanthine oxidase generates H_2O_2 when it catalyzes two reactions in the pathway that converts purine bases into the nitrogenous waste molecule uric acid. Peroxisomes use H_2O_2 to oxidize toxic molecules such as formaldehyde or alcohol. If not used in such reactions, this highly reactive molecule is detoxified by the enzyme catalase.

$$H_2O_2 \rightarrow 2H_2O + O_2$$

Peroxisome biogenesis occurs by two distinct pathways. In the *de novo* process preperoxisomal vesicles bud off from a specialized region of the ER. These vesicles then fuse together to form mature peroxisomes. As many as 32 peroxins,

Visit the companion website at www.oup.com/us/mckee to read the Biochemistry in Perspective essay on organelles and human disease.

proteins required for peroxisome assembly, have been identified. Preexisting peroxisomes can grow larger with the transfer of membrane from the ER and then divide to produce new peroxisomes.

Chloroplasts

Chloroplasts are specialized chromoplasts that convert light energy into chemical energy. (A chromoplast is a cellular organelle in plants that accumulates the pigments responsible for the colors of leaves, flower petals, and fruits.) In this process, called **photosynthesis**, which will be described in Chapter 13, light energy is used to drive the synthesis of carbohydrate from CO_2. The structure of chloroplasts (**Figure 2.26**) is similar in several respects to that of mitochondria. For example, the outer membrane is highly permeable, whereas the relatively impermeable inner membrane contains special carrier proteins that control molecular traffic into and out of the organelle. In addition, chloroplasts undergo fission.

An intricately folded internal membrane system, called the **thylakoid membrane**, is responsible for the metabolic function of chloroplasts. For example, chlorophyll molecules, which capture light energy during photosynthesis, are bound to thylakoid membrane proteins. Certain portions of thylakoid membrane form tightly stacked structures called **grana** (singular: granum), whereas the entire membrane encloses a compartment known as the *thylakoid lumen*. Surrounding the thylakoid membrane is the **stroma**, a dense enzyme-filled substance, analogous to the mitochondrial matrix. In addition to enzymes, the stroma contains DNA, RNA, and ribosomes. Membrane segments that connect adjacent grana are referred to as *stroma lamellae* (singular: lamella).

Cytoskeleton

The **cytoskeleton** is a network of fibers, filaments, and associated proteins called the **cytoskeleton** (**Figure 2.27**). Its components include microtubules, microfilaments, and intermediate fibers.

Microtubules, the largest constituent of the cytoskeleton (outer diameter 25 nm, inner diameter 12 nm), are girderlike hollow cylinders composed of protofilaments. Each protofilament is formed by the reversible polymerization of the protein tubulin. *Tubulin* is a dimer that consists of two polypeptides: α-tubulin and β-tubulin, which is a GTP-binding molecule. Microtubules are polar; that is, their ends are different. At the plus (+) end, polymerization can occur rapidly. The minus (−) end grows more slowly. As the microtubule grows at the plus end, it extends toward the cell's periphery. Microtubule dynamics are regulated by microtubule-associated proteins (MAPs), a series of molecules that

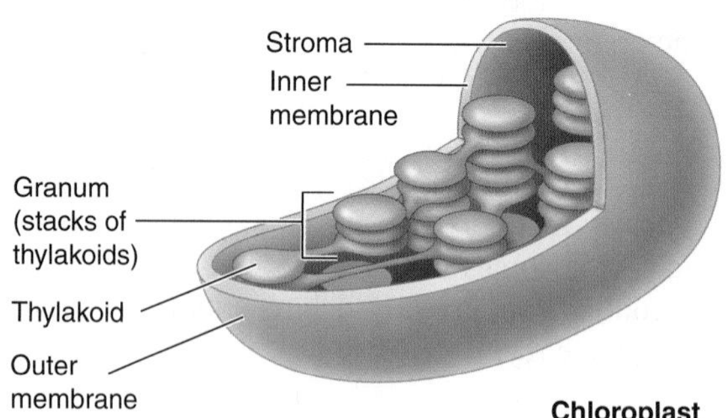

FIGURE 2.26
The Chloroplast
Chloroplasts convert light energy into the chemical bond energy of organic biomolecules.

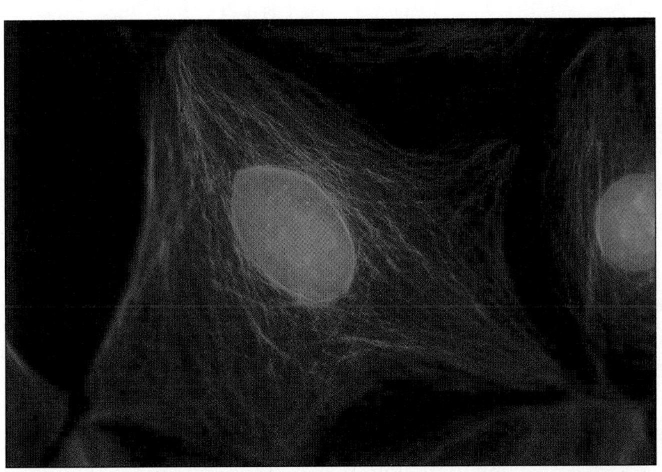

(a)

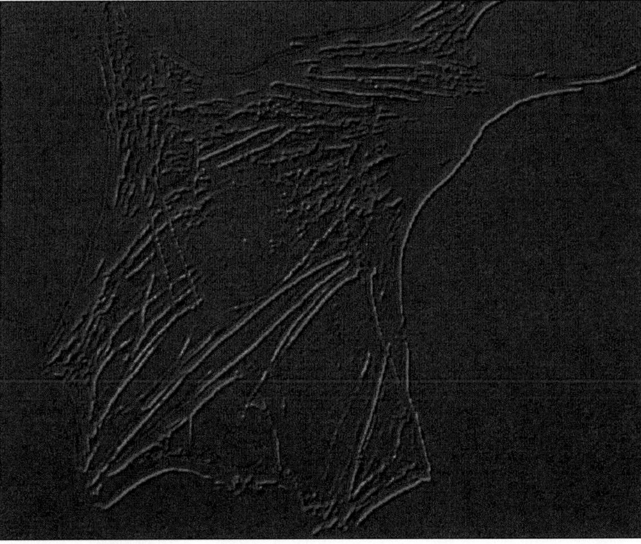

(b)

(c)

FIGURE 2.27

The Cytoskeleton

The major components of the cytoskeleton are (a) microtubules, (b) microfilaments, and (c) intermediate filaments. The intracellular distribution of each type of cytoskeletal component is visualized by staining with fluorescent dyes.

control microtubule stability by promoting or preventing the assembly process. Other functions of MAPs include guiding microtubules toward specific cellular locations and cross-linking that creates microtubule bundles. The ATP-dependent motor proteins kinesin and dynein move along microtubules. In general, kinesin moves cargo such as vesicles or organelles toward the plus end and dynein moves toward the minus end. Although found in many cellular regions, microtubules are most prominent in long, thin structures that require support (e.g., the extended axons and dendrites of nerve cells). They are also found in the *mitotic spindle* (the structure formed in dividing cells that is responsible for the equal dispersal of chromosomes into daughter cells) and the slender, hairlike organelles of locomotion known as cilia and flagella (**Figure 2.28**).

Cilia and flagella, whiplike appendages encased in plasma membrane, are highly specialized for their roles in propulsion. The most prominent examples include the motile cilia on the surface of tracheal cells that move debris-laden mucus away from the lungs and the flagellum of sperm cells that seek out egg cells. The microtubules in the inner core of flagella and cilia, referred to as the *axoneme*, form a ring of nine fused pairs with a centrally located unfused pair (a 9 + 2 pattern). The undulating motion of cilia and flagella is the result of the outer microtubule pairs sliding relative to each other. Bending occurs as the ATP-driven structural changes in the dynein molecules (called "arms") cause them to alternately attach to and "walk along" the adjacent microtubule and

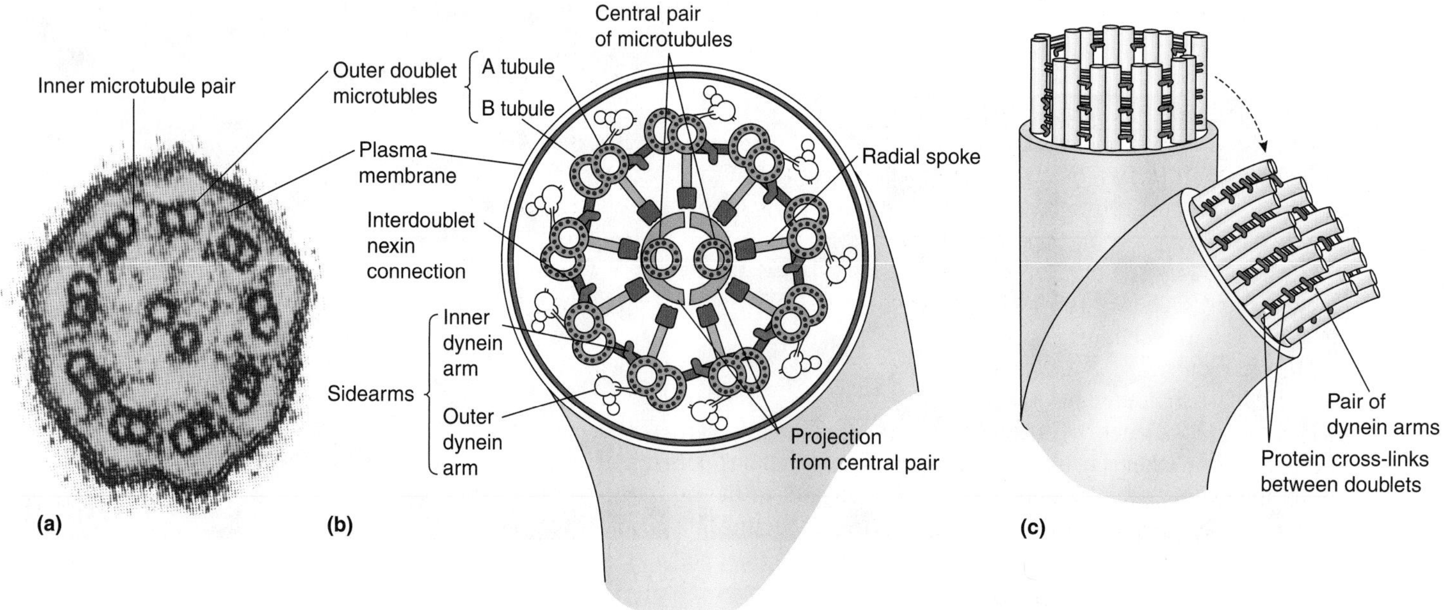

FIGURE 2.28

The Axoneme of Eukaryotic Cilia and Flagella

The axoneme is microtubule-based cytoskeletal structure. In whiplike motile cilia and flagella the axoneme has a classic $9 \times 2 + 2$ microtubule pattern (i.e., nine outer pairs [doublet microtubules] and an inner microtubule pair). (a) Transmission electron micrograph of a cross-section of a motile eukaryoutic flagellum. (b) This diagram of an axoneme in cross-section illustrates the classic axoneme pattern. The dynein arms are ATP-hydrolizing motor proteins that "walk along" the microtubules to produce axoneme bending. The nexin linkages connect the outer doublet microbules. Radial spokes are believed to regulate axoneme motion. (c) This diagram illustrates axoneme bending, the result of the sliding of the microtubule doublets relative to each other.

then detach. The microtubules also transport cargo (e.g., newly synthesized axonemal proteins) within cilia and flagella. In a process called *intraflagellar transport (IFT)*, kinesins move particles containing molecules required for ciliary or flagellar assembly and maintenance along the outer pair of microtubules toward the cell periphery. Dyneins move substances (e.g., kinesins that have discharged their cargo) in the opposite direction. A nonmotile version of cilia, referred to as *primary cilia*, is an important structural feature of most vertebrate cells. Their impact on human health is described in the Biochemistry in Perspective essay Primary Cilia and Human Disease on p. 67.

Microfilaments are small fibers (5–7 nm in diameter) composed of polymers of globular actin (G-actin). The filamentous or polymeric form (F-actin) exists as a coil of two actin polymers with a plus end and a minus end. Polymerization, driven by ATP hydrolysis, occurs more rapidly at the plus end. The individual filaments, being highly flexible, are usually cross-linked into bundles of different sizes. A large variety of actin-binding proteins regulate the structure and functional properties of microfilaments. They cross-link, stabilize, sever (cut into fragments), or cap (block polymerization) microfilaments. Microfilaments can exert force simply by polymerizing or depolymerizing. Together with the myosins, a large family of ATP-dependent motor proteins, microfilaments generate contractile forces that create tension. Important roles of microfilaments include involvement in cytoplasmic streaming (a process that is most easily observed in plant cells in which cytoplasmic currents rapidly displace organelles such as chloroplasts), ameboid movement (a type of locomotion created by the formation of temporary cytoplasmic protrusions), and muscle contraction.

Intermediate filaments (8–12 nm in diameter) are a large group of flexible, strong, and relatively stable polymers. They provide cells with significant mechanical support. A network of intermediate filaments (IFs) extends from a ringlike meshwork around the nucleus to attachment points on the plasma membrane. There are six classes of IF proteins, which differ in their amino acid sequences. Well-known examples are the keratins found in skin and hair cells and the lamins that reinforce the nuclear envelope. Despite this diversity, each IF type consists of a rodlike domain flanked by globular head and tail domains. IF polypeptides assemble into dimers (two polypeptides), tetramers (four polypeptides), and higher-order structures. IFs are especially prominent in cells that are subjected to mechanical stress.

The cytoskeleton, a dynamic mechanical system, is an integral feature of most cell activities. The unique functional properties of the cytoskeleton are made possible by a balance of mechanical forces between compression-resistant microtubules and tension generated by contractile microfilaments. IFs connect microtubules and microfilaments to each other and to the nucleus and plasma membrane. As a result of this functional "cytoarchitecture," opposing forces are continuously equilibrated throughout all cytoskeletal elements (**Figure 2.29**). Living cells are, therefore, in a constant state of dynamic instability. Cytoskeletal reorganization, triggered by a vast array of chemical and physical signals, is a principal feature of most cellular processes.

Among the most important functions made possible by the properties of the cytoskeleton are the following.

1. **Cell shape**. Eukaryotic cells come in a vast variety of shapes including the bloblike amoeba, columnar epithelial cells, and neurons with complex branching architecture. Changes in cell shape result from responses to external signals. Amoebas, for example, rapidly change shape as they move closer to a source of nutrient molecules.
2. **Large- and small-scale cell movement**. Large-scale cellular movements are made possible by a dynamic cytoskeleton that can rapidly assemble and disassemble its structural elements according to the cell's immediate needs. Organelles are moved around within cells by being attached to cytoskeletal structures. For example, after cell division, the extension of the endoplasmic reticulum membrane from the newly formed nuclear membrane out to the cell's periphery and the re-formation of the Golgi complex is accomplished by attachment to microtubules. Movement occurs as specific motor proteins linked to microtubules and to the membrane cargo undergo ATP hydrolysis–dependent conformational changes.
3. **Solid state biochemistry**. Many of the biochemical reactions previously believed to occur within the liquid phase of the cytoplasm instead proceed in large measure on a cytoskeletal platform. Biochemical pathways are both more efficient and more easily controlled when enzymes assemble into complexes on a solid surface. Prominent examples are the reactions of glycolysis, an ATP-generating pathway in carbohydrate metabolism. The binding of glycolytic enzymes to cytoskeletal filaments has been observed to vastly increase reaction rates. Drugs that disrupt cytoskeletal structure cause glycolytic enzyme detachment and a rapid decrease in cytoplasmic ATP production.
4. **Signal transduction**. Cells are information-processing systems that respond to a wide range of external chemical and physical stimuli. Examples include the binding of hormones and growth factors to cell surface receptors, action potentials in nerve and muscle cell membranes, and mechanical forces such as tensional force and hydrostatic pressure. Cells possess a constellation of signal transduction mechanisms (voltage- and

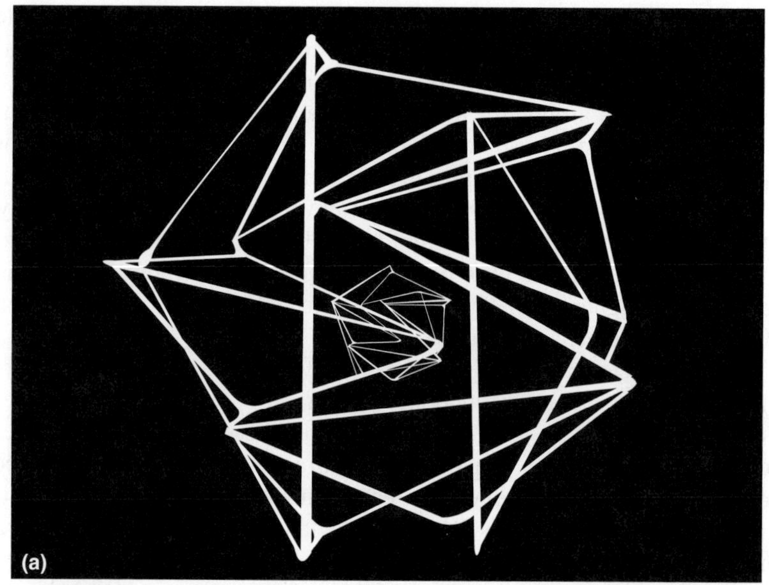

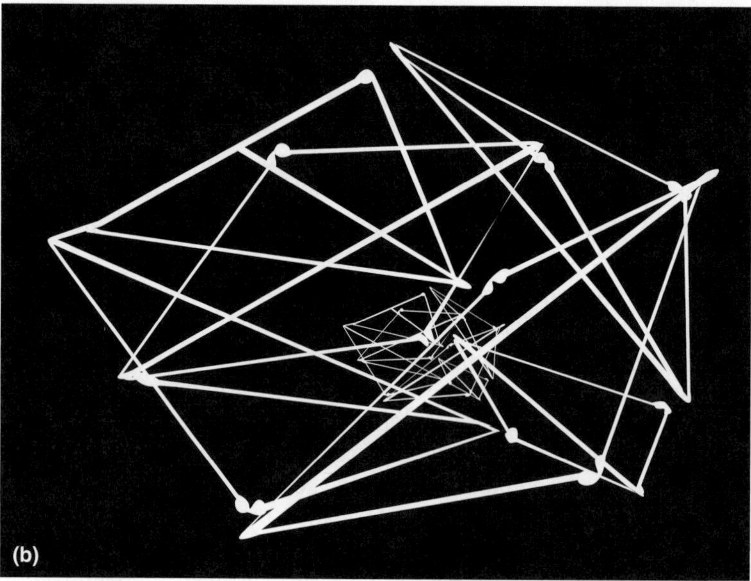

FIGURE 2.29
Cytoskeletal Reorganization Model

Both (a) and (b) are stable structures that are held together by balanced mechanical stresses, namely, tensional strings and rigid struts. The "cell" here is composed of aluminum struts and thin elastic cord; the "nucleus," a geodesic sphere, is constructed of wooden sticks and white elastic thread. When an external force is applied to structure (a), it rearranges to form structure (b).

KEY CONCEPT

The cytoskeleton, a highly structured network of proteinaceous filaments, is responsible for maintenance of cell shape, large- and small-scale cell movement, solid state biochemistry, and signal transduction.

stretch-sensitive ion channels, signal complexes, biochemical pathways, and gene expression devices) that can be thought of as resembling the *integrated circuits* (microchips) in computers: information-processing devices composed of transistors and capacitors, connected by wires, and driven by electricity. It is the filaments of the cytoskeleton and nucleoskeleton that facilitate and support signal transduction processes.

QUESTION 2.4

Cancer is a group of diseases characterized by unregulated cell division. Taxol, a drug used to treat ovarian cancer, attaches to and stabilizes microtubules. Briefly, what is the basis of Taxol's anticancer action?

Biochemistry IN PERSPECTIVE

Primary Cilia and Human Disease

What effects do nonmotile cilia have on human health?

Most differentiated vertebrate cells possess a single nonmotile cilium called the **primary cilium**. In contrast to motile cilia, primary cilia lack the central microtubule pair within the axoneme (a 9 + 0 pattern) and the dynein arms and radial spokes that are required for motility. The primary cilium functions as a sensory organelle; that is, it acts as a cellular antenna. A large number of receptor molecules and other proteins embedded in the ciliary membrane facilitate the sensing of environmental cues, such as mechanical pressure, signal molecules, and light. For example, the primary cilium of kidney tubule cells protrudes into the tubule, where it senses the flow of urine. Mechanical bending of the cilium caused by urine flow results in the inward flow of Ca^{2+}. One of the consequences of inward calcium flow is cell division suppression. Other examples of primary cilia functions include wound healing (fibroblasts migrate toward a wound when primary cilium plasma membrane receptors bind to platelet-derived growth factor, PDGF), olfaction (the primary cilia of olfactory sensory neurons detect odorant molecules), and sight (the outer segment of rod cells in the retina is essentially a highly modified primary cilium with an enlarged tip packed with visual pigments).

The confined space within a primary cilium enables the tight integration of several signaling systems. Examples include the hedgehog and Wnt signaling pathways, both of which have important roles in animal development. Efficient transport by intraflagellar transport of ciliary components and intermediate signal molecules back and forth between the cilium and the cytoplasm also facilitates signal transduction. Signal molecules transported by IFT ultimately result in gene expression changes within the nucleus. A number of human diseases, referred to as *ciliopathies*, are attributed to defects in primary cilia.

Human Ciliopathies

Considering that primary cilia are present on most cells in the human body, the wide spectrum of primary cilia-related human diseases is not surprising. Some ciliopathies appear to affect one or a small number of cell types or organs, whereas others affect many of the body's systems and can vary greatly in their symptoms. Retinitis pigmentosa (RP) is a group of more than 30 different progressive genetic eye disorders that lead to blindness. One form of RP is caused by a defective version of RP1, a gene that codes for a MAP in the outer segment of rod cells. In polycystic kidney disease, loss of function in the kidneys and several other organs is caused by cyst formation that is linked to defects in either of two genes that code for the primary cilium proteins polycystin 1 (PC1) and polycystin 2 (PC2). Together PC1 and PC2 (a cation channel) act as a mechanoreceptor that monitors fluid flow in the kidney tubule cells. When this function is disrupted by mutation of either protein, cell division, which is controlled in part by primary cilium function, is stimulated. The resulting increased cell division leads to the formation of thousands of cysts (fluid-filled sacs) and eventually results in kidney failure.

Bardet-Biedl syndrome (BBS) is an example of a pleiotropic disease (a condition in which a genetic defect results in numerous and seemingly unrelated symptoms). In BBS, which can be caused by mutations in any of 12 genes, retinal degeneration and kidney and liver cysts occur in addition to an array of clinical symptoms that include several of the following: obesity, hearing loss, olfactory deficits, diabetes, mental retardation, polydactyly (extra fingers or toes on either or both hands or feet), and situs inversus (left-to-right reversal of the internal organs). The genes linked to BBS form the BBSome, a protein complex that plays an important role in IFT in primary cilia.

SUMMARY: Nonmotile primary cilia have vital roles in the health of vertebrate cells. Defects in primary cilia result in numerous human diseases.

Biochemistry IN THE LAB

Cell Technology

During the past 50 years, our understanding of the functioning of living organisms has undergone a revolution. Much of our current knowledge of biochemical processes is a direct result of technological innovations. Three of the most important cellular techniques used in biochemical research are briefly described: cell fractionation, electron microscopy, and autoradiography.

Cell Fractionation

Cell fractionation techniques (**Figure 2A**) allow the study of organelles in a relatively intact form outside of cells. For example, functioning mitochondria can be used to study cellular energy generation. In these techniques, cells are gently disrupted and separated into several organelle-containing fractions. Cells may be disrupted by several methods, but homogenization is the most commonly used. In this process a cell suspension is placed either in a glass tube fitted with a specially designed glass pestle or into an electric blender. The resulting homogenate is separated into several fractions by means of **differential centrifugation**. In this procedure, a refrigerated instrument, the *ultracentrifuge*, generates enormous centrifugal forces that separate cell components on the basis of size, surface area, and relative density. (Forces as large as 500,000 times the force of gravity, or 500,000 *g*, can be generated in unbreakable test tubes placed in the rotor of an ultracentrifuge.) The homogenate is first spun in the ultracentrifuge at low speed (700–1000 *g*) for 10 to 20 minutes. The heavier particles, such as the nuclei, form a sediment, or pellet. Lighter particles, such as mitochondria and lysosomes, remain suspended in the *supernatant*, the liquid above the pellet. The supernatant is then transferred to another centrifuge tube and spun at a higher speed (15,000–20,000 *g*) for 10 to 20 minutes. The resulting pellet contains mitochondria, lysosomes, and peroxisomes. The supernatant, which contains **microsomes** (small closed vesicles formed from ER during homogenization), is transferred to another tube and spun at 100,000 *g* for 1 to 2 hours. Microsomes are deposited in the pellet, and the supernatant contains ribosomes, various cellular membranes, and granules such as glycogen, a carbohydrate polymer. After this latest supernatant has been recentrifuged at 200,000 *g* for 2 to 3 hours, ribosomes and large macromolecules are recovered from the pellet.

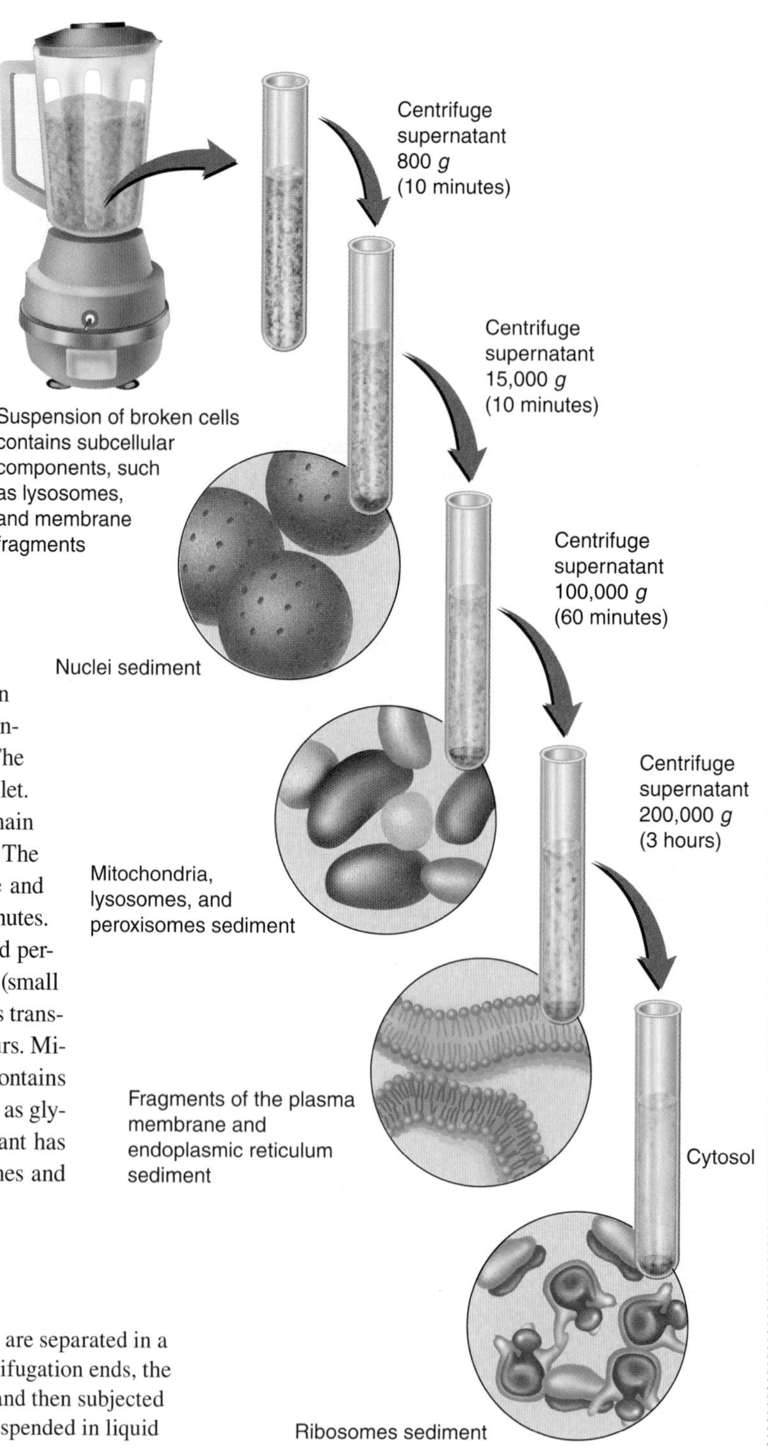

FIGURE 2A
Cell Fractionation

After homogenization of cells in a blender, cell components are separated in a series of centrifugations at increasing speeds. As each centrifugation ends, the supernatant is removed, placed into a new centrifuge tube, and then subjected to greater centrifugal force. The collected pellet can be resuspended in liquid and examined by microscopy or biochemical tests.

Often, the organelle fractions obtained with this technique are not sufficiently pure for research purposes. One method often employed to further purify cell fractions is **density-gradient centrifugation** (**Figure 2B**). In this procedure the fraction of interest

Biochemistry IN THE LAB cont.

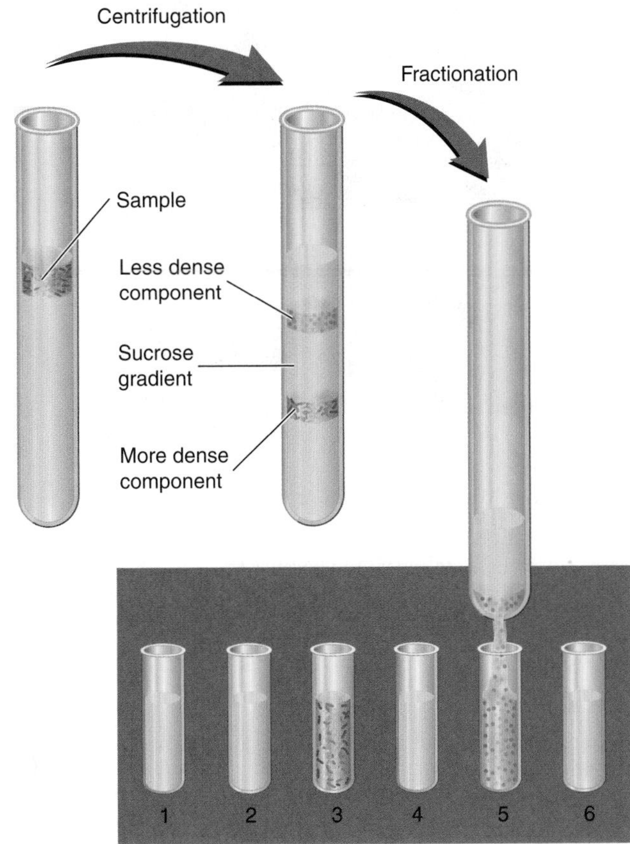

FIGURE 2B
Density-Gradient Centrifugation

The sample is gently layered onto the top of a preformed gradient of an inert substance such as sucrose. As centrifugal force is applied, particles in the sample migrate through the gradient bands according to their densities. After centrifugation, the bottom of the tube is punctured and the individual bands are collected in separate tubes.

is layered on top of a solution that consists of a dense substance such as sucrose. (In the centrifuge tube containing the solution, the sucrose concentration increases from the top to the bottom.) During centrifugation at high speed for several hours, particles move downward in the gradient until they reach a level that has a density equal to their own. Then the plastic centrifuge tube is punctured, and the cell components are collected in drops from the bottom. The purity of the individual fractions can be assessed by visual inspection (electron microscopy). However, assays for **marker enzymes** (enzymes that are known to be present in especially high concentration in specific organelles) are more commonly used. For example, glucose-6-phosphatase, the enzyme responsible for converting glucose-6-phosphate to glucose in the liver, is a marker for liver microsomes. DNA polymerase, which is involved in DNA synthesis, is a marker for nuclei.

Electron Microscopy

The electron microscope (EM) permits a view of cell ultrastructure not possible with the more common light microscope. Direct magnifications as high as 1,000,000× have been obtained with the EM. Electron micrographs may be enlarged photographically to 10,000,000×. The light microscope, in contrast, magnifies an image to about 1,000×. The greater resolving power of EM (0.5 nm) compared with that of the light microscope (0.2 μm) is a result of the wavelength sizes of their illumination sources. EM uses a stream of electrons, which have much shorter wavelengths than those of visible light. As a result, more detailed images can be obtained. In general, shorter wavelengths allow greater resolution.

There are two types of EM. In transmission electron microscopy (TEM) electrons pass through thin specimens. Images are formed because of variations in the absorption of electrons by the specimen. Scanning electron microscopy (SEM) is used to form three-dimensional images by detecting electrons emitted from specimen surfaces coated with a thin layer of a heavy metal. Although only surface features can be examined with the SEM, this form of microscopy provides very useful information about cell structure and function.

Autoradiography

Autoradiography is used to study the intracellular location and behavior of cellular components. It has been an invaluable tool in biochemistry. Radioactively labeled molecules have been used in the investigation of nucleic acid and protein synthesis, gene expression, signal transduction, and metabolic pathways. Commonly used radioisotopes include ^{3}H (tritium), ^{32}P, and ^{35}S. The tritiated nucleotide thymidine, for example, is used to study DNA synthesis because thymidine is incorporated only into DNA molecules. After exposure to the radioactive precursor, the cells are processed for light or electron microscopy. The resulting slides are then dipped in photographic emulsion. After storage in the dark, the emulsion is developed by standard photographic techniques. The location of radioactively labeled molecules is indicated by the developed pattern of silver grains.

Live Cell Imaging

The dynamic activities of living cells are best observed by live cell imaging using light microscopy. Two examples include phase contrast microscopy and fluorescence microscopy.

Phase contrast microscopy takes advantage of variations in the refraction of light as it passes through substances with different densities. Phase contrast microscopes are fitted with an

Biochemistry IN THE LAB cont.

annulus aperture (limits the angle of the incoming light rays) and a phase plate (contains a phase ring that shifts light wavelengths (λ) along the horizontal axis). Although living cells are translucent, a phase shift of 90° relative to the background light results in sufficient contrast that cell structures can be observed. Phase contrast microscopy is used in investigations such as chemotaxis when low-resolution level is sufficient. (Chemotaxis is the movement of an organism such as an amoeba toward or away from a chemical stimulus.)

Cell biologists and biochemists use *fluorescence microscopy* to investigate cell function with *fluorophores* (molecules that absorb photons of light and then re-emit photons of lower energy). (Fluorescence is described on p. 485.) Fluorophores can be small molecules such as DAPI, TRITC, and FITC or fluorescent proteins such as green fluorescent protein (GFP) linked to a protein of interest. Fluorophores are used in the time-lapse investigation of a wide variety of cell functions including signal transduction mechanisms.

A fluorescence microscope is fitted with a light source, an excitation filter (transmits only the wavelengths that excite a specific fluorophore), a dichroic mirror (a beamsplitter composed of coated glass that reflects the excitation light and transmits the emitted fluorescence), an emission filter (transmits peak emission light waves), and a camera.

Several fluorophores can be used simultaneously. In **Figure 2C** cell nuclei are stained blue with DAPI, a fluorescent molecule that binds to DNA, actin filaments appear red because they are labeled with TRITC bound to the actin filament–specific binding protein phalloidin, and microtubules are green because they are bound to antibody proteins linked to FITC.

Note that the use of fluorescence microscopy in the observation of living cells is limited to short periods of time by the toxic nature of fluorescent stains.

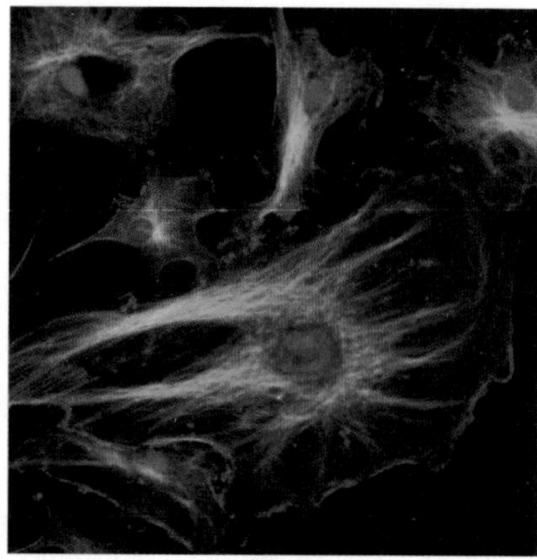

FIGURE 2C
Fluorescence Micrograph of an Arterial Endothelial Cell
Nuclei are stained blue because the fluorophore DAPI binds to DNA; actin filaments appear red because they are labeled with TRITC; microtubules are green because microbule-specific antibodies are linked to FITC.

Chapter Summary

1. Cells are the structural units of all living organisms. Within each living cell are hundreds of millions of densely packed biomolecules. The unique chemical and physical properties of water are a crucial determining factor in the behavior of all other biomolecules. Biological membranes are thin, flexible, and relatively stable sheetlike structures that enclose cells and organelles. They are formed from biomolecules such as phospholipids and proteins that together form a selective physical barrier.
2. Self-assembly of supramolecular structures occurs within living cells because of the steric information encoded into the intricate shapes of biomolecules that allows numerous weak, noncovalent interactions between complementary surfaces. Many of the multisubunit complexes involved in cellular processes are now known to function as molecular machines; that is, they are mechanical devices composed of moving parts that convert energy into directed motion. Macromolecular crowding, created by the density of proteins within the cell, is an important factor in the wide variety of cellular phenomena. Signal transduction mechanisms allow cells to process internal and external information. Proteostasis (protein homeostasis) exists in cells as the result of a dynamic equilibrium between protein synthesis and folding and protein degradation.
3. All currently existing organisms contain either prokaryotic or eukaryotic cells. Prokaryotes are simpler in structure than eukaryotes. They also have a vast biochemical diversity across species lines because almost any organic molecule can be used as a food source by some species of prokaryote. Unlike the prokaryotes, the eukaryotes carry out their metabolic functions in membrane-bound compartments called organelles.
4. DNA molecules in prokaryotic cells are located in an irregularly shaped region called the nucleoid. Many bacteria contain additional small circular DNA molecules called plasmids. Plasmids may carry genes for special function proteins that provide protection, metabolic specialization, or reproductive advantages to the organism.

5. The plasma membrane of both prokaryotes and eukaryotes performs several vital functions. The most important of these is controlled molecular transport, which is facilitated by carrier and channel proteins.
6. The ER is a system of interconnected membranous tubules, vesicles, and large flattened sacs found in eukaryotic cells. There are two forms of ER. The RER, which is primarily involved in protein synthesis, is so named because of the numerous ribosomes that stud its cytoplasmic surface. The second form lacks attached ribosomes and is called SER. Functions of the SER include lipid synthesis and biotransformation.
7. Formed from relatively large, flattened, saclike membranous vesicles that resemble a stack of plates, the Golgi apparatus is involved in the modification, packaging, and release of cell products into the vesicular compartment for delivery to target locations in the cell.
8. The cell contains a system of vesicular organelles involved in processing both endogenous and exogenous materials around, into, and out of the cell and performing specialized biochemical functions.
9. The nucleus of any eukaryote contains DNA, the cell's genetic information. Ribosomal RNA is synthesized in the nucleolus, found within the nucleus. Separating DNA replication and transcription processes from the cytoplasm is the nuclear envelope; it is composed of two membranes that fuse at structures called the nuclear pores.
10. Aerobic respiration, a process by which cells use oxygen to generate energy, takes place in mitochondria. Each mitochondrion is bounded by two membranes. The smooth outer membrane is permeable to most molecules with masses less than 10,000 Da. The inner membrane, which is impermeable to ions and a variety of organic molecules, projects inward into folds that are called cristae. Embedded in this membrane are mitochondrial respiratory chains, molecular complexes that are responsible for the synthesis of ATP.
11. Peroxisomes are small spherical membranous organelles that contain a variety of oxidative enzymes. These organelles are most noted for their involvement in the generation and breakdown of peroxides.
12. Chromoplasts accumulate the pigments that are responsible for the color of leaves, flower petals, and fruits. Chloroplasts are a type of chromoplast that are specialized to convert light energy into chemical energy.
13. The cytoskeleton, a supportive network of fibers and filaments, is involved in the maintenance of cell shape, large- and small-scale cellular movement, solid state biochemistry, and signal transduction.

Take your learning further by visiting the **companion website** for Biochemistry at **www.oup.com/us/mckee** where you can complete a multiple-choice quiz on water to help you prepare for exams.

Suggested Readings

Bell, L., Mitochondria Gone Bad, *Sci. News* 175(5):20–23, 2009.

Clemente, J. C. *et al.*, The Impact of the Gut Microbiota on Human Health: An Integrative View, *Cell* 148(6):1258–1270, 2012.

Fang, S., and Evans, R. M., Wealth Management in the Gut, *Nature* 500:538–539, 2013.

Fischer, F., Hamann, A., and Osiewacz, D., Mitochondrial Quality Control: An Integrated Network of Pathways, *Trends Biochem. Sci.* 37(7):284–292, 2012.

Goodsell, D. S. *The Machinery of Life*, 2nd ed., Springer Verlag, New York, 2009.

Ingber, D. E., The Architecture of Life, *Sci. Am.* 278(1):48–57, 1998.

Lane, N., *Power, Sex, Suicide: Mitochondria and the Meaning of Life*, Oxford University Press, New York, 2005.

Platta, H. W., and Stenmark, H., Endocytosis and Signaling, *Curr. Opinion Cell Biol.* 23:393–403, 2011.

Powers, E. T. and Balch, W. E., Diversity in the Origins of Proteostasis Networks—A Driver for Protein Function in Evolution, *Nat. Rev. Mol. Cell Biol.* 14(4):237–248, 2013.

Saey, T. H., Nouveaux Antennas, *Sci. News* 182(9):16–19, 2012.

Satir, P., Pederson, L. B., and Christensen, S. T., The Primary Cilia at a Glance, J. Cell. Sci. 123:499–503, 2010.

Wickstead, B. and Gull, K., The Evolution of the Cytoskeleton, *J. Cell Biol.* 194(4):513–525, 2011.

Key Words

aerobic metabolism, *57*
apoptosis, *58*
biotransformation reaction, *48*
carrier protein, *36*
caveolae, *53*
caveolar endoeytosis, *53*
cell fractionation, *68*
channel protein, *36*
chloroplast, *62*
chromatin fiber, *55*
chromosome, *43*
clathrin, *51*
clathrin-dependent endocytosis, *51*
cytoskeleton, *62*
density-gradient centrifugation, *68*
differential centrifugation, *68*
endocytic cycle, *53*
endocytosis, *51*
endomembrane system, *44*
endoplasmic reticulum (ER), *47*

ER-associated protein degradation, 47
ER stress, 47
exocytosis, 50
extracellular matrix (ECM), 47
glycocalyx, 44
Golgi apparatus (Golgi complex), 49
granum, 62
hormone, 39
hydrophilic, 36
hydrophobic, 36
inner mitochondrial membrane, 55
inner nuclear membrane, 55
integral protein, 36
intermediate filament, 65
ligand, 39
lysosome, 51
marker enzyme, 69
membrane skeleton, 47
microbiota, 33
microfilament, 64
microsome, 68
microtubule, 62
mitochondrion, 57
motor protein, 38
neurotransmitter, 39
nuclear envelope, 55
nuclear lamina, 56
nuclear matrix, 57
nuclear pore complexes, 55
nucleoid, 43
nucleolus, 57
nucleoplasm, 54
nucleus, 54
organelle, 35
outer mitochondrial membrane, 58
outer nuclear membrane, 55
perinuclear space, 55
peripheral protein, 36
peroxisome, 61
photosynthesis, 43
plasma membrane, 42
plasmid, 43
primary cilium, 67
proteome, 38
proteostasis, 39
proteostasis network, 39
receptor, 36
respiration, 43
rough ER (RER), 47
signaling cascade, 39
signal transduction, 39
smooth ER (SER), 47
stroma, 62
thylakoid membrane, 62
unfolded protein response, 48
vesicles, 44

Review Questions

These questions are designed to test your knowledge of the key concepts discussed in this chapter before moving on to the next chapter. You may like to compare your answers to the solutions provided in the back of the book and in the accompanying Study Guide.

1. Define the following terms:
   a. prokaryote
   b. eukaryote
   c. organelle
   d. hydrophilic
   e. hydrophobic
2. Define the following terms:
   a. lipid bilayer
   b. polar head group
   c. hydrocarbon tail
   d. integral protein
   e. peripheral protein
3. Define the following terms:
   a. supramolecular complex
   b. ribosome
   c. channel protein
   d. carrier protein
   e. receptor
4. Define the following terms:
   a. ligand
   b. motor protein
   c. GTP
   d. macromolecular crowding
   e. excluded volume
5. Define the following terms:
   a. signal transduction
   b. neurotransmitter
   c. hormone
   d. cytokine
   e. LPS
6. Define the following terms:
   a. endotoxin
   b. periplasmic space
   c. biofilm
   d. slime layer
   e. bacterial capsule
7. Define the following terms:
   a. plasma membrane
   b. photosynthesis
   c. respiration
   d. nucleoid
   e. chromosome
8. Define the following terms:
   a. pilus
   b. conjugation
   c. flagellum
   d. endomembrane system
   e. vesicle
9. Define the following terms:
   a. glycocalyx
   b. extracellular matrix
   c. cell cortex
   d. endoplasmic reticulum
   e. ER lumen
10. Define the following terms:
   a. rough ER
   b. smooth ER
   c. clathrin
   d. unfolded protein response
   e. caveolae
11. Define the following terms:
   a. biotransformation reaction
   b. sarcoplasmic reticulum
   c. Golgi apparatus
   d. Golgi cisterna
   e. exocytosis
12. Define the following terms:
   a. nucleoplasm
   b. chromatin fiber
   c. nuclear matrix
   d. nucleolus
   e. nuclear envelope

13. Define the following terms:
   a. nuclear pore complex
   b. endocytosis
   c. proteome
   d. endocytic cycle
   e. lysosome
14. Define the following terms:
   a. acid hydrolase
   b. autophagy
   c. clathrin-dependent endocytosis
   d. caveolar endocytosis
   e. proteostasis
15. Define the following terms:
   a. mitochondrion
   b. aerobic respiration
   c. apoptosis
   d. outer mitochondrial membrane
   e. inner mitochondrial membrane
16. Define the following terms:
   a. intermembrane space
   b. mitochondrial matrix
   c. peroxisome
   d. peroxin
17. Define the following terms:
   a. outer nuclear membrane
   b. inner nuclear membrane
   c. chloroplast
   d. photosynthesis
   e. thylakoid membrane
18. Define the following terms:
   a. cytoskeleton
   b. microtubule
   c. MAP
   d. IFT
   e. primary cilium
19. Define the following terms:
   a. microfilament
   b. F-actin
   c. G-actin
   d. ameboid movement
   e. intermediate filament
   f. keratin
20. Define the following terms:
   a. ciliopathy
   b. retinitis pigmentosum
   c. polycystic kidney disease
   d. Bardet-Biedl syndrome
   e. anterograde transport
21. List four physiological functions required for human health that are performed by microorganisms.
22. Approximately 70% of immune system cells are located in the wall of the lower digestive tract (intestines). Can you suggest a reason for this phenomenon?
23. Why is it difficult for the colon to reestablish a beneficial flora after several antibiotic treatments?
24. What factors can promote protein misfolding in cells?
25. Draw a diagram of a bacterial cell. Label and explain the function of each of the following components:
   a. nucleoid
   b. plasmid
   c. cell wall
   d. pili
   e. flagella
26. Indicate whether the following structures are present in prokaryotic or eukaryotic cells:
   a. nucleus
   b. plasma membrane
   c. endoplasmic reticulum
   d. mitochondria
   e. nucleoid
   f. cytoskeleton
27. Explain why the term "crowded" rather than "concentrated" is used to describe the densely packed molecules in the interior of living cells.
28. What are the functions of nuclear lamina?
29. Describe the four phases of signal transduction in living organisms.
30. What are the components of the endomembrane system? How are these components functionally connected?
31. Outline the role of cytoskeleton in intracellular signal transduction.
32. How do lysosomes participate in the life of a cell?
33. What purpose does mitochondrial fusion and fission serve?
34. List and describe six diseases linked to organelles.
   [*Hint*: Visit the companion website at www.oup/us/mckee to read the Biochemistry in Perspective essay for Chapter 2, Organelles and Human Disease.]
35. What functions does the cytoskeleton perform in living cells?
36. What are the two essential functions of the nucleus?
37. What major roles do plasma membrane proteins play in cells?
38. Name the two forms of endoplasmic reticulum. What functions do they serve in the cell?
39. Describe the functions of the Golgi apparatus.
40. Distinguish among the terms ER stress, unfolded protein response, and ER-associated protein degradation.
41. List three environmental signals detected by primary cilia.
42. Describe the structural and functional properties of the nuclear pore complex.
43. Describe the functions of intraflagellar transport.
44. Eukaryotic cells possess a system of vesicular organelles. Name and describe three specific examples.
45. Peroxisomes are not included in the population of vesicular organelles because of their specialized function and unique biogenesis. Describe the function of this organelle and explain how it is formed.
46. What is the function of COP II in cells?
47. Describe the intracellular and extracellular structures that protect the eukaryotic plasma membrane.
48. Describe the functions of the smooth endoplasmic reticulum in hepatocytes and muscle cells.
49. Describe the structure and functional properties of the nuclear envelope.
50. What is proteostasis? How important is it in the life of cells?

Fill in the Blank

51. Techniques that the body uses to protect itself from the microbe members of the human superorganism are impenetrable tissue barriers and ________________ system cells.
52. ________________ are phagocytic cells that engulf and digest foreign cells.
53. There are two types of living cells: ________________ and prokaryotic cells.
54. There are two types of prokaryotes: bacteria and ________________.
55. Common features of prokaryotes and eukaryotes include similar chemical composition and ________________.
56. ________________ compounds exclude water.
57. The most basic and critical function of membranes is to serve as a ________________.
58. The two types of membrane proteins are peripheral and ________________.
59. Living organisms require both information and ________________ to create order.
60. In contrast to the eukaryotic cell, the prokaryotic cell is characterized by the lack of a ________________.

Short Answer

61. Why are eukaryotic cells so much larger than prokaryotic cells?
62. How does soap kill bacteria?
63. What would happen if the cell membrane was covalently linked rather than held together by relatively weak van der Waals forces?
64. Suggest a reason why phospholipids are constituents of cell membranes rather than carboxylic acids.
65. Suggest a reason why some eukaryotic cells lack cell walls.

Thought Questions

These questions are designed to reinforce your understanding of all of the key concepts discussed in the book so far, including this chapter and the chapter before it. They may not have one right answer! The authors have provided possible solutions to these questions in the back of the book and in the accompanying Study Guide for your reference.

66. Cyst formation causes a catastrophic loss of function in polycystic kidney disease. Genetic research has linked this disease to defects in genes that code for primary cilium proteins. Describe in general terms how malfunctioning primary cilia cause the formation of kidney cysts.
67. Primary cilia have evolved as primary sensory organelles for vertebrate cells. What structural features of these cilia make them ideal for this purpose?
68. Several pathogenic bacteria (e.g., *Bacillus anthracis*, the cause of anthrax) produce an outermost mucoid layer called a capsule. Capsules may be composed of polysaccharide or protein. What effect do you think this "coat" would have on a bacterium's interactions with a host animal's immune system?
69. In addition to providing support, the cytoskeleton immobilizes enzymes and organelles in the cytoplasm. What advantage does this immobilization have over allowing the cell contents to freely diffuse in the cytoplasm?
70. Familial hypercholesterolemia (FH) is an inherited disease characterized by high blood levels of cholesterol, xanthomas (lipid-laden nodules that develop under the skin near tendons), and early-onset atherosclerosis (the formation of yellowish plaques within arteries). In the milder form of this disease, patients have half the plasma membrane low-density lipoprotein (LDL) receptors needed for cells to bind to and internalize LDL (a plasma lipoprotein particle that transports cholesterol and other lipids to tissues). These patients have their first heart attacks in young adulthood. In the severe form of FH, in which patients have no functional LDL receptors, heart attacks begin at about age 8, with death occurring a few years later. Based on what you have learned in this chapter, briefly describe the cellular processes that are defective in FH.
71. Mycoplasmas are unusual bacteria that lack cell walls. With a diameter of 0.3 μm, they are believed to be the smallest known free-living organisms. Some species are pathogenic to humans. For example, *Mycoplasma pneumoniae* causes a very serious form of pneumonia. Assuming that mycoplasmas are spherical, calculate the volume of an individual cell. Compare the volume of a mycoplasma with that of *E. coli.*
72. The dimensions of prokaryotic ribosomes are approximately 14 nm by 20 nm. If ribosomes occupy 20% of the volume of a bacterial cell, calculate how many ribosomes are in a typical cell such as *E. coli*. Assume that the shape of a ribosome is approximately that of a cylinder.
73. The *E. coli* cell is 2 μm long and 1 μm in diameter, whereas a typical eukaryotic cell is 20 μm in diameter. Assuming that the *E. coli* cell is a perfect cylinder and the eukaryotic cell is a perfect sphere, calculate the surface-to-volume ratio for each cell type (cylinder volume, $V = \pi r^2 h$; cylinder area $A = 2\pi r^2 + 2\pi rh$; sphere volume, $V = 4/3(\pi r^3)$; sphere area, $A = 4\pi r^2$). What do these numbers tell you about the evolutionary changes that would have to occur to generate an efficient eukaryotic cell, considering that most biochemical processes depend on membrane-bound transport processes?

CHAPTER

3 Water: The Matrix of Life

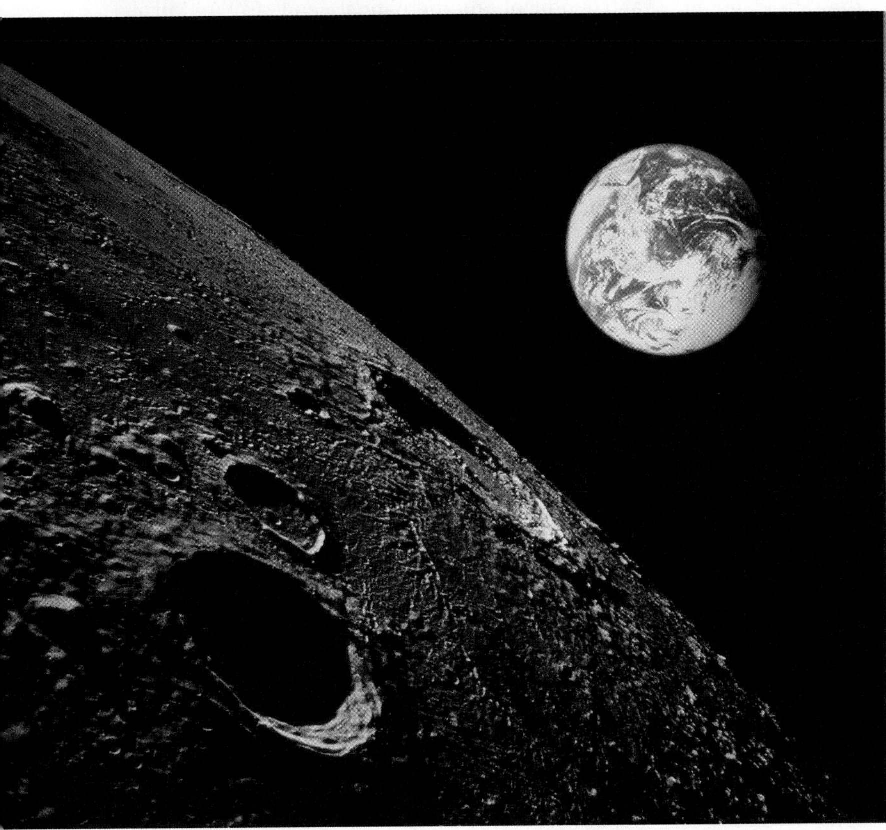

The Water Planet Unique among the planets in the solar system, the earth is an oceanic world. Water's properties make life on earth possible.

OUTLINE

Water, Water, Everywhere

Water, water, everywhere,
And all the boards did shrink;
Water, water, everywhere,
Nor any drop to drink.

The Rime of the Ancient Mariner
Samuel Taylor Coleridge (1772–1834)

Coleridge's mariner laments the fate (dying of thirst) of himself and his shipmates as their becalmed wooden sailing ship is surrounded by an endless vista of undrinkable ocean water. The plight of these fictional characters is also our own. We are also threatened by our dependence on clean, unsalted water to maintain our health when usable, accessible water resources are vanishingly small. Our bodies are, on average, approximately 60% water because it is the principal component of every body fluid (e.g., blood, saliva, lymph, and gastric juices). Among water's numerous roles are nutrient absorption and transport, waste product excretion, body temperature regulation, and joint lubrication. Our requirements for water are so stringent that even mild dehydration causes tiredness, headaches, and loss of concentration. However, of an estimated total world water supply of 366×10^{18} gallons, approximately 97% (355×10^{18} gallons) is ocean water. Only 3% of the world's water is fresh (drinkable), and most of that is locked away in glaciers and polar icecaps. Together, fresh water in rivers, lakes, underground aquifers, and the atmosphere accounts for less than 1% of the world's water!

In the 10,000 years since the Neolithic agricultural revolution, humans have usually settled near convenient sources of water (dependable rainfall and/or large rivers) so that agriculture could flourish. The first civilizations in the West occurred between the Tigris and Euphrates rivers in Mesopotamia and the Nile River in Egypt. In Asia, the Yellow River basin has been described as the "cradle of Chinese civilization." As human populations in towns and then cities increased, survival depended on water resource management. In ancient Rome by the third century CE 11 aqueducts provided sufficient water for an estimated 1 million people. Inadequate water results in catastrophe. The Old Kingdom in Egypt in the third millennium BCE built the great pyramids, but it collapsed in the midst of a 300-year drought.

The Mayan civilization, a series of city-states in the Yucatan peninsula, is an exceptionally well-researched example of the effect of water availability on human survival because of climate data (temperature, volcanic eruptions, forest fires, and precipitation) obtained from Greenland ice cores. During 250–950 CE, the Mayans developed a culture with a written language and astronomical, mathematical, and architectural achievements that rivaled those of other contemporary civilizations (**Figure 3.1**). Despite these magnificent accomplishments, beginning around 750 CE and lasting until about 950 CE, Mayan cities began to decline and were eventually abandoned. Despite sophisticated rain collection and storage facilities and irrigation systems, the Mayan population (at its highest about 13 million) was severely stressed by a serious dry period that lasted more than 250 years. The drought, at its most severe in the ninth century, became especially dire when it was punctuated with 3- to 9-year periods of little or no rainfall. The collapse of the Mayan civilization, the result of severe malnutrition and disease, chronic warfare, and social chaos, was the inevitable consequence of inadequate food production caused by insufficient water. However, the Mayan collapse is not just a cautionary tale of geography and climate. It also illustrates that human-caused environmental destruction can have catastrophic consequences.

The Yucatan peninsula is a seasonal desert. Before the Mayans arrived, this land was a dense rainforest growing on top of a porous, limestone bedrock and was dependent almost entirely on summer rains. Analysis of pollen and mineral concentrations of modern lake sediments reveals that as the Mayan population grew, the land gradually became deforested for agricultural purposes. Deforestation disrupts the water cycle, the process by which water continuously moves on, above and below the surface of the Earth. *Evapotranspiration*, in which water is transpired by plants and evaporated from soil, is an essential feature of the water cycle. The water vapor created by these processes is carried into the atmosphere by air currents and then returns to the Earth as rain. The immediate effect of deforestation in the Yucatan peninsula in the ninth century was erosion as rainwater washed away soil previously held in place by tree roots. In many areas of the Yucatan peninsula, this unimpeded water percolated through porous limestone down into

Water, Water, Everywhere cont.

water tables, many of which were too deep for wells. For the Mayans it was a tragedy that deforestation, with its resultant erosion, water runoff, and decreased rainfall (caused by water cycle disruption), occurred in the midst of a regional drought. Together these circumstances contributed to the failure of the Mayans to sustain their civilization.

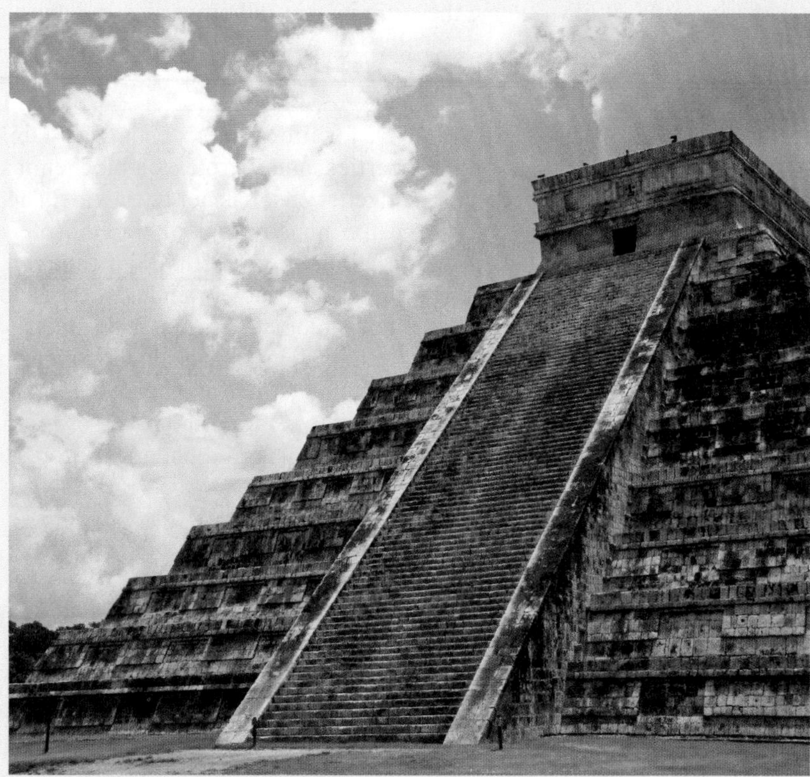

FIGURE 3.1
A Mayan Pyramid
The Mayans developed an elaborate and sophisticated culture with remarkable achievements in architecture, mathematics, and astronomy. The Mayan civilization eventually collapsed because its population grew so large that it surpassed water and food resources.

Overview

EARTH IS UNIQUE AMONG THE PLANETS IN OUR SOLAR SYSTEM, PRIMARILY BECAUSE OF ITS VAST OCEANS OF WATER. OVER BILLIONS OF YEARS, WATER was produced during high-temperature interactions between atmospheric hydrocarbons and the silicate and iron oxides in the earth's mantle. Moisture reached the planet's surface as steam emitted during volcanic eruptions. Oceans formed as the steam condensed and fell back to Earth as rain.

Over millions of years, water has profoundly affected our planet. Whether falling as rain or flowing in rivers, water has eroded the hardest rocks and transformed the mountains and continents. Many scientists today believe that life arose in a *primordial pudding* of clay and water. Shallow clay pools can promote the synthesis of macromolecules and accumulate the building blocks of life. In another scenario, life arose in close proximity to hydrothermal vents, openings in the sea floor out of which flows heated mineral-rich water. Whatever its origin, it is not an accident that life arose in association with water because this substance has several unusual properties that suit it to be the matrix of life. Among these are its thermal properties and unusual solvent characteristics. Water's properties are directly related to its molecular structure.

Why is water so vital for life? Water's chemical stability, its remarkable solvent properties, and its role as a biochemical reactant have long been recognized. What has not been widely appreciated is the critical role that *hydration* (the noncovalent interaction of water molecules with solutes) plays in the architecture, stability, and functional dynamics of macromolecules such as proteins and nucleic acids. Water is now known to be an indispensable component of biological processes as diverse as protein folding and biomolecular recognition in signal transduction mechanisms, the self-assembly of supramolecular structures such as ribosomes, and gene expression. Understanding how essential water is in living processes requires a review of its molecular structure and the physical and chemical properties that are the consequences of that structure.

3.1 MOLECULAR STRUCTURE OF WATER

The water molecule (H_2O) is composed of two atoms of hydrogen and one of oxygen. Water has a tetrahedral geometry because its oxygen atom is sp^3 hybridized and at the center of the tetrahedron is the oxygen atom. Two of the corners are occupied by hydrogen atoms, each of which is linked to the oxygen atom by a single covalent bond (**Figure 3.2**). This arrangement gives the water molecule an overall bent geometry. The other two corners are occupied by the unshared electron pairs of the oxygen. Oxygen is more electronegative than hydrogen (i.e., oxygen has a greater capacity to attract electrons when bonded to hydrogen). Consequently, the larger oxygen atom bears a partial negative charge (δ^-) and each of the two hydrogen atoms bears a partial positive charge (δ^+) (**Figure 3.3**). The electron distribution in oxygen–hydrogen bonds is displaced toward the oxygen and, therefore, the bond is **polar**. If water molecules were linear, like those of carbon dioxide (O=C=O), then the bond polarities would balance each other and water would be nonpolar. However, water molecules are bent (the bond angle is 104.5°, slightly less than the symmetrical tetrahedral angle of 109°) because the lone pair electrons occupy more space than the bonding electron pairs of the O—H bonds (**Figure 3.4**).

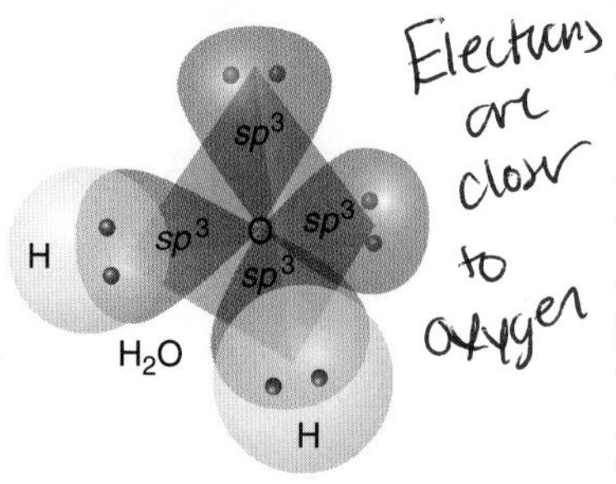

FIGURE 3.2

Tetrahedral Structure of Water

In water, two of the four sp^3 orbitals of oxygen are occupied by two lone pairs of electrons. Each of the other two half-filled sp^3 orbitals is filled by the addition of an electron from hydrogen.

Electrons are closer to oxygen

Molecules such as water, in which charge is separated, are called **dipoles**. When molecular dipoles are subjected to an electric field, they orient themselves in the direction opposite to that of the field (**Figure 3.5**).

The electron-deficient hydrogens of one water molecule are attracted to the unshared pairs of electrons of another water molecule because of the large difference in electronegativity of hydrogen and oxygen. (Hydrogens attached to nitrogen and fluorine also behave the same way.) In this interaction, called a **hydrogen bond** (**Figure 3.6**), the hydrogen is unequally shared by the two electronegative centers: oxygen nuclei in the case of a pair of water molecules.

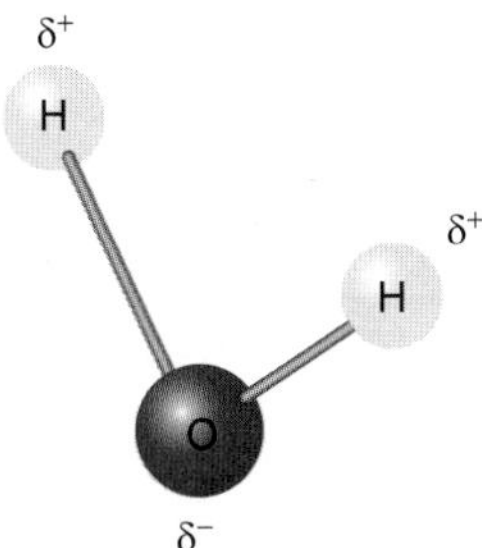

FIGURE 3.3
Charges on a Water Molecule

The two hydrogen atoms in each molecule carry partial positive charges. The oxygen atom carries a partial negative charge.

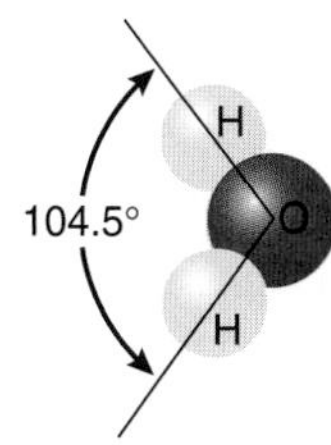

FIGURE 3.4
Space-Filling Model of a Water Molecule

Because the water molecule has a bent geometry, the distribution of charge within the molecule is asymmetric. Water is therefore polar.

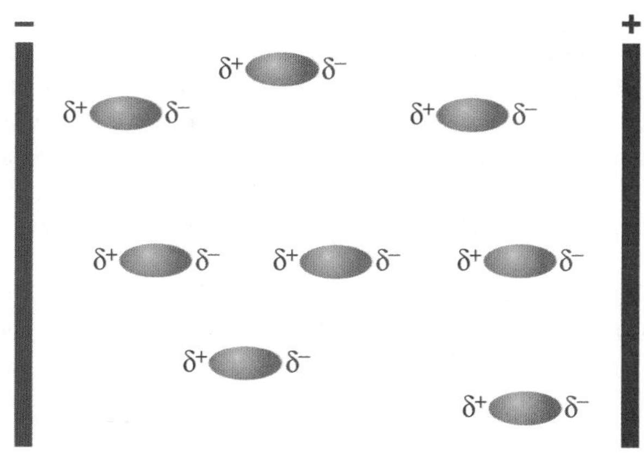

FIGURE 3.5
Molecular Dipoles in an Electric Field

When polar molecules are placed between charged plates, they line up in opposition to the field.

The bond has both electrostatic (ionic) and covalent character. **Electrostatic interactions** occur between any two opposite partial charges (polar molecules) or full charges (ions or charged molecules). **Covalent bonds** involve electron sharing with orbital overlap or mixing. Covalent character confers directionality to the bond or interaction, unlike the uniformly spherical force field around an ion.

3.2 NONCOVALENT BONDING

Noncovalent interactions are usually electrostatic; that is, they occur between the positive nucleus of one atom and the negative electron clouds of another nearby atom. Unlike the stronger covalent bonds, individual noncovalent interactions are relatively weak and are therefore easily disrupted (**Table 3.1**). Nevertheless, they play a vital role in determining the physical and chemical properties of water and the structure and function of biomolecules because the cumulative effect of many weak interactions can be considerable. Large numbers of noncovalent interactions stabilize macromolecules and supramolecular structures, whereas the capacity of these bonds to rapidly form and break endows biomolecules with the flexibility required for the rapid flow of information that occurs

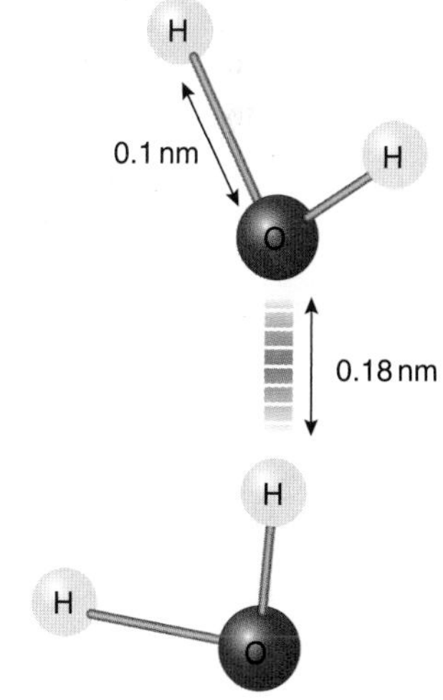

FIGURE 3.6
The Hydrogen Bond

A hydrogen bond results when the electronegative oxygen atoms of two water molecules compete for the same electron-deficient hydrogen atom. The hydrogen bond is represented by short parallel lines designating the weak covalent character and directionality of the bond.

TABLE 3.1 Bond Strengths of Bonds Typically Found in Living Organisms

	Bond Strength*	
Bond Type	**kcal/mol**†	**kJ/mol**
Covalent	>50	>210
Noncovalent		
Ionic interactions	1–20	4–80
Van der Waals forces	<1–2.7	<4–11.3
Mixed: hydrogen bonds	3–7	12–29

* The actual strength varies considerably with the identity of the interacting species.
† 1 cal = 4.184 J.

in dynamic living processes. In living organisms, the most important noncovalent interactions include ionic interactions, hydrogen bonds, and van der Waals interactions.

Ionic Interactions

The ionic interactions that occur between charged atoms or groups are nondirected (i.e., they are felt uniformly in space around the center of charge). Oppositely charged ions such as sodium (Na^+) and chloride (Cl^-) are attracted to each other. In contrast, ions with like charges, such as Na^+ and K^+ (potassium), repel each other. In proteins, certain amino acid side chains contain ionizable groups. For example, the side chain of the amino acid glutamic acid ionizes at physiological pH as $—CH_2CH_2COO^-$. The side chain group of the amino acid lysine ($—CH_2CH_2CH_2CH_2—NH_2$) ionizes as $—CH_2CH_2CH_2CH_2NH_3^+$ at physiological pH. The attraction of positively and negatively charged amino acid side chains forms **salt bridges** ($—COO^-\ ^+H_3N—$), and the repulsive forces created when similarly charged species come into close proximity are an important feature in many biological processes, such as protein folding, enzyme catalysis, and molecular recognition. It should be noted that stable salt bridges rarely form between biomolecules in the presence of water; this is because the hydration of ions is preferred, and the attraction between the biomolecules decreases significantly. Most salt bridges in biomolecules occur in relatively water-free depressions or at biomolecular interfaces where water is excluded.

Hydrogen Bonds

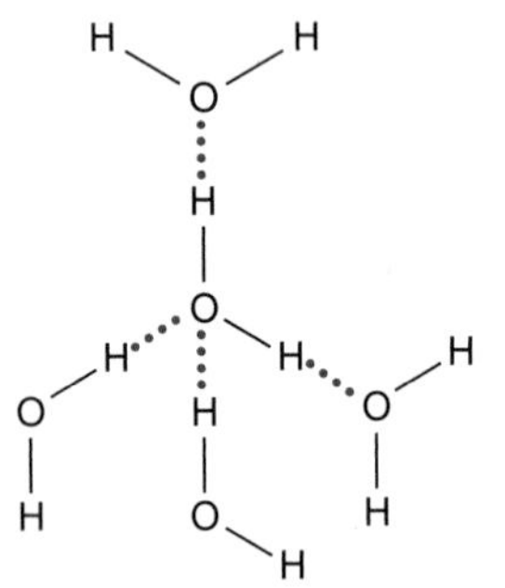

FIGURE 3.7
Tetrahedral Aggregate of Water Molecules

In water, each molecule can form hydrogen bonds with four other water molecules.

Covalent bonds between hydrogen and oxygen or nitrogen are sufficiently polar that the hydrogen nucleus is weakly attracted to the lone pair of electrons of an oxygen or nitrogen on a neighboring molecule. In the water molecule, each of oxygen's unshared electron pairs can form a weak electrostatic attraction to a hydrogen atom in an interaction, referred to as a *hydrogen bond*, with nearby water molecules (**Figure 3.7**). Because this interaction is partially covalent, the force of attraction has directionality. Maximum attraction occurs when the two O—H bonds of the participating water molecules are colinear. The resulting intermolecular "bonds" act as a bridge between water molecules. Neither hydrogen bond is especially strong (about 20 kJ/mol) in comparison to covalent bonds (e.g., 393 kJ/mol for N—H bonds and 460 kJ/mol for O—H bonds). However, when large numbers of intermolecular hydrogen bonds can be formed (e.g., in the liquid and solid states of water), the molecules effectively become large, dynamic, three-dimensional aggregates. In water, the substantial amounts of energy that are required to break up this aggregate explain the high values for its boiling and melting points, heat of vaporization, and heat capacity. Other properties of water, such as surface tension and viscosity, are also largely a result of its capacity to form large numbers of hydrogen bonds.

Van der Waals Forces

Van der Waals forces are relatively weak electrostatic interactions that arise when biomolecules containing neutral permanent dipoles approach each other or an inducible dipole (such as a π cloud). The more polar and colinear the groups involved, the stronger the van der Waals force. Even in pure hydrocarbons (no polar bonds as in hydrophobic regions of proteins and hydrocarbon tails of lipids), close approach will induce charge delocalization (electrons shift and the charges spread out) that results in cohesiveness. The attraction between molecules is greatest at a distance called the van der Waals radius. If molecules approach more closely, a repulsive force develops. In biological systems, the sum total of repulsive and attractive forces creates the stable, functional structure of large biomolecules and biomolecular complexes.

There are three types of van der Waals force:

1. **Dipole-dipole interactions**. These forces, which occur between molecules containing electronegative atoms, cause molecules to orient themselves so that the positive end of one polar group is directed toward the negative end of another (**Figure 3.8a**). Hydrogen bonds are an especially strong type of dipole–dipole interaction.
2. **Dipole–induced dipole interactions**. A permanent dipole induces a transient dipole in a nearby molecule by distorting its electron distribution (**Figure 3.8b**). For example, a carbonyl-containing molecule is weakly attracted to an aromatic ring because of the ability of the permanent dipole of the carbonyl group to delocalize (shift) the electrons of the π electron cloud of the aromatic ring. Dipole–induced dipole interactions are weaker than dipole–dipole interactions.
3. **Induced dipole–induced dipole interactions**. The motion of electrons in nearby nonpolar molecules results in transient charge imbalance in adjacent molecules (**Figure 3.8c**). A transient dipole in one molecule polarizes the electrons in a neighboring molecule. This attractive interaction, often called **London dispersion forces**, is extremely weak. The stacking of the base rings in a DNA molecule, a classic example of this type of interaction, is made possible because of the ability of the loosely held π electrons to distribute unequally above and below closely spaced parallel rings. Although individually weak, these interactions extending over the length of the DNA molecule provide significant stability.

KEY CONCEPTS

- Noncovalent bonds (i.e., ionic interactions and van der Waals forces) are important in determining the physical and chemical properties of living systems.
- Hydrogen bonds, with both dipole-dipole and covalent character, play a critical role in the properties of water and its place in the structure and function of cells.

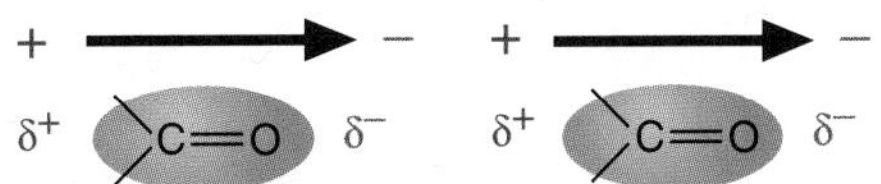

(a) Dipole-dipole interactions

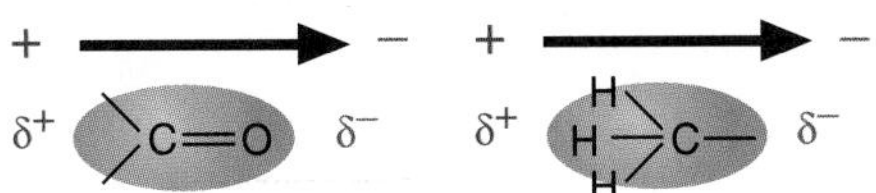

(b) Dipole–induced dipole interactions

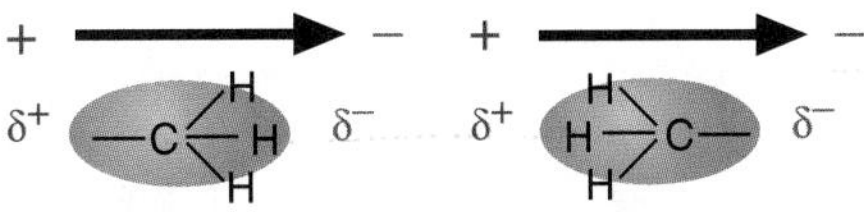

(c) Induced dipole–induced dipole interactions

FIGURE 3.8
Dipolar Interactions

The three types of electrostatic interaction involving dipoles are (a) dipole-dipole interactions, (b) dipole–induced dipole interactions, and (c) induced dipole–induced dipole interactions. The relative ease with which electrons respond to an electric field determines the magnitude of van der Waals forces. Dipole-dipole interactions are the strongest; induced dipole–induced dipole interactions are the weakest.

3.3 THERMAL PROPERTIES OF WATER

Perhaps the oddest property of water is that it is a liquid at room temperature. Compared with related molecules of similar molecular weight, water's melting and boiling points are exceptionally high (**Table 3.2**). If water followed the pattern of compounds such as hydrogen sulfide, it would melt at –100°C and boil at –91°C. Under these conditions, most of Earth's water would be steam, making life unlikely. However, water actually melts at 0°C and boils at +100°C. Consequently, it is a liquid over most of the wide range of temperatures typically found on Earth's surface. Hydrogen bonding is responsible for this anomalous behavior.

Each water molecule can form hydrogen bonds with four other water molecules that, in turn, can form hydrogen bonds with other water molecules. The maximum number of hydrogen bonds form when water has frozen into ice (**Figure 3.9**). Energy is required to break these bonds. When ice is warmed to its melting point, approximately 15% of the hydrogen bonds break. The energy required to melt ice (*heat of fusion*) is substantially higher than expected (Table 3.3). Liquid water consists of icelike clusters of molecules whose hydrogen bonds are continuously breaking and forming. As the temperature rises, the movement and vibrations of the water molecules accelerate, and additional hydrogen bonds are broken. At the boiling point, the water molecules break free from one another and vaporize.

Water is an effective modulator of climatic temperature because of its high *heat of vaporization* (the energy required to vaporize one mole of a liquid at a pressure of one atmosphere) and high *heat capacity* (the energy that must be

TABLE 3.2 Melting and Boiling Points of Water and Three Other Group VI Hydrogen-Containing Compounds

Name	Formula	Molecular Weight (Da)*	Melting Point (°C)	Boiling Point (°C)
Water	H_2O	18	0	100
Hydrogen sulfide	H_2S	34	−85.5	−60.7
Hydrogen selenide	H_2Se	81	−50.4	−41.5
Hydrogen telluride	H_2Te	129.6	−49	−2

* I dalton (Da) = 1 atomic mass unit (amu).

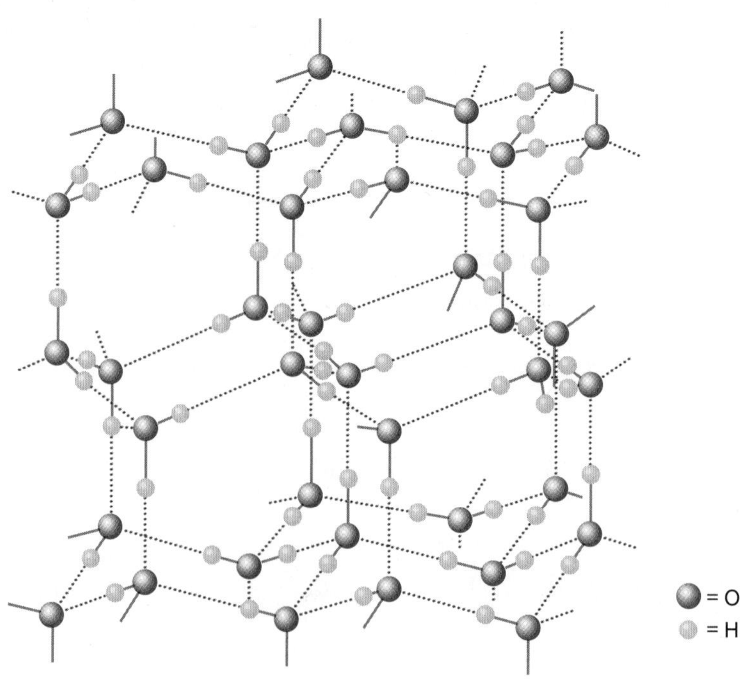

FIGURE 3.9
Hydrogen Bonding between Water Molecules in Ice

Hydrogen bonding in ice produces a very open structure. Ice is less dense than water in its liquid state.

TABLE 3.3 Heat of Fusion of Water and Two Other Group VI Hydrogen-Containing Compounds

Name	Formula	Molecular Weight (Da)	Heat of Fusion* cal/g	J/g
Water	H_2O	18	80	335
Hydrogen sulfide	H_2S	34	16.7	69.9
Hydrogen selenide	H_2Se	81	7.4	31

* The heat of fusion is the amount of heat required to change 1 g of a solid into a liquid at its melting point; 1 cal = 4.184 J.

added or removed to change the temperature by one degree Celsius). Water also plays an important role in the thermal regulation of living organisms. Its high heat capacity, coupled with the high water content found in most organisms (between 50% and 95%, depending on the species), helps maintain an organism's internal temperature. The evaporation of water serves as a cooling mechanism. An adult human may eliminate as much as 1200 g of water daily in expired air, sweat, and urine. The associated heat loss may amount to approximately 20% of the total heat generated by metabolic processes.

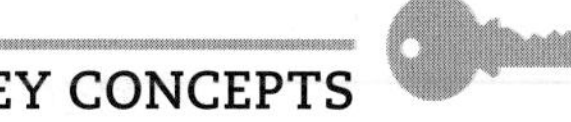

KEY CONCEPTS

- Hydrogen bonding is responsible for water's unusually high freezing and boiling points.
- Because water has a high heat capacity, it can absorb and release heat slowly. Water plays an important role in regulating body temperature in living organisms.

QUESTION 3.1

Water (H_2O), ammonia (NH_3), and methane (CH_4) have approximately the same molecular weight: 18, 17, and 16 g/mol, respectively. Although these molecules are all structurally in the tetrahedral family, they differ significantly in physical properties. For example, the heat of fusion decreases somewhat from water (6.01 kJ/mol) to ammonia (5.66 kJ/mol) and significantly from water to methane (0.94 kJ/mol). Draw the structure of these molecules and explain the difference in properties based on what you know about hydrogen bonding in the solid state. If it were realistic to generate NH_3 ice (melting point −97.8°C), would you expect it to be more or less dense than liquid ammonia?

WORKED PROBLEM 3.1

Water's capacity to absorb large amounts of energy (heat capacity) with only minimal increases in its temperature is an important factor in the success of life on Earth. Heat capacity (energy absorbed or liberated by a substance when its temperature changes) is given by $q = g \cdot C \cdot \Delta T$, where

q = energy in joules
g = mass in grams
C = heat capacity
ΔT = change in temperature

Calculate how much energy is required to raise the temperature of 10 g of water 10°C where $C_{H_2O} = 4.178$ J/g · °C. Then calculate the amount of energy needed to raise the temperature of 10 g of sand (SiO_2) ($C_{sand} = 0.74$ J/g · °C) by 10°C.

SOLUTION

The energy absorbed by 10 g of water is determined by substituting the values given for water into the formula:

$$q = (10\text{ g})(4.178\text{ J/g}\cdot{}^\circ\text{C})(10^\circ\text{C}) = 418\text{ J.}$$

The energy absorbed by sand is calculated as

$$q = (10\text{ g})(0.74\text{ J/g}\cdot{}^\circ\text{C})(10^\circ\text{C}) = 56\text{ J.}$$

■

Visit the companion website at www.oup.com/us/mckee to read the Biochemistry in Perspective essay on water, abiotic stress, and compatible solutes.

3.4 SOLVENT PROPERTIES OF WATER

Water is the ideal biological solvent. It easily dissolves a wide variety of the constituents of living organisms. Examples include ions (e.g., Na^+, K^+, and Cl^-), sugars, and many of the amino acids. Supramolecular structures (e.g., membranes) and numerous biochemical processes (e.g., protein folding) are possible because water cannot dissolve other substances, such as lipids and certain amino acids. This section describes the behavior of hydrophilic and hydrophobic substances in water. The discussion is followed by a brief review of osmotic pressure, one of the colligative properties of water. Colligative properties are physical properties that are affected not by the specific structure of dissolved solutes, but by their numbers.

Hydrophilic Molecules, Cell Water Structuring, and Sol-Gel Transitions

A dipolar structure and the capacity to form hydrogen bonds with electronegative atoms enable water to dissolve both ionic and polar substances. An important aspect of all ionic interactions in aqueous solution is the hydration of ions. Salts such as sodium chloride (NaCl) are held together by ionic forces. Because water molecules are polar, they are attracted to charged ions such as Na^+ and Cl^-. Shells of water molecules, referred to as **solvation spheres**, cluster around both positive and negative ions (**Figure 3.10**). The size of the solvation sphere depends upon the charge density of the ion (i.e., size of charge per unit volume). As ions become hydrated, the attractive force between them is reduced, and the charged species dissolves in the water. Organic molecules with ionizable groups and many neutral organic molecules with polar functional groups also dissolve in water, primarily because of the solvent's hydrogen bonding capacity. Such associations form between water and the carbonyl groups of aldehydes and

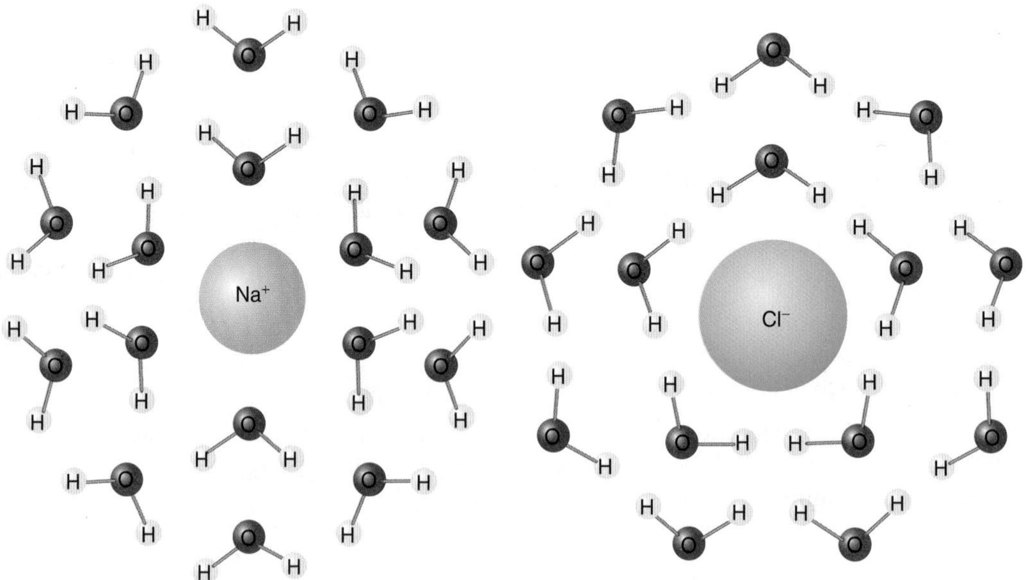

FIGURE 3.10

Solvation Spheres of Water Molecules around Na^+ and Cl^- Ions

When an ionic compound such as NaCl is dissolved in water, its ions separate because the polar water molecules attract the ions more than the ions attract each other. In reality, the solvation sphere of Na^+ has four times the volume of that of Cl^- because of the higher charge density of the sodium ion (the same unit charge distributed over a smaller volume).

ketones and the hydroxyl groups of alcohols. The capacity of a solvent to reduce the electrostatic attraction between charges is indicated by its *dielectric constant.* Water, sometimes referred to as the *universal solvent* because of the large variety of ionic and polar substances it can dissolve, has a very large dielectric constant.

STRUCTURED WATER The arrangement of water molecules in living organisms is distinctive. Although organismal water is in liquid form, most water molecules are not in the "bulk water" state (i.e., they do not flow freely). At any given time, most of a cell's billions of water molecules are noncovalently associated with macromolecules and membrane surfaces throughout its densely packed interior. The surfaces of proteins, for example, are studded with positive and negative charges and polar functional groups. Dipolar water molecules readily form hydrogen bonds with such species (**Figure 3.11**). Recall that water molecules are tetrahedral and that each one can form hydrogen bonds with four other water molecules. For this reason, a single layer of water molecules attracts additional water molecules and an extended three-dimensional network of water molecules forms. In crowded cells, numerous water layers bridge the space between adjacent macromolecules. Moreover, the water molecules in these layers, referred to as *structured water,* are in perpetual motion and constantly rearranging. They exchange with bulk water molecules, farther away from the protein's surface, on time scales that range from femtoseconds (10^{-15} s) to picoseconds (10^{-12} s). The pace of the exchange for an individual water molecule depends on how restricted its motion is. In other words, the closer a water molecule is to a polar surface, the slower its motion. The dynamics of structured water contribute to the structural stability of macromolecules such as proteins. They also facilitate the flexibility required for function.

SOL-GEL TRANSITIONS Cytoplasm, like any water-based material that contains polymers, has the properties of a gel. A *gel* is a colloidal mixture, which is a type of mixture in which small particles are evenly distributed throughout another substance. In cells, the gelatinous cytoplasm is a semisolid composed of

(a)

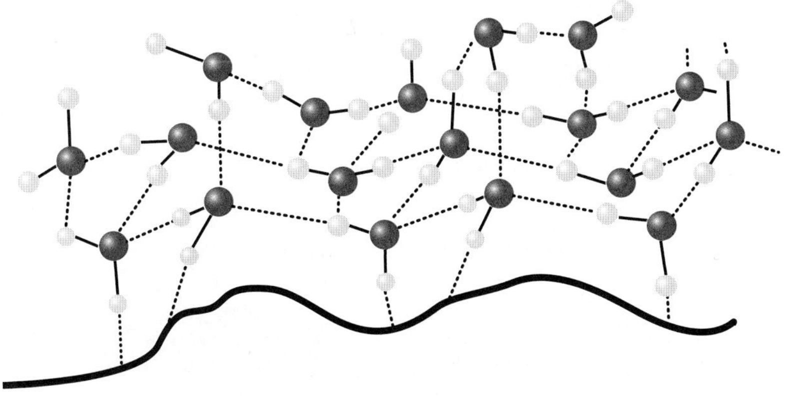

(b)

FIGURE 3.11

Diagrammatic View of Structured Water

Polar surfaces of macromolecules attract water molecules: (a) a short segment of a polar macromolecular surface with a single water layer, which attracts additional water molecules that form an extended network (b).

biopolymers with polar surfaces in association with adsorbed water. Gelatin desserts are well-known examples of gels with fibers of the protein collagen suspended and hydrated in a large quantity of water. The highly structured solvation layers on a matrix of protein give the viscoelastic properties we associate with Jell-O. The stability of a gel depends on the length and cross-linking of the polymer and the continuity of the adsorbed water. The freedom of solutes to move within this meshwork or gel matrix varies with the trabecular (resembling a sponge) arrangement of the protein polymers. If you punch wells in a petri dish filled with solidified gelatin and pour a solution of inorganic ions into the wells, the ions will migrate out into the gel at rates related to their size and degree of hydration. You can observe the results of this sieving effect in just a few minutes. In addition, the water solvating the surface of the gelatin (collagen polymers) is for all practical purposes fixed in position; that is, diffusion is limited.

Changes in temperature (and therefore molecular motion), matrix architecture, and inclusion of solutes can lead to a transition from the gel to a "sol" or liquid state. Cells behave in a similar way because of the highly structured solvation surfaces of the polymeric proteins. Transitions from gel to sol (from more solid to less solid) contribute to many aspects of cell function, most notably cell movement. These transitions are caused by the reversible polymerization of G-actin to form F-actin and the subsequent cross-linking of actin filaments. These transitions are carefully regulated by signal transduction mechanisms that affect the concentrations and functions of a group of proteins called *actin-binding proteins* (p. 64). Various actin-binding proteins can inhibit polymerization or they can cross-link or sever actin filaments.

Amoeboid motion provides an example of the highly regulated nature of cellular sol-gel transitions and the forces that such transformations create. The principal feature of amoeboid motion is the protrusion of a cellular extension called a *pseudopodium* (**Figure 3.12**). The pseudopodium moves forward because polymerizing actin filaments in the cell cortex (the *ectoplasm*) in this isolated part of the cell undergo further cross-linking (a sol-to-gel transition). Once this has occurred, actin filaments in the cell's interior (the *endoplasm*) depolymerize, effecting a gel-to-sol transition. Simultaneously, a contractile force is created by the binding of actin filaments to myosin (a motor protein) in the trailing end of the cell. This force squeezes the freely flowing endoplasm, causing it to stream forward into the pseudopodium.

Hydrophobic Molecules and the Hydrophobic Effect

Small amounts of nonpolar substances mixed with water are excluded from the solvation network of the water; that is, they coalesce into droplets. This process is called the *hydrophobic effect*. **Hydrophobic** ("water-hating") molecules, such as the hydrocarbons, are virtually insoluble in water. Their association into droplets (or, in larger amounts, into a separate layer) results from the solvent properties of water, not from the relatively weak attraction between the associating nonpolar molecules. When nonpolar molecules enter an aqueous environment, the water molecules organize into a cagelike structure that drives the

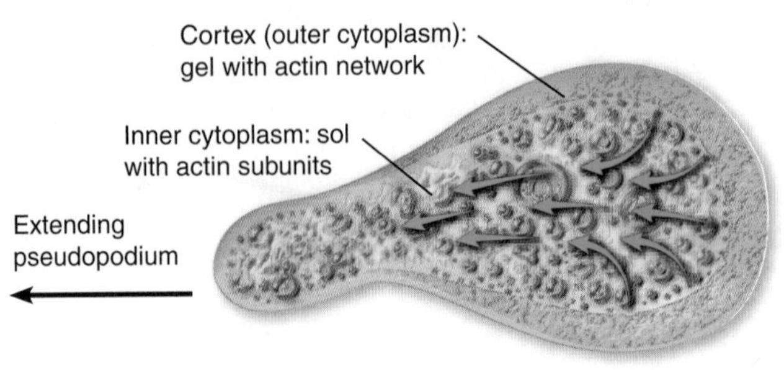

FIGURE 3.12
Amoeboid Motion and Sol-Gel Transitions
The cell moves forward because of the coordination of sol-gel transitions in the cell cortex (ectoplasm) and cytoplasm in the cell's interior (endoplasm). A contractile force in the rear of the cell squeezes the fluid endoplasm forward.

QUESTION 3.2

Proteins are amino acid polymers. Noncovalent bonding plays an important role in determining the three-dimensional structures of proteins. The noncovalent interactions indicated here by shaded areas are typical of the bonding that occurs between amino acid side chains.

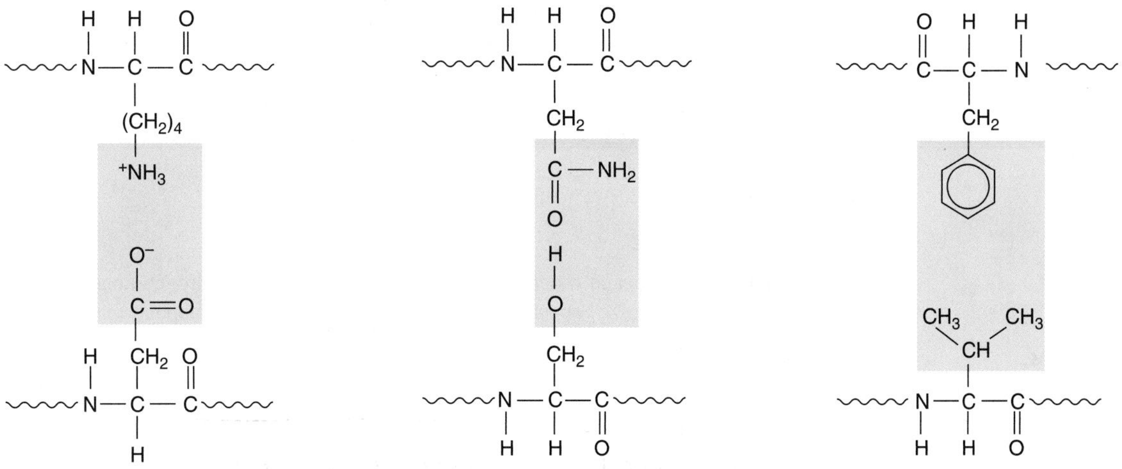

Which noncovalent bond is primarily responsible for the interactions indicated in the figure?

QUESTION 3.3

Collagen, a large fiberlike protein, combined with other molecules, forms a gel-like material found in shock-absorbing body components (e.g., tendons and ligaments). Explain the role of structured water in the function of these tissues. [*Hint*: Water is an incompressible substance.]

hydrophobic region in on itself (a partitioning or exclusion process). The excluded hydrophobic phase is ultimately stabilized by van der Waals interactions between closely spaced nonpolar regions (**Figure 3.13**). The water-caged structure, or *clathrate*, is stabilized when exposure of water to the hydrophobic material is minimized. The hydrophobic effect is responsible for the generation of stable lipid membranes and contributes to the fidelity of protein folding.

Amphipathic Molecules

A large number of biomolecules, referred to as **amphipathic**, contain both polar and nonpolar groups. This property significantly affects their behavior in water. For example, ionized fatty acids are amphipathic molecules because they contain hydrophilic carboxylate groups and hydrophobic hydrocarbon groups. When they are mixed with water, amphipathic molecules form structures called **micelles** (**Figure 3.14**). In micelles, the charged species (the carboxylate groups), called *polar heads*, orient themselves so that they are in contact with water. The nonpolar hydrocarbon "tails" become sequestered in the hydrophobic interior. The tendency of amphipathic biomolecules to spontaneously rearrange themselves in water is an important feature of numerous cell components. For example, a group of bilayer-forming phospholipid molecules is the basic structural feature of biological membranes (see Chapter 11).

KEY CONCEPTS

- Water's dipolar structure and its capacity to form hydrogen bonds enable water to dissolve many ionic and polar substances.
- Nonpolar molecules cannot form hydrogen bonds with water and are excluded via clathrate formation.
- Amphipathic molecules, such as fatty acid salts, spontaneously rearrange themselves in water to form micelles.

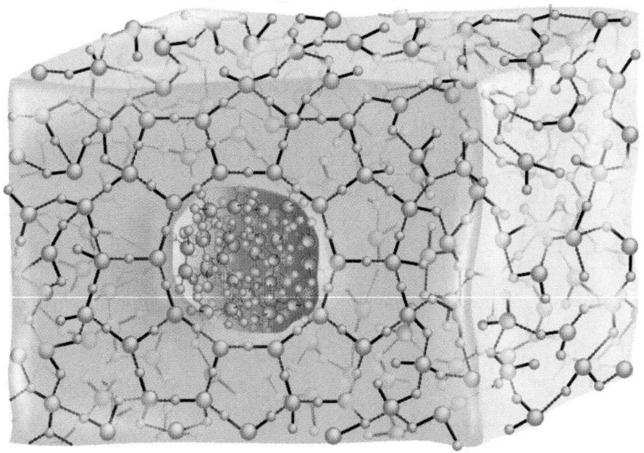

FIGURE 3.13
The Hydrophobic Effect

When nonpolar molecules and water are mixed, a cage of organized hydrogen-bonded water molecules forms to minimize exposure to the hydrophobic substance. Nonpolar molecules, when in close proximity, are attracted to each other by van der Waals forces. However, the driving force in the formation of the cage and exclusion of the hydrophobic substance is the strong tendency of water molecules to form hydrogen bonds among themselves. Nonpolar molecules are excluded because they cannot form hydrogen bonds.

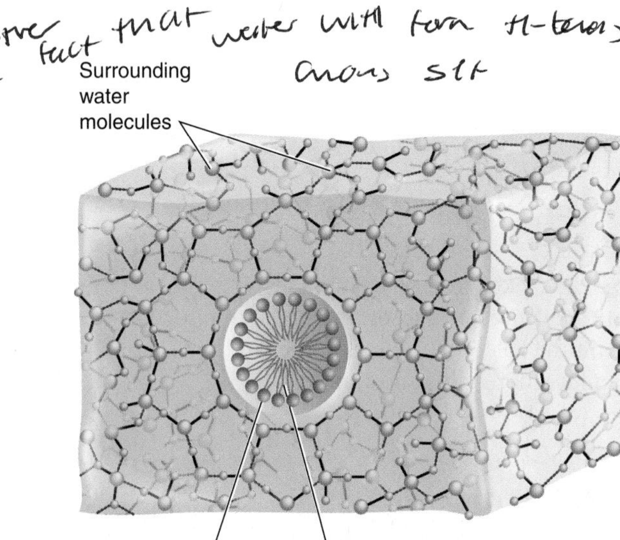

FIGURE 3.14
Formation of Micelles

The polar heads of amphipathic molecules orient themselves so that they are hydrogen-bonded to water. The nonpolar tails aggregate in the center, away from water.

Osmotic Pressure

Osmosis is the spontaneous passage of solvent molecules through a semipermeable membrane that separates a solution of lower solute concentration from a solution of higher solute concentration. Pores in the membrane are wide enough to allow solvent molecules to pass through in both directions but too narrow for the larger solute molecules or ions to pass. **Figure 3.15** illustrates the movement of solvent across a membrane. As the process begins, there are fewer water molecules on the high solute concentration side of the membrane. Over time, more water moves from side A (lower solute concentration) to side B (higher solute concentration). The higher the concentration of water in a solution (i.e., the lower the solute concentration), the greater the rate of water flow through the membrane.

Osmotic pressure is the pressure required to stop the net flow of water across the membrane. As the principal cause of water flow across cellular membranes, osmotic pressure is a driving force in numerous living processes. For example, osmotic pressure appears to be a significant factor in the formation of sap in trees.

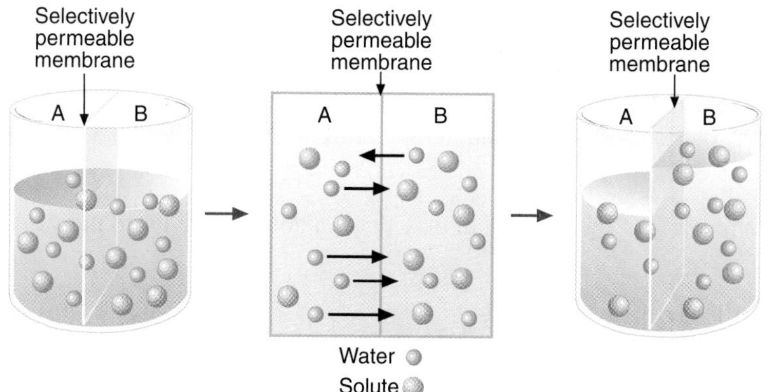

FIGURE 3.15

Osmotic Pressure

Over time water diffuses from side A (more dilute) to side B (more concentrated). Equilibrium between the solutions on both sides of a semipermeable membrane is attained when there is no net movement of water molecules from side A to side B. Osmotic pressure stops the net flow of water across the membrane.

Cell membranes are not, strictly speaking, osmotic membranes because they permit molecules other than solvent (water) to move across the membrane. The term *dialyzing membrane* would be more accurate.

Osmotic pressure depends on solute concentration. A device called an *osmometer* (**Figure 3.16**) measures osmotic pressure. Osmotic pressure can also be calculated using the following equation, keeping in mind that the final osmotic pressure reflects the contribution of all solutes present.

$$\pi = iMRT$$

where π = osmotic pressure (atm)
i = van't Hoff factor (reflects the extent of ionization of solutes)
M = molarity (mol/L)
R = gas constant (0.082 L·atm/K·mol)
T = temperature (K)

The concentration of a solution can be expressed in terms of *osmolarity*. The unit of osmolarity is osmoles (osmol) per liter. In the equation $\pi = iMRT$, the osmolarity is equal to iM, where i (the van't Hoff factor) represents the degree of ionization of the solute species, which varies with temperature. The degree of ionization of a 1 M NaCl solution is 90%, with 10% of the NaCl existing as ion pairs. Thus if we write

$$i = [Na^+] + [Cl^-] + [NaCl]_{\text{un-ionized}} = 0.9 + 0.9 + 0.1 = 1.9$$

the value of i for this solution is 1.9. The value of i approaches 2 for NaCl solutions as they become increasingly more dilute. The value of i for a 1 M solution of a weak acid that undergoes a 10% ionization is 1.1. The value of i for a nonionizable solute is always 1.0. Problems 3.1, 3.2 and 3.3 use the concept of osmotic pressure.

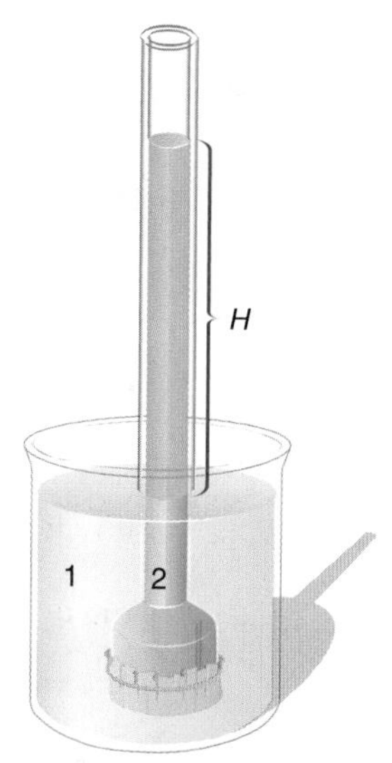

FIGURE 3.16

The Measurement of Osmosis Using an Osmometer

Volume 1 contains pure water. Volume 2 contains a solution of sucrose. The membrane is permeable to water but not to the sucrose. Therefore there will be a net movement of water into the osmometer. The osmotic pressure is proportional to the height H of the solution in the tube.

Osmotic pressure creates some critical problems for living organisms. Cells typically contain fairly high concentrations of solutes, that is, small organic molecules and ions, as well as lower concentrations of macromolecules. Consequently, cells may gain or lose water because of the concentration of solute in their environment. If cells are placed in an **isotonic solution** (i.e., the concentration of solute and water is the same on both sides of the selectively permeable plasma membrane) there is no net movement of water in either direction across the membrane (**Figure 3.17**). For example, red blood cells are isotonic to a 0.9% NaCl solution. When cells are placed in a solution with a lower solute concentration (i.e., a **hypotonic solution**), water moves into the cells. When red blood cells are immersed in pure water, for example, they swell and rupture in a process called *hemolysis*. In solutions with higher solute concentrations (i.e., **hypertonic solutions**), cells shrivel because there is a net movement of water out of the cell. The shrinkage of red blood cells in a hypertonic solution (e.g., a 3% NaCl solution) is referred to as *crenation*.

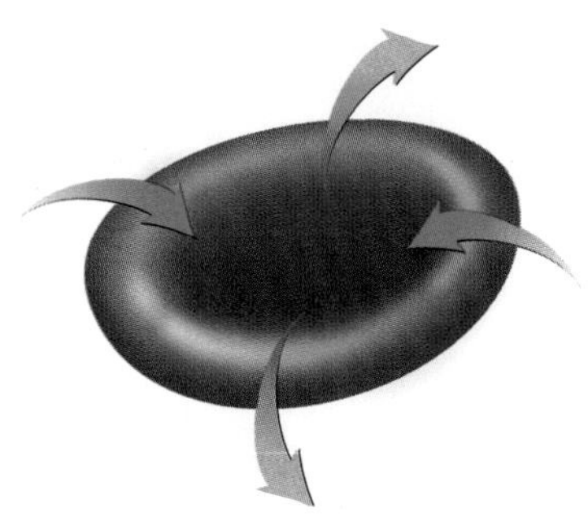

(a)

(b)

(c)

FIGURE 3.17

The Effect of Hypertonic and Hypotonic Solutions on Animal Cells

(a) Isotonic solutions do not change cell volume because water is entering and leaving the cell at the same rate.
(b) Hypotonic solutions cause cell rupture.
(c) Hypertonic solutions cause cell shrinkage (crenation).

WORKED PROBLEM 3.2

When 0.1 g of urea (MW 60) is diluted to 100 mL with water, what is the osmotic pressure of the solution? [Assume room temperature, i.e., 25°C (298 K).]

SOLUTION

Calculate the molarity of the urea solution. Urea is a nonelectrolyte, so the van't Hoff factor (*i*) is 1.

$$\text{Molarity} = \frac{0.10\text{ g urea} \times 1.0\text{ mol}}{60} \times \frac{1}{0.10\text{ L}} = 1.7 \times 10^{-2}\text{ mol/L}$$

The osmotic pressure at room temperature is given by

$$\pi = iMRT$$

$$\pi = (1)\frac{1.7 \times 10^{-2}\text{ mol}}{\text{L}}\ \frac{0.0821\text{L} \cdot \text{atm}}{\text{K} \cdot \text{mol}}\ (298\text{ K})$$

$$\pi = 0.4\text{ atm}$$

■

WORKED PROBLEM 3.3

Estimate the osmotic pressure of a solution of 0.1 M NaCl at 25°C. Assume 100% ionization of solute.

SOLUTION

A solution of 0.10 M NaCl produces 0.2 mol of particles per liter (0.10 mol of Na^+ and 0.10 mol of Cl^-). The osmotic pressure at room temperature is

$$\pi = \frac{2 \times 0.10\text{ mol}}{\text{L}}\ \frac{0.0821\text{L} \cdot \text{atm}}{\text{K} \cdot \text{mol}} \times 298\text{ K}$$

$$\pi = 4.9\text{ atm}$$

■

WORKED PROBLEM 3.4

Osmotic pressure can be used as a method to estimate the molecular mass of a biomolecule. Determine the molecular mass (m) of the nonionic compound *X* ($i = 1$). When 1.0 g of compound *X* is dissolved in 100 ml of water, the solution has an osmotic pressure of 0.2 atm at 25°C.

SOLUTION

Calculate the molarity (M) of the solution using the formula for osmotic pressure.

$$\pi = iMRT$$

$$M = \pi/iRT$$

$$M = (0.20\text{ atm})/[(1)(0.0821\text{ L} \cdot \text{atm/mol} \cdot \text{K})\ (298\text{ K})]$$

$$M = 8.2 \times 10^{-3}\text{ mol/L}$$

Use the formula for molarity and the given mass and volume (V) to calculate the molecular mass (m) of compound *X*.

$$\text{M} = \text{mass}/[\text{m/V}]$$

$$\text{m} = \text{mass/MV}$$

$$\text{m} = 1.0\text{ g}/[8.2 \times 10^{-3}\text{ mol/L }(0.1\text{ L})]$$

$$\text{m} = 1.2 \times 10^{3}\text{ g/mol}$$

■

ION DISTRIBUTION ACROSS CELL MEMBRANES Macromolecules have little direct effect on cellular osmolarity because their cellular molar concentrations are relatively low. However, macromolecules such as the proteins contain a large number of ionizable groups. The ions of opposite charge that are attracted to these groups have a substantial effect on intracellular osmolarity. The nature of this effect is determined by the interaction of the structured water associated with proteins with hydrated ions. The size of an ion's solvation sphere is inversely related to its charge density (size of charge per unit volume). For example, sodium and potassium ions have nonhydrated diameters of 1.96 and 2.66 Å, respectively. The hydrated diameters of sodium and potassium ions are 9.0 and 6.0 Å, respectively. Consequently, the hydrated volume of Na^+ is 3.4 times that of K^+. In addition, since the solvation sphere of K^+ is much smaller than that of Na^+, the potassium ion is easier to remove to form ion pairs with the surface anions of proteins. It costs significantly more energy to remove the solvation sphere of Na^+, a step that must occur for the sodium ion to move through ion channels. As a result, the ion distribution across the cell membrane is unequal, with the tendency to accumulate inside the cell being much greater for K^+ (159 mM) than for Na^+ (10 mM). This inequality would occur even if specific ion pumps were not present. (Refer to the Biochemistry in Perspective essay Cell Volume Regulation and Metabolism for an insight into how osmotic pressure affects cell volume.)

Unlike most inorganic ions, the ionizable groups of cellular proteins are fixed within the cell, conferring a significant net negative charge to the intracellular environment. As a consequence, there exists an electronegative gradient across the cell membrane: that is, the ions and negative charge are distributed unequally. The cytoplasmic side of the membrane is strongly negative, an effect that is partly offset by potassium ions. The outside of the membrane is positive because of the relatively large number of extracellular sodium ions. The existence of this asymmetry on the surfaces of cell membrane results in the establishment of an electrical gradient, called a **membrane potential**, which provides the means for electrical conduction, active transport, and even passive transport.

Although hydrated sodium ions tend to be excluded from structured water inside cells, leakage of these ions back across the plasma membrane does occur. Small intracellular increases in $[Na^+]$ cause the cytoplasm to be slightly less negative. As a result, small amounts of K^+ move down their concentration gradient out of the cell. Animals and bacteria control cell volume by opposing this process with ATP-driven Na^+-K^+ pumps. Ion pumping in these cells requires substantial amounts of energy.

Several species, such as some protozoa and algae, control cell volume by periodically expelling water from special contractile vacuoles. Because plant cells have rigid cell walls, plants use osmotic pressure to create an internal hydrostatic pressure, called *turgor pressure*. This process drives cellular growth and expansion and makes many plant structures rigid.

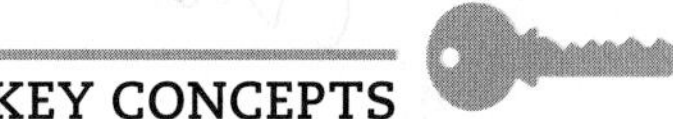

KEY CONCEPTS

- Osmosis is the movement of water across a semipermeable membrane from a dilute solution to a more concentrated solution.
- Osmotic pressure is the pressure exerted by water on a semipermeable membrane as a result of a difference in the concentration of solutes on either side of the membrane.

3.5 IONIZATION OF WATER

Liquid water molecules have a limited capacity to ionize to form a proton, or hydrogen ion (H^+), and a hydroxide ion (OH^-). Protons do not actually exist in aqueous solution. In water a proton combines with a water molecule to form H_3O^+, commonly referred to as *hydronium ion* (**Figure 3.18**). For convenience, H^+ will be used to represent the ionization reactions of water.

The disassociation of water

$$H_2O(l) \rightleftharpoons H^+ + OH^-$$

may be expressed as

$$K_{eq} = \frac{[H^+][OH^-]}{[H_2O]}$$

where K_{eq} is the equilibrium constant for the reaction.

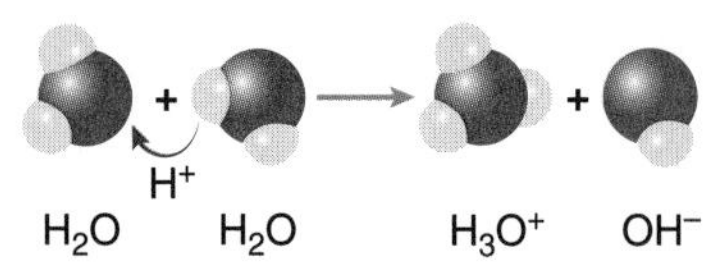

FIGURE 3.18
Hydronium Ion

The hydronium ion (H_3O^+) is formed by the protonation of a water molecule. When a water molecule dissociates to yield a proton (H^+) and a hydroxide ion (OH^-), the electronegative oxygen atom of a nearby water molecule is attracted to the positively charged proton.

The concentration of water is essentially unchanged and can be thought of as a constant. The equilibrium expression can be rewritten combining the two constants as

$$K_{eq}[H_2O] = [H^+]\,[OH^-]$$

The term $K_{eq}[H_2O]$ is referred to as the *ion product of water* or K_w. After substitution of the term K_w, the preceding equation may be rewritten as

$$K_w = [H^+]\,[OH^-]$$

The K_w for H_2O at 25°C and 1 atm pressure is 1.0×10^{-14} and is a fixed characteristic of water at this temperature and pressure. In pure water, where there are no other contributors of H^+ or OH^-, the concentrations of these ions are equal:

$$[H^+] = [OH^-] = (K_w)^{1/2} = (1.0 \times 10^{-14})^{1/2} = 1.0 \times 10^{-7}\ \text{M}$$

A solution that contains equal amounts of H^+ and OH^- is said to be *neutral*. When an ionic or polar substance is dissolved in water, it may change the relative numbers of H^+ and OH^-. Solutions with an excess of H^+ are *acidic*, whereas those with a greater number of OH^- are *basic*. Hydrogen ion concentration varies over a wide range: commonly between 10^0 and 10^{-14} M, which provides the basis of the pH scale (pH = −log [H^+]).

Acids, Bases, and pH

The concentration of the hydrogen ion, one of the most important ions in biological systems, affects most cellular and organismal processes. For example, the structure and function of proteins and the rates of most biochemical reactions are strongly affected by hydrogen ion concentration. Additionally, hydrogen ions play a major role in processes such as energy generation (see Chapter 10) and endocytosis.

Many biomolecules have acidic and/or basic properties. Large polymers and macromolecular complexes usually have amphoteric surfaces; that is, they possess both acidic and basic groups. A side group of a molecule is said to be an **acid** if it is a proton donor and a **base** if it is a proton acceptor.

Strong acids (e.g., HCl) and bases (e.g., NaOH) ionize almost completely in water:

$$HCl \rightarrow H^+ + Cl^-$$

$$NaOH \rightarrow Na^+ + OH^-$$

Many acids and bases, however, do not dissociate completely. Organic acids (compounds with carboxyl groups) are referred to as **weak acids** because of their partial dissociation in water. Organic bases have a small but measurable capacity to combine with hydrogen ions. Many common **weak bases** contain amino groups.

The dissociation of an organic acid is described by the following reaction:

$$\underset{\text{Weak acid}}{HA} \rightleftharpoons H^+ + \underset{\text{Conjugate base of HA}}{A^-}$$

Note that the deprotonated product of the dissociation reaction is referred to as a **conjugate base**. For example, acetic acid (CH_3COOH) dissociates to form the conjugate base acetate (CH_3COO^-).

The strength of a weak acid (i.e., its capacity to release hydrogen ions) may be determined using the following expression:

$$K_a = \frac{[H^+][A^-]}{[HA]}$$

TABLE 3.4 Dissociation Constants and pK_a Values for Common Weak Acids*

Acid	HA	A^-	K_a	pK_a
Acetic acid	CH_3COOH	CH_3COO^-	1.76×10^{-5}	4.76
Carbonic acid	H_2CO_3	HCO_3^-	4.5×10^{-7}	6.35
Bicarbonate	HCO_3^-	CO_3^{2-}	5.61×10^{-11}	10.33
Lactic acid	$CH_3CH(OH)COOH$	$CH_3CH(OH)COO^-$	1.38×10^{-4}	3.86
Phosphoric acid	H_3PO_4	$H_2PO_4^-$	7.25×10^{-3}	2.14
Dihydrogen phosphate	$H_2PO_4^-$	HPO_4^{2-}	6.31×10^{-8}	7.20

* Equilibrium constants should be expressed in terms of activities rather than concentrations (activity is the effective concentration of a substance in a solution). However, in dilute solutions, concentrations may be substituted for activities with reasonable accuracy.

where K_a is the acid dissociation constant. The larger the value of K_a, the stronger the acid is. Because K_a values vary over a wide range, they are expressed using a logarithmic scale:

$$pK_a = -\log K_a$$

The lower the pK_a, the stronger the acid. Dissociation constants and pK_a values for several common weak acids are given in **Table 3.4**.

The **pH scale** (**Figure 3.19**) can be used to determine hydrogen ion concentration $[H^+]$:

$$pH = -\log[H^+]$$
$$[H^+] = \text{antilog}(-pH)$$

On the pH scale, neutrality is defined as pH 7; that is, $[H^+]$ is equal to 1×10^{-7} M. Acidic solutions have pH values less than 7; that is, $[H^+]$ is greater than 1×10^{-7} M. A pH value greater than 7 indicates a solution that is basic, or alkaline.

It is important to note that although pK_a and pH appear to be similar mathematical expressions, they are in fact different. At constant temperature, the pK_a value of a substance is a constant. In contrast, the pH values of a system may vary.

Buffers

The regulation of pH is a universal and essential activity of living organisms. Hydrogen ion concentration must typically be kept within narrow limits. For example, normal human blood has a pH of 7.4. It may vary between 7.35 and 7.45, depending on the concentrations of acidic and basic waste products and metabolites. Certain disease processes cause pH changes that, if not corrected, can be disastrous. **Acidosis**, a condition that occurs when human blood pH falls below 7.35, results from an excessive production of acid in the tissues, loss of base from body fluids, or the failure of the kidneys to excrete acidic metabolites. Acidosis occurs in certain diseases (e.g., diabetes mellitus) and during starvation. If blood pH drops below 7, the central nervous system becomes depressed, resulting in coma and eventually death. When blood pH rises above 7.45, **alkalosis**

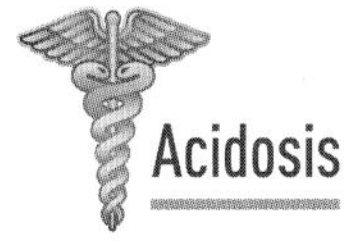

FIGURE 3.19

The pH Scale and the pH Values of Common Fluids

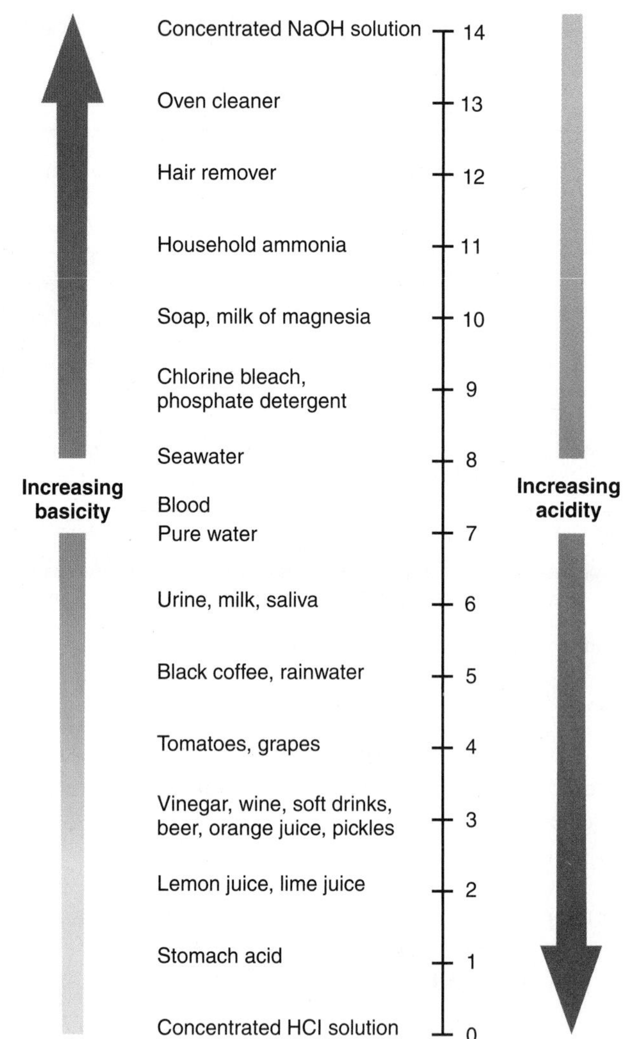

results. This condition, brought on by prolonged vomiting or by ingestion of excessive amounts of alkaline drugs, overexcites the central nervous system. Muscles then go into a state of spasm. If this situation is uncorrected, convulsions and respiratory arrest develop.

Buffers help maintain a relatively constant hydrogen ion concentration. The most common buffers consist of mixtures of weak acids and their conjugate bases. A buffered solution can resist pH changes because an equilibrium between the buffer's components is established. Therefore buffers obey **Le Chatelier's principle**, which states that if a stress is applied to a reaction at equilibrium, the equilibrium will be displaced in the direction that relieves the stress. Consider a solution containing acetate buffer, which consists of acetic acid and sodium acetate (**Figure 3.20**). The buffer is created by mixing a solution of sodium acetate with a solution of acetic acid to create an equilibrium mixture of the correct pH and ionic strength.

$$CH_3-\overset{\displaystyle O}{\overset{\|}{C}}-OH \rightleftharpoons CH_3-\overset{\displaystyle O}{\overset{\|}{C}}-O^- + H^+$$

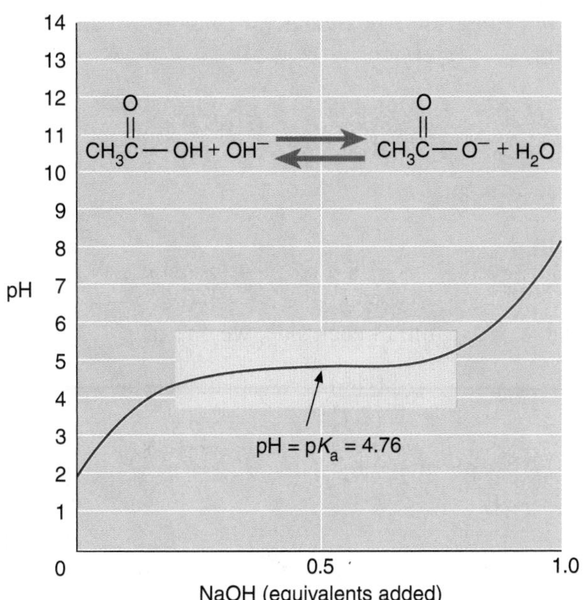

FIGURE 3.20

Titration of Acetic Acid with NaOH

The shaded band indicates the pH range over which acetate buffer functions effectively. A buffer is most effective at or near its pK_a value.

If hydrogen ions are added, the equilibrium shifts toward the formation of acetic acid with the $[H^+]$ changing little:

$$H^+ + CH_3COO^- \rightarrow CH_3COOH$$

If hydroxide ions are added, they react with the free hydrogen ions to form water, the equilibrium shifts to the acetate ion, and the pH changes little.

$$CH_3—C(=O)—OH \rightleftharpoons CH_3—C(=O)—O^- + H^+$$

$$H^+ + OH^- \rightarrow H_2O$$

BUFFERING CAPACITY The capacity of a buffer to maintain a specific pH depends on two factors: (1) the molar concentration of the acid–conjugate base pair and (2) the ratio of their concentrations. Buffering capacity is directly proportional to the concentration of the buffer components. In other words, the more molecules of buffer present, the more H^+ and OH^- ions can be absorbed without changing the pH. The concentration of the buffer is defined as the sum of the concentration of the weak acid and its conjugate base. For example, a 0.2 M acetate buffer may contain 0.1 mol of acetic acid and 0.1 mol of sodium acetate in 1 L of H_2O. Such a buffer may also consist of 0.05 mol of acetic acid and 0.15 mol of sodium acetate in 1 L of H_2O. The most effective buffers are those that contain equal concentrations of both components or the pH is equal to the pK_a. Biological systems generate acids during metabolism, and buffer capacity for acid neutralization must be maximized. Consequently, biological buffers often contain a higher concentration of the conjugate base. Bicarbonate buffer (p. 100) is an example of such a buffering system.

HENDERSON-HASSELBALCH EQUATION In choosing or making a buffer, the pH and pK_a concepts are useful. The relationship between these two quantities is expressed in the Henderson-Hasselbalch equation, which is derived from the following equilibrium expression:

$$K_a = \frac{[H^+][A^-]}{[HA]}$$

Solving for $[H^+]$ results in

$$[H^+] = K_a \frac{[HA]}{[A^-]}$$

Taking the negative logarithm of each side, we obtain

$$-\log [H^+] = -\log K_a - \log \frac{[HA]}{[A^-]}$$

Defining $-\log [H^+]$ as pH and $-\log K_a$ as pK_a gives

$$pH = pK_a - \log \frac{[HA]}{[A^-]}$$

If the log term is inverted, thereby changing its sign, the *Henderson-Hasselbalch equation* is obtained:

$$pH = pK_a + \log \frac{[A^-]}{[HA]}$$

Note that when $[A^-] = [HA]$, the equation becomes

$$pH = pK_a + \log 1$$
$$= pK_a + 0$$

Under this circumstance, pH is equal to pK_a. **Figure 3.20** illustrates that buffers are most effective when they are composed of equal amounts of weak acid and conjugate base. The most effective buffering occurs in the portion of the titration curve that has a minimum slope, that is, 1 pH unit above and below the value of pK_a. In the graph, the abscissa displays the equivalents added. Here, an equivalent is the mass of base that can accept 1 mol of H^+ ions; an acid equivalent gives the mass of acid that can accept a mole of protons.

Problems 3.5 through 3.11 are typical buffer problems.

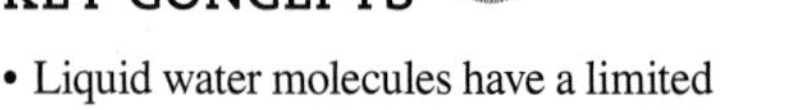

KEY CONCEPTS

- Liquid water molecules have a limited capacity to ionize to form H^+ and OH^- ions.
- The concentration of hydrogen ions is a crucial feature of biological systems primarily because of their effects on biochemical reaction rates and protein structure.
- Buffers, which consist of weak acids and their conjugate bases, prevent changes in pH (a measure of $[H^+]$).

WORKED PROBLEM 3.5

Calculate the pH of a mixture of 0.25 M acetic acid and 0.10 M sodium acetate. The pK_a of acetic acid is 4.76.

SOLUTION

$$pH = pK_a + \log \frac{[acetate]}{[acetic\ acid]}$$

$$pH = 4.76 + \log \frac{0.10}{0.25} = 4.76 - 0.40 = 4.36$$ ■

WORKED PROBLEM 3.6

What is the pH in the preceding problem if the mixture consists of 0.10 M acetic acid and 0.25 M sodium acetate?

SOLUTION

$$\text{pH} = 4.76 + \log \frac{0.25}{0.10} = 4.76 + 0.40 = 5.16$$

■

WORKED PROBLEM 3.7

Calculate the ratio of lactic acid and lactate required in a buffer system of pH 5.00. The pK_a of lactic acid is 3.86.

SOLUTION

The equation

$$\text{pH} = \text{p}K_a + \log \frac{[\text{lactate}]}{[\text{lactic acid}]}$$

can be rearranged to

$$\log \frac{[\text{lactate}]}{[\text{lactic acid}]} = \text{pH} - \text{p}K_a$$

$$= 5.00 - 3.86 = 1.14$$

Therefore the required ratio is

$$\frac{[\text{lactate}]}{[\text{lactic acid}]} = \text{antilog } 1.14$$

$$= 13.8$$

For a lactate buffer to have a pH of 5, the lactate and lactic acid components must be present in a ratio of 13.8:1. A good buffer is a mixture of a weak acid and its conjugate base present in near equal concentrations, and the buffered pH should be within 1 pH unit of the pK_a. Thus lactate buffer is a poor choice in this situation. With a pK_a of 4.76, the acetate buffer would be a better choice. ■

WORKED PROBLEM 3.8

What is the pH of a solution of 100 ml of 0.01 M H_3PO_4 and 100 ml of 0.01 M Na_3PO_4?

SOLUTION

First, determine what species are present in the solution. The two reagents will react to give a mixture composed of 0.01 mol NaH_2PO_4 (weak acid) and 0.01 mol Na_2HPO_4 (conjugate base).

WORKED PROBLEM 3.8 CONT

Using the Henderson-Hasselbalch equation,

$$\begin{aligned} pH &= pK_a + \log [A^-]/[HA] \\ pH &= pK_a + \log 0.01 \text{ mol}/0.01 \text{ mol} \\ pH &= pK_a + \log 1 \\ pH &= pK_a + 0, \text{ therefore} \\ pH &= pK_a. \end{aligned}$$

Because the pK_a of Na_2PO_4 is 7.2 (refer to Table 3.4), this is also the pH of the solution. ■

WORKED PROBLEM 3.9

During the fermentation of wine, a buffer system consisting of tartaric acid and potassium hydrogen tartrate is produced by a biochemical reaction. Assuming that at some time the concentration of potassium hydrogen tartrate is twice that of tartaric acid, calculate the pH of the wine. The pK_a of tartaric acid is 2.96.

SOLUTION

$$\begin{aligned} pH &= pK_a + \log \frac{[\text{hydrogen tartrate}]}{[\text{tartaric acid}]} \\ &= 2.96 + \log 2 \\ &= 2.96 + 0.30 = 3.26 \end{aligned}$$

■

WORKED PROBLEM 3.10

What is the pH of a solution prepared by mixing 150 mL of 0.10 M HCl with 300 mL of 0.10 M sodium acetate (NaOAc) and diluting the mixture to 1 L? The pK_a of acetic acid is 4.76.

SOLUTION

The amount of acid present in the solution is found by multiplying the volume of the solution, in milliliters, by M, the molarity of the solution; it is expressed in millimoles (mmol):

$$150 \text{ mL} \times 0.10 \text{ M} = 15 \text{ mmol acid}$$

The amount of sodium acetate is found using the same equation:

$$300 \text{ mL} \times 0.10 \text{ M} = 30 \text{ mmol base}$$

Each mole of HCl will consume 1 mol of sodium acetate and produce 1 mol of acetic acid. This will give 15 mmol of acetic acid with 15 mmol remaining of sodium acetate (i.e., 30 mmol – 15 mmol). Substituting these values into the Henderson-Hasselbalch equation gives

$$\begin{aligned} pH &= 4.76 + \log \frac{15}{15} \\ &= 4.76 + \log 1 \\ &= 4.76 \end{aligned}$$

Because the log term is a ratio of two concentrations, the volume factor can be eliminated and the molar amounts can be used directly. ■

WORKED PROBLEM 3.11

What would be the effect of adding an additional 50 mL of 0.10 M HCl to the solution in Problem 3.10 before dilution to 1 L?

SOLUTION

Using the same equation as in Problem 3.7, the amount of HCl would be

$$200 \text{ mL} \times 0.10 \text{ M} = 20 \text{ mmol}$$

which is also equal to the concentration of acetic acid.

The amount of sodium acetate would be

$$30 \text{ mmol} - 20 \text{ mmol} = 10 \text{ mmol}$$

Substituting into the Henderson-Hasselbalch equation gives

$$\begin{aligned} \text{pH} &= 4.76 + \log \frac{10}{20} \\ &= 4.76 + \log 0.50 \\ &= 4.76 - 0.30 \\ &= 4.46 \end{aligned}$$

■

WEAK ACIDS WITH MORE THAN ONE IONIZABLE GROUP Some molecules contain more than one ionizable group. Phosphoric acid (H_3PO_4) is a weak polyprotic acid; that is, it can donate more than one hydrogen ion (in this case, three hydrogen ions). During titration with NaOH (**Figure 3.21**) these ionizations occur in a stepwise fashion with 1 proton being released at a time:

$$H_3PO_4 \underset{}{\overset{pK_1 = 2.1}{\rightleftharpoons}} H^+ + H_2PO_4^- \underset{}{\overset{pK_2 = 7.2}{\rightleftharpoons}} H^+ + HPO_4^{2-} \underset{}{\overset{pK_3 = 12.3}{\rightleftharpoons}} H^+ + PO_4^{3-}$$

The pK_a for the most acidic group is referred to as pK_1. The pK_a for the next most acidic group is pK_2. The third most acidic pK_a value is pK_3.

At low pH most molecules are fully protonated. As NaOH is added, protons are released in the order of decreasing acidity, with the least acidic proton (with the largest pK_a value) ionizing last. When the pH is equal to pK_1, equal amounts of H_3PO_4 and $H_2PO_4^-$ exist in the solution.

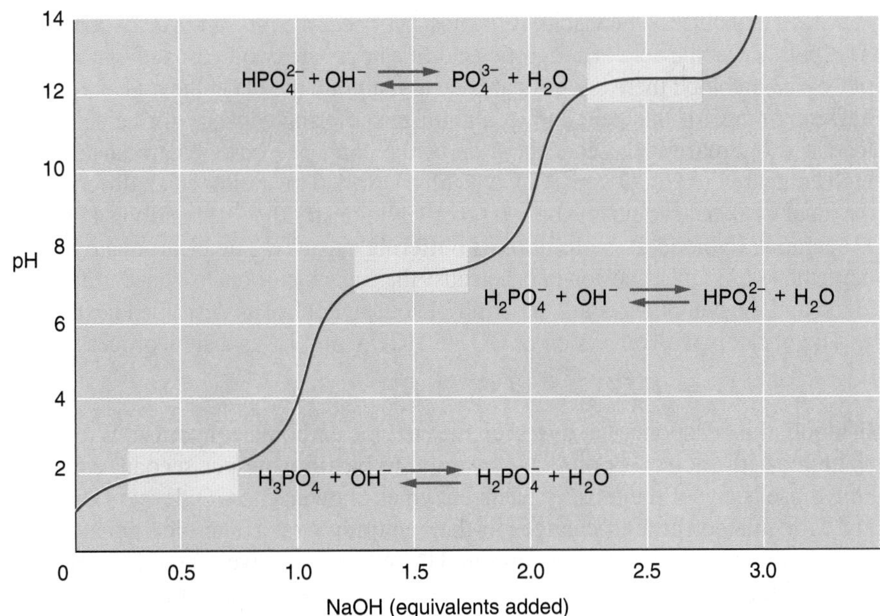

FIGURE 3.21
Titration of Phosphoric Acid with NaOH
Phosphoric acid (H_3PO_4) is a polyprotic acid that releases three protons sequentially upon titration with NaOH.

Amino acids are biomolecules that contain several ionizable groups. Like all amino acids, alanine contains both a carboxyl group and an amino group. At low pH, both of these groups are protonated. As the pH rises during a titration with NaOH, the acidic carboxyl group (COOH) loses its proton to form a carboxylate group (COO^-). The addition of more NaOH eventually causes the ionized amino group to release its proton:

$$H_3\overset{+}{N}-CH(CH_3)-COOH \overset{pK_1 = 2.3}{\rightleftharpoons} H^+ + H_3\overset{+}{N}-CH(CH_3)-COO^- \overset{pK_1 = 9.7}{\rightleftharpoons} H^+ + H_2N-CH(CH_3)-COO^-$$

Certain amino acids also possess side chains with ionizable groups. For example, the side chain of lysine possesses an ionizable amino group. Because of their structures, alanine, lysine, and the other amino acids can act as effective buffers at or near their respective pK_a values [e.g., lysine: $pK_1 = 2.0$, $pK_a = 9.0$, pK_3 (R group) $= 10.7$]. See Chapter 5 for further descriptions of the titration and buffering capacity of amino acids.

Physiological Buffers

The three most important buffers in the body are the bicarbonate buffer, the phosphate buffer, and the protein buffer. Each is adapted to solve specific physiological problems in the body.

BICARBONATE BUFFER Bicarbonate buffer, one of the more important buffers in blood, has three components. The first of these, carbon dioxide, reacts with water to form carbonic acid:

$$CO_2 + H_2O \rightleftharpoons \underset{\text{Carbonic acid}}{H_2CO_3}$$

Carbonic acid then rapidly dissociates to form H^+ and HCO_3^- ions:

$$H_2CO_3 \rightleftharpoons H^+ + \underset{\text{Bicarbonate}}{HCO_3^-}$$

Because the concentration of H_2CO_3 is very low in blood, the preceding equations may be simplified to

$$CO_2 + H_2O \rightleftharpoons H^+ + HCO_3^-$$

Recall that buffering capacity is greatest at or near the pK_a of the acid–conjugate base pair. Carbonic acid is a diprotic acid (it can donate two hydrogen ions) with a pK_1 of 6.3. In blood there is a critical need to maintain the pH at the high end of the buffering range of this acid and to maximize buffering capacity for acid. Therefore, it is optimal for the concentration of the conjugate base, bicarbonate, to be high compared to H_2CO_3 (or CO_2), typically 11 to 1. This ratio, which differs from the ideal weak acid:conjugate base ratio of 1:1, indicates that the bicarbonate buffer is operating in blood at the limit of its buffering capacity. The bicarbonate buffer, nevertheless, is effective for two reasons: the high bicarbonate concentration in blood and the components are under physiological control as described next.

The uncatalyzed conversion of CO_2 to HCO_3^- and H^+ is a slow process:

$$CO_2 + H_2O \rightleftharpoons H_2CO_3 \rightleftharpoons HCO_3^- + H^+$$

In blood, this reaction is catalyzed by the enzyme carbonic anhydrase. With rates as high as 10^6 molecules of CO_2 converted to bicarbonate per second, carbonic anhydrase is one of the most efficient enzymes known. The CO_2 level is kept low and is regulated through changes in the respiratory rate. The bicarbonate level stays high because the kidneys excrete H^+. When excessive amounts of HCO_3^- are produced, the kidney excretes the bicarbonate. As acid, a metabolic waste

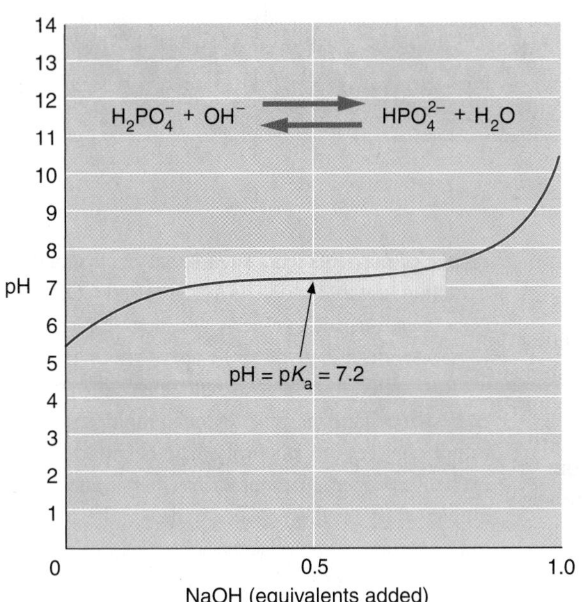

FIGURE 3.22

Titration of $H_2PO_4^-$ by Strong Base

The shaded band indicates the pH range over which the weak acid–conjugate base pair $H_2PO_4^-/HPO_4^{2-}$ functions effectively as a buffer.

product, is added to the body's bicarbonate system, the concentration of HCO_3^- decreases and CO_2 is formed. Because the excess CO_2 is exhaled, the ratio of HCO_3^- to CO_2 remains essentially unchanged.

PHOSPHATE BUFFER Phosphate buffer consists of the weak acid–conjugate base pair $H_2PO_4^-/HPO_4^{2-}$ (**Figure 3.22**):

$$\underset{\text{Dihydrogen phosphate}}{H_2PO_4^-} \rightleftharpoons H^+ + \underset{\text{Hydrogen phosphate}}{HPO_4^{2-}}$$

With pK_a 7.2, it would appear that phosphate buffer is an excellent choice for buffering the blood. Although the blood pH of 7.4 is well within this buffer system's capability, the concentrations of $H_2PO_4^-$ and HPO_4^{2-} in blood are too low to have a major effect. Instead, the phosphate system is an important buffer in intracellular fluids where its concentration is approximately 75 milliequivalents (mEq) per liter. Phosphate concentration in extracellular fluids such as blood is about 4 mEq/L. Because the normal pH of cell fluids is approximately 7.2 (the range is from 6.9 to 7.4), an equimolar mixture of $H_2PO_4^-$ and HPO_4^{2-} is typically present. Although cells contain other weak acids, these substances are unimportant as buffers. Their concentrations are quite low, and their pK_a values are significantly lower than intracellular pH. For example, lactic acid has a pK_a of 3.86.

PROTEIN BUFFER Proteins are a significant source of buffering capacity. Composed of amino acids linked together by peptide bonds, proteins contain several types of ionizable groups in side chains that can donate or accept protons. Protein molecules are powerful buffers because they are present in significant concentration in living organisms. For example, the oxygen-carrying protein hemoglobin is the most abundant biomolecule in red blood cells. Hemoglobin plays a major role in maintaining blood pH because of its structure and high cellular concentration. Also present in blood in high concentrations and buffering capacity are the serum albumins and other proteins.

KEY CONCEPT

The most important buffers in the body are the bicarbonate buffer (blood), the phosphate buffer (intracellular fluids), and the protein buffer.

QUESTION 3.4

Severe diarrhea is one of the most common causes of death in young children. One of the principal effects of diarrhea is the excretion of large quantities of sodium bicarbonate. In which direction does the bicarbonate buffer system shift under this circumstance? What is the resulting condition called?

Biochemistry IN PERSPECTIVE

Cell Volume Regulation and Metabolism

Is there a relationship between metabolism and cell volume? Living cells are in constant danger. Even the smallest changes in the balance of solutes between their interiors and their surroundings make cells vulnerable to potentially damaging changes in osmotic pressure. Any inability to manage osmotic balance can lead to distortions in shape and volume that compromise cell function. In multicellular organisms such as animals, however, individual cells are usually not exposed to significant fluctuations in the osmolarity of their surroundings. Instead, it is now realized, they are continuously challenged by internal variations that are created by normal metabolic processes. Routine tasks such as the uptake of nutrients (e.g., sugars, fatty acids, and amino acids), the excretion of waste products (e.g., H^+ and CO_2), and metabolic processes such as the synthesis and degradation of macromolecules (e.g., proteins and glycogen) cause osmotic imbalances.

Research efforts have revealed that cells possess several sophisticated mechanisms that together rapidly correct even the most minor changes in osmolarity. The best understood of these is the exchange of inorganic ions across membranes (**Figure 3A**). For example, when a cell is engaged in the synthesis of protein, the resulting reduction in the concentration of amino acids causes water to flow out of the cell. The cell responds by importing K^+, Na^+, and Cl^- (in exchange for HCO_3^-) through specialized membrane

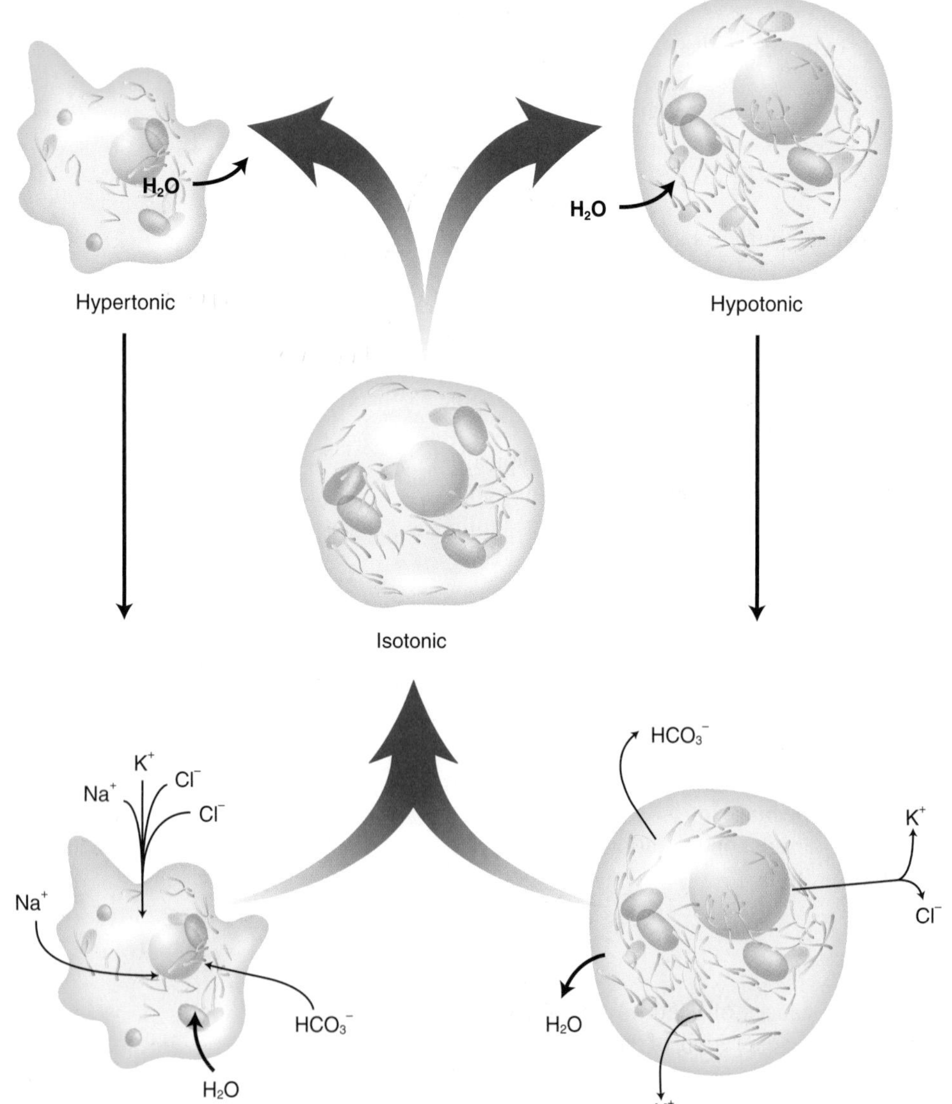

FIGURE 3A
Osmotic Pressure and Cell Volume Changes

Cells shrink when they are exposed to a hypertonic medium or when biochemical processes reduce the number of osmotically active particles. The cell's osmotic balance is restored when inorganic ions such as Na^+, K^+, and Cl^- enter via anion and cation channels and pumps. Ions can exchange with like-charged ions or be carried along a Na^+ or K^+ gradient. The cell returns to its normal volume as water then flows back into the cell. Cells swell when they are placed in a hypotonic medium or they increase their concentration of osmotically active particles through transport or the degradation of macromolecules. Osmotic balance is restored with the expulsion of inorganic ions, followed by the outflow of water.

Biochemistry IN PERSPECTIVE cont.

channel complexes. The osmotic gradient created by this process results in the flow of water into the cell, thus restoring the cell's normal volume. When protein is degraded, the opposite process occurs. The increased concentration of osmotically active amino acids causes the cell to swell. Ions (e.g., K^+, Cl^-, and HCO_3^-) followed by water then move across the plasma membrane out of the cell, and cell volume is restored.

Cell volume can also be controlled by the synthesis of quantities of osmotically active substances called **osmolytes**. For example, when confronted with osmotic stress the cells of some organisms produce large amounts of alcohols (e.g., sorbitol, see p. 246), amino acids, or amino acid derivatives such as taurine (see p. 461). Cells have also been observed to restore osmotic balance by synthesizing or degrading macromolecules such as glycogen. The precise means by which cells manage osmotic balance are not yet resolved. It is known that cell volume changes signaled by distortions of the cytoskeleton cause alterations in the expression of genes, some of which code for the synthesis of membrane channel proteins and osmolytes.

SUMMARY: Living cells are constantly engaged in managing the balance of solutes across their membranes. Nutrient uptake, waste product excretion, and metabolic processes such as macromolecule synthesis affect this balance. Any significant failure to correct imbalances can cause potentially lethal cell volume changes.

Chapter Summary

1. Water molecules (H_2O) are composed of two atoms of hydrogen and one of oxygen. Each hydrogen atom is linked to the oxygen atom by a single covalent bond. The oxygen-hydrogen bonds are polar, and water molecules are dipoles. One consequence of water's polarity is that water molecules are attracted to each other by the electrostatic force between the oxygen of one molecule and the hydrogen of another. This attraction is called a hydrogen bond.
2. Noncovalent bonds are relatively weak and, therefore, easily disrupted. They play a vital role in determining the physical and chemical properties of water and biomolecules. Ionic interactions occur between charged atoms or groups. Although each hydrogen bond is not especially strong in comparison to covalent bonds, large numbers of them have a significant effect on the molecules involved. Van der Waals forces, either attractive or repulsive, occur between permanent and/or induced dipoles.
3. Water has an exceptionally high heat capacity. Its boiling and melting points are significantly higher than those of compounds of comparable structure and molecular weight. Hydrogen bonding is responsible for this anomalous behavior.
4. Water is a remarkable solvent. Water's dipolar structure and its capacity to form hydrogen bonds enable it to dissolve many ionic and polar substances.
5. Most water molecules in living organisms are structured; that is, they are noncovalently associated with macromolecules and membrane surfaces. The networks of water molecules that form act as bridges between macromolecules in the densely crowded cytoplasm. Cytoplasm has the properties of a gel, a semisolid viscoelastic substance that resists flow and stores mechanical energy. Gels can undergo reversible transitions to a liquid or sol state.
6. Hydrophobic molecules are virtually insoluble in water. When nonpolar molecules enter an aqueous environment, they form droplets surrounded by water molecules that rearrange into their most energetically favorable configuration.
7. Amphipathic molecules contain both polar and nonpolar groups. Fatty acids are amphipathic molecules that form structures called micelles when they are placed in water.
8. Several physical properties of liquid water change when solute molecules are dissolved. The most important of these for living organisms is osmotic pressure, the pressure that prevents the flow of water across cellular membranes. Macromolecules have little direct effect on cellular osmolarity. The large number of ionizable groups on these molecules attracts ions of opposite charge. The structured water network that surrounds macromolecules such as proteins tends to exclude Na^+ because of its relatively large hydrated volume. The charge asymmetry across the cell membrane (negative on the inside and positive on the outside) creates an electrical gradient called a membrane potential.
9. Liquid water molecules have a limited capacity to ionize to form a hydrogen ion (H^+) and a hydroxide ion (OH^-). When a solution contains equal amounts of H^+ and OH^- ions, it is said to be neutral. Solutions with an excess of H^+ are acidic, whereas those with a greater number of OH^- are basic. Because organic acids do not completely dissociate in water, they are referred to as weak acids. The acid dissociation constant K_a is a measure of the strength of a weak acid. Because K_a values vary over a wide range, pK_a values ($-\log K_a$) are used instead.
10. The hydrogen ion is one of the most important ions in biological systems. The pH scale conveniently expresses

hydrogen ion concentration. pH is defined as the negative logarithm of the hydrogen ion concentration.

11. Because hydrogen ion concentration affects living processes so profoundly, it is not surprising that regulating pH is a universal and essential activity of living organisms. Hydrogen ion concentration is typically kept within narrow limits. Because buffers combine with H^+ ions, they help maintain a relatively constant hydrogen ion concentration. The ability of a solution to resist pH changes is called buffering capacity. Most buffers consist of mixtures of a weak acid and its conjugate base.

Take your learning further by visiting the **companion website** for Biochemistry at **www.oup.com/us/mckee** where you can complete a multiple-choice quiz on water to help you prepare for exams.

Suggested Readings

Ball, P., Water as an Active Constituent in Cell Biology, *Chem. Rev.* 108(1):74–108.

Diamond, J., *Collapse: How Societies Choose to Fail or Succeed*, Penguin, New York, 2005.

Gerstein, M., and Levitt, M., Simulating Water and the Molecules of Life, *Sci. Am.* 279(5):100–105, 1998.

Ho, M-W., *Living Rainbow H_2O*, World Scientific, Singapore, 2012.

Kemsley, J., Hydrogen Bonds Visualized, *Chem. Engineer. News* 91(39):5, 2013.

Kollipara, P., Diving Deeper into Souring Oceans, *Chem. Engineer. News* 91(41):33–35, 2013.

Lang, F., and Waldegger, S., Regulating Cell Volume, *Am. Sci.* 85:456–463, 1997.

Leterrier, J.-F., Water and the Cytoskeleton, *Cell Mol. Biol. (Noisy-le-Grand)* 47(5):901–923, 2001.

Pollack, G. H., *Cells, Gels and the Engine of Life*, Ebner and Sons, Seattle, Washington, 2001.

Key Words

acid, *92*
acidosis, *93*
alkalosis, *93*
amphipathic molecule, *87*
base, *92*
buffer, *94*
conjugate base, *92*
covalent bond, *79*
dipole, *78*
electrostatic interaction, *79*
hydrogen bond, *78*
hydrophobic interaction, *86*
hypertonic solution, *89*
hypotonic solution, *89*
isotonic solution, *89*
Le Chatelier's principle, *94*
London dispersion force, *81*
membrane potential, *91*
micelle, *87*
osmolyte, *103*
osmosis, *88*
osmotic pressure, *88*
pH scale, *93*
polar, *78*
salt bridge, *80*
solvation sphere, *84*
van der Waals force, *81*
weak acid, *92*
weak base, *92*

Review Questions

These questions are designed to test your knowledge of the key concepts discussed in this chapter before moving on to the next chapter. You may like to compare your answers to the solutions provided in the back of the book and in the accompanying Study Guide.

1. Define the following terms:
   a. polar
   b. hydrogen bond
   c. electrostatic interaction
   d. salt bridge
   e. dipole
2. Define the following terms:
   a. heat of fusion
   b. solvation sphere
   c. amphipathic
   d. micelle
   e. hydrophobic effect
3. Define the following terms:
   a. osmosis
   b. osmotic pressure
   c. isotonic solution
   d. membrane potential
   e. hydronium ion
4. Define the following terms:
   a. acid
   b. base
   a. weak acid
   d. weak base
   e. conjugate base

5. Define the following terms:
   a. buffer
   b. acidosis
   c. alkalosis
   d. pH
   e. pK_a
6. Which of the following are acid–conjugate base pairs?
   a. H_2CO_3, CO_3^{2-}
   b. $H_2PO_4^-$, PO_4^{3-}
   c. HCO_3^-, CO_3^{2-}
   d. H_2O, OH^-
7. What is the hydrogen ion concentration in a solution at pH 8.3?
8. Describe how you would prepare a 0.1 M phosphate buffer with a pH of 7.2. What ratio of conjugate base to acid would you use?
9. What is the osmolarity of a 1.3 M solution of sodium phosphate (Na_3PO_4)? Assume 85% ionization for this solution.
10. A dialysis bag containing a 3 M solution of the sugar fructose is placed in the following solutions. In each case, give the direction in which water flows.
    a. 1 M sodium lactate
    b. 3 M sodium lactate
    c. 4.5 M sodium lactate

H O
CH3—C—C—O⁻ Na⁺
OH

11. A water molecule can form hydrogen bonds with a maximum of four other water molecules. Draw this interaction.
12. How many hydrogen bonds can form when methyl alcohol is substituted for water? Draw the interaction.
13. A solution of 0.10 M acetic acid and 0.10 M sodium acetate has a pH of 4.76. Determine the pK_a of acetic acid. What is the K_a for acetic acid?
14. What interactions occur between the following molecules and ions?
    a. water and ammonia
    b. lactate and ammonium ion
    c. benzene and octane
    d. carbon tetrachloride and chloroform
    e. chloroform and diethylether
15. A solution containing 56 mg of a protein in 30 mL of distilled water exerts an osmotic pressure of 0.01 atm at $T = 25°C$. Determine the molecular weight of the unknown protein.
16. Which of the following molecules would you expect to have a dipole moment?
    a. CCl_4
    b. $CHCl_3$
    c. H_2O
    d. CH_3OCH_3
    e. CH_3CH_3
    f. H_2
17. Which of the following molecules would you expect to form micelles?
    a. NaCl
    b. CH_3COOH
    c. $CH_3COO^-NH_4^+$
    d. $CH_3(CH_2)_{10}COO^-Na^+$
    e. $CH_3(CH_2)_{10}CH_3$
18. Bicarbonate is one of the main buffers of the blood, and phosphate is the main buffer of the cells. Why might this be?
19. Describe how you can increase the buffering capacity of a 0.1 M acetate buffer.
20. If the total concentration of a buffer is known (and not the individual concentrations of the weak acid and its conjugate base), can the pH of the buffer be calculated?
21. Carbonic acid has the following pK_a values: $pK_{a1} = 6.4$ and $pK_{a2} = 10.2$. Indicate which species are present at each of the following pH values: 6.4, 8, 13.
22. Which of the following molecules or ions are weak acids? Explain.
    a. HCl
    b. $H_2PO_4^-$
    c. CH_3COOH
    d. HNO_3
    e. HSO_4^-
23. Which of the following species can form buffer systems?
    a. NH_4^+ Cl^-
    b. CH_3COOH, HCl
    c. CH_3COOH, $CH_3COO^-Na^+$
    d. H_3PO_4, PO_4^{3-}
24. What effect does hyperventilation have on blood pH?
25. Is it possible to prepare a buffer consisting of only carbonic acid and sodium carbonate?
26. Calculate the ratio of dihydrogen phosphate to hydrogen phosphate in blood at pH 7.4. The K_a is 6.3×10^{-8}.
27. Calculate the pH of a solution prepared by mixing 300 mL of 0.25 M sodium hydrogen ascorbate and 150 mL of 0.2 M HCl. The pK_{a1} of ascorbic acid is 4.04.
28. What is the pH of a solution that is 1×10^{-8} M in HCl?
29. Compounds such as the sugar trehalose are used as compatible solutes (i.e., water replacements) in desiccated organisms. What are the requirements for a substance that is to be used in this manner?
30. Calculate the pH for a mixture of one mole of benzoic acid and one mole of sodium benzoate. The pK_a of benzoic acid is 4.2.
31. Detergents that are good micelle formers often have strong antibacterial properties. Considering the structure of the cell membrane, suggest a reason for this antibacterial action.
32. Determine the pH of a solution composed of 1 M acetic acid and 1 M sodium acetate.
33. What would be the pH of the solution in Question 32 if 1 mL of 1 M HCl is added?
34. What would be the pH of 1 L of water if 1 mL of 1 M HCl is added?
35. Many molecules are polar, yet they do not form significant hydrogen bonds. What is so unusual about water that hydrogen bonding becomes possible?

Fill in the Blank

36. ____________________ is responsible for the unusual thermal properties of ammonia.
37. Bicarbonate buffer in blood involves the dissociation of ____________________.
38. ____________________ is a rule concerning chemical processes that states that stress applied to a reaction at equilibrium results in the displacement of the equilibrium so as to relieve the stress.
39. The conjugate acid of water is ____________________.
40. When blood pH rises above 7.45, ____________________ results.
41. A buffer is composed of a weak acid and its ____________________ base.
42. Osmotically active substances are called ____________________.
43. A semisolid viscoelastic substance that resists flow and stores mechanical energy is called a ____________________.
44. The most important ion in biological systems is the ____________________ ion.
45. The ability of a solution to resist pH changes is called its ____________________.

Short Answer

46. Some of the water on earth was formed by the interaction of iron and silicon oxides and methane. Write the balanced equation for the reaction of ferric oxide and methane.
47. Examine the elements in the second row of the periodic table. Note that HF, H_2O, NH_3, and CH_4 all are sp^3 hybridized. The first three are all capable of hydrogen bonding. Methane, on the other hand, is not capable of hydrogen bonding. Explain.
48. Why is ice less dense than water?
49. Chloroform ($HCCl_3$) is capable of weak hydrogen bonding. Explain.
50. Considering the hydration spheres of sodium and potassium, would you expect lithium ions to easily penetrate the cell or remain outside the cell?

Thought Questions

These questions are designed to reinforce your understanding of all of the key concepts discussed in the book so far, including this chapter and all of the chapters before it. They may not have one right answer! The authors have provided possible solutions to these questions in the back of the book and in the accompanying Study Guide for your reference.

51. Gelatin consists of collagen that is suspended and hydrated by water. What happens to gelatin when NaCl is sprinkled over its surface?
52. A well is punched into a slab of gelatin and a solution of equimolar amounts of NaCl and KCl is placed in it. After an hour the concentration of Na^+ and K^+ present in the well is measured. Which ion, Na^+ or K^+, would be present in greater abundance in the well? Explain your answer.
53. The potassium and sodium ions have identical charges, yet the hydrated volume of Na^+ is 3.4 times that of K^+. Explain.
54. A cell's ATP-driven sodium-potassium pump fails. Will the cell undergo crenation of hemolysis?
55. Many fruits can be preserved by candying. The fruit is immersed in a highly concentrated sugar solution, and then the sugar is allowed to crystallize. How does the sugar preserve the fruit?
56. Explain why ice is less dense than water. If ice were not less dense than water, how would the oceans be affected? How would the development of life on Earth be affected?
57. Why can't seawater be used to water plants?
58. Explain how the acids produced in metabolism are transported to the liver without greatly affecting the pH of the blood.
59. The pH scale is valid only for water. Why is this so?
60. Gelatin is a mixture of protein and water that is mostly water. Explain how the water-protein mixture becomes a solid.
61. Water has been described as the universal solvent. If this statement were strictly true, could life have arisen in a water medium? Explain.
62. Alcohols (ROH) are structurally similar to water. Why are alcohols not as powerful a solvent as water for ionic compounds? [*Hint*: Methanol is a better solvent for ionic compounds than is propanol.]
63. During stressful situations, some cells in the body convert glycogen to glucose. What effect does this conversion have on cellular osmotic balance? Explain how cells handle this situation.
64. Would you expect a carboxylic acid group within the water-free interior of a protein to have a higher or lower K_a than it would have if it occurred on the protein's surface, where it is hydrated?
65. The strength of ionic interactions is weaker in water than in an anhydrous medium. Explain how water weakens these interactions.
66. In many cells that can survive severe dehydration, certain sugars replace water. These sugars interact with and protect membrane surfaces and prevent protein aggregation. What

structural feature of the sugar molecules is responsible for this phenomenon?

67. Consider the following ion series:

$$Mg^{2+} > Ca^{2+} > Na^{+} > K^{+} > Cl^{-} > NO_3^{-}$$

Ions to the left are more strongly hydrated than ions to the right. Indicate whether ions such as Mg^{2+} and Cl^- would move easily into the structured water that is associated with cellular macromolecules.

68. Suggest a structure for the micelle formed by dissolving the following molecule in water.

$$^{+}Na\ ^{-}OOC—(CH_2)_{16}—COO^{-}\ Na^{+}$$

69. Pure sugars are often crystalline solids. Frequently, the process used to concentrate aqueous solutions of sugars produces syrups rather than crystals. Explain.

70. Sketch the titration curve of the amino acid tyrosine starting with the following structure.

H—O—(C₆H₄)—CH$_2$—CH(+NH$_3$)—C(=O)—OH

The pK_a values are as follows: amino group = 9.11, carboxyl group = 2.2, and side chain hydroxyl group = 10.07.

71. The heat absorbed or liberated by a substance (q) can be calculated using the following equation: $q = mc\Delta T$, where m is the mass in grams, c is the heat capacity per unit mass, and ΔT is the change in temperature. Use the following values to calculate the energy required to convert one gram of ice at 0°C to one gram of steam at 100°C. The heat capacity of water is 4.25 J/g · °C. The heat of fusion (the amount of heat required to change a solid into a liquid at its melting point) of ice is 335 J/g. The heat of vaporization of water is 2258 J/g.

72. Calculate the energy required to convert solid hydrogen sulfide (H_2S) to a gas. The heat capacity of H_2S is 1.03 J/g · °C. Its heats of fusion and vaporization are 69.9 and 549 J/g, respectively. Compare your answer with the value for water.

73. Potassium chloride (KCl) is slightly soluble in methyl alcohol. Draw the hydration sphere of the potassium ion.

74. Consider the following compound:

$$Cl—CH_2—(CH_2)_{10}—CH_2—COOH$$

Would this molecule form into a lipid bilayer and, if so, what would it look like?

75. Water forms stronger hydrogen bonds than ammonia. Suggest a reason for this.

76. When acetic acid ionizes under normal conditions, both the departing proton and the acetate anion are solvated by water molecules. In the absence of water, would you expect the pK_a of acetic acid to be larger or smaller? Explain your answer.

CHAPTER

4 Energy

Energy Transformation Wild horses are herbivores that consume a diet of grass. They transform the chemical bond energy in food molecules into the energy needed to escape predators such as mountain lions. Horses gallop at speeds that average 25 to 30 mph, with a top speed of 55 mph during short sprints.

OUTLINE

Energy and Life's Deep, Dark Secrets

Chris and Steve are microbiologists on a mission: to retrieve rock from the sizzling depths of a South African gold mine (**Figure 4.1**). Wearing coveralls, boots, and hard hats fitted with lamps, and carrying their sterilized tools and specimen bags and a 4-L supply of water, the two scientists descend along with the miners 3 km into the hot, oppressive darkness of the deepest excavation on Earth. Chris and Steve then work quickly to remove rock samples. They are limited, by the threat of heat stress, to only 4 hours in the mine because mine temperatures often exceed 49°C. Work is often interrupted by the removal of liters of sweat from boots and gloves. Their goal is to find microbes with unique biochemical capabilities (e.g., radiation resistance), most notably for use in *bioremediation* (microbes used to remove toxic substances from contaminated soil or water).

How and why did Chris and Steve find their way into such a hostile work place? Recent research efforts have overturned the longstanding conviction that the biosphere is limited largely to Earth's surface. The hot and toxic planet's crust is not only suffused with life, but also its biochemically diverse microbial inhabitants have vital roles in the biogeochemical cycles that maintain the health of the entire biosphere.

Biogeochemical cycles are pathways driven by solar and geothermal energy in which chemical elements move throughout Earth's biotic (biosphere) and abiotic compartments [*lithosphere* (crust and upper mantle), *hydrosphere* (surface water, such as lakes, rivers, and oceans), and atmosphere]. Because Earth's supply of elements is fixed and finite, the cyclic transport and chemical transformations of elements are critical for living organisms. In the carbon cycle, for example, CO_2 is incorporated into organic biomolecules by light-driven photosynthesis and animals use O_2-based respiration to convert consumed plant biomolecules back into CO_2, which is released (along with CO_2 generated from the decomposition of dead organisms and fossil fuel combustion) into the atmosphere. Carbon is also incorporated into carbonate minerals in a solar energy–driven process in which CO_2 and water react with silicate minerals

FIGURE 4.1
To Hell and Back

In their investigations of some of the Earth's most inaccessible microorganisms, scientists must endure searing heat in deep mines where rock temperatures can reach 60°C (140°F) and the ever-present threat of dehydration. Research into the unique biochemistry of organisms that survive these conditions has revealed that some microbes generate energy by reducing geologically produced sulfate with hydrogen released when water molecules are split by radiation emitted from radioactive elements such as uranium and thorium.

Energy and Life's Deep, Dark Secrets cont.

(e.g., $CaSiO_3$) to form calcium carbonate ($CaCO_3$) and silicon dioxide (SiO_2). As the result of erosion, $CaCO_3$ and other minerals are then washed into the oceans where they become buried in sediments that also contain the calcium carbonate-containing shells of a variety of dead marine organisms along with organic detritus formed from decomposing organisms. With life spans of thousands of years, subseafloor microbes slowly decompose this organic matter, releasing CO_2 into ocean water for reuse in photosynthesis or evaporation into the atmosphere.

Although it is well known that CO_2 is also released into the atmosphere via volcanism, the role of subsurface microorganisms has only been appreciated recently. Volcanoes are the inevitable result of the slow recycling of oceanic crust tectonic plates that is driven by geothermal energy (heat energy partially generated by radioactive element decay). New oceanic crust is formed at a mid-ocean ridge (the boundary between crust and continental plates) by the upwelling of magma (molten rock) from the underlying mantle. As the magma seeps through the cracks between the two plates, it solidifies and displaces existing oceanic crust. At the opposite end of each of the two enlarging oceanic plates, there is a collision with a continental plate in which the denser crust slab and any surface sediment will sink (subduct) beneath the continental plate. Trapped by the subducting slab, the sediment microbes continue to release CO_2 from any remaining organic matter until the slab's temperature becomes incompatible with life. As subduction proceeds, crust melts to form magma as it is forced deeper into the Earth. Some of the melted rock flows into magma chambers of volcanoes near crustal plate edges. The intense heat within the mantle converts $CaCO_3$ back into $CaSiO_3$ and CO_2, which then escapes with other volcanic gases into the atmosphere, thereby completing the cycle.

Overview

ENERGY! IT IS CERTAINLY ESSENTIAL TO LIFE, BUT WHAT IS IT AND WHY IS IT SO VITAL TO LIVING ORGANISMS? ENERGY IS *THE* BASIC CONSTITUENT OF the universe. The relationship between matter and its energy equivalent is defined by Einstein's famous equation $E = mc^2$. In other words, energy and matter are interconvertible: matter is condensed energy. The total energy (E) in joules ($kg{\cdot}m^2/s^2$) in a particle is equal to the mass (m) in kilograms of the particle multiplied by the speed of light ($c = 3.0 \times 10^8$ m/s) squared. Energy is more commonly defined, however, as the capacity to do work. **Work** is organized molecular motion that causes the displacement or movement of an object by the application of force and results in a specific physical change (e.g., the force of flowing water turns the blades of a turbine in a hydroelectric plant). Energy comes in many interconvertible forms: gravitational, nuclear, radiant, chemical, mechanical, electrical, and thermal (heat).

Electromagnetic radiation, electrical energy, and chemical energy are high-quality energy sources, whereas heat is lower-quality energy. Compare, for example, the work potential of electricity as it enters a building carried by electrical wire and the heat energy radiating from a lightbulb. In the right circumstances, however, heat can be useful energy. For example, heat flow from the Earth's core to the mantle and within the mantle by the movement of magma is an important contributing force in biogeochemical cycles.

Energy flows continuously through the biosphere. Originating as either solar or geothermal energy, it drives the flow of matter (e.g., nutrients) and biochemical processes in living organisms. There are three energy-generating mechanisms used by living organisms: photosynthesis, chemoorganotrophy, and chemolithotrophy. Photosynthesis (Chapter 13) is a process that converts light energy into chemical bond energy (ATP). Chemoorganotrophs and chemolithotrophs are heterotrophs that generate ATP by oxidizing organic and inorganic compounds, respectively. All of these methods of capturing and transforming energy involve oxidation-reduction (redox) reactions (described in Chapter 9) in which electrons are transferred from an **electron donor** to an **electron acceptor**. Living organisms use the energy provided by ATP to power thousands of molecular machines. Work performed by these machines includes maintenance of concentration gradients and the synthesis of biomolecules.

The investigation of energy transformations that accompany physical and chemical changes in matter is called **thermodynamics**. **Bioenergetics**, a branch of thermodynamics, is the study of energy transformations in living organisms. It is especially useful in determining the direction and extent to which specific biochemical reactions occur. These reactions are affected by three factors. Two of these, **enthalpy** (total heat content) and **entropy** (disorder), are related to the first and second laws of thermodynamics, respectively. The third factor, called **free energy** (energy available to do chemical work and a measure of the spontaneity of chemical reactions), is explained by a mathematical relationship between enthalpy and entropy.

The chapter begins with some basic thermodynamic concepts and their relationship to biochemical reactions. This is followed by a discussion of free energy, a useful measure of the spontaneity of chemical reactions. The chapter ends with the structure and function of ATP and other high-energy compounds.

4.1 THERMODYNAMICS

The modern concept of energy is an invention of the Industrial Revolution. In the nineteenth century, investigations of the relationship between mechanical work and heat led to the discovery of a set of rules, called the *laws of thermodynamics*, that describe energy transformations:

1. **The first law of thermodynamics:** The total amount of energy in the universe is constant. Energy can be neither created nor destroyed, but it can be transformed from one form into another.
2. **The second law of thermodynamics:** The disorder of the universe always increases. Chemical and physical processes occur spontaneously only when the disorder of the universe increases.
3. **The third law of thermodynamics:** As the temperature of a perfect crystalline solid approaches absolute zero (0 K), disorder approaches zero.

The first two laws are powerful tools that biochemists use to investigate the energy transformations in living systems.

Thermodynamics considers heat and energy transformations in a "universe" composed of a system and its surroundings (**Figure 4.2**). A system is defined according to the interests of the investigator: an entire organism, or a single cell, or a reaction occurring in a flask. In an *open system* matter and energy are exchanged between the system and its surroundings. If only energy can be

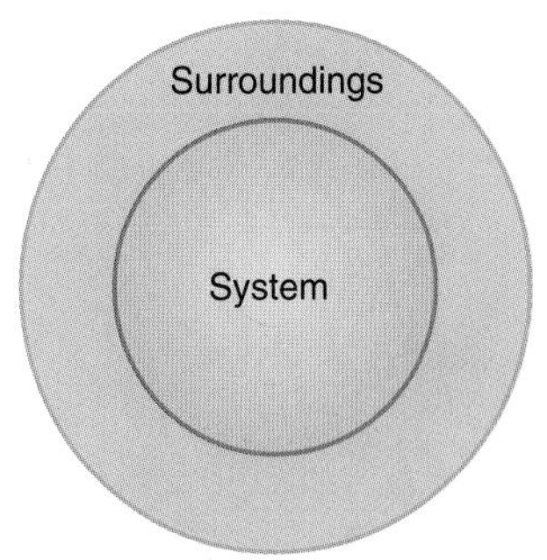

FIGURE 4.2
A Thermodynamic Universe
A universe consists of a system and its surroundings.

exchanged with the surroundings, then the system is said to be *closed*. Living organisms, which consume nutrients from their surroundings and release waste products into it, are open systems.

Thermodynamic functions include enthalpy, entropy, and free energy. Knowledge of these functions enables biochemists to predict whether a process is spontaneous (*thermodynamically favorable*). Spontaneity alone does not indicate that a reaction will occur, only that it can occur under the right set of conditions. Reactions occur only if there is sufficient energy available to the system. These reactions are described as *kinetically favorable.*

Several thermodynamic properties are *state functions*, which depend only on their initial and final states. State functions are independent of the pathway taken to get from the initial state to the final state. For example, the energy contained in a glucose molecule is inherent in its molecular structure. How the energy of glucose is distributed when it is degraded, however, is not fixed but is governed by the system or pathway undergoing change. For example, living cells use some of the energy in glucose molecules to perform cellular work such as muscle contraction. The remainder is released as disordered heat energy. If glucose molecules are instead ignited in a laboratory dish, the overall reaction is the same. However, all of the chemical bond energy in the glucose is transformed directly into heat, and little or no measurable work is performed. The energy content of the glucose molecules is the same in each process. The work accomplished by each process is different. In other words, work and heat are not state functions; their values vary with the pathway.

The exchange of energy between a system and its surroundings can occur in only two ways. Heat (q), random molecular motion, may be transferred to or from the system. Alternatively, the system may do work (w) on its surroundings or have work done on it by its surroundings. Energy is transferred as heat when the system and its surroundings are at different temperatures. Energy is transferred as work when an object is moved by force.

First Law of Thermodynamics

The first law of thermodynamics expresses the relationship between the internal energy (E) of a closed system and the heat (q, or disorganized motion) and work (w, or organized motion) transferred between the system and its surroundings (**Figure 4.2**). It is an alternative statement of the law of *conservation of energy*: the total energy of an isolated system (e.g., our universe) is constant. In other words, for a closed system,

$$\Delta E = q + w \tag{1}$$

where ΔE = the change in energy of the system
q = the heat absorbed or released by the system
w = the work done by or to the system

Chemists define enthalpy (H), a measure of the system's internal energy:

$$H = E + PV \tag{2}$$

where PV = pressure-volume work, that is, the work done on or by a system that involves changes in pressure and volume

In biochemical systems pressure is nearly constant and volume changes are negligible. Changes in enthalpy are then essentially equal to changes in internal energy:

$$\Delta H = \Delta E \tag{3}$$

If ΔH is negative ($\Delta H < 0$), the reaction or process gives off heat and is referred to as **exothermic**. If ΔH is positive ($\Delta H > 0$), heat is absorbed from the surroundings, and the process by which it is emitted is called **endothermic**. In **isothermic** processes ($\Delta H = 0$), heat is not exchanged with the surroundings.

Equation (3) indicates that the total energy change of a biological system is equivalent to the heat evolved or absorbed by the system. Because the enthalpy of a reactant or product is a state function (independent of pathway), then the enthalpy change for a particular reaction forming that substance can be used to calculate the ΔH of any other reaction involving that substance. If the sum of the ΔH values ($\Sigma\Delta H$) for both the reactants and the products is known, then the enthalpy change for the reaction can be calculated using the following equation:

$$\Delta H_{\text{reaction}} = \Sigma\Delta H_{\text{products}} - \Sigma\Delta H_{\text{reactants}} \tag{4}$$

The standard enthalpy of formation per mole (25°C, 1 atm), symbolized by ΔH_f°, is commonly used in enthalpy calculations; H_f° is the energy evolved or absorbed when 1 mol of a substance is formed from its most stable elements under standard conditions. Note that Equation (4) cannot predict the direction of any chemical reaction. It determines only the heat flow. Problems 4.1 and 4.2 give standard enthalpy calculations.

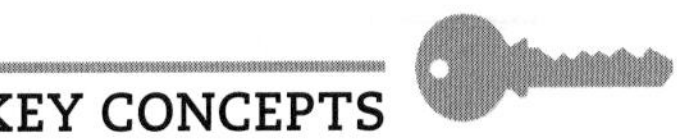

KEY CONCEPTS

- At constant pressure, a system's enthalpy change ΔH is equal to the flow of heat energy.
- If ΔH is negative, the reaction or process is exothermic. If ΔH is positive, the reaction or process is endothermic. In isothermic processes no heat is exchanged with the surroundings.

WORKED PROBLEM 4.1

Given the following ΔH_f° values, where ΔH_f° is the energy change required to produce a compound from its elements, calculate ΔH_f° for the reaction

$$6CO_2 + 6H_2O \rightarrow C_6H_{12}O_6 + 6O_2$$

	ΔH_f°	
	kcal/mol	kJ/mol
$C_6H_{12}O_6$	−304.7	−1274.9
CO_2	−94.0	−393.3
H_2O	−68.4	−286.2
O_2	0	0

The units in the table have the following definitions: 1 kcal is the energy required to raise the temperature of 1000 g of water 1°C; the joule (J) is a unit of energy that is replacing the calorie (cal) in scientific usage (1 cal = 4.184 J).

SOLUTION

The total enthalpy for a reaction is equal to the sum of enthalpy values of the products minus those of the reactants.

$$\begin{array}{lcl} 6CO_2 + \quad 6H_2O & \rightarrow & C_6H_{12}O_6 + 6O_2 \\ 6(-393.3) + 6(-286.2) & & -1274.9 + 6(0) \\ -2359.8 + (-1717.2) & & -1274.9 \end{array}$$

$$\Delta H = -1274.9 - (-4077.0) = 2802.1 \text{ kJ/mol}$$

The positive ΔH indicates that the reaction is endothermic. ■

WORKED PROBLEM 4.2

Given the following data, calculate the change in mileage of a car when it is converted from using gasoline (*n*-octane) as fuel to ethanol. Assume that your car has a mileage rating of 7.92 mi/L (30 mi/gal) of gasoline and that it burns gasoline with 100% efficiency to yield only CO_2 and H_2O.

ΔH_f (kj/mol)		Density (g/mL)		Mass (g/mol)
CO_2	−393.5	CH_3CH_2OH	0.80	46.1
H_2O	−285.8	Octane	0.70	114.2
CH_3CH_2OH	−277.7			
Octane	−250.1			

SOLUTION

1. Calculate the energy evolved per mole for each molecule.

$$\begin{array}{lllll} CH_3CH_2OH & + \ O_2 & \rightarrow & 2CO_2 & + \ 3H_2O \\ -277.7 & 0 & & 2(-3935) & 3(-285.8) \\ -277.7 & & & -787.0 & -857.4 \end{array}$$

$$\Delta H = -1644.4 - (-277.7) = -1366.7 \text{ kj/mol}$$

$$\begin{array}{lllll} C_8H_{18} & + \ O_2 & \rightarrow & 8CO_2 & + \ 9\,H_2O \\ -250.1 & 0 & & 8(-393.5) & 9(-285.8) \\ -250.1 & & & -3148.0 & -2572.2 \end{array}$$

$$\Delta H = -5720.2 - (-250.1) = -5470.1 \text{ kJ/mol}$$

2. Calculate the number of moles per liter for each fuel burned.

CH_3CH_2OH (0.80 g/mL)(1000 mL)/46.1 g/mol = 17.4 mol
n-Octane (0.70 g/ml)(1000 ml)/114.2 g/mol = 6.1 mol

3. Calculate the amount of energy produced per liter of each fuel and the change in mileage when the fuel is switched from gasoline to ethanol.

CH_3CH_2OH (17.4 mol)(1366.7 kJ/mol) = −23,780.6 kJ
n-Octane (6.1 mol)(5470.1 kJ/mol) = −33,367.6 kJ
Relative heat production (ethanol/*n*-octane) = 23,780.6kJ/−33,367.6kJ
= 0.7

Assuming that a car using gasoline has a mileage rating of 7.9 mi/L, then switching to ethanol would give (0.7)(7.9 mi/L) = 5.5 mi/L or 21.0 mpg. ■

Second Law of Thermodynamics

The first law accounts for the energy changes that can occur during a process, but it cannot predict to what extent the process will occur. In some circumstances, whether processes occur appears obvious: for example, the behavior of ice at room temperature or gasoline in an internal combustion engine. Experience tells us that ice melts at temperatures above 0°C and that gasoline molecules can be converted to energy in the presence of oxygen to form CO_2 and H_2O. Physical or chemical changes that occur with the release of energy are said to be **spontaneous**. Nonspontaneous processes are those that occur when a constant input of energy is required to support a change. Experience convinces us that certain processes will not occur: ice will not form at temperatures above 0°C, and gasoline molecules are not formed from an engine's exhaust fumes. In other words, we intuitively understand that there is a direction to these processes and that we can easily predict their outcome. When experience cannot be relied on to allow us to make predictions concerning spontaneity and direction, the second law can be used. According to the second law, all spontaneous processes occur in the direction that increases the total disorder of the universe (a system and its surroundings) (**Figure 4.3**). As a result of spontaneous processes, matter and energy become more disorganized. Gasoline molecules, for example, are hydrocarbons in which carbon atoms are linked in an orderly arrangement. When gasoline burns, the carbon atoms in the gaseous products are randomly dispersed (**Figure 4.4**). Similarly, the energy that is released as gasoline burns becomes more disordered; it becomes less concentrated and less useful. In a car engine, increased gas pressure in the cylinders drives the pistons and causes the car to move. When we compare the chemical energy in the gasoline molecules and the kinetic energy that moves the car, it becomes apparent that a significant amount of energy does no useful work. Rather, it is dissipated (dispersed) into the surroundings, producing a hot engine and exhaust fumes.

The degree of disorder of a system is measured by the state function called **entropy** (S). The more disordered a system is, the greater is its entropy value. According to the second law, the entropy change of the universe is positive for every spontaneous process. The increase may take place in any part of the universe, either the system or its surroundings (ΔS_{sys} or ΔS_{surr}):

$$\Delta S_{univ} = \Delta S_{sys} + \Delta S_{surr}$$

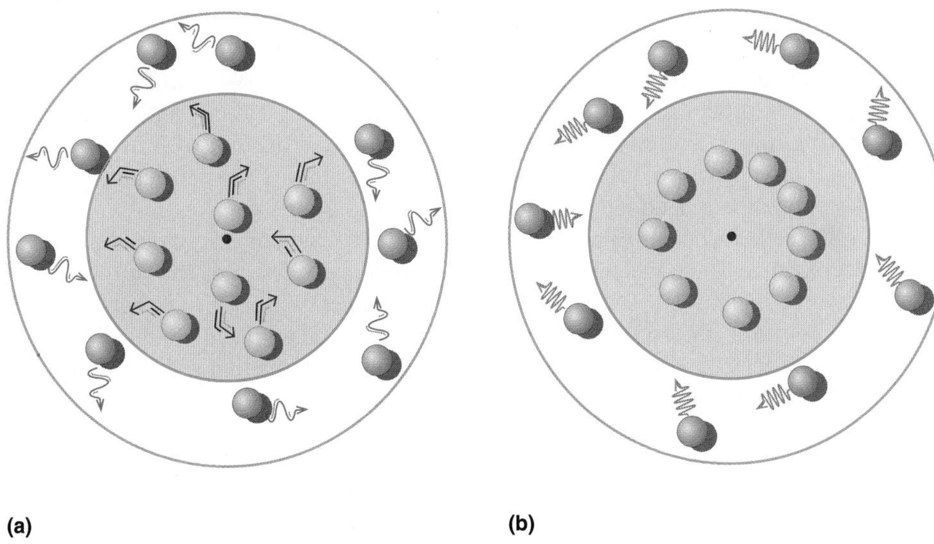

FIGURE 4.3
A Living Cell as a Thermodynamic System

(a) The molecules of the cell and its surroundings are in a relatively disordered state. (b) Heat is released from the cell as a consequence of reactions that create order among the molecules inside the cell. This energy increases the random motion, and therefore the disorder, of the molecules outside the cell (indicated by tighter springs on the outer molecules). This process causes a net positive entropy change. The cell's decrease in entropy is more than offset by an increase in the entropy of the surroundings.

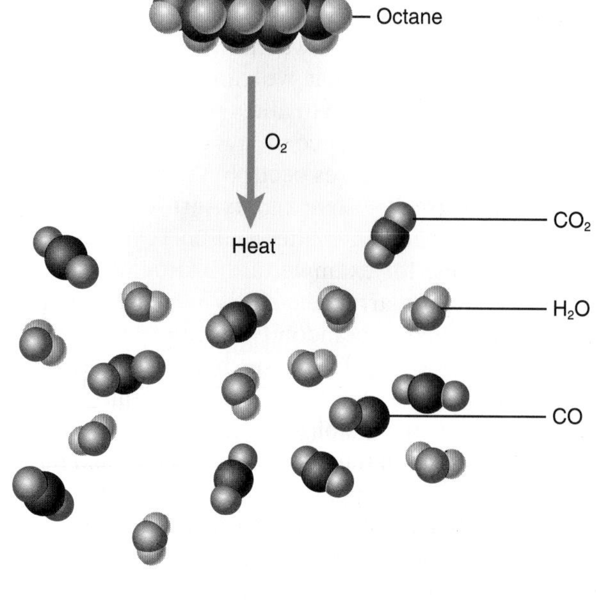

FIGURE 4.4
Gasoline Combustion

When hydrocarbons such as octane are burned, the release of energy is accompanied by the conversion of highly ordered reactant molecules into relatively disorganized gaseous products such as CO_2 and H_2O. However, gasoline combustion is inefficient; that is, other substances such as the environmental pollutant carbon monoxide (CO) are also released.

Living cells do not increase their internal disorder when they consume and metabolize nutrients. The organism's surroundings increase in entropy instead. For example, humans consume food to provide the energy and structural material needed to maintain their complex bodies. The food molecules are converted into vast amounts of disordered waste products (e.g., CO_2, H_2O, and heat) that are discharged into their surroundings.

Although entropy may be considered unusable energy, the formation of entropy is not a useless activity. Some reactions are said to be entropy-driven because the increase in entropy in the system overrides a gain in enthalpy to result in a spontaneous reaction. (By definition, a spontaneous process will occur. The rate at which it occurs, however, may be very rapid or very slow.) In irreversible processes, processes that proceed in only one direction, entropy and enthalpy are driving forces. Entropy directs a system toward equilibrium with its surroundings. Once a process has reached equilibrium (i.e., there is no net change in either direction), there is no longer any driving force to propel it.

KEY CONCEPTS

- The second law of thermodynamics states that the universe tends to become more disorganized.
- Entropy increases may take place anywhere in the system's universe.
- For processes in living organisms, the increase in entropy takes place in the surroundings.

To predict whether a process is spontaneous, the sign of ΔS_{univ} must be known. For example, if the value of ΔS_{univ} for a process is positive (i.e., the entropy of the universe increases), then the process is spontaneous. If ΔS_{univ} is negative, the process does not occur, but the reverse process takes place spontaneously. If ΔS_{univ} is zero, neither process tends to occur. Organisms that are at equilibrium with their surroundings are dead.

4.2 FREE ENERGY

Although the entropy of the universe always increases in a spontaneous process, measuring it is often impractical because both ΔS_{sys} and ΔS_{surr} must be known. A more convenient thermodynamic function for predicting the spontaneity of a process is free energy, which can be derived from the expression for ΔS_{univ}:

$$\Delta S_{univ} = \Delta S_{surr} + \Delta S_{sys}$$

The ΔS_{surr} is defined as the quantity of heat exchanged per kelvin (K) of temperature in the course of a specific chemical or physical change. The ΔS_{surr} can therefore be defined as

$$\Delta S_{surr} = -\Delta H/T$$

By substitution

$$\Delta S_{univ} = -\Delta H/T + \Delta S_{sys}$$

Multiply both sides by $-T$:

$$-T\Delta S_{univ} = \Delta H - T\Delta S_{sys}$$

Josiah Gibbs defined the state function $-T\Delta S_{univ}$, now known as *Gibbs free energy change* or ΔG:

$$\Delta G = \Delta H - T\Delta S_{sys}$$

At constant temperature and pressure, the change in free energy is negative when ΔS_{univ} is positive, which reflects a spontaneous reaction said to be **exergonic** (**Figure 4.5**). If the ΔG is positive, the process is said to be **endergonic** (nonspontaneous). When the ΔG is zero, the process is at equilibrium. As with other thermodynamic functions, ΔG provides no information about reaction rates. Reaction rates depend on the precise mechanism by which a process occurs and are dealt with under the study of kinetics (Chapter 6).

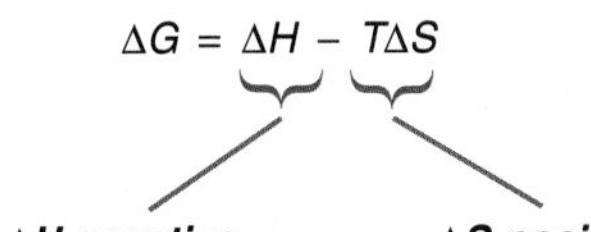

ΔH negative
Energy released during the reaction

ΔS positive
Randomness or disorder in the system increases. If $T\Delta S$ is sufficiently large, then ΔG will be negative and ΔS_{univ} will increase (favorable reaction).

FIGURE 4.5

The Gibbs Free Energy Equation

At constant pressure, enthalpy (H) is essentially equal to the total energy content of the system. A process is spontaneous if it decreases free energy. At constant temperature and pressure, free energy changes (ΔG) are negative if enthalpy decreases or if the entropy term $T\Delta S$ is sufficiently large.

ΔG = − when ΔS = + (Spontaneous) Exergonic
ΔG = 0, process is at equilibrium.

Standard Free Energy Changes

The *standard state* provides a uniform basis for free energy calculations. The standard free energy, $\Delta G°$, is defined for reactions at 25°C (298 K) and 1.0 atm pressure with all solutes at a concentration of 1.0 M.

The standard free energy change is related to the reaction's equilibrium constant, K_{eq}. This is the value of the reaction quotient at equilibrium when the forward and reverse reaction rates are equal. For a reaction

$$aA + bB \rightleftharpoons cC + dD$$

the equilibrium constant is related to the reaction's equilibrium concentrations:

$$K_{eq} = \frac{[C]^c[D]^d}{[A]^a[B]^b}$$

Based on the observations that the free energy of an ideal gas depends on its pressure (concentration) and that the state function G can be manipulated in the same way as the state function H, the following equation was derived:

$$\Delta G = \Delta G° + RT \ln \frac{[C]^c[D]^d}{[A]^a[B]^b}$$

If the reaction is allowed to go to equilibrium, the ΔG is 0 and the expression reduces simply to

$$\Delta G° = -RT \ln K_{eq}$$

This equation allows the calculation of $\Delta G°$ if the K_{eq} is known. Because most biochemical reactions take place at or near pH 7 ($[H^+] = 1.0 \times 10^{-7}$ M), this exception is made in the 1.0 M solute rule in bioenergetics and the free energy change is expressed as $\Delta G°'$.

WORKED PROBLEM 4.3

For the reaction $HC_2H_3O_2 \rightleftharpoons C_2H_3O_2^- + H^+$, calculate $\Delta G°$ and $\Delta G°'$. Assume that $T = 25°C$. The ionization constant for acetic acid is 1.8×10^{-5}. Is this reaction spontaneous? Recall that

$$K_{eq} = \frac{[C_2H_3O_2^-][H^+]}{[HC_2H_3O_2]}$$

SOLUTION

1. Calculate $\Delta G°$.

$$\begin{aligned}\Delta G° &= -RT \ln K_{eq} \\ &= -(8.315\ J/mol{\cdot}K)(298\ K)\ln(1.8 \times 10^{-5}) \\ &= 27{,}071 = 27.1\ kJ/mol\end{aligned}$$

The $\Delta G°$ indicates that under these conditions the reaction is not spontaneous.

2. Calculate $\Delta G°'$. Use the relation between free energy change and standard free energy change.
For this example, the expression becomes

$$\Delta G°' = \Delta G° + RT \ln[H^+]$$

Substituting values, we have

$$\begin{aligned}\Delta G°' &= 27{,}071\ J/mol + (8.315\ J/mol.K)(298\ K)(\ln 10^{-7}) \\ &= 27{,}071 - 39{,}939 \\ &= -12867.54 = -12.9\ kJ/mol\end{aligned}$$

Under the conditions specified for $\Delta G°$ (i.e., 1 M concentrations for all reactants including H^+) the ionization of acetic acid is not spontaneous, as indicated by the positive $\Delta G°$. When the pH value is 7, however, the reaction becomes spontaneous. A low $[H^+]$ makes the ionization of a weak acid such as acetic acid a more likely process, as indicated by the negative $\Delta G°'$. ■

Coupled Reactions

Many chemical reactions within living organisms have positive $\Delta G°'$ values. Fortunately, free energy values are additive in any reaction sequence.

$$A + B \rightleftharpoons C + D \qquad \Delta G°'_{\text{reaction 1}} \qquad (1)$$

$$C + E \rightleftharpoons F + G \qquad \Delta G°'_{\text{reaction 2}} \qquad (2)$$

$$A + B + E \rightleftharpoons D + F + G \qquad \Delta G°'_{\text{overall}} = \Delta G°'_{\text{reaction 1}} + \Delta G°'_{\text{reaction 2}} \qquad (3)$$

Note that reactions (1) and (2) are coupled (i.e., they have a common intermediate, C). If the net $\Delta G°'$ value ($\Delta G°'_{\text{overall}}$) is sufficiently negative, forming the products F and G is an exergonic process.

The conversion of glucose-6-phosphate to fructose-1,6-bisphosphate illustrates the principle of coupled reactions (**Figure 4.6**). The common intermediate in this reaction sequence is fructose-6-phosphate. Because the formation of fructose-6-phosphate from glucose-6-phosphate is endergonic ($\Delta G°'$ is +1.7 kJ/mol), the reaction is not expected to proceed as written (at least under standard conditions). The conversion of fructose-6-phosphate to fructose-1,6-bisphosphate is strongly exergonic because it is coupled to the cleavage of the phosphoanhydride bond of ATP. (The cleavage of ATP's phosphoanhydride bond to form ADP yields approximately −30.5 kJ/mol. ATP in living organisms is discussed in Section 4.3.) Because $\Delta G°'_{\text{overall}}$ for the coupled reactions is negative, the reactions do proceed in the direction written at standard conditions.

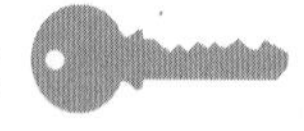

KEY CONCEPT

Free energy is a thermodynamic function that can be used to predict the spontaneity of a process. Spontaneous reactions are exergonic ($-\Delta G$). Nonspontaneous reactions are endergonic ($+\Delta G$).

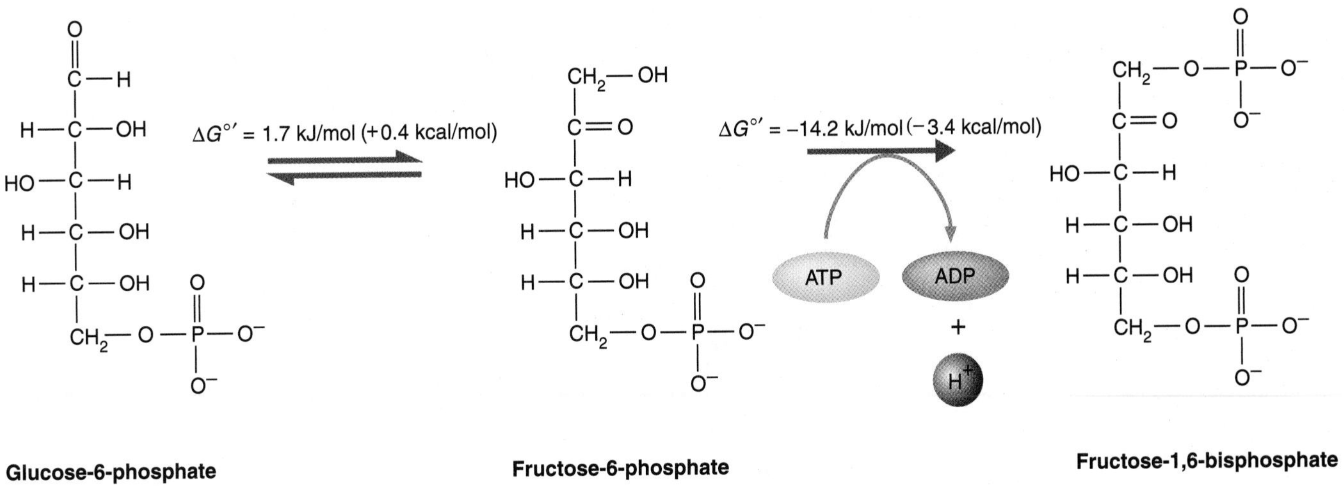

FIGURE 4.6
A Coupled Reaction
The net $\Delta G°'$ value for the two reactions is −12.5 kJ/mol (−3.0 kcal/mol).

WORKED PROBLEM 4.4

Glycogen is synthesized from glucose-1-phosphate. To be incorporated into glycogen, glucose-1-phosphate is converted to a derivative of the nucleotide uridine diphosphate (UDP). The UDP serves as an excellent leaving group in the condensation reaction to form the glycogen polymer. The reaction is

$$\text{Glucose-1-phosphate} + \text{UTP} + H_2O \rightarrow \text{UDP-glucose} + PP_i$$

where PP_i is the inorganic compound pyrophosphate.

If the $\Delta G°'$ value for this reaction is approximately zero, is this reaction favorable? If PP_i is hydrolyzed, then

$$PP_i + H_2O \rightarrow 2P_i$$

where P_i is the inorganic compound orthophosphate.

The loss in free energy ($\Delta G°'$) is −33.5 kJ. How does this second reaction affect the first one? What is the overall reaction? Determine the $\Delta G°'_{\text{overall}}$ value.

SOLUTION

The overall reaction is

$$\text{Glucose-1-phosphate} + \text{UTP} \rightarrow \text{UDP-glucose} + 2P_i$$

$$\begin{aligned}\Delta G°'_{\text{overall}} &= \Delta G°'_{\text{reaction 1}} + \Delta G°'_{\text{reaction 2}} \\ &= 0 + (-33.5\ \text{kJ}) \\ &= -33.5\ \text{kJ}\end{aligned}$$

The hydrolysis of PP_i drives the formation of UDP-glucose to the right. ■

WORKED PROBLEM 4.5

Consider the following reaction in which the sugar fructose reacts with ATP to form fructose-6-phosphate:

ATP + Fructose → ADP + Fructose-6-phosphate

Calculate the equilibrium constant for the reaction given the following free energy values for the two half reactions:

	$\Delta G^{\circ\prime}$(kJ/mol)
ATP → ADP + P_i	−30.5
Fructose + P_i → Fructose-6-phosphate	+15.9

SOLUTION

1. Sum the free energy values for the two reactions

ATP + Fructose → Fructose-6-phosphate + ADP
−30.5 kJ/mol + 15.9 kJ/mol = −14.6 kJ/mol

2. Determine the value of K_{eq} using the equation

$$\Delta G^\circ = -\text{RT}\ln K_{eq}$$

$$-14{,}600 \text{ J/mol} = (-8.315 \text{ J/mol K})(298 \text{ K})(2.303) \log K_{eq}$$
$$-14{,}600 \text{ J/mol} = -(5706.5 \text{ J/mol}) \log K_{eq}$$

$$\log K_{eq} = 2.6$$
$$K_{eq} = 398.1$$

■

QUESTION 4.1

In living cells, the concentrations of ATP and the products of its hydrolysis (ADP and P_i) are significantly lower than the standard 1 M concentrations. Therefore, the actual free energy of hydrolysis of ATP ($\Delta G'$) differs from the standard free energy ($\Delta G^{\circ\prime}$). Unfortunately, it is difficult to obtain an accurate measure of the concentrations of cellular components. For this reason, only estimates can be made. The following equation includes a correction for nonstandard concentrations:

$$\Delta G' = \Delta G^{\circ\prime} + RT \ln \frac{[\text{ADP}][\text{P}_i]}{[\text{ATP}]}$$

The temperature is 37°C. Assume that the pH is 7. In a liver cell, the concentrations (mM) are as follows:

ATP = 4.0, ADP = 1.35, P_i = 4.65
$\Delta G^{\circ\prime}$ = −30.5 kJ/mol

What is the actual $\Delta G'$ for the hydrolysis of ATP under these conditions?

The Hydrophobic Effect Revisited

Understanding the spontaneous aggregation of nonpolar substances in water is enhanced by consideration of thermodynamic principles. When nonpolar molecules are mixed with water, they disrupt water's energetically favorable hydrogen-bonded interactions. The hydrogen bonds that stabilize the highly ordered cagelike structures around clusters of nonpolar molecules restrict the

Biochemistry IN PERSPECTIVE

Nonequilibrium Thermodynamics and the Evolution of Life

How does thermodynamic theory relate to energy flow in living organisms? The thermodynamic concepts described in this chapter, referred to as classical thermodynamics, were discovered during investigations of internal combustion engines in the nineteenth century. Classical thermodynamics explains energy flow in ideal systems in or near equilibrium. Living organisms, however, are open systems that are never at equilibrium until they die. In contrast to stable systems that are in thermodynamic equilibrium, systems that are far from equilibrium are inherently unstable. Thus, a critical question arises: How can an organized living system (a living organism) not in equilibrium remain structurally stable for an extended period of time?

The properties of a Benard cell provide a clue to the phenomenon of order within a universe that favors disorder. A Benard cell is a fluid-filled insulated container that is fitted with a cold reservoir on top and a heat source at the bottom. The liquid placed in the container at the beginning of the experiment has a uniform temperature. As the temperature of the liquid at the bottom of the container gradually increases, a temperature gradient is created. Warm (less dense) liquid begins to rise and cooler (more dense) liquid moves downward. As a specific temperature threshold is reached, convection currents that are organized, rotating, and dynamically stable form spontaneously. The term used to describe the capacity of far-from-equilibrium systems such as the Benard cell to form ordered structures under the influence of an energy gradient is *dissipative*.

Living organisms are **dissipative systems** that facilitate the reduction in the enormous energy gradient between the sun and the earth. Energy dissipation begins when photosynthetic organisms capture a fraction of the total solar radiation (about 10^{18} kJ/day). Phototrophs dissipate a portion of the captured energy by producing their own ordered structure. The dissipation process continues as animals and other heterotrophs consume phototrophs. Eventually all of the energy captured from the sun is released as heat (disorganized energy). Living organisms can be compared to the phenomenon of the Benard cell only in the limited sense that the energy-driven creation of ordered structures is reminiscent of the organized convection currents in the Benard cell.

The critical property of energy flow in living organisms is that equilibrium is never reached because the key feature of the living state is the ability to dissipate energy. The ordered structure of living systems, made possible by the uninterrupted flow of high-quality energy, is maintained by the capacity to release heat and more disordered waste products into the surroundings. The maintenance of dissipative systems requires that continuous work be done on the system because otherwise all natural processes will proceed toward equilibrium. In living organisms this far-from-equilibrium state is maintained by transport, chemical, and mechanical work.

The evolution of living systems is driven by the size of the energy gradient to be dissipated and the dictates of the second law of thermodynamics, namely, the required increase in entropy in the universe. But although the second law determines the direction of living processes, it is insufficient to explain the precise molecular mechanisms that sustain life. The mechanisms by which energy flow is coupled to the performance of work to build and maintain living organisms evolved over several billion years, and the precise details of the mechanism of energy dissipation by living organisms have yet to be resolved. Through trial and error, living organisms, taking advantage of the physical and chemical properties of elements such as carbon, nitrogen, and oxygen, have developed gigantic and complex energy-dissipating biochemical and information-processing networks. The vast species diversification observed on Earth can, therefore, be viewed as a means to provide the greatest number of pathways for energy dissipation. The region of the planet that exhibits the greatest number of species of living organisms is at the equator, where the energy gradient between the sun and Earth is at its largest. Unsurprisingly, nonequilibrium thermodynamics is an active area of research.

SUMMARY: Living organisms are far-from-equilibrium dissipative structures. They create internal organization via a continuous flow of energy.

motion of the water molecules, thus resulting in a decrease in entropy. Consequently, the free energy of dissolving nonpolar molecules is unfavorable (i.e., ΔG is positive because ΔH is positive and $-T\Delta S$ is strongly positive). The decrease in entropy, however, is proportional to the surface area of contact between nonpolar molecules and water. The aggregation of nonpolar molecules significantly decreases the surface area of their contact with water, and thus the water becomes less ordered (i.e., the entropy change, ΔS, is now positive). Because $-T\Delta S$ becomes negative, the free energy of the process is negative, and therefore, it proceeds spontaneously. The spontaneous exclusion of nonpolar groups or molecules by the interaction of water with other water molecules or nearby polar groups is a major factor in biological processes such as protein folding and the assembly of supramolecular structures such as membranes.

4.3 THE ROLE OF ATP

Adenosine triphosphate is a nucleotide that plays an extraordinarily important role in living cells. The hydrolysis of ATP (**Figure 4.7**) immediately and directly provides the free energy to drive an immense variety of endergonic biochemical reactions. Produced from ADP and P_i with energy released by the breakdown of food molecules and the light reactions of photosynthesis, ATP drives processes of several types (**Figure 4.8**). These include (1) biosynthesis of biomolecules, (2) active transport of substances across cell membranes, and (3) mechanical work such as muscle contraction.

ATP is ideally suited to its role as universal energy currency because of its structure (**Figure 4.9**). ATP is a nucleotide composed of adenine, ribose, and a triphosphate unit. Its two terminal phosphoryl groups ($—PO_3^{2-}$) are linked by phosphoanhydride bonds. Although anhydrides are easily hydrolyzed, the phosphoanhydride bonds of ATP are sufficiently stable under mild intracellular conditions. Specific enzymes facilitate ATP hydrolysis.

The tendency of ATP to undergo hydrolysis, also referred to as its **phosphoryl group transfer potential**, is not unique. A variety of biomolecules can transfer phosphate groups to other compounds. **Table 4.1** lists several important examples.

Phosphorylated compounds with high negative $\Delta G°'$ values of hydrolysis have larger phosphoryl group transfer potentials than those compounds with smaller, negative values. Because ATP has an intermediate phosphoryl group transfer

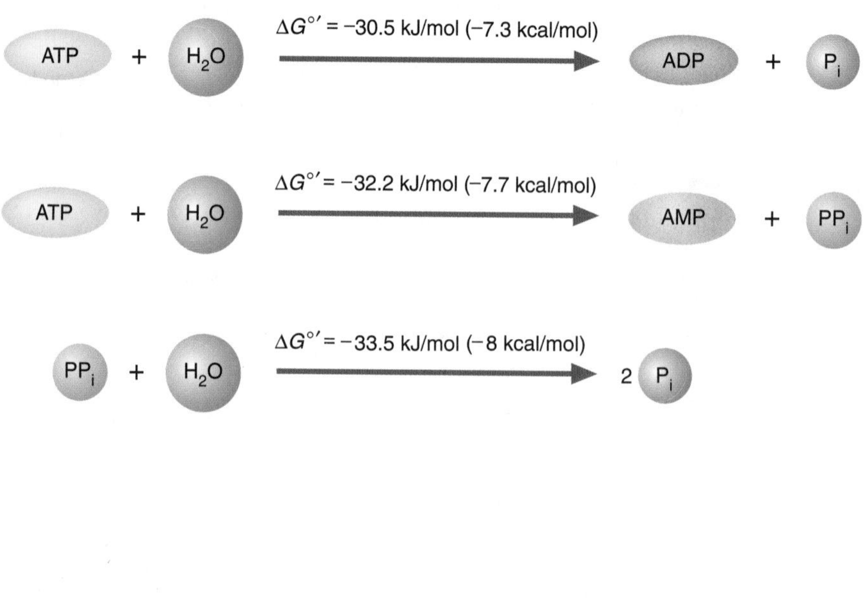

FIGURE 4.7
Hydrolysis of ATP

ATP may be hydrolyzed to form ADP and P_i (orthophosphate) or AMP (adenosine monophosphate) and PP_i (pyrophosphate). Pyrophosphate may be subsequently hydrolyzed to orthophosphate, releasing additional free energy. The hydrolysis of ATP to form AMP and pyrophosphate is often used to drive reactions with high positive $\Delta G°'$ values or to ensure that a reaction goes to completion.

$$\text{ATP} + H_2O \xrightarrow{\Delta G°' = -30.5\ \text{kJ/mol}\ (-7.3\ \text{kcal/mol})} \text{ADP} + P_i$$

$$\text{ATP} + H_2O \xrightarrow{\Delta G°' = -32.2\ \text{kJ/mol}\ (-7.7\ \text{kcal/mol})} \text{AMP} + PP_i$$

$$PP_i + H_2O \xrightarrow{\Delta G°' = -33.5\ \text{kJ/mol}\ (-8\ \text{kcal/mol})} 2\ P_i$$

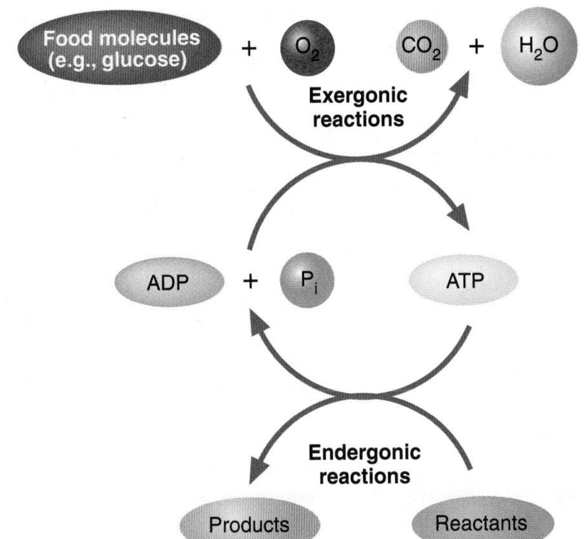

FIGURE 4.8
The Role of ATP

ATP is an intermediate in the flow of energy from food molecules to the biosynthetic reactions of metabolism.

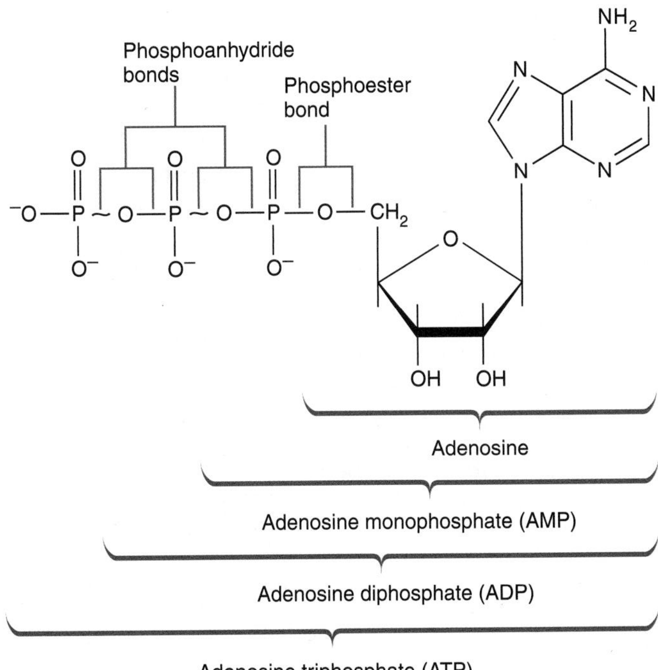

FIGURE 4.9
The Structure of ATP

The squiggles (~) in ATP indicate that the bonds so connected are easily hydrolyzed.

potential, it can be an intermediate carrier of phosphoryl groups from higher-energy compounds such as phosphoenolpyruvate to low-energy compounds (**Figure 4.10**). ATP is therefore the "energy currency" for living systems because cells usually transfer phosphate by coupling reactions to ATP hydrolysis. The two phosphoanhydride bonds of ATP are often referred to as "high energy." The term *high-energy bond* is now considered inappropriate, however, because it denotes instability of the bond and, therefore, its ability to participate in reactions rather than the quantitative value of the bond energy. To understand why ATP hydrolysis is so exergonic, several factors must be considered.

Pyrophosphate
2-PO_3^-

TABLE 4.1 Standard Free Energy of Hydrolysis of Selected Phosphorylated Biomolecules

Molecule	$\Delta G^{\circ\prime}$ kcal/mol	$\Delta G^{\circ\prime}$ kJ/mol
Glucose-6-phosphate	−3.3	−13.8
Fructose-6-phosphate	−3.8	−15.9
Glucose-1-phosphate	−5	−20.9
ATP → ADP + P_i	−7.3	−30.5
ATP → AMP + PP_i	−7.7	−32.2
PP_i → $2P_i$	−8.0	−33.5
Phosphocreatine	−10.3	−43.1
Glycerate-1,3-bisphosphate	−11.8	−49.4
Carbamoyl phosphate	−12.3	−51.5
Phosphoenolpyruvate	−14.8	−61.9

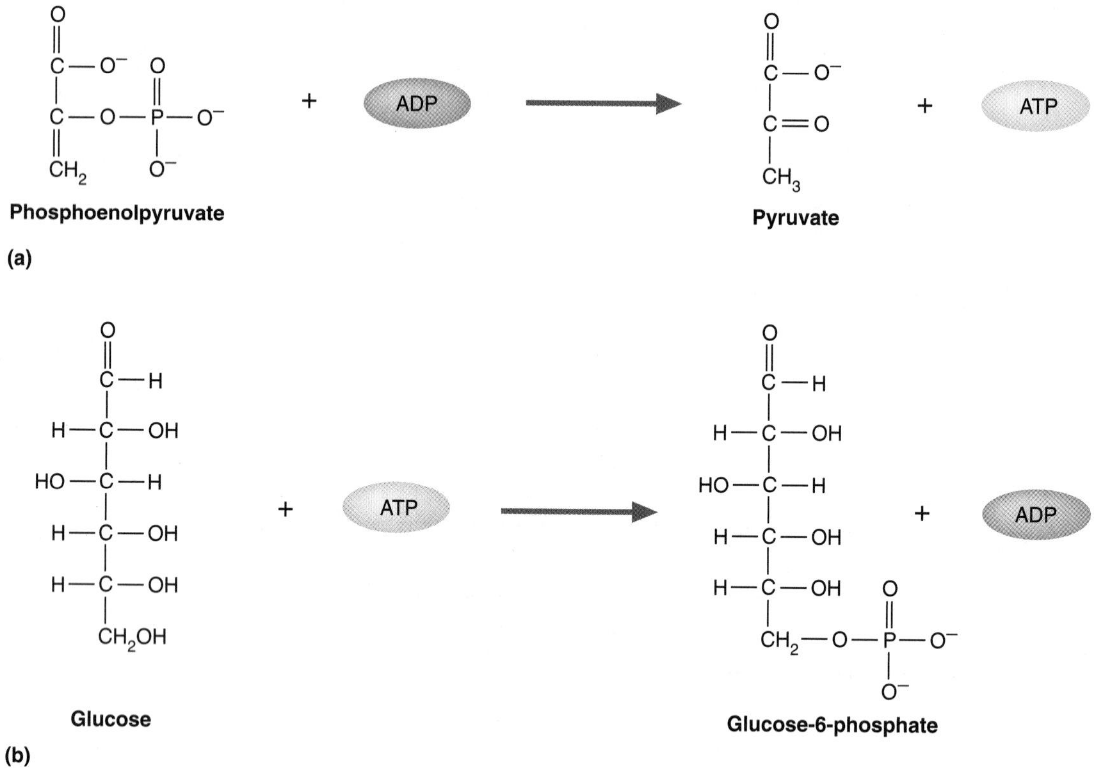

FIGURE 4.10
Transfer of Phosphoryl Groups

(a) Transfer of a phosphoryl group from phosphoenolpyruvate to ADP. As discussed in Chapter 8, this reaction is one of two steps that form ATP during glycolysis, a reaction pathway that breaks down glucose. (b) Transfer of a phosphoryl group from ATP to glucose. The product of this reaction, glucose-6-phosphate, is the first intermediate formed during glycolysis.

FIGURE 4.11

Contributing Structure of the Resonance Hybrid of Phosphate

At physiological pH, orthophosphate is HPO_4^{2-}. In this illustration, H^+ is not assigned permanently to any of the four oxygen atoms.

1. At typical intracellular pH values, the triphosphate unit of ATP carries three or four negative charges that repel each other. Hydrolysis of ATP reduces electrostatic repulsion.
2. Because of *resonance stabilization*, the products of ATP hydrolysis are more stable than a resonance-restricted ATP. When a molecule has two or more alternative structures that differ only in the position of electrons, the result is called a **resonance hybrid**. The electrons in a resonance hybrid with several contributing structures possess much less energy than those with fewer contributing structures. The contributing structures of the phosphate resonance hybrid are illustrated in **Figure 4.11**.
3. The hydrolyzed products of ATP, either ADP and P_i or AMP and PP_i, are more easily solvated than ATP. Recall that the water molecules that form the solvation spheres around ions shield them from one another. The resulting decrease in the repulsive force between phosphoryl groups drives the hydrolytic reaction.
4. There is an increase in disorder caused by an increase in the number of molecules. ATP is converted into two molecules (ADP and P_i), both of which now move randomly.

KEY CONCEPTS

- The hydrolysis of ATP immediately and directly provides the free energy to drive an immense variety of endergonic biochemical reactions.
- Because ATP has an intermediate phosphoryl group transfer potential, it can carry phosphoryl groups from high-energy compounds to low-energy compounds.
- ATP is the energy currency for living systems.

QUESTION 4.2

Walking consumes approximately 100 kcal/mi. In the hydrolysis of ATP (ATP → ADP + P_i), the reaction that drives muscle contraction, $\Delta G°'$ is −7.3 kcal/mol (−30.5 kJ/mol). Calculate how many grams of ATP must be produced to walk a mile. ATP synthesis is coupled to the oxidation of glucose ($\Delta G°' = -686$ kcal/mol). How many grams of glucose are actually metabolized to produce this amount of ATP? (Assume that only glucose oxidation is used to generate ATP and that 40% of the energy generated from this process is used to phosphorylate ADP. The gram molecular weight of glucose is 180 g and that of ATP is 507 g.)

Chapter Summary

1. All living organisms unrelentingly require energy. Bioenergetics, the study of energy transformations, can be used to determine the direction and extent to which biochemical reactions proceed. Enthalpy (a measure of heat content) and entropy (a measure of disorder) are related to the first and second laws of thermodynamics, respectively. Free energy (the portion of total energy that is available to do work) is related to a mathematical relationship between enthalpy and entropy.
2. Energy and heat transformations take place in a "universe" composed of a system and its surroundings. In an open system, matter and energy are exchanged between the system and its surroundings. If energy but not matter can be exchanged with the surroundings, then the system is said to be closed. Living organisms are open systems.
3. Several thermodynamic quantities are state functions; that is, their value does not depend on the pathway used to make or degrade a specific substance. Examples of state functions are total energy, free energy, enthalpy, and entropy. Quantities such as work and heat depend on the pathway and thus are not state functions.
4. Free energy, a state function that relates the first and second laws of thermodynamics, represents the maximum useful work obtainable from a process. Exergonic processes, that is, processes in which free energy decreases ($\Delta G < 0$), are spontaneous. If the free energy change is positive ($\Delta G > 0$), the process is called endergonic. A system is at equilibrium when the free energy change is zero. The standard free energy (ΔG°) is defined for reactions at 25°C, 1 atm pressure, and 1 M solute concentrations. The standard pH in bioenergetics is 7. The standard free energy change $\Delta G^{\circ\prime}$ at pH 7 is used in this textbook.
5. ATP hydrolysis provides most of the free energy required for living processes. ATP is ideally suited to its role as universal energy currency because it is a relatively unstable phosphoanhydride and has a phosphoryl group transfer potential that is an intermediate in the production of other phosphorylated biomolecules.

Take your learning further by visiting the **companion website** for Biochemistry at **www.oup.com/us/mckee** where you can complete a multiple-choice quiz on energy to help you prepare for exams.

Suggested Readings

Bustamante, C., Liphardt, J., and Ritort, F., The Nonequilibrium Thermodynamics of Small Systems, *Phys. Today* 58(7):43–48, 2005 [http://www.physicstoday.org].

Falkowski, P. G., Fenchel, T., and Delong, E. F., The Microbial Engines That Drive Earth's Biogeochemical Cycles, *Science* 320:1034–1039, 2008.

Harold, F. M., Molecules into Cells: Specifying Spatial Architecture, *Microbiol. Mol. Biol. Rev.* 69(4):544–564, 2005.

Hanson, R. W., The Role of ATP in Metabolism, *Biochem. Educ.* 17:86–92, 1989.

Ho, M. W., *The Rainbow and the Worm: The Physics of Organisms*, 3rd ed., World Scientific, Singapore, 2008.

Kleidon, A., Non-equilibrium Thermodynamics, Maximum Entropy Production and Earth-System Evolution, *Phil. Trans. R. Soc.* 368(1910):181–196, 2010.

Mascarelli, A. L., Low Life, *Nature* 459:770–773, 2009.

Pross, A., The Driving Force for Life's Emergence: Kinetic and Thermodynamic Considerations, *J. Theor. Biol.* 220:393–406, 2003.

Rubi, J. M., The Long Arm of the Second Law, *Sci. Am.* 299(5):62–67, 2008.

Schneider, E. D., and Sagan, D., *Into the Cool: Energy Flow, Thermodynamics and Life*, University of Chicago Press, Chicago, 2005.

Schrödinger, E., *What Is Life?* Cambridge University Press, Cambridge, 1944.

Key Words

bioenergetics, *111*
biogeochemical cycle, *109*
chemolithotrophs, *111*
chemoorganotrophs, *111*
dissipative system, *121*
electron acceptor, *111*
electron donor, *111*
endergonic process, *117*
endothermic reaction, *113*
enthalpy, *111*
entropy, *111, 115*
exergonic process, *117*
exothermic reaction, *113*
free energy, *111*
isothermic reaction, *113*
phosphoryl group transfer potential, *122*
resonance hybrid, *125*
spontaneous changes, *114*
thermodynamics, *111*
work, *110*

Review Questions

These questions are designed to test your knowledge of the key concepts discussed in this chapter before moving on to the next chapter. You may like to compare your answers to the solutions provided in the back of the book and in the accompanying Study Guide.

1. Define the following terms:
   a. thermodynamics
   b. bioenergetics
   c. enthalpy
   d. entropy
   e. free energy
2. Define the following terms:
   a. work
   b. exothermic reaction
   c. endothermic reaction
   d. isothermic process
   e. spontaneous process
3. Define the following terms:
   a. exergonic reaction
   b. endergonic reaction
   c. phosphoryl group transfer potential
   d. dissipative system
   e. phosphoanhydride bond
4. Define the following terms:
   a. redox reaction
   b. resonance hybrid
   c. chemolithotroph
   d. electron donor
   e. biogeochemical cycle
5. Which of the following thermodynamic quantities are state functions? Explain.
   a. work
   b. entropy
   c. enthalpy
   d. free energy
6. Which of the following reactions could be driven by coupling to the hydrolysis of ATP? (The $\Delta G°'$ value for each reaction is indicated in parentheses in units of kilojoules per mole.)

   ATP + H_2O → ADP + P_i (−30.5)

   a. Pyruvate + P_i → phosphoenolpyruvate (+31.7)
   b. Glucose + P_i → glucose-6-phosphate (+13.8)
   c. Fructose-6-phosphate → Fructose + P_i (−15.9)
   d. Maltose + H_2O → 2 glucose (–15.5)
   e. Glycerol + P_i → glycerol phosphate (+9.2)
7. The K_a for the ionization of formic acid is 1.8×10^{-4}. Calculate $\Delta G°$ for this reaction.
8. At what temperature is $\Delta G°' = \Delta H$ for all conditions?
9. Reactions are coupled when the product of one is a reactant in another. What principle is involved in this phenomenon?
10. What are the first and second laws of thermodynamics and what are their defining equations?
11. The equilibrium constant for the dissociation of acetic acid is 1.8×10^{-5}. What is the free energy change for this reaction?
12. The following reaction is catalyzed by the enzyme glutamine synthase:

    ATP + glutamate + NH_3 → ADP + P_i + glutamine

    Use the following equations (with $\Delta G°'$ values given in kJ/mol) to calculate $\Delta G°'$ for the overall reaction.

    ATP + H_2O → ADP + P_i (− 30.5)
    Glutamine + H_2O → glutamate + NH_3 (−14.2)
13. $\Delta G°'$ values (kJ/mol) for the following reactions are indicated in parentheses.

    Ethyl acetate + water → ethyl alcohol + acetic acid (−19.7) (i)
    Glucose-6-phosphate + water → glucose + P_i (−13.8) (ii)

    Indicate whether each of the following statements is true, false, or undetermined. Explain your answers.
    a. The rate of reaction (i) is greater than the rate of reaction (ii).
    b. The rate of reaction (ii) is greater than the rate of reaction (i).
    c. Neither reaction is spontaneous.
    d. Reaction (ii) can be used to synthesize ATP from ADP and P_i.
14. Define the thermodynamic term work. Provide two physiological examples of work.
15. Under standard conditions, which statements are true?
    a. $\Delta G = \Delta G°$
    b. $\Delta H = \Delta G$
    c. $\Delta G = \Delta G° + RT \ln K_{eq}$
    d. $\Delta G° = \Delta H - T\Delta S$
    e. $P = 1$ atm
    f. $T = 273$ K
    g. [reactants] = [products] = 1 M
16. Which of the following compounds would you expect to liberate the least free energy when hydrolyzed? Explain.
    a. ATP
    b. ADP
    c. AMP
    d. phosphoenolpyruvate
    e. phosphocreatine
17. Which statements are true and which are false? Modify each false statement so that it reads correctly.
    a. In a closed system, neither energy nor matter is exchanged with the surroundings.
    b. State functions are independent of the pathway.
    c. A process is isothermic if $\Delta H = 0$.
    d. The sign and magnitude of ΔG give important information about the direction and rate of a reaction.
    e. At equilibrium, $\Delta G = \Delta G°$.
    f. For two reactions to be coupled, they must have a common intermediate.

18. Which statements concerning free energy change are true?
    a. Free energy change is a measure of the rate of a reaction.
    b. Free energy change is a measure of the maximum amount of work available from a reaction.
    c. Free energy change is a constant for a reaction under any conditions.
    d. Free energy change is related to the equilibrium constant for a specific reaction.
    e. Free energy change is equal to zero at equilibrium.
19. Consider the following reaction:

    Glucose-1-phosphate → glucose-6-phosphate

    $\Delta G° = -7.1$ kJ/mol

    What is the equilibrium constant for this reaction at 25°C?
20. Describe why ATP, the molecule that serves as the energy currency for the body, has an intermediate phosphoryl group transfer potential.
21. *Methanococcus janaschii* obtains energy by converting carbon dioxide to methane:

    $CO_2 + 4H_2 \rightarrow CH_4 + 2H_2O$

    Considering that CO_2 has a lower energy content than CH_4, how does the organism accomplish this conversion?
22. The $\Delta G°'$ value for glucose-1-phosphate is −20.9 kJ/mol. If glucose and phosphate are both at 4.8 mM, what is the equlibrium concentration of glucose-1-phosphate?
23. Given the following equation

    Glycerol-3-phosphate → glycerol + P_i $\Delta G°' = -9.7$ kJ/mol

    At equilibrium the concentrations of both glycerol and inorganic phosphate are 1 mM. Under these conditions, calculate the final concentration of glycerol-3-phosphate.
24. Magnesium ion (Mg^{2+}) forms complexes with the negative charges of the phosphate in ATP. In the absence of Mg^{2+}, would ATP have more, less, or the same stability as when the ion is present?
25. The free energy of hydrolysis ($\Delta G°'$) of pyrophosphate (PP_i) at pH 7 is 19.2 kJ/mol. Does the value of $\Delta G°'$ change with pH? If so, why? If not, why not?
26. The $\Delta G°'$ value for the hydrolysis of glucose-6-phosphate is −13.8 kJ/mol. Assuming a concentration of 4 mM for this molecule, what would the phosphate concentration be at equilibrium ($\Delta G = 0$)?
27. If glucose, phosphate, and glucose-6-phosphate are combined in concentrations of 4.8, 4.8, and 0.25 mM, respectively, what is the equilibrium constant for the hydrolysis of glucose-6-phosphate at a temprature of 25°C?
28. Using the data for the reaction in Question 26 calculate the ΔG value ($\Delta G°' = -13.8$ kJ/mol).
29. Consider the following reaction:

    ATP → AMP + $2P_i$

    Calculate the equilibrium constant (K_{eq}) given the following $\Delta G°'$ values:

    ATP → AMP + PP_i (−32.2 kJ/mol)
    PP_i → $2P_i$ (−33.5 kJ/mol)

Fill in the Blank

30. An ____________________ reaction liberates heat.
31. Energy is defined as the capacity to perform ____________________.
32. Entropy is a measure of the ____________________ of the system.
33. According to the ____________________ law of thermodynamics, the disorder of the universe always increases.
34. The exchange of energy between a system and its surroundings can occur in two ways: ____________________ and ____________________.
35. The state function that measures the disorder of the system is ____________________.
36. A ____________________ occurs when there are two or more alternate structures of a molecule that differ in the position of electrons.
37. Spontaneous processes occur with the release of ____________________.
38. Absolute zero is ____________________ K.
39. $\Delta H = 0$ in an ____________________ process.

Short Answer

40. It is thermodynamically favorable for methane to autoignite and burn at room temperature, but this does not occur. Explain.
41. The first and second laws of thermodynamics are useful for biochemists who investigate chemical reactions in living organisms. Explain why the third law is not useful.
42. A spontaneous process usually involves an increase in temperature (an exothermic reaction), yet when salt (NaCl) dissolves in water, the solution cools. Explain.
43. What is the difference between an exothermic and an exergonic reaction?
44. How does a common intermediate couple two chemical reactions?

Thought Questions

These questions are designed to reinforce your understanding of all of the key concepts discussed in the book so far, including this chapter and all of the chapters before it. They may not have one right answer! The authors have provided possible solutions to these questions in the back of the book and in the accompanying Study Guide, for your reference.

45. In many ways arsenate (AsO_4^{3-}) is very similar to phosphate (PO_4^{3-}), yet it does not substitute for phosphate in biomolecules. After reviewing the essential atomic characteristics of the element arsenic, explain this phenomenon.
46. Many salts produce heat when dissolved in water and the solution becomes warm, reflecting the fact that the ΔH term is positive. In other cases the solution becomes quite cold and the enthalpy term is negative. What must be true about the Gibbs equation for these reactions to be spontaneous?
47. Pyruvate oxidizes to form carbon dioxide and water and liberates energy at the rate of 1142.2 kJ/mol. If electron transport also occurs, approximately 12.5 ATP molecules are produced. The free energy of hydrolysis for ATP is −30.5 kJ/mol. What is the apparent efficiency of ATP production?
48. In the reaction

    ATP + glucose → ADP + glucose-6-phosphate

    $\Delta G°$ is −16.7 kJ/mol. Assume that both ATP and ADP have a concentration of 1 M and T = 25°C. What ratio of glucose-6-phosphate to glucose would allow the reverse reaction to begin?
49. Thermodynamics is based on the behavior of large numbers of molecules. Yet within a cell there may only be a few molecules of a particular type at a time. Do the laws of thermodynamics apply under these circumstances?
50. Frequently, when salts dissolve in water, the solution becomes warm. Such a process is exothermic. When other salts, such as ammonium chloride, dissolve in water, the solution becomes cold, indicating an endothermic process. Because endothermic processes are usually not spontaneous, why does ammonium chloride dissolve in water?
51. Of the three thermodynamic quantities ΔH, ΔG, and ΔS, which provides the most useful criterion of spontaneity in a reaction? Explain.
52. What factors make ATP suitable as an "energy currency" for the cell?
53. Glucose-1-phosphate has a higher phosphoryl group transfer potential than does glucose-6-phosphate. Review the structures of these molecules and suggest a reason for this phenomenon.
54. Matter has been described as condensed energy. Using Einstein's equation, calculate the amount of energy in 1 mg of dust. How much coal would have to be burned (yielding 393.3 kJ/mol) to produce an equivalent amount of energy? Assume that coal is pure carbon.
55. Given the following data, calculate K_{eq} for the denaturation reaction of β-lactoglobin at 25°C:

    $\Delta H° = -88$ kJ/mol
    $\Delta S° = 0.3$ kJ/mol
56. The free energy of hydrolysis of ATP in systems free of Mg^{2+} is −35.7 kJ/mol. When the concentration of this ion is 5 mM, $\Delta G°_{observed}$ is approximately −31 kJ/mol at pH 7 and 38°C. Suggest a possible reason for this effect.
57. Balance the following reaction and calculate its ΔH value:

    $C_{17}H_{35}COOH + O_2 \rightarrow CO_2 + H_2O$

    where the ΔH values (kcal/mol) are as follows:

    $C_{17}H_{35}COOH$ (−211.4)
    O_2 (0)
    CO_2 (−94)
    H_2O (−68.4)
58. The free energy of hydrolysis for acetic anhydride is −21.8 kJ/ mol. The conversion of ATP to ADP also involves the cleavage of an anhydride bond. Its free energy of hydrolysis is −30 kJ/ mol. Explain the difference in these values.
59. Calculate the temperature for the reaction in Question 55 at which $\Delta G°$ is zero.
60. Consider the following reactions and their $\Delta G°'$ values:

    Ethyl acetate + H_2O → ethanol + acetate (−19.6 kJ/mol)
    Acetyl-S–CoA → Acetate + CoASH (−31 kJ/mol)

    Explain why thioester hydrolysis reactions have more negative $\Delta G°'$ values than those of esters.
61. Glucose-1-phosphate has a $\Delta G°'$ value of −20.9 kJ/mol, whereas that for glucose-6-phosphate is −12.5 kJ/mol. After reviewing the molecular structures of these compounds, explain why there is such a difference in these values.
62. After reviewing the discussion on pp. 109–110, sketch the phases of the carbon cycle that are described.

CHAPTER

5 Amino Acids, Peptides, and Proteins

A Spider's Web Constructed with Silk Fiber The amino acid sequence of spider silk protein and the spider's silk fiber spinning process combine to make spider silk, one of the strongest materials on earth.

OUTLINE

Spider Silk: A Biosteel Protein

Spiders have evolved over 400 million years into exceptionally successful predators. These invertebrate animals are a class of arthropods, called the arachnids, that have an exoskeleton, a segmented body, and jointed appendages. Although spiders possess an efficient venom injection system, their most impressive feature is the production of silk, a multiuse protein fiber. Silk, which is spun through spinnerets at the end of the spider's abdomen, is used in locomotion, mating, and offspring protection. The most prominent use of spider silk, however, is prey capture. The spiral, wheel-shape orb web, which is oriented vertically to intercept fast-moving flying prey, is the best known method. Spider silk's mechanical properties ensure that the web readily absorbs impact energy so that prey is retained until the spider can subdue it. Orb webs (and the species that produce them) have fascinated humans for many thousands of years because of their dramatic visual impact. Ancient Greeks and Romans, for example, explained the occurrence of spiders and orb webs with the myth of Arachne, in which the mortal woman Arachne, an extraordinarily gifted weaver, offended Minerva (Athena in the Greek version), the goddess of weaving and other crafts, with her arrogant acceptance of a challenge to a weaving contest with the goddess. When confronted with Arachne's flawless work, an enraged Minerva transformed her into a spider, doomed to forever weave webs.

Humans have also long appreciated spider webs for their physical properties. Examples range from the ancient Greeks, who used spider webs to treat wounds, to the Australian aborigines who used spider silk to make fishing lines. In modern times spider silk has served as crosshairs in scientific equipment and gun sights. In the past several decades, spider silk and orb webs have attracted the attention of life scientists, bioengineers, and material scientists as they began to appreciate the unique mechanical properties of this remarkable protein.

There are eight different types of spider silk, although no spider makes all of them. Dragline silk, a very strong fiber, is used for frame and radial lines in orb webs and as a safety line (to break a fall or escape other predators). Capture silk, an elastic and sticky fiber, is used in the spiral of webs.

Spider silk is a lightweight fiber with impressive mechanical properties. *Toughness*, a combination of stiffness and strength, is a measure of how much energy is needed to rupture a fiber. Spider silk is about five times as tough as high-grade steel wire of the same weight and about twice as tough as synthetic fibers such as Kevlar (used in body armor). Spider silk's *tensile strength*, the resistance of a material to breaking when stretched, is as great as that of Kevlar and greater than that of high-grade steel wire. *Torsional resistance*, the capacity of a fiber to resist twisting (an absolute requirement for draglines used as safety lines), is higher for spider silk than for all textile fibers, including Kevlar. It also has superior *elasticity* and *resilience*, the capacity of a material when it is deformed elastically to absorb and then release energy. Scientists estimate that a 2.54 cm (1 in)–thick rope made of spider silk could be substituted for the flexible steel arresting wires used on aircraft carriers to rapidly stop a jet plane as it lands.

Overview

PROTEINS ARE MOLECULAR TOOLS THAT PERFORM AN ASTONISHING VARIETY OF FUNCTIONS. IN ADDITION TO SERVING AS STRUCTURAL materials in all living organisms (e.g., actin and myosin in animal muscle cells), proteins are involved in such diverse functions as catalysis, metabolic regulation, transport, and defense. Proteins are composed of one or more polypeptides, unbranched polymers of 20 different amino acids. The genomes of most organisms specify the amino acid sequences of thousands or tens of thousands of proteins.

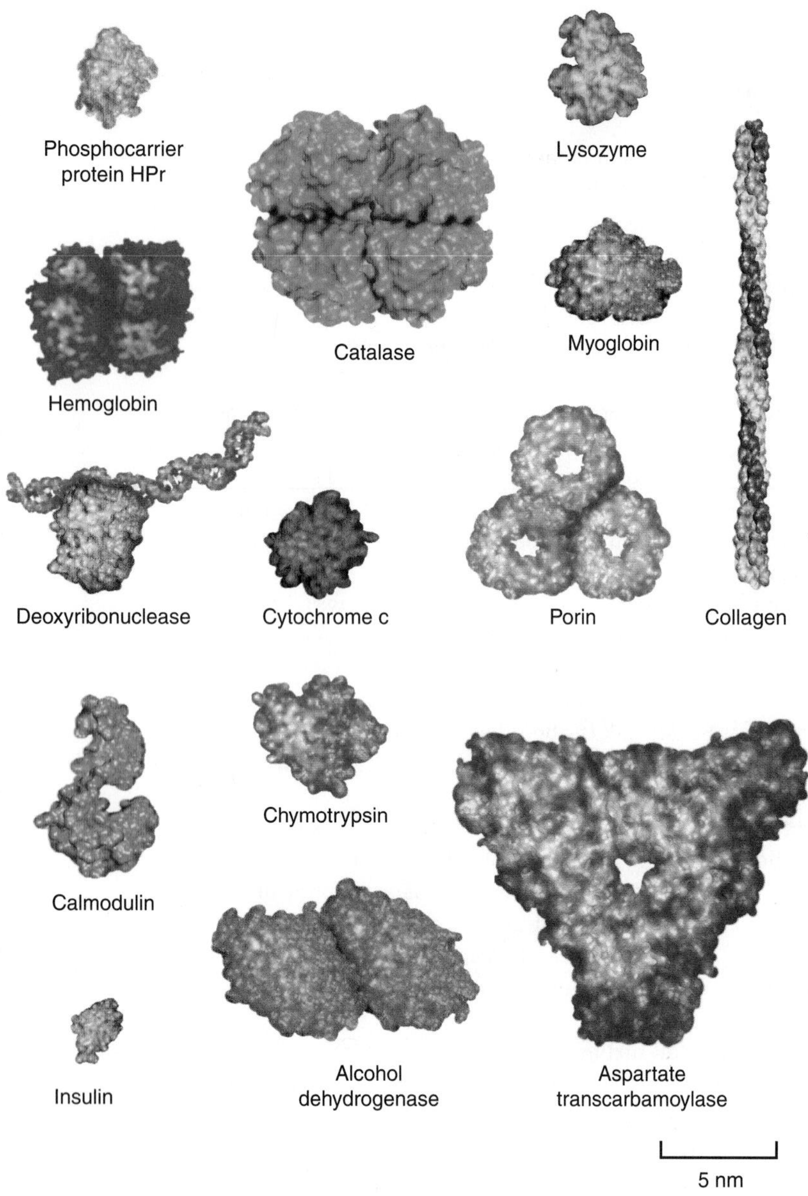

FIGURE 5.1
Protein Diversity

Proteins occur in an enormous diversity of sizes and shapes.

Proteins are a diverse group of macromolecules (**Figure 5.1**). This diversity is directly related to the combinatorial possibilities of the 20 amino acid monomers. Amino acids can be theoretically linked to form protein molecules in any imaginable size or sequence. Consider, for example, a hypothetical protein composed of 100 amino acids. The total possible number of combinations for such a molecule is an astronomical 20^{100}. However, of the trillions of possible protein sequences, only a small fraction (possibly no more than 2 million) is actually produced in all living organisms. An important reason for this remarkable discrepancy is demonstrated by the complex set of structural and functional properties of naturally occurring proteins that have evolved over billions of years in response to selection pressure. Among these are (1) structural features that make protein folding a relatively rapid and successful process, (2) the presence of binding sites that are specific for one or a small group of molecules, (3) an appropriate balance of structural flexibility and rigidity so that function is maintained, (4) surface structure that is appropriate for a protein's immediate environment (i.e., hydrophobic in membranes and hydrophilic in cytoplasm), and (5) vulnerability of proteins to degradation reactions when they become damaged or no longer useful.

Proteins can be distinguished based on their number of amino acids (called **amino acid residues**), their overall amino acyl composition, and their amino acid sequence. Selected examples of the diversity of proteins are illustrated in **Figure 5.1**. Molecules with molecular weights ranging from several thousand to several million daltons are called **polypeptides**. Those with low molecular weights, typically consisting of fewer than 50 amino acids, are called **peptides**. The term **protein** describes molecules with more than 50 amino acids. Each protein consists of one or more polypeptide chains.

This chapter begins with a review of the structures and chemical properties of the amino acids. This is followed by descriptions of the structural and functional features of peptides and proteins and the protein folding process. The emphasis throughout is on the intimate relationship between the structure and function of polypeptides. In Chapter 6 the functioning of the enzymes, an especially important group of proteins, is discussed. Protein synthesis is covered in Chapter 19.

5.1 AMINO ACIDS

The hydrolysis of each polypeptide yields a set of amino acids, referred to as the molecule's *amino acid composition*. The structures of the 20 amino acids that are commonly found in naturally occurring polypeptides are shown in **Figure 5.2**.

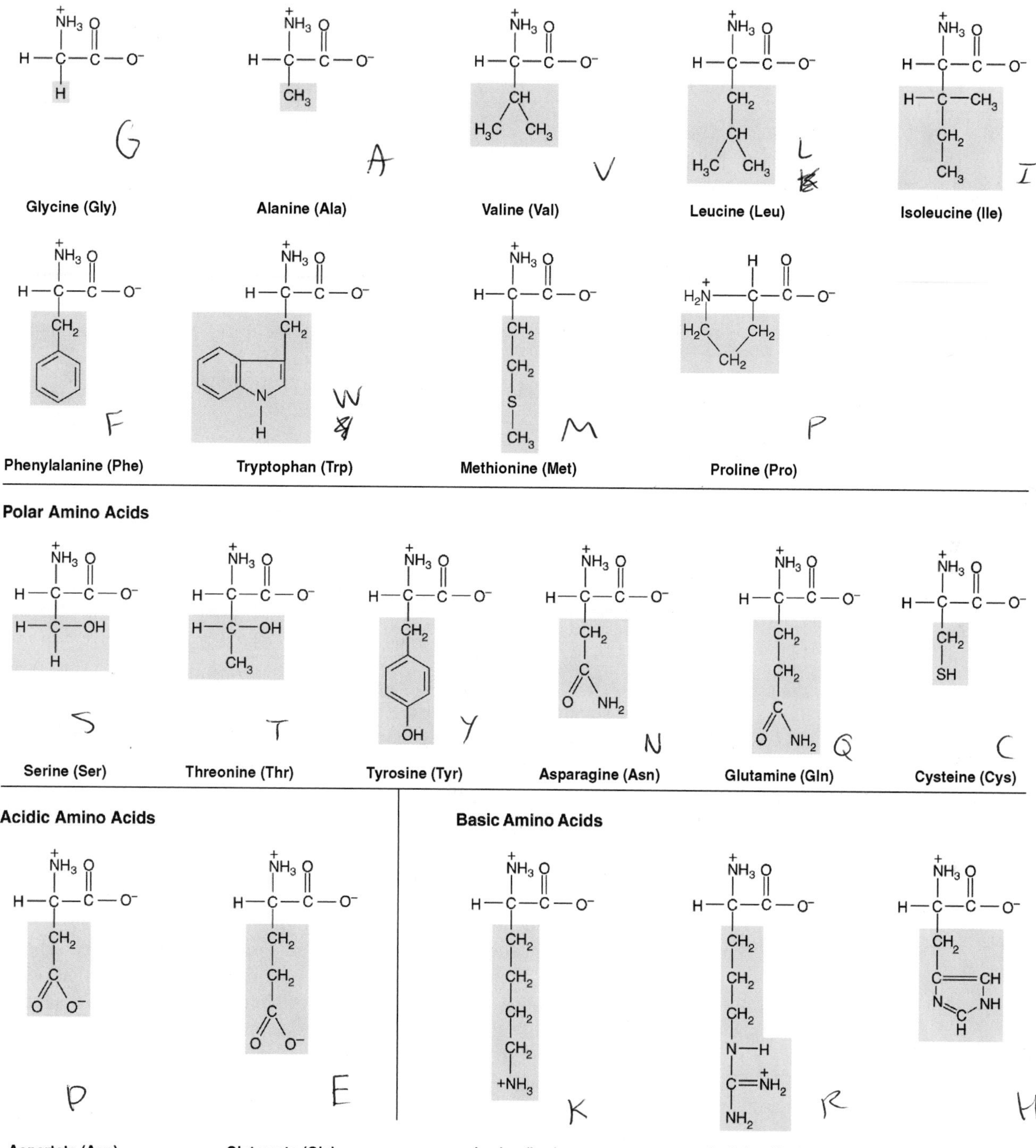

FIGURE 5.2

The Standard Amino Acids

The ionization state of the amino acid molecules in this illustration represents the dominant species that occur at a pH of 7. The side chains are indicated by shaded boxes.

TABLE 5.1 Names and Abbreviations of the Standard Amino Acids

Amino Acid	Three-Letter Abbreviation	One-Letter Abbreviation
Alanine	Ala	A
Arginine	Arg	R
Asparagine	Asn	N
Aspartate	Asp	D
Cysteine	Cys	C
Glutamate	Glu	E
Glutamine	Gln	Q
Glycine	Gly	G
Histidine	His	H
Isoleucine	Ile	I
Leucine	Leu	L
Lysine	Lys	K
Methionine	Met	M
Phenylalanine	Phe	F
Proline	Pro	P
Serine	Ser	S
Threonine	Thr	T
Tryptophan	Trp	W
Tyrosine	Tyr	Y
Valine	Val	V

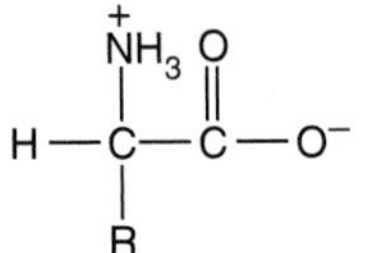

FIGURE 5.3
General Structure of the α-Amino Acids

These amino acids are referred to as *standard* amino acids. Common abbreviations for the standard amino acids are listed in **Table 5.1**. Note that 19 of the standard amino acids have the same general structure (**Figure 5.3**). These molecules contain a central carbon atom (the α-carbon) to which an amino group, a carboxylate group, a hydrogen atom, and an R (side chain) group are attached. The exception, proline, differs from the other standard amino acids in that its amino group is secondary, formed by ring closure between the R group and the amino nitrogen. Proline confers rigidity to the peptide chain because rotation about the α-carbon is not possible. This structural feature has significant implications in the structure and, therefore, the function of proteins with a high proline content.

Nonstandard amino acids consist of amino acid residues that have been chemically modified after incorporation into a polypeptide or amino acids that occur in living organisms but are not found in proteins. Nonstandard amino acids found in proteins are usually the result of *posttranslational modifications* (chemical changes that follow protein synthesis). Selenocysteine, an exception to this rule, is discussed in Chapter 19.

At a pH of 7, the carboxyl group of an amino acid is in its conjugate base form ($—COO^-$), and the amino group is in its conjugate acid form ($—NH_3^+$). Thus each amino acid can behave as either an acid or a base. The term **amphoteric** is used to describe this property. Molecules that bear both positive and negative charges are called **zwitterions**. The R group gives each amino acid its unique properties.

Amino Acid Classes

The sequence of amino acids determines the three-dimensional configuration of each protein. Their structures are therefore examined carefully in the next four subsections. Amino acids are classified according to their capacity to interact with water. Using this criterion, four classes may be distinguished: (1) nonpolar, (2) polar, (3) acidic, and (4) basic.

NONPOLAR AMINO ACIDS The nonpolar amino acids contain mostly hydrocarbon R groups that do not bear positive or negative charges. Nonpolar (i.e., hydrophobic) amino acids play an important role in maintaining the three-dimensional structures of proteins because they interact poorly with water. Two types of hydrocarbon side chains are found in this group: aromatic and aliphatic. **Aromatic** hydrocarbons contain cyclic structures that constitute a class of unsaturated hydrocarbons with planar conjugated π electron clouds. Benzene is one of the simplest aromatic hydrocarbons (p. P-25). The term **aliphatic** refers to nonaromatic hydrocarbons such as methane and cyclohexane (p. 5). Phenylalanine and tryptophan contain aromatic ring structures. Glycine, alanine, valine, leucine, isoleucine, and proline have aliphatic R groups. A sulfur atom appears in the thioether-containing aliphatic side chain (—S—CH_3) of methionine. Its derivative *S*-adenosyl methionine (SAM) is an important metabolite that serves as a methyl donor in numerous biochemical reactions. It should be noted that glycine with a hydrogen atom side chain instead of a hydrocarbon side chain is slightly hydrophilic. Its small side chain introduces structural flexibility into proteins.

POLAR AMINO ACIDS The functional groups of polar amino acids easily interact with water through electrostatic interactions, such as hydrogen bonding. Serine, threonine, tyrosine, asparagine, and glutamine belong to this category. Serine, threonine, and tyrosine contain a polar hydroxyl group, which enables them to participate in hydrogen bonding, an important factor in protein structure. The hydroxyl groups serve other functions in proteins. For example, the formation of the phosphate ester of tyrosine is a common regulatory mechanism. Additionally, the —OH groups of serine and threonine are points for attaching carbohydrates. Asparagine and glutamine are amide derivatives of the acidic amino acids aspartic acid and glutamic acid, respectively. Because the amide functional group is highly polar, the hydrogen-bonding capability of asparagine and glutamine has a significant effect on protein stability. The sulfhydryl group (—SH) of cysteine is highly reactive and is an important component of many enzymes. It also binds metals (e.g., iron and copper ions) in proteins. Additionally, the sulfhydryl groups of two cysteine molecules oxidize easily in the extracellular compartment to form a disulfide compound called cystine. (See p. 144 for a discussion of this reaction.)

ACIDIC AMINO ACIDS Two standard amino acids have side chains with carboxylate groups. Aspartic acid and glutamic acid are often referred to as aspartate and glutamate because carboxylate groups are negatively charged at physiological pH.

QUESTION 5.1

Classify these standard amino acids according to whether their structures are nonpolar, polar, acidic, or basic.

(a) $H_3\overset{+}{N}$—CH(COO^-)—CH_2SH

(b) $H_3\overset{+}{N}$—CH(COO^-)—CH_2—C_6H_5

(c) $H_3\overset{+}{N}$—CH(COO^-)—CH_2—CH_2—CH_2—CH_2—$^+NH_3$

(d) $H_3\overset{+}{N}$—CH(COO^-)—CH_2—CH_2—COO^-

KEY CONCEPT

Amino acids are classified according to their capacity to interact with water. This criterion may be used to distinguish four classes: nonpolar, polar, acidic, and basic.

BASIC AMINO ACIDS Basic amino acids bear a positive charge at physiological pH. They can therefore form ionic bonds with acidic amino acids. Lysine, which has a side chain amino group, accepts a proton from water to form the conjugate acid (—NH_3^+). When lysine side chains in collagen fibrils, a vital structural component of ligaments and tendons, are oxidized and subsequently condensed, strong intramolecular and intermolecular cross-linkages are formed. Because the guanidino group of arginine has a pK_a range of 11.5 to 12.5 in proteins, it is permanently protonated at physiological pH and, therefore, does not function in acid-base reactions. The imidazole side chain of histidine, on the other hand, is a weak base that is only partially ionized at pH 7 because its pK_a is approximately 6. Histidine's capacity under physiological conditions to accept or donate protons in response to small changes in pH plays an important role in the catalytic activity of numerous enzymes.

Biologically Active Amino Acids

In addition to their primary function as components of protein, amino acids have several other biological roles.

1. Several α-amino acids or their derivatives act as chemical messengers (**Figure 5.4**). For example, glycine, glutamate, γ-amino butyric acid (GABA, a derivative of glutamate), and serotonin and melatonin (derivatives of tryptophan) are **neurotransmitters**, substances released from one nerve cell that influence the function of a second nerve cell or a muscle cell. Thyroxine (a tyrosine derivative produced in the thyroid gland of animals) is a **hormone**—a chemical signal molecule produced in one cell that regulates the function of other cells.
2. Amino acids are precursors of a variety of complex nitrogen-containing molecules. Examples include the nitrogenous base components of nucleotides and the nucleic acids, heme (the iron-containing organic group required for the biological activity of several important proteins), and chlorophyll (a pigment of critical importance in photosynthesis).
3. Several standard and nonstandard amino acids act as metabolic intermediates. For example, arginine (**Figure 5.2**), citrulline, and ornithine (**Figure 5.5**) are components of the urea cycle (Chapter 15). The synthesis of urea, a molecule formed in vertebrate livers, is the principal mechanism for the disposal of nitrogenous waste.

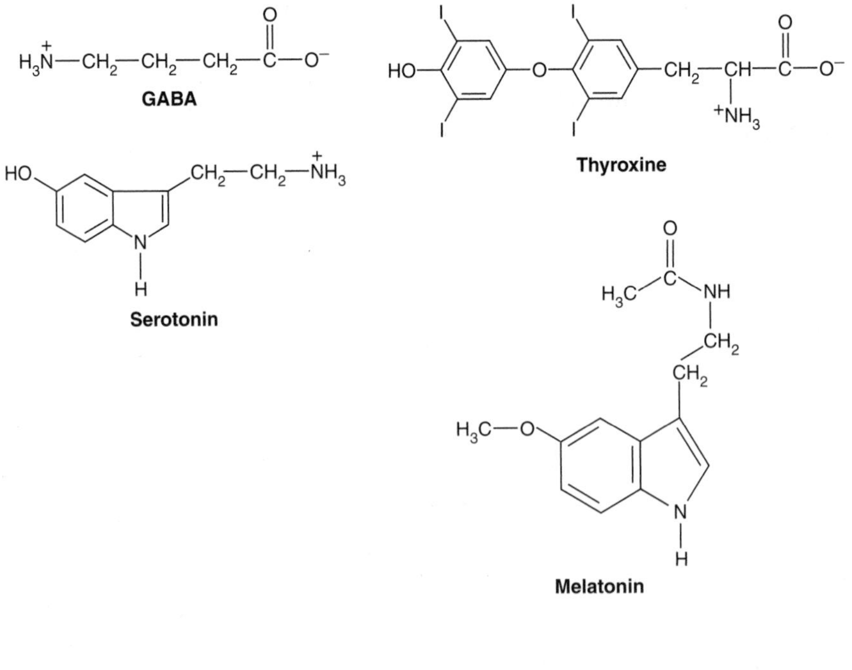

GABA

Thyroxine

Serotonin

Melatonin

FIGURE 5.4
Some Derivatives of Amino Acids

Modified Amino Acids in Proteins

Several proteins contain amino acid derivatives that are formed after a polypeptide chain has been synthesized. Among these modified amino acids is γ-carboxyglutamic acid (**Figure 5.6**), a calcium-binding amino acid residue found in the blood-clotting protein prothrombin. Both 4-hydroxyproline and 5-hydroxylysine are important structural components of collagen, the most abundant protein in connective tissue. Phosphorylation of the hydroxyl-containing amino acids serine, threonine, and tyrosine is often used to regulate the activity of proteins. For example, the synthesis of glycogen is significantly curtailed when the enzyme glycogen synthase is phosphorylated. Two other modified amino acids, selenocysteine and pyrolysine, are discussed in Chapter 19.

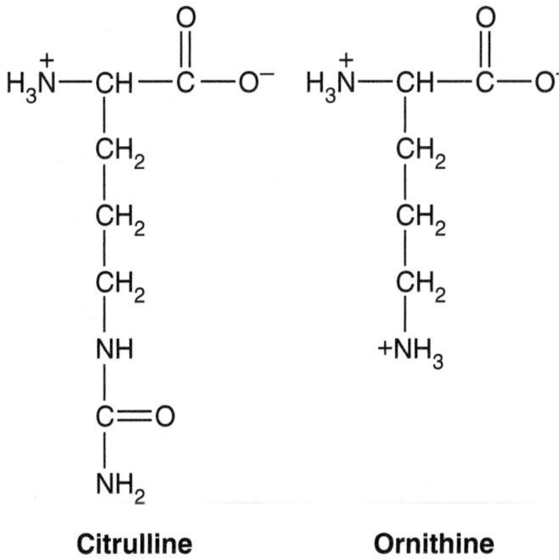

Citrulline

Ornithine

FIGURE 5.5
Citrulline and Ornithine

Amino Acid Stereoisomers

Because the α-carbons of 19 of the 20 standard amino acids are attached to four different groups (i.e., a hydrogen, a carboxyl group, an amino group, and an R group), they are referred to as **asymmetric**, or **chiral**, **carbons**. Glycine is a symmetrical molecule because its α-carbon is attached to two hydrogens. Molecules with chiral carbons can exist as **stereoisomers**, molecules that differ only in the spatial arrangement of their atoms. Three-dimensional representations of amino acid stereoisomers are illustrated in **Figure 5.7**. Notice in the figure that the atoms of the two isomers are bonded together in the same pattern except for the position of the ammonium group and the hydrogen atom. These two isomers are mirror images of each other. Such molecules, called **enantiomers**, cannot be superimposed on each other. Enantiomers have identical physical properties except that they rotate plane-polarized light in opposite directions. Plane-polarized light is produced by passing unpolarized light through a special filter; the light waves vibrate in only one plane. Molecules that possess this property are called **optical isomers**.

Glyceraldehyde is the reference compound for optical isomers (**Figure 5.8**). One glyceraldehyde isomer rotates the light beam in a clockwise direction and is said to be dextrorotatory (designated by +). The other glyceraldehyde isomer, referred to as levorotatory (designated by −), rotates the beam in the opposite direction to an equal degree. Optical isomers are often designated as D or L (e.g., D-glucose, L-alanine) to indicate the similarity of the arrangement of atoms around a molecule's asymmetric carbon to the asymmetric carbon in either of the glyceraldehyde isomers.

Most biomolecules have more than one chiral carbon. As a result, the letters D and L refer only to a molecule's structural relationship to either of the glyceraldehyde isomers, not to the direction in which it rotates plane-polarized light. Most asymmetric molecules found in living organisms occur in only one stereoisomeric form, either D or L. For example, with few exceptions, only L-amino acids are found in proteins.

Chirality has had a profound effect on the structural and functional properties of biomolecules. For example, the right-handed helices observed in proteins result from the exclusive presence of L-amino acids. Polypeptides synthesized in the laboratory from a mixture of both D- and L-amino acids do not form helices. In addition, because the enzymes are chiral molecules, most bind substrate (reactant) molecules in only one enantiomeric form. Proteases,

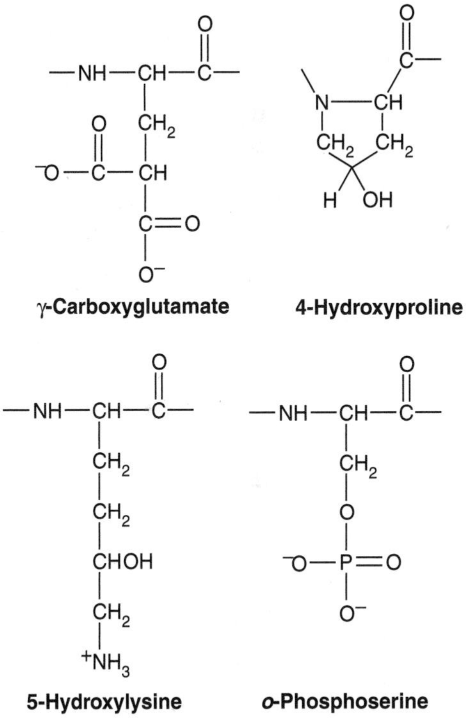

γ-Carboxyglutamate

4-Hydroxyproline

5-Hydroxylysine

o-Phosphoserine

FIGURE 5.6
Some Modified Amino Acid Residues Found in Polypeptides

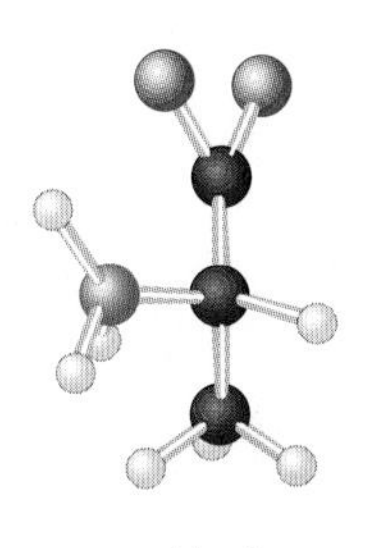

L-Alanine

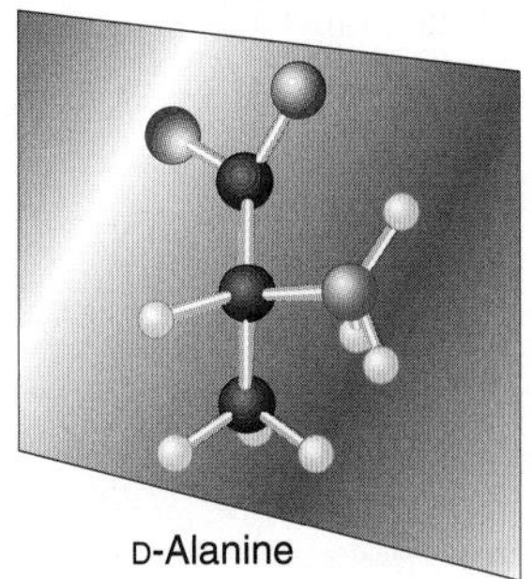

D-Alanine

FIGURE 5.7
Two Enantiomers

L-Alanine and D-alanine are mirror images of each other. (Nitrogen = large blue ball; hydrogen = small gray ball; carbon = black ball; oxygen = red balls)

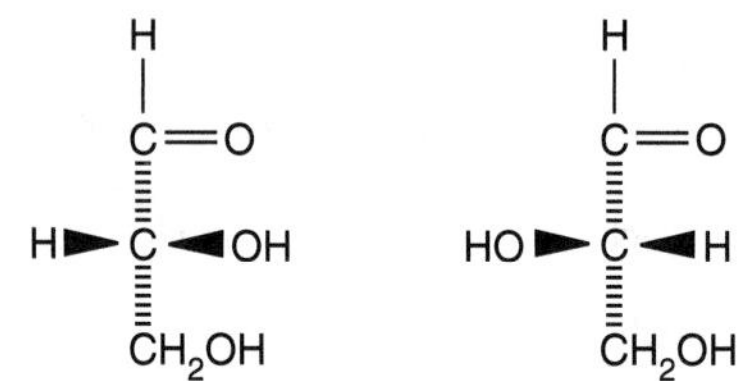

D-Glyceraldehyde L-Glyceraldehyde

FIGURE 5.8

D- and L-Glyceraldehyde

These molecules are mirror images of each other.

KEY CONCEPTS

- Molecules with an asymmetric or chiral carbon atom differ only in the spatial arrangement of the atoms attached to the carbon.
- The mirror-image forms of a molecule are called enantiomers.
- Most asymmetric molecules in living organisms occur in only one stereoisomeric form.

enzymes that degrade proteins by hydrolyzing peptide bonds, cannot degrade artificial polypeptides composed of D-amino acids.

QUESTION 5.2

Certain bacterial species have outer layers composed of polymers made of D-amino acids. Immune system cells, whose task is to attack and destroy foreign cells, cannot destroy these bacteria. Suggest a reason for this phenomenon.

Titration of Amino Acids

The predominant ionic form of the ionizable groups of amino acids in solution depends on pH (**Table 5.2**). Titration of an amino acid illustrates the effect of pH on amino acid structure (**Figure 5.9a**). Titration is also a useful tool in determining the reactivity of amino acid side chains. Consider alanine, a simple amino acid, which has two titratable groups. During titration with a strong base such as NaOH, alanine loses two protons in stepwise fashion. In a strongly acidic solution (e.g., at pH 0), alanine is in the form in which the carboxyl group is uncharged. In this circumstance the molecule's net charge is +1 because the ammonium group is protonated. If the H^+ concentration is lowered, the carboxyl group loses its proton to become a negatively charged carboxylate group. (In a polyprotic acid, the protons are first lost from the group with the lowest pK_a.) Once the carboxyl group has lost its proton, alanine has no net charge and is electrically neutral. The pH at which this occurs is called the **isoelectric point** (pI). The isoelectric point for alanine may be calculated as follows:

$$pI = \frac{pK_1 + pK_2}{2}$$

TABLE 5.2 pK_a Values for the Ionizing Groups of the Amino Acids

Amino Acid	pK_1 (—COOH)	pK_2 ($—NH_3^+$)	pK_R
Glycine	2.34	9.60	
Alanine	2.34	9.69	
Valine	2.32	9.62	
Leucine	2.36	9.60	
Isoleucine	2.36	9.60	
Serine	2.21	9.15	
Threonine	2.63	10.43	
Methionine	2.28	9.21	
Phenylalanine	1.83	9.13	
Tryptophan	2.83	9.39	
Asparagine	2.02	8.80	
Glutamine	2.17	9.13	
Proline	1.99	10.60	
Cysteine	1.71	10.78	8.33
Histidine	1.82	9.17	6.00
Aspartic acid	2.09	9.82	3.86
Glutamic acid	2.19	9.67	4.25
Tyrosine	2.20	9.11	10.07
Lysine	2.18	8.95	10.79
Arginine	2.17	9.04	12.48

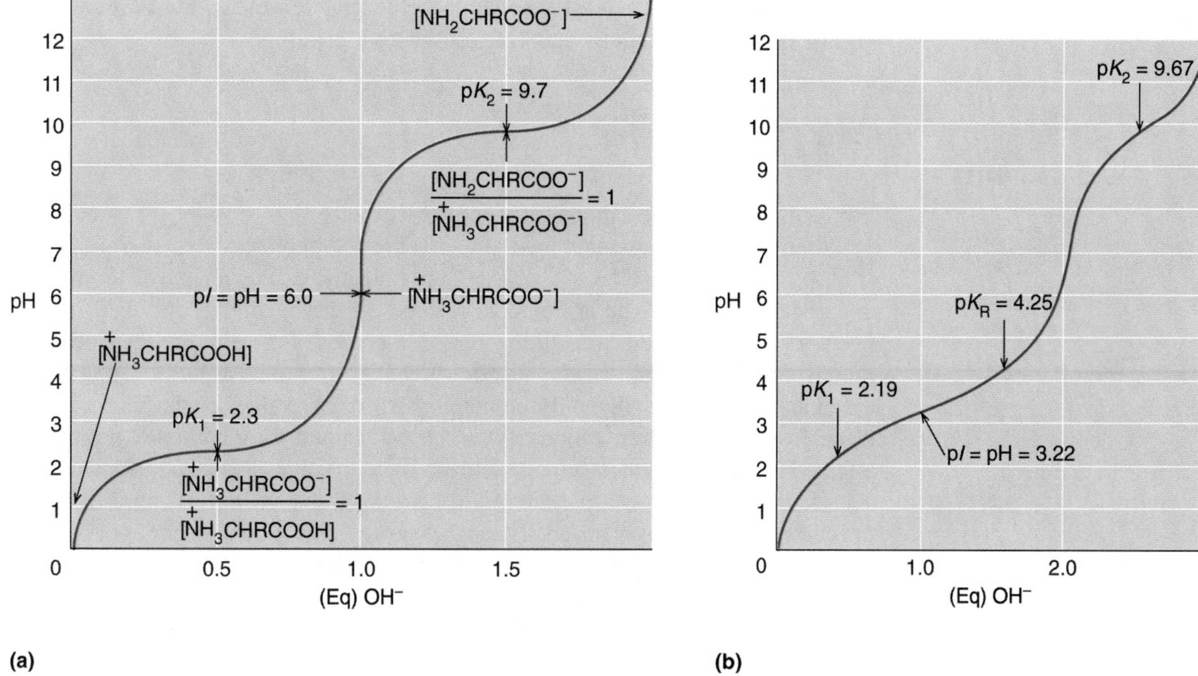

(a) **(b)**

FIGURE 5.9

Titration of Two Amino Acids

(a) Alanine and (b) glutamic acids. The ionized forms of glutamic acid are illustrated below.

The pK_1 and pK_2 values for alanine are 2.34 and 9.9, respectively (see **Table 5.2**). The pI value for alanine is therefore

$$pI = \frac{2.34 + 9.69}{2} = 6.02$$

As the titration continues, the ammonium group loses its proton, leaving an uncharged amino group. The molecule then has a net negative charge because of the carboxylate group.

Amino acids with ionizable side chains have more complex titration curves. Glutamic acid, for example, has a carboxyl side chain group (**Figure 5.9b**). At low pH, glutamic acid has a net charge of +1. As base is added, the α-carboxyl group loses a proton to become a carboxylate group. Glutamate now has no net charge.

$H_3\overset{+}{N}CH(CH_2CH_2COOH)COOH \overset{pK_1}{\rightleftharpoons} H_3\overset{+}{N}CH(CH_2CH_2COOH)COO^- \overset{pK_R}{\rightleftharpoons} H_3\overset{+}{N}CH(CH_2CH_2COO^-)COO^- \overset{pK_2}{\rightleftharpoons} H_2NCH(CH_2CH_2COO^-)COO^-$

Net charge: +1, 0, −1, −2

Titration of glutamic acid

As more base is added, the second carboxyl group loses a proton, and the molecule has a –1 charge. Adding additional base results in the ammonium ion losing its proton. At this point, glutamate has a net charge of –2. The pI value for glutamate is the pH halfway between the pK_a values for the two carboxyl groups (i.e., the pK_a values that bracket the zwitterion):

$$pI = \frac{2.19 + 4.25}{2} = 3.22$$

Problems 5.1 to 5.3 are sample titration problems.

When amino acids are incorporated in polypeptides, the α-amino and α-carboxyl groups lose their charges. Consequently, except for the α-amino and α-carboxyl groups of the amino acid residues at the beginning and end, respectively, of a polypeptide chain, all the ionizable groups of proteins are the side chain groups of seven amino acids: histidine, lysine, arginine, aspartate, glutamate, cysteine, and tyrosine. It should be noted that the pK_a values of these groups can differ from those of free amino acids. The pK_a values of individual R groups are affected by their positions within protein microenvironments. For example, when the side chain groups of two aspartate residues are in close proximity, the pK_a of one of the carboxylate groups is raised. The significance of this phenomenon will become apparent in the discussion of enzyme catalytic mechanisms (Section 6.4).

WORKED PROBLEM 5.1

Consider the following amino acid and its pK_a values:

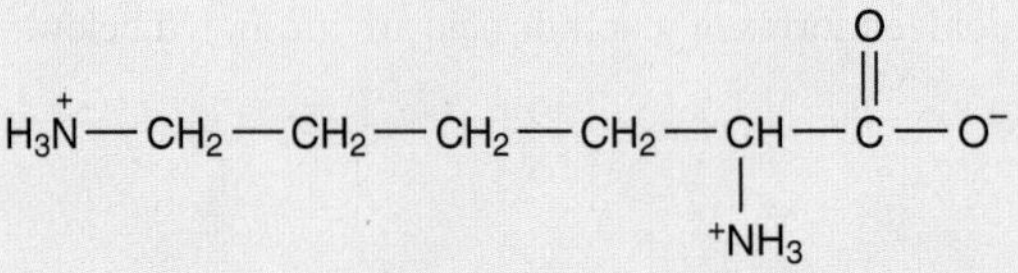

$$H_3\overset{+}{N}-CH_2-CH_2-CH_2-CH_2-\underset{\underset{^{+}NH_3}{|}}{CH}-\overset{\overset{O}{\|}}{C}-O^-$$

pK_{a1} = 2.18 pK_{a2} = 8.95 pK_{aR} = 10.79

a. Draw the structure of the amino acid as the pH of the solution changes from highly acidic to strongly basic.

SOLUTION (A)

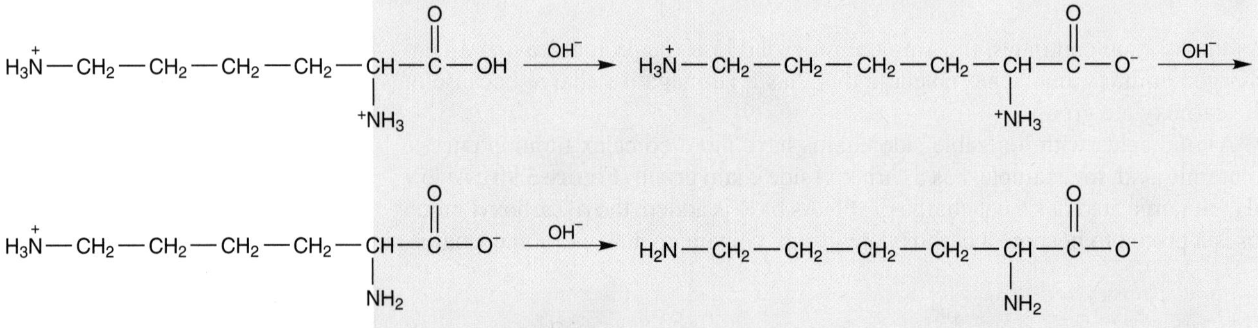

$$H_3\overset{+}{N}-CH_2-CH_2-CH_2-CH_2-\underset{\underset{^{+}NH_3}{|}}{CH}-\overset{\overset{O}{\|}}{C}-OH \xrightarrow{OH^-} H_3\overset{+}{N}-CH_2-CH_2-CH_2-CH_2-\underset{\underset{^{+}NH_3}{|}}{CH}-\overset{\overset{O}{\|}}{C}-O^- \xrightarrow{OH^-}$$

$$H_3\overset{+}{N}-CH_2-CH_2-CH_2-CH_2-\underset{\underset{NH_2}{|}}{CH}-\overset{\overset{O}{\|}}{C}-O^- \xrightarrow{OH^-} H_2N-CH_2-CH_2-CH_2-CH_2-\underset{\underset{NH_2}{|}}{CH}-\overset{\overset{O}{\|}}{C}-O^-$$

The ionizable hydrogens are lost in the order of acidity, with the most acidic ionizing first.

b. Which form of the amino acid is present at the isoelectric point?

SOLUTION (B)

The form present at the isoelectric point is electrically neutral:

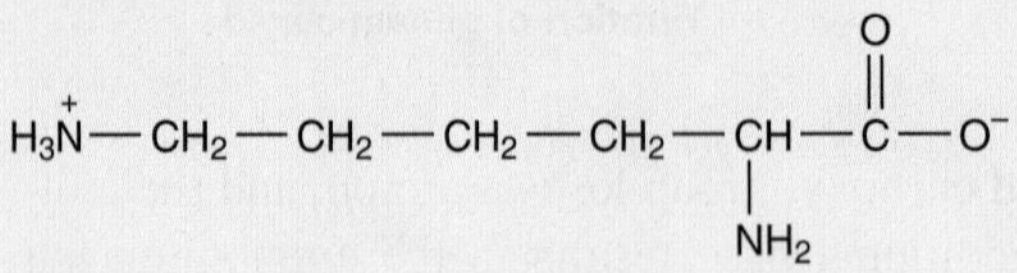

$$H_3\overset{+}{N}-CH_2-CH_2-CH_2-CH_2-\underset{\underset{NH_2}{|}}{CH}-\overset{\overset{O}{\|}}{C}-O^-$$

KEY CONCEPTS

- Titration is useful in determining the relative ionization potential of acidic and basic groups in an amino acid or peptide.
- The pH at which an amino acid has no net charge is called its isoelectric point.

WORKED PROBLEM 5.1 CONT.

c. Calculate the isoelectric point.

SOLUTION (C)

The isoelectric point is the average of the two pK_as bracketing the zwitterion.

$$pI = \frac{pK_2 + pK_R}{2} = \frac{8.95 + 10.79}{2} = 9.87$$

■

WORKED PROBLEM 5.2

a. Sketch the titration curve for the amino acid lysine.

SOLUTION (A)

Plateaus appear at the pK_a and are centered about 0.5 equivalent (Eq), 1.5 Eq, and 2.5 Eq of base. There is a sharp rise at 1 Eq, 2 Eq, and 3 Eq. The isoelectric point is midway on the sharp rise between pK_{a1} and pK_{aR}.

b. In what direction does the amino acid move when placed in an electric field at the following pH values: 1, 3, 5, 7, 9, and 12? *Choice 1*: does not move, *Choice 2*: toward the *cathode* (negative electrode), *Choice 3*: toward the *anode* (positive electrode).

SOLUTION (B)

At pH values below the p*I* (in this case 9.87), the amino acid is positively charged and moves to the cathode. Therefore, the amino acid in this problem will move to the cathode at the pH values of 1, 3, 5, 7, and 9. The amino acid will be negatively charged at a pH value of 12. Under this condition, the amino acid will move to the anode. ■

WORKED PROBLEM 5.3

Consider the following dipeptide:

$$H_3\overset{+}{N}-CH_2-\overset{O}{\overset{\|}{C}}-\overset{H}{\overset{|}{N}}-\underset{CH_2-C_6H_5}{\underset{|}{CH}}-\overset{O}{\overset{\|}{C}}-O^-$$

a. What is its isoelectric point?

SOLUTION (A)

The isoelectric point is the average of the pK_as of the amino group of glycine and the carboxyl group of phenylalanine (obtained from Table 5.2).

$$pI = (9.60 + 1.83)/2 = 5.72$$

b. In which direction will the dipeptide move at pH 1, 3, 5, 7, 9, and 12?

SOLUTION (B)

At pH values below that of the p*I* the dipeptide will move to the cathode (i.e., 1, 3, and 5). At pH values above the p*I* the dipeptide will move to the anode. These are 7, 9, and 12. ■

QUESTION 5.3

Calculate the isoelectric point of the following tripeptide:

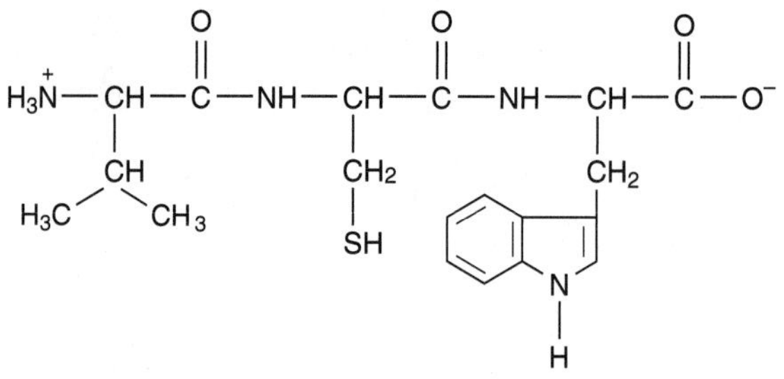

$$H_3\overset{+}{N}-CH(CH(CH_3)_2)-C(=O)-NH-CH(CH_2SH)-C(=O)-NH-CH(CH_2\text{-indolyl})-C(=O)-O^-$$

Assume that the pK_a values listed for the amino acids in Table 5.2 are applicable to this problem.

Amino Acid Reactions

The functional groups of organic molecules determine which reactions they may undergo. Amino acids with their carboxyl groups, amino groups, and various R groups can undergo numerous chemical reactions. Peptide bond and disulfide bridge formation, however, are of special interest because of their effect on protein structure. Schiff base formation is another important reaction.

PEPTIDE BOND FORMATION Polypeptides are linear polymers composed of amino acids linked together by peptide bonds. **Peptide bonds (Figure 5.10)** are amide linkages formed when the unshared electron pair of the α-amino nitrogen atom of one amino acid attacks the α-carboxyl carbon of another in a nucleophilic acyl substitution reaction (see pp. P-28–P-29). A generalized acyl substitution reaction is shown:

$$R-\overset{O}{\overset{\|}{C}}-X + Y^- \longrightarrow R-\overset{O}{\overset{\|}{C}}-Y + X^-$$

The linked amino acids in a polypeptide are referred to as *amino acid residues* because peptide bond formation is a dehydration reaction (i.e., a water molecule is removed). When two amino acid molecules are linked, the product is called a dipeptide. For example, glycine and serine can form the dipeptides glycylserine or serylglycine. As amino acids are added and the chain lengthens, the prefix reflects the number of residues: a tripeptide contains three amino acid residues, a tetrapeptide four, and so on. By convention, the amino acid residue with the free amino group is called the *N-terminal* residue and is written to the left. The free carboxyl group on the *C-terminal* residue appears on the right. Peptides are named using their amino acid sequences, beginning from their N-terminal residue. For example,

$$H_3\overset{+}{N}\text{—Tyr—Ala—Cys—Gly—}COO^-$$

is a tetrapeptide named tyrosylalanylcysteinylglycine.

Large polypeptides often have well-defined, three-dimensional structures. This structure, referred to as the molecule's native conformation, is a direct consequence of its *amino acid sequence* (the order in which the amino acids are linked together). Because all the linkages connecting the amino acid residues consist of single bonds, each polypeptide might be expected to undergo constant conformational changes caused by rotation around the single bonds. However, many polypeptides spontaneously fold into a single biologically active form. In the early 1950s, Linus Pauling (1901–1994, 1954 Nobel Prize in Chemistry) and his colleagues proposed an explanation. Using X-ray diffraction studies, they characterized the peptide bond (1.33 Å) as rigid and planar (flat) (**Figure 5.11**).

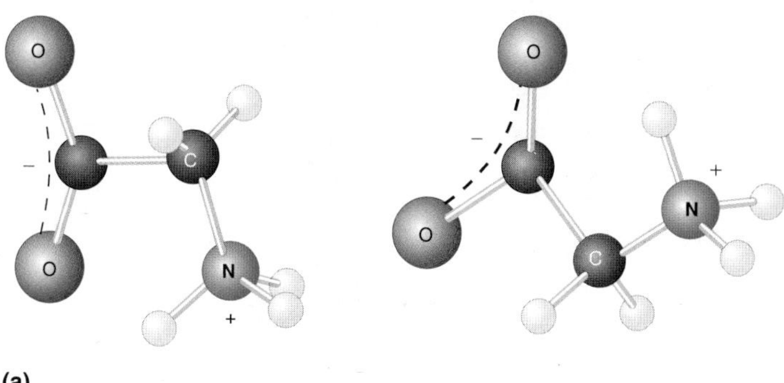

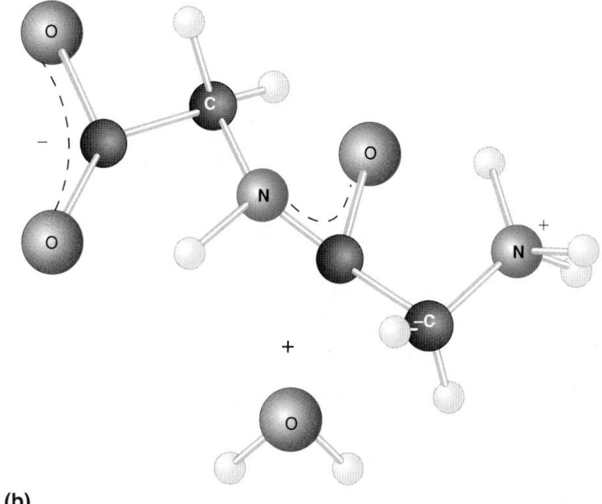

FIGURE 5.10

Formation of a Dipeptide

(a) A peptide bond forms when the α-carboxyl group of one amino acid reacts with the amino group of another. (b) A water molecule is formed in the reaction.

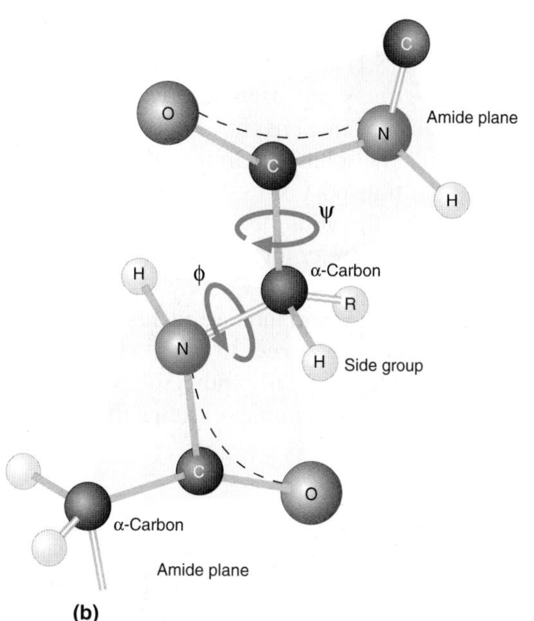

FIGURE 5.11

The Peptide Bond

(a) Resonance forms of the peptide bond. (b) Dimensions of a dipeptide. Because peptide bonds are rigid, the conformational degrees of freedom of a polypeptide chain are limited to rotations around the Cα—C and Cα—N bonds. The corresponding rotations are represented by Ψ and ϕ, respectively.

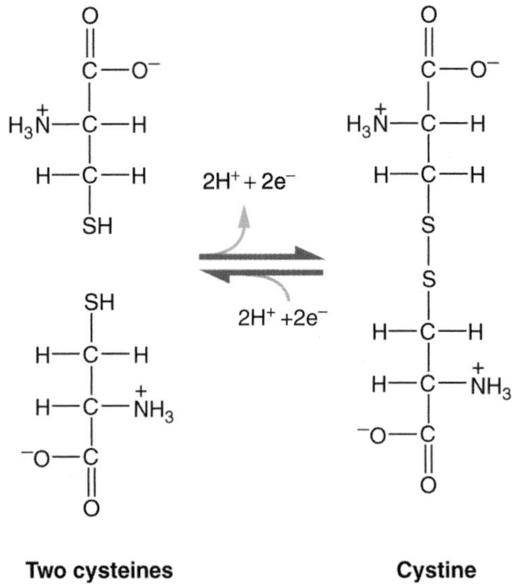

Two cysteines ⇌ Cystine ($2H^+ + 2e^-$)

FIGURE 5.12
Oxidation of Two Cysteine Molecules to Form Cystine
The disulfide bond in a polypeptide is called a disulfide bridge.

Having discovered that the C—N bonds joining each two amino acids are shorter than other types of C—N bonds (1.45 Å), Pauling deduced that peptide bonds have a partial double-bond character. (This indicates that peptide bonds are resonance hybrids.) Because of the rigidity of the peptide bond, fully one-third of the bonds in a polypeptide backbone chain cannot rotate freely. Consequently, there are limits to the number of conformational possibilities.

CYSTEINE OXIDATION The sulfhydryl group of cysteine is highly reactive. The most common reaction of this group is a reversible oxidation that forms a disulfide. Oxidation of two molecules of cysteine forms cystine, a molecule that contains a disulfide bond (**Figure 5.12**). When two cysteine residues form such a bond in peptides or polypeptides, it is referred to as a **disulfide bridge**. This bond can occur in a single chain to form a ring or between two separate chains to form an intermolecular bridge. Disulfide bridges help stabilize many polypeptides and proteins.

QUESTION 5.4

In extracellular fluids such as blood (pH 7.2–7.4) and urine (pH 6.5), the sulfhydryl groups of cysteine (pK_a 8.1) are subject to oxidation to form cystine. In peptides and proteins thiol groups are used to advantage in stabilizing protein structure and in thiol transfer reactions, but free cystine in tissue fluids can be problematic because of its low solubility. In a genetic disorder known as *cystinuria*, defective membrane transport of cystine results in excessive excretion of cystine into the urine. Crystallization of the amino acid results in formation of calculi (stones) in the kidney, ureter, or urinary bladder. The stones may cause pain, infection, and blood in the urine. Cystine concentration in the kidney is reduced by massively increasing fluid intake and administering D-penicillamine. It is believed that penicillamine (**Figure 5.13**) is effective because penicillamine–cysteine disulfide, which is substantially more soluble than cystine, is formed. What is the structure of the penicillamine–cysteine disulfide?

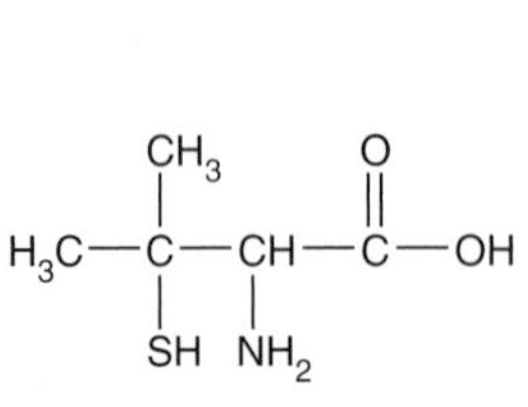

FIGURE 5.13
Structure of Penicillamine

SCHIFF BASE FORMATION Molecules such as amino acids that possess primary amine groups can reversibly react with carbonyl groups. The imine products of this reaction are often referred to as **Schiff bases**. In a *nucleophilic addition reaction*, an amine nitrogen attacks the electrophilic carbon of a carbonyl group to form an alkoxide product. The transfer of a proton from the amine group to the oxygen to form a carbinolamine, followed by the transfer of another proton from an acid catalyst, converts the oxygen into a good leaving group (OH_2^+). The subsequent elimination of a water molecule followed by loss of a proton from the nitrogen yields the imine product. The most important examples of Schiff base formation in biochemistry occur in amino acid metabolism. Schiff bases, referred to as **aldimines**, formed by the reversible reaction of an amino group with an aldehyde group, are *intermediates* (species formed during a reaction) in transamination reactions (pp. 517–519).

$$R{-}NH_2 + R'{-}C(=O){-}R'' \rightleftharpoons R{-}\overset{+}{N}H_2{-}C(O^-)(R'){-}R'' \rightleftharpoons$$

Amine — Alkoxide

$$R{-}\ddot{N}H{-}C(OH)(R'){-}R'' \xrightarrow{H^+} R{-}NH{-}C(\overset{+}{O}H_2)(R'){-}R'' \xrightarrow{-H_2O}$$

Carbinolamine

$$R{-}\overset{+}{N}H{=}C(R')(R'') \rightleftharpoons R{-}\ddot{N}{=}C(R')(R'')$$

Aldimine
(Schiff Base)

KEY CONCEPTS

- Polypeptides are polymers composed of amino acids linked by peptide bonds. The order of the amino acids in a polypeptide is called the amino acid sequence.
- Disulfide bridges, formed by the oxidation of cysteine residues, are an important structural element in polypeptides and proteins.
- Schiff bases are imines that form when amine groups react reversibly with carbonyl groups.

5.2 PEPTIDES

Although less structurally complex than the larger protein molecules, peptides have significant biological activities. The structure and function of several interesting examples, presented in **Table 5.3**, are now discussed.

TABLE 5.3 Selected Biologically Important Peptides

Name	Amino Acid Sequence
Glutathione	$^-O{-}C(=O){-}CH(\overset{+}{N}H_3){-}CH_2{-}CH_2{-}C(=O){-}NH{-}CH(CH_2{-}SH){-}C(=O){-}NH{-}CH_2{-}C(=O){-}O^-$
Oxytocin	Cys—Tyr—Ile—Gln—Asn—Cys—Pro—Leu—Gly—NH_2 (Cys1 and Cys6 linked by S—S)
Vasopressin	Cys—Tyr—Phe—Gln—Asn—Cys—Pro—Arg—Gly—NH_2 (Cys1 and Cys6 linked by S—S)
Atrial natriuretic factor	Ser^{1}—Leu—Arg—Arg—Ser—Ser—Cys—Phe—Gly—Gly^{10}—Arg—Met—Asp—Arg—Ile—Gly—Ala—Gln—Ser—Gly—Leu—Gly—Cys—Asn—Ser—Phe—Arg—Tyr^{28}

The tripeptide glutathione (γ-glutamyl-L-cysteinylglycine) contains an unusual γ-amide bond. (Note that the γ-carboxyl group of the glutamic acid residue, not the α-carboxyl group, contributes to the peptide bond.) Found in almost all organisms, glutathione (GSH) (pp. 533–534) is involved in protein and DNA synthesis, drug and environmental toxin metabolism, amino acid transport, and other important biological processes. One group of glutathione's functions exploits its effectiveness as a reducing agent. Glutathione protects cells from the destructive effects of oxidation by reacting with substances such as peroxides (R–O–O–R), by-products of O_2 metabolism. For example, in red blood cells, hydrogen peroxide (H_2O_2) oxidizes the iron of hemoglobin to its ferric form (Fe^{3+}). Methemoglobin, the product of this reaction, is incapable of binding O_2. Glutathione protects against the formation of methemoglobin by reducing H_2O_2 in a reaction catalyzed by the enzyme glutathione peroxidase. In the oxidized product GSSG, two tripeptides are linked by a disulfide bond:

$$2\text{ GSH} + H_2O_2 \rightarrow \text{GSSG} + 2H_2O$$

Because of the high GSH:GSSG ratio normally present in cells, glutathione is an important intracellular antioxidant. The abbreviation GSH is used because the reducing component of the molecule is the —SH group of the cysteine residue.

Peptides are one class of signal molecules that multicellular organisms use to regulate their complex activities. The dynamic interplay between opposing processes maintains a stable internal environment, a condition called *homeostasis*. Peptide molecules with opposing functions are now known to affect numerous processes (e.g., blood pressure regulation). The roles of selected peptides are briefly described.

Blood pressure, the force exerted by blood against the walls of blood vessels, is influenced by two peptides called vasopressin and atrial natriuretic factor. Vasopressin, also called antidiuretic hormone, contains nine amino acid residues. It is synthesized in the hypothalamus, a small structure in the brain that regulates a wide variety of functions including water balance, appetite, body temperature, and sleep. In response to low blood pressure or a high blood Na^+ concentration, osmoreceptors in the hypothalamus trigger vasopressin secretion. Vasopressin stimulates water reabsorption in the kidneys by initiating a signal transduction mechanism that inserts aquaporins (water channels) into kidney tubule membrane. Blood pressure rises as water then flows down its concentration gradient through the tubule cells and back into the blood. Atrial natriuretic factor (ANF), a peptide produced by specialized cells in the heart in response to stretching and in the nervous system, stimulates the production of a dilute urine, an effect opposite to that of vasopressin. ANF exerts its effect, in part, by increasing the excretion of Na^+, a process that causes increased excretion of water, and by inhibiting the secretion of renin by the kidney. (Renin is an enzyme that catalyzes the formation of angiotensin, a hormone that constricts blood vessels.)

The structure of vasopressin is remarkably similar to that of another peptide produced in the hypothalamus called oxytocin, the signal molecule that stimulates the ejection of milk by mammary glands during lactation. Oxytocin produced in the uterus stimulates the contraction of uterine muscle during childbirth. Because vasopressin and oxytocin have similar structures, it is not surprising that the functions of the two molecules overlap. Oxytocin has mild antidiuretic activity and vasopressin has some oxytocinlike activity.

QUESTION 5.5

Write out the complete structure of oxytocin. What would be the net charge on this molecule at the average physiological pH of 7.3? At pH 4? At pH 9? Indicate which atoms in oxytocin can potentially form hydrogen bonds with water molecules.

QUESTION 5.6

The structural features of vasopressin that allow binding to vasopression receptors are the rigid hexapeptide ring and the amino acid residues at positions 3 (Phe) and 8 (Arg). The aromatic phenylalanine side chain, which fits into a hydrophobic pocket in the receptor, and the large positively charged arginine side chain are especially important structural features. Compare the structures of vasopressin and oxytocin and explain why their functions overlap. Can you suggest what will happen to the binding properties of vasopressin if the arginine at position 8 is replaced by lysine?

KEY CONCEPT

Although small in comparison to larger protein molecules, peptides have significant biological activity. They are involved in a variety of signal transduction processes.

5.3 PROTEINS

Of all the molecules encountered in living organisms, proteins have the most diverse functions, as the following list suggests.

1. **Catalysis**. Catalytic proteins called the *enzymes* accelerate thousands of biochemical reactions in such processes as digestion, energy capture, and biosynthesis. These molecules have remarkable properties. Enzymes can increase reaction rates by factors of between 10^6 and 10^{12}. They can perform this feat under mild conditions of pH and temperature because they can induce or stabilize strained reaction intermediates. For example, ribulose bisphosphate carboxylase is an important enzyme in photosynthesis, and the protein complex nitrogenase is responsible for nitrogen fixation.
2. **Structure**. Structural proteins often have specialized properties. For example, collagen (the major component of connective tissues) and fibroin (silkworm protein) have significant mechanical strength. Elastin, the rubberlike protein found in elastic fibers, is found in blood vessels and skin that must be elastic to function properly.
3. **Movement**. Proteins are involved in all cell movements. Cytoskeletal proteins such as actin, tubulin, and their associated proteins, for example, are active in cell division, endocytosis, exocytosis, and the ameboid movement of white blood cells.
4. **Defense**. A wide variety of proteins are protective. In vertebrates, keratin, a protein found in skin cells, aids in protecting the organism against mechanical and chemical injury. The blood-clotting proteins fibrinogen and thrombin prevent blood loss when blood vessels are damaged. The immunoglobulins (or antibodies), produced by lymphocytes, protect against invasion of the body by foreign organisms such as bacteria. Binding antibodies to an invading organism is the first step in its destruction.
5. **Regulation**. Binding a hormone molecule or a growth factor to cognate receptors on its target cell changes cellular function. For example, insulin and glucagon are peptide hormones that regulate blood glucose levels. Growth hormone stimulates cell growth and division. Growth factors are polypeptides that control animal cell division and differentiation. Examples include platelet-derived growth factor (PDGF) and epidermal growth factor (EGF).
6. **Transport**. Many proteins function as carriers of molecules or ions across membranes or between cells. Examples of membrane transport proteins include the enzyme Na^+-K^+ ATPase and the glucose transporter. Other transport proteins include hemoglobin, which carries O_2 to the tissues from the lungs, and the lipoproteins LDL and HDL, which transport water-insoluble lipids in the blood. Transferrin and ceruloplasmin are serum proteins that transport iron and copper, respectively.

Visit the companion website at www.oup.com/us/mckee to read the Biochemistry in Perspective essay on protein poisons.

7. **Storage**. Certain proteins serve as a reservoir of essential nutrients. For example, ovalbumin in bird eggs and casein in mammalian milk are rich sources of organic nitrogen during development. Plant proteins such as zein perform a similar role in germinating seeds.
8. **Stress response.** The capacity of living organisms to survive abiotic stresses is mediated by a variety of proteins. For example, cytochrome P_{450} is a diverse group of enzymes found in animals and plants that usually convert a variety of toxic organic contaminants into less toxic derivatives. Metallothionein is another example. Found in virtually all mammalian cells, metallothionein is a cysteine-rich intracellular protein that binds to and sequesters toxic metals such as cadmium, mercury, and silver. Excessively high temperatures and other stresses result in the synthesis of a class of proteins called the **heat shock proteins** (hsps), which promote the correct refolding of damaged proteins. When proteins are severely damaged, hsps promote their degradation. (Certain hsps function in the normal process of protein folding.) Cells are protected from radiation by DNA repair enzymes.
9. **Toxins.** Many organisms produce protein toxins, which in general are used in predation or in defense. For example, snakes, scorpions, and some spiders use neurotoxins to subdue prey animals. The bacterium *Clostridium botulinum* secretes botulinum toxin (p. 418), which causes muscle paralysis. Bees protect their hives by injecting pain- and inflammation-producing apitoxin via a bee's stinger. Mellitin is a major peptide toxin component of apitoxin.

Protein research efforts in recent years have revealed that numerous proteins have multiple and often unrelated functions. Once thought to be a rare phenomenon, **multifunction proteins** (sometimes referred to as *moonlighting proteins*) are a diverse class of molecules. Glyceraldehyde-3-phosphate dehydrogenase (GAPD) is a prominent example. As the name suggests, GAPD (p. 278) is an enzyme that catalyzes the oxidation of glyceraldehyde-3-phosphate, an intermediate in glucose catabolism. The GAPD protein is now known to have roles in such diverse processes as DNA replication and repair, membrane fusion events, and microtubule bundling.

Other Protein Classifications

In addition to their functional classifications, proteins are categorized on the basis of amino acid sequence similarities and overall three-dimensional shape. **Protein families** are composed of protein molecules that are related by amino acid sequence similarity. Such proteins share an obvious common ancestry. Classic protein families include the hemoglobins (blood oxygen transport proteins, pp. 170–175) and the immunoglobulins, the antibody proteins produced by the immune system in response to foreign antigens (substances that elicit an immune response). Proteins more distantly related are often classified into **superfamilies**. For example, the globin superfamily includes a variety of heme-containing proteins that serve in the binding and/or transport of oxygen. In addition to the hemoglobins and myoglobins (oxygen-binding proteins in muscle cells), the globin superfamily includes neuroglobin and cytoglobin (oxygen-binding proteins in brain and other tissues, respectively) and the leghemoglobins (oxygen-sequestering proteins in the root nodules of leguminous plants).

Proteins are often classified in two additional ways: shape and composition. There are two major groups of protein shape. As the name suggests, **fibrous proteins** are long, rod-shape molecules that are insoluble in water and physically tough. Fibrous proteins, such as the keratins found in skin, hair, and nails, have structural and protective functions. **Globular proteins** are compact spherical molecules that are usually water-soluble. Typically, globular proteins have dynamic functions. For example, nearly all enzymes have globular structures. Other examples include the immunoglobulins and the transport proteins hemoglobin and albumin (a carrier of fatty acids in blood).

On the basis of composition, proteins are classified as simple or conjugated. Simple proteins, such as serum albumin and keratin, contain only amino acids. Each **conjugated protein** consists of a simple protein combined with a nonprotein component. The nonprotein component is called a **prosthetic group**. (A protein without its prosthetic group is called an **apoprotein**. A protein molecule combined with its prosthetic group is referred to as a **holoprotein**.) Prosthetic groups typically play an important, even crucial, role in the function of proteins. Conjugated proteins are classified according to the nature of their prosthetic groups. For example, **glycoproteins** contain a carbohydrate component, **hemoproteins** contain heme groups (p. 170), and **lipoproteins** contain lipid molecules. Examples of proteins with inorganic cofactors include **metalloproteins**, which contain metal ions, and **phosphoproteins**, which possess phosphate groups.

Protein Structure

Proteins are extraordinarily complex molecules. Complete models depicting even the smallest of the polypeptide chains are almost impossible to comprehend. Simpler images that highlight specific features of a molecule are useful. Two methods of conveying structural information about proteins are presented in **Figure 5.14**. Another structural representation, referred to as a ball-and-stick model, is presented later (**Figures 5.36** and **5.39**).

Biochemists have distinguished several levels of the structural organization of proteins. **Primary structure**, the amino acid sequence, is specified by genetic information. As the *nascent* (newly synthesized) polypeptide chain folds, it forms certain localized arrangements of adjacent (but not necessarily contiguous) amino acids that constitute **secondary structure**. The overall three-dimensional shape that a polypeptide assumes is called the **tertiary structure**. Proteins that consist of two or more polypeptide chains (or subunits) are said to have a **quaternary structure**.

PRIMARY STRUCTURE Every polypeptide has a specific amino acid sequence. The interactions between amino acid residues determine the protein's three-dimensional structure and its functional role and relationship to other proteins. Polypeptides that have similar amino acid sequences and have arisen from the same ancestral gene are said to be **homologous**. Sequence comparisons among homologous polypeptides have been used to trace the genetic relationships of different species. For example, the sequence homologies of the mitochondrial redox protein cytochrome c have been used extensively in the study of evolution of species. Sequence comparisons of cytochrome c, an essential molecule in energy production, among numerous species reveal a significant amount of sequence conservation. The amino acid residues that are identical in all homologues of a protein, referred to as *invariant*, are presumed to be essential for the protein's function. (In cytochrome c the invariant residues interact with heme, a prosthetic group, or certain other proteins involved in energy generation.)

PRIMARY STRUCTURE, EVOLUTION, AND MOLECULAR DISEASES Over time, as the result of evolutionary processes, the amino acid sequences of polypeptides change. These modifications are caused by random and spontaneous alterations in DNA sequences called mutations. A significant number of primary sequence changes do not affect a polypeptide's function. Some of these substitutions are said to be *conservative* because an amino acid with a chemically similar side chain is substituted. For example, at certain sequence positions leucine and isoleucine, which both contain hydrophobic side chains, may be

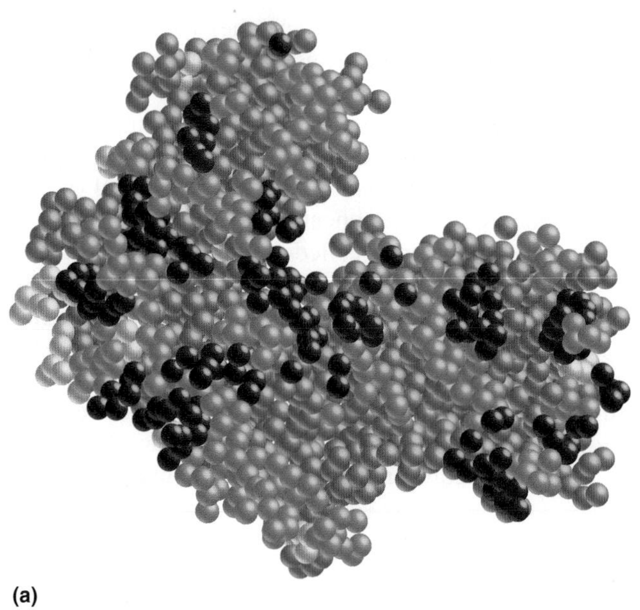

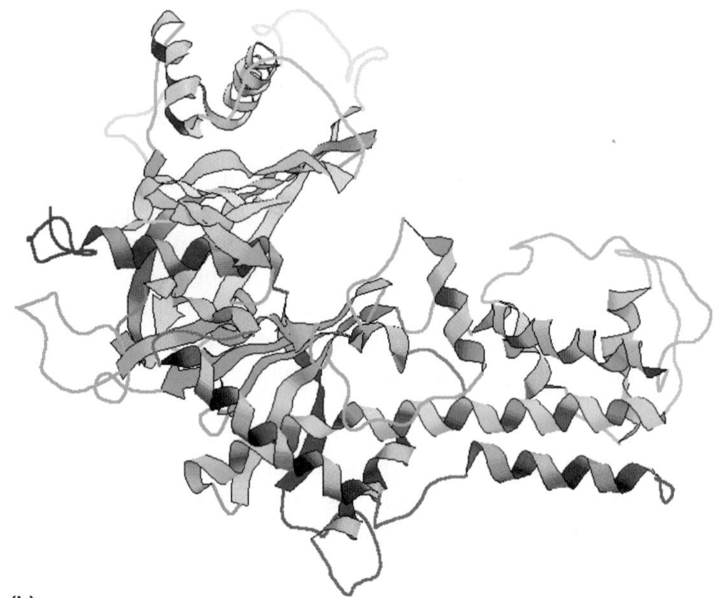

FIGURE 5.14
The Enzyme Adenylate Kinase

(a) This space-filling model illustrates the volume occupied by molecular components and overall shape. (b) In a ribbon model β-pleated strands are represented by flat arrows. The α-helices appear as spiral ribbons. α-helices and β-pleated strands are described on p. 152.

KEY CONCEPTS

- The primary structure of a polypeptide is its amino acid sequence. The amino acids are connected by peptide bonds.
- Amino acid residues that are essential for the molecule's function are referred to as invariant.
- Proteins with similar amino acid sequences and functions and a common origin are said to be homologous.

substituted for each other without affecting function. Some sequence positions are significantly less stringent. These residues, referred to as *variable*, apparently perform nonspecific roles in the polypeptide's function.

Substitutions at conservative and variable sites have been used to trace evolutionary relationships. These studies assume that the longer the time since two species diverged from each other, the larger the number of differences in a certain polypeptide's primary structure. For example, humans and chimpanzees are believed to have diverged relatively recently (perhaps only 4 million years ago). This presumption, based principally on fossil and anatomical evidence, is supported by cytochrome c primary sequence data indicating that the protein is identical in both species. Kangaroos, whales, and sheep, whose cytochrome c molecules each differ by 10 residues from the human protein, are believed to have evolved from a common ancestor that lived more than 50 million years ago. It is interesting to note that quite often the overall three-dimensional structure does not change despite numerous amino acid sequence changes. The shape of proteins coded for by genes that diverged millions of years ago may show a remarkable resemblance to each other.

Mutations, however, can also be deleterious. Such random changes in gene sequence can range from moderate to severe. Individual organisms with nonconservative, variable amino acid substitutions at the conservative, invariant residues of cytochrome c, for example, are not viable. Mutations can also have a profound effect without being immediately lethal. Sickle-cell anemia, which is caused by mutant hemoglobin, is a classic example of a group of maladies that Linus Pauling and his colleagues referred to as **molecular diseases**. (Dr. Pauling first demonstrated that sickle-cell patients have a mutant hemoglobin through the use of electrophoresis, a technique described on p. 182.) Human adult hemoglobin (HbA) is composed of two identical α-chains and two identical β-chains. Sickle-cell anemia results from a single amino acid substitution in the β-chain of HbA. Analysis of the hemoglobin molecules of sickle-cell patients reveals that the only difference between HbA and sickle-cell hemoglobin (HbS) is at amino acid residue 6 in the β-chain (**Figure 5.15**). Because of the substitution of a hydrophobic valine for a negatively charged glutamic acid, HbS molecules aggregate to form rigid rodlike structures in the oxygen-free state (**Figure 5.16**). The patient's red blood cells become sickle shaped and are susceptible to hemolysis, resulting in severe anemia. These red blood cells have an abnormally low oxygen-binding capacity. Intermittent clogging of capillaries by sickled cells also causes tissues to be deprived of oxygen. Sickle-cell anemia is characterized by excruciating pain, eventual organ damage, and earlier death.

Until recently, because of the debilitating nature of sickle-cell disease, affected individuals rarely survived beyond childhood. Thus one might predict that the deleterious mutational change that causes this affliction would be rapidly eliminated from human populations. However, the sickle-cell gene is not as rare

Hb A Val—His—Leu—Thr—Pro—Glu—Glu—Lys—

Hb S Val—His—Leu—Thr—Pro—Val—Glu—Lys—

1 2 3 4 5 6 7 8

FIGURE 5.15
Segments of β-Chain in HbA and HbS

Individuals possessing the gene for sickle-cell hemoglobin produce β-chains with valine instead of glutamic acid at residue 6.

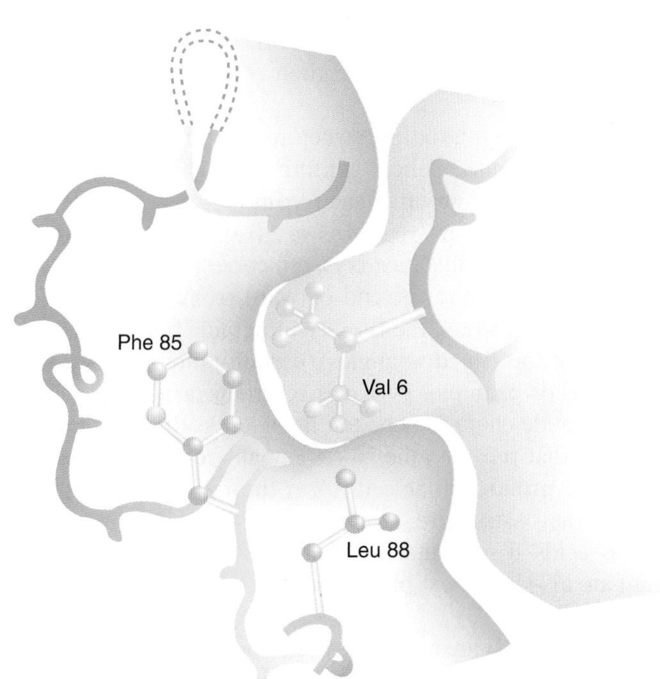

FIGURE 5.16
Sickle-Cell Hemoglobin

HbS molecules aggregate into rodlike filaments because the hydrophobic side chain of valine, the substituted amino acid in the β-chain, interacts with a hydrophobic pocket in a second hemoglobin molecule.

as would be expected. Sickle-cell disease is a homozygous recessive illness, that is, it occurs only in individuals who have inherited two copies of the sickle-cell gene. The term *homozygous* indicates that the affected individual has inherited one copy of the defective gene from each parent. Each of the parents is said to have the *sickle-cell trait*. Such people, referred to as *heterozygous* because they have one normal HbA gene and one defective HbS gene, are relatively sympton-free, even though about 40% of their hemoglobin is HbS. The incidence of sickle-cell trait is especially high in some regions of Africa. In these areas malaria, caused by the *Anopheles* mosquito–borne parasite *Plasmodium*, is a serious health problem. Individuals with the sickle-cell trait are less vulnerable to malaria because their red blood cells are a less favorable environment for the growth of the parasite than are normal cells. Because sickle-cell trait carriers are more likely to survive malaria than normal individuals, the incidence of the sickle-cell gene has remained high. (In some areas, the sickle-cell trait is present in as much as 40% of the native population.)

QUESTION 5.7

A genetic disease called *glucose-6-phosphate dehydrogenase deficiency* is inherited in a manner similar to that of sickle-cell anemia. The defective enzyme cannot keep erythrocytes supplied with sufficient amounts of the antioxidant molecule NADPH (Chapter 8). NADPH protects cell membranes and other cellular structures from oxidation. Describe in general terms the inheritance pattern of this molecular disease. Why do you think that the antimalarial drug primaquine, which stimulates peroxide formation, results in devastating cases of hemolytic anemia in carriers of the defective gene? Does it surprise you that this genetic anomaly is commonly found in African and Mediterranean populations?

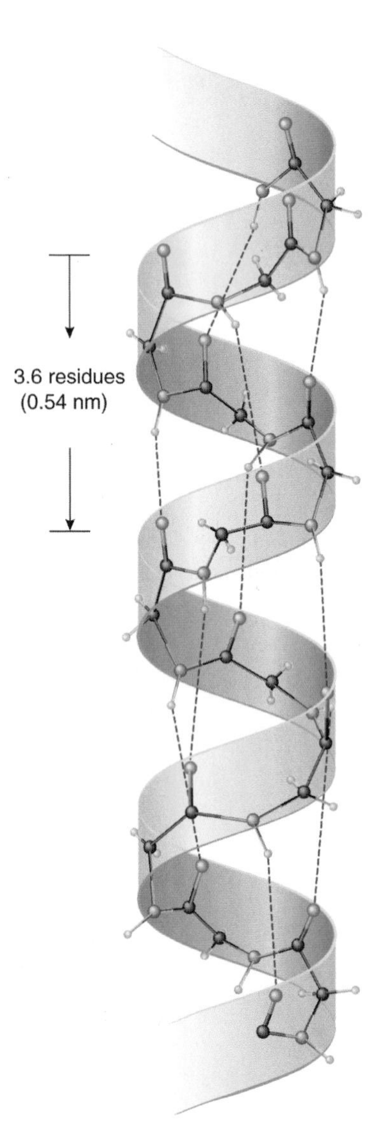

FIGURE 5.17
The α-Helix

Hydrogen bonds form between carbonyl and N—H groups along the long axis of the α-helix. Note that there are 3.6 residues per turn of the helix, which has a pitch of 0.54 nm.

SECONDARY STRUCTURE The secondary structure of polypeptides consists of several repeating patterns. The most commonly observed types of secondary structure are the α-helix and the β-pleated sheet. Both α-helix and β-pleated sheet patterns are stabilized by localized hydrogen bonding between the carbonyl and N—H groups in the polypeptide's backbone. Because peptide bonds are rigid, the α-carbons are swivel points for the polypeptide chain. Several properties of the R groups (e.g., size and charge, if any) attached to the α-carbon influence the ϕ and Ψ angles. Certain amino acids foster or inhibit specific secondary structural patterns. Many fibrous proteins are composed almost entirely of secondary structural patterns.

α-Helix. The *α-helix* is a rigid, rodlike structure that forms when a polypeptide chain twists into a right-handed helical conformation (**Figure 5.17**). Hydrogen bonds form between the N—H group of each amino acid and the carbonyl group of the amino acid four residues away. There are 3.6 amino acid residues per turn of the helix, and the pitch (the distance between corresponding points per turn) is 0.54 nm. Amino acid R groups extend outward from the helix. Because of several structural constraints (i.e., the rigidity of peptide bonds and the allowed limits on the values of the ϕ and ψ angles), certain amino acids do not foster α-helical formation. For example, glycine's R group (a hydrogen atom) is so small that the polypeptide chain may be too flexible. Proline, on the other hand, contains a rigid ring that prevents the N—C_α bond from rotating. In addition, proline has no N—H group available to form the intrachain hydrogen bonds that are crucial in α-helix structure. Amino acid sequences with large numbers of charged amino acids (e.g., glutamate and aspartate) and bulky R groups (e.g., tryptophan) are also incompatible with α-helix structures.

β-Pleated Sheets. *β-Strands* are a second type of secondary structure. *β-Pleated sheets* form when two or more β-strands line up side by side (**Figure 5.18**). Rather

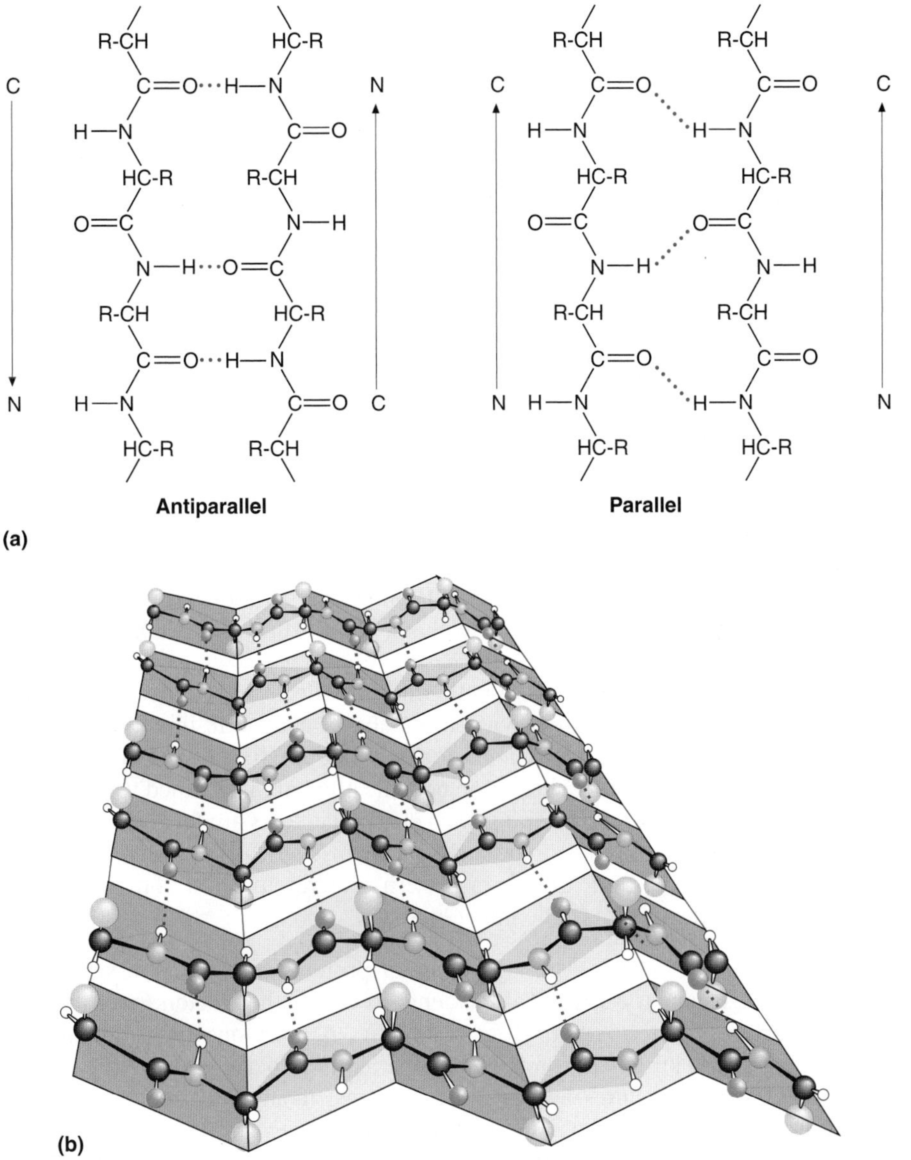

FIGURE 5.18
β-Pleated Sheet

(a) Two forms of β-pleated sheet: antiparallel and parallel. Hydrogen bonds are represented by dotted lines. (b) A more detailed view of antiparallel β-pleated sheet. Note that the hydrogen bonds in antiparallel β-pleated sheets are perpendicular to the β-strands and those in parallel β-pleated sheets are evenly spaced but slanted.

than being coiled, each β-strand is fully extended. β-Pleated sheets are stabilized by hydrogen bonds that form between the polypeptide backbone N—H and carbonyl groups of adjacent chains. β-Pleated sheets are either parallel or antiparallel. In *parallel* β-pleated sheet structures, the hydrogen bonds in the polypeptide chains are arranged in the same direction; in antiparallel chains these bonds are arranged in opposite directions. Occasionally, mixed parallel-antiparallel β-sheets are observed.

Supersecondary Motifs. Many globular proteins contain combinations of α-helix and β-pleated sheet secondary structures (**Figure 5.19**). These patterns are called **supersecondary structures** or **motifs**. In the *βαβ unit*, two parallel β-pleated sheets are connected by an α-helix segment. The structure of βαβ units is stabilized by hydrophobic interactions between nonpolar side chains projecting from the interacting surfaces of the β-strands and the α-helix. Abrupt changes in direction of a polypeptide involve structural elements called loops. The *β-turn*, a commonly observed type of loop, is a 180° turn involving four residues. The carbonyl oxygen of the first residue in the loop forms a hydrogen bond with the amide hydrogen of the fourth residue. Glycine and proline residues often occur in β-turns. Glycine's lack of an organic side group permits a contiguous

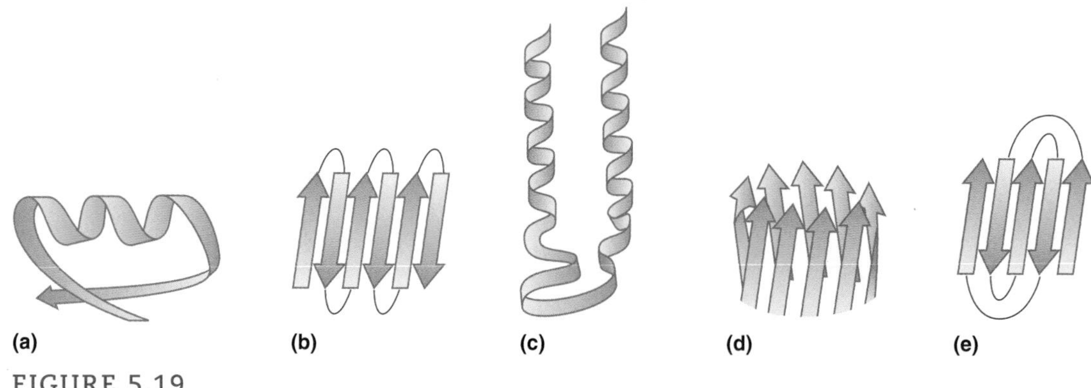

FIGURE 5.19

Selected Supersecondary Structures

(a) $\beta\alpha\beta$ units, (b) β-meander, (c) $\alpha\alpha$ unit, (d) β-barrel, and (e) Greek key. Note that β-strands are depicted as arrows. Arrow tips point toward the C-terminus.

proline to assume a cis orientation (same side of the peptide plane), and a tight turn can form in a polypeptide strand. Proline is a helix-breaking residue that alters the direction of the polypeptide chain. The β-turn is common in proteins rich in α-helical segments.

In the *β-meander* pattern, two antiparallel β-sheets are connected by polar amino acids and glycines to effect a more abrupt change in direction called a reverse or *hairpin turn.* In *$\alpha\alpha$ units* (or helix-loop-helix units), two α-helical regions separated by a nonhelical loop become aligned in a defined way because of interacting side chains. Several *β-barrel* arrangements are formed when various β-sheet configurations fold back on themselves. When an antiparallel β-sheet doubles back on itself in a pattern that resembles a common Greek pottery design the motif is called the *Greek key.*

TERTIARY STRUCTURE Although globular proteins often contain significant numbers of secondary structural elements, several other factors contribute to their structure. The term *tertiary structure* refers to the unique three-dimensional conformations that globular proteins assume as they fold into their native (biologically active) structures and prosthetic groups, if any, are inserted. **Protein folding**, a process in which an unorganized, *nascent* (newly synthesized) molecule acquires a highly organized structure, occurs as a consequence of the interactions between the molecule's side chains. Tertiary structure has several important features:

1. Many polypeptides fold in such a fashion that amino acid residues that are distant from each other in the primary structure come into close proximity.
2. Globular proteins are compact because of efficient packing as the polypeptide folds. During this process, most water molecules are excluded from the protein's interior making interactions between both polar and nonpolar groups possible.
3. Large globular proteins (i.e., those with more than 200 amino acid residues) often contain several compact units called structural domains. *Domains* (**Figure 5.20**) are typically structurally independent segments that have specific functions (e.g., binding an ion or small molecule). The core three-dimensional structure of a domain is called a **fold**. Well-known examples of folds include the nucleotide-binding Rossman fold, often found in nucleotide binding proteins, and the globin fold, found in the oxygen-binding globins. Domains are classified on the basis of their core motif structure. Examples include α, β, α/β, and $\alpha + \beta$. α-Domains are composed exclusively of α-helices, and β-domains consist of antiparallel β strands. α/β-Domains contain various combinations of an α-helix alternating with β-strands ($\beta\alpha\beta$ motifs). $\alpha + \beta$ Domains are primarily

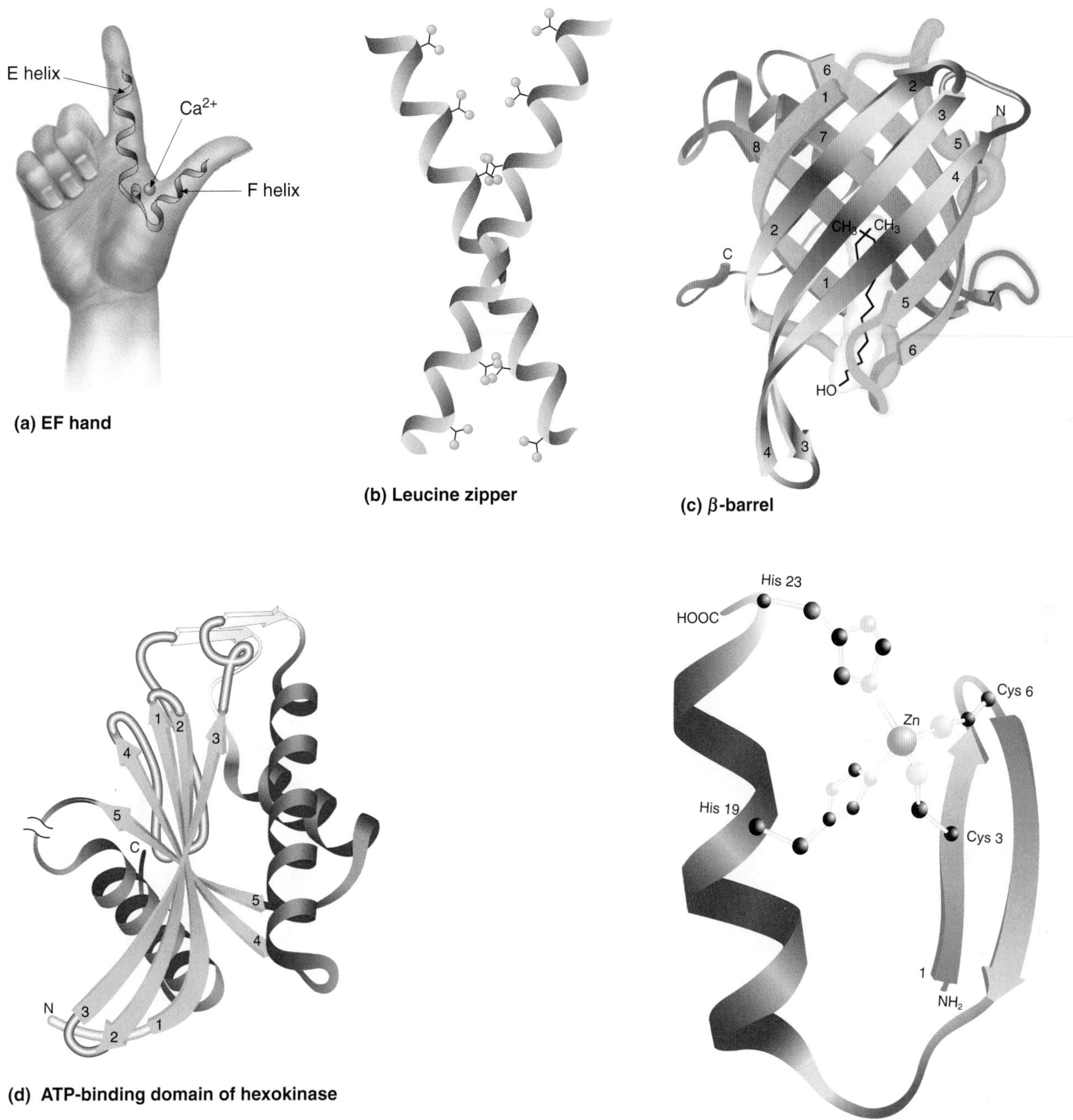

FIGURE 5.20

Selected Domains Found in Large Numbers of Proteins

(a) The EF hand, a helix-loop-helix that binds specifically to Ca^{2+}, and (b) the leucine zipper, a DNA-binding domain, are two examples of α-domains. (c) Human retinol-binding protein, a type of β-barrel domain (retinol, a visual pigment molecule, is shown in yellow). (d) The ATP-binding domain of hexokinase, a type of α/β-domain. (e) The α/β zinc-binding motif, a core feature of numerous DNA-binding domains.

β-sheets with one or more outlying α-helices. Most proteins contain two or more domains.

4. A number of eukaryotic proteins, referred to as **modular** or **mosaic proteins**, contain numerous duplicate or imperfect copies of one or more domains that are linked in series. Fibronectin (**Figure 5.21**) contains three repeating domains: Fl, F2, and F3. All three domains, which are found in a variety of extracellular matrix (ECM) proteins, contain binding sites for

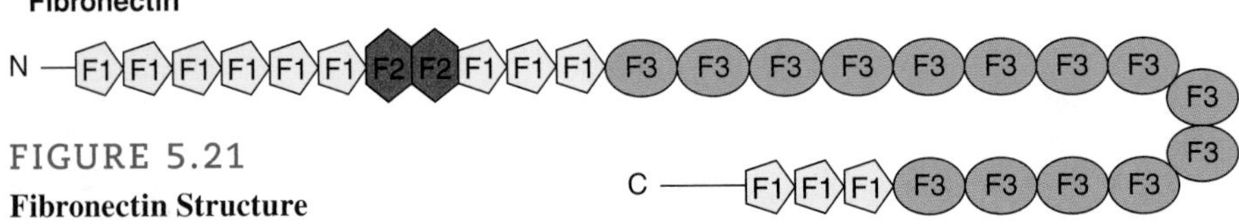

FIGURE 5.21
Fibronectin Structure

Fibronectin is a mosaic protein that is composed of multiple copies of F1, F2, and F3 modules.

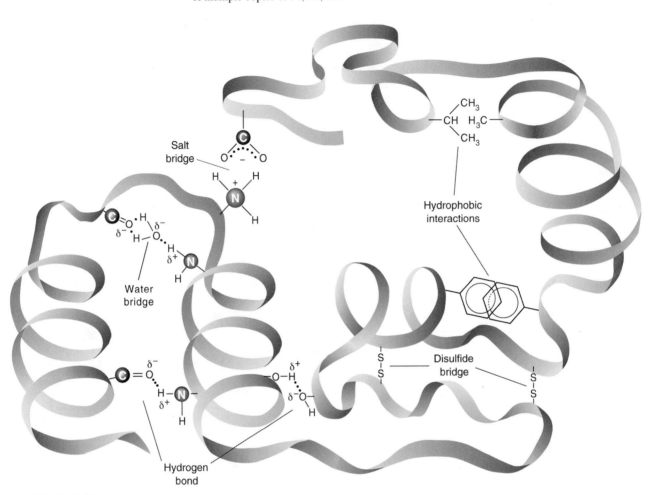

FIGURE 5.22
Interactions That Maintain Tertiary Structure

other ECM molecules such as collagen (p. 167) and heparan sulfate (p. 248), as well as certain cell surface receptors. Domain modules are coded for by genetic sequences created by gene duplications (extra gene copies that arise from errors in DNA replication). Such sequences are used by living organisms to construct new proteins. For example, the immunoglobulin structural domain is found not only in antibodies, but also in a variety of cell surface proteins.

Several types of interactions stabilize tertiary structure (**Figure 5.22**):

1. **Hydrophobic interactions**. As a polypeptide folds, hydrophobic R groups are brought into close proximity because they are excluded from water. Then the highly ordered water molecules in solvation shells are released from the interior, increasing the disorder (entropy) of the water molecules. The favorable entropy change is a major driving force in protein folding. It should be noted that a few water molecules remain within the core of folded proteins, where each forms as many as four hydrogen bonds with the

polypeptide backbone. The stabilization contributed by small "structural" water molecules may free the polypeptide from some of its internal interactions. The resulting increased flexibility of the polypeptide chain is believed to play a critical role in the binding of molecules called **ligands** to specific sites. Ligand binding is an important protein function.

2. **Electrostatic interactions**. The strongest electrostatic interaction in proteins occurs between ionic groups of opposite charge. Referred to as **salt bridges**, these noncovalent bonds are significant only in regions of the protein where water is excluded because of the energy required to remove water molecules from ionic groups near the surface. Salt bridges have been observed to contribute to the interactions between adjacent subunits in complex proteins. The same is true for the weaker electrostatic interactions (ion-dipole, dipole-dipole, van der Waals). They are significant in the interior of the folded protein and between subunits or in protein-ligand interactions. (In proteins that consist of more than one polypeptide chain, each polypeptide is called a **subunit**.) Ligand-binding pockets are water-depleted regions of the protein.
3. **Hydrogen bonds**. A significant number of hydrogen bonds form within a protein's interior and on its surface. In addition to forming hydrogen bonds with one another, the polar amino acid side chains may interact with water or with the polypeptide backbone. Again, the presence of water precludes the formation of hydrogen bonds with other species.
4. **Covalent bonds**. Covalent linkages are created by chemical reactions that alter a polypeptide's structure during or after its synthesis. (Examples of these reactions, referred to as posttranslational modifications, are described in Section 19.2.) The most prominent covalent bonds in tertiary structure are the disulfide bridges found in many extracellular proteins. In extracellular environments, these strong linkages partly protect protein structure from adverse changes in pH or salt concentrations. Intracellular proteins do not contain disulfide bridges because of high cytoplasmic concentrations of reducing agents.
5. **Hydration**. As described previously (p. 85) structured water is an important stabilizing feature of protein structure. The dynamic hydration shell that forms around a protein (**Figure 5.23**) also contributes to the flexibility required for biological activity.

The precise nature of the forces that promote the folding of proteins (described on pp. 163–167) has not been completely resolved. It is clear, however, that protein folding is a thermodynamically favorable process with an overall negative free energy change. According to the free energy equation

$$\Delta G^\circ = \Delta H^\circ - T\Delta S^\circ$$

a negative free energy change in a process is the result of a balance between favorable and unfavorable enthalpy and entropy changes (pp. 116–117). As a polypeptide folds, favorable (negative) ΔH values are the result in part of the sequestration of hydrophobic side chains within the interior of the molecule and the optimization of other noncovalent interactions. Opposing these factors is the unfavorable decrease in entropy that occurs as the disorganized polypeptide folds into its highly organized native state. The change in entropy of the water that surrounds the protein is positive because of the decreased organization of the water in going from the unfolded to the folded state of the protein. For most polypeptide molecules the net free energy change between the folded and unfolded state is relatively modest (the energy equivalent of several hydrogen bonds). The precarious balance between favorable and unfavorable forces allows proteins the flexibility they require for biological function.

QUATERNARY STRUCTURE Many proteins, especially those with high molecular weights, are composed of several polypeptide chains. As mentioned, each polypeptide component is called a subunit. Subunits in a protein complex

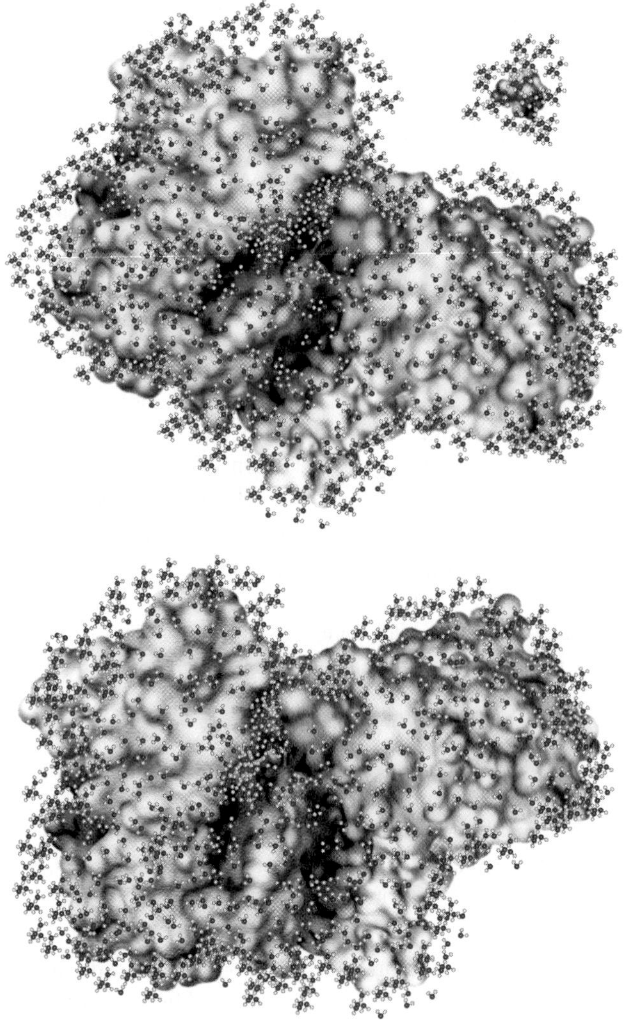

FIGURE 5.23
Hydration of a Protein

Three layers of structured water molecules surround a space-filling model of the enzyme hexokinase before and after binding the sugar glucose. Hexokinase (p. 276) is an enzyme that catalyzes the nucleophilic attack of the carbon–6 hydroxyl group of glucose on the phosphorus in the terminal phosphate of ATP. As the hydrated glucose molecule enters its binding site in a cleft in the enzyme, it sheds its water molecules and displaces water molecules occupying the binding site. The water exclusion process promotes the conformation change that moves the domains together to create the catalytic site. Water exclusion from this site also prevents the unproductive hydrolysis of ATP.

may be identical or quite different. Multisubunit proteins in which some or all subunits are identical are referred to as **oligomers**. Oligomers are composed of **protomers**, which may consist of one or more subunits. A large number of oligomeric proteins contain two or four subunit protomers, referred to as dimers and tetramers, respectively. There appear to be several reasons for the common occurrence of multisubunit proteins:

Efficiency

1. Synthesis of separate subunits may be more efficient than substantially increasing the length of a single polypeptide chain.
2. In supramolecular complexes such as collagen fibers, replacement of smaller worn-out or damaged components can be managed more effectively.
3. The complex interactions among multiple subunits help regulate a protein's biological function.

Polypeptide subunits assemble and are held together by noncovalent interactions such as the hydrophobic and electrostatic interactions and hydrogen bonds, as well as covalent cross-links. As with protein folding, the hydrophobic effect is clearly the most important because the structures of the complementary interfacing surfaces between subunits are similar to those observed in the interior of globular protein domains. Although they are less numerous, covalent cross-links significantly stabilize certain multisubunit proteins. Prominent examples include the disulfide bridges in the immunoglobulins (**Figure 5.24**) and the desmosine and lysinonorleucine linkages in certain connective tissue proteins. *Desmosine*

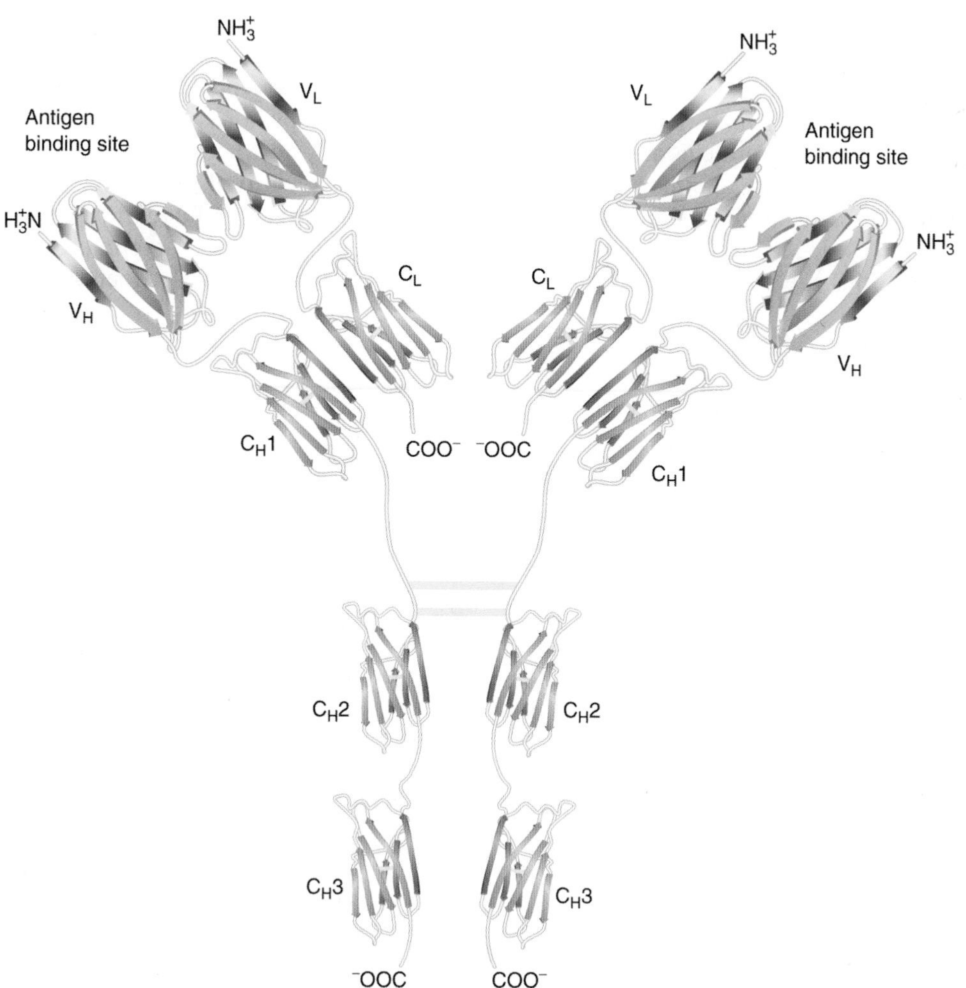

FIGURE 5.24
Structure of Immunoglobulin G (IgG)

IgG is an antibody molecule composed of two heavy chains (H) and two light chains (L) that together form a Y-shaped molecule. Each of the heavy and light chains contains constant (C) and variable (V), β-barrel domains (the classic immunoglobulin fold). The chains are held together by disulfide bridges (yellow lines) and noncovalent interactions. The variable domains of the H and L chains form the site that binds to antigens (foreign molecules). Many antigenic proteins bind to the external surface of these sites. Note that disulfide bridges are also a structural feature within each constant domain.

(**Figure 5.25**) cross-links connect four polypeptide chains in the rubberlike connective tissue protein elastin. They are formed as a result of a series of reactions involving the oxidation and condensation of lysine side chains. A similar process results in the formation of *lysinonorleucine*, a cross-linking structure that is found in elastin and collagen.

Quite often the interactions between subunits are affected by the binding of ligands. In **allostery**, which is the control of protein function through ligand binding, binding a ligand to a specific site in a protein triggers a conformational change that alters its affinity for other ligands. Ligand-induced conformational changes in such proteins are called **allosteric transitions**, and the ligands that trigger them are called **effectors** or **modulators**. Allosteric effects can be positive or negative, depending on whether effector binding increases or decreases the protein's affinity for other ligands. One of the best understood examples of allosteric effects, the reversible binding of O_2 and other ligands to hemoglobin, is described on pp. 172–175. (Because allosteric enzymes play a key role in the control of metabolic processes, allostery is discussed further in Sections 6.3 and 6.5.)

Desmosine

Lysinonorleucine

FIGURE 5.25
Desmosine and Lysinonorleucine Linkages

UNSTRUCTURED PROTEINS In the traditional view of proteins, a polypeptide's function is determined by its specific and relatively stable three-dimensional structure. In recent years, however, as a result of new genomic methodologies and new applications of various forms of spectroscopy, it has become apparent that many proteins are in fact partially or completely unstructured. Unstructured proteins are referred to as **IUPs (intrinsically unstructured proteins)**. If there is a complete lack of ordered structure, the term **natively unfolded proteins** is used. Most IUPs are eukaryotic. Amazingly, estimates of eukaryotic proteins that are partially or completely disordered are as high as 45%, whereas only

QUESTION 5.8

Review the following illustrations of globular proteins. Identify examples of secondary and supersecondary structure.

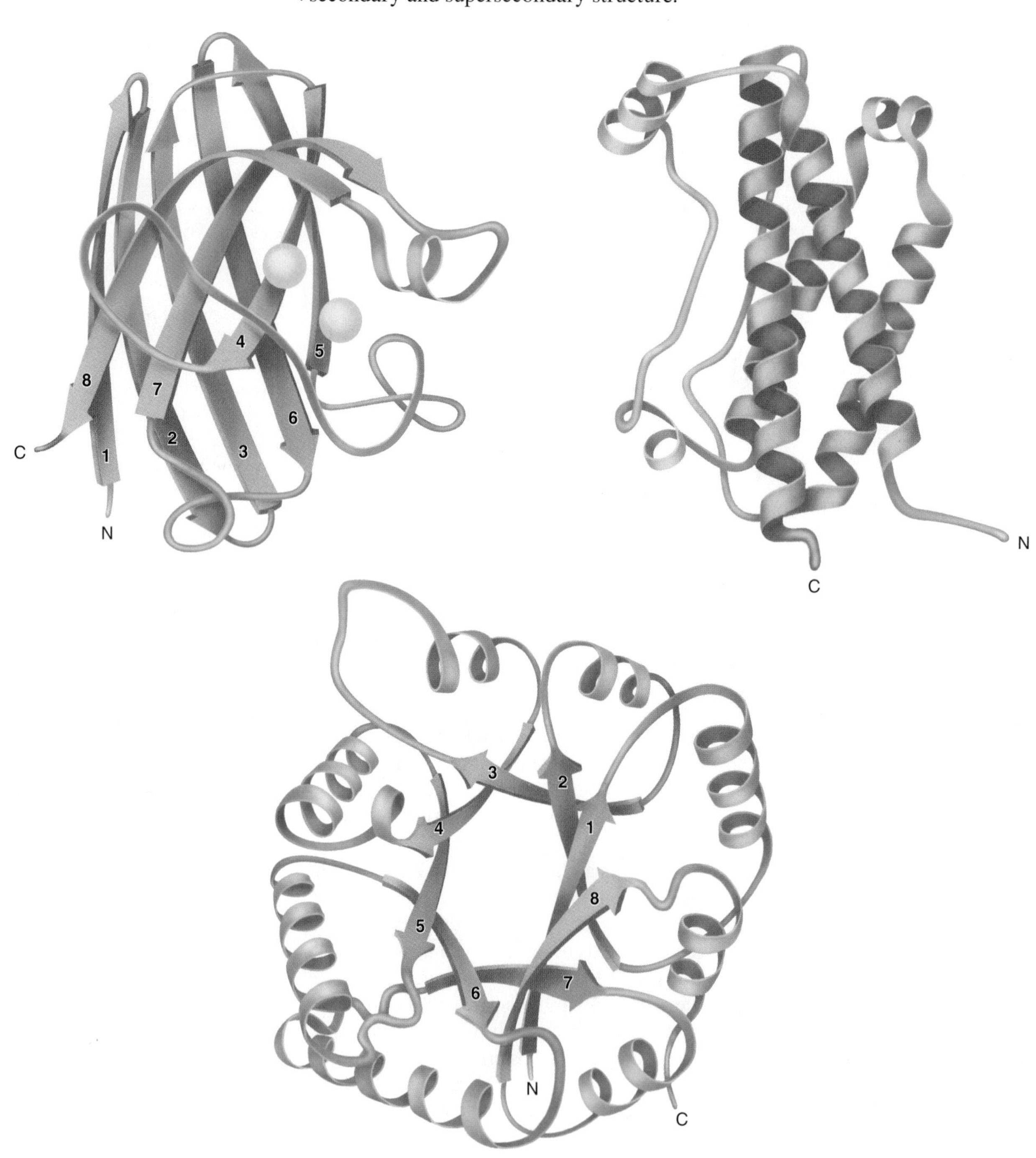

QUESTION 5.9

Illustrate the noncovalent interactions that can occur between the following side chain groups in folded polypeptides: (a) serine and glutamate, (b) arginine and aspartate, (c) threonine and serine, (d) glutamine and aspartate, and (e) phenylalanine and tryptophan.

about 2 and 4% of archaean and bacterial proteins, respectively, can be described as unstructured. The folding of IUPs into stable three-dimensional conformations is prevented by biased amino acid sequences that contain high percentages of polar and charged amino acids (e.g., Ser, Gln, Lys, and Glu) and low quantities of hydrophobic amino acids (e.g., Leu, Val, Phe, and Trp).

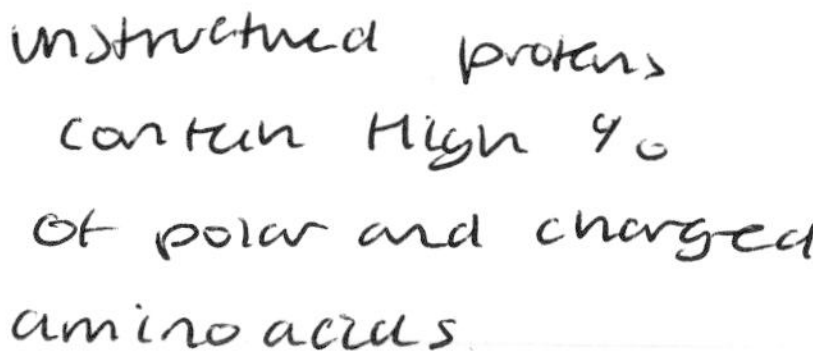

IUPs have a diversity of functions. Many are involved in the regulation of such processes as signal transduction, transcription, translation, cell proliferation, and multiprotein complex assembly. Highly extended and malleable disordered segments enable the molecule to "search" for binding partners. For example, CREB, a transcription regulatory protein discussed later (p. 591), binds to CRE, one type of DNA sequence called a *response element*. When the KID (*k*inase *i*nducible *d*omain) of CREB is phosphorylated by a kinase (an enzyme that attaches phosphate groups to specific amino acid side chains) it becomes unstructured. The unstructured phosphorylated KID (pKID) domain is then able to search out and bind to a domain of CREB-binding protein (CBP) called KIX (KID-binding domain) (**Figure 5.26**). As often happens with IUPs, the disordered pKID domain transitions into a more ordered conformation as it binds to the KIX domain of CBP. As a result of CREB-CBP binding, CREB forms a dimer that alters the expression of certain genes when it binds to its response element.

p53 provides an example of the utility of unstructured protein domains. p53 is a major tumor suppressor (p. 717), as indicated by the fact that p53 mutations occur in least 50% of all cancers. Active as a homotetramer, p53 regulates the expression of hundreds of genes, most notably those involved in cancer-suppression processes such as DNA repair, cell cycle arrest, autophagy, apoptosis, and energy metabolism. p53 is activated in response to a variety of stimuli such as DNA damage, ER stress, ultraviolet (UV) light, and hypoxia. Each p53 polypeptide possesses four domains: an N-terminal unstructured transactivation domain (a peptide sequence that recruits and binds to numerous transcription factors), a DNA-binding domain, a tetramerization domain, and a C-terminal domain that is involved in nuclear localization. p53 can integrate information from multiple signaling pathways because covalent modifications (e.g., phosphorylation and acetylation) of its unstructured N-terminal domain result in structural variations that allow interactions with a wide variety of proteins.

KEY CONCEPTS

- Biochemists distinguish four levels of the structural organization of proteins.
- In primary structure, the amino acid residues are connected by peptide bonds.
- The secondary structure of polypeptides is stabilized by hydrogen bonds. Prominent examples of secondary structure are α-helices and β-pleated sheets.
- Tertiary structure is the unique three-dimensional conformation that a protein assumes because of the interactions between amino acid side chains. Several types of interaction stabilize tertiary structure: hydrophobic and electrostatic interactions, hydrogen bonds, and certain covalent bonds.
- Proteins that consist of several separate polypeptide subunits exhibit quaternary structure.
- Both noncovalent and covalent bonds hold the subunits together. Some proteins are partially or completely unstructured.

LOSS OF PROTEIN STRUCTURE Considering the small differences in the free energy of folded and unfolded proteins, it is not surprising that protein structure

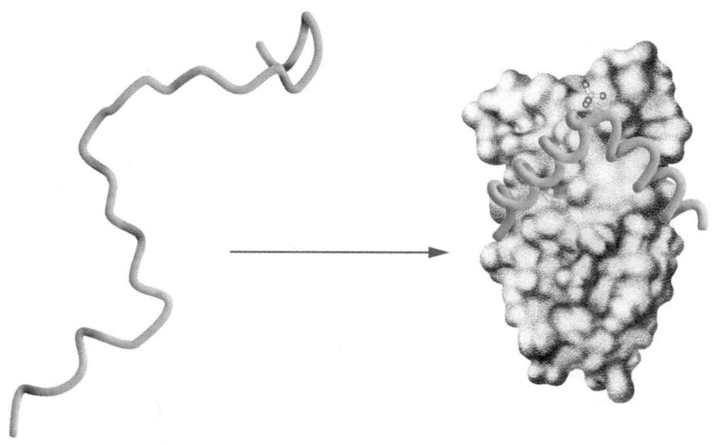

FIGURE 5.26
Disordered Protein Binding

The intrinsically disordered phosphorylated KID domain (pKID) (left) of the transcription regulatory protein CREB searches out and binds to the KIX domain of the transcription coactivator protein CBP (right). As pKID binds to KIX, it undergoes a disorder-to-order transition as it folds into a pair of helices.

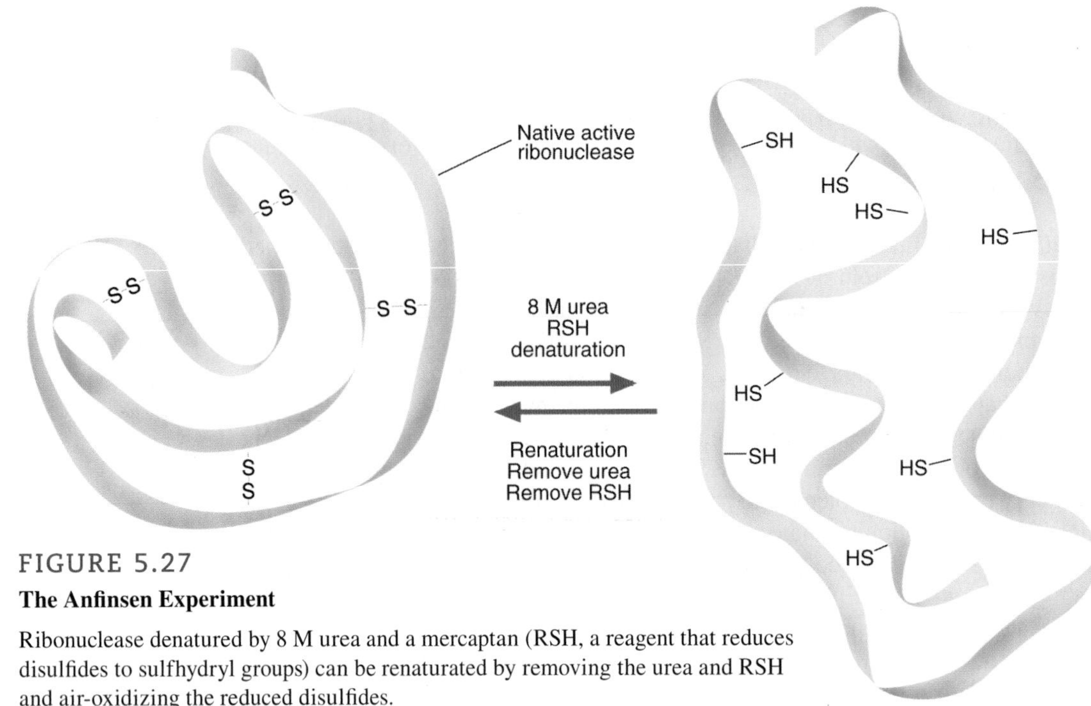

FIGURE 5.27

The Anfinsen Experiment

Ribonuclease denatured by 8 M urea and a mercaptan (RSH, a reagent that reduces disulfides to sulfhydryl groups) can be renaturated by removing the urea and RSH and air-oxidizing the reduced disulfides.

is especially sensitive to environmental factors. Many physical and chemical agents can disrupt a protein's native conformation. The process of structure disruption, which may or may not involve protein unfolding, is called **denaturation**. (Denaturation is not usually considered to include the breaking of peptide bonds.) Depending on the degree of denaturation, the molecule may partially or completely lose its biological activity. Denaturation often results in easily observable changes in the physical properties of proteins. For example, soluble and transparent egg albumin (egg white) becomes insoluble and opaque on heating. Like many denaturations, cooking eggs is an irreversible process.

The following example of a reversible denaturation was demonstrated in the 1950s by Christian Anfinsen, who shared the Nobel Prize in Chemistry in 1972. Bovine pancreatic ribonuclease (a digestive enzyme from cattle that degrades RNA) is denatured when treated with β-mercaptoethanol and 8 M urea (**Figure 5.27**). During this process, ribonuclease, composed of a single polypeptide with four disulfide bridges, completely unfolds and loses all biological activity. Careful removal of the denaturing agents with dialysis results in a spontaneous and correct refolding of the polypeptide and re-formation of the disulfide bonds. Anfinsen's experimental treatment resulting in a full restoration of the enzyme's catalytic activity provided an important early insight into the roles of different forces and primary structure in protein folding. However, most proteins treated similarly do not renature.

Denaturing conditions include the following:

1. **Strong acids or bases**. Changes in pH alter the protonation state of certain protein side chain groups, which in turn alters hydrogen bonding and salt bridge patterns. As a protein approaches its isoelectric point, it becomes less soluble and may precipitate from solution.
2. **Organic solvents**. Water-soluble organic solvents such as ethanol interfere with hydrophobic interactions because they interact with nonpolar R groups and form hydrogen bonds with water and polar protein groups. Nonpolar solvents also disrupt hydrophobic interactions.
3. **Detergents**. Detergents are substances that disrupt hydrophobic interactions, causing proteins to unfold into extended polypeptide chains. These molecules are called **amphipathic** because they contain both hydrophobic and hydrophilic components.

4. **Reducing agents**. In the presence of reagents such as urea, reducing agents (e.g., β-mercaptoethanol) convert disulfide bridges to sulfhydryl groups. Urea disrupts hydrogen bonds and hydrophobic interactions.
5. **Salt concentration**. When there is an increase in the salt concentration of an aqueous solution of protein, some of the water molecules that interact with the protein's ionizable groups are attracted to the salt ions. As the number of solvent molecules available to interact with these groups decreases, protein-protein interactions increase. If the salt concentration is high enough, there are so few water molecules available to interact with ionizable groups that the solvation spheres surrounding the protein's ionized groups are removed. The protein molecules aggregate and then precipitate. This process is referred to as *salting out*. Because salting out is usually reversible and different proteins salt out at different salt concentrations, it is often used as an early step in protein purification.
6. **Heavy metal ions**. Heavy metals such as mercury (Hg^{2+}) and lead (Pb^{2+}) affect protein structure in several ways. They may disrupt salt bridges by forming ionic bonds with negatively charged groups. Heavy metals also bond with sulfhydryl groups, a process that may result in significant changes in protein structure and function. For example, Pb^{2+} binds to sulfhydryl groups in two enzymes in the heme synthetic pathway. The resultant decrease in hemoglobin synthesis causes severe anemia. (In anemia the number of red blood cells or the hemoglobin concentration is lower than normal.) Anemia is one of the most easily measured symptoms of lead poisoning.
7. **Temperature changes**. As the temperature increases, the rate of molecular vibration increases. Eventually, weak interactions such as hydrogen bonds are disrupted and the protein unfolds. Some proteins are more resistant to heat denaturation, and this fact can be used in purification procedures.
8. **Mechanical stress**. Stirring and grinding actions disrupt the delicate balance of forces that maintain protein structure. For example, the foam formed when egg white is beaten vigorously contains denatured protein.

Visit the companion website at www.oup.com/us/mckee to read the Biochemistry in Perspective essays on lead poisoning and heme biosynthesis (Chapter 14).

The Folding Problem

The direct relationship between a protein's primary sequence and its final three-dimensional conformation, and by extension its biological activity, is among the most important assumptions of modern biochemistry. One of the principal underpinnings of this paradigm has already been mentioned: the series of experiments reported by Christian Anfinsen in the late 1950s. Working with bovine pancreatic RNase, Anfinsen demonstrated that under favorable conditions a denatured protein could refold into its native and biologically active state (**Figure 5.27**). This discovery suggested that the three-dimensional structure of any protein could be predicted if the physical and chemical properties of the amino acids and the forces that drive the folding process (e.g., bond rotations, free energy considerations, and the behavior of amino acids in aqueous environments) were understood. Unfortunately, several decades of painstaking research with the most sophisticated tools available (e.g., X-ray crystallography and nuclear magnetic resonance (NMR) in combination with site-directed mutagenesis and computer-based mathematical modeling) resulted in only limited progress. However, this work did reveal that protein folding is a stepwise process in which secondary structure formation (i.e., α-helix and β-pleated sheet) is an early feature. Hydrophobic interactions are an important force in folding. In addition, amino acid substitutions experimentally introduced into certain proteins reveal that changes in surface amino acids rarely affect the protein's structure. In contrast, substitutions of amino acids within the hydrophobic core often lead to serious structural changes in conformation.

Visit the companion website at www.oup.com/us/mckee to read the Biochemistry in Perspective essay on protein folding and human health.

In recent years important advances have been made by biochemists in protein-folding research. Protein-folding researchers have determined that the process does not consist, as was originally thought, of a single pathway. Instead, there are

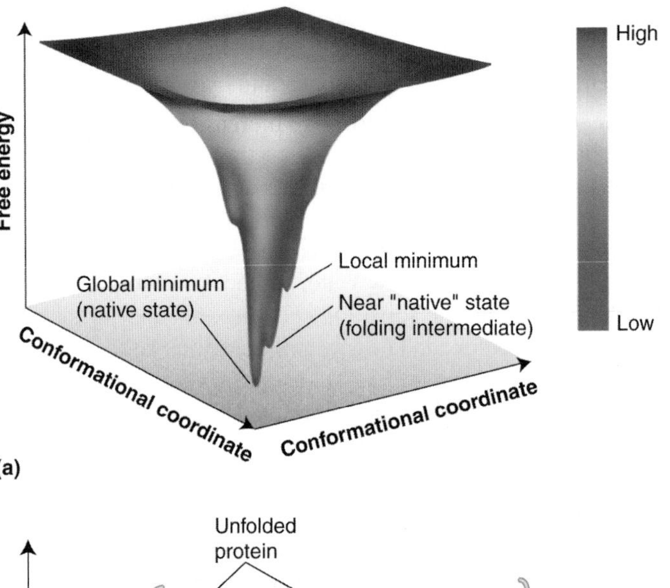

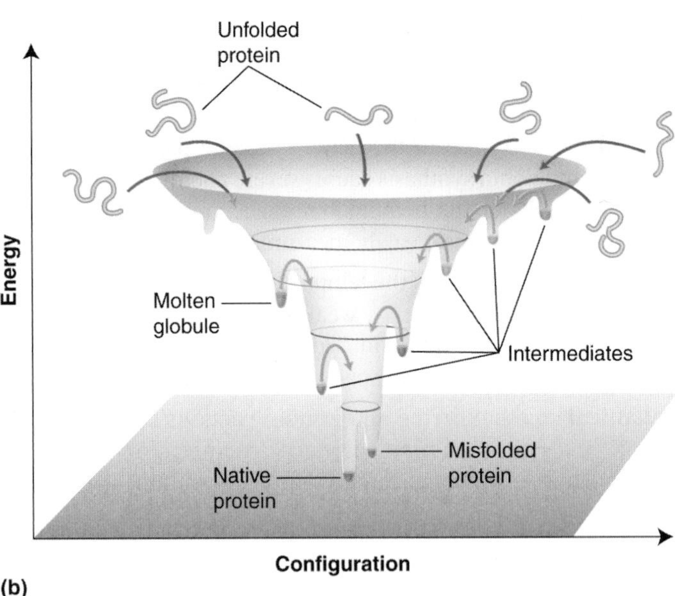

FIGURE 5.28

The Energy Landscape for Protein Folding

(a) Color is used to indicate the entropy level of the folding polypeptide. As folding progresses the polypeptide moves from a disordered state (high entropy, red) toward a progressively more ordered conformation until its unique biologically active conformation is achieved (lower entropy, blue). (b) A depiction of the conformational state of a polypeptide during folding: polypeptides can fold into their native states by several different pathways. Many molecules form transient intermediates, whereas others may become trapped in a misfolded state.

numerous routes that a polypeptide can take to fold into its native state. As illustrated in **Figure 5.28a**, an energy landscape with a funnel shape appears to best describe how an unfolded polypeptide with its own unique set of constraints (e.g., its amino acid sequence and posttranslational modifications, as well as environmental features within the cell such as temperature, pH, and molecular crowding) negotiates its way to a low-energy folded state. Depending largely on its size, a polypeptide may or may not form intermediates (species existing long enough to be detected) that are momentarily trapped in local energy wells (**Figure 5.28b**). Small molecules (fewer than 100 residues) often fold without intermediate formation (**Figure 5.29a**). As these molecules begin emerging from the ribosome, a rapid and cooperative folding process begins in which side chain interactions facilitate the formation and alignment of secondary structures. The folding of larger polypeptides typically involves the formation of several intermediates (**Figure 5.29b**, c). In many of these molecules or the domains within a molecule, the hydrophobically collapsed shape of the intermediate is referred to as a molten globule. The term **molten globule** refers to a partially organized globular state of a folding polypeptide that resembles the molecule's native state. Within the interior of a molten globule, tertiary interactions among amino acid side chains are fluctuating; that is, they have not yet stabilized.

Many proteins require assistance in folding into their native conformations. The **molecular chaperones** are a network of proteins that play a central role in

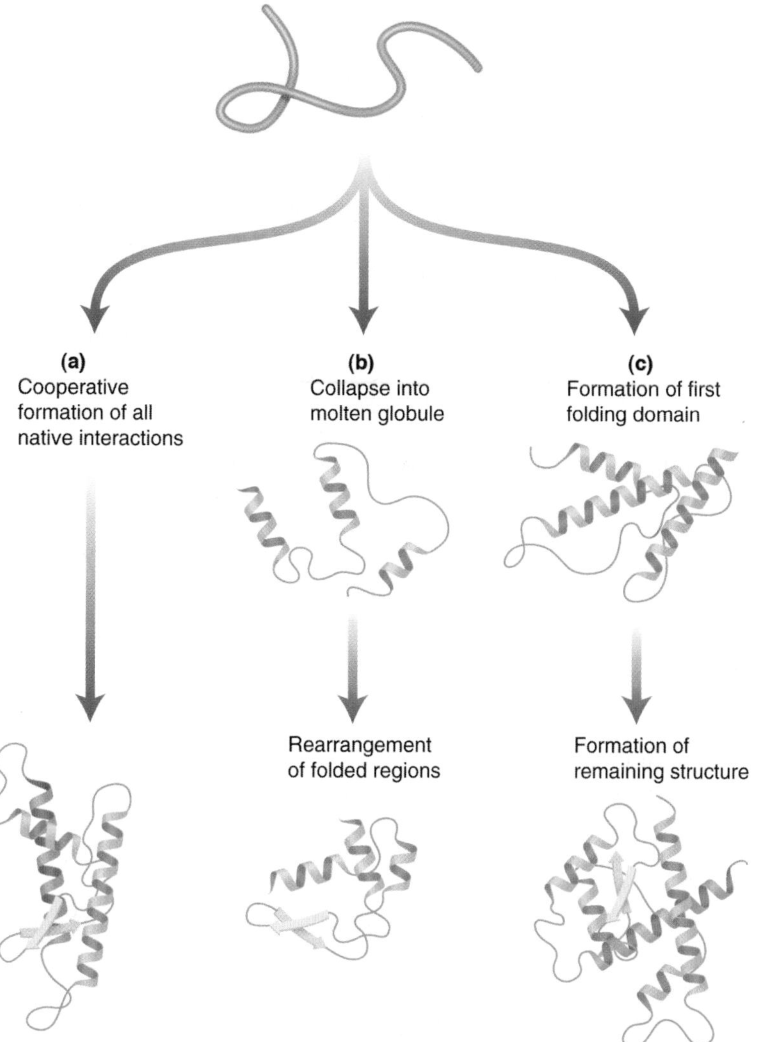

FIGURE 5.29
Protein Folding

(a) In many small proteins, folding is cooperative with no intermediates formed. (b) In some larger proteins, folding involves the initial formation of a molten globule followed by rearrangement into the native conformation. (c) Large proteins with multiple domains follow a more complex pathway, with each domain folding separately before the entire molecule progresses to its native conformation.

cellular protein quality control by facilitating the folding of *nascent* (newly synthesized) proteins and the refolding of preexisting proteins. In addition, they also assist in the assembly of multisubunit proteins and other protein-containing structures (e.g., chromatin) and, if necessary, target misfolded proteins to the cell's degradation pathways (pp. 556–559). Many molecular chaperones are hsps, so named because they were originally characterized in cells exposed to high temperatures. Found in organisms ranging from bacteria to animals and plants, the molecular chaperones have a high degree of sequence homology. In eukaryotes, they are found in cytoplasm and several organelles (mitochondria, ER, and chloroplasts). The properties of these important molecules are described next.

MOLECULAR CHAPERONES Molecular chaperones apparently assist unfolded proteins by protecting them from inappropriate hydrophobic protein-protein interactions that can result in misfolding or aggregation. There are four groups of molecular chaperones involved in *de novo* (new) protein folding:

1. **Ribosome-Associated Chaperones.** Chaperone action begins as the nascent polypeptide begins to emerge from the ribosome's exit tunnel. Proteins such as *trigger factor* in bacteria (p. 743), RAC (*r*ibosome-*a*ssociated *c*omplex) in yeast, and *NAC* (*n*ascent polypeptide-*a*ssociated *c*omplex) in eukaryotes bind to the ribosome and the emerging polypeptide. This binding process prevents folding until an entire domain or polypeptide has emerged from the tunnel.

2. **Hsp70s**. The hsp70s bind to, stabilize, and promote the folding of nascent polypeptides. They are also involved in the refolding of misfolded and aggregated proteins and in the transmembrane transfer of organelle (e.g., ER, mitochondria and chloroplast) or secretory polypeptides. Each hsp70 possesses two domains connected by a short interdomain linker: an N-terminal ATP-binding domain and a domain containing a substrate-binding pocket with an affinity for peptides with hydrophobic amino acid residues. There is also a C-terminal α-helical structure that acts as a lid for the substrate-binding pocket. Hsp70 ATPase activity is stimulated when a peptide segment enriched in hydrophobic amino acid residues binds to the binding pocket. Hsp70s interact with *co-chaperones*, accessory proteins that assist hsps in individual steps in chaperone activities. For example, hsp40 assists hsp70 by regulating ATP hydrolysis and mediating binding to unfolded or oxidized protein substrates. Hsp100 assists hsp70 in the disassembly of protein aggregates. When ATP in the N-terminal domains of an hsp70 dimer are hydrolyzed, the lid closes, thus trapping a peptide segment of the substrate protein. Peptide segments are released from Hsp70s as a result of a conformational change initiated by the exchange of ADP for ATP. On its release, a polypeptide either folds into its functional conformation, or Hsp70 and its associated cochaperones transfer it to other downstream chaperones.
3. **Hsp90s**. The hsp90s, found in organisms ranging from bacteria to mammals, have roles in diverse cellular pathways. In eukaryotes the hsp90s are found in cytoplasm, nucleus, mitochondria, and chloroplasts where they do not bind to nascent polypeptides. Instead Hsp90 proteins finalize the folding of a limited, but diverse, set of partially unfolded molecules, referred to as *client proteins*. Reversibly linked to hsp70 by the co-chaperone HOP (hsp70-hsp90 organizing protein), hsp90 completes the folding process. Hsp70 and hsp90 also work together to identify proteins damaged by oxidative or heat stress and either refold them or target them for proteasome-mediated destruction. In addition, hsp90 also coordinates the assembly of protein complexes such as RNA polymerase II (p. 703), RNA-induced silencing complex (p. 716), and 26S proteasome (p. 557). Hsp90 functions as a dimer. Each hsp90 molecule is composed of three domains: the N-terminal domain, containing an ATP binding site, a middle domain that is involved in client protein binding and ATP hydrolysis regulation, and the C-terminal domain, which contains the interaction site for dimer formation. In an ATP-regulated cycle, an open hsp90 dimer (V-shaped) binds a client protein and closes in a clamplike motion driven by ATP hydrolysis. Following ATP hydrolysis, the dimer opens, releasing the now folded client protein.
4. **Chaperonins**. The chaperonins are large, double-ring complexes that promote faster, more efficient polypeptide refolding within an internal compartment. They are classified into two groups. Group I chaperonins, found in bacteria, mitochondria, and chloroplasts, are composed of two stacked seven-membered rings. Known as GroEL (**Figure 5.30**) in bacteria and hsp60s in eukaryotes, their function requires lid-shaped cochaperones. TRIC, a group II chaperonin in the cytoplasm of eukaryotic cells, is composed of two eight-member rings with a built-in lid. Protein folding commences after lid closure entraps a substrate protein. ATP hydrolysis converts the cavity within GroEL into a hydrophilic microenvironment that facilitates the collapse of the hydrophobic core of the folding protein into the molten globule form. It takes 15–20 s for all seven ATPs to hydrolyze in the ring subunits and to complete the folding process. In the $(ADP)_7$ state, the hydrophobic character of the cavity returns, the chamber opens, and the folded protein or domain is released. A new unfolded protein can now bind to repeat the cycle. Protein folding proceeds with two cycles

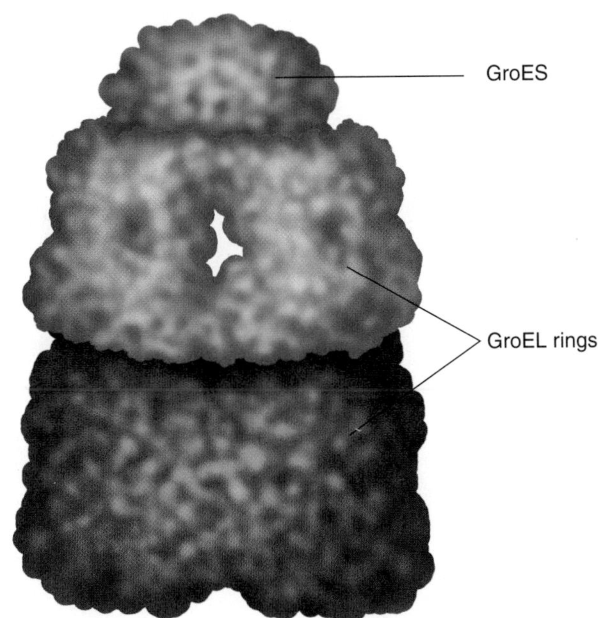

FIGURE 5.30

Space-Filling Model of the *E. coli* Chaperonin Called the GroES-GroEL Complex

GroES (a *co-chaperonin*, or *hsp10*) is a seven-subunit ring that sits on top of GroEL. GroEL (a *chaperonin*, or *hsp60*) is composed of two stacked, seven-subunit rings with a cavity in which ATP-dependent protein folding takes place.

occurring in an overlapping fashion depending on the ATP/ADP binding status of the two cavities. The structure of GroEL and its lid-shape co-chaperone GroES is illustrated in **Figure 5.31**. TRiC is a eukaryotic group II chaperonin composed of two eight-member rings that contains its own lid. Along with its co-chaperone hsp70, TRiC facilitates the cotranslational folding of proteins with complex domain folds, most notably actin and tubulin.

In addition to promoting the folding of nascent protein, molecular chaperones direct the refolding of protein that was partially unfolded as a consequence of stressful conditions. If refolding is not possible, molecular chaperones promote protein degradation. A diagrammatic view of protein folding involving GroEL and its lid co-chaperone, GroES, is presented in **Figure 5.31**. The effects of protein misfolding on human health can be considerable. Both Alzheimer's and Huntington's diseases are neurodegenerative diseases caused by accumulations of insoluble protein aggregates. (See the online Biochemistry in Perspective essay Protein Folding and Human Disease.)

KEY CONCEPTS

- All the information required for each newly synthesized polypeptide to fold into its biologically active conformation is encoded in the molecule's primary sequence.
- Some relatively simple polypeptides fold spontaneously into their native conformations.
- Other larger molecules require the assistance of proteins called molecular chaperones to ensure correct folding.

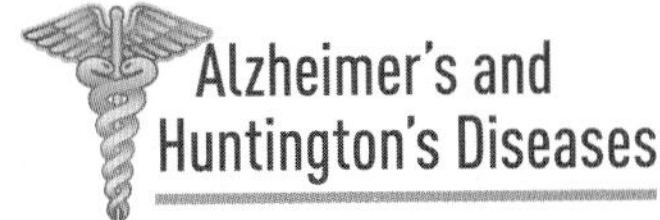

Fibrous Proteins

Fibrous proteins typically contain high proportions of regular secondary structures, such as α-helices or β-pleated sheets. As a consequence of their rodlike or sheetlike shapes, many fibrous proteins have structural rather than dynamic roles. Keratin (**Figure 5.32**) is a fibrous protein composed of bundles of α-helices, whereas the polypeptide chains of the silkworm silk protein fibroin (**Figure 5.33**) are arranged in antiparallel β-pleated sheets. The structural features of collagen, the most abundant protein in vertebrates, are described in some detail.

COLLAGEN Collagen is synthesized by connective tissue cells and then secreted into the extracellular space to become part of the connective tissue matrix. The 28 major families of collagen molecules include many closely related proteins that have diverse functions. The genetically distinct collagen molecules in skin, bones, tendons, blood vessels, and corneas impart to these structures many of their special properties (e.g., the tensile strength of tendons and the transparency of corneas).

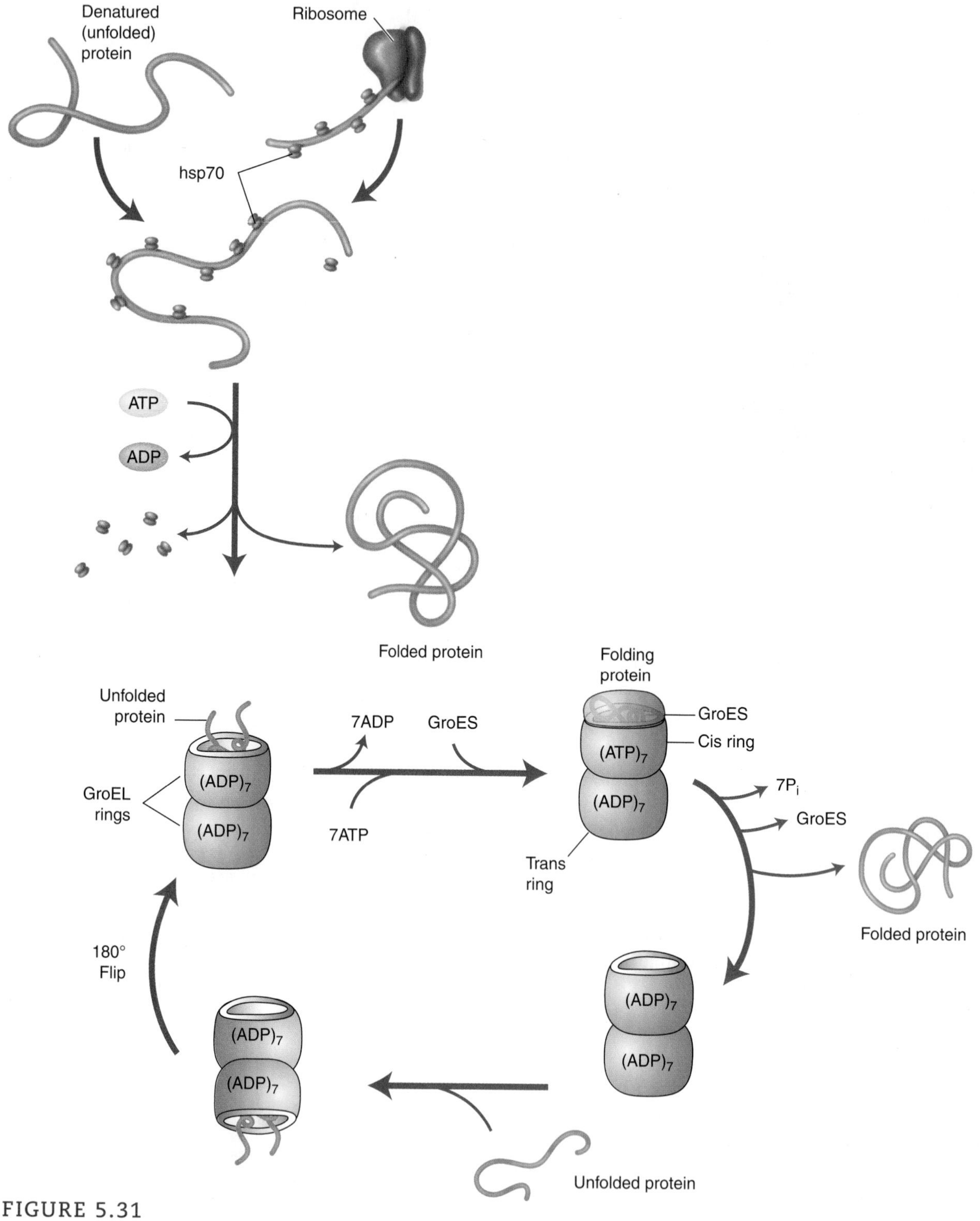

FIGURE 5.31

Molecular Chaperone-Assisted Protein Folding

Molecular chaperones bind transiently to both nascent proteins and unfolded proteins (i.e., those denatured by stressful conditions). The members of the hsp70 family stabilize nascent proteins and reactivate some denatured proteins. Many proteins also require hsp60 protein complexes to achieve their final conformations. In *E. coli*, cellular proteins that do not fold spontaneously require processing by the GroEL-GroES complex. [Note that during a folding cycle the GroEL ring capped by GroES is referred to as the *cis*-ring. The other attached ring, the one that has not yet initiated a protein folding cycle, is called the *trans*-ring.] At the beginning of a folding cycle, an unfolded protein (or protein domain) is loosely bound via hydrophobic interactions to the cavity entrance of one of the GroEL-$(ADP)_7$ rings. ADP/ATP exchange converts the cavity to a hydrophobic, expanded microenvironment that then traps the protein substrate under a GroES lid. Subsequently, sequential hydrolysis of the seven ATPs converts the cavity to a hydrophilic microenvironment, driving both the formation of the molten globule state of the protein substrate and the progression of the folding process. When all seven ATPs have been hydrolyzed, the hydrophobic surface of the cavity is reestablished and GroES and the newly folded protein leave the GroEL ring. Meanwhile, the trans-GroEL-$(ADP)_7$ ring is already beginning the loading, trapping, and folding process for another unfolded protein or domain.

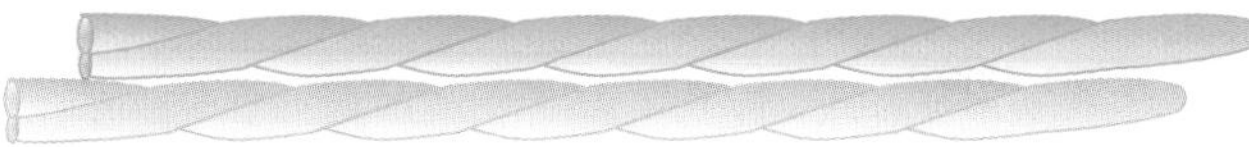

FIGURE 5.32
α-Keratin

The α-helical rodlike domains of two keratin polypeptides form a coiled coil. Two staggered antiparallel rows of these dimers form a supercoiled protofilament. Hydrogen bonds and disulfide bridges are the principal interactions between subunits. Hundreds of filaments, each containing four protofilaments, form a macrofibril. Each hair cell, also called a fiber, contains several macrofibrils. Each strand of hair consists of numerous dead cells packed with keratin molecules. In addition to hair, the keratins are also found in wool, skin, horns, and fingernails.

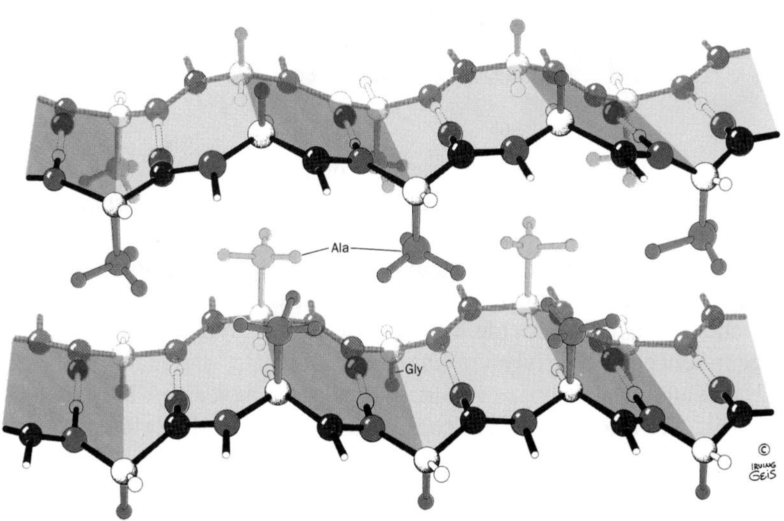

FIGURE 5.33
Molecular Model of Silk Fibroin

In fibroin, the silk fibrous protein produced by silkworms, the polypeptide chains are arranged in fully extended antiparallel β-pleated sheet conformations. Note that the R groups of alanine on one side of each β-pleated sheet interdigitate with similar residues on the adjacent sheet. Silk fibers (fibroin embedded in an amorphous matrix) are flexible because the pleated sheets are loosely bonded to each other (primarily with weak van der Waals forces) and slide over each other easily.

Collagen is composed of three left-handed polypeptide helices that are twisted around each other to form a right-handed triple helix (**Figure 5.34**). Type I collagen molecules, found in teeth, bone, skin, and tendons, are about 300 nm long and 1.5 nm wide. Approximately 90% of the collagen found in humans is type I.

The amino acid composition of collagen is distinctive. Glycine constitutes approximately one-third of the amino acid residues. Proline and 4-hydroxyproline may account for as much as 30% of a collagen molecule's amino acid composition. Small amounts of 3-hydroxyproline and 5-hydroxylysine also occur. Specific proline and lysine residues in collagen's primary sequence are hydroxylated within the rough ER after the polypeptides have been synthesized. These reactions, which are discussed in Chapter 19, require ascorbic acid (p. 754).

Collagen's amino acid sequence primarily consists of large numbers of repeating triplets with the sequence of Gly—X—Y, in which X and Y are often proline and hydroxyproline. Hydroxylysine is also found in the Y position. Simple carbohydrate groups are often attached to the hydroxyl group of hydroxylysine residues. It has been suggested that collagen's carbohydrate components are required for *fibrilogenesis*, the assembly of collagen fibers in their extracellular locations, such as tendons and bone.

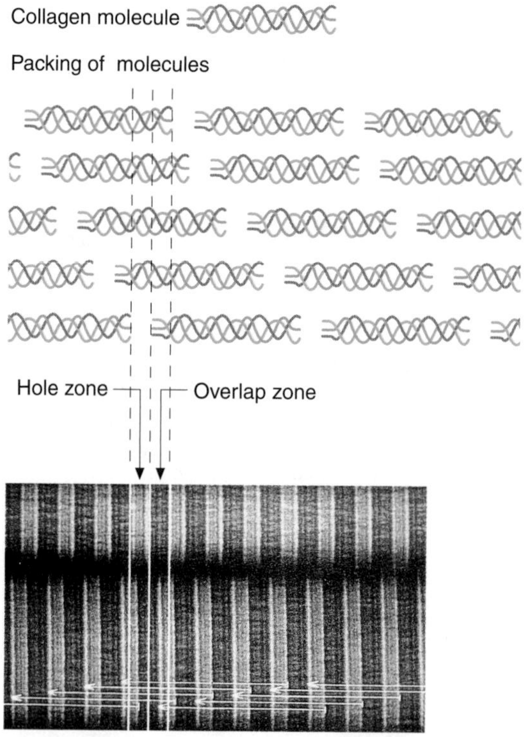

FIGURE 5.34

Collagen Fibrils

The bands are formed by staggered collagen molecules. Cross-striations are about 680 Å apart. Each collagen molecule is about 3000 Å long.

Collagen molecule

Packing of molecules

Hole zone

Overlap zone

The enzyme lysyl oxidase converts some of the lysine and hydroxylysine side groups to aldehydes through oxidative deamination, and this facilitates the spontaneous nonenzymatic formation of strengthening aldimine and aldol cross-links. (An aldol cross-link is formed in a reaction, called an **aldol condensation**, in which two aldehydes form an α, β-unsaturated aldehyde linkage. In condensation reactions, a small molecule, in this case H_2O, is removed.) Cross-linkages also occur between hydroxylysine-linked carbohydrates and the amino group of other lysine and hydroxylysine residues on adjacent molecules. Increased cross-linking with age leads to the brittleness and breakage of the collagen fibers that occur in older organisms.

Glycine is prominent in collagen sequences because the triple helix is formed by interchain hydrogen bonding involving the glycine residues. Therefore every third residue is in close contact with the other two chains. Glycine is the only amino acid with an R group sufficiently small for the space available. Larger R groups would destabilize the superhelix structure. The triple helix is further strengthened by hydrogen bonding between the polypeptides (caused principally by the large number of hydroxyproline residues) and lysinonorleucine linkages that stabilize the orderly arrays of triple helices in the final collagen fibril.

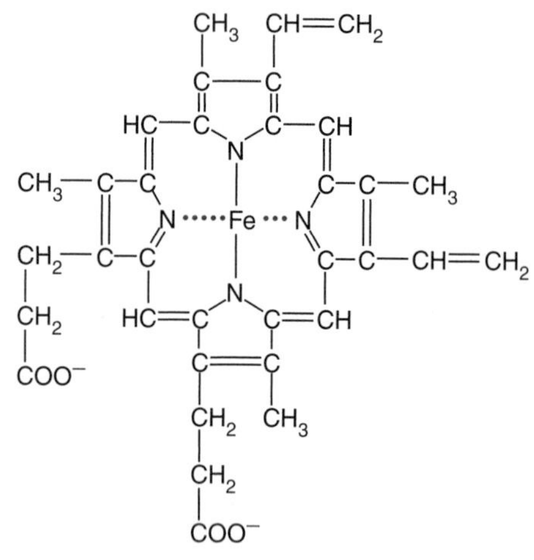

FIGURE 5.35

Heme

Heme consists of a porphyrin ring (composed of four pyrroles) with Fe^{2+} in the center.

Globular Proteins

The biological functions of globular proteins usually involve the precise binding of small ligands or large macromolecules such as nucleic acids or other proteins. Each protein possesses one or more unique cavities or clefts whose structure is complementary to a specific ligand. After ligand binding, a conformational change occurs in the protein that is linked to a biochemical event. For example, the binding of ATP to myosin in muscle cells is a critical event in muscle contraction.

The oxygen-binding proteins myoglobin and hemoglobin are interesting and well-researched examples of globular proteins. They are both members of the hemoproteins, a specialized group of proteins that contain the prosthetic group heme. Although the heme group (**Figure 5.35**) in both proteins is responsible for

QUESTION 5.10

Covalent cross-links contribute to the strength of collagen. The first reaction in cross-link formation is catalyzed by the copper-containing enzyme lysyl oxidase, which converts lysine residues to the aldehyde allysine:

Allysine then reacts with other side chain aldehyde or amino groups to form cross-linkages. For example, two allysine residues react to form an aldol cross-linked product:

In a disease called *lathyrism*, which occurs in humans and several other animals, a toxin (β-aminopropionitrile) found in sweet peas *(Lathyrus odoratus)* inactivates lysyl oxidase. Consider the abundance of collagen in animal bodies and suggest some likely symptoms of this malady.

the reversible binding of molecular oxygen, the physiological roles of myoglobin and hemoglobin are significantly different. The chemical properties of heme are dependent on the Fe^{2+} ion in the center of the prosthetic group. Fe^{2+}, which forms six coordinate bonds, is bound to the four nitrogens in the center of the protoporphyrin ring. Two other coordinate bonds are available, one on each side of the planar heme structure. In myoglobin and hemoglobin, the fifth coordination bond is to the nitrogen atom in a histidine residue, and the sixth coordination bond is available for binding oxygen. In addition to serving as a reservoir for oxygen within muscle cells, myoglobin also facilitates the intracellular diffusion of oxygen. The role of hemoglobin, the primary protein of red blood cells, is to deliver oxygen to cells throughout the body. A comparison of the structures of these two proteins illustrates several important principles of protein structure, function, and regulation.

MYOGLOBIN Myoglobin, found in high concentration in skeletal and cardiac muscle, gives these tissues their characteristic red color. The muscles of diving mammals such as whales, which remain submerged for long periods, have high myoglobin concentrations. Because of the extremely high concentrations of

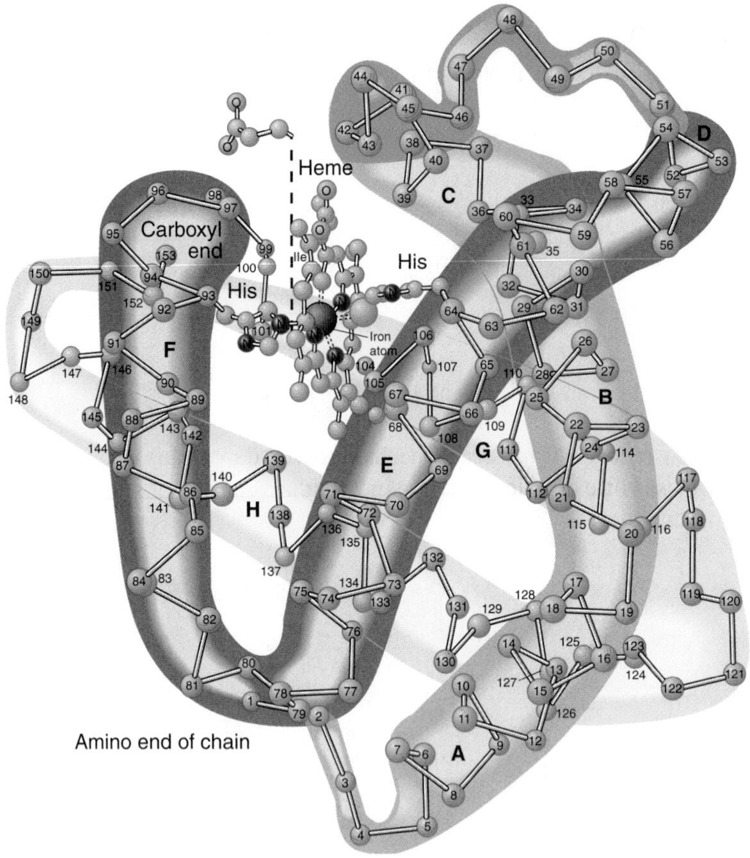

FIGURE 5.36
Myoglobin

With the exception of the side chain groups of two histidine residues, only the α-carbon atoms of the globin polypeptide are shown. Myoglobin's eight helices are designated A through H. The heme group has an iron atom that binds reversibly with oxygen. To improve clarity, one of heme's propionic acid side chains has been displaced.

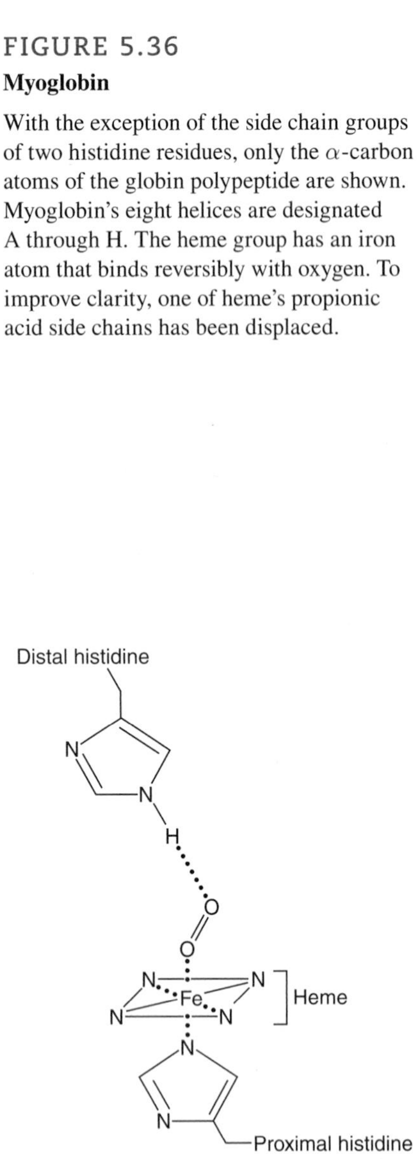

FIGURE 5.37
The Oxygen-Binding Site of Heme Created by a Folded Globin Chain

myoglobin, such muscles are typically brown. The protein component of myoglobin, called globin, is a single polypeptide chain that contains eight segments of α-helix (**Figure 5.36**). The folded globin chain forms a crevice that almost completely encloses a heme group. Free heme [Fe^{2+}] has a high affinity for O_2 and is irreversibly oxidized to form hematin [Fe^{3+}]. Hematin cannot bind O_2. Noncovalent interactions between amino acid side chains and the nonpolar porphyrin ring within myoglobin's oxygen-binding crevice decrease heme's affinity for O_2. The decreased affinity protects Fe^{2+} from oxidation and allows for the reversible binding of O_2. All of the heme-interacting amino acids are nonpolar except for two histidines, one of which (the proximal histidine) binds directly to the iron (**Figure 5.37**). The other (the distal histidine) stabilizes the oxygen-binding site.

Myoglobin's oxygen-dissociation curve (**Figure 5.38**) is hyperbolic where O_2 affinity is high and O_2 saturation occurs at very low partial pressures of O_2 (pO_2). This reflects its structure as a single subunit protein and its role as an O_2 storage protein in muscle tissue. Myoglobin gives up its O_2 only when the muscle cell's oxygen concentration is very low (i.e., during strenuous exercise).

HEMOGLOBIN Hemoglobin is a roughly spherical molecule found in red blood cells, where its primary function is to transport oxygen from the lungs to every tissue in the body. The HbA molecule (**Figure 5.39**) is designated $\alpha_2\beta_2$ (HbA_2 is a variant designated $\alpha_2\delta_2$). Before birth, variants of β-chains are produced: ε in embryonic life and γ in the fetus. Because $\alpha_2\varepsilon_2$ and $\alpha_2\gamma_2$ hemoglobin both have greater affinities for O_2 than $\alpha_2\beta_2$, the fetus can preferentially absorb oxygen from the maternal bloodstream.

Hemoglobin's four chains are arranged in two identical $\alpha\beta$-dimers. Each globin chain has a heme-binding pocket similar to that of myoglobin. Whereas myoglobin

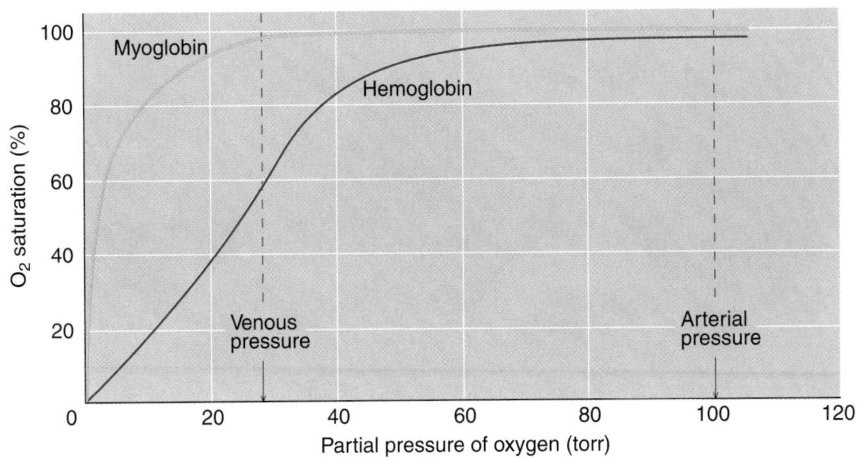

FIGURE 5.38
Dissociation Curves Measure the Affinity of Hemoglobin and Myoglobin for Oxygen.

Arterial blood, enriched in O_2, delivers it to the tissues. Venous blood, which drains from tissues, is O_2 depleted.

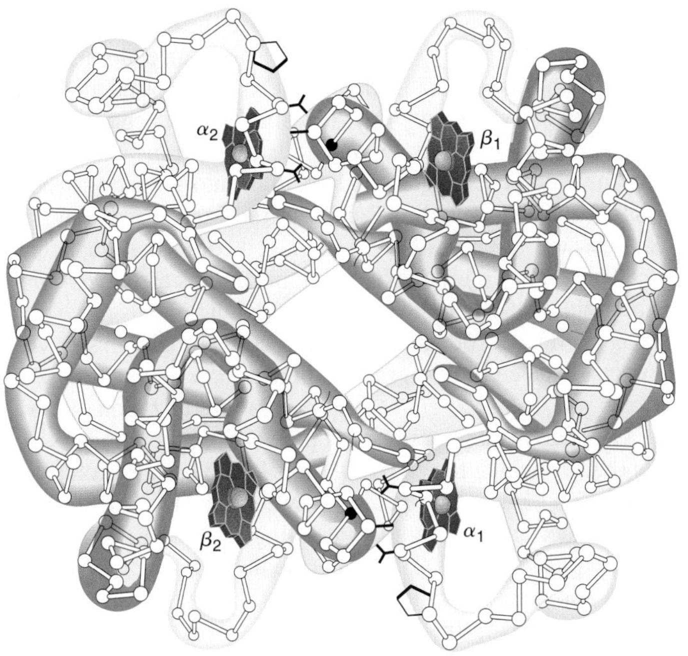

FIGURE 5.39
Hemoglobin

The protein contains four subunits, designated α and β. Each subunit contains a heme group that binds reversibly with oxygen.

shows hyperbolic O_2-binding kinetics, hemoglobin exhibits a sigmoidal curve suggestive of **cooperative binding** (ligand binding by one subunit affects the binding behavior of other subunits) and **allostery** (ligand binding affected by effector molecules). (Cooperativity and allosteric regulation are described in Chapter 6.) When hemoglobin is oxygenated, specific salt bridges and hydrogen bonds between the $\alpha\beta$-dimers are disrupted and the dimers slide past each other and rotate 15° relative to each other (**Figure 5.40**). The deoxygenated conformation (deoxyHb) is referred to as the *T(aut) state* and the oxygenated conformation (oxyHb) is referred to as the *R(elaxed)* state. The oxygen-induced readjustment in the interdimer contacts is almost simultaneous. In other words, a conformational change in one subunit is rapidly propagated to the other subunits. Consequently, hemoglobin alternates between two stable conformations, the T and R states.

The oxygen dissociation curve of hemoglobin (**Figure 5.38**) has a sigmoidal shape because of T to R transitions. When an oxygen molecule binds to one subunit in the low-affinity T state, that subunit shifts to the high-affinity configuration and induces shifts in conformation in the remaining three subunits. Subsequent oxygen molecules bind with high affinity. This shift in binding

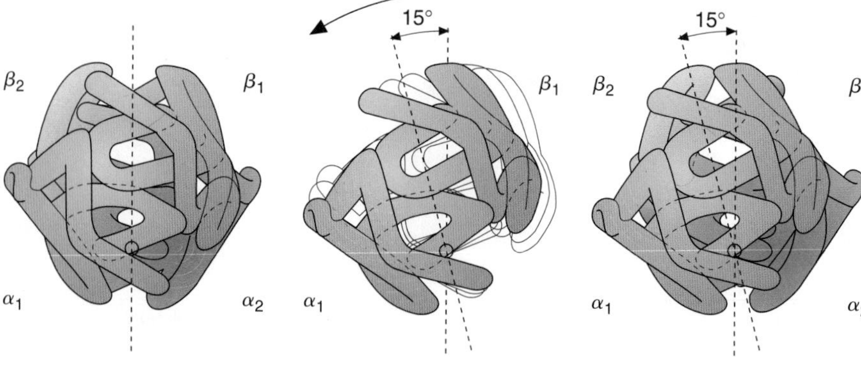

FIGURE 5.40

The Hemoglobin Allosteric Transition

When hemoglobin is oxygenated, the $\alpha_1\beta_1$ and $\alpha_2\beta_2$ dimers slide by each other and rotate 15°.

properties, called cooperative binding, results in a shift from the T to R conformational forms of hemoglobin. In the lungs where pO_2 is high, and the pH is high, hemoglobin is quickly saturated (converted to the R state). Hemoglobin also binds nitric oxide (NO) (p. 537), produced by vascular endothelial cells in the lungs, via thiol groups on the globin chains. NO is transported along with oxygen to the tissues. In tissues where the pO_2 is low and CO_2 concentration is higher and pH is lower, hemoglobin releases O_2 and NO. NO is transported out of the red blood cell via the anion exchanger AE-1 (p. 408) into the bloodstream, where it facilitates changes in the endothelial cells to increase blood flow to the target tissue. O_2 is released as a result of proton-induced conformational changes in the Hb molecule. The α-amino groups of the globin chains are *carbamoylated* (CO_2 attached), stabilizing the T state for return to the lungs.

The binding of ligands other than O_2 affects hemoglobin's oxygen-binding properties, as can be seen in its sigmoidal binding curve when these ligands are present. For example, the *Bohr effect* describes the stabilization of the T state and unloading of O_2 when pH decreases ([H^+] increases). Metabolically active tissue produces lots of CO_2, which forms HCO_3^- and H^+ in blood. Protonation of the Hb subunits reduces salt bridge stabilization between the Hb dimers and induces an R- to T-state transition facilitating the unloading of O_2 and NO.

2,3-Bisphosphoglycerate (BPG), an intermediate of glycolysis (glucose breakdown), binds to and stabilizes the T-state, increasing the amount of O_2 that is unloaded in the tissue (**Figure 5.41**). Although most cells contain only trace

KEY CONCEPTS

- Globular protein function usually involves binding to small ligands or to other macromolecules.
- The oxygen-binding properties of myoglobin and hemoglobin are determined in part by the number of subunits they contain.

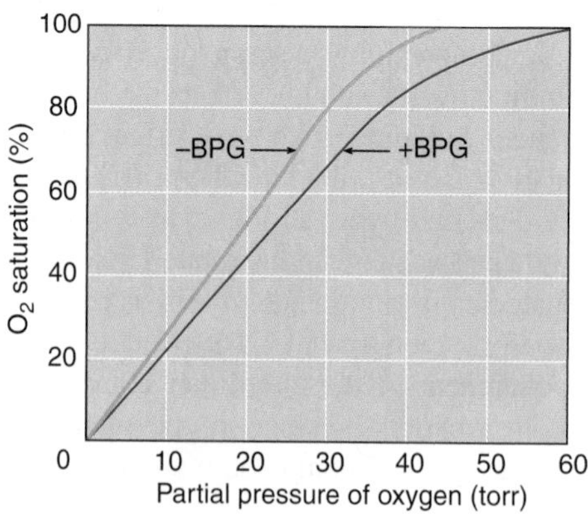

FIGURE 5.41

The Effect of 2,3-Bisphosphoglycerate (BPG) on the Affinity between Oxygen and Hemoglobin

In the absence of BPG (–BPG), hemoglobin has a high affinity for O_2; where BPG is present and binds to hemoglobin (+BPG), its affinity for O_2 decreases.

amounts of BPG, red blood cells and the brain generate a considerable amount. BPG binds to the T-state because a cavity lined with positively charged amino acid residue side chains is exposed in this conformation. Why is this important? BPG stabilizes deoxyHb, thus providing a mechanism for increasing O_2 unloading when energy demand is high.

In the lungs the process is reversed. A high oxygen concentration drives the conversion from the deoxyHb configuration to that of oxyHb. The change in the protein's three-dimensional structure initiated by the binding of the first oxygen molecule releases bound CO, H^+, and BPG. The H^+ recombines with HCO_3^- to form carbonic acid, which then dissociates to form CO_2 and H_2O. Afterward, CO_2 diffuses from the blood into the alveoli.

Carbon monoxide (CO_2) is a competitive inhibitor of hemoglobin because it binds to hemoglobin with an affinity 250 times that of O_2. Cyanide (CN^-) and NO also inhibit O_2 transport, but at higher concentration than CO. HbCO is has a bright red color, so cherry-red skin is a symptom of CO poisoning. In addition, any chemical that oxidizes the Fe^{2+} results in the formation of Fe^{3+}-Hb or methemoglobin that does not bind O_2, although the oxidation state of iron in oxyHb is near 3. The extra electron in Fe^{2+} is required to coordinate with an unpaired electron in the O_2 molecule to generate oxyHb.

QUESTION 5.11

Fetal hemoglobin binds to BPG to a lesser extent than does HbA because His 143 in the BPG binding pocket in β-globin has been replaced with a serine residue in γ-globin. As a result of the loss of two positive charges (one for each of two γ-globins), the binding pocket binds BPG less avidly. What are the consequences of this phenomenon for mother and fetus?

QUESTION 5.12

Myoglobin stores O_2 in muscle tissue to be used by the mitochondria only when the cell is in oxygen debt, whereas hemoglobin can effectively transport O_2 from the lungs and deliver it discriminately to cells in need of O_2. Describe the structural features that allow these two proteins to accomplish separate functions.

5.4 MOLECULAR MACHINES

Purposeful movement is the hallmark of living organisms. This behavior takes myriad forms that range from the record-setting 110 km/h chasing sprint of the cheetah to more subtle movements such as the migration of white blood cells in the animal body, cytoplasmic streaming in plant cells, intracellular transport of organelles, and the enzyme-catalyzed unwinding of DNA. The multisubunit proteins responsible for these phenomena (e.g., the muscle sarcomere and various other types of cytoskeletal components and DNA polymerase) function as biological machines. Machines are defined as mechanical devices with moving parts that perform work (the product of force and distance). When machines are used correctly, they permit the accomplishment of tasks that would often be impossible without them. Although biological machines are composed of relatively fragile proteins that cannot withstand the physical conditions associated with human-made machines (e.g., heat and friction), the two types do share important features. In addition to having moving parts, all machines require energy-transducing mechanisms; that is, they convert energy into directed motion.

Despite the wide diversity of motion types in living organisms, in all cases, energy-driven changes in protein conformations result in the accomplishment of work. Protein conformation changes occur when a ligand is bound. When a specific ligand binds to one subunit of a multisubunit protein complex, the change in its conformation will affect the shapes of adjacent subunits. These changes are reversible; that is, ligand dissociation from a protein causes it to revert to its previous conformation. The work performed by complex biological machines requires that the conformational and, therefore, functional changes occur in an orderly and directed manner. In other words, an energy source (usually provided by the hydrolysis of ATP or GTP) drives a sequence of conformational changes of adjacent subunits in one functional direction. The directed functioning of biological machines is possible because nucleotide hydrolysis is irreversible under physiological conditions.

Motor Proteins

Despite their functional diversity, all biological machines possess one or more protein components that bind nucleoside triphosphates (NTPs). These subunits, called NTPases, function as mechanical transducers or **motor proteins**. The NTP hydrolysis–driven changes in the conformation of a motor protein trigger ordered conformational changes in adjacent subunits in the molecular machine. NTP-binding proteins perform a wide variety of functions in eukaryotes, most of which occur in one or more of the following categories.

1. **Classical motors**. Classical motor proteins are ATPases that move a load along a protein filament, as shown earlier (**Figure 2.5**). The best-known examples include the **myosins**, which move along actin filaments, and the kinesins and dyneins, which move vesicles and organelles along microtubules. **Kinesins** walk along the microtubules toward the (+) end, away from the centrosome (the microtubule organizing center). **Dyneins** walk along the microtubules toward the (–) end, toward the centrosome.
2. **Timing devices**. The function of certain NTP-binding proteins is to provide a delay period during a complex process that ensures accuracy. The prokaryotic protein synthesis protein EF-Tu (online Biochemistry in Perpective essay EF-Tu: A Motor Protein, Chapter 19) is a well-known example. The relatively slow rate of GTP hydrolysis by EF-Tu when it is bound to an aminoacyl-tRNA allows sufficient time for the dissociation of the complex from the ribosome if the tRNA-mRNA base sequence binding is not correct.
3. **Microprocessing switching devices**. A variety of GTP-binding proteins act as on-off molecular switches in signal transduction pathways. Examples include the β-subunits of the trimeric G proteins. Numerous intracellular signal control mechanisms are regulated by G proteins.
4. **Assembly and disassembly factors**. Numerous cellular processes require the rapid and reversible assembly of protein subunits into larger molecular complexes. Among the most dramatic examples of protein subunit polymerization are the assembly of tubulin and actin into microtubules and microfilaments, respectively. The slow hydrolysis of GTP by tubulin and ATP by actin monomers, after the incorporation of these molecules into their respective polymeric filaments, promotes subtle conformational changes that later allow disassembly.

The best-characterized motor protein is myosin. A brief overview of the structure and function of myosin in the molecular events in muscle contraction is provided online in the Biochemistry in Perspective essay Myosin: A Molecular Machine.

Visit the companion website at www.oup.com/us/mckee to read the Biochemistry in Perspective essay on myosin: a molecular machine.

Biochemistry IN PERSPECTIVE

Spider Silk and Biomimetics

What properties of spider silk have made it the subject of research worth hundreds of millions of dollars? The female golden orb-web spider of Madagascar (*Nephila madagascariensis*) is a large spider (length = 12.7 cm or 5 in) that is so named because of its bright yellow silk. In a recent and astonishing effort two Madagascar businessmen oversaw the creation of a hand-woven 11-foot brocaded spider silk textile (Figure 5A). The large amount of silk required in this endeavor was obtained by literally harnessing thousands of orb-web spiders. (After gentle hand pulling of the dragline silk, the spiders were then released.) The silk fiber used in the weaving process was twisted into 96-ply thread. Amazingly, when the Madagascar tapestry was on display in a New York museum, the owners challenged an onlooker to break a thread in one of the tassels. Unable to do so, he compared its strength with that of a chain in a bicycle lock.

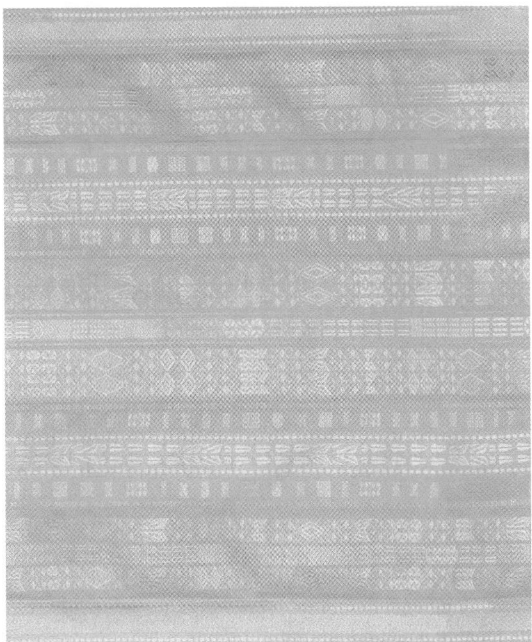

FIGURE 5A
Detail from the Madagascar Spider Silk Tapestry
The design is a classic Madagascan weaving pattern.

In addition to textiles, biodegradable and lightweight spider silk is preferable to artificial fibers for a variety of applications, such as artificial tendon and ligament components, surgical thread (e.g., eye sutures), and lightweight armor (e.g., bulletproof vests and helmets). This fiber is desirable not only because of its remarkable mechanical properties, but also because spiders produce it at ambient temperature and pressure with water as the solvent. In contrast, Kevlar, an aramid (aromatic amide) polymer derived from petroleum, is synthesized by forcing an almost boiling mixture of monomers dissolved in sulfuric acid through the small holes of an industrial spinneret (a multipored device used to convert a polymer into individual fibers). However, despite enormous effort and millions of dollars of investment, most applications of spider silk have not materialized for the simple reason that there is no adequate source. The obvious solution would be spider farming, similar to the more than 5000-year-old practice (originating in China) of cultivating the domesticated silkworm moth (*Bombyx mori*). (Silkworm silk is less tough and elastic than spider silk.) Unfortunately, spider silk farming is not commercially feasible because spiders are cannibals. (In close quarters these aggressive organisms proceed to eat one another.) Handling each of thousands of spiders separately is also unworkable because of its expense. (The creation of the Madagascar tapestry cost $500,000.) An alternative strategy is the biomimetic industrial synthesis of artificial spider silk. *Biomimetics* solves engineering problems by emulating biological processes such as the spinning of spider silk. Success, unfortunately, has been limited. For example, attempts to use recombinant DNA technology to synthesize spider silk by inserting silk genes into bacteria and yeast and then recovering silk proteins have been disappointing. There has been limited success with transgenic goats, animals into which a spider silk protein gene has been inserted. Efforts at spinning the goat silk protein, purified from the animal's milk, resulted in fibers with inferior mechanical properties with diameters (10–60 μm) that were considerably thicker than that of natural spider silk (2.5–5 μm). However, the research efforts continue because the goal of industrial engineers will certainly be worth the investment: biodegradable, environmentally safe artificial spider silk will offer an alternative to petroleum-based fibers. Throughout this effort, scientists will further probe both spider silk structure and the biological spinning process.

▶

Biochemistry IN PERSPECTIVE cont.

Spider Silk Structure

Dragline silk is composed of two proteins, spidroin 1 and spidroin 2, that have molecular masses that range from 200 to 350 kDa. Spidroin amino acid composition is distinctive because the majority of residues are glycine (42%) and alanine (25%), with smaller amounts of amino acids with bulkier side chains (Arg, Tyr, Gln, Ser, Leu, and Pro). Both spidroin proteins contain two major types of repeating units: polyalanine sequences (5 to 10 residues) and glycine-enriched motifs such as polyGlyAla, GlyProGlyGly*X* (where *X* is often Gln), and GlyGly*X* (where *X* can be various amino acids). In mature silk protein the polyalanine and polyGlyAla sequences form antiparallel β-pleated sheets, the microcrystalline structures that give silk its tensile strength (Figure 5B). The β-pleated sheets are connected by glycine-enriched sequences that form random coil, β-spirals (similar to β-turns), and GlyGly*X* helical structures that together constitute an amorphous and elastic matrix.

Spider Silk Fiber Assembly

Dragline spider silk fiber production provides a rare opportunity to observe protein folding as it occurs in a living organism. The spinning process (Figure 5C) begins in the ampullate gland in the spider's abdomen where epithelial cells secrete the spidroin into the gland's lumen. Silk protein, referred to as the silk feedstock or *dope*, is highly concentrated (as high as 50%). At this stage spidroin's globular conformation (about 30% α-helices) ensures its solubility in water and prevents aggregation. The silk dope is squeezed through the ampullate gland and the narrow funnel that connects to the spinning duct. Here the flowing dope begins to assume the properties of a liquid crystal as long axes of the protein molecules are forced into parallel orientation. The tapered S-shape spinning duct has three segments. As the protein moves through the segments, nascent silk polymer forms as a result of increasing shear stress (force applied by the parallel duct wall) and several biochemical environment changes. Within the duct, Na^+ and Cl^- are extracted and phosphate and K^+ are pumped in. An increase in the K^+:Na^+ ratio, combined with the secretion of phosphate and H^+, is believed to cause the conversion of α-helical conformations to β-pleated sheets. At first randomly oriented, the β-pleated sheets are eventually forced into parallel alignment with the long axis of the filament. In the third segment of the duct, large amounts of water, released from the silk protein as hydrophobic interactions increase, are pumped out by epithelial cells. The valve at the end of the duct is believed to act as a clamp that grips the silk and a means of restarting the spinning process if the silk breaks. The silk polymer then enters one of numerous spigots within a spinneret (Figure 5D). As the silk filament emerges and the remaining

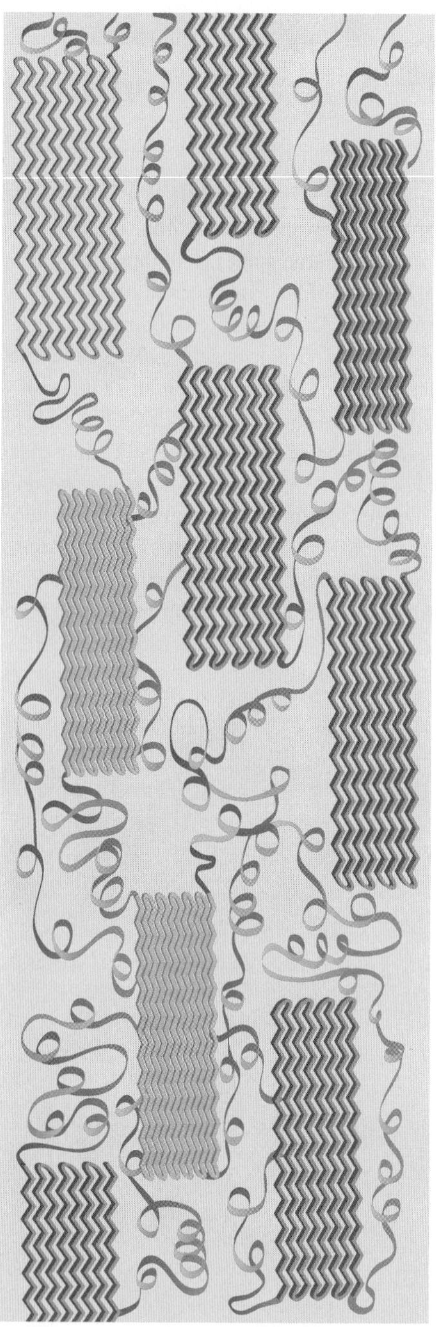

FIGURE 5B

Diagrammatic View of a Spider Silk Filament Segment

The structure of spider silk is not known with precision. It is known that two types of β-pleated sheet (highly ordered and less ordered) are responsible for spider silk strength. They are linked to each other by polypeptide sequences that form the random coil, left-hand helices, and β-spirals that provide elasticity. Silks differ in their β-pleated sheet and random coil content. For example, dragline silk has a higher content of β-pleated sheet than does capture silk.

Biochemistry IN PERSPECTIVE cont.

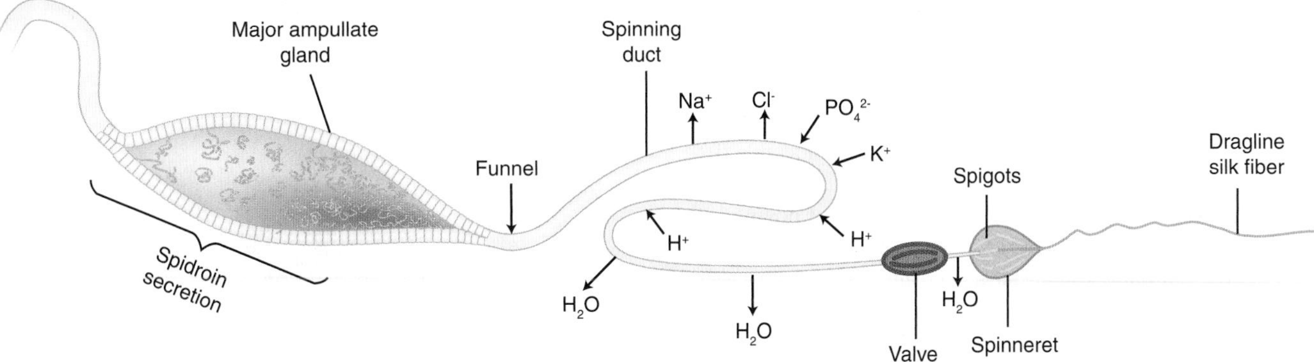

FIGURE 5C

Processing of Spider Dragline Silk

After the spidroins are secreted into the lumen of the major ampullate gland they move toward the funnel, where they exit into the beginning of the spinning duct. As a result of shear stress and other forces (e.g., squeezing of the wall of the ampullate gland and the pulling of the silk fiber out of the spinneret by the spider), the spidroins in the silk dope are compressed and forced to align along their long axes. As the silk polymer progresses down the tapering duct, biochemical changes (e.g., Na^+, K^+, and H^+) cause the conversion of α-helices into hydrophobic β-pleated sheets that expel H_2O. After passing through the valve, the polymer is forced through one of several spigots. Several emerging filaments are twisted together to form a silk fiber that is pulled out of the spinneret by the spider.

water evaporates, it is solid. The filaments from numerous spigots wrap around each other to form a cablelike fiber. The diameter and strength of the fiber depend on the muscular tension within the spinneret valve and how fast the spider draws it out.

FIGURE 5D

Illustration of the Silk Spinning Spigots of a Spider Spinneret

Note that emerging filaments are twisting together to form a fiber.

SUMMARY: Biodegradable, lightweight, strong spider silk has an enormous number of potential applications. Intense, and as yet unsuccessful, research efforts have focused on duplicating the natural process by which spiders produce this remarkable fiber.

Biochemistry IN THE LAB

Protein Technology

Living organisms produce a stunning variety of proteins. Consequently, it is not surprising that considerable time, effort, and funding have been devoted to investigating their properties. Since the amino acid sequence of bovine insulin was determined by Frederick Sanger in 1953, the structures of several thousand proteins have been elucidated.

In contrast to the 10 years required for insulin, current technologies allow protein sequence determination within a few days by mass spectrometry. The amino acid sequence of a protein can be generated from its DNA or mRNA sequence if this information is available. After a brief review of protein purification methods, mass spectrometry is described. An older means of determining the primary sequence of polypeptides, the Edman degradation method, is described in an online Biochemistry in the Lab box Protein Sequencing Analysis: The Edman Degradation. Note that all the techniques for isolating, purifying, and characterizing proteins exploit differences in charge, molecular weight, and binding affinities. Many of these technologies apply to the investigation of other biomolecules.

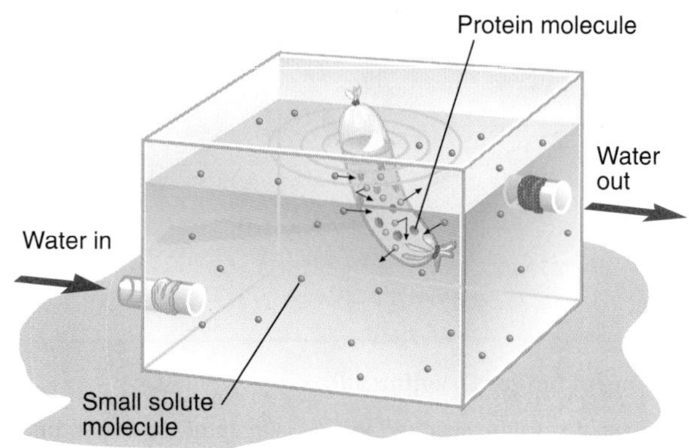

FIGURE 5E
Dialysis

Proteins are routinely separated from low-molecular-weight impurities by dialysis. When a dialysis bag (an artificial semipermeable membrane) containing a cell extract is suspended in water or a buffered solution, small molecules pass out through the membrane's pores. If the solvent outside the bag is continually renewed, all low-molecular-weight impurities are removed from the inside.

Purification

Protein analysis begins with isolation and purification. Extraction of a protein requires cell disruption and homogenization (see Biochemistry in the Lab, Cell Technology, Chapter 2). This process is often followed by differential centrifugation and, if the protein is a component of an organelle, by density gradient centrifugation. After the protein-containing fraction has been obtained, several relatively crude methods may be used to enhance purification. In **salting out**, high concentrations of salts such as ammonium sulfate [$(NH_4)_2SO_4$] are used to precipitate proteins. Because each protein has a characteristic salting-out point, this technique removes many impurities. (Unwanted proteins that remain in solution are discarded when the liquid is decanted.) When proteins are tightly bound to membrane, organic solvents or detergents often aid in their extraction. Dialysis (Figure 5E) is routinely used to remove low-molecular-weight impurities such as salts, solvents, and detergents.

As a protein sample becomes progressively more pure, more sophisticated methods are used to achieve further purification. Among the most commonly used techniques are chromatography and electrophoresis.

Chromatography

Originally devised to separate low-molecular-weight substances such as sugars and amino acids, chromatography has become an invaluable tool in protein purification. A wide variety of chromatographic techniques are used to separate protein mixtures on the basis of molecular properties such as size, shape, and weight or certain binding affinities. Often several techniques must be used sequentially to obtain a demonstrably pure protein.

In all chromatographic methods the protein mixture is dissolved in a liquid known as the **mobile phase**. As the protein molecules pass across the **stationary phase** (a solid matrix), they separate from each other because they are differently distributed between the two phases. The relative movement of each molecule results from its capacity to remain associated with the stationary phase while the mobile phase continues to flow.

Three chromatographic methods commonly used in protein purification are gel-filtration chromatography, ion-exchange chromatography, and affinity chromatography. **Gel-filtration chromatography** (Figure 5F) is a form of size-exclusion chromatography in which particles in an aqueous solution flow through a column (a hollow tube) filled with gel and are separated according to size. Molecules that are larger than the gel pores are excluded and therefore move through the column quickly. Molecules that are smaller than the gel pores diffuse in and out of the pores, so their movement through the column is retarded. Differences in the rates of particle movement separate the protein mixture into bands, which are then collected separately.

Ion-exchange chromatography separates proteins on the basis of their charge. Anion-exchange resins, which consist of positively charged materials, bind reversibly with a protein's negatively charged groups. Similarly, cation-exchange resins bind positively charged groups. After proteins that do not bind to the resin have been removed, the protein of interest is recovered by

Biochemistry IN THE LAB cont.

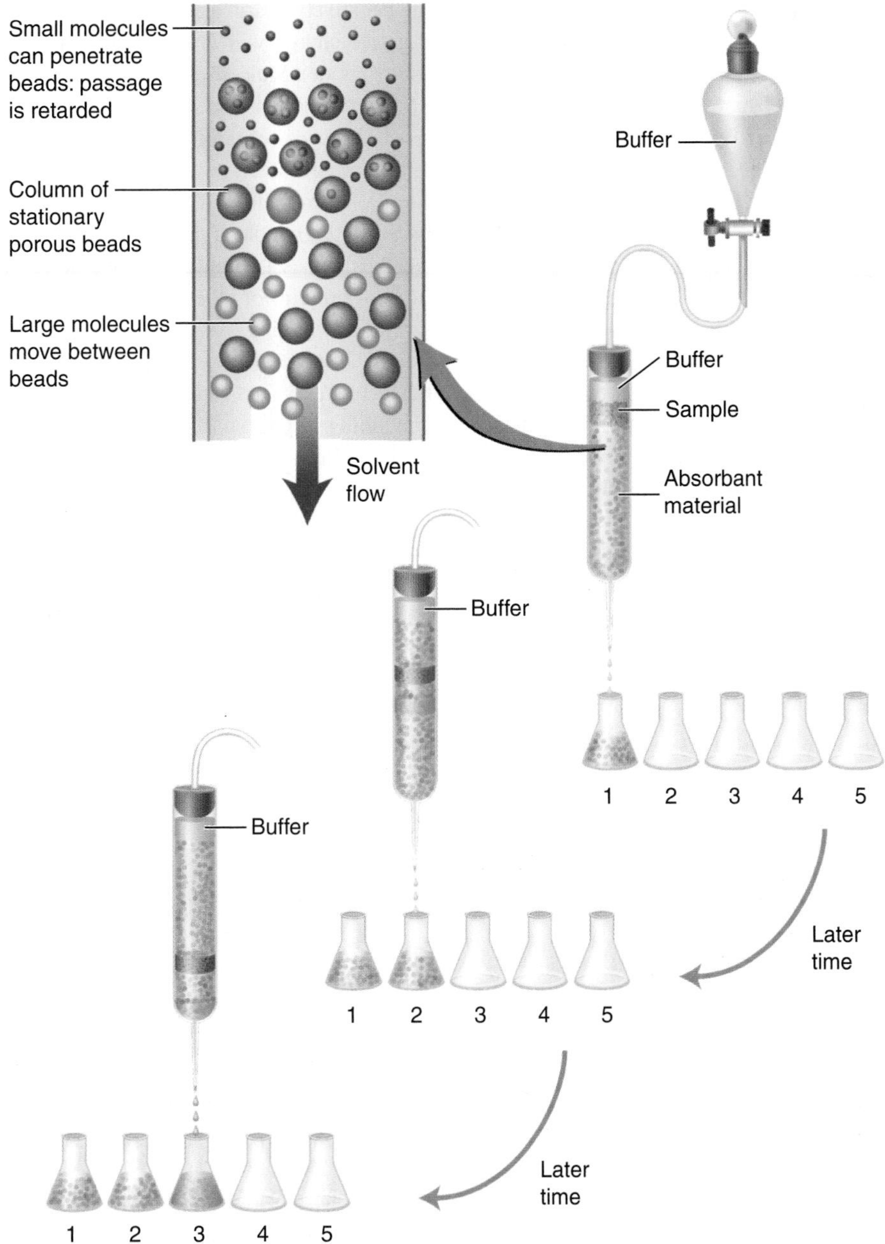

FIGURE 5F
Gel-Filtration Chromatography

In gel-filtration chromatography the stationary phase is a gelatinous polymer with pore sizes selected by the experimenter to separate molecules according to their sizes. The sample is applied to the top of the column and is eluted with buffer (the mobile phase). As elution proceeds, larger molecules travel faster through the gel than smaller molecules, whose progress is slowed because they can enter the pores. If fractions are collected, the larger molecules appear in the earlier fractions and later fractions contain smaller molecules.

an appropriate change in the solvent pH and/or salt concentration. (A change in pH alters the protein's net charge.)

Affinity chromatography is an efficient method of protein purification that takes advantage of the unique biological properties of proteins. That is, it uses a reversible noncovalent binding affinity between a specific protein and a special molecule (the ligand). The ligand is covalently bound to an insoluble matrix, which is placed in a column. After nonbinding protein molecules have passed through the column, the protein of interest is removed by altering the conditions that affect protein-ligand binding (i.e., pH or salt concentration). Affinity chromatography is commonly used in the purification of recombinant proteins.

Biochemistry IN THE LAB cont.

(A recombinant protein is produced in a host organism by the translation of a recombinant DNA sequence. Recombinant DNA, described on pp. 690–692, is synthesized in the laboratory by joining together DNA from different organisms.) The recombinant protein is typically a fusion protein that contains an affinity tag such as GST (glutathione-S-transferase, see p. 533) that is grafted either to the N- or C-terminal of the protein of interest. GST is a small protein (26 kDa) that has a high binding affinity for GSH. (GSH is an important antioxidant in cells; i.e., it protects biomolecules from oxygen radical damage, see pp. 372–373.) When a protein mixture is mixed with beads with GSH attached, the GST fusion protein will bind to the beads. After the beads are gently washed to remove other proteins, the beads are then washed with free GSH, which causes the detachment of the fusion protein. The protein of interest is recovered by the enzyme-catalyzed hydrolysis of a specific site between the protein of interest and the GST tag.

Electrophoresis

Because proteins are electrically charged, they move in an electric field. In this process, called **electrophoresis**, molecules separate from each other because of differences in their net charge. For example, molecules with a positive net charge migrate toward the negatively charged electrode (cathode). Molecules with a net negative charge will move toward the positively charged electrode (anode). Molecules with no net charge will not move at all.

Electrophoresis, one of the most widely used techniques in biochemistry, is usually carried out using gels such as polyacrylamide or agarose. The gel, functioning much as it does in gel-filtration chromatography, also acts to separate proteins on the basis of their molecular weight and shape. Consequently, gel electrophoresis is highly effective at separating complex mixtures of proteins or other molecules.

Bands resulting from a gel electrophoretic separation may be treated in several ways. Specific bands may be excised from the gel after visualization with ultraviolet light. Each protein-containing slice is then eluted with buffer and prepared for further analysis. Because of its high resolving power, gel electrophoresis is also used to assess the purity of protein samples. Staining gels with a dye such as Coomassie brilliant blue is a common method for quickly assessing the success of a purification step.

SDS–polyacrylamide gel electrophoresis (SDS-PAGE) is a widely used variation of electrophoresis that can be used to determine molecular weight (Figure 5G). SDS, a negatively charged detergent, binds to the hydrophobic regions of protein molecules, causing the proteins to denature and assume rodlike shapes. Because most molecules bind SDS in a ratio roughly

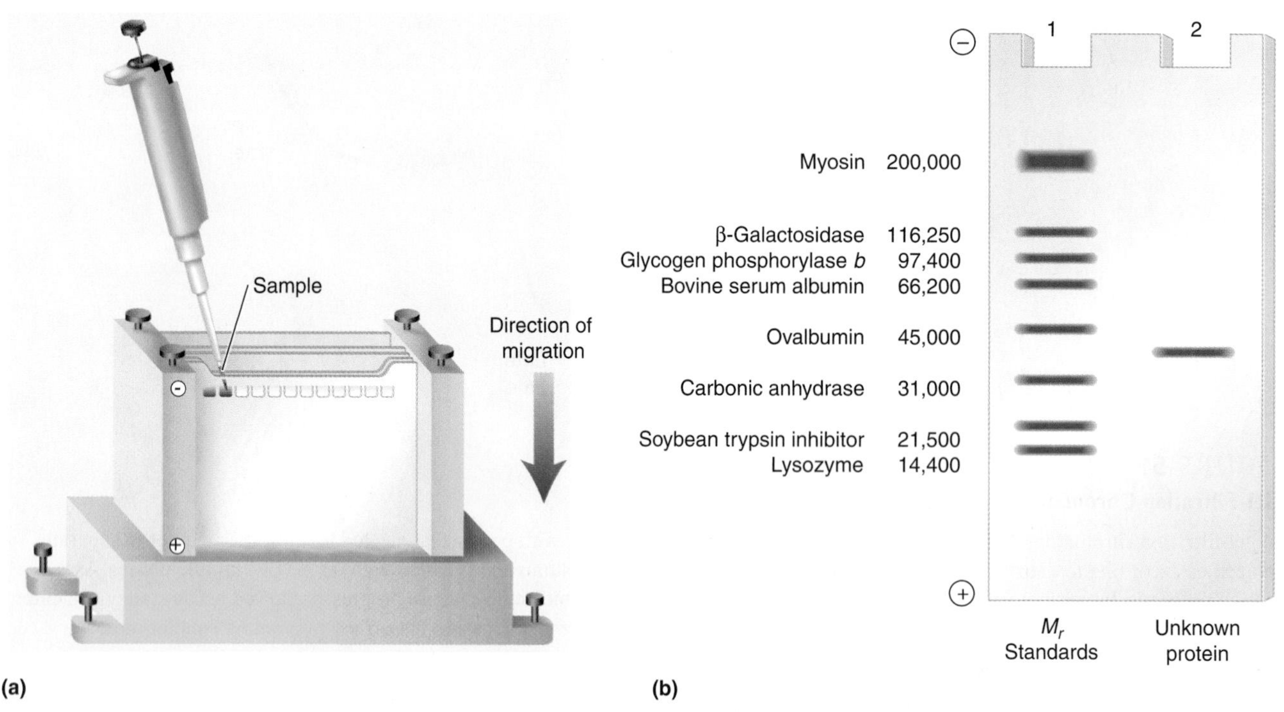

FIGURE 5G
Gel Electrophoresis

(a) Gel apparatus. The samples are loaded into wells. After an electric field is applied, the proteins move into the gel. (b) Molecules separate and move in the gel as a function of molecular weight and shape.

Biochemistry IN THE LAB cont.

proportional to their molecular weights, during electrophoresis SDS-treated proteins migrate toward the anode (+ pole) only in relation to their molecular weight.

Mass Spectrometry

Mass spectrometry (MS) is a powerful and sensitive technique for separating, identifying, and determining the mass of molecules. It exploits differences in their mass-to-charge (*m*/*z*) ratios. In a mass spectrometer, ionized molecules flow through a magnetic field (Figure 5H). The magnetic field force deflects the ions depending on their *m*/*z* ratios with lighter ions being more deflected from a straight-line path than heavier ions. A detector measures the deflection of each ion. In addition to protein identity and mass determinations, MS is also used to detect bound cofactors and protein modifications. Because MS analysis involves the ionization and vaporization of the substances to be investigated, its use in the analysis of thermally unstable macromolecules such as proteins and nucleic acids did not become feasible until methods such as electrospray ionization and matrix-assisted laser desorption ionization (MALDI) had been developed. In electrospray ionization a solution containing the protein of interest is sprayed in the presence of a strong electrical field into a port in the spectrometer. As the protein droplets exit the injection device, typically an ultrafine glass tube, the protein molecules become charged. In MALDI, a laser pulse vaporizes the protein, which is embedded in a solid matrix. Once the sample has been ionized, its molecules, now in the gas phase, are separated according to their individual *m*/*z* ratios. A detector within the mass spectrometer produces a peak for each ion. In a computer-assisted process, information concerning each ion's mass is compared against data for ions of known structure and used to determine the sample's molecular identity.

Protein sequencing analysis makes use of tandem MS (two mass spectrometers linked in series, MS/MS). A protein of interest, often extracted from a band in a gel, is then digested by a

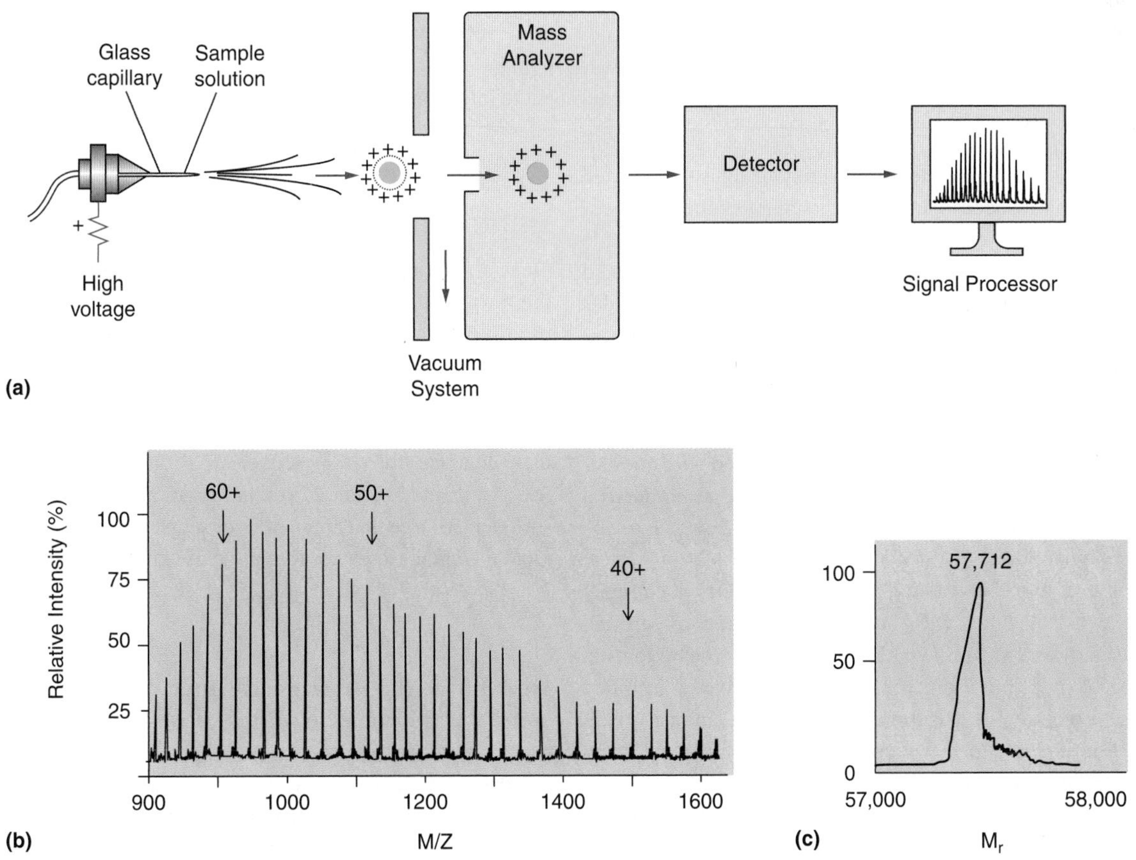

FIGURE 5H
Mass Spectrometry

(a) The principal steps in electrospray ionization. The sample (a protein dissolved in a solvent) is injected via a glass capillary into the ionization chamber. The voltage difference between the electrospray needle and the injection port results in the creation of protein ions. The solvent evaporates during this phase. The ions enter the mass spectrometer, which then measures their *m*/*z* ratios. (b) An electrospray mass spectrum showing the *m*/*z* ratios for several peaks. (c) A computer analysis of the data showing the molecular mass of the sample protein (M_r = molecular weight).

Biochemistry IN THE LAB cont.

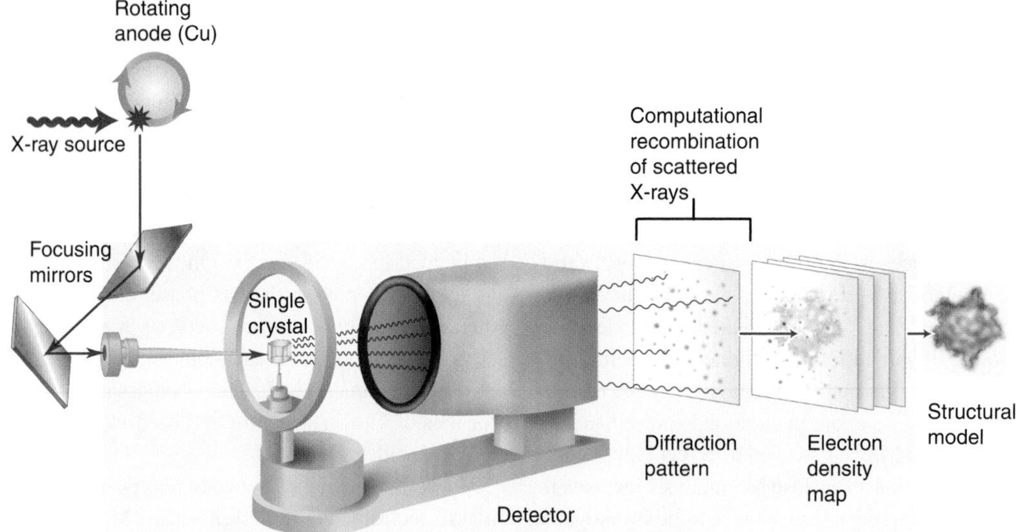

FIGURE 5I

Schematic Diagram of X-Ray Crystallography

X-rays are useful in the analysis of biomolecules because their wavelength range is similar to the magnitude of chemical bonds. Consequently, the resolving power of X-ray crystallography is equivalent to interatomic distances.

proteolytic enzyme. Subsequently, the enzyme digest is injected into the first mass spectrometer, which separates the oligopeptides according to their m/z ratios. One by one, each oligopeptide ion is directed into a collision chamber, where it is fragmented by collisions with hot inert gas molecules. Product ions, peptides that differ from each other in size by one amino acid residue, are then sequentially directed into the second mass spectrometer. A computer identifies each peak and automatically determines the amino acid sequence of the peptides. The process is then repeated with oligopeptides derived from digestion with another enzyme. The computer uses the sequence information derived from both digests to determine the amino acid sequence of the original polypeptide.

Recently, a higher-resolution mass spectrometry technique, referred to as top-down MS, has been developed for protein structure analysis. In top-down MS (as opposed to the bottom-up technique of digesting a protein before MS analysis), an intact protein or protein mixture is directly injected into an instrument called a *Fourier transform ion cyclotron resonance mass spectrometer* (FTICR-MS). In FTICR-MS signals generated by the accelerated movement of charged particles in a static magnetic field and a rapidly varying electric field are analyzed by performing a mathematical transformation called a Fourier transform to yield a mass spectrum. FRICR-MS has been used successfully to obtain high-resolution sequencing of proteins with masses under 100,000 kDa and the characterization of posttranslational modifications and protein-protein interactions.

Protein Sequence-Based Function Prediction

Once a polypeptide has been isolated, purified, and sequenced, the next logical step is to determine its function. This endeavor usually begins with a database search of known protein sequences. BLAST (*B*asic *L*ocal *A*lignment *S*earch *T*ool) is a computer program (www.ncbi.nim.nih.gov/blast) that allows fast searches of known sequences for matches to the unknown protein sequence (the query sequence). Protein sequence databases (e.g., UniProt [*U*niversal *P*rotein resource] www.uniprot.org) are sufficiently large that about 50% of sequence comparison queries yield matched sequences that are close enough to infer function.

X-Ray Crystallography

Much of the three-dimensional structural information about proteins was obtained by X-ray crystallography. Because the bond distances in proteins are approximately 0.15 nm, the electromagnetic radiation used to resolve protein structure must have a short wavelength. Visible light wavelengths [(λ) = 400–700 nm] clearly do not have sufficient resolving power for biomolecules. X-rays, however, have very short wavelengths (0.07–0.25 nm).

In X-ray crystallography, highly ordered crystalline specimens are exposed to an X-ray beam (Figure 5I). As the X-rays hit the crystal, they are scattered by the atoms in the crystal. The

Biochemistry IN THE LAB cont.

diffraction pattern that results is recorded on charge-coupled device (CCD) detectors. The diffraction patterns are used to construct an electron density map. Because there is no objective lens to recombine the scattered X-rays, the three-dimensional image is reconstructed mathematically. Computer programs now perform these extremely complex and laborious computations. The three-dimensional structure of a polypeptide can also be determined using *homologous modeling*, a method that is based on the observation that three-dimensional protein structure is more conserved than protein sequences. A structural model is constructed from X-ray diffraction data of one or more homologous proteins in the Protein Data Bank (www.pdb.org).

NMR Spectroscopy

NMR is a commonly used form of spectroscopy based on the absorption of electromagnetic radiation (radio waves) by atomic nuclei with magnetic properties (i.e., isotopes that have nuclear spin such as ^{1}H, ^{13}C, and ^{15}N) that are aligned with a strong magnetic field. (*Nuclear spin* is a form of angular momentum that occurs in atomic nuclei with an odd number of protons or neutrons. A spinning nucleus is similar to a miniature bar magnet.) Unlike X-ray crystallography, which is limited to proteins that crystallize easily, NMR can be used to investigate the structure of moderately sized proteins (40–60 kDa), as well as other macromolecules, without regard to their crystallization capacity. The three-dimensional structure of protein molecules is determined using short pulses of radio waves, which disrupt the alignment of the nuclei along the external magnetic field. Since the total magnetic field experienced by each nucleus includes local magnetic fields of nearby atoms, each NMR-active nucleus will react differently to the radio waves. An NMR spectrum is a computer program–derived plot of radio pulse frequency against the energy absorbed by each of the molecule's nuclei. Once a protein's amino acid sequence is known, its folded structure is determined by the analysis of several NMR experiments in which the properties and relative location of each atom in the molecule is resolved. NMR has another important advantage over X-ray crystallography. X-ray crystallography can only provide a snapshot of a crystallized protein, whereas NMR can be used in the investigation of protein dynamics (e.g., internal structural changes that are essential for a molecule's function). NMR can also provide information about a protein's responses to interactions with other molecules or changes in conditions such as temperature and pH.

Chapter Summary

1. Polypeptides are amino acid polymers. Proteins may consist of one or more polypeptide chains.
2. Each amino acid contains a central carbon atom (the α-carbon) to which an amino group, a carboxylate group, a hydrogen atom, and an R group are attached. In addition to comprising protein, amino acids have several other biological roles. According to their capacity to interact with water, amino acids may be separated into four classes: nonpolar, polar, acidic, and basic.
3. Titration of amino acids and peptides illustrates the effect of pH on their structures. The pH at which a molecule has no net charge is called its isoelectric point.
4. Amino acids undergo several chemical reactions. Three reactions are especially important: peptide bond formation, cysteine oxidation, and Schiff base formation.
5. Proteins have a vast array of functions in living organisms. In addition to serving as structural materials, proteins are involved in metabolic regulation, transport, defense, and catalysis. Some proteins are multifunctional; that is, they have two or more seemingly unrelated functions. Proteins can also be classified into families and superfamilies, according to their sequence similarities. Fibrous proteins (e.g., collagen) are long, rod-shape molecules that are insoluble in water and physically tough. Globular proteins (e.g., hemoglobin) are compact, spherical molecules that are usually soluble in water.
6. Biochemists have distinguished four levels of protein structure. Primary structure, the amino acid sequence, is specified by genetic information. As the polypeptide chain folds, local folding patterns constitute the protein's secondary structure. The overall three-dimensional shape that a polypeptide assumes is called the tertiary structure. Proteins that consist of two or more polypeptides have quaternary structure. The functions of numerous proteins, especially molecules that participate in eukaryotic regulatory processes, are partially or completely unstructured. Many physical and chemical conditions disrupt protein structure. Denaturing agents include strong acids or bases, reducing agents, organic solvents, detergents, high salt concentrations, heavy metals, temperature changes, and mechanical stress.
7. One of the most important aspects of protein synthesis is the folding of polypeptides into their biologically active conformations. Despite decades of investigation into the physical and chemical properties of polypetide chains, the mechanism by which a primary sequence dictates the molecule's final conformation is unresolved. Many proteins require molecular chaperones to fold into their final three-dimensional conformations. Protein misfolding is now known to be an important feature of several human diseases, including Alzheimer's disease and Huntington's disease.

8. Fibrous proteins (e.g., α-keratin and collagen), which contain high proportions of α-helices or β-pleated sheets, have structural rather than dynamic roles. Despite their varied functions, most globular proteins have features that allow them to bind to specific ligands or sites on certain macromolecules. These binding events involve conformational changes in the globular protein's structure.
9. The biological activity of complex multisubunit proteins is often regulated by allosteric interactions in which small ligands bind to the protein. Any change in the protein's activity is caused by changes in the interactions among the protein's subunits. Effectors can increase or decrease the function of a protein.

Take your learning further by visiting the **companion website** for Biochemistry at **www.oup.com/us/mckee** where you can complete a multiple-choice quiz on amino acids, peptides, and proteins to help you prepare for exams.

Suggested Readings

Arnaud, C. H., Top-Down Proteomics Becomes Reality, *Chem. Eng. News* 91(20):11–17, 2013.

Bustamonte, C., Of Torques, Forces, and Protein Machines, *Protein Sci.* 13:3061–3065, 2004.

Chouard, T., Breaking the Protein Rules, *Nature* 471:151–153, 2011.

Dunker, A. K., and Kriwacki, R. W., The Orderly Chaos of Proteins, *Sci. Am.* 30:68–73, 2011.

Everts, S., Mirror Molecules, *Sci. Am.* 306(5):79–81, 2013.

Heim, M., Keerl, D., and Scheibel, T., Spider Silk: From Soluble Protein to Extraordinary Fiber, *Angewandte Chem. Int. Ed.* 48:3584–3596, 2009.

Kim, Y. E., *et al.,* Molecular Chaperone Function in Protein Folding and Proteostasis, *Annu. Rev. Biochem.* 82:323–355, 2013.

Lewis, T., Disorder at Work, *Sci. News* 183(3):26–29, 2013.

Schnabel, J., Protein Folding: The Dark Side of Proteins, *Nature* 484:828–829, 2010.

Tompa, P., Szasz, C., and Buday, L., Structural Disorder Throws New Light on Moonlighting. *Trends Biochem. Sci.* 30(9):484–489, 2005.

Key Words

affinity chromatography, *181*
aldimine, *145*
aldol condensation, *170*
aliphatic hydrocarbon, *135*
allosteric transition, *159*
allostery, *159*
Alzheimer's disease, *167*
amino acid residue, *132*
amphipathic molecule, *162*
amphoteric molecule, *134*
antigen, *148*
apoprotein, *149*
aromatic hydrocarbon, *135*
asymmetric carbon, *137*
chaperonins, *166*
chiral carbon, *137*
conjugated protein, *149*
cooperative binding, *173*
denaturation, *162*
disulfide bridge, *144*
dynein, *176*
effector, *159*
electrophoresis, *182*
enantiomer, *137*
fibrous protein, *148*
fold, *154*
gel-filtration chromatography, *180*
globular protein, *148*
glycoprotein, *149*
heat shock protein, *148*
hemoprotein, *149*
holoprotein, *149*
homologous polypeptide, *149*
hormone, *136*
hsp70, *166*
hsp90, *166*
Huntington's disease, *167*
intrinsically unstructured protein, *160*
ion-exchange chromatography, *180*
isoelectric point, *138*
kinesin, *176*
ligand, *157*
lipoprotein, *149*
mass spectrometry, *183*
metalloprotein, *149*
mobile phase, *180*
modular protein, *155*
modulator, *159*
molecular chaperone, *164*
molecular disease, *151*
molten globule, *164*
motif, *153*
motor protein, *176*
multifunction protein, *148*
myosin, *176*
natively unfolded protein, *160*
neurotransmitter, *136*
oligomer, *158*
optical isomer, *137*
peptide, *132*
peptide bond, *142*
phosphoprotein, *149*
polypeptide, *132*
primary structure, *149*
prosthetic group, *149*
protein, *132*
protein family, *148*
protein folding, *154*
protein superfamily, *148*
protomer, *158*
quaternary structure, *149*
response element, *161*
salt bridge, *157*
salting out, *180*
Schiff base, *144*
SDS–polyacrylamide gel electrophoresis, *182*
secondary structure, *149*
stationary phase, *180*
stereoisomer, *137*
subunit, *157*
supersecondary structure, *153*
tertiary structure, *149*
zwitterion, *134*

Review Questions

These questions are designed to test your knowledge of the key concepts discussed in this chapter, before moving on to the next chapter. You may like to compare your answers to the solutions provided in the back of the book and in the accompanying Study Guide.

1. Define the following terms:
   a. supersecondary structure
   b. protomer
   c. phosphoprotein
   d. denaturation
   e. ion exchange chromatography
2. Define the following terms:
   a. protein motif
   b. conjugated protein
   c. dynein
   d. zwitterion
   e. electrophoresis
3. Define the following terms:
   a. metalloprotein
   b. hormone
   c. holoprotein
   d. intrinsically unstructured protein
   e. kinesin
4. Define the following terms:
   a. aliphatic hydrocarbon
   b. neurotransmitter
   c. asymmetric carbon
   d. chiral carbon
   e. stereoisomer
5. Define the following terms:
   a. optical isomer
   b. isoelectric point
   c. peptide bond
   d. disulfide bridge
   e. Schiff base
6. Define the following terms:
   a. aldimine
   b. heat shock protein
   c. multifunction protein
   d. protein family
   e. protein superfamily
7. Define the following terms:
   a. moonlighting protein
   b. fibrous protein
   c. globular protein
   d. prosthetic group
   e. apoprotein
8. Define the following terms:
   a. homologous polypeptide
   b. molecular disease
   c. protein fold
   d. mosaic protein
   e. ligand
9. Define the following terms:
   a. salt bridge
   b. oligomer
   c. allosteric transition
   d. protein denaturation
   e. amphipathic molecule
10. Define the following terms:
    a. molecular chaperone
    b. chaperonin
    c. cooperative binding
    d. motor protein
    e. primary structure
11. Define the following terms:
    a. aldol condensation
    b. α-carbon
    c. hydrophobic amino acid
    d. secondary structure
    e. β-pleated sheet
12. Distinguish among proteins, peptides, and polypeptides.
13. Indicate whether each of the following amino acids is polar, nonpolar, acidic, or basic:
    a. glycine
    b. tyrosine
    c. glutamic acid
    d. histidine
    e. proline
    f. lysine
    g. cysteine
    h. asparagine
    i. valine
    j. leucine
14. Arginine has the following pK_a values:
    pK_1 = 2.17, pK_2 = 9.04, pK_R = 12.48
    Give the structure and net charge of arginine at the following pH values: 1, 4, 7, 10, 12
15. Consider the following molecule.

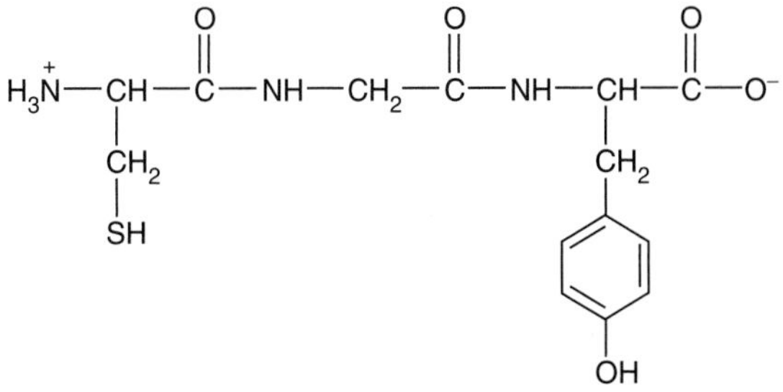

    $H_3\overset{+}{N}$—CH(CH_2—SH)—C(=O)—NH—CH_2—C(=O)—NH—CH(CH_2—C_6H_4—OH)—C(=O)—O^-

    a. Name it.
    b. Use the three-letter symbols for the amino acids to represent this molecule.
16. Rotation about the peptide bond in glycylglycine is hindered. Draw the resonance forms of the peptide bond and explain why.
17. List six functions of proteins in the body.
18. Indicate the level(s) of protein structure to which each of the following contributes:
    a. amino acid sequence
    b. β-pleated sheet
    c. hydrogen bond
    d. disulfide bond

19. What type of secondary structure would the following amino acid sequence be *most* likely to have?
    a. polyproline
    b. polyglycine
    c. Ala—Val—Ala—Val—Ala—Val—
    d. Gly—Ser—Gly—Ala—Gly—Ala
20. List three factors that do not foster α-helix formation.
21. Denaturation is the loss of protein function from structural change or chemical reaction. At what level of protein structure or through what chemical reaction does each of the following denaturation agents act?
    a. heat
    b. strong acid
    c. saturated salt solution
    d. organic solvents (e.g., alcohol or chloroform)
22. A polypeptide has a high p*I* value. Suggest which amino acids might comprise it.
23. Outline the steps to isolate a typical protein. What is achieved at each step?
24. Outline the steps to purify a protein. What criteria are used to evaluate purity?
25. List the types of chromatography used to purify proteins. Describe how each separation method works.
26. Describe the problems associated with using a polypeptide's primary sequence to determine its final three-dimensional shape.
27. Describe the forces involved in protein folding.
28. What are the characteristics of motor proteins? How do organisms use them?
29. Briefly outline the roles of ribosome-associated chaperones, hsp70s, hsp90s, and chaperonins in protein folding.
30. What is the p*I* value for the dipeptide Gly-Ala?
31. The muscles of deep diving mammals such as whales contain large amounts of myoglobin. How does this circumstance contribute to prolonged dives?

Fill in the Blank

32. Amino acid polymers consisting of more than 50 amino acids are called _______.
33. Molecules that can behave as an acid or a base are called __________________.
34. Molecules with both a positive and a negative charge, but an overall neutral charge, are called __________________.
35. Nonsuperimposable mirror-image stereoisomers are referred to as _______________.
36. The amide linkages of proteins are referred to as ______________________.
37. Cystine is unique among the amino acids because it contains a _______________ bond.
38. Primary amine groups of amino acids react with aldehydes or ketones to form __________________.
39. Keratin and fibrinogen are examples of __________________ proteins.
40. Collagen and elastin are examples of _______________ proteins.
41. Multisubunit proteins in which some or all of the subunits are identical are referred to as ______________________.

Short Answer

42. With the exception of glycine, all of the standard amino acids have one chiral center. Are there any amino acids that have two chiral centers?
43. What would be the effect on the pK_a of a carboxylic acid if it exists in the water-free interior of a protein?
44. There are three possible levels of structure for a polypeptide: primary, secondary, and tertiary. Which of these has only one possible structure?
45. An emulsifying agent is a molecule that promotes the dispersion of one substance in another. Emulsifying agents are used to cause the interactions of hydrophobic and hydrophilic substances. They do so because their structures contain both hydrophobic and hydrophilic groups. Taurocholic acid is a bile salt, an emulsifying agent produced in the liver that promotes fat digestion in the intestine. Its structure contains the amino acid taurine. Suggest a reason why taurine makes taurocholic acid a better emulsifying agent than glycocholic acid, a molecule in which glycine is substituted for taurine.

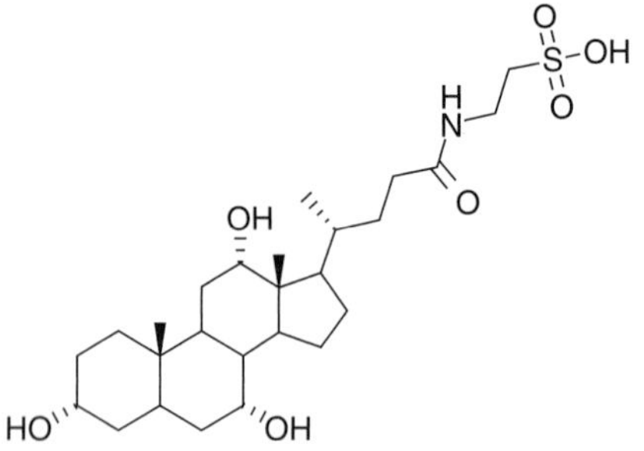

Taurocholic acid

46. Review the structures of the amino acids in Figure 5.2 and determine which amino acid serves as the precursor of taurine, whose structure forms part of taurocholic acid shown in Question 45.

Thought Questions

These questions are designed to reinforce your understanding of all of the key concepts discussed in the book so far, including this chapter and all of the chapters before it. They may not have one right answer! The authors have provided possible solutions to these questions in the back of the book and in the accompanying Study Guide for your reference.

47. Glucose is a polyhydroxyaldehyde. Determine the structure of the product of the reaction of glycine with glucose.
48. What effect would hyperventilation (rapid breathing) have on the concentration of oxyhemoglobin in the bloodstream?
49. Complete the following reactions of penicillamine:

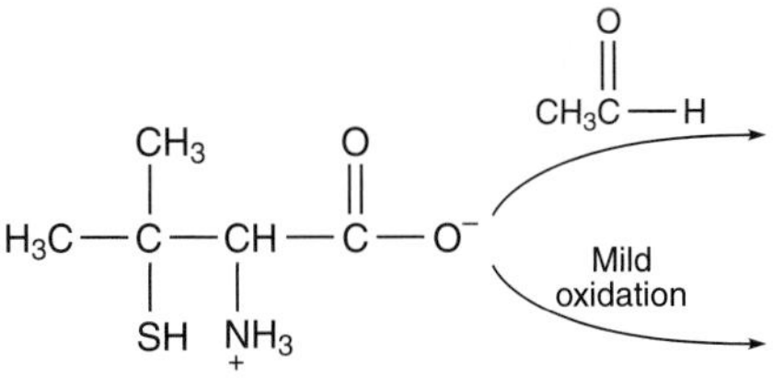

$$H_3C-\underset{SH}{\overset{CH_3}{\underset{|}{\overset{|}{C}}}}-\underset{\overset{+}{N}H_3}{\underset{|}{CH}}-\overset{O}{\overset{\|}{C}}-O^- \xrightarrow{CH_3\overset{O}{\overset{\|}{C}}-H} \qquad \xrightarrow{\text{Mild oxidation}}$$

50. The folding of intrinsically unstructured proteins is prevented by the presence of Ser, Gly, Lys, and Glu residues. What do these amino acids have in common? Why do they disrupt the folding of IUPs?
51. Why do some proteins require molecular chaperones to fold into their active conformations, whereas others do not?
52. Residues such as valine, leucine, isoleucine, methionine, and phenylalanine are often found in the interior of proteins, whereas arginine, lysine, aspartic acid, and glutamic acid are often found on the surface of proteins. Suggest a reason for this observation. Where would you expect to find glutamine, glycine, and alanine?
53. Proteins that are synthesized by living organisms adopt a biologically active conformation. Yet when such molecules are prepared in the laboratory, they usually fail to spontaneously adopt their active conformations. Can you suggest why?
54. The active site of an enzyme contains sequences that are conserved because they participate in the protein's catalytic activity. The bulk of an enzyme, however, is not part of the active site. Because a substantial amount of energy is required to assemble enzymes, why are they usually so large?
55. A structural protein may incorporate large amounts of immobilized water as part of its structure. Can you suggest how protein molecules "freeze" the water in place and make it part of the protein structure?
56. The peptide bond is a stronger bond than that of esters. What structural feature of the peptide bond gives it additional bond strength?
57. Because of their tendency to avoid water, nonpolar amino acids play an important role in forming and maintaining the three-dimensional structure of proteins. Can you suggest how these molecules accomplish this feat?
58. Consider the following tripeptide:

    Gly—Ala—Val

    a. What is the approximate isoelectric point?

    b. In which direction will the tripeptide move if placed in an electric field at pH 1, 5, 10, and 12?
59. When the multifunction protein glyceraldehyde-3-phosphate dehydrogenase (GAPD) catalyzes a key reaction in glycolysis (a metabolic pathway in cytoplasm), it does so as a homotetramer (four identical subunits). The GAPD monomer is a nuclear DNA repair enzyme. Describe in general terms what structural properties of multifunction proteins allow this phenomenon.
60. Many proteins have several functions. Provide examples. What natural forces are responsible for this phenomenon?
61. Why are multifunctional proteins necessary and/or desirable?
62. Given the following decapeptide sequence, which amino acids would you expect to be on the surface of this molecule once it folds into its native conformation?

    Gly—Phe—Tyr—Asn—Tyr—Met—Ser—His—Val—Leu
63. What amino acid residues of the decapeptide in Question 62 would tend to be found on the interior of the molecule?
64. What would be the products of the acid hydrolysis for 3 hours of the decapeptide in Question 62?
65. A mutational change alters a polypeptide by substituting three adjacent prolines for three glycines. What possible effect will this event have on the protein's structure?
66. As a genetic engineer, you have been given the following task: alter a protein's structure by converting a specific amino acid sequence that forms an extended α-helix to one that forms a β-barrel. What types of amino acid are probably in the α-helix, and which ones would you need to substitute?
67. Why is it advantageous for p53 to be activated by factors such as ER stress, UV light, and hypoxia (low oxygen concentrations)?
68. The synthesis of 2,3-BPG from the glycolytic intermediate 1,3-BPG (glycerate-1,3-bisphosphate) is catalyzed by the enzyme bisphosphoglycerate mutase, an iron-requiring enzyme. What effect on oxygen delivery to the body's tissues would be expected in the event of an iron-deficient diet?
69. Amino acids are the precursors of a vast number of biologically active nitrogen-containing molecules. Which amino acids are the precursors of the following molecules?

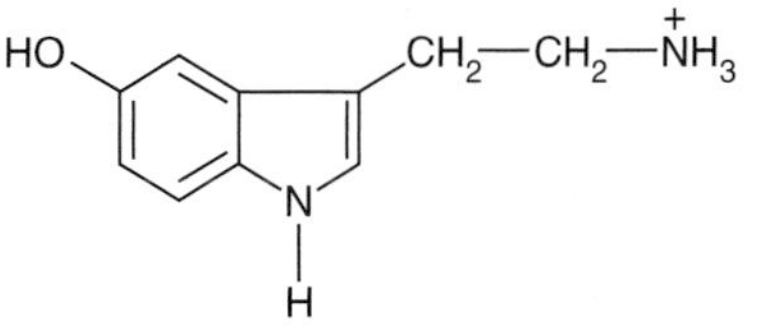

HO– (indole ring, N–H) –$CH_2-CH_2-\overset{+}{N}H_3$

Serotonin

(benzene ring with OH, OH) –$CH_2-CH_2-{}^+NH_3$

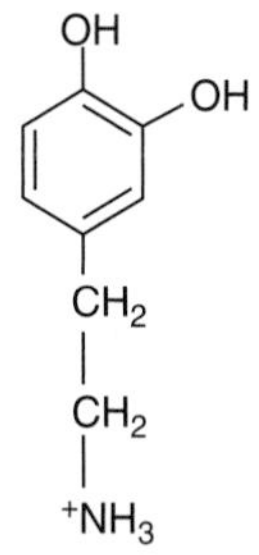

Dopamine

70. Suggest a protocol for separating oxytocin and vasopressin from an extract of the posterior pituitary gland.

CHAPTER

6 Enzymes

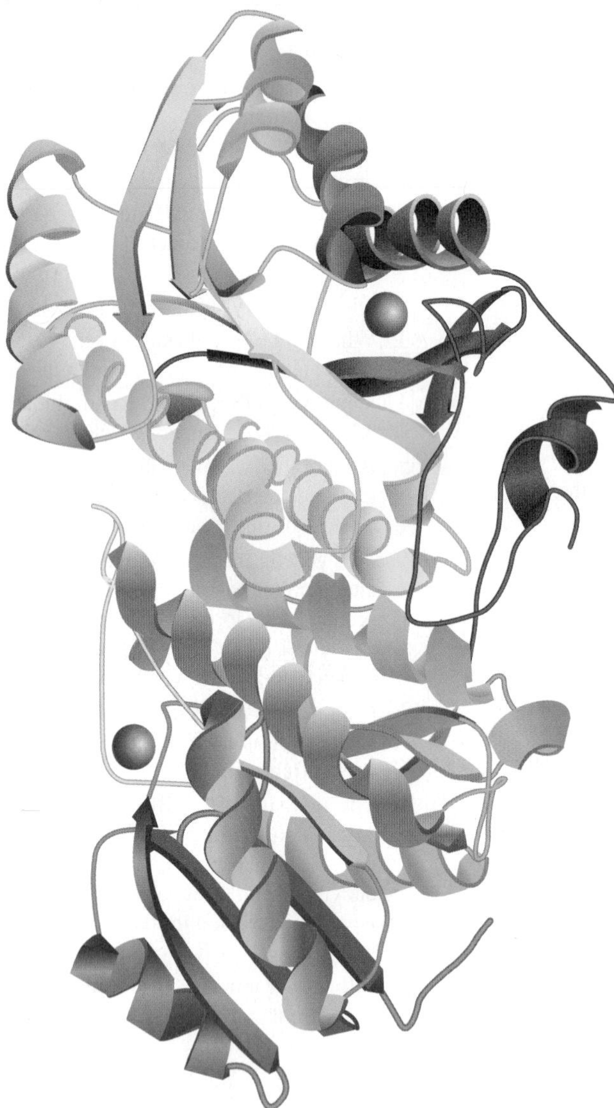

Alcohol Dehydrogenase The alcohol dehydrogenases (ADHs) are a class of enzymes that protect organisms from the toxic effects of endogenous and exogenous alcohols. A large proportion of ADHs require zinc ions for catalytic and structural purposes.

OUTLINE

Humans and Enzymes: A Brief History

Humans have taken advantage of **enzymes** (catalytic proteins) from other species since the Neolithic period. Soon after the domestication of animals such as sheep, goats (estimated at between 9000 and 11000 BCE), and cattle (7000 to 8000 BCE), humans made discoveries that allowed the conversion of excess milk into cheese and yogurt. According to ancient legends, milk carried in containers made from the stomachs of young calves or goats was observed to transform into curds (insoluble coagulated protein) and whey (a liquid containing soluble protein). Eventually, milk curd (now known to be the result of rennin, a protein-degrading enzyme in the mammalian stomach enzyme complex called rennet) was converted into cheese by pressing with large flat stones (to remove the remaining water) and salting. Another ancient means of making cheese, mentioned by Homer, the legendary Greek poet and author of the *Iliad*, involved the use of fig juice (containing the enzyme ficin) to coagulate milk. Yogurt resulted from the storage of milk in goatskin bags. As nomads and traders traveled, bacterial enzymes, in combination with ambient heat and the motion of pack animals, transformed milk into tangy custard. The Neolithic period also introduced wine, beer, and, eventually, leavened bread, all products of enzyme-catalyzed reactions of the yeast *Saccharomyces cerevisiae*.

The existence of enzymes was unknown until the nineteenth century when Louis Pasteur, a French chemist and microbiologist, proved in 1857 that live yeast cells were required in the conversion of grape extracts into wine. He referred to the yeast cell components responsible for alcohol production as "ferments." In 1897, the German chemist Eduard Buchner discovered "zymase," an enzymatic activity in a cell-free extract of yeast that converted sucrose into ethanol. Questions about the molecular nature of enzymes were resolved in the twentieth century. Together, James B. Sumner in his work with urease, which catalyses the conversion of urea into ammonia and water, and John H. Northrup and Wendell M. Stanley in their work on the pancreatic protein-digesting enzymes trypsin, chymotrypsin, and pepsin proved that enzymes are proteins. Sumner, Northrup, and Stanley shared the 1946 Nobel Prize in Chemistry for their groundbreaking work. Research efforts to reveal the molecular properties of enzymes have continued to the present day as biochemists probe the catalytic proteins that make life possible.

Overview

BIOCHEMISTS HAVE INVESTIGATED ENZYMES (BIOLOGICAL CATALYSTS) FOR MORE THAN 140 YEARS. LONG BEFORE THEY HAD ANY REALISTIC understanding of the physical basis of the living state, biochemists instinctively appreciated the importance of enzymes. Using the technologies devised by biochemists, life scientists gradually determined the properties of biological systems. This work eventually demonstrated that almost every event in living organisms occurs because of enzyme-catalyzed reactions. Until recently all known enzymes were proteins, but groundbreaking research led to the revelation that RNA molecules also have catalytic properties. This chapter is devoted to catalytic proteins. The characteristics of catalytic RNA molecules are described in Chapter 18.

Without enzymes, most of the thousands of biochemical reactions that sustain living processes would occur at imperceptible rates. Recent determinations of uncatalyzed (unenhanced) reaction rates in water range from 5 s for CO_2 hydration to 1.1 billion years for glycine decarboxylation. In contrast, enzyme-catalyzed reactions typically occur within time frames ranging from micro- to milliseconds. Enzymes are in fact the means by which living organisms channel the flow of energy and matter. Today, as a result of accumulating evidence derived from protein dynamics (conformational motion studies) and macromolecular crowding analysis, enzyme research is undergoing revolutionary changes. For example, according to long-held views, enzyme function depends almost entirely on the complementary shapes and catalytic interactions between reactant molecules and their more or less flexible binding sites. Recently, however, investigators have demonstrated that the catalytic function of certain enzymes can be linked to internal motions that extend throughout the protein molecule. Similarly, it is now recognized that enzymes function in conditions that are vastly different from those traditionally studied (i.e., purified molecules in dilute concentration). Instead, the *in vivo* ("in life") milieu of enzymes is a crowded gel-like environment. As a result of recent investigations, models of enzyme kinetics are evolving, and methods of experimentation, data collection, and computer simulations are becoming more sophisticated and closer to realistic *in vivo* conditions. This chapter reviews the structural and functional properties of enzymes.

TABLE 6.1 Key Characteristics of Enzymes

• Increase reaction rates
• Obey the laws of thermodynamics (i.e., no effect on K_{eq} values)
• Catalyze the forward and backward reactions of reversible reactions
• Usually present in low concentrations because they are not consumed by reactions
• Controlled via regulatory mechanisms
• Transition state of reacting substrates bound in enzyme active sites

6.1 PROPERTIES OF ENZYMES

Enzymes have several remarkable properties (**Table 6.1**). First, the rates of enzymatically catalyzed reactions are often phenomenally high. Rate increases of 10^7 to 10^{19} have been observed. Second, in marked contrast to inorganic catalysts, the enzymes are highly specific to the reactions they catalyze and side products are rarely formed. Finally, because of their relatively large and complex structures, enzymes can be regulated. This is an especially important consideration in living organisms, which must conserve energy and raw materials.

Enzyme Catalysts: The Basics

How do enzymes work? The answer to this question requires a review of the role of catalysts. By definition, a **catalyst** enhances the rate of a chemical reaction but is not permanently altered by the reaction. Catalysts perform this feat because they decrease the activation energy required for a chemical reaction. In other words, catalysts provide an alternative reaction pathway that requires less energy (**Figure 6.1**). The free energy of activation, $\Delta G^{\ddagger}$, is defined as the amount of energy required to convert 1 mol of **substrate** (reactant) molecules from the ground state (the stable, low-energy form of a molecule) to the **transition state**, which is the structural form of the reactant molecule that has the highest energy along the reaction coordinate. Since the transition state possesses more energy than the reactant or the product, it is also the least stable. In the reaction in which ethanol is oxidized to form acetaldehyde

$$CH_3—CH_2—OH \xrightarrow[2\,H]{[O]} CH_3—\overset{\overset{\displaystyle O}{\|}}{C}—H$$

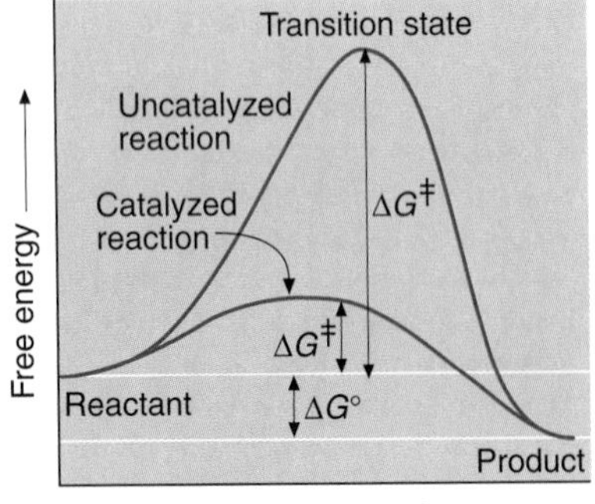

FIGURE 6.1
A Catalyst Reduces the Activation Energy of a Reaction

A catalyst alters the free energy of activation $\Delta G^{\ddagger}$, not the standard free energy ΔG° of the reaction. The transition state occurs at the apex of both reaction pathways.

this transition state might look like

$$H_3C-\overset{\displaystyle H}{\overset{|}{\underset{\underset{\displaystyle H^{\delta-}}{\vdots}}{C}}}\cdots O\cdots H^{\delta+}$$

Note that the alcoholic H is beginning to leave as a H^+ ion and the methylene H is beginning to leave as a hydride ion ($H:^-$).

Enzymes: Activation Energy and Reaction Equilibrium

To proceed at a viable rate, most chemical reactions require an initial input of energy. At temperatures above absolute zero (0 K, or –273.1°C) all molecules possess vibrational energy, which increases as the molecules are heated. Consider the following spontaneous reaction:

$$A + B \rightarrow C$$

As the temperature rises, vibrating molecules A and B are more likely to collide. A chemical reaction occurs when the colliding molecules possess a minimum amount of energy called the **activation energy** (E_a) or, more commonly in biochemistry, the *free energy of activation* ($\Delta G^{\ddagger}$). Not all collisions result in chemical reactions because only a fraction of the molecules have sufficient energy or the correct orientation to react (i.e., to break bonds or rearrange atoms into product molecules). Increasing collisions by raising the temperature or increasing reactant concentrations can improve product formation rates. In living systems, however, elevated temperature is unrealistic because of structural damage to biomolecules, and the concentration of most reactants is relatively low. Living organisms use enzymes to circumvent these restrictions. Each type of enzyme contains a unique, intricately shaped binding surface called an **active site**. Each active site is a cleft or crevice in a large protein molecule into which substrate molecules can bind in a catalysis-promoting orientation. The active site is more than a binding site, however. Several of the amino acid side chains that line the active site actively participate in the catalytic process.

The shape and charge distribution of an enzyme's active site constrains the motions and allowed conformations of the substrate, forcing it to adopt a conformation more like that of the transition state. In other words, the structure of the active site is used to optimally orient the substrate. As a result, the enzyme-substrate complex converts to product and free enzyme without the high-energy requirement of the constrained transition state. Consequently, the reaction rate increases significantly over that of the uncatalyzed reaction. Several other factors (described in Section 6.4) also contribute to rate enhancement.

Enzymes, like all catalysts, cannot alter the equilibrium of the reaction, but they can increase the rate toward equilibrium. Consider the following reversible reaction:

$$A \rightleftharpoons B$$

Without a catalyst, the reactant A is converted into the product B at a certain rate. Because this is a reversible reaction, B is also converted into A. The rate expression for the forward reaction is $k_F[A]^n$, and the rate expression for the reverse reaction is $k_R[B]^m$. The superscripts n and m represent the order of a reaction. Reaction order reflects the mechanism by which A is converted to B and vice versa. A reaction order of 2 for the conversion of A to B indicates that it is a bimolecular process and two molecules of A must collide for the reaction to occur

(Section 6.3). At equilibrium, the rates for the forward and reverse reactions must be equal:

$$k_F[A]^n = k_R[B]^m \tag{1}$$

which rearranges to

$$\frac{k_F}{k_R} = \frac{[B]^m}{[A]^n} \tag{2}$$

The ratio of the forward and reverse constants is the equilibrium constant:

$$K_{eq} = \frac{[B]^m}{[A]^n} \tag{3}$$

For example, in Equation (3), if $m = n = 1$ and $k_F = 1 \times 10^{-3}\ s^{-1}$ and $k_R = 1 \times 10^{-6}\ s^{-1}$, then

$$K_{eq} = \frac{10^{-3}}{10^{-6}} = 10^3$$

At equilibrium, therefore, the ratio of products to reactants is 1000 to 1.

In a catalyzed reaction, both the forward rate and the backward rate are increased, but the K_{eq} (in this case, 1000) remains unchanged. If the catalyst increases both the forward and the reverse rates by a factor of 100, then the forward rate becomes 100,000 and the reverse rate becomes 100. Because of the dramatic increase in the rate of the forward reaction made possible by the catalyst, equilibrium is approached in seconds or minutes instead of hours or days.

Enzymes and Macromolecular Crowding Effects

Enzymes operate in living organisms in crowded conditions (p. 38). However, enzyme-catalyzed reactions have been traditionally investigated using dilute buffered solutions. This strategy is based on the simplifying assumption of an ideal solution. Ideal solutions, for example, contain solutes in such low concentration that interactions such as steric repulsion or attractive forces are nonexistent. Reactions that occur in living organisms, however, deviate from ideality. In such circumstances, equilibrium constants are based not on solute concentrations, but on activities, quantities called *effective concentrations* that take intermolecular interactions into account. The effective concentration or *activity* (a) of a solute is equal to

$$a = \gamma c \tag{4}$$

where γ is a correction factor called the *activity coefficient*, which depends on the size and charge of the species and on the ionic strength of the solution in which the species is reacting, and c is the concentration in moles per liter. The impact of this phenomenon can be considerable. For example, the oxygen-binding capacity of hemoglobin, the predominant red blood cell protein, differs by several orders of magnitude depending on whether it is measured within red blood cells or in dilute buffer. The equilibrium constant for a reaction under nonideal conditions is given by

$$K_{eq}^{\circ} = \gamma_B[B]/\gamma_A[A] = K_{eq}^{i}\Gamma \tag{5}$$

where K_{eq}^{i} is the ideal constant and Γ is the nonideality factor, the ratio of the activity coefficients of the products and reactants. Over the past two decades it has become increasingly apparent that the assumption of ideal conditions needs to be reevaluated. Consequently, many investigators now

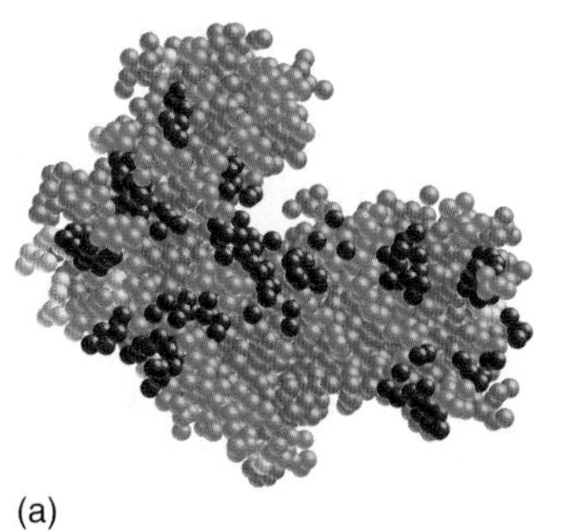

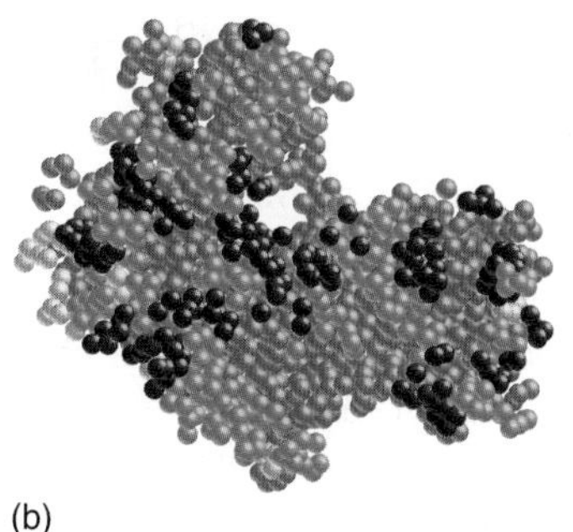

FIGURE 6.2
The Induced-Fit Model

Substrate binding causes enzymes to undergo conformational change. Hexokinase, a single polypeptide with two domains, is shown (a) before and (b) after glucose binding. The domains move relative to each other to close around a glucose molecule (not shown).

use high-molecular-weight "crowding agents" such as dextran (a glucose polymer produced by some bacteria) or serum albumin to simulate intracellular conditions in enzyme studies. Enzyme assays in the presence of crowding agents are closer to those of direct *in vivo* measurements, but the environment of the assay is still too homogeneous and differs significantly from the crowded heterogeneous conditions *in vivo*. It is a challenge for today's biochemists to construct assays and models that duplicate *in vivo* conditions.

Enzyme Specificity

Enzyme specificity is an enzyme property that is partially accounted for by the lock-and-key model, introduced by Emil Fischer in 1890. Each enzyme binds to a single type of substrate because the active site and the substrate have complementary structures. The substrate's overall shape and charge distribution allow it to enter and interact with the enzyme's active site. In a modern variation of the lock-and-key model, Daniel Koshland's *induced-fit model*, the flexible structure of proteins is taken into account (**Figure 6.2**). In this model, substrate does not fit precisely into a rigid active site. Instead, noncovalent interactions between the enzyme and substrate change the three-dimensional structure of the active site, conforming the shape of the active site to the shape of the substrate in its transition state conformation.

Although the catalytic activity of some enzymes depends only on interactions between active site amino acids and the substrate, other enzymes require nonprotein components for their activities. Enzyme **cofactors** may be ions, such as Mg^{2+} or Zn^{2+}, or complex organic molecules, referred to as **coenzymes**. The protein component of an enzyme that lacks an essential cofactor is called an **apoenzyme**. Intact enzymes with their bound cofactors are referred to as **holoenzymes**.

The activities of some enzymes can be regulated. Adjustments in the rates of enzyme-catalyzed reactions allow cells to respond effectively to environmental changes. Organisms may control enzyme activities directly, principally through the binding of activators or inhibitors, the covalent modification of enzyme molecules, or indirectly, by regulating enzyme synthesis. (Control of enzyme synthesis requires gene expression changes, a topic covered in Chapters 18 and 19.)

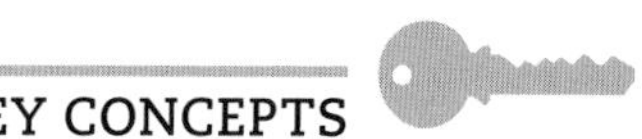

KEY CONCEPTS

- Enzymes are catalysts.
- Catalysts modify the rate of a reaction because they provide an alternative reaction pathway that requires less activation energy than the uncatalyzed reaction.
- Most enzymes are proteins.

QUESTION 6.1

The hexokinases are a class of enzymes that catalyze the ATP-dependent phosphorylation of hexoses (sugars with six carbons). The hexokinases will bind only D-hexose sugars and not their L-counterparts. In general terms, describe the features of enzyme structure that make this specificity possible.

6.2 CLASSIFICATION OF ENZYMES

In the early days of biochemistry, enzymes were named at the whim of their discoverers. Often enzyme names provided no clue to their function (e.g., trypsin), and sometimes several names were used for the same enzyme. Enzymes were often named by adding the suffix *-ase* to the name of the substrate. For example, urease catalyzes the hydrolysis of urea. To eliminate confusion, the International Union of Biochemistry (IUB) instituted a systematic naming scheme for enzymes. Each enzyme is now classified and named according to the type of chemical reaction it catalyzes. In this scheme, an enzyme is assigned a four-number classification and a two-part name called a *systematic name*. In addition, a shorter version of the systematic name, called the *recommended name*, is suggested by the IUB for everyday use. For example, alcohol:NAD^+ oxidoreductase (EC 1.1.1.1) is usually referred to as alcohol dehydrogenase. (The letters EC are an abbreviation for the Enzyme Commission of the IUB.) Because many enzymes were discovered before the institution of the systematic nomenclature, the old well-known names have been retained in quite a few cases.

The following are the six major enzyme categories:

1. **Oxidoreductases**. **Oxidoreductases** catalyze oxidation-reduction reactions in which the oxidation state of one or more atoms in a molecule is altered. Oxidation-reduction in biological systems involves one- or two-electron transfer reactions accompanied by the compensating change in the amount of hydrogen and oxygen in the molecule. Prominent examples include the redox reactions facilitated by the dehydrogenases and the reductases. For example, alcohol dehydrogenase catalyzes the oxidation of ethanol and other alcohols, and ribonucleotide reductase catalyzes the reduction of ribonucleotides to form deoxyribonucleotides. The oxygenases, oxidases, and peroxidases are among the enzymes that use O_2 as an electron acceptor.
2. **Transferases**. **Transferases** are enzymes that transfer molecular groups from a donor molecule to an acceptor molecule. Such groups include amino, carboxyl, carbonyl, methyl, phosphoryl, and acyl (RC=O). Common trivial names for the transferases often include the prefix trans; the transcarboxylases, transmethylases, and transaminases are examples.
3. **Hydrolases**. **Hydrolases** catalyze reactions in which the cleavage of bonds such as C—O, C—N, and O—P is accomplished by the addition of water. The hydrolases include esterases, phosphatases, and proteases.
4. **Lyases**. **Lyases** catalyze reactions in which groups (e.g., H_2O, CO_2, and NH_3) are removed by elimination to form a double bond or are added to a double bond. Decarboxylases, hydratases, dehydratases, deaminases, and synthases are examples of lyases.
5. **Isomerases**. A heterogeneous group of enzymes, the **isomerases** catalyze several types of intramolecular rearrangements. The sugar isomerases interconvert *aldoses* (aldehyde-containing sugars) and *ketoses* (ketone-containing sugars). The epimerases catalyze the inversion of asymmetric carbon atoms and the mutases catalyze the intramolecular transfer of functional groups.
6. **Ligases**. **Ligases** catalyze bond formation between two substrate molecules. For example, DNA ligase links DNA strand fragments together. The names of many ligases include the term *synthetase*. Several other ligases are called carboxylases.

Table 6.2 presents an example from each enzyme class.

TABLE 6.2 Selected Examples of Enzymes

Enzyme Class	Example	Reaction Catalyzed
Oxidoreductase	Alcohol dehydrogenase	$CH_3{-}CH_2{-}OH + NAD^+ \longrightarrow CH_3{-}\overset{O}{\overset{\Vert}{C}}H + NADH + H^+$
Transferase	Hexokinase	α-D-Glucose + ATP ⟶ α-D-Glucose-6-Phosphate + ADP
Hydrolase	Chymotrypsin	Polypeptide + H_2O ⟶ Peptides
Lyase	Pyruvate decarboxylase	$CH_3{-}\overset{O}{\overset{\Vert}{C}}{-}\overset{O}{\overset{\Vert}{C}}{-}O^- + H^+ \longrightarrow CH_3{-}\overset{O}{\overset{\Vert}{C}}{-}H + CO_2$ (Pyruvate ⟶ Acetaldehyde + Carbon Dioxide)
Isomerase	Alanine racemase	D-Alanine ⇌ L-Alanine
Ligase	Pyruvate carboxylase	$CH_3{-}\overset{O}{\overset{\Vert}{C}}{-}\overset{O}{\overset{\Vert}{C}}{-}O^- + HCO_3^- \xrightarrow{ATP \;\; ADP + P_i} {}^-O{-}\overset{O}{\overset{\Vert}{C}}{-}CH_2{-}\overset{O}{\overset{\Vert}{C}}{-}\overset{O}{\overset{\Vert}{C}}{-}O^-$ (Pyruvate ⟶ Oxaloacetate)

QUESTION 6.2

Which type of enzyme catalyzes each of the following reactions?

(a) $HO{-}\overset{O}{\overset{\Vert}{C}}{-}CH{=}CH{-}\overset{O}{\overset{\Vert}{C}}{-}OH$ (trans) ⟶ $HO{-}\overset{O}{\overset{\Vert}{C}}{-}CH{=}CH{-}\overset{O}{\overset{\Vert}{C}}{-}OH$ (cis)

(b) $CH_3{-}S{-}CH_2{-}\overset{O}{\overset{\Vert}{C}}{-}OH + H_2N{-}CH(CH_3){-}CH_3 \longrightarrow HS{-}CH_2{-}\overset{O}{\overset{\Vert}{C}}{-}OH + CH_3{-}NH{-}CH(CH_3){-}CH_3$

$CH_3—CH(OH)—CH_3 \longrightarrow CH_3—CH{=}CH_2 + H_2O$

(c)

$CH_3—CH_2—CH_2—C(=O)—OH \xrightarrow{NAD^+ \rightarrow NADH + H^+} CH_3—CH{=}CH—C(=O)—OH$

(d)

$2\ H_2N—CH_2—C(=O)—OH \xrightarrow{ATP \rightarrow ADP + P_i} H_2N—CH_2—C(=O)—N(H)—CH_2—C(=O)—OH + H_2O$

(e)

$CH_3—C(=O)—O—CH_3 + H_2O \longrightarrow CH_3—C(=O)—OH + CH_3—OH$

(f)

QUESTION 6.3

Aspartame, an artificial sweetener, has the following structure:

$H_3\overset{+}{N}—CH(CH_2—C(=O)—O^-)—C(=O)—N(H)—CH(CH_2—C_6H_5)—C(=O)—OCH_3$

Once consumed in food or beverages, aspartame is degraded in the digestive tract to its component molecules. Predict what the products of this process are. What classes of enzymes are involved?

6.3 ENZYME KINETICS

Recall from Chapter 4 that the principles of thermodynamics can predict whether a reaction is spontaneous but cannot predict its rate. The rate or **velocity** of a biochemical reaction is defined as the change in the concentration of a reactant or product per unit time. The initial velocity v_0 of the reaction $A \rightarrow P$, where A and P are substrate and product molecules, respectively, is

$$v_0 = \frac{-\Delta[A]}{\Delta t} = \frac{\Delta[P]}{\Delta t} \tag{6}$$

where [A] = concentration of substrate
[P] = concentration of product
t = time

Initial velocity (ν_0) is the velocity of a reaction when [A] greatly exceeds the concentration of enzyme E, [E], and the reaction time is very short. Measurements of ν_0 are made immediately after the mixing of enzyme and substrate because it can be assumed that the reverse reaction (i.e., conversion of product into substrate) has not yet occurred to any appreciable extent.

The quantitative study of enzyme catalysis, referred to as **enzyme kinetics**, provides information about reaction rates. Kinetic studies also measure the affinity of enzymes for substrates and inhibitors and provide insight into reaction mechanisms. Enzyme kinetics, in turn, provides insight into the forces that regulate metabolic pathways. The rate of the reaction A → P is proportional to the frequency with which the reacting molecules form product. The reaction rate is

$$\nu_0 = k[A]^x \qquad (7)$$

where ν_0 = initial rate
k = a rate constant that depends on the reaction conditions (e.g., temperature, pH, and ionic strength)
x = the order of the reaction

Combining Equations (6) and (7), we have

$$\frac{\Delta[A]}{\Delta t} = k[A]^x \qquad (8)$$

Order, defined as the sum of the exponents on the concentration terms in the rate expression, is determined empirically, that is, by experimentation (Figure 6.3). Determining the order of a reaction allows an experimenter to draw certain conclusions regarding the reaction's mechanism. A reaction is said to follow *first-order kinetics* if the rate depends on the first power of the concentration of a single reactant and suggests that the rate-limiting step is a unimolecular reaction (i.e., no molecular collisions are required). In such a reaction (A → P), it is assumed that

$$\text{Rate} = k[A]^1 \qquad (9)$$

If [A] is doubled, the rate is observed to double. Reducing [A] by half results in halving the observed reaction rate. In first-order reactions the concentration of the reactant is a function of time, so k is expressed in units of s^{-1}. In any reaction the time required for one-half of the reactant molecules to be consumed is called a *half-life* ($t_{1/2}$).

In the reaction A + B → P, if the order of A and B is 1 each, then the reaction is said to be *second-order*, and A and B must collide for product to form (a biomolecular reaction):

$$\text{Rate} = k[A]^1[B]^1 \qquad (10)$$

In this circumstance the reaction rate depends on the concentrations of the two reactants. In other words, both A and B take part in the reaction's rate-determining step. Second-order rate constants are measured in units of $M^{-1}s^{-1}$.

Sometimes second-order reactions involve reactants such as water that are present in great excess:

$$A + H_2O \rightarrow P$$

The second-order rate expression is

$$\text{Rate} = k[A]^1\,[H_2O]^1 \qquad (11)$$

Because water is present in excess and $[H_2O]$ is essentially constant, however, the reaction appears to be first-order. Such reactions are said to be *pseudo-first-order*. Hydrolysis reactions in biochemical systems are assumed to be pseudo-first-order

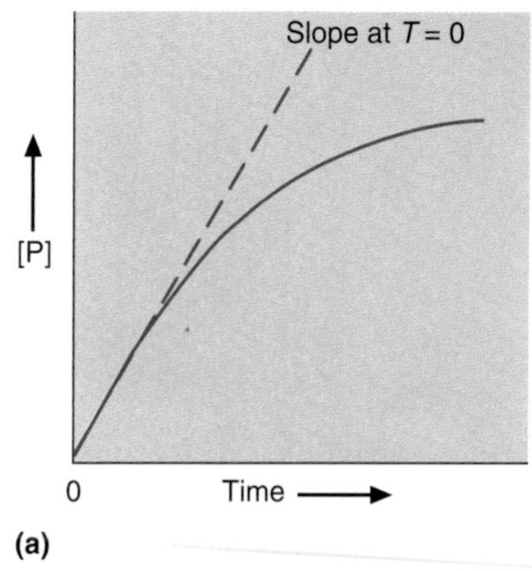

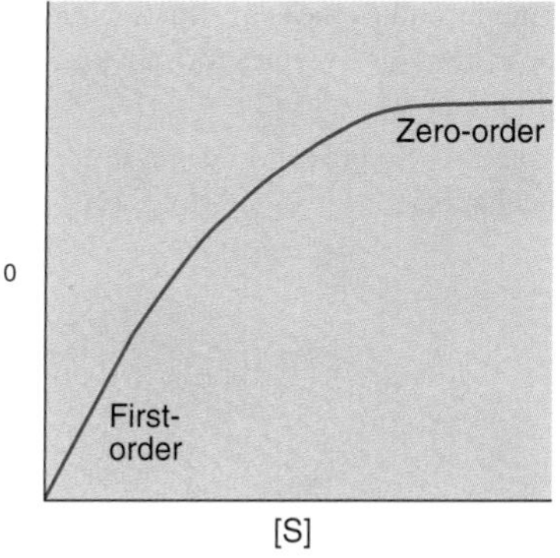

FIGURE 6.3
Enzyme Kinetic Studies

(a) Conversion of substrate to product per unit time. The slope of the curve at $t = 0$ equals the initial rate of the reaction. (b) Plot of initial velocity v versus substrate concentration [S]. The rate of the reaction is directly proportional to substrate concentration only when [S] is low. When [S] becomes high enough to saturate the enzyme, the rate of the reaction is zero-order with respect to substrate. At intermediate substrate concentrations, the reaction has a mixed order (i.e., the effect of substrate on reaction velocity is in transition).

because of the ready availability of the second reactant, H_2O, in aqueous environments.

Another possibility is that only one of the two reactants is involved in the rate-determining step and it alone appears in the rate expression. For the above reaction, if rate = $k[A]^2$, then the rate-limiting step involves collisions between A molecules. The water is involved in a fast, non-rate-limiting step in the reaction mechanism.

When the addition of a reactant does not alter a reaction rate, the reaction is said to be *zero-order* for that reactant. For the reaction A → P, the experimentally determined rate expression under such conditions is

$$\text{Rate} = k[\text{A}]^0 = k \quad (12)$$

KEY CONCEPTS

- Enzyme kinetics is the quantitative study of enzyme catalysis.
- Kinetic studies measure reaction rates and the affinity of enzymes for substrates and inhibitors.
- Kinetics also provides insight into reaction mechanisms.

The rate is constant because the reactant concentration is high enough to saturate all the catalytic sites on the enzyme molecules. An example of an order determination is given in Problem 6.1.

Reaction order can also be characterized in another way. A theoretical term can be used to characterize simple reactions: *molecularity* is defined as the number of colliding molecules in a single-step reaction. A *unimolecular* reaction A → B has a molecularity of one, whereas the *bimolecular* reaction A + B → C + D has a molecularity of two.

Michaelis-Menten Kinetics

Leonor Michaelis and Maud Menten proposed one of the most useful models in the systematic investigation of enzyme rates in 1913. The concept of the

WORKED PROBLEM 6.1

Consider the following reaction:

$$CH_3{-}CH_2{-}OH + NAD^+ \longrightarrow CH_3\overset{\displaystyle O}{\overset{\|}{C}}{-}H + NADH + H^+$$

Given the following rate data, determine the order in each reactant and the overall order of the reaction.

Initial Concentrations (mol/L)		
Ethanol	**NAD^+**	**Rate (mmol/s)**
0.1	0.1	1×10^2
0.2	0.1	2×10^2
0.1	0.2	2×10^2
0.2	0.2	4×10^2

SOLUTION

The overall initial rate expression is

$$\text{Rate} = k[\text{ethanol}]^x[\text{NAD}^+]^y$$

To evaluate x and y, determine the effect on the rate of the reaction of increasing the concentration of one reactant while keeping the concentration of the other constant.

For this experiment, doubling the concentration of ethanol doubles the rate of the reaction; therefore, x is 1. Doubling the concentration of NAD^+ doubles the rate of the reaction. So y is also 1. The rate expression then is

$$\text{Rate} = k[\text{ethanol}]^1[\text{NAD}^+]^1$$

The reaction is first-order in both reactants and second-order overall. ■

enzyme-substrate complex, first enunciated by Victor Henri in 1903, is central to Michaelis-Menten kinetics. When the substrate S binds in the active site of an enzyme E, an intermediate complex (ES) is formed. ES complex formation lowers the energy of the transition state and facilitates the product formation state. After a brief time, the product dissociates from the enzyme. This process can be summarized:

$$\mathrm{E} + \mathrm{S} \underset{k_{-1}}{\overset{k_1}{\rightleftharpoons}} \mathrm{ES} \overset{k_2}{\rightarrow} \mathrm{E} + \mathrm{P} \tag{13}$$

where k_1 = rate constant for ES formation
k_{-1} = rate constant for ES dissociation
k_2 = rate constant for product formation and release from the active site

Equation (13) ignores the reversibility of the step in which the ES complex is converted into enzyme and product. This simplifying assumption is allowed if the reaction rate is measured while [P] is still very low. Recall that initial velocities are measured in most kinetic studies. In addition, many enzymes have little affinity for the product, so the reverse reaction is not possible.

According to the Michaelis-Menten model, as currently conceived, it is assumed that (1) k_{-1} is negligible compared with k_1 and (2) the rate of formation of ES is equal to the rate of its degradation over most of the course of the reaction (i.e., the [ES] remains the same throughout the reaction). The latter premise is referred to as the *steady state assumption*. The general expression for the velocity of the reaction is

$$\text{Rate} = \frac{\Delta \mathrm{P}}{\Delta t} = k_2[\mathrm{ES}] \tag{14}$$

To be useful, a reaction rate must be defined in terms of [S] and [E]. The rate of formation of ES is equal to k_1[E][S], whereas the rate of ES dissociation is equal to $(k_{-1} + k_2)$[ES]. The steady state assumption equates these two rates:

$$k_1[\mathrm{E}][\mathrm{S}] = (k_{-1} + k_2)[\mathrm{ES}] \tag{15}$$

$$[\mathrm{ES}] = \frac{[\mathrm{E}][\mathrm{S}]}{(k_{-1} + k_2)/k_1} \tag{16}$$

Michaelis and Menten introduced a new constant, K_m (now referred to as the *Michaelis constant*):

$$k_m = \frac{k_{-1} + k_2}{k_1} \tag{17}$$

They also derived the equation

$$v = \frac{V_{max}[\mathrm{S}]}{[\mathrm{S}] + K_m} \tag{18}$$

where V_{max} = maximum velocity that the reaction can attain. This equation, now referred to as the *Michaelis-Menten equation*, has proven to be very useful in defining certain aspects of enzyme behavior. For example, when [S] is equal to K_m, the denominator in Equation (18) is equal to 2[S], and v is equal to $V_{max}/2$ (**Figure 6.4**). The experimentally determined value K_m (measured in moles per liter of substrate) is considered a constant that is characteristic of the enzyme and the substrate under specified conditions. It may reflect the affinity of the enzyme for its substrate. (If k_2 is much smaller than k_{-1}, that is, $k_2 \ll k_{-1}$, then the K_m value approximates k_1. In this circumstance, K_m is the dissociation constant for the ES complex.) The lower the value of K_m, the lower the [S] required to reach 1/2 V_{max} and, therefore, the greater the "affinity" of the enzyme for the substrate.

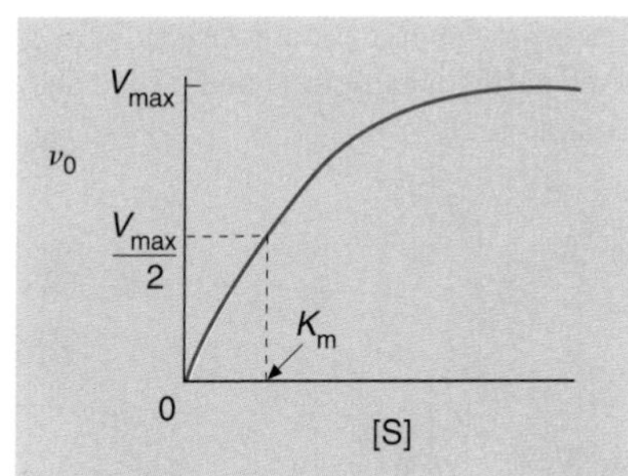

FIGURE 6.4

Initial Reaction Velocity v_0 and Substrate Concentration [S] for a Typical Enzyme-Catalyzed Reaction

The enzyme has half-maximal velocity at substrate concentration K_m.

An enzyme's kinetic properties can also be used to determine its catalytic efficiency. The **turnover number** (k_{cat}) of an enzyme is defined as

$$k_{cat} = \frac{V_{max}}{[E_t]} \tag{19}$$

where k_{cat} = number of substrate molecules converted to product per unit time by an enzyme molecule under saturating conditions

$[E_t]$ = total enzyme concentration

Under physiological conditions, [S] is usually significantly lower than K_m. A more useful measure of catalytic efficiency is obtained by rearranging Equation (19) as

$$V_{max} = k_{cat}[E_t] \tag{20}$$

and substituting this function in the Michaelis-Menten equation (Equation 18):

$$v = \frac{k_{cat}[E_t][S]}{k_m + [S]} \tag{21}$$

When [S] is very low, $[E_t]$ is approximately equal to [E] and Equation (21) reduces to

$$v = (k_{cat}/K_m)[E][S] \tag{22}$$

In Equation (22) the term k_{cat}/K_m, also referred to as the **specificity constant**, is the second-order rate constant for a reaction in which [S] << K_m, a common occurence in biological systems. In this reaction the [S] is sufficiently low that the value of k_{cat}/K_m reflects the relationship between catalytic rate and substrate binding affinity. A substrate with a high specificity constant will have a low K_m (high affinity) and a high kinetic efficiency (high turnover number). The specificity constant can be useful when comparing different substrates for the same enzyme. When [S] is low and the $[E_t]$ can be accurately measured, Equation (22) holds. Examples of k_{cat}, K_m, and k_{cat}/K_m values of selected enzymes are provided in Table 6.3. It should be noted that the upper limit for an enzyme's k_{cat}/K_m value cannot exceed the maximal value of the rate at which the enzyme can bind to substrate molecules (k_1). This limit is imposed by the rate of diffusion of substrate into an enzyme's active site. The *diffusion control limit* on enzymatic reactions is approximately 10^8 to 10^9 $M^{-1}s^{-1}$. Several enzymes, such as those listed in **Table 6.3**, have k_{cat}/K_m values that approach the diffusion control limit. Because such enzymes convert substrate to product virtually every time the substrate diffuses into the active site, they are said to have achieved *catalytic perfection*. Living organisms overcome the diffusion control limit for the enzymes in biochemical pathways that do not achieve this high degree of catalytic efficiency by organizing them into multienzyme complexes. In these complexes, the active sites of the enzymes are in such close proximity to each other that diffusion is not a factor in the transfer of substrate and product molecules.

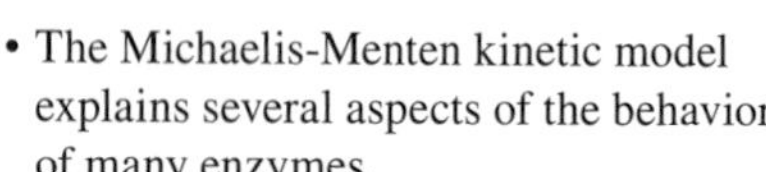

KEY CONCEPTS

- The Michaelis-Menten kinetic model explains several aspects of the behavior of many enzymes.
- Each enzyme has a characteristic K_m for a particular substrate under specified conditions.

WORKED PROBLEM 6.2

Consider the Michaelis-Menten plot illustrated in **Figure 6.5**. Identify the following points on the curve:

a. V_{max}
b. K_m

SOLUTION

Refer to **Figure 6.4**.

a. The maximum rate the enzyme can attain is V_{max}. Further increases in substrate concentration do not increase the rate.
b. K_m = [S] at $V_{max}/2$ ■

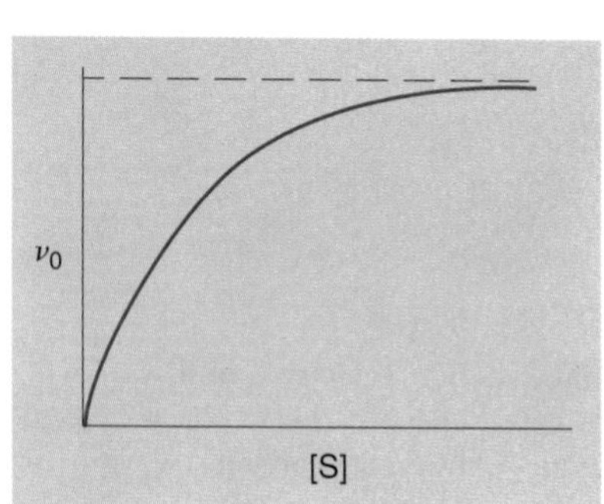

FIGURE 6.5
A Michaelis-Menten Plot

TABLE 6.3 The Values of k_{cat}, K_m, and k_{cat}/K_m for Selected Enzymes

Enzyme	Reaction Catalyzed	k_{cat} (s^{-1})	K_m (M)	k_{cat}/K_m ($M^{-1}s^{-1}$)
Acetylcholinesterase	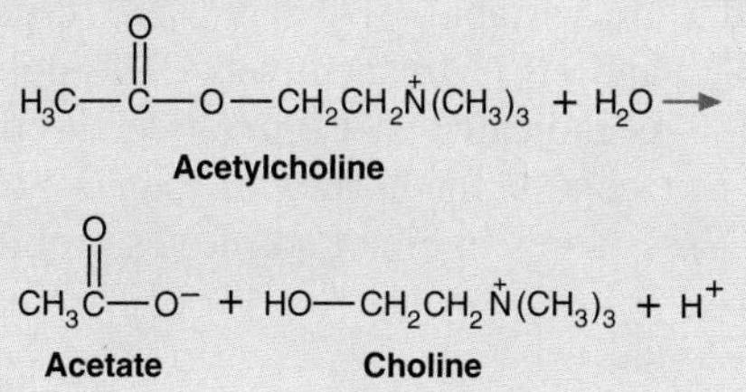 $H_3C-C(=O)-O-CH_2CH_2\overset{+}{N}(CH_3)_3 + H_2O \longrightarrow$ Acetylcholine $CH_3C(=O)-O^- + HO-CH_2CH_2\overset{+}{N}(CH_3)_3 + H^+$ Acetate, Choline	1.4×10^4	9×10^{-5}	1.6×10^8
Carbonic anhydrase	$HCO_3^- + H^+ \rightleftharpoons CO_2 + H_2O$	4×10^5	0.026	1.5×10^7
Catalase	$2\ H_2O_2 \longrightarrow 2\ H_2O + O_2$	4×10^7	1.1	4×10^7
Fumarase	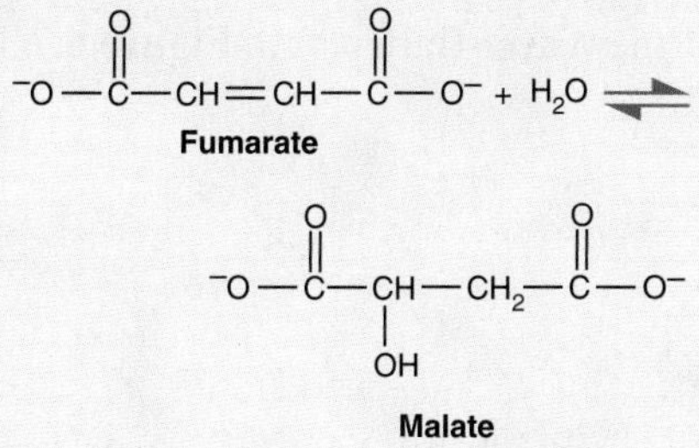 $^-O-C(=O)-CH=CH-C(=O)-O^- + H_2O \rightleftharpoons$ Fumarate $^-O-C(=O)-CH(OH)-CH_2-C(=O)-O^-$ Malate	8×10^2	5×10^{-6}	1.6×10^8
Triosephosphate isomerase	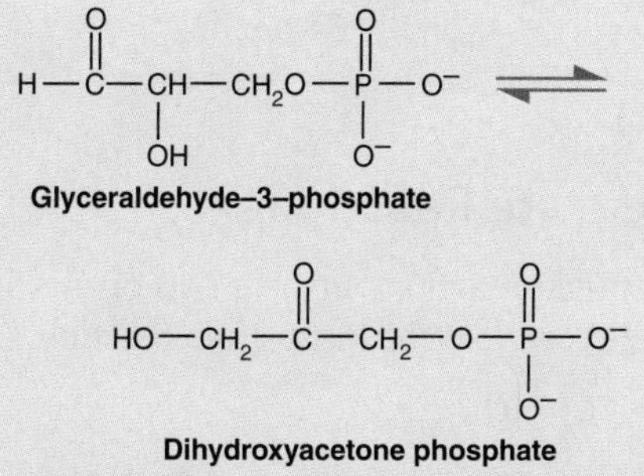 $H-C(=O)-CH(OH)-CH_2O-P(=O)(O^-)-O^- \rightleftharpoons$ Glyceraldehyde-3-phosphate $HO-CH_2-C(=O)-CH_2-O-P(=O)(O^-)-O^-$ Dihydroxyacetone phosphate	4.3×10^3	4.7×10^{-4}	2.4×10^8

Source: Adapted from A. Fersht, Structure and Mechanism in Protein Science: A Guide to Enzyme Catalysis and Protein Folding, 2nd ed., W. H. Freeman, New York, 1999.

Enzyme activity is measured in *international units* (IU). One IU is defined as the amount of enzyme that produces 1 μmol of product per minute. An enzyme's *specific activity*, a quantity that is used to monitor enzyme purification, is defined as the number of international units per milligram of protein. [A new unit for measuring enzyme activity called the *katal* has recently been introduced. One katal (kat) indicates the amount of enzyme that transforms 1 mole of substrate per second. One katal is equal to 6×10^7 IU.]

Lineweaver-Burk Plots

The K_m and V_{max} values for an enzyme are determined by measuring initial reaction velocities at various substrate concentrations. Approximate values of K_m and V_{max} can be obtained by constructing a graph, as shown in **Figure 6.4**. A more accurate determination of these values results from an algebraic transformation of the data. The Michaelis-Menten equation, whose graph is a hyperbola,

$$v = \frac{V_{max}[S]}{[S] + K_m}$$

can be rearranged by taking its reciprocal:

$$\frac{1}{v_0} = \frac{K_m}{V_{max}}\frac{1}{[S]} + \frac{1}{V_{max}}$$

The reciprocals of the initial velocities are plotted as functions of the reciprocals of substrate concentrations. In such a graph, referred to as a *Lineweaver-Burk double-reciprocal plot*, the straight line that is generated has the form $y = mx + b$, where y and x are variables ($1/v$ and $1/[S]$, respectively) and m and b are constants (K_m/V_{max} and $1/V_{max}$, respectively). The slope of the straight line is K_m/V_{max} (**Figure 6.6**). As indicated in **Figure 6.6**, the intercept on the vertical axis is $1/V_{max}$. The intercept on the horizontal axis is $-1/K_m$.

Problem 6.3 is an example of a kinetics problem using the Lineweaver-Burk plot.

FIGURE 6.6

Lineweaver-Burk, or Double-Reciprocal, Plot

If an enzyme obeys Michaelis-Menten kinetics, a plot of the reciprocal of the reaction velocity $1/v_0$ as a function of the reciprocal of the substrate concentration $1/[S]$ will fit a straight line. The slope of the line is K_m/V_{max}. The intercept on the vertical axis is $1/V_{max}$. The intercept on the horizontal axis is $-1/K_m$.

WORKED PROBLEM 6.3

Consider the Lineweaver-Burk plot in **Figure 6.7**. Identify:

a. $-1/K_m$
b. $1/V_{max}$
c. K_m/V_{max}

SOLUTION

a. $A = -1/K_m$
b. $B = 1/V_{max}$
c. K_m/V_{max} = slope ■

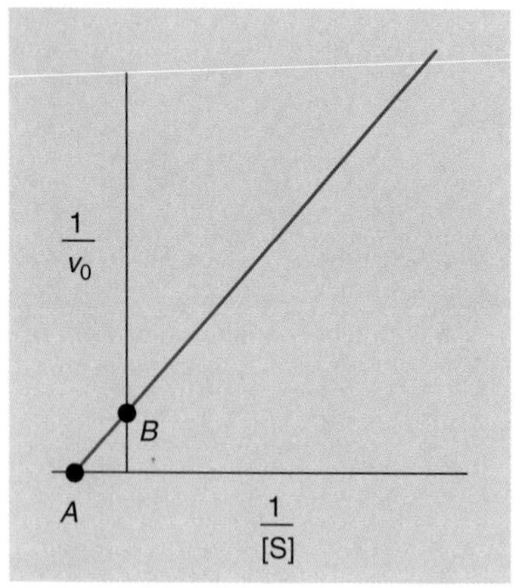

FIGURE 6.7

A Lineweaver-Burk Plot

Multisubstrate Reactions

Most biochemical reactions involve two or more substrates. The most common multisubstrate reaction, the bisubstrate reaction, is represented as

$$A + B \rightleftharpoons C + D$$

In most of these reactions there is a transfer of a specific functional group from one substrate to the other (e.g., phosphate or methyl groups) or oxidation-reduction reactions in which redox coenzymes (e.g., NAD^+/NADH or FAD/$FADH_2$) are substrates. The kinetic analysis of bisubstrate reactions is, by necessity, more complicated than that of one-substrate reactions. Often, however, the determination of the K_m for each substrate (when the other substrate is present in a saturating amount) is sufficient for most purposes. On the basis of kinetic analysis, multisubstrate reactions can be divided into two classes: sequential and double displacement.

SEQUENTIAL REACTIONS A sequential reaction cannot proceed until all substrates have been bound in the enzyme's active site. There are two possible sequential mechanisms: ordered and random. In a bisubstrate ordered mechanism, the first substrate must bind to the enzyme before the second for the reaction to proceed to product formation. The notation for such a reaction, introduced by W. W. Cleland, is

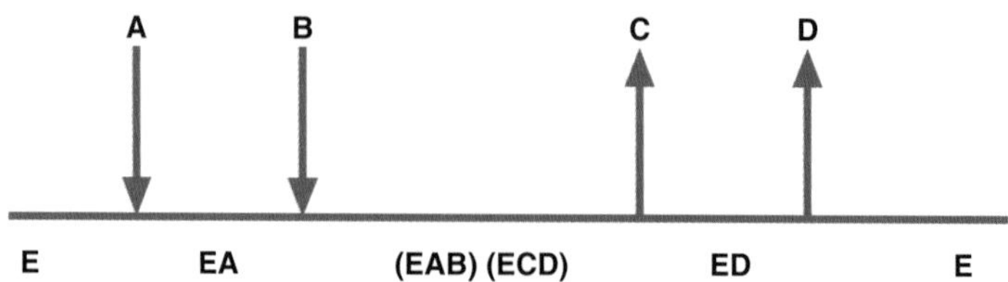

In a random sequential mechanism, the substrates can bind in any order and the products can be released in any order.

A B C D

EA ED

E EB (EAB) (ECD) E

EC

B A D C

DOUBLE-DISPLACEMENT REACTIONS In a double displacement, or "ping-pong" mechanism, the first product is released before the second substrate binds.

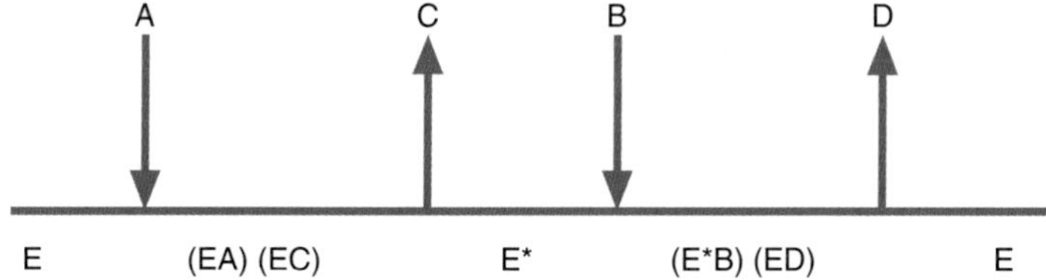

where E* = modified enzyme

In such a reaction, the enzyme is altered by the first phase of the reaction. The enzyme is then restored to its original form during the second phase, in which the second substrate is converted to product.

Enzyme Inhibition

The activity of enzymes can be inhibited. Molecules that reduce an enzyme's activity, called **inhibitors**, include many drugs, antibiotics, food preservatives, poisons, and metabolites of normal biochemical processes. The investigations of enzyme inhibition and inhibitors carried out by biochemists are important for several reasons. First, in living systems enzyme inhibition is an important means by which metabolic pathways are regulated. Numerous small biomolecules are used routinely to modulate the rates of specific enzymatic reactions so that the needs of the organism are consistently met. Second, numerous clinical therapies are based on enzyme inhibition. For example, many antibiotics and other drugs reduce or eliminate the activity of specific enzymes. The most effective AIDS treatment is a multidrug therapy that includes protease inhibitors, molecules that disable a viral enzyme required to make new virus. Finally, investigations of enzyme inhibition have enabled biochemists to develop techniques for probing the physical and chemical architecture, as well as the functional properties, of enzymes.

Enzyme inhibition can be reversible or irreversible. **Reversible inhibition** occurs when the inhibitory effect of a compound can be counteracted by increasing substrate levels or removing the inhibitor compound while the enzyme remains intact. Reversible inhibition will be **competitive** if the inhibitor and the substrate bind to the same site, **noncompetitive** if the inhibitor binds to a site other than the active site, and **uncompetitive** if the inhibitor binding site is created after the substate is bound to the enzyme. **Irreversible inhibition** occurs when inhibitor binding permanently impairs the enzyme, usually through a covalent reaction that chemically modifies the enzyme.

COMPETITIVE INHIBITORS Competitive inhibitors bind reversibly to free enzyme, not the ES complex, to form an enzyme-inhibitor (EI) complex.

$$\mathrm{E} + \mathrm{S} \underset{k_{-1}}{\overset{k_1}{\rightleftharpoons}} \mathrm{ES} \xrightarrow{k_2} \mathrm{P} + \mathrm{E}$$

$$+\ \mathrm{I}$$

$$k_{\mathrm{I}-1} \upharpoonleft\downharpoonright k_{\mathrm{I}1}$$

$$\mathrm{EI} + \mathrm{S} \longrightarrow \text{No reaction}$$

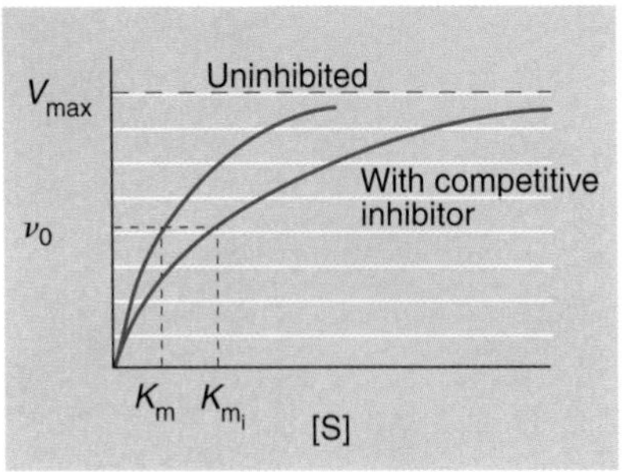

FIGURE 6.8

Michaelis-Menten Plot of Uninhibited Enzyme Activity versus Competitive Inhibition

Initial velocity v_0 is plotted against substrate concentration [S]. With competitive inhibition, V_{max} stays constant and K_m increases.

The substrate and the inhibitor compete for the same site on the enzyme. As soon as the inhibitor binds, substrate access to the active site is blocked. The concentration of EI complex depends on the concentration of free inhibitor and on the dissociation constant K_I:

$$K_I = \frac{[E][I]}{[EI]} = \frac{k_{I-1}}{k_{I1}}$$

where K_I is a measure of the enzyme's binding affinity for the inhibitor.

Because the EI complex readily dissociates, the enzyme is again available for substrate binding. The enzyme's activity declines with an increased K_m and unchanged V_{max} (**Figure 6.8**) in the presence of a competitive inhibitor because the EI complex does not participate in the catalytic process. Because V_{max} is unchanged, the effect of a competitive inhibitor on activity is reversed by increasing the concentration of substrate. At high [S], all the active sites are filled with substrate, and reaction velocity reaches the value observed without an inhibitor. The Michaelis-Menten equation for competitive inhibition that takes the formation of [EI] into account is obtained by incorporating a term for the effect of the inhibitor on K_m:

$$v = \frac{V_{max}\,[S]}{[S] + \alpha K_m}$$

where $\alpha = 1 + [I]/K_i$. The term α is a function of both the competitive inhibitor's concentration and its affinity for the enzyme active site. The K_m value (the [S] at which $v = 1/2\ V_{max}$) increases in the presence of the inhibitor by the factor α. The term αK_m is often referred to as the apparent K_m (K_m^{app}).

Substances that behave as competitive inhibitors (i.e., reduce an enzyme's apparent affinity for substrate) are often similar in structure to the substrate. Such molecules include reaction products or unmetabolizable analogues or derivatives of substrate molecules. Succinate dehydrogenase, an enzyme in the Krebs citric acid cycle (Chapter 9), catalyzes a redox reaction that converts succinate to fumarate.

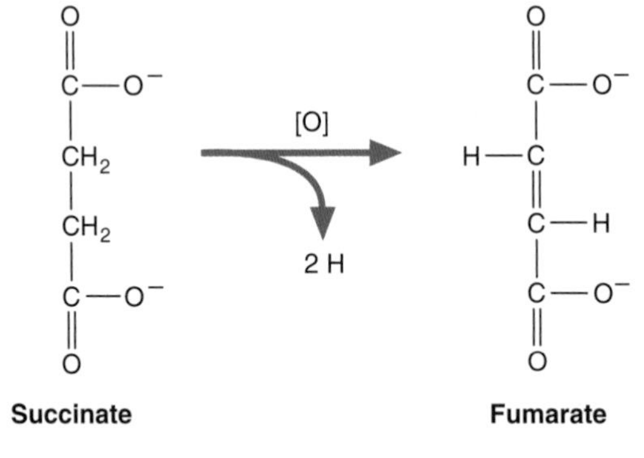

$$^{-}OOC{-}CH_2{-}CH_2{-}COO^{-} \xrightarrow{[O]} {}^{-}OOC{-}CH{=}CH{-}COO^{-} + 2\,H$$

Succinate **Fumarate**

This reaction is inhibited by malonate (**Figure 6.9**). Malonate binds to the enzyme's active site but cannot be converted to product. Molecules that resemble the structure of a substrate's transition state, referred to as *transition state analogs*, can be especially efficient competitive inhibitors. They bind in the enzyme active site with a higher affinity than does the substrate. Oseltamivir (Tamiflu), used to prevent and treat influenza, is converted in the liver into a transition state analog of sialic acid (a nine-carbon sugar). Tamiflu's binding in the active site of the viral enzyme neuraminidase inhibits cleavage of sialic acid from certain host cell proteins, thereby preventing the release of new virus from infected cells.

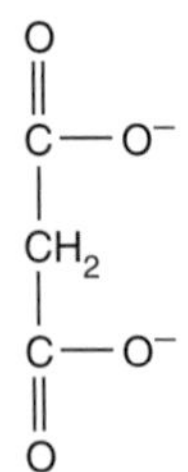

FIGURE 6.9

Malonate

NONCOMPETITIVE INHIBITORS In some enzyme-catalyzed reactions an inhibitor can bind to both the enzyme and the enzyme-substrate complex:

$$\begin{array}{ccccc} E + S & \rightleftharpoons & ES & \longrightarrow & P + E \\ + & & + & & \\ I & & I & & \\ k_{I-1} \uparrow\downarrow k_{I1} & & k_{I-1} \uparrow\downarrow k_{I1} & & \\ EI + S & \underset{k_{I-1}}{\overset{k_{I1}}{\rightleftharpoons}} & EIS & & \end{array}$$

In such circumstances, referred to as **noncompetitive** inhibition, the inhibitor binds to a site other than the active site. Inhibitor binding results in a modification of the enzyme's conformation that prevents product formation (**Figure 6.10**). Noncompetitive inhibitors have little or no structural resemblance to substrate, but they may influence substrate binding if their binding sites are in close proximity to the substrate binding site. In some cases, noncompetitive inhibition may be partially reversed by increasing the substrate concentration. Analysis of reactions inhibited by noncompetitive inhibitors is often complex because usually two or more substrates are involved. Thus the characteristics of the inhibition that are observed may depend in part on factors such as the order in which the different substrates bind. In addition, there are two determinations of K_I for the binding of noncompetitive inhibitors

$$K_{Ia} = \frac{[E][I]}{[EI]}$$

and

$$K_{Ib} = \frac{[E][S][I]}{[EIS]}$$

Depending on the inhibitor being considered, the values of these dissociation constants may or may not be equivalent. There are two forms of noncompetitive inhibition: pure and mixed. In pure noncompetitive inhibition (the inhibitor binds far from the active site), a rare phenomenon, both K_I values are equivalent. Mixed noncompetitive inhibition (the inhibitor binds close to the active site) is typically more complicated because the K_I values are different. The Michaelis-Menten equation that describes mixed noncompetitive inhibition is

$$v = \frac{V_{max}\,[S]}{\alpha'[S] + \alpha K_m}$$

where $\alpha = 1 + [I]/K_{Ia}$ and $\alpha' = (1 + [I]/K_{Ib})$. In pure noncompetitive inhibition (i.e., $\alpha = \alpha'$), the values of V_{max} change, but K_m remains unaffected. In mixed noncompetitive inhibition the values of both K_m ($K_m^{app} = \alpha/\alpha' K_m$) and V_{max} ($V_{max}^{app} = V_{max}/\alpha'$) are altered.

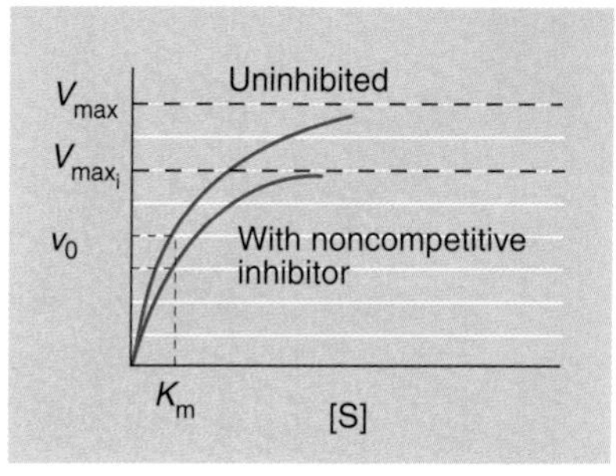

FIGURE 6.10

Michaelis-Menten Plot of Uninhibited Enzyme Activity versus Noncompetitive Inhibition

Initial velocity v_0 is plotted against substrate concentration [S]. With noncompetitive inhibition V_{max} decreases and K_m remains unchanged if access to the active site is uncompromised. If the active site is partially blocked when the inhibitor is bound, the K_m will increase.

UNCOMPETITIVE INHIBITORS In uncompetitive inhibition, which is considered a rare type of noncompetitive inhibition, the inhibitor binds only to the enzyme-substrate complex, not to the free enzyme. As a consequence, the inhibitor is ineffective at low substrate concentrations because very little of the ES complex is present.

$$\begin{array}{ccccc} E + S & \rightleftharpoons & ES & \longrightarrow & P + E \\ & & + & & \\ & & I & & \\ & & k_{I-1} \uparrow\downarrow k_{I1} & & \\ & & EIS & & \end{array}$$

The dissociation constant for the binding step of an uncompetitive inhibitor to an enzyme is

$$K_I' = \frac{[ES][I]}{[EIS]}$$

Uncompetitive inhibition is most easily observed at high [S]. When I binds to ES, the substrate is not free to dissociate from the enzyme and the K_m ($K_m^{app} = K_m/\alpha'$)

decreases, giving the appearance of higher substrate affinity. The Michaelis-Menten equation that describes uncompetitive inhibition is

$$v = \frac{V_{max}\,[S]}{\alpha'[S] + K_m}$$

where $\alpha' = 1 + [I]/K_I'$. Because an uncompetitive inhibitor binds only to ES, the equilibrium $E + S \rightleftharpoons ES$ is shifted to the right with the result of increasing the amount of ES available to bind to the inhibitor. The value of V_{max} ($V_{max}^{app} = V_{max}/\alpha'$) is lowered by a factor of α'. Uncompetitive inhibition is most commonly observed for enzymes that bind more than one substrate.

KINETIC ANALYSIS OF ENZYME INHIBITION Competitive, noncompetitive, and uncompetitive inhibition can be distinguished with double-reciprocal plots (**Figure 6.11a–e**). In two sets of rate determinations, enzyme concentration is held constant. The first experiment establishes the velocity and kinetic parameters (K_m and V_{max}) of the uninhibited enzyme. In the second experiment, a constant amount of inhibitor is included in each enzyme assay. **Figure 6.11** illustrates the different effects that inhibitors have on enzyme activity. Competitive inhibition increases the K_m of the enzyme but the V_{max} is unchanged. (This is shown in the double-reciprocal

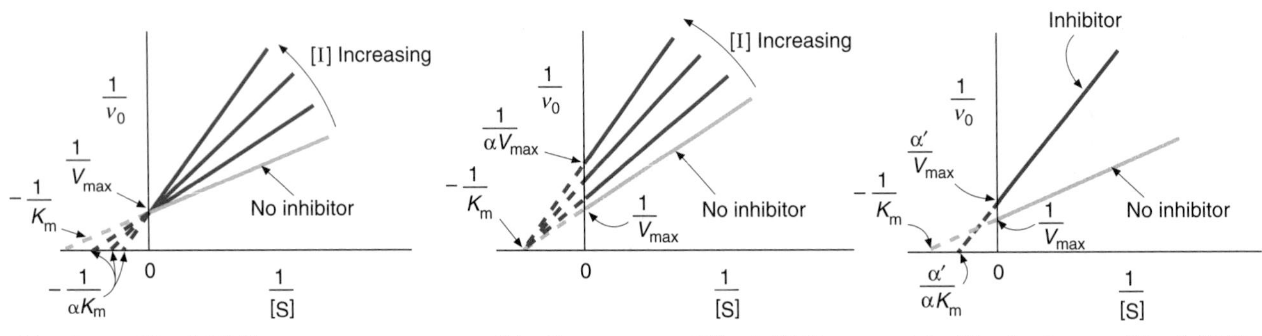

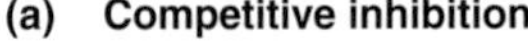
(a) Competitive inhibition

(b) Pure noncompetitive inhibition

(c) Mixed noncompetitive inhibition

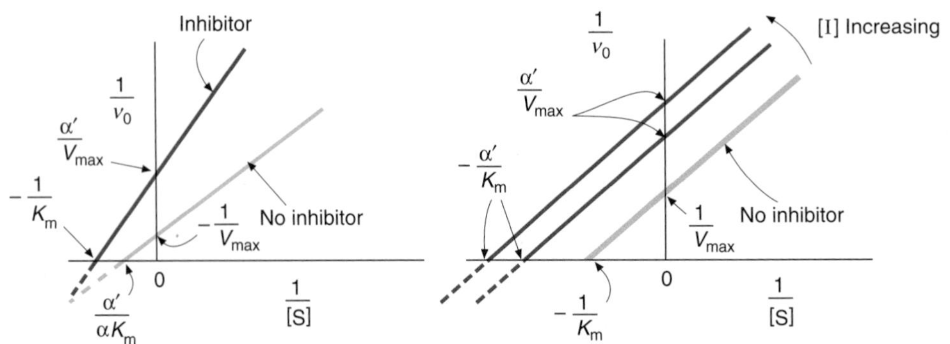

(d) Mixed noncompetitive inhibition

(e) Uncompetitive inhibition

FIGURE 6.11
Kinetic Analysis of Enzyme Inhibition

(a) Competitive inhibition. Plots of $1/v$ versus $1/[S]$ in the presence of several concentrations of the inhibitor result in no change in the vertical intercept (i.e., $1/V_{max}$, so V_{max} is unchanged) and a decrease in the vertical intercept (i.e., $-1/K_m$, so K_m is increased). The Lineweaver-Burk equation for competitive inhibition is $1/v = (\alpha K_m/V_{max})\,(1/[S]) + 1/V_{max}$. (b) Noncompetitive inhibition. Plots of $1/v$ versus $1/[S]$ in the presence of several concentrations of the inhibitor intersect at the same point on the horizontal axis, $-1/K_m$. In pure noncompetitive inhibition the constants for ES and EIS are assumed to stay the same. In mixed noncompetitive inhibition the plots of $1/v$ versus $1/[S]$ will intersect the vertical axis above the horizontal axis (c) if K_{Ib}' is greater than K_{Ia} or below the horizontal axis (d) if K_{Ib}' is less than K_{Ia}. The Lineweaver-Burk equation for mixed noncompetitive inhibition is $1/v = (\alpha K_m/V_{max})\,(1/[S]) + \alpha'/V_{max}$. (e) Uncompetitive inhibition. Plots of $1/v$ versus $1/[S]$ in the presence of several concentrations of the inhibitor result in a decrease in K_m (α'/K_m) and a decrease in V_{max} (α'/V_{max}) such that the slope of the line is unchanged but shifts up and to the left. The Lineweaver-Burk equation for uncompetitive inhibition is $1/v = (K_m/V_{max})\,(1/[S]) + \alpha'/V_{max}$.

plot as a shift in the horizontal intercept.) In pure noncompetitive inhibition (i.e., $K_{Ia} = K_{Ib}$) V_{max} is lowered (i.e., the vertical intercept is shifted); the K_m is unchanged in this rare case because the k values for the binding to E and ES are the same. In mixed noncompetitive inhibition these k values are different because the inhibitor binding site and the active site are close together, and their binding activities interfere with each other. Consequently, both V_{max} and K_m change. In such cases, the plots of $1/v$ versus 1/[S] will intersect to the left of the vertical axis either above the horizontal axis if K_I' is greater than K_I or below the axis if K_I' is less than K_I. In uncompetitive inhibition both K_m and V_{max} are changed, although their ratio (i.e., the slope K_m/V_{max}) remains the same.

IRREVERSIBLE INHIBITION In reversible inhibition the inhibitor can dissociate from the enzyme because it binds through noncovalent bonds. Irreversible inhibitors usually bond covalently to the enzyme, often to a side chain group in the active site. For example, enzymes containing free sulfhydryl groups can react with alkylating agents such as iodoacetate:

$$\text{Enzyme—CH}_2\text{—SH} + \text{I—CH}_2\text{—}\overset{\overset{\text{O}}{\|}}{\text{C}}\text{—O}^- \longrightarrow \text{Enzyme—CH}_2\text{—S—CH}_2\text{—}\overset{\overset{\text{O}}{\|}}{\text{C}}\text{—O}^- + \text{HI}$$

Glyceraldehyde-3-phosphate dehydrogenase, an enzyme in the glycolytic pathway (Chapter 8), is inactivated by alkylation with iodoacetate. Enzymes that use sulfhydryl groups to form covalent bonds with metal cofactors are often irreversibly inhibited by heavy metals (e.g., mercury and lead). The anemia that is symptomatic of lead poisoning is caused in part because lead binds to a sulfhydryl group of ferrochelatase. Ferrochelatase catalyzes the insertion of Fe^{2+} into the heme prosthetic group of hemoglobin.

Problem 6.4 is concerned with enzyme inhibition.

KEY CONCEPTS

- Reversible inhibition of enzymes can be competitive, noncompetitive, or uncompetitive.
- Competitive inhibitors reversibly compete with substrate for the same site on free enzyme.
- Noncompetitive inhibitors can bind to both the enzyme and the enzyme-substrate complex because it binds to a site outside the active site.
- Uncompetitive inhibitors bind only to the enzyme-substrate complex, not the free enzyme.
- Irreversible inhibitors usually bind covalently to the enzyme.

WORKED PROBLEM 6.4

Consider the Lineweaver-Burk plot illustrated in **Figure 6.12**.

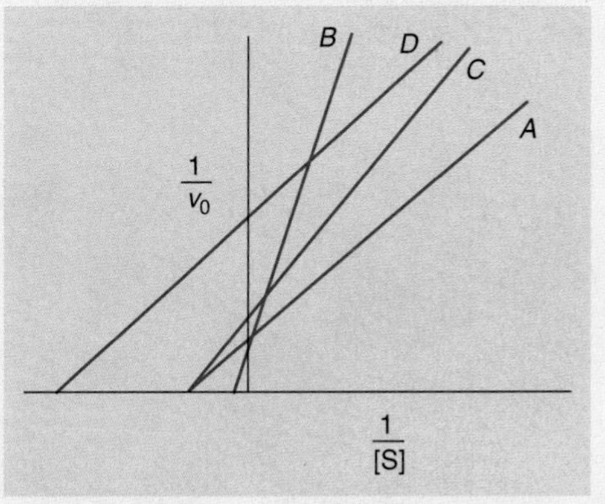

FIGURE 6.12
A Lineweaver-Burk Plot

Line A = normal enzyme-catalyzed reaction
Line B = compound B added
Line C = compound C added
Line D = compound D added

Identify the type of inhibitory action shown by compounds B, C, and D.

SOLUTION

Compound B is a competitive inhibitor because the K_m only has changed. Compound C is a pure noncompetitive inhibitor because the V_{max} only has changed. Compound D is an uncompetitive inhibitor because both K_m and V_{max} have changed. ■

WORKED PROBLEM 6.5

An enzyme has a K_m of 10 μM. When 5 μM of a competitive inhibitor is added, K_I is determined to be 2.5 μM. Determine (a) the value of α and (b) K_m^{app}.

SOLUTION

(a) The value of α is determined as follows:

$$\alpha = 1 + [I]/K_I = 1 + 5\ \mu M/2.5\ \mu M = 1 + 2 = 3$$

(b) K_m^{app}, the measured K_m value when the enzyme is inhibited, is calculated as follows:

$$\alpha = K_m^{app}/K_m$$
$$K_m^{app} = \alpha K_m = (3)\ (10\ mM) = 30\ mM.$$

■

QUESTION 6.4

Iodoacetamide is an irreversible inhibitor of several enzymes that have a cysteine residue in their active sites. After examining its structure, predict the product of the reaction of iodoacetamide with such an enzyme.

$$I{-}CH_2{-}\overset{\overset{O}{\|}}{C}{-}NH_2$$

ALLOSTERIC ENZYMES Although the Michaelis-Menten model is an invaluable tool, it does not explain the kinetic properties of many enzymes. For example, plots of reaction velocity versus substrate concentration for many enzymes with multiple subunits are often sigmoidal rather than hyperbolic, as predicted by the Michaelis-Menten model (**Figure 6.13**). Such effects are seen in an important group of enzymes called the **allosteric enzymes**. Note that the substrate-binding curve in **Figure 6.13** resembles the oxygen-binding curve of hemoglobin (p. 173).

The properties of allosteric enzymes are discussed on pp. 227–229.

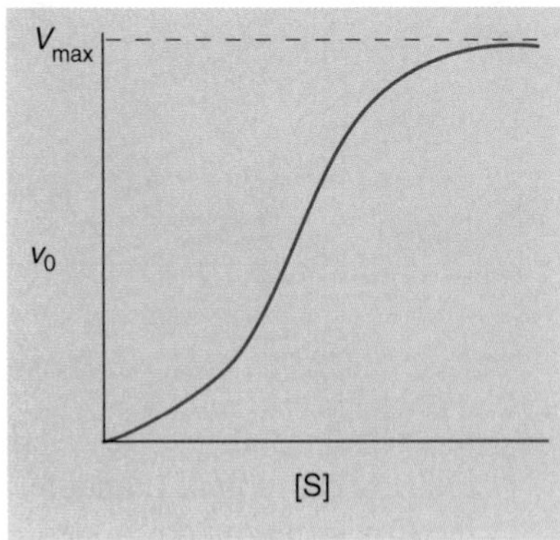

FIGURE 6.13
The Kinetic Profile of an Allosteric Enzyme
The sigmoidal binding curve displayed by many allosteric enzymes resembles the curve for the cooperative binding of O_2 to hemoglobin.

QUESTION 6.5

Drinking methanol can cause blindness in humans, as well as a severe acidosis that may be life-threatening. Methanol is toxic because it is converted in the liver to formaldehyde and formic acid by the enzymes alcohol dehydrogenase and aldehyde dehydrogenase. Methanol poisoning is treated with dialysis and infusions of bicarbonate and ethanol. Explain why each treatment is used.

Enzyme Kinetics, Metabolism, and Macromolecular Crowding

The ultimate goal of enzyme kinetic investigations is the development of a realistic understanding of metabolism in living organisms. Despite a wealth of knowledge, gathered over many decades, about biochemical pathways and the *in vitro* ("in glass," i.e., measured in the laboratory) kinetic properties of enzymes, insight into metabolic processes in "live" cells is elusive. An important reason for this lack of success is now believed to lie in the assumptions of the law of mass action. According to this rule, when an equilibrium constant for a reaction such as

$$A + B \underset{k_R}{\overset{k_F}{\rightleftharpoons}} C$$

is calculated, it is assumed that the forward rate is *linear* (i.e., is directly proportional) to the concentrations of A and B, and the rate of the reverse reaction (k_R) is linear with respect to the concentration of C. For the K_{eq} value derived for this or any reaction to be valid, it is further assumed that (1) the system in which the reaction occurs is homogeneous (i.e., there is uniform mixing of its contents) and (2) interacting molecules move randomly and independently of each other. It is now understood that under crowded *in vivo* conditions, neither of these assumptions appears to hold true. The crowded interior of cells is highly heterogeneous (i.e., an enormous variety of different molecular species are present) and filled with macromolecules, membranes, cytoskeletal components, and other obstacles that impede molecular movement. This phenomenon is called **macromolecular crowding**. Among its observed effects within cells are increased values for the effective concentrations of macromolecules, enhanced protein folding and binding affinities, changes in reaction rates and equilibrium constants, and decreased diffusion rates. These effects are *nonlinear*; that is, they are difficult to predict. For example, the *in vivo* kinetic values (i.e., K_m, V_{max}, and K_{eq}) of each biochemical reaction depend on the activity coefficient of its enzyme and the diffusion rates of substrate and effector molecules, neither of which are easy to describe in the heterogeneous environment of the cell. Slow substrate diffusion can be overcome by the segregation of reactions into microcompartments where substrate concentration is relatively high or the recruitment of some or all of a pathway's enzymes into complexes called *metabolons* in which pathway intermediates are directly transferred, or "channeled," from one active site to the next. These microenvironments have their own unique ionic composition that promotes the efficient operation of the resident enzymes. It is a significant challenge to biochemists to identify these local conditions and duplicate them *in vitro* to facilitate the investigation of specific enzymes and determine realistic kinetic parameters. Discrepancies between results derived *in vivo*, *in vitro*, and in silico (by computer simulation) indicate that an enzyme isolated from its metabolic pathway and other intersecting pathways does not function with the same kinetic parameters.

Despite the daunting complexities of living cells, systems biologists have endeavored to develop new strategies to investigate metabolic processes. This work is often done by metabolic engineers, industrial scientists who improve the metabolic capacities of microorganisms used to produce commercial products. Among the most important tools used by metabolic engineers are dynamic computer simulations of metabolic processes that are based on both *in vivo* and *in vitro* data. The mathematical model upon which the computer simulation is based is constructed from kinetic data and equations that define *metabolic flux*, the rate of flow of metabolites, such as substrates, products, and intermediates along biochemical pathways. (Despite the limitations of *in vitro* kinetic values just described, they are recognized as invaluable baseline data during the early phases of model development.) The ultimate goal of a computer simulation is to approximate the *in vivo* steady state behavior of metabolic processes and predict how they will respond to parameter changes such as nutrient availability or mutant forms of enzymes. Insights already gained from computer simulations (e.g., see the discussion of glycolysis and jet engines in Chapter 8) guarantee that in silico modeling and pathway simulation will form an increasingly significant part of future enzyme kinetics investigations.

6.4 CATALYSIS

However valuable kinetic studies are, they do not reveal any information about the catalytic mechanisms of enzymes. [A *mechanism* describes how bonds are broken and new ones formed in the conversion of substrate(s) to product(s).] Enzyme mechanism investigations seek to relate enzyme activity to the structure and function of the active site. Scientists who study enzyme mechanisms use such methods as X-ray crystallography, chemical inactivation of active site side chains, and modeling with simple model compounds as substrates and as inhibitors.

Organic Reactions and the Transition State

Biochemical reactions follow the same set of rules as the reactions studied by organic chemists. The essential features of both are the reaction between electron-deficient atoms (electrophiles) and electron-rich atoms (nucleophiles) and the formation of transition states. Each is discussed briefly.

Chemical bonds form when a nucleophile donates an electron pair to an electrophile. For example, in the following reaction

$$R_2C{=}CH_2 + H_3O^+ \longrightarrow R{-}C(OH)(R){-}CH_2{-}H + H^+$$

the π electrons of the nucleophilic double bond will react with a partially positive hydrogen atom of an electrophilic hydronium ion. As in all organic reactions, product formation takes place in several steps.

$$R_2C{=}CH_2 + H_3O^+ \longrightarrow R_2C^+{-}CH_3 + H_2O \longrightarrow$$

Carbocation intermediate

$$R_2C(OH_2^+){-}CH_3 + H_2O \longrightarrow R{-}C(OH)(R){-}CH_3 + H_3O^+$$

This step-by-step description is referred to as a **reaction mechanism**. The curved arrows illustrate the flow of electrons away from a nucleophile and toward an electrophile. During the course of the reaction, one or more intermediates form. An **intermediate** is a species that exists for a finite length of time (10^{-13} s or less) and then is transformed into product. In the example reaction, a carbocation intermediate forms as π electrons in the double bond attack an electrophilic hydronium ion. (A **carbocation**, or carbonium ion, contains an electron-deficient, positively charged carbon atom.) This type of intermediate is stabilized by the

R groups that decrease the positive charge on the carbon atom. In the reaction's next step, an electron pair of the oxygen atom in a water molecule forms a σ bond with the positively charged carbon atom. In the final step the alcohol product results from a proton transfer to another water molecule. Other examples of reactive intermediates observed in biochemical reactions include **carbanions** (nucleophilic carbon anions with three bonds and an unshared electron pair) and **free radicals** (highly reactive species with at least one unpaired electron).

Transition State Stabilization

In any chemical reaction, only molecules that reach the activated condition known as the transition state (see p. 192 and **Figure 6.1**) can convert into product molecules. The conversion of a stable reactant molecule into an activated one is analogous to the energy-requiring process of rolling a boulder up a hill. Once the boulder has reached the top, only a slight push will cause it to slide down the other side, releasing energy as it does so. The rate of a reaction is determined by the relative number of reactant molecules that possess sufficient energy to overcome the activation energy barrier (E_a) (see **Figure 6.1**). Reaction rates increase if E_a can be lowered.

The capacity to lower E_a has been attributed to transition state stabilization. In *transition state theory* it is assumed that the enzyme binds more strongly to the transition state than to the original substrate molecule. Chemical reactions occur in the following manner:

$$\mathrm{E + S \rightleftharpoons ES \rightleftharpoons ETS \rightarrow E + P}$$

where ETS = enzyme-transition state complex.

The dissociation constants for the ES complex and the ETS complex are K_S = [E][S]/[ES] and K_T = [E][TS]/[ETS], respectively. The rate accelerations achieved by enzymes requires that $K_T < K_S$. In other words, an enzyme stabilizes the transition state by binding it more tightly than to the substrate. Enzymes can bind more effectively to the transition state when the active site is complementary in its shape and electrostatic structure to the transition state's structure.

Triose phosphate isomerase (TPI) is an example of an enzyme whose kinetic and mechanistic properties have been researched in some detail. TPI is an exceptionally efficient enzyme that is composed of two identical subunits. It catalyzes the reversible conversion of dihydroxyacetone phosphate (DHAP) to glyceraldehyde-3-phosphate (G-3-P), which is a reaction in glycolysis (Chapter 8), the biochemical pathway that degrades the sugar glucose. The enzyme's reaction mechanism involves the formation of an enediol intermediate and two transition states. (An *enediol* is a molecule containing the atomic arrangement –C(OH)=C(OH) –. It is produced by proton migration from the CH of a CHOH group to the oxygen of a carbonyl group to yield a carbon-carbon double bond where each carbon bears an OH group. See p. 247.)

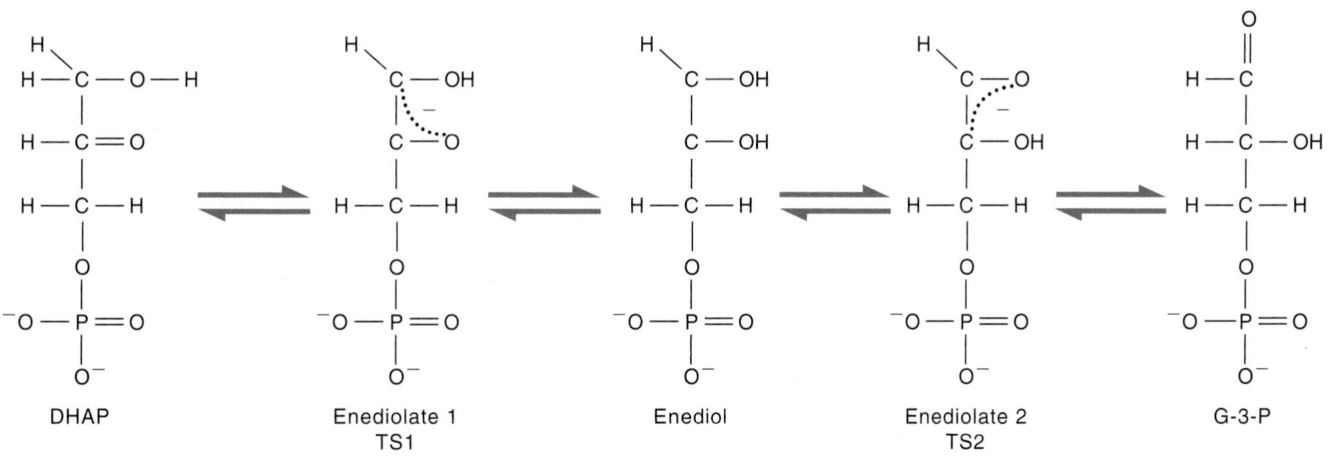

DHAP

Enediolate 1
TS1

Enediol

Enediolate 2
TS2

G-3-P

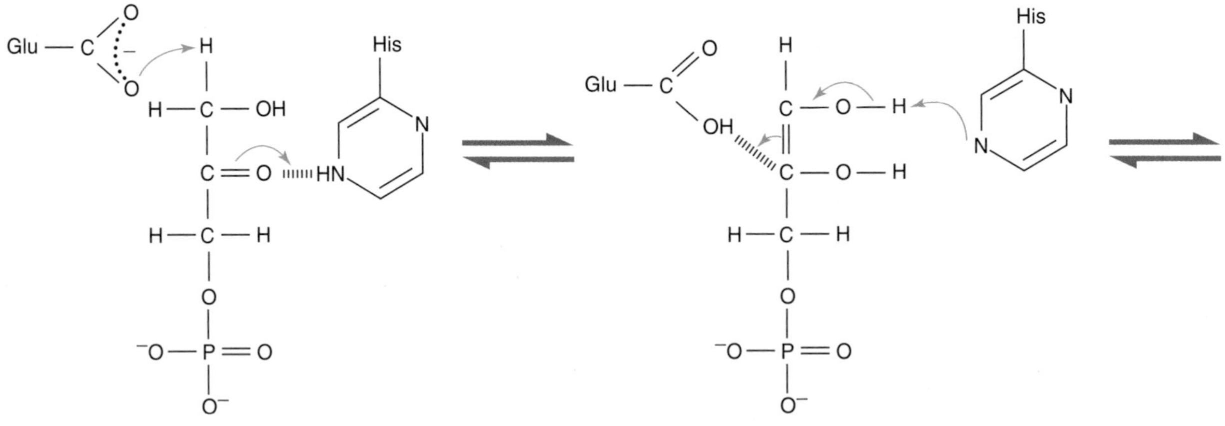

FIGURE 6.14

The Triose Phosphate Isomerase Mechanism

The isomerization reaction in which dihydroxyacetone phosphate (DHAP) is converted to glyceraldehyde-3-phosphate (G-3-P) is an example of acid-base catalysis that involves a glutamate and a histidine within the active site. The reaction begins when the nucleophilic glutamate carboxyl group abstracts a proton from the substrate and a histidine residue donates a proton to yield the enediol intermediate. The enediol collapses as the histidine removes a proton from the C-1 hydroxyl group and C-2 abstracts a proton from the protonated glutamate to form the G-3-P product.

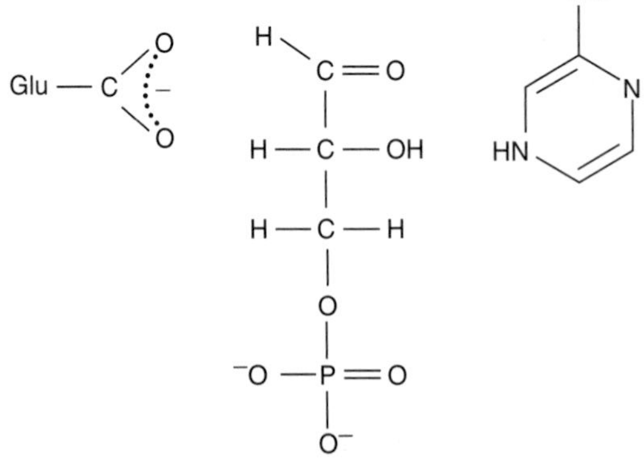

The active site of TPI possesses numerous precisely oriented amino acid side chains that either participate directly in the reaction mechanism (Glu 165 and His 95) or stabilize the substrate, transition states, or the enediol intermediate (**Figure 6.14**). The conversion of DHAP into the enediol involves the deprotonation of C-1 by a nucleophilic glutamate residue and the donation of a proton to the C-2 carbonyl oxygen by an electrophilic histidine residue. Once the enediol is formed, its C-2 atom abstracts a proton from the protonated glutamate residue and the removal of a proton from the C-1 hydroxyl group to yield the product G-3-P. The energy diagram for a two-step reaction in **Figure 6.15** illustrates the difference between

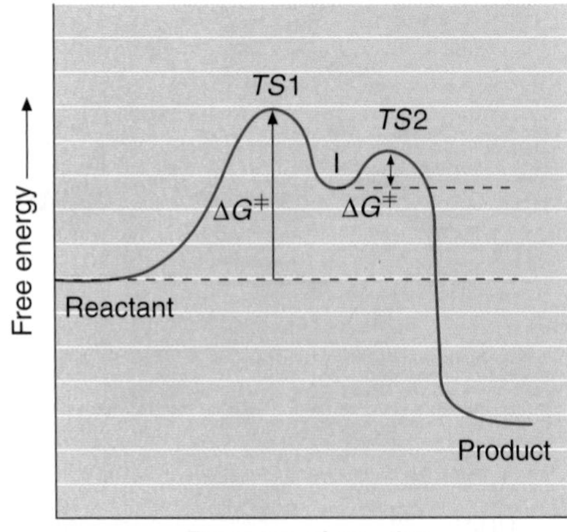

FIGURE 6.15

Energy Diagram for a Two-Step Reaction

There are two transition states (*TS*1 and *TS*2) in a two-step reaction. The overall rate of the reaction is determined by the rate of the step with the highest Ea (ΔG), in this case the first step. The reaction intermediate (I), a molecular entity formed from the reactant molecule(s) that reacts further to form the product, exists in the valley between the transition state peaks.

an intermediate (a reactive species with a finite lifetime) and a transition state (an unstable species with maximum free energy).

Evidence that enzyme mechanisms involve tight binding of transition states has been acquired with the use of transition state analogs, molecules that so closely resemble a transition state that they bind to an enzyme's active site more tightly than substrate molecules. The molecule 2-phosphoglycolic acid (PGA) resembles TPI's enediolate 1 transition state. PGA was found to bind so tightly to the active site of TPI, most notably to Glu 165 and His 65 side chains, that it functions as a potent inhibitor of the enzyme.

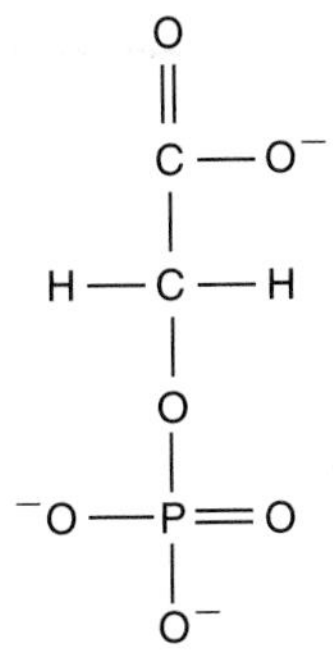

O
||
C—O⁻
|
H—C—H
|
O
|
⁻O—P=O
|
O⁻

2-Phosphoglycolic acid

Catalytic Mechanisms

Despite extensive research, the mechanisms of only a few enzymes are known in significant detail. It is known that enzymes achieve significantly higher catalytic rates than other catalysts because their active sites possess structures that are uniquely suited to promote catalysis. Several factors contribute to enzyme catalysis. The most important of these are proximity and orientation effects, electrostatic effects, acid-base catalysis, and covalent catalysis. Quantum tunneling, which occurs when hydrogen is transferred, is discussed in an online Biochemistry in Perspective essay. It should be noted that none of these catalytic factors is mutually exclusive. In varying degrees, they all participate in each type of catalytic mechanism.

Visit the companion website at www.oup.com/us/mckee to read the Biochemistry in Perspective essay on quantum tunneling.

PROXIMITY AND ORIENTATION EFFECTS For a biochemical reaction to occur, the substrate must come into close proximity to catalytic functional groups (side chain groups involved in a catalytic mechanism) within the active site. In addition, the substrate must be precisely oriented to the catalytic groups. Reactions occur faster when substrates are correctly positioned.

ELECTROSTATIC EFFECTS Recall that the strength of electrostatic interactions is inversely related to the hydration of participating species (Chapter 3). Hydration shells increase the distance between charge centers and reduce electrostatic attraction. The local dielectric constant is often low within active sites because of water exclusion. For example, water exclusion results from a conformational change in TPI that occurs when an enediol intermediate is in the active site. As a result, a lidlike flexible loop in the enzyme moves so as to shield the active site from solvent. The shutting of the lid also traps and stabilizes the intermediate. The charge distribution in the relatively anhydrous active site facilitates the optimum positioning of substrate molecules and influences their chemical reactivity. In addition, weak electrostatic interactions, such as those between permanent and induced dipoles in both the active site and the substrate, are believed to contribute to catalysis.

ACID-BASE CATALYSIS Acid-base catalysis (proton transfer) is an important factor in chemical reactions. For example, consider the hydrolysis of an ester:

$$R-\overset{O}{\overset{\|}{C}}-O-R' + H_2O \longrightarrow R-\overset{O}{\overset{\|}{C}}-OH + R'-OH$$

Because water is a weak nucleophile, ester hydrolysis is relatively slow in neutral solution. Ester hydrolysis takes place much more rapidly if the pH is raised. As the hydroxide ion attacks the polarized carbon atom of the carbonyl group (**Figure 6.16a**), a tetrahedral intermediate is formed. As the intermediate breaks down, a proton is transferred from a nearby water molecule. The reaction is complete when the alcohol is released. However, hydroxide ion catalysis is not

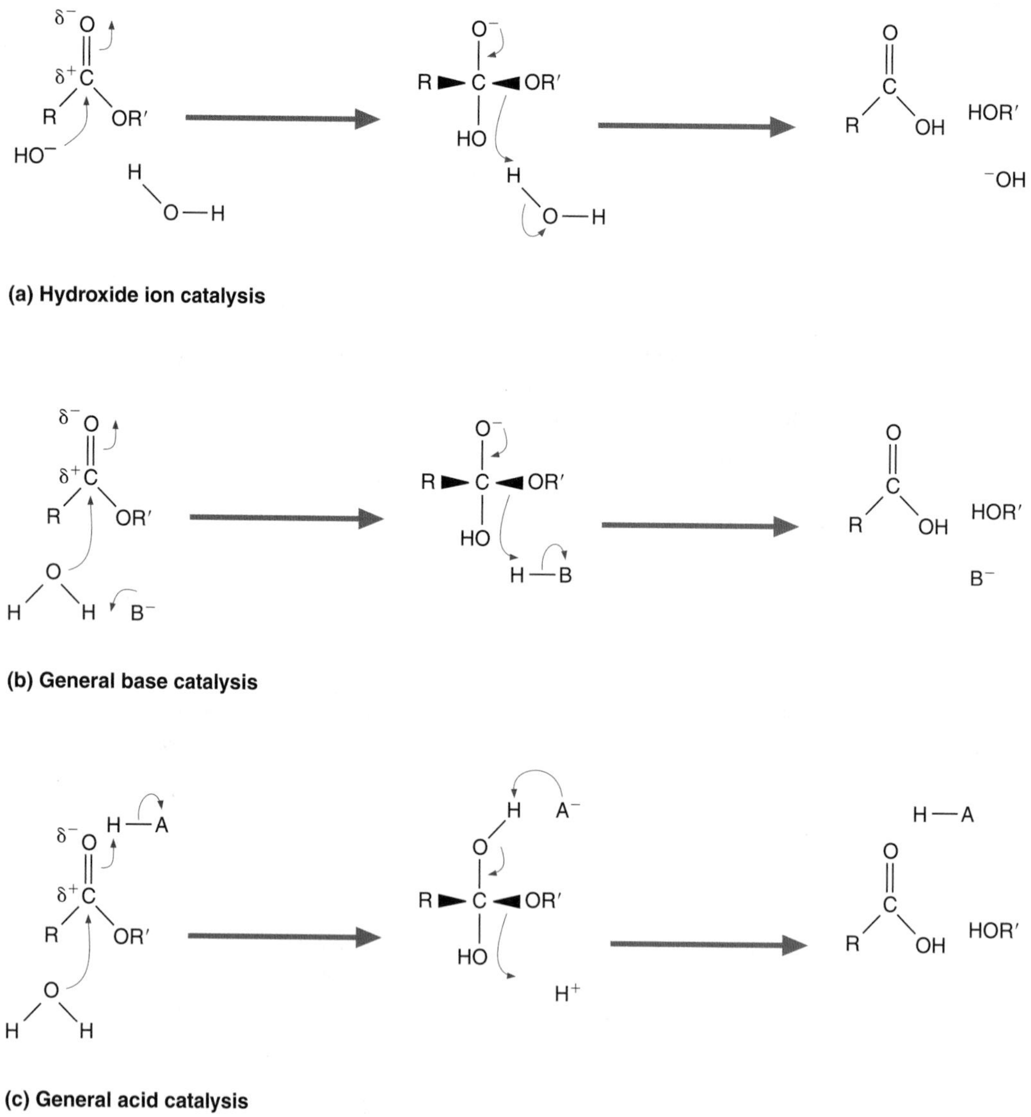

(a) Hydroxide ion catalysis

(b) General base catalysis

(c) General acid catalysis

FIGURE 6.16
Ester Hydrolysis

Esters can be hydrolyzed in several ways: (a) catalysis by free hydroxide ion, (b) general base catalysis, and (c) general acid catalysis. A colored arrow represents the movement of an electron pair during each mechanism.

practical in living systems. Enzymes use several functional groups that behave as general bases to transfer protons efficiently. Such groups can be precisely positioned in relation to the substrate (**Figure 6.16b**). Ester hydrolysis can also be catalyzed by a general acid (**Figure 6.16c**). As the oxygen of the ester's carbonyl group binds to the proton, the carbon atom becomes more electrophilic. The ester then becomes more susceptible to the nucleophilic attack of a water molecule.

Within enzyme active sites, the functional groups on the side chains of histidine (imidazole), aspartate and glutamate (carboxylate), tyrosine (hydroxyl), cysteine (sulfhydryl), and lysine (amine) can act either as proton donors (called general acids) or as proton acceptors (called general bases) depending on their state of protonation. Each of these side chain groups has a characteristic pK_a value that may be influenced by nearby charged or polar atoms. For example, the side chain of histidine often participates in concerted acid-base catalysis because its pK_a range is close to physiological pH. The protonated imidazole ring can serve as a general acid, and the deprotonated imidazole ring can serve as a general base:

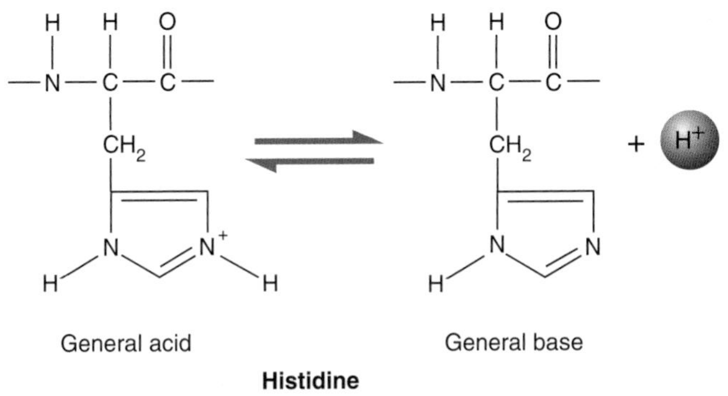

General acid ⇌ General base + H^+

Histidine

Within the active site of the serine proteases (a class of proteolytic enzymes such as trypsin and chymotrypsin, see pp. 223–225), the close proximity of an aspartate carboxylate group to histidine raises the latter's pK_a. As a consequence, histidine, acting as a general base, abstracts a proton from a nearby serine side chain, thus converting the oxygen of serine into a better nucleophile.

COVALENT CATALYSIS In some enzymes a nucleophilic side chain group forms an unstable covalent bond with an electrophilic group on the substrate. As just described, the serine proteases use the —CH_2—OH group of serine as a nucleophile to hydrolyze peptide bonds. During the first step, the nucleophile (i.e., the oxygen of serine) attacks the carbonyl group of the peptide substrate. As the ester bond is formed, the peptide bond is broken. The resulting acyl-enzyme intermediate is hydrolyzed by water in a second reaction:

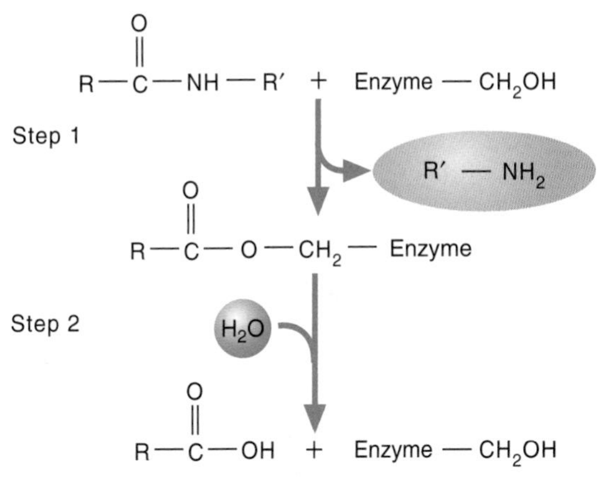

$$R{-}C(=O){-}NH{-}R' + \text{Enzyme}{-}CH_2OH$$

Step 1 → $R'{-}NH_2$

$$R{-}C(=O){-}O{-}CH_2{-}\text{Enzyme}$$

Step 2 ← H_2O

$$R{-}C(=O){-}OH + \text{Enzyme}{-}CH_2OH$$

Several other amino acid side chains may act as nucleophiles. The sulfhydryl group of cysteine, and the carboxylate groups of aspartate and glutamate, can play this role.

The Roles of Amino Acids in Enzyme Catalysis

The active sites of enzymes are lined with amino acid side chains that are in close proximity as a result of the protein folding process. Together these side chains create a microenvironment that is conducive to catalysis. The functions of active site side chains fall into two major categories: catalytic and noncatalytic. Catalytic residues directly participate in the catalytic mechanism, whereas noncatalytic residues have support functions. Of the 20 amino acids found in proteins (**Figure 5.2**), only those with polar and charged side chains actually participate in catalysis. These amino acids (and their side chain groups) are as follows: serine, threonine, and tyrosine (hydroxyl); cysteine (thiol); glutamine and asparagine (amide); glutamate and aspartate (carboxylate); lysine (amine); arginine (guanidinium); and histidine (imidazole).

Extensive research has revealed that catalytic mechanisms require the precise positioning of one or more catalytic units that are composed of either two or three amino acid side chains called dyads or triads, respectively. Although the number of enzymes is vast, the catalytic units are composed of relatively few combinations of amino acids. Commonly observed examples include the arginine-arginine, carboxylate-carboxylate, and carboxylate-histidine dyads.

An arginine-arginine dyad is a catalytic unit in adenylate kinase, an enzyme that catalyzes the transfer of phosphoryl groups from ATP to other nucleotides. The polarizing effect of the two arginines on the phosphate group's oxygens has the effect of converting phosphate into a good leaving group.

Carboxylate-carboxylate dyads occur in the active sites of the aspartic proteases, a family of proteolytic enzymes such as pepsin, which animals use to digest dietary protein. The close proximity of the two negatively charged aspartate carboxyl groups raises the pK_a of one of the aspartates, making it less acidic and more basic. Its enhanced capacity to accept a proton initiates an acid-base hydrolytic mechanism.

The functional properties of aspartate-histidine dyads result from a polarized imidazolium ring caused by the close proximity of the aspartate's negatively charged carboxylate group. In the active site of aconitase, the enzyme that catalyzes the isomerization of citrate to form isocitrate (p. 331), histidine acts as a general acid and protonates the –OH group of the citrate, making it a better leaving group.

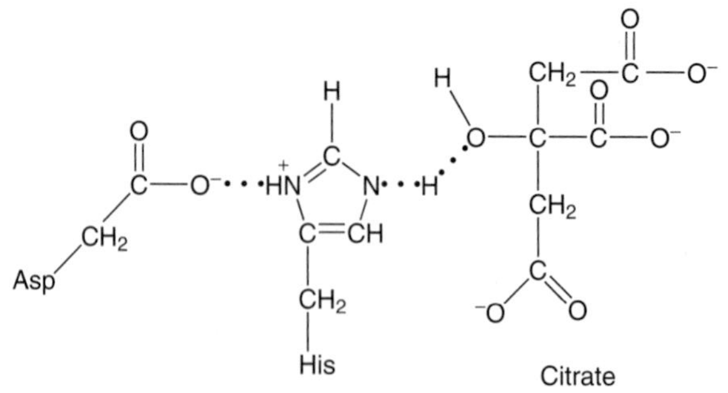

The resulting HOH is held in the active site by histidine long enough to be able to attack the intermediate and generate the isomer of citrate (isocitrate). An aspartate-histidine dyad is also, most notably, a component of the well-researched serine protease triad (Asp-His-Ser) (p. 223). The close proximity of the aspartate carboxylate group to the imidazole group of histidine raises the latter group's pK_a, thus facilitating its ability to remove a proton from serine. The deprotonated serine is thus converted into a better nucleophile.

The functions of noncatalytic side groups, which include substrate orientation and transition state stabilization, are more subtle than those of catalytic residues. For example, the substrate specificity of chymotrypsin (pp. 223–225), manifested in the cleavage of peptide bonds on the C-terminal side of the aromatic amino acids tryptophan, tyrosine, and phenylalanine, is made possible by the relatively large size of a hydrophobic pocket within the active site that both accommodates and orients the aromatic side chains. The **oxyanion** intermediate (a molecular species with a negatively charged oxygen, see p. 223 and **Figure 6.19**) that forms during the catalytic mechanism is stabilized by interactions between the substrate's peptide bond carbonyl group and the backbone amide hydrogens of serine and glycine residues.

The Role of Cofactors in Enzyme Catalysis

In addition to active site amino acid side chains, many enzymes require nonprotein cofactors, that is, metal cations and the coenzymes. Each group has distinctive structural properties and chemical reactivities.

METALS The important metals in living organisms fall into two classes: the alkali and alkaline earth metals (e.g., Na^+, K^+, Mg^{2+}, and Ca^{2+}) and the transition metals (e.g., Zn^{2+}, Fe^{2+}, and Cu^{2+}). The alkali and alkaline earth metals in enzymes are loosely bound and usually have structural roles. In contrast, the transition metals play key roles in catalysis either bound to functional groups such as carboxylate, imidazole, or hydroxyl groups, or as components of prosthetic groups such as Fe^{2+} in heme.

Several properties of transition metals make them useful in catalysis. Metal ions provide a high concentration of positive charge that is especially useful in binding small molecules. Because transition metals act as *Lewis acids* (electron pair acceptors), they are effective electrophiles. (Amino acid side chains are poor electrophiles because they cannot accept unshared pairs of electrons.) Because the metal's directed d shell valences allow them to interact with two or more ligands, metal ions help orient the substrate within the active site. As a consequence, the enzyme–metal ion complex polarizes the substrate and promotes catalysis. For example, carbonic anhydrase is the enzyme that catalyzes the reversible hydration of CO_2 to form bicarbonate (HCO_3^-). Its active site contains a zinc (Zn^{2+}) cofactor that is coordinated with three histidine side chains. The zinc ion polarizes a water molecule resulting in a Zn^{2+}-bound OH group. The OH group (acting as a nucleophile) attacks CO_2, converting it into HCO_3^-:

$$\text{Enzyme—Zn}^{2+} + H_2O \rightleftharpoons \text{Enzyme—Zn}^{2+}\cdots{}^{-}OH + H^+$$

$$\text{Enzyme—Zn}^{2+}\cdots{}^{-}OH + H_2O + O{=}C{=}O \rightleftharpoons \text{Enzyme—Zn}^{2+}\cdots O(H)\text{—}C({=}O)\text{—}OH$$

$$\rightleftharpoons \text{Enzyme—Zn}^{2+} + H_2CO_3 \rightleftharpoons \text{Enzyme—Zn}^{2+} + HCO_3^- + H^+$$

Finally, because transition metals have two or more valence states, they can mediate oxidation-reduction reactions by reversibly gaining or losing electrons. For example, the reversible oxidation of Fe^{2+} to form Fe^{3+} is important in the function of cytochrome P_{450}, a type of enzyme that processes toxic substances in animals. [See Biochemistry in Perspective essays in Chapters 10 (Ischemia and Reperfusion) and 12 (Biotransformation).]

QUESTION 6.6

Copper is a cofactor in several enzymes, including lysyl oxidase and superoxide dismutase. Ceruloplasmin, a deep-blue glycoprotein, is the principal copper-containing protein in blood. It is used to transport Cu^{2+} and maintain appropriate levels of Cu^{2+} in the body's tissues. Ceruloplasmin also catalyzes the oxidation of Fe^{2+} to Fe^{3+}, an important reaction in iron metabolism. Because the metal is widely found in foods, copper deficiency is rare in humans. Deficiency symptoms include anemia, leukopenia (reduction in blood levels of white blood cells), bone defects, and weakened arterial walls. The body is partially protected from exposure to excessive copper (and several other metals) by metallothionein, a small, metal-binding protein that possesses a large proportion of cysteine residues. Certain metals (most notably zinc and cadmium) induce the synthesis of metallothionein in the intestine and liver.

In *Menkes' syndrome* intestinal absorption of copper is defective. How can affected infants be treated to avoid the symptoms of the disorder, which include seizures, retarded growth, and brittle hair?

QUESTION 6.7

In another rare inherited disorder, called *Wilson's disease*, excessive amounts of copper accumulate in liver and brain tissue. A prominent symptom of the disease is the deposition of copper in greenish-brown layers surrounding the cornea, called Kayser-Fleischer rings. Wilson's disease is now known to be caused by a defective ATP-dependent protein that transports copper across cell membranes. Apparently, the copper transport protein is required to incorporate copper into ceruloplasmin and to excrete excess copper. In addition to a diet low in copper, Wilson's disease is treated with zinc sulfate and the chelating agent penicillamine (p. 144). Describe how these treatments work. [*Hint:* Metallothionein has a greater affinity for copper than for zinc.]

COENZYMES Coenzymes are organic molecules that provide enzymes with chemical versatility because they either possess reactive groups not found on amino acid side chains or can act as carriers for substrate molecules. Some coenzymes are only transiently bound to the enzyme and are essentially cosubstrates, whereas others are tightly bound by covalent or noncovalent bonds. Unlike ordinary catalysts, coenzyme structures are changed by the reactions in which they participate. Their catalytically active forms must be regenerated before another catalytic cycle can occur. Transiently bound coenzymes are usually regenerated by a reaction catalyzed by another enzyme, whereas tightly bound coenzymes are regenerated during a step in the catalytic cycle. Most coenzymes are derived from vitamins. **Vitamins** (organic nutrients required in small amounts in the human diet) are divided into two classes: water-soluble and lipid-soluble. In addition, there are certain vitamin-like substances that organisms can synthesize in sufficient amounts to facilitate enzyme-catalyzed reactions. Examples include lipoic acid, carnitine, coenzyme Q, biopterin, *S*-adenosylmethionine, and *p*-aminobenzoic acid.

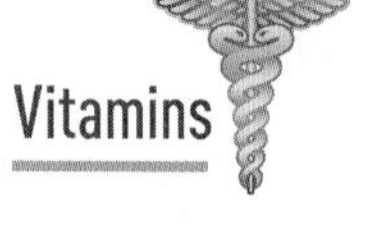

Coenzymes can be classified according to function into three groups: electron transfer, group transfer, and high-energy transfer potential. Coenzymes involved in redox (electron or hydrogen transfer) reactions include nicotinamide adenine dinucleotide (NAD^+), nicotinamide adenine dinucleotide phosphate ($NADP^+$), flavin adenine dinucleotide (FAD), flavin mononucleotide (FMN), coenzyme Q (CoQ), and tetrahydrobiopterin (BH_4). Coenzymes such as thiamine pyrophosphate (TPP), coenzyme A (CoASH), and pyridoxal phosphate are involved in the transfer of aldehyde, acyl, and amino groups, respectively. One-carbon transfers, a diverse set of reactions in which carbon atoms are transferred in various oxidation states between substrates, require biotin, tetrahydrofolate (TH_4) or *S*-adenosylmethionine (SAM). Nucleotides, known for their high-energy transfer potential, function as coenzymes in that they activate metabolic intermediates and/or serve as phosphate donors or as carriers for small molecules. Examples of the latter include UDP-glucose, an intermediate in glycogen synthesis (p. 305), and CDP (cytidine dephosphate)-ethanolamine, an intermediate in the synthesis of certain lipids (p. 453). In such cases the nucleotide serves as a molecular carrier and a good leaving group in subsequent reactions. The numerous noncovalent bonds that form when it binds within the enzyme's active site assure that the attached metabolite is correctly positioned. Note that the structures of coenzymes such as NAD^+, $NADP^+$, FAD, FMN, and CoASH also contain adenine nucleotide components. **Table 6.4** lists the vitamins and vitaminlike substances, their coenzyme forms, and the reactions they facilitate, along with the pages on which their structural and functional properties are described.

KEY CONCEPTS

- The amino acid side chains in the active site of enzymes catalyze proton transfers and nucleophilic substitutions. Other reactions require nonprotein cofactors, that is, metal cations and the coenzymes.
- Metal ions are effective electrophiles, and they help orient the substrate within the active site. In addition, certain metal cations mediate redox reactions.
- Coenzymes are organic molecules that have a variety of functions in enzyme catalysis.

QUESTION 6.8

Identify each of the following as a cofactor, coenzyme, apoenzyme, holoenzyme, or none of these. Explain each answer.

a. Zn^{2+}
b. active alcohol dehydrogenase
c. alcohol dehydrogenase lacking Zn^{2+}
d. FMN
e. NAD^+
f. CDP-ethanolamine
g. biotin

TABLE 6.4 Vitamins, Vitaminlike Molecules, and Their Coenzyme Forms

Molecules	Coenzyme Form	Reaction or Process Promoted	See Page
Water-Soluble Vitamins			
Thiamine (B_1)	Thiamine pyrophosphate	Decarboxylation, aldehyde group transfer	329
Riboflavin (B_2)	FAD and FMN	Redox	322
Pyridoxine (B_6)	Pyridoxal phosphate	Amino group transfer	517
Nicotinic acid (niacin)	NAD and NADP	Redox	322
Pantothenic acid	Coenzyme A	Acyl transfer	328
Biotin	Biotin	Carboxylation	442
Folic acid	Tetrahydrofolic acid	One-carbon group transfer	529
Vitamin B_{12}	Deoxyadenosylcobalamin, methylcobalamin	Intramolecular rearrangements	514
Ascorbic acid (vitamin C)	Unknown	Hydroxylation	375
Lipid-Soluble Vitamins			
Vitamin A	Retinal	Vision, growth, and reproduction	375
Vitamin D	1, 25-Dihydroxycholecalciferol	Calcium and phosphate metabolism	400
Vitamin E	Unknown	Lipid antioxidant	375
Vitamin K	Unknown	Blood clotting	399
Vitaminlike Molecules			
Coenzyme Q	Coenzyme Q	Redox	350
Biopterin	Tetrahydrobiopterin	Redox	534
S-Adenosylmethionine	*S*-Adenosylmethionine	Methylation	531
Lipoic acid	Lipoamide	Redox	329

Effects of Temperature and pH on Enzyme-Catalyzed Reactions

Any environmental factor that disturbs protein structure may change enzymatic activity. Enzymes are especially sensitive to changes in temperature and pH.

TEMPERATURE All chemical reactions are affected by temperature. In general, the higher the temperature, the higher the reaction rate; that is, the greater the increase in the number of collisions. The reaction velocity increases because more molecules have sufficient energy to enter into the transition state. The rates of enzyme-catalyzed reactions also increase with increasing temperature. However, enzymes are proteins that become denatured at high temperatures. An enzyme's *optimum temperature*, the temperature at which it operates at maximum efficiency (**Figure 6.17**), is determined in the laboratory under specified conditions of pH, ionic strength, and solute concentrations. In living organisms there is no single definable optimal temperature for enzymes.

pH Hydrogen ion concentration, expressed as a pH value, affects enzymes in several ways. First, catalytic activity is related to the ionic state of the active site. Changes in hydrogen ion concentration can affect the ionization of active site groups. For example, the catalytic activity of a certain enzyme requires the protonated form of a side chain amino group. If the pH becomes so alkaline that the group loses its proton, the enzyme's activity may be depressed. In addition, substrates may be affected. If a substrate contains an ionizable group, a change in pH may alter its capacity to bind to the active site. Second, changes in ionizable groups may change the tertiary structure of the enzyme. Drastic changes in pH often lead to denaturation.

Although a few enzymes tolerate large changes in pH, most enzymes are active only within a narrow pH range. For this reason, living organisms employ buffers to closely regulate pH. The pH value at which an enzyme's activity is maximal is called the **pH optimum** (**Figure 6.18**). The pH optima of enzymes, determined in the laboratory, vary considerably. For example, the optimum pH of pepsin, a proteolytic enzyme produced in the chief cells of the stomach, is approximately 2. The stomach's low pH is the result of the secretion of hydrochloric acid (HCl) by parietal cells. The optimum pH for the proteolytic enzyme chymotrypsin is 8. Produced in the pancreas as a larger molecule (p. 227), chymotrypsin is one of several digestive enzymes that are secreted into the small intestine as a component of pancreatic juice. Pancreatic juice also contains sodium bicarbonate ($NaHCO_3$), which neutralizes the acidity of the partially digested food entering the small intestine from the stomach, thereby raising its pH to about 8.

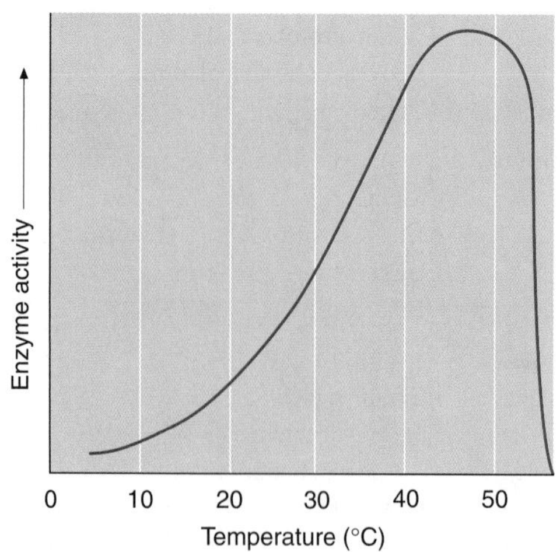

FIGURE 6.17
Effect of Temperature on Enzyme Activity

Modest increases in temperature increase the rate of enzyme-catalyzed reactions because of an increase in the number of collisions between enzyme and substrate. Eventually, increasing the temperature decreases the reaction velocity. Catalytic activity is lost because heat denatures the enzyme.

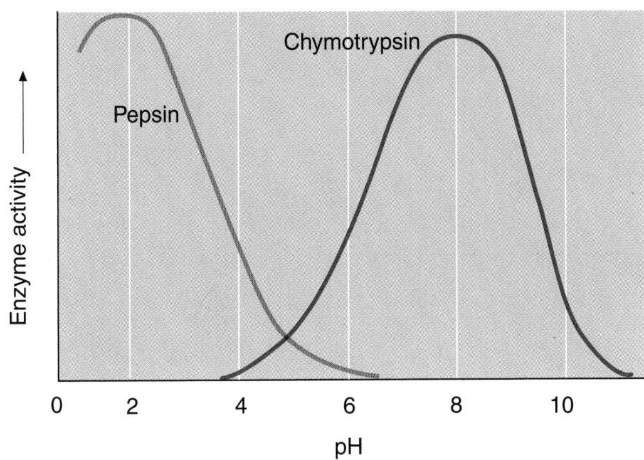

FIGURE 6.18

Effect of pH on Two Enzymes

Each enzyme has a certain pH at which it is most active. A change in pH can alter the ionizable groups within the active site or affect the enzyme's conformation.

Detailed Mechanisms of Enzyme Catalysis

Each of the thousands of enzymes that have been investigated has a unique structure, substrate specificity, and reaction mechanism. The mechanisms of a variety of enzymes have been investigated intensively over the past several decades. The subsections that follow describe the catalytic mechanisms of two well-characterized enzymes.

CHYMOTRYPSIN Chymotrypsin is a 27,000 Da protein that belongs to the serine proteases. The active sites of all serine proteases contain a characteristic set of amino acid residues, often referred to as the serine protease triad (p. 218). In the chymotrypsin numbering system these are Asp 102, His 57, and Ser 195. Studies of crystallized enzyme bound to substrate analogues reveal that these residues are close to each other in the active site. The active site serine residue plays an especially important role in the catalytic mechanisms of this group of enzymes. Serine proteases are irreversibly inhibited by diisopropylfluorophosphate (DFP). In DFP-inhibited enzymes, the inhibitor is covalently bound only to Ser 195 and not to any of the other 29 serines. The special reactivity of Ser 195 is attributed to the proximity of His 57 and Asp 102. The imidazole ring of His 57 lies between the carboxyl group of Asp 102 and the —CH_2OH group of Ser 195. The carboxyl group of Asp 102 polarizes His 57, thus allowing it to act as a general base (i.e., the abstraction of a proton by the imidazole group is facilitated):

$$—CH_2—\overset{\overset{O}{\|}}{C}—O^- \cdots\cdots \text{HN} \quad \text{N:} \quad HO—CH_2—$$

Removing the proton from the serine OH group converts it into a more effective nucleophile.

Chymotrypsin catalyzes the hydrolysis of peptide bonds adjacent to aromatic amino acids. The probable mechanism for this reaction is illustrated in **Figure 6.19**. Step (a) of the figure shows the initial enzyme-substrate complex. The alignment of the amino acid residues within the active site, including the catalytic triad of Asp 102, His 57, and Ser 195, creates a temporarily unoccupied position called the oxyanion hole. In addition, the active site provides a hydrophobic binding pocket for the substrate's phenylalanine side chain. The nucleophilic hydroxyl oxygen of Ser 195 launches a nucleophilic attack on the carbonyl carbon of the substrate. The tetrahedral intermediate that forms (step b) is sufficiently distorted that the newly created oxyanion is stabilized by hydrogen bonds to the amide hydrogens of Ser 195 and Gly 193 and enters the oxyanion

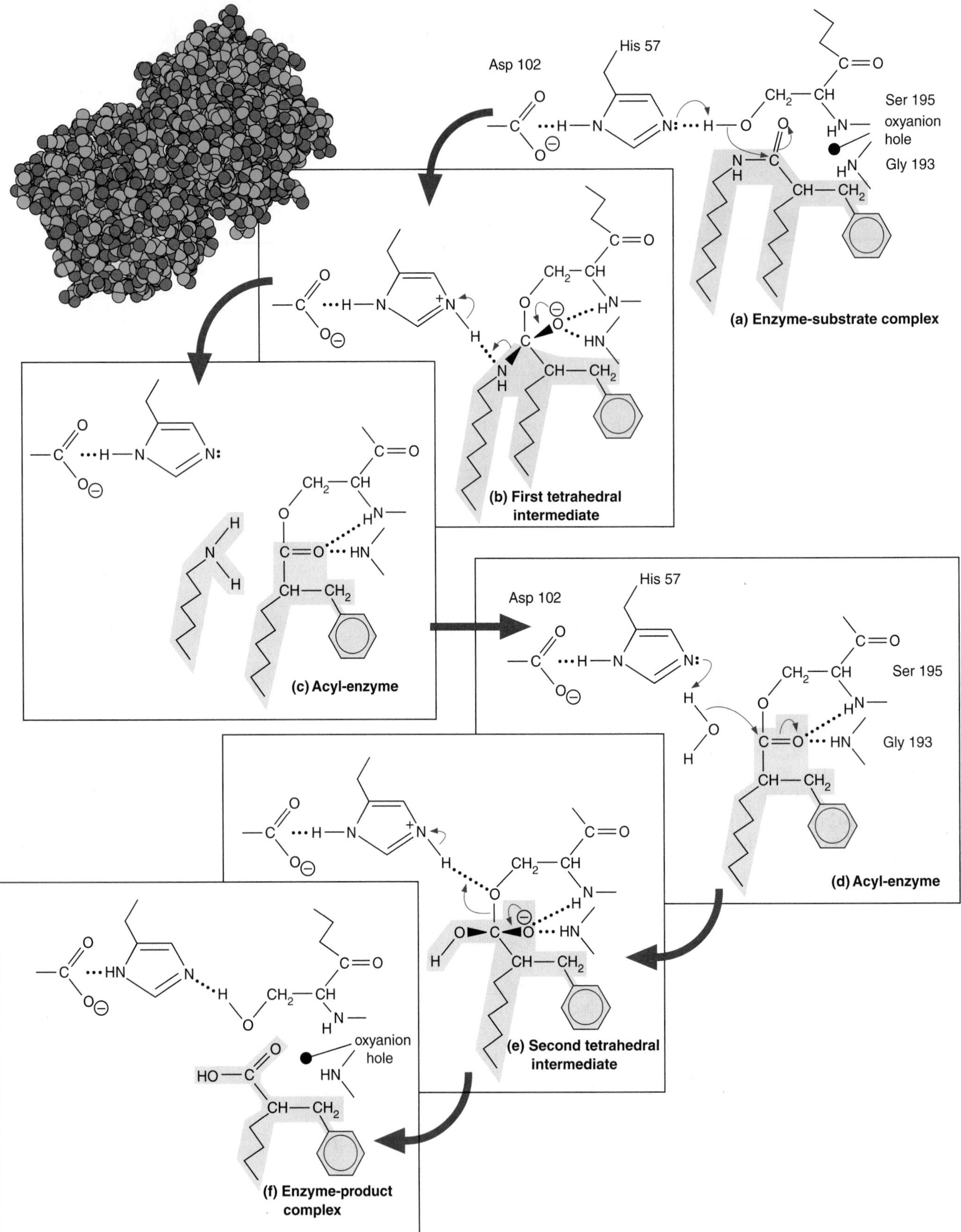

FIGURE 6.19

The Probable Mechanism of Action of Chymotrypsin

There is a fast acylation step during which the carbonyl end of the target peptide bond of the substrate is transferred to the enzyme to form an acyl-enzyme adduct and the amino end of the substrate leaves the active site. A second slow deacylation of the enzyme follows, releasing the carbonyl end of the substrate.

hole. It is believed that the formation of the tetrahedral intermediate, which resembles the transition state, and its preferential binding by the enzyme, is responsible for the catalytic efficiency of the serine proteases.

The tetrahedral intermediate subsequently decomposes to form the covalently bound acyl-enzyme intermediate (step c). The residue His 57, acting as a general acid, is believed to facilitate this decomposition. In steps (d) and (e), the two previous steps are reversed. With water acting as an attacking nucleophile, a tetrahedral (oxyanion) intermediate is formed. By step (f) (the final enzyme-product complex), the bond between the serine oxygen and the carbonyl carbon has been broken. Serine is again hydrogen-bonded to His 57.

ALCOHOL DEHYDROGENASE The alcohol dehydrogenases (ADHs) are a diverse group of enzymes that are distributed throughout the three domains of living organisms (Bacteria, Archaea, and Eukarya). They catalyze the reversible oxidation of alcohols to form aldehydes or ketones. In the following reaction, the oxidation of ethanol, two electrons and two protons are removed from the alcohol molecule. The coenzyme NAD acts as a hydride ion ($H:^-$) acceptor.

$$CH_3{-}CH_2{-}OH + NAD^+ \rightleftharpoons CH_3{-}\overset{\overset{\displaystyle O}{\|}}{C}{-}H + NADH + H^+$$

Most ADHs consist of two or four subunits, each of which contains two zinc ions. One zinc ion, located in the active site, is crucial for catalysis, whereas the other has a structural function. The active site of alcohol dehydrogenase also contains two cysteine residues (Cys 48 and Cys 174) and a histidine residue (His 67), all of which are coordinated to a zinc ion (**Figure 6.20a**). After NAD^+ binds to the active site, the substrate ethanol enters and binds to the Zn^{2+} as the alcoholate anion (**Figure 6.20b**). The electrostatic effect of Zn^{2+} stabilizes the transition state. As the intermediate decomposes, the hydride ion is transferred from the substrate to the nicotinamide ring of NAD^+. After the aldehyde product is released from the active site, NADH also dissociates.

KEY CONCEPTS

- Each enzyme has a unique structure, substrate specificity, and reaction mechanism.
- Each mechanism is affected by catalysis-promoting factors that are determined by the structure of the substrate and the enzyme's active site.

6.5 ENZYME REGULATION

Living organisms have sophisticated mechanisms for regulating their vast networks of biochemical pathways. Regulation is essential for several reasons:

1. **Maintenance of an ordered state**. Regulation of each pathway results in the production of the substances required to maintain cell structure and function in a timely fashion and without wasting resources.
2. **Conservation of energy**. Cells ensure that they consume just enough nutrients to meet their energy requirements by constantly controlling energy-generating reactions.
3. **Responsiveness to environmental changes**. Cells can make relatively rapid adjustments to changes in temperature, pH, ionic strength, and nutrient concentrations because they can increase or decrease the rates of specific reactions.

The regulation of biochemical pathways is achieved primarily by adjusting the concentrations and activities of certain enzymes. Control is accomplished by (1) genetic control, (2) covalent modification, (3) allosteric regulation, and (4) compartmentation.

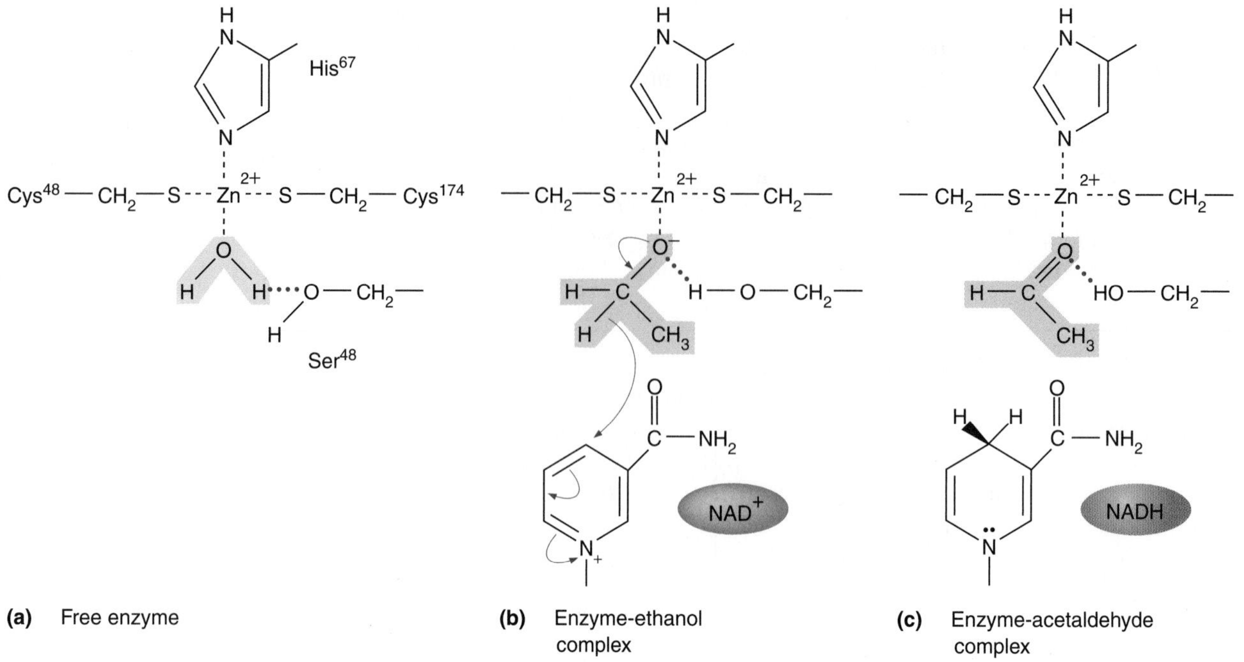

FIGURE 6.20

Functional Groups of the Active Site of Alcohol Dehydrogenase

(a) Without a substrate, a molecule of water is one of the ligands of the Zn^{2+} ion. (b) The substrate ethanol probably binds to the Zn^{2+} as the alcoholate anion, displacing the water molecule. (c) NAD^+ accepts a hydride ion from the substrate and the aldehyde product is formed.

Genetic Control

The synthesis of enzymes in response to changing metabolic needs, a process referred to as **enzyme induction**, allows cells to respond efficiently to changes in their environment. For example, *E. coli* cells grown without the sugar lactose initially cannot metabolize this nutrient when it is introduced into the bacterium's growth medium. The introduction of lactose in the absence of glucose, however, activates the genes that code for enzymes needed to utilize lactose as an energy source. After all the lactose has been consumed, synthesis of these enzymes is terminated.

Covalent Modification

Some enzymes are regulated by the reversible interconversion between their active and inactive forms. Several covalent modifications of enzyme structure cause these changes in function. Many such enzymes have specific residues that may be phosphorylated and dephosphorylated. For example, glycogen phosphorylase (Chapter 8) catalyzes the first reaction in the degradation of glycogen, a carbohydrate energy storage molecule. In a process controlled by hormones, the inactive form of the enzyme (glycogen phosphorylase b) is converted to the active form (glycogen phosphorylase a) by the addition of a phosphate group to a specific serine residue. In the nonphosphorylated enzyme, the peptide segment directly surrounding the serine residue is disordered. Phosphorylation of the serine side chain triggers the conversion of the disordered segment into α-helices, thereby resulting in a more active enzyme. Other types of reversible covalent modification include methylation, acetylation, and nucleotidylation (the covalent addition of a nucleotide).

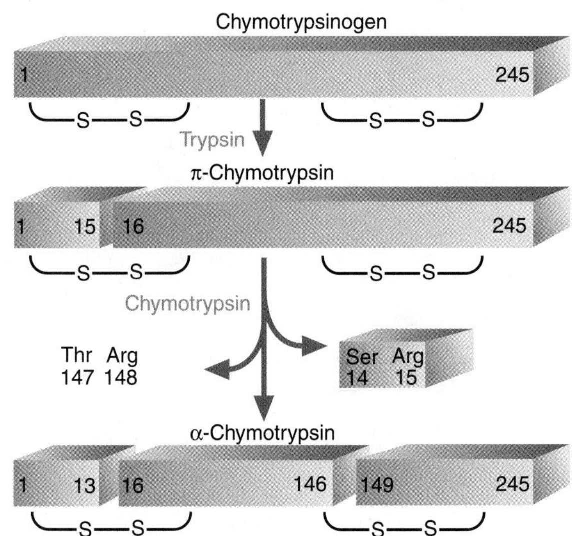

FIGURE 6.21

The Activation of Chymotrypsinogen

The inactive zymogen chymotrypsinogen is activated in several steps. After its secretion into the small intestine, chymotrypsinogen is converted into π-chymotrypsin when trypsin, another proteolytic enzyme, cleaves the peptide bond between Arg 15 and Ile 16. Later cleaved chymotrypsin molecules activate each other by removing two depeptide segments to cause the formation of α-chymotrypsin.

Several enzymes are produced and stored as inactive precursors called **proenzymes** or **zymogens**. Zymogens are converted into active enzymes by the irreversible cleavage of one or more peptide bonds. For example, chymotrypsinogen is produced in the pancreas. After chymotrypsinogen is secreted into the small intestine, it is converted to its active form that is responsible for digesting dietary protein in several steps (**Figure 6.21**).

Allosteric Regulation

In each biochemical pathway there are one or more enzymes whose catalytic activity can be modulated (i.e., increased or decreased) by the binding of effector molecules. The binding of ligands to *allosteric sites* on such an enzyme triggers rapid conformational changes that can increase or decrease its rate of substrate binding. A plot of a reaction catalyzed by an allosteric enzyme differs from those of enzymes that observe Michaelis-Menten kinetics. Instead the rate curve is sigmoidal (**Figure 6.22**). If the ligand is the same as the substrate (i.e., if the binding of a substrate influences the binding of additional substrate), the allosteric effects are referred to as *homotropic. Heterotropic effects* involve modulating ligands, which are different from the substrate. Most allosteric enzymes are multisubunit enzymes with multiple substrate and effector binding sites.

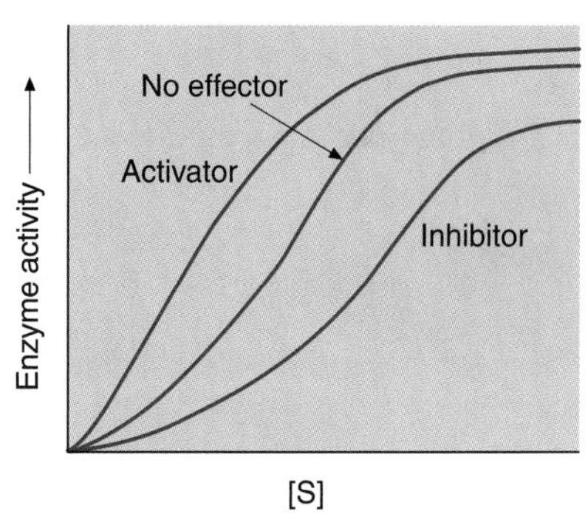

FIGURE 6.22

The Rate of an Enzyme-Catalyzed Reaction as a Function of Substrate Concentration

The activity of allosteric enzymes is affected by positive effectors (activators) and negative effectors (inhibitors).

Two theoretical models that attempt to explain the behavior of allosteric enzymes are the concerted model and the sequential model (**Figure 6.23**). In the *concerted* (or *symmetry*) model, it is assumed that the enzyme exists in only two states (Section 5.3): T(aut) and R(elaxed). Activators bind to and stabilize the R conformation, whereas inhibitors bind to and stabilize the T conformation. The term *concerted* is applied to this model because the conformations of all the protein's subunits are believed to change simultaneously when the first effector binds. (This rapid concerted change in conformation maintains the protein's overall symmetry.) The binding of an activator shifts the equilibrium in favor of the R form. An inhibitor shifts the equilibrium toward the T conformation.

The concerted model explains some properties of allosteric enzymes, but not others. For example, the concerted model accounts for **positive cooperativity**, in which the first ligand increases subsequent ligand binding, but not **negative cooperativity**, a phenomenon observed in enzymes in which the

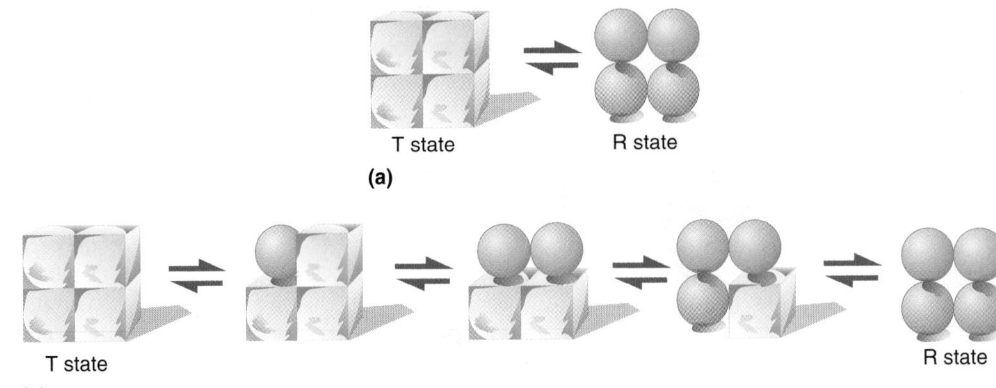

FIGURE 6.23
Allosteric Interaction Models

(a) In the concerted model, the enzyme exists in only two conformations. Substrates and activators have a greater affinity for the R state. Inhibitors favor the T state. (b) In the sequential model, one subunit assumes an R conformation as it binds to substrate. As the first subunit changes its conformation, the affinity of nearby subunits for ligand is affected.

binding of the first ligand reduces the affinity of the enzyme for similar ligands. In addition, the concerted model makes no allowances for hybrid conformations. Examples of enzymes whose behavior appears to be consistent with the concerted model are aspartate transcarbamoylase (ATCase) in *E. coli* and phosphofructokinase (PFK).

ATCase catalyzes the first step in a reaction pathway that leads to the synthesis of the pyrimidine nucleotide cytidine triphosphate (CTP) (**Figure 6.24**). CTP acts as an inhibitor of ATCase activity. It shifts the rate curve to the right, indicating an increase in the apparent K_m of ATCase that corresponds to a lower velocity of enzyme activity at a given concentration of substrate. CTP inhibition of ATCase is an example of negative feedback inhibition (p. 24), the process in which the product of a pathway inhibits the activity of an enzyme at or near the beginning of a biochemical pathway or at a branch point. The purine nucleotide ATP acts as an activator. ATP activation of ATCase makes sense because nucleic acid biosynthesis requires relatively equal amounts of purine and pyrimidine nucleotides. When ATP concentration is higher than that of CTP, ATCase is activated. When ATP concentration is lower than that of CTP, the net effect on ATCase is inhibitory.

Phosphofructokinase (PFK) catalyzes an important reaction in glycolysis (pp. 276–277): the transfer of a phosphate group from ATP to the OH group on C-1 of fructose-6-phosphate.

$$\text{Fructose-6-phosphate} + \text{ATP} \rightarrow \text{fructose-1,6-bisphosphate} + \text{ADP}$$

PFK is composed of four identical subunits, each of which has an active site and a number of allosteric sites. The R state of the enzyme is stabilized by ADP and AMP (a sensitive indicator of ATP depletion and the cell's need for energy). The T state is stabilized by molecules such as ATP and other molecules that are indicators that cellular energy is high.

In the sequential model, first proposed by Daniel Koshland, the binding of a ligand to one flexible subunit in a multisubunit protein triggers a conformational change that is sequentially transmitted to adjacent subunits. The more sophisticated sequential model allows for the intermediate conformations that may be more realistic representations of the operation of some enzymes. The model can also accommodate negative cooperativity (ligand binding to one subunit induces conformational changes in adjacent subunits that make ligand binding less likely). Consequently, the sequential model can be considered a general model that accounts for all allosteric possibilities. In this view the concerted model is a simple example of the sequential model.

There are two important aspects of allosteric regulation that should be noted. First, the concerted and sequential models are theoretical models; that is, the behavior of many allosteric proteins appears to be more complex than can be accounted for by either model. For example, the cooperative binding of O_2 by hemoglobin (the most thoroughly researched allosteric protein) appears to exhibit features of both models. The binding of the first O_2 initiates a concerted $T \rightarrow R$ transition that involves small changes in the conformation of each subunit (a feature of the sequential model; see p. 173). In addition, hemoglobin species with only one or two bound O_2 have been observed. A second and more important issue is that there are no simple rules that explain metabolic regulation. For example, attempts to increase the metabolic flux of pathways in organisms such as yeast by engineering increased synthesis of allosteric enzymes failed. (*Flux* is the rate of turnover of molecules through a pathway.) Flux increases were observed only when all of the enzymes in the pathway were increased. It appears that pathway regulation is the result of regulatory contributions to a greater or lesser degree by all or most of the enzymes.

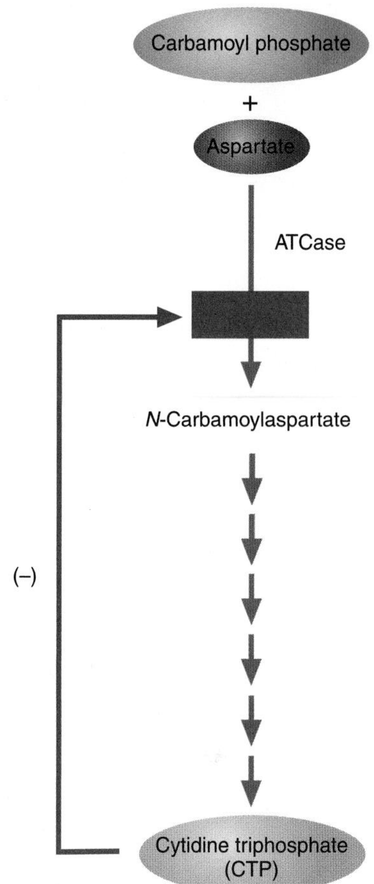

FIGURE 6.24
Feedback Inhibition

ATCase (aspartate transcarbamoylase) catalyzes the committed step in the synthesis of cytidine triphosphate (CTP). The binding of CTP, the product of the pathway, to ATCase inhibits the enzyme.

Compartmentation

Compartmentation, created by cellular infrastructure, is a significant means for regulating biochemical reactions because physical separation makes separate control possible. The intricate internal architecture of cells contains compartments (e.g., eukaryotic organelles) and microcompartments of diverse types (e.g., individual enzymes or multiprotein complexes attached to membranes or cytoskeletal filaments). Cellular compartmentation solves several interrelated problems:

1. **Divide and control**. The physical separation of competing reactions (i.e., those that would undo the other, such as kinases and phosphatases) allows for coordinated regulation that prevents the wasteful dissipation of resources.
2. **Diffusion barriers**. Within crowded cells, diffusion of substrate molecules is a potentially limiting factor in reaction rates. Cells circumvent this problem by creating microenvironments in which enzymes and their substrates are concentrated, as well as by metabolite channeling, the transfer of product molecules from one enzyme to the next in a multienzyme complex.
3. **Specialized reaction conditions**. Certain reactions require an environment with unique properties. For example, the low pH within lysosomes facilitates hydrolytic reactions.

KEY CONCEPTS

- All biochemical pathways are regulated to maintain the ordered state of living cells.
- Regulation is accomplished by genetic control, covalent modification of enzymes, allosteric regulation, and cell compartmentation.

4. **Damage control**. The segregation of potentially toxic reaction products protects other cellular components.

Overall metabolic control requires the integration of all the cell's biochemical pathways, which is accomplished, in part, by transport mechanisms that transfer metabolites and signal molecules between compartments.

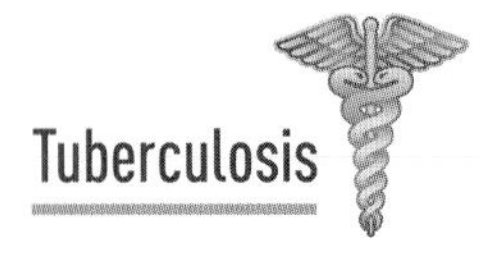

QUESTION 6.9

Certain enzymes are involved in the metabolism of the medicinal chemicals that alter or enhance physiological processes. For example, aspirin suppresses pain, and antibiotics kill infectious organisms. Once such a drug has been consumed, it is absorbed and distributed to the tissues, where it performs its function. Eventually, drug molecules are processed (primarily in the liver) and excreted. The dosage of each drug that physicians prescribe is based on the amount required to achieve a therapeutic effect and the drug's average rate of excretion from the body. Various reactions prepare drug molecules for excretion. Examples include oxidation, reduction, and conjugation reactions. (In conjugation reactions, small polar or ionizable groups are attached to a drug molecule to improve its solubility.) The amount of certain enzymes directly affects a patient's ability to metabolize a specific drug. For example, isoniazid is used to treat tuberculosis, a highly infectious, chronic, disease caused by *Mycobacterium tuberculosis*. This drug is metabolized by N-acetylation, in which an amide bond forms between the substrate and an acetyl group. The rate at which isoniazid is acetylated determines its clinical effectiveness.

Two tuberculosis patients with similar body weights and symptoms are given the same dose of isoniazid. Although both take the drug as prescribed, one patient fails to show a significant clinical improvement. The other patient is cured. Genetic factors appear to be responsible for the differences in drug metabolism. Can you suggest a reason why the two patients reacted so differently to isoniazid? How can physicians improve the percentage of patients who are cured?

Visit the companion website at www.oup.com/us/mckee to read the Biochemistry in Perspective essay on enzymes and clinical medicine.

Biochemistry IN PERSPECTIVE

Alcohol Dehydrogenase: A Tale of Two Species

Why can humans make and consume alcoholic beverages?

The alcohol dehydrogenases have diverse functions in living organisms. Many single-celled organisms generate energy by fermentation, a set of *anaerobic* ("without oxygen") biochemical pathways that yield organic waste products such as alcohols and acids. In the bacterium *Clostridium acetobutylicum,* the conversion of butyraldehyde and acetone into the end products butanol and isopropanol, respectively, is catalyzed by two different ADHs. In World War I *C. acetobutylicum* was used as an alternative source of these solvents when petroleum was in short supply. Another ADH, cinnamyl alcohol dehydrogenase, converts cinnamyl aldehydes into the cinnamyl alcohols, the precursors of lignin, the rigid cell wall polymer in the woody stems of angiosperms. Of all the ADHs, however, the enzymes that most interest researchers are those of *Saccharomyces cerevisiae* and *Homo sapiens*.

Yeast Alcohol Dehydrogenases

S. cerevisiae, a type of yeast (unicellular fungus), has a unique property that humans have exploited for thousands of years: a robust capacity for converting sugar (either glucose or fructose) into ethanol, either in the presence or in the absence of molecular oxygen (O_2). (Unlike most fermenting organisms, yeast are protected by several enzymes from the toxic effects of O_2.) In fermentation pathways a six-carbon sugar molecule is split in half and converted into two molecules each of pyruvate, ATP, and NADH (**Figure 6A**). For fermentation to continue, NAD^+ (the oxidized form of NADH) must be regenerated. Fermenting organisms solve this problem by converting pyruvate into one or more reduced molecules (e.g., alcohols, ketones, or acids). In ethanol-producing yeast, pyruvate decarboxylase removes a carboxylate group from pyruvate to form acetaldehyde and alcohol dehydrogenase (a zinc-containing tetramer) converts acetaldehyde into ethanol. The highly efficient ethanol-producing process of modern *S. cerevisiae*, which humans use in beer brewing and wine making, is believed to have originated about 80 million years ago with the emergence of angiosperm species producing fleshy fruits. As a result of fierce competition with other microorganisms for fruit sugar molecules, ancestors of *S. cerevisiae* developed the capacity to rapidly produce large amounts of ethanol as a means of killing its competitors. *S. cerevisiae* possesses about 250 genes that code for molecules that protect against ethanol toxicity (e.g., detoxification enzymes and membrane-stabilizing molecules). *S. cerevisiae* possesses two alcohol dehydrogenases: ADH1 (converts acetaldehyde to ethanol) and ADH2 (converts ethanol to acetaldehyde). The role of ADH2 is described in Chapter 8.

Human Alcohol Dehydrogenases

Humans can consume moderate amounts of ethanol because of ADH. Altogether humans have seven ADHs, each of which is a zinc-containing dimer. Ethanol is primarily detoxified by three ADH isoenzymes (ADH1A, ADH1B, and ADH1C). These molecules have a high affinity for ethanol even when it is present in low concentrations. The ADH1 isoenzymes are called the liver ADHs because they are present in relatively high quantities in liver, the major site of ethanol detoxification. (The small intestine, colon, lung, and adrenal gland also possess ADH1s, but in lower amounts.) The other human ADHs differ from ADH1 in their substrate specificities and/or kinetic properties. ADH2, also found in the liver and the GI

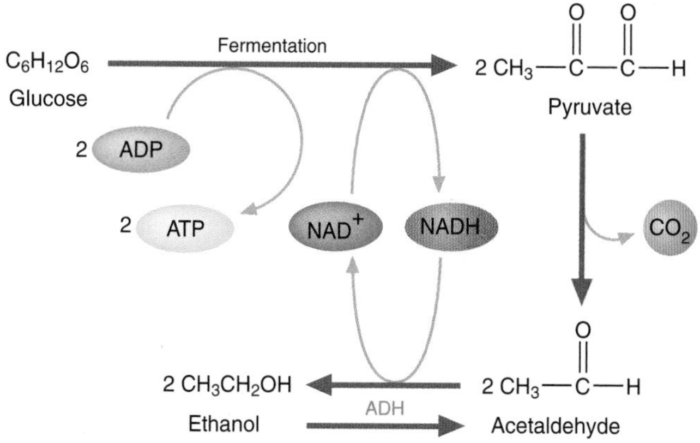

FIGURE 6A

Synthesis of Ethanol from Glucose by *S. cerevisiae*

An anaerobic biochemical pathway called fermentation yields two ATP for each glucose molecule that is converted into two pyruvate molecules. To regenerate NAD^+, yeast cells first convert pyruvate into acetaldehyde in a reaction catalyzed by pyruvate decarboxylase. Acetaldehyde is then reversibily reduced to form ethanol and NAD^+ in a reaction catalyzed by ADH.

Biochemistry IN PERSPECTIVE cont.

tract, has a significantly lower affinity for ethanol and therefore plays only a minor role in ethanol metabolism. Among its many functions, ADH3 (the ancestral vertebrate ADH) is involved in formaldehyde detoxification and the synthesis of retinoic acid, derived from retinol (vitamin A). ADH4 oxidizes a wide variety of substrates, which include retinol, aliphatic alcohols, and hydroxylated steroids. The functions of ADH5 are unresolved.

SUMMARY: The production of alcoholic beverages by humans is possible because *S. cerevisiae* can rapidly and efficiently convert large amounts of sugar into ethanol. Humans can consume moderate amounts of toxic ethanol molecules because of the detoxifying reaction catalyzed by the liver ADH isoenzymes.

Chapter Summary

1. Enzymes are biological catalysts. They enhance reaction rates because they provide an alternative reaction pathway that requires less energy than an uncatalyzed reaction. In contrast to some inorganic catalysts, most enzymes catalyze reactions at mild temperatures. In addition, enzymes are specific to the types of reaction they catalyze. Each type of enzyme has a unique, intricately shaped binding surface called an active site. Substrate binds to the enzyme's active site, which is a small cleft or crevice in a large protein molecule. In the lock-and-key model of enzyme action, the structures of the enzyme's active site and the substrate transition state are complementary. In the induced-fit model, the protein molecule is assumed to be flexible.
2. Enzymes are classified and named according to the type of reaction each one catalyzes. There are six major enzyme categories: oxidoreductases, transferases, hydrolases, lyases, isomerases, and ligases.
3. Enzyme kinetics is the quantitative study of enzyme catalysis. According to the Michaelis-Menten model, when the substrate S binds in the active site of an enzyme E, an ES transition state complex is formed. During the transition state, the substrate is converted into product. After a time the product dissociates from the enzyme. The symbol V_{max} represents the maximal velocity for the reaction, and K_m is a rate constant called the Michaelis constant. Experimental determinations of K_m and V_{max} are made with Lineweaver-Burk double-reciprocal plots.
4. The turnover number (k_{cat}) is a measure of the number of substrate molecules converted to product per unit time by an enzyme when it is saturated with substrate. Because [S] is relatively low under physiological conditions ($[S] \ll K_m$), the specificity constant k_{cat}/K_m is a more reliable gauge of the catalytic efficiency of enzymes.
5. Enzyme inhibition may be reversible or irreversible. In reversible inhibition, the inhibitor can dissociate from the enzyme. The most common types of reversible inhibition are competitive, noncompetitive, and uncompetitive. The assumptions of the law of mass action upon which *in vitro* enzyme analysis are based do not appear to hold true in the crowded and heterogeneous interior space of living organisms. Irreversible inhibitors usually bind covalently to enzymes.
6. The kinetic properties of allosteric enzymes are not explained by the Michaelis-Menten model. Most allosteric enzymes are multisubunit proteins. The binding of substrate or effector to one subunit affects the binding properties of other subunits.
7. Enzymes use the same catalytic mechanisms as nonenzymatic catalysts. Several factors contribute to enzyme catalysis: proximity and orientation effects, electrostatic effects, acid-base catalysis, and covalent catalysis. Combinations of these factors affect enzyme mechanisms.
8. Active site amino acid side chains can facilitate proton transfer and nucleophilic substitutions and stabilize the transition state. Many enzymes use nonprotein cofactors (metals and coenzymes) to facilitate reactions.
9. Enzymes are sensitive to environmental factors such as temperature and pH. Each enzyme has an optimum temperature and an optimum pH.
10. The chemical reactions in living cells are organized into a series of biochemical pathways. The pathways are controlled primarily by adjusting the concentrations and activities of enzymes through genetic control, covalent modification, allosteric regulation, and compartmentation.

Take your learning further by visiting the **companion website** for Biochemistry at **www.oup.com/us/mckee** where you can complete a multiple-choice quiz on water to help you prepare for exams.

Suggested Readings

Gramser, S., Alcohol and Science: The Party Gene, *Nature* 438:1068–1069, 2005.

Gutteridge, A., and Thornton, J. M., Understanding Nature's Toolkit, *Trends Biochem. Sci.* 30(11):622–629, 2005.

Palson, B., The Challenges of In Silico Biology, *Nature Biotechnol.* 18:1147–1150, 2000.

Schnell, S., and Turner, T. E., Reaction Kinetics in Intracellular Environments with Macromolecular Crowding: Simulations and Rate Laws, *Prog. Biophys. Mol. Biol.* 85:234–260, 2004.

Stitt, M., and Gibon, Y., Why Measure Enzyme Activities in the Era of Systems Biology? *Trends Plant Sci.* 19(4):256–265, 2014.

Storey, K. B. (ed.), *Functional Metabolism: Regulation and Adaptation*, Wiley-Liss, Hoboken, New Jersey, 2004.

Thomson, J. M., *et al.*, Resurrecting Ancestral Alcohol Dehydrogenases from Yeast, *Nat. Genet.* 37:630–635, 2005.

Wolfenden, R., and Snider, M. J., The Depth of Chemical Time and the Power of Enzymes as Catalysts, *Acc. Chem. Res.* 34(12):938–945, 2001.

Woofit, M., and Wolfe, K., The Gene Duplication That Greased Society's Wheels, *Nat. Genet.* 37:566–567, 2005.

Key Words

activation energy, *193*
active site, *193*
activity coefficent, *194*
allosteric enzyme, *210*
apoenzyme, *195*
carbanion, *213*
carbocation, *212*
catalyst, *192*
coenzyme, *195*
cofactor, *195*
competitive inhibition, *205*
enzyme, *192*
enzyme induction, *226*
enzyme kinetics, *199*
free radical, *213*
holoenzyme, *195*
hydrolase, *195*
inhibitor, *205*
intermediate, *212*
irreversible inhibition, *205*
isomerase, *196*
ligase, *196*
lyase, *196*
macromolecular crowding, *211*
negative cooperativity, *227*
noncompetitive inhibition, *205, 207*
oxidoreductase, *196*
oxyanion, *219*
pH optimum, *222*
positive cooperativity, *227*
proenzyme, *227*
reaction mechanism, *212*
reversible inhibition, *205*
specificity constant, *202*
substrate, *195*
transferase, *196*
transition state, *192*
turnover number, *202*
uncompetitive inhibition, *205*
velocity, *198*
vitamin, *220*
zymogen, *227*

Review Questions

These questions are designed to test your knowledge of the key concepts discussed in this chapter before moving on to the next chapter. You may like to compare your answers to the solutions provided in the back of the book and in the accompanying Study Guide.

1. Define the following terms:
   a. catalyst
   b. transition state
   c. substrate
   d. activation energy
   e. active site
2. Define the following terms:
   a. cofactor
   b. coenzyme
   c. apoenzyme
   d. holoenzyme
   e. velocity
3. Define the following terms:
   a. reaction order
   b. turnover number
   c. double-displacement reaction
   d. inhibitor
   e. reaction mechanism
4. Define the following terms:
   a. competitive inhibitor
   b. uncompetitive inhibitor
   c. noncompetitive inhibitor
   d. reversible inhibition
   e. irreversible inhibition
5. Define the following terms:
   a. Lineweaver-Burk plot
   b. allosteric enzyme
   c. macromolecular crowding
   d. reaction intermediate
   e. carbocation
6. Define the following terms:
   a. proenzyme
   b. positive cooperativity
   c. negative cooperativity
   d. zymogen
   e. free radical
7. What are four important properties of enzymes?
8. Living things must regulate the rate of catalytic processes. Explain how the cell regulates enzymatic reactions.
9. Several factors contribute to enzyme catalysis. What are they? Briefly explain the effect of each.
10. Why is the regulation of biochemical processes important? List three reasons.
11. Describe negative feedback inhibition.

12. Define the following terms:
    a. oxidoreductase
    b. lyase
    c. ligase
    d. transferase
    e. hydrolase
    f. isomerase
13. Describe the two models that explain the binding of allosteric enzymes. Use either model to explain the binding of oxygen to hemoglobin.
14. What are the major coenzymes? Briefly describe the function of each.
15. What properties of transition metals make them useful as enzyme cofactors?
16. The change in enthalpy ΔH for the following reaction is −28.2 kJ/mol:

    $$C_6H_{12}O_6 + 6O_2 \rightarrow 6CO_2 + 6H_2O$$

    Explain why glucose ($C_6H_{12}O_6$) is stable in an oxygen atmosphere for appreciable periods of time.
17. Enzymes act by reducing the activation energy of a reaction. Describe several ways in which this is accomplished.
18. In enzyme kinetics, why are measurements made at the start of a reaction?
19. Histidine is frequently used as a general acid or general base in enzyme catalysis. Consider the pK_a values of the side groups of the amino acids listed in **Table 5.2** to suggest a reason why this is so.
20. Enzymes are stereochemically specific; that is, they often convert only one stereoisomeric form of substrate into product. Why is such specificity inherent in their structure?
21. The uncatalyzed reaction rate for the conversion of substrate X to product Y is one year. The enzyme-catalyzed rate is one millisecond. Describe the features of the enzyme that are probably responsible for this rate difference.
22. Describe the relationship between the law of mass action, solute concentration, and effective concentration.
23. Define the following terms.
    a. oxygenase
    b. epimerase
    c. protease
    d. hydroxylase
    e. oxidase
24. Define the following terms:
    a. metabolon
    b. *in vivo*
    c. *in vitro*
    d. in silico
    e. metabolic flux
25. Draw the structures of the amino acids whose side chains most commonly participate in catalytic mechanisms when they occur in the active sites of enzymes.
26. Use specific examples to describe the roles played in enzyme catalysis by the amino acids side chains from Question 25.
27. Describe compartmentation within eukaryotic cells. What problems does this phenomenon solve for living organisms?
28. Explain the role that compartmentation plays in the regulation of metabolic pathways. Provide several examples.
29. What are the two types of enzyme inhibitors? Give an example of each.
30. Review the structure of the standard amino acids and list those that are capable of acting as acids or bases in enzyme catalysis. Are there any that can function as either acids or bases?
31. In any given enzyme, the active site is only a small portion of the entire molecule. Synthesis of such a relatively large molecular machine requires an enormous amount of cellular energy. Explain why this inefficiency is tolerated.
32. Catalysts are believed to lower the activation energy of the transition state in a chemical reaction. How do they accomplish this task?
33. What term describes k_{cat}/K_m? What is the maximum value that this term can have? Explain.
34. How does an enzyme attain catalytic perfection?
35. List three effects of macromolecular crowding on the properties of enzymes and the reactions they catalyze.
36. Transition metals can act as Lewis acids. Explain.
37. Is a reaction mechanism altered by the presence of a catalyst? Explain.
38. When calculating the order of a reaction, the enzyme does not appear in the equation. Explain.
39. Quantum tunneling appears to play a significant role in the facilitation of efficient enzyme catalysis in reactions involving the transfer of protons and hydride ions. Provide an energy diagram for a hypothetical reaction in the presence and absence of an enzyme that illustrates the tunneling process. (*Hint*: Refer to **Figure 6.1** and the online Biochemistry in Perspective essay Quantum Tunneling and Catalysis.)
40. Mercuric ion and methanol are inhibitors of alcohol dehydrogenase. Explain.
41. Explain the difference between the energy of a reaction and energy of activation.
42. What is the structure of the intermediate formed during the conversion of dihydroxyacetone phosphate to glyceraldehyde-3-phosphate?

Fill in the Blank

43. Four important properties of enzymes are high catalytic rate, high degree of substrate specificity, negligible formation of side products, and ____________________.
44. Each type of enzyme contains a unique, intricately shaped binding surface called an ____________________.
45. ____________________ are enzymes that catalyze oxidation-reduction reactions.
46. Molecules that reduce an enzyme's activity are called ____________________.
47. Reversible inhibition will be ____________________ if the inhibitor and substrate bind to the same active site.
48. An ____________________ is a species that exists for a finite period of time during a reaction before it is converted to product.
49. Species with unpaired electrons are called ____________________.
50. The synthesis of enzymes in response to changing metabolic needs is referred to as ____________________.
51. Hexokinase is an example of a general class of enzyme known as the ____________________.
52. The ____________________ model states that substrate binding causes enzymes to undergo conformational change.

Short Answer

53. Until recently all known enzymes were proteins. Several RNA molecules have been shown to have catalytic properties. Explain why RNA molecules can have catalytic activity.
54. What is meant by the term "activation energy"?
55. In many cases an enzyme will attack one particular stereoisomer but not its enantiomer. Explain.
56. Reactions taking place in dilute solution obey the rules of simple reaction kinetics because their rates depend on diffusion. How does this approach relate to investigations of enzymes in living cells?
57. Suicide substrates are substances that resemble the substrate, but when incorporated into the active site they form a covalent bond with the enzyme, permanently inhibiting it. These molecules have been used to label the amino-acid side chains that are involved in the catalytic reaction occurring within the active site. This is usually accomplished by hydrolyzing the inhibited enzyme. How could an investigator identify the amino acid bound to the suicide substrate molecule?

Thought Questions

These questions are designed to reinforce your understanding of all of the key concepts discussed in the book so far, including this chapter and all of the chapters before it. They may not have one right answer! The authors have provided possible solutions to these questions in the back of the book and in the accompanying Study Guide for your reference.

58. Consider the following reaction:

$$CH_3-\overset{\overset{O}{\|}}{C}-\overset{\overset{O}{\|}}{C}-O^- + \text{ADP} + P_i \longrightarrow CH_3-\overset{\overset{O}{\|}}{C}-H + CO_2 + \text{ATP}$$

Pyruvate

Using the following data, determine the order of the reaction for each substrate and the overall order of the reaction.

Experiment	Concentration (mol/L) Pyruvate	ADP	P_i	Rate (mol L^{-1} s^{-1})
1	0.1	0.1	0.1	8×10^{-4}
2	0.2	0.1	0.1	1.6×10^{-3}
3	0.2	0.2	0.1	3.2×10^{-3}
4	0.1	0.1	0.2	3.2×10^{-3}

59. Consider the following data for an enzyme-catalyzed hydrolysis reaction in the presence and absence of inhibitor I:

[Substrate] (M)	v_0 (μmol/min)	v_{0I} (μmol/min)
6×10^{-6}	20.8	4.2
1×10^{-5}	29	5.8
2×10^{-5}	45	9
6×10^{-5}	67.6	13.6
1.8×10^{-4}	87	16.2

Using a Michaelis-Menten plot, determine K_m for the uninhibited reaction and the inhibited reaction.

60. Use the data in Question 59 to do the following.
    a. Generate Lineweaver-Burk plots for the data.
    b. Explain the significance of the horizontal intercept, the vertical intercept, and the slope.
    c. Identify the type of inhibition being measured.

61. Two experiments were performed with the enzyme ribonuclease. In experiment 1 the effect of increasing substrate concentration on reaction velocity was measured. In experiment 2 the reaction mixtures were identical to those in experiment 1 except that 0.1 mg of an unknown compound was added to each tube. Plot the data according to the Lineweaver-Burk method. Determine the effect of the unknown compound on the enzyme's activity. (Substrate concentration is measured in millimoles per liter. Velocity is measured in the change in optical density per hour.)

Experiment 1		Experiment 2	
[S]	v	[S]	v
0.5	0.81	0.5	0.42
0.67	0.95	0.67	0.53
1	1.25	1	0.71
2	1.61	2	1.08

62. The mechanism for the exothermic hydrolysis of *t*-butyl chloride to form *t*-butyl alcohol and chloride ion is as follows:

$$(CH_3)_3\,CCl \rightarrow (CH_3)_3C+ \;+\; Cl^-$$

$$(CH_3)_3C+ \;+\; H_2O \rightarrow (CH_3)_3COH + H^+$$

Draw the transition state for each step.

63. Alcohol dehydrogenase (ADH) is inhibited by numerous alcohols. Using the data given in the following table, calculate the k_{cat}/K_m values for each of the alcohols. Which of the listed alcohols is most easily metabolized by alcohol dehydrogenase?
Kinetic parameters for hamster testes ADH.

Substrate	K_m (μM)	k_{cat}(min^{-1})
Ethanol	960	480
1-Butanol	440	450
1-Hexanol	69	182
12-Hydroxydodecanoate	50	146
All-*trans*-retinol	20	78
Benzyl alcohol	410	82
2-Butanol	250,000	285
Cyclohexanol	31,000	122

64. 4-Methyl pyrazole (4-MP) has been developed as a long-acting and less toxic alternative to ethanol in the treatment of ethylene glycol poisoning. Shown is a Lineweaver-Burk plot of the inhibition of alcohol dehydrogenase by various concentrations of 4-methyl pyrazole. What type of inhibition appears to be exhibited by this molecule?

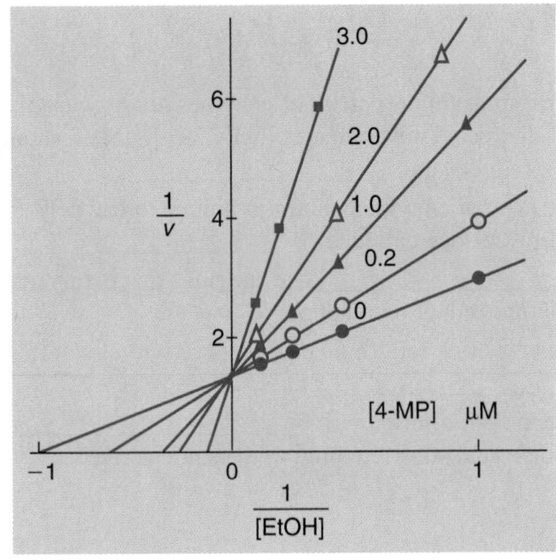

65. The table presents the rates of reaction at specific substrate concentrations for an enzyme that displays classical Michaelis-Menten kinetics. Two sets of inhibitor data are also included. Determine the K_m and V_{max} for the uninhibited enzyme.

	v_o (mM/s)		
[S] (mM)	**Without inhibitor**	**With inhibitor A**	**With inhibitor B**
1.3	2.50	1.17	0.62
2.6	4.00	2.10	1.42
6.5	6.30	4.00	2.65
13.0	7.60	5.70	3.12
26.0	9.00	7.20	3.58

66. Determine the type of inhibition exhibited by each inhibitor in Question 65. What kind of affinity would these inhibitors have for ES versus E?

67. Catalase has a K_m of 25 mM and a k_{cat} of 4.0×10^7 s^{-1} with H_2O_2 as a substrate. Carbonic anhydrase has a K_m of 26 mM and a k_{cat} of 4.0×10^5 s^{-1}. What do these data tell you about these two enzymes?

68. The K_m and k_{cat} for fumarase with fumarate as a substrate are 5×10^{-6} M and 8×10^2 s^{-1}, respectively. When malate is the substrate the K_m and k_{cat} are 2.5×10^{-5} M and 9×10^2 s^{-1}, respectively. What do the data tell you about the operation of this enzyme in the citric acid cycle?

69. What possible effect would macromolecular crowding have on activity coefficients? Explain how macromolecular crowding can increase the effective concentration of a substance.

70. Consider the following reaction along with its rate information.

A + B → C

[A] (mM)	[B] (mM)	Rate (mM/s)
0.05	0.05	2×10^7
0.10	0.05	4×10^7
0.05	0.1	4×10^7
0.1	0.1	8×10^7

What is the overall rate expression for this reaction? What is the order of the reaction?

71. In the serine protease triad, the proximity of an aspartate carboxylate group to the imidazole group of histidine raises the latter's pK_a. Explain.

72. Review the electronic configuration of the magnesium ion and explain why Mg^{2+} can function as a Lewis acid.

73. Provide the reaction between the residue Ser 195 of chymotrypsin and diisopropylfluorophosphate. Why is this particular serine so reactive?

74. Ethylene glycol (HO—CH_2—CH_2—OH) is frequently used as antifreeze in automobile engines. Every year children and pets are poisoned because they tasted this sweet-tasting material. Ethylene glycol is metabolized in the liver by alcohol dehydrogenase. Suggest a possible medical treatment for ethylene glycol intoxication.

75. Given the following rate expression, complete the following table.

Rate = $k[A]^2[B]$

[A] (mM)	[B] (mM)	Rate (mM/s)
0.1	0.01	1×10^6
0.1	0.02	———
0.2	0.01	———
0.2	0.02	———

76. Write out a mechanism for the serine triad hydrolysis of glycylglycine.

77. When free water molecules are excluded from an enzyme's catalytic pocket, is a catalytic —OH group a stronger or weaker nucleophile? Explain.

78. Show how the arginine-histidine dyad within an active site could catalyze the dehydration of *t*-butyl alcohol.

79. Within an enzyme active site, how does the aspartate carboxyl group activate the histidine nitrogen and convert it into a stronger base?

80. Review the following approximate K_m values for ethanol and acetaldehyde for ADH1 in *Saccharomyces cerevisiae* and explain why these data support the observed role of this enzyme in the organism's ethanol metabolism.

	K_m (ethanol)	K_m (acetaldehyde)
ADH1	20,000 μM	1500 μM

81. Given the equation for competitive inhibition:

$$v = \frac{V_{max}[S]}{\alpha K_m + [S]}$$

a. Can the value of α ever be less than 1?
b. What would happen if the value of α is less than 1?

CHAPTER

7 Carbohydrates

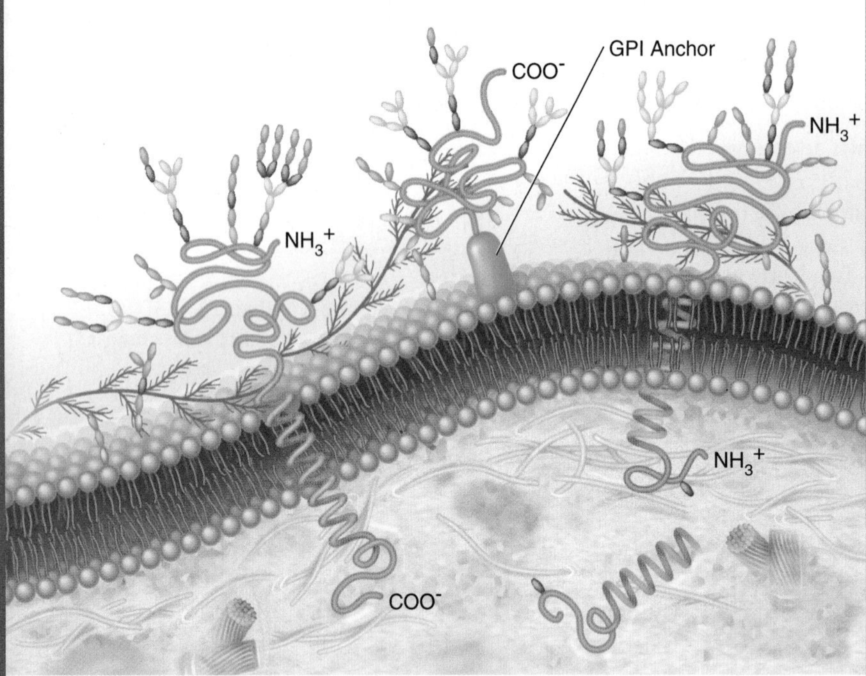

Sugars and Cells Sugar molecules shape the molecular landscape of living organisms. Carbohydrates linked to membrane proteins and lipids are especially prominent on the external surface of cells.

OUTLINE

AVAILABLE ONLINE

Biochemistry in Perspective
Scurvy and Ascorbic Acid

Biochemistry in Perspective
Sweet Medicine

Biochemistry in Perspective
Conversion of Fischer Structures into Haworth Structures

Biochemistry in the Lab
Glycomics

Sweet and Bitter Taste: The Roles of Sugar Molecules

For several hundred million years, plants and plant-eating animals (called herbivores) have been engaged in a struggle for survival. Herbivores derive energy and nutrients from plant biomass, the product of photosynthesis. To protect themselves, many plant species synthesize toxic molecules that either deter or kill herbivores. Examples include oxalate ($C_2O_4^{2-}$) crystals, which cause intense burning and irritation in the mouth, and strychnine, which causes muscular convulsions and asphyxiation by blocking a chloride channel in the spinal cord and brain. Although many animals have evolved several types of defenses against plant toxins (e.g., inactivating biotransformation reactions), their principal means of protection is the sense of taste. Taste, mediated by taste receptor cells within taste buds, allows an animal to quickly evaluate what it is eating. The binding of food molecules to taste receptors initiates the depolarization of taste receptor cell membranes. Taste signals, carried via action potentials in the sensory nerve fibers in close proximity to taste receptor cells, are ultimately transmitted to several areas of the brain. Chemosensory neurons within the cerebral cortex are responsible for taste perception.

Most animals have five primary types of taste perception: sweet, savory, bitter, sour, and salty. Each allows animals to identify specific nutrients or physiological threat. Sweet and savory tastes, for example, indicate that a food is a rich source of sugars or amino acids, respectively. These tastes are pleasurable and promote feeding behavior. Avoidance behavior is triggered by bitter taste, which indicates toxicity, and high levels of salty and sour tastes, which indicate the presence of electrolytes and acids in food. Of the five taste types, sweet and bitter are the most important in evaluating food energy content and safety. Sweet receptor cells detect sugars at 0.01 M or above so that only energy-rich foods will elicit eating behavior. In contrast, the threshold for bitter taste is very low. Quinine, an example of the alkaloids (a group of nitrogen-containing plant molecules with potent physiological properties) from the Cinchona tree, is detected at 0.8 μM.

One of the more intriguing features of sweet and bitter taste perception is that sugar molecules are not only sweet, energy-rich nutrients; they can also be components of bitter, toxic molecules. In two prominent classes of potentially lethal plant toxins, the cyanogenic and cardiac glycosides, one or more sugar molecules are covalently linked to a toxic aglycone (the noncarbohydrate portion of a molecule). (The sugar residues improve the water solubility of the aglycone.) *Cyanogenic glycosides* are rapidly hydrolyzed to form hydrogen cyanide (HCN; a potent inhibitor of aerobic respiration, see p. 357) when consumed by animals. *Cardiac glycosides* inhibit the Na^+/ K^+-ATPase in the membrane of cells, resulting in nausea, dizziness, confusion, and cardiac arrest. Humans have learned how to render some plants safe to eat by inactivating toxins. For example, roasting the roots of the cassava plant, an important food source for many humans, destroys cyanogenic glycosides such as linamarin. Some plant toxins have been used for their medicinal properties. In small and carefully regulated amounts, digitoxin, a cardiac glycoside from the foxglove plant, has been used for hundreds of years to treat congestive heart failure.

Not all bitter plant molecules are toxic. In fact, numerous bitter molecules in various plant-derived foods and beverages have beneficial effects on human health. Examples include bioflavonoids (polyhydroxypolyphenols) in tea and citrus fruits that protect against cell-damaging free radicals and organosulfur compounds in cruciferous vegetables such as broccoli that have anticancer properties.

Overview

CARBOHYDRATES ARE NOT JUST AN IMPORTANT SOURCE OF RAPID ENERGY PRODUCTION FOR LIVING CELLS. THEY ARE ALSO STRUCTURAL building blocks of cells and components of numerous metabolic pathways. Sugar polymers linked to proteins and lipids are now recognized as a high-density coding system. Their vast structural diversity is exploited by living organisms to produce the informational capacity required for living processes. Chapter 7 describes the structures and chemistry of typical carbohydrate molecules found in living organisms. An introduction to glycomics, the investigation of the sugar code, is available online.

Carbohydrates, the most abundant biomolecules in nature, are a direct link between solar energy and the chemical bond energy of living organisms. (More than half of all "organic" carbon is found in carbohydrates.) They are formed during *photosynthesis* (Chapter 13), a biochemical process in which light energy is captured and used to drive the biosynthesis of energy-rich organic molecules from the energy-poor inorganic molecules CO_2 and H_2O. Most carbohydrates contain carbon, hydrogen, and oxygen in the ratio $(CH_2O)_n$, hence the name "hydrate of carbon." They have been adapted for a wide variety of biological functions, which include energy sources (e.g., glucose), structural elements (e.g., cellulose and chitin in plants and insects, respectively), cellular communication and identity, and precursors in the production of other biomolecules (e.g., amino acids, lipids, purines, and pyrimidines). Carbohydrates are classified as monosaccharides, disaccharides, oligosaccharides, and polysaccharides according to the number of simple sugar units they contain. Carbohydrate moieties also occur as components of other biomolecules. A vast array of *glycoconjugates* (protein and lipid molecules with covalently linked carbohydrate groups) are distributed among all living species, most notably among the eukaryotes. Certain carbohydrate molecules (the sugars ribose and deoxyribose) are structural elements of nucleotides and nucleic acids.

Chapter 7 provides a foundation for understanding the complex processes in living organisms by reviewing the structure and function of the most common carbohydrates and glycoconjugates. The chapter ends with a discussion of the *sugar code*, the mechanism by which carbohydrate structure is used to encode biological information.

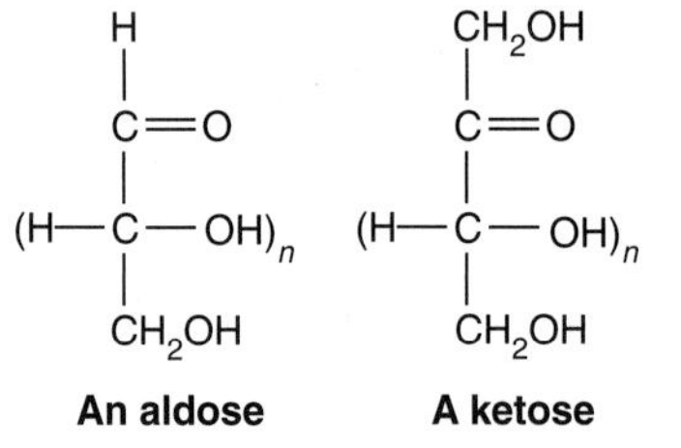

H
|
C=O
|
(H—C—OH)$_n$
|
CH_2OH

An aldose

CH_2OH
|
C=O
|
(H—C—OH)$_n$
|
CH_2OH

A ketose

FIGURE 7.1

General Formulas for the Aldose and Ketose Forms of Monosaccharide

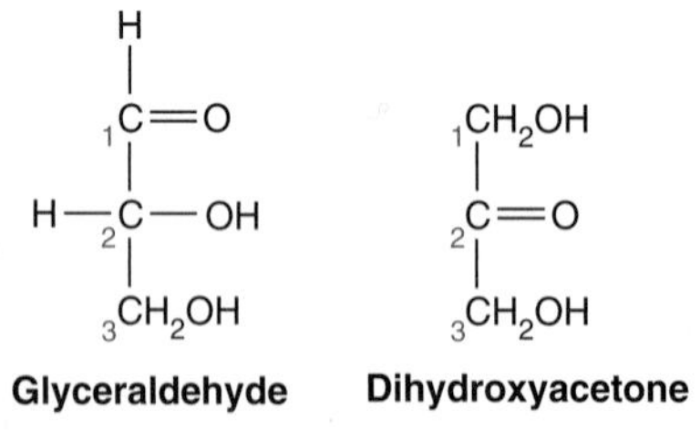

H
|
$_1$C=O
|
H—$_2$C—OH
|
$_3CH_2OH$

Glyceraldehyde

$_1CH_2OH$
|
$_2$C=O
|
$_3CH_2OH$

Dihydroxyacetone

FIGURE 7.2

Glyceraldehyde (an Aldotriose) and Dihydroxyacetone (a Ketotriose)

7.1 MONOSACCHARIDES

Monosaccharides, or simple sugars, are polyhydroxy aldehydes or ketones. Recall from Chapter 1 that monosaccharides with an aldehyde functional group are called **aldoses**, whereas those with a ketone group are called *ketoses* (**Figure 7.1**). The simplest aldose and ketose are glyceraldehyde and dihydroxyacetone, respectively (**Figure 7.2**). Sugars are also classified according to the number of carbon atoms they contain. For example, the smallest sugars, called *trioses*, contain three carbon atoms. Four-, five-, and six-carbon sugars are called *tetroses, pentoses*, and *hexoses*, respectively. The most abundant monosaccharides found in living cells are the pentoses and hexoses. Often, class names such as aldohexoses and ketopentoses, which combine information about carbon number and functional groups, describe monosaccharides. For example, glucose, a six-carbon, aldehyde-containing sugar, is referred to as an aldohexose.

The sugar structures shown in Figures 7.1 and 7.2 are known as Fischer projections (in honor of the German chemist Emil Fischer). In these structures the carbohydrate backbone is drawn vertically with the most highly oxidized carbon usually shown at the top. The horizontal lines are understood to project toward the viewer; the vertical lines recede from the viewer.

QUESTION 7.1

Identify the class of each of the following sugars. For example, glucose is an aldohexose.

(a) H–C=O; H–C–OH; HO–C–H; CH_2OH

(b) CH_2OH; C=O; H–C–OH; HO–C–H; CH_2OH

(c) CH_2OH; C=O; H–C–OH; H–C–OH; HO–C–H; CH_2OH

Monosaccharide Stereoisomers

When the number of chiral carbon atoms increases in optically active compounds, the number of possible optical isomers also increases. The total number of possible isomers can be determined using van't Hoff's rule: A compound with n chiral carbon atoms has a maximum of 2^n possible stereoisomers. For example, when n is 4, there are 2^4 or 16 stereoisomers (8 D-stereoisomers and 8 L-stereoisomers).

In optical isomers the reference carbon is the asymmetric carbon that is most remote from the carbonyl carbon. Its configuration is similar to that of the asymmetric carbon in either D- or L-glyceraldehyde. Almost all naturally occurring sugars have the D-configuration. They can be considered to be derived from either the triose D-glyceraldehyde (the aldoses) or the nonchiral triose dihydroxyacetone (the ketoses). (Note that although dihydroxyacetone does not have an asymmetric carbon, it clearly is the parent compound for the ketoses.) In the D-aldose family of sugars (**Figure 7.3**), which contains most biologically important monosaccharides, the hydroxyl group is to the right on the chiral carbon atom in the Fischer model farthest from the most oxidized carbon (in this case the aldehyde group) in the molecule (e.g., carbon 5 in a six-carbon sugar).

Stereoisomers that are not enantiomers (mirror-image isomers) are called **diastereomers**. For example, the aldopentoses D-ribose and L-ribose are enantiomers, as are D-arabinose and L-arabinose (**Figure 7.4**). The sugars D-ribose and D-arabinose, which are isomers but not mirror images, are diastereomers. Diastereomers that differ in the configuration at a single asymmetric carbon atom are called **epimers**. For example, D-glucose and D-galactose are epimers because their structures differ only in the configuration of the OH group at carbon 4 (**Figure 7.3**). D-Mannose and D-galactose are not epimers because their configurations differ at more than one carbon.

Cyclic Structure of Monosaccharides

Sugars that contain four or more carbons exist primarily in cyclic forms. Ring formation occurs in aqueous solution because aldehyde and ketone groups react reversibly with hydroxyl groups present in the sugar to form cyclic **hemiacetals** and **hemiketals**, respectively. Ordinary hemiacetals and hemiketals, which form when molecules containing an aldehyde or ketone functional group react with an alcohol, are unstable and easily revert to the aldehyde or ketone forms (**Figure 7.5**). When the aldehyde or ketone group and the alcohol functional group are part of the same molecule, however, an intramolecular cyclization

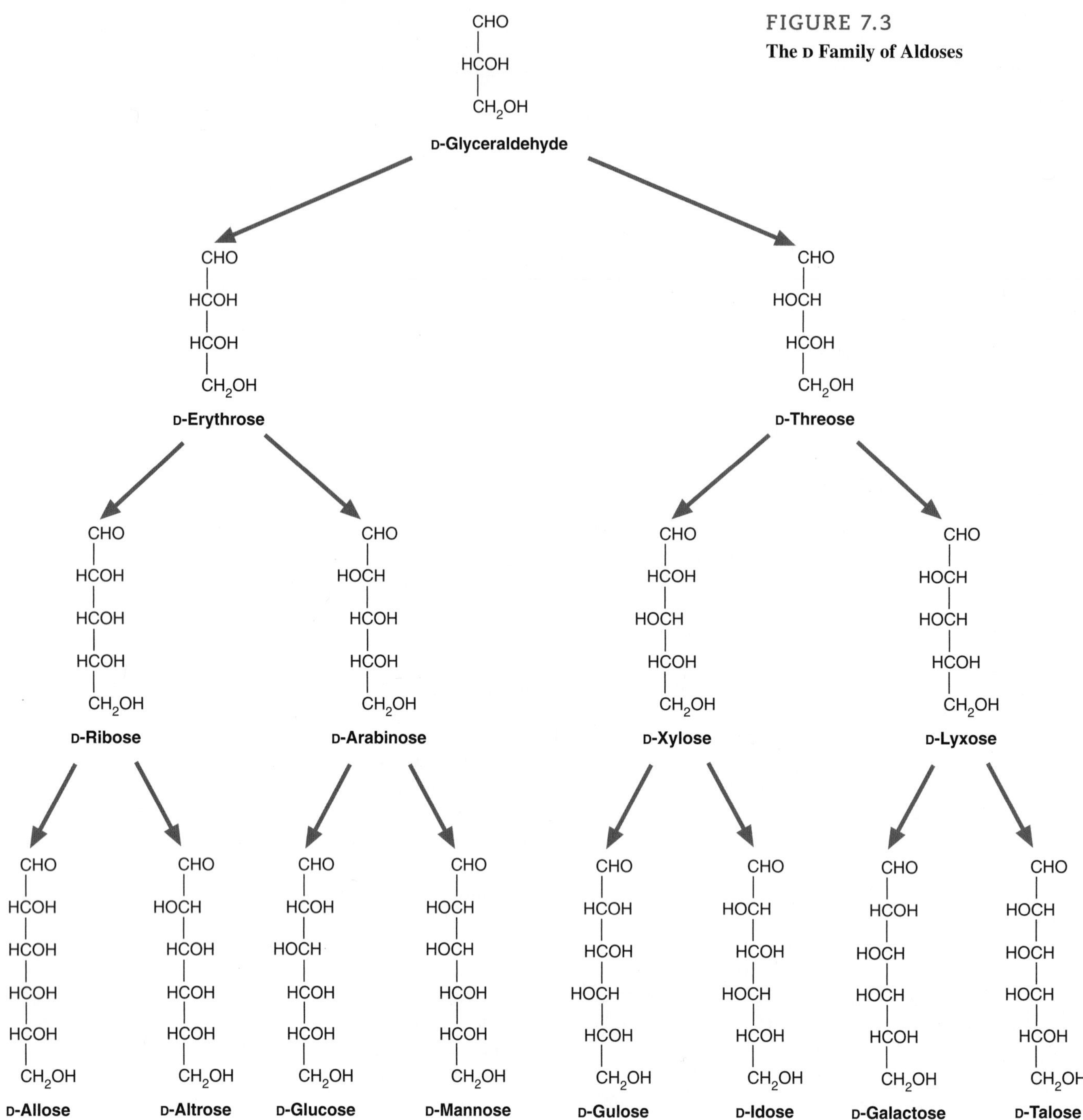

FIGURE 7.3
The D Family of Aldoses

reaction occurs that can form stable products. The most stable cyclic hemiacetal and hemiketal rings contain five or six atoms. As cyclization occurs, the carbonyl carbon becomes a new chiral center. This carbon is called the *anomeric carbon atom*. The two possible diastereomers that may form during the cyclization reaction are called **anomers**.

In aldose sugars the hydroxyl group of the newly formed hemiacetal occurs on carbon 1 (the anomeric carbon). Because the anomeric carbon is chiral, two stereoisomers of the aldose can form: either the α-anomer or the β-anomer. In Fischer projections, the α-anomeric hydroxyl occurs on the right and the β-hydroxyl occurs on the left (**Figure 7.6**). It is important to note that the anomers are defined relative to the D- and L-classification of sugars. These rules apply only to D-sugars, the most common ones found in nature. In the L-sugars

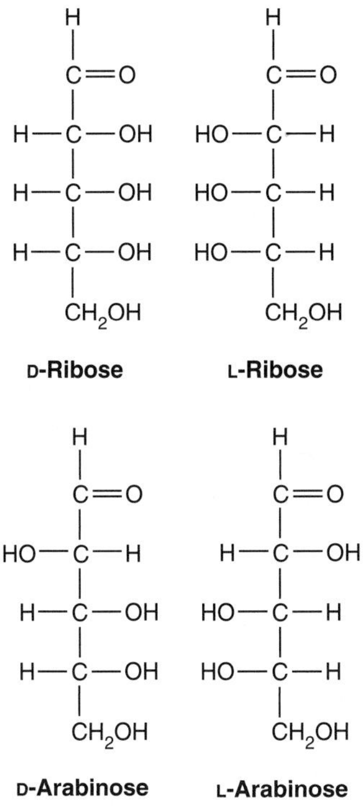

FIGURE 7.4

The Optical Isomers D- and L-Ribose and D- and L-Arabinose

D-Ribose and D-arabinose are diastereomers; that is, they are not mirror images.

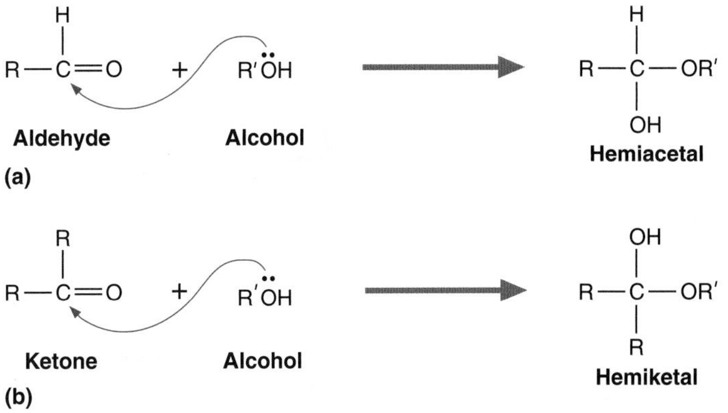

FIGURE 7.5

Formation of Hemiacetals and Hemiketals

(a) From an aldehyde. (b) From a ketone.

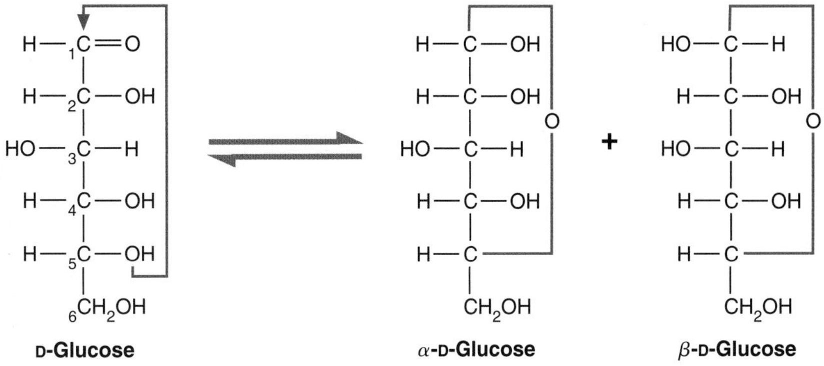

FIGURE 7.6

Monosaccharide Structure

Formation of the hemiacetal structure of glucose. Both the α- and the β-anomers of glucose form. Note that the right angles in the hemiacetal linkage in Fischer models of monosaccharides do not represent methylene groups.

the α-anomeric OH group occurs on the left. The cyclization of sugars is more easily visualized using Haworth structures.

HAWORTH STRUCTURES Fischer representations of cyclic sugar molecules use a long bond to indicate ring structure. A more accurate picture of carbohydrate structure was developed by the English chemist W. N. Haworth (**Figure 7.7**). Haworth structures more closely depict proper bond angles and lengths than do Fischer representations. In Haworth depictions of aldoses, the hydroxyl group on the anomeric carbon occurs either above the ring on the same side as the CH_2OH group (the "up" position) or below the ring on the side opposite the CH_2OH group (the "down" position). For D-sugars, when the hydroxyl is down the structure is in the α-anomeric form. If the hydroxyl is up, the structure is in the β-anomeric form. In the L-sugars, this rule is reversed: the α-anomeric OH group is above the ring and the β-anomeric OH group is below the ring.

Five-membered hemiacetal rings are called *furanoses* because of their structural similarity to furan (**Figure 7.8**). For example, the cyclic form of fructose depicted in **Figure 7.9** is called fructofuranose. Six-membered rings are called *pyranoses* because of their similarity to pyran. Glucose, in the pyranose form, is called glucopyranose.

Visit the companion website at www.oup.com/us/mckee to read the Biochemistry in Perspective essay on conversion of Fischer structures into Haworth structures.

KEY CONCEPTS

- Monosaccharides, which may be polyhydroxy aldehydes or ketones, are either aldoses or ketoses.
- Sugars that contain four or more carbons primarily have cyclic forms.
- Cyclic aldoses or ketoses are hemiacetals and hemiketals, respectively.

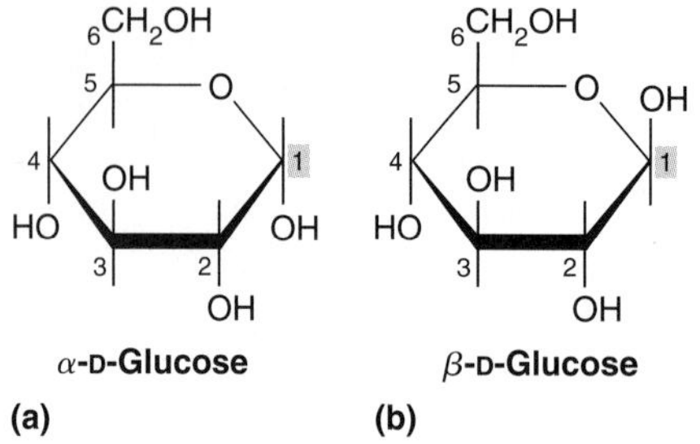

α-D-Glucose (a)

β-D-Glucose (b)

FIGURE 7.7
Haworth Structures of the Anomers of D-Glucose

(a) α-D-Glucose. (b) β-D-Glucose. Note that in carbohydrate chemistry, the hydrogens bonded to carbons in sugar rings can be represented by single lines.

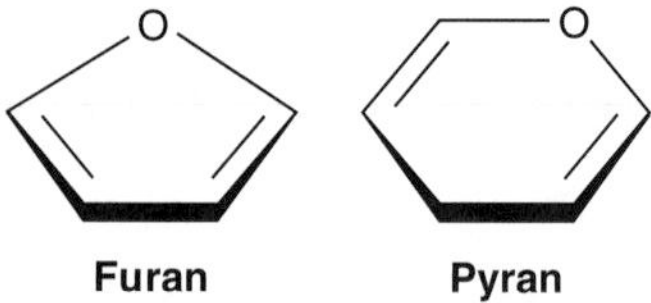

FIGURE 7.8
Furan and Pyran

FIGURE 7.9
Fischer and Haworth Representations of D-Fructose

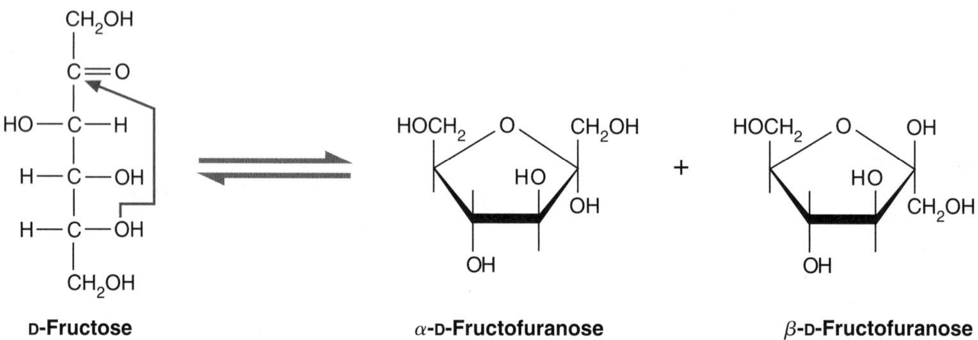

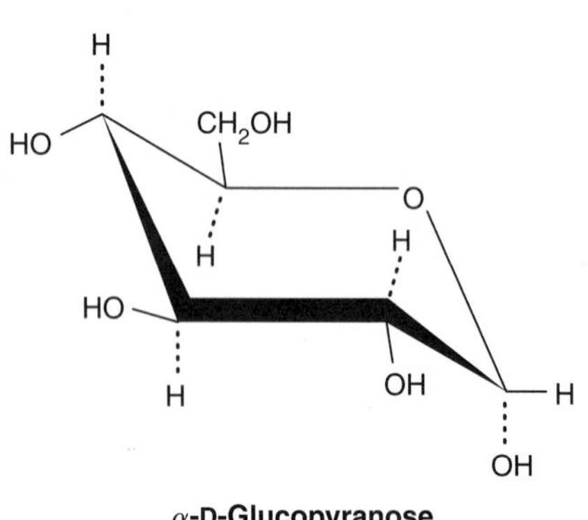

CONFORMATIONAL STRUCTURES Although Haworth projection formulas are often used to represent carbohydrate structure, they are oversimplifications. Bond angle analysis and X-ray analysis demonstrate that *conformational formulas* are more accurate representations of monosaccharide structure (**Figure 7.10**) because they illustrate the puckered nature of sugar rings.

Space-filling models, whose dimensions are proportional to the van der Waals radius of the atoms, also give useful structural information (see later: Figures 7.21, 7.22, and 7.23).

Monosaccharides undergo most of the reactions that are typical of aldehydes, ketones, and alcohols. The most important of these reactions in living organisms are described.

MUTAROTATION The α- and β-forms of monosaccharides are readily interconverted when dissolved in water. This spontaneous process, called **mutarotation**, produces an equilibrium mixture of α- and β-forms in both furanose and pyranose ring structures. The proportion of each form differs according to sugar type. Glucose, for example, exists primarily as a mixture of α- (38%) and β- (62%) pyranose forms (**Figure 7.11**). Fructose is predominantly found in the α- and β-furanose forms. The open chain formed during mutarotation can participate in oxidation-reduction reactions.

β-D-Glucopyranose

FIGURE 7.10
Conformational Representations of α- and β-D-Glucose

Reactions of Monosaccharides

The carbonyl and hydroxyl groups of sugars can undergo several chemical reactions. Among the most important are oxidation, reduction, isomerization, esterification, glycoside formation, and glycosylation reactions.

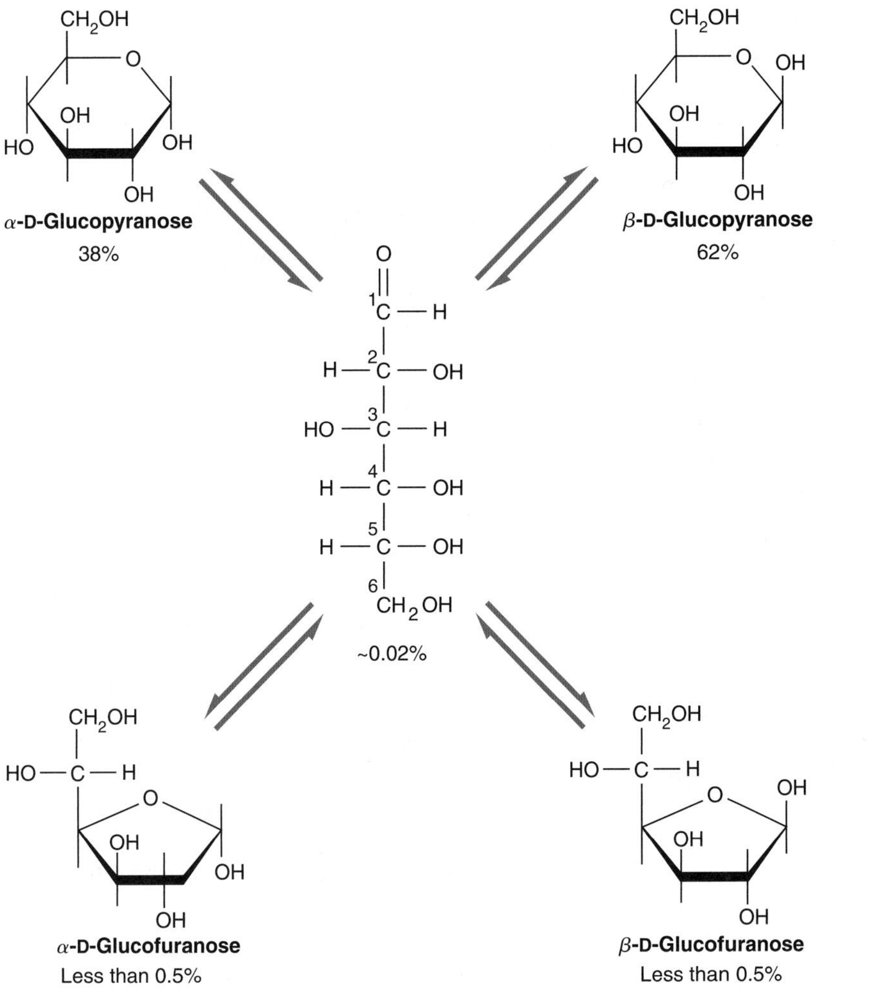

FIGURE 7.11

Equilibrium Mixture of D-Glucose

When glucose is dissolved in water at 25°C, the anomeric forms of the sugar undergo very rapid interconversions. When equilibrium is reached (i.e., there is no net change in the occurrence of each form), the glucose solution contains the percentages shown.

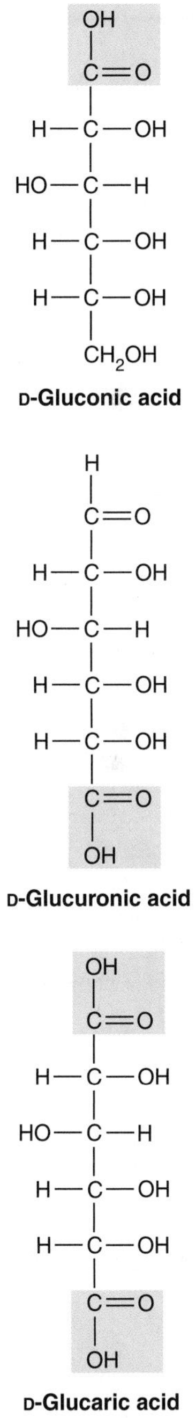

OXIDATION In the presence of oxidizing agents, metal ions such as Cu^{2+}, and certain enzymes, monosaccharides readily undergo several oxidation reactions. Oxidation of an aldehyde group yields an **aldonic acid**, whereas oxidation of a terminal CH_2OH group (but not the aldehyde group) gives a **uronic acid**. Oxidation of both the aldehyde and CH_2OH gives an **aldaric acid** (**Figure 7.12**).

The carbonyl groups in both aldonic and uronic acids can react with an OH group in the same molecule to form a cyclic ester known as a **lactone**:

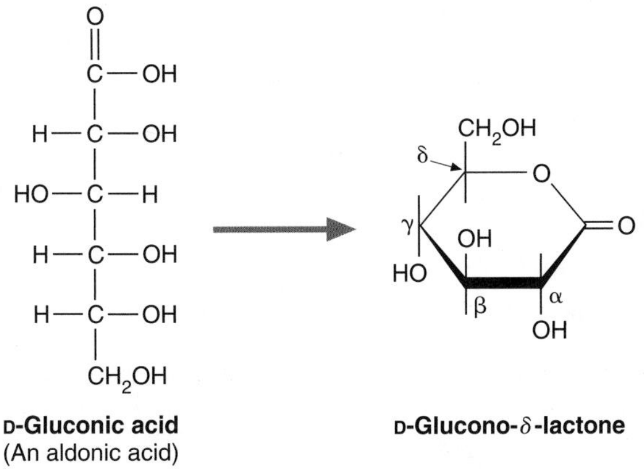

FIGURE 7.12

Oxidation Products of Glucose

The newly oxidized groups are highlighted.

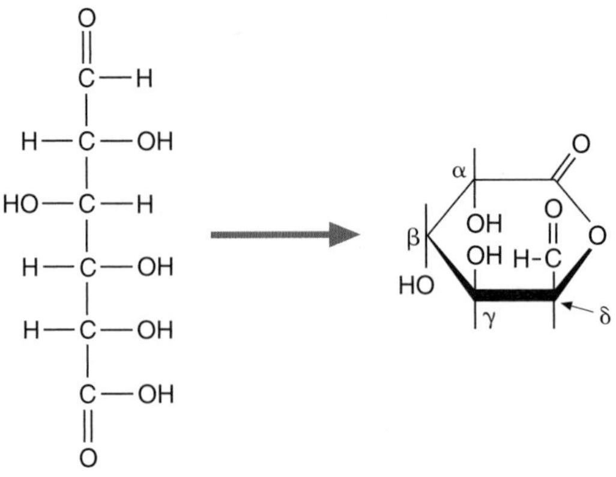

D-Glucuronic acid
(A uronic acid)

D-Glucurono-δ-lactone

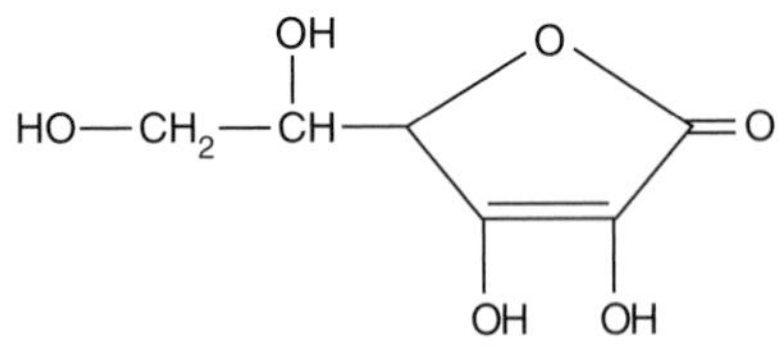

Ascorbic acid

FIGURE 7.13
Structure of Ascorbic Acid

Humans and guinea pigs cannot synthesize ascorbic acid because they lack gluconolactone oxidase, one of the three enzymes required to synthesize the acid from its precursor glucuronate.

Visit the companion website at www.oup.com/us/mckee to read the Biochemistry in Perspective essay on scurvy and ascorbic acid.

Lactones are commonly found in nature. For example, L-ascorbic acid, also known as vitamin C (**Figure 7.13**), is a lactone derivative of D-glucuronic acid. It is synthesized by all mammals except guinea pigs, apes, fruit-eating bats, and, of course, humans. These species must obtain ascorbic acid in their diet, hence the name vitamin C. Ascorbic acid is a powerful reducing agent; that is, it protects cells from highly reactive oxygen and nitrogen species (see p. 375). In addition, it is required in the hydroxylation reactions of collagen.

Sugars that can be oxidized by weak oxidizing agents such as Benedict's reagent are called **reducing sugars** (**Figure 7.14**). The reaction occurs only with sugars that reform an aldehyde group when they revert to the open chain form. All aldoses are, therefore, reducing sugars. Ketoses such as fructose are reducing sugars because they convert to aldoses via isomerization reactions (see below).

REDUCTION Reduction of the aldehyde and ketone groups of monosaccharides yields the sugar alcohols (**alditols**). Reduction of D-glucose, for example, yields D-glucitol, also known as D-sorbitol (**Figure 7.15**). Sugar alcohols are used commercially in processing foods and pharmaceuticals. Sorbitol, for example, improves the shelf life of candy because it helps prevent moisture loss. Adding sorbitol syrup to artificially sweetened canned fruit reduces the unpleasant aftertaste of the artificial sweetener saccharin. Sorbitol is converted into fructose in the liver.

ISOMERIZATION Monosaccharides undergo several types of isomerization. For example, after several hours an alkaline solution of D-glucose also contains

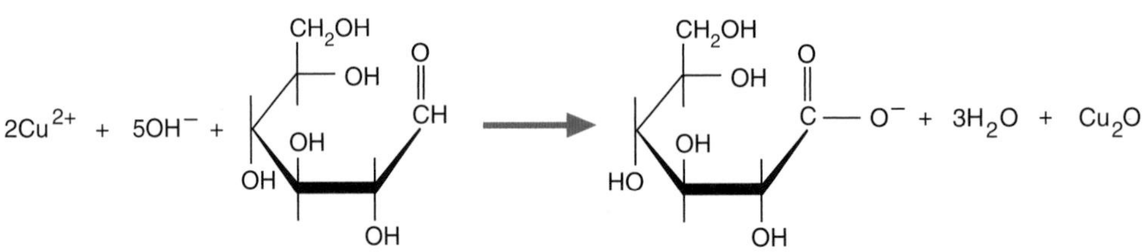

$2Cu^{2+} + 5OH^- +$ glucose $\longrightarrow$ gluconate $+ 3H_2O + Cu_2O$

FIGURE 7.14
Reaction of Glucose with Benedict's Reagent

Benedict's reagent, copper(II) sulfate in a solution of sodium carbonate and sodium citrate, is reduced by the monosaccharide glucose. Glucose is oxidized to form the salt of gluconic acid. The reaction also forms the reddish-brown precipitate Cu_2O and other oxidation products (not shown).

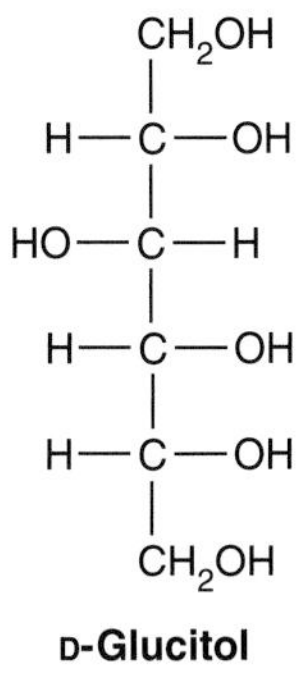

D-mannose and D-fructose. Both isomerizations involve an intramolecular shift of a hydrogen atom and a relocation of a double bond (**Figure 7.16**). The intermediate formed is called an **enediol**. The reversible transformation of glucose to fructose is an aldose-ketose interconversion. Because the configuration at a single asymmetric carbon changes, the conversion of glucose to mannose is referred to as an **epimerization**. Several enzyme-catalyzed reactions involving enediols occur in carbohydrate metabolism (Chapter 8).

ESTERIFICATION Like all free OH groups, those of carbohydrates can be converted to esters by reactions with acids. Esterification often dramatically changes a sugar's chemical and physical properties. Phosphate and sulfate esters of carbohydrate molecules are among the most common ones found in nature.

Phosphorylated derivatives of certain monosaccharides are metabolic components of living cells that are frequently formed during reactions with ATP. They are important because many biochemical transformations use nucleophilic substitution reactions. Such reactions require a leaving group. In a carbohydrate molecule this group is most likely to be an OH group. However, because OH groups are poor leaving groups, any substitution reaction is unlikely. The problem is solved by converting an appropriate OH group to a phosphate ester, which can then be displaced by an incoming nucleophile. As a consequence, a slow reaction now occurs much more rapidly.

Sulfate esters of carbohydrate molecules are found predominantly in the proteoglycan components of connective tissue. Sulfate esters are charged so they bind large amounts of water and small ions. They also participate in forming salt bridges between carbohydrate chains.

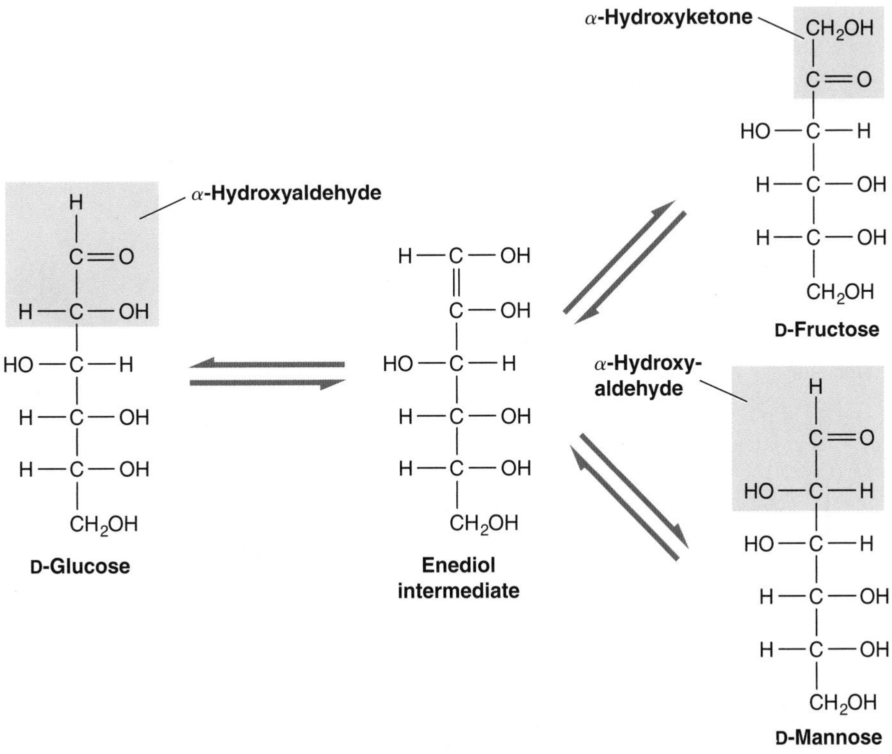

CH_2OH
H—C—OH
HO—C—H
H—C—OH
H—C—OH
CH_2OH
D-Glucitol

FIGURE 7.15
Structure of D-Glucitol (Sorbitol)

α-Hydroxyaldehyde
H
C=O
H—C—OH
HO—C—H
H—C—OH
H—C—OH
CH_2OH
D-Glucose

H—C—OH
C—OH
HO—C—H
H—C—OH
H—C—OH
CH_2OH
Enediol intermediate

α-Hydroxyketone
CH_2OH
C=O
HO—C—H
H—C—OH
H—C—OH
CH_2OH
D-Fructose

α-Hydroxy-aldehyde
H
C=O
HO—C—H
HO—C—H
H—C—OH
H—C—OH
CH_2OH
D-Mannose

FIGURE 7.16
Isomerization of D-Glucose to Form D-Mannose and D-Fructose

An enediol intermediate is formed in this process.

QUESTION 7.2

Draw the following compounds:

a. α- and β-anomers of D-galactose
b. aldonic acid, uronic acid, and aldaric acid derivatives of galactose
c. galactitol
d. δ-lactone of galactonic acid

GLYCOSIDE FORMATION Hemiacetals and hemiketals react with alcohols to form the corresponding **acetal** or **ketal** (**Figure 7.17**). When the cyclic hemiacetal or hemiketal form of the monosaccharide reacts with an alcohol, the new linkage is called a **glycosidic linkage**, and the compound is called a **glycoside**. The name of the glycoside specifies the sugar component. For example, the acetals of glucose and the ketals of fructose are called *glucoside* and *fructoside*, respectively. Additionally, glycosides derived from sugars with five-membered rings are called *furanosides*; those from six-membered rings are called *pyranosides*. A relatively simple example shown in **Figure 7.18** illustrates the reaction of glucose with methanol to form two anomeric types of methyl glucosides. Because glycosides are acetals, they are stable in basic solutions. Carbohydrate molecules that contain only acetal groups do not test positive with Benedict's reagent. (Acetal formation "locks" a ring so it cannot undergo oxidation or mutarotation.) Only hemiacetals act as reducing agents.

If an acetal linkage is formed between the hemiacetal hydroxyl group of one monosaccharide and a hydroxyl group of another monosaccharide, the resulting glycoside is called a **disaccharide**. A molecule containing a large number of monosaccharides linked by glycosidic linkages is called a **polysaccharide**.

FIGURE 7.17
Formation of Acetals and Ketals

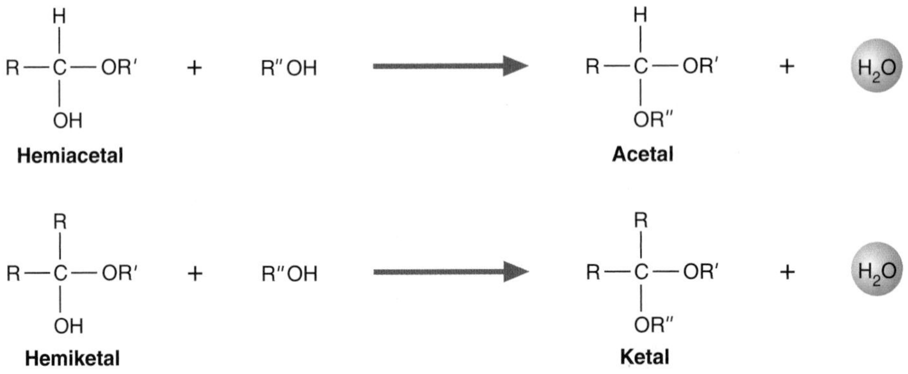

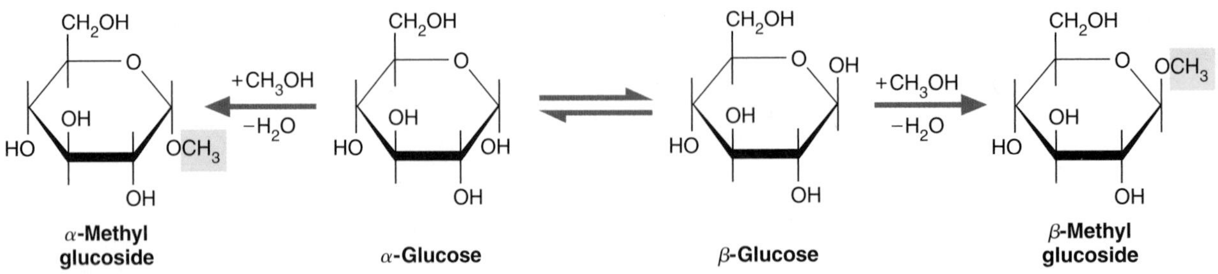

FIGURE 7.18
Methyl Glucoside Formation

Noncarbohydrate components of glycosides are called aglycones. The highlighted methyl groups are aglycones.

QUESTION 7.3

Draw the structure of a D-glucose molecule linked to threonine via a β-glycosidic linkage.

QUESTION 7.4

Glycosides are commonly found in nature. One example is salicin (Figure 7.19), a compound found in willow tree bark that has antipyretic (fever-reducing) and analgesic (pain-relieving) properties. Can you identify the carbohydrate and aglycone (noncarbohydrate) components of salicin?

FIGURE 7.19
Salicin

GLYCOSYLATION REACTIONS Glycosylation reactions attach sugars or glycans (sugar polymers) to proteins or lipids. Analogous to glycoside formation between sugar molecules, the glycosylation reactions, catalyzed by the glycosyl transferases, form glycosidic bonds between anomeric carbons in certain sugars and nitrogen or oxygen atoms in other types of molecules. For example, both N- and O-glycosidic bonds are prominent structural features of glycoproteins.

Reducing sugars can also react with nucleophilic nitrogen atoms in nonenzymatic reactions. These so-called *glycation* reactions occur rapidly in the presence of heat (e.g., during the cooking or baking of sugar-containing food) or slowly within the body when excess sugar molecules are present. The best-researched example is the reaction of glucose with the side chain amino nitrogen of lysine residues in proteins. The nonenzymatic glycation of protein, called the Maillard reaction (named for the French chemist who discovered it in 1912), begins with the nucleophilic attack of the amino nitrogen on the anomeric carbon of the reducing sugar (**Figure 7.20**). The Schiff base that forms rearranges to yield a stable ketoamine called the *Amadori product.* Both the protein-bound Schiff base and the Amadori product can undergo further reactions (e.g., oxidations, rearrangements, and dehydrations) to produce additional protein-bound products, referred to collectively as *a*dvanced *g*lycation *e*nd products (AGEs). Reactive carbonyl-containing products such as the dicarbonyl compound glyoxal (CHOCHO) cause rapid protein cross-linkage and adduct formation. (An **adduct** is the product of an addition reaction, that is, the reactions of two molecules to form a third molecule.) Consequently, glycation alters the structural and functional properties of proteins. For example, the glycation of long-lived proteins such as collagen and elastin disrupts the structure of vascular and connective tissues. In addition, AGEs trigger the production of molecules such as the cytokines that promote inflammatory processes. The accumulation of AGEs has been linked to such age-related conditions as vascular and neurodegenerative diseases and arthritis. In one vascular disease, *atherosclerosis*, cells lining the arterial blood vessels are damaged by AGE formation. AGE-mediated damage initiates a repair process involving macrophages and growth factors that triggers an inflammatory process leading to the formation of artery-clogging deposits called *plaque.* The capacity of affected blood vessels to nourish nearby tissue is

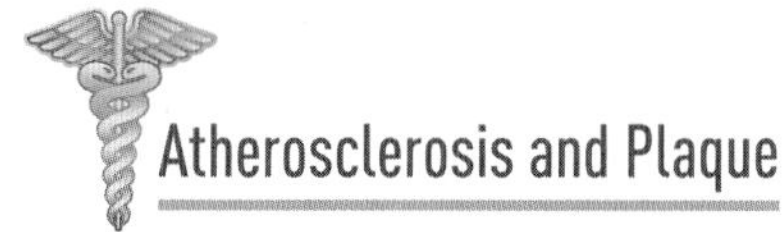

FIGURE 7.20

The Maillard Reaction

Any molecule that contains an amino group can undergo the Maillard reaction, so nucleotides and amines also react with glucose molecules. Since proteins have greater exposure to elevated circulating simple sugars, they are more compromised by the process. The amino nitrogen of a protein side chain reacts with the carbonyl carbon of an aldose or ketose (1) to produce an intermediate (2) that then undergoes dehydration to form an imine bond (Schiff base) (3). The imine undergoes a tautomerization (p. 281) to yield an intermediate (4), which then undergoes a second tautomerization to yield the Amadori product (5) with a ketone and an amine bond. Amadori products undergo further reactions (oxidation and cleavage) to yield highly reactive carbonyl-containing products that form adducts with the amine groups of other proteins.

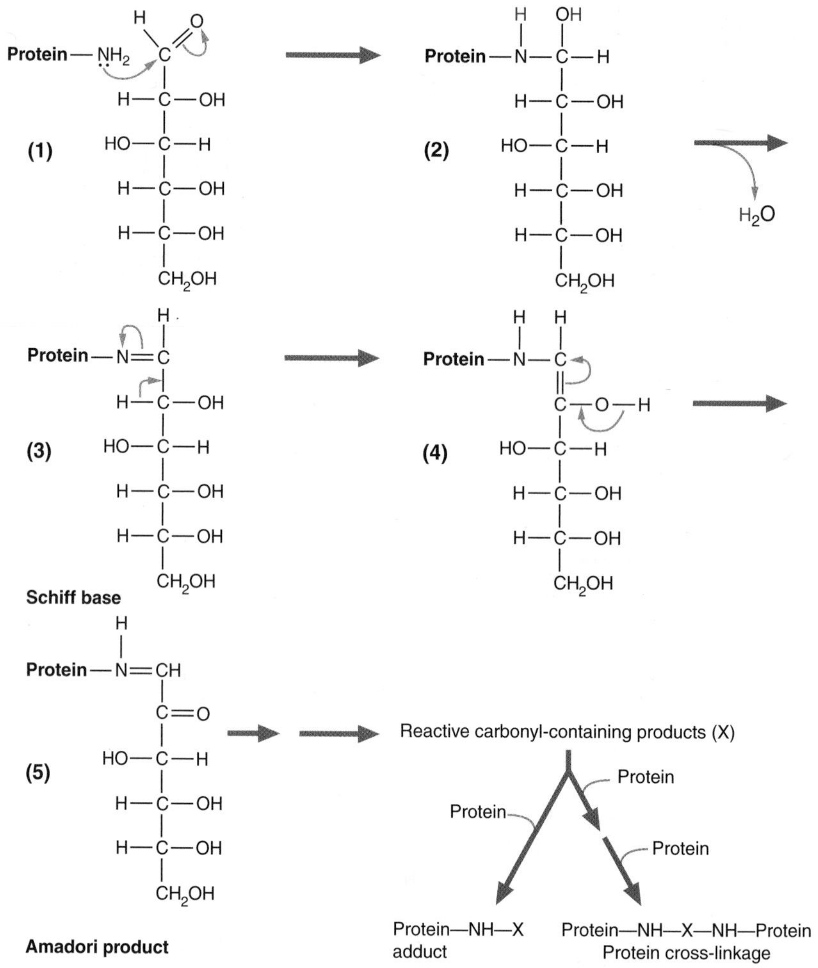

eventually compromised. The excessively high blood glucose levels that occur in diabetes mellitus (see Biochemistry in Perspective essay, Diabetes Mellitus, in Chapter 16) cause an accelerated form of atherosclerosis as well as numerous other AGE-related pathological changes.

Important Monosaccharides

Among the most important monosaccharides found in living organisms are glucose, fructose, and galactose. The principal functional roles of these molecules are briefly described.

GLUCOSE D-Glucose, originally called dextrose, is found in large quantities throughout the living world (**Figure 7.21**). It is the primary fuel for living cells. In animals, glucose is the preferred energy source of brain cells and cells that have few or no mitochondria, such as erythrocytes. Cells that have a limited oxygen supply, such as those in the eyeball, also use large amounts of glucose to generate energy. Dietary sources include plant starch and the disaccharides lactose, maltose, and sucrose.

FRUCTOSE D-Fructose, originally called levulose, is often referred to as fruit sugar because of its high content in fruit. It is also found in some vegetables and in honey (**Figure 7.22**). This molecule is an important member of the ketose family of sugars. On a per-gram basis, fructose is twice as sweet as sucrose. It can therefore be used in smaller amounts. For this reason, fructose is often

FIGURE 7.21
α-D-Glucopyranose

Compare the information provided by these two representations. (a) The space-filling model, with carbon, oxygen, and hydrogen atoms in green, red, and white, respectively, and (b) the Haworth structure.

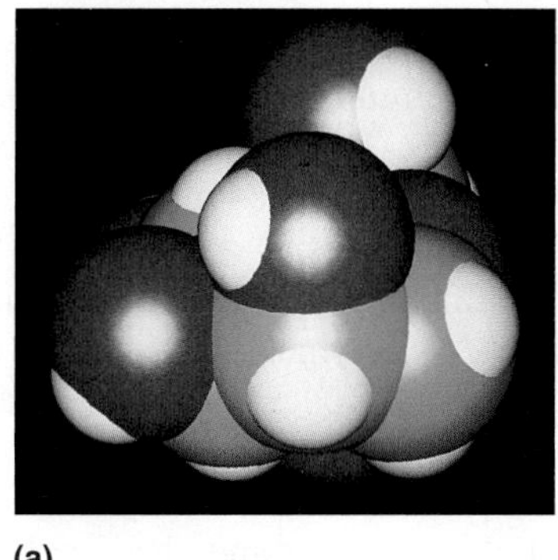

FIGURE 7.22
β-D-Fructofuranose

(a) Space-filling model and (b) Haworth structure.

FIGURE 7.23
α-D-Galactopyranose

(a) Space-filling model and (b) Haworth structure.

used as a sweetening agent in processed food products. Large amounts of fructose are used in the male reproductive tract. It is synthesized in the seminal vesicles and then incorporated into semen. Sperm use the sugar as an energy source.

GALACTOSE Galactose is necessary to synthesize a variety of biomolecules (**Figure 7.23**). These include lactose (in lactating mammary glands), glycolipids, certain phospholipids, proteoglycans, and glycoproteins. Synthesis of these substances is not diminished by diets that lack galactose or the disaccharide lactose (the principal dietary source of galactose) because the sugar is readily synthesized from glucose-1-phosphate.

In *galactosemia*, a genetic disorder, an enzyme required to metabolize galactose is missing. Galactose, galactose-1-phosphate, and galactitol (a sugar alcohol derivative) accumulate and cause liver damage, cataracts, and severe mental retardation. The only effective treatment is early diagnosis and a diet free of galactose.

KEY CONCEPT

Glucose, fructose, and galactose are among the most important monosaccharides in living organisms.

Monosaccharide Derivatives

Simple sugars may be converted to closely related chemical compounds. Several of these are important metabolic and structural components of living organisms.

URONIC ACIDS Recall that uronic acids are formed when the terminal CH_2OH group of a monosaccharide is oxidized. Two uronic acids are important in animals: D-glucuronic acid and its epimer, L-iduronic acid (α-D-glucuronate and β-L-iduronate in **Figure 7.24**). In liver cells, glucuronic acid is combined with molecules such as steroids, certain drugs, and bilirubin (a degradation product of heme in the oxygen-carrying protein hemoglobin) to improve water solubility. This process helps remove waste products from the body. Both D-glucuronic acid and L-iduronic acid are abundant in connective tissue carbohydrate components.

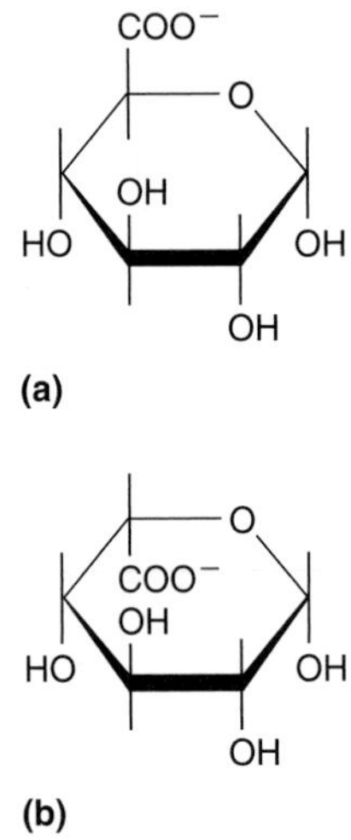

FIGURE 7.24
Uronic Acids
(a) α-D-Glucuronate and (b) β-L-iduronate.

AMINO SUGARS In amino sugars a hydroxyl group (most commonly on carbon 2) is replaced by an amino group (**Figure 7.25**). These compounds are common constituents of the complex carbohydrate molecules found attached to cellular proteins and lipids. The most common amino sugars of animal cells are D-glucosamine and D-galactosamine. Amino sugars are often acetylated. One such molecule is *N*-acetylglucosamine. *N*-Acetylneuraminic acid (the most common form of sialic acid) is a condensation product of D-mannosamine and pyruvic acid, a 2-ketocarboxylic acid. Sialic acids are ketoses containing nine carbon atoms that may be amidated with acetic or glycolic acid (hydroxyacetic acid). They are common components of glycoproteins and glycolipids.

DEOXY SUGARS Monosaccharides in which an —H has replaced an —OH group are known as *deoxy sugars*. Two important deoxy sugars found in cells are L-fucose (formed from D-mannose by reduction reactions) and 2-deoxy-D-ribose (**Figure 7.26**). Fucose is often found among the carbohydrate components of glycoproteins, such as those of the ABO blood group determinants on the surface

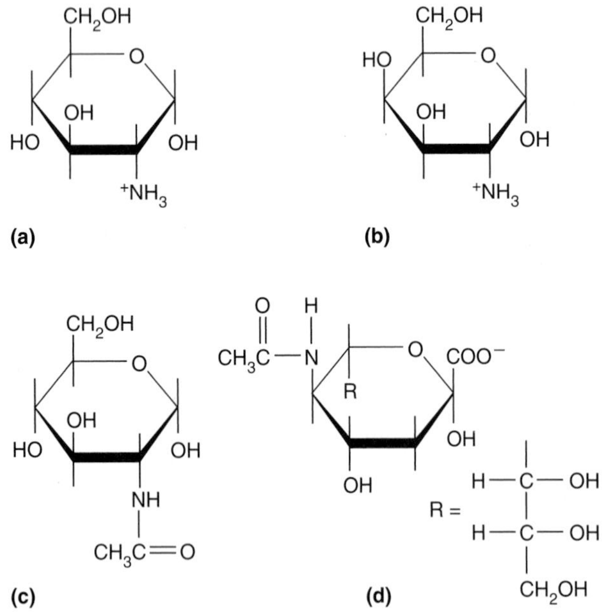

FIGURE 7.25
Amino Sugars
(a) α-D-Glucosamine, (b) α-D-galactosamine, (c) *N*-acetyl-α-D-glucosamine, and (d) *N*-acetylneuraminic acid (sialic acid).

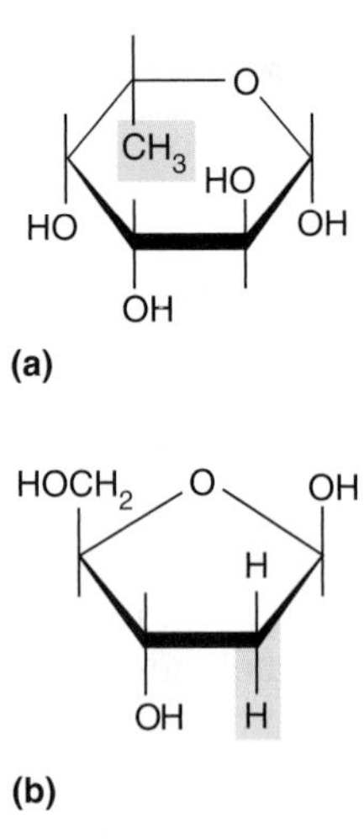

FIGURE 7.26
Deoxy Sugars
(a) β-L-Fucose (6-deoxygalactose) and (b) 2-deoxy-β-D-ribose. The carbon atoms that have —OH groups replaced by —H are highlighted.

of red blood cells. 2-Deoxyribose, the pentose sugar component of DNA, was shown earlier (**Figure 1.8**).

7.2 DISACCHARIDES

Disaccharides are molecules composed of two monosaccharides that are linked by a glycosidic bond. If one monosaccharide molecule is linked through its anomeric carbon atom to the hydroxyl group on carbon 4 of another monosaccharide, the glycosidic linkage is designated as 1,4. Because the anomeric hydroxyl group may be in either the α- or the β-configuration, two possible disaccharides may form when two sugar molecules are linked: α(1,4) or β(1,4). Other varieties of glycosidic linkages [i.e., α or β(1,1), (1,2), (1,3), and (1,6) linkages] also occur (**Figure 7.27**).

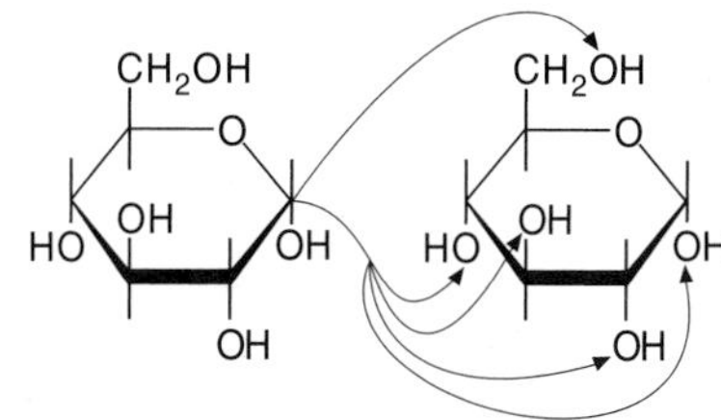

FIGURE 7.27
Glycosidic Bonds

Several types of glycosidic bonds can form between monosaccharides. The sugar α-D-glucopyranose (left) can theoretically form glycosidic linkages with any of the alcoholic functional groups of another monosaccharide, in this case another molecule of α-D-glucopyranose.

Digestion of disaccharides and other carbohydrates is mediated by enzymes synthesized by cells lining the small intestine. Deficiency of any of these enzymes causes unpleasant symptoms when the undigestible disaccharide sugar is ingested. Because carbohydrates are absorbed principally as monosaccharides, any undigested disaccharide molecules pass into the large intestine, where osmotic pressure draws water from the surrounding tissues (diarrhea). Bacteria in the colon digest the disaccharides (fermentation), thus producing gas (bloating and cramps). The most commonly known deficiency is *lactose intolerance*, which may occur in most human adults except those with ancestors from northern Europe and/or certain African groups. Caused by the greatly reduced synthesis of the enzyme lactase following childhood, lactose intolerance is treated by eliminating the sugar from the diet or (in some cases) by treating food with the enzyme lactase.

Lactose (milk sugar) is a disaccharide found in milk. It is composed of one molecule of galactose linked through the hydroxyl group on carbon 1 in a β-glycosidic linkage to the hydroxyl group of carbon 4 of a molecule of glucose (**Figure 7.28**). Because the anomeric carbon of galactose is in the β-configuration, the linkage between the two monosaccharides is designated as β(1,4). Because the glucose component contains a hemiacetal group, lactose is a reducing sugar.

Maltose, also known as malt sugar, is an intermediate product of starch hydrolysis and does not appear to exist freely in nature. Maltose is a disaccharide with an α(1,4) glycosidic linkage between two D-glucose molecules. In solution the free anomeric carbon undergoes mutarotation, which results in an equilibrium mixture of α- and β-maltoses (**Figure 7.29**).

Cellobiose, a degradation product of cellulose, contains two molecules of glucose linked by a β(1,4) glycosidic bond (**Figure 7.30**). Like maltose, whose

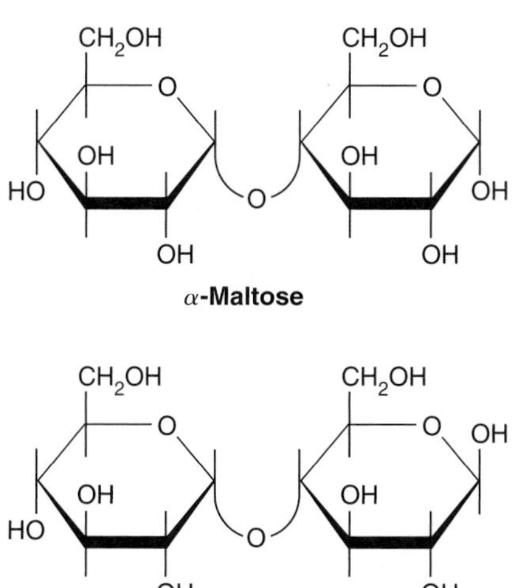

FIGURE 7.29
α- and β-Maltose

α-Lactose

β-Lactose

FIGURE 7.28
α- and β-Lactose

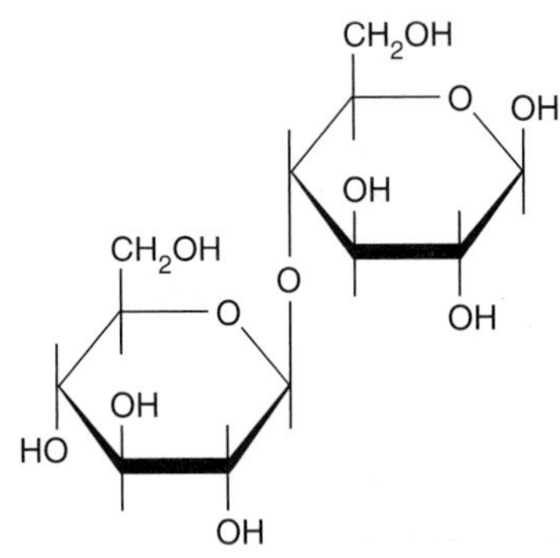

FIGURE 7.30
β-Cellobiose

KEY CONCEPTS

- Disaccharides are glycosides composed of two monosaccharide units.
- Maltose, lactose, cellobiose, and sucrose are disaccharides.

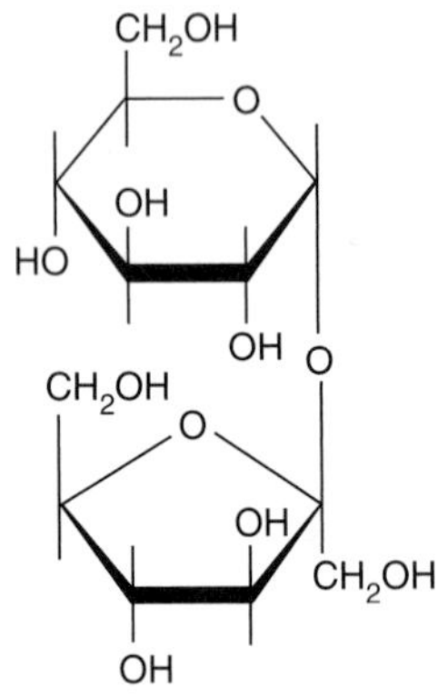

FIGURE 7.31
Sucrose

The glucose and fructose residues are linked by an α, β(1,2) glycosidic bond.

structure is identical except for the direction of the glycosidic bond, cellobiose does not occur freely in nature.

Sucrose (common table sugar: cane sugar or beet sugar) is produced in the leaves and stems of plants. It is a transportable energy source throughout the entire plant. Containing both α-glucose and β-fructose residues, sucrose differs from the previously described disaccharides in that the monosaccharides are linked through a glycosidic bond between both anomeric carbons (**Figure 7.31**). Because neither monosaccharide ring can revert to the open-chain form, sucrose is a nonreducing sugar.

QUESTION 7.5

Which of the following sugars or sugar derivatives are reducing sugars?
a. glucose
b. fructose
c. α-methyl-D-glucoside
d. sucrose
Which of these compounds are capable of mutarotation?

7.3 POLYSACCHARIDES

Polysaccharides, also referred to as **glycans**, are composed of large numbers of monosaccharide units connected by glycosidic linkages. Smaller glycans, called **oligosaccharides**, are polymers containing up to about 10 or 15 monomers, most often attached to polypeptides in glycoproteins (p. 260) and some glycolipids (p. 395). Among the best-characterized oligosaccharide groups are those attached to membrane and secretory proteins. There are two broad classes of oligosaccharides: N-linked and O-linked. The N-linked oligosaccharides are attached to polypeptides by an N-glycosidic bond with the side chain amide nitrogen of the amino acid asparagine. There are three major types of asparagine-linked oligosaccharide: high-mannose, hybrid, and complex (**Figure 7.32**). The O-linked oligosaccharides are attached to polypeptides by the side chain hydroxyl group of the amino acids serine or threonine in polypeptide chains or the hydroxyl group of membrane lipids. Larger glycans may contain from hundreds to thousands of sugar units. These molecules may have a linear structure or they may have branched shapes. Polysaccharides may be divided into two classes: **homoglycans**, which are composed of one type of monosaccharide, and **heteroglycans**, which contain two or more types of monosaccharides.

Homoglycans

The **homoglycans** found in abundance in nature are starch, glycogen, cellulose, and chitin. Starch, glycogen, and cellulose all yield D-glucose when they are hydrolyzed. Starch and glycogen are energy storage molecules in plants and animals, respectively. Cellulose is the primary structural component of plant cells. **Chitin**, the principal structural component of the exoskeletons of arthropods such as insects and crustaceans and the cell walls of many fungi, yields the glucose derivative *N*-acetylglucosamine when it is hydrolyzed.

Polysaccharides such as starch and glycogen, unlike proteins and nucleic acids, have no fixed molecular weights. The size of such molecules reflects the metabolic state of the cell producing them. For example, when blood sugar levels are high (e.g., after a meal), the liver synthesizes glycogen. Glycogen molecules in a well-fed animal may have molecular weights as high as 2×10^7 Da. When blood sugar levels fall, the liver enzymes begin breaking down the glycogen molecules, releasing glucose into the bloodstream. If the animal continues to fast, the process continues until glycogen reserves are almost used up.

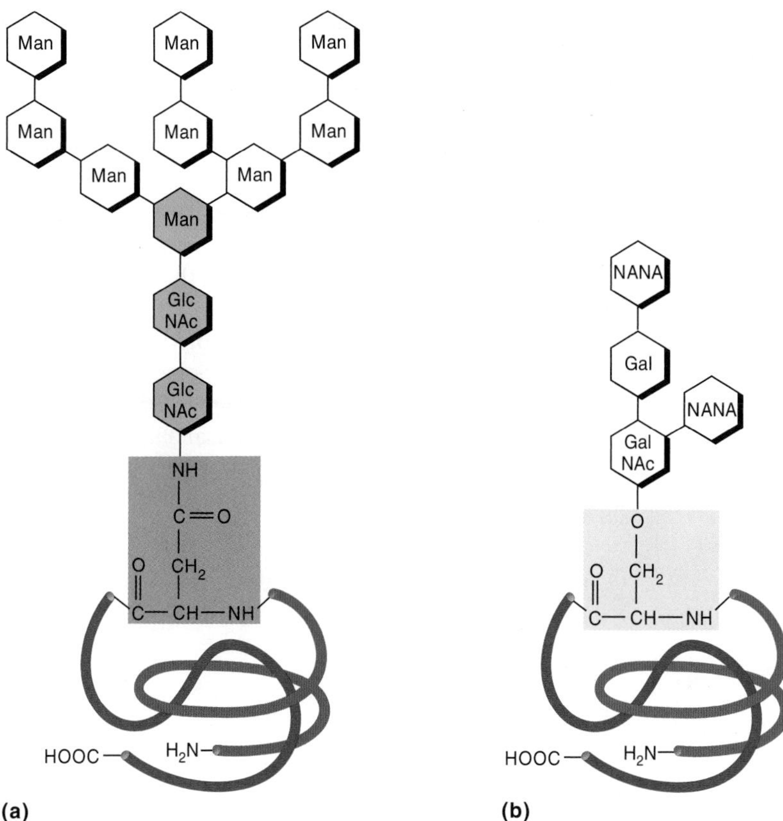

FIGURE 7.32

Oligosaccharides Linked to Polypeptides

Two classes of oligosaccharides are linked to proteins: (a) N-linked and (b) O-linked. High mannose oligosaccharides are an example of N-linked oligosaccharides, which are linked to polypeptides via an asparagine residue. Complex and hybrid N-linked oligosaccharides (not shown) contain other sugar residues [e.g., sialic acid (SA), and/or galactose (Gal)] in addition to mannose (Man) and *N*-acetylglucosamine (GlcNAc). O-linked oligosaccharides contain *N*-acetylneuraminic acid (NANA) in addition to mannose, galactose, and *N*-acetylglucosamine residues.

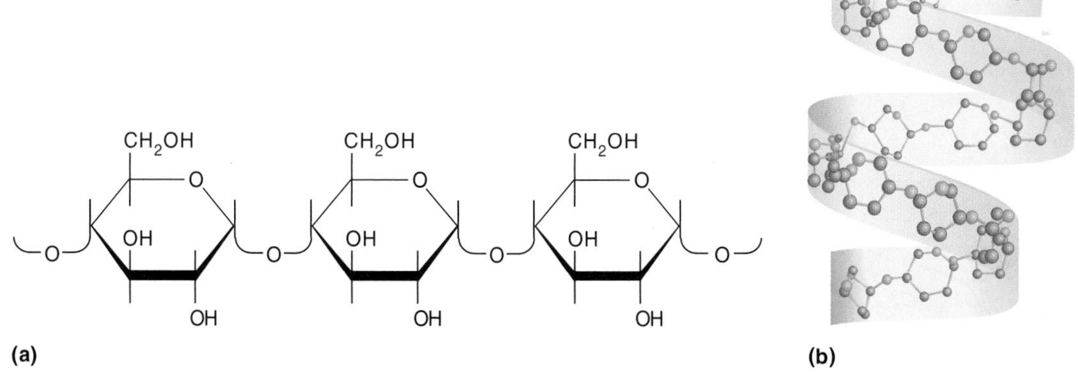

FIGURE 7.33

Amylose

(a) The D-glucose residues of amylose are linked through $\alpha(1,4)$ glycosidic bonds.
(b) The amylose polymer forms a left-handed helix.

STARCH Starch, the energy reservoir of plant cells, is a significant source of carbohydrate in the human diet. Much of the nutritional value of the world's major foodstuffs (e.g., potatoes, rice, corn, and wheat) comes from starch. Two polysaccharides occur together in starch: amylose and amylopectin.

Amylose is composed of long, unbranched chains of D-glucose residues that are linked with $\alpha(1,4)$ glycosidic bonds (**Figure 7.33**). A number of polysaccharides, including both types of starch, have one *reducing end* in which the ring can open to form a free aldehyde group with reducing properties. The internal anomeric carbons in these molecules are involved in acetal linkages and are not free to act as reducing agents.

Amylose molecules, which typically contain several thousand glucose residues, vary in molecular weight from 150,000 to 600,000 Da. Because the linear amylose molecule forms long, tight helices, its compact shape is ideal for its storage function. The common iodine test for starch works because molecular iodine inserts itself into these helices. (The intense blue color of a positive test comes from electronic interactions between iodine molecules and the helically arranged glucose residues of the amylose.)

The other form of starch, **amylopectin**, is a branched polymer containing both $\alpha(1,4)$ and $\alpha(1,6)$ glycosidic linkages. The $\alpha(1,6)$ branch points may occur every 20 to 25 glucose residues and prevent helix formation (**Figure 7.34a**). The number of glucose units in amylopectin may vary from a few thousand to a million.

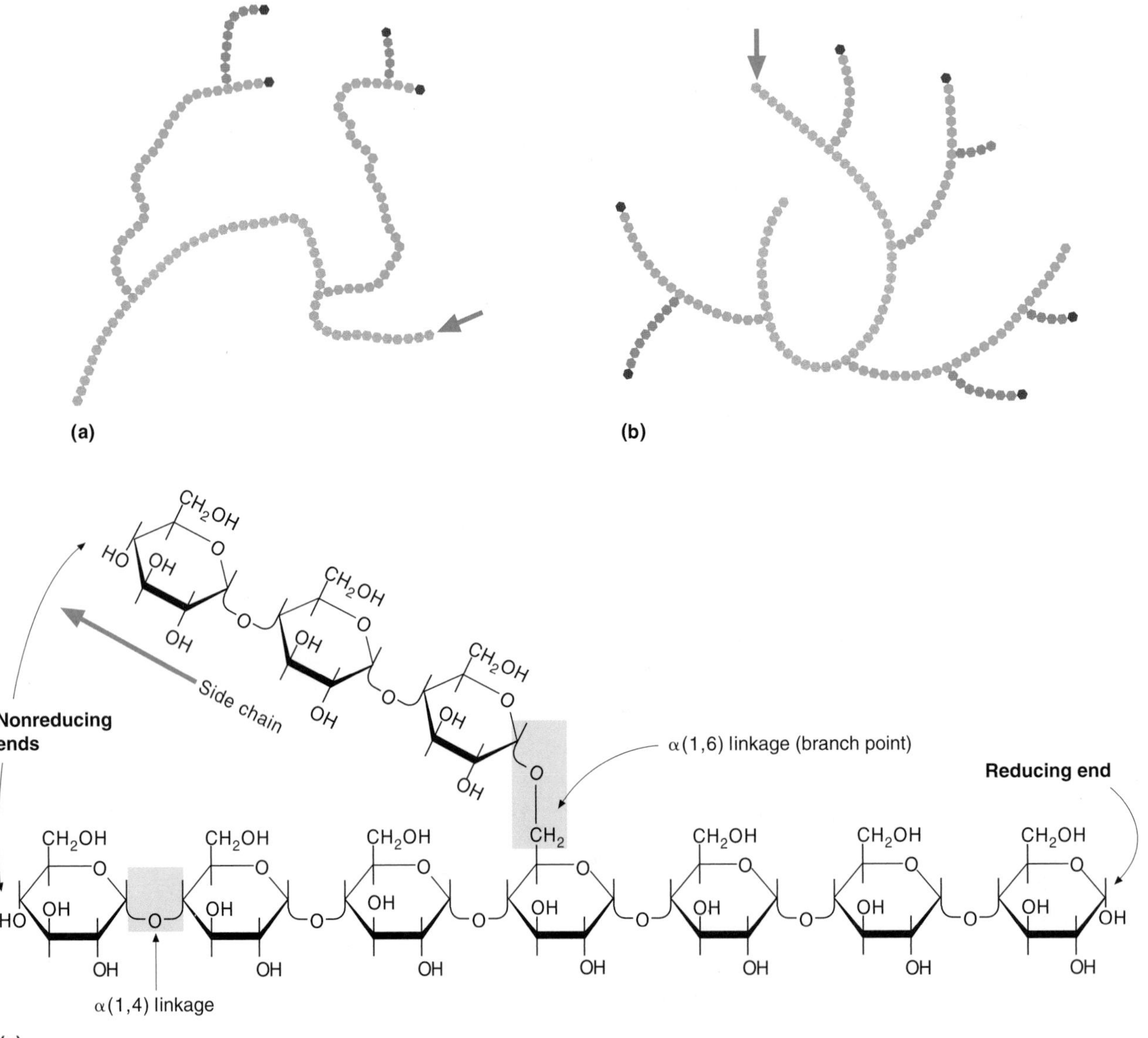

FIGURE 7.34

(a) Amylopectin and (b) Glycogen

Each hexagon represents a glucose molecule. Notice that each molecule has only one reducing end (arrow) and numerous nonreducing ends. (c) Detail from (a) or (b).

Starch digestion begins in the mouth, where the salivary enzyme α-amylase initiates hydrolysis of the glycosidic linkages. Digestion continues in the small intestine, where pancreatic α-amylase randomly hydrolyzes all the α(1,4) glycosidic bonds except those next to the branch points. The products of α-amylase are maltose, the trisaccharide maltotriose, and the α-limit dextrins [oligosaccharides that typically contain eight glucose units with one or more α(1,6) branch points]. Several enzymes secreted by cells that line the small intestine convert these intermediate products into glucose. Glucose molecules are then absorbed into enterocytes, the cells that line the small intestine. After passage into the bloodstream, they are transported to the liver and then to the rest of the body.

GLYCOGEN **Glycogen** is the carbohydrate storage molecule in vertebrates. It is found in greatest abundance in liver and muscle cells. (Glycogen may make up as much as 8–10% of the wet weight of liver cells and 2–3% of that of muscle cells.) Glycogen (**Figure 7.34b**) is similar in structure to amylopectin except that it has more branch points, possibly at every fourth glucose residue in the core of the molecule. In the outer regions of glycogen molecules, branch points are not so close together (approximately every 8–12 residues). Because the molecule is more compact than other polysaccharides, it takes up little space, an important consideration in mobile animal bodies. Because hydrolysis of glucose monomers occurs from the many nonreducing ends of the glycogen molecule, energy mobilization can be rapid.

QUESTION 7.6

It has been estimated that two high-energy phosphate bonds must be expended to incorporate one glucose molecule into glycogen. Why is glucose stored in muscle and liver in the form of glycogen, not as individual glucose molecules? In other words, why is it advantageous for a cell to expend metabolic energy to polymerize glucose molecules? [*Hint*: Besides the reasons given in Section 7.3, refer to Chapter 3 for another problem that glucose polymerization solves.]

CELLULOSE **Cellulose** is a polymer composed of D-glucopyranose residues linked by β(1,4) glycosidic bonds (**Figure 7.35**). It is the most important structural polysaccharide of plants. Because cellulose comprises about one-third of plant biomass, it is the most abundant organic substance on earth. Approximately 100 trillion kilograms of cellulose are produced each year.

Unbranched cellulose molecules, which may contain as many as 12,000 glucose units each, are held together by hydrogen bonding to form tough and inflexible sheetlike strips called *microfibrils* (**Figure 7.36**). With a tensile strength comparable to that of steel wire, cellulose microfibrils are components of both plant primary and secondary cell walls, where they provide a structural framework that both protects and supports cells.

The ability to digest cellulose is found only in microorganisms that possess the enzyme cellulase. Certain animal species (e.g., termites and cows) use such

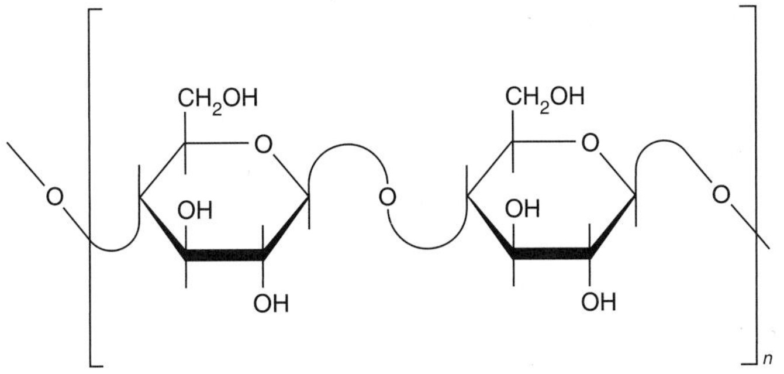

CH_2OH CH_2OH O O O O O OH OH OH OH n

FIGURE 7.35
The Disaccharide Repeating Unit of Cellulose

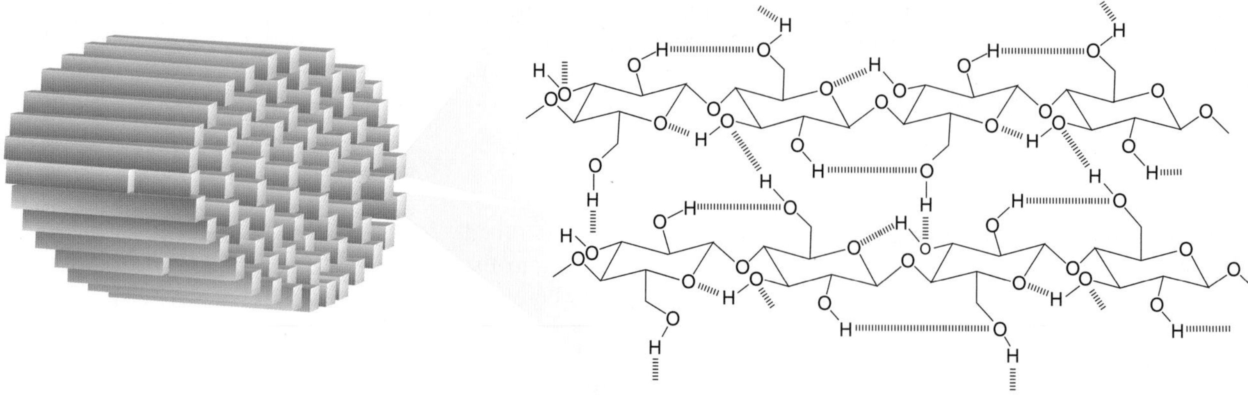

FIGURE 7.36
Cellulose Microfibrils
Intermolecular hydrogen bonds between adjacent cellulose molecules are largely responsible for the great strength of cellulose.

organisms in their digestive tracts to digest cellulose. The breakdown of the cellulose makes glucose available to both the microorganisms and their host. Although many animals cannot digest cellulose-containing plant materials, these substances play a vital role in nutrition. Cellulose is one of several plant products that make up the dietary fiber that is now believed to be important for good health.

Because of its structural properties, cellulose has enormous economic importance. Products such as wood, paper, and textiles (e.g., cotton, linen, and ramie) owe many of their unique characteristics to their cellulose content.

Heteroglycans

Heteroglycans are high-molecular-weight carbohydrate polymers that contain more than one kind of monosaccharide. The major classes of heteroglycans found in mammals are N- and O-linked heteropolysaccharides (N- and O-glycans) attached to proteins, the glycosaminoglycans of the extracellular matrix, and the glycan components of glycolipids and GPI (glycosylphosphatidylinositol) anchors. The structure and properties of the N- and O-glycans and the glycosaminoglycans are described next. Discussion of glycolipids and GPI anchors, a means for attaching peripheral proteins to membrane, is deferred to Chapter 11.

N-AND O-GLYCANS Many proteins have both N- and O-linked oligosaccharides that can comprise a significant proportion of the molecule's molecular weight. The N-linked oligosaccharides (**N-glycans**) are linked via a β-glycosidic bond between the core *N*-acetylglucosamine anomeric carbon and a side chain amide nitrogen of an asparagine residue. In addition to *N*-acetylglucosamine, the most commonly observed sugars among the N-glycans include mannose, galactose, *N*-acetylneuraminic acid, and glucose. Note that all N-glycans have the same core structure:

$^+NH_3$

Man—β 1,4—GlcNac—β 1,4—GlcNAc—Asn

C═O
|
O^-

N-Glycan Core Structure

The O-linked oligosaccharides **(O-glycans)** have a disaccharide core of galactosyl-β-(1,3)-*N*-acetylgalactosamine linked to the protein via an α-glycosidic bond to the hydroxyl oxygen of serine or threonine residues. In the collagens, the core β-linked disaccharide may be Gal-Gal or Glc-Gal and is linked to the side chain hydroxyl oxygen of 5-hydroxylysine. Other sugars found in O-glycans are *N*-acetylneuraminic acid and sialic acid.

GLYCOSAMINOGLYCANS **Glycosaminoglycans** (GAGs) are linear polymers with disaccharide repeating units. Many of the sugar residues are amino derivatives. There are five GAG classes: hyaluronic acid, chondroitin sulfate, dermatan sulfate, heparin and heparan sulfate, and keratan sulfate. The repeating units contain a hexuronic acid (a uronic acid containing six carbon atoms), except for keratan sulfate, which contains galactose. Usually an *N*-acetylhexosamine sulfate is also present, except in hyaluronic acid, which contains *N*-acetylglucosamine. Many disaccharide units contain both carboxyl and sulfate groups. GAGs are classified according to their sugar residues, the linkages between these residues, and the presence and location of sulfate groups.

GAGs have many negative charges at physiological pH. The charge repulsion keeps GAGs separated from each other. Additionally, the relatively inflexible polysaccharide chains are strongly hydrophilic. GAGs occupy a huge volume relative to their mass because they attract large volumes of water. For example, hydrated hyaluronic acid may occupy a volume 1000 times greater than its dry state.

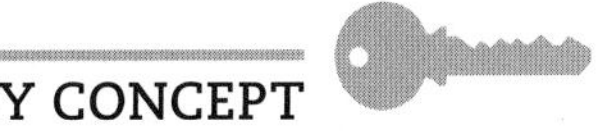

KEY CONCEPT

Polysaccharide molecules, composed of large numbers of monosaccharide units, are used in energy storage and as structural materials.

7.4 GLYCOCONJUGATES

The compounds that result from the covalent linkages of carbohydrate molecules to both proteins and lipids are collectively known as the **glycoconjugates**. These substances have profound effects on the function of individual cells, as well as the cell-cell interactions of multicellular organisms. There are two classes of carbohydrate-protein conjugate: proteoglycans and glycoproteins. Although both molecular types contain carbohydrate and protein, their structures and functions appear, in general, to be substantially different. The *glycolipids* (Chapter 11), which are oligosaccharide-containing lipid molecules, are found predominantly on the outer surface of plasma membranes.

Proteoglycans

Proteoglycans are distinguished from the more common glycoproteins by their extremely high carbohydrate content, which may constitute as much as 95% of the dry weight of such molecules. These molecules occur on cell surfaces or are secreted into the extracellular matrix. All proteoglycans contain GAG chains that are linked to protein molecules (known as *core proteins*) by N- and O-glycosidic linkages. Proteoglycans are produced in the Golgi apparatus where GAG chains are synthesized and then covalently linked to a core protein previously synthesized in the RER. The diversity of proteoglycans results from both the number of different core proteins and the large variety of classes and lengths of the carbohydrate chains. Examples include the syndecans, glypicans, and aggrecan. The syndecans are a class of heparan sulfate and chondroitin sulfate containing proteoglycans in which the core protein is a transmembrane protein. The glypicans are proteoglycans that contain heparan sulfate and are linked to the cell membrane by GPI anchors (p. 394). Aggrecan, a proteoglycan found in abundance in cartilage, consists of a core protein to which are attached more than 100 chondroitin sulfate and about 40 keratan sulfate chains. Up to 100 aggrecan monomers are in turn attached to hyaluronic acid to form a proteoglycan aggregate (**Figure 7.37**).

In addition to their roles in organizing extracellular matrices, proteoglycans participate in all cellular processes that involve events at cell surfaces. For

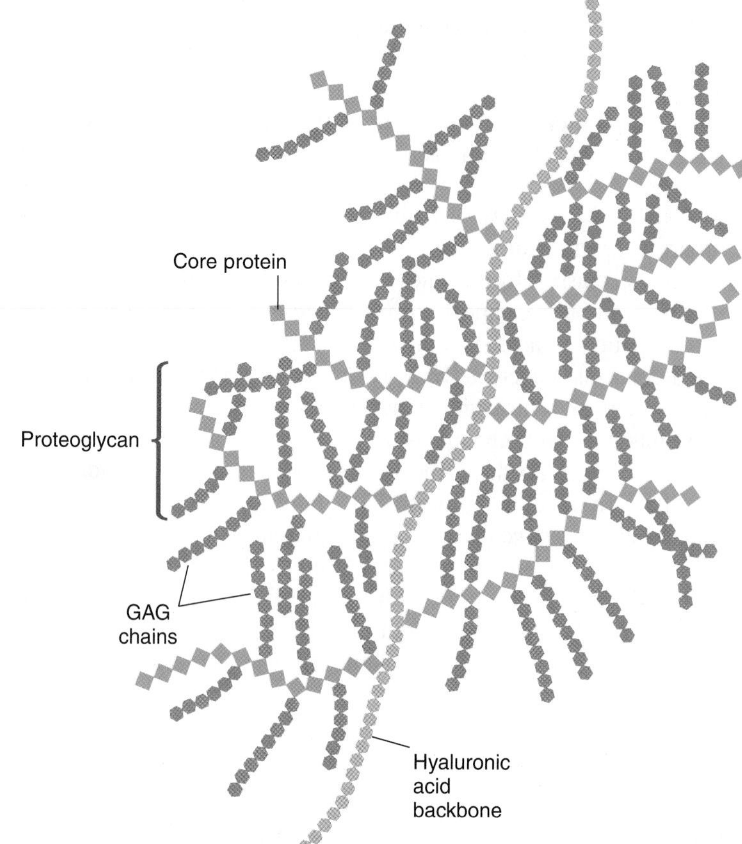

FIGURE 7.37

Proteoglycan Aggregate

Proteoglycan aggregates are typically found in the extracellular matrix of connective tissue. The noncovalent attachment of each aggrecan monomer to hyaluronic acid via the core protein is mediated by two linker proteins (not shown). Proteoglycans interact with numerous fibrous proteins in the extracellular matrix such as collagen, elastin, and fibronectin (a glycoprotein involved in cell adhesion).

example, the membrane-bound syndecans and glypicans that bind to specific signal molecules (e.g., growth factors) are components in several signal transduction pathways that regulate the cell cycle. Because of their vast numbers of polyionic GAG chains, the aggrecans trap large volumes of water. Consequently, these molecules occupy thousands of times as much space as a densely packed molecule of the same mass. The strength, flexibility, and resilience of cartilage are made possible by the combination of the compressive stiffness contributed by repulsion between the negatively charged GAGs and the tensile strength of collagen fibers.

A number of genetic diseases associated with proteoglycan metabolism, known as *mucopolysaccharidoses*, have been identified. Because proteoglycans are constantly being synthesized and degraded, their excessive accumulation (caused by missing or defective lysosomal enzymes) has serious consequences. For example, in *Hurler's syndrome*, deficiency of a specific enzyme causes dermatan sulfate to accumulate. Symptoms include mental retardation, skeletal deformity, and death in early childhood. Like Tay-Sachs disease, Hurler's syndrome is an autosomal recessive disorder (i.e., one copy of the defective gene is inherited from each parent).

Glycoproteins

Glycoproteins are commonly defined as proteins that are covalently linked to carbohydrate through N- or O-linkages. The carbohydrate composition of glycoprotein varies from 1% to more than 85% of total weight. The carbohydrates found include monosaccharides and disaccharides such as those attached to the structural protein collagen and the branched oligosaccharides on plasma glycoproteins. Although the glycoproteins are sometimes considered to include the proteoglycans, for structural reasons they are examined separately. There is a relative absence in glycoproteins of uronic acids, sulfate groups, and the disaccharide repeating units that are typical of proteoglycans.

As described previously, the carbohydrate groups of glycoproteins are linked to the polypeptide by either an N-glycosidic linkage between *N*-acetylglucosamine (GlcNAc) and the amino acid asparagine (Asn) or an O-glycosidic linkage between *N*-acetylgalactosamine (GalNAc) and the hydroxyl group of serine (Ser) or threonine (Thr). The former glycoprotein class is sometimes referred to as *asparagine-linked*; the latter is often called *mucin-type*. The asparagine-linked oligosaccharides are constructed on a membrane-bound lipid molecule and covalently linked to asparagine residues during ongoing protein synthesis (Chapter 19). Several additional reactions, in the lumen of the endoplasmic reticulum and the Golgi complex, form the final N-linked oligosaccharide structures. Examples of proteins with asparagine-linked oligosaccharides include the iron transport protein transferrin and ovalbumin, a nutritional storage protein in chicken eggs. Mucin-type carbohydrate (O-glycan) units vary considerably in size and structure, from disaccharides such as Gal-1,3-GalNAc, found in the antifreeze glycoprotein of antarctic fish, to the complex oligosaccharides of blood groups such as those of the ABO system.

GLYCOPROTEIN FUNCTIONS Glycoproteins are a diverse group of molecules that are ubiquitous constituents of most living organisms (**Table 7.1**). They occur in cells, in both soluble and membrane-bound form, and in extracellular fluids. Vertebrate animals are particularly rich in glycoproteins. Examples include the metal-transport proteins transferrin and ceruloplasmin, the blood-clotting factors, and many of the components of complement (proteins involved in cell destruction during immune reactions). A number of hormones are glycoproteins. Consider, for example, follicle-stimulating hormone, produced by the anterior pituitary gland. Follicle-stimulating hormone stimulates the development of both eggs and sperm. Additionally, many enzymes are glycoproteins. Ribonuclease (RNase), the enzyme that degrades ribonucleic acid, is a well-researched example. Other glycoproteins are integral membrane proteins (Chapter 11). Of these, Na^+-K^+-ATPase (an ion pump found in the plasma membrane of animal cells) and the major histocompatibility antigens (cell surface markers used to cross-match organ donors and recipients) are especially interesting examples.

Recent research has focused on how carbohydrate stabilizes protein molecules and functions in recognition processes in multicellular organisms. The presence of carbohydrate on protein molecules protects them from denaturation. For example, bovine RNase A is more susceptible to heat denaturation than its glycosylated counterpart RNase B. Several other studies have shown that sugar-rich glycoproteins are relatively resistant to proteolysis (enzyme-catalyzed hydrolysis of polypeptides). Because the carbohydrate is on the molecule's surface, it may shield the polypeptide chain from proteolytic enzymes.

TABLE 7.1 Glycoproteins

Type	Example	Source	Molecular Mass (Da)
Enzyme	Ribonuclease B	Bovine	14,700
Immunoglobulin	Immunoglobulin A	Human	160,000
	Immunoglobulin M	Human	950,000
Hormone	Chorionic gonadotropin	Human placenta	38,000
	Follicle-stimulating hormone	Human	34,000
Membrane protein	Glycophorin	Human red blood cells	31,000
Lectin	Potato lectin	Potato	50,000
(carbohydrate-	Soybean agglutinin	Soybean	120,000
binding proteins)	Ricinus lectin	Castor bean	120,000

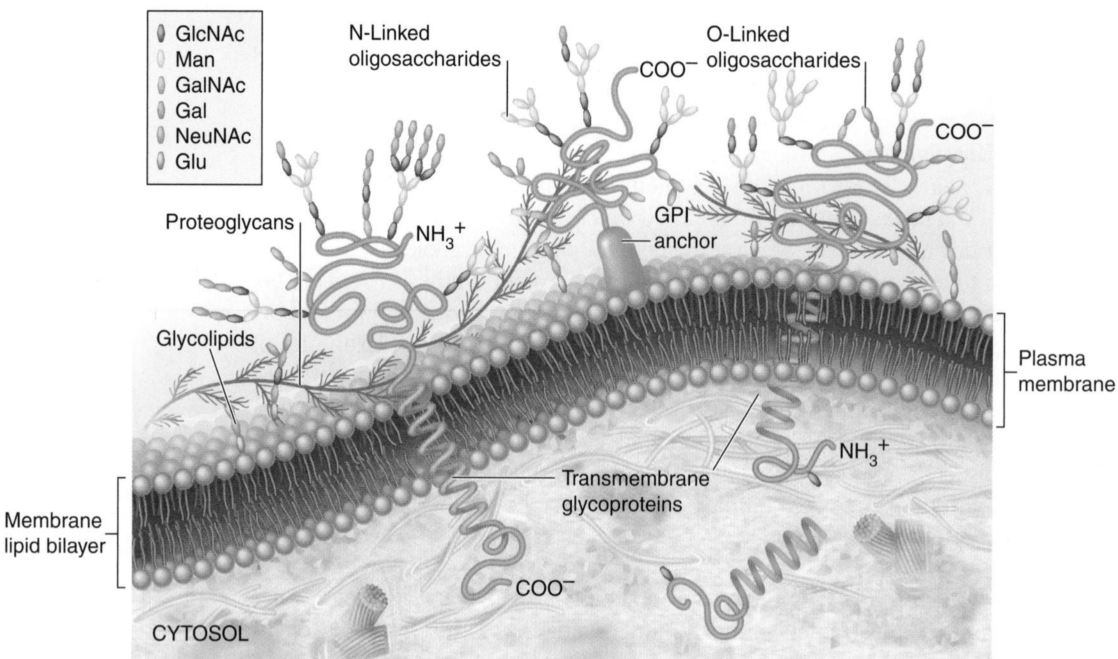

FIGURE 7.38
The Glycocalyx

The glycocalyx is made up of the carbohydrate groups attached to the glycoprotein, proteoglycan, and glycolipid components on the external surface of eukaryotic cells. A glycosylphosphatidylinositol (GPI) anchor is a specialized structure that attaches several diverse types of oligosaccharide to the plasma membrane of some cells.

The carbohydrates in glycoproteins also affect biological function. In some glycoproteins this contribution is more easily discerned than in others. For example, a large content of sialic acid residues is responsible for the high viscosity of salivary mucins (the lubricating glycoproteins of saliva). The disaccharide residues of the antifreeze glycoproteins of antarctic fish form hydrogen bonds with water molecules. This process retards the growth of ice crystals.

Glycoproteins, as components of the glycocalyx (p. 46), are now known to be important in complex recognition phenomena. A prime example is the insulin receptor, whose binding to insulin facilitates the transport of glucose into numerous cell types. It does so, in part, by recruiting glucose transporters to the plasma membrane. A variety of cell surface glycoproteins are also involved in cellular adhesion, a critical event in the cell-cell interactions of growth and differentiation (**Figure 7.38**). The best-characterized of these substances are called cell adhesion molecules (CAMs). Examples include the *selectins* (transient cell-cell interactions), the *integrins* (cell attachment to the components of the extracellular matrix), and the *cadherins* (Ca^{2+}-dependent binding of cells to each other within a tissue). The roles of glycoconjugates in living processes are explored further in the next section.

KEY CONCEPTS

- Glycoconjugates are biomolecules in which carbohydrate is covalently linked to either proteins or lipids.
- Proteoglycans are composed of relatively large amounts of carbohydrate (GAG units) covalently linked to small polypeptide components.
- Glycoproteins are proteins covalently linked to carbohydrate through N- or O-linkages.

7.5 THE SUGAR CODE

Living organisms require extraordinarily large coding capacities. Each information transfer event, whether it is the binding of a substrate within an enzyme's active site, the transduction of a hormonal signal, or the engulfment of a bacterial cell by a macrophage, is initiated by the specific binding of one unique molecule by another that has been selected from millions of other nearby molecules. In other words, the functioning of systems as profoundly complicated as living organisms requires a correspondingly large repertoire of molecular codes. To succeed as a coding mechanism, a class of molecules must provide a large capacity for variations in shape because the number of different messages that must be quickly and unambiguously deciphered is tremendous.

For more than 60 years research efforts to understand information flow in biosystems focused primarily on the nucleic acids DNA and RNA. As a result of this monumental work, life scientists fully expected to find that approximately 100,000 genes existed to code for proteins in humans. Instead, analysis of the data generated by the Human Genome Project (pp. 694–695) produced a much lower number of about 21,000 genes.

Whatever the reasons for this surprisingly low number, it is known that living organisms have two strategies for expanding the coding capacity of their genes: alternative splicing and covalent modification. *Alternative splicing* (described in Chapter 18) is a mechanism whereby eukaryotes produce several polypeptides from the same gene by cutting RNA transcripts and then splicing together various combinations of the RNA fragments. Each type of spliced mRNA product is translated into a unique polypeptide. *Posttranslational modifications* (described in Chapter 19) are enzyme-catalyzed changes in a protein's structure that occur after its synthesis.

Of all the types of posttranslational modification (e.g., phosphorylation, acetylation, and proteolytic cleavage), glycosylation is the most important in terms of coding capacity, as the following examples illustrate. Recall that just 20 amino acids account for the enormous diversity of proteins observed in living organisms. The total number of hexapeptides that can be synthesized from these amino acids is an impressive 20^6 (6.4×10^7). Carbohydrates have structural properties (e.g., glycosidic linkage variations, branching, and anomeric isomers) that provide them with significant coding capacity. In contrast to the peptide linkages that form exclusively between the amino and carboxyl groups in amino acid residues to create a linear peptide molecule, the glycosidic linkages between monosaccharides can be considerably more variable. Consequently, the potential number of permutations in oligosaccharides is substantially higher than that predicted for peptides. For example, the total number of possible linear and branched hexasaccharides that can form from 20 simple or modified monosaccharides is 1.44×10^{15}. In addition to their immense combinatorial possibilities, oligosaccharides, whether they are attached to proteins or lipids, have one other property: their relative inflexibility (in comparison to peptides), which allows them to bind more precisely with ligands.

Lectins: Translators of the Sugar Code

Once information has been encoded, it must be translated. The translation of the sugar code is accomplished by lectins. **Lectins** are carbohydrate-binding proteins (CBPs) that are not antibodies and have no enzymatic activity. Originally discovered in plants, they are now known to exist in all organisms. Lectins, which usually consist of two or four subunits, possess recognition domains that bind to specific carbohydrate groups via hydrogen bonds, van der Waals forces, and hydrophobic interactions. Biological processes that involve lectin binding include an array of cell-cell interactions (**Figure 7.39**). Prominent examples include infections by microorganisms, the mechanisms of many toxins, and physiological processes such as leukocyte rolling.

Infection by many bacteria is initiated when the microorganisms become firmly attached to host cells. Often, attachment is mediated by the binding of bacterial lectins to oligosaccharides on the cell's surface. *Helicobacter pylori*, the causative agent of gastritis and stomach ulcers, possesses several lectins that allow it to establish a chronic infection in the mucous lining of the stomach. One of these lectins binds with high affinity to a portion of the type O blood group determinant, an oligosaccharide. This circumstance explains the observation that humans with type O blood are at considerably greater risk of developing ulcers

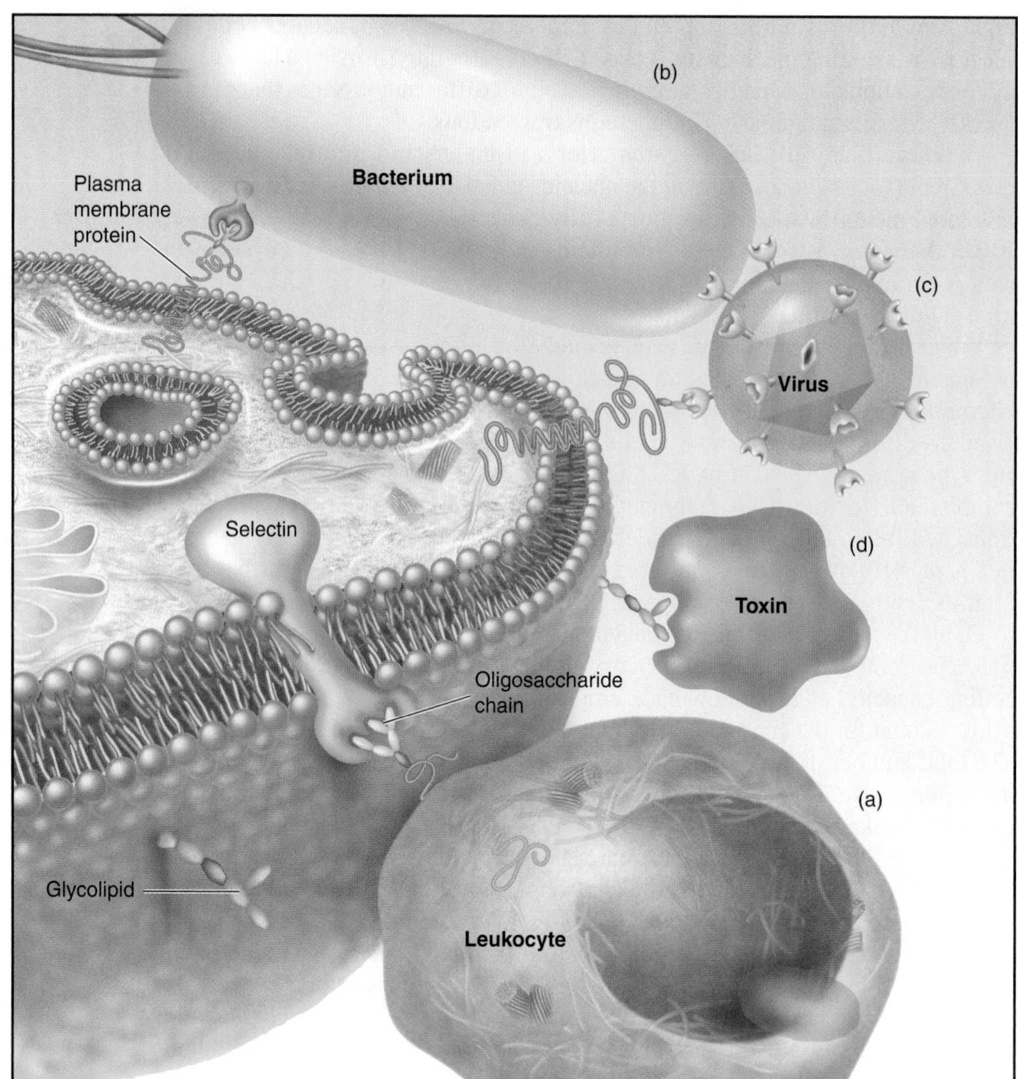

FIGURE 7.39

Role of Oligosaccharides in Biological Recognition

The specific binding of lectins (carbohydrate-binding proteins) to the oligosaccharide groups of glycoconjugate molecules is an essential feature of many biological phenomena. (a) Cell-cell interactions (e.g., leukocyte rolling), (b and c) cell-pathogen infections, and (d) the binding of toxins (e.g., cholera toxin) to cells.

than those with other blood types. Individuals with type A or B blood are not immune to infection, however, since the bacterium can also use other lectins to achieve adhesion.

The damaging effects of many bacterial toxins occur only after endocytosis into the host cell, a process that is initiated by lectin-ligand binding. The binding of the B subunit of cholera toxin to a glycolipid on the surface of intestinal cells results in the uptake of the toxic A subunit. Once internalized, the A subunit proceeds to disrupt the mechanism that regulates chloride transport, a process that results in a life-threatening diarrhea.

Leukocyte rolling is a well-known example of cell-cell interaction mediated by lectin binding. When a tissue becomes damaged in an animal either by infection with a pathogenic organism or by physical trauma, it emits signal molecules that trigger an inflammatory process. In response to certain of these molecules, the endothelial cells that line nearby blood vessels produce and insert into their plasma membranes a protein called selectin. The selectins, a family of lectins that act as cell adhesion molecules, are displayed on the surface of the endothelial cells. They bind transiently to the *selectin* ligand (an oligosaccharide) on white blood cells such as neutrophils. These relatively weak binding events serve to slow the rapid motion of the neutrophils as they flow in blood, causing them to appear to be rolling along the luminal surface of the blood vessel. Once rolling has been initiated and white blood cells approach the inflammation site, they encounter other signal molecules that cause them to express another lectin called *integrin* on their surfaces. The binding of integrin with its oligosaccharide ligand on the endothelial surface of the blood vessel causes the neutrophils to stop rolling. Subsequently, the neutrophils undergo changes that allow them to squeeze between the cells of the endothelium and migrate to the infected site, where they proceed to consume and degrade bacteria or cellular debris.

Visit the companion website at www.oup.com/us/mckee to read the Biochemistry in Perspective essay on sweet medicine.

The Glycome

The term **glycome**, derived from *glyco* (sweet) and *-ome* (as in genome), was created to describe the total set of sugars and glycans that a cell or organism produces. Glycomes are constantly in flux because cells responding to environmental signals fine-tune biological responses by altering the glycan structures attached to proteins and lipids. This capacity exists in large part because there is no template for glycan biosynthesis. In contrast to nucleic acid and protein biosynthesis, which are template-driven processes (i.e., multiple identical copies are produced using a nucleotide base sequence), glycans are constructed stepwise, on an assembly line within the ER and Golgi complex. Because of factors such as variations in sugar nucleotide precursor concentrations and the localization of glycan-processing enzymes, the glycan components of each type of glycoprotein may be produced in a series of slightly different forms called **glycoforms**. It has been suggested that this phenomenon, referred to as **microheterogeneity**, may be a means by which cells can generate cell- or tissue-specific signal transduction ligands and/or a mechanism whereby cells elude pathogens whose binding to certain glycan structures initiates an infective process.

KEY CONCEPTS

- The covalent modification of biomolecules such as proteins and lipids provides living organisms with enormous coding capacity.
- The entire set of sugars and glycans produced by a cell or organism is referred to as the glycome.
- Glycoproteins are often produced in slightly different versions called glycoforms.

QUESTION 7.7

The sugar code, with its diverse and subtle nontemplate mechanism for encoding information, has been described as an "analog" system, whereas genetic information processing (DNA- and RNA-directed protein synthesis) is considered "digital." Explain.

Chapter Summary

1. Carbohydrates, the most abundant organic molecules in nature, are classified as monosaccharides, disaccharides, oligosaccharides, and polysaccharides according to the number of simple sugar units they contain. Carbohydrate moieties also occur as components of other biomolecules. Glycoconjugates are protein and lipid molecules with covalently linked carbohydrate groups. They include proteoglycans, glycoproteins, and glycolipids.
2. Monosaccharides with an aldehyde functional group are called aldoses; those with a ketone group are known as ketoses. Aldoses belong to either the D or the L family, according to whether the configuration of the asymmetric carbon farthest from the aldehyde group resembles the D- or L-isomer of glyceraldehyde. The D family of aldoses contains most biologically important sugars.
3. Monosaccharides containing five or six carbons exist in cyclic forms that result from reactions between hydroxyl groups and either aldehyde (hemiacetal product) or ketone groups (hemiketal product). In both five-membered rings (furanoses) and six-membered rings (pyranoses), the hydroxyl group attached to the anomeric carbon lies either below (α) or above (β) the plane of the ring for D-sugars. The spontaneous interconversion between α- and β-forms is called mutarotation.
4. Simple sugars undergo a variety of chemical reactions. Derivatives of these molecules, such as uronic acids, amino sugars, deoxy sugars, and phosphorylated sugars, have important roles in cellular metabolism. Glycosylation reactions attach sugars to proteins or lipids. Glycation reactions are nonenzymatic reactions in which reducing sugars react with nucleophilic nitrogens.
5. Hemiacetals and hemiketals react with alcohols to form acetals and ketals, respectively. When the cyclic hemiacetal or hemiketal form of a monosaccharide reacts with an alcohol, the new linkage is called a glycosidic linkage. Molecules in which a sugar is linked via a glycosidic linkage to a noncarbohydrate moiety (aglycone) are called glycosides.
6. Glycosidic bonds form between the anomeric carbon of one monosaccharide and one of the free hydroxyl groups of another monosaccharide. Disaccharides are carbohydrates composed of two monosaccharides. Oligosaccharides, carbohydrates that typically contain as many as 10 to 15 monosaccharide units, are often attached to proteins and lipids. Polysaccharide molecules, which are composed of large numbers of monosaccharide units, may have a linear structure like cellulose and amylose or a branched structure like glycogen and amylopectin. Oligosaccharides and polysaccharides are now referred to as glycans. Glycans may consist of only one sugar type (homoglycans) or multiple types (heteroglycans).
7. The three most common homoglycans found in nature (starch, glycogen, and cellulose) all yield D-glucose when hydrolyzed. Cellulose is a plant structural material; starch and glycogen are storage forms of glucose in plant and animal cells, respectively. Chitin, the principal structural material in insect exoskeletons, is composed of *N*-acetylglucosamine residues linked in unbranched chains. The major classes of heteroglycans, carbohydrate polymers that contain more than one kind of monosaccharide, are N- and O-glycans, glycosaminoglycans, and the glycan components of glycolipids and GPI anchors.
8. The enormous heterogeneity of proteoglycans, which are found predominantly in the extracellular matrix of animal tissues, allows them to play diverse, but as yet poorly understood, roles in living organisms. Glycoproteins occur in cells, in both soluble and membrane-bound forms, and in extracellular fluids. The diverse structures of the glycoconjugates, which include proteoglycans, glycoproteins, and glycolipids, allow them to play important roles in information transfer in living organisms. The glycome is the total set of sugars and glycans that a cell or organism produces.

Take your learning further by visiting the **companion website** for Biochemistry at **www.oup.com/us/mckee** where you can complete a multiple-choice quiz on carbohydrates to help you prepare for exams.

Suggested Readings

Andrews, Z. B., and Horvath, Why Calories Taste Delicious: Eating and the Brain, *Scientific American.com/Mind Matters*, Sept. 2008.

Freeze, H. H, and Ng, B. G., Golgi Glycosylation and Human Inherited Diseases, *Cold Spring Harbor Perspect. Biol.* 3(9):a005371, 2011.

Gabius, H.-J., Biological Information Transfer beyond the Genetic Code: The Sugar Code, *Naturwissenschaften* 87(3):108–121, 2000.

Maeder, T., Sweet Medicines, *Sci. Am.* 287(1):40–47, 2002.

Meyers, B., and Brewer, M. S., Sweet Taste in Man: A Review, *J. Food Sci.* 73(6):R81–R90, 2008.

Reed, D. R., Tanaka, T., and McDaniel, A. H., Diverse Tastes: Genetics of Sweet and Bitter Taste Perception, *Physiol. Behav.* 88:215–226, 2006.

Rillahan, C. D., and Paulson, J. C., Glycan Microarrays for Decoding the Glycome, *Ann. Rev. Biochem.* 89:797–823, 2011.

Sasisekharan, R., and Myette, J. R., The Sweet Science of Glycobiology, *Am. Sci.* 91:432–441, 2003.

Seeberger, P. H., Exploring Life's Sweet Spot, *Nature* 437:1239, 2005.

Key Words

acetal, *248*
adduct, *249*
aldaric acid, *245*
alditol, *246*
aldonic acid, *245*
aldose, *240*
amylopectin, *256*
amylose, *255*
anomer, *242*
cellobiose, *253*
cellulose, *257*
chitin, *254*
diastereomer, *241*
disaccharide, *248*
enediol, *241*
epimer, *241*
epimerization, *247*
glycan, *254*
glycoconjugate, *259*
glycoform, *265*
glycogen, *257*
glycome, *265*
glycosaminoglycan, *259*
glycoside, *248*
glycosidic linkage, *248*
hemiacetal, *241*
hemiketal, *241*
heteroglycan, *258*
homoglycan, *254*
ketal, *248*
lactone, *245*
lactose, *253*
lectin, *263*
maltose, *253*
microheterogeneity, *265*
monosaccharide, *252*
mutarotation, *244*
N-glycan, *258*
O-glycan, *259*
oligosaccharide, *254*
polysaccharide, *248*
proteoglycan, *259*
reducing sugar, *246*
sucrose, *254*
uronic acid, *245*

Review Questions

These questions are designed to test your knowledge of the key concepts discussed in this chapter before moving on to the next chapter. You may like to compare your answers to the solutions provided in the back of the book and in the accompanying Study Guide.

1. Define the following terms:
   a. monosaccharide
   b. aldose
   c. ketose
   d. diastereomer
   e. epimer
2. Define the following terms:
   a. hemiacetal
   b. hemiketal
   c. anomer
   d. furan
   e. pyran
3. Define the following terms:
   a. aldonic acid
   b. uronic acid
   c. aldaric acid
   d. lactone
   e. reducing sugar
4. Define the following terms:
   a. alditol
   b. enediol
   c. epimerization
   d. acetal
   e. ketal
5. Define the following terms:
   a. glycosidic linkage
   b. glycoside
   c. disaccharide
   d. oligosaccharide
   e. polysaccharide
6. Define the following terms:
   a. Maillard reaction
   b. Schiff base
   c. Amadori product
   d. adduct
   e. reactive carbonyl-containing product
7. Define the following terms:
   a. adduct
   b. galactosemia
   c. cellobiose
   d. glycan
   e. chitin
8. Define the following terms:
   a. homoglycan
   b. heteroglycan
   c. amylose
   d. amylopectin
   e. enterocyte
9. Define the following terms:
   a. glycogen
   b. cellulose
   c. *N*-glycan
   d. *O*-glycan
   e. glycosaminoglycan
10. Define the following terms:
    a. glycoconjugate
    b. glycolipid
    c. proteoglycan
    d. glycoprotein
    e. sugar code
11. Define the following terms:
    a. lectin
    b. glycoform
    c. microheterogeneity
    d. glycome
    e. glycomics
12. Give an example of each of the following:
    a. epimer
    b. glycosidic linkage
    c. reducing sugar
    d. monosaccharide
    e. anomer
    f. diastereomer

13. What structural relationship is indicated by the term D-sugar? Why are (+) glucose (shifts polarized light to the right) and (−) fructose (shifts polarized light to the left) both classified as D-sugars?
14. Name an example of each of the following classes of compounds:
    a. glycoprotein
    b. proteoglycan
    c. disaccharide
    d. glycosaminoglycan (GAG)
15. What is the difference between a heteroglycan and a homoglycan? Give examples.
16. Which of the following carbohydrates are reducing and which are nonreducing?
    a. starch
    b. cellulose
    c. fructose
    d. sucrose
    e. ribose
17. What structural differences characterize starch, cellulose, and glycogen?
18. Draw the structure of a disaccharide unit in a polysaccharide composed of D-glucose linked $\alpha(1,4)$ to D-galactosamine.
19. Raffinose, the most abundant trisaccharide found in nature, occurs in whole grains and numerous vegetables (e.g., asparagus, cabbage, and beans). Hydrolysis of raffinose yields galactose and sucrose.
    a. Provide the systematic name for this trisaccharide.
    b. Is raffinose a reducing or nonreducing sugar?
    c. Is raffinose capable of mutarotation?

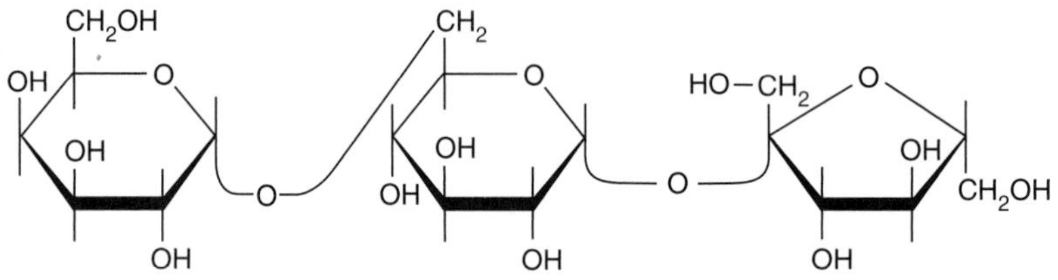

CH_2OH OH O OH O OH CH_2 O OH OH O OH $HO-CH_2$ O OH CH_2OH O OH

20. Give at least one function of each of the following:
    a. glycogen
    b. glycosaminoglycans
    c. glycoconjugates
    d. proteoglycans
    e. glycoproteins
    f. polysaccharides
21. The polymer chains of glycosaminoglycans are widely spread apart and bind large amounts of water.
    a. What two functional groups of the polymer make this binding of water possible?
    b. What type of bonding is involved?
22. In glycoproteins, what are the three amino acids to which the carbohydrate groups are most frequently linked? To what functional group is the glycan linked in each case?
23. Chondroitin sulfate chains have been likened to a large fishnet, passing small molecules through their matrix but excluding large ones. Use the structure of chondroitin sulfate and proteoglycans to explain this analogy.
24. Define the term *reducing sugar*. What structural feature does a reducing sugar have?
25. Compare the structures of proteoglycans and glycoproteins. How are structural differences related to their functions?
26. What role is carbohydrate thought to play in maintaining glycoprotein stability?
27. Classify each of the following sugar pairs as enantiomers, diastereomers, epimers, or an aldose-ketose pair.
    a. D-erythrose and D-threose
    b. D-glucose and D-mannose
    c. D-ribose and L-ribose
    d. D-allose and D-galactose
    e. D-glyceraldehyde and dihydroxyacetone

Fill in the Blank

28. Sugars that differ in one chiral center of many are called ____________________.
29. Aldehyde sugars are also referred to as ________________.
30. ____________________ are stereoisomers of sugars that differ only in the configuration at an acetal or hemiacetal carbon.
31. ____________________ are nonsuperimposable, non-mirror-image stereoisomers.
32. ____________________ is the structural glucose-containing polysaccharide of plants.
33. Sugars that react with Tollens' reagent are called ____________________ sugars.
34. The product of the reaction between an aldehyde or ketone and an amine is called a ____________________ base.
35. D-Erythrose and D-threose are referred to as ____________________ because their structures only differ in the configuration at one asymmetric carbon atom.
36. Oxidation of the aldehyde group of a sugar produces an ____________________ acid.
37. Polysaccharides that contain more than one type of monosaccharide are called ____________________.

Short Answer

38. Do all sugars have at least one chiral center?
39. Pure α-D-glucopyranose has a specific rotation of +19°, whereas that of β-D-glucopyranose is +112°. The specific rotation of the equilibrating mixture of α- and β-D-glucopyranose produced during mutarotation is +53°. What is the composition of this equilibrating mixture?
40. Why can aldoses and ketoses both behave as reducing sugars?
41. What do the proteoglycan and collagen components of cartilage contribute to this tissue?
42. Fish living in environments such as the Arctic Ocean, where water temperatures are below the freezing point of their blood, are able to survive because of so-called antifreeze glycoproteins. Suggest how these glycoproteins work.

Thought Questions

These questions are designed to reinforce your understanding of all of the key concepts discussed in the book so far, including this chapter and all of the chapters before it. They may not have one right answer! The authors have provided possible solutions to these questions in the back of the book and in the accompanying Study Guide for your reference.

43. β-Galactosidase is an enzyme that hydrolyses only β(1,4) linkages of lactose. An unknown trisaccharide is converted by β-galactosidase into maltose and galactose. Draw the structure of the trisaccharide.
44. Steroids are cholesterol-derived, polycyclic, lipid-soluble molecules that are very insoluble in water. Reaction with glucuronic acid makes a steroid much more water soluble and enables transport through the blood. What structural feature of the glucuronic acid increases the solubility?
45. Many bacteria are surrounded by a proteoglycan coat. Use your knowledge of the properties of this substance to suggest a function for such a coat.
46. It has long been recognized that breast milk protects infants from infectious diseases, especially those that affect the digestive tract. The main reason for this protection appears to be a large group of oligosaccharides that are components of human milk. Suggest a rationale for the protective effect of these oligosaccharides.
47. Ripe fruit has a high carbohydrate content and tastes sweet. In contrast, grain, which also has a high carbohydrate content, does not elicit a sweet sensation when consumed. Suggest a reason for this disparity from the plant's point of view.
48. A polysaccharide is found in the shells of arthropods (e.g., lobsters and grasshoppers) and of mollusks (e.g., oysters and snails). It can be obtained from these sources by soaking the shells in cold dilute hydrochloric acid to dissolve the calcium carbonate. The threadlike substance formed is composed of linear long-chain molecules. Hydrolysis with boiling acid gives D-glucosamine and acetic acid in equimolar amounts. Milder enzymatic hydrolysis gives *N*-acetyl-D-glucosamine as the sole product. The polysaccharide's linkages are identical to those of cellulose. What is the structure of this polymer?
49. Why is it advantageous for a plant toxin to elicit a bitter taste when an animal eats the plant rather than a bland or sweet taste?
50. Proteoglycan aggregates in tissues form hydrated, viscous gels. Can you think of any obvious mechanical reason why their capacity to form gels is important to cell function? [*Hint:* Liquid water is virtually incompressible.]
51. Alginic acid, isolated from seaweed and used as a thickening agent for ice cream and other foods, is a polymer of D-mannuronic acid with β(1,4) glycosidic linkages.
    a. Draw the structure of alginic acid.
    b. Why does this substance act as a thickening agent?

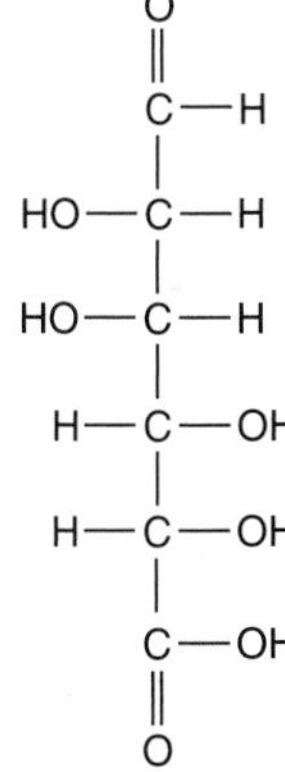

D-Mannuronic acid

52. Cellulose is virtually insoluble in water, whereas amylose is relatively soluble. Compare their structures and explain this disparity in solubility.
53. What is the maximum number of stereoisomers for mannuronic acid?
54. The ABO blood group antigens are the terminal sugars covalently linked to the end of the glycolipid in the red blood cell membrane. The H antigen is the precursor of the A and B antigens. Individuals with type A blood produce a gene that codes for an enzyme that adds *N*-acetylgalactosamine in an α(1,3) linkage to the Gal* residue in the H antigen. Type B blood requires that an enzyme add a D-α galactose in an

$\alpha(1,3)$ linkage to the Gal*. Draw the structures of the A and B antigens.

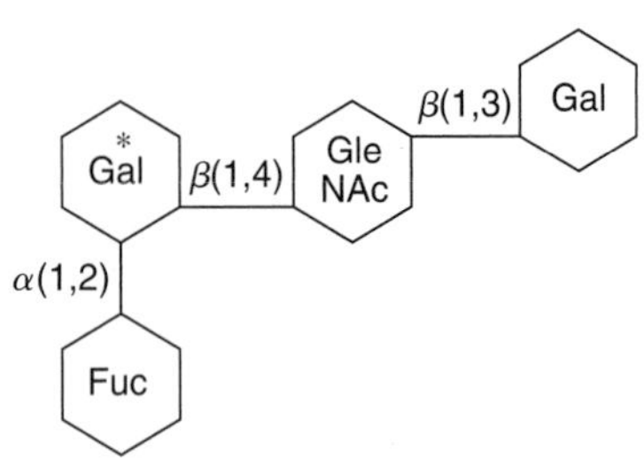

H Antigen

55. What would happen to the H antigen precursor (see Question 54) if an individual has both A and B genes? Keep in mind that the substrate of the respective enzymes is the H antigen.
56. Phosphate esters can form at positions 2 to 6 of an aldohexose but not at position 1. Explain.
57. In strong base, glucose converts to fructose. Explain how this conversion occurs.
58. Treatment of fructose with methyliodide produces 1,3,4,6-tetra-*O*-methylfructose. What does this information tell you about the fructose ring structure?
59. When glucose is reduced, only one alditol is produced. When fructose undergoes the same reaction, however, two diasteriomeric sugars are produced. Draw their structures.
60. In the liver, the water solubility of some hydrophobic molecules (e.g., drug molecules and steroid hormones) is improved by converting them into sulfate derivatives. Once such molecules are water soluble, they can be easily excreted. How does sulfate improve solubility?
61. Sucrose does not undergo mutarotation. Explain.
62. An oligosaccharide isolated from an organism is found to contain two glucose residues and one galactose residue. Exhaustive methylation followed by hydrolysis produced two glucoses with methoxy groups at positions 2, 3, and 6 and galactose with methoxy groups at positions 2, 3, 4, and 6. What is the structure of the original oligosaccharide?
63. A newly isolated aldohexose is oxidized to produce the corresponding aldaric acid that has an internal plane of symmetry; that is, it is a symmetrical molecule. What is the structure of the original aldohexose?
64. What sugar is produced by the epimerization of galactose?
65. Before a sugar can be analyzed by gas chromatography (GLC) or GLC/MS, it must be converted to a volatile derivative. Why can't sugars be analyzed directly?
66. The carbohydrate molecule 3-ketoglucose can exist in several ring forms. Draw them and determine which is the most stable.
67. Suggest a reason why sorbitol prevents moisture loss in candy.
68. Glyceraldehyde, the simplest aldose, does not have a cyclic form when tetroses with only one more carbon can form a ring readily. Can you suggest a reason for this phenomenon?
69. Olestra has been used in certain snack foods as an alternative to fats and oils. Its structure consists of a sucrose molecule in which all free hydroxyl groups have formed esters with oleic acid (an 18-carbon monounsaturated fatty acid). Olestra molecules contain no calories because they are exceptionally large and cannot be digested. Draw the structure of olestra. Use R—COOH as an abbreviation for oleic acid.
70. For a sugar to behave as a reducing sugar, it must have a free aldehyde group. Fructose is a ketose, yet it behaves like a reducing sugar. Explain.

CHAPTER

8 Carbohydrate Metabolism

Wine: A Product of Fermentation
Humans use microorganisms, in this case yeast, to metabolize sugar in the absence of oxygen. The aging of wine in oak barrels improves its taste and aroma.

OUTLINE

Metabolism and Jet Engines

Can systems biology improve our understanding of biochemical pathways such as glycolysis? Modern species are the result of billions of years of rigorous natural selection, which has adapted organisms to their various environments. This selection process also governs the metabolic pathways that manage the biochemical transformations that sustain life. As systems biologists analyzed metabolic processes, it became apparent that evolution, operating under thermodynamic and kinetic constraints, has converged again and again into a relatively small set of designs. Catabolic pathways, those that degrade organic molecules and release energy, provide an important example. These pathways typically have two characteristics: optimal ATP production and kinetic efficiency (i.e., minimal response time to changes in cellular metabolic requirements). Living organisms have optimized catabolic pathways, in part, through highly exergonic reactions at the beginning of a pathway. The early "activation" of nutrient molecules thus makes subsequent ATP-producing reactions (usually near the end of the pathway) run thermodynamically downhill. As a result, the pathway can produce ATP under varying substrate and product concentrations. The term "turbo design," inspired by the turbo engines in jet aircraft, describes this phenomenon. A good example is glycolysis, the energy-capturing reaction pathway that converts the hexose glucose into pyruvate (**Figure 8.1**).

A jet engine creates propulsion by mixing air with fuel to create hot, expanding, fast-moving exhaust gases, which blast out the back. Some of the air drawn in at the front of the engine is diverted into compressors, where its pressure increases substantially before flowing into combustion chambers and mixing with fuel molecules. As the burning fuel molecules expand, they flow through turbines fitted with fan blades that drive the compressors. An important feature of this process is that hot exhaust gases are also fed back into the engine to accelerate the fuel input step. This chapter reveals how carbohydrate fuel is burned by living cells with the same remarkable efficiency.

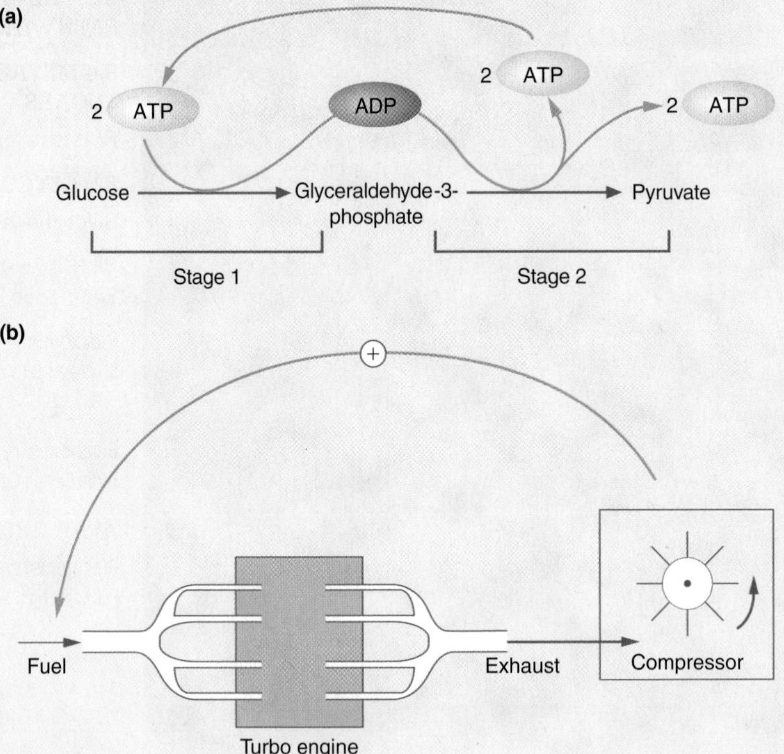

FIGURE 8.1

Comparison of Glycolysis and the Turbo Jet Engine

(a) Glycolysis is a two-stage catabolic pathway. Two of the four ATPs produced in stage 2 are used to activate an incoming glucose molecule (stage 1). The ADPs used in stage 2 are generated from the two ATPs used in stage 1 and in ATP-requiring reactions throughout the cell. (b) The schematic representation of a turbo jet engine illustrates how energy generated by an engine can be used to improve efficiency. Before exiting the engine, hot exhaust gases are diverted around the compressor turbines, raising the temperature and increasing the efficiency of fuel combustion.

Overview

CARBOHYDRATES PLAY SEVERAL CRUCIAL ROLES IN THE METABOLIC PROCESSES OF LIVING ORGANISMS. THEY SERVE AS ENERGY SOURCES and as structural elements in living cells. This chapter looks at the role of carbohydrates in energy production. Because the monosaccharide glucose is a prominent energy source in almost all living cells, major emphasis is placed on its synthesis, degradation, and storage.

Living cells are in a state of ceaseless activity. To maintain its "life," each cell depends on highly coordinated biochemical reactions. Carbohydrates are an important source of the energy that drives these reactions. This chapter discusses pathways of carbohydrate metabolism. During **glycolysis**, an ancient pathway found in almost all organisms, a small amount of energy is captured as a glucose molecule is converted to two molecules of pyruvate. Glycogen, a storage form of glucose in vertebrates, is synthesized by **glycogenesis** when glucose levels are high and degraded by **glycogenolysis** when glucose is in short supply. Glucose can also be synthesized from noncarbohydrate precursors by reactions referred to as **gluconeogenesis**. The **pentose phosphate pathway** enables cells to convert glucose-6-phosphate, a derivative of glucose, to ribose-5-phosphate (the sugar used to synthesize nucleotides) and other types of monosaccharides. NADPH, an important cellular reducing agent, is also produced by this pathway. In Chapter 9, the *glyoxylate cycle*, used by some organisms (primarily plants) to manufacture carbohydrate from fatty acids, is considered. *Photosynthesis*, a process in which light energy is captured to drive carbohydrate synthesis, is described in Chapter 13.

Any discussion of carbohydrate metabolism focuses on the synthesis and usage of glucose, a major fuel for most organisms. In vertebrates, glucose is transported throughout the body in the blood. If cellular energy reserves are low, glucose is degraded by the glycolytic pathway. Glucose molecules not required for immediate energy production are stored as glycogen in liver and muscle. The energy requirements of many tissues (e.g., brain, red blood cells, and exercising skeletal muscle cells) depend on an uninterrupted flow of glucose. Depending on a cell's metabolic requirements, glucose can also be used to synthesize, for example, other monosaccharides, fatty acids, and certain amino acids. **Figure 8.2** summarizes the major pathways of carbohydrate metabolism in animals.

8.1 GLYCOLYSIS

Glycolysis occurs, at least in part, in almost every living cell. This series of reactions is believed to be among the oldest of all the biochemical pathways. Both the enzymes and the number and mechanisms of the steps in the pathway are highly conserved in prokaryotes and eukaryotes. Also, glycolysis is an anaerobic process, which would have been necessary in the oxygen-poor atmosphere of pre-eukaryotic Earth.

In glycolysis, also referred to as the *Embden-Meyerhof-Parnas pathway*, each glucose molecule is split and converted to two three-carbon units (pyruvate). During this process several carbon atoms are oxidized. The small amount of energy captured during glycolytic reactions (about 5% of the total available) is stored temporarily in two molecules each of ATP and NADH (the reduced form of the coenzyme NAD^+). The subsequent metabolic fate of pyruvate depends on the organism being considered and its metabolic circumstances. In **anaerobic organisms**

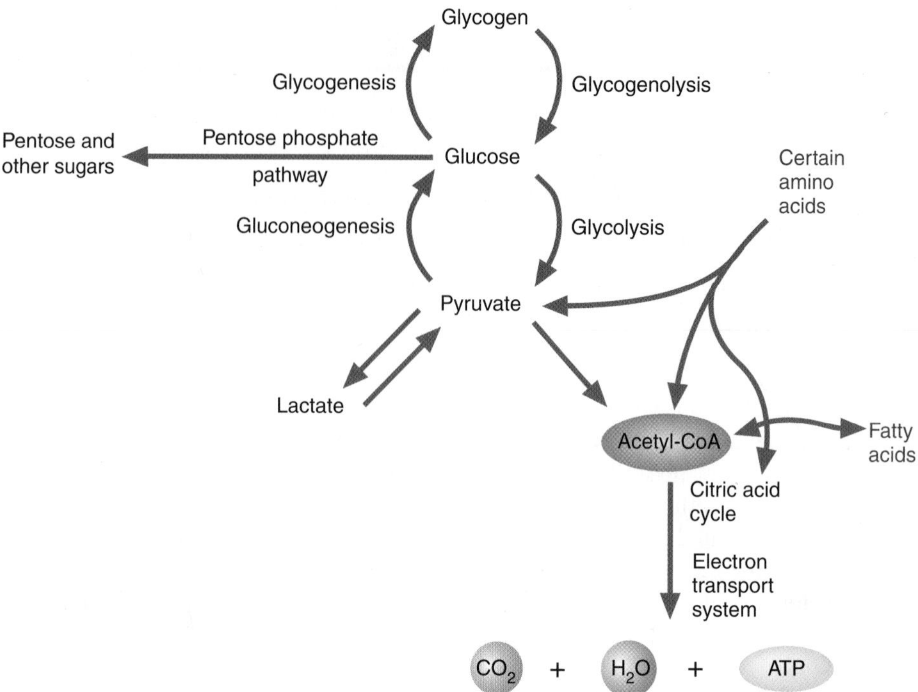

FIGURE 8.2

Major Pathways in Carbohydrate Metabolism

In animals, excess glucose is converted to its storage form, glycogen, by glycogenesis. When glucose is needed as a source of energy or as a precursor molecule in biosynthetic processes, glycogen is degraded by glycogenolysis. Glucose can be converted to ribose-5-phosphate (a component of nucleotides) and NADPH (a powerful reducing agent) by means of the pentose phosphate pathway. Glucose is oxidized by glycolysis, an energy-generating pathway that converts it to pyruvate. In the absence of oxygen, pyruvate is converted to lactate. When oxygen is present, pyruvate is further degraded to form acetyl-CoA. Significant amounts of energy in the form of ATP can be extracted from acetyl-CoA by the citric acid cycle and the electron transport system. Note that carbohydrate metabolism is inextricably linked to the metabolism of other nutrients. For example, lactate and certain amino acids can be used to synthesize glucose (gluconeogenesis). Acetyl-CoA is also generated from the breakdown of fatty acids and certain amino acids. When acetyl-CoA is present in excess, a different pathway converts it into fatty acids.

(those that do not use oxygen to generate energy), pyruvate may be converted to waste products such as ethanol, lactic acid, acetic acid, and similar molecules. Using oxygen as a terminal electron acceptor, aerobic organisms such as animals and plants completely oxidize pyruvate to form CO_2 and H_2O in an elaborate stepwise mechanism known as **aerobic respiration** (Chapters 9 and 10).

Glycolysis (**Figure 8.3**), which consists of 10 reactions, occurs in two stages:

1. Glucose is phosphorylated twice and cleaved to form two molecules of glyceraldehyde-3-phosphate (G-3-P). The two ATP molecules consumed during this stage are like an investment because this stage creates the actual substrates for oxidation.
2. G-3-P is converted to pyruvate. Four ATP and two NADH molecules are produced. Because two ATP were consumed in stage 1, the net production of ATP per glucose molecule is 2.

The glycolytic pathway can be summed up in the following equation:

$$\text{D-Glucose} + 2\ \text{ADP} + 2\ \text{P}_i + 2\ \text{NAD}^+ \rightarrow 2\ \text{pyruvate} + 2\ \text{ATP} + 2\ \text{NADH} + 2\text{H}^+ + 2\text{H}_2\text{O}$$

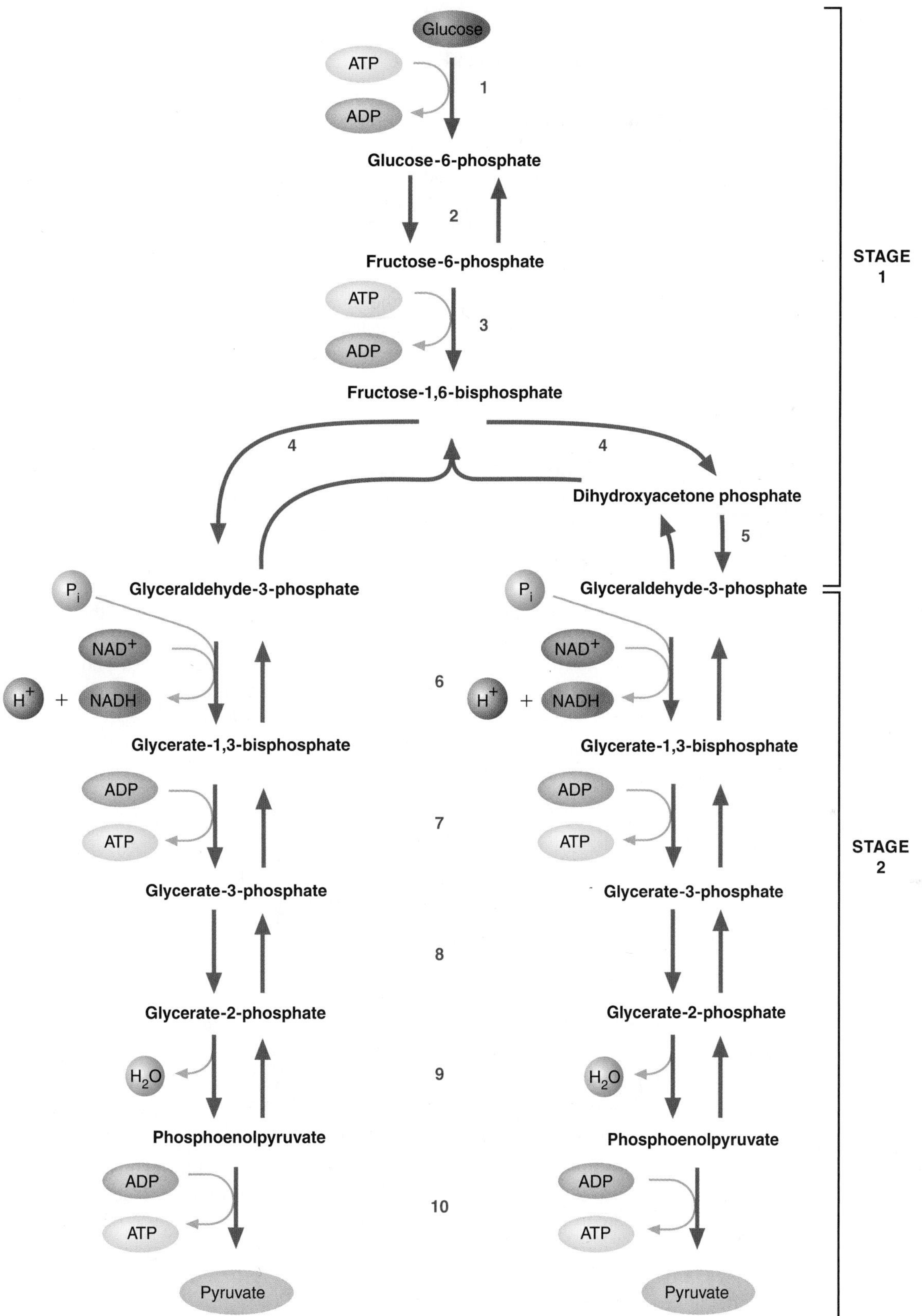

FIGURE 8.3

The Glycolytic Pathway

In glycolysis, a pathway with 10 reactions, each glucose molecule is converted into two pyruvate molecules. In addition, two molecules each of ATP and NADH are produced. Reactions with double arrows are reversible reactions and those with single arrows are irreversible reactions that serve as control points in the pathway.

The Reactions of the Glycolytic Pathway

Glycolysis is summarized in **Figure 8.3**. The 10 reactions of the glycolytic pathway are as follows.

1. **Synthesis of glucose-6-phosphate**. Immediately after entering a cell, glucose and other sugar molecules are phosphorylated by hexokinase (Hk). Phosphorylation prevents transport of glucose out of the cell and increases the reactivity of the oxygen in the resulting phosphate ester. Several hexokinases catalyze the phosphorylation of hexoses in all cells in the body. ATP, a cosubstrate in the reaction, is complexed with Mg^{2+}. (ATP-Mg^{2+} complexes are common in kinase-catalyzed reactions.) The carbon-6 OH group attacks the phosphorus-oxygen double bond to yield ADP and the 6-hydroxyphosphate ester of glucose.

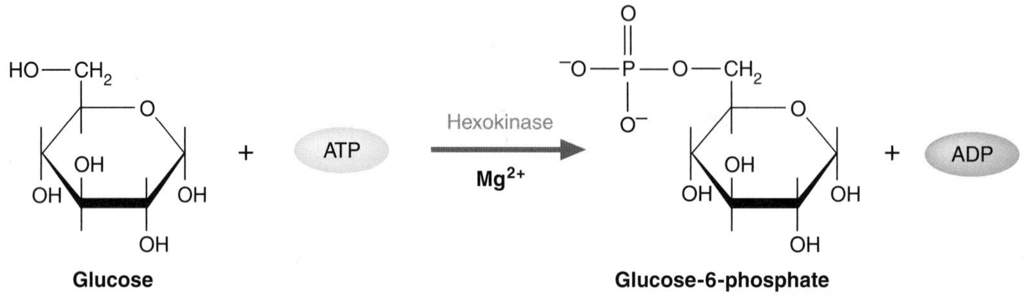

Glucose + ATP —Hexokinase, Mg^{2+}→ Glucose-6-phosphate + ADP

Under intracellular conditions the reaction is irreversible; that is, the enzyme has no ability to retain or accommodate the product of the reaction in its active site, regardless of the concentration of glucose-6-phosphate.

2. **Conversion of glucose-6-phosphate to fructose-6-phosphate**. During reaction 2 of glycolysis, the open chain form of the aldose glucose-6-phosphate is converted to the open chain form of the ketose fructose-6-phosphate by phosphoglucose isomerase (PGI) in a readily reversible reaction:

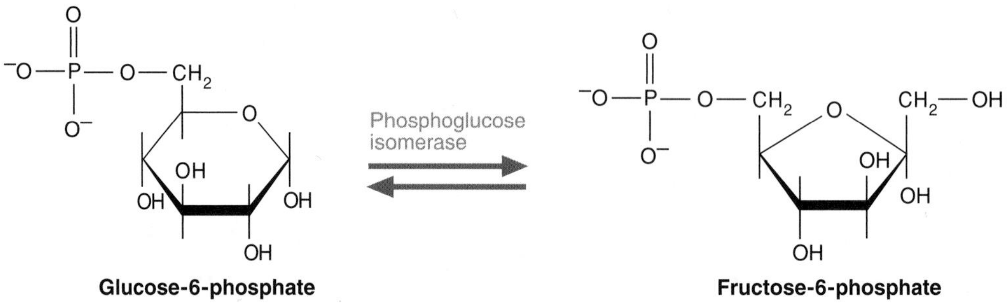

Glucose-6-phosphate ⇌ (Phosphoglucose isomerase) Fructose-6-phosphate

Recall that the isomerization reaction of glucose and fructose involves an enediol intermediate (**Figure 7.16**). This transformation makes C-1 of the fructose product available for phosphorylation. The hemiacetal hydroxy group of glucose-6-phosphate is more difficult to phosphorylate.

3. **The phosphorylation of fructose-6-phosphate**. Phosphofructokinase-1 (PFK-1) irreversibly catalyzes the phosphorylation of fructose-6-phosphate to form fructose-1,6-bisphosphate:

Fructose-6-phosphate + ATP —PFK-1, Mg^{2+}→ Fructose-1,6-bisphosphate + ADP

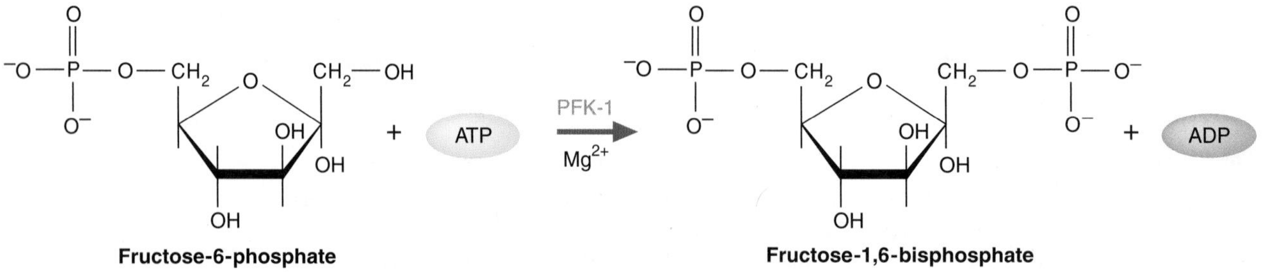

The PFK-1-catalyzed reaction is irreversible under cellular conditions. It is, therefore, the first committed step in glycolysis. Unlike glucose-6-phosphate and fructose-6-phosphate, the substrate and product, respectively, of the previous reaction, fructose-1,6-bisphosphate cannot be diverted into other pathways. Investing a second molecule of ATP serves several purposes. First, because ATP is used as the phosphorylating agent, the reaction proceeds with a large decrease in free energy. After fructose-1,6-bisphosphate has been synthesized, the cell is committed to glycolysis. Because fructose-1,6-bisphosphate eventually splits into two trioses, another purpose for phosphorylation is to prevent any later product from diffusing out of the cell because charged molecules cannot easily cross membranes.

4. **Cleavage of fructose-1,6-bisphosphate**. Stage 1 of glycolysis ends with the cleavage of fructose-1,6-bisphosphate into two three-carbon molecules: G-3-P and dihydroxyacetone phosphate (DHAP). The hydroxyl hydrogen on carbon 4 is removed by the enzyme to produce a carbonyl group, simultaneously breaking the C-3–C-4 bond. This reaction is an **aldol cleavage**, hence the name of the enzyme: aldolase. Aldol cleavages are the reverse of aldol condensations, described on p. 170. In aldol cleavages an aldehyde and a ketone are products.

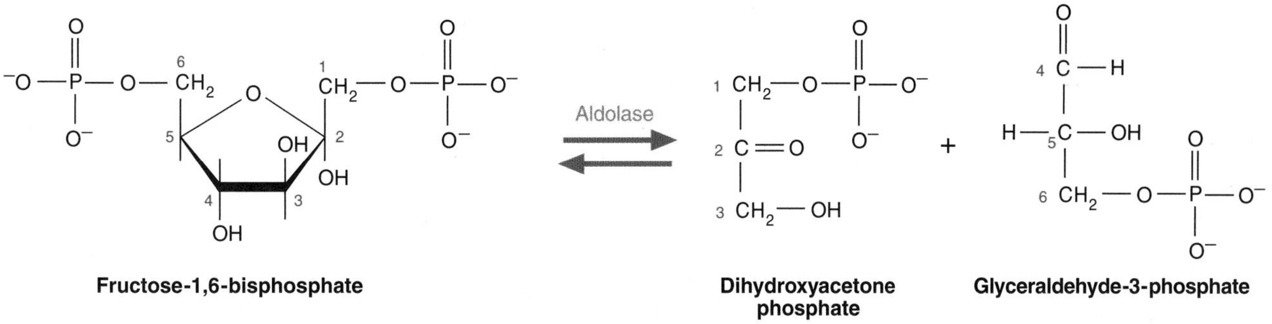

Aldolase

Fructose-1,6-bisphosphate ⇌ **Dihydroxyacetone phosphate** + **Glyceraldehyde-3-phosphate**

The open chain form of fructose-1,6-bisphosphate is initially bound to the enzyme active site by a Schiff base with carbon-2 of the sugar. A general base within the active site removes a proton from the C-4 hydroxy group. The formation of a carbonyl group at C-4 results in cleavage of the C-3–C-4 bond to yield G-3-P. In a second step, the cleavage of the C–N bond releases DHAP. Although the cleavage of fructose-1,6-bisphosphate is thermodynamically unfavorable ($\Delta G^{\circ\prime} = +23.8$ kJ/mol), the reaction proceeds because the products are rapidly removed.

5. **The interconversion of glyceraldehyde-3-phosphate and dihydroxyacetone phosphate**. Of the two products of the aldolase reaction, only G-3-P serves as a substrate for the next reaction in glycolysis. To prevent the loss of the other three-carbon unit from the glycolytic pathway, triose phosphate isomerase catalyzes the reversible conversion of DHAP to G-3-P (**Figure 8.4**):

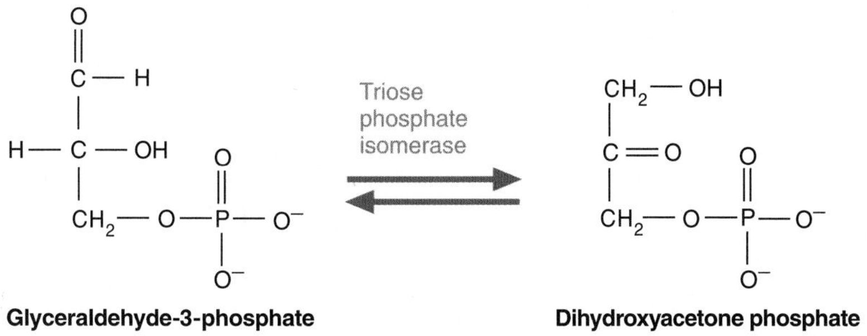

Triose phosphate isomerase

Glyceraldehyde-3-phosphate ⇌ **Dihydroxyacetone phosphate**

FIGURE 8.4

Reversible Interconversion of Dihydroxyacetone Phosphate to Glyceraldehyde-3-Phosphate

The Glu165 residue of the enzyme, acting as a nucleophile, deprotonates the substrate (DHAP) as the electrophilic His95 donates a proton to yield the enediol intermediate. The enediol is converted to the product (G-3-P) by the removal of a proton by His65 and the abstraction of a proton from Glu195.

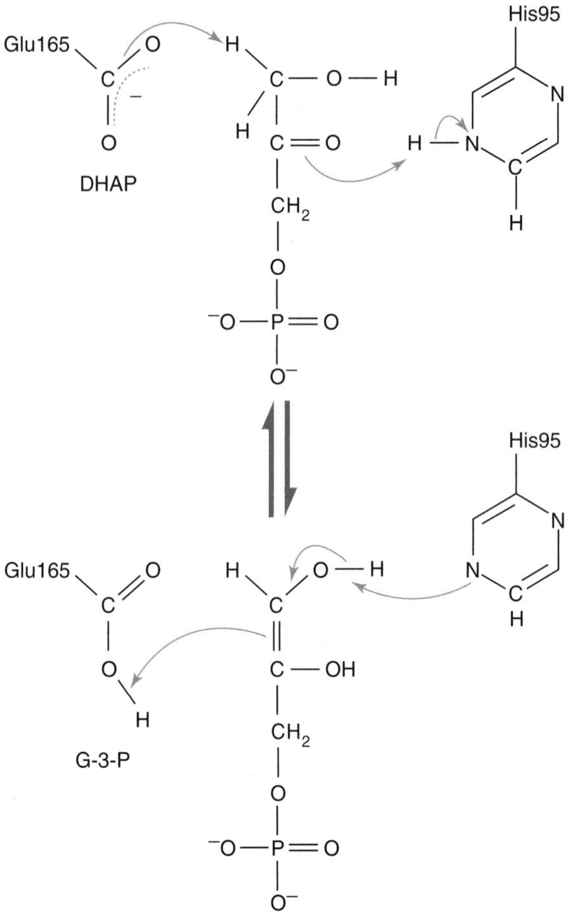

After this reaction, the original molecule of glucose has been converted to two molecules of G-3-P.

6. **Oxidation of glyceraldehyde-3-phosphate.** During reaction 6 of glycolysis, G-3-P undergoes oxidation and phosphorylation. The product, glycerate-1,3-bisphosphate, contains a high-energy phosphoanhydride bond, which is used in the next reaction to generate ATP:

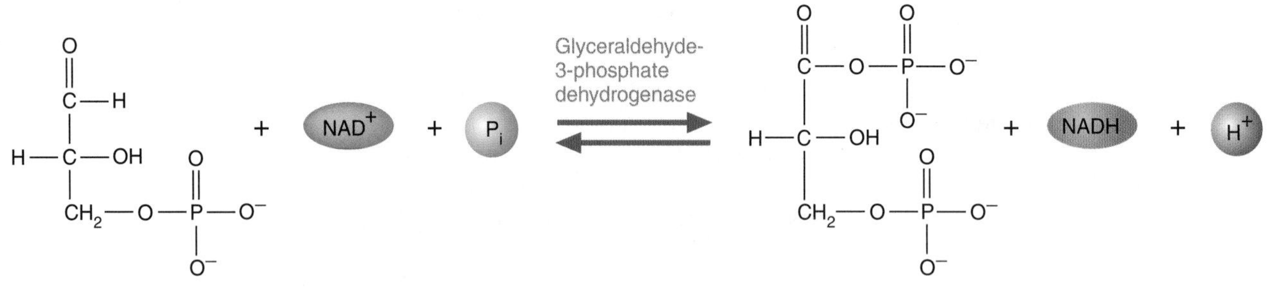

Glyceraldehyde-3-phosphate **Glycerate-1,3-bisphosphate**

This complex process is catalyzed by glyceraldehyde-3-phosphate dehydrogenase, a tetramer composed of four identical subunits. Each subunit contains one binding site for G-3-P and another for NAD^+, an oxidizing agent. As the enzyme forms a covalent thioester bond with the substrate (**Figure 8.5**), a hydride ion ($H:^-$) is transferred to NAD^+ in the active site. NADH, the reduced form of NAD^+, then leaves the active site and is replaced by an incoming NAD^+. The acyl enzyme adduct is attacked by inorganic phosphate and the product leaves the active site.

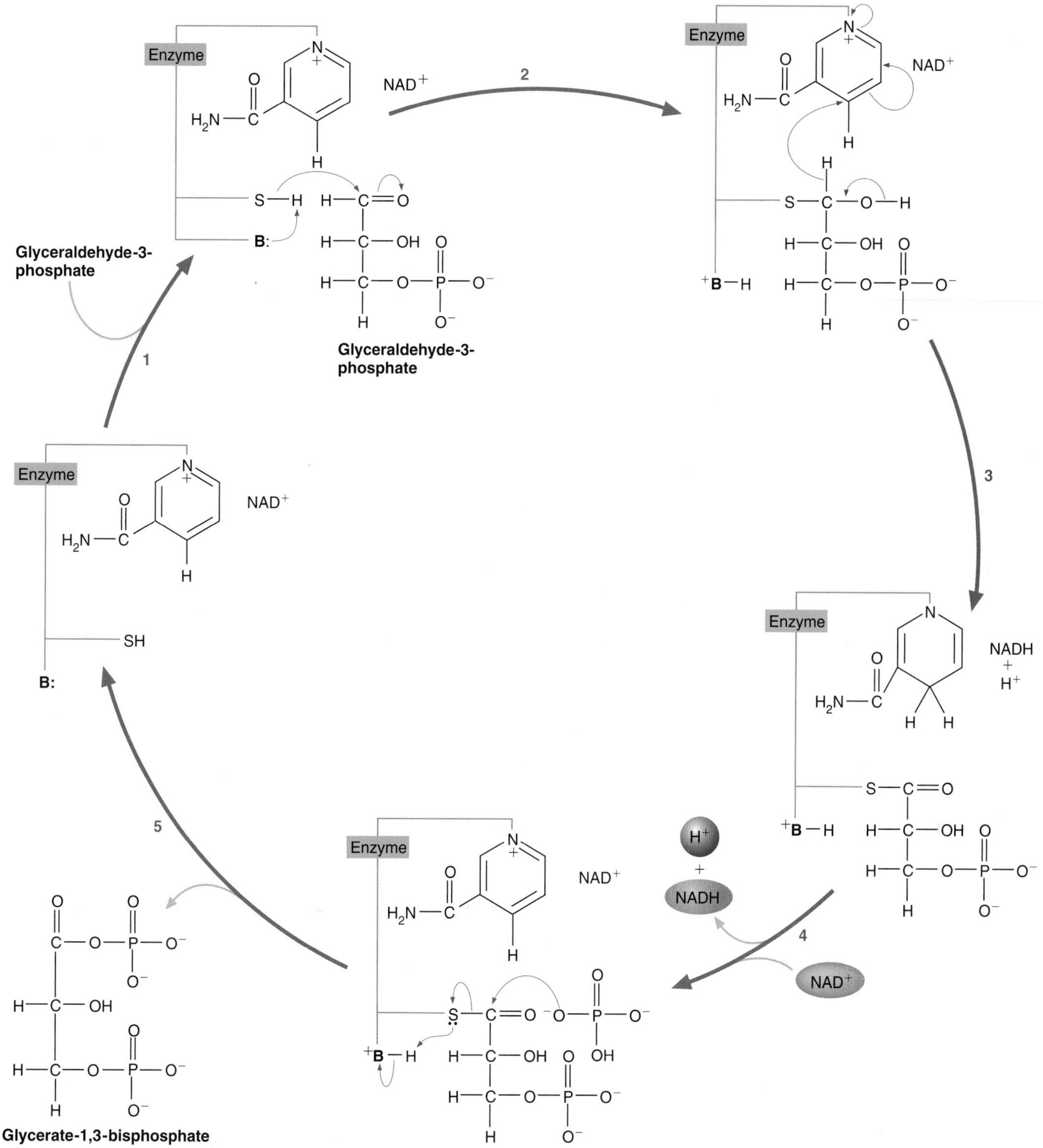

Enzyme
NAD$^+$
Glyceraldehyde-3-phosphate
Glyceraldehyde-3-phosphate
NADH + H^+
H^+ + NADH
NAD$^+$
Glycerate-1,3-bisphosphate
1
2
3
4
5

FIGURE 8.5
Glyceraldehyde-3-Phosphate Dehydrogenase Reaction

In the first step the substrate, glyceraldehyde-3-phosphate, enters the active site. As the enzyme catalyzes the reaction of the substrate with a sulfhydryl group within the active site (step 2), the substrate is oxidized (step 3). The noncovalently bound NADH is exchanged for a cytoplasmic NAD$^+$ (step 4). Displacement of the enzyme by inorganic phosphate (step 5) liberates the product, glycerate-1,3-bisphosphate, thus returning the enzyme to its original form.

7. **Phosphoryl group transfer.** In this reaction ATP is synthesized as phosphoglycerate kinase catalyzes the transfer of the high-energy phosphoryl group of glycerate-1,3-bisphosphate to ADP:

Glycerate-1,3-bisphosphate + ADP ⇌ (Phosphoglycerate kinase, Mg^{2+}) Glycerate-3-phosphate + ATP

The terminal phosphoryl group of ADP acting as a nucleophile attacks the phosphorus of the phosphoanhydride of glycerate-1,3-bisphosphate to yield glycerate-3-phosphate. Reaction 7 is an example of a substrate-level phosphorylation. Because the synthesis of ATP is endergonic, it requires an energy source. In **substrate-level phosphorylations,** ATP is produced by the transfer of a phosphoryl group from a substrate with a high phosphoryl transfer potential (glycerate-1,3-bisphosphate) (refer to **Table 4.1**) to produce a compound with a lower transfer potential (ATP) and therefore $\Delta G < 0$. Because two molecules of glycerate-1,3-bisphosphate are formed for every glucose molecule, this reaction produces two ATP molecules, and the investment of phosphate bond energy is recovered. ATP synthesis later in the pathway represents a net gain.

8. **The interconversion of glycerate-3-phosphate and glycerate-2-phosphate.** Glycerate-3-phosphate has a low phosphoryl group transfer potential. As such, it is a poor candidate for further ATP synthesis ($\Delta G^{\circ\prime}$ for ATP synthesis is –30.5 kJ/mol). Cells convert glycerate-3-phosphate with its energy-poor phosphate ester to phosphoenolpyruvate (PEP), which has an exceptionally high phosphoryl group transfer potential. (The standard free energies of hydrolysis of glycerate-3-phosphate and PEP are −12.6 and −61.9 kJ/mol, respectively.) In the first step in this conversion (reaction 8), phosphoglycerate mutase catalyzes the conversion of a C-3 phosphorylated compound to a C-2 phosphorylated compound through a two-step addition/elimination cycle.

Glycerate-3-phosphate + P_i ⇌ (Phosphoglycerate mutase) [Glycerate-2,3-bisphosphate] ⇌ (Phosphoglycerate mutase; H_2O in, P_i out) Glycerate-2-phosphate

9. **Dehydration of glycerate-2-phosphate.** Enolase catalyzes the dehydration of glycerate-2-phosphate to form PEP:

Glycerate-2-phosphate ⇌ (Enolase) Phosphoenolpyruvate (PEP) + H_2O

The enzyme removes a proton from C-2 while simultaneously a carboxylic acid side chain group protonates the OH group. As a result, a water molecule leaves and a carbon-carbon double bond is formed. PEP has a higher phosphoryl group transfer potential than does glycerate-2-phosphate because it contains an enol-phosphate group instead of a simple phosphate ester. The reason for this difference is made apparent in the next reaction. Aldehydes and ketones have two isomeric forms. The *enol* form contains a carbon-carbon double bond and a hydroxyl group. Enols exist in equilibrium with the more stable carbonyl-containing *keto* form. The interconversion of keto and enol forms, also called **tautomers**, is referred to as **tautomerization**:

Enol form ⇌ Keto form

This tautomerization is restricted by the presence of the phosphate group, as is the resonance stabilization of the free phosphate ion. As a result, phosphoryl transfer to ADP in reaction 10 is highly favored.

10. **Synthesis of pyruvate.** In the final reaction of glycolysis, pyruvate kinase (PK) catalyzes the transfer of a phosphoryl group from PEP to ADP. The terminal phosphoryl group of ADP attacks the phosphorus of PEP to produce the enol form of pyruvate, which then spontaneously transforms into the keto form. Two molecules of ATP are formed for each molecule of glucose.

PEP + ADP ⇌ (Mg^{2+}; ATP) Pyruvate (enol form) → (H^+) Pyruvate (keto form)

This reaction is irreversible because of the high phosphoryl group transfer potential of PEP (**Table 4.1**). There is an exceptionally large free energy loss associated with the spontaneous conversion (tautomerization) of the enol form of pyruvate to the more stable keto form. The 10 reactions of glycolysis are illustrated in **Figure 8.6**.

The Fates of Pyruvate

In terms of energy, the result of glycolysis is the production of two ATPs and two NADHs per molecule of glucose. Pyruvate, the other product of glycolysis, is still an energy-rich molecule, which can yield a substantial amount of ATP. Whether further energy can be produced, however, depends on the cell type and the

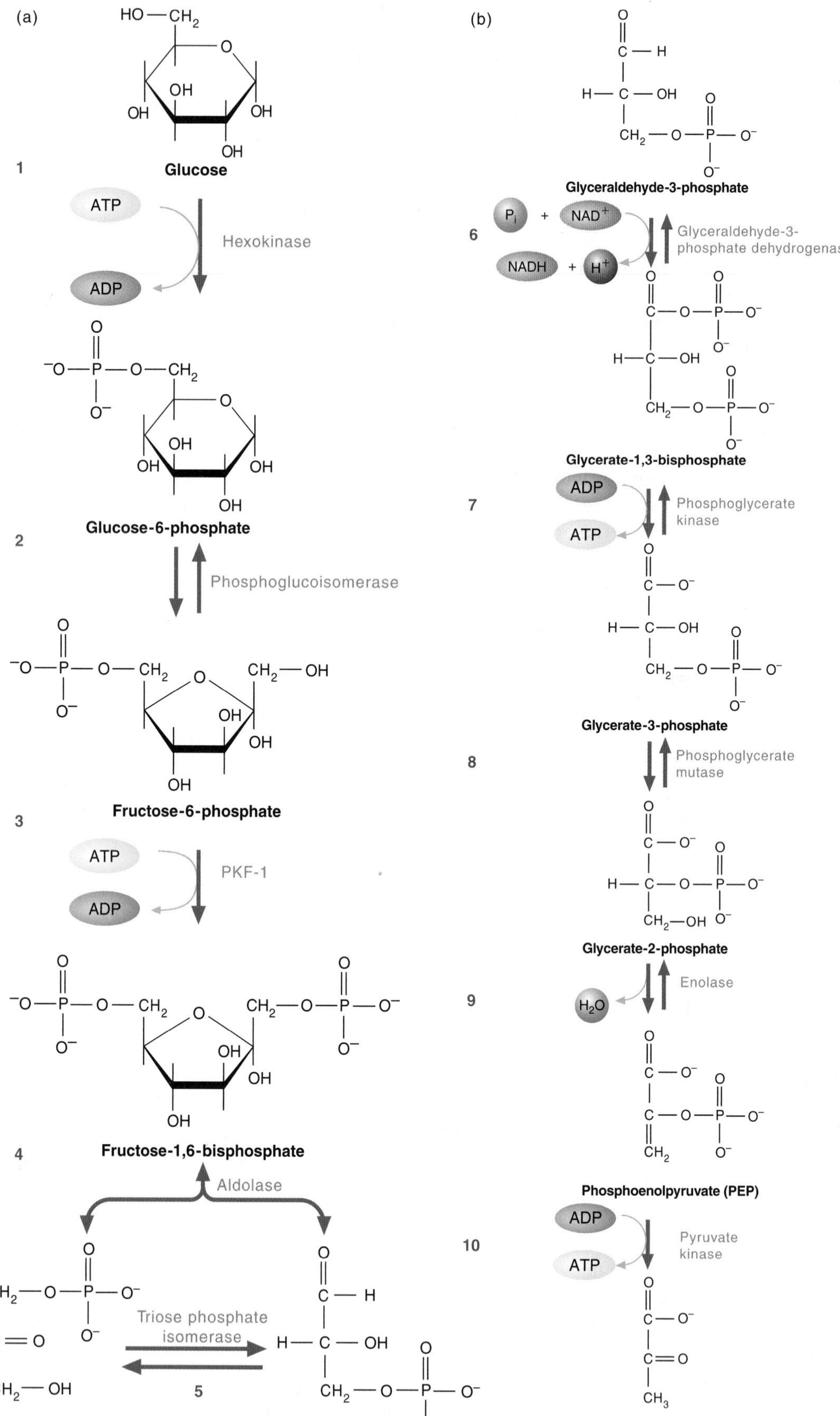

FIGURE 8.6
The Reactions of Glycolysis

In all there are 10 reactions in the glycolytic pathway. (a) In stage 1, reactions 1 through 5 convert glucose into glyceraldehyde-3-phosphate. Two ATP are consumed in stage 1 for each glucose molecule. (b) In stage 2, reactions 6 through 10 convert glyceraldehyde-3-phosphate into pyruvate. In addition to pyruvate, the reactions of stage 2 also produce 4 ATP and 2 NADH per glucose molecule.

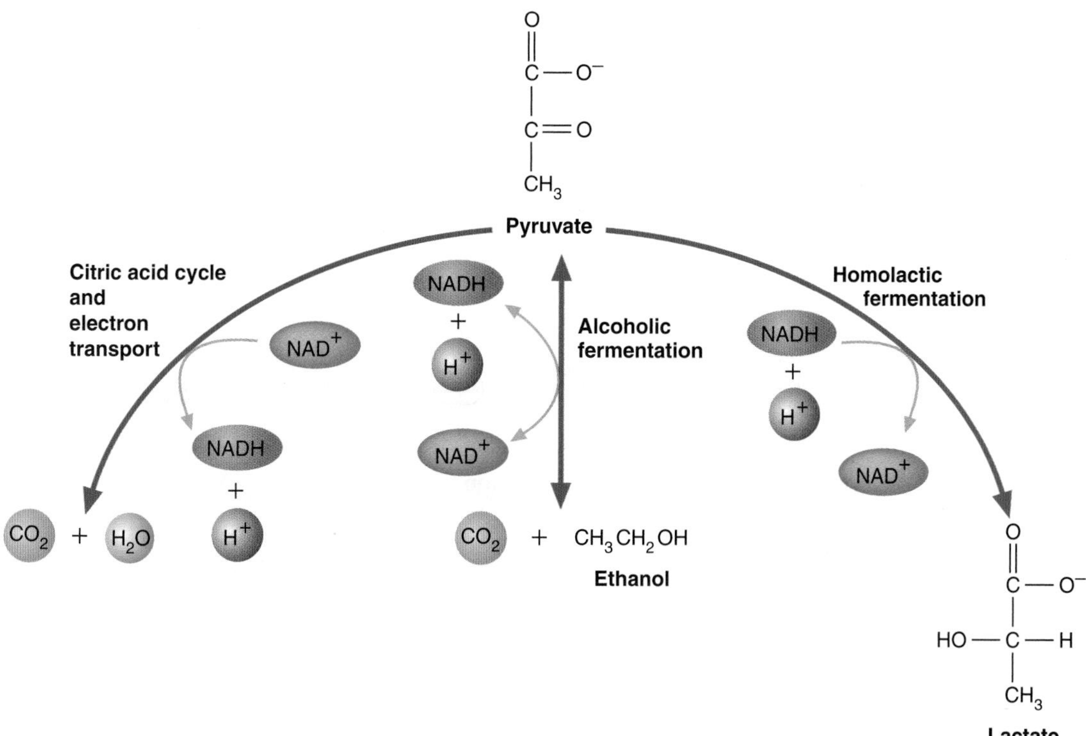

FIGURE 8.7
The Fates of Pyruvate

When oxygen is available (*left*), aerobic organisms completely oxidize pyruvate to CO_2 and H_2O. In the absence of oxygen, pyruvate can be converted to several types of reduced molecules. In some cells (e.g., yeast), ethanol and CO_2 are produced (*middle*). In others (e.g., muscle cells), homolactic fermentation occurs in which lactate is the only organic product (*right*). Some microorganisms use heterolactic fermentation reactions (not shown) that produce other acids or alcohols in addition to lactate. In all fermentation processes, the principal purpose is to regenerate NAD^+ so that glycolysis can continue.

availability of oxygen. Under aerobic conditions, most cells in the body convert pyruvate into acetyl-CoA, the entry-level substrate for the **citric acid cycle**, an amphibolic pathway that completely oxidizes the two acetyl carbons to form CO_2 and the reduced molecules NADH and $FADH_2$. (An **amphibolic pathway** functions in both anabolic and catabolic processes.) The **electron transport system**, a series of oxidation-reduction reactions, transfers electrons from NADH and $FADH_2$ to O_2 to form water. The energy that is released during electron transport is coupled to a mechanism that synthesizes ATP. Under anaerobic conditions, further oxidation of pyruvate is impeded. A number of cells and organisms compensate by converting this molecule to a more reduced organic compound and regenerating the NAD^+ required for glycolysis to continue (**Figure 8.7**). (Recall that the hydride ion acceptor molecule NAD^+ is a cosubstrate in the reaction catalyzed by glyceraldehyde-3-phosphate dehydrogenase.) This process of NAD^+ regeneration is referred to as **fermentation**. Muscle cells, red blood cells, and certain bacterial species (e.g., *Lactobacillus*) regenerate NAD^+ by transforming pyruvate into lactate:

$$\underset{\text{Pyruvate}}{CH_3{-}CO{-}COO^-} + \text{NADH} + \text{H}^+ \xrightleftharpoons{\text{Lactate dehydrogenase}} \underset{\text{Lactate}}{CH_3{-}CH(OH){-}COO^-} + \text{NAD}^+$$

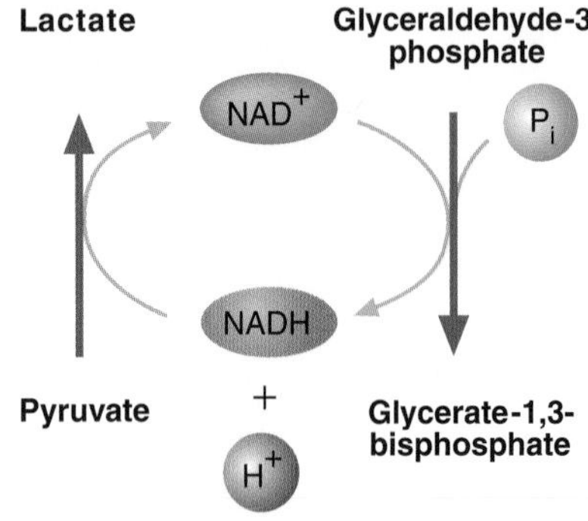

FIGURE 8.8

Recycling of NADH during Anaerobic Glycolysis

The NADH produced during the conversion of glyceraldehyde-3-phosphate to glycerate-1, 3-bisphosphate is oxidized when pyruvate is converted to lactate. This process allows the cell to continue producing ATP under anaerobic conditions as long as glucose is available.

In rapidly contracting muscle cells, the demand for energy is high. After the O_2 supply is depleted, *lactic acid fermentation* provides sufficient NAD^+ to allow glycolysis (with its low level of ATP production) to continue for a short time (**Figure 8.8**).

QUESTION 8.1

Most molecules of ethanol are detoxified in the liver by two reactions. In the first, ethanol is oxidized to form acetaldehyde. This reaction, catalyzed by ADH, produces large amounts of NADH:

$$CH_3-CH_2-OH + NAD^+ \xrightarrow{\text{ADH}} CH_3\overset{O}{\overset{\|}{C}}-H + NADH + H^+$$

Soon after its production, acetaldehyde is converted to acetate by aldehyde dehydrogenase, which catalyzes a reaction that also produces NADH:

$$CH_3-\overset{O}{\overset{\|}{C}}-H + NAD^+ + H_2O \xrightarrow[\text{dehydrogenase}]{\text{Aldehyde}} CH_3-\overset{O}{\overset{\|}{C}}-O^- + NADH + 2\,H^+$$

One common effect of alcohol intoxication is the accumulation of lactate in the blood. Can you explain why this effect occurs?

In yeast and certain bacterial species, pyruvate is decarboxylated to form acetaldehyde, which is then reduced by NADH to form ethanol. (In a **decarboxylation** reaction, an organic acid loses a carboxyl group as CO_2.)

$$\underset{\text{Pyruvate}}{CH_3-\overset{O}{\overset{\|}{C}}-COO^-} \xrightarrow[\;CO_2\;]{\text{Pyruvate decarboxylase}} \underset{\text{Acetaldehyde}}{CH_3-\overset{O}{\overset{\|}{C}}-H} \xrightarrow[NADH + H^+ \;\to\; NAD^+]{\text{Alcohol dehydrogenase}} \underset{\text{Ethanol}}{CH_3-CH_2-OH}$$

This process, called alcoholic fermentation, is used commercially to produce wine, beer, and bread. Certain bacterial species produce organic molecules other than ethanol. For example, *Clostridium acetobutylicum*, an organism related to the causative agents of botulism and tetanus, produces butanol. Until recently, this organism was used commercially to synthesize butanol, an alcohol used in the production of detergents and synthetic fibers. A petroleum-based synthetic process has now replaced microbial fermentation.

Visit the companion website at www.oup.com/us/mckee to read the Biochemistry in Perspective essay on fermentation.

KEY CONCEPTS

- During glycolysis, glucose is converted to two molecules of pyruvate. A small amount of energy is captured in two molecules each of ATP and NADH.
- In anaerobic organisms, pyruvate is converted to waste products in a process called fermentation.
- In the presence of oxygen the cells of aerobic organisms convert pyruvate into CO_2 and H_2O.

The Energetics of Glycolysis

During glycolysis, the energy released as glucose is converted to pyruvate is coupled to the phosphorylation of ADP with a net yield of 2 ATP. However, evaluation of the standard free energy changes of the individual reactions does not explain the efficiency of this pathway. A more useful method for evaluating free energy changes takes into account the conditions (e.g., pH and metabolite concentrations) under which cells actually operate. As illustrated in **Figure 8.9**, free energy changes measured in red blood cells indicate that only three reactions

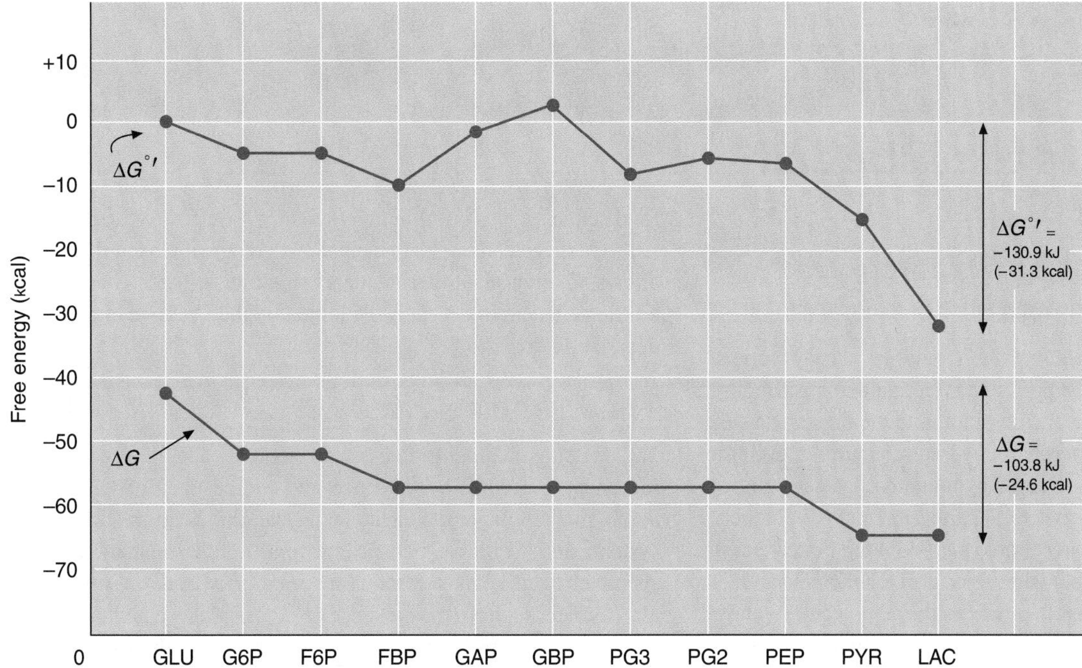

FIGURE 8.9

Free Energy Changes during Glycolysis in Red Blood Cells

The standard free energy changes ($\Delta G^{\circ\prime}$) for the reactions in glycolysis show no consistent pattern (upper plot). In contrast, actual free energy values (ΔG) based on metabolite concentrations measured in red blood cells (lower plot) clearly illustrate why reactions 1, 3, and 10 (the conversions of glucose to glucose-6-phosphate, fructose-6-phosphate to fructose-1,6-bisphosphate, and phosphoenolpyruvate to pyruvate, respectively) are irreversible. The ready reversibility of the remaining reactions is indicated by their near-zero ΔG values. (GLU = glucose, G6P = glucose-6-phosphate, F6P = fructose-6-phosphate, FBP = fructose-1,6-bisphosphate, GAP = glyceraldehyde phosphate, PG3 = glycerate-3-phosphate, PG2 = glycerate-2-phosphate, PEP = phosphoenolpyruvate, PYR = pyruvate, LAC = lactate.) Note that the conversion of DHAP to GAP is not counted in this list since FBP is broken into GAP and DHAP, which is reconverted into GAP.

(1, 3, and 10, see **Figure 8.6**) have significantly negative ΔG values. These reactions, catalyzed by hexokinase, PFK-1, and pyruvate kinase, respectively, are for all practical purposes irreversible; that is, each goes to completion as written. The values for the remaining reactions (2, 4–9) are so close to zero that they operate near equilibrium. Consequently, these latter reactions are easily reversible; small changes in substrate or product concentrations can alter the direction of each reaction. Not surprisingly, in gluconeogenesis (Section 8.2), the pathway by which glucose can be generated from pyruvate and certain other substrates, all of the glycolytic enzymes are involved except for those that catalyze reactions 1, 3, and 10. Gluconeogenesis uses different enzymes to bypass the irreversible steps of glycolysis.

Regulation of Glycolysis

The rate at which the glycolytic pathway operates in a cell is directly controlled primarily by the kinetic properties of its hexokinase isoenzymes and the allosteric regulation of the enzymes that catalyze the three irreversible reactions: hexokinase, PFK-1, and pyruvate kinase.

THE HEXOKINASES The animal liver has four hexokinases. Three of these enzymes (hexokinases I, II, and III) are found in varying concentrations in other body tissues, where they bind reversibly to an anion channel (called a *porin*) in the outer membrane of mitochondria. As a result, ATP is readily available.

Biochemistry IN PERSPECTIVE

Saccharomyces cerevisiae and the Crabtree Effect

What unique properties of *Saccharomyces cerevisiae* make it so useful in the production of wine, beer, and bread? The yeast *Saccharomyces cerevisiae,* a eukaryotic microorganism, has had a profound effect on humans. Beginning in the Neolithic age and continuing to the present day, this organism is used to convert carbohydrate-containing food into the wine, beer, and bread that many humans consider indispensible to life. What properties of *S. cerevisiae* make it uniquely suited for the oldest of biotechnologies? Although many yeast species can ferment carbohydrates to form ethanol and CO_2, only *S. cerevisiae* efficiently produces these molecules in large quantities. A simple experiment helps explain why. If fleshy fruits such as grapes are crushed and placed in a vat, they will begin to ferment. At the beginning of fermentation, a survey of the microorganisms present reveals many different types, but relatively few *S. cerevisiae.* As the fermentation process proceeds (and the ethanol content increases), however, *S. cerevisiae* cells become a larger proportion of the microbes until eventually they virtually become the only microbes present. How *S. cerevisiae* performs this feat has been the subject of considerable research—and not only because of the economic importance of traditional fermentation biotechnologies. The goal of producing ethanol biofuel from cellulose (not from a food staple such as corn) in an efficient, cost-effective manner still remains elusive. The principal physiological reason that allows *S. cerevisiae* to ferment carbohydrates efficiently and dominate its environment is explained by the Crabtree effect, described next.

The Crabtree Effect

S. cerevisiae is a *facultative anaerobe*: it is capable of generating energy in both the presence and the absence of O_2 using aerobic metabolism (the citric acid cycle, the electron transport system, and oxidative phosphorylation) and fermentation, respectively. Unlike most fermenting organisms, *S. cerevisiae* can also ferment sugar in the presence of O_2. As glucose and/or fructose levels rise, pyruvate is diverted away from the citric acid cycle (the first phase of aerobic energy generation) into ethanol synthesis by conversion to acetaldehyde and CO_2 by pyruvate decarboxylase. This phenomenon, in which glucose represses aerobic metabolism, is the **Crabtree effect.** (In most organisms that use oxygen the **Pasteur effect** is observed: glycolysis is depressed when the gas is available.) In *S. cerevisiae* cells high glucose levels result in changes in gene expression. These changes promote the insertion of hexose transporters into the plasma membrane (resulting in rapid transport of glucose into the cell), the synthesis of glycolytic enzymes, and the inhibition of aerobic respiration enzyme synthesis. The diversion of

FIGURE 8A

Ethanol Metabolism in *S. cerevisiae*

When glucose levels are high, yeast cells shift into the "make, accumulate, consume" ethanol pathway. At first, glucose is converted by ADH1 into ethanol molecules to regenerate NAD^+. Ethanol is then released into the environment, where it kills competing microbes. Once glucose is depleted, glucose repression ends. Among the results of derepression is that ADH2, the enzyme that converts ethanol back into acetaldehyde, is synthesized. Acetaldehyde is subsequently converted into acetyl-CoA, the substrate for the citric acid cycle. Note that although the "make, accumulate, consume" strategy is expensive (i.e., the energy expended to synthesize the enzymes needed to convert ethanol into acetyl-CoA and the ATPs used to synthesize acetyl phosphate), yeast cells manage to kill off the competition and retrieve a waste product that they then use as an energy source.

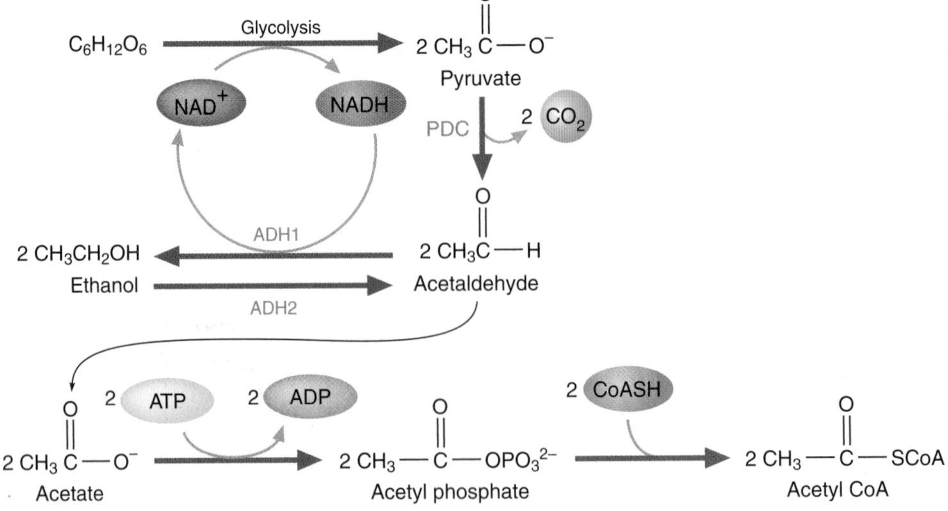

Biochemistry IN PERSPECTIVE cont.

pyruvate into ethanol production is also believed to result from an overflow phenomenon. There are too few pyruvate dehydrogenase molecules to convert pyruvate entirely into acetyl-CoA. Consequently, pyruvate decarboxylase converts excess pyruvate molecules into acetaldehyde.

At first glance the glucose-induced repression of aerobic metabolism appears inefficient because the production of ATP in fermentation (2 ATP per glucose) is minor compared with that of oxidative phosphorylation (approximately 30 ATP per glucose). However, rapid synthesis of ethanol along with its release into the environment by ethanol-tolerant *S. cerevisiae* has the effect of eliminating microbial competitors and predators. Once glucose levels are depleted and O_2 is available, there is a significant change in gene expression, called the *diauxic shift*, which alters yeast energy metabolism. Glucose repression ends, and the yeast cells proceed to reabsorb ethanol and reconvert it to acetaldehyde using ADH2 (p. 231). Acetaldehyde is then converted to acetyl-CoA, the citric acid cycle substrate. In effect, as a result of its unique energy metabolism, referred to as "make, accumulate, consume," *S. cerevisiae* cells can use their waste product as an energy source (**Figure 8A**). Humans take advantage of the first phase of yeast energy metabolism in the commercial production of wine and beer. Early in the process, aerobic fermentation is accompanied by O_2-requiring reactions that facilitate cell division, thereby increasing the number of yeast. As the oxygen in the fermentation vessel is depleted, ethanol production accelerates. Eventually the sugar level drops and fermentation slows. By excluding O_2 from the fermentation at this stage of the process, the ethanol content of the product is maximized because the shift to aerobic degradation of ethanol is prevented.

SUMMARY: A metabolic adaptation in the ancestors of *S. cerevisiae* allowed them to produce large quantities of ethanol, a toxic molecule that eliminated microbial competitors. Humans take advantage of this adaptation when they use *S. cerevisiae* in the production of alcoholic beverages and bread.

These isozymes have high affinities for glucose relative to its concentration in blood; that is, they are half-saturated at concentrations of less than 0.1 mM, although blood glucose levels are approximately 4 to 5 mM. In addition, hexokinases I, II, and III are inhibited by glucose-6-phosphate, the product of the reaction. When blood glucose levels are low, these properties allow cells such as those in brain and muscle to obtain sufficient glucose. When blood glucose levels are high, cells do not phosphorylate more glucose molecules than required to meet their immediate needs.

Hexokinase IV (or glucokinase) catalyzes the same reaction but has significantly different kinetic properties. Glucokinase (GK), found in liver as well as in certain cells in pancreas, intestine, and brain, requires much higher glucose concentrations for optimal activity (about 10 mM), and it is not inhibited by glucose-6-phosphate. In liver, GK diverts glucose into storage as glycogen. This capacity provides the resources used to maintain blood glucose levels, a major role of the liver. Consequently, after a carbohydrate meal the liver does not remove large quantities of glucose from the blood for glycogen synthesis until other tissues have satisfied their requirements for this molecule. In cell types where it occurs, GK is believed to be a *glucose sensor.* Because GK does not usually work at maximum velocity, it is highly sensitive to small changes in blood glucose levels. Its activity is linked to a signal transduction pathway. For example, the release of insulin (the hormone that promotes the uptake of glucose into muscle and adipose tissue cells) by pancreatic β-cells in response to rising blood levels of glucose is initiated by GK. GK regulation involves its binding to GK regulator protein (GKRP), a process that is triggered by high fructose-6-phosphate levels. GKRP/GK then moves into the nucleus. When blood glucose levels rise after a meal, GKRP releases GK (caused by exchange with fructose-1-phosphate), and GK moves back through the nuclear pores and can again phosphorylate glucose.

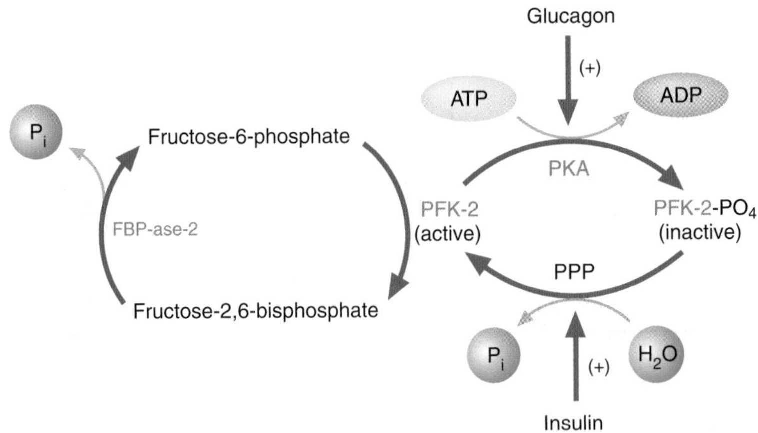

FIGURE 8.10
Fructose-2,6-Bisphosphate Level Regulation

Glycolysis is stimulated when fructose-2,6-bisphosphate, the activator of PFK-1, is synthesized by PFK-2. PFK-2 is, in turn, activated by a dephosphorylation reaction catalyzed by phosphoprotein phosphatase (PPP), an enzyme activated by insulin. PFK-2 is a bifunctional protein with two enzymatic activities: PFK-2 and fructose-2-6-bisphosphatase-2 (FBPase-2). As a result, the dephosphorylation reaction also inhibits FBP-ase-2, the enzymatic activity that converts fructose-2,6-bisphosphate to fructose-6-phosphate. Glycolysis is inhibited when fructose-2-6-bisphosphate levels are low, which is the result of a glucagon-stimulated and protein kinase A (PKA)-catalyzed phosphorylation reaction that inactivates PFK-2.

ALLOSTERIC REGULATION OF GLYCOLYSIS The reactions catalyzed by hexokinases I, II, and III, PFK-1, and PK can be switched on and off by allosteric effectors. In general, allosteric effectors are molecules whose cellular concentrations are sensitive indicators of a cell's metabolic state. Some allosteric effectors are product molecules. For example, hexokinases I, II, and III are inhibited by excess glucose-6-phosphate. Several energy-related molecules also act as allosteric effectors. For example, a high AMP concentration (an indicator of low energy production) activates PK. In contrast, PK is inhibited by a high ATP concentration (an indicator that the cell's energy requirements are being met). Acetyl-CoA, which accumulates when ATP is in rich supply, inhibits PK.

Of the three key enzymes in glycolysis, PFK-1 is the most carefully regulated. Its activity is allosterically inhibited by high levels of ATP and citrate, which are indicators that the cell's energy charge is high and that the citric acid cycle, a major component of the cell's energy-generating capacity, has slowed down. AMP is an allosteric activator of PFK-1. AMP levels, which increase when the energy charge of the cell is low, are a better predictor of energy deficit than ADP levels. Fructose-2,6-bisphosphate, an allosteric activator of PFK-1 activity in the liver, is synthesized by phosphofructokinase-2 (PFK-2) in response to hormonal signals correlated to blood glucose levels (**Figure 8.10**). When serum glucose levels are high, hormone-stimulated increase in fructose-2,6-bisphosphate coordinately increases the activity of PFK-1 (activates glycolysis) and decreases the activity of the enzyme that catalyzes the reverse reaction, fructose-1,6-bisphosphatase (inhibits gluconeogenesis, Section 8.2). AMP is an allosteric inhibitor of fructose-1,6-bisphosphatase. PFK-2 is a bifunctional enzyme that behaves as a phosphatase when phosphorylated in response to the hormone glucagon (released into blood in reponse to low blood sugar; see later). It functions as a kinase when dephosphorylated in response to the hormone insulin (high blood sugar). Fructose-2,6-bisphosphate, produced via hormone-induced covalent modification of PFK-2, is an indicator of high levels of available glucose and allosterically activates PFK-1. Accumulated fructose-1,6-bisphosphate activates PK, providing a feed-forward mechanism of control (i.e., fructose-1,6-bisphosphate is an allosteric activator). The allosteric regulation of glycolysis is summarized in **Table 8.1**.

HORMONAL REGULATION Glycolysis is also regulated by the peptide hormones glucagon and insulin. **Glucagon**, released by pancreatic α-cells when blood glucose is low, activates the phosphatase function of PFK-2, thereby reducing the level of fructose-2,6-bisphosphate in the cell. As a result, PFK-1 activity and flux through glycolysis are decreased. In liver, glucagon also inactivates PK. Glucagon's effects, triggered by binding to its receptor on target cell surfaces, are mediated by cyclic AMP (cAMP). cAMP (p. 309) is synthesized from ATP in a reaction catalyzed by adenylate cyclase, a plasma

TABLE 8.1 Allosteric Regulation of Glycolysis

Enzyme	Activator	Inhibitor
Hexokinase		Glucose-6-phosphate, ATP
PFK-1	Fructose-2,6-bisphosphate, AMP	Citrate, ATP
Pyruvate kinase	Fructose-1,6-bisphosphate, AMP	Acetyl-CoA, ATP

membrane protein. Once synthesized, cAMP binds to and activates protein kinase A (PKA). PKA then initiates a signal cascade of phosphorylation/dephosphorylation reactions that alter the activities of a diverse set of enzymes and transcription factors. **Transcription factors** are proteins that regulate or initiate RNA synthesis by binding to specific DNA sequences called **response elements**.

Insulin is a peptide hormone released from pancreatic β-cells when blood glucose levels are high. The effects of insulin on glycolysis include activation of the kinase function of PFK-2, which increases the level of fructose-2,6-bisphosphate in the cell, in turn increasing glycolytic flux. In cells containing insulin-sensitive glucose transporters (muscle and adipose tissue but not liver or brain) insulin promotes the translocation of glucose transporters to the cell surface. When insulin binds to its cell-surface receptor, the receptor protein undergoes several autophosphorylation reactions, which trigger numerous intracellular signal cascades that involve phosphorylation and dephosphorylation of target enzymes and transcription factors. Many of insulin's effects on gene expression are mediated by the transcription factor SREBP1c, a sterol regulatory element binding protein (p. 452). As a result of SREBP1c activation, there is increased synthesis of GK and PK, and key enzymes in *de novo* ("new") **lipogenesis** (synthesis of "new" fatty acids)

GLUCOSE-INDUCED GENE EXPRESSION Glucose acts as a signaling molecule in lipogenic (i.e., fat-synthesizing) organs such as liver and adipose tissue. When there are high concentrations of glucose, the molecule promotes the transcription of genes involved in *de novo* lipogenesis (synthesis of new fatty acids from carbohydrate molecules). Target genes encode enzymes in glycolysis (e.g., aldolase and PK), the pentose phosphate pathway (e.g., glucose-6-phosphate dehydrogenase and transketoase), and fatty acid synthesis (e.g., fatty acid synthase). Glucose-stimulated transcription of target genes is mediated by a heterotetramer composed of ChREBP (carbohydrate response element binding protein) and Mlx. In the liver when blood glucose concentrations are low, *ChREBP*, located in the cytoplasm in its inactive form (phosphorylated at Ser196), is activated when (1) glucose is phosphorylated by glucokinase to yield glucose-6-phosphate and (2) two other sugars are available: xylulose-5-phosphate (a pentose phosphate pathway intermediate) and fructose-2,6-bisphosphate (the allosteric activator of PFK-1). Xylulose-5-phosphate activates a protein phosphatase that dephosphorylates ChREBP, which can then move into the nucleus. Once the ChREBP/Mlx heterodimer forms, it binds to the *carbohydrate response elements* (ChREs) that occur in the promoters of glucose responsive genes. (Promoter sequences control where and if transcription will occur.) Note that both glucose and insulin regulate *de novo* fat synthesis from excess glucose molecules.

AMPK: A METABOLIC MASTER SWITCH AMP-activated protein kinase (AMPK) is an enzyme that plays a central role in energy metabolism. First discovered as a regulator of lipid metabolism, AMPK is now known to affect glucose metabolism as well. Once AMPK is activated, as a result of an increase in a cell's AMP:ATP ratio, it phosphorylates target proteins (enzymes and transcription factors). AMPK switches off anabolic pathways (e.g., protein and lipid synthesis) and switches on catabolic pathways (e.g., glycolysis and fatty acid oxidation). AMPK's regulatory effects on glycolysis include the following. In cardiac and skeletal muscle, AMPK promotes glycolysis by facilitating the stress- or exercise-induced recruitment of glucose transporters to the plasma membrane. In cardiac cells, AMPK stimulates glycolysis by activating PFK-2. AMPK structure and its functional properties are described in Chapter 12.

QUESTION 8.2

Insulin is a hormone secreted by the pancreas when blood sugar increases. Its most easily observable function is to reduce the blood sugar level to normal. The binding of insulin to its target cells promotes the transport of glucose across the plasma membrane. The capacity of an individual to respond to a carbohydrate meal by reducing blood glucose concentration quickly is referred to as *glucose tolerance*. Chromium-deficient animals show a decreased glucose tolerance; that is, they cannot remove glucose from blood quickly enough. The metal is believed to facilitate the binding of insulin to cells. Do you think the chromium is acting as an allosteric activator or cofactor?

QUESTION 8.3

Louis Pasteur, the great nineteenth-century French chemist and microbiologist, was the first scientist to observe that cells that can oxidize glucose completely to CO_2 and H_2O use glucose more rapidly in the absence of O_2 than in its presence. The oxygen molecule seems to inhibit glucose consumption. Explain in general terms the significance of this finding, now referred to as the Pasteur effect.

8.2 GLUCONEOGENESIS

Gluconeogenesis, the formation of new glucose molecules from noncarbohydrate precursors, occurs primarily in the liver. Precursor molecules include lactate, pyruvate, glycerol, and certain α-keto acids (molecules derived from amino acids). Under certain conditions (i.e., metabolic acidosis or starvation) the kidney can make small amounts of new glucose. Between meals adequate blood glucose levels are maintained by the hydrolysis of liver glycogen. When liver glycogen is depleted (e.g., because of prolonged fasting or vigorous exercise), the gluconeogenesis pathway provides the body with adequate glucose. Brain and red blood cells rely exclusively on glucose as their energy source.

Gluconeogenesis Reactions

The reaction sequence in gluconeogenesis is largely the reverse of glycolysis. Recall, however, that three glycolytic reactions (the reactions catalyzed by hexokinase, PFK-1, and PK) are irreversible. In gluconeogenesis, alternate reactions catalyzed by different enzymes are used to bypass these obstacles. The reactions unique to gluconeogenesis are listed next. The entire gluconeogenic pathway and its relationship to glycolysis are illustrated in Figure 8.11. The bypass reactions of gluconeogenesis are as follows:

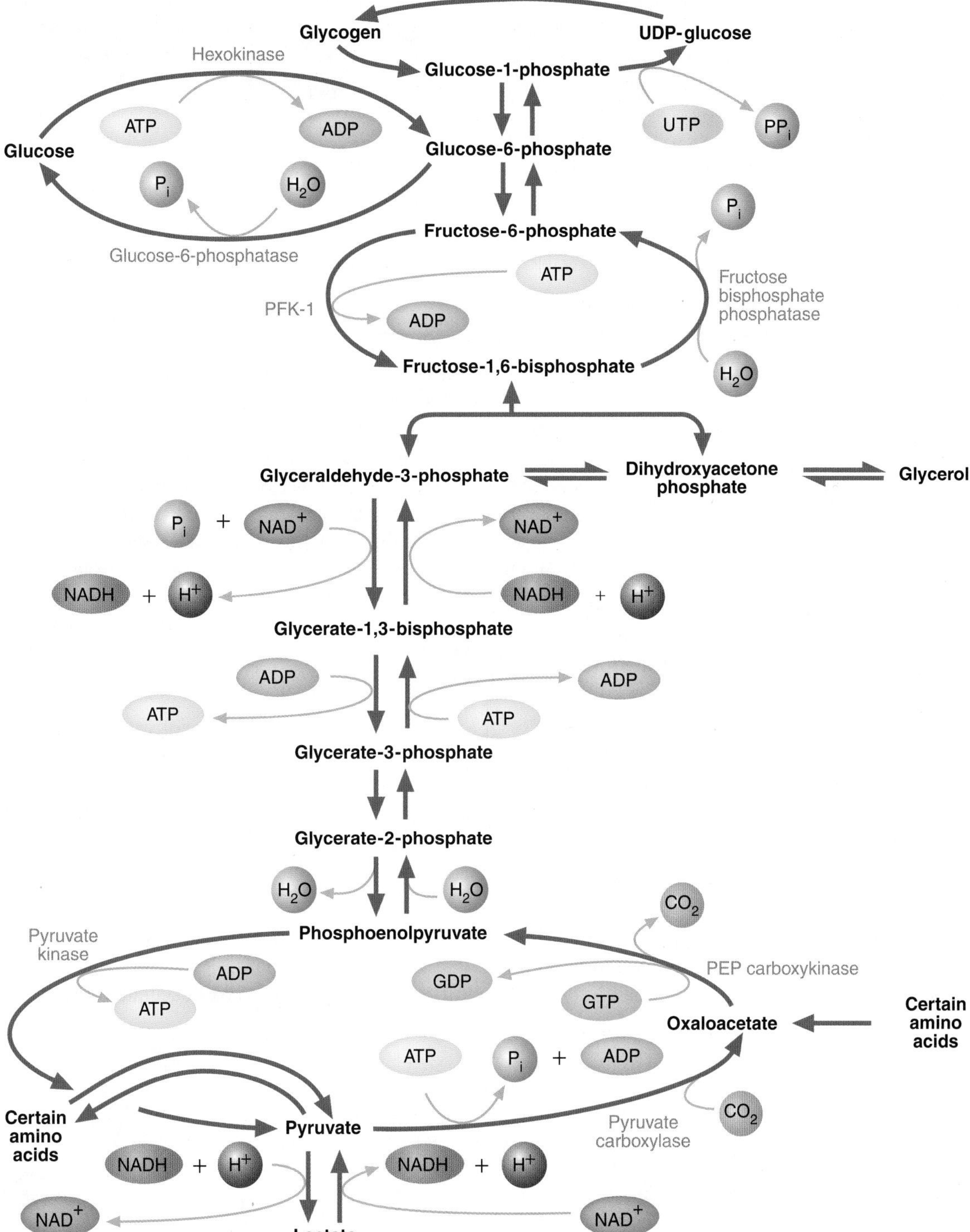

FIGURE 8.11

Carbohydrate Metabolism: Gluconeogenesis and Glycolysis

In gluconeogenesis, which occurs when blood sugar levels are low and liver glycogen is depleted, 7 of the 10 reactions of glycolysis are reversed. Three irreversible glycolytic reactions are bypassed by alternative reactions. The major substrates for gluconeogenesis are certain amino acids (derived from muscle), lactate (formed in muscle and red blood cells), and glycerol (produced from the degradation of triacylglycerols). In contrast to the reactions of glycolysis, which occur only in cytoplasm, the gluconeogenesis reactions catalyzed by pyruvate carboxylase and, in some species, PEP carboxykinase occur within the mitochondria. The reaction catalyzed by glucose-6-phosphatase takes place in the endoplasmic reticulum. Note that gluconeogenesis and glycolysis do not occur simultaneously. Pyruvate, the product of glycolysis, is converted either to acetyl-CoA (not shown) or to lactate.

Biochemistry IN PERSPECTIVE

Turbo Design Can Be Dangerous

Why must turbo design pathways be rigorously controlled? Catabolic pathways with a turbo design are optimized and efficient. In glycolysis, for example, two of the four ATPs produced from each glucose molecule are fed back to the fuel input stage of the pathway to drive the pathway forward. As indicated by its use by most modern living organisms, glycolysis has been a tremendously successful energy-generating strategy. However, the early phases of such pathways must be negatively regulated to prevent buildup of intermediates and overuse of fuel. For example, in certain circumstances, the turbo design of glycolysis makes some cells vulnerable to a phenomenon called "substrate-accelerated death."

Certain types of mutant yeast cells, for example, are unable to grow anaerobically on glucose despite having a completely functional glycolytic pathway. These mutants die when exposed to large concentrations of glucose. Amazingly, research efforts have revealed that defects in *TPS1*, the gene that codes for the catalytic subunit of trehalose-6-phosphate synthase, are responsible. Trehalose-6-phosphate (Tre-6-P), an α-(1,1)-disaccharide of glucose, is a compatible solute (see the online Chapter 3 Biochemistry in Perspective essay entitled Water, Abiotic Stress, and Compatible Solutes) used by yeast and various other organisms to resist several forms of abiotic stress. Apparently, Tre-6-P is a normal inhibitor of HK (and possibly a glucose transporter). In the absence of a functional TPS1 protein and when glucose becomes available, glycolytic flux in the mutant cells rapidly accelerates. In a relatively short time, as a result of the turbo design of the pathway, most available phosphate has been incorporated into glycolytic intermediates, and the cell's ATP level is too low to sustain cellular processes. This and other similar examples of substrate-accelerated cell death in other species provide insight into the importance of the intricate regulatory mechanisms observed in living organisms.

SUMMARY: Defects in the intricate regulatory mechanism that controls a turbo design pathway can make an organism vulnerable to substrate-accelerated death through uncontrolled pathway flux.

1. **Synthesis of PEP**. PEP synthesis from pyruvate requires two enzymes: pyruvate carboxylase and PEP carboxykinase. Pyruvate carboxylase, found within mitochondria, converts pyruvate to oxaloacetate (OAA):

$$\underset{\text{Pyruvate}}{CH_3-CO-COO^-} + CO_2 + H_2O + ATP \xrightarrow[\text{(biotin)}]{\text{Pyruvate carboxylase}} \underset{\text{Oxaloacetate (OAA)}}{^-OOC-CH_2-CO-COO^-} + ADP + P_i + H^+$$

The transfer of CO_2 to form the product OAA is mediated by the coenzyme *biotin* (p. 442), which is covalently bound within the enzyme's active site. The keto form of pyruvate is converted to the enol form by a general base in the enzyme active site. As the carbonyl on carbon-2 reforms, the electron pair of the double bond attacks CO_2, released from biotin, to yield OAA.

OAA is then decarboxylated and phosphorylated by PEP carboxykinase in a reaction driven by the hydrolysis of GTP:

OAA + GTP → (PEP carboxykinase) → PEP + GDP + CO_2

PEP carboxykinase is found within the mitochondria of some species and in the cytoplasm of others. In humans this enzymatic activity is found in both compartments. Because the inner mitochondrial membrane is impermeable to OAA, cells that lack mitochondrial PEP carboxykinase transfer OAA into the cytoplasm using the **malate shuttle**. In this process, OAA is converted into malate by mitochondrial malate dehydrogenase. After the transport of malate across mitochondrial membrane, the reverse reaction (to form OAA and NADH) is catalyzed by cytoplasmic malate dehydrogenase. The malate shuttle allows gluconeogenesis to continue because it provides the NADH required for the reaction catalyzed by glyceraldehyde-3-phosphate dehydrogenase.

OAA + NADH + H^+ ⇌ (Malate dehydrogenase) ⇌ Malate + NAD^+

2. **Conversion of fructose-1,6-bisphosphate to fructose-6-phosphate.** The irreversible PFK-1–catalyzed reaction in glycolysis is bypassed by fructose-1,6-bisphosphatase:

Fructose-1,6-bisphosphate + H_2O → (Fructose-1,6-bisphosphatase) → Fructose-6-phosphate + P_i

This exergonic reaction ($\Delta G^{\circ\prime} = -16.7$ kJ/mol) is also irreversible under cellular conditions. ATP is not regenerated, and inorganic phosphate (P_i) is also produced. Fructose-1,6-bisphosphatase is an allosteric enzyme. Its activity is stimulated by citrate and inhibited by AMP and fructose-2,6-bisphosphate.

3. **Formation of glucose from glucose-6-phosphate**. Glucose-6-phosphatase, found only in liver and kidney, catalyzes the irreversible hydrolysis of glucose-6-phosphate to form glucose and P_i. Glucose is subsequently released into the blood.

   Each of the foregoing reactions is matched by an opposing irreversible reaction in glycolysis. Each set of such paired reactions is referred to as a *substrate cycle*. Because they are coordinately regulated (an activator of the enzyme catalyzing the forward reaction serves as an inhibitor of the enzyme catalyzing the reverse reaction), little energy is wasted, even though both enzymes may be operating at some level at the same time. *Flux control* (regulation of the flow of substrate and removal of product) is more effective if transient accumulation of product is funneled back through the cycle. The catalytic velocity of the forward enzyme will remain high if the concentration of the substrate is maximized. The gain in catalytic efficiency more than makes up for the small energy loss in recycling the product.

   Gluconeogenesis is an energy-consuming process. Instead of generating ATP (as in glycolysis), gluconeogenesis requires the hydrolysis of six high-energy phosphate bonds.

QUESTION 8.4

Malignant hyperthermia is a rare, inherited disorder triggered during surgery by certain anesthetics. A dramatic (and dangerous) rise in body temperature (as high as 112°F) is accompanied by muscle rigidity and acidosis. The excessive muscle contraction is initiated by a large release of calcium from the sarcoplasmic reticulum, a calcium-storing organelle in muscle cells. Acidosis results from excessive lactic acid production. Prompt treatment to reduce body temperature and to counteract the acidosis is essential to save the patient's life. A probable contributing factor to this disorder is wasteful cycling between glycolysis and gluconeogenesis. Explain why this is a reasonable explanation.

QUESTION 8.5

After examining the gluconeogenic pathway reaction summary illustrated here, account for each component in the equation. [*Hint:* The hydrolysis of each nucleotide releases a proton.]

$$\underset{\text{Pyruvic acid}}{2C_3H_4O_3} + 4\,\text{ATP} + 2\,\text{GTP} + 2\,\text{NADH} + 2\,H^+ + 6\,H_2O \longrightarrow$$

$$\underset{\text{Glucose}}{1C_6H_{12}O_6} + 4\,\text{ADP} + 2\,\text{GDP} + 2\,\text{NAD}^+ + 6\,HPO_4^{2-} + 6\,H^+$$

QUESTION 8.6

Patients with *von Gierke's disease* (a glycogen storage disease) lack glucose-6-phosphatase activity. Two prominent symptoms of this disorder are fasting hypoglycemia and lactic acidosis. Can you explain why these symptoms occur?

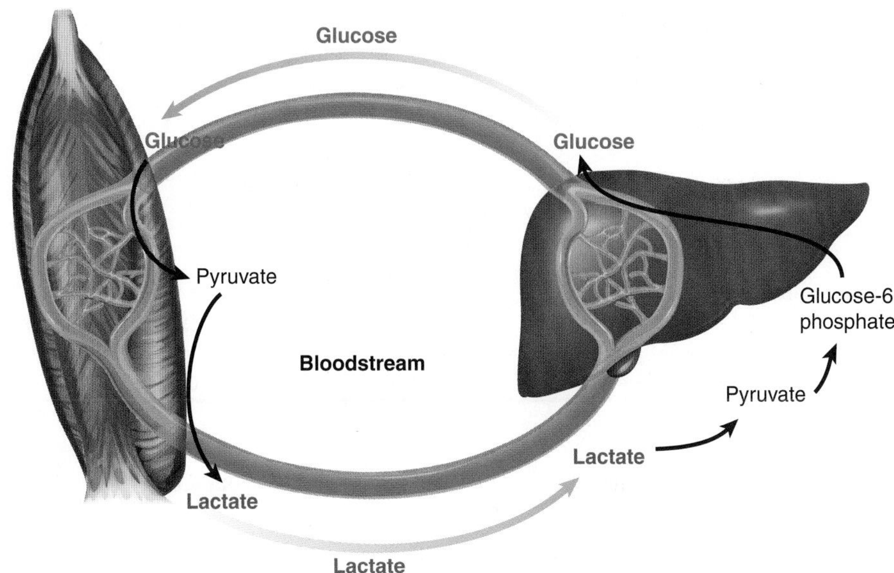

FIGURE 8.12
The Cori Cycle

During strenuous exercise, lactate is produced anaerobically in muscle cells. After passing through blood to the liver, lactate is converted to glucose by gluconeogenesis.

Gluconeogenesis Substrates

As previously mentioned, several metabolites are gluconeogenic precursors. Three of the most important substrates are described briefly.

Lactate is released by red blood cells and other cells that lack mitochondria or have low oxygen concentrations. In the **Cori cycle**, lactate is released by skeletal muscle during exercise (**Figure 8.12**). After lactate is transferred to the liver, it is reconverted to pyruvate by lactate dehydrogenase and then to glucose by gluconeogenesis.

Glycerol, a product of fat metabolism in adipose tissue, is transported to the liver in the blood and then converted to glycerol-3-phosphate by glycerol kinase. Oxidation of glycerol-3-phosphate to form DHAP occurs when cytoplasm NAD^+ concentration is relatively high.

$$\underset{\textbf{Glycerol}}{HOCH_2{-}CH(OH){-}CH_2OH} \xrightarrow[ATP \;\to\; ADP]{\text{Glycerol kinase}} \underset{\textbf{Glycerol-3-phosphate}}{HOCH_2{-}CH(OH){-}CH_2{-}O{-}PO_3^{2-}} \xrightarrow[NAD^+ \;\to\; NADH + H^+]{\text{Glycerol phosphate dehydrogenase}} \underset{\textbf{DHAP}}{HOCH_2{-}C({=}O){-}CH_2{-}O{-}PO_3^{2-}}$$

Of all the amino acids that can be converted to glycolytic intermediates (molecules referred to as *glucogenic*), alanine is perhaps the most important. When exercising muscle produces large quantities of pyruvate, some of these molecules are converted to alanine by a transamination reaction involving glutamate:

$$\underset{\textbf{Pyruvate}}{CH_3-\overset{\overset{\displaystyle O}{\|}}{C}-\overset{\overset{\displaystyle O}{\|}}{C}-O^-} + \underset{\textbf{L-Glutamate}}{^-O-\overset{\overset{\displaystyle O}{\|}}{C}-CH_2-CH_2-\overset{\overset{\displaystyle H}{|}}{\underset{\underset{\displaystyle +NH_3}{|}}{C}}-\overset{\overset{\displaystyle O}{\|}}{C}-O^-}$$

$$\Big\Updownarrow \text{ Alanine transaminase}$$

$$\underset{\textbf{L-Alanine}}{CH_3-\overset{\overset{\displaystyle H}{|}}{\underset{\underset{\displaystyle +NH_3}{|}}{C}}-\overset{\overset{\displaystyle O}{\|}}{C}-O^-} + \underset{\boldsymbol{\alpha}\textbf{-Ketoglutarate}}{^-O-\overset{\overset{\displaystyle O}{\|}}{C}-CH_2-CH_2-\overset{\overset{\displaystyle O}{\|}}{C}-\overset{\overset{\displaystyle O}{\|}}{C}-O^-}$$

After it has been transported to the liver, alanine is reconverted to pyruvate and then to glucose. The **glucose-alanine cycle** (**Figure 8.13**) serves several purposes. In addition to its role in recycling α-keto acids between muscle and liver, the glucose-alanine cycle is a mechanism for transporting amino nitrogen to the liver. In α-keto acids, sometimes referred to as carbon skeletons, a carbonyl group is directly attached to the carboxyl group. Once alanine reaches the liver, it is reconverted to pyruvate. The amino nitrogen is then incorporated into urea or transferred to other α-keto acids to restore the amino acid balance in the liver (Chapter 15).

Gluconeogenesis Regulation

As with other metabolic pathways, the rate of gluconeogenesis is affected primarily by substrate availability, allosteric effectors, and hormones. Not surprisingly, gluconeogenesis is stimulated by high concentrations of lactate,

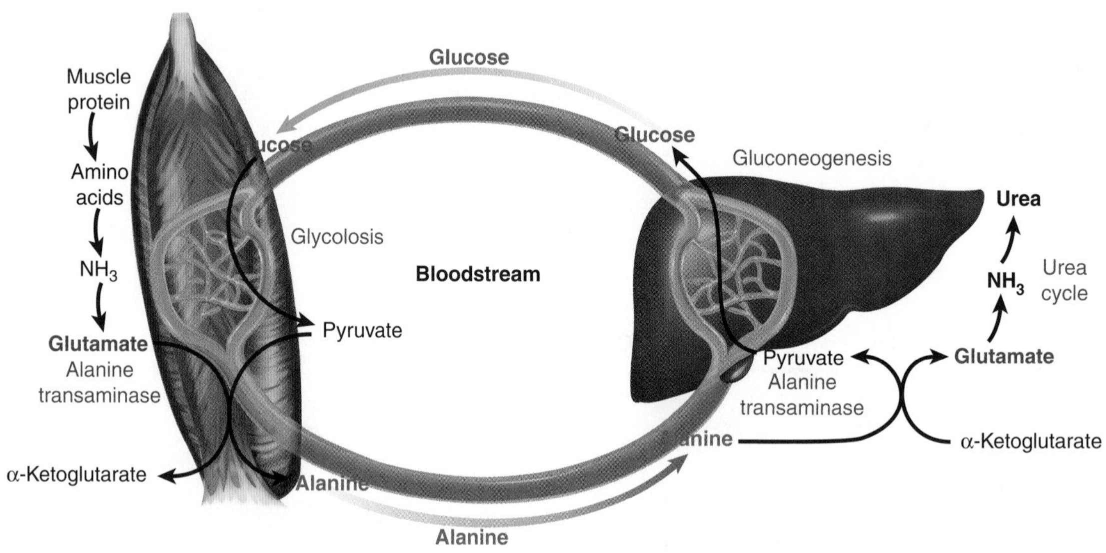

FIGURE 8.13
The Glucose-Alanine Cycle

Alanine is formed from pyruvate in muscle. After it has been transported to the liver, alanine is reconverted to pyruvate by alanine transaminase. Eventually pyruvate is used in the synthesis of new glucose. Because muscle cannot synthesize urea from amino nitrogen, the glucose-alanine cycle is used to transfer amino nitrogen to the liver.

glycerol, and amino acids. A high-fat diet, starvation, and prolonged fasting make large quantities of these molecules available.

ALLSOTERIC MODULATION The four key enzymes in gluconeogenesis (pyruvate carboxylase, PEP carboxykinase, fructose-1,6-bisphosphatase, and glucose-6-phosphatase) are affected to varying degrees by allosteric modulators. For example, fructose-1,6-bisphosphatase is activated by citrate and inhibited by AMP and fructose-2,6-bisphosphate. Acetyl-CoA activates pyruvate carboxylase. (The concentration of acetyl-CoA, a product of fatty acid degradation, is especially high during starvation.) **Figure 8.14** provides an overview of the allosteric regulation of glycolysis and gluconeogenesis.

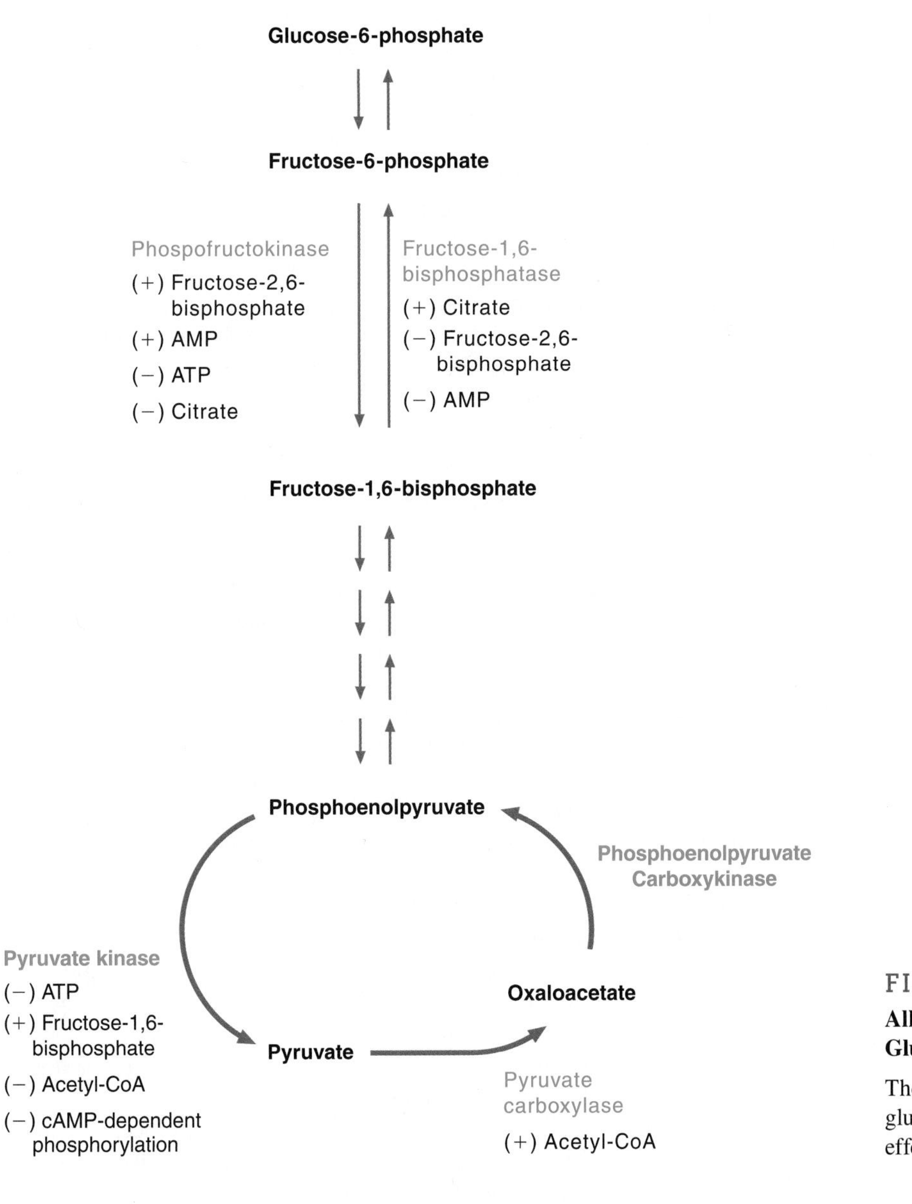

FIGURE 8.14

Allosteric Regulation of Glycolysis and Gluconeogenesis

The key enzymes in glycolysis and gluconeogenesis are regulated by allosteric effectors. Activator, +; inhibitor, −.

HORMONAL REGULATION As with other biochemical pathways, hormones affect gluconeogenesis by altering the concentrations of allosteric effectors and the key rate-determining enzymes. As mentioned previously, glucagon depresses the synthesis of fructose-2,6-bisphosphate, which releases the inhibition of fructose-1,6-bisphosphatase, and inactivates the glycolytic enzyme PK, via a cAMP-triggered phosphorylation reaction (p. 288). Hormones also influence gluconeogenesis by altering enzyme synthesis. For example, the synthesis of gluconeogenic enzymes is stimulated by cortisol, a steroid hormone produced in the cortex of the adrenal gland that facilitates the body's adaptation to stressful situations. Finally, insulin action leads to the synthesis of new molecules of GK, PFK-1 (SREBP1c-induced), and PFK-2 (glycolysis favored). Insulin also depresses the synthesis (also via SREBP1c) of PEP carboxykinase, fructose-1,6-bisphosphatase, and glucose-6-phosphatase. Glucagon action leads to the synthesis of additional molecules of PEP carboxykinase, fructose-1,6-bisphosphatase, and glucose-6-phosphatase (gluconeogenesis favored).

The hormones that regulate glycolysis and gluconeogenesis alter the phosphorylation state of certain target proteins in the liver cell, which in turn modifies gene expression. The key point to remember is that insulin and glucagon have opposing effects on carbohydrate metabolism. The direction of metabolite flux, (i.e., whether either glycolysis or gluconeogenesis is active) is largely determined by the ratio of insulin to glucagon. After a carbohydrate meal, the insulin/glucagon ratio is high and glycolysis in the liver predominates over gluconeogenesis. After a period of fasting or following a high-fat, low-carbohydrate meal, the insulin:glucagon ratio is low and gluconeogenesis in the liver predominates over glycolysis. The availability of ATP is the second important regulator in the reciprocal control of glycolysis and gluconeogenesis in that high levels of AMP, the low-energy hydrolysis product of ATP, increase the flux through glycolysis at the expense of gluconeogenesis, and low levels of AMP increase the flux through gluconeogenesis at the expense of glycolysis. Although control at the PFK-1/fructose-1,6-bisphosphatase cycle would appear to be sufficient for this pathway, control at the PK step is key because it permits the maximal retention of PEP, a molecule with a high phosphate transfer potential.

KEY CONCEPTS

- Gluconeogenesis, the synthesis of new glucose molecules from noncarbohydrate precursors, occurs primarily in the liver.
- The reaction sequence is the reverse of glycolysis except for three reactions that bypass irreversible steps in glycolysis.

8.3 THE PENTOSE PHOSPHATE PATHWAY

The pentose phosphate pathway is an alternative metabolic pathway for glucose oxidation in which no ATP is generated. Its principal products are NADPH, a reducing agent required in several anabolic processes, and ribose-5-phosphate, a structural component of nucleotides and nucleic acids. The pentose phosphate pathway occurs in the cytoplasm in two phases: oxidative and nonoxidative. In the oxidative phase of the pathway, the conversion of glucose-6-phosphate to ribulose-5-phosphate is accompanied by the production of two molecules of NADPH. The nonoxidative phase involves the isomerization and condensation of a number of different sugar molecules. Three intermediates in this process that are useful in other pathways are ribose-5-phosphate, fructose-6-phosphate, and glyceraldehyde-3-phosphate.

The oxidative phase of the pentose phosphate pathway consists of three reactions (**Figure 8.15a**). In the first reaction, glucose-6-phosphate dehydrogenase (G-6-PD) catalyzes the oxidation of glucose-6-phosphate. 6-Phosphogluconolactone and NADPH are products in this reaction. 6-Phospho-D-glucono-δ-lactone is then hydrolyzed to produce 6-phospho-D-gluconate. A second molecule of NADPH is produced during the oxidative decarboxylation of 6-phosphogluconate, a reaction that yields ribulose-5-phosphate.

A substantial amount of the NADPH required for reductive processes (i.e., lipid biosynthesis) is supplied by these reactions. For this reason this pathway is most active in cells in which relatively large amounts of lipids are synthesized

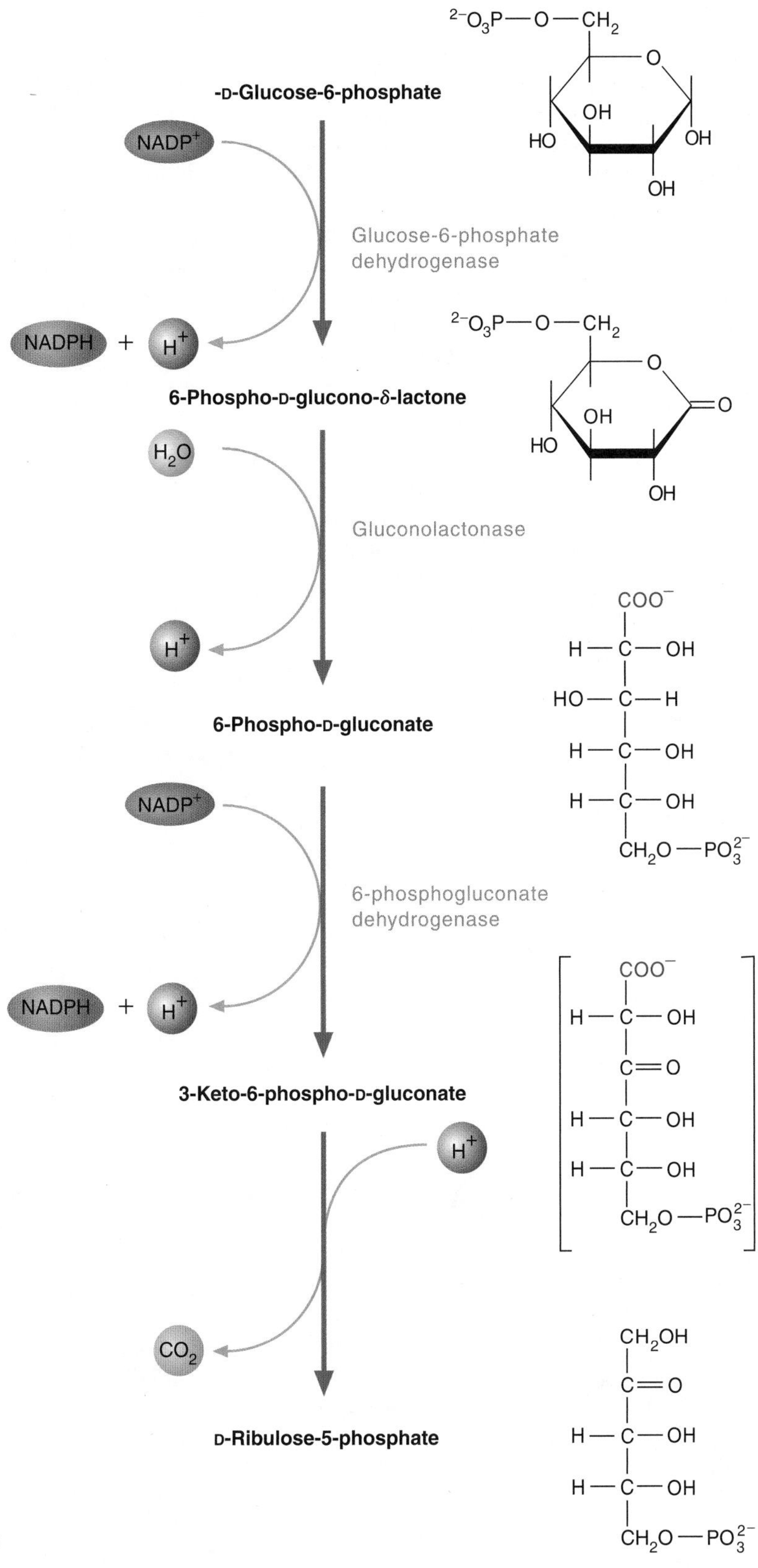

FIGURE 8.15a

The Pentose Phosphate Pathway

(a) The oxidative phase. NADPH is an important product of these reactions.

(e.g., adipose tissue, adrenal cortex, mammary glands, and the liver). NADPH is also a powerful antioxidant. (**Antioxidants** are substances that prevent the oxidation of other molecules. Their roles in living processes are described in Chapter 10.) Consequently, the oxidative phase of the pentose phosphate pathway is also quite active in cells that are at high risk for oxidative damage, such as red blood cells.

The nonoxidative phase of the pathway begins with the conversion of ribulose-5-phosphate to ribose-5-phosphate by ribulose-5-phosphate isomerase or to xylulose-5-phosphate by ribulose-5-phosphate epimerase. During the remaining reactions of the pathway (**Figure 8.15b**), transketolase and transaldolase catalyze the interconversions of trioses, pentoses, and hexoses. *Transketolase* is a

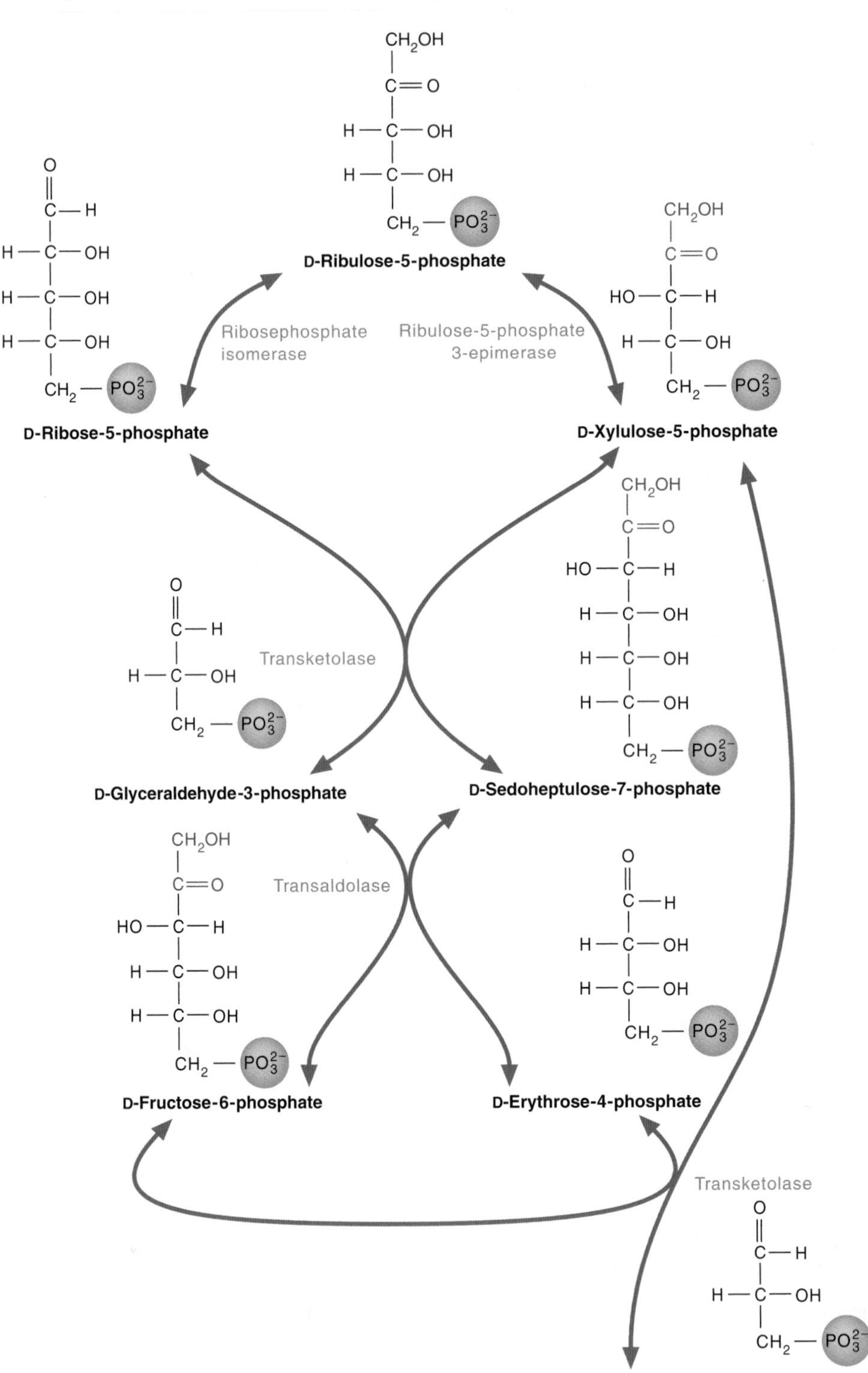

FIGURE 8.15b
The Pentose Phosphate Pathway

(b) The nonoxidative phase. When cells require more NADPH than pentose phosphates, the enzymes in the nonoxidative phase convert ribose-5-phosphate into the glycolytic intermediates fructose-6-phosphate and glyceraldehyde-3-phosphate.

thiamine pyrophosphate (TPP)–requiring enzyme that transfers two-carbon units from a ketose to an aldose. TPP is the coenzyme form of thiamine, also known as vitamin B_1. Transketolase catalyzes two reactions. In the first reaction, the enzyme transfers a two-carbon unit from xylulose-5-phosphate to ribose-5-phosphate, yielding G-3-P and sedoheptulose-7-phosphate. In the second transketolase-catalyzed reaction, a two-carbon unit from another xylulose-5-phosphate molecule is transferred to erythrose-4-phosphate to form a second molecule of G-3-P and fructose-6-phosphate. *Transaldolase* transfers three-carbon units from a ketose to an aldose. In the reaction catalyzed by transaldolase, a three-carbon unit is transferred from sedoheptulose-7-phosphate to G-3-P. The products formed are fructose-6-phosphate and erythrose-4-phosphate. The result of the nonoxidative phase of the pathway is the synthesis of ribose-5-phosphate and the glycolytic intermediates G-3-P and fructose-6-phosphate.

When pentose sugars are not required for biosynthetic reactions, the metabolites in the nonoxidative portion of the pathway are converted into glycolytic intermediates that can then be further degraded to generate energy or converted into precursor molecules for biosynthetic processes (**Figure 8.16**).

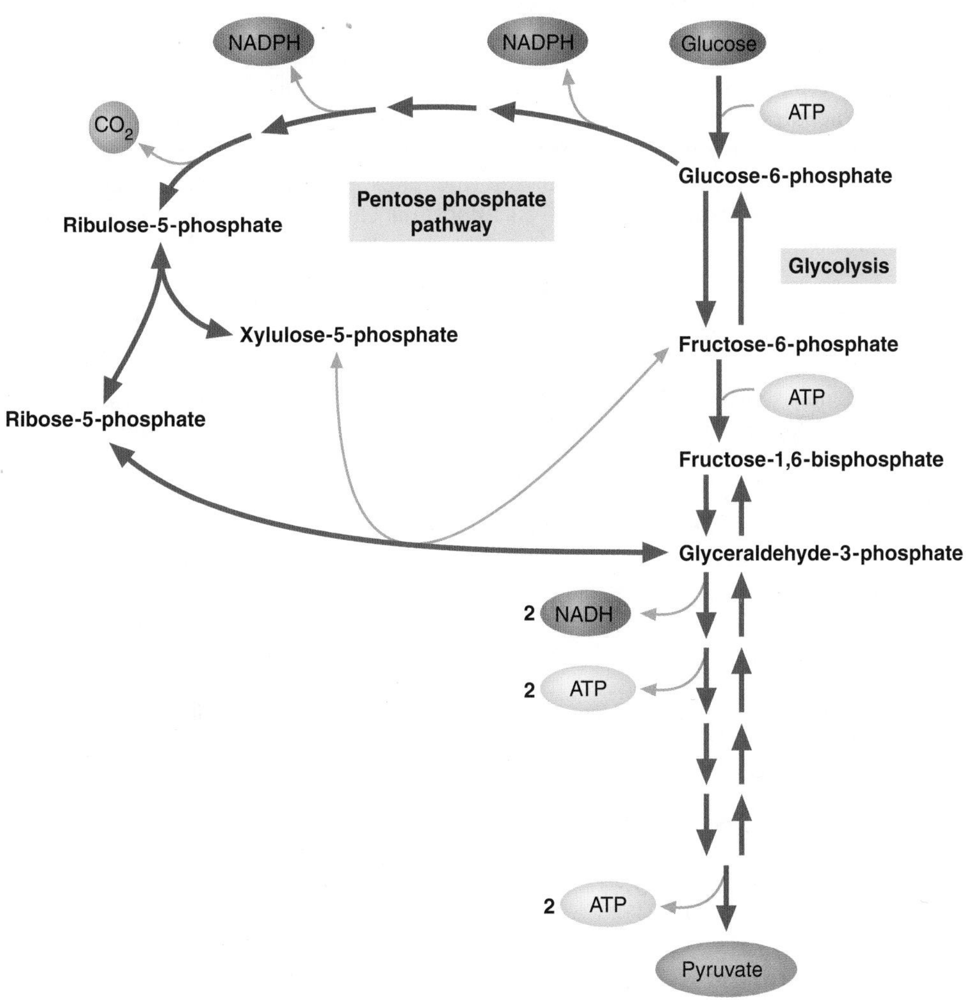

FIGURE 8.16

Carbohydrate Metabolism: Glycolysis and the Pentose Phosphate Pathway

If the cell requires more NADPH than ribose molecules, it can channel the products of the nonoxidative phase of the pentose phosphate pathway into glycolysis. As this overview of the two pathways illustrates, excess ribulose-5-phosphate can be converted into the glycolytic intermediates fructose-6-phosphate and G-3-P.

For this reason the pentose phosphate pathway is also referred to as the *hexose monophosphate shunt*. In plants, the pentose phosphate pathway is involved in the synthesis of glucose during the dark reactions of photosynthesis (Chapter 13).

The pentose phosphate pathway is regulated to meet the cell's moment-by-moment requirements for NADPH and ribose-5-phosphate. The oxidative phase is very active in cells such as red blood cells or hepatocytes in which demand for NADPH is high. (Lipid synthesis is a major consumer of NADPH.) In contrast, the oxidative phase is virtually absent in cells (e.g., muscle cells) that synthesize little or no lipid. G-6-PD, which catalyzes a key regulatory step in the pentose phosphate pathway, is inhibited by NADPH and stimulated by GSSG, the oxidized form of the antioxidant glutathione (p. 533), and glucose-6-phosphate. In addition, diets high in carbohydrate increase the synthesis of both G-6-PD and phosphogluconate dehydrogenase.

KEY CONCEPT

The pentose phosphate pathway produces NADPH, ribose-5-phosphate, and the glycolytic intermediates fructose-6-phosphate and G-3-P.

8.4 METABOLISM OF OTHER IMPORTANT SUGARS

Several sugars other than glucose are important in vertebrates. The most notable of these are fructose, galactose, and mannose. Besides glucose, these molecules are the most common sugars found in oligosaccharides and polysaccharides. They are also energy sources. The reactions by which these sugars are converted into glycolytic intermediates are illustrated in **Figure 8.17**. The metabolism of fructose, an important component of the human diet, is discussed.

Fructose Metabolism

Dietary sources of fructose include fruit, honey, sucrose, and high-fructose corn syrup, an inexpensive sweetener used in a wide variety of processed foods and beverages.

Fructose, second only to glucose as a source of carbohydrate in the modern human diet, enters the glycolytic pathway in the liver, where it is converted to fructose-1-phosphate by fructokinase:

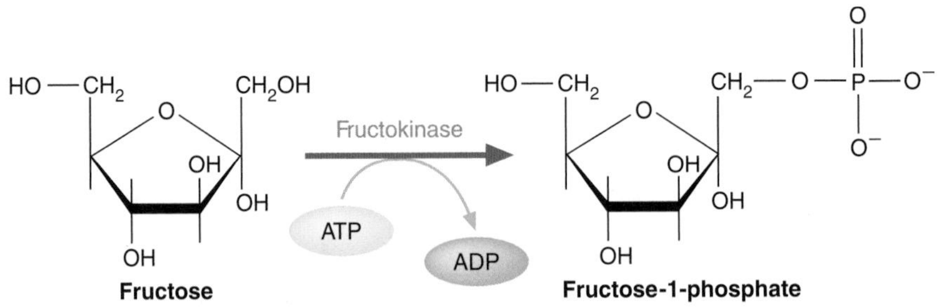

HO—CH_2 O CH_2OH OH OH OH
Fructose

Fructokinase
ATP → ADP

HO—CH_2 O CH_2—O—P(=O)(O$^-$)—O$^-$ OH OH OH
Fructose-1-phosphate

When fructose-1-phosphate enters the glycolytic pathway, it is first split into DHAP and glyceraldehyde by fructose-1-phosphate aldolase. DHAP is then converted to G-3-P by triose phosphate isomerase. G-3-P is generated from glyceraldehyde and ATP by glyceraldehyde kinase.

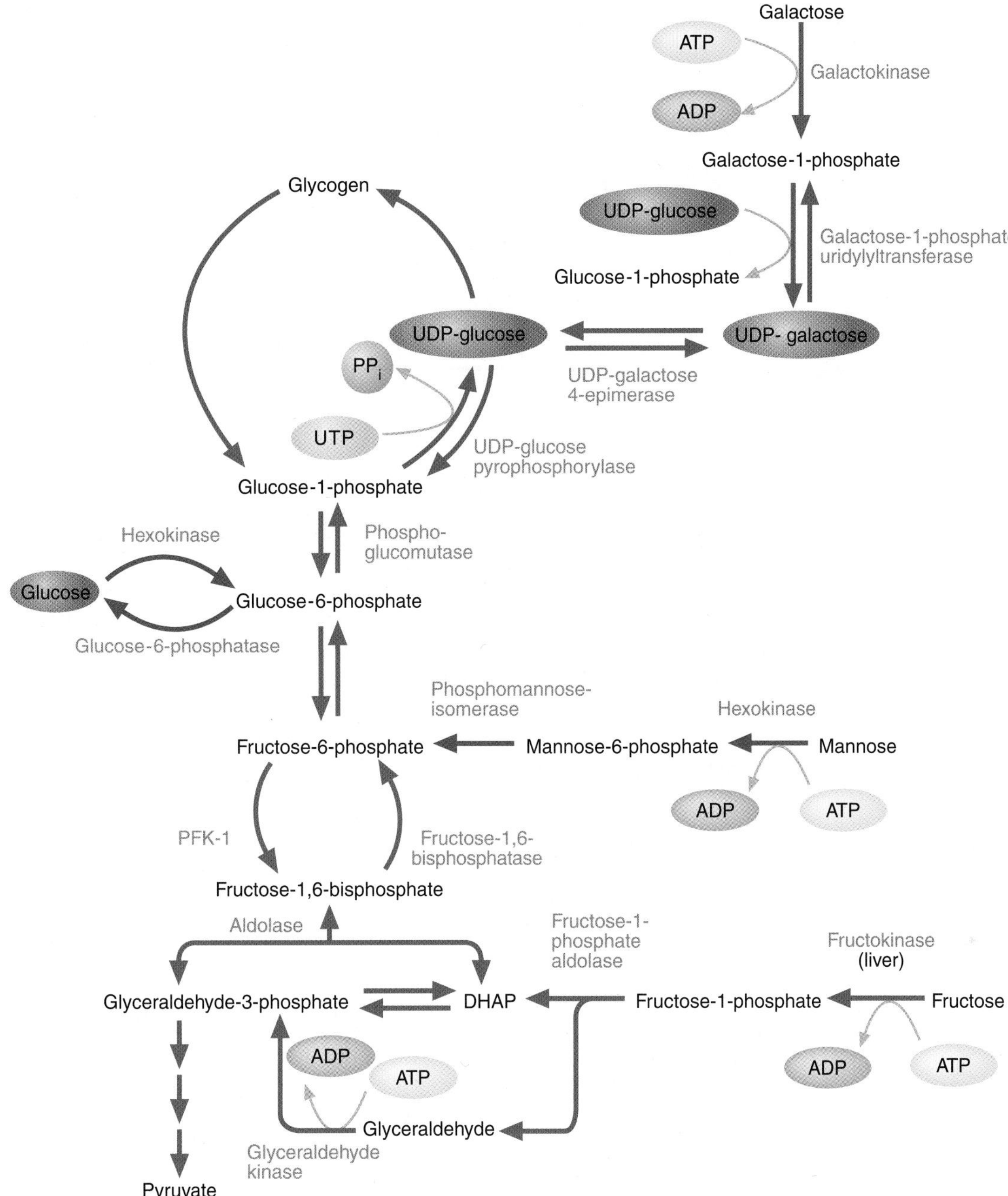

FIGURE 8.17

Carbohydrate Metabolism: Other Important Sugars

Fructose enters the glycolytic pathway in liver cells. Fructokinase converts the sugar to fructose-1-phosphate that is then split into DHAP and glyceraldehyde. Glyceraldehyde is phosphorylated by glyceraldehyde kinase to yield glyceraldehyde-3-phosphate. Galactose is converted into galactose-1-phosphate, which then reacts with UDP-glucose to form UDP-galactose. UDP-galactose is converted to its epimer, UDP-glucose, the substrate for glycogen synthesis. Mannose is phosphorylated by hexokinase to form mannose-6-phosphate, which is then isomerized to fructose-6-phosphate.

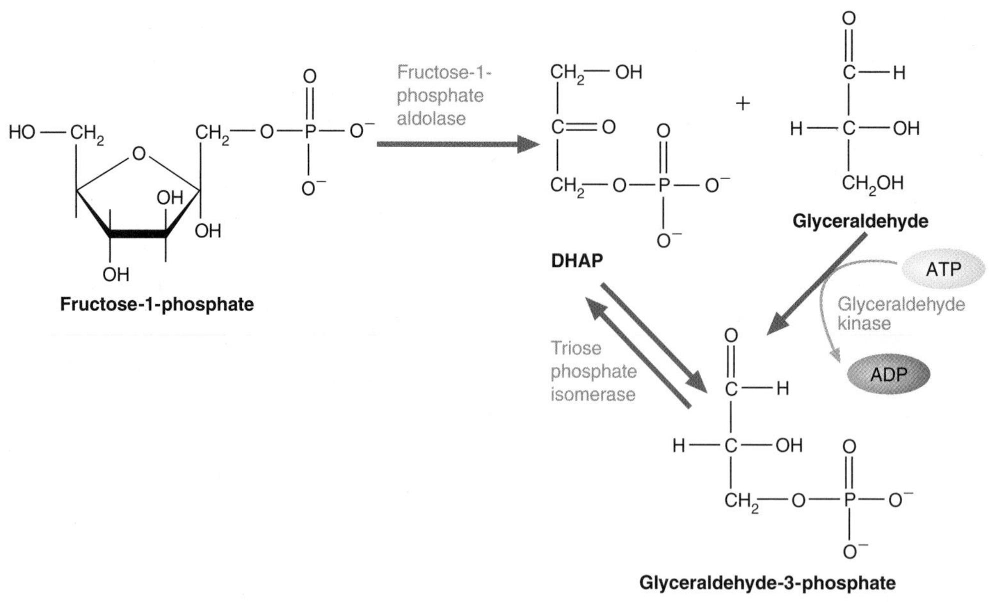

Fructose-1-phosphate

Glyceraldehyde

DHAP

Glyceraldehyde-3-phosphate

The conversion of fructose-1-phosphate into glycolytic intermediates bypasses two regulatory steps (the reactions catalyzed by hexokinase and PFK-1); thus in comparison to glucose, the entrance of fructose into the glycolytic pathway is essentially unregulated.

8.5 GLYCOGEN METABOLISM

Glycogen is the storage form of glucose. The synthesis and degradation of glycogen are carefully regulated so that sufficient glucose is available for the body's energy needs. Both glycogenesis and glycogenolysis are controlled primarily by three hormones: insulin, glucagon, and epinephrine.

Glycogenesis

Glycogen synthesis occurs after a meal, when blood glucose levels are high. It has long been recognized that the consumption of a carbohydrate meal is followed promptly by liver glycogenesis. The synthesis of glycogen from glucose-6-phosphate involves the following set of reactions.

1. **Synthesis of glucose-1-phosphate**. Glucose-6-phosphate is reversibly converted to glucose-1-phosphate by phosphoglucomutase, an enzyme that contains a phosphoryl group attached to a reactive serine residue:

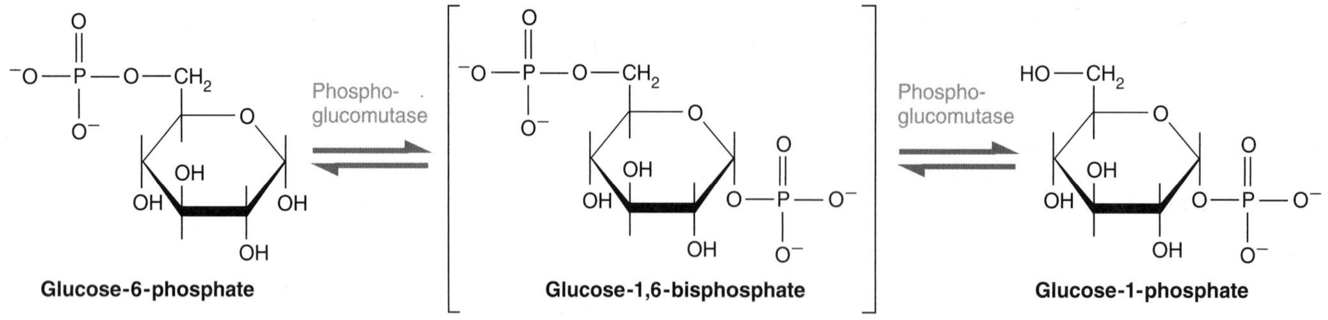

Glucose-6-phosphate

Glucose-1,6-bisphosphate

Glucose-1-phosphate

The enzyme's phosphoryl group is transferred to glucose-6-phosphate, forming glucose-1,6-bisphosphate. As glucose-1-phosphate forms, the phosphoryl group attached to C-6 is transferred to the enzyme's serine residue.

2. **Synthesis of UDP-glucose**. Glycosidic bond formation is an endergonic process. Derivatizing the sugar with a good leaving group provides the driving force for most sugar transfer reactions. For this reason, sugar-nucleotide synthesis is a common reaction preceding sugar transfer and polymerization processes. UDP-glucose is more reactive than glucose and is held more securely in the active site of the enzymes catalyzing transfer reactions (referred to as a group as glycosyl transferases). Formation of UDP-glucose, whose $\Delta G^{\circ\prime}$ value is near zero, is a reversible reaction catalyzed by UDP-glucose pyrophosphorylase:

$$\text{Glucose-1-phosphate} + \text{UTP} \rightleftharpoons \text{UDP-glucose} + PP_i$$

However, the reaction is driven to completion because pyrophosphate (PP_i) is immediately and irreversibly hydrolyzed by pyrophosphatase with a large loss of free energy ($\Delta G^{\circ\prime} = -33.5$ kJ/mol):

$$PP_i + H_2O \longrightarrow 2\ P_i$$

(Recall that removing product shifts the reaction equilibrium to the right. This cellular strategy is common.)

3. **Synthesis of glycogen from UDP-glucose**. The formation of glycogen from UDP-glucose requires two enzymes: (a) glycogen synthase, which catalyzes the transfer of the glucosyl group of UDP-glucose to the nonreducing ends of glycogen (**Figure 8.18a**), and (b) amylo-α(1,4 $\rightarrow$ 1,6)-glucosyl transferase (branching enzyme), which creates the α(1,6) linkages for branches in the molecule (**Figure 8.18b**).

   Glycogen synthesis requires a preexisting oligosaccharide. The first glucosyl residue is linked to Tyr-194 in a primer protein called *glycogenin* by its autocatalytic glucosyltransferase activity. Once the oligosaccharide chain consists of about 12 α(1,4)-linked glucosyl residues, the glycogen chain is then extended by glycogen synthase and branching enzyme. Large glycogen granules, each consisting of a single highly branched glycogen molecule, can be observed in the cytoplasm of liver and muscle cells of well-fed animals. The enzymes responsible for glycogen synthesis and degradation coat each granule's surface.

Glycogenolysis

Glycogen degradation requires the following two reactions.

1. **Removal of glucose from the nonreducing ends of glycogen**. Glycogen phosphorylase uses P_i to cleave the α(1,4) linkages on the outer branches of glycogen to yield glucose-1-phosphate. Glycogen phosphorylase stops when it comes within four glucose residues of a branch point (**Figure 8.19**). (A glycogen molecule that has been degraded to its branch points is called a *limit dextrin*.)

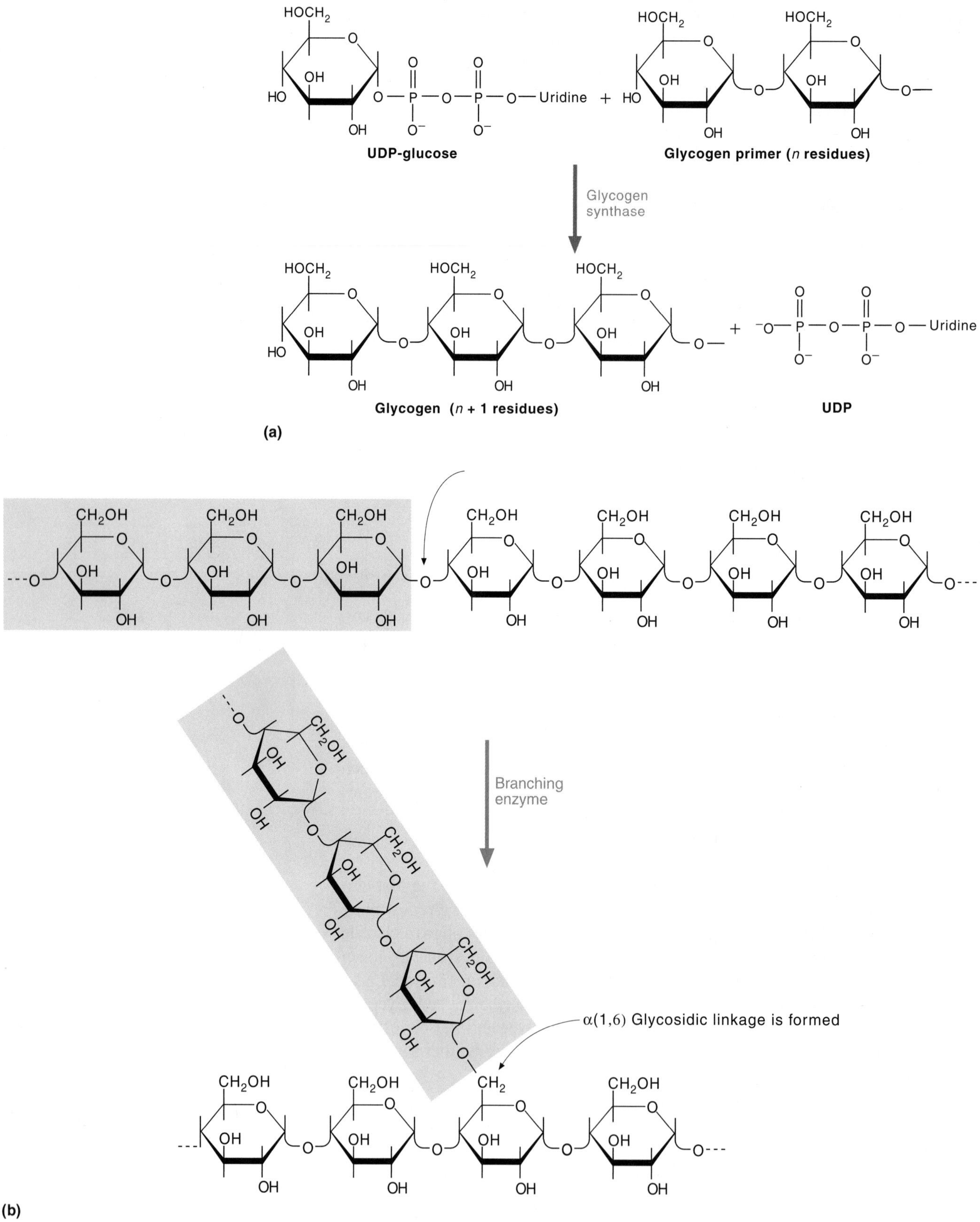

UDP-glucose

Glycogen primer (*n* residues)

Glycogen synthase

Glycogen (*n* + 1 residues)

UDP

(a)

Branching enzyme

α(1,6) Glycosidic linkage is formed

(b)

FIGURE 8.18

Glycogen Synthesis

(a) The enzyme glycogen synthase breaks the ester linkage of UDP-glucose and forms an $\alpha(1,4)$ glycosidic bond between glucose and the growing glycogen chain. (b) Branching enzyme is responsible for the synthesis of $\alpha(1,6)$ linkages in glycogen.

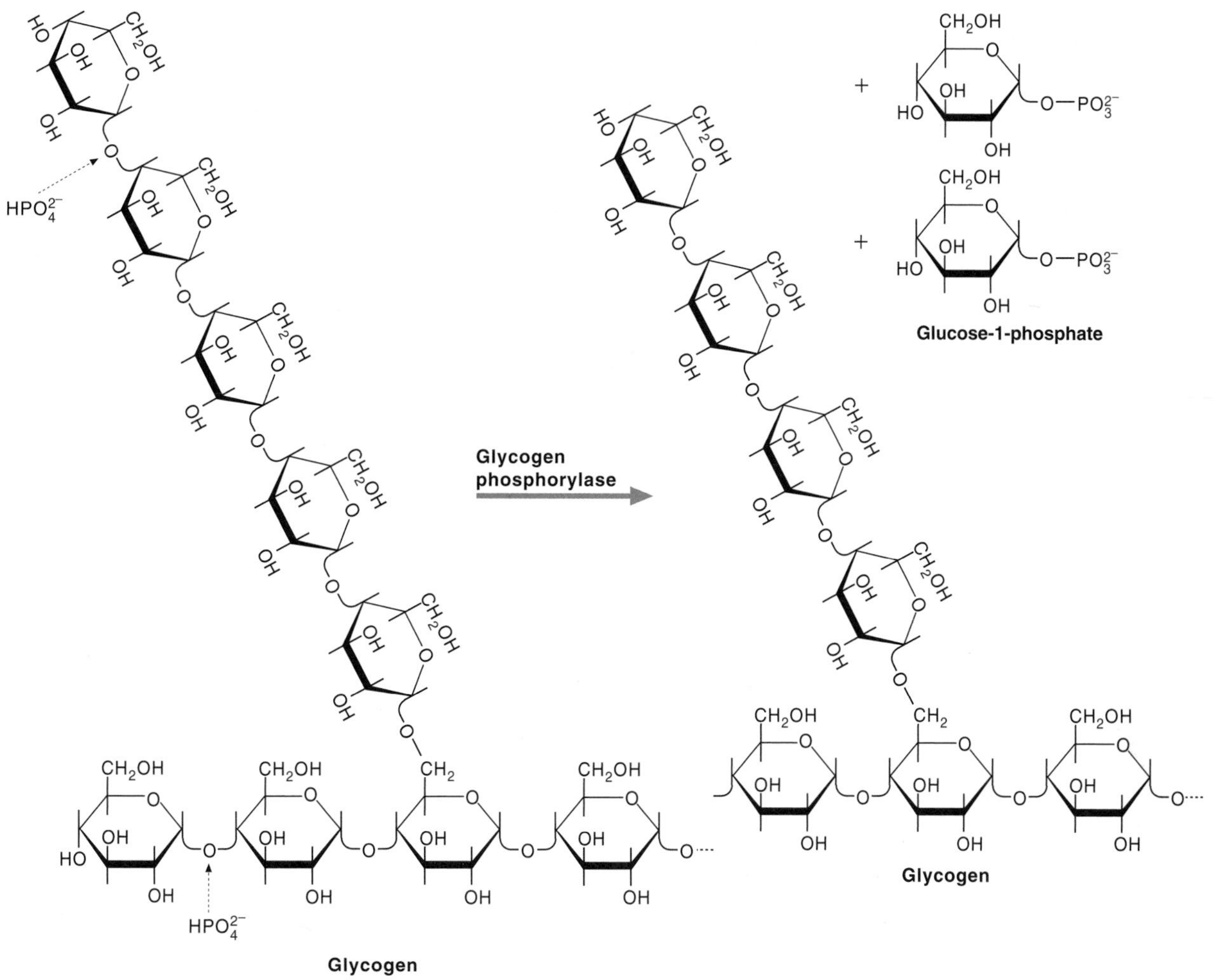

FIGURE 8.19

Glycogen Degradation

Glycogen phosphorylase catalyzes the removal of glucose residues from the nonreducing ends of a glycogen chain to yield glucose-1-phosphate. In this illustration one glucose residue is removed from each of two nonreducing ends. Removal of glucose residues continues until there are four residues at a branch point.

2. **Hydrolysis of the α(1,6) glycosidic bonds at branch points of glycogen.** Amylo-α(1,6)-glucosidase, also called debranching enzyme, begins the removal of α(1,6) branch points by transferring the outer three of the four glucose residues attached to the branch point to a nearby nonreducing end. It then removes the single glucose residue attached at each branch point. The product of this latter reaction is free glucose (**Figure 8.20**).

Glucose-1-phosphate, the major product of glycogenolysis, is diverted to glycolysis in muscle cells to generate energy for muscle contraction. In hepatocytes, glucose-1-phosphate is converted to glucose by phosphoglucomutase and glucose-6-phosphatase, which is then released into the blood. A summary of glycogenolysis is shown in **Figure 8.21**.

Regulation of Glycogen Metabolism

Glycogen metabolism is carefully regulated to avoid wasting energy. Both synthesis and degradation are controlled through a complex mechanism involving insulin, glucagon, and epinephrine, as well as allosteric regulators. Glucagon is released from the pancreas when blood glucose levels drop in the hours after a meal. It binds

FIGURE 8.20

Glycogen Degradation via Debranching Enzyme

Branch points in glycogen are removed by the debranching enzyme amylo-α(1,6)-glucosidase. After transferring the three-residue unit that precedes the branch point to a nearby nonreducing end of the glycogen molecule, the enzyme cleaves the α(1,6) linkage, releasing a glucose molecule.

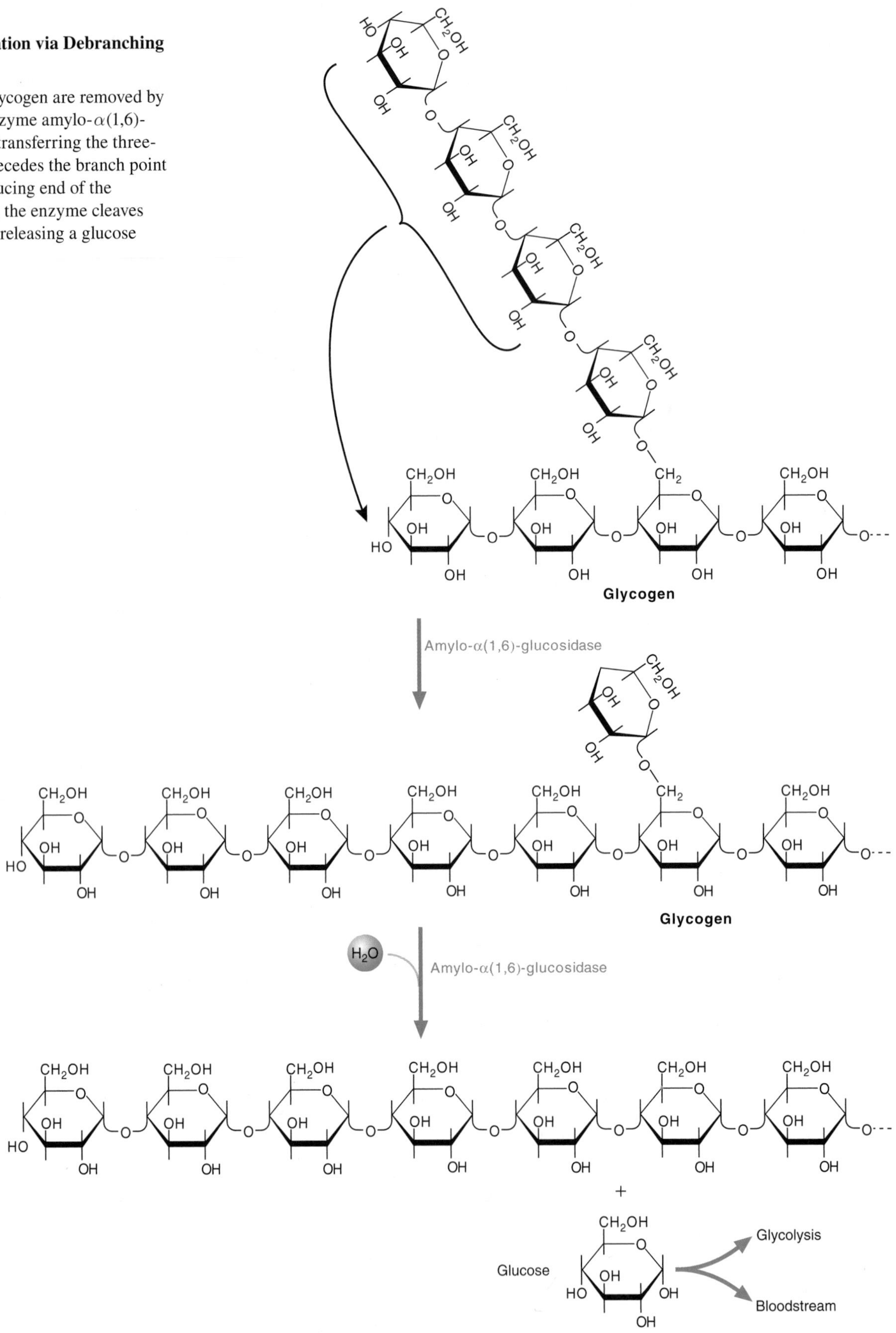

Glycogen

Amylo-α(1,6)-glucosidase

Glycogen

H_2O

Amylo-α(1,6)-glucosidase

+

Glucose

Glycolysis

Bloodstream

to receptors on hepatocytes and initiates a signal transduction process that elevates intracellular cAMP levels. cAMP amplifies the original glucagon signal and initiates a phosphorylation cascade that leads to the activation of glycogen phosphorylase along with a number of other proteins. Within seconds, glycogenolysis leads to the release of glucose into the bloodstream.

When occupied, the insulin receptor becomes an active tyrosine kinase enzyme that causes a phosphorylation cascade that ultimately has the opposite effect of the glucagon/cAMP system: the enzymes of glycogenolysis are inhibited and the enzymes of glycogenesis are activated. Insulin also increases the rate of glucose uptake into several types of target cells, but not liver or brain cells.

Emotional or physical stress releases the hormone *epinephrine* from the adrenal medulla. Epinephrine promotes glycogenolysis and inhibits glycogenesis. In emergency situations, when epinephrine is released in relatively large quantities, massive production of glucose provides the energy required to manage the situation. This effect is referred to as the flight-or-fight response. Epinephrine initiates the process by activating adenylate cyclase in liver and muscle cells. Calcium ions and inositol trisphosphate (Chapter 16) are also believed to be involved in epinephrine's action.

Glycogen synthase (GS) and glycogen phosphorylase have both active and inactive conformations that are interconverted by covalent modification. The active form of GS, known as the I (independent) form, is converted to the inactive or D (dependent) form by phosphorylation. The activity of GS can be finely modulated in response to a range of signal intensities because it is inactivated by phosphorylation reactions catalyzed by a large number of kinases. Physiologically, the most important kinases are glycogen synthase kinase 3 (GSK3) and casein kinase 1 (CS1). In contrast to GS, the inactive form of glycogen phosphorylase (phosphorylase b) is converted to the active form (phosphorylase a) by the phosphorylation of a specific serine residue. The phosphorylating enzyme is called phosphorylase kinase. Phosphorylation of both glycogen synthase (inactivating) and phosphorylase kinase (activating) is catalyzed by PKA, a protein kinase activated by cAMP. (The role of cAMP in signal transduction is described in Chapter 16 on pp. 589–591.) Glycogen synthesis occurs when GS and glycogen phosphorylase have been dephosphorylated. This conversion is catalyzed by phosphoprotein phosphatase 1 (PP1), which also inactivates phosphorylase kinase. It is noteworthy that PP1 is linked to both glycogen synthase and glycogen phosphorylase by an anchor protein called PTG (protein targeting to glycogen). The effects of glucagon, insulin, and epinephine on glycogen metabolism are summarized in **Figure 8.22**.

Several allosteric regulators also regulate glycogen metabolism. In muscle cells, both calcium ions released during muscle contraction and AMP bind to sites on glycogen phosphorylase b and promote its conversion to phosphorylase a. The reverse process, the conversion of glycogen phosphorylase a to phosphorylase b, is promoted by high levels of ATP and glucose-6-phosphate. GS activity is stimulated by glucose-6-phosphate. In hepatocytes, glucose is an allosteric regulator that promotes the inhibition of glycogen phosphorylase.

KEY CONCEPTS

- During glycogenesis, glycogen synthase catalyzes the transfer of the glucosyl group of UDP-glucose to the nonreducing ends of glycogen, and glycogen branching enzyme catalyzes the formation of branch points.
- Glycogenolysis requires glycogen phosphorylase and debranching enzyme. Glycogen metabolism is regulated by the actions of three hormones: glucagon, insulin, and epinephrine and several allosteric regulators.

QUESTION 8.7

Glycogen storage diseases are caused by inherited defects of one or more enzymes involved in glycogen synthesis or degradation. Patients with *Cori's disease,* caused by a deficiency of debranching enzyme, have enlarged livers (*hepatomegaly*) and low blood sugar concentrations (**hypoglycemia**). Can you suggest what causes these symptoms?

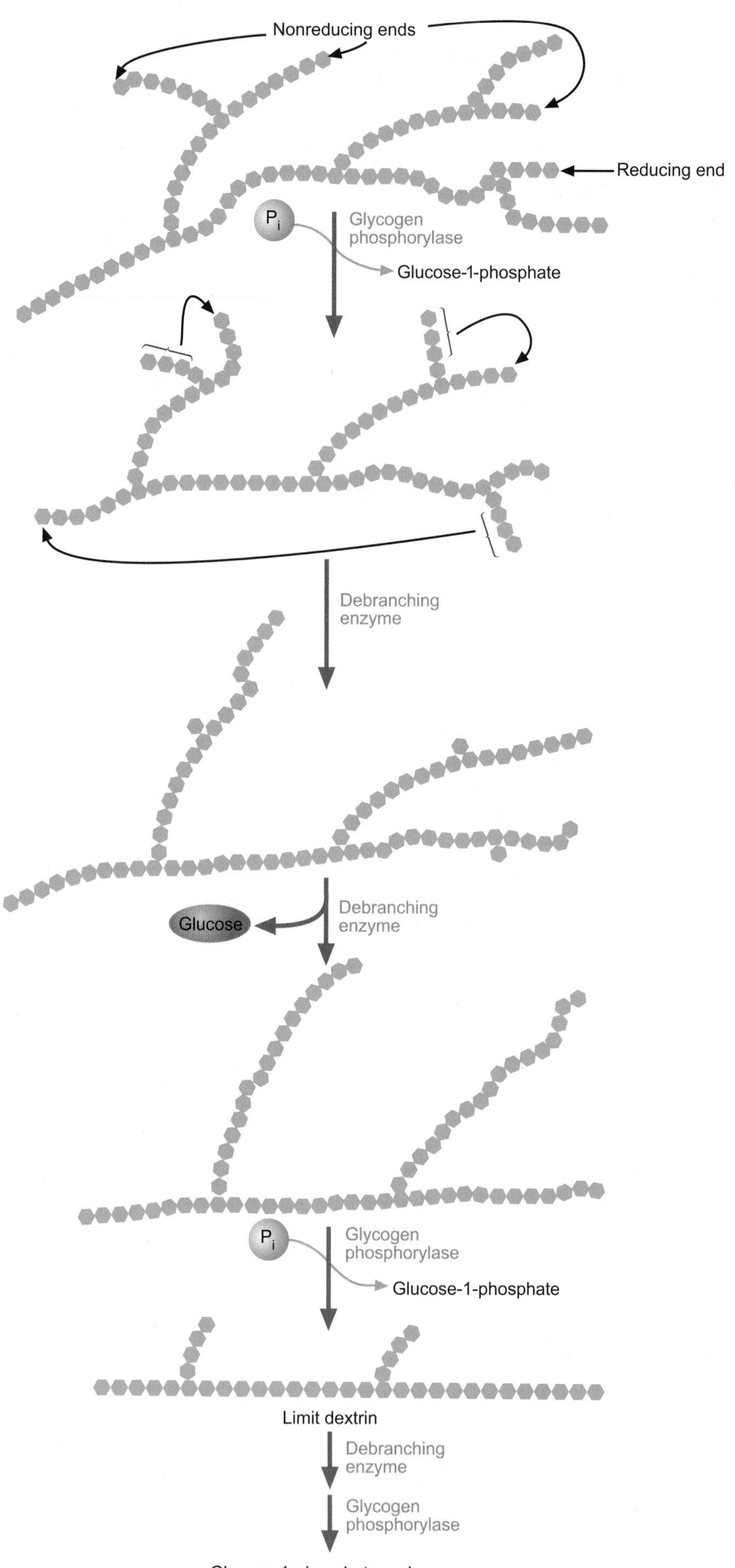

FIGURE 8.21
Glycogen Degradation: Summary

Glycogen phosphorylase cleaves the $\alpha(1,4)$ linkages of glycogen to yield glucose-1-phosphate until it comes within four glucose residues of a branch point. Debranching enzyme transfers three of these residues to a nearby nonreducing end and releases the fourth residue as free glucose. The repeated actions of both enzymes can lead to the complete degradation of glycogen.

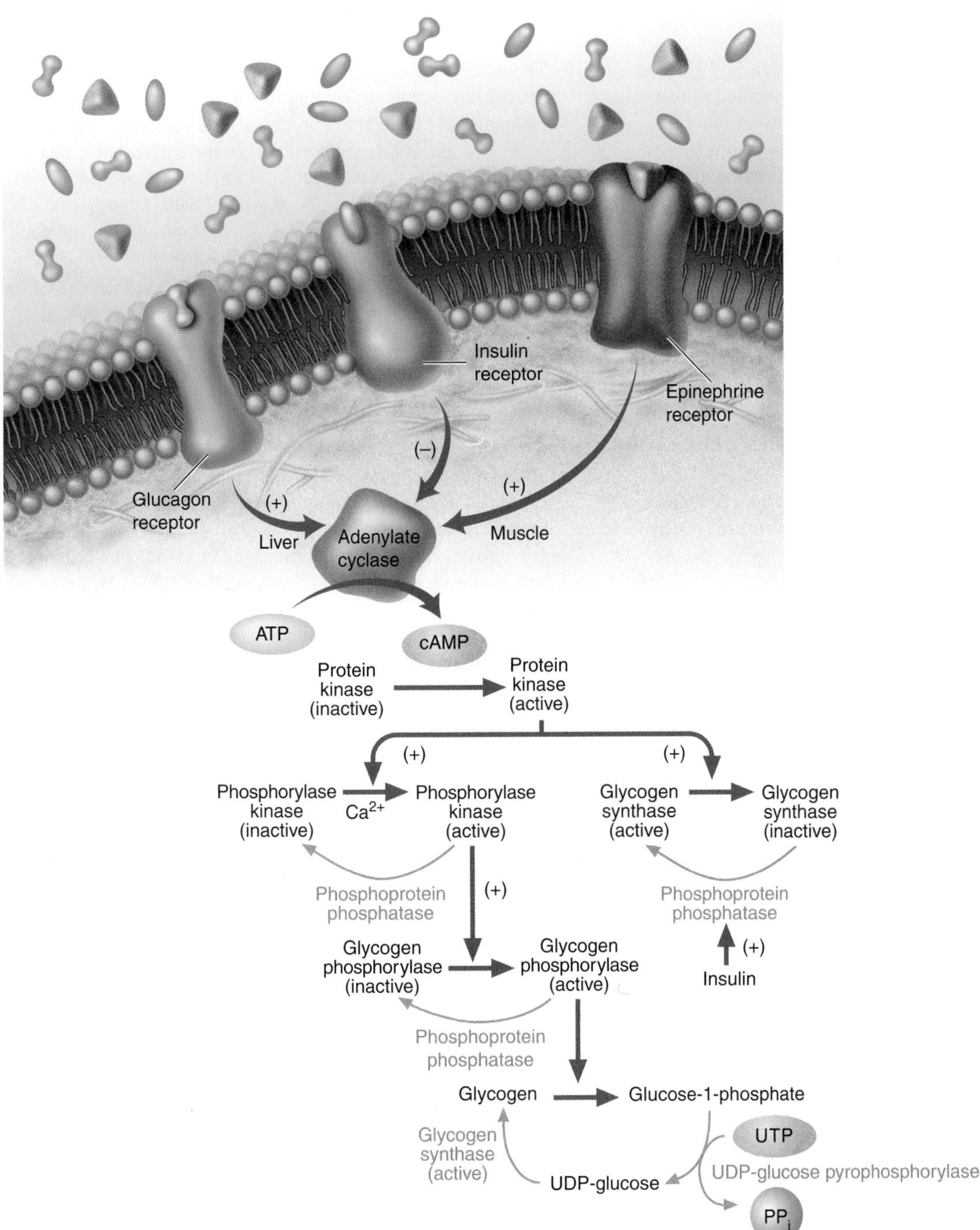

FIGURE 8.22

Major Factors Affecting Glycogen Metabolism

The binding of glucagon (released from the pancreas in response to low blood sugar) and/or epinephrine (released from the adrenal glands in response to stress) to their cognate receptors on the surface of target cells initiates a reaction cascade that converts glycogen to glucose-1-phosphate and inhibits glycogenesis. Insulin inhibits glycogenolysis and stimulates glycogenesis in part by decreasing the synthesis of cAMP and activating phosphoprotein phosphatase. Note that adenylate cyclase is a transmembrane protein with its functional domains protruding into the cytoplasm. In this illustration, for the sake of clarity, adenylate cyclase appears to be a cytoplasmic protein.

Chapter Summary

1. The metabolism of carbohydrates is dominated by glucose because this sugar is an important fuel molecule in most organisms. If cellular energy reserves are low, glucose is degraded by the glycolytic pathway. Glucose molecules that are not required for immediate energy production are stored as either glycogen (in animals) or starch (in plants).
2. During glycolysis, glucose is phosphorylated and cleaved to form two molecules of G-3-P. Each G-3-P is then converted to a molecule of pyruvate. A small amount of energy is captured in two molecules each of ATP and NADH. In anaerobic organisms, pyruvate is converted to waste products. During this process, NAD^+ is regenerated so that glycolysis can continue. In the presence of O_2, aerobic organisms convert pyruvate to acetyl-CoA and then to CO_2 and H_2O. Glycolysis is controlled primarily by allosteric regulation of three enzymes—hexokinase, PFK-1, and PK—and by the hormones glucagon and insulin.
3. During gluconeogenesis, molecules of glucose are synthesized from noncarbohydrate precursors (lactate, pyruvate, glycerol, and certain amino acids). The reaction sequence in gluconeogenesis is largely the reverse of glycolysis. The three irreversible glycolytic reactions (the synthesis of pyruvate, the conversion of fructose-1,6-bisphosphate to fructose-6-phosphate, and the formation of glucose from glucose-6-phosphate) are bypassed by alternate energetically favorable reactions.
4. The pentose phosphate pathway, in which glucose-6-phosphate is oxidized, occurs in two phases. In the oxidative phase, two molecules of NADPH are produced as glucose-6-phosphate is converted to ribulose-5-phosphate. In the nonoxidative phase, ribose-5-phosphate and other sugars are synthesized. If cells need more NADPH than ribose-5-phosphate, a component of nucleotides and the nucleic acids, then metabolites of the nonoxidative phase are converted into glycolytic intermediates.
5. Several sugars other than glucose are important in vertebrate carbohydrate metabolism. These include fructose, galactose, and mannose.
6. The substrate for glycogen synthesis is UDP-glucose, an activated form of the sugar. UDP-glucose pyrophosphorylase catalyzes the formation of UDP-glucose from glucose-1-phosphate and UTP. Glucose-6-phosphate is converted to glucose-1-phosphate by phosphoglucomutase. Glycogen synthesis requires two enzymes: glycogen synthase and branching enzyme. Glycogen degradation requires glycogen phosphorylase and debranching enzyme. The balance between glycogenesis (glycogen synthesis) and glycogenolysis (glycogen breakdown) is carefully regulated by several hormones (insulin, glucagon, and epinephrine) and allosteric regulators.

Take your learning further by visiting the **companion website** for Biochemistry at **www.oup.com/us/mckee** where you can complete a multiple-choice quiz on carbohydrate metabolism to help you prepare for exams.

Suggested Readings

Bakker, B. M., Mensonides, F. I. C., Teusink, B., van Hoek, P., Michels, P. A. M., and Westerhoff, H. V., Compartmentation Protects Trypanosomes from the Dangerous Design of Glycolysis, *Proc. Nat. Acad. Sci. USA* 97(5):2087–2092, 2000.

Fothergill-Gilmore, L. A., and Michels, P. A., Evolution of Glycolysis, *Prog. Biophys. Mol. Biol.* 59:105–135, 1993.

Frommer, W. B., Schulze, W. X., and LaLonde, S., Hexokinase, Jack-of-All-Trades, *Science* 300:261, 263, 2003.

Malik, V. S., Schulze, M. B., and Hu, F. B., Intake of Sugar-Sweetened Beverages and Weight Gain: A Systematic Review, *Am. J. Clin. Nutr.* 84:274–288, 2006.

Polaskova, P., Herszage, J., and Ebeler, E. E., Wine Flavor: Chemistry in a Glass, *Chem. Soc. Rev.* 37:2478–2489, 2008.

Piskur, J., et al., How Did *Saccharomyces cerevisiae* Evolve to Become a Good Brewer? *Trends Genet.* 22(4):183–186, 2006.

Teusink, B., Walsh, M. C., van Dam, K., and Westerhoff, H. V., The Danger of Metabolic Pathways with Turbo Design, *Trends Biochem. Sci.* 23(5):162–169, 1998.

Key Words

aerobic respiration, *262*
aldol cleavage, *265*
amphibolic pathway, *270*
anaerobic organism, *261*
antioxidant, *287*
citric acid cycle, *270*
Cori cycle, *282*
Crabtree effect, *273*
decarboxylation, *271*
electron transport system, *270*
epinephrine, *295*
fermentation, *270*
glucagon, *276*
gluconeogenesis, *261*
glucose-alanine cycle, *283*
glycogenesis, *261*
glycogenolysis, *261*
glycolysis, *261*
hypoglycemia, *298*
insulin, *276*
malate shuttle, *280*
Pasteur effect, *273*
pentose phosphate pathway, *261*
response element, *276*
substrate-level phosphorylation, *266*
tautomer, *268*
tautomerization, *268*
transcription factor, *276*

Review Questions

These questions are designed to test your knowledge of the key concepts discussed in this chapter before moving on to the next chapter. You may like to compare your answers to the solutions provided in the back of the book and in the accompanying Study Guide.

1. Define the following terms:
   a. glycolysis
   b. pentose phosphate pathway
   c. gluconeogenesis
   d. glycogenolysis
   e. glycogenesis
2. Define the following terms:
   a. anaerobic organism
   b. aerobic organism
   c. aerobic respiration
   d. aldol cleavage
   e. substrate-level phosphorylation
3. Define the following terms:
   a. tautomerization
   b. tautomer
   c. amphibolic pathway
   d. electron transport system
   e. decarboxylation reaction
4. Define the following terms:
   a. Crabtree effect
   b. transcription factor
   c. response element
   d. insulin
   e. malate shuttle
5. Define the following terms:
   a. Cori cycle
   b. glucagon
   c. glucose-alanine cycle
   d. hypoglycemia
   e. antioxidant
6. Upon entering a cell, glucose is phosphorylated. Give two reasons why this reaction is required.
7. Describe the functions of the following molecules:
   a. insulin
   b. glucagon
   c. fructose-2,6-bisphosphate
   d. UDP-glucose
   e. cAMP
   f. GSSG
   g. NADPH
8. Designate the reactions catalyzed by the following enzymes:
   a. aldolase
   b. enolase
   c. hexokinase
   d. amylo-α(1,6)-glucosidase
   e. phosphoglucomutase
9. Why is "turbo design" used in catabolic pathways in living organisms?
10. Identify in which reactions in glycolysis the following functions occur.
    a. ATP consumption
    b. ATP synthesis
    c. NADH synthesis
11. Name the glycolytic enzymes that are allosterically regulated.
12. Name the three unique reactions in gluconeogenesis.
13. Explain and contrast the Pasteur and Crabtree effects.
14. Explain the role of glycogenin in glycogen synthesis.
15. List the three principal hormones that regulate glucose metabolism. Briefly explain the effects these molecules have on carbohydrate metabolism.
16. In which locations in the eukaryotic cell do the following processes occur?
    a. gluconeogenesis
    b. glycolysis
    c. pentose phosphate pathway
17. Compare the entry-level substrates, products, and metabolic purposes of glycolysis and gluconeogenesis.
18. Define substrate-level phosphorylation. Which two reactions in glycolysis are in this category?
19. What is the principal reason that organisms such as yeast produce alcohol?
20. Why is pyruvate not oxidized to CO_2 and H_2O under anaerobic conditions?
21. Describe how epinephrine promotes the conversion of glycogen to glucose.
22. Glycolysis occurs in two stages. Describe what is accomplished in each stage.
23. What effects do the following molecules have on gluconeogenesis?
    a. lactate
    b. ATP
    c. pyruvate
    d. glycerol
    e. AMP
    f. acetyl-CoA
24. Describe the physiological conditions that activate gluconeogenesis.
25. The following two reactions constitute a wasteful cycle:

    $$\text{Glucose} + \text{ATP} \rightarrow \text{glucose-6-phosphate}$$
    $$\text{Glucose-6-phosphate} + H_2O \rightarrow \text{glucose} + P_i$$

    Suggest how such wasteful cycles are prevented or controlled.
26. Describe the central role of glucose in carbohydrate metabolism.
27. Draw the structure of glucose with the carbons numbered 1 through 6. Number these carbons again as they appear in the two molecules of pyruvate formed during glycolysis.
28. Draw the reactions that convert glucose into ethanol.
29. After reviewing **Figure 8.17**, draw the reactions that convert galactose into a glycolytic intermediate. Include the Haworth formulas in your work.
30. Draw the reactions that convert fructose into glycolytic intermediates.

31. Why is severe hypoglycemia so dangerous?
32. Explain why ATP hydrolysis occurs so early in glycolysis, an ATP-producing pathway.
33. In which reaction in glycolysis does a dehydration occur?
34. Describe the Cori cycle. What is its physiological function?
35. Describe the effects of insulin and glucagon on glycogen metabolism.
36. Describe the effects of insulin and glucagon on blood glucose.
37. What cells produce insulin, glucagon, epinephrine, and cortisol?
38. Describe the different functions of glycogen in liver and muscle.
39. Describe the fate of pyruvate under anaerobic and aerobic conditions.

Fill in the Blank

40. Rapidly equilibrating isomers that differ in the position of a double bond and a hydrogen are called ________________.
41. A metabolic pathway that functions in both anabolic and catabolic processes is called an ________________ pathway.
42. The cleavage of fructose-1,6-bisphosphate to glyceraldehyde-3-phosphate and dihydroxyacetone phosphate is an example of an ________________ reaction.
43. The ___________ effect is a phenomenon in which glucose represses aerobic respiration.
44. In the _____________ effect, glycolysis is repressed in the presence of oxygen.
45. ________________________ factors are proteins that regulate or initiate RNA synthesis by binding directly or indirectly to specific DNA sequences called response elements.
46. Cells that lack mitochondrial PEP carboxykinase transfer oxaloacetate into the cytoplasm using the _____________ shuttle.
47. Lactose released by skeletal muscle during physical exercise is transported to the liver in the _____________ cycle.
48. _____________ are substances that prevent oxidation of other substances.
49. ________________ is the substrate for glycogen synthesis.

Short Answer

50. During which reaction(s) in glycolysis does oxidation occur?
51. Glucose fermentation in yeast produces ethanol as a waste product. In contrast, *Clostridium acetobutylicum* produces butanol. Suggest a possible sequence of reactions that would produce butyl alcohols.
52. What is a diauxic shift, a term that is used in reference to microorganisms in cell culture? Where does it occur in glucose metabolism?
53. Glucose is converted to pyruvate in glycolysis, yielding a net synthesis of 2 ATP. In certain cells pyruvate can be reconverted to glucose during gluconeogenesis. How many ATPs are required to convert pyruvate back to glucose?
54. What is the function of the glucose-alanine cycle in normal metabolism?

Thought Questions

These questions are designed to reinforce your understanding of all of the key concepts discussed in the book so far, including this chapter and all of the chapters before it. They may not have one right answer! The authors have provided possible solutions to these questions in the back of the book and in the accompanying Study Guide for your reference.

55. In the first stage of glycolysis, fructose-1,6-bisphosphate is cleaved to form glyceraldehyde-3-phosphate and dihydroxyacetone phosphate. The latter molecule can then be converted to glyceraldehyde-3-phosphate. Illustrate the mechanisms whereby these reactions occur.
56. After a carbohydrate-rich meal is consumed and the glucose requirements of all tissues have been met, the liver begins to store excess glucose in glycogen molecules. Explain the role of the hexokinases in this phenomenon.
57. Glucokinase acts as a glucose sensor in hepatocytes (liver cells), α- and β-cells in the pancreas, enterocytes (intestinal wall cells), and the hypothalamus (a control center in the brain of numerous physiological processes). Explain why glucokinase can perform this role.
58. An individual has a genetic deficiency that prevents the production of glucokinase. Following a carbohydrate meal, do you expect blood glucose levels to be high, low, or about normal? What organ accumulates glycogen under these circumstances?
59. Glycogen synthesis requires a short primer chain. Explain how new glycogen molecules are synthesized given this limitation.
60. Why is fructose metabolized more rapidly than glucose?

61. What is the difference between an enol-phosphate ester and a normal phosphate ester that gives PEP such a high phosphoryl group transfer potential?
62. In aerobic oxidation, oxygen is the ultimate oxidizing agent (electron acceptor). Name two common oxidizing agents in anaerobic fermentation.
63. Why is it important that gluconeogenesis is not the exact reverse of glycolysis?
64. Compare the structural formulas of ethanol, acetate, and acetaldehyde. Which molecule is the most oxidized? Which is the most reduced? Explain your answers.
65. *Trypanosoma brucei* is a parasitic protozoan that causes sleeping sickness in humans, Transmitted by the tsetse fly, sleeping sickness is a fatal disease characterized by fever, anemia, inflammation, lethargy, headache, and convulsions. When trypanosomes are present in the human bloodstream, they depend on glycolysis entirely for energy generation. The first seven glycolytic enzymes in these organisms are localized in peroxisome-like organelles called glycosomes, which are only regulated weakly by allosteric regulator molecules. Glycosomes take up glucose and export glycerate-3-phosphate. There are two pools of ADP and ATP (cytoplasmic and glycosomal), and the glycosomal membrane is impermeable to both nucleotides as well as most other glycolytic intermediates. If the glycosomal membrane is compromised, the concentration of phosphoylated glycolytic intermediates rises and the cells die. Explain.
66. The consumption of large amounts of soft drink beverages and processed foods sweetened with high-fructose corn syrup has been linked to obesity. After reviewing Figures 8.1 and 8.17, suggest a likely reason for this phenomenon.
67. How does phosphorylation increase the reactivity of glucose?
68. Examine the structure of phosphoenolpyruvate and explain why it has such a high phosphoryl group transfer potential.
69. Both glycogen and triacylglycerols are energy sources used by the body. Suggest a reason why both are required.
70. Severe dieting results in both the reduction of fat stores and the loss of muscle mass. Use biochemical reactions to trace the conversion of muscle protein to glucose production.
71. Suggest a reason why glycolysis produces NADH and the pentose phosphate pathway produces NADPH.
72. Cells in culture are fed glucose molecules labeled with ^{14}C at carbon 2. Trace the radioactive label through one pass through the pentose phosphate pathway.
73. Ethanol is especially toxic in children for several reasons. For example, ethanol consumption results in elevated levels of NADH in the liver. Suggest a mechanism that explains this phenomenon.

CHAPTER

9 Aerobic Metabolism I: The Citric Acid Cycle

Aerobic Cells In aerobic cells most energy is generated within the mitochondrion. Dioxygen (O_2) is the final electron acceptor in the oxidation of nutrient molecules.

OUTLINE

Oxygen and Evolution: Chance and Necessity

Aerobic organisms use oxygen (O_2) to extract energy from nutrient molecules. *Aerobic respiration,* the oxygen-requiring energy-generating mechanism, generates significantly more energy than does fermentation, an anaerobic (without oxygen) process. When it first appeared about 2 billion years ago (bya), aerobic respiration was a critical turning point in the evolution of life. The large increases in energy it supplied to organisms with the molecular equipment to exploit oxygen provided the resources for evolutionary innovation. It is not coincidental that the origin of eukaryotic cells (about 1.5 bya) and multicellular organisms (about 1 bya) occurred after aerobic respiration became a common means of energy generation. Oxygen accumulated in the primordial atmosphere because it is a waste product of *oxygenic photosynthesis* (Chapter 13), a 3 billion-year-old process initiated by cyanobacteria that uses sunlight to drive the synthesis of biomolecules from CO_2 and the hydrogen atoms of water. Oxygenic photosynthesis is so complex and thermodynamically challenging, however, that it might never have come about at all.

Why Oxygen?

Why is oxygen so useful in energy generation? The answer lies in the nature of energy generation and the chemistry of oxygen and water. Energy is captured when electrons are transferred from an electron donor to an electron acceptor. In aerobic respiration oxygen acts as the terminal acceptor of electrons removed from organic nutrients. As O_2 combines with these electrons as well as protons, water molecules are formed. Oxygen is an excellent electron acceptor for two major reasons. First, the element oxygen is abundant, which makes it far more useful as an oxidizing agent than less commonly available elements such as sulfur. Second, oxygen is a powerful oxidizing agent because it is highly electronegative (i.e., it has a considerable affinity for electrons). Consequently, there is a correspondingly large energy release with every electron that is transferred from carbon to oxygen.

Why Is Oxygenic Photosynthesis Such a Challenge?

The origin of oxygenic photosynthesis is improbable for two major reasons. First, because oxygen atoms are powerful electron acceptors, water is a very poor reducing agent. In contrast, nonoxygenic photosynthesis, the precursor of the oxygen-generating process, used more powerful reducing agents such as H_2 and H_2S. Significantly more light energy is required to oxidize water molecules than for H_2 and H_2S. As a result, the photosynthetic pigment molecules that absorb light energy had to be adapted to absorbing higher-energy photons (electromagnetic particles). Second, photosynthetic electron transport (one electron at a time) and water oxidation (a sequential and concerted four-electron process) are seemingly incompatible. The cyanobacteria solved this extraordinarily complex set of problems with a multisubunit pigment-enzyme complex now referred to as *photosystem II* (PSII).

The One and Only Oxygen-Evolving Complex

Within PSII, water molecules are split into electrons, protons, and O_2 by the *oxygen-evolving complex* (OEC), a protein complex that contains a unique inorganic cubelike cofactor: the Mn_3CaO_4 cubane. Of all of the problems solved by the cyanobacteria, it is the OEC that provides insight into the tenuous origin of oxygenic photosynthesis. All modern photosynthesizing organisms contain an identical water-splitting mechanism. Unlike all other biochemical devices, the OEC was invented only once. The emergence of the water-splitting OEC of PSII was the most pivotal event in the history of life on Earth. Without the OEC, all modern organisms would be anaerobic prokaryotes.

Overview

MODERN AEROBIC ORGANISMS TRANSDUCE THE CHEMICAL BOND ENERGY OF FOOD MOLECULES INTO THE BOND ENERGY OF ATP. THEY PERFORM THIS feat because oxygen is used as the terminal acceptor of the electrons extracted from food molecules. The capacity to use oxygen to oxidize nutrients such as glucose and fatty acids yields a substantially greater amount of energy than does fermentation.

As atmospheric O_2 began accumulating on earth, about 2 bya, existing organisms were confronted with a serious problem: molecular oxygen forms toxic oxygen ions and peroxides referred to as *reactive oxygen species* (ROS). ROS react with and damage or destroy biomolecules. Consequently, exposure to O_2 acted as a severe selection pressure. Species in existence during this period either evolved a means of adapting to O_2 or became extinct. Modern organisms are classified based on the strategies they use to cope with ROS or use O_2 in energy generation:

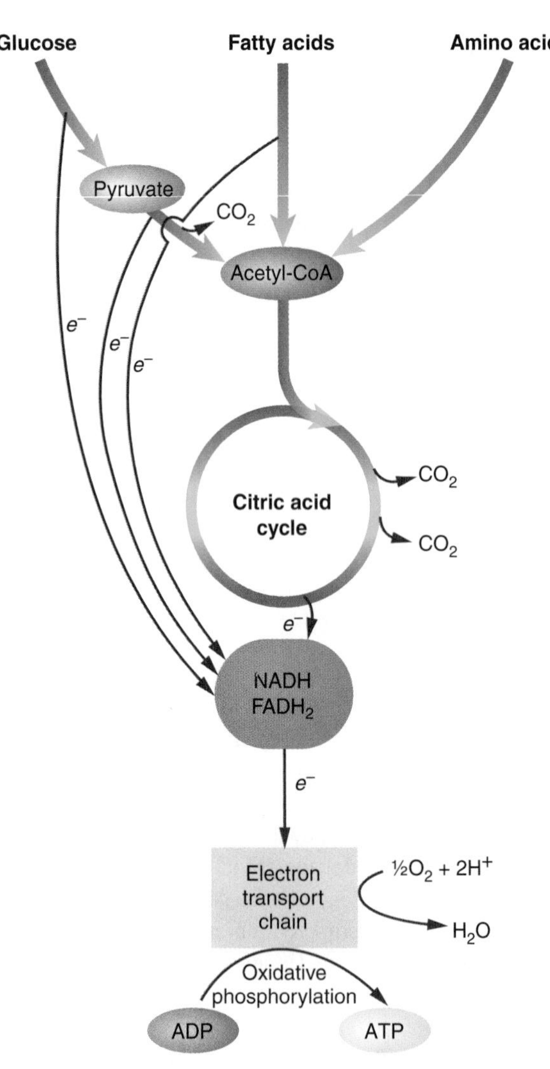

FIGURE 9.1
Overview of Aerobic Metabolism

In aerobic metabolism, the nutrient molecules glucose, fatty acids, and some amino acids are degraded to form acetyl-CoA. Acetyl-CoA then enters the citric acid cycle. Electron carriers (NADH and $FADH_2$) produced by glucose and fatty acid degradation and several citric acid cycle reactions donate electrons (e^-) to the electron transport chain. Energy captured by the electron transport chain is then used to synthesize ATP in a process referred to as oxidative phosphorylation. Note that O_2, the terminal electron acceptor in aerobic metabolism, combines with protons to form water molecules.

1. **Obligate anaerobes** are organisms that grow only in the absence of O_2 (i.e., they live in highly reduced environments such as soil) and use fermentation to generate energy.
2. **Aerotolerant anaerobes**, which also depend on fermentation, possess detoxifying enzymes and antioxidant molecules that detoxify ROS.
3. **Facultative anaerobes** not only possess the biochemical mechanisms required for detoxifying ROS, but also can use O_2 as an electron acceptor when the gas is available.
4. **Obligate aerobes** are highly dependent on O_2 for energy production. They protect themselves from ROS with elaborate detoxifying mechanisms that are composed of enzymes and numerous endogenous and exogenous antioxidant molecules.

Aerobic metabolism consists of the following biochemical processes: the citric acid cycle, the electron transport pathway, and oxidative phosphorylation (**Figure 9.1**). In eukaryotes these processes occur within the mitochondrion (**Figure 9.2**). The citric acid cycle is a metabolic pathway in which two-carbon fragments derived from organic fuel molecules are oxidized to form CO_2 and the coenzymes NAD^+ and FAD are reduced to form NADH and $FADH_2$, respectively. The electron transport pathway, also referred to as the electron transport chain (ETC), is a mechanism by which electrons are transferred from NADH and $FADH_2$ to a series of electron carriers that are sequentially reduced and then oxidized. The terminal electron acceptor is O_2. In oxidative phosphorylation, the energy released by electron transport is captured in the form of a proton gradient that drives the synthesis of ATP, the energy currency of living organisms.

Chapter 9 begins with a review of oxidation-reduction reactions and the relationship between electron flow and energy transduction. This is followed by a detailed discussion of the citric acid cycle, the central pathway in aerobic metabolism, and its roles in energy generation and biosynthesis. In Chapter 10, the discussion of aerobic metabolism continues with an examination of electron transport and oxidative phosphorylation, the means by which aerobic organisms use oxygen to generate significant amounts of ATP. It ends with a review of *oxidative stress*, a series of reactions in which toxic oxygen species are

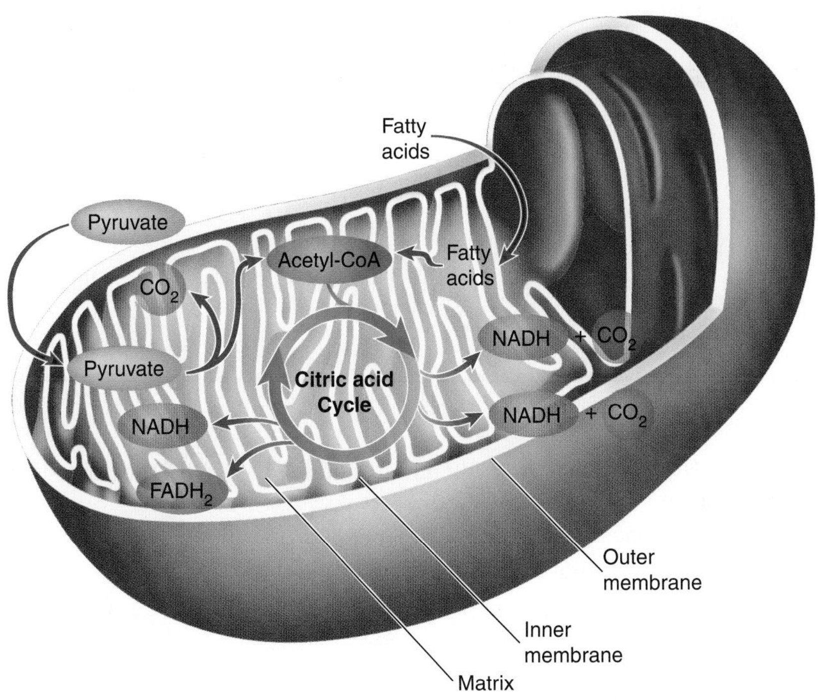

FIGURE 9.2

Aerobic Metabolism in the Mitochondrion

In eukaryotic cells aerobic metabolism occurs within the mitochondrion. Acetyl-CoA, the oxidation product of pyruvate, fatty acids, and certain amino acids (not shown), is oxidized by the reactions of the citric acid cycle within the mitochondrial matrix. The principal products of the cycle are the reduced coenzymes, NADH and $FADH_2$, and CO_2. The high-energy electrons of NADH and $FADH_2$ are subsequently donated to the electron transport chain (ETC), a series of electron carriers in the inner membrane. The terminal electron acceptor for the ETC is O_2. The energy derived from the electron transport mechanism drives ATP synthesis by creating a proton gradient across the inner membrane. The large folded surface of the inner membrane is studded with ETC complexes, transport proteins of numerous types, and ATP synthase, the enzyme complex responsible for ATP synthesis.

created and subsequently damage cells. Chapter 10 also describes the principal mechanisms used by living organisms to protect themselves from oxidative stress.

9.1 OXIDATION-REDUCTION REACTIONS

In living organisms, both energy-capturing and energy-releasing processes consist largely of redox reactions. Recall that redox reactions occur when electrons are transferred between an electron donor (reducing agent) and an electron acceptor (oxidizing agent). In some redox reactions, only electrons are transferred. For example, in the reaction

$$Cu^{+} + Fe^{3+} \rightleftharpoons Cu^{2+} + Fe^{2+}$$

an electron is transferred from Cu^{+} to Fe^{3+}. Cu^{+}, the reducing agent, is oxidized to form Cu^{2+}. Meanwhile, Fe^{3+} is reduced to Fe^{2+}. In many reactions, however, both electrons and protons are transferred. For example, the reaction catalyzed by lactate dehydrogenase begins with the transfer of a hydride ion ($H:^{-}$), that is, a hydrogen nucleus and two electrons, from NADH to pyruvate.

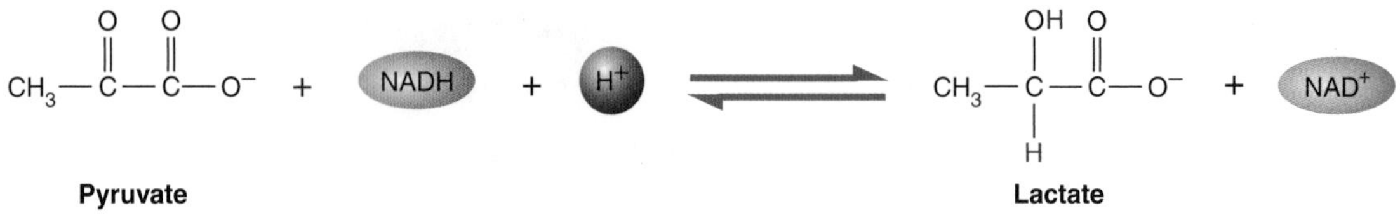

$$\underset{\textbf{Pyruvate}}{CH_3{-}\overset{\overset{\displaystyle O}{\|}}{C}{-}\overset{\overset{\displaystyle O}{\|}}{C}{-}O^-} + NADH + H^+ \rightleftharpoons \underset{\textbf{Lactate}}{CH_3{-}\underset{\underset{\displaystyle H}{|}}{\overset{\overset{\displaystyle OH}{|}}{C}}{-}\overset{\overset{\displaystyle O}{\|}}{C}{-}O^-} + NAD^+$$

FIGURE 9.3

Reduction of Pyruvate by NADH

In this redox reaction, a hydride ion ($H:^-$) is transferred from NADH to pyruvate and the product is protonated from the surrounding medium to form lactate.

A proton (H^+) is gained from the environment (Figure 9.3) to form the final products lactate and NAD^+.

Redox reactions are more easily understood if they are separated into half-reactions. In the reaction between copper and iron the Cu^+ ion loses an electron to become Cu^{2+}:

$$Cu^+ \rightleftharpoons Cu^{2+} + e^-$$

This equation indicates that Cu^+ is the electron donor. (Together Cu^+ and Cu^{2+} constitute a **conjugate redox pair**.) As Cu^+ loses an electron, Fe^{3+} gains an electron to form Fe^{2+}:

$$Fe^{3+} + e^- \rightleftharpoons Fe^{2+}$$

In this half-reaction, Fe^{3+} is an electron acceptor. The separation of redox reactions emphasizes that electrons are always the common intermediates between half-reactions.

The constituents of half-reactions may be observed in an electrochemical cell (**Figure 9.4**). Each half-reaction takes place in a separate container or *half-cell*. The movement of electrons generated in the half-cell undergoing oxidation (e.g., $Cu^+ \rightarrow Cu^{2+} + e^-$) creates a voltage (or potential difference) between the two half-cells. The sign of the voltage (measured by a voltmeter) is positive or negative according to the direction of the electron flow. The magnitude of the potential difference is a measure of the energy that drives the reaction.

The tendency for a specific substance to gain electrons is called its **reduction potential**. The **standard reduction potential** ($E°$) of a substance is measured in a galvanic cell relative to a standard hydrogen electrode. A standard cell has all solutes at 1.0 M concentration, all gases at 1 atm pressure, and the

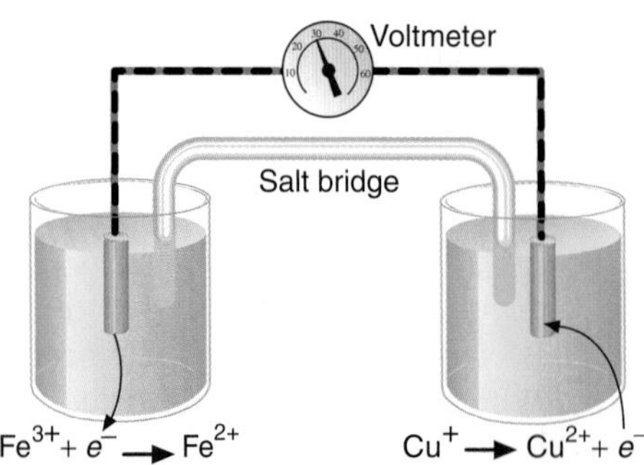

FIGURE 9.4

An Electrochemical Cell

Electrons flow from the Cu^{2+}/Cu^+ half-cell (cathode) through the voltmeter to the Fe^{3+}/Fe^{2+} half-cell (anode). The salt bridge containing KCl completes the electrical circuit. The voltmeter measures the electrical potential, which drives electrons from one half-cell to the other.

temperature at 25°C. The reduction potential for the half-reaction, $2H^+ + 2e^- \rightarrow H_2(g)$, against the standard hydrogen electrode is set at 0.00 V.

In biochemistry the reference half-reaction is

$$2H^+ + 2\,e^- \leftrightharpoons H_2$$

$$\begin{aligned} \text{when pH} &= 7 \\ \text{temperature} &= 25°\text{C} \\ \text{pressure} &= 1 \text{ atm} \end{aligned}$$

Under these conditions the reduction potential ($E^{\circ\prime}$) of the hydrogen electrode is –0.42 V when measured against the standard hydrogen electrode in which the hydrogen ion concentration is 1 M. Substances with reduction potentials lower than –0.42 V (i.e., those with more negative values) have a lower affinity for electrons than does H^+. Substances with higher reduction potentials (i.e., those with more positive values) have a greater affinity for electrons (**Table 9.1**). The pH in the test electrode is 7.0 for each of the redox half-reactions, and the pH of the reference standard electrode is 0 or the $[H^+]$ is 1.0 M.

A substance with a more negative (less positive) reduction potential will receive electrons from a substance with a more positive reduction potential, and the overall cell potential ($\Delta E^{\circ\prime}$) will be positive. The relationship between $\Delta E^{\circ\prime}$ and $\Delta G^{\circ\prime}$ is

$$\Delta G^{\circ\prime} = -nF\,\Delta E^{\circ\prime}$$

where $\Delta G^{\circ\prime}$ = the standard free energy at pH 7
n = the number of electrons transferred
F = the Faraday constant (96,485 J/V·mol)

TABLE 9.1 Standard Reduction Potentials*

Redox Half-Reaction	Standard Reduction Potentials ($E^{\circ\prime}$) (V)
$2H^+ + 2e^- \rightarrow H_2$	−0.42
α-Ketoglutarate + $CO_2 + 2H^+ + 2e^- \rightarrow$ isocitrate	−0.38
$NADP^+ + H^+ + 2e^- \rightarrow$ NADPH	−0.324
$NAD^+ + H^+ + 2e^- \rightarrow$ NADH	−0.32
$S + 2H^+ + 2e^- \rightarrow H_2S$	−0.23
FAD + $2H^+ + 2e^- \rightarrow FADH_2$	−0.22
Acetaldehyde + $2H^+ + 2e^- \rightarrow$ ethanol	−0.20
Pyruvate+ $2H^+ + 2e^- \rightarrow$ lactate	−0.19
Oxaloacetate + $2H^+ + 2e^- \rightarrow$ malate	−0.166
$Cu^{2+} + e^- \rightarrow Cu^+$	+0.16
Fumarate + $2H^+ + 2e^- \rightarrow$ succinate	+0.031
Cytochrome b (Fe^{3+}) + $e^- \rightarrow$ cytochrome b (Fe^{2+})	+0.075
Cytochrome c_1 (Fe^{3+}) + $e^- \rightarrow$ cytochrome c_1 (Fe^{2+})	+0.22
Cytochrome c (Fe^{3+}) + $e^- \rightarrow$ cytochrome c (Fe^{2+})	+0.235
Cytochrome a (Fe^{3+}) + $e^- \rightarrow$ cytochrome a (Fe^{2+})	+0.29
$NO_3^- + 2H^+ + 2e^- \rightarrow NO_2^- + H_2O$	+0.42
$NO_2^- + 8H^+ + 6e^- \rightarrow NH_4^+ + 2H_2O$	+0.44
$Fe^{3+} + e^- \rightarrow Fe^{2+}$	+0.77
$\frac{1}{2}O_2 + 2H^+ + 2e^- \rightarrow H_2O$	+0.82

* By convention, redox reactions are written with the reducing agent to the right of the oxidizing agent and the number of electrons transferred. In this table the redox pairs are listed in order of increasing $E^{\circ\prime}$ values. The more negative the $E^{\circ\prime}$ value is for a redox pair, the lower the affinity of the oxidized component for electrons. The more positive the $E^{\circ\prime}$ value is, the greater the affinity of the oxidized component of the redox pair for electrons. Under appropriate conditions, a redox half-reaction reduces any of the half-reactions below it in the table.

$\Delta E^{\circ\prime}$ = the difference in reduction potential between the electron donor and the electron acceptor under standard conditions

Living organisms utilize redox coenzymes as high-energy electron carriers. The most prominent examples are described next.

Redox Coenzymes

The coenzyme forms of the vitamin molecules nicotinic acid and riboflavin are universal electron carriers. Their structural and functional properties are as follows.

NICOTINIC ACID There are two coenzyme forms of nicotinic acid: **nicotinamide adenine dinucleotide (NAD)** and **nicotinamide adenine dinucleotide phosphate (NADP)**. These coenzymes occur in oxidized forms (NAD^+ and $NADP^+$) and reduced forms (NADH and NADPH). The structures of NAD^+ and $NADP^+$ both contain adenosine and the *N*-ribosyl derivative of nicotinamide, which are linked together through a pyrophosphate group (**Figure 9.5a**). $NADP^+$ has an additional phosphate attached to the 2′ OH group of adenosine. (The ring atoms of the sugar in a nucleotide are designated with a prime to distinguish them from atoms in the base.) Both NAD^+ and $NADP^+$ carry electrons for several enzymes in a group known as the dehydrogenases. (Dehydrogenases catalyze hydride transfer reactions. Many dehydrogenases that catalyze reactions involved in energy generation use the coenzyme NADH. The enzymes that require NADPH usually catalyze biosynthetic reactions. A small number of dehydrogenases can use either NADH or NADPH.)

Recall that alcohol dehydrogenase catalyzes the reversible oxidation of ethanol to form acetaldehyde (p. 225).

$$CH_3—CH_2—OH + NAD^+ \rightleftharpoons CH_3—\overset{\overset{\displaystyle O}{\|}}{C}—H + NADH + H^+$$

During this reaction NAD^+ accepts a hydride ion from ethanol, the substrate molecule undergoing oxidation. The product deprotonates to form the acetaldehyde molecule. The reversible reduction of NAD^+ is illustrated in **Figure 9.5b**.

In most reactions catalyzed by dehydrogenases, the NAD^+ (or $NADP^+$) is bound only transiently to the enzyme. After the reduced version of the coenzyme is released from the enzyme, it donates the hydride ion to another molecule, called an *electron acceptor*, with a more positive reduction potential than NADH.

RIBOFLAVIN Riboflavin (vitamin B_2) is a component of two coenzymes: **flavin mononucleotide (FMN)** and **flavin adenine dinucleotide (FAD)** (Figure 9.6). FMN and FAD function as tightly bound prosthetic groups in a class of enzymes known as the **flavoproteins**. Flavoproteins are a diverse group of redox enzymes; they function as dehydrogenases, oxidases, and hydroxylases. These enzymes use the isoalloxazine group of FAD or FMN as

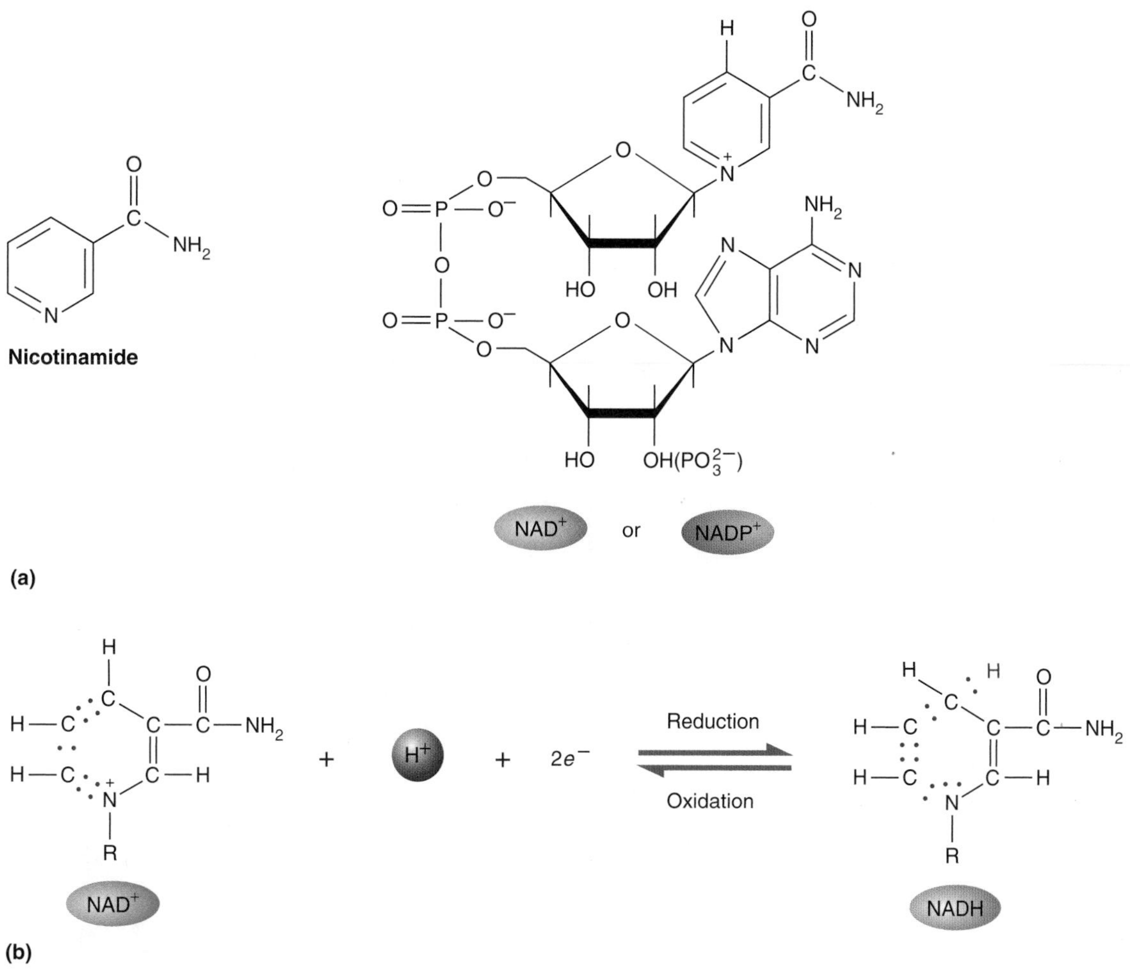

FIGURE 9.5

Nicotinamide Adenine Dinucleotide (NAD)

(a) Nicotinamide and $NAD(P)^+$. (b) Reversible reduction of NAD^+ to NADH. To simplify the equation, only the nicotinamide ring is shown. The rest of the molecule is designated R.

a donor or acceptor of two hydrogen atoms. FMN plays a key role in the link between two-electron transfer reactions in the mitochondrial matrix and the one-electron transfer reactions of the electron transport chain because it can transfer one hydrogen atom at a time. Succinate dehydrogenase is a prominent example of a flavoprotein. It catalyzes the oxidation of succinate to form fumarate, an important reaction in the citric acid cycle.

QUESTION 9.1

Use **Table 9.1** to determine which of the following reactions will proceed as written:

$$CH_3CH_2OH + 2 \text{ cyt b } (Fe^{3+}) \rightarrow CH_3CHO + 2 \text{ cyt b } (Fe^{2+}) + 2H^+$$

$$NO_2^- + H_2O + 2 \text{ cyt b } (Fe^{3+}) \rightarrow 2 \text{ cyt b } (Fe^{2+}) + NO_3^- + 2H^+$$

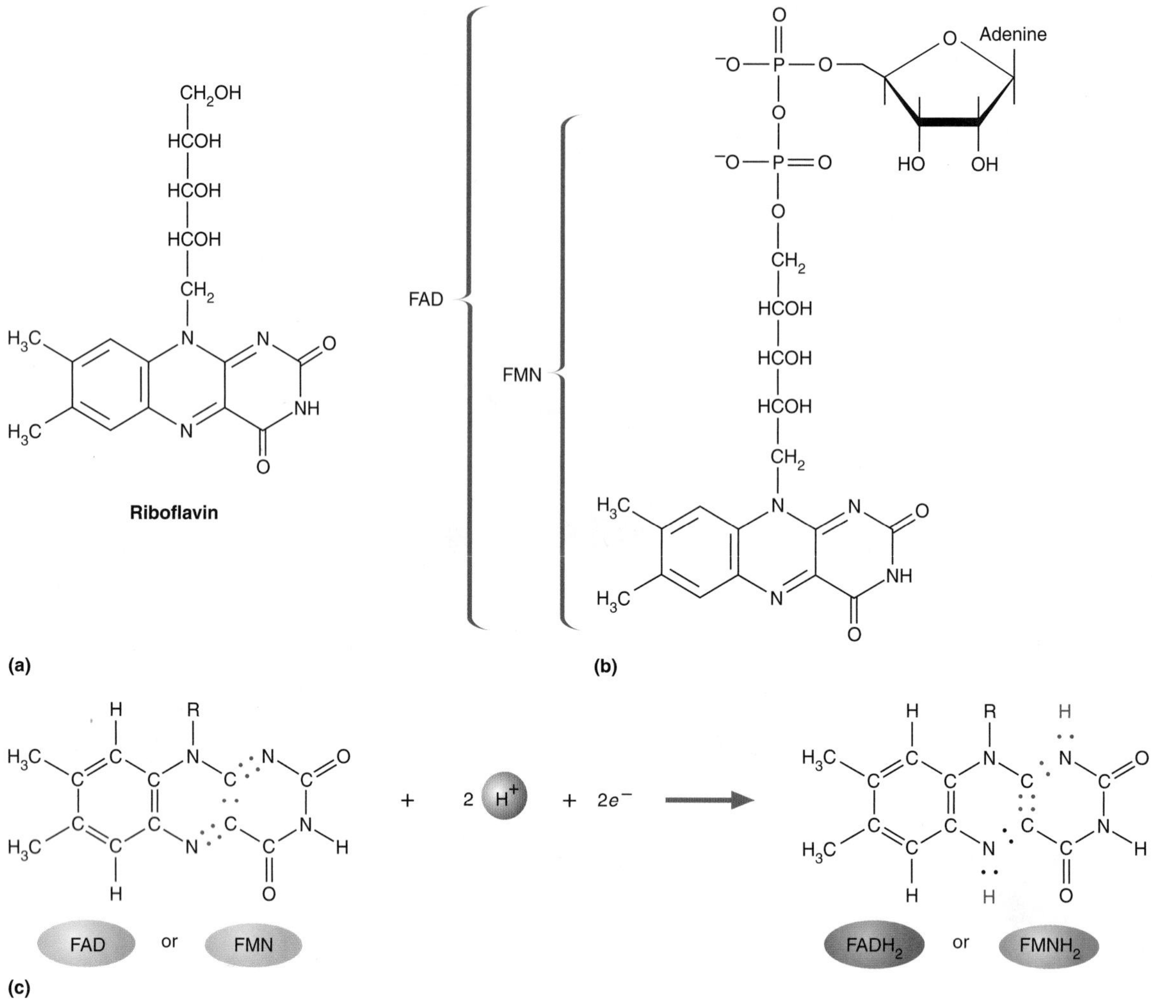

FIGURE 9.6

Flavin Coenzymes

(a) The vitamin riboflavin consists of an isoalloxazine ring system linked to ribotol (an alcohol formed by the reduction of ribose). (b) Structure of FAD and FMN. (c) Reversible reduction of flavin coenzymes: to simplify the equation, only the isoalloxazine ring system is shown. The rest of the coenzyme is designated R.

QUESTION 9.2

Which of the following reactions are redox reactions? For each redox reaction identify the oxidizing and reducing agents.

1. Glucose + ATP → glucose-1-phosphate + ADP
2.

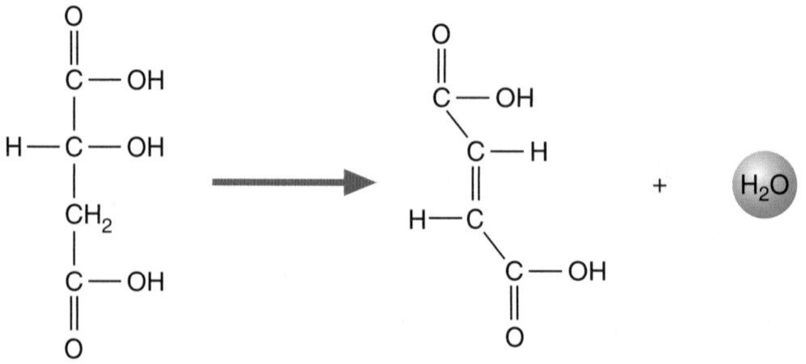

3. Lactate + NAD^+ → pyruvate + NADH + H^+
4. NO_2^- + $8H^+$ + 6 cyt b (Fe^{2+}) → NH_4^+ + $2H_2O$ + 6 cyt b (Fe^{3+})
5. CH_3CHO + NADH + H^+ → CH_3CH_2OH + NAD^+

WORKED PROBLEM 9.1

Use the following half-cell potentials to calculate (a) the overall cell potential and (b) $\Delta G^{\circ\prime}$.

Succinate + $\frac{1}{2}O_2$ → fumarate + H_2O

The half-reactions are

Succinate → fumarate + $2H^+ + 2e^-$ $(E^{\circ\prime} = -0.031\ V)$
$\frac{1}{2}O_2 + 2H^+ + 2e^- \rightarrow H_2O$ $(E^{\circ\prime} = +0.82\ V)$

SOLUTION

Write the fumarate reaction as an oxidation (lower reduction potential), balance if the number of electrons transferred differs between the two half-reactions (not necessary here), and add the two reactions to get the net reaction.

Succinate + $\frac{1}{2}O_2$ → fumarate + H_2O

a. The overall potential is defined by the following:

$\Delta E^{\circ\prime} = E^{\circ\prime}$ (electron acceptor) $- E^{\circ\prime}$ (electron donor)
$\Delta E^{\circ\prime} = (+0.82\ V) - (-0.031\ V)$
$\Delta E^{\circ\prime} = +0.85\ V$

b. Use the formula to find $\Delta G^{\circ\prime}$.

$$\begin{aligned}\Delta G^{\circ\prime} &= -nF\Delta E^{\circ\prime}\\ &= -(2)(96.5\ kJ/V{\cdot}mol)(0.85\ V)\\ &= -164.05\ kJ/mol\\ &= -164\ kJ/mol\end{aligned}$$

■

QUESTION 9.3

Because redox reactions play an important role in living processes, biochemists need to determine the oxidation state of the atoms in a molecule. In one method, the oxidation state of an atom is determined by assigning numbers to carbon atoms based on the type of groups attached to them. For example, a bond to a hydrogen is assigned the value −1. A bond to another carbon atom is valued at 0, and a bond to an electronegative atom such as oxygen or nitrogen is valued at +1. The values of a single carbon atom in a molecule may range from −4 (e.g., CH_4) to +4 (CO_2). Note that methane is a high-energy molecule and carbon dioxide is a low-energy molecule. As carbon changes its oxidation state from −4 to +4, a large amount of energy is released. This process is therefore highly exothermic.

Ethanol is degraded in the liver by a series of redox reactions. Identify the oxidation state of the indicated carbon atom in each molecule in the following reaction sequence:

$CH_3—CH_2—OH$ (Ethanol) → $CH_3—C(=O)—H$ (Acetaldehyde) → $CH_3—C(=O)—OH$ (Acetic acid)

QUESTION 9.4

As CO_2 is incorporated into organic molecules during photosynthesis, is it being oxidized or reduced?

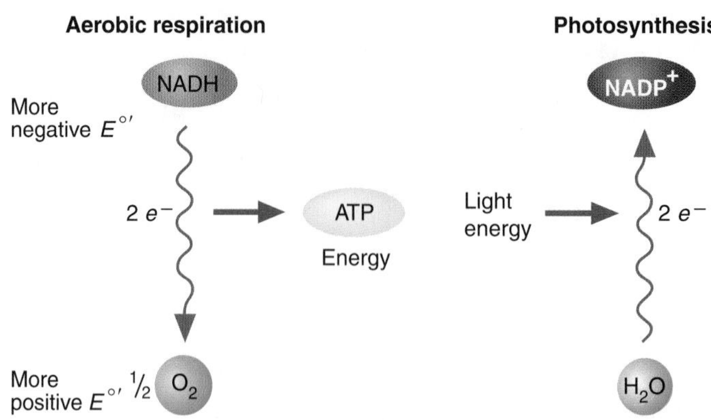

FIGURE 9.7
Electron Flow and Energy

Electron flow is used to generate and capture energy in aerobic respiration. Energy is also used to drive electron flow in photosynthesis. Note that the energy captured by photosynthesis in the chemical bonds of sugars and other biomolecules is released by aerobic respiration and used to synthesize ATP.

Aerobic Metabolism

Most of the aerobic cell's free energy is captured by the mitochondrial ETC. During this process, electrons are transferred from a redox pair with a more negative reduction potential ($NADH/NAD^+$) to those with more positive reduction potentials. The last component in the system is the $H_2O/{}^1\!/_2O_2$ pair:

$$ {}^1\!/_2O_2 + NADH + H^+ \rightarrow H_2O + NAD^+ $$

The free energy released as a pair of electrons passes from NADH to O_2 under standard conditions is calculated as follows:

$$\begin{aligned} \Delta G^{\circ\prime} &= -nF\Delta E^{\circ\prime} \\ &= -2(96.5\ \text{kJ/V}\cdot\text{mol})[0.815 - (-0.32)] \\ &= -220\ \text{kJ/mol} \end{aligned}$$

A significant portion of the free energy generated as electrons move from NADH to O_2 in the ETC is used to drive ATP synthesis.

It should be noted that in several metabolic processes, electrons move from redox pairs with more positive reduction potentials to those with more negative reduction potentials. Of course, energy is required. The most prominent example of this phenomenon is photosynthesis (Chapter 13). Photosynthetic organisms use captured light energy to drive electrons from electron donors, such as water, to electron acceptors with more negative reduction potentials (**Figure 9.7**). The energized electrons eventually flow back to acceptors with more positive reduction potentials, thereby providing energy for ATP synthesis and CO_2 reduction to form carbohydrate.

In Section 9.2 the citric acid cycle is examined. In this pathway, which is the first phase of aerobic metabolism, the energy released by the oxidation of two-carbon fragments derived from glucose, fatty acids, and some amino acids is captured by and carried in the reduced coenzymes NADH and $FADH_2$.

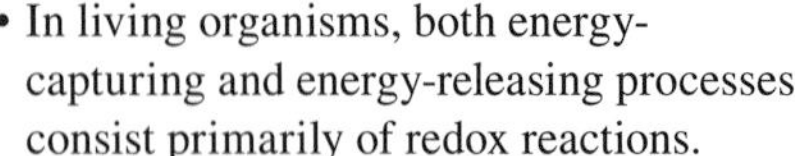

KEY CONCEPTS

- In living organisms, both energy-capturing and energy-releasing processes consist primarily of redox reactions.
- In redox reactions, electrons move between an electron donor and an electron acceptor.
- In many reactions, both electrons and protons are transferred.
- In biological systems, most redox reactions involve hydride ion transfer ($NADH/NAD^+$) or hydrogen atom transfer ($FADH_2/FAD$).

9.2 CITRIC ACID CYCLE

The citric acid cycle (**Figure 9.8**) is a series of biochemical reactions aerobic organisms use to release chemical energy stored in the two-carbon acetyl group in acetyl-CoA. Acetyl-CoA is composed of an acetyl group derived from the breakdown of carbohydrates, lipids, and some amino acids that is linked to the acyl carrier molecule **coenzyme A** (**Figure 9.9**). Acetyl-CoA is synthesized from pyruvate and is also the product of fatty acid catabolism (described in Chapter 11) and certain reactions in amino acid metabolism (Chapter 15). In the

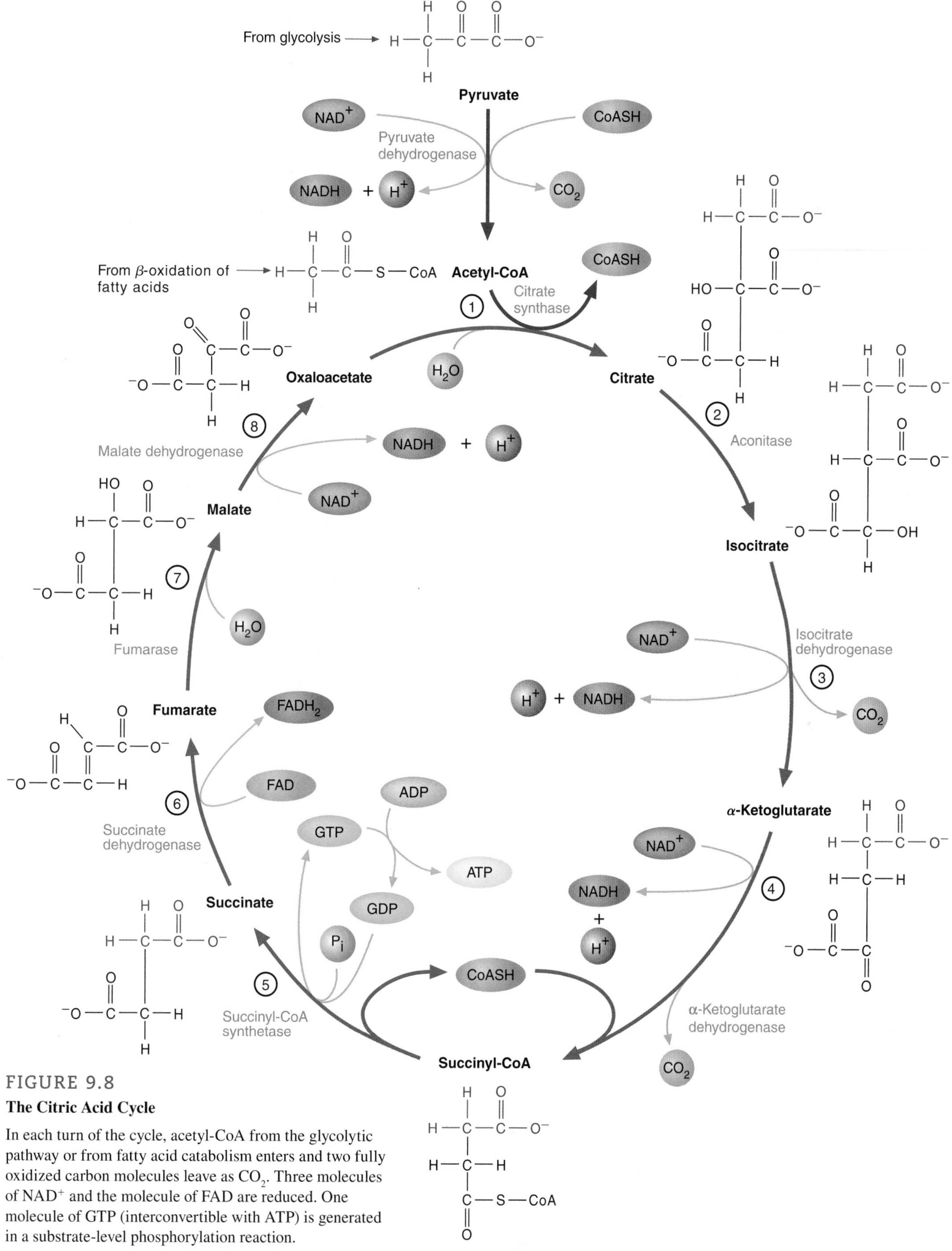

FIGURE 9.8

The Citric Acid Cycle

In each turn of the cycle, acetyl-CoA from the glycolytic pathway or from fatty acid catabolism enters and two fully oxidized carbon molecules leave as CO_2. Three molecules of NAD^+ and the molecule of FAD are reduced. One molecule of GTP (interconvertible with ATP) is generated in a substrate-level phosphorylation reaction.

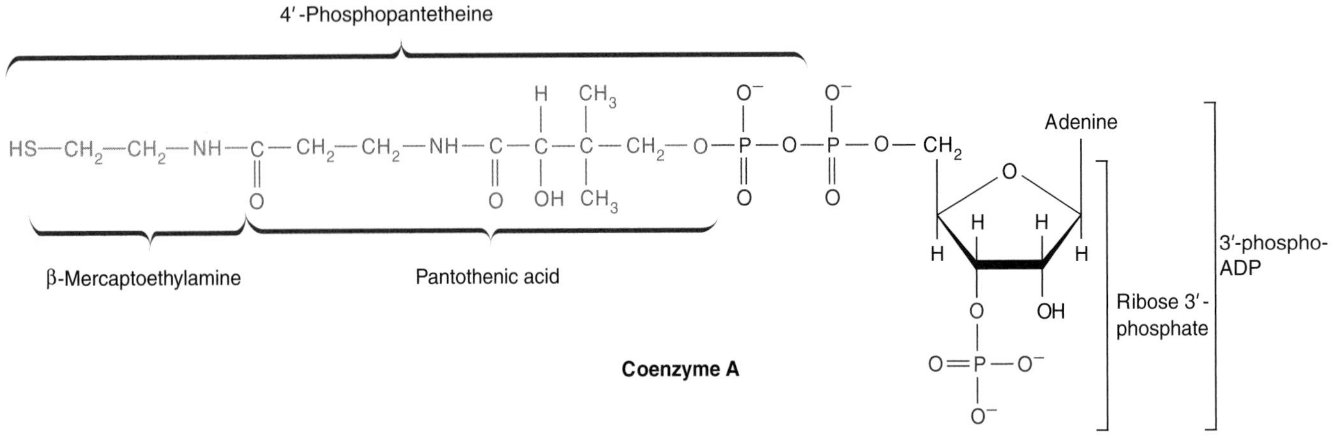

FIGURE 9.9

Coenzyme A

In coenzyme A a 3′-phosphate derivative of ADP is linked to pantothenic acid via a phosphate ester bond. The β-mercaptoethylamine group of coenzyme A is attached to pantothenic acid by an amide bond. Coenzyme A is a carrier of acyl groups that range in size from the acetyl group to long-chain fatty acids. Because the reactive SH group forms a thioester bond with acyl groups, coenzyme A is often abbreviated as CoASH. Note that sulfur is a better leaving group than oxygen. Consequently, the carbon-sulfur linkage of a thioester is a high-energy bond that is more easily cleaved than the carbon-oxygen bond of an ester.

citric acid cycle, the acetyl group's carbon atoms are eventually oxidized to CO_2 and the electrons are transferred to NAD^+ and FAD.

In the first reaction of the citric acid cycle, a two-carbon acetyl group condenses with a four-carbon molecule (oxaloacetate) to form a six-carbon molecule (citrate). During the subsequent seven reactions, in which two CO_2 molecules are produced and four pairs of electrons are removed from carbon compounds, citrate is reconverted to oxaloacetate. During one step in the cycle, the high-energy molecule GTP is produced during a substrate-level phosphorylation. The net reaction for the citric acid cycle is as follows:

$$\text{Acetyl-CoA} + 3\,\text{NAD}^+ + \text{FAD} + \text{GDP} + \text{P}_i + 2\text{H}_2\text{O} \rightarrow$$
$$2\text{CO}_2 + 3\,\text{NADH} + \text{FADH}_2 + \text{CoASH} + \text{GTP} + 2\text{H}^+$$

In addition to its role in energy production, the citric acid cycle plays another important role in metabolism. Cycle intermediates are substrates in a variety of biosynthetic reactions. **Table 9.2** provides a summary of the roles of coenzymes in the citric acid cycle.

Conversion of Pyruvate to Acetyl-CoA

After its transport into the mitochondrial matrix, pyruvate is converted to acetyl-CoA in a series of reactions catalyzed by the enzymes in the pyruvate

TABLE 9.2 Summary of the Coenzymes in the Citric Acid Cycle

Coenzyme	Functions
Thiamine pyrophosphate (TPP)	Decarboxylation and aldehyde group transfer
Lipoic acid	Carrier of hydrogens or acetyl groups
NADH	Electron carrier
$FADH_2$	Electron carrier
Coenzyme A (CoASH)	Acetyl group carrier

TABLE 9.3 *E. coli* Pyruvate Dehydrogenase Complex

Enzyme Activity	Function	Copies per Complex*	Coenzymes
Pyruvate dehydrogenase (E_1)	Decarboxylates pyruvate	24 (20–30)	TPP
Dihydrolipoyl transacetylase (E_2)	Catalyzes transfer of acetyl group to CoASH	24 (60)	Lipoic acid, CoASH
Dihydrolipoyl dehydrogenase (E_3)	Reoxidizes dihydrolipoamide	12 (20–30)	NAD^+, FAD

* The number of molecules of each enzyme activity found in mammalian pyruvate dehydrogenase is shown in parentheses.

dehydrogenase complex (PDHC). The net reaction, an oxidative decarboxylation, is as follows:

$$\text{Pyruvate} + NAD^+ + \text{CoASH} \rightarrow \text{Acetyl-CoA} + \text{NADH} + CO_2 + H^+$$

Despite the apparent simplicity of this highly exergonic reaction ($\Delta G^{\circ\prime} = -33.5$ kJ/mol), its mechanism is one of the most complex known. The PDHC is a large multienzyme structure that contains multiple copies of three enzyme activities: pyruvate dehydrogenase (E_1), dihydrolipoyl transacetylase (E_2), and dihydrolipoyl dehydrogenase (E_3). **Table 9.3** summarizes the number of copies of each enzyme and the required coenzymes of *E. coli* pyruvate dehydrogenase.

In the first step, pyruvate dehydrogenase catalyzes the decarboxylation of pyruvate (**Figure 9.10**). A nucleophile is formed when a basic residue of the enzyme extracts a proton from the thiazole ring of **thiamine pyrophosphate** (TPP). (TPP is the coenzyme form of thiamine, also called vitamin B_1.) The intermediate, hydroxyethyl-TPP (HETPP), forms after the nucleophilic thiazole ring has attacked the carbonyl group of pyruvate with the resulting loss of CO_2.

In the next several steps, the hydroxyethyl group of HETPP is converted to acetyl-CoA by dihydrolipoyl transacetylase. **Lipoic acid** (**Figure 9.11**) an acyl transfer coenzyme that contains two thiol groups that can be reversibly oxidized, plays a crucial role in this transformation. Lipoic acid is bound to the enzyme through an amide linkage with the -amino group of a lysine residue. Lipoamide reacts with HETPP to form an acetylated lipoamide and free TPP. The acetyl group is then transferred to the sulfhydryl group of coenzyme A. Subsequently, the reduced lipoamide is reoxidized by dihydrolipoyl dehydrogenase. The $FADH_2$ product is reoxidized by NAD^+ (with its more negative reduction potential) to form the FAD required for the oxidation of the next reduced lipoamide residue. The mobile NADH can deliver its electrons to the ETC and is replaced in the enzyme by another NAD^+ molecule so that the cycle can begin again.

KEY CONCEPT

Pyruvate is converted to acetyl-CoA by the enzymes in the pyruvate dehydrogenase complex. TPP, FAD, NAD^+, Coenzyme A, and lipoic acid are required coenzymes.

PDHC is stringently regulated because of its central role in energy metabolism, most notably by linking glycolysis to the citric acid cycle. Its activity is controlled largely through allosteric effectors and covalent modification. The enzyme complex is allosterically activated by NAD^+, CoASH, and AMP. It is inhibited by high concentrations of ATP and the reaction products acetyl-CoA and NADH. In mammals, acetyl-CoA and NADH also activate a kinase, which converts the active PDHC to an inactive phosphorylated form. High concentrations of the substrates pyruvate, CoASH, and NAD^+ inhibit the activity of the kinase. The PDHC is reactivated by a dephosphorylation reaction catalyzed by the phosphoprotein phosphatase pyruvate dehydrogenase phosphatase (PDP). PDP is activated when the mitochondrial ATP concentration is low. PDP is also activated by Ca^{2+} and insulin.

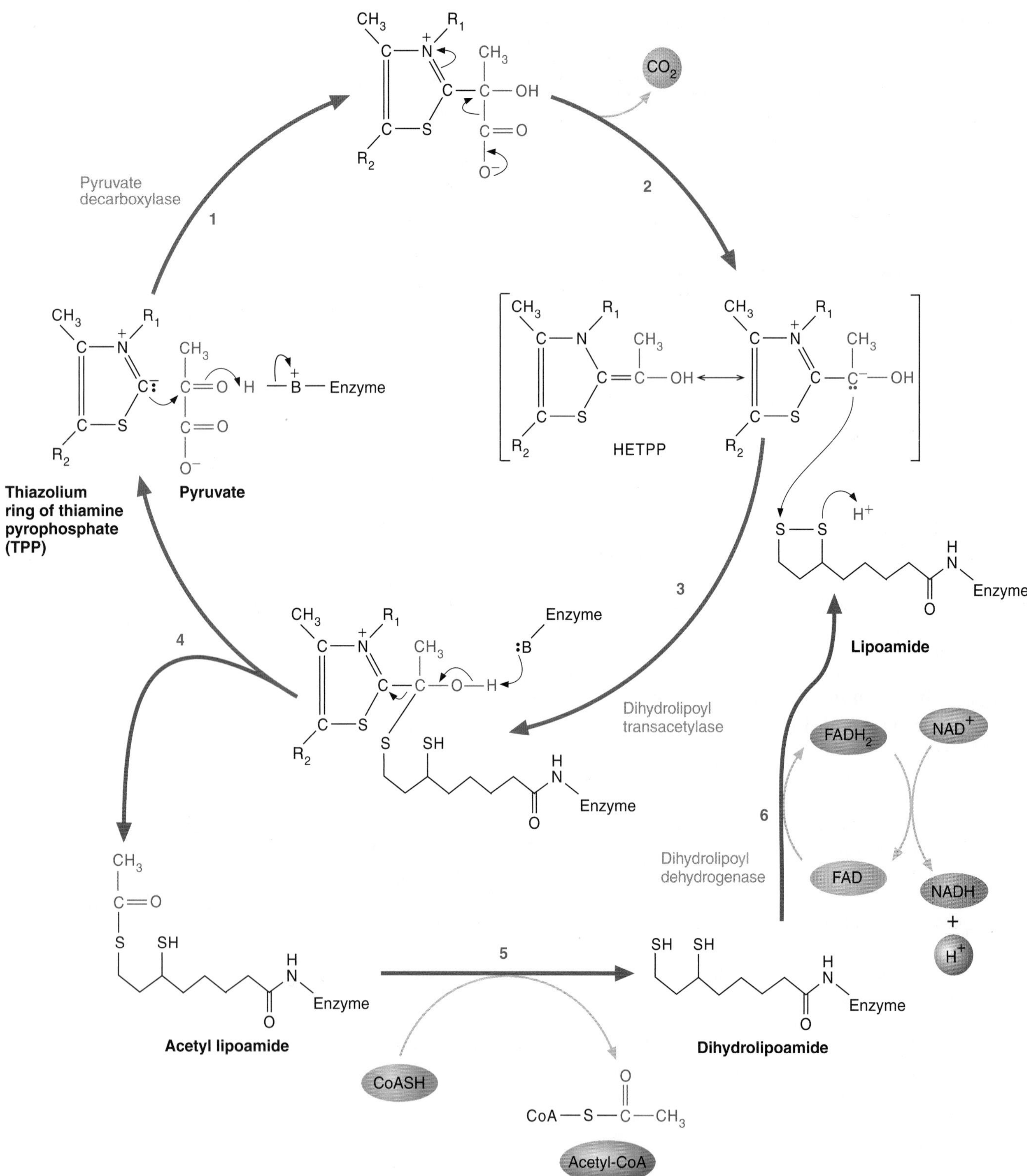

FIGURE 9.10

The Reactions Catalyzed by the Pyruvate Dehydrogenase Complex

Pyruvate decarboxylase, which contains TPP, catalyzes the formation of HETPP. (1) TPP attacks pyruvate to produce a reactive intermediate that (2) decarboxylates to produce HETPP. Dihydrolipoyl transacetylase uses lipoic acid as a cofactor to convert the hydroxyethyl group of HETPP to acetyl-CoA. In step (3) a resonance-stabilized carbanion attacks the oxidized lipoamide of dihydrolipoyl transacetylase. (4) The enzyme converts the hydroxyethyl group into an acetyl group. Note that the acetyl group is linked to lipoamide through a thiol ester bond. (5) A nucleophilic attack of CoASH on the acetyl group's carbonyl carbon yields acetyl-CoA and dihydrolipoamide. In step (6) dihydrolipoyl dehydrogenase reoxidizes the reduced lipoamide. FAD is regenerated when $FADH_2$ donates a hydride ion to NAD^+. (Refer to Figure 9.11 for the structure of lipoamide.)

Reactions of the Citric Acid Cycle

The citric acid cycle is composed of eight reactions that occur in two stages:

1. The two-carbon acetyl group of acetyl-CoA enters the cycle by reacting with the four-carbon compound OAA and two molecules of CO_2 are subsequently liberated (reactions 1–4).
2. OAA is regenerated so it can react with another acetyl-CoA (reactions 5–8).

The enzymes that catalyze these reactions associate through noncovalent interactions into metabolons (p. 211), multienzyme complexes that ensure efficient channeling of the product of each reaction to the next enzyme in the pathway.

The reactions of the citric acid cycle are as follows.

1. **Introduction of two carbons as acetyl-CoA**. The citric acid cycle begins with the condensation of acetyl-CoA with oxaloacetate to form citrate:

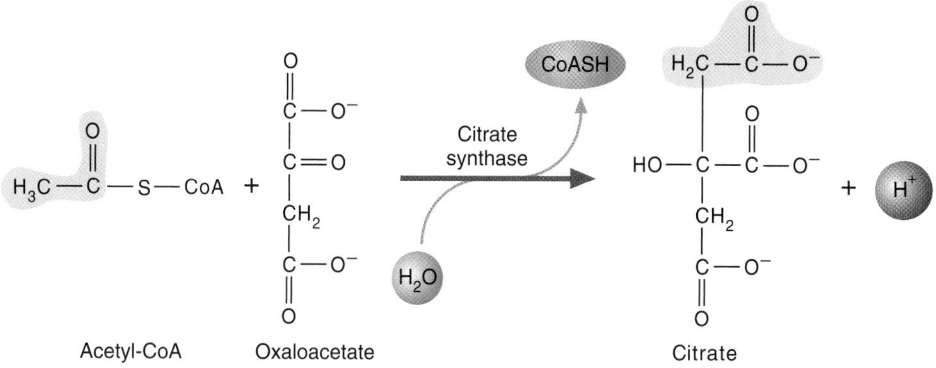

Acetyl-CoA + Oxaloacetate + H_2O —(Citrate synthase)→ Citrate + CoASH + H^+

Note that this reaction is an aldol condensation. In this reaction (**Figure 9.12**) the enzyme removes a proton from the methyl group of acetyl-CoA, thereby converting it to an enol. The enol subsequently attacks the carbonyl carbon of oxaloacetate. The product, citroyl-CoA, rapidly hydrolyzes to form citrate and CoASH. Because of the hydrolysis of the high-energy thioester bond, the overall standard free energy change is –33.5 kJ/mol, and citrate formation is highly exergonic.

2. **Citrate is isomerized to form a secondary alcohol that can be easily oxidized**. In the next reaction of the cycle, citrate, which contains a tertiary alcohol, is reversibly converted to isocitrate by aconitase. During this isomerization reaction, an intermediate called *cis*-aconitate is formed by dehydration. The carbon-carbon double bond of *cis*-aconitate is then rehydrated in a nucleophilic addition reaction to form the more reactive secondary alcohol, isocitrate. Although the standard free energy change of citrate isomerization is positive ($\Delta G^{\circ\prime}$ = 13.3 kJ), the reaction is pulled forward by the rapid removal of isocitrate by the next reaction.

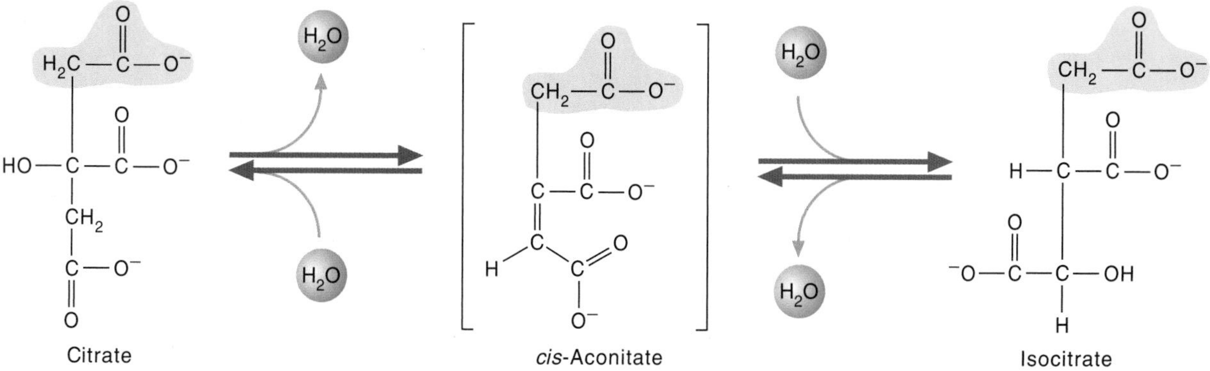

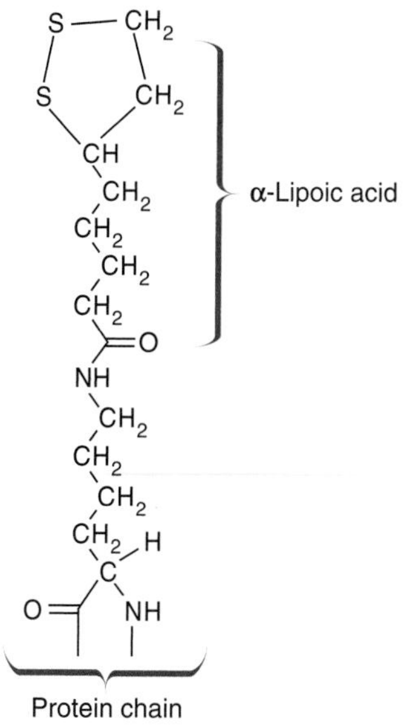

Citrate ⇌ ($-H_2O$ / $+H_2O$) [*cis*-Aconitate] ⇌ ($+H_2O$ / $-H_2O$) Isocitrate

α-Lipoic acid

Protein chain

FIGURE 9.11

Lipoamide

Lipoic acid is covalently bonded to the enzyme through an amide linkage with the ε-amino group of a lysine residue.

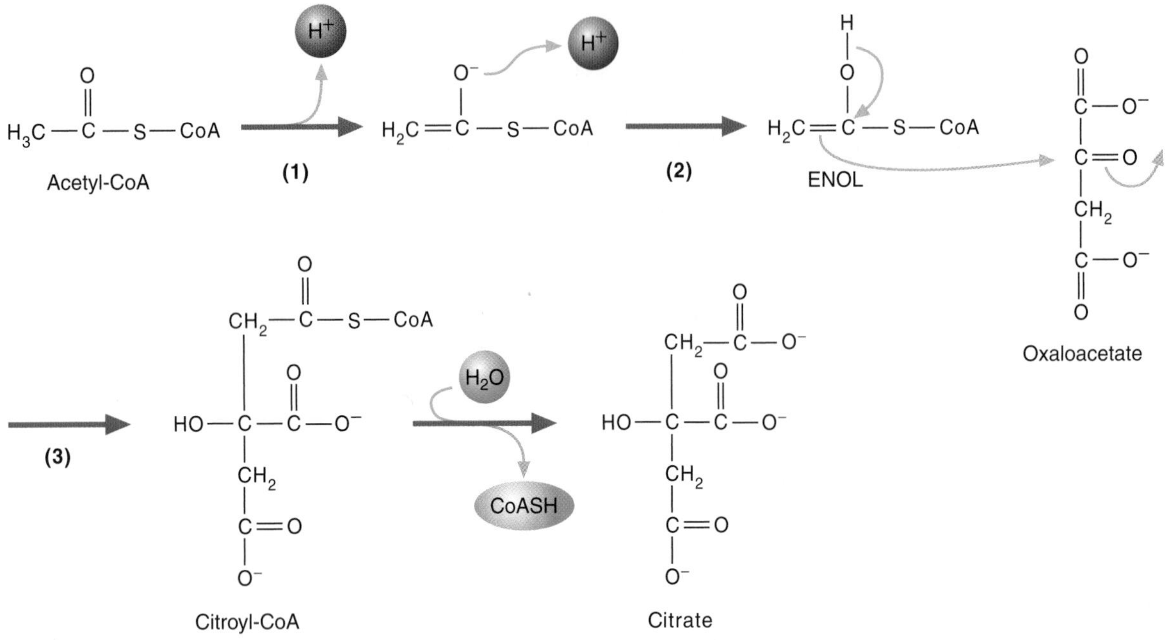

FIGURE 9.12
Citrate Synthesis

(1) A side chain carboxylate group of the enzyme citrate synthase removes a proton from the methyl group of acetyl-CoA. (2) Simultaneously, a side chain NH group of a histidine residue donates a proton to the carbonyl oxygen, thus generating an enol intermediate. (3) The same histidine side chain then deprotonates the enol to produce an enolate anion that attacks the carbonyl carbon of oxaloacetate. (4) The product, citroyl-CoA, is then hydrolyzed in a nucleophilic acyl substitution reaction to yield citrate and CoASH.

3. **Isocitrate is oxidized to form NADH and CO_2.** The oxidative decarboxylation of isocitrate, catalyzed by isocitrate dehydrogenase, occurs in three steps. First, isocitrate is oxidized to form a transient ketone-group-containing intermediate called oxalosuccinate. A manganese (Mn^{2+}) cofactor facilitates the polarization of the newly formed ketone group.

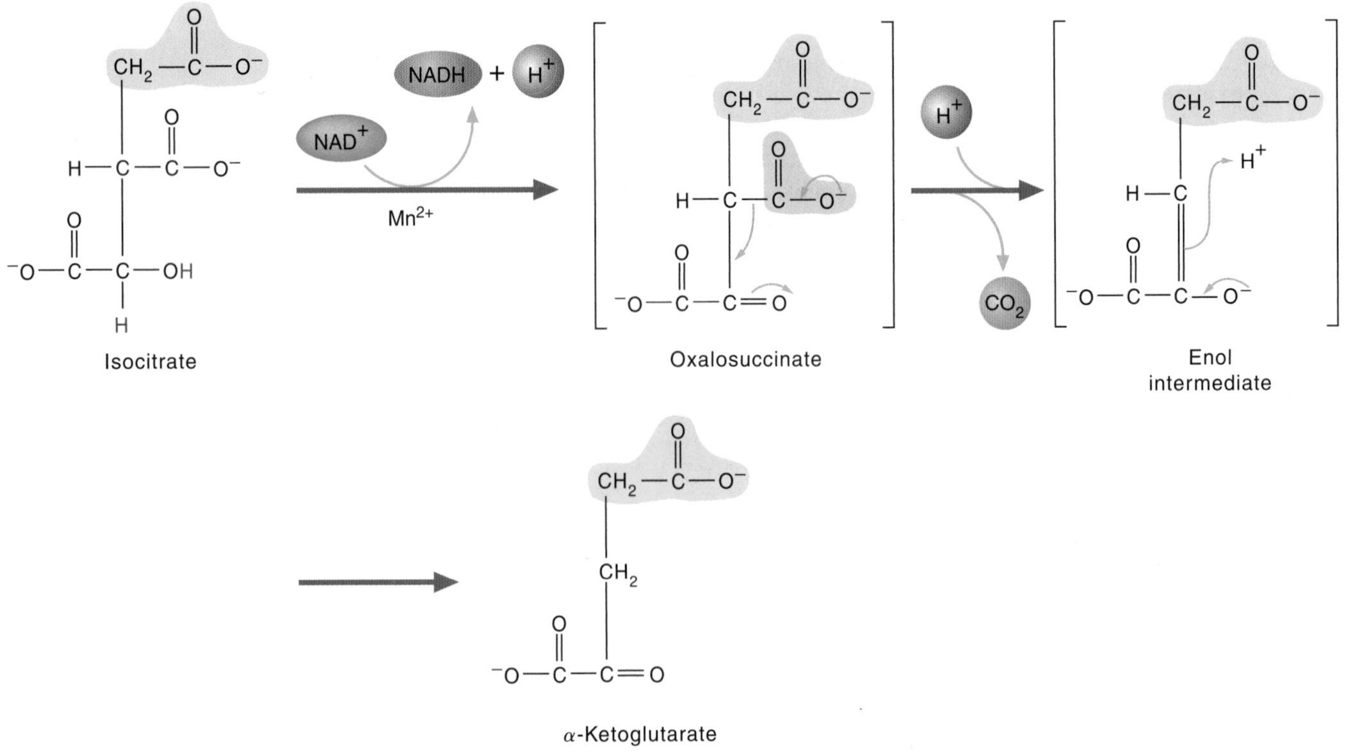

Immediate decarboxylation of oxalosuccinate results in the formation of an enol intermediate that rearranges to form α-ketoglutarate, an α-keto acid. There are three forms of isocitrate dehydrogenase in mammals. The NAD^+-requiring isozyme (IDH3) is found only within mitochondria. The other isozymes, IDH1 (cytoplasm) and IDH2 (mitochondria), use $NADP^+$ as a cofactor. NADPH is required in biosynthetic processes. Note that the NADH produced in the conversion of isocitrate to α-ketoglutarate is the first link between the citric acid cycle and the ETC and oxidative phosphorylation.

4. **α-Ketoglutarate is oxidized to form a second molecule each of NADH and CO_2.** The conversion of α-ketoglutarate to succinyl-CoA is catalyzed by the enzyme activities in the α-ketoglutarate dehydrogenase complex: α-ketoglutarate dehydrogenase, dihydrolipoyl transsuccinylase, and dihydrolipoyl dehydrogenase.

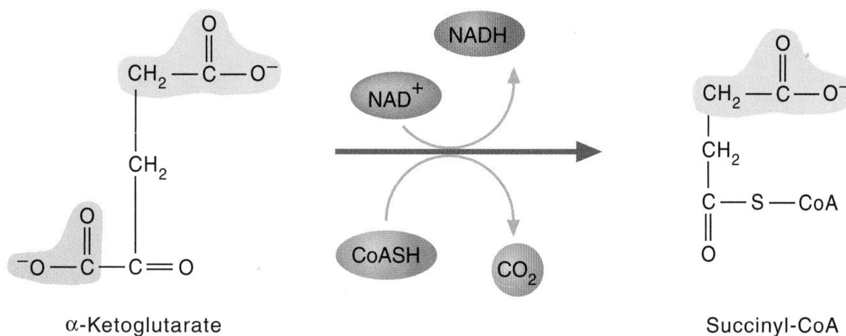

This highly exergonic reaction ($\Delta G^{\circ\prime} = -33.5$ kJ/mol), an oxidative decarboxylation, is analogous to the conversion of pyruvate to acetyl-CoA catalyzed by pyruvate dehydrogenase. Both reactions have energy-rich thioester molecules as products, that is, acetyl-CoA and succinyl-CoA. Other similarities between the two multienzyme complexes are that the same cofactors (TPP, CoASH, lipoic acid, NAD^+, and FAD) are required, and the same or similar allosteric effectors are inhibitors. α-Ketoglutarate dehydrogenase is inhibited by succinyl-CoA, NADH, ATP, and GTP. An important difference between the two complexes is that the control mechanism of the α-ketoglutarate dehydrogenase complex does not involve covalent modification.

5. **The cleavage of succinyl-CoA is coupled to a substrate-level phosphorylation.** The cleavage of the high-energy thioester bond of succinyl-CoA to form succinate is a reversible reaction catalyzed by succinyl-CoA synthetase (also called succinate thiokinase). In mammals the reaction is coupled to the substrate-level phosphorylation of ADP or GDP. There are two forms of succinyl-CoA synthetase: one is specific for ATP and the other for GTP. In many tissues both enzymes are produced, although their relative amounts vary.

Succinyl-CoA + ADP or GDP + P_i $\rightleftharpoons$ Succinate + ATP or GTP + CoASH

The direction of the reaction depends on the relative concentrations of nucleoside diphosphates (ADP and/or GDP) and nucleotide triphosphates (ATP and/or GTP). GTP is used for RNA, DNA, and protein synthesis within mitochondria. The phosphoryl group of GTP can also be donated to ADP in a reversible reaction catalyzed by nucleoside diphosphate kinase.

GTP + ADP ⇌ GDP + ATP

6. **The four-carbon molecule succinate is oxidized to form fumarate and $FADH_2$.** Succinate dehydrogenase catalyzes the oxidation of succinate to form fumarate:

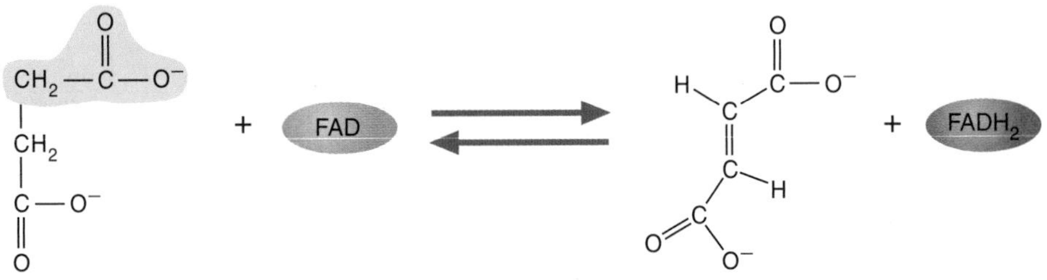

Succinate + FAD ⇌ Fumarate + $FADH_2$

A general base within the enzyme's active site removes a proton from C-2 to form a C-2–C-3 double bond followed by the release of a C-3 hydride to FAD to yield $FADH_2$ and fumarate. Unlike the other citric acid cycle enzymes, succinate dehydrogenase is not found within the mitochondrial matrix. Instead, it is tightly bound to the inner mitochondrial membrane. Succinate dehydrogenase is a flavoprotein that uses FAD to drive the oxidation of succinate to fumarate because the oxidation of an alkane requires a stronger oxidizing agent than NAD^+. Succinate dehydrogenase is composed of four subunits. ShdA contains the succinate binding site and a covalently bound FAD. ShdB possesses three iron-sulfur clusters that function as electron carriers between $FADH_2$ and coenzyme Q, a component of the ETC. Subunits ShdC and ShdD are hydrophobic molecules that anchor the enzyme complex into the inner membrane. The $\Delta G^{\circ\prime}$ for succinate oxidation is −5.6 kJ/mol. Succinate dehydrogenase is activated by high concentrations of succinate, ADP, and P_i and inhibited by OAA. The enzyme is also inhibited by malonate, a structural analogue of succinate, whose accumulation redirects energy to fatty acid synthesis. This inhibitor was used by Hans Krebs in his pioneering work on the citric acid cycle.

Visit the companion website at www.oup.com/us/mckee to read the Biochemistry in Perspective essay on Hans Krebs and the citric acid cycle.

7. **Fumarate is hydrated.** Fumarate is converted to L-malate in a reversible stereospecific hydration reaction catalyzed by fumarase (also referred to as fumarate hydratase):

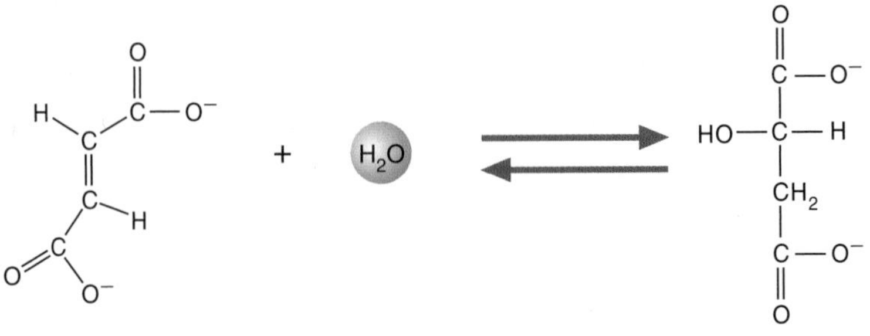

Fumarate + H_2O ⇌ L-Malate

A general base of the enzyme activates a water molecule that then attacks the double bond to form a hydroxyl group at C-2 and a carbanion at C-3. A general acid then protonates the carbanion to yield L-malate.

8. **Malate is oxidized to form OAA and a third NADH.** Finally, OAA is regenerated with the oxidation of L-malate by malate dehydrogenase:

$$\mathrm{^{-}O{-}C({=}O){-}CH(OH){-}CH_2{-}C({=}O){-}O^{-}} + NAD^{+} \rightleftharpoons \mathrm{^{-}O{-}C({=}O){-}C({=}O){-}CH_2{-}C({=}O){-}O^{-}} + NADH + H^{+}$$

L-Malate L-Oxaloacetate

A histidine side chain in the active site of the enzyme removes a hydrogen from the hydroxyl group at C-2 of L-malate. Simultaneously, with the formation of the carbonyl group, a hydride is transferred to NAD^+ to yield NADH. Malate dehydrogenase uses NAD^+ as the oxidizing agent in a highly endergonic reaction ($\Delta G^{\circ\prime} = +29$ kJ/mol). The reaction is pulled to completion because of the removal of oxaloacetate in the next round of the cycle.

Fate of Carbon Atoms in the Citric Acid Cycle

In each turn of the citric acid cycle, two carbon atoms enter as the acetyl group of acetyl-CoA and two molecules of CO_2 are released. A careful review of Figure 9.8 reveals that the two carbon atoms released as CO_2 molecules are not the same two carbons that just entered the cycle. Instead, the released carbon atoms are derived from OAA that reacted with the incoming acetyl-CoA. The incoming carbon atoms subsequently form one-half of succinate. Because of the symmetric structure of succinate, the carbon atoms derived from the incoming acetyl group become evenly distributed in all of the molecules derived from succinate. Consequently, incoming carbon atoms are released as CO_2 only after two or more turns of the cycle.

KEY CONCEPTS

- The citric acid cycle begins with the condensation of a molecule of acetyl-CoA with oxaloacetate to form citrate, which is eventually reconverted to oxaloacetate.
- During this process, two molecules of CO_2, three molecules of NADH, one molecule of $FADH_2$, and one molecule of GTP are produced.

QUESTION 9.5

Trace the labeled carbon in $CH_3{}^{14}C(=O){-}SCoA$ through one round of the citric acid cycle. After examining **Figure 9.8**, show why more than two turns of the cycle are required before all the labeled carbon atoms are released as $^{14}CO_2$.

QUESTION 9.6

A mutated IDH1 isoenzyme is found in a high percentage of a type of brain cancer called glioblastoma. Instead of converting isocitrate to α-ketoglutarate, mutated IDH1 converts its substrate to 2-hydroxyglutarate, a circumstance that disrupts the citric acid cycle. Review the structures of isocitrate and α-ketoglutarate and determine the structure of 2-hydroxyglutarate.

The Amphibolic Citric Acid Cycle

Amphibolic pathways can function in both anabolic and catabolic processes. The citric acid cycle is obviously catabolic: acetyl groups are oxidized to form CO_2, and energy is conserved in reduced coenzyme molecules. The citric acid cycle is also anabolic, since several citric acid cycle intermediates are precursors in biosynthetic pathways (**Figure 9.13**). For example, OAA is a gluconeogenesis substrate (Chapter 8) and a precursor in the synthesis of the amino acids lysine, threonine, isoleucine, and methionine (Chapter 14). α-Ketoglutarate also plays an important role in amino acid synthesis as a precursor of glutamate, glutamine, proline, and arginine. The synthesis of porphyrins such as heme requires succinyl-CoA (Chapter 14). Finally, excess citrate molecules are transported into the cytoplasm, where they are cleaved to form OAA and acetyl-CoA. The latter molecule is used to synthesize fatty acids and steroid molecules such as cholesterol (Chapter 12).

Anabolic processes drain the citric acid cycle of the molecules required to sustain its role in energy generation. Several reactions, referred to as **anaplerotic** reactions, replenish them. One of the most important anaplerotic reactions is catalyzed by pyruvate carboxylase (p. 292). A high concentration of acetyl-CoA, an indicator of an insufficient OAA concentration, activates pyruvate carboxylase. As a result, OAA concentration increases. Other anaplerotic reactions include the synthesis of succinyl-CoA from certain fatty acids (Chapter 12) and the α-keto acids α-ketoglutarate and OAA from the amino acids glutamate and aspartate, respectively, via transamination reactions (Chapter 14).

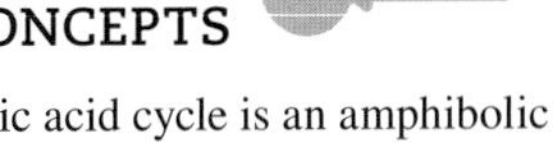

KEY CONCEPTS

- The citric acid cycle is an amphibolic pathway; that is, it plays a role in both anabolism and catabolism.
- The citric acid cycle intermediates used in anabolic processes are replenished by several anaplerotic reactions.

QUESTION 9.7

Pyruvate carboxylase deficiency, a disease that is usually fatal, is caused when the enzyme that converts pyruvate to OAA is missing or defective. It is characterized by varying degrees of mental retardation and disturbances in several metabolic pathways, especially those involving amino acids and their degradation products. A prominent symptom of this malady is *lactic aciduria* (lactic acid in the urine). After reviewing the function of pyruvate carboxylase, explain why this symptom occurs.

Citric Acid Cycle Regulation

The citric acid cycle is precisely regulated to meet the cell's energy and biosynthetic requirements (**Figure 9.14**). Regulation is achieved via control of three irreversible enzymes within the cycle: citrate synthase, isocitrate dehydrogenase, and α-ketoglutarate dehydrogenase. These three enzymes operate far from equilibrium (i.e., with highly negative $\Delta G^{\circ\prime}$ values), and they also catalyze reactions that represent important metabolic branch points. Strategies of control include substrate availability, product inhibition, and competitive feedback inhibition. Increased matrix levels of Ca^{2+} also activate all three enzymes.

CITRATE SYNTHASE Citrate synthase, the first enzyme in the cycle, catalyzes the formation of citrate from acetyl-CoA and OAA. The concentrations of acetyl-CoA and OAA are low in mitochondria in relation to the amount of the enzyme. Therefore, any increase in substrate availability stimulates citrate synthesis. In most eukaryotes citrate synthase has no allosteric regulators. Its rate is controlled primarily by the availability of the substrate molecule OAA. Because OAA is the

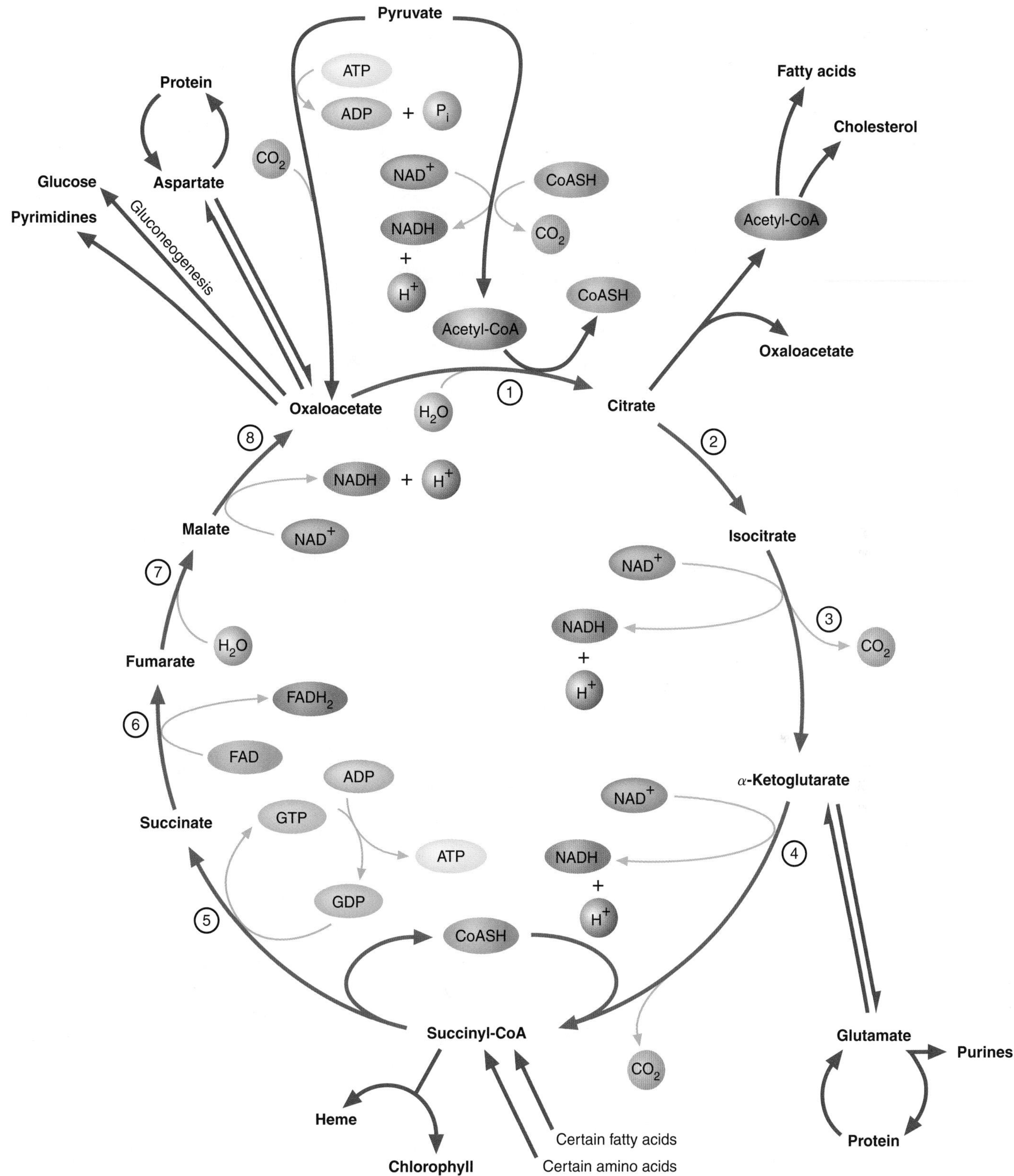

FIGURE 9.13

The Amphibolic Citric Acid Cycle

The citric acid cycle operates in both anabolic processes (e.g., the synthesis of fatty acids, cholesterol, heme, and glucose) and catabolic processes (e.g., amino acid degradation and energy production).

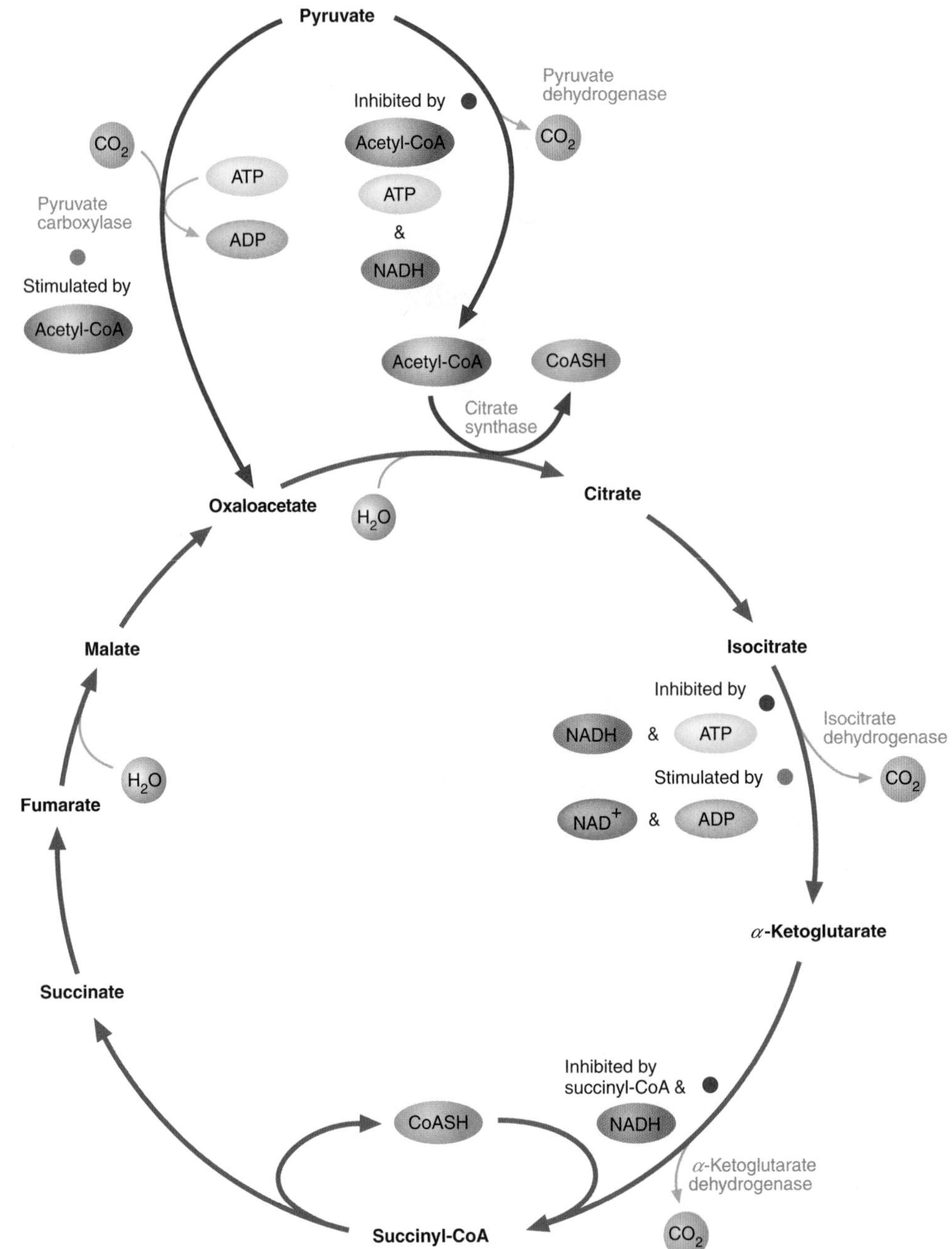

FIGURE 9.14

Control of the Citric Acid Cycle

The major regulatory sites of the cycle are indicated. Activators and inhibitors of regulated enzymes are shown in color.

product of an endergonic reaction, its concentration in mitochondria is quite low relative to malate unless the NADH/NAD$^+$ ratio is low. (In many Gram-negative bacteria such as *E. coli*, ATP, NADH, and succinyl-CoA allosterically inhibit citrate synthase.) Citrate synthase is also inhibited by its product, citrate.

ISOCITRATE DEHYDROGENASE Isocitrate dehydrogenase catalyzes the second regulated reaction in the cycle. Its activity is stimulated by relatively high concentrations of ADP and NAD$^+$ and inhibited by ATP and NADH. Isocitrate

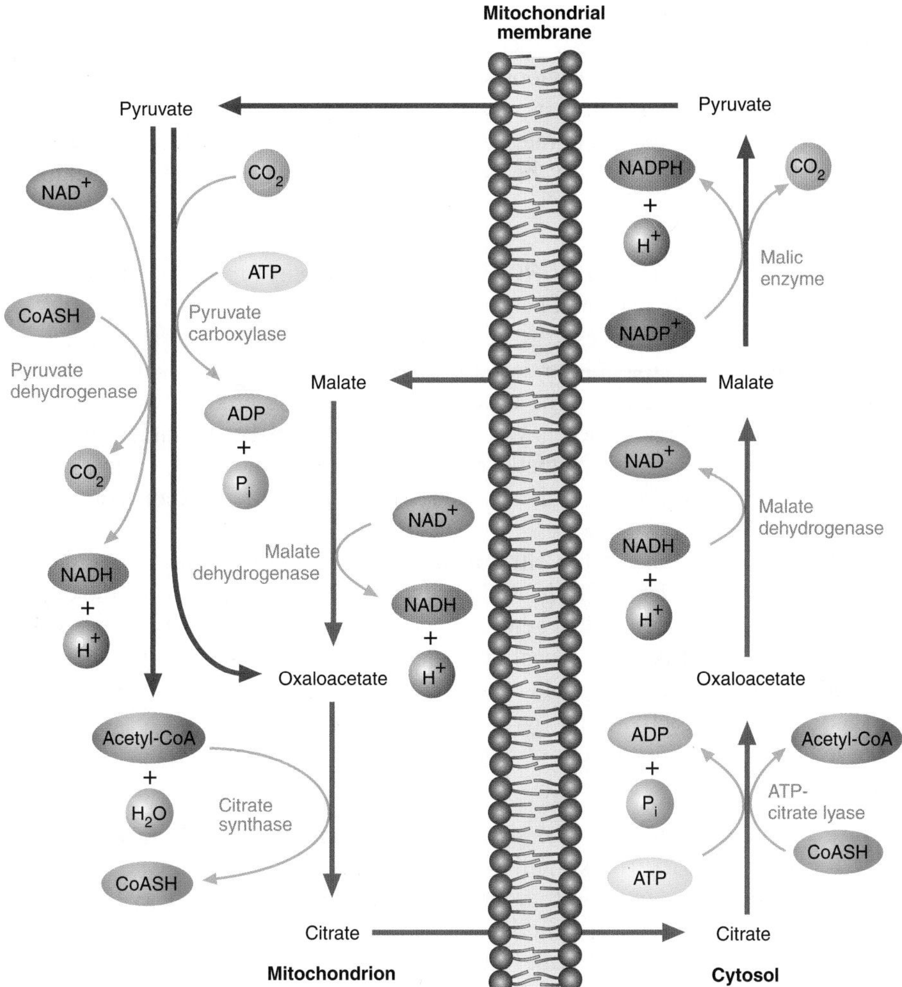

FIGURE 9.15

Citrate Metabolism

When citrate, a citric acid cycle intermediate, moves from the mitochondrial matrix into the cytoplasm, it is cleaved to form acetyl-CoA and oxaloacetate by citrate lyase. The citrate lyase reaction is driven by ATP hydrolysis. Most of the oxaloacetate is reduced to malate by malate dehydrogenase. Acetyl-CoA molecules can be used in biosynthetic pathways such as fatty acid synthesis. Malate may then be oxidized to pyruvate and CO_2 by malic enzyme. The NADPH produced in this reaction is used in cytoplasmic biosynthetic processes, such as fatty acid synthesis. Pyruvate enters the mitochondria, where it may be converted to oxaloacetate or acetyl-CoA. Malate may also reenter the mitochondria, where it is reoxidized to form oxaloacetate.

dehydrogenase is closely regulated because of its important role in citrate metabolism (**Figure 9.15**). As described earlier, the conversion of citrate to isocitrate is reversible. An equilibrium mixture of the two molecules consists largely of citrate. (The reaction is driven forward because isocitrate is rapidly transformed to α-ketoglutarate.) Of the two molecules, only citrate can penetrate the mitochondrial inner membrane. When cellular energy demands are met, excess citrate molecules are transported out of mitochondria and into the cytoplasm. Citrate is then cleaved by ATP-citrate lyase to yield acetyl-CoA and oxaloacetate. The acetyl-CoA formed is used in fatty acid synthesis. (Citrate transport is used to transfer acetyl-CoA out of mitochondria because acetyl-CoA cannot penetrate the inner mitochondrial membrane.) OAA is used in biosynthetic reactions, or it can be converted to malate. Malate either reenters the mitochondrion, where it is reconverted to OAA, or is converted in the cytoplasm to pyruvate by malic enzyme. Pyruvate then reenters the mitochondrion. In addition to being a precursor of acetyl-CoA and OAA in the cytoplasm, citrate also acts directly to regulate several cytoplasmic processes. Citrate is an allosteric activator of the first reaction of fatty acid synthesis. In addition, citrate metabolism provides some of the NADPH used in fatty acid synthesis. Finally, because citrate is an inhibitor of PFK-1, it inhibits glycolysis.

α-KETOGLUTARATE DEHYDROGENASE The activity of α-ketoglutarate dehydrogenase is strictly regulated because of the important role of α-ketoglutarate in several metabolic processes (e.g., amino acid metabolism). When a cell's energy stores are low, α-ketoglutarate dehydrogenase is activated and α-ketoglutarate is retained within the cycle at the expense of biosynthetic processes. As the cell's supply of NADH rises, the enzyme is inhibited, and α-ketoglutarate molecules become available for biosynthetic reactions. The enzyme is also inhibited by its product succinyl-CoA and activated by AMP, a critical indicator of low energy charge.

KEY CONCEPTS

- The citric acid cycle is closely regulated, thus ensuring that the cell's energy and biosynthetic needs are met.
- Allosteric effectors and substrate availability primarily regulate the enzymes citrate synthase, isocitrate dehydrogenase, α-ketoglutarate dehydrogenase, pyruvate dehydrogenase, and pyruvate carboxylase.

Two enzymes outside the citric acid cycle profoundly affect its regulation. The relative activities of PDHC and pyruvate carboxylase determine the degree to which pyruvate is used to generate energy and biosynthetic precursors. For example, if a cell is using a cycle intermediate such as α-ketoglutarate in biosynthesis, the concentration of OAA falls and acetyl-CoA accumulates. Because acetyl-CoA is an activator of pyruvate carboxylase (and an inhibitor of PDHC), more OAA is produced from pyruvate, thus replenishing the cycle.

CALCIUM REGULATION The signal transduction mechanisms by which cells respond to a variety of stimuli (e.g., hormones, growth factors, and neurotransmitters) often involve transient increases in cytoplasm [Ca^{2+}], followed rapidly by increases in [Ca^{2+}] in the mitochondrial matrix (p. 59). A principal role of Ca^{2+} in the matrix is the stimulation of ATP synthesis by activating the enzymes that regulate the pace of the citric acid cycle. Calcium ions stimulate PDHC activity by activating the dephosphorylating enzyme PDP (p. 329). Both isocitrate dehydrogenase and α-ketoglutarate dehydrogenase are activated directly by Ca^{2+} when the ion binds to a regulatory site on each enzyme. The linkage of the cell's response to a stimulus-driven signal transduction pathway with the uptake of Ca^{2+} into the mitochondrial matrix thus serves to match energy demand with energy production.

The Citric Acid Cycle and Human Disease

Encephalopathy and Cancers

Although rare, several human diseases have been attributed to deficits in citric acid cycle enzymes. Because of the brain's high energy requirements, the most commonly observed illnesses are severe forms of *encephalopathy* (brain dysfunction characterized by cognitive deficits, tremor, and seizures). For example, encephalopathies have been linked to mutations in the genes that code for α-ketoglutarate dehydrogenase, the A subunit of succinate dehydrogenase, fumarase, and succinyl-CoA synthetase. Several rare cancers are also caused by citric acid cycle enzyme deficits (p. 343). SHB and SHD mutations can cause a pheochromocytoma, an adrenal tumor that secretes excessive amounts of the hormone/neurotransmitter molecules epinephrine and norepinephrine. Symptoms include excessive heart rate and sweating, high blood pressure, and anxiety. A form of renal cell cancer is caused by mutations in fumarase.

The Glyoxylate Cycle

Plants and some fungi, algae, protozoans, and bacteria can grow using two-carbon compounds. (Molecules such as ethanol, acetate, and acetyl-CoA, derived from fatty acids, are the most common substrates.) The series of reactions responsible for this capability, referred to as the **glyoxylate cycle**, is a modified version of the citric acid cycle. In plants the glyoxylate cycle occurs in organelles called glyoxysomes, a type of peroxisome (p. 61) found in germinating seeds. In the absence of photosynthesis, growth in germinating seed is supported by the conversion of oil reserves (triacylglycerol) to carbohydrate. In other eukaryotic organisms and in bacteria, glyoxylate enzymes occur in cytoplasm.

The glyoxylate cycle (**Figure 9.16**) consists of five reactions. The first two reactions (the synthesis of citrate and isocitrate) are familiar because they also occur in the citric acid cycle. However, the formation of citrate from OAA and acetyl-CoA and the isomerization of citrate to form isocitrate are catalyzed by glyoxysome-specific isozymes. The next two reactions are unique to the glyoxylate cycle. Isocitrate is split into two molecules (succinate and glyoxylate) by isocitrate lyase. (This reaction is an aldol cleavage.) Succinate, a four-carbon molecule, is eventually converted to malate by mitochondrial enzymes (**Figure 9.17**). The two-carbon molecule glyoxylate reacts with a second molecule of acetyl-CoA to form malate in a reaction catalyzed by malate synthase. The cycle is completed as malate is converted to OAA by malate dehydrogenase.

The glyoxylate cycle allows for the net synthesis of larger molecules from two-carbon molecules. The decarboxylation reactions of the citric acid cycle, in which two molecules of CO_2 are lost, are bypassed. Using two molecules of acetyl-CoA, the glyoxylate cycle produces one molecule each of succinate and OAA. The succinate product is used in the synthesis of metabolically

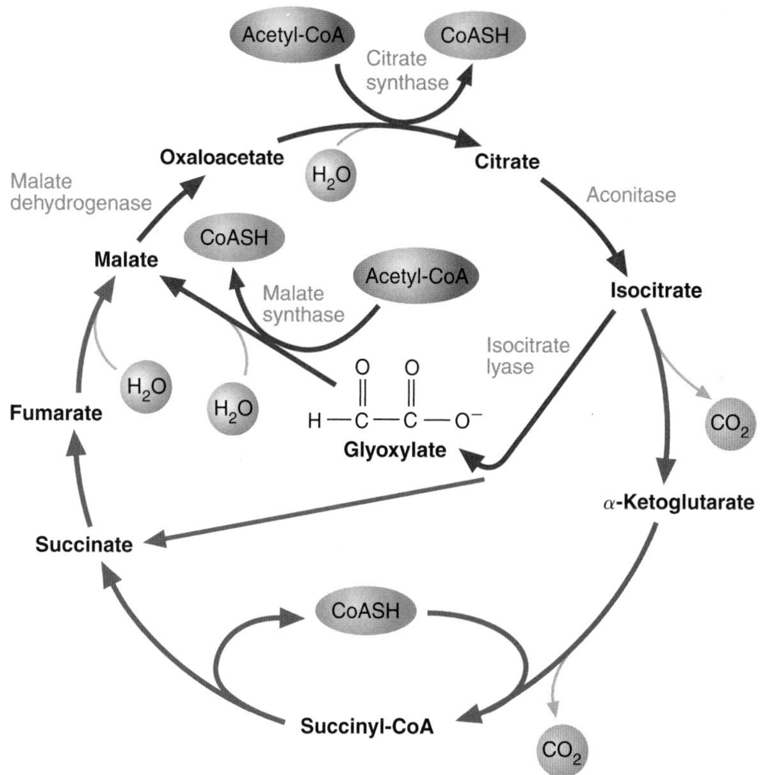

FIGURE 9.16
The Glyoxylate Cycle

Using some of the enzymes of the citric acid cycle, the glyoxylate cycle converts two molecules of acetyl-CoA to one molecule of oxaloacetate. Both decarboxylation reactions of the citric acid cycle are bypassed.

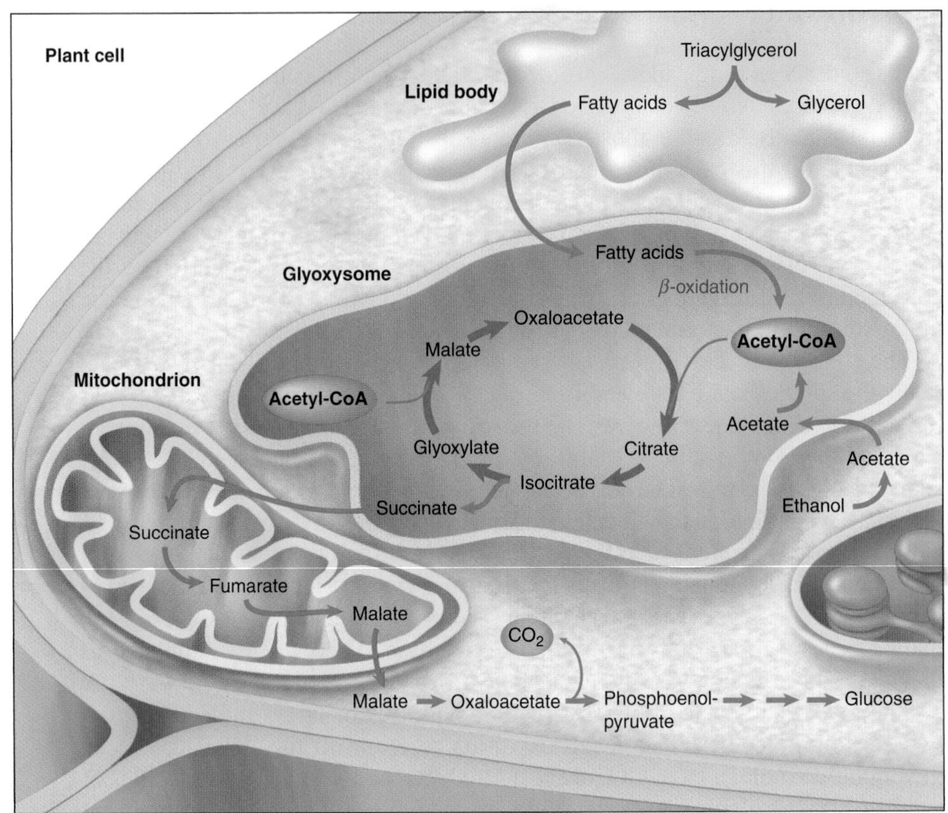

FIGURE 9.17

Role of the Glyoxylate Cycle in Gluconeogenesis

The acetyl-CoA used in the glyoxylate cycle is derived from the breakdown of fatty acids (β-oxidation, see Chapter 12). In organisms with the appropriate enzymes, glucose can be produced from two-carbon compounds such as ethanol and acetate. In plants, the reactions are localized in lipid bodies, glyoxysomes, mitochondria, and the cytoplasm.

KEY CONCEPTS

- Organisms in which the glyoxylate cycle occurs can use two-carbon molecules to sustain growth.
- In plants, the glyoxylate cycle is found in organelles called glyoxysomes.

important molecules such as glucose. (In organisms, such as animals, that do not possess isocitrate lyase and malate synthase, gluconeogenesis substrates are always molecules with at least three carbon atoms. In these organisms there is no net synthesis of glucose from fatty acids.) The oxaloacetate product is used to sustain the glyoxylate cycle.

Biochemistry IN PERSPECTIVE

Carcinogenesis: The Warburg Effect and Metabolic Reprogramming

What is aerobic glycolysis and how does it impact cancerous cells? In one of the earliest biochemical investigations of cancer, Otto Warburg (1883–1970) observed that most tumor cells generate energy principally by glycolysis instead of aerobic respiration even if O_2 is plentiful. He believed that this exception to the Pasteur effect (p. 286) in cancerous cells, now referred to as the Warburg effect, is caused by mitochondrial damage. Although there is mitochondrial damage in some forms of cancer, emerging research has revealed that "aerobic glycolysis" is more closely associated with altered cell signaling pathways that cause a reprogramming of metabolic processes. Such alterations give rise to selective advantages to cancerous cells in energy generation and biomolecule synthesis.

Cancer is a large group of diseases in which there is unregulated cell growth. Cancerous cells require a large amount of rapidly generated energy and a large supply of metabolic precursor molecules to sustain rapid growth. Although glycolytic ATP synthesis is inefficient, the process of carcinogenesis involves metabolic changes that include markedly increased glucose uptake and glycolytic activity in "transformed" cells. Accomplishing these transformative changes involves alterations in gene expression and a reconfiguration of some pathways. Considering the complexity of cell processes, it is not a coincidence that cancer genome sequencing has revealed an astonishing level of heterogeneity in tumors arising in different tissues and even within the same tumor. Despite this complexity, a relatively small number of genes play pivotal roles in carcinogenic aerobic glycolysis. Examples include c-myc, HIF-1, PKB, and p53. Although most cancers have normal oxidative phosphorylation, increased reliance on aerobic glycolysis in some cells can also be caused by tumor cell mitochondria with mutations in certain citric acid cycle enzymes.

C-MYC The c-myc gene codes for a transcription factor that regulates about 15% of all genes, including many that are involved in cell division and apoptosis (p. 58). Among c-myc's target genes are those that code for most of the glycolytic enzymes and several glucose transporters (GLUTs, p. 600). Tumors with unregulated c-myc have increased lactic acid production and diversion of glycolytic intermediates into pathways that synthesize nucleotides, amino acids, and lipids. Unregulated c-myc also promotes glutaminolysis, an alternative energy-generating pathway in which glutamine is transported into mitochondria and converted into the citric acid cycle intermediate α-ketoglutarate. Both the glutamine transporter and glutaminase, the enzyme that converts glutamine to glutamate, are upregulated by c-myc.

HIF-1 HIF-1 (hypoxia inducible factor 1) is a transcription factor that responds to decreases in cellular oxygen levels by inducing the transcription of genes that promote survival under low-oxygen conditions. The most prominent examples are genes for glycolytic enzymes, glucose transporters, and pyruvate dehydrogenase kinase 1 (PDK1), an enzyme that inactivates pyruvate dehydrogenase, thereby diverting pyruvate away from mitochondrial oxidation. HIF-1 is composed of a constitutively expressed β-subunit and an O_2-regulated α-subunit. When oxygen is available, HIF-1 is inactivated, and the α-subunit is degraded by the ubiquitin-proteasome system in response to oxygen- and α-ketoglutarate-dependent prolyl hydroxylation. Under the hypoxic conditions of transformed cells, a stabilized HIF-1 reduces oxygen requirements in part by inhibiting c-myc and increasing the rate of c-myc degradation by proteasomes. HIF-1 is stabilized in some tumors by mutations in succinate dehydrogenase and fumarase. Succinate or fumarate accumulations result in leakage of these molecules into cytoplasm, where they act as competitive inhibitors of prolyl hydroxylase.

PKB Protein Kinase B (PKB) (p. 593), also referred to as Akt, is a serine/threonine protein kinase that regulates glucose metabolism, among several other processes. Activated PKB enhances glucose uptake and glycolysis by inducing increased glucose transporter insertion into the plasma membrane, hexokinase activation, and PFK2-dependent activation of PFK-1 (p. 288). PKB also promotes the diversion of glucose carbon into the biosynthetic pathways that produce fatty acids and cholesterol and other lipids required for rapid growth. The PKB facilitated transport of pyruvate into mitochondria and its rapid conversion into acetyl-CoA. The subsequent increase in citrate results in its export into the cytoplasm where acetyl-CoA, the substrate for lipid synthesis, is regenerated by the PKB-activated enzyme ATP-citrate lyase.

P53 With more than 100 target genes, p53 is one of the most important tumor suppressors. In addition to its well-known roles in DNA repair, cell cycle regulation, and apoptosis induction, p53 is now known as an important regulator of energy metabolism. P53 coordinates the regulation of glycolysis and oxidative phosphorylation. It does so by maintaining oxidative phosphorylation and down- regulating glycolysis (via decreases in glucose transporters and glycolytic enzymes). P53-induced regulation of energy metabolism is exerted through inducing the synthesis and/or activation of enzymes such as PKB, mTOR (p. 593), and TIGAR (TP53-induced glycolysis and apoptosis regulator), which decreases glycolytic flux by lowering the intracellular concentrations of fructose 2,6-bisphosphate (the PFK-1 allosteric activator). The Warburg effect is one of many consequences of p53 inactivation.

Biochemistry IN PERSPECTIVE cont.

SUMMARY: Aerobic glycolysis is a process in tumor cells in which there is rapid glycolysis-generated ATP synthesis that occurs even when O_2 is present. Loss of glycolysis regulation is one facet of carcinogenesis, the set of mechanisms whereby normal cells are gradually transformed into cancerous cells that are no longer responsive to the body's regulatory signals.

Chapter Summary

1. Aerobic organisms have an enormous advantage over anaerobic organisms, that is, a greater capacity to obtain energy from organic food molecules. To use oxygen to generate energy requires the following biochemical pathways: the citric acid cycle, the electron transport pathway, and oxidative phosphorylation.
2. Most reactions that capture or release energy are redox reactions. In these reactions, electrons are transferred between an electron donor (reducing agent) and an electron acceptor (oxidizing agent). In most biochemical reactions, hydride ions are transferred to NAD^+ or $NADP^+$ or hydrogen atoms are transferred to FAD or FMN. The tendency for a substance to gain electron(s) is called its reduction potential. Electrons flow spontaneously from a substance with a less positive (more negative) reduction potential to a substance with a more positive (less negative) reduction potential. In favorable redox reactions $\Delta E^{\circ\prime}$ is positive and $\Delta G^{\circ\prime}$ is negative.
3. The citric acid cycle is a series of biochemical reactions that eventually completely oxidize organic substrates, such as glucose and fatty acids, to form CO_2, H_2O, and the reduced coenzymes NADH and $FADH_2$. Pyruvate, the product of the glycolytic pathway, is converted to acetyl-CoA, the citric acid cycle substrate.
4. In addition to its role in energy generation, the citric acid cycle plays an important role in several biosynthetic processes, such as gluconeogenesis, amino acid synthesis, and porphyrin synthesis.
5. The glyoxylate cycle, found in plants and some fungi, algae, protozoans, and bacteria, is a modified version of the citric acid cycle in which two-carbon molecules, such as acetate, are converted to precursors of glucose.

Take your learning further by visiting the **companion website** for Biochemistry at **www.oup.com/us/mckee** where you can complete a multiple-choice quiz on the citric acid cycle to help you prepare for exams.

Suggested Readings

Bilgen, T., Metabolic Evolution and the Origin of Life, in Storey, K. B. (ed.), *Functional Metabolism: Regulation and Adaptation*, pp. 557–582, Wiley-Liss, Hoboken, New Jersey, 2004.

Koch, L. G., and Britton, S. L., Aerobic Metabolism Underlies Complexity and Capacity, *J. Physiol.* 586:83–95, 2008.

Kornberg, H., Krebs and His Trinity of Cycles, *Nat. Rev. Mol. Cell Biol.* 1(3):225–227, 2000.

Krebs, H. A., The History of the Tricarboxylic Cycle, *Perspect. Biol. Med.* 14:154–170, 1970.

Lane, N., *Oxygen: The Molecule That Made the World*, Oxford University Press, Oxford, 2002.

Lane, N., *Power, Sex and Suicide and the Meaning of Life*, Oxford University Press, Oxford, 2005.

Ward, P. S., and Thompson, C. B., Metabolic Reprogramming: A Cancer Hallmark Even Warburg Did Not Anticipate, *Cancer Cell* 21:297–308, 2012.

Wu, W., and Zhao, S., Metabolic Changes in Cancer: Beyond the Warburg Effect, *Acta Biochim. Biophys. Sin. (Shanghai)* 45(1):18–26, 2013.

Key Words

aerotolerant anaerobe, *318*
amphibolic pathway, *337*
anaplerotic, *337*
coenzyme A, *326*
conjugate redox pair, *320*
facultative anaerobe, *318*
flavin adenine dinucleotide (FAD), *322*
flavin mononucleotide (FMN), *322*
flavoprotein, *322*
glyoxylate cycle, *341*
lipoic acid, *329*
nicotinamide adenine dinucleotide (NAD), *322*
nicotinamide adenine dinucleotide phosphate (NADP), *322*
obligate aerobe, *318*
obligate anaerobe, *318*
redox potential, *326*
reduction potential, *320*
standard reduction potential, *320*
thiamine pyrophosphate, *329*

Review Questions

These questions are designed to test your knowledge of the key concepts discussed in this chapter before moving on to the next chapter. You may like to compare your answers to the solutions provided in the back of the book and in the accompanying Study Guide.

1. Define the following terms:
   a. obligate anaerobe
   b. aerotolerant anaerobe
   c. facultative anaerobe
   d. obligate aerobe
   e. reactive oxygen species
2. Define the following terms:
   a. thiamine pyrophosphate
   b. lipoic acid
   c. PHDC
   d. HETPP
   e. nucleoside diphosphate kinase
3. Define the following terms:
   a. amphibolic pathway
   b. anaplerotic reaction
   c. glyoxylate cycle
   d. reduction potential
   e. conjugate redox pair
4. Define the following terms:
   a. glyoxosome
   b. coenzyme A
   c. $NAD(P)^+$
   d. FAD
   e. FMN
5. Describe in general terms how the appearance of molecular oxygen in Earth's atmosphere about 3 billion years ago affected the history of living organisms.
6. A runner needs a tremendous amount of energy during a race. Explain how the use of ATP by contracting muscle affects the citric acid cycle.
7. Describe two important roles of the citric acid cycle.
8. Acetyl-CoA is manufactured in the mitochondria and used in the cytoplasm to synthesize fatty acids. However, acetyl-CoA cannot penetrate the mitochondrial membrane. How is this problem solved?
9. Outline the steps of the glyoxylate cycle.
10. How does the glyoxylate cycle differ from the citric acid cycle?
11. Which of the following conditions indicates a low cell energy status? What impact does each of the following conditions have on flux through the citric acid cycle?
    a. high $NADH/NAD^+$ ratio
    b. high ATP/ADP ratio
    c. high acetyl-CoA concentration
    d. low citrate concentration
    e. high succinyl-CoA concentration
12. Write balanced equations for each of the reactions of the citric acid cycle.
13. What are the coenzymes required in each of the reactions in Question 12?
14. Discuss the mechanisms of control of the irreversible steps in the citric acid cycle.
15. List the biochemical processes required to obtain energy from glucose using O_2 as an electron acceptor.
16. Calculate the free energy changes that occur in the following reactions:
    a. $\frac{1}{2} O_2 + NADH + H^+ \rightarrow H_2O + NAD^+$
    b. $S + NADH + H^+ \rightarrow H_2S + NAD^+$
17. Referring to Question 16, calculate the difference in energy produced by the oxidation of NADH by O_2 and sulfur.
18. The citric acid cycle operates only when O_2 is present, yet O_2 is not a substrate for the cycle. Explain.
19. If a small amount of [1-^{14}C]glucose is added to an aerobic yeast culture, where will the ^{14}C label initially appear in citrate molecules?
20. Describe in detail the structure of the pyruvate dehydrogenase complex.
21. Describe the role played by each enzyme, cofactor, and coenzyme of the pyruvate dehydrogenase complex.
22. Each of the following genes has a role in carcinogenic aerobic glycolysis: c-myc, HIF-1, PKB, and p53. Describe the function of each gene.
23. Outline the synthesis of glucose from acetyl-CoA in germinating seed.
24. Explain why animals cannot produce glucose from two-carbon molecules such as acetate or ethanol.
25. Write the net equation for the citric acid cycle.
26. What steps in the citric acid cycle are regulated? Why are they regulated?
27. Provide examples of biosynthetic pathways that utilize citric acid cycle intermediates as precursor molecules.

Fill in the Blank

28 ______________ pathways can function in both anabolic and catabolic processes.

29. ______________ reactions replenish the intermediate molecules of the citric acid cycle.

30. The acetyl group of acetyl-CoA is derived from the breakdown of lipids, carbohydrates, and certain ______________.

31. Pyruvate is converted to acetyl-CoA by the enzymes of the ______________ enzyme complex.

32. In the glyoxylate cycle, two molecules of acetyl-CoA are converted into ______________ and malate.

33. ______________ produced in the citric acid cycle is an important carbon source for gluconeogenesis.

34. The most pivotal event in the history of life on earth, allowing for the emergence of eukaryotic cells, was the ______________.

35. The ______________ branch of the primitive citric acid cycle provides a source of oxidized NAD.

36. In eukaryotic cells the citric acid cycle occurs in the ______________.

37. The tendency for a substance to gain electrons is called the ______________ potential.

Short Answer

38. One of the consequences of ethanol addiction is fatty liver disease, an illness in which liver cells accumulate triacylglycerols, the esters derived from glycerol and fatty acids. Ethanol is oxidized in the cytoplasm of liver cells by alcohol dehydrogenase and aldehyde dehydrogenase to yield acetate and 2 NADH. Acetate is then transported into the mitochondrion, where it is converted to acetyl-CoA and metabolized by the citric acid cycle. When alcohol is consumed in excessive quantities, the resulting high levels of NADH cause metabolic abnormalities, one of which is high levels of fatty acid synthesis. Fatty acid synthesis, also a cytoplasmic process, uses acetyl-CoA as a substrate and NADPH as a reducing agent. Determine how a high level of cytoplasmic NADH provides a source of NADPH for fatty acid synthesis.

39. How does calcium regulate the citric acid cycle?

40. How do germinating seeds convert their triacylglycerol reserves to the glucose molecules required in the synthesis of complex carbohydrate such as cellulose?

41. What enzymes outside of the citric acid cycle affect its regulation?

42. How does pyruvate carboxylase deficiency result in lactic aciduria, an illness in which lactate appears in the urine?

Thought Questions

These questions are designed to reinforce your understanding of all of the key concepts discussed in the book so far, including this chapter and all of the chapters before it. They may not have one right answer! The authors have provided possible solutions to these questions in the back of the book and in the accompanying Study Guide for your reference.

43. The plant toxin fluoroacetate ($F\text{-}CH_2COO^-$) is easily converted to fluorocitrate when an animal ingests the plant. The 2-fluoro-3-hydroxy citrate molecule is a suicide substrate for the enzyme aconitase. Consider the overall goal of this enzyme (to remove the OH group from the tertiary carbon and place another OH group on a secondary carbon) and speculate as to why the reaction to produce isocitrate does not occur as planned. (*Hint*: Fluorine is more electronegative than oxygen.)

44. In answering Question 43, provide the chemical structure of fluorocitrate and the suspected suicide inhibitor that is produced by the enzyme citrate synthase.

45. What is the significance of substrate-level phosphorylation reactions? Which of the reactions in the citric acid cycle involve a substrate-level phosphorylation? Name another example from a biochemical pathway with which you are familiar.

46. You have just consumed a piece of fruit. Trace the carbon atoms in the glucose in the fruit through the biochemical pathways between their uptake into tissue cells and their conversion to CO_2.

47. Determine the standard free energy ($\Delta G^{\circ\prime}$) for the following reactions:
    a. $NADH + H^+ + \frac{1}{2}O_2 \rightarrow NAD^+ + H_2O$
    b. Cytochrome c (Fe^{2+}) + $\frac{1}{2}O_2$ → cytochrome c (Fe^{3+}) + H_2O

48. One of the many effects of chronic alcohol abuse is thiamine deficiency, caused by impaired absorption of the vitamin through the intestinal wall and diminished storage in a damaged liver. When thiamine levels are inadequate, cellular energy generation is diminished. What are the three enzymes involved in cellular metabolism that require thiamine? Describe the metabolic consequences of inadequate thiamine levels.

49. Despite the absence of the glyoxylate cycle in animals, when ^{14}C-labeled acetate is fed to lab animals, small amounts of the radioactive label later appear in glycogen stores. Explain.

50. Malonate (p. 206) poisons the citric acid cycle because it inhibits succinate dehydrogenase. After reviewing its

structure, describe how the inhibitory effect of malonate can be overcome.

51. Fatty acid degradation stimulates the citric acid cycle through the activation of pyruvate carboxylase by acetyl-CoA. Why would the activation of pyruvate carboxylase increase energy generation from fatty acids?

52. Shock (failure of the circulatory system) is an abnormal condition in which blood flow is inadequate. The most common causes of shock are massive blood loss and obstruction of blood flow. As a result of shock there is a failure to provide cells with oxygen and nutrients. In this circumstance cells swell and lysosomal membranes rupture, among other effects. Describe how energy is generated during shock and why cell structure becomes destabilized.

53. Lactic acidosis occurs as a result of shock. Explain why low oxygen levels promote lactate production.

54. Because dichloroacetate inhibits the enzyme pyruvate dehydrogenase kinase, this compound has been used, with limited results, to treat lactic acidosis. The phosphorylation of the α-subunit of the pyruvate dehydrogenase component of the pyruvate dehydrogenase complex by pyruvate dehydrogenase kinase causes complete loss of enzymatic activity. Describe the theory behind the clinical use of dichloroacetate.

55. Pyruvate dehydrogenase deficiency is a fatal disease usually diagnosed in children. Symptoms include severe neurological damage. Elevated blood levels of lactate, pyruvate, and alanine are also seen. Explain how the deficiency of pyruvate dehydrogenase causes these elevated values.

56. The large amount of energy used during aerobic exercise (e.g., running) requires large amounts of oxaloacetate. Explain why acetyl-CoA cannot be used to produce oxaloacetate in this circumstance. What is the source of oxaloacetate molecules during aerobic activity?

57. Using the data in **Table 9.1** and the equation given here, calculate the free energy ($\Delta G°$) produced by the reduction of sulfur to hydrogen sulfide and oxygen to water by NADH. How much more free energy is produced by the reduction of oxygen compared to that of sulfur?

$$\Delta G° = -nF\Delta E°$$

58. Systemic fungal infections (e.g., *Candida* and *Cryptococcus*) and tuberculosis (*Mycobacterium*) are on the rise. One of the reasons these organisms have high virulence is that they fare better following macrophage phagocytosis than other microorganisms. Phagocytosis activates the glyoxylate cycle in fungi and mycobacteria, allowing them to use two-carbon substrates to sustain growth in an otherwise nutrient-poor environment of the phagolysosome. Suggest some likely molecules that could be processed through the glyoxylate cycle in this circumstance.

59. Calculate the standard free energy change ($\Delta G°'$) for the following reaction:

$$\tfrac{1}{2}\,O_2 + FADH_2 \rightarrow H_2O + FAD$$

Compare the $\Delta G°'$ for this reaction with that for NADH (see Question 16a).

Aerobic Metabolism II: Electron Transport and Oxidative Phosphorylation

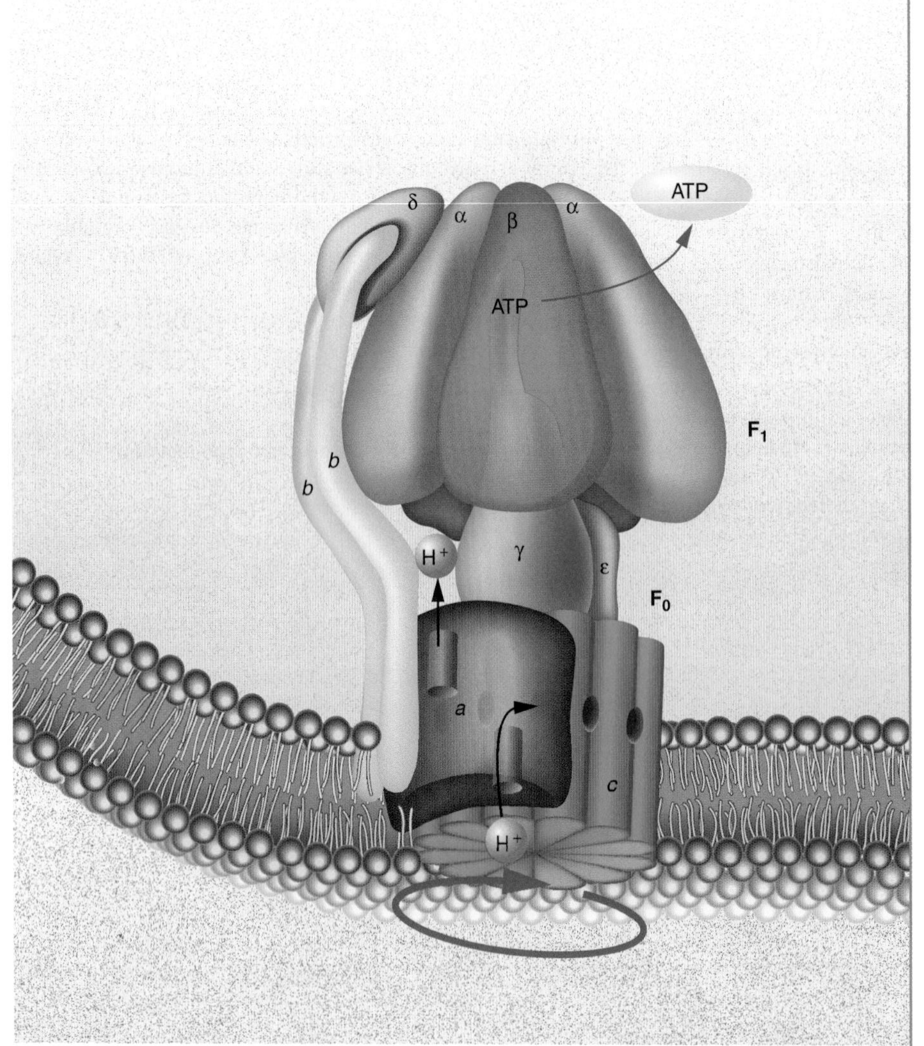

The ATP Synthase The ATP synthase is a rotating molecular machine that synthesizes ATP. This multiprotein complex is composed of two principal domains: the membrane-spanning F_0 component and the ATP-synthesizing F_1 component. The flow of protons through F_0, made possible by a gradient created by electron transport, generates a torque that forces the shaft (the γ subunit) to rotate. Rotational force within F_1 then triggers conformational changes that result in ATP synthesis.

OUTLINE

Oxygen: A Molecular Paradox

Oxygen is a very dangerous molecule! Oxygen is a *diradical* (i.e., it has two unpaired electrons) that combines with most other elements to form unstable and highly reactive intermediates. It can cause significant damage. Oxygen's reaction with iron to form rust, for example, costs the U.S. economy several tens of billions of dollars every year. But oxygen is also extremely useful. It is such a powerful oxidizing agent that liquefied oxygen and hydrogen served as fuel in the main engines of the Space Shuttle. Similarly, hydrocarbon fuel combustion is an oxygen-requiring process that under controlled conditions is used to heat homes and move vehicles. Considering oxygen's dangerous properties, why can living organisms use it to generate energy? The answer to this question begins with a comparison between combustion and aerobic respiration.

Combustion

The reaction of oxygen with hydrocarbons (RH) is a rapid and extremely exothermic radical chain reaction. This reaction is controllable because there is an energy barrier that prevents the spontaneous oxidation of most organic molecules when they are in contact with air. Combustion begins only after the introduction of an energy source (e.g., a lighted match). The chain reaction begins with the abstraction of a hydrogen atom by the oxygen diradical to form a hydroxyperoxide radical (HOO·) and an alkyl radical (R·), both of which proceed to react with other hydrocarbon molecules. The combustion process generates heat as chemical bonds break. The chain reaction accelerates as newly formed radicals react with other fuel olecules. Other radicals formed during the chain reaction include ROO·, RO·, and HO·. Because radicals react with the first molecule in their path, the process is unstoppable until either the hydrocarbon fuel is expended or oxygen is excluded from the reaction. Complete combustion of hydrocarbons produces CO_2 and H_2O.

Aerobic Respiration: Controlled Combustion and the Oxygen Paradox

Aerobic respiration is similar to hydrocarbon combustion in that organic molecules (carbohydrates and fats) are converted into CO_2 and H_2O. Instead of releasing energy in a rapid burst of flames, however, aerobic organisms have evolved a rigorously controlled step-by-step process that extracts hydrogen atoms from fuel molecules while simultaneously protecting themselves from ROS (Section 10.3). As with hydrocarbon combustion, the diradical molecule O_2 accepts electrons one at a time: $O_2 + 4e^- + 4H^+ \rightarrow 2H_2O$. Released energy is stored in ATP molecules. Although aerobic organisms usually keep ROS formation to a minimum, reactive intermediates can leak out and react with cell constituents. Despite antioxidant enzyme systems and repair processes, eventually the damage becomes so great that function is compromised. Oxidative damage has been linked to the aging process and numerous degenerative diseases.

So this is the paradox of oxygen: aerobic organisms use oxygen to generate the vast amounts of energy required to maintain their metabolic processes, as they risk damage caused by this highly reactive molecule.

Overview

THE AEROBIC LIFESTYLE DEPENDS ON THE LARGE QUANTITIES OF ENERGY MADE POSSIBLE BY OXYGEN. OXYGEN IS ALSO REQUIRED DIRECTLY OR indirectly for 1000 biochemical reactions that cannot occur under anaerobic conditions. A high price is paid for oxygen's enormous benefits, however. Research efforts have revealed that aerobic organisms have evolved an array of mechanisms that provide protection from the toxic by-products of oxygen metabolism. Numerous enzymes and antioxidant molecules usually prevent

most oxidative cell damage. Despite this protection, however, injury does occur. Oxygen metabolites are now known to contribute to an array of human disorders that include cancer and heart and neurological diseases.

Oxygen has several properties that in combination have made possible a highly favorable mechanism for extracting energy from organic molecules. First, oxygen is found almost everywhere on the Earth's surface. In contrast, most other electron acceptors are relatively rare. Second, oxygen diffuses easily across cell membranes. This is not true of several other electron acceptors such as the charged species sulfate and nitrate. Finally, its diradical structure allows oxygen to readily accept electrons. This capacity is also responsible for its tendency to form highly destructive metabolites called *reactive oxygen species* (ROS).

Chapter 10 describes the basic principles of *oxidative phosphorylation*, the complex mechanism by which modern aerobic cells manufacture ATP. The discussion begins with a review of the electron transport system in which electrons are donated by reduced coenzymes to the ETC. The ETC is a series of electron carriers in the inner membrane of the mitochondria of eukaryotes and the plasma membrane of aerobic prokaryotes. The chapter's next topic is *chemiosmosis*, the means by which the energy extracted from electron flow is captured and used to synthesize ATP. Chapter 10 ends with a discussion of the formation of toxic oxygen products and the strategies that cells use to protect themselves.

10.1 ELECTRON TRANSPORT

The mitochondrial ETC, also referred to as the electron transport system, is a series of electron carriers arranged in the inner membrane in order of increasing electron affinity; it is these molecules that transfer the electrons derived from the reduced coenzymes NADH and $FADH_2$ to oxygen. During this transfer, a decrease in reduction potential ($\Delta E^{\circ\prime}$) occurs. When NADH is the electron donor and oxygen is the electron acceptor, the change in standard reduction potential is +1.14 V (i.e., +0.82V − (−0.32V); see **Table 9.1**). The energy released during electron transfer is coupled to several endergonic processes, the most prominent of which is ATP synthesis. Other processes driven by electron transport transfer Ca^{2+} into the mitochondrial matrix via MAMs (p. 59) and generate heat in brown adipose tissue (described on p. 366). Reduced coenzymes, derived from glycolysis, the citric acid cycle, and fatty acid oxidation, are the principal sources of electrons.

Electron Transport and Its Components

The components of the ETC in eukaryotes, located in the inner mitochondrial membrane (**Figure 10.1**), are organized into four complexes. Each of these complexes, which function as oxidoreductase enzymes, consists of several proteins and prosthetic groups. The structural and functional properties of each complex are briefly described. The roles of coenzyme Q (ubiquinone, UQ) and the cytochromes are also described. **Figure 10.1** provides an overview of electron flow in the ETC.

Complex I, also referred to as the *NADH dehydrogenase complex*, catalyzes the transfer of electrons from NADH to UQ. It is the largest protein component in the inner membrane. The bovine heart mitochondrial enzyme, for example, has at least 45 subunits. In addition to one molecule of FMN (**Figure 9.6**), the complex contains seven iron-sulfur clusters (**Figure 10.2**). Iron-sulfur clusters

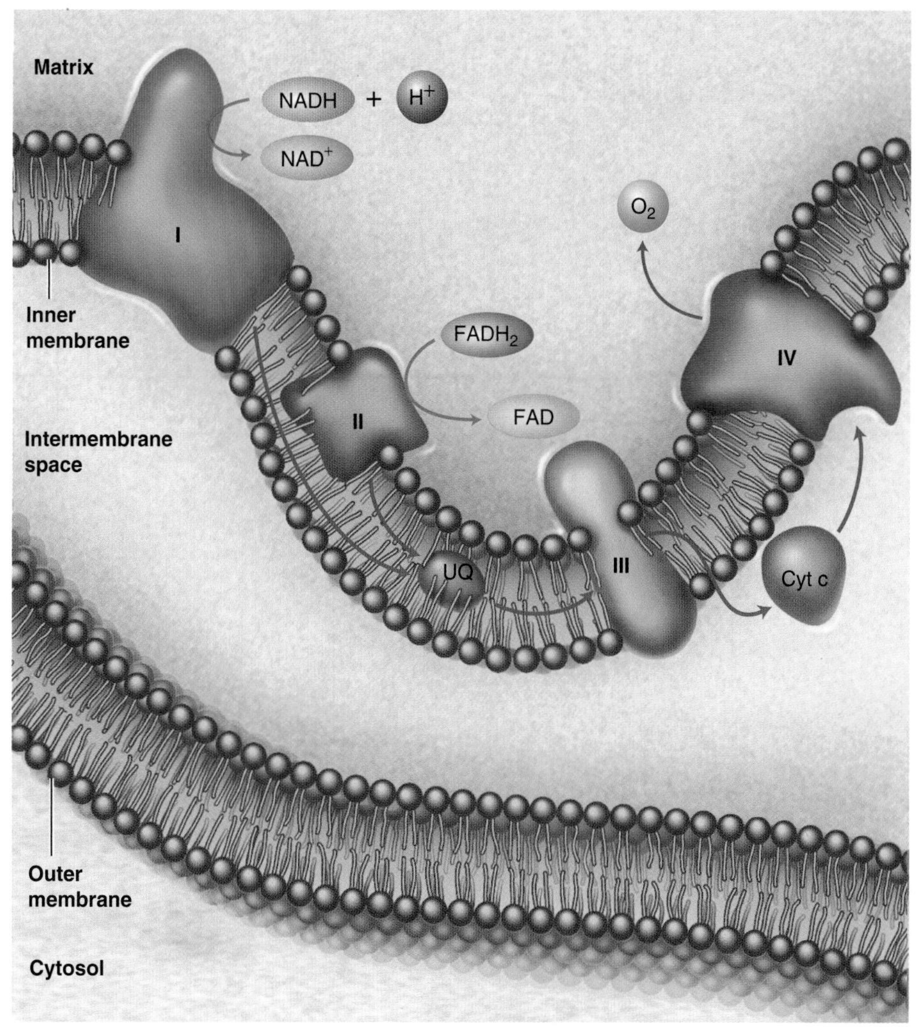

FIGURE 10.1

The Electron Transport Chain

Complexes I and II transfer electrons from NADH and succinate (via $FADH_2$), respectively, to UQ. Complex III transfers electrons from UQH_2 to cytochrome c. Complex IV transfers electrons from cytochrome c to O_2. The arrows represent the flow of electrons. The function of electron transport is the transfer of protons from the matrix across the inner membrane into the inner membrane space. ATP synthesis is linked to proton transport.

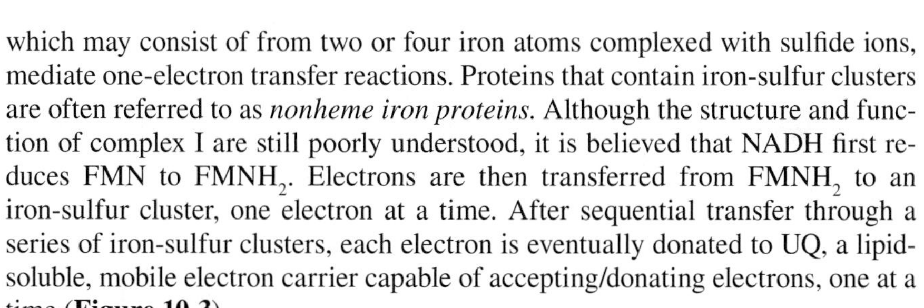

which may consist of from two or four iron atoms complexed with sulfide ions, mediate one-electron transfer reactions. Proteins that contain iron-sulfur clusters are often referred to as *nonheme iron proteins.* Although the structure and function of complex I are still poorly understood, it is believed that NADH first reduces FMN to $FMNH_2$. Electrons are then transferred from $FMNH_2$ to an iron-sulfur cluster, one electron at a time. After sequential transfer through a series of iron-sulfur clusters, each electron is eventually donated to UQ, a lipid-soluble, mobile electron carrier capable of accepting/donating electrons, one at a time (**Figure 10.3**).

Figure 10.4 illustrates the transfer of electrons through complex I. Electron transport is accompanied by the net movement of protons from the matrix across the inner membrane and into the intermembrane space. The significance of this phenomenon for ATP synthesis will be discussed.

The *succinate dehydrogenase complex* (complex II), also called succinate ubiquinol reductase, consists of four protein subunits. Both ShdA, a flavoprotein with a succinate binding site and a covalently bound FAD, and ShdB, an iron-sulfur protein with three iron-sulfur clusters, extend into the matrix (p. 334). Subunits C and D are integral membrane proteins. The hydrophobic UQ binding site is in a crevice that is lined with side chains from ShdC and ShdD. The CD dimer also contains a binding site for a heme group. The role of the heme group is to suppress electron leakage from complex II, a process that can result in oxygen radical formation.

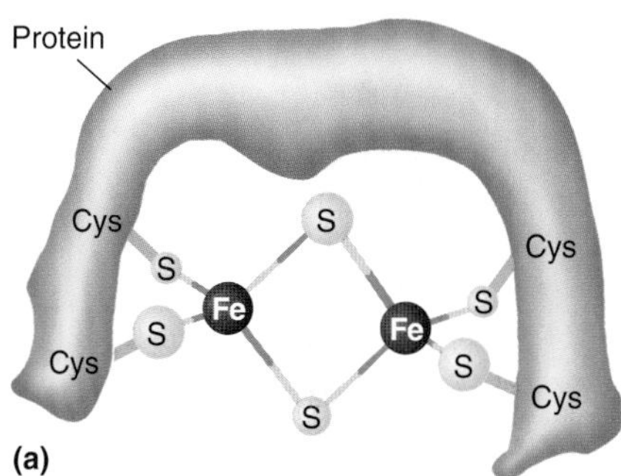

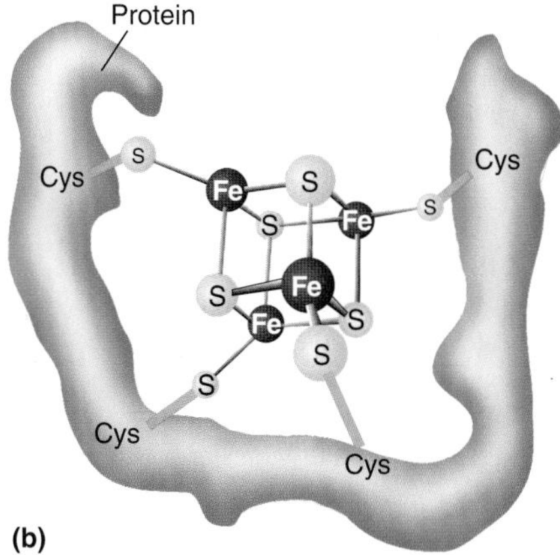

FIGURE 10.2

Two Iron-Sulfur Clusters

(a) Protein whose iron-sulfur cluster contains 2 irons and 2 sulfurs. (b) A 4 Fe-4S iron-sulfur cluster. In both cases, the cysteine residues are part of a polypeptide.

FIGURE 10.3

Structure and Oxidation States of Coenzyme Q

The length of the side chain varies among species. For example, some bacteria have six isoprene units. For mammals, however, $n = 10$, and such molecules are referred to as Q_{10}.

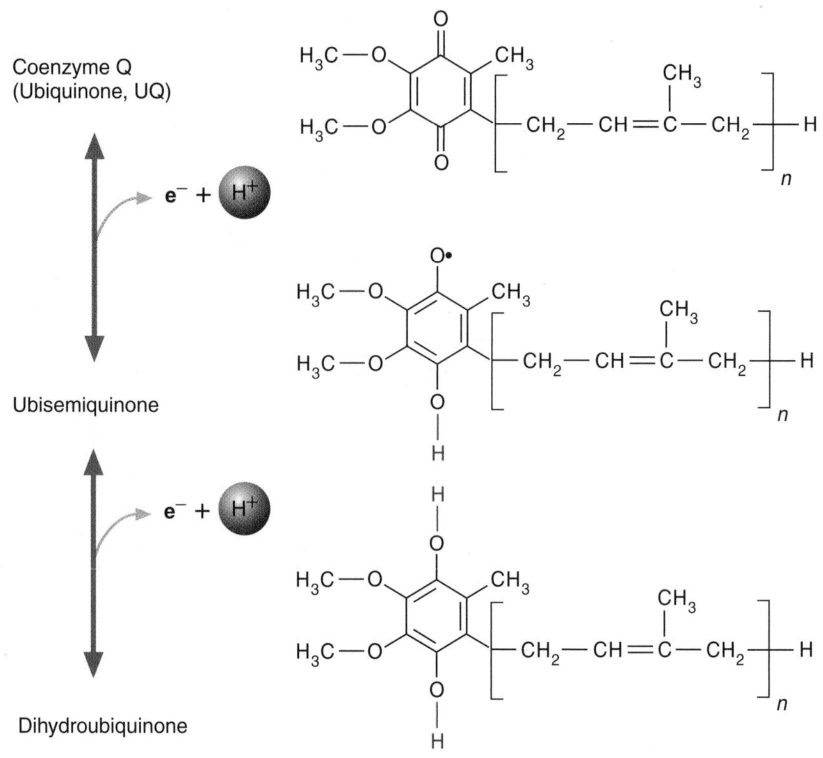

Coenzyme Q
(Ubiquinone, UQ)

$e^- + H^+$

Ubisemiquinone

$e^- + H^+$

Dihydroubiquinone

FIGURE 10.4

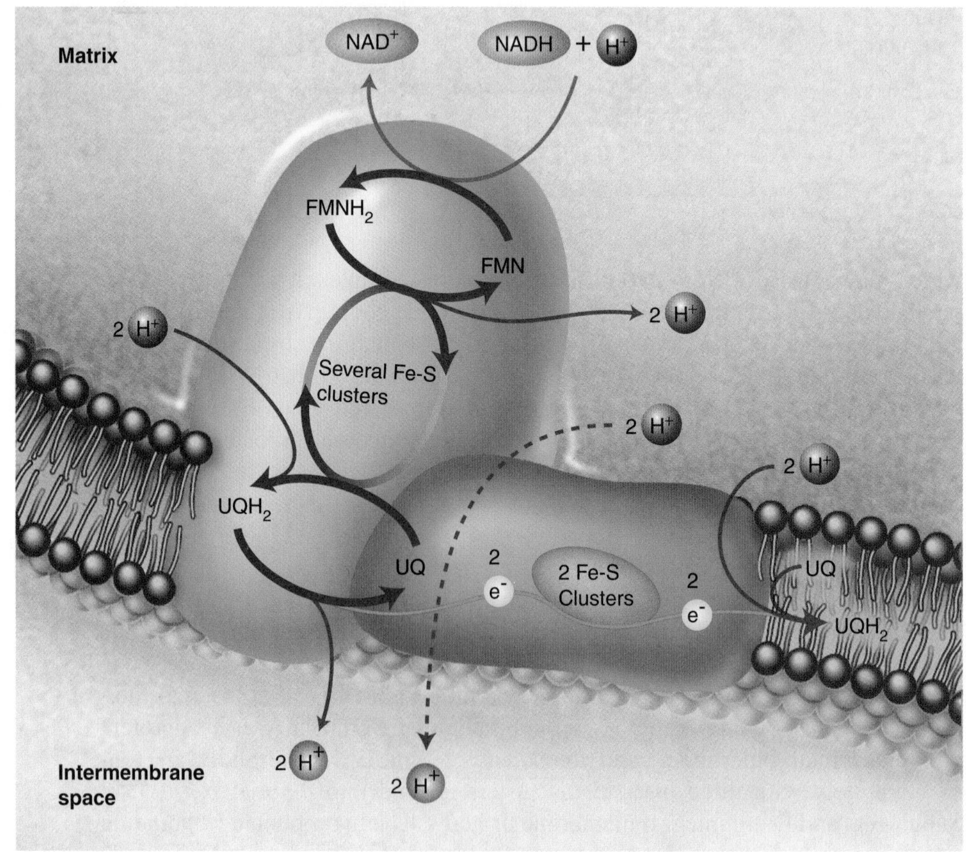

Transfer of Electrons through Complex I of the Mitochondrial Electron Transport Chain

Electron transfer begins with the reduction of FMN by NADH, a process that requires one proton from the matrix. $FMNH_2$ subsequently transfers two electrons, one at a time, to six to eight Fe-S clusters. The sequential transfer of the two electrons to the first Fe-S cluster eventually releases four protons into the intermembrane space. The mechanism by which these protons are transferred across the membrane is still unclear. However, an internal UQ is believed to be involved in the transfer of two of the protons. The second pair of protons is transferred to an external UQ as the two electrons are sequentially transferred from the internal UQ through a series of Fe-S clusters.

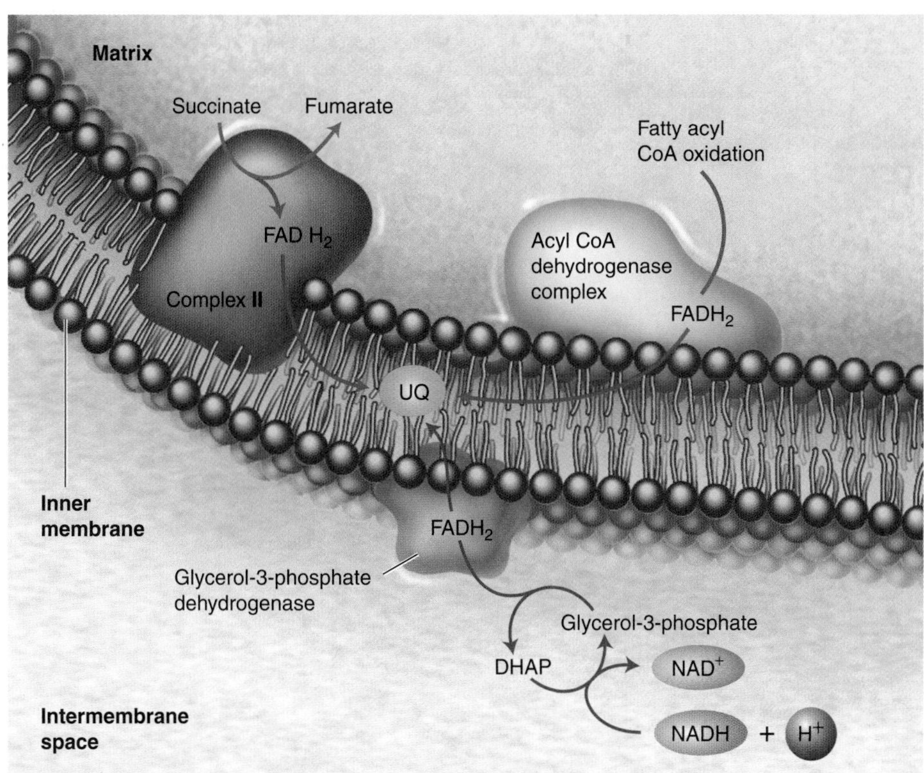

FIGURE 10.5

Path of Electrons from Succinate, Glycerol-3-Phosphate, and Fatty Acids to UQ

Electrons from succinate are transferred to FAD in complex II and several Fe-S clusters and then to UQ. Electrons from cytoplasmic NADH are transferred to UQ via a pathway involving glycerol-3-phosphate and the flavoprotein glycerol-3-phosphate dehydrogenase (see p. 295). Fatty acids are oxidized as coenzyme A derivatives. Acyl-CoA dehydrogenase, one of several enzymes in fatty acid oxidation, transfers two electrons to FAD. They are then donated to UQ.

Complex II mediates the transfer of electrons from succinate to UQ. After FAD is reduced (in the oxidation of the citric acid cycle intermediate succinate, p. 334), its electrons are transferred to a series of three iron-sulfur clusters and then to UQ. Unlike the other ETC complexes, succinate dehydrogenase does not translocate protons from the matrix to the inner membrane space. Other flavoproteins also donate electrons to UQ (**Figure 10.5**). In some cell types glycerol-3-phosphate dehydrogenase, an enzyme located on the outer face of the inner mitochondrial membrane, transfers electrons from cytoplasmic NADH to the ETC (see p. 295). Acyl-CoA dehydrogenase, the first enzyme in fatty acid oxidation (Chapter 12), transfers electrons to UQ from the matrix side of the inner membrane.

Complex III, also referred to as *cytochrome bc$_1$ complex*, is a dimer with each monomer containing 11 subunits. Among these are three cytochromes (cyt b_L, cyt b_H, and cyt c_1) and one iron-sulfur cluster. The *cytochromes* are a series of electron transport proteins that contain a heme prosthetic group similar to those found in hemoglobin and myoglobin. Electrons are transferred by cytochromes one at a time in association with a reversible change in the oxidation state of a heme iron (i.e., between a reduced Fe^{2+} and an oxidized Fe^{3+}). The function of complex III is the transfer of electrons from reduced coenzyme Q (UQH_2) to a protein called cytochrome c (cyt c). Cyt c (**Figure 10.6**) is a mobile electron carrier that is loosely associated with the outer face of the inner mitochondrial membrane.

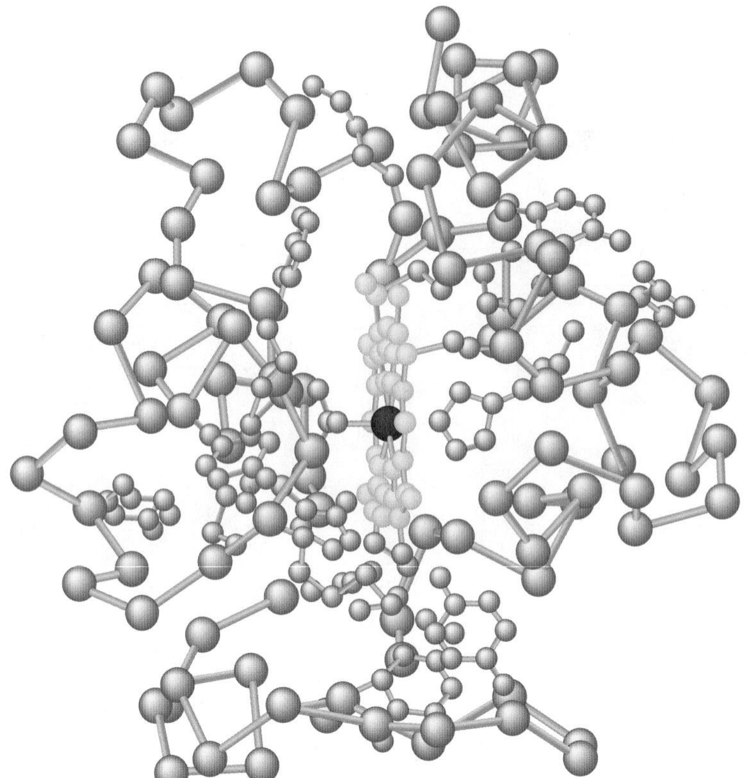

FIGURE 10.6
Structure of Cytochrome c

Cytochrome c is a member of the class of small proteins called cytochromes, each of which possesses a heme prosthetic group. During the electron transport process, the iron in the heme is alternately oxidized and reduced.

The passage of electrons through complex III (**Figure 10.7**), referred to as the **Q cycle**, is complicated. However, the overall reaction for the process whereby each UQH_2 donates two electrons to cyt c is straightforward:

$$UQH_2 + 2\ \text{cyt}\ c_{ox}(Fe^{3+}) + 2H^+_{matrix} \rightarrow UQ + 2\ \text{cyt}\ c_{red}(Fe^{2+}) + 4H^+_{cytosol} \quad (1)$$

During the Q cycle, coenzyme Q molecules diffuse within the inner membrane between the electron donors in complex I or II and the electron acceptor in complex III. UQH_2 donates its two electrons one at a time. One electron flows to the Fe-S protein, which transfers it to cyt c_1 and, subsequently, to cyt c (**Figure 10.7**). The products of this transfer are $UQ^{\overline{\bullet}}$ and two protons from UQH_2 that are transferred to the intermembrane space. The $UQ^{\overline{\bullet}}$ then transfers its second electron first to cyt b_L and then to cyt b_H. The product of the first round of the Q cycle is UQ, with one electron transferred to cyt c and the second electron to cyt b_H; two protons are transferred into the intermembrane space. The second round of the Q cycle involves a second molecule of UQH_2, which transfers one electron to cyt c in the same manner as in the first round. The product $UQ^{\overline{\bullet}}$ then accepts the electron from cyt b_H and two protons from the mitochondrial matrix to form UQH_2. The net effect is that two electrons and four protons are fed off the intermembrane space side of the inner mitochondrial membrane to oxidize cyt c and contribute to the proton gradient. One molecule each of UQ and UQH_2 is formed within the membrane with two protons contributed from the mitochondrial matrix.

Cytochrome oxidase (complex IV) is a protein complex that catalyzes the four-electron reduction of O_2 to form H_2O. The membrane-spanning complex (**Figure 10.8**) in mammals contains 13 subunits, depending on species. Complex IV contains cytochromes a and a_3 and three copper ions. Two copper ions form Cu_A/Cu_A, a binuclear Cu-Cu center, and heme a_3 and Cu_B form a binuclear Fe-Cu center. Both centers accept electrons, one at a time. Electrons flow from cytochrome c to Cu_A/Cu_A to cytochrome a and then to a_3-Cu_B and, finally, to O_2. Four protons and four electrons are shuttled through complex IV from the outer face

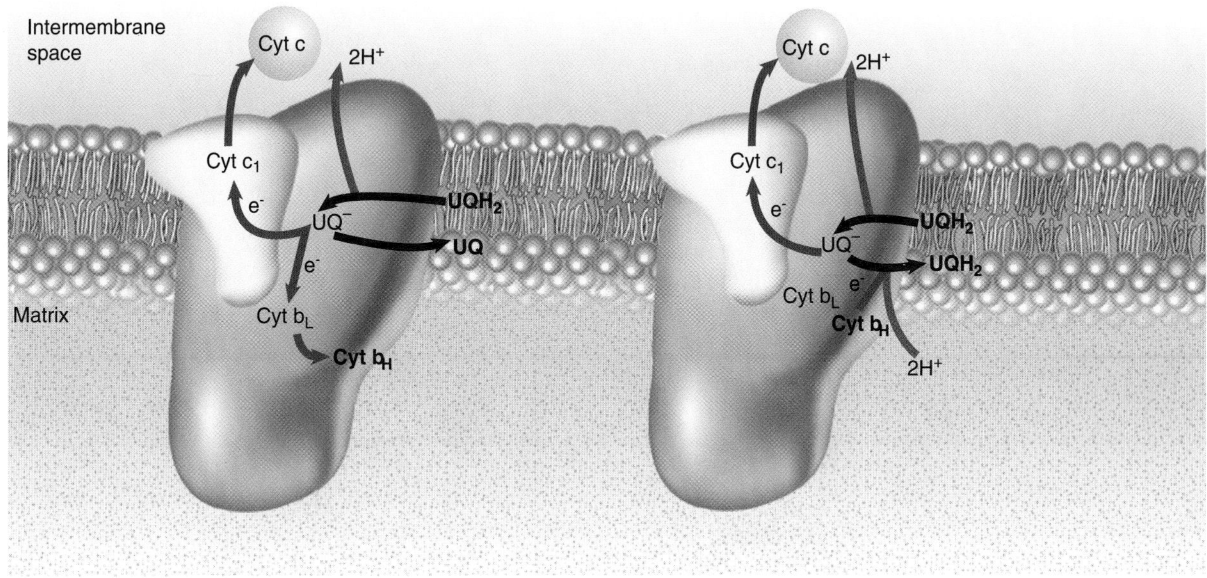

(a) Round 1 of Q cycle (b) Round 2 of Q cycle

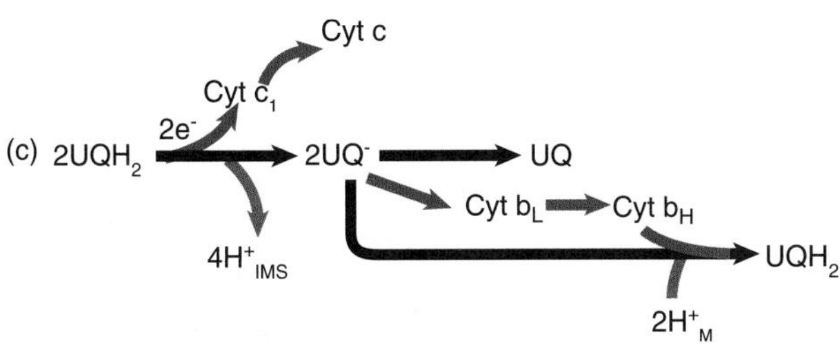

FIGURE 10.7

Electron Transport through Complex III

Two molecules of UQH_2 are oxidized sequentially to supply two electrons (e^-) to cyt c. The first electron from each UQH_2 is transferred to the Fe-S protein (not shown) and then to cyt c_1 and cyt c as two protons each are transferred to the inner membrane space. (a) The second electron from one $UQ^{\bullet}$ is transferred to cyt b_L and then to cyt b_H. UQ is released into the inner mitochondrial membrane. (b) The second $UQ^{\bullet}$ picks up the electron from cyt b_H and two protons from the mitochondrial matrix to form UQH_2. (c) The summary reaction shows that two UQH_2 molecules enter complex III and UQ and UQH_2 are released from the complex. (Black arrows = reaction arrows; red arrows = proton transfer; blue arrows = electron flow.)

of the inner mitochondrial membrane to the matrix for delivery to the cytochrome a_3-Fe(II)-bound dioxygen. Two water molecules are formed and leave the site:

$$O_2 + 4H^+ + 4\,e^- \rightarrow 2H_2O \qquad (2)$$

There are ATP-binding regulatory sites on cyt c and complex IV. When ATP concentrations are high, ATP, acting as an allosteric inhibitor, binds to these sites and causes decreased electron transport activity.

NADH oxidation results in the release of a substantial amount of energy, measured by decreased reduction potentials ($\Delta E^{\circ\prime}$), as electrons flow through complexes I, III, and IV (**Figure 10.9**). Approximately 2.5 molecules of ATP are synthesized for each pair of electrons transferred between NADH and O_2 in the ETC. Approximately 1.5 molecules of ATP result from the transfer of each pair donated by the $FADH_2$ produced by succinate oxidation. The mechanism by which ATP synthesis is coupled to electron transport, the proton gradient, is described on page 358. The components of the ETC are summarized in Table 10.1.

KEY CONCEPT

The electron transport chain is a series of complexes consisting of molecular electron carriers located in the inner mitochondrial membrane of eukaryotic cells.

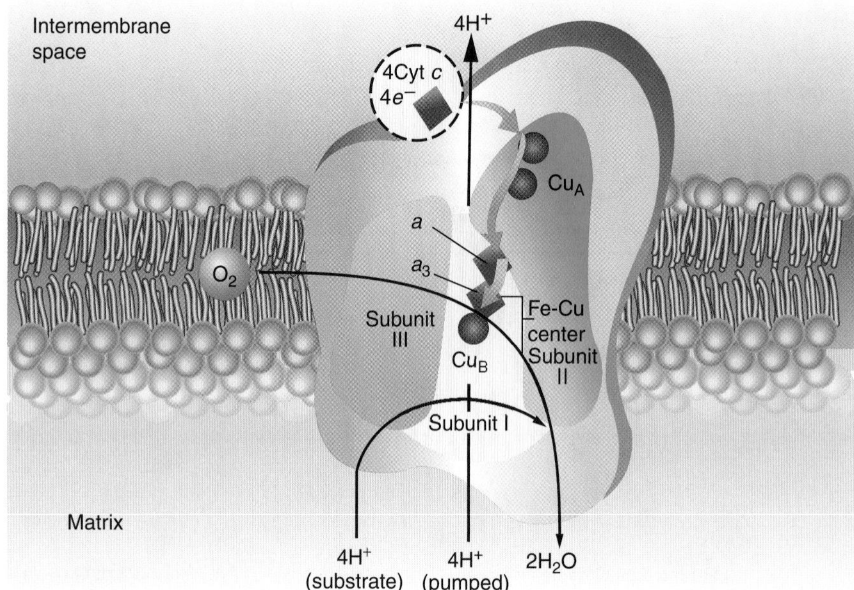

FIGURE 10.8

Electron Transport through Complex IV

Each of two reduced cyt c molecules donates two electrons, one at a time, to Cu_A on subunit II. The electrons are then transferred to cyt a, cyt a_3, and Cu_B on subunit I. The O_2 binds to the heme group in cyt a_3 and is converted to the protonated form of O_2^{-2}, its peroxy derivative, by two electrons transferred from the Fe-Cu center (cyt a_3 and Cu_B) in subunit I. Transfer of two additional electrons from cyt c and four protons from the matrix results in the formation of two water molecules. Simultaneously, four more protons are pumped from the matrix into the intermembrane space.

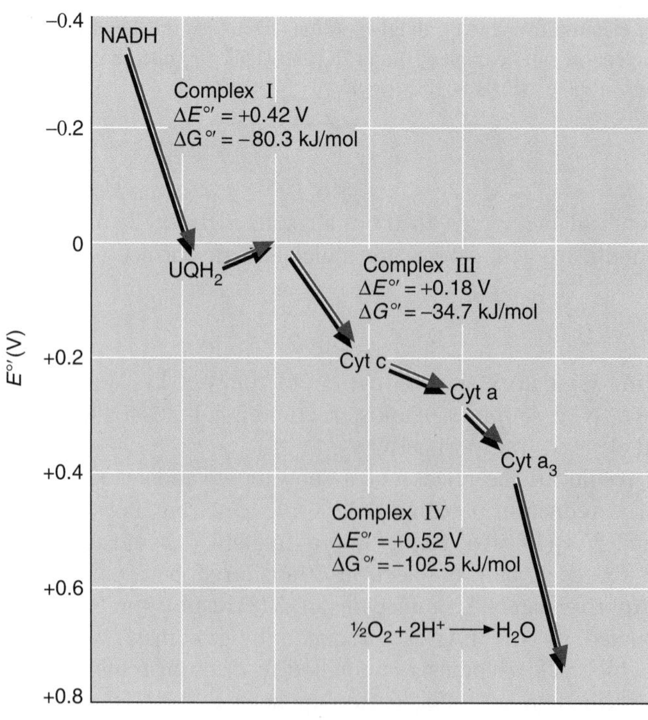

FIGURE 10.9

The Energy Relationships in the Mitochondrial Electron Transport Chain

Relatively large decreases in free energy occur in three steps. During each of these steps, sufficient energy is released to account for the synthesis of ATP.

TABLE 10.1 Supramolecular Components of the Electron Transport Chain

Enzyme Complex	Prosthetic Groups
Complex I (NADH dehydrogenase)	FMN, FeS
Complex II (succinate dehydrogenase)	FAD, FeS
Complex III (cytochrome bc_1 complex)	Hemes, FeS
Cytochrome c	Heme
Complex IV (cytochrome oxidase)	Hemes, Fe, Cu

Electron Transport: The Fluid State and Solid State Models

The fluid state model has been the conventional view of mitochondrial electron transport for several decades. Based in part on the experimental extraction and purification of the four ETC complexes in enzymatically active forms, the fluid-state model describes electron transfer between ETC components. Electron transfer is seen as the result of random collisions among the ETC complexes and the mobile electron carriers UQ and cytochrome c. Within the past two decades, however, significant evidence supports a solid state organization of the electron transport process instead. As a result of more gentle methods of extraction and purification, several supercomplexes composed of combinations of complexes I, III, and IV (e.g., I, III and III, IV_{1-4}, but not complex II) have been isolated and identified. The I, III_2, IV_{1-2} supercomplex has been found in several animals, plants, and fungi. Now referred to as the **respirasome**, it is believed to be the functional respiration unit because it is the most active and stable version of the ETC. (The symbol III_2 represents the complex III dimer.)

According to the solid state model, electron transfer is efficient because of short diffusion distances for the mobile electron carriers. Structural studies of the bovine cardiac I, III_2, IV supercomplex indicate that the binding sites of the mobile electron carriers in the ETC complexes are in close proximity. For example, the UQ binding site in complex I directly faces the complex III binding site for reduced UQ. Another feature of the inner mitochondrial membrane also supports the solid state model. This membrane is exceptionally protein dense (a 75%:25% protein:lipid ratio). Consequently, random protein movement, as envisioned in the fluid state model, is severely restricted in the heavily folded inner membrane.

Electron Transport Inhibitors

Several molecules specifically inhibit the electron transport process (**Figure 10.10**). Used in conjunction with reduction potential measurements, inhibitors have been invaluable in determining the correct order of ETC components. In such experiments, electron transport is measured with an oxygen electrode. (Oxygen consumption is a sensitive measure of electron transport.) When electron transport is inhibited, oxygen consumption is reduced or eliminated. Oxidized ETC components accumulate on the O_2-reducing side of the site of inhibition. Reduced ETC components accumulate on the nonoxygen side of the site of inhibition. For example, antimycin A inhibits cyt b in complex III. If this inhibitor is added to a suspension of mitochondria, NAD^+, the flavins, and cyt b molecules become more reduced. The cytochromes c_1, c, and a become more oxidized. Other prominent examples of ETC inhibitors include rotenone and amytal, which inhibit NADH dehydrogenase (complex I). Carbon monoxide (CO), azide (N_3^-), and cyanide (CN^-) inhibit cytochrome oxidase.

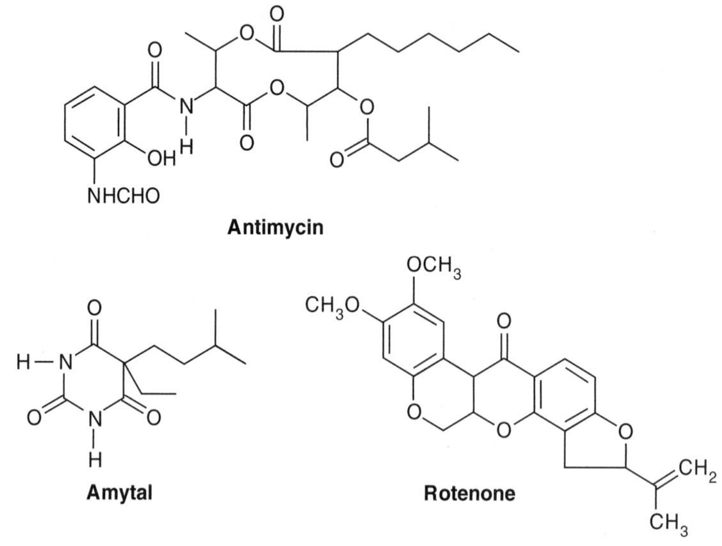

FIGURE 10.10

Several Inhibitors of the Mitochondrial Electron Transport Chain

Antimycin blocks the transfer of electrons from the b cytochromes. Amytal and rotenone block NADH dehydrogenase.

Antimycin

Amytal

Rotenone

QUESTION 10.1

Which compound in each of the following pairs is the better reducing agent?

a. $NADH/H_2O$
b. $UQH_2/FADH_2$
c. Cyt c (reduced)/cyt b (reduced)
d. $FADH_2/NADH$
e. $NADH/FMNH_2$

10.2 OXIDATIVE PHOSPHORYLATION

Oxidative phosphorylation, the process whereby the energy generated by the ETC is conserved by the phosphorylation of ADP to yield ATP, is explained by the **chemiosmotic coupling theory** proposed by Peter Mitchell in 1961. In *chemiosmosis* the energy released by electron transport generates an electrochemical gradient across a membrane, which in turn drives ATP synthesis.

The Chemiosmotic Theory

The chemiosmotic coupling theory has the following features:

1. As electrons pass through the ETC, protons are transported from the matrix and released into the intermembrane space. As a result, an electrical potential Ψ and a proton gradient ΔpH arise across the inner membrane. The electrochemical proton gradient is sometimes referred to as the **protonmotive force** Δp.
2. Protons, which are present in the intermembrane space in great excess as a result of the electron transport process, can pass through the inner membrane and back into the matrix down their concentration gradient only through special channels. (The inner membrane itself is impermeable to ions such as protons.) ATP is synthesized from ADP and P_i by a molecular machine called *ATP synthase*. Also referred to as complex V, it contains a proton channel. ATP synthesis occurs as a result of a thermodynamically favorable flow of protons through the channel.

Mitchell used the term *chemiosmotic* to emphasize that chemical reactions can be coupled to osmotic gradients. An overview of the chemiosmotic model as it operates in the mitochrondion is illustrated in **Figure 10.11**.

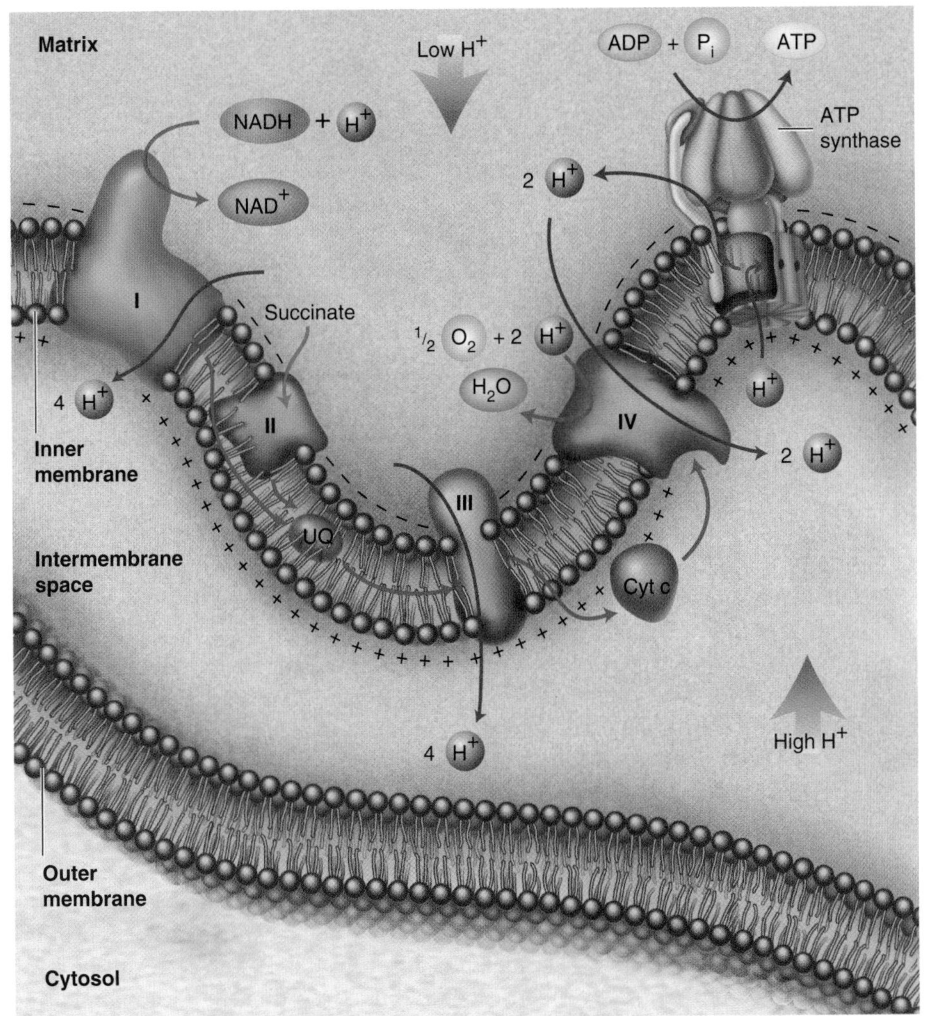

FIGURE 10.11

Overview of the Chemiosmotic Model

In Mitchell's model, protons are driven from the mitochondrial matrix across the inner membrane and into the intermembrane space by the electron transport mechanism. The energy captured from electron transport is used to create an electrical potential and a proton gradient. Because the inner membrane is impermeable to protons, they can traverse the membrane only by flowing through specific proton channels. The flow of protons through the ATP synthase drives the synthesis of ATP. (See **Figure 10.1** for brief descriptions of the roles of complexes I, II, III, and IV in electron transport.) Note that as a result of proton flow across the inner membrane, the pH of the matrix rises and the matrix side of the membrane becomes more negatively charged.

Examples of the evidence that supports the chemiosmotic theory include the following:

1. Actively respiring mitochondria expel protons. The pH of a weakly buffered suspension of mitochondria measured by an electrode drops when O_2 is added. The typical pH gradient across the inner membrane is approximately 0.05 pH unit.
2. ATP synthesis stops when the inner membrane is disrupted. For example, although electron transport continues, ATP synthesis stops in mitochondria placed in a hypertonic solution. Mitochondrial swelling results in proton leakage across the inner membrane.
3. A variety of ATP synthesis inhibitor molecules are now known to specifically dissipate the proton gradient (**Figure 10.12**). According to the chemiosmotic theory, a disrupted proton gradient dissipates the energy derived from food molecules as heat. **Uncouplers** such as dinitrophenol collapse the proton gradient by equalizing the proton concentration on both sides of membranes. (As they diffuse across the membrane, uncouplers pick up protons from one side and release them on the other.) **Ionophores** are hydrophobic molecules that dissipate osmotic gradients by inserting themselves into a membrane and forming a channel. For example, gramicidin is an antibiotic that forms a channel in membranes that allows the passage of H^+, K^+, and Na^+.

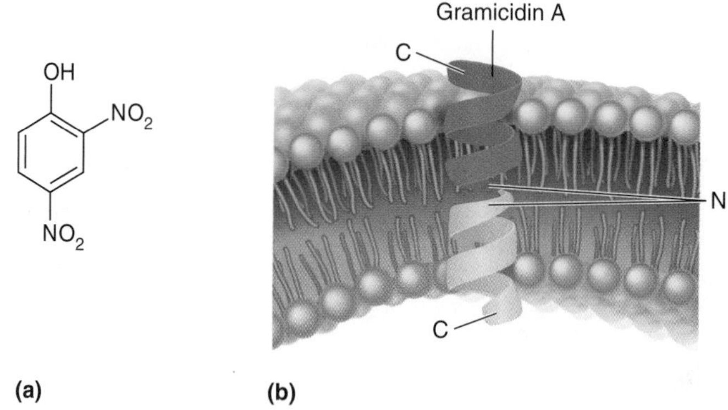

FIGURE 10.12
Uncouplers

(a) Dinitrophenol, which diffuses across the membrane, picks up protons on one side and releases them on the other. (b) Gramicidin A, an 11-residue peptide, forms an end-to-end dimer that creates a proton-permeable pore in the membrane: C = carboxy terminus; N = amino terminus.

The proton gradients generated by electron transport systems are dissipated for two general purposes: ATP is synthesized as protons flow through the ATP synthase and regulated proton leakage is used to drive several other types of biological work. Examples include heat generation and transport of substances such as phosphate and the adenine nucleotides ADP and ATP across the inner membrane. In the next section, descriptions of ATP synthesis and its regulation within mitochondria are followed by a brief overview of energy generation from glucose. Section 10.2 ends with a discussion of nonshivering thermogenesis, the mechanism by which the mitochondrial proton gradient in certain animal cells regulates body temperature.

QUESTION 10.2

Dinitrophenol (DNP) is an uncoupler that was used as a diet aid in the 1920s, until several people who had been taking it died. Suggest why DNP consumption results in weight loss. The deaths caused by DNP were a result of liver failure. Explain. [*Hint:* Liver cells contain an extraordinarily large number of mitochondria.]

ATP Synthesis

Early electron microscopic studies of mitochondria showed numerous lollipop-shape structures studding the inner membrane on its inner surface (**Figure 10.13**). Experiments begun in the early 1960s revealed that each lollipop is a proton translocating ATP synthase. These investigations made use of *submitochondrial particles*, which are small membranous vesicles formed when mitochondria are subjected to sonication. Further work showed that the ATP synthase (**Figure 10.14**) consists of two major components. The F_1 unit, the active ATPase, possesses five different subunits present in the ratio α_3:β_3:γ:δ:ε. There are three nucleotide-binding catalytic sites on F_1. The F_0 unit, a transmembrane channel for protons, has three subunits present in the ratio a:b_2:c_{10-12}. The F_0 subunit is so named because its function is inhibited by oligomycin, which is an antibiotic produced by *Streptomyces*, a group of Gram-positive bacteria. Oligomycin blocks the proton channel when it binds to subunit a.

The ATP synthase consists of two rotors (rotary motors) linked together by a strong, flexible *stator* (a stationary component of a motor). In respiring organisms the F_0 motor converts the protonmotive force into a rotational force that drives ATP synthesis catalyzed by the F_1 unit. The revolving component, the c ring (formed from the c subunits), which is attached to a central shaft composed of the ε and γ subunits, rotates within the α,β hexamer of the F_1 unit. The stator (the b and δ subunits) prevents the α,β hexamer from rotating.

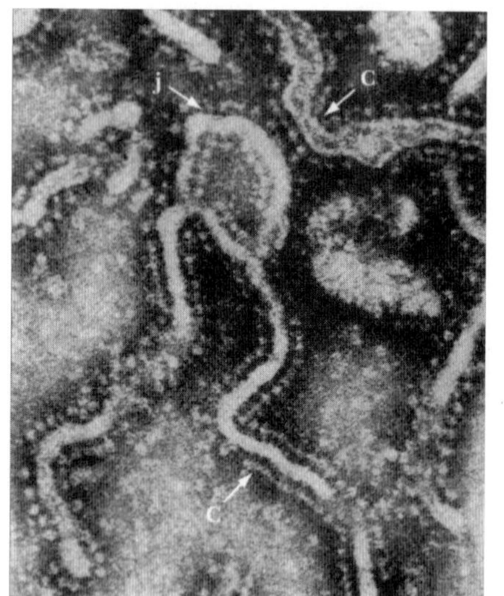

FIGURE 10.13
The ATP Synthase

An early electron micrograph of submitochondrial particles revealing the "lollipop"-like structures that would eventually be identified as ATP synthase.

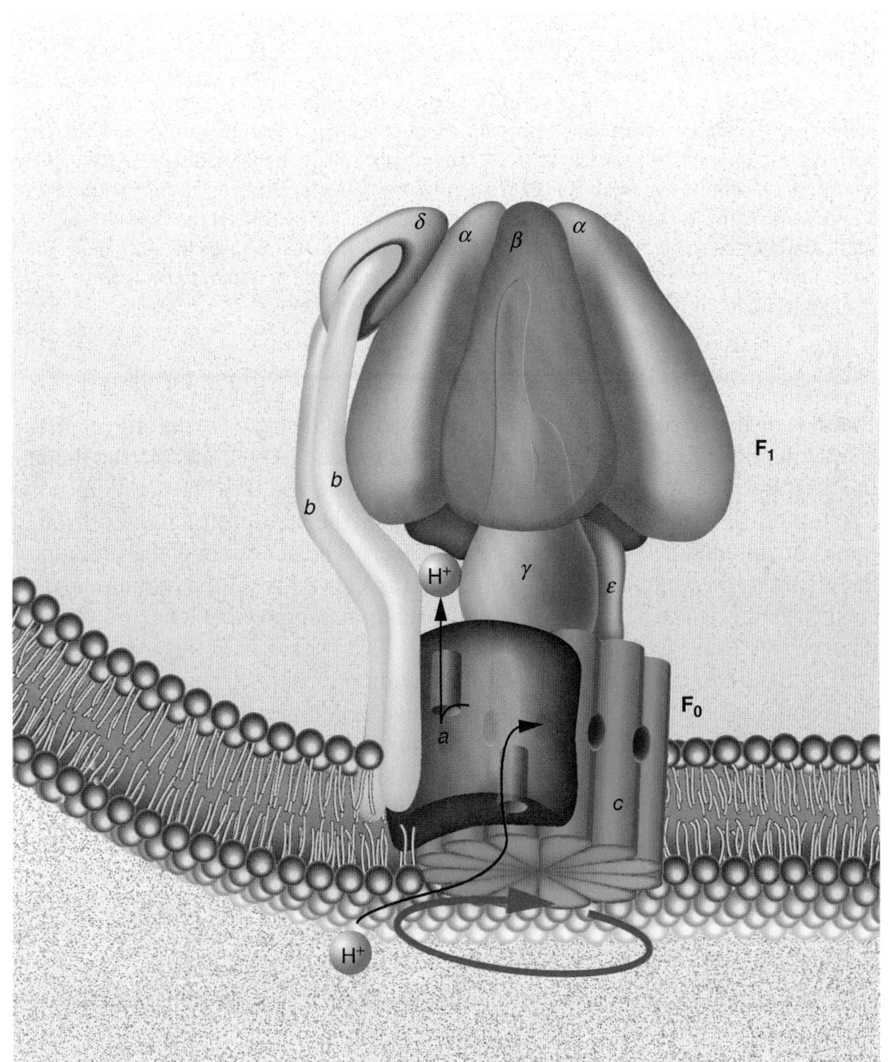

FIGURE 10.14

ATP Synthase from *Escherichia coli*

The rotor consists of ε, γ, and c_{12} subunits. The stator consists of a, b_2, δ, α_3, and β_3 subunits. The molecular components of ATP synthase are well conserved among bacteria, plants, and animals.

The synthesis of each ATP requires the translocation of three protons through the ATP synthase. (The transfer of an additional proton is required for the transport of ATP and OH^- out of the matrix in exchange for ADP and P_i.) As protons flow through F_0, the rotation of the proton channel (c_{12}, also referred to as the c ring) is transmitted to the γ subunit that projects into the core of the F_1 unit. The rotation of the central shaft puts it in three possible positions relative to each α,β dimer. In effect, the protonmotive force induces three sequential 120° rotations of the α,β hexamer. As rotation proceeds, each of three nucleotide-binding sites undergoes a series of conformational changes that result in ATP synthesis. In certain circumstances (e.g., *E. coli* in anaerobic conditions and fermentative lactic acid bacteria) the F_1 unit, acting as a motor, works in the opposite direction so that ATP is hydrolyzed. As a result, protons are pumped outward. The outward-moving proton gradient is then used to perform cellular work such as flagella rotation and nutrient transport.

The mechanism of ATP synthesis by the ATP synthase is as follows. In the F_0 unit each c subunit in the c ring consists of two transmembrane antiparallel helices. The C-terminal helix of c subunits contains an essential aspartate residue that upon protonation causes a swiveling motion that in turn triggers the rotation of the entire subunit. Deprotonation of this residue causes the C-terminal helix to return to its original conformation. Protons enter the c ring through a

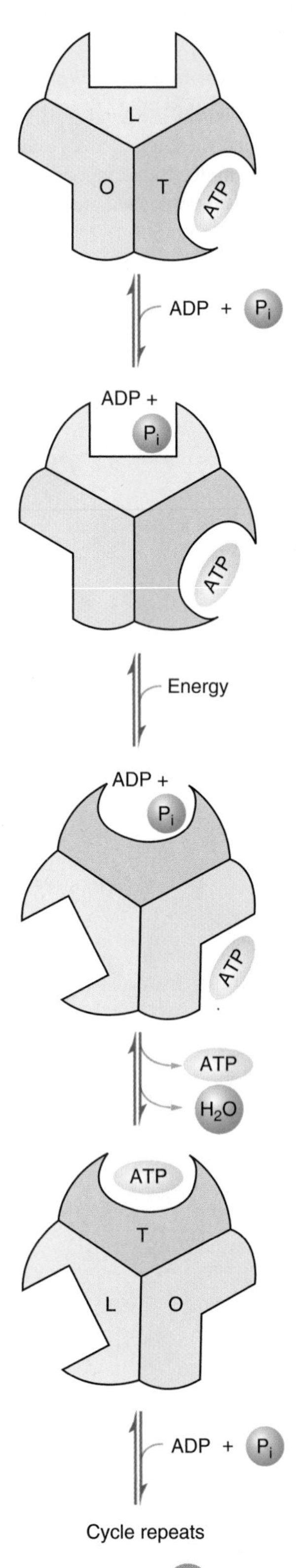

FIGURE 10.15
ATP Synthesis Model

The catalytic sites in the β subunits of ATP synthase are believed to have three conformations: open (O: inactive with low affinity for adenine nucleotides), tight (T: active with high affinity), and loose (L: also inactive). ATP synthesis begins with the binding of ADP and P_i to an L site. A conformational change driven by the rotation of the central shaft within the α,β hexamer converts the L site into a T conformation. ATP is then synthesized. As the shaft continues to rotate, the T site is converted to an O site that releases the ATP molecule. ATP cannot be released from the O site unless ADP and P_i are bound to the adjacent β-subunit site in the T conformation.

channel in the *a* subunit. At the end of this channel, at the interface of the *a* subunit and the proximal *c* subunit, a basic residue (Arg) transfers the incoming proton to the Asp residue in the *c* subunit. As a result of the rotation of the c subunit, it is displaced from the a subunit. As the next proton moves through the *a* subunit channel, the process repeats itself. The net effect is the rotation of the c ring in a counterclockwise direction (as viewed from the membrane). As each *c* subunit reaches a channel on the *a* subunit, deprotonation occurs, and the proton exits into the mitochondrial matrix. The *torque* (twisting force) generated by the *c* subunit rotation causes the asymmetric central shaft (composed of the ε and γ subunits) to rotate within a sleeve inside the α,β hexamer. The α,β hexamer catalytic sites are within the β subunits. They occur in three conformations in terms of affinity for adenine nucleotide ligands: *open* (O) (inactive with low affinity), *tight* (T) (active with high affinity), and *loose* (L) (also inactive) (**Figure 10.15**). Interconversion between these conformations is caused by interaction with the rotating γ subunit. There are essentially three steps in the ATP synthesizing process: (1) ADP and P_i bind to an L site, (2) ATP is synthesized when the L conformation is transformed to a T conformation, and (3) ATP is released as the T conformation converts to an O conformation. As the γ subunit rotates and sequentially interacts with each β subunit, each active site is forced through the O, T, and L conformations.

QUESTION 10.3

A suspension of inside-out submitochondrial particles is placed in a solution that contains ADP, P_i, and NADH. Will increasing the proton concentration of the solution result in ATP synthesis? Explain.

Control of Oxidative Phosphorylation

Control of oxidative phosphorylation allows cells to adapt to ever-fluctuating energy requirements. Recall that in normal circumstances, electron transport and ATP synthesis are tightly coupled. The value of the *P:O ratio* (the number of moles of P_i consumed for each oxygen atom reduced to H_2O) reflects the degree of coupling observed between the protonmotive force, created by electron transport, and ATP synthesis. The measured maximum ratio for the oxidation of NADH is 2.5. The maximum P:O ratio for $FADH_2$ is 1.5. If isolated mitochondria are provided with an oxidizable substrate (e.g., succinate), all of the ADP is eventually converted to ATP. At this point, oxygen consumption becomes greatly depressed. Oxygen consumption increases dramatically when ADP is supplied. The control of aerobic respiration by ADP is referred to as **respiratory control**.

The formation of ATP appears to be strongly related to the ATP mass action ratio ([ATP]:[ADP][P_i]). In other words, ATP synthase is inhibited by a high

KEY CONCEPTS

- In aerobic organisms, the energy used to drive the synthesis of most ATP molecules is the protonmotive force.
- The protonmotive force is generated as free energy is released when electrons flow through the electron transport chain.

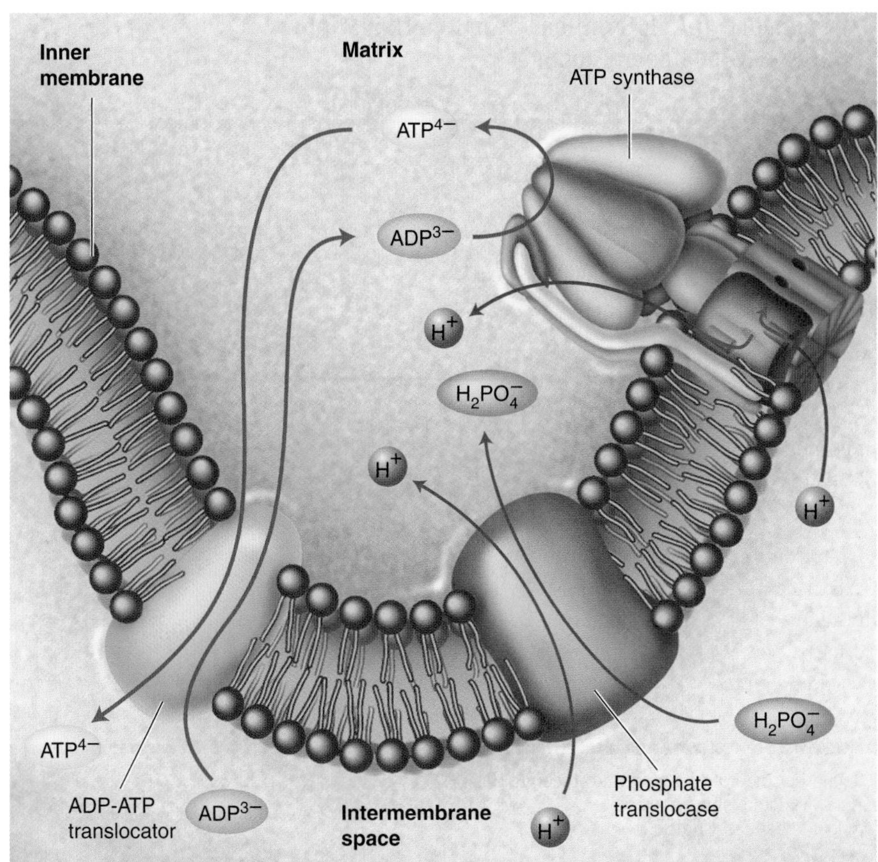

FIGURE 10.16

The ADP-ATP Translocator and the Phosphate Translocase

The transport of $H_2PO_4^-$ across the inner mitochondrial membrane by the phosphate translocase is driven by the proton gradient. For every four protons that are transported out of the matrix, three drive the ATP synthase rotor and one drives the inward transport of phosphate. The simultaneous exchange of ADP^{3-} and ATP^{4-}, required for continuing ATP synthesis and mediated by the ADP-ATP translocator, is driven by the potential difference across the inner membrane.

concentration of its product (ATP) and activated when ADP and P_i concentrations are high. The relative amounts of ATP and ADP within mitochondria are controlled largely by two transport proteins in the inner membrane: the ADP-ATP translocator and the phosphate carrier. The *ADP-ATP translocator* (**Figure 10.16**) is a dimeric protein responsible for the 1:1 exchange of intramitochondrial ATP for the ADP produced in the cytoplasm. As previously described, there is a potential difference across the inner mitochondrial membrane (negative inside). Because ATP molecules have one more negative charge than ADP molecules, the outward transport of ATP and inward transport of ADP are favored. The transport of $H_2PO_4^-$ along with a proton is mediated by the phosphate translocase, also referred to as the $H_2PO_4^-/H^+$ symporter. (*Symporters* are transmembrane transport proteins that move solutes across a membrane in the same direction. See Section 11.2 for a discussion of membrane transport mechanisms.) The inward transport of 4 protons is required for the synthesis of each ATP molecule: 3 to drive the ATP synthase rotor and 1 to drive the inward transport of phosphate.

The Complete Oxidation of Glucose

Table 10.2 summarizes the sources of ATP produced from one molecule of glucose. ATP production from fatty acids, the other important energy source, is discussed in Chapter 12. Recall that two molecules of NADH are produced during glycolysis. When oxygen is available, the oxidation of this NADH by the ETC is preferable (in terms of energy production) to lactate formation. The inner mitochondrial membrane, however, is impermeable to NADH. Animal cells have evolved several shuttle mechanisms to transfer electrons from cytoplasmic NADH to the mitochrondrial ETC. The most prominent examples are the glycerol phosphate shuttle and the malate-aspartate shuttle.

KEY CONCEPTS

- The P/O ratio reflects the coupling between electron transport and ATP synthesis.
- The measured maximum P:O ratios for NADH and $FADH_2$ are 2.5 and 1.5, respectively.

TABLE 10.2 Summary of ATP Synthesis from the Oxidation of One Molecule of Glucose

	NADH	$FADH_2$	ATP
Glycolysis (cytoplasm)			
Glucose → glucose-6-phosphate			−1
Fructose-6-phosphate → fructose-1,6-bisphosphate			−1
Glyceraldehyde-3-phosphate → glycerate-1,3-bisphosphate	+2		
Glycerate-1,3-bisphosphate → glycerate-3-phosphate			+2
Phosphoenolpyruvate → pyruvate			+2
Mitochondrial Reactions			
Pyruvate → acetyl-CoA	+2		
Citric acid cycle			
Oxidation of isocitrate, α-ketoglutarate, and malate	+6		
Oxidation of succinate		+2	
GDP → GTP			+1.5*
Oxidative Phosphorylation			
2 Glycolytic NADH			+4.5† (3)‡
2 NADH (pyruvate to acetyl-CoA)			+5
6 NADH (citric acid cycle)			+15
2 $FADH_2$ (citric acid cycle)			+3
			31 (29.5)

* This number reflects the price of transport into the cytoplasm.
† Assumes the malate-aspartate shuttle.
‡ Assumes the glycerol phosphate shuttle.

In the **glycerol phosphate shuttle** (**Figure 10.17a**), DHAP, a glycolytic intermediate, is reduced by NADH to form glycerol-3-phosphate. This reaction is followed by the oxidation of glycerol-3-phosphate by mitochondrial glycerol-3-phosphate dehydrogenase. (The mitochondrial enzyme uses FAD as an electron acceptor.) Because glycerol-3-phosphate interacts with the mitochondrial enzyme on the outer face of the inner membrane, the substrate does not actually enter the matrix. The $FADH_2$ produced in this reaction is then oxidized by the ETC. FAD as an electron acceptor produces only 1.5 ATP per molecule of cytoplasmic NADH.

Although the **malate-aspartate shuttle** (**Figure 10.17b**) is a more complicated mechanism than the glycerol phosphate shuttle, it is more energy efficient. The shuttle begins with the reduction of cytoplasmic OAA to malate by NADH. After its transport into the mitochondrial matrix, malate is reoxidized. The NADH produced is then oxidized by the ETC. For the shuttle to continue, OAA must be returned to the cytoplasm. Because the inner membrane is impermeable to OAA, it is converted to aspartate in a transamination reaction (Chapter 14) involving glutamate.

The aspartate is transported to the cytoplasm in exchange for glutamate (via the glutamate-aspartate transport protein), where it can be converted to OAA. The α-ketoglutarate is transported to the cytoplasm in exchange for malate (via the malate-α-ketoglutarate transport protein), where it can be converted to glutamate. The glutamate-aspartate transporter requires moving a proton into the matrix. Therefore, the net ATP synthesis using this mechanism is somewhat reduced. Instead of generating 2.5 molecules of ATP for each NADH molecule, the yield is approximately 2.25 molecules of ATP.

One final issue concerned with ATP synthesis from glucose remains. Recall that two molecules of ATP are produced in the citric acid cycle (from GTP). The price for their transport into the cytoplasm, where they will be used, is the uptake of two protons into the matrix. Therefore the total amount of ATP produced from a molecule of glucose is reduced by about half a molecule of ATP.

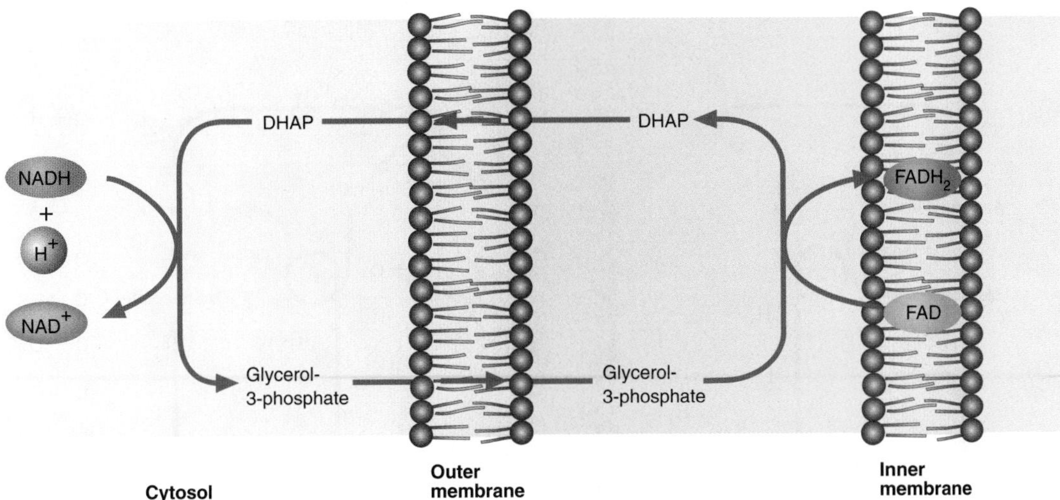

FIGURE 10.17a

Shuttle Mechanisms That Transfer Electrons from Cytoplasmic NADH to the Respiratory Chain

(a) The glycerol-3-phosphate shuttle. Dihydroxyacetone phosphate (DHAP) is reduced to form glycerol-3-phosphate. Glycerol-3-phosphate is reoxidized by mitochondrial glycerol-3-phosphate dehydrogenase and FAD is reduced to $FADH_2$. (b) The aspartate-malate shuttle. Oxaloacetate is reduced by NADH to form malate. Malate is transported into the mitochondrial matrix, where it is reoxidized to form oxaloacetate and NADH. Because oxaloacetate cannot penetrate the inner membrane, it is converted to aspartate in a transamination involving glutamate. Two inner membrane carriers are required for this shuttle mechanism: the glutamate-aspartate transport protein and the malate–α-ketoglutarate transport protein.

Depending on the shuttle used, the total number of molecules of ATP produced per molecule of glucose varies (approximately) from 29.5 to 31. Assuming that the average amount of ATP produced is 30 molecules, the net reaction for the complete oxidation of glucose is as follows:

$$C_6H_{12}O_6 + 6O_2 + 30\ ADP + 30\ P_i \rightarrow 6CO_2 + 6H_2O + 30\ ATP \qquad (3)$$

The number of ATP molecules generated during the complete oxidation of glucose is in sharp contrast to the two molecules of ATP formed by glycolysis. Obviously, organisms that use oxygen to oxidize glucose have a substantial advantage.

KEY CONCEPT

The aerobic oxidation of glucose yields between 29.5 and 31 ATP molecules.

QUESTION 10.4

Traditionally, the oxidation of each NADH and $FADH_2$ by the ETC was believed to result in the synthesis of three molecules of ATP and two molecules of ATP, respectively. As noted, recent measurements, which have considered such factors as proton leakage across the inner membrane, have reduced these values somewhat. Use the earlier values to calculate the number of ATP molecules generated by the aerobic oxidation of a glucose molecule. First assume that the glycerol phosphate shuttle is operating; then assume that the malate-aspartate shuttle is transferring reducing equivalents into the mitochondrion.

QUESTION 10.5

Calculate the maximum number of ATP that can be generated from a mole of sucrose.

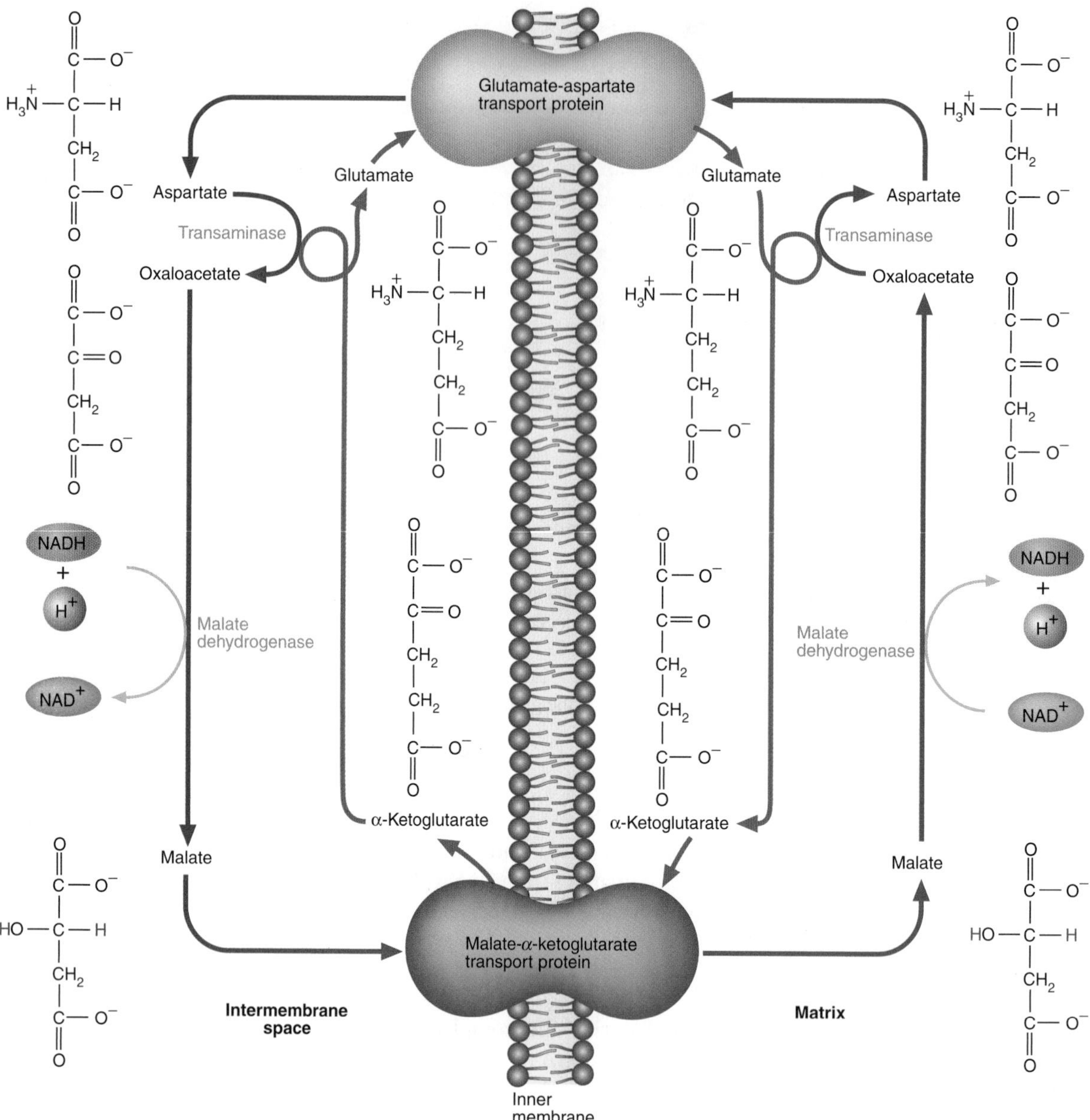

FIGURE 10.17b

Uncoupled Electron Transport

Certain proteins, called **uncoupling proteins**, partially dissipate oxidative energy by translocating protons across the mitochondrial inner membrane without involving ATP synthesis. Uncoupling protein 1 (UCP1), the best-characterized example, is a dimer that forms a proton channel. UCP1, or *thermogenin*, is found exclusively in the mitochondria of brown fat, a specialized form of adipose tissue present in newborn babies (to help normalize body temperature) and hibernating mammals (to warm their bodies at the end of winter). (The characteristic color of brown adipose tissue results from the large number of mitochondria it contains.) The functions of UCP_2 and UCP_3, which occur in a variety of different tissues, are unresolved.

UCP1, which constitutes about 10% of the protein in the mitochondrial inner membrane, is activated when it is bound to fatty acids. As a result of the reduction of the proton gradient by UCP1, the energy captured during electron transport is dissipated as heat. The entire process of heat generation from brown fat, called *nonshivering thermogenesis*, is regulated by norepinephrine (p. 534).

(In shivering thermogenesis, heat is produced by nonvoluntary muscle contraction.) Norepinephrine, a neurotransmitter released from specialized neurons that terminate in brown adipose tissue, initiates a cascade mechanism that ultimately hydrolyzes fat molecules. The fatty acid products of fat hydrolysis activate the uncoupler protein. Fatty acid oxidation continues until the norepinephrine signal is terminated or the cell's fat reserves are depleted.

Increased expression of uncoupler proteins (especially UCP2) has been observed in several cancers (e.g., colon, breast, and liver). UCP2 promotes carcinogenesis by decreasing ROS formation, thereby protecting tumor cells from oxidative damage, and by increasing glycolytic flux since ATP inhibits several glycolytic enzymes.

10.3 OXYGEN, CELL FUNCTION, AND OXIDATIVE STRESS

All living processes take place within a redox environment that is defined as the sum of the reduction potential and the reducing capacity of linked redox pairs such as NAD(P)H/NAD(P)$^+$ and GSH/GSSG (reduced and oxidized forms, respectively, of glutathione, a key cellular reducing agent, p. 533). Each cell's "redox state" is regulated within a narrow range because of the redox-sensitive nature of many metabolic and signaling pathways. These processes contain numerous proteins whose functional properties (activation or inactivation) are altered when the redox status of critical thiols (–SH groups) is altered. For example, the oxidation of sulfhydryl groups in proteins to form sulfenic (R-SOH), sulfinic (R-SO_2H), and sulfonic (R-SO_3H) acids can change the functional properties of these molecules. Limited redox changes do occur during certain normal cellular processes. For example, the cytoplasm of cells entering into cell division becomes more reduced, whereas that of differentiated cells becomes relatively more oxidized. Intracellular compartments also have distinctive redox conditions. The nucleus and mitochondria are more reduced than the cytoplasm (GSH/GSSG is high) and the ER is more oxidized (GSH/GSSG is low).

Redox regulation is of paramount importance because of the nature of molecular oxygen. As previously mentioned, the advantages of using oxygen are linked to a dangerous property: oxygen can accept single electrons to form unstable derivatives, referred to as **reactive oxygen species (ROS)**. Examples of ROS include the superoxide radical, hydrogen peroxide, the hydroxyl radical, and singlet oxygen. ROS are so reactive that they can seriously damage living cells if formed in significant amounts. In living organisms, ROS formation is usually kept to a minimum by antioxidant defense mechanisms. **Antioxidants** are substances that react more easily with ROS than with critical biomolecules and, therefore, mitigate the tissue-damaging effects of these highly reactive metabolic by-products. Despite their potentially toxic properties, however, ROS function in small amounts as cell signaling devices by altering the redox status of target proteins such as metabolic enzymes, cytoskeletal components, cell cycle regulator proteins, transcription factors, and translation regulators. Their small size, easy diffusibility, and short half-lives allow ROS, in controlled amounts, to act as important integrators of cellular function.

Under certain conditions, referred to collectively as **oxidative stress**, antioxidant mechanisms are overwhelmed, ROS levels rise, and some damage may occur. Damage results primarily from enzyme inactivation, polysaccharide depolymerization, DNA breakage, and membrane destruction. Examples of circumstances that may cause serious oxidative damage include infection, inflammation, certain metabolic abnormalities, the overconsumption of certain drugs or exposure to intense radiation, and repeated contact with certain environmental

contaminants (e.g., tobacco smoke). In addition to contributing to the aging process, oxidative damage has been linked to at least 100 human diseases. Examples include cancer, cardiovascular disorders such as atherosclerosis, myocardial infarction, and hypertension, as well as neurological disorders such as amyotrophic lateral sclerosis (ALS, or Lou Gehrig's disease), Parkinson's disease, and Alzheimer's disease.

Several types of cells deliberately produce ROS. In animal bodies scavenger cells such as macrophages and neutrophils continuously search for microorganisms and damaged cells. In an oxygen-consuming process referred to as the **respiratory burst**, ROS are generated and used to kill and dismantle these cells.

Reactive Oxygen Species

The properties of oxygen are of course directly related to its molecular structure. Dioxygen is a diradical because it possesses two unpaired electrons. (A **radical** is an atom or group of atoms that contains one or more unpaired electrons.) For this and other reasons, when it reacts, dioxygen can accept only one electron at a time.

Recall that during mitochondrial electron transport, H_2O is formed as a consequence of the sequential transfer of four electrons to O_2. During this process, several ROS are formed. Cytochrome oxidase (like other oxygen-activating proteins) traps these reactive intermediates within its active site until all four electrons have been transferred to oxygen. However, electrons may leak out of the electron transport pathway and react with O_2 to form ROS (Figure 10.18).

In normal circumstances, cellular antioxidant defense mechanisms minimize any damage. ROS are also formed during nonenzymatic processes. For example, exposure to UV light and ionizing radiation causes ROS formation.

The first ROS formed during the reduction of oxygen is the superoxide radical $O_2^{\bar{\cdot}}$. Most superoxide radicals are produced by electrons derived from the flavoprotein NADH dehydrogenase (complex I) and the Q cycle in complex III. However, $O_2^{\bar{\cdot}}$ acts as a nucleophile and (in specific circumstances) as either an oxidizing agent or a reducing agent. Because of its solubility properties, $O_2^{\bar{\cdot}}$ causes considerable damage to the phospholipid components of membranes. When it is generated in an aqueous environment, $O_2^{\bar{\cdot}}$ reacts with itself to produce O_2 and hydrogen peroxide (H_2O_2):

$$2H^+ + 2O_2^{\bar{\cdot}} \rightarrow O_2 + H_2O_2 \qquad (4)$$

Since H_2O_2 does not have any unpaired electrons, it is not a radical. The limited reactivity of H_2O_2 allows it to cross membranes and become widely dispersed. The subsequent reaction of H_2O_2 with Fe^{2+} (or other transition metals) results in the production of the hydroxyl radical (•OH), a highly reactive species:

$$Fe^{2+} + H_2O_2 \rightarrow Fe^{3+} + \bullet OH + OH^- \qquad (5)$$

The hydroxyl radical diffuses only a short distance before it reacts with whatever biomolecule it collides with. Radicals such as the hydroxyl radical are especially dangerous because they can initiate an autocatalytic radical chain reaction (**Figure 10.19**). Singlet oxygen ($^1O_2$), an excited state of dioxygen in which the unpaired electrons have become paired, can form from superoxide:

$$2O_2^{\bar{\cdot}} + 2H^+ \rightarrow H_2O_2 + {}^1O_2 \qquad (6)$$

or from peroxides:

$$2ROOH \rightarrow 2ROH + {}^1O_2 \qquad (7)$$

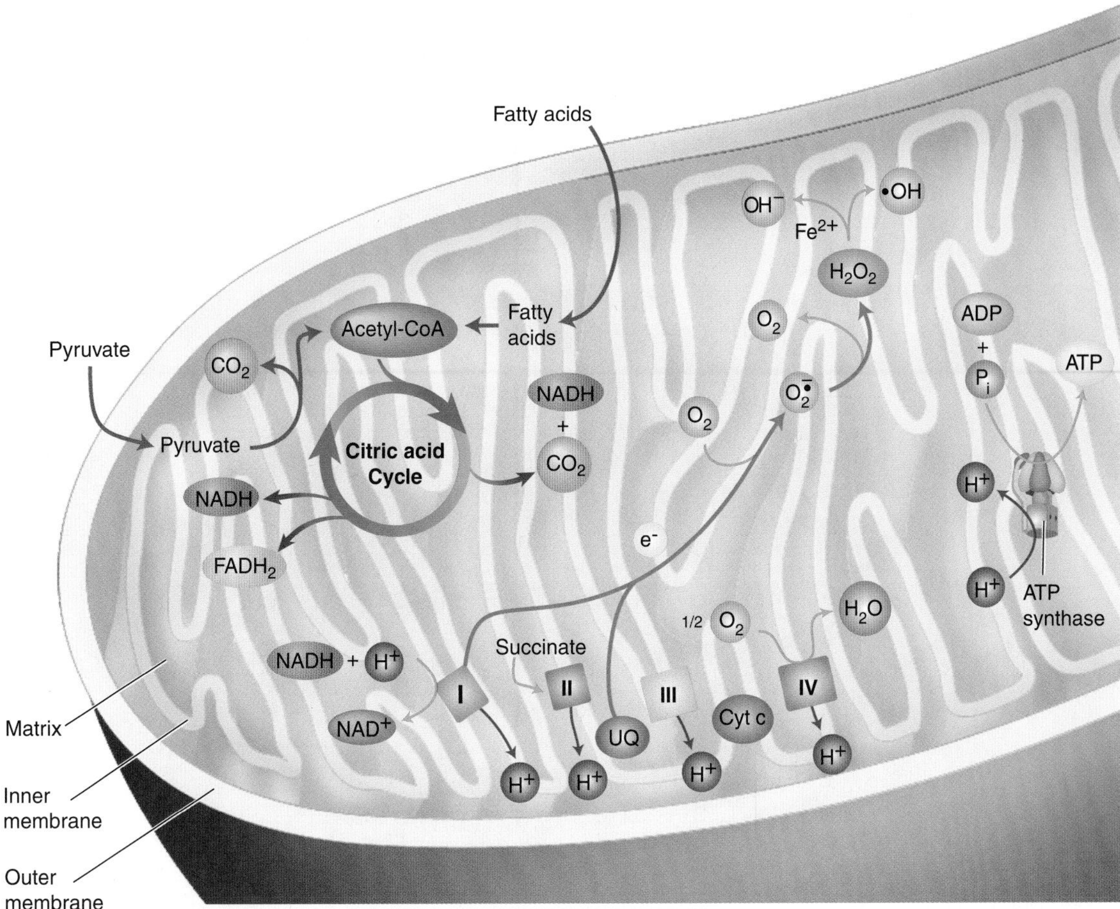

FIGURE 10.18

Overview of Oxidative Phosphorylation and ROS Formation in the Mitochondrion

Oxidative phosphorylation involves five multiprotein complexes: complexes I, II, III, and IV (the principal components of the ETC) and the ATP synthase. Pyruvate and fatty acids, the major fuel molecules, are transported into the mitochondrion, where they are oxidized by the citric acid cycle. The hydrogen atoms liberated during this process are carried by NADH and $FADH_2$ to the ETC. The energy that is released by the electron transport system is used to pump protons from the matrix through the inner membrane into the intermembrane space. The electrochemical gradient across the inner membrane created by proton pumping is used to synthesize ATP as protons flow through the ATP synthase. However, no system is perfect. Electrons leak from the ETC, especially complexes I and III, and react with O_2 to form superoxide ($O_2^{\bullet-}$). When respiratory chain activity is low (e.g., in ischemia; see p. 377), an increase in the NADH:NAD^+ ratio may lead to superoxide formation by α-ketoglutarate and pyruvate dehydrogenases. In the presence of Fe^{2+}, superoxide is converted into the hydroxyl radical $(\bullet OH)_2$. Superoxide is also converted to hydrogen peroxide. ROS react with and damage any molecule they encounter. Accumulating oxidative damage within mitochondria can compromise their capacity to generate energy.

Singlet oxygen formed from certain reactions of H_2O_2 and during photosynthetic light harvesting can react with double bonds in biomolecules. It is particularly damaging to aromatics and conjugated alkenes.

As mentioned (see p. 368), ROS are generated during several other cellular activities besides the reduction of O_2 to form H_2O. These include the biotransformation of xenobiotics and the respiratory burst (**Figure 10.20**) in white blood cells. In addition, electrons often leak from the electron transport pathways in the endoplasmic reticulum (e.g., the cytochrome P_{450} electron transport system) to form superoxide by combining with O_2. Reactions involving ROS include

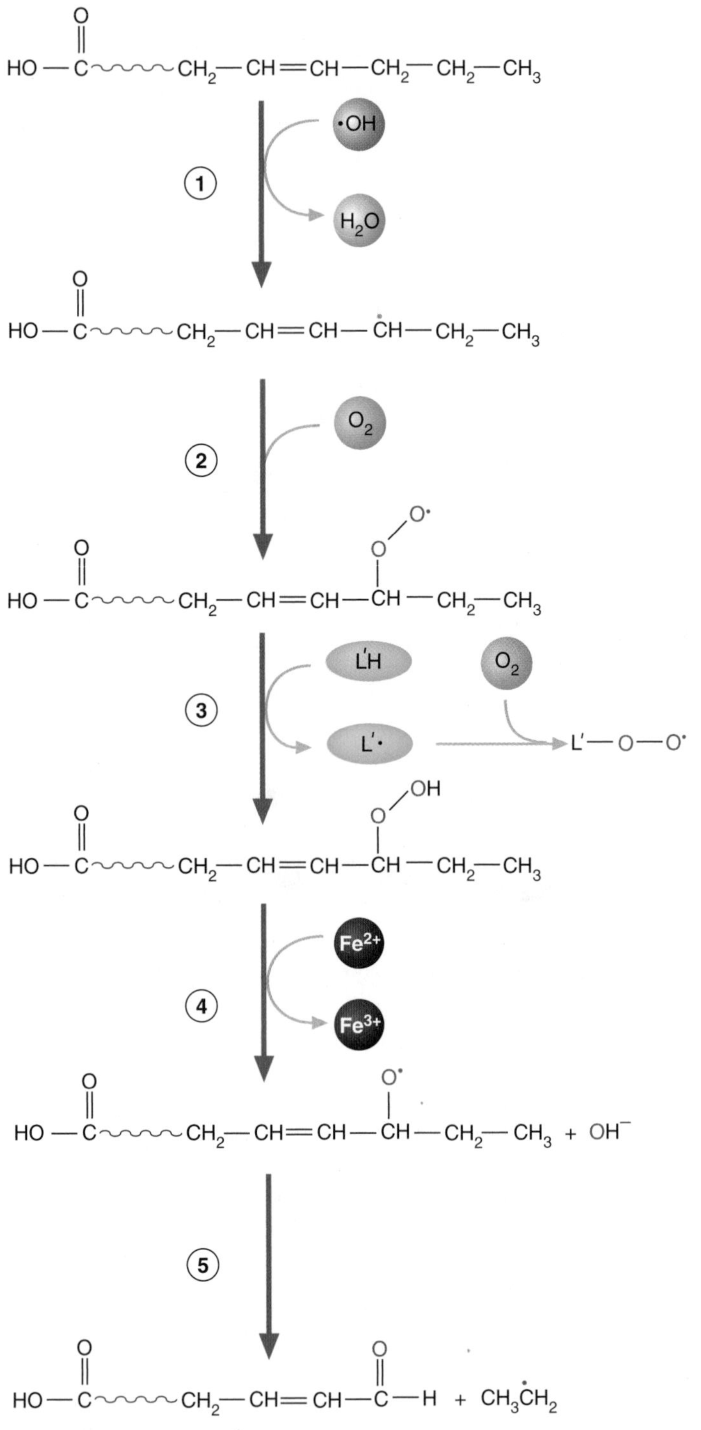

FIGURE 10.19

Radical Chain Reaction

Step 1: Lipid peroxidation reactions begin after the extraction of a hydrogen atom from an unsaturated fatty acid (LH → L•). Step 2: The lipid radical (L•) then reacts with O_2 to form a peroxyl radical (L• + O_2 → L–O—O•). Step 3: The radical chain reaction begins when the peroxyl radical extracts a hydrogen atom from another fatty acid molecule (L—O—O• + L′H → L—O—OH + L′•). Step 4: The presence of a transition metal such as Fe^{2+} initiates further radical formation (L—O—O—H + Fe^{2+} → LO• + HO^- + Fe^{3+}). Step 5: One of the most serious consequences of lipid peroxidation is the formation of α,β-unsaturated aldehydes, which involves a radical cleavage reaction. The chain reaction continues as the free radical product then reacts with a nearby molecule. Reactive carbonyl products are also products of this process.

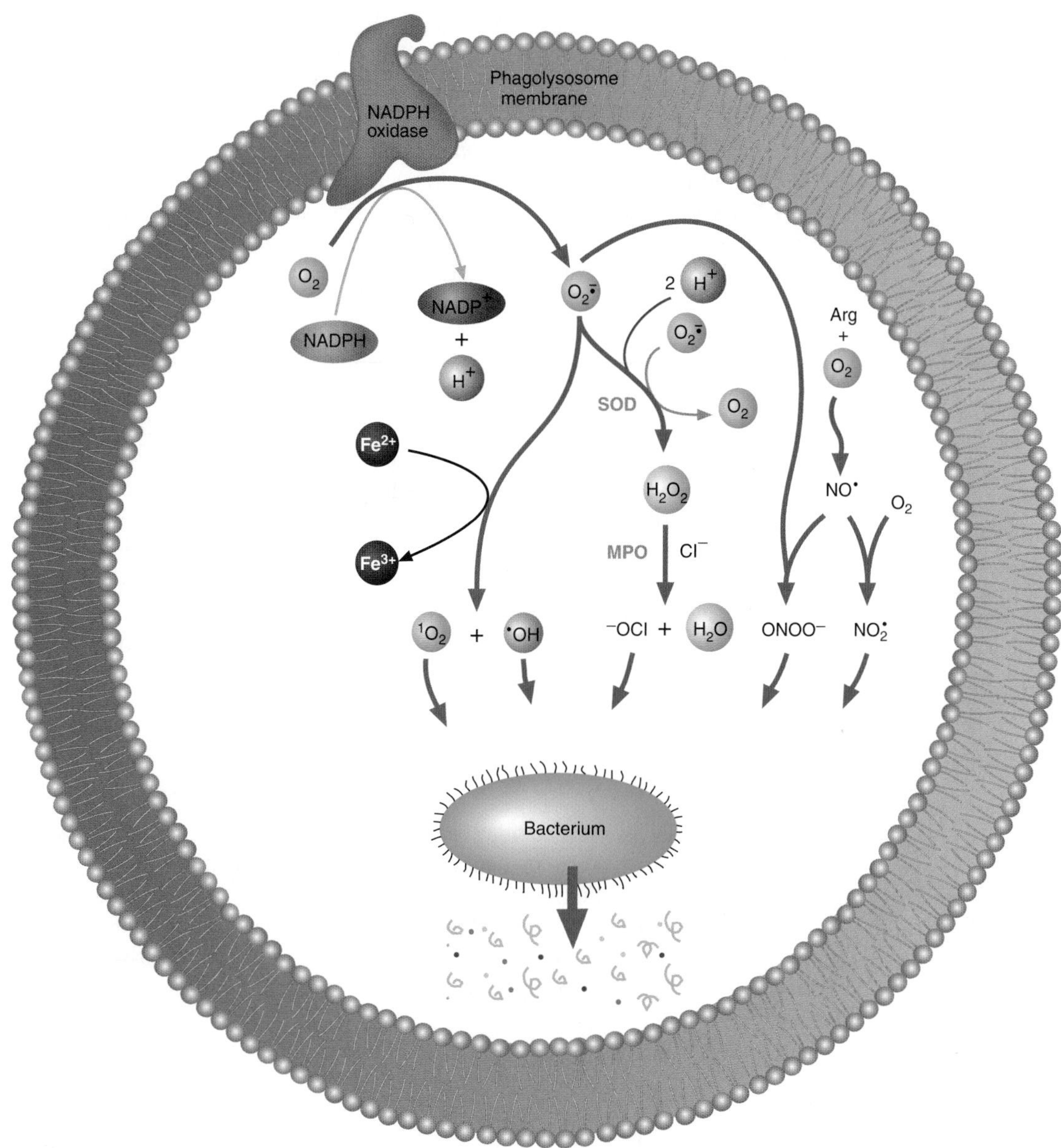

FIGURE 10.20

The Respiratory Burst

The respiratory burst provides a dramatic example of the destructiveness of ROS. Within seconds after a phagocytic cell binds to a bacterium (or other foreign structure), its oxygen consumption increases nearly 100-fold. During endocytosis the bacterium is incorporated into a large vesicle called a phagosome. Phagosomes then fuse with lysosomes to form phagolysosomes. Two destructive processes then ensue: the respiratory burst and digestion by lysosomal enzymes. The respiratory burst is initiated when NADPH oxidase converts O_2 to $O_2^{\bullet}$. Two molecules of $O_2^{\bullet}$ combine in a reaction catalyzed by SOD (superoxide dismutase) to form H_2O_2. H_2O_2 is next converted to several types of bactericidal (bacteria-killing) molecules by myeloperoxidase (MPO), an enzyme found in abundance in phagocytes. For example, MPO catalyzes the oxygenation of halide ions (e.g., Cl^-) to form hypohalides. Hypochlorite (the active ingredient in household bleach) is extremely bactericidal. In the presence of Fe^{2+}, $O_2^{\bullet}$ and H_2O_2 react to form •OH and $^1O_2$ (singlet oxygen), both of which are extremely reactive. Nitric oxide, synthesized by nitric oxide synthase from arginine and O_2, reacts with superoxide to form peroxynitrite and with molecular oxygen to form nitrogen dioxide. In addition to the damage inflicted by various types of ROS, these species, along with MPO products, activate proteases, which degrade microbial proteins. Proteases are themselves protected from oxidative injury by MPO, which also possesses a catalase activity. After the disintegration of the bacterial cell, lysosomal enzymes digest the fragments that remain.

hydroxylation and peroxidation. Another such reaction, carbonylation, is a nonenzymatic protein modification resulting from the oxidation of amino acid side chains (i.e., Thr, Lys, Arg, or Pro) or the reaction of Cys, Lys, or His side chains with reactive carbonyl radicals.

There are also several nitrogen-containing radicals. Because their synthesis is often linked to that of ROS, **reactive nitrogen species** (RNS) are often classified as ROS. Among the most important examples are nitric oxide (·NO), nitrogen dioxide ($^{\bullet}NO_2$), and peroxynitrite ($ONOO^-$). Nitric oxide (·NO) is a highly reactive gas. Because of its free radical structure ·NO was regarded, until recently, primarily as a contributing factor in the destruction of the ozone layer in Earth's atmosphere and as a precursor of acid rain. Recent research has revealed, however, that ·NO is an important signal molecule that is produced throughout the mammalian body.

Physiological functions in which ·NO is now believed to play a role include blood pressure regulation, blood clotting inhibition, and macrophage-induced destruction of foreign, damaged, or cancerous cells. The disruption of the normally precise regulation of ·NO synthesis has been linked to numerous pathological conditions that include stroke, migraine headache, male impotence, septic shock, and several neurodegenerative diseases such as Parkinson's disease. ·NO can damage proteins with sulfhydryl groups, such as glyceraldehyde-3-phosphate dehydrogenase (p. 278), by converting SH groups into nitrosothiol (—SNO) derivatives. ·NO also damages iron-sulfur proteins. Some of the damage attributed to ·NO is, in fact, caused by its oxidation products, $^{\bullet}NO_2$ ($2\ {}^{\bullet}NO + O_2 \rightarrow 2\ {}^{\bullet}NO_2$) and peroxynitrite ($^{\bullet}NO + O_2 \rightarrow ONOO^-$).

KEY CONCEPTS

- Reactive oxygen species form because oxygen is reduced by accepting one electron at a time.
- ROS formation is a normal by-product of metabolism and the result of conditions such as exposure to radiation.
- RNS are often classified as ROS because the synthesis of these species is often linked.

Antioxidant Enzyme Systems

To protect themselves from oxidative stress, living organisms have developed several antioxidant defense mechanisms. These mechanisms employ several metalloenzymes and antioxidant molecules.

The major enzymatic defenses against oxidative stress are provided by four enzymes: superoxide dismutase, glutathione peroxidase, peroxiredoxin, and catalase. The wide distribution of these enzymatic activities underscores the ever-present problem of oxidative damage.

The superoxide dismutases (SOD) are a class of enzymes that catalyze the formation of H_2O_2 and O_2 from the superoxide radical:

$$2O_2^{\bullet-} + 2H^+ \rightarrow H_2O_2 + O_2 \qquad (8)$$

There are three major forms of SOD in humans. SOD1 is a Cu-Zn isoenzyme that occurs in cytoplasm. SOD3, also requiring Cu and Zn, is an extracellular enzyme. A manganese-containing isozyme, SOD2, is found in the mitochondrial matrix. About 20% of inherited ALS, also known as Lou Gehrig's disease, are caused by a mutation in the gene that codes for SOD1. ALS is a fatal degenerative disease in which motor neurons are destroyed.

Glutathione peroxidase (GPx), a selenium-containing enzyme, is a key component in an enzymatic system most responsible for controlling cellular peroxide levels. Recall that this enzyme catalyzes the reduction of a variety of substances by the tripeptide reducing agent GSH (**Table 5.3**). In addition to reducing H_2O_2 to form water, glutathione peroxidase transforms organic peroxides into alcohols:

$$2\ GSH + R{-}O{-}O{-}H \rightarrow G{-}S{-}S{-}G + R{-}OH + H_2O \qquad (9)$$

In humans, GPx1 reduces H_2O_2 and GPx4 acts on lipid hydroperoxides. Several ancillary enzymes support glutathione peroxidase function (**Figure 10.21**). GSH is regenerated from GSSG by glutathione reductase:

$$G{-}S{-}S{-}G + NADPH + H^+ \rightarrow 2\ GSH + NADP^+ \qquad (10)$$

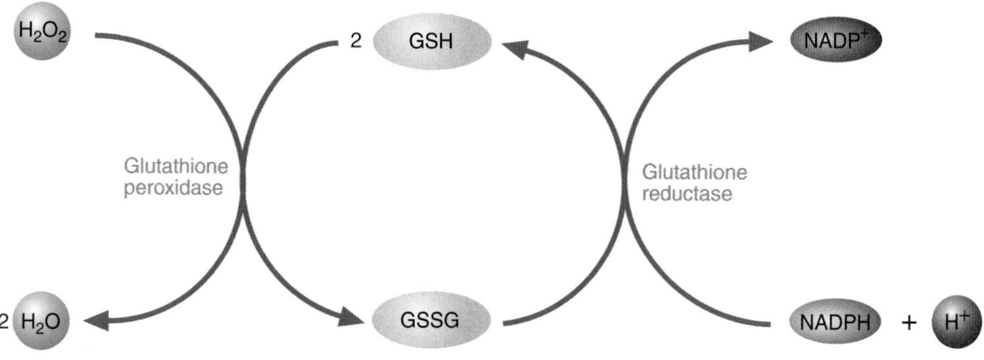

FIGURE 10.21

The Glutathione-Centered System

Glutathione peroxidase utilizes GSH to reduce the peroxides generated by cellular aerobic metabolism. GSH is regenerated from its oxidized form, GSSG, by glutathione reductase. NADPH, the reducing agent in this reaction, is supplied by the pentose phosphate pathway and several other reactions.

The NADPH required in the reaction is provided primarily by several reactions of the pentose phosphate pathway (Chapter 8). Recall that NADPH is also produced by the reactions catalyzed by isocitrate dehydrogenase (p. 333) and malic enzyme (p. 339).

The *peroxiredoxins* (PRX) are a class of thiol-containing enzymes that detoxify peroxides. Their catalytic mechanism involves the oxidation of a redox-active cysteine side chain sulfhydryl group by the peroxide substrate to form sulfenic acid (RSOH). Sulfenic acid is subsequently reduced by a thiol-containing protein such as thioredoxin. *Thioredoxin* (TRX) is involved in redox reactions mediated by the peroxiredoxin/thioredoxin reductase (TR) system (sometimes referred to as the TRX-centered system) (**Figure 10.22**). TR and TRX also return other oxidized cellular proteins, including many transcription factors, to their functional reduced sulfhydryl form; this change is accomplished by means of an enzyme-catalyzed shift of electrons from reduced thioredoxin [TRX-$(SH)_2$] to the target protein. Thioredoxin reductase reduces oxidized thioredoxin [TRX(S_2)] with electrons delivered to it through a mobile NADPH and a bound $FADH_2$. TRX also serves as an electron shuttle for other nonantioxidant enzyme systems such as that of ribonucleotide reductase (p. 545).

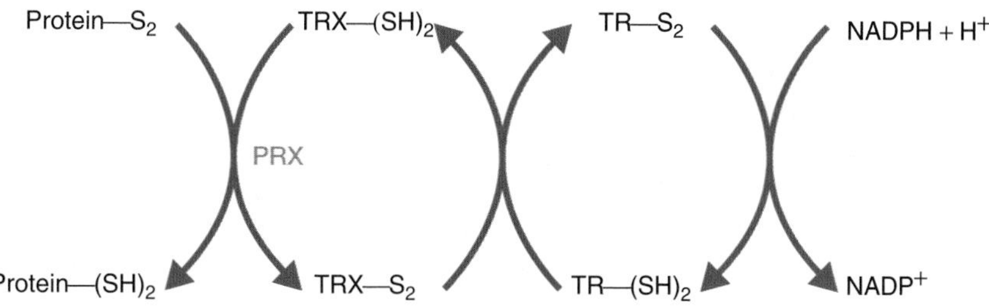

FIGURE 10.22

The Thioredoxin-Centered System

A substrate protein that contains function-altering disulfides is reduced by electron transfer from thioredoxin [TRX-$(SH)_2$] in a reaction mediated by the peroxiredoxin enzyme (PRX). The oxidized thioredoxin (TRX-S_2) is returned to its reduced form by thioredoxin reductase (TR). The electrons required to restore the TR to its reduced form come from NADPH via an enzyme-bound FAD.

Catalase is an enzyme whose primary role is to catalyze the conversion of H_2O_2 to water and dioxygen. (For every two molecules of H_2O_2 that are detoxified, one is oxidized to form O_2 and the other is reduced to H_2O.) Catalase has a heme-Fe(III) prosthetic group with the following mechanism:

$$\text{Step 1: } H_2O_2 + \text{Fe(III)—enzyme} \rightarrow H_2O + \text{O=Fe(IV)—enzyme}$$

$$\text{Step 2: } H_2O_2 + \text{O=Fe(IV)—enzyme} \rightarrow O_2 + H_2O + \text{Fe(III)—enzyme} \quad (11)$$

Two isozymes of catalase have been identified: HPI and HPII. HPI is a bifunctional enzyme in that the second H_2O_2 can be replaced by an oxygen-containing organic group (phenols, aldehydes, acids, and alcohols). This peroxidative reaction converts the organic molecule to an often less toxic form:

$$\text{R—CH}_2\text{—OH} + \text{O=Fe(IV)—enzyme} \rightarrow \text{R—CH}_3 + O_2 + \text{Fe(III)—enzyme} \quad (12)$$

KEY CONCEPTS

- The major enzymatic defenses against oxidative stress are superoxide dismutase, glutathione peroxidase, peroxiredoxin, and catalase.
- The pentose phosphate pathway produces the reducing agent NADPH.

The respiratory burst of macrophages, for example, generates H_2O_2 as a microbicide primarily through the incomplete oxidation of fatty acids. Unlike HPI, HPII has only one substrate, H_2O_2. It is found in high quantity in erythrocytes and peroxisomes of phagocytic white blood cells. Outside the peroxisome, the primary H_2O_2 generator is SOD1.

ROS level reduction can also be mediated by the tumor suppressor protein p53 (p. 343) in response to certain cellular stress signals by upregulating the synthesis of several antioxidant enzymes. Among these are sestrins 1 and 2, which regenerate peroxiredoxins, and TIGAR [TP53-induced glycolysis and apoptosis regulator], which diverts glucose into the NADPH-producing pentose phosphate pathway. When cells are under severe stress (e.g., irreparable DNA damage), p53 upregulates the transcription of pro-oxidant genes such as PUMA (p53 upregulated modulator of apoptosis) and depresses antioxidant gene transcription, a circumstance that results in a massive increase in ROS levels, one component of p53-mediated apoptosis.

QUESTION 10.6

Selenium is generally considered a toxic element. (It is the active component of loco weed.) However, there is growing evidence that selenium is also an essential trace element. Because glutathione peroxidase activity is essential to protect red blood cells against oxidative stress, a selenium deficiency can damage red blood cells. Although sulfur is in the same chemical family as selenium, it cannot be substituted. Can you explain why? [*Hint*: Selenium is more easily oxidized than sulfur.] Is sulfur or selenium a better scavenger for oxygen when this gas is present in trace amounts?

QUESTION 10.7

Ionizing radiation is believed to damage living tissue by producing hydroxyl radicals. Drugs that protect organisms from radiation damage, which must be taken *before* radiation exposure usually have —SH groups. How do such drugs protect against radiation damage? Can you suggest any type of nonsulfhydryl group–containing molecule that would protect against hydroxyl radical–induced damage?

Antioxidant Molecules

Antioxidants in Diet

Living organisms use antioxidant molecules to protect themselves from radicals. Humans obtain α-tocopherol (vitamin E), ascorbic acid (vitamin C), and β-carotene (**Figure 10.23**) from the diet.

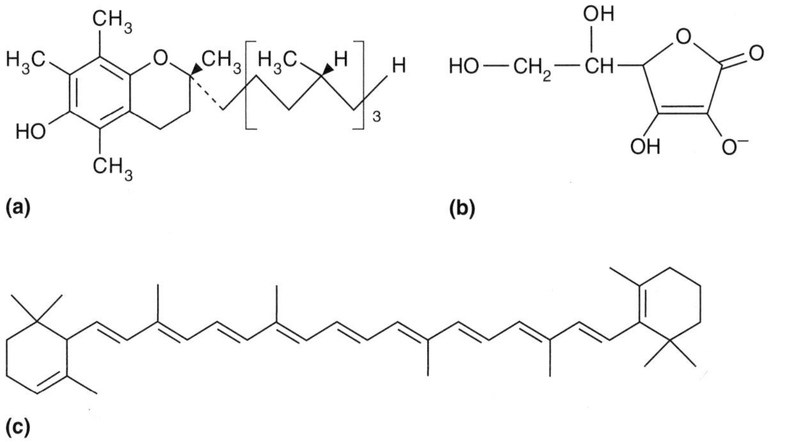

(a) (b) (c)

FIGURE 10.23

Selected Antioxidant Molecules

(a) α-Tocopherol (vitamin E). (b) Ascorbate (vitamin C). (c) β-Carotene.

α-Tocopherol, a potent radical scavenger, belongs to a class of compounds referred to as *phenolic antioxidants.* Phenols are effective antioxidants because the radical products of these molecules are resonance stabilized and thus relatively unreactive:

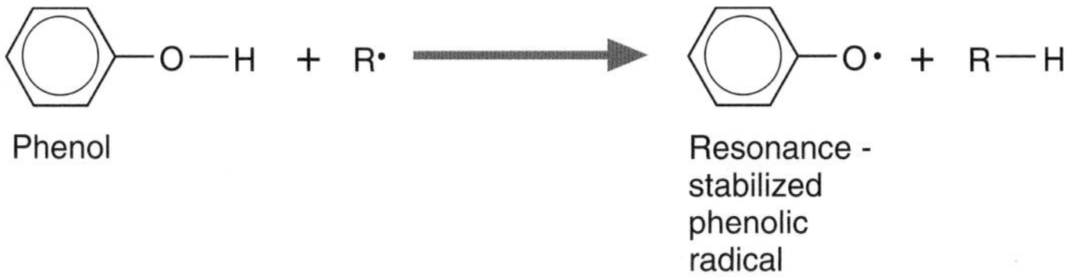

Phenol–O–H + R• → Phenol–O• + R–H (13)

Phenol; Resonance-stabilized phenolic radical

Because vitamin E (found in vegetable and seed oils, whole grains, and green, leafy vegetables) is lipid soluble, it plays an important role in protecting membranes from lipid peroxyl radicals.

β-Carotene, found in yellow-orange and dark green fruits and vegetables such as carrots, sweet potatoes, broccoli, and apricots, is a member of a class of plant pigment molecules referred to as the *carotenoids.* In plant tissue the carotenoids absorb some of the light energy used to drive photosynthesis and protect against the ROS that form at high light intensities. In animals, β-carotene is a precursor of retinol (vitamin A) and an important antioxidant in membranes. (Retinol is a precursor of retinal, the light-absorbing pigment found in the rod cells of the retina.)

Ascorbic acid is an efficient antioxidant. Present largely as ascorbate, this water-soluble molecule scavenges a variety of ROS within the aqueous compartments of cells and in extracellular fluids. Ascorbate is reversibly oxidized as shown:

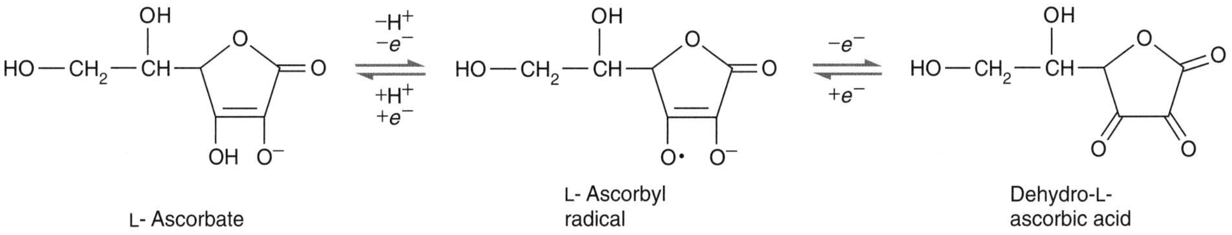

L-Ascorbate $\underset{+H^+, +e^-}{\overset{-H^+, -e^-}{\rightleftharpoons}}$ L-Ascorbyl radical $\underset{+e^-}{\overset{-e^-}{\rightleftharpoons}}$ Dehydro-L-ascorbic acid (14)

Ascorbate protects membranes through two mechanisms. First, ascorbate prevents lipid peroxidation by reacting with peroxyl radicals formed in the cytoplasm before they can reach the membrane. Second, ascorbate enhances the antioxidant activity of vitamin E by regenerating reduced α-tocopherol from the α-tocopheroxyl radical (**Figure 10.24**). Ascorbate is then regenerated by reacting with GSH.

LOOH

Tocopheroxyl radical

L-Ascorbate

LOO•

α-Tocopherol

Ascorbyl radical

FIGURE 10.24

Regeneration of α-Tocopherol by L-Ascorbate

L-Ascorbate, a water-soluble molecule, protects membranes from oxidative damage by regenerating α-tocopherol from α-tocopheroxyl radical. The ascorbyl radical formed in this process is reconverted to L-ascorbate during a reaction with GSH.

It is noteworthy that in well-nourished individuals, the consumption of excessive amounts of antioxidant supplements renders the body's cells more vulnerable to oxidative stress. In small quantities, ROS act as signal molecules. When cells are experiencing oxidative stress (e.g., infection or inflammation), ROS levels begin to rise. Early in this process, ROS oxidize and/or covalently modify the sulfur groups of transcription factors, thereby triggering the expression of dozens of genes that strengthen the antioxidant defenses of the cell. In addition to increased concentrations of catalase, SOD, and other antioxidant enzymes, other stress proteins are produced. If cells contain excessive amounts of antioxidant molecules, obtained from dietary supplements, ROS-triggered defense mechanisms are compromised.

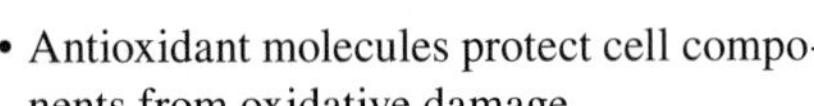

KEY CONCEPTS

- Antioxidant molecules protect cell components from oxidative damage.
- Prominent antioxidants include GSH and the dietary components α-tocopherol, β-carotene, and ascorbic acid.

QUESTION 10.8

The antioxidant BHT (butylated hydroxytoluene) is widely used as a food preservative. Quercitin is a member of a large group of potent antioxidants found in fruits and vegetables called the flavonoids.

BHT

Quercitin

What structural characteristic of these molecules is responsible for their antioxidant properties?

Biochemistry IN PERSPECTIVE

Myocardial Infarct: Ischemia and Reperfusion

How are heart cells damaged by the inadequate nutrient and oxygen flow caused by blood clots, and why does the reintroduction of O_2 cause further damage? The initial tissue damage that occurs during a myocardial infarction (heart attack) is caused by *ischemia*, a condition in which there is inadequate blood flow. Heart attacks are usually caused by atherosclerosis accompanied by blood clot formation in the carotid artery. In atherosclerosis, soft masses of fatty material called *plaques* are formed in the linings of blood vessels. Unlike skeletal muscle, which is fairly resistant to ischemic injury, the heart is extremely sensitive to hypoxic (low-oxygen) conditions. The shift from oxygen-requiring fatty acid oxidation to anaerobic glycolysis, which leads to lactate production and acidosis, is an early response of cells to ischemia. Under hypoxic conditions fatty acid oxidation, which normally provides at least half of the heart's energy, is depressed. As a result, flux through glycolysis is increased because of lower levels of modulators such as citrate (pp. 288 and 326). Because energy production by glycolysis is inefficient, ATP levels begin to fall. As they do so, adenine nucleotides are degraded to form hypoxanthine (Chapter 15).

Without sufficient ATP, cardiac cells cannot maintain appropriate intracellular ion concentrations. For example, cytoplasmic calcium levels rise as a result of the depressed activity of the plasma membrane's Ca^{2+} ATPase. One of the consequences of this circumstance is the activation of calcium-dependent enzymes such as proteases and phospholipases (enzymes that degrade the phospholipids in membranes). As osmotic pressure increases, affected cells swell and leak their contents. The blood supply is further compromised as neutrophils, attracted to the damaged site via chemotaxis, clog the blood vessels. Eventually, lysosomal enzymes begin to leak from lysosomes. Because lysosomal enzymes are active only at low pH values, their presence in an increasingly acidic cytoplasm eventually results in the hydrolysis of cell components.

ER stress is another important feature of hypoxic tissue. In normal circumstances, the ER is an oxidizing environment that promotes protein folding and disulfide bond formation. In addition, a high level of calcium ions supports the functions of the ER molecular chaperones. Under hypoxic conditions protein folding is compromised and the unfolded protein response (UPR, p. 48) is triggered. If oxygen deprivation is not prolonged, UPR-triggered stress genes may allow an adaptive response that protects affected cells from some of the damage caused by the reintroduction of O_2. Severe or prolonged ER stress, however, may result in cell death.

The reoxygenation of an ischemic tissue, a process referred to as *reperfusion*, can be a life-saving therapy. For example, the use of streptokinase to digest artery-occluding clots in heart attack patients, accompanied by administration of oxygen, has saved many lives. However, depending on the duration of the hypoxic episode, the reintroduction of oxygen to ischemic tissue may also result in further damage.

Reperfusion injury, cellular damage caused by the reestablishment of the blood supply, results from a combination of numerous factors. Among the most important of these are ROS production and the opening of the mitochondrial permeability transition pore (mPTP). Upon the reintroduction of O_2 and nutrients, ROS are first generated by reenergized mitochondrial ETC and soon by NADPH oxidase and xanthine oxidase (from neutrophils attracted to the injury site). In addition, the release of iron from cell components such as myoglobin, which can result from ROS-inflicted damage, can cause additional production of •OH. Finally, the acidosis caused by lactate accumulation in compromised heart muscle cells unloads abnormally high amounts of oxygen from hemoglobin. This latter condition greatly facilitates further ROS synthesis. Reperfusion also promotes the synthesis of large amounts of •NO, which react with superoxide to produce damaging peroxynitrite. NO• also modifies oxidizable cysteine residues in the Ca^{2+} channels of the SR (smooth ER in muscle cells), another factor contributing to ER stress. mPTP, a nonspecific channel that forms where the inner and outer mitochondrial membranes meet, allows the passage of molecules smaller than 1500 Da. mPTP opening is triggered by ROS in combination with low ATP levels and high calcium levels (caused in part by ROS-inflicted damage of the SR Ca^{2+}-ATPase). mPTP opening leads to mitochondrial membrane potential collapse and mitochondrial swelling (caused by osmotic pressure). The release of cyt c from damaged mitochondria leads to apoptosis (p. 58).

SUMMARY: Damage to heart cells as a result of oxygen deprivation originates with inefficient energy production, followed by osmotic pressure increases, lysosomal breakage, ER stress, and high cytoplasmic calcium levels. The reperfusion of damaged cells with O_2 leads to ROS formation, causing further damage.

Chapter Summary

1. Dioxygen (O_2), generally referred to as oxygen, is used by aerobic organisms as a terminal electron acceptor in energy generation. Several physical and chemical properties of oxygen make it suitable for this role. In addition to its ready availability (it occurs almost everywhere on the Earth's surface), oxygen diffuses easily across cell membranes. Oxygen is a reactive diradical and an excellent oxidizing agent, readily accepting electrons from other species.
2. The NADH and $FADH_2$ molecules produced in glycolysis, the β-oxidation pathway, and the citric acid cycle generate usable energy in the electron transport pathway. The pathway consists of a series of redox carriers that receive electrons from NADH and $FADH_2$. At the end of the pathway the electrons, along with protons, are donated to oxygen to form H_2O.
3. During the oxidation of NADH, there are three steps in which the energy loss is sufficient to account for ATP synthesis. These steps occur within complexes I, III, and IV of the ETC.
4. Oxidative phosphorylation is the mechanism by which electron transport is coupled to the synthesis of ATP. According to the chemiosmotic theory, the creation of a proton gradient that accompanies electron transport is coupled to ATP synthesis.
5. The complete oxidation of a molecule of glucose results in the synthesis of 29.5 to 31 molecules of ATP, depending on whether the glycerol phosphate shuttle or the malate-aspartate shuttle transfers electrons from cytoplasmic NADH to the mitochondrial ETC.
6. The use of oxygen by aerobic organisms is linked to the production of ROS. These species form because the diradical oxygen molecule accepts electrons one at a time. Examples of ROS include the superoxide radical, hydrogen peroxide, the hydroxyl radical, and singlet oxygen. Prominent RNS include nitric oxide, nitrogen dioxide, and peroxynitrite.

Take your learning further by visiting the **companion website** for Biochemistry at **www.oup.com/us/mckee** where you can complete a multiple-choice quiz on electron transport and oxidative phosphorylation to help you prepare for exams.

Suggested Readings

Mailloux, R. J., McBride, S. L., and Harper, M-L., Unearthing the Secrets of Mitochondrial ROS and Glutathione in Bioenergetics, *Trends Biochem. Sci.* 38(12):592–602, 2013.

Minamino, T., and Kitakaze, M., ER Stress in Cardiovascular Disease, *J. Mol. Cell. Cardiol.* 48(6):1105–1110, 2010.

Nicholls, D. G., and Ferguson, S. J., *Bioenergetics 4*, Academic Press, London, 2013.

Parlakpinar, H., and Orum, M. H., Pathophysiology of Myocardial Ischemia Reperfusion Injury: A Review, *Med-Science* 2(4):935–954, 2013.

Vijg, J. , Aging Genomes: A Necessary Evil in the Logic of Life, *Bioessays* 36(3):282–292, 2014.

Winyard, P. G., Moody, C. J., and Jacob, C., Oxidative Activation of Antioxidant Defense, *Trends Biochem. Sci.* 30(8):453–461, 2005.

Key Words

antioxidant, *367*
β-carotene, *375*
chemiosmotic coupling theory, *358*
glycerol phosphate shuttle, *364*
ionophore, *359*
malate-aspartate shuttle, *364*
oxidative phosphorylation, *358*
oxidative stress, *367*
protonmotive force, *358*
Q cycle, *354*
radical, *368*
reactive nitrogen species (RNS), *372*
reactive oxygen species (ROS), *367*
respirasome, *357*
respiratory burst, *368*
respiratory control, *362*
α-tocopherol, *375*
uncoupler, *359*
uncoupling protein, *366*

Review Questions

These questions are designed to test your knowledge of the key concepts discussed in this chapter before moving on to the next chapter. You may like to compare your answers to the solutions provided in the back of the book and in the accompanying Study Guide.

1. Define the following terms:
   a. electron transport
   b. ETC
   c. Q cycle
   d. proton motive force
   e. UQH_2
2. Define the following terms:
   a. Complex I
   b. Complex II
   c. Complex III
   d. Complex IV
   e. respirasome
3. Define the following terms:
   a. chemiosmotic coupling theory
   b. uncoupler
   c. ionophore
   d. submitochondrial particle
   e. α,β hexamer
4. Define the following terms:
   a. stator
   b. rotor
   c. torque
   d. ATP synthase
   e. respiratory control
5. Define the following terms:
   a. cytochrome
   b. glycerol phosphate shuttle
   c. malate-aspartate shuttle
   d. uncoupling protein
   e. oligomycin
6. Define the following terms:
   a. reactive oxygen species
   b. antioxidant
   c. oxidative stress
   d. respiratory burst
   e. reactive nitrogen species
7. Define the following terms:
   a. superoxide dismutase
   b. peroxiredoxin
   c. thioredoxin
   d. β-carotene
   e. ascorbate
8. Define the following terms:
   a. ischemia
   b. reperfusion injury
   c. MPTP
   d. α-tocopherol
   e. glutathione
9. Define the following terms:
   a. radical
   b. ROS
   c. RNS
   d. GSH
   e. SOD
10. Describe the principal features of the chemiosmotic theory.
11. The chemical coupling hypothesis (substrate-level phosphorylation) failed to explain why mitochondrial membrane must be intact during ATP synthesis. How does the chemiosmotic theory account for this phenomenon?
12. What are the principal sources of electrons for the electron transport pathway?
13. Describe the processes that are driven by mitochondrial electron transport.
14. How does dinitrophenol inhibit ATP synthesis?
15. Four protons are required to drive the phosphorylation of ADP. Account for the function of each proton in this process.
16. List several reasons why oxygen is widely used in energy production.
17. Compare the amount of energy captured from a mole of glucose from glycolysis alone with that of oxidative phosphorylation.
18. When taken in appropriate amounts, vitamin E protects the body from ROS. However, when it is taken in excessive amounts vitamin E can potentially make the body more susceptible to ROS. Explain.
19. Which of the following are reactive oxygen species? Why is each ROS dangerous?
   a. O_2
   b. OH^-
   c. RO•
   d. $O_2^{\cdot-}$
   e. CH_3OH
   f. $^1O_2$
20. Describe the types of cellular damage produced by ROS.
21. Describe the enzymatic activities used by cells to protect themselves from oxidative damage.
22. Explain how a defect in the gene for glucose-6-phosphate dehydrogenase can provide a survival advantage. [*Hint*: Visit the companion website at www.oup.com/us/mckee to read the Biochemistry in Perspective essay for Chapter 10 entitled Glucose-6-Phosphate Dehydrogenase Deficiency.]
23. What would be the end products when the following substances are final electron acceptors in an electron transport system: nitrate, ferric ion, carbon dioxide, sulfate, and sulfur?
24. What advantage does dioxygen have over the oxidizing agents in Question 23?
25. Provide the reaction equations that illustrate the production of ROS from electrons leaking from the electron transport system.
26. List some of the causes of reperfusion-triggered cardiac cell damage.
27. List the protein complexes in the mitochondrial ETC and describe their functions.

28. Valinomycin is an ionophore antibiotic that renders biological membranes permeable to K^+. Its side effects in patients with bacterial infections include a rise in body temperature and sweating. Explain.
29. Describe the mechanism whereby uncoupling agents disrupt phosphorylation.
30. The electron transport system consists of a series of oxidations rather than one reaction. Why is this an important feature of energy capture?
31. What metabolites accumulate when azide (N_3^-) is added to actively respiring mitochondria?
32. What is the minimum voltage drop for individual electron transfer events in the mitochondrial electron transport systems that is necessary for ATP synthesis?
33. When rotenone is added to actively respiring mitochondria, the ratio of NADH to NAD^+ increases, but the $FADH_2$/FAD ratio remains unchanged. What step in the system is being inhibited?
34. Describe the role of UCP1 in nonshivering thermogenesis.

Fill in the Blank

35. Complexes I and II of the electron transport system transfer electrons from NADH and succinate from ______________, respectively, to UQ.
36. The functional respiration unit within the mitochondria of plants and animals is called the ______________.
37. An electrochemical proton gradient across a membrane is referred to as the ______________ force.
38. Hydrophobic molecules that dissipate osmotic gradients by inserting themselves into a membrane and forming a channel are called ______________.
39. ______________ collapse a proton gradient by equalizing the proton gradient on both sides of a membrane.
40. The control of aerobic respiration by ADP is referred to as ______________.
41. Under conditions of ______________, the antioxidant mechanisms of cells are overwhelmed.
42. Macrophages utilize an oxygen-consuming process called the ______________ in which reactive oxygen species are generated to kill microorganisms or damaged cells.
43. A chemical species with an unpaired electron is called a ______________.
44. The major enzymatic defense against oxidative stress is provided by superoxide dismutase, glutathione, peroxidase, peroxiredoxin, and ______________.

Short Answer

45. Hypochlorite ion is a strong oxidizing agent that is produced during the respiratory burst. Explain how it is synthesized from O_2 and Cl^-.
46. What is nonshivering thermogenesis?
47. The antifungal drug nystatin kills fungal cells in part through the formation of membrane pores that cause K^+ leakage. To what class of membrane-inserting molecule does nystatin belong?
48. Explain why the combustion of a fuel such as methane is a series of free radical reactions.
49. One of the effects of low oxygen levels in cardiac cells is a rise in cytoplasmic calcium levels as a result of the depressed activity of the plasma membrane's Ca^{2+}-ATPase. What are some of the consequences of this phenomenon?

Thought Questions

These questions are designed to reinforce your understanding of all of the key concepts discussed in the book so far, including this chapter and all of the chapters before it. They may not have one right answer! The authors have provided possible solutions to these questions in the back of the book and in the accompanying Study Guide for your reference.

50. During an experiment, $^{14}CH_3$—COOH is fed to microorganisms. Trace the ^{14}C label through the citric acid cycle. How many ATP molecules can be generated from 1 mol of this substance? (The conversion of acetate to acetyl-CoA requires the consumption of 2 ATP.)
51. In some regions where malaria is endemic (e.g., the Middle East), fava beans are a staple food. Fava beans are now known to contain two β-glycosides called vicine and convicine:

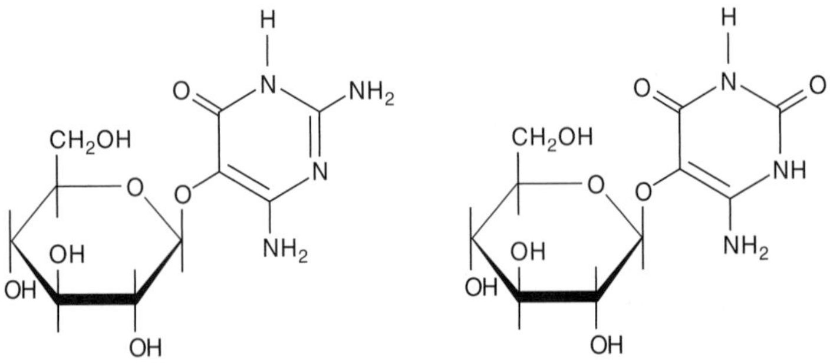

Vicine

Convicine

It is believed that the aglycone components of these substances, called divicine and isouramil, respectively, can oxidize GSH. Individuals who eat fresh fava beans are protected to a certain extent from malaria. A condition known as *favism* results when some glucose-6-phosphate dehydrogenase–deficient individuals develop a severe hemolytic anemia after eating the beans. Explain why.

52. Ethanol is oxidized in the liver to form acetate, which is converted to acetyl-CoA. Determine how many molecules of ATP are produced from 1 mol of ethanol. Note that 2 mol of NADH are produced when ethanol is oxidized to form acetate.
53. Glutamine is degraded to form NH_4^+, CO_2, and H_2O. How many molecules of ATP can be generated from 1 mol of this amino acid? (Removal of the amino group yields one molecule of NADH.)
54. Consumption of dinitrophenol by animals results in an immediate increase in body temperature. Explain the phenomenon. Why should this decoupler not be used as a diet aid?
55. The glutaridoxins are a class of small antioxidant enzymes that use GSH as a reducing agent. They have functions similar to those of the thioredoxins. Describe the pathway whereby an organic peroxide is reduced by a glutaridoxin. Include the redox cycle of GSH in your answer.
56. According to the chemiosmotic theory, what would be the effect on oxidative phosphorylation of allowing other positive ions to diffuse across the inner mitochondrial membrane?
57. Cyanide causes an irreversible inhibition of electron transport that prevents ATP synthesis, whereas the inhibitory effect of small amounts of dinitrophenol on ATP synthesis is reversible. Explain the difference.
58. The reduction potentials of the iron in each of the cytochromes in the electron transport system vary from −0.1 V to −0.39 V. Explain why these different values are necessary for the operation of the system.
59. Explain why an inhibitor of complex I will cause not only an increase in the ratio of NADH to NAD^+ but also an increase in the UQ/UQH_2 ratio.
60. Suppose that the cytochrome complexes were not embedded within the mitochondrial inner membrane. According to the chemiosmotic theory, what would be the consequences?
61. Explain why rotenone inhibits oxidative phosphorylation when the substrate is pyruvate, but not when succinate is used.
62. In a major class of peroxiredoxins, referred to as the 2-cys Prxs, two cysteines in the active site are involved in the catalytic mechanism. During the reaction, one of the cysteines is oxidized to a sulfenic acid by the substrate peroxide to yield an alcohol product. The sulfenic acid group then reacts with the sulfhydryl group of the other cysteine residue to yield a disulfide bond and a molecule of water. Provide the pathway of this reaction. Include the regeneration of the two reduced cysteine residues in your answer.
63. If nitrate is used as the terminal electron acceptor in the electron transport system, how many ATPs could be synthesized? [*Hint*: Refer to the reduction potential differences between NADH and nitrate (**Table 9.1**).]
64. Dehydroascorbate is unstable at pH values greater than 6 and decomposes to form tartrate and oxalate. Cells use GSH to reduce the loss of ascorbate. What is the reaction pathway for the regeneration of ascorbate?
65. Among the many destructive consequences of oxidative stress are reactions of •OH with polypeptide backbone atoms. The process begins with the abstraction of α-hydrogen atoms to form carbon radicals.

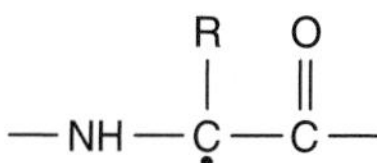

Subsequently, such radicals react with O_2 to form alkylperoxyl radicals (ROO•). Starting with the undamaged backbone atoms and O_2, describe the pathway that results in the formation of an alkylperoxyl radical.

66. Polypeptides under oxidative attack can form intra- and intermolecular cross-links. Referring to Question 65, suggest one mechanism whereby a cross-link could form.

CHAPTER

11 Lipids and Membranes

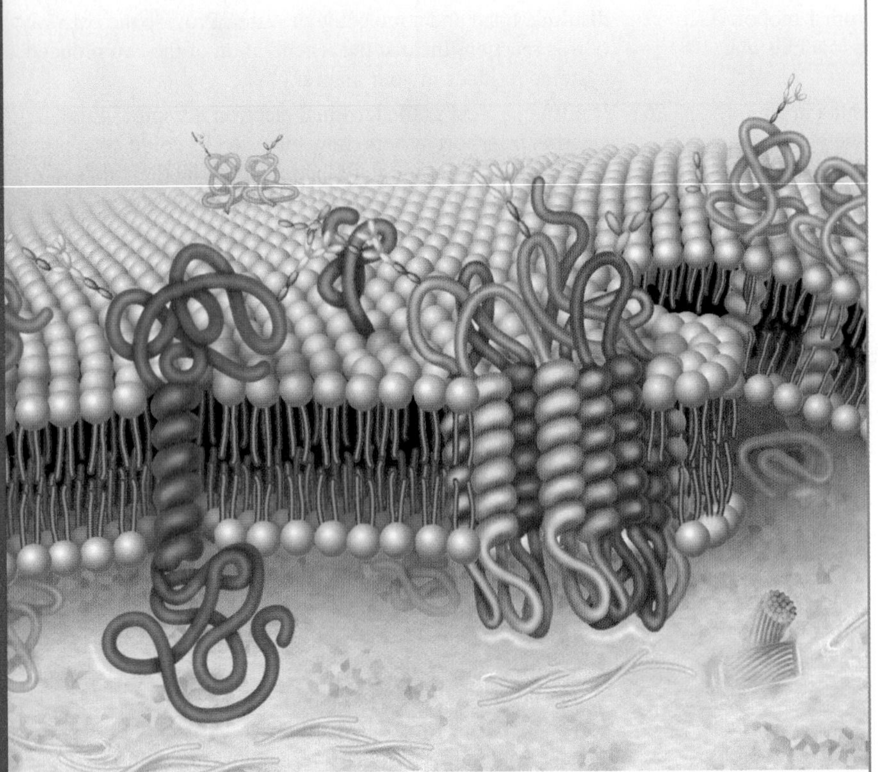

Biological Membrane
A biological membrane is a dynamic compartmental barrier composed of a lipid bilayer noncovalently complexed with proteins, glycoproteins, glycolipids, and cholesterol.

OUTLINE

The Low-Fat Diet

Obesity is a worldwide phenomenon. The World Health Organization estimates that there are currently more than 300 million clinically obese adults, with 700 million more described as overweight. These are remarkable statistics considering that obesity was relatively rare before the twentieth century. The obesity rate in the United States was stable at about 12–14% until the late 1970s, when the rate began to rise. By the late 1980s approximately 25% of American adults were obese or significantly overweight. This number is now over 50%. The current epidemic, and the chronic diseases associated with it (e.g., diabetes mellitus, cardiovascular disease, and certain forms of cancer), has been linked to the ready availability of cheap, energy-dense, and nutrient-deficient processed food and a shift toward physically less demanding lifestyles. Throughout this period there has been an overwhelming amount of dietary advice, most notably the promotion, beginning in the late 1970s, of low-fat diets. Unfortunately, because of the unpalatable nature of low-fat food, carbohydrates (usually sucrose or high-fructose corn syrup) soon replaced the fat content. Apparently, many consumers believed that low-fat foods could be eaten in excess without considering that the caloric content of these foods equals that of the fat-containing foods they replaced.

Although low-fat diets are still popular, they have proven to be ineffective for achieving sustained weight loss. Health problems associated with low-fat diets, however, do not end with weight gain in obesity-prone individuals. The original nutritional recommendation that diets should be low in saturated fats was often interpreted as low in all fats. Consequently, without realizing the diverse functions of dietary fats, many weight-onscious individuals reduced or eliminated almost all fats from their diets. Among the consequences of an extreme low-fat diet (15% or less) are deficiencies in the fat-soluble vitamins (A, D, E, and K) and the essential fatty acids linoleic and linolenic acids. The fat-soluble vitamins are important for numerous physiological processes (e.g., growth, immunity, cell repair, and blood clotting). The essential fatty acids (EFAs), so called because the body's cells cannot make them, perform a wide variety of functions. In addition to being important constituents of cell membranes, the EFAs are also converted into biologically active derivatives with roles in immunity, inflammatory responses, and nervous system function. EFA deficiencies in adults have been associated with dry scaly skin, brittle hair, fatigue, high blood pressure, atherosclerosis, depressed immunity, poor wound healing, depression, and cravings for fatty foods. In addition to inadequate growth, EFA deficiencies in children have been linked to impaired brain development with symptoms that include hyperactivity, attention deficit, and aggressive behavior.

Ironically, recent research indicates that modest amounts of healthy dietary fat not only provide the body with required nutrients, but also promote satiety (the perception of fullness that inhibits further eating behavior). Among the most intriguing satiety-inducing mechanisms is the dietary fat–induced release of endorphins by the central nervous system. Endorphins are endogenous opiate-like proteins that are partly responsible for the pleasure that results from the consumption of delicious food.

The substitution of sugar for fat in processed foods, the other unforeseen result of the low-fat recommendation, had even more serious consequences for human health. The metabolic mechanisms whereby dietary sugar contributes to obesity, hypertension, and atherosclerosis are outlined in Chapter 16.

Overview

LIPIDS ARE NATURALLY OCCURRING WATER-INSOLUBLE SUBSTANCES THAT PERFORM A STUNNING ARRAY OF FUNCTIONS IN LIVING ORGANISMS. Some lipids are vital energy reserves. Others are the primary structural components of biological membranes. Still other lipid molecules act as hormones, antioxidants, pigments, or vital growth factors and vitamins. This chapter describes the structures and properties of the major lipid classes found in living organisms as well as the structural and functional properties of biomembranes.

Lipids are a diverse group of biomolecules. Because of this diversity, the term **lipid** has an operational rather than a structural definition. Lipids are defined as substances from living organisms that dissolve in nonpolar solvents such as ether, chloroform, and acetone but not appreciably in water. The functions of lipids are also diverse. Several types of lipid molecules are important structural components in cell membranes. Another type, the fats and oils, store energy efficiently. Other lipid molecules are chemical signals, vitamins, or pigments. Finally, some lipid molecules that occur in the outer surfaces of various organisms have protective or waterproofing functions.

Chapter 11 describes the structure and function of each major type of lipid and discusses the lipoproteins, complexes of protein and lipid that transport lipids in animals. Chapter 11 ends with an overview of membrane structure and function. In Chapter 12 the metabolism of several major lipids is described.

11.1 LIPID CLASSES

Lipids may be classified in many different ways. For this discussion, lipids can be subdivided into the following classes:

1. Fatty acids
2. Triacylglycerols
3. Wax esters
4. Phospholipids (phosphoglycerides and sphingomyelin)
5. Sphingolipids (molecules other than sphingomyelin that contain the amino alcohol sphingosine)
6. Isoprenoids (molecules made up of repeating isoprene units, a branched five-carbon hydrocarbon)

Each class is discussed.

Fatty Acids

Fatty acids are monocarboxylic acids that typically contain hydrocarbon chains of variable lengths (between 12 and 20 or more carbons) (**Figure 11.1**). Fatty acids are numbered from the carboxylate end. Greek letters are used to designate certain carbon atoms. The α-carbon in a fatty acid is adjacent to the carboxylate group, the β-carbon is two atoms removed from the carboxylate group, and so forth. The terminal methyl carbon atom is designated the omega (ω) carbon. **Table 11.1** gives the structures, names, and standard abbreviations of several common fatty acids. Fatty acids are important components of several types of lipid molecules. They occur primarily in triacylglycerols and several types of membrane-bound lipid molecules.

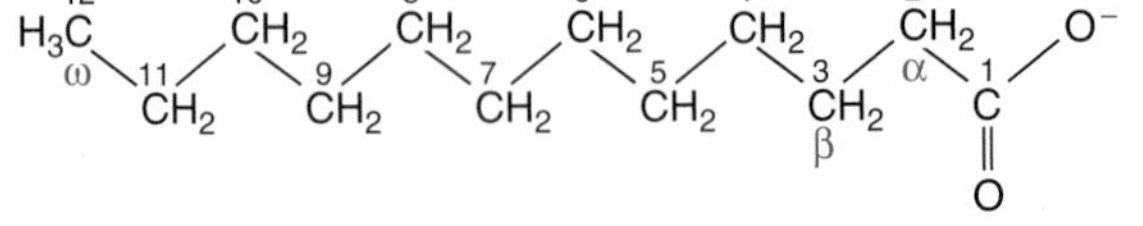

FIGURE 11.1

Fatty Acid Structure

Fatty acids consist of a long-chain hydrocarbon covalently bonded to a carboxylate group. The lipid shown is dodecanoic acid (common name, lauric acid), a 12-carbon saturated fatty acid (12:0).

TABLE 11.1 Examples of Fatty Acids

Common Name	Structure	Abbreviation
Saturated Fatty Acids		
Myristic acid	$CH_3(CH_2)_{12}COOH$	14:0
Palmitic acid	$CH_3(CH_2)_{12}CH_2CH_2COOH$	16:0
Stearic acid	$CH_3(CH_2)_{12}CH_2CH_2CH_2CH_2COOH$	18:0
Arachidic acid	$CH_3(CH_2)_{12}CH_2CH_2CH_2CH_2CH_2CH_2COOH$	20:0
Lignoceric acid	$CH_3(CH_2)_{12}CH_2CH_2CH_2CH_2CH_2CH_2CH_2CH_2CH_2CH_2COOH$	24:0
Cerotic acid	$CH_3(CH_2)_{12}CH_2CH_2CH_2CH_2CH_2CH_2CH_2CH_2CH_2CH_2CH_2CH_2COOH$	26:0
Unsaturated Fatty Acids		
Palmitoleic acid	$CH_3(CH_2)_5CH{=}CH(CH_2)_7COOH$	$16{:}1^{\Delta 9}$
Oleic acid	$CH_3(CH_2)_7CH{=}CH(CH_2)_7COOH$	$18{:}1^{\Delta 9}$
Linoleic acid	$CH_3(CH_2)_4CH{=}CH-CH_2-CH{=}CH(CH_2)_7COOH$	$18{:}2^{\Delta 9,12}$
α-Linolenic acid	$CH_3CH_2CH{=}CH-CH_2-CH{=}CH-CH_2-CH{=}CH(CH_2)_7COOH$	$18{:}3^{\Delta 9,12,15}$
γ-Linolenic acid	$CH_3(CH_2)_3-(CH_2-CH{=}CH)_3-(CH_2)_4-COOH$	$18{:}3^{\Delta 6,9,12}$
Arachidonic acid	$CH_3(CH_2)_3-(CH_2-CH{=}CH)_4-(CH_2)_3COOH$	$20{:}4^{\Delta 5,8,11,14}$

Most naturally occurring fatty acids have an even number of carbon atoms that form an unbranched chain. (Unusual fatty acids with branched or ring-containing chains are found in some species.) Fatty acid chains that contain only carbon-carbon single bonds are referred to as *saturated*. Those molecules that contain one or more double bonds are said to be *unsaturated*. Double bonds are rigid structures so molecules that contain them can occur in two isomeric forms: *cis* and *trans*. In *cis* isomers, similar or identical groups are on the same side of a double bond (**Figure 11.2a**). When such groups are on opposite sides of a double bond, the molecule is said to be a *trans*-isomer (**Figure 11.2b**). The double bonds in most naturally occurring fatty acids are in a *cis* configuration. The presence of a *cis* double bond causes an inflexible "kink" in a fatty acid chain (Figure 11.3). Because of this structural feature, unsaturated fatty acids do not pack as closely together as saturated fatty acids. Less energy is required to disrupt the intermolecular forces between unsaturated fatty acids. Therefore, they have lower melting points and are liquids at room temperature. For example, a sample of palmitic acid (16:0), a saturated fatty acid, melts at 63°C, whereas palmitoleic acid ($16{:}1^{\Delta 9}$) melts at 0°C. Note that in the abbreviations for specific fatty acids, the number to the left of the colon is the total number of carbon atoms, and the number to the right is the number of double bonds. A superscript denotes the placement of a double bond. For example, Δ9 signifies that there are eight carbons between the carboxyl group and the double bond; that is, the double bond occurs between carbons 9 and 10.

Unsaturated fatty acids are also classified according to the location of the first double bond relative to the terminal methyl (omega, ω) end of the molecule.

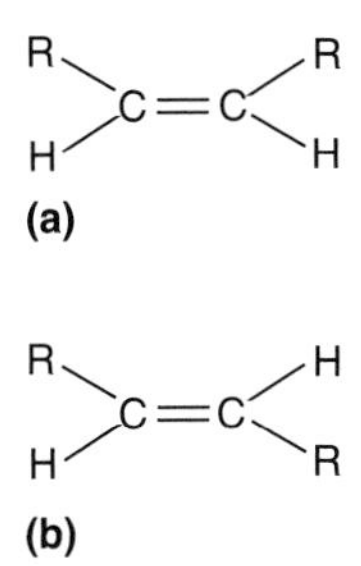

(a) R(H)C=C(H)R (cis)

(b) R(H)C=C(R)H (trans)

FIGURE 11.2

Isomeric Forms of Unsaturated Molecules

In *cis*-isomers (a) both R groups are on the same side of the carbon-carbon double bond. In *trans*-isomers (b), the R groups are on different sides.

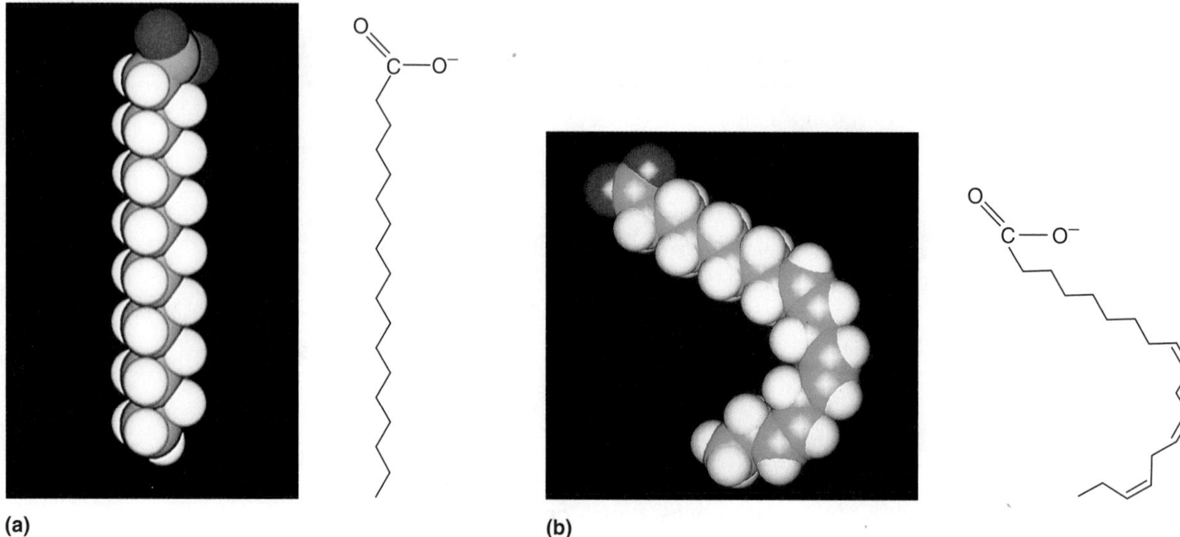

FIGURE 11.3

Space-Filling and Conformational Models

(a) A saturated fatty acid (stearic acid) and (b) an unsaturated fatty acid (α-linolenic acid). (Green spheres = carbon atoms; white spheres = hydrogen atoms; red spheres = oxygen atoms.)

For example, linoleic acid and α-linolenic acid can be designated as 18:2ω-6 (equivalent to $18{:}2^{\Delta 9,12}$) and 18:3ω-3 (equivalent to $18{:}3^{\Delta 9,12,15}$), respectively. (The number to the right of ω designates the carbon at which the first double bond occurs, counting from the methyl end of the fatty acid. Sequential double bonds are always three carbons apart.) It is noteworthy that fatty acids with *trans* double bonds have three-dimensional structures similar to those of saturated fatty acids. In addition, the presence of one or more double bonds in a fatty acid makes it susceptible to oxidation (**Figure 10.19**). The consequences include the effects of oxidative stress on cell membranes and the tendency of oils to become rancid (i.e., they contain unpleasant smelling or tasting short-chain organic acids).

Fatty acids with one double bond are referred to as **monounsaturated** molecules. When two or more double bonds occur in fatty acids, usually separated by methylene groups (—CH_2—), they are referred to as **polyunsaturated**. The monounsaturated fatty acid oleic acid ($18{:}1^{\Delta 9}$) and the polyunsaturated linoleic acid ($18{:}2^{\Delta 9,12}$) are among the most abundant fatty acids in living organisms.

Organisms such as plants and bacteria can synthesize all the fatty acids they require from acetyl-CoA (Chapter 12). Mammals obtain most of their fatty acids from dietary sources, but they can synthesize saturated fatty acids and some monounsaturated fatty acids. Mammals can also modify some dietary fatty acids by adding two-carbon units and introducing some double bonds. Fatty acids that can be synthesized are called **nonessential fatty acids**. Because mammals do not possess the enzymes required to synthesize linoleic ($18{:}2^{\Delta 9,12}$) and α-linolenic ($18{:}2^{\Delta 9,12,15}$) acids, these **essential fatty acids** must be obtained from the diet.

Linoleic acid ($18{:}2^{\Delta 9,12}$ or 18:2ω-6) is the precursor for numerous derivatives, formed by elongation and/or desaturation reactions. Prominent examples include γ-linolenic acid ($18{:}3^{\Delta 6,9,12}$ or 18:3ω-6), arachidonic acid ($20{:}4^{\Delta 5,8,11,14}$ or 20:4ω-6), and docosapentaenoic acid ($22{:}5^{\Delta 4,7,10,13,16}$ or 22:5ω-6) (DPA). Together, linoleic acid and its derivatives are referred to as the **omega-6 fatty acids**. Food sources include various vegetable oils (e.g., sunflower and soybean oils), eggs, and poultry. α-Linolenic acid ($18{:}3^{\Delta 9,12,13}$ or 18:3ω-3) and its

derivatives such as eicosapentaenoic acid (20:5$^{\Delta 5,8,11,14,17}$ or 20:5ω-3) (EPA) and docosahexaenoic acid (22:6$^{\Delta 4,7,10,13,16,19}$ or 22:6ω-3) (DHA) are referred to as the **omega-3 fatty acids**. Sources of α-linolenic include flaxseed and soybean oils, and walnuts. EPA and DHA, also found in fish and fish oils (e.g., salmon, tuna, and sardines), are now believed to promote cardiovascular health. Effects associated with diets with adequate amounts of these two fatty acids include lower blood levels of triacylglycerols (triglycerides), lower blood pressure, and decreased platelet aggregation. Essential fatty acids are used as structural components (e.g., phospholipids in membranes) and as precursors for several important metabolites. Prominent examples of the latter include the eicosanoids and anandamine.

The eicosanoids are hormonelike molecules derived from omega-6 or omega-3 fatty acids. In general, omega-6-derived eicosanoids promote inflammation, whereas omega-3 derivatives are less inflammatory. The ratio of omega-6 to omega-3 fatty acids in the diet influences the relative amounts of inflammatory and anti-inflammatory eicosanoids that are synthesized. It is currently believed that 1:1 to 1:4 ratios are healthy. In many developed countries typical diets provide 10:1 to 30:1 ratios. These ratios favor a net increase in unfavorable inflammatory reactions in the body, an undesirable condition that increases the risk of chronic disease.

Anandamine (*N*-arachidonyl ethanolamine), a derivative of arachidonic acid, is a Δ9-*endocannabinoid*, a substance produced in the body that binds to the same receptor as tetrahydrocannabinol, a psychoactive drug. Anandamine acts as a neurotransmitter in the central and peripheral nervous systems, where it affects feeding and sleep behavior, short-term memory, and pain relief.

Fatty acids have several important chemical properties. The reactions that they undergo are typical of short-chain carboxylic acids. For example, fatty acids react with alcohols to form esters:

$$R{-}\overset{\overset{\displaystyle O}{\|}}{C}{-}OH + R'{-}OH \rightleftharpoons R{-}\overset{\overset{\displaystyle O}{\|}}{C}{-}O{-}R' + H_2O \quad (1)$$

This reaction is reversible; that is, under appropriate conditions a fatty acid ester can react with water to produce a fatty acid and an alcohol. Unsaturated fatty acids with double bonds can undergo hydrogenation reactions to form saturated fatty acids. Finally, as described (**Figure 10.19**), unsaturated fatty acids are susceptible to oxidation.

Certain fatty acids are covalently attached to a wide variety of eukaryotic proteins. Such proteins are referred to as *acylated* proteins. Fatty acid groups (called **acyl groups**) clearly facilitate the interactions between membrane proteins and their hydrophobic environment. Myristoylation and palmitoylation, the most common forms of protein acylation, are now known to influence a variety of structural and functional properties of proteins. Promotion of protein binding to membranes is a prominent example. In addition, hydrophobic fatty acid molecules are transported from fat cells to body cells by means of the acylation of water-soluble serum proteins.

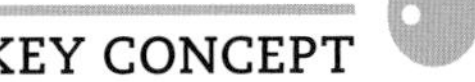

KEY CONCEPT

Fatty acids are monocarboxylic acids, most of which are found in triacylglycerol molecules, several types of membrane-bound lipid molecules, or acylated membrane proteins.

The Eicosanoids

The **eicosanoids**, produced in most mammalian tissues, include the prostaglandins, thromboxanes, and leukotrienes (**Figure 11.4**). Together the eicosanoids mediate a wide variety of physiological processes, including smooth muscle contraction, inflammation, pain perception, and blood flow regulation. Eicosanoids are also implicated in several diseases such as myocardial infarction and

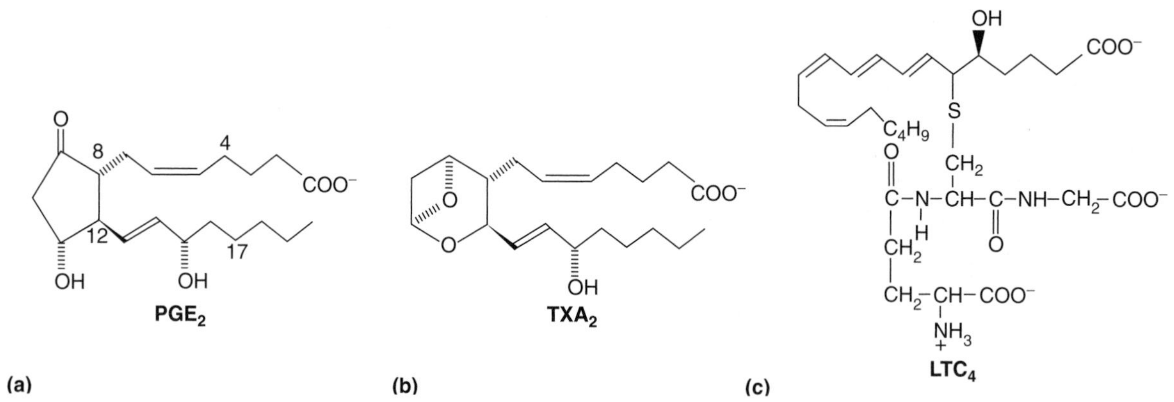

FIGURE 11.4
Eicosanoids

(a) Prostaglandin E_2. (b) Thromboxane A_2. (c) Leukotriene C_4. Note that LTC_4 has a glutathione substituent.

rheumatoid arthritis. Because they are generally active within the cell in which they are produced, the eicosanoids are called **autocrine** regulators.

Eicosanoids, which are usually designated by their abbreviations, are named according to the following system. The first two letters indicate the type of eicosanoid (PG = prostaglandin, TX = thromboxane, LT = leukotriene). The third letter identifies the type of modification made to the parent compound of the eicosanoid (e.g., A = hydroxyl group and an ether ring, B = two hydroxyl groups). The number in an eicosanoid name indicates the number of double bonds in the molecule. Eicosanoids are extremely difficult to study because they are active for short periods (often measured in seconds or minutes). In addition, they are produced only in small amounts.

Eicosanoids are derived from either arachidonic acid or EPA. Production of eicosanoids begins after arachidonic acid or EPA is released from membrane phospholipid molecules by the enzyme phospholipase A_2. A brief overview of each class of eicosanoid is provided next.

Prostaglandins contain a cyclopentane ring and hydroxy groups at C-11 and C-15. Molecules belonging to the E series of prostaglandins have a carbonyl group at C-9, whereas F series molecules have an OH group at the same position. The 2 series, derived from arachidonic acid, appears to be the most important group of prostaglandins in humans. EPA is the precursor of the 3 series of prostaglandins. Prostaglandins are involved in inflammation, an infection-fighting process that produces pain and fever, reproduction (e.g., ovulation and uterine contractions during conception and labor), and digestion (e.g., inhibition of gastric secretion). Biosynthesis of the 2 series of prostaglandins begins once arachidonic acid has been released from a membrane phosphoglyceride. Arachidonic acid is first converted into PGH_2, a precursor of several prostaglandins, by cyclooxygenase and then a peroxidase. (Aspirin relieves minor pain and reduces fever and inflammation because it inactivates cyclooxygenase by acetylating a critical serine residue in the enzyme's active site.) PGH_2 is converted to PGE_2 (**Figure 11.4a**), a fever-inducing prostaglandin, by prostaglandin endoperoxidase isomerase.

The **thromboxanes** are also derivatives of arachidonic acid or EPA. They differ from other eicosanoids in that their structures have a cyclic ether. TXA_2 (**Figure 11.4b**) is produced from arachidonic acid primarily by platelets. Thromboxane A synthase converts PGH_2 to TXA_2. Once platelets are activated, they release TXA_2, which promotes platelet aggregation and vasoconstriction following tissue injury. TXA_2 is rapidly converted to its inactive metabolite TXB_2 by an isomerase.

The **leukotrienes** are linear (noncyclic) molecules whose synthesis is initiated by a peroxidation reaction catalyzed by lipoxygenase. The leukotrienes differ in the position of this peroxidation step and the nature of the thioether group attached near the site of peroxidation. The leukotrienes are powerful chemotactic agents (i.e., they attract immune system cells to damaged tissue); they also induce vasoconstriction and bronchoconstriction (caused by smooth muscle contraction in blood vessels and air passages in the lungs, respectively). Leukotrienes LTC_4, LTD_4, and LTE_4 have been identified as components of slow-reacting substance of anaphylaxis (SRS-A). *Anaphylaxis*, an unusually severe allergic reaction, is triggered when an allergen binds to an IgE antibody on mast cells that occur in connective tissue all over the body. The mast cells then release granules that contain leukotrienes as well as other inflammatory substances (heparin, histamine, and prostaglandins) into the surrounding tissue. Symptoms of anaphylaxis may include itching, hives, and swelling. In anaphylactic shock, the inflammatory process is so severe that circulatory system collapse and suffocation caused by bronchial swelling may occur. The SRS-A molecules are derived from LTA_4, which is produced when 5-lipo-oxygenase introduces a peroxide group into arachidonic acid. The product of this reaction is then converted into an epoxide by LTA_4 synthase. LTA_4 is then converted to LTC_4 (**Figure 11.4c**) with the addition of the γ-glutamylcysteinylglycine tripeptide GSH (p. 533) to carbon 6. Removal of the γ-glutamyl group from the tripeptide constituent of LTC_4 yields LTD_4. LTE_4 is produced by the removal of glycyl residue from LTD_4.

QUESTION 11.1

Rheumatoid arthritis is an autoimmune disease in which the joints are chronically inflamed. In **autoimmune diseases** the immune system fails to distinguish between self and nonself. For reasons that are not understood, specific lymphocytes are stimulated to produce antibodies, referred to as *autoantibodies*. These molecules bind to surface antigens on the patient's own cells as if they were foreign. In rheumatoid arthritis, the binding of an autoantibody called rheumatoid factor (RF) to F_c portion of IgG promotes inflammation because it stimulates the infiltration of joint tissue by several types of white blood cells. The leakage of lysosomal enzymes from actively phagocytosing cells (neutrophils and macrophages) leads to further tissue damage. The inflammatory response is perpetuated by the release of several eicosanoids. For example, macrophages are known to produce PGE_4, TXA_2, and several leukotrienes.

Currently, the treatment of rheumatoid arthritis consists of suppressing pain and inflammation. Despite treatment, however, the disease continues to progress. Aspirin plays an important role in the treatment of rheumatoid arthritis and other types of inflammation because of its low cost and relative safety. Certain steroids, which inhibit phospholipase A_2 (see p. 388), are more potent than aspirin in reducing inflammation; that is, they immediately and dramatically reduce painful symptoms. However, steroids have serious side effects. For example, prednisone may depress the immune system, cause fat redistribution to the neck ("buffalo hump"), and cause serious behavioral changes. For these and other reasons, prednisone is used to treat rheumatoid arthritis only when a patient does not respond to aspirin or similar drugs.

Review the effects of aspirin and steroids on eicosanoid metabolism and suggest a reason why this information is relevant to the treatment of rheumatoid arthritis. Does it explain the difference between the effectiveness of aspirin and steroids in treating inflammation?

$$
\begin{array}{l}
\mathrm{H-\overset{\displaystyle H}{\overset{|}{C}}-O-\overset{\displaystyle O}{\overset{\|}{C}}-(CH_2)_{16}-CH_3} \\
\mathrm{H-\underset{|}{\overset{|}{C}}-O-\overset{\displaystyle O}{\overset{\|}{C}}-(CH_2)_7-CH{=}CH-(CH_2)_7-CH_3} \\
\mathrm{H-\underset{\displaystyle H}{\underset{|}{C}}-O-\overset{\displaystyle O}{\overset{\|}{C}}-(CH_2)_{14}-CH_3}
\end{array}
$$

FIGURE 11.5
Triacylglycerol

Each triacylglycerol molecule is composed of glycerol esterified to three (usually different) fatty acids.

Triacylglycerols

Triacylglycerols are esters of glycerol with three fatty acid molecules (**Figure 11.5**). Glycerides with one or two fatty acid groups, called monoacylglycerols and diacylglycerols, respectively, are metabolic intermediates. They are normally present in small amounts. Because triacylglycerols have no charge (i.e., the carboxyl group of each fatty acid is joined to glycerol through a covalent bond), they are sometimes referred to as **neutral fats**. Most triacylglycerol molecules contain fatty acids of varying lengths; the acids themselves may be unsaturated, saturated, or a combination (Figure 11.6). Depending on their fatty acid compositions, triacylglycerol mixtures are referred to as fats or oils. *Fats*, which are solid at room temperature, contain a large proportion of saturated fatty acids. *Oils* are liquid at room temperature because of their relatively high unsaturated fatty acid content. (Recall that unsaturated fatty acids do not pack together as closely as do saturated fatty acids.)

In animals, triacylglycerols (usually referred to as *fat*) have several roles. First, they are the major storage and transport form of fatty acids. Triacylglycerol molecules store energy more efficiently than glycogen for several reasons:

1. Triacylglycerols are hydrophobic, and therefore they coalesce into compact, anhydrous droplets within cells. A specialized type of cell called the *adipocyte*, found in adipose tissue, stores triacylglycerols. The anhydrous triacylglycerols store an equivalent amount of energy in about one-eighth of the volume of glycogen (the other major energy storage molecule), which binds a substantial amount of water.
2. Triacylglycerols are more reduced and can thus release more electrons when oxidized than an equivalent amount of carbohydrate. Therefore, triacylglycerols release more energy (38.9 kJ/g of fat compared with 17.2 kJ/g of carbohydrate) when they are degraded.

A second important function of fat is to provide insulation in low temperatures. Fat, a poor conductor of heat, prevents heat loss. Adipose tissue, with its high triacylglycerol content, is found throughout the body (especially underneath the skin). Finally, in some animals fat molecules secreted by specialized glands make fur or feathers water-repellent.

In plants, triacylglycerols constitute an important energy reserve in fruits and seeds. Because these molecules contain relatively large amounts of unsaturated fatty acids (e.g., oleic and linoleic), they are referred to as plant oils. Seeds rich in oil include peanut, corn, palm, safflower, soybean, and flax. Avocados and olives are fruits with a high oil content.

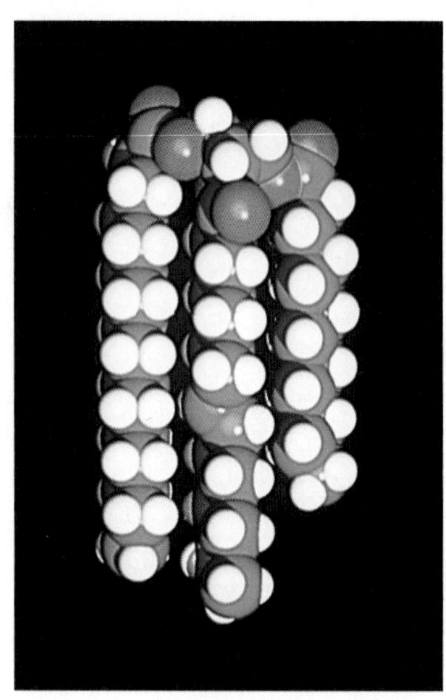

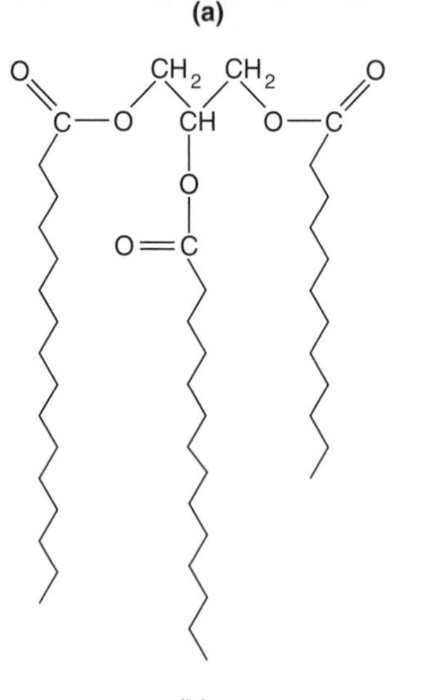

FIGURE 11.6
Space-Filling (a) and Conformational (b) Models of a Triacylglycerol

Triacylglycerols are molecules that serve as a rich source of chemical bond energy.

QUESTION 11.2

Oils can be converted to fats in a commercial nickel-catalyzed process referred to as *partial hydrogenation*. Under relatively mild conditions (180°C and pressures of about 1013 torr or 1.33 atm) enough double bonds are hydrogenated for liquid oils to solidify. This solid material, oleomargarine, has a consistency like butter. However, oils are not completely hydrogenated during commercial hydrogenation processes. Propose a practical reason for this.

QUESTION 11.3

Soapmaking is an ancient process. The Phoenicians, a seafaring people who dominated trade in the Mediterranean area about 3000 years ago, are believed to have been the first to manufacture soap. Traditionally, soap has been made by heating animal fat with potash. (Potash is a mixture of potassium hydroxide (KOH) and potassium carbonate (K_2CO_3) obtained by mixing wood ash with water.) Currently, soap is made by heating beef tallow or coconut oil with sodium or potassium hydroxide. During this reaction, which is a *saponification*, triacylglycerol molecules are hydrolyzed to give glycerol and the sodium or potassium salts of fatty acids:

$$\underset{\text{Triacylglycerol}}{\begin{matrix} CH_2{-}O{-}\overset{O}{\overset{\|}{C}}{-}R \\ |\\ CH{-}O{-}\overset{O}{\overset{\|}{C}}{-}R' \\ | \\ CH_2{-}O{-}\overset{O}{\overset{\|}{C}}{-}R'' \end{matrix}} + 3\,NaOH \longrightarrow \underset{\text{Soap}}{\begin{matrix} Na^+\ {}^-O{-}\overset{O}{\overset{\|}{C}}{-}R \\ + \\ Na^+\ {}^-O{-}\overset{O}{\overset{\|}{C}}{-}R' \\ + \\ Na^+\ {}^-O{-}\overset{O}{\overset{\|}{C}}{-}R'' \end{matrix}} + \underset{\text{Glycerol}}{\begin{matrix} CH_2{-}OH \\ | \\ CH{-}OH \\ | \\ CH_2{-}OH \end{matrix}}$$

Fatty acid salts (soaps) are amphipathic molecules (i.e., they possess polar and nonpolar domains) that spontaneously form into micelles (Figure 3.14). Soap micelles have negatively charged surfaces that repel each other. Soap is used to remove dirt mixed with grease because it is an *emulsifying agent*; that is, it promotes the dispersion of one substance in another. The mixing of soap and grease results in an emulsion—specifically, a system in which the oil molecules are dispersed in the soap micelles. Complete the diagram in Figure 11.7 and explain how this process occurs.

KEY CONCEPTS

- Triacylglycerols are molecules consisting of glycerol esterified to three fatty acids.
- In both animals and plants they are a rich energy source.

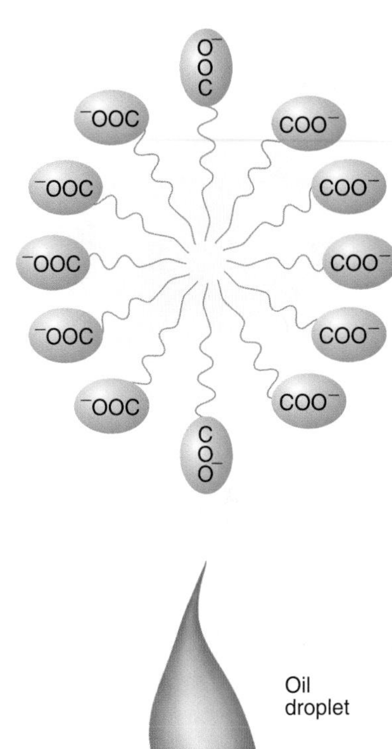

FIGURE 11.7
Soap: An Emulsifying Agent

How does soap interact with an oil droplet? [*Hint*: Recall that "like dissolves like."]

Wax Esters

Waxes are complex mixtures of nonpolar lipids. They are protective coatings on the leaves, stems, and fruits of plants and on the skin and fur of animals. **Wax esters** composed of long-chain fatty acids and long-chain alcohols are prominent constituents of most waxes. Well-known waxes include carnauba wax, produced by the leaves of the Brazilian wax palm, and beeswax. The predominant constituent of carnauba wax is the wax ester melissyl cerotate (**Figure 11.8**). Triacontyl hexadecanoate is one of several important wax esters in beeswax. Waxes also contain hydrocarbons, long-chain alcohols and aldehydes, fatty acids, and sterols (steroid alcohols).

Phospholipids

Phospholipids have several roles in living organisms. They are first and foremost structural components of membranes. In addition, several phospholipids are emulsifying agents and surface active agents. (A *surface active agent* is a substance that lowers the surface tension of a liquid, usually water, so that it spreads out over a surface.) Phospholipids are suited to these roles because, like fatty acid salts, they are amphipathic molecules. The hydrophobic domain of a phospholipid is composed largely of the hydrocarbon chains of fatty acids; the hydrophilic domain, called a **polar head group**, contains phosphate and other charged or polar groups.

$$CH_3{-}(CH_2)_{24}{-}\overset{O}{\overset{\|}{C}}{-}O{-}(CH_2)_{29}{-}CH_3$$

FIGURE 11.8
The Wax Ester Melissyl Cerotate

Found in carnauba wax, melissyl cerotate is an ester formed from melissyl alcohol and cerotic acid.

When phospholipids are suspended in water, they spontaneously rearrange into ordered structures (**Figure 11.9**). As these structures form, phospholipid hydrophobic groups are buried in the interior to exclude water. Simultaneously, hydrophilic polar head groups are oriented so that they are exposed to water. When phospholipid molecules are present in sufficient concentration, they form bimolecular layers. This property of phospholipids (and other amphipathic lipid molecules) is the basis of membrane structure (see pp. 404–417).

There are two types of phospholipid: phosphoglycerides and sphingomyelins. **Phosphoglycerides** are molecules that contain glycerol, fatty acids, phosphate, and an alcohol (e.g., choline). **Sphingomyelins** differ from phosphoglycerides in that they contain sphingosine instead of glycerol. Because sphingomyelins are also classified as sphingolipids, their structures and properties are discussed separately.

Phosphoglycerides are the most numerous phospholipid molecules found in cell membranes. The simplest phosphoglyceride, phosphatidic acid, is the precursor for all other phosphoglyceride molecules. Phosphatidic acid is composed of glycerol-3-phosphate that is esterified with two fatty acids. Phosphoglyceride molecules are classified according to which alcohol becomes esterified to the phosphate group. For example, if the alcohol is choline, the molecule is called phosphatidylcholine (PC) (also referred to as *lecithin*). Other types of phosphoglyceride include phosphatidylethanolamine (PE), phosphatidylserine (PS), diphosphatidylglycerol (dPG) (also called cardiolipin), and phosphatidylinositol (PI). (Refer to **Table 11.2** for the structures of the common types of phosphoglyceride.) The most common fatty acids in the phosphoglycerides have between 16 and 20 carbons. Saturated fatty acids usually occur at C-1 of glycerol. The fatty acid substituent at C-2 is usually unsaturated.

Both phosphatidylethanolamine and cardiolipin have relatively small polar heads. Usually found in the inner leaflet of membranes, PE (25% of all phospholipids in humans) stabilizes membrane curvature and has a role in membrane fusion. Cardiolipin is found almost entirely in membranes whose function is to generate the electrochemical potential that drives ATP synthesis (i.e., in bacterial plasma membrane and mitochondrial inner membrane, where it constitutes 20% of total lipid). Composed of a lipid dimer with two phosphatidyl groups and a total of four hydrocarbon chains, cardiolipin stabilizes ETC supercomplexes (p. 357).

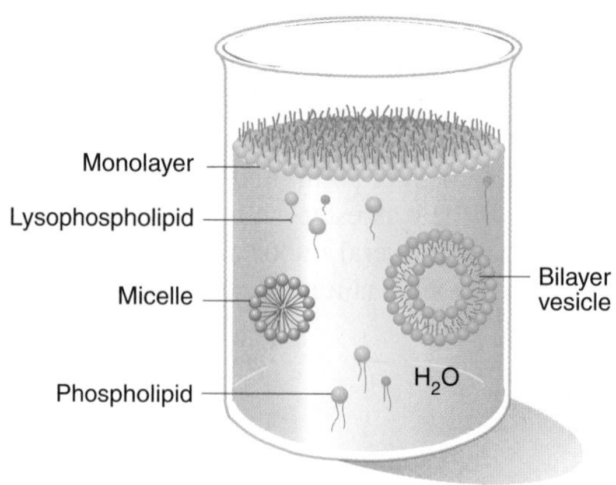

FIGURE 11.9

Phospholipid Molecules in Aqueous Solution

Each molecule is represented as a polar head group attached to one or two fatty acyl chains. (Lysophospholipid molecules possess only one fatty acyl chain.) The monolayer on the surface of the water forms first. As the phospholipid concentration increases, bilayer vesicles begin to form. Because of their wedge shape (compared to the cylindrical shape of phospholipids that contain two fatty acid chains), lysophospholipid molecules form micelles.

TABLE 11.2 Major Classes of Phosphoglycerides

$$R_2\overset{O}{\overset{\|}{C}}O-CH(CH_2O-\overset{O}{\overset{\|}{C}}-R_1)-CH_2O-\overset{O}{\overset{\|}{P}}(O^-)-O-X$$

X Substituent		
Name of X-OH	**Formula of X**	**Name of Phospholipid**
Water	—H	Phosphatidic acid
Choline	$-CH_2CH_2\overset{+}{N}(CH_3)_3$	Phosphatidylcholine (lecithin)
Ethanolamine	$-CH_2CH_2\overset{+}{N}H_3$	Phosphatidylethanolamine (cephalin)
Serine	$-CH_2-CH(\overset{+}{N}H_3)(COO^-)$	Phosphatidylserine
Glycerol	$-CH_2CH(OH)CH_2OH$	Phosphatidylglycerol
Phosphatidylglycerol	$-CH_2CH(OH)-CH_2-O-\overset{O}{\overset{\|}{P}}(O^-)-O-CH_2-CH(O\overset{O}{\overset{\|}{C}}R)-CH_2O\overset{O}{\overset{\|}{C}}R$	Diphosphatidylglycerol (cardiolipin)
Inositol	OH, OH, OH, OH, OH	Phosphatidylinositol

A derivative of phosphatidylinositol, namely, phosphatidylinositol-4,5-bisphosphate (PIP_2), is found in only small amounts in plasma membranes. PIP_2 is now recognized as an important component of intracellular signal transduction. The *phosphatidylinositol cycle*, initiated when certain hormones bind to membrane receptors, is described in Section 16.2.

Phosphatidylinositol is also a prominent structural component of GPI anchors. **GPI anchors** (**Figure 11.10**), which are also composed of a trimannosylglucosamine group and phosphoethanolamine, attach certain proteins to the external surface of the plasma membrane. Proteins are attached to the anchor molecule via an amide linkage between the carboxyl terminal of the protein and the amino nitrogen of ethanolamine. The two fatty acids of the phosphatidylinositol component are embedded in the plasma membrane.

KEY CONCEPTS

- Phospholipids are amphipathic molecules that play important roles in living organisms as membrane components, emulsifying agents, and surface active agents.
- There are two types of phospholipid: phosphoglycerides and sphingomyelins.

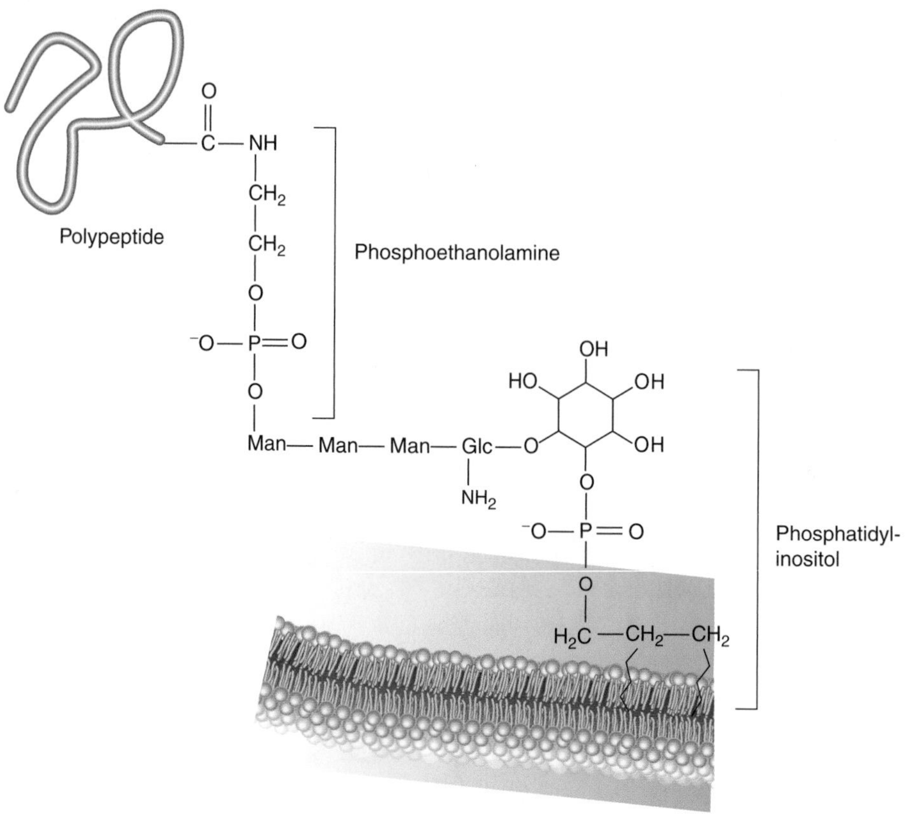

FIGURE 11.10
GPI Anchor

GPI-anchored proteins are attached to the external surface of the membrane through a linker element, phosphoethanolamine-Man_3-$GlcNH_2$, connecting the polypeptide at its carboxy terminus via an amide bond to a membrane phosphatidylinositol via an ether bond. Note that there are variations of this structure. For example, $GlcNH_2$ can be acetylated, and the phosphate of phosphatidic acid can be linked to either C-2 or C-3 of inositol.

QUESTION 11.4

Respiratory Distress Syndrome

Dipalmitoylphosphatidylcholine is the major component of *surfactant*, or surface active agent (an amphipathic molecule), that is secreted into lung alveoli to reduce the surface tension of the primarily aqueous extracellular fluid of the alveolar epithelia. Alveoli, also referred to as alveolar sacs, are the functional units of respiration. Oxygen and carbon dioxide diffuse across the walls of alveolar sacs, which are one cell thick. The water on alveolar surfaces has a high surface tension because of the attractive forces between the molecules. If the water's surface tension is not reduced, the alveolar sac tends to collapse, making breathing extremely difficult. If premature infants lack sufficient surfactant, they are likely to die of suffocation. This condition is called *respiratory distress syndrome.* Draw the structure of dipalmitoylphosphatidylcholine. Considering the general structural features of phospholipids, propose a reason why surfactant is effective in reducing surface tension.

Phospholipases

Phospholipases hydrolyze ester bonds in glycerophospholipid molecules (**Figure 11.11**). They are classified according to the specific bond that they cleave. Phospholipases A_1 (PLA_1) and A_2 (PLA_2) hydrolyze the ester bonds at C_1 and C_2 of glycerol, respectively. The products of PLA_1 and PLA_2 are a fatty acid and a *lysophosphatide* (a glycerophospholipid from which one fatty acid has been removed). Phospholipase B (PLB) can hydrolyze both the C-1 and the C-2 ester bonds. Phospholipases C (PLC) and D (PLD) are phosphodiesterases that yield diacylglycerol and phosphatidic acid, respectively. Phospholipases have three major functions: membrane remodeling, signal transduction, and digestion. They are also used by some organisms as biological weapons.

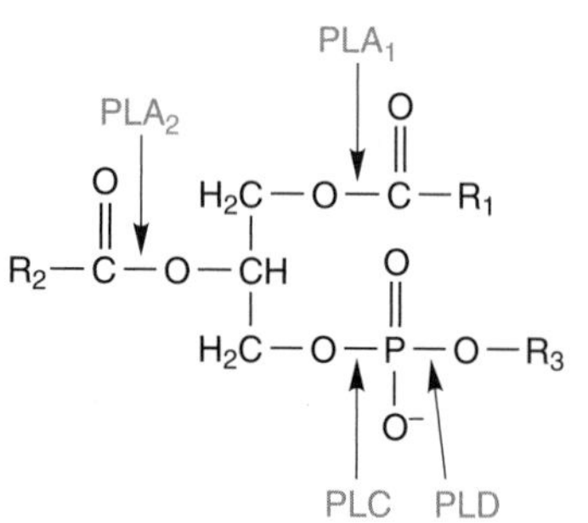

FIGURE 11.11
Phospholipases

Phospholipases hydrolyze ester bonds in phospholipids. Note that PLB has both PLA_1 and PLA_2 activities.

Membrane Remodeling Cells use phospholipases to alter the flexibility of membranes (p. 405) by adjusting the ratio of saturated and unsaturated fatty acids or to replace a damaged fatty acid. Fatty acid removal from a phospholipid is followed by a reacylation reaction catalyzed by an acyltransferase.

Signal Transduction Numerous hormones initiate signal transduction mechanisms that involve phospholipid hydrolysis (pp. 591–592). For example, the PLC-catalyzed cleavage of PIP_2, a phosphorylated derivative of phosphatidylinositol, yields the signal molecules inositol-1,4,5-trisphosphate (IP_3) and diacylglycerol (DAG). Synthesis of eicosanoids (p. 387) is initiated with the PLA_2-catalyzed release of arachidonic acid.

Digestion In mammals, fat digestion occurs in the small intestine where bile salts convert large fat globules into smaller droplets that can be acted on by enzymes. Pancreatic phospholipases, delivered to the small intestine along with other digestive enzymes, degrade dietary phospholipids. Lysosomal phospholipases degrade the phospholipid components of cellular membranes.

Toxic Phospholipases Various organisms use membrane-degrading phopholipases as a means of inflicting damage on other species. PLA_2 in snake venoms, for example, not only digests cell membranes at the site of a bite, but also causes diverse forms of systemic damage (e.g., necrosis of skeletal and heart muscle, neurotoxicity, and red blood cell lysis). *Clostridium perfringins* is an anaerobic, Gram-positive bacterium that causes *gas gangrene* (tissue death accompanied by gas formation). A phospholipase called α-toxin facilitates the organism's penetration into the tissue surrounding a wound.

Sphingolipids

Sphingolipids are important components of animal and plant membranes. All sphingolipid molecules contain a long-chain amino alcohol. In animals this alcohol is primarily sphingosine (**Figure 11.12**). Phytosphingosine is found in plant sphingolipids. The core of each type of sphingolipid is *ceramide*, a fatty acid amide derivative of sphingosine. In *sphingomyelin*, the 1-hydroxyl group of ceramide is esterified to the phosphate group of phosphorylcholine or phosphorylethanolamine (**Figure 11.13**). Sphingomyelin is found in most animal cell membranes. However, as its name suggests, sphingomyelin is found in greatest abundance in the myelin sheath of nerve cells. The myelin sheath is formed by successive wrappings of the cell membrane of a specialized myelinating cell around a nerve cell axon. Its insulating properties facilitate the rapid transmission of nerve impulses.

The **glycolipids** are lipid molecules with carbohydrate groups attached. They include the *glycosphingolipids* (**Figure 11.14**) and GPI anchors (p. 394). The glycosphingolipids contain ceramide, but they differ from sphingomyelin in that

$CH_3(CH_2)_{12}CH{=}CH{-}CH(OH){-}CH(\overset{+}{N}H_3){-}CH_2OH$

Sphingosine

$CH_3(CH_2)_{12}CH_2CH(OH){-}CH(OH){-}CH(\overset{+}{N}H_3){-}CH_2OH$

Phytosphingosine

$CH_3(CH_2)_{12}CH{=}CH{-}CH(OH){-}CH(NH{-}C({=}O){-}(CH_2)_{12}{-}CH_3){-}CH_2OH$

A ceramide

FIGURE 11.12
Sphingolipid Components

Note that the *trans*-isomer of sphingosine occurs in sphingolipids.

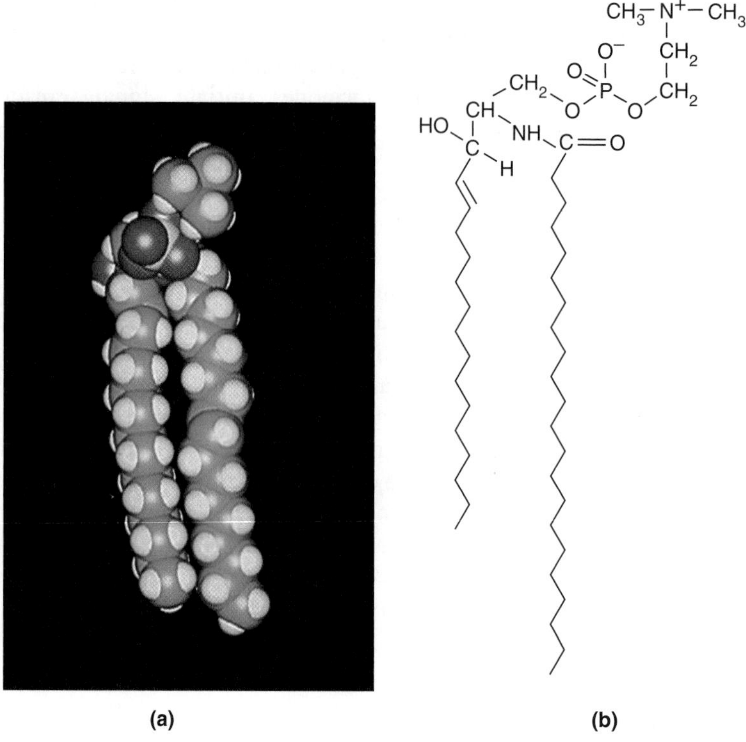

FIGURE 11.13
(a) Space-Filling and Conformational (b) Models of Sphingomyelin

The fatty acid component of sphingomyelins can be saturated or monounsaturated and 16 to 24 carbons in length, depending on the species and tissue of origin. The sphingosine base can be replaced by sphinganine (no double bond) and other C-20 homologues, although sphingosine is by far the most abundant.

they contain no phosphate. The most important glycolipid classes are the cerebrosides, the sulfatides, and the gangliosides.

Cerebrosides are sphingolipids in which the head group is a monosaccharide. (These molecules, unlike phospholipids, are nonionic.) Galactocerebrosides, the most common example of this class, are found almost entirely in the cell membranes of the brain. If a cerebroside is sulfated, it is referred to as a *sulfatide*. Sulfatides are negatively charged at physiological pH.

Sphingolipids that possess oligosaccharide groups with one or more sialic acid residues are called *gangliosides*. Although gangliosides were first isolated from nerve tissue, they also occur in most other animal tissues. The names of gangliosides include subscript letter and numbers. The letters M, D, and T indicate whether the molecule contains one, two, or three sialic acid residues (see **Figure 7.25d**), respectively. The numbers designate the sequence of sugars that are attached to ceramide. The Tay-Sachs ganglioside G_{M2} is illustrated in **Figure 11.14**.

The normal role of glycolipids is still unclear. Certain glycolipid molecules may bind bacterial toxins, as well as bacterial cells, to animal cell membranes. For example, the toxins that cause cholera, tetanus, and botulism bind to glycolipid cell membrane receptors. Bacteria that have been shown to bind to glycolipid receptors include *E. coli*, *Streptococcus pneumoniae*, and *Neisseria gonorrhoeae*, the causative agents of urinary tract infections, pneumonia, and gonorrhea, respectively.

KEY CONCEPTS

- Sphingolipids, important membrane components of animals and plants, contain a complex long-chain amino alcohol (either sphingosine or phytosphingosine).
- The core of each sphingolipid is ceramide, a fatty acid amide derivative of the alcohol molecule. Glycolipids are derivatives of ceramide that possess a carbohydrate component.

Sphingolipid Storage Diseases

Sphingolipidoses

Each lysosomal storage disease (see p. 54) is caused by hereditary deficiency of an enzyme required for the degradation of a specific metabolite. Several lysosomal storage diseases are associated with sphingolipid metabolism. Most of these diseases, also referred to as the *sphingolipidoses*, are fatal. The most common sphingolipid storage disease, Tay-Sachs disease, is caused by a deficiency of β-hexosaminidase A, the enzyme that degrades the ganglioside G_{M2}. As cells

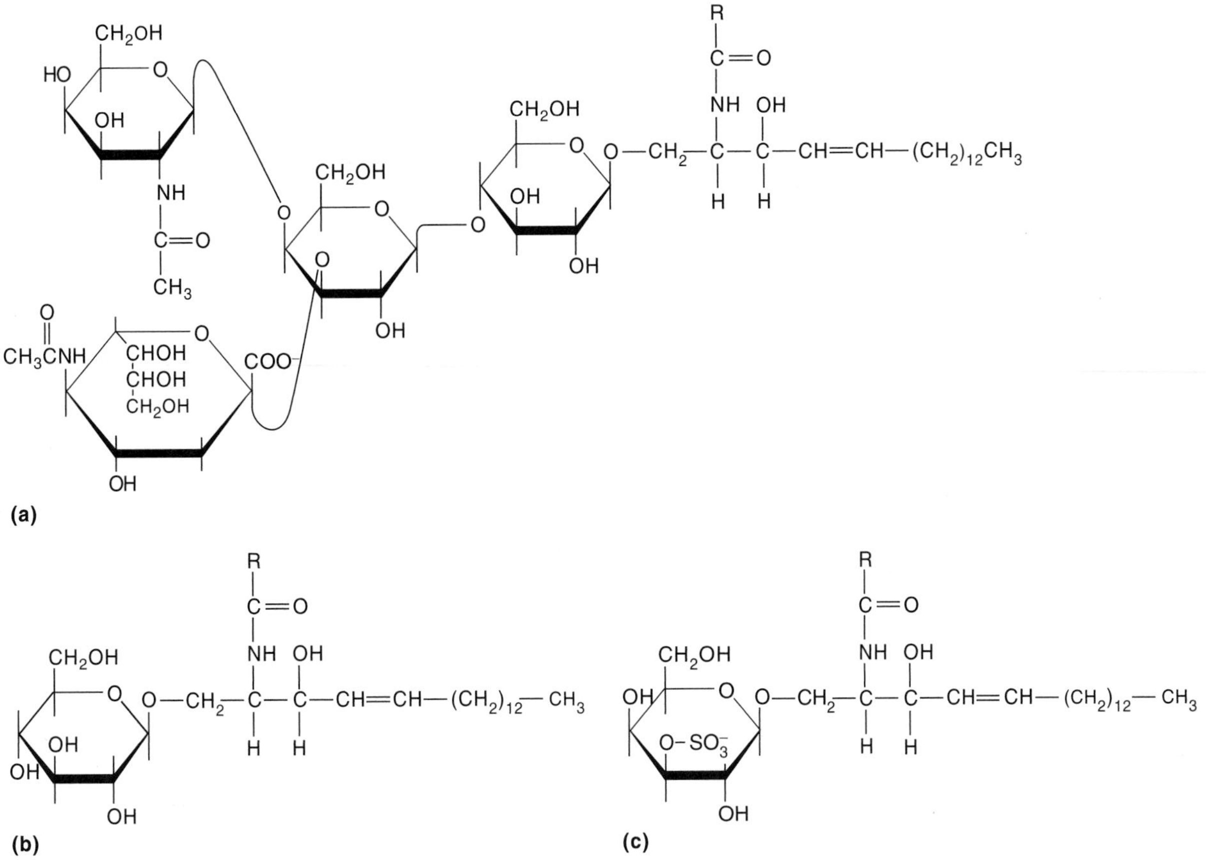

(a)

(b)

(c)

FIGURE 11.14

Selected Glycolipids

(a) Tay-Sachs ganglioside (G_{M2}), (b) glucocerebroside, and (c) galactocerebroside sulfate (a sulfatide).

accumulate this molecule, they swell and eventually die. Tay-Sachs symptoms (blindness, muscle weakness, seizures, and mental retardation) usually appear several months after birth. Because there is currently no therapy for Tay-Sachs disease or for any other of the sphingolipidoses, the condition is always fatal (usually by age 3). Examples of the sphingolipidoses are summarized in **Table 11.3**.

TABLE 11.3 Selected Sphingolipid Storage Diseases*

Disease	Symptom	Accumulating Sphingolipid	Enzyme Deficiency
Tay-Sachs disease	Blindness, muscle weakness, seizures, mental retardation	Ganglioside G_{M2}	β-Hexosaminidase A
Gaucher's disease	Mental retardation, liver and spleen enlargement, erosion of long bones	Glucocerebroside	β-Glucosidase
Krabbe's disease	Demyelination, mental retardation	Galactocerebroside	β-Galactosidase
Niemann-Pick disease	Mental retardation	Sphingomyelin	Sphingomyelinase

*Many diseases are named for the physicians who first described them. Tay-Sachs disease was reported by Warren Tay (1843–1927), a British ophthalmologist, and Bernard Sachs (1858–1944), a New York neurologist. Phillipe Gaucher (1854–1918), a French physician, and Knud Krabbe (1885–1961), a Danish neurologist, first described Gaucher's disease and Krabbe's disease, respectively. Niemann-Pick disease was first characterized by the German physicians Albert Niemann (1880–1921) and Ludwig Pick (1868–1944).

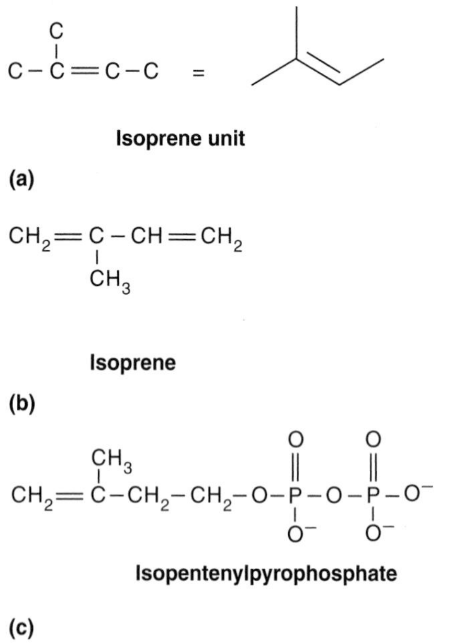

C–C(C)=C–C = [skeletal structure]

Isoprene unit

(a)

$CH_2{=}C(CH_3){-}CH{=}CH_2$

Isoprene

(b)

$CH_2{=}C(CH_3){-}CH_2{-}CH_2{-}O{-}P(=O)(O^-){-}O{-}P(=O)(O^-){-}O^-$

Isopentenylpyrophosphate

(c)

FIGURE 11.15
Isoprene

(a) Basic isoprene structure.
(b) The organic molecule isoprene.
(c) Isopentenylpyrophosphate.

Isoprenoids

The **isoprenoids** are a vast array of biomolecules that contain repeating five-carbon structural units known as *isoprene units* (**Figure 11.15**). Isoprenoids are not synthesized from isoprene (methylbutadiene). Instead, their biosynthetic pathways all begin with the formation of isopentenyl pyrophosphate from acetyl-CoA (Chapter 12).

The isoprenoids consist of terpenes and steroids. **Terpenes** are an enormous group of molecules that are found largely in the essential oils of plants. Steroids are derivatives of the hydrocarbon ring system of cholesterol.

TERPENES The terpenes are classified according to the number of isoprene units they contain (**Table 11.4**). *Monoterpenes* are composed of two isoprene units (10 carbon atoms). Geraniol is a monoterpene found in the *essential oils*, volatile hydrophobic liquid mixtures extracted from plants, fruits, or flowers (e.g., roses, lemon, and geranium). Each essential oil has a characteristic odor, and some are used to make perfumes.

Terpenes that contain three isoprenes (15 carbons) are referred to as *sesquiterpenes*. Farnesene, an important constituent of oil of citronella, which is used in soap and perfumes, is a sesquiterpene. Phytol, a plant alcohol, is a *diterpene*, a molecule composed of 4 isoprene units. Squalene is a prominent example of the *triterpenes*; this intermediate in the synthesis of the steroids is found in large quantities in shark liver oil, olive oil, and *yeast*. **Carotenoids**, the orange

TABLE 11.4 Examples of Terpenes

Type	Number of Isoprene Units	Example: Name	Example: Structure
Monoterpene	2	Geraniol	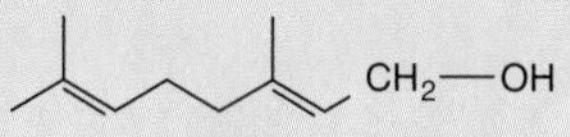 [skeletal structure] CH_2—OH
Sesquiterpene	3	Farnesene	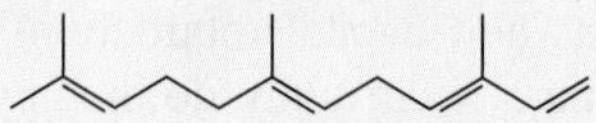 [skeletal structure]
Diterpene	4	Phytol	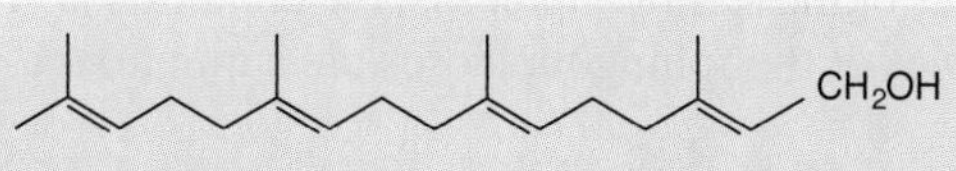 [skeletal structure] CH_2OH
Triterpene	6	Squalene	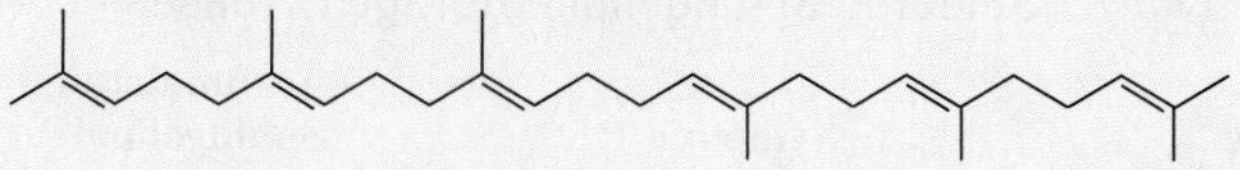 [skeletal structure]
Tetraterpene	8	β-Carotene	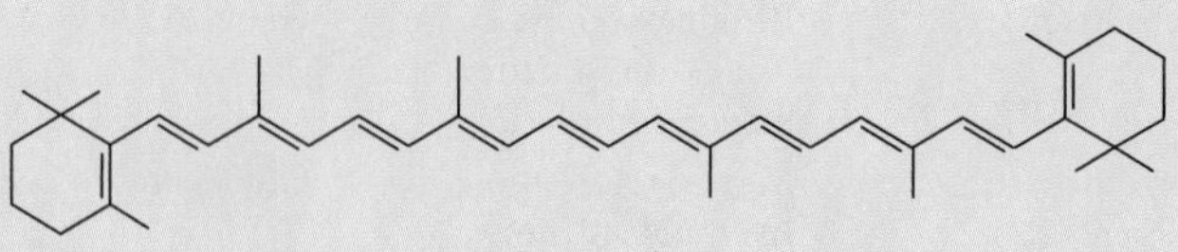 [skeletal structure]
Polyterpene	9–24	Dolichol	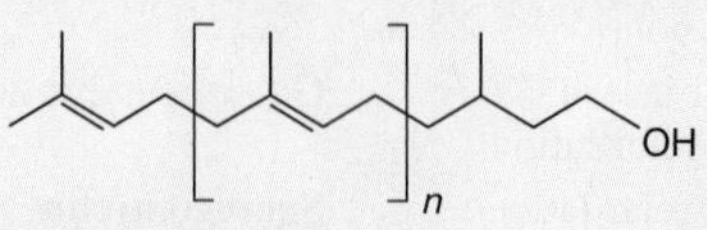 [skeletal structure, repeating unit]$_n$ OH
	Thousands	Rubber	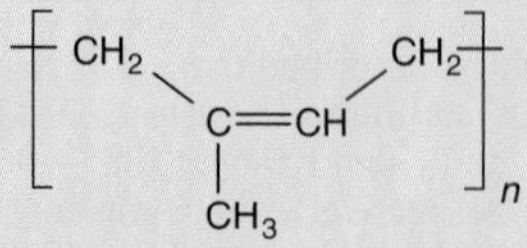 $[-CH_2-C(CH_3){=}CH-CH_2-]_n$

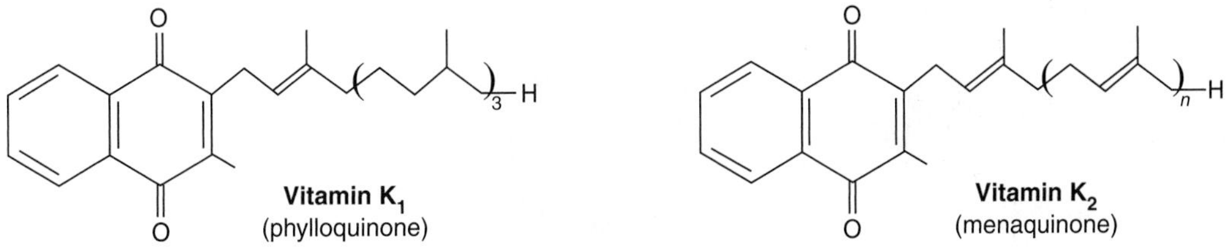

Vitamin K_1 (phylloquinone)

Vitamin K_2 (menaquinone)

FIGURE 11.16

Vitamin K, a Mixed Terpenoid

Vitamin K_1 (phylloquinone) is found in plants, where it acts as an electron carrier in photosynthesis. Vitamin K_2 (menaquinone) is synthesized by intestinal bacteria and plays an important role in blood coagulation.

pigments found in most plants, are the only *tetraterpenes* (molecules composed of 8 isoprene units). The *carotenes* are hydrocarbon members of this group. The *xanthophylls* are oxygenated derivatives of the carotenes. *Polyterpenes* are high-molecular-weight molecules composed of up to thousands of isoprene units. Natural rubber is a polyterpene composed of between 3000 and 6000 isoprene units. *Dolichols* are polyisoprenoid alcohols (16–19 isoprene units) that function as sugar carriers in glycoprotein synthesis.

Several important biomolecules are composed of nonterpene components attached to isoprenoid groups (often referred to as *prenyl* or *isoprenyl* groups). Examples of these biomolecules, referred to as **mixed terpenoids**, include vitamin E (α-tocopherol) (**Figure 10.23a**), ubiquinone (**Figure 10.3**), vitamin K (**Figure 11.16**), and plastoquinone (p. 480).

A variety of proteins in eukaryotic cells are now known to be covalently attached to prenyl groups after their biosynthesis on ribosomes. The prenyl groups most often involved in this process, referred to as **prenylation**, are farnesyl and geranylgeranyl groups (**Figure 11.17**). Farnesyl and geranylgeranyl groups are intermediates on the cholesterol biosynthesis pathway (**Figure 12.29**). The function of protein prenylation is not clear. There is some evidence that it plays a role in the control of cell growth. For example, *Ras proteins*, a group of cell growth regulators, are activated by prenylation reactions.

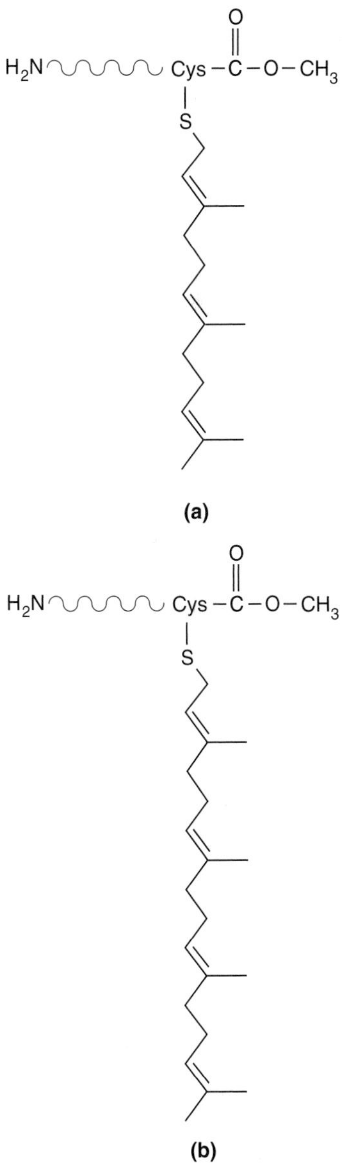

FIGURE 11.17

Prenylated Proteins

Prenyl groups are covalently attached at the SH group of C-terminal cysteine residues. Many prenylated proteins are also methylated at this residue. (a) Farnesylated protein. (b) Geranylgeranylated protein.

QUESTION 11.5

The majority of terpenes contain one or more ring structures. Consider the following examples. Determine which terpene class they belong to, and outline the positions of the isoprene units.

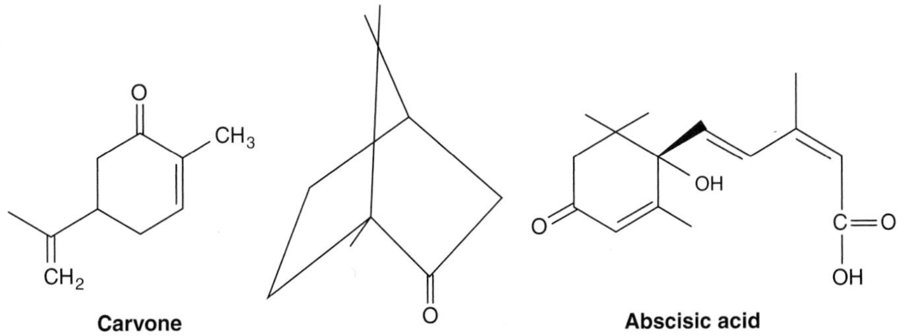

Carvone (spearmint oil) **Camphor** **Abscisic acid (plant growth regulator)**

STEROIDS Steroids are derivatives of triterpenes with four fused rings. They are found in all eukaryotes and a small number of bacteria. Steroids are distinguished

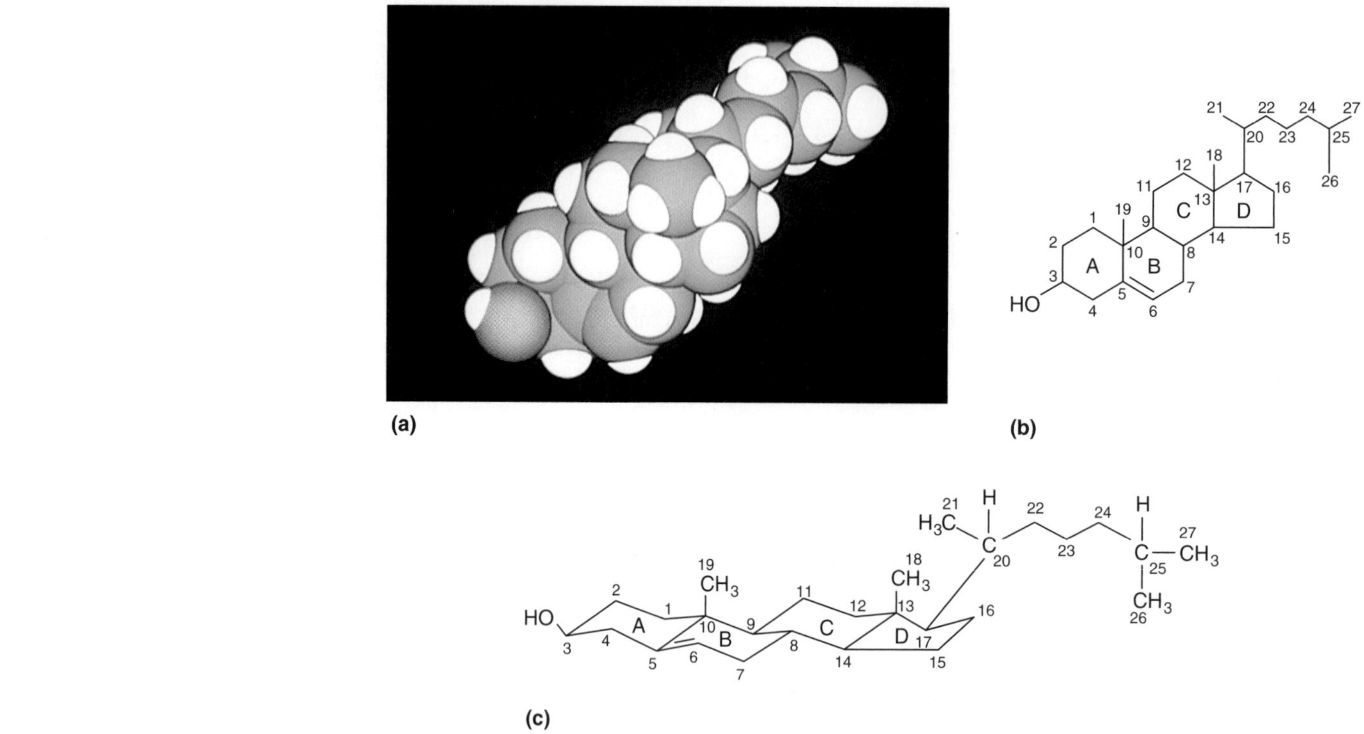

(a)

(b)

(c)

FIGURE 11.18
Structure of Cholesterol

(a) Space-filling model, (b) conventional view, and (c) conformational model. Space-filling models and conformational models represent molecular structure more accurately than the conventional view.

from each other by the placement of carbon-carbon double bonds and various substituents (e.g., hydroxyl, carbonyl, and alkyl groups).

Cholesterol, an important molecule in animals, is an example of a steroid (**Figure 11.18**). In addition to being an essential component in animal cell membranes, cholesterol is a precursor in the biosynthesis of all steroid hormones, vitamin D, and bile salts (**Figure 11.19**). Cholesterol (C-27) is formed from the linear triterpene squalene (C-30) by intramolecular ring closure, oxidation, and cleavage. The only double bond retained migrates to the Δ5 position, and C-3 is oxidized to a hydroxyl group, which justifies its classification as a *sterol*. (Although the term *steroid* is most properly used to designate molecules that contain one or more carbonyl or carboxyl groups, it is often used to describe all derivatives of the steroid ring structure.) Cholesterol is usually stored within cells as a fatty acid ester. The esterification reaction is catalyzed by the enzyme *acyl-CoA:cholesterol acyltransferase* (ACAT), located on the cytoplasmic face of the ER.

QUESTION 11.6

Bile salts are emulsifying agents; that is, they promote the formation of mixtures of hydrophobic substances and water. Produced in the liver, bile salts assist in the digestion of fats in the small intestine. They are formed by linking bile acids to hydrophilic substances such as the amino acid glycine. After reviewing the structure of cholic acid in Figure 11.19, suggest how the structural features of bile salts contribute to their function.

(a) Progesterone, Testosterone, 17-β-Estradiol

(b) Aldosterone (c) Cortisol (d) Cholic acid

FIGURE 11.19
Animal Steroids

(a) Sex hormones (molecules that regulate the development of primary and secondary sex characteristics and various reproductive behaviors). (b) A mineralocorticoid (a molecule produced in the adrenal cortex that regulates plasma concentrations of several ions, especially sodium). (c) A glucocorticoid (a molecule that regulates the metabolism of carbohydrates, fats, and proteins). (d) A bile acid. (Bile acids are converted into bile salts. Produced in the liver, bile salts aid the absorption of dietary fats and fat-soluble vitamins in the intestine).

Cardiac glycosides, molecules that increase the force of cardiac muscle contraction, are among the most interesting steroid derivatives. Glycosides are carbohydrate-containing acetals (see p. 248). Although several cardiac glycosides are extremely toxic (e.g., *ouabain*, obtained from the seeds of the plant *Strophanthus gratus* or climbing oleander), others have valuable medicinal properties (**Figure 11.20**). For example, *digitalis*, an extract of the dried leaves of *Digitalis purpurea* (the foxglove plant), is a time-honored stimulator of cardiac muscle contraction. *Digitoxin*, the major "cardiotonic" glycoside in digitalis, is used to treat congestive heart failure, an illness in which the heart is so damaged by disease processes (e.g., myocardial infarcts) that pumping is impaired. In higher than therapeutic doses, digitoxin is extremely toxic. Both ouabain and digitoxin inhibit Na^+–K^+ ATPase (see p. 412).

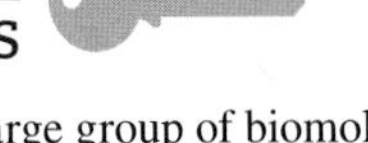

KEY CONCEPTS

- Isoprenoids are a large group of biomolecules with repeating units derived from isopentenyl pyrophosphate.
- There are two types of isoprenoids: terpenes and steroids.

Lipoproteins

Although the term *lipoprotein* can describe any protein that is covalently linked to lipid groups (e.g., fatty acids or prenyl groups), it is most often applied to a group of molecular complexes found in the blood plasma of mammals (especially humans). Plasma lipoproteins transport lipid molecules (triacylglycerols, phospholipids, and cholesterol) through the bloodstream from one organ to another. Lipoproteins also contain several types of lipid-soluble antioxidant molecules (e.g., α-tocopherol and several carotenoids). (The function of *antioxidants*, substances that protect biomolecules from free radicals, is described in Chapter 10.) The protein components of lipoproteins, called *apolipoproteins* or *apoproteins*, are synthesized in the liver or intestine. There are five major classes of apolipoproteins: A, B, C, D, and E. A generalized lipoprotein is shown in **Figure 11.21**. The relative amounts of lipid and protein components of the major types of lipoprotein are summarized in **Figure 11.22**.

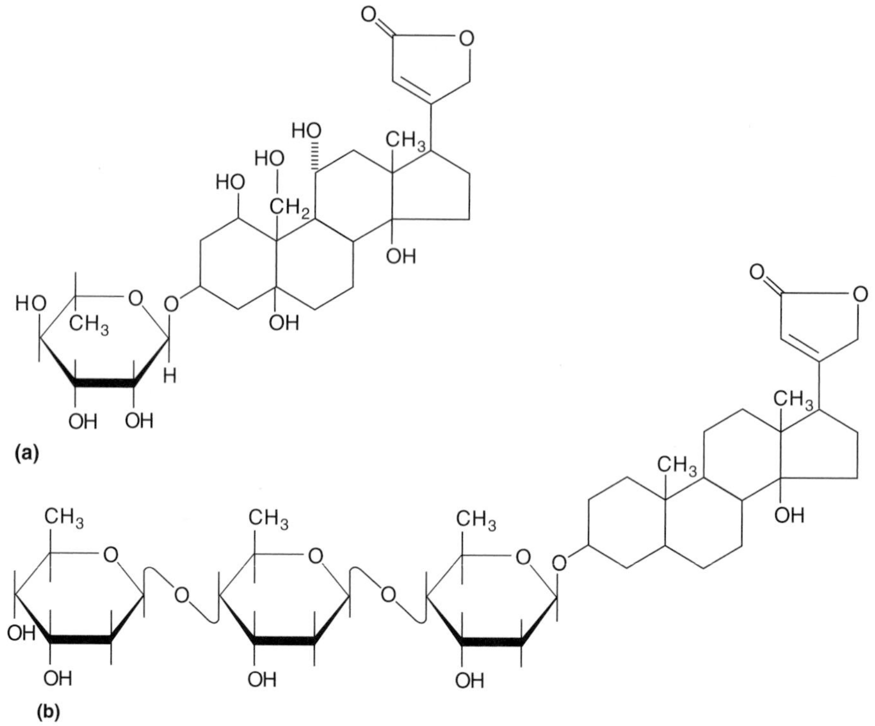

FIGURE 11.20

Cardiac Glycosides

Each cardiac glycoside possesses a glycone (carbohydrate) and an aglycone component. (a) In ouabain the glycone is one rhamnose residue. The steroid aglycone of ouabain is called ouabagenin. (b) The glycone of digitoxin is composed of three digitoxose residues. The aglycone of digitoxin is called digitoxigenin.

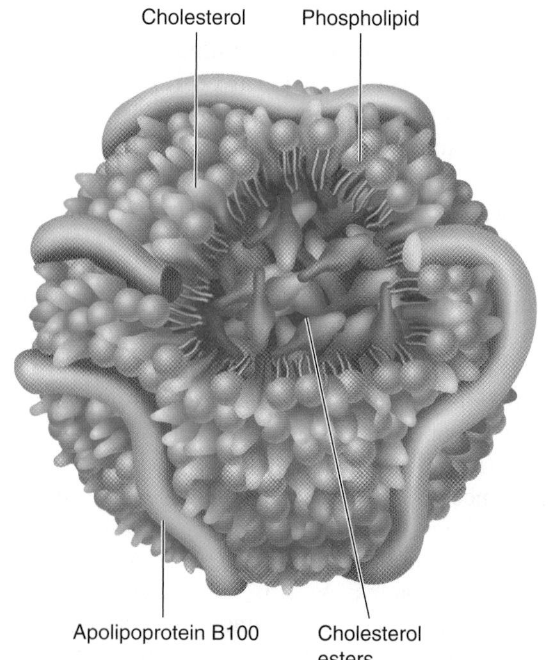

FIGURE 11.21

Plasma Lipoproteins

Lipoproteins vary in diameter from 5 to 1000 nm. Each type of lipoprotein contains a neutral lipid core composed of cholesteryl esters and/or triacylglycerols. This core is surrounded by a layer of phospholipid, cholesterol, and protein. Charged and polar residues on the surface of a lipoprotein enable it to dissolve in blood. In a low-density lipoprotein (LDL), as illustrated in this figure, each particle is composed of a core of cholesteryl esters surrounded by a monolayer that consists of hundreds of cholesterol and phospholipid molecules and several apolipoproteins, including apolipoprotein B100, the ligand for the LDL receptor (p. 416).

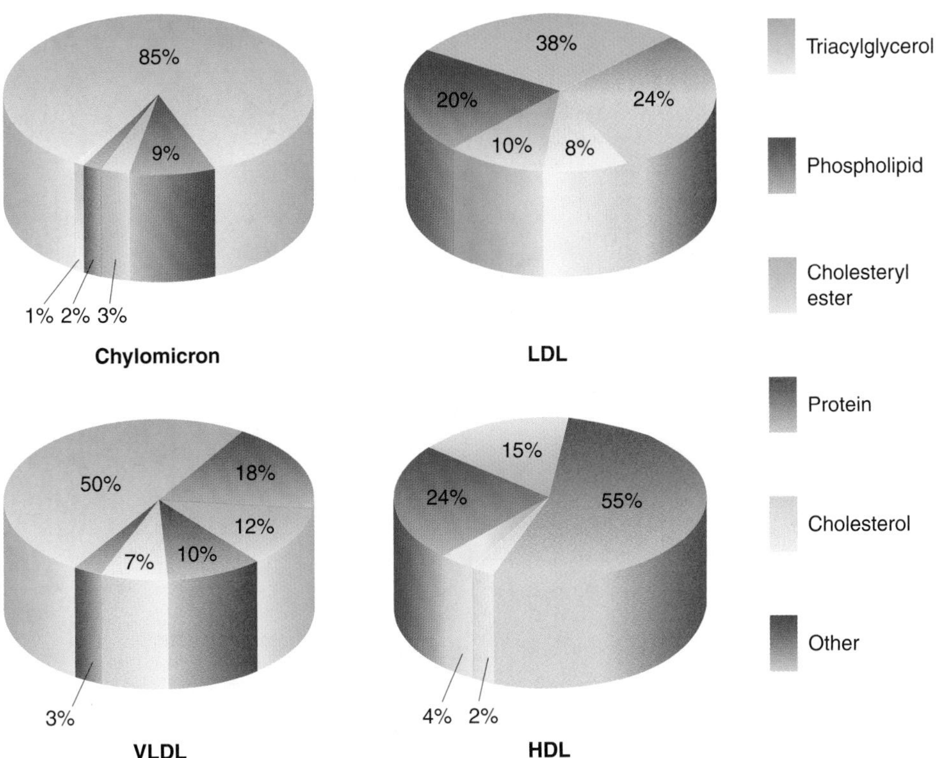

FIGURE 11.22

Proportional (Relative) Mass of Cholesterol, Cholesteryl Ester, Phospholipid, and Protein Molecules in Four Major Classes of Plasma Lipoproteins

Chylomicrons are the largest but least dense of the plasma lipoproteins because of their high content of triacylglycerols. In contrast, high-density lipoprotein (HDL) is a smaller dense particle that contains a high percentage by mass of protein and low triacylglycerol content.

Lipoproteins are classified according to their density. **Chylomicrons**, which are large lipoproteins (diameter $\leqq$ 1000 nm) of extremely low density (<0.95 g/cm^3), transport dietary triacylglycerols and cholesteryl esters from the intestine to muscle and adipose tissues. Chylomicron remnants are then taken up by the liver via endocytosis. **Very-low-density lipoproteins** (VLDLs) (0.98 g/cm^3, D = 30–90 nm), synthesized in the liver, transport lipids to tissues. As VLDLs are depleted of triacylglycerol and some apolipoprotein and phospholipids, they shrink in size, become more dense, and are referred to as **intermediate-density lipoproteins** (IDLs) or VLDL remnants (1 g/cm^3, D = 40 nm). IDLs may continue to lose triacylglycerol to form a higher-density lipoprotein called **low-density lipoprotein** (LDL), or they may be removed from the bloodstream by the liver. LDL, or IDL remnants (1.04 g/cm^3, D = 20 nm), are the principal transporters of cholesterol and cholesteryl esters to tissues. In an intricate process (p. 416) elucidated by Michael Brown and Joseph Goldstein (recipients of the 1985 Nobel Prize in Physiology or Medicine), LDL particles bind to LDL receptors and then are engulfed by cells. LDLs are also classified according to their diameters. LDLs with diameters less than 25 nm are referred to as small dense LDLs (sdLDLs). Those with diameters greater than 25 nm are called large buoyant LDL. sdLDLs are more atherogenic (i.e., prone to promotion of fatty plaques in arteries) than buoyant LDLs because they easily enter artery walls, where they are susceptible to oxidation. Risk factors for high blood sdLDL levels include genetic predisposition, high carbohydrate diet, physical inactivity, and insulin resistance (p. 588). The role of **high-density lipoproteins** (HDLs) (1.2 g/cm^3, D = 9 nm), protein-rich particles produced in the liver and the intestine, appears to be the scavenging of excessive cholesterol from cell membranes and cholesteryl esters

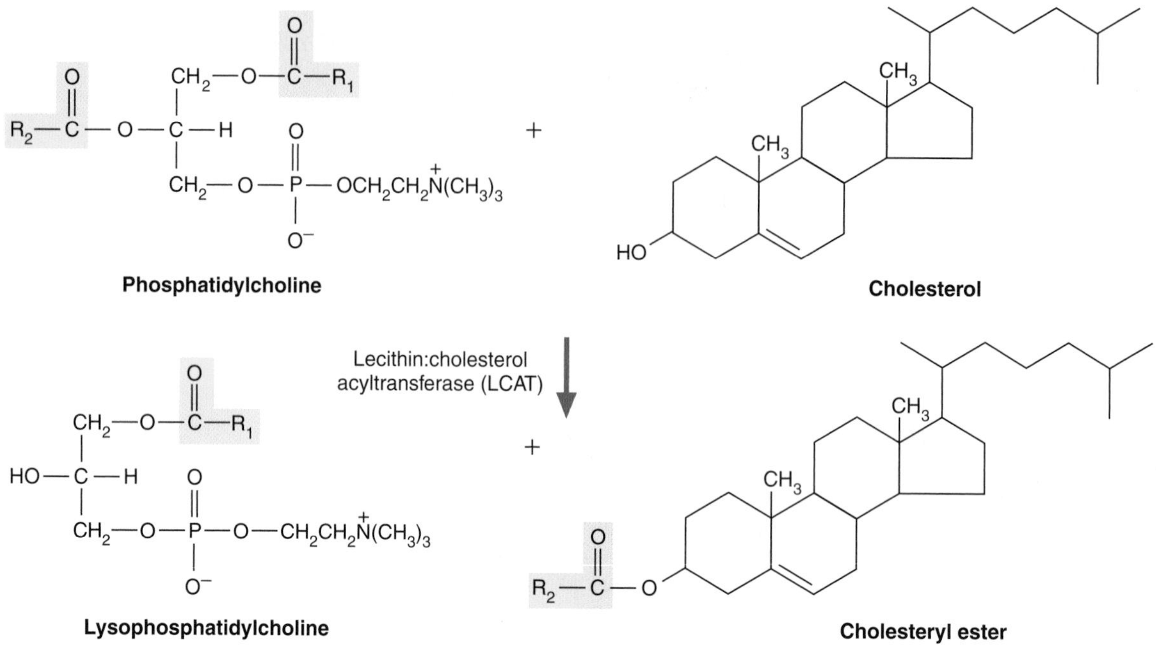

Phosphatidylcholine + Cholesterol

Lecithin:cholesterol acyltransferase (LCAT)

Lysophosphatidylcholine + Cholesteryl ester

FIGURE 11.23

Reaction Catalyzed by Lecithin:Cholesterol Acyltransferase (LCAT)

Cholesteryl ester transfer protein, a protein associated with the LCAT–high-density lipoprotein (HDL) complex, transfers cholesteryl esters from HDL to very-low-density lipoprotein and LDL. The acyl groups are highlighted in color.

KEY CONCEPTS

- Plasma lipoproteins transport lipids through the bloodstream.
- On the basis of density, lipoproteins are classified into five major classes: chylomicrons, VLDL, IDL, LDL, and HDL.

from VLDL and LDL. Cholesteryl esters are formed when the plasma enzyme lecithin:cholesterol acyltransferase (LCAT) transfers a fatty acid residue from lecithin to cholesterol (**Figure 11.23**). (Apolipoprotein A1, a component of HDL, is a cofactor of LCAT.) HDL transports cholesteryl esters to the liver and steroid-producing organs such as adrenal glands, ovary, and testes. The liver, the only organ that can dispose of excess cholesterol, converts most of it to bile acids (Chapter 12). The role of lipoproteins in atherosclerosis, a chronic disease of the cardiovascular system, is discussed in Chapter 12.

11.2 MEMBRANES

Most of the properties attributed to living organisms (e.g., movement, growth, reproduction, and metabolism) depend, either directly or indirectly, on membranes. All biological membranes have the same general structure. As previously mentioned (Chapter 2), membranes contain lipid and protein molecules. In the currently accepted concept of membranes, referred to as the **fluid mosaic model**, or the Singer-Nicholson model, a membrane is a noncovalent heteropolymer of a **lipid bilayer** and associated proteins. The nature of these molecules determines each membrane's biological functions and mechanical properties. Because of the importance of membranes in biochemical processes, the remainder of this chapter is devoted to a discussion of their structure and functions.

Membrane Structure

Each type of living cell has a unique set of functions. Consequently, the membranes of each cell type have unique features. Not surprisingly, the proportions of lipid and protein vary considerably among cell types and among organelles within each cell (**Table 11.5**). The types of lipid and protein found in each membrane also vary.

TABLE 11.5 Chemical Composition of Some Cell Membranes

Membrane	Protein (%)	Lipid (%)	Carbohydrate (%)
Human erythrocyte plasma membrane	49	43	8
Mouse liver cell plasma membrane	46	54	2–4
Amoeba plasma membrane	54	42	4
Mitochondrial inner membrane	76	24	1–2
Spinach chloroplast lamellar membrane	70	30	6
Halobacterium purple membrane	75	25	0

Source: G. Guidotti, Membrane Proteins, *Annu. Rev. Biochem.* 41:731, 1972.

MEMBRANE LIPIDS When amphipathic molecules are suspended in water, they spontaneously rearrange into ordered structures (**Figure 11.9**). As these structures form, hydrophobic groups become buried in the water-depleted interior. Simultaneously, hydrophilic groups become oriented so that they are exposed to water. Phospholipids form into bimolecular layers at relatively low concentration. This property of phospholipids (and other amphipathic lipid molecules) is the basis of membrane structure. Membrane lipids are largely responsible for several other important features of biological membranes as well.

Membrane Fluidity. The term *fluidity* refers to the viscosity of the lipid bilayer (i.e., the degree of resistance of membrane components to movement). Membrane fluidity is largely determined by the percentage of unsaturated fatty acids in its phospholipid molecules. (Recall that the hydrocarbon chains of unsaturated fatty acids pack less densely than saturated chains; see p. 385). A membrane's fluidity increases as its percentage of unsaturated fatty acids increases. Cholesterol contributes stability to animal cell membranes because of its rigid ring system and ability to form van der Waals interactions with contiguous hydrocarbon chains. However, fluidity remains high because of cholesterol's incomplete penetration into the membrane and the flexibility of its hydrocarbon tail (**Figure 11.24**). Membrane fluidity is an important feature of biological membranes because rapid lateral movement (**Figure 11.25**) of lipid molecules is apparently responsible for the proper functioning of many membrane proteins. The movement of lipid molecules from one side of a lipid bilayer to the other occurs only during membrane synthesis or under conditions of lipid imbalance and requires the function of ATP-requiring mediator proteins, in the process called facilitated diffusion. *Flippase* transfers phospholipids from the outer to the inner membrane leaflet, whereas *floppase* transfers them in the opposite direction. *Scramblase* is a nonspecific, energy-independent redistributor of phospholipids across membranes. One measure of membrane fluidity, the ability of membrane components to diffuse laterally, can be demonstrated when two different cell types are fused to form a *heterokaryon*. (Certain viruses or chemicals are used to promote cell-cell fusion.) The plasma membrane proteins of each cell type can be tracked when they are labeled with different fluorescent markers. Initially, the proteins are confined to their own side of the heterokaryon membrane. As time passes, the two fluorescent markers intermix, indicating that proteins move freely in the lipid bilayer.

Selective Permeability. Hydrophobic hydrocarbon chains in lipid bilayers provide a virtually impenetrable barrier to the transport of ionic and polar substances. Specific membrane proteins regulate the movement of such substances into and out of cells. To cross a lipid bilayer, a polar substance must shed some or all of its hydration sphere and bind to a carrier protein for membrane translocation or pass through an aqueous protein channel. Both methods shield the hydrophilic molecule from the hydrophobic core of the membrane. Water crosses the membrane through protein channels called *aquaporins* (p. 412), which exhibit cell type variation in their permeability to both water and accompanying ions. Nonpolar substances simply diffuse through the lipid bilayer down their concentration gradients. Each membrane exhibits its own transport capability or selectivity based on its protein component.

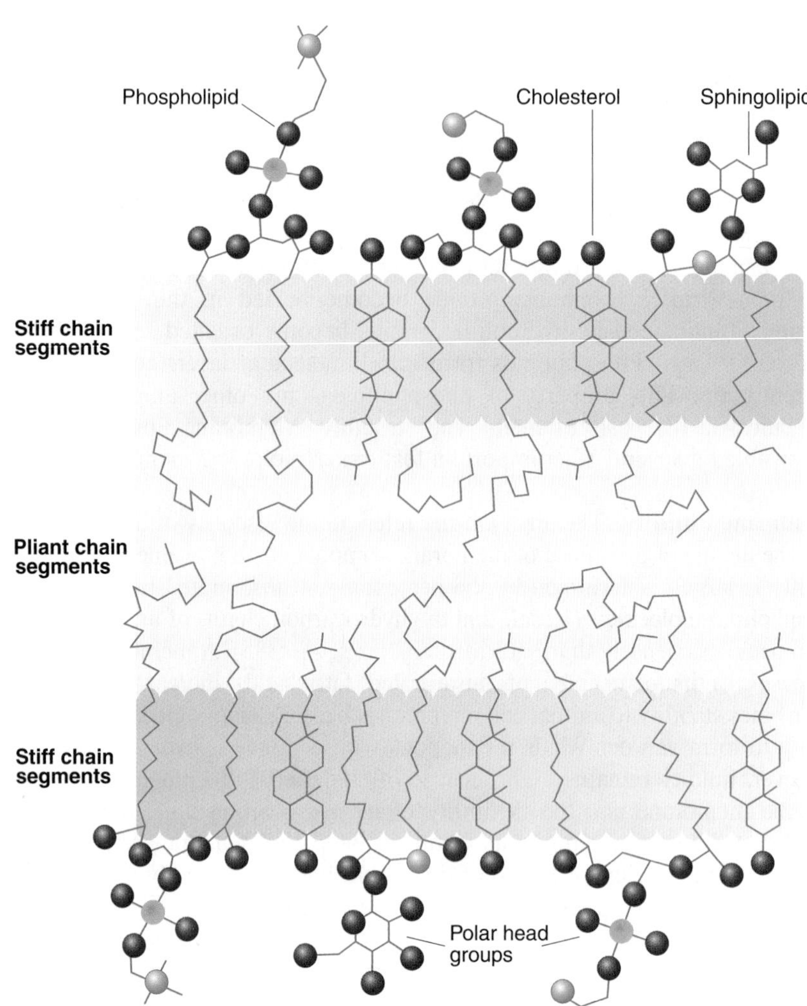

FIGURE 11.24

Diagrammatic View of a Lipid Bilayer

The flexible hydrocarbon chains in the hydrophobic core (lightly shaded area in the middle) make the membrane fluid. The phospholipids in the membrane have different levels of unsaturation and vary in the nature of the polar head group. Cholesterol's compact and rigid ring system creates structural stability in the outer region of each leaflet. Red blood cells and other cells that are subjected to mechanical stress have a high content of cholesterol and cardiolipin (two phospholipids linked by a glycerol). Cell membranes are about 7 to 9 nm thick. (Nitrogen atoms are blue, oxygen atoms are red, and phosphorus atoms are orange.)

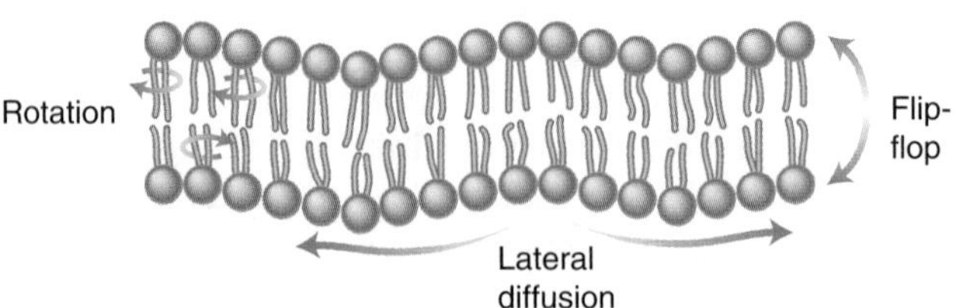

FIGURE 11.25

Lateral Diffusion in Biological Membranes

Lateral movement of phospholipid molecules is usually relatively rapid. "Flip-flop," the transfer of a lipid molecule from one side of a lipid bilayer to the other, occurs during new-membrane synthesis and membrane remodeling. Rotation of phospholipids within cell membranes is rapid.

Self-Sealing. When lipid bilayers are disrupted, they soon reseal. Small plasma membrane breaks spontaneously seal via the lateral flow of lipid molecules. Repair of larger lesions caused by mechanical stress, however, is an energy-requiring, Ca^{2+}-dependent process. A tear in the plasma membrane results in the inward flow of Ca^{2+} down its concentration gradient. Calcium ions trigger the movement of nearby endomembrane-derived vesicles to the lesion site. In an exocytosis-like process that involves cytoskeletal rearrangements, motor proteins such as dynein and kinesin, and membrane fusion proteins, the vesicles fuse with the plasma membrane to form a membrane patch. The repair process is rapid, usually occurring within a few seconds of the traumatic event.

Asymmetry. Biological membranes are asymmetric; that is, the lipid composition of each half of a bilayer is different. New membrane is synthesized by insertion of additional phospholipid molecules from the cytoplasmic face of existing membranes. Lipid molecules are transferred by mediator proteins to the opposite leaflet until membrane stability has been attained. Because the two faces of the resulting membrane are not chemically equivalent, the resulting leaflets are not identical in lipid composition. For example, the human red blood cell membrane possesses substantially more phosphatidylcholine and sphingomyelin on its outside surface. Most of the membrane's phosphatidylserine and phosphatidylethanolamine are on the inner side. The protein components of membranes (discussed next) also exhibit considerable asymmetry, with distinctly different functional domains within membrane and on the cytoplasmic and extracellular faces of membrane.

MEMBRANE PROTEINS Most of the functions associated with biological membranes require protein molecules. Membrane proteins are often classified by the function they perform: structure, transport, catalysis, signal transduction, or immunological identity. Membrane proteins are also classified according to their structural relationship to membrane. Proteins that are embedded in and/or extend through a membrane are referred to as ***integral proteins*** (**Figure 11.26**). Such molecules can be extracted only by disrupting the membrane with organic solvents or detergents. ***Peripheral proteins*** are bound to membrane primarily through noncovalent interactions with integral membrane proteins or covalent bonds to myristic, palmitic, or prenyl groups. GPI anchors link a wide variety of cell-surface proteins (e.g., lipoprotein lipase, folate receptor, alkaline phosphatase, and the core proteins of glypicans) to plasma membranes. Some peripheral proteins interact directly with the lipid bilayer.

The band 3 anion exchanger protein (AE1) (**Figure 11.27**), found in red blood cell membrane, is a well-researched example of integral membrane protein. AE1 is composed of two identical subunits, each consisting of 929 amino acids. With

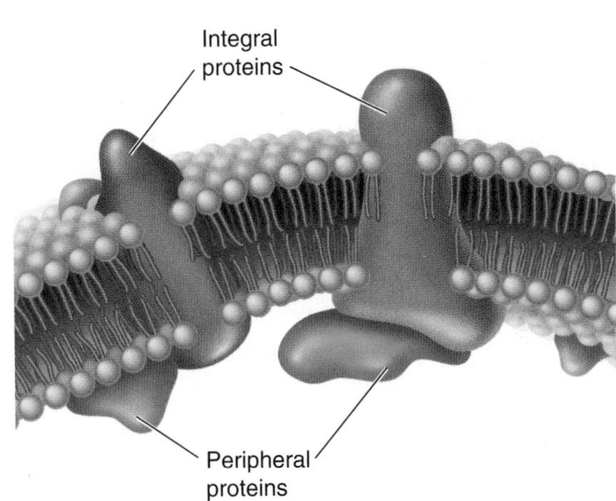

FIGURE 11.26
Integral and Peripheral Membrane Proteins

Integral membrane proteins are released only if the membrane is disrupted with detergents. Many peripheral membrane proteins can be removed with mild reagents, such as high salt concentration.

FIGURE 11.27
Red Blood Cell Integral Membrane Proteins

The integral membrane proteins glycophorin and anion-exchange protein are components in a network of linkages that connect the plasma membrane to structural elements of the cytoskeleton (e.g., actin, spectrin, protein band 4.1, and ankyrin). Note that the oligosaccharides on glycophorin are the ABO and MN blood group antigens.

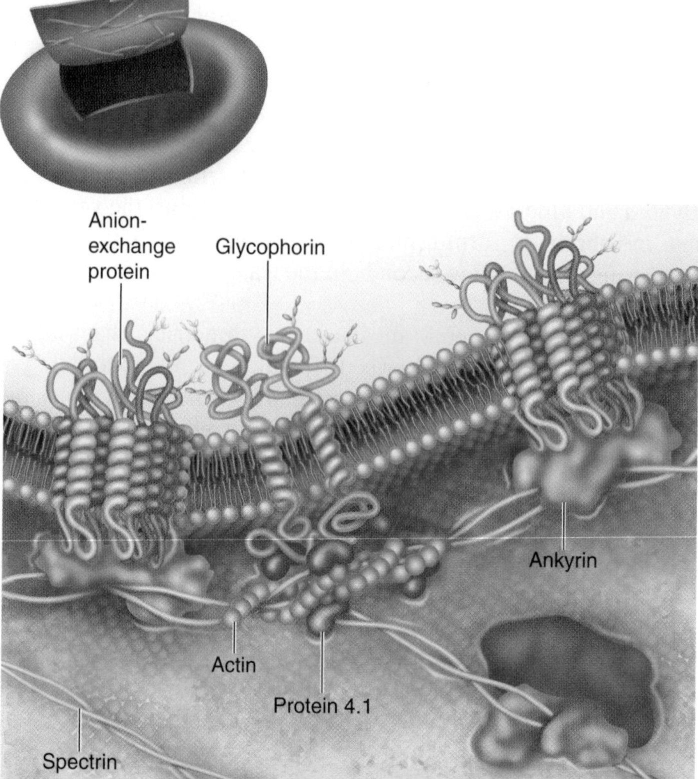

more than 1 million copies per cell, this protein channel plays an important role in CO_2 transport in blood. The HCO_3^- ion formed from CO_2 with the aid of carbonic anhydrase diffuses into and out of the red blood cell through the anion channel in exchange for chloride ion Cl^-. (The exchange of Cl^- for HCO_3^-, called the *chloride shift*, preserves the electrical potential of the red blood cell membrane.)

Red blood cell membrane peripheral proteins are composed largely of spectrin, ankyrin, and band 4.1. They are primarily involved in preserving the cell's unique biconcave shape, which maximizes the surface area:volume ratio and exposure of diffusing O_2 to intracellular hemoglobin. *Spectrin* is a tetramer, composed of two $\alpha\beta$ dimers, that binds to ankyrin and band 4.1. *Ankyrin* is a large globular polypeptide (215 kDa) that links spectrin to the anion channel protein. (This is a connecting link between the red blood cell's cytoskeleton and its plasma membrane.) *Band 4.1* binds to both spectrin and *actin filaments* (a cytoskeletal component found in many cell types). Because band 4.1 also binds to glycophorin, it too links the cytoskeleton and the membrane.

KEY CONCEPTS

- The basic structural feature of membrane structure is the lipid bilayer, which consists of phospholipids and other amphipathic lipid molecules.
- Membrane proteins, embedded in or associated with the phospholipid bilayer, contribute specialized functions to the membrane depending on its cell type and its role in biological processes.

THE BAND 3 ANION EXCHANGER MACROCOMPLEX AND THE CHLORIDE SHIFT The red blood cell protein AE1 contains three domains: a 12-segment membrane-spanning domain that carries out anion exchange, an extended cytoplasmic N-terminal domain, and a short cytoplasmic C-terminal domain. In its tetramer form AE1 is a macrocomplex with numerous integral and peripheral membrane proteins. The transmembrane domains bind to glycophorin and AQP1 (a water channel; see p. 412). The C-terminal domains bind carbonic anhydrase (CA), and the N-terminal domains bind several glycolytic enzymes (e.g., phosphofructokinase, aldolase, and glyceraldehyde-3-phosphate dehydrogenase), deoxyhemoglobin, and ankyrin, among others.

As freshly oxygenated blood reaches the tissues, carbon dioxide molecules released by respiring cells enter red blood cells, where CA rapidly converts them into HCO_3^- and H^+. Bicarbonate ions then leave the cell via the AE1 channel in exchange for chloride ions. The water molecules required for the CA-catalyzed reaction are supplied by AQP1. Excess protons cause the intracellular pH to drop, which decreases the affinity of hemoglobin for oxygen (the Bohr effect, p. 174). As deoxyHb forms, a few molecules displace the glycolytic enzymes from the N-terminal domains of AE1 in a process that increases their catalytic activity. Two products of glycolysis, BPG (p. 174) and ATP, have unique roles in erythrocytes. As described previously, BPG binds to and stabilizes deoxyHb, thereby promoting oxygen release to the tissues. As red blood cells squeeze through narrow capillaries, it is ATP that drives the membrane pumps that restore the ion concentrations disrupted by leakage caused by mechanical stress.

When blood flows through the lungs, its pH and oxygen level rise, and the process reverses. The T-to-R shift in hemoglobin conformation (p. 173) occurs as oxygen binds to hemoglobin and deoxyHb is destabilized by the release of BPG, protons, and other allosteric effectors. The resulting shift in the chemical equilibrium of the CA-catalyzed reaction is in favor of the conversion of HCO_3^- to CO_2. Bicarbonate enters the cell via AE1 in exchange for chloride ions and CO_2 flows out of the blood down its concentration gradient and into the lung's alveolar cells. The release of newly oxygenated hemoglobin from the N-terminal domain of AE1 facilitates the rebinding of the glycolytic enzymes, depressing their activity. As a result, more glucose molecules are diverted into the pentose phosphate pathway, causing the synthesis of higher levels of NADPH. NADPH is required in oxygenated red blood cells for the reduction of the ferrous iron in methemoglobin (p. 146) and protection of the red blood cell membrane from oxidative stress.

MEMBRANE MICRODOMAINS The lipids and proteins in membranes are not distributed uniformly. A notable example is provided by "lipid rafts," specialized microdomains in the external leaflet of eukaryotic plasma membranes (Figure 11.28). The components of lipid rafts are primarily cholesterol and sphingolipids and certain membrane proteins. The rigid fused rings of cholesterol pack tightly alongside the more saturated acyl chains of sphingolipid molecules. Consequently, the lipid molecules in these microdomains are more ordered

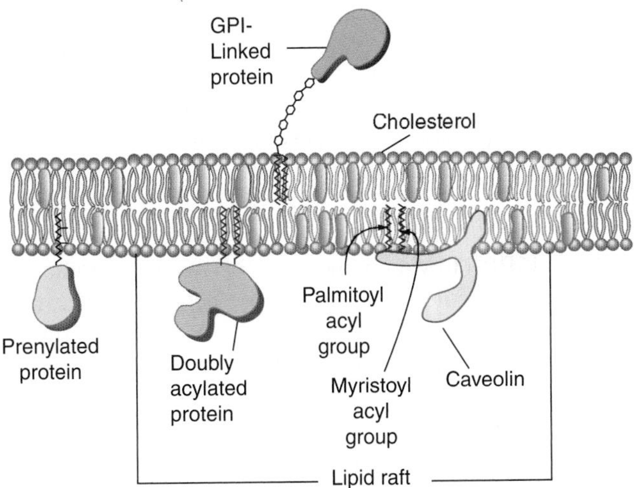

FIGURE 11.28

Lipid Rafts

As a result of stable associations between cholesterol and sphingolipids, slightly thicker membrane microdomains called lipid rafts form. Lipid rafts are also enriched in specific types of membrane protein. (Caveolin is a membrane protein found in caveolae, curved lipid rafts involved in clathrin-independent endocytosis and other processes.)

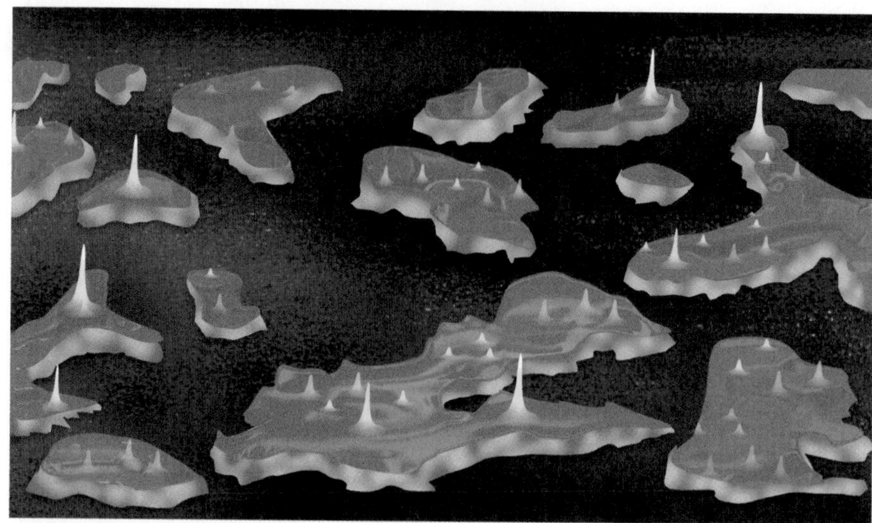

FIGURE 11.29
The Lipid Raft Environment
Atomic force microscopy provides a view of thick lipid rafts surrounded by a more fluid bilayer. The peaks in this micrograph represent GPI-linked proteins.

(i.e., less fluid) than those in nonraft regions, where unsaturated acyl chains (in phospholipids) are more common. As their name suggests, lipid rafts seemingly float in a sea of more loosely packed membrane lipids (**Figure 11.29**).

Lipid rafts are enriched in some classes of proteins and they exclude others. Lipid-raft-associated proteins include GPI-anchored proteins, doubly acetylated tyrosine kinases, and certain transmembrane proteins. Some proteins are always present in lipid rafts, whereas others enter lipid rafts only as a result of an activation process. Lipid rafts have been implicated in a variety of cellular processes. Examples include exocytosis, endocytosis, and signal transduction. Lipid rafts are believed to function as platforms where the molecules that drive these processes are spatially organized. Caveolae (p. 410) are a special type of lipid raft.

Membrane Function

Among the vast array of membrane functions are the transport of polar and charged substances into and out of cells and organelles and the relay of signals that initiate change in metabolic and developmental aspects of cell function. Each of these topics is discussed briefly. A description of membrane receptors follows.

MEMBRANE TRANSPORT Membrane transport mechanisms are vital to living organisms. Ions and molecules constantly move across cell plasma membranes and across the membranes of organelles. This flux must be carefully regulated to meet each cell's metabolic needs. Additionally, the plasma membrane regulates intracellular ion concentrations. Because lipid bilayers are generally impenetrable to ions and polar substances, specific transport components must be inserted into cellular membranes. Several examples of these structures, referred to as transport proteins, are discussed.

Biological transport mechanisms are classified according to whether they require energy. Major types of biological transport are illustrated in Figure 11.30. In **passive transport** (simple and facilitated diffusion), substances are moved across the membrane with their concentration gradient and require no direct input of energy. In contrast, **active transport** requires energy to transport molecules against a concentration gradient.

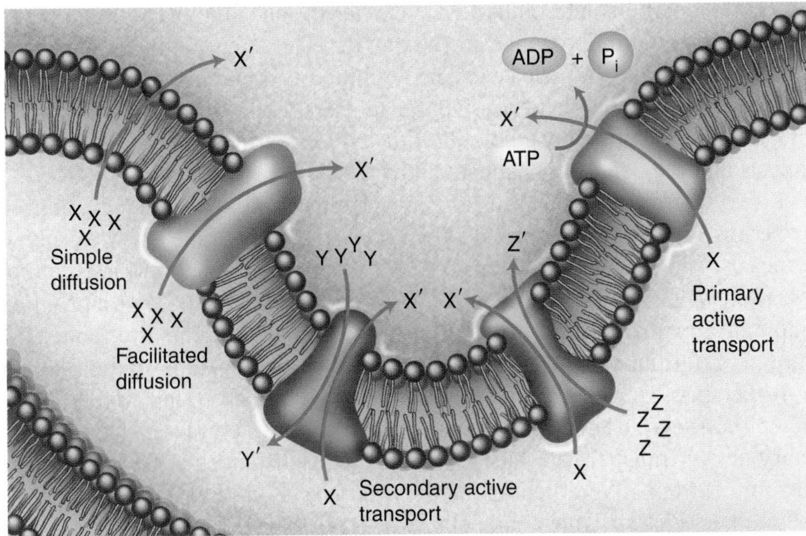

FIGURE 11.30

Transport across Membranes

The major transport processes are simple and facilitated diffusion and primary and secondary active transport. In simple diffusion the spontaneous transport of a specific solute is driven by its concentration gradient. Facilitated diffusion, the movement of a solute down its concentration gradient across a membrane, occurs through protein channels or carriers. Both primary and secondary active transport require energy to move solutes across a membrane against their concentration gradients. In primary active transport, this energy is usually provided directly by ATP hydrolysis. In secondary active transport, solutes (X) are moved across a membrane by energy stored in a concentration gradient of a second substance (Y or Z) that has been created by ATP hydrolysis or another energy-requiring mechanism.

In **simple diffusion**, each solute, propelled by random molecular motion, moves down its concentration gradient (i.e., from an area of high concentration to an area of low concentration). In this spontaneous process, there is a net movement of solute until an equilibrium is reached. A system reaching equilibrium becomes more disordered: that is, entropy increases. Because there is no input of energy, transport occurs with a negative change in free energy. In general, the higher the concentration gradient, the faster the rate of solute diffusion. The diffusion of gases such as O_2 and CO_2 across membranes is proportional to their concentration gradients. The diffusion of nonpolar organic molecules (such as steroid hormones) also depends on molecular weight and lipid solubility.

In **facilitated diffusion**, certain large or charged molecules are moved through special channels or carriers. *Channels* are tunnel-like transmembrane proteins. Each type usually transports a specific solute. Many channels are regulated. Chemically regulated channels open or close in response to a specific chemical signal. For example, a *chemically gated Na^+ channel* in the nicotinic acetylcholine receptor complex (found in muscle cell plasma membranes) opens when the neurotransmitter molecule acetylcholine binds. Then Na^+ rushes into the cell, and the membrane potential falls. Because membrane potential is an electrical gradient across the membrane (see p. 91), a decrease in membrane potential is membrane *depolarization*. Local depolarization caused by acetylcholine leads to the opening of nearby Na^+ channels (these are referred to as *voltage-gated Na^+ channels*). *Repolarization*, the reestablishment of the membrane potential, begins with the diffusion of K^+ ions out of the cell through *voltage-gated K^+ channels*. (The diffusion of K^+ ions out of the cell makes the inside less positive, that is, more negative.)

Another form of facilitated diffusion involves membrane proteins called *carriers* (sometimes referred to as *passive transporters*). In carrier-mediated

Visit the companion website at www.oup.com/us/mckee to read the Biochemistry in Perspective essay on cell volume regulation and metabolism in Chapter 3.

transport, a specific solute binds to the carrier on one side of a membrane and causes a conformational change in the carrier. The solute is then translocated across the membrane and released. The red blood cell *glucose transporter* is the best characterized example of passive transporters. It allows D-glucose to diffuse across the red blood cell membrane with its concentration gradient for use in glycolysis and the pentose phosphate pathway.

The two forms of active transport are primary and secondary. In **primary active transport**, energy is provided by ATP. Transmembrane ATP-hydrolyzing enzymes use the energy derived from ATP to drive the transport of ions or molecules. The Na^+-K^+ pump (also referred to as the Na^+-K^+ ATPase) is a prominent example of a primary transporter. (The Na^+ and K^+ gradients are required for maintaining normal cell volume and membrane potential. Refer to the Biochemistry in Perspective essay entitled Cell Volume Regulation and Metabolism in Chapter 3). In **secondary active transport**, concentration gradients generated by primary active transport are harnessed to move substances across membranes. For example, the Na^+ gradient created by the Na^+-K^+ ATPase pump is used in kidney tubule cells and intestinal cells to transport D-glucose against its concentration gradient (**Figure 11.31**). Two examples of membrane transport proteins, the aquaporins and cystic fibrosis–related chloride channel, are described next.

THE AQUAPORINS A basic characteristic of living cells is the capacity to rapidly move water across cell membranes in response to changes in osmotic pressure. For years, many researchers assumed that simple diffusion was responsible for most water flow. In a wide variety of cell types such as red blood cells and certain kidney cells, it became apparent that water flow is extraordinarily rapid. In the early 1990s, investigators characterized the first of a series of water channel proteins, now called the aquaporins. Found initially in red blood cell membrane and then in kidney tubule cells, **aquaporin** 1 (AQP1) is an intrinsic membrane protein complex that facilitates water flow, about 3×10^9 water molecules/s/ channel. Aquaporins have been found in almost all living organisms, with at least 10 different forms in mammals, with different water and ion permeability characteristics.

Recent experimental evidence suggests that water flow through aquaporin channels is regulated. For example, three mammalian aquaporins appear to be regulated by pH. Others are regulated by phosphorylation reactions or by the binding of specific signal molecules. In 1993, the cause of a rare inherited form of **nephrogenic diabetes insipidis** (a disease in which the kidneys of

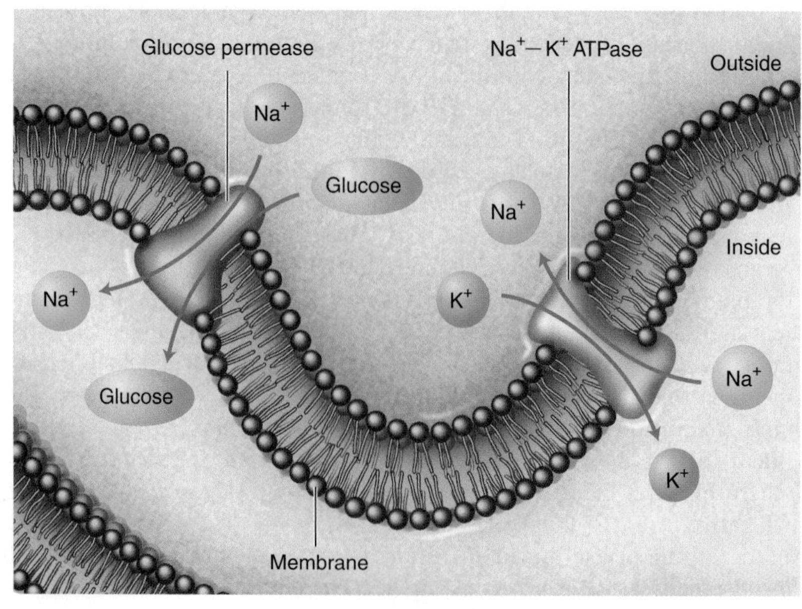

FIGURE 11.31
The Na^+-K^+ ATPase and Glucose Transport

The Na^+-K^+ ATPase preserves the Na^+ gradient essential to maintain membrane potential. In certain cells, glucose transport depends on the Na^+ gradient. Glucose permease transports both Na^+ and glucose. Only when both substrates are bound does the protein change its conformation, thus initiating transport.

affected individuals cannot produce concentrated urine) was discovered to be a mutation in the gene for AQP2. The mutated AQP2 does not respond to the antidiuretic hormone vasopressin (**Table 5.3**, p. 145).

Of all the aquaporins, AQP1, a homotetramer that has high permeability to water only, is the best characterized. Each subunit is a polypeptide containing 269 amino acid residues that form a water-transporting pore with six α-helical membrane-spanning domains connected by five loops. Although each monomer is an independent water channel, the formation of the tetramer is required for full function. In the functional monomer, the two loops that both possess an Asn-Pro-Ala (NPA) sequence meet in the middle to form the water-binding site. The pore, which has been measured at 3 Å, is only slightly larger than a water molecule (2.8 Å). The movement through the channel of water molecules only, and not smaller species such as H^+, is believed to be made possible by the formation of hydrogen bonds between water molecules and the Asn residues of the two NPA sequences (**Figure 11.32**). The hydrophobic environment created by the amino acid residues on the other helices that comprise the pore causes the hydrogen bonds between water molecules to break as they move in single file toward the narrowest portion of the pore. It also forces the oxygen atom of each water molecule to orient itself toward the Asn residues. When the water molecule approaches the 3 Å constriction in the pore, its oxygen atom sequentially forms and breaks hydrogen bonds with the side chains of the two Asn residues. The absence of other hydrogen bonding partners prevents the ionization of H_2O and the generation of protons. Several aquaporins, namely AQP3, AQP7, AQP9, and AQP10, are referred to as aquaglyceroporins because they are permeable to small solutes such as glycerol in addition to water. Peter Agre received the Nobel Prize in Chemistry in 2003 for his discovery and investigation of aquaporins.

THE CYSTIC FIBROSIS TRANSMEMBRANE CONDUCTANCE REGULATOR

Impaired membrane transport mechanisms can have serious consequences. One of the best understood examples of dysfunctional transport occurs in cystic fibrosis. **Cystic fibrosis** (CF), a fatal autosomal recessive disease, is caused by

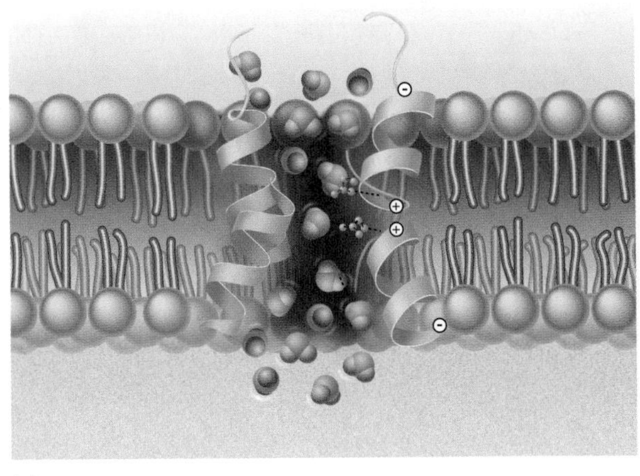

(a)

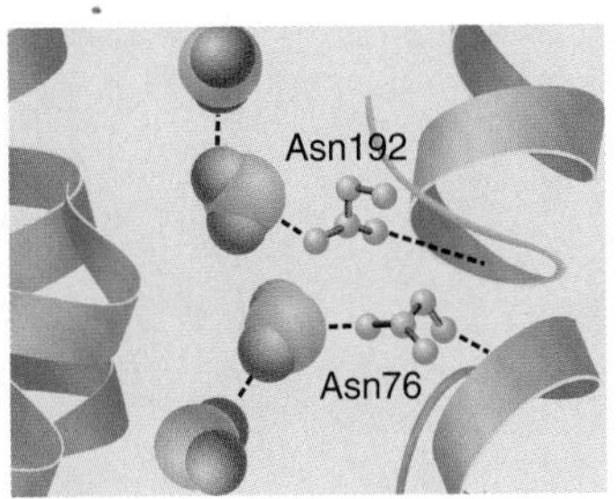

(b)

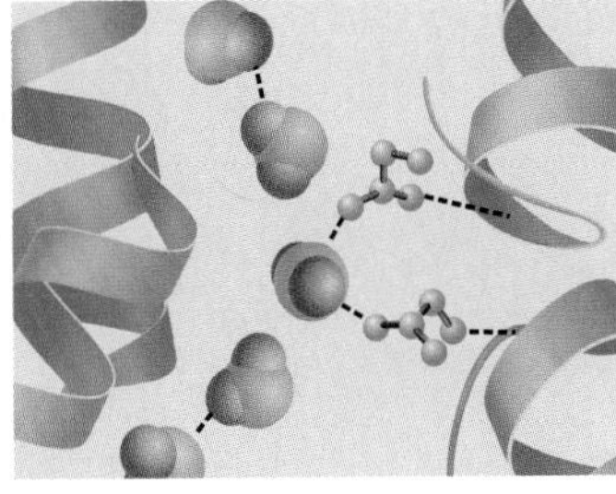

(c)

FIGURE 11.32

Water Transport through the AQP1 Monomer

Water molecules move through the pore in single file. As each molecule approaches the constriction in the pore, it is forced to orient its oxygen atom so it can then form and break hydrogen bonds with the side chains of the two Asn residues. (a) Within the pore of the aquaporin monomer, there is a positive electrostatic environment in which the oxygen atom of each water molecule is oriented toward the two Asn residues. (b) and (c) The sequential formation and breakage of hydrogen bonds between the oxygen of water molecules and the two Asn residue side chains mediate the movement of water through the pore.

KEY CONCEPTS

- Membrane transport mechanisms are classified as passive or active according to whether they require energy.
- In passive transport, solutes moving across membranes move down their concentration gradient.
- In active transport, energy derived directly or indirectly from ATP hydrolysis or another energy source is required to move an ion or molecule against its concentration gradient.

a missing or defective plasma membrane glycoprotein called **cystic fibrosis transmembrane conductance regulator (CFTR)**. CFTR (**Figure 11.33**), which functions as a chloride channel in epithelial cells, is a member of a family of proteins referred to as ABC transporters. (ABC transporters are so named because each contains a polypeptide segment called an *ATP-binding cassette*.) The CFTR gene on chromosome 7 codes for the CFTR protein, which contains five domains. Two domains (MSD1 and MSD2), each containing six membrane-spanning helices, form the Cl^- channel pore. Chloride transport through the pore is controlled by the other three domains (all of which occur on the cytoplasmic side of the plasma membrane). Two are nucleotide-binding domains (NBD_1 and NBD_2) that bind and hydrolyze ATP and use the released energy to drive conformational changes in the pore. The regulatory (R) domain contains several amino acid residues that must be phosphorylated by cAMP-dependent protein kinase (PKA) for chloride transport to occur.

The chloride channel is vital for proper absorption of salt (NaCl) and water across the apical (top) membrane surface of epithelial cells that line ducts and tubes in tissues such as lungs, liver, small intestine, and sweat glands. Chloride channel opening occurs in response to a signal molecule, cAMP. The cAMP-dependent kinase PKA then phosphorylates specific residues in the R domain, causing a change in its conformation that triggers the binding of ATP molecules to NBD_1 and NBD_2. The two nucleotide-binding domains then form a head-to-tail heterodimer-like structure with the ATP-binding sites on the inside surfaces. As a result of these intramolecular rearrangements, the chloride channel gate opens and chloride ions flow down their concentration gradient. Hydrolysis of one of the NBD-bound ATP molecules causes dimer disruption that results in channel closing. The NBD dimer acts as a timing device in that the rate of ATP hydrolysis determines the length of time that the channel is open.

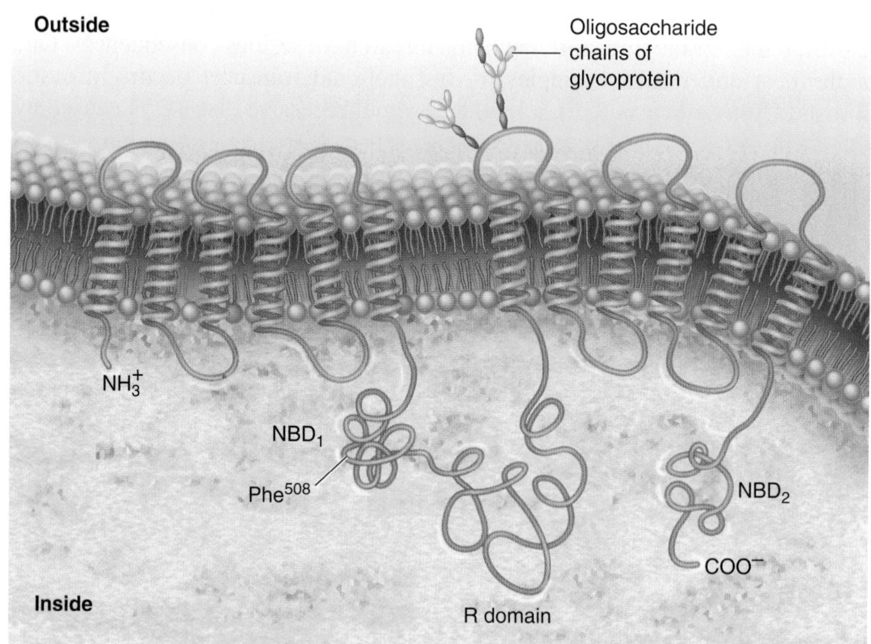

FIGURE 11.33

The Cystic Fibrosis Transmembrane Conductance Regulator (CFTR)

CFTR is a chloride channel composed of two domains (each with six membrane-spanning helices) that constitute the Cl^- pore, two nucleotide-binding domains (NBD), and a regulatory (R) domain. Transport of Cl^- through the pore, driven by ATP hydrolysis, occurs when specific amino acid residues on the R domains are phosphorylated. The most commonly observed CF-causing mutation is a deletion of Phe^{508} in NBD_1, which prevents proper targeting of CFTR-containing vesicles to the plasma membrane. The precise structural relationships among the pore-forming helices remain unclear.

In cystic fibrosis, the failure of CFTR channels results in the retention of Cl^- within the cells. A thick mucus or other secretion forms because osmotic pressure causes the excessive uptake of water from the mucus. The most obvious features of CF are lung disease (obstructed air flow and chronic bacterial infections) and pancreatic insufficiency (impaired production of digestive enzymes that can result in severe nutritional deficits). In the majority of CF patients, CFTR is defective because of a deletion mutation at Phe^{508}, which causes protein misfolding that prevents the processing and insertion of the mutant protein into the plasma membrane. Less common causes of CF (more than 100 have been described) include defective formation of CFTR mRNA molecules, mutations in the nucleotide-binding domains that result in ineffective binding or hydrolysis of ATP, and mutations in the pore-forming domains that cause reduced chloride transport.

Before the development of modern therapies, CF patients rarely survived childhood. It is only because of antibiotics (used principally in the treatment of lung infections) and commercially available digestive enzymes (replacing the enzymes normally produced in the pancreas) that many CF patients can now expect to live into their 30s. Yet as with the sickle-cell gene (see p. 151), defective CF genes are not rare. With an approximate incidence of 1 in every 2500 Caucasians, CF is the most common fatal genetic disorder in this population. Recent experiments with "knockout" mice indicate that carriers of the mutant gene are protected from diseases that kill because of diarrhea. (Knockout animals are bred to contain one copy of a defective gene in all of their cells.) The experimental animals used in CF research lose significantly less body fluid because they have a reduced number of functional chloride channels. It is suspected that CF carriers (individuals having only one copy of a defective CF gene) are also less susceptible to fatal diarrhea (e.g., in cholera) for the same reason. The CF gene did not spread beyond western Europe (the incidence among East Asians is approximately 1 in 100,000) because CF carriers secrete slightly more salt in their sweat than noncarriers and the epithelial cells that line sweat gland ducts cannot reabsorb chloride efficiently. In warmer climates, where sweating is a common feature of daily life, chronic excessive salt loss is far more dangerous than intermittent exposure to diarrhea-causing microorganisms. In recent years, researchers have attempted to solve mutation-caused CFTR dysfunction with small molecules called correctors (molecules that improve folding in the ER and trafficking to the cell surface) and potentiators (molecules that enhance the function of plasma membrane–bound CFTR). Several such molecules have completed clinical trials. For example, the corrector molecule VX809 partially restores CFTR function in individuals with the Phe^{508} deletion mutation. VX809 acts as a molecular chaperone by stabilizing the folded MSD1 in a process that results in some CFTR molecules reaching the cell surface. VX770 (also called Kalydeco) is a potentiator that improves chloride transport in individuals with a $G_{551}D$ mutation (glycine is replaced by aspartate). The $G_{551}D$ mutation results in a CFTR protein that reaches the cell surface but cannot transport Cl^- through the ion channel.

QUESTION 11.7

Suggest the mechanism(s) by which each of the following substances is (are) transported across cell membranes:

a. CO_2

b. Glucose

c. Cl^-

d. K^+

e. Fat molecules

f. α-Tocopherol

QUESTION 11.8

Describe the types of noncovalent interaction that promote the stability and the functional properties of biological membranes.

QUESTION 11.9

Transport mechanisms are often categorized according to the number of transported solutes and the direction of solute transport.

1. *Uniporters* transport one solute.
2. *Symporters* transport two different solutes simultaneously in the same direction.
3. *Antiporters* transport two different solutes simultaneously in opposite directions.

After examining the examples of transport discussed in this chapter, determine the category to which each belongs.

MEMBRANE RECEPTORS Membrane receptors provide mechanisms by which cells monitor and respond to changes in their environment. In multicellular organisms the binding to membrane receptors of chemical signals, such as the hormones and neurotransmitters of animals, is a vital link in intercellular communication. Other receptors are engaged in cell-cell recognition or adhesion. For example, lymphocytes perform a critical function of the immune system by transiently binding to the cell surface of virus-infected cells. This binding event triggers the lymphocyte-induced death of the infected cell. Similarly, the capacity of cells to recognize and adhere to other appropriate cells in a tissue is of crucial importance in many organismal processes, such as embryonic and fetal development.

The binding of a ligand to a membrane receptor results in a conformational change, which then causes a specific programmed response. Sometimes, receptor responses appear to be relatively straightforward. For example, the binding of acetylcholine to an acetylcholine receptor opens a cation channel. However, most responses are complex. The most intensively researched example of membrane receptor function is LDL receptor–mediated endocytosis, which is discussed next.

LDL RECEPTORS The low density lipoprotein receptor is responsible for the uptake of cholesterol-containing lipoproteins into cells. The LDL receptor is a glycoprotein found on the surface of many cells. When cells need cholesterol for the synthesis of membrane or steroid hormones, they produce LDL receptors and insert them into discrete regions of plasma membrane. (These membrane regions usually constitute about 2% of a cell's surface). The protein *clathrin* forms a latticelike polymer on the cytosolic side of the membrane during the initial stages of endocytosis. The number of receptors per cell varies from 15,000 to 70,000, depending on cell type and cholesterol requirements.

The process of LDL receptor–mediated endocytosis occurs in several steps (**Figure 2.17**). It begins within several minutes after LDLs have bound to LDL receptors clustered in *coated* pits. Coated pits are concave regions of the membrane with a clathrin cage on the intracellular face that are high in LDL receptor proteins. The LDL-occupied coated pit pinches off inside the cell to become a coated vesicle. Subsequently, *uncoated vesicles* are formed as clathrin depolymerizes. Before uncoated vesicles fuse with early endosomes (p. 53), LDLs are uncoupled from LDL receptors as the pH changes from 7 to 5. (This change is created by ATP-driven proton pumps in the vesicle membrane.) LDL receptors are recycled to the plasma membrane, and LDL-containing late endosomes fuse with lysosomes. Proteins associated with the LDL particles are degraded to amino acids, and cholesteryl esters are hydrolyzed to cholesterol and fatty acids.

Under normal circumstances, LDL receptor–mediated endocytosis is a highly regulated process. In liver cells, transcription factors called SREBPs (sterol regulatory element binding proteins) have been characterized. SREBP precursors are membrane-bound ER proteins. When liver cell cholesterol levels are low, the precursor proteins are transported to the Golgi complex, where they are cleaved to form the active transcription factors. The SREBPs then migrate into the nucleus and bind to sterol regulatory elements (SRE); afterward, they together activate up to 30 genes involved in lipid metabolism, including the gene for the LDL receptor. Cellular cholesterol levels then rise in response to a combination of LDL uptake and increased endogenous cholesterol synthesis. High-fat diets block LDL receptor synthesis because the accumulation of ingested cholesterol in ER membranes prevents SREBP processing reactions.

The LDL receptor was discovered during an investigation of an inherited disease, *familial hypercholesterolemia* (FH). The LDL receptor was discovered as Brown and Goldstein were investigating the uptake of LDLs into the fibroblasts of FH patients. The biochemical defect that causes FH was then identified as mutations in the LDL receptor gene.

Patients with FH have elevated levels of plasma cholesterol because they have missing or defective LDL receptors. Heterozygous individuals (also referred to as *heterozygotes*) inherit one defective LDL receptor gene. Consequently, they possess half the number of functional LDL receptors. With blood cholesterol values of 300 to 600 mg/100 mL, heterozygotes have heart attacks as early as the age of 40. They also develop disfiguring *xanthomas* (cholesterol deposits in the skin) in their 30s. With a population frequency of 1 in 500, heterozygous FH is one of the most common human genetic anomalies. In contrast, *homozygotes* (individuals who have inherited a defective LDL receptor gene from both parents) are rare (approximately 1 in 1 million). These patients have plasma cholesterol values of 650 to 1200 mg/100 mL. Both xanthomas and heart attacks occur during childhood or early adolescence. Death usually occurs before the age of 20. The genetic defects that cause FH prevent affected cells from obtaining sufficient cholesterol from LDLs. The most common defect is failure to synthesize the receptor. Other defects include ineffective intracellular processing of newly synthesized receptor, defects in the receptor's binding of LDL, and the inability of receptors to cluster in coated pits.

Biochemistry IN PERSPECTIVE

Botulism and Membrane Fusion

What is the biochemical mechanism whereby botulism toxin causes an often-fatal muscle paralysis? *Botulism* is a form of muscle paralysis caused by a protein toxin produced by the anaerobic bacterium *Clostridium botulinum*. The toxin usually enters the body either by the consumption of improperly sterilized canned foods or by the contamination of a wound by the bacterium. Muscle paralysis begins in facial muscles, causing symptoms such as loss of facial expression and difficulty with swallowing and speech. Death may be caused by respiratory failure, the result of paralysis of the intercostal muscles that move the chest wall. The botulinum toxin prevents the release of the neurotransmitter acetylcholine (ACH) from the presynaptic axon of motor neurons. It does so by interfering with the fusion of ACH-containing synaptic vesicles with the nerve cell membrane. After a brief overview of the membrane fusion process, botulinum toxin mechanism is described.

Membrane Fusion

Membrane reorganization is a constant operational feature of eukaryotic cells. In the endomembrane system (p. 44), biosynthetic and secretory processes involve the movement of biomolecules through a series of transfers from one membrane–bound compartment to another. Molecules such as digestive enzymes and hormones, for example, are processed through the ER and Golgi complex and are eventually secreted from the cell by exocytosis (Figure 2.15). At each transfer step, these substances are passed from a donor compartment to a target compartment in a process that involves the fusion of membranes. Membranous vesicles pinch off from the donor compartment, and after transport by the cytoskeleton to their destination, the vesicles deliver their cargo by fusing with the membrane of a target compartment.

Membrane fusion, the merging of two lipid bilayers, is a precise and highly regulated process that is made possible by a specialized set of proteins that function as a fusion machine. The **SNAREs** (*s*oluble *N*-ethylmaleimide sensitive factor *a*ttachment protein *re*ceptors), the most essential components of the fusion machinery, are a large class of small transmembrane proteins (18–42 kDa), each consisting of a membrane bound C-terminal domain and a helical domain that extends into the cytoplasm. (*N*-Ethylmaleimide is a reagent used to elucidate the functional properties of the fusion proteins.) There are two categories of SNAREs: v-SNAREs (vesicle-specific proteins) and t-SNAREs (target membrane proteins). The principal features of the fusion mechanism are as follows.

As a vesicle (the donor membrane) approaches the target membrane, the helices of the v-SNARE and the t-SNARE interact, forming four helix bundles (the core complex) from relatively unstructured helices (Figure 11A). The "zippering" of each set of coiled coil structures creates a torsional force that

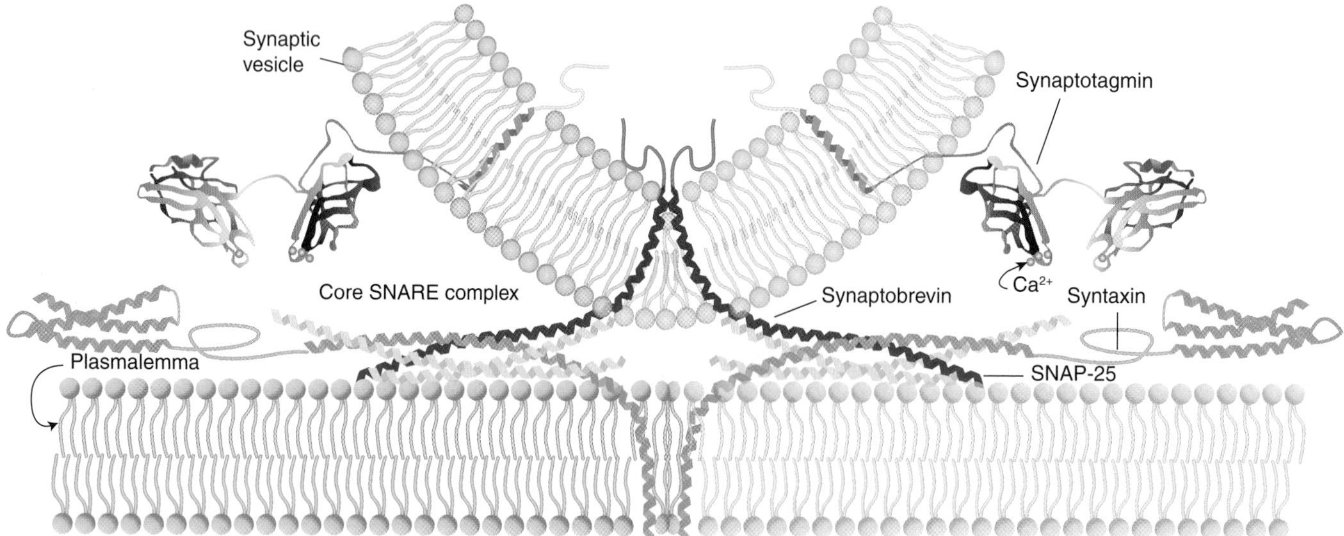

FIGURE 11A
Membrane Fusion

The fusion of neurotransmitter vesicles with the presynaptic membrane of neurons begins with the formation of the core SNARE complex from synaptobrevin (a v-SNARE), syntaxin (a t-SNARE), and two helices of SNAP-25, a peripheral membrane protein component of the t-SNARE complex. Synaptotagmin is a Ca^{2+} sensor that triggers the late stages of membrane fusion when local Ca^{2+} levels are high.

Biochemistry IN PERSPECTIVE cont.

draws the opposing membranes into intimate contact in a process that expels water molecules (i.e., hydrostatic pressure is overcome). Once assembled, the fusion machinery is activated when synaptotagmin, a Ca^{2+}-sensor protein in the vesicle membrane, undergoes a conformational change that is triggered by a localized rise in calcium levels. The now-active fusion machinery promotes the rearrangement of the two lipid bilayers to form the fused membrane. After fusion is complete, the SNARE complexes are disassembled by *N*-ethylmaleimide sensitive factor (NSF). NSF, an ATPase, contains a clamp-like module that in combination with α-SNAP (*s*oluble *N*SF *a*ttachment *p*rotein) exerts the mechanical force necessary to pry the stable SNARE complex apart so that their components can be recycled.

The Botulinum Toxin Mechanism

Botulinum toxin consists of a heavy chain (100 kDa) linked by a disulfide bridge to a light chain (50 kDa). All seven toxin types (A, B, C, D, E, F, and G) inhibit the release of ACH from motor neurons. The toxin enters the cell via endocytosis triggered by the binding of the heavy chain to a plasma membrane receptor. The light chain exits the endocytotic vesicles and migrates to the presynaptic membrane, where it cleaves a SNARE protein, thereby disabling the fusion machinery. Each toxin type disables a specific membrane fusion protein. For example, toxins A and B cleave SNAP-25 (a t-SNARE) and synaptobrevin (a v-SNARE), respectively.

SUMMARY: Botulinum toxin causes muscle paralysis by preventing the membrane fusion event that releases the neurotransmitter ACH into the neuromuscular junction.

Chapter Summary

1. Lipids are a diverse group of biomolecules that dissolve in nonpolar solvents. They can be separated into the following classes: fatty acids and their derivatives, triacylglycerols, wax esters, phospholipids, lipoproteins, sphingolipids, and the isoprenoids.
2. Fatty acids are monocarboxylic acids that occur primarily in triacylglycerols, phospholipids, and sphingolipids. The eicosanoids are a group of powerful hormonelike molecules derived from long-chain fatty acids. The eicosanoids include the prostaglandins, the thromboxanes, and the leukotrienes.
3. Triacylglycerols are esters of glycerol with three fatty acid molecules. Triacylglycerols that are solid at room temperature (i.e., possess mostly saturated fatty acids) are called fats. Those that are liquid at room temperature (i.e., possess a high unsaturated fatty acid content) are referred to as oils. Triacylglycerols, the major storage and transport form of fatty acids, are an important energy storage form in animals. In plants they store energy in fruits and seeds.
4. Phospholipids are structural components of membranes. There are two types of phospholipid: phosphoglycerides and sphingomyelins.
5. Sphingolipids are also important components of animal and plant membranes. Like the sphingomyelins, they contain a ceramide base (*N*-acylsphingosine) but do not contain phosphate. Their polar head group is one or more sugar residues.
6. Isoprenoids are molecules that contain repeating five-carbon isoprene units. The isoprenoids consist of the terpenes and the steroids.
7. Plasma lipoproteins transport lipid molecules through the bloodstream from one organ to another. They are classified according to their density. Chylomicrons are large lipoproteins of extremely low density that transport dietary triacylglycerols and cholesteryl esters from the intestine to adipose tissue and skeletal muscle. VLDLs, which are synthesized in the liver, transport lipids to tissues. As VLDLs unload some of their lipid molecules, they are converted to LDLs. LDLs bind to LDL receptors on the plasma membrane and then are engulfed by cells. HDLs, also produced in the liver, scavenge cholesterol from cell membranes and other lipoprotein particles. LDLs play an important role in the development of atherosclerosis.
8. According to the fluid mosaic model, the basic structure of membranes is a lipid bilayer in which proteins float. Membrane lipids (the majority of which are phospholipids) are primarily responsible for the fluidity, regional partitioning (lipid rafts), and sealing and fusion properties of membranes. Membrane proteins usually define the biological functions of specific membranes. Depending on their location, membrane proteins can be classified as integral or peripheral. Membranes are involved in transport and in the binding of hormones and other extracellular metabolic signals.
9. Membranes exhibit heterogeneity with respect to their lipid, protein, and carbohydrate components. New lipid bilayer is synthesized from the cytoplasmic face, and lipids are distributed throughout the bilayer by specific mediator proteins. The compartments on either side of the membrane are chemically different, and the membrane surfaces reflect that difference.
10. The movement of substances across a cell membrane can be accomplished by non-energy-requiring passive transport

with the concentration gradient, (the passive diffusion of small nonpolar molecules and the facilitated diffusion of large or polar molecules or ions via carrier or channel proteins), primary active transport (ATP energy used to concentrate a substance on one side of the membrane), secondary active transport (ion gradient generated by primary active transport used to concentrate a substance on one side of the membrane), or receptor-mediated transport (receptor and ligand in coated pits engulfed by endocytosis). Some transport channels are gated; that is, they open only in the presence of a certain gating substance (neurotransmitter, ions, etc.) or membrane condition (voltage, pH).

Take your learning further by visiting the **companion website** for Biochemistry at **www.oup.com/us/mckee** where you can complete a multiple-choice quiz on Lipids and Membranes to help you prepare for exams.

Suggested Readings

Calder, P. C., Omega-3 Fatty Acids and Inflammatory Processes, *Nutrients* 2(3):355–374, 2010.

Hanraham, J. W., Sampson, H. M., and Thomas, D. Y., Novel Pharmacological Strategies to Treat Cystic Fibrosis, *Trends Pharmacol. Sci.* 34(2):119–125, 2013.

Jahn, R., and Fasshauer, D., Molecular Machines Governing Exocytosis of Synaptic Vesicles, *Nature* 490: 201–207, 2012.

Janmey, P. A., and Kinnunen, P. K. J., Biophysical Properties of Lipids and Dynamic Membranes, *Trends Cell Biol.* 16(10):538–546, 2006.

Los, F. C. O., *et al.*, Role of Pore-forming Toxins in Bacterial Infectious Diseases, *Microbiol. Mol. Biol. Rev.* 77(2):173–207, 2013.

Rodriguez, A. et al., Physiology and Pathophysiology of Aquaporins, *Adipobiology* 2:9–22, 2010.

Trivedi, B. P., Doorway to a Cure, *Discover* 34(7):42–51, 2013.

Wu, X., and Schauss, A. G., Mitigation of Inflammation with Foods, *J. Agri. Food Chem* 60(27):6703–6717, 2012.

Key Words

active transport, *410*
acyl group, *387*
aquaporin, *412*
autocrine, *388*
autoimmune disease, *389*
carotenoid, *398*
chylomicron, *403*
cystic fibrosis, *413*
cystic fibrosis transmembrane conductance regulator, *414*
eicosanoid, *387*
essential fatty acid, *386*
facilitated diffusion, *411*
fluid mosaic model, *404*
glycolipid, *395*
GPI anchor, *393*
high-density lipoprotein, *403*
integral protein, *407*
intermediate-density lipoprotein, *403*
isoprenoid, *398*
leukotriene, *389*
lipid, *384*
lipid bilayer, *404*
low-density lipoprotein, *403*
mixed terpenoid, *399*
monounsaturated, *386*
nephrogenic diabetes insipidis, *412*
neutral fat, *390*
nonessential fatty acid, *386*
omega-3 fatty acid, *387*
omega-6 fatty acid, *386*
passive transport, *410*
peripheral protein, *407*
phosphoglyceride, *392*
phospholipid, *394*
polar head group, *391*
polyunsaturated, *386*
prenylation, *399*
prostaglandin, *388*
simple diffusion, *411*
sphingolipid, *395*
sphingomyelin, *392*
steroid, *399*
terpene, *398*
thromboxane, *388*
very-low-density lipoprotein, *403*
wax, *391*
wax ester, *391*

Review Questions

These questions are designed to test your knowledge of the key concepts discussed in this chapter before moving on to the next chapter. You may like to compare your answers to the solutions provided in the back of the book and in the accompanying Study Guide.

1. Define the following terms:
   a. fatty acid
   b. monunsaturated fatty acid
   c. polyunsaturated fatty acid
   d. saturated fatty acid
   e. acyl group

2. Define the following terms:
   a. essential fatty acid
   b. nonessential fatty acid
   c. omega-3 fatty acid
   d. omega-6 fatty acid
   e. eicosanoid

3. Define the following terms:
   a. prostaglandin
   b. thromboxane
   c. leukotriene
   d. autocrine
   e. anaphylaxis
4. Define the following terms:
   a. triacylglycerol
   b. neutral fat
   c. partial hydrogenation
   d. autoimmune disease
   e. soap
5. Define the following terms:
   a. saponification
   b. wax ester
   c. wax
   d. phospholipid
   e. polar head group
6. Define the following terms:
   a. phosphoglyceride
   b. sphingolipid
   c. GPI anchor
   d. glycolipid
   e. sphingomyelin
7. Define the following terms:
   a. membrane remodeling
   b. phospholipase
   c. cerebroside
   d. ganglioside
   e. sphingolipidoses
8. Define the following terms:
   a. isoprenoid
   b. isoprene unit
   c. terpene
   d. carotenoid
   e. mixed terpenoid
9. Define the following terms:
   a. prenylation
   b. steroid
   c. digitalis
   d. lipoprotein
   e. apolipoprotein
10. Define the following terms:
    a. chylomicron
    b. VLDL
    c. IDL
    d. LDL
    e. HDL
11. Define the following terms:
    a. lipid bilayer
    b. membrane fluidity
    c. flippase
    d. floppase
    e. AE1
12. Define the following terms:
    a. peripheral protein
    b. integral protein
    c. lipid raft
    d. passive transport
    e. active transport
13. Define the following terms:
    a. CFTR
    b. simple diffusion
    c. facilitated diffusion
    d. Na^+-K^+ pump
    e. aquaporin
14. Define the following terms:
    a. nephrogenic diabetes insipidis
    b. LDL receptor
    c. coated pit
    d. clathrin
    e. familial hypercholesterolemia
15. Define the following terms:
    a. botulism
    b. botulinum toxin
    c. t-SNARE
    d. v-SNARE
    e. membrane fusion
16. List a major function of each of the following classes of lipid:
    a. phospholipids
    b. sphingolipids
    c. oils
    d. waxes
    e. steroids
    f. carotenoids
17. What role do plasma lipoproteins play in the human body?
18. Why do plasma lipoproteins require a protein component to accomplish their role?
19. Describe several factors that influence membrane fluidity.
20. Which of the following statements concerning ionophores are true?
    a. Ionophores form channels through which ions flow.
    b. They require energy.
    c. Ions may diffuse in either direction.
    d. Ionophores may cause voltage gates.
    e. They transport all ions with equal ease.
21. What role do bile salts play in the body?
22. Describe the possible consequences of a low-fat diet.
23. What do the abbreviations ACAT and LCAT stand for? What functions do these molecules serve in lipid metabolism?
24. Explain the differences in the ease of lateral movement and bilayer translocation movement of phospholipids.
25. Explain how potassium moves across a nerve cell membrane during both depolarization and repolarization.
26. From what fatty acid are most of the eicosanoids derived? List several medical conditions in which the suppression of eicosanoid synthesis may appear to be advantageous.
27. In which of the following processes do the prostaglandins not have a major recognized role?
    a. reproduction
    b. blood clotting
    c. respiration
    d. inflammation
    e. blood pressure modification
28. What is the difference between an autocrine regulator and a hormone?
29. How do water molecules move through hydrophobic cell membranes?

30. Classify each of the following as a monoterpene, diterpene, triterpene, tetraterpene, sesquiterpene, or polyterpene:

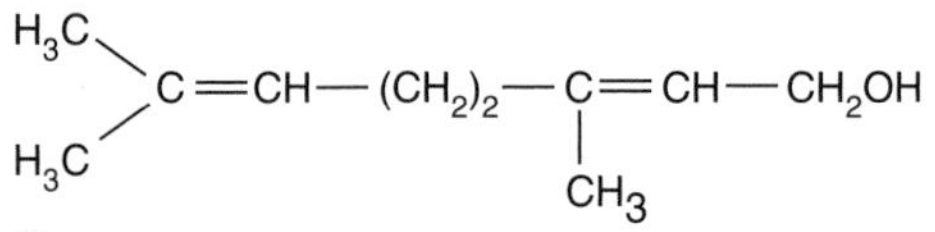

a. $(H_3C)_2C{=}CH{-}(CH_2)_2{-}C(CH_3){=}CH{-}CH_2OH$

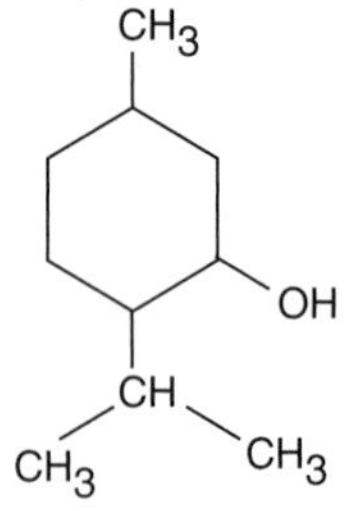

b.

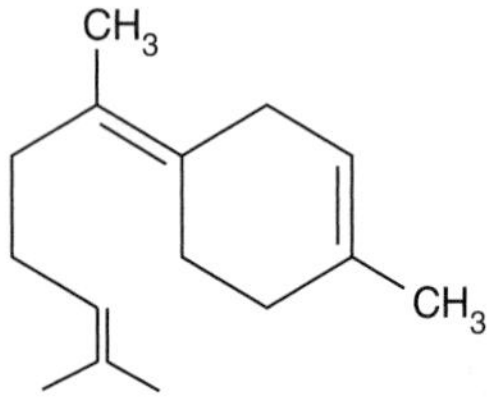

c.

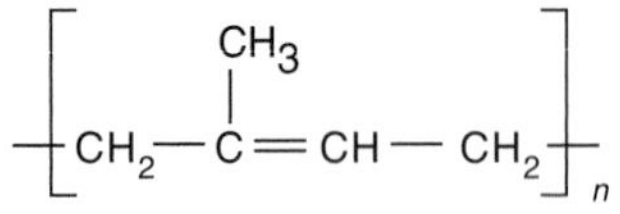

d. $-[CH_2{-}C(CH_3){=}CH{-}CH_2]_n-$

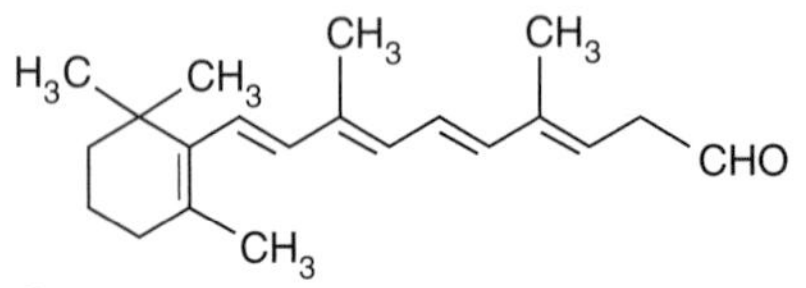

e.

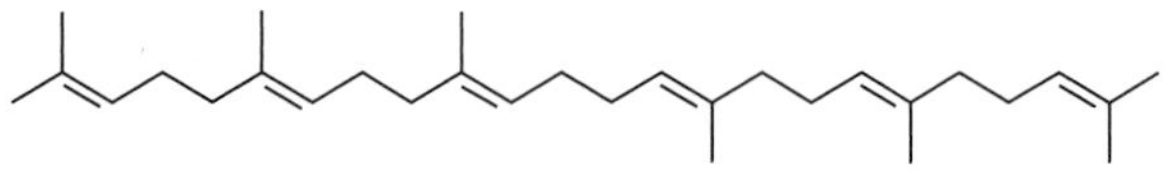

f.

31. Discuss the functions of triacylglycerols.
32. Sphingomyelins are amphipathic molecules. Review the structure of a typical sphingomyelin. Which regions are hydrophilic and which are hydrophobic?
33. What changes might you make in the structure of a cell membrane to increase the cell's resistance to mechanical stress?
34. Low density lipoproteins referred to as sdLDLs are more atherogenic than large, buoyant LDLs. Explain.
35. Describe the structural and functional properties of lipid rafts.
36. List three mediator proteins that move phospholipid molecules from one side of a membrane to the other. Why are these proteins required?
37. Describe how glucose is transported across membranes in the kidney. What type of transport is involved?
38. Describe and give a specific example of a facilitated diffusion process.
39. Compare and contrast the following processes. Give an example of each to illustrate.
    a. primary vs secondary active transport
    b. passive vs facilitated diffusion
    c. carrier-mediated vs channel-mediated transport
40. How do lipoproteins transport water-insoluble lipid molecules in the bloodstream?
41. Suggest a reason why trans fatty acids have melting points similar to analogous saturated fatty acids.
42. Detergents are synthetic soaplike substances that are used to disrupt membranes and extract membrane proteins. Explain how this process works.
43. Explain why triacylglycerols are not components of lipid bilayers.

Fill in the Blank

44. A derivative of a triterpene that contains four fused rings is referred to as a ____________________.
45. ____________________ proteins are embedded in a membrane.
46. Bile salts emulsify dietary ____________.
47. Transport of a substance across membranes that requires no direct input of energy is referred to as ____________ transport.
48. Three proteins that facilitate the movement of phospholipid molecules across a membrane are flippase, floppase, and ____________________.
49. ____________ is a form of membrane transport that requires no direct energy input but does require a channel or a carrier.
50. The term ____________ is used to describe the viscosity of a membrane.
51. In active transport ATP provides energy to transport a substance ____________ a concentration gradient.
52. Prostaglandins have recognized roles in reproduction, respiration, and ____________.
53. ____________ is a type of cardiac glucoside that increases the force of cardiac contraction.

Short Answer

54. What appears to be the role of HDLs?
55. What are terpenes? Name three examples.
56. What are the sphingolipid storage diseases?
57. What is a toxic phospholipase?
58. What are endorphins and what is their connection to dietary fat?

Thought Questions

These questions are designed to reinforce your understanding of all of the key concepts discussed in the book so far, including this chapter and all of the chapters before it. They may not have one right answer! The authors have provided possible solutions to these questions in the back of the book and in the accompanying Study Guide for your reference.

59. Animal cells are enclosed in a cell membrane. According to the fluid mosaic model, this membrane is held together by hydrophobic interactions. Consider the shear forces involved. Why does this membrane not break every time an animal moves?
60. Suggest a reason why elevated LDL levels are a risk factor for coronary artery disease.
61. Glycolipids are nonionic lipids that can orient themselves into bilayers as phospholipids do. They accomplish this feat although they lack an ionic group like that of the phospholipids. Suggest a reason why this is possible.
62. Explain why spontaneous phospholipid translocation (the movement of a molecule from one leaflet of a bilayer to the other) is so slow.
63. Mammals in the Arctic (e.g., reindeer) have higher levels of unsaturated fatty acids in their legs than in the rest of their bodies. Suggest a reason for this phenomenon. Does it have a survival advantage?
64. Explain why entropy increases when a lipid bilayer forms from phospholipid molecules.
65. The fluid mosaic model of membrane structure has been very useful in explaining membrane behavior. However, the description of membrane as proteins floating in a phospholipid sea is oversimplified. Describe some components of membrane that are restricted in their lateral motion.
66. Plants often produce waxes on the surface of their leaves to prevent dehydration and protect against insects. What structural feature of waxes makes them more suitable for this task than carbohydrates or proteins?
67. Changes in temperature affect membrane properties. How would you expect the lipid composition of thermophile membranes to differ from those of prokaryotes that live at more normal temperatures?
68. Men on calorie-restricted diets frequently experience an intensified growth of facial hair, which correlates with a rise in blood steroid levels. How would you explain this phenomenon?
69. Boric acid is a potent insecticide that dissipates the wax coat of insects. Explain how this substance kills insects.
70. The myelin sheath insulates the axons of certain neurons in the body. What structural feature of this covering makes it a good insulator?

CHAPTER

12 Lipid Metabolism

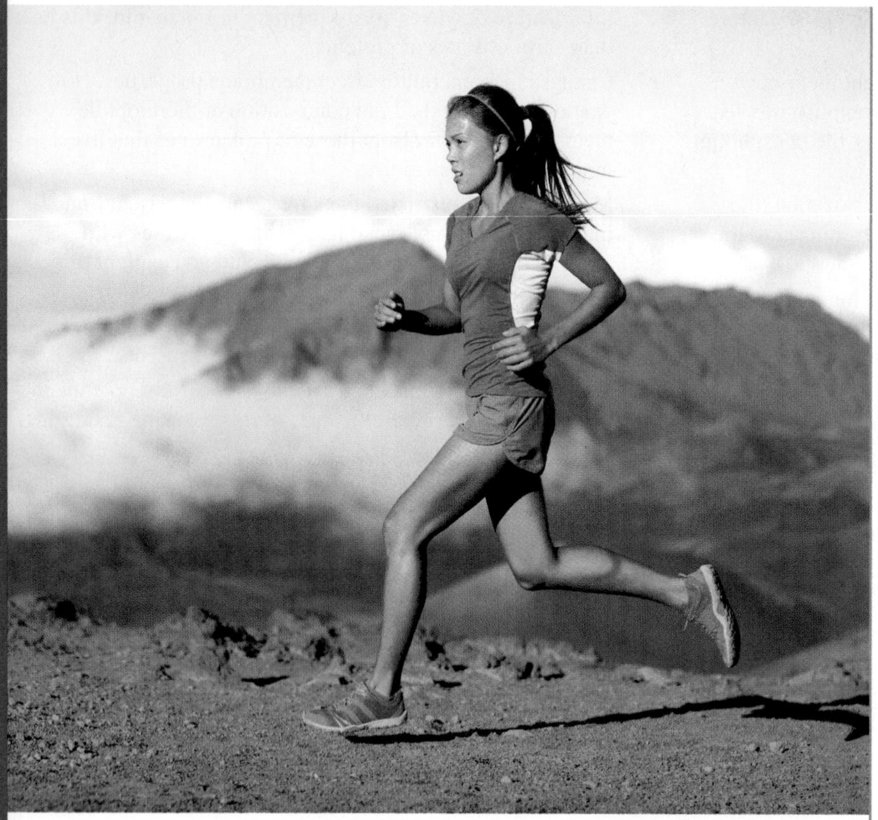

A Long-Distance Runner
Long-distance running requires physical strength and stamina. Among the most important characteristics of long-distance runners is the long-term capacity to deliver adequate O_2 and calorie-dense fatty acids to contracting muscle.

OUTLINE

Abetalipoproteinemia

After months of chronic diarrhea and symptoms described as "failure to thrive," a 10-month-old boy was taken to a gastroenterologist (a physician specializing in disorders of the stomach and intestines) by his increasingly desperate parents. Although he had no feeding difficulties, the child's height and weight were lower than the fifth percentile. Laboratory exams were normal except for a fasting lipid profile, which revealed exceptionally low values for triacylglycerols (less than 10 mg/dl; normal is less than 150 mg/dl) and cholesterol (less than 50 mg/dl; normal is 170 mg/dl). In addition, the values for blood hemoglobin and hematocrit (proportion of blood volume occupied by red blood cells) were also low because of malabsorption of iron and folic acid. Visual examination of blood showed the presence of *acanthocytes*, abnormal red blood cells with spiny projections caused by abnormal lipid content in the red blood cell membrane. The use of an endoscope, an instrument fitted with a camera that allows viewing of the inside of the body, showed that much of the duodenum, the first part of the small intestine immediately beyond the stomach, was almost white in color and not a normal pink color. Microscopic examination of retrieved enterocytes revealed that the cytoplasm contained large lipid droplets, which, combined with steatorrhea (excessive fats in the feces), indicated an inability to absorb dietary fat.

Diagnosis and Cause

This constellation of symptoms led the child's physician to suspect *abetalipoproteinemia*, an exceptionally rare autosomal recessive metabolic disorder (a frequency of 1 in 1 million humans). This disorder affects the absorption of dietary fat, cholesterol, and the fat-soluble vitamins. The diagnosis was confirmed by a blood test for lipoproteins, which showed that the patient's blood contained virtually no chylomicrons, VLDL, and LDL (see pp. 401–404). In recent years the molecular defect that causes alipoproteinemia was finally identified. Apo-B-containing lipoproteins fail to form as a result of mutations in the gene that codes for a molecular chaperone called microsomal triglyceride transfer protein (MTTP). MTTP facilitates the transfer of lipids onto apoB during the production of apo-B-containing lipoproteins.

Prognosis and Treatment

Clinical research over the past several decades has improved the lives of individuals with abetalipoproteinemia. Early diagnosis and treatment can now help patients manage their symptoms and delay or prevent serious long-term damage. Just a few examples of the damage observed in patients not diagnosed until adulthood include ataxia (progressive neurological damage that causes uncoordinated muscle movement), muscle weakness, and loss of peripheral vision. Treatment of affected children consists of several years of intravenous feeding with a low-fat formula supplemented with medium-chain triacylglycerols and fat-soluble vitamins. This is followed by a rigorous low-fat diet with oral doses of medium-chain fatty acids and fat-soluble vitamins. (Medium-chain fatty acids are more easily absorbed than are long-chain fatty acids, and they are transported by albumin in the blood stream instead of chylomicrons.)

Overview

LIPIDS PLAY A UNIQUE ROLE IN LIVING ORGANISMS LARGELY BECAUSE OF THEIR HYDROPHOBIC STRUCTURES. LIPIDS SERVE AS (1) HIGHLY EFFICIENT and compact energy storage molecules (triacylglycerols), (2) essential components of biological membranes (phospholipids, sphingolipids, and cholesterol), and (3) diverse membrane-associated molecules that have signaling (e.g., steroid hormones and prostaglandins) or protective (e.g., α-tocopherol) functions. Chapter 12 focuses on the metabolism of the major classes of lipids, that is, how they are synthesized and degraded and how these processes are regulated.

Major emphasis is placed on the central metabolite in lipid metabolism: acetyl-coenzyme A. The metabolism of cholesterol is also discussed because of its prominent role in cardiovascular disease.

The structural and functional diversity of lipids is certainly impressive. All lipids are derived in whole or in part from acetyl-CoA (**Figure 9.9**). For example, acetyl-CoA is the substrate in the synthesis of fatty acids, the terpenes (e.g., β-carotene), and the steroids (e.g., cholesterol). When cells require energy, fatty acids are degraded to yield acetyl-CoA, which is then diverted into the citric acid cycle. In Chapter 12 the metabolism of the major classes of lipid is discussed: fatty acids, triacylglycerols, phospholipids, sphingolipids, and isoprenoids. In addition, the synthesis of the ketone bodies is reviewed. Several metabolic control mechanisms are discussed throughout the chapter.

12.1 FATTY ACIDS, TRIACYLGLYCEROLS, AND THE LIPOPROTEIN PATHWAYS

Triacylglycerols (fat molecules) are an important and efficient energy source in animals. For example, in the average U.S. diet, between 30 and 40% of calories ingested are provided by fat. Triacylglycerol molecules (TGs) are digested within the lumen of the small intestine (**Figure 12.1**). The absorption of TGs and other lipid nutrients and their distribution to the body's tissues via lipoproteins is referred to as the *exogenous pathway.* (The *endogenous pathway* in which lipoproteins transport lipids produced in the liver to the body's cells is described on p. 453.)

Dietary Fat: Digestion, Absorption, and Transport

This section describes the digestion and absorption of dietary fat, TG metabolism in adipose tissue, energy-yielding fatty acid degradation reactions, and fatty acid biosynthesis. The section ends with an overview of fatty acid metabolism regulation and a short review of the endogenous lipoprotein pathway, the mechanism whereby the liver packages lipids into lipoproteins for distribution throughout the body. After dietary fat mixes with **bile salts**, amphipathic molecules with detergent properties (see p. 464), are digested by pancreatic lipase to form fatty acids and monoacylglycerol. These latter molecules are then transported across the plasma membrane of intestinal wall cells (enterocytes). Short- (C4 to C6) and medium- (C6 to C12) chain fatty acids are transferred to the blood stream, where they bind to serum albumin, which carries them to the liver. Long-chain fatty acids are delivered to the enterocyte SER, where they are incorporated into TGs. Enterocytes combine triacylglycerols with dietary cholesterol, newly synthesized phospholipids, and lipoprotein B-48 to form *nascent* (newly made) chylomicrons (large, low-density lipoproteins; see p. 403). (Apolipoprotein B-48, the main lipoprotein component of nascent chylomicrons, is synthesized from an mRNA that is a truncated version of the apoliproprotein B-100 mRNA, the protein found on lipoprotein produced in the liver.) After their secretion into the lymph (tissue fluid derived from blood), chylomicrons pass from the lymph into the bloodstream at the thoracic duct. Nascent chylomicrons are converted into mature chylomicrons as they circulate in blood and lymph when HDLs transfer two lipoprotein molecules. Apolipoprotein C-II activates lipoprotein lipase (LPL), and lipoprotein E binds to a specific receptor on the surface of hepatocytes.

Most of the triacylglycerol content of circulating chylomicrons is removed from blood by muscle and the adipose tissue cells (adipocytes), which comprise the body's primary lipid storage depot. Lipoprotein lipase, synthesized by cardiac and skeletal muscle, lactating mammary gland, and adipose tissue, is transferred

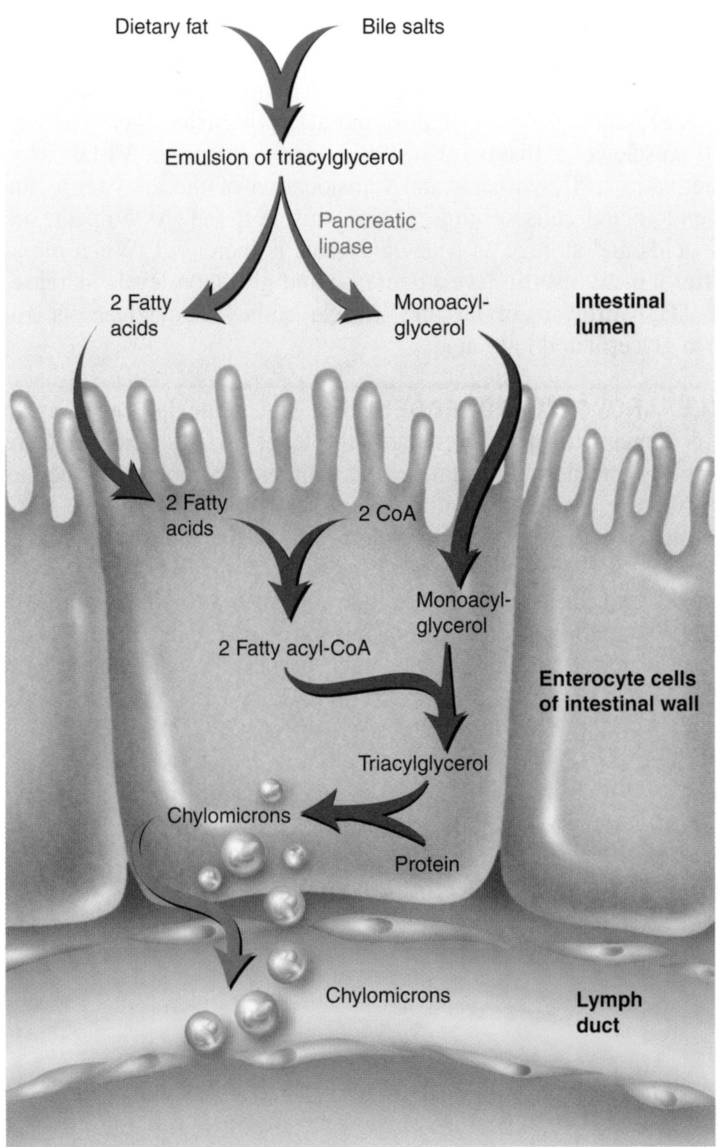

FIGURE 12.1

Digestion and Absorption of Triacylglycerols in the Small Intestine

After triacylglycerols have been emulsified (solubilized) by mixing with bile salts, they are digested by intestinal lipases, the most important of which is pancreatic lipase. The products, fatty acids and monoacylglycerol, are transported into enterocytes and resynthesized to form triacylglycerol. The triacylglycerol molecules, along with newly synthesized phospholipid and protein, are then incorporated into chylomicrons. After the chylomicrons have been transported into lymph, via exocytosis, and then into blood, the triacylglycerols are drawn off by muscle and fat cells. Chylomicron remnants are removed from the blood by the liver.

to the endothelial surface of the capillaries. Once it is activated by apolipoprotein CII, lipoprotein lipase converts the triacylglycerol in chylomicrons into fatty acids and glycerol.

The fatty acids are taken up by cells, whereas glycerol is carried in the blood to the liver, where the enzyme glycerol kinase converts it to glycerol-3-phosphate. Glycerol-3-phosphate is then used in the synthesis of triacylglycerols, phospholipids, or glucose. When LPL has removed about 90% of TGs in chylomicrons, the **chylomicron remnants** are removed from the blood by liver cells via the binding of apolipoprotein E to chylomicron remnant receptors. Hydrolysis of the remaining TGs within lysosomes releases fatty acids and glycerol that can be either metabolized by liver cells immediately or stored for later use. Cholesterol molecules released from chylomicron remnants have several fates. Some are esterified with fatty acids and then packaged into nascent lipoproteins, whereas others are either converted into bile acids or secreted directly into bile.

Triacylglycerol Metabolism in Adipocytes

Fatty acids, stored in TGs primarily in adipocytes, are the body's most concentrated energy source. Depending on an animal's current metabolic needs, fatty acids may be released from triacylglycerols to be degraded to generate energy or used in membrane synthesis. Immediately after a meal, for example, insulin is released in response to high blood glucose levels. Insulin promotes triacylglycerol storage by inactivating *hormone-sensitive lipase* (an enzyme that hydrolyzes the ester bonds of fat molecules) and activating triacylglycerol synthesis in fat and muscle cells. Insulin also stimulates the release of VLDL from the liver and activates LPL synthesis and translocation of the enzyme to the surface of the endothelial cells serving fat and muscle tissue. As a result, the uptake of fatty acids and storage as triacylglycerols is increased. When blood glucose falls after a meal, insulin levels decrease and glucagon levels increase, promoting net TG hydrolysis in fat and muscle cells. Triacylglycerols are degraded to form glycerol and fatty acids.

THE TG CYCLE AND GLYCERONEOGENESIS The **triacylglycerol cycle** (**Figure 12.2**) is a mechanism that regulates the level of fatty acids that are available to the body for energy generation and synthesis of molecules such as phospholipids. TGs are constantly both synthesized and hydrolyzed to fatty acids and glycerol. This seemingly futile cycle occurs at the cellular level (e.g., in adipocytes) and at the whole body level. **Figure 12.2** illustrates TG cycling between adipocytes and the liver. TGs are hydrolyzed in adipocytes with the release of a relatively small fraction of fatty acids into blood. Once in the blood, fatty acids are transported to the body's other tissues. In the liver a high

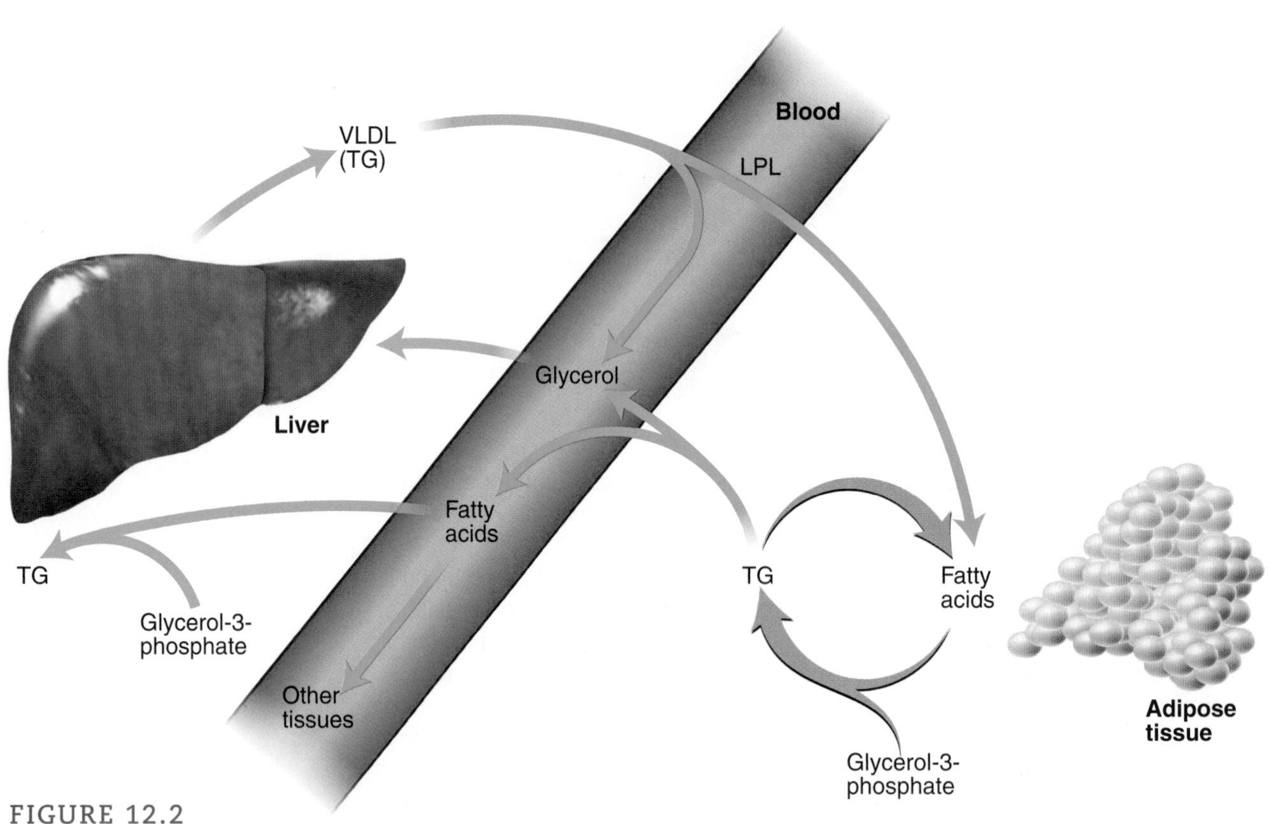

FIGURE 12.2

The Triacylglycerol Cycle

In adipocytes, the body's primary energy storage depot, TGs are synthesized from fatty acids obtained from the blood and glycerol-3-phosphate. The rate at which fatty acids are released into blood to meet current energy needs of other tissues is increased by glucagon and epinephrine and depressed by insulin. Under all metabolic conditions, however, the percentage of adipocyte fatty acids (about 75%) that are reesterified is remarkably constant. In the liver most fatty acids removed from blood are used to synthesize TGs that are incorporated into VLDL. Once VLDL are secreted into blood, they travel to tissues such as adipose tissue where TGs are hydrolyzed by lipoprotein lipase. Fatty acids are then transported into adipocytes. Glycerol, the other product of TG hydrolysis, is removed from blood by the liver.

proportion of the fatty acids are reincorporated into TGs, most of which are packaged into VLDL. The net result of TG cycling is that a flexible system ensures that sufficient fatty acids are available for the body's energy and biosynthetic requirements. Excess fatty acids, which can have toxic effects on cells, are efficiently reesterified into TGs. The pathways by which TGs are synthesized and hydrolyzed are described next.

TRIACYLGLYCEROL BIOSYNTHESIS Triacylglycerol synthesis (referred to as **lipogenesis**) is illustrated in **Figure 12.3**. Glycerol-3-phosphate or DHAP reacts sequentially with three molecules of acyl-CoA (fatty acid esters of CoASH). Acyl-CoA molecules are produced in the following reaction:

$$R{-}\overset{\overset{\large O}{\|}}{C}{-}O^- + \text{ATP} + \text{CoASH} \longrightarrow R{-}\overset{\overset{\large O}{\|}}{C}{-}S{-}\text{CoA} + \text{PP}_i + \text{AMP} \qquad (1)$$

Note that the reaction is driven to completion by the hydrolysis of pyrophosphate by pyrophosphatase.

In the synthesis of triacylglycerols, phosphatidic acid is formed by two sequential acylations of glycerol-3-phosphate or by a pathway involving the direct acylation of DHAP. In the latter pathway, acyldihydroxyacetone phosphate is later reduced to form lysophosphatidic acid. Depending on the pathway used, lysophosphatidic acid synthesis utilizes either an NADH or an NADPH cofactor. Phosphatidic acid is produced when lysophosphatidic acid reacts with a second acyl-CoA. Once formed, phosphatidic acid is converted to diacylglycerol by phosphatidic acid phosphatase. A third acylation reaction forms triacylglycerol. Fatty acids derived from both the diet and *de novo* synthesis are incorporated into triacylglycerols. (The term *de novo* is used by biochemists to indicate new synthesis.) *De novo* synthesis of fatty acids is discussed on p. 441. Glyceroneogenesis, the principal means of producing glycerol-3-phosphate required in TG synthesis, is described next.

Glyceroneogenesis (**Figure 12.4**) is an abbreviated version of gluconeogenesis in which glycerol-3-phosphate (required for TG synthesis) is synthesized from substrates other than glucose or glycerol. The key enzymes for glyceroneogenesis are pyruvate carboxylase (PC) and the cytoplasmic isoform of phosphoenolpyruvate carboxykinase (PEPCK-C). Both enzymes are found in large amounts in lipogenic (TG-producing) tissues, such as adipose tissue and lactating mammary glands, and organs involved in gluconeogenesis (i.e., liver and kidney). PC and PEPCK-C are also found in moderate amounts in brain, heart, and adrenal glands.

TRIACYLGLYCEROL HYDROLYSIS When energy reserves are low, the body's fat stores are mobilized in a process referred to as **lipolysis** (**Figure 12.5**). Lipolysis in adipose tissue occurs during fasting or vigorous exercise and in response to stress. Several hormones (e.g., the catecholamines epinephrine and norepinephrine) bind to specific adipocyte plasma membrane receptors and begin a reaction sequence similar to the activation of glycogen phosphorylase. Hormone binding to the receptor elevates cytoplasmic cAMP levels, which, in turn, activates the phosphorylation of a protein called perilipin A. Perilipin A regulates lipid storage by coating the adipocyte lipid droplet, thereby preventing lipase access. Once perilipin A is phosphorylated, its conformational change exposes TGs to lipase-catalyzed hydrolysis. The first and committed step in TG hydrolysis is catalyzed by adipose triglyceride lipase (ATGL), which requires a coactivator called CGI-58. The products are diacylglycerol and a fatty acid. The hydrolysis of diacylglycerol to yield monoacylglycerol and a fatty acid is catalyzed by hormone-sensitive lipase (HSL). The final reaction, the conversion of monoacylglycerol to glycerol and a fatty acid, is catalyzed by monoacylglycerol

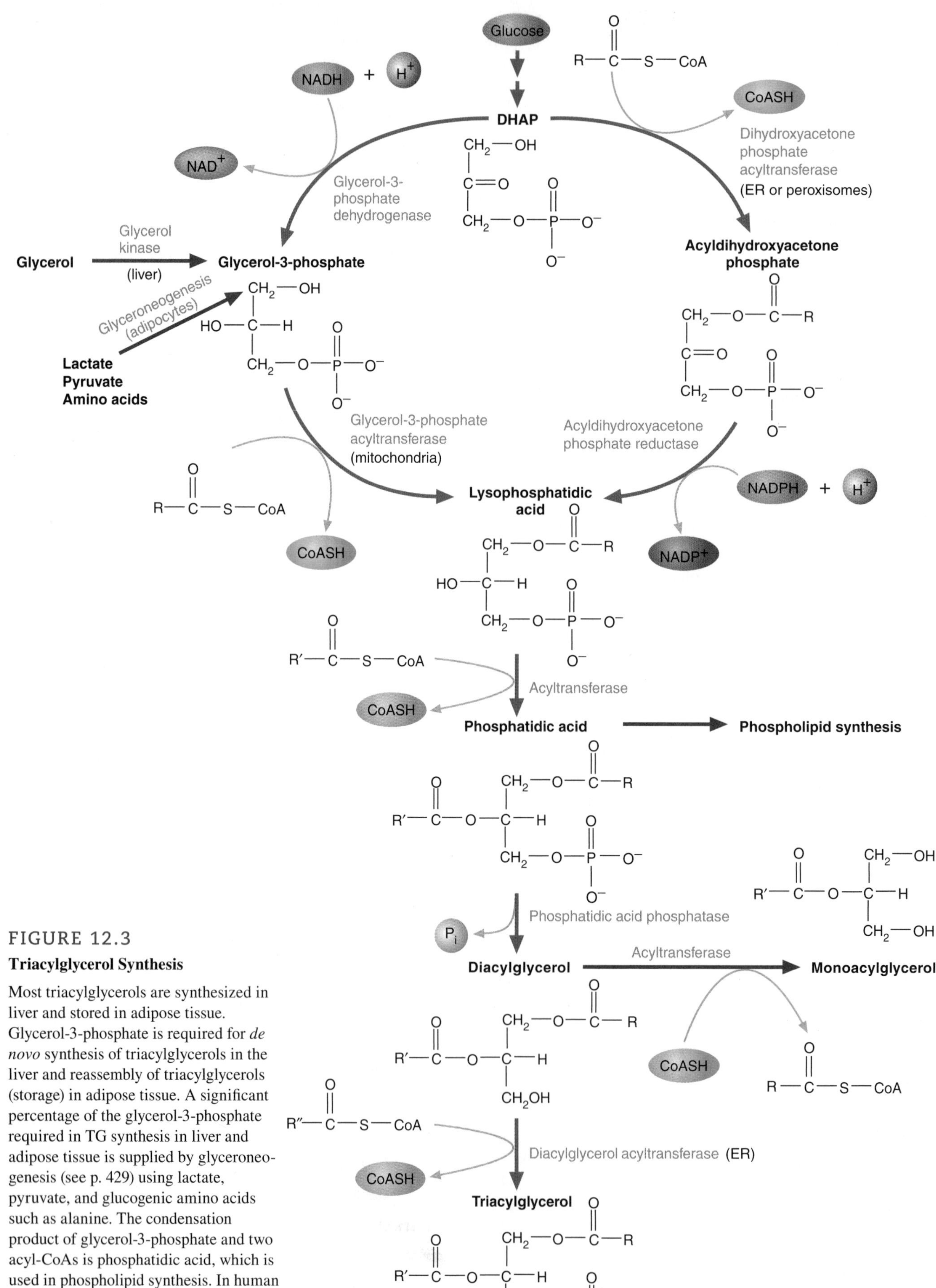

FIGURE 12.3

Triacylglycerol Synthesis

Most triacylglycerols are synthesized in liver and stored in adipose tissue. Glycerol-3-phosphate is required for *de novo* synthesis of triacylglycerols in the liver and reassembly of triacylglycerols (storage) in adipose tissue. A significant percentage of the glycerol-3-phosphate required in TG synthesis in liver and adipose tissue is supplied by glyceroneogenesis (see p. 429) using lactate, pyruvate, and glucogenic amino acids such as alanine. The condensation product of glycerol-3-phosphate and two acyl-CoAs is phosphatidic acid, which is used in phospholipid synthesis. In human triacylglycerols, palmitate is often attached at C-1 and oleate at C-2.

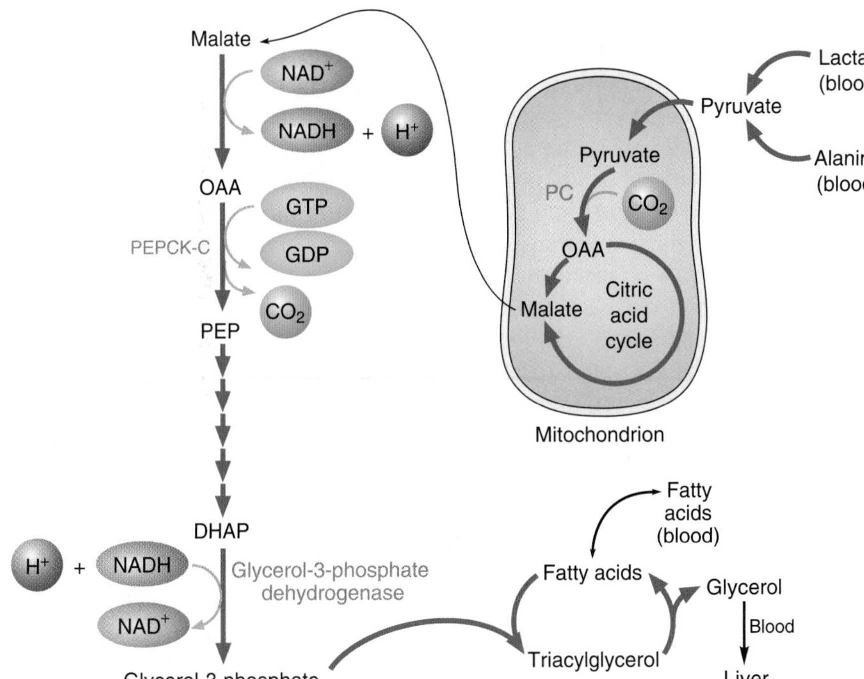

FIGURE 12.4

Glyceroneogenesis in Adipocytes

Glyceroneogenesis, an abbreviated form of gluconeogenesis, is a major source of glycerol-3-phosphate required for TG synthesis. Substrates for this pathway include lactate, pyruvate, and glucogenic amino acids such as alanine. Pyruvate is converted to OAA within the mitochondrion by PC (pyruvate carboxylase). After OAA is reduced by NADH, the product, malate, is transported out of the mitochondrion where the reaction is reversed to form OAA. OAA is then phosphorylated and decarboxylated by phosphoenolpyruvate carboxykinase (PEPCK-C) in a GTP-requiring reaction to form phosphoenolpyruvate (PEP). PEP is then converted via gluconeogenesis to DHAP. DHAP is reduced by glycerol-3-phosphate dehydrogenase to glycerol-3-phosphate, which is then utilized in TG synthesis.

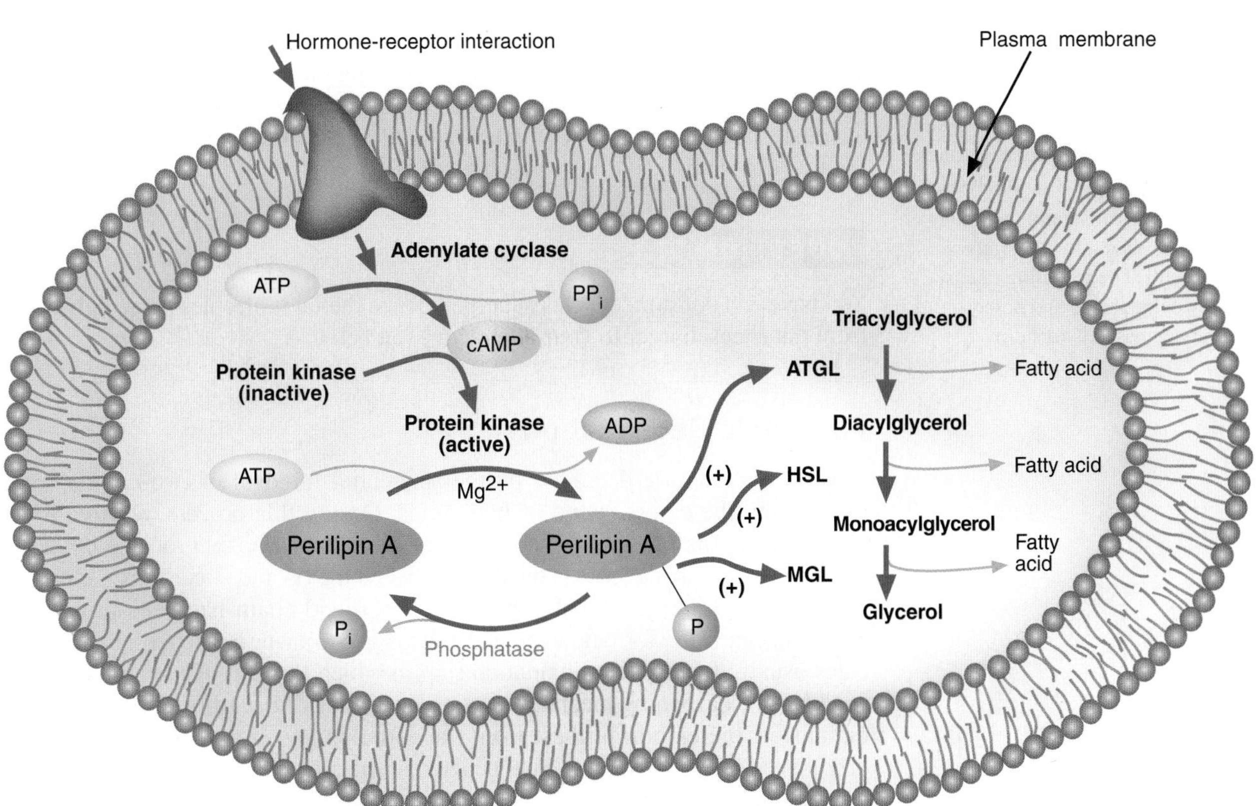

FIGURE 12.5

Diagrammatic View of Lipolysis in Adipocyte

The binding of certain hormones to their receptors on the surface of adipocytes initiates a cAMP-mediated mechanism that activates perilipin A, a protein that coats lipid droplets. Phosphorylation changes the conformation of perilipin A so as to expose the triacylglycerols to lipases. TGs are then hydrolyzed by a series of three lipolytic enzymes, each with its own substrate preferences. The initial reaction is catalyzed by ATGL (adipose triglyceride lipase), which requires a coactivator called CG1-58, to yield a diacylglycerol and a fatty acid. The product of ATGL lipolysis, diacylglycerol, is hydrolyzed by HSL (hormone sensitive lipase) to yield monoacylglycerol and a fatty acid. The final lipase, MGL (monoacylglycerol lipase), catalyzes the conversion of monocylglycerol to glycerol and a fatty acid. Note that phosphorylated perilipin A facilitates (+) TG hydrolytic enzymes by allowing their access to substrate molecules.

lipase (MGL). The products of lipolysis (i.e., fatty acids and glycerol) are released into the blood. The synthesis of ATGL is promoted by PPARγ and suppressed by insulin. Insulin also suppresses lipolysis by decreasing intracellular cAMP.

After their transport across the adipocyte plasma membrane, fatty acids become bound to serum albumin. The albumin-bound fatty acids are carried to tissues throughout the body, where they are released from albumin and taken up by cells. Fatty acids are transported into cells by a protein in the plasma membrane in a process linked to the active transport of sodium. The amount of fatty acid that is transported depends on its concentration in blood and the relative activity of the fatty acid transport mechanism. Cells vary widely in their capacity to transport and use fatty acids. Some cells (e.g., brain and red blood cells) cannot use fatty acids as fuel, although others (e.g., cardiac muscle) rely on them for a significant portion of their energy requirements. Once they enter a cell, fatty acids must be transported to their destinations (mitochondria, ER, and other organelles). Several **fatty acid–binding proteins** (water-soluble proteins whose sole function is to bind and transport hydrophobic fatty acids) are responsible for this transport.

Most fatty acids are degraded to form acetyl-CoA within mitochondria in a process referred to as β-oxidation. β-Oxidation also occurs in peroxisomes. Other oxidative mechanisms are also available to degrade certain nonstandard fatty acids.

Fatty acids are synthesized when an organism has met its energy needs and nutrient levels are high. (Glucose and several amino acids are substrates for fatty acid synthesis.) Fatty acids are synthesized from acetyl-CoA in a process that is similar to the reverse of β-oxidation. Although most fatty acids are supplied in the diet, most animal tissues can synthesize some saturated and unsaturated fatty acids. In addition, animals can elongate and desaturate dietary fatty acids. For example, arachidonic acid is produced by adding a two-carbon unit and introducing two double bonds to linoleic acid.

KEY CONCEPTS

- In the exogenous pathway, triacylglycerols and other lipid nutrients are absorbed into the body and distributed to the tissues by chylomicrons.
- When energy reserves are high, triacylglycerols are stored in the process called lipogenesis.
- When energy reserves are low, triacylglycerols are degraded to form fatty acids and glycerol. This process is called lipolysis.
- Triacylglycerols are constantly being synthesized and hydrolyzed to yield fatty acids and glycerol. The recycling rate is stimulated by epinephrine and norepinephrine and depressed by insulin.

QUESTION 12.1

You have just consumed a cheeseburger. Trace the fat molecules (triacylglycerol) from the cheeseburger to your adipocytes (fat cells).

Fatty Acid Degradation

Most fatty acids are degraded by the sequential removal of two-carbon fragments from the carboxyl end of fatty acids. During this process, referred to as **β-oxidation**, the β-carbon (second carbon from the carboxyl group) is oxidized and acetyl-CoA is released as the bond between the α- and β-carbons is cleaved. This process is repeated until the entire fatty acid chain has been processed. Other mechanisms for degrading fatty acids are known. Branched-chain molecules usually require an α-oxidation step in which the fatty acid chain is shortened by one carbon by means of a stepwise oxidative decarboxylation. In some organisms, the carbon farthest from the carboxyl group may be oxidized by a process called ω-oxidation that generates short-chain dicarboxylic acids. In ω-oxidation the terminal methyl group is converted to an alcohol by an O_2- and NADPH-requiring ER enzyme called *cytochrome* P_{450} (p. 468). The alcohol is subsequently converted to a carboxyate group by two sequential reactions catalyzed by ADH and aldehyde dehydrogenase. The resulting dicarboxylic acids are then shortened by β-oxidation in mitochondria to short-chain water-soluble dicarboxylic acids such as succinate and adipic acid (p. 441). In humans, ω-oxidation is a minor pathway that becomes relevant only when β-oxidation is impaired. β-Oxidation is discussed next. The degradation of odd-chain, branched-chain, and unsaturated fatty acids is also described.

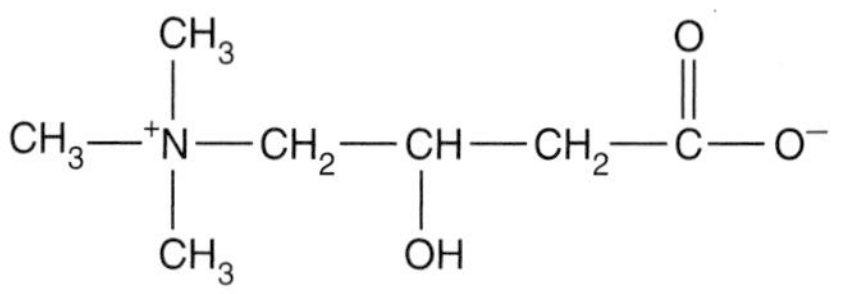

β-Oxidation occurs primarily within mitochondria. Before β-oxidation begins, each fatty acid is activated in a reaction with ATP and CoASH (see p. 429). The enzyme that catalyzes this reaction, acyl-CoA synthetase, is found in the outer mitochondrial membrane. Because the mitochondrial inner membrane is impermeable to most acyl-CoA molecules, a special carrier called *carnitine* is used to transport acyl groups into the mitochondrion (**Figure 12.6**). Carnitine-mediated transfer of acyl groups into the mitochondrial matrix is accomplished through the following mechanism (**Figure 12.7**):

$$CH_3-\overset{CH_3}{\underset{CH_3}{\overset{|}{\underset{|}{N^+}}}}-CH_2-\underset{OH}{\underset{|}{CH}}-CH_2-\overset{O}{\overset{\|}{C}}-O^-$$

FIGURE 12.6
Structure of Carnitine

1. Each acyl-CoA molecule is converted to an acylcarnitine derivative:

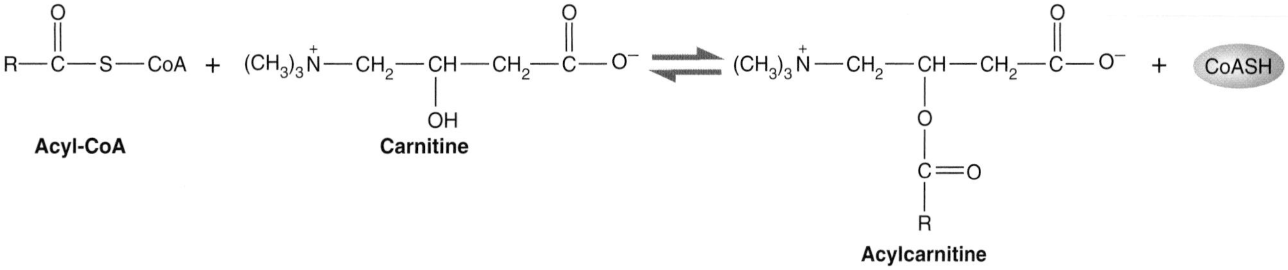

$$\underset{\textbf{Acyl-CoA}}{R-\overset{O}{\overset{\|}{C}}-S-CoA} + \underset{\textbf{Carnitine}}{(CH_3)_3\overset{+}{N}-CH_2-\underset{OH}{\underset{|}{CH}}-CH_2-\overset{O}{\overset{\|}{C}}-O^-} \rightleftharpoons \underset{\textbf{Acylcarnitine}}{(CH_3)_3\overset{+}{N}-CH_2-\underset{\underset{R}{\underset{|}{\underset{C=O}{\underset{|}{O}}}}}{\underset{|}{CH}}-CH_2-\overset{O}{\overset{\|}{C}}-O^-} + CoASH \qquad (2)$$

   This reaction is catalyzed by carnitine acyltransferase I (CAT-I).

2. A carrier protein within the mitochondrial inner membrane transfers acylcarnitine into the mitochondrial matrix.
3. Acyl-CoA is regenerated by carnitine acyltransferase II (CAT-II).
4. Carnitine is transported back into the intermembrane space by the carrier protein. It then reacts with another acyl-CoA.

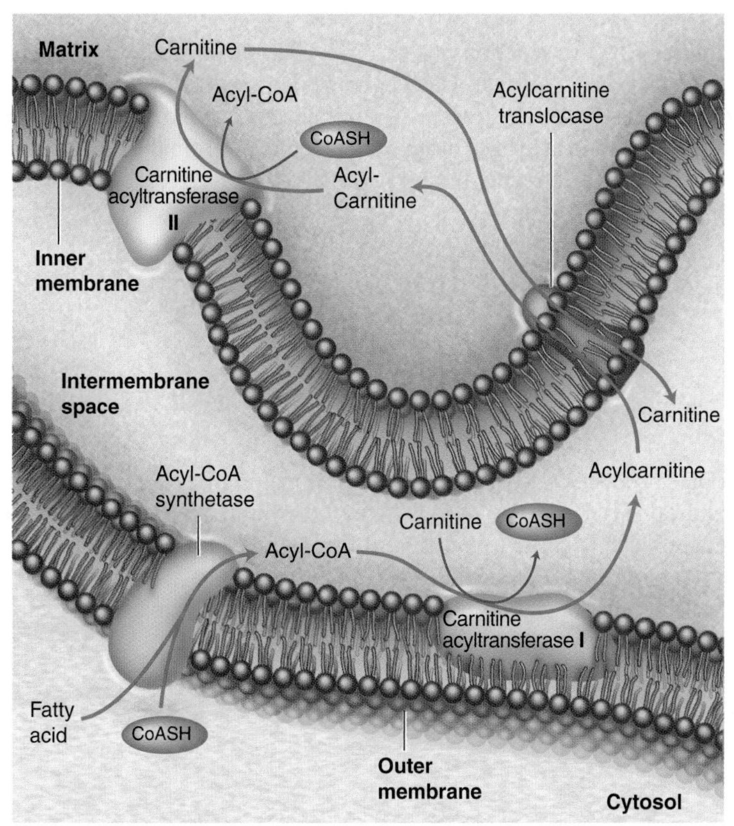

FIGURE 12.7
Fatty Acid Transport into the Mitochondrion

Fatty acids are activated to form acyl-CoA by acyl-CoA synthetase, an enzyme in the outer mitochrondrial membrane. Acyl-CoA then reacts with carnitine to form an acylcarnitine derivative. Carnitine acyltransferase I catalyzes this reaction. Acylcarnitine is transported across the inner membrane by acylcarnitine translocase and then reconverted to carnitine and acyl-CoA by carnitine acyltransferase II. Note that acylcarnitine translocase is an antiport protein (i.e., it transports one acylcarnitine into the mitochondrial matrix for every carnitine that is transported into the intermembrane space).

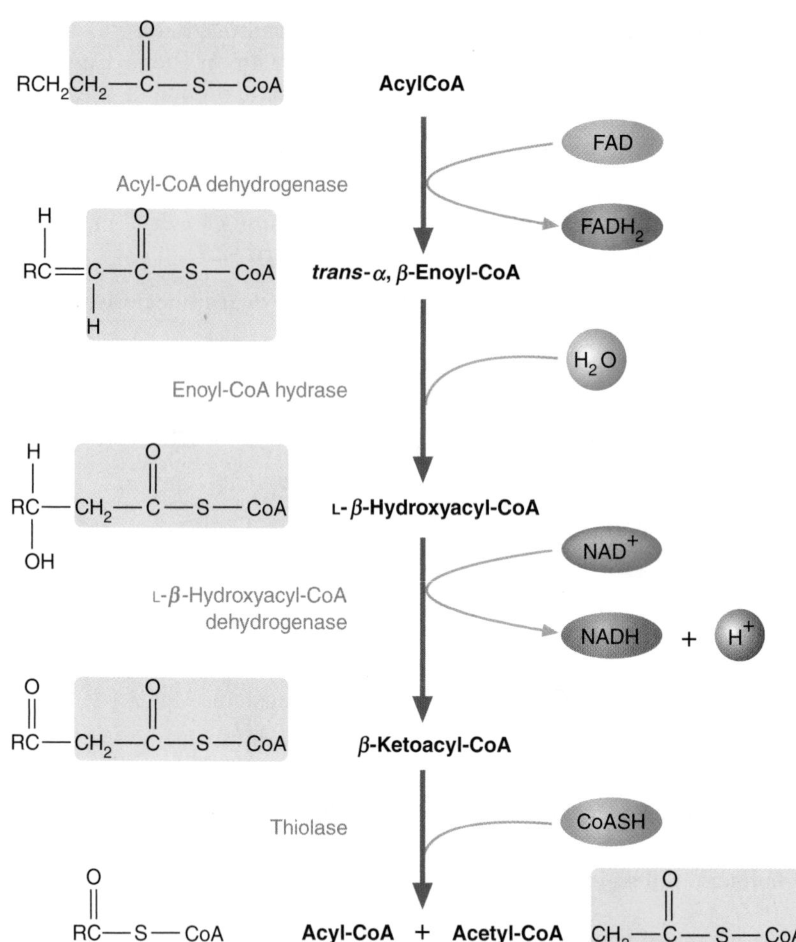

FIGURE 12.8
β-Oxidation of Acyl-CoA

The β-oxidation of acyl-CoA molecules consists of four reactions that occur in the mitochondrial matrix. Each cycle of reactions forms acetyl-CoA and an acyl-CoA that is shorter by two carbons.

A summary of the reactions of the β-oxidation of saturated fatty acids is shown in **Figure 12.8**. The pathway begins with an oxidation-reduction reaction, catalyzed by acyl-CoA dehydrogenase (an inner mitochondrial membrane flavoprotein), in which one hydrogen atom each is removed from the α- and β-carbons and transferred to the enzyme-bound FAD:

$$\underset{\text{Acyl-CoA}}{R-\underset{\beta}{CH_2}-\underset{\alpha}{CH_2}-\overset{O}{\overset{\|}{C}}-S-CoA} \xrightarrow{FAD \rightarrow FADH_2} \underset{\textit{trans-}\alpha,\beta\text{-Enoyl-CoA}}{R-\overset{H}{\overset{|}{C}}=\underset{H}{\underset{|}{C}}-\overset{O}{\overset{\|}{C}}-S-CoA} \quad (3)$$

The $FADH_2$ produced in this reaction then donates two electrons to UQ in the mitochondrial ETC (**Figure 10.5**). There are several isozymes of acyl-CoA dehydrogenase, each specific to a different fatty acid chain length. The product of this reaction is *trans*-α,β-enoyl-CoA.

The second reaction, catalyzed by enoyl-CoA hydrase, involves a hydration of the double bond between the α- and β-carbons:

$$\underset{\textit{trans-}\alpha,\beta\text{-Enoyl-CoA}}{R-\overset{H}{\overset{|}{C}}=\underset{H}{\underset{|}{C}}-\overset{O}{\overset{\|}{C}}-S-CoA} + H_2O \longrightarrow \underset{\beta\text{-Hydroxyacyl-CoA}}{R-\underset{H}{\underset{|}{\overset{OH}{\overset{|}{C}}}}-\underset{H}{\underset{|}{\overset{H}{\overset{|}{C}}}}-\overset{O}{\overset{\|}{C}}-S-CoA} \quad (4)$$

The β-carbon is now hydroxylated. In the next reaction this hydroxyl group is oxidized. The production of a β-ketoacyl-CoA is catalyzed by β-hydroxyacyl-CoA dehydrogenase:

$$\mathrm{R{-}C_{\beta}H(OH){-}C_{\alpha}H_2{-}C(=O){-}S{-}CoA} + NAD^+ \longrightarrow \mathrm{R{-}C_{\beta}(=O){-}C_{\alpha}H_2{-}C(=O){-}S{-}CoA} + NADH + H^+ \quad (5)$$

3-Hydroxyacyl-CoA → **β-Ketoacyl-CoA**

The electrons transferred to NAD^+ are later donated to complex I of the ETC. Finally, thiolase (sometimes referred to as β-ketoacyl-CoA thiolase) catalyzes a C_α-C_β cleavage:

$$\mathrm{R{-}C_{\beta}(=O){-}C_{\alpha}H_2{-}C(=O){-}S{-}CoA} + CoASH \longrightarrow \mathrm{R{-}C(=O){-}S{-}CoA} + \mathrm{CH_3{-}C(=O){-}S{-}CoA} \quad (6)$$

β-Ketoacyl-CoA → **Acyl-CoA** + **Acetyl-CoA**

In this reaction, sometimes called a **thiolytic cleavage**, an acetyl-CoA molecule is released. The other product, an acyl-CoA, now contains two fewer C atoms.

The four steps just outlined constitute one cycle of β-oxidation. During each later cycle, a two-carbon fragment is removed. In a process, called the *β-oxidation spiral*, the β-oxidation cycle is repeated until, in the last cycle, a four-carbon acyl-CoA is cleaved to form two molecules of acetyl-CoA.

The following equation summarizes the oxidation of palmitoyl-CoA:

$$\mathrm{CH_3(CH_2)_{14}C(=O){-}S{-}CoA} + 7\,FAD + 7\,NAD^+ + 7\,CoASH + 7\,H_2O \longrightarrow 8\,\mathrm{CH_3C(=O){-}S{-}CoA} + 7\,FADH_2 + 7\,NADH + 7\,H^+ \quad (7)$$

In muscle, the rate of β-oxidation depends on the availability of its substrate (i.e., the concentration of fatty acids in blood) and the tissue's current energy requirements. When NADH/NAD^+ ratios are high, β-hydroxyacyl–CoA dehydrogenase is inhibited. High acetyl-CoA levels depress the activity of thiolase. In liver, where fatty acids are also used in the synthesis of triacylglycerols and phospholipids, the rate of β-oxidation depends on how quickly these molecules are transported into mitochondria. When glucose levels are high and excess glucose molecules are being converted into fatty acids, malonyl-CoA, the product of the first committed step in fatty acid synthesis, prevents a futile cycle by inhibiting CAT-I.

The acetyl-CoA molecules produced by fatty acid oxidation are converted via the *citric acid cycle* to CO_2 and H_2O as additional NADH and $FADH_2$ are formed. A portion of the energy released as NADH and $FADH_2$ are oxidized by the ETC is later captured in ATP synthesis via *oxidative phosphorylation*. The complete oxidation of acetyl-CoA is discussed in Chapter 10. The calculation of the total number of ATP that can be generated from palmitoyl is reviewed next.

QUESTION 12.2

Medium-chain acyl-CoA dehydrogenase (MCAD) is a mitochondrial enzyme that catalyzes the first reaction in the β-oxidation cycle. Its substrates are 12-carbon acyl-CoAs. MCAD deficiency is an autosomal recessive disorder caused by a mutated version of the MCAD gene. Symptoms of this disorder, which include fatigue, vomiting, and hypoglycemia, are triggered by fasting. Apparently, the accumulation of toxic amounts of acyl-CoAs depresses gluconeogenesis. Considering that liver glycogen is depleted quickly in children, suggest a treatment that would help MCAD-deficient patients manage their symptoms.

QUESTION 12.3

Identify each of the following biomolecules:

(a) $R'—C(=O)—O—CH(CH_2—O—C(=O)—R)(CH_2—O—P(=O)(O^-)—O^-)$

(b) $R—C(=O)—S—CoA$

(c) $(CH_3)_3\overset{+}{N}—CH_2—CH(OH)—CH_2—C(=O)—O^-$

QUESTION 12.4

In the absence of oxygen, cells can produce small amounts of ATP from the anaerobic oxidation of glucose. This is not true for fatty acid oxidation. Explain.

The Complete Oxidation of a Fatty Acid

The aerobic oxidation of a fatty acid generates a large number of ATP molecules. As previously described (see p. 262), the oxidation of each $FADH_2$ during electron transport and oxidative phosphorylation yields approximately 1.5 molecules of ATP. Similarly, the oxidation of each NADH yields approximately 2.5 molecules of ATP. The yield of ATP from the oxidation of palmitoyl-CoA, which generates 7 $FADH_2$, 7 NADH, and 8 acetyl-CoA molecules to form CO_2 and H_2O, is calculated as follows:

$$\begin{aligned} 7\ \text{FADH}_2 \times 1.5\ \text{ATP/FADH}_2 &= 10.5\ \text{ATP} \\ 7\ \text{NADH} \times 2.5\ \text{ATP/NADH} &= 17.5\ \text{ATP} \\ 8\ \text{Acetyl-CoA} \times 10\ \text{ATP/acetyl-CoA} &= \underline{80\ \ \text{ATP}} \\ &\ 108\ \text{ATP} \end{aligned}$$

The formation of palmitoyl-CoA from palmitic acid uses two ATP equivalents. The net synthesis of ATP per molecule of palmitoyl-CoA is therefore 106 molecules of ATP.

The yield of ATP from the oxidations of palmitic acid and glucose can be compared. Recall that the total number of ATP molecules produced per glucose molecule is approximately 31. If glucose and palmitic acid molecules are compared in terms of the number of ATP molecules produced per carbon atom, palmitic acid is a superior energy source. The ratio for glucose is 31/6 or 5.2 ATP molecules per carbon atom. Palmitic acid yields 106/16 or 6.6 ATP molecules per carbon atom. The oxidation of palmitic acid generates more energy than that of glucose because palmitic acid is a more reduced molecule. (Glucose with its six oxygen atoms is a partially oxidized molecule.)

QUESTION 12.5

Determine the number of moles of NADH, $FADH_2$, and ATP molecules that can be synthesized from 1 mol of stearic acid.

β-OXIDATION IN PEROXISOMES β-Oxidation of fatty acids also occurs within peroxisomes. In animals, peroxisomal β-oxidation shortens very-long-chain fatty acids without ATP synthesis. The resulting medium-chain fatty acids are further degraded within mitochondria. Peroxisomal membrane possesses an acyl-CoA synthetase activity that is specific for very-long-chain fatty acids. Mitochondria apparently cannot activate long-chain fatty acids such as tetracosanoic (24:0) and hexacosanoic (26:0) acids. Peroxisomal carnitine acyltransferases catalyze the transfer of these molecules into peroxisomes, where they are oxidized to form acetyl-CoA and medium-chain acyl-CoA molecules that are further degraded via β-oxidation within mitochondria.

Although the reactions of peroxisomal β-oxidation are similar to those in mitochondria, there are some notable differences. First, the initial reaction in the peroxisomal pathway is catalyzed by an acyl-CoA oxidase. The reduced coenzyme $FADH_2$ then donates its electrons directly to O_2 instead of UQ. The H_2O_2 produced when $FADH_2$ is oxidized is converted to H_2O by catalase. Second, the next two reactions in peroxisomal β-oxidation are catalyzed by two enzyme activities (enoyl-CoA hydrase and 3-hydroxyacyl CoA dehydrogenase) found on the same protein molecule. Finally, the last enzyme in the pathway (β-ketoacyl-CoA thiolase) and its mitochondrial version have different substrate specificities. The former does not efficiently bind medium-chain acyl-CoAs.

THE KETONE BODIES Most of the acetyl-CoA produced during fatty acid oxidation is used by the citric acid cycle or in isoprenoid synthesis (Section 12.3). Under normal conditions, fatty acid metabolism is so carefully regulated that only small amounts of excess acetyl-CoA are produced. In a process called **ketogenesis**, excess acetyl-CoA molecules are converted to acetoacetate, β-hydroxybutyrate, and acetone, a group of molecules called the **ketone bodies** (**Figure 12.9**).

Ketone body formation, which occurs within the matrix of liver mitochondria, begins with the condensation of two acetyl-CoAs to form acetoacetyl-CoA. Then acetoacetyl-CoA condenses with another acetyl-CoA to form β-hydroxy-β-methylglutaryl-CoA (HMG-CoA). In the next reaction, HMG-CoA is cleaved to form acetoacetate and acetyl-CoA. Acetoacetate is then reduced to form β-hydroxybutyrate. Acetone is formed by the spontaneous decarboxylation of acetoacetate when the latter molecule's concentration is high. (This condition, referred to as **ketosis**, occurs during starvation and in uncontrolled diabetes, a metabolic disease discussed in Chapter 16: see the Biochemistry in Perspective essay entitled Diabetes Mellitus. In both conditions there is a heavy reliance on fat stores and β-oxidation of fatty acids to supply energy.)

Several tissues, most notably cardiac and skeletal muscle, use ketone bodies to generate energy. During prolonged starvation (i.e., in the absence of sufficient glucose) the brain uses ketone bodies as an energy source, thereby reducing its dependence on glucose. Ketone body oxidation also spares skeletal muscle protein, a source of substrates for gluconeogenesis (glucose-alanine cycle, p. 296). Other cells that use ketone bodies to generate energy during starvation include enterocytes and adipocytes. The mechanism by which acetoacetate and β-hydroxybutyrate are converted to acetyl-CoA is illustrated in **Figure 12.10**.

KEY CONCEPTS

- In β-oxidation, fatty acids are degraded by breaking the bond between the α- and β-carbon atoms.
- The ketone bodies are produced from excess molecules of acetyl-CoA.

Fatty Acid Oxidation: Double Bonds and Odd Chains

The β-oxidation pathway degrades saturated fatty acids with an even number of carbon atoms. Certain additional reactions are required to degrade unsaturated, odd-chain, and branched-chain fatty acids.

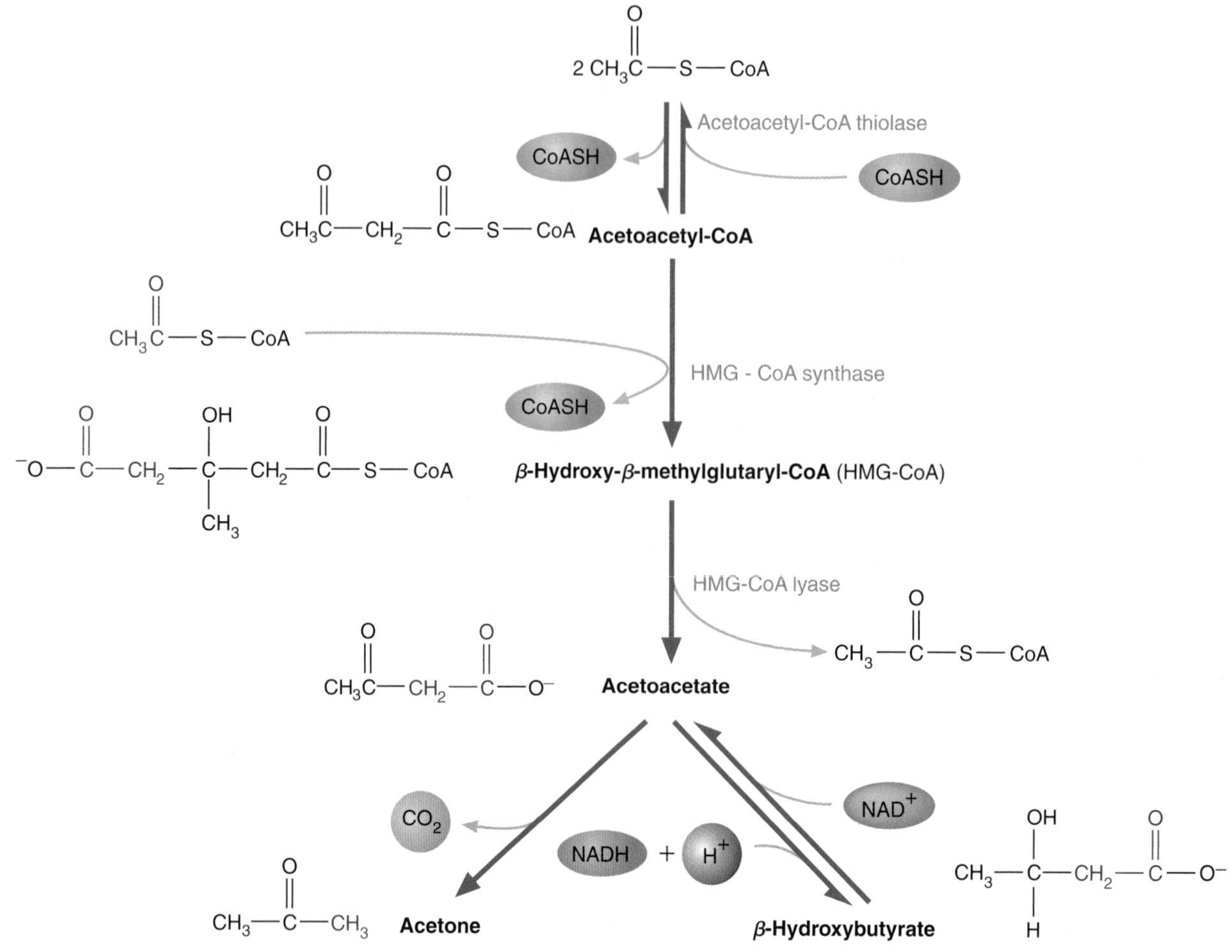

FIGURE 12.9

Ketone Body Formation

Ketone bodies (acetoacetate, acetone, and β-hydroxybutyrate) are produced within the mitochondria when excess acetyl-CoA is available. Under normal circumstances, only small amounts of ketone bodies are produced.

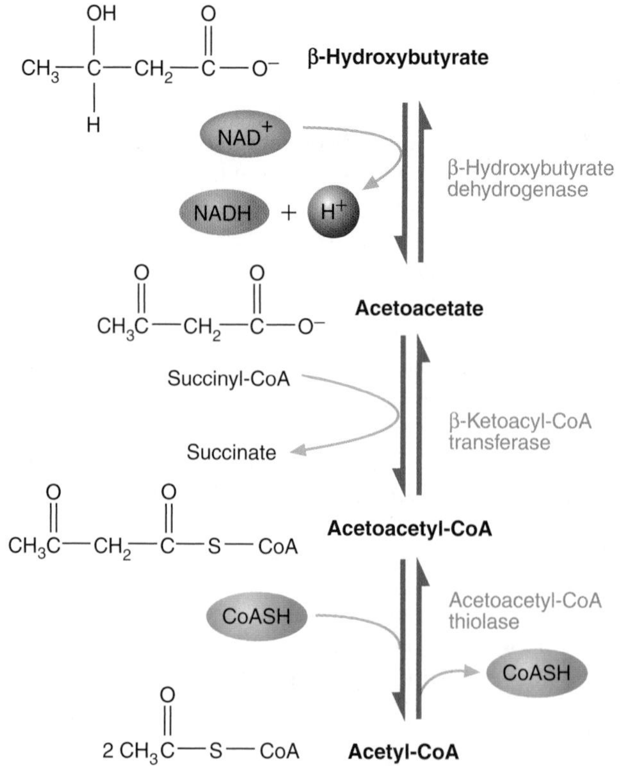

FIGURE 12.10

Conversion of Ketone Bodies to Acetyl-CoA

Some organs (e.g., heart and skeletal muscle) can use ketone bodies (β-hydroxybutyrate and acetoacetate) as an energy source under normal conditions. During starvation, however, ketone bodies become an important fuel source for the brain. Because liver does not have β-ketoacid-CoA transferase, it cannot use ketone bodies as an energy source. These reactions are reversible. The energy yield from the catabolism of β-hydroxybutyrate, 21.5 ATPs, is calculated as follows. Two acetyl CoA products yield 20 ATPs in the citric acid cycle, electron transport, and oxidative phosphorylation. An additional NADH, produced by the oxidation of β-hydroxybutyrate to form acetoacetate, yields 2.5 ATPs. One ATP equivalent is subtracted from the sum of 22.5 ATPs to account for the activations of acetoacetate by succinyl CoA.

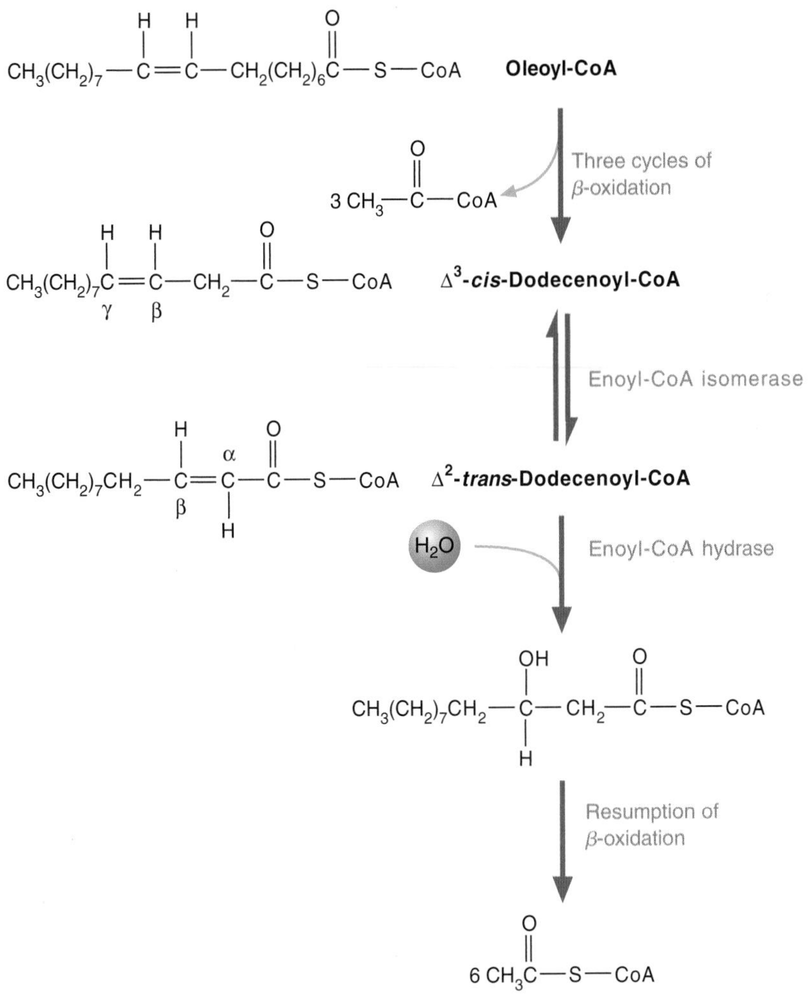

$CH_3(CH_2)_7—CH{=}CH—CH_2(CH_2)_6C(=O)—S—CoA$ **Oleoyl-CoA**

Three cycles of β-oxidation → $3\ CH_3—C(=O)—CoA$

$CH_3(CH_2)_7CH{=}CH—CH_2—C(=O)—S—CoA$ **Δ^3-*cis*-Dodecenoyl-CoA** (γ, β)

Enoyl-CoA isomerase

$CH_3(CH_2)_7CH_2—CH{=}CH—C(=O)—S—CoA$ **Δ^2-*trans*-Dodecenoyl-CoA** (β, α)

H_2O Enoyl-CoA hydrase

$CH_3(CH_2)_7CH_2—CH(OH)—CH_2—C(=O)—S—CoA$

Resumption of β-oxidation

$6\ CH_3C(=O)—S—CoA$

FIGURE 12.11
β-Oxidation of Oleoyl-CoA

β-Oxidation of the CoA derivative of oleic acid progresses until Δ^3-*cis*-dodecenoyl-CoA is produced. This molecule is not a suitable substrate for β-oxidation because it contains a *cis* double bond. After conversion of the β,γ-*cis* double bond to an α,β-*trans* double bond, β-oxidation recommences.

UNSATURATED FATTY ACID OXIDATION The oxidation of unsaturated fatty acids such as oleic acid requires additional enzymes. They are needed because, unlike the *trans* double bonds introduced during β-oxidation, the double bonds of most naturally occurring unsaturated fatty acids have a *cis* configuration. The enzyme enoyl-CoA isomerase converts the *cis*-β,γ double bond to a *trans*-α,β double bond. **Figure 12.11** illustrates the β-oxidation of oleic acid.

ODD-CHAIN FATTY ACID OXIDATION Although most fatty acids contain an even number of carbon atoms, certain organisms (e.g., some plants and microorganisms) produce odd-chain fatty acid molecules. β-Oxidation of such fatty acids proceeds normally until the last β-oxidation cycle, which yields one acetyl-CoA molecule and one propionyl-CoA molecule. Propionyl-CoA is then converted to succinyl-CoA, a citric acid cycle intermediate (**Figure 12.12**). Ruminant animals such as cattle and sheep derive a substantial amount of energy from the oxidation of odd-chain fatty acids. These molecules are produced by microbial fermentation processes in the rumen (stomach).

α-OXIDATION *α-Oxidation* is a mechanism for degrading branched-chain fatty acid molecules such as phytanic acid. Phytanic acid, a 20-carbon fatty acid, is an oxidation product of phytol, a diterpene alcohol esterified to chlorophyll, the photosynthetic pigment. Phytol, found in green vegetables, is converted to phytanic acid after ingestion. Phytanic acid is a component of dairy products and other foods derived from herbivorous (plant-eating) animals. In humans, α-oxidation takes place in peroxisomes.

β-Oxidation of phytanic acid is blocked by the methyl group substituent on C-3 (the β-position). Consequently, the first step in phytanic acid catabolism is an

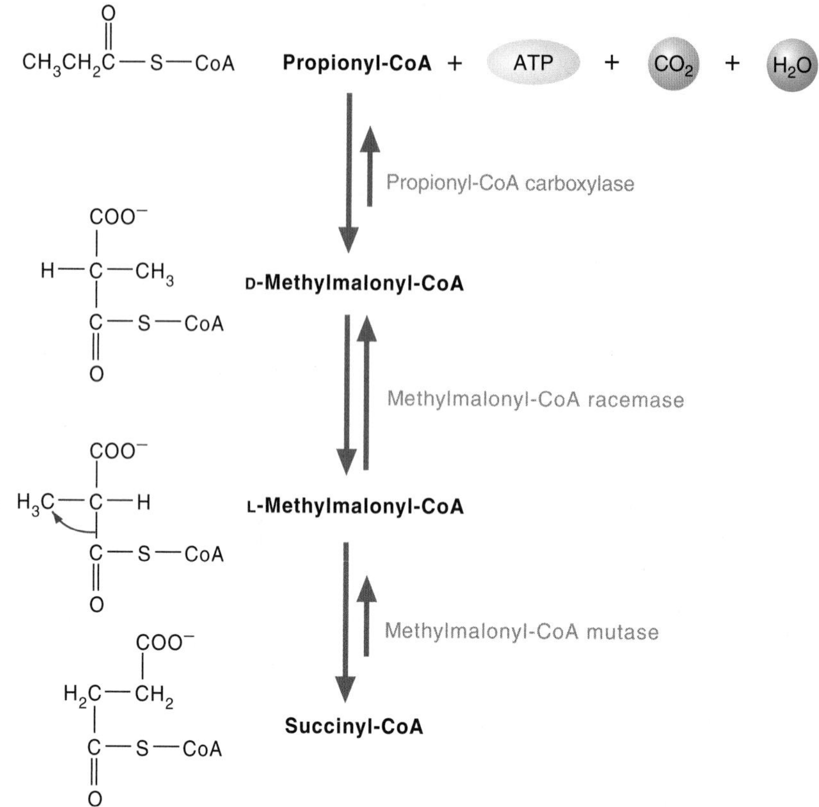

FIGURE 12.12

Conversion of Propionyl-CoA to Succinyl-CoA

In the first step, propionyl-CoA is carboxylated by propionyl-CoA carboxylase, an enzyme with a biotin cofactor (see p. 444). The product, D-methylmalonyl-CoA, is isomerized by methylmalonyl-CoA racemase to form L-methylmalonyl-CoA. In the last step, a hydrogen atom and the carbonyl-CoA group exchange positions. This unusual reaction is catalyzed by methylmalonyl-CoA mutase, an enzyme that requires 5′-deoxyadenosylcobalamin, usually designated as vitamin B_{12}.

α-oxidation in which the molecule is converted to a α-hydroxy fatty acid. This reaction is followed by the removal of the carboxyl group (Figure 12.13). After activation to a CoA derivative, the product, pristanic acid, can be further degraded by β-oxidation. All subsequent side chain methyl groups will now be in the α-position, which is not a problem for β-oxidation enzymes. Phytanic acid oxidation is critical because large quantities of this molecule are found in the diet. In *Refsum's disease* (also referred to as *phytanic acid storage syndrome*) a buildup of phytanic acid causes serious neurological problems. In this rare autosomal recessive condition, nerve damage is caused by a missing or defective gene that codes for phytanoyl-CoA hydroxylase. Phytanic acid accumulation interferes with myelination (myelin sheath formation, see p. 395). Eating less phytanic acid–containing foods (i.e., dairy products) can significantly reduce nerve damage.

KEY CONCEPT

Several reactions in addition to β-oxidation are required to degrade unsaturated, odd-chain, and branched-chain fatty acids.

QUESTION 12.6

In the past, mammals were believed to be unable to use fatty acids in gluconeogenesis. (Acetyl-CoA cannot be converted to pyruvate because the reaction catalyzed by pyruvate dehydrogenase is irreversible.) Recent experimental evidence indicates that certain unusual fatty acids (i.e., those with odd chains or two carboxylic acid groups) can be converted to glucose in small but measurable quantities. One molecule of propionyl-CoA is produced when an odd-carbon chain fatty acid is oxidized. Describe a possible biochemical pathway by which a liver cell might synthesize glucose from propionyl-CoA. [*Hint*: Refer to **Figure 12.12**.]

QUESTION 12.7

One of the products of the β-oxidation of dicarboxylic acids is succinyl-CoA. Propose a biochemical pathway for the conversion of the molecule illustrated in **Figure 12.14** to glucose.

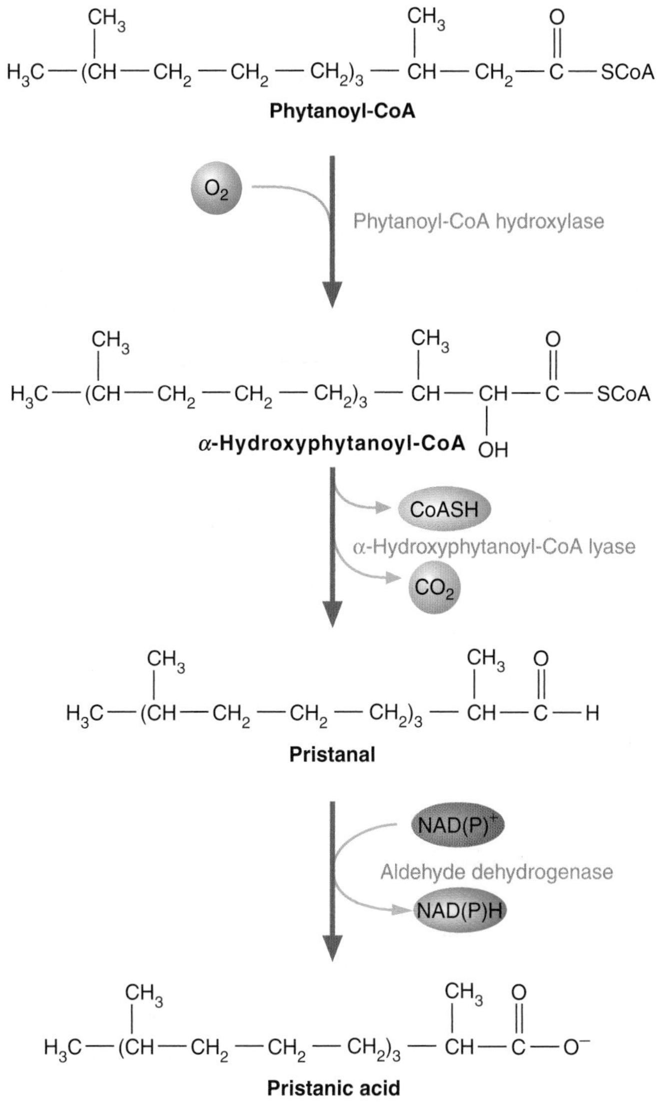

FIGURE 12.13

α-Oxidation of Phytanic Acid in Peroxisomes

In α-oxidation, phytanoyl-CoA is converted to α-hydroxyphytanoyl- CoA in an O_2-requiring reaction. α-Hydroxyphytanoyl-CoA is converted to pristanal in a TPP-requiring decarboxylation reaction in which the α-carbon is oxidized. The thioester bond of the other product, formyl-CoA, is subsequently cleaved to form CoASH and formic acid (HCOOH), which is then oxidized to form CO_2. Pristanal is oxidized in an $NAD(P)^+$-requiring reaction to form pristanic acid. Pristanic acid is then further degraded by β-oxidation after esterification with CoASH. Products of this process are three acetyl-CoAs, three propionyl-CoAs, and one isobutyl-CoA.

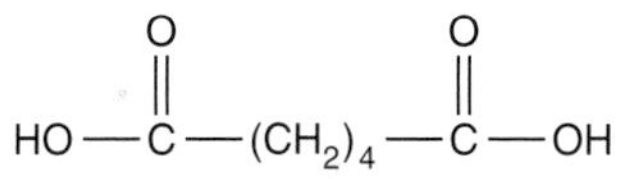

FIGURE 12.14

Adipic Acid

Fatty Acid Biosynthesis

Although fatty acid synthesis occurs within the cytoplasm of most animal cells, liver is the major site for this process. (Recall, for example, that liver produces VLDL. See p. 403.) Fatty acids are synthesized when the diet is low in fat and/or high in carbohydrate or protein. Most fatty acids are synthesized from excess dietary carbohydrate. As discussed, glucose is converted to pyruvate in the cytoplasm. After entering the mitochondrion, pyruvate is converted to acetyl-CoA, which condenses with OAA, a citric acid cycle intermediate, to form citrate. When mitochondrial citrate levels are sufficiently high (i.e., cellular energy requirements are low), citrate enters the cytoplasm, where it is cleaved to form acetyl-CoA and OAA. Acetyl-CoA is then used in fatty acid biosynthesis. The net reaction for the synthesis of palmitic acid from acetyl-CoA is as follows:

$$8\ \text{Acetyl-CoA} + 14\ \text{NADPH} + 14\ H^+ + 7\ \text{ATP} \longrightarrow$$

$$\text{Palmitate} + 14\ \text{NADP}^+ + 7\ \text{ADP} + 7\ P_i + 8\ \text{CoASH} + 6\ H_2O \quad (8)$$

A relatively large quantity of NADPH is required in fatty acid synthesis. A substantial amount of NADPH is provided by the pentose phosphate pathway (see p. 298). Reactions catalyzed by isocitrate dehydrogenase (see p. 333) and malic enzyme (see p. 340) provide smaller amounts.

The biosynthesis of fatty acids is outlined in **Figure 12.15**. At first glance, fatty acid synthesis appears to be the reverse of the β-oxidation pathway. For example, fatty acids are constructed by the sequential addition of two-carbon groups supplied by acetyl-CoA. Additionally, the same intermediates (i.e., β-ketoacyl-, β-hydroxyacyl-, and α,β-unsaturated acyl groups) are found in both pathways. A closer examination, however, reveals several notable differences between fatty acid synthesis and β-oxidation. First, fatty acid synthesis occurs predominantly in the cytoplasm. (Recall that β-oxidation occurs within mitochondria and peroxisomes.) Second, the enzymes that catalyze fatty acid synthesis are significantly different in structure from those in β-oxidation. In eukaryotes, most of these enzymes are components of a multienzyme complex referred to as fatty acid synthase. Third, the intermediates in fatty acid synthesis are linked through a thioester linkage to **acyl carrier protein** (ACP), a component of fatty acid synthase. (Recall that acyl groups are attached to CoASH through a thioester linkage during β-oxidation.) Note that acyl groups are attached to both ACP and CoASH via a phosphopantetheine prosthetic group (**Figure 12.16**). And finally, in contrast to β-oxidation, which produces NADH and $FADH_2$, fatty acid synthesis consumes NADPH. Fatty acid synthesis occurs in two phases: the carboxylation of acetyl-CoA to form malonyl-CoA by acetyl-CoA carboxylase and the synthesis of palmitate by the sequential addition of two carbon units to a growing fatty acyl chain by fatty acid synthase.

ACETYL-COA CARBOXYLASE The carboxylation of acetyl-CoA to form malonyl-CoA is an irreversible reaction that is catalyzed by acetyl-CoA carboxylase (ACC) (**Figure 12.17**). The first phase of this reaction is the ATP-dependent carboxylation of biotin to give carboxybiotin. Subsequent decarboxylation results in the transfer of an activated CO_2 from biotin to acetyl-CoA. Acetyl-CoA carboxylation, the rate-limiting step in fatty acid synthesis, is an activating reaction that is necessary because this carbon-carbon condensation is thermodynamically unfavorable. As a result of resonance stabilization, free carboxylate groups are insufficiently reactive.

Acetyl-CoA carboxylase is found in most organisms. In eukaryotes, ACC contains three domains: BCCP (biotin carboxyl carrier protein), BC (biotin carboxylase), and CT (carboxyltransferase). Biotin, a coenzyme that carries carboxyl groups, is bound to BCCP via an amide linkage to the side chain of a lysine residue. The flexible lysine side chain is, in effect, a swinging arm that transfers the newly carboxylated biotin from the active site of the BC domain to the active site of the CT domain (a 7-Å distance). CT then catalyzes the transfer of the carboxyl group from biotin to acetyl-CoA to form the product malonyl-CoA. In mammals, there are two forms of ACC. A cytoplasmic enzyme, ACC1, is expressed in lipogenic tissues such as liver, adipose tissue, and lactating mammary gland. A mitochondrial enzyme, ACC2, occurs in oxidative tissues such as cardiac and skeletal muscle, where its product, malonyl-CoA, acts as a strong inhibitor of carnitine acyltransferase I. ACC2, therefore, serves a regulatory function in fatty acid oxidation. The liver, which both oxidizes and synthesizes fatty acids, contains both forms of ACC.

Mammalian ACC contains two subunits, each with a bound biotin cofactor. ACC becomes active when ACC dimers aggregate to form high-molecular-weight polymers (4 million to 8 million daltons) composed of 10 to 20 dimers. ACC, a key enzyme in fatty acid metabolism, is highly regulated by allosteric modulators and phosphorylation reactions (**Figure 12.18**). The allosteric effects of

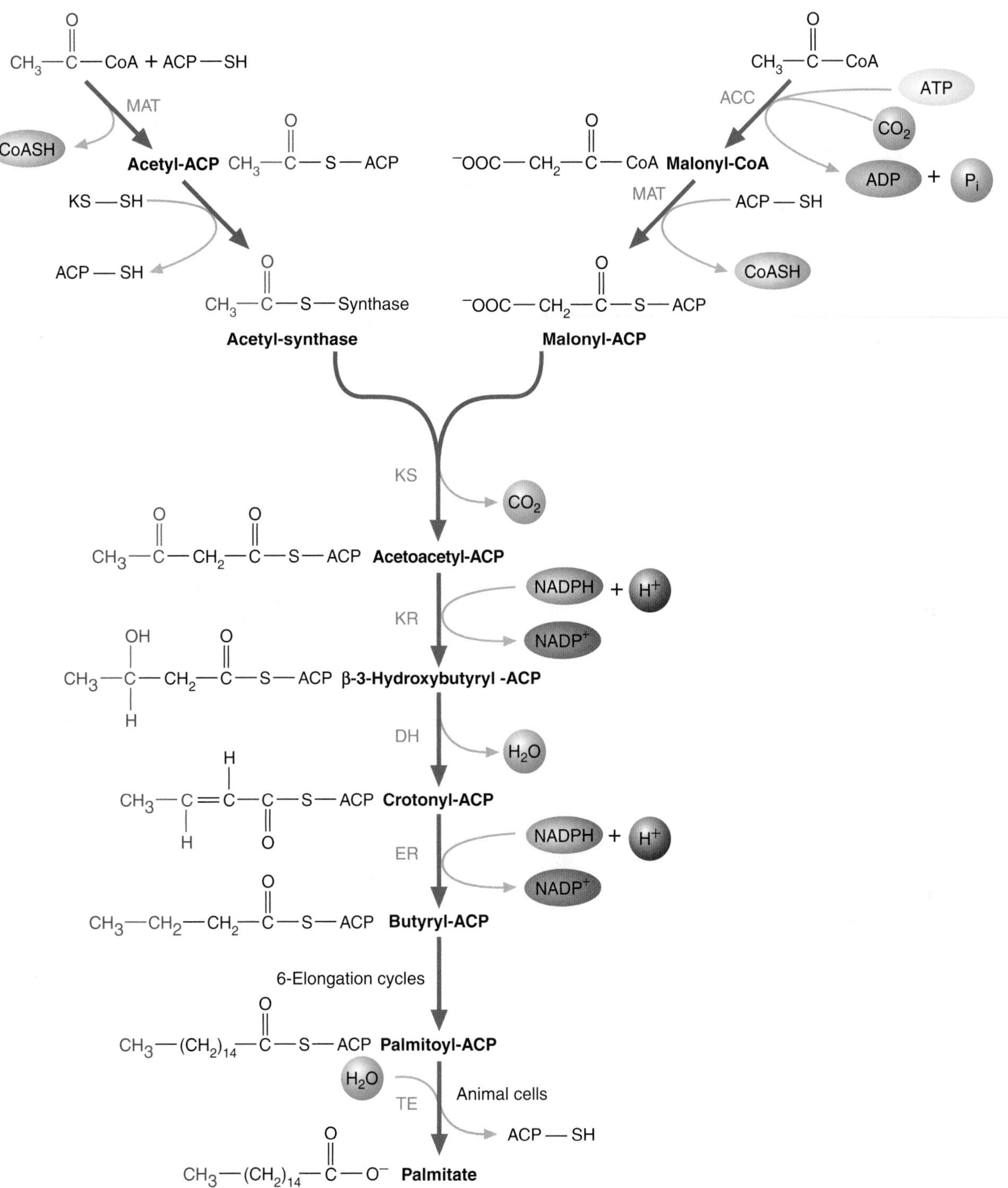

FIGURE 12.15

Fatty Acid Biosynthesis

The substrates for fatty acid synthesis are acetyl-ACP and malonyl-ACP, formed from ACP and acetyl-CoA and malonyl-CoA, respectively. Both reactions are catalyzed by malonyl/acetyl transferase (MAT). Formation of a fatty acid chain begins with a condensation reaction. Catalyzed by β-ketoacyl synthase (KS), the acetyl group (linked to KS via a thioester linkage) is transferred to the malonyl group to form acetoacetyl-ACP. The reduction of the β-carbonyl group, catalyzed by β-ketoacyl-ACP reductase (KR), forms an alcohol. The removal of water to form a carbon-carbon double bond is catalyzed by β-hydroxyacyl-ACP dehydratase (DH). Reduction by enoyl-ACP reductase (ER) yields a saturated four-carbon acyl group. This acyl group is then transferred from ACP to the SH group of KS to begin a new elongation cycle. The acyl chain lengthens by two carbons as it condenses with another ACP-linked malonyl group. In animal cells, fatty acid synthesis ends with the release of palmitate from ACP, catalyzed by thioesterase (TE).

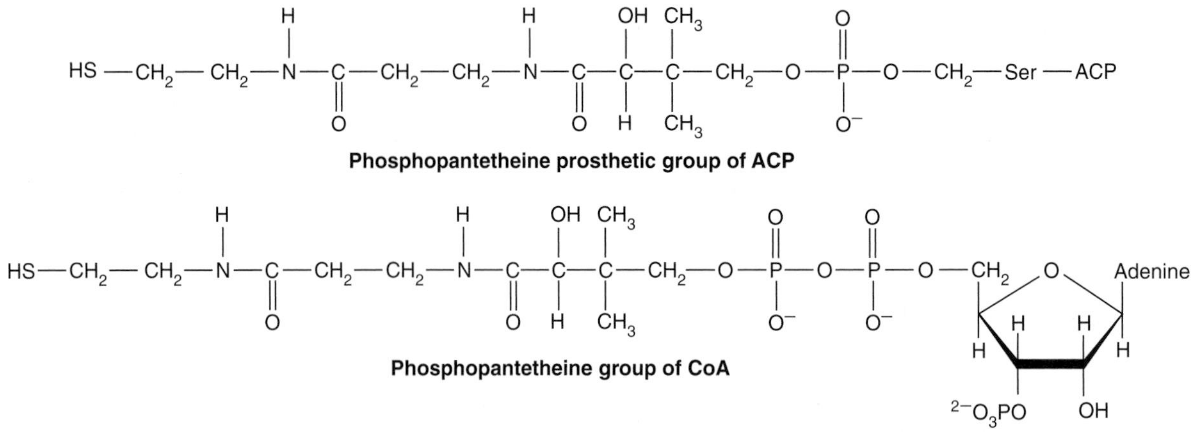

FIGURE 12.16

Comparison of the Phosphopantetheine Group in Acyl Carrier Protein (ACP) and in Coenzyme A (CoASH)

Fatty acids are attached to this prosthetic group on ACP during fatty acid biosynthesis and on CoASH during β-oxidation.

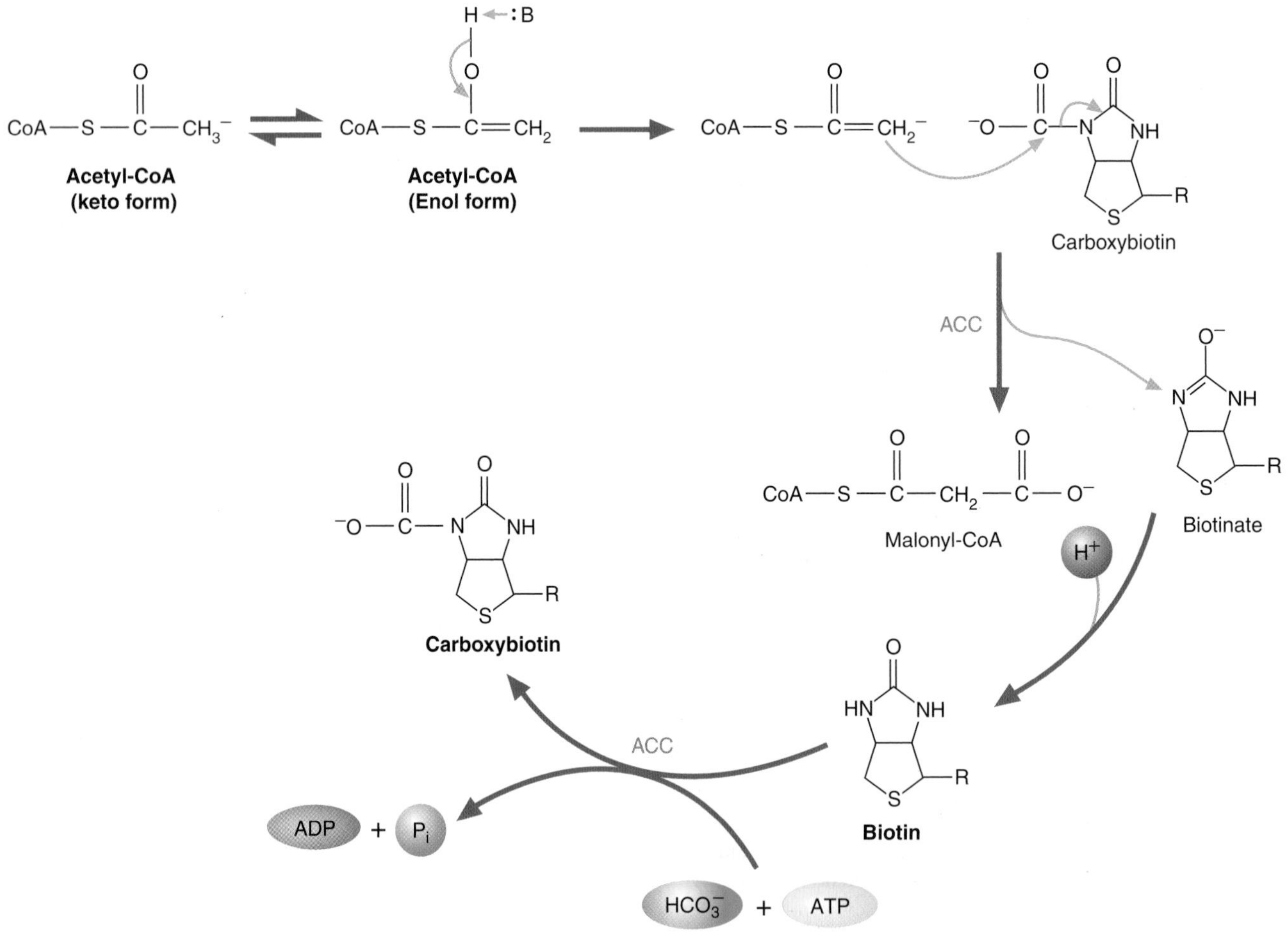

FIGURE 12.17

Synthesis of Malonyl-CoA

The reaction begins with the ATP-dependent carboxylation of the biotin cofactor of acetyl-CoA carboxylase (ACC). The carboxylase abstracts a proton from the α-carbon of the enol form of acetyl-CoA to generate a reactive carbanion. The carbanion attacks the carbonyl carbon of carboxybiotin to yield malonyl-CoA and biotinate. The biotinate is protonated by the enzyme to regenerate its biotin form. ACC then regenerates carboxybiotin, which reacts with another acetyl-CoA.

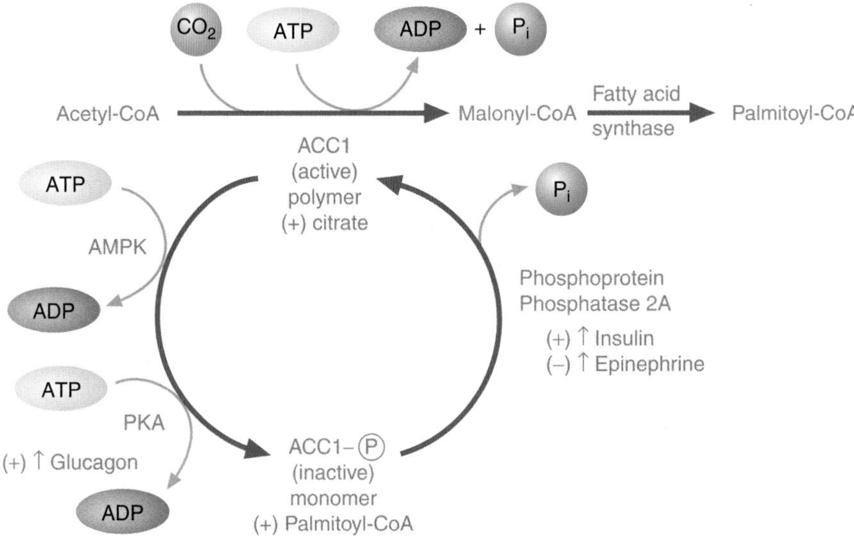

FIGURE 12.18

Regulation of Acetyl-CoA Carboxylase 1

ACC1 is inactivated (depolymerized) by phosphorylation reactions catalyzed by AMPK (as a result of high AMP levels), PKA (stimulated by glucagon), and palmitoyl-CoA accumulation. ACC1 is activated (polymerized) by phosphoprotein phosphatase 2A, which is activated by high insulin levels (glucose readily available) and deactivated by high glucagon levels (low blood glucose) and/or the presence of epinephrine (stress demand for energy mobilization).

citrate, a feed-forward activator that promotes polymerization, and palmitoyl-CoA, an end product inhibitor that causes depolymerization, are dependent on the phosphorylation state of the enzyme. ACC is phosphorylated and therefore inhibited (depolymerized) by AMP-activated protein kinase (AMPK), an important regulatory enzyme in energy metabolism. (AMP is a more sensitive indicator of a cell's energy charge than either ADP or ATP.) Phosphorylation by the c-AMP-activated protein kinase PKA, stimulated by glucagon, also plays a role in ACC inhibition. Depolymerization is also favored by the presence of palmitoyl-CoA, which binds to and stabilizes the dimeric form of the enzyme. Glucagon and epinephrine help to maintain the phosphorylated inactive form of ACC by inactivating phosphoprotein phosphatase-2A (PP-2A), the enzyme that mediates the dephosphorylation of a number of target proteins including ACC. Insulin activates PP-2A, which dephosphorylates ACC and permits its polymerization and, therefore, activation. The polymerized form of ACC is stabilized by binding to citrate, which accumulates when acetyl-CoA levels are high.

FATTY ACID SYNTHASE In humans, the remaining reactions in fatty acid synthesis take place on the fatty acid synthase multienzyme complex (FAS) (**Figure 12.19**). FAS is an X-shaped head-to-head homodimer composed of two identical 272-kDa polypeptides. Each polypeptide has seven catalytic domains and ACP. As a result, FAS synthesizes two fatty acids simultaneously. During fatty acid synthesis the acyl intermediates are covalently bound to the 2-nm-long phosphopantetheine group of ACP through a thiol ester linkage. The flexibility of ACP, a relatively unstructured domain of FAS, and the phosphopantetheine group allow the transfer of attached acyl intermediates from one active site to another in the complex.

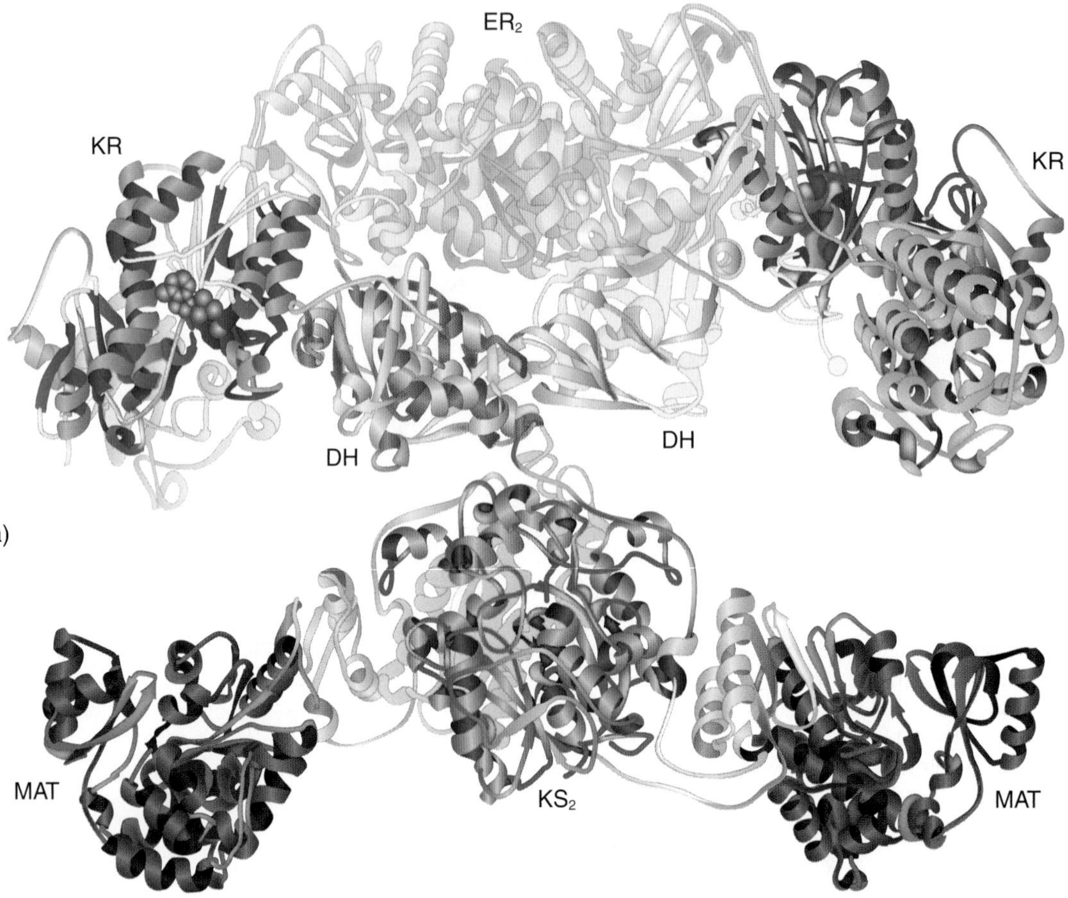

FIGURE 12.19

Fatty Acid Synthase Structure

(a) The structure of mammalian FAS is based on X-ray crystal studies. Note that because ACP is unstructured, its structure (not shown) remains unresolved. Each ACP domain is probably located near a KR domain. (b) This model illustrates the domain organization of the FAS homodimer. [KS = β-ketoacyl synthase; MAT = malonyl/acetyltransferase; DH = β-hydroxyacyl-ACP dehydratase; ER = enoyl-ACP reductase; KR = β-ketoacyl reductase; ACP = acyl carrier protein; TE = thioesterase.]

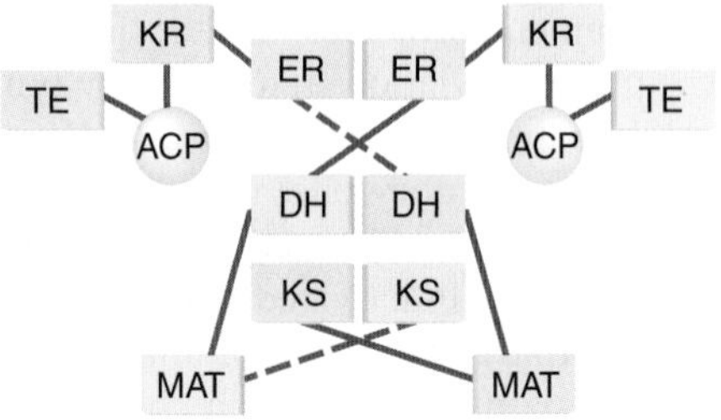

The synthesis of a fatty acid (**Figure 12.15**) is initiated with the transfer of the acetyl group of acetyl-CoA and the malonyl group of malonyl-CoA to ACP. Both reactions are catalyzed by malonyl/acetyl transferase (MAT). The acetyl group is then transferred from acetyl-ACP to a cysteinyl side chain of β-ketoacyl synthase (KS). KS then catalyzes a condensation reaction (**Figure 12.20**) in which the decarboxylation of the malonyl group creates a carbanion. The carbanion attacks the carbonyl carbon of the acetyl group to yield the product acetoacetyl-ACP.

During the next three steps, consisting of two reductions and a dehydration, the acetoacetyl group is converted to a butyryl group. β-Ketoacyl-ACP reductase (KR) catalyzes the reduction of acetoacetyl-ACP to form β-hydroxybutyryl-ACP. β-Hydroxyacyl-ACP dehydratase (DH) later catalyzes a dehydration, thus forming crotonyl-ACP. Butyryl-ACP is produced when 2,3-*trans*-enoyl-ACP reductase (ER) reduces the double bond in crotonyl-ACP. In the last step of the first cycle of fatty acid synthesis, the butyryl group is transferred from the phosphopantetheine group to the cysteine residue of KS. The newly freed ACP-SH group now binds to another malonyl group. The process is then repeated until, eventually, palmitoyl-ACP is synthesized. The palmitoyl group is released from fatty

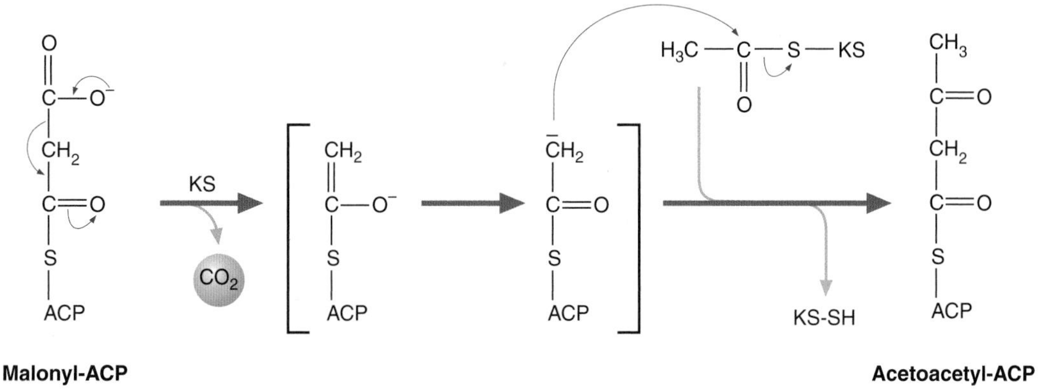

Malonyl-ACP **Acetoacetyl-ACP**

FIGURE 12.20
Formation of Acetoacetyl-ACP

The decarboxylation of malonyl-ACP, catalyzed by β-ketoacyl-ACP synthase (KS), results in the formation of a carbanion. The attack of the carbanion on the carbonyl carbon of an acetyl group linked via a thioester bond to the enzyme yields acetoacetyl-ACP.

acid synthase when thioesterase cleaves the thioester bond. Depending on cellular conditions, palmitate can be used directly in the synthesis of several types of lipid (e.g., triacylglycerol or phospholipids), or it can enter the mitochondrion, where several enzymes catalyze elongating and desaturating reactions. ER possesses similar enzymes.

FATTY ACID ELONGATION AND DESATURATION Elongation and desaturation of fatty acids synthesized in cytoplasm or obtained from the diet are accomplished primarily by ER enzymes. Fatty acid elongation and desaturation (the formation of double bonds) are especially important in the regulation of membrane fluidity and the synthesis of the precursors for a variety of fatty acid derivatives, such as the eicosanoids. For example, myelination (p. 440) depends especially on the ER fatty acid synthetic reactions. Very-long-chain saturated and monounsaturated fatty acids are important constituents of the cerebrosides and sulfatides found in myelin. Cells regulate membrane fluidity by adjusting the types of fatty acids that are incorporated into membrane lipids. For example, in cold weather, more unsaturated fatty acids are incorporated. (Recall that unsaturated fatty acids have a lower freezing point than do saturated fatty acids. See p. 385.) If the diet does not provide a sufficient number of these molecules, fatty acid synthetic pathways are activated. Although elongation and desaturation are closely integrated processes, for the sake of clarity they are discussed separately.

ER fatty acid chain elongation, which uses two-carbon units provided by malonyl-CoA, is a cycle of condensation, reduction, dehydration, and reduction reactions similar to those observed in cytoplasmic fatty acid synthesis. In contrast to the cytoplasmic process, the intermediates in the ER elongation process are CoA esters. These reactions can lengthen both saturated and unsaturated fatty acids. Reducing equivalents are provided by NADPH.

Acyl-CoA molecules are desaturated in ER membrane in the presence of NADH and O_2. Cytochrome b_5 reductase (a flavoprotein), cytochrome b_5, and oxygen-dependent desaturases, functioning together as an electron transport system, efficiently introduce double bonds into long-chain fatty acids (**Figure 12.21**). Both the flavoprotein and cytochrome b_5 (found in a ratio of approximately 1:30) have hydrophobic peptides that anchor the proteins into the ER membrane. Animals typically have Δ^9, Δ^6, and Δ^5 desaturases that use electrons supplied by NADH to activate the oxygen needed to create the double bond. Because elongation and desaturation systems are in close proximity to each other, a variety of long-chain polyunsaturated acids are typically produced. One example of this interaction is the synthesis of arachidonic acid ($20{:}4^{\Delta 5,8,11,14}$) from linoleic acid ($18{:}2^{\Delta 9,12}$).

KEY CONCEPTS

- In animals, fatty acids are synthesized in the cytoplasm from acetyl-CoA and malonyl-CoA.
- Mitochondrial and ER enzymes elongate and desaturate newly synthesized fatty acids as well as those obtained in the diet.

H^+ + NADH → NAD^+

Cytochrome b_5 reductase (F) → Cytochrome b_5 reductase (FH_2)

2 H^+

2 Cytochrome b_5 (Fe^{2+}) (reduced) → 2 Cytochrome b_5 (Fe^{3+}) (oxidized)

2 H^+ + O_2 → Desaturase → 2 H_2O

$CH_3(CH_2)_{16}C(=O)—S—CoA$ **Stearoyl-CoA** → $CH_3(CH_2)_7CH{=}CH(CH_2)_7C(=O)—S—CoA$ **Oleoyl-CoA**

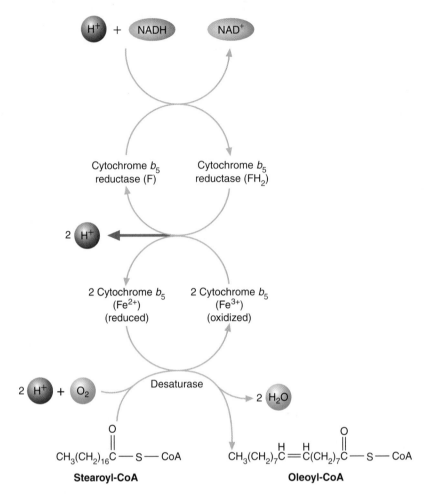

FIGURE 12.21
Desaturation of Stearoyl-CoA

The desaturase uses electrons provided by an electron transport system composed of cytochrome b_5 reductase and cytochrome b_5 to activate the oxygen (not shown) needed to create the double bond. NADH is the electron donor.

COMPARISON OF FATTY ACID OXIDATION AND FATTY ACID SYNTHESIS The functions of β-oxidation and fatty acid synthesis are clearly different. β-Oxidation degrades fatty acids to yield acetyl-CoA, the substrate of the energy-generating citric acid cycle. In contrast, energy is stored when acetyl-CoA is converted into fatty acids. Although the cellular locations, the redox coenzymes, the acyl group carriers, and the enzymes involved in the β-oxidation and fatty acid synthetic pathways are quite different, the reactions are similar enough to cause confusion. **Table 12.1** outlines the differences between the two processes.

QUESTION 12.8

Excessive consumption of fructose has been linked to obesity and to the condition referred to as *hypertriglyceridemia* (high blood levels of triacylglycerols). Common sources of fructose for most Americans are sucrose and high-fructose corn syrup. Over the past several decades, high-fructose corn syrup has replaced sucrose in many processed foods and beverages because it is inexpensive in comparison to sucrose. It now comprises at least 40% of caloric sweeteners.

TABLE 12.1 Comparisons of Fatty Acid β-Oxidation and Fatty Acid Synthesis

	Fatty Acid β-Oxidation	Fatty Acid Synthesis
Subcellular location	Mitochondrion	Cytoplasm
Substrates	Acyl-CoA, FAD, NAD^+, CoASH	Acetyl-CoA, NADPH
Products	Acetyl-CoA, $FADH_2$, NADH	Palmatate, $NADP^+$, CO_2, CoASH
Reaction cycle	$RCH_2CH_2C(=O)CoA$	$CH_3C(=O)$—ACP
	① Redox reaction: FAD → $FADH_2$	① Condensation reaction: Malonyl-CoA → CoASH + CO_2
	$RCH{=}CHC(=O)CoA$	$CH_3C(=O)$—$CH_2C(=O)$—ACP
	② Hydration: H_2O	② Redox reaction: NADPH + H^+ → $NADP^+$
	$RCH(OH)CH_2C(=O)CoA$	$CH_3CH(OH)$—$CH_2C(=O)$—ACP
	③ Redox reaction: NAD^+ → NADH + H^+	③ Dehydration: H_2O
	$RC(=O)CH_2C(=O)CoA$	$CH_3CH{=}CHC(=O)$—ACP
	④ Cleavage reaction: CoASH	④ Redox reaction: NADPH + H^+ → $NADP^+$
	$RC(=O)CoA + CH_3C(=O)CoA$	$CH_3CH_2CH_2C(=O)$—ACP

(The fructose content of fresh fruits and vegetables is so low that it would be difficult to consume sufficient quantities to induce hypertriglyceridemia.) Sucrose is digested in the small intestine by the enzyme sucrase to yield one molecule each of fructose and glucose. Digestion is so rapid that the blood concentrations of these sugars can become quite high. Whatever its source, once fructose reaches the liver, it is converted to fructose-1-phosphate (see p. 302). In addition, whereas high blood levels of glucose trigger the release of insulin and *leptin* (a hormone secreted by adipose tissue), both of which curb appetite, this does not happen with fructose. After reviewing fructose metabolism and fatty acid and triacylglycerol synthesis, suggest how hypertriglyceridemia and obesity might result from a diet that is rich in sucrose and high-fructose corn syrup.

Regulation of Fatty Acid Metabolism in Mammals

Animals have such varying requirements for energy that the metabolism of fatty acids (*the* major energy source in animals) is carefully regulated. Both short- and long-term regulatory mechanisms are used. In short-term regulation (measured in minutes) the activities of existing molecules of key regulatory enzymes are modified by allosteric regulators (**Figure 12.22**), covalent modification, and

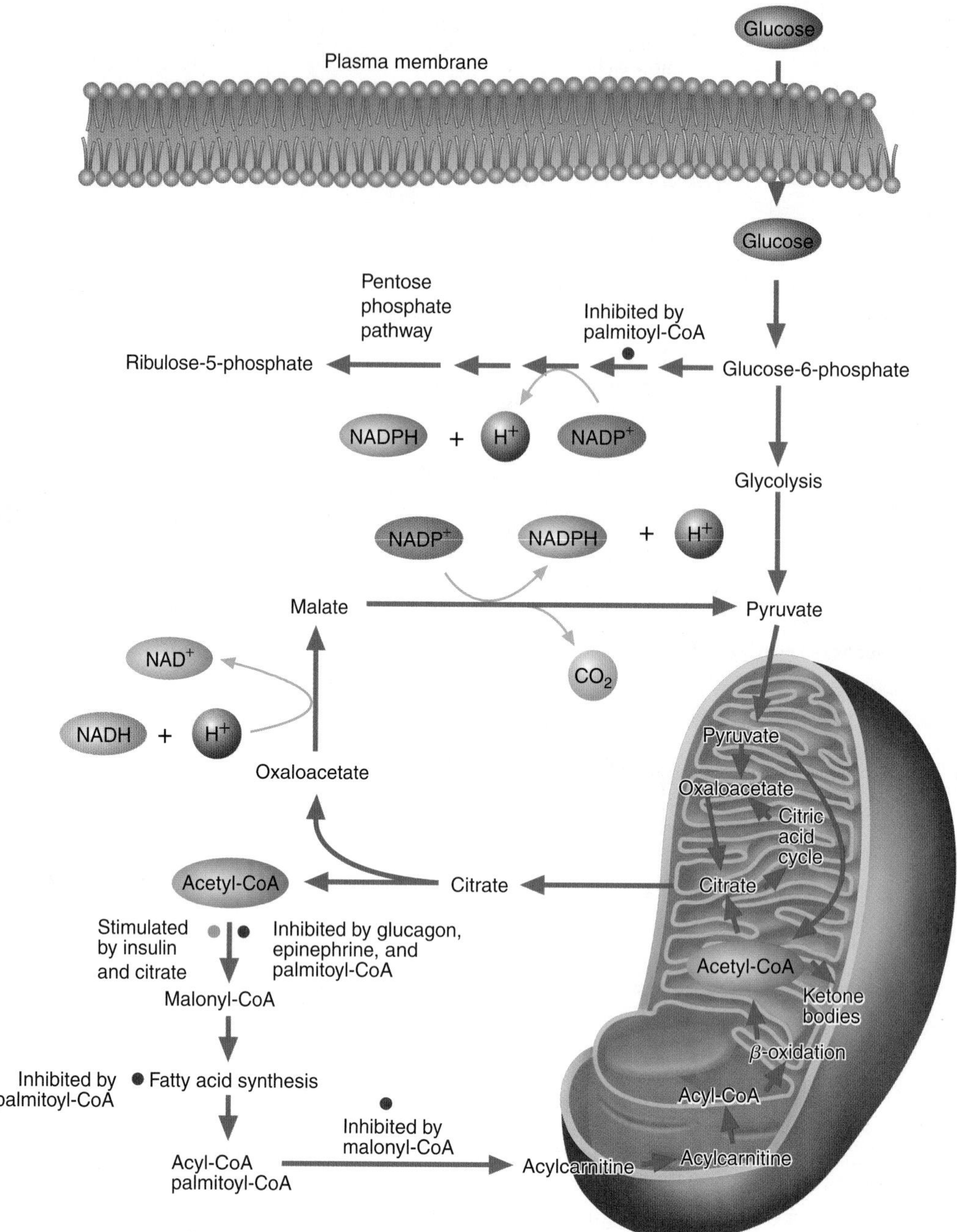

FIGURE 12.22

Regulation of Intracellular Fatty Acid Metabolism

Fatty acids are synthesized in cytoplasm from acetyl-CoA, which is formed within the mitochondrion. Because the inner membrane is impermeable to acetyl-CoA, it is transferred out as citrate. Citrate is produced from acetyl-CoA and oxaloacetate in the citric acid cycle, a reaction pathway in the mitochondrial matrix. Citrate is transferred to cytoplasm when β-oxidation is suppressed, that is, when the cell needs little energy. It is then cleaved to form oxaloacetate and acetyl-CoA. When the cell needs more energy, fatty acids are transported into the mitochondrion as acylcarnitine derivatives. Then acyl-CoA is degraded to acetyl-CoA via β-oxidation. Note that glucagon facilitates fatty acid oxidation, possibly by stimulating CAT-1. AMPK, the hormones glucagon and epinephrine and the substrates citrate, malonyl-CoA, and palmitoyl-CoA are important regulators of fatty acid metabolism.
A portion of the NADPH, the reducing agent required in fatty acid synthesis, is generated by several reactions in the pentose phosphate pathway. NADPH is also produced by converting malate, formed by the reduction of oxaloacetate, to pyruvate.

hormones. When energy levels are high, β-oxidation is depressed by the binding of the allosteric modulators NADH and acetyl-CoA to β-hydroxyacyl–CoA dehydrogenase and thiolase, respectively. Similarly, malonyl-CoA, the product of ACC, is an allosteric regulator of CAT-I. In liver when insulin:glucagon ratios are high, malonyl-CoA levels rise and cause the inhibition of β-oxidation, thus preventing a futile cycle. High cellular concentrations of long-chain fatty acyl-CoA esters inhibit ACC by promoting its depolymerization.

AMPK Fatty acid synthesis and β-oxidation are also rapidly regulated by changes in energy demand by AMPK. **AMPK** (pp. 289–290) is a trimeric enzyme composed of an α subunit (catalytic) and β and γ subunits (regulatory). As AMP:ATP ratios begin to rise, AMPK is activated by upstream AMPK kinases and by its allosteric modulator, AMP. In addition to acting as an allosteric activator, AMP promotes the activating phosphorylation reactions and inhibits dephosphorylation by protein phosphatases. AMP levels are a sensitive indicator of cellular energy status, and they rise in response to stresses that deplete cellular ATP levels (e.g., nutrient deprivation, hypoxia, heat shock, prolonged exercise). Once activated, AMPK switches off anabolic pathways (e.g., fatty acid and triacylglycerol synthesis by phosphorylating ACC1 (p. 442) and glycerol-3-phosphate acyltransferase, respectively) and switches on catabolic pathways (e.g., β-oxidation is stimulated by AMPK-induced activation of malonyl-CoA decarboxylase (MCD), the enzyme that decreases the concentration of malonyl-CoA and inhibition of ACC2). AMPK's influence over the body's major metabolic processes is outlined in **Figure 12.23**.

HORMONES Hormones play an important role in both short- and long-term regulation of fatty acid metabolism. Short-term effects of insulin that promote fat synthesis are caused by rapid signal transduction mechanisms. Insulin activates phosphoprotein phosphatase 2A, which dephosphorylates and activates ACC1, in combination with allosteric regulators and covalent modifications (p. 445).

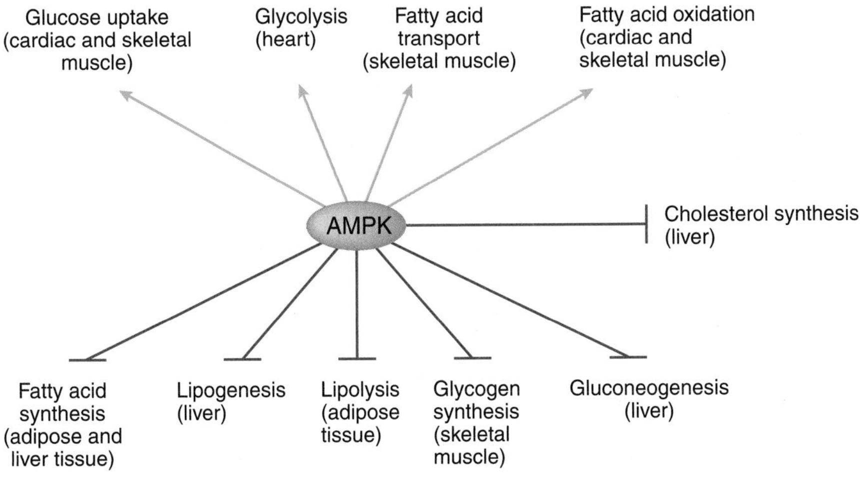

FIGURE 12.23

AMPK-Regulated Pathways in Lipid and Carbohydrate Metabolism

Elevated cellular AMP levels activate AMPK, a major metabolic switch that regulates numerous biochemical pathways. AMPK shifts metabolism from energy-consuming processes to energy-generating processes by phosphorylating target proteins. The effects of AMPK in the principal metabolic pathways in cardiac and skeletal muscle, liver, and adipose tissue are indicated in the diagram. AMPK also regulates the body's metabolism via its effects on insulin secretion from pancreatic β-cells (inhibition) and the appetite center in the brain (stimulates feeding behavior, see p. 606).

Insulin also activates ATP-citrate lyase (p. 339) and pyruvate dehydrogenase (in adipocytes). Insulin promotes fat synthesis in adipocytes by triggering the movement of the transporter GLUT-4 to the cell surface, thus facilitating the entry into the cell of glucose (the precursor of glycerol-3-phosphate and fatty acids). Insulin simultaneously depresses fat mobilization in adipocytes by stimulating the phosphorylation of hormone-sensitive lipase. Epinephrine increases lipolysis by stimulating the dephosphorylation and inactivation of hormone-sensitive lipase via signal transduction–mediated phosphorylation reactions. Glucagon increases fatty acid oxidation by an unresolved mechanism, possibly by activating CAT-I.

TRANSCRIPTION FACTORS Changes in the long-term regulation of fatty acid metabolism, which occur in response to fluctuating nutrient availability and energy demand, are effected by alterations in gene expression. Two classes of transcription factors are prominent components of an intricate regulatory process: the SREBPs and peroxisome proliferator–activated receptors (PPARs). Each type of transcription factor, when activated, binds to a regulatory element near target genes, a process that triggers the binding of coactivator molecules and, subsequently, transcription.

The **SREBPs** are a group of three proteins, SREBPla, SREBP lc, and SREBP 2, which are coded for by two genes. SREBPla and (more prominently) SREBPlc regulate the expression of genes involved in fatty acid metabolism. (SREBPla, which activates all SREBP-responsive genes, is continuously expressed at low levels in most animal tissues.) SREBP2 regulates genes in cholesterol metabolism (see p. 462). In liver and adipose tissue, the activation of SREBPlc in response to insulin upregulates the transcription of the genes that code for enzymes in fatty acid synthetic pathway and NADPH synthesis. Glucagon and high levels of long-chain fatty acids inhibit SREBPlc. **PPARs**, named for the ability of certain synthetic compounds to cause the proliferation of liver cell peroxisomes, are ligand-activated transcription factors that bind to PPAR response elements associated with target genes. PPARα, a transcription factor originally discovered because of the peroxisome proliferation phenomenon, controls the expression of several genes in lipid metabolism. In liver and adipose tissues, under fasting conditions, it stimulates fatty acid catabolism and ketogenesis. PPARγ, primarily expressed in adipose tissue, in combination with insulin and SREBP1, promotes fat storage by stimulating glucose uptake and fatty acid and triacylglycerol synthesis. The activity of PPARs is stimulated by the binding of several lipid molecules (e.g., saturated and unsaturated fatty acids, prostaglandins, and leukotrienes).

When there are high blood glucose levels, the glucose-responsive transcription factor ChREBP (p. 289) in both liver and adipose tissue is activated. ChREBP/Mlx heterodimer promotes the synthesis of enzymes that convert excess sugar molecules into fatty acids (e.g., ACC and fatty acid synthase).

KEY CONCEPTS

- The metabolism of fatty acids, the major energy source in animals, is regulated in the short term by allosteric modulators, covalent modification, and hormones.
- Long-term regulation, which occurs in response to fluctuating nutrient availability and energy demand, is effected by changes in gene expression.

QUESTION 12.9

Identify each of the following biomolecules:

(a) $CH_3-CH(OH)-CH_2-C(=O)-O^-$

(b) $CoA-S-C(=O)-CH_2-C(=O)-O^-$

(c) O, HN, NH, S, R

(d) $CH_3-C(=O)-S-ACP$

What is the function of each?

Lipoprotein Metabolism: The Endogenous Pathway

The endogenous lipoprotein pathway, which transports recently synthesized lipids throughout the body, begins in the liver, where VLDLs are assembled on the cytoplasmic surface of hepatocyte ER. Nascent VLDLs contain apolipoprotein B-100, TGs, phospholipids, cholesterol, and cholesterol esters. Once VLDLs are secreted into blood, they are transformed into mature VLDL with the acquisition of apolipoproteins C-II and E, which are transferred from HDL. VLDLs then proceed to unload TGs as they encounter LPL (activated by apolipoprotein C-II) located primarily near the surface of target cells. Fatty acids transported into adipocytes are reconverted into TGs that coalesce into fat droplets. The fatty acids that are transported into muscle cells are oxidized to generate the energy required to drive muscle contraction. Once the TG content of VLDLs has been depleted, they are referred to as IDLs. The removal of IDLs from the blood by endocytosis is mediated by the binding of lipoprotein E to its receptor on the surface of hepatocytes. The TG content of IDLs is further reduced by hepatic lipase. Once the cholesterol content of IDLs exceeds that of TGs, the lipoproteins are referred to as LDLs. LDLs are released from the liver into the bloodstream that carries them to target tissues. After the binding of apolipoproitein B-100 to LDL receptors, LDLs are internalized via endocytosis into cells, where they release their content, primarily cholesterol.

12.2 MEMBRANE LIPID METABOLISM

The lipid bilayer of cell membranes is composed primarily of phospholipids and sphingolipids. The metabolism of each lipid class is briefly described.

Phospholipid Metabolism

The membrane of the eukaryotic cell's endomembrane system (p. 44) originates in the SER with the synthesis of phospholipid at the interface of SER and cytoplasm. The fatty acid composition of the SER membrane subsequently changes, with unsaturated fatty acids replacing the original saturated fatty acids. This remodeling, which is accomplished by phospholipases and acyltransferases, allows cells to adjust the fluidity of their membranes.

The syntheses of PE and PC are similar (**Figure 12.24**). PE synthesis begins in the cytoplasm when ethanolamine enters the cell and is immediately phosphorylated. Subsequently, phosphoethanolamine reacts with CTP (cytidine triphosphate) to form the activated intermediate CDP-ethanolamine. (CDP derivatives have an important role in the transfer of polar head groups in phosphoglyceride synthesis.) CDP-ethanolamine is converted to PE when it reacts with diacylglycerol. As noted, the biosynthesis of phosphatidylcholine is similar to that of PE. The choline required in this pathway is obtained in the diet. However, PC is also synthesized in the liver from PE. PE is methylated in three steps by the enzyme phosphatidylethanolamine-*N*-methyltransferase to form the trimethylated product phosphatidylcholine. *S-Adenosylmethionine* (SAM) is the methyl donor in this set of reactions. (The role of SAM in cellular methylation processes is discussed in Chapter 14.)

In animals, PS is generated by the reversible polar head group exchange with PE mediated by PE–serine transferase. The decarboxylation of PS to yield PE is an important reaction in many eukaryotes.

Phospholipid turnover is rapid. (**Turnover** is the rate at which all molecules in a structure are degraded and replaced with newly synthesized molecules.) For example, in animal cells, approximately two cell divisions are required for

Biochemistry IN PERSPECTIVE

Atherosclerosis

What is the biochemical basis of arterial damage in the disease process called atherosclerosis? **Atherosclerosis** is a chronic disease in which soft masses called atheromas accumulate within arterial walls, eventually compromising their functional structure. Normal arterial walls are strong and flexible. They consist of three well-defined layers: the *intima* (a single layer of endothelial cells attached to an underlying extracellular matrix), the *media* (layers of smooth muscle cells embedded in an extracellular matrix consisting of elastic fibers, collagen, and proteoglycan), and the *adventitia* (the outermost layer, which consists of fibroblasts, smooth muscle cells, collagen, and elastin).

Once believed to be little more than passive conduits of blood, blood vessels are now known to be physiologically active. The endothelial cells that line arteries perform several vital functions. Among these are providing a barrier that prevents toxic substances from penetrating into the vessel wall and regulating the response of arteries to shear stress (the fluctuating mechanical force created by blood flow). Nitric oxide (NO), the vasodilator produced by endothelial NO synthase, relaxes smooth muscle cells. Endothelial cells also produce molecules that provide arterial linings with smooth Teflon-like surfaces that prevent white blood cells from adhering. In the course of aging, the endothelial barrier becomes leaky and vulnerable, a result that is accelerated by poor diet, smoking, and a sedentary lifestyle.

The atherosclerotic process is initiated by injury to the surface of endothelial cells, for example, AGE formation caused by high blood glucose levels (p. 249) or the binding of AGEs inhaled in tobacco smoke. Endothelial cells respond by secreting chemokines that attract white blood cells to the injured site and by producing surface adhesion molecules that permit the binding of monocytes, cells that differentiate into macrophages. Monocytes/macrophages bind to AGEs via the AGE receptor (RAGE), a protein in the immunoglobulin superfamily. Oxidative stress ensues as the result of a RAGE-ligand initiated signaling pathway. The inflammatory process continues as sdLDLs (p. 403) enter the damaged site where they become ensnared by interactions with proteoglycans. Trapped sdLDLs soon become depleted of antioxidant molecules, allowing oxidative damage to accumulate. The attempt of macrophages to clear the site of oxidized LDL is overwhelmed, and the phagocytes become so filled with lipoprotein that they transform into "foam cells." Together, damaged endothelial cells, macrophages, and smooth muscle cells release molecules that promote cell division and cell migration, which has the effect of reorganizing the vessel wall architecture. In effect, the attempt to heal the injury results in the formation of a fibrous cap that walls off the damaged tissue.

The atherosclerotic process usually causes the formation of atheromas that extend sideways and not outward, which would cut off blood flow. Atherosclerotic lesions (*plaque*) usually do not cause obvious problems for long periods of time, perhaps decades. Eventually the inflammatory process weakens the fibrous cap, which may suddenly rupture. It is the subsequent formation of a thrombus (blood clot) that prevents blood flow, especially in smaller vessels such as the coronary arteries. Sudden death, the most common symptom of coronary artery damage, results from such an event. In a small number of myocardial infarctions (heart attacks), plaque formation grows outward, slowly occluding blood flow. In these cases, decreased blood flow through one or more coronary arteries causes angina pectoris (pain and tightness in the chest) by preventing O_2 and nutrients from reaching the myocardial cells. Prompt medical attention to life-threatening occlusions can prevent death.

SUMMARY: Atherosclerosis, which may lead to myocardial infarction, is initiated by damage to the endothelial cells that line arteries. The formation of atherosclerotic lesions begins with the accumulation of LDL and progresses to an inflammatory process that degrades arterial structure and function.

FIGURE 12.24
Phospholipid Synthesis

After ethanolamine or choline has entered a cell, it is phosphorylated and converted to a CDP derivative. Diacylglycerol then reacts with the CDP derivative, and phosphatidylethanolamine or phosphatidylcholine is formed. Triacylglycerol is produced if diacylglycerol reacts with acyl-CoA. CDP-diacylglycerol, formed from phosphatidic acid and CTP, is a precursor of several phospholipids (e.g., phosphatidylglycerol and phosphatidylinositol).

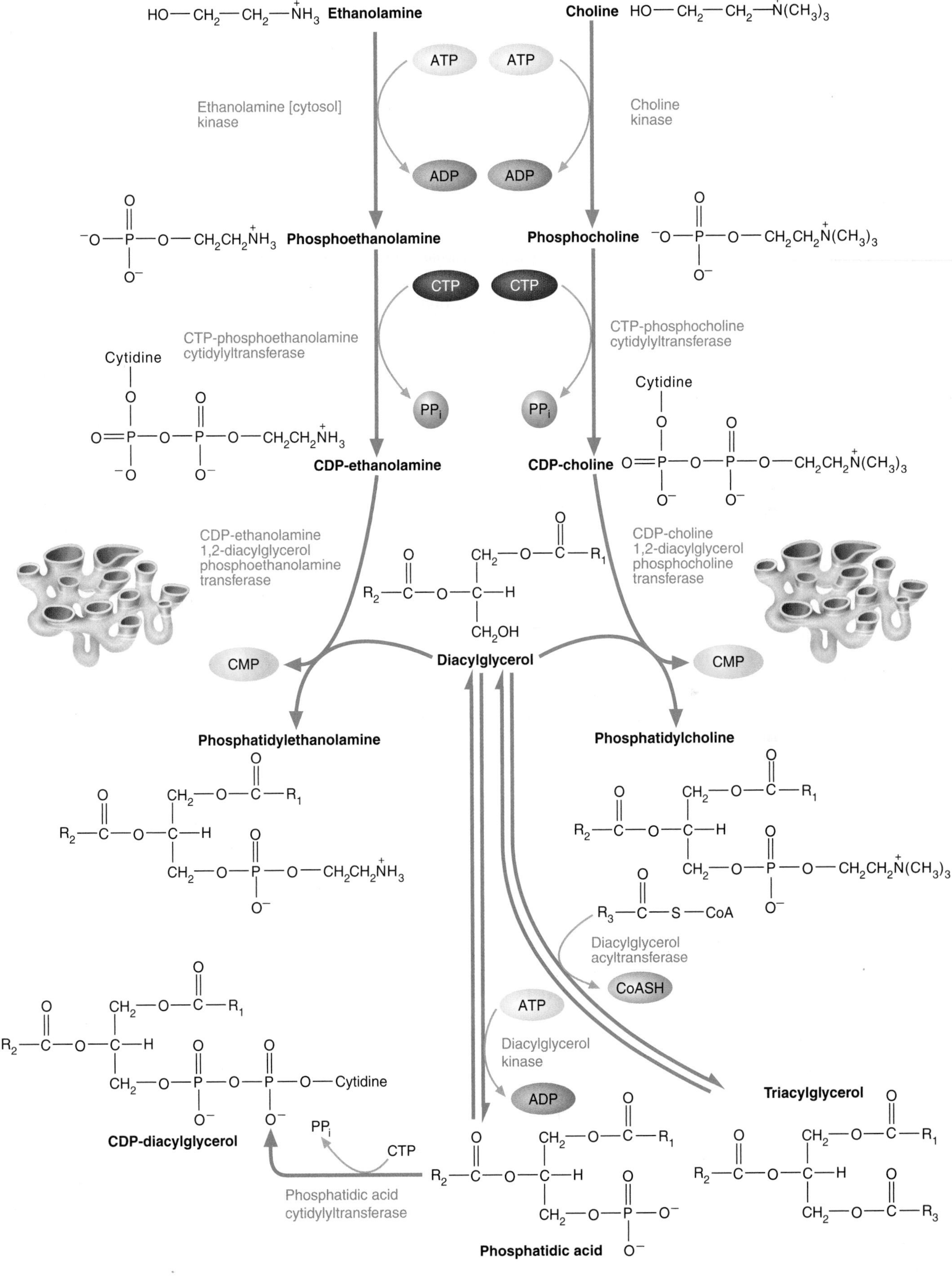
Ethanolamine
Choline
ATP
ATP
Ethanolamine [cytosol] kinase
Choline kinase
ADP
ADP
Phosphoethanolamine
Phosphocholine
CTP
CTP
CTP-phosphoethanolamine cytidylyltransferase
CTP-phosphocholine cytidylyltransferase
Cytidine
PPi
PPi
CDP-ethanolamine
CDP-choline
CDP-ethanolamine 1,2-diacylglycerol phosphoethanolamine transferase
CDP-choline 1,2-diacylglycerol phosphocholine transferase
CMP
Diacylglycerol
CMP
Phosphatidylethanolamine
Phosphatidylcholine
R3—C—S—CoA
Diacylglycerol acyltransferase
CoASH
ATP
Diacylglycerol kinase
ADP
Triacylglycerol
CDP-diacylglycerol
PPi
CTP
Phosphatidic acid cytidylyltransferase
Phosphatidic acid

KEY CONCEPTS

- Phospholipid synthesis occurs in the membrane of the SER. After phospholipids have been synthesized, they are remodeled by altering their fatty acid composition.
- Phospholipid degradation is catalyzed by several phospholipases.

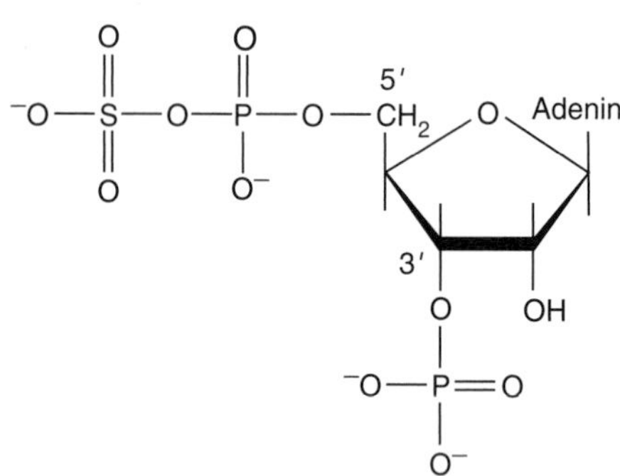

FIGURE 12.25

3′-Phosphoadenosine-5′-Phosphosulfate (PAPS)

PAPS is a high-energy sulfate donor.

KEY CONCEPT

The synthesis of all sphingolipids begins with the production of ceramide. Sphingolipids are degraded within lysosomes by specific hydrolytic enzymes.

the replacement of one-half of the total number of phospholipid molecules. Phosphoglycerides are degraded by the phospholipases, each of which catalyzes the cleavage of a specific bond in phosphoglyceride molecules. Phospholipases A_1 and A_2 hydrolyze the ester bonds of phosphoglycerides at C-1 and C-2, respectively (see **Figure 11.11**, p. 394).

Sphingolipid Metabolism

Animal sphingolipids possess ceramide, a derivative of the amino alcohol sphingosine. Ceramide synthesis begins with the condensation of palmitoyl-CoA with serine to form 3-ketosphinganine. This reaction is catalyzed by 3-ketosphinganine synthase, a pyridoxal-5′-phosphate–dependent enzyme. (Because pyridoxal-5′-phosphate plays an important role in amino acid metabolism, the biochemical function of this coenzyme is discussed in Chapter 14.) 3-Ketosphinganine is subsequently reduced by NADPH to form sphinganine. In a two-step process involving acyl-CoA and $FADH_2$, sphinganine is converted to ceramide. Sphingomyelin is formed when ceramide reacts with phosphatidylcholine. (In an alternative reaction, CDP-choline is used in place of phosphatidylcholine.) When ceramide reacts with UDP-glucose, glucosylceramide (a common cerebroside, sometimes referred to as glucosylcerebroside) is produced. Galactocerebroside, a precursor of other glycolipids, is synthesized when ceramide reacts with UDP-galactose. The sulfatides are synthesized when the galactocerebrosides react with the sulfate donor molecule **3′-phosphoadenosine-5′-phosphosulfate** (PAPS) (**Figure 12.25**). The transfer of sulfate groups is catalyzed by the ER enzyme sulfotransferase. Sphingolipids are degraded within lysosomes. Recall that specific diseases, called the sphingolipidoses (p. 396), result when enzymes required for degrading these molecules are missing or defective. The synthesis of sphingomyelin and glycosphingolipids is shown in **Figure 12.26**).

12.3 ISOPRENOID METABOLISM

Isoprenoids occur in all eukaryotes. Despite the astonishing diversity of isoprenoid molecules, the mechanisms by which different species synthesize them are similar. In fact, the initial phase of isoprenoid synthesis (the synthesis of isopentenyl pyrophosphate) appears to be identical in all of the species in which this process has been investigated. **Figure 12.27** illustrates the relationships among the isoprenoid classes.

Because of its importance in human biology, cholesterol has received enormous attention from researchers. For this reason the metabolism of cholesterol is better understood than that of any other isoprenoid molecule.

Cholesterol Metabolism

Cholesterol is derived from two sources: diet and *de novo* synthesis. When the diet provides sufficient cholesterol, usually about 400 mg per day, the synthesis of this molecule is depressed. In normal individuals cholesterol delivered by LDL suppresses the synthesis of both cholesterol and LDL receptors. Cholesterol biosynthesis, which averages about 900 mg per day, and LDL receptor synthesis are stimulated when the diet is low in cholesterol. As described previously, cholesterol is a vital cell membrane component and a precursor in the synthesis of important metabolites. Cholesterol is also used to form bile salts.

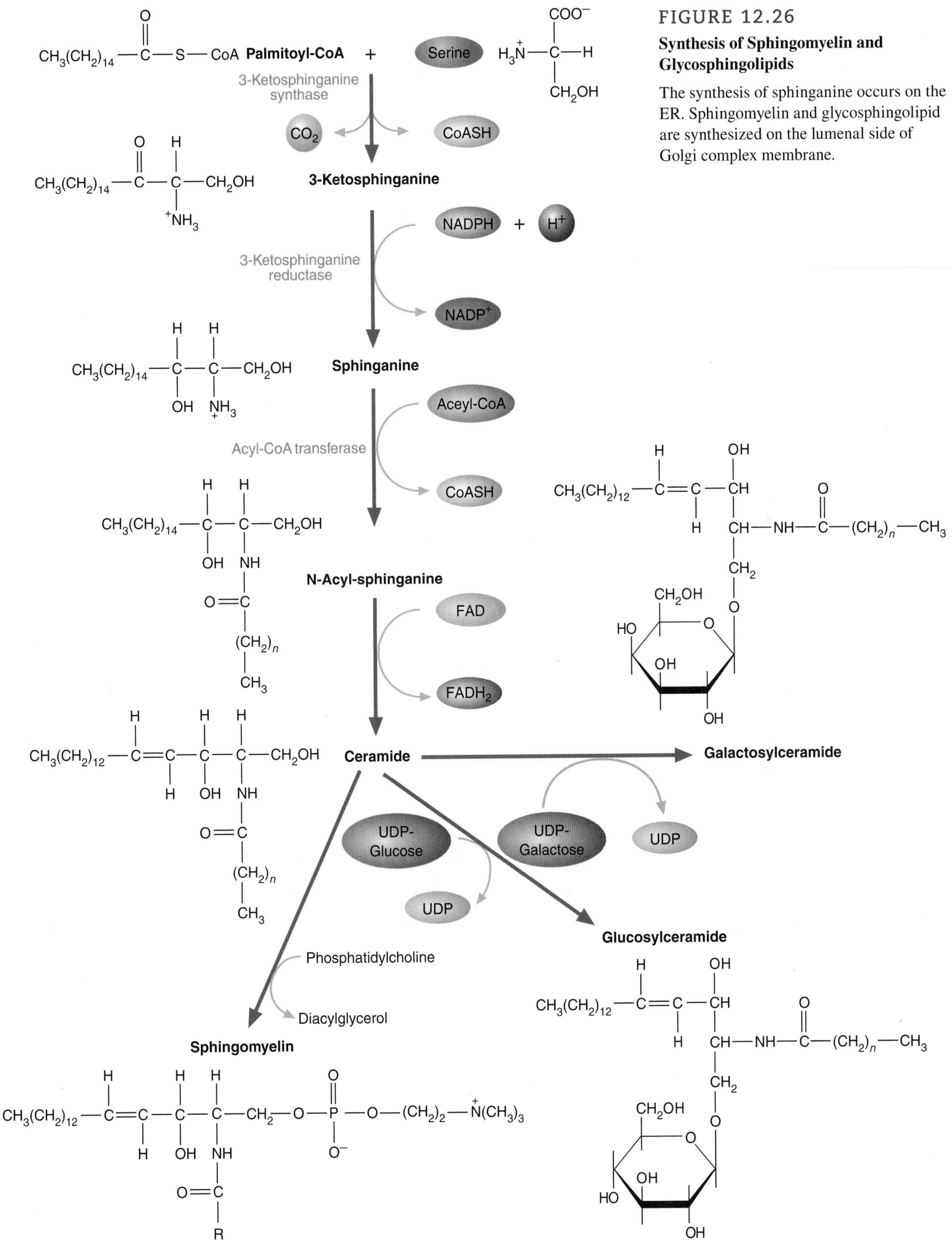

FIGURE 12.26

Synthesis of Sphingomyelin and Glycosphingolipids

The synthesis of sphinganine occurs on the ER. Sphingomyelin and glycosphingolipid are synthesized on the lumenal side of Golgi complex membrane.

Palmitoyl-CoA + Serine → (3-Ketosphinganine synthase; CO_2, CoASH released) → 3-Ketosphinganine → (3-Ketosphinganine reductase; NADPH + H^+ → $NADP^+$) → Sphinganine → (Acyl-CoA transferase; Aceyl-CoA → CoASH) → N-Acyl-sphinganine → (FAD → $FADH_2$) → Ceramide

Ceramide → (UDP-Galactose → UDP) → Galactosylceramide

Ceramide → (UDP-Glucose → UDP) → Glucosylceramide

Ceramide → (Phosphatidylcholine → Diacylglycerol) → Sphingomyelin

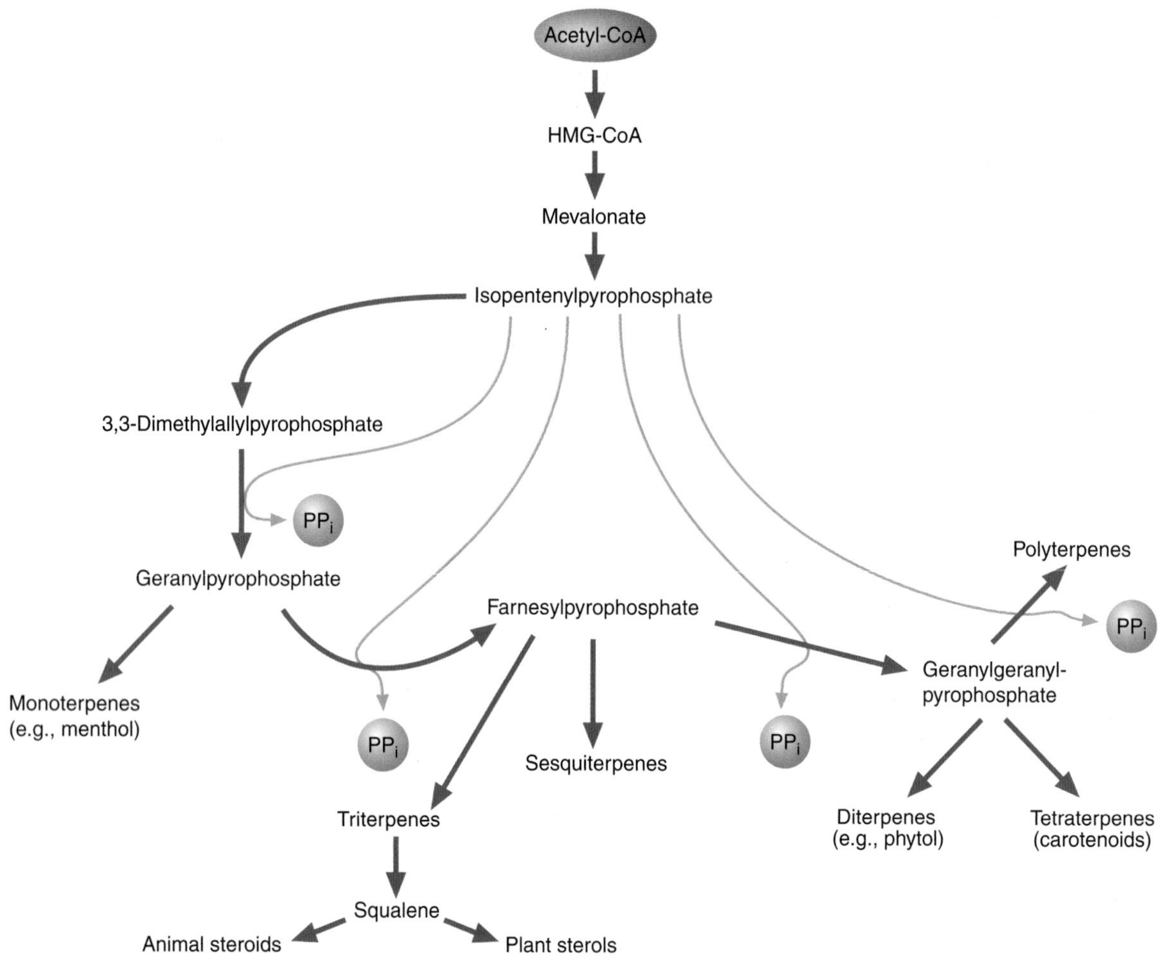

FIGURE 12.27

Isoprenoid Biosynthesis

Isoprenoid biosynthetic pathways produce an astonishing variety of products in different cell types and in different species. Despite this diversity, the beginning of isoprenoid biosynthesis appears to be identical in the species investigated (e.g., yeast, mammals, and plants). (HMG-CoA = β-hydroxy-β-methylglutaryl-CoA)

CHOLESTEROL SYNTHESIS Although all tissues (e.g., adrenal glands, ovaries, testes, skin, and intestine) can make cholesterol, most cholesterol molecules are synthesized in the liver. Cholesterol synthesis can be divided into three phases:

1. Formation of HMG-CoA (β-hydroxy-β-methylglutaryl-CoA) from acetyl-CoA
2. Conversion of HMG-CoA to squalene
3. Conversion of squalene to cholesterol

The first phase of cholesterol synthesis is a cytoplasmic process. (Recall that the initial substrate, acetyl-CoA, is produced in mitochondria from fatty acids or pyruvate. Also observe the similarity of the first phase of cholesterol synthesis to ketone body synthesis. Refer to **Figure 12.9**.) The condensation of two acetyl-CoA molecules to form β-ketobutyryl-CoA (also referred to as acetoacetyl-CoA) is catalyzed by thiolase.

$$2\,CH_3{-}\overset{O}{\overset{\|}{C}}{-}S{-}CoA \rightleftharpoons CH_3{-}\overset{O}{\overset{\|}{C}}{-}CH_2{-}\overset{O}{\overset{\|}{C}}{-}S{-}CoA + CoASH \qquad (9)$$

Acetyl-CoA $\qquad$ **β-Ketobutyryl-CoA**

In the next reaction, β-ketobutyryl-CoA condenses with another acetyl-CoA to form β-hydroxy-β-methylglutaryl-CoA (HMG-CoA). This reaction is catalyzed by β-hydroxy-β-methylglutaryl-CoA synthase (HMG-CoA synthase).

$$\underset{\beta\text{-Ketobutyryl-CoA}}{CH_3-\overset{O}{\overset{\|}{C}}-CH_2-\overset{O}{\overset{\|}{C}}-S-CoA} + \underset{\text{Acetyl-CoA}}{CH_3-\overset{O}{\overset{\|}{C}}-S-CoA} \longrightarrow \underset{\text{HMG-CoA}}{{}^{-}O-\overset{O}{\overset{\|}{C}}-CH_2-\overset{CH_3}{\underset{OH}{C}}-CH_2-\overset{O}{\overset{\|}{C}}-S-CoA} + CoASH \qquad (10)$$

The second phase of cholesterol synthesis begins with the reduction of HMG-CoA to form mevalonate. This reaction is catalyzed by HMG-CoA reductase (HMGR), the rate-limiting enzyme in cholesterol synthesis. NADPH is the reducing agent.

$$\underset{\text{HMG-CoA}}{{}^{-}O-\overset{O}{\overset{\|}{C}}-CH_2-\overset{CH_3}{\underset{OH}{C}}-CH_2-\overset{O}{\overset{\|}{C}}-S-CoA} \xrightarrow{2\,NADPH\ +\ 2\,H^+ \ \to\ 2\,NADP^+} \underset{\text{Mevalonate}}{{}^{-}O-\overset{O}{\overset{\|}{C}}-CH_2-\overset{CH_3}{\underset{OH}{C}}-CH_2-CH_2-OH} + CoASH \qquad (11)$$

The HMGR polypeptide consists of three major domains: an N-terminal transmembrane anchor domain, a catalytic domain, and a linker that connects the membrane and catalytic domains. The enzyme, located on the cytoplasmic face of the SER, consists of two HMGR dimers that associate to form a tetramer. Each dimer has an active site at the interface between the two dimers. The reaction (**Figure 12.28**) begins with the nucleophilic acyl substitution in which there is a hydride transfer from NADPH to the thioester carbonyl group of HMG-CoA. This transfer is assisted by a hydrogen bond between a lysine and the thioester carbonyl oxygen. The C—S bond in the product mevaloyl-CoA is hydrolyzed to form mevaldehyde. The CoA-thiolate anion product is protonated by a histidine residue and then released. Protonation of the carbonyl oxygen of mevaldehyde by the lysine residue facilitates a second hydride transfer from NADPH to form mevalonate.

In a series of cytoplasmic reactions, mevalonate is converted to farnesylpyrophosphate. Mevalonate kinase catalyzes the synthesis of phosphomevalonate. A second phosphorylation reaction catalyzed by phosphomevalonate kinase produces 5-pyrophosphomevalonate.

$$\underset{\text{Mevalonate}}{{}^{-}O-\overset{O}{\overset{\|}{C}}-CH_2-\overset{CH_3}{\underset{OH}{C}}-CH_2-CH_2OH} \xrightarrow{ATP\ \to\ ADP} \underset{\text{Phosphomevalonate}}{{}^{-}O-\overset{O}{\overset{\|}{C}}-CH_2-\overset{CH_3}{\underset{OH}{C}}-CH_2-CH_2-O-\overset{O}{\underset{O^-}{P}}-O^-}$$

$$\xrightarrow{ATP\ \to\ ADP} \underset{\text{5-Pyrophosphomevalonate}}{{}^{-}O-\overset{O}{\overset{\|}{C}}-CH_2-\overset{CH_3}{\underset{OH}{C}}-CH_2-CH_2-O-\overset{O}{\underset{O^-}{P}}-O-\overset{O}{\underset{O^-}{P}}-O^-} \qquad (12)$$

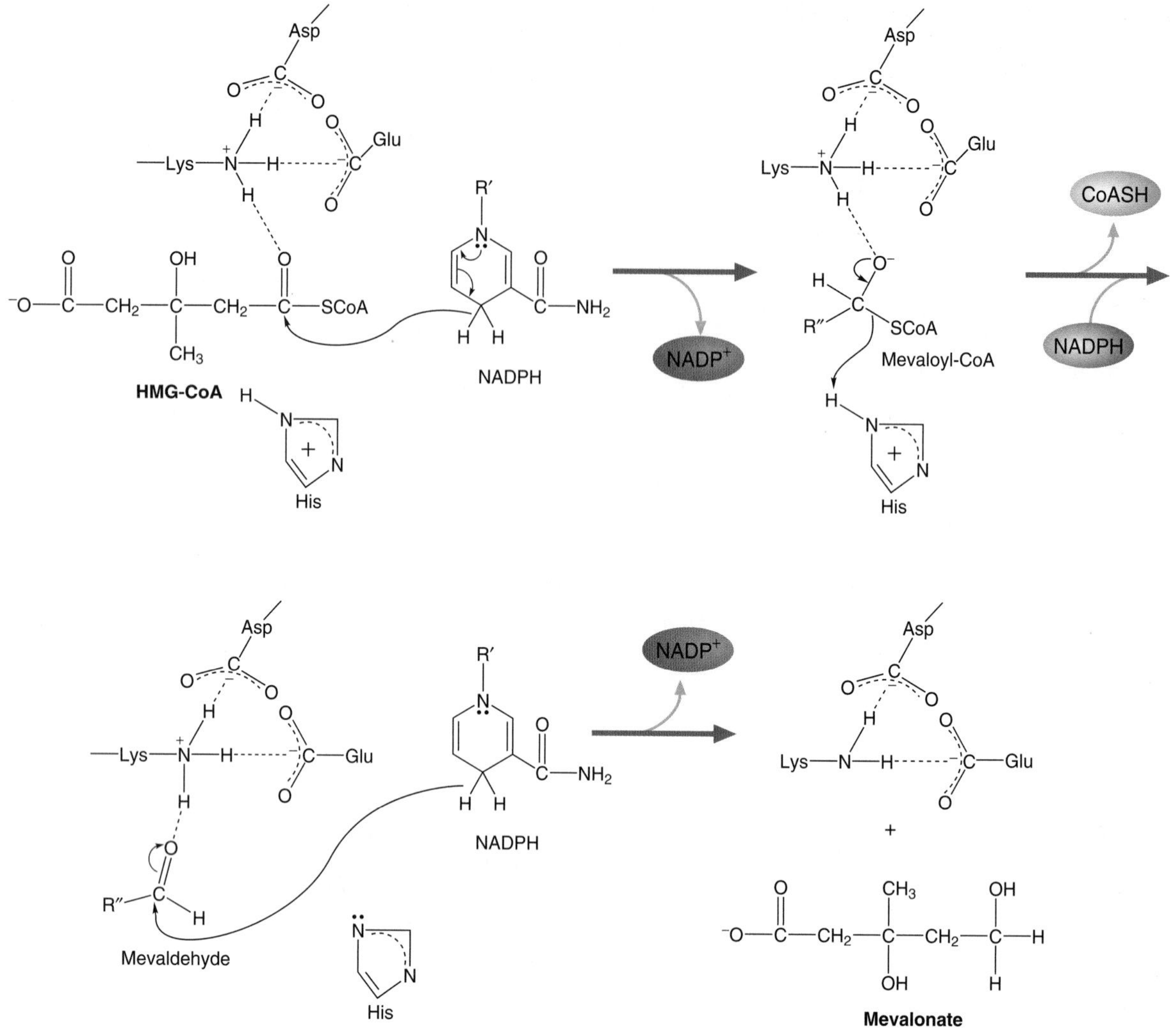

FIGURE 12.28

The HMGR-Catalyzed Reaction

In the HMGR-catalyzed reaction, there is an initial hydride transfer from NADPH to the thioester carbonyl of the substrate HMG-CoA. Subsequently, the C—S bond of mevaloyl-CoA is hydrolyzed to form mevaldehyde, and a histidine residue protonates the CoA-thiolate anion. A second hydride transfer from NADPH, facilitated by the protonation of the carbonyl oxygen of mevaldehyde, results in the formation of mevalonate, the product of the reaction. Note that carboxylate side chain groups of Asp and Glu residues orient the Lys side chain amino group within the active site.

(Phosphorylation reactions significantly increase the solubility of these hydrocarbon molecules in the cytoplasm.) 5-Pyrophosphomevalonate is converted to isopentenylpyrophosphate in a process involving a decarboxylation and a dehydration:

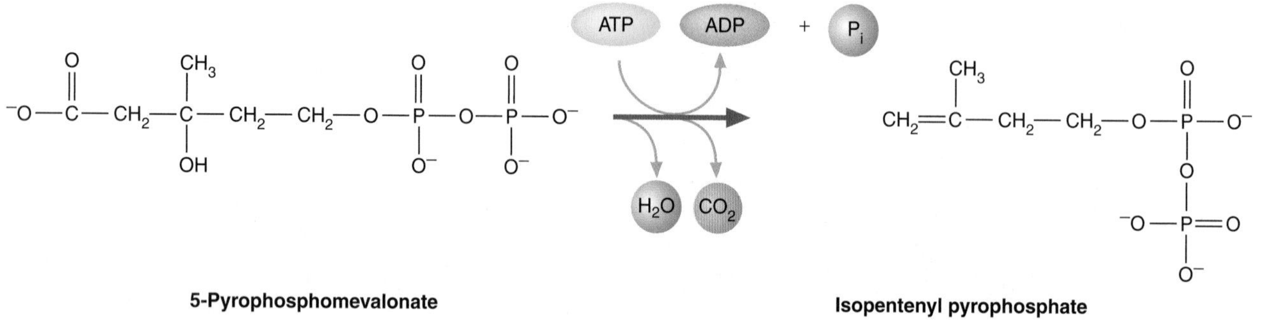

(13)

Isopentenyl pyrophosphate is next converted to its isomer dimethylallylpyrophosphate by isopentenylpyrophosphate isomerase. (A CH_2=CH—CH_2— group on an organic molecule is sometimes referred to as an *allyl group*.)

$$CH_2{=}C(CH_3){-}CH_2{-}CH_2{-}O{-}PO(O^-){-}O{-}PO(O^-){-}O^- \rightleftharpoons (CH_3)_2C{=}CH{-}CH_2{-}O{-}PO(O^-){-}O{-}PO(O^-){-}O^- \quad (14)$$

Isopentenylpyrophosphate **Dimethylallylpyrophosphate**

Geranylpyrophosphate is generated during a condensation reaction between isopentenylpyrophosphate and dimethylallylpyrophosphate (**Figure 12.29**). Pyrophosphate is also a product of this reaction and two subsequent reactions. (Recall that reactions in which pyrophosphate is released are irreversible because of subsequent pyrophosphate hydrolysis.) Geranyl transferase catalyzes the head-to-tail condensation reaction between geranylpyrophosphate and isopentenylpyrophosphate that forms farnesylpyrophosphate. Squalene is synthesized when farnesyl transferase catalyzes the head-to-head condensation of two farnesylpyrophosphate molecules. (Farnesyl transferase is sometimes referred to as squalene synthase.) This reaction requires NADPH as an electron donor.

The last phase of the cholesterol biosynthetic pathway (**Figure 12.30**) begins by binding squalene to a specific cytoplasmic protein carrier called **sterol carrier protein**. The conversion of squalene to lanosterol occurs while the intermediates are bound to this protein. The enzyme activities required for the oxygen-dependent epoxide formation (squalene monooxygenase) and subsequent cyclization (2,3-oxidosqualene lanosterol cyclase) that result in lanosterol synthesis have been localized in microsomes. Squalene monooxygenase requires NADPH and FAD for activity. After its synthesis, lanosterol binds to a second carrier protein, to which it remains attached during the remaining reactions. All of the enzyme activities that catalyze the remaining 20 reactions needed to convert lanosterol to cholesterol are embedded in SER membranes. In a series of transformations involving NADPH and oxygen, lanosterol is converted to 7-dehydrocholesterol. This product is then reduced by NADPH to form cholesterol.

CHOLESTEROL DEGRADATION Unlike many other types of biomolecules, cholesterol and other steroids cannot be degraded to smaller molecules. Instead, downregulation of synthesis along with loss caused by bile acid synthesis and excretion and steroid hormone biotransformation combine to lower circulating cholesterol levels. About one-half of the cholesterol synthesized daily is used to produce bile acids. Bile acid synthesis occurs in the liver. The synthesis of cholic acid, one of the principal bile acids, is outlined in **Figure 12.31**. The conversion of cholesterol to 7-α-hydroxycholesterol, catalyzed by cholesterol-7-hydroxylase (a SER enzyme), is the rate-limiting reaction in bile acid synthesis. Cholesterol-7-hydroxylase is a cytochrome P_{450} enzyme. In later reactions, the double bond at C-5 is rearranged and reduced, and an additional hydroxyl group is introduced. The products of this process, cholic acid and deoxycholic acid, are converted to bile salts by SER enzymes that catalyze conjugation reactions. (In **conjugation reactions** a molecule's solubility is increased by converting it into a derivative that contains a water-soluble group. Amides and esters are common examples of these conjugated derivatives.) Most bile acids are conjugated with glycine or taurine ($H_3N^+CH_2CH_2SO^-_3$).

The bile salts are important components of *bile*, a yellowish green liquid produced by hepatocytes that aids in the digestion of lipids. In addition to bile salts, bile contains cholesterol, phospholipids, and bile pigments (bilirubin and biliverdin). The bile pigments are degradation products of heme. After bile is secreted into the bile ducts and stored in the gall bladder, it is used in the small intestine

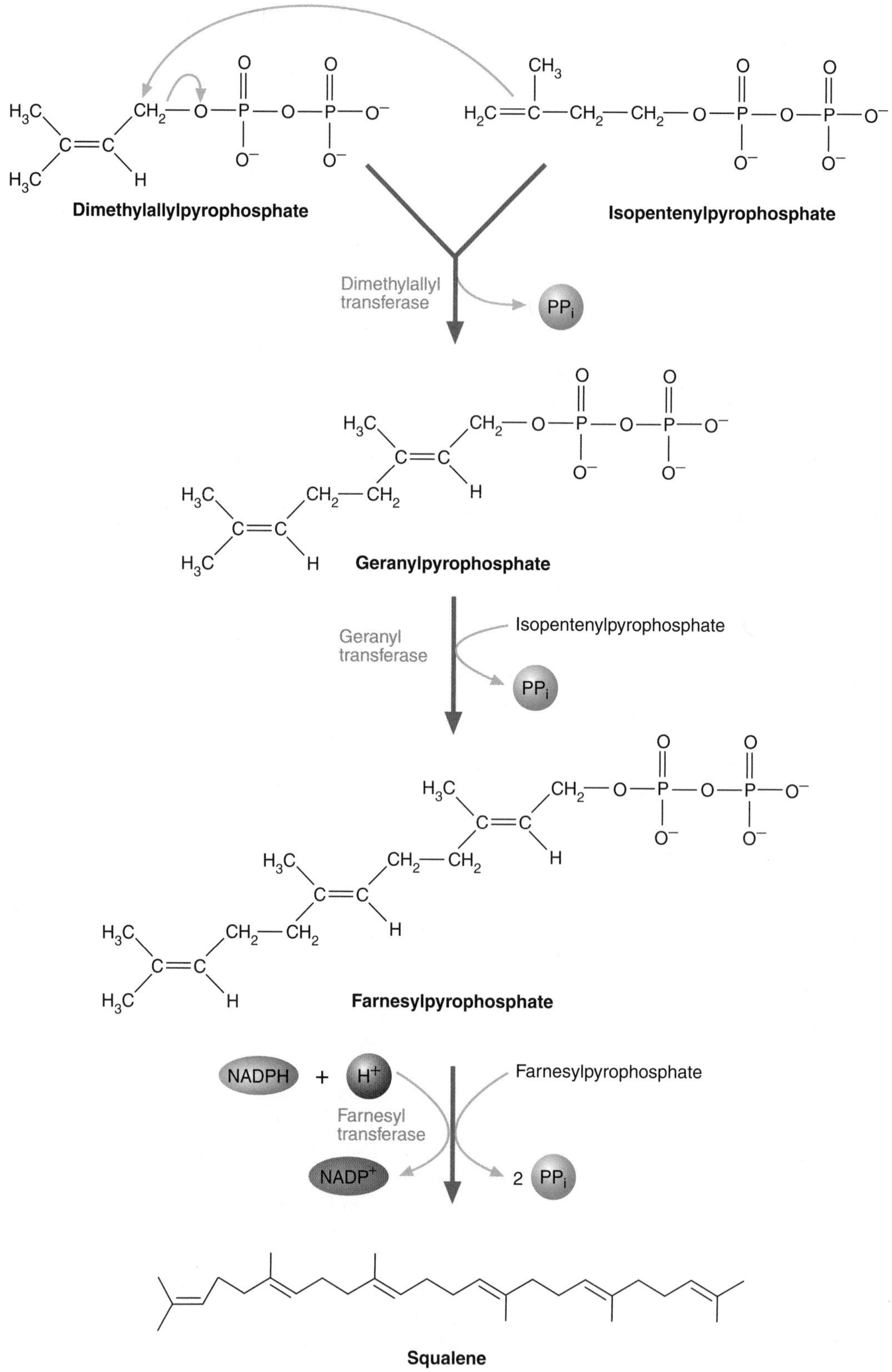

FIGURE 12.29

Synthesis of Squalene

The head-to-tail condensation of dimethylallylpyrophosphate and isopentenylpyrophosphate generates the terpene geranylpyrophosphate. A subsequent head-to-tail condensation with another isopentenylpyrophosphate generates the C_{15} farnesylpyrophosphate. Head-to-head condensation of two molecules of farnesylpyrophosphate produces the C_{30} triterpene, squalene. The geranylgeranyl groups used in prenylation reactions are synthesized in a reaction between farnesyl pyrophosphate and isopentenyl pyrophosphate.

Squalene

O_2 NADPH + H^+

Squalene monooxgenase

$NADP^+$

Squalene-2, 3-epoxide

2,3-Oxidosqualene lanosterol cyclase

H^+

Lanosterol

19 reactions

7-Dehydrocholesterol

NADPH + H^+

$NADP^+$

Cholesterol

FIGURE 12.30

Synthesis of Cholesterol from Squalene

This is the major route in mammals. In an alternative minor route, squalene is converted to desmosterol, which is then reduced to form cholesterol. The details of these and many reactions in the major route are poorly understood. (Desmosterol differs from cholesterol because of a C=C bond between C-24 and C-25.)

as an emulsifying agent to form biliary micelles and enhance the absorption of dietary fat and fat-soluble vitamins (A, D, E, and K). Most bile salts (about 90%) are reabsorbed in the distal ilium (near the end of the small intestine). They enter the blood and are transported back to the liver, where they are resecreted into the bile ducts with other bile components. Bile acid conjugation reactions prevents premature absorption of bile acids in the biliary tract (the duct system and gallbladder) and small intestine. The reabsorption of bile salts in the distal ilium of the small intestine (necessary for effective recycling) is apparently triggered by the glycine or taurine signal. (It has been estimated that bile salt molecules are recycled 18 times before they are finally eliminated.)

QUESTION 12.10

The formation in the gallbladder or bile ducts of gallstones (crystals usually composed of cholesterol and inorganic salts) afflicts millions of people. Predisposing factors for this excruciatingly painful disorder include obesity and infection of the gallbladder (*cholecystitis*). Because cholesterol is virtually insoluble in water, it is solubilized in bile by its incorporation into micelles composed of bile salts and phospholipids. Gallstones tend to form when cholesterol is secreted into bile in excessive quantities. Suggest a reason why an obese person is prone to gallstone formation. [*Hint*: HMG-CoA reductase activity is higher in obese individuals.]

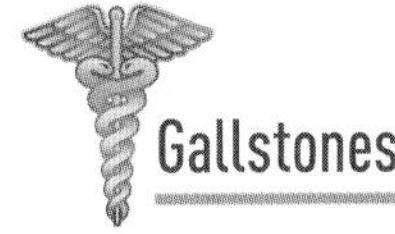

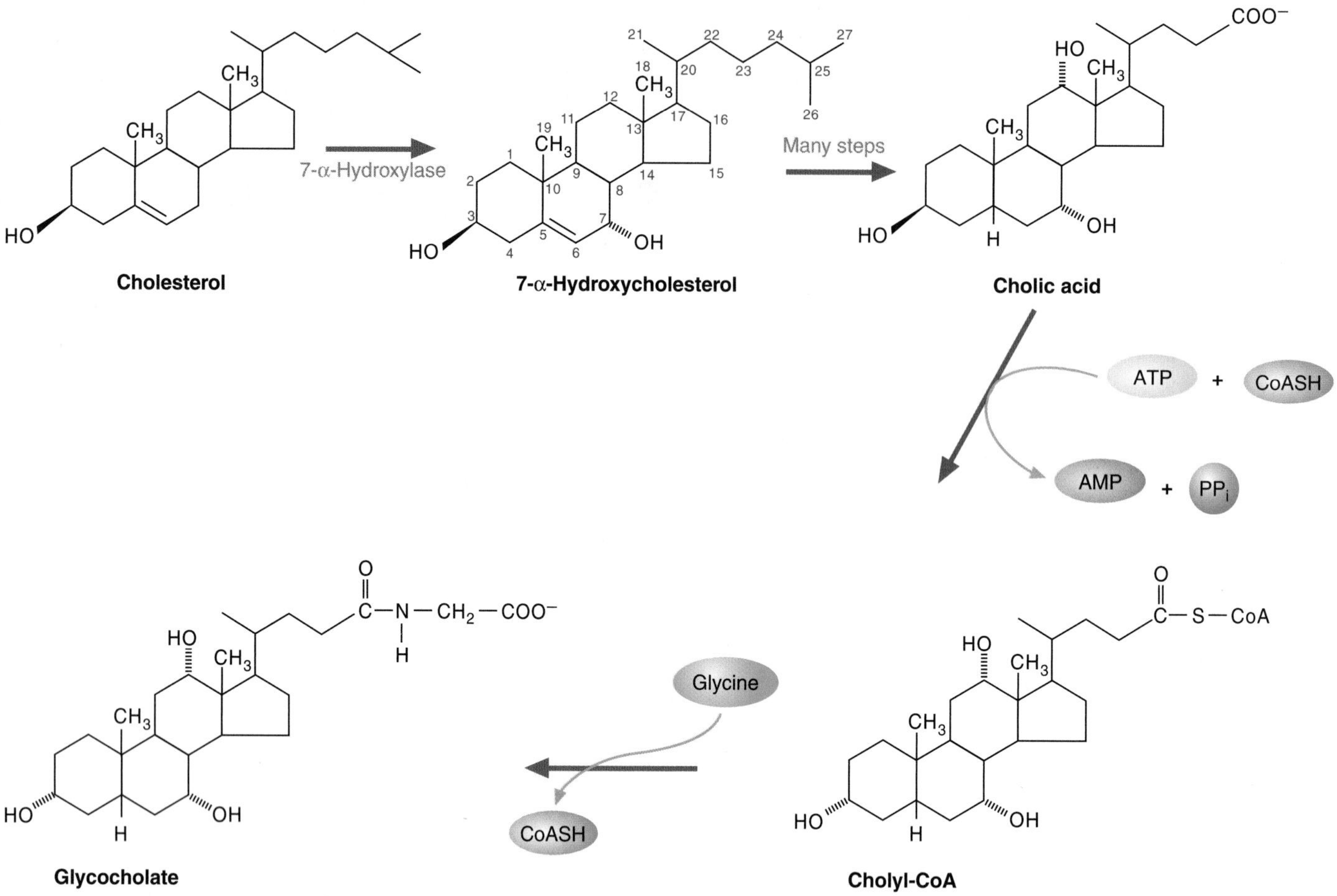

FIGURE 12.31
Bile Salt Synthesis

Bile salts are emulsifying agents that facilitate dietary fat digestion in the small intestine. They are synthesized in the liver from cholesterol. The bile acid cholic acid (cholate) is produced from cholesterol in a series of reactions, two of which are hydroxylation reactions catalyzed by cytochrome P_{450} enzymes: 7-α-hydroxylase and 12-α-hydroxylase (not shown). The bile salt glycocholate is produced when cholyl-CoA reacts with glycine to form an amide bond linkage.

CHOLESTEROL HOMEOSTASIS The critical roles of cholesterol in animal bodies, combined with the potentially toxic properties it exhibits when present in excessive amounts, require that its concentration be maintained within normal limits. Cholesterol homeostasis is achieved through intricate mechanisms that regulate its biosynthetic pathway, LDL receptor activity, and bile acid biosynthesis. The regulation of cholesterol biosynthesis is accomplished primarily by modulation of existing HMGR molecules, gene expression changes, and enzyme degradation.

The principal means by which the activity of existing HMGR molecules is regulated is downregulation via phosphorylation reactions (**Figure 12.32**). HMGR activity is depressed by phosphorylation by AMPK in response to high cellular concentrations of AMP, which have the effect of integrating cholesterol biosynthesis, a metabolically expensive process, into the cell's energy metabolism. cAMP, which is regulated by hormones such as glucagon and epinephrine, also depresses HMGR by activating phosphoprotein phosphatase inhibitor 1 (PPI-1) by means of a phosphorylation reaction catalyzed by PKA (protein kinase A). Activated PPI-1 inhibits several phosphatases that can increase HMGR activity by removing phosphate groups. Insulin increases HMGR activity, in part by inhibiting cAMP synthesis. HMGR is also regulated by a negative feedback mechanism involving various sterols, including cholesterol derived from the endocytosis of LDL receptors, and nonsterol derivatives of mevalonate.

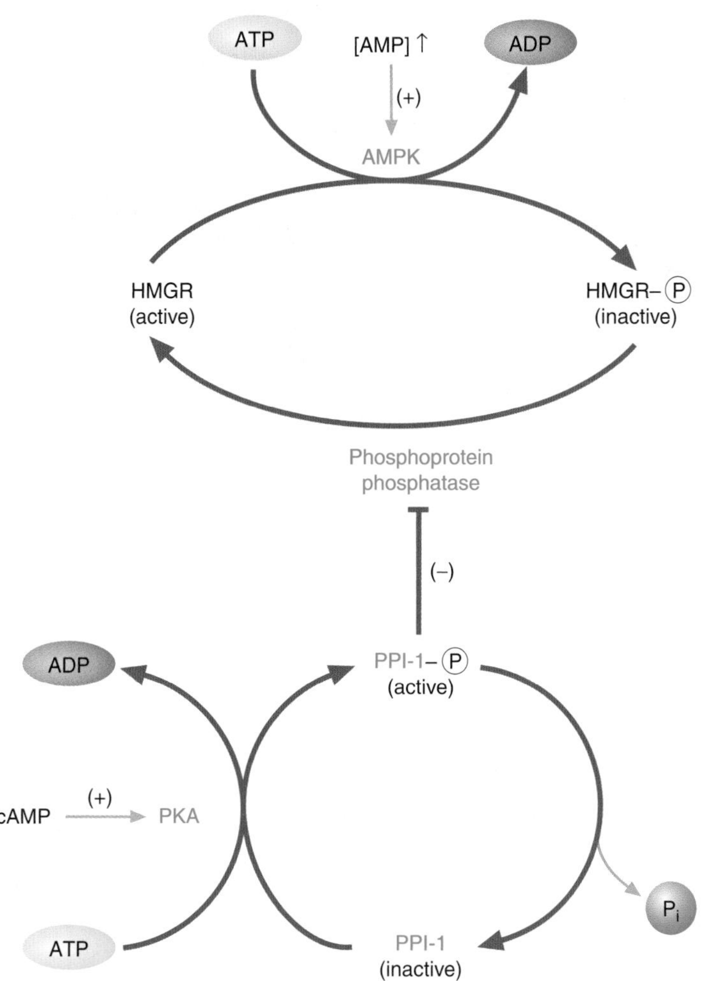

FIGURE 12.32

Regulation of HMGR by Covalent Modification

HMGR is inactivated by phosphorylation reactions catalyzed by AMPK in response to rising AMP levels. Activation of HMGR is effected by phosphoprotein phosphatase. When cAMP levels are high, phosphoprotein phosphatase inhibitor 1 (PPI-1), activated by a phosphorylation reaction, inhibits phosphoprotein phosphatase. An inhibited phosphoprotein phosphatase ensures the inactive status of HMGR.

Sterol-mediated changes in gene expression are a major feature of cholesterol homeostasis. The ER membrane protein SREBP2 is the predominant regulator of cholesterol biosynthesis. In addition to stimulating the expression of cholesterol biosynthesis genes, SREBP2 activates the LDL receptor gene and three genes required for NADPH synthesis (G-6-PD, 6-phosphogluconate dehydrogenase, and malic enzyme). SREBP2 is bound to SREBP cleavage-activating protein (SCAP). When cells have sufficient cholesterol, the sterol-sensing domain (SSD) of SCAP binds cholesterol and an ER retention protein called Insig (*ins*ulin-*i*nduced *g*ene). In cholesterol-depleted cells, Insig no longer binds to SCAP, which now escorts SREBP-2 from the ER to the Golgi (**Figure 12.33**). When the SREBP-SCAP complex reaches the Golgi, two proteases release the N-terminal domain of SREBP2, the active transcription factor, from the membrane. The activated SREBP2 product then translocates to the nucleus, where it binds to the SREs (sterol regulatory elements) of the target genes. In addition to SREBP, the expression of some target genes requires the binding of coregulatory transcription factors. For example, transcription of the genes for HMGR and HMG-CoA synthase also requires the binding of nuclear factor-1 and CREB (cAMP response element binding protein).

As cellular cholesterol levels begin to rise as a result of newly synthesized enzymes and the internalization of LDL from the bloodstream via LDL receptors, a concentration threshold is reached and a few molecules bind to the sterol-sensing domains of SCAP. As the result of a conformation change in the SCAP-SREBP complex, the ER protein Insig replaces cholesterol in the sterol-sensing domain of SCAP and the complex is retained in the ER, thus ending

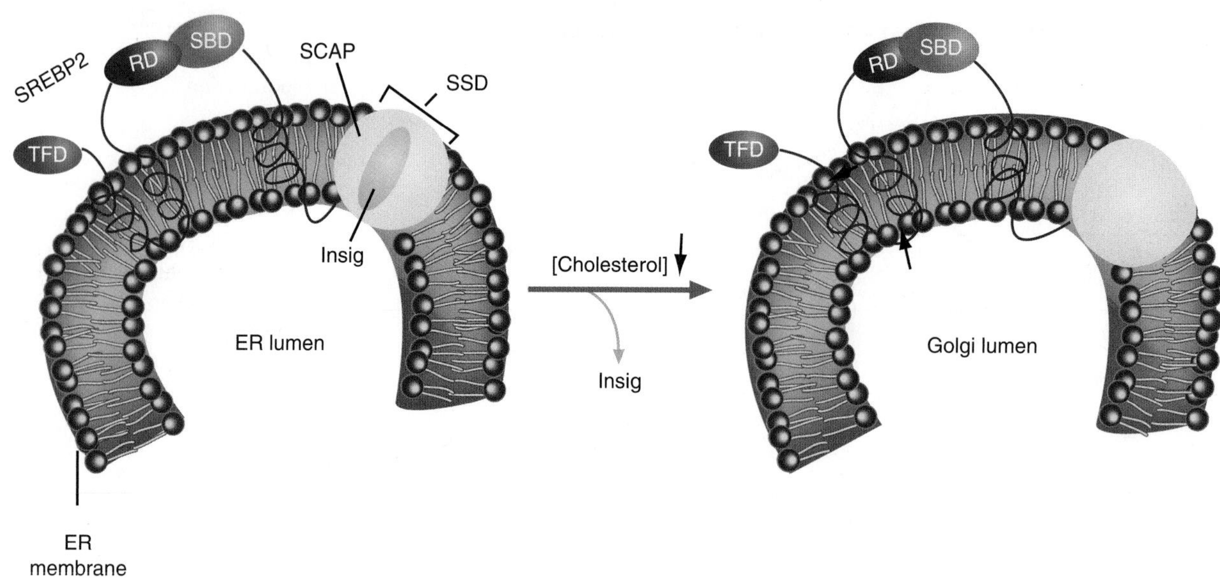

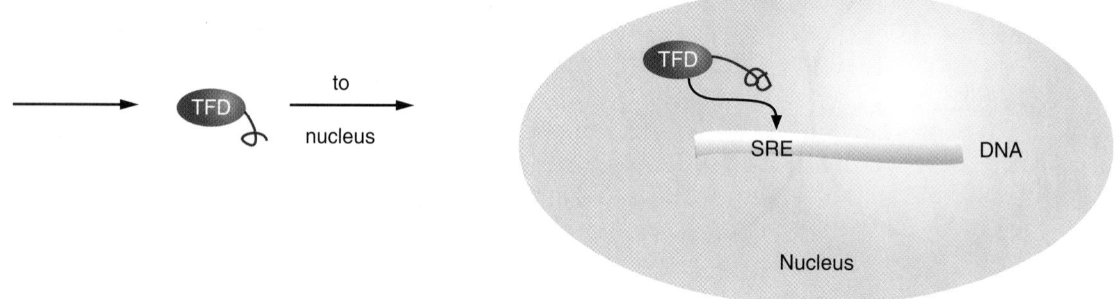

FIGURE 12.33

SREBP2 Regulation

SREBP2, an ER protein, is complexed to SCAP (SREBP cleavage-activating protein) via the binding of the regulatory domain (RD) of SREBP2 with the SREBP-binding domain (SBD) of SCAP. The SSD (sterol-sensing domain) of SCAP binds cholesterol. When cholesterol levels are high and cholesterol is bound to SSD, the ER retention protein Insig is also bound. When cholesterol levels are depleted and cholesterol is no longer bound to SSD, Insig is released, and the SREBP/SCAP complex is transferred to the Golgi complex. When SREBP2 is present in the Golgi complex, two proteases cleave it at two sites (arrows) to release the now active TFD (transcription factor domain). The SREBP transcription factor then moves into the nucleus, where it binds to SREs (sterol regulatory elements) that are associated with sterol-related genes.

SREBP2 processing and the transcription of target genes. High levels of cholesterol and other metabolites of mevalonate also depress further cholesterol synthesis by inhibiting the translation of existing HMGR mRNA. In liver, excess cholesterol activates ACAT, the enzyme that catalyzes the transfer of a fatty acid from a fatty acyl–CoA molecule to the hydroxyl group of cholesterol to generate a cholesterol ester storage molecule.

HMGR degradation, another means by which cholesterol levels are controlled, is mediated by Insig. When cholesterol levels are high, sterols bind to the enzyme's sterol-sensing N-terminal domain. Subsequently HMGR binds to Insig, which in turn is associated with ubiquitin ligase, an enzyme that initiates a major proteolytic mechanism (described in Chapter 15).

KEY CONCEPTS

- Cholesterol is synthesized from acetyl-CoA in a multistep pathway that occurs primarily in the liver.
- Cholesterol is degraded primarily by conversion to the bile salts, which facilitate the emulsification and absorption of dietary fat.

High cholesterol concentrations in the liver also trigger bile acid biosynthesis. When cholesterol begins to accumulate, some molecules become oxidized to form oxysterols (e.g., 25-hydroxycholesterol). Oxysterols bind to and activate LXR (liver X receptor), which then forms an active heterodimer transcription factor with RXR (retinol X receptor). The heterodimer then causes the transcription of 7-α-hydroxylase, the rate-limiting enzyme in bile synthesis.

The Cholesterol Biosynthetic Pathway and Drug Therapy

A majority of medical researchers believe that high blood serum levels of total cholesterol (the sum of cholesterol in VLDL, LDL, and HDL) combined with high LDL levels are strongly associated with cardiovascular disease. Currently, a class of drugs called the *statins* is routinely used to lower serum cholesterol in an attempt to reduce the risk of heart attack and stroke. Statins such as lovastatin are competitive inhibitors of HMG-CoA reductase, the rate-limiting enzyme in cholesterol biosynthesis. Because most cholesterol synthesis occurs in the liver, cholesterol serum levels are reduced when liver cells compensate for reduced synthesis by removing cholesterol from serum LDLs. Because most of the body's cholesterol is synthesized at night, statins are usually taken in the evening. Statin drugs also interfere with the synthesis of the mixed terpenoid coenzyme Q (ubiquinone), a critical molecule in energy generation. Consequently, statin therapy must be accompanied by CoQ supplementation.

Bisphosphonates are a class of drugs that are used to treat *osteoporosis*, a bone disease in which bone mineral density is reduced. Because bisphosphonates bind Ca^{2+}, they are readily absorbed into bone tissue where they kill osteoclasts. *Osteoclasts* are bone-remodeling cells that break down and remove old bone tissue. Bisphosphonates such as alendronate trigger osteoclast apoptosis by inhibiting the enzymatic activities that convert isopentenylpyrophosphate to farnesylpyrophosphate (see **Figure 12.29**). Osteoclast cell death occurs as a result of the absence of geranylpyrophosphate and farnesylpyrophosphate, the substrates for the prenylation reactions that link several cell signal proteins (e.g., Ras) to the plasma membrane.

Biochemistry IN PERSPECTIVE

Biotransformation

How are potentially toxic hydrophobic molecules metabolized by the body? In **biotransformation**, a series of enzyme-catalyzed processes in which toxic substances are converted into less toxic metabolites, the enzymes generally possess broad specificities. In mammals, biotransformation is used principally to convert toxic molecules, which are usually hydrophobic, into water-soluble derivatives so that they may be more easily excreted. The enzymes that catalyze the biotransformation of xenobiotics (foreign molecules) are similar to several of the enzymes that dispose of hydrophobic endogenous molecules. Although biotransformation reactions occur in several locations within the cell (e.g., the cytoplasm and mitochondria), most occur within the SER. Cell types also differ in their biotransforming potential. In general, cells located near the major points of xenobiotic entry into the body (e.g., liver, lung, and intestine) possess greater concentrations of biotransforming enzymes than others.

Biotransformation processes have been differentiated into two major types. During **phase I**, reactions involving oxidoreductases and hydrolases convert hydrophobic substances into more polar molecules. **Phase II** consists of reactions in which metabolites containing appropriate functional groups are conjugated with substances such as glucuronate, glutamate, sulfate, or glutathione. In general, conjugation dramatically improves solubility, which then promotes rapid excretion. (In animals, excretion of biotransformed molecules is sometimes referred to as phase III.) Although many substances undergo these phases sequentially, a significant number do not. For example, some molecules are excreted as phase I metabolites, whereas others undergo only phase II reactions. Moreover, variations in enzyme concentrations, availability of cosubstrates, and the order in which the reactions occur may cause certain substances to be converted into more than one end product. However, despite these and other complications, basic biotransformation patterns have emerged. Several well-researched examples of phase I and phase II reactions are described below. In the following discussions, the term **detoxication** refers to the process by which a toxic molecule is converted to a more soluble (and usually less toxic) product. The more familiar term **detoxification** implies the correction of a state of toxicity, that is, the chemical reactions that produce sobriety in an intoxicated person.

Phase I reactions usually convert substrates to more polar forms by introducing or unmasking a functional group (e.g., —OH, —NH_2, or —SH). Many phase I enzymes are located in the SER membrane, but others such as the dehydrogenases (e.g., alcohol dehydrogenases and peroxidases) occur in the cytoplasm, whereas still others (e.g., monoamine oxidase) are localized in mitochondria. The predominant enzymes of SER oxidative metabolism are the monooxygenases, sometimes referred to as *mixed function oxidases.* They are so named because in a typical reaction, one molecule of oxygen is consumed (reduced) per substrate molecule: one oxygen atom appearing in the product and the other in a molecule of water. Monooxygenases can carry out an immense variety of chemical reactions. Some of these reactions form highly unstable (and therefore toxic) intermediates.

There are two major types of SER monooxygenase, both of which require NADPH as an external reductant: the cytochrome P_{450} (cyt P_{450}) system and flavin-containing monooxygenases. The **cytochrome P_{450} system**, which consists of two enzymes (NADPH–cytochrome P_{450} reductase and cytochrome P_{450}), is involved in the oxidative metabolism of many endogenous substances (e.g., steroids and bile acids), as well as the detoxication of a wide variety of xenobiotics. **Flavin-containing monooxygenases** catalyze an NADPH- and an oxygen-requiring oxidation of substances (primarily xenobiotics) bearing functional groups containing nitrogen, sulfur, or phosphorus. The properties of the cyt P_{450} electron transport systems are described.

Cytochrome P_{450} Electron Transport Systems

In cytochrome P_{450} electron transport systems, found in SER and inner mitochondrial membranes, two electrons are transferred one at a time from NADPH to a cytochrome P_{450} protein by NADPH–cytochrome P_{450} reductase (**Figure 12A**). The latter enzyme is a flavoprotein that contains both FAD and FMN in a ratio of 1:1 per mole of enzyme. In addition to its role in the cytochrome P_{450} system, the reductase is also believed to be involved in the function of heme oxygenase. (See the Chapter 15 online Biochemistry in Perspective essay Heme Biotransformation.)

The hemoproteins referred to as cytochrome P_{450} are so named because of the complexes they form with carbon monoxide. In the presence of the gas, light is strongly absorbed at a wavelength of 450 nm. More than 6000 cyt P_{450} genes have been identified in species as diverse as mammals and bacteria. In humans, 63 cyt P_{450} genes, classified into 18 families, have been identified. Each gene codes for a protein with a unique specificity range. Cytochrome P_{450} proteins found in the liver have broad and overlapping specificities. For example, molecules as diverse as alkanes, aromatics, ethers, and sulfides are routinely oxidized. In contrast, cytochrome P_{450} proteins in the adrenal glands, ovaries, and testes that add hydroxyl groups to steroid molecules have narrow specificities. Despite this diversity, all cytochrome P_{450} isozymes contain one molecule of heme and are similar in their physical properties and catalytic mechanisms.

Despite an enormous variety of substrates, all of the oxidative reactions catalyzed by cytochrome P_{450} may be viewed as

Biochemistry IN PERSPECTIVE cont.

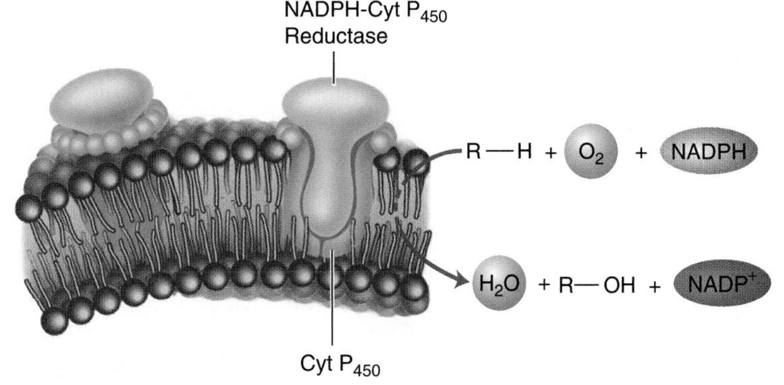

FIGURE 12A

The Cytochrome P_{450} Electron Transport System

Cytochrome P_{450} and cytochrome P_{450} reductase are components of an electron transport system used to oxidize both endogenous and exogenous molecules.

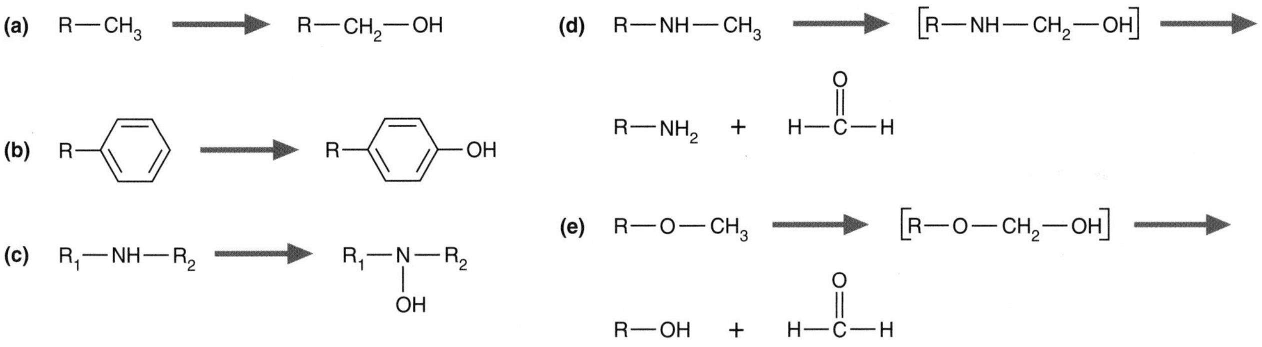

(a) $R—CH_3 \longrightarrow R—CH_2—OH$

(b) R–(benzene ring) ⟶ R–(benzene ring)–OH

(c) $R_1—NH—R_2 \longrightarrow R_1—N(OH)—R_2$

(d) $R—NH—CH_3 \longrightarrow [R—NH—CH_2—OH] \longrightarrow R—NH_2 + H—C(=O)—H$

(e) $R—O—CH_3 \longrightarrow [R—O—CH_2—OH] \longrightarrow R—OH + H—C(=O)—H$

FIGURE 12B

Diverse Substrates Oxidized by Cytochrome P_{450} Isozymes

Among the reactions catalyzed by cytochrome P_{450} are (a) aliphatic oxidation, (b) aromatic hydroxylation, (c) N-hydroxylation, (d) N-dealkylation, and (e) O-dealkylation.

hydroxylation reactions (i.e., an OH group appears in each reaction) (**Figure 12B**). The general reaction is as follows:

$$R—H + O_2 + NADPH + H^+ \rightarrow ROH + H_2O + NADP^+$$

where R—H is the substrate.

The oxygenation reaction is initiated when the substrate binds to oxidized cytochrome P_{450} (Fe^{3+}). This binding promotes a reduction of the enzyme substrate complex by an electron transferred from NADPH via cytochrome P_{450} reductase (Fe^{2+}—substrate). After reduction, cytochrome P_{450} can bind O_2. Then the electron from heme iron is transferred to the bound O_2, thus forming a transient $Fe^{3+}—O_2^-$—substrate species. (If the bound substrate is easily oxidized, it can be converted into a peroxy radical.) A second electron transferred from the flavoprotein results in the generation of a $Fe^{3+}—O_2^{2-}$—substrate complex. This brief association ends when the oxygen-oxygen bond is broken. One atom is released in a water molecule, while the other remains bound to heme. After a hydrogen atom or electron has been abstracted from the substrate, the oxygen species (now a powerful oxidant) is transferred to the substrate. The cycle ends with release of the product from the active site. Depending on the nature of the substrate, the product is either an **epoxide** (a highly reactive ether in which the oxygen is incorporated into a three-membered ring) or an alcohol. The function of conjugation (phase II) reactions is to inactivate biologically active substances and/or to form more polar (and therefore more easily excretable) derivatives. During this process, lipophilic metabolites, bearing functional groups that can act as acceptors, undergo enzyme-catalyzed reactions along with second (or donor) substrates. Among the most frequently used donor substrates are glucuronic acid, glutathione (see later, Section 14.3), sulfate, and amino acids.

SUMMARY: Biotransformation is the enzyme-catalyzed process in which toxic, hydrophobic molecules are converted to less toxic, water-soluble molecules.

Chapter Summary

1. Acetyl-CoA plays a central role in most lipid-related metabolic processes. For example, acetyl-CoA is used in the synthesis of fatty acids. When fatty acids are degraded to generate energy, acetyl-CoA is the product.
2. In the exogenous lipoprotein pathway dietary lipid is distributed to the body's tissues. The process begins within enterocytes where TGs, cholesterol, and other lipid molecules and apolipoprotein B-48 are packaged into chylomicrons.
3. Depending on the body's current energy requirements, newly digested fat molecules are used to generate energy or are stored within adipocytes. When the body's energy reserves are low, fat stores are mobilized in a process referred to as lipolysis in which triacylglycerols are hydrolyzed to fatty acids and glycerol. Glycerol is transported to the liver, where it can be used in glucose synthesis. Most fatty acids are degraded to form acetyl-CoA in the β-oxidation pathway, a series of four reactions in which the β-carbon is oxidized, followed by the breakage of the bond between the α- and β-carbons. Peroxisomal β-oxidation appears to shorten very long fatty acids. Additional reactions are required to degrade odd-chain and unsaturated fatty acids and dicarboxylic acids. When the acetyl-CoA product is present in excess, ketone bodies are produced and are used as an energy source in some tissues.
4. Fatty acid synthesis begins with the carboxylation of acetyl-CoA to form malonyl-CoA. ACC, a key enzyme in fatty acid metabolism, is regulated by allosteric modulators and phosphorylation reactions. The remaining reactions of fatty acid synthesis take place on the fatty acid synthase multienzyme complex. Several enzymes are available to elongate and desaturate dietary and newly synthesized fatty acids.
5. Both short- and long-term regulatory mechanisms are used to control fatty acid metabolism. Short-term regulation includes the use of malonyl-CoA as an inhibitor of CAT-I, and AMPK-catalyzed phosphorylation of ACC1 and glycerol-3-phosphate acyltransferase. Hormones such as insulin, glucagon, epinephrine, and cortisol have roles in short- and long-term regulation. Long-term regulation of fatty acid metabolism involves changes in gene expression triggered by transcription factors. Two prominent examples are the SREBPs and the PPARs.
6. After phospholipids have been synthesized at the interface of the SER and the cytoplasm, they are often "remodeled"; that is, their fatty acid composition is adjusted. The turnover (i.e., the degradation and replacement) of phospholipids, mediated by the phospholipases, is rapid.
7. Synthesis of the ceramide component of sphingolipids begins with the condensation of palmitoyl-CoA with serine to form 3-ketosphinganine. In a two-step process involving acyl-CoA and $FADH_2$, sphinganine (formed when 3-ketosphinganine is reduced by NADPH) is converted to ceramide. Sphingolipids are degraded within lysosomes.
8. Cholesterol synthesis can be divided into three phases: formation of HMG-CoA from acetyl-CoA, conversion of HMG-CoA to squalene, and conversion of squalene to cholesterol. Cholesterol is the precursor for all steroid hormones and the bile salts. Bile salts are used to emulsify dietary fat. Cholesterol homeostasis is achieved through regulation of cholesterol biosynthesis, LDL receptor activity, and bile acid biosynthesis.

Take your learning further by visiting the **companion website** for Biochemistry at **www.oup.com/us/mckee** where you can complete a multiple-choice quiz on lipid metabolism to help you prepare for exams.

Suggested Readings

Gurr, M. I., Frayn, K. N., and Harwood, J. L., *Lipid Biochemistry: An Introduction*, 5th ed., Blackwell Scientific, Oxford, 2002.

Hardie, D. G., and Sakamoto, K., AMPK: A Key Sensor of Fuel and Energy Status in Skeletal Muscle, *Physiology* 21:48–60, 2006.

Jorgensen, S. B., Richter, E. A., and Wojtaszewski, F. P., Role of AMPK in Skeletal Muscle Metabolic Regulation and Adaptation in Relation to Exercise, *J. Physiol.* 574:17–31, 2006.

Mestek, M. L., Physical Activity, Blood Lipids, and Lipoproteins, *Am. J. Lifestyle Med.* 3(4):279–283, 2009.

Neels, J. G., and Olefsky, J. M., A New Way to Burn Fat, *Science* 312:1756–1758, 2006.

Nye, C., *et al.*, Reassessing Triglyceride Synthesis in Adipose Tissue, *Trends Endocr. Metab.* 19(10):356–361, 2008.

Santamarina-Fojo, S., Gonzalez-Navarro, H., Freeman, L., Wagner, E., and Nong, Z., Hepatic Lipase, Lipoprotein Metabolism, and Atherosclerosis, *Arterioscler. Thromb. Vasc. Biol.* 24:1750–1754, 2004.

Scott, J., The Liver X Receptor and Atherosclerosis, *N. Engl. J. Med.* 357:2195–2197, 2007.

Vance, D. E., and Vance, J. E., *Biochemistry of Lipids, Lipoproteins, and Membranes*, 4th ed., Elsevier, Amsterdam, 2008.

Yang, J., Kalhan, S. C., and Hanson, R. W., What Is the Metabolic Role of Phosphoenolpyruvate Carboxylase? *J. Biol. Chem.* 284(40):27025–27029, 2009.

Key Words

acyl carrier protein, *442*
AMPK, *451*
atherosclerosis, *454*
bile salts, *426*
biotransformation, *468*
chylomicron remnants, *427*
conjugation reaction, *461*
cytochrome P_{450} system, *468*
detoxication, *468*
detoxification, *468*
epoxide, *469*
fatty acid–binding protein, *432*
flavin-containing monooxygenase, *468*
glyceroneogenesis, *429*
ketogenesis, *437*
ketone body, *437*
ketosis, *437*
lipogenesis, *429*
lipolysis, *429*
β-oxidation, *432*
phase I reaction, *468*
phase II reaction, *468*
3′-phosphoadenosine-5′-phosphosulfate, *456*
PPAR, *452*
SREBP, *452*
sterol carrier protein, *461*
thiolytic cleavage, *435*
triacylglycerol cycle, *428*
turnover, *453*

Review Questions

These questions are designed to test your knowledge of the key concepts discussed in this chapter before moving on to the next chapter. You may like to compare your answers to the solutions provided in the back of the book and in the accompanying Study Guide.

1. Define the following terms:
   a. chylomicron remnants
   b. glyceroneogenesis
   c. exogenous lipoprotein pathway
   d. apolipoprotein B-48
   e. enterocyte
2. Define the following terms:
   a. β-oxidation
   b. carnitine
   c. ketogenesis
   d. ketone bodies
   e. ketosis
3. Define the following terms:
   a. adipocyte
   b. MCAD
   c. ACP
   d. α-oxidation
   e. odd-chain oxidation
4. Define the following terms:
   a. sterol carrier protein
   b. fatty acid binding protein
   c. biotransformation
   d. *de novo*
   e. turnover
5. Define the following terms:
   a. SREBP1
   b. SREBP2
   c. PPAR
   d. hypertriglyceridemia
   e. atheroma
6. Define the following terms:
   a. cytochrome P_{450}
   b. mixed-function oxidase
   c. flavin-containing monooxygenase
   d. detoxication
   e. detoxification
7. Define the following terms:
   a. allyl group
   b. epoxide
   c. SAM
   d. PAPS
   e. phase I reaction
8. Define the following terms:
   a. MTTP
   b. abetalipoproteinemia
   c. endogenous lipoprotein pathway
   d. PEPCK-C
   e. triacylglycerol cycle
9. What are the differences between β-oxidation in mitochondria and in peroxisomes? What similarities are there between these processes?
10. List three differences between fatty acid synthesis and β-oxidation.
11. Explain how hormones act to modify the metabolism of fatty acids in both the short and the long term. Give examples.
12. Show how the following fatty acid is oxidized:

$$CH_3CH_2CH_2CH(CH_3){-}CH_2{-}C({=}O){-}OH$$

13. What are the products of the oxidation of the molecule in Question 12?
14. Describe the glyceroneogenesis pathway. What molecules are its substrates?
15. Compare the energy content of a stearic acid molecule compared with glucose.
16. Define the term thiolytic cleavage. In what biochemical process does it occur?
17. How do cells adjust the fluidity of their membranes?
18. Describe the difference between IDLs and LDLs.
19. Explain why low-fat diets supplemented with medium-chain fatty acids are prescribed for patients with abetalipoproteinemia.
20. Outline the reactions in the α-oxidation pathway.
21. β-Oxidation of naturally occurring monounsaturated fatty acids requires an additional enzyme. What is this enzyme, and how does it accomplish its task?

22. Identify the hydrophobic and hydrophilic regions in the following molecule. How does it orient itself in a membrane?

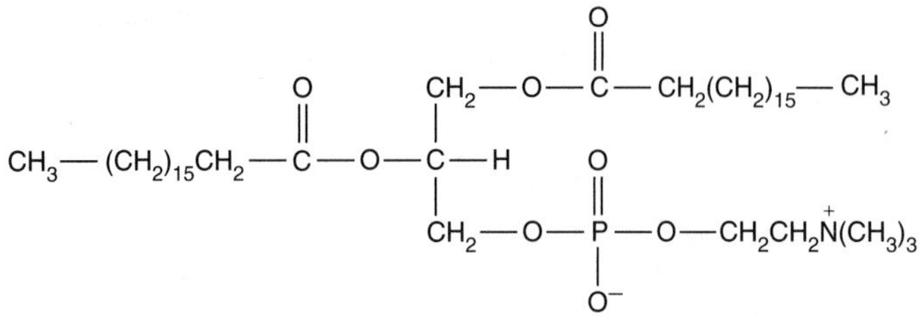

23. What is the function of each of the following?
    a. glycocorticoid
    b. ketone bodies
    c. biotransformation
    d. phase I reaction
    e. ACP

24. Gaucher's disease is an inherited deficiency of β-glucocerebrosidase. Glucocerebroside is deposited in macrophages that die, releasing their contents into the tissues. Some affected individuals experience neurologic disorders while quite young; others do not show ill effects until much later in life. The disease may be detected by assaying white blood cells for the ability to hydrolyze the β-glycosidic bond of artificial substrates. Examine the following glucocerebroside and determine which bond is cleaved by glucocerebrosidase.

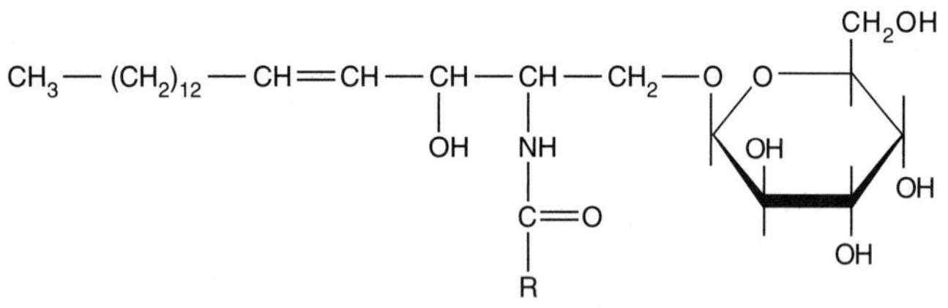

25. Determine the number of moles of ATP that can be generated from the fatty acids in 1 mol of tristearin, a triacylglycerol composed of glycerol esterified to three stearic acid molecules. What is the fate of the glycerol?

26. How would you describe the three phases of biotransformation?

27. Outline the biosynthesis of bile salts. What are the functions of these substances?

28. How are lipid molecules such as animal steroid molecules like estrogen and β-carotene related to each other? What biosynthetic reactions do these specific molecules have in common?

29. What is the function of each of the following substances?
    a. AMPK
    b. sterol carrier protein
    c. lipoprotein lipase
    d. GLUT-4
    e. hormone-sensitive lipase

30. List and describe the components of the cytochrome P_{450} electron transport system. What is the role of each component?

31. What is the function of conjugation reactions in the biotransformation process?

32. The absorption of triacylglycerols in the small intestine is an energy-requiring process that involves hydrolytic reactions to yield monoacylglycerol and fatty acids. After their transport into enterocytes, fatty acids and monoacylglycerol are reconverted into triacylglycerols. Suggest a reason why triacylglycerols are not absorbed directly without the hydrolytic reactions.

33. The peroxisomal enzyme β-ketoacyl–CoA-thiolase does not bind medium-chain acyl-CoA, in contrast to the analogous mitochondrial enzyme. Explain why this phenomenon is an advantage to the cell.

34. Under severe starvation conditions, people develop "acetone breath." Explain.

35. Provide an explanation for the intracellular separation of fatty acid metabolic processes (i.e., fatty acid biosynthesis in cytoplasm and degradation in mitochondria and peroxisomes).

36. Describe the mechanism in which NADH is involved in the activation of oxygen molecules in fatty acid desaturation processes.

37. Describe the role of insulin in lipid metabolism.

38. Membranes that contain a significant proportion of cis unsaturated fatty acids are more fluid than similar membranes with higher levels of saturated fatty acids. Explain.

39. When $^{14}CO_2$ is used in the synthesis of malonyl-CoA from acetyl-CoA, no label appears in the eventual fatty acid products. Explain.

40. Draw a cholesterol molecule and indicate the isopentenyl units.

41. Determine how many ATP equivalents are obtained by the oxidation of oleic acid.

42. Describe the fate of glycerol generated from triacylglycerol hydrolysis in adipocytes.

43. Many processed and fast foods contain trans fatty acids. Explain why these molecules are a problem for the body.

44. Before β-oxidation, fatty acids are converted to their CoASH derivatives. Explain why this reaction is necessary.

45. Review the steps in β-oxidation and determine which ones are actually oxidation reactions.

Fill in the Blank

46. Oxidation of phytanic acid produces acetyl-CoA and ____________________ as the end products.

47. Of the four reactions of the β-oxidation cycle, which ones are redox reactions?

48. The carrier of carbon dioxide in fatty acid synthesis is ____________________.

49. β-Hydroxybutyrate is a product of ____________________ metabolism.

50. After the triacylglycerol content of VLDL has been depleted, the lipoprotein is referred to as ____________________.

51. Glycerol generated from TG hydrolysis in adipocytes is converted by the liver into ______________________, which serves as a substrate for the synthesis of glucose, among other molecules.
52. ______________ is the steroid precursor of the bile salts.
53. The oxidation of 1 mol of palmitic acid yields a total of ____________ mol of metabolic water.
54. SAM is the abbreviation for ________________.
55. ________________ is a carrier molecule required for the transport of fatty acids into the mitochondria.

Short Answer

56. Why do gallstones form in susceptible individuals?
57. What are the conjugation reactions that the bile acids undergo and what is their function?
58. What are the roles of SREBPs and PPARs in fatty acid metabolism?
59. Why is the synthesis of desaturated fatty acids called an electron transport system?
60. What function does the enzyme ACC1 serve and how is it regulated?

Thought Questions

These questions are designed to reinforce your understanding of all of the key concepts discussed in the book so far, including this chapter and all of the chapters before it. They may not have one right answer! The authors have provided possible solutions to these questions in the back of the book and in the accompanying Study Guide for your reference.

61. When blood levels of glucose are higher than normal, glucose molecules react with protein side chains such as that of lysine residues to produced glycated side chains. Explain why this process is dangerous. What is the name of the first functional group that is formed during glycation?
62. Describe the possible effects of low levels of carnitine on a person's metabolism.
63. Phospholipases show an enhanced activity for a substrate above the critical micelle concentration. (The critical micelle concentration, or cmc, is that concentration of a lipid above which micelles begin to form.)
    a. What type of noncovalent interactions are possible between the lipid and the enzyme at this stage?
    b. What do these interactions suggest about the structure of phospholipases?
64. When the production of acetyl-CoA exceeds the body's capacity to oxidize it, acetoacetate, β-hydroxybutyrate, and acetone accumulate. When generated in large amounts, these substances can exceed the blood's buffering capacity. As the blood pH falls, the ability of red blood cells to carry oxygen is affected. Subsequently, the brain can be starved for oxygen, and a fatal coma can result. Explain how severe dieting can produce this condition.
65. The acyl-CoA dehydrogenase deficiency diseases are a group of inherited defects that impair the β-oxidation of fatty acids. Symptoms of the disease range from nausea and vomiting to frequent comas. Symptoms may be alleviated by eating regularly and avoiding periods of starvation (12 hours or more). Why does this simple procedure alleviate the symptoms?
66. There is an unusually high concentration of phosphatidylcholine on the lumenal side of the ER. What structural feature of phosphatidylcholine is responsible for this? Explain how this structural feature produces this effect.
67. During periods of stress or fasting, blood glucose levels fall. In response, fatty acids are released by adipocytes. Explain how the drop in blood glucose triggers fatty acid release.
68. Pharmaceuticals in a class called the statins inhibit the enzyme HMG-CoA reductase. What is the primary effect of this drug on patients?
69. The adaptations of desert animals to their environment include water conservation mechanisms. A number of these organisms conserve water so successfully that they never actually drink it. They depend instead on water generated during metabolism. Determine how much water can be obtained by the oxidation of 1 mole of palmitic acid.
70. Consider the structure of the following fatty acid and determine the reactions by which it is oxidized.

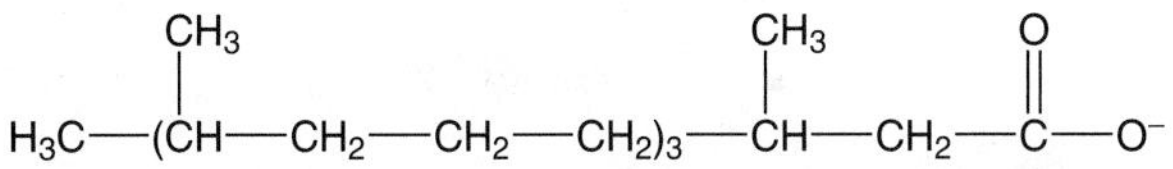

$$H_3C\text{—}(\underset{|}{\overset{CH_3}{CH}}\text{—}CH_2\text{—}CH_2\text{—}CH_2)_3\text{—}\overset{CH_3}{\underset{}{CH}}\text{—}CH_2\text{—}\overset{O}{\overset{\|}{C}}\text{—}O^-$$

71. Cholestyramine is an anion-exchange resin used by clinicians to lower serum cholesterol in patients. This oral medication forms insoluble complexes with bile salts in the intestine, thus preventing their reabsorption. Explain the mechanism by which cholestyramine lowers serum cholesterol levels.

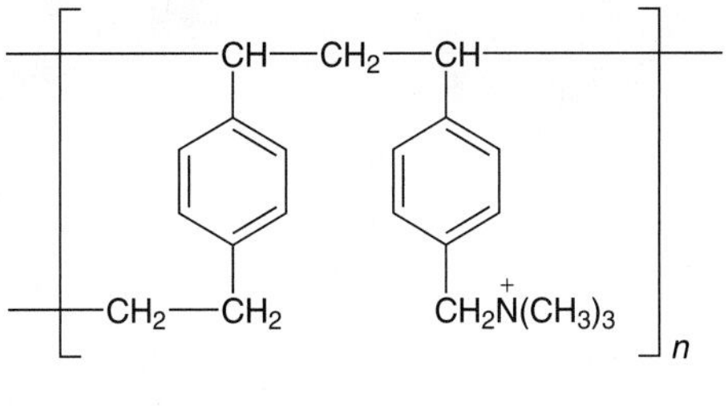

$$\left[\begin{array}{c} \text{—CH—CH}_2\text{—CH—} \\ \text{(C}_6\text{H}_4\text{)} \quad \text{(C}_6\text{H}_4\text{)} \\ \text{—CH}_2\text{—CH}_2 \quad \text{CH}_2\overset{+}{N}(\text{CH}_3)_3 \end{array}\right]_n$$

CHOLESTYRAMINE RESIN

72. Provide an explanation for the fact that most fatty acids are 16 or 18 carbons long.
73. An experimenter using acetyl-CoA with a ^{14}C label on the carbonyl group traces the label in a cell synthesizing cholesterol. On what atoms of mevalonate will the label appear?
74. Determine the position of ^{14}C (see Question 73) in isopentenylpyrophosphate.

CHAPTER

13 Photosynthesis

A Light in the Forest Light has a profound impact on most living organisms on Earth. Its energy, captured by a molecular mechanism called photosynthesis, is used to manufacture organic biomolecules.

OUTLINE

Climate Change, Renewable Energy, and Photosynthesis

Ever-accelerating glacier melting and iceberg formation in both the Arctic Circle and Antarctica combined with rising sea levels are impressive indicators that the Earth's climate is warming up! During the Earth's long history there have been numerous transitions between glacial "ice ages" and interglacial warm periods. These climate changes often lasted from thousands to tens of thousands of years. Abrupt transitions lasting only a few decades have occurred, but were always associated with dramatic natural events. Significant changes in the Earth's orbit or the sun's intensity can alter the amounts of solar radiation that reach Earth. Volcanic eruptions of unusual duration and intensity can emit massive amounts of CO_2 and other greenhouse gases that trap solar heat, and altered ocean current patterns can redistribute heat around the Earth. Since the beginning of the Industrial Revolution about 150 years ago, however, a steady increase in human activities, most notably fossil fuel burning and deforestation, has resulted in the release of massive amounts of greenhouse gases into the atmosphere. Their impact on the Earth's climate is significant.

Global warming is a difficult and seemingly intractable problem, with consequences that include an increased incidence and intensity of drought and flooding that will cause catastrophic food and water shortages. Other effects include economic damage from more powerful hurricanes and an increased prevalence of insect-borne diseases. Slowing the progress of global warming will require substantial investments. New technologies are needed to replace fossil fuels with economically competitive fuels that are carbon neutral (i.e., with no net release of CO_2 into the atmosphere). *Biofuels* are renewable energy sources synthesized by living organisms directly or indirectly via photosynthesis, the light energy–driven process that converts CO_2 and H_2O into organic molecules. Currently produced biofuels include ethanol (from corn kernels), biodiesel (long-chain alkyl esters derived from vegetable oil or animal fat), and microbe-generated inorganic molecules such as hydrogen gas (H_2) or methane (CH_4).

Sustainable biofuel production requires that the following three criteria be met. First, biofuels must be economically generated in large enough quantities—about 10 terrawatts (TW) of energy per year (1 TW = 10^{12} watts or 3.2×10^{18} J/year). Second, biofuels must not compete with food production. The use of arable farmland for biomass-derived energy production drives up the cost and availability of food. Finally, biofuel production must not adversely affect the environment. Most biofuel production methods currently in use fail to meet these criteria. For example, corn-based ethanol production is cost ineffective (energy costs alone account for at least 90% of output), and the diversion of corn, a food staple, into energy generation has caused increased food prices. Palm oil-derived biodiesel, made possible by new plantations in Southeast Asia, has resulted in deforestation that not only destroyed the habitat of endangered animals (e.g., tigers, gibbons, and orangutans), but also removed a critical "sink" for atmospheric CO_2 removal.

Considering the drawbacks of current biofuel technologies, is there any hope for the development of a successful fossil fuel substitute? Current research demonstrates that biodiesel production involving photosynthetic algae and cyanobacteria, which does not use arable land, can be up to 100 times that obtained from current methods. It is hoped that the combination of this promising technology with improvements in methane production from agricultural, industrial, and municipal waste can contribute substantially to solving the Earth's human-caused climate change problem.

Overview

WITHOUT QUESTION, OXYGENIC PHOTOSYNTHESIS IS THE MOST IMPORTANT BIOCHEMICAL PROCESS ON EARTH. WITH A FEW MINOR EXCEPTIONS, PHOTOSYNTHESIS is the only mechanism by which an external abiotic source of energy is harnessed by the living world and it is the source of O_2, which sustains all aerobic organisms.

Chapter 13, which is devoted to a discussion of the principles of photosynthetic processes, emphasizes the relationship between photosynthetic reactions and the structure of chloroplasts and the relevant properties of light.

KEY CONCEPTS

- Incorporating CO_2 into organic molecules requires energy and reducing power.
- In photosynthesis, both these requirements are provided by a complex process driven by light energy.

Photosynthesis is the light-driven biochemical mechanism whereby CO_2 is incorporated into organic molecules such as glucose (**Figure 13.1**). Captured light energy is used to synthesize ATP and NADPH, which drive this process. The reducing power of NADPH is necessary because a strong electron donor is required to reduce the fully oxidized, low-energy carbon atoms in CO_2 to the carbon units of organic molecules.

Photosynthesis is performed by biochemical mechanisms, referred to as **photosystems**, which are membrane-bound protein complexes found in chloroplasts (plants and algae) or the cell membrane of photosynthetic bacteria. There are two photosystems in chloroplasts: photosystem I (PSI) and photosystem II (PSII). Each photosystem is composed of two functional components. The **light-harvesting antenna** captures solar energy and transfers it to the **reaction center**, which uses captured light energy to drive transmembrane electron transport. In PSI light-driven electron transport is used to synthesize NADPH. Electron flow from PSII to cytochrome b_6f complex pumps protons across a membrane to subsequently drive ATP synthesis. The electrons of water molecules replace the PSII reaction center electrons, yielding the waste product molecule O_2.

Chapter 13 describes the principles of oxygenic photosynthesis. The discussion begins with a detailed view of chloroplast structure. After a brief review of the relevant properties of light, the reactions that constitute modern photosynthesis will be described. These include the light reactions and the light-independent reactions. It is during the light reactions that electrons are energized and eventually used in the synthesis of both ATP and NADPH. Both molecules are then used in the light-independent reactions (often referred to as the dark reactions or Calvin cycle) to drive the synthesis of carbohydrate. Several photosynthetic

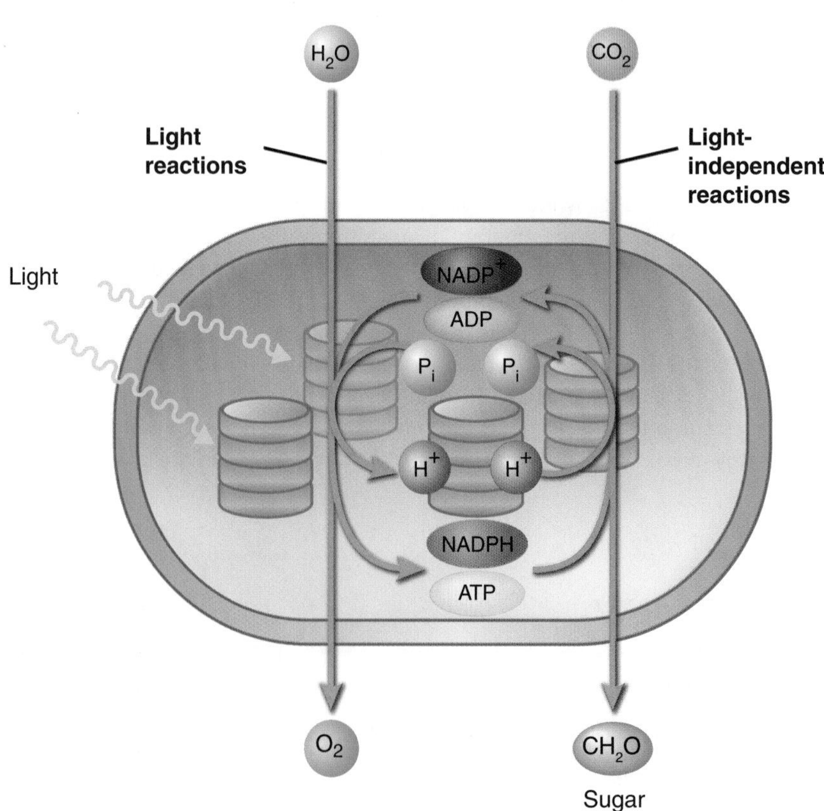

FIGURE 13.1
Overview of Oxygenic Photosynthesis

Oxygenic (oxygen-producing) photosynthesis occurs in two sets of reactions: the light reactions and the light-independent reactions. In chloroplasts (**Figure 2.26**) the light reactions occur within the thylakoid membrane, and the light-independent reactions (the Calvin cycle) occur within the stroma. The light reactions use light energy to drive the synthesis of NADPH and ATP. Water molecules are the source of the electrons and protons used to synthesize these molecules. O_2 is released as a waste product. The light-independent reactions use NADPH and ATP to convert CO_2 into sugar molecules.

variations, referred to as C4 metabolism and crassulacean acid metabolism, are also discussed. Chapter 13 ends with a discussion of several mechanisms that control photosynthesis in plants.

13.1 CHLOROPHYLL AND CHLOROPLASTS

Photosynthesis begins with the absorption of light energy by specialized pigment molecules (**Figure 13.2**). The **chlorophylls** are green pigment molecules that resemble heme. *Chlorophyll a* plays the principal role in oxygenic photosynthesis because its absorption of light energy directly drives photochemical events. Chlorophyll a is also involved in **light harvesting**, the process whereby absorbed energy is channeled to a reaction center. *Chlorophyll b* is a light-harvesting pigment that passes absorbed energy on to chlorophyll a. The orange-colored **carotenoids** are isoprenoid molecules that either function as light-harvesting pigments (e.g., lutein, a xanthophyll, see p. 399) or protect against overexcitation and ROS (e.g., β-carotene).

Chloroplasts, the photosynthetic organelle in plants and algae, resemble mitochondria in several respects. First, both organelles have an outer and an inner membrane with different permeability characteristics (**Figure 13.3**). As with mitochondria, the outer membrane of chloroplasts is highly permeable, and the inner membrane possesses specialized carrier molecules that regulate molecular traffic. Second, the chloroplast inner membrane encloses an inner space, referred to as the **stroma**, that resembles the mitochondrial matrix. The stroma possesses a variety of enzymes (e.g., those that catalyze the light-independent reactions and starch synthesis), DNA, and ribosomes.

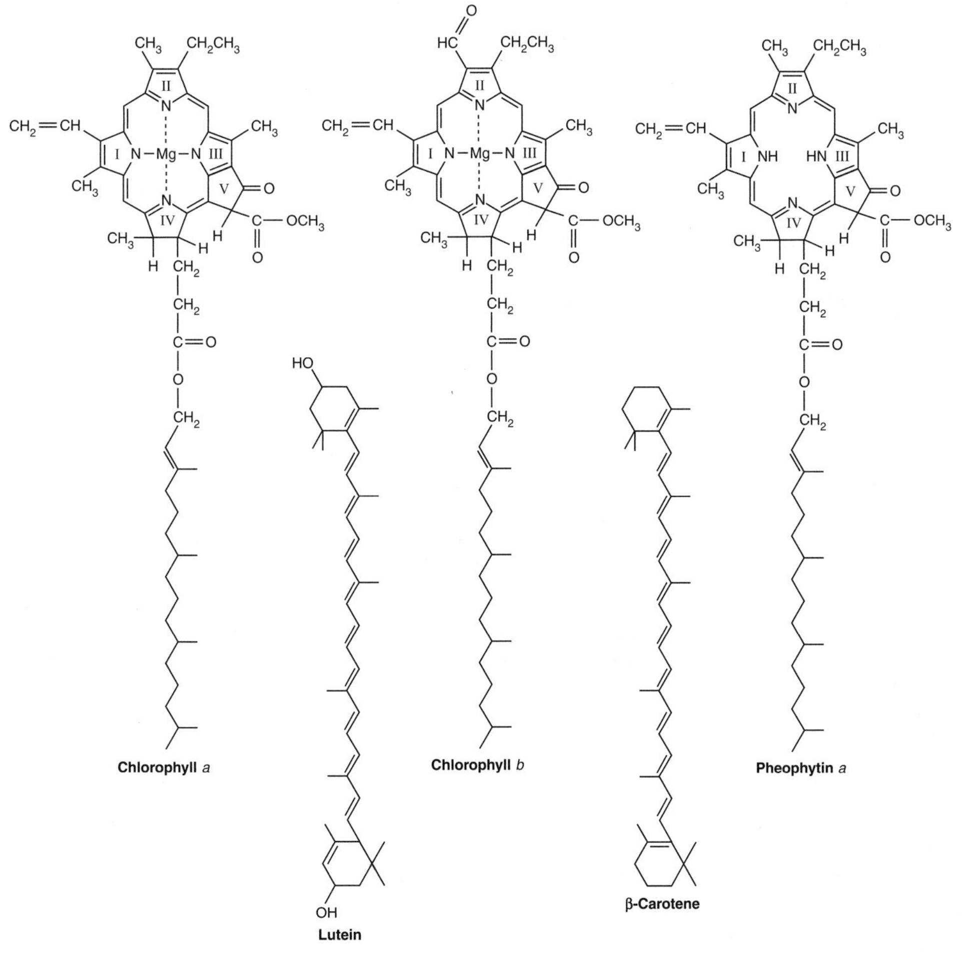

Chlorophyll *a* Lutein Chlorophyll *b* β-Carotene Pheophytin *a*

FIGURE 13.2

Pigment Molecules Used in Photosynthesis

Chlorophylls a and b are found in almost all oxygenic photosynthesizing organisms. They possess a complex cyclic structure (called a porphyrin) with a Mg (II) ion at its center. Chlorophyll a possesses a methyl group attached to ring II of the porphyrin, whereas chlorophyll b has an aldehyde group attached to the same site. Pheophytin a is similar in its structure to chlorophyll a. The Mg (II) ion is replaced by two protons. Chlorophylls a and b and pheophytin a all possess a phytol chain esterified to the porphyrin. The phytol chain extends into and anchors the molecule to the membrane. Lutein and β-carotene are the most abundant carotenoids in thylakoid membranes.

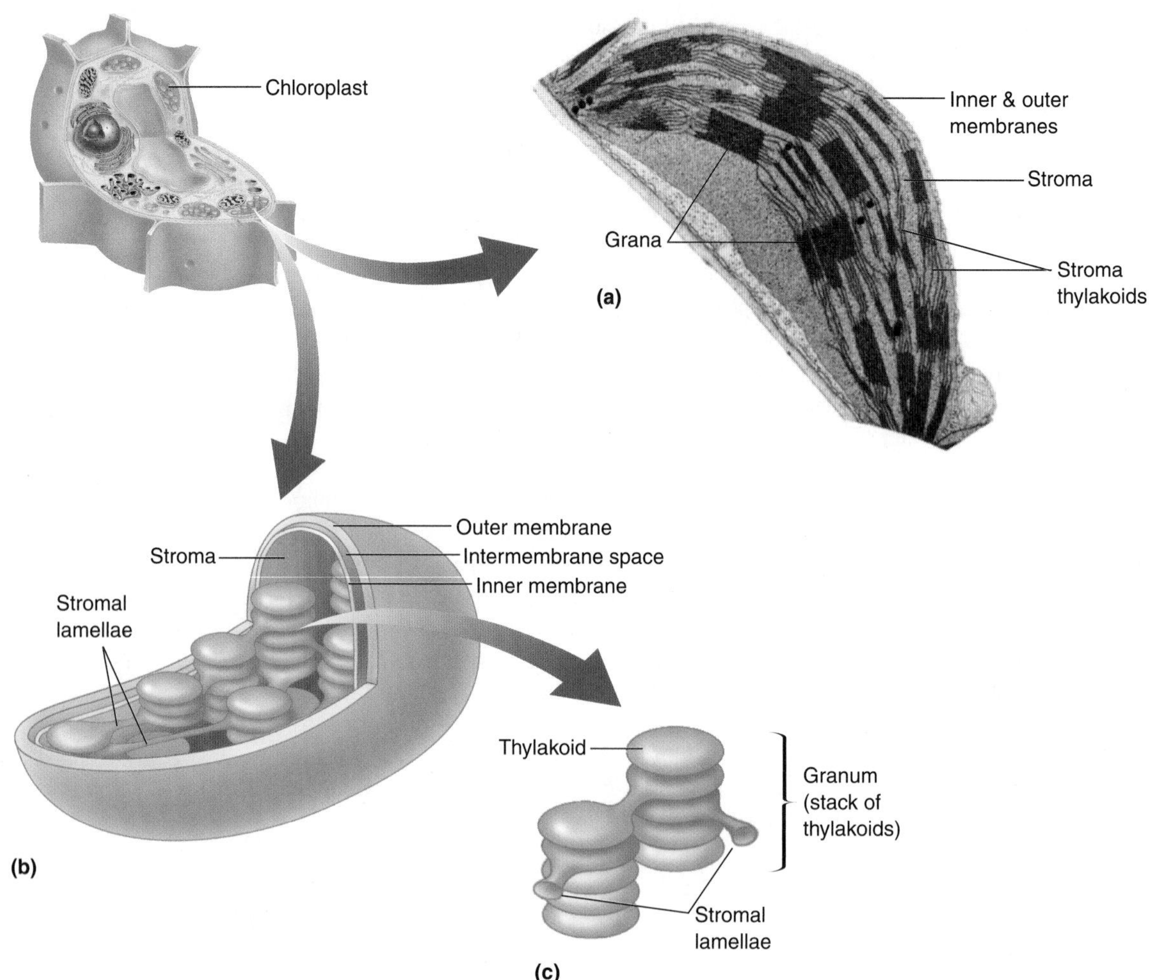

FIGURE 13.3
Chloroplast Structure

Chloroplasts have inner and outer membranes. A third membrane forms within the aqueous, enzyme-rich stroma into flattened sacs called thylakoids. A stack of thylakoids is called a granum. Unstacked, connecting thylakoid membrane is referred to as stromal lamellae. (a) Electron micrograph of a chloroplast. (b) Diagrammatic view of a chloroplast. (c) Cutaway view of a granum.

There are also notable differences between the organelles. For example, chloroplasts are substantially larger than plant mitochondria, rod-shape structures approximately 1500 nm long and 500 nm wide. Chloroplasts are spheroidal, from 4000 to 6000 nm long and approximately 2000 nm wide. In addition, chloroplasts possess a distinct third membrane, referred to as the **thylakoid membrane** that is folded into a series of disc-like vesicular structures called **grana**. Each ***granum*** is a stack of several flattened vesicles. The internal compartment created by the formation of grana is referred to as the **thylakoid lumen** (or *space*). The thylakoid membranes that interconnect the grana are called **stromal lamellae**. Adjacent layers of membrane that fit closely together within each granum are said to be appressed. The stromal lamellae are nonappressed.

The pigments and proteins responsible for the light-dependent reactions of photosynthesis are found within thylakoid membrane (**Figure 13.4**). Most of these molecules are organized into the working units of photosynthesis.

1. **Photosystem I**. PSI (**Figure 13.5**) is a large membrane-spanning multisubunit complex that energizes and transfers the electrons that eventually are donated to $NADP^+$. The essential role of PSI, the donation of energized

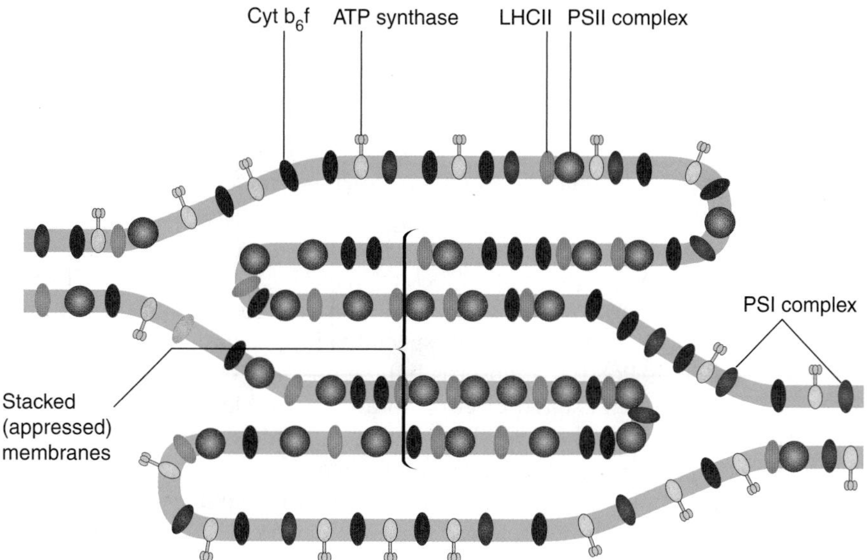

FIGURE 13.4

The Working Units of Photosynthesis

PSI complexes are most abundant in the unstacked stromal lamellae. In contrast, PSII complexes are located primarily in the stacked regions of thylakoid membrane. Cytochrome b_6f is found in both areas of thylakoid membrane. The ATP synthase is found only in thylakoid membrane that is directly in contact with the stroma.

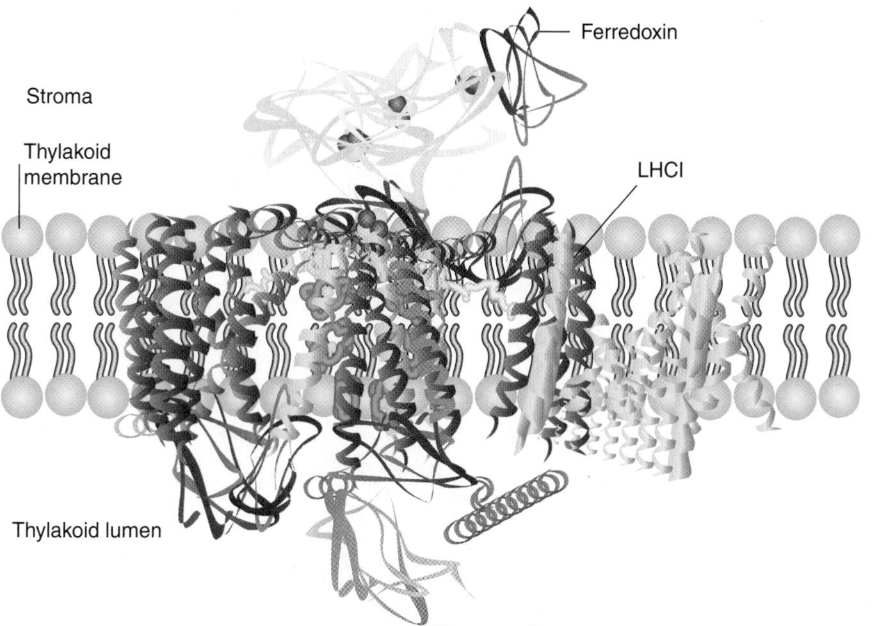

FIGURE 13.5

Structure of Photosystem I in Plants

PSI is a multisubunit protein complex composed of a reaction center formed largely by a PsaA and PsaB heterodimer and the peripheral light-harvesting complex (LHCI), shown here as green ribbon structures. Two other electron transport proteins are also shown: plastocyanin (orange ribbon structure) and ferredoxin (dark pink ribbon structure).

electrons to a series of electron carriers within thylakoid membrane, is performed by two special chlorophyll a molecules that reside within the reaction center. These molecules, referred to as a *special pair*, are located in the core complex of PSI, the PsaAB dimer. The special pair in PSI is sometimes referred to as P700 because it absorbs light at 700 nm. In addition to the special pair, the PsaAB dimer contains a series of single electron carriers: A_0, A_1, and F_x. A_0 is a specific chlorophyll a molecule that accepts an energized electron from P700 and transfers it to A_1, which has been identified as phylloquinone (vitamin K_1). The electron is then transferred from A_1 to a series of iron-sulfur clusters (F_x, F_A, and F_B). Ultimately, the electron is donated to $NADP^+$ to form NADPH. PSI also contains a large number of chlorophyll a molecules other than the special pair, as well as chlorophyll b and carotenoids that act as antenna pigments. **Antenna pigments** absorb light energy and transfer it to the reaction center. Additional antenna pigment molecules in a peripheral light-harvesting complex (LHCI) associated with PSI also contribute to efficient absorption of light energy. This phenomenon is described more fully in Section 13.2. Most PSI complexes are located in nonappressed thylakoid membrane, that is, membrane that is directly exposed to the stroma.

2. **Photosystem II**. The function of PSII is to oxidize water molecules and donate energized electrons to electron carriers that eventually reduce PSI. PSII is a large membrane-spanning protein-pigment complex located in appressed grana membrane (**Figure 13.6**). The most active form of PSII is a dimer. The PSII reaction center is a protein-pigment complex composed of two polypeptide subunits known as D_1 (33 kDa) and D_2 (31 kDa) (the D_1/D_2 dimer), two core subunits, CP47 and CP43, and cytochrome b_{559}. The D_1/D_2 complex binds a special pair of chlorophyll a molecules (referred to as P680) that absorbs light at 680 nm. Once it absorbs light, P680 transfers an excited electron to a series of electron acceptors and eventually to *plastoquinone* (PQ or Q), a molecule similar to ubiquinone. The electron donated by the reaction center is replaced by the oxygen-evolving complex (OEC), also known as the water-splitting complex, and a tyrosine residue, often referred to as Y_z, located on D_1. The water splitting site is a cubelike Mn_4CaO_5 cluster surrounded by amino acid side chains of D_1 and CP43 that form direct ligands to the metals. Several hundred antenna pigment molecules are also associated with the reaction center. The preponderance of accessory pigment molecules and several proteins belong to a detachable unit referred to as *light-harvesting-complex II (LHCII)*. LHCII is a trimer of light-harvesting proteins, each of which binds 12 to 14 chlorophyll a and b molecules as well as several carotenoid molecules. In plants, LHCII trimers are tightly packed in appressed grana membranes.

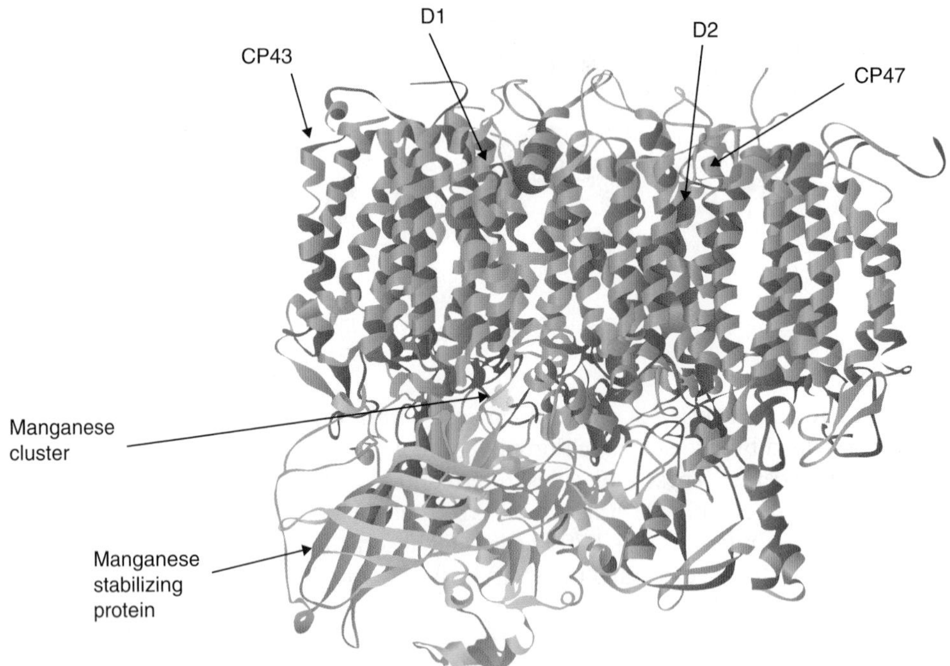

FIGURE 13.6

Photosystem II Monomer Structure

The PSII monomer is composed of about 20 subunits. The most important of these are the reaction pair subunits D_1 and D_2 and CP43 and CP47, antenna subunits that bind chlorophyll (not shown). Manganese stabilizing protein is a peripheral membrane protein component of PSII that maintains the efficiency of oxygen production by stabilizing the manganese cluster.

3. **Cytochrome b_6f complex**. Cytochrome b_6f complex, found throughout the thylakoid membrane, is similar in structure and function to the cytochrome bc_1 complex (**Figure 10.7**) in mitochondrial inner membrane. The cytochrome b_6f complex plays a critical role in the transfer of photoexcited electrons from PSII to PSI. An iron-sulfur protein in the complex, referred to as the Rieske Fe-S protein, accepts electrons from the membrane-soluble electron carrier PQ and donates them to a small water-soluble, copper-containing protein called plastocyanin (PC). The mechanism (**Figure 13.7**) that transports electrons from PQH_2 through the cytochrome b_6f complex is similar to the Q cycle in mitochondria.
4. **ATP synthase**. The chloroplast ATP synthase (**Figure 13.8**), also referred to as *CF_0CF_1ATP synthase*, is structurally and functionally similar to the

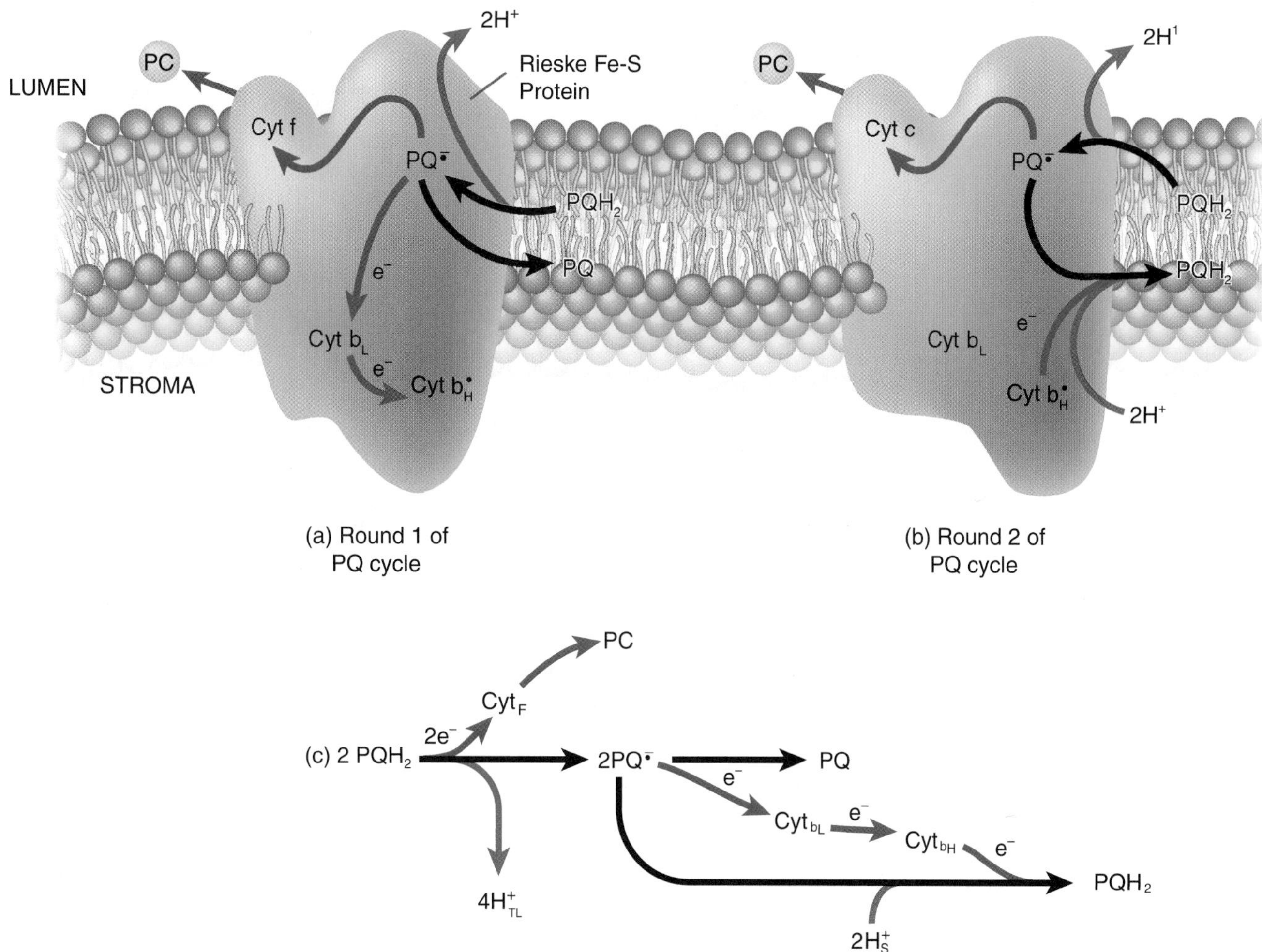

FIGURE 13.7

Electron Transport through Cytochrome b_6f Complex

Two molecules of plastoquinol (PQH_2) are oxidized sequentially to supply two electrons (e^-) to plastocyanin (PC). PC, a water-soluble copper-containing protein, then transfers each electron to the reaction center (P700) of PSI. The first electron from each PQH_2 is transferred to the Rieske iron-sulfur protein and then to cyt f and PC as two protons each are transferred to the thylakoid lumen (TL). (a) An electron from one $PQ^{\bar{\cdot}}$ is transferred to cyt b_L and then to cyt b_H. PQ is released into the thylakoid membrane. (b) The second $PQ^{\bar{\cdot}}$ picks up the electron from cyt b_H and 2 protons from the stroma (S) to form PQH_2. (c) The summary reaction shows that 2 PQH_2 molecules enter cyt b_6f and PQ and PQH_2 are released from the protein complex. (Black arrows = reaction arrows; red arrows = proton transfer; blue arrows = electron flow.)

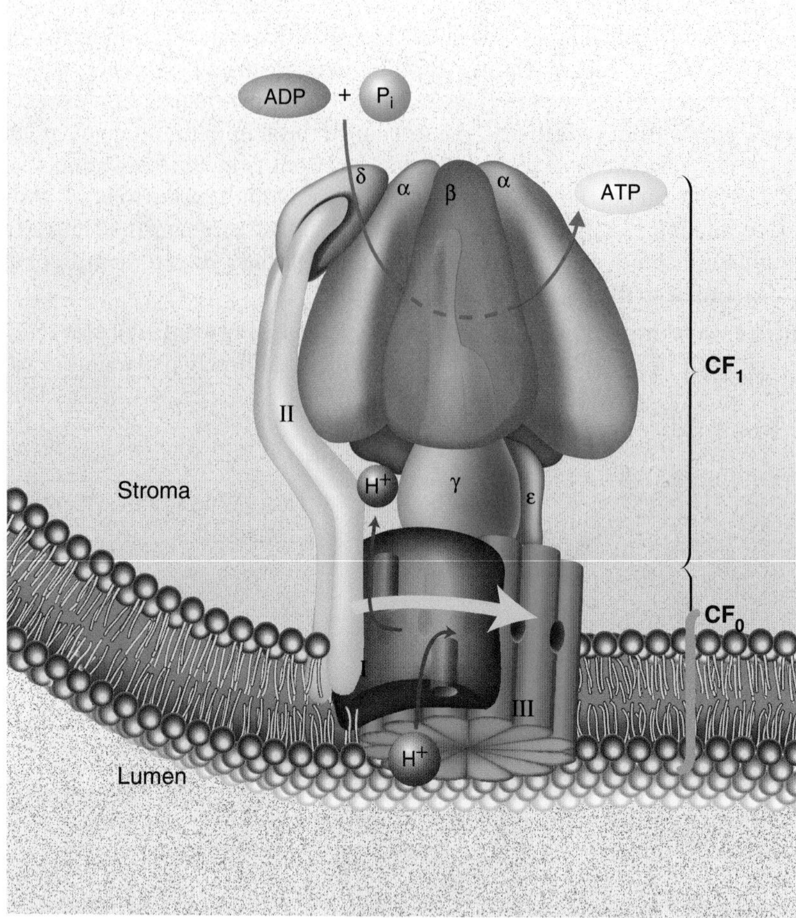

FIGURE 13.8

Diagrammatic View of the Chloroplast ATP Synthase

The ATP synthase is composed of two components: an integral membrane protein complex (CF_0) that contains a proton pore and an extrinsic protein complex (CF_1) that synthesizes ATP. CF_0 contains four different types of subunits: I, II, III, and IV. The proton pore is composed of 12–14 copies of subunit III. Subunit IV (not shown) binds CF_0 to CF_1. CF_1 consists of five different subunits: α, β, γ, δ, and ε. The mitochondrial ATP synthase has a similar structure (Chapter 10).

mitochondrial ATP synthase. The CF_0 component is a membrane-spanning protein complex that contains a proton-conducting channel. The CF_1 head piece, which projects into the stroma, possesses an ATP-synthesizing activity. A transmembrane proton gradient produced during light-driven electron transport causes the rotation of the CF_0 proton channel complex, which in turn drives ADP phosphorylation. The synthesis of each ATP molecule requires pumping approximately four protons across the membrane into the thylakoid space. Thylakoid membrane that is directly in contact with the stroma contains the ATP synthase.

KEY CONCEPTS

- In chloroplasts a double membrane encloses an inner space called the stroma. The stroma contains the enzymes that catalyze the light-independent reactions of photosynthesis. The third membrane forms into flattened sacs called thylakoids.
- Thylakoid membrane contains the pigments and proteins of the light-dependent reactions.

QUESTION 13.1

Draw a sketch of a chloroplast and indicate where each of the following is located: CF_0CF_1, P700, P680, and Calvin cycle reactions. Describe the function of each.

13.2 LIGHT

The sun emits energy in the form of electromagnetic radiation, which propagates through space as waves, some of which impinge on Earth. Visible light, the energy source that drives photosynthesis, occupies a small part of the electromagnetic radiation spectrum (**Figure 13.9**). Of the approximately 178,000 TW of solar energy that reach Earth per year, only a fraction (100 TW) are captured by photosynthetic organisms. Many of the properties of light are explained by its wave behavior (**Figure 13.10**). Energy waves are described by the following terms:

1. **Wavelength**. Wavelength (λ) is the distance from the crest of one wave to the crest of the next wave.
2. **Amplitude**. Amplitude (a) is the height of a wave. The intensity of electromagnetic radiation (e.g., the brightness of light) is proportional to a^2.
3. **Frequency**. Frequency (v) is the number of waves that pass a point in space per second.

For each type of radiation the wavelength multiplied by the frequency equals the velocity (c) of the radiation:

$$\lambda v = c$$

This equation rearranges to

$$\lambda = c/v$$

The wavelength therefore depends on both the frequency and the velocity of the wave.

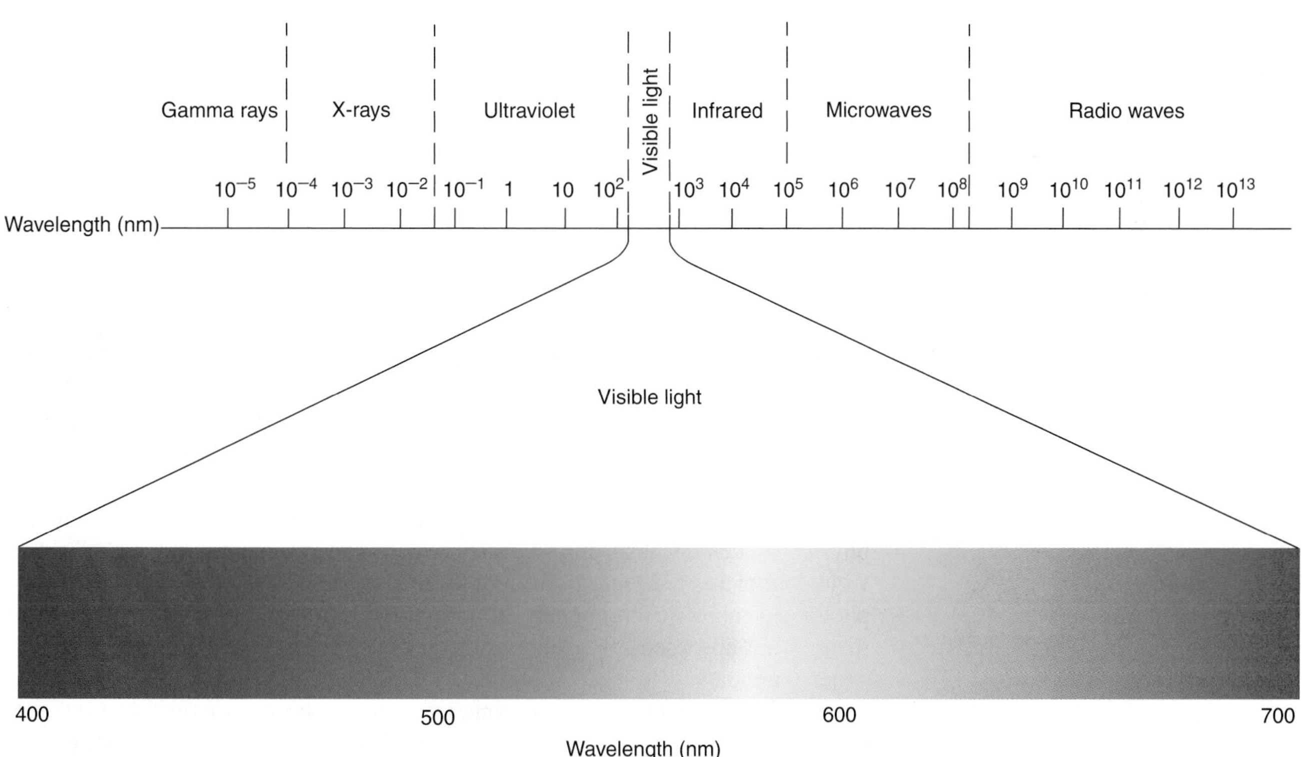

FIGURE 13.9

The Electromagnetic Spectrum

Gamma rays, which have short wavelengths, have high energy. At the other end of the spectrum, the radio waves (long wavelengths) have low energy. Visible light is the portion of the spectrum to which the visual pigments in the retina of the eye are sensitive. Pigment molecules in chloroplasts are also sensitive to portions of the visible spectrum. UV light near the visible spectrum is subdivided into UVA (400–315 nm), UVB (315–280 nm), and UVC (280–100 nm).

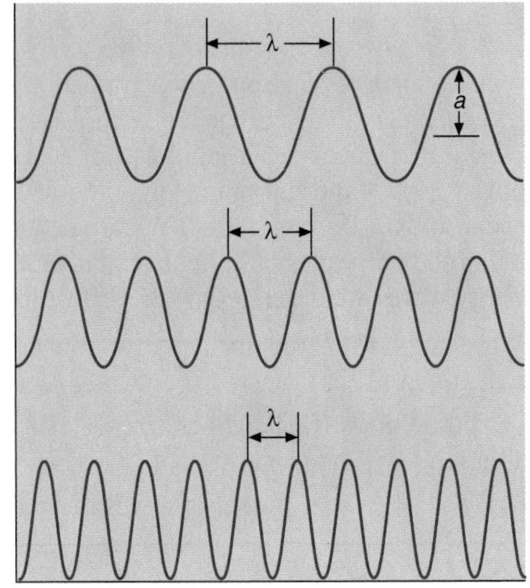

FIGURE 13.10
Properties of Waves

A wavelength (λ) is the distance between two consecutive peaks in a wave. The amplitude (*a*) or height of a wave is related to the intensity of electromagnetic radiation. Frequency is the number of waves that pass a point in space per second. Radiation with the shortest wavelength has the highest frequency.

The wavelengths of visible light range from 400 nm (violet light) to 700 nm (red light). In comparison, highly energetic X-rays and γ-rays have wavelengths that are 10^4 to 10^7 times shorter. On the other end of the spectrum are low-energy radio waves; these have wavelengths on the order of meters to kilometers.

QUESTION 13.2

Why do green light waves have less energy than blue light waves?

In addition to behaving like a wave, visible light (and other types of electromagnetic radiation) exhibits the properties of particles such as mass and acceleration. Einstein's observation that energy has mass, or $E = mc^2$, applies to the photon. When light interacts with matter, it does so in discrete packets of energy called photons. The energy (E) of a photon is proportional to the frequency of the radiation:

$$E = hv$$

where h is Planck's constant (6.63×10^{-34} J • s).

According to quantum theory, radiant energy can be absorbed or emitted only in specific quantities called quanta. When a molecule absorbs a quantum of energy, an electron is promoted from its ground state orbital (lowest energy level) to a higher energy state. For absorption to occur, the energy difference between the two energy states must exactly equal the energy of the absorbed photon. Complex molecules often absorb at several wavelengths. For example, chlorophyll produces an absorption spectrum with broad and multiple peaks (blue-violet region and red region). Both of these facts suggest that chlorophyll absorbs photons of many different energies with varying probabilities. The wavelengths that are not absorbed are visible to us, and so a chlorophyll solution (or a leaf) appears green.

Molecules that absorb electromagnetic energy have structural components called chromophores. Electrons in **chromophores** move easily to higher energy levels when energy is absorbed. Visible chromophores typically possess extended chains of conjugated double bonds and aromatic rings. For example, the anthocyanins, a class of water-soluble pigments, contain chromophores that protect plants from light-induced damage by absorbing blue-green and UV light. Chromophores undergo electronic transitions wherein an electron moves from a ground state occupied orbital to a higher-energy unoccupied orbital (**Figure 13.11**).

Visit the companion website at www.oup.com/us/mckee to read the Biochemistry in the Lab box on photosynthetic studies for a discussion of the absorption spectrum of chlorophyll and several other pigment molecules.

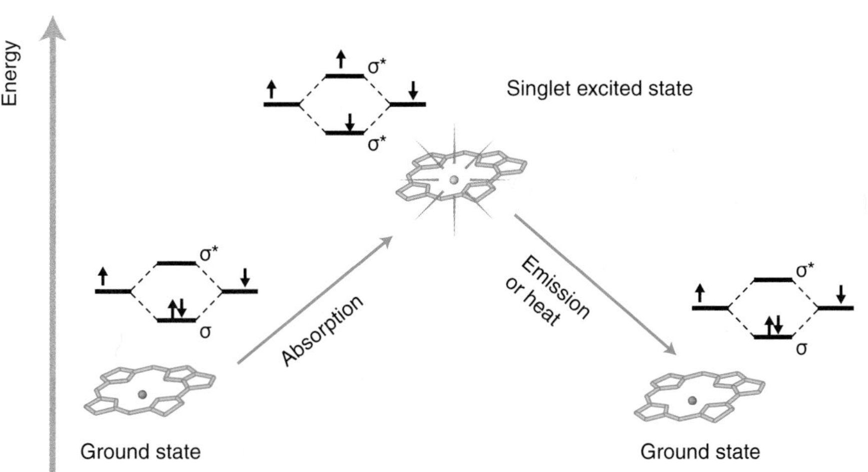

FIGURE 13.11

Excitation of a Chromophore

If a chromophore molecule (e.g., chlorophyll) absorbs a photon of visible light, its energy increases as a result of the movement of an electron to a higher molecular orbital. An excited molecule can return to its ground state by releasing energy as emission (fluorescence) or heat. The excited molecule can also donate its excitation energy to a neighbor molecule (excitation energy acceptor) or give away its electron to an electron acceptor.

Molecules with a small number of conjugated double bonds or isolated double bonds absorb energy in the UV portion of the electromagnetic spectrum. The extensive conjugation of the chromophores of photosynthetic and accessory pigments allows electronic transitions to occur at longer wavelengths (lower energy) across the visible spectrum. The π electrons of these systems require less energy to make the transition into a higher-energy orbital.

Once excited, an electron can return to its ground state in several ways:

1. **Fluorescence**. In **fluorescence** a molecule's excited state decays as it emits a photon. Because the excited electron loses some energy initially by relaxing to a lower vibrational (energy) state, a transition resulting in the emission of a photon has lower energy than the photon originally absorbed. Fluorescent decay can occur as quickly as 10^{-15} s. (Although various chlorophylls absorb light energy throughout the visible spectrum, they emit only photons with low energy at or beyond the red end of the visible spectrum.)
2. **Resonance energy transfer**. In **resonance energy transfer**, the excitation energy is transferred to a neighboring chromophore through interaction between adjacent molecular orbitals. A chromophore whose absorption spectrum overlaps the emission spectrum of the target chromophore can absorb the quantum of energy released when that chromophore returns to its ground state.
3. **Oxidation-reduction**. An excited electron is transferred to a neighboring molecule. An excited electron occupies a normally unoccupied orbital and is bound less tightly than when it occupies a ground state orbital. A molecule with an excited electron is a strong reducing agent. It returns to its ground state by reducing another molecule.
4. **Radiationless decay**. The excited molecule decays to its ground state by converting the excitation energy into heat.

Of all these responses to energy absorption, the most important in photosynthesis are resonance energy transfer and oxidation-reduction. Resonance energy transfer plays a critical role in the harvesting of light energy by accessory pigment molecules (**Figure 13.12a**). Eventually, the energy absorbed and transferred by light-harvesting complexes reaches the special chlorophyll a molecules P700 and P680, in the reaction centers of PSI and PSII, respectively. When these molecules are excited, they are referred to as P700* and P680*. Both P700* and 680* are electron donors because they can easily lose an electron to a specific acceptor molecule. P700* passes an electron to the electron acceptor A_0, a chlorophyll a molecule, and P680* transfers its electron to a molecule of pheophytin a. The electron hole left in the oxidized P700 and P680 is filled by an electron from

KEY CONCEPTS

- The light energy absorbed by chromophores causes electrons to move to higher energy levels.
- In photosynthesis, it is energy absorption that drives electron flow.

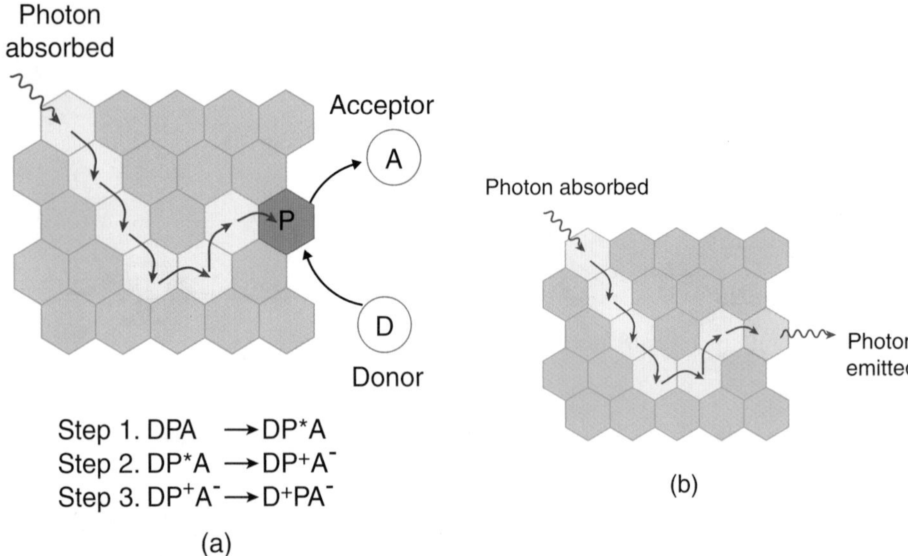

FIGURE 13.12

Energy Transfer in Photosystems

A photon absorbed by a chlorophyll molecule in the light-harvesting antenna promotes it to a singlet excited state. The excited chlorophyll molecule donates its energy to neighboring molecules by resonance energy transfer. The excitation randomly migrates through the antenna molecules (yellow hexagons) until it is trapped by the reaction center (dark green hexagon, P). (a) The excitation trap in the reaction center is a special chlorophyll molecule called the primary donor (P700 or P680), whose lowest excited state is lower than the antenna molecules. A molecule of the primary donor can be excited either by direct absorption of light energy or by transfer of the excitation from the nearest antenna molecule (step 1). The molecule of the primary donor in its excited state (P*) initiates the electron transfer in the reaction center by reducing the acceptor molecule (A) (step 2). The oxidized primary donor (P^+) extracts an electron from the nearest electron donor (D) (step 3). The electron from the reduced acceptor molecule (A^-) is transferred further along the chain of electron transfer carriers in the reaction center. (b) Under high-stress conditions such as high light intensity, reduced numbers of available reaction centers result in the loss of excitations through fluorescence.

a donor molecule. Plastocyanin and water play this role in PSI and PSII, respectively. Fluorescence also plays a role in photosynthesis when light absorption exceeds the capacity of the photosystems to transfer energy (**Figure 13.12b**). Then photons are reemitted by a protective mechanism.

QUESTION 13.3

Explain the observation that the different absorption spectra of antenna pigment molecules are different from those of P680 and P700.

13.3 LIGHT REACTIONS

In photosynthesis the **light reactions** are a mechanism by which electrons are energized and subsequently used in ATP and NADPH synthesis. Species that evolve O_2 require both PSI and PSII. Species that live without oxygen can use either PSI- or PSII-like complexes. Working in series, the two photosystems couple the light-driven oxidation of water molecules to the reduction of $NADP^+$. The overall reaction is

$$2\,NADP^+ + 2H_2O \rightleftharpoons 2\,NADPH + O_2 + 2H^+ \quad (1)$$

The standard reduction potentials for the half-reactions are

$$O_2 + 4e^- + 4H^+ \quad 2H_2O \qquad E^{\circ\prime} = +0.816\text{ V} \qquad (2)$$

and

$$NADP^+ + H^+ + 2e^- \quad NADPH \quad E^{\circ\prime} = -0.320\text{ V} \qquad (3)$$

Therefore, the coupled process has a standard redox potential of -1.136 V. The minimum free energy change for this process (calculated from $\Delta G^{\circ\prime} = -nF\,\Delta E^{\circ\prime}$; Section 9.1) is approximately 438 kJ (104.7 kcal) per mole of O_2 generated. In comparison, a mole of photons of 700-nm light provides approximately 170 kJ (40.6 kcal). Experimental observations have revealed that the absorption of 8 or more photons (i.e., 2 photons per electron) is required for each O_2 generated. Consequently, a total of 1360 kJ (325 kcal) (i.e., 8 times 170 kJ) is absorbed for each mole of O_2 produced. This energy is more than sufficient to account for reducing $NADP^+$ and to establish the proton gradient for ATP synthesis.

The process of light-driven photosynthesis begins with the excitation of PSII by light energy. One electron at a time is transferred to a chain of electron carriers that connects the two photosystems. As electrons are transferred from PSII to PSI, protons are pumped across the thylakoid membrane from the stroma into the thylakoid space. ATP is synthesized as protons flow back into the stroma through the ATP synthase. When P700 absorbs an additional photon, it releases an energized electron. This electron is immediately

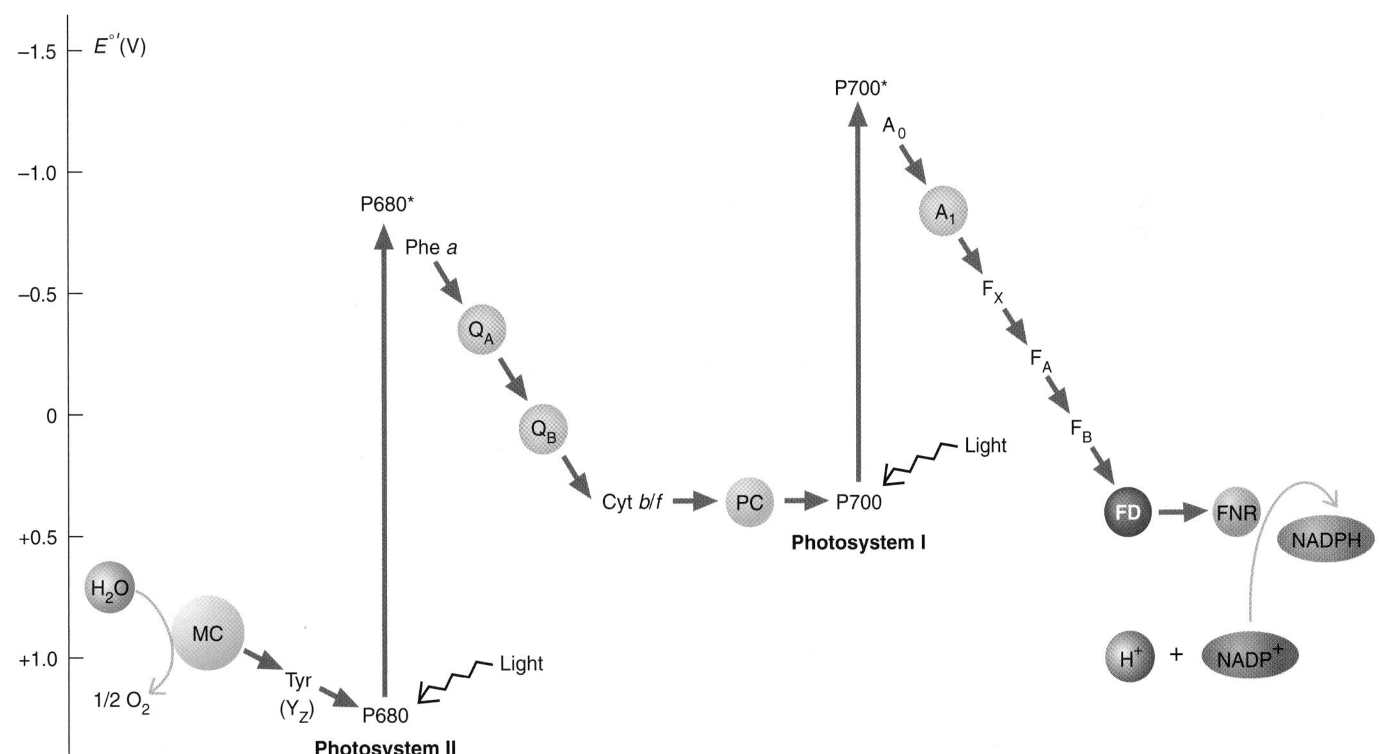

FIGURE 13.13
The Z Scheme

The flow of electrons from photosystem II to photosystem I drives the transport of protons into the thylakoid lumen. Electron transfer through the iron-sulfur proteins Fe_A and Fe_B is not understood. The $E^{\circ\prime}$ values are approximate. (MC = manganese cluster, tyr (Y_z) = tyr^{161} in D_1, PC = plastocyanin, A_0 = chlorophyll a, A_1 = phylloquinone, F_X, F_A, and F_B = a series of iron-sulfur clusters, FD = ferredoxin, FNR = ferredoxin-NADP oxidoreductase)

replaced by an electron provided by PSII. The newly energized PSI electron is passed through a series of iron-sulfur proteins and a flavoprotein to $NADP^+$, the final electron acceptor. This entire sequence, referred to as the **Z scheme**, is outlined in **Figure 13.13**.

Photosystem II and Water Oxidation

When LHCII absorbs a photon, its energy is transferred to P680 in PSII; the newly energized electron is ejected and subsequently donated to *pheophytin a* (**Figure 13.14**), a molecule similar to chlorophyll in its structure. Reduced pheophytin a passes this electron to a chain of two electron carriers Q_A and Q_B (**Figure 13.14**). Although both molecules are plastoquinones, they perform different functions in PSII. Q_A is tightly bound to the protein. This is a single electron carrier and it never binds protons. Q_A transfers its electron to Q_B, which is loosely bound to the protein and can be doubly reduced receiving its electrons from Q_A one at a time upon binding two stromal protons. Reaction of a double reduction of Q_B is shown below:

$$\underset{Q_B}{\text{H}_3\text{C},\ \text{H}_3\text{C},\ \text{H-quinone}-[\text{CH}_2-\text{CH}=\text{C}(\text{CH}_3)-\text{CH}_2]_9-\text{H}} \underset{}{\overset{2\,H^+,\ 2\,e^-}{\rightleftharpoons}} \underset{Q_BH_2}{\text{H}_3\text{C},\ \text{H}_3\text{C},\ \text{H-hydroquinone (OH, OH)}-[\text{CH}_2-\text{CH}=\text{C}(\text{CH}_3)-\text{CH}_2]_9-\text{H}} \tag{4}$$

The reduced Q_B (plastoquinol, Q_BH_2) is then released to the membrane pool of plastoquinones that donate electrons to cytochrome b_6f complex forming a transmembrane proton gradient.

An *OEC*, composed in part of the Mn_4CaO_5 cluster on the lumenal side of PSII and the tyrosine residue located on D_1, is responsible for the transfer of electrons from H_2O to oxidized P680 ($P680^+$). Recall that the excited state of P680 (P680*) reduces pheophytin a, thereby resulting in the formation of $P680^+$. The very high redox potential of this ion (+1.25 V) enables it to oxidize the tyrosine residue Y_z in D_1 and subsequently water. Tyrosine is effective in electron transfer because the tyrosyl radical formed is resonance stabilized.

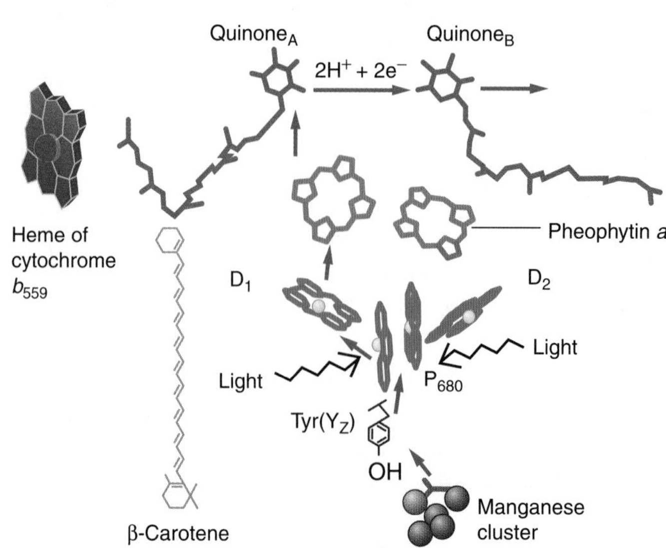

FIGURE 13.14
Photosystem II Electron Transport

Arrows indicate the electron transport pathway in PSII. A photon of light energizes an electron in P680. The electron donated by P680 to pheophytin a is then transferred to the quinones. Note that the manganese cluster and the Tyr(Y_Z) side chain replace the electron in oxidized P680. Cytochrome b_{559} plays a role in PSII photodamage protection.

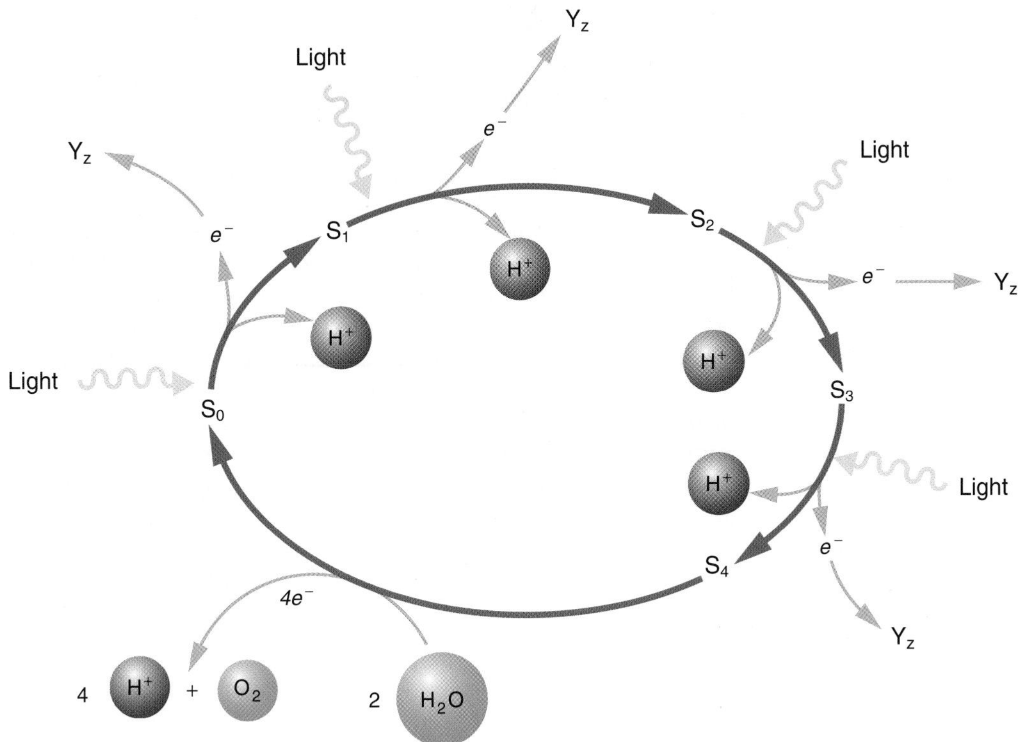

FIGURE 13.15

The Water-Oxidizing Clock

The absorption of four photons causes the abstraction of four electrons and four protons from two water molecules to yield O_2. The O_2-evolving complex has five oxidation states (S_0, S_1, S_2, S_3, and S_4), which represent different oxidation states of the Mn cluster. The sequential absorption of each of four photons drives the removal of an electron from a water molecule. Each electron is donated first to Y_Z, the tyrosine residue in D_1, and then to $P680^+$. Four protons created by the oxidation of the two water molecules are released into the thylakoid lumen.

The evolution of one O_2 requires splitting two H_2O molecules, which releases 4 protons and 4 electrons. Experimental evidence indicates that H_2O is converted to O_2 by a mechanism referred to as the *water-oxidizing clock* (**Figure 13.15**). The O_2-evolving complex has five oxidation states: S_0, S_1, S_2, S_3, and S_4. In the OEC, S_0 is the most reduced state and S_4 is the most oxidized state. It is now believed that the Mn_4CaO_5 cluster, near the PSII reaction center, is responsible for these transitions. Oxygen–oxygen bond formation is the rate-limiting step in water oxidation. The oxygen-evolving complex also abstracts protons from H_2O as it cycles through the oxidation states. The protons are released into the thylakoid lumen, where they contribute to the pH gradient that drives ATP synthesis.

QUESTION 13.4

Excessive amounts of light can depress photosynthesis. Recent research indicates that PSII is extremely vulnerable to light damage. Plants often survive this damage because they possess efficient repair systems. It now appears that cells delete and resynthesize damaged components and recycle undamaged ones. For example, the D_1 polypeptide, apparently the most vulnerable component of PSII, is rapidly replaced after it is damaged. Review the role of PSII and suggest the proximate cause of light-induced damage of D_1. [*Hint*: The D_1/D_2 dimer binds two molecules of β-carotene.]

Photosystem I and NADPH Synthesis

The absorption of a photon by P700 leads to the release of an energized electron that is passed through a series of electron carriers (**Figure 13.16**). The first electron carrier is a chlorophyll a molecule (A_0). As the electron is donated sequentially to phylloquinone (A_1) and to several iron-sulfur proteins (the last of which is ferredoxin), it is moved from the lumenal surface of the thylakoid membrane to its stromal surface. Ferredoxin, a mobile, water-soluble protein, then donates each electron to a flavoprotein called ferredoxin-NADP oxidoreductase (FNR). The flavoprotein uses a total of two electrons and a stromal proton to reduce $NADP^+$ to NADPH. The transfer of electrons from ferredoxin to $NADP^+$ is referred to as the *noncyclic electron transport pathway* (**Figure 13.17**). In this pathway the absorption of eight photons yields an ATP:NADPH ratio of 3:2. In *cyclic electron transport* (**Figure 13.18**), reduced ferredoxin donates its electrons to plastoquinone, which then passes them to the cyt b_f complex, plastocyanin, and eventually P700. In this process, which typically occurs when a chloroplast has a high NADPH:$NADP^+$ ratio, no NADPH is produced. Instead, electron transport results in the transfer of additional protons across the thylakoid membrane. As a result, additional molecules of ATP are synthesized.

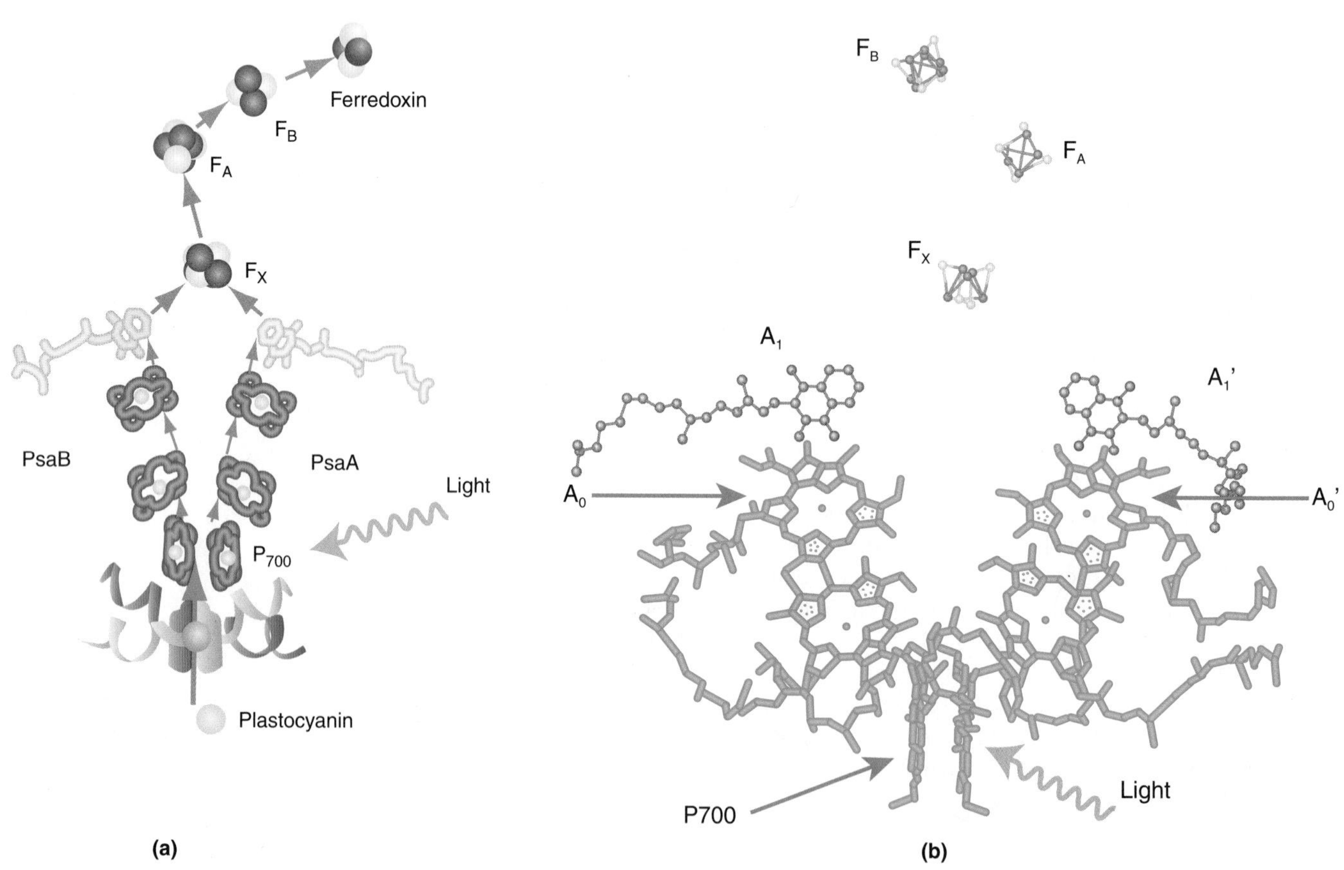

FIGURE 13.16

Two Views of Electron Transport in Photosystem I

(a) Electron carriers in the reaction center are bound to the PsaA and PsaB protein subunits of PSI and are in the same positions as **Figure 13.5**. (b) Excited P700 transfers an electron to the acceptor A_0, a chlorophyll a molecule, which then passes the electron to an electron carrier A_1. This is a molecule of plastoquinone that transfers the electron to a series of iron-sulfur clusters (F_X, F_A, and F_B) and finally to ferredoxin. Note that there are two potential branches of the electron transfer from the reaction center toward F_X. Oxidized P700 obtains its missing electron from the nearby plastocyanin in a process that is facilitated by two tryptophan residues in the two polypeptide segments shown in (a) as pink and blue ribbon structures.

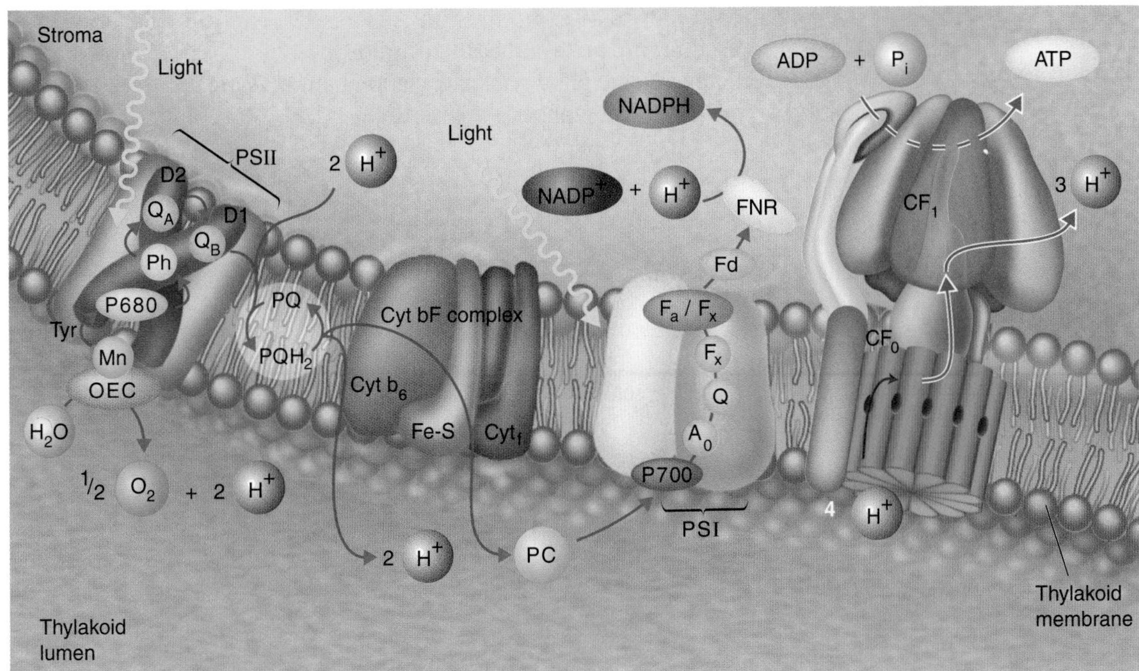

FIGURE 13.17

Membrane Organization of the Light Reactions in Chloroplasts: The Noncyclic Electron Transport Chain and the ATP Synthase Complex

As two electrons move from each water molecule to $NADP^+$ (blue arrows), about two H^+ are pumped from the stroma into the thylakoid lumen. Two additional H^+ are generated within the lumen by the oxygen-evolving complex. The flow of protons through the proton pore in CF_0 drives the synthesis of ATP in CF_1. (OEC = oxygen evolving complex, Ph = pheophytin; Fd = ferredoxin; FNR = the flavoprotein ferredoxin-NADP oxidoreductase)

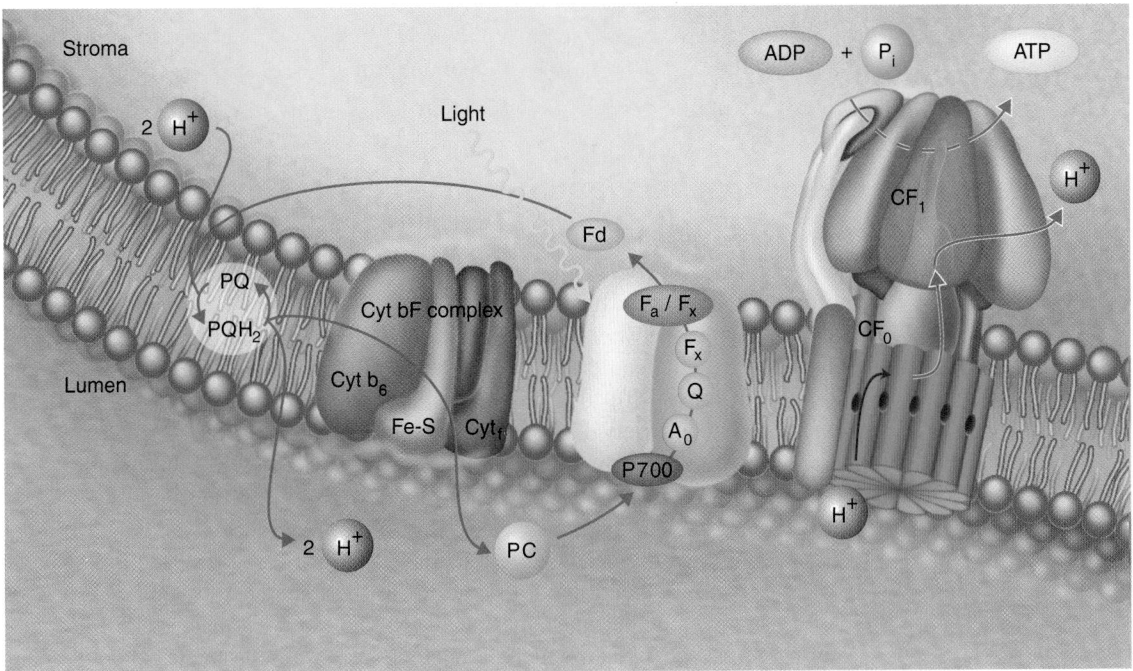

FIGURE 13.18

The Cyclic Electron Transport Pathway

A Q cycle similar to that observed in the electron transport pathway that links PSI and PSII is believed to be responsible for pumping two H^+ across the thylakoid membrane for each electron transported. The proton flow drives ATP synthesis. No NADPH is produced.

KEY CONCEPTS

- Eukaryotic photosynthesizing cells possess two photosystems, PSI and PSII, which are connected in series in a mechanism referred to as the Z scheme.
- The water-oxidizing clock component of PSII generates O_2.
- The protons are used in the synthesis of ATP in a chemiosmotic mechanism.
- PSI is responsible for the synthesis of NADPH.

The Calvin cycle (p. 493) requires an ATP:NADPH ratio of 3:2. However, ATP is also used for processes other than carbohydrate synthesis. Consequently, both noncyclic and cyclic photophosphorylation pathways are required for sufficient ATP synthesis during photosynthesis.

WORKED PROBLEM 13.1

Calculate $\Delta G^{\circ\prime}$ for the four-electron oxidation of H_2O by $NADP^+$ in the light reactions.

SOLUTION

The overall reaction is

$$2\,H_2O + 2\,NADP^+ \rightarrow O_2 + 2\,NADPH + 2\,H^+$$

The reduction potentials ($\Delta E^{\circ\prime}$) for the two half reactions are

$$1/2\,O_2 + 2\,H^+ + 2e^- \rightarrow H_2O\ (\Delta E^{\circ\prime} = +\,0.82\text{ V})$$

$$NADP^+ + H^+ + 2e^- \rightarrow NADPH + H^+\ (\Delta E^{\circ\prime} = -0.32\text{ V})$$

$\Delta G^{\circ\prime}$ is calculated using the equation $\Delta G = -nF\Delta E^{\circ\prime}$
Substituting the $\Delta E^{\circ\prime}$ values for the two half reactions

$$\begin{aligned}\Delta G^{\circ\prime} &= -4\,(96.5\text{ kJ/V}\cdot\text{mol})\,[-0.32\text{ V}-(0.82\text{ V})]\\ &= (386\text{ kJ/V}\cdot\text{mol})\,(-1.14\text{ V})\\ &= -440\text{ kJ/mol}\end{aligned}$$

■

QUESTION 13.5

Describe the role of each of the following molecules in photosynthesis:

a. plastocyanin
b. β-carotene
c. ferredoxin
d. plastoquinone
e. pheophytin a
f. lutein

Photophosphorylation

During photosynthesis light energy captured by an organism's photosystems is transduced into ATP phosphate bond energy. This conversion is referred to as **photophosphorylation**. It is apparent from the preceding discussions that there are many similarities between mitochondrial and chloroplast ATP synthesis. For example, many of the same molecules and terms that are encountered in aerobic respiration (Chapter 10) are also relevant to discussions of photosynthesis. Although there are a variety of differences between aerobic respiration and photosynthesis, the essential difference between the two processes is the conversion of light energy into redox energy by chloroplasts. (Recall that mitochondria produce redox energy by extracting high-energy electrons from food molecules.) Another critical difference involves the permeability characteristics of mitochondrial inner membrane and thylakoid membrane. In contrast to the inner membrane of mitochondria, the thylakoid membrane is permeable to Mg^{2+} and Cl^-. Therefore, Mg^{2+} and Cl^- move across the thylakoid membrane, thereby dissipating electrical potential as protons are transported across the membrane during the light reaction. The electrochemical gradient across the thylakoid membrane that drives ATP synthesis therefore consists mainly of a proton gradient that may be as great as 3.5 pH units.

Experimental measurements of H^+:ATP ratios indicate that the movement across the thylakoid membrane of about 12 protons in noncyclic photophosphorylation yields three molecules of ATP. The synthesis of these ATPs is made possible by the absorption of eight photons, one for each of the electrons from two water molecules. Proton transport occurs as these electrons are transported down the non-cyclic electron transport system (**Figure 13.17**). (The measured difference between the proton:ATP ratio of chloroplasts (4 H^+:ATP) and mitochondria (3 H^+:ATP) is explained in part by a structural difference in the ATP synthase proton channels of the two organelles. More protons are required for a 360ϒ turn of the chloroplast C_0 proton channel complex because it consists of a larger number of subunits than does the mitochondrial proton channel complex.) In cyclic photophosphorylation, the pumping of eight protons by the cyt b_6f complex, as the result of the absorption of four photons, yields two molecules of ATP.

QUESTION 13.6

A variety of herbicides kill plants by inhibiting photosynthetic electron transport. Atrazine, a triazine herbicide, blocks electron transport between Q_A and Q_B in PSII. The compound 3-(3,4-dichlorophenyl)-1,1-dimethylurea (DCMU) also blocks electron flow between the two molecules of plastoquinone. Paraquat is a member of a family of compounds called bipyridylium herbicides. Paraquat is reduced by PSI but is easily reoxidized by O_2 in a process that produces superoxide and hydroxyl radicals. Plants die because their cell membranes are destroyed by radicals. Of the herbicides just discussed, determine which, if any, are most likely to be toxic to humans and other animals. What specific damage may occur?

13.4 THE LIGHT-INDEPENDENT REACTIONS

The incorporation of CO_2 into carbohydrate by eukaryotic photosynthesizing organisms, a process that occurs within chloroplast stroma, is often referred to as the **Calvin cycle**. Because the reactions of the Calvin cycle can occur without light if sufficient ATP and NADPH are supplied, they have often been called the *dark reactions*. The term is somewhat misleading, however. The Calvin cycle reactions typically occur only when the plant is illuminated because ATP and NADPH are produced by the light reactions. Therefore, **light-independent reactions** is a more appropriate term. Because of the types of reaction that occur in the Calvin cycle, it is also referred to as the *reductive pentose phosphate cycle* (RPP cycle) and the *photosynthetic carbon reduction cycle* (PCR cycle).

The Calvin Cycle

The net equation for the Calvin cycle (**Figure 13.19**) is

$$3CO_2 + 6\ NADPH + 9\ ATP \rightarrow \text{glyceraldehyde-3-phosphate} + 6\ NADP^+ + 9\ ADP + 8\ P_i \quad (5)$$

For every three molecules of CO_2 that are incorporated into carbohydrate molecules, there is a net gain of one molecule of glyceraldehyde-3-phosphate. The fixation of six CO_2 into glucose occurs at the expense of 12 NADPH and 18 ATP. The reactions of the cycle can be divided into three phases: carbon fixation, reduction, and regeneration.

Carbon fixation. **Carbon fixation**, the mechanism by which inorganic CO_2 is incorporated into organic molecules, consists of a single reaction. Ribulose-1,5-bisphosphate carboxylase (Rubisco) is a Mg^{2+}-requiring enzyme that catalyzes the carboxylation of ribulose-1,5-bisphosphate to form two molecules of glycerate-3-phosphate (reaction 1). The reaction mechanism is illustrated in **Figure 13.20**. Plants that produce glycerate-3-phosphate as the first stable product of photosynthesis are referred to as **C3 plants**. (Notable exceptions are described on pages 498–501.) Rubisco, composed of eight large (L 54 kDa) subunits and eight small (S 14 kDa) subunits, is the pacemaker enzyme of the Calvin

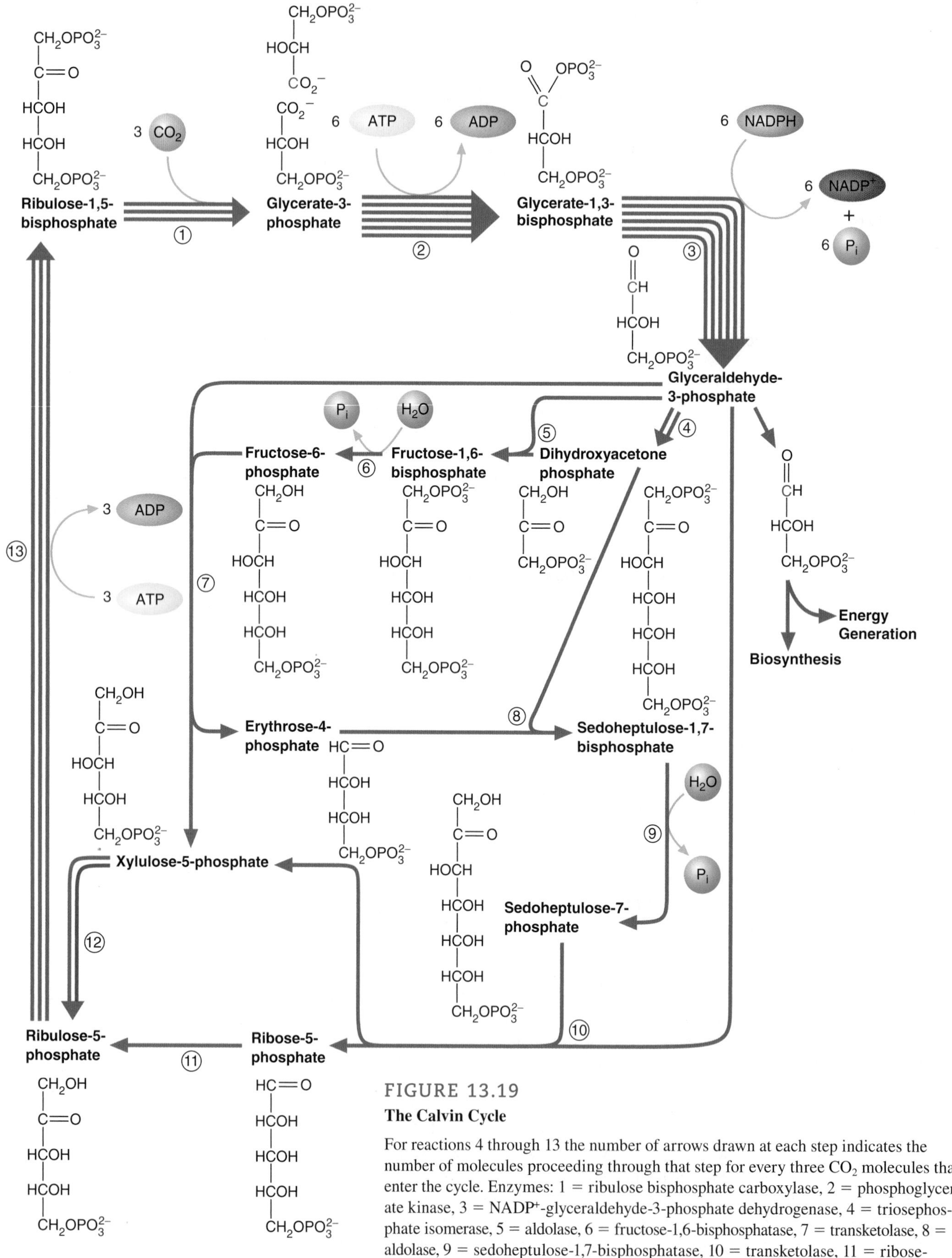

FIGURE 13.19

The Calvin Cycle

For reactions 4 through 13 the number of arrows drawn at each step indicates the number of molecules proceeding through that step for every three CO_2 molecules that enter the cycle. Enzymes: 1 = ribulose bisphosphate carboxylase, 2 = phosphoglycerate kinase, 3 = $NADP^+$-glyceraldehyde-3-phosphate dehydrogenase, 4 = triosephosphate isomerase, 5 = aldolase, 6 = fructose-1,6-bisphosphatase, 7 = transketolase, 8 = aldolase, 9 = sedoheptulose-1,7-bisphosphatase, 10 = transketolase, 11 = ribose-5-phosphate isomerase, 12 = ribulose-5-phosphate epimerase, 13 = ribulose-5-phosphate kinase.

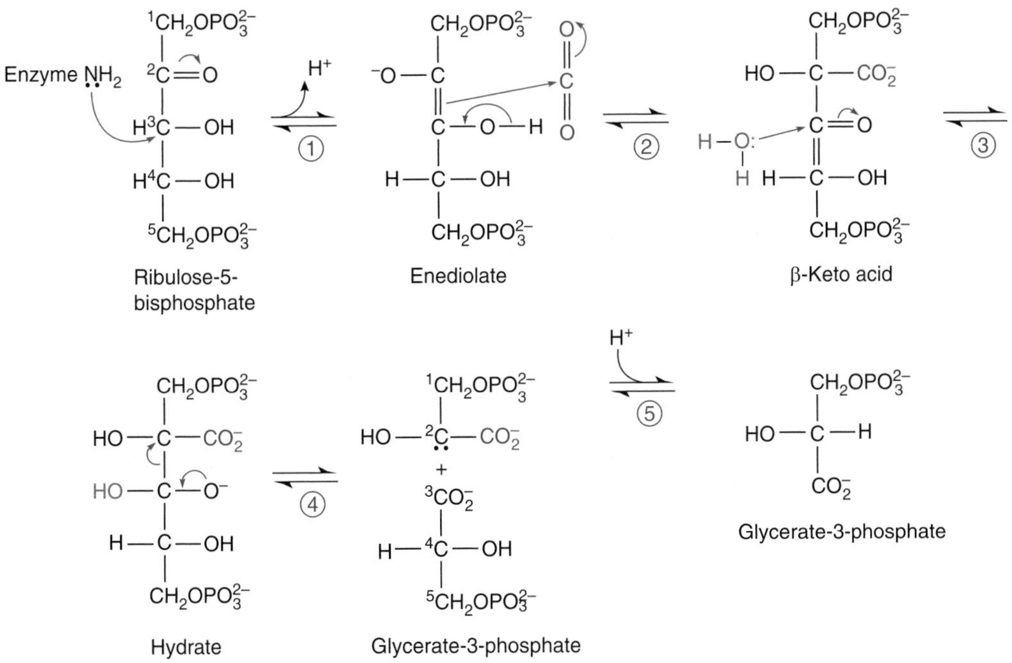

FIGURE 13.20

The Rubisco Carboxylation Mechanism

(1) The C-3 proton, made more acidic by its proximity to Mg^{2+}, is removed by a lysine side chain to yield an enediolate. (2) The enediolate attacks a CO_2 polarized by Mg^{2+} to form a six-carbon β-keto acid. (3) A water molecule then attacks the carbonyl carbon of the β-keto acid to yield a hydrated intermediate. (4) The hydrated intermediate is rapidly cleaved into two three-carbon products: a glycerate-3-phosphate anion and glycerate-3-phosphate. (5) The protonation of the anion by another lysine side chain (not shown) yields a second molecule of glycerate-3-phosphate.

cycle. Each L subunit contains an active site that binds substrate. The catalytic activity of the L subunits is enhanced by the S subunits. Because the CO_2 fixation reaction is extremely slow, plants compensate by producing a large number of copies of the enzyme, which often constitutes approximately half of a leaf's soluble protein. For this reason, rubisco is often described as the world's most abundant enzyme.

Reduction. In the reduction phase of the Calvin cycle, glycerate-3-phosphate is reduced to glyceraldehyde-3-phosphate. In the first of two reactions, six molecules of glycerate-3-phosphate are phosphorylated at the expense of six ATP molecules to form six glycerate-1,3-bisphosphate (reaction 2). The latter molecules are then reduced by $NADP^+$-glyceraldehyde-3-phosphate dehydrogenase to form six molecules of glyceraldehyde-3-phosphate (reaction 3). These reactions are similar to reactions encountered in gluconeogenesis. Unlike the dehydrogenase in gluconeogenesis, the Calvin cycle enzyme uses NADPH as a reducing agent.

Regeneration. Several regenerative phase reactions are similar to those of other biochemical pathways. Two reactions each are catalyzed by aldolase (glycolysis) and transketolase (pentose phosphate pathway). Fructose-1,6-bisphosphatase is a gluconeogenic enzyme. As noted previously, the net production of fixed carbon in the Calvin cycle is one molecule of glyceraldehyde-3-phosphate. The other five glyceraldehyde-3-phosphate molecules are processed in the remainder of the Calvin cycle reactions to regenerate three molecules of ribulose-1,5-bisphosphate. The regeneration of ribulose-1,5-bisphosphate begins with reactions involving glyceraldehyde-3-phosphate. In reaction 4 two molecules of glyceraldehyde-3-phosphate are isomerized to form two dihydroxyacetone phosphate. Aldolase catalyzes the condensation of one of these dihydroxyacetone molecules with a third glyceraldehyde-3-phosphate to form fructose-1,6-bisphosphate (reaction 5).

The latter molecule is then hydrolyzed by fructose-1,6-bisphosphatase (reaction 6) to fructose-6-phosphate. Fructose-6-phosphate subsequently combines with a fourth molecule of glyceraldehyde-3-phosphate in a transketolase-catalyzed reaction to form xylulose-5-phosphate and erythrose-4-phosphate (reaction 7). In reaction 8 aldolase catalyzes the condensation of erythrose-4-phosphate with the second molecule of DHAP to form sedoheptulose-1,7-bisphosphate, which is then hydrolyzed to form sedoheptulose-7-phosphate (reaction 9). Transketolase catalyzes the reaction of the fifth molecule of glyceraldehyde-3-phosphate with sedoheptulose-7-phosphate to form ribose-5- phosphate and a second molecule of xylulose-5-phosphate (reaction 10). Ribose-5-phosphate and both molecules of xylulose-5-phosphate are separately isomerized (reactions 11 and 12) to ribulose-5-phosphate. In the last step, three molecules of ribulose-5-phosphate are phosphorylated at the expense of three ATP molecules by ribulose-5-phosphate kinase (reaction 13) to form three molecules of ribulose-1,5-bisphosphate. The remaining molecule of glyceraldehyde-3-phosphate is either used within the chloroplast in starch synthesis or exported to the cytoplasm, where it may serve in the synthesis of sucrose or other metabolites.

Visit the companion website at www.oup.com/us/mckee to read the Biochemistry in Perspective essay on starch and sucrose metabolism.

QUESTION 13.7

When plant cells are illuminated, their cytoplasmic ATP:ADP and NADH:NAD^+ ratios rise significantly. The following shuttle mechanism is believed to contribute to the transfer of ATP and reducing equivalents from the chloroplast into the cytoplasm. Once DHAP has been transported from the stroma into the cytoplasm, it is converted to glyceraldehyde-3-phosphate and then to glycerate-1,3-bisphosphate. (This reaction is the reverse of the reaction in which glyceraldehyde-3-phosphate is formed during carbon fixation.) In the cytoplasmic reaction, the reducing equivalents are donated to NAD^+ to form NADH. In a later reaction, glycerate-1,3-bisphosphate is converted to glycerate-3-phosphate with the concomitant production of one molecule of ATP. Glycerate-3-phosphate is then transported back into the chloroplast, where it is reconverted to glyceraldehyde-3-phosphate.

This shuttle somewhat depresses mitochondrial respiration processes. Review the regulation of aerobic respiration (Chapter 9) and suggest how photosynthesis suppresses this aspect of mitochondrial function.

WORKED PROBLEM 13.2

Glyceraldehyde-3-phosphate, the first product of the Calvin cycle, is used to synthesize the energy storage molecules starch and sucrose. Outline the pathway by which two glyceraldehyde-3-phosphate molecules are incorporated into starch and calculate the cost in ATP molecules. Note that the precursor molecule in starch synthesis is ADP-glucose. Compare the cost of incorporating a glucose molecule into starch with that of its degradation into CO_2 and H_2O. Assume that the net production of glucose catabolism is 30 ATP.

SOLUTION

One molecule each of glyceraldehyde-3-phosphate and its isomer DHAP are converted by aldolase into fructose-1,6-bisphosphate. The latter molecule is converted via the gluconeogenic enzyme fructose-bisphosphate phosphatase into fructose-6-phosphate, which is isomerized to form glucose-6-phosphate. Glucose-1-phosphate, formed from glucose-6-phosphate by phosphoglucomutase, is then converted into ADP-glucose by ADP-glucose pyrophosphorylase. Therefore, only one ATP is used to synthesize ADP-glucose from two glyceraldehyde-3-phosphate. The cost of the incorporation of these two molecules into starch compared with the energy released by glucose is 1/30 or 3.3% of the total number of ATPs that can be generated by glucose catabolism. ■

Photorespiration

Photorespiration (**Figure 13.21**) is perhaps the most curious feature of photosynthesis. In this light-dependent process, oxygen is consumed and CO_2 is liberated by plant cells that are actively engaged in photosynthesis. Photorespiration is a multistep mechanism initiated by ribulose bisphosphate carboxylase, which also possesses an oxygenase activity. (For this reason the name *ribulose-1, 5-bisphosphate carboxylase-oxygenase*, or *rubisco*, is sometimes used.) Both CO_2 and O_2 compete for rubisco's active site.

In the oxygenation reaction, ribulose-1,5-bisphosphate is converted to glycolate-2-phosphate (**Figure 13.22**) and glycerate-3-phosphate. Glycolate-2-phosphate is hydrolyzed to form glycolate, which is then oxidized by O_2 to form glyoxylate and H_2O_2. Glyoxylate is converted through a series of reactions (outlined in **Figure 13.21**)

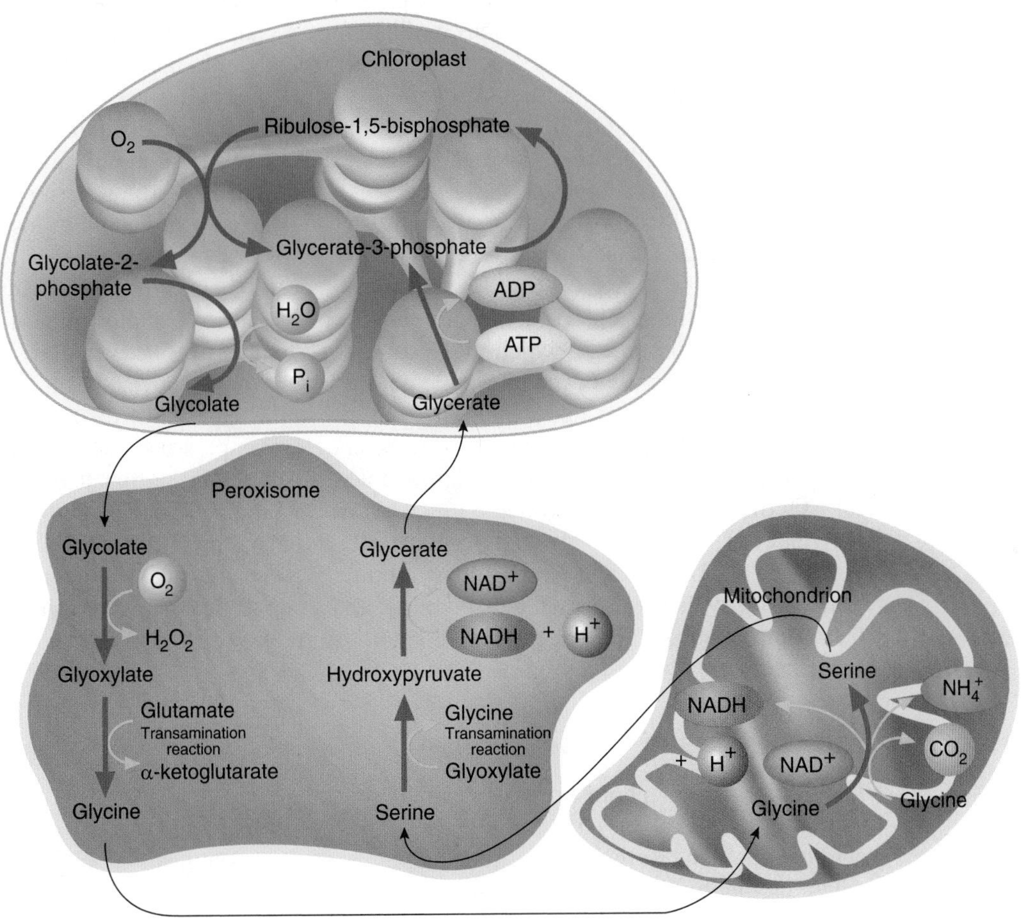

FIGURE 13.21
Photorespiration

Photorespiration is a wasteful process that occurs because rubisco can link ribulose-1,5-bisphosphate under certain conditions to O_2 to form glycolate-2-phosphate. This multistep pathway, which is a mechanism for salvaging fixed carbon from glycolate-2-phosphate, is catalyzed by enzymes in three organelles: chloroplasts, peroxisomes, and mitochondria. After glycolate-2-phosphate is hydrolyzed in chloroplast stroma, the product glycolate is transferred to a peroxisome, where it reacts with O_2 to form glyoxylate and H_2O_2. Glyoxylate then undergoes a transamination reaction to form glycine, which is then transferred out of the peroxisome and into a mitochondrion. Within the mitochondrial matrix two glycine molecules are converted into serine, CO_2, and NH_3, and NAD^+ is reduced to form NADH in a series of reactions catalyzed by enzymes in the glycine decarboxylase complex. Serine then returns to the peroxisome, where it undergoes a transamination reaction involving glyoxylate to form hydroxypyruvate. Hydroxypyruvate is subsequently reduced by NADH to yield glycerate. Once glycerate enters a chloroplast it reacts with ATP to yield glycerate-3-phosphate, the Calvin cycle intermediate.

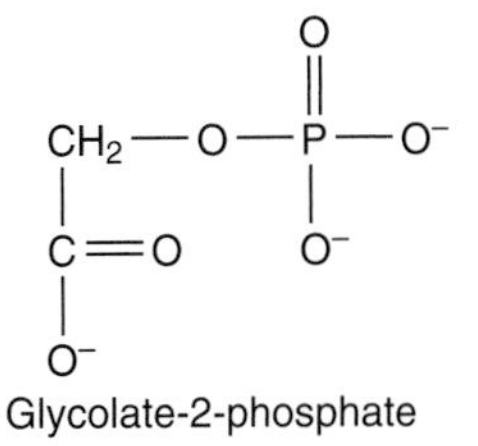

Glycolate-2-phosphate

FIGURE 13.22
Structure of Glycolate-2-Phosphate

to glycerate-3-phosphate. Glycerate-3-phosphate then enters the Calvin cycle where it is converted to ribulose-1,5-bisphosphate. Photorespiration is a wasteful process. It loses fixed carbon (as CO_2), and consumes both ATP and NADH.

The rate of photorespiration depends on several parameters. These include the concentrations of CO_2 and O_2 to which photosynthesizing cells are exposed. Photorespiration is depressed by CO_2 concentrations above 0.2%. (Because photorespiration and photosynthesis occur concurrently, CO_2 is released during CO_2 fixation. When the rates of CO_2 release and fixation are equal, the *CO_2 compensation point* has been reached. The lower the CO_2 compensation point, the less photorespiration takes place. Many C3 plants have CO_2 compensation points between 0.02 and 0.03% of CO_2 in the air near photosynthesizing cells.) In contrast, high O_2 concentrations and high temperatures promote photorespiration. Consequently, this process is favored when plants are exposed to high temperatures and any condition that causes low CO_2 and/or high O_2 concentrations. For example, photorespiration is a serious problem for C3 plants in hot, dry environments. To conserve water, these plants close their stomata, thus reducing the CO_2 concentration within leaf tissue. (*Stomata* are pores on the surface of leaves. When they are open, CO_2, O_2, and H_2O vapor can readily diffuse down the concentration gradients between the leaf's interior and the external environment.) As photosynthesis continues, O_2 levels increase. Depending on the severity of the circumstances, from 30 to 50% of a plant's yield of fixed carbon may be lost. This effect can be serious because several C3 plants (e.g., soybeans and oats) are major food crops.

Photorespiration is an artifact of the evolutionary history of photosynthesis. In the early atmosphere in which the first photosystem evolved, oxygen levels were very low. Thus, over the long time period before oxygen levels became a problem, there was no selection pressure to improve the capacity of the rubisco active site to distinguish between CO_2 and O_2. Selection pressure could occur only when oxygen levels increased significantly. It is noteworthy that selection for CO_2 over O_2 is higher in modern green plants than in bacteria. The pathway that evolved to convert glycolate-2-phosphate to glycerate-3-phosphate, although costly in ATP and NADH consumption, is viewed as a salvage operation that recovers previously fixed and partly reduced carbon. C4 plants, which have developed an elaborate mechanism to suppress photorespiration, are described next.

KEY CONCEPTS

- The Calvin cycle is a series of light-independent reactions in which CO_2 is incorporated into organic molecules.
- The Calvin cycle reactions occur in three phases: carbon fixation, reduction, and regeneration.
- Photorespiration is a wasteful process in which photosynthesizing cells evolve CO_2.

Alternatives to C3 Metabolism

In addition to C3 photosynthesis, which is used by most plants, there are two other mechanisms for fixing CO_2: C4 metabolism and crassulacean acid metabolism. Both improve the efficiency of photosynthesis in climates where temperatures are high and water is scarce.

C4 METABOLISM C4 plants include sugarcane and maize (corn); they thrive in the tropics and can successfully tolerate conditions of drought and high temperatures. The name **C4 plants** indicates the prominent role of a four-carbon molecule (OAA) in a biochemical pathway that avoids photorespiration. This pathway is called **C4 metabolism**, the *C4 pathway*, or the *Hatch-Slack pathway* (after its discoverers).

The leaves of C4 plants possess two types of photosynthesizing cells: mesophyll cells and bundle sheath cells. (In C3 plants, photosynthesis occurs in mesophyll cells.) Most mesophyll cells in both plant types are positioned so that they are in direct contact with air when the leaf's stomata are open. In C4 plants, CO_2 is captured in specialized mesophyll cells where it is converted into bicarbonate and then incorporated into oxaloacetate (**Figure 13.23**). PEP carboxylase

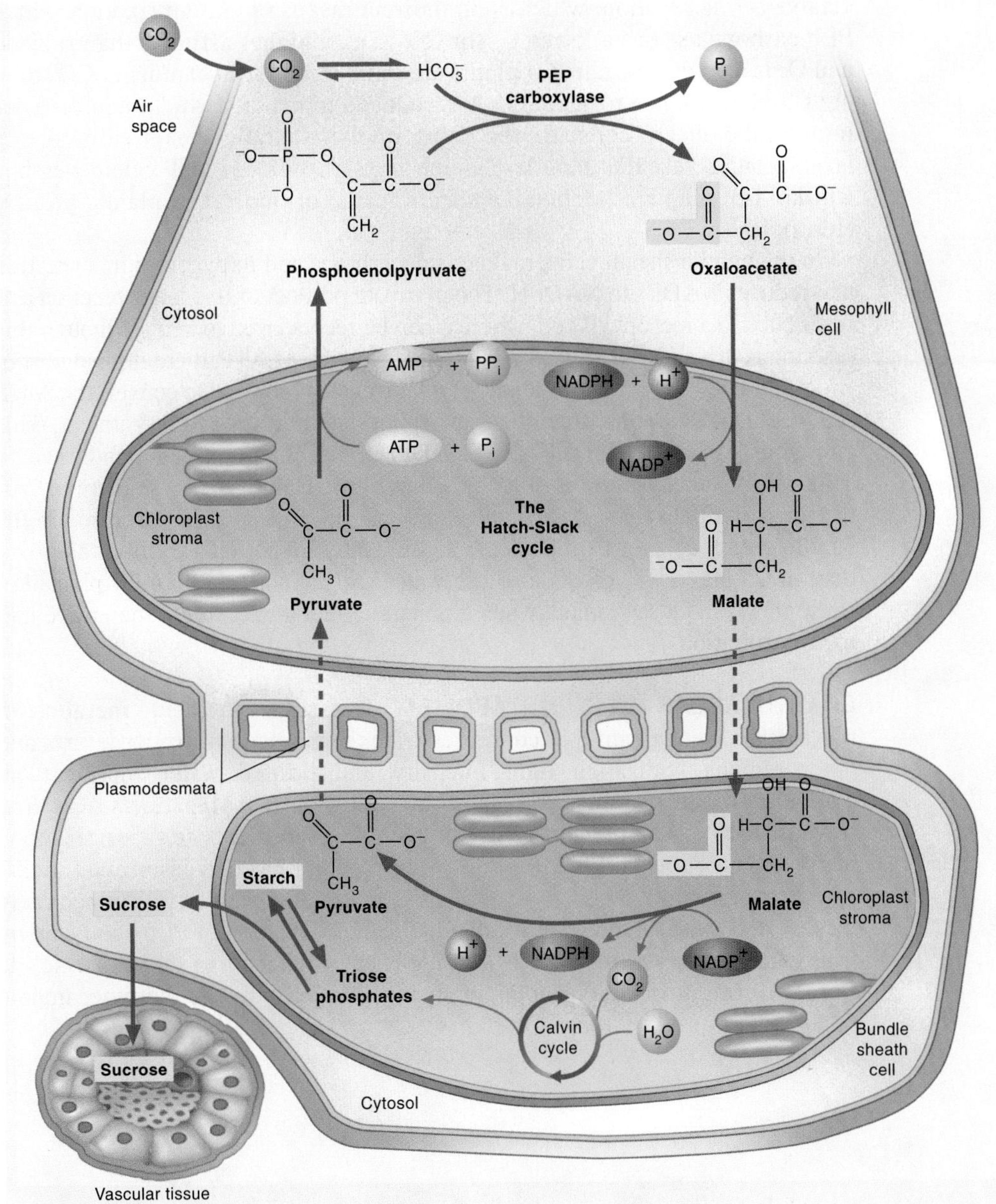

FIGURE 13.23

C4 Metabolism

In the C4 pathway mesophyll cells, which are in direct contact with the air space in the leaf, take up CO_2 and use it to synthesize oxaloacetate, which is then reduced to malate. (Some C4 plants synthesize aspartate instead of malate.) Malate then diffuses to bundle sheath cells, where it is reconverted to pyruvate. The CO_2 released in this reaction is used in the Calvin cycle, eventually yielding triose phosphate molecules. Triose phosphate is subsequently converted to starch or sucrose. Pyruvate returns to the mesophyll cell. Starch is synthesized from glucose-1-phosphate, which is subsequently converted to ADP-glucose by ADP-glucose pyrophosphorylase. ADP-glucose molecules are then incorporated into a preexisting polysaccharide chain by starch synthase. Sucrose-6-phosphate is synthesized from UDP-glucose and fructose-6-phosphate by sucrose phosphate synthase. Sucrose phosphatase catalyzes the hydrolysis of sucrose-6-phosphate to form sucrose and P_i. Note that glyceraldehyde-3-phosphate and dihydroxyacetone phosphate are referred to as triose phosphates.

catalyzes this reaction, which is an indirect means of carbon fixation. Since PEP carboxylase has a lower K_m for CO_2 (i.e., a higher affinity) than rubisco and O_2 is a poor substrate, C4 plants are more effective at capturing CO_2 than are C3 plants. Once formed, OAA is reduced to malate, which then diffuses into bundle sheath cells. As the name implies, bundle sheath cells form a layer around vascular bundles, which contain phloem and xylem vessels. Unlike C3 plants, the bundle sheath cells of most C4 plants possess chloroplasts.

In the bundle sheath cells, malate is decarboxylated to pyruvate in a reaction that reduces $NADP^+$ to NADPH. The pyruvate product of this latter reaction diffuses back to a mesophyll cell, where it can be reconverted to PEP. Although this reaction is driven by the hydrolysis of one molecule of ATP, there is a net cost of two ATP molecules. An additional ATP molecule is required to convert the AMP product to ADP so that it can be rephosphorylated during photosynthesis. This circuitous process delivers CO_2 and NADPH to the chloroplasts of bundle sheath cells, where rubisco and the other enzymes of the Calvin cycle use them to synthesize triose phosphates. The concentrations of CO_2 available to rubisco in the bundle sheath cells of C4 plants are significantly higher (10–20 times as great) than in C3 plants. C4 plants also use water more efficiently than C3 plants because they can close stomata when ambient temperature is high, thereby reducing transpiration.

CRASSULACEAN ACID METABOLISM **Crassulacean acid metabolism** (CAM) is a mechanism that conserves water in plants that live in deserts and other regions with high light intensity and limited water supply. (The Crassulaceae are a family of plants in which the CAM pathway was first investigated.) CAM plants, most of which are succulents (e.g., cacti), open their stomata only at night after the air temperature has decreased and the risk of water loss is low. The CO_2 enters mesophyll cells, where it is immediately incorporated into an OAA molecule via the carboxylation of PEP catalyzed by PEP carboxylase (**Figure 13.24**). OAA is then reduced to malate, which is stored overnight in the vacuoles of mesophyll cells. In the daytime, malate

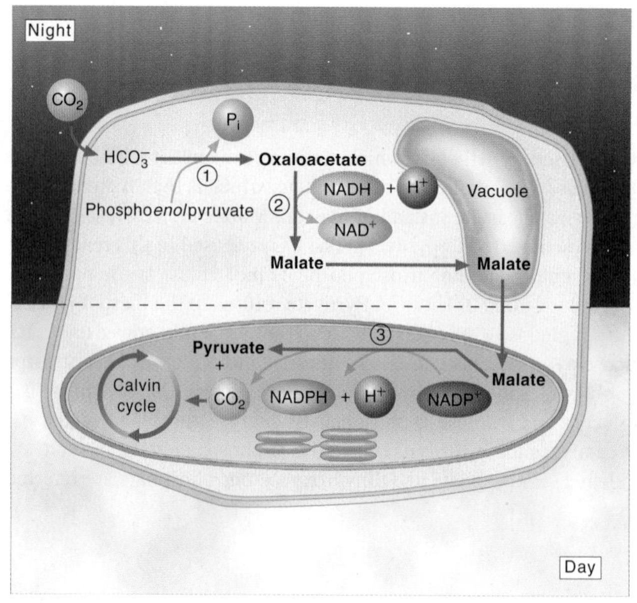

FIGURE 13.24
Crassulacean Acid Metabolism (CAM)

At night the stomata of CAM plants open to allow CO_2 to enter. Within mesophyll cells, PEP carboxylase (1) incorporates CO_2 (as HCO_3^-) into oxaloacetate. Afterward, oxaloacetate is reduced by malate dehydrogenase (2) to form malate. Malate is stored in the cell's vacuole until daylight. Light stimulates the decarboxylation of malate by malic enzyme (3) to form pyruvate and CO_2. As a result of this temporal separation of reactions, CO_2 can be incorporated into sugar molecules via the Calvin cycle during the day, when the plant's stomata are closed to avoid water loss.

molecules are broken down to pyruvate and the rubisco substrate CO_2. The temporal separation of carbon fixation and the Calvin cycle allows CAM plants to close their stomata during daytime, thus minimizing water loss through transpiration.

13.5 REGULATION OF PHOTOSYNTHESIS

Plants must adapt to a wide variety of environmental conditions. The regulation of photosynthesis is, therefore, complex. Although the control of most photosynthetic processes is far from being completely understood, several control features are well established. Most of these processes are directly or indirectly controlled by light. After a brief description of general light-related effects, the control of the activity of rubisco, the key regulatory enzyme in photosynthesis, is discussed.

Light Control of Photosynthesis

Investigations of photosynthesis are complicated by several factors. The most prominent of these is that the photosynthetic rate depends on temperature and cellular CO_2 concentration, as well as on light. Nevertheless, numerous investigations have firmly established light as an important regulator of most aspects of photosynthesis.

Many of the effects of light on plants are mediated by changes in the activities of key enzymes. Because plant cells possess enzymes that operate in several competing pathways (i.e., glycolysis, pentose phosphate pathway, and the Calvin cycle), careful metabolic regulation is critical. Light assists in this regulation by activating certain photosynthetic enzymes and deactivating several enzymes in degradative pathways. Among the light-activated enzymes are ribulose-1,5-bisphosphate carboxylase, $NADP^+$-glyceraldehyde-3-phosphate dehydrogenase, fructose-1,6-bisphosphatase, sedoheptulose-1,7-bisphosphatase, and ribulose-5-phosphate kinase. Light-inactivated enzymes include phosphofructokinase and glucose-6-phosphate dehydrogenase (G-6-PD).

Light affects enzymes by indirect mechanisms. Among the best researched are the following.

1. **pH**. During the light reactions, protons are pumped across the thylakoid membrane from the stroma into the thylakoid lumen. As the pH of the stroma increases from 7 to approximately 8, the activities of several enzymes are affected. For example, the pH optimum of ribulose-1,5-bisphosphate carboxylase is 8.
2. **Mg^{2+}**. Several photosynthetic enzymes (e.g., fructose-1,6-bisphosphatase) are activated by Mg^{2+}. Light induces an increase in the stromal Mg^{2+} concentration from 1 to 3 mM to about 3 to 6 mM. (Recall that Mg^{2+} moves across the thylakoid membrane into the stroma during the light reactions.)
3. **The ferredoxin-thioredoxin system**. Thioredoxins are small proteins that transfer electrons from reduced ferredoxin to certain enzymes (**Figure 13.25**). (Recall that ferredoxin is an electron donor in PSI.) When exposed to light, PSI reduces ferredoxin, which then reduces ferredoxin-thioredoxin reductase (FTR), an iron-sulfur protein that mediates the transfer of electrons between ferredoxin and thioredoxin. Reduced thioredoxins alter the activities of several enzymes. For example, the Calvin cycle enzymes fructose-1,6-bisphosphatase, sedoheptulose-1,7-bisphosphatase,

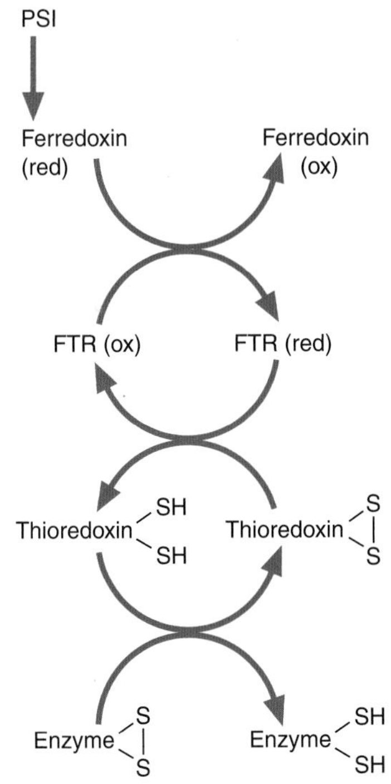

FIGURE 13.25
The Ferredoxin-Thioredoxin System
Using the light energy captured by PSI, energized electrons are donated to ferredoxin. Electrons donated by ferredoxin to ferredoxin-thioredoxin reductase (FTR) are used to reduce the disulfide bridge of thioredoxin. Thioredoxin then reduces the disulfide bridges of susceptible enzymes. Some enzymes are activated by this process, whereas others are inactivated.

KEY CONCEPT

Light is the principal regulator of photosynthesis. In photosynthetic processes, light affects the activities of regulatory enzymes such as rubisco by means of indirect mechanisms, which include changes in pH, Mg^{2+} concentration, the ferredoxin-thioredoxin system, and phytochrome. Rubisco is also regulated by covalent modification in that carbamoylation of an active site lysine residue is required for optimal activity.

Visit the companion website at www.oup.com/us/mckee to read the Biochemistry in the Lab box on photosynthetic studies for a brief discussion of the technologies used to investigate photosynthesis.

$NADP^+$-glyceraldehyde-3-phosphate dehydrogenase, and ribulose-5-phosphate kinase are activated and G-6-PD, the pentose phosphate pathway enzyme, is inhibited. By depressing the activity of the pentose phosphate pathway, light-driven reduced theoredoxin production prevents carbon flow through an opposing pathway.

4. **Phytochrome**. Phytochrome is a 120 kDa protein that possesses a red light-sensitive chromophore that exists in two forms: P_r and P_{fr}. The inactive blue form, P_r, absorbs red light (670 nm). The absorption of longer wavelengths (i.e., far red light, 720 nm) converts P_r to P_{fr}, the active green form. In the dark, P_{fr} decays back to P_r. Phytochrome activation triggers several signal transduction pathways that mediate hundreds of plant responses to light. In addition to phytochrome's effects on plant processes such as seed germination, it also has specific effects on photosynthetic processes. These include controlling the rate of synthesis of the small subunit of rubisco, regulation of light absorption by LHCIIb, a component of LHCII, and positioning chloroplasts within photosynthesizing cells.

Control of Ribulose-1,5-Bisphosphate Carboxylase

Rubisco is regulated by light-stimulated changes in gene expression, ion concentrations, and covalent modification. Rubisco genes, found within the chloroplast (the L subunit) and the nucleus (the S subunit), are activated by increased light intensity. Once the S subunit has been transported from the cytoplasm into the chloroplast, both subunits assemble to form the L_8S_8 holoenzyme. When illumination is low, the synthesis of both subunits is rapidly depressed.

The activity of rubisco is also modified by several metabolic signals. When photosynthesis is active, the pH in the stroma increases (protons are being pumped out of the stroma into the thylakoid lumen) and the Mg^{2+} concentration increases (Mg^{2+} moves into the stroma as H^+ moves out). Both changes increase the activity of rubisco. An important consideration in this process is whether the stomata are open or closed (see the discussion of photorespiration on p. 497). Although CO_2 is the preferred substrate for rubisco, under physiological conditions both the carboxylase activity and the oxidase activity compete significantly with each other. If the stomata are closed, as they would be on a hot, dry day, O_2 accumulation in the leaf tissue greatly decreases the proportional participation of the carboxylase activity of rubisco. Recall that C4 plants diminish this competition by trapping the CO_2 in a four-carbon intermediate and delivering the CO_2 via decarboxylation directly to a rubisco molecule that is protected from exposure to O_2.

Rubisco is subject to covalent modification. The active site of the L subunit must be carbamoylated at a specific lysine residue to be active. Carbamoylation is the nonenzymatic carboxylation of a free primary amino group, in this case the ε-amino group of a certain lysine in the active site of rubisco (**Figure 13.26**). The rate of carbamoylation is dependent on the CO_2 concentration and an alkaline pH, which ensures that CO_2 fixation occurs at an appreciable rate only when the CO_2 levels and available energy are high. Ribulose-1,5-bisphosphate can and does bind to the active site in both its modified and its unmodified forms, but catalysis can occur only when rubisco is carbamoylated. The level of activation is cooperative and increases as more of the eight subunits are modified. An enzyme called rubisco activase mediates an ATP-dependent removal of ribulose-1,5-bisphosphate from the active site so that carbamoylation can occur, followed by enzyme activation. In the absence of light, photosynthesis is

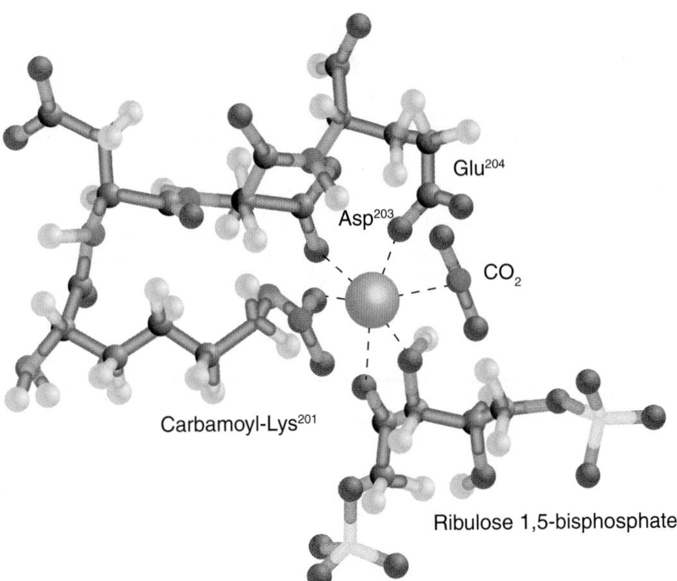

FIGURE 13.26

The Carbamoylated Active Site of Rubisco

Within the active site of rubisco, a magnesium ion orients the substrates ribulose-1,5-bisphosphate and CO_2 with oxygen atoms in aspartate and glutamate side chains and the carbamoyl group of a lysine residue.

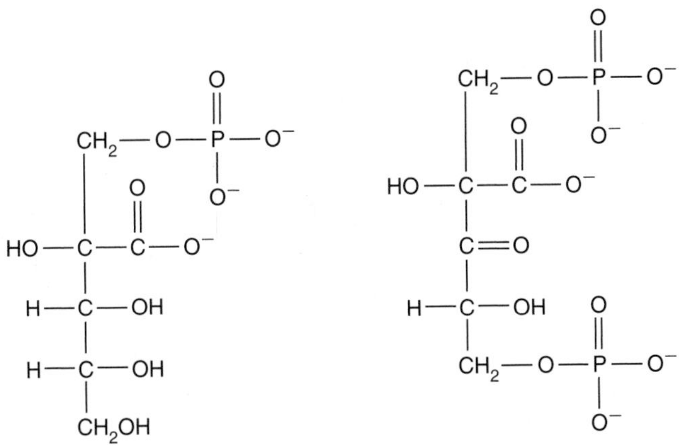

D-2-Carboxyarabinitol-1-phosphate

(a)

Rubisco transition state

(b)

FIGURE 13.27

2-Carboxyarabinitol-1-Phosphate

(a) 2-Carboxyarabinitol-1-phosphate (CA1P) is a competitive inhibitor of rubisco. It is a substrate analog that has a similar structure to the (b) transition state intermediate of the enzyme's substrate ribulose-1,5-bisphosphate.

depressed and the ATP required for this activation process is greatly reduced, as is the NADPH required for the Calvin cycle. In some plants rubisco activase also releases a competitive inhibitor molecule, 2-carboxyarabinitol-1-phosphate (CA1P), that binds to the active site of rubisco in the dark (**Figure 13.27**).

Biochemistry IN PERSPECTIVE

The Artificial Leaf: Biomimetic Photosynthesis

Can humans successfully produce solar fuels as a substitute for environmentally damaging fossil fuels? As human use of dwindling fossil fuels increases and the Earth's atmospheric concentration of the greenhouse gas CO_2 moves ever higher, scientists and engineers search for alternative sources of energy. Solar energy, radiant energy derived from nuclear fusion reactions within the sun, is an attractive possibility since the annual total power requirement of humans is currently about 16 TW, just a small fraction of the estimated 120,000 TW of solar energy that reaches the Earth's land masses every year. (A watt is a unit of power equal to 1 joule per second. A TW is 1 trillion watts.) Solar power is the attempt to use of manufactured photovoltaic cells, composed of monocrystalline silicon wafers, to convert sunlight into electrical energy. Although solar power is a clean and environmentally safe energy source, current solar technology is expensive. For example, many solar cells are manufactured with rare transition metals such as platinum and iridium to increase energy conversion efficiency. In addition, solar cells must be used in large arrays, called solar panels, to be economically viable, and they require constant maintenance.

Artificial photosynthesis research is an attempt to imitate the essential feature of the natural process, photoinduced electron transfer creating a charge-separated state, to harness solar power in the form of a synthesized fuel. The principal components of any successful artificial photosynthetic system are the same as those observed in natural photosynthesis: (1) a chromophore-containing light-harvesting mechanism associated with a reaction center complex that transduces photon energy into electrochemical energy (i.e., a charge-separated state is created when an electron is released from the reaction center complex, thereby creating an electron hole); (2) a water-oxidation catalyst that converts a water molecule to four electrons (which one at a time fill the electron hole in the reaction center complex), four hydrogen ions (H^+), and O_2; and (3) a catalyst that uses the water-derived electrons (reducing equivalents) and hydrogen ions to make fuel molecules, most notably H_2 ($2\ H^+ + 2e^- \rightarrow H_2$) or methanol (via reduction of CO_2). A model of artificial photosynthesis based on a light-harvesting supermolecular complex is outlined in **Figure 13A**.

Progress has been made in the development and assembly of a variety of artificial photosynthetic components, but the goal of producing a robust, photochemically stable and environmentally friendly means of solar-driven fuel synthesis has not yet been achieved. Of the numerous problems that must be solved, researchers currently consider the water-splitting mechanism the most challenging. The reaction $2\ H_2O \rightarrow 4\ e^- + 4\ H^+ + O_2$ is endothermic and, therefore, requires a catalyst to achieve or surpass the electron-transfer efficiency of the natural water-oxidizing complex (the manganese cluster and the tyrosine amino acid side chain). Among the best candidates for this role are manganese-containing complexes, dye-sensitized titanium dioxide (TiO_2), and cobalt oxide (CoO). However, these and

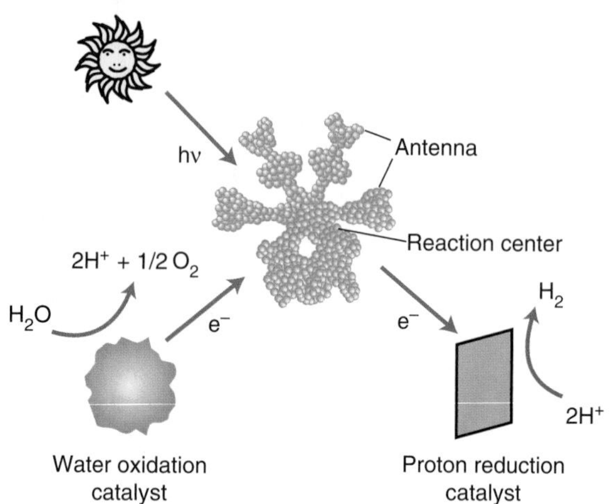

FIGURE 13A
An Artificial Photosynthesis Model

As with natural photosynthesis, all artificial photosynthetic devices must include a photoinduced electron transfer mechanism that creates a charge-separated state. Light energy is harvested by the chromophore components of a supermolecular complex such as the one illustrated in the center of the figure. This complex, referred to as a molecular hexad, is composed of three types of chromophores [bis(phenylethynyl) anthracene, borondipyrromethene, and the centrally located zinc tetraarylporphyrin] that together absorb energy across the visible region of the electromagnetic spectrum. After energy transfer from the other antennas, the zinc porphyrin donates an electron to the attached spherical fullerene to yield a charge-separated state. Despite the success of such molecular complexes in creating a charge-separated state, artificial photosynthesis is not yet feasible. In addition to the current lack of effective water oxidizing and proton reduction catalysts, current research has yet to provide a mechanism that can coordinate the four-electron water-oxidizing process (yielding O_2), the one-electron charge-separating reaction center process, and the two-electron H_2-generating reaction.

Biochemistry IN PERSPECTIVE cont.

other water-splitting catalysts so far investigated have electron transfer properties that are too slow. In addition, they lack coordination in the four electron–releasing process of water oxidation with the one-electron charge separation mechanism in the reaction center complex. As a result, water splitting remains a significant barrier to reaching the goal of competitively priced solar-driven energy production and reduction in the use of global warming–linked fossil fuels.

SUMMARY: Artificial photosynthesis, the use of solar energy to drive the synthesis of storable fuel such as H_2, is a goal that scientists and engineers are currently working toward. Although progress in the creation of light-driven charge separation has been made in the laboratory, the goal of cost-effective solar-driven fuel synthesis remains elusive. The seemingly intractable problems encountered in artificial photosynthesis research were solved by living organisms several billions of years ago.

Chapter Summary

1. In plants, photosynthesis takes place in chloroplasts. Chloroplasts possess three membranes. The outer membrane is highly permeable, whereas the inner membrane possesses a variety of carrier molecules that regulate molecular traffic into and out of the chloroplast. A third membrane, called the thylakoid membrane, forms an intricate series of flattened vesicles called grana connected by stromal lamellae that house the photosynthetic machines.
2. Photosynthesis consists of two major phases: the light reactions and the light-independent reactions. During the light reactions, water is oxidized, O_2 is evolved, and the ATP and NADPH required to drive carbon fixation are produced. The major working units of the light reactions are photosystems I and II, the cytochrome b_6f complex, and the ATP synthase. In noncyclic electron transport, electrons from water molecules are transferred from photosystem II to photosystem I to $NADP^+$ with the production of O_2, ATP and NADPH. Cyclic electron transport involves only PSI and generates additional ATP but no NADPH. During the light-independent reactions, CO_2 is incorporated into organic molecules. The first stable product of carbon fixation is glycerate-3-phosphate. The Calvin cycle is composed of three phases: carbon fixation, reduction, and regeneration.
3. Most of the carbon incorporated during the Calvin cycle is used initially to synthesize starch and sucrose, both of which are important energy sources. Sucrose is also important because it is used to translocate fixed carbon throughout the plant.
4. Photorespiration is an apparently wasteful process whereby O_2 is consumed and CO_2 is released from plants. Its role in plant metabolism is not understood. C4 plants and crassulaceans, which must tolerate hot, dry environments, have developed biochemical and anatomical mechanisms for suppressing photorespiration.
5. Light is an important regulator of most aspects of photosynthesis. Many of the effects of light are mediated by changes in the activities of key enzymes. The mechanisms by which light effects these changes include changes in pH, Mg^{2+} concentration, the ferredoxin-thioredoxin system, and phytochrome. The most important enzyme in photosynthesis is ribulose-1,5-bisphosphate carboxylase. Its activity is highly regulated. Light activates the synthesis of both types of the enzyme's subunits. Its activity is affected by allosteric effectors as well as covalent modification. Carbamoylation of an active site lysine residue is required for activation of rubisco.

Take your learning further by visiting the **companion website** for Biochemistry at **www.oup.com/us/mckee** where you can complete a multiple-choice quiz on water to help you prepare for exams.

Suggested Readings

Gross, M., *Light and Life,* Oxford University Press, Oxford, 2002.

Jones, M. R., and Fyfe, P. K., Photosynthesis: A New Step in Oxygen Evolution, *Current Biol.* 14:R320–R322, 2004.

Melkozernov, A. N., Barber, J., and Blankenship, R. E., Light Harvesting in Photosystem I Supercomplexes, *Biochemistry* 45(2):331–345, 2006.

Merchant, S., and Sawaya, M. R., The Light Reactions: A Guide to Recent Acquisitions for the Picture Gallery, *Plant Cell,* 17:648–663, 2005.

Nelson, N., and Ben-Shem, A., The Structure of Photosystem I and Evolution of Photosynthesis, *BioEssays* 27:914–922, 2005.

Rittman, B. E., Opportunities for Renewable Bioenergy Using Microorganisms, *Biotechnol. Bioeng.* 100(2):203–211, 2008.

Key Words

antenna pigment, *479*
C3 plant, *493*
C4 metabolism, *498*
C4 plant, *498*
Calvin cycle, *493*
carbon fixation, *493*
carotenoid, *477*
chlorophyll, *477*
chromophore, *484*
crassulacean acid metabolism, *500*
fluorescence, *485*
granum, grana, *478*
light-harvesting, *477*
light-harvesting antenna, *476*
light-independent reaction, *493*
light reaction, *486*
photophosphorylation, *482*
photorespiration, *497*
photosystem, *476*
reaction center, *476*
resonance energy transfer, *485*
stroma, *477*
stromal lamella, *478*
thylakoid membrane, *478*
thylakoid lumen, *478*
Z scheme, *488*

Review Questions

These questions are designed to test your knowledge of the key concepts discussed in this chapter before moving on to the next chapter. You may like to compare your answers to the solutions provided in the back of the book and in the accompanying Study Guide.

1. Define the following terms:
   a. photosynthesis
   b. photosystem
   c. reaction center
   d. PSI
   e. PSII
2. Define the following terms:
   a. chloroplast
   b. thylakoid
   c. stroma
   d. granum
   e. stromal lamellae
3. Define the following terms:
   a. phylloquinone
   b. lutein
   c. Q_A
   d. PQH_2
   e. carotenoids
4. Define the following terms:
   a. light-dependent reactions
   b. light-independent reactions
   c. chlorosome
   d. P680
   e. P700
5. Define the following terms:
   a. Rieske protein
   b. Psa dimer
   c. D_1/D_2 dimer
   d. A_0
   e. A_1
6. Define the following terms:
   a. cytochrome b_6f complex
   b. CF_0
   c. CF_1
   d. LCHII
   e. Mn_4CaO_5
7. Define the following terms:
   a. wavelength
   b. chromophore
   c. fluorescence
   d. radiationless decay
   e. resonance energy transfer
8. Define the following terms:
   a. C3 metabolism
   b. C4 metabolism
   c. CAM
   d. phytochrome
   e. CA1P
9. What was the most significant contribution of early photosynthetic organisms to Earth's environment?
10. List the three primary photosynthetic pigments and describe the role each plays in photosynthesis.
11. List five ways in which chloroplasts resemble mitochondria.
12. Excited molecules can return to the ground state by several means. Describe each briefly. Which of these processes are important in photosynthesis? Describe how they function in a living organism.
13. What is the final electron acceptor in photosynthesis when the $NADPH/NADP^+$ ratio is low? Does your answer change if the $NADPH/NADP^+$ ratio is high?

14. What reactions occur during the light and light-independent reactions of photosynthesis?
15. What are the potential advantages of promoting artificial photosynthesis research?
16. The overall equation for photosynthesis is

$$6CO_2 + 6H_2O \xrightarrow{\text{light}} C_6H_{12}O_6 + 6O_2$$

From which of the two substrates are the atoms of molecular oxygen derived?
17. Why is the oxygen-evolving system referred to as a clock?
18. If the rate of photosynthesis is plotted versus the incident wavelength of light, an action spectrum is obtained. How can the action spectrum provide information about the nature of the light-absorbing pigments involved in photosynthesis? [*Hint*: Refer to the Biochemistry in Perspective essay entitled Photosynthetic Studies at the companion website, www.oup.com/us/mckee.]

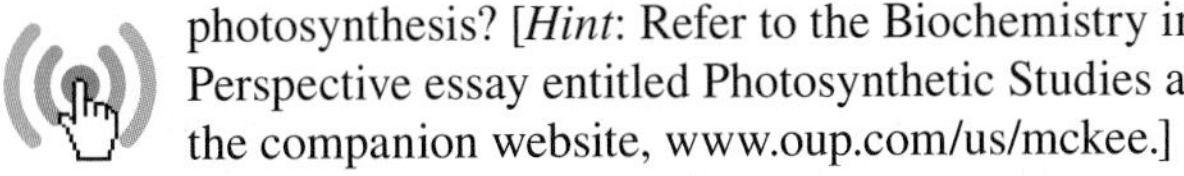

19. Review the net reaction for photosynthesis and explain the origin of the oxygen atoms in glucose molecules.
20. What is the function of the phytol chain in chlorophyll molecules?
21. What is a special pair and how does it function?
22. Why is the term dark reactions misleading?
23. Plants actively engaged in photosynthesis also evolve carbon dioxide. Explain.
24. Deforestation greatly decreases the plant biomass in an area. What impact will deforestation have on the quality of life in the region?
25. List four consequences of global warming.
26. List three criteria for sustainable biofuel production.
27. Using the action spectrum for photosynthesis, determine what wavelengths of light appear to be optimal for photosynthesis. [*Hint*: Refer to the Biochemistry in Perspective essay entitled Photosynthetic Studies at the companion website, www.oup.com/us/mckee.]

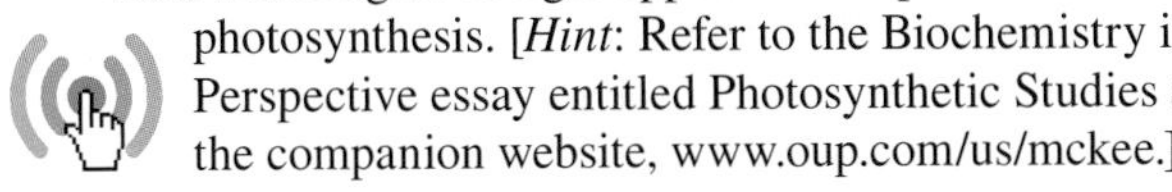

28. List the types of metal that are components of the photosynthesis mechanism. What functions do they serve?
29. Explain the following observation. When a photosynthetic system is exposed to a brief flash of light, no oxygen is evolved. Only after several bursts of light is oxygen evolved.
30. Describe the Z scheme of photosynthesis. How are the products of this reaction used to fix carbon dioxide?
31. Where does carbon dioxide fixation take place in the cell with reference to the light-dependent reaction?
32. The chloroplast has a highly organized structure. How does this structure help make photosynthesis possible?
33. Visit the companion website at www.oup.com/us/mckee to read the Biochemistry in Perspective essay entitled Photosynthetic Studies. Then describe the Emerson enhancement effect. How was it used to demonstrate the existence of two different photosystems?
34. Why does CO_2 depress photorespiration?
35. After one turn of the Calvin cycle in a C3 plant, where would you expect to find the radioactive label of $^{14}CO_2$?
36. What are triose phosphates? Where are they formed in plant carbohydrate metabolism?
37. A C4 plant is exposed to $^{15}O_2$. Where will this radioactive label first appear?
38. If corn, a C4 plant, is exposed to $^{14}CO_2$, in what molecule will it initially appear? Indicate the portion of the molecule in which the label will be found.
39. What is the end product of photosynthesis when H_2S is the source of hydrogen atoms?
40. What effect would you expect dinitrophenol to have on photosynthesis?

Fill in the Blank

41. Two ways in which chloroplasts resemble mitochondria are the presence of inner and outer membranes, which have different permeability characteristics, and ______________________.
42. The photosynthetic mechanisms referred to as ______________________ are membrane-bound protein complexes found in chloroplasts.
43. Unlike mitochondria, chloroplasts contain a distinct third membrane called the ______________________ membrane.
44. In photosynthesis, ______________ pigments absorb light energy and transfer it to the reaction center.
45. Molecules that absorb electromagnetic energy have structural components called ______________.
46. ______________________ is a process in which an excited molecule decays to its ground state by converting excess energy to heat.
47. In ______________________ the exited molecule decays to its ground state by emitting a photon.
48. The products of the light reactions of photosynthesis are __________ and NADPH.
49. Both carbon dioxide and ______________ compete for the active site of rubisco.
50. In addition to C3 photosynthesis (the most common form), there are two other mechanisms for fixing carbon dioxide. These are C4 metabolism and ______________.

Short Answer

51. What reactions occur during the reduction phase of the Calvin cycle?
52. Why is photorespiration considered an evolutionary artifact?
53. What is blackbody radiation, and how can photosynthetic organisms near hydrothermal vents take advantage of it?
54. What is the function of the ferredoxin-thioredoxin system in photosynthesis control?
55. A chromophore is the component of a molecule that gives it its color. What properties of chromophores are responsible for this phenomenon?

Thought Questions

These questions are designed to reinforce your understanding of all of the key concepts discussed in the book so far, including this chapter and all of the chapters before it. They may not have one right answer! The authors have provided possible solutions to these questions in the back of the book and in the accompanying Study Guide for your reference.

56. The burning of fossil fuels releases CO_2 into the atmosphere and is detrimental to the Earth's ecosystems. Explain why the use of biofuels, which also release CO_2, is an improvement over fossil fuels.
57. H_2S oxidation was one of the earliest mechanisms of photosynthesis on the ancient Earth. Because this mechanism requires less energy input than H_2O oxidation, why was there a major shift to the use of water?
58. Without carbon dioxide, chlorophyll fluoresces. How does carbon dioxide prevent this fluorescence?
59. The statement has been made that the more extensively conjugated a chromophore, the less energy a photon needs to excite it. What is conjugation, and how does it contribute to this phenomenon?
60. Increasing the intensity of the incident light but not its energy increases the rate of photosynthesis. Why is this so?
61. Both oxidative phosphorylation and photophosphorylation trap energy in high-energy bonds. How are these processes different? How are they the same?
62. In C3 plants, high concentrations of oxygen inhibit photosynthesis. Why is this so?
63. Generally, increasing the concentration of carbon dioxide increases the rate of photosynthesis. What conditions could prevent this effect?
64. It has been suggested that chloroplasts, like mitochondria, evolved from living organisms. What features of the chloroplast suggest that this is true?
65. How would biofuels generated by photosynthetic algae be an improvement over current methods?
66. Why does exposing C3 plants to high temperatures raise the carbon dioxide compensation point?
67. Herbicides that act by promoting photorespiration are lethal to C3 plants but do not affect C4 plants. Why is this so?
68. Corn, a grain of major economic importance, is a C4 plant, and many weeds in temperate climates are C3 plants. Therefore, the herbicides described in Question 67 are widely used. What effect is likely if these toxic materials are not degraded before they wash into the ocean?
69. Although both organelles originated as free-living prokaryotes, mitochondria are significantly smaller than chloroplasts. Suggest a reason for this discrepancy. [*Hint*: Consider the energy sources for both organelles.]
70. Suggest a reason why photosynthetic pigments readily absorb in the blue region of the visible spectrum, but do so with low probability in the ultraviolet.
71. If a C3 plant and a C4 plant are placed in separate sealed containers and provided with adequate light and water, both will survive. If both plants are placed in a single sealed container, the C3 plant will die. Explain.
72. Which plants expend more energy per molecule of glucose produced, C3 or C4?
73. Triazine herbicides are effective C3 plant toxins that are used to suppress weed growth. They apparently bind to a plastocyanin-binding protein. Suggest a mechanism whereby triazines undermine C3 growth.
74. Photosynthesizing organisms of the deep ocean capture long-wave radiation. Determine how many quanta of light at 1000 nm are required to provide as much energy as plants that absorb at 700 nm. [*Hint*: Recall that $\Delta E = hc/\lambda$, where h is Planck's constant (1.58×10^{-37} kcal/s), c is the speed of light (3×10^8 m/s), and λ is wavelength.]

CHAPTER

14 Nitrogen Metabolism I: Synthesis

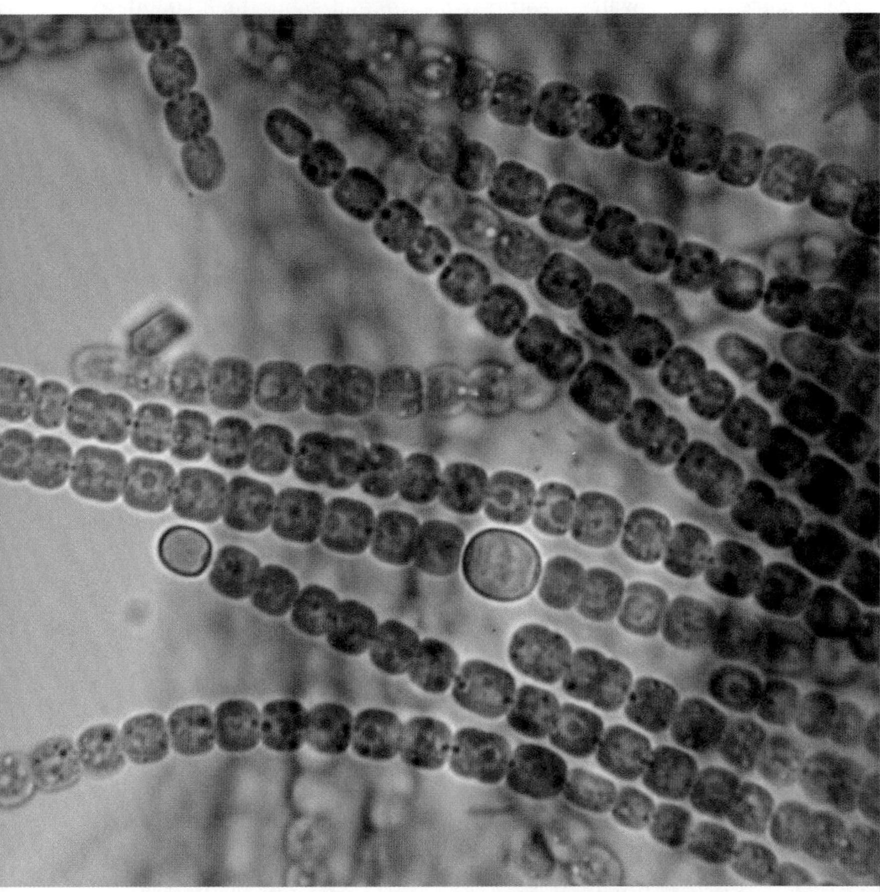

Nitrogen Fixation by Cyanobacteria Nitrogen-fixing organisms such as cyanobacteria convert atmospheric nitrogen (N_2) to ammonia (NH_3), a biologically useful form. Nitrogen fixation takes place in heterocysts, large, specialized cells with several thick cell walls that exclude O_2, the waste product released by the more numerous photosynthesizing vegetative cells.

OUTLINE

Nitrogen and the Gulf of Mexico Dead Zone

Every summer for at least three decades, there has been an expanding phytoplankton overgrowth in the coastal water of the Gulf of Mexico near Louisiana and Texas. This overgrowth, referred to as a phytoplankton "bloom," triggers the formation of a zone in which no aerobic organisms can exist. In 2014, this *dead zone* (**Figure 14.1**) covered almost 14,245 km^2 (about 5500 mi^2), an area the size of Massachusetts. The dead zone not only causes habitat destruction, but also is an obvious threat to both human health and a multibillion-dollar commercial and recreational fishing industry.

Phytoplankton (e.g., cyanobacteria and diatoms) are microorganisms that produce more than half of the Earth's oxygen. Consumed by zooplankton (microscopic animals) and fish, phytoplankton serve as the foundation of aquatic food webs. Their growth requires light and CO_2 and nutrient availability. Ordinarily, phytoplankton growth is limited only by temperature and low levels of nutrients, especially those of nitrogen, and, to a lesser extent, phosphorus. Small-scale, transient blooms, which occur in ocean water when air and surface water temperatures are high, are typically linked to upwelling. In this process, caused by high winds, cold nutrient-dense ocean water moves from the bottom to the surface.

The Gulf of Mexico dead zone, one of several hundred now observed on Earth, has been definitively linked to excess nitrogen discharged into the Gulf by the Mississippi River. The major causes of the massive Gulf phytoplankton bloom are now known to be a combination of nitrogen-based agricultural fertilizers (ammonia and nitrate), un- or undertreated sewage release from water treatment plants, and septic tank runoff into the rivers and streams in the Mississippi River watershed that drain water from two-thirds of the continental United States. The concentration of phytoplankton is so high (thousands of cells per milliliter) in the Gulf phytoplankton bloom that, depending on the dominant species, the water changes color to green, brown, or red.

Phytoplankton blooms trigger dead zone formation because once a bloom ends, dead phytoplankton and the zooplankton that fed on them drift to the bottom, where decomposition is carried out by aerobic bacteria. So much O_2 is used in this decaying process that the water becomes hypoxic (dissolved oxygen <2 mg/L). (In a healthy ocean environment the concentration of dissolved oxygen is about 5 mg/L.) Soon fish and other oxygen-requiring organisms (e.g., shrimp, clams, and oysters) suffocate. Once the bottom sediments become anoxic (no oxygen), foul-smelling substances [e.g., hydrogen sulfide (H_2S)] released by anaerobic bacteria begin accumulating.

Can the Gulf of Mexico dead zone be reversed? The answer to this question is a qualified yes.

FIGURE 14.1

Satellite Images of the Northern Gulf of Mexico in Winter and Summer

In these images, red and orange represent high concentrations of phytoplankton and sediment that originated in the Mississippi River. Note that the size of the dead zone off the coasts of Louisiana and Texas is larger and extends farther from land in summer (right) than in winter (left).

Nitrogen and the Gulf of Mexico Dead Zone cont.

Reductions in nitrogen pollution have been shown to result in the disappearance of dead zones. The Black Sea dead zone, once the largest on Earth, began shrinking in 1991 with the collapse of the Soviet Union, which had previously subsidized massive fertilizer use in the surrounding Soviet block countries. Subsequently, several European Union–financed sewage treatment facilities reduced nitrogen pollution even further. By 2001 the dead zone had completely disappeared, and Black Sea commercial fishing has begun to recover.

The Gulf of Mexico dead zone recovery will be difficult and expensive because of the size of the Mississippi River watershed. In addition to substantial improvements in the efficient and targeted use of fertilizers, recovery will require significantly upgraded water treatment facilities throughout the Mississippi River watershed. It will also depend on the restoration of wetlands, ecosystems that can reduce nutrient pollution before runoff water reaches the river.

Overview

NITROGEN IS FOUND IN AN ASTONISHINGLY VAST ARRAY OF BIOMOLECULES, INCLUDING THE AMINO ACIDS AND THE NITROGENOUS BASES, USED IN THE synthesis of proteins and the nucleic acids, respectively. Other essential nitrogen-containing biomolecules include the porphyrins (e.g., heme and chlorophyll), certain membrane lipids, and a diverse group of metabolically important biomolecules that are synthesized in smaller amounts (e.g., several neurotransmitters and glutathione). This chapter traces nitrogen from nitrogen fixation, the process that converts inert N_2 to biologically useful ammonia (NH_3), through the synthesis of the major nitrogen-containing biomolecules.

The nitrogen cycle is the biogeochemical cycle in which nitrogen atoms flow through the biosphere. Several biochemical processes convert nitrogen from one form to another. Nitrogen fixation, the incorporation of nitrogen into organic molecules, begins with the fixation (reduction) of N_2 by prokaryotic microorganisms to form ammonia (NH_3). Plants such as corn depend on absorbing NH_3 and NO_3^- (nitrate), the oxidation product of NH_3, synthesized by soil bacteria or provided by artificial fertilizers. Nitrogen supply is often the limiting factor in plant growth and development because the amount of fixed nitrogen available to plants is usually small.

Whether plants acquire NH_3 by nitrogen fixation, by absorption from the soil, or by reduction of absorbed NO_3^-, it is assimilated by conversion into the amide group of glutamine. This "organic nitrogen" is then transferred to other carbon-containing compounds to produce the amino acids used by the plant to synthesize nitrogenous molecules (e.g., proteins, nucleotides, and heme). Organic nitrogen, primarily in the form of amino acids, then flows throughout the ecosystem as plants are consumed by animals and decomposing microorganisms. When organisms die, organic nitrogen is *mineralized* (i.e., it is converted through the actions of numerous types of microbes into NH_3, NO_3^-, NO_2^- (nitrite), and eventually N_2).

After a discussion of nitrogen fixation, the essential features of amino acid biosynthesis are described. This is followed by descriptions of the biosynthesis of selected nitrogen-containing molecules. A special emphasis is placed on the anabolic pathways of the nucleotides. Chapter 15 traces the flow of nitrogen atoms through several catabolic pathways to the nitrogenous waste products excreted by animals.

14.1 NITROGEN FIXATION

Several circumstances limit the amount of usable nitrogen available in the biosphere. The most notable are the limited number of species that can convert N_2 into NH_3, a more chemically reactive molecule, and the high energy requirements of this process, which is referred to as **nitrogen fixation**. Among the most prominent nitrogen-fixing species are free-living bacteria (e.g., *Azotobacter vinelandii* and *Clostridium pasteurianum*), the cyanobacteria (e.g., *Nostoc muscorum* and *Anabaena azollae*), and symbiotic bacteria (e.g., several species of *Rhizobium*). Symbiotic organisms form mutualistic, that is, mutually beneficial, relationships with host plants or animals. *Rhizobium* species, for example, infect the roots of leguminous plants such as soybeans and alfalfa.

Nitrogen fixation requires a large energy input because reduction of N_2 to form NH_3 involves the breaking of the nonpolar triple bond of atmospheric dinitrogen gas. In commercial nitrogen fixation, NH_3 is the product of the Haber-Bosch reaction, where H_2 and N_2 are heated at 400 to 650°C under 200 to 400 atm pressure in the presence of an iron catalyst. Unlike the Haber-Bosch process, nitrogen fixing species convert N_2 to NH_3 at ambient temperature and atmospheric pressure. The energy requirements for the biological process, however, are also high, with a minimum of 16 ATP required to reduce one N_2 to form two ammonia molecules. The overall reaction of nitrogen fixation is $N_2 + 8e^- + 16\text{ ATP} + 10H^+ \rightarrow 2NH_4^+ + 16\text{ ADP} + 16\text{ P}_i + H_2$.

The Nitrogen Fixation Reaction

All species that can fix nitrogen possess the *nitrogenase complex*. Its structure, similar in all species so far investigated, consists of two proteins called dinitrogenase and dinitrogenase reductase (**Figure 14.2**). Dinitrogenase (240 kDa), also referred to as MoFe *protein*, is an $\alpha_2\beta_2$ heterotetramer. Each $\alpha\beta$ dimer is a catalytic unit that contains two unique metal prosthetic groups: a *P cluster* [8Fe-7S] and *molybdenum-iron cofactor* (MoFe cofactor or M cluster), which contains a carbide atom (a carbon atom at the center of a metal cluster). The FeMo cofactor is linked to the tricarboxylic acid homocitrate [7Fe-9S-Mo-C-homocitrate]. The MoFe protein catalyzes the reaction $N_2 + 8H^+ + 8e^- \rightarrow 2NH_3 + H_2$. (Note that NH_3 is the initial product. Under cellular pH conditions it will be in equilibrium with NH_4^+.). Dinitrogenase reductase (60 kDa) (also referred to as *Fe protein*) is a dimer containing identical subunits, each of which has a MgATP-binding site. A 4Fe-4S cluster is bound at the interface of the two subunits, 15 Å from the MgATP-binding site and close to the docking site for the dinitrogenase tetramer (MoFe protein). The Fe protein transfers electrons, ultimately derived from NAD(P)H, one at a time to the Mo-Fe protein. Both proteins in the nitrogenase complex are irreversibly inactivated by O_2.

The first step in nitrogen fixation (**Figure 14.3**) is the transfer of electrons from NAD(P)H to ferredoxin, a powerful reducing agent that in turn donates the electrons to the Fe protein FeS cluster, one at a time. Each electron transfer begins with the reduction of the oxidized [4Fe-4S] cluster in the Fe protein with two bound ADP [$Fe^{ox}(ADP)_2$] to yield [$Fe^{red}(ADP)_2$]. This reduction triggers the replacement of ADP with ATP in both subunits. The product [$Fe^{red}(ATP)_2$] then docks with the MoFe protein to form the active complex [$Fe^{red}(ATP)_2$:MoFe]. The docking event is followed by the transfer of an electron from the P cluster of

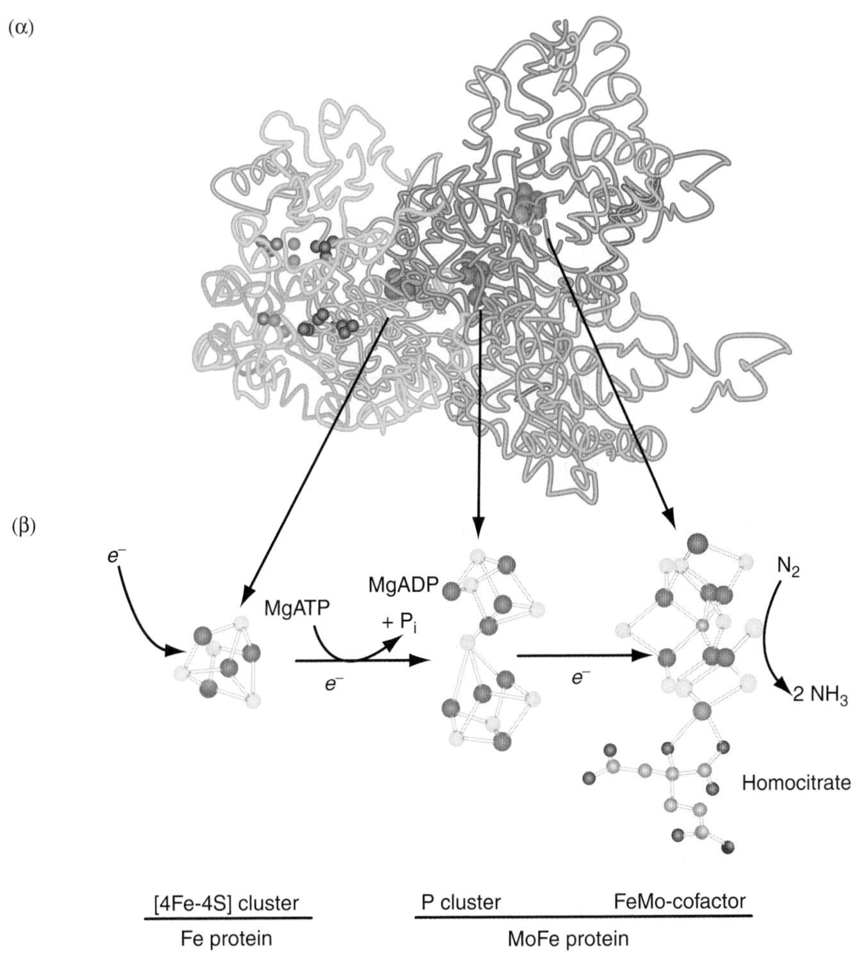

FIGURE 14.2

Nitrogenase Complex Structure

(a) Left: Fe protein dimer (pink and red); right: an $\alpha\beta$ dimer of the MoFe protein, with the α subunit in blue and the β subunit in green. (b) The metal clusters of the Fe protein [4Fe-4S] and the P cluster and FeMo cofactor of the FeMo protein are illustrated as ball-and-stick models. Atom colors: carbon, gray; oxygen, red; phosphorus, dark green; sulfur, yellow; iron, green; molybdenum, pink.

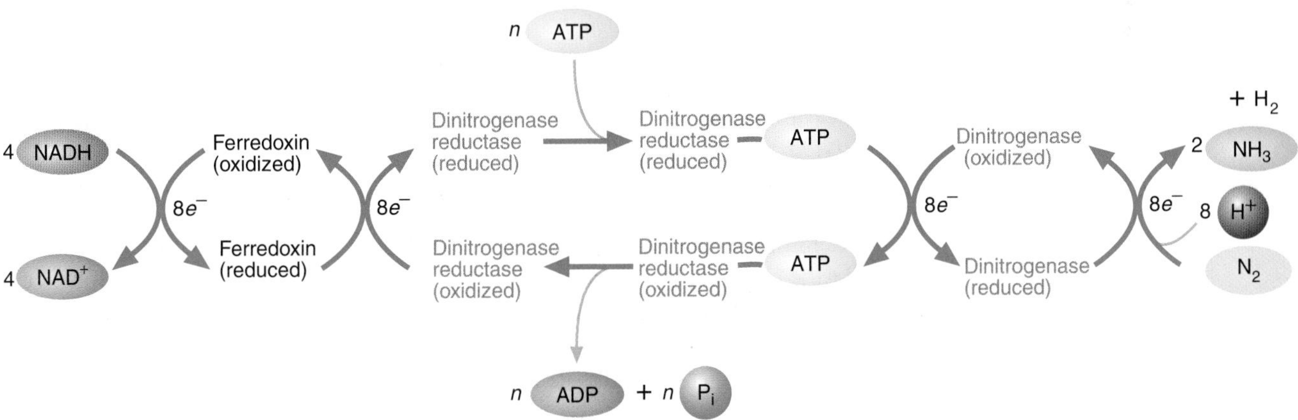

FIGURE 14.3

Schematic Diagram of the Nitrogenase Complex Illustrating the Flow of Electrons and Energy in Enzymatic Nitrogen Fixation

The high energy of activation of nitrogen fixation is overcome by a large number of ATP molecules (about 16 ATP per N_2 molecule). Both the binding of ATP to dinitrogenase reductase (Fe protein) and its subsequent hydrolysis cause conformational changes in the protein that facilitate the transfer of electrons to dinitrogenase (P cluster/MoFe cofactor).

the MoFe protein to the M cluster. An electron from the Fe protein is then transferred to the oxidized P cluster to yield $[Fe^{ox}(ATP)_2{:}MoFe^{red}]$. ATP hydrolysis and P_i release cause a conformational change that allows the dissociation of $[Fe^{ox}(ADP)_2]$ from the reduced MoFe protein. The electron transfer process is then repeated until eight electrons have been delivered to the MoFe cofactor. (In its oxidized electron-deficient form, the Fe protein has a higher affinity for MgATP.) The reduction potentials for the Fe protein and the P cluster have been recorded in the –400- and –300-mV ranges, respectively. The transfer of the first two electrons leads to the reduction of H^+ (the $MoFeH_2$ state of the enzyme). This stage of the process occurs with or without the presence of N_2. The incoming N_2 exchanges with H_2 in the active site (between the molybdenum ion and four coordinating Fe centers) to form a stable intermediate ($MoFeN_2$). Subsequently, six electrons and six protons are transferred to the active site to form the diimine (HN = NH, two electrons added), then the hydrazine (H_2N—NH_2, four electrons added) and, finally, two NH_3 products (six electrons added). Therefore, two electrons are used to reduce $2H^+$ to H_2 (essential to the catalytic process) and six electrons are used to reduce N_2 in three steps to yield two NH_3 molecules.

In addition to the large quantities of ATP that are required to drive the reduction of N_2 (a minimum of 16 ATP), a substantial number of copies of nitrogenase complex proteins must be synthesized because of the enzyme's slow turnover time: about 6 NH_3 produced per second per molecule of enzyme. (Nitrogenase complex protein may constitute as much as 20% of cellular protein in diazotrophs, the bacteria that fix nitrogen.) As a result, the regulation of nitrogen fixation by diazotrophs, which must respond to numerous environmental variations (e.g., fixed nitrogen and oxygen levels and carbon source availability), is both intricate and stringent. The major form of regulation is transcriptional control of the approximately 20 nitrogen fixation (*nif*) genes. In addition to coding for dinitrogenase reductase (Fe protein) and the α and β subunits of dinitrogenase (MoFe protein), *nif* genes code for a variety of enzymes that synthesize components of the nitrogen fixation process, such as metal clusters, homocitrate, or ferredoxin, and regulatory proteins that mediate responses to environmental cues.

WORKED PROBLEM 14.1

How expensive is nitrogen fixation? Determine the number of ATPs required to reduce a nitrogen atom to NH_4^+. Assume that one NADH is the equivalent of 2.5 ATPs.

SOLUTION

Refer to the overall equation for nitrogen fixation on p. 512. The reduction of a molecule of N_2 to 2 NH_4^+ requires 8 electrons and 16 ATP. Once the number of ATP equivalents for the 4 NADHs is calculated ($4 \times 2.5 = 10$), the number of ATPs required for the reduction of N_2 rises to 26. The number of ATPs required for the reduction of one nitrogen atom is half of this number, or 13 per NH_4^+ formed. ■

Nitrogen Assimilation

Nitrogen assimilation is the incorporation of inorganic nitrogen compounds into organic molecules. In plants, nitrogen assimilation begins in the roots, whether by the transfer of NH_4^+ (ammonium ions) from symbiotic bacteria in the root nodules of leguminous plants or by the absorption of NH_4^+ or NO_3^- (nitrate) from soil. Nitrate is produced by nitrifying soil bacteria. Organisms such as *Nitrosomonas* oxidize NH_4^+ to form NO_2^- (nitrite), which is then further oxidized to NO_3^- by bacteria such as *Nitrobacter.*

The assimilation of inorganic nitrogen into biomolecules in plants is effected by the incorporation of ammonium nitrogen into amino acids. Glutamine synthetase, the most important enzyme in nitrogen assimilation, catalyzes the ATP-dependent reaction of glutamate with NH_4^+ to form glutamine. In the next step, glutamine reacts with α-ketoglutarate to form glutamate. (The synthesis of other amino acids via the transfer of the amino group of glutamate is described in Section 14.2). When nitrate is the nitrogen source, it must first be converted to NH_4^+ in a two-step process. After nitrate has been converted to nitrite by nitrate reductase, ammonia is produced by the reduction of nitrite, which is catalyzed by nitrite reductase.

QUESTION 14.1

The nitrogenase complex can reduce molecules other than N_2. Provide the structures for the products for each of the following substrates (real and hypothetical) that contain triple bonds: hydrogen cyanide, dinitrogen, and acetylene.

QUESTION 14.2

A red heme-containing protein called leghemoglobin is found in the nitrogen-fixing root nodules of leguminous plants. The protein component is produced by the plant, whereas bacterial cells produce the precursor of heme. Can you deduce the function of leghemoglobin? [*Hint*: Review the function of the hemoglobins.]

14.2 AMINO ACID BIOSYNTHESIS

Living organisms differ in their capacity to synthesize the amino acids required for protein synthesis. Although plants and many microorganisms can produce all their amino acids from readily available precursors, other organisms must obtain some preformed amino acids from their environment. Animals can synthesize only about half the amino acids they require. The **nonessential amino acids (NAA)** are synthesized from readily available metabolites. The amino acids that must be provided in the diet are referred to as **essential amino acids (EAA)**. Mammalian tissues can synthesize NAA (**Table 14.1**) by relatively simple reaction pathways. In contrast, EAA must be obtained from the diet because mammals lack the complex reaction pathways required for their synthesis.

TABLE 14.1 The Essential and Nonessential Amino Acids in Humans

Essential	Nonessential
Isoleucine	Alanine
Leucine	Arginine*
Lysine	Asparagine
Methionine	Aspartate
Phenylalanine	Cysteine*
Threonine	Glutamate
Tryptophan	Glutamine
Valine	Glycine
	Histidine*
	Proline
	Serine
	Tyrosine*

* Referred to as the *semi-essential amino acids*, these molecules are essential for infants and children up to 5 years of age. The pathways that produce the semi-essential amino acids are not fully functional in young children.

Amino Acid Metabolism Overview

Amino acids serve a number of functions. In addition to their most important role in the synthesis of proteins, amino acids are the principal source of the nitrogen atoms required in various synthetic reaction pathways. The nonnitrogen components of amino acids (referred to as carbon skeletons) are a source of energy, as well as precursors in several reaction pathways. Therefore an adequate intake of amino acids, in the form of dietary protein, is essential for an animal's proper growth and development.

DIETARY AMINO ACIDS Dietary protein sources differ widely in their proportions of the EAA. In general, complete proteins (those containing sufficient quantities of EAA) are of animal origin (e.g., meat, milk, and eggs). Plant proteins often lack one or more EAA. For example, gliadin (wheat protein) has insufficient amounts of lysine, and zein (corn protein) is low in both lysine and tryptophan. Because plant proteins differ in their amino acid compositions, plant foods can provide a high-quality source of essential amino acids only if they are eaten in appropriate combinations. One such combination consists of beans (low in methionine) and cereal grains (low in lysine). Following the digestion of dietary protein in the body's digestive tract, free amino acids are transported across intestinal enterocytes and into the blood. Most diets do not provide amino acids in the proportions that the body requires. Their concentrations, therefore, must be adjusted by metabolic mechanisms. The amino acids released to the blood in the intestine already show some changes in their relative concentrations. The intestinal mucosa is a very active and constantly replaced tissue that sustains its structure and function by means of incoming nutrients other than glucose. The amino acid glutamine, for example, is a primary energy source for enterocytes. The blood from the GI tract goes first to the liver, another active tissue. The liver synthesizes the serum proteins, among others, and draws amino acids from the blood for this purpose. It also preferentially uses amino acids (especially alanine and serine) to synthesize glucose for export.

BRANCHED CHAIN AMINO ACIDS AND THE AMINO ACID POOL The blood that leaves the liver to nourish the rest of the body has a much higher concentration of **branched-chain amino acids (BCAA**: leucine, isoleucine, and valine) than that leaving the GI tract. The BCAA are EAAs and provide critical hydrophobic side chains in protein structure (e.g., leucine zipper motifs in DNA-binding proteins). BCAA also represent a major transport form of amino nitrogen from the liver to other tissues, where they are used in the synthesis of the NAAs required for protein synthesis, as well as various amino acid derivatives.

Amino acid metabolism is a complex series of reactions in which the amino acid molecules required for the syntheses of proteins and metabolites are continuously being synthesized and degraded. Depending on current metabolic requirements, certain amino acids are synthesized or interconverted and then transported to tissue, where they are used. The amino acid molecules that are immediately available for use in metabolic processes are referred to as the **amino acid pool**. In animals, amino acids in the pool are derived from the breakdown of both dietary and tissue proteins.

When nitrogen intake (primarily amino acids) equals nitrogen loss, the body is said to be in *nitrogen balance*. This is the condition of healthy adults. In *positive nitrogen balance*, a condition that is characteristic of growing children, pregnant women, and recuperating patients, nitrogen intake exceeds nitrogen loss. The excess of nitrogen is retained because the amount of tissue proteins being synthesized exceeds the amount being degraded. *Negative nitrogen balance* exists when an individual cannot replace nitrogen losses with dietary sources. *Kwashiorkor* ("the disease the first child gets when the second is on the way") is a form of malnutrition caused by a prolonged insufficient intake of protein. Its symptoms include growth failure, apathy, ulcers, liver enlargement, and diarrhea, as well as decreased mass and function of the heart and kidneys. Prevalent in Africa, Asia, and Central and South America, kwashiorkor can be treated or prevented by feeding children a high-energy protein-containing peanut-based food called Plumpy'nut, which is the equivalent of a glass of milk and a multiple vitamin. Plumpy'nut is inexpensive, requires no refrigeration, and can be shipped to remote clinics in poor areas of the world.

AMINO ACID TRANSPORT Transport of amino acids into cells is mediated by specific membrane-bound transport proteins, several of which have been identified in mammalian cells. They differ in their specificity for the types of amino acid transported and in whether the transport process is linked to the movement of Na^+ across the plasma membrane. (Recall that the gradient created by the active transport of Na^+ can move molecules across the membrane. Na^+-dependent amino acid transport is similar to that observed in the glucose transport process illustrated in **Figure 11.31**.) For example, several Na^+-dependent transport systems have been identified in the lumenal plasma membrane of enterocytes. Na^+-independent transport systems are responsible for transporting amino acids across the portion of enterocyte plasma membrane in contact with blood vessels. The γ-glutamyl cycle (see **Figure 14.20** later in this chapter) is believed to assist in transporting some amino acids into specific tissues (i.e., brain, intestine, and kidney).

Reactions of Amino Groups

Once amino acid molecules have entered cells, the amino groups are available for numerous synthetic reactions. This metabolic flexibility is effected by transamination reactions and reactions in which NH_4^+ or the amide nitrogen of glutamine is used to supply the amino group or the amide nitrogen of certain amino acids. Both reaction types are discussed next.

TRANSAMINATION **Transamination** reactions dominate amino acid metabolism. In these reactions, catalyzed by a group of enzymes referred to as the *aminotransferases* or *transaminases*, α-amino groups are transferred from an α-amino acid to an α-keto acid:

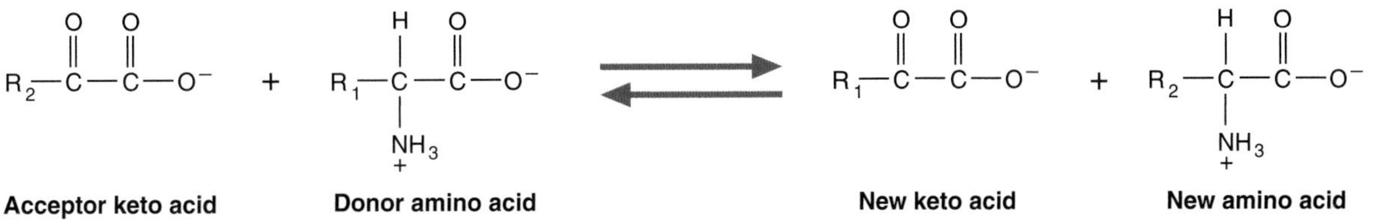

$$R_2-\overset{O}{\overset{\|}{C}}-\overset{O}{\overset{\|}{C}}-O^- + R_1-\underset{\overset{+}{N}H_3}{\underset{|}{\overset{H}{\overset{|}{C}}}}-\overset{O}{\overset{\|}{C}}-O^- \rightleftharpoons R_1-\overset{O}{\overset{\|}{C}}-\overset{O}{\overset{\|}{C}}-O^- + R_2-\underset{\overset{+}{N}H_3}{\underset{|}{\overset{H}{\overset{|}{C}}}}-\overset{O}{\overset{\|}{C}}-O^- \quad (1)$$

Acceptor keto acid **Donor amino acid** **New keto acid** **New amino acid**

Transamination reactions, which are readily reversible, play important roles in both the synthesis and the degradation of the amino acids.

Eukaryotic cells possess a large variety of aminotransferases. Found within both the cytoplasm and the mitochondria, these enzymes possess two types of specificity: (1) the type of α-amino acid that donates the α-amino group and (2) the α-keto acid that accepts the α-amino group. Although the aminotransferases vary widely in the type of amino acids they bind, most of them use glutamate as the amino group donor. Because glutamate is produced when α-ketoglutarate (a citric acid cycle intermediate) accepts an amino group, these two molecules, the *α-ketoglutarate/glutamate pair*, have a strategically important role in both amino acid metabolism and metabolism in general. Two other such pairs have important functions in metabolism. In addition to its role in transamination reactions, the *OAA/aspartate pair* is involved in the disposal of nitrogen in the urea cycle (Chapter 15). One of the most important functions of the *pyruvate/alanine pair* is in the glucose-alanine cycle (**Figure 8.12**). α-Ketoglutarate and OAA are citric acid cycle intermediates. Consequently, transamination reactions often represent an important mechanism for meeting the energy requirements of cells. Recall, for example, that in the glucose-alanine cycle transamination reactions are used to recycle the carbon skeleton of pyruvate between muscle and liver.

Transamination reactions require the coenzyme pyridoxal-5′-phosphate (PLP), which is derived from pyridoxine (vitamin B_6). PLP is also required in numerous other reactions of amino acids. Examples include racemizations, decarboxylations, and several side chain modifications. (**Racemizations** are reactions in which mixtures of L- and D-amino acids are formed.) The structures of the vitamin and its coenzyme form are illustrated in **Figure 14.4**.

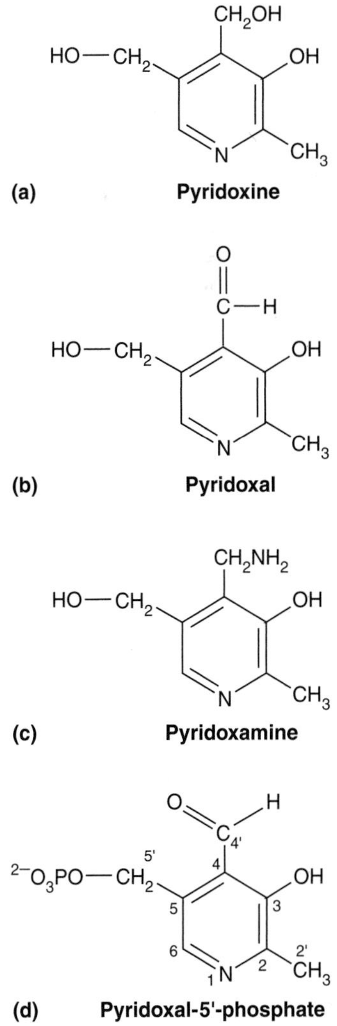

FIGURE 14.4
Vitamin B_6

Vitamin B_6 includes (a) pyridoxine, (b) pyridoxal, and (c) pyridoxamine. (Pyridoxine is found in leafy green vegetables. Pyridoxal and pyridoxamine are found in animal foods such as fish, poultry, and red meat.) The biologically active form of vitamin B_6 is (d) pyridoxal-5′-phosphate.

PLP is bound in the enzyme active site as a Schiff base (R′—CH = N—R, an aldimine) formed by the condensation of the aldehyde group of PLP and the ε-amino group of a lysine residue.

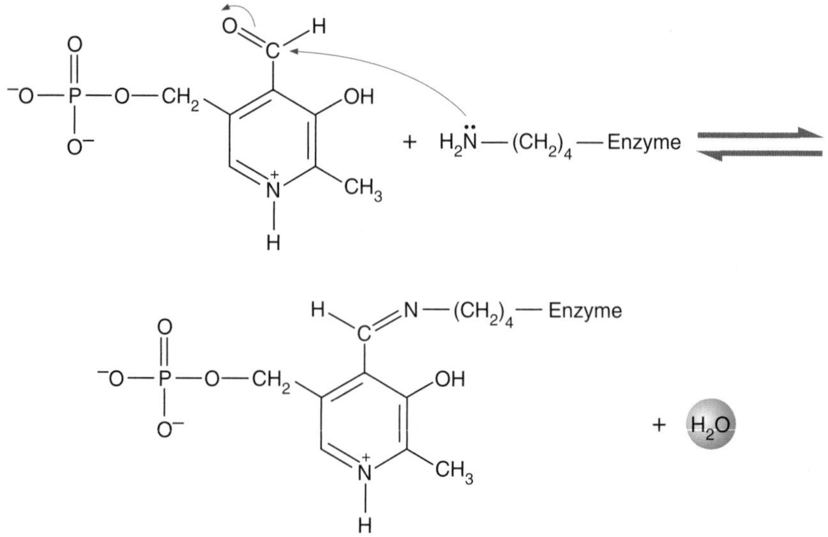

$$^{-}O-P(=O)(O^{-})-O-CH_2\text{-pyridoxal} + H_2\ddot{N}-(CH_2)_4-\text{Enzyme} \rightleftharpoons \text{PLP}-CH=N-(CH_2)_4-\text{Enzyme} + H_2O$$

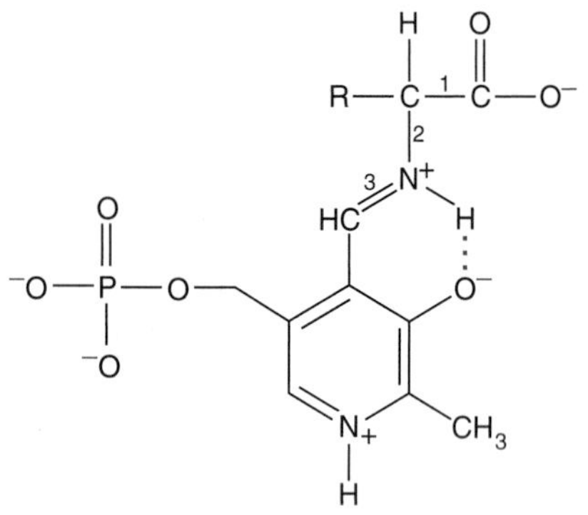

FIGURE 14.5

Intermediate Schiff Base Formed between Pyridoxine and an Amino Acid

When an amino acid binds to pyridoxine in the active site of a PLP-dependent enzyme, one of three bonds will break. This selectivity depends on the presence or absence of a nearby base catalyst and the orientation of the amino acid in the active site. If an initial deprotonation of the α-carbon of the amino group donor occurs, then transamination (bond 2 broken) or racemization or elimination (bond 3 broken) may occur. If the initial deprotonation does not occur, then decarboxylation results (bond 1 broken).

Additional stabilizing forces include ionic interactions between amino acid side chains and PLP's pyridinium ring and phosphate group. The positively charged pyridinium ring also functions as an electron sink, stabilizing negatively charged reaction intermediates.

Amino acid substrates become bound to PLP via the α-amino group in an imine exchange reaction. Then one of three bonds in the substrate is selectively broken in the active sites in each type of PLP-dependent enzyme (**Figure 14.5**).

THE TRANSAMINATION MECHANISM The transamination reaction begins with the formation of a Schiff base between PLP and the α-amino group of an α-amino acid (**Figure 14.6**). When the α-hydrogen atom is removed by a general base in the enzyme active site, a resonance-stabilized intermediate forms. With the donation of a proton from a general acid and a subsequent hydrolysis, the newly formed α-keto acid is released from the enzyme. A second α-keto acid then enters the active site and is converted into an α-amino acid in a reversal of the reaction process that has just been described. Transamination reactions are examples of a reaction mechanism referred to as a double displacement, or *ping-pong reaction* (p. 205). The mechanism is so named because the first substrate must leave the active site before the second one can enter.

Transamination reactions are reversible. It is, therefore, theoretically possible for all amino acids to be synthesized by transamination. However, experimental evidence indicates that there is no net synthesis of an amino acid if its α-keto acid precursor is not independently synthesized by the organism. For example, alanine, aspartate, and glutamate are nonessential for animals because their α-keto acid precursors (i.e., pyruvate, OAA, and α-ketoglutarate) are readily available metabolic intermediates. Because the reaction pathways for synthesizing molecules such as phenylpyruvate, α-keto-β-hydroxybutyrate, and imidazole pyruvate do not occur in animal cells, phenylalanine, threonine, and histidine must be provided in the diet. (Reaction pathways that synthesize amino acids from metabolic intermediates, and not by transamination, are referred to as *de novo* pathways.)

DIRECT INCORPORATION OF AMMONIUM IONS INTO ORGANIC MOLECULES There are two principal means by which ammonium ions are incorporated into amino acids and eventually other metabolites: (1) reductive

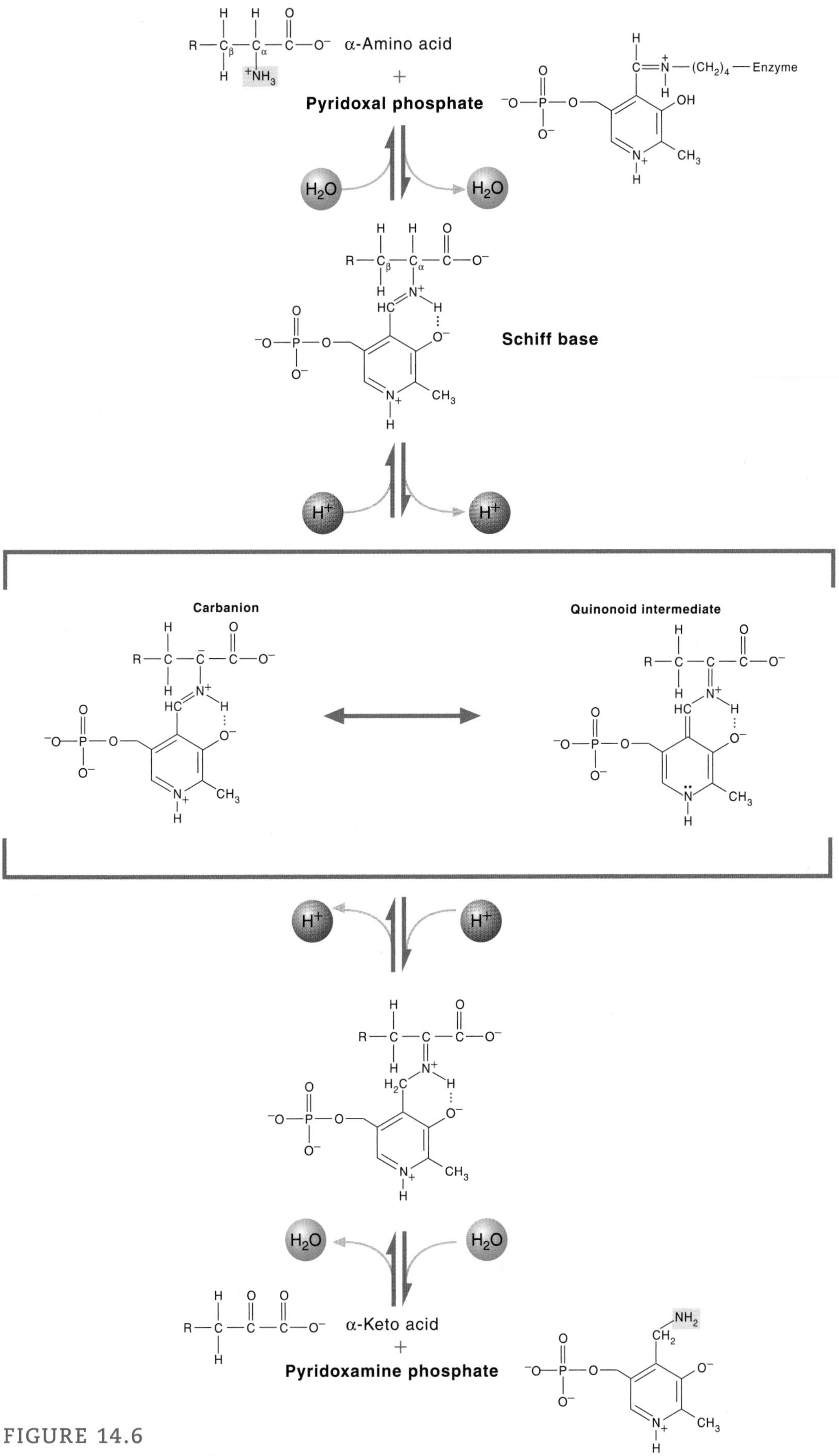

α-Amino acid
+
Pyridoxal phosphate

H_2O H_2O

Schiff base

H^+ H^+

Carbanion

Quinonoid intermediate

H^+ H^+

H_2O H_2O

α-Keto acid
+
Pyridoxamine phosphate

FIGURE 14.6

The Transamination Mechanism

The donor amino acid forms a Schiff base with the coenzyme pyridoxal phosphate within the enzyme's active site. A proton is lost, and a carbanion forms the coenzyme, which is resonance-stabilized by interconversion to a quinonoid intermediate. After an enzyme-catalyzed proton transfer and a hydrolysis, the α-keto product is released. A second α-keto acid then enters the active site. This acceptor α-keto acid is converted to an α-amino acid product as the mechanism just described is reversed. Note that the chirality of the donor amino acid is preserved in the α-amino acid product. Within the active site, the orientation of the quinonoid intermediate allows the proton to be added in a manner that confers on the resulting product, the Schiff base, an L-configuration.

amination of α-keto acids and (2) formation of the amides of aspartic and glutamic acid with subsequent transfer of the amide nitrogen to form other amino acids.

Glutamate dehydrogenase, an enzyme found in both the mitochondria and the cytoplasm of eukaryotic cells and in some bacterial cells, catalyzes the direct amination of α-ketoglutarate:

$$^{-}O{-}\overset{O}{\overset{\|}{C}}{-}CH_2{-}CH_2{-}\overset{O}{\overset{\|}{C}}{-}\overset{O}{\overset{\|}{C}}{-}O^{-} + NH_4^{+} + NADH + H^{+} \rightleftharpoons$$

α-Ketoglutarate

$$^{-}O{-}\overset{O}{\overset{\|}{C}}{-}CH_2{-}CH_2{-}\underset{+NH_3}{\underset{|}{\overset{H}{\overset{|}{C}}}}{-}\overset{O}{\overset{\|}{C}}{-}O^{-} + NAD^{+} + H_2O \qquad (2)$$

Glutamate

The primary function of this enzyme in eukaryotes appears to be catabolic (i.e., a means of producing NH_4^+ in preparation for nitrogen excretion). However, the reaction is reversible. When excess ammonia is present, the reaction is driven toward glutamate synthesis.

Ammonium ions are also incorporated into cell metabolites by the formation of glutamine, the amide of glutamate:

$$^{-}O{-}\overset{O}{\overset{\|}{C}}{-}CH_2{-}CH_2{-}\underset{+NH_3}{\underset{|}{\overset{H}{\overset{|}{C}}}}{-}\overset{O}{\overset{\|}{C}}{-}O^{-} + ATP + NH_4^{+} \longrightarrow$$

Glutamate

$$H_2N{-}\overset{O}{\overset{\|}{C}}{-}CH_2{-}CH_2{-}\underset{+NH_3}{\underset{|}{\overset{H}{\overset{|}{C}}}}{-}\overset{O}{\overset{\|}{C}}{-}O^{-} + ADP + P_i \qquad (3)$$

Glutamine

KEY CONCEPTS

- In transamination reactions, amino groups are transferred from one carbon skeleton to another.
- In reductive amination, amino acids are synthesized by the incorporation of free NH_4^+ or the amide nitrogen of glutamine or asparagine into α-keto acids.
- Ammonium ions are also incorporated into cellular metabolites by the amination of glutamate to form glutamine.

The brain, a rich source of the enzyme glutamine synthetase, is especially sensitive to the toxic effects of NH_4^+. Brain cells convert NH_4^+ and glutamate to glutamine, a neutral, nontoxic molecule. Glutamine is then transported to the liver, where the amide nitrogen is released as NH_4^+ (p. 560). Ammonium ion is then disposed of by incorporation into urea, the principal nitrogenous waste product in mammals.

Biosynthesis of the Amino Acids

Despite the tremendous diversity of amino acid synthetic pathways, they have one common feature. The carbon skeleton of each amino acid is derived from a commonly available metabolic intermediate. Thus in animals, all NAA molecules are derivatives of glycerate-3-phosphate, pyruvate, α-ketoglutarate, or OAA. Tyrosine, synthesized from the EAA phenylalanine, is an exception to this rule.

On the basis of the similarities in their synthetic pathways, the amino acids can be grouped into six families: glutamate, serine, aspartate, pyruvate, the aromatics, and histidine (**Figure 14.7**). The amino acids in each family are ultimately derived from one precursor molecule. In the discussions of amino acid synthesis that follow, the intimate relationship between amino acid metabolism

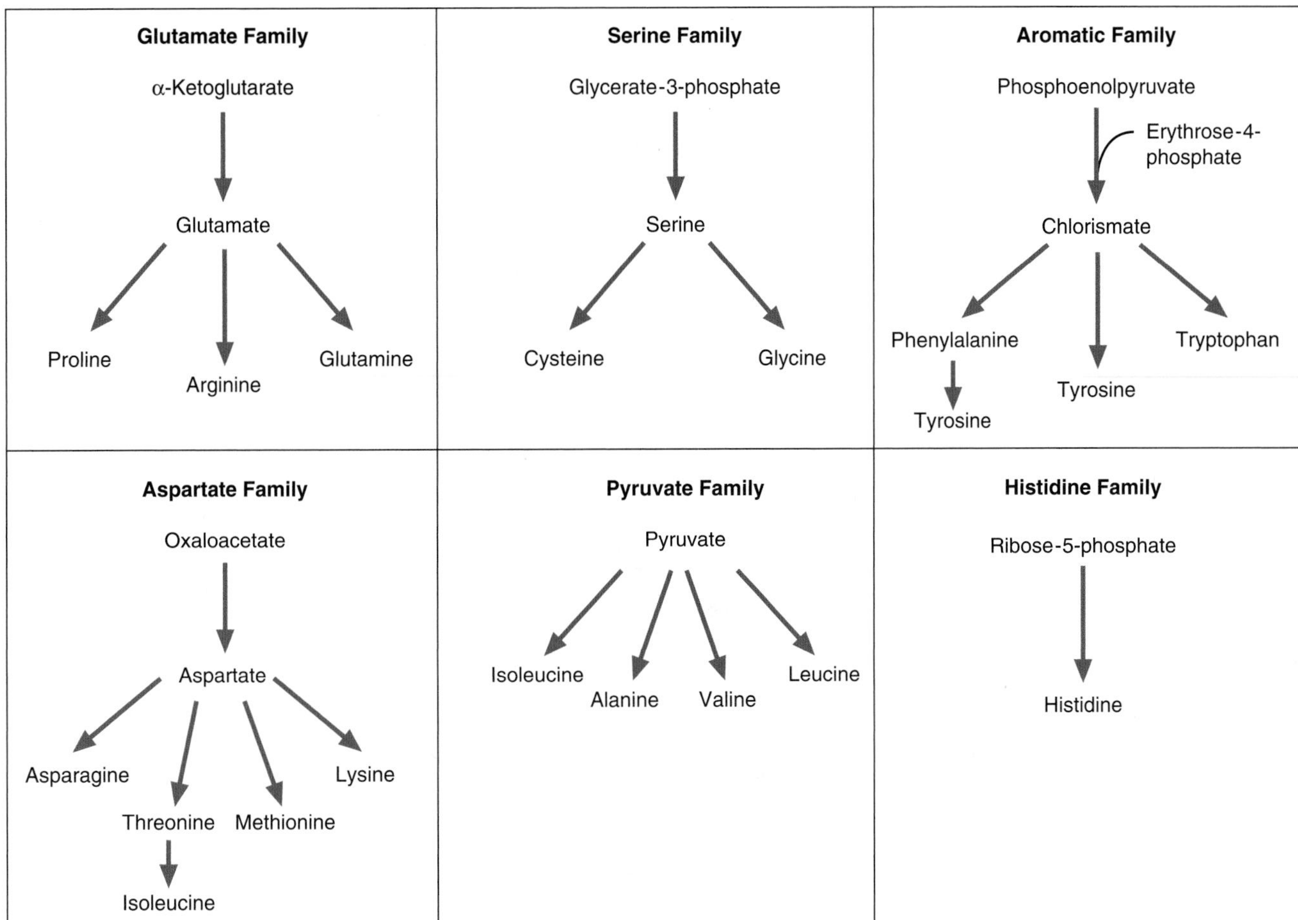

FIGURE 14.7
The Amino Acid Biosynthetic Families
Each family of amino acids is derived from a common precursor molecule.

and several other metabolic pathways is apparent. Amino acid biosynthesis is outlined in **Figure 14.8**.

THE GLUTAMATE FAMILY The glutamate family includes—in addition to glutamate—glutamine, proline, and arginine. As described, α-ketoglutarate may be converted to glutamate by reductive amination and by transamination reactions involving a number of amino acids. Although the relative contribution of these reactions to glutamate synthesis varies with cell type and metabolic circumstances, transamination appears to play a major role in the synthesis of most glutamate molecules in eukaryotic cells.

The conversion of glutamate to glutamine, catalyzed by glutamine synthetase, takes place in liver, brain, kidney, muscle, and intestine. BCAA are an important source of amino groups in glutamine synthesis. As mentioned, blood that leaves the liver is selectively enriched in BCAA. Many more BCAA are taken up by peripheral tissues than are needed for protein synthesis. BCAA amino groups may be used primarily for the synthesis of OAA. In addition to its role in protein synthesis, glutamine is the amino group donor in numerous biosynthetic reactions (e.g., purine, pyrimidine, and amino sugar syntheses) and, as previously mentioned, as a safe storage and transport form of NH_4^+. Glutamine is therefore a major metabolite in living organisms. Other functions of glutamine vary, depending on the cell type being considered. For example, in the kidney and small intestine, glutamine is a major source of energy. In the small intestine, approximately 55% of glutamine carbon is oxidized to CO_2.

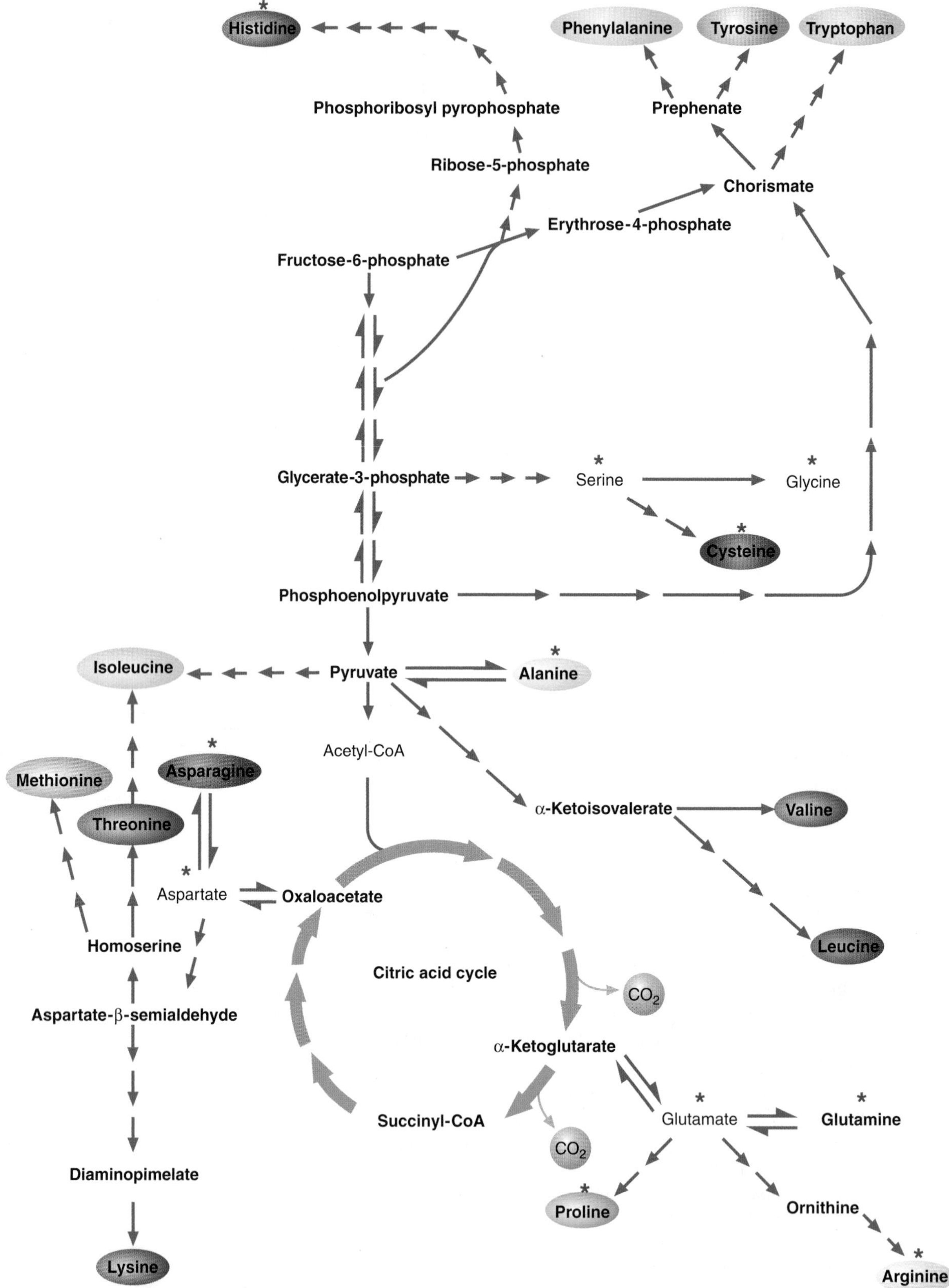

FIGURE 14.8

Biosynthesis of the Amino Acids

Intermediates in the central metabolic pathways provide the carbon skeleton precursor molecules required for the synthesis of each amino acid. The number of reactions in each pathway is indicated. The nonessential amino acids for mammals are indicated by asterisks. (In mammals, tyrosine can be synthesized from phenylalanine.)

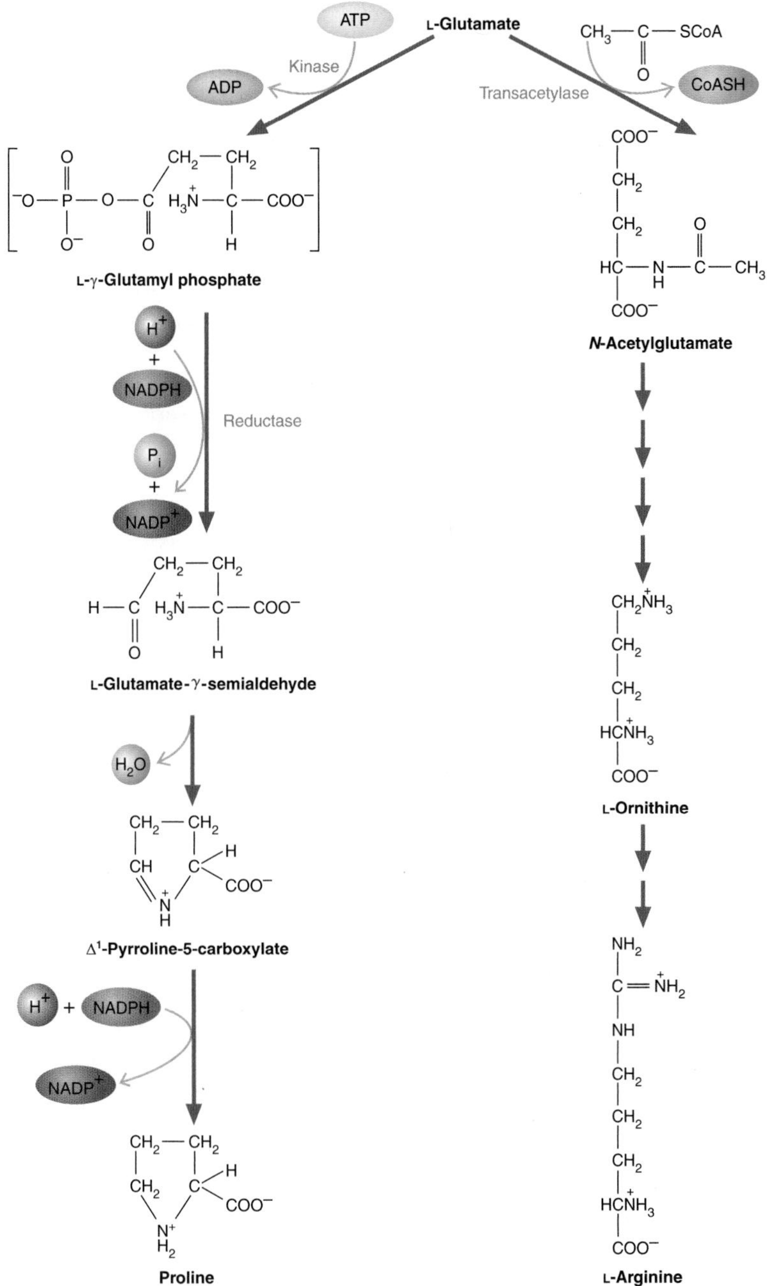

FIGURE 14.9

Biosynthesis of Proline and Arginine from Glutamate

Proline is synthesized from glutamate in four steps. The third step is a spontaneous cyclization reaction. In arginine synthesis the acetylation of glutamate prevents the cyclization reaction. In mammals the reactions that convert ornithine to arginine are part of the urea cycle.

Proline is a cyclized derivative of glutamate. As shown in **Figure 14.9**, a γ-glutamyl phosphate intermediate is reduced to glutamate-γ-semialdehyde. The enzyme catalyzing the phosphorylation of glutamate (γ-glutamyl kinase) is regulated by negative feedback inhibition by proline. Glutamate-γ-semialdehyde cyclizes spontaneously to form Δ^1-pyrroline-5-carboxylate. Then Δ^1-pyrroline-5-carboxylate reductase catalyzes the reduction of Δ^1-pyrroline-5-carboxylate to form proline. The interconversion of Δ^1-pyrroline-5-carboxylate and proline may act as a shuttle mechanism to transfer reducing equivalents derived from the pentose phosphate pathway into mitochondria. This process may partially explain the high turnover of proline in many cell types. Proline can also be synthesized from ornithine, a urea cycle intermediate. The enzyme catalyzing ornithine's conversion to glutamate-γ-semialdehyde, ornithine aminotransferase, is found in relatively high concentration in fibroblasts where the demand for proline incorporation into collagen is high.

Arginine synthesis begins with the acetylation of the α-amino group of glutamate. *N*-Acetylglutamate is then converted to ornithine in a series of reactions that include a phosphorylation, a reduction, a transamination, and a deacetylation (removal of an acetyl group). The subsequent reactions in which ornithine is converted to arginine are part of the urea cycle (p. 560). In infants, in whom the urea cycle is insufficiently functional, arginine is an EAA.

THE SERINE FAMILY The members of the serine family—serine, glycine, and cysteine—derive their carbon skeletons from the glycolytic intermediate glycerate-3-phosphate. The members of this group play important roles in numerous anabolic pathways. Serine is a precursor of ethanolamine and sphingosine. Glycine is used in the purine, porphyrin, and glutathione synthetic pathways. Together, serine and glycine contribute to a series of biosynthetic pathways that are referred to collectively as one-carbon metabolism (p. 528). Cysteine plays a significant role in sulfur metabolism (p. 569).

Serine is synthesized in a direct pathway from glycerate-3-phosphate that involves dehydrogenation, transamination, and hydrolysis by a phosphatase (**Figure 14.10**). Cellular serine concentration controls the pathway through feedback inhibition of phosphoglycerate dehydrogenase and phosphoserine phosphatase. The latter enzyme catalyzes the only irreversible step in the pathway.

The conversion of serine to glycine consists of a single complex reaction catalyzed by serine hydroxymethyltransferase, a pyridoxal phosphate–requiring enzyme. During the reaction, which is an aldol cleavage, serine binds to pyridoxal phosphate. The reaction yields glycine and a chemically reactive formaldehyde group that is transferred to the coenzyme tetrahydrofolate (THF) (p. 529) to form N^5,N^{10}-methylene tetrahydrofolate. Serine is the major source of glycine. Smaller amounts of glycine can be derived from choline when the latter molecule is present in excess. The synthesis of glycine from choline consists of two dehydrogenations and a series of demethylations.

Cysteine synthesis is a primary component of sulfur metabolism (p. 569). The carbon skeleton of cysteine is derived from serine (**Figure 14.11**). In animals the sulfhydryl group is transferred from methionine by way of a demethylated derivative homocysteine. Both enzymes involved in the conversion of serine to cysteine (cystathionine β-synthase, or CBS, and γ-cystathionase, or CSE) require pyridoxal phosphate.

THE ASPARTATE FAMILY Aspartate, the first member of the aspartate family of amino acids, is derived from OAA in a transamination reaction:

$$\underset{\textbf{Glutamate}}{{}^{-}O-\overset{\overset{O}{\|}}{C}-CH_2-CH_2-\underset{\underset{+NH_3}{|}}{\overset{\overset{H}{|}}{C}}-\overset{\overset{O}{\|}}{C}-O^-} + \underset{\textbf{Oxaloacetate}}{{}^{-}O-\overset{\overset{O}{\|}}{C}-CH_2-\overset{\overset{O}{\|}}{C}-\overset{\overset{O}{\|}}{C}-O^-} \rightleftharpoons$$

$$\underset{\alpha\textbf{-Ketoglutarate}}{O-\overset{\overset{O}{\|}}{C}-CH_2-CH_2-\overset{\overset{O}{\|}}{C}-\overset{\overset{O}{\|}}{C}-O^-} + \underset{\textbf{Aspartate}}{{}^{-}O-\overset{\overset{O}{\|}}{C}-CH_2-\underset{\underset{+NH_3}{|}}{\overset{\overset{H}{|}}{C}}-\overset{\overset{O}{\|}}{C}-O^-} \quad (4)$$

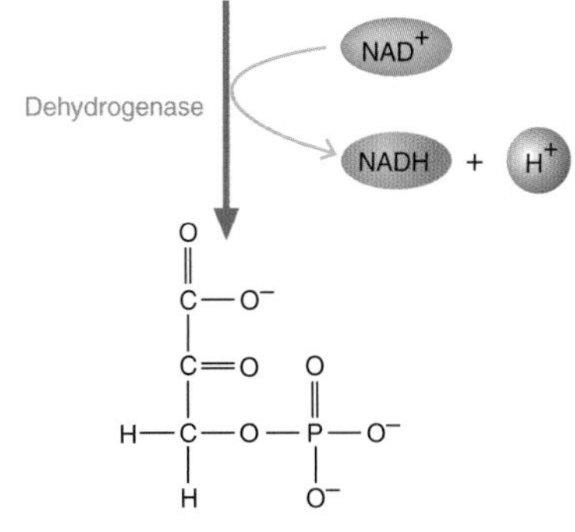

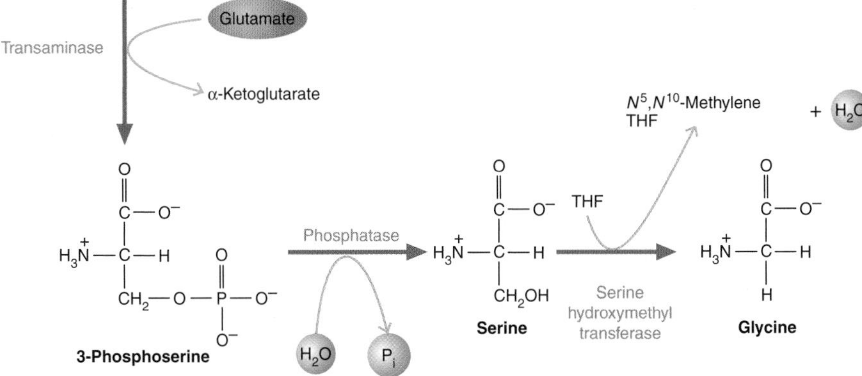

FIGURE 14.10

Biosynthesis of Serine and Glycine

Serine biosynthesis begins with the oxidation of glycerate-3-phosphate. The carbonyl-containing product 3-phosphohydroxypyruvate then undergoes a transamination reaction with glutamate to yield 3-phosphoserine. The hydrolysis of 3-phosphoserine yields serine. Serine hydroxymethyl transferase catalyzes the conversion of serine to glycine and N^5, N^{10}-methylene THF. Serine inhibits glycerate-3-phosphate dehydrogenase, the first reaction in the pathway.

Aspartate transaminase (AST) (also known as glutamic oxaloacetic transaminase or GOT), the most active of the aminotransferases, is found in most cells. AST isozymes occur in both mitochondria and the cytoplasm, and the reaction that it catalyzes is reversible. This enzymatic activity, therefore, significantly influences the flow of carbon and nitrogen within the cell. For example, excess glutamate is converted via AST to aspartate. Aspartate is then used as a source of both nitrogen (for urea formation) and the citric acid cycle intermediate OAA. Aspartate is also an important precursor in nucleotide synthesis.

Visit the companion website at www.oup.com/us/mckee to read the Biochemistry in Perspective essay on the essential amino acids.

The aspartate family also contains asparagine, lysine, methionine, and threonine. Threonine contributes to the reaction pathway in which isoleucine is synthesized. The synthesis of isoleucine, often considered a member of the pyruvate family, is discussed in the online Biochemistry in Perspective essay The Essential Amino Acids.

Asparagine, the amide of aspartate, is not formed directly from aspartate and NH_4^+. Instead, the amide group of glutamine is transferred by amide group transfer during an ATP-requiring reaction catalyzed by asparagine synthase:

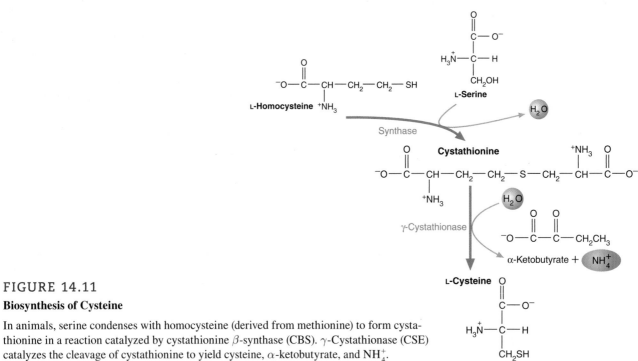

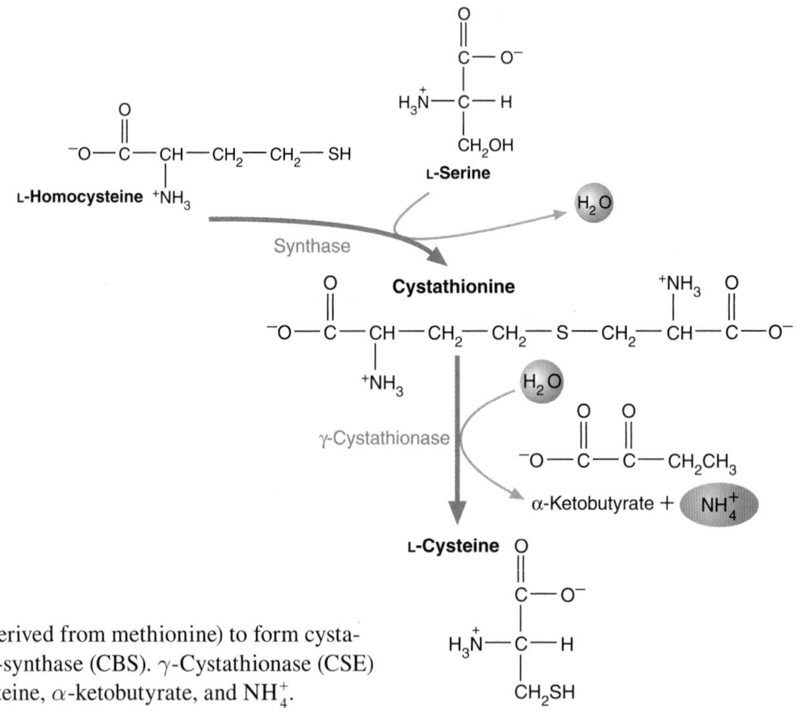

FIGURE 14.11

Biosynthesis of Cysteine

In animals, serine condenses with homocysteine (derived from methionine) to form cystathionine in a reaction catalyzed by cystathionine β-synthase (CBS). γ-Cystathionase (CSE) catalyzes the cleavage of cystathionine to yield cysteine, α-ketobutyrate, and NH_4^+.

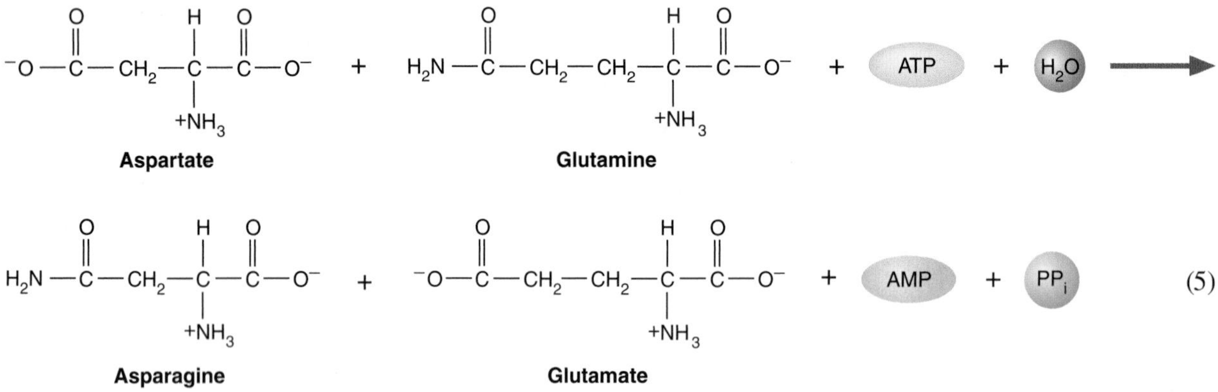

$$\text{Aspartate} + \text{Glutamine} + \text{ATP} + H_2O \longrightarrow \text{Asparagine} + \text{Glutamate} + \text{AMP} + PP_i \quad (5)$$

The biosynthesis of the EAA in the aspartate family (lysine, methionine, and threonine) is described briefly in the online reading The Essential Amino Acids.

THE PYRUVATE FAMILY The pyruvate family consists of alanine, valine, leucine, and isoleucine. Alanine is synthesized from pyruvate in a single step:

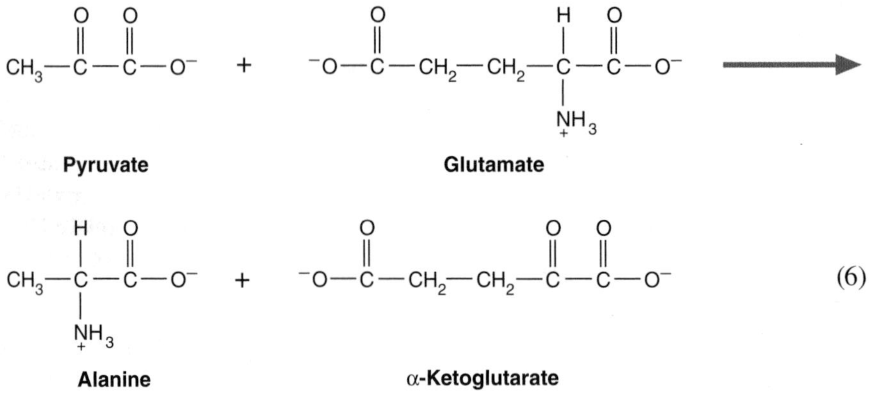

$$\text{Pyruvate} + \text{Glutamate} \longrightarrow \text{Alanine} + \alpha\text{-Ketoglutarate} \quad (6)$$

Although the enzyme that catalyzes this reaction, alanine aminotransferase, has cytoplasmic and mitochondrial forms, the majority of its activity has been found in the cytoplasm. Recall that the alanine cycle (Chapter 8) contributes to the maintenance of blood glucose. BCAA are the ultimate source of many of the amino groups transferred from glutamate in the alanine cycle. Their biosynthesis is described in the online reading The Essential Amino Acids.

THE AROMATIC FAMILY The aromatic family of amino acids includes phenylalanine, tyrosine, and tryptophan. Of these, only tyrosine is considered nonessential in mammals. Either phenylalanine or tyrosine is required for the synthesis of dopamine, epinephrine, and norepinephrine, an important class of biologically potent molecules referred to as the *catecholamines* (pp. 534–536). Tryptophan is a precursor in the synthesis of NAD^+, $NADP^+$, and the neurotransmitter serotonin.

The benzene ring of the aromatic amino acids is formed by the *shikimate pathway* (**Figure 14.12**). The carbons in the benzene ring are derived from erythrose-4-phosphate and PEP. These two molecules condense to form a molecule that is subsequently converted to chorismate. Chorismate is the branch point in the syntheses of various aromatic compounds. For example, the aromatic rings in the mixed terpenoids (e.g., the tocopherols, the ubiquinones, and plastoquinone) are derived from chorismate.

Tyrosine is not an EAA in animals because it is synthesized from phenylalanine in a hydroxylation reaction. The enzyme involved, phenylalanine-4-monooxygenase, requires the coenzyme tetrahydrobiopterin (BH_4, pp. 534 and 568), a folic acid–like molecule derived from GTP. Because this reaction also is a first step in phenylalanine catabolism, it is discussed further in Chapter 15. The synthesis of the aromatic family of amino acids is described in the online reading The Essential Amino Acids.

HISTIDINE Histidine is considered nonessential in healthy human adults. In human infants and many animals, histidine must be provided by the diet. Histidine contributes substantially to protein structure and function because of its unique chemical properties. Recall, for example, that histidine residues bind heme prosthetic groups in hemoglobin. In addition, histidine often acts as a general acid during enzyme-catalyzed reactions.

Of all the amino acids, histidine's biosynthesis is the most unusual. Histidine is synthesized from phosphoribosylpyrophosphate (PRPP), ATP, and glutamine (**Figure 14.13**). Synthesis begins with the condensation of PRPP with ATP to form phosphoribosyl-ATP, which is then hydrolyzed to phosphoribosyl-AMP. Subsequently, a hydrolysis reaction opens the adenine ring. After an isomerization and the transfer of an amino group from glutamine, imidazole glycerol phosphate is synthesized. (The other product of the latter reaction, 5′-phosphoribosyl-4-carboxamide-5-aminoimidazole, is used in the synthesis of purine nucleotides. See Section 14.3.) Histidine is produced from imidazole glycerol phosphate in a series of reactions that include a dehydration, a transamination, a phosphorolysis, and an oxidation.

Phosphoenolpyruvate

+

Erythrose-4-phosphate

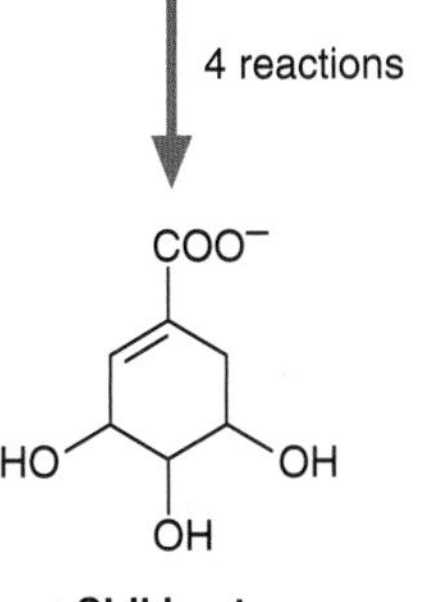

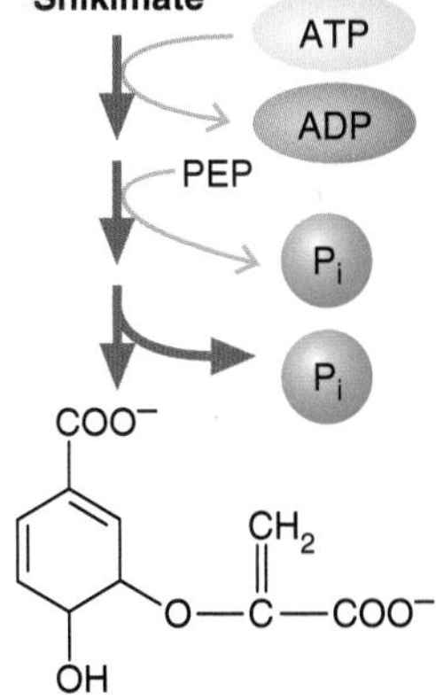

FIGURE 14.12

Chorismate Biosynthesis

Chorismate is an intermediate in the shikimate pathway. The formation of chorismate involves the ring closure of an intermediate (not shown) and the subsequent creation of two double bonds. The side chain of chorismate is derived from phosphoenolpyruvate (PEP).

14.3 BIOSYNTHETIC REACTIONS INVOLVING AMINO ACIDS

As described, amino acids are precursors of many physiologically important nitrogen-containing molecules, in addition to serving as building blocks for polypeptides. In the following discussion the syntheses of several examples of these molecules (neurotransmitters, glutathione, and nucleotides) are described. The synthesis of heme and chlorophyll is described in a reading, available online.

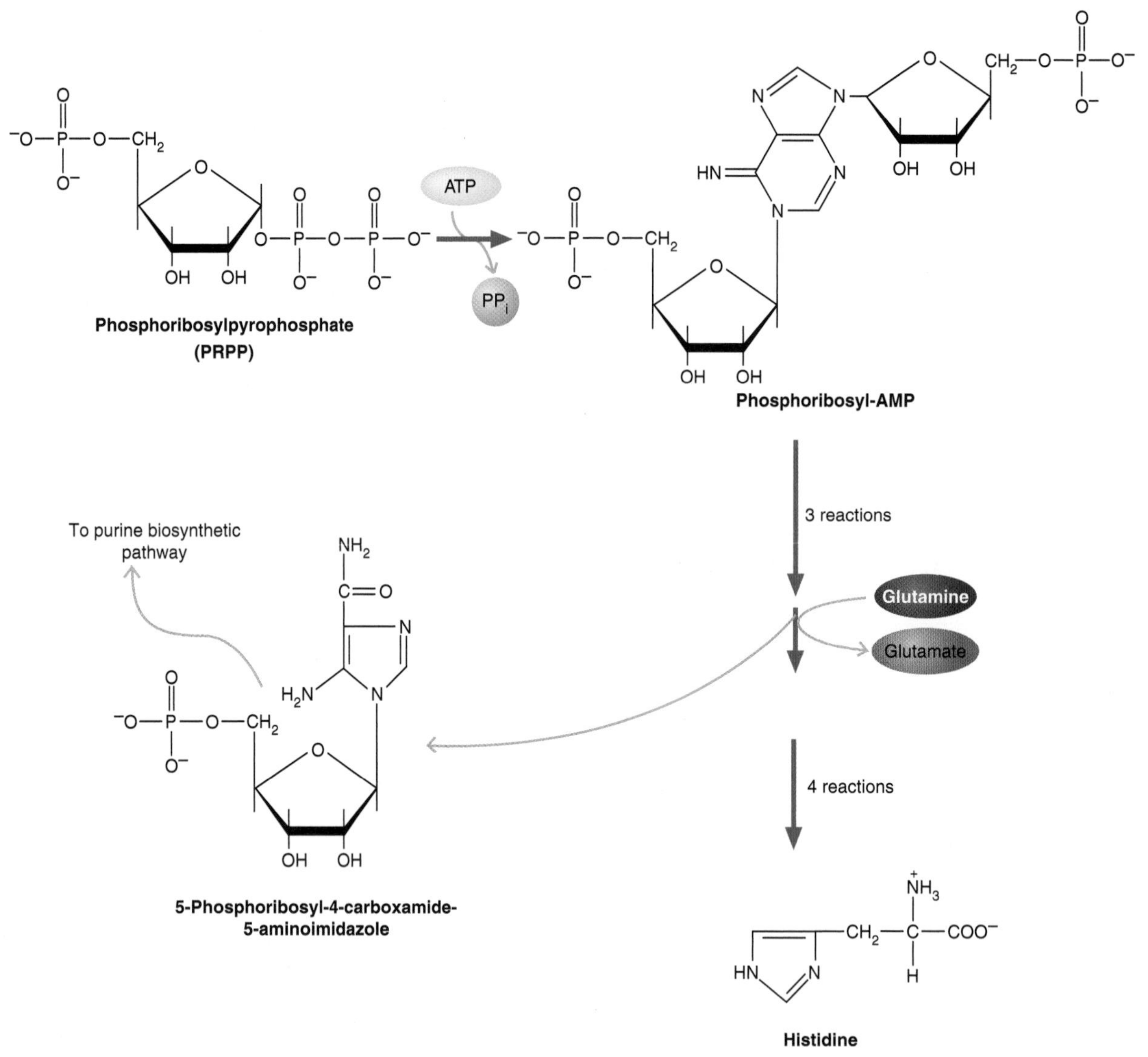

FIGURE 14.13

Histidine Biosynthesis

Histidine is derived from three biomolecules: phosphoribosylpyrophosphate (PRPP) (five carbons), the adenine ring from ATP (one nitrogen and one carbon), and glutamine (one nitrogen). The ATP used in the first reaction in the pathway is regenerated when 5-phosphoribosyl-4-carboxamide-5-aminoimidazole (released in a subsequent reaction) is diverted into the purine nucleotide biosynthetic pathway.

KEY CONCEPTS

- There are six families of amino acids: glutamate, serine, aspartate, pyruvate, the aromatics, and histidine.
- The nonessential amino acids are derived from precursor molecules available in many organisms.
- The essential amino acids are synthesized from metabolites produced only in plants and some microorganisms.

Because many biosynthetic processes involve the transfer of carbon groups, this section begins with a brief description of one-carbon metabolism.

One-Carbon Metabolism

One-carbon metabolism is a set of reactions in which single carbon atoms are transferred from one molecule to another. Carbon atoms have several oxidation states. Those of biological interest are found in methanol (+1), formaldehyde (+2), and formate (+3). **Table 14.2** lists the equivalent one-carbon groups that are actually involved in biosynthetic reactions. The most important carriers of one-carbon groups in biosynthetic pathways are folic acid and *S*-adenosylmethionine.

TABLE 14.2 One-Carbon Groups

Oxidation Level	Methanol (most reduced)	Formaldehyde	Formate (most oxidized)
One-carbon group	Methyl (—CH_3)	Methylene (—CH_2—)	Formyl (—CHO) Methenyl (—CH=)

The metabolism of each is described briefly. (The function of biotin, a carrier of CO_2 groups, is discussed in Section 8.2.)

FOLIC ACID Folic acid, also known as folate or folacin, is a B vitamin found in beans, peas, broccoli, beets, spinach, and sunflower seeds. Folic acid's structure consists of a pteridine nucleus and *para*-aminobenzoic acid, linked to one or more glutamic acid residues (**Figure 14.14**). Once absorbed by the body, folic acid is converted by dihydrofolate reductase to the biologically active form, **tetrahydrofolate (THF)**. The carbon units carried by THF (i.e., methyl, methylene, methenyl, and formyl groups) are bound to N^5 of the pteridine ring and/or N^{10} of the para-aminobenzoate ring. **Figure 14.15** illustrates the interconversions of the one-carbon units carried by THF, as well as their origin and metabolic fate. A substantial number of one-carbon units enter the THF pool as N^5,N^{10}-methylene THF, produced during the conversion of serine to glycine and the cleavage of glycine (catalyzed by glycine synthase).

Methylcobalamin, a coenzyme form of vitamin B_{12}, is required for the N^5-methyl THF–dependent conversion of homocysteine to methionine (**Figure 14.15**). **Vitamin B_{12}** (cobalamin) is a complex, cobalt-containing molecule synthesized only by microorganisms (**Figure 14.16**). (During the purification of cobalamin, a cyanide group attaches to cobalt.) Another coenzyme form of vitamin B_{12}, 5′-deoxyadenosylcobalamin, is required for the isomerization of methylmalonyl-CoA to succinyl-CoA, which is catalyzed by methylmalonyl-CoA

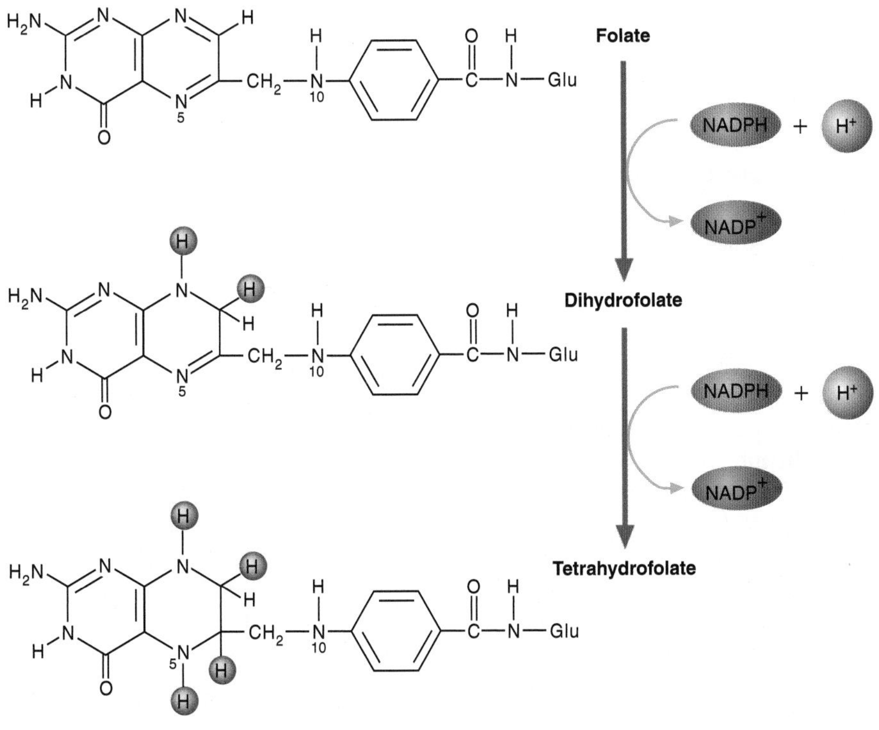

Folate

NADPH + H+

NADP+

Dihydrofolate

NADPH + H+

NADP+

Tetrahydrofolate

FIGURE 14.14
Biosynthesis of Tetrahydrofolate (THF)
The vitamin folic acid (folate) is converted to its biologically active form by two successive reductions of the pteridine ring. Both reactions are catalyzed by dihydrofolate reductase.

FIGURE 14.15

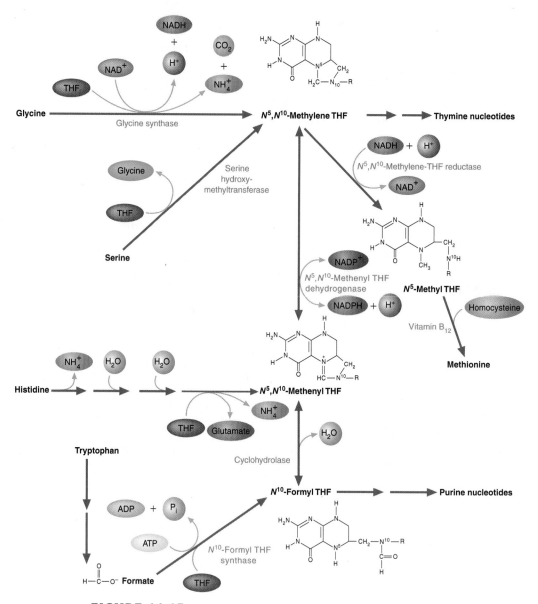

Structures and Enzymatic Interconversions of THF Coenzymes

The THF coenzymes play a critical role in one-carbon metabolism. The interconversions of the coenzymes are reversible except for the conversion of N^5, N^{10}-methylene THF to N^5-methyl THF.

mutase (refer to **Figures 12.12** and **15.11**). Animals obtain cobalamin from intestinal flora and by consuming foods derived from other animals (e.g., liver, eggs, shrimp, chicken, and pork).

A deficiency of vitamin B_{12} results in *pernicious anemia*. In addition to low red blood cell counts, the symptoms of this malady include weakness and various neurological disturbances. Pernicious anemia is most often caused by decreased secretion of intrinsic factor, a glycoprotein secreted by stomach cells, which is required for the absorption of the vitamin in the intestine. Vitamin B_{12} absorption can also be inhibited by several gastrointestinal disorders, such as celiac disease and tropical sprue, both of which damage the lining of the intestine. A reduction in vitamin B_{12} absorption has also been observed in the presence of intestinal overgrowths of microorganisms induced by antibiotic treatments.

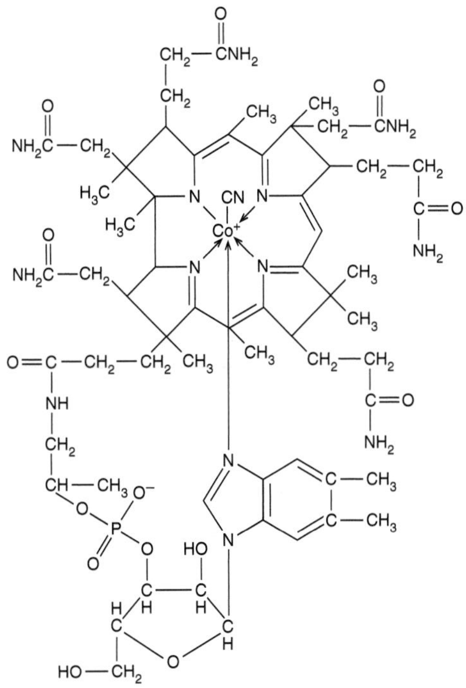

FIGURE 14.16

Structure of Cyanocobalamin, a Derivative of Vitamin B_{12}

During the purification of cobalamin, a cyanide group attaches to cobalt.

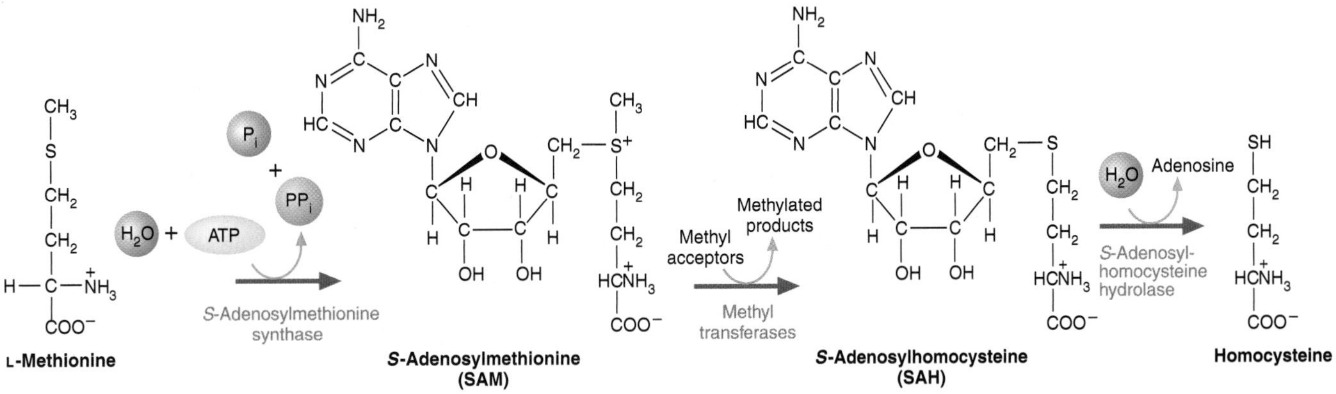

FIGURE 14.17

The Formation of *S*-Adenosylmethionine (SAM)

One of the principal functions of SAM is to serve as a methylating agent. The SAM product of these reactions is then hydrolyzed to form homocysteine.

TABLE 14.3 Examples of Transmethylation Acceptors and Products

Methyl Acceptors	Methylated Product
Phosphatidylethanolamine	Phosphatidylcholine (p. 392)
Norepinephrine	Epinephrine (p. 309)
Guanidinoacetate	Creatine (p. 533)
γ-Aminobutyric acid	Carnitine (p. 433)

***S*-ADENOSYLMETHIONINE** ***S*-Adenosylmethionine (SAM)** is the major methyl group donor in one-carbon metabolism. Formed from methionine and ATP (**Figure 14.17**), SAM contains an "activated" methyl thioether group, which can be transferred to a variety of acceptor molecules (**Table 14.3**). *S*-Adenosylhomocysteine (SAH) is a product in these reactions. The loss of free

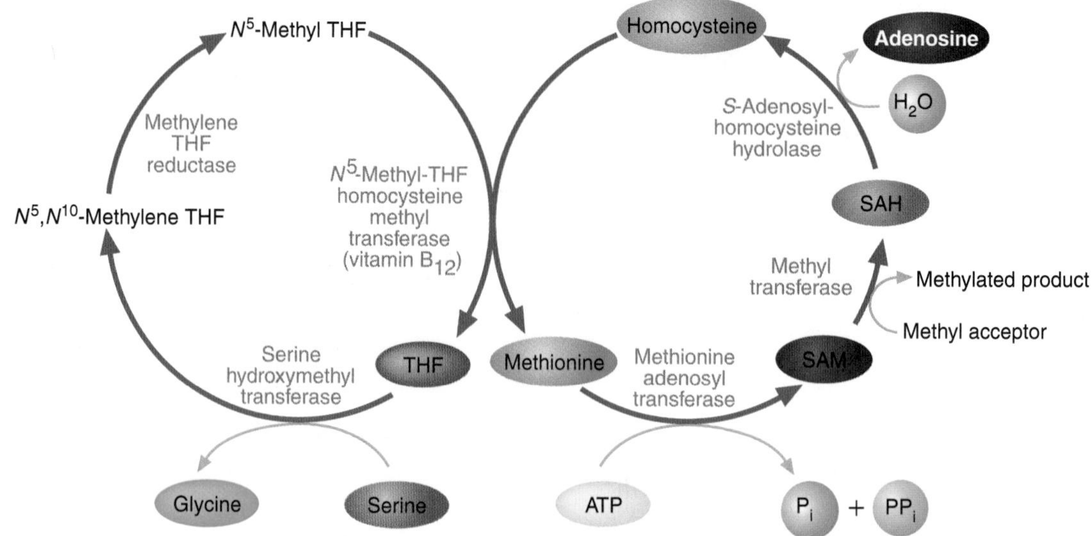

FIGURE 14.18

The Tetrahydrofolate (THF) and *S*-Adenosylmethionine (SAM) Pathways

The THF and SAM pathways intersect at the reaction, catalyzed by N^5-methyl THF homocysteine methyltransferase, in which homocysteine is converted to methionine.

energy that accompanies SAH formation makes the methyl transfer irreversible. SAM acts as a methyl donor in at least 115 transmethylation reactions, some of which occur in the synthesis of phospholipids, several neurotransmitters, and glutathione. *DNA* methylation reactions play a significant role in gene expression. SAM also plays a role in the synthesis of the polyamines (p. 631) by donating aminopropyl groups. Polyamines are polycationic molecules that bind to DNA when it is compressed into chromosomes.

The importance of SAM in metabolism is reflected in the several mechanisms that provide for the synthesis of sufficient amounts of its precursor, methionine, when the latter molecule is temporarily absent from the diet. For example, choline is used as a source of methyl groups to convert homocysteine into methionine. Homocysteine can also be methylated in a reaction utilizing N^5-methyl THF. This latter reaction is a bridge between the THF and SAM pathways (**Figure 14.18**).

QUESTION 14.3

Amethopterin, also referred to as methotrexate, is a structural analogue of folate. Methotrexate has been used to treat several types of cancer and autoimmune diseases. It has been especially successful in childhood leukemia. Autoimmune diseases treated with methotrexate include rheumatoid arthritis, Crohn's disease, and psoriasis (excessive skin cell production and inflammation).

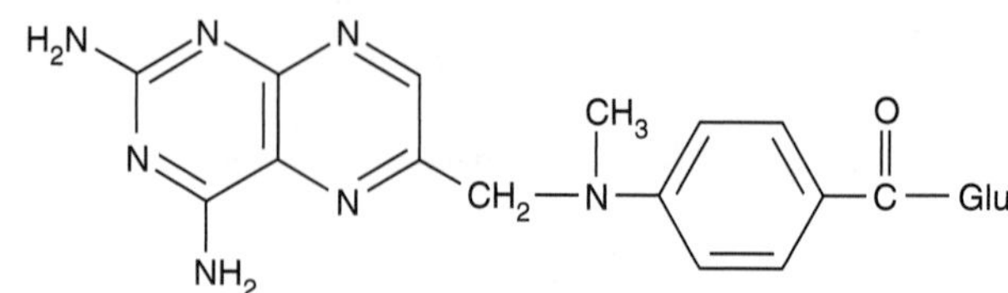

Amethopterin (methotrexate)

Using your knowledge of cell biology and biochemistry, suggest a biochemical mechanism that explains why amethopterin is effective against cancer. [*Hints*: Compare the structures of folate and methotrexate. Review **Figure 14.14**.]

QUESTION 14.4

Melatonin is a hormone derived from serotonin. It is produced in the brain's light-sensitive pineal gland. The pineal's secretion of melatonin is depressed by nerve impulses that originate in the retina of the eye and other light-sensitive tissue in the body in response to light. Pineal function is involved in *circadian rhythms*, patterns of activity associated with light and dark, such as sleep/wake cycles. Melatonin is also a powerful antioxidant, especially in the central nervous system. After serotonin (5-hydroxytryptamine) is produced in the pineal gland, it is converted to 5-hydroxy-*N*-acetyltryptamine by *N*-acetyltransferase. 5-Hydroxy-*N*-acetyltryptamine is then methylated by *O*-methyltransferase. SAM is the methylating agent. With this information, draw the synthetic pathway of melatonin.

QUESTION 14.5

Creatine is a nitrogen-containing organic acid found primarily in muscle and brain. Both of these tissues experience large and fluctuating energy demands. Phosphocreatine, the product of the reaction of creatine and ATP, serves as a short-term storage form of high-energy phosphate. (Refer to **Table 4.1**.) When energy demands are high and available ATP molecules are hydrolyzed, phosphocreatine donates its phosphoryl group to ADP to yield ATP. Most creatine molecules are synthesized in the body in a two-reaction pathway. In the first step arginine and glycine are converted in the kidney to guanidinoacetate and ornithine (p. 523) by L-arginine:glycine amidinotransferase (AGAT).

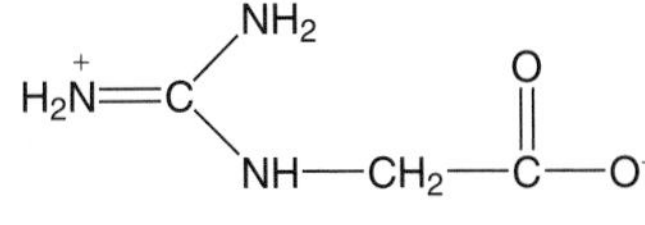

Guanidinoacetate

In liver, guanidinoacetate reacts with SAM to form creatine and SAH in a reaction catalyzed by *S*-adenosyl-L-methionine:*N*-guanidinoacetate methyltransferase (GAMT). With the information provided, write out the creatine biosynthetic pathway.

KEY CONCEPT

Tetrahydrofolate, the biologically active form of folic acid, and *S*-adenosylmethionine are important carriers of single carbon atoms in a variety of synthetic reactions.

Glutathione

With concentrations that average about 5 mM in mammalian cells, glutathione (γ-glutamylcysteinylglycine, see **Table 5.3**) is the most common intracellular reducing agent. Glutathione (GSH) performs several critical functions. First, GSH is a major endogenous antioxidant molecule that quenches superoxide ($O_2^{\bullet-}$), hydroxyl radicals (·OH), and peroxynitrite anions ($ONOO^-$). GSH is a powerful antioxidant because the thiol group of its cysteine residue donates a reducing equivalent (H^+ and e^-) to unstable groups or radicals. GSH is also a cofactor for several antioxidant enzymes. Examples of enzymes that utilize GSH as a reducing agent include glutathione peroxidase (p. 372) and dehydroascorbate reductase, the enzyme that maintains vitamin C in its reduced form. GSH also protects cells from oxidative stress by maintaining the sulfhydryl groups of enzymes and other molecules in a reduced state. Second, GSH has roles in diverse biochemical processes. Examples include DNA synthesis and repair, protein synthesis, and leukotriene synthesis (LTC_4, **Figure 11.4c**). Third, GSH protects cells from xenobiotics. GSH conjugates of these molecules are formed either spontaneously or in reactions catalyzed by the GSH-S-transferases (**Figure 14.19**). Before their excretion, GSH conjugates are usually converted to mercapturic acids by a series of reactions initiated by γ-glutamylpeptidase.

KEY CONCEPTS

- Glutathione (GSH), the most common intercellular thiol, is involved in many cellular activities.
- In addition to reducing sulfhydryl groups, GSH protects cells against toxins and promotes the transport of some amino acids.

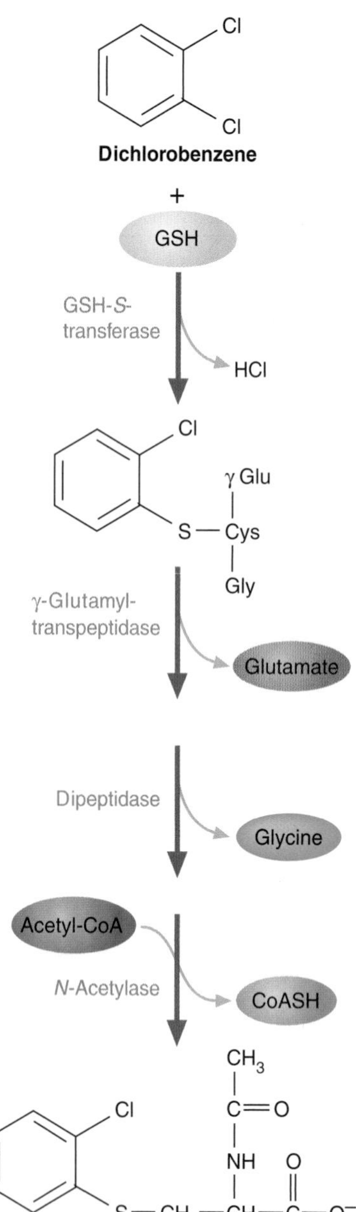

FIGURE 14.19

Formation of a Mercapturic Acid Derivative of a Typical Organic Contaminant

GSH-*S*-transferase catalyzes the synthesis of a GSH derivative of dichlorobenzene.

Visit the companion website at www.oup.com/us/mckee to read the Biochemistry in Perspective essay on the amine neurotransmitters.

GSH is synthesized in a pathway composed of two reactions. In the first reaction, γ-glutamylcysteine synthase catalyzes the condensation of glutamate with cysteine (**Figure 14.20**). γ-Glutamylcysteine, the product of this reaction, then combines with glycine to form GSH in a reaction catalyzed by glutathione synthase.

In certain tissues an additional enzyme, γ-glutamyl transpeptidase, is synthesized that facilitates amino acid transport across the plasma membrane. In this process, called the γ-glutamyl cycle (**Figure 14.20**), the γ-glutamyl transpeptidase catalyzes the reaction between an extracellular amino acid and GSH to yield an intracellular γ-glutamylamino acid and glycine and cysteine. The γ-glutamylamino acid is then hydrolyzed by γ-glutamylcyclotransferase to yield 5-oxoproline and the amino acid. 5-Oxoproline is then converted to glutamate by 5-oxoprolinase (**Figure 14.21**).

GSH synthesis is regulated, in part, by cellular cysteine levels. Because homocysteine is a precursor in cysteine synthesis (**Figure 14.11**), GSH synthesis is linked to both the SAM (**Figure 14.17**) and the transsulfuration pathways. (The transulfuration pathway, described in Chapter 15, is a biochemical pathway that controls cell concentrations of sulfur-containing biomolecules.)

Neurotransmitters

More than 30 different substances, including several amino acids, function as neurotransmitters. **Neurotransmitters**, signal molecules released from neurons, are either excitatory or inhibitory. *Excitatory neurotransmitters* (e.g., glutamate and acetylcholine) open sodium channels and promote the depolarization of the membrane in another cell (either another neuron or an effector cell, such as a muscle cell). If the second (postsynaptic) cell is a neuron, the wave of depolarization (referred to as an action potential) triggers the release of neurotransmitter molecules as it reaches the end of the axon. (Most neurotransmitter molecules are stored in numerous membrane-enclosed *synaptic vesicles*.) When the action potential reaches the nerve ending, the neurotransmitter molecules are released by exocytosis into the synapse. If the postsynaptic cell is a muscle cell, sufficient release of the excitatory neurotransmitter acetylcholine results in muscle contraction. *Inhibitory neurotransmitters* (e.g., glycine) open chloride channels and make the membrane potential in the postsynaptic cell even more negative, that is, they inhibit the formation of an action potential.

A significant percentage of neurotransmitter molecules are either amino acids or amino acid derivatives (**Table 14.4**). The latter class, referred to as the **biogenic amines**, includes γ-aminobutyric acid (GABA), the catecholamines, serotonin, and histamine. The synthesis and inactivation of the inhibitory neurotransmitters γ-aminobutyric acid (GABA, derived from glutamate) and serotonin (derived from tryptophan), are described in the online reading The Amine Neurotransmitters.

THE CATECHOLAMINES The *catecholamines* (dopamine, norepinephrine, and epinephrine) are derivatives of tyrosine. Dopamine (D) and norepinephrine (NE) are used in the brain as excitatory neurotransmitters. Outside the central nervous system, NE and epinephrine (E) are released primarily from the adrenal medulla, as well as the peripheral nervous system. Because both NE and E regulate aspects of metabolism, they are often considered hormones.

The first, and rate-limiting, step in catecholamine synthesis is the hydroxylation of tyrosine to form 3,4-dihydroxyphenylalanine (L-DOPA) (**Figure 14.22**). Tyrosine hydroxylase, the mitochondrial enzyme that catalyzes the reaction, requires a cofactor known as *tetrahydrobiopterin* (BH_4). A folic acid–like molecule, BH_4 (**Figure 15.9** on p. 568) is an essential cofactor in the hydroxylation of aromatic amino acids; it is regenerated from its oxidized metabolite, BH_2, by reduction with NADPH.

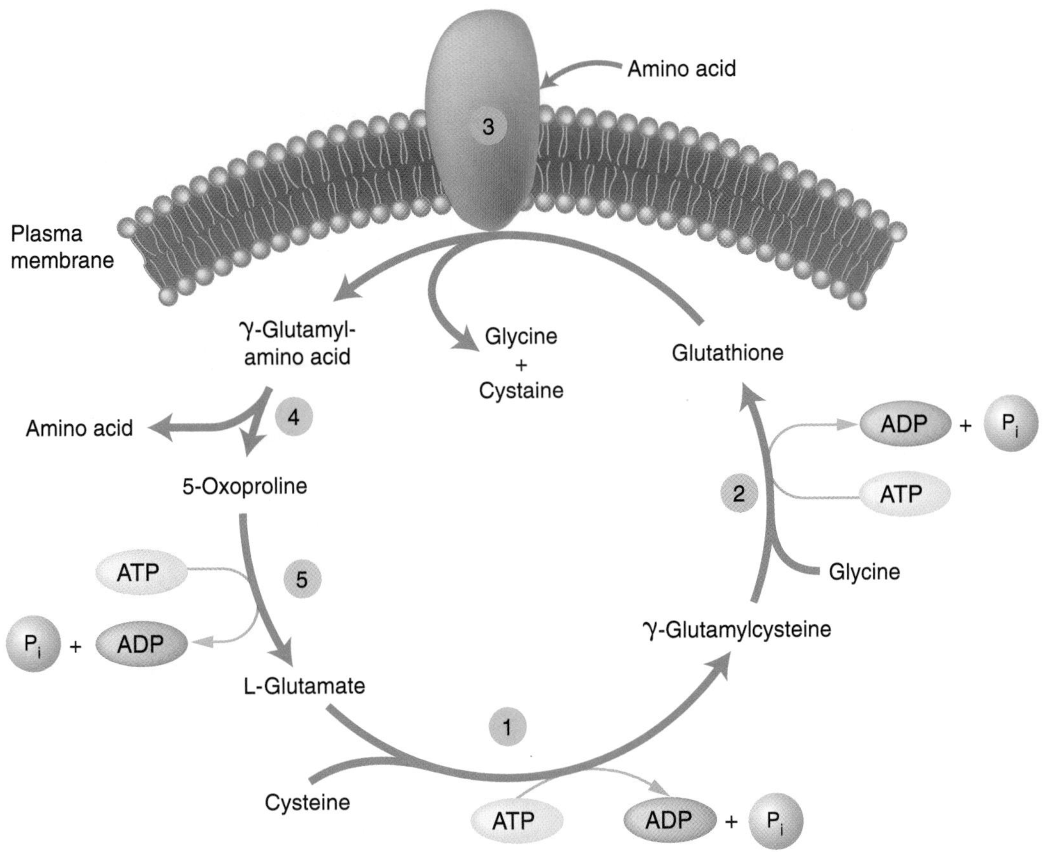

FIGURE 14.20

Glutathione Biosynthesis and the γ-Glutamyl Cycle

GSH is synthesized from glutamate, cysteine, and glycine in two ATP-requiring reactions. γ-Glutamylcysteine synthetase (1) converts glutamate and cysteine into γ-glutamylcysteine. Glutathione synthetase (2) then catalyzes the formation of a peptide bond that links γ-glutamylcysteine to glycine, yielding glutathione. In some tissues (e.g., intestinal and kidney cells) amino acids are transported across the plasma membrane because an additional enzyme, γ-glutamyl transpeptidase (3), is produced that catalyzes a reaction between GSH and an amino acid to yield a γ-glutamylamino acid and cysteinylglycine. Cysteinylglycine is subsequently hydrolyzed to cysteine and glycine. The γ-glutamylamino acid is hydrolyzed by γ-glutamyl cyclotransferase (4) to yield 5-oxoproline and the amino acid. Glutamate is regenerated from 5-oxoproline by 5-oxoprolinase (5), an ATP-requiring enzyme.

TABLE 14.4 Amino Acid and Amine Neurotransmitters

Amino Acids	Amines
Glycine	Norepinephrine*
Glutamate	Epinephrine*
γ-Aminobutyric acid (GABA)	Dopamine*
	Serotonin
	Histamine

* A catecholamine.

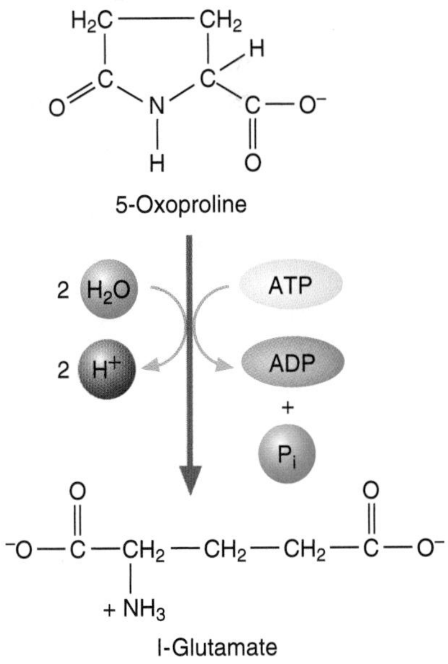

FIGURE 14.21

Conversion of 5-Oxyproline to L-Glutamate

The ATP-driven hydrolysis of 5-oxoproline to yield L-glutamate is catalyzed by 5-oxoprolinase.

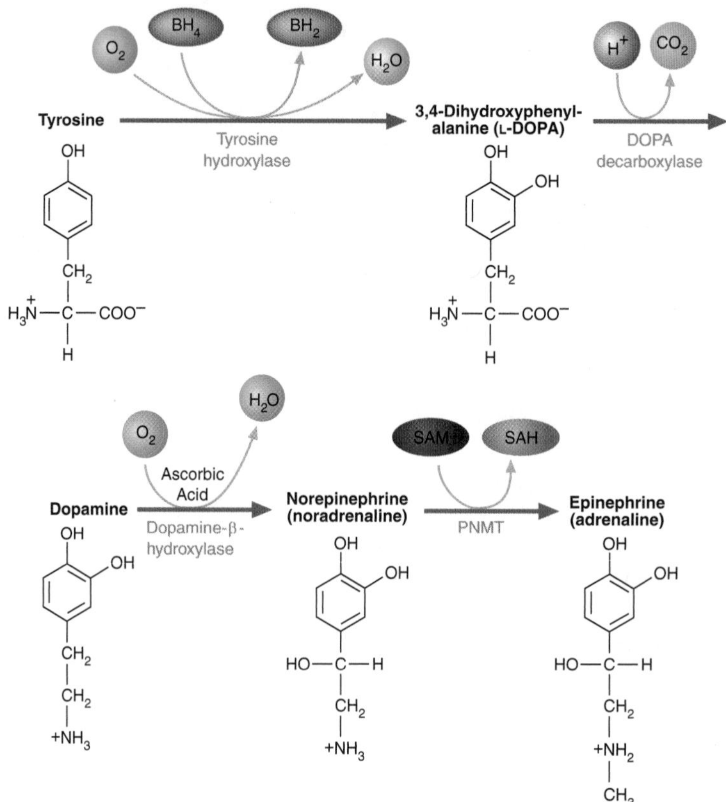

FIGURE 14.22
Biosynthesis of the Catecholamines

Dopamine, norepinephrine, and epinephrine act as neurotransmitters and/or hormones. (PNMT = phenylethanolamine-*N*-methyltransferase)

Tyrosine hydroxylase uses BH_4 to activate O_2. One oxygen atom is attached to tyrosine's aromatic ring, while the other atom oxidizes the coenzyme. DOPA, the product of the reaction, is used in the synthesis of the other catecholamines.

DOPA decarboxylase, a pyridoxal phosphate–requiring enzyme, catalyzes the synthesis of dopamine from L-DOPA. Dopamine is produced in neurons found in certain structures in the brain. It is believed to exert an inhibitory action in the central nervous system. Deficiency in dopamine production has been found to be associated with Parkinson's disease, a serious degenerative neurological disorder. The precursor L-DOPA is used to alleviate the symptoms of Parkinson's disease because dopamine cannot penetrate the blood-brain barrier. (Although most lipid-soluble substances readily pass across the blood-brain barrier, many polar molecules and ions cannot move from blood capillaries.) Once L-DOPA has been transported into appropriate nerve cells, it is converted to dopamine.

Visit the companion website at www.oup.com/us/mckee to read the Biochemistry in Perspective essay on Parkinson's disease and dopamine.

Norepinephrine is synthesized from tyrosine in the chromaffin cells of the adrenal medulla in response to fright, cold, and exercise, as well as low levels of blood glucose. NE acts to stimulate the degradation of triacylglycerol and glycogen. It also increases cardiac output and blood pressure. The hydroxylation of dopamine to produce NE is catalyzed by the copper-containing enzyme dopamine-β-hydroxylase, an oxidase that requires ascorbic acid, acting as a reducing agent, for full activity.

As described, the secretion of epinephrine in response to stress, trauma, extreme exercise, or hypoglycemia causes a rapid mobilization of energy stores, that is, glucose from the liver and fatty acids from adipose tissue. The reaction in which NE is methylated to form E is catalyzed by the enzyme phenylethanolamine-*N*-methyltransferase (PNMT). Although the enzyme occurs predominantly in the chromaffin cells of the adrenal medulla, it is also found in certain portions of the brain, where E functions as a neurotransmitter. Recent evidence indicates that both E and NE are present in several other organs (e.g., liver, heart, and lung). Bovine PNMT is a monomeric protein (30 kDa) that uses SAM as a source of methyl groups.

Biochemistry IN PERSPECTIVE

Gasotransmitters

How do gas molecules, previously thought to be toxic at any concentration, act as signal molecules? **Gasotransmitters** are endogenous gaseous molecules that have been recently recognized as a class of signal molecules. These molecules, which are synthesized by highly regulated enzymes, have several characteristics in common: (1) they are lipid-soluble gases that can diffuse through cellular membranes; (2) their biological effects at physiological concentrations (usually μmolar) are mediated by second messenger molecules and/or ion channels; and (3) signals are not terminated by enzymes but largely by discontinued synthesis and diffusion away from cellular targets. Gasotransmitters include nitric oxide (NO), carbon monoxide (CO), hydrogen sulfide (H_2S), and certain ROS. It is noteworthy that each of these substances is extremely toxic and/or potentially lethal at high concentrations.

Nitric Oxide

NO is a highly reactive gas. Because of its free radical structure, NO (symbolized as NO•) was regarded until recently only as a contributing factor in the destruction of the ozone layer in the Earth's atmosphere and as a precursor of acid rain. NO• is an important signal molecule, produced throughout the mammalian body. Physiological functions such as the regulation of blood pressure, the inhibition of blood clotting, and the destruction of foreign, damaged, or cancerous cells by macrophages are triggered when NO• binds to guanylate cyclase. The product of guanylate cyclase is the second messenger molecule cGMP (cyclic GMP, see p. 592). NO• is also a neurotransmitter and is produced in many areas of the brain where its formation has been linked to the excitatory neurotransmitter glutamate. When glutamate is released from a neuron and binds to a certain class of glutamate receptor, a transient flow of Ca^{2+} through the postsynaptic membrane stimulates NO• synthesis.

Once synthesized, NO• diffuses back to the presynaptic cell, where it signals further release of glutamate. Thus NO• acts as a *retrograde neurotransmitter*; that is, it promotes a cycle in which glutamate is released from the presynaptic neuron and then binds to and promotes action potentials in the postsynaptic neuron. This potentiating mechanism is now believed to play a role in learning and memory formation, as well as other functions in mammalian brain. The disruption of the normally precise regulation of NO• synthesis has been linked to numerous pathological conditions that include stroke, migraine headache, male impotence, septic shock, and inflammatory conditions such as multiple sclerosis and insulin-dependent diabetes (p. 596).

NO• is synthesized by NO synthase (NOS), a heme-containing metalloenzyme that catalyzes a two-step oxidation of L-arginine to L-citrulline (**Figure 14A**). In this complex reaction electrons are transferred from NADPH to O_2 by an electron transport chain that involves several redox components. The functional enzyme is a homodimer (**Figure 14B**). Each monomer has two major domains. The reductase domain possesses binding sites for NADPH, FAD, and FMN. The oxygenase domain, binds BH_4 and the substrates arginine and O_2. Between the two major domains is the binding site for calmodulin (CAM), a small calcium-binding protein that regulates a variety of enzymes. During NO• synthesis CAM accelerates the rate of electron transfer from the reductase domain to the heme group.

The biosynthesis of NO• begins with the hydroxylation of L-arginine. NADPH donates two electrons to FAD, which in turn reduces FMN. BH_4 is essential for the activation of O_2 by the electrons donated by NADPH. The product of this reaction, L-hydroxyarginine, remains bound to NOS. The steps in the subsequent reaction have not yet been resolved. It is believed that L-hydroxyarginine reacts with a heme-peroxy complex (R—O—OH) to give citrulline and NO•.

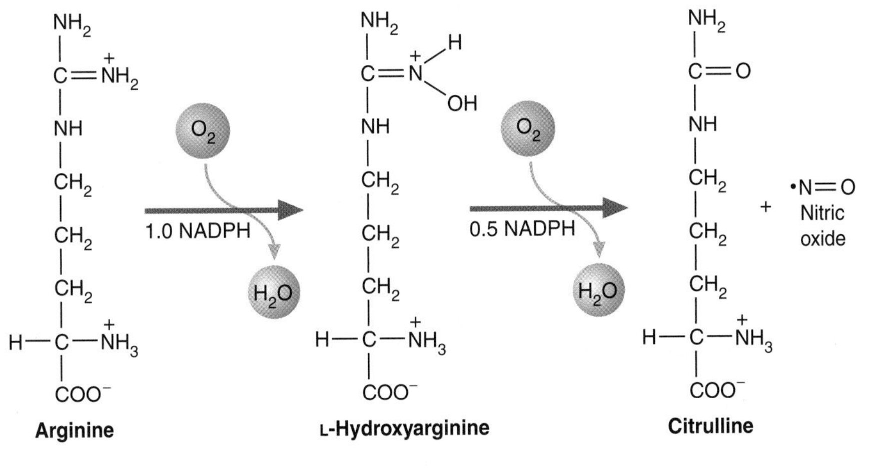

FIGURE 14A
The NOS-Catalyzed Reaction

The biosynthesis of NO• is a two-step oxidation of arginine to form citrulline. During the reaction, 2 mol of O_2 and 1.5 mol of NADPH are consumed per mole of citrulline formed.

Biochemistry IN PERSPECTIVE cont.

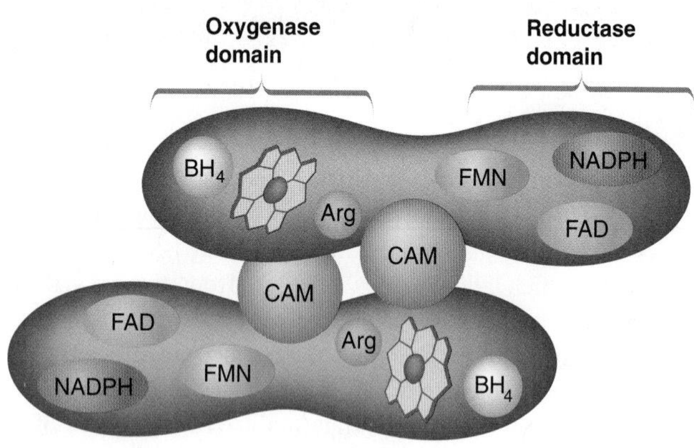

FIGURE 14B

Diagrammatic Structure of NOS

The catalytically active NOS is a homodimer. Each monomer binds NADPH, FAD, FMN, BH_4, and CAM in addition to the substrates arginine and O_2.

Carbon Monoxide

The most important physiological roles played by CO, a colorless and odorless gas, include neuromodulation (i.e., regulation of neurotransmitter release and other neural activities related to learning, memory, and thermal regulation), cardiac protection during hypoxia against ischemia-reperfusion damage (see the Biochemistry in Perspective essay in Chapter 10 entitled Myocardial Infarct: Ischemia and Reperfusion, p. 377), and vascular relaxation (i.e., promotion of smooth muscle relaxation and, therefore, blood vessel dilation). CO is produced by the catabolism of heme in a reaction described in the online Chapter 15 Biochemistry in Perspective essay on heme biotransformation. CO synthesis is catalyzed by heme oxygenase (HO), an ER enzyme. In addition to CO, the products of this reaction are biliverdin and free Fe(II) ions.

The catabolism of heme is an ongoing activity in all tissues that is particularly persistent in spleen (hemoglobin breakdown) and liver (cytochrome P_{450} turnover). Heme itself and its oxidized form, hemin, are strong oxidants and have the potential to confer significant tissue damage. Biliverdin and its reduced form, bilirubin, have much lower oxidant potential than the

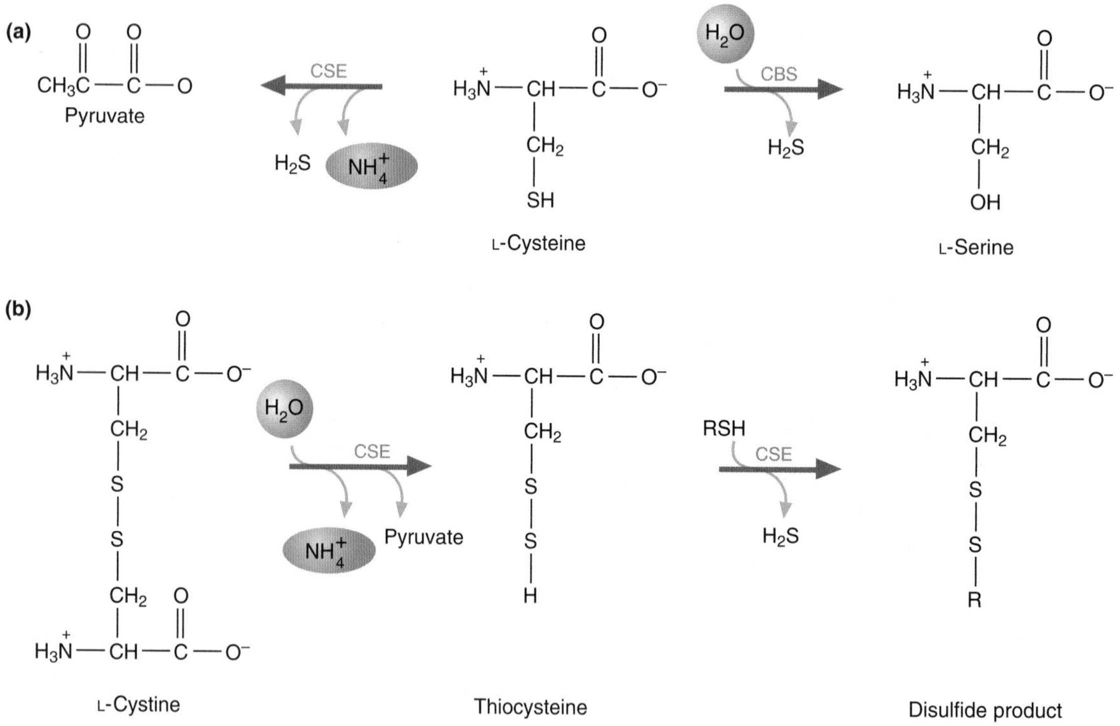

(a)
Pyruvate
CSE
H_2S NH_4^+
L-Cysteine
H_2O
CBS
H_2S
L-Serine

(b)
L-Cystine
H_2O
CSE
NH_4^+ Pyruvate
Thiocysteine
RSH
CSE
H_2S
Disulfide product

FIGURE 14C

H_2S Biosynthesis

(a) Most H_2S-producing reactions are catalyzed by either CBS or CSE with cysteine as the substrate. CBS hydrolyzes cysteine to form serine and H_2S and CSE catalyzes the conversion of cysteine to pyruvate, NH_4^+, and H_2S. (b) CSE also catalyzes a two-reaction sequence in which cystine is converted to thiocysteine, pyruvate, and NH_4^+. Thiocysteine then reacts with a thiol (e.g., GSH or another cysteine) to yield H_2S and a cysteine disulfide product.

Biochemistry IN PERSPECTIVE cont.

parent heme. All cells must have a pool of available free heme to serve as a resource for the synthesis of heme-containing proteins, but that level is kept in a safe range by the action of HO. The free iron is sequestered by ferritin to prevent redox damage. Many actions of CO are mediated by its binding to heme-containing proteins or activation of K^+ channels. CO, at physiologically safe levels, has also been shown to reduce inflammation and suppress apoptosis (cell death).

Hydrogen Sulfide

H_2S is a toxic gas with an unpleasant, foul odor. Within the body, where it is produced in small quantities, H_2S is known to affect numerous physiological processes. The most prominent of these are its functions as a neuromodulator in brain and as a vascular relaxant. H_2S is synthesized from cysteine by several enzymes (**Figure 14C**). The most important of these are the pyridoxal phosphate–requiring enzymes cystathione β-synthetase (CBS) and γ-cystathionase (CSE), both of which catalyze other reactions in the transulfuration pathway (see later: Section 15.1). With cystine used as an alternate substrate, CSE catalyzes a two-reaction pathway that yields H_2S. CSE is the dominant H_2S-producing enzyme in the liver and the cardiovascular system. CBS appears to be dominant in the brain. H_2S dilates arterioles by triggering the opening of ATP-sensitive K^+ channels, a process that results in the hyperpolarization of smooth muscle cell membrane.

H_2S exerts its neuromodulatory effects by enhancing the activation of receptors for NMDA (*N*-methyl-D-aspartate). *NMDA receptors* are a type of glutamate receptor; when activated, they open an ion channel that allows the inward flow of Ca^{2+} and other ions. NMDA receptors are believed to play an important role in *synaptic plasticity*, the capacity of a synapse (the connection between two neurons) to change in chemical strength. Changes in synaptic strength are believed to be the basis for memory formation.

SUMMARY: At very low concentrations, NO, CO, and H_2S are signal molecules that diffuse easily through cell membranes and whose synthesis is rigorously controlled.

Nucleotides

Nucleotides are complex nitrogen-containing molecules required for cell growth and development. Not only are nucleotides the building blocks of the nucleic acids, they also play several essential roles in energy transformation and regulate many metabolic pathways. As described, each nucleotide is composed of three parts: a nitrogenous base, a pentose sugar, and one or more phosphate groups. The nitrogenous bases are derivatives of either purine or pyrimidine, which are planar heterocyclic aromatic compounds.

Common naturally occurring **purines** include adenine, guanine, xanthine, and hypoxanthine; thymine, cytosine, and uracil are common **pyrimidines** (**Figure 14.23**). Because of their aromatic structures, the purines and pyrimidines absorb UV light. At pH 7, this absorption is especially strong at 260 nm. Purine and pyrimidine bases have tautomeric forms; that is, they undergo spontaneous shifts in the relative position of a hydrogen atom and a double bond in a three-atom sequence involving heteroatoms. This property is especially important because the precise location of hydrogen atoms on the oxygen and nitrogen atoms affects the interaction of bases in nucleic acid molecules. Adenine and cytosine have both amino and imino forms; guanine, thymine, and uracil have both keto (lactam) and enol (lactim) forms. At physiological pH the amino and keto forms are the most stable. The amino and imino forms of adenine and the keto and enol forms of thymine are illustrated in **Figure 14.24**.

When a purine or pyrimidine base is linked through a β-N-glycosidic linkage to C-1 of a pentose sugar, the molecule is called a **nucleoside** (**Figure 14.25**). and it contains one of two types of sugar: ribose or deoxyribose. Ribose-containing nucleosides with adenine, guanine, cytosine, and uracil are referred to as adenosine, guanosine, cytidine, and uridine, respectively. When the sugar

7 N 5 6 N1 8 2 9N 4 N 3 H

Purine

4 5 N3 6 2 N1

Pyrimidine

component is deoxyribose, the prefix *deoxy* is used. For example, the deoxy nucleoside with adenine is called deoxyadenosine. Deoxythymidine is called thymidine because the base thymine usually occurs only in deoxyribonucleosides. Possible confusion in the identification of atoms in the base and sugar components of nucleosides is avoided by using a superscript prime to denote the atoms in the sugar (**Figure 14.25**). Rotation around the N-glycosidic bond of nucleosides creates two conformations: *syn* and *anti*. Purine nucleosides occur as either *syn* or *anti* forms. In pyrimidine nucleosides the *anti* conformation predominates because of steric hinderance between the pentose sugar and the carbonyl oxygen at C-2.

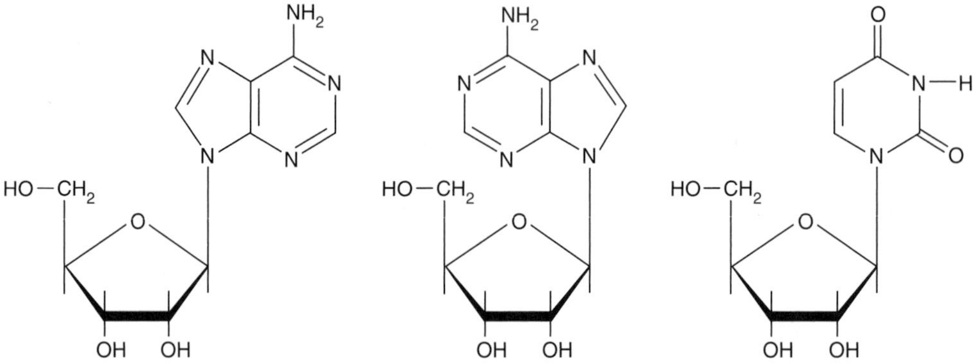

anti-Adenosine *syn*-Adenosine *anti*-Uridine

Nucleotides are nucleosides in which one or more phosphate groups are bound to the sugar (**Figure 14.26**). Most naturally occurring nucleotides are 5′-phosphate esters. If one phosphate group is attached at the 5′-carbon of the sugar, the molecule is named as a nucleoside monophosphate (e.g., adenosine-5′-monophosphate: AMP). Nucleoside di- and triphosphates contain two and three phosphate groups, respectively. Phosphate groups make nucleotides strongly acidic. (Protons dissociate from the phosphate groups at physiological pH.) Because of their acidic nature, nucleotides may also be named as acids. For example, AMP is often referred to as adenylic acid or adenylate. Nucleoside di- and triphosphates form complexes with Mg^{2+}. In nucleoside triphosphates such as ATP, Mg^{2+} can form α,β (shown) and β,γ complexes:

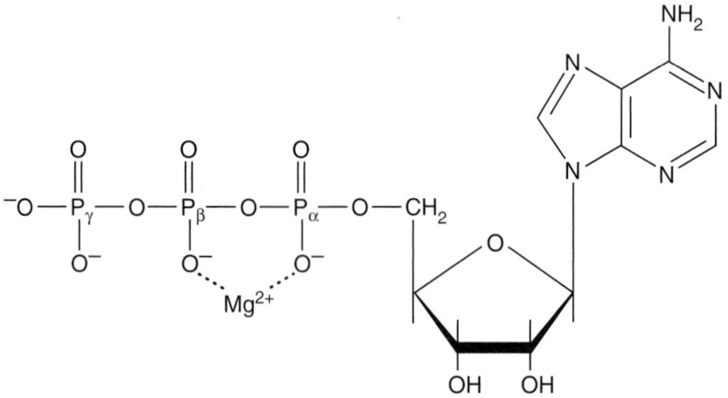

Purine and pyrimidine nucleotides can be synthesized in *de novo* and salvage pathways. Both types of pathway are described.

PURINE NUCLEOTIDE BIOSYNTHESIS The *de novo* synthesis of purine nucleotides begins with the formation of 5-phospho-α-D-ribosyl-1-pyrophosphate (PRPP) (**Figure 14.27**) catalyzed by ribose-5-phosphate pyrophosphokinase (PRPP synthetase).

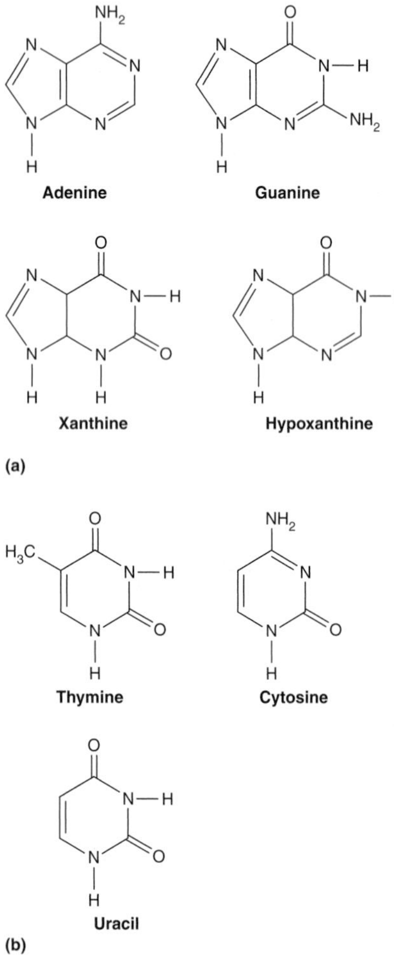

Adenine Guanine

Xanthine Hypoxanthine

(a)

Thymine Cytosine

Uracil

(b)

FIGURE 14.23

The Most Common Naturally Occurring Purines (a) and Pyrimidines (b)

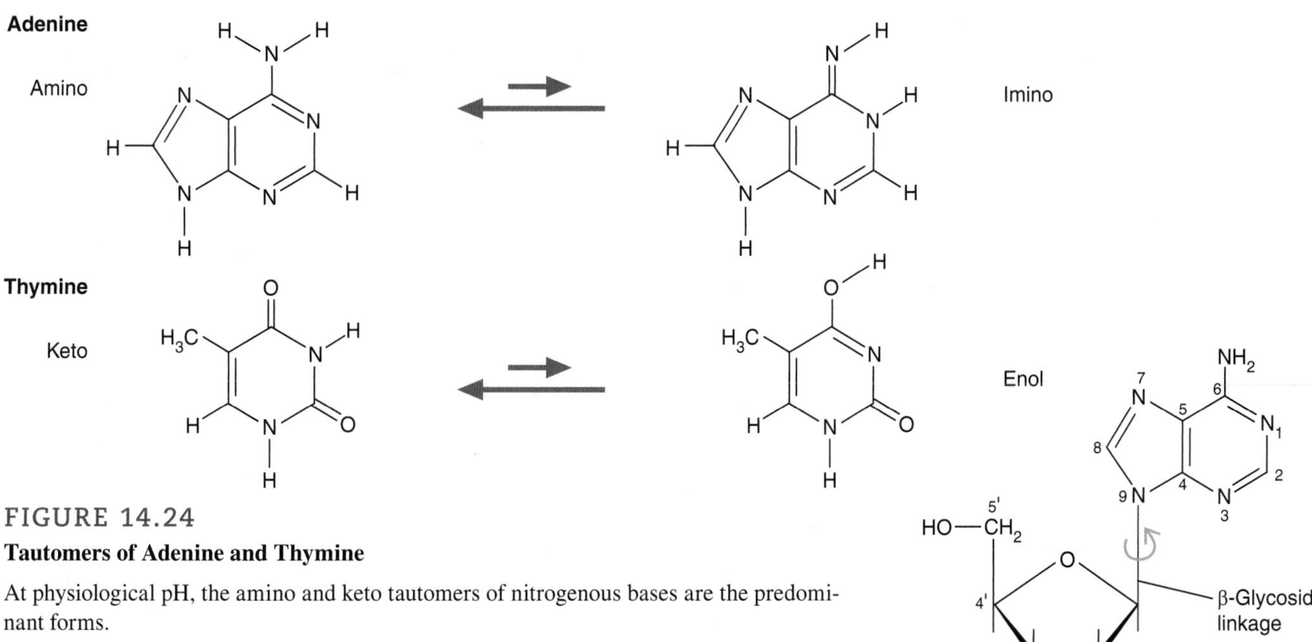

FIGURE 14.24

Tautomers of Adenine and Thymine

At physiological pH, the amino and keto tautomers of nitrogenous bases are the predominant forms.

FIGURE 14.25

Nucleoside Structure

Nucleosides are molecules in which nitrogenous bases such as adenine (shown) are linked via a β-glycosidic bond to C-1 of a pentose sugar, in this case ribose.

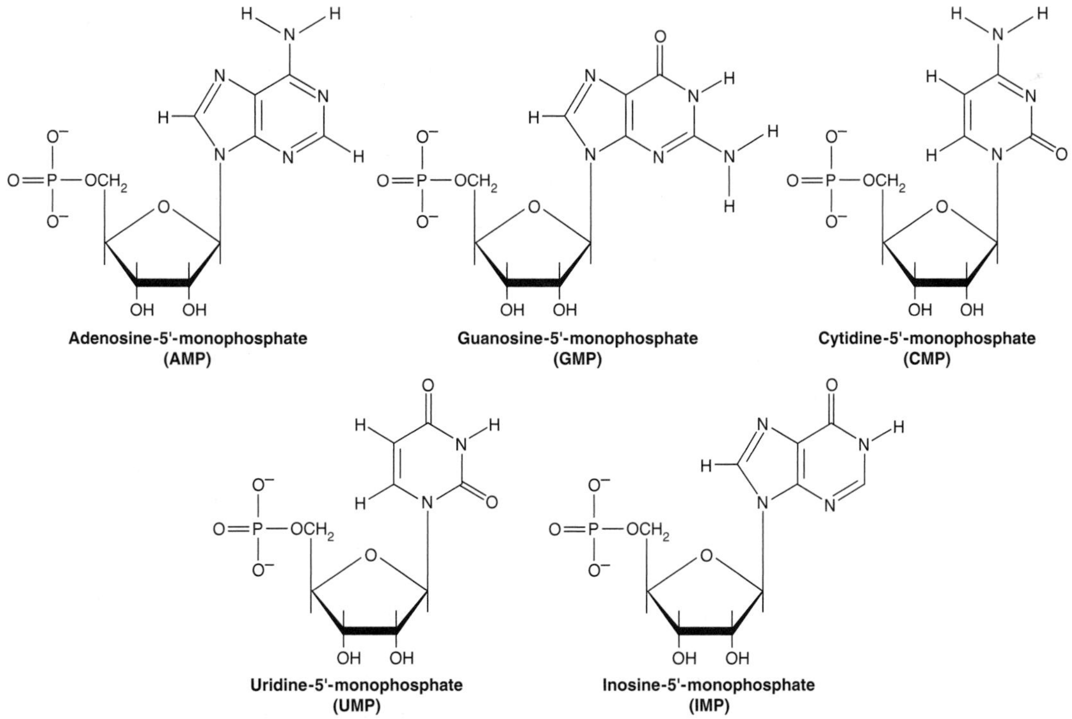

Adenosine-5'-monophosphate (AMP)

Guanosine-5'-monophosphate (GMP)

Cytidine-5'-monophosphate (CMP)

Uridine-5'-monophosphate (UMP)

Inosine-5'-monophosphate (IMP)

FIGURE 14.26

Common Ribonucleotides

Ribonucleotides contain ribose. When nucleotides contain deoxyribose instead of ribose, their names include the prefix *deoxy.* Inosine-5′-monophosphate (IMP) is an intermediate in purine nucleotide synthesis. The base component of IMP is hypoxanthine.

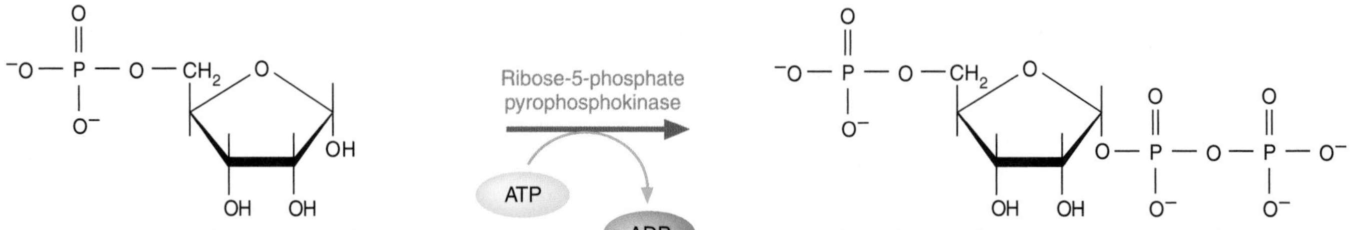

α-D-ribose-5-phosphate

Ribose-5-phosphate pyrophosphokinase

ATP

ADP

5-Phospho-α-D-ribosyl-1-pyrophosphate (PRPP)

FIGURE 14.27
PRPP Synthesis

The *de novo* purine nucleotide pathway begins with the synthesis of PRPP (5-phospho-α-D-ribosyl-1-pyrophosphate), which is catalyzed by ribose-5-phosphate pyrophosphatase.

Visit the companion website at www.oup.com/us/mckee to read the related Biochemistry in Perspective essay on the nucleotides.

(The substrate for this reaction, α-D-ribose-5-phosphate, is a product of the pentose phosphate pathway.) The pathway by which PRPP is converted to inosine monophosphate (inosinate), the first purine nucleotide, is described in the online reading The Nucleotides. The origin of the ring atoms of purines is illustrated in **Figure 14.28**.

The conversion of inosine-5′-monophosphate (IMP) to either AMP (or adenylate) or guanosine monophosphate (GMP or guanylate) requires two reactions (**Figure 14.29**).

AMP differs from IMP in only one respect: an amino group replaces a keto oxygen at position 6 of the purine base. The amino nitrogen provided by aspartate becomes linked to IMP in a GTP-requiring reaction catalyzed by adenylosuccinate synthetase. In this step the product adenylosuccinate eliminates fumarate to form AMP. (The enzyme that catalyzes this reaction also catalyzes a similar step in IMP synthesis.) The conversion of IMP to GMP begins with a dehydrogenation utilizing NAD^+, which is catalyzed by IMP dehydrogenase. The product, referred to as xanthosine monophosphate (XMP), is then converted to GMP by the donation of an amino nitrogen by glutamine in an ATP-requiring reaction catalyzed by GMP synthase.

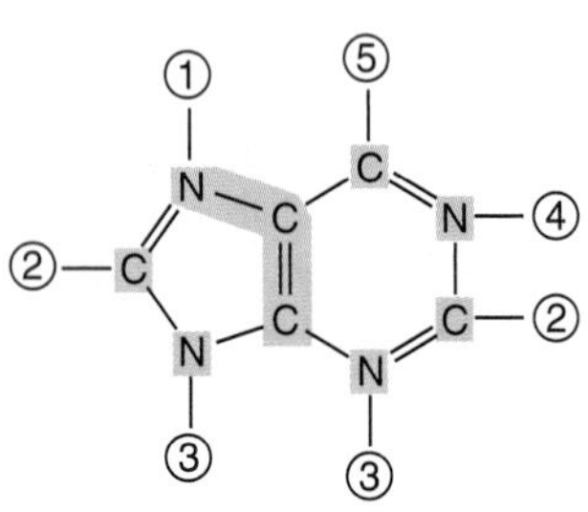

FIGURE 14.28
Origin of Purine Ring Atoms

The sources of the nitrogen and carbon atoms in purine rings, determined by isotopic labeling experiments, are as follows: 1 = glycine; 2 = formate; 3 = glutamine's amide nitrogen; 4 = aspartate; 5 = CO_2.

Nucleoside triphosphates are the most common nucleotides used in metabolism. They are formed in the following manner. Recall that ATP is synthesized from ADP and P_i during certain reactions in glycolysis and aerobic metabolism. ADP is synthesized from AMP and ATP in a reaction catalyzed by adenylate kinase:

$$\text{AMP} + \text{ATP} \rightleftharpoons 2\ \text{ADP} \tag{7}$$

Other nucleoside triphosphates are synthesized in ATP-requiring reactions catalyzed by a series of nucleoside monophosphate kinases:

$$\text{NMP} + \text{ATP} \rightleftharpoons \text{NDP} + \text{ADP} \tag{8}$$

Nucleoside diphosphate kinase catalyzes the formation of nucleoside triphosphates,

$$N_1DP + N_2TP \rightleftharpoons N_1TP + N_2DP \tag{9}$$

where N_1 and N_2 are purine or pyrimidine bases.

In the purine salvage pathway, purine bases obtained from the normal turnover of cellular nucleic acids or (to a lesser extent) from the diet are reconverted into nucleotides. Because the *de novo* synthesis of nucleotides is metabolically expensive (i.e., relatively large amounts of phosphoryl bond energy are used), many cells have mechanisms to retrieve purine bases. Hypoxanthine-guaninephosphoribosyltransferase (HGPRT) catalyzes nucleotide synthesis using PRPP and either hypoxanthine or guanine. The hydrolysis of pyrophosphate makes these reactions irreversible.

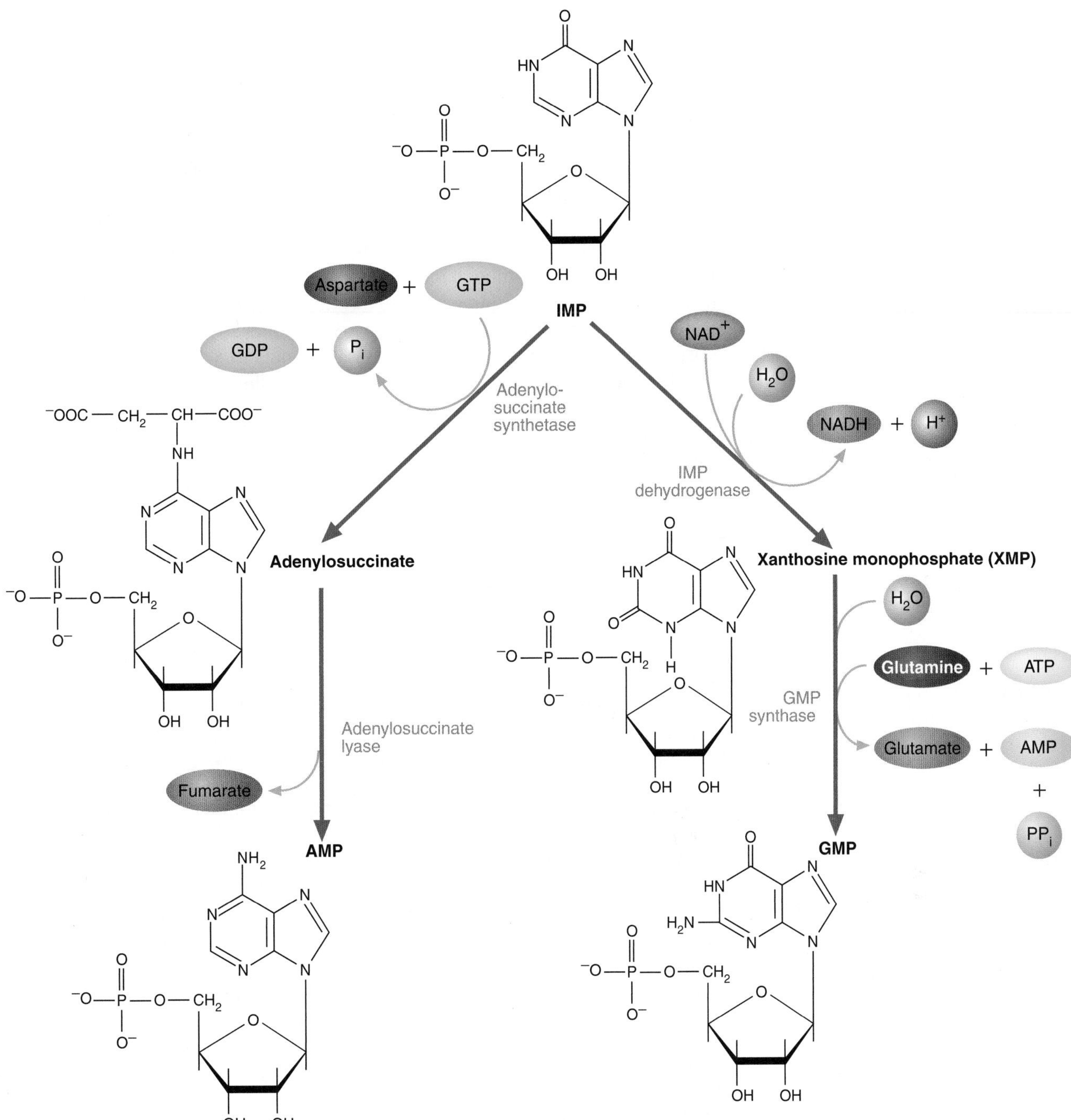

FIGURE 14.29

Biosynthesis of AMP and GMP from IMP

In the first step of AMP synthesis, the C-6 keto oxygen of the hypoxanthine base moiety of IMP is replaced by the amino group of aspartate. In the second step the product of the first reaction, adenylosuccinate, is hydrolyzed to form AMP and fumarate. GMP synthesis begins with the oxidation of IMP to form XMP. GMP is produced as the amide nitrogen of glutamine replaces the C-2 keto oxygen of XMP. Note that AMP formation requires GTP and GMP formation requires ATP.

$$\text{Hypoxanthine} + \text{PRPP} \longrightarrow \text{IMP} + \text{PP}_i \quad (10)$$

$$\text{Guanine} + \text{PRPP} \longrightarrow \text{GMP} + \text{PP}_i \quad (11)$$

Deficiency of HGPRT causes *Lesch-Nyhan syndrome*, a devastating X-linked disease that occurs primarily in males. It is characterized by excessive production of uric acid, the degradation product of purine nucleotides (Section 15.3), and certain neurological symptoms (self-mutilation, involuntary movements, and mental retardation). Although a powerful antioxidant, uric acid can act as a pro-oxidant when present in large amounts. Oxidative stress is now believed to contribute to the symptoms of Lesch-Nyhan syndrome. Affected children appear normal at birth but begin to deteriorate at about 3 to 4 months of age. Death, usually caused by renal failure, occurs in childhood. A partially defective HGPRT enzyme causes one form of *gout* (a condition in which high blood uric acid concentrations result in the accumulation of sodium urate crystals in joints, especially those in feet).

Visit the companion website at www.oup.com/us/mckee to read the Biochemistry in Perspective essay on gout in Chapter 15.

Adenine phosphoribosyltransferase (ARPT) catalyzes the transfer of adenine to PRPP, thus forming AMP:

$$\text{Adenine} + \text{PRPP} \longrightarrow \text{AMP} + \text{PP}_i \quad (12)$$

The relative importance of the *de novo* and salvage pathways is unclear. However, the severe symptoms of hereditary HGPRT deficiency indicate that the purine salvage pathway is vitally important. In addition, investigations of purine nucleotide synthesis inhibitors for treating cancer indicate that both pathways must be inhibited for significant tumor growth suppression.

The regulation of purine nucleotide biosynthesis is summarized in **Figure 14.30**. The pathway is controlled to a considerable degree by PRPP availability. Several products of the pathway inhibit both ribose-5-phosphate pyrophosphokinase and glutamine-PRPP amidotransferase (the rate-limiting enzyme in IMP synthesis). The combined inhibitory effect of the end products is synergistic (i.e., the net inhibition is greater than the inhibition of each nucleotide acting alone). At the IMP branch point, both AMP and GMP regulate their own syntheses by feedback inhibition of adenylosuccinate synthetase and IMP dehydrogenase. The hydrolysis of GTP drives the synthesis of adenylosuccinate, whereas ATP drives XMP synthesis. This reciprocal arrangement facilitates the maintenance of appropriate cellular concentrations of adenine and guanine nucleotides.

PYRIMIDINE NUCLEOTIDE BIOSYNTHESIS Pyrimidine nucleotide synthesis occurs in the cytoplasm where the pyrimidine ring is assembled first and then linked to ribose phosphate. The carbon and nitrogen atoms in the pyrimidine ring are derived from bicarbonate, aspartate, and glutamine. Synthesis begins with the formation of carbamoyl phosphate in an ATP-requiring reaction catalyzed by carbamoyl phosphate synthetase II (CPSII) (**Figure 14.31**). (Carbamoyl phosphate synthetase I is a mitochondrial enzyme involved in the urea cycle, described in Chapter 15.) One molecule of ATP provides a phosphate group, whereas the hydrolysis of another ATP drives the reaction. Aspartate transcarbamoylase (ATCase) catalyzes the reaction of carbamoyl phosphate with aspartate to form carbamoyl aspartate. The closure of the pyrimidine ring is then catalyzed by dihydroorotase. The product, dihydroorotate, is oxidized to form orotate. Dihydroorotate dehydrogenase, the enzyme that catalyzes this reaction, is a flavoprotein associated with the inner mitochondrial membrane. (The NADH produced in this reaction donates its electrons to the ETC.) Once

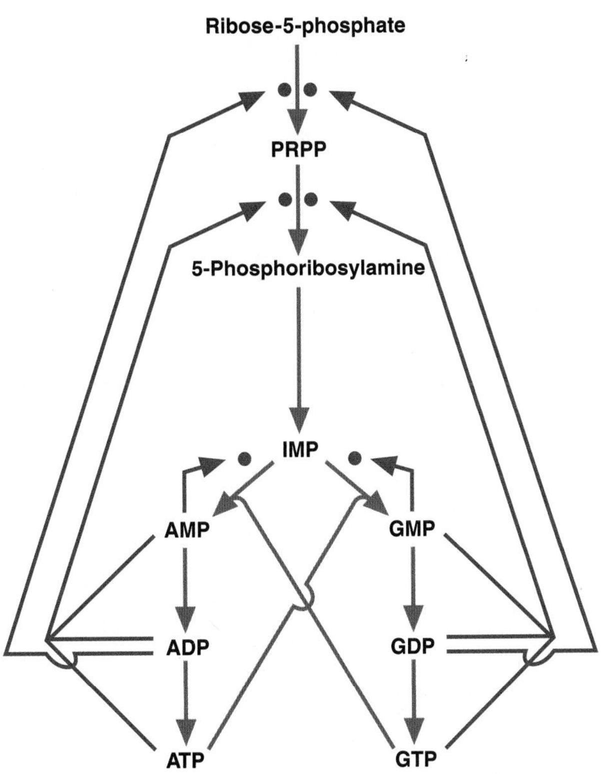

FIGURE 14.30
Purine Nucleotide Biosynthesis Regulation

Feedback inhibition is indicated by red arrows. The stimulation of AMP synthesis by GTP and GMP synthesis by ATP ensures a balanced synthesis of both purine nucleotide families.

synthesized on the cytoplasmic face of the inner mitochondrial membrane, orotate is converted by orotate pyrophosphoribosyl transferase to orotidine-5′-phosphate (OMP), the first nucleotide in the pathway, by reacting with PRPP. Uridine-5′-phosphate (UMP) is produced when OMP is decarboxylated in a reaction catalyzed by OMP decarboxylase. Both orotate pyrophosphoribosyl transferase and OMP decarboxylase activities occur on a protein referred to as UMP synthase. UMP is a precursor for the other pyrimidine nucleotides. Two sequential phosphorylation reactions form UTP, which then accepts an amide nitrogen from glutamine to form CTP. The origin of the ring atoms in pyrimidines is illustrated in **Figure 14.32**.

In a rare genetic disease called *orotic aciduria*, there is excessive urinary excretion of orotic acid because UMP synthase is defective. Symptoms include anemia and retardation in growth. Treatment with a combination of pyrimidine nucleotides, which inhibit the production of orotate and provide the building blocks for nucleic acid synthesis, reverses the disease process.

DEOXYRIBONUCLEOTIDE BIOSYNTHESIS All the nucleotides discussed so far are ribonucleotides, molecules that are principally used as the building blocks of RNA, as nucleotide derivatives of molecules such as sugars, or as energy sources. The nucleotides required for DNA synthesis, the 2′-deoxyribonucleotides, are produced by reducing ribonucleoside diphosphates in a reaction catalyzed by ribonucleotide reductase (**Figure 14.33**). The electrons used in the synthesis of 2′-deoxyribonucleotides are ultimately donated by NADPH. Thioredoxin mediates the transfer of hydrogen atoms from NADPH to ribonucleotide reductase. The regeneration of reduced thioredoxin is catalyzed by thioredoxin reductase.

Ribonuclease reductase I, found in mammals, is a tetramer of two different subunits. Subunit 1 has a number of reactive thiols required for catalysis plus the allosteric sites involved in regulation. Subunit 2 possesses a critical binuclear Fe(III) center that generates and stabilizes a tyrosyl radical essential for enzyme function. The interface of the four subunits forms the active site. The tyrosyl radical initiates the radical-mediated reduction of substrate NDPs by abstracting an H atom from one of the thiols in subunit 1, generating a transient thiyl radical.

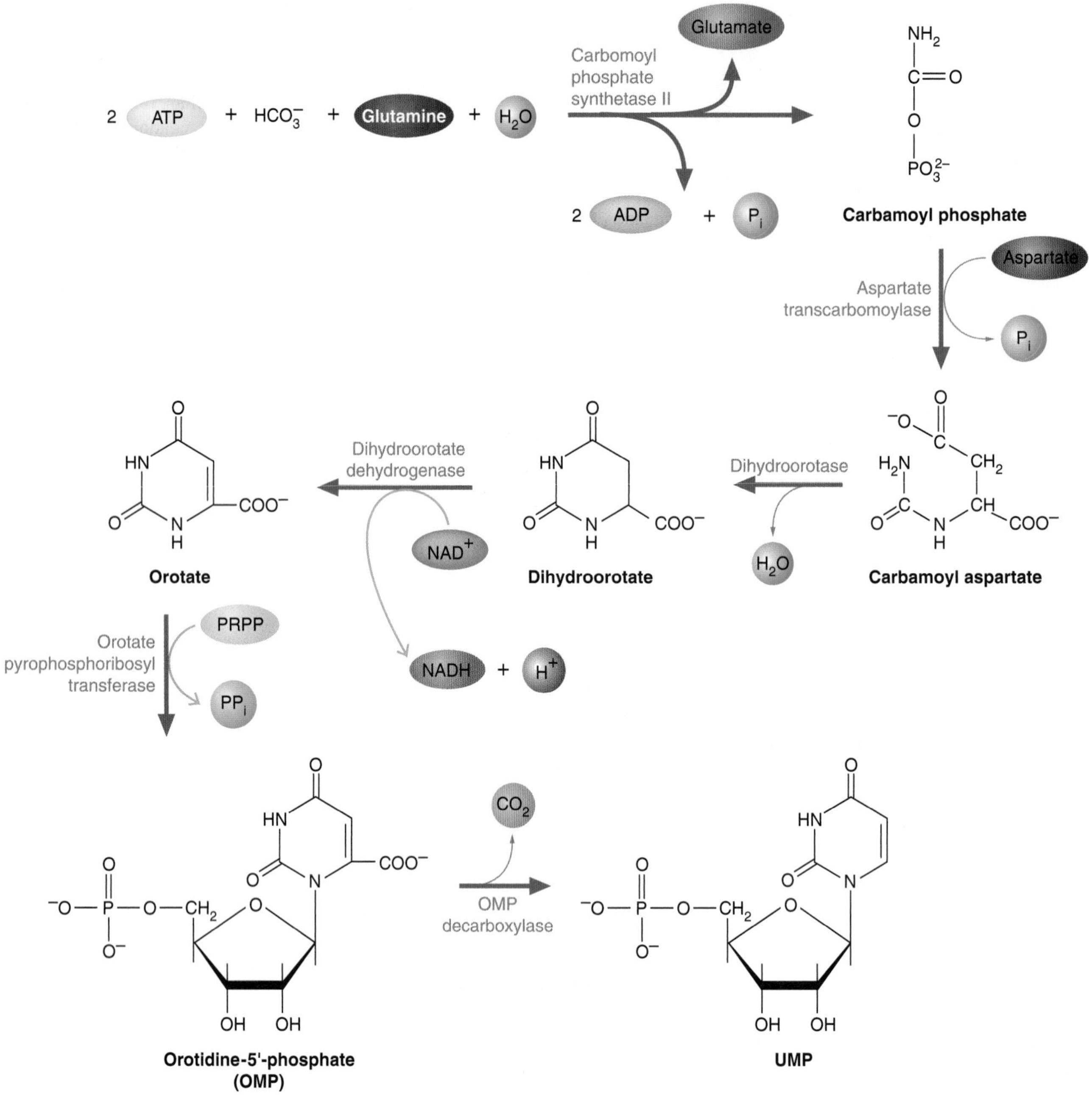

FIGURE 14.31

Pyrimidine Nucleotide Synthesis

The metabolic pathway in which UMP is synthesized is composed of six enzyme-catalyzed reactions. In mammals, the first three enzymatic activities in the pathway (**c**arbamoyl phosphate synthetase II, **a**spartate transcarbamoylase, and **d**ihydroorotate dehydrogenase) are located on a single polypeptide, referred to as CAD. Unlike the other enzymes in pyrimidine biosynthesis, which are cytoplasmic, CAD is located on the outer face of the inner mitochondrial membrane. The CPSII domain is activated by ATP and inhibited by UTP and CTP. In *E. coli* the rate-limiting reaction in pyrimidine synthesis is catalyzed by the 12-subunit complex aspartate transcarbamoylase (**Figure 6.25**). Bacterial ATCase is stimulated by ATP and inhibited by UTP and CTP.

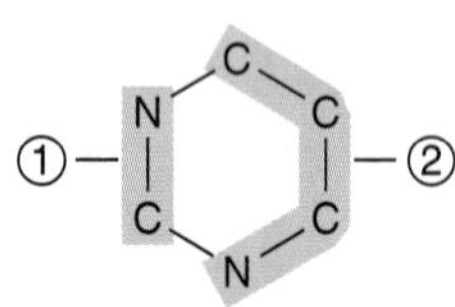

FIGURE 14.32

Origin of Pyrimidine Ring Atoms

The sources of the nitrogen and carbon atoms in pyrimidine rings are as follows: 1 = carbamoyl phosphate; 2 = aspartate.

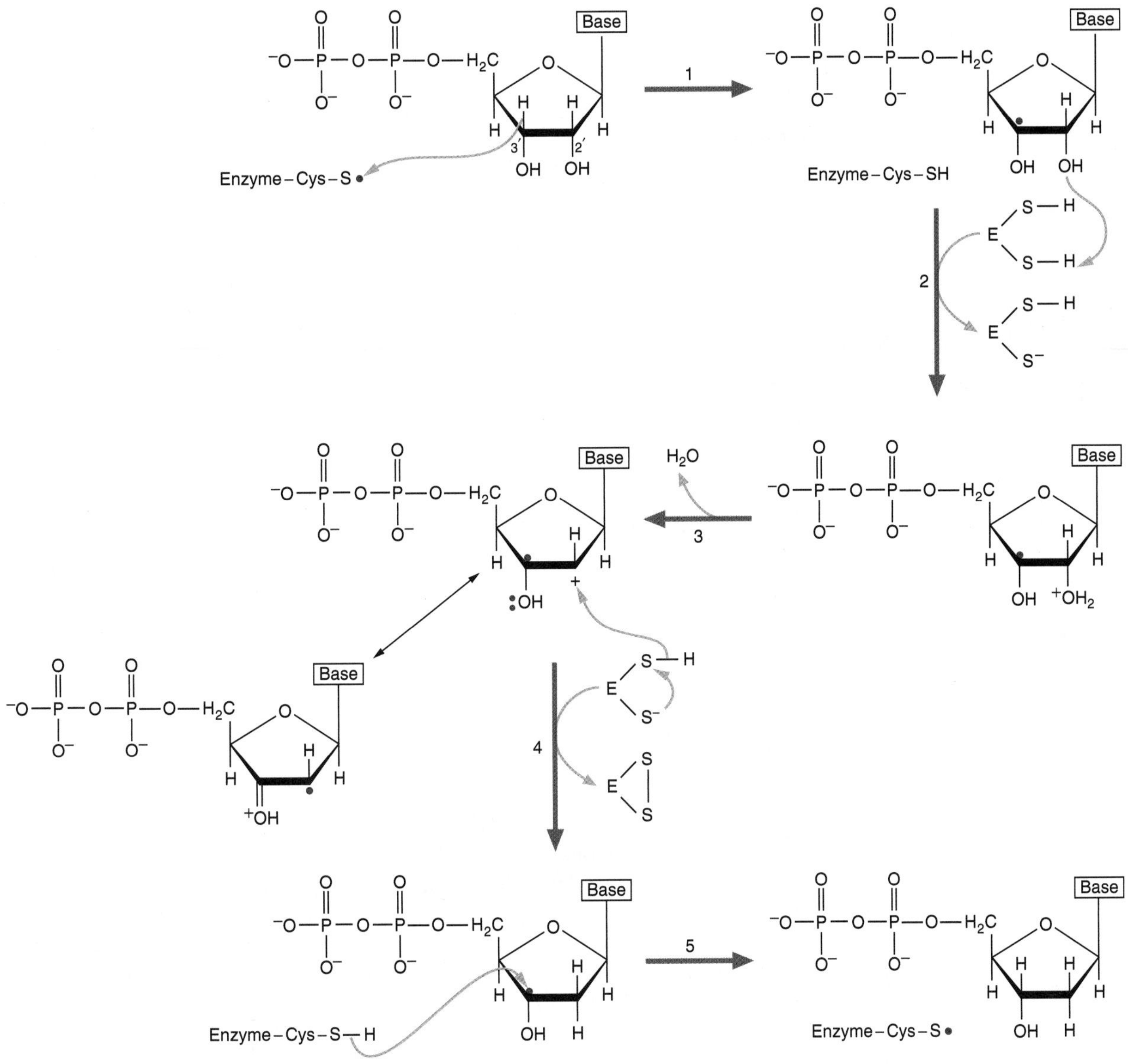

FIGURE 14.33

Mechanism of Ribonucleotide Reductase

The reaction begins with the tyrosyl radical–induced formation of a transient thiyl radical and the binding of NDP. (1) The thiyl radical abstracts a H atom from C3′. The C2′—OH (2) is protonated by a reactive thiol and H_2O is eliminated (3) to generate a carbocation. A dithiol reduces the cation radical (4) and (5) an H atom is transferred from the initiating thiyl S to the C3′ and the product dNDP leaves the active site. The subsequent reduction of a disulfide–mediated by thioredoxin/NADPH and regeneration of the tyrosyl radical returns the enzyme to its ready state to receive new substrate.

The thiyl radical abstracts an H atom from C3′ of the substrate, generating a radical. A nearby thiol protonates the C2′—OH group and it departs as H_2O, leaving behind a carbocation. Hydride ion shift from an active site thiol resolves the carbocation and a disulfide bridge forms in the active site. The H atom abstracted by the thiyl radical is returned to C3′, and tyrosine transfers an H atom to resolve the thiyl radical. The product dNDP leaves the active site, and the enzyme is returned to its reduced free thiol state by electron transfer from NADPH mediated through thioredoxin (**Figure 14.34**).

Regulation of ribonucleotide reductase is intricate. The binding of deoxyadenosine triphosphate to a regulatory site on the enzyme decreases catalytic activity. The binding of deoxyribonucleoside triphosphates to several other

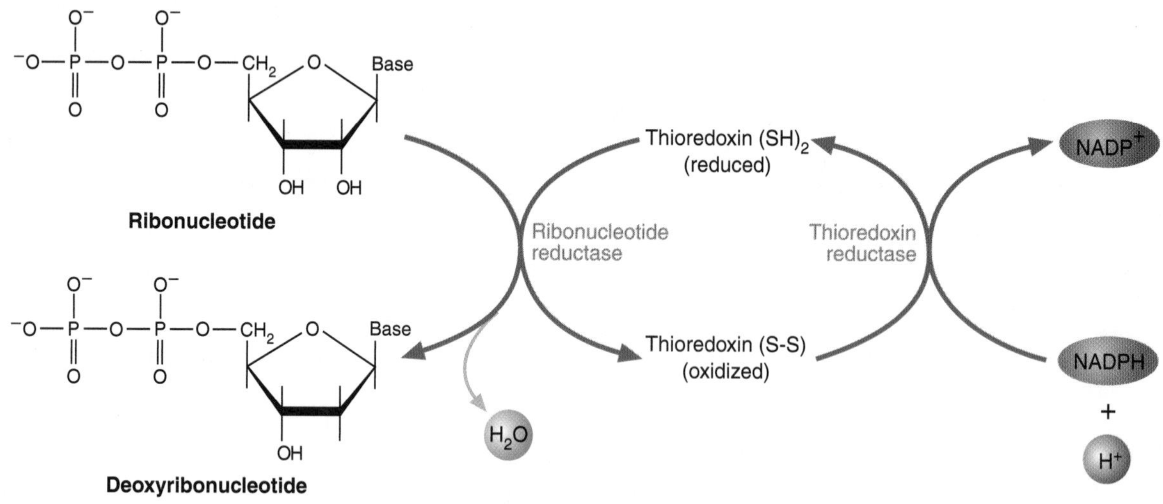

FIGURE 14.34
Deoxyribonucleotide Biosynthesis

Electrons for the reduction of ribonucleotides ultimately come from NADPH. Thioredoxin, a small protein with two thiol groups, mediates the transfer of electrons from NADPH to ribonucleotide reductase.

enzyme sites alters substrate specificity so that there are differential increases in the concentrations of each of the deoxyribonucleotides. This latter process balances the production of the 2′-deoxyribonucleotides required for cellular processes, especially that of DNA synthesis.

The deoxyuridylate (dUMP) produced by dephosphorylation of the dUDP product of ribonucleotide reductase is not a component of DNA, but its methylated derivative deoxythymidylate (dTMP) is. The methylation of dUMP is catalyzed by thymidylate synthase, which utilizes N^5,N^{10}-methylene THF. As the methylene group is transferred, it is reduced to a methyl group, while the folate coenzyme is oxidized to form dihydrofolate. THF is regenerated from dihydrofolate by dihydrofolate reductase and NADPH. (This reaction is the site of action of some anticancer drugs, such as methotrexate.) Deoxyuridylate can also be synthesized from dCMP by deoxycytidylate deaminase.

KEY CONCEPTS

- Nucleotides are the building blocks of the nucleic acids. They also regulate metabolism and transfer energy.
- The purine and pyrimidine nucleotides are synthesized in both *de novo* and salvage pathways.

In mammals, carbamoyl phosphate synthetase II is the key regulatory enzyme in the biosynthesis of pyrimidine nucleotides. The enzyme is inhibited by UTP, the product of the pathway, and stimulated by purine nucleotides. In many bacteria, aspartate carbamoyl transferase is the key regulatory enzyme. It is inhibited by CTP and stimulated by ATP. The pyrimidine salvage pathway, which uses preformed pyrimidine bases from dietary sources or from nucleotide turnover, is of minor importance in mammals.

Heme

Heme (**Figure 5.35**), one of the most complex molecules synthesized by mammalian cells, has an iron-containing porphyrin ring. Heme is an essential structural component of hemoglobin (**Figure 5.39**), myoglobin (**Figure 5.36**), peroxidase, and the cytochromes (**Figure 10.6**). Although it occurs in almost all aerobic cells, the heme biosynthetic pathway is especially prominent in liver and reticulocytes (the nucleus-containing precursor cells of red blood cells in bone marrow). Both heme and chlorophyll are synthesized from the relatively simple components glycine and succinyl-CoA.

The pathway that produces these two substances is outlined in a Biochemistry in Perspective essay available online. Note that heme biosynthesis is inhibited by heavy metal poisoning, a topic discussion in a Biochemistry in Perspective essay on lead poisoning.

Chapter Summary

1. Nitrogen, found in proteins, nucleic acids, and myriad other biomolecules, is an essential element in living systems. Biologically useful nitrogen, a scarce resource, is produced in a process referred to as nitrogen fixation. The nitrogenase enzyme complex that converts N_2 to NH_3 resides in some free soil bacteria, cyanobacteria, and symbiotic root nodule bacteria.
2. Organisms vary widely in their ability to synthesize amino acids. Some organisms (e.g., plants and some microorganisms) can produce all required amino acid molecules from fixed nitrogen. Animals can produce only some amino acids. Nonessential amino acids are produced from readily available precursor molecules, whereas essential amino acids must be acquired in the diet.
3. Two types of reaction play prominent roles in amino acid metabolism. In transamination reactions, new amino acids are produced when α-amino groups are transferred from donor α-amino acids to acceptor α-keto acids. Because transamination reactions are reversible, they play an important role in both amino acid synthesis and degradation. Ammonium ions or the amide nitrogen of glutamine can also be directly incorporated into amino acids and eventually other metabolites.
4. On the basis of the biochemical pathways in which they are synthesized, the amino acids can be divided into six families: glutamate, serine, aspartate, pyruvate, aromatics, and histidine.
5. Amino acids are precursors of many physiologically important biomolecules. Many of the processes that synthesize these molecules involve the transfer of carbon groups. Because many of these transfers involve one-carbon groups (e.g., methyl, methylene, methenyl, and formyl), the overall process is referred to as one-carbon metabolism. *S*-Adenosylmethionine (SAM) and tetrahydrofolate (THF) are the most important carriers of one-carbon groups.
6. Molecules derived from amino acids include several neurotransmitters (e.g., GABA, the catecholamines, serotonin, and histamine) and hormones (e.g., melatonin). Glutathione is a tripeptide that plays an essential role in cells. The nucleotides, molecules that serve as the building blocks of the nucleic acids (as well as energy sources and metabolic regulators), possess heterocyclic nitrogenous bases as part of their structures. These bases, called the purines and the pyrimidines, are derived from various amino acid molecules. Heme is an example of a complex heterocyclic ring system that is derived from glycine and succinyl-CoA. The biosynthetic pathway that produces heme is similar to the one that produces chlorophyll in plants.

Take your learning further by visiting the **companion website** for Biochemistry at **www.oup.com/us/mckee** where you can complete a multiple-choice quiz on synthesis of nitrogen-containing biomolecules to help you prepare for exams.

Suggested Readings

Bedoya, F. J., *et al.*, Regulation of Pancreatic β-Cell Survival by Nitric Oxide, *Islets* 4(2):108–118, 2012.

Diaz, R. J., and Rosenberg, R., Spreading Dead Zones and Consequences for Marine Ecosystems, *Science* 321:926–929, 2008.

Flora, G., Gupta, D., and Tiwari, A., Toxicity of Lead: A Review with Recent Updates, *Interdisciplinary Toxicology* 5(2):47–58, 2012.

Gotoh, T., and Mori, M., Nitric Oxide and Endoplasmic Reticulum Stress, *Arterioscler. Thromb. Vasc. Biol.* 26:1439–1446, 2006.

Parle-McDermott, A., and Ozaki, M., The Impact of Nutrition on Differential Methylated Regions of the Genome, *Adv. Nutr.* 2:463–471, 2011.

Paul, B. D., and Snyder, S., H_2S Signaling Through Protein Sulfhydration and Beyond, *Nat. Rev. Mol. Cell Biol.* 13:499–507, 2012.

Stec, D. E., Drummond, H. A., and Vera, T., Role of Carbon Monoxide in Blood Pressure Regulation, *Hypertension* 51:597–604, 2008.

Wu, G., Fang, Y.-Z., Yang, S., Lupton, J. R., and Turner, N. D., Glutathione Metabolism and Its Implications for Health, *J. Nutr.* 134:489–492, 2004.

Key Words

S-adenosylmethionine, *531*
amino acid pool, *516*
biogenic amine, *534*
branched-chain amino acid, *516*
catecholamine, *535*
essential amino acid, *515*
gasotransmitter, *537*
neurotransmitter, *534*
nitrogen fixation, *462*
nonessential amino acid, *515*
nucleoside, *539*
one-carbon metabolism, *528*
purine, *539*
pyrimidine, *539*
racemization, *517*
tetrahydrobiopterin, *534*
tetrahydrofolate, *529*
transamination, *517*
vitamin B_{12}, *529*

Review Questions

These questions are designed to test your knowledge of the key concepts discussed in this chapter before moving on to the next chapter. You may like to compare your answers to the solutions provided in the back of the book and in the accompanying Study Guide.

1. Define the following terms:
   a. nitrogen fixation
   b. nitrogenase
   c. nitrogen assimilation
   d. amino acid pool
   e. nitrogen cycle
2. Define the following terms:
   a. nonessential amino acid
   b. essential amino acid
   c. branched chain amino acid
   d. nitrogen balance
   e. transamination
3. Define the following terms:
   a. biogenic amine
   b. catecholamine
   c. pyridoxal phosphate
   d. urea
   e. L-DOPA
4. Define the following terms:
   a. vitamin B_{12}
   b. blood-brain barrier
   c. neurotransmitter
   d. serotonin
   e. pernicious anemia
5. Define the following terms:
   a. excitatory neurotransmitter
   b. inhibitory neurotransmitter
   c. retrograde neurotransmitter
   d. dopamine
   e. epinephrine
6. Define the following terms:
   a. synaptic vesicles
   b. thioredoxin
   c. orotic aciduria
   d. *anti*-adenosine
   e. PRPP
7. Define the following terms:
   a. GSH
   b. mercapturic acid
   c. CSE
   d. CBS
   e. γ-glutamyl cycle
8. Define the following terms:
   a. SAM
   b. SAH
   c. THF
   d. homocysteine
   e. one-carbon metabolism
9. Define the following terms:
   a. methotrexate
   b. circadian rhythms
   c. calmodulin
   d. Lesch-Nyhan syndrome
   e. thioredoxin reductase
10. Why are transamination reactions important in both the synthesis and degradation of amino acids?
11. Why are nitrogen compounds limited in the biosphere? Give two reasons.
12. Nitrogenase complexes are irreversibly inactivated by oxygen. Explain how nitrogen-fixing bacteria solve this problem.
13. Use reaction equations to illustrate how α-ketoglutarate is converted to glutamate. Name the enzymes and cofactors required.
14. In PLP-catalyzed reactions, the pyridinium ring acts as an electron sink. Describe this process.
15. The concentrations of the following amino acids differ significantly in blood flowing from the liver to the rest of the body and in the nutrient pool entering the enterocyte. Explain why the concentration changes in each case.
    a. isoleucine
    b. glutamine
    c. serine
    d. valine
    e. alanine
16. What are the two major classes of neurotransmitter? How do their modes of action differ? Give an example of each type of neurotransmitter.
17. Illustrate the pathways to synthesize the following amino acids:
    a. glutamine
    b. serine
    c. arginine
    d. glycine
    e. cysteine
18. Determine the synthetic family to which each of the following amino acids belongs:
    a. alanine
    b. phenylalanine
    c. methionine
    d. tryptophan
    e. histidine
    f. serine
19. What are the two most important carriers in one-carbon metabolism? Give two examples of processes in which each one participates.
20. Glutathione is an important intracellular thiol. List five functions of glutathione in the body.
21. List 10 essential amino acids in humans. Why are they essential?
22. In pyrimidine nucleosides the *anti* conformation predominates because of steric interactions with pentose. Do the purine nucleosides have similar interactions?
23. Describe the steric interactions that determine the conformations that pyrimidine nucleosides assume.
24. Through what intermediates do nitrogen atoms pass during its reduction by nitrogenase?

25. Vegans (individuals who do not consume meat or meat products such as milk and eggs) may develop pernicious anemia if they take antibiotics. Explain.
26. Explain how the γ-glutamyl cycle acts to transport amino acids across a membrane. How does the location of the γ-glutamyl transpeptidase help drive this process?
27. The amino acids glutamine and glutamate are central to amino acid metabolism. Explain.
28. Transamination reactions have been described as ping-pong reactions. Use the reaction of alanine with α-ketoglutarate to indicate how this ping-pong reaction works.
29. Pyridoxal phosphate acts as an intermediate carrier of amino groups during transamination reactions. Write a series of reactions to show the role of pyridoxal phosphate in the reaction of alanine and α-ketoglutarate.
30. What is the biologically active form of folic acid? How is it formed?
31. Identify the following biomolecules. Describe the metabolic role of each.

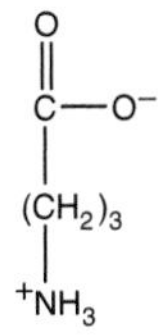

(a) 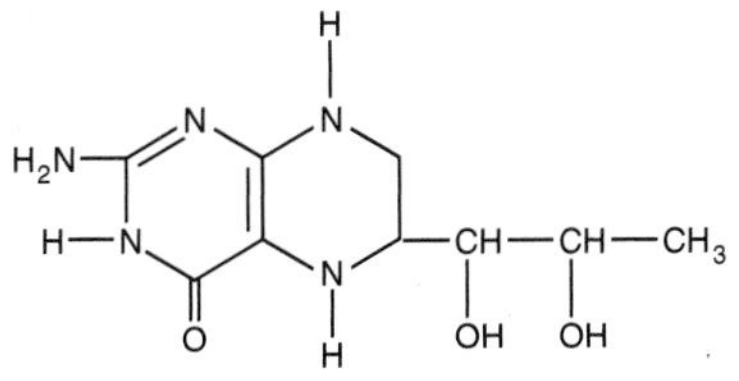 (b) 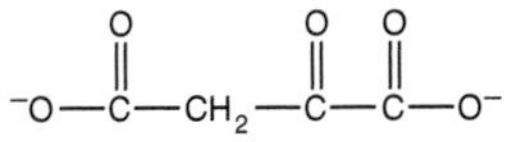 (c) 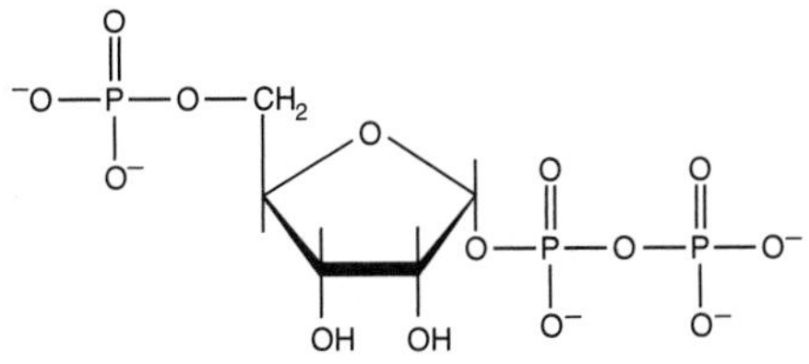 (d) 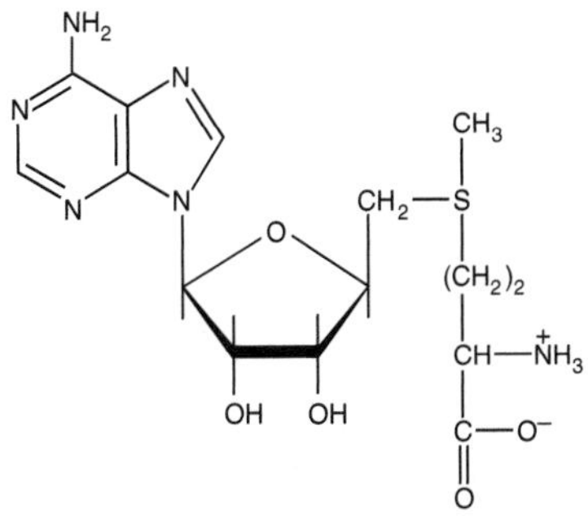 (e)

32. Long before living organisms developed the nitrogenase system, there was ammonia in Earth's atmosphere. Suggest a naturally occurring method of fixing nitrogen.
33. What carbon in uracil is derived from carbon dioxide? Draw the structure to illustrate your answer.
34. What is the function of ATP in the conversion of glutamate to glutamine?
35. Individuals with inadequate tyrosine metabolism are often light-sensitive and easily develop severe cases of sunburn. Explain. [*Hint*: The skin pigment melanin is derived from L-DOPA.]
36. In the nitrogen reductase system, for every molecule of ammonia released, one molecule of hydrogen gas is also produced. What is the source of the hydrogen gas?

Fill in the Blank

37. Two factors affecting the amount of usable nitrogen in the soil are a limited number of nitrogen-fixing bacteria and ____________________.
38. The incorporation of inorganic nitrogen into organic molecules is called ____________________.
39. Amino acids that must be provided in the diet are called ____________________ amino acids.
40. The reactions in which single carbon atoms are transferred from one molecule to another are called ____________________.
41. Neurotransmitters that are either amino acids or derived from amino acids are called ____________________ amines.
42. Dopamine, norepinephrine, and epinephrine are all derived from the amino acid ____________________.
43. Adenine and guanine are examples of the ____________________ class of nitrogenous bases.
44. Uracil is an example of the ____________________ class of nitrogenous bases.
45. When a purine or pyrimidine is linked through a β-*N*-glycosidic link to C-1 of a pentose, the molecule is called a ____________________.
46. The most common intercellular thiol is ____________________.

Short Answer

47. Describe the cause and symptoms of Lesch-Nyhan syndrome.
48. Orotic aciduria is an autosomal recessive disease caused by a deficiency in the enzyme UMP synthase (a bifunctional protein composed of orotate pyrophosphoribosyl transferase and OMP decarboxylase activities). What symptom accounts for this disease's name? How is this illness treated?
49. Describe the functions of the neurotransmitter carbon monoxide.
50. What is an NMDA receptor?
51. How is nitric oxide (NO$^\bullet$) synthesized?

Thought Questions

These questions are designed to reinforce your understanding of all of the key concepts discussed in the book so far, including this chapter and all of the chapters before it. They may not have one right answer! The authors have provided possible solutions to these questions in the back of the book and in the accompanying Study Guide for your reference.

52. Both oxygen and nitrogen are present in the atmosphere as gases. Oxygen is reactive and nitrogen is relatively inert. What features of the two molecules account for this difference?
53. Although they are consumed by animals in the diet, the purine and pyrimidine bases (unlike fatty acids and sugars) are not used to generate energy. Explain.
54. Tyrosine is a nonessential amino acid in humans. Under what circumstance would it become an essential amino acid?
55. The antibiotic sulfanilamide is an analogue of the para-aminobenzoic acid component of folate. Why would certain microorganisms die when exposed to sulfanilamide?
56. Why do marathon runners prefer beverages with sugar instead of amino acids during a long run?
57. When susceptible people consume monosodium glutamate, they experience several extremely unpleasant symptoms, such as increased blood pressure and body temperature. Use your knowledge of glutamate activity to explain these symptoms.
58. Radiation exerts part of its damaging effect by causing the formation of hydroxyl radicals. Write a reaction equation to explain how glutathione acts to protect against this form of radiation damage.
59. In purine nucleotide synthesis, the carbon and nitrogen atoms are derived from bicarbonate, aspartate, and glutamine. Devise a simple experiment to prove the source of the nitrogen atoms. Do not forget to take into account the nitrogen exchange in amino acids.
60. By definition, essential amino acids are not synthesized by an organism. Arginine is classified as an essential amino acid in infants even though it is part of the urea cycle. Explain.
61. In PLP-catalyzed reactions, the bond broken in the substrate molecule must be perpendicular to the plane of the pyridinium ring. Considering the bonds present in this ring, describe why this arrangement stabilizes the carbanion.
62. During your experimental investigation of the conversion of glutamine to proline, you have labeled the γ-carbonyl group of glutamine with ^{14}C. What carbon in the proline product will be labeled?
63. In water, cytosine will gradually convert to uracil. Using reactions you have learned in this and previous chapters, show how the conversion takes place.

CHAPTER

15 Nitrogen Metabolism II: Degradation

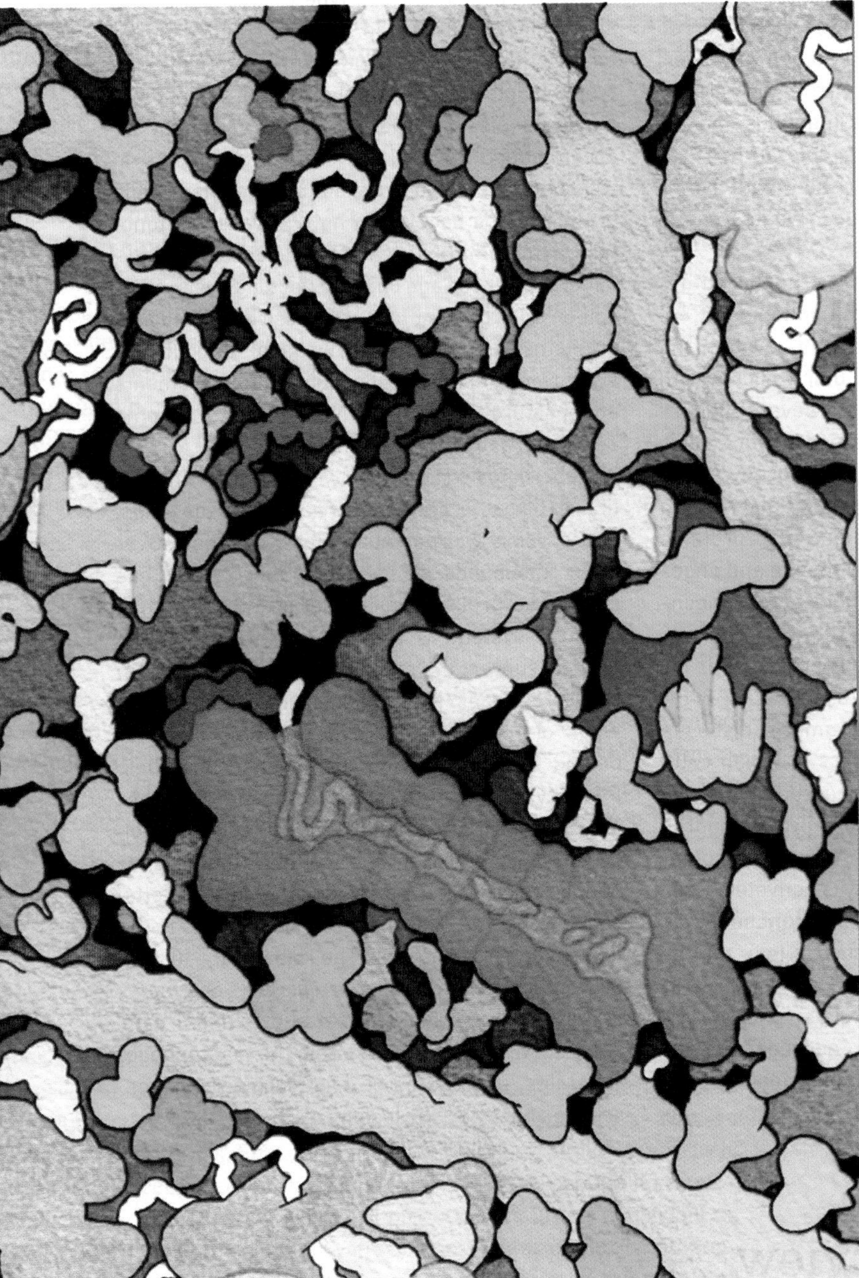

The Proteasome The proteasome is a large multisubunit molecular machine that contains several proteolytic activities and ATPases. It destroys worn-out or abnormal proteins and short-lived signal proteins. In this illustration p53 (shown in yellow), a cell cycle regulator protein, is the target protein to be degraded. Before its destruction, p53 is tagged with several ubiquitin molecules (shown in red). Ubiquitin chains are required for a protein's entry into the proteasome.

OUTLINE

TABLE 15.1 Human Protein Half-Lives

Protein	Approximate Value of Half-Life (h)
Ornithine decarboxylase	0.5
Tyrosine aminotransferase	2
Tryptophan oxygenase	2
PEP carboxykinase	5
Arginase	96
Aldolase	118
Glyceraldehyde-3-phosphate dehydrogenase	130
Cytochrome c	150
Hemoglobin	2880

Proteins differ significantly in their turnover rates, which are measured in half-lives. (A *half-life* is the time required for 50% of a specified amount of a protein to be degraded.) Proteins that play structural roles typically have long half-lives. For example, some connective tissue proteins (e.g., the collagens) often have half-lives that are measured in years. In contrast, the half-lives of regulatory enzymes are typically measured in minutes. **Table 15.1** lists several examples.

Although some proteins are degraded by proteolytic enzymes in cytoplasm (e.g., Ca^{2+}-activated calpains), most cellular proteins are degraded by two major systems: the ubiquitin proteasomal system and the autophagy lysosomal system (**Figure 15.1**).

Ubiquitin Proteasomal System

In the **ubiquitin proteasomal system** (UPS) (**Figure 15.1a**), protein degradation is initiated with a covalent modification referred to as ubiquitination in which **ubiquitin**, an 8.5-kDa protein, is attached to substrate proteins. UPS degrades most short-lived proteins (e.g., regulatory proteins such as transcription factors). UPS is also triggered by ERAD (ER associated protein degradation, p. 47).

The mechanisms that target protein for destruction by ubiquitination are not fully understood. It is known that many proteins have sequence motifs, called *degrons*, which mark them for proteolytic destruction:

1. **N-terminal residues**. Very short-lived proteins often have basic (Arg, Lys, His) or bulky hydrophobic (Leu, Phe, Tyr, Trp, Ile) N-terminal residues. More stable proteins characteristically have sulfur-containing, hydroxyl-containing, or nonbulky hydrophobic amino acids at the N-terminus.
2. **Peptide motifs**. Proteins with certain homologous sequences are rapidly degraded. For example, proteins that have extended sequences containing proline, glutamate, serine, and threonine have half-lives of less than 2 hours. (PEST sequences are named for the one-letter abbreviations for these amino acids. See **Table 5.1**.) Ensuring rapid ubiquitination is the *cyclin destruction box*, a set of homologous nine-residue sequences near the N-terminus of cyclins.

Once ubiquitinated, target proteins are transferred to massive proteolytic molecular machines called **proteasomes** that cleave them into peptide fragments with an average of seven to eight amino acid residues. Such fragments are further degraded by cytoplasmic proteases to amino acids that can be recycled into new protein molecules. **Ubiquitination**, the attachment of a small, highly conserved 76-residue protein called ubiquitin to worn-out or damaged proteins, or short-lived regulatory proteins, occurs in several stages and involves three enzyme classes: E1, E2, and E3. In the first step (**Figure 15.1a**), E1 (*ubiquitin-activating enzyme*) activates a ubiquitin molecule via adenylation and transfers it to an active site thiol of E1 to form a high-energy thioester. The C-terminal glycine carboxyl group of the ubiquitin molecules participates in this reaction. Ubiquitin is then transferred to an active site thiol of E2 (*ubiquitin-conjugating enzyme*) via a transthiolation reaction. There is only one E1, but dozens of E2 enzymes in mammalian cells. The E2 enzymes vary in their specificity for association with E3 (*ubiquitin ligase*). (Thirty-five different E2 enzymes have been discovered in humans.) The E3 enzymes determine the substrate specificity of ubiquitination because they interact with E2 and the target protein and transfer ubiquitin to a specific internal lysine side chain of the target protein via a thioester-to-amide transition. An E3 enzyme recognizes its target protein by binding to a degron, which may be revealed by posttranslational modification (e.g., acetylation, hydroxylation, or proteolytic cleavage) or they may be revealed by a change in a protein's conformation that exposes hydrophobic residues. Subsequent ubiquitination lengthens the ubiquitin tag on the protein from 4 to 50 units. The specificity

KEY CONCEPTS

- Protein turnover, the continuous synthesis and degradation of proteins, provides living organisms with metabolic flexibility and protects cells from the accumulation of abnormal proteins.
- Most short-lived cellular proteins are degraded by the ubiquitin proteasomal system to yield short peptides. Long-lived proteins and organelles are degraded by the autophagy lysosomal system.
- The amino acid products of the peptides that are cleaved by cytoplasmic proteases enter the amino acid pool and are available for incorporation into new protein molecules.

of ubiquitination and therefore regulated proteolysis derives in part from the substrate specificities of the large number of E2s (35 in humans) and E3s (more than 600 in humans). It is noteworthy that proteasomal digestion of highly oxidized proteins does not require ubiquitination.

The proteasome (**Figure 15.2**) is a large (2000-kDa) multisubunit complex in the form of a hollow cylinder with dimensions of 1500 by 1150 nm. Referred to as the 26S proteasome, this structure consists of a 20S core particle and a 900-kDa 19S regulatory particle. [The Svedberg unit (S) is a measure of the rate at which a particle sediments in an ultracentrifuge. Since S values are related to both the mass and the shape of particles, they are not additive.] The 20S particle consists of four heptameric protein rings ($\alpha_7\beta_7\beta_7\alpha_7$). The two inner β rings

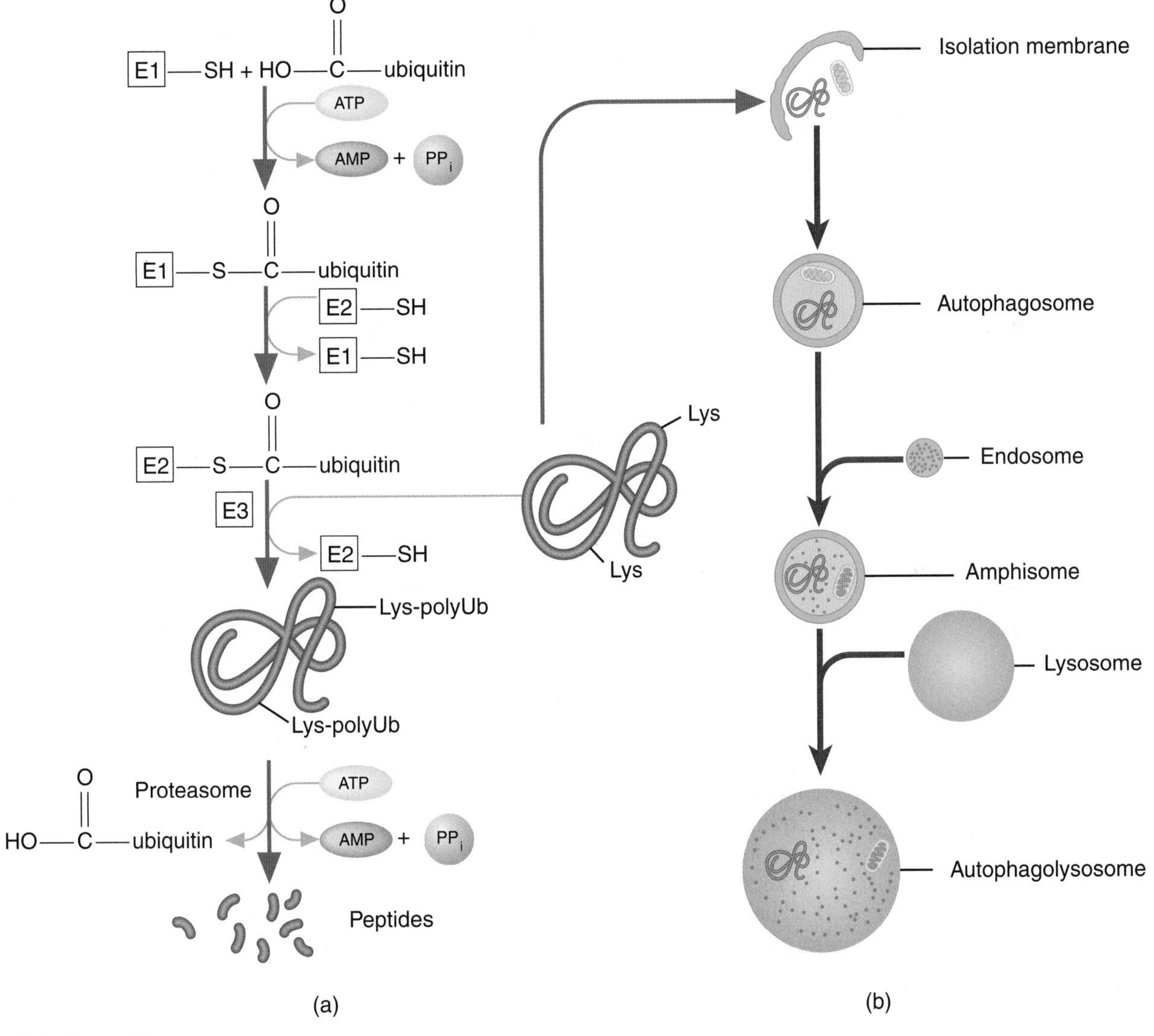

FIGURE 15.1

The Ubiquitin Proteasomal and the Autophagy Lysosomal Systems

(a) Protein ubiquitination begins with the ATP hydrolysis-driven formation of a thiol ester bond between E1 (ubiquitin-activating enzyme) and ubiquitin. Ubiquitin is then transferred to E2 (ubiquitin-conjugating enzyme). E3 (ubiquitin ligase) transfers ubiquitin from E2 to a lysine side chain on the target protein and subsequently to ubiquitin moieties already on the target protein. Polyubiquitination continues until there are 4 to 50 ubiquitins attached to the target lysine residues. The polyubiquitinated protein is then degraded by a proteasome. (b) In autophagy, a seemingly random process, an isolation membrane begins to form that surrounds and sequesters cytoplasmic components. The expanding isolation membrane eventually seals to form the autophagosome. When the autophagosome fuses with an endosome, forming an amphisome, the internal pH begins to drop. Fusion of the amphisome with a lysosome results in the formation of the autophagolysosome. Lysosomal enzymes then degrade the cargo. The products are recycled or degraded to generate energy. Once thought to be independent pathways, UPS and autophagy are interrelated. They share some regulatory proteins and under certain conditions either pathway may degrade some protein substrates.

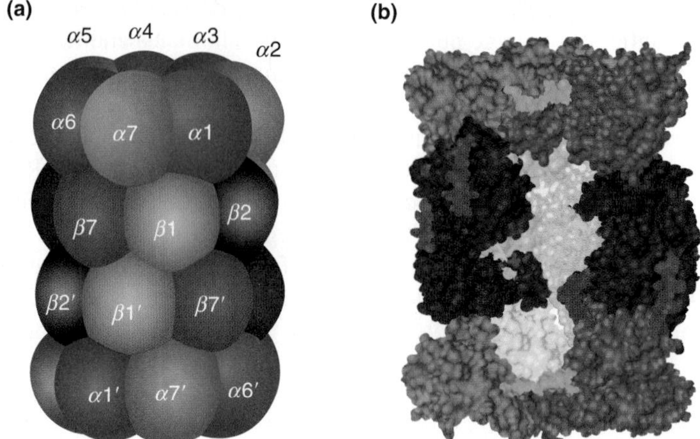

FIGURE 15.2
The Proteasome

The 20S proteasome (core particle) (a) is a 700-kDa barrel-shape structure that contains 28 proteins: two α rings with seven subunits each and two β rings with seven subunits each. A cutaway view (b) reveals the inner catalytic chamber.

possess three different types of proteolytic activity that face the inner chamber. The α-ring components have an N-terminal segment that limits access to the narrow proteolytic chamber to unfolded peptide chains. The 19S particles consist of a 9-subunit *lid* that binds directly to the α-ring of the core particle and a 10-subunit *base*. Six of the base subunits possess ATPase activity. ATP binding to the 19S ATPase subunits allows the association of the 19S and 20S particles and facilitates the protein unfolding that is critical for threading substrate polypeptides into the catalytic chamber. The lid subunits participate in substrate selection and ubiquitin removal. This gatekeeping function ensures that only appropriate target proteins are translocated into the catalytic chamber for degradation. One protein is degraded at a time, and the 6- to 10-residue peptide products are released from the proteasome for hydrolysis to free amino acids by cytoplasmic proteases.

Ubiquitination also has nonproteolytic functions. The reversible covalent attachment of ubiquitin to target proteins is a pervasive mechanism used to control diverse cellular processes such as DNA repair and cell signaling.

Autophagy-Lysosomal System

Autophagy ("self-eating") is a cellular degradation pathway in which cell components, most notably long-lived proteins and organelles, are degraded by the hydrolytic enzymes within lysosomes. Autophagy has several roles in cells. In addition to its obvious role in degrading worn-out or damaged cell components, autophagy provides a recycling mechanism that maintains vital functions when nutrient levels are low (during fasting). Autophagy is also involved in the regulation of development (cellular remodeling) and the destruction of invading microorganisms.

There are three forms of autophagy: chaperone-mediated autophagy, microautophagy, and macroautophagy. **Chaperone-mediated autophagy** is a receptor-mediated process in which specific proteins that are bound to a chaperone complex are unfolded and then translocated into a lysosome, where they are degraded by lysosomal proteases. In **microautophagy**, small amounts of cytoplasm are directly engulfed by lysosomes. **Macroautophagy** (**Figure 15.1b**), often referred to as autophagy, uses a lysosomal pathway for bulk degradation of cytoplasmic components. It is the major catabolic mechanism used by eukaryotic cells to maintain optimal function and respond to changing environmental conditions.

Autophagy is induced by a large number of stressors including ER stress (e.g., stress caused by the unfolded protein response, see p. 48), hypoxia, oxidative stress, nutrient deprivation, high temperature, and viral infections. Autophagy begins with the formation of the *autophagosome* from a double membrane structure called the *isolation membrane* (probably originating from a ribosome-free zone of the RER). The isolation membrane surrounds and sequesters cytoplasmic material as it expands and eventually closes to form the autophagosome. The autophagosome then fuses with an endosome to form an *amphisome*. Endosomal membrane-bound ATPases begin pumping protons into the amphisome lumen. Eventually, the outer membrane of the amphisome fuses with a lysosome to form an autophagolysosome. Lysosomal enzymes then proceed to digest the cytoplasmic cargo and the inner amphisome membrane. The degradation products (e.g., amino acids and sugars) are exported to the cytoplasm where they may be used in biosynthesis and/or degraded to generate energy. Any autophagolysosomes that contain substances that resist digestion, referred to as residual bodies, remain in the cytoplasm. For example, *lipofuscin granules*, which contain indigestible brown-pigmented debris, are found in aging nerve, heart, kidney, and adrenal cells.

Autophagy is a housekeeping process that operates at a basal level in almost all eukaryotic cells. It can be rapidly upregulated when cells are under stress or when energy and nutrient levels are low. For example, the protein synthesis initiation factor 2α (eIF2α) and AMPK stimulate autophagy when the levels of nutrients and energy, respectively, are low. Autophagy is inhibited by the serine-threonine kinase mammalian target of rapamycin (mTOR) when nutrient and energy levels are high. (Rapamycin is a bacterial molecule used clinically to prevent the rejection of transplanted organs.) mTOR, which integrates intracellular signals (e.g., nutrient and energy levels and redox status) with extracelluar signals (e.g., hormones and growth factors), is a central regulator of cell metabolism.

15.2 AMINO ACID CATABOLISM

The catabolism of the amino acids usually begins with the removal of the amino group. Amino groups can then be disposed of in urea synthesis. The resulting carbon skeletons are then degraded to form one or more of seven possible metabolic products: acetyl-CoA, acetoacetyl-CoA, pyruvate, α-ketoglutarate, succinyl-CoA, fumarate, or OAA. Depending on the animal's current metabolic requirements, these molecules are used to synthesize fatty acids or glucose or to generate energy. Amino acids degraded to form acetyl-CoA or acetoacetyl-CoA are referred to as **ketogenic** because they can be converted to either fatty acids or ketone bodies. The carbon skeletons of the **glucogenic** amino acids, which are degraded to pyruvate or a citric acid cycle intermediate, can then be used in gluconeogenesis. Most amino acids are glucogenic. Discussions of deamination pathways and urea synthesis are followed by descriptions of the pathways that degrade carbon skeletons.

Deamination

The removal of the α-amino group from amino acids involves two types of biochemical reaction: transamination and oxidative deamination. Both reactions have been described (Section 14.2). Because these reactions are reversible, amino groups are easily shifted from abundant amino acids and used to synthesize those that are scarce. Amino groups become available for urea synthesis when amino acids are in excess. Urea is synthesized in especially large amounts when the diet is high in protein or when there is massive breakdown of protein, for example, during starvation.

In muscle, excess amino groups are transferred to α-ketoglutarate to form glutamate:

$$\alpha\text{-Ketoglutarate} + \text{L-Amino acid} \rightleftharpoons \text{L-Glutamate} + \alpha\text{-Keto acid} \qquad (1)$$

The amino groups of glutamate molecules are transported in blood to the liver by the alanine cycle (**Figure 8**.12):

$$\text{Pyruvate} + \text{L-Glutamate} \rightleftharpoons \text{L-Alanine} + \alpha\text{-Ketoglutarate} \qquad (2)$$

In the liver, glutamate is formed as the reaction catalyzed by alanine transaminase is reversed. The oxidative deamination of glutamate yields α-ketoglutarate and NH_4^+.

In most extrahepatic tissues, the amino group of glutamate is released via oxidative deamination as NH_4^+. Ammonia is carried to the liver as the amide group of glutamine. The ATP-requiring reaction in which glutamate is converted to glutamine is catalyzed by glutamine synthetase:

$$\text{L-Glutamate} + NH_4^+ + \text{ATP} \rightarrow \text{L-Glutamine} + \text{ADP} + P_i \qquad (3)$$

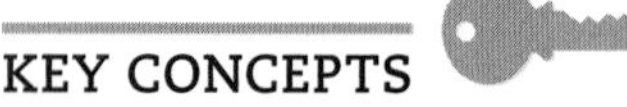

KEY CONCEPTS

- The degradation of most amino acids begins with removal of the α-amino group.
- Two types of biochemical reaction are involved in amino group removal: transamination and oxidative deamination.

After its transport to the liver, glutamine is hydrolyzed by glutaminase to form glutamate and NH_4^+. Additional NH_4^+ is generated as glutamate dehydrogenase converts glutamate to α-ketoglutarate:

$$\text{L-Glutamine} + H_2O \rightarrow \text{L-Glutamate} + NH_4^+ \qquad (4)$$

$$\text{L-Glutamate} + H_2O + NAD^+ \rightarrow \alpha\text{-Ketoglutarate} + \text{NADH} + H^+ + NH_4^+ \qquad (5)$$

Most of the ammonia generated in amino acid degradation is produced by the oxidative deamination of glutamate. Additional ammonia is produced in several other reactions catalyzed by the following enzymes.

The *L-amino acid oxidases* are FMN-requiring liver and kidney enzymes that convert some of the amino acids to α-keto acids, NH_4^+ and H_2O_2. The *serine and threonine dehydratases* are hepatic pyridoxal-requiring enzymes that convert serine and threonine to pyruvate and α-ketobutyrate, respectively. Large quantities of ammonia are produced by intestinal bacterial *urease*, which hydrolyzes urea circulating in the bloodstream. Afterward, ammonia diffuses into the blood and is transported to the liver. *Adenosine deaminase* (p. 575) releases NH_4^+ from the adenine ring of AMP in a nucleotide catabolic pathway.

Urea Synthesis

The urea cycle disposes of approximately 90% of surplus nitrogen in ureotelic organisms (i.e., those that convert ammonia to urea). As shown in **Figure 15.3**, urea is formed from ammonia, CO_2, and aspartate in a cyclic pathway referred to as the **urea cycle**. Because the urea cycle was discovered by Hans Krebs and Kurt Henseleit, it is often referred to as the **Krebs urea cycle** or the *Krebs-Henseleit cycle*. The overall equation for urea synthesis is

$$CO_2 + NH_4^+ + \text{Aspartate} + 3\,\text{ATP} + 2\,H_2O \rightarrow$$
$$\text{Urea} + \text{Fumarate} + 2\,\text{ADP} + 2\,P_i + \text{AMP} + PP_i + 5\,H^+ \qquad (6)$$

Urea synthesis, which occurs in hepatocytes, begins with the formation of carbamoyl phosphate in the matrix of mitochondria. The substrates for this reaction, catalyzed by carbamoyl phosphate synthetase (CPSI), are NH_4^+ and HCO_3^-.

HO—C(=O)—O⁻ + ⁻O—P(=O)(O⁻)—O—ADP → (ADP) → HO—C(=O)(NH3)—O—P(=O)(O⁻)—O⁻ → (Pi)

H_2N—C(=O)—O⁻ → (ATP → ADP) → H_2N—C(=O)—O—P(=O)(O⁻)—O⁻ (7)

Carbamate — Carbamoyl phosphate

Carbamoyl phosphate synthesis is essentially irreversible because two molecules of ATP are consumed. (One is used to activate HCO_3^-. The second

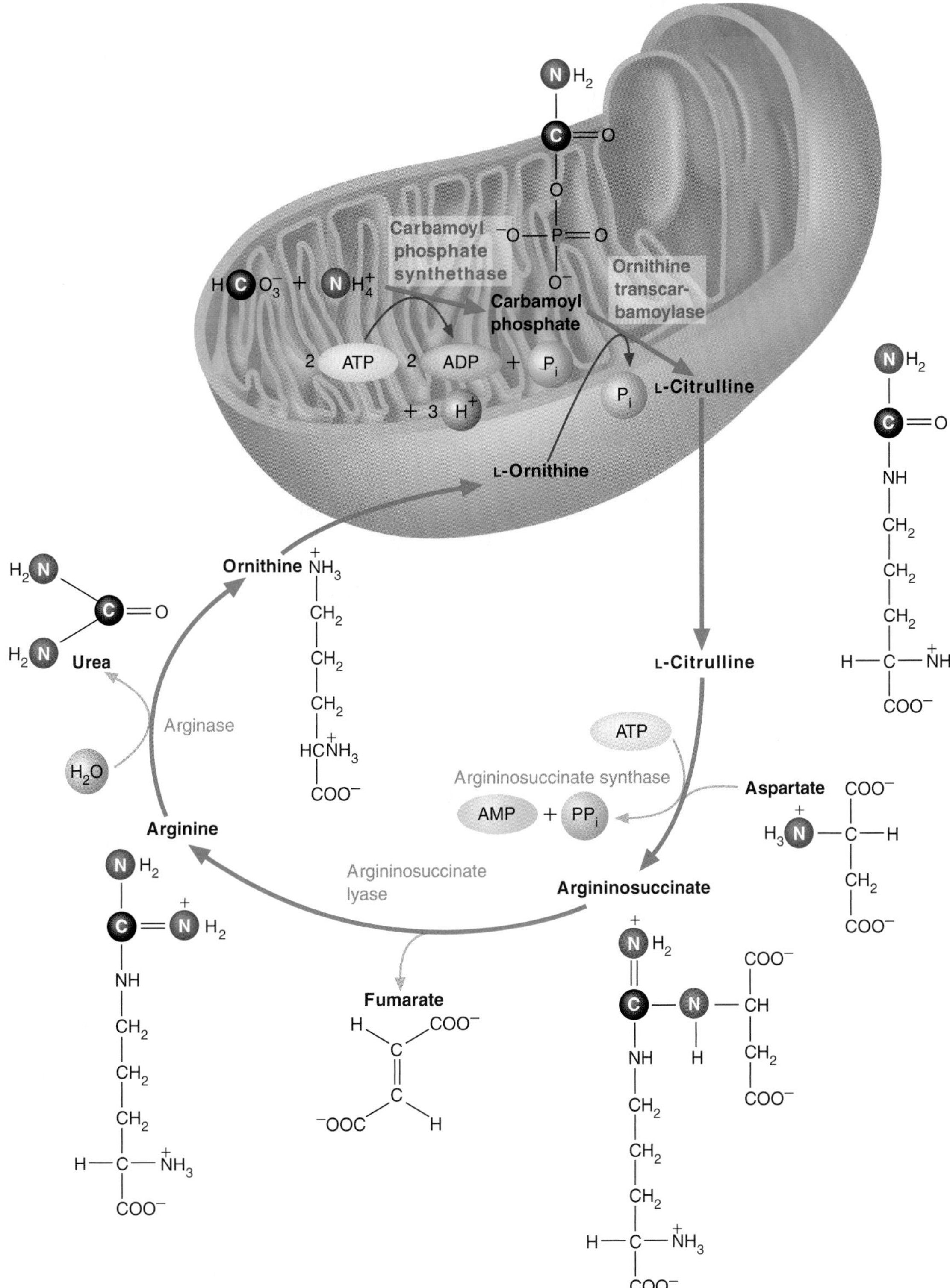

FIGURE 15.3
The Urea Cycle

The urea cycle converts NH_4^+ to urea, a less toxic molecule. The sources of the atoms in urea are shown in color. Citrulline is transported across the inner membrane by a carrier for neutral amino acids. Ornithine is transported by ornithine translocase in exchange for H^+ and citrulline. Fumarate is transported back into the mitochondrial matrix (for reconversion to malate) by carriers for α-ketoglutarate or tricarboxylic acids.

molecule is used to phosphorylate carbamate.) Carbamoyl phosphate subsequently reacts with ornithine to form citrulline. Citrulline is synthesized in a nucleophilic acyl substitution reaction in which the side chain amino group of ornithine is the nucleophile and phosphate is the leaving group.

Ornithine + Carbamoyl phosphate → Citrulline + P_i (8)

This reaction, catalyzed by ornithine transcarbamoylase, is driven to completion because phosphate is released from carbamoyl phosphate. (Recall from **Table 4.1** that carbamoyl phosphate has a high phosphoryl group transfer potential.) Once formed, citrulline is transported to the cytoplasm, where it reacts with aspartate to form argininosuccinate. (The α-amino group of aspartate, formed from OAA by transamination reactions in the liver, provides the second nitrogen that is ultimately incorporated into urea.) In this reaction, which is catalyzed by argininosuccinate synthase, citrulline is activated by reacting with ATP to form a citrulline-AMP intermediate and pyrophosphate. The amino nitrogen of aspartate, acting as a nucleophile, displaces the AMP to form argininosuccinate.

Citrulline-AMP + Aspartate → Argininosuccinate + AMP (9)

This acyl substitution reaction is driven forward by the cleavage of pyrophosphate by pyrophosphatase. Argininosuccinate lyase subsequently cleaves argininosuccinate to form arginine (the immediate precursor of urea) and fumarate.

Argininosuccinate → (BH) → Arginine + Fumarate (10)

A histidine residue within the enzyme's active site, acting as a base (B:), removes a proton from the substrate to form a carbanion. The carbanion then expels the nitrogen to form a C = C bond. The nitrogen accepts a proton from a proton donor (HA) (perhaps the protonated histidine).

In the final reaction of the urea cycle, arginase catalyzes the hydrolysis of arginine to form ornithine and urea.

$$\left[{}^{-}O{-}\overset{O}{\overset{\|}{C}}{-}\underset{{}^{+}NH_3}{\overset{H}{C}}{-}CH_2CH_2\right]CH_2{-}\overset{H}{N}{-}\overset{\overset{+}{N}H_2}{\overset{\|}{C}}{-}NH_2 + H_2\ddot{O} \longrightarrow R{-}CH_2{-}\overset{H}{\ddot{N}}{-}\underset{:OH}{\overset{\overset{H\ \ H}{N}}{C}}{-}NH_2 \longrightarrow$$

Arginine

$${}^{-}O{-}\overset{O}{\overset{\|}{C}}{-}\underset{{}^{+}NH_3}{\overset{H}{C}}{-}CH_2CH_2CH_2NH_2 + H_2N{-}\overset{O}{\overset{\|}{C}}{-}NH_2 \qquad (11)$$

Ornithine **Urea**

Once it forms, urea diffuses out of the hepatocytes and into the bloodstream. It is ultimately eliminated in urine by the kidney. Ornithine returns to the mitochondria for condensation with carbamoyl phosphate to begin the cycle again. Because arginase is found in significant amounts only in the ureotelic animal liver, urea is produced only in this organ.

After its transport back into the mitochondrial matrix, fumarate is hydrogenated to form malate, a component of the citric acid cycle. The OAA product of the citric acid cycle can be used in energy generation or it can be converted to glucose or aspartate. The relationship between the urea cycle and the citric acid cycle, often referred to as the **Krebs bicycle**, is outlined in **Figure 15.4**.

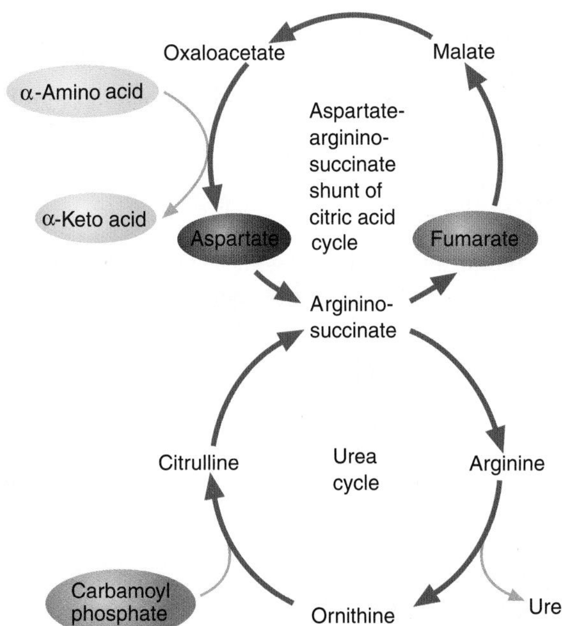

FIGURE 15.4
The Krebs Bicycle

The aspartate used in urea synthesis is generated from oxaloacetate, a citric acid cycle intermediate. This transamination reaction removes amino nitrogen from many amino acids.

WORKED PROBLEM 15.1

Review the urea cycle and then determine the number of ATP molecules used to synthesize one urea molecule.

SOLUTION

Two ATP are required for the synthesis of carbamoyl phosphate from NH_4^+ and CO_2. The synthesis of argininosuccinate involves the conversion of one ATP to an AMP product. Two ATP equivalents are required to convert AMP to ATP. The total number of ATP equivalents used to synthesize one urea molecule is therefore four. ■

Hyperammonemia, a potentially fatal condition in which blood levels of NH_4^+ become excessive when the liver's capacity to synthesize urea is compromised, is discussed in the Biochemistry in Perspective essay on that topic.

Visit the companion website at www.oup.com/us/mckee to read the Biochemistry in Perspective essay on hyperammonemia.

KEY CONCEPTS

- Urea is synthesized from ammonia, CO_2, and aspartate.
- The urea cycle is carefully regulated.

Control of the Urea Cycle

Urea is a toxic molecule. Its synthesis is, therefore, stringently regulated. There are long- and short-term regulatory mechanisms. The levels of all five urea cycle enzymes are altered by variations in dietary protein consumption. Within several days after a significant dietary change, there are twofold to threefold changes in enzyme levels. Several hormones are involved in the altered rates of enzyme synthesis. Glucagon and the glucocorticoids activate the transcription of urea cycle enzymes, whereas insulin represses their synthesis.

The urea cycle enzymes are controlled in the short term by the concentrations of their substrates. For example, urea synthesis is stimulated by a high-protein diet or by fasting. Carbamoyl phosphate synthetase I (CPSI) is also allosterically activated by *N-acetylglutamate*. This latter molecule is a sensitive indicator of the cell's concentration of glutamate, a source of NH_4^+. *N*-Acetylglutamate (NAG) is produced from glutamate and acetyl-CoA in a reaction catalyzed by *N*-acetylglutamate synthetase, which is allosterically activated by arginine. CPSI activation by NAG is a positive-feedback regulatory process because an increase in arginine concentration results in an increase in NAG synthesis. Substrate channeling (p. 211) also enhances the efficiency of the urea cycle. Of all the urea cycle metabolites, only urea, the product of the pathway, has been observed to mix freely with other cytoplasmic metabolites.

QUESTION 15.1

Although arginine is an intermediate in the urea cycle, it is an EAA in young animals. Suggest a reason for this phenomenon.

QUESTION 15.2

In some clinical circumstances, patients with hyperammonemia are treated with antibiotics. Suggest a rational basis for this therapy.

Catabolism of Amino Acid Carbon Skeletons

The α-amino acids can be grouped into classes according to their end products: acetyl-CoA, acetoacetyl-CoA, pyruvate, and several citric acid cycle intermediates. Each group is briefly discussed. The degradation pathways for the 20 α-amino acids found in proteins are outlined in **Figure 15.5**.

AMINO ACIDS FORMING ACETYL-CoA In all, 10 α-amino acids yield acetyl-CoA. This group is further divided according to whether pyruvate is an intermediate in acetyl-CoA formation. The amino acids whose degradation involves pyruvate are

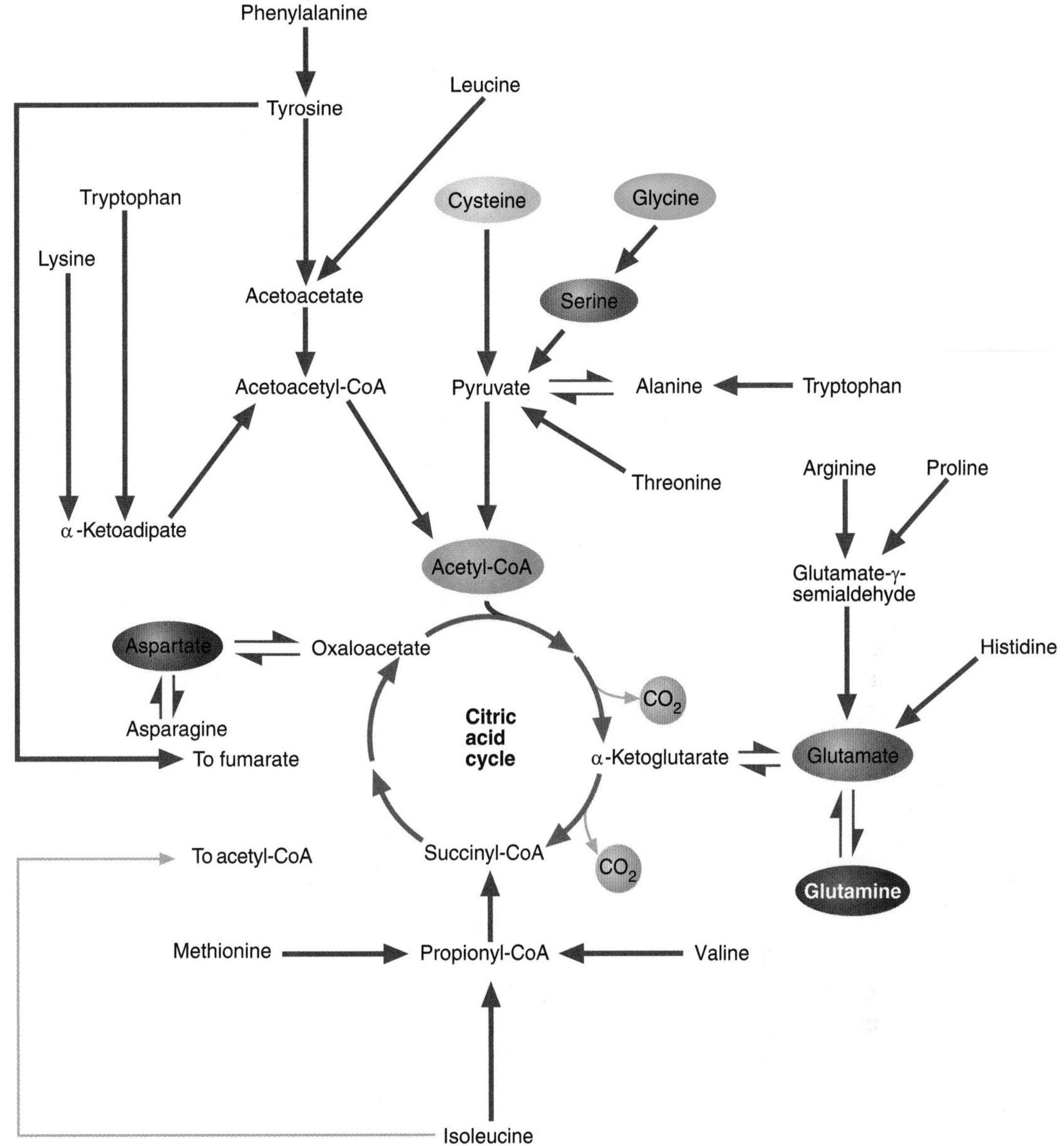

FIGURE 15.5

Degradation of the 20 α-Amino Acids Found in Proteins

The α-amino groups are removed early in the catabolic pathways. Carbon skeletons are converted to common metabolic intermediates.

alanine, serine, glycine, cysteine, and threonine. These amino acids can be ketogenic or glucogenic depending on the relative activities of pyruvate dehydrogenase and pyruvate carboxylase. Depending on a cell's metabolic requirements, pyruvate may be converted to acetyl-CoA to be oxidized or used to synthesize fatty acids, or it can be converted to OAA, which can be diverted into gluconeogenesis. The other five amino acids converted to acetyl-CoA by pathways not involving pyruvate are lysine, tryptophan, tyrosine, phenylalanine, and leucine. The two reaction sequences are outlined in Figures 15.6 and 15.7.

The individual catabolic pathways for these molecules are as follows:

1. **Alanine**. Recall that the reversible transamination reaction involving alanine and pyruvate is an important component of the alanine cycle discussed previously (Section 8.2).
2. **Serine**. As described, serine is converted to pyruvate by the pyridoxal phosphate-requiring enzyme serine dehydratase.

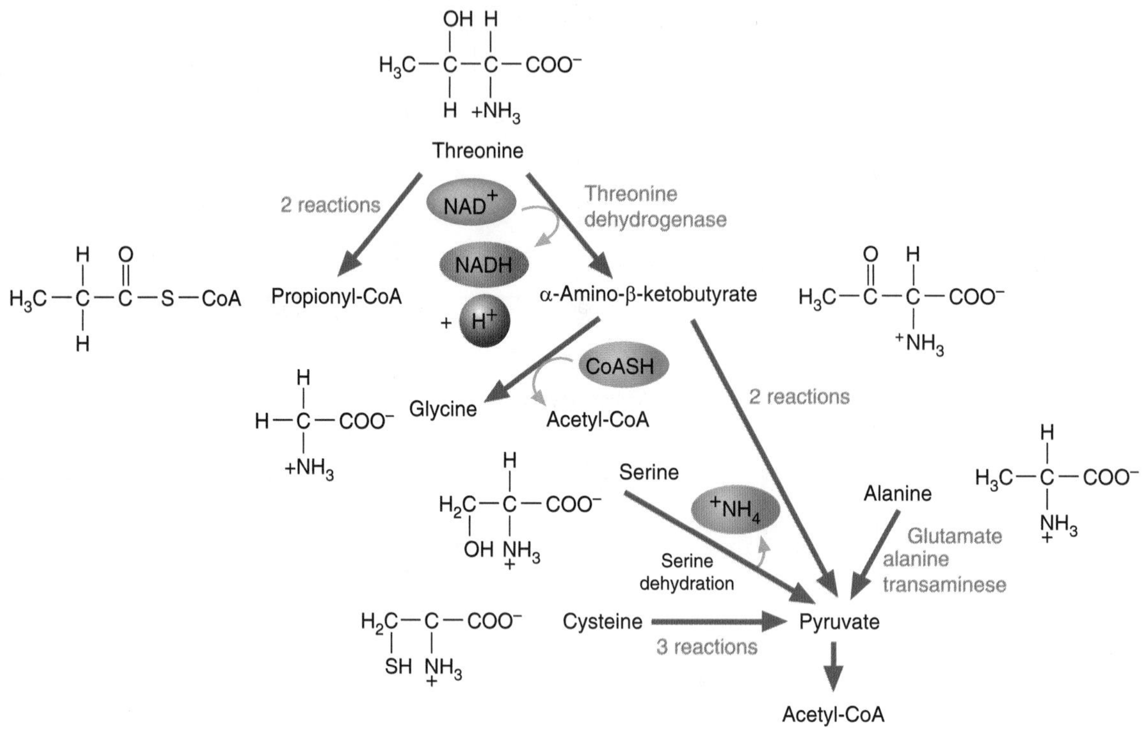

FIGURE 15.6

The Catabolic Pathways of Threonine, Glycine, Serine, Cysteine, and Alanine

Pyruvate is an intermediate in the conversion of these amino acids to acetyl-CoA. Threonine can be degraded by two pathways. In one pathway threonine is converted to pyruvate by a three-reaction pathway. α–Amino-β-ketobutyrate, an intermediate in this pathway, can also react with CoASH to yield acetyl-CoA and glycine. Note that glycine is also degraded by glycine synthase to form CO_2, NH_4^+, and N^5,N^{10}-methylene THF in an NAD^+-requiring reaction. In primates, most threonine molecules are degraded to propionyl-CoA, a molecule, which is then converted to the citric acid cycle intermediate succinyl-CoA (**Figure 12.12**). Serine is converted to pyruvate with the release of NH_4^+ by serine dehydratase. The major pathway for cysteine catabolism is a three-reaction pathway in which the oxidized intermediate cysteine sulfinate undergoes a transamination and then a desulfuration to yield pyruvate and bisulfite (HSO_3^-). A transamination reaction converts alanine to pyruvate.

3. **Glycine**. Glycine can be converted to serine by serine hydroxymethyltransferase. (The hydroxymethyl group is donated by N^5,N^{10}-methylene THF as described in Section 14.3.) Then serine is converted to pyruvate, as previously described. Most glycine molecules, however, are degraded by glycine cleavage enzyme to CO_2, NH_4^+, and N^5,N^{10}-methylene THF.
4. **Cysteine.** In animals, cysteine is converted to pyruvate by several pathways. In the principal pathway cysteine is oxidized to cysteine sulfinate. Pyruvate is then produced after a transamination and a desulfuration reaction.
5. **Threonine**. In the major degradative pathway, threonine is oxidized by threonine dehydrogenase to form α-amino-β-ketobutyrate. The latter molecule is metabolized further to form pyruvate, or it can be cleaved by α-amino-β-ketobutyrate lyase to form acetyl-CoA and glycine. Alternatively, threonine can be degraded to α-ketobutyrate by threonine dehydratase and subsequently to propionyl-CoA. Propionyl-CoA is then converted to succinyl-CoA (see p. 439).
6. **Lysine**. Lysine is converted to α-ketoadipate in a series of reactions that include two oxidations, removal of the side chain amino group, and a transamination. Acetoacetyl-CoA is produced in a reaction pathway that includes several oxidations, a decarboxylation, and a hydration. Acetoacetyl-CoA can be converted to acetyl-CoA in a reaction that is the reverse of a step in ketone body formation.

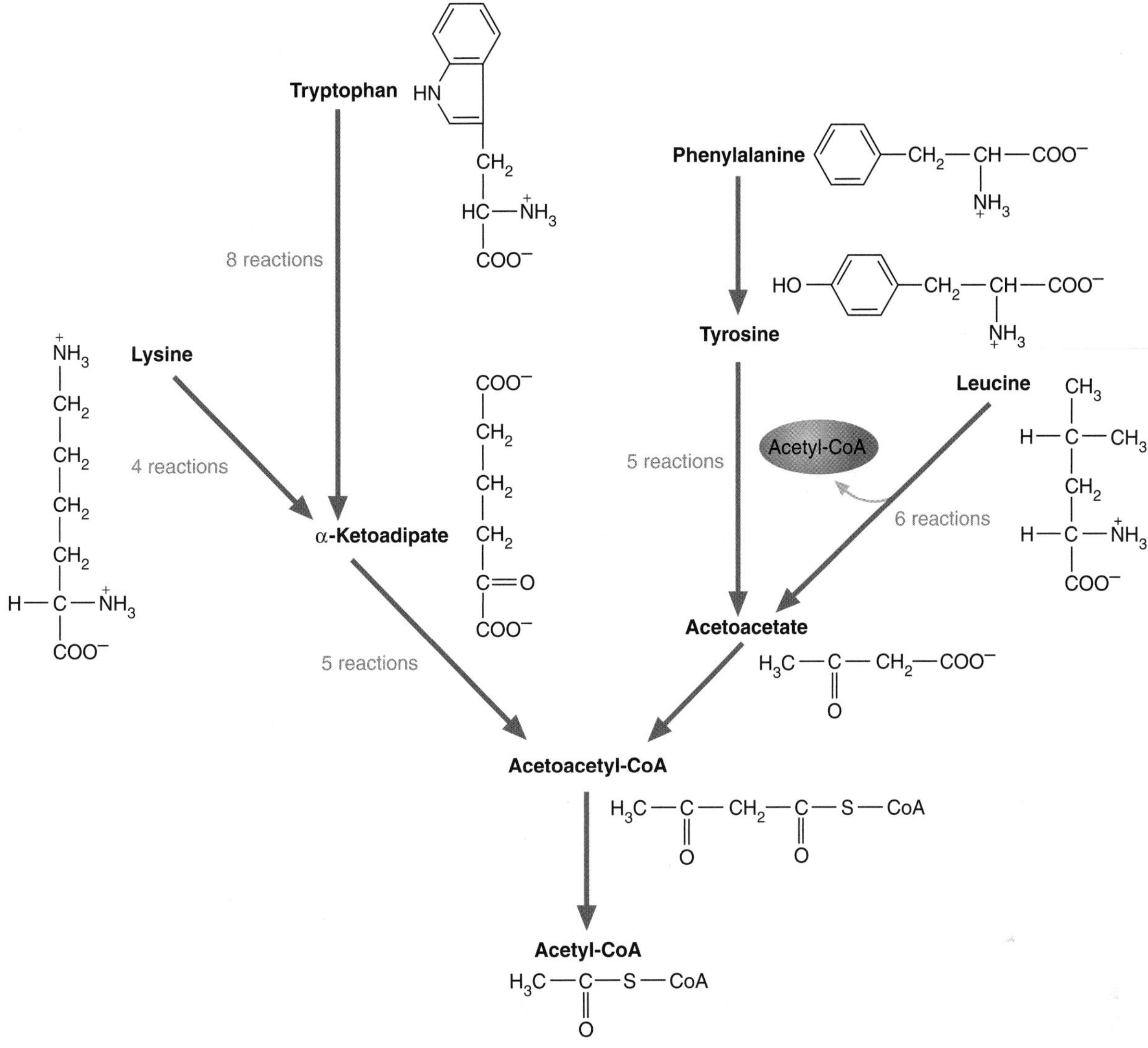

FIGURE 15.7

The Catabolic Pathways of Lysine, Tryptophan, Phenylalanine, Tyrosine, and Leucine

These pathways are long and complex. The number of reactions in each segment is indicated.

7. **Tryptophan**. Tryptophan is converted to α-ketoadipate in eight reactions, which also yield formate and alanine. Acetyl-CoA is synthesized from α-ketoadipate as described for lysine. The alanine produced in this pathway is converted to acetyl-CoA via pyruvate.
8. **Tyrosine**. Tyrosine catabolism begins with a transamination and a dehydroxylation. Homogentisate is synthesized in the latter reaction, catalyzed by the ascorbate-requiring enzyme parahydroxyphenylpyruvate dioxygenase. Homogentisate is converted to maleylacetoacetate by homogentisate oxidase. Acetoacetate and fumarate are then generated in isomerization and hydration reactions.
9. **Phenylalanine**. Phenylalanine is hydroxylated to form tyrosine by phenylalanine-4-monooxygenase (**Figure 15.8**) in a reaction requiring O_2 and tetrahydrobiopterin (BH_4), a folic acid-like molecule (**Figure 15.9**). Tyrosine is degraded to form acetoacetate and fumarate.
10. **Leucine**. Leucine, one of the branched-chain amino acids, is converted to HMG-CoA in a series of reactions that include a transamination, two oxidations, a carboxylation, and a hydration. HMG-CoA is then converted to acetyl-CoA and acetoacetate by HMG-CoA lyase.

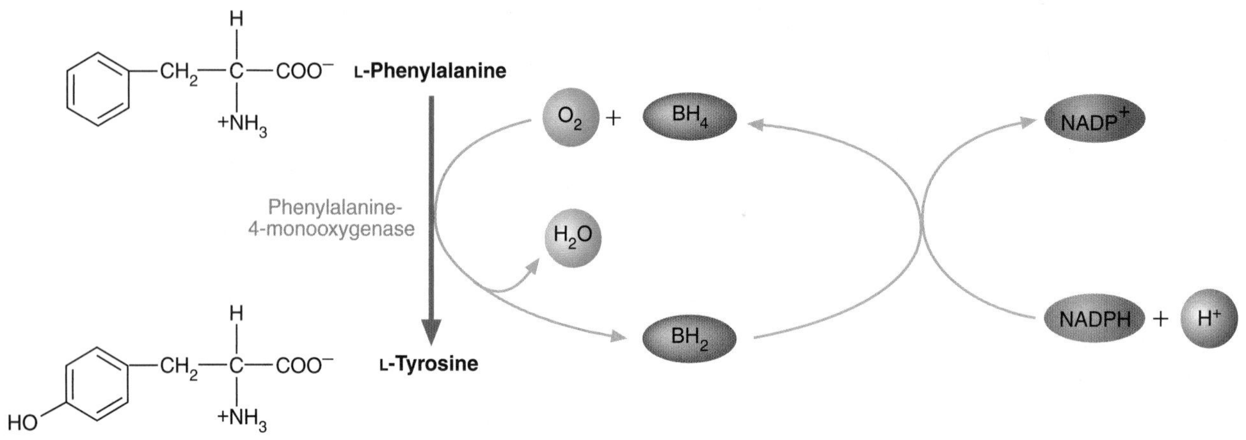

L-Phenylalanine

Phenylalanine-4-monooxygenase

O_2 + BH_4

H_2O

BH_2

$NADP^+$

NADPH + H^+

L-Tyrosine

FIGURE 15.8

The Conversion of Phenylalanine to Tyrosine

The reaction catalyzed by phenylalanine-4-monooxygenase is irreversible. The electrons required for the hydroxylation of phenylalanine are carried to O_2 from NADPH by tetrahydrobiopterin.

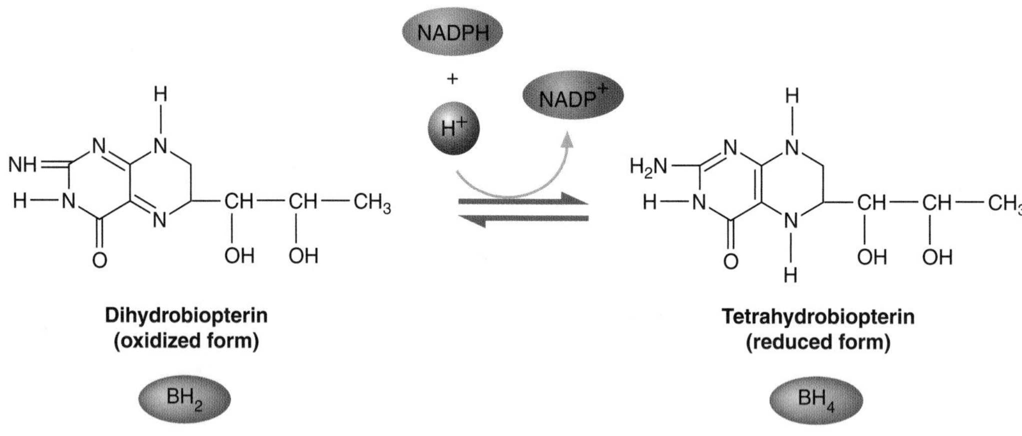

FIGURE 15.9

Tetrahydrobiopterin

Tetrahydrobiopterin (BH_4), derived from GTP, is an essential cofactor in the biosynthesis of several neurotransmitters (the catecholamines and serotonin) as well as melatonin (p. 533) and nitric oxide (p. 537). BH_4 (the reduced form) is regenerated from dihydrobiopterin (BH_2) (the oxidized form) by reduction with NADPH.

AMINO ACIDS FORMING α-KETOGLUTARATE Five amino acids (glutamate, glutamine, arginine, proline, and histidine) are degraded to α-ketoglutarate. An outline of their catabolism is illustrated in **Figure 15.10**. Each pathway is briefly described.

1. **Glutamate and glutamine**. Glutamine is converted to glutamate and NH_4^+ by glutaminase. As described previously, glutamate is converted to α-ketoglutarate by glutamate dehydrogenase or by transamination.
2. **Arginine**. Recall that arginine is cleaved by arginase to form ornithine and urea. In a subsequent transamination reaction, ornithine is converted to glutamate-γ-semialdehyde. Glutamate is then produced as glutamate-γ-semialdehyde is hydrated and oxidized. α-Ketoglutarate is produced by a transamination reaction or by oxidative deamination.
3. **Proline**. Proline catabolism begins with an oxidation reaction that produces Δ^1-pyrroline. The latter molecule is converted to glutamate-γ-semialdehyde by a hydration reaction. Glutamate is then formed by another oxidation reaction.

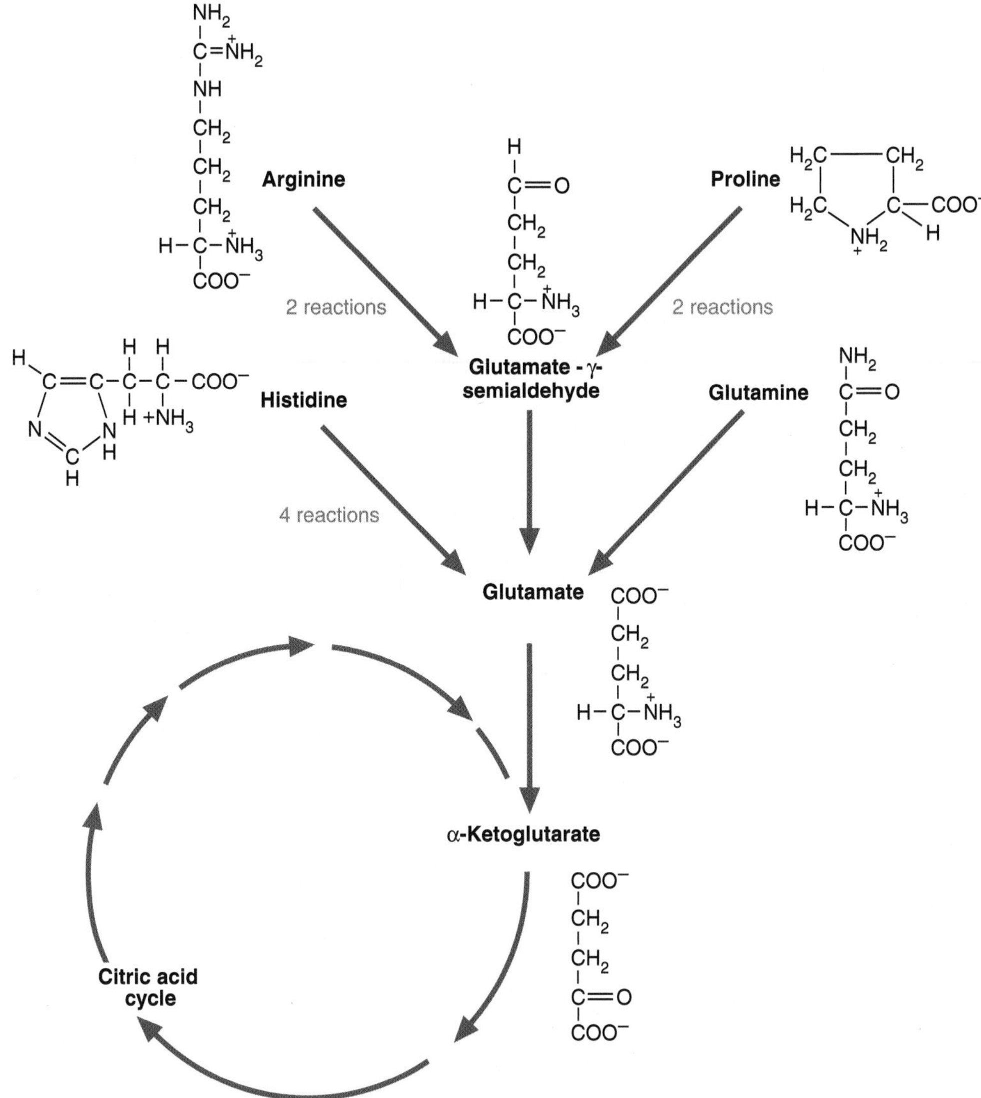

FIGURE 15.10

The Catabolic Pathways of Glutamate, Glutamine, Arginine, Proline, and Histidine

All these amino acids are eventually converted to α-ketoglutarate.

4. **Histidine**. Histidine is converted to glutamate in four reactions: a nonoxidative deamination, two hydrations, and the removal of a formamino group (NH = CH—) by THF.

AMINO ACIDS FORMING SUCCINYL-CoA Succinyl-CoA is formed from the carbon skeletons of methionine, isoleucine, valine, and threonine (as already discussed). **Figure 15.11** outlines the reactions that degrade the first three of these amino acids.

1. **Methionine**. Methionine degradation begins with the formation of SAM, which is followed by a demethylation reaction, as described (**Figure 14.20**). The product SAH is hydrolyzed to adenosine and homocysteine. Homocysteine is metabolized to yield α-ketobutyrate, cysteine, and NH_4^+. α-Ketobutyrate is then converted to propionyl-CoA by α-keto acid dehydrogenase. Propionyl-CoA is converted to succinyl-CoA in three steps (**Figure 12.12**). The conversion of methionine to cysteine is sometimes referred to as the **transsulfuration pathway** (**Figure 15.12**). A substantial amount of the sulfate produced from cysteine degradation is excreted in urine. Sulfate in the form of PAPS (p. 456) is also used in the synthesis of

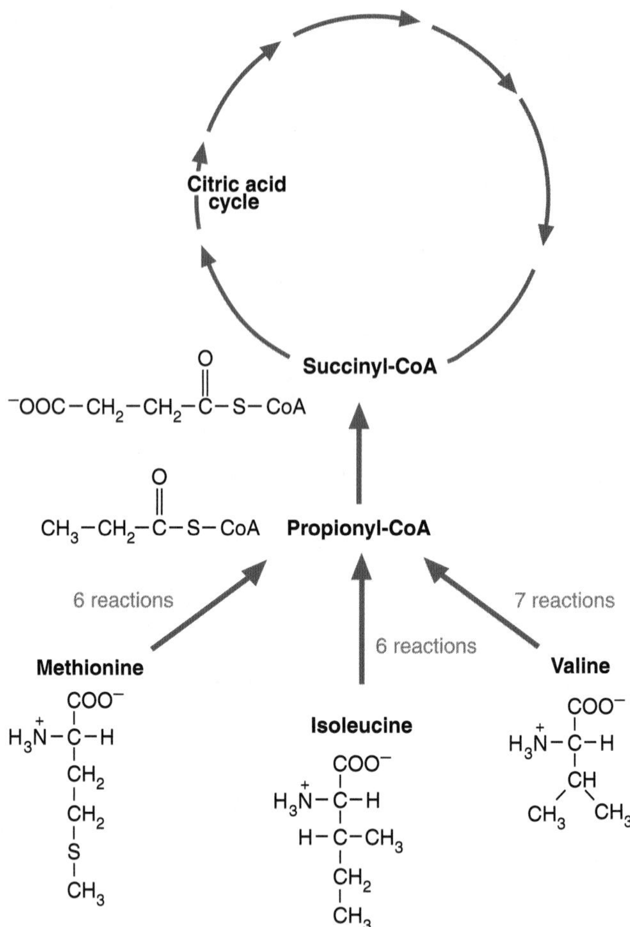

FIGURE 15.11

The Catabolic Pathways of Methionine, Isoleucine, and Valine

Propionyl-CoA is a common intermediate in the degradation of methionine, isoleucine, and valine. Methionine is first converted to homocysteine (**Figure 14.18**), which in turn yields cysteine and α-ketobutyrate (**Figure 14.11**). α-Ketobutyrate is decarboxylated to generate propionyl-CoA. Products of the conversion of isoleucine to propionyl-CoA include acetyl-CoA, three NADH, and one CO_2. Valine degradation products include three NADH, one $FADH_2$, and two CO_2. The conversion of propionyl-CoA to succinyl-CoA, a citric acid cycle intermediate, is outlined in **Figure 12.12**. Note that threonine is also degraded via the propionyl-CoA/succinyl-CoA pathway (see **Figure 15.6**).

sulfatides and proteoglycans. Additionally, molecules such as steroids and certain drugs are excreted as sulfate esters. Also recall that the gasotransmitter H_2S is synthesized from cysteine in reactions catalyzed by CBS or CSE.

2. **Isoleucine and valine**. The first four reactions in the degradation of isoleucine and valine are catalyzed by the same four enzymes (**Figure 15.13**). Several reactions in both pathways are similar to the β-oxidation reactions that yield NADH and $FADH_2$. The products of the isoleucine pathway are acetyl-CoA and propionyl-CoA, which is subsequently converted to succinyl-CoA. Hence, isoleucine is both a ketogenic and a glucogenic amino acid. The valine degradative pathway is similar but yields only succinyl-CoA. Valine is, therefore, a glucogenic amino acid. Numerous tissues can use valine, isoleucine, and leucine (p. 567), the BCAA, to generate energy. Most BCAA oxidation, however, occurs in skeletal muscle during exercise. BCAA are an important source of energy because in addition to their high concetration in muscle protein, BCAA degradation generates NADH and $FADH_2$ and their end products (acetyl-CoA, succinyl-CoA, and acetoacetate) are oxidized by the citric acid cycle.

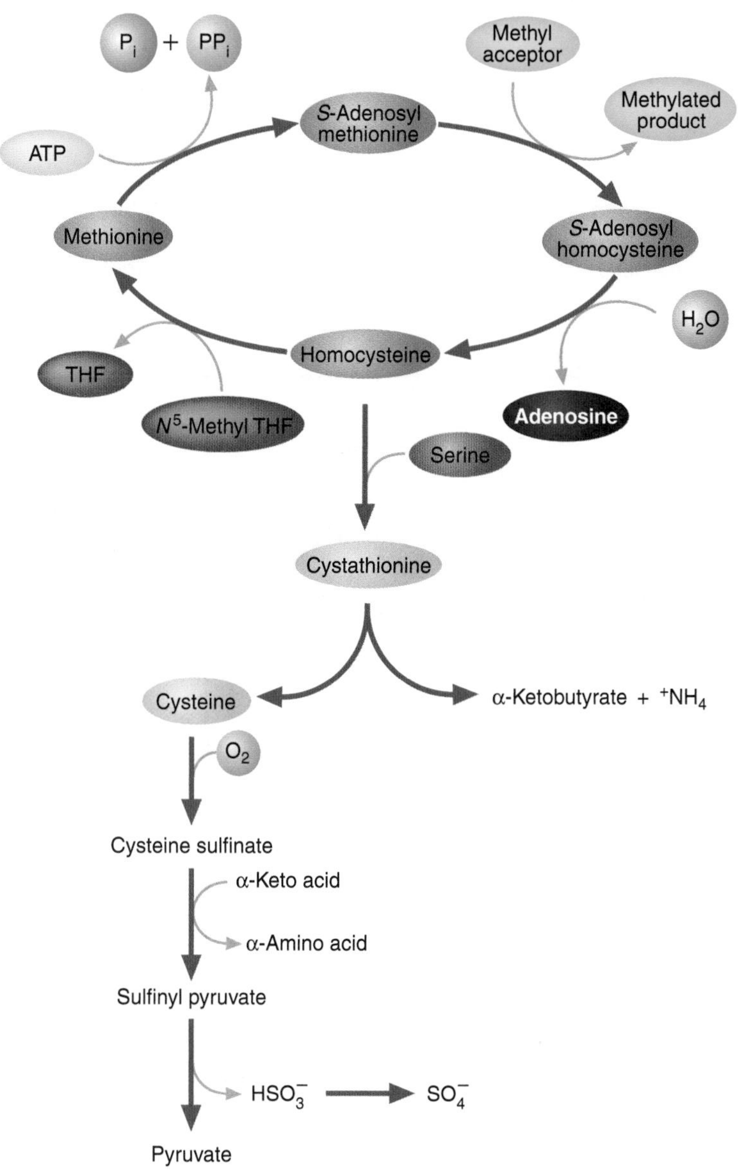

FIGURE 15.12
The Transsulfuration Pathway

The sulfur atom of methionine becomes the sulfur atom of cysteine in two reactions. Cystathionine β-synthase (CBS) converts homocysteine and serine into cystathionine. (CBS activity is depressed by cysteine when the latter molecule's cellular concentration is high.) Cystathionine is converted to cysteine, α-ketobutyrate, and NH_4^+ by γ-cystathionase (CSE). Cysteine may then be incorporated into glutathione, coenzyme A, or proteins; or it may be oxidized by cysteine dioxygenase to form cysteine sulfinate. Cysteine sulfinate can undergo a transamination reaction followed by desulfuration to yield pyruvate and sulfite (HSO_3^-). Sulfite is subsequently converted to sulfate (SO_4^-) by sulfite oxidase. The sulfate generated in cysteine catabolism is excreted or used in several biosynthetic or catabolic pathways. Note that the transsulfuration and methylation pathways are intimately related.

AMINO ACIDS FORMING OXALOACETATE Both aspartate and asparagine are degraded to form OAA. Aspartate is converted to OAA with a single transamination reaction. Asparagine is initially hydrolyzed to yield aspartate and NH_4^+ by asparaginase.

KEY CONCEPT 

Amino acid carbon skeletons can be degraded into one or more of several metabolites. These include acetyl-CoA, acetoacetyl-CoA, α-ketoglutarate, succinyl-CoA, and oxaloacetate.

QUESTION 15.3

Taurine is a sulfur-containing amine synthesized from cysteine. Although taurine is present in high concentrations in mammalian cells, except for its incorporation in bile salts, the physiological role of this amine is still poorly understood. However, several pieces of information suggest that taurine is an important metabolite. For example, taurine is found in brain tissue in large amounts. In addition, domestic cats have been observed to develop congestive heart failure if fed a taurine-free diet. (Cats cannot synthesize taurine. For this reason they must consume meat in their diet. Cats that are fed vegetarian diets soon become listless and will die prematurely.) In most animals, taurine is synthesized from cysteine sulfinate (the oxidation product of cysteine) in two reactions: a decarboxylation followed by an oxidation of the sulfinate group ($—SO_2^-$) to form sulfonate ($—SO_3^-$). With this information, determine the biosynthetic pathway for taurine. [*Hint*: The structure of taurine is illustrated in Chapter 12 on p. 461. Also refer to **Figure 15.12.**]

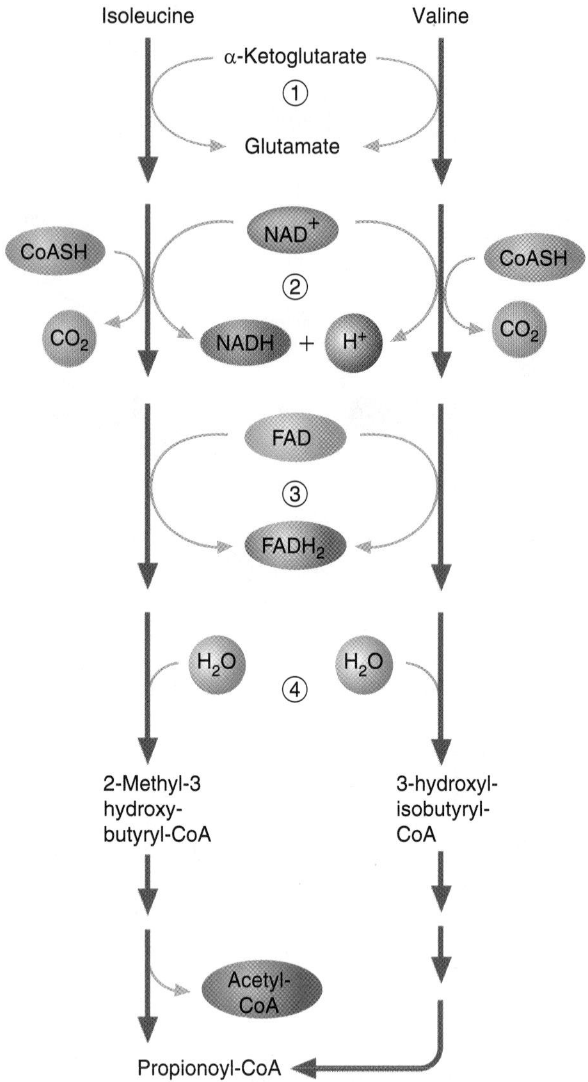

FIGURE 15.13
Degradation of Isoleucine and Valine

The degradation of both isoleucine and valine begins with the same four reactions: a transamination catalyzed by *branched-chain amino acid transaminase* (enzyme 1), an oxidative decarboxylation catalyzed by *branched-chain α-keto acid dehydrogenase* (enzyme 2), an oxidation catalyzed by an FAD-requiring acyl-CoA dehydrogenase (enzyme 3), and a hydration catalyzed by *enoyl-CoA hydratase* (enzyme 4). Isoleucine degradation continues in three reactions to yield acetyl-CoA and propionyl-CoA. The latter molecule is then converted to succinyl-CoA. Valine degradation continues in four reactions to yield propionyl-CoA.

QUESTION 15.4

Taurine is the most abundant amino acid in white blood cells, where it reacts with HOCl produced during the respiratory burst catalyzed by myeloperoxidase. The product of this reaction, taurine monochloramine (Tau-Cl), is relatively nontoxic and stable compared with HOCl. Tau-Cl modulates the inflammation process by downregulating the production of $NO^\bullet$ and proinflammatory proteins such as tumor necrosis factor α (TNF-α). Provide the reactions in which HOCl and Tau-Cl are produced. [*Hint*: Review the respiratory burst in **Figure 10.20**.]

Biochemistry IN PERSPECTIVE

Disorders of Amino Acid Catabolism

What are the effects on human health of deficiency of a single enzyme in amino acid metabolism? *Defects in amino acid catabolism were among* the first genetic diseases to be recognized and investigated by medical scientists. These "inborn errors of metabolism" result from **mutations** (permanent changes in genetic information, i.e., DNA structure). Most commonly, in the genetic diseases related to amino acid metabolism, the defective gene codes for an enzyme. The metabolic blockage that results from such a deficit disrupts what are ordinarily highly coordinated cellular and organismal processes, producing abnormal amounts and/or types of metabolites. Because these metabolites (or their heightened concentrations) are often toxic, permanent damage or death ensues. Several of the most commonly observed inborn errors of amino acid metabolism are discussed below.

Alkaptonuria, the first disease to be linked to genetic inheritance involving a single enzyme, is caused by a deficiency of homogentisate oxidase, an enzyme required for the catabolism of the aromatic ring of phenylalanine and tyrosine. In 1902 Archibald Garrod proposed that a single inheritable unit (later called a gene) was responsible for the urine in alkaptonuric patients turning black. Large quantities of homogentisate, the substrate for the defective enzyme, are excreted in urine. Homogentisate turns black when it is oxidized as the urine is exposed to air. Although black urine appears to be an essentially benign (if somewhat disconcerting) condition, alkaptonuria is not innocuous because alkaptonuric patients develop arthritis in later life. In addition, pigment accumulates gradually and unevenly darkens the skin.

Albinism is an example of a genetic defect with serious consequences. The enzyme tyrosinase is deficient. Consequently, *melanin*, a black pigment found in skin, hair, and eyes and formed from tyrosine in several cell types, is not produced. In such cells, tyrosinase converts tyrosine to L-DOPA and L-DOPA to dopaquinone. A large number of molecules of the latter product, which is highly reactive, condense to form melanin. Because of the lack of pigment, affected individuals (called albinos) are extremely sensitive to sunlight. In addition to their susceptibility to skin cancer and sunburn, they often have poor eyesight.

Phenylketonuria (PKU), caused by a deficiency of phenylalanine hydroxylase, is one of the most common genetic diseases associated with amino acid metabolism. If this condition is not identified and treated immediately after birth, mental retardation and other forms of irreversible brain damage occur. This damage results mostly from the accumulation of phenylalanine. High phenylalanine blood levels result in the saturation of the transport mechanism for large neutral amino acids across the blood-brain barrier. Brain damage results from decreased levels of protein and neurotransmitter synthesis. When present in excess, phenylalanine undergoes transamination to form phenylpyruvate, which is also converted to phenyllactate and phenylacetate. Large amounts of these molecules are excreted in the urine. Phenylacetate gives the urine its characteristic musty odor. PKU is treated with a low-phenylalanine diet.

In *maple syrup urine disease*, also called *branched-chain ketoaciduria*, the α-keto acids derived from leucine, isoleucine, and valine accumulate in large quantities in blood. Their presence in urine imparts a characteristic odor that gives the malady its name. All three α-keto acids accumulate because of a deficient branched-chain α-keto acid dehydrogenase complex. (This enzymatic activity is responsible for the conversion of the α-keto acids to their acyl-CoA derivatives.) If left untreated, affected individuals experience vomiting, convulsions, severe brain damage, and mental retardation. They often die before 1 year of age. As with phenylketonuria, treatment consists of rigid dietary control.

Deficiency of methylmalonyl-CoA mutase (an enzyme involved in the conversion of propionyl-CoA into succinyl-CoA) results in *methylmalonic acidemia*, a condition in which methylmalonate accumulates in blood. The symptoms are similar to those of maple syrup urine disease. Methylmalonate may also accumulate because of a deficiency of adenosylcobalamin or weak binding of this coenzyme by a defective enzyme. Some affected individuals respond to injections of large daily doses of vitamin B_{12}.

SUMMARY: The deficiency in humans of a single enzyme in amino acid metabolism has widespread effects that typically include brain damage.

FIGURE 15.14
Inactivation of the Catecholamines

Monoamine oxidase is a flavoprotein that catalyzes the oxidative deamination of amines to form the corresponding aldehydes; O_2 is the electron acceptor, and NH_3 and H_2O_2 are the other products. (PNMT = phenylethanolamine-*N*-methyltransferase)

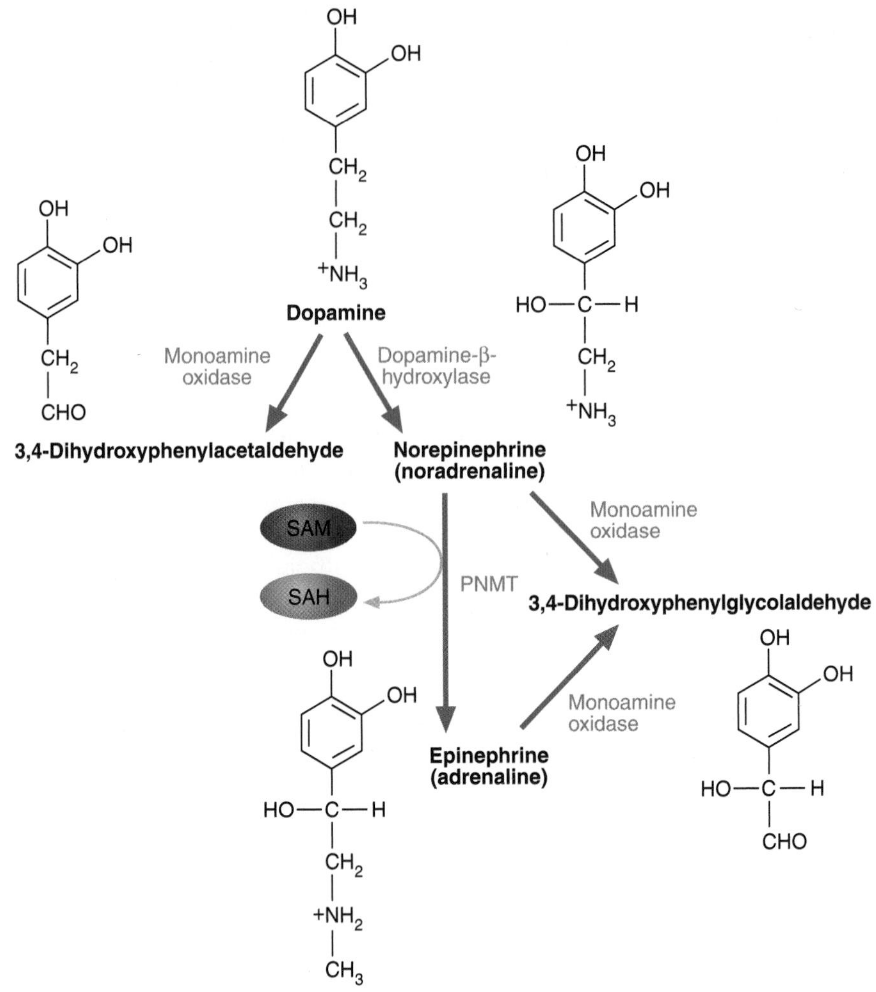

15.3 NEUROTRANSMITTER DEGRADATION

The previous discussion of amino acid catabolic disorders indicates that catabolic processes are just as important for the proper functioning of cells and organisms as are anabolic processes. This is no less true for neurotransmitters such as the catecholamines.

The catecholamines epinephrine, norepinephrine, and dopamine are inactivated by oxidation reactions catalyzed by monoamine oxidase (MAO) (**Figure 15.14**). Catecholamines must be transported out of the synaptic cleft before inactivation because MAO is located in nerve endings. (The process by which neurotransmitters are transported back into nerve cells so that they can be reused or degraded is referred to as *reuptake*.) Epinephrine, released as a hormone from the adrenal gland, is carried in the blood and is catabolized in nonneural tissue (perhaps the kidney). Catecholamines are also inactivated in methylation reactions catalyzed by catechol-*O*-methyltransferase (COMT). These two enzymes (MAO and COMT) work together to produce a large variety of oxidized and methylated metabolites of the catecholamines.

KEY CONCEPT

Information transfer in animals requires that after their release, neurotransmitters be quickly degraded or removed from the synaptic cleft.

15.4 NUCLEOTIDE DEGRADATION

In most living organisms, purine and pyrimidine nucleotides are constantly degraded and/or recycled. During digestion, nucleic acids are hydrolyzed to oligonucleotides by enzymes called **nucleases**. (Short nucleic acid segments containing fewer than 50 nucleotides are called **oligonucleotides**.) Enzymes that are specific for breaking internucleotide bonds in DNA are called *deoxyribonucleases* (DNases); those that degrade RNA are called *ribonucleases* (RNases). Once formed, oligonucleotides are further hydrolyzed by various *phosphodiesterases* in a process that produces a mixture of mononucleotides. *Nucleotidases* remove phosphate groups from nucleotides, yielding nucleosides. These latter molecules are hydrolyzed by *nucleosidases* to free bases and ribose or deoxyribose, which are then absorbed.

Generally speaking, dietary purine and pyrimidine bases are not used in significant amounts to synthesize cellular nucleic acids. Instead, they are degraded within enterocytes. Purines are degraded to uric acid in humans and birds. Pyrimidines are degraded to β-alanine or β-aminoisobutyric acid, as well as NH_3 and CO_2. In contrast to the catabolic processes for other major classes of biomolecules (e.g., sugars, fatty acids, and amino acids), purine and pyrimidine catabolism does not result in ATP synthesis. The major pathways for the degradation of purine and pyrimidine bases are described next.

Purine Catabolism

Purine nucleotide catabolism is outlined in **Figure 15.15**. There is some variation in the specific pathways used by different organisms or tissues to degrade AMP. In most tissues, AMP is hydrolyzed by 5′-nucleotidase to form adenosine. Adenosine is then deaminated by adenosine deaminase (also called adenosine aminohydrolase) to form inosine. In muscle, AMP is initially converted to IMP by AMP deaminase (also referred to as adenylate aminohydrolase). IMP is subsequently hydrolyzed to inosine by 5′-nucleotidase. The AMP deaminase reaction is also a component of the purine nucleotide cycle (**Figure 15.16**). In this pathway IMP reacts with aspartate to yield adenylosuccinate. This GTP-requiring reaction is

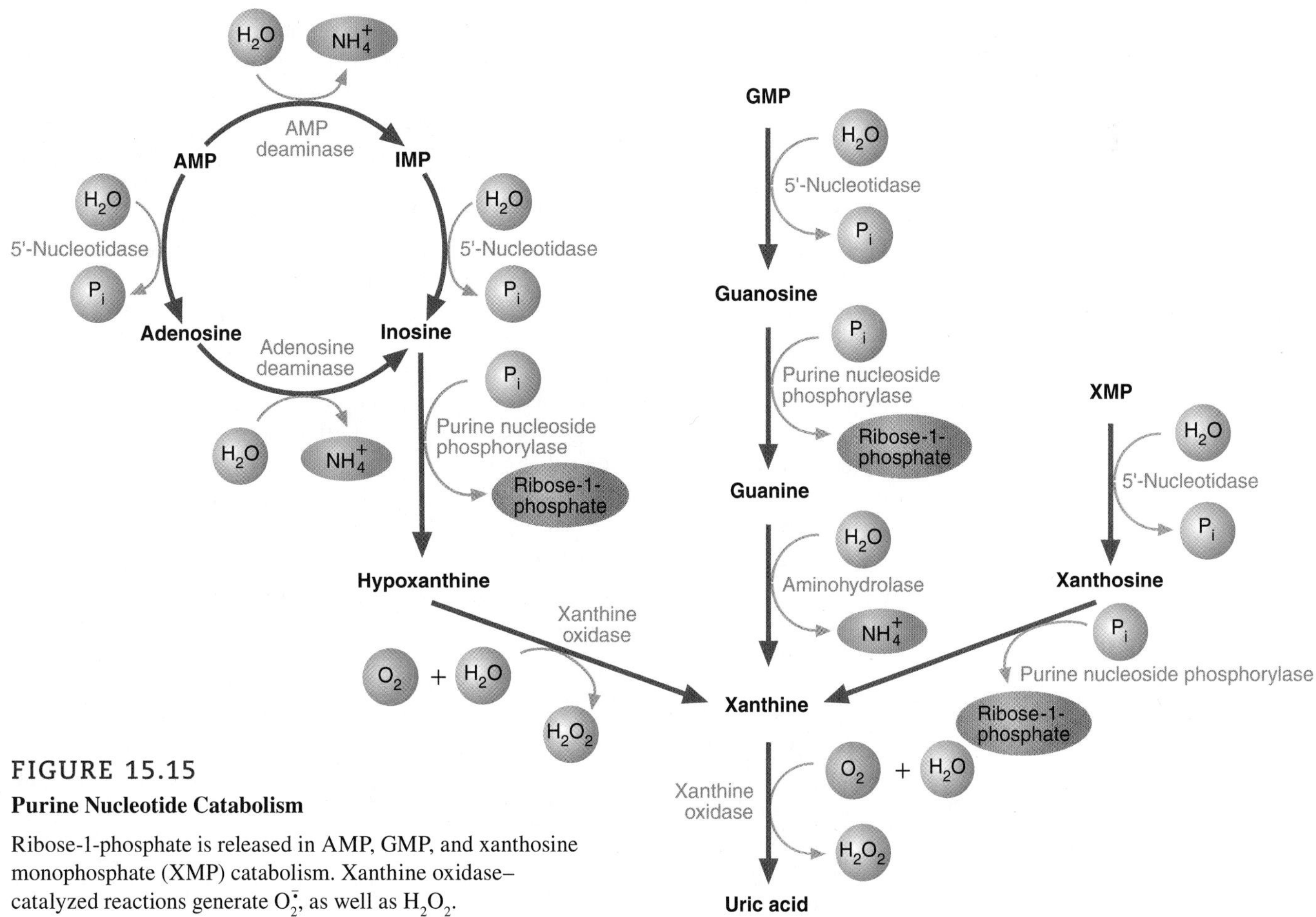

FIGURE 15.15

Purine Nucleotide Catabolism

Ribose-1-phosphate is released in AMP, GMP, and xanthosine monophosphate (XMP) catabolism. Xanthine oxidase–catalyzed reactions generate $O_2^{\cdot-}$, as well as H_2O_2.

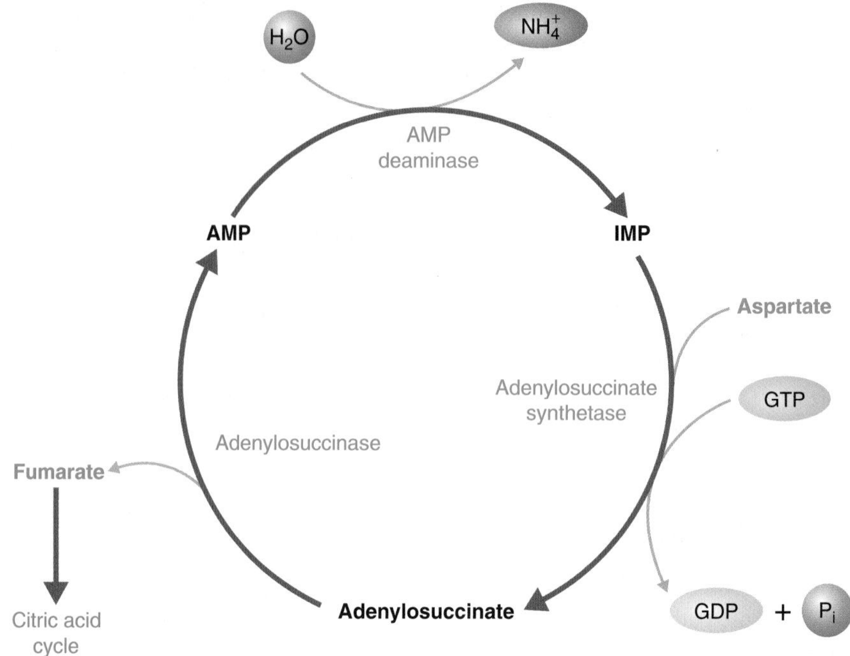

FIGURE 15.16
The Purine Nucleotide Cycle

In skeletal muscle, the purine nucleotide cycle is an anaplerotic process that replenishes citric acid cycle intermediates by producing fumarate from aspartate.

catalyzed by adenylosuccinate synthetase. Adenylosuccinate is then converted by adenylosuccinase to AMP and fumarate. The purine nucleotide cycle is a means of converting amino acids (via aspartate) to citric acid cycle intermediates (via fumarate). In skeletal muscle, AMP deaminase activity is exceptionally high. In contrast to other tissues, skeletal muscle can replenish citric acid cycle intermediates only with the fumarate produced by the purine nucleotide cycle. Activation of muscle AMP deaminase (called myoadenylate deaminase) and increased flux through the purine nucleotide cycle (**Figure 15.16**) occur during intense exercise.

Purine nucleoside phosphorylase converts inosine, guanosine, and xanthosine to hypoxanthine, guanine, and xanthine, respectively. (The ribose-1-phosphate formed during these reactions is reconverted to PRPP by ribose-5-phosphate pyrophosphokinase.) Hypoxanthine is oxidized to xanthine by xanthine oxidase, an enzyme that contains molybdenum, FAD, and two different Fe-S clusters. (Xanthine oxidase–catalyzed reactions produce $O_2^{\cdot-}$ in addition to forming H_2O_2.) Guanine is deaminated to xanthine by guanine deaminase (also called guanine aminohydrolase). Xanthine molecules are further oxidized to uric acid by xanthine oxidase. Xanthine oxidase is inhibited by allopurinol, a structural analogue of hypoxanthine.

Many animals degrade uric acid further (**Figure 15.17**). Urate oxidase converts uric acid to allantoin, an excretory product in many mammals. Allantoinase catalyzes the hydration of allantoin to form allantoate, which is excreted by bony fish. Other fish, as well as amphibians, produce allantoicase, which splits allantoic acid into glyoxylate and urea. Finally, marine invertebrates degrade urea to NH_4^+ and CO_2 in a reaction catalyzed by urease.

Several diseases result from defects in purine catabolic pathways. *Gout*, which is often characterized by high blood levels of uric acid and recurrent attacks of arthritis, is caused by several metabolic abnormalities. Allopurinol is used to treat gout. Two different immunodeficiency diseases are now known to result from defects in purine catabolic reactions. *Adenosine deaminase deficiency* results in high levels of deoxyadenosine, which is toxic, especially in the T and B lymphocytes. (*T lymphocytes*, or **T cells**, bear antibody-like molecules on their surfaces. They bind to and destroy foreign cells in a process referred to as **cellular immunity**. *B lymphocytes*, or **B cells**, secrete antibodies that bind to foreign substances, thereby initiating their destruction by other immune system cells. The production of antibodies by B cells is referred to as the **humoral immune response**.) Children with adenosine deaminase deficiency usually die before the

Visit the companion website at www.oup.com/us/mckee to read the Biochemistry in Perspective essay on gout.

age of 2 because of massive infections. In *purine nucleoside phosphorylase deficiency*, levels of purine nucleotides are high and synthesis of uric acid decreases. High levels of dGTP are apparently responsible for the impairment of T cells that is characteristic of this malady. Individuals with *myoadenylate deaminase deficiency* exhibit exercise-induced muscle fatigue.

QUESTION 15.5

Unlike primates and birds, many animals possess the enzyme urate oxidase. Suggest a reason why these organisms do not suffer from gout.

Pyrimidine Catabolism

In humans the purine ring cannot be degraded. This is not true for the pyrimidine ring. An outline of the pathway for pyrimidine nucleotide catabolism is illustrated in **Figure 15.18**.

Before they can be degraded, cytidine and deoxycytidine are converted to uridine and deoxyuridine, respectively, by deamination reactions catalyzed by cytidine deaminase. Similarly, deoxycytidylate (dCMP) is deaminated to form deoxyuridylate (dUMP). The latter molecule is then converted to deoxyuridine by 5′-nucleotidase. Uridine and deoxyuridine are then further degraded by nucleoside phosphorylase to form uracil. Thymine is formed from thymidylate (dTMP) by the sequential actions of thymidine kinase and thymidine phosphorylase.

Uracil and thymine are converted to their end products, β-alanine and β-aminoisobutyrate, respectively, in parallel pathways. In the first step, uracil and thymine are reduced by dihydropyrimidine dehydrogenase to their corresponding dihydro derivatives. As these latter molecules are hydrolyzed, the rings open, yielding β-ureidopropionate and β-ureidoisobutyrate, respectively. Finally, β-alanine and β-aminoisobutyrate are produced in deamination reactions catalyzed by β-ureidopropionase.

In several conditions, β-aminoisobutyrate is produced in such large quantities that it appears in urine. Among these are a genetic predisposition for slow β-aminoisobutyrate conversion to succinyl-CoA and diseases that cause massive cell destruction, such as leukemia. Because it is soluble, excess β-aminoisobutyrate does not cause problems comparable to those observed in gout.

QUESTION 15.6

Identify each of the following biomolecules. Explain how they are produced.

(a) $NH_2—C(=O)—NH_2$

(b) [uric acid structure]

(c) $H_3\overset{+}{N}—CH_2—CH_2—C(=O)—O^-$

Uric acid (primates, birds)

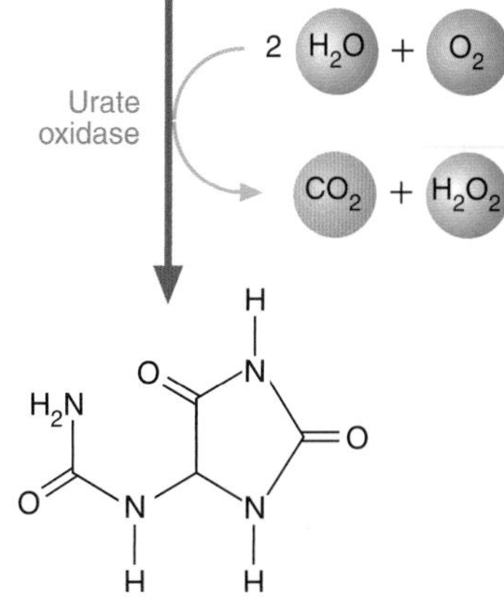

Allantoin (some mammals)

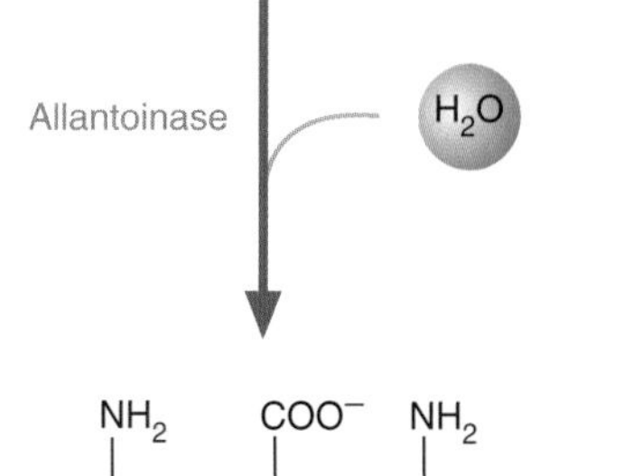

Allantoate (bony fish)

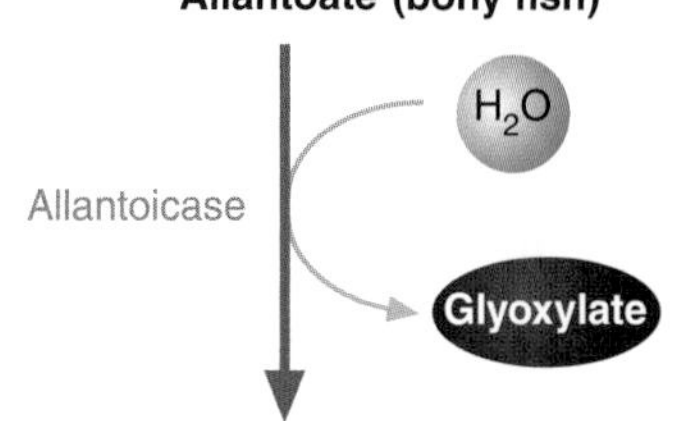

$2\ NH_2—C(=O)—NH_2$

Urea (amphibians)

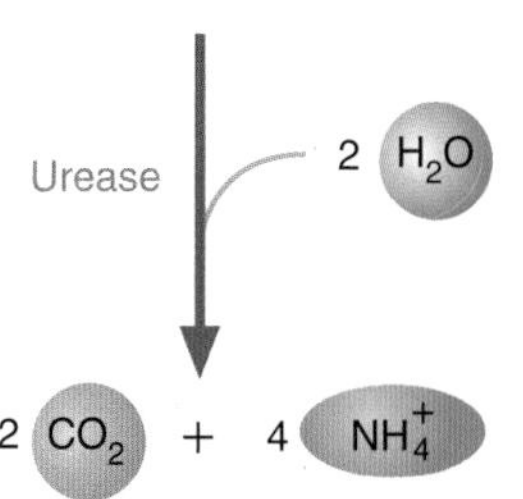

(Marine invertebrates)

FIGURE 15.17
Uric Acid Catabolism

Many animals possess enzymes that allow them to convert uric acid to other excretory products. The final excretory products of specific animal groups are indicated.

2'-Deoxycytidine
H_2O
Cytidine deaminase
NH_4^+
dCMP
H_2O
Cytidine deaminase
NH_4^+
dUMP
dTMP
H_2O
Thymidine kinase
P_i
2'-Deoxythymidine
Cytidine
H_2O
Cytidine deaminase
NH_4^+
Uridine
H_2O
P_i
2'-Deoxyuridine
P_i
Thymidine phosphorylase
Deoxyribose-1-phosphate
P_i
P_i
Nucleoside phosphorylase
Ribose-1-phosphate
Uracil
Deoxyribose-1-phosphate
Thymine
NADPH + H^+
NADPH + H^+
Dihydropyrimidine dehydrogenase
$NADP^+$
$NADP^+$
Dihydrouracil
Dihydrothymine
H_2O
Dihydropyrimidinase
H_2O
β-Ureidopropionate
β-Ureidoisobutyrate
H_2O
NH_4^+
β-Ureidopropionase
H_2O
NH_4^+
CO_2
CO_2
β-Alanine
β-Aminoisobutyrate
3 reactions
3 reactions
Acetyl-CoA
Succinyl-CoA

FIGURE 15.18

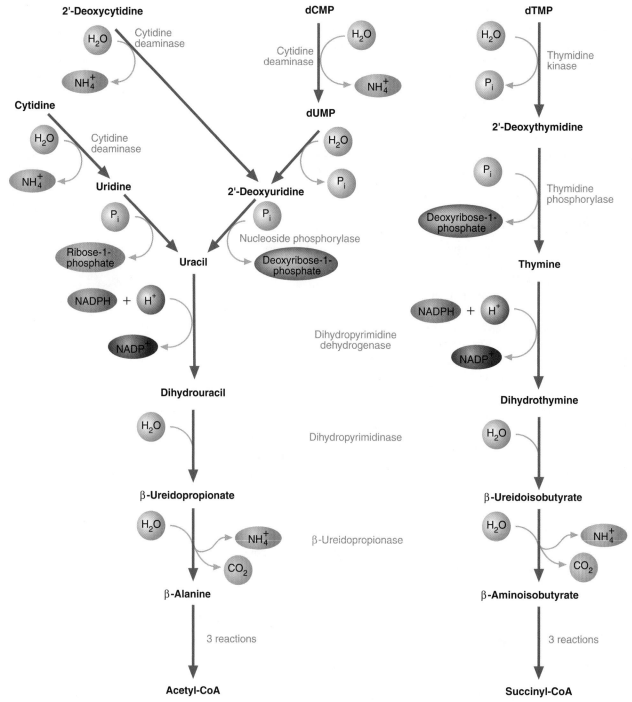

Degradation of Pyrimidine Bases

Uracil and thymine are degraded to β-alanine and β-aminoisobutyrate, respectively, in parallel pathways. The entire pathway is present in mammalian liver.

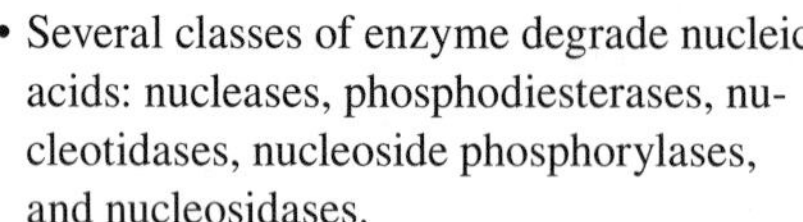

KEY CONCEPTS

- Several classes of enzyme degrade nucleic acids: nucleases, phosphodiesterases, nucleotidases, nucleoside phosphorylases, and nucleosidases.
- The bases of purine nucleotides are degraded to form the nitrogenous waste product uric acid.
- β-Alanine and β-aminoisobutyrate are the nitrogenous waste products of pyrimidine base catabolism.

QUESTION 15.7

The products of pyrimidine base catabolism, β-alanine and β-aminoisobutyrate, can be further degraded to acetyl-CoA and succinyl-CoA, respectively. Can you suggest the types of reaction required to accomplish these transformations?

Chapter Summary

1. Animals are constantly synthesizing and degrading nitrogen-containing molecules such as proteins and nucleic acids. Protein turnover is believed to provide cells with metabolic flexibility, protection from accumulations of abnormal proteins, and the timely destruction of proteins during developmental processes. Most short-lived cellular proteins are degraded by the ubiquitin proteasomal system. The process begins with the covalent modification of target proteins called ubiquitination. Ubiquitin, a small, highly conserved protein is linked to worn-out or damaged proteins or short-lived regulatory proteins. The autophagy-lysosomal system degrades long-lived protiens and organelles.
2. In general, amino acid degradation begins with deamination. Most deamination is accomplished by transamination reactions, which are followed by oxidative deaminations that produce ammonia. Although most deaminations are catalyzed by glutamate dehydrogenase, other enzymes also contribute to ammonia formation. Ammonia is prepared for excretion by the enzymes of the urea cycle. Aspartate and CO_2 also contribute atoms to urea.
3. Amino acids are classified as ketogenic or glucogenic on the basis of whether their carbon skeletons are converted to fatty acids or to glucose. Several amino acids can be classified as both ketogenic and glucogenic because their carbon skeletons are precursors for both fat and carbohydrates.
4. The degradation of neurotransmitters is critical to the proper functioning of information transfer in animals. The amine neurotransmitters such as acetylcholine, the catecholamines, and serotonin are among the best-researched examples.
5. The turnover of nucleic acids is accomplished by enzymes of several types. The nucleases degrade the nucleic acids to oligonucleotides. (The deoxyribonucleases degrade DNA; the ribonucleases degrade RNA.) The phosphodiesterases convert the oligonucleotides to mononucleotides. By removing phosphate groups, the nucleotidases convert nucleotides to nucleosides. The nucleosidases hydrolyze nucleosides to form free bases and ribose or deoxyribose. The nucleoside phosphorylases convert ribonucleosides to free bases and ribose-1-phosphate. Dietary nucleic acids are generally degraded in the intestine and are not used in salvage pathways. Cellular purines are converted to uric acid. Many animals degrade uric acid further because they produce enzymes that are not present in primates. Pyrimidine bases are degraded to either β-alanine (UMP, CMP, dCMP) or β-aminoisobutyrate (dTMP).

Take your learning further by visiting the **companion website** for Biochemistry at **www.oup.com/us/mckee** where you can complete a multiple-choice quiz on degradation of nitrogen-containing biomolecules to help you prepare for exams.

Suggested Readings

Bhattacharyya, S., Regulated Protein Turnover: Snapshots of the Proteasome in Action, *Nat. Rev. Cell Mol. Biol.* 15(2):122–133, 2014.

Biswas, S., Chida, A. S., and Rahman, I., Redox Modifications of Protein-Thiols: Emerging Roles in Cell Signaling, *Biochem. Pharmacol.* 71:551–564, 2006.

Dormandy, T., *The White Death: A History of Tuberculosis*, New York University Press, New York, 2000.

Finkelstein, J. D., Inborn Errors of Sulfur-Containing Amino Acid Metabolism, *J. Nutr.* 136:1750S–1754S, 2006.

Greenayre, J. T., Huntington's Disease: Making Connections, *New Engl. J. Med.* 356(5):518–520, 2007.

Lamb, C. A., Yoshimori, T., and Tooze, S. A., The Autophagosome: Origins Unknown, Biogenesis Complex, *Nat. Rev. Mol. Cell Biol.* 14(12):759–774, 2013.

Shen, H.-M., and Mizushima, N., At the End of the Autophagic Road: An Emerging Understanding of Lysosomal Functions in Autophagy, *Trends Biochem. Sci.* 39(2):61–71, 2014.

Taylor, R. C., Berendzen, K. M., and Dillin, A., Systemic Stress Signaling: Understanding the Cell Non-Autonomous Control of Proteostasis, *Nat. Rev. Mol. Cell Biol.* 15(3):211–217, 2014.

Warner, D. F., and Mizrahi, V., The Survival Kit of *Mycobacterium tuberculosis, Nat. Med.* 13(3): 282–284, 2007.

Key Words

autophagy, *558*
B cell, *576*
cellular immunity, *576*
chaperone-mediated autophagy, *558*
glucogenic, *559*
humoral immune response, *576*
hyperammonemia, *564*
ketogenic, *559*
Krebs bicycle, *563*
Krebs urea cycle, *560*
macro autophagy, *558*
micro autophagy, *558*
mutation, *573*
nuclease, *574*
oligonucleotide, *574*
protein turnover, *555*
proteasome, *542*
T cell, *576*
transsulfuration pathway, *569*
ubiquitin, *556*
ubiquitination, *556*
ubiquitin proteasomal system, *556*
urea cycle, *560*

Review Questions

These questions are designed to test your knowledge of the key concepts discussed in this chapter before moving on to the next chapter. You may like to compare your answers to the solutions provided in the back of the book and in the accompanying Study Guide.

1. Define the following terms:
   a. protein turnover
   b. proteasome
   c. ubiquitin
   d. ubiquitination
   e. autophagy
2. Define the following terms:
   a. autophagosome
   b. endosome
   c. amphisome
   d. lysosome
   e. autophagolysosome
3. Define the following terms:
   a. L-amino acid oxidase
   b. serine dehydratase
   c. bacterial urease
   d. adenosine deaminase
   e. glutaminase
4. Define the following terms:
   a. urea cycle
   b. Krebs bicycle
   c. carbamoyl phosphate
   d. ureotelic organism
   e. ornithine translocase
5. Define the following terms:
   a. glucogenic
   b. ketogenic
   c. *N*-acetylglutamate
   d. hyperammonemia
   e. BH_4
6. Define the following terms:
   a. transsulfuration pathway
   b. cystathionine
   c. homocysteine
   d. PAPS
   e. *S*-adenosylmethionine
7. Define the following terms:
   a. chaperone-mediated autophagy
   b. microautophagy
   c. macroautophagy
   d. ubiquitin proteasomal system
   e. autophagy lysosomal system
8. Define the following terms
   a. albinism
   b. maple syrup urine disease
   c. alkaptonuria
   d. methylmalonic acidemia
   e. phenylketonuria
9. Define the following terms:
   a. MAO
   b. PNMT
   c. COMT
   d. NPC
   e. TB
10. Define the following terms:
    a. oligonucleotide
    b. nuclease
    c. phosphodiesterase
    d. nucleosidase
    e. nucleotidase
11. Define the following terms:
    a. uric acid
    b. allantoin
    c. allantoate
    d. urate oxidase
    e. allantoicase
12. Define the following terms:
    a. T cells
    b. B cells
    c. cellular immunity
    d. humoral immunity
    e. myoadenylate deaminase deficiency
13. What are the major molecules that serve in the excretion of nitrogen?
14. What are three purposes served by protein turnover?
15. What are the structural features of proteins that mark them for destruction?
16. What are the seven metabolic products produced by the degradation of amino acids?
17. Describe how aspartate is degraded.
18. In humans the purine ring cannot be degraded. How is it excreted? What reactions are involved?
19. Indicate which of the following amino acids are ketogenic and which are glucogenic:
    a. tyrosine
    b. lysine
    c. glycine
    d. alanine
    e. valine
    f. threonine
20. Describe how glutamate is degraded.
21. The urea cycle occurs partially in the cytoplasm and partially in the mitochondria. Discuss the urea cycle reactions with reference to their cellular locations.
22. Describe how the glucose-alanine cycle acts to transport ammonia to the liver.
23. Describe how alanine is degraded.
24. In individuals with PKU, is tyrosine an essential amino acid? Explain.
25. Urea formation is energetically expensive, requiring the expenditure of 4 mol of ATP per mole of urea formed. However, NADH is produced when fumarate is reconverted to aspartate. How many ATP molecules are produced by the mitochondrial oxidation of the NADH? What is the net ATP requirement for urea synthesis?
26. Describe the Krebs bicycle. What compound links the citric acid and urea cycles?

27. Most amino acids are degraded in the liver. This is not true of the branched-chain amino acids, most of which are degraded in extrahepatic tissues with high protein turnover. Suggest some examples of these tissues.
28. Describe how a protein is targeted for degradation in a proteasome.
29. Describe how lysine is degraded.
30. Provide the names of the organisms that form the following substances as nitrogenous waste molecules:
    a. uric acid
    b. urea
    c. allantoate
    d. NH_4^+
    e. allantoin
31. Which of the following molecules yields uric acid when degraded?
    a. DNA
    b. FAD
    c. CTP
    d. PRPP
    e. β-alanine
    f. urea
    g. NAD^+
32. Describe how tyrosine is degraded.
33. Explain why providing domestic cats with a vegetarian diet is a bad idea.
34. Trace the origin of the nitrogen atoms in urea.
35. Explain why the carbon skeletons of ketogenic amino acids yielding acetylCoA only cannot be converted into glucose.
36. Describe how the urea cycle is activated after a high-protein meal.
37. Foods high in purines can trigger a gout attack. Provide several examples of these foods and explain how they contribute to this malady. (Visit the companion website at www.oup.com/us/mckee to read the Biochemistry in Perspective essay on gout.)
38. Which nitrogen atom in uracil eventually ends up in urea molecules?

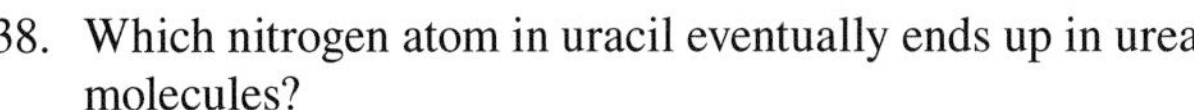

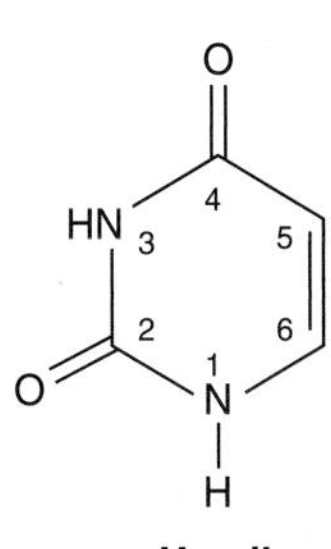

Uracil

39. Explain why the amino acid tryptophan is both ketogenic and glucogenic.
40. Describe the symptoms of maple syrup urine disease and the metabolic basis of this malady.
41. What molecules cause the urine odor that is characteristic of the disease mentioned in Question 40?
42. Why can't humans simply excrete waste nitrogen atoms as ammonia rather than utilize the energetically expensive process of urea synthesis?
43. Explain how tyrosinase deficiency causes albinism.
44. Why can't humans degrade purine rings?

Fill in the Blank

45. The process of continuous synthesis and degradation of protein is referred to as ______________________________.
46. Amino acids that degrade, yielding acetyl-CoA or acetoacetyl-CoA, are referred to as ________________.
47. ______________ amino acids are degraded to pyruvic acid and citric acid cycle intermediates.
48. ______________________________ are short nucleic acid segments containing fewer than 50 nucleotides.
49. The production of antibodies by B lymphocytes is referred to as __________________ immunity.
50. The purine nucleotide cycle produces fumarate, ______________ and, ___________.
51. T cells bind to and destroy foreign cells in a process referred to as ________________ immunity.
52. The bases of purine nucleotides are degraded to ________________.
53. β-Alanine is one of the waste products of ___________________________ metabolism.
54. In the urea cycle, urea is synthesized from carbon dioxide and ____________________.

Short Answer

55. Where is taurine monochloride (Tau-Cl) synthesized and what role does it play in the inflammatory process?
56. Purine nucleoside phosphorylase (PNP) deficiency disease is an autosomal recessive disease that results in severe immunodeficiency. Explain.
57. What is the end product of catabolism of the pyrimidine base thymine? Unlike uric acid, the end product of purine catabolism, excess amounts of this molecule do not cause problems comparable to gout. What circumstances cause excess amounts of the end product and why doesn't a gout-like illness result?
58. What is methylmalonic acidemia and how is it treated?
59. What is the Krebs bicycle?

Thought Questions

These questions are designed to reinforce your understanding of all of the key concepts discussed in the book so far, including this chapter and all of the chapters before it. They may not have one right answer! The authors have provided possible solutions to these questions in the back of the book and in the accompanying Study Guide for your reference.

60. Mammals excrete most nitrogen atoms as urea. The urea cycle itself is costly, requiring considerable amounts of ATP energy. What mechanism does the cell have to compensate for that energy input?
61. Create a diagram that illustrates how the purine nucleotide cycle contributes to energy metabolism in skeletal muscle. Include glycolysis and the citric acid cycle in your drawing. What enzymatic activities would you expect to be absent in skeletal muscle?
62. Describe how increasing concentrations of ammonia stimulate the formation of *N*-acetylglutamate and turn on the urea cycle.
63. Phenylketonuria can be caused by deficiencies in phenylalanine hydroxylase and by enzymes catalyzing the formation and regeneration of 5,6,7,8-tetrahydrobiopterin. How can this second defect cause the symptoms of PKU?
64. In their *in vitro* studies using liver slices, Krebs and Henseleit observed that urea formation was stimulated by the addition of ornithine, citrulline, and arginine. Other amino acids had no effect. Explain these observations.
65. Specify the type of carbon unit that is transferred in one-carbon metabolism by each of the following compounds:
    a. N^5,N^{10}-methylene THF
    b. serine
    c. choline
    d. *S*-adenosylmethionine
66. Caffeine, a methylated xanthine found in chocolate, coffee, and tea, is excreted as uric acid. Use your knowledge of the metabolism of other purine compounds to suggest how caffeine is metabolized.
67. Some animals living in a fluid medium excrete nitrogen as ammonia. Land animals, which conserve water, excrete urea and uric acid. Why does the excretion of these molecules aid in water conservation?
68. Parkinson's disease is a devastating, progressive disorder of the central nervous system. Damage to dopaminergic nerve tracts in the brain causes tremors and impaired muscle coordination. An early-onset (inherited) form of the disease has been linked to a defective protein, now called parkin. Parkin has E3 activity. Describe, in general terms, how neurons with this defective protein are damaged.
69. Dihydrouracil and β-uredidopropionate (*N*-carbamoyl-β-alanine) are intermediates in the conversion of uracil to β-alanine. Provide the structures of the molecules in this pathway. (Refer to the numbered illustration of uracil in Question 38.)
70. Refer to Question 38 and outline the pathway by which the nitrogen of uracil is used in the synthesis of urea.
71. Why do primates suffer from gout whereas most other animals do not?
72. Diabetes is a complex set of metabolic diseases with the common symptom of an inability to transport glucose into target cells (muscle cells and adipocytes). The body compensates in part by degrading muscle protein to generate energy. Explain how this process works.
73. Individuals with hyperammonemia are given α-keto acids as a treatment. Explain. (Visit the companion website at www.oup.com/us/mckee to read the Biochemistry in Perspective essay on hyperammonemia.)
74. Review the Krebs bicycle and calculate the ATPs degraded or synthesized in this pathway. Is urea biosynthesis a net energy-requiring or energy-generating process?

CHAPTER

16 Integration of Metabolism

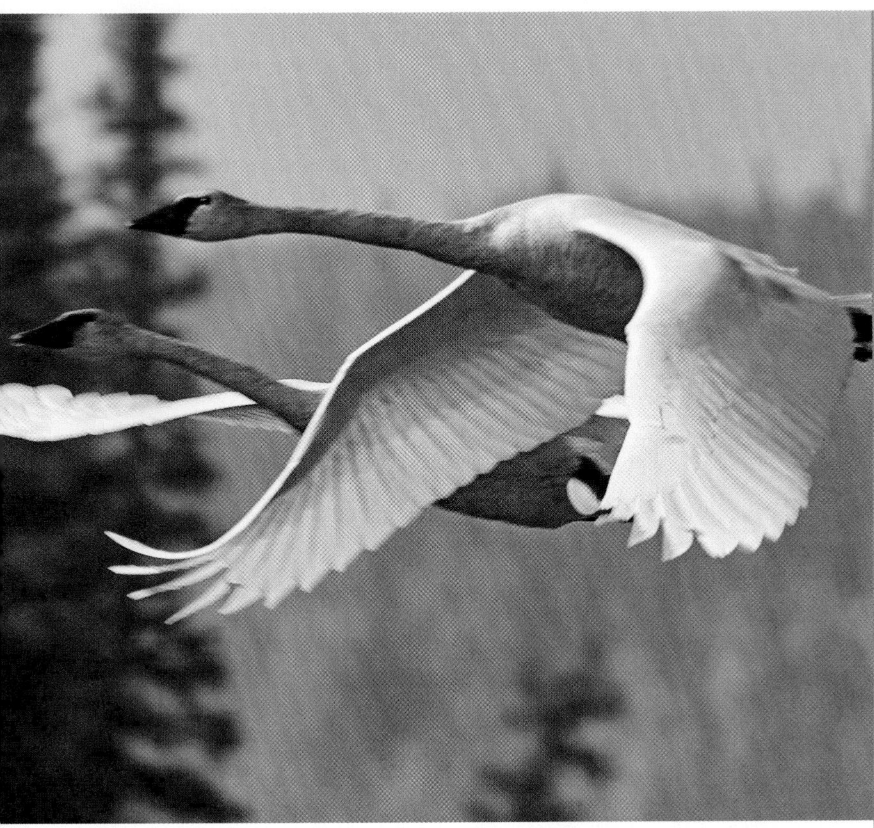

Integration of Metabolism
Swans fly at speeds of 25 to 50 mph as a result of metabolic processes, most notably the conversion of food molecules derived from aquatic plants into skeletal muscle contraction.

OUTLINE

Hypertension and Uric Acid: A Diet Connection?

Jake is a 15 year old with a problem. He is an overweight teenager newly diagnosed with hypertension at the hypertension clinic at a local university medical center. Jake and several dozen other overweight teens with mild untreated hypertension (blood pressures of 135/80 [systolic/diastolic]; normal values = 110–120/60–70) were selected for a clinical trial because their hypertension had recently been diagnosed. They also had serum uric acid (SUA) levels of 6 mg/dL (normal values are 5.5 mg/dL or lower) and no kidney damage or other diseases related to hypertension. Jake's clinical trial was designed to ascertain the relationship of high blood uric acid levels (hyperuricemia) to hypertension. The experimental drug used in the trial, allopurinol, is an inhibitor of xanthine oxidase (p. 576), the enzyme that converts the purine derivatives hypoxanthine and xanthine to uric acid, a nitrogenous waste product in humans. Although hypertension has long been linked to hyperuricemia (and assigned a causative role by early researchers), modern biomedical scientists have regarded high SUA levels as a consequence of hypertension (decreased uric acid excretion caused by hypertensive kidney damage). Recent research by epidemiologists (public health researchers that investigate disease risk factors) and experimental studies using lab animals have challenged this view.

In Jake's clinical trial allopurinol lowered SUA levels in two-thirds of the patients to 5 mg/dL or less. Lower SUA levels in these patients were associated with statistically significant reductions in blood pressure. In other investigations, vascular damage has been demonstrated in hyperuricemic lab animals as well. Together, these results support the hypothesis that hypertension is the result and not the cause of high SUA levels. In other words, hypertension-related damage to renal blood vessels is caused by toxic amounts of uric acid.

High serum uric acid levels are statistically linked not only to hypertension and kidney disease, but also to obesity, diabetes, and cardiovascular diseases (pp. 607–608). SUA values have been steadily rising over the past century. Long thought to be the result of the sedentary lifestyles typical of industrialized societies, these disease conditions have recently been associated by epidemiologists with excessive fructose consumption. Major increases in hypertension and obesity beginning about three decades ago parallel the introduction of increased quantities of fructose (via sucrose and high-fructose corn syrup) into processed foods and beverages. Indeed, administration of fructose-rich food to humans in clinical trials results in pronounced blood pressure increases. Recall that fructose metabolism bypasses the key regulatory step in glycolysis, catalyzed by PFK-1 (p. 304). After a high-fructose meal, fructose, metabolized primarily in the liver, is rapidly converted into phosphorylated trioses that then enter pathways involved in lipogenesis and VLDL synthesis and secretion. This unregulated process depletes hepatic ATP levels and increases AMP, the substrate for uric acid synthesis. High SUA levels are believed to damage blood vessels, including those of the kidney.

Overview

PREVIOUS CHAPTERS DEAL WITH THE METABOLISM OF CARBOHYDRATES, LIPIDS, AND OTHER MOLECULES. HOWEVER, THE WHOLE IS NOT JUST THE sum of its parts. Multicellular organisms are extraordinarily complex, more so than their components would suggest. Chapter 16 takes a wider view of functioning of the mammalian body. A review of the mechanisms of action of hormones and other signal molecules that make the sophisticated regulation possible is followed by a description of the feeding-fasting cycle, a physiological process that ensures that adequate energy and nutrient resources are available. The contributions of the major organs in mammals to this process is described in an online reading.

The most distinctive characteristic of living organisms is their capacity to sustain adequate, if not always optimal, operating conditions despite changes in their internal and external environments. And if this feat is not amazing enough, consider that they must also simultaneously repair damaged components and, when possible, undergo cell divisions and other forms of growth. To accomplish these functions, the anabolic and catabolic reaction pathways that use carbohydrates, lipids, and proteins as energy sources and biosynthetic precursors must be precisely regulated. For multicellular organisms such as animals, this endeavor is astonishingly complicated. The operation of such a complex system as the body is maintained by a continuous flow of information among its parts. A simple system for information transfer is composed of a primary signal (e.g., a hormone), a target (a specific receptor), and a transducer system (that converts the signal to a cellular response). Considering the complexities of multicellular organisms, the need for a large number of primary signals, specific receptors, and transducer systems is not surprising. In the mammalian body, much information transfer is accomplished by hormones, molecules produced in specific cells that affect cells in other parts of the body. Hormones allow for a high degree of sophisticated regulation.

The chapter focuses on the integration of the major metabolic processes in mammals. The chapter begins with an overview of metabolic processes and descriptions of the major classes of signal molecules and their mechanisms of action. The chapter ends with an overview of the hormone- and neurotransmitter-regulated feeding-fasting cycle, which has great physiological importance because of its role in energy acquisition. The chapter also provides descriptions of disorders of metabolic regulation: diabetes, obesity, and the metabolic syndrome.

16.1 OVERVIEW OF METABOLISM

The central metabolic pathways are common to most organisms. Throughout the life of an organism, a precise balance is struck between anabolic (synthetic) and catabolic (degradative) processes. An overview of the principal anabolic and catabolic pathways in heterotrophs such as animals is illustrated in **Figure 16.1**. Except during youth, illness, or pregnancy, the animal's tissues exist in a metabolic steady state throughout the remainder of its life. In a **steady state**, the rate of anabolic processes is approximately equal to that of catabolic processes.

How are animals (or other multicellular organisms) able to maintain a balance between anabolic and catabolic processes as they respond and adapt to changes in their environment? Various forms of intercellular communication are believed to play an important role. Most intercellular communication occurs by means of biochemical signals. Once released into the extracellular environment, each signal molecule is recognized by specific cells (called **target cells**), which then respond in a specific manner. Most signals are modified amino acids, fatty acid derivatives, peptides, proteins, or steroids.

In animals the nervous and endocrine systems are primarily responsible for coordinating metabolism. The nervous system provides a rapid and efficient mechanism for acquiring and processing environmental information. Nerve cells, called neurons, release neurotransmitters (Section 14.3) at the end of long cell extensions called axons into tiny intercellular spaces called synapses. The neurotransmitter molecules bind to nearby cells, evoking specific responses from those cells.

Metabolic regulation by the endocrine system is achieved by secretion of signals called hormones directly into the blood. The endocrine system is composed of specialized cells, many of which are found in glands. After these hormone molecules, referred to as **endocrine** hormones, have been secreted, they

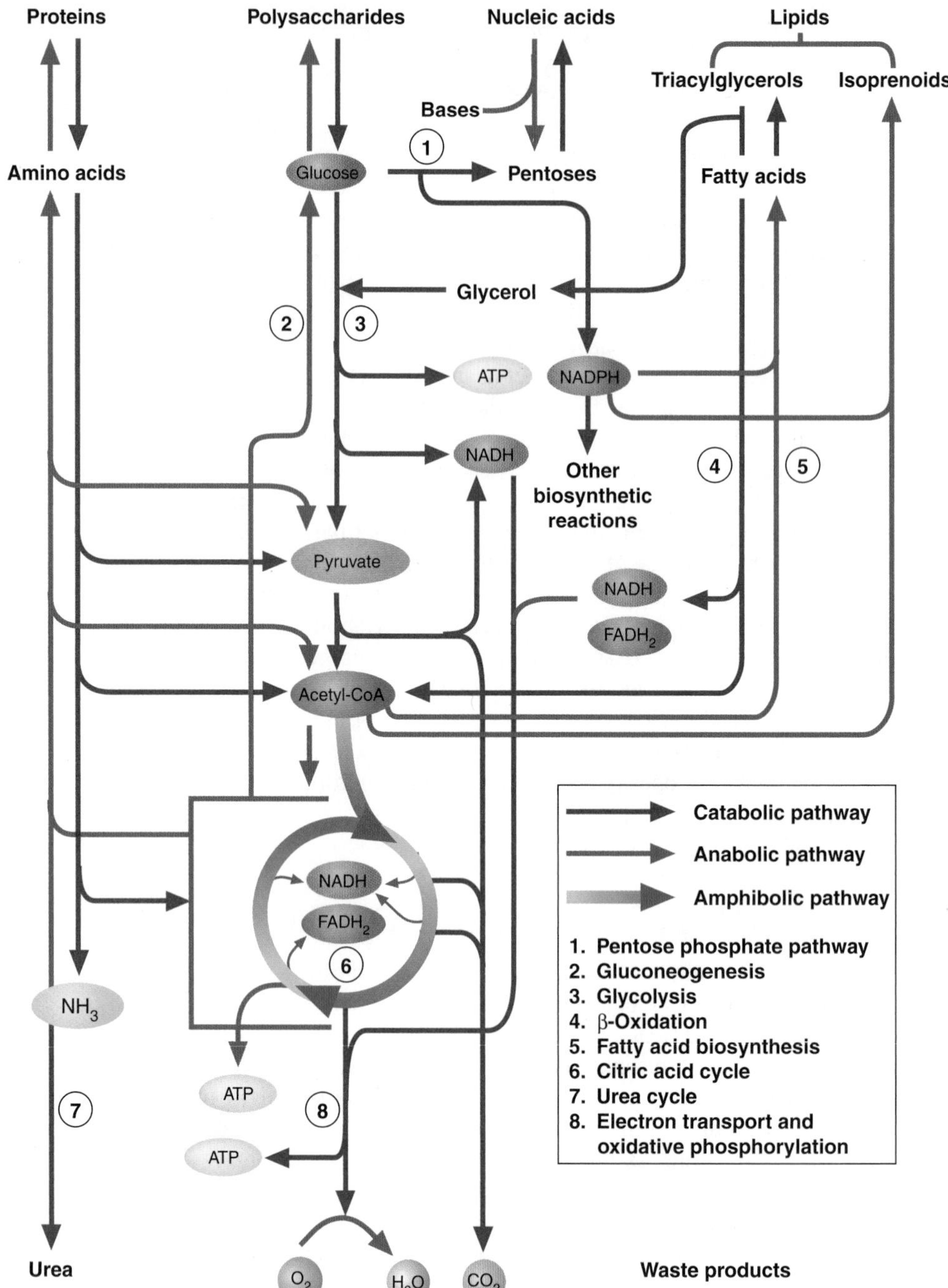

FIGURE 16.1
Overview of Metabolism

This simplified overview of metabolism illustrates the anabolic and catabolic pathways of the major biomolecules in heterotrophs (i.e., the biochemical pathways that synthesize, degrade, or interconvert important biomolecules and generate energy).

travel through the blood until they reach a target cell. Some hormones exert very specific effects on one type of target cell; other hormones act on a variety of target cells. For example, thyroid-stimulating hormone (TSH), released from the pituitary gland in the brain, stimulates follicular cells in the thyroid gland to release T_3 (triiodothyronine) and T_4 (thyroxine) (**Figure 16.2**). In contrast, T_3 (the most active form of thyroid hormone) and T_4 stimulate a variety of cellular reactions in numerous cell types (e.g., T_3 stimulates glycogenolysis in liver cells and glucose absorption in the small intestine).

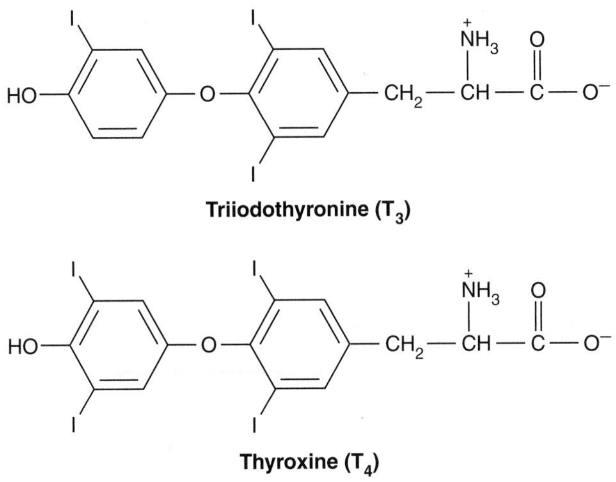

FIGURE 16.2
Structure of the Thyroid Hormones T_3 and T_4

16.2 HORMONES AND INTERCELLULAR COMMUNICATION

A wide variety of biomolecules regulate the body's metabolic activities. Most hormone-induced changes in cell function result from alterations in the activity, concentration, or location of enzymes that are triggered by hormone-receptor binding events. Hormones can be classified as water-soluble hormones (peptides, polypeptides, and amino acid derivatives such as epinephrine) and growth factors (proteins that regulate cell growth and cell division) and lipid-soluble hormones (e.g., steroid and thyroid hormones).

Peptide Hormones

In animals, the vast majority of water-soluble hormones are peptides or polypeptides. They initiate their actions by binding to receptors on the outer surface of the target cell's plasma membrane. In mammals, the synthesis and secretion of many of these hormones are regulated by a complex cascade mechanism and are ultimately controlled by the central nervous system. (An overview of mammalian hormones is provided in an online Biochemistry in Perspective essay on mammalian hormones and the hormone cascade system.) Sensory signals are received by the hypothalamus, an area in the brain that integrates the nervous and endocrine systems. For example, hypothalamic osmoreceptor cells trigger the secretion of the antidiuretic peptide hormone vasopressin (p. 146), in response to high blood Na^+ levels, into the posterior pituitary where it is then released into the bloodstream.

The receptors for most water-soluble hormones are located on the surface of target cells. The binding of these hormones to membrane-bound receptors triggers an intracellular response. The intracellular actions of many hormones are mediated by a group of molecules referred to as **second messengers**. (The hormone molecule is the first messenger.) Several second messengers have been identified. These include the nucleotides cAMP and cGMP, calcium ions, and the inositol-phospholipid system. Most second messengers act to modulate enzymes, often by an enzyme cascade. In an *enzyme cascade*, a powerful amplification device (**Figure 16.3**), enzymes undergo conformational transitions that switch the enzymes from their inactive forms to their active forms, or vice versa. The sequentially expanding array leads to a substantial amplification of the original signal. This process is often initiated when a second messenger binds to a specific enzyme. For example, the binding of cAMP to inactive PKA converts it to active PKA, which in turn modifies the activity of many target enzymes through phosphorylation. The original signal generates an amplified and diversified response,

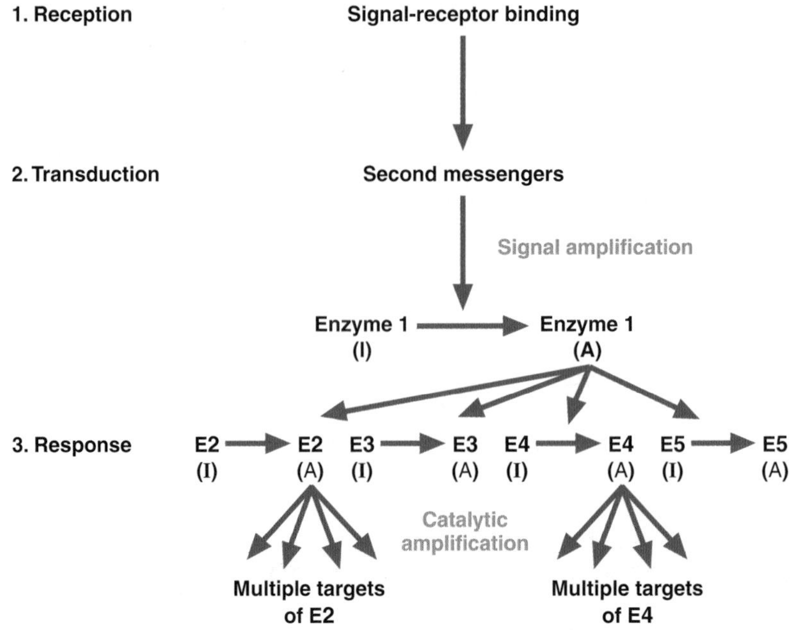

FIGURE 16.3
Signal Transduction

Signal transduction mechanisms occur in three phases. (1) Reception: a signal molecule binds to its receptor. (2) Transduction: as a result of receptor binding, an enzyme cascade is initiated, typically by a second messenger molecule. The enzyme activated by the second messenger modifies multiple copies of a number of different target enzymes. Target enzymes that are activated may also modify multiple copies of a second set of target proteins. (3) Response: cellular functions are altered by changes in the activities of existing enzymes, rearrangements of the cytoskeleton, and/or altered gene expression. Signal amplification occurs at each step of the cascade. (I = inactive, A = active)

via a second messenger (at the signal level) in some cases and an enzyme cascade (at the catalytic level) in most cases. A cAMP system accomplishes amplification at both levels.

Animals employ several mechanisms to prevent excessive hormone synthesis and release. The most prominent of these is feedback inhibition. The hypothalamus and anterior pituitary are controlled by the target cells they regulate. For example, TSH release by the anterior pituitary is inhibited when blood levels of T_3 and T_4 rise. The thyroid hormones inhibit the responsiveness of TSH-synthesizing cells to TRH (thyrotropin releasing hormone). In addition, several tropic hormones inhibit the synthesis of their releasing factors.

Target cells also possess mechanisms that protect against overstimulation by hormones. In a process referred to as **desensitization**, target cells adjust to changes in stimulation levels by decreasing the number of cell-surface receptors or by inactivating those receptors. The reduction in cell-surface receptors in response to stimulation by specific hormone molecules is called **downregulation**. In downregulation, receptors are internalized by endocytosis. Depending on cell type and several metabolic factors, the receptors may eventually be recycled to the cell surface or be degraded. If degraded, new receptor proteins must be synthesized to replace old receptors. Some disease states are caused by or associated with target cell insensitivity to specific hormones. For example, some cases of diabetes mellitus are associated with **insulin resistance**, caused by a decrease in functional insulin receptors. (Diabetes is discussed in a Biochemistry in Perspective essay later in this chapter: pp. 596–597.)

Sensitive techniques are now available to detect and measure hormones. The most common of these are enzyme-linked immunosorbent assays (ELISA).

KEY CONCEPT

Many of the hormones in the mammalian body are controlled by a complex cascade mechanism and ultimately regulated by the central nervous system.

Visit the companion website at www.oup.com/us/mckee to read the Biochemistry in the Lab box on hormone methods.

QUESTION 16.1

Review the epinephrine-stimulated activation of glycogen breakdown presented earlier (**Figure 8.22**). Identify the following signal transduction components in this biochemical process: Primary signal, receptor, transducer, and response.

There are two major types of cell-surface receptor: G-protein-coupled receptors and receptor tyrosine kinases. Guanylate cyclase receptors are also described.

G-PROTEIN-COUPLED RECEPTORS **G-protein-coupled receptors (GPCRs)** are the largest known protein receptor family (800 genes in humans). They are composed of seven membrane-spanning helices that are arranged into a three-dimensional barrel-like shape. An extracellular N-terminal segment forms part of the ligand-binding site and an intracellular C-terminal segment interacts with G-proteins, also known as heterotrimeric guanosine nucleotide-binding proteins. GPCRs transduce a wide variety of stimuli into intracellular signals. In addition to hormones such as glucagon, TSH, the catecholamines, and the endocannabinoids (e.g., the arachidonic acid derivative anandamide), GPCRs also respond to neurotransmitters (e.g., glutamate, dopamine, and GABA), neuropeptides (e.g., vasopressin and oxytocin), odorants and tastants (molecules that stimulate the senses of odor and taste, respectively), and light (rhodopsin).

G proteins are the molecular switches that transduce ligand binding to GPCRs into intracellular signals (**Figure 16.4**). G proteins are composed of α, β, and γ subunits. The α subunit binds GTP and GDP.

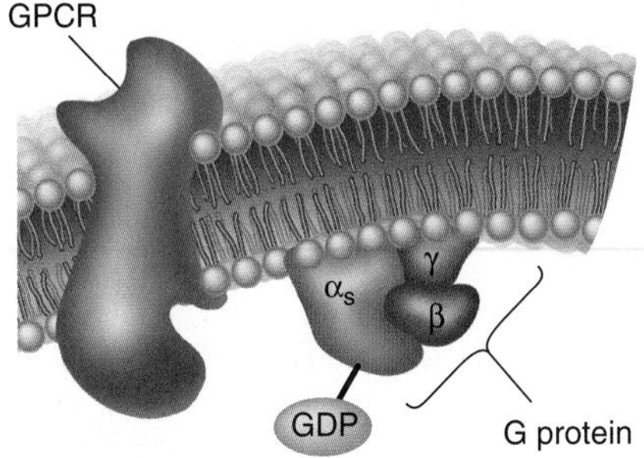

FIGURE 16.4
G-Protein-Coupled Receptor and G Protein

When a ligand binds to a G-protein-coupled receptor, a signaling mechanism is initiated that is mediated by a G protein. Each G protein consists of three subunits: α, β, and γ. Before a G protein is activated, it binds GDP.

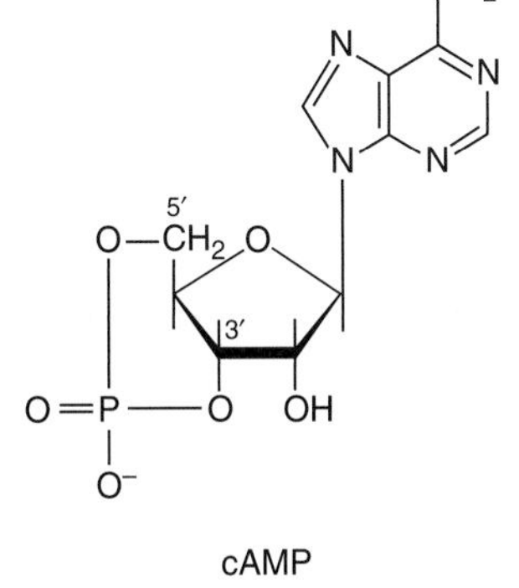

Of the 20 known α subunits, the best researched examples include those of G_S and G_i, which stimulate and inhibit adenylate cyclase, respectively, The $\beta\gamma$ complex, composed of β and γ subunits, binds to and inhibits the α subunit. G proteins are anchored to the plasma membrane by myristoyl or palmitoyl groups attached to the α subunits and farnesyl or geranylgeranyl groups attached to γ subunits. The $\beta\gamma$ dimer promotes the association of the α subunit to the GPCR and, in the absence of receptor activation, prevents GDP/GTP exchange (discussed shortly). It also facilitates the anchoring of the α subunit to the membrane and plays a role in effector signaling downstream (postreceptor activation).

G-protein activation occurs when ligand binds to the GPCR. A conformational change in the transmembrane region of the receptor leads to GDP/GTP exchange mediated by a **guanine nucleotide exchange factor (GEF)** followed by the release of the GTP-α subunit. The activated α subunit moves over the cytoplasmic surface of the membrane to activate a second messenger–generating enzyme. The GTP-α_s subunit activates adenylate cyclase and increases the intracellular synthesis of cAMP that triggers the signal transduction cascade that follows. GTP-α_i inhibits adenylate cyclase, reducing the intracellular concentration of cAMP. Other molecules in addition to cAMP are involved in the transduction of some GPCR signaling mechanisms. These include components of the phosphatidylinositol cycle and calcium ions. cGMP, a second messenger molecule synthesized by guanylate cylase works via a similar mechanism.

cAMP cAMP (**Figure 16.5**) is generated from ATP by adenylate cyclase when hormones such as glucagon, TSH, and epinephrine bind to their receptors. G_s bound to an occupied receptor (signal initiation) undergoes GDP/GTP exchange, and the GTP-α_s subunit dissociates and activates adenylate cyclase (**Figure 16.6**). GTP hydrolysis terminates this association, and GDP-α_s recombines with GPCR-$\beta\gamma$ (primary signal termination). (In effect, α_s-GTP is a timing device. The rate at which the α subunit hydrolyzes GTP determines the signal's duration.) The activated adenylate cyclase synthesizes a number of cAMP molecules (signal amplification), which diffuse into the cytoplasm, where they bind to and activate the regulatory subunits of the heterotetramer cAMP-dependent protein kinase (PKA). The now-active PKA subunits then phosphorylate and thereby alter the

cAMP

FIGURE 16.5
Structure of the Second Messenger Molecule cAMP

Note that cAMP, also referred to an adenosine 3′,5′-phosphate, is a diester that links the 3′ and 5′ carbons.

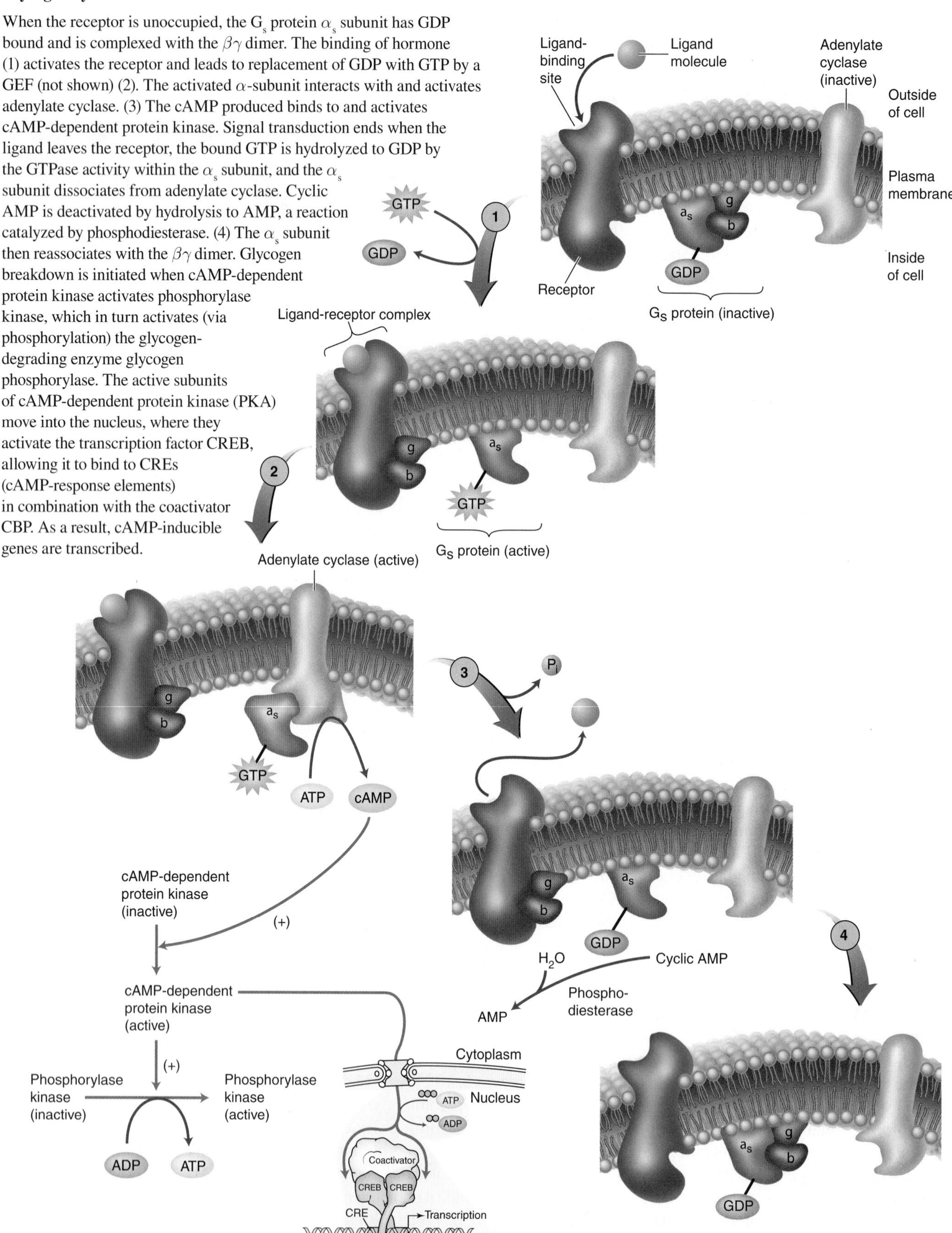

FIGURE 16.6

The Adenylate Cyclase Second Messenger System That Controls Glycogenolysis

When the receptor is unoccupied, the G_s protein α_s subunit has GDP bound and is complexed with the $\beta\gamma$ dimer. The binding of hormone (1) activates the receptor and leads to replacement of GDP with GTP by a GEF (not shown) (2). The activated α-subunit interacts with and activates adenylate cyclase. (3) The cAMP produced binds to and activates cAMP-dependent protein kinase. Signal transduction ends when the ligand leaves the receptor, the bound GTP is hydrolyzed to GDP by the GTPase activity within the α_s subunit, and the α_s subunit dissociates from adenylate cyclase. Cyclic AMP is deactivated by hydrolysis to AMP, a reaction catalyzed by phosphodiesterase. (4) The α_s subunit then reassociates with the $\beta\gamma$ dimer. Glycogen breakdown is initiated when cAMP-dependent protein kinase activates phosphorylase kinase, which in turn activates (via phosphorylation) the glycogen-degrading enzyme glycogen phosphorylase. The active subunits of cAMP-dependent protein kinase (PKA) move into the nucleus, where they activate the transcription factor CREB, allowing it to bind to CREs (cAMP-response elements) in combination with the coactivator CBP. As a result, cAMP-inducible genes are transcribed.

catalytic activity of key regulatory enzymes. The cAMP is quickly hydrolyzed by phosphodiesterase (secondary signal termination). The activated PKA subunits migrate into the nucleus where they phosphorylate CREB (cAMP response element binding protein, p. 161), thereby creating the unstructured segment that allows binding to a coactivator called CBP (CREB-binding protein). CRB promotes transcription in part by acetylating histones, which results in increasing the accessibility of DNA to transcription factors. Examples of metabolic enzymes synthesized in response to activated CREB include the gluconeogenic enzymes PEPCK and glucose-6-phosphatase, as well as tyrosine hydroxylase (p. 536).

The target proteins affected by cAMP depend on the cell type. In addition, several hormones may activate the same G protein. Therefore, different hormones may elicit the same effect. For example, glycogen degradation in liver cells is initiated by both epinephrine and glucagon.

Some hormones inhibit adenylate cyclase activity. Such molecules depress cellular protein phosphorylation reactions because their receptors interact with G_i protein. When G_i is activated, its α_i subunit dissociates from the $\beta\gamma$ dimer and prevents the activation of adenylate cyclase. For example, because its receptors in adipocytes are associated with G_i, PGE_1 depresses lipolysis. (Recall that lipolysis is stimulated by epinephrine.)

THE PHOSPHATIDYLINOSITOL CYCLE, IP_3, DAG, AND CALCIUM The phosphatidylinositol cycle (**Figure 16.7**) mediates the actions of hormones and growth

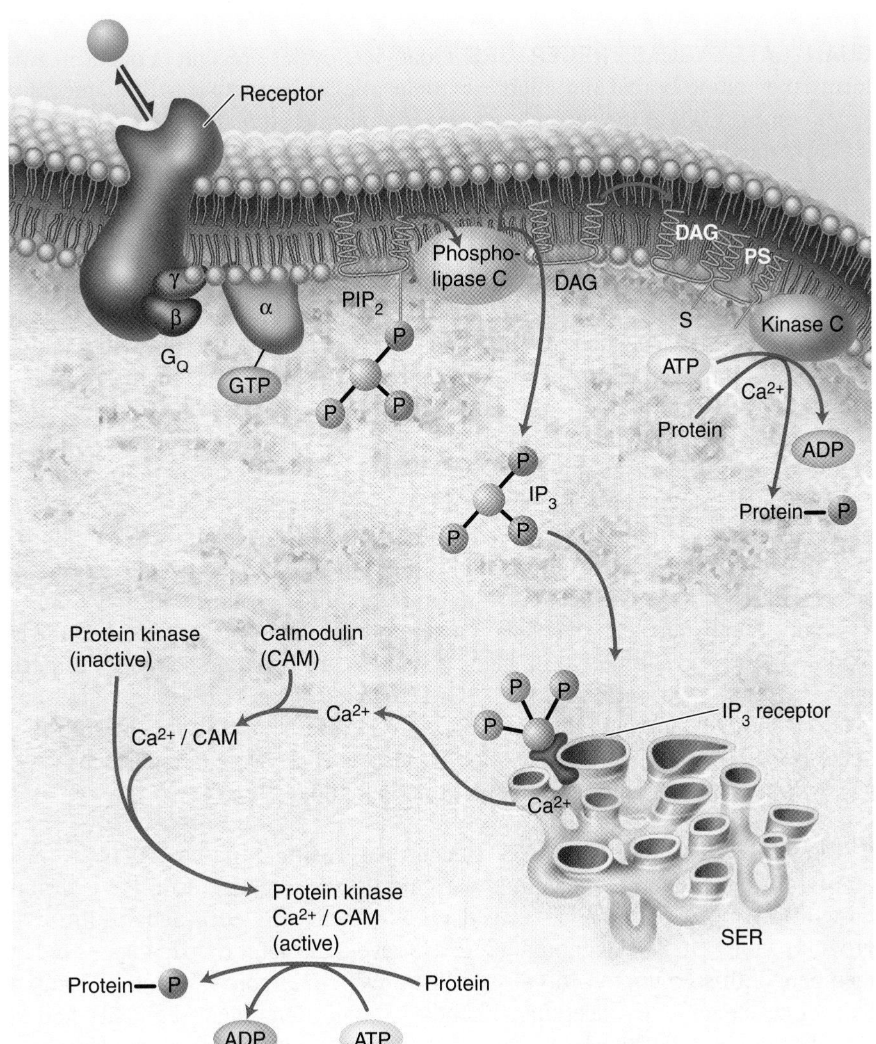

FIGURE 16.7
The Phosphatidylinositol Pathway

The binding of certain hormones to their receptor activates the α subunit of a G protein. The α subunit then activates phospholipase C, which cleaves IP_3 from PIP_2, leaving DAG in the membrane. DAG, acting with phosphatidylserine (PS) and Ca^{2+}, activates protein kinase C, which subsequently phosphorylates key regulators. IP_3 binds to receptors on the SER, opening Ca^{2+} channels. Then Ca^{2+} moves into the cytoplasm and activates additional targets.

factors. Examples include acetylcholine (e.g., insulin secretion in pancreatic cells), vasopressin, and epinephrine (α_i receptors). Phosphatidylinositol-4,5-bisphosphate (PIP_2) is cleaved by phospholipase C to form the second messengers **DAG (diacylglycerol)** and **IP_3 (inositol-1,4,5-trisphosphate)**. A hormone-receptor complex induces the activation of a G protein, which in turn activates phospholipase C. Several types of G protein may be involved in the phosphatidylinositol cycle. For example, G_Q (shown in **Figure 16.7**) mediates the actions of vasopressin.

The DAG product of the phospholipase C–catalyzed reaction activates protein kinase C (PKC). Several PKC activities have been identified. Depending on the cell, activated PKC phosphorylates specific regulatory enzymes, thereby activating or inactivating them.

Once generated, IP_3 diffuses to the calcisome (SER), where it binds to the IP_3 receptor (a calcium channel). Cytoplasmic calcium levels then rise as calcium ions flow through the activated open channel. Calcium ions are involved in the regulation of a large number of cellular processes including contributing to the activation of plasma membrane–associated PKC. Because calcium levels are still relatively low even when the calcium release mechanism has been activated (approximately 10^{-6} M), the calcium-binding sites on calcium-regulated proteins must have a high affinity for the ion. Several calcium-binding proteins modulate the activity of other proteins in the presence of calcium. Calmodulin, a type of calcium-binding protein, mediates many calcium-regulated reactions. In fact, calmodulin is a regulatory subunit for some enzymes (e.g., phosphorylase kinase, which converts phosphorylase b to phosphorylase a in glycogen metabolism).

GUANYLATE CYCLASE RECEPTORS Guanylate cyclase receptors occur in two forms: membrane-bound and soluble (cytoplasmic). When activated, both receptor types convert GTP to the second messenger molecule *cyclic GMP* (cGMP).

Two types of molecule are now known to activate membrane-bound guanylate cyclase: atrial natriuretic peptide and bacterial enterotoxin. *Atrial natriuretic factor* (ANF), a peptide that is released from heart atrial cells in response to increased blood volume, lowers blood pressure via vasodilation and causes diuresis (increased urine production). The biological effects of ANF are mediated by cGMP. ANF activates guanylate cyclase in several cell types, resulting in the activation of a cGMP-dependent protein kinase (PKG). In one type, those in the kidney's collecting tubules, ANF-stimulated cGMP synthesis increases renal excretion of Na^+ and water.

Enterotoxin (produced by several bacterial species) causes diarrhea by binding to another type of guanylate cyclase found in the plasma membrane of intestinal cells. For example, one form of traveler's diarrhea is caused by a strain of *E. coli* that produces *heat-stable enterotoxin*. This toxin binds to an enterocyte plasma membrane receptor linked to guanylate cyclase triggering excessive secretion of electrolytes and water into the lumen of the small intestine.

Soluble guanylate cyclase (sGC) is a heterodimer that is activated when nitric oxide (·NO) (p. 537) binds to its heme prosthetic groups (one heme per dimer). Once it is activated, sGC regulates cGMP-dependent protein kinases and ion-gated channels. For example, blood pressure decreases as a result of ·NO-triggered cGMP synthesis in arterial smooth muscle cells. Vasodilation is caused by a cGMP-mediated reduction in intracellular calcium.

RECEPTOR TYROSINE KINASES **Receptor tyrosine kinases (RTKs)** are a family of transmembrane receptors that bind ligands such as insulin, epidermal growth factor (EGF), platelet-derived growth factor (PDGF), and insulin-like growth factor I (IGF-I). Although there are several structural differences among members of this group, they do possess features in common: an external domain that binds specific extracellular ligands, a transmembrane segment, and a cytoplasmic catalytic domain with tyrosine kinase activity. When a ligand binds to the external domain, a conformational change in the receptor protein activates the tyrosine kinase domain. The tyrosine kinase activity initiates a phosphorylation cascade that begins with an autophosphorylation of the tyrosine kinase domain. Most research efforts have been devoted to the insulin receptor.

The *insulin receptor* (**Figure 16.8**) is an example of a receptor with tyrosine kinase activity. It is a transmembrane glycoprotein composed of two types of subunit connected by disulfide bridges. Two large α subunits (130 kDa) extend extracellularly, where they form the insulin-binding site. Each of the two β subunits (90 kDa) contains a transmembrane segment and a tyrosine kinase domain.

The binding of an insulin molecule to each of the α- subunits activates receptor tyrosine kinase activity, which in turn causes several phosphorylation cascades that modulate the activities of numerous intracellular proteins. Insulin alters the expression of more than 150 genes. Insulin receptor substrate 1 (IRS-1), one of six IRS proteins, is among the most important proteins that are directly phosphorylated. Activated IRS-1 then binds to and activates several proteins, including phosphatidylinositol-3-kinase (PI3K) (**Figure 16.9a**). PI3K subsequently phosphorylates PIP_2, a minor constituent of cell membrane, to form PIP_3 (phosphatidy 1-3,4,5-trisphosphate). Once PIP_3-dependent protein kinase (PDK1) has bound PIP_3, it then activates several kinases. Among the most important of these is PKB (also known as Akt), a serine/threonine kinase that has a central role in cell signal transduction mechanisms (**Figure 16.9b**). Activities of PKB include stimulation of glycogen synthesis (via inhibition of GSK3, see p. 309) and inhibition of lipolysis (via inhibition of PKA activity). PKB also facilitates glucose transport into adipocytes and muscle cells by stimulating the translocation of the GLUT4 transporter (p. 600) to the plasma membrane.

The most prominent role of PKB is the activation of mTOR, a component of mTORC1 (mTOR complex 1), which is a central kinase sensor that integrates hormonal activity, nutrient availability, energy status, and cell responses to various forms of stress (e.g., osmotic, oxidative, and inflammatory). Examples of activated mTORC1-affected gene expression changes include increased ribosome and protein synthesis and decreased autophagy (p. 558). SREBP-1c and PPARγ (p. 452) are two major transcription factors that are activated by mTORC1. SREBP-1c, in combination with ChREBP (carbohydrate response element binding protein) (p. 452), causes the expression of lipogenic genes such as those coding for FAS and ACC (p. 442). SREBP-1c also suppresses gluconeogenic enzyme synthesis (p. 298). Activated PPARγ (p. 452) stimulates the expression of lipogenic genes.

mTORC1/insulin-driven processes are modulated by at least two mechanisms. First, mTORC1 is part of an autoregulatory pathway whereby IRS-1 is phosphorylated so as to inhibit PIK3 activity. Second, AMPK, activated in response to low energy and nutrient levels or other cell stressors, represses mTORC1 via phosphorylation of an mTOR regulatory protein.

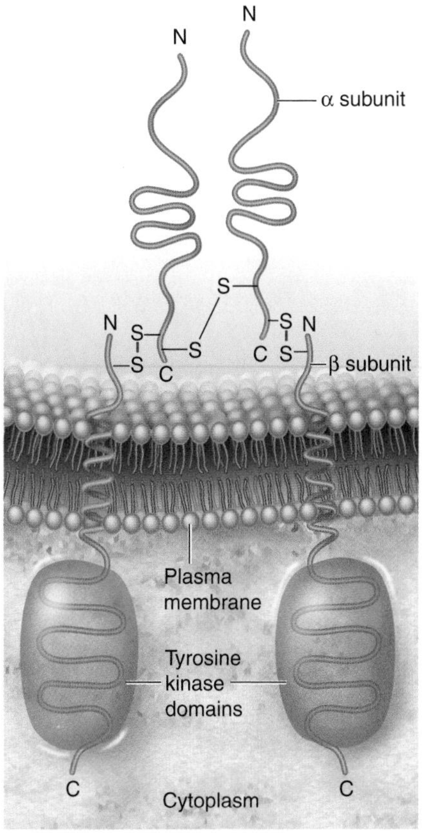

FIGURE 16.8

The Insulin Receptor

The insulin receptor is a tetramer composed of two pairs of α and β subunits. The subunits are connected to each other by disulfide bridges.

QUESTION 16.2

In a hypothetical cAMP-mediated signal transduction cascade, the GTP-α_s/adenylate cyclase interaction following a single hormone-receptor binding event lasts for 2.3 seconds. The catalytic rate (turnover number) for the adenylate cyclase in question is 350 cAMP molecules produced per second. How many cAMP molecules would be produced if five hormone-receptor binding events were to occur before the hormone molecule dissipates in the bloodstream? What is the amplification effect of this step in the signaling pathway?

QUESTION 16.3

Explain the sequence of events that occurs when epinephrine triggers the synthesis of cAMP. Once formed, cAMP breaks down rapidly. Why is this an important feature for a second messenger in a signal transduction process?

KEY CONCEPTS

- There are two major types of cell-surface receptor that bind to hormone molecules: G-protein-linked receptors and receptor tyrosine kinases.
- The G proteins activated by GPCRs utilize one or more second messenger molecules to transduce the original signal into a signal cascade.
- Receptor tyrosine kinases activate phosphorylation cascades when they undergo autophosphorylation triggered by the binding of signal molecules.

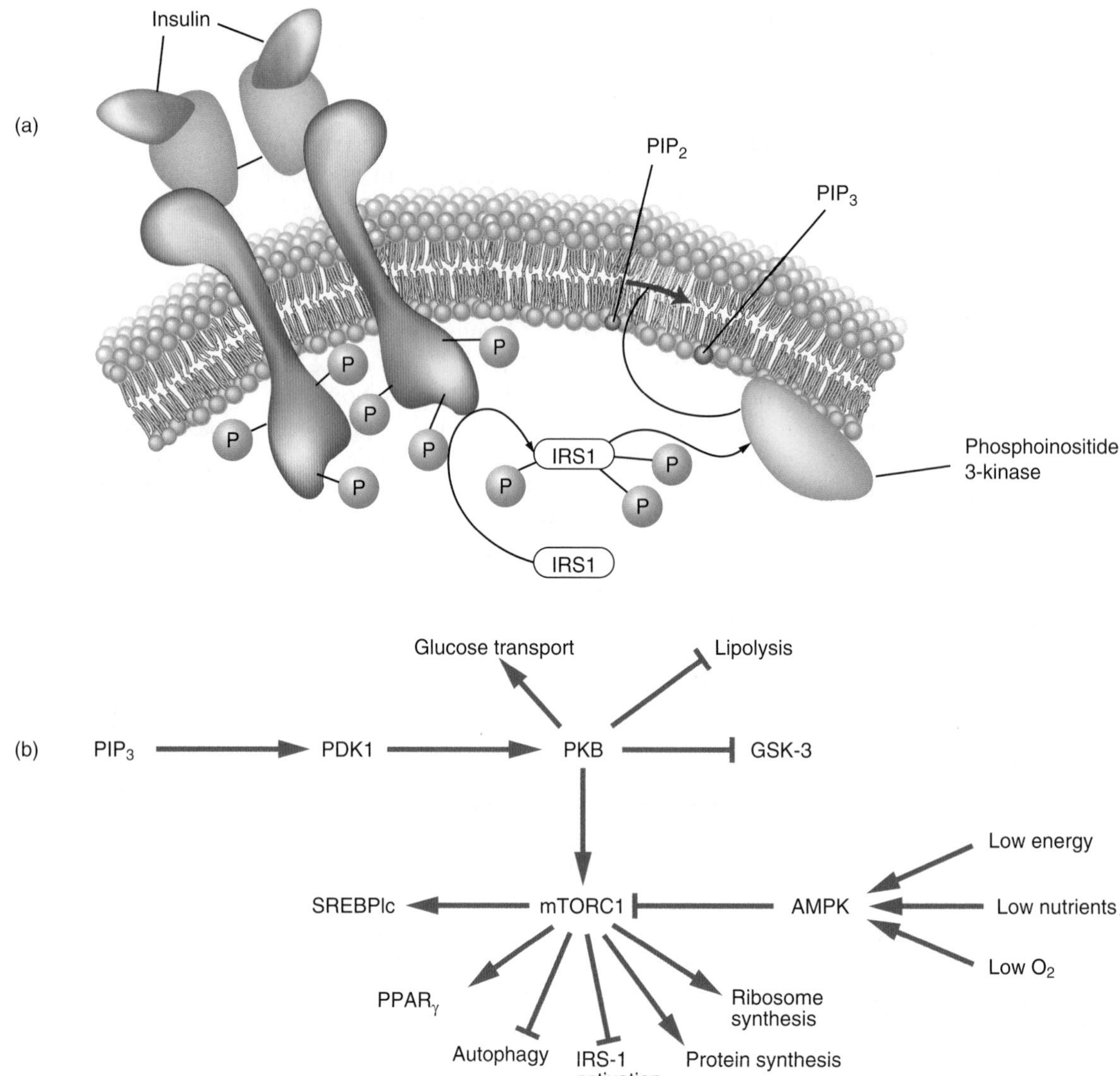

FIGURE 16.9

Simplified Model of Insulin Signaling

(a) The binding of insulin to its tyrosine kinase receptor causes autophosphorylation reactions. The activated insulin receptor subsequently phosphorylates an initial set of substrate molecules. Only one is illustrated in this figure: insulin receptor substrate 1 (IRS1). The newly phosphorylated IRS1 then binds to and activates phosphoinositide-3-kinase, which then phosphorylates PIP_2 to form PIP_3. PIP_3 subsequently binds to and activates PIP_3-dependent protein kinase (PDK1), which in turn activates, via phosphorylation, various kinases (e.g., PKB and PKC). These latter molecules continue the signal cascade, ending with alterations in gene expression. (b) PKB stimulates glycogen synthesis by phosphorylating (and thereby inactivating) GSK3, an enzyme that inactivates glycogen synthase. PKB also inhibits lipolysis and activates glucose transport and the activity of mTOR in the mTORC1 complex, which regulates numerous cellular processes. Examples of mTOR-affected processes shown in this figure include protein and ribosome synthesis (stimulated) and autophagy (inhibited). mTOR is also known to affect the expression of certain genes. For example, mTOR promotes, via the transcription factors SREBP1c and PPARγ (p. 452), the expression of genes that code for proteins involved in lipid and cholesterol synthesis. mTOR activity is inhibited by AMPK when the levels of ATP, nutrients, and O_2 are low. When nutrient levels are high, mTOR modulates its own activity by activating an IRS1 and PI3K inhibitor, which reduces the activity of the insulin-IRS1-PI3K signal cascade.

QUESTION 16.4

The A subunit of cholera toxin causes the cAMP-mediated opening of chloride channels. A massive diarrhea results because the GTP hydrolysis of GTP-α_s is prevented. Describe why this inhibition leads to the diarrhea.

QUESTION 16.5

Cancer often results from a multistage process involving an initiating event (mediated by a viral infection or a carcinogenic chemical) followed by exposure to tumor promoters. Tumor promoters, a group of molecules that stimulate cell proliferation, cannot induce tumor formation by themselves. The phorbol esters, found in croton oil (obtained from the seeds of the croton plant, *Croton tiglium*), are potent tumor promoters. (Other examples of tumor promoters include asbestos and several components of tobacco smoke.) In one of the tumor-promoting actions of the phorbol esters, these molecules mimic the actions of DAG. In contrast to DAG, the phorbol esters are not easily disposed of. Explain the possible biochemical consequences of phorbol esters in an "initiated" cell. What enzyme is activated by both DAG and phorbol esters?

QUESTION 16.6

The term *diabetes*, derived from the Greek word *diabeinein* ("to go to excess"), was first used by Aretaeus (AD 81–138) to identify a group of symptoms that included intolerable thirst and "a liquefaction of the flesh and limbs into urine." After reviewing the accompanying Biochemistry in Perspective essay (pp. 596–597), explain the physiological and biochemical basis for Aretaeus's findings.

Growth Factors

The survival of multicellular organisms requires that cell growth and cell division (mitosis) be rigorously controlled. A variety of hormonelike polypeptides and proteins, the **growth factors**, and some cytokines regulate the growth, differentiation, and proliferation of various cells. Often, the actions of several growth factors are required to promote cellular responses. Growth factors differ from hormones in that they are synthesized by a variety of cell types rather than by specialized glandular cells. Examples of mammalian growth factors include epidermal growth factor, platelet-derived growth factor, and insulin-like growth factor 1 and insulin-like growth factor 2. The term **cytokine** has traditionally described proteins produced by blood-forming cells and immune system cells. Cytokines may stimulate or inhibit cell growth or proliferation. Interleukins and interferons are examples of cytokines.

Epidermal growth factor (EGF) (6.4 kDa), one of the first cellular growth factors identified, is a **mitogen** (a stimulator of cell division) for a large number of epithelial cells, such as epidermal and gastrointestinal lining cells. EGF triggers cell division when it binds to plasma membrane EGF receptors, which are transmembrane tyrosine kinases structurally similar to insulin receptors.

Platelet-derived growth factor (PDGF) (31 kDa) is secreted by blood platelets during the clotting reaction. Acting with EGF, PDGF stimulates mitosis in fibroblasts and other nearby cells during wound healing. PDGF also promotes collagen synthesis in fibroblasts.

Insulin-like growth factor 1 (IGF-1) and **insulin-like growth factor 2** (IGF-2) are polypeptides that mediate the growth-promoting actions of growth hormone (GH). Produced in the liver and in a variety of other tissue cells (e.g., muscle, fibroblasts, bone, and kidney) when GH binds to its cell surface receptor, IGF-1 (highest levels in childhood and young adulthood) and IGF-2 (highest levels during gestation) are the major stimulators of growth in animals. As their names

(Continued on p. 598)

Biochemistry IN PERSPECTIVE

Diabetes Mellitus

Why does diabetes mellitus, a disease in which glucose transport is compromised, damage the entire body?

Diabetes mellitus is a group of devastating metabolic diseases caused by insufficient insulin synthesis, increased insulin destruction, or ineffective insulin action. There are two major forms of diabetes: type 1 and type 2. **Type 1 diabetes** (previously referred to as *juvenile-onset* or *insulin-dependent diabetes*) is an autoimmune disease caused by the destruction of the insulin-producing pancreatic β-cells. **Type 2 diabetes** (previously referred to as *adult-onset* or *non-insulin-dependent diabetes*) is caused by insensitivity of target tissues to insulin. Once quite rare, diabetes is now a leading cause of death in the United States, where it afflicts at least 7% of the population. The worldwide incidence of diabetes is 2.8%.

The basic feature of diabetes is dysfunctional fuel metabolism. Insulin deficiency or the insensitivity of insulin's target tissues (muscle, adipose tissue, and liver) results in **hyperglycemia** (high blood glucose levels) and **dyslipidemia** (abnormal blood lipid and lipoprotein levels). Without effective insulin action, blood glucose levels are higher than normal because muscle and adipose tissue fail to absorb it. This circumstance is exacerbated in the liver by gluconeogenesis and glycogenolysis (ordinarily suppressed by insulin), which produce additional glucose for delivery into an already hyperglycemic bloodstream. Without effective insulin action, lipid metabolism also becomes impaired, especially in liver and adipose tissue.

Hyperglycemia is the proximate cause of the acute symptoms seen in all forms of diabetes. Extreme thirst and frequent urination are caused by **glucosuria** (glucose in urine), which leads to **osmotic diuresis** [excessive loss of water and electrolytes (Na^+, K^+, and Cl^-)]. Extreme fatigue results from the inability of cells to generate sufficient energy. The body's glucose-starved cells trigger a hunger response in the brain's appetite center (p. 604), causing extreme hunger (polyphagia). Hyperglycemia also activates several processes that lead to long-term damage to the body. The formation of advanced glycation end products via the glycation of proteins (p. 249) damages blood vessels and contributes to atherosclerosis and other degenerative diseases.

High blood glucose also stimulates the sorbitol pathway. In some cells in which glucose uptake is insulin dependent (e.g., peripheral nerves and the lens of the eye), excess glucose molecules are converted into sorbitol (p. 246) by NADPH-requiring *aldose dehydrogenase*. Sorbitol accumulation causes glycation of intracellular proteins. Excess $NADP^+$ can result in lower cellular levels of GSH (p. 372) and NO (p. 537). The oxidation of some sorbitol molecules is coupled to NAD^+ reduction. In addition to increasing lactate synthesis (from pyruvate), excess NADH also stimulates superoxide production by the mitochondrial electron transport system by activation of the superoxide-producing enzyme *NADH oxidase*. In diabetics, the accumulation of sorbitol and the redox changes are associated with nerve damage and cataract formation. Finally, hyperglycemia is one of several factors that cause a chronic systemic inflammatory process, mediated by proinflammatory cytokines [e.g., tumor necrosis factor-α (TNF-α)] that activate a network of inflammatory signaling pathways.

In the absence of effective insulin action, there is increased lipolysis (Section 12.1) in adipose tissue (caused by the unopposed action of glucagon), which releases large quantities of fatty acids into blood. In the liver, because these molecules are degraded by β-oxidation in combination with low concentrations of OAA (caused by excessive gluconeogenesis), large amounts of acetyl-CoA, the substrate for forming ketone bodies, are produced. Fatty acids not used to generate energy or ketone bodies are used in VLDL synthesis. This process causes *hyperlipoproteinemia* (high blood concentrations of lipoproteins) because lipoprotein lipase synthesis is depressed when insulin is lacking.

Type 1 Diabetes

Most cases of type 1 diabetes are caused by the autoimmune destruction of the insulin-producing β-cells in the pancreas. The symptoms tend to appear abruptly when almost all insulin-producing capacity has been destroyed, the result of an ongoing inflammatory process over several months or years. As in other inflammatory and autoimmune processes, β-cell destruction is initiated when an autoantibody binds to a cell-surface antigen. One of the most common autoantibodies found in type I diabetes is now believed to bind specifically to an antigen with glutamate decarboxylase activity. Autoantibodies to insulin and the tyrosine phosphatase IA-2 have also been detected.

The most serious acute symptom of type I diabetes is **ketoacidosis**, which is the result of unrestrained fatty acid oxidation. Ketone bodies are released in such large amounts that the body's capacity to oxidize them is exceeded. Elevated concentrations of ketones in the blood (**ketosis**) and low blood pH along with hyperglycemia cause excessive water losses. (The odor of acetone on a patient's breath is characteristic of ketoacidosis because the major mechanism for the removal of this volatile molecule is through the lungs.) Ketoacidosis and dehydration, if left untreated, can lead to coma and death. Type 1 patients are treated with injections of insulin obtained from animals or from recombinant DNA technology. Before Frederick Banting and Charles Best discovered insulin in 1922, most type 1 diabetics died within a year after being diagnosed. Although exogenous insulin prolongs life, it is not a cure. Most diabetics have a shortened life span because of the long-term complications of their disease.

Many researchers now believe that type 1 diabetes is caused by both genetic and environmental factors. Although the precise

Biochemistry IN PERSPECTIVE cont.

cause is still unknown, individuals who have inherited certain genetic markers are at high risk for developing the disease. Certain HLA antigens, that is, specific variants of HLA-DR3 and HLA-DR4, are found in a large majority of type I diabetics. The *HLA* or *histocompatibility antigens*, present on the surface of most of the body's cells, play an important role in determining how the immune system will react to foreign substances or cells.

Type 2 Diabetes

Type 2 diabetes appears to be a milder disease than type 1 in that onset is gradual, but its long-term effects on the body are devastating. In contrast to type 1 diabetics, most individuals with type 2 have elevated blood levels of insulin at the time of diagnosis. For various reasons type 2 diabetics are resistant to insulin. The most common cause of insulin resistance is disruption of the insulin receptor catalyzed tyrosine phosphorylation of IRS (insulin receptor substrate) proteins. Insulin signaling is inhibited by kinases (e.g., JNK) that phosphorylate IRS serine residues. JNK, activated by ER stress and high levels of free fatty acids, is one of several IRS-modifying enzymes that trigger the expression of genes that promote inflammation. Inflammation is increasingly recognized as a central feature of type 2 diabetes pathology. Patients with type 2 diabetes have high blood levels of inflammatory proteins such as C-reactive protein, interleukin-1β (IL-1β), and IL-6.

The insensitivity of target tissues to insulin causes blood glucose levels to rise, which leads to increased release of insulin from pancreatic β-cells. The increased exposure of target cell insulin receptors to insulin promotes receptor internalization and decreased synthesis of the receptor and several downstream signaling proteins. One of the eventual consequences of high blood insulin levels (**hyperinsulinemia**) is compromised function and reduced mass of the insulin-secreting β-cells. The formation of amyloid deposits, aggregates of a misfolded protein, in islet cells may be a contributing factor to β-cell apoptosis. (Disease-related protein aggregation is described on p. 764.)

The onset of type 2 diabetes occurs when the insulin response falls to a level where fasting blood glucose levels exceed 126 mg/dl (normal = <100 mg/dl). Approximately 85% of type 2 diabetics are obese. Because obesity itself promotes tissue insensitivity to insulin, individuals who are prone to this form of diabetes are at risk for the disease when they gain weight.

Treatment of type 2 diabetes usually consists of diet control and exercise. Often, obese patients become more sensitive to insulin (i.e., there is an upregulation of insulin receptors) when they lose weight. Because sustained muscular activity increases the uptake of glucose without requiring insulin, exercise also decreases hyperglycemia. In some cases, medications are prescribed. Oral hypoglycemic agents such as the sulfonylureas stimulate insulin release and, therefore, reduce gluconeogenesis and glycogenolysis in the liver and increase transport of glucose into insulin-sensitive body cells. Metformin, a biguanide antidiabetic drug, inhibits hepatic gluconeogenesis and promotes peripheral glucose uptake and fatty acid oxidation by activating AMPK.

When the failure of type 2 diabetic patients to control hyperglycemia is accompanied by other serious medical conditions (e.g., renal insufficiency, myocardial infarction, or infections), a serious metabolic state referred to as **hyperosmolar hyperglycemic nonketosis** (HHNK) can result. (Ketoacidosis is rare in type 2 diabetes.) Because of the additional metabolic stress, insulin resistance is exacerbated, and blood glucose levels rise. The patient may then become severely dehydrated. The resulting lower blood volume depresses renal function, which causes further increases in blood glucose concentrations. Eventually, the patient becomes comatose. Because the onset is slow, it may not be recognized until the dehydration is severe. (This is especially true for elderly diabetics, who often have a depressed thirst mechanism.) For this reason, HHNK is often more life-threatening than ketoacidosis.

Long-Term Complications of Diabetes

Despite the efforts of physicians and patients to control the symptoms of diabetes, few diabetics avoid the long-term consequences of their disease. Diabetics are especially prone to develop kidney failure, myocardial infarction, stroke, blindness, and neuropathy. In *diabetic neuropathy*, nerve damage causes the loss of sensory and motor functions. In addition, circulatory problems often cause gangrene, which leads to tens of thousands of amputations annually.

Most diabetic complications stem from damage to the vascular system. For example, damaged capillaries in the eye and kidney lead to blindness and kidney damage, respectively. Similarly, the accelerated form of atherosclerosis found in diabetics leads to serious cases of myocardial infarction and stroke. Much of this damage results from glycation (p. 249), which initiates atherosclerosis. One facet of this disease process is hyperglycemia-initiated ROS formation, which causes mitochondria in endothelial cells to produce large quantities of superoxide. Some superoxide species react with ·NO to form peroxynitrite (ONOO-, see p. 372). The subsequent nitrosylation of proteins, most notably antioxidant enzymes and NO synthase, leads to increased oxidative stress and a decrease in ·NO, a critical vasodilator molecule.

SUMMARY: Diabetes is an example of how a single defect (the inability to synthesize or respond to insulin) in a complex biological system can cause devastating damage.

suggest, IGF-1 and IGF-2 promote (but to a lesser degree) the same metabolic processes as does the hormone insulin (e.g., glucose transport and fat synthesis). Like other polypeptide growth factors, IGF-1 and IGF-2 trigger intracellular processes by binding to cell surface receptors. Not surprisingly, these receptors are similar in structure to the insulin receptor. For example, the intracellular β-chains have tyrosine kinase activity.

Interleukin 2 (IL-2) (13 kDa) is a member of a group of cytokines that regulate the immune system in addition to promoting cell growth and differentiation. IL-2 is secreted as a result of binding by activated T cells to a specific antigen-presenting cell. These cells are also stimulated to produce IL-2 receptors. The binding of IL-2 to these receptors stimulates cell division so that numerous identical T cells are produced. This process, as well as other aspects of the immune response, continues until the antigen has been eliminated from the body.

Several cytokines are growth inhibitors. The **interferons** are a group of polypeptides produced by a variety of cells in response to several stimuli, such as antigens, mitogens, viral infections, and certain tumors. The type I interferons protect cells from viral infection by stimulating the phosphorylation and inactivation of a protein factor (eIF2α) required to initiate protein synthesis. Type II interferons, produced by T lymphocytes, inhibit the growth of cancerous cells in addition to having several immunoregulatory effects. As the name implies, the **tumor necrosis factors (TNF)** are toxic to tumor cells. Both TNF-α (produced by antigen-activated phagocytic white blood cells) and TNF-β (produced by activated T cells) suppress cell division. TNF-α is an important regulator of immune system cells.

KEY CONCEPT

Growth factors and cytokines are a group of hormonelike polypeptides and proteins that influence cell growth, proliferation, and differentiation.

Steroid and Thyroid Hormone Mechanisms

The signal transduction mechanisms of the hydrophobic steroid and thyroid hormones result in changes in gene expression that in turn cause changes in the pattern of proteins that a target cell produces. Steroid and thyroid hormones are transported in the blood to their target cells bound to several types of protein. Examples of steroid transport proteins include corticosteroid-binding globulin, sex hormone–binding protein, and albumin. In addition to albumin, the thyroid hormones are transported by thyroid-binding globulin.

Once they have reached their target cells, hydrophobic hormone molecules dissociate from their transport proteins, diffuse through the plasma membrane, and bind to their intracellular receptors (**Figure 16.10**). These receptors are high-affinity ligand-binding molecules that belong to a large family of structurally similar DNA-binding proteins. Depending on the type of hormone involved, initial binding to receptors may occur within the cytoplasm (e.g., glucocorticoid) or the nucleus (e.g., estrogen, androgens, and thyroid hormone). In the absence of hormone, several types of receptor have been observed to form complexes with other proteins. For example, unoccupied glucocorticoid receptors are found in the cytoplasm bound to chaperone proteins, such as hsp90. (Recall that hsps are so named because of their increased synthesis in response to cellular stressors such as elevated temperature.) The chaperone proteins block the receptor's DNA-binding site when the hormonal ligand is not present. When the hormone binds to its receptor, the chaperones dissociate and the receptor-ligand complex migrates to the nucleus as a homodimer.

Within the nucleus, each hormone-receptor complex binds to specific DNA segments called **hormone response elements** (HRE). The binding of the hormone-receptor complex to the base sequence of an HRE via zinc finger domains (see p. 155) in the receptor either enhances or diminishes the transcription of a specific gene. The same hormone-receptor complex can bind to and influence the transcription of as many as 50 to 100 different genes, therefore inducing global changes in cellular function.

KEY CONCEPT

Hydrophobic hormones such as the steroids and thyroid hormones diffuse across cellular membranes and bind to intracellular receptors.

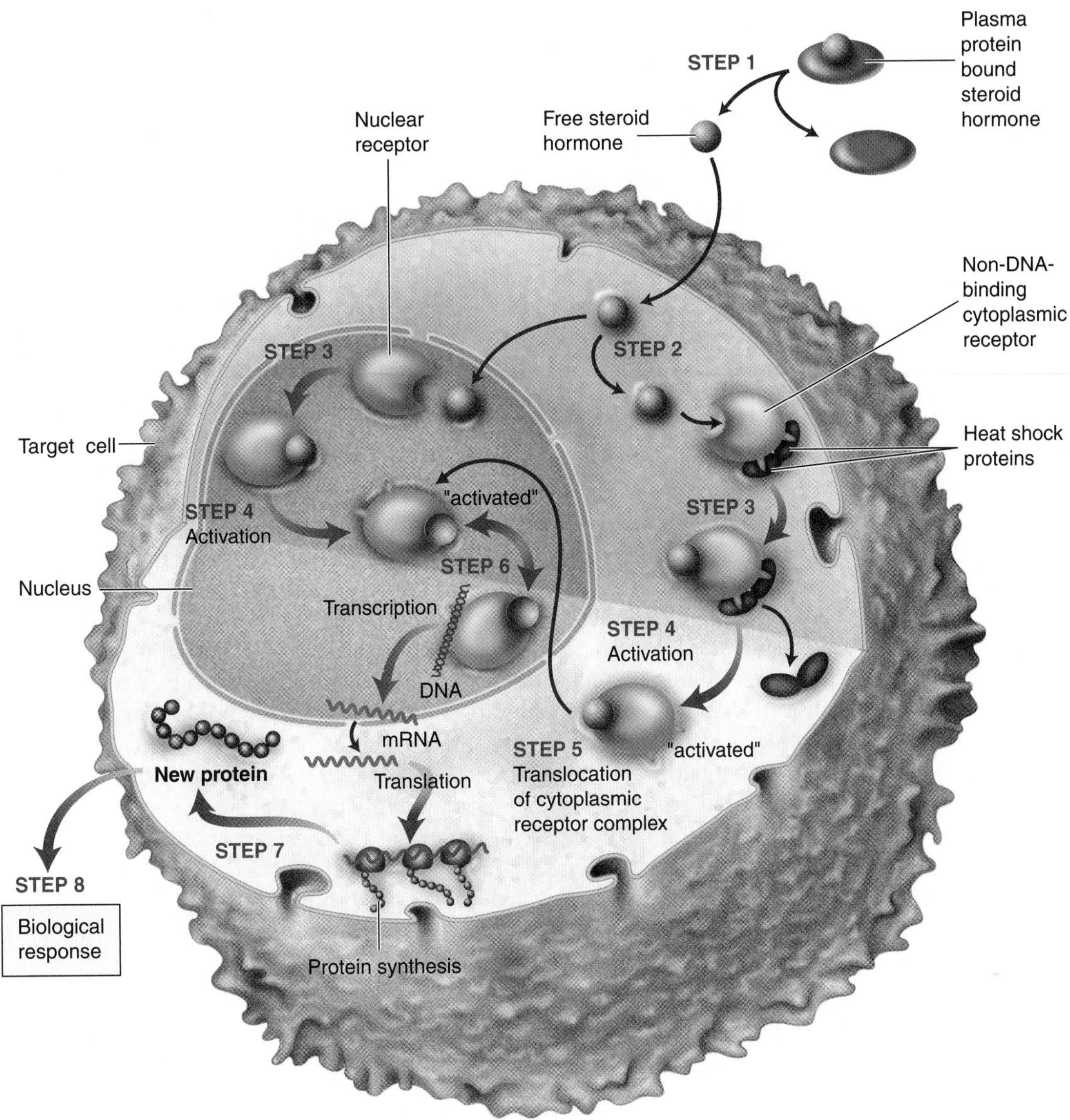

FIGURE 16.10

Model of Steroid Hormone Action within a Target Cell

Steroid hormones are transported in blood associated with plasma proteins. When they reach a cell and are released (1), the hormone molecules diffuse through the plasma membrane, where they bind to receptor molecules in cytoplasm (2) or nucleus (3). After activation (4), a cytoplasmic hormone-receptor complex, a dimer (not shown), migrates to the nucleus (5). The binding of an activated hormone-receptor complex to HRE sequences within DNA (6) results in a change in the rate of transcription of specific genes and therefore, in the pattern of proteins (7) that the cell produces. The net effect of the steroid hormone (8) is a change in the metabolic functioning of the cell.

16.3 METABOLISM IN THE MAMMALIAN BODY: THE FEEDING-FASTING CYCLE

Each of the major metabolic pathways that sustain life in multicellular organisms has now been covered. Any true understanding of metabolism, however, requires a more integrated approach. The feeding-fasting cycle, the self-regulating mechanism by which the mammalian body extracts energy and nutrients from food, is a well-understood process. A brief review of its operation provides an opportunity to observe biochemical reactions as they actually occur.

Each organ in the mammalian body contributes to the individual's function in several ways. For example, some organs are consumers of energy so that they may perform certain energy-driven tasks (e.g., brain and skeletal, cardiac, and smooth muscle). Other organs, such as those in the digestive tract, are responsible for efficiently supplying energy-rich nutrient molecules for use elsewhere. Information in the form of signal molecules (e.g., hormones and neurotransmitters) is used to regulate the balance between energy generation and energy expenditure. Prominent examples of protein hormones are ghrelin (hgr), peptide YY (PYY), cholecystokinin (CCK), and glucagon-like peptide 1 (GLP-1). **Ghrelin** (ghr), produced by cells in the stomach and small intestine, stimulates appetite (food intake), whereas insulin, PYY, CCK, and GLP-1 promote satiety (i.e., inhibit food intake). Nutrient transport across cell plasma membranes is also an important feature of organ function. Glucose transport is a well-researched example. Active transport of glucose by the Na^+/glucose transporter is linked to a Na^+ gradient established by the ATP-driven Na^+-K^+ pump (p. 412). The facilitated diffusion of glucose across cell membranes occurs by glucose carriers called GLUTs: GLUT1 (most cells), GLUT2 (liver, β-pancreatic cells, and intestinal enterocytes), GLUT3 (neurons), and GLUT4 (insulin-sensitive muscle and adipose tissue cells). GLUT5, principally found in intestinal enterocyte and liver cell plasma membranes, transports fructose.

Despite their consistent requirements for energy and biosynthetic precursor molecules, mammals consume food only intermittently. This is possible because of elaborate mechanisms for storing and mobilizing energy-rich molecules derived from food (**Figure 16.11**). The changes in the status of various biochemical pathways during transitions between feeding and fasting illustrate metabolic integration and the profound regulatory influence of hormones. Substrate concentrations are also an important factor in metabolism. In the **postprandial** state, which occurs directly after a meal has been digested or absorbed, blood nutrient levels are elevated above those in the fasting phase. During the **postabsorptive** state, for example, after an overnight fast, nutrient levels in blood are low.

The Feeding Phase

As the feeding phase begins, food is propelled along the GI tract by muscle contractions initiated and controlled by the enteric (intestinal) nervous system. As it moves through the organs, food is broken into smaller particles and exposed to enzymes. Ultimately, the products of digestion (consisting largely of sugars, fatty acids, glycerol, and amino acids) are absorbed by the small intestine and transported into the blood and lymph. This phase is regulated by interactions between enzyme-producing cells of the digestive organs, the nervous system, and several hormones. The enteric nervous system, which is influenced by the parasympathetic and sympathetic nerves, is responsible for the waves of smooth muscle contraction that propel food along the tract, as well as for regulating the secretions of several digestive structures (e.g., from salivary and gastric glands). Hormones such as gastrin, secretin, and CCK also contribute to the digestive process. They do so by stimulating the secretion of enzymes or digestive aids such as bicarbonate and bile.

The early postprandial state is illustrated in **Figure 16.12**. Sugars and amino acids are absorbed from the small intestine and transported by the portal blood to the liver. The portal blood also contains a high level of lactate, a product of enterocyte metabolism. Most lipid molecules are transported from the small intestine in lymph as chylomicrons. Chylomicrons pass into the bloodstream, which carries them to tissues such as muscle and adipose tissue. After most triacylglycerol molecules have been removed from chylomicrons, these structures, now referred to as *chylomicron remnants*, are then taken up by the liver. The phospholipid, protein, cholesterol, and few remaining triacylglycerol molecules

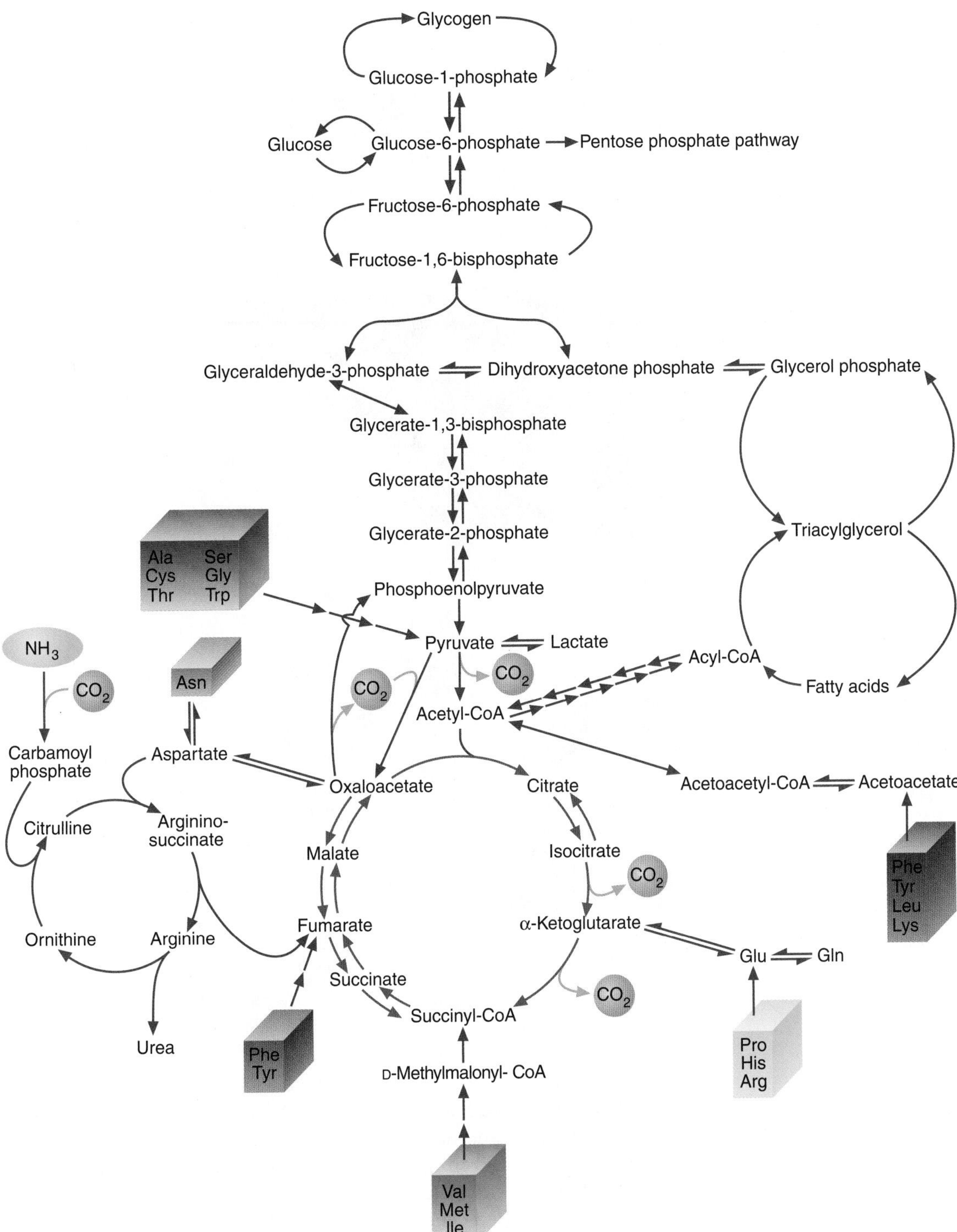

FIGURE 16.11

Nutrient Metabolism in Mammals

Despite the variability of the mammalian diet, these organisms usually provide their cells with adequate nutrients. Control mechanisms that regulate biochemical pathways are responsible for this phenomenon.

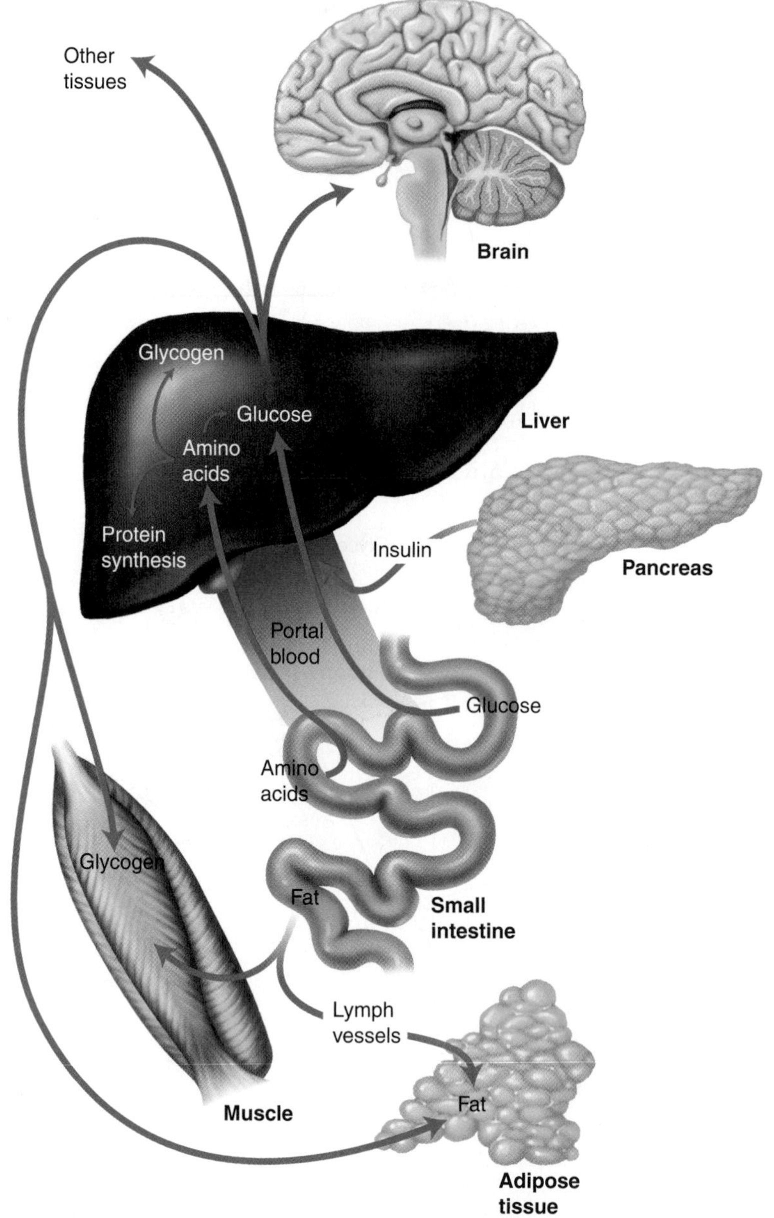

FIGURE 16.12
The Early Postprandial State

The primary substrates for glycogen synthesis in liver are amino acids and lactate (not shown) derived from portal blood. Note that the normal use for glucose in fat cells is as the precursor of glycerol. Fat cells do not carry out significant *de novo* fatty acid synthesis, but instead obtain most from the diet. Chylomicrons (not shown) carry lipids from the small intestine to the body's tissues, especially muscle and adipose tissue. Note that brain uses glucose as its sole fuel.

are then degraded or reused. For example, cholesterol is used to synthesize bile acids, and fatty acids are used in new phospholipid synthesis. Phospholipids, as well as other newly synthesized lipid and protein molecules, are then incorporated into lipoproteins for export to other tissues.

As glucose moves through the blood from the small intestine to the liver, the β cells within the pancreas are stimulated to release insulin. (High blood glucose and insulin levels depress glucagon secretion by the pancreatic α cells.) Insulin release triggers several processes that ensure the storage of nutrients. These include glucose uptake by muscle and adipose tissue, glycogenesis in liver and muscle, fat synthesis in liver, and fat storage in adipocytes. Insulin also represses lipolysis in adipocyes and gluconeogenesis and glycogenolysis in liver. In addition, insulin influences amino acid metabolism. For example, insulin promotes the transport of amino acids into the cells (especially liver and muscle cells). In general, insulin stimulates protein synthesis in most tissues.

Although the effects of insulin on postprandial metabolism are profound, other factors (e.g., substrate supply and allosteric effectors) also affect the rate and degree to which these processes occur. For example, elevated levels of fatty acids in blood promote *lipogenesis* (TG synthesis) in adipose tissue. Regulation by several allosteric effectors further ensures that competing pathways do not occur simultaneously; in many cell types, for example, fatty acid synthesis is promoted by citrate (an activator of acetyl-CoA carboxylase), whereas fatty acid oxidation is depressed by malonyl-CoA (an inhibitor of carnitine acyltransferase I activity).

The Fasting Phase

The early postabsorptive state (**Figure 16.13**) of the feeding-fasting cycle begins as the nutrient flow from the intestine diminishes. As blood glucose and insulin levels fall, glucagon is released. Glucagon acts to prevent hypoglycemia by promoting glycogenolysis and gluconeogenesis in liver. Decreased insulin reduces energy storage in several tissues and leads to increased lipolysis and the release of amino acids such as alanine and glutamine from muscle. Recall that several tissues use fatty acids in preference to glucose. Glycerol and alanine (i.e., the glucose-alanine

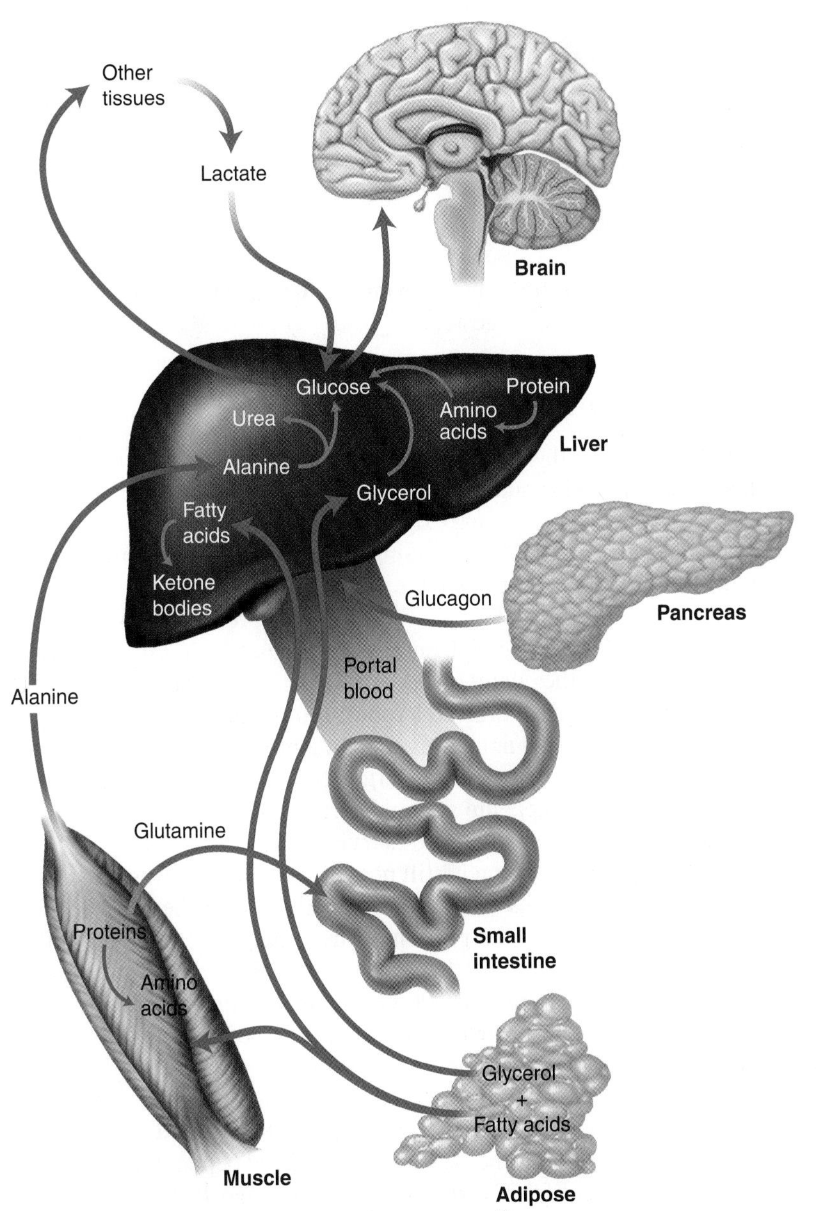

FIGURE 16.13
The Postabsorptive State

Between meals the body, under the influence of hormones, obtains nutrients from skeletal muscle (e.g., alanine for gluconeogenesis in liver and glutamine for energy generation in enterocytes) and adipose tissue (fatty acids). See the text for further details.

cycle, p. 296) are substrates for gluconeogenesis, and glutamine is an energy source for enterocytes.

When a fast becomes prolonged (e.g., overnight), several metabolic strategies maintain blood glucose levels. Increased mobilization of fatty acids from adipose tissue during the postabsorptive state is stimulated by epinephrine. These fatty acids provide an alternative to glucose for muscle. (Reduced skeletal muscle consumption of glucose spares its use for brain.) In addition, the action of glucagon increases gluconeogenesis, using amino acids derived from muscle.

Under conditions of extraordinarily prolonged fasting (starvation), the body makes metabolic changes to ensure that adequate amounts of blood glucose are available to sustain energy production in the brain and other glucose-requiring cells. Additionally, fatty acids from adipose tissue and ketone bodies from liver are mobilized to sustain the other tissues. Because glycogen is depleted after several hours of fasting, gluconeogenesis plays a critical role in providing sufficient glucose. During early starvation, large amounts of amino acids from muscle are used for this purpose. However, after several weeks, the breakdown of muscle protein declines significantly because the brain is using ketone bodies as a fuel source.

KEY CONCEPTS

- During the feeding phase, food is consumed, digested, and absorbed.
- Absorbed nutrients are then transported to the organs, where they are either used or stored.
- During fasting, several metabolic strategies maintain blood glucose levels.

QUESTION 16.7

Explain the metabolic changes that occur during starvation. What appears to be the principal purpose for the preferential degradation of muscle tissue during starvation?

QUESTION 16.8

Explain the changes in liver metabolism that occur when blood glucose levels drop after a meal has been digested.

Feeding Behavior

Feeding behavior is the complex mechanism by which animals, including humans, seek out and consume food. In mammals, regulation of feeding behavior involves hormonal and neural signals from peripheral organs (e.g., the GI tract and adipose tissue) and sensory input from the external environment (e.g., the sight, smell, and taste of palatable food) that together are integrated in the brain to regulate appetite and the body's metabolic processes (**Figure 16.14**).

For mammals, which have high energy requirements, the consumption of sufficient food to ensure the energy needed to sustain life is of critical importance. To this end, mammals have evolved a robust food-seeking system involving several neuronal pathways and numerous signaling molecules. In addition to providing a mechanism for balancing energy consumption and utilization, the mammalian brain links appetite systems to taste, olfaction, and reward systems to create a powerful drive that ensures survival.

Although appetite regulation is still not completely understood, it is clear that the principal neural circuits that control appetite are in the hypothalamus, located in the ventral (underside) portion of the vertebrate brain, and in the brain stem. Despite its small size, the hypothalamus, one of the most evolutionarily conserved regions of the brain, has a wide array of functions, including control of body temperature, electrolyte balance, monitoring of nutrient levels (e.g., blood glucose), and several aspects of emotional behavior.

The primary neurons that control feeding behavior are in the *arcuate nucleus* (ARC) of the hypothalamus (**Figure 16.15**). Activation of ARC neurons that produce NPY (neuropeptide Y) and AgRP (agouti-related peptide) stimulate appetite, whereas stimulation of POMC (pro-opiomelanocortin) cells that produce α-MSH (α-melanocyte-stimulating hormone) suppress appetite. Hormones that affect

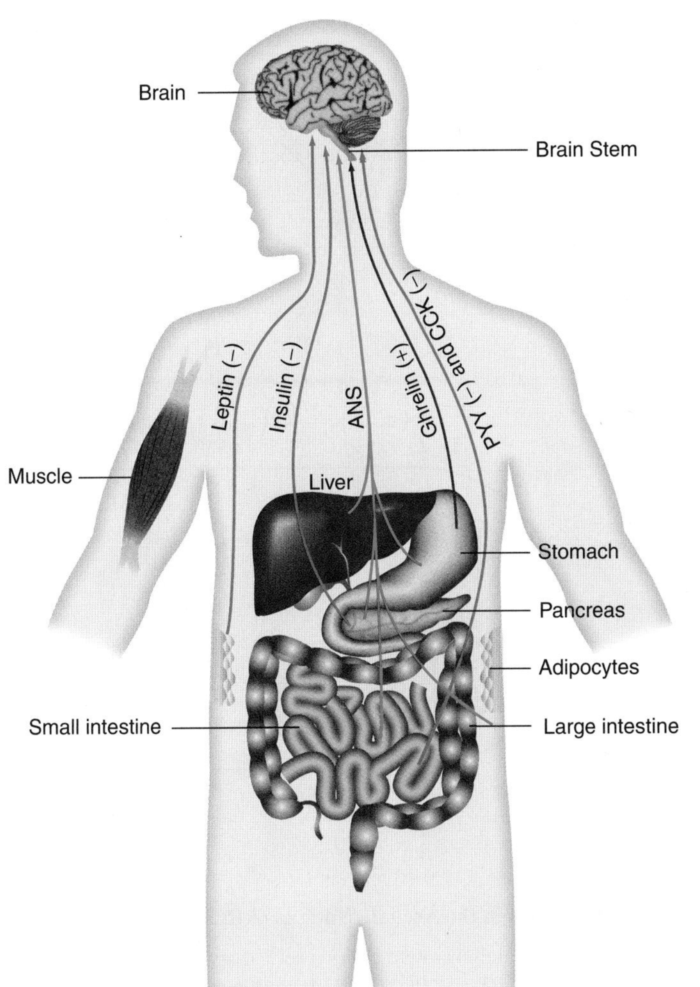

FIGURE 16.14

Feeding Behavior in Humans

Appetite and satiety in humans are regulated by hormonal and neural signals from peripheral organs. Peptide hormones such as PYY and CCK (produced by cells in the GI tract), insulin (produced by pancreatic β cells), and leptin (produced by adipose tissue) inhibit appetite (–); that is, they promote satiety. Ghrelin (produced by cells in the stomach and small intestine) stimulates appetite (+). Neuronal pathways of the autonomic nervous system (ANS) such as the vagus nerve continuously supply the brain with information related to the status of the body's internal organs.

ARC-regulated feeding behavior include **leptin** (a satiety-including protein secreted by adipose tissue in proportion to adipose tissue mass), insulin, ghr, and PYY. Under normal conditions, when leptin levels rise, indicating that the body's energy resources are sufficient, NPY/AgRP neurons are inhibited and POMC neurons are activated. In this circumstance appetite for food is depressed. In contrast, increased food consumption, triggered by the activation of NPY/AgRP neurons and the inhibition of POMC neurons, results from falling leptin levels (caused by weight loss). All these neurons send signals to other neurons in the hypothalamus (referred to collectively as second-order neurons) that in turn relay them to other parts of the brain. Among these targets is the *nucleus tractus solitarius* (NTS) within the brain stem, which integrates this information with appetite-regulating signals from the GI tract via the *vagus nerve.*

The results of feeding behavior-related signaling, depending on circumstances, range from increased appetite to a sense of satiety. Insulin also reduces food intake via the NPY/AgRP and POMC neurons, although to a lesser extent than leptin. Insulin regulates leptin synthesis. During fasting or calorie restriction (dieting) decreased leptin levels, caused by decreased adipose tissue mass, contribute to increased hunger and subsequent weight gain. Other important appetite-regulating hormones include ghr and PYY. Ghr, an orexigenic (appetite stimulating) molecule released by cells within the stomach, as well as the intestine, activates NPY/AgRP neurons. PYY, an anorexigenic (appetite inhibiting) molecule produced by cells in the small intestine and colon, inhibits NPY/AgRP neurons. The ARC neurons are also sensitive to local levels of glucose, fatty acids, and the amino acid leucine.

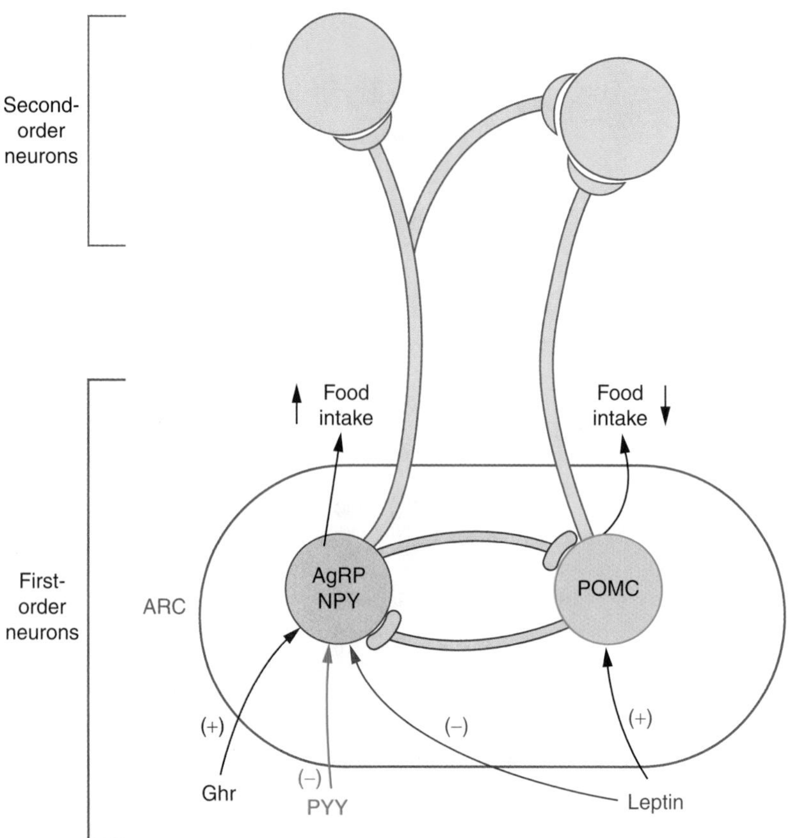

FIGURE 16.15

Appetite-Regulating Neurons in the Arcuate Nucleus (ARC)

Within the ARC of the hypothalamus, there are two sets of appetite-regulating neurons with opposing effects: AgRP/NPY and POMC. Both neuron types are first-order neurons (i.e., they respond to peripheral signal molecules). Activation of AgRP/NPY neurons by Ghr increases appetite and energy-requiring metabolic processes. When POMC neurons are activated by leptin (and insulin to a lesser extent), appetite is depressed. Appetite is also negatively affected by the actions of leptin and PYY on AgRP/NPY neurons. Appetite-regulating signals generated by AgRP/NPY and POMC neurons are communicated via second-order neurons to other parts of the hypothalamus and then other brain centers. Signals from these centers are subsequently sent to the NTS in the brain stem, where they are integrated with neural signals from the GI tract and other organs.

The integration of disparate appetite-regulating hormonal and nutrient signals received within the hypothalamus appears to be mediated by AMPK and mTOR. The signal transduction events that cause neurons to fire and release NPY and AgRP neurotransmitter molecules are triggered by activation of AMPK by appetite-promoting molecules such as ghr. These molecules bind to cell-surface receptors in neurons in ARC and other hypothalamic regions, in combination with low blood glucose levels in the process that leads to appetite stimulation. When appetite-inhibiting hormones such as leptin and insulin bind to their cell-surface receptors, AMPK activity is inhibited, with the result that NPY/ArGP neurons are inhibited. ARC nutrient-sensing neurons also utilize mTOR (p. 593) to regulate feeding behavior. mTOR activity, which varies inversely with that of AMPK, is stimulated by leptin, insulin, and nutrients. As a result of the mTOR-triggered signal transduction pathway, appetite is depressed. Low energy availability (i.e., a high AMP:ATP ratio) results in AMPK (p. 451) activation. Activated AMPK inhibits mTOR with the result that appetite increases.

Biochemistry IN PERSPECTIVE

Obesity and the Metabolic Syndrome

Why are so many humans predisposed to obesity in the modern world, especially in the past 30 years? Obesity, excess body weight that results from an imbalance between energy intake and energy expenditure, is a worldwide epidemic. If the human brain is so adept at balancing appetite, satiety, and physical activity, why has obesity become such a threat to health in the past 30 years? The answer lies in the long human struggle for survival and a significant recent change to the modern human diet. For most of its existence, *Homo sapiens* has had physically challenging hunter-gatherer and agrarian lifestyles and has been threatened by famines. Robust appetites ensured sufficient calories required to sustain life and, when combined with the physiological capacity to store large quantities of fat, had significant survival value. During a famine, individuals with large fat stores can go without food for many weeks and have a better chance to survive than lean people.

In the modern world, in contrast, an increasing proportion of human populations live under vastly different circumstances. The availability of inexpensive, calorie-dense food is combined with sedentary lifestyles. As a result, there has been significant weight gain among individuals genetically predisposed to what have now become problems with body weight regulation. For many people, a calorie-restricted environment combined with vigorous physical work would mask such vulnerability. Several rare mutations, linked to monogenic disorders of body weight regulation (**Figure 16A**), provide insight into human feeding behavior. For example, obese patients with leptin or leptin receptor defects have insatiable appetites. In Prader-Willi syndrome, another rare disorder, exceptionally high blood levels of ghrelin drive an insatiable hunger. By one estimate, at least 40% of the factors causing obesity can be attributed to genetics. Obesity may result from a combination of a calorie-dense environment, sedentary lifestyles, and mutations in one or more genes coding for appetite and satiety signal transduction pathways components. Recent research has linked the obesity epidemic over the past several decades to the introduction in the 1970s of large amounts of fructose into processed foods. Fructose promotes weight gain because, unlike glucose, it does not suppress the release of ghr, the appetite-stimulating hormone, or stimulate the release of leptin and insulin, both of which promote satiety.

FIGURE 16A
Obesity and Genetics

These mice have identical genomes except that the leptin gene was deleted from the mouse on the left.

In addition to the discomfort and severe social stigma associated with obesity, excessive body weight is a serious risk factor for a variety of illnesses such as hypertension, heart disease, several forms of cancer, osteoarthritis, and diabetes mellitus. Obesity is now recognized as a contributing factor in metabolic syndrome.

Metabolic Syndrome

Metabolic syndrome is the term used to describe a cluster of clinical disorders that include, in addition to obesity, hypertension, dyslipidemia (high blood levels of total cholesterol and triacylglycerol and low HDL levels), and insulin resistance. Insulin resistance, one of the earliest manifestations of the metabolic syndrome, originates in part from excess levels of free fatty acids (FFA) in blood. Expanding adipocytes, especially those of visceral (abdominal) adipose tissue, release fatty acids into the bloodstream. As the FFA levels in blood rise, these molecules begin accumulating in other cells. In insulin-sensitive tissue (muscle, liver, and pancreatic β cells) FFA disrupt signal transduction pathways. Among the consequences of this process, called *lipotoxicity*, are dyslipidemia and hyperinsulinemia caused by FFA-stimulated insulin secretion. Other effects include excess glucose production in liver (uninhibited gluconeogenesis and glycogenolysis) and inhibited insulin-mediated glucose uptake by muscle. When adipocytes also become insulin-resistant, there is an increase in lipolysis that results in the release of more FFA into the bloodstream.

As obesity progresses, adipose tissues develop a chronic state of low-level inflammation. Apparently, overstimulated adipocytes undergo physiological changes that cause ER stress and oxidative stress, both of which activate inflammatory signaling pathways. Eventually, adipose tissue, now infiltrated with macrophages, releases inflammatory cytokines. (e.g., TNF-α, IL-6, and C-reactive protein). These and other cytokines also contribute to vascular disease by increasing insulin resistance. In genetically predisposed individuals, the metabolic syndrome may develop into type 2 diabetes (see the earlier Biochemistry in Perspective essay in this chapter, pp. 596–597).

In addition to promoting obesity and hypertension (p. 584), excessive fructose consumption contributes to the metabolic syndrome by other means. First, fructose has a greater propensity than glucose toward glycation reactions (p. 249) and

Biochemistry IN PERSPECTIVE cont.

AGE formation (linked to atherosclerosis and other inflammation-based diseases) because it spends proportionally more time in its open-chain form. Second, hepatic fructose metabolism is highly lipogenic. Consumption of large amounts of fructose has been linked not only to dyslipidemia, but also to nonalcoholic fatty liver disease, a disorder that compromises liver function. (Fructose has such a negative effect on the liver because most dietary fructose molecules are metabolized by this organ because of the presence of GLUT5 in the hepatocyte plasma membrane.) Third, fructose consumption preferentially results in increased abdominal fat accumulation, a major cause of inflammation. Finally, fructose-induced lipogenesis and uric acid production contribute to insulin resistance.

SUMMARY: Natural selection in response to the rigors of chronic food scarcity has left many humans with the propensity to gain weight when food is plentiful. The body's inability to cope with lipotoxicity, caused by excessive body weight and fructose-caused metabolic stress, can result in metabolic syndrome.

WORKED PROBLEM 16.1

Body mass index (BMI) is a measure of a person's body composition that is based both on weight and on height. It is defined as weight in kg/(height in m)^2. A normal healthy person's BMI falls within the range of 18.5 to 24.9. Individuals who have BMI values between 25 and 29.9 or above 30 are designated as overweight and obese, respectively. Calculate the BMI values for the following individuals and determine how they would be classified: three 6-foot (1.829 m) men with weights of 150 lb (68 kg), 200 lb (91 kg), and 250 lb (115 kg).

SOLUTION

Substitute the given values into the equation BMI = weight/(height in m)^2. For the first man, his BMI is calculated

$$\text{BMI} = 68\text{ kg}/(1.829\text{ m})^2 = 68\text{ kg}/3.349\text{ m}^2 = 20.3\text{ (normal)}.$$

For the second man, his BMI is calculated

$$\text{BMI} = 91\text{ kg}/3.349\text{ m}^2 = 27\text{ (overweight)}.$$

For the third man, his BMI is calculated

$$\text{BMI} = 115\text{ kg}/3.349\text{ m}^2 = 34\text{ (obese)}.$$

■

Chapter Summary

1. Multicellular organisms require sophisticated regulatory mechanisms to ensure that all their cells, tissues, and organs cooperate.
2. Hormones are molecules organisms use to convey information between cells. When target cells are distant from the hormone-producing cell, such molecules are called endocrine hormones. To ensure proper control of metabolism, the synthesis and secretion of many mammalian hormones are ultimately controlled by the central nervous system. In addition, a negative feedback mechanism precisely controls various hormone syntheses. A variety of diseases are caused by either overproduction or

underproduction of a specific hormone or by the insensitivity of target cells.

3. Growth factors and some cytokines are polypeptides that regulate the growth, differentiation, or proliferation of various cells. They differ from hormones in that they are often produced by a variety of cell types rather than by specialized glandular cells.
4. Signal molecules initiate their effects in the target cell by binding to a specific receptor. Polar molecules, such as amines and peptides, bind to cell-surface receptors. They alter the activities of several enzymes and/or transport mechanisms in the cell. G-protein-coupled receptors use one or more of the second messengers (cAMP, cGMP, IP_3, DAG, and Ca^{2+}) to mediate the primary signal's effect on the target cell, and this has a significant amplifying effect on the signal transduction pathway. Tyrosine kinase receptors do not directly involve the generation of a second messenger. The nonpolar steroid and thyroid hormones diffuse through the lipid bilayer and bind to intracellular receptors. The hormone-receptor complex subsequently binds to a DNA sequence referred to as a hormone response element (HRE). The binding of a hormone-receptor complex to an HRE enhances or diminishes the expression of specific genes.
5. The feeding-fasting cycle illustrates how a variety of organs contribute via hormones and neurotransmitters to the acquisition of food molecules and their use. Feeding behavior is a mechanism by which animals seek out and consume food. The goal is to maintain balance between energy acquisition and energy expenditure. The hypothalamus contains the critical neural circuits that control appetite and satiety.

Take your learning further by visiting the **companion website** for Biochemistry at **www.oup.com/us/mckee** where you can complete a multiple-choice quiz on integration of metabolism to help you prepare for exams.

Suggested Readings

Duncan, D. E., The Covert Plague, *Discover* 26(12):60–66, 2005.

Edwards, N. L., The Role of Hyperuricemia in Vascular Disorder, *Curr. Opin. Rheumatol.* 21:132–137, 2009.

Erion, D. M., and Shulman, G. J., Diacylglycerol-Mediated Insulin Resistance, *Nat. Med.* 16(4):400–402, 2010.

Feig, D. I., Kang, D.-H., and Johnson, R. J., Uric Acid and Cardiovascular Risk, *N. Engl. J. Med.* 359:1811–1821, 2008.

Johnson, R. J., *et al.*, Hypothesis: Could Excessive Fructose Intake and Uric Acid Cause Type 2 Diabetes? *Endocr. Rev.* 30(1):96–116, 2009.

Laplante, M., and Sabatini, D. M., mTOR Signaling at a Glance, *J. Cell Sci.*, 122:3589–3594, 2009.

Paneni, F., *et al.*, Diabetes and Vascular Disease: Pathophysiology, Clinical Consequences, and Medical Therapy, Part I, *Eur. Heart J.* 34(31):2436–2443, 2013.

Schwartz, M. W., *et al.*, Cooperation between Brain and Islet in Glucose Homeostasis and Diabetes, *Nature* 503:59–66, 2013.

Trivedi, B. P., The Bypass Cure, *Discover* 33(10):52–60, 2012.

Woods, S. C., Seeley, R. J., and Cota, D., Regulation of Food Intake through Hypothalamic Signaling Networks Involving mTOR, *Annu. Rev. Nutr.* 28:295–311, 2008.

Key Words

Review Questions

These questions are designed to test your knowledge of the key concepts discussed in this chapter, before moving on to the next chapter. You may like to compare your answers to the solutions provided in the back of the book and in the accompanying Study Guide.

1. Define the following terms:
   a. second messenger
   b. desensitization
   c. target cell
   d. insulin resistance
   e. adenylate cyclase
2. Define the following terms:
   a. ketosis
   b. ketoacidosis
   c. osmotic diuresis
   d. GLUT4
   e. body mass index
3. Define the following terms:
   a. G protein
   b. GPCR
   c. RTK
   d. growth factor
   e. cytokine
4. Define the following terms:
   a. hyperinsulinemia
   b. dyslipidemia
   c. hyperglycemia
   d. glucosuria
   e. hyperosmolar hyperglycemic nonketosis
5. Define the following terms:
   a. insulin-like growth factor
   b. interferon
   c. interleukin
   d. hormone response element
   e. histocompatability antigen
6. Define the following terms:
   a. metabolic syndrome
   b. hyperurisemia
   c. hypothalamus
   d. anorexigenic
   e. orexigenic
7. Define the following terms:
   a. postprandial
   b. postabsorptive
   c. ARC
   d. NPY
   e. POMC
8. Define the following terms:
   a. mitogen
   b. phorbol ester
   c. enteric
   d. SUA
   e. endocrine
9. Define the following terms:
   a. tumor promoter
   b. guanine nucleotide exchange factor
   c. DAG
   d. steady state
   e. IP_3
10. Which organ carries out each of the following activities?
    a. urea synthesis
    b. gluconeogenesis
    c. nutrient absorption
    d. neural and endocrine integration
    e. lipogenesis
11. State the action of each of the following hormones:
    a. triiodothyronine
    b. insulin
    c. glucagon
    d. growth hormone
    e. cholecystokinin
12. NADH is an important reducing agent in cellular catabolism, whereas NADPH is an important reducing agent in anabolism. Review previous chapters and show how the synthesis and degradation of these two molecules are interconnected.
13. Briefly discuss the major classes of second messenger that are now recognized.
14. How do phorbol esters promote tumor growth?
15. After about 6 weeks of fasting, the production of urea is decreased. Explain.
16. Extreme thirst is a characteristic symptom of diabetes. Explain.
17. State the action(s) of each of the following signal molecules:
    a. vasopressin
    b. PYY
    c. leptin
    d. ghrelin
    e. adiponectin
18. During periods of prolonged exercise, muscles burn fat released from adipocytes in addition to glucose. Explain how the need for additional fatty acids by muscle is communicated to the adipocytes.
19. Bodybuilders often take anabolic steroids to increase their muscle mass. How do these steroids achieve this effect? (Common side effects of anabolic steroid abuse include heart failure, violent behavior, and liver cancer.)
20. State the action(s) of each of the following signal molecules:
    a. IGF-1
    b. PDGF
    c. epidermal growth factor
    d. IRS-1
    e. tumor necrosis factor
21. In Alzheimer's disease, nerve cell death is associated with the accumulation of aggregates of misfolded protein. Compare this process with the onset of diabetes mellitus.
22. During periods of fasting, some muscle protein is depleted. How is this process initiated, and what happens to the amino acids in these proteins?
23. The kidney has an unusually large demand for glutamine and glutamate. How does the metabolism of these compounds help maintain pH balance?
24. Hemoglobin molecules exposed to high levels of glucose are converted to glycated products. The most common, referred

to as hemoglobin A_{1C} (HbA_{1C}), contains a β-chain glycated adduct. Because red blood cells last about 3 months, HbA_{1C} concentration is a useful measure of a patient's blood sugar control. In general terms, describe why and how HbA_{1C} forms.

25. State the action of each of the following signal molecules:
    a. mTOR
    b. SREBP-1C
    c. PDK1
    d. TSH
    e. interleukin
26. What are the most common sites on proteins that are phosphorylated during signal transduction cascades?
27. Ketoacidosis is a common feature of insulin-dependent diabetes mellitus, but not of insulin-independent diabetes mellitus. Explain.
28. Type 2 diabetics are often obese. Explain how obesity contributes to the onset of diabetes.
29. Because the prime early symptom of diabetes is a high level of blood glucose, insulin is often associated primarily with carbohydrate metabolism. List several other processes that are insulin-dependent.
30. During the first week of a prolonged diet there is a relatively rapid weight loss. In addition to water loss, what other factor contributes to this phenomenon?
31. Hormones can be assigned to what three general classes based on the molecules they are derived from?
32. List three examples of steroid transport proteins.
33. Describe two functions concerned with nutrient metabolism for each of the following organs:
    a. intestine
    b. liver
    c. muscle
    d. adipose tissue
    e. kidney
    f. brain

Fill in the Blank

34. The reduction of cell-surface receptors in response to stimulation by a specific hormone molecule is called ____________________.
35. The two major types of cell-surface receptors that bind to hormone molecules are G-protein linked receptors and ____________________.
36. A group of molecules called ____________________ mediate the action of many hormones.
37. ____________________ is a peptide hormone released from heart atrial cells in response to increased blood volume.
38. Type 1 diabetes is caused by the destruction of ____________________.
39. Type 2 diabetes is caused by an insensitivity to ____________________.
40. Glucose in the urine is called ____________________.
41. Excessive loss of water and electrolytes in diabetes is called ____________________.
42. The most serious symptom of type 1 diabetes is ____________________.
43. Elevated concentration of ketones in the blood is called ____________________.

Short Answer

44. The binding of insulin to its receptor on the surface of a target cell activates the receptor's tyrosine kinase activity, which in turn causes several phosphorylation cascades that alter the activity of numerous enzymes and alters the expression of genes. Describe how one of these phosphorylation cascades stimulates glycogen synthesis.
45. Why is a diet dominated by fructose consumption a major contributing factor for cardiovascular disease?
46. What do leptin deficiency and Prader-Willi syndrome have in common?
47. Adiponectin, a hormone secreted by adipose tissue cells, suppresses several metabolic derangements that are associated with metabolic syndrome. For example, low adiponectin levels correlate with obesity. Provide another reason for linking adiponectin to metabolic syndrome.
48. Peptide YY (PYY) is a 36-amino-acid peptide that is so named because it contains two tyrosines. (The one-letter symbol for tyrosine is Y. See **Table 5.1**.) Describe the function of PYY and where it is synthesized.

Thought Questions

These questions are designed to reinforce your understanding of all of the key concepts discussed in the book so far, including this chapter and all of the chapters before it. They may not have one right answer! The authors have provided possible solutions to these questions in the back of the book and in the accompanying Study Guide, for your reference.

49. In severely diabetic patients, the blood glucose level is so high that it appears in the urine. Before the development of blood tests by modern medical research, diabetics could often be recognized by the appearance of flies around their feet. Suggest a reason for this observation.
50. Explain why obese individuals are often insulin-resistant.
51. Explain how a second messenger works. Why use a second messenger rather than simply relying on the original hormone to produce the desired effect?

52. Dieters frequently fast in an attempt to reduce their weight. During these fasts, they often lose considerable muscle mass rather than fat. Why is this so?
53. You are being stalked by a large tiger. Explain how your metabolism responds to help you escape.
54. During fasting in humans, virtually all the glucose reserves are consumed in the first day. The brain requires glucose to function and adjusts only slowly to other energy sources. Explain how the body supplies the glucose required by the brain.
55. In uncontrolled diabetes, levels of hydrogen ions are elevated. Explain how these ions are generated.
56. Why is it important for hormones to act at low concentrations and be degraded quickly?
57. Hormones are often synthesized and stored in an inactive form within secretory vesicles. Secretion usually occurs only when the hormone-producing cell is stimulated. Explain the advantages that this process has over making the hormone molecules as they are needed.
58. Steroid hormones are often present in cells in low concentrations. This makes them difficult to isolate and identify. It is sometimes easier to isolate the proteins to which they bind by using affinity chromatography. (Refer to the Biochemistry in the Lab box entitled Protein Technology in Chapter 5, pp. 180–185). Explain how you would use this technique to isolate a protein suspected of steroid hormone binding.
59. Skeletal muscle cannot synthesize fatty acids, yet it produces the enzyme acetyl-CoA carboxylase. Explain the role of this enzyme.
60. Long-term weight loss, as the result of dieting (calorie restriction), has a failure rate of about 95%. Review the basic principles of systems biology and explain how the body resists conscious efforts to lose weight.
61. Animals can convert glucose to fat, but not the reverse. Explain.
62. In controlled clinical trials the consumption of soft drinks that contain artificial sweeteners was observed to result in an increased appetite for carbohydrate-containing food. Based on your knowledge of appetite control by the brain, speculate as to why this phenomenon occurs.
63. Describe the relationship between AMPK and mTORC1.
64. Describe the effects of starvation on urea production.
65. Describe the effect of leptin on feeding behavior and weight control.
66. The binding of small amounts of hormones to target cell receptors triggers an intricate signal cascade. Why is the signal cascade necessary? Why not just have a simple molecular mechanism between the hormone and cellular effect?
67. Several hormones may activate the same G protein. Therefore, different hormones may have the same effect. For example, glycogen degradation is initiated by both epinephrine and glucagon. Why is overlap of function an advantage?
68. As a result of social and economic changes after World War II, the Pima Indians of Arizona began to adopt a "Westernized" lifestyle that included high-calorie diets and sedentary occupations. Soon afterward, obesity and type 2 diabetes became common. (A genetically related group, Pima Indians in Mexico, who live in remote mountain villages and maintain the diet and activity levels of their ancestors, have a low rate of diabetes and obesity.) Researchers have discovered that IRS1 levels are reduced in obese individuals from the Arizona Pima population, as compared with lean individuals. It was also shown that specific mutations in the IRS1 gene in combination with obesity led to a 50% reduction in insulin sensitivity. Explain this phenomenon.
69. Pima Indians are encouraged to exercise regularly to delay the onset of diabetes and/or improve diabetic symptoms. What impact does vigorous physical exercise have on their health?
70. The following are the energy reserves of the body: blood glucose, liver glycogen, muscle glycogen, adipose fat, and muscle protein. Comment on their importance in the body's energy economy, their calorie content, and their relative levels in an average body.
71. The fat store of a normal 150-pound (about 68 kg) man is about 1.5 kg or 141,000 cal. Assuming that during a prolonged fast such a person "burns" 2000 cal/day, determine how long this reserve will last.
72. Under starvation conditions, death usually occurs before the body's energy reserves are totally exhausted. Suggest a reason for this phenomenon.
73. Hyperinsulinism, the result of the injection of an excessive insulin dose or an insulin-secreting tumor, can result in brain damage. Explain.
74. Explain the rationale that links hypertension to excess fructose consumption.
75. Describe the effect of excess fructose consumption on hepatic lipid metabolism.

CHAPTER

17 Nucleic Acids

Genetic Inheritance The traits of human parents, such as eye and hair color, are inherited by their children. DNA is the biomolecule that transmits genetic information from one generation to the next.

OUTLINE

What Makes Us Human?

Chimpanzees (*Pan troglodytes*) are our closest living relatives. In addition to obvious similarities between humans and chimpanzees in anatomy, physiology, and social behavior, we share an astonishing 99.5% of our genome with them (after taking inserted and deleted DNA sequences into account). And yet humans differ in significant ways from chimpanzees. In the 6 million years since the last common ancestor of chimpanzees and humans lived, we have evolved such uniquely human traits as bipedalism (upright walking), opposable thumbs, and the cognitive power and flexibility that make tool making, abstract thought, complex language, and art possible.

In 2005 biostatisticians began the process of determining exactly how we differ from chimpanzees. Equipped with complete chimpanzee and human genomes and powerful clustered computers, they compared the DNA sequences of both species and ascertained which sequences are exclusively human (altogether about 15 million bases). Although there are some sequence changes in protein-coding genes, most are found in noncoding DNA, formerly referred to as "junk DNA." Acting as genetic switches, these fast-evolving human regulatory sequences are involved in the activation or repression of nearby genes. Instead of having a lot of new protein-coding genes, humans regulate their genes somewhat differently than other primates. Not surprisingly, many human genetic switch sequences are active in the brain. Specific human changes have also occurred as copy number variations (CNVs), or changes in the number of copies of a specific DNA sequence. CNVs are often the result of deletions or duplications. Among the most distinctive human sequences are the following:

HARs

HARs (human accelerated regions) are DNA sequences that are highly conserved in vertebrates, but have changed significantly during human evolution. Of the 202 HARs that have been identified, HAR1 and HAR2 are the most notable. HAR1 codes for a 106 RNA-base sequence, which differs from the chimpanzee sequence by 18 bases; It is located on chromosome 20 with two overlapping ncRNA genes, HAR1A and HAR1B, which are also known as HACNS1. HAR1A is active in the fetal and adult brain where it is involved in cerebral cortex development. With 12 human-specific base substitutions in its 119-base sequence, HAR2 codes for a human-specific developmental enhancer sequence that contributes to human wrist and thumb formation. (An *enhancer* is a short DNA sequence that can increase the transcription of one or more specific genes.)

ASPM

ASPM (*a*bnormal *s*pindle-like *p*rotein *m*icrocephaly-associated) is a gene, located on chromosome 1, that codes for a protein that contributes to enlarged brain size. When human ASPM is mutated, babies are born with very small heads, a disorder called microcephaly.

FOXP2

The FOXP2 (*fo*rkhead bo*x* protein P2) gene, located on chromosome 7, codes for a highly conserved transcription factor that is most active in brain and lung. In the developing human brain, FOXP2 regulates the expression of genes involved in several aspects of language and speech production. These include the motor control of the mouth and tongue that makes human vocalization possible and cognitive skills including those required for language comprehension (i.e., processing of words according to grammatical and syntax rules). When compared with the chimpanzee version, human FOXP2 has two specific amino acid changes, which allow the transcription factor to play a key role in synaptic plasticity. The brain regions most affected are the *basal ganglia*, which is associated with motor control and learning, and the *inferior frontal cortex*, which contains *Broca's area*, linked to speech production. (*Synaptic plasticity*, required in such functions as learning and memory, is the capacity to change the strength of a synaptic connection, often by altering the amount of neurotransmitter molecules that are released.) Mutations in human FOXP2 are exceedingly rare. People with this anomaly are not only incapable of intelligible speech; they also have cognitive deficits with lower verbal and nonverbal IQ values.

What Makes Us Human? cont.

AMY1

AMY1 on chromosome 1 codes for *salivary amylase*, the enzyme that initiates starch degradation in the mouth. Although many mammals have multiple copies, humans have an especially large number (an average of seven copies). Consequently, starch is a large component in the human diet. With only two copies, chimpanzees consume a low-starch diet.

LCT

The LCT gene located on chromosome 2 codes for the enzyme lactase that hydrolyzes lactose to yield glucose and galactose. Mammals synthesize lactase so that newborn animals can digest milk sugar until weaning when the enzyme is no longer produced. Some humans whose ancestors lived in select areas in Europe and Africa in which cattle were domesticated possess a trait, called *lactose persistence*, that allows them to produce lactase throughout life. The single base change mutations in both human populations that allowed continuous lactase synthesis occurred in a regulatory sequence embedded in a nearby protein-coding gene.

As researchers continue to identify and characterize other DNA sequence differences between our closest relatives and ourselves, this work will reveal the genetic events that have shaped our species.

Overview

THE NUCLEIC ACIDS DNA AND RNA ARE POLYNUCLEOTIDES THAT ENCODE THE GENETIC INFORMATION USED TO CONSTRUCT AND MAINTAIN LIVING organisms. Double-stranded DNA is, in effect, the blueprint used to direct cell processes. Cells then convert DNA's operating instructions into the nucleotide sequence of single-stranded RNA molecules. RNAs have numerous functions, which include polypeptide synthesis, regulation of gene expression (control of when or if a specific polypeptide is synthesized), and protection from foreign nucleic acids introduced by viral infections. Investigations of nucleic acid structure and function, now more than 60 years old, have given humans a previously unimagined understanding of biological processes and a powerful tool used in such diverse fields as disease diagnosis and treatment and forensic investigations.

For countless centuries, humans have observed inheritance patterns without understanding the mechanisms that transmit physical traits and developmental processes from parent to offspring. Many human cultures have used such observations to improve their economic status, as in the breeding of domesticated animals or seed crops. It was not until the nineteenth century that the scientific investigation of inheritance, now referred to as **genetics**, began. By the beginning of the twentieth century, scientists generally recognized that physical traits are inherited as discrete units (later called "genes") and that chromosomes within the nucleus are the repositories of genetic information. Eventually, the chemical composition of chromosomes was elucidated and, after many decades of investigation, deoxyribonucleic acid (DNA) was identified as the genetic information. In the decades that followed the 1953 discovery of DNA structure by James Watson and Francis Crick (**Figure 17.1**), a new science emerged. (Refer to the online Biochemistry in

FIGURE 17.1

The First Complete Structural Model of DNA

When James Watson (left) and Francis Crick discovered the structure of DNA in 1953 using Rosalind Franklin's data, they were research students at the Henry Cavendish Laboratory of Cambridge University.

Perspective essay A Short History of DNA Research: The Early Years for a brief overview of the work of other scientists that led to the discovery of DNA structure by Watson and Crick.) **Molecular biology** is devoted to the investigation of gene structure and genetic information processing. Using the technologies developed by molecular biologists and biochemists, life scientists have studied the processes by which living organisms organize and process genetic information. This work has revealed the following principles.

1. DNA directs the functioning of living cells and is transmitted to offspring. DNA is composed of two polydeoxynucleotide strands that form a double helix (**Figure 17.2**). The information in DNA is encoded in the form of the sequence of purine and pyrimidine bases (Refer to **Figure 14.23**). A **gene** is a DNA sequence that contains the base sequence information necessary to code for a gene product (a polypeptide or several types of RNA molecules) and regulatory sequences that control the production of the gene. The complete DNA base sequence in an organism is referred to as the **genome**. DNA synthesis, referred to as **replication**, involves the complementary pairing of purine and pyrimidine bases between the old parental strand and the newly synthesized strand. The physiological and genetic function of DNA requires the synthesis of error-free copies. Consequently, most organisms employ several DNA repair mechanisms.
2. The mechanism by which genetic information is decoded and used to direct cellular processes begins with the synthesis of another type of nucleic acid, ribonucleic acid (RNA). Its synthesis, referred to as **transcription** (**Figure 17.3a**), involves the complementary pairing of ribonucleotide bases with the bases in a DNA molecule. Each newly synthesized RNA molecule is called a **transcript**. The term **transcriptome** designates the complete set of RNA molecules that are transcribed from a cell's genome.

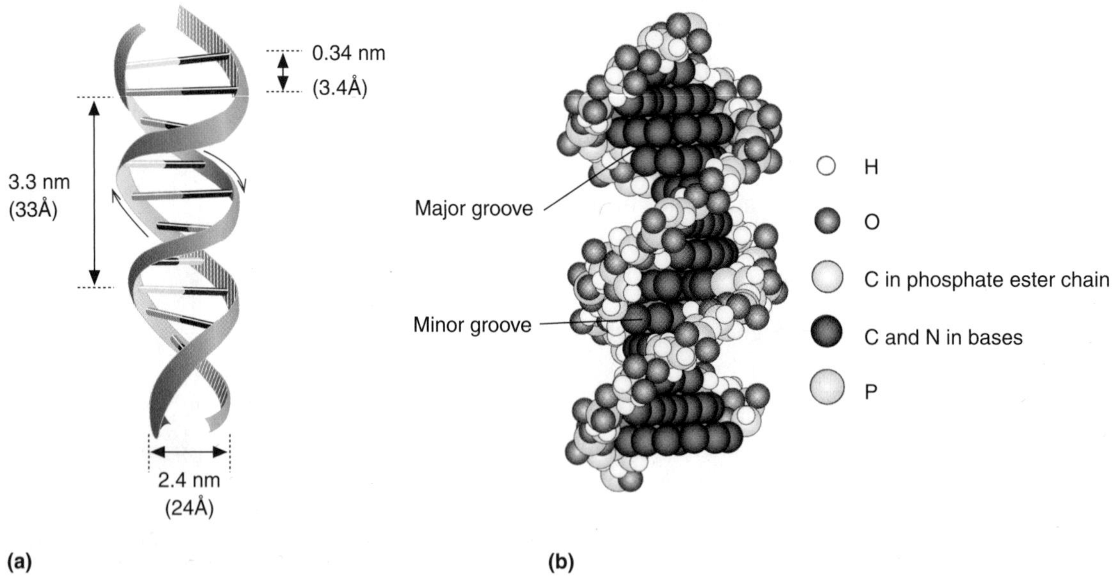

FIGURE 17.2

Two Models of DNA Structure

(a) The DNA double helix is represented as a spiral ladder; this is the conformation originally proposed by Watson and Crick and now known as B-DNA. (For the structural properties of three forms of DNA, see **Table 17.1**, later.) The sides of the spiral ladder represent the sugar-phosphate backbones. The rungs represent the base pairs. (b) In a space-filling model, the sugar-phosphate backbones are represented by colored spheres. The base pairs consist of horizontal arrangements of dark blue spheres. Wide and narrow grooves are created by twisting the two strands around each other in a right-handed sense.

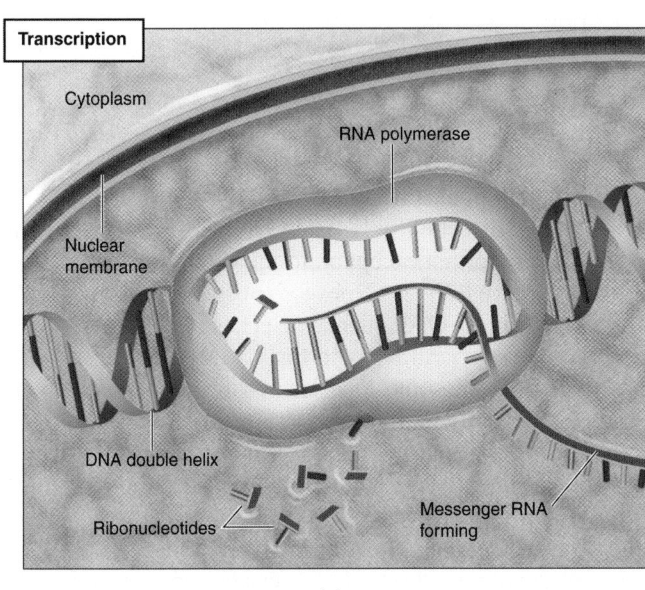

(a)

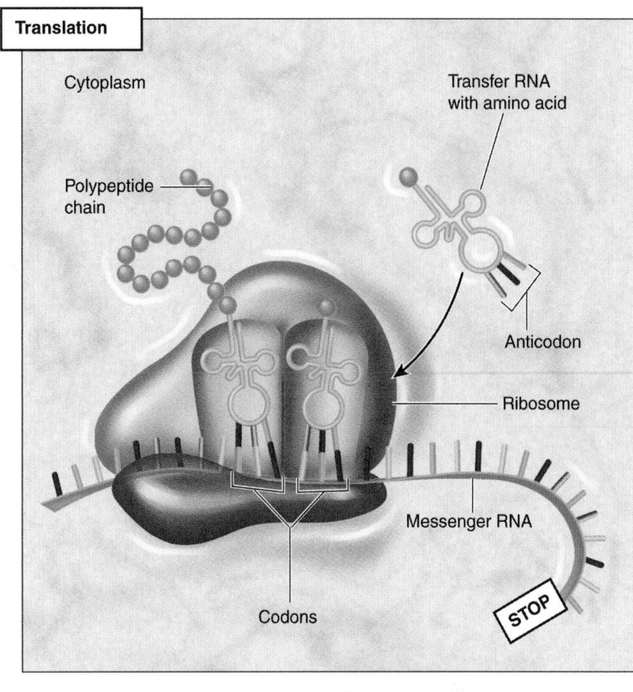

(b)

FIGURE 17.3

An Overview of Genetic Information Flow

The genetic information in DNA is converted into the linear sequence of amino acids in polypeptides in a two-phase process. During *transcription* (a), RNA molecules are synthesized from a DNA strand through complementary base pairing between the bases in DNA and the bases in free ribonucleoside triphosphate molecules. During the second phase, called *translation* (b), mRNA molecules bind to ribosomes that are composed of rRNA and ribosomal proteins. Transfer RNA–aminoacyl complexes position their amino acid cargo in the catalytic site within the ribosome in a process that involves complementary base pairing between the mRNA base triplets called codons and tRNA base triplets called anticodons. When the amino acids are correctly positioned within the catalytic site, a peptide bond is formed. After the mRNA molecule moves relative to the ribosome, a new codon enters the ribosome's catalytic site and base-pairs with the appropriate anticodon on another aminoacyl-tRNA complex. As a stop codon in the mRNA enters the catalytic site, the newly formed polypeptide is released from the ribosome.

3. Several types of RNA participate directly in the synthesis of the enzymes and other proteins required for the regulated manufacture of all other biomolecules needed in organismal function. The base sequence of each messenger RNA (mRNA) specifies the primary sequence of a specific polypeptide. Ribosomal RNA (rRNA) molecules are components of the ribosomes. Each transfer RNA (tRNA) molecule is covalently bound to a specific amino acid and delivers it to the ribosome for incorporation into a polypeptide chain. Protein synthesis, called **translation** (**Figure 17.3b**), occurs within ribosomes, the ribonucleoprotein molecular machines that translate the base sequences of mRNAs into the amino acid sequences of polypeptides. The entire set of proteins synthesized by a cell is referred to as the **proteome**.
4. **Gene expression** is the set of mechanisms whereby cells control the timing of gene product synthesis in response to environmental or developmental cues. A vast array of proteins, called **transcription factors**, and RNA molecules, called *noncoding RNA molecules* (ncRNAs), regulate gene expression when they bind to specific DNA sequences. The term **metabolome** refers to the sum total of all the low-molecular-weight metabolite molecules produced by a cell as the result of its gene expression pattern.

The flow of genetic information can be summarized by a sequence called the *central dogma*:

This diagram illustrates that the genetic information encoded in DNA base sequences flows from DNA, a molecule that is replicated during cell division (indicated by the arrow encircling DNA), to RNA, which specifies the primary structure of proteins. As originally conceived, the central dogma asserted that genetic information flows in one direction only: from DNA to RNA to protein. Several years ago, however, an important exception to the central dogma was revealed. Some of the viruses that have RNA genomes also possess an enzyme activity referred to as *reverse transcriptase*. Once such a virus has infected a host cell, the reverse transcriptase acts to copy viral RNA to form a DNA copy. The viral DNA is then inserted into a host chromosome. One such virus is HIV, discussed in the Biochemistry in Perspective essay HIV Infection.

Chapter 17 focuses on the structure of the nucleic acids. It begins with a description of DNA structure and how that structure can be altered by mutations. This is followed by a discussion of current knowledge of genome and chromosome structure, as well as the structure and roles of the several forms of RNA. A Biochemistry in Perspective essay provides a brief overview of epigenetics, the covalent modification of DNA that adds another layer of gene regulation and inheritance but is not bound by the Mendelian laws of inheritance. Chapter 17 ends with the description of viruses, macromolecular complexes composed of nucleic acid and proteins that are cellular parasites. Chapter 18 discusses several aspects of nucleic acid synthesis and function (i.e., DNA replication and transcription). Protein synthesis (translation) is described in Chapter 19. The strategies and techniques that are routinely used to isolate, purify, characterize, and manipulate nucleic acids are described in Biochemistry in the Lab boxes in Chapters 17 and 18 (Nucleic Acid Methods, pp. 644–648, and Genomics, pp. 690–696).

17.1 DNA

DNA consists of two polydeoxynucleotide strands that wind around each other to form a right-handed double helix (**Figure 17.2**). The structure of DNA is so distinctive that this molecule is often referred to as *the double helix*. As described earlier (Sections 1.3 and 14.3), each nucleotide monomer in DNA is composed of a nitrogenous base (either a purine or a pyrimidine), a deoxyribose sugar, and phosphate. The mononucleotides are linked to each other by 3′,5′-phosphodiester bonds. These bonds join the 5′-hydroxyl group of the deoxyribose of one nucleotide to the 3′-hydroxyl group of the sugar unit of another nucleotide through a phosphate group (**Figure 17.4**). The antiparallel orientation of the two polynucleotide strands allows hydrogen bonds to form between the nitrogenous bases that are oriented toward the helix interior (**Figure 17.5**). There are two types of base pair (bp) in DNA: (1) adenine (a purine) pairs with thymine (a pyrimidine) (AT pair), and (2) the purine guanine pairs with the pyrimidine cytosine (GC pair). The overall structure of DNA resembles a twisted staircase because each base pair is oriented at an angle to the long axis of the helix. The average dimensions of crystalline B-DNA have been measured.

1. One turn of the double helix spans 3.32 nm and consists of approximately 10.3 base pairs. (Changes in pH and salt concentrations affect these values slightly.)
2. The diameter of the double helix is 2.37 nm. Note that the interior space of the double helix is only suitable for base pairing a purine and a pyrimidine.

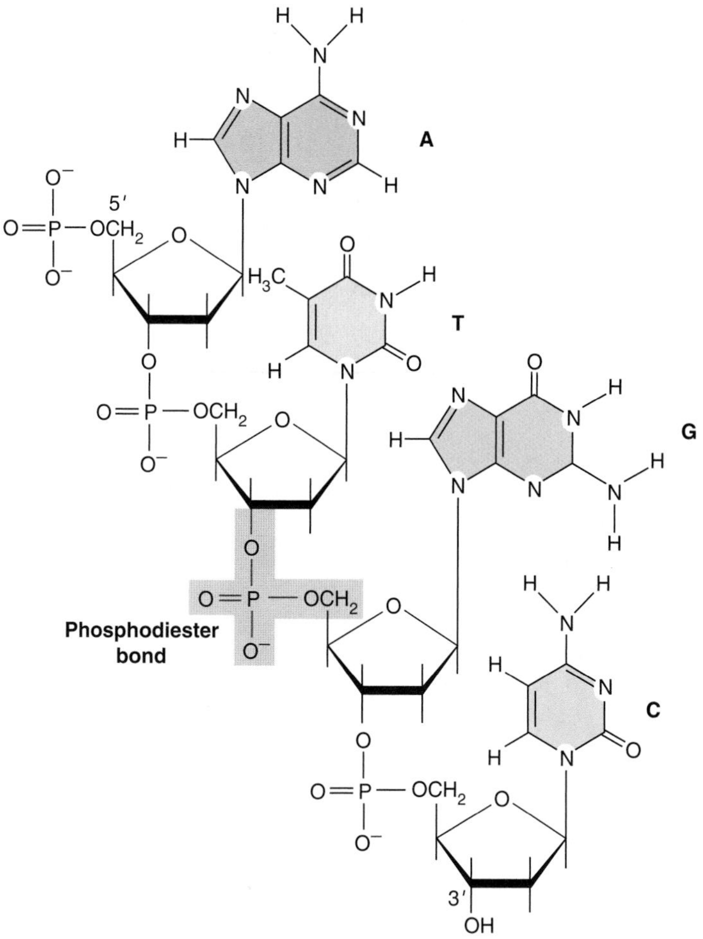

FIGURE 17.4

DNA Strand Structure

In each DNA strand the deoxyribonucleotide residues are connected to each other by 3′, 5′-phosphodiester linkage. The sequence of the strand section illustrated in this figure is 5′-ATGC-3′. Refer to p. 539 for the numbering system of the atoms in nucleotides. (Note that in sugar molecules, the hydrogens bonded to carbon can be represented by single lines.)

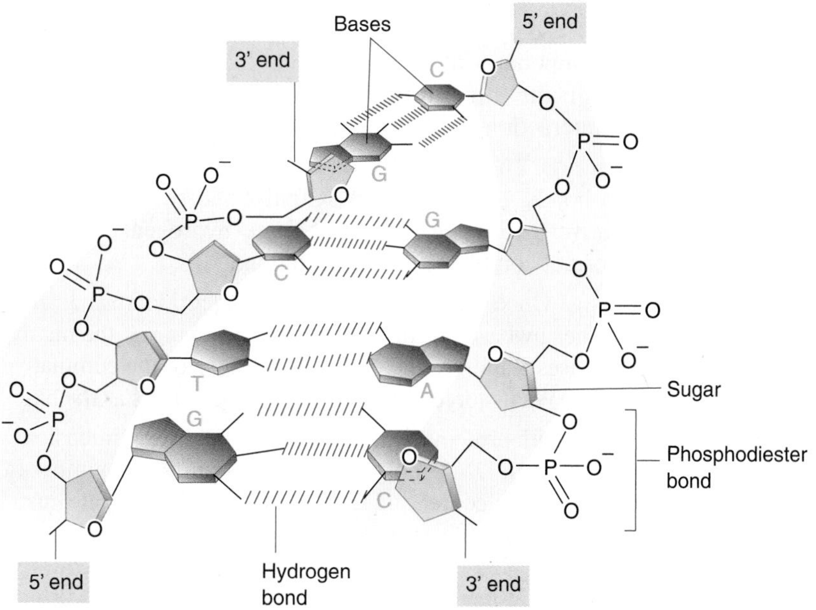

FIGURE 17.5

DNA Structure

In this short segment of DNA, the bases are shown in orange and the sugars are blue. Each base pair is held together by either two or three hydrogen bonds. The two polynucleotide strands are antiparallel. Because of base pairing, the order of bases in one strand determines the order of bases along the other.

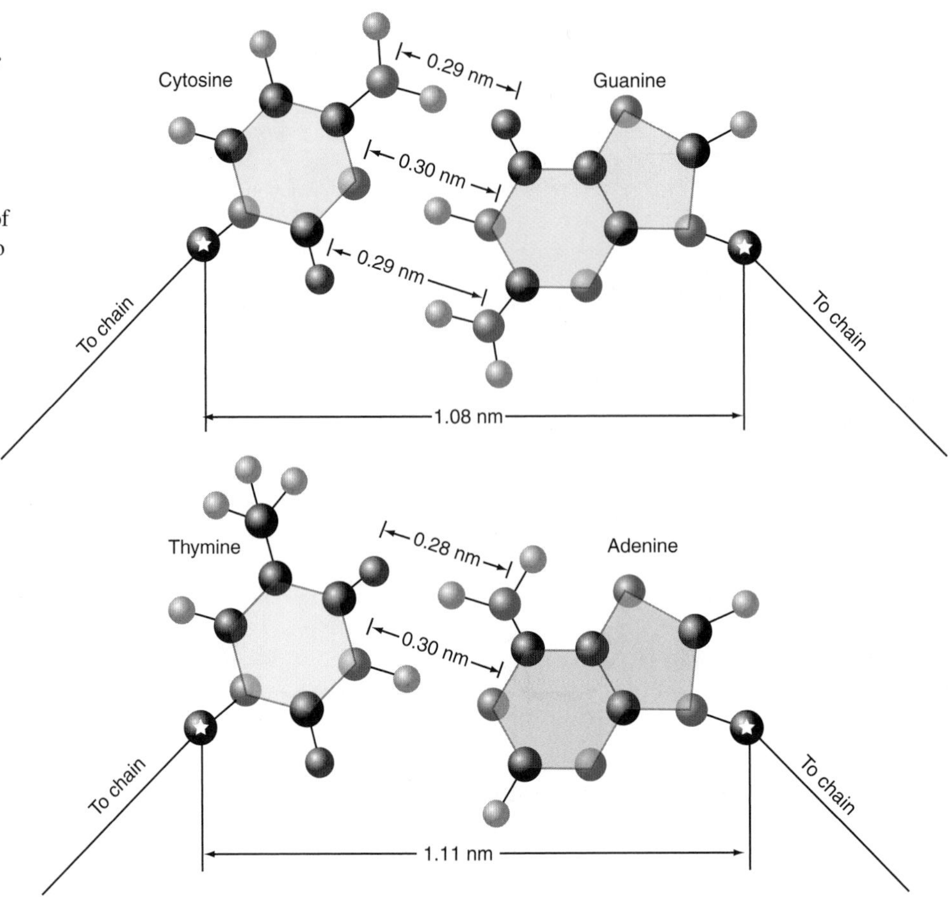

FIGURE 17.6
DNA Structure: AT and GC Base Pair Dimensions

Two hydrogen bonds are formed in each AT base pair and three in each GC base pair. The near equal dimensions of base pairs of both types allow the formation of uniform helical conformations of the two polynucleotide strands.

Pairing two pyrimidines would create a gap, and two purines would not fit in the interior space of the double helix. The relative dimensions of both types of base pair are illustrated in **Figure 17.6**.

As befits its genetic information storage role in living processes, DNA is a relatively stable molecule. Several types of noncovalent interactions contribute to the stability of its helical structure.

1. **Hydrophobic interactions**. The base ring π cloud of electrons between stacked purine and pyrimidine bases is relatively nonpolar. The clustering of the base components of nucleotides within the double helix is a stabilizing factor in the three-dimensional macromolecule because it minimizes their interactions with water, thereby increasing overall entropy.
2. **Hydrogen bonds**. The base pairs, on close approach, form a preferred set of hydrogen bonds, three between GC pairs and two between AT pairs. The cumulative "zippering" effect of these hydrogen bonds keeps the strands in correct complementary orientation.
3. **Base stacking**. Once the antiparallel polynucleotide strands have been brought together by base pairing, the parallel stacking of the nearly planar heterocyclic bases stabilizes the molecule because of the cumulative effect of weak van der Waals forces generated by π cloud shifts as the bases stack.
4. **Hydration**. As with proteins, water stabilizes the three-dimensional structure of nucleic acids. DNA molecules bind a significant number of water molecules. The water content of B-DNA, the conformation illustrated in **Figure 17.2**, is about 30% by weight. Water molecules bind to phosphate groups, ribose 3′- and 5′-oxygen atoms, and electronegative atoms in the nucleotide bases. When measured under laboratory conditions, each

nucleotide in B-DNA binds about 18 to 19 water molecules. Each phosphate can bind a maximum of 6 water molecules.

5. **Electrostatic interactions**. DNA's external surface, referred to as the *sugar-phosphate backbone*, possesses negatively charged phosphate groups. Repulsion between nearby phosphate groups, a potentially destabilizing force, is minimized by the shielding effects of water and divalent cations such as Mg^{2+} and polycationic molecules such as the polyamines (p. 631) and histones (see pp. 631–632).

KEY CONCEPT

DNA is a relatively stable molecule composed of two antiparallel polynucleotide strands wound around each other to form a right-handed double helix.

WORKED PROBLEM 17.1

Deoxynucleotides (deoxynucleoside triphosphates) are the substrates for DNA synthesis. Calculate the number of glucose molecules that must be degraded to generate the energy required for the synthesis of 1000 deoxynucleotides used in DNA synthesis. Assume that glucose is oxidized completely to CO_2 and H_2O, the malate-aspartate shuttle is in operation, and NADPH is the equivalent of 4 ATP.

SOLUTION

The biosynthesis of each deoxynucleotide requires two reactions. Ribonucleoside diphosphate molecules containing the bases guanine, thymine, and cytosine are converted to their corresponding nucleoside triphosphates by ATP-requiring nucleoside kinase (p. 542). Each ribonucleoside triphosphate is then converted to a dexoyribonucleoside triphosphate in an NADPH-requiring reaction catalyzed by ribonucleotide reductase (p. 545). Because each NADPH is the equivalent of 4 ATP, the total number of ATPs required to synthesize a deoxyribonucleoside triphosphate is 5 ATP. As a result, the cost of synthesizing 1000 deoxyribonucleotides (1000 × 5) is 5000. Because each completely oxidized glucose molecule yields 31 ATPs, the total number of glucose molecules required to synthesize 5000 nucleotides is approximately 161 (5000/31). ■

QUESTION 17.1

When DNA is heated, it denatures; that is, the strands separate because hydrogen bonds are broken and some base-stacking and hydrophobic interactions are disrupted. The higher the temperature, the larger the number of hydrogen bonds that are broken. After reviewing DNA base pair structure, determine which of the following molecules will denature first as the temperature is raised. Explain your reasoning.

a. 5′-GCATTTCGGCGCGTTA-3′
   3′-CGTAAAGCCGCGCAAT-5′
b. 5′-ATTGCGCTTATATGCT-3′
   3′-TAACGCGAATATACGA-5′

DNA Structure: The Nature of Mutation

Although its structure makes DNA eminently suited for information storage, it is not a static molecule. DNA is vulnerable to several types of disruptive forces that can cause mutations, permanent structural base sequence changes. Mutations range from small-scale alterations (e.g., single base changes) to large-scale chromosomal abnormalities. Although most mutations are either deleterious or neutral (having no discernible effect on an organism's fitness), on rare occasions a mutation can enhance the adaptation of an organism to its environment. In other words, mutational change is the raw material of evolution. It should also be noted that the mutation rates of most organisms (mutation frequency per cell division or generation) are usually low because of two factors: the accuracy of the DNA replication process (p. 667) and the efficiency of DNA repair (p. 676).

MUTATION TYPES The most commonly observed mutations include single base changes, insertions, deletions, duplications, and genome rearrangements. As their name implies, a single base change involves a change in the identity of a single base in one strand of a DNA sequence. Also referred to as **point mutations**, single base changes can result, if the damage is unrepaired, in transition or transversion mutations. In **transition mutations** (**Figure 17.7**), caused by deamination reactions or tautomerization, a pyrimidine is substituted for another pyrimidine or a purine is substituted for another purine. **Transversion mutations**, caused by alkylating agents or ionizing radiation, occur when a pyrimidine is substituted for a purine or vice versa. Point mutations that occur in a population to any extent are referred to as **single nucleotide polymorphisms**. Point mutations can also result from insertions or deletions of an individual base pair. Point mutations within DNA sequences that code for gene products (e.g., polypeptides) are classified as silent, missense, or nonsense according to their impact on the gene product's structure and/or function. In **silent mutations**, a base change has no discernible effect (e.g., coding for the same or a different amino acid in a polypeptide results in no functional difference), whereas in a **missense mutation**, there is an observable effect (e.g., coding for a different amino acid that causes a change in a polypeptide's structure and function). In a **nonsense mutation**, a point mutation converts the code for an amino acid into a premature stop signal. The polypeptide product of the transcribed sequence containing the nonsense mutation will be incomplete and probably nonfunctional.

Insertions and deletions, referred to as **indels**, are mutations that occur when from one to thousands of bases are either inserted or removed from a DNA

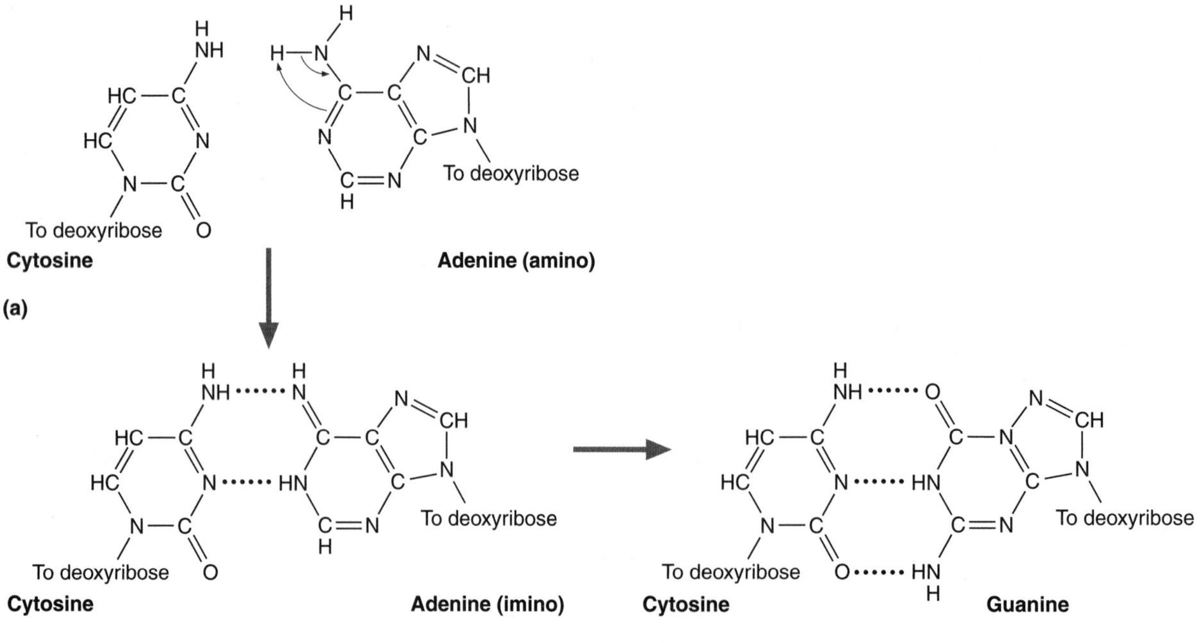

To deoxyribose
Cytosine
To deoxyribose
Adenine (amino)

(a)

To deoxyribose
Cytosine
To deoxyribose
Adenine (imino)

(b)

To deoxyribose
Cytosine
To deoxyribose
Guanine

(c)

FIGURE 17.7

A Tautomeric Shift Causes a Transition Mutation

As adenine (a) undergoes a tautomeric shift, its imino form (b) can base-pair with cytosine. The transition shows up in the second generation of DNA replication when cytosine base-pairs with guanine. In this manner an A-T base pair is replaced by a C-G base pair (c). Refer to pp. 539–541 for a description of tautomeric shifts of purine and pyrimidine bases.

sequence. If the bases that are inserted or deleted into a polypeptide coding sequence are not divisible by 3 (the length of an individual amino acid coding sequence), a *frameshift mutation* (p. 733) can occur that results in either an altered or a truncated polypeptide. Indels are the result of either unequal crossing over of misaligned homologous chromosomes in meiosis or *slipped strand mispairing*, a mistake made during DNA replication in which there is displacement and then mispairing of the DNA strands.

Genome rearrangements such as inversions, translocations, and duplications can cause disruptions in gene structure or regulation. They can result from double-stranded DNA breaks caused by a variety of circumstances including errors in meiosis and exposure to mutagens or radiation. An **inversion** results when a deleted DNA fragment is reinserted into its original position, but in the opposite orientation. **Translocation** is a chromosomal abnormality observed in eukaryotes in which a DNA fragment from one chromosome inserts into a different position on the same chromosome or into a different (nonhomologous) chromosome. **Gene duplication**, the creation of duplicate genes or parts of genes, can result from unequal crossing over in meiosis or from retrotransposition, a process in which genetic elements called retrotransposons (p. 639) insert themselves into a DNA sequence. In rare instances, gene duplication is an important feature in evolution because duplicates can, as the result of mutation over long periods of time, take on different functions.

CAUSES OF DNA DAMAGE DNA damage can result from endogenous and exogenous disruptive forces. Endogenous causes of mutations include spontaneous events such as tautomeric shifts, depurination, deamination, and ROS-induced oxidative damage. Exogenous factors such as radiation and xenobiotic exposure can also be mutagenic.

Tautomeric Shifts and Spontaneous Hydrolytic Reactions. Tautomeric shifts (**Figure 14.24**) are spontaneous changes in nucleotide base structure that result in amino to imino and keto to enol changes in configuration. Usually tautomeric shifts have little effect on overall, three-dimensional DNA structure. However, if tautomers form during DNA replication, base mispairings may result. For example, the imino form of adenine will not base-pair with thymine. Instead, it forms a base pair with cytosine (**Figure 17.7**). If this pairing is not corrected immediately, a transition mutation results because cytosine has been incorporated during the replication process in a position that should carry thymine.

Several spontaneous hydrolytic reactions also cause DNA damage. For example, it has been estimated that several thousand purine bases are lost daily from the DNA in each human cell. In depurination reactions the N-glycosyl linkage between a purine base and deoxyribose is cleaved. The protonation of N-3 and N-7 of guanine promotes hydrolysis. A point mutation results if repair mechanisms do not replace the purine nucleotide. Similarly, bases can be spontaneously deaminated. For example, the deaminated product of cytosine converts to uracil via a tautomeric shift. Eventually, what should be a CG base pair is converted to an AT base pair. (Uracil is similar in structure to thymine.)

Ionizing Radiation. Some types of ionizing radiation (e.g., UV, X-rays, and γ-rays) can alter DNA structure. Low radiation levels may cause mutation; high levels can be lethal. UV light-induced damage caused by a free radical mechanism (either abstraction of hydrogen atoms or the creation of •OH and other ROS) includes strand breaks, DNA-protein cross-linking (e.g., via thymine-tyrosine linkages), ring openings, and base modifications. The hydroxyl radical, formed by the radiolysis of water, as well as oxidative stress is known to cause strand breakage and numerous base modifications (e.g., thymine glycol, 5-hydroxymethyluracil, and 8-hydroxyguanine). The enzyme superoxide dismutase (p. 372) plays a key role in preventing ROS-induced damage to both nuclear and mitochondrial DNA.

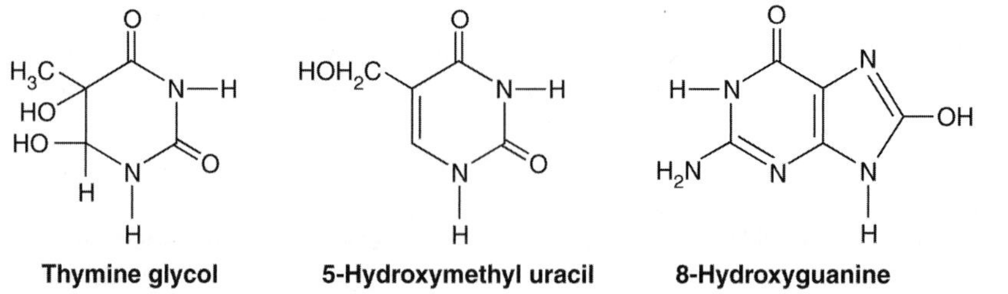

Thymine glycol   5-Hydroxymethyl uracil   8-Hydroxyguanine

8-Hydroxydeoxyguanosine (8-OHdG) levels in urine are used to measure the body's production of unquenched ROS. Smoking, for example, results in increases of 8-OHdG excretion by as much as 50%.

The most common UV-induced products, created by UV-B energy absorption by double bonds, are thymine dimers (**Figure 17.8**). The helix distortion that results from dimer formation stalls DNA replication machinery.

Xenobiotics. A large number of xenobiotics can damage DNA. The most important of these molecules are base analogues, alkylating agents, nonalkylating agents, and intercalating agents. The structures of **base analogues** are so similar to normal nucleotide bases that they can be incorporated into DNA. For example, 5-bromouracil (5-BU) is a base analog of thymine. Because the enol tautomer (p. 539) of 5-BU forms a base pair with guanine, an AT pair can be converted into a GC pair in the next round of replication. **Alkylating agents** are electrophiles that attract molecules that possess an unshared pair of electrons, adding alkyl groups. Adenine and guanine are especially susceptible to alkylation, although thymine and cytosine can also be affected. Alkylated bases often pair incorrectly (e.g., methylguanine with thymine instead of with cytosine), leading to possible transition mutations on subsequent rounds of replication. In the case of methylguanine, a GC pair becomes an AT pair. Transversion mutations may also

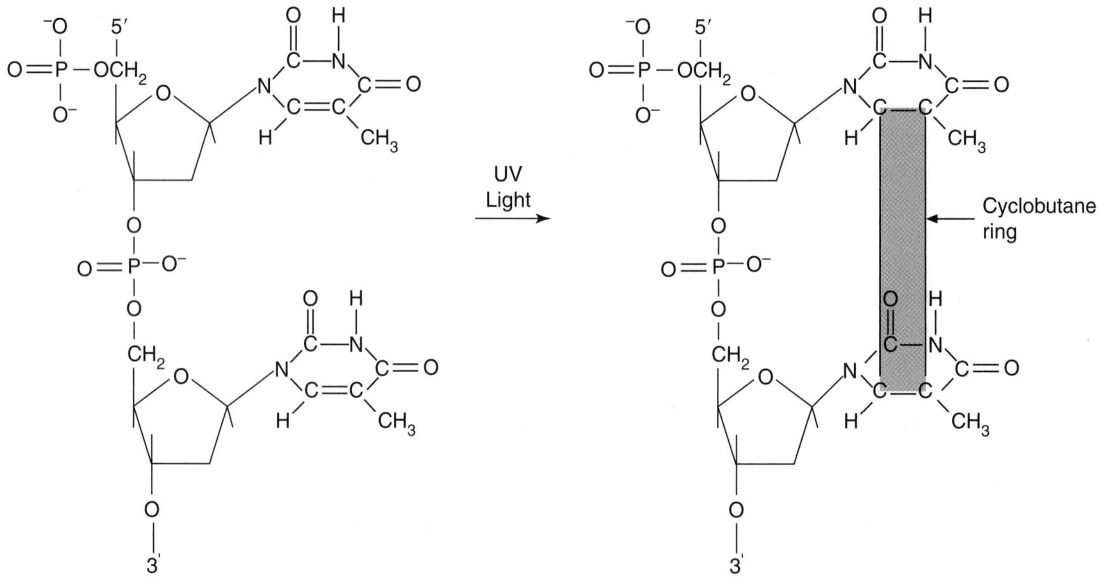

UV Light

Cyclobutane ring

FIGURE 17.8

Thymine Dimer Formation

UV light induces the formation of a covalent linkage in the form of a cyclobutane ring (red) between adjacent thymine bases in a strand of DNA.

occur when the alkylating group is bulky. The polycyclic aromatic hydrocarbon benzo[*a*]pyrene, found in cigarette smoke, causes transversion mutations because it is converted to a highly reactive epoxide derivative by several biotransformation reactions including those catalyzed by cytochrome P_{450} (p. 468). (An *epoxide* is a cyclic ether with three ring atoms.) The benzo[*a*]pyrene epoxide then forms an adduct of guanine, causing distortion of DNA structure that results in G-to-T transversions as well as disrupted DNA replication. Alkylations can also promote tautomer formation, which may result in transition mutations. Examples of alkylating agents include dimethylsulfate and dimethylnitrosamine.

A variety of **nonalkylating agents** can modify DNA structure. Nitrous acid (HNO_2), derived from the nitrosamines and from sodium nitrite ($NaNO_2$), deaminates bases. (Both nitrosamines and $NaNO_2$ are found in processed meats and in any foodstuffs preserved with nitrite pickling salt.) HNO_2 oxidatively deaminates adenine, guanine, and cytosine to hypoxanthine, xanthine, and uracil, respectively. Certain planar polycyclic aromatic molecules are referred to as **intercalating agents** because they can distort DNA by inserting themselves (intercalating) between the stacked base pairs of the double helix. The resulting distortion of local DNA structure can disrupt DNA replication. Depending on the level of exposure, intercalating agents can cause damage that ranges from deletions or insertions to cell death. Doxorubicin and ethidium bromide are examples of intercalating agents. The chemotherapeutic agent doxorubicin inhibits DNA replication in fast-growing cancer cells. Ethidium bromide is a fluorescent tag molecule used as a nucleic acid stain in a variety of molecular biology lab techniques.

Mutations caused by DNA replication machinery errors and transposition (the movement of DNA sequences within a genome) are described in Chapter 18.

KEY CONCEPT

DNA is vulnerable to certain types of disruptive force that can result in mutations, permanent changes in its base sequence.

QUESTION 17.2

How will each of the following substances or conditions affect DNA structure? a. ethanol b. heat c. dimethylsulfate d. nitrous acid e. 5-BU

QUESTION 17.3

The accumulation of oxidative DNA damage now appears to be a major cause of aging in mammals. Animals that have high metabolic rates (i.e., use large amounts of oxygen) or excrete large amounts of modified bases in the urine typically have shorter life spans. The excretion of relatively large amounts of oxidized bases indicates a reduced capacity to prevent oxidative damage. Despite substantial evidence that oxygen radicals damage DNA, the actual radicals that cause the damage are still not clear. Suggest possible culprits in addition to the hydroxyl radical. Some tissues sustain more oxidative damage than others. For example, the human brain is believed to sustain more oxidative damage than most other tissues during an average life span. Suggest two reasons for this phenomenon.

DNA Structure: The Genetic Material

The publication of James Watson and Francis Crick's model of DNA structure in the April 25, 1953, issue of *Nature* was both an end and a beginning. Their work was the culmination of research over the course of nearly a century. Among the most important discoveries were (1) the deduction in 1928 by the British bacteriologist Frederick Griffith that bacteria can transfer genetic information between cells, (2) the identification of Griffith's "transforming factor" as DNA by Avery, MacLeod, and McCarty in 1944, and (3) the experiment by Hershey and Chase in 1952 that confirmed that DNA is the genetic material of living organisms. The publication of the Watson-Crick model marked the beginning of a new field

Visit the companion website to read the Biochemistry in Perspective essay A Short History of DNA Research: The Early Years. This essay describes the critical experiments that led to the discovery of DNA as the genetic material.

called **molecular biology**, which deals with the biosynthesis of DNA, RNA, and proteins and the mechanisms that regulate these processes.

The information used by Watson and Crick to construct their DNA model included the following:

1. The chemical structures and molecular dimensions of deoxyribose, the nitrogenous bases, and phosphate.
2. The 1:1 ratios of adenine to thymine and guanine to cytosine in the DNA isolated from a wide variety of species investigated by Erwin Chargaff between 1948 and 1952. (These 1:1 relationships are sometimes referred to as **Chargaff's rules**.)
3. Superb X-ray diffraction studies performed by Rosalind Franklin (**Figure 17.9**) indicating that DNA is a symmetrical molecule and probably a helix.
4. The diameter and pitch of the helix estimated by Maurice Wilkins and his colleague Alex Stokes from other X-ray diffraction studies.
5. The demonstration by Linus Pauling that protein, another class of complex molecule, could exist in a helical conformation.

KEY CONCEPT

The model of DNA structure proposed by James Watson and Francis Crick in 1953 was based on information derived from the efforts of many individuals.

The 1962 Nobel Prize in Medicine or Physiology was awarded to James Watson, Francis Crick, and Maurice Wilkins.

DNA Structure: Variations on a Theme

The structure discovered by Watson and Crick, referred to as **B-DNA**, represents the sodium salt of DNA under highly humid conditions. DNA can assume different conformations because deoxyribose is flexible and the C^1-N-glycosidic linkage rotates. (Recall that furanose rings have a puckered conformation.)

A-DNA When DNA becomes partially dehydrated—that is, when the number of water molecules bound to each of the nucleotides drops to about 13 to 14—the molecule assumes the A form (**Figure 17.10** and **Table 17.1**). In **A-DNA**, the base pairs are no longer at right angles to the helical axis. Instead, they tilt 20° away from the horizontal. In addition, the distance between adjacent base pairs is slightly reduced, with 11 bp per helical turn instead of the 10.5 bp found in the

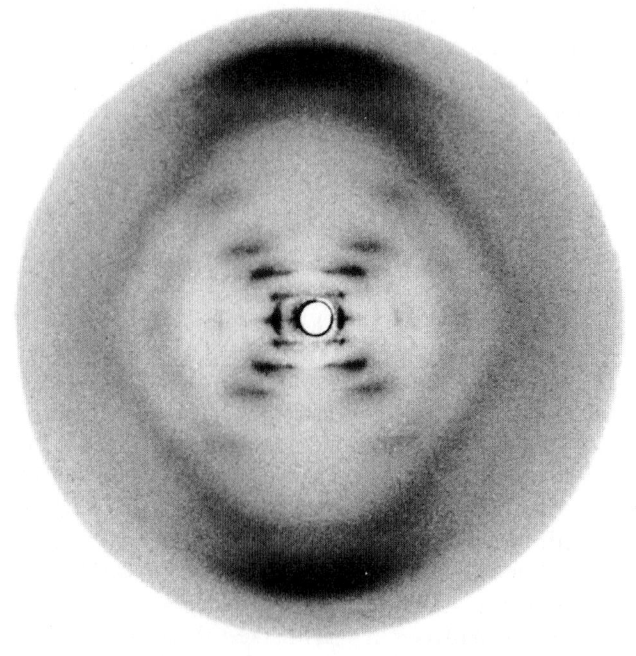

FIGURE 17.9
X-Ray Diffraction Study of DNA by Rosalind Franklin and R. Gosling

The symmetry of the X-ray diffraction pattern indicates a helical structure.

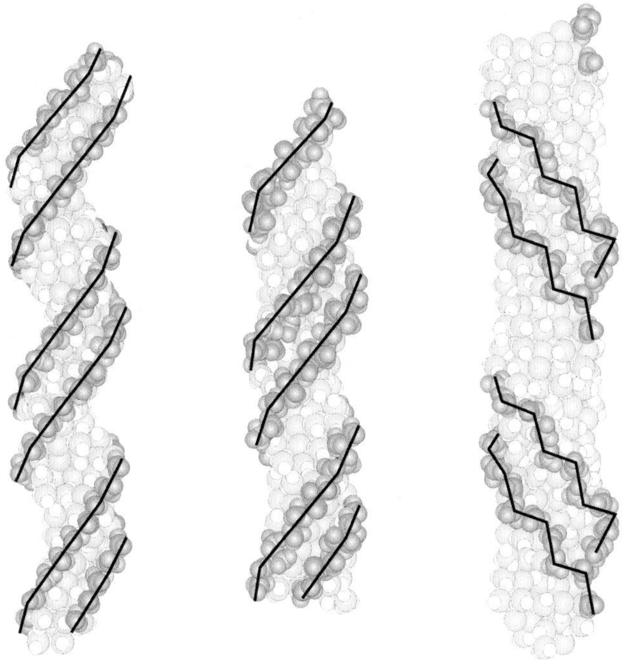

B form A form Z form

FIGURE 17.10
B-DNA, A-DNA, and Z-DNA

Because DNA is a flexible molecule, it can assume different conformational forms depending on its base sequence and/or isolation conditions. Each molecular form in the figure possesses the same number of base pairs. Refer to **Table 17.1** for the dimensions of these three DNA structures.

TABLE 17.1 Selected Structural Properties of B-, A-, and Z-DNA

	B-DNA (Watson-Crick Structure)	A-DNA	Z-DNA
Helix diameter	2.37 nm	2.55 nm	1.84 nm
Base pairs per helical turn	10.5	11	12
Helix rise per helical turn	3.32 nm	2.46 nm	4.56 nm
Helix rise per base pair	0.34 nm	0.24 nm	0.37 nm
Helix rotation	Right-handed	Right-handed	Left-handed

B form. Each turn of the double helix occurs in 2.46 nm, instead of 3.32 nm, and the molecule's diameter swells to approximately 2.55 nm from the 2.37 nm observed in B-DNA. The A form of DNA is observed when it is extracted with solvents such as ethanol. The significance of A-DNA under cellular conditions is that the structure of RNA duplexes and RNA/DNA duplexes formed during transcription resembles the A-DNA structure.

Z-DNA The Z form of DNA (named for its "zigzag" conformation) radically departs from the B form. **Z-DNA** (D = 1.84 nm), which is considerably slimmer than B-DNA (D = 2.37 nm), is twisted into a left-handed spiral with 12 bp per turn. Each turn of Z-DNA occurs in 4.56 nm, compared with 3.32 nm for B-DNA. DNA segments with alternating purine and pyrimidine bases (especially CGCGCG) are most likely to adopt a Z configuration. In Z-DNA, the bases stack in a left-handed staggered dimeric pattern, which gives the DNA a zigzag appearance and its flattened, nongrooved surface. Regions of DNA rich in GC repeats are often regulatory, binding specific proteins that initiate or block transcription. Although the physiological significance of Z-DNA is unclear, it is known that certain physiologically relevant processes such as methylation and negative supercoiling (discussed on p. 628) stabilize the Z form. In addition, short segments have been observed to form as the result of torsional strain during transcription.

Certain segments of DNA have been observed to have higher-order structures. Cruciforms are an important example. As their name implies, *cruciforms* are cross-like structures. They are likely to form when a DNA sequence contains a palindrome, a sequence that provides the same information whether it is read forward or backward (e.g., "MADAM, I'M ADAM."). In contrast to language palindromes, the nucleotide base "letters" are read in one direction on one of the complementary strands of DNA and in the opposite direction on the other strand. One-half of the palindrome on each strand is complementary to the other half. The DNA sequences that form palindromes, which may consist of several bases or thousands of bases, are called *inverted repeats.* Cruciform formation, which occurs during DNA recombination (pp. 681–689) and DNA repair (pp. 676–681), begins with a small bubble, or *protocruciform*, and progresses as intrastrand base pairing occurs.

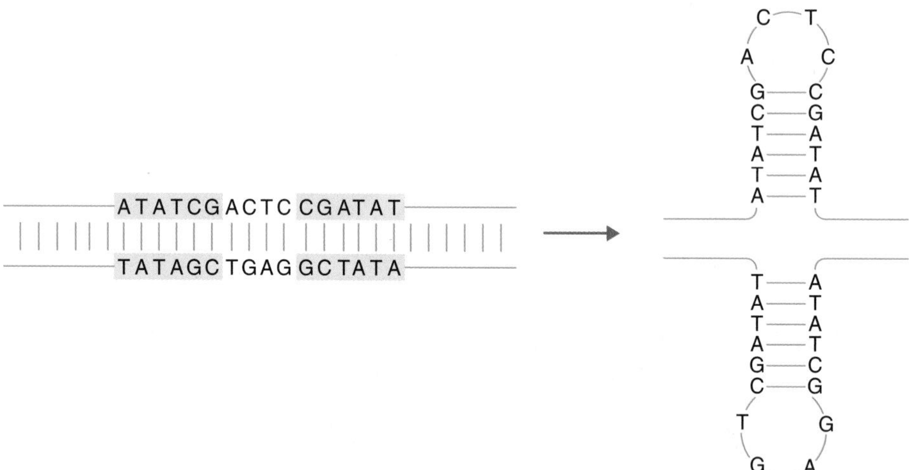

Packaging large DNA molecules to fit into cells requires DNA supercoiling. To undergo supercoiling, a single strand of double-helical DNA must be broken, or "nicked," and then either overwound or underwound before resealing. (A "nick" is a break in a single strand of double-helical DNA.) Small changes in DNA shape depend on sequence. For example, four sequential AT pairs produce a bend in the molecule. Significant bending or wrapping around associated proteins, however, requires supercoiling.

WORKED PROBLEM 17.2

The haploid human genome consists of approximately 3.2×10^9 bp (or 6.4×10^9 bp per diploid genome). Assuming that DNA is B-DNA, calculate the total length of a single cell's DNA.

SOLUTION

Given that the helix rise per base pair of B-DNA is 0.34 nm (**Table 17.1**), the total length of the DNA in a diploid cell is

$$6.4 \times 10^9 \text{ bp} \times 0.34 \times 10^{-9} \text{ m/bp} = 2.2 \text{ m}$$ ■

DNA Supercoiling

DNA supercoiling facilitates several biological processes. Examples include packaging DNA into a compact form, as well as replicating and transcribing DNA (Chapter 18). Because DNA supercoiling is a dynamic three-dimensional

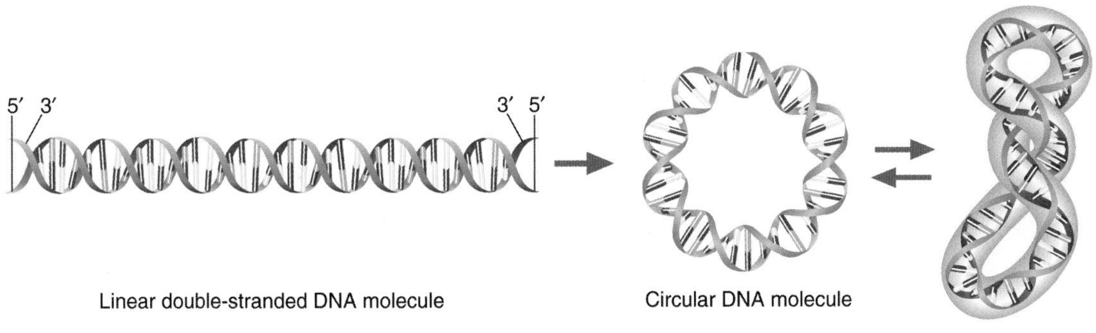

5′ 3′ 3′ 5′

Linear double-stranded DNA molecule

Circular DNA molecule

FIGURE 17.11
Linear and Circular DNA and DNA Winding

A linear DNA molecule circularizes to form a relaxed circular DNA molecule. When a relaxed circular DNA molecule is twisted, it reverts to its flat structure upon release.

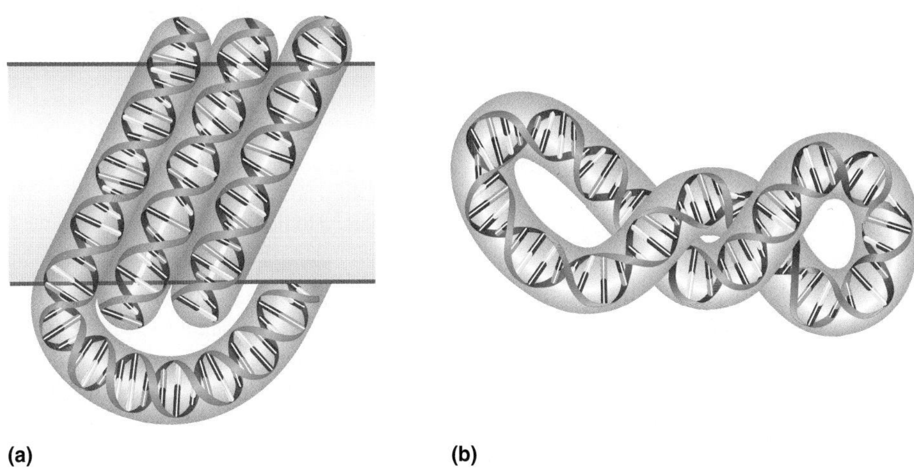

FIGURE 17.12
Supercoils

Supercoils occur in two major forms: (a) toroidal (spiral) and (b) plectonemic. In plectonemic DNA, the DNA coils are wrapped aound each other.

process, the information that two-dimensional illustrations can convey is limited. To understand supercoiling, therefore, consider the following thought experiment. A long, linear DNA molecule is laid on a flat surface. Then the ends are brought together and sealed to form an unpuckered circle (**Figure 17.11a**). Because this molecule is sealed without under- or overwinding, the helix is said to be relaxed, and it remains flat on a surface. If the relaxed circular DNA molecule is held and twisted a few times, it takes the shape shown in **Figure 17.11b**. When this twisted molecule is returned to the flat surface and made to lie on the plane, it spontaneously rotates to eliminate the twist. Note that supercoiling is a mathematical concept borrowed from knot theory. The supercoiling of a DNA molecule is the sum of *twist* (the number of helical turns) and *writhe* (the number of times a DNA molecule crosses over itself).

When a linear DNA molecule is *underwound* (i.e., the right-handed DNA helix is twisted in a left-handed direction) and then sealed, the circular molecule twists to the right to relieve strain, with negative supercoiling as the result. Negatively supercoiled DNA (most naturally occurring DNA molecules) can form either of two interconvertible shapes: a toroidal supercoil or a plectonemic (interwound) supercoil (**Figure 17.12**). A negatively supercoiled DNA molecule stores potential energy in the form of torque (force that causes rotation). The stored energy, in turn, facilitates strand separation during processes such as DNA replication and transcription (**Figure 17.13**). An *overwound* DNA molecule (i.e., twisted in the right-handed direction before it is sealed to form a circle) twists to the left to relieve stress and is positively supercoiled. Supercoils that form during strand separation in DNA replication interfere with the replication machinery. They are removed by enzymes called topoisomerases (e.g., DNA gyrase in *E. coli*), which make reversible cuts that allow the supercoiled DNA segments to relax.

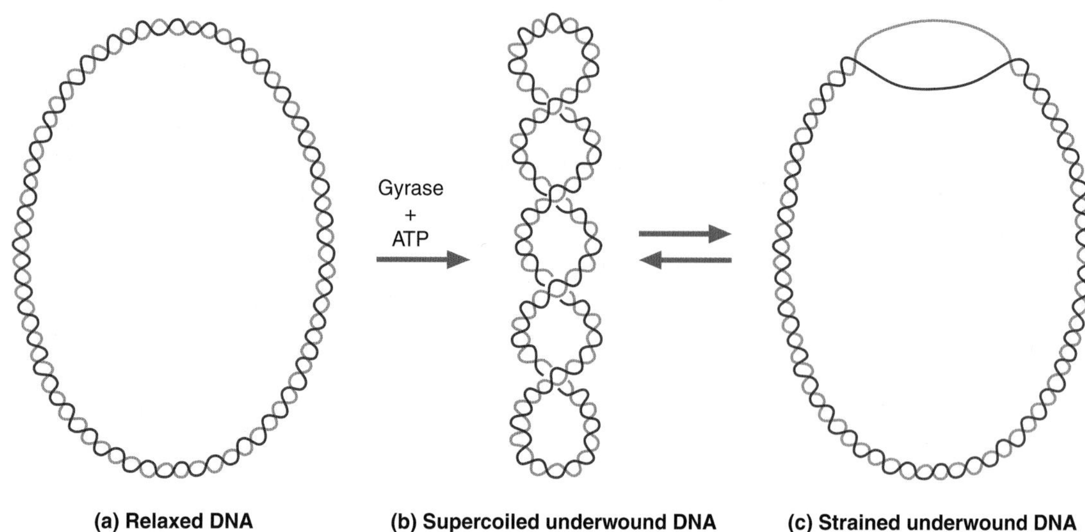

FIGURE 17.13
Effect of Strain on a Circular DNA Molecule

When a negatively supercoiled DNA molecule is forced to lie in a plane, the strain relieved by the formation of the negative supercoiling is reintroduced. Breakage and re-formation of a phosphodiester linkage allows the conversion of a relaxed circular form (a) to the negatively supercoiled (underwound) form (b). The strain relieved by the supercoiling process will be reintroduced when the underwound molecule is forced to lie in a plane (c).

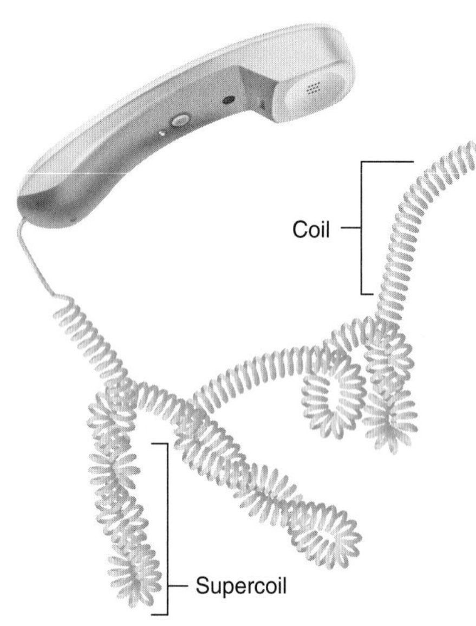

FIGURE 17.14
Supercoiling

The telephone cord resembles a coiled DNA molecule in that it is a right-handed coil that can, through rotations introduced over time, form supercoils.

DNA coiling can be compared with a coiled telephone cord (**Figure 17.14**). Through extensive use a coiled telephone cord, which connects the phone to its receiver, is rotated with the result of introducing supercoils. The coiled cord can only lie flat if it is rotated so as to undo the supercoils. Underwinding and overwinding can be observed if the hands of two individuals grasp each end of an unconnected telephone cord. One end is held stationary, while the second end is twisted. If twisting occurs in the same direction of the cord's coil (e.g., a right-handed coil twisted to the right), then the coil becomes overwound (positively supercoiled). Twisting in the opposite direction (e.g., a right-handed coil twisted to the left), the coil becomes underwound (negatively supercoiled).

Chromosomes and Chromatin

DNA is packaged into structures called chromosomes. The term **chromosome** originally referred only to the dense, dark-staining structures visible within eukaryotic cells during meiosis or mitosis. However, this term is now also used to describe the DNA molecules that occur in prokaryotic cells. The physical structure and genetic organization of prokaryotic and eukaryotic chromosomes are significantly different.

PROKARYOTES In prokaryotes such as *E. coli*, a chromosome is a circular DNA molecule that is extensively looped and coiled so that it can be compressed into a relatively small space (1 μm × 2 μm). Yet the information in this highly condensed molecule must be readily accessible. The *E. coli* chromosome (circumference 1.6 μm) consists of a supercoiled DNA that is complexed with a protein core (**Figure 17.15**).

In this structure, called the *nucleoid*, the chromosome is attached to the protein core in at least 40 places. This structural feature produces a series of loops that limit the unraveling of supercoiled DNA if a strand break is introduced. Compression is further enhanced by packaging with architectural proteins that bind bacterial DNA and facilitate bending and supercoiling. In archaeal prokaryotes, DNA is packaged with histones that have structures similar to eukaryotic histones.

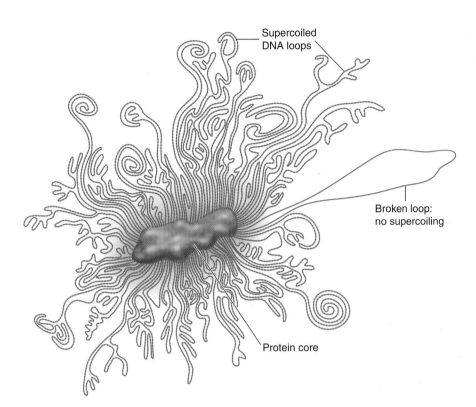

FIGURE 17.15

An *E. coli* Chromosome Removed from a Cell

The circular *E. coli* chromosome is complexed with a protein core. Because the chromosome (3×10^6 bp) is highly supercoiled within the cell, the entire chromosome complex ordinarily measures only 2 μm across. The attachment of each of the DNA loops to the protein core may prevent the unraveling of the entire supercoiled chromosome when a strand breaks.

In addition, the polyamines (polycationic molecules such as spermidine and spermine) assist in attaining the chromosome's highly compressed structure.

$$H_3\overset{+}{N}-CH_2-CH_2-CH_2-CH_2-\overset{+}{N}H_2-CH_2-CH_2-CH_2-\overset{+}{N}H_3$$

Spermidine

$$H_3\overset{+}{N}-CH_2-CH_2-CH_2-\overset{+}{N}H_2-CH_2-CH_2-CH_2-CH_2-\overset{+}{N}H_2-CH_2-CH_2-CH_2-\overset{+}{N}H_3$$

Spermine

The positively charged polyamines bind to the negatively charged DNA backbone, thus overcoming the charge repulsion between adjacent DNA coils.

EUKARYOTES In comparison to prokaryotes, the eukaryotes possess genomes that are extraordinarily large. Depending on species, the chromosomes of eukaryotes vary in both length and number. For example, humans possess 23 pairs of chromosomes and have a haploid genome of approximately 3 billion bp. The fruit fly *Drosophila melanogaster* has four chromosome pairs with 180 million bp, and corn (*Zea mays*) has 10 chromosome pairs with a total of 2.4 billion bp.

Each eukaryotic chromosome consists of a single linear DNA molecule complexed with histone proteins, which is referred to as **chromatin**. Several types of nonhistone protein are also associated with chromatin. Examples include DNA replication and repair enzymes, chromatin remodeling proteins, and a vast number of transcription factors.

HISTONES The histones are a group of small basic proteins found in all eukaryotes that bind to DNA, resulting in the formation of **nucleosomes**, the structural units of eukaryotic chromosomes. Consisting of five major classes (H1, H2A, H2B, H3, and H4), the histones are similar in their primary structure among eukaryotic species. They are particularly rich in the basic (i.e., positively charged) amino acids lysine and arginine. In electron micrographs, chromatin has a beaded appearance (**Figure 17.16**). Each of the "beads" is a nucleosome,

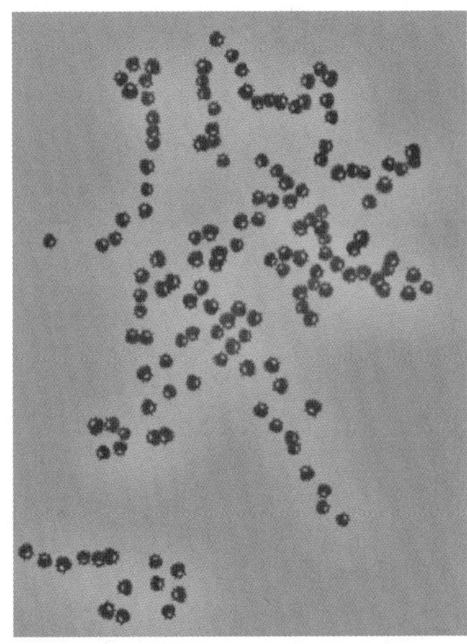

FIGURE 17.16

Electron Micrograph of Chromatin

A chromatin specimen spread out for electron microscopy. Note the "bead on a necklace" structure of this DNA-containing material.

which is composed of a positively supercoiled segment of DNA forming a toroidal coil around an octameric histone core (two copies each of H2A, H2B, H3, and H4).

Each of the highly conserved core histones (**Figure 17.17a**) contains a common structural feature called the *histone fold*: three α-helices separated by two short unstructured segments. It is the structure of the histone fold that allows the formation of histone octomers. The flexible N-terminal tails of core histones consist of between 25 and 40 amino acid residues that extend away from nucleosomes. Covalent modifications of tail residues of several types (e.g., acetylation and methylation) alter their interaction with nearby nucleosomes to facilitate either the compaction or the unfolding of nearby chromatin or modify the accessibility of DNA to proteins such as transcription factors. Such modifications are referred to as *epigenetic* modifications. (See the Biochemistry in Perspective essay Epigenetics and the Epigenome on pp. 641–643.)

The histone core forms when two sets of H2A and H2B form two head-to-tail heterodimers and H3 and H4 histones form two sets of head-to-tail heterodimers. The H3•H4 heterodimers then associate to form a $H3_2$•$H4_2$ tetramer (**Figure 17.17b**). The assembly of the nucleosome begins when the $H3_2$•$H4_2$ tetramer binds to DNA. When the two H2A•H2B dimers associate with the tetramer, nucleosome assembly is complete (**Figure 17.17c**). One molecule of histone H1, a protein that does not contain the histone fold, binds to the nucleosome where the DNA enters and exits and acts as a clamp that prevents nucleosome unraveling (**Figure 17.18**). Approximately 146 bp (1.7 helical turns) are in contact with each histone octamer. Connection between adjacent nucleosomes is by means of linker DNA, which may vary between 20 and 70 bp, depending on species and tissue and even within the same cell.

Eukaryotic cells also synthesize several types of *histone variants*, which are molecules with sequence homology and structure similar to specific classes of the core histones. Although they are synthesized in fewer numbers than the core histones, histone variants play important roles in chromatin function. For example, histones H2AX and H3.3 have roles in DNA repair and recombination, and transcriptional activation, respectively. Histone variants are inserted into nucleosomes by ATP-dependent remodeling enzymes with the aid of histone chaperones.

The histones can also be classified according to when they are synthesized during the cell cycle. Most histones are synthesized during the S, or synthesis phase, of the cell cycle (p. 675), in coordination with DNA replication, and are therefore referred to as *replication-dependent histones*. In multicellular animals, the mRNAs that code for these molecules have several unique properties that will be described in Chapter 18. *Replication-independent histones* consist of replacement variant histones that are synthesized at low levels throughout the cell cycle.

Chromatin Structure. In anticipation of cell division, chromatin is compacted (about 10,000-fold) to form chromosomes. The nucleosomes are coiled into a higher-order of structure referred to as the *30 nm fiber* (**Figure 17.19**). Despite extensive research, the precise internal structure of the 30-nm chromatin fiber is still unresolved, possibly because of irregularities caused by variation in linker length and histone posttranslational modifications. Currently there are two classes of 30-nm structural models. In one-start helix or solenoid models, nucleosomes connected by linker DNA form into a helix with 6 to 8 nucleosomes per turn. In two-start helix models, a zig-zag pattern of two nucleosomes connected by straight linker DNA are arranged into a helical structure. The 30-nm fiber is further coiled to form higher-order structures that form supercoiled loops attached to a central protein complex referred to as a nuclear scaffold (**Figure 17.20**). During interphase of the cell cycle, chromatin is observed in

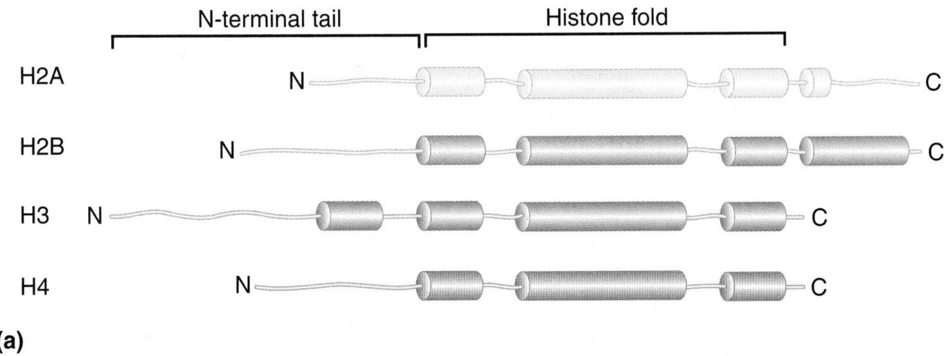

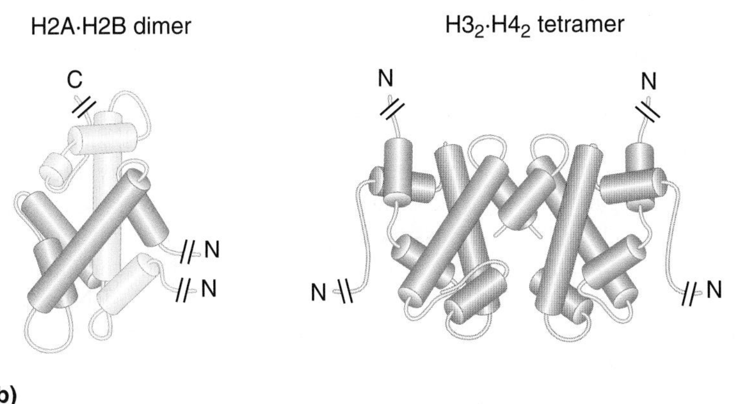

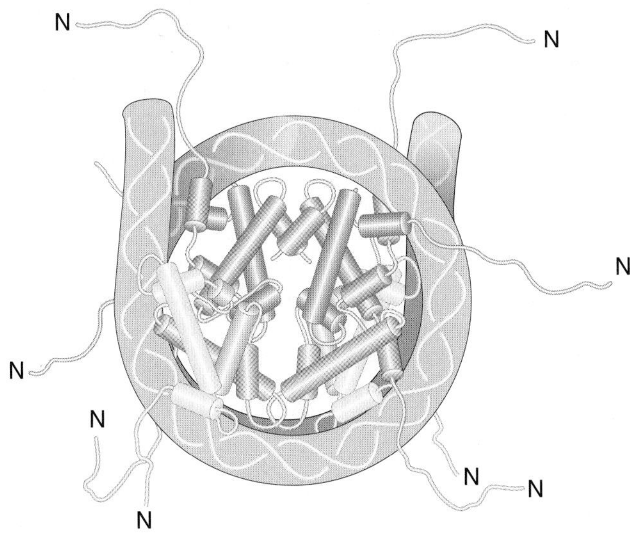

FIGURE 17.17
The Core Histones

Each nucleosome contains eight histone molecules: two each of H2A, H2B, H3, and H4. Each of these molecules contains a globular domain referred to as the histone fold and a long unstructured N-terminal domain. (a) These domains are illustrated as linear molecules with cylinders representing α-helices. (b) The formation of the histone core begins with the association of two sets of H2A and H2B to form two heterodimers and two molecules each of H3 and H4 combine to form a tetramer. The $H3_2 \bullet H4_2$ tetramer then binds to DNA. The nucleosome structure is complete (c) when 2 H2A•H2B heterodimers bind to the tetramer.

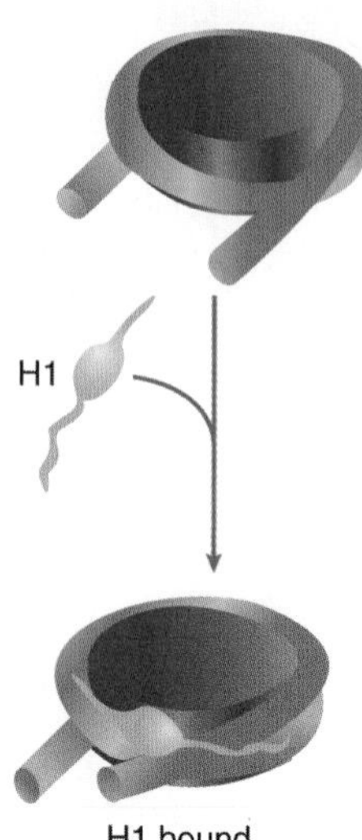

FIGURE 17.18
Histone H1

The binding of H1 to two different sites on the nucleosome DNA stablizes the wrapping of DNA around the histone octomer.

two forms: heterochromatin and euchromatin. **Constitutive heterochromatin**, composed of repetitive DNA sequences (pp. 638–640) such as those found in transposons (mobile genetic elements), centromeres, and telomeres, is so highly condensed that it is transcriptionally inactive. Other portions of heterochromatin, referred to as **facultative heterochromatin**, do not contain repetitive sequences. As its name suggests, facultative heterochromatin can lose its condensed structure and become transcriptionally active in response to specific signaling mechanisms. As a result of developmental processes, the pattern of silenced genes is different in each type of differentiated cell. **Euchromatin**, a less condensed form of chromatin, has varying levels of transcriptional activity. Transcriptionally active euchromatin is the least condensed. Inactive euchromatin is somewhat more condensed, but not as much as heterochromatin. The mechanism by which chromatin reversibly condenses is unresolved, but DNA methylation, several histone covalent modifications, and certain chromatin remodeling enzymes are known to play significant roles.

ORGANELLE DNA Mitochondria and chloroplasts are semiautonomous organelles; that is, they possess DNA and their own version of protein-synthesizing machinery. These organelles, both of which are descended from free-living prokaryotes that reproduce by binary fission, require a substantial contribution of proteins and other molecules that are coded for by the nuclear genome. For example, mitochondrial DNA (mtDNA) codes for 2 rRNAs, 22 tRNAs, and several proteins, most of which are used for electron transport. The remainder of mitochondrial proteins are synthesized in the cytoplasm and transported into the mitochondria (p. 759). Similarly, the chloroplast genome codes for several types of RNA and certain proteins, many of which are directly associated with photosynthesis. The activities of nuclear and organelle genomes are highly coordinated. Consequently, their individual contributions to organelle function are often difficult to discern. Because of the origins of mitochondria and chloroplasts, it is not surprising that they are susceptible to the actions of antibiotics (e.g., molecules such as chloramphenicol and erythromycin that inhibit bacterial genome function) if their concentrations are sufficiently high.

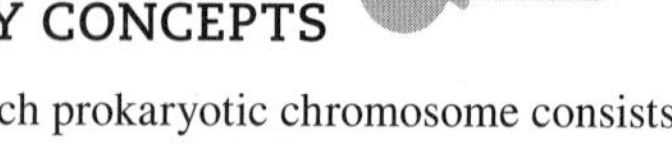

KEY CONCEPTS

- Each prokaryotic chromosome consists of a supercoiled circular DNA molecule complexed to a protein core.
- Each eukaryotic chromosome consists of a single linear DNA molecule that is complexed with histones and other proteins to form chromatin.

QUESTION 17.4

Compare the structural features that distinguish B-DNA from A-DNA and Z-DNA. What is known about the functional properties of these variants of B-DNA, the Watson-Crick structure?

QUESTION 17.5

Explain the hierarchical relationships among the following: genomes, genes, nucleosomes, chromosomes, and chromatin.

Genome Structure

The genome is an organism's operating system, the full set of inherited instructions required to sustain all living processes. Within each genome are the genes, the units of inheritance that determine the primary structure of gene products (polypeptides and RNA molecules). Genomes differ in size, shape, and sequence complexity. Genome size—the number of base-paired nucleotides, which is loosely related to organismal complexity—varies over an enormous range from less than 10^6 bp in some species of *Mycoplasma* (the smallest known bacteria) to greater than 10^{10} bp in certain plants. Most

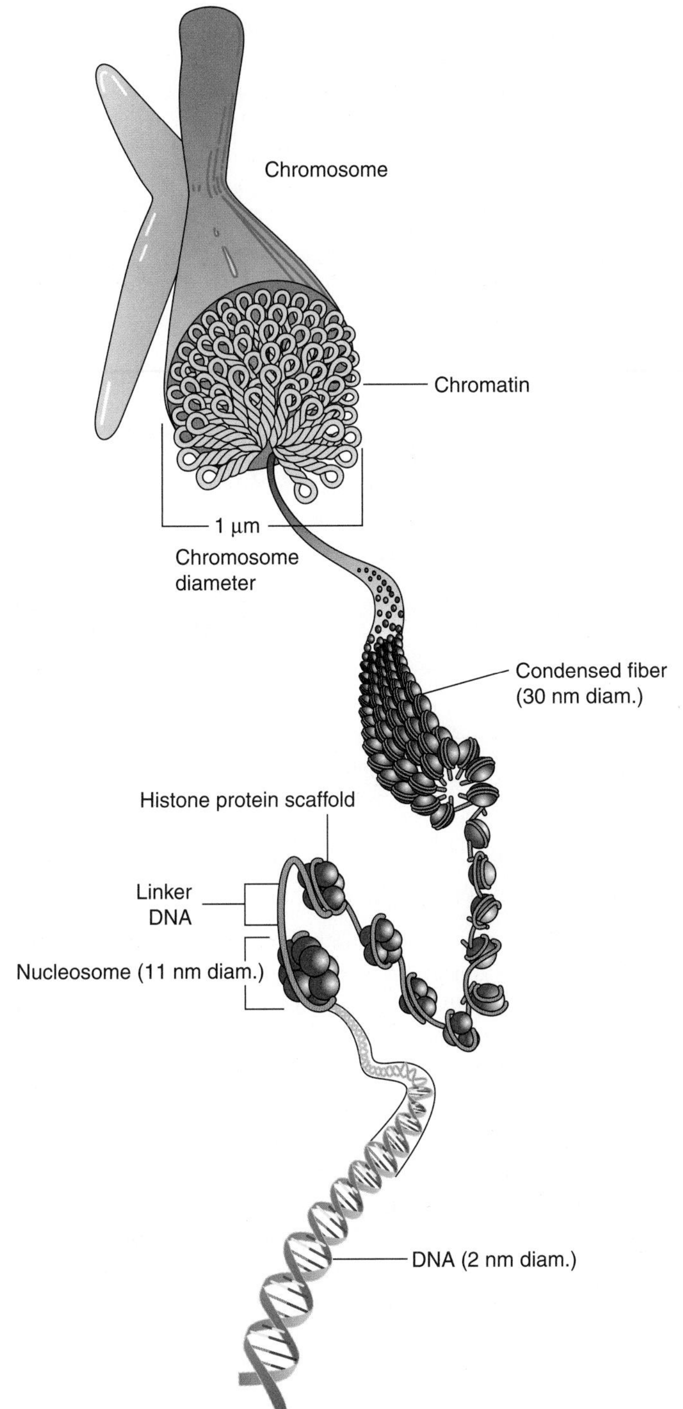

FIGURE 17.19
Chromatin

Nuclear chromatin contains many levels of coiled structure. Despite decades of research efforts, higher-order chromatin structure is largely unresolved.

prokaryotic genomes are smaller than those of eukaryotes. In contrast to prokaryotic genomes, which typically consist of single circular DNA molecules, eukaryotic genomes are divided into two or more linear DNA molecules. The most significant difference between prokaryotic and eukaryotic genomes, however, is the vastly larger information-coding capacity of eukaryotic DNA. Amazingly, the majority of eukaryotic sequences do not code for gene products. For this reason, each type of genome will be considered separately. Short segments of the genomes of several eukaryotes are compared with that of *E. coli* in **Figure 17.21**.

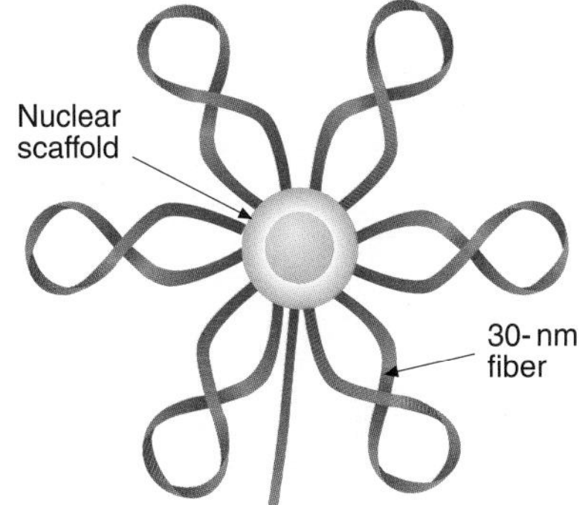

FIGURE 17.20
Chromatin

In one proposal for chromatin structure, the 30-nm fiber is looped and attached to a nuclear scaffold composed of protein.

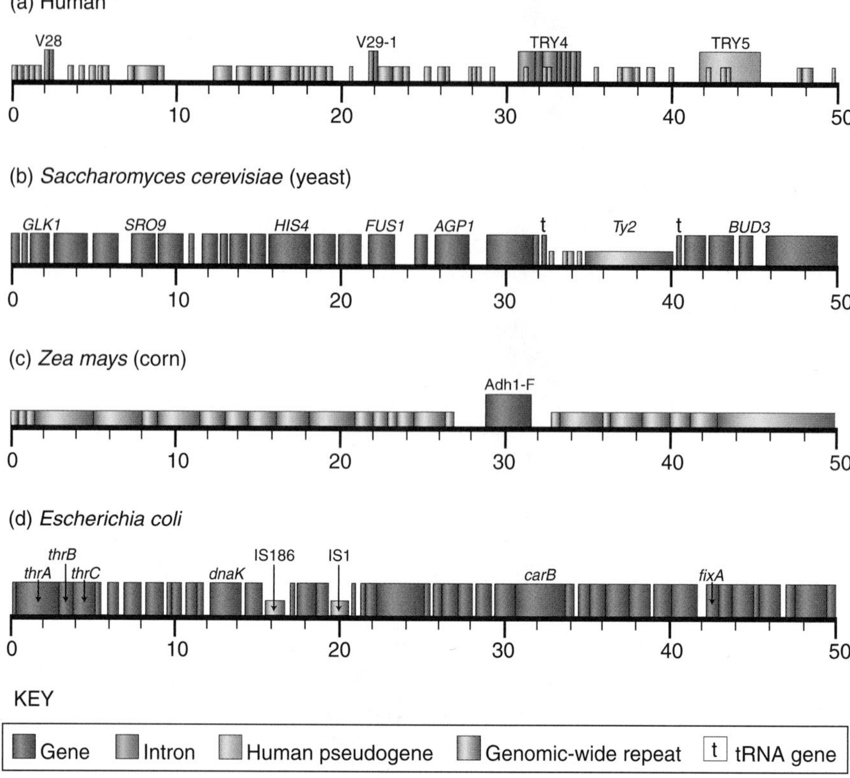

FIGURE 17.21

Comparison of 50-kb Segments of the Genomes of Selected Eukaryotes with the Prokaryote *E. coli* Genome

As indicated, the genomes of organisms such as (a) humans, (b) *Saccharomyces cerevisiae*, (c) maize, and (d) *E coli* can vary considerably in their complexity and gene density. Genes are indicated by letters and/or numbers. Humans and other complex eukaryotes have genes that are interrupted with sequences such as introns and nonfunctional sequences called pseudogenes that resemble true genes. Bacteria have few if any genome-wide repeats (repetitive, noncoding segments).

PROKARYOTIC GENOMES Investigations of prokaryotic chromosomes, especially those of several strains of *E. coli*, have revealed the following features.

1. **Genome size**. As described, most prokaryotic genomes are relatively small, having considerably fewer genes than those of eukaryotes. The K12 *E. coli* chromosome contains about 4.6 megabases, which code for 4377 protein-coding genes and at least 109 ncRNAs [1 megabase (Mb) = 1×10^6 bases].
2. **Coding capacity**. The genes of prokaryotes are compact and continuous; that is, they contain only about 15% noncoding DNA sequences. This is in sharp contrast to human DNA, in which more than 80% of DNA can be in noncoding form. Prokaryotic genomes, however, do contain numerous ncRNA-coding sequences, which function as regulators of cellular processes. During nutrient deprivation, for example, the small RNA (sRNA) molecule 6S binds to and inhibits RNA polymerase, the DNA-dependent enzyme complex that converts DNA sequences into RNA. Other bacterial ncRNAs function as riboswitches (p. 712), noncoding segments of mRNAs that control gene expression in response to cell conditions.
3. **Gene expression**. The regulation of many functionally related genes is enhanced by organizing them into operons. An **operon** is a set of linked genes, the transcription of which is regulated as a unit. About one-fourth of the genes of *E. coli* are organized into operons (see pp. 712–722).

Recall that prokaryotes also often possess additional small pieces of DNA (see p. 43) called *plasmids*, which are usually, but not always, circular. Plasmids typically have genes that are not present on the main chromosome and are seldom essential for bacterial growth and survival. They may, however, code for biomolecules that provide the cell with a growth or survival advantage: antibiotic resistance, unique metabolic capacities (e.g., nitrogen fixation; degradation of unique energy sources such as aromatic compounds) or virulence (e.g., toxins or other factors that undermine host defense mechanisms).

EUKARYOTIC GENOMES The organization of genetic information in eukaryotic chromosomes has proven to be substantially more complex than that observed in prokaryotes. Eukaryotic nuclear genomes possess the following unique features:

1. **Genome size**. Eukaryotic genomes tend to be substantially larger than those of prokaryotes. However, in the higher eukaryotes genome size is not necessarily a measure of the complexity of the organism. For example, the haploid genome of humans is 3200 Mb. The genomes of peas and the salamander are 4800 and 40,000 Mb, respectively.
2. **Coding capacity**. Although eukaryotic genomes have relatively enormous capacity, only a small fraction is devoted to coding for proteins (e.g., only 1.5% of DNA sequences in the human genome). Until recently, much of the remaining DNA sequences were thought to be nonfunctional "junk." ENCODE (*Enc*yclopedia *o*f *D*NA *E*lements), a decade-long research project begun in 2003 and funded by the U.S. Human Research Institute, revealed that 80% of the human DNA sequences have biological functions. In addition to regulatory DNA sequences that control protein-coding genes, an enormous number of DNA sequences code for diverse types of ncRNAs that regulate every facet of genome function.
3. **Coding continuity**. Most eukaryotic genes are discontinuous. Noncoding sequences (called **introns** or intervening sequences) are interspersed between sequences called **exons** (expressed sequences), which code for part of a gene product (e.g., a polypeptide, or an entire ncRNA). Intron sequences, which may together be significantly longer than the exons in a protein coding gene, are removed from primary RNA transcripts by a splicing mechanism (Section 18.2) to produce a functional RNA molecule.

The existence of exons and introns enables eukaryotes to produce more than one polypeptide from each protein-coding gene. By utilizing a process called *alternative splicing* (p. 713), various combinations of exons can be joined together to form a series of mRNAs. For example, the random rearrangements of gene sequences that encode the antigen receptors of the immunoglobulins have a major role in generating the millions of antibodies produced by the mammalian immune system.

THE HUMAN GENOME Of the approximately 3200 Mb of the human genome, approximately 1.5% comprise protein-coding genes. Humans have approximately 21,000 protein-coding genes. The remainder of the genome (98%) is referred to as noncoding DNA (ncDNA). The functions of almost one-quarter of known protein-coding genes (**Figure 17.22**) are related to DNA synthesis and repair and gene expression. Signal transduction proteins are coded for by about 21% of genes and about 17% code for general biochemical functions of cells (i.e., metabolic enzymes). The remaining genes, about 38%, code for an array of proteins involved in transport processes (e.g., ion channels), protein folding (molecular chaperones and proteosomal subunits), structural proteins (e.g., actin, myosin, tubulin, and cell adhesion proteins), and immunological proteins (e.g., antibodies and surfactant proteins that modulate inflammatory responses). The functions of almost one-quarter of protein-coding genes are as yet unknown. Protein coding genes are not distributed evenly among the chromosomes. For example, chromosomes 1, 11, and 19 contain about 22% of all human protein-coding genes.

About 80 to 90% of the human genome consists of intergenic or noncoding sequences. Noncoding DNA sequences are diverse in their functions. Examples include ncRNA genes (pp. 651–652), introns, and UTRs (untranslated regions before and after the coding sequence in an mRNA). Noncoding DNA also includes regulatory DNA sequences, pseudogenes, and repetitive DNA.

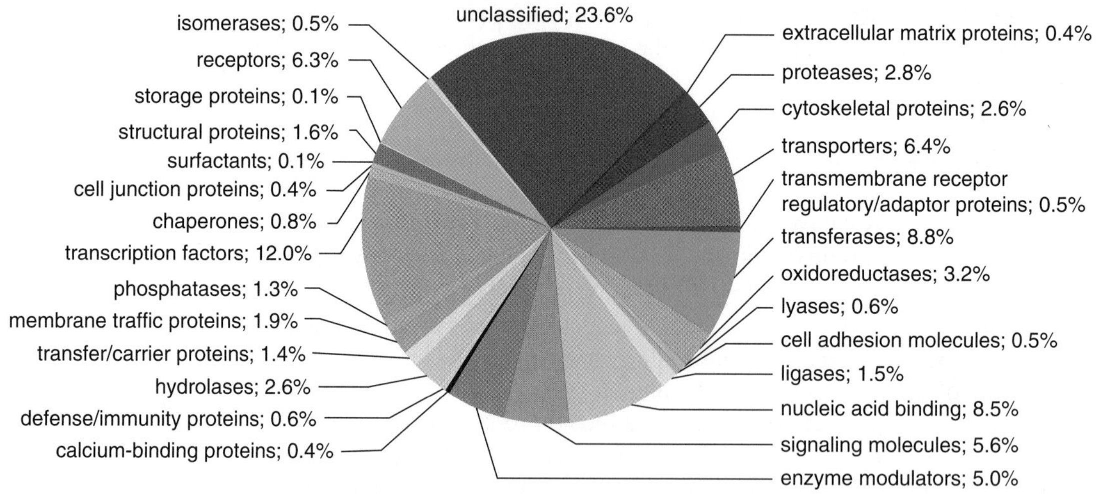

FIGURE 17.22
Human Protein-Coding Genome
Human genes are annotated for function and their percentage of all protein-coding genes.

DNA REGULATORY SEQUENCES The major types of DNA regulatory sequences are promoters, enhancers, silencers, and insulators. **Promoters** are DNA sequences (100–1000 bp) in close proximity to a transcription start site of a specific gene. For a gene to be transcribed, certain transcription factors bind to the gene's promoter and then recruit an RNA polymerase complex to the site. An **enhancer** is a DNA sequence (50–1500 bp) that, when bound to an activator transcription factor, interacts with and stimulates the activity of an RNA polymerase complex. When it is bound to a repressor protein, a **silencer** sequence inhibits the transcription of its gene by preventing an RNA polymerase from binding to the promoter. An **insulator** sequence binds to an insulator-binding protein, which in turn blocks the interaction between enhancers and the promoters of neighboring genes. Insulators also prevent the spread of heterochromatin to active genes.

PSEUDOGENES A **pseudogene** is defined as a nonfunctional DNA sequence that is homologous to a known protein or RNA gene. It is estimated that humans have at least 17,000 pseudogenes, many of which we share with other primates. There are three principal types of pseudogenes: nonprocessed, processed, and disabled. A nonprocessed pseudogene is the result of a gene duplication, followed by a series of disabling mutations that eventually render it nonfunctional. A processed pseudogene is the result of a process called retrotransposition, in which a segment of an mRNA is reverse transcribed into a DNA copy that is then inserted into a chromosome. Processed pseudogenes are so called because they have the characteristics of processed mRNAs (i.e., the absence of promoter sequences and introns and the presence of a poly A tail). A disabled pseudogene is a gene that is not expressed because of disabling mutations, but has not been duplicated. The classic example is the human gene for the enzyme L-gulono-γ-lactone oxidase, which catalyzes the conversion of L-gulono-γ–lactone to L-ascorbate (p. 246).

REPETITIVE DNA The term **repetitive DNA** refers to patterns of DNA that occur in multiple copies thoughout a genome. There are two general classes: tandem repeats and interspersed genome-wide repeats. Each is briefly described.

Tandem repeats are DNA sequences in which one or more nuclestides is repeated and multiple copies are arranged next to each other. These sequences were originally referred to as **satellite DNA** because they form a separate or "satellite" band when genomic DNA is broken into pieces and centrifuged to separate the fragments by density gradient centrifugation (see the box entitled Biochemistry in the Lab: Nucleic Acid Methods, pp. 644–648). Total lengths of the tandem repeats often vary between 10^5 and 10^7 bp. Certain types of tandem repeats apparently play structural roles in **centromeres** (the structures that contain kinetochores, which attach chromosomes to the mitotic spindle during mitosis and meiosis) and **telomeres** (structures at the ends of chromosomes that buffer the loss of critical coding sequences after a round of DNA replication). Two relatively small repetitive sequence types are referred to as minisatellites and microsatellites. **Minisatellites** have tandemly repeated sequences of 10 to 100 bp with total lengths between 10^2 and 10^5 bp. Telomeres contain minisatellite clusters. In **microsatellites**, also referred to as single-sequence repeats, there is a core sequence of 1 to 4 bp that is tandemly repeated from 10 to 100 times. The functions of these repetitive sequences are for the most part unknown. Because of their large number in genomes and because they are pleomorphic (i.e., they vary with each individual organism), minisatellites and microsatellites are used as markers in genetic disease diagnosis, in kinship and population studies, and in forensic investigations (see the Biochemistry in Perspective essay entitled Forensic Investigations, pp. 649–650).

As their name implies, **interspersed genome-wide repeats** are repetitive sequences that are scattered around the genome. Most of these sequences are the result of **transposition** (Section 18.1), a mechanism whereby certain DNA sequences, referred to as **mobile genetic elements**, can be duplicated and enabled to move within the genome. **Transposable DNA elements**, referred to as **transposons**, excise themselves and then insert at another site. More commonly, however, transposition mechanisms involve an RNA transcript intermediate. These latter DNA elements are called **RNA transposons** or **retrotransposons**. Retrotransposons can be classified into two groups based on the presence or absence of long terminal repeats (LTRs), sequences that are involved in reverse transcription.

LTR retrotransposons, which are believed to be decayed viruses, are often referred to as **endogenous retroviruses**. Human endogenous retroviruses (HERVs) comprise about 8% of the genome and are believed to be the result of ancient infections of germ cells (eggs and sperm). Although HERVs are in general inactive, several functional HERV sequences have been identified. For example, the fusion protein syncytin 1, a product of a member of the HERV-W family, is expressed during the development of the placenta. Inadequate synthesis of syncytin 1 is one of several factors that cause *pre-eclampsia*, pregnancy-induced hypertension that may lead to maternal mortality.

Non-LTR retrotransposons with lengths greater than 5 kb are called **LINEs** (long interspersed nuclear elements). LINE sequences contain a strong *promoter* (a base sequence upstream of a gene that is required for transcription initiation), an integration sequence (a base sequence required for insertion into another DNA molecule), and the coding sequences for transposition enzymes. LINEs, which constitute 17% of the human genome, have undergone duplication and mutation over time, and only a small percentage of them (about 0.1%) are at all functional. One in every 1200 mutations in humans is estimated to be the result of a LINE insertion. One example is hemophilia A (a blood-clotting disorder), which results when a LINE sequence inserts into the gene for clotting factor VIII.

SINEs (short interspersed nuclear elements) are non-LTR retrotransposons with less than 500 bp. Although SINEs contain insertional sequences, they cannot undergo transposition without the aid of a functional LINE sequence. SINEs have greatly expanded over time to comprise 11% of the human genome with more than a million copies. The only SINE insertion linked to human disease involves the *Alu element* that mediates chromosome rearrangements, insertions, deletions, and recombinations. Mutations mediated by the *Alu* subfamily of SINEs have accounted for more than 20 different human genetic diseases.

KEY CONCEPTS

- In each organism's genome, the information required to direct living processes is organized for efficient storage and use.
- Genomes from different types of organism differ in their sizes and levels of complexity.

Alu-mediated insertions have been observed to cause hemophilia B (defective clotting factor IX), and unequal recombinations resulting from Alu-mediated insertions have been linked to Lesch-Nyhan disease (p. 544) and Tay-Sachs disease (p. 396). *Recombination* (Section 18.1) is the shuffling of DNA sequences via crossing over of homologous chromosomes in the cells that give rise to egg or sperm cells. Unequal recombination is an aberrant process that causes insertion or deletion mutations.

QUESTION 17.6

Compare the sizes and coding capacity of prokaryotic genomes with those of eukaryotes. What other features distinguish them?

17.2 RNA

RNA is an extraordinarily large and diverse group of molecules that are synthesized by the transcription of DNA sequences. In addition to their well-known roles in protein synthesis, RNAs are amazingly versatile molecules that perform functions previously thought to be exclusively performed by proteins. Examples include gene expression regulation, cell differentiation, and catalysis. The primary structure of polyribonucleotides is similar to that of their DNA counterparts, but there are several differences.

1. The sugar moiety of RNA is ribose instead of deoxyribose in DNA. The presence of the 2′-OH group of ribose makes RNA more reactive than DNA.
2. The nitrogenous bases in RNA differ somewhat from those observed in DNA. Instead of thymine, RNA molecules use uracil. In addition, the bases in some RNA molecules are modified by a variety of enzymes (e.g., methylases, thiolases, and deaminases).
3. In contrast to the double helix of DNA, RNA exists as a single strand. For this reason, RNA can coil back on itself and form unique and often quite complex three-dimensional structures (**Figure 17.23**). The shape of these structures is determined by complementary base pairing by specific RNA sequences, as well as by base stacking and interactions between double-stranded regions (formed from single-stranded regions of the same molecule or between single-stranded regions of neighboring molecules) and free loops of RNA. Base-pairing rules apply in double-stranded regions where A-U, G-U, and G-C pairing occurs. Loop or single-stranded RNA (ssRNA) regions contain a number of modified bases in mature non-mRNA molecules. The base composition of RNA does not follow Chargaff's rules because RNA molecules are single-stranded.
4. RNA molecules have catalytic properties because of the complex three-dimensional structures with binding pockets that they can form. The majority of catalytic RNA molecules, which are referred to as **ribozymes**, catalyze self-cleavages or the cleavage of other RNAs. The most notable example of ribozyme activity, however, is peptide bond formation within ribosomes. A magnesium ion cofactor is usually required in RNA catalysis because Mg^{2+} stabilizes transition states.

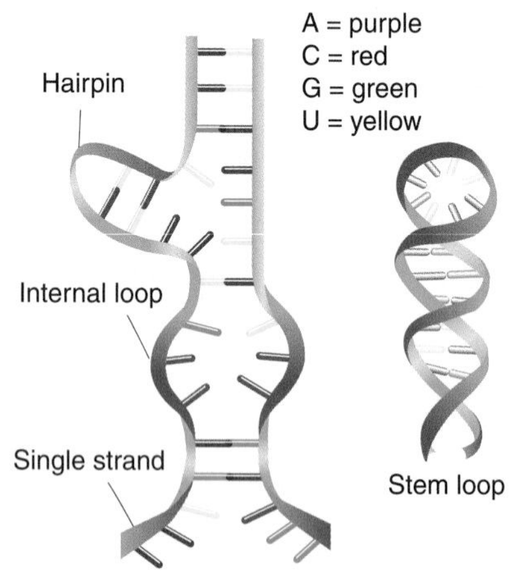

FIGURE 17.23

Secondary Structure of RNA

Three of the many different types of secondary structure that occur in RNA molecules. Hairpin loops form when there are at least four bases that form base pairs. Internal loops form when both sides of a double-stranded segment cannot form base pairs. Stable stem loops are between four and eight bases long. Note that RNA hairpins and stem loops form because of the inverted repeat sequences of DNA palindromes (see p. 628).

The most prominent types of RNA directly involved in protein synthesis are transfer RNA, ribosomal RNA and messenger RNA. The structure and function of each of these molecules is discussed next. Examples of several other types of RNA, referred to as ncRNAs, are also provided.

(Continued on p. 643)

Biochemistry IN PERSPECTIVE

Epigenetics and the Epigenome: Genetic Inheritance beyond DNA Base Sequences

How do covalent modifications of DNA and histones affect the functions of multicellular organisms? How do the more than 200 cell types in humans arise from a fertilized egg? Life scientists have known for many years that the transformation of a single cell into a multicellular organism is the result of cell specialization effected by gene expression changes that occur during the developmental process. Early signal mechanisms must "instruct" cells, each with an identical genetic blueprint, to progress down separate developmental pathways to yield terminally differentiated cells such as red blood cells, neurons, or skeletal muscle cells. In recent years it has become apparent that this process, the result of sequential, programmed changes in the pattern of expressed and silenced genes in each cell type, does not depend on genetic information (DNA base sequences) alone. Rather, development is the result of chromatin remodeling that is effected by two mechanisms: DNA methylation and histone covalent modifications. Because covalent modification–induced gene activations and repressions are heritable but do not change DNA base sequences, this phenomenon is referred to as **epigenetics** [epi (Gk) = over or above]. Epigenetic modifications convert affected DNA sequences within facultative heterochromatin into transcriptionally active euchromatin or vice versa. DNA methylation is also a means whereby cells silence transposable elements. Each differentiated cell type has unique epigenetic modifications that are referred to as its **epigenome**. After a brief description of epigenetic modifications, the role of epigenetics is discussed as an interface between genomes and the environment.

DNA Methylation

In DNA methylation reactions a methyl group is donated by SAM (p. 531) to carbon-5 of cytosine residues (**Figure 17A**). In mammals, methylated cytosines occur predominantly in 5′-CG-3′ sequences, which are referred to as CpG dinucleotides or **CpGs**. The C-5 methyl groups of cytosine residues protrude into the major groove where they prevent binding of certain DNA-binding proteins (i.e., transcription factors). They also enable the binding of proteins with methyl-CpG binding domains, called *methyl-CpG-binding proteins (MeCPs)*, that promote heterochromatin formation. CpGs are relatively rare in mammalian genomes. However, there are CpG-rich regions, called **CpG islands**, in which CpGs are typically about 50% of bases. The methylation of CpG islands, which are located upstream of most constitutively expressed (continuously produced) genes and some regulated genes, represses gene expression.

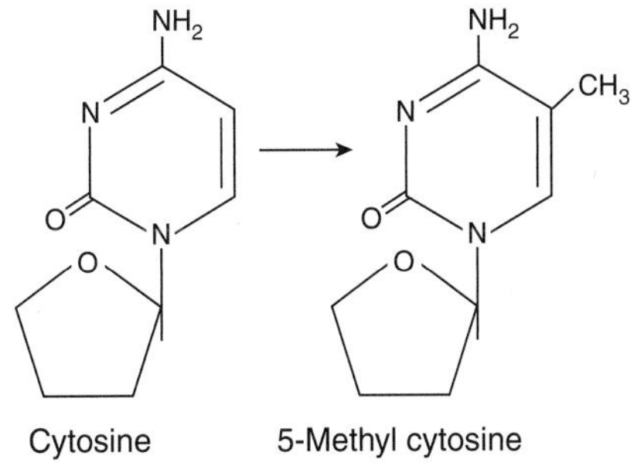

FIGURE 17A
Cytosine Methylation

Cytosine residues in CpG dinucleotides are methylated by specific methyltransferases.

There are two classes of CpG methylating enzymes: maintenance methyltransferases and *de novo* methyltransferases. *Maintenance methyltransferases* recognize methylated CpGs in the parental DNA strand and then catalyze the methylation of cytosines in the corresponding CpGs in the newly synthesized strand. It is this process that is responsible for the stable inheritance of DNA methylation patterns between cell generations. The addition of methyl groups to previously unmodified CpGs is catalyzed by *de novo methyltansferases*, usually in response to various signal transduction mechanisms. The mechanism whereby CpGs are demethylated is still obscure.

Histone Modifications

Histones have a featured role in epigenetic gene expression regulation. Covalent modification of histone N-terminal tails (**Figure 17B**) can occur at specific amino acid residues because the unstructured tails protrude outward from the nucleosome where they are accessible to modifying enzymes. The most commonly observed modifications are methylation, acetylation, and ubiquitinylation of lysine, methylation of arginine, and phosphorylation of serine. (Histone modifications are designated by histone type followed by a one-letter symbol of the modified amino acid [**Table 5.1**, p. 134] and an abbreviation of the modification type. For example, mono- and dimethyl modifications of lysine 4 on histone 3 are referred to as H3K4me and H3K4me2, respectively.) According to the *histone code hypothesis*, the pattern of histone modifications within each DNA sequence regulates gene expression by serving as a platform for the binding of specific accessory proteins. Once they are bound

Biochemistry IN PERSPECTIVE cont.

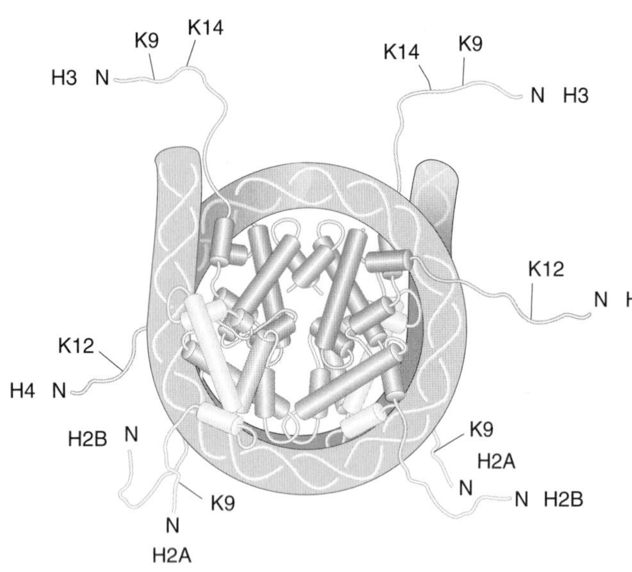

FIGURE 17B
Histone Modifications in a Nucleosome

The acetylation of the N-terminal tail lysine residues (K) in this illustration, in combination with several other types of covalent modifications (not shown), triggers chromatin remodeling that results in gene transcription.

to modified histones, the accessory protein complex initiates processes that alter chromatin structure such that transcription is inhibited or facilitated.

Acetylation and methylation are the best characterized histone modifications. The acetylation of certain lysine residues in histone tails (e.g., H3K9ac) by *histone acetyltransferases* (HAT) promotes transcription by facilitating access of DNA to transcription factors. Acetylation does so by reducing histone affinity for DNA, which causes loosening of the DNA wrap around the histone octomer and reducing interactions between adjacent nucleosomes. Lysine acetylation, in combination with other histone modifications, also creates binding sites for proteins that initiate chromatin remodeling. Lysine deacetylation, catalyzed by *histone deacetylases* (HDAC), impedes transcription because deacetylated histones have an increased affinity for DNA, a circumstance that results in tighter chromatin coiling. Most histone methylation reactions repress transcription. MeCPs mediate gene silencing by preferentially binding to 5-Me CpGs and recruiting HDAC to the site along with histone methylases. The conversion of acetylated lysine residues on histone H3 tails to methylated lysines is especially effective in gene silencing.

Epigenetics: The Connection between Gene Expression and the Environment

Life scientists have long suspected that environmental factors can influence heredity. For example, monozygotic (identical) twins often have different susceptibilities to diseases such as diabetes or cancer. In 2000 a genetic experiment involving agouti mice unambiguously linked inheritance to the environment. Agouti mice are yellow, obese, and prone to diabetes and cancer. A seemingly subtle change in diet was observed to cause a profound alteration in the health and appearance of agouti offspring.

Agouti mice are obese animals with a yellow coat color because they express the agouti gene that codes for AgRP. AgRP is an antagonist of melanocortin receptors (MCRs). MCRs are G-protein-coupled receptors that bind α-MSH, a peptide hormone that in some cells stimulates the synthesis of melanin pigment. Yellow coat color results from AgRP inhibition of the hair follicle MCR. Obesity in agouti mice is caused by MCR inhibition in the hypothalamic appetite control center (p. 604). When pregnant yellow, obese agouti mice were fed a diet supplemented with methyl donors such as folic acid, methionine, and choline, their offspring were slender with a dark coat color (**Figure 17C**). Subsequent research revealed that agouti mice have a spontaneous insertion of a virus-like genetic element near the AgRP gene sequence. Hypomethylation of the inserted genetic element results in continuous synthesis of AgRP. Diet-induced methylation of the genetic element results in vastly reduced AgRP synthesis.

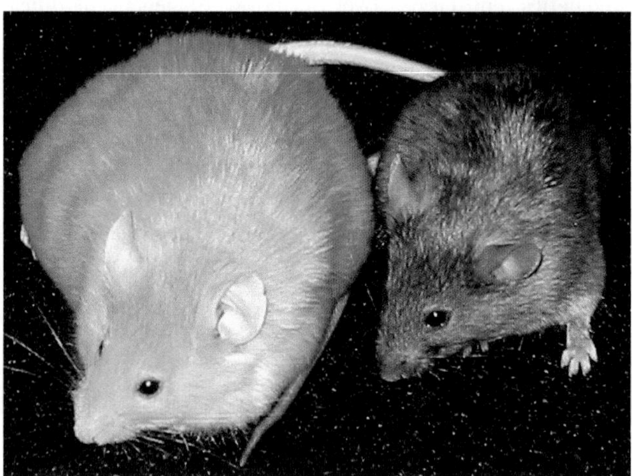

FIGURE 17C
Agouti Mice and Epigenetics

Both mice in this photograph have identical DNA sequences. The mouse on the left has yellow color and obesity, both typical agouti traits. The mouse on the right is the offspring of a yellow, obese mother fed a diet enriched in methyl donors such as folic acid and methionine.

Biochemistry IN PERSPECTIVE cont.

Over the past decade research efforts have revealed numerous examples of abnormal epigenome changes, referred to as **epimutations**, linked to environmental factors such as diet, toxins, pathogens, and behavior. Folate-deficient diets, which cause global and gene-specific DNA hypomethylation, have been observed to increase risk for cardiovascular disease and several forms of cancer (e.g., colorectal cancer). Toxins such as those in cigarette smoke cause epimutations that result in several cancers. Stomach cancer, induced by hypermethylation of a tumor suppressor gene, is caused by infection by the bacterium *Heliobacter pylori.*

Behavior can also have an affect on health. Experiments performed with rats indicate that maternal neglect has a lifelong impact on the health and behavior of offspring. Rats raised by non-nurturing mothers (measured by grooming behavior) are anxious and prone to stress-related disease. They have increased DNA methylation and decreased histone acetylation at the promoter region of a glucocorticoid receptor gene in the hippocampus (a brain region that contributes to brain functions such as emotion and behavior). Maternal nurturing behavior causes the release of serotonin in the brain, which initiates epigenetic-induced receptor synthesis that in turn results in stress hormone release inhibition.

SUMMARY: The heritable covalent modification of certain cytosine bases in DNA and histone tail residues is a sophisticated mechanism that regulates gene expression by determining the accessibility of DNA sequences to transcription machinery.

Transfer RNA

Transfer RNA (tRNA) molecules transport amino acids to ribosomes for assembly into proteins. The length of tRNA molecules varies from 75 to more than 90 nucleotides. Each tRNA molecule becomes bound to a specific amino acid. Consequently, cells possess at least one type of tRNA for each of the 20 standard amino acids. The three-dimensional structure of tRNA molecules, which resembles a warped cloverleaf (**Figure 17.24**), results primarily from extensive intrachain base pairing. tRNA molecules contain a variety of modified bases. Examples include pseudouridine, 4-thiouridine, 1-methylguanosine, and dihydrouridine:

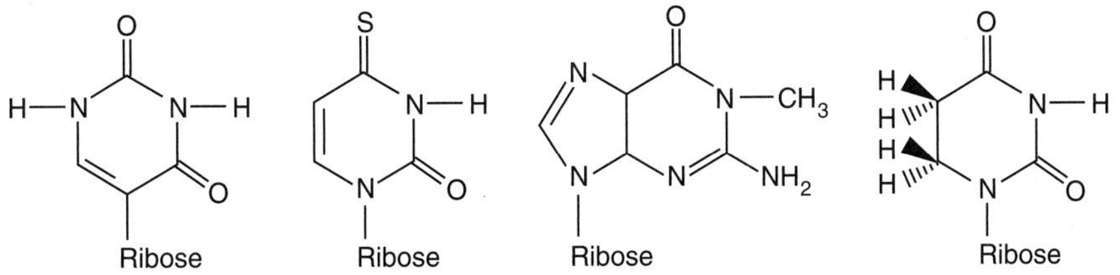

Ribose **Pseudouridine** | Ribose **4-Thiouridine** | Ribose **1-Methylguanosine** | Ribose **Dihydrouridine**

The structure of tRNA allows it to perform two critical functions involving the 3′-terminus and the anticodon loop. The 3′-terminus forms a covalent bond to a specific amino acid. (This specificity is achieved because the set of enzymes called the *aminoacyl-tRNA synthetases* link each amino acid to its tRNA.) The *anticodon loop* contains a three-base sequence that is complementary to the DNA triplet code for the specific amino acid. The conformational relationship between the 3′-terminus and the anticodon loop allows the tRNA to align its attached amino acid properly during protein synthesis. (This process is discussed in Chapter 19.)

tRNAs also possess three other prominent structural features, referred to as the D loop, the TψC loop, and the variable loop. (The Greek letter psi, ψ, stands for the modified base pseudouridine.) These structures facilitate the specific binding to the appropriate aminoacyl-tRNA synthetase and the appropriate alignment of the aminoacyl-tRNA within the nucleoprotein scaffold of the ribosome.

(Continued on p. 648)

Biochemistry IN THE LAB

Nucleic Acid Methods

The techniques used in the isolation, purification, and characterization of biomolecules take advantage of their physical and chemical properties. Most of the techniques used in nucleic acid research are based on differences in molecular weight or shape, base sequences, or complementary base pairing. Techniques such as chromatography, electrophoresis, and ultracentrifugation, which have been used successfully in protein research, have also been adapted to use with nucleic acids. In addition, other techniques have been developed that exploit the unique properties of nucleic acids. For example, under certain conditions DNA duplexes reversibly melt (separate) and reanneal (base-pair to form a duplex again). One of several techniques that exploit this phenomenon, called *Southern blotting*, is often used to locate specific (and often rare) nucleic acid sequences. After brief descriptions of several techniques used to purify and characterize nucleic acids, the common method for determining DNA sequences is outlined. More complex techniques are described in Chapter 18 in the Biochemistry in the Lab box entitled Genomics (p. 690).

Techniques Adapted from Use with Other Biomolecules

Many of the techniques used in protein purification procedures have also been adapted for use with nucleic acids. For example, several types of chromatography (e.g., ion-exchange, gel filtration, and affinity) have been used in several stages of nucleic acid purification and in the isolation of individual nucleic acid sequences. Because of its speed, HPLC (high-performance liquid chromatography) has replaced many slower chromatographic separation techniques applied to small samples.

The movement of nucleic acid molecules in an electric field depends on both their molecular weight and their three-dimensional structure. However, because DNA molecules often have relatively high molecular weights, their capacity to penetrate some gel preparations (e.g., polyacrylamide) is limited. Although DNA sequences with less than 500 bp can be separated by polyacrylamide gels with especially large pore sizes, more porous gels must be used with larger DNA molecules. Agarose gels, which are composed of a cross-linked polysaccharide, are used to separate DNA molecules with lengths between 500 bp and approximately 150 kilobases (kb). Larger sequences are now isolated by pulsed-field electrophoresis (PFGE), a variation of agarose gel electrophoresis in which two electric fields (perpendicular to each other) are alternately turned on and off. DNA molecules reorient themselves each time the electric field alternates, resulting in a very efficient and precise separation of heterogeneous groups of DNA molecules. The reorientation times decrease with the size of the DNA.

Density gradient centrifugation with cesium chloride (CsCl) has been widely used in nucleic acid research (see the Biochemistry in the Lab box entitled Cell Technology, p. 68). At high speeds, a linear gradient of CsCl is established. Mixtures of DNA, RNA, and protein migrating through this gradient separate into discrete bands at positions where their densities are equal to the density of the CsCl. DNA molecules with high guanine and cytosine content are more dense than those with a higher proportion of adenine and thymine. This difference helps separate heterogeneous mixtures of DNA fragments.

Techniques That Exploit the Unique Structural Features of the Nucleic Acids

Several unique properties of the nucleic acids are exploited in nucleic acid research. These properties include absorption of UV light at specific wavelengths and their tendency to reversibly form double-stranded complexes.

Because of their aromatic structures, the purine and pyrimidine bases absorb UV light. At pH 7 this absorption is especially strong at 260 nm. However, when the nitrogenous bases are incorporated into polynucleotide sequences, various noncovalent forces promote close interactions between them. This decreases their absorption of UV light. This **hypochromic effect** is an invaluable aid in studies involving nucleic acid. For example, absorption changes are routinely used to detect the disruption of the double-stranded structure of DNA or the hydrolytic cleavage of polynucleotide strands by enzymes.

The binding forces that hold the complementary strands of DNA together can be disrupted. This process, referred to as **denaturation** (**Figure 17D**), is promoted by heat, low salt concentrations, and extremes in pH. (Because it is easily controlled, heating is the most common denaturing method in nucleic acid investigations.) When a DNA solution is slowly heated, absorption at 260 nm remains constant until a threshold temperature is reached. Then the sample's absorbance increases. The absorbance change is caused by the unstacking of bases and the disruption of base pairing. The temperature at which one-half of a DNA sample is denatured, referred to as the melting temperature (T_m), varies among DNA molecules according to their base compositions. (Recall that DNA stability is affected by the number of hydrogen bonds between GC and AT pairs and base stacking interactions [see p. 620]. More energy is required, therefore, to "melt" DNA molecules with high G and C content.) If the separated DNA strands are held at a temperature approximately 25°C below the T_m for an extended time, renaturation is possible. Renaturation, or reannealing, does not occur

Biochemistry IN THE LAB cont.

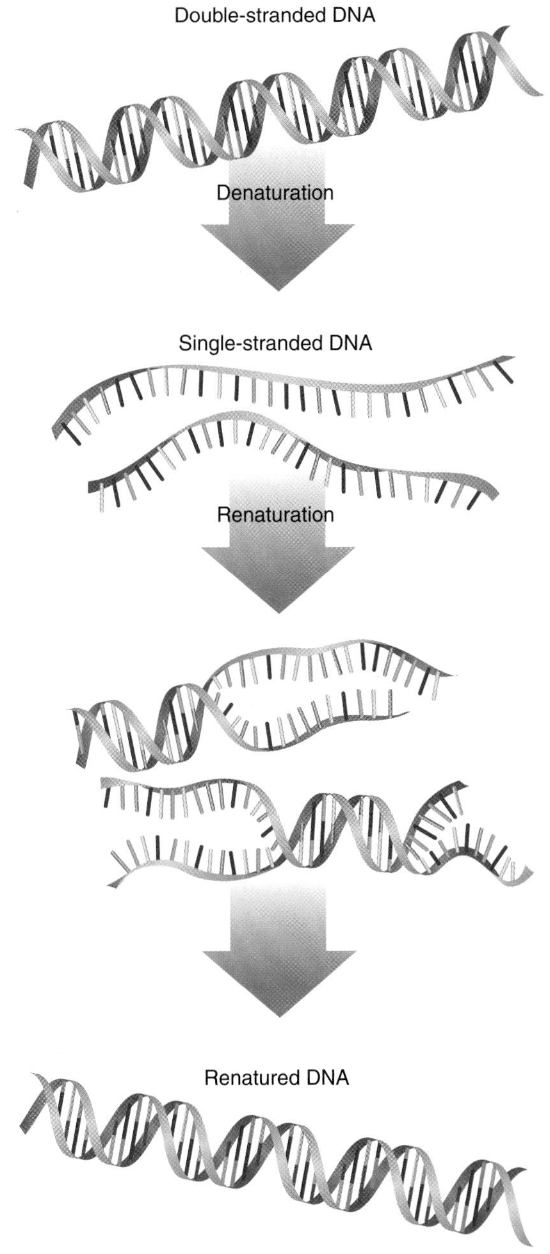

FIGURE 17D
Denaturation and Renaturation of DNA

Under appropriate conditions, DNA that has been denatured can renature; that is, strands with complementary sequences will re-form into a double helix.

instantaneously because the strands explore various configurations until they achieve the most stable one (i.e., the one having paired complementary regions).

Hybridization can also be used to locate and/or identify specific genes or other DNA sequences. For example, ssDNA from two different sources (e.g., tumor cells and normal cells) can be screened for sequence differences. If one set of ssDNA is biotinylated, then the double-stranded hybrids bind to an avidin column (avidin is a protein that binds with high affinity to biotin.). If any unhybridized sequence is present, it passes through the column. Then it can be isolated and identified. In **Southern blotting** (**Figure 17E**) radioactively labeled DNA or RNA probes (sequences with known identities) locate a complementary sequence in the midst of a DNA digest, which typically contains a large number of heterogeneous DNA fragments. A DNA digest is obtained by treating a DNA sample with restriction enzymes that cut at specific nucleotide sequences (**Figure 17F**). (Produced by bacterial cells, restriction enzymes protect bacteria against viral infection by cleaving viral DNA at specific sequences.) Once the DNA sample has been digested, the fragments are separated by agarose gel electrophoresis according to their sizes. After the gel has been soaked in 0.5 M NaOH, a process that converts dsDNA to ssDNA, the DNA fragments are transferred to nitrocellulose filter paper by placing them on a wet sponge in a tray with a high-salt buffer. (Nitrocellulose has the unique property of binding strongly to ssDNA.) Absorbent dry filter paper is placed in direct contact with the nitrocelluose filter/agarose gel sandwich. As buffer is drawn through the gel and filter paper by capillary action, the DNA is transferred and becomes permanently bound to the nitrocellulose filter. (The transfer of DNA to the filter is the "blotting" referred to in the name of this technique.) Subsequently, the nitrocellulose filter is exposed to the radioactively labeled probe, which binds to any ssDNA with a complementary sequence. For example, an mRNA that codes for β-globin binds specifically to the β-globin gene, even though β-globin mRNA lacks the introns present in the gene. Apparently, there is sufficient base pairing between the two single-stranded molecules to locate the gene.

Detection techniques other than autoradiography are also used. Nucleic acid probes can be labeled with ethidium bromide or luminol, for example. The intercalating agent ethidium bromide (p. 625) fluoresces when the gel is exposed to UV light. Probes covalently linked to luminol, a chemiluminescent molecule, emit a blue light when luminol is exposed to an oxidizing agent (H_2O_2) and a catalyst (iron atoms).

DNA Sequencing

The determination of DNA nucleotide sequences has provided valuable insights in biochemistry, medical science, and evolutionary biology. The analysis of long DNA sequences may be accomplished with the use of multiple primers or it may begin with the formation of smaller fragments by means of one type of restriction enzyme. Each fragment is then sequenced independently by the chain-terminating method. As with protein primary structure determinations, these steps are repeated with a

Biochemistry IN THE LAB cont.

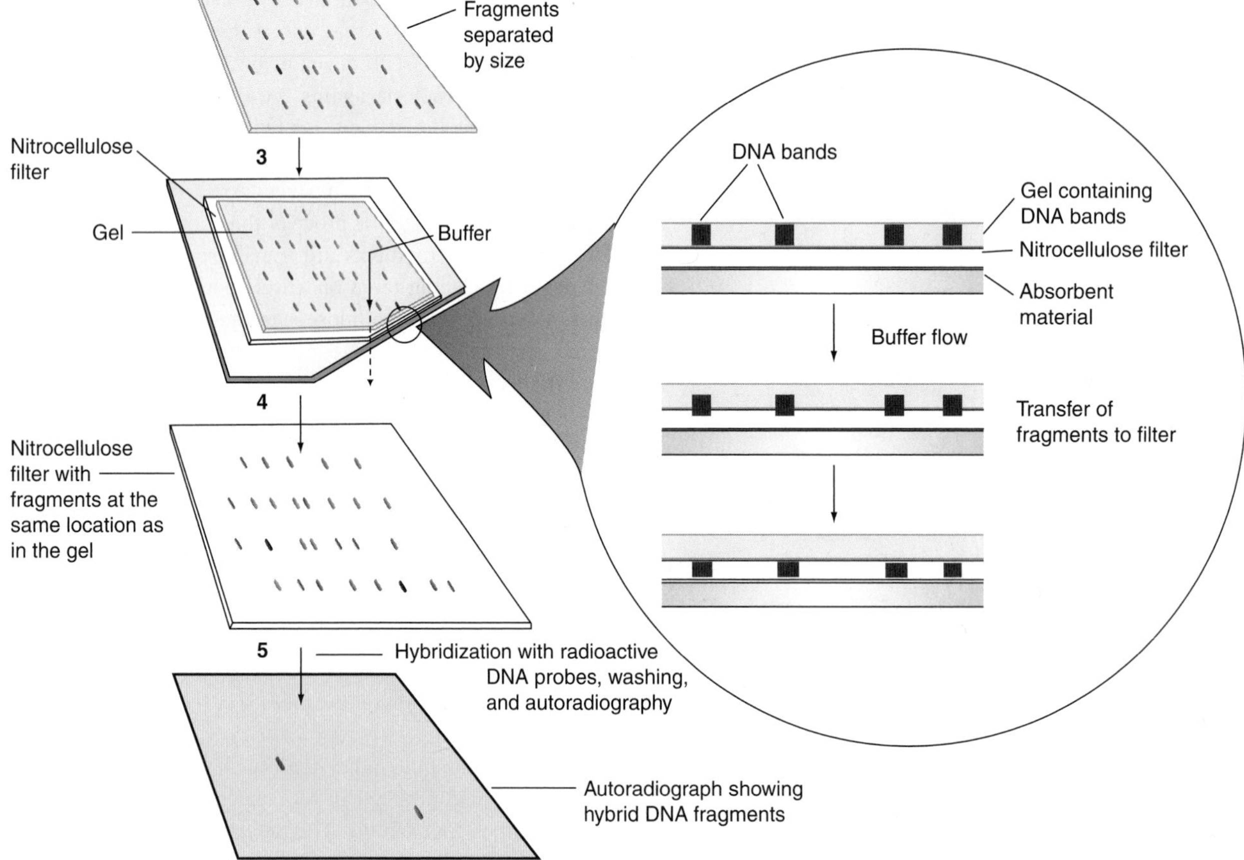

FIGURE 17E
Southern Blotting

(1) DNA analysis begins with its digestion by a restriction enzyme. (2) DNA fragments are separated by agarose gel electrophoresis. (3) The DNA fragments are transferred to nitrocellulose filter paper under denaturing conditions. (4) The ssDNA on the nitrocellulose filter paper is hybridized with a radioactively labeled ssDNA probe. (5) Any hybridized DNA can be visualized by autoradiography or other detection techniques.

FIGURE 17F
Restriction Enzymes

Restriction endonucleases are enzymes isolated from bacteria that cut DNA at specific sequences. In this example the enzyme EcoRI (obtained from *E. coli*) makes staggered cuts that result in the formation of "sticky ends." A "sticky end" is a single-stranded terminus on a double-stranded DNA fragment. Sticky ends facilitate the formation of recombinant DNA because a single-stranded segment on one DNA fragment can anneal with a complementary sticky end on another DNA fragment. Some restriction enzymes make what are called "blunt cuts," For a discussion of recombinant DNA, see box in Chapter 18 (Biochemistry in the Lab, Genomics, pp. 690–696).

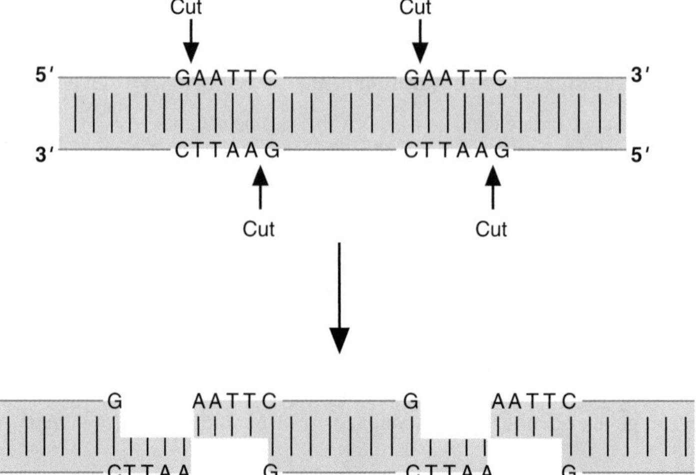

Biochemistry IN THE LAB cont.

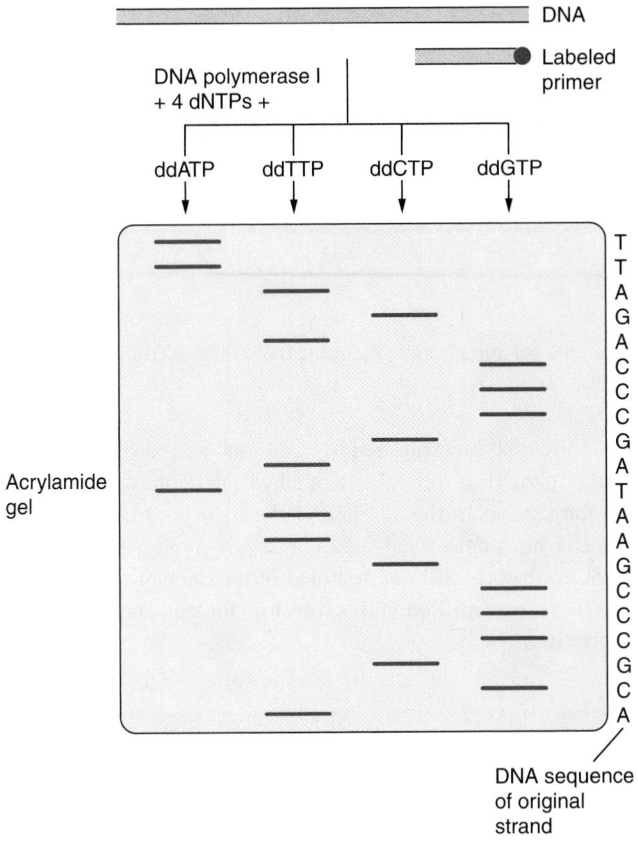

FIGURE 17G

The Sanger Chain Termination Method

A specific ^{32}P-labeled primer is chosen so that DNA synthesis will begin at the point of interest on the template strand to be replaced. DNA synthesis continues in the presence of a mixture of dNTPs and a trace of one of four dideoxynucleotides (ddNTPs). Chain termination occurs when the ddNTP is incorporated into the growing DNA chain. The mixture of truncated fragments is separated by gel electrophoresis and analyzed by autoradiography. Four samples are run in separate lanes, each with a different ddNTP. The DNA sequence is "read" from the bottom to the top of the gel.

different set of polynucleotide fragments (generated by another type of restriction enzyme) that overlap the first set. Sequence information from both sets of experiments then orders the fragments into a complete sequence.

In DNA sequencing by the **chain-terminating method** (**Figure 17G**), developed by Frederick Sanger, restriction enzymes cleave large DNA segments into smaller fragments. Each fragment is separated into two strands, one of which is used as a template to produce a complementary copy. The sample is further divided into four test tubes. To each tube is added the substances required for DNA synthesis (e.g., the enzyme DNA polymerase and the four deoxyribonucleotide triphosphates and a ^{32}P-labeled primer (a short segment of a complementary DNA strand). The investigator determines the site at which sequencing is to start by selecting the appropriate primer, usually beginning with the recognition sequence of the restriction enzyme used to generate the DNA fragment of interest.

In addition to a template, DNA synthesis materials, and a primer, each of the four tubes contains a different 2′, 3′-dideoxynucleotide derivative. (The dideoxy derivatives are synthetic nucleotide analogues in which the hydroxy groups on the 2′- and 3′-carbons have been replaced with hydrogens.) Dideoxynucleotides can be incorporated into a growing polynucleotide chain, but they cannot form a phosphodiester linkage with another nucleotide. Consequently, when dideoxynucleotides are incorporated, they terminate the chain. Because small amounts of the dideoxynucleotides are used, they are randomly incorporated into growing polynucleotide strands. Each tube, therefore, contains a mixture of DNA fragments containing strands of different lengths. Each newly synthesized strand ends in a dideoxynucleotide residue. The reaction products in each tube are separated by gel electrophoresis and analyzed together by autoradiography. Each band in the autoradiogram corresponds to a polynucleotide that differs in length by one nucleotide from the one that precedes it in any of the four lanes of the autoradiogram. Note that the smallest polynucleotide appears on the bottom of the gel because it moves more quickly than larger molecules.

Eventually, a more rapid and efficient automated version of the Sanger method that used fluorescent-tagged dideoxynucleotides was introduced (**Figure 17H**). However, in the mid-2000s, the demand for low-cost sequencing led to the development of a series of next-generation high-throughput methods such as Illumina sequencing and ion torrent sequencing. In Illumina sequencing, which uses a reversible dye terminator method, DNA molecules and primers are attached to a slide and then amplified by a polymerase to produce DNA clusters. First, four types of fluorescently labeled base (adenine, guanine, cytosine, and thymine) are introduced, each with a blocking group attached. Next, each DNA cluster is monitored by a laser-scanning confocal microscope, which acquires images point by point and then reconstructs them via computer. The four bases, each with a specific fluorescent color, compete with each other for binding sites on the template DNA molecules. During each sequencing cycle, a single nucleotide is added to the DNA chain. After non-incorporated bases are washed away, the fluorescent dye bound to the incorporated base is identified. The cycle ends with the removal of the terminal blocking group. The cycles are repeated until the DNA molecule is completely sequenced. With an estimated cost of \$0.10 per million bases, Illumina sequencing is significantly cheaper than Sanger sequencing (\$2400/million bases).

In ion torrent sequencing, DNA synthesis monitoring by computer is used in combination with a high-density array

Biochemistry IN THE LAB cont.

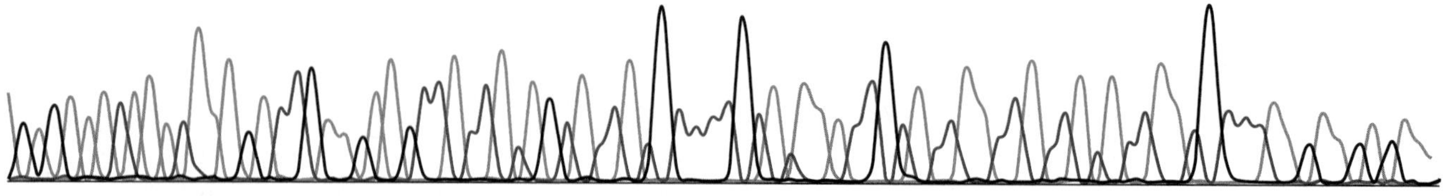

FIGURE 17H

Automated DNA Sequencing

With the use of fluorescent tags on the dideoxynucleotides, a detector can scan a gel quickly and determine the sequence from the order of the colors in the bands.

of micromachined wells in a semiconductor chip. The ion torrent chip is so designed that a hypersensitive ion-sensitive layer below the wells acts as a miniature pH meter. Each well contains many copies of a different DNA template and DNA polymerase. All four unmodified nucleotides are then sequentially introduced into the wells. A cycle begins with adding one type of nucleotide to the wells. If an introduced nucleotide is incorporated (i.e., there is a covalent bond formed), the resulting release of a hydrogen ion is detected by the ion sensor. In the absence of the incorporating reaction, there is no biochemical reaction and hydrogen ion release. Each cycle ends with the removal of unattached nucleotides. The cost per million bases for ion torrent sequencing is approximately $1.

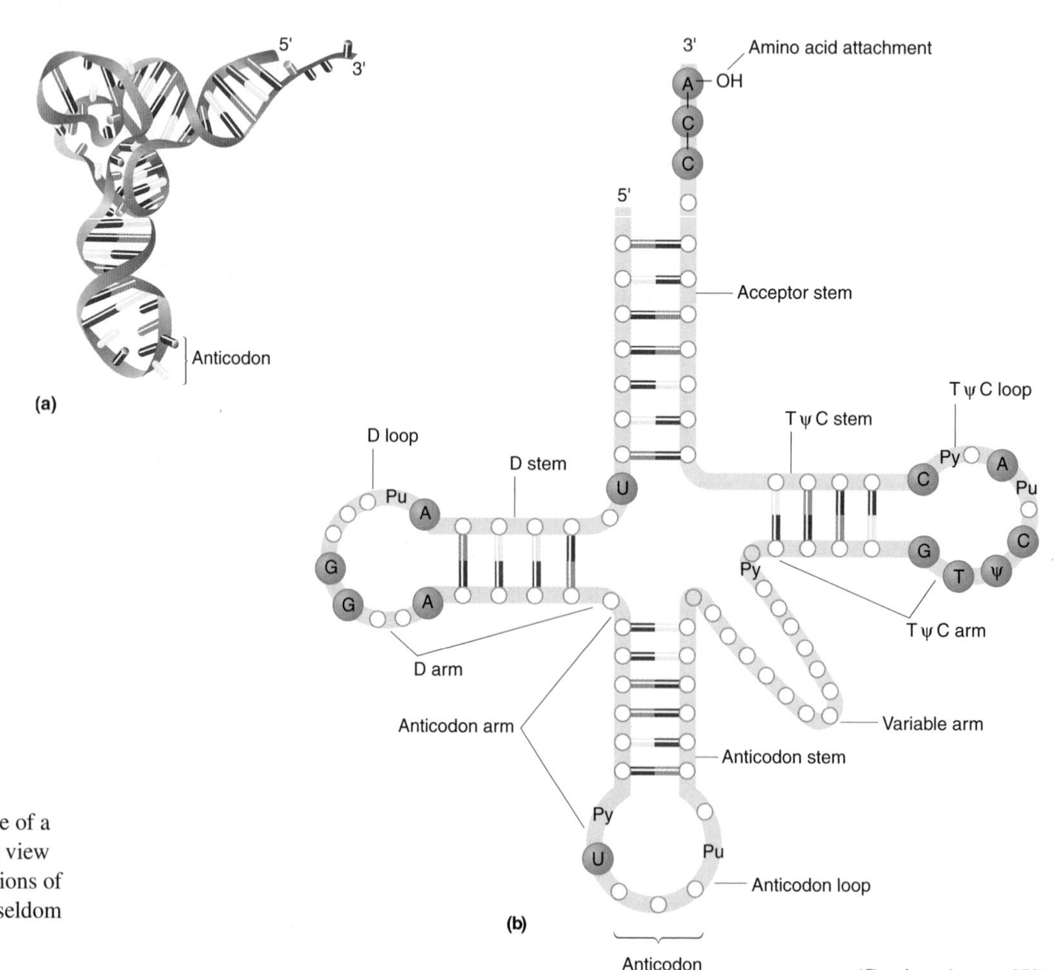

FIGURE 17.24

Transfer RNA

(a) Three-dimensional structure of a tRNA molecule. (b) Schematic view of a tRNA molecule. The positions of invariant bases and bases that seldom vary are indicated.

(Continued on p. 650)

Biochemistry IN PERSPECTIVE

Forensic Investigations

How is DNA analysis used in the investigation of violent crime? DNA persists for many years in dried biological specimens (e.g., blood, saliva, hair, and semen) and in bone. Consequently, DNA can be used as evidence in any type of forensic investigation in which such specimens are available. DNA analysis techniques that are typically used to ascertain the identity of victims and/or perpetrators of violent crimes are referred to as **DNA typing**, or profiling. DNA typing involves the analysis of variable sequences called markers. Markers include tandem repeat variations (p. 638) and single nucleotide polymorphisms (SNPs). *SNPs* are single nucleotide variations or point mutations that occur in at least 1% of the human population. With several million SNPs identified in both coding and noncoding DNA, they represent the majority of genetic variation among humans. Although 99.9% of the DNA humans share is identical, variations in the remaining 0.1% allow investigators to generate identifying genetic profiles for each individual human.

In numerous court cases since the 1990s, DNA typing has provided decisive information concerning defendants' presence at a crime scene or their absence. The techniques now available differ in their capacity to differentiate between individuals and in the speed with which results can be obtained. **DNA fingerprinting**, introduced in 1985 by the British geneticist Alec Jeffreys, is a variation of Southern blotting. In this technique, referred to as restrction fragment polymorphisms, or RFLPs, the banding characteristics of DNA minisatellites (see p. 639) from different individuals are compared—for example, crime scene specimen DNA with samples from suspects. When the quantity of DNA extracted from a crime scene sample is too minute to analyze, it is amplified by means of the *polymerase chain reaction* (PCR), a technique that is used to amplify the number of copies of DNA in a tiny sample. Up to 10^9 copies can be obtained. (Refer to p. 692.) Consequently, DNA from a single cell is now sufficient for DNA fingerprint analysis. The entire genome in each sample is isolated and treated with a restriction enzyme. (See p. 646.)

Although RFLP testing is an accurate method, it does have limitations. Among these are the substantial amounts of time (6–8 weeks), labor, and expertise required to obtain DNA profiles. A newer methodology that analyzes **short tandem repeats (STR)**, (DNA sequences with between 2 and 4 bp repeats, called microsatellites, see p. 639), has significantly greater discriminating power than RFLP and is relatively rapid (several hours).

Moreover, STR sequences are sufficiently robust that STR analysis can often be successfully used to analyze degraded specimens. After DNA has been extracted from a specimen, several target STR sequences are amplified by PCR and linked to fluorescent dye molecules.

In the United States, 13 core polymorphic (highly variable) DNA markers, called *loci*, are used to generate genetic profiles and to distinguish between individuals. A **DNA profile** (**Figure 17I**), which results when the PCR products are separated in an electrophoretic gel, consists of the pattern and the number of repeats of each target sequence visible on the gel. Fluorescence detection increases the sensitivity of the technique. Unlike RFLP, STR-based DNA typing is easily automated. If the DNA profiles from individual samples are compared and determined to be identical, the samples are said to be a match. If compared profiles are not identical, they are said to have come from different sources. The results are reported in terms of the probabilities of a random match (the chance that a randomly selected person from the population will have an identical DNA profile to that of the specimen of interest such as that left at a crime scene). The use of multiple markers and the sensitivity of the methodology reduce the random match probability to at least one in several billion.

When DNA is too degraded for nuclear DNA STR analysis, mitochondrial DNA (mtDNA) and SNPs can often yield results. Both types of DNA analysis are used as a last resort because they are expensive and labor intensive. MtDNA analysis,

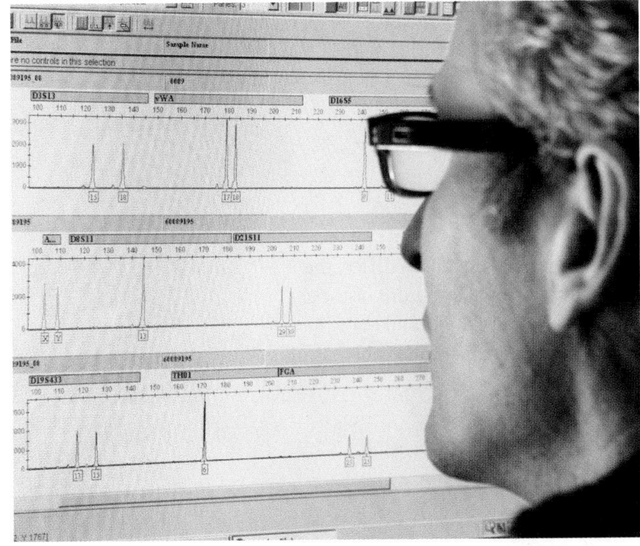

FIGURE 17I
A DNA STR Profile

STR DNA analysis is often used to analyze evidence collected at a crime scene.

Biochemistry IN PERSPECTIVE cont.

which involves sequencing of the entire mtDNA, can be successful when nuclear DNA analysis fails because most cells have hundreds to thousands of mitochondria. SNP markers are small enough that they can often be identified in degraded samples, but their discriminating power is not as high as those used in STR analysis. In contrast to the 13 loci used in STR analysis, SNP analysis requires at least 50 loci to discriminate between individuals.

SUMMARY: Forensic scientists use PCR and other technologies to amplify crime scene DNA to generate the unique genetic profile that distinguishes one individual from all others.

The *D loop* is so named because it contains dihydrouridine. Similarly, the TψC loop contains the base sequence thymine, pseudouridine, and cytosine. tRNAs can be classified on the basis of the length of their *variable loop*. The majority (approximately 80%) of tRNAs have variable loops with four to five nucleotides, whereas the others have variable loops with as many as 20 nucleotides.

Ribosomal RNA

Ribosomal RNA (rRNA) is the most abundant form of RNA in living cells. Since rRNA is present in all living organisms, its sequences have been used to characterize the evolutionary relationships between organisms. rRNA has an extraordinarily complex structure (**Figure 17.25**). Although there are species differences in the primary nucleotide sequences of rRNA, the overall three-dimensional structure of this class of molecules is conserved. As its name suggests, rRNA is a component of ribosomes.

Ribosomes, the cytoplasmic ribonucleoprotein complexes that synthesize proteins, are similar in shape and function in prokaryotes and eukaryotes, although they differ in size and in their chemical composition. Both types of ribosome consist of two subunits of unequal size, which are usually referred to in terms of their S values. (The Svedberg [or sedimentation] unit, S, is a measure of sedimentation velocity in a centrifuge. Because sedimentation velocity depends on the molecular weight and the shape of a particle, S values are not necessarily additive.) Prokaryotic ribosomes (70S) are composed of a 50S subunit and a

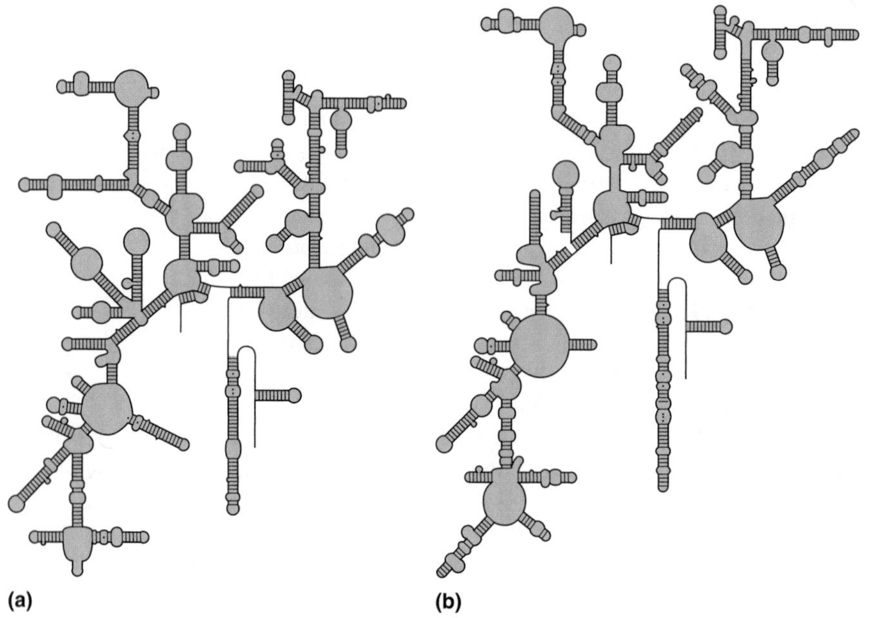

FIGURE 17.25
rRNA Structure
Although their sequences differ, the three-dimensional structure of these rRNAs from (a) *E. coli* and (b) *S. cerevisiae* (yeast) appears remarkably similar.

30S subunit, whereas ribosomes of eukaryotes (80S) contain a 60S subunit and a 40S subunit. It has been estimated that there are approximately 15,000 ribosomes per prokaryotic cell and as many as 1 million ribosomes in eukaryotic cells.

Several different kinds of rRNA and protein are found in each type of ribosomal subunit. The large ribosomal subunit of *E. coli*, for example, contains 5S and 23S rRNAs and 34 polypeptides. The small ribosomal subunit of *E. coli* contains a 16S rRNA and 21 polypeptides. The bacterial rRNA genes are organized in an operon in the following order: 3′-16S, 23S, 5S-5′. *E. coli* has seven copies in its genome. A typical large eukaryotic ribosomal subunit contains three rRNAs (5S, 5.8S, and 28S) and 49 polypeptides; the small subunit contains an 18S rRNA and approximately 30 polypeptides. Eukaryotic rRNA gene copy numbers vary from 50 to more than 5000. Humans are estimated to have about 350 copies present in clusters on chromosomes 13, 14, 15, 21, and 22. The rRNA serves as a scaffold for the self-assembly of proteins to form the native ribosomal subunit. Peptidyl transferase, the enzymatic activity responsible for peptide bond formation, resides in the 23S rRNA in prokaryotic ribosomes and 28S rRNA in eukaryotic ribosomes.

Messenger RNA

As its name suggests, **messenger RNA** (mRNA) is the carrier of genetic information from DNA for the synthesis of protein. mRNA molecules contain three-base sequences, called *codons*, that dictate specific amino acids in the subsequently synthesized protein. The base sequence within an mRNA that codes for a polypeptide is called an **open reading frame (ORF)**. An ORF begins with an initiation codon and ends with a termination or stop codon. mRNAs can vary widely in length. For example, the number of bases in mRNA from *E. coli* varies from 500 to 6000. ORFs are flanked by untranslated sequences that are referred to as the 5′UTR (upstream of the ORF) and the 3′UTR.

Prokaryotic mRNA and eukaryotic mRNA differ in several respects. First, many prokaryotic mRNAs are *polycistronic*; that is, they contain coding information for several polypeptide chains. In contrast, eukaryotic mRNA typically codes for a single polypeptide and is therefore referred to as *monocistronic*. (A **cistron** is a DNA sequence that contains polypeptide coding information and several signals that are required for ribosome function.) Second, prokaryotic and eukaryotic mRNAs are processed differently. In contrast to prokaryotic mRNAs, which are translated into protein by ribosomes during or immediately after they are synthesized, eukaryotic mRNAs are modified extensively. These modifications include capping (linkage of 7-methylguanosine to the 5′-terminal residue), splicing (removal of introns), and the attachment of an adenylate polymer referred to as a poly (A) tail. (Each of these processes is described in Chapter 18.)

Noncoding RNA

RNAs that do not directly code for polypeptides are called **noncoding RNAs (ncRNAs)**. In addition to tRNAs and rRNAs, eukaryotes also produce a diverse array of other ncRNAs. These molecules have roles in numerous cell processes that include DNA replication, gene expression, transcription, translation, stress management, genome structure and defense, and epigenetic regulation. Noncoding RNAs are classified according to their lengths: small ncRNAs (sncRNAs) have lengths less than 200 nt and long ncRNAs (lncRNAs) have lengths greater than 200 nt. Among the most important of the ncRNAs are microRNA, small interfering RNA, small nucleolar RNA, and small nuclear RNA. Note that these ncRNAs often perform their functions as components of ribonucleoprotein complexes.

The **microRNAs (miRNAs)**, between 22 and 26 nt in length, are involved in gene expression regulation. After binding to several proteins to form the RNA-induced silencing complex (RISC), each type of miRNA binds to complementary base sequences on the 3′UTR of target mRNAs, thereby preventing their translation or enhancing their degradation. Each miRNA that is expressed in a cell may

KEY CONCEPTS

- RNA is a nucleic acid that is involved in various aspects of protein synthesis and in the regulation of gene expression.
- The most abundant types of RNA are transfer RNA, ribosomal RNA, and messenger RNA.
- Important noncoding RNAs include miRNAs, siRNAs, snoRNAs, and snRNAs.

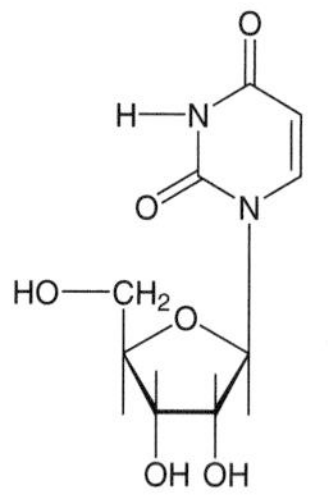

Uridine

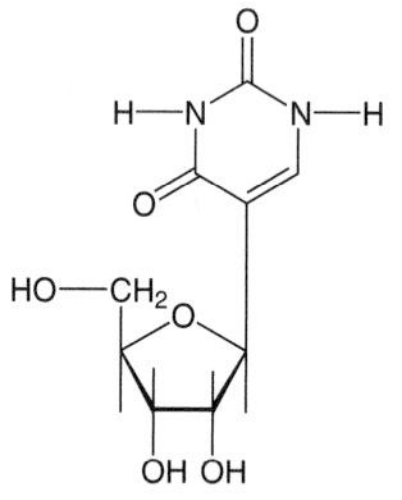

Pseudouridine

FIGURE 17.26
Structures of uridine and pseudouridine.

target as many as 200 mRNAs. MiRNAs are believed to regulate about 60% of human protein-coding genes.

Small interfering RNAs (siRNAs) are 21- to 23-nt double-stranded RNAs (dsRNAs) with 2-nt-long 3′ overhangs that play a crucial role in *RNA interference* (RNAi) (see p. 716). RNAi is an RNA-degrading process that defends cells from RNA-containing viruses and any inadvertently transcribed transposons. siRNAs are also used to interfere with the expression of specific genes by initiating the degradation of mRNAs.

Small nucleolar RNAs (snoRNAs) are single-stranded RNAs containing 70 to 300 nucleotides; they facilitate chemical modifications of rRNA, tRNA and snRNA within the nucleolus. Encoded within the introns of rRNA genes, snoRNAs are a component of the small nucleolar ribonucleoprotein (snoRNP). The function of snoRNAs (more than 100 in humans) is to guide the snoRNP via base pairing to the specific sequence site on a target rRNA. Modifications that occur during rRNA processing include methylation of the 2′OH of ribose and the isomerization of uridine to form pseudouridine (**Figure 17.26**). Several snoRNAs are involved in certain tRNA and snRNA base modifications.

The primary function of the **small nuclear RNAs** (snRNAs) is the processing of pre-mRNA. Composed of an average of 150 nt, the snRNAs U1, U2, U4, U5, and U6 combine with several proteins to form **small nuclear ribonucleoproteins (snRNPs)**, often called "snurps." Together with several other proteins, the snRNPs form a molecular machine called the spliceosome because of its function: splicing is a key step in the processing of eukaryotic mRNAs. **Spliceosomes** excise introns from pre-mRNA and then join exons together. Other snRNAs have been shown to regulate the activity of transcription factors and of RNA polymerase II (p. 703).

Long ncRNAs (IncRNAs) often have structures resembling mRNAs because many are spliced, polyadenylated, and 5′-capped. The roles of most of the approximately 9000 lncRNAs in humans are not known. Those whose functions have been defined are involved in regulation of transcription and posttranscriptional modification reactions. lncRNAs have also been shown to encode miRNAs. Cell processes involving lncRNAs include cell cycle regulation and epigenetic regulation such as imprinting (the differential expression of a gene depending on whether it was paternally or maternally inherited).

QUESTION 17.7

When a gene is transcribed, only one DNA strand acts as a template for the synthesis of the RNA molecule. This strand is referred to as an **antisense** (or noncoding) strand; the nontranscribed DNA strand is called the **sense** (or coding) strand. The base sequence of the sense strand is the DNA version of the mRNA used to synthesize the polypeptide product of the gene. The antisense RNA (the transcript of the antisense DNA strand) plays a role in transcriptional and translational regulation. An antisense RNA can anneal specifically to a corresponding mRNA and prevent translation.

Because mRNA-antisense RNA binding is so specific, antisense molecules are considered promising research tools. Numerous investigators are using antisense RNA molecules to study eukaryotic function by selectively turning on and off the activities of specific genes. This so-called *reverse genetics* is also useful in medical research. Although serious problems have been encountered in antisense research (e.g., the inefficiency of inserting oligonucleotides into living cells and high manufacturing costs), antisense technology has already provided valuable insight into the mechanisms of cancer and viral infections.

Consider the following sense DNA sequence:

5′-GCATTCGAATTGCAGACTCCTGCAATTCGGCAAT-3′

Determine the sequence of its complementary strand. Then determine the mRNA and antisense RNA sequences. (Recall that in RNA structure, U is substituted for T. So A in a DNA strand is paired with a U as RNA is synthesized.)

17.3 VIRUSES

Viruses lack most of the properties that distinguish life from nonlife, the most important of which is the inability to carry on metabolic process. Yet under appropriate conditions they can wreak havoc on living organisms. Often described as obligate intracellular parasites, viruses can also be viewed as mobile genetic elements because of their structure, a piece of nucleic acid enclosed in a protective coat. Once a virus has infected a host cell, its nucleic acid can hijack the cell's nucleic acid and protein-synthesizing machinery. As viral components accumulate, complete new viral particles are produced and then released from the host cell. In many circumstances, so many new viruses are produced that the host cell lyses (ruptures). Alternatively, the viral nucleic acid may insert itself into a host chromosome, resulting in transformation of the cell.

Viruses occur in a bewildering array of sizes and shapes. Virions (complete viral particles) range from 10 nm to approximately 400 nm in diameter. Although most viruses are too small to be seen with the light microscope, a few (e.g., the pox viruses) can be visualized because they are as large as the smallest bacteria.

Simple virions are composed of a *capsid* (a protein coat made of interlocking protein molecules called capsomeres), which encloses nucleic acid. (The term *nucleocapsid* is often used to describe the complex formed by the capsid and the nucleic acid.) Most capsids are either helical or icosahedral (20-sided structures composed of triangular capsomeres). The nucleic acid component of virions is either DNA or RNA. Although most viruses possess double-stranded DNA (dsDNA) or single-stranded RNA (ssRNA), examples with ssDNA and dsRNA genomes have also been observed. There are two types of ssRNA genome. A *positive-sense* RNA genome [(+)-ssRNA] acts as a giant mRNA; that is, it directs the synthesis of a long polypeptide that is cleaved and processed into smaller molecules. A *negative-sense* RNA genome [(−)-ssRNA] is complementary in base sequence to the mRNA that directs the synthesis of viral proteins. Viruses that employ (−)-ssRNA genomes, called **retroviruses**, must provide an enzyme, referred to as a reverse transcriptase, that synthesizes the mRNA.

In more complex viruses, the nucleocapsid is surrounded by a membrane envelope, which usually arises from the host cell nuclear or plasma membranes. Envelope proteins, coded for by the viral genome, are inserted into the envelope membrane during virion assembly. Proteins that protrude from the surface of the envelope, called spikes, are believed to mediate the attachment of the virus to the host cell. Human immunodeficiency virus (HIV) is an example of an enveloped virus.

Bacteriophage T4: A Viral Lifestyle

The T4 bacteriophage (**Figure 17.27**) is a large virus with an icosahedral head and a long, complex tail. The head contains dsDNA, and the tail attaches to the host cell and injects the viral DNA into the host cell.

The life cycle of T4 begins with adsorbing the virion to the surface of an *E. coli* cell. The entire virion cannot penetrate into the cell's interior because the bacterial cell wall is rigid. Instead, the DNA is injected by flexing and constricting the tail apparatus. Once the viral DNA has entered the cell, the infective process is complete, and the next phase (replication) begins.

Within 2 minutes after the injection of T4 phage DNA into an *E. coli* cell, synthesis of host DNA, RNA, and protein stops and phage mRNA synthesis begins. Phage mRNA codes for the synthesis of capsid proteins and some of the enzymes required for the replication of the viral genome and the assembly of virion components. In addition, other enzymes are synthesized that weaken the cell wall of the host, allowing release of the new phage for new rounds of infection. Approximately 22 minutes after the injection of viral DNA (vDNA), the host cell, now filled with several hundred new virions, lyses. Upon release, the virions attach to nearby bacteria, thus initiating new infections.

KEY CONCEPTS

- Viruses are composed of nucleic acid enclosed in a protective coat. The nucleic acid may be a single- or double-stranded DNA or RNA.
- In simple viruses the protective coat, called a capsid, is composed of protein.
- In more complex viruses the nucleocapsid, composed of nucleic acid and protein, is surrounded by a membranous envelope derived from host cell membrane.

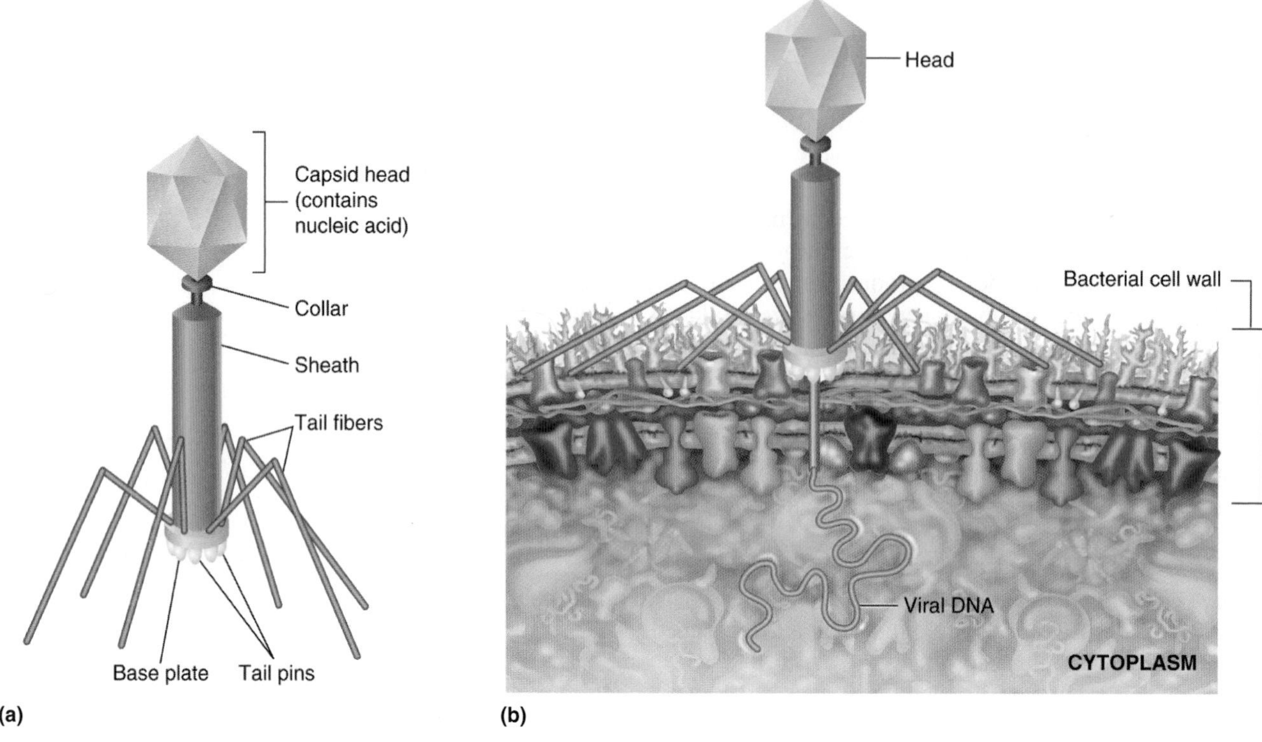

FIGURE 17.27
The T4 Baceteriophage

(a) The structure of an intact T4 bacteriophage. (b) Penetration of the cell wall and injection of viral DNA (vDNA) into the host cell by the bacteriophage. The vDNA directs the host cell to synthesize about 30 proteins that facilitate new virion synthesis.

Bacteriophage that initiate this so-called **lytic cycle** are referred to as *virulent* because they destroy their host cells. Many phage, however, do not initially kill their hosts. So-called *temperate* or *lysogenic* phage integrate their genome into that of the host cell. (The term **lysogeny** describes a condition in which the phage genome is integrated into a host chromosome.) The integrated viral genome, called the **prophage**, is copied along with host DNA during cell division for an indefinite time. Occasionally, lysogenic phage enter a *lytic* phase. Certain external conditions, such as UV or ionizing radiation, activate the prophage, which directs the synthesis of new virions. Sometimes, a lysing bacterial cell releases a few virions that contain some bacterial DNA along with the phage DNA. When such a virion infects a new host cell, this DNA is introduced into the host genome. This process is referred to as **transduction**.

QUESTION 17.8

Recall that according to the central dogma, the flow of genetic information is from DNA to RNA and then to protein. Retroviruses are an exception to this rule. The alterations of the central dogma that are observed in retroviruses and other RNA viruses can be illustrated as follows:

Compare this with the original central dogma (p. 618). Describe in your own words the implications of each component of these figures.

Biochemistry IN PERSPECTIVE

HIV Infection

How does HIV infect human cells? The human immunodeficiency virus (HIV) is the causative agent of acquired immune deficiency syndrome (AIDS). Left untreated, AIDS is a lethal condition because HIV destroys the body's immune system, rendering it defenseless against disease-causing organisms (e.g., bacteria, protozoa, and fungi, as well as other viruses) in addition to some forms of cancer.

HIV (**Figure 17J**) is a lentivirus, a member of a group of RNA viruses called the retroviruses. Retroviruses are so named because they contain an enzymatic activity reverse transcriptase, which catalyzes the synthesis of a DNA copy of an ssRNA genome. A typical retrovirus consists of an RNA genome enclosed in a protein capsid. Wrapped around the capsid is a membranous envelope that is formed from a host cell lipid bilayer.

HIV Infection: An Overview

In the reproductive cycle of HIV (**Figure 17K**), the infective process begins when the virus binds to a host cell. Binding, which occurs between viral surface glycoproteins and specific plasma membrane receptors, initiates a fusion process between host cell membrane and viral membrane. Subsequently, the viral capsid is released into the cytoplasm and the viral reverse transcriptase catalyzes the synthesis of DNA strands complementary to two copies of the viral ssRNA. This enzymatic activity also catalyzes the conversion of the single-stranded DNA into a double-stranded molecule. The double-stranded DNA version of the viral genome is then translocated into the nucleus, where it integrates into a host chromosome. The integrated proviral genome, acting like a prophage, is replicated each time the cell undergoes DNA synthesis. The mRNA transcripts produced when the viral

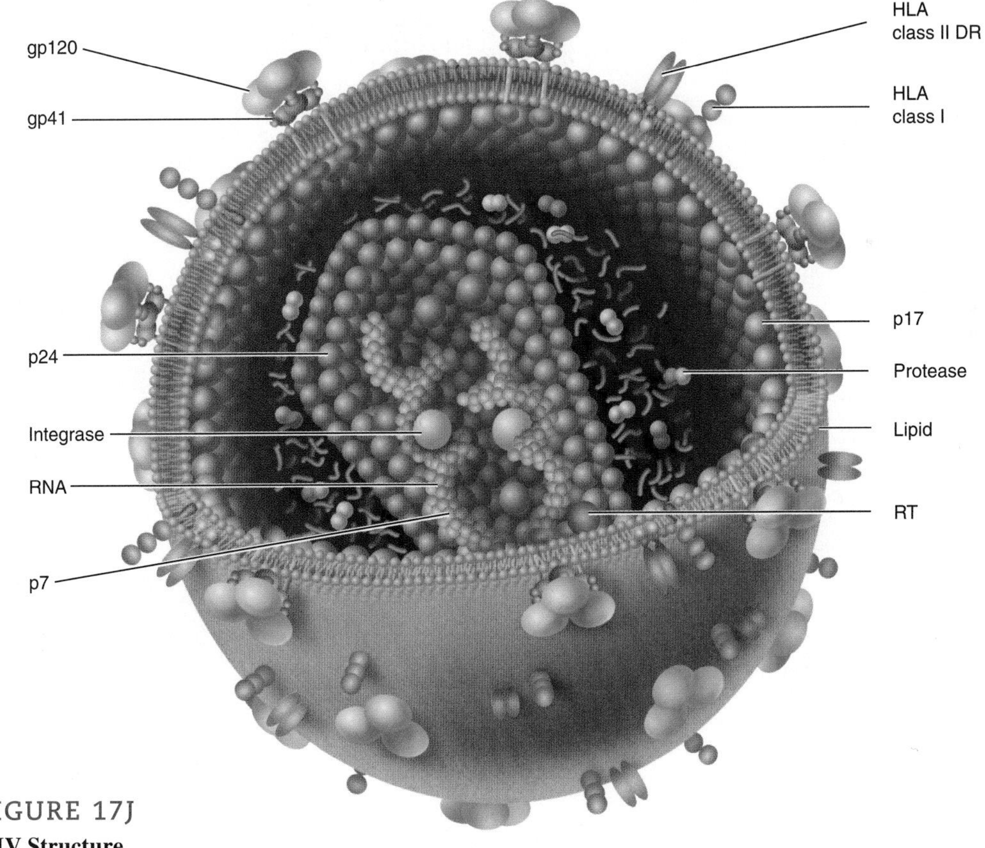

FIGURE 17J
HIV Structure

The surface of the virus is a lipid bilayer in which are embedded the viral glycoproteins gp120 and gp41, as well as HLA (human leukocyte antigens) membrane proteins taken from host cells. (HLA proteins are signals that protect the viral particle from the immune system, which ordinarily searches out and destroys foreign invaders.) Lining the inside of the envelope are hundreds of copies of the matrix protein p17. Two copies of the RNA genome are contained in a bullet-shape capsid composed of the core protein p24. The nucleocapsid protein p7 coats the RNA genome. Enzymes associated with viral genome are reverse transcriptase (RT), integrase, and protease.

Biochemistry IN PERSPECTIVE cont.

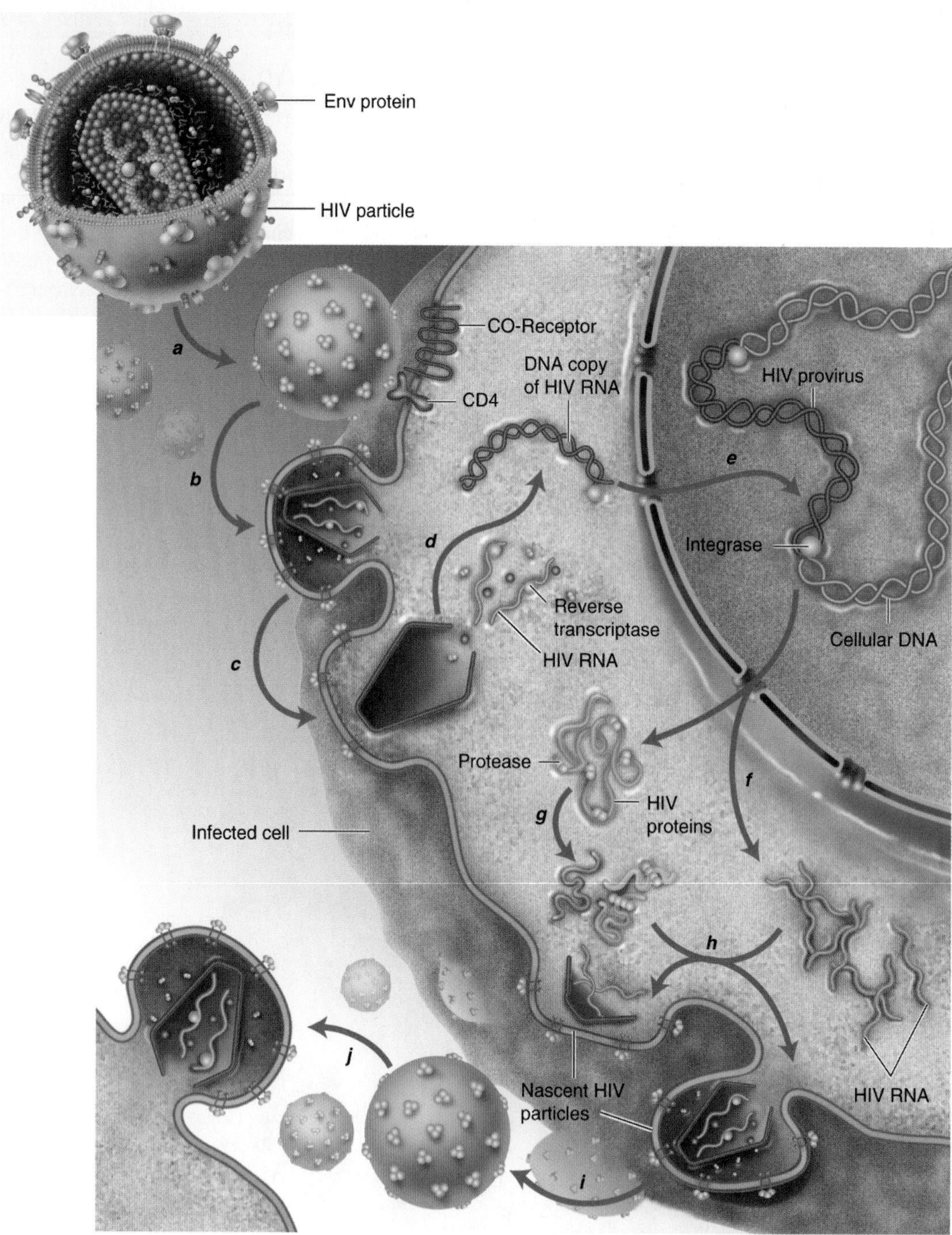

FIGURE 17K
Reproductive Cycle of HIV, a Retrovirus

After the viral particle binds to surface receptors on the host cell (a), its envelope fuses with the cell's plasma membrane (b), thus releasing the capsid and its contents (vRNA and several viral enzymes) into the cytoplasm (c). The viral enzyme reverse transcriptase catalyzes the synthesis of a DNA strand complementary to the vRNA (d) and then proceeds to form a second DNA strand that is complementary to the first. Subsequently the double-stranded viral DNA (vDNA) transfers to the nucleus, where it integrates itself into a host chromosome with the aid of a viral integrase (e). The provirus (the integrated viral genome) is replicated every time the cell synthesizes new DNA. Transcription of viral DNA results in the formation of two types of RNA transcript: RNA molecules that function as the viral genome (f) and molecules that code for the synthesis of viral protein (e.g., reverse transcriptase, capsid proteins, envelope proteins, and viral integrase) (g). The protein molecules are combined with the vRNA genome during the creation of new virus (h) that buds from the surface of the host cell (i) and then proceeds to infect other cells (j).

Biochemistry IN PERSPECTIVE cont.

genome is transcribed direct the synthesis of numerous copies of viral proteins. New virus, created as copies of the viral RNA genome are packaged with viral proteins, is released from the host cell by a "budding" process.

HIV Structure

HIV (diameter = 120 nm) is an enveloped virus (i.e., its external surface, called an envelope, is derived from the plasma membrane of a host cell). The principal component of the HIV envelope is the glycoprotein *Env*, a complex of three gp120/gp41 heterodimers. (The numbers indicate the size of the protein. For example, gp120 is a glycoprotein with a mass of 120 kDa.) The overall structure of the Env trimer, also called the gp160 spike, consists of a cap composed of three gp120 molecules. These are noncovalently linked to a stem, composed of three gp41 molecules, that anchors Env into the viral envelope. Env is a fusion machine that enables HIV to attach to and fuse with target cells.

HIV contains a cylindrical core within its capsid. Inside the core are two copies of its ssRNA genome. The RNA molecules are coated with a nucleocapsid protein p7, essential for vRNA translation and packaging. The bullet-shape core itself is composed of the core protein p24. Copies of the matrix protein p17 form the inner lining of the viral envelope. The core also contains several enzymes: reverse transcriptase (RT), integrase, and protease. The RT has RNA-directed DNA polymerase activity, which converts the viral ssRNA into a ssDNA and DNA-directed DNA polymerase activity, which converts the viral ssDNA into dsDNA. The RT subsequently degrades the viral ssRNA. Later, when new viral particles are being assembled, the protease cleaves newly synthesized polypeptides to create the protein components of infectious HIV.

The HIV Infection Cycle

The first step in viral entry into target cells is the high-affinity binding of gp120 with the CD4 receptor, a glycoprotein on the surface of target cells, most notably the T-4 helper lymphocytes. T-4 helper cells play a critical role in regulating the activities of other immune system cells. T-cell infection requires the interaction of the gpl20-CD4 complex with a chemokine receptor, which acts as a co-receptor. (The immune system chemotactic agents called chemokines stimulate T cells by binding to receptors on the T-cell plasma membrane.) In the early stages of infection, the co-receptor CCR5 (or less often CXCR4) helps HIV enter T cells; other cells that are known to be infected by HIV include some intestinal and nervous system cells. Recent evidence suggests that humans with two copies of a gene for a variant of the CCR5 receptor, CCR5-Δ32, are resistant to HIV infection. CCR5-Δ32 apparently provides protection against plague and smallpox as well, but for all three diseases, the portion of the population that benefits is relatively small.

The binding of gp120 to the CD4 receptor and the co-receptor causes a conformational change in the Env complex that converts it into a fusion-active state with the exposure of the gp41 fusion protein. Gp41 then proceeds to insert itself into the target cell plasma membrane. Further conformational changes in gp41 cause the viral and cell membranes to be pulled close enough for fusion to take place. Once the viral envelope has fused with the cell's plasma membrane, HIV's RNA and enzymes (RT, integrase, and protease) are injected into the cytoplasm and then transported via microtubules into the nucleus. Along the way, RT catalyzes the synthesis of double-stranded vDNA using vRNA as a template. After the vDNA is integrated into a host cell chromosome, proviral DNA remains latent until the specific infected T cell is activated in an immune response. The proviral DNA then directs the cell to synthesize viral components. Newly synthesized viruses bud from the infected cell.

Within 30 minutes of an active infection in a cell, the expression of 500 cellular genes has been suppressed and 200 have been activated. Within hours, host cell mRNA has largely been replaced by viral mRNA. The virus has crippled the cell's capacity to generate energy and repair virally inflicted DNA damage.

Cell death is triggered by several mechanisms that include the following:

1. A virus activates the genes that induce apoptosis, a normal cell mechanism by which cells respond to external signals such as those that occur in developmental processes.
2. The simultaneous budding of numerous viral particles from the cell membrane may tear the membrane and cause leakages that cannot be repaired, or the massive release of new virus from a cell may so deplete the cell that it disintegrates.
3. The binding of cell surface gpl20 molecules to CD4 receptors on nearby healthy cells leads to the formation of large, nonfunctional multinucleated cell masses called *syncytia*.

HIV Infection and AIDS

HIV infection occurs because of direct exposure of an individual's bloodstream to the body fluids of an infected person. Most HIV is transmitted through sexual contact, blood transfusions, and perinatal transmission from mother to child. Once HIV has entered the body, it infects cells that bear the CD4 antigen on their plasma membranes.

HIV infection progresses through several stages, which may vary considerably in length among individuals. Initial symptoms, which usually occur soon after the initial exposure to the virus and last for several weeks, include fever, lethargy, headache and other neurological complaints, diarrhea, and lymph node enlargement. (Antibodies to HIV are detectable during this period.) Exaggerated versions of these symptoms, referred to as the AIDS-related complex, may often recur. Eventually, the immune system becomes so compromised that

▶

Biochemistry IN PERSPECTIVE cont.

the individual becomes susceptible to serious opportunistic diseases and is said to have developed AIDS. The time required for the development of AIDS may vary from 2 years to 8 or 10 years. For reasons that are not understood, a few patients do not develop AIDS even after 15 years of HIV infection. (It has recently been suggested that some of these individuals are infected with attenuated HIV variants.) Among the most common AIDS-related diseases are *Pneumocystis carinii* pneumonia, cryptococcal meningitis (inflammation of membranes that cover the brain and spinal cord), toxoplasmosis (brain lesions, heart and kidney damage, and fetal abnormalities), cytomegalovirus infections (pneumonia, kidney and liver damage, and blindness), and tuberculosis. HIV infection is also associated with several types of cancer, the most common of which is a rare skin cancer called Kaposi's sarcoma.

There is no cure for AIDS. Treatment seeks to suppress symptoms (e.g., antibiotics for the infections) and to slow viral reproduction. Mortality rates have decreased since 1995 because of the introduction of a treatment protocol called highly active antiretroviral therapy that consists of combinations of drugs from the following categories: (1) entry (fusion) inhibitors (maraviroc and enfuvirtide), (2) nucleoside reverse transcriptase inhibitors (e.g., azidothymidine, also called zidovudine or AZT, and abacavir), (3) nonnucleoside reverse transcriptase inhibitors (NNRTIs: e.g., efavirenz and rilpivirine), (4) protease inhibitors (e.g., indinavir and lopinavir), and (5) integrase inhibitors (raltegravir and elvitegravir).

The development of an AIDS vaccine is problematic because the viral genome mutates frequently (i.e., its surface antigens become altered).

SUMMARY: HIV infection disrupts cell function. By suppressing some cellular genes and activating others, the HIV genome directs the host cell to produce new HIV particles that proceed to infect other cells.

Chapter Summary

1. The information required for directing all living processes is stored in the nucleotide sequences of DNA. DNA is composed of two antiparallel polynucleotide strands that wind around each other to form a right-handed double helix. The deoxyribose-phosphodiester bonds form the backbones of the double helix, and the nucleotide bases project to its interior. The nucleotide base pairs form because of hydrogen bonding between the following bases: adenine with thymine, and cytosine with guanine.
2. Several types of noncovalent interaction contribute to the stability of DNA's structure: hydrophobic and van der Waal's interactions between stacked heterocyclic bases, hydrogen bonds between GC and AT base pairs, and hydration with water molecules. Decoding of the genetic information in DNA requires molecular machinery largely composed of proteins.
3. Mutations are changes in DNA structure, which may be caused by collisions with solvent molecules, thermal fluctuations, ROS, radiation, or xenobiotics. In a transition mutation, one pyrimidine base is substituted for another pyrimidine base or a purine base is substituted for another purine. In a transversion mutation, a pyrimidine base is substituted for a purine base or vice versa. Other mutation types include indels (insertions and deletions), inversions, translocations, and duplications.
4. DNA can have several conformations depending on the nucleotide sequence and the degree of hydration of the double helix. In addition to the classical structure determined by Watson and Crick (B-DNA), A-DNA and Z-DNA have also been observed. DNA supercoiling is a critical feature of several biological processes, such as DNA packaging, replication, and transcription.
5. Each eukaryotic chromosome is composed of nucleohistone, a complex formed by winding a DNA molecule around histone octamers to form nucleosomes. DNA methylation and several types of histone covalent modification (e.g., acetylation and methylation) alter chromatin structure and gene expression. The DNA of mitochondria and chloroplasts is similar to the chromosomes found in prokaryotes.
6. A genome is the full set of the inherited instructions required to sustain an organism's living processes. Although there are some similarities, the genomes of prokaryotes and eukaryotes are substantially different in size, coding capacity, coding continuity, and gene expression mechanisms.
7. The majority of the DNA sequences in humans do not code for proteins or functional RNAs. There are two general classes of intergenic sequences: tandem repeats and interspersed genome-wide repeats. Mobile genetic elements can be duplicated and then move within the genome. Retrotransposons can cause disease by inserting into genes or regulatory sequences.
8. RNA differs from DNA in that it contains ribose (instead of deoxyribose), has a somewhat different base composition, and is usually single-stranded. The forms of RNA involved in protein synthesis are transfer, ribosomal, and messenger RNAs. Transfer RNA molecules have specific amino acids attached to them by specific enzymes and transport these to the ribosome for incorporation into newly

synthesized protein, where they are properly aligned during protein synthesis. The ribosomal RNAs are components of ribosomes, where they are the sites of catalytic activity. Messenger RNA contains within its nucleotide sequence the coding instructions for synthesizing a specific polypeptide. There are several classes of noncoding RNAs that have diverse roles in genome regulation and protection. Examples include miRNAs, siRNAs, snoRNAs, and snRNAs.

9. Viruses are obligate intracellular parasites. Although they are acellular and cannot carry out metabolic activities on their own, viruses can wreak havoc on living organisms. Each type of virus infects a specific type of host (or small set of hosts). This is possible because a virus can either inject its genome into the host cell or gain entrance for the entire viral particle. Each virus has the capacity to use the host cell's metabolic processes to manufacture new copies of itself, called virions. Viruses possess dsDNA, ssDNA, dsRNA, or ssRNA genomes.

10. Retroviruses are a class of RNA viruses that possess a reverse transcriptase activity that converts their RNA genome to a DNA molecule. The retrovirus HIV causes AIDS.

Take your learning further by visiting the **companion website** for Biochemistry at **www.oup.com/us/mckee** where you can complete a multiple-choice quiz on nucleic acids to help you prepare for exams.

Suggested Readings

Cech, T. R., and Steitz, J. A., The Noncoding RNA Revolution: Trashing Old Rules to Forge New Ones, *Cell* 157(1):77–94, 2014.

Cohen, J., Bound for Glory: The Discovery of Antibodies That Foil Almost Every HIV Varient Has Transformed the AIDS Vaccine Search, *Science* 341:1168–1171, 2013.

Cooke, M. S., and Evans, D. D., Reactive Oxygen Species: From DNA Damage to Disease, *Sci. Med.* 10(2):98–111, 2005.

Eliscovich, C., *et al.*, mRNA on the Move: The Road to Its Biological Destiny, *J. Biol. Chem.* 288(28):20361–20368, 2013.

Geisler, S., and Coller, J., RNA in Unexpected Places: Long Noncoding RNA Functions in Diverse Cellular Contexts, *Nat. Rev. Cell Mol. Biol.* 14(11):699–712, 2013.

Gomes, A. Q., Nolasco, S., and Soares, H., Non-Coding RNAs: Multi-Tasking Molecules in the Cell, *Int. J. Mol. Sci.* 14:16010–16039, 2013.

Grigoryev, S. A., Nucleosome Spacing and Chromatin Higher-order Folding, *Nucleus* 3(6):493–499, 2012.

Grigoryev, S. A. and Woodcock, C. L., Chromatin Organization: The 30 nm Fiber, *Exp. Cell Res.* 318:1448–1455, 2012.

Luger, K., Dechassa, L., and Tremethick, D. J., New Insights into Nucleosome and Chromatin Structure: An Ordered State or a Disordered Affair?, *Nat. Rev. Mol. Cell Biol.*, 13(7)436–437, 2012.

O'Bleness, M., *et al.*, Evolution of Genetic and Genomic Features Unique to the Human Lineage, *Nat. Rev. Genet.* 13(12): 853–866, 2012.

Pennisi, E., ENCODE Project Writes Eulogy for Junk DNA, *Science* 337:1159–1161, 2012.

Peschansky, V. J., and Wahlestedt, C., Non-coding RNAs as Direct and Indirect Modulators of Epigenetic Regulation, *Epigenetics* 9(1):3–12, 2014.

Pollard, K. S., What Makes Us Human? *Sci. Am.* 300(5):44–49, 2009.

Van Steensel, B., Chromatin: Constructing the Big Picture, *EMBO J.* 30(10):1885–1895, 2011.

Weiss, R. A., and Stoye, J. P. Our Viral Inheritance, *Science* 340:820–821, 2013.

Yang, L., Froberg, J. E., and Lee, J. T., Long Coding RNAs: Fresh Perspectives into RNA World, *Trends. Biochem. Sci.* 39(1):35–43, 2014.

Key Words

A-DNA, *626*
alkylating agent, *624*
antisense, *651*
base analogues, *624*
B-DNA, *626*
centromere, *639*
chain-terminating method, *647*
Chargaff's rules, *626*
chromatin, *631*
chromosome, *630*
cistron, *651*
CpG, *641*
CpG island, *641*
denaturation, *644*
DNA fingerprinting, *649*
DNA profile, *649*
DNA typing, *649*
endogenous retrovirus, *639*
epigenetics, *641*
epigenome, *641*
epimutation, *643*
euchromatin, *634*
exon, *637*
genes, *616*
gene duplication, *623*
genetics, *602*
genome, *621*
heterochromatin, *634*
hybridization, *645*
hypochromic effect, *644*
indel, *622*
intercalating agent, *625*
interspersed genome-wide repeat, *639*
intron, *637*
inversion, *623*
LINE, *639*
lysogeny, *654*
lytic cycle, *654*
messenger RNA, *651*
metabolome, *617*
methyl-CpG, *641*
microRNA, *651*

Review Questions

These questions are designed to test your knowledge of the key concepts discussed in this chapter before moving on to the next chapter. You may like to compare your answers to the solutions provided in the back of the book and in the accompanying Study Guide.

1. Define the following terms:
   a. molecular biology
   b. genetics
   c. replication
   d. transcription
   e. transcriptome
2. Define the following terms:
   a. transcript
   b. proteome
   c. metabolome
   d. double helix
   e. base stacking
3. Define the following terms:
   a. point mutation
   b. transition mutation
   c. transversion mutation
   d. silent mutation
   e. missense mutation
4. Define the following terms:
   a. single nucleotide polymorphism
   b. nonsense mutation
   c. indel
   d. inversion
   e. translocation
5. Define the following terms:
   a. alkylating agent
   b. base analogue
   c. nonalkylating agent
   d. intercalating agent
   e. ethidium bromide
6. Define the following terms:
   a. Chargaff's rules
   b. constitutive heterochromatin
   c. bacteriophage
   d. replication-dependent histones
   e. replication-independent histones
7. Define the following terms:
   a. A-DNA
   b. B-DNA
   c. Z-DNA
   d. cruciform
   e. palindrome
8. Define the following terms:
   a. positive supercoiling
   b. negative supercoiling
   c. polyamines
   d. chromatin
   e. nucleosome
9. Define the following terms:
   a. histones
   b. heterochromatin
   c. euchromatin
   d. intergenic sequences
   e. tandem repeats
10. Define the following terms:
    a. satellite DNA
    b. transposition
    c. transposon
    d. retrotransposon
    e. endogenous retrovirus
11. Define the following terms:
    a. epigenetics
    b. CpG island
    c. epimutation
    d. DNA methylation
    e. histone acetylation
12. Define the following terms:
    a. DNA typing
    b. short tandem repeats
    c. DNA profile
    d. ribozyme
    e. noncoding RNA

13. Define the following terms:
    a. hypochromic effect
    b. DNA denaturation
    c. restriction endonucleases
    d. DNA hybridization
    e. Southern blotting
14. Define the following terms:
    a. cistron
    b. operon
    c. miRNA
    d. siRNA
    e. snoRNA
15. List three differences between eukaryotic and prokaryotic DNA.
16. List three biological properties facilitated by supercoiling.
17. Describe the structure of a nucleosome.
18. Describe the structural differences between RNA and DNA.
19. Describe the roles of promoters, enhancers, silencers, and insulators in gene expression.
20. Z-DNA derives its name from the zig-zag conformation of phosphate groups. What features of the DNA molecule allow this distinctive structure to form?
21. There is one base pair for every 0.34 nm of DNA and the total contour length of all the DNA in a single human cell is 2 m. Calculate the number of base pairs in a single cell. Assuming that there are 10^{14} cells in the human body, calculate the total length of DNA. How does this estimate compare to the distance from the earth to the sun (1.5×10^8 km)?
22. A DNA sample contains 21% adenine. What is its complete percentage base composition?
23. The melting temperature of a DNA molecule increases as the GC content increases. Explain.
24. What was the impact of the 1953 publication of the Watson-Crick paper on genetic research?
25. Provide the complementary strand and the RNA transcription product for the following DNA template segment: 5′-AGGGGCCGTTATCGTT-3′
26. Describe the structural features that stabilize RNA molecules.
27. Describe how DNA methylation patterns are retained from one cell generation to the next.
28. Which error would cause more damage to a cell: A DNA replication or a DNA transcription error?
29. When an aromatic hydrocarbon intercalates between two stacked base pairs, what effect may there be on DNA structure?
30. What role does the Env trimer play in HIV infection?
31. Chargaff's rules apply to DNA, but not RNA. Explain.
32. In ion torrent DNA sequencing, covalent bond formation is detected by an ion sensor sensitive to protons. What is the source of these protons?
33. Describe how water affects DNA structure.
34. Describe the types of noncovalent interactions that stabilize DNA structure.
35. How do polyamines contribute to DNA structure?
36. Describe in general terms the histone code hypothesis.
37. Describe the evidence that Watson and Crick used to construct their DNA model.
38. Compare the sizes and coding capacity of prokaryotic genomes with those of eukaryotes. What other features distinguish them?
39. The xenobiotic molecule ethyl chloride (CH_3CH_2Cl) is mutagenic. To what class of mutagenic substances does it belong?
40. What are Alu elements? How can they adversely affect human health?

Fill in the Blank

41. The term ____________________ refers to the sum total of all the low-molecular-weight metabolite molecules produced by a cell.
42. The set of proteins produced by a cell is called the ____________________.
43. A base change that has no discernible effect is called a ____________________ mutation.
44. A ____________________ is a point mutation that converts a codon for an amino acid to a premature stop signal.
45. DNA sequences in which multiple copies are arranged next to each other are called ____________________.
46. ____________________ are structures at the end of chromosomes that buffer DNA from the critical loss of coding sequences after a round of replication.
47. Noncoding DNA sequences are called ____________________.
48. Expressed sequences of DNA are called ____________________.
49. Abnormal epigenome changes are referred to as ____________________.
50. Enzymes isolated from bacteria that cut DNA at specific sequences are called ____________________.

Short Answer

51. Pre-eclampsia is the most common dangerous complication of pregnancy. Characterized by high blood pressure and the occurrence of protein in urine, pre-eclampsia occurs most commonly after the 32nd week of pregnancy. Without treatment, patients with pre-eclampsia may progress to eclampsia, a life-threatening disorder, characterized by placental rupture, seizures, coma, and the risk of maternal death. The causes of pre-eclampsia and eclampsia remain uncertain. However, there may be a connection between these disorders and an endogenous retrovirus sequence located on chromosome 7. Explain.

52. Agouti mice are heterozygous for the agouti gene (A^{vy}) (i.e., their genotype is $A^{vy}a$). When an agouti female mouse is mated with a male mouse with the aa genotype, the coat colors of the $A^{vy}a$ offspring vary considerably from yellow to pseudoagouti (brown). If trace amounts of bisphenol A (a chemical used in plastics manufacturing) are added to the diet of pregnant $A^{vy}a$ mice, the coat color of the offspring is shifted toward yellow. Bisphenol A (BPA) is believed to be an endocrine disrupter (a chemical that interferes with the normal action of a hormone in the development of animals, including humans) with estrogen-like activity. Estrogen, like other steroid hormones, alters the expression of genes controlled by hormone response elements. Provide an explanation of how BPA could cause the coat color distribution change in the $A^{vy}a$ offspring.
53. An experiment was conducted with two groups of genetically identical pregnant $A^{vy}a$ mice that were mated with aa male mice. Both groups ate lab chow with BPA (50 mg/Kg). The lab chow of one of the groups was supplemented with methyl donors (e.g., folate, choline, and vitamin B12). The effects of BPA were abolished in the $A^{vy}a$ offspring of the mice fed with the methyl donor–supplemented chow (i.e., the coat color distribution observed was shifted toward brown). Explain.
54. Because the mitochondrial electron transport system is the major source of ROS production in cells, mtDNA molecules are especially vulnerable to oxidative damage because they are attached to the inner membrane on the matrix side. Currently, the mitochondrial free radical theory of aging proposes that oxidative damage to mitochondrial DNA is a major factor in aging. Suggest several DNA-protecting mechanisms that you would expect to find in mitochondria.
55. The polyamines spermine and spermidine (see p. 631) have numerous effects on both prokaryotic and eukaryotic cells. Examples include chromatin condensation, transcription, translation, and apotosis. They are best known for their role in promoting DNA stability. Explain how polyamines enhance DNA structure and promote supercoiling.

Thought Questions

These questions are designed to reinforce your understanding of all of the key concepts discussed in the book so far, including this chapter and all of the chapters before it. They may not have one right answer! The authors have provided possible solutions to these questions in the back of the book and in the accompanying Study Guide for your reference.

56. Under physiological conditions, DNA ordinarily forms B-DNA. However, RNA hairpins and DNA-RNA hybrids adopt the structure of A-DNA. Considering the structural differences between DNA and RNA, explain this phenomenon.
57. What structural features of DNA cause the major groove and the minor groove to form?
58. In contrast to the double helix of DNA, RNA exists as a single strand. What effects does this have on the structure of RNA?
59. Jerome Vinograd found that when circular DNA from a polyoma virus is subjected to cesium chloride gradient centrifugation, it separates into two distinct bands, one consisting of supercoiled DNA and the other of relaxed DNA. Explain how you would identify each band.
60. 5-Bromouracil is an analogue of thymine that usually pairs with adenine. However, 5-bromouracil frequently pairs with guanine. Explain.
61. The flow of genetic information is from DNA to RNA to protein. In certain viruses, the flow of information is from RNA to DNA. Does it appear possible for that information flow to begin with proteins? Explain.
62. You wish to isolate mitochondrial DNA without contamination with nuclear DNA. Describe how you would accomplish this task.
63. Unlike linker DNA and deproteinized DNA, DNA segments wrapped around histone cores are relatively resistant to the hydrolytic actions of nucleases. Explain.
64. The set of mRNAs present within a cell changes over time. Explain.
65. DNA and RNA are information-rich molecules. Explain the significance and implications of this statement.
66. A 10-year-old murder case in a small town has finally been solved, thanks to the efforts of a detective who took advantage of a new statewide DNA database containing samples of DNA from convicted felons. Explain how this case might have been solved, starting with the arrival of a crime scene unit at the murder scene. What technological advances made this case solvable?
67. Some living organisms are under considerable pressure to streamline their genomes for the sake of more efficient operation. As a result, the mitochondria of eukaryotic species have lost, to one degree or another, the overwhelming majority of their genes. During this process, several hundred mitochondrial genes were transferred to the nuclear genome. Yet mitochondria still retain a genome with the capacity to produce several electron transport proteins. Review mitochondrial electron transport and suggest a reason why these energy-generating organelles retained the genes to produce this set of molecules.
68. During the infective process, viruses may incorporate segments of the host genome into their structures. Explain how this phenomenon can contribute to the formation of new diseases or exacerbate existing pathologies.
69. It has been estimated that each phosphate group in B-DNA can form hydrogen bonds with six water molecules. Draw a

diagram of a DNA phosphodiester linkage with its associated water molecules.

70. Alkylating agents are compounds that react with hydroxyl and amino groups to form alkylated molecules. For example:

$$\text{Cytosine} \xrightarrow{CH_3I} N\text{-Methylaminocytosine}$$

Cytosine **N-Methylaminocytosine**

When an individual is accidentally exposed to these substances, a variety of physiological systems are affected via genome alterations. Cancer (uncontrolled cell proliferation caused by genome damage) is not an uncommon result, in part, of a series of such exposures. Using your knowledge of the genome, suggest in general terms how cancer might occur.

71. Identical twin brothers begin life with identical genomes and epigenomes. How will this circumstance change with age? Suggest how these changes could be used as a forensic tool.

72. The suggestion has been made that DNA extracted from ancient fossils could be extracted, cloned, and used to resurrect an extinct species. Comment on the practicality of this idea.

73. Fluorouracil is a structural analogue of thymine. The fluorine promotes enolization. How is this effect used in the treatment of cancer?

CHAPTER

18 Genetic Information

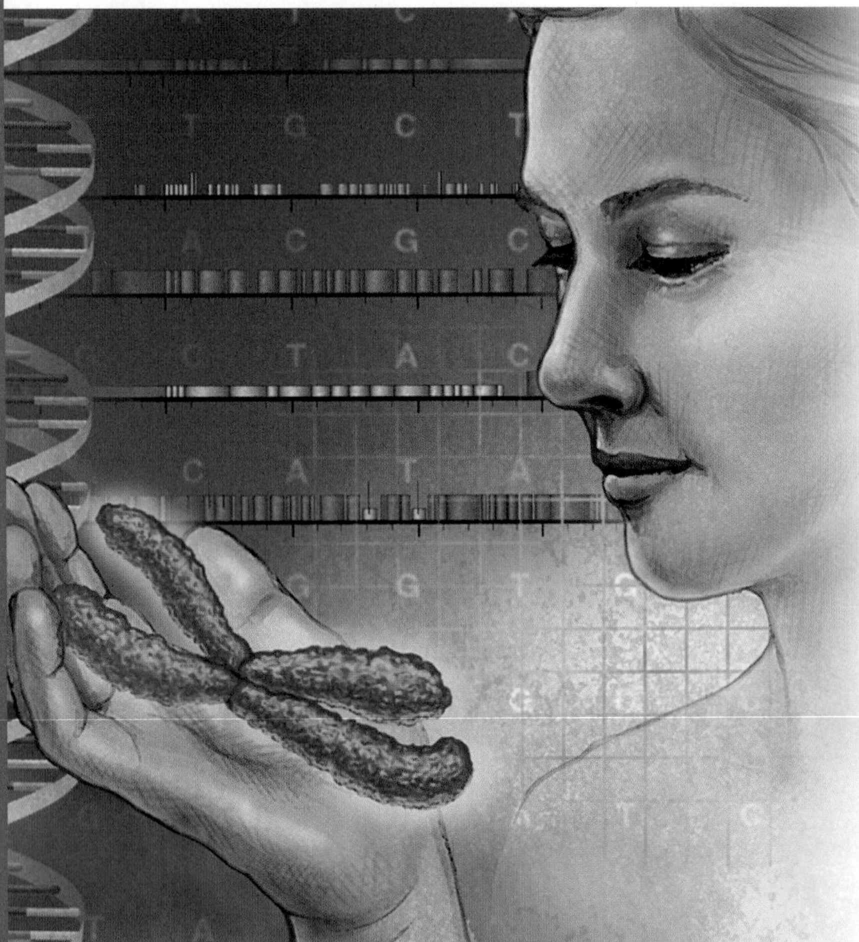

The Human Genome and ENCODE (*Enc*yclopedia *o*f *D*NA *E*lements) Projects The Human Genome Project, performed by thousands of scientists in public and private laboratories around the world, took more than 15 years to accomplish. Researchers are just beginning to interpret and utilize the resulting tidal wave of biological information to solve the medical and biological problems of humans. The ENCODE project, intended as a follow-up to the Human Genome Project, aims to identify all functional elements in the human genome.

OUTLINE

DNA and Chimeras: A Biological and Legal Mystery

In 2002, the lives of two women intersected in a very surprising and unexpected way when their status as mothers was challenged. Living on opposite coasts of the United States, both women, L.F. in Washington State and K.K. in Boston, were leading ordinary lives until they underwent genetic testing. In December 2002 L.F., the pregnant mother of two other children, received the results of the DNA testing performed in support of her welfare application. Separated from her children's father and currently unemployed, L.F. could only receive temporary public support if the father's paternity was confirmed. To her utter astonishment, the test confirmed the father's paternity, but revealed that she could not be her children's mother. Despite hospital birth records and assurances from family members and her obstetrician who witnessed the births, state prosecutors proceeded, solely on the basis of DNA test results, to charge L.F. with welfare fraud and threaten her with the loss of her children.

DNA profile analysis, the result of research spanning several decades, is considered the most accurate means of identifying individuals. Used in parental testing and forensic investigations, DNA profiling technology takes advantage of two facts: DNA is inherited from parents and, with the exception of identical twins, the DNA profile of each human is unique. How could well-accepted DNA profiling science be reconciled with the puzzling DNA profile results of L.F. and her children?

As a consequence of L.F.'s vehement protests, a trial judge appointed a court representative who witnessed the birth of the third child and the taking of blood samples from L.F. and the newborn infant. Subsequent DNA testing yielded a bewildering result: L.F. was not the mother of this child either. Fortunately, a prosecution lawyer found a precedent for this remarkable phenomenon in a recently published article in the *New England Journal of Medicine* that reported a similar case.

K.K., a Boston teacher, was a *chimera*, an individual with two genetically distinct, intermingled cell lines. This discovery was the result of research triggered by a 1998 search for a suitable kidney donor that involved the testing of K.K. and her family members for major histocompatability complex (MHC) antigens. MHC antigens, the cell surface proteins coded for by genes on chromosome 6, allow immune system T cells to distinguish between self and nonself (foreign cells). Although organ rejection is suppressed by antirejection drugs, transplant success is enhanced if the MHC antigens of the donor and the recipient are either identical or a close match. Genetic testing of blood samples from K.K. and her family (husband and three sons) revealed that K.K. was not the biological mother of two of her three sons.

Several years of intense research by genetic investigators at a Boston medical center eventually revealed that K.K. is a *tetragametic chimera*, an individual who developed from the fusion of two nonidentical zygotes (fertilized eggs) formed from four genetically distinct gametes (two eggs and two sperm). If fusion had not occurred, the two zygotes would have developed into fraternal female twins. The genetic markers found in two of her sons (and later in her brother's genetic profile) that differed from those in blood and saliva were subsequently found in several other tissues (e.g., skin and thyroid cells). The researchers concluded that K.K. had two separate egg cell types in her ovaries.

As a result of the 2002 paper detailing the chimera research, several tissues of L.F. were tested. She was subsequently declared to be the mother of her children when the DNA markers missing in her blood were located in cervical epithelial cells.

Overview

ALL LIVING ORGANISMS ARE INFORMATION-PROCESSING SYSTEMS. THEIR ULTIMATE SOURCE OF INFORMATION IS ENCODED IN THE NUCLEOTIDE BASE sequence of DNA. As biochemists have searched ever more deeply into the mysteries of genetic information storage and transmission, of how DNA is replicated and gene expression is controlled, they have transformed all of the life sciences. The knowledge and technologies acquired during this pursuit have provided us with an understanding of the intricacies of living processes that is still unfolding.

In any successful information-based system the instructions required to produce a certain type of organization (e.g., for building a house or for reproducing a living organism) must be stably stored to safeguard their accuracy and availability for use. Information must also be converted into a form that can be utilized. Living organisms have partitioned these functions as follows. DNA is a relatively stable molecule with structural features that maximize information storage and facilitate duplication. RNA molecules, more reactive than DNA, have numerous roles in protein synthesis and gene expression regulation. Finally, the proteins have diverse and flexible three-dimensional structures that perform most of the tasks that sustain the living state.

Although DNA contains the genetic information that drives living processes, it does not directly control cellular processes. The decoding of DNA base sequences requires molecular machinery, largely composed of proteins and ncRNAs and powered by cellular energy resources. These machines bend, twist, unwind, and unzip DNA during replication and transcription. At first glance, the seemingly repetitious and regular structure of DNA makes it an unlikely partner for the productive binding of specific base sequences with appropriate proteins. However, numerous contacts (often about 20 or so) involving hydrophobic interactions, hydrogen bonds, and ionic bonds between amino acid residues and the edges of bases within the major groove (and to a lesser extent the minor groove) of DNA result in highly specific DNA-protein binding. The three-dimensional structures of most DNA-binding proteins analyzed thus far have surprisingly similar features. In addition to usually possessing a twofold axis of symmetry, many of these molecules can be separated into families (**Figure 18.1**) on the basis of the following structures: (1) helix-turn-helix, (2) helix-loop-helix, (3) leucine

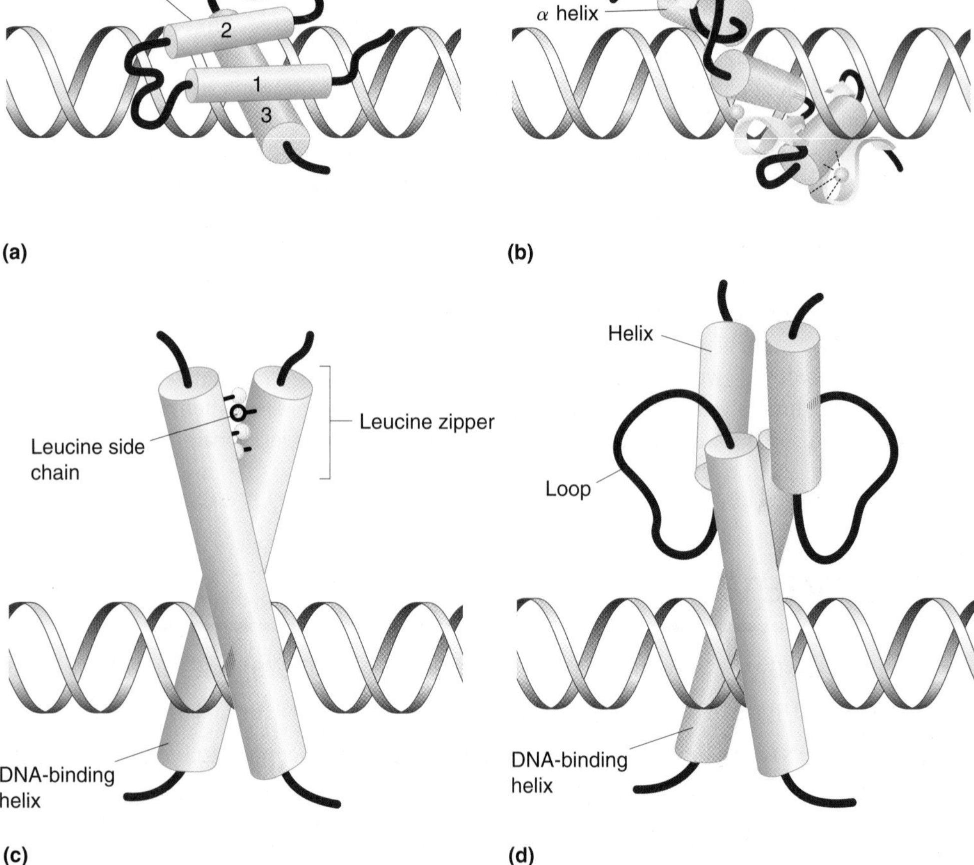

FIGURE 18.1
DNA-Protein Interactions

DNA-binding proteins contain specific structural motifs for interacting with DNA: (a) helix-turn-helix, (b) zinc fingers, (c) leucine zipper, and (d) helix-loop-helix.

zipper, and (4) zinc finger. DNA-binding proteins, many of which are transcription factors, often form dimers. For example, a variety of transcription factors with leucine zipper motifs form dimers as their leucine-containing α-helices associate via van der Waals interactions.

Chapter 18 provides an overview of the mechanisms that living organisms use to synthesize the nucleic acids DNA and RNA that direct cellular processes. The chapter begins with a discussion of several aspects of DNA replication, repair, and **recombination** (the reassortment of DNA sequences). This is followed by descriptions of the synthesis and processing of RNA. Also included is an overview of several of the basic tools of biotechnology that biochemists use to investigate living processes. Chapter 18 ends with a section devoted to gene expression, the mechanisms cells use to produce gene products in an orderly and timely manner.

18.1 GENETIC INFORMATION: REPLICATION, REPAIR, AND RECOMBINATION

All viable living organisms possess the following features: rapid and accurate DNA synthesis and genetic stability maintained by effective DNA repair mechanisms. Paradoxically, the long-term survival of species also depends on genetic variations that allow adaptation to changing environments. In most species these variations arise predominantly from genetic recombination, although mutation also plays a role. The following sections discuss the mechanisms that prokaryotes and eukaryotes use to achieve these goals.

DNA Replication

DNA replication occurs before every cell division. The mechanism by which DNA copies are produced is similar in all living organisms. After the two strands have separated, each serves as a template for the synthesis of a complementary strand (**Figure 18.2**). (In other words, each of the two new DNA molecules contains one old strand and one new strand.) This process, referred to as **semiconservative replication**, was first demonstrated in an elegant experiment reported in 1958 by Matthew Meselson and Franklin Stahl. Refer to the Online Biochemistry in Perspective essay The Meselson-Stahl Experiment for a description of the historic work that proved that DNA replication is a semiconservative process.

In the years since the Meselson and Stahl experiment, many of the details of DNA replication have been discovered. Until recently it was assumed that DNA replication machinery moved along a DNA "track" that is for the most part stationary. Recent research efforts have revealed that DNA replication occurs in specific nuclear or nucleoid compartments called **replication factories**, which are relatively stationary during the process of replication. The replication machinery within these factories performs as an energy-driven DNA pump. In prokaryotes, replication factories are attached to the cell membrane. In eukaryotes, replication factories assemble during the synthesis phase of the cell cycle (S. phase: see later, **Figure 18.12**) in association with the nuclear matrix.

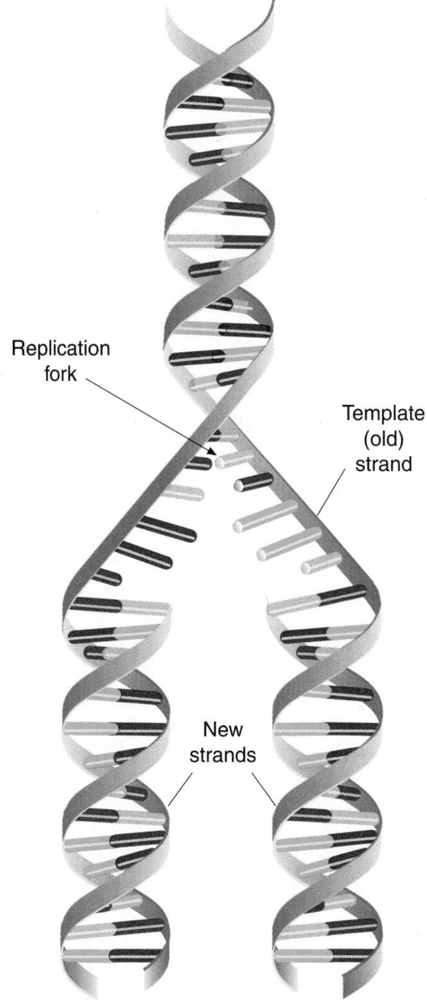

FIGURE 18.2
Semiconservative DNA Replication
As the double helix unwinds at the replication fork, each old strand serves as a template for the synthesis of a new strand.

DNA SYNTHESIS IN PROKARYOTES DNA replication in *E. coli* consists of several basic steps, each of which requires enzymatic activities associated with the following processes: DNA unwinding, primer synthesis, and DNA polynucleotide synthesis. DNA fragment ligation and supercoiling control are also described.

DNA unwinding Helicases, as the name implies, are ATP-requiring motor proteins that separate the annealed strands of duplex DNA. The principal helicase in *E. coli* is DnaB, a ring-shape hexamer that opens the double helix.

Primer synthesis The formation of short RNA segments complementary to a single-stranded DNA template called **primers** is required for the initiation of DNA replication. Primer synthesis is catalyzed by **primase**, an RNA polymerase. Primase is a 60- kDa polypeptide product of the dnaG gene. A multienzyme complex containing primase and several auxiliary proteins is called the **primosome.**

DNA synthesis The synthesis of a complementary DNA strand in the $5' \rightarrow 3'$ direction by forming phosphodiester linkages between nucleotides base-paired to a template strand is catalyzed by large multienzyme complexes referred to as the DNA polymerases (**Figure 18.3**). In the current model for the catalytic mechanism of DNA polymerases, illustrated in **Figure 18.4**, the 3′-hydroxyl oxygen is a nucleophile that attacks the α-phosphate of the incoming nucleotide to form a new P—O bond. With a polymerization rate of 1000 nt/s, DNA polymerase III (pol III) is the major DNA-synthesizing enzyme in prokaryotes. The pol III holoenzyme is composed of at least 10 subunits (Table 18.1). The core polymerase is formed from three subunits: α, ε, and θ. The β-protein (also called the sliding clamp protein, or the β_2-clamp) is composed of two subunits (**Figure 18.5**). It forms a donut-shape ring around the template DNA strand. The γ-complex is composed of τ, γ, δ, δ', χ, and ψ. Of these, τ, γ, δ, and δ' contain a motor ATPase domain that uses the energy released by ATP hydrolysis to catalyze the mechanical function of DNA clamp loading. The γ-complex recognizes single DNA strands with primer and, acting as a **clamp loader**, transfers the β_2-clamp dimer to the core polymerase, where it forms a closed ring around the DNA strand. The inside diameter of the β_2-clamp is about 3.5 Å larger than dsDNA, large enough for hydrated DNA strands to slide through easily. The β_2-clamp promotes **processivity**; that is, it prevents frequent dissociation of the polymerase from the DNA template. The γ-complex is ejected in a process driven by ATP hydrolysis, and the pol III holoenzyme can proceed to replicate DNA. Note that the subunit τ allows two core enzyme complexes to form a dimer, which also improves processivity. The DNA replicating machine, called the **replisome**, is composed of two copies of the pol III holoenzyme, the primosome, and DNA unwinding proteins.

E. coli also possesses four other DNA polymerases: I, II, IV, and V. DNA polymerase I (pol I), the first DNA polymerase to be discovered (Arthur Kornberg, Nobel Prize in Physiology or Medicine, 1959), is a versatile enzyme with several roles in DNA replication and repair. It has three distinctly different catalytic activities: $5' \rightarrow 3'$ exonuclease activity, $5' \rightarrow 3'$ template-directed polymerase activity, and $3' \rightarrow 5'$ exonuclease activity. (An **exonuclease** is an enzyme that removes nucleotides from the end of a polynucleotide strand.) Pol I's $5' \rightarrow 3'$ exonuclease activity removes (depolymerizes) mispaired segments of DNA during replication. This same activity also removes RNA primers. Once the RNA primer is removed, the $5' \rightarrow 3'$ template directed polymerase activity of pol I replaces ribonucleotides with deoxyribonucleotides. (Note that pol I is a slow enzyme (18 nt/s) with low processivity, in contrast to Pol III, the major replicating enzyme.) As pol I synthesizes short DNA segments, it also uses its proofreading $3' \rightarrow 5'$ exonuclease activity to ensure accuracy. Pol I also has important roles in several types of postreplication damage repair (p. 676–681).

Pols II and IV are repair enzymes normally present in cells where they repair low levels of DNA damage. Exposure of cells to high levels of UV light or mutagenic chemicals activates an emergency repair mechanism called the *SOS response*. SOS response–induced enzymes and other proteins prevent cell death caused by high levels of DNA damage that prevent replication. Early in the SOS response, pol II and pol IV gene expression is upregulated, to be followed later

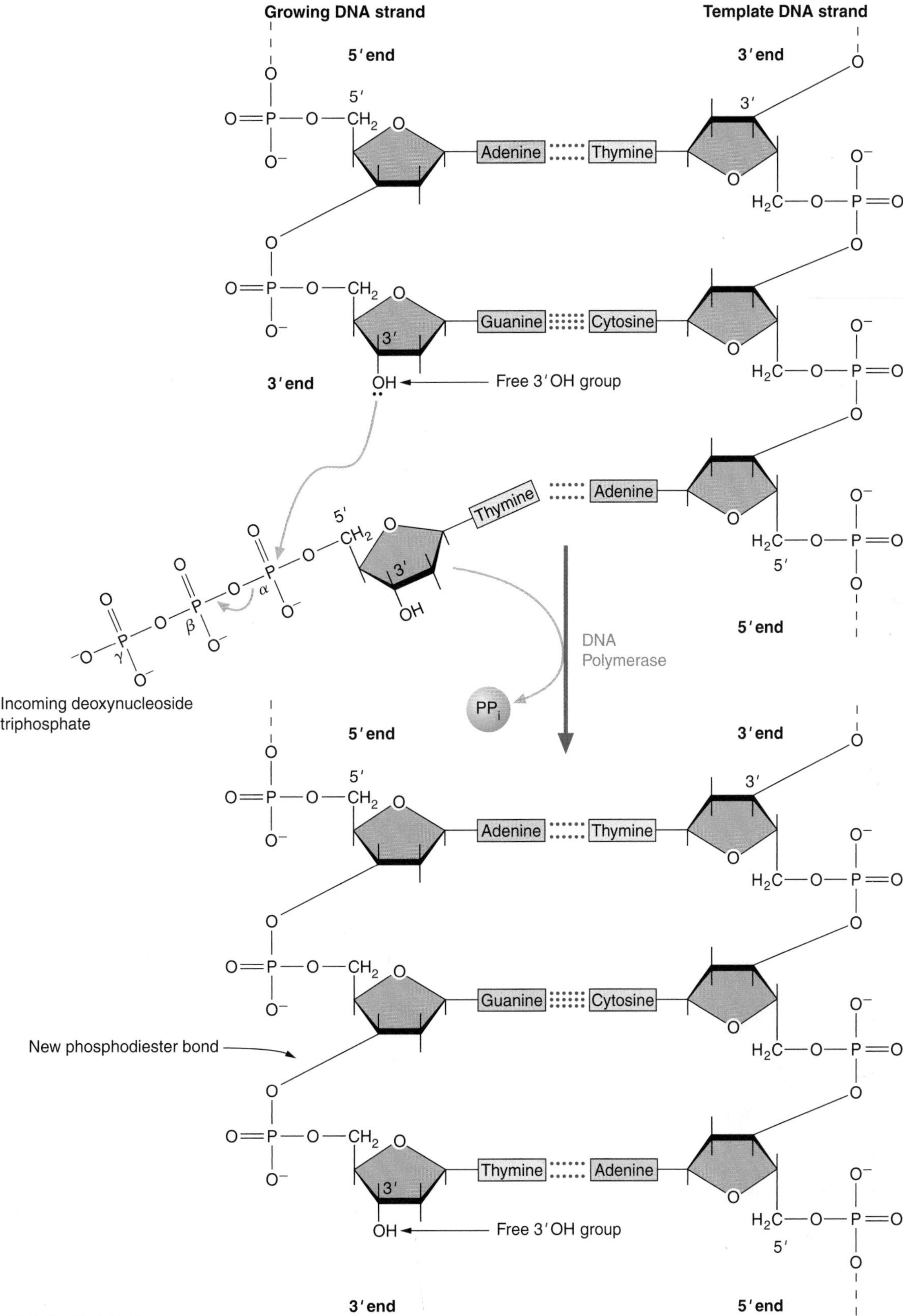

FIGURE 18.3

The DNA Polymerase Reaction

The essential feature of DNA synthesis is the formation of a phosphodiester linkage between a growing 5′ → 3′ DNA strand and an incoming dNTP (deoxyribonucleoside triphosphate). The bond is created by a nucleophilic attack of the 3′-hydroxyl group of the terminal residue on the α-phosphate of the dNTP. Pyrophosphate (PP_i), formed as the leaving group, is then hydrolyzed. The energy released by pyrophosphate hydrolysis drives the overall process forward.

FIGURE 18.4

Mechanism of DNA Polymerases

All DNA (and RNA) polymerases apparently use the same mechanism for template-driven nucleotide polymerization: an in-line nucleotidyl transfer. In this example from *E. coli* pol I, two Mg^{2+} ions (labeled A and B) are coordinated with the α-phosphate group of the incoming nucleotide (dNTP) and are themselves bridged by two aspartate side chain carboxylate groups. One ion, Mg^{2+} A, lowers the affinity of the 3′-hydroxyl oxygen for its hydrogen atom. The 3′-oxygen is a nucleophile that attacks the α-phosphate to form a new P—O bond. Both metal ions stabilize the negative charge of the transition state. By stabilizing the negatively charged pyrophosphate, Mg^{2+} B facilitates its departure.

TABLE 18.1 Subunits of DNA Polymerase III

		Subunit	Function
Holoenzyme	Core polymerase	α	5′ → 3′ polymerase
		ε	3′ → 5′ exonuclease
		θ	Assists ε
	γ-Complex (clamp loader)	τ	ATPase, assists in dimerization of core
		γ	ATPase
		δ and δ'	ATPases, stimulates clamp loading
		χ and ψ	Stabilize the clamp loader complex and promate replication initiation
		β	Sliding clamp as β_2

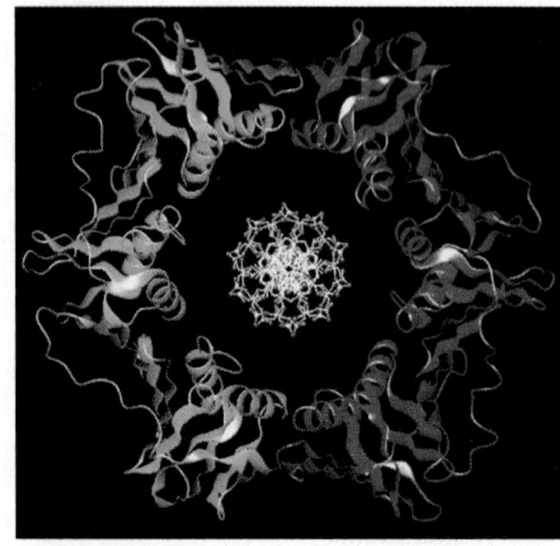

FIGURE 18.5

Cross-Section of the β_2-Clamp of DNA Polymerase III

The β-protein is a dimer (shown in red and orange) that encircles the DNA and acts like a clamp.

by expression of the pol V gene. These repair enzymes are referred to as the *translesion polymerases* because their structures allow the use of damaged DNA as a template. When replication stalls because pol III has encountered a lesion (e.g., base adducts, base loss, or thymine dimers), then pol III is replaced with one of the repair enzymes. Once the lesion has been bypassed by the incorporation of nucleotides opposite to the lesion, the translesion polymerase is removed and replaced with pol III, which continues replication until the next lesion is encountered. Unfortunately, repair by the SOS system comes at a high cost because it is error prone. The structural features of pol IV and V, and to a lesser extent pol II, that allow binding to DNA lesions also decrease accuracy. Mutations result if the errors introduced by the translesion polymerases are not corrected by postreplication repair (pp. 676–681).

Joining DNA fragments Frequently, during DNA replication (as well as DNA repair and recombination processes) DNA strand segments must be joined together. An enzyme called **DNA ligase** catalyzes the formation of a covalent phosphodiester bond between the 3′-OH end of one segment and the 5′-phosphate end of an adjacent segment.

Supercoiling control DNA *topoisomerases* prevent tangling of DNA strands. They function ahead of the replication machinery to relieve *torque* (rotary force) that can slow down the replication process. The generation of torque is a very real problem because the double helix unwinds rapidly (as many as 50 revolutions per second during bacterial DNA replication). Topoisomerases are enzymes that change the supercoiling of the DNA (see p. 628) by breaking one or both strands, which is followed by passing the DNA through the break and rejoining the strands. The terms *topoisomerase* and *topoisomers* (DNA molecules that differ only in their degree of supercoiling) are derived from *topology*, a branch of mathematics that examines changes in shape or position that can be achieved without cutting. When appropriately controlled, supercoiling can facilitate the unzipping of DNA molecules. Type I topoisomerases produce transient single-strand breaks in DNA; type II topoisomerases produce transient double-strand breaks. In prokaryotes, a type II topoisomerase called DNA gyrase helps to separate (decatanate) the replication products (i.e., linked circular chromosomes) and to create the negative (−) supercoils required for genome packaging.

THE PROKARYOTIC REPLICATION PROCESS The replication of the circular *E. coli* chromosome (**Figure 18.6**) begins when there is a high ATP:ADP ratio and sufficient copies of the DNA-binding protein DnaA (53 kDa) that serve to initiate the replication process by melting the dsDNA and attracting other replication proteins to the site (**Figure 18.7**). The initiation site on the *E. coli* chromosome is referred to as oriC. Once it has been initiated, replication proceeds in two directions. Helicases unwind the DNA duplex, two replisomes assemble, and replication proceeds outward in both directions. As the two sites of active DNA synthesis (referred to as **replication forks**) move farther away from each other, a "replication eye" forms. Because an *E. coli* chromosome contains one initiation site, it is considered a single replication unit. A replication unit, or **replicon**, is a DNA molecule (or DNA segment) that contains an initiation site and appropriate regulatory sequences.

When DNA replication was first observed experimentally (using electron microscopy and autoradiography), investigators were confronted with a paradox. The bidirectional synthesis of DNA as it appeared in their research seemed to indicate that continuous synthesis occurs in the 5′ → 3′ direction on one strand and in the 3′ → 5′ direction on the other strand. However, all the enzymes that catalyze DNA synthesis do so in the 5′ → 3′ direction only. It is now known that only one strand, referred to as the *leading strand*, is continuously synthesized in the 5′ → 3′ direction. The other strand, the *lagging strand*, is also synthesized in the 5′ → 3′ direction but in small segments (**Figure 18.8**). Reiji Okazaki and his

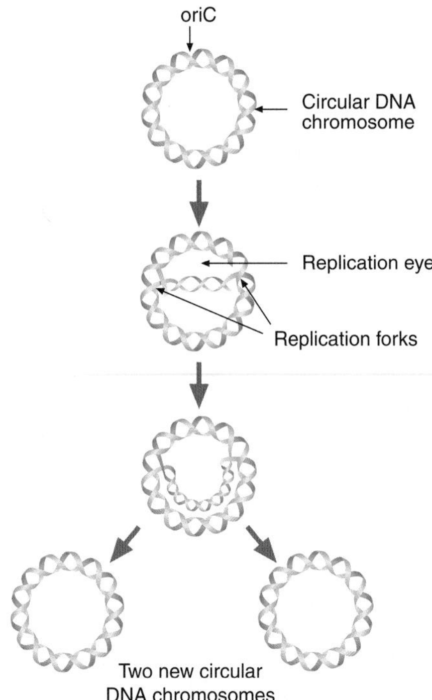

FIGURE 18.6
Replication of Prokaryotic DNA

As DNA replication of a circular chromosome proceeds, two replication forks can be observed using autoradiography. The structure that forms is called a replication eye.

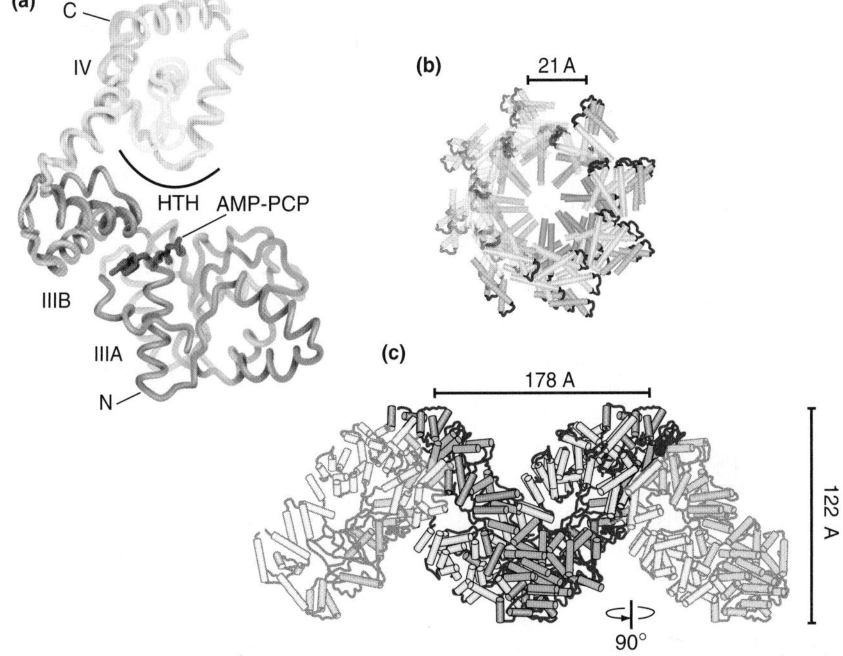

FIGURE 18.7
DnaA Structure

(a) DnaA consists of four domains: III (red and green) and IV (yellow) are illustrated. In this illustration AMP-PCP, an analogue of ATP (designated in dark blue), is bound in the ATP-binding site of DnaA. When ATP binds to the ATP-binding site within domain IIIA of DnaA monomers, they undergo a conformational change that facilitates the formation of oligomers that bind, via domain IV, to highly conserved 9-bp sequences called DnaA boxes (HTH = helix-turn-helix motif) (b) and (c) Top and side views, respectively, of DnaA oligomers. The architecture of the DnaA oligomer (an open right-handed helix) causes DNA that contains DnaA boxes to wrap around its exterior.

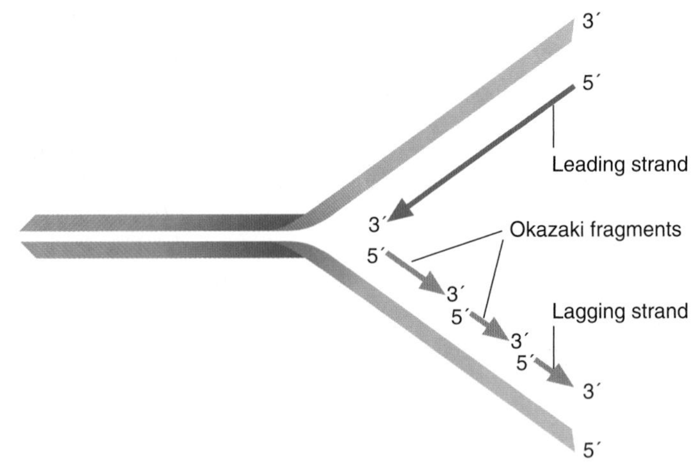

FIGURE 18.8
DNA Replication at a Replication Fork

The 5′ → 3′ synthesis of the leading strand is continuous. The lagging strand is also synthesized in the 5′ → 3′ direction but in small segments (Okazaki fragments).

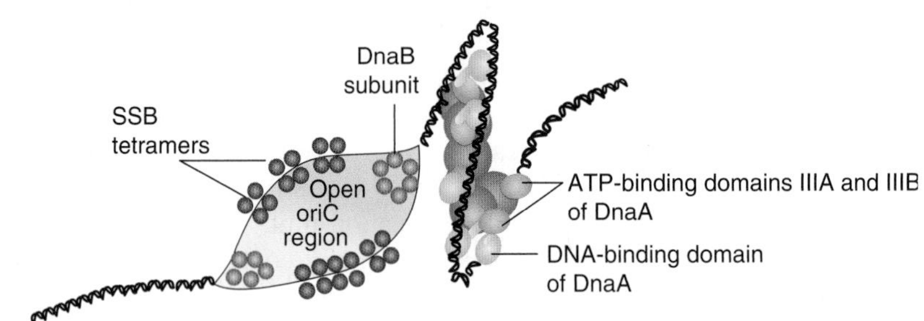

FIGURE 18.9
Replication Fork Formation

After DnaA and DnaB binding, the DnaB helicase separates the duplex DNA strands at two replication forks. Each DNA-binding domain (yellow) of DnaA binds to a 9-bp DnaA box. The binding of SSB to newly formed ssDNA prevents reassociation of the single strands. The helicase loader DnaC (not shown) facilitates the binding of DnaB helicase to DNA.

colleagues provided the experimental evidence for discontinuous DNA synthesis. Once the RNA primer is removed from each segment and replaced by DNA, these lagging strand segments, now called **Okazaki fragments**, are covalently linked together by DNA ligase. In prokaryotes such as *E. coli*, Okazaki fragments possess approximately 1000 nucleotides.

Replication begins when DnaA binds to five to eight 9-bp sites, referred to as *DnaA boxes*, within the oriC sequence. Prokaryotic organisms vary in the number of DnaA boxes. *E. coli* has five DnaA boxes. The oligomerization of DnaA, which results in a nucleosomelike structure, requires ATP and the histonelike protein HU. As the DnaA-DNA complex forms, localized "melting" of the DNA duplex in a nearby region containing three 13-bp repeats causes a small segment of the double helix to open up (**Figure 18.9**). *DnaB* (a 300-kDa helicase composed of six subunits), complexed with *DnaC* (29 kDa), then enters the open oriC region. When DnaB is loaded onto the DNA, the DnaC is released. The replication fork moves forward as DnaB unwinds the helix. Topoisomerases relieve torque ahead of the replication machinery. As DNA unwinding proceeds, DnaA is displaced. The hydrolysis of bound ATP molecules causes DnaA to revert to an inactive conformation that is incapable of binding DNA. The single strands are kept apart by the binding of numerous copies of *single-stranded DNA-binding protein* (SSB). SSB, a tetramer, may also protect vulnerable ssDNA segments from attack by nucleases.

A model of DNA synthesis at a replication fork is illustrated in **Figure 18.10**. For pol III to initiate DNA synthesis, an RNA primer must be synthesized. On the leading strand, where DNA synthesis is continuous, primer formation occurs only once per replication fork. In contrast, the discontinuous synthesis on the lagging strand requires primer synthesis for each of the Okazaki fragments. The primosome travels along the lagging strand and stops and reverses direction at intervals to synthesize a short RNA primer. Subsequently, pol III synthesizes DNA beginning at the 3′ end of the primer. As lagging strand synthesis continues, the RNA primers are removed by pol I, which also synthesizes a complementary DNA segment. DNA ligase then joins the Okazaki fragments.

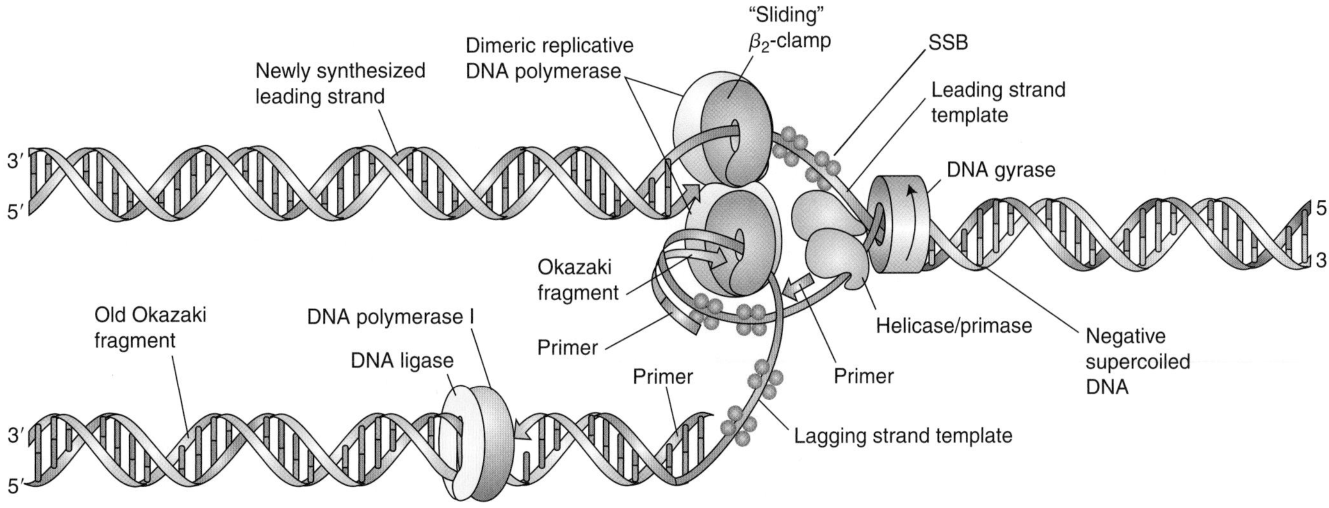

FIGURE 18.10

***E. coli* DNA Replication Model**

The DNA duplex at the replication fork is unwound by DnaB and DNA gyrase (a topoisomerase). DNA gyrase uses ATP hydrolysis to introduce negative supercoils into DNA just ahead of the replication fork, which relieves torque, thereby returning the DNA molecule to a more relaxed state. As the helix unwinds, an RNA primer is synthesized on the lagging strand. New strand synthesis is catalyzed by two pol III holoenzymes tethered to each other by the τ-complex (not shown). Each pol III complex is linked to a β_2-clamp that tethers the polymerase to the template strand. Both leading and lagging strand synthesis move in the same direction because the lagging strand is looped out before it enters the pol III holoenzyme. When the lagging strand pol III complex completes an Okazaki fragment, it releases the strand. (Because of the alternate lengthening and shortening of the lagging strand, this replication model is referred to as the trombone model.) The primosome then synthesizes a new primer. Working together, pol I and DNA ligase remove the primer and fill and seal the gaps between the Okazaki fragments. Pol III rebinds the lagging strand at a new primer and begins the synthesis of a new Okazaki fragment.

As illustrated in **Figure 18.10**, the synthesis of both the leading and the lagging strands is coupled. The tandem operation of two pol III complexes requires that one strand (the lagging strand) be looped around the replisome. When the lagging strand pol III complex completes an Okazaki fragment, it releases the duplex DNA by severing its connection to the sliding clamp. It then reassociates with a new sliding clamp assembled on the newly synthesized RNA primer by γ-complex that is directly adjacent to the replication fork. This allows the primosome to move in and synthesize the next RNA primer.

Despite the complexity of DNA replication in *E. coli*, as well as its high processivity rate, this process is amazingly accurate: approximately one error per 10^9 to 10^{10} base pairs per generation. This low error rate is largely a consequence of the precise nature of the copying process itself (i.e., complementary base pairing). Within the active site of pol III and pol I there is a pocket that is precisely shaped to fit a nucleotide base pair in which a purine and a pyrimidine are properly aligned by hydrogen bonds and van der Waals interactions. If the nucleotide bases are mismatched, they do not fit into the pocket, and the incoming nucleotide usually leaves the site before the reaction occurs. Both pol III and pol I also proofread newly synthesized DNA. Most mispaired nucleotides are removed (by the $3' \rightarrow 5'$ exonuclease activities of pol III and pol I) and then replaced. Several postreplication repair mechanisms also contribute to the low error rate in DNA replication.

Replication ends when the replication forks meet on the other side of the circular chromosome at the termination site, the *ter* (τ) region. The ter region in *E. coli* is composed of six 20-bp terminator sites. When ter binds to *tus,* an antihelicase that halts DNA polymerase movement, replication arrest results. The asymmetric tus-ter complex prevents a replication fork from traveling in one

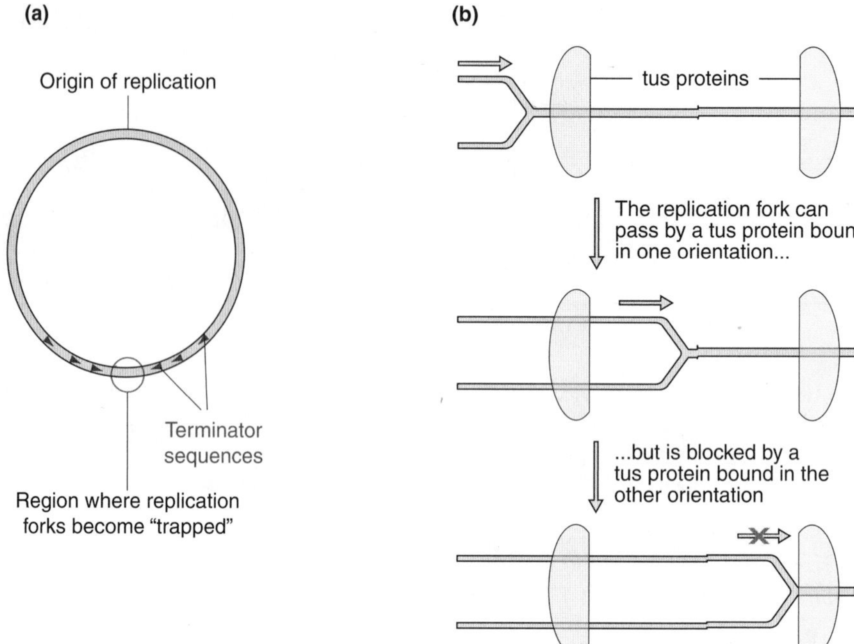
(a)
(b)
Origin of replication
Terminator sequences
Region where replication forks become "trapped"
tus proteins
The replication fork can pass by a tus protein bound in one orientation...
...but is blocked by a tus protein bound in the other orientation

FIGURE 18.11

Role of tus in DNA Replication Termination in *E. coli*

(a) Within the ter region of the *E. coli* chromosome, there are six termination sequences. The arrowheads that designate these sequences indicate the direction that each sequence can be passed by a replication fork. (b) A ter sequence binds a pair of tus proteins, oriented in reverse directions. Depending on its binding orientation on a dsDNA ter segment, tus can prevent unwinding. In this diagram a replication fork is passing by the left-handed tus. The tus bound at this site is disrupted by the DnaB helicase, so the replication fork passes through unobstructed until it encounters the second tus protein. The second tus protein (on the right), oriented on the ter sequence in the opposite direction from the left-handed tus, blocks the replication fork by inhibiting DnaB.

direction but not the other via direction-dependent inhibition of DnaB helicase (**Figure 18.11**). After replication ends, the replisomes disassemble and the two daughter molecules are separated by a type II topoisomerase.

DNA SYNTHESIS IN EUKARYOTES Although the principles of DNA replication in prokaryotes and eukaryotes have a great deal in common (e.g., semiconservative replication, the DNA polymerase mechanism, and bidirectional replicons), they also have significant differences. Not surprisingly, these differences appear to be related to the size and complexity of eukaryotic genomes.

DNA polymerases There are 15 eukaryotic DNA polymerases. Of these, 3 (α, δ, and ε) are involved in nuclear DNA replication. DNA polymerase α (pol α) is a primase that initiates DNA replication by synthesizing a short 10-nt RNA segment followed by a 10- to 20-nt DNA segment. After primer synthesis on the leading strand, DNA synthesis is continued by pol ε. Pol δ is the lagging strand polymerase. Both pol δ and pol ε are highly accurate and processive polymerases with $3' \rightarrow 5'$-exonuclease proofreading activity. Pol δ corrects errors made by pol α. Polymerases β, ζ (zeta), and η (eta) function in nuclear DNA repair. Pol γ replicates and repairs mitochondrial DNA. Unlike the prokaryotic DNA polymerases, none of the eukaryotic enzymes removes RNA primers. Instead, the enzymes Dna2 with nuclease and helicase activities and FEN1, an endonuclease, remove the primers.

Timing of replication In contrast to rapidly growing bacterial cells, in which replication occurs throughout most of the cell division cycle, eukaryotic replication is limited to a specific period referred to as the *S phase* (**Figure 18.12**). Eukaryotic cells produce certain proteins (Section 18.3) that regulate phase transitions within the cell cycle.

Replication rate DNA replication is significantly slower in eukaryotes than in prokaryotes. The eukaryotic rate is approximately 50 nucleotides per second per replication fork. (Recall that the rate in prokaryotes is about 20 times higher.) This discrepancy is presumably a result, in part, of the complex structure of chromatin.

Replicons Despite the relative slowness of eukaryotic DNA synthesis, the replication process is relatively brief, considering the large sizes of eukaryotic genomes. For example, on the basis of the replication rate, the replication of an average eukaryotic chromosome (approximately 150 million bp) should take more than a month to complete. Instead, this process usually is completed in several hours. Eukaryotes use multiple replicons to compress the replication of their large genomes into short periods (**Figure 18.13**). About every 40 kb along eukaryotic chromosomes, there is a site where replication machinery assembles. Humans have about 30,000 origins of replication.

Okazaki fragments From 100 to 200 nucleotides long, the Okazaki fragments of eukaryotes are significantly shorter than those in prokaryotes.

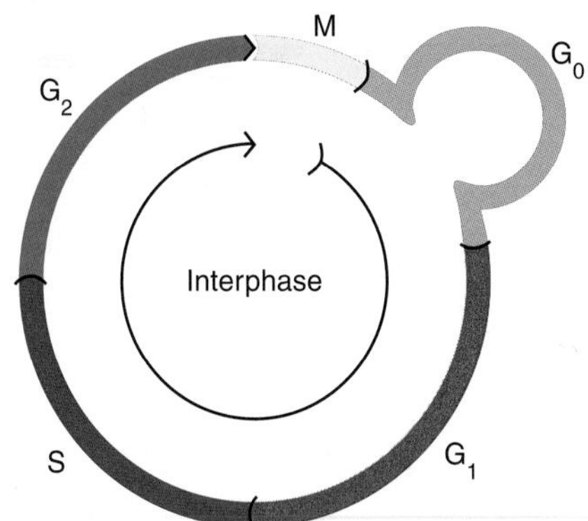

FIGURE 18.12
The Eukaryotic Cell Cycle

Interphase (the period between mitotic divisions) is divided into several phases. DNA replication occurs during the synthesis or S phase. The G_1 (first gap) phase is the time between the previous mitosis and the beginning of the next S phase. During the G_2 phase, protein synthesis increases as the cell readies itself for mitosis (M phase). After mitosis, many cells enter a resting phase (G_0).

THE EUKARYOTIC REPLICATION PROCESS In higher eukaryotes, replication begins with the sequential assembly of the **preinitiation replication complex (preRC)** (**Figure 18.14**). The formation of this complex involves a process that starts in early G_1 phase of the cell cycle when levels of cyclin-dependent kinase (Cdk) (p. 717) and cell division cycle (Cdc) proteins are low. The process, called licensing, limits DNA replication to once per cell cycle.

The components of preRC are assembled beginning when the **origin of replication complex (ORC)**, the subunits of which are analogues of DnaA, binds to the DNA initiation region or origin. Cdc6/Cdc18 and Cdt1 bind to ORC and recruit the **MCM complex** to the site. MCM is thought to be the major DNA helicase in eukaryotes. The recruitment of additional proteins of the cell cycle control machinery (e.g., cdk2-cyclin E and cdc45) allows for the DNA replication proteins to be loaded onto the replication fork and DNA synthesis to begin (**Figure 18.15**). The conversion of licensed preRC to an active initiation complex requires the addition of pol α/primase, pol ε, and a number of accessory proteins, which occurs only at the onset of S phase. The cell cycle regulating kinases then phosphorylate and activate the components of the preRC. The proteins that bind to the ORC and complete the structure of the preRC are referred to as **replication licensing factors**.

When the initiation complex is active, newly phosphorylated MCM separates the DNA strands, each of which is then stabilized by **replication protein A (RPA)** (**Figure 18.16**). Replication is begun by primase, which synthesizes the RNA primers of the leading strand and each Okazaki fragment on the lagging strand. Pol α/primase extends each primer by a short DNA strand about 20 nt long. Pol α is then displaced and pols δ and ε continue the replication process. The attachment of pol ε to the leading strand and pol δ to the lagging strand is controlled by **replication factor C (RFC)**, which is a clamp loader protein. After binding ATP, RFC then binds to PCNA, a processivity

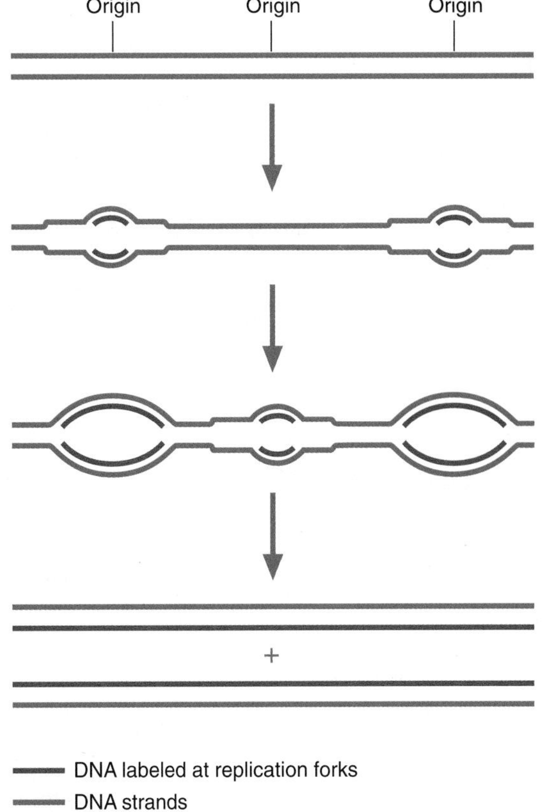

FIGURE 18.13
Multiple-Replicon Model of Eukaryotic Chromosomal DNA Replication

Short segment of a eukaryotic chromosome during replication.

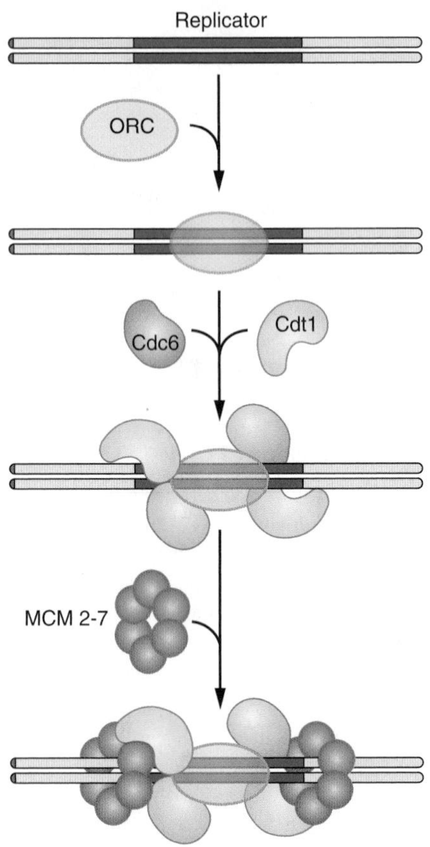

FIGURE 18.14
Formation of a Preinitiation Replication Complex

Assembly of the preRC begins (during G_1 phase of the cell cycle) when the ORC binds to the replication origin sequence, sometimes referred to as the *replicator*. ORC proceeds to recruit Cdc6 and Cdt1. When the MCM complex subsequently binds, preRC formation is complete.

KEY CONCEPTS

- DNA is replicated by a semiconservative mechanism involving several enzymes.
- The leading strand is synthesized continuously in the $5' \rightarrow 3'$ direction.
- The lagging strand is synthesized in pieces in the $5' \rightarrow 3'$ direction; the pieces are then covalently linked.

factor. The RFC/PCNA complex, which converts pols δ and ε into processive enzymes, then loads either polymerase onto the DNA, triggering ATP hydrolysis.

DNA replication of each chromosome continues until the replicons meet and fuse. When the replication machinery reaches the 3′ end of the lagging strand, there is insufficient space to synthesize a new RNA primer. Incomplete lagging strand synthesis leaves the template strand without its complementary base pairs at the end of the chromosome. Chromosomes with 3′-ssDNA overhangs are very susceptible to nuclease digestion and tend to fuse with each other, leading to chromosomal breakage during mitosis. Eukaryotic cells can compensate for this problem with **telomerase**, a ribonucleoprotein with reverse transcriptase activity, and an RNA molecule with a base sequence that is complementary to the TG-rich sequence of telomeres. Recall that telomeres (p. 639) are minisatellite sequences that occur at the ends of linear chromosomes. Telomerase uses the RNA base sequence to synthesize a single-stranded DNA to extend the 3′ strand of the telomere (**Figure 18.17**). Afterward, the normal replication machinery synthesizes a primer and a new Okazaki fragment. The chromosome ends are then sequestered and stabilized by **telomere end-binding proteins (TEBPs)** that bind to GT-rich telomere sequences and **telomere repeat-binding factors (TRFs)** that secure the 3′ overhang (now further away from critical coding sequences) into a knotlike T-loop.

In most multicellular eukaryotes, telomerase is active only in germ cells (cells that give rise to eggs and sperm). In the human body during normal aging, the telomeres of somatic cells (differentiated cells not including eggs and sperm) shorten over time. Once telomeres are reduced to a critical length, chromosomes can no longer replicate. As a result, somatic cells eventually die. It is noteworthy that the fibroblasts (connective tissue cells) of patients with Hutchinson-Guilford progeria syndrome have abnormally short telomeres. These patients age rapidly, with death occurring in the preteen years. It is also known that telomerase is overexpressed in approximately 90% of all cancers.

QUESTION 3.1

Compare the replication processes of prokaryotes and eukaryotes.

DNA Repair

Cells continuously monitor for DNA damage. The effects of normal metabolic activities and environmental exposures on DNA are considerable. Estimates of DNA lesions in human cells, for example, vary between tens and hundreds of thousands per cell per day. And yet, only a small fraction of these lesions are passed on to daughter cells as mutations. The average natural mutation rate for both animal and plant gametes (reproductive cells) is 1 mutation per 100,000 genes per generation. This low rate is the result of a genome maintenance network that detects DNA lesions and then repairs them. There are several classes of DNA repairs. Direct repairs eliminate chemical damage to DNA (e.g., pyrimidine dimers) by reversing it. When damage is localized to one of the two DNA strands, several forms of excision repair (base excision repair, nucleotide excision repair, and mismatch repair) can be used. Double-strand DNA breaks are repaired by either nonhomologous end joining or homologous recombination repair. In humans and other mammals DNA damage responses are regulated by ATM and ATR, members of a superfamily of serine-threonine kinases that also includes mTOR (p. 593). Of these, ATM (*a*taxia *t*elangiectasia *m*utated) and ATR (*a*taxia *t*elangiectasia and *R*ad3-related protein) initiate global responses to DNA damage that activate large numbers of DNA repair and cell cycle regulatory proteins. (*Ataxia telangiectasia* is a rare human disease characterized by radiation sensitivity, a genomic instability that predisposes affected individuals to neurodegeneration and cancer.) A few types of DNA damage can be repaired

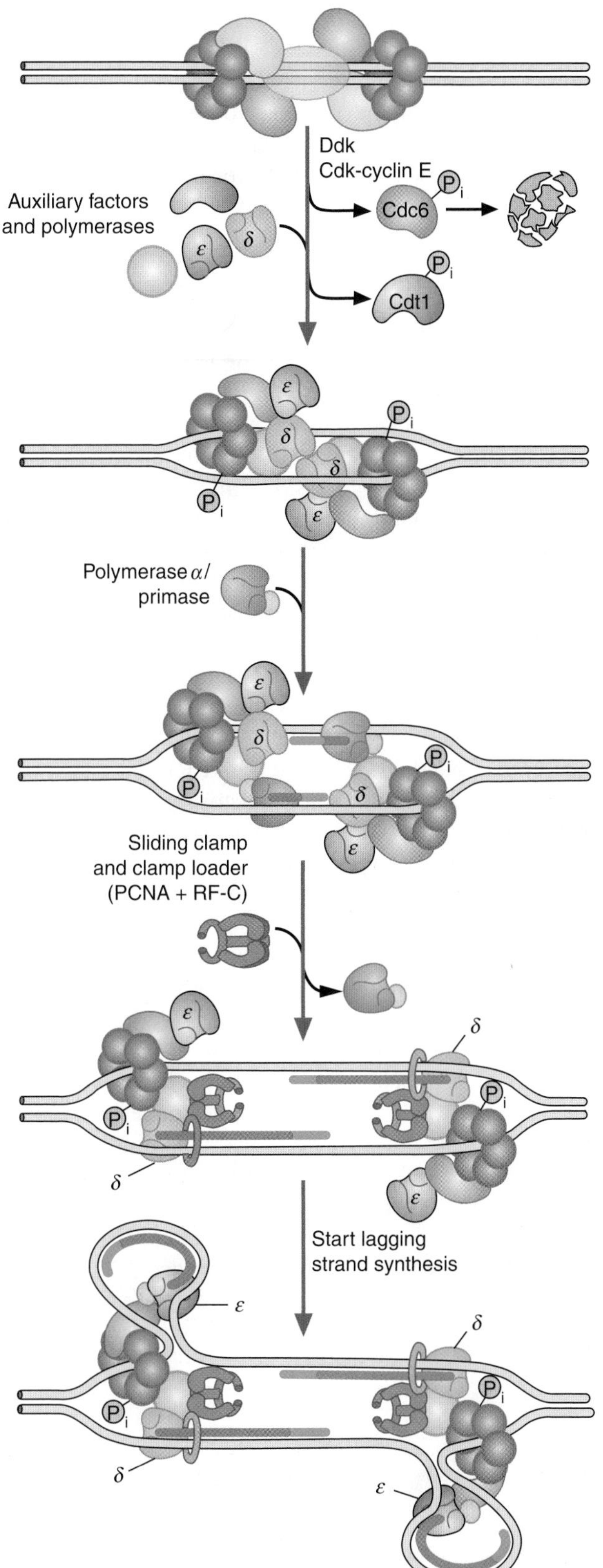

FIGURE 18.15

Eukaryotic Replication Fork Formation

Ddk and Cdk-cyclin E trigger replication initiation by phosphorylating several proteins. Among the results is the release of Cdc6 and Cdtl from ORC. Upon recruitment of DNA polymerase δ and ε the initiation complex is complete. Polymerase α/primase is then recruited. Once the RNA primers have been synthesized on the leading strand and the primer sequence briefly extended by polymerase α, the clamp loader (RFC) binds to the sliding clamp (PCNA). The sliding clamp/clamp loader complex then transforms polymerase δ into a processive enzyme. After DNA has unwound sufficiently, lagging strand synthesis is initiated. In this illustration DNA polymerase ε is bound to the lagging strand. The sliding clamp/clamp loader complex then binds to both the template and the newly synthesizing strands, which transforms pol δ into a processive enzyme.

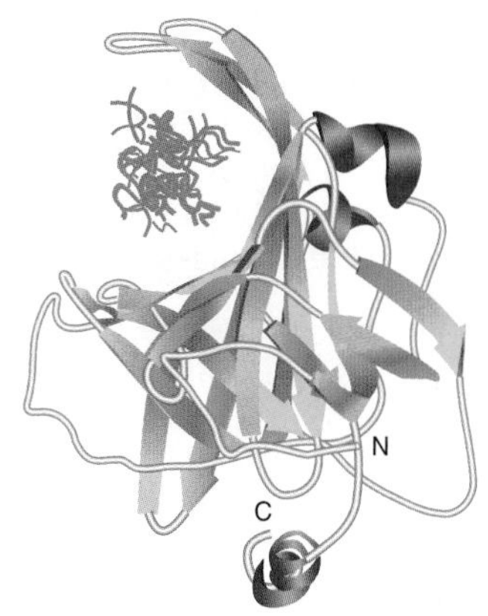

FIGURE 18.16

Replication Protein A Structure

Eukaryotes use RPA, a single-stranded DNA-binding protein, to prevent DNA strands from reannealing or being degraded by nucleases. The β-sheet in RPA forms a channel in which DNA (dark orange) binds.

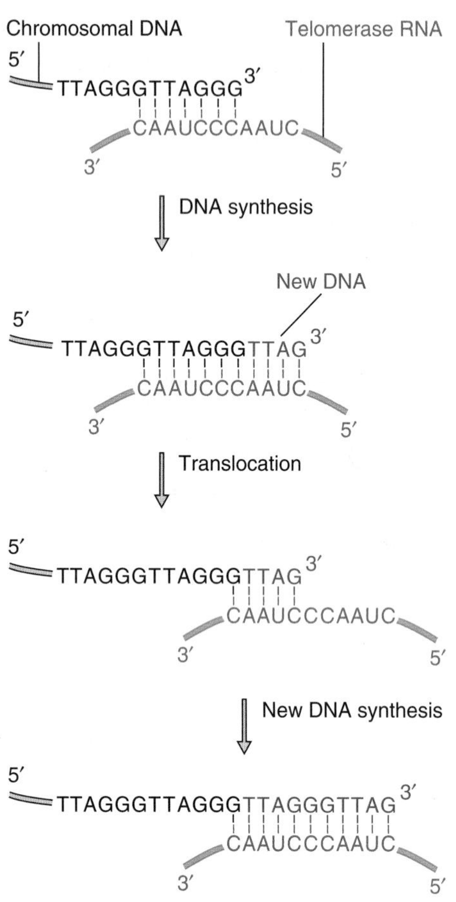

FIGURE 18.17

Telomerase-Catalyzed Extension of a Chromosome

The GT-rich, 3′-ssDNA end of a chromosome (the telomere): a portion of the RNA component of telomerase base pairs with this region, and the reverse transcriptase activity further extends the template strand in the 5′ to 3′ direction. The telomerase translocates to the end of the new segment, and the process is repeated until the ssDNA is long enough to accommodate the replication machinery and a primer. A new Okazaki fragment is then synthesized.

without the removal of nucleotides. For example, breaks in the phosphodiester linkages can be repaired by DNA ligase. In **photoreactivation repair**, or **light-induced repair**, pyrimidine dimers are restored to their original monomeric structures (**Figure 18.18**). In the presence of visible light, DNA photolyase cleaves the dimer, leaving the phosphodiester bonds intact. Light energy captured by the enzyme's flavin and pterin chromophores breaks the cyclobutane ring.

Direct Repairs

SINGLE-STRAND DNA REPAIRS Several repair mechanisms repair damage limited to a single-strand DNA segment using the complementary, undamaged strand as a template. **Base excision repair** is a mechanism that removes and then replaces individual nucleotides whose bases have undergone damage (e.g., alkylation, deamination, or oxidation). One of several enzymes called **DNA glycosylases** cleaves the N-glycosidic linkage between the damaged base and the deoxyribose component of the nucleotide (**Figure 18.19**). The resulting **apurinic** or **apyrimidinic** (AP) **sites** are resolved through the action of nucleases that remove the deoxyribose residue and several additional nucleotides. The gap is repaired by a DNA polymerase (pol I in bacteria and DNA polymerase β in mammals) and DNA ligase.

In **nucleotide excision repair** (NER), a large variety of bulky, helix-destabilizing (2–30 nt) lesions (e.g., pyrimidine dimers and benzo[*a*]pyrene adducts) are removed, and the resulting gap is filled. There are two forms of nucleotide excision repair: *global genomic repair* (GG-NER) and *transcription coupled repair* (TC-NER). They differ in the mechanisms by which DNA damage is recognized. In both forms the excision enzymes appear to recognize the physical distortion rather than specific base sequences. In *E. coli* GG-NER (**Figure 18.20**), the excision nuclease (exinuclease), composed of Uvr A, B, and C, cuts the damaged DNA and removes a 12- to 13-nt ssDNA sequence containing the lesion. A $UvrA_2UvrB$ (A_2B) complex scans DNA for damage (e.g., a thymine dimer). Once UvrA senses a helix distortion, it partially unwinds the affected segment. UvrB further destabilizes the segment by inserting a β-hairpin domain. Next A_2B bends the DNA, and Uvr A dissociates from Uvr B–DNA. Uvr C then binds to Uvr B, cutting the damaged DNA strand 4 or 5 nucleotides to the 3′ side of the thymine dimer. Then Uvr C cuts the strand 8 nucleotides to the 5′ side. Uvr D, a helicase, releases Uvr C and the thymine dimer–containing oligonucleotide. The excision gap is repaired by pol I and DNA ligase.

Transcription coupled repair occurs only on a strand actively being transcribed. The damage is recognized when the transcribing enzyme, RNA polymerase, is stalled. In *E. coli,* Mfd (mutation frequency decline), a transcription-repair coupling factor, then proceeds to displace the polymerase and recruit the $UvrA_2B$ complex that will initiate damage removal.

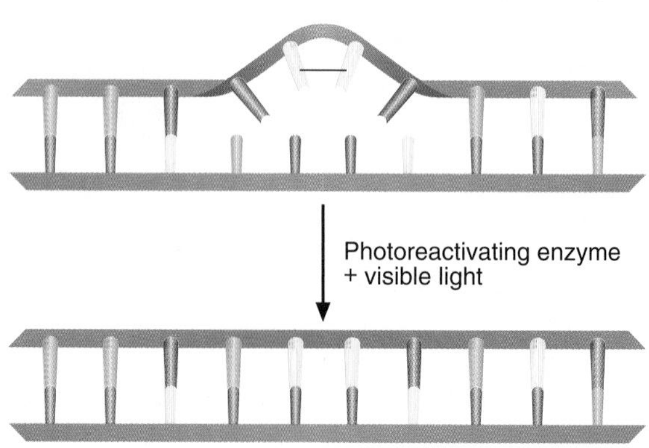

FIGURE 18.18

Photoreactivation Repair of Thymine Dimers

Light provides the energy for converting the dimer to two thymine monomers. $FADH_2$ is the electron donor in this reaction. No nucleotides are removed in this repair mechanism. Photoreactivation repair occurs widely but unevenly in bacteria, archaea, protozoa, fungi, plants, and animals (but not humans).

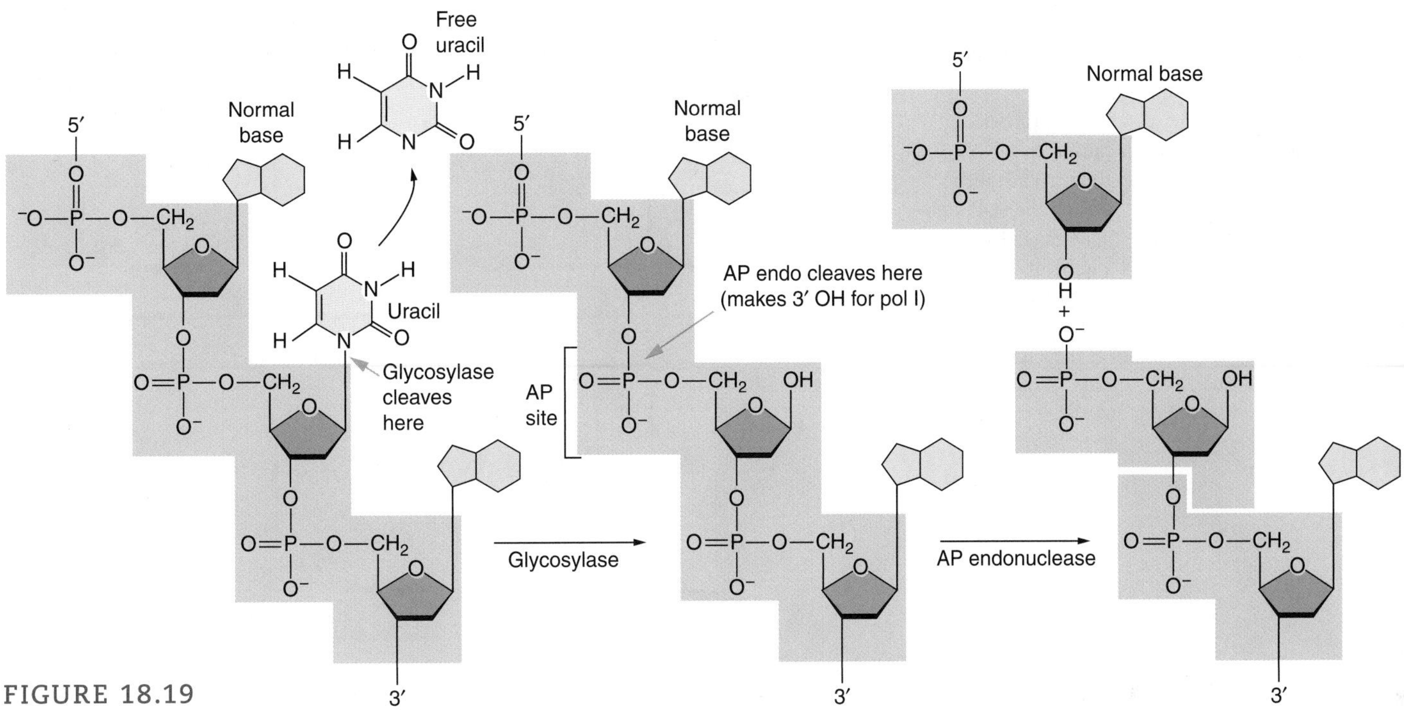

FIGURE 18.19
Base Excision Repair

DNA glycosylase hydrolyzes the N-glycosidic linkage to release the base (in this case, uracil). AP endonuclease cleaves the DNA backbone at the 5′ position of the AP (apyrimidic) site. An endonuclease (not shown) removes the AP residue and several additional nucleotides in the 5′ → 3′ direction. A DNA polymerase then fills in the gap and DNA ligase repairs the nick.

Nucleotide excision repair in humans, like that of other eukaryotes, is more complicated than the prokaryotic process, involving numerous proteins (e.g., 30 in mammals). Most of these proteins are named for their association with two diseases caused by their deficiency: *xeroderma pigmentosum* (an inherited disorder characterized by extreme sensitivity to light) and *Cockayne syndrome* (premature aging and hearing and eye abnormalities). In GG-NER, XPA, XPC, and XPE are involved in damage recognition and the subsequent recruitment of other proteins. Examples include helicases (XPB and XPD), which unwind the DNA segment containing the damage, and repair proteins such as XPFG and XPG, which excise the oligonucleotide strand. A DNA polymerase uses the undamaged strand as a template to synthesize the complementary strand. GG-NER is complete when DNA ligase repairs the single strand breaks.

In eukaryotic TC-NER, the stalled RNA polymerase serves as the damage recognition signal. The polymerase binds to CSB (an ATP-dependent chromatin remodeling enzyme) and CSA (a component of an E3 ubiquitin ligase complex). Following this DNA damage detection step, CSB triggers the recruitment of the same NER proteins used in GG-NER.

Mismatch repair (MMR) is a single-strand repair mechanism that corrects helix-distorting base mispairings that are the result of replication proofreading errors or as the result of replication slippage. (*Replication slippage* is a type of error that occurs when repeat sequences are skipped or copied twice, causing the formation of bubbles that require repair.) The replication process often results in deletions (skipped sequences) or insertions (recopied sequences). Some recombination errors and several forms of chemical damage (e.g., 8-oxoguanine and carcinogen adducts) can also be repaired. MMR has been estimated to cause a 100-fold increase in replication fidelity. A key feature of MMR is the capacity to distinguish between old and newly synthesized strands. In *E. coli*, this is accomplished by the methylation of both strands of parental DNA. DNA methyl transferase (Dam) methylates N-6 of adenine residues in 5′-GATC-3′ sequences, and

KEY CONCEPTS

- DNA is constantly exposed to chemical and physical processes that alter its structure.
- Each organism's survival depends on its capacity to repair this structural damage.
- Examples include light-induced repair, base excision repair, nucleotide excision repair, mismatch repair, nonhomologous end joining, and homologous recombination repair.

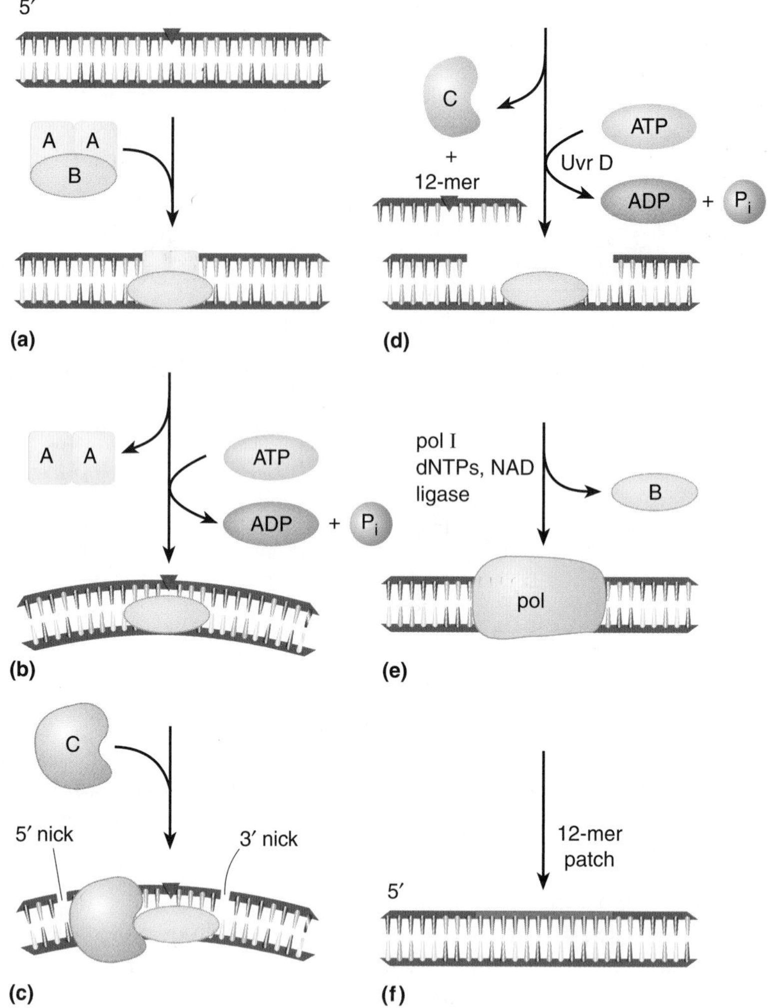

FIGURE 18.20
Excision Repair of a Thymine Dimer in *E. coli*

Uvr A, a damage recognition protein, detects helical distortion caused by DNA adducts such as thymine dimers (a). It then associates with Uvr B to form the A_2B complex. After binding to the damaged segments, A_2B forces DNA to bend. Uvr A then dissociates (b). The binding of the nuclease Uvr C to Uvr B (c) and the action of the helicase Uvr D (d) results in the excision of a 12-nucleotide DNA strand (12-mer). Then Uvr B is released (e), and the excision gap is repaired by pol I (f).

DNA cytosine methylase (Dcm) converts the cytosines in 5′CCTGG-3′ and 5-CCAGG-3′ sequences to 5-methycytosine. Because these sequences form palindromes, both strands are methylated equally. For a finite amount of time after replication, each daughter DNA is hemimethylated (i.e., it consists of one methylated strand and one nonmethylated strand). It is during this brief period that DNA is scanned for mismatched base pairs. The MMR system consists of three proteins: MutS, MutL, and MutH. A MutS homodimer recognizes and binds to a site on the newly synthesized strand containing a mispaired base. MutH then locates and binds the nearest GATC site. After ATP-dependent activation by MutS and MutL, MutH proceeds to nick the unmethylated strand, thus marking it for repair. A DNA helicase separates the two strands and several exonucleases proceed to degrade the unmethylated strand from the nicked site to a few nucleotide residues past the mismatch site. Pol III then proceeds to resynthesize the missing segment using the methylated strand as a template.

The human homologs of MutS and MutL, MSH2 and MLH1, respectively, cause microsatellite instability if mutated. Most cases of hereditary nonpolyposis colorectal cancers (HNPCCs) are linked to mutations in the genes for these two proteins. The risk for other cancers (e.g., endometrial, ovarian, stomach, and small intestine) is also increased by MMR protein mutations.

DOUBLE-STRAND BREAKS DNA double strand-breaks (DSBs) are especially dangerous for cells because they can result in genome rearrangements or a lethal breakdown of chromosomes. Caused by radiation, ROS, or DNA-damaging chemicals (e.g., asbestos or cisplatin, an anticancer drug) or as a result of errors in DNA replication, DSBs are repaired by two mechanisms. One mechanism ligates two DNA ends together (nonhomologous end joining), while the other takes advantage of the base sequence information on a homologous chromosome (homologous recombination). In humans and other mammals, nonhomologous end joining (NHEJ) is the preferred pathway for DSB repair. Both processes begin with sensing that a DSB has occurred. In mammals DSB sensing involves several molecules. Among these are ATM and ATR (p. 676), which initiate global responses to DNA damage that activate large numbers of DNA repair and cell cycle regulatory proteins, and DNA-PK, which is involved in DSB repair.

NHEJ begins with the binding of DNA-PK to the two DNA ends. DNA-PK is a heterotrimer of DNA-PK_{CS} (protein kinase DNA-activated catalytic polypeptide) and a Ku dimer. DNA-PK_{CS} undergoes autophosphorylation and then recruits and phosphorylates a protein called artemis. *Artemis* is a nuclease that converts the two ends into ligation substrates. A trimeric complex containing DNA ligase IV and two accessory proteins finalizes the repair by ligating the broken DNA ends. Because there is no requirement for sequence homology, NHEJ is an error- prone pathway. For example, the occurrence of several DSBs in a cell can lead to the inadvertent joining of DNA ends from different chromosomes and the loss of nucleotides at the break site can result in deletions. NHEJ proteins are used by immune system cells to generate antibody diversity (estimated at 10 billion different antibodies in humans) by recombination of V, D, and J gene segments. Homologous recombination repair, which closely resembles general recombination, is discussed in the next section.

QUESTION 18.2

List six types of DNA repair. Explain the basic features of each.

DNA Recombination

Recombination can be defined as the rearrangement of DNA sequences by exchanging segments from different molecules. The process of recombination, which produces new combinations of genes and gene fragments, is primarily responsible for diversity among living organisms. More important, the large number of variations made possible by recombination gives species opportunities to adapt to changing environments. In other words, recombination is a principal source of the variations that make evolution possible. Recombination is also used to repair (ds)DNA breaks.

There are two forms of recombination: general and site-specific. **General** or **homologous recombination**, which occurs between homologous DNA molecules, most notably occurs during meiosis. (Meiosis is the form of eukaryotic cell division in which haploid gametes are produced.) In **site-specific recombination**, the exchange of sequences from different molecules requires only short regions of DNA homology. These regions are flanked by extensive nonhomologous sequences. Site-specific recombinations, which depend more on protein-DNA interactions than on sequence homology, occur throughout nature. For example, this mechanism is used by bacteriophage to integrate its genome into the *E. coli* chromosome. In eukaryotes, site-specific recombination is responsible for a wide variety of developmentally controlled gene rearrangements that are partially responsible for cell differentiation in complex multicellular organisms. One of the most interesting examples of gene rearrangement is the generation of antibody diversity in mammals. In the variation of site-specific recombination referred to as **transposition**, DNA sequences called **transposable elements** (p. 639) move from one chromosome or chromosomal region to another.

FIGURE 18.21

General Recombination: The Holliday Model

Once one strand in each duplex is nicked, each broken strand invades the other duplex (a). Covalent bonds are formed (b), cross-linking the two duplexes. Branch migration then occurs (c). The bending of the chi structure in (d) and (e) makes later events easier to understand. In (f) the same original strands are nicked. The resulting heteroduplex (g) contains a patch. Nicking the opposite strands (h) results in the formation (i) of a spliced heteroduplex.

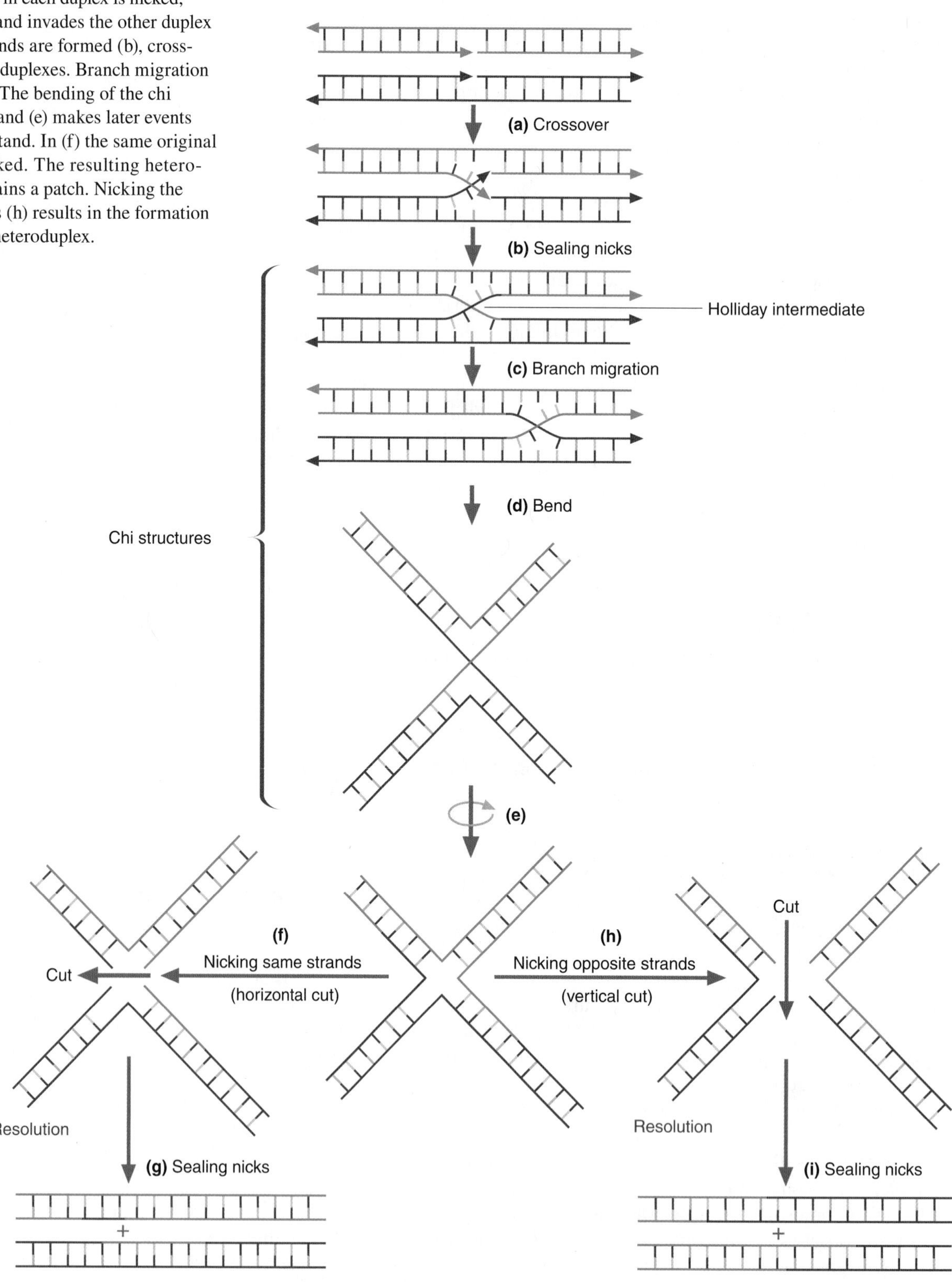

GENERAL RECOMBINATION General recombination occurs in all living organisms, but has been investigated primarily in *E. coli* and fungi such as *S. cervisiae* and *Aspergillus nidulans*. In addition to generating genetic diversity, general recombination is also an important DNA damage repair process. Several models explaining recombination have been proposed. Examples include the Holliday, Meselson-Radding, double-strand break repair, and synthesis-dependent strand annealing models.

The first model explaining general recombination was based on Robin Holliday's work with fungi. The Holliday model (**Figure 18.21**) involves the following steps:

1. Two homologous DNA molecules become paired.
2. Two of the DNA strands, one in each molecule, are nicked (cleaved) at identical (if not similar) locations.
3. The two nicked strand segments cross over, thus forming a Holliday intermediate.
4. DNA ligase seals the cut ends.
5. Branch migration caused by base-pairing exchange leads to the transfer of a segment of DNA from one homologue to the other.
6. A second series of DNA strand cuts occurs.
7. DNA polymerase fills any gaps, and DNA ligase seals the cut strands.

The **Meselson-Radding model** is the result of efforts to account for several laboratory observations not explained by the Holliday model. Among these is the fact that recombination sometimes results in only one of the homologous DNA molecules having a recombinant strand. According to the Meselson-Radding model (**Figure 18.22**) recombination occurs as follows:

1. One strand of one of the two homologous DNA molecules is nicked.
2. The extension of the newly created 3′-end by DNA polymerase causes the displacement of the strand on the other side of the nick. As the growing strand becomes longer, the displaced strand invades the double helix of a homologous segment of the second chromosome to form a D-loop structure.
3. The D-loop is cleaved and the invading strand is ligated to the newly created 3′-end of the homologous strand.
4. The 3′-end of the newly synthesized strand and the 5′-end of a homologous strand are ligated to form a Holliday intermediate.
5. Branch migration may occur.
6. Strand nicks and resolution of the Holliday junction (refer to **Figure 18.21f–i**) result in either a crossover product (vertical cut) or a noncrossover product (horizontal cut).

In eukaryotes, general recombination has a featured role in *meiosis*, a type of cell division necessary for sexual reproduction. In a mechanism that accurately aligns homologous maternal and paternal chromosomes, the general

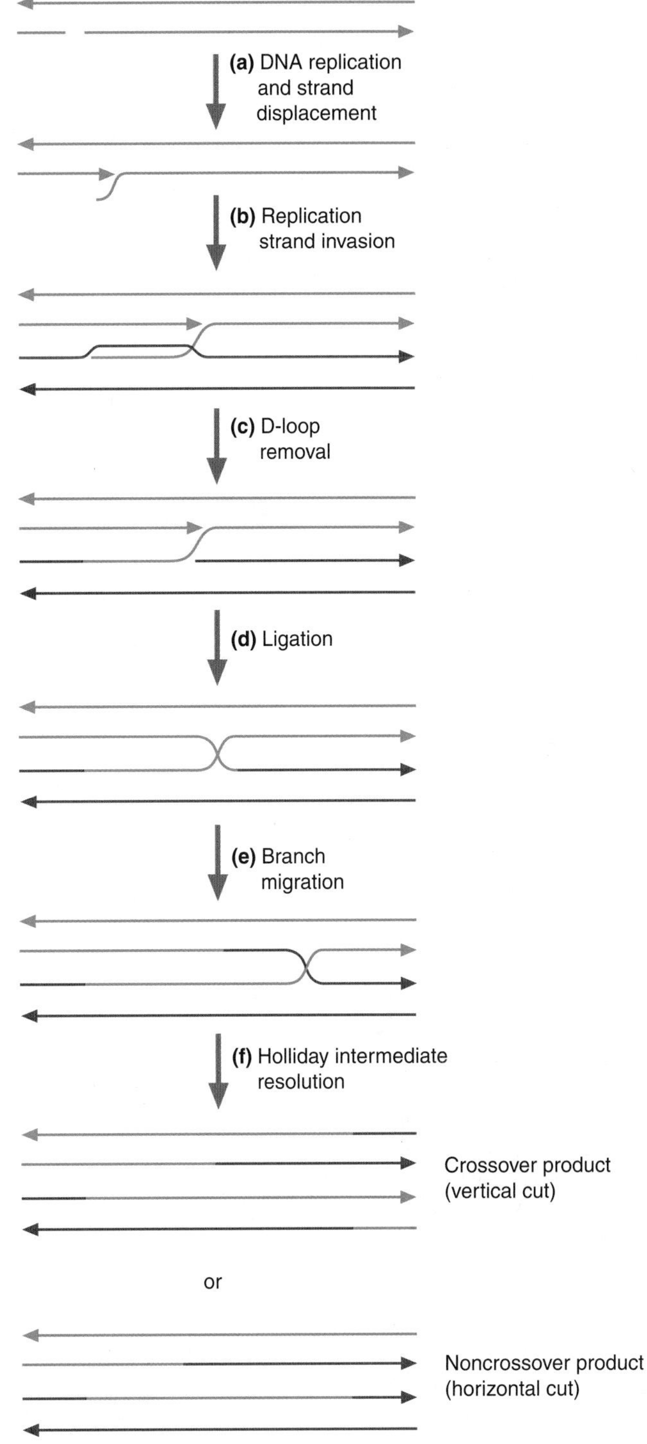

FIGURE 18.22

General Recombination: The Meselson-Radding Model

The recombination process is initiated by a single-strand nick on a DNA molecule. As DNA polymerase proceeds to synthesize a strand extension at the 3′-end, the DNA strand on the other side of the nick (i.e., the 5′-end) is displaced (a). It is this 5′-tail that invades into a homologous DNA molecule (b) to form a D-loop. The D-loop is subsequently removed (c) by a nuclease. A Holliday intermediate forms as the two free ends are ligated (d). Branch migration may occur (e). As in the Holliday model (**Figure 18.21 e–i**) the Holliday intermediate resolves (f) into either noncrossover or crossover products.

recombination machinery facilitates the exchange of chromosomal segments that creates genetic diversity. The **double-strand break repair model** (DSBR) (**Figure 18.23a–d**) explains many of the features of meiotic recombination not accounted for by the Holliday and Meselson-Radding models. DSBR mechanisms also protect prokaryotic and eukaryotic cells from DSBs (p. 680) caused by mutagenic factors such as ionizing radiation and ROS. The major steps in DSBR are as follows:

1. An endonuclease induces a DSB in one of a pair of homologous DNA molecules.
2. Exonucleases proceed to degrade the 5′-ends, leaving 3′-tails.
3. One of the 3′-tails invades a homologous DNA molecule and after a successful homology search forms a D-loop.
4. DNA polymerase extends both the invading and the noninvading 3′-tails.
5. Branch migration combined with the ligation between the invading 3′-tail by the 5′-end on the other side of the break results in the formation of a double Holliday junction.
6. Resolution of the two Holliday structures can give noncrossover or crossover products.

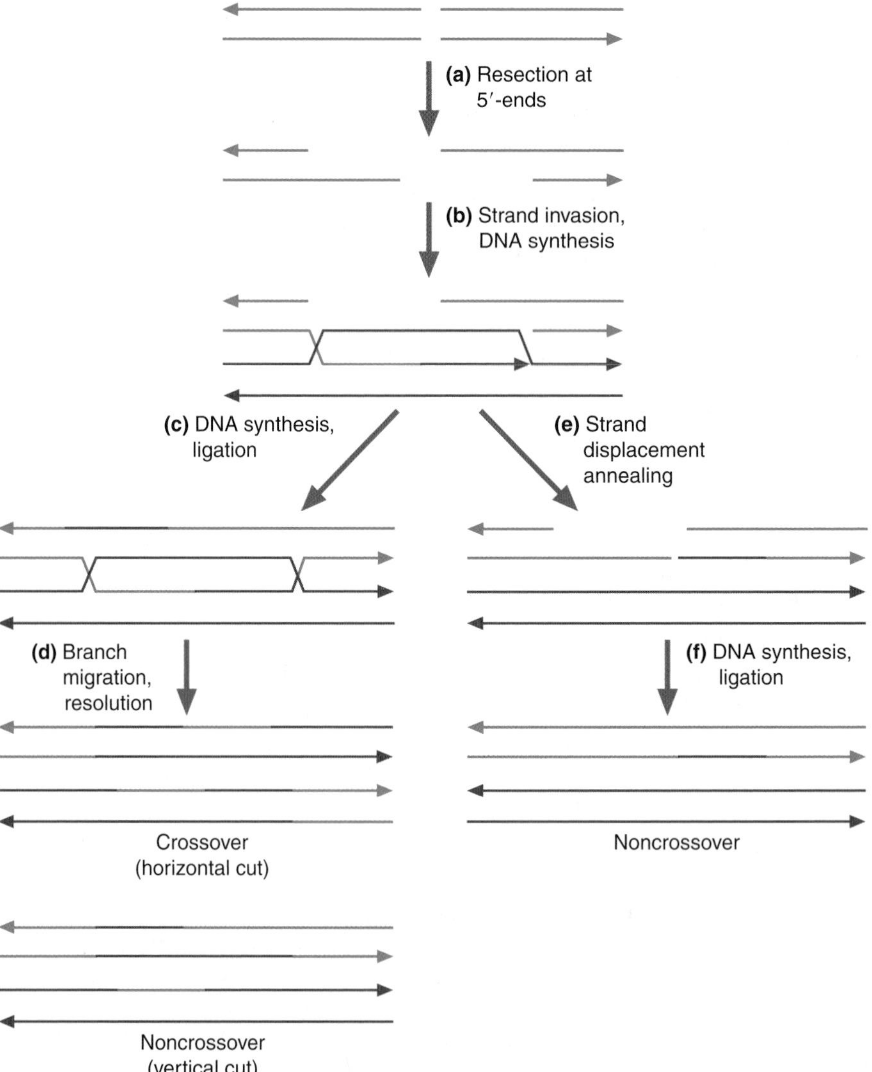

FIGURE 18.23

Double-Strand Break Repair (DSBR) and Synthesis-Dependent Strand Annealing (SDSA) Models of General Recombination

DSBR begins with a double strand cut of one of the homologous DNA molecules. An exonuclease then resects (trims) the two 5′-ends (a) to yield two 3′-tails. (b) One of the 3′-tails invades an equivalent segment of the second homologous DNA molecule to form a D-loop. As the 3′-end of this tail is extended by DNA synthesis, using the homologous DNA strand as a template, a Holliday intermediate forms. (c) The other 3′-tail is extended by DNA synthesis using the complementary strand segment in the D-loop as a template. The DNA synthesis phase of the process ends with the action of DNA ligase. (d) The double Holliday junction undergoes branch migration and is then resolved by an endonuclease into crossover or noncrossover products. In SDSA after 5′-end resection (a), strand invasion, and DNA synthesis (b), a second Holliday intermediate does not form. Instead, (e) the invading strand is displaced, thus allowing a reannealing process that provides a template for the DNA synthesis repair (f) of the other broken strand. SDSA ends with ligation of the single-stranded gaps.

DSBR also occurs in mitosis, usually without crossover products. The **synthesis-dependent strand annealing model** (SDSA) explains how this phenomenon occurs. SDSA diverges from DSBR when the strand in the D-loop is displaced (**Figure 18.23e** and **f**). The invading strand proceeds to anneal with the complementary single strand of the other DSB end. The process ends with further DNA synthesis, followed by ligation.

PROKARYOTIC RECOMBINATION Recombination in *E. coli* is initiated when RecBCD, an enzyme complex that possesses both exonuclease and helicase activities, encounters a Chi (*c*rossover *h*otspot *i*nstigator) sequence site. After binding to a DNA molecule, RecBCD cleaves one of the strands and proceeds to unwind the double helix until it reaches 5′-GCTGGTGG-3′ (the chi site), a sequence that occurs frequently in *E. coli* DNA. Strand exchange begins when a nucleoprotein filament formed when monomers of RecA, an ATPase, coats one of the strands. Powered by ATP hydrolysis, the RecA-coated strand segment then searches for homology in nearby dsDNA. Once a homologous segment is located, DNA synthesis causes strand displacement, which is followed by D-loop cleavage, strand capture, and Holliday junction formation. Subsequent branch migration (**Figure 18.24**) is initiated as RuvA recognizes and binds to the Holliday junction. Two copies of RuvB, a hexamer with ATPase and helicase activities, then form a ring on either side of the junction. Branch migration is catalyzed by the RuvAB complex. This molecular machine separates, rotates, and pulls the strands in the two sets of helices, even after RecA dissociates. The migration ends when a specific sequence [5′-(A or T)TT (G or C)-3′] is reached. As RuvAB detaches, two RuvC proteins bind to the junction. Recombination ends as RuvC cleaves the crossover strands and the Holliday structure resolves itself to form two separate double-helical DNA molecules.

In bacteria, general recombination appears to be involved in several forms of intermicrobial DNA transfer.

1. **Transformation**. In **transformation**, naked DNA fragments enter a bacterial cell through a small opening in the cell wall and are introduced into

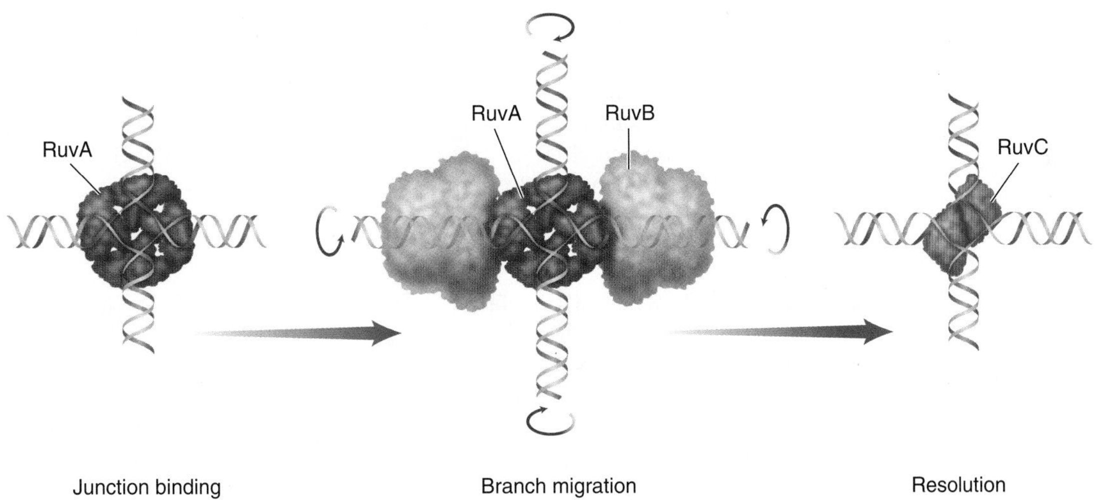

FIGURE 18.24

Model of the Association of Ruv Proteins with a Holliday Junction

RuvA, a tetramer, first binds to the Holliday junction point. Two hexameric RuvB rings then form on both sides of the DNA/RuvA complex, with the DNA passing through the rings. Branch migration occurs as ATP hydrolysis drives the two RuvB rings to rotate the DNA helices in opposite directions. After branch migration, RuvA and RuvB detach and two RuvC proteins bind to the junction. RuvC, a nuclease, proceeds to cut the crossover strands, thus resolving the Holliday structure.

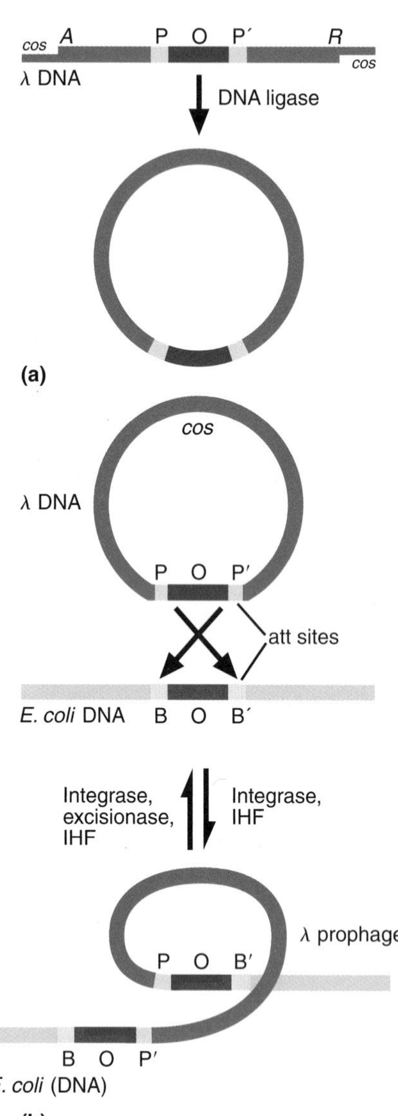

FIGURE 18.25

Insertion of the Bacteriophage λ Genome into the *E. coli* Chromosome

(a) The λ DNA circularizes as the single-stranded *cos* sequences anneal. The viral att site sequence consists of POP′ where O is a 15-bp sequence at the center. The bacterial att sequence BOB′ contains an identical O sequence. (b) Insertion occurs through site-specific recombination between short homologous O sequences within the phage and bacterial att sites.

the bacterial genome. (Recall Fred Griffith's experiment, see the online Biochemistry in Perspective essay A Short History of DNA Research: The Early Years.)

2. **Transduction**. **Transduction** occurs when bacteriophage inadvertently carry bacterial DNA to a recipient cell. After a suitable recombination, the cell uses the transduced DNA.
3. **Conjugation**. Certain bacterial species are known to engage in **conjugation**, an unconventional sexual mating that involves a donor cell and a recipient cell. The donor cell possesses a specialized plasmid (p. 43) that allows it to synthesize a sex pilus, a filamentous appendage that functions in a DNA exchange process. After the pilus has attached to the surface of the recipient cell, a fragment of the donor's genetic material is transferred. The transferred DNA segment can be integrated into the recipient's chromosome by recombination, or it may exist outside it in plasmid form.

EUKARYOTIC RECOMBINATION In eukaryotes, general recombination occurs during the first phase of meiosis to ensure accurate homologous chromosome pairing and *crossing over*, the mechanism for introducing genetic variation. DNA damage repair of DSBs using homologous DNA molecules occurs during the S and G2 phases of the cell cycle (p. 675) when newly replicated DNA is available. (In the other phases of the cycle DSBs are repaired by NHEJ.)

The basic mechanism of eukaryotic general recombination is believed to be similar to the process in prokaryotes. However, because of vastly more complex genomes, the number of proteins involved in eukaryotic recombination is substantially larger. Rad52, a multifunction protein, is believed to be the initial sensor of DSBs. In mammals the MRN complex (*M*re11, *R*ad50, and *N*bs1) creates a scaffold that stabilizes the DNA ends at DSBs either caused by exogenous damage or introduced by the meiotic enzyme Spo11. MRN also catalyzes 5′-end resection. When exogenous agents cause DSBs, MRN recruits and activates ATM (p. 676), which in turn activates several DNA repair proteins and cell cycle regulatory proteins. Examples of ATM-activated DNA repair proteins include Rad51, BRCA1, and BRCA2. Rad51 is a RecA homolog that binds to ssDNA and facilitates strand invasion. BRCA1 (breast cancer type 1 susceptibility protein) functions in a number of repair-associated pathways including cell cycle regulation and chromatin remodeling. BRCA1, linked to BRCA2, interacts with Rad51 during DSB repair. BRCA1 also promotes genome stability by repressing satellite DNA transcription.

SITE-SPECIFIC RECOMBINATION AND TRANSPOSITION Site-specific recombination relies on short segments of homologous DNA called **attachment (att) sites** or **insertional (IS) elements**. Recombination at these sites can lead to insertions, inversions, deletions, and translocations that may benefit or harm the cell. Examples of site-specific recombination include insertion of viral DNA into a host cell genome, acquisition of antibiotic resistance, phenotypic variation in plants, and antibody maturation in mammals. A simple case of insertion is illustrated by the integration of bacteriophage λ DNA into the *E. coli* chromosome (**Figure 18.25**). The process requires homologous att sites in the phage and bacterial genomes, a viral recombinase called λ integrase, and a bacterial gene product, the integration host factor. The mechanism with Holliday junction resolution results in the insertion of the λ genome into the bacterial chromosome.

QUESTION 18.3

Bacterial conjugation has medical consequences. For example, certain plasmids contain genes that code for toxins. The causative agent of a deadly form of food poisoning, *E. coli* 0157, synthesizes a toxin that causes massive bloody diarrhea and kidney failure. This toxin is now believed to have originated in *Shigella*, another bacterium that causes dysentery. Similarly, the growing problem of antibiotic resistance is partly attributable to the spread of antibiotic-resistant genes among bacterial populations. Antibiotic resistance develops because antibiotics are overused in medical practice and in livestock feeds. Suggest a mechanism by which this extensive use promotes antibiotic resistance. [*Hint*: The high-level use of antibiotics acts as a selection pressure.]

Barbara McClintock, a geneticist working with Indian corn (maize), reported in the 1940s that mobile genetic elements were responsible for the phenotypic variation in corn kernel color. Acceptance of this idea was slow in coming because it required a paradigm shift away from the view of the genome as a static component of the cell. In 1967 transposable elements (**transposons** or "jumping genes") (see p. 639) were confirmed in *E. coli*, and the concept of genome plasticity was finally accepted. Dr. McClintock received the Nobel Prize in Physiology or Medicine in 1983.

The IS elements of simple prokaryotic transposons consist of the gene for the transposition enzyme transposase, flanked by short inverted terminal repeats that define the boundary of the transposon (**Figure 18.26**). (An **inverted repeat** is a sequence that is the reverse of another downstream sequence. Inverted repeats without intervening sequences constitute a **palindrome**, a DNA sequence whose

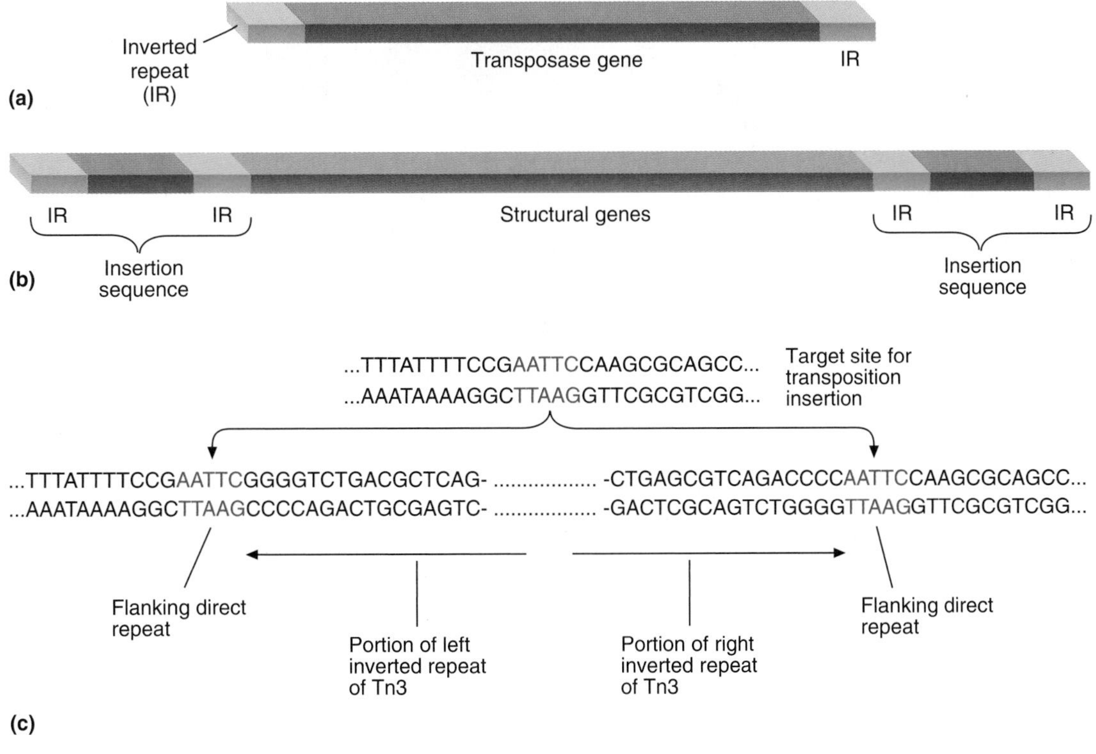

FIGURE 18.26
Bacterial Insertion Elements

(a) An insertion sequence. (b) A composite transposon. (c) Insertion of a transposon (Tn3) into bacterial DNA. The insertion process involves the duplication of the target site.

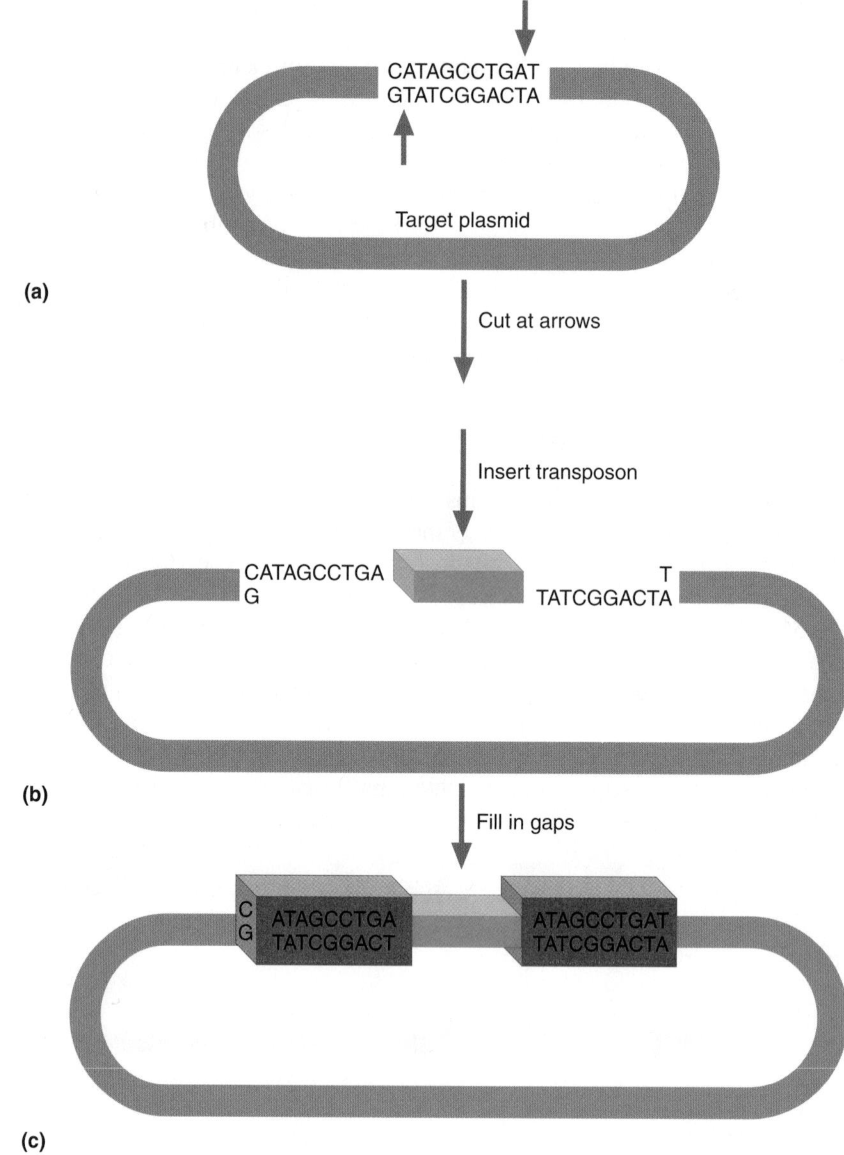

FIGURE 18.27
Nonreplicative Transposition

(a) Host DNA is cut (arrows) in a staggered fashion. (b) Each end of the transposon (blue) is covalently attached to an overhanging end of host DNA. (c) After the gaps have been filled by a DNA polymerase, there are nine base pair repeats of host DNA (red) flanking the transposon.

complement reads the same in reverse. See p. 628 for an example of an inverted repeat.) More complicated bacterial transposons, called **composite transposons**, are made up of two separate transposons that are linked by the DNA between them. The two transposons (IS elements) become linked when inactivating mutations (e.g., in the inner inverted repeats or one of the IS transposase genes) prevent independent jumping of the two IS elements. The presence of DNA sequences between the two IS elements that have useful properties (e.g., antibiotic resistance genes) also promotes retention of a composite transposon. Two transposition mechanisms have been observed, nonreplicative and replicative.

Nonreplicative transposition The transposase makes a double-stranded cut in the donor DNA and splices it into the staggered ssDNA cut ends of the target site (a "cut-and-paste" mechanism). The cell's DNA repair system fills the gaps in the target DNA, resulting in a short direct repeat on either side of the transposon insertion (**Figure 18.27**). An unrepaired gap can be lethal to the cell.

KEY CONCEPTS 

- In general recombination, the exchange of DNA sequences takes place between homologous sequences.
- In site-specific recombination, DNA-protein interactions are principally responsible for the exchange of nonhomologous sequences.
- The DNA sequences called transposons can move from one site in a genome to another.

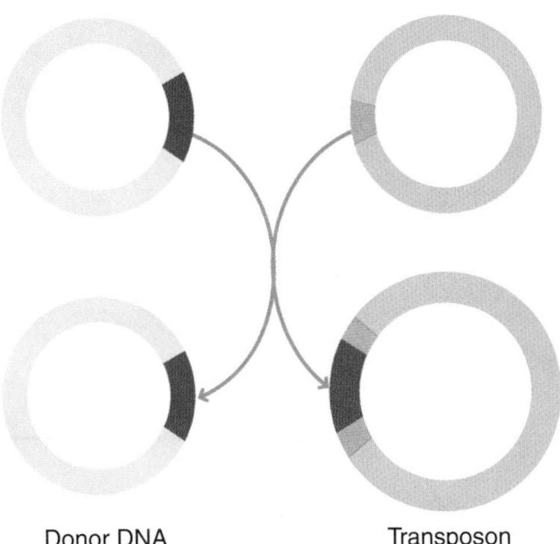

FIGURE 18.28
Replicative Transposition

The transposition of a transposable element (red) to a new location (pink) does not involve its loss from the original location. Note that in this type of transposition process, which occurs via replication of the entire transposon followed by site-specific recombination, the donor DNA is undamaged.

Replicative transposition In replicative transposition only one strand of the donor DNA is transferred to the target position, and replication followed by site-specific recombination results in the duplication of the transposon rather than insertion at the new site (**Figure 18.28**). An intermediate (the cointegrate) then forms that consists of the donor segment covalently linked to the target DNA. An additional enzyme, called resolvase, catalyzes a site-specific recombination that allows the resolution of the cointegrate into two separate molecules.

Some transposons found in eukaryotes resemble those found in bacteria. For example, the Ac element, the maize transposon that was first described by McClintock, is composed of a transposase gene flanked by short inverted repeats. McClintock referred to the Ac transposon as a "controlling element" because it appeared to control the synthesis of the pigment anthocyanin in corn kernels (i.e., there was a change in gene expression). Many other eukaryotic transposons, however, have somewhat different structures than those observed in bacteria. As described previously (p. 639), retrotransposons, also referred to as **retroposons** or **retroelements**, are a significant feature of eukaryotic genomes. Depending on the changes and their location, the effects of transposons can be viewed either as disruptive and damaging or as providing opportunities for genetic diversity. Some effects of transposition are observed as changes in gene expression, a topic that is discussed in Section 18.3.

QUESTION 18.4

One of the fascinating aspects of complex organisms such as mammals is the existence of gene families, groups of genes that code for the synthesis of a series of closely related proteins. For example, several different types of collagen are required for the proper structure and function of connective tissues. Similarly, there are several types of globin gene. It is currently believed that gene families originate from a rare event in which a DNA sequence is duplicated. Some gene duplications provide a selective advantage by providing larger quantities of important gene products. Alternatively, the two duplicate genes evolve independently. One copy continues to serve the same function, while the other eventually evolves to serve another function. Can you speculate about how gene duplications occur? Once a gene has been duplicated, what mechanisms introduce variations?

Biochemistry IN THE LAB

Genomics

Genomics and functional genomics have accelerated research efforts in all of the life sciences. **Genomics** is the large-scale analysis of entire genomes. **Functional genomics** is a methodology used to analyze the functional properties of genes and proteins and how these molecules interact within living organisms. As a result of major breakthroughs in DNA technology, genomics has yielded genome sequence information for more than several thousand organisms and a variety of functional genome technologies. For example, with DNA chips (glass or plastic wafers to which thousands of different DNA sequence probes are attached) researchers can simultaneously monitor the expression of thousands of genes in cultured cells. The following discussion describes the basic tools used in genomic technology.

The isolation, characterization, and manipulation of DNA sequences, now considered commonplace, is made possible by a series of techniques referred to as **recombinant DNA technology**. The essential feature of this technology is that DNA molecules obtained from various sources can be cut and spliced together. These techniques have made genomes more accessible to investigation than ever before because the large number of DNA copies required in DNA sequencing methods can be obtained through molecular cloning and (more recently) PCR. Commercial applications of recombinant DNA techniques have revolutionized medical practice. For example, human gene products such as insulin and growth hormone, as well as certain vaccines (e.g., hepatitis B vaccine) and diagnostic tests (e.g., HIV diagnosis), are now produced in large quantities by bacterial cells into which recombinant genes have been inserted. Currently, several research groups are investigating the use of recombinant techniques in human gene therapy, a process in which (it is hoped) defective genes can be replaced by their normal counterparts.

Figure 18A illustrates the basic features of recombinant DNA construction. The process begins by using a restriction enzyme (see the Biochemistry in the Lab box entitled Nucleic Acid Methods in Chapter 17, pp. 644–648) that generates sticky ends to cleave DNA from two different sources. The DNA fragments are then mixed under conditions that allow annealing (base pairing) between the sticky ends. Once base pairing has occurred, the fragments are covalently bonded together by DNA ligase. After recombinant DNA molecules have been isolated and purified, it is usually necessary to reproduce them so that sufficient quantities are available for further investigation. Molecular cloning, a commonly used method for increasing the number of copies of DNA, is discussed next.

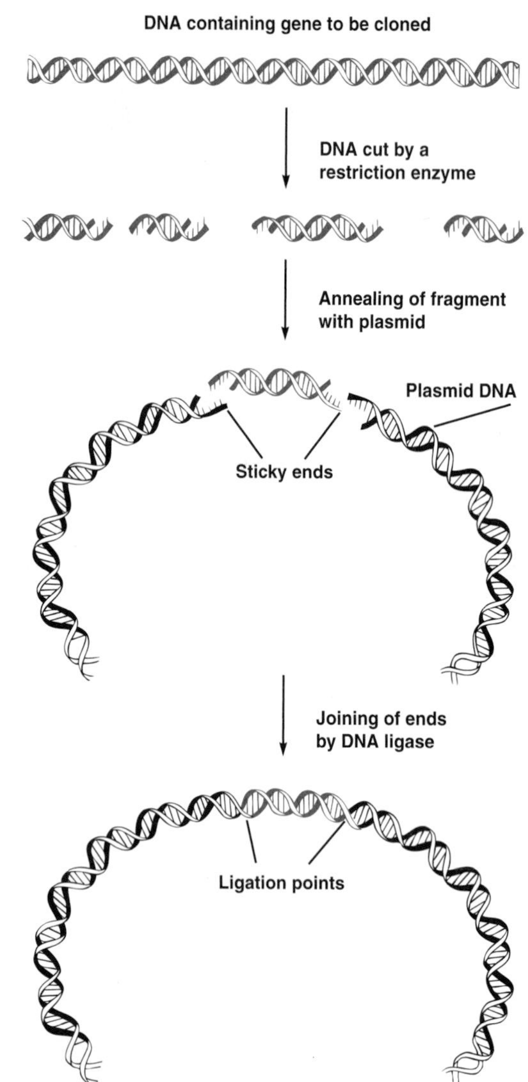

FIGURE 18A
Recombinant DNA Construction

Recombinant DNA molecules are created by treating DNA from two sources with restriction enzymes. Under hybridizing conditions, DNA fragments with sticky ends anneal together. Once base pairing has occurred, DNA ligase joins the fragments together.

Molecular Cloning

The term *molecular cloning* refers to the experimental methods used to create recombinant DNA molecules. In molecular cloning, a piece of DNA isolated from a donor cell (e.g., any animal or plant cell) is spliced into a vector. A **vector** is a DNA molecule capable of replication that is used to transport a foreign DNA sequence, often a gene, into a host cell.

The choice of vector depends on the size of donor DNA. For example, bacterial plasmids are often used to clone small pieces (15 kb) of DNA. Somewhat larger pieces (24 kb) are incorporated

Biochemistry IN THE LAB cont.

into bacteriophage λ vectors, whereas cosmid vectors are used for DNA fragments as large as 50 kb. Bacteriophage λ can be used as a vector because a substantial portion of its genome does not code for phage production and can therefore be removed. The removed viral DNA can then be replaced by foreign DNA. **Cosmids** are cloning vehicles that contain λ bacteriophage *cos* sites incorporated into plasmid DNA sequences with one or more selectable markers. The *cos* sites allow delivery to the host cell in a phage head, the plasmid DNA facilitates independent replication of the recombined unit, and the selectable markers permit detection of successful recombinants. Still larger pieces can be inserted into bacterial artificial chromosomes and yeast artificial chromosomes. **Bacterial artificial chromosomes** (BAC), derived from a large *E. coli* plasmid called the F-factor, are used to clone DNA sequences as long as 300 kb. **Yeast artificial chromosomes** (YAC), which can accommodate up to 1000 kb, are constructed by using yeast DNA sequences that are autonomously replicating (i.e., contain a eukaryotic DNA replication origin).

As noted, forming recombinant DNA requires a restriction enzyme, which cuts the vector DNA (e.g., a plasmid) open (**Figure 18B**). After the sticky ends of the plasmid have annealed with those of the donor DNA, a DNA ligase activity joins the two molecules covalently. Then the recombinant vector is inserted into host cells.

In some circumstances the introduction of a cloning vector into a host cell is trivial. For example, phage vectors are designed to introduce recombinant DNA in an infective process called **transfection**, and some bacteria take up plasmids unaided. However, most host cells must be induced to take up foreign DNA. Several methods are used. In some prokaryotic and eukaryotic cells, the addition of Ca^{2+} to the medium promotes uptake. In others, a process called **electroporation**, in which cells are treated with an electric current, is used. One of the most effective methods for transforming animal and plant cells is the direct microinjection of genetic material. Transgenic animals, for example, are created by the microinjection of recombinant DNA into fertilized ova.

Once introduced, each type of cell replicates the recombinant DNA along with its own genome. Note that recombinant vectors must contain regulatory regions recognized by host cell enzymes.

As host cells that have been successfully transformed proliferate, they rapidly amplify the recombinant DNA. For example, under favorable conditions of nutrient availability and temperature, a single recombinant plasmid introduced into an *E. coli* cell can be replicated a billion times in about 11 hours. However, transformed and untransformed cells usually look exactly alike. Consequently, researchers often design cloning protocols that use vectors with selectable **marker genes** (genes whose presence can be detected) to facilitate the identification of transformed cells. Antibiotic resistance genes or color selection markers (**Figure 18C**), for example, are usually incorporated into the plasmid vectors introduced into bacteria. When bacteria exposed to plasmids with antiboitic resistance genes are plated out on a medium containing the antibiotic, only the transformed cells will grow. With eukaryotic organisms such as

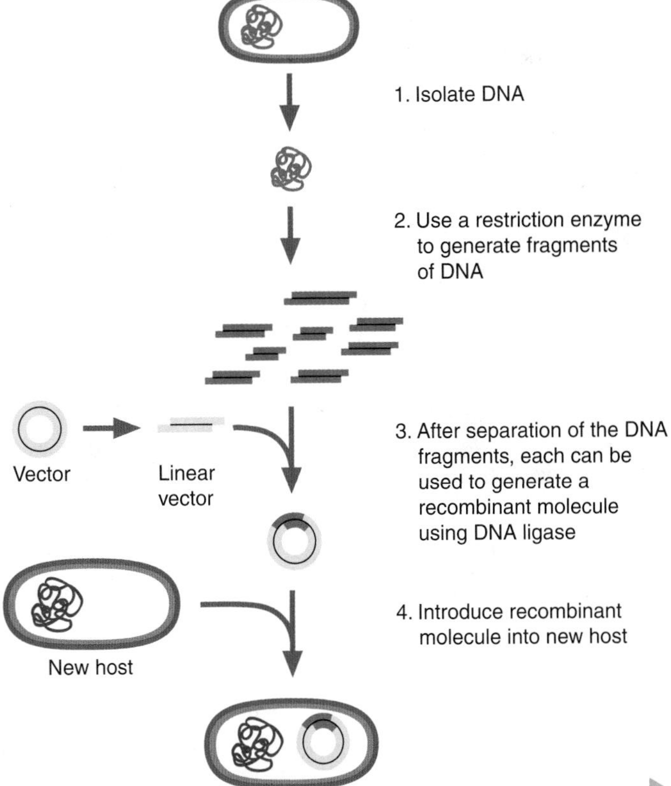
1. Isolate DNA
2. Use a restriction enzyme to generate fragments of DNA
Vector
Linear vector
3. After separation of the DNA fragments, each can be used to generate a recombinant molecule using DNA ligase
New host
4. Introduce recombinant molecule into new host

FIGURE 18B
DNA Cloning

In the cloning process, each clone is produced by introducing a recombinant molecule into a host cell, which then replicates the vector along with its own genome. The same restriction enzyme is used to create the linear vector and the DNA fragments.

Biochemistry IN THE LAB cont.

yeast, cells that lack an enzyme required to synthesize a nutrient may be used. For example, vectors containing the LEU2 gene are used to transform mutant yeast cells that lack a specific enzyme in the leucine biosynthetic pathway. Only cells that have successfully transformed are able to grow in a leucine-deficient medium.

In another approach, the **colony hybridization technique** (**Figure 18D**), bacteria are screened by using a radioactively labeled nucleic acid probe, an RNA molecule or a single-stranded DNA molecule with a sequence complementary to that of a specific sequence within the recombinant DNA. Bacterial cells are plated out onto solid media in petri dishes and allowed to grow into colonies. Each plate is then blotted with a nitrocellulose filter. (Some cells from each of the original colonies remain on the petri dishes.) The cells on the nitrocellulose filter are lysed, and the released DNA is treated so that hybridization with the probe can occur. Once nonhybridized probe molecules have been washed away, autoradiography (see the Biochemistry in the Lab box in Chapter 2 entitled Cell Technology, pp. 68–70) is used to identify the colonies on the master plate that possess the recombinant DNA of interest.

Polymerase Chain Reaction

Although cloning has been immensely useful in molecular biology, **PCR** is a more convenient method for obtaining large numbers of DNA copies. Using a heat-stable DNA polymerase from *Thermus aquaticus* (Taq polymerase), PCR can amplify any DNA sequence, provided the flanking sequences are known

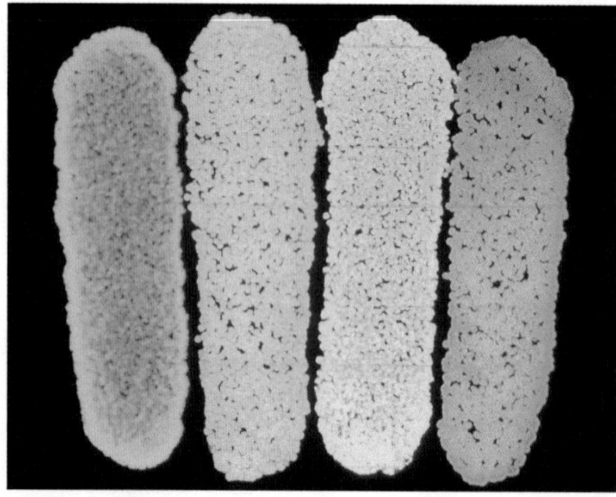

FIGURE 18C

Four Recombinant *E. coli* Cells, Each with a Different Variant of the Gene for Luciferase

Luciferase is an enzyme found in fireflies, mollusks, and several types of deep-sea fish. When in the presence of ATP and luciferin, luciferase catalyzes a light-emitting reaction. The luciferins are a group of bioluminescent compounds that emit light when they are oxidized by O_2 in a reaction catalyzed by a luciferase.

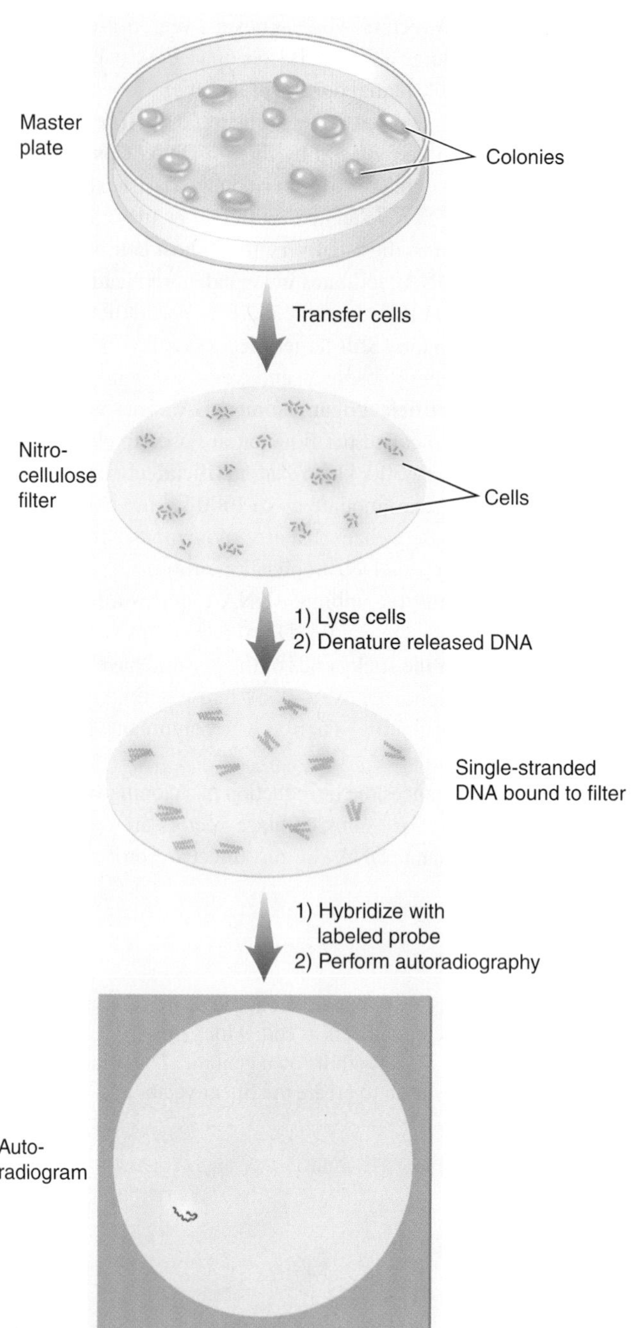

FIGURE 18D

Colony Hybridization

Bacterial cells are plated onto a solid medium that allows the growth of transformed cells only. When the colonies become visible, the plate is blotted with a nitrocellulose filter. The cells clinging to the filter are lysed, and the released DNA is denatured and deproteinized. Then a labeled probe is added and unhybridized probe molecules are washed away. Cells that possess DNA sequences that hybridize with the probe are identified by comparing the autoradiogram of the filter with the master plate.

Biochemistry IN THE LAB cont.

(**Figure 18E**). Flanking sequences must be known because PCR amplification requires primers. Priming sequences are produced by automated DNA-synthesizing machines.

PCR begins by adding *Taq* polymerase, the primers, and the ingredients for DNA replication to a heated sample of the target DNA. (Recall that heating DNA separates its strands.) As the mixture cools, the primers attach to their complementary sequences on either side of the target sequence. Each strand then serves as a template for DNA replication. At the end of this process, referred to as a *cycle*, the copies of the target sequence have been doubled. The process can be repeated indefinitely, synthesizing an extraordinary number of copies. By the end of 30 cycles, for example, a single DNA fragment has been amplified 1 billion times.

Genomic Libraries

Genomic libraries, also called clones or gene banks, are collections of clones derived from fragments of entire chromosomes or genomes. They are used for a variety of purposes, the most important of which are the isolation of specific genes whose chromosomal location is unknown and in genome-wide sequencing efforts (gene mapping). Genomic libraries are produced in a process, referred to as **shotgun cloning**, in which a genome is randomly digested (**Figure 18F**). The range of fragment sizes, which is determined by the type of restriction enzyme and the experimental conditions chosen, must be compatible with the vector. To ensure that all sequences of interest are represented in the library, DNA samples are often only partially digested. The location of any gene can be identified if an appropriate probe is available.

In a variation of genomic libraries, collections of complementary DNA molecules called **cDNA libraries** are produced from mRNA molecules by reverse transcription. This technique can be used to evaluate the transcriptome of certain cell types under specified circumstances. In other words, it is a method for determining which genes are expressed in a particular cell type. For example, with the use of DNA chip technology (DNA microarrays), gene expression in normal and diseased cells can be investigated and compared. cDNA libraries are especially useful when eukaryotic DNA is cloned because mRNA molecules lack noncoding or intron sequences. Consequently, gene products can be more easily identified, and large amounts of gene product can be generated in bacteria, which cannot process introns.

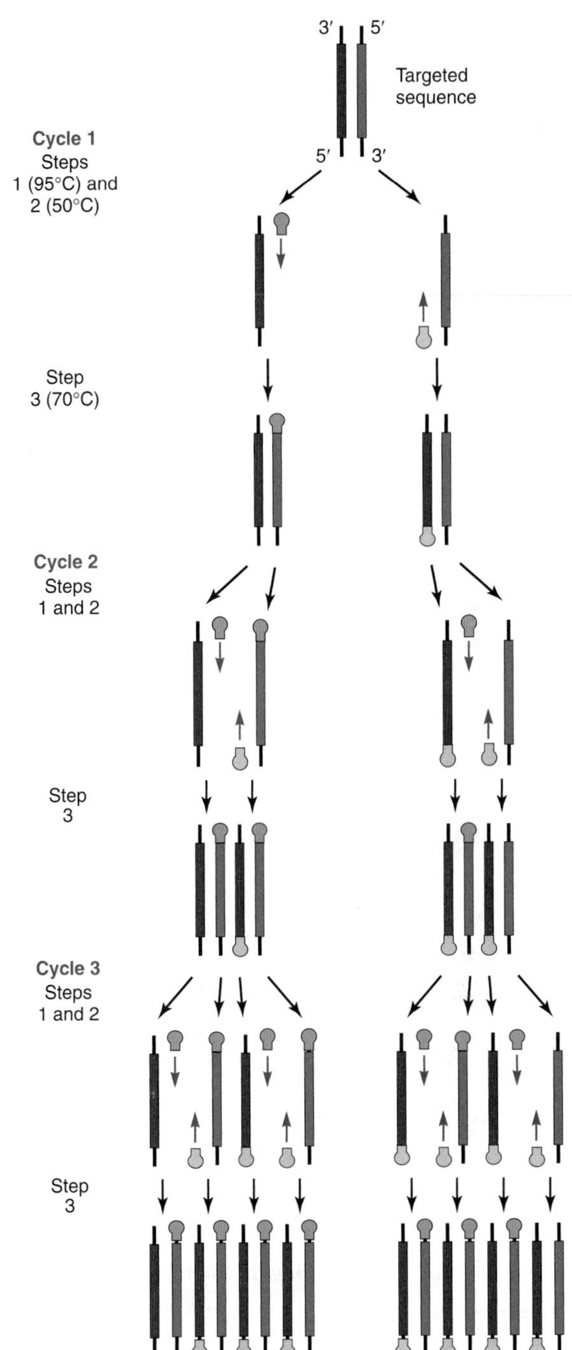

FIGURE 18E
Polymerase Chain Reaction

A single DNA molecule can be amplified millions of times by replicating a three-step cycle. In the first step the dsDNA sample is denatured by heating to 95°C. In step 2 the temperature is quickly lowered to 50°C and an oligonucleotide primer is added. The primer hybridizes to complementary sequences on the ends of the two strands. During step 3, DNA synthesis occurs as the temperature is raised to 70°C, the optimal temperature of *Taq* polymerase. The cycle is then repeated with both old and new strands serving as templates.

Chromosome Walking

Chromosome walking is used when a DNA sequence (a clone) in a genomic library is too large to sequence. The cloned DNA is fragmented and subcloned. One of the subclones is picked and sequenced and a small fragment at one end is used as a probe to select one of the remaining subclones that contains that

Biochemistry IN THE LAB cont.

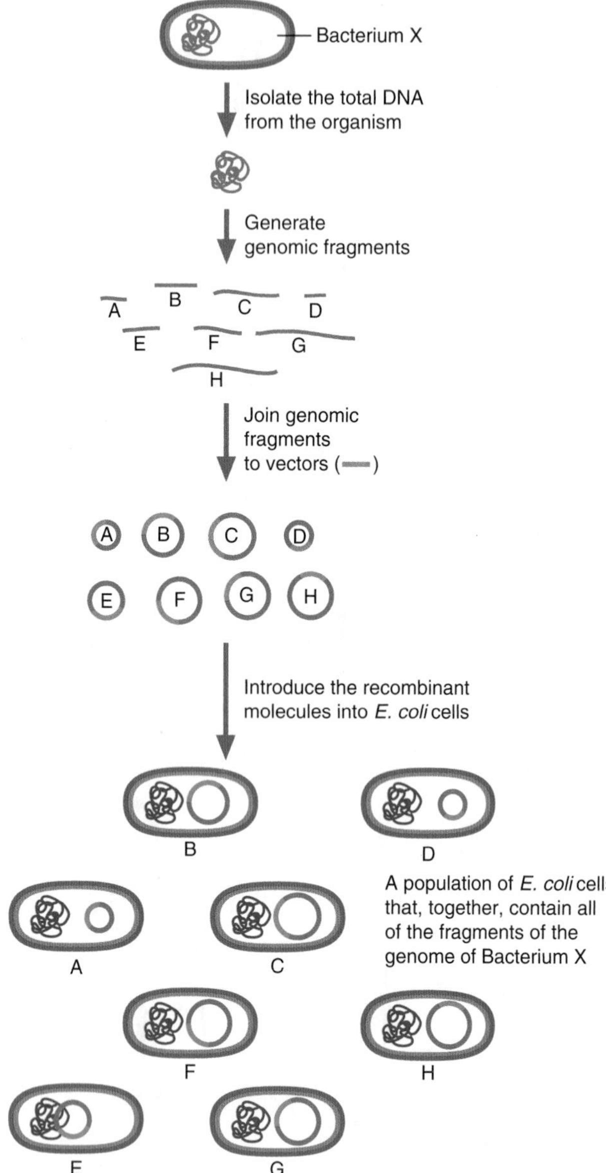

FIGURE 18F

Creation of a DNA Library Using the Shotgun Method

After the organism's DNA is isolated and purified, it is cleaved into fragments with a restriction enzyme. The fragments are then randomly incorporated into vectors, and the recombinant molecules are introduced into cells such as *E. coli*. The collection of these cells is called a genomic library.

sequence (**Figure 18G**). The new fragment is sequenced and a portion is used as a probe to select other overlapping clones. In this way, contiguous sequences can be mapped. A set of overlapping sequences is referred to as a **contig**. When eukaryotic genomes are analyzed, their large size often requires the use of large cloning vectors such as YACs and a technique called chromosomal jumping. In **chromosomal jumping** the overlapping clones contain DNA sequences of several hundred kb that are generated using restriction enzymes that cut infrequently.

DNA Microarrays

DNA microarrays, or DNA "chips," are used to analyze the expression of thousands of genes simultaneously (**Figure 18H**). Often no larger than a postage stamp, a DNA microarray consists of thousands, or hundreds of thousands, of oligonucleotides or ssDNA fragments attached to a glass or plastic support. At each position in the microarray the attached sequence, acting as a DNA probe, is designed to hybridize with a target, a specific DNA or RNA sequence.

In investigations of gene expression, an entire set of mRNA molecules from the cells of interest (i.e., the transcriptome) are reverse-transcribed into cDNA. The cDNA molecules, labeled with a fluorescent dye, are then incubated with a microarray under hybridizing conditions. After incubation, the microarray is washed to remove unhybridized molecules. Researchers then determine which genes are being expressed by identifying the positions on the microarray that are fluorescing. Scientists have used microscopes, photomultiplier tubes, and computer software to observe changes in gene expression in a variety of circumstances. Examples include comparisons of normal and cancerous cells and cells exposed to different nutrients or signal molecules.

Genome Projects

Each genome project determines the entire set of DNA base sequences of a particular organism. The process entails taking a large number of sequence fragments obtained by fracturing the genome and then determining their base sequences with an automatic sequencing method. The sequence data for each of the fragments are then assembled using computer methods to yield the entire genome sequence.

The Human Genome Project was an intensive international effort to determine the nucleotide sequence of the human genome. As this goal was reached, the attention of researchers shifted to the **annotation** (i.e., functional identification) of approximately 21,000 human genes. Just as scientists have historically used structural and functional comparisons of other organisms in anatomy, biochemistry, physiology, and medicine to better understand human biology, the current effort to interpret human genome data is being aided immensely by comparisons with the information obtained in other genome projects. The genomes of well-researched organisms as diverse as bacteria (e.g., *E. coli*), yeast (e.g., *S. cerevisiae*), the worm *Caenorhabditis elegans*, the fruit fly *Drosophila*, and various mammals (e.g., the mouse) have been used in genome structure analysis and in the assignment of recently discovered genes in other organisms.

Biochemistry IN THE LAB cont.

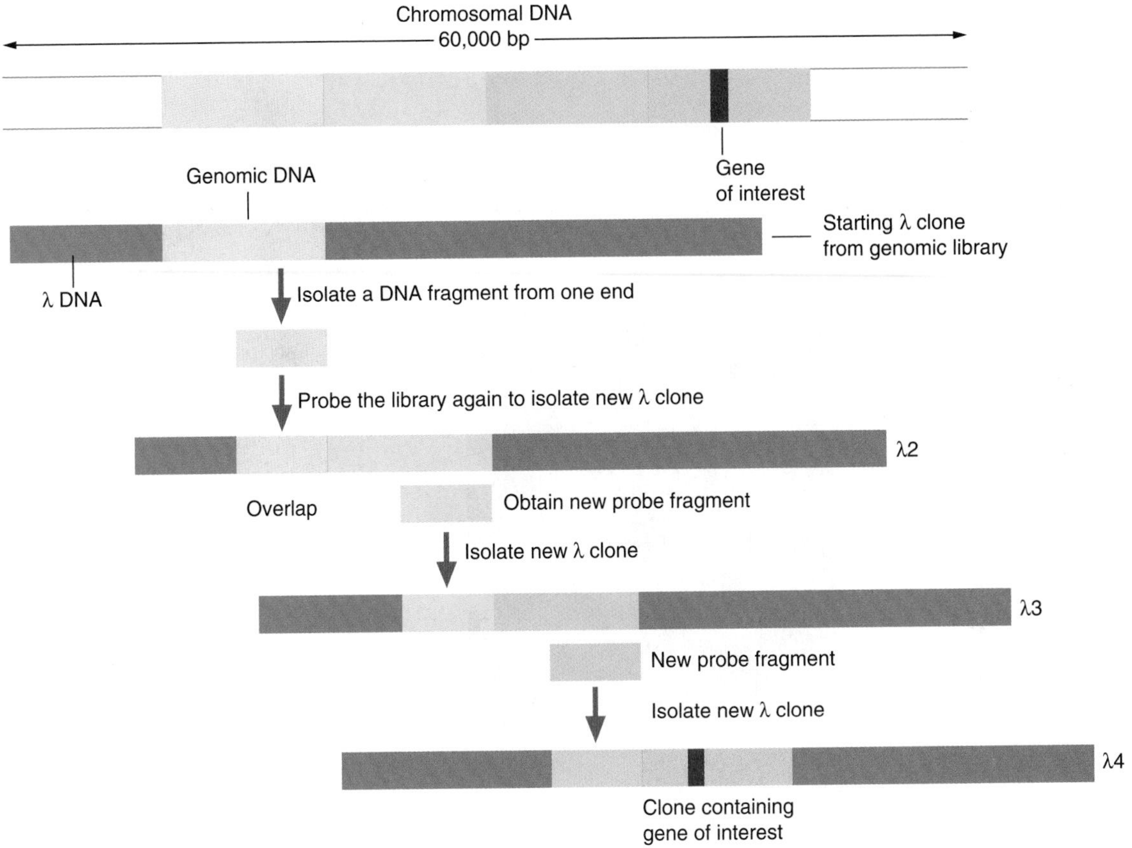

FIGURE 18G
Chromosome Walking

In chromosome walking, DNA clones, which contain overlapping sequences, are systematically identified. They may then be mapped and sequenced. Unknown genes may also be searched for. The process begins when DNA is cleaved into pieces and cloned. (In this example, bacteriophage λ vectors are used.) One end of the starting clone is labeled and used as a probe to identify the clone in the λ library that contains both that sequence and an adjacent sequence. Repeating this step results in the labeling of the end of the second clone, which is used as a probe to identify yet another overlapping clone. The process continues until a collection of clones that together contain all the sequences in the original DNA fragment has been sequenced and mapped.

ENCODE (the Encyclopedia of DNA Elements) is a research project, funded by the U.S. National Human Genome Research Institute (a division of the U.S. National Institutes of Health), that involves a worldwide group of research laboratories. Its goal, made possible by ever more powerful DNA sequencing technologies, is two-fold. First, it is to determine the function of the 98.5% of the human genome that does not code for proteins. Second, it is to stimulate the development of knowledge that can be used to develop DNA-based therapies to prevent and treat human diseases. Examples of ENCODE's preliminary results include the surprising fact that the human genome is pervasively transcribed and that the vast majority (about 80%) of human DNA sequences are in close proximity of ENCODE-identified regulatory sequences.

Bioinformatics

The emergence of the *high-throughput* (i.e., rapid, high-volume, automated) technologies to analyze living organisms has created a vast amount of data on nucleic acid and polypeptide sequences. The information, which is collected from genome and proteome sequencing projects and from microarray analysis of cell processes such as transcription, is placed in databases that are available to the scientific community. How do scientists analyze such enormous volumes of raw data? As a result of technological advances in computer science, applied mathematics, and statistics, **bioinformatics** has provided researchers with a powerful investigational tool. The use of computer algorithms has made the investigation of a wide variety of

Biochemistry IN THE LAB cont.

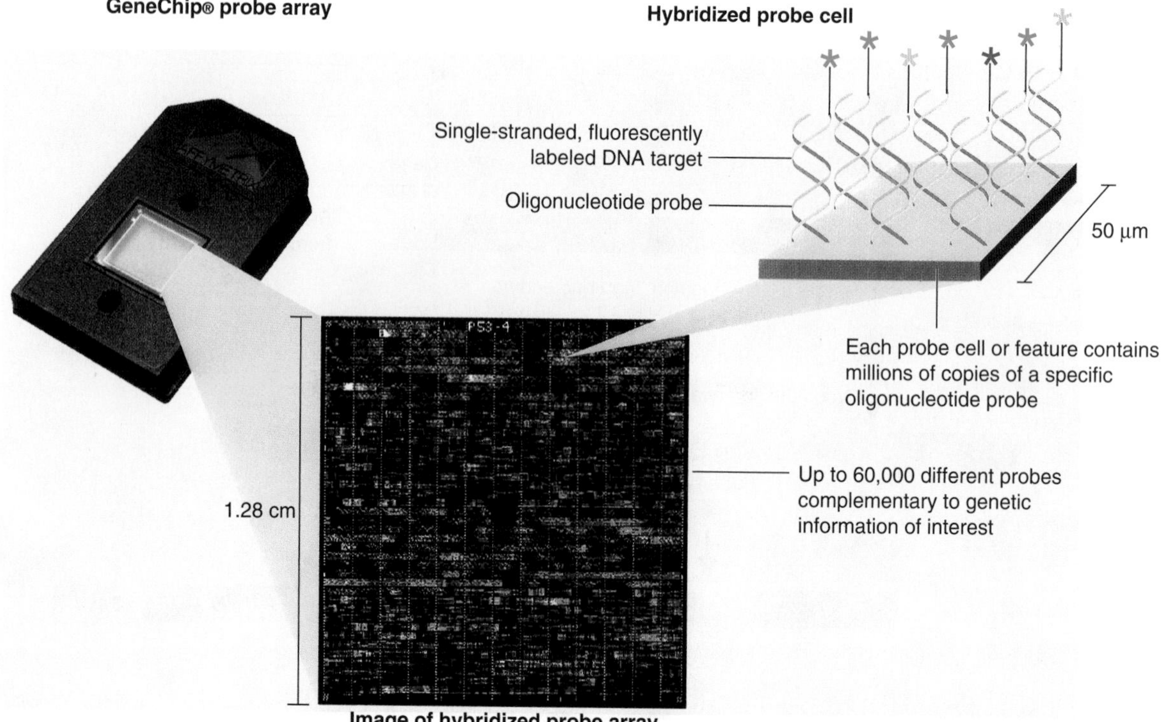

FIGURE 18H
DNA Microarray Technology

DNA microarrays can be used to determine which genes are expressed in a specific cell type because each "chip" can accommodate from thousands to millions of DNA probes. (Oligonucleotide probes are synthesized on the chip surface by means of photolithographic techniques similar to those used in the manufacture of computer chips.) The microarray is incubated under hybridizing conditions with fluorescently labeled cDNA. The cDNA molecules are derived from mRNA extracted from the cells of interest.

previously intractable problems feasible, as the following examples illustrate.

1. Genes can be located by a process referred to as sequence inspection. Gene prediction programs utilize several clues to locate sequences that can potentially code for polypeptides called open reading frames (ORFs) (p. 651). ORFs are extended DNA sequences that could potentially code for a polypeptide. They begin with the three-base sequence AUG called a *start codon* and end with a *stop codon* of UAA, UAG, or UGA. Eukaryotic ORF scanning is complicated by the presence of introns, some of which are longer than the polypeptide domain–coding exons.
2. Alignment of DNA sequences allows researchers to search the genomes of hundreds of organisms for similarities among gene or regulatory sequences and has provided invaluable insight into the relatedness of living organisms and the mechanisms used to sustain living processes.
3. Protein structure prediction has been facilitated by a method called homology modeling. Once a new protein-coding gene has been discovered, bioinformatic analysis is used to search among homologous or near-homologous molecules whose structure is already known.
4. Bioinformatic analysis of the massive data derived from protein microarrays and MS-derived cell proteome data provides an invaluable means of analyzing cell protein synthesis patterns. For example, this type of data analysis allows medical scientists to compare how normal cell proteins are altered in disease states.
5. In the field of evolutionary biology, bioinformatic programs have been used to trace the lineages of organisms based on rare events such as gene duplications and lateral gene transfer. (Lateral gene transfer is the transfer of genes between species.)
6. High-throughput gene expression analysis is now used to identify the genes involved in medical disorders (e.g., in comparisons of the transcription products of normal and cancerous cells).
7. Systems biologists' use of complex mathematical modeling combined with an ever-increasing source of biological data promises to significantly improve our understanding of life's operating systems.

18.2 TRANSCRIPTION

Transcription, the creation of RNA copies of DNA sequences, is a highly regulated process that transforms environmental cues (e.g., nutrient availability in bacteria and developmental signals in multicellular eukaryotes) into gene expression changes. As with all aspects of nucleic acid function, the synthesis of RNA molecules is a complex process involving a variety of enzymes and associated proteins. Recall that RNA molecules are transcribed from the cell's genes. As RNA synthesis proceeds, the incorporation of ribonucleotides is catalyzed by RNA polymerase, sometimes referred to as DNA-dependent RNA polymerase, or RNAP. The reaction catalyzed by all RNA polymerases is

$$\text{NTP} + (\text{NMP})_n \longrightarrow (\text{NMP})_{n+1} + \text{PP}_\text{i}$$

Because the nontemplate or plus (+) strand has the same base sequence as the RNA transcription product (except for the substitution of U for T), it is also called the **coding strand** (**Figure 18.29**). By convention, the direction of the gene, a segment of double-stranded DNA, is the same as the direction of the coding strand. Polymerization proceeds from the 5′ end to the 3′ end of the gene because the template DNA strand, also called the minus (−) strand, and the newly made RNA molecule are antiparallel. As noted, transcription generates several types of RNA, of which rRNA, tRNA, and mRNA are directly involved in protein synthesis (Chapter 19).

Transcription in Prokaryotes

The RNA polymerase (**Figure 18.30**) in *E. coli* catalyzes the synthesis of all RNA classes. The core enzyme (α_2, β, and β'), with a molecular weight of 370 kDa, catalyzes RNA synthesis. Another protein, the ϖ subunit, promotes the assembly of the RNAP holoenzyme. The transient binding of the σ (sigma) factor to the core enzyme allows it to bind both the correct template strand and the proper site to initiate transcription. A variety of σ factors have been identified. For example, in *E. coli*, σ^{70} is involved in the transcription of most genes, whereas σ^{32} and σ^{28} promote the transcription of heat shock genes and the flagellin gene, respectively. (As its name suggests, flagellin is a protein component of bacterial flagella.) The superscript indicates the protein's molecular weight in kilodaltons.

The transcription of an *E. coli* gene is outlined in **Figure 18.31**. The process consists of three stages: initiation, elongation, and termination. Each is discussed briefly.

The initiation of transcription involves the binding of RNA polymerase to a **promoter**, a regulatory DNA sequence that is located upstream (i.e., toward the 5′ end of a polynucleotide) of a gene. Although prokaryotic promoters are variable in size (from 20 to 200 bp), two short sequences at positions about 10 and 35 bp away from the transcription initiation site are remarkably similar among

DNA (+)
5′— TTTGGACAACGTCCAGCGATC —3′ Nontemplate strand (coding strand)
3′— AAACCTGTTGCAGGTCGCTAG —5′ Template strand (noncoding strand) (−)

RNA
5′—UUUGGACAACGUCCAGCGAUC —3′

FIGURE 18.29
DNA Coding Strand

One of the two complementary DNA strands, referred to as the template (−) strand, is transcribed. The RNA transcript is identical in sequence to the nontemplate (+) or coding strand, except for the substitution of U for T.

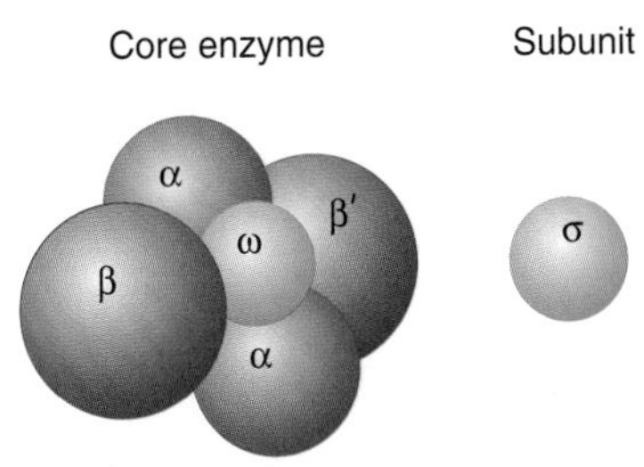

FIGURE 18.30
***E. coli* RNA Polymerase**

The *E. coli* RNA polymerase consists of two α subunits and one each of β, and β' subunits. The transient binding of a σ subunit allows binding of the core enzyme to appropriate DNA sequences. Note that ω promotes the assembly of the core enzyme.

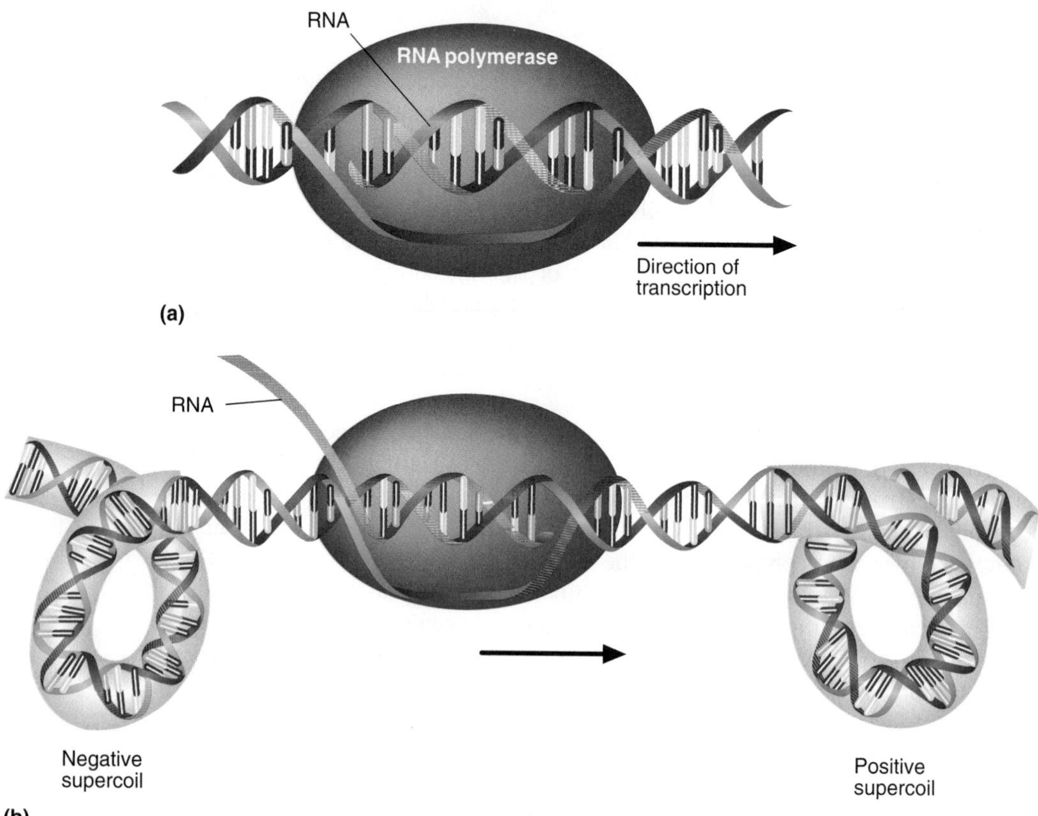

FIGURE 18.31
Transcription Initiation in *E. coli*

(a) A transcription bubble forms as a short DNA segment unwinds. An RNA-DNA hybrid forms as transcription progresses. The bubble moves to keep up with transcription as DNA unwinds ahead of it and rewinds behind it. (b) Transcription induces coiling. Positive supercoils form ahead of the bubble, while negative supercoils form behind it.

various bacterial species. There is a set of base sequences, called consensus sequences, at each of these sites. A **consensus sequence** represents an average of a number of closely related but nonidentical sequences. The sequences shown in **Figure 18.32** are named in relation to the transcription starting point, the −35 region and the −10 region. (The −10 region is also called the *Pribnow box*, after its discoverer.) RNA polymerase slides along the DNA until it reaches a promoter sequence. Promoters vary widely in the efficiency with which they productively bind RNA polymerase. Transcription initiation rates between "strong" promoters (close to the consensus sequence) and "weak" promoters (far from the consensus sequence) may vary by as much as a thousandfold. Mutations within a promoter sequence usually weaken the promoter but can also convert a weak promoter into a stronger one. Neither possibility is favorable.

RNAP and the Prokaryotic Transcription Process

The RNAP holoenzyme (i.e., the core enzyme and its associated σ factor) binds to the promoter region. A short DNA segment then unwinds as the β' and σ subunits break the hydrogen bonds between 13 bp in the Pribnow box. Because the DNA strands are now separated, the enzyme-promoter complex is referred to as "open."

Transcription begins with the binding of the first nucleoside triphosphate (usually ATP or GTP) to the RNA polymerase complex. A nucleophilic attack by the 3′-OH group of the first nucleoside triphosphate on the α-phosphate of a second nucleoside triphosphate (also positioned by base pairing in the adjacent site)

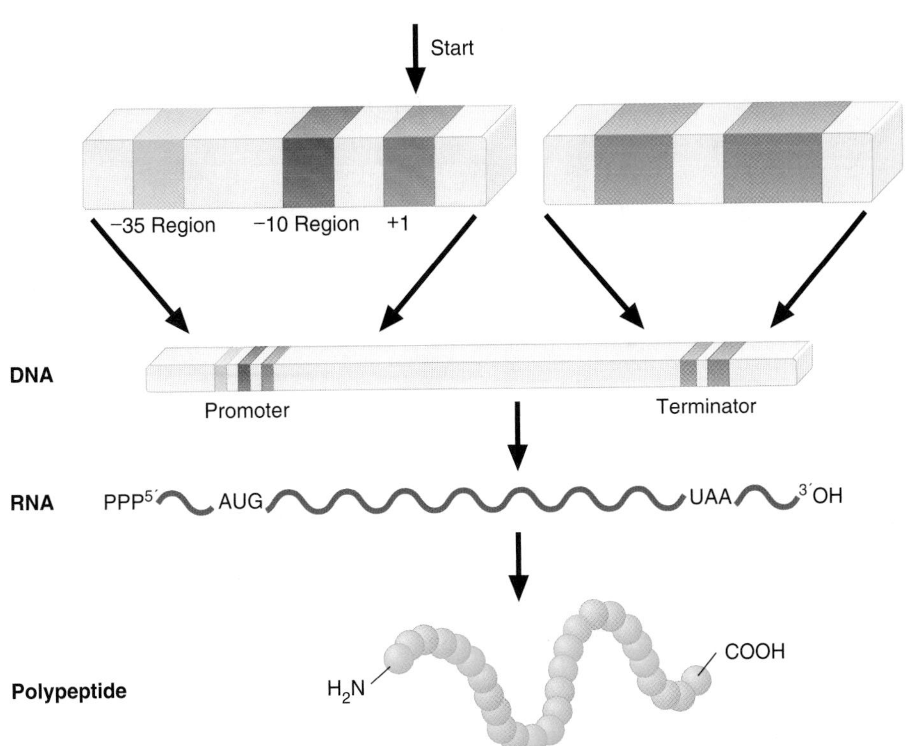

FIGURE 18.32

Typical *E. coli* Transcription Unit

If RNA polymerase can bind to the promoter, DNA transcription begins at the start site +1. Translation of mRNA begins as soon as the ribosome-binding site on the mRNA transcript is available.

causes the first phosphodiester bond to form. (Because the phosphate groups of the first molecule are not involved in this reaction, the 5′ end of prokaryotic transcripts possesses a triphosphate group.) RNAP proceeds to catalyze additional phosphodiester bonds between ribonucleotides that are base-paired to the DNA template strand within the promoter. For the initiation phase to end successfully (i.e., for the RNAP holoenzyme to move away from the promoter), the growing RNA chain must reach a length of about 10 nt. Several attempts at "promoter clearance" usually fail, and the truncated transcripts are released and then degraded. When the transcribed sequence reaches a length of 10 nucleotides, the conformation of the RNA polymerase complex changes, the σ factor is released, and the initiation phase ends. As soon as one RNA polymerase has moved beyond the promoter site, another RNA polymerase can move in, bind to the site, and start another round of RNA synthesis.

The elongation phase begins once the σ factor has detached and the affinity of the RNAP complex for the promoter site has decreased. The core RNA polymerase converts to an active transcription complex as it binds several accessory proteins. As RNA synthesis proceeds in the $5' \rightarrow 3'$ direction (**Figure 18.31**), the DNA unwinds ahead of the *transcription bubble* (the transiently unwound DNA segment, composed of 12 to 14 bp, in which an RNA-DNA hybrid has formed). At any one time there are the equivalent of about 30 bp of DNA within RNAP. The active site of the enzyme complex lies between the β and β' subunits. (**Figure 18.33**). As the dsDNA enters the enzyme and is separated into two strands, the template strand enters the active site through a channel. The nontemplate strand is looped away from the active site and travels in its own channel. When the template and nontemplate strands emerge from their separate channels, they re-form a double helix. Meanwhile, the growing RNA transcript exits through its own channel formed in the β and β' subunits. The unwinding action of RNA polymerase creates positive supercoils ahead of the transcription bubble and negative supercoils behind the bubble, which are resolved by topoisomerases. As the bubble moves down the gene, it is said to move "downstream." The incorporation of ribonucleotides continues until a termination signal is reached.

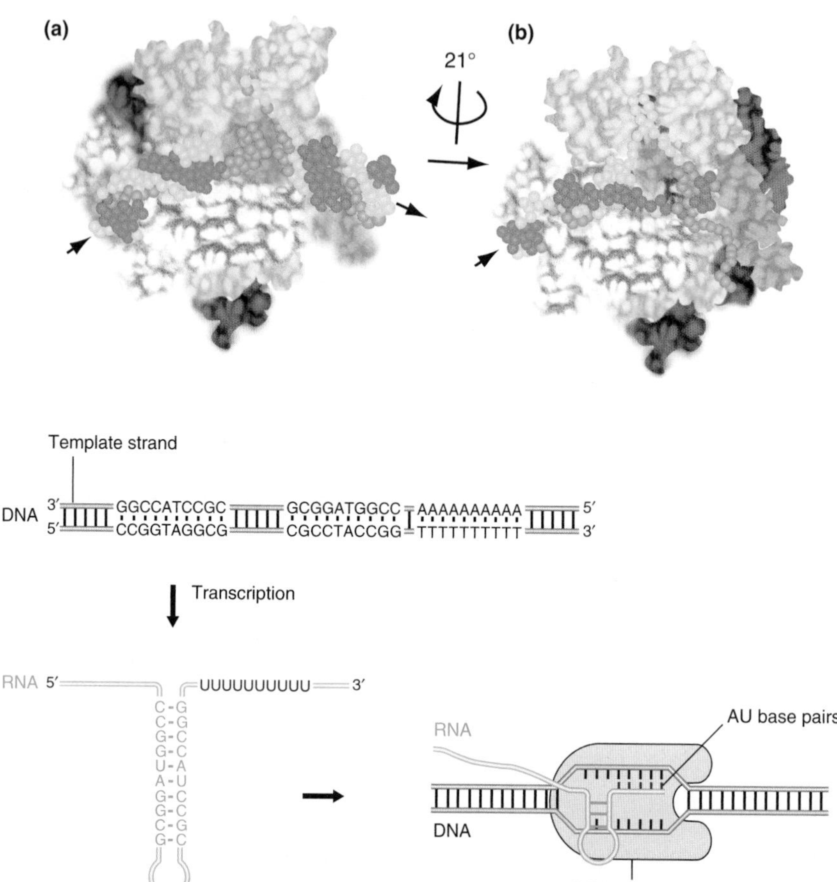

FIGURE 18.33

Bacterial RNA Polymerase Model

In this model the components of RNAP are color-coded as follows: β (blue), β' (pink). The DNA template strand is red and the nontemplate strand is yellow. In (a) and (b) the double helix lies within a horizontal trough between the β and β' subunits. The black arrows indicate the direction of DNA movement within the enzyme. Rotation of the model by 21° reveals the growing RNA transcript (gold) as it exits the RNAP.

FIGURE 18.34

Intrinsic Termination

When the termination sequence (an inverted repeat followed by a series of As) in the template strand has been transcribed, the resulting RNA sequence forms into a stable hairpin (stem-loop) structure. After the disruption of DNA-RNA interactions by the hairpin, the RNA transcript is held on the template strand only by a short sequence of AU base pairs. The RNA molecule quickly dissociates because AU interactions are very weak.

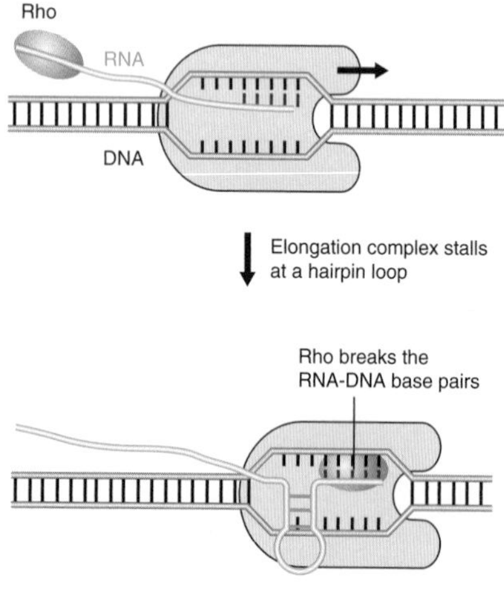

FIGURE 18.35

Rho-Dependent Termination

Rho factor is an ATP-dependent helicase that binds to a C-rich sequence in the RNA transcript. Once bound, rho moves along the transcript in the 5′ → 3′ direction until it reaches the termination site now inside the transcription bubble. The RNA polymerase has stalled because, it is believed, of the formation of a weak hairpin. Rho disrupts DNA-RNA interactions by using ATP-derived energy to unwind the DNA-RNA hybrid, and the RNA transcript is released.

There are two types of transcription termination in bacteria: intrinsic termination and rho-dependent termination. In **intrinsic termination** (also referred to as **rho-independent termination**) RNA synthesis is terminated as the result of the transcription of the termination sequence that consists of an inverted repeat sequence followed by from 6 to 8 As (**Figure 18.34**). When the termination sequence is transcribed, the inverted repeat forms a stable hairpin that causes the RNA polymerase to slow or stop. The RNA transcript is then released because the RNA-DNA hybrid dissociates owing to weak base pair interactions between the As in a short polyadenylate [poly(A)] sequence that follows the inverted repeat and the complementary Us in the transcript. In **rho-dependent termination**, strong hairpins do not form and termination requires the aid of a protein known as the rho factor, an ATP-dependent helicase (**Figure 18.35**). Rho factor binds to a specific recognition sequence on the nascent mRNA strand upstream from the termination site. It then proceeds to unwind the RNA-DNA helix to release the transcript and dislodge the polymerase.

In prokaryotes, mRNA is translated as soon as a ribosome-binding site is exposed (cotranscriptional translation). However, mature rRNA and tRNA molecules are produced from larger transcripts by posttranscriptional processing. The RNA processing reactions for *E. coli* rRNA are outlined in **Figure 18.36**. The *E. coli* genome contains several sets of the rRNA genes 16S, 23S, and 5S. In the primary processing step, the polycistronic 30S transcript is methylated and then cleaved by several RNases into a number of smaller segments. Further cleavage by different RNases produces mature rRNAs. A few tRNAs are also produced. The other tRNAs are produced from primary transcripts in a series of processing reactions in which they are trimmed down by several RNases. In the last step of tRNA processing, a large number of bases are altered by several modification reactions (e.g., deamination, methylation, and reduction).

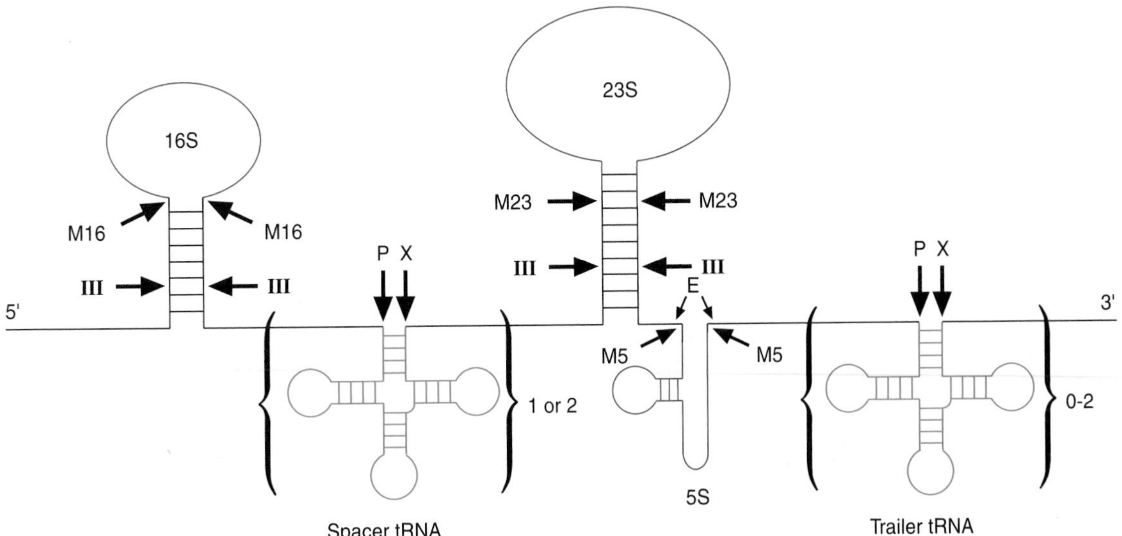

FIGURE 18.36

Ribosomal RNA Processing in *E. coli*

Each rRNA operon encodes a primary transcript that contains one copy each of 16S, 23S, and 5S rRNAs. Each transcript also encodes one or two spacer tRNAs and as many as two trailer tRNAs. Posttranscriptional processing involves numerous cleavage reactions catalyzed by various RNases and splicing reactions. (Individual RNases are identified by letters and/or numbers, e.g., M5, X, III.) RNase P is a ribozyme.

Transcription in Eukaryotes

KEY CONCEPTS

- During transcription, an RNA molecule is synthesized from a DNA template.
- In prokaryotes this process involves a single RNA polymerase activity.
- Transcription is initiated when the RNA polymerase complex binds to a promoter sequence.

DNA transcription in prokaryotes and eukaryotes is similar in several respects. For example, the bacterial RNA polymerases and their eukaryotic counterparts are structurally similar and use a common transcription mechanism (e.g., promoter recognition, truncated transcripts, and promoter clearance). Also, although the initiation factors in bacteria and eukaryotes are only distantly related, they perform similar functions. These two types of organisms differ significantly, however, in the regulatory mechanisms that control gene expression. One of the most prominent examples of these differences is the limited access of the eukaryotic transcription machinery to DNA. Chromatin is usually at least partially condensed. Yet for transcription to occur, DNA must be sufficiently exposed and accessible for RNA polymerase activity. For DNA to be permissive to transcription, the histone tails must be acetylated by histone acetyltransferases, and the histone-DNA contacts must be weakened by **chromatin-remodeling complexes.** There are two classes of chromatin-remodeling complex. The SWI and SNF proteins (as a SWI/SNF complex) facilitate the release of the histone particle, whereas the NURF proteins release the contacts only enough to allow the histone particle to slide out of the way (**Figure 18.37**).

Principally because of the complexity of eukaryotic genomes, eukaryotic DNA transcription is not understood as completely as the prokaryotic process. However, the eukaryotic process is known to possess several unique features.

RNA POLYMERASE ACTIVITY Eukaryotes possess three nuclear RNA polymerases, each of which differs in the type of RNA synthesized, subunit structure, and relative amounts. RNA polymerase I (RNAP I), which is localized within the nucleolus, transcribes the larger rRNAs (28S, 18S, and 5.8S). The precursors of mRNA and most miRNAs and snRNAs are transcribed by RNA polymerase II (RNAP II), and RNA polymerase III (RNAP III) is responsible for transcribing the precursors of the tRNAs, 5S rRNA, U6 snRNA, and the snoRNAs. Each polymerase possesses two large subunits and several (8–12) smaller subunits. For example, the two large subunits of RNAP II have molecular

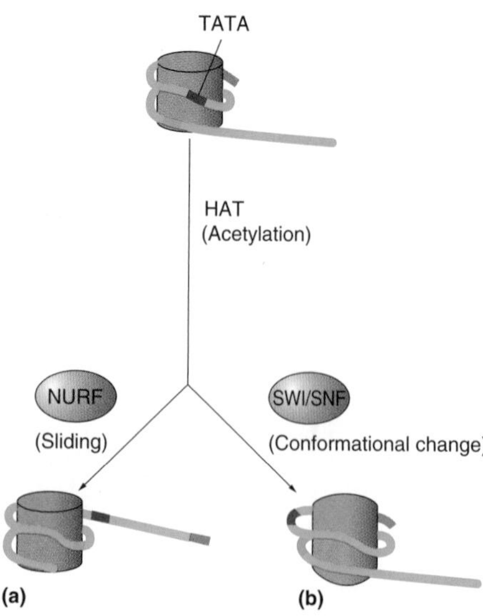

FIGURE 18.37
Chromatin Remodeling

The acetylation by histone acetyltransferase (HAT) of the histone tails breaks their contacts with DNA. The core histones are then released from DNA contacts by sliding out of the way through the action of NURF proteins (a) to expose the promoter region on the DNA to the transcriptional machinery or by a localized conformational change in chromatin (b) promoted by the SWI/SNF chromatin-remodeling complex.

weights of 215 and 139 kDa. The number of smaller subunits varies among species; for example, plants possess eight, whereas vertebrates have six. Some of the smaller subunits are also present in the other two RNA polymerases. In contrast to the prokaryotic RNA polymerase, the eukaryotic enzymes cannot initiate transcription themselves. Various transcription factors must be bound at the promoter before transcription can begin.

RNA polymerases function within discrete clusters where active genes come together via chromatin looping to be transcribed. These clusters, called *transcription factories,* are located throughout the nucleus. The number of transcription factories varies with species and cell type. The nuclei of HeLa cells (epithelial cells from an immortal cell line used in research), for example, contain about 10,000 factories, of which 8000 are RNAPII factories and 2000 are RNAPIII factories. The nucleolus is considered a large RNAPI factory. RNAPII transcription factories, each of which contains an estimated 8 polymerase molecules, have diameters ranging from 40 to 180 nm (depending on the level of transcriptional activity). Each RNAPII-associated factory has a protein-rich core with active RNA polymerase enzyme complexes attached to its surface.

EUKARYOTIC PROMOTERS Eukaryotic promoters are more complex and variable than those of prokaryotes. Each consists of a *core promoter,* the minimum DNA sequence required for RNA polymerase binding and transcription initiation, and additional proximal (nearby) and distal (distant) DNA sequences that contribute to transcription regulation.

There are two classes of eukaryotic core promoters: focused and dispersed. Focused promoters contain a *transcription start site* (TSS) that consists of a single nucleotide or a very narrow range of nucleotides and a variable set of sequence motifs, referred to as *core promoter elements* (CPEs), that allow binding of the transcription machinery. The TATA box, originally identified in yeast, is the best researched eukaryotic CPE. Found only in about 10% of mammalian core promoters, the AT-rich TATA box is recognized and bound by TATA-binding protein (TBP), a subunit of the transcription factor TFIID. Other examples of CPEs found in focused core promoters include Inr (initiator), BRE (B recognition element), MTE (motif ten element), and DPE (downstream promoter element) (**Figure 18.38**). Dispersed core promoters are so named because they possess many TSSs (usually one strong and several weak start sites), which are distributed seemingly at random over a broad region of 50 to 100 bp. Commonly found in vertebrates and especially in mammals, dispersed core promoters typically occur within CpG islands (p. 641) and are mainly

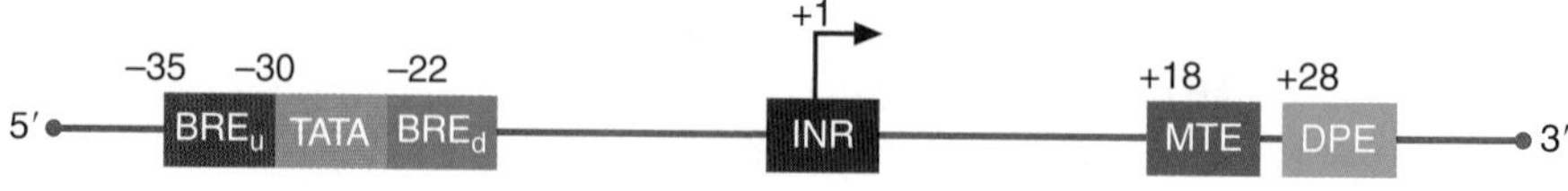

FIGURE 18.38
The Eukaryotic RNAPII Core Promoter

The core promoter, the minimum DNA sequence required for RNA polymerase binding and transcription initiation for a specific gene, often contains sequence motifs that researchers have identified as core promoter elements (CPE). The TATA box is a well-researched CPE. BRE_u and BRE_d are CPEs that when present are located immediately upstream and downstream of the TATA box, respectively. Together or separately BRE_u and BRE_d modulate transcription activation via their interactions with TFIIB. The majority of genes lack a TATA box. Instead, they may contain Inr (initiator element) or DPE (downstream core promoter element). MTE (motif 10 element), acting in conjunction with Inr, promotes the binding of TFIID to the core promoter. No CPE is present in all eukaryotes core promoters, and some core promoters have no known CPEs. Note that when it is present in a core promoter, Inr contains the transcription start sequence.

associated with housekeeping genes. (*Housekeeping genes* code for proteins such as glycolytic enzymes and ribosomal proteins that are constantly required by cells for essential functions.) CpGs are now believed to facilitate transcription by destabilizing nucleosomes.

Proximal promoter elements are transcription factor binding sites up to about 250 bp upstream of the TSS. Examples of proximal promoter elements include the GC and CCAAT boxes, both of which enhance transcription when bound to certain transcription factors. For example, transcription rates increase when the transcription factor Spl binds to GC box elements upstream of genes that code for proteins such as the insulin receptor (p. 593) and HMG-CoA reductase (p. 459).

Distal regulatory sequences are distance-independent DNA elements that can be located upstream or downstream or within the introns of their target genes. Distal elements can increase (*enhancers*) or decrease (*silencers*) transcription. Transcription is activated when an activator protein binds to an enhancer element, and it is inhibited when a repressor protein binds to a silencer element.

When enhancer or silencer elements come into contact with an RNA polymerase complex at a promoter, the intervening DNA sequences (as long as 100 kbp) bend into loops. Such loops are stabilized by a ring-shape protein called *cohesin*. The interactions of enhancers and promoters can be blocked by boundary DNA elements called *insulators*. In gene- and promoter-dense DNA regions in vertebrates, CTCF, an insulator protein dimer, interacts with cohesin at insulator elements to prevent enhancer interaction with other promoters.

RNA POLYMERASE II AND THE EUKARYOTIC TRANSCRIPTION PROCESS

DNA-directed RNA polymerase II (RNAP II) is the most investigated type of RNA polymerase in eukaryotes, largely because of its role in mRNA synthesis. The structural and functional properties of yeast RNAP II were determined by Roger Kornberg (**Figure 18.39**), who received the 2006 Nobel Prize in Chemistry for this work.

The core RNAP II enzyme contains 12 subunits in humans. RBP1, the largest subunit, forms part of the enzyme's active site, the groove that binds DNA. RBP1 also contains a C-terminal domain (CTD) with between 25 and 52 repeats of the heptad YSPTSPS (52 in humans). The CTD's capacity to act as a docking site for proteins involved in different phases of transcription is regulated by reversible covalent modifications of various heptad residues (primarily phosphorylation of Ser2 and Ser5). RNAP II can bind to promoters when the heptads are unphosphorylated. As transcription progresses, a pattern of phosphorylated Ser2 and Ser5 residues created by kinases and phosphatases allows the recruitment of

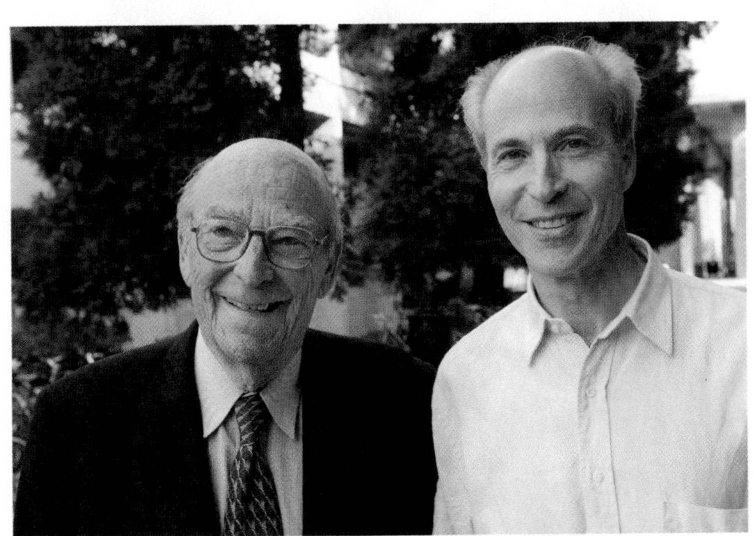

FIGURE 18.39

Roger Kornberg (right) won the 2006 Nobel Prize in Chemistry for his elucidation of the molecular mechanism of eukaryotic transcription. His father, Arthur Kornberg, who died in 2007, received the 1959 Nobel Prize in Physiology or Medicine for his work on prokaryotic DNA polymerases.

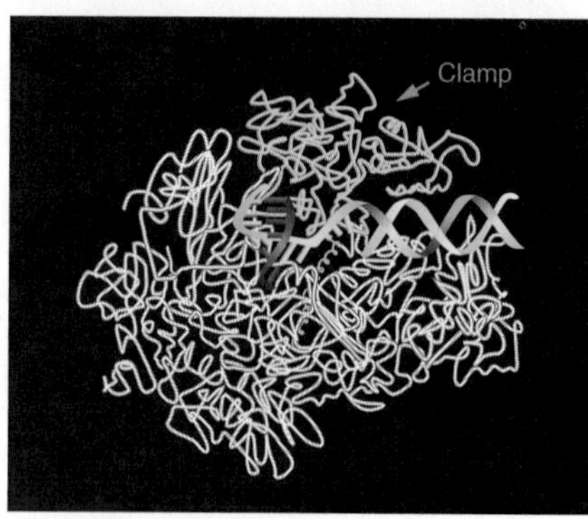

FIGURE 18.40
Yeast RNAP II Core Enzyme Structure

This view of the core RNAP II shows the polypeptide chains in white and orange (clamp protein), DNA (blue = template strand and green = nontemplate strand), and RNA (lavender).

specific RNA processing enzymes. During transcription the pattern of serine phosphorylation changes from high Ser5P and low Ser2P levels near the beginning to the reverse further downstream. Examples of processes affected by CTD serine phosphorylation patterns include the conversion of RNAP II into a processive enzyme, histone remodeling, and mRNA 5′-cap formation and splicing initiation (p. 706). By the time that RNAP II begins synthesizing the poly(A) tail (p. 706), Ser5 residues are unphosphorylated. Complete CTD dephosphorylation allows RNAP II disengagement and its subsequent recycling in another round of transcription.

In addition to the core enzyme (**Figure 18.40**), the RNAP II transcription machinery (3000 kDa) consists of a set of six transcription factors referred to as *general transcription factors*, and a multisubunit protein complex called Mediator. The general transcription factors (GTFs) TFIIA, TFIIB, TFIID, TFIIE, TFIIF, and TFIIH are the minimum number of additional proteins that are necessary for accurate transcription. They facilitate construction of the transcription **preinitiation complex** (PIC), recognition of the promoter, and the ATP-dependent unwinding of DNA.

Mediator is a protein complex that is required for the transcription of almost all RNAP II promoters. Human mediator is a 26-subunit complex with a molecular mass of 1.2 MDa. Its structure consists of three domains: head, middle, and tail (**Figure 18.41**). Mediator is essentially a signal integration platform that acts as an adaptor between RNAP II and the transcription factors that are bound at positive (enhancers) and negative (silencers) regulatory DNA sequences, which may be some distance away from the gene(s) they modulate (**Figure 18.42**). **Figure 18.43** illustrates the PIC bound to mediator in the context of a DNA loop.

Eukaryotic transcription occurs in several phases: PIC assembly, initiation, elongation, and termination. PIC assembly (**Figure 18.44**) begins with the binding of the TATA-binding protein (TBP) subunit of TFIID to the TATA box, a process that is facilitated by TFIIA. TBP is a saddle-shape protein that causes strand separation by distorting DNA structure. The insertion of four phenylalanine side chains between base pairs in the minor groove causes the formation of a bend in the DNA segment. As the DNA bends its interactions with TBP increase, thus enhancing the shape of the bend to approximately 90°. The subsequent binding of the other GTFs followed by Mediator yields the transcriptionally active PIC. Transcription bubble formation is catalyzed by the

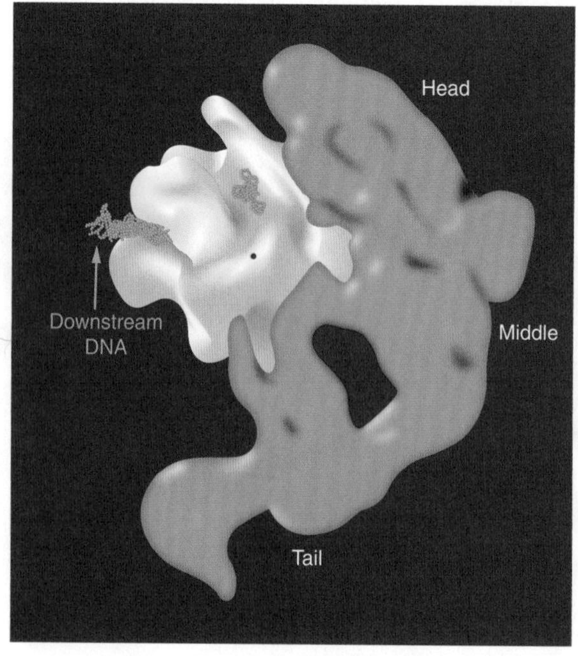

FIGURE 18.41
The Yeast Mediator–RNA Polymerase II Holoenzyme Complex

The RNAP II preinitiation complex (RNAP II with associated transcription factors), shown in white, interacts with the crescent-shape Mediator (shown in blue) via its head and middle domains. The tail domain of Mediator interacts with a regulatory protein that is bound to DNA.

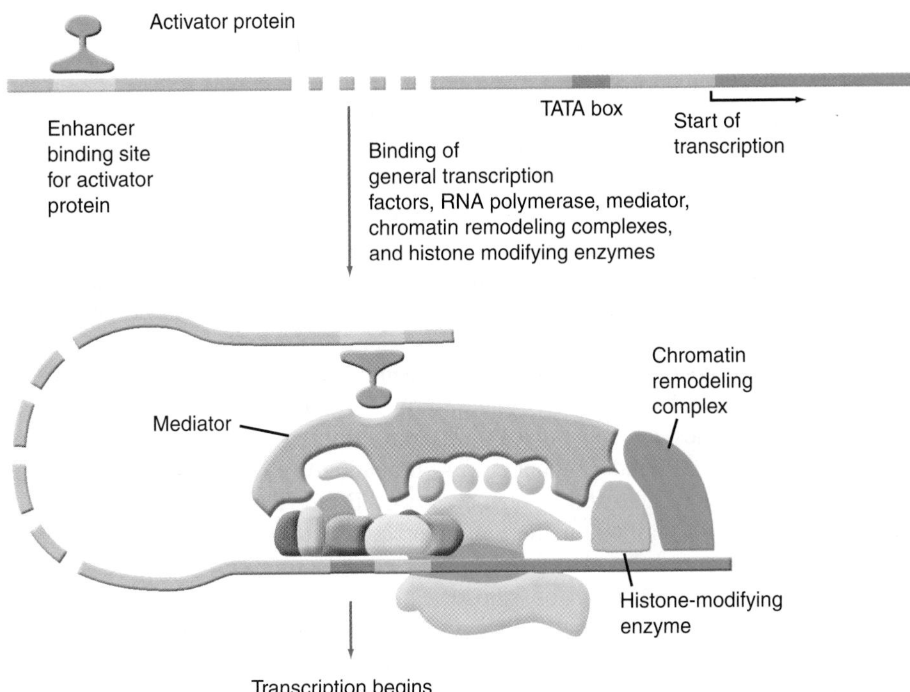

FIGURE 18.42

Transcription Activation by Activator Proteins

The interaction between an activating transcription factor bound to an enhancer element distal from the core promoter (indicated by the dashed lines) and Mediator facilitates the recruitment of RNAP II and the GTFs to the core promoter. Mediator binds to the CTD of RNAP II. Chromatin remodeling complexes and histone-modifying enzymes are also recruited to the PIC. Although only one activator-mediator interaction is illustrated, the transcription of most genes involves several activator proteins that together specify the rate of the process. Mediator can also act to repress transcription when it interacts with repressor proteins (not shown) bound to silencers.

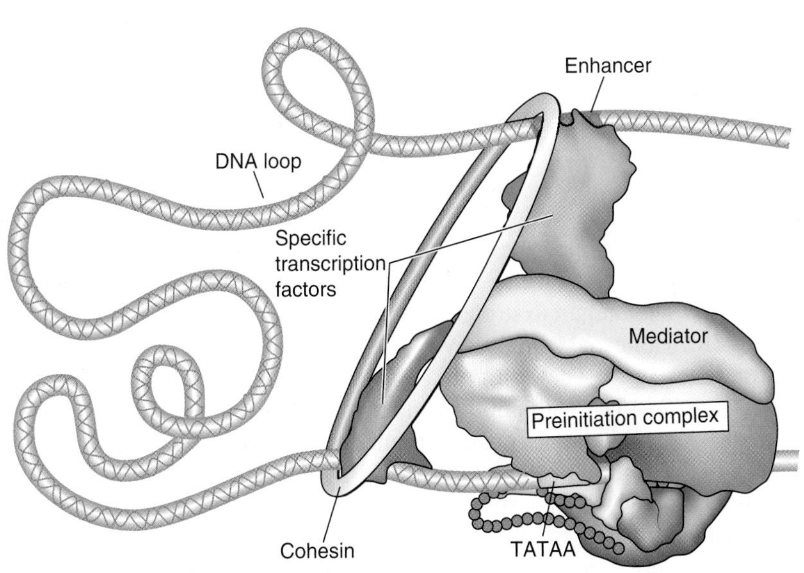

FIGURE 18.43

DNA Looping and Gene Activation

A DNA loop forms between an enhancer element bound to an activator transcription factor and a distant core promoter site via linkage to a PIC/Mediator complex. Cohesin is a ring-like complex that stabilizes the loop. In some cases of enhancer-promoter DNA loops, the insulator protein CTCF (not shown) binds in close proximity to cohesin to block the interaction of enhancers in one DNA domain with promoters in another. The DNA loops created by Mediator and cohesin are different in different cells because they co-occupy different enhancers and promoters as a result of the unique gene expression pattern of each cell type.

ATPase and helicase activities of TFIIH subunits, which create negative supercoil tension. The bubble is maintained in the open position by TFIIF, which binds to the coding strand. The noncoding strand then enters the active site of RNAPII. Once the first bond is synthesized, the transcription process continues until the mRNA reaches about 23 nt, where it often enters a paused state. The binding of cohesin at the promoter is believed to stimulate the transition to the elongation phase of transcription. This phase begins with promoter clearance and the release of the GIFs TFIIE and TFIIH.

Once promoter clearance has been achieved, the RNAP II holoenzyme dissociates from Mediator and the elongation phase of transcription begins. The RNAP II complex continues the transcription process well past the functional end of the nascent transcript until a signal sequence (5′-AAUAAA-3′) called a poly(A) signal is reached. As soon as the poly(A) signal is transcribed, several proteins now linked to the RNAP II CTD bind to it and cause termination by cleaving the transcript about 10–30 nt downstream of the poly(A) signal.

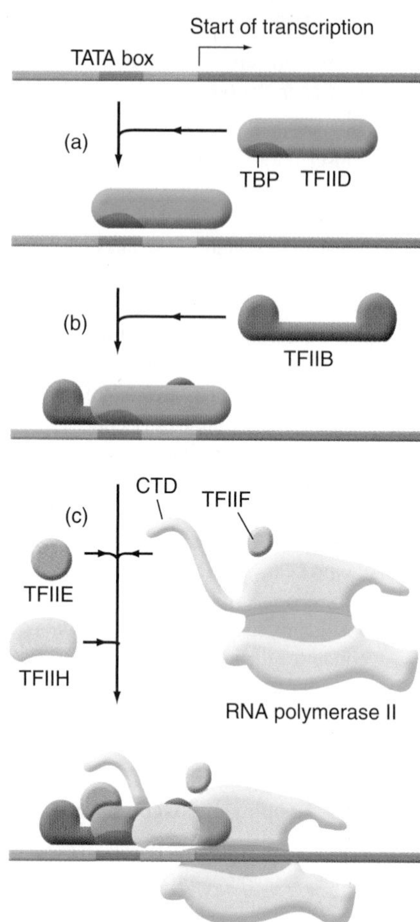

FIGURE 18.44

Preinitiation Complex Formation at a TATA Box

(a) The DNA-distorting TBP subunit of the GTF TFIID recognizes and binds to the TATA box. TFIIA (not shown) stabilizes the interaction between TFIID and the TATA box. (b) The binding of TFIID enables the subsequent binding of TFIIB. (c) The other GTFs (TFIIE, TFIIF, and TFIIH) and RNAP II assemble at the promoter to form an active preinitiation complex. Note that the mediator complex (not shown) controls PIC formation.

RNA PROCESSING In contrast to prokaryotic mRNA, which usually requires little or no processing, eukaryotic mRNAs are the products of extensive posttranscriptional processing. Throughout processing, which begins shortly after transcription initiation, the nascent mRNA transcripts, called pre-mRNAs, become associated with about 20 different types of nuclear protein in ribonucleoprotein particles (hnRNP). Shortly after the transcription of the primary transcript begins, a modification of the 5′ end called *capping* occurs. The cap structure (**Figure 18.45**), which consists of 7-methylguanosine linked to the mRNA through a triphosphate linkage, is synthesized when the pre-mRNA is about 30 nt long. It protects the 5′ end from exonucleases, facilitates transfer into the cytoplasm, and promotes mRNA translation by ribosomes.

With the exception of the replication-dependent histone mRNAs of multicellular animals, all eukaryotic mRNAs are linked to a polyadenylate strand (100–250 adenylate residues) called a *poly(A) tail*. Transcript termination by RNAP II and poly(A) addition are independent events, although the enzyme complex involved in the process is located at the polymerase's CTD tail. The poly(A) tail is synthesized by a poly(A) tail polymerase and then covalently linked to the 3′-terminus of the mRNA transcript. The poly(A) tail facilitates transport of mRNAs out of the nucleus and translation on ribosomes. In most cells, the poly(A) tails of mRNAs in the cytoplasm gradually get shorter, with the result that they are translated less and become vulnerable to degradation. The mRNAs of many protein-coding genes have more than one poly(A) signal and therefore differ in their 3′-UTR sequences. Consequences of alternative polyadenylation include the synthesis of different proteins (if the coding region is altered) and different binding sites for miRNAs.

One of the more remarkable features of eukaryotic RNA processing is the removal of introns from RNA transcripts. In this process, called **RNA splicing**, introns are cut out of the primary transcript and the exons are linked together to form a functional product. Most research efforts have been concerned with the splicing of pre-mRNAs, which is now described. The number of introns in the protein-coding eukaryotic genes varies widely, from one in the intron-containing genes in lower eukaryotes such as yeast to dozens or even hundreds in some mammalian genes. The human dystrophin gene, the largest in humans, is 2.4 million base pairs (Mb) long and contains 79 exons. (Synthesis of the dystrophin primary transcript takes 16 h!) The dystrophin-spliced mRNA (14 kb) codes for a structural protein (427 kDa) with 3600 amino acid residues that connects the muscle cell cytoskeleton with the plasma membrane and the extracellular matrix. Several mutations in the dystrophin gene cause the X-linked recessive disorder, Duchenne muscular dystrophy, which presents as progressive muscle degeneration.

RNA splicing occurs in close proximity to transcription in a 4.8-megadalton (MDa) RNA-protein complex called a **spliceosome** formed from the snRNPs U1, U2, U4, U5, and U6 (p. 652). In eukaryotic nuclear pre-mRNA transcripts most splicing events occur at GU-AG sequences. In GU-AG introns, 5′-GU-3′ and 5′-AG-3′ are the first and last dinucleotide sequences, respectively, of the intron. The splicing event is initiated by complementary binding between the intron's 5′-splice site sequence and the 5′ terminal sequence of the snRNA U1. This is followed by the binding of a U2 sequence in the U2 snRNP to the "branch site," a 2′-OH of an adenosine nucleotide within the intron. The other spliceosome components (U4, U5, and U6 RNPs) then bind to form the entire spliceosome with the U5 sequence binding upstream of the 5′ splice site. Splicing (**Figure 18.46**) is composed of two reactions:

1. A 2′-OH of the "branch site" attacks a phosphate in the 5′ splice site in a transesterification reaction. This reaction, which is accompanied by the displacement of the U1 and U6 RNPs and the release of the U1 and U4 RNP, results in the formation of a loop called a *lariat* because of the newly created 5′→2′ phosphodiester bond (**Figure 18.47**).

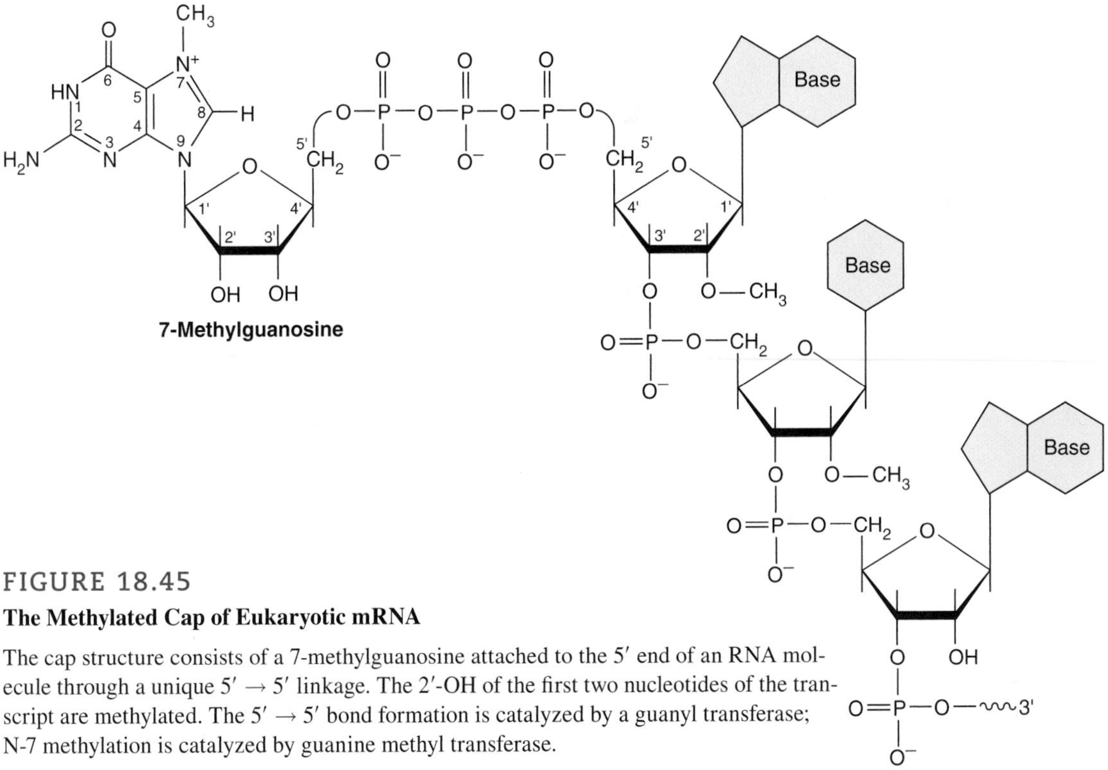

7-Methylguanosine

FIGURE 18.45

The Methylated Cap of Eukaryotic mRNA

The cap structure consists of a 7-methylguanosine attached to the 5′ end of an RNA molecule through a unique 5′ → 5′ linkage. The 2′-OH of the first two nucleotides of the transcript are methylated. The 5′ → 5′ bond formation is catalyzed by a guanyl transferase; N-7 methylation is catalyzed by guanine methyl transferase.

2. The lariat is cleaved and degraded and the two exons are joined when the 3′-OH, which is bound to U5 of the upstream exon, attacks a phosphate that is adjacent to the lariat.

As a splicing event ends, an *exon junction complex* (EJC) binds to each splice site 20 nt upstream from the exon-exon junction to form the mature messenger ribonucleoprotein (mRNP). EJCs remain associated with mRNAs during nuclear export and cytoplasmic localization until the first round of translation by a ribosome. EJCs play a role in *nonsense-mediated decay* (NMD), an mRNA surveillance mechanism that detects premature termination signals (stop codons) that result in the synthesis of truncated and possibly toxic polypeptides. Caused by RNA splicing errors (as well as random mutations and faulty DNA rearrangements), premature stop codons are distinguished from normal stop codons by their position relative to an EJC. As the mRNA is translated by a ribosome, detection of a stop codon upstream of an EJC (indicating there is another stop codon downstream) causes the surveillance mechanism to initiate the mRNA decay process. NMD occurs within *P bodies*, cytoplasmic structures that consist of enzymes involved in several mRNA degradation or silencing processes (see p. 715).

The majority of mammalian pre-mRNAs, most of which contain multiple introns, are processed by **suprapliceosomes** (21.1 MDa), each formed from four active spliceosomes (**Figure 18.48**). It has been suggested that the supraspliceosome increases the speed and efficiency of transcript splicing and provides opportunities for intron excision proofreading.

Processing of the replication-dependent histone mRNAs of multicellular animals is different in several respects from that of other mRNAs. The histone mRNA transcript has a 5′-cap but no introns or poly(A) tail. Instead of a poly(A) tail, there is an evolutionarily conserved 25-nt stem-loop at the 3′-end that binds to SLBP (stem-loop binding protein). SLBP facilitates mRNA transport, translation, and degradation. The histone mRNA stem-loop is essential for the rapid degradation of histone mRNAs when DNA synthesis ends at the conclusion of the S phase of the cell cycle.

KEY CONCEPTS

- Transcription in eukaryotes is significantly more complex than its counterpart in prokaryotes.
- In addition to chromatin-remodeling and RNA-processing reactions, gene transcription requires the binding of unique sets of transcription factors to promoter sequences.
- Eukaryotic RNA transcripts undergo several processing reactions.

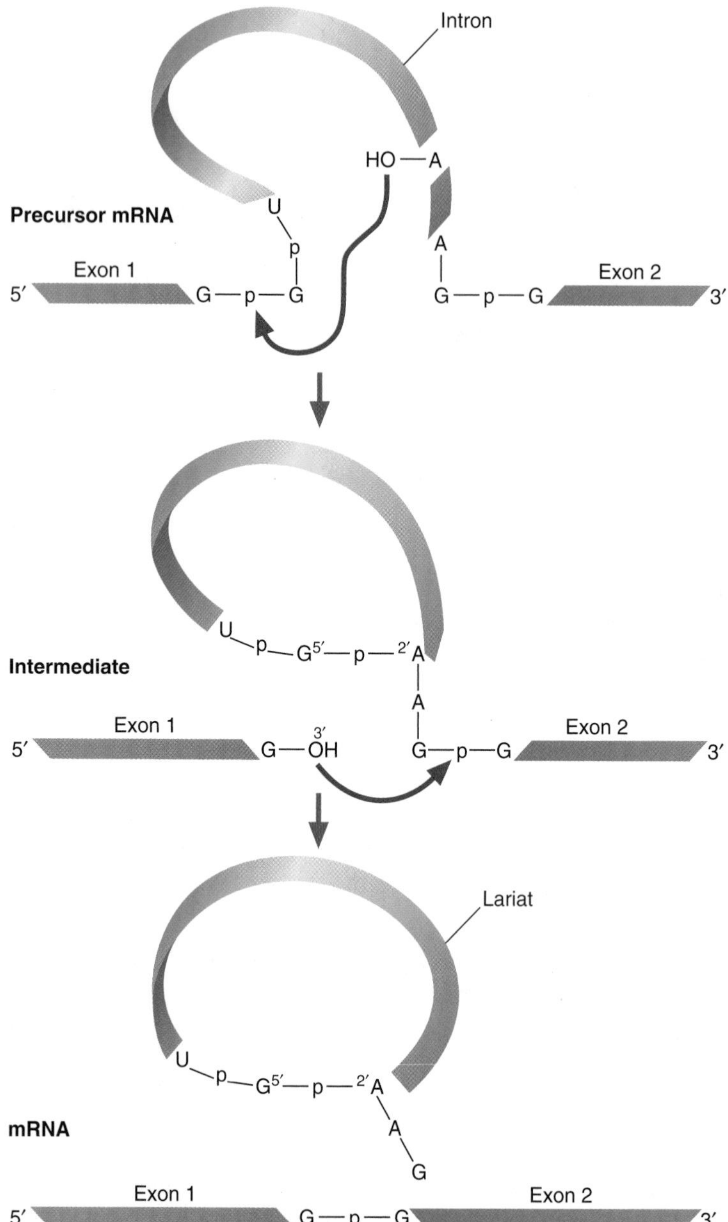

FIGURE 18.46
RNA Splicing

mRNA splicing, which occurs in the nucleus, begins with the nucleophilic attack of the 2′-OH of a specific adenosine on a phosphate in the 5′-splice site. A lariat is formed by a 2′,5′-phosphodiester bond. In the next step, 3′-OH of exon 1 (acting as a nucleophile) attacks a phosphate adjacent to the lariat. This reaction releases the intron and ligates the two exons.

18.3 GENE EXPRESSION

Ultimately, the internal order most essential to living organisms requires the precise and timely regulation of gene expression. It is, after all, the capacity to switch genes on and off that enables cells to respond efficiently to a changing environment. In multicellular organisms, complex programmed patterns of gene expression are responsible for cell differentiation and intercellular cooperation.

The regulation of genes, as measured by their transcription rates, is the result of a complex hierarchy of control elements that coordinate the cell's metabolic activities. Some genes, referred to as **constitutive** or housekeeping **genes** (p. 703), are routinely transcribed because they code for gene products required for cell function. In the differentiated cells of multicellular organisms, certain specialized protein products of transcribed genes are produced that cannot be detected elsewhere (e.g., hemoglobin in red blood cells). Genes that are expressed only in certain circumstances are referred to as *inducible*. For example, the enzymes that are required for lactose metabolism in *E. coli* are synthesized

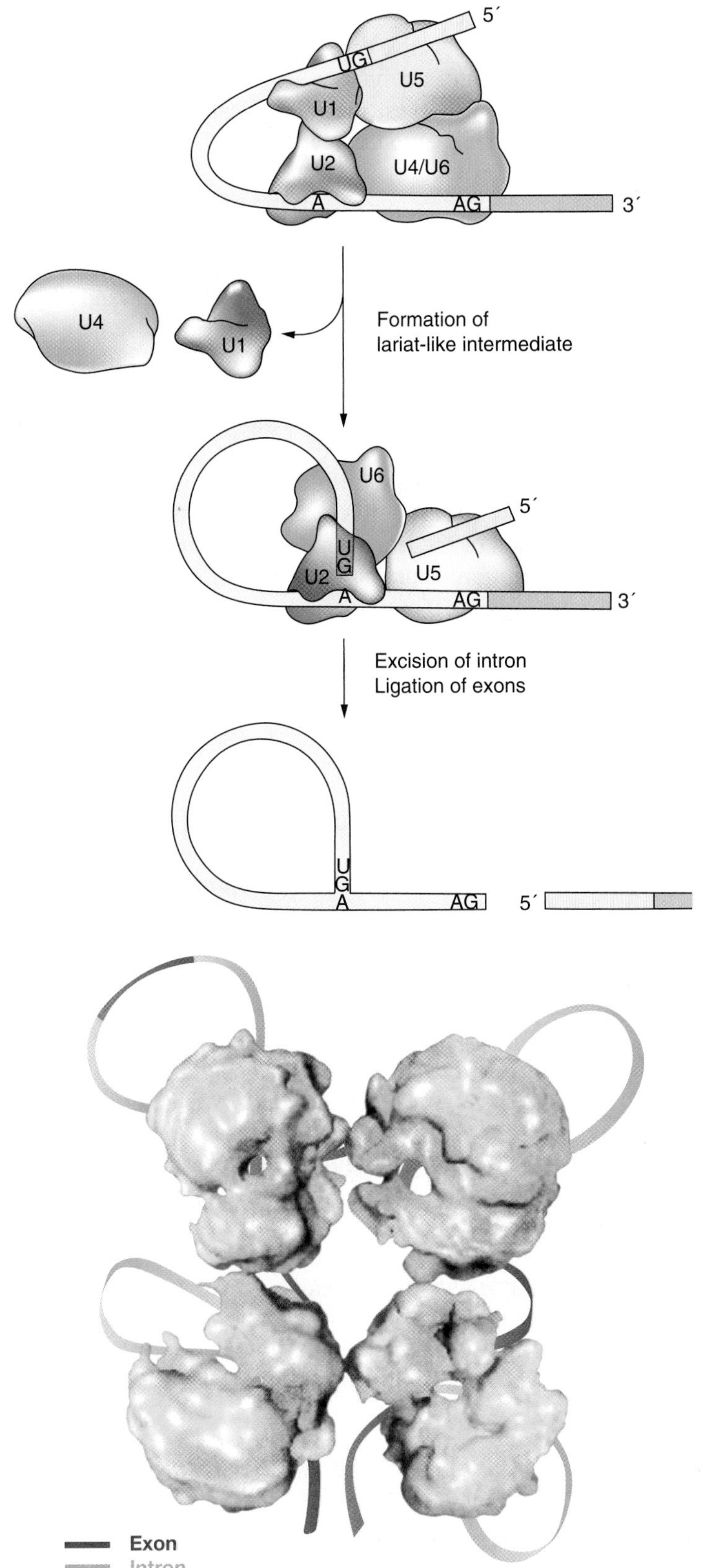

FIGURE 18.47

The Spliceosome

The lariat intermediate forms as a result of the cut made at the 5′-splice site and the 5′-end of the intron is then connected to the conserved adenine. During this process U6 dissociates from U4, both U1 and U4 are released from the splicing complex, and a U6/U2 interaction occurs. U5 bound to the 3′-splice site positions 5′-exon close to the 3′-exon so they can be readily ligated.

FIGURE 18.48

The Supraspliceosome

In this model of supraspliceosome structure and function, a pre-mRNA is being processed by a complex formed by four spliceosomes. Loops of the introns (blue) being spliced extend outward. Note that an alternative exon (red) (see p. 713) is illustrated in the upper left-hand corner.

only when lactose is actually present and glucose, the bacterium's preferred energy source, is absent.

In general, prokaryotic gene expression involves the interaction of specific proteins with DNA in the immediate vicinity of a transcription start site. Such interactions may have either a positive effect (i.e., transcription is initiated or increased) or a negative effect (i.e., transcription is blocked). In an interesting variation, the inhibition of a negative regulator (called a *repressor*) activates affected genes. (The inhibition of a repressor gene is referred to as derepression.) Eukaryotic gene expression uses these mechanisms and several others, including gene rearrangement and amplification, epigenetic mechanisms, and various complex transcriptional, RNA processing, and translational controls. In addition, the spatial separation of transcription and translation inherent in eukaryotic cells provides another opportunity for regulation: mRNA transport control.

This section describes several examples of gene expression control. The discussion of prokaryotic gene expression focuses on operons and riboswitches. An **operon** is a set of genes under the regulation of the same operator and promoter(s). The **operator** is a regulatory sequence that binds to specific repressor or activator proteins that modulate gene expression. The *lac* operon in *E. coli*, originally investigated by François Jacob and Jacques Monod in the 1950s, is the most thoroughly researched example. Bacteria also employ an RNA-based control mechanism called the **riboswitch**, made up of a specific untranslated sequence within the mRNA along with the small metabolite to which it binds. The action of the riboswitch usually represses gene expression by terminating transcription, blocking translation, or causing mRNA self-destruction. Eukaryotic gene expression is less understood than that of prokaryotes such as *E. coli* because of larger genomes and more varied and complex regulatory mechanisms. The section ends with a brief overview of growth factor–triggered gene expression.

Gene Expression in Prokaryotes

The highly regulated metabolism of prokaryotes such as *E. coli* allows these organisms to manage limited resources and to respond to a changing environment. The timely synthesis of enzymes and other gene products only when needed, combined with the rapid destruction of mRNAs by ribonucleases (RNases), prevents dissipation of energy and nutrients. This flexibility is made possible at the genetic level where the control of inducible genes is often effected by groups of linked structural genes and regulatory genes called operons. Investigations of operons, especially the *lac* operon, have provided substantial insight into how gene expression can be altered by environmental conditions. The *lac* operon and several types of riboswitch, a more recent discovery, are described.

THE *LAC* OPERON The *lac* operon (**Figure 18.49**) consists of a control element and structural genes that code for the enzymes of lactose metabolism. The control element contains the promoter site, which overlaps the operator site. The catabolite gene activator protein (CAP) site (described shortly), a 16-bp DNA sequence upstream of the promotor, is another regulatory element. The structural genes Z, Y, and A specify the primary structure of β-galactosidase, lactose permease, and thiogalactoside transacetylase, respectively. β-Galactosidase catalyzes the hydrolysis of lactose, which yields the monosaccharides galactose and glucose,

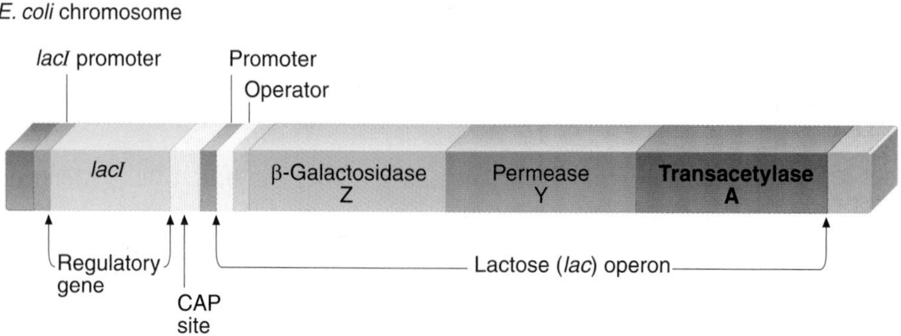

FIGURE 18.49
The *lac* Operon in *E. coli*

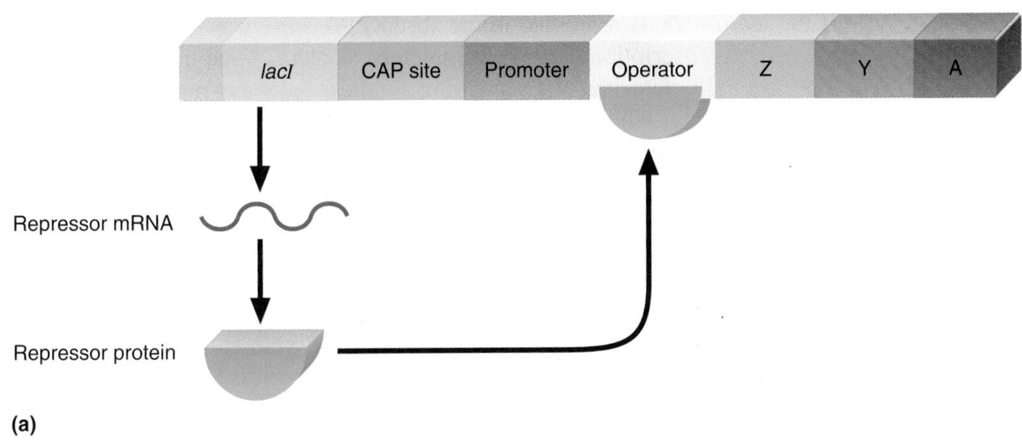

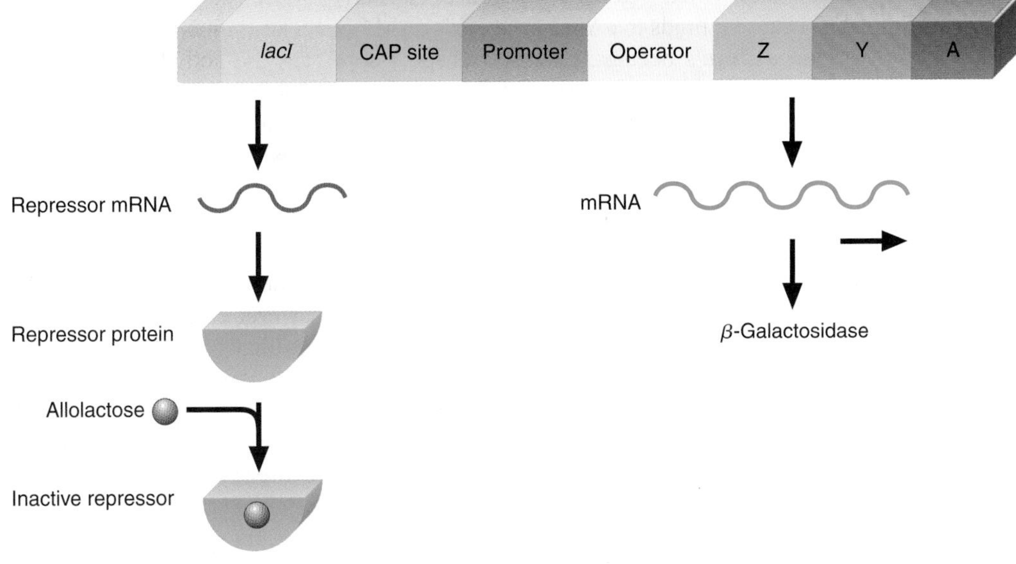

FIGURE 18.50

Function of the *lac* Operon

(a) The repressor gene *lacI* encodes a repressor that binds to the operator when lactose (the inducer) is not present. (b) When lactose is present, its isomer allolactose binds to the repressor protein, thereby inactivating it. (Not shown: the effect of glucose on the *lac* operon. Refer to the text for a discussion of this topic.)

whereas lactose permease promotes lactose transport into the cell. The role of thiogalactose transacetylase is unclear, since lactose metabolism proceeds normally without it. A repressor gene *lacI*, directly adjacent to the *lac* operon, codes for the *lac* repressor protein, a tetramer that binds to the operator site with high affinity. (There are about 10 copies of *lac* repressor protein per cell.) The binding of the *lac* repressor to the operator prevents the functional binding of RNA polymerase to the promoter (**Figure 18.50**).

Without its inducer (allolactose, a β-1,6-isomer of lactose) the *lac* operon remains repressed because the *lac* repressor binds to the operator. When lactose becomes available, a few molecules are converted to allolactose by β-galactosidase. Allolactose then binds to the repressor, changing its conformation and promoting dissociation from the operator. Once the inactive repressor has diffused away from the operator, the transcription of the structural genes begins. The *lac* operon remains active until the lactose supply is consumed or glucose, the preferred energy source, becomes available. Then the repressor reverts to its active form and rebinds to the operator.

If the *lac* operon is repressed in the absence of lactose, what is the source of allolactose? Transcription is never completely blocked, since the repressor protein occasionally detaches, with resulting synthesis of a low number of operon-coded proteins. So when the bacterial cell encounters lactose, there are a few molecules of lactose permease available to facilitate the transport of lactose into the cell, where it is converted to allolactose.

Glucose is the preferred carbon and energy source for *E. coli*. If both glucose and lactose are available, the glucose is metabolized first. Synthesis of the *lac* operon enzymes is induced only after the glucose has been consumed. (This makes sense because glucose is more commonly available and has a central role in cellular metabolism. Why expend the energy to synthesize the enzymes required for the metabolism of other sugars if glucose is also available?)

The delay in activating the *lac* operon is mediated by a CAP. CAP is an allosteric homodimer that binds to the chromosome at a site directly upstream of the lac promoter when glucose is absent. CAP is an indicator of glucose concentration because it binds to cAMP. The cell's cAMP concentration is inversely related to glucose levels because glucose transport depresses the activity of adenylate cyclase. The binding of cAMP to CAP, which occurs only when glucose is absent and cAMP levels are high, causes a conformational change that allows the protein to bind to the *lac* promoter. CAP binding promotes transcription by increasing the affinity of RNA polymerase for the *lac* promoter. In other words, CAP exerts a positive or activating control on lactose metabolism.

RIBOSWITCHES A **riboswitch** is a metabolite-sensing domain in the 5′-untranslated region of mRNAs. Found mostly in bacteria, riboswitches monitor cellular concentrations of specific metabolites. Genes that contain riboswitches typically code for proteins that are involved either in the transport or in the synthesis of molecules that are expensive to produce, such as TPP or FMN. Riboswitches are toggle switch–like devices that act as feedback inhibitors to prevent the wasteful acquisition of molecules that are already present in sufficient concentrations. They are composed of two structural elements: an *aptamer*, which directly binds the metabolite, and an *expression platform*, the gene expression regulator. When the aptamer binds the metabolite, it undergoes a structural change that in turn alters the structure of the expression platform. The most common results of this process are transcription and translation inhibition.

When TPP binds to its aptamer (**Figure 18.51a**) the riboswitch is converted from a structure that has an open translation initiation site to one in which the initiation site is sequestered in a hairpin loop, effectively blocking translation. The FMN riboswitch illustrated in **Figure 18.51b** causes premature transcription termination when it reconfigures to form a terminator hairpin that causes RNA polymerase to no longer be able to bind. Examples of other riboswitches include those for cobalamin (coenzyme B_{12}), purines, and lysine. In rare instances, the riboswitch catalyzes a self-cleavage reaction (**Figure 18.51c**) that results in a decrease in mRNA copy number.

KEY CONCEPTS

- Constitutive genes are routinely transcribed, whereas inducible genes are transcribed only under appropriate circumstances.
- In prokaryotes, inducible genes and their regulatory sequences are grouped into operons.
- Riboswitches are metabolite-sensing domains in mRNAs that regulate the transport or synthesis of certain types of molecules.

Gene Expression in Eukaryotes

Eukaryotic genomes are also substantially more intricately regulated than those of prokaryotes, as the following example confirms. With the exception of mature red blood cells, human cells contain the same genome. (As red blood cells develop from precursor cells called reticulocytes, they lose their nuclei.) Yet each of more than 200 highly differentiated cell types expresses a unique subset of genes that changes in response to intercellular signaling mechanisms and/or environmental cues. Eukaryotic diversity is made possible by vastly larger groups of transcription factors and ncRNA than occurs in prokaryotes. Most eukaryotic transcription factors can be classified into the following categories, which are based on DNA binding specificity or effects on chromatin structure: DNA

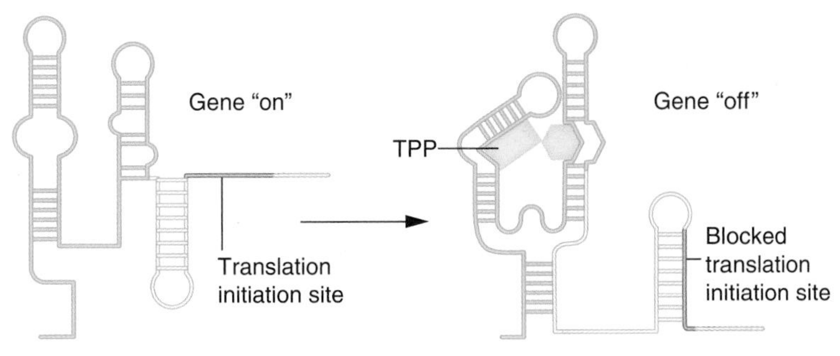

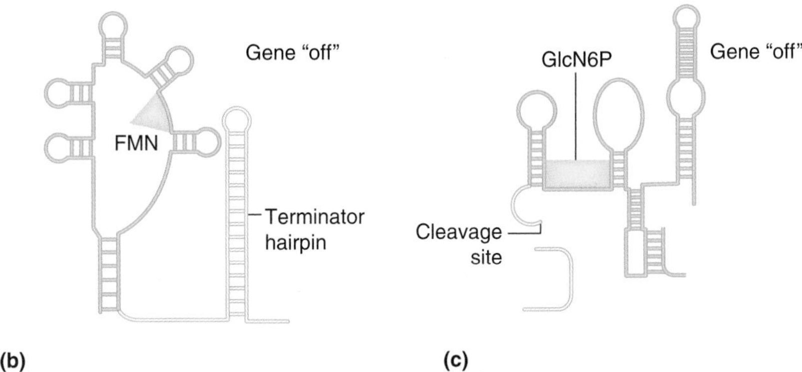

FIGURE 18.51
Riboswitches

(a) Translation prevention. When TPP binds to the aptamer, the riboswitch rearranges so that the expression platform forms a hairpin that blocks translation initiation. (b) Transcription termination. FMN binding causes the formation of a hairpin structure that halts transcription. (c) Self-cleavage. In rare instances, the binding of a metabolite (in this case, the sugar GlcN6P) triggers a self-cleavage reaction.

sequence-specific and general or nonspecific chromatin structure factors, and chromatin remodeling factors, and histone methyltransferases. Eukaryotic gene expression is also regulated at more levels than the prokaryotic version. These levels include genomic control, transcriptional control (see pp. 701–705), RNA processing, RNA editing, RNA transport, and translational control. A brief description of these topics is followed by an overview of signal transduction-triggered gene expression.

GENOMIC CONTROL As described previously, there appear to be two major influences on eukaryotic transcription initiation: chromatin structure and transcription factor–regulated RNA polymerase complex formation. Gene expression is affected by changes in the structural organization of the genome (i.e., chromatin remodeling induced by DNA methylation and histone covalent modification). A significant amount of gene regulation also occurs through transcription initiation control.

Other examples of genomic control include gene rearrangements and gene amplification. The differentiation of certain cells involves gene rearrangements, for example, the rearrangements of antibody genes in B lymphocytes. Transposition (see pp. 686–689) is also believed to affect gene regulation. During certain stages in development, the requirement for specific gene products may be so great that the genes that code for their synthesis are selectively amplified. Amplification occurs via repeated rounds of replication within the amplified region. For example, the rRNA genes in various animals (most notably amphibians, insects, and fish) are amplified within immature egg cells (called oocytes).

RNA PROCESSING Several types of eukaryotic RNA processing reaction have already been described (pp. 706–707). Among the most important of these is *alternative splicing*, the joining of different combinations of exons to form cell-specific proteins (**Figure 18.52**). Tropomyosin is a protein found in a wide variety

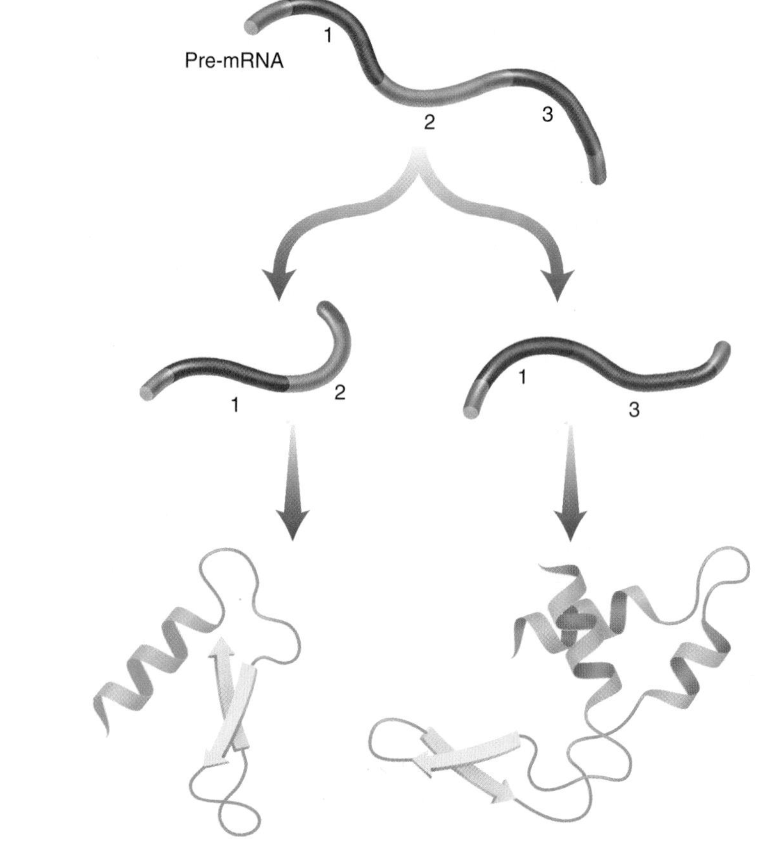

FIGURE 18.52
RNA Processing

The coding properties of an mRNA molecule depend on the types of processing event its precursor undergoes. Different polypeptides can be synthesized from splicing different combinations of exons from the same pre-mRNA transcript.

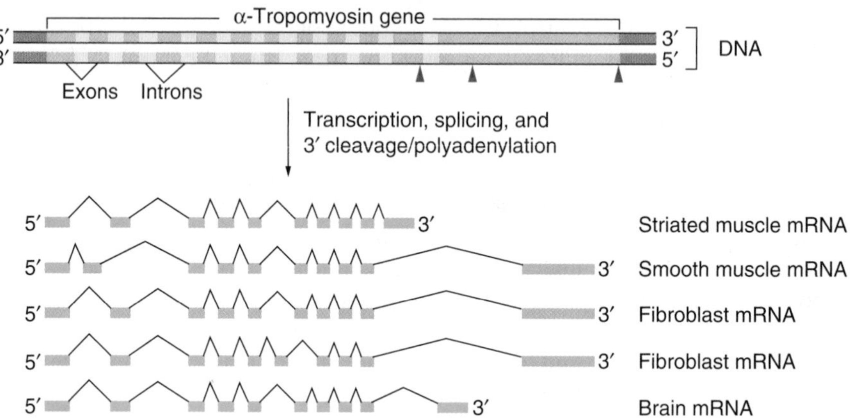

FIGURE 18.53
Alternate Splicing of the Tropomyosin Gene

Tropomyosin is a regulatory protein in muscle contraction. Alternate splicing of the primary transcript in other cells results in alternate versions of tropomyosin that are utilized for different purposes.

of cells (e.g., skeletal, smooth, and cardiac muscle, fibroblasts, and brain) (**Figure 18.53**). The vertebrate tropomyosin gene consists of 13 to 15 exons. Five of the exons are common to all isoforms of the protein, while the remaining exons are alternately used in different tropomyosin mRNAs. For example, rat striated muscle tropomyosin mRNA contains exons 3, 11, and 12, but not exons 2 or 13, while the smooth muscle isoform contains exons 2 and 3, but not exons 11 and 12. This variation accounts in part for the differences in contractile fiber structure and function in these two muscle types. Rat brain tropomyosin mRNA lacks exons 2, 7, and 11 through 13 and functions in the actin-myosin cytoskeletal system in these noncontractile cells.

The selection of alternative sites for polyadenylation also affects mRNA function. In addition to altering mRNA binding sites for miRNAs, changes in polyadenylation sites also affect mRNA longevity (see p. 706). Polyadenylation site changes can also alter an mRNA's structural and functional properties. The mRNA that codes for the heavy chain of the antibody IgM is a well-researched example. There are two forms of IgM: the membrane-bound antibody and the secreted antibody. During the early phase of B-lymphocyte differentiation, the cell produces the plasma membrane IgM because heavy-chain transcript polyadenylation occurs at a site downstream from two exons that code for membrane anchor domains. Later, the cell synthesizes both the membrane-bound and the secreted versions of IgM. The heavy chain of secreted IgM is a truncated molecule that lacks the membrane anchor because the mRNA transcript is polyadenylated upstream of the membrane domain exons. Selection of the heavy-chain mRNA polyadenylation site depends on the expression of the protein CstF, which determines the site that is processed.

After transcription, base changes are effected by means of **RNA editing**. Alterations in mRNA base sequence can have several consequences. When they occur in the 5′ and 3′ UTRs, for example, translation initiation and RNA stability, respectively, may be affected. Other possibilities include the alteration of intron splice sites and changes in the amino acid sequence of the polypeptide product. Among the best-researched examples of RNA editing are C → U and A → I conversions, where I is an abbreviation for inosine. The mRNA for the apolipoprotein B-100 gene codes for a polypeptide with 4563 amino acid residues, which is a component of VLDL. Intestinal cells produce a shorter version of this molecule, namely, apolipoprotein B-48 (2153 amino acid residues), which becomes incorporated into the chylomicron particles produced by these cells. The cytosine in a CAA codon that specifies glutamine is converted by cytidine deaminase into a uracil. The new codon, UAA, is a stop signal in translation; hence a truncated polypeptide is produced during translation of the edited mRNA.

Among the best examples of an A → I transition occurs in some brain neurons, where a specific adenosine residue in the mRNA for a glutamate receptor subunit is deaminated by adenosine deaminase acting on RNA. When an edited mRNA is read by a ribosome, the I base is read as a G. As a result, an arginine residue (codon CGA) is substituted for a glutamate (codon CAA), and the ion channel in the receptor becomes less permeable to Ca^{2+}. Transgenic mice in which the A → I transition was prevented were observed to develop a severe form of epilepsy.

POSTTRANSCRIPTIONAL GENE SILENCING **Gene silencing**, a form of posttranscriptional gene regulation in higher eukaryotes, involves short 22-nt RNAs called microRNAs (p. 651). First discovered in the nematode worm *Caenorhabditis elegans*, miRNAs inhibit the translation of target mRNAs by binding to partially complementary sequences in their 3′ UTRs, although examples of 5′ UTR and coding region binding have been observed. The first miRNA to be investigated, lin-4, regulates an early phase in the worm's larval development. By preventing the translation of lin-14 and lin-28 mRNAs, lin-4 facilitates the progression of developing larvae from an early stage to a later one. MiRNAs, each of which may have hundreds of target mRNAs, have been discovered in organisms as diverse as fruit flies, frogs, rice, and maize. Humans are estimated to have as many as 1500 miRNAs, which are believed to regulate about one-third of human genes. Defects in miRNA regulation have been associated with diseases such as cancer (e.g., leukemia) and heart disease.

Gene silencing (**Figure 18.54**) begins with the RNAP II–catalyzed synthesis of a 70-nt-long ss-RNA precursor called a primary miRNA (pri-miRNA). Each pri-miRNA, which may be thousands of nts in length, is a stem-loop structure that is processed in the nucleus into one or more miRNAs by a protein complex called microprocessor. *Microprocessor*, composed of *pasha* and a RNase called *drosha*, cleaves pri-miRNA 11 nt away from the base of each hairpin structure to release pre-miRNA. After GTP-driven export from the nucleus via the carrier protein exportin-5, each pre-miRNA is cleaved by an RNase called *dicer* to yield a double-stranded miRNA. One strand of the miRNA duplex, called the guide strand,

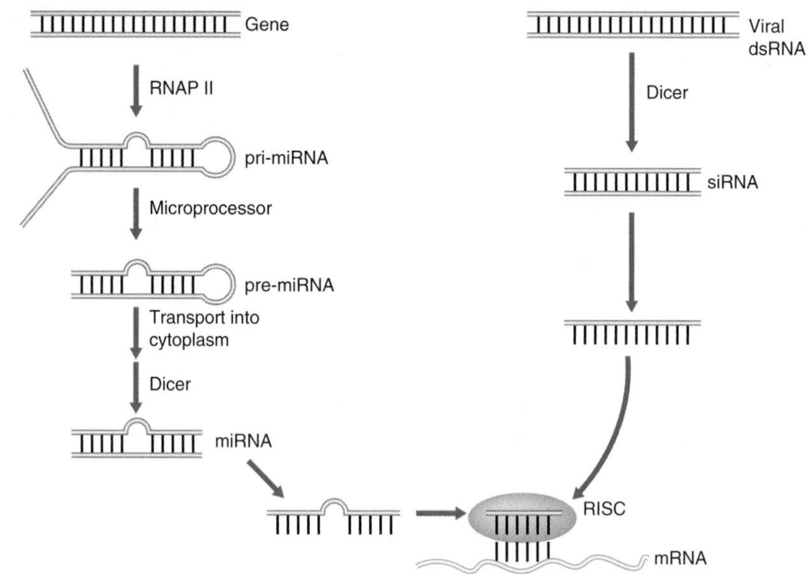

FIGURE 18.54
mi-RNA and si-RNA Processing

In posttranscriptional gene silencing, the primary transcript of a miRNA gene, pri-miRNA, is processed by microprocessor, a protein complex containing pasha and drosha, and dicer to form miRNA. The miRNA guide strand is then incorporated into the RISC ribonucleoprotein complex where it binds a complementary sequence in the 3′ UTR of its target mRNA. Because these two sequences are not perfectly complementary, the mRNA is silenced, but not degraded. In RNA interference, a foreign dsRNA is cleaved by dicer to yield the ds-RNA molecule siRNA. Once the guide strand of the siRNA has been positioned within the RISC, it binds to its complementary sequence on the viral mRNA. Because these two sequences are perfectly complementary, the slicer activity of the RISC proceeds to cleave the mRNA into pieces.

is incorporated into a ribonucleoprotein complex called *RISC*. The other miRNA strand (the passenger strand) is degraded. A RISC protein called *argonaute* positions the miRNA so it can bind the target mRNA, thereby inactivating it.

MiRNA-mediated gene silencing utilizes components of **RNA interference**, a process originally believed to be limited to protection against viruses and transposons. Cells use double-stranded si-RNAs to recognize and then degrade target mRNAs. siRNAs are the products of dicer-induced cleavage of larger RNA molecules (e.g., a viral RNA genome). Once the guide siRNA is incorporated into RISC (**Figure 18.54**), it binds to its complementary sequence on the target mRNA. Because the sequences match exactly, *slicer* (an enzymatic activity in a domain of argonaute) cleaves the mRNA into pieces.

RNA TRANSPORT mRNA transport out of the nucleus, a highly regulated process, occurs in three phases: processing reactions, docking and passage through NPC (p. 55), and release into the cytoplasm. In the first phase pre-mRNA molecules are simultaneously processed into mRNAs and packaged into ribonucleoprotein complexes (mRNPs). mRNP proteins (e.g., cap binding protein, EJCs, and poly(A)-binding protein) recruit export factors that allow NPC targeting. The capping and splicing proteins allow binding to TREX, an export protein complex. Once mRNPs are linked via a TREX subunit to Nxf1-Nxt1, a heterodimer nuclear export receptor, they move through the NPC. When an mRNP complex reaches the cytoplasm, the release of export proteins triggers the remodeling of the complex that in turn directs transport to its final destination where translation will occur.

TRANSLATIONAL CONTROL Eukaryotic cells can respond to various stimuli (e.g., heat shock, viral infections, and cell cycle phase changes) by selectively altering protein synthesis. The covalent modification of several translation factors (nonribosomal proteins that assist in the translation process) has been observed to alter the overall protein synthesis rate and/or enhance the translation of specific mRNAs. For example, when cellular iron levels are low, a repressor protein binds to mRNAs coding for the iron storage protein ferritin. When iron levels rise sufficiently the binding of iron to the repressor protein triggers a conformational change that causes it to dissociate from mRNA. The ferrritin mRNA can then be translated.

SIGNAL TRANSDUCTION AND GENE EXPRESSION All cells respond to signals from their environment in part by altering gene expression patterns.

Signal transduction–triggered changes in gene expression are initiated by the binding of a ligand to either a cell surface receptor or an intracellular receptor. The mechanisms by which signal molecules switch certain genes on or off are an intricate series of reactions and protein conformation changes that transmit information from the cell's environment to specific DNA sequences in the nucleus. Considering the enormous research efforts devoted to the investigation of cancer (see the Biochemistry in Perspective essay entitled Carcinogenesis at the end of this chapter), the best understood examples of such signal transduction pathways are those that affect cell division.

In contrast to single-cell organisms in which cell growth and cell division are governed largely by nutrient availability, the proliferation of cells in multicellular organisms is regulated by an elaborate intercellular network of growth-promoting and growth-inhibiting signal molecules. Several complicating features of intracellular signal transduction mechanisms have been revealed by research efforts in cell proliferation. The mechanisms by which signal molecules alter gene expression often involve the simultaneous activation of several different pathways. Depending on circumstances, the activation of several types of receptor may result in overlapping responses.

In the eukaryotic cell cycle, cells repeatedly progress through each of the four phases (M, G_1, S, and G_2: refer to **Figure 18.12**). Checkpoints occur in G_1 (in yeast cells it is referred to as START), G_2, and M phases. The cell is prevented from entering the next phase until the conditions are optimal (e.g., sufficient cell growth in G_2 or alignment of chromosomes in M). The fixed, rhythmic activities observed in cell division are regulated so that each phase is completed before the next one starts. Progression is accomplished by the alternating synthesis and degradation of the cyclins, a group of regulatory proteins that bind to and activate the *cyclin-dependent protein kinases* (Cdks). The Cdks phosphorylate a variety of proteins that control the passage of cells though checkpoints in the cell cycle.

Cell division regulation involves both positive and negative controls. Positive control is exerted largely by growth factors that bind to specialized cell receptors. The initiation of cell division typically requires binding a variety of such factors. Cell proliferation is inhibited by the protein products of *tumor suppressor genes*. Well-known examples of these genes include *Rb* (so named because of the role played by the loss of Rb gene function in the childhood eye cancer *retinoblastoma*) and the p53 gene (a cyclin chaperone that arrests cell cycle progression). The arrest of the cell cycle is prolonged when a certain amount of DNA damage has occurred, as in overexposure to radiation. If DNA repair mechanisms are incomplete, a complex mechanism involving p53 leads to programmed cell death, or **apoptosis** (**Figure 2.22**).

The positive effects exerted by growth factors are now believed to include gene expression that specifically overcomes the inhibitions at the cell cycle checkpoints, especially the G_1 checkpoint. The binding of growth factors to their cell surface receptors initiates a cascade of reactions that induces two classes of genes: early response genes and delayed response genes.

Early response genes These genes, which usually code for transcription factors, are rapidly activated, usually within 15 minutes. Among the best-characterized early response genes are the *jun, fos*, and *myc* protooncogenes. **Protooncogenes** are normal genes that, if mutated, can promote carcinogenesis. (Refer to the Biochemistry in Perspective essay entitled Carcinogenesis). Each of the *jun* and *fos* protooncogene families codes for a series of transcription factors containing leucine zipper domains. Both Jun and Fos proteins form dimers that can bind DNA. Among the best-characterized of these is a Jun-Fos heterodimer, referred to as AP-1, which forms through a leucine zipper interaction. One member of a large class of transcription factors that possess the basic helix-loop-helix-leucine zipper (bHLHZip) DNA-protein binding motif is Myc. Members of the Myc protein family form homo- and heterodimers with themselves and with members of certain other transcription factor families. When Myc forms a heterodimer with a protein called Max, the expression of a large number of genes is affected. The products of some of Myc/Max target genes have a stimulatory effect on the cell cycle.

Delayed response genes These genes are induced by the activities of the transcription factors and other proteins produced or activated during the early response phase. Among the products of the delayed response genes are the Cdks, the cyclins, and other components required for cell division.

As mentioned earlier (see p. 592), many growth factors bind to tyrosine kinase receptors, and some of these are linked via G-protein-like mechanisms to DAG and IP_3 generation (**Figure 16.7**). EGF is one growth factor of this type, and **Figure 18.55** gives a brief overview of its role in the activation of the transcription factor AP-1. The *ras* oncogene, first isolated from rat sarcomas, serves in its protooncogene form as the G-protein component of this system. Activation of ras is achieved when it binds to SOS/GRB2, a protein complex that has bound to the tyrosine kinase domain of the EGF receptor. SOS is a type of **GEF** that causes ras to release GDP and bind a GTP when it is in turn bound to the protein called GRB2. Ras becomes inactive when GTP is hydrolyzed, in a reaction catalyzed by **GTPase-activating proteins**, or GAPs.

PLCγ is also activated when it binds to the EGF receptor. Like its counterpart in G-protein–linked receptors, active PLCγ hydrolyzes PIP_2 to form DAG and IP_3. DAG activates PKC, the enzyme that in turn activates a variety of proteins involved in cell growth and proliferation.

Phosphorylation cascades are induced by both the activation of ras and increased levels of DAG and IP_3 in the cell following EGF binding. One of the key enzymes activated in the phosphorylation cascade is MAPKK (mitogen-activated protein kinase kinase). Activated MAPKK then phosphorylates both a tyrosine and a threonine of MAP kinase (MAPK). (This unusual reaction appears to ensure that MAP kinase is activated only by MAPKK.) Active MAPK (a serine/threonine kinase) then phosphorylates a variety of cellular proteins. Among those are Jun, Fos, and Myc. Phosphorylated jun and fos proteins then combine to form AP-1, which promotes the transcription of several delayed response genes that promote cell division.

QUESTION 18.5

The mechanism by which light influences plant gene expression is referred to as *photomorphogenesis*. Because of serious technical problems with plant cell culture, relatively little is known about plant gene expression. However, certain DNA sequences, referred to as *light-responsive elements* (LRE), have been identified. On the basis of the gene expression patterns observed in animals, can you suggest (in general terms) a mechanism whereby light induces gene expression? [*Hint*: Recall that phytochrome is an important component of light-induced gene expression.]

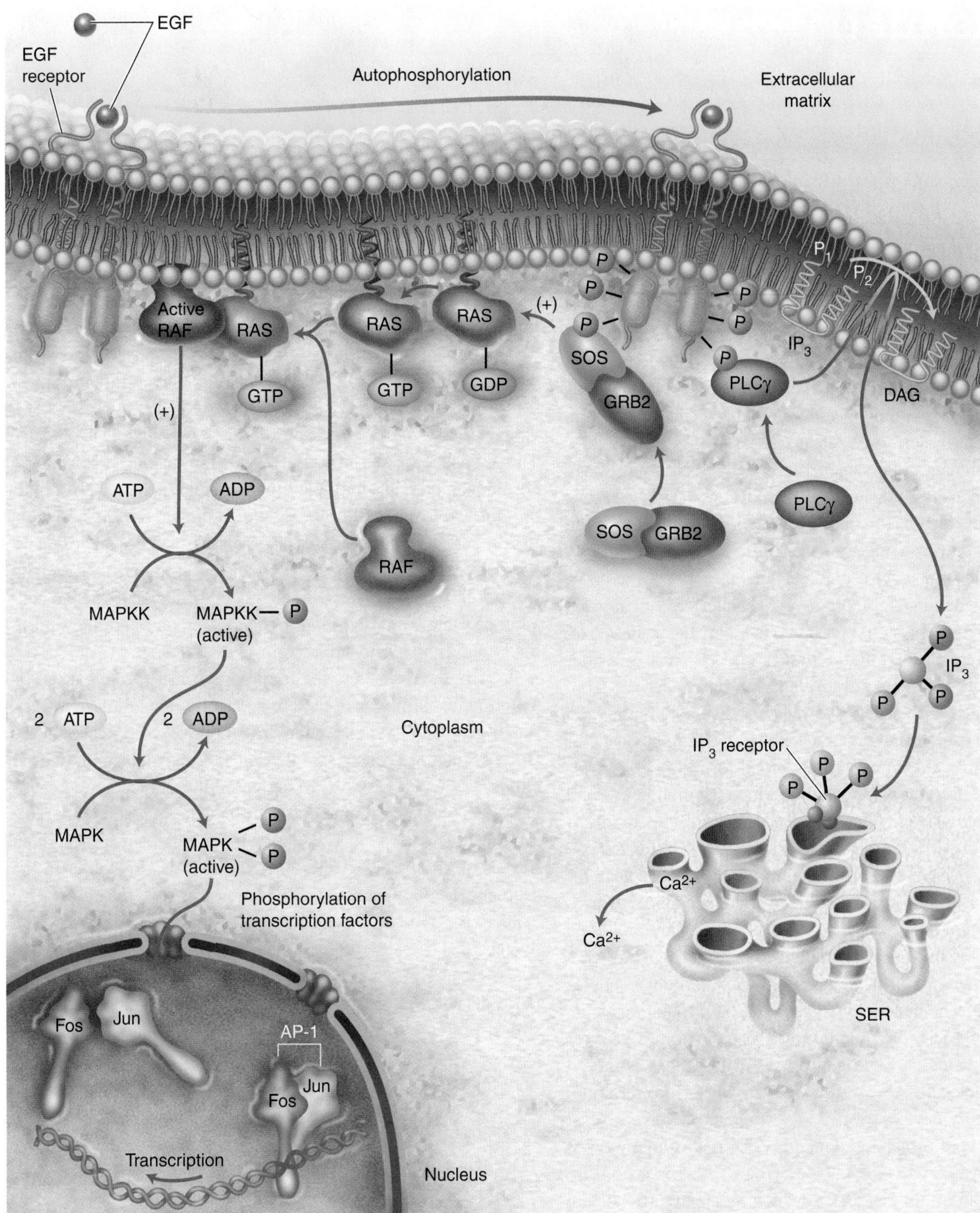

FIGURE 18.55

A Simplified Example of Eukaryotic Gene Expression Triggered by Growth Factor Binding

Signal transduction pathway illustrating selected events that are triggered when a growth factor (e.g., EGF) binds to its plasma membrane receptor. Subsequent changes in growth factor– or hormone-triggered gene expression that occur are typically mediated by several different mechanisms. This example illustrates only Ras and PLCγ activation. When EGF binds, it promotes dimerization of the receptor and autophosphorylation of the tyrosine residues on its cytoplasmic domain. As a result, the receptor binds a variety of cytoplasmic proteins. When one of these proteins, GRB2, binds to the receptor, SOS (a GEF) is activated, and it in turn activates Ras by promoting the exchange of GDP for GTP. Activated Ras initiates a phosphorylation cascade by activating the protein kinase RAF, which in turn activates MAPKK. MAPKK activates MAPK, which migrates to the nucleus and activates a number of transcription factors (e.g., fos and jun that form AP-1). When PLCγ binds to the EGF receptor, it catalyzes the cleavage of PIP_2 into IP_3 and DAG. IP_3 stimulates the release of Ca^{2+} into the cytoplasm. As illustrated in **Figure 16.7**, in the presence of Ca^{2+} and DAG, protein kinase C activates yet another series of protein kinases that in turn affect the function of several cell growth regulatory proteins.

Biochemistry IN PERSPECTIVE

Carcinogenesis

What is cancer, and what are the biochemical processes that facilitate the transformation of normal cells into those with cancerous properties? Cancer is a group of diseases in which genetically damaged cells proliferate autonomously. Such cells cannot respond to normal regulatory mechanisms that ensure the intercellular cooperation required in multicellular organisms. Consequently, they continue to proliferate, thereby robbing nearby normal cells of nutrients and eventually crowding surrounding healthy tissue. Depending on the damage they have sustained, abnormal cells may form either benign or malignant tumors. Benign tumors, which grow slowly and are limited to a specific location, are not considered cancerous and rarely cause death. In contrast, *malignant* tumors are often fatal because they can undergo *metastasis*, migration through blood or lymph vessels to distant locations throughout the body. Wherever new malignant tumors arise, they interfere with normal functions. When life-sustaining processes fail, patients die.

Cancers are classified by the tissues affected. The vast majority of cancerous tumors are *carcinomas* (tumors derived from epithelial tissue cells such as skin, various glands, breasts, and the lining of most internal organs). *Leukemias* (lymphoblastic or myelogenic) are cancers of various blood-forming cells. *Lymphomas* are solid tumors of lymphoid tissue (lymph nodes and spleen). Tumors arising in connective tissue are called *sarcomas*. Despite the differences among the diseases in this diverse class, they also have several common characteristics that include genomic instability, growth factor-independence, growth suppression signal insensitivity, and apoptosis evasion. Each tumor originates from a single damaged cell. In other words, a tumor is a clone derived from a cell in which heritable changes have occurred. The genetic damage consists of mutations (e.g., point mutations, deletions, and inversions) and chromosomal rearrangements or losses. Such changes result in the loss or altered function of molecules involved in cell growth or proliferation. Tumors typically develop over a long time and involve several independent types of genetic damage. (The risk of many types of cancer increases with age.)

Transformation, the process in which an apparently normal cell is converted or "transformed" into a malignant cell, is a complex process that may involve numerous genes in diverse cellular pathways. Although a cancerous cell may contain thousands of mutations, typically only a small set (e.g., 9 in breast cancer and 12 in colorectal cancer) is required for transformation. There are three phases in transformation: initiation, promotion, and progression.

During the *initiation* phase of carcinogenesis, a destabilizing change in a cell's genome provides it with a growth advantage over its neighbors. Most initiating mutations affect protooncogenes or tumor suppressor genes. Protooncogenes code for a variety of growth factors, growth factor receptors, enzymes, or transcription factors that promote cell growth and/or cell division. Mutated versions of protooncogenes that promote abnormal cell proliferation are called **oncogenes** (Table 1). **Tumor suppressor genes** code for proteins that actively protect cells from progressing toward cancer. For example, in cells with damaged DNA the tumor suppressor RB1 codes for a protein (Rb) that inhibits a transcription factor complex required for cell cycle progression. Other tumor suppressors are DNA repair genes such as BRCA1 and BRCA 2 (p. 686), which are involved in the repair of double-stranded DNA breaks. The tumor suppressor protein p53, coded for by the TP53 gene on chromosome 17, has multiple functions that include DNA repair protein activation, cell cycle regulation, and apoptosis initiation. In response to cellular stresses such as DNA damage and oxidative stress, p53 is activated by phosphorylation of its N-terminal domain by specific kinases. The subsequent conformational change converts p53 into a transcription regulatory protein and allows it to bind transcription activator proteins.

In hereditary cancers the initiation event is often caused by genomic instability attributed to a germline mutation in a DNA repair gene. For example, about 5% of breast cancer cases in the United States are caused by inherited heterozygous mutations in either BRCA1 or BRCA2. Many hereditary cancers are genetically recessive (i.e., both copies of the gene must be inoperable for the cancer to occur). Because mutations occur throughout life, as an individual grows older the risk of a mutation in the second gene copy increases. By some estimates, the cumulative cancer risk of a carrier of one copy of mutated BRCA1 is 80% by age 70. The cause of genomic instability in sporadic cancers is less clear. However, it is believed that several factors contribute to destabilizing DNA damage. Examples include oxidative stress, proteotoxic stress (an inadequate unfolded protein response), and chronic inflammation. Initiating events, caused by

TABLE 1 Selected Oncogenes*

Oncogene	Protooncogene Function
sis	Platelet-derived growth factor
erbB	Epidermal growth factor receptor
src	Tyrosine-specific protein kinase
raf	Serine/threonine-specific protein kinase
ras	GTP-binding protein
jun	Transcription factor
fos	Transcription factor
myc	Transcription factor

*Abnormal versions of protooncogenes that mediate cancerous transformations.

Biochemistry IN PERSPECTIVE cont.

damage that alters the expression or function of protooncogenes and/or tumor suppressors, include exposure to carcinogenic chemicals, radiation, and viral infections. The damage that alters the function of protooncogenes and tumor suppressor genes is caused by the following.

1. **Carcinogenic chemicals**. Most cancer-causing chemicals are mutagenic; that is, they alter DNA structure. Some carcinogens (e.g., nitrogen mustard) are highly reactive electrophiles that attack electron-rich groups in DNA (as well as RNA and protein). Other carcinogens (e.g., benzo[*a*]pyrene) are actually procarcinogens, which are converted to active carcinogens by one or more enzyme-catalyzed reactions.
2. **Radiation**. Some radiation (UV, X-rays, and γ-rays) is carcinogenic. As noted, the damage inflicted on DNA includes single- and double-strand breaks, pyrimidine dimer formation, and the loss of both purine and pyrimidine bases. Radiation exposure causes the formation of ROS, which may be responsible for most of radiation's carcinogenic effects.
3. **Viruses**. Viruses appear to contribute to the transformation process in several ways. Some introduce oncogenes into a host cell chromosome as they insert their genome. Viruses can also affect the expression of cellular protooncogenes through insertional mutagenesis, a random process in which viral genome insertion inactivates a regulatory site or alters the protooncogene's coding sequence. Most virus-associated cancers have been detected in animals. Only a few human cancers have been proven to be associated with viral infection. Examples include human papilloma virus and hepatitis B virus, which induce cervical cancer and liver cancer, respectively.

Tumor development can also be promoted by chemicals that do not alter DNA structure. So-called **tumor promoters** contribute to carcinogenesis by two principal methods. By activating components of intracellular signaling pathways, some molecules (e.g., the phorbol esters) provide the cell a growth advantage over its neighbors. (Recall that phorbol esters, see p. 595, activate PKC because they mimic the actions of DAG.) The effects of many other tumor promoters are unknown but may involve transient effects such as increasing cellular Ca^{2+} levels or increasing synthesis of the enzymes that convert procarcinogens into carcinogens. Unlike initiating agents, the effects of tumor promoters are reversible. They produce permanent damage only with prolonged exposure after an affected cell has undergone an initiating mutation.

Following initiation and promotion, cells go through a process referred to as progression. During *progression*, genetically vulnerable precancerous cells, which already possess significant growth advantages over normal cells, are further damaged. Eventually, the continued exposure to carcinogens and promoters makes further random mutations inevitable. If these mutations affect cellular proliferative or differentiating capacity, then an affected cell may become sufficiently malignant to produce a tumor.

Cancer is not only a genetic disease; it is also currently appreciated as an epigenetic disease. Chromatin-associated gene silencing has been implicated in the progress of a number of cancers. Epigenetic analysis of DNA methylation utilizes DNA sequencing methods on bisulfite-treated genomic DNA. (Bisulfite [HSO_3^-] converts unmethylated cytosines to uracil, but does not affect methylated cytosines.) This work has revealed that DNA in cancer cells is, in general, hypomethylated in comparison to normal cells. Hypomethylation of repetitive sequences, which becomes more pronounced as normal cells are transformed into cancerous cells, results in genomic instability, transposable element activation, and the expression of normally repressed genes. Lung cancer caused by smoking provides an example of cancerous epigenetic change. Cigarette smoke has been definitively proven to cause demethylation of the γ-synuclein gene, which codes for an unfolded, soluble protein. Normally expressed only in neurons, the γ-synuclein gene is an oncogene in lung tumors where it facilitates metastasis. Epimutations also include hypermethylation of CpG islands near normally expressed gene sequences that render them inactive. For example, hypermethylation silences the WRN gene that codes for a protein with DNA helicase and exonuclease activity. In the absence of the WRN protein, affected cells have unstable chomosomes with large numbers of deletions, inversions, and translocations.

An Ounce of Prevention . . .

Because of the enormous cost and limited success of cancer therapy, it has become increasingly recognized that cancer prevention is cost-effective since the majority of cancer cases are preventable. For example, more than one-third of cancer mortality is directly caused by tobacco use, and another one-third of cancer deaths have been linked to inadequate diets. Tobacco smoke, which contains thousands of chemicals, many of which are either carcinogens or tumor promoters, is responsible for most cases of lung cancer and contributes to cancers of the pancreas, bladder, and kidneys, among others. Diets that are high in fat and low in fiber content have been associated with increased incidence of cancers of the large bowel, breast, pancreas, and prostate. Other dietary risk factors include low consumption of fresh vegetables and fruit. Folic acid deficiency is also a risk factor for cancer. Found in leafy vegetables, legumes, sunflower seeds, and fortified grains, folic acid in its biologically active form, tetrahydrofolate, is a key component of one-carbon metabolism (p. 528) that provides the methyl groups used by epigenetic methyltransferases. Although folic acid is known to protect against cancer initiation, excessive supplementation has been linked to the progression of several types of already existing tumors.

In addition to providing sufficient antioxidant vitamins, many vegetables (and fruits to a lesser extent) contain numerous nonnutritive components that actively inhibit carcinogenesis. Some carcinogenesis inhibitors (e.g., organosulfides),

Biochemistry IN PERSPECTIVE cont.

referred to as *blocking agents*, prevent carcinogens from reacting with DNA or inhibit the activity of tumor promoters. Other inhibitors, referred to as *suppressing agents* (e.g., inositol hexaphosphate), prevent the further development of neoplastic processes that are already in progress. Many nonnutritive food components (e.g., tannins and protease inhibitors) possess both blocking and suppressing effects. In general, these molecules effectively protect against cancer because many of them inhibit the arachidonic acid cascade and oxidative damage. Apparently low-fat, high-fiber diets that are rich in raw or fresh vegetables that are leafy green (e.g., spinach), cruciferous (e.g., broccoli), and members of the allium family (e.g., onions), as well as fresh fruits, are a prudent choice for individuals seeking to reduce their risk of cancer.

SUMMARY: Carcinogenesis is the process whereby cells with a growth advantage over their neighbors are transformed by mutations in the genes that control cell division into cells that no longer respond to regulatory signals.

Chapter Summary

1. Living organisms must possess efficient mechanisms for rapid and accurate DNA synthesis. DNA replication occurs by a semiconservative mechanism; that is, each of the two parental strands serves as a template to synthesize a new strand. Enzymatic activities required in DNA replication are used in DNA unwinding, primer synthesis, polynucleotide synthesis, supercoiling control, and ligation. Although the basic features of DNA replication are similar in prokaryotes and eukaryotes, there are significant differences (e.g., replication time and rate, replication origin numbers, Okazaki fragment size, and replication machinery structure).
2. There are several types of DNA repair mechanism. These include photoreactivation repair, base excision and nucleotide excision repair, double-strand break repair, and recombinational repair.
3. Genetic recombination, a process in which DNA sequences are exchanged between different DNA molecules, occurs in two forms. In general recombination, the exchange occurs between sequences in homologous chromosomes. In site-specific recombination, the exchange of sequences requires only short homologous sequences. DNA-protein interactions are principally responsible for the exchange of largely nonhomologous sequences.
4. Transposition, the movement of genetic elements (transposons) from one location to another within a genome, can cause genetic changes such as insertions, deletions, and translocations. The movement of retrotransposons, found in large numbers in eukaryotic genomes, can cause disease or provide opportunities for genetic diversity.
5. The synthesis of RNA, referred to as DNA transcription, requires a variety of proteins. Transcription initiation involves binding an RNA polymerase to a specific DNA sequence called a promoter. Regulation of transcription differs significantly between prokaryotes and eukaryotes. Examples of eukaryotic transcription processes not observed in prokaryotes include RNA-processing features such as capping, poly(A) tail synthesis, and RNA splicing.
6. Prokaryotic gene expression regulation involves regulatory units called operons and RNA-based structures called riboswitches. Eukaryotes have a wide variety of mechanisms that control gene expression. Prominent examples include DNA methylation, histone covalent modification and chromatin remodeling, RNA-processing reactions such as alternative splicing, and RNA editing, RNA transport, posttranscriptional gene silencing, and translational controls.

Take your learning further by visiting the **companion website** for Biochemistry at **www.oup.com/us/mckee** where you can complete a multiple-choice quiz on genetic information to help you prepare for exams.

Suggested Readings

Bell, L., Catching a Cancer: Viral Culprits May Explain a Host of Tumors with as-yet Unknown Triggers, *Sci. News* 182(2): 22–25, 2012.

Carlsten, J. O. P., Zhu, X., and Gustafsson, C. M., The Multitalented Mediator Complex, *Trends Biochem. Sci.* 38(11):531–537, 2013.

Egloff, S., and Murphy, S., Cracking the RNA Polymerase II CTD Code, *Trends Genet.* 24(6):280–288, 2008.

Goosen, N., Scanning the DNA for Damage by the Nucleotide Excision Repair Machinery, *DNA Repair* 9:593–596, 2010.

Lai, F., and Shiekhatter, R., Enhancer RNAs: The New Molecules of Transcription, *Curr. Opin. Genet. Dev.* 25:38–42, 2014.

Lavelle, C., Pack, Unpack, Bend, Twist, Pull, Push: The Physical Side of Gene Expression, *Curr. Opin. Genet. Dev.* 25:74–84, 2014.

Lee, J. T., Epigenetic Regulation by Long Noncoding RNAs, *Science* 338:1435–1439, 2012.

Misteli, T., The Inner Life of the Genome, *Sci. Am.* 304(2):66–73, 2011.

Negrini, S., Gorgoulis, V. G., and Halazonetis, T. D., Genomic Instability—An Evolving Hallmark of Cancer, *Nat. Rev. Mol. Cell Biol.* 11:220–228, 2010.

O'Blessness, M., et al., Evolution of Genetic and Genomic Features Unique to the Human Lineage, *Nat. Rev. Genet.* 13(12):853–856, 2012.

Rieder, D., Trajanoski, Z., and McNally, J. G., Transcription Factories, *Frontiers Genet.* 3:221, 2012.

Scharer, O. D., Nucleotide Excision Repair in Eukaryotes, *Cold Spring Harbor Perspect. Biol.* 5:a012609, 2013.

Tammen, S. A., Friso, S., and Choi, S.-W., Epigenetics: The Link between Nature and Nurture, *Mol. Aspects Med.* 34(4): 753–764, 2013.

Taubes, G., Double-Edged Genes, *Discover* 34(3):60–66, 2013.

Tian, B., and Manley, J. L., Alternative Cleavage and Polyadenylation: The Long and Short of It, *Trends Biochem. Sci.* 38(6): 312–320, 2013.

Watson, J. D., Baker, T. A., Bell, S. P., Gann, A., Levine, M., and Losick, R., *Molecular Biology of the Gene*, Pearson Benjamin Cummings, San Francisco, 7th ed., 2014.

Key Words

annotation, *694*
apoptosis, *717*
apurinic site, *678*
apyrimidinic site *678*
attachment site, *686*
bacterial artificial chromosome, *691*
base excision repair, *678*
bioinformatics, *695*
cDNA library, *693*
chromosomal jumping, *694*
chromatin-remodeling complexes, *701*
clamp loader, *668*
coding strand, *697*
colony hybridization technique, *692*
composite transposon, *688*
conjugation, *686*
consensus sequence, *698*
constitutive gene, *708*
contig, *694*
cosmid, *691*
DNA glycosylase, *678*
DNA ligase, *670*
DNA microarray, *694*
double-strand break repair model, *684*
electroporation, *691*
exonuclease, *668*
functional genomics, *690*
general recombination, *681*
gene silencing, *715*
genomics, *690*
GTPase-activating protein, *718*
guanine nucleotide exchange facter, *718*
helicase, *668*
insertional element, *686*
intrinsic termination *700*
inverted repeat, *687*
light-induced repair, *678*
marker gene, *691*
MCM complex, *675*
mediator, *704*
Meselson-Radding model, *683*
mismatch repair, *679*
nucleotide excision repair, *678*
Okazaki fragment, *672*
oncogene, *720*
operator, *710*
operon, 710
origin of replication complex, *675*
palindrome, *687*
photoreactivation repair, *678*
polymerase chain reaction, *693*
preinitiation complex, *704*
preinitiation replication complex, *675*
primase, *668*
primer, *668*
primosome, *668*
processivity, *668*
promoter, *697*
protooncogene, *717*
recombinant DNA technology, *690*
recombination, *667*
replication, *667*
replication factor C, *675*
replication factory, *667*
replication fork, *671*
replication licensing factor, *675*
replication protein A, *675*
replicon, *671*
replisome, *668*
retroelement, *689*
retroposon, *689*
rho-dependent termination, *700*
rho-independent termination, *700*
riboswitch, *710*
RNA editing, *715*
RNA interference, *716*
RNA splicing, *706*
semiconservative replication, *667*
shotgun cloning, *693*
site-specific recombination, *681*
spliceosome, *706*
supraspliceosome, *707*
synthesis-dependent strand annealing model, *676*
telomerase, *676*
telomere-end binding protein, *676*
telomere repeat-binding factors, 676
transcription coupled repair, *678*
transduction, *686*
transfection, *691*
transformation, *685*
transposable element, *681*
transposition, *681*
transposon, *687*
tumor promoter, *721*
tumor suppressor gene, *720*
vector, *690*
yeast artificial chromosome, *677*

Review Questions

These questions are designed to test your knowledge of the key concepts discussed in this chapter before moving on to the next chapter. You may like to compare your answers to the solutions provided in the back of the book and in the accompanying Study Guide.

1. Define the following terms:
   a. replication
   b. semiconservative
   c. replication factory
   d. primosome
   e. clamp loader
2. Define the following terms:
   a. processivity
   b. replisome
   c. exonuclease
   d. DNA ligase
   e. replication fork
3. Define the following terms:
   a. replicon
   b. Okazaki fragment
   c. ter region
   d. tus protein
   e. preinitiation complex
4. Define the following terms:
   a. ORC
   b. licensing factors
   c. RPA
   d. TEBP
   e. TRF
5. Define the following terms:
   a. RFC
   b. DNA glycosylase
   c. apurinic site
   d. apyrimidinic site
   e. mismatch repair
6. Define the following terms:
   a. transposition
   b. transposable element
   c. bacterial transformation
   d. transduction
   e. conjugation
7. Define the following terms:
   a. nonreplicative transposition
   b. replicative transposition
   c. composite transposon
   d. retrotransposon
   e. insertional element
8. Define the following terms:
   a. transfection
   b. cosmid
   c. electroporation
   d. transgenic animal
   e. colony hybridization technique
9. Define the following terms:
   a. PCR
   b. DNA microarray
   c. chromosomal jumping
   d. genome project
   e. bioinformatics
10. Define the following terms:
    a. promoter
    b. consensus sequence
    c. operon
    d. chromatin-remodeling complex
    e. general transcription factors
11. Define the following terms:
    a. RNA splicing
    b. spliceosome
    c. operator
    d. riboswitch
    e. *lac* operon
12. Define the following terms:
    a. gene silencing
    b. RNA interference
    c. tumor suppressor gene
    d. protooncogene
    e. GEF
13. Define the following terms:
    a. cell transformation
    b. oncogene
    c. apoptosis
    d. early response gene
    e. delayed response gene
14. Define the following terms:
    a. cohesin
    b. transcription factory
    c. P bodies
    d. CTCF
    e. ENCODE
15. What role does the poly(A) tail play in mRNA function?
16. List and describe the steps in prokaryotic DNA replication. How does this process appear to differ from eukaryotic DNA replication?
17. Indicate the stage of DNA replication when each of the following enzymes is active:
    a. helicase
    b. primase
    c. DNA polymerases
    d. ligase
    e. topoisomerase
    f. DNA gyrase

18. DNA is polymerized in the 5′ → 3′ direction. Demonstrate with the incorporation of three nucleotides into a single strand of DNA how the 5′ → 3′ directionality is derived.
19. Mutations are caused by chemical and physical phenomena. Indicate the type of mutation that each of the following reactions or molecules might cause:
    a. ROS
    b. intercalating agents
    c. a small alkylating agent
    d. a large alkylating agent
    e. nitrous acid
20. How can viruses cause mutations?
21. Explain the significance of "jumping gens."
22. Describe two forms of genetic recombination. What functions do they fulfill?
23. Although genetic variation is required for species to adapt to changes in their environment, most genetic changes are detrimental. Explain why genetic mutations are rarely beneficial.
24. General recombination occurs in bacteria, where it is involved in several types of intermicrobial DNA transfer. What are these types of transfer, and by what mechanisms do they occur?
25. Compare and contrast the mechanisms of replicative and non-replicative transposition.
26. Within cells, cytosine slowly converts to uracil. To what type of mutation would this lead in DNA molecules? What impact would the same modification have if it occurred at the RNA or gene product level?
27. A correlation has been found among species between life span and the efficiency of DNA repair systems. Suggest a reason why this is so.
28. What are the similarities and differences between cellular DNA replication and PCR?
29. Describe the purpose of marker genes in recombinant DNA technology.
30. Define and describe the roles of the following in transcription:
    a. transcription factors
    b. RNA polymerase
    c. promoter
    d. sigma factor
    e. enhancer
    f. TATA box
31. List the steps in the processing of a typical eukaryotic mRNA precursor that prepare for its functional role.
32. Describe the advantages and disadvantages for organisms that arrange genes in operons.
33. Determine the magnitude of amplification of a single DNA molecule that can be attained with PCR during five cycles.
34. Many genes generate different products depending on the type of cell expressing the gene. How is this phenomenon accomplished?
35. Provide an example for the process described in Question 34.
36. During base excision repair, DNA glycosylase cleaves the N-glycosidic link between the altered base and the deoxyribose component of the nucleotide. Draw a typical nucleotide and indicate which bond is cleaved.
37. During certain stages of development, the requirement for certain gene products may require gene amplification. What purpose does gene amplification serve?
38. How are the genes referred to in Question 37 amplified?
39. Cells exposed to ultraviolet light develop thymine dimers and visible light reverses this damage. Explain how this form of DNA repair occurs.
40. RNA molecules are more reactive than DNA. Explain.
41. Describe how LINE 1 element transposition occurs.
42. Describe the function of telomere end-binding proteins.
43. Describe the function of telomere repeat-binding factors.
44. Describe the role of mismatch repair (MMR) in DNA repair. Why do mutations in MMR proteins increase the risk of human cancer?
45. What is a tetragametic chimera and how does it occur?
46. What role does a stalled RNA polymerase play in transcription coupled nucleotide excision repair?

Fill in the Blank

47. The process whereby DNA molecules are cut and rejoined in new combinations is called ____________________.
48. A ____________________ is a multienzyme complex that synthesizes RNA primers during *E. coli* DNA replication.
49. An enzyme called ____________________ catalyzes the formation of a covalent phosphodiester bond between the 3′ OH end of one DNA strand and the 5′ phosphate end of an adjacent strand during DNA replication.
50. In prokaryotes Okazaki fragments are approximately ____________________ nucleotides long.
51. ____________________ is a mechanism that removes and replaces individual nucleotides in DNA when bases have been damaged.
52. The two forms of recombination are general and ____________________.
53. General recombination occurs between ____________________ DNA molecules.
54. DNA sequences called ____________________ can move from one site in a genome to another.
55. DNA ____________________ are enzymes that prevent tangling of DNA strands during DNA replication.
56. The average of a number of closely related but nonidentical sequences is referred to as a ____________________.

Short Answer

57. Telomerase activity in cancer cells has been found to be 10 to 20 times more active than in somatic normal cells. What is the significance of this circumstance?
58. What would you expect the functional properties of a riboswitch specific for the amino acid lysine to be?
59. The retinoblastoma tumor suppressor gene Rb (RB1) codes for the retinoblastoma protein (pRB). pRB prevents the progression of the cell cycle through G1 if DNA has been damaged. It does so in part because it binds a transcription-activating dimer referred to as E2F-DP. The pRB-E2F/DP complex recruits a histone deacetylase to chromatin. Explain.
60. The tumor suppressor pRB also binds to and suppresses the activity of retinoblastoma binding protein 2 (RBP2), a histone demethylase that removes methyl groups from di- and trimethylated lysines in histone 3. What is the possible consequence of an inactivating mutation in RB1 that causes an inability of pRB to bind RBP2?
61. All eukaryotes possess a surveillance pathway referred to as non-sense-mediated mRNA decay (NMD). Its principal function is to eliminate mRNA transcripts with premature stop codons. Such faulty transcripts are detected during translation and subsequently destroyed by removal of the 5′ cap followed by degradation by a nuclease. Describe how premature stop codons are detected and what type of error causes them.

Thought Questions

These questions are designed to reinforce your understanding of all of the key concepts discussed in the book so far, including this chapter and all of the chapters before it. They may not have one right answer! The authors have provided possible solutions to these questions in the back of the book and in the accompanying Study Guide, for your reference.

62. In eukaryotes the DNA replication rate is 50 nucleotides per second. How long does the replication of a chromosome of 150 million base pairs take? If eukaryotic chromosomes were replicated like those of prokaryotes, the replication of a genome would take months. Actually, eukaryotic replication takes only several hours. How do eukaryotes achieve this high rate?
63. There appears to be insufficient genetic material to direct all the activities of certain types of eukaryotic cell (e.g., such as T lymphocytes). Explain how genetic recombination, gene splicing, and alternative RNA splicing help solve this problem.
64. Mustard gas is an extremely toxic substance that severely damages lung tissue when inhaled in large amounts. In small amounts, mustard gas is a mutagen and carcinogen. Considering that mustard gas is a bifunctional alkylating agent, explain how it mutates genes and impacts DNA replication.
65. Infection caused by a rare, virulent strain of group A streptococcus has appeared relatively recently. In approximately 25 to 50% of these cases (reported in Great Britain and the United States), infection resulted in necrotizing fasciitis, a rapidly spreading destruction of flesh, often accompanied by hypotension (low blood pressure), organ failure, and toxic shock. If antibiotic treatment is not initiated within 3 days of exposure to the bacterium, gangrene and death may result. Similar cases were reported in the 1920s. However, these earlier cases had a significantly lower fatality rate, although antibiotics were not then available. (Physicians reported treating affected areas by washing with acidic solutions.) Group A streptococci are converted into the pathogenic form by becoming infected with a certain virus. This virus's genome contains a gene that codes for a tissue-destroying toxin. Can you describe in general terms how a viral infection might cause a permanent change in the pathogenicity of a group A streptococcus bacterium? Considering the apparent difference in virulence between the bacterium in the 1920s and the present, is there any method for determining whether the same strain of group A streptococcus is responsible for both sets of cases? Preserved specimens of infected tissue from these early cases are available.
66. Adjacent pyrimidine bases in DNA form dimers with high efficiency after exposure to UV light. If these dimers are not repaired, skin cancers can result. Melanin is a natural sunscreen produced by melanocytes, a type of skin cell, when the skin is exposed to sunlight. Individuals who spend long periods over many years developing a tan eventually acquire thick and highly wrinkled skin. Such individuals are also at high risk for skin cancer. Can you explain, in general terms, why these phenomena are related?
67. Phorbol esters have been observed to induce the transcription of AP-1–influenced genes. Explain how this process could occur. What are the consequences of AP-1 transcription? What role does intermittent exposure to phorbol esters have on an individual's health?
68. Because of overuse of antibiotics and/or weakened governmental surveillance of infectious disease, several diseases that had been thought to be no longer a threat to human health (e.g., pneumonia and tuberculosis) are rapidly becoming unmanageable. In several instances, so-called superbugs (microorganisms that are resistant to almost all known antibiotics) have been detected. How did this circumstance arise? What will happen if this process continues?
69. Retinoblastoma is a rare cancer in which tumors develop in the retina of the eye. The tumors arise because of the loss of the *Rb* gene, which codes for a tumor suppressor. Hereditary retinoblastoma usually appears in children who have inherited only one functional copy of *Rb*. Explain why the nonhereditary form of retinoblastoma usually occurs later in life.

70. Explain the difference between the potential effects on an individual organism of errors made during replication and those made during transcription.
71. Explain how a reverse transcriptase activity within a cell can result in gene amplification.
72. Use the techniques described in Chapter 18 to describe in general terms how a researcher can map the genome of a newly discovered organism.
73. It was once thought that the DNA polymerase machinery moves along DNA in a manner analogous to a train on a track. Current evidence indicates that the polymerizing machinery is instead stationary, and DNA strands are pumped through the complex. What advantages does this stationary mechanism have?
74. Describe the features of a riboswitch that is activated by excess metabolite concentrations.
75. DNA is the cellular repository of genetic information, and proteins are synthesized from RNA transcripts. Why don't cells simply skip the transcription step and use DNA directly in protein synthesis?

CHAPTER

19 Protein Synthesis

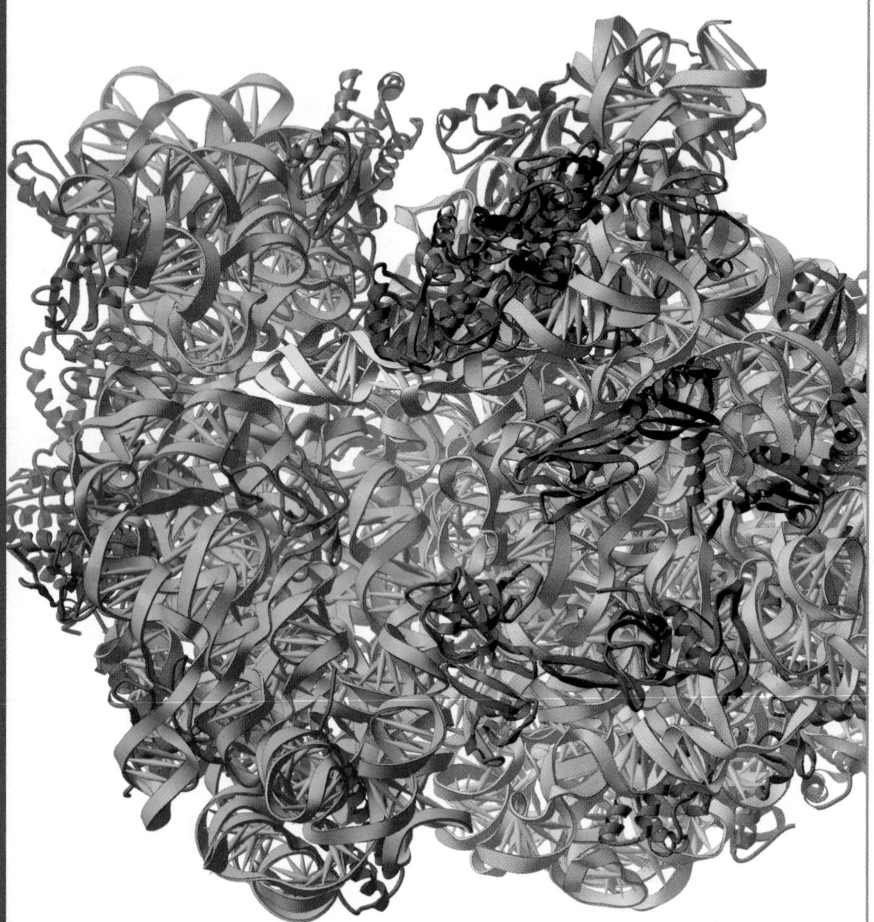

The Ribosome Ribosomes are ribonucleoprotein molecular machines that synthesize proteins in all living cells. In this illustration of the high-resolution structure of a complete bacterial 70S ribosome, proteins are illustrated in dark blue and magenta, and rRNA molecules are shown in cyan and gray. The tRNAs are orange and yellow. Note that the ribosome is primarily composed of rRNA, which perform most catalytic activities. Protein molecules largely serve supporting roles.

OUTLINE

MRSA: The Superbug

If you were to ask Sam about his leg infection, it is unlikely that he could tell you how or even exactly when it began. Skin abrasions are very frequent with physically active teenagers, especially on high school football teams. The torn skin on his left lower leg probably occurred during a practice game. After a quick toweling off in the locker room, this busy, stressed high school student was off to his next activity. The leg didn't become bothersome until several days later when the abrasion became red and itchy, followed later by slight swelling and warmth. Still, Sam was not overly concerned: his experiences with cuts and scrapes had never amounted to anything more troubling than a few Band-Aids. About a week later when Sam mentioned it to his parents, the now abscessed (pus-releasing) lesion had become very painful. When an antibiotic prescription failed to heal the leg and red streaks just below the skin began to appear, it was apparent to everyone that Sam had a very serious infection. The morning of Sam's first appointment with an infection specialist, 4 days after the initial visit to his family doctor, his symptoms (fever, chills, weakness, and nausea), although still relatively mild, indicated that he was in the beginning stages of septicemia (infection of the bloodstream by microorganisms). After the specialist sent a specimen to be cultured, Sam was diagnosed with an MRSA-caused infection.

MRSA (methicillin-resistant *Staphylococcus aureus*) is a strain of the Gram-positive bacterium *S. aureus* that is resistant to the β-lactams, a class of β-lactam ring–containing antibiotics that includes the penicillins and penicillin-related antibiotics (e.g., the cephalosporins). The β-lactams kill susceptible bacteria by stopping transpeptidation, a key reaction in cell wall synthesis. Resistant bacteria synthesize β-lactamase, an enzyme that hydrolyzes the β-lactam ring, thus inactivating the antibiotic.

Antibiotic resistance is not restricted to the β-lactams. It is an evolutionary phenomenon caused by a selection pressure, in this case the unrestricted use of antibiotics. The widespread use of a specific antibiotic provides any bacterium with a plasmid that contains an antibiotic resistance gene, allowing it to evade the antibiotic's action against the pathogen with a significant growth advantage. In other words, the antibiotic removes the resistant bacterium's competition for resources. Some strains of bacteria, called superbugs, are resistant to several antibiotics because they possess several resistance genes.

In most cases of MRSA infection, the drug of choice is vancomycin. Vancomycin kills Gram-positive bacteria by interfering with transglycosylation, yet another reaction in cell wall formation. To the horror of the hospital physicians and Sam's parents, lab cultures of the organism replicating in Sam's bloodstream proved it to be insensitive to vancomycin as well. One of the few remaining antibiotic treatments for MRSA infections is linezolid, a molecule that disrupts the initiation phase of bacterial protein synthesis. Within a few days after beginning intravenous therapy with linezolid, Sam began to recover. After 1 week he was released from the hospital with a prescription for linezolid tablets. Both Sam and his parents were warned that he had to adhere rigorously to the medication schedule (one tablet every 12 hours for 7 days). Failure to do so would create the risk of a relapse, a serious threat to Sam's health. Any further treatment with linezolid could result in toxic side effects such as an irreversible peripheral neuropathy. It could even lead to resistance of the organism to linezolid. Without any antibiotics proven effective against this particular strain of MRSA, Sam's life would be threatened.

Sam's parents waited until he recovered from the infection to tell him that Jake, a fellow member of his team, had also been infected with MRSA. For unknown reasons Jake's infection was more aggressive than Sam's. Despite the heroic efforts of his doctors, Jake died after 2 weeks in the hospital. Afterward, the school district temporarily closed the high school so that it could be thoroughly sanitized. The students were also informed about prevention strategies. In addition to routine cleaning of sports equipment, students were warned that such simple behaviors as hand washing, the covering of open cuts and abrasions, and not sharing towels or razors could prevent this tragedy from happening again.

Overview

PROTEINS ARE THE MOST DYNAMIC, NUMEROUS, AND VARIED CLASS OF BIOMOLECULES. THE UNIQUENESS OF EACH CELL TYPE IS CAUSED ALMOST entirely by the proteins it produces. It is not surprising, therefore, that a relatively large amount of cellular energy is used in protein synthesis. Because of their strategic importance in the cellular economy, the synthesis of proteins is a regulated process. Although control is also of major importance at the transcriptional level, control of the translation of genetic messages allows for additional opportunities for regulation. This is especially true in multicellular eukaryotes, whose complex lifestyles require amazingly diverse regulatory mechanisms.

Protein synthesis is the process in which genetic information encoded in the nucleic acids is translated into the 20 standard amino acid "alphabet'" of polypeptides. In addition to translation (the mechanism by which a nucleotide base sequence directs the polymerization of amino acids), protein synthesis includes the processes of posttranslational modification and targeting. *Posttranslational modification* consists of the chemical alterations cells use to prepare polypeptides for their functional roles. Several modifications assist in *targeting*, which directs newly synthesized molecules to a specific intracellular or extracellular location.

In all, at least 100 different molecules are involved in protein synthesis. Among the most important of these are the components of the ribosomes, large ribonucleoprotein machines that synthesize polypeptides. Each ribosome "reads" the base sequence of an mRNA and, fueled by GTP, rapidly and precisely converts this information into the amino acid sequence of a polypeptide. Speed is required because organisms must respond expeditiously to ever-changing environmental conditions. In prokaryotes such as *E. coli*, for example, a polypeptide of 100 residues is synthesized in less than 6 s. Eukaryotes are slower, at about 2 residues per second. Precision in mRNA translation is critical because the proper functioning of each polypeptide is determined not only by the molecule's primary sequence but also by its accurate folding.

As the result of decades of intense work, an increasingly more detailed understanding of ribosome structure and function is emerging. One of the unexpected findings was that rRNA performs the critical functions of the ribosome. For example, the catalytic activity that forms peptide bonds resides in an RNA molecule. In addition, rRNA molecules also have roles in tRNA-mRNA docking, ribosomal subunit association, proofreading, and the binding of translation factors. Ribosomal proteins, for the most part, have supporting roles. In recent years it has also become apparent that the ribosome is a major component of the cellular protein quality control process.

Chapter 19 provides an overview of protein synthesis. The chapter begins with a discussion of the genetic code, the mechanism by which nucleic acid-base sequences specify the amino acid sequences of polypeptides. This is followed by discussions of protein synthesis as it occurs in both prokaryotes and eukaryotes and a description of the mechanisms that convert polypeptides into their biologically active conformations. A critical feature of this process, protein folding, was described in Section 5.3. Chapter 19 ends with a section devoted to the role of the proteostasis network in protein folding and an introduction to **proteomics**, a relatively new technology being developed to characterize the protein products of the genome.

19.1 THE GENETIC CODE

Translation is fundamentally different from the transcription process that precedes it. During transcription, the language of DNA sequences is converted to the closely related dialect of RNA sequences. During protein synthesis, however, a nucleic acid-base sequence is converted to a clearly different language (i.e., an amino acid sequence), hence the term *translation*. Researchers were at first at a loss to explain how an mRNA base code could be converted into an amino acid polymer. Then Francis Crick realized that a series of adaptor molecules must mediate the translation process. This role was eventually assigned to tRNA molecules (**Figure 17.24**).

Before adaptor molecules could be identified, however, a more important problem had to be solved: deciphering the genetic code. The **genetic code** can be described as a coding dictionary that specifies a meaning for each base sequence. Once the importance of the genetic code was recognized, investigators speculated about its dimensions. Only four different bases (G, C, A, and U) occur in mRNA, but 20 amino acids must be specified. It therefore appeared reasonable that a combination of bases codes for each amino acid. A sequence of two bases would specify only a total of 16 amino acids (i.e., $4^2 = 16$). However, a three-base sequence, or **codon**, provides more than sufficient base combinations for translation (i.e., $4^3 = 64$).

The codon assignments for the 64 possible trinucleotide sequences, determined by Marshall Nirenberg, Heinrich Matthaei, and Har Gobind Khorana, are presented in Table 19.1. Of these, 61 code for amino acids. Four codons serve as punctuation signals. UAA, UAG, and UGA are *stop* (polypeptide chain terminating) signals. AUG, the codon for methionine, also serves as a *start* signal (sometimes referred to as the *initiating codon*). The genetic code possesses the following properties.

1. **Specific**. Each codon is a signal for a specific amino acid.
2. **Degenerate**. Any coding system in which several signals have the same meaning is said to be degenerate. The genetic code is partially degenerate because most amino acids are coded for by several codons. For

TABLE 19.1 The Genetic Code

First position (5′ end)	Second Position: U		C		A		G		Third position (3′ end)
U	UUU	Phe	UCU	Ser	UAU	Tyr	UGU	Cys	U
U	UUC	Phe	UCC	Ser	UAC	Tyr	UGC	Cys	C
U	UUA	Leu	UCA	Ser	UAA	STOP	UGA*	STOP	A
U	UUG	Leu	UCG	Ser	UAG*	STOP	UGG	Trp	G
C	CUU	Leu	CCU	Pro	CAU	His	CGU	Arg	U
C	CUC	Leu	CCC	Pro	CAC	His	CGC	Arg	C
C	CUA	Leu	CCA	Pro	CAA	Gln	CGA	Arg	A
C	CUG	Leu	CCG	Pro	CAG	Gln	CGG	Arg	G
A	AUU	Ile	ACU	Thr	AAU	Asn	AGU	Ser	U
A	AUC	Ile	ACC	Thr	AAC	Asn	AGC	Ser	C
A	AUA	Ile	ACA	Thr	AAA	Lys	AGA	Arg	A
A	AUG	Met	ACG	Thr	AAG	Lys	AGG	Arg	G
G	GUU	Val	GCU	Ala	GAU	Asp	GGU	Gly	U
G	GUC	Val	GCC	Ala	GAC	Asp	GGC	Gly	C
G	GUA	Val	GCA	Ala	GAA	Glu	GGA	Gly	A
G	GUG	Val	GCG	Ala	GAG	Glu	GGG	Gly	G

*The stop codons UGA and UAG are also used by some organisms and under specific conditions (described later in the Biochemistry in Perspective essay entitled Context-Dependent Coding Reassignment) to insert selenocysteine or pyrrolysine, respectively, into a polypeptide sequence.

example, leucine is coded for by six different codons (UUA, UUG, CUU, CUC, CUA, and CUG). In fact, methionine (AUG) and tryptophan (UGG) are the only amino acids that are coded for by a single codon.

3. **Nonoverlapping**. The mRNA coding sequence is "read" by a ribosome starting from the initiating codon (AUG) as a continuous sequence taken three bases at a time until a stop codon is reached. The set of contiguous triplet codons in an mRNA that code for the amino acids in a polypeptide is called an **open reading frame**.
4. **Almost universal**. The genetic code is used by the vast majority of organisms, but there are a few exceptions. Minor deviations have been observed in mitochondria. For example, instead of six arginine codons, there are only four. The remaining two codons (AGA and AGG) are instead used as stop codons. Since UGA, normally a stop codon, instead codes for tryptophan, mitochondria have four stop codons: UAG, UAA, AGA, and AGG. Similar minor changes have been observed in a few species of prokaryotes and eukaryotes such as protozoa and yeast.

It appears that the degenerate genetic code evolved to diminish the deleterious effects of point mutations; that is, base substitutions often result in the incorporation of the same or similar amino acids, as the following two examples involving leucine illustrate. Since all the codons with CU in the first two positions code for leucine, base substitutions at position 3 will have no effect. (Refer to the discussion of the wobble hypothesis, p. 733, for another aspect of this phenomenon.) An analysis of the genetic code also reveals that all of the codons with U in the second position code for hydrophobic amino acids. If the first base in the CUU codon (leucine) is replaced with A, another hydrophobic amino acid residue, isoleucine, will be substituted.

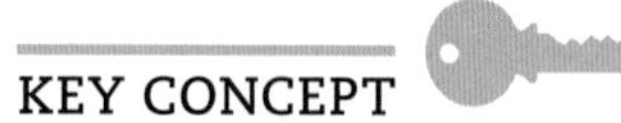

KEY CONCEPT

The genetic code is a mechanism by which ribosomes translate nucleotide base sequences into the primary sequence of polypeptides.

In certain circumstances, living organisms are not limited to the codon assignments of the genetic code. In two recently recognized examples, the nonstandard amino acids selenocysteine and pyrrolysine are coded for by stop codons. This context-dependent codon reassignment is described later in the Biochemistry in Perspective essay on pp. 752–753.

Codon Usage Bias

Bioinformatic analysis of the transcriptomes of numerous organisms has revealed that *synonymous* codons, which code for the same amino acid, are not used uniformly. Each organism has a distinct preference for particular codons. For example, *E. coli* prefers to use the UUU codon for phenylalanine, whereas humans use UUC most often. Codon usage can also vary between genes in the same organism. The human α-globin mRNA uses the phenylalanine codon UUU almost exclusively. In contrast, dystrophin mRNA uses both UUU and UUC. A correlation between frequently used codons and cellular levels of matching tRNAs has also been observed. It has been suggested that codon usage bias is the result of evolutionary pressures for translational efficiency because codon bias is especially common in fast-growing unicellular organisms such as bacteria and yeast. In slower growing cells in multicellular organisms, codon bias frequently occurs in highly expressed genes, such as α-globin in reticulocytes.

Codon-Anticodon Interactions

Transfer RNA molecules are the "adaptors" that are required for the translation of the genetic message. Each type of tRNA carries a specific amino acid (at the 3′ terminus) and possesses a three-base sequence called the **anticodon**. Codon-anticodon base pairing is responsible for the actual translation of mRNAs. Although codon-anticodon pairings are antiparallel, both sequences are given in

QUESTION 19.1

As described earlier, DNA damage can cause deletion or insertion of base pairs. If a nucleotide base sequence of a coding region changes by any number of bases other than three base pairs, or multiples of 3, a *frameshift mutation* occurs. Depending on the location of the sequence change, such mutations can have serious effects. The following synthetic mRNA sequence codes for the beginning of a polypeptide:

5′-AUGUCUCCUACUGCUGACGAGGGAAGGAGGUGGCUUAUCAU-GUUU-3′

First, determine the amino acid sequence of the polypeptide. Then determine the types of mutation that have occurred in the following altered mRNA segments. What effect do these mutations have on the polypeptide products?

a. 5′-AUGUCUCCUACUUGCUGACGAGGGAAGGAGGUGGCUUAU-CAUGUUU-3′
b. 5′-AUGUCUCCUACUGCUGACGAGGGAGGAGGUGGCUUAUCAU-GUUU-3′
c. 5′-AUGUCUCCUACUGCUGACGAGGGAAGGAGGUGGCCCUUAU-CAUGUUU-3′
d. 5′-AUGUCUCCUACUGCUGACGGAAGGAGGUGGCUUAUCAU-GUUU-3′

the 5′ → 3′ direction. For example, the codon UGC binds to the anticodon GCA (**Figure 19.1**).

Once the genetic code had been determined, researchers anticipated the identification of 61 types of tRNAs in living cells. Instead, they discovered that cells often operate with substantially fewer tRNAs than expected. Most cells possess about 50 tRNAs, although lower numbers have been observed. Further investigation of tRNAs revealed that the anticodon in some molecules contains inosinate (I), which typically occurs at the third anticodon position. (The base adenine is deaminated to form hypoxanthine; see **Figure 14.26** on p. 541.) As tRNAs were investigated further, it became increasingly clear that some molecules recognize several codons. In 1966, after reviewing the evidence, Crick proposed a rational explanation, the **wobble hypothesis**.

KEY CONCEPTS

- The genetic code is translated through base-pairing interactions between mRNA codons and tRNA anticodons.
- The wobble hypothesis explains why cells usually have fewer tRNAs than expected.

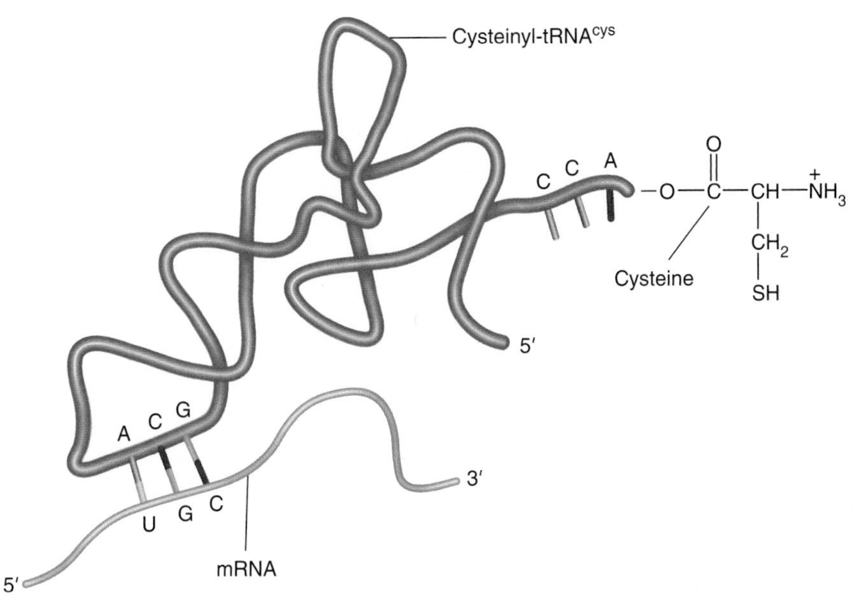

FIGURE 19.1
Codon-Anticodon Base Pairing of Cysteinyl-tRNA^{cys}

The pairing of the codon UGC with the anticodon GCA ensures that the amino acid cysteine will be incorporated into a growing polypeptide chain.

The wobble hypothesis, which allows for multiple codon-anticodon interactions by individual tRNAs, is based principally on the following observations.

1. The first two base pairings in a codon-anticodon interaction confer most of the specificity required during translation since most redundant codons specifying a certain amino acid possess identical nucleotides in the first two positions. These interactions are standard (i.e., Watson-Crick) base pairings.
2. The interactions between the third codon and anticodon nucleotides are less stringent. In fact, nontraditional base pairs (i.e., non-Watson-Crick) often occur. For example, tRNAs containing G in the 5′ (or "wobble") position of the anticodon can pair with two different bases; that is, G can interact with either C or U. The same is true for U, which can interact with A or G (**Figure 19.2a**). When I is in the wobble position of an anticodon, a tRNA can base-pair with three different codons, because I can interact with U or A or C (**Figure 19.2b**).

   A careful examination of the genetic code and the "wobble rules" indicates that all 61 codons can be translated with a minimum of 31 tRNAs. An additional tRNA for initiating protein synthesis brings the total to 32 tRNAs.

QUESTION 19.2

The sequence of a DNA segment is GGTTTA. What is the sequence of the tRNA anticodons?

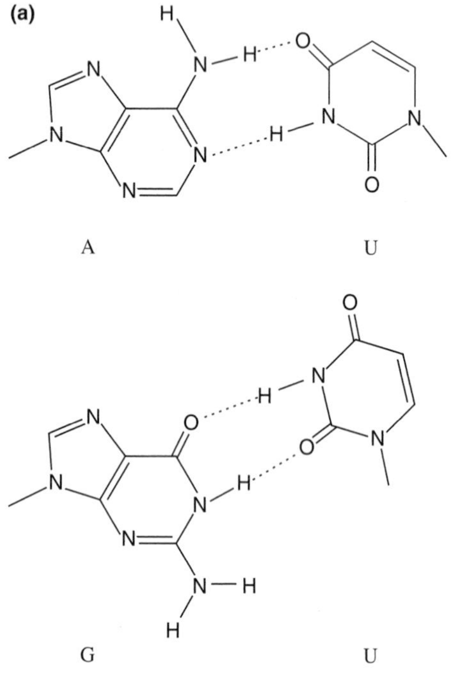

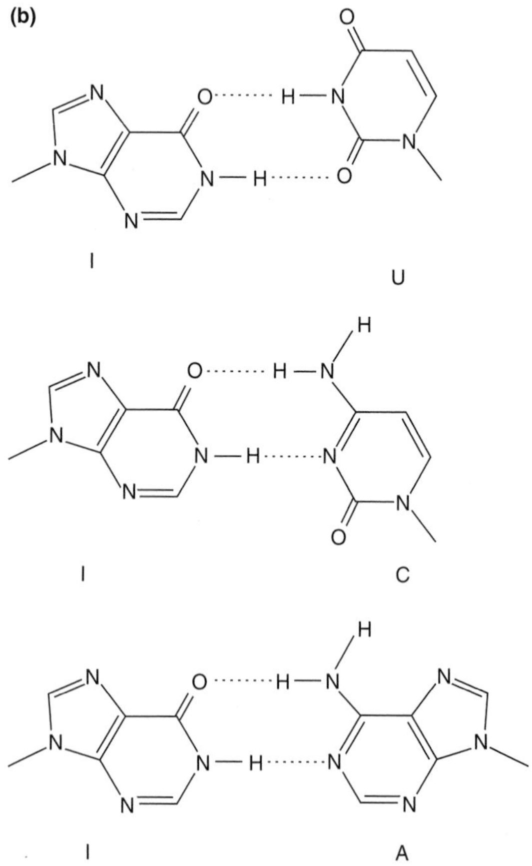

FIGURE 19.2
Wobble Base Pairs

Nonstandard base pairing is critical for the translation of the genetic code. Examples of wobble base pairs include (a) AU and GU base pairs and (b) IU, IC, and IA base pairs.

QUESTION 19.3

The amino acid sequence for a short peptide is Tyr-Leu-Thr-Ala. What are the possible base sequences of the mRNA and the transcribed DNA strand that code for it? What are the anticodons?

The Aminoacyl-tRNA Synthetase Reaction

Although the accuracy of translation (approximately one error per 10^4 amino acids incorporated) is lower than that of DNA replication and transcription, it is remarkably higher than one would expect of such a complex process. The principal reasons for the accuracy with which amino acids are incorporated into polypeptides include codon-anticodon base pairing and the mechanism by which amino acids are attached to their cognate tRNAs. The attachment of amino acids to tRNAs, considered the first step in protein synthesis, is catalyzed by a group of enzymes called the aminoacyl-tRNA synthetases.

In most organisms there is at least one aminoacyl-tRNA synthetase for each of the 20 amino acids. Each enzyme links its specific amino acid to an appropriate tRNA. The process that links an amino acid to the 3′ terminus of the correct tRNA consists of two sequential reactions (**Figure 19.3**), both of which occur within the active site of the synthetase.

1. **Activation**. The synthetase first catalyzes the formation of aminoacyl-AMP. This reaction, which activates the amino acid by forming a high-energy mixed anhydride bond, is driven to completion through the hydrolysis of its other product, pyrophosphate. (An **anhydride** is a molecule containing two carbonyl groups linked through an oxygen atom. The term **mixed anhydride** describes an anhydride formed from two different acids, e.g., a carboxylic acid and phosphoric acid.)
2. **tRNA linkage**. A specific tRNA, also bound in the active site of the synthetase, becomes covalently bound to the aminoacyl group through an ester linkage. (Depending on the synthetase, the ester linkage may be through the 2′-OH or 3′-OH of the ribose moiety of the tRNA's 3′-terminal nucleotide. Subsequently, the aminoacyl group can migrate between the 2′-OH and 3′-OH groups. Only the 3′-aminoacyl esters are used during translation.) Although the aminoacyl ester linkage to the tRNA is lower in energy than the mixed anhydride of aminoacyl AMP, it still possesses sufficient energy to participate in acyl transfer reactions (peptide bond formation).

The sum of the reactions catalyzed by the aminoacyl-tRNA synthetases is as follows:

$$\text{Amino acid} + \text{ATP} + \text{tRNA} \rightarrow \text{aminoacyl-tRNA} + \text{AMP} + \text{PP}_i$$

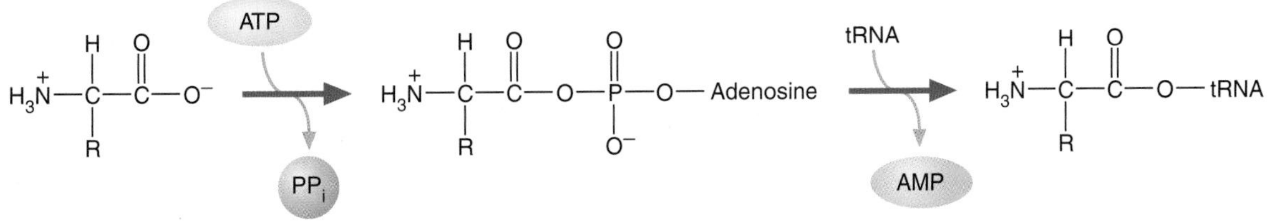

$$H_3\overset{+}{N}{-}CH(R){-}C(=O){-}O^- \xrightarrow[\text{PP}_i]{\text{ATP}} H_3\overset{+}{N}{-}CH(R){-}C(=O){-}O{-}P(=O)(O^-){-}O{-}\text{Adenosine} \xrightarrow[\text{AMP}]{\text{tRNA}} H_3\overset{+}{N}{-}CH(R){-}C(=O){-}O{-}\text{tRNA}$$

FIGURE 19.3
Formation of Aminoacyl-tRNA

Each aminoacyl-tRNA synthetase catalyzes two sequential reactions in which an amino acid is linked to the 3′-terminal ribose residue of the tRNA molecule.

The product PP_i is immediately hydrolyzed with a large loss of free energy. Consequently, tRNA charging is irreversible. Because AMP is a product of this reaction, the metabolic price for the linkage of each amino acid to its tRNA is the equivalent of the hydrolysis of two molecules of ATP to ADP and P_i.

The aminoacyl-tRNA synthetases are a diverse group of enzymes that vary in molecular weight, primary sequence, and number of subunits. Each enzyme efficiently produces a specific aminoacyl-tRNA product relatively accurately. The specificity with which each of the synthetases binds the correct amino acid and its cognate tRNA is crucial for the fidelity of the translation process. Some amino acids can easily be differentiated by their size (e.g., tryptophan vs glycine) or the presence of positive or negative charges in their side chains (e.g., lysine and aspartate). Other amino acids, however, are more difficult to discriminate because their structures are similar. For example, isoleucine and valine differ only by a methylene group. Despite this difficulty, isoleucyl-$tRNA^{ile}$ synthetase usually synthesizes the correct product. However, this enzyme occasionally also produces valyl-$tRNA^{ile}$. Isoleucyl-$tRNA^{ile}$ synthetase, as well as several other synthetases, can correct such a mistake because it possesses a separate *proofreading* site. Because of its size, this site binds valyl-$tRNA^{ile}$ and excludes the larger isoleucyl-$tRNA^{ile}$. After its binding in the proofreading site, the ester bond of valyl-$tRNA^{ile}$ is hydrolyzed.

19.2 PROTEIN SYNTHESIS

An overview of protein synthesis is illustrated in **Figure 19.4**. Despite its complexity and the variations among species, the translation of a genetic message into the primary sequence of a polypeptide can be divided into three phases: initiation, elongation, and termination.

1. **Initiation**. Translation begins with **initiation**, when the small ribosomal subunit binds an mRNA. The anticodon of a specific tRNA, referred to as an *initiator tRNA*, then base-pairs with the initiation codon AUG on the mRNA. Initiation ends as the large ribosomal subunit combines with the small subunit. There are two sites on the complete ribosome for the codon-anticodon interactions involved in translation: the P (peptidyl) site (now occupied by the initiator tRNA) and the A (aminoacyl) site. Bacterial ribosomes also contain an E or exit site. The E site is occupied by an uncharged tRNA before it is released from the ribosome. In both prokaryotes and eukaryotes, mRNAs are read simultaneously by numerous ribosomes. An mRNA with several ribosomes bound to it is referred to as a **polysome**. In actively growing prokaryotes, for example, the ribosomes attached to an mRNA molecule may be separated from each other by as few as 80 nucleotides.
2. **Elongation**. During the **elongation** phase the polypeptide is synthesized according to the specifications of the genetic message. The mRNA base sequence is read in the $5' \rightarrow 3'$ direction, and polypeptide synthesis proceeds from the N-terminal to the C-terminal. The elongation cycle is composed of three steps: codon-anticodon pairing in the A site, peptide bond formation, and the transfer of the peptidyl-tRNA to the P site. Elongation begins as the next aminoacyl-tRNA binds to the A site as the result of codon-anticodon pairing. Peptide bond formation is then catalyzed by peptidyl transferase. In this *transpeptidation* reaction, the α-amino nitrogen of the A site amino acid (the nucleophile) attacks the carbonyl group of the P site amino acid (**Figure 19.5**). As a result of peptide bond formation, the growing peptide chain is now attached to the A site tRNA. Finally, ***translocation*** occurs. As the ribosome moves along the mRNA one triplet length, the peptidyl chain linked to the A site tRNA is shifted to the P site, and the uncharged tRNA in the P site is

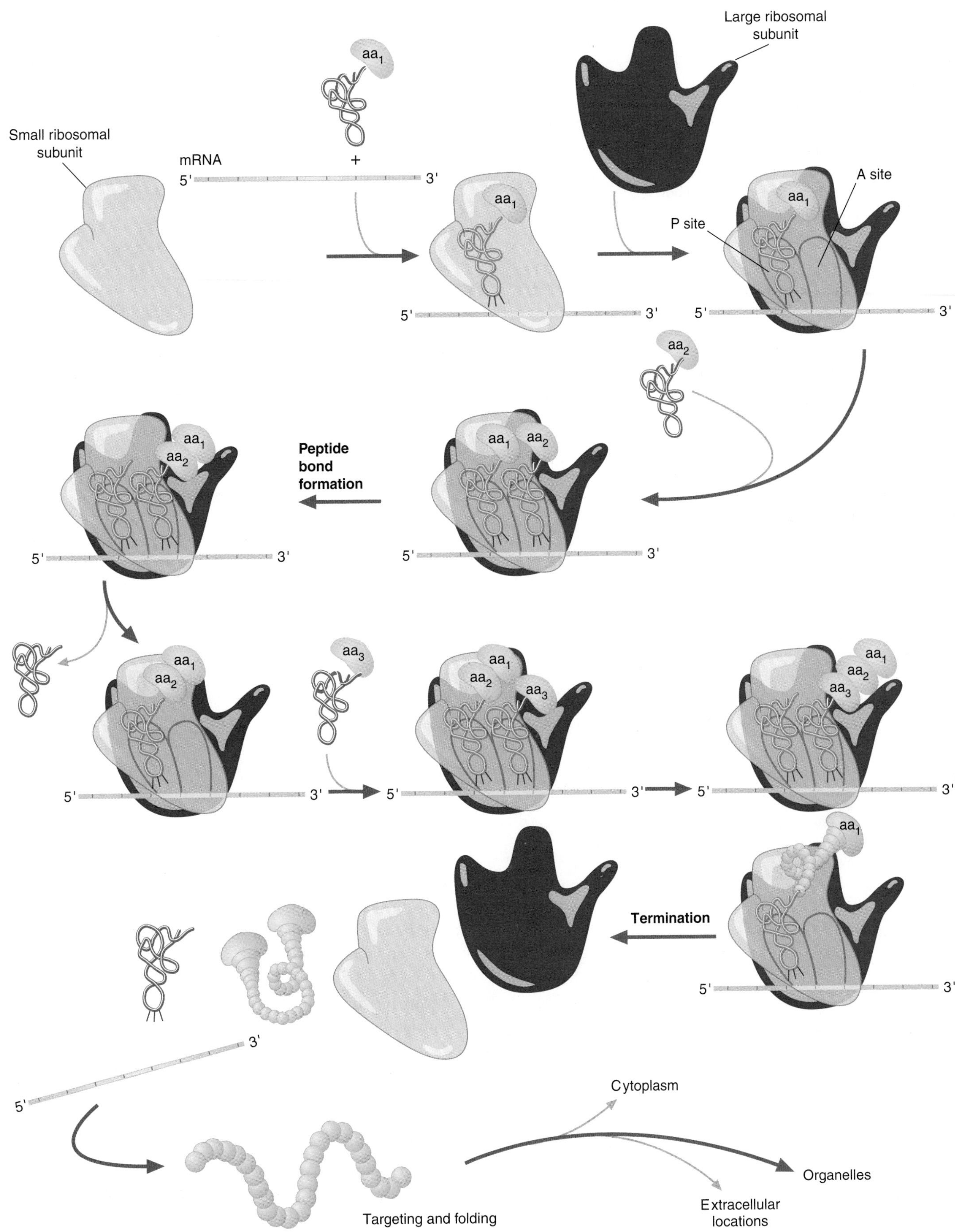

FIGURE 19.4

Protein Synthesis

No matter what the organism, translation consists of three phases: initiation, elongation, and termination. The elongation reactions, which include peptide bond formation and translocation, are repeated many times until a stop codon is reached. The numerous protein factors that facilitate each step in protein synthesis are different in prokaryotes and eukaryotes. Posttranslational reactions and targeting processes vary according to cell type.

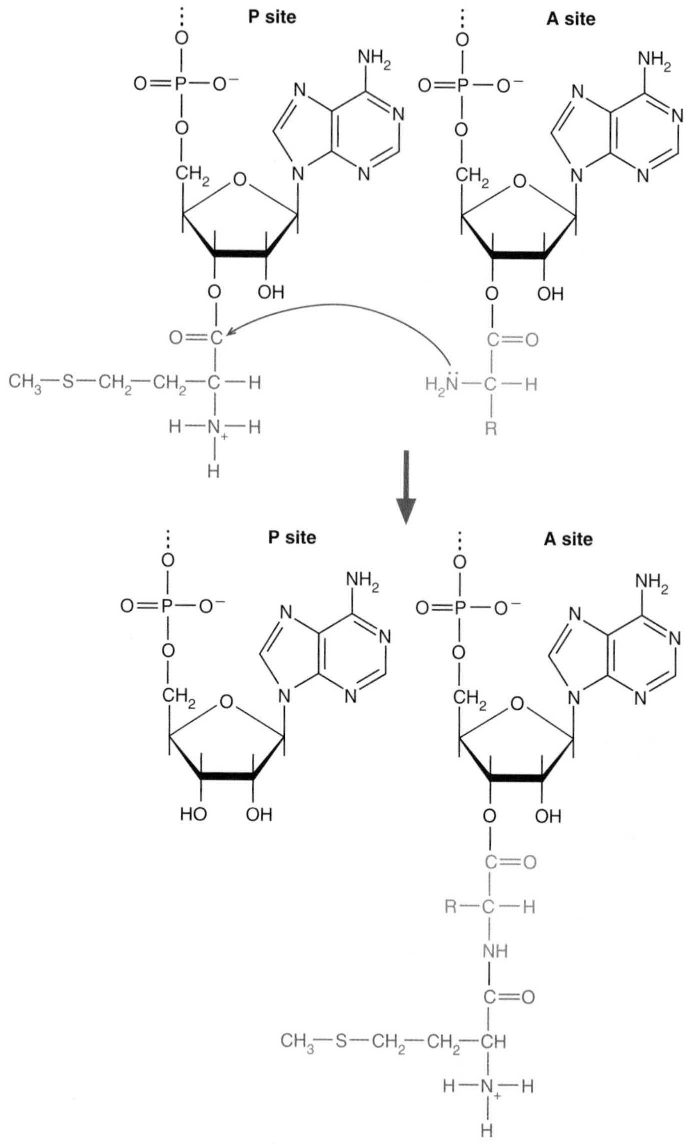

FIGURE 19.5
Peptide Bond Formation

Elongation begins when a peptide bond forms because of the nucleophilic attack of the A-site amino acid's amino group on the carbonyl carbon of the methionine residue linked to the initiating tRNA in the P site. Because a peptide bond has formed, both amino acids are now attached to the A site tRNA.

released from the ribosome (eukaryotes) or shifted to the E or exit site (prokaryotes) where it is subsequently released from the ribosome. As the A site is vacated, the next codon, now positioned in the A site, binds to its cognate tRNA anticodon. This elongation cycle is repeated until a stop codon enters the A site.

3. **Termination**. During **termination** the polypeptide chain is released from the ribosome. Translation terminates because a stop codon cannot bind an aminoacyl-tRNA. Instead, a protein-releasing factor binds to the A site. Subsequently, peptidyl transferase (acting as an esterase) hydrolyzes the bond connecting the now-completed polypeptide chain and the tRNA in the P site. Translation ends as the ribosome releases the mRNA and dissociates into the large and small subunits.

In addition to the ribosomal subunits, mRNA, and aminoacyl-tRNAs, translation requires an energy source (GTP) and a wide variety of protein factors. These factors perform several roles. Some have catalytic functions; others stabilize specific structures that form during translation. Translation factors are classified according to the phase of the translation process they affect, that is, initiation, elongation, or termination. The major differences between prokaryotic and eukaryotic translation appear to be largely a result of the identity and functioning of these protein factors.

Regardless of the species, immediately after translation, some polypeptides fold into their final form without further modifications. Frequently, however, newly synthesized polypeptides are modified. These alterations, referred to as **posttranslational modifications**, can be considered the fourth phase of translation. They may include removal of portions of the polypeptide by proteases, chemical modification of the side chains of certain amino acid residues, and insertion of cofactors. Often, individual polypeptides then combine to form multisubunit proteins.

Posttranslational modifications appear to serve two general purposes: (1) to prepare a polypeptide for its specific function and (2) to direct a polypeptide to a specific location, a process referred to as **targeting**. Targeting is an especially important process in eukaryotes because proteins must be precisely directed to a vast array of possible destinations. In addition to cytoplasm and the plasma membrane (the principal destinations in prokaryotes), eukaryotic proteins may be sent to a variety of organelles (e.g., mitochondria, chloroplasts, lysosomes, or peroxisomes).

Although there are many similarities between prokaryotic and eukaryotic protein synthesis, there are also notable differences. Consequently, the details of prokaryotic and eukaryotic processes are discussed separately. Each discussion is followed by a brief description of mechanisms that control translation.

KEY CONCEPTS

- Translation consists of three phases: initiation, elongation, and termination.
- After synthesis, many proteins are chemically modified and targeted to specific cellular or extracellular locations.

Prokaryotic Protein Synthesis

Protein synthesis in bacteria takes place on 2.4-MDa ribosomes capable of polymerizing amino acids at a rate of approximately 20 per second. The 70S bacterial ribosome is composed of a large 50S subunit and a small 30S subunit (**Figure 19.6**). The large subunit (about 1.5 MDa) consists of 23S and 5S rRNAs and 34 proteins. The small subunit (about 0.8 MDa) contains a 16S rRNA and 21 proteins. In addition to the P, A, and E sites, there are three other functional centers: the decoding center, the peptidyl transferase center, and the GTPase-associated region.

The **decoding center**, located at the A site on the 30S subunit, is where an mRNA codon is matched to an incoming tRNA anticodon. Three highly conserved 16S rRNA bases (A1492, A1493, and G 530) are contiguous with the codon-anticodon base pair triplet. When correct Watson-Crick base pairs have formed between the first two base pairs of the codon-anticodon base pair triplet, a conformational change occurs in A1492 and A1493 that accelerates the tRNA selection phase of the elongation cycle.

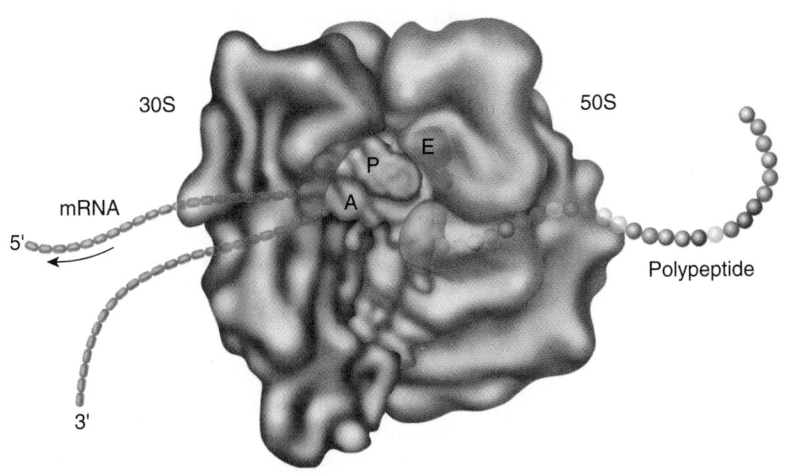

FIGURE 19.6
The Functional Ribosome

In this three-dimensional reconstruction of an *E. coli* ribosome during protein synthesis, the large and small subunits are shown in pink and orange, respectively. The relative positions of the mRNA, tRNAs, and the growing polypeptide chain are also illustrated. The tRNAs are identified as A and P to indicate their positions within the acyl and peptidyl sites where peptide bond formation occurs. The tRNA labeled E is in the exit position; that is, having discharged its amino acid during ongoing protein synthesis, it is in the process of leaving the ribosome. The movement of the mRNA through the ribosome is indicated by an arrow.

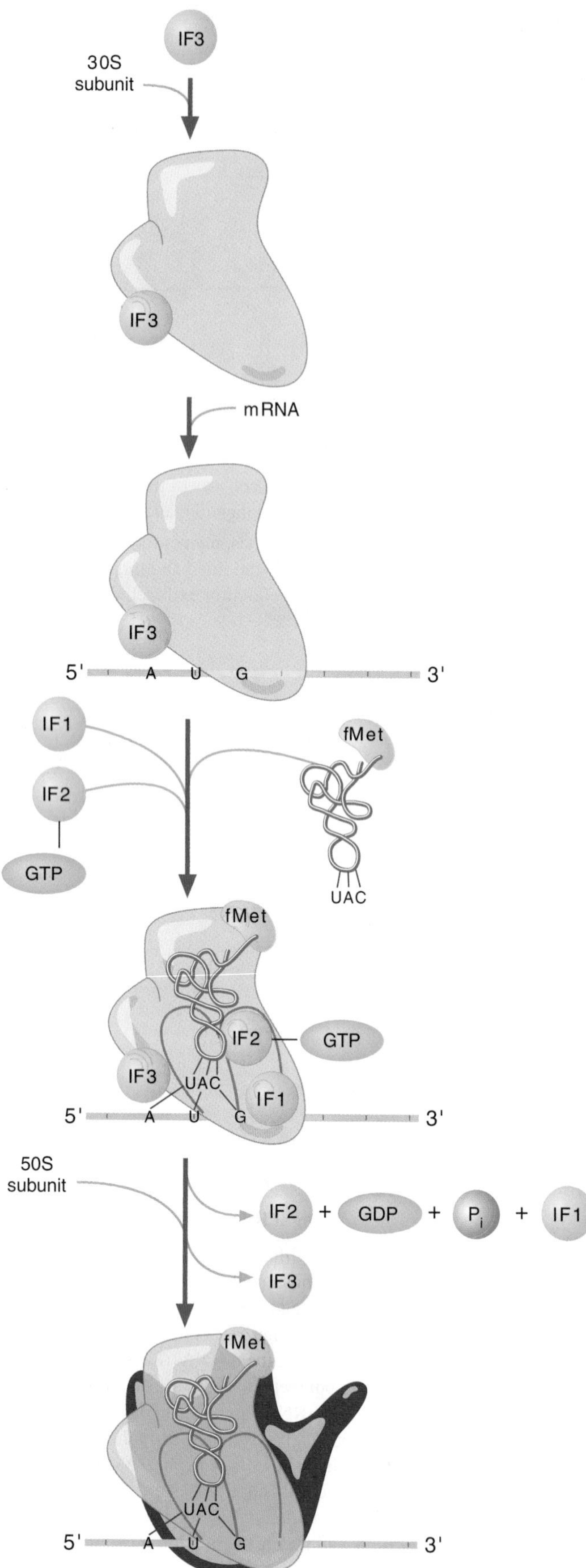

The **peptidyl transferase center** (PTC), where peptide bond formation occurs, is located in a cleft in the large subunit containing a domain of 23S rRNA. The core of the PTC, which consists of five conserved bases (A2451, U2505, U2585, C2452, and A2602), binds to the 3′ ends of the aminoacyl- and peptidyl-tRNAs. Peptide bond formation is the result of a concerted proton shuttling mechanism (pp. 741–742) that is triggered when aminoacyl-tRNA binding occurs in the precisely organized rRNA nucleotides within the PTC active site.

The **GTPase associated region** (GAR) is a set of overlapping binding sites on the 50S subunit composed of 23S rRNA structural elements. GAR acts as a GAP (p. 718). When translation factors with GTPase activity interact with GAR, the resulting GTP hydrolysis drives a conformational change in the protein that affects a translation event. The L12 stalk of the large subunit recruits the GTPase translation factors and facilitates their binding to GAR.

INITIATION Translation begins with the formation of an initiation complex (**Figure 19.7**). In prokaryotes this process requires three initiation factors: IF1, IF2, and IF3. IF3 binding to the 30S subunit prevents that subunit from binding prematurely to the 50S subunit and promotes mRNA binding. IF1 then binds to the A site of the 30S subunit, thereby blocking it during initiation. As an mRNA binds to the 30S subunit, it is guided into a precise location (so that the initiation codon AUG is correctly positioned) by a purine-rich sequence referred to as the **Shine-Dalgarno sequence**. The Shine-Dalgarno sequence, which occurs a short distance upstream from AUG, binds to a complementary sequence, referred to as the anti–Shine-Dalgarno sequence, which is contained in the 16S rRNA component of the 30S subunit. Base pairing between the Shine-Dalgarno sequence and the anti–Shine-Dalgarno sequence provides a mechanism for distinguishing a start codon from an internal methionine codon. It also increases the stability of the ribosome-mRNA complex.

Each gene on a polycistronic mRNA possesses its own Shine-Dalgarno sequence and an initiation codon. The translation of each gene appears to occur independently; that is, translation of the first gene in a polycistronic message may or may not be followed by the translation of subsequent genes.

FIGURE 19.7
Formation of the Prokaryotic Initiation Complex

The initiation phase of translation begins with the binding of the initiation factor IF3 to the 30S subunit. After the mRNA binds to the 30S subunit, the GTPase IF2 binds to fmet-tRNA^{fmet} and subsequently promotes the association of the small and large subunits. GTP hydrolysis triggers the release of the initiation factors, GDP, and P_i. The fully functional 70S ribosome is now ready to enter the elongation phase.

In the next step in initiation, IF2 (a GTPase with a bound GTP) binds to the initiating tRNA and facilitates its entry into the P site. The initiating tRNA in bacteria is a *N*-formylmethionine-tRNA (fmet-$tRNA^{fmet}$), which is synthesized in the following process. After an initiator tRNA ($tRNA^{fmet}$) is charged with methionine, the amino acid residue is formylated by an N^{10}-formyl THF-requiring enzyme. The initiation phase ends as IF2-GTP promotes the joining of the two subunits via its binding to the GAR site of the 50S subunit. The subsequent GAR-initiated GTP hydrolysis causes a conformational change that results in the simultaneous joining of the two subunits and the release of the three initiation factors, GDP, and P_i. The 70S ribosome is now primed for the elongation phase of protein synthesis.

ELONGATION As noted, the elongation cycle consists of three steps, referred to collectively as an elongation cycle: (1) positioning an aminoacyl-tRNA in the A site, (2) peptide bond formation, and (3) translocation.

The prokaryotic elongation process begins when an aminoacyl-tRNA, specified by the next codon, binds to the now-empty A site. Before aminoacyl-tRNAs can enter the A site they must bind the elongation factor EF-Tu-GTP. EF-Tu-GTP is a motor protein that positions its cargo within the A site so that the tRNA anticodon is free to interact with an mRNA codon. EF-Tu-GTP prevents unregulated peptide bond formation and serves to protect the aminoacyl linkage to the tRNA from hydrolysis. Entry of the aminoacyl-tRNA into the A site requires the binding of the EF-Tu-GTP-aa-tRNA complex to the 50S subunit GAR, a process that is assisted by the subunit's L12 stalk.

After the anticodon of the aminoacyl-tRNA has been correctly paired to the mRNA codon, GTP hydrolysis releases EF-Tu-GDP from the ribosome. Then another elongation factor (called EF-Ts), acting as a **guanine nucleotide exchange factor (GEF)**, promotes EF-Tu regeneration by displacing its GDP moiety. EF-Ts is then itself displaced by an incoming GTP molecule (**Figure 19.8**). The newly formed EF-Tu-GTP can then bind to a new aminoacyl-tRNA. (The structural and functional properties of EF-Tu are described in an online Biochemistry in Perspective essay entitled EF-Tu: A Motor Protein.)

After EF-Tu has delivered an aminoacyl-tRNA to the A site, the formation of a peptide bond is catalyzed by the PTC within the 23S rRNA that is located in a cleft in the 50S ribosomal subunit on the side facing the 30S subunit. The mechanism whereby the peptide bond is formed occurs via intrasubstrate proton shuttling (**Figure 19.9**). The ribosome facilitates the reaction in several ways. These include precise positioning of the substrates within the PTC (i.e., the acceptor ends of the A-site and P-site tRNAs become fixed in place via interactions with nucleotide residues of 23S rRNA) and an electrostatic environment that assists the proton shuttle process. The ribosome also reduces the free energy requirements of the reaction by providing a relatively anhydrous environment in the active site that is essential for the formation of a highly polar transition state. The energy required to drive this reaction is provided by the high-energy ester bond linking the P-site amino acid to its tRNA.

Immediately after peptide bond formation, the tRNA in the A site (referred to as the peptidyl tRNA because it is linked to the growing peptide chain) is still in the A site and the now-deacylated tRNA is in the P site. This phase of elongation is referred to as the *pretranslocation state*. Translocation, the shifting of the base-paired tRNAs by one codon position (putting the uncharged tRNA in the E site, the peptidyl-tRNA in the P site, and the new codon in the empty A site), requires the binding of another GTPase, referred to as EF-G. When EF-G-GTP binds near the A site assisted by L12 stalk motion, GTP hydrolysis is triggered. The resulting conformational change causes the two subunits to rotate in opposite directions in relation to each other. The rotational movement, which creates a space between the two subunits, accommodates the

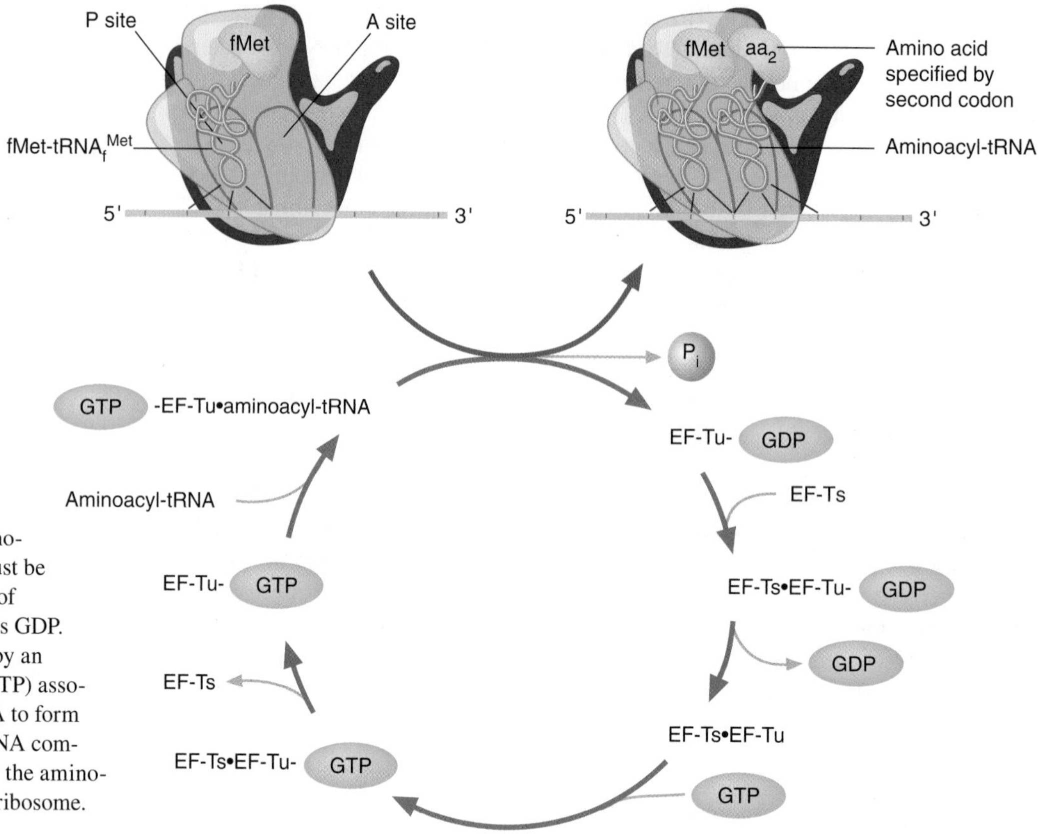

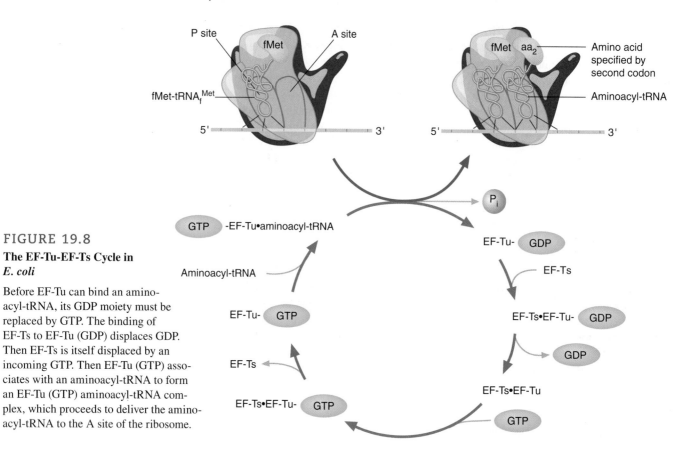

FIGURE 19.8
The EF-Tu-EF-Ts Cycle in *E. coli*

Before EF-Tu can bind an aminoacyl-tRNA, its GDP moiety must be replaced by GTP. The binding of EF-Ts to EF-Tu (GDP) displaces GDP. Then EF-Ts is itself displaced by an incoming GTP. Then EF-Tu (GTP) associates with an aminoacyl-tRNA to form an EF-Tu (GTP) aminoacyl-tRNA complex, which proceeds to deliver the aminoacyl-tRNA to the A site of the ribosome.

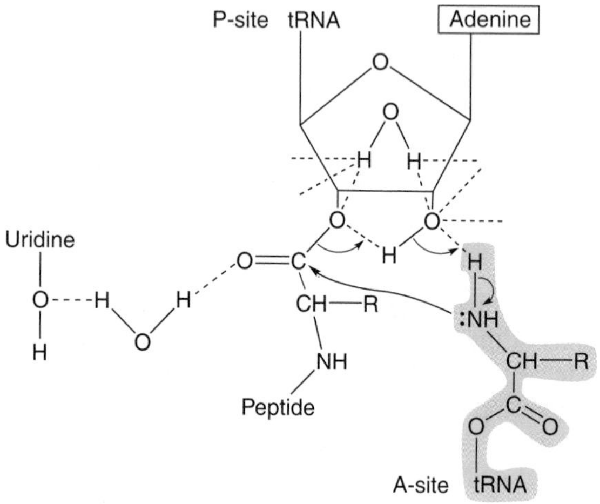

FIGURE 19.9
The Peptidyl Transferase Proton Shuttle Mechanism

The reaction begins with the nucleophilic attack of the α-amino nitrogen on the A-site aminoacyl group carbonyl carbon of the peptidyl chain (linked to P-site tRNA by an ester bond to 3′-OH group of the ribose of residue 76). During the first elongation cycle, a single *N*-formylmethionyl aminoacyl group is linked to the P-site tRNA. A six-membered transition state is thus created in which the 2′-OH group of the A-site ribose donates its proton to the adjacent 3′-oxygen atom as the latter receives an amino proton. The precise alignment of the substrates within the active site is fostered by hydrogen bonds between the ribose oxygens and a bridging water molecule (black dashed lines) and hydrogen bonds between the ribose 2′-hydroxyl oxygen and the bridging water molecule and certain cytosine and adenosine residues (not shown) of 23S rRNA (dashed lines). The carbonyl oxygen of the peptidyl group of the P-site tRNA is stabilized by a hydrogen bond, via a water molecule bridge, to the hydroxyl oxygen of a uridine residue of 23S rRNA.

rachet-like shift of the tRNAs along the mRNA strand. In the *posttranslocation state*, the deacylated tRNA is in the E site. The release of the deacylated tRNA from the E site occurs as the incoming EF-Tu aminoacyl-tRNA complex enters the A site and codon-anticodon interactions are initiated. After the release of EF-G-GDP, the ribosome is ready for the next elongation cycle. As the polypeptide elongates, it passes through a 10- to 20-Å-wide and 100-Å-long exit tunnel in the 50S subunit. Each **nascent** (newly synthesized) polypeptide emerges from the tunnel when it is between 30 and 50 amino acid residues in length, depending on the polypeptide's propensities for forming secondary structures in the 20-Å-wide vestibule near the subunit's surface. Elongation continues until a stop codon enters the A site.

TERMINATION The termination phase begins when a termination codon (UAA, UAG, or UGA) enters the A site. Three **release factors** (RF1, RF2, and RF3) are involved in termination. Both RF1 and RF2 resemble tRNAs in shape and size. RF1 recognizes the stop codons UAA and UAG, and RF2 recognizes UAA and UGA. RF3 is a GTPase that promotes the binding of RF1 and RF2 to the ribosome. The binding of RF3-GDP to the RF1- or RF2-ribosome complex induces RF3 to exchange GDP for GTP. RF binding alters ribosomal function. Hydrolysis of the GTP-bound to RF3 triggers the release of RF1 or RF2. Release factor binding causes a change in the orientation of the decoding center bases (A1492, A1493, and G530), which triggers a conformational change within the PTC that transiently transforms peptidyl transferase into an esterase. The subsequent hydrolysis reaction cleaves the bond linking the completed polypeptide and the P-site tRNA. Following the polypeptide's release from the ribosome, the mRNA and tRNA also dissociate. The termination phase ends when the ribosome dissociates into its constituent subunits in a process that requires **ribosome recycling factor**, a tRNA-shaped protein that binds within the A site. The energy required to separate the two ribosomal subunits is supplied by EF-G-GTP. Then IF3 binds to the small subunit to prevent premature rebinding by the large subunit.

POSTTRANSLATIONAL MODIFICATIONS The folding process begins as each nascent polypeptide emerges from the exit tunnel, where it first encounters a molecular chaperone called trigger factor. *Trigger factor* (TF), a 48 kDa protein, is transiently linked to the ribosome via an attachment between its N-terminal domain and ribosomal protein L23. Trigger factor's C-terminal domain, an elongated, narrow, and flexible flaplike structure, is positioned at the ribosome exit site. TF provides a specialized surface that guides the early steps of the folding process. Downstream chaperones (e.g., DNAK (hsp70) and GroES-GroEL, p. 167) further assist the folding process if needed. As mentioned, most polypeptides also undergo a series of modifying reactions that prepare them for their functional role. Most of the information concerning posttranslational modifications has been obtained through research on eukaryotes. However, prokaryotic polypeptides are known to undergo several types of covalent modifications.

The best-researched examples are proteolytic processing reactions. These include removing the formylmethionine residue and signal peptide sequences. **Signal peptides**, or leader peptides, are short peptide sequences, typically near the amino terminal, that determine a polypeptide's destination. In bacteria, for example, a signal peptide is required to insert a polypeptide into the plasma membrane.

Posttranslational chemical modifications of prokaryotic proteins include methylation, phosphorylation, and covalent linkage to lipid molecules. In *E. coli* chemotaxis is regulated by methylation and phosphorylation of signal transduction proteins. (*Chemotaxis* is the process in which cells alter their movements in response to certain chemicals in their environment.)

Lipoproteins are fairly common in prokaryotes. Blc, a lipoprotein found in the outer membrane of *E. coli*, is a type of lipocalin (a protein that binds hydrophobic ligands) that is produced under stressful conditions such as starvation. Blc, which forms covalent linkages with fatty acids and phospholipids, plays a role in membrane biogenesis and repair.

TRANSLATIONAL CONTROL MECHANISMS Protein synthesis is an exceptionally expensive process, costing four high-energy phosphate bonds per peptide bond (i.e., two bonds expended during tRNA charging and one each during A-site–tRNA binding and translocation). It is perhaps not surprising that enormous quantities of energy are involved. For example, approximately 90% of *E. coli* energy production used in the synthesis of macromolecules may be devoted to the manufacture of proteins. Although the speed and accuracy of translation require a high energy input, the cost would be even higher without

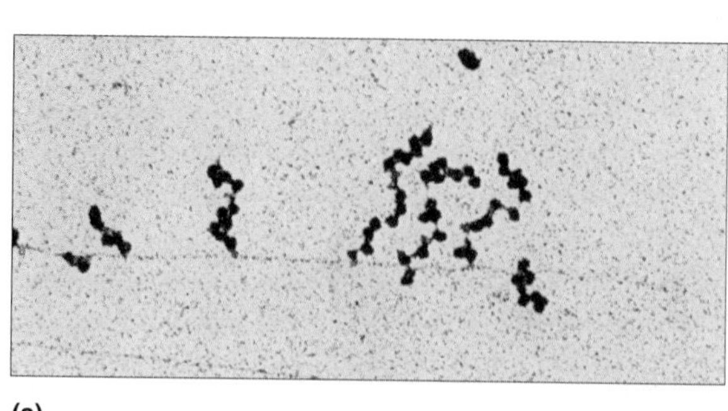

(a)

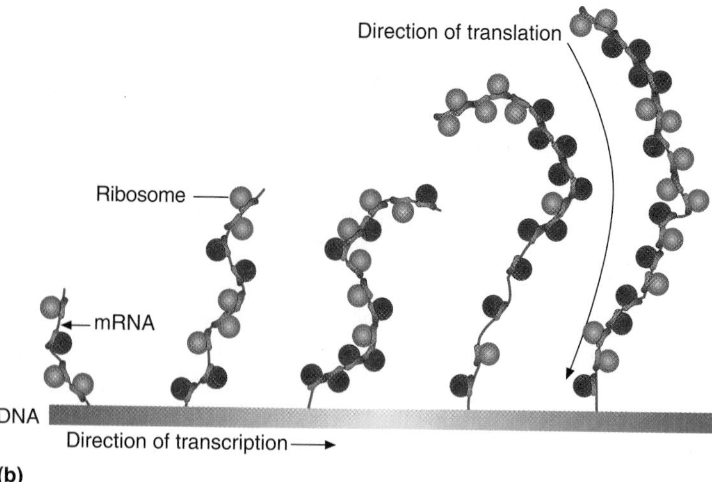

(b)

FIGURE 19.10

Transcription and Translation in *E. coli*

(a) An electron micrograph of *E. coli* transcription and translation. In *E. coli*, as in other prokaryotes, transcription and translation are directly coupled. (b) Diagram of (a). Note polyribosomes.

metabolic control mechanisms. These mechanisms allow prokaryotic cells to compete with each other for limited nutritional resources.

In prokaryotes such as *E. coli*, most of the control of protein synthesis occurs at the level of transcription initiation. (Refer to Section 18.3 for a discussion of the principles of prokaryotic transcriptional control.) This circumstance makes sense for several reasons. First, transcription and translation are spatially and temporally coupled; that is, translation is initiated shortly after transcription begins (**Figure 19.10**). Second, the lifetime of prokaryotic mRNA is usually relatively short. With half-lives of between 1 and 3 minutes, the types of mRNA produced in a cell can be quickly altered as environmental conditions change. Most mRNA molecules in *E. coli* are degraded by two exonucleases, referred to as RNase II and polynucleotide phosphorylase.

Despite the preeminence of transcriptional control mechanisms, the rates of prokaryotic mRNA translation also vary. A large portion of this variation is attributed to differences in Shine-Dalgarno sequences. Because Shine-Dalgarno sequences facilitate the selection of the initiation codon, sequence variations may affect the rate of translating genetic messages. For example, the gene products of the *lac* operon (β-galactosidase, galactose permease, and galactoside transacetylase) are not produced in equal quantities. Thiogalactoside transacetylase is produced at approximately one-fifth the rate of β-galactosidase.

The structural and functional differences between prokaryotic and eukaryotic protein synthesis are the basis of the therapeutic and research uses of antibiotics, antimicrobial molecules used to treat infections. The actions of several antibiotics are listed in Table 19.2.

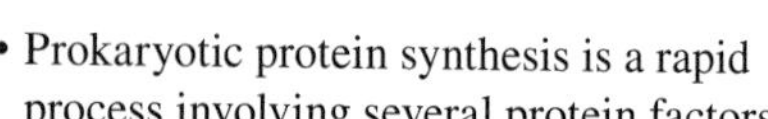

KEY CONCEPTS

- Prokaryotic protein synthesis is a rapid process involving several protein factors.
- Although most prokaryotic gene expression appears to be regulated by transcription initiation, several types of translational regulation have been detected.

TABLE 19.2 Selected Antibiotic Inhibitors of Protein Synthesis

Antibiotic	Action
Chloramphenicol	Blocks prokaryotic A site
Cycloheximide	Inhibits eukaryotic peptidyl transferase
Erythromycin	Blocks prokaryotic exit site
Lincosamide	Binding to 23S rRNA of the 50S subunit
Streptomycin	Blocks binding of fmet-tRNA$_i$ to 30S P site
Streptogramins	Premature release of polypeptide chain
Tigecycline	Binding to 16S rRNA prevents entry of aa-tRNA to A site

Eukaryotic Protein Synthesis

Although eukaryotic protein synthesis resembles bacterial translation, there are distinct differences in the two processes. At 4.3 MDa the eukaryotic ribosome, also known as the 80S ribosome, is larger than the bacterial version. The large 60S ribosomal subunit is composed of 28S, 5S, and 5.8S rRNAs and 47 proteins. The small 40S subunit is composed of an 18S rRNA and 32 proteins. In addition, a significantly larger number of protein factors assist a vastly more sophisticated translation process. The posttranslational modifications of eukaryotic polypeptides are remarkably more numerous than those observed in prokaryotes. Eukaryotic polypeptide targeting mechanisms are also quite intricate.

INITIATION Many of the major differences between the prokaryotic and eukaryotic versions of protein synthesis can be observed during the initiation phase. Among the reasons for the additional complexity of eukaryotic initiation are the following.

1. **mRNA secondary structure**. Eukaryotic mRNA is processed with the addition of a methylguanosine cap (**Figure 18.45**) and a poly(A) tail and the removal of introns. In addition, eukaryotic mRNA does not associate with a ribosome until it has left the nucleus complexed with several proteins.
2. **mRNA scanning**. In contrast to prokaryotic mRNA, eukaryotic molecules lack Shine-Dalgarno sequences, which allow for the identification of the initiating AUG sequence. Instead, eukaryotic ribosomes "scan" each mRNA. This scanning is a complex process in which ribosomes bind to the capped 5′ end of the molecule and migrate in a 5′ $\rightarrow$ 3′ direction searching for a translation start site.

Eukaryotes use a more complex spectrum of initiation factors than prokaryotes. There are at least 12 eukaryotic initiating factors (eIFs), several of which possess numerous subunits. The functional roles of most of these factors are still under investigation.

Eukaryotic initiation begins with the assembly of the preinitiation complex (PIC) (**Figure 19.11**), composed of the small (40S) subunit and several eIFs. The small subunit is prevented from binding to the large (60S) subunit during this phase because it is associated with eIF3 and the 60S subunit is bound to eIF6. The process begins when the small subunit binds to eIF1 and eIF1A, a molecule with the same function as the bacterial factor IF; that is, it blocks the A site during initiation. eIF3 and eIF5 also bind to the 40S subunit. eIF3 is a multisubunit protein complex that facilitates the mRNA-binding process in addition to preventing premature binding to the large subunit. eIF5 is a GAP (p. 718) that is specific for the GTPase eIF2. Once these proteins have bound to the small subunit, the initiating methionyl-tRNA (met-$tRNA_i$), complexed with eIF2-GTP, binds at the P site. The **43S preinitiation complex**—composed of the 40S subunit, eIF1A, eIF2-GTP, eIF3, eIF5, and met-$tRNA_i$—is now ready to bind an mRNA. Most mRNAs cannot perform this step until they have bound to a cap-binding complex.

The **cap-binding complex** (CBC), also referred to as eIF4F, consists of eIF4E (cap-binding protein), eIF4A (a helicase), and eIF4G (a scaffold protein). eIF4E is the rate-limiting factor in eukaryotic protein synthesis. Some cellular proteins and viruses can evade the necessity of using eIF4E, a protein regulated by mTOR (p. 593) and MAPK (pp. 718–719) signal transduction pathways. The mRNAs for proteins such as the hsps and amyloid precursor protein and viruses such as poliovirus possess internal ribosome entry site sequences (IRES) that allow eIF4E-independent initiation.

CBC binds to the mRNA cap region after an ATP-dependent helicase (eIF4A), assisted by eIF4B, removes any secondary structures in the 5′-UTR that can interfere with the mRNA scanning process. The 3′-poly(A) tail of the mRNA is

Biochemistry IN PERSPECTIVE

Trapped Ribosomes: RNA to the Rescue!

How are ribosomes bound to damaged mRNAs retrieved so they can be recycled? What happens when a ribosome translates an mRNA that is either truncated or missing a stop codon? The answer is that translation stalls and ribosomes attempting to translate such defective mRNAs become trapped. They cannot move the process forward nor can they eject the mRNA. Because of the critical importance of efficient protein synthesis, bacteria solve this seemingly intractable problem with a unique 10S RNA molecule called tmRNA. tmRNA is so named because it possesses a tRNA-like domain (TLD) and an mRNA-like domain (MLD). The TLD, formed by the 5′ and 3′ ends of tmRNA, mimics the acceptor stem, T-stem, variable stem, and 3′CCA end of an alanyl-tRNA. The MLD contains an ORF that specifies the amino acid sequence (about 10 residues long) of the peptide tag that the ribosome will eventually add to the stalled polypeptide.

Ribosome rescue (**Figure 19A**) begins with the charging of the TLD with an alanine, catalyzed by alanyl-tRNA synthetase. The alanyl-charged TLD then binds to EF-Tu and SmpB, a small protein that mimics the anticodon arm of a tRNA. The TLD of

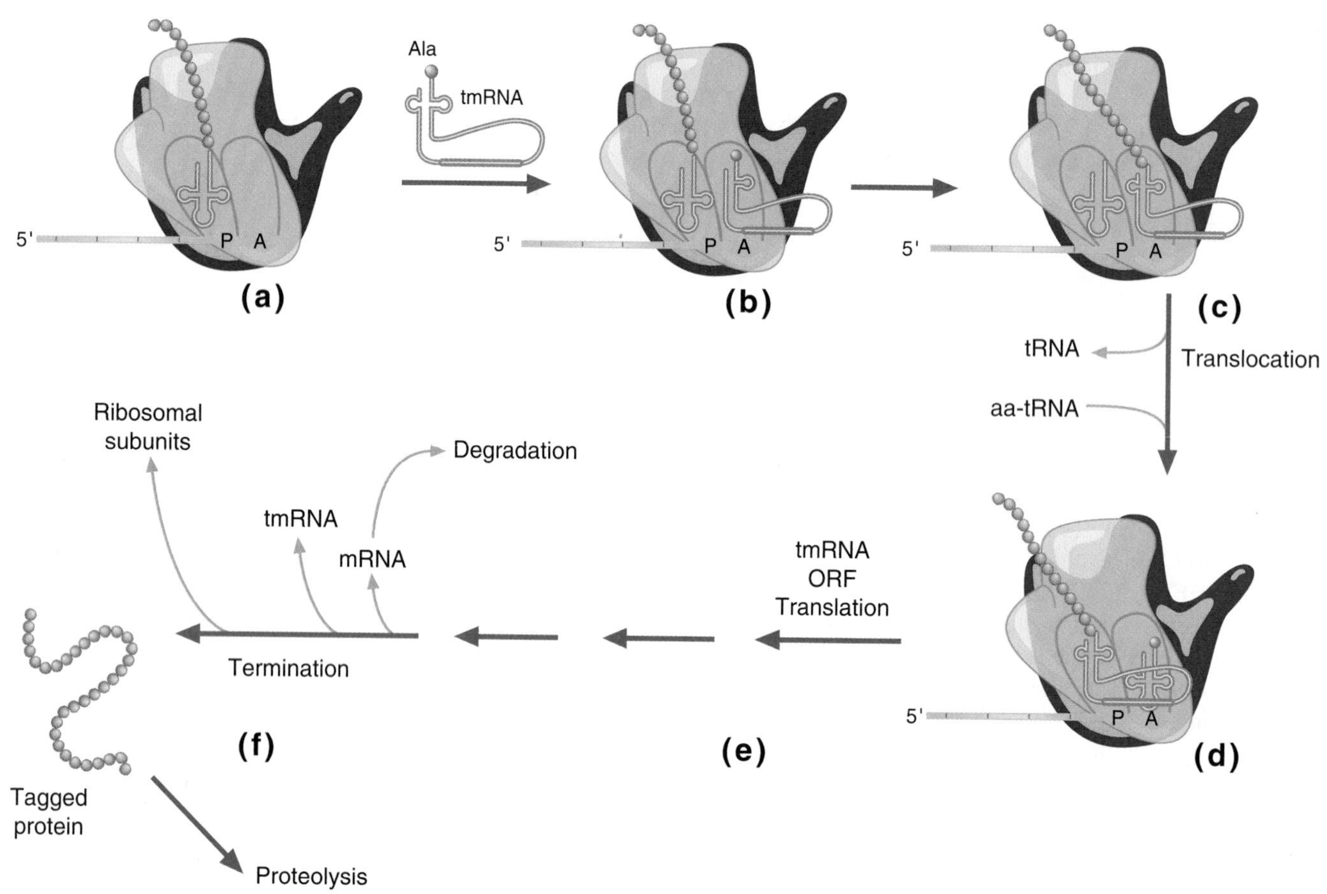

FIGURE 19A

Trapped Ribosome Rescue

(a) A tmRNA linked to an alanine residue binds via codon-anticodon interactions in the A site of the stalled ribosome. (b) The ribosome-catalyzed addition of the alanine to the C-terminus of the nascent polypeptide is followed by translocation (c). (d) The MLD of the tmRNA moves into position so that the first codon in its ORF is in the A site. (e) The MLD ORF is then translated by the ribosome until the stop codon is reached. (f) Release factor-mediated termination releases the tmRNA and mRNA and causes ribosomal subunit separation. The defective mRNA and the polypeptide are then degraded.

Biochemistry IN PERSPECTIVE cont.

the tmRNA-protein complex then binds at the A site, thus positioning its bound alanine within the PTC. The alanyl residue is then covalently bound to the nascent polypeptide. The nearby MLD is poised for the ribosome to switch from the damaged mRNA to its ORF base sequence. Although this transition is structurally awkward, translation soon resumes and continues until the MLD stop codon (UAA) is reached. Soon after its release from the ribosome, the damaged mRNA is degraded in a process that is facilitated by tmRNA. The newly released polypeptide is also degraded, having been targeted for destruction by protease binding sites within the newly synthesized peptide tag.

Eukaryotes do not have a ribosome rescue operation comparable to the tm-tRNA system. When eukaryotic mRNAs lack a stop codon, the ribosome translates the poly(A) tail, one AAA codon at a time, until the 3′ end is reached, thereby producing a C-terminal polylysine segment. An mRNA surveillance system called *nonstop-mediated mRNA decay* detects and rescues the stalled ribosome. A protein called Ski7 binds to the vacant A site, thereby causing the ribosome to release the mRNA and the polypeptide. The defective mRNA is then destroyed. The polypeptide is degraded because the polylysine peptide tag on the C-terminal end of the polypeptide is a target for proteases.

SUMMARY: Living organisms, faced with the high metabolic cost of protein synthesis, have evolved the means of ensuring the efficiency of the process. Both prokaryotes and eukaryotes have methods for rescuing ribosomes trapped by association with damaged mRNAs.

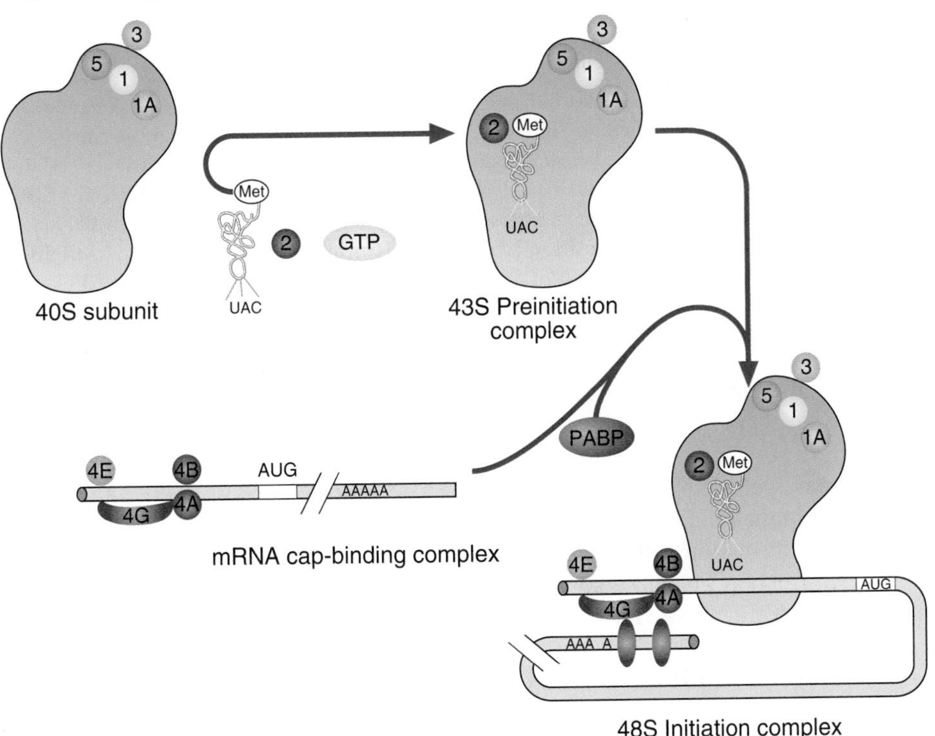

FIGURE 19.11

Eukaryotic Initiation: Assembly of the 43S Preinitiation Complex and the 48S Initiation Complex

In preparation for its role in polypeptide synthesis, the 40S ribosomal subunit, previously bound to eIF1, eIF1A, eIF3, and eIF5, now binds a complex of eIF2-GTP and the initiation tRNA, met-tRNA$_i$. Once the 40S subunit has bound to the met-tRNA$_i$-eIF2-GTP complex, it is referred to as the 43S preinitiation complex. Once eIF4B, a helicase, has removed secondary structure from the 5′-UTR of the mRNA, the cap-binding complex, composed of eIF4A, eIF4B, eIF4E, and eIF4G, binds to the 5′-cap structure. The 48S initiation complex forms as the 3′-poly(A) tail of the mRNA is linked to the 5-capped end by interactions between eIF4G and multiple copies of PABP. Note that the eukaryotic initiation process involves at least 30 proteins. Only the most important are mentioned in the text and illustrated in this figure.

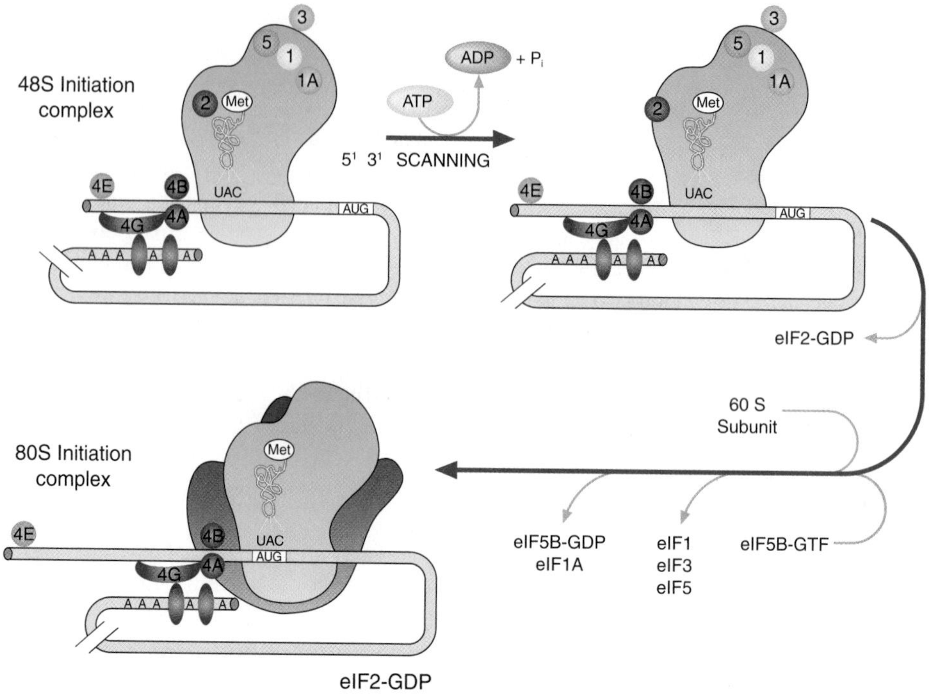

FIGURE 19.12

Eukaryotic Initiation: Assembly of the 80S Initiation Complex

The newly formed 48S complex moves in an ATP-requiring scanning process along the mRNA in the 5′ ⟶ 3′ direction in search for the start codon. Once correct base pairing between the AUG codon and the anticodon of met-tRNA$_i$ has occurred eIF5 triggers GTP hydrolysis and the release of eIF2-GDP. The subsequent binding of eIF5B-GTP to the CBC-mRNA complex facilitates the joining of the complex with the 60S subunit, an event that involves the displacement of eIF1, eIF3, and eIF5. Hydrolysis of the GTP linked to eIF5B causes the dissociation of eIF5B-GDP and eIF1A from the elongation competent 80S ribosome.

then brought into close proximity to the 5′capped end by interactions between eIF4G and multiple copies of **poly(A)-binding protein (PABP)** to form a circular mRNA. The completed **48S initiation complex** proceeds to scan the mRNA in search of the 5′-AUG-3′ near the 5′ end (**Figure 19.12**). Both eIF1 and eIF1A assist in the scanning process. When the initiation codon is reached, a change in the conformation of the scanning complex causes it to lock on to the mRNA. The conformation change also induces the eIF5-mediated hydrolysis of the GTP bound to eIF2. Once the GTP is hydrolyzed, eIF2-GTP is regenerated by eIF2B, a GEF (p. 589). The subsequent binding of eIF5B, a ribosome dependent GTPase (homologous to IF2), promotes the joining of the 60S subunit (now devoid of eIF6) to the 48S initiation complex and displacement of eIF2-GDP, eIF3, and eIF5. Hydrolysis of the GTP bound to eIF5B allows the dissociation of eIF5B-GDP and eIF1A from the active 80S ribosome.

ELONGATION **Figure 19.13** illustrates the eukaryotic elongation cycle. Several elongation factors (eEFs) are required during this phase of translation. *eEF1α*, a 50-kDa polypeptide, is the eukaryotic equivalent of EF-Tu, that is, it is a GTPase that binds aminoacyl-tRNAs and delivers them to the A site. If correct codon-anticodon pairing occurs, eEF1α hydrolyzes its bound GTP and exits the ribosome, leaving its aminoacyl-tRNA behind. If correct pairing does not occur, the complex leaves the A site, thereby preventing incorrect amino acid residues from being incorporated.

During the next elongation step (peptide bond formation), the peptidyl transferase activity of the large ribosomal subunit catalyzes the nucleophilic attack of the

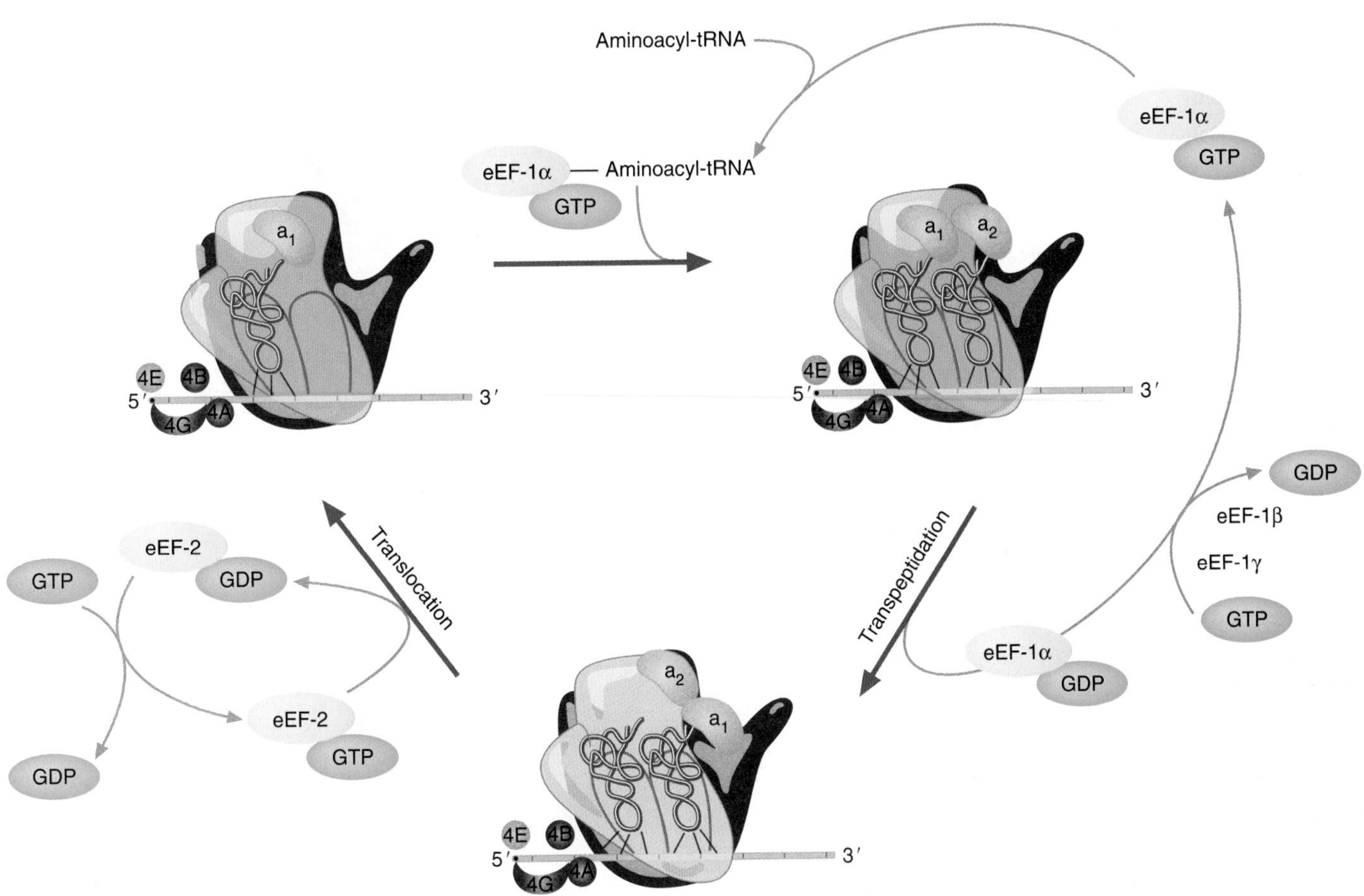

FIGURE 19.13

The Elongation Cycle in Eukaryotic Translation

Elongation comprises three phases: binding of an aminoacyl-tRNA to the A site, transpeptidation, and translocation.

A-site α-amino group on the carboxyl carbon of the P-site amino acid residue. eEF1α-GDP dissociates from the ribosome immediately before transpeptidation. eEF1β and eEF1γ are GEFs that mediate the regeneration of eEF-1α-GTP by promoting an exchange of GDP for GTP.

Translocation in eukaryotes requires a 100-kDa polypeptide referred to as *eEF2*, which is also a GTPase. During translocation, eEF2-GTP binds to the ribosome. GTP is then hydrolyzed to GDP, and eEF2-GDP is released. As noted, GTP hydrolysis provides the energy needed to physically move the ribosome along the mRNA. At the end of translocation, a new codon is exposed in the A site.

TERMINATION In eukaryotic cells, two release factors mediate the termination process: eRF1 (a molecule that resembles the size and overall shape of a tRNA and recognizes and binds to stop codons) and eRF3 (a GTPase). When a stop codon (UAG, UGA, or UAA) enters the active site, eRF1 binds to it (**Figure 19.14**). As a result of this binding process, peptidyl transferase catalyzes the hydrolysis of the ester linkage between the polypeptide and the P-site tRNA. The hydrolysis of the GTP bound to eRF3 is believed to then trigger the dissociation of eRF1 from the ribosome. In eukaryotes, the dissociation of ribosomal subunits has been attributed to eIF3, eIF1, and eIF1A.

The efficiency of eukaryotic translation (i.e., the number of polypeptides that can be synthesized per unit of time) is made possible to a large extent by the circular conformation of eukaryotic polysomes (**Figure 19.15**), As soon as ribosomal subunits and their associated protein factors are released, they are optimally positioned for recruitment into new ribosomes.

KEY CONCEPTS

- Eukaryotic protein synthesis, like its prokaryotic counterpart, has three phases: initiation, elongation, and termination.
- Features that are unique to eukaryotic translation include an abundance of protein factors that facilitate each step, cap-binding protein, and the formation of circular mRNA polysomes.

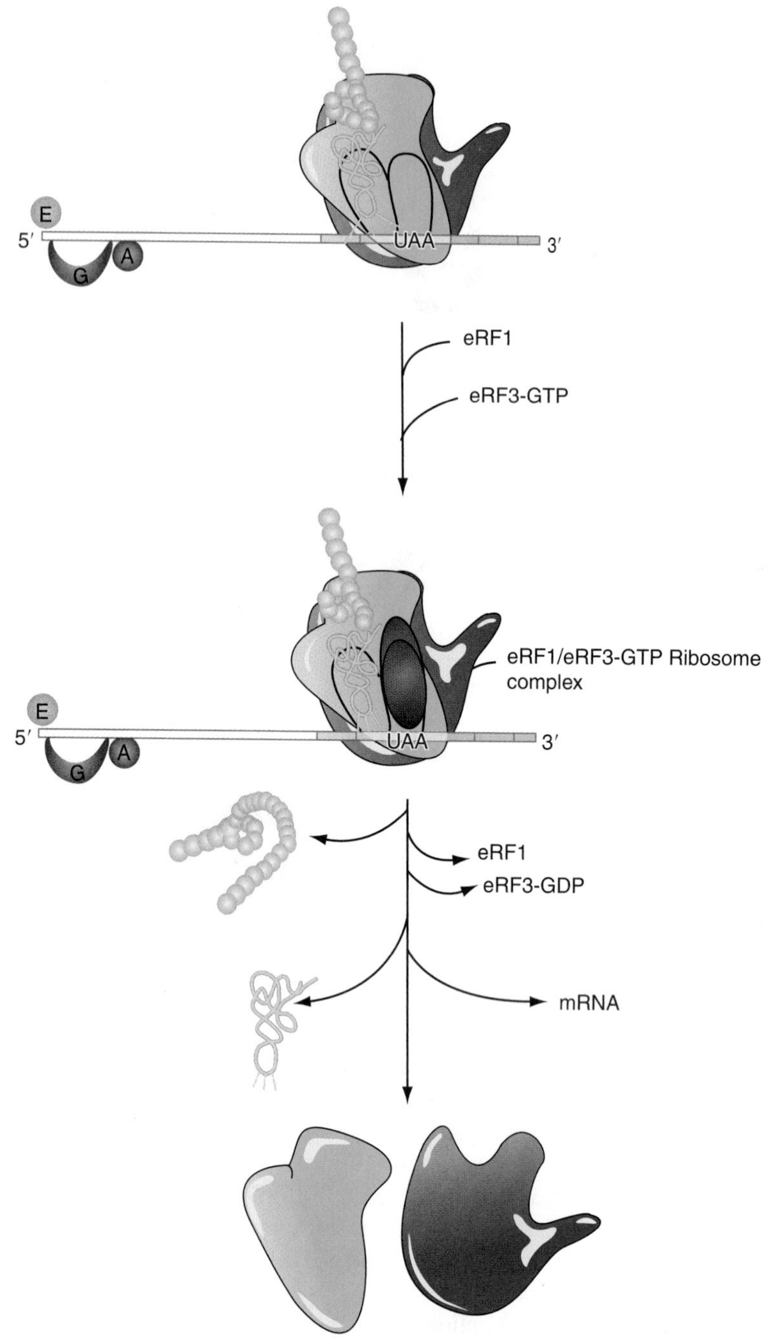

FIGURE 19.14
Eukaryotic Protein Synthesis Termination

As a stop codon (UAG, UGA, or UAA) enters the A site, release factor eRF1 recognizes the stop codon and binds to the A site. eRF1 in association with eRF3-GTP promotes the conversion of the peptidyl transferase to a hydrolase that catalyzes the hydrolysis of the ester bond that links the now complete polypeptide chain to the P-site tRNA. The release of the polypeptide is followed by the dissociation of eRF1 and eRF3-GDP from the ribosome. The dissociation of the ribosomal subunits is mediated principally by eIF3, eIF1, and eIF1A.

QUESTION 19.4

Diphtheria is a highly contagious and potentially fatal respiratory tract illness. A throat lesion called a *pseudomembrane*, which is formed from bacterial cells and damaged throat epithelial cells, causes suffocation. Once considered a serious threat to children (e.g., the 1735–1740 epidemic in colonial New England killed a substantial proportion of children under the age of 16), diphtheria is now completely preventable because of a highly effective vaccine. Diphtheria is caused by an exotoxin-producing pathogenic strain of the bacterium *Corynebacterium diphtheriae*. The exotoxin is a protein that is coded for by a genetic element introduced into the bacterial cell by a bacteriophage. After the diphtheria exotoxin is released by the bacterium, it kills host cells by forming diphthamide, a specific ADP-ribosylated histidine residue in eEF2.

ADP–O–CH_2 (ribose, OH OH) nicotinamide ring (C=O, NH_2, N^+) + eEF-2 $\xrightarrow{\text{Toxin}}$ ADP–O–CH_2 (ribose, OH OH)–eEF-2 + nicotinamide ring (C=O, NH_2, N)

NAD$^+$ Nicotinamide

Cells die because they cannot synthesize proteins. The mechanism by which eEF2 function is affected by ADP-ribosylation is unknown. Can you suggest any possibilities?

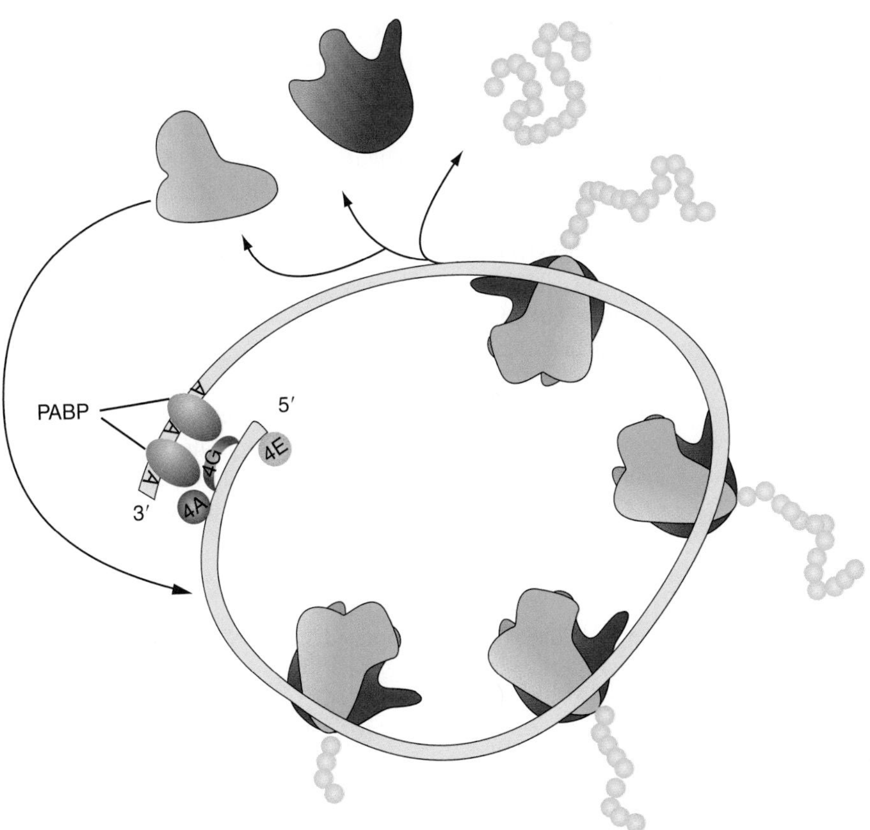

FIGURE 19.15

The Eukaryotic mRNA Polysome

Eukaryotic mRNA is circularized via an interaction between the 5′ UTR and the 3′-poly(A) tail that is mediated by PABP, the poly(A)-binding protein that links the poly(A) sequence to the CBC. As a result of this structural feature, when a polypeptide's synthesis is completed and the ribosomal subunits are released, the close proximity of these subunits to the 5′ cap facilitates immediate recruitment for another round of protein synthesis.

Biochemistry IN PERSPECTIVE

Context-Dependent Coding Reassignment

How are selenocysteine and pyrrolysine, two nonstandard amino acids, incorporated into polypeptides during protein synthesis? There are two variations of the genetic code in which nonstandard amino acids are incorporated into polypeptides during protein synthesis. Selenocysteine and pyrrolysine (**Figure 19B**), now referred to as the 21st and 22nd amino acids in proteins, are coded for by codons that are ordinarily stop signals. The indirect mechanisms used, called *context-dependent codon reassignment*, for both amino acids are described.

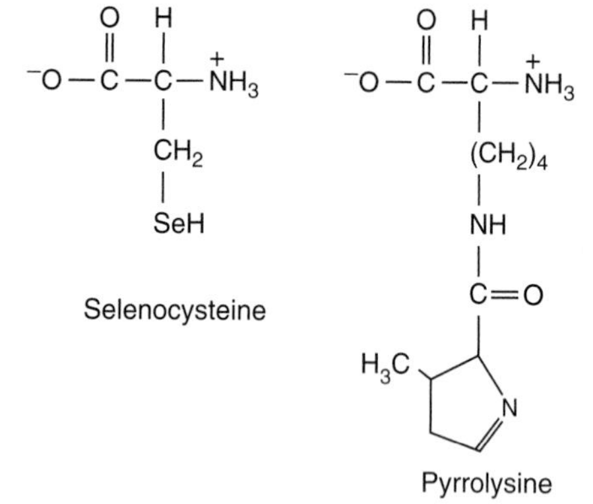

Selenocysteine

Pyrrolysine

FIGURE 19B
The 21st and 22nd Amino Acids

Selenocysteine

The incorporation of selenocysteine into selenoproteins (e.g., glutathione peroxidase and thioredoxin reductase) is widespread and has been observed in Eubacteria, Archaea, and Eukarya. The following discussion is limited to mammals.

Selenocysteine (sec) codon reassignment involves several molecules: a specific tRNA ($tRNA^{[ser]sec}$), a seryl-tRNA synthetase with sec-generating ability, an mRNA-binding protein (SPB2), and a specialized elongation factor (EFsec). The seryl-tRNA synthetase binds $tRNA^{sec}$, and serine is attached to its acceptor arm. The seryl moiety is then converted to selenocysteine by a pyridoxal phosphate–containing enzyme called selenocysteine synthase to form sec-$tRNA^{sec}$. The coding of selenocysteine requires a **SECIS (*s*elenocysteine *i*nsertion *se*quence) element** in the 3′ UTR of the mRNAs for all selenoproteins. The SECIS element rearranges into a loop-stem-bubble structure that recruits SBP2 to form a SECIS/SBP2 complex (**Figure 19C**). An AGU/AG sequence in the stem forms a nucleotide quartet that is the primary conserved sequence within the SECIS. One of the suggested functions of the sequence is stop codon suppression. Once formed, the SECIS/SBP2 complex binds to EFsec, which has previously bound sec $tRNA^{sec}$. When the ribosomal A site becomes vacant (i.e., the UGA codon has moved into position), the SECIS element complex donates the sec $tRNA^{sec}$, after which a peptide bond forms. The selenocysteine residue is now incorporated into the polypeptide.

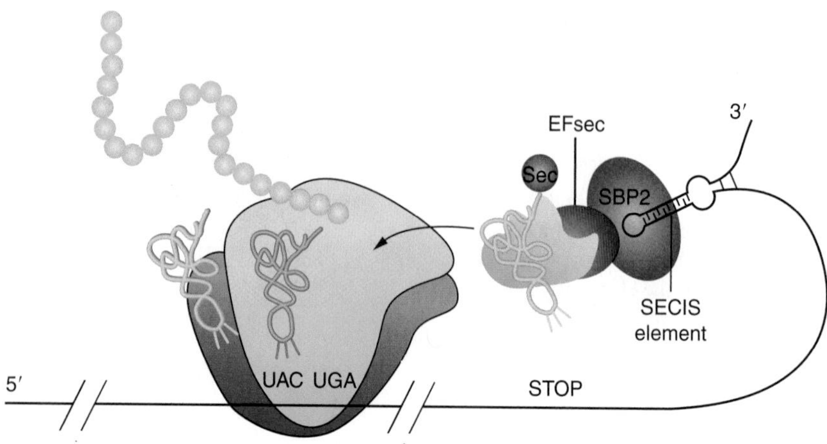

FIGURE 19C
Mechanism of Selenocysteine Incorporation into Eukaryotic Selenoproteins

The stop codon UGA is reassigned for selenocysteine incorporation by the interaction of a complex composed of a SECIS element bound to SBP2 (red) and EFsec (blue and gray) that is charged with sec-$tRNA^{sec}$ with the ribosome. The sec-$tRNA^{sec}$ complex is shown approaching the A site of the 80S ribosome. Once the sec-$tRNA^{sec}$ has been donated to the A site, a peptide bond forms between it and the nascent polypeptide.

Biochemistry IN PERSPECTIVE cont.

Pyrrolysine

Pyrrolysine is found in methyltransferases used by some methane-producing Archaea. A naturally occurring dipeptide composed of lysine linked via an ε-N-amide bond to 4-methylpyrroline-5-carboxylate, pyrrolysine is coded for by the stop codon UAG. The context-dependent coding of pyrrolysine involves a $tRNA^{Pyl}$ with a CUA anticodon and a tRNA synthetase that specifically binds pyrrolysine to its tRNA. In contrast to selenocysteine, pyrrolysine is synthesized before it is linked to a tRNA molecule. The precise mechanism whereby pyrrolysine is incorporated into protein is unknown. It has been noted that UAG codons are used significantly less often than other stop codons. A stem-loop PYLIS (*pyrrolysine insertion sequence*) element that is downstream of the UAG codon is believed to promote pyrrolysine insertion into methyltransferase polypeptides.

SUMMARY: In context-dependent codon reassignment, a specific tRNA, a tRNA synthetase, and other molecules are used to transform a stop codon into one that codes for the incorporation of a nonstandard amino acid.

POSTTRANSLATIONAL MODIFICATIONS IN EUKARYOTES Most nascent polypeptides undergo one or more types of covalent modification. These alterations, which may occur either during ongoing polypeptide synthesis or afterward, consist of reactions that modify the side chains of specific amino acid residues or break specific bonds. In general, posttranslational modifications prepare each molecule for its functional role and/or for folding into its native (i.e., biologically active) conformation. More than 200 different types of posttranslational processing reaction have been identified. Most of them occur in one of the following classes.

Proteolytic Cleavage The proteolytic processing of proteins is a common regulatory mechanism in eukaryotic cells. Typical examples of proteolytic cleavage (hydrolysis by proteases) include removal of the N-terminal methionine and signal peptides (see p. 757). Proteolytic cleavage is also used to convert inactive precursor proteins, called **proproteins**, to their active forms. Recall, for example, that certain enzymes, referred to as proenzymes or zymogens, are transformed into their active forms by cleavage of specific peptide bonds. The proteolytic processing of insulin (**Figure 19.16**) provides a well-researched example of the conversion of a polypeptide hormone into its active form. The inactive insulin precursor produced by removing the signal peptide is referred to as proinsulin. Inactive precursor proteins with removable signal peptides are called **preproproteins**. The insulin precursor containing a signal peptide is referred to as preproinsulin.

Glycosylation A wide variety of eukaryotic proteins are glycosylated for structural and informational purposes (p. 262). Glycosylation reactions begin in the ER where N-linked oligosaccharides are synthesized in association with phosphorylated dolichol (**Figure 19.17**). The product of this process, $Glc_3Man_9GlcNAc_2$, is transferred to a nascent polypeptide at an asparagine residue in either of the tripeptide sequences Asn-X-Ser or Asn-X-Thr by oligosaccharyltransferase. (X is any amino acid residue except proline.) Two of the three terminal glucoses are then removed, generating $Glc_1Man_9GlcNAc_2$, the binding site for the Ca^{2+}-requiring molecular chaperones calnexin and calreticulin. Calnexin (a membrane-bound protein) and calreticulin (a lumenal protein) are lectin-like molecules that promote protein folding. (Lectins,

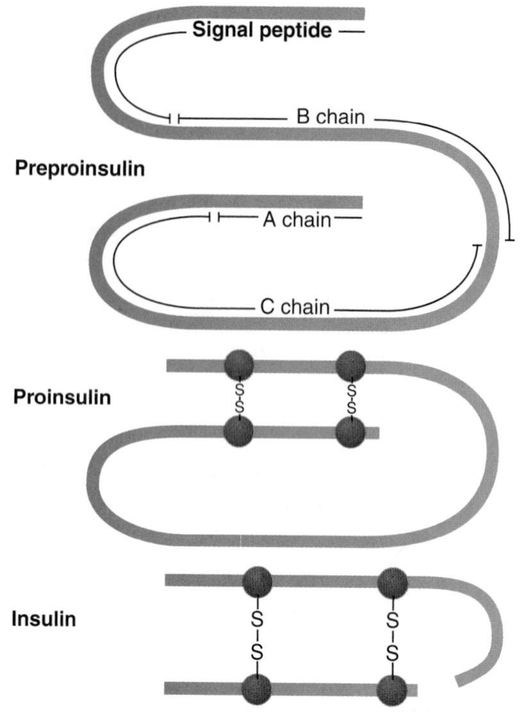

FIGURE 19.16
Proteolytic Processing of Insulin

After the removal of the signal peptide, the peptide segment referred to as the C chain is removed by a specific proteolytic enzyme. Two disulfide bonds are also formed during insulin's posttranslational processing.

described on p. 263, are carbohydrate-binding proteins.) When the glycoprotein is properly folded, the terminal glucose residue is removed, which causes release by the molecular chaperones. The now properly folded glycoprotein is then transferred to the Golgi apparatus where the *N*-oligosaccharide is modified further to yield a high mannose, complex, or hybrid product (p. 254).

N-Linked protein glycosylation plays a vital role in protecting the ER from misfolded glycoproteins. If a protein is incompletely folded, it enters the *calnexin-calreticulin cycle*. A glucose residue is attached to an N-linked oligosaccharide ($Man_9GlcNAc_2$) in close proximity to the misfolded segment by UGGT1 (UDP-glucose/glycoprotein glucosyltransferase). Calnexin/calreticulin can then reassociate with the glycoprotein for another folding attempt. If a glycoprotein is permanently misfolded, a mannose residue in the middle branch of the oligosaccharide is removed by α-mannosidase I. Glycoprotein molecules with one or more $Man_8GlcNAc_2$ oligosaccharides are targeted for ER-associated protein degradation (p. 47), a pathway in which misfolded proteins are translocated into the cytoplasm where they are degraded by the ubiquitin proteasomal system (p. 256).

Hydroxylation Hydroxylation of the amino acids proline and lysine is required for the structural integrity of the connective tissue proteins collagen (Section 5.3) and elastin. Additionally, 4-hydroxyproline is found in acetylcholinesterase (the enzyme that degrades the neurotransmitter acetylcholine) and complement (a series of serum proteins involved in the immune response). Three RER mixed-function oxygenases (prolyl-4-hydroxylase, prolyl-3-hydroxylase, and lysyl hydroxylase) are responsible for hydroxylating certain proline and lysine residues. Substrate requirements are highly specific. For example, prolyl-4-hydroxylase hydroxylates only proline residues in the Y position of peptides containing Gly-X-Y sequences, whereas prolyl-3-hydroxylase requires Gly-Pro-4-Hyp sequences (Hyp stands for hydroxyproline; X and Y represent other amino acids). Hydroxylation of lysine occurs only when the sequence Gly-X-Lys is present. (Polypeptide hydroxylation by prolyl-3-hydroxylase and lysyl hydroxylase occurs only before helical structure forms.) **Figure 19.18** illustrates the synthesis of

Visit the companion website at www.oup.com/us/mckee to read the related Biochemistry in Perspective essay on scurvy for Chapter 7.

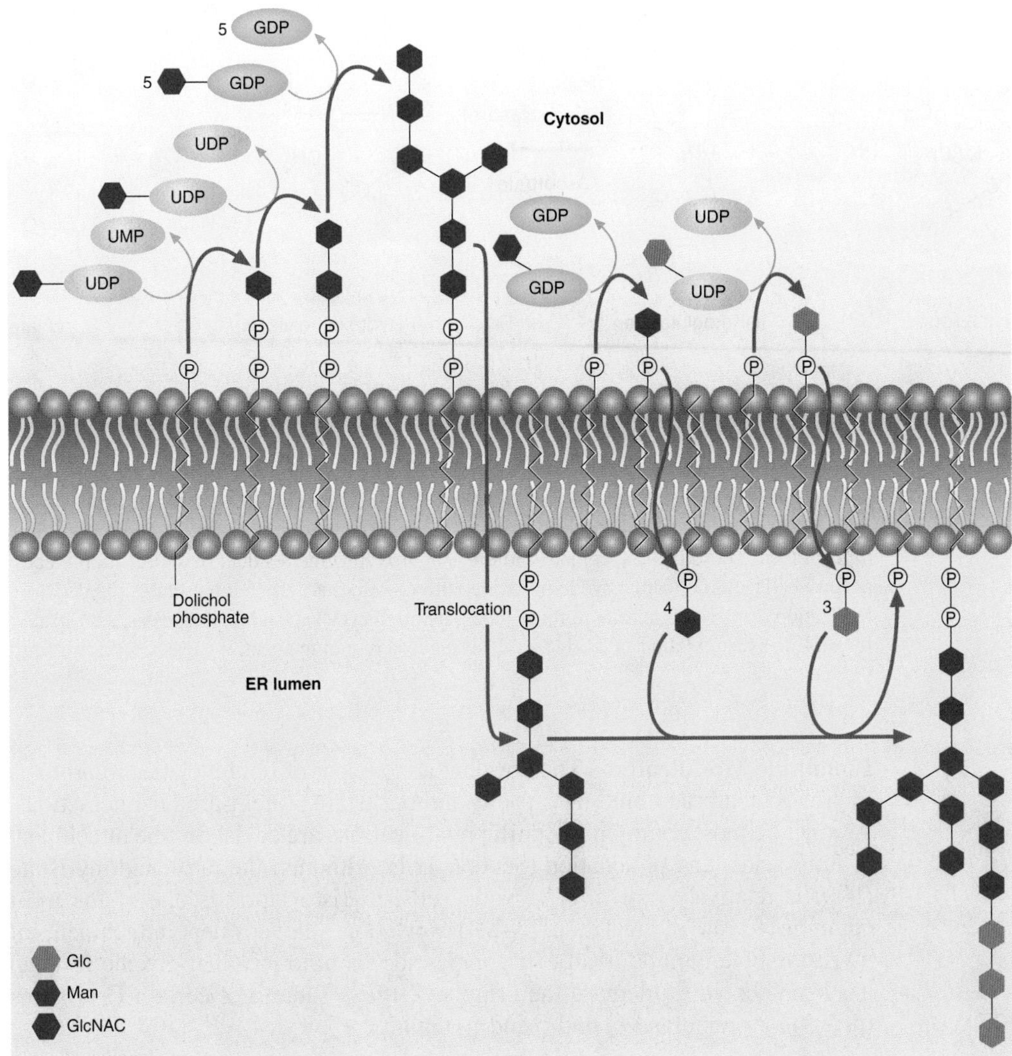

FIGURE 19.17

Synthesis of N-Linked Oligosaccharide

In the first step, GlcNAc-1-P is transferred from UDP-GlcNAc to dolichol phosphate (Dol-P). (Dolichol is a polyisoprenoid found within all cell membranes. Phosphorylated dolichol is found predominantly in ER membrane.) The next GlcNAc and the following five mannose residues are then transferred from nucleotide-activated forms. After the entire structure has flipped to the lumenal side of the membrane, each of the remaining sugars (four mannoses and three glucoses) is transferred first to Dol-P and then to the growing oligosaccharide. N-glycosylation of protein takes place in the ER in a one-step reaction catalyzed by a membrane-bound enzyme called glycosyl transferase.

4-Hyp. Ascorbic acid (vitamin C) is required to hydroxylate proline and lysine residues in collagen. Inadequate dietary intake of vitamin C can result in scurvy. The symptoms of scurvy (e.g., blood vessel fragility and poor wound healing) are effects of weak collagen fiber structure.

Phosphorylation Examples of the roles of protein phosphorylation in metabolic control and signal transduction have already been discussed. Protein phosphorylation may also play a critical (and interrelated) role in protein-protein interactions. For example, the autophosphorylation of tyrosine residues in PDGF receptors precedes the binding of cytoplasmic target proteins.

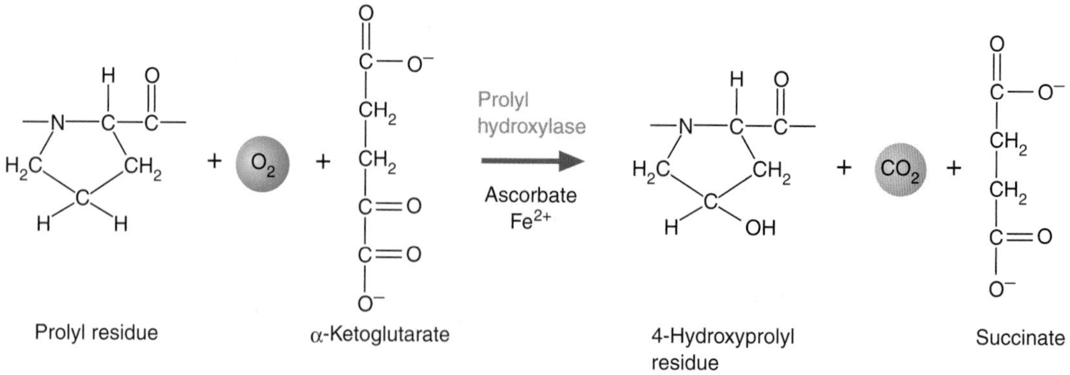

Prolyl residue + O_2 + α-Ketoglutarate → (Prolyl hydroxylase; Ascorbate, Fe^{2+}) 4-Hydroxyprolyl residue + CO_2 + Succinate

FIGURE 19.18

Hydroxylation of Proline

Prolyl-4-hydroxylase, the enzyme that catalyzes the hydroxylation of the C-4 position of certain prolyl residues in nascent polypeptides, is an example of a dioxygenase. Both oxygen atoms of O_2 are incorporated into the two substrates, α-ketoglutarate and the prolyl residue, to form the two products, succinate and the 4-hydroxyproline residue. In the reaction mechanism Fe(II) and O_2 form a cyclic peroxide with α-ketoglutarate that facilitates the decarboxylation of α-ketoglutarate to form succinate and Fe(IV)=O, which serves as a substrate for proline hydroxylation. Ascorbic acid, acting as a reducing agent, restores cofactor iron to the ferrous state.

Lipophilic Modifications The covalent attachment of lipid moieties to proteins improves membrane-binding capacity and/or certain protein-protein interactions. Among the most common lipophilic modifications are acylation (the attachment of fatty acids) and prenylation (Section 11.1). Although the fatty acid myristate (14:0) is relatively rare in eukaryotic cells, myristoylation is one of the most common forms of acylation. N-Myristoylation (the covalent attachment of myristate by an amide bond to a polypeptide's amino terminal glycine residue) has been shown to increase the affinity of the α subunit of certain G proteins (p. 589) for membrane-bound β and γ subunits.

Methylation Protein methylation serves several purposes in eukaryotes. The methylation of altered aspartate residues by a specific type of methyltransferase promotes either the repair or the degradation of damaged proteins. Other methyltransferases catalyze reactions that alter the cellular roles of certain proteins. For example, methylated lysine residues have been found in such disparate proteins as ribulose-2,3-bisphosphate carboxylase, calmodulin, histones, certain ribosomal proteins, and cytochrome c. Other amino acid residues that may be methylated include histidine (e.g., histones, rhodopsin, and eEF2) and arginine (e.g., hsps and ribosomal proteins).

Carboxylation Vitamin K–dependent carboxylation of glutamyl residues to form γ-carboxyglutamyl residues increases a protein's sensitivity to Ca^{2+}-dependent modulation. The carboxylation requires an NADPH-dependent reductase to convert phylloquinone, a vitamin K quinone (**Figure 11.16**), to its hydroquinone form, a carboxylase that adds a carboxyl group to the γ-carbon of a glutamyl residue, and an epoxide reductase that converts the vitamin K-2,3-epoxide product of the carboxylation reaction to the original vitamin K quinone. The target protein must contain an appropriate signal sequence that binds to the carboxylase enzyme and a $(GluXXX)_n$ repeat (n = 3–12, X = other amino acids). Many of the known target proteins are involved in blood clotting (factors VII, IX, and X, and prothrombin). The anticoagulant coumadin (warfarin), used to prevent blood clot formation, acts by inhibiting the two reductases required in protein carboxylation.

Disulfide Bond Formation Disulfide bonds are generally found only in secretory proteins (e.g., insulin) and certain membrane proteins. (Recall that disulfide bridges are favored in the oxidizing environment outside the cell and confer considerable structural stability on the molecules that contain them.) As described earlier (Section 5.3), cytoplasmic proteins generally do not possess disulfide bonds because of the reducing conditions within cytoplasm. The ER has a nonreducing environment, so disulfide bonds form spontaneously in the RER as the nascent polypeptide emerges into the lumen. Although some proteins have disulfide bridges that form sequentially as the polypeptide enters the lumen (i.e., the first cysteine pairs with the second, the third residue pairs with the fourth, etc.), this is not true for many other molecules. Proper disulfide bond formation for these latter proteins is now presumed to be facilitated by **disulfide exchange**. During this process, disulfide bonds rapidly migrate from one position to another until the most stable structure is achieved. An ER enzymatic activity, referred to as protein disulfide isomerase (PDI), is a thioredoxin-like enzyme that catalyzes this process. Protein disulfide isomerase also acts as a chaperone by rescuing wrongly folded polypeptides.

TARGETING Despite the vast complexities of eukaryotic cell structure and function, each newly synthesized polypeptide is normally directed to its proper destination. There appear to be two principal mechanisms by which polypeptides are directed to their correct locations: transcript localization and signal peptides. Each is briefly discussed.

It is generally recognized that cells often have asymmetrical protein distributions within the cytoplasm. For example, mature *Drosophila* eggs contain a gradient of bicoid, a protein that plays a critical role during development by influencing the translation of certain genes. A high concentration of bicoid in the anterior portion of the egg is required for the normal development of anterior body parts (i.e., head segments), whereas the low bicoid concentration in the posterior portion of the egg cytoplasm promotes the development of posterior body parts. If posterior cytoplasm is removed from one egg and substituted for anterior cytoplasm in a second egg, two sets of posterior body parts appear in the larva that develops from the recipient egg.

It is now believed that cytoplasmic protein gradients are created by **transcript localization**, that is, the binding of specific mRNA to receptors in certain cytoplasmic locations. mRNA cargo in the form of mRNP (messenger ribonucleoprotein) complexes is directed to its final location by localization elements or "zip codes" in the 3′ UTR of the transcript. RNA-binding proteins facilitate this process. Once the mRNP exits the nucleus it may lose or gain proteins, depending on its final destination. For example, some mRNPs are moved along cytoskeletal filaments via attachment to motor proteins such as kinesin, dynein, or myosin. It is known that bicoid mRNA is transported from nearby nurse cells into the developing oocyte (an immature egg cell) where it is carried along microtubules via a connection between a motor protein and its 3′ UTR to the anterior ends of the cell. Once in the oocyte, bicoid mRNA binds via its 3′ end to certain components of the anterior cytoskeleton. After the mature egg has been fertilized, translation of bicoid mRNA, coupled with protein diffusion, gives rise to the concentration gradient.

Polypeptides destined for secretion or for use in the plasma membrane or any of the membranous organelles are specifically targeted to their proper location by sorting signals referred to as signal peptides. Each signal peptide sequence facilitates the insertion of the polypeptide that contains it into an appropriate membrane. Signal peptides generally consist of a positively charged region followed by a central hydrophobic region and a more polar region. Although numerous signal peptides occur at the amino terminal, they may also occur elsewhere along the polypeptide.

The **signal hypothesis**, proposed by Gunter Blobel in 1975, explains the translocation of polypeptides across RER membrane (**Figure 19.19**). As soon as about 70 amino acids have been incorporated into the polypeptide that emerges from a

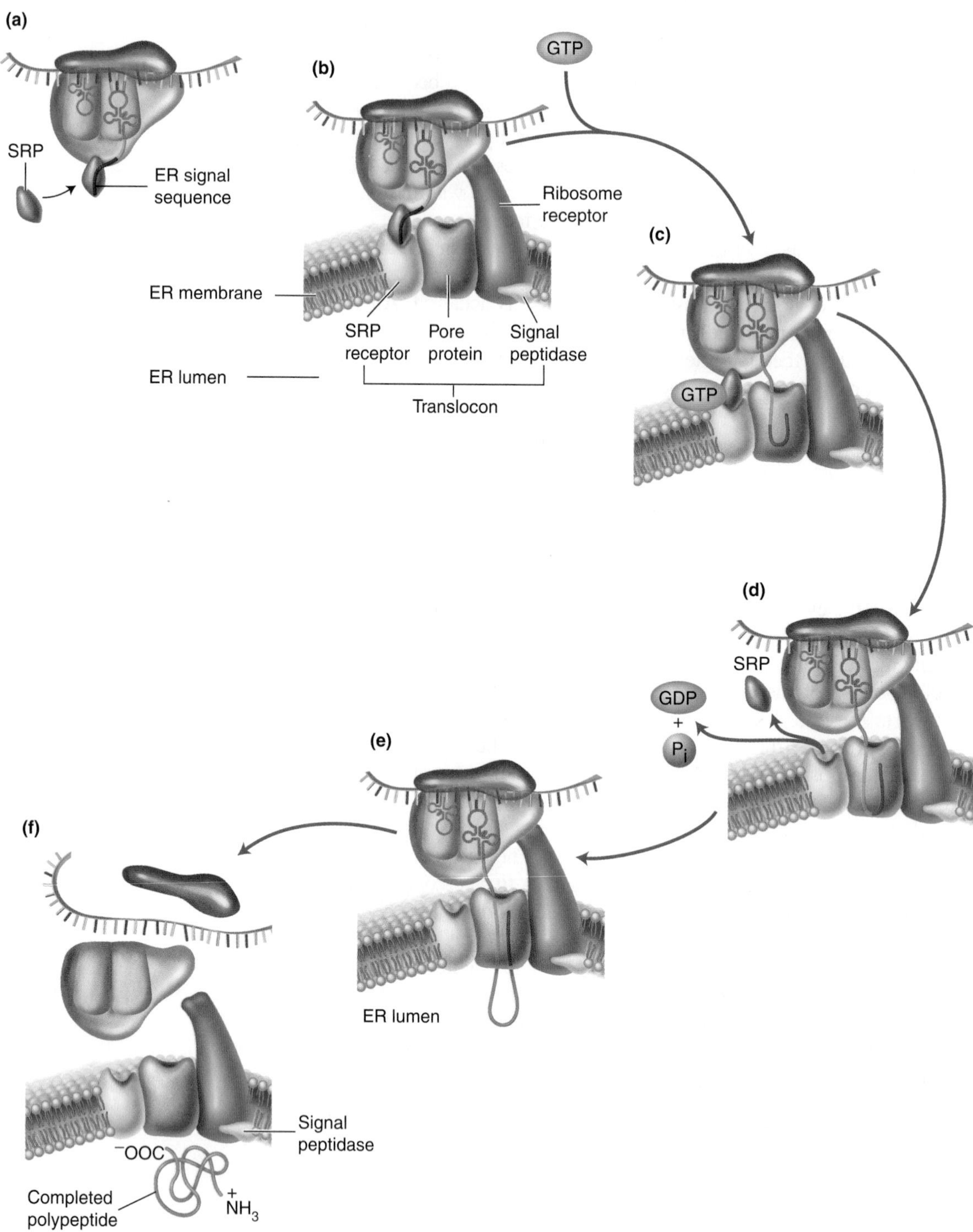

FIGURE 19.19

Cotranslational Transfer across the RER Membrane

(a) When the nascent polypeptide is long enough to protrude from the ribosome, the SRP binds to the signal sequence, causing a transient cessation of translation. (b) The subsequent binding of SRP to the SRP receptor results in the binding of the ribosome to the translocon complex in the RER membrane. (c) Polypeptide synthesis begins again as GTP binds to the SRP–SRP receptor complex. GTP hydrolysis accompanies the binding of the signal sequence to the translocon and (d) the dissociation of SRP from its receptor. (e) The polypeptide continues to elongate until (f) translation is completed. The signal peptide is removed by signal peptidase in the RER lumen. The polypeptide is released into the lumen.

ribosome, a rodlike ribonucleoprotein complex called the **signal recognition particle** (SRP) binds to the ribosome. The SRP, which consists of six proteins and a 7S RNA, is a GTPase that recognizes and transiently binds short RER signal sequences with about eight nonpolar amino acid residues. As a result of SRP-ribosome binding, the EF2 binding site is blocked and translation is arrested. The SRP mediates binding of the ribosome to the ER via *docking protein*, a heterodimer composed of two GTPases, also referred to as ***SRP receptor protein***. Once ribosome binding to the RER has occurred, both docking protein GTPs are hydrolyzed. Subsequent SRP release allows translation to restart. The growing polypeptide then inserts into the **translocon**, a protein complex composed of a transmembrane pore and several associated proteins that facilitate translocation and processing. When polypeptide synthesis and translocation occur simultaneously, the process is referred to as **cotranslational transfer**. In **posttranslational translocation**, previously synthesized polypeptides are transported across the RER membrane by ATP-binding translocon-associated molecular chaperones hsp40 and hsp70.

The fate of a targeted polypeptide depends on the location of the signal peptide and any other signal sequences. As illustrated in **Figure 19.19**, soluble secretory protein transmembrane transfer is usually followed by removal of an N-terminal signal peptide by signal peptidase, a process that releases the protein into the ER lumen. Such molecules usually undergo further posttranslational processing. The initial phase of the translocation of transmembrane proteins is similar to that of secretory proteins. For these molecules, the amino-terminal signal peptide serves as a *start signal* that remains bound in the membrane as the remaining polypeptide sequence is threaded through the membrane. So-called "single-pass" transmembrane proteins possess a *stop transfer signal* (or stop signal), which prevents further transfer across the membrane (**Figure 19.20a**). Membrane proteins with multiple membrane-spanning segments (multipass) possess a series of alternating start and stop signals (**Figure 19.20b**).

Most proteins that are translocated into the RER are directed to other destinations. After undergoing initial posttranslational modifications, both soluble and membrane-bound proteins are transferred to the Golgi complex via transport vesicles that bud off from the ER and fuse with the *cis* face of the Golgi membrane (**Figure 19.21**). Proteins that ultimately reside in the ER possess retention signals. In most vertebrate cells this signal consists of the carboxy-terminal tetrapeptide Lys-Asp-Glu-Leu (KDEL) sequence.

Within the Golgi complex, proteins undergo further modifications. For example, N-linked oligosaccharides are processed further, and O-linked glycosylation of certain serine and threonine residues occurs. Lysosomal proteins are targeted to the lysosomes by adding a mannose-6-phosphate residue. It is still unclear what signals direct secretory proteins to the cell surface (via exocytosis) or promote the delivery of plasma membrane proteins to their destination, although a "default mechanism" has been proposed. (In default mechanisms, the absence of a signal results in a specific sequence of events.) When protein modification is complete, transport vesicles exit from the trans face of the Golgi and move to their target locations.

Proteins targeted to mitochondria are synthesized on cytoplasmic ribosomes as preproteins and then bind to a multichaperone complex composed of the ATPases hsp70 and hsp90. Transport into a mitochondrion begins with the docking of the molecular chaperone-bound preprotein to receptors of the TOM complex (*t*ranslocase of the *m*itochondrial *o*uter *m*embrane). The delivery of preprotein to the TOM complex protein-conducting channel is an ATP-dependent process. About half of imported mitochondrial proteins are translocated into the mitochondrial matrix. The translocation of a matrix protein (**Figure 19.22**) begins with the transfer of the preprotein through the TOM channel. As the N-terminal signal sequence emerges in the intermembrane space it binds to a receptor of the

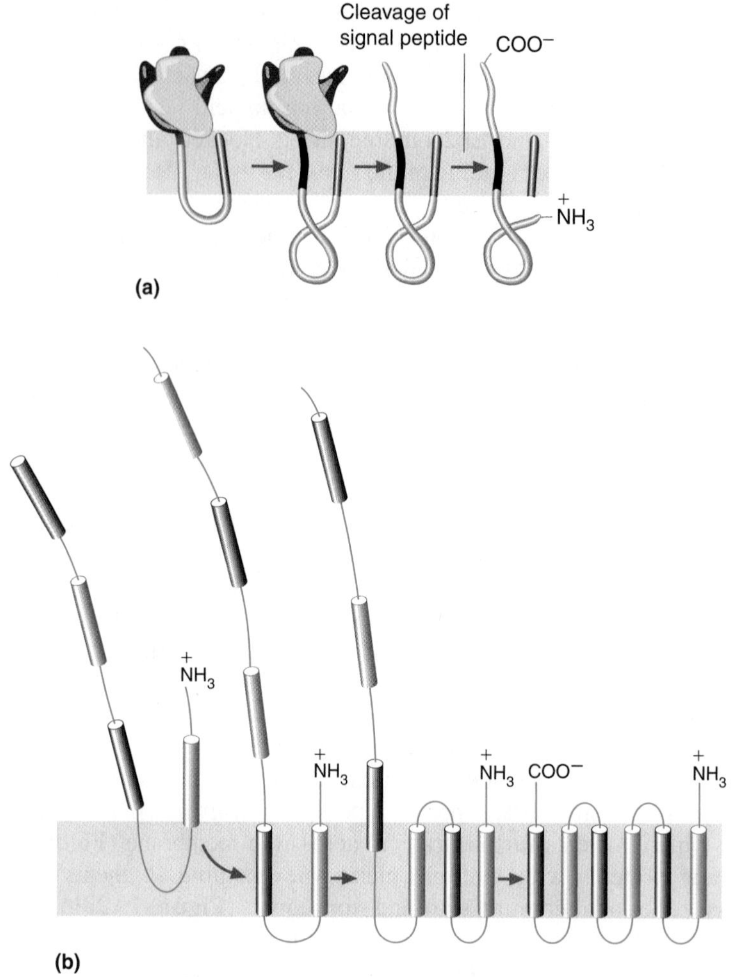

FIGURE 19.20
Cotranslational Transfer of Integral Membrane Proteins

(a) Transfer of a single-pass transmembrane protein. (b) Transfer of a multipass membrane protein. For the sake of clarity, the transfer apparatus has been omitted from the diagrams. In addition, the ribosome has been omitted from (b). The shaded segment is a signal peptide. The black segment is a stop transfer signal.

TIM (*t*ranslocase of the mitochondrial *i*nner *m*embrane) complex. The preprotein is then transported through the TIM channel in a process driven by the electrochemical proton gradient of the inner membrane. Once it is in the matrix, the preprotein's signal sequence is cleaved by a signal peptidase. ATP-dependent mitochondrial matrix chaperones, mthsp70 and mthsp60, assist in the folding of the protein into its biologically active conformation.

TRANSLATION CONTROL MECHANISMS Eukaryotic translation control mechanisms are proving to be exceptionally complex, substantially more so than those observed in prokaryotes. In eukaryotes these mechanisms appear to occur on a continuum, from *global* controls (i.e., the translation of a wide variety of mRNAs is altered) to *specific* controls (i.e., the translation of a specific mRNA or small group of mRNAs is altered). Although most aspects of eukaryotic translational control are currently unresolved, the following features are believed to be important.

mTOR-Mediated Translational Control Recall that the mTORC1 signaling pathway (p. 593) integrates nutrient availability, energy levels, and hormone and growth factor signals. Because amino acid polymerization consumes a substantial portion of cellular resources, it is not surprising that mTORC1 has a significant effect on the rate of protein synthesis. mTORC1 is a nutrient, energy, and redox sensor. It activates translation in a process that involves three proteins: eIF4E, ribosomal protein S6, and eEF2. The availability of eIF4E (cap-binding protein)

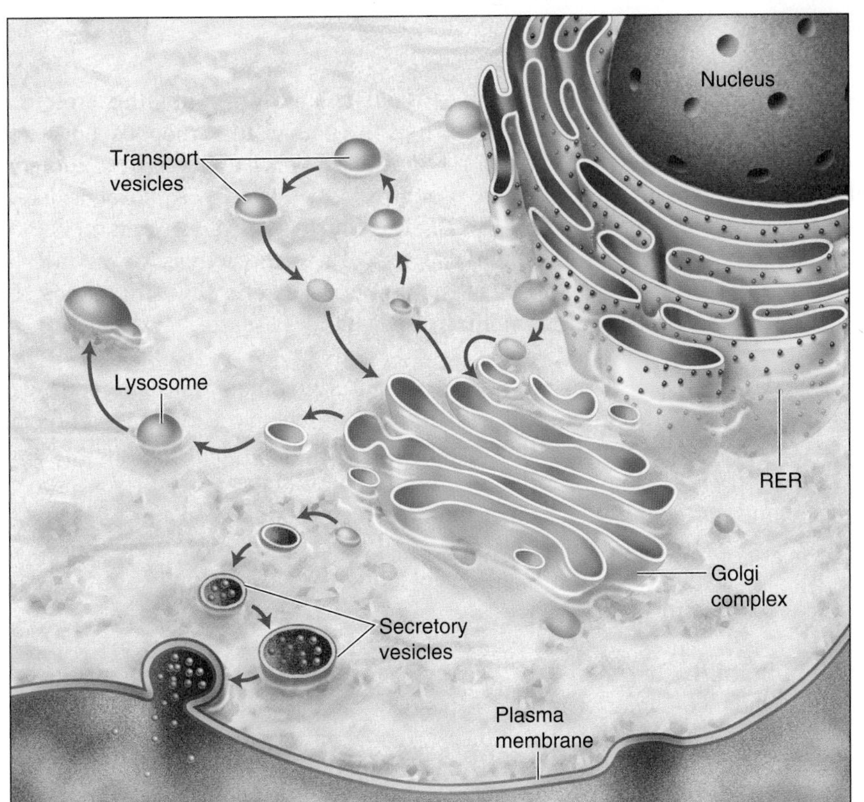

FIGURE 19.21

The ER, Golgi, and Plasma Membrane

Transport vesicles transfer new membrane components (protein and lipids) and secretory products from the ER to the Golgi complex, from one Golgi cisterna to another, and from the trans-Golgi network to other organelles (e.g., lysosomes) or to the plasma membrane.

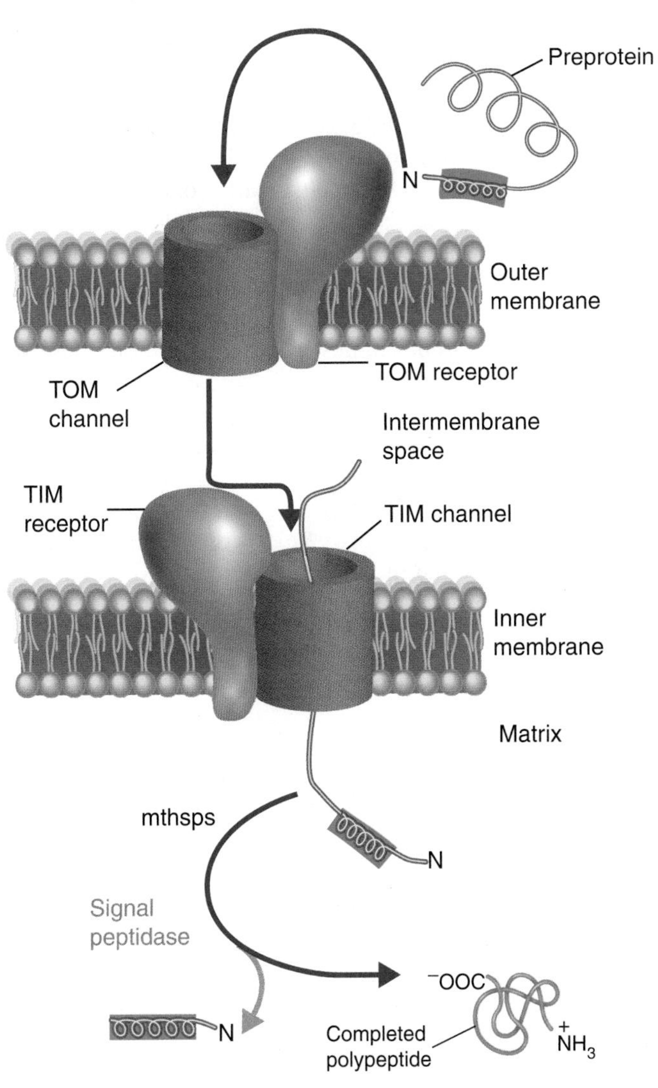

FIGURE 19.22

Import of a Mitochondrial Matrix Protein

The preprotein, which is complexed with a multichaperone complex (not shown), is recognized by a TOM receptor protein via its N-terminal signal sequence (an α-helix in which there are hydrophobic residues on one side and positively changed residues on the other). Preprotein translocation through the TOM transmembrane channel is assisted by ATP hydrolysis catalyzed by the hsp70 and hsp90 components of the multichaperone complex. Once the N-terminal signal sequence enters the intermembrane space and is recognized by a TIM receptor, the proprotein enters the TIM channel. Translocation of the preprotein across the inner membrane is driven by the electrochemical proton gradient created by the electron transport process. Once the preprotein enters the matrix it is processed by a signal peptidase and mitochondrial chaperones to yield the final biologically active protein.

for cap-binding complex formation is regulated by eIF4E-binding protein 1 (eIF4E-BP1). In its inactive state eIF4E is bound to hypophosphorylated eIF4-BP1. eIF4E is activated when mTORC1 phosphorylates eI4E-BP1, allowing it to separate from the IF. Ribosomal protein S6 is activated by a phosphorylation reaction catalyzed by S6 kinase 1 (SK1). SK1, which is activated by an mTORC1-mediated phosphorylation reaction, also upregulates the activity of the mRNA helicase eIF4B. Finally, mTORC1 signaling activates eEF2 by stimulating the phosphorylation of the eEF2-inhibiting enzyme eEF2 kinase.

mRNA Export The spatial separation of transcription and translation afforded by the nuclear membrane appears to provide eukaryotes with significant opportunities for gene expression regulation. Export through the nuclear pore complex (p. 716) is known to be a carefully controlled, energy-driven process whose minimum requirements include the presence of a 5′-cap and a 3′ poly(A) tail. mRNAs are exported as mRNPs. In addition to exon junction complexes (p. 707) and export proteins (p. 716), an elaborate group of targeting proteins are associated with mRNAs.

mRNA Stability In general, the translation rate of any mRNA species is related to its abundance, which is in turn dependent on its rates of both synthesis and degradation. mRNA half-lives range from about 20 minutes to more than 24 hours. Several features of mRNA structure are known to affect its stability, that is, its capacity to avoid degradation by various nucleases. The presence of certain sequences may confer resistance to nuclease action (e.g., palindromes that create hairpins), whereas other sequences may increase the likelihood of nuclease action, particularly if present in multiple copies. The binding of specific proteins to certain sequences can also affect mRNA stability. Finally, reversible adenylation and deadenylation of the 3′ end of mRNA strongly influence both its stability and its translational activity. As described previously, most mRNAs transported into the cytoplasm have poly(A) tails containing between 100 and 200 nucleotides. As time passes, many poly(A) tails progressively shorten, bearing no fewer than 30 residues when the entire mRNA has been degraded. In certain circumstances the poly(A) tail of some mRNAs is selectively elongated or shortened. For example, mRNAs in mature oocytes are "masked" by removal of most of their poly(A) tail nucleotides. After fertilization, these mRNAs are reactivated by adding adenine nucleotides.

KEY CONCEPTS

- Eukaryotic protein synthesis is slower and more complex than that of its prokaryotic counterpart. In addition to requiring a larger number of translation factors and a more complex initiation mechanism, the eukaryotic process involves vastly more complicated posttranslational processing and targeting mechanisms.
- Eukaryotes use a wide spectrum of translational control mechanisms.

Negative Translational Control The translation of certain specific mRNAs is known to be blocked by binding repressor proteins to sequences near their 5′ ends. Recall that ferritin mRNA contains an iron response element (IRE) that binds an iron-binding repressor protein (p. 716). When cellular iron concentrations are high, the iron atoms binding in large numbers to the repressor protein cause it to dissociate from the IRE. Then ferritin mRNA is translated.

QUESTION 19.5

The mechanism involved in the posttranslational transport of proteins into chloroplasts has so far received only limited attention. However, the import of plastocyanin into the thylakoid lumen has been determined to require two import signals near the N-terminal of the newly synthesized protein. Assuming that chloroplast protein import resembles the import process for mitochondria, suggest a reasonable hypothesis to explain how plastocyanin (a lumen protein associated with the inner surface of the thylakoid membrane) is transported and processed. What enzymatic activities and transport structures do you expect are involved in this process?

19.3 THE PROTEOSTASIS NETWORK

Within the highly crowded and dynamic interior of living cells, millions of proteins perform a vast array of functions such as DNA replication and transcription, cell signal transduction, immune responses, cell cycle control, and molecular transport. Life depends on the proper function of proteins, which in turn requires that these linear macromolecules fold into their "native states" yet retain some degree of conformational flexibility. As a result, many proteins, especially those that are composed of 100 or more amino acids or are completely or even partially unstructured, are marginally stable and therefore prone to misfolding. Misfolded or partially folded proteins often have exposed hydrophobic patches that may interact with other molecules to form amorphous aggregates. In addition, some misfolded molecules may rearrange to form the β-strands of amyloid fibrillar aggregates. The proteome is also challenged by a constant barrage of metabolic and environmental stresses (e.g., heat or heavy metals exposure, amino acid side chain oxidation, hypoxia, and toxins) that can damage them. When combined with the incidence of random errors in protein synthesis, proteotoxic stress-related protein misfolding and other types of damage are a severe threat to cell function.

Healthy young cells maintain proteostasis with a robust and highly conserved interconnected network of pathways, referred to as the **proteostasis network** (PN) (**Figure 19.23**). Using stress-responsive signaling pathways, the PN monitors proteins from their synthesis by ribosomes, through folding, refolding, transport, and degradation when their useful life is over or they are damaged. PN processes are accomplished with the aid of molecular chaperones (p. 165), stress-response transcription factors, detoxifying enzymes, and degradation processes such as the ubiquitin-proteosomal system (p. 556) and autophagy (p. 558). The resources that are devoted to proteome protection indicate the importance of the PN. For example, the human PN involves about 2000 genes. Under stressful conditions PN processes can be activated throughout the cell, that is, cytoplasm and the

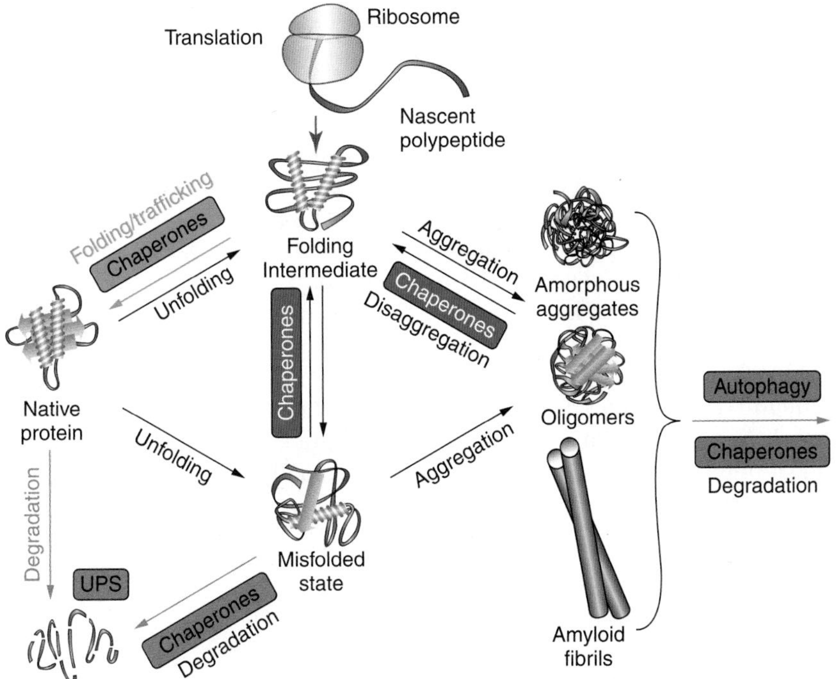

FIGURE 19.23
The Proteostasis Network

The proteostasis network consists of molecular chaperones that assist proteins in *de novo* folding and in maintaining them in their native states. The network also includes enzymes and protein complexes that degrade misfolded, damaged, and obsolete proteins. As each nascent polypeptide emerges from the exit tunnel, it encounters ribosome-associated chaperones. If necessary, additional folding assistance is provided by downstream molecular chaperones such as the hsp70s and hsp90s and their associated proteins. Misfolded proteins are degraded by a combination of chaperones and E3 ubiquitin ligases that together recognize and target them for destruction by the UPS (ubiquitin-proteosome system). Aggregated proteins that resist digestion by proteasomes are removed by autophagy.

nucleus and other organelles (e.g., the unfolded protein response in the ER [erUPR] (p. 48) and mitochondria [mtUPR]).

The Heat Shock Response

The best understood stress response is the *heat shock response* (HSR). As with other stress responses, the *HSR* works to protect an affected cell and its proteome from heat-induced damage (as well as other stresses such as oxidative stress and heavy metals). It does so by rapid and global changes in gene expression that inhibit nonessential protein synthesis on ribosomes and increase the concentration of PN components. In *E. coli*, HSR is mediated by σ^{32}, the bacterial transcription IF that directs RNA polymerase to the promoters of HSR target genes. Synthesis of σ^{32} is initiated by heat-induced melting of the secondary structure of the 5′ UTR of the σ^{32}-coding mRNA, revealing the molecule's Shine-Dalgarno sequence (p. 740). In eukaryotes, HSR, which repairs heat-induced damage to cytoplasmic proteins, is initiated by heat shock factors (HSFs), most notably heat shock factor 1 (HSF1). In unstressed cells HSF1 exists in cytoplasm as a monomer bound to hsp90. Heat stress triggers the assembly of an HSF1 trimer that relocates to the nucleus, where it stimulates the activity of chromatin remodeling proteins and the subsequent transcription of HSR genes. HSF1 activation is the result of hsp90 displacement as the molecular chaperone attempts to refold heat-damaged proteins. An RNA thermometer mechanism is also believed to play a role. HSF1 trimer formation is stimulated by a complex formed by the heat shock–induced binding of the ncRNA molecule heat shock RNA1 (HSR1) to the translation elongation factor eEF1A.

The Proteostasis Network and Human Disease

Defective protein folding is responsible for a large number of human diseases, which are referred to as protein-folding diseases or protein conformational disorders. Some protein-folding diseases involve a single genetic mutation that causes a change in a polypeptide's sequence, which in turn results in its improper folding. Examples of such loss-of-function disorders include cystic fibrosis (pp. 413–415) and CBS deficiency (refer to **Figure 15.12**). Mutations in the gene coding for p53, a major tumor-suppressor protein, occur in more than 50% of all cancers. Normal p53 regulates the cell cycle, monitors the genome for damage, and can initiate apoptosis if DNA damage is irreparable. A mutated and improperly folded p53 results in inadequately repaired DNA damage, a circumstance that eventually allows cells to grow in an uncontrolled manner.

In a large number of protein-folding disorders, there is chronic proteostasis dysfunction that arises from adverse interactions between aggregated proteins and proteosomal components. Examples include Alzheimer's, Huntington's, and Parkinson's diseases and type 2 diabetes. A major risk factor for these and other protein-folding diseases is aging, since the PN progressively declines in effectiveness as individuals grow older. A common mechanism in these disorders is the misfolding of specific proteins, such as β-amyloid (Alzheimer's disease), huntingtin (Huntington's disease), α-synuclein (Parkinson's disease), and amylin (type 2 diabetes). When aging cells fail to degrade these molecules, a small number of them form oligomers. Oligomers eventually associate to form insoluble aggregates that may entangle other denatured proteins and components of the now seriously compromised PN. The aggregated proteins are described as having a toxic gain of function.

Biochemistry IN THE LAB

Proteomics

The technology of proteomics is being developed to investigate the proteome, the functional output of the genome. The proteome of each organism or cell type includes not only the entire set of the protein products of mRNA translations, but also all of their covalent modifications, few of which can be predicted from mRNA sequences. The goals of proteomics are primarily twofold: to study the global changes in the expression of cellular proteins and to determine the identity and functions of all the proteins in the proteomes of organisms. The potential applications for proteomic research are many and varied. In addition to providing opportunities to resolve basic biological problems (e.g., ascertaining the precise mechanisms by which cellular processes such as neuronal transport or mRNA splicing occur), proteomics-based technology has obvious uses in biomedical research. Examples of the latter include investigations of the causes and diagnosis of genetic and infectious diseases and the development of effective drugs.

Proteomic Tools

The investigation of proteomes is currently focused primarily on developing fast, automated methods for identifying and characterizing the proteins produced by normal and diseased cells. Laser capture dissection microscopy (LCDM), which uses a laser beam to harvest single cells from a tissue, is providing scientists with a powerful new tool. Combined with increasingly more sophisticated proteomic technologies and bioinformatic analysis, LCDM will significantly impact the pace of discovery. Among the most useful proteomic technologies have been two-dimensional gel electrophoresis and mass spectrometry. Protein-protein interactions, the identification of the functional partners of proteins (e.g., protein kinase substrates or transcription factor activators), are investigated with a number of technologies, including protein microarrays and the yeast two-hybrid screening method.

Two-Dimensional Gel Electrophoresis Protein expression is currently analyzed with two-dimensional (2-D) gels. Proteins are separated according to charge (within a pH gradient) in the first dimension of the gel and on the basis of molecular mass in the second (**Figure 19D**). As many as 3000 individual proteins can be visualized on a 2-D gel. Protein analysis of multiple gels (e.g., determinations of presence, absence, or relative concentration of specific proteins in healthy and diseased cells) is accomplished by comparisons of 2-D gel images with proteome databases with the aid of specialized computer software. Although 2-D gel technology has been improved in both speed and capacity, it also has limitations. In addition to being labor intensive, 2-D gels are not useful in the evaluation of proteins of certain types, such as membrane proteins or proteins found in very low concentrations. Newer technologies are being developed to overcome these problems.

Mass Spectrometry As described previously (p. 183), MS is a technique in which molecules are vaporized and then bombarded by a high-energy electron beam, which causes them to fragment as cations. As the ionized fragments enter the spectrometer, they pass through a strong magnetic field that separates them according to their mass-to-charge (*m/z*) ratio. Each type of molecule is identified by the pattern of fragments that is generated, each pattern or "fingerprint" being unique. Because proteins do not vaporize, they are instead digested and then dissolved in a volatile solvent and sprayed into the vacuum chamber of the mass spectrometer. The electron beam ionizes these peptide fragments, and the positively charged peptides are passed through the magnetic field. The peptide mass fingerprint that results is then compared to fragmentation information in protein databases. Although MS is highly accurate and automated, it is usually insufficient for identifying all the proteins in a sample. To improve protein identification, tandem MS (MS/MS), a

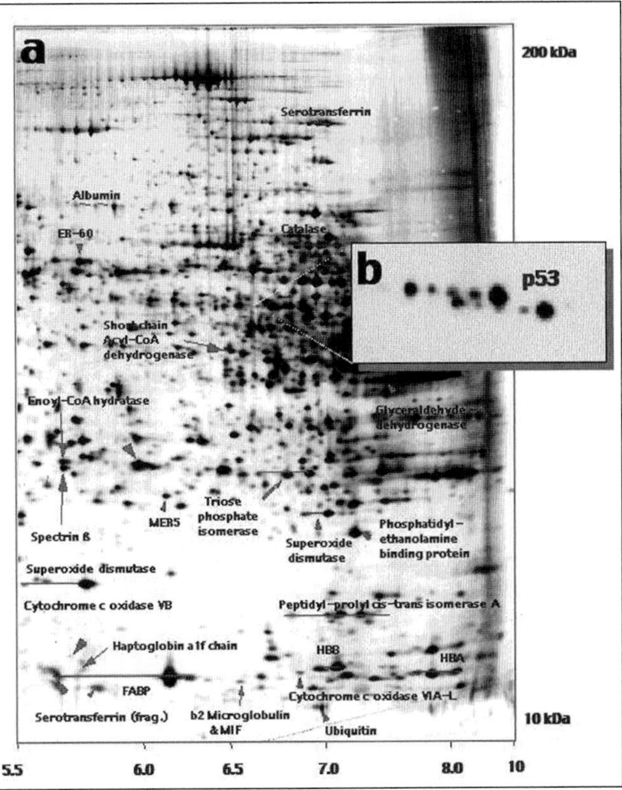

FIGURE 19D
Two-Dimensional Gel Pattern

A gel to which the extract of a liver sample had been added was run first in a pH gradient (isoelectric focusing, p. 138) and then in SDS-PAGE (p. 182) to separate the proteins according to molecular mass. The insert shows an enlargement of multiple versions of p53, a protooncogene product, that was revealed by treatment with antibodies.

Biochemistry IN THE LAB cont.

method in which two mass spectrometers are linked in tandem, has been developed. In this technique the oligopeptide fragments produced in the first MS are transferred to the second MS, where they are further fragmented and analyzed. MS/MS is used to rapidly sequence proteins.

Protein Microarrays

Protein microarrays, also called protein chips, are manufactured arrays of specific molecules, called capture molecules, on a solid support (e.g., a glass slide). Capture molecules can include antibodies, receptors, or enzymes, among others. The most commonly used detection method involves the use of fluorescent tags. An extracted protein mixture (labeled with fluorescent dyes) is first applied to a slide containing a library of immobilized antibodies. Binding between specific antibodies and proteins in the mixture is detected by a fluorescence scan.

Yeast Two-Hybrid Screening

Two-hybrid screening is a technique that uses genetically modified yeast cells to detect physical interactions between proteins that may indicate a functional relationship. A productive binding of two proteins is identified as follows. One of the proteins, called the "bait protein," is fused using DNA recombinant technology to the DNA-binding domain fragment of a specific transcription factor. The other protein is fused to the "prey protein," an activation domain fragment of the same transcription factor. Plasmids containing the DNA sequences coding for the two fused proteins are then simultaneously introduced into mutant yeast cells. The binding of the bait and prey segments of the two fused proteins results in the formation of a functional transcription factor. This in turn causes a change in the cell's phenotype. Thus, the transcription of a reporter gene results in the synthesis of an easily detectable protein.

Despite the recent advances in proteomic research techniques, several problems are a serious barrier to accomplishing the enormous task of characterizing entire proteomes. Among the most important of these is the inefficiency of the methods available and the lack of a technique equivalent to PCR for amplification of proteins found in very small amounts.

Chapter Summary

1. Protein synthesis is a complex process in which information encoded in nucleic acids is translated into the primary sequence of proteins. During the translation phase of protein synthesis, the incorporation of each amino acid is specified by one or more triplet nucleotide base sequences, referred to as codons.
2. The genetic code consists of 64 codons: 61 codons that specify the amino acids and three stop codons. Two of the stop codons can be used by some organisms to code for the nonstandard amino acids selenocysteine and pyrrolysine.
3. Translation involves the tRNAs, a set of molecules that act as carriers of the amino acids. The base-pairing interactions between codons and the anticodon base sequence of tRNAs result in the accurate translation of genetic messages.
4. Translation consists of three phases: initiation, elongation, and termination. Each phase requires several types of protein factor. Although prokaryotic and eukaryotic translational mechanisms bear a striking resemblance to each other, they differ in several respects. One of the most notable differences is the identity, quantity, and function of the translation factors.
5. A unique set of posttranslational modifications can occur that prepare the polypeptide for its functional role, assist in folding, or target it to a specific destination. These covalent alterations include proteolytic processing, modification of certain amino acid side chains, and insertion of cofactors.
6. Prokaryotes and eukaryotes differ in their usage of translational control mechanisms. Prokaryotes use variations in Shine-Dalgarno sequences and negative translational control (the repression of the translation of a polycistronic mRNA by one of its products). In contrast, a wide variety of eukaryotic translational controls have been observed. These mechanisms range from global controls in which the translation rate of a large number of mRNAs is altered to specific controls in which the translation of a specific mRNA or small group of mRNAs is altered.
7. The proteostasis network (PN) is a system of interconnected pathways that maintain the structural integrity of cellular proteins. The PN, with the aid of molecular chaperones, monitors proteins from their synthesis by ribosomes, through folding, refolding, transport, and degradation (via the ubiquitin-proteasomal system or autophagy). The heat shock response protects cells from stress-induced damage (e.g., heat, oxidative stress, and heavy metals) by inducing rapid and global changes in gene expression that can repair damage if it is not too severe.
8. Proteomics is a technology that is used to investigate the proteome, the complete set of proteins produced from an organism's genome. The goals of proteomics are to study the global changes in the expression of cellular proteins over time and to determine the identity and the functions of all the proteins produced by organisms.

Take your learning further by visiting the **companion website** for Biochemistry at **www.oup.com/us/mckee** where you can complete a multiple-choice quiz on protein synthesis to help you prepare for exams.

Suggested Readings

Akerfelt, M., Morimoto, R. I., and Sistonen, L. Heat Shock Factors: Integrators of Cell Stress, Development and Lifespan, *Nat. Rev. Cell Mol. Biol.* 11:545–555, 2010.

Arner, E. S. J., Selenoproteins: What Unique Properties Can Arise with Selenocysteine in Place of Cysteine, *Exp. Cell Res.* 316:1296–1303, 2010.

Back, S. H., and Kaufman, R. J., Endoplasmic Reticulum Stress and Type 2 Diabetes, *Annu. Rev. Biochem.* 81:767–793, 2012.

Jackson, R. J., Hellen, C. U. T., and Pestova, T. V., The Mechanism of Eukaryotic Translation Initiation and Principles of Its Regulation, *Nat. Rev. Mol. Cell Biol.* 11:113–127, 2010.

Kim, Y. E., *et al.* Molecular Chaperone Functions in Protein Folding and Proteostasis, *Annu. Rev. Biochem.* 82:323–355, 2013.

LaPlante, M., and Sabatini, D. M., mTOR at a Glance, *J. Cell Sci.* 122:3589–3584, 2009.

Lykke-Anderson, J., and Bennett, E. J., Protecting the Proteome: Eukaryotic Cotranslational Quality Control Pathways, *J. Cell Biol.* 204(4):467–476, 2014.

McKenna, M., Antibiotic Resistance: The Last Resort, *Nature* 499:394–396, 2013.

Pechmann, S., Wilmund, F., and Frydman, J., The Ribosome as a Hub for Protein Quality Control, *Mol. Cell* 49(3):411–421, 2013.

Preissler, S., and Deuerling, E., Ribosome-associated Chaperones as Key Players in Proteostasis, *Trends Biochem. Sci.* 37(7): 274–283, 2012.

Rodnina, M. V., Beringer, M., and Wintermeyer, W., How Ribosomes Make Peptide Bonds, *Trends Biochem. Sci.* 32(1):20–26, 2007.

Schmeing, T. M., and Ramakrishnan, V., What Recent Ribosome Structures Have Revealed about the Mechanism of Translation, *Nature* 461:1234–1242, 2009.

Sherman, M. Y., and Qian, S.-B., Less Is More: Improving Proteostasis by Translation Slow Down, *Trends Biochem. Sci.* 38(12):585–591, 2013.

Wilson, D. N., and Cate, J. H. D., The Structure and Function of the Eukaryotic Ribosome, *Cold Spring Harbor Perspec. Biol.* 4(5): a01536, 2012.

Zimmerman, E., and Yonath, A., Biological Implications of the Ribosome's Stunning Stereochemistry, *ChemBioChem* 10: 63–72, 2009.

Key Words

anhydride, *735*
anticodon, *732*
cap-binding complex, *745*
codon, *731*
cotranslational transfer, *759*
decoding center *739*
disulfide exchange, *757*
elongation, *736*
genetic code, *731*
GTPase associated region, *740*
initiation, *736*
48S initiation complex, *748*
mass spectrometry, *765*
mixed anhydride, *735*
nascent, *742*
open reading frame, *732*
peptidyl tranferase center, *740*
poly(A)-binding protein, *748*
polysome, *736*
posttranslational modification, *739*
posttranslational translocation, *759*
43S preinitiation complex, *745*
preproprotein, *753*
proprotein, *753*
proteomics, *730*
proteostasis network, 763
release factor, *743*
ribosome recycling factor, *743*
SECIS element, *752*
Shine-Dalgarno sequence, *740*
signal hypothesis, *757*
signal peptide, *743*
signal recognition particle, *759*
SRP receptor protein, *759*
targeting, *739*
termination, *738*
transcript localization, *757*
translocation, *736*
translocon, *759*
wobble hypothesis, *733*

Review Questions

These questions are designed to test your knowledge of the key concepts discussed in this chapter before moving on to the next chapter. You may like to compare your answers to the solutions provided in the back of the book and in the accompanying Study Guide.

1. Define the following terms:
   a. codon
   b. anticodon
   c. genetic code
   d. open reading frame
   e. codon usage bias

2. Define the following terms:
   a. wobble hypothesis
   b. cognate tRNA
   c. AUG sequence
   d. Shine-Dalgarno sequence
   e. aminoacyl-tRNA synthetase

3. Define the following terms:
   a. decoding center
   b. peptidyl transferase center
   c. GTPase-associated center
   d. guanine nucleotide exchange factor
   e. proton shuttle mechanism
4. Define the following terms:
   a. signal peptide
   b. preinitiation complex
   c. initiation complex
   d. PABP
   e. cap-binding complex
5. Define the following terms:
   a. mRNA scanning
   b. transcript localization
   c. glycosylation
   d. targeting
   e. lipophilic modification
6. Define the following terms:
   a. proprotein
   b. preproprotein
   c. disulfide exchange
   d. proline hydroxylation
   e. proteolytic cleavage
7. Define the following terms:
   a. SRP
   b. translocon
   c. docking protein
   d. SRP receptor protein
   e. signal peptidase
8. Define the following terms:
   a. cotranslational transfer
   b. posttranslational transfer
   c. TOM receptor protein
   d. TIM complex
   e. mthsp70
9. Define the following terms:
   a. tmRNA
   b. SECIS element
   c. initiation
   d. elongation
   e. termination
10. Define the following terms:
   a. nascent
   b. signal hypothesis
   c. posttranslational modification
   d. context-dependent codon reassignment
   e. diphthamide
11. Define the following terms:
   a. proteomics
   b. protein chip
   c. yeast two-hybrid screening
   d. mass spectroscopy
   e. mixed anhydride
12. Define the following terms:
   a. proteostasis network
   b. heat shock response
   c. mRNP
   d. mTORC1
   e. NMD
13. What two observations prompted the wobble hypothesis?
14. Describe the two sequential reactions that occur in the active site of aminoacyl-tRNA synthetases.
15. What are the major differences between eukaryotic and prokaryotic translation?
16. What are the major prokaryotic translation control mechanisms?
17. Describe the steps in the elongation cycle.
18. Explain the differences among preproproteins, proproteins, and proteins.
19. Describe the structure and function of the signal recognition particle.
20. Describe the function of the translocon in cotranslational transfer.
21. Describe how eukaryotic mRNA structure can affect translational control.
22. In general terms, describe the intracellular processing of a typical glycoprotein that is destined for secretion from a cell.
23. Why are tRNAs described as adaptor molecules?
24. What steps in the elongation cycle of protein synthesis require GTP hydrolysis? What role does it play in each step?
25. Name and explain the roles of the protein factors that participate in the initiation phase of prokaryotic protein synthesis.
26. Compare the RNA and protein components of prokaryotic and eukaryotic ribosomes.
27. Explain the roles of the large and small subunits of ribosomes.
28. The term *translation* refers to which of the following?
   a. DNA → RNA
   b. RNA → DNA
   c. proteins → RNA
   d. RNA → proteins
29. List and describe the major classes of eukaryotic posttranslational modifications.
30. Explain the importance of the proper targeting of nascent polypeptides.
31. Explain the critically important role of aminoacyl-tRNA synthetases in protein synthesis.
32. A peptide sequence is composed of 10 serine residues. Determine how many different mRNA sequences could code for the peptide.
33. Describe each step in the eukaryotic elongation process.
34. Describe the process whereby proteins are directed to their final destinations.
35. List the major translational control mechanisms of eukaryotes.
36. Name the misfolded protein associated with each of the following diseases: Alzheimer's, Huntington's, Parkinson's, and type 2 diabetes.
37. Indicate the phase of protein synthesis during which each of the following processes occurs:
   a. A ribosomal subunit binds to a messenger RNA.
   b. The polypeptide is actually synthesized.
   c. The ribosome moves along the codon sequence.
   d. The ribosome dissociates into its subunits.

38. Determine the total amount of nucleotide bond energy that is required in the synthesis of the following tetrapeptide: Lys-Ala-Ser-Val.
39. Provide a DNA base sequence that could code for the following peptide: Ala-Ser-Phe-Tyr-Ser-Lys-Lys-Leu-Ala-Asp-Val-Ile.
40. What is the mRNA base sequence for the peptide in Question 39?
41. What would be the effect of a single base deletion in the codon for the second Ser residue in the DNA sequence that codes for the peptide in Question 39?
42. Eukaryotic protein synthesis is considerably slower than synthesis of prokaryotes. Explain.
43. Determine the codon sequence for the peptide sequence glycylserylcysteinylarginylalanine. How many possibilities are there?
44. Discuss the role of GTP in the functioning of translation factors.

Fill in the Blank

45. Four basic properties of the genetic code are specificity, degenerate, almost universal, and ____________________.
46. The three steps in protein synthesis are initiation, elongation, and ____________________.
47. Posttranslational modifications serve two purposes: to prepare a polypeptide for a specific function and ____________________.
48. A posttranslational modification that directs a polypeptide to a specific site is called a ____________________ modification.
49. Inactive precursor proteins with removable signal peptides are called ____________________.
50. The major differences between eukaryotic and prokaryotic translation are speed, location, complexity, and the variety of ____________________.
51. The chemical bond that links amino acids in a polypeptide is an ____________________.
52. The chemical bond involved between the nucleotides in a polynucleotide strand is a ____________________.
53. The ____________________ is an explanation for why there are fewer tRNAs than expected in living cells.
54. A ____________________ is a purine-rich sequence close to AUG (the initiation codon) on a prokaryotic mRNA that binds to a complementary sequence on the 30S ribosome subunits, thereby promoting the formation of the correct preinitiation complex.

Short Answer

55. How do proteins targeted to the mitochondrial matrix reach their destination?
56. What is the signal hypothesis?
57. What feature of eukaryotic translation is especially responsible for its efficiency?
58. What function does a guanine nucleotide exchange factor (GEF) serve?
59. In the first step of aminoacyl group linkage to a tRNA, an amino acid reacts with ATP to yield an aminoacyl-AMP and pyrophosphate. The bond between the amino acyl group and AMP is described as a mixed anhydride. Explain the difference between an anhydride bond and a mixed anhydride bond. What makes this reaction irreversible?

Thought Questions

These questions are designed to reinforce your understanding of all of the key concepts discussed in the book so far, including this chapter and all of the chapters before it. They may not have one right answer! The authors have provided possible solutions to these questions in the back of the book and in the accompanying Study Guide, for your reference.

60. Selenocysteine and pyrolysine differ in the way that they are converted to their cognate charged tRNAs. Explain.
61. The three-dimensional structures of ribosomal RNA and ribosomal protein are remarkably similar among species. Suggest reasons for these similarities.
62. Explain the significance of the following statement: The functioning of the aminoacyl-tRNA synthetases is referred to as the second genetic code.
63. Although aminoacyl-tRNA synthetases make few errors, occasionally an error does occur. How can these errors be detected and corrected?
64. What specific roles do translation factors play in both prokaryotic and eukaryotic translation processes?
65. Estimate the minimum number of ATP and GTP molecules required to polymerize 200 amino acids.
66. Posttranslational modifications serve several purposes. Discuss and give examples.

67. Describe how the base pairing between the Shine-Dalgarno sequence and the 30S subunit provides a mechanism for distinguishing a start codon from a methionine codon. What is the eukaryotic version of this mechanism?
68. Given an amino acid sequence for a polypeptide, can the base sequence for the mRNA that codes for it be predicted?
69. Because of the structural similarity between isoleucine and valine, the aminoacyl-tRNA synthetases that link them to their respective tRNAs possess proofreading sites. Examine the structures of the other α-amino acids and determine other sets of amino acids whose structural similarities might also require proofreading.
70. What advantages are there for synthesizing an inactive protein that must subsequently be activated by posttranslational modifications?
71. What factors ensure accuracy in protein synthesis? How does the level of accuracy usually attained in protein synthesis compare with that of replication or transcription?
72. Can you suggest a reason why ribosomes in all living organisms consist of two subunits and not one supramolecular complex?
73. A prokaryotic species is facing a new environmental stress that can be ameliorated by a catalytic activity that requires the side chain of a unique amino acid derivative called pyrovaline. How would such an organism develop a mechanism for the incorporation of this nonstandard amino acid into an enzyme molecule? What would be the properties of the molecules required to solve this problem?

Appendix

Solutions

Chapter 1: End-of-Chapter Questions

Review Questions

3

a. functional group—group of atoms within a molecule with distinct chemical properties
b. R-group—group of atoms that make up a side chain in amino acids
c. carboxyl group—functional group consisting of R-C(=O)OH or R-COOH.
d. amino group—functional group consisting of $R\text{-}NH_2$
e. hydroxyl group—functional group consisting of R-OH

6

a. fatty acid—monocarboxylic acids represented by the formula R-COOH in which R is an alkyl group that contains carbon and hydrogen
b. saturated fatty acid—fatty acid that contains no carbon-carbon double bonds
c. unsaturated fatty acid—fatty acid that contains one or more carbon-carbon double bonds
d. triacylglycerol—esters containing glycerol and three fatty acids
e. phosphoglyceride—esters containing glycerol and two fatty acids

9

a. mRNA—messenger RNA
b. tRNA—transfer RNA
c. rRNA—ribosomal RNA
d. siRNA—small interfering RNA
e. miRNA—micro RNA

12

a. elimination reaction—a double bond is formed when atoms in a molecule are removed
b. hydrolysis—nucleophilic substitution reactions in which the oxygen of water serves as the nucleophile
c. addition reaction—two molecules combine to form a single product
d. dehydration reaction—removal of water from biomolecules containing alcohol functional groups
e. hydration reaction—water is added to an alkene producing an alcohol

15

a. autotroph—organisms that transform the energy of the sun into energy
b. chemoautotroph—organisms that transform the energy of the various chemicals into energy
c. photoautotroph—organisms that transform the energy of the sun into energy
d. chemoheterotroph—consume preformed food molecules as their energy source
e. photoheterotroph—use both light and organic biomolecules as their energy source

18

A few examples of life science fields that require a solid understanding of biochemistry include agronomy, forensics, marine biology, plant biology, pharmacology, plant or animal genetics, environmental science, and wildlife biology.

21

Amino acids occur in peptides and proteins. Sugars occur in oligosaccharides and polysaccharides. Nucleotides are the components of nucleic acids. Fatty acids are components of several types of lipid molecules, e.g., triacylglycerols and phospholipids.

24

Cells use oxidation-reduction reactions to convert bond energy in biomolecules into higher energy ATP bonds. Energy is captured as electrons are transferred from reduced molecules to more oxidized ones.

27

Catabolic pathways convert nutrients to small-molecule starting materials. Anabolic pathways utilize the small-molecule precursors to synthesize complex structure and function.

30

Examples of reactions involving acyl transfer nucleophilic substitution include the formation of thioesters such as acyl-SCoA molecules from acyl monophosphate and coenzyme A, the reaction of a carboxylic acid with an alcohol to form an ester (e.g., ethyl acetate is the product of the reaction between acetic acid and ethanol), and the hydrolysis of the amide bonds in protein molecules to yield amino acids.

33

The molecules belong to the following classes:
a. amino acid
b. sugar
c. fatty acid
d. nucleotide

36

Important ions found in living organisms are Na^+, K^+, Ca^{2+}, and Cl^-. Many polyatomic icons are also common, such as NH_4^+, PO_4^{3-}, and CO_3^{2-}.

39
The functions of polypeptides include transport, structural composition, defense, storage, regulation, movement, and catalysis.

42
Examples of waste products produced by animals include carbon dioxide, ammonia, urea, uric acid, and water.

45
Complex control mechanisms and protective systems allow living organisms to withstand various physical and/or chemical challenges, e.g., fluctuations in temperature, availability of nutrients, and energy needs. As such, living organisms are robust, yet they are fragile in their vulnerability to unusual or rare events that cause irreparable damage. For example, a bleeding cut will clot and heal, but extended exposure to high levels of carbon monoxide causes death.

Fill in the Blank

48. Anhydride
51. Metabolism
54. Sugars
57. Emergent

Short Answer

60
Organisms that capture energy by combining photosynthesis with a capacity to utilize organic molecules as energy and carbon sources are referred to as photoheterotrophs.

Thought Questions

63
The leaving group in nucleophilic acyl substitution reactions involving thioesters is RS^-. When the thioester involved contains coenzyme A, the leaving group is $CoAS^-$.

66
The smaller the pK_a value of a chemical group, the better it functions as a leaving group. In other words, its anion is more stable.

69
In the normally functioning system the lipid molecule is not particularly toxic (i.e., its presence at normal levels does not interfere with brain function). In individuals that lack the key enzyme, this lipid begins to accumulate within neurons and other brain cells. Its physical presence interferes with the way nerve tracts are laid down. Neurological systems that require integration of brain functions (e.g., motor control and intelligence) begin to fail.

72
Both normal and tumor cells are robust, i.e., they remain alive despite perturbations. The genetic instability of tumor cells facilitates the survival of the tumor. Out of the millions of cells in a tumor, there is a distinct possibility that one or more will express the P-glycoprotein. Under the selection pressure of the drug, these cells will survive. Cells not expressing P-glycoprotein or another means of detoxifying the drug will die.

Chapter 2: In-Chapter Questions

2.1
The volume of a prokaryotic cell is calculated as follows:

$$\pi r^2 h = 3.14 \times (0.5\ \mu\text{m})^2 \times 2\ \mu\text{m} = 1.57\ \mu\text{m}^3$$

The volume of a eukaryotic cell is calculated as follows:

$$4/3\ \pi r^3 = 4 \times (3.14 \times 10^3)/3 = 4200\ \mu\text{m}^3$$

By dividing the volume of the hepatocyte by the volume of the prokaryotic cell ($4200\ \mu m^3/1.57\ \mu m^3$) the number of prokaryotic cells that would fit within the heptocyte is obtained: 2700.

2.2
Without a means of disposal, the lipid molecules will accumulate in the cells. Cell function is eventually compromised and the cells die.

2.3
In response to neural signals originating in the retina of the chameleon's eyes, granules containing the appropriate pigments are released from melanosome-like organelles in the organism's skin. The change in skin color is the result of the ingestion of the pigment granules by keratinocytes.

2.4
Ten percent of 70 kg is 7 kg, which can be expressed as 7×10^6 mg. To calculate the weight of a single "average" mitochondrion, divide the estimated total weight of mitochondria by the estimated number of mitochondria (1×10^{16}):

$$\frac{7 \times 10^6 \text{ mg}}{1 \times 10^{16}}$$

The answer (i.e., the weight of an average mitochondrion) is approximately

$$7 \times 10^{-10} \text{ mg, or } 7 \times 10^{-7} \text{ mg. } (1 \text{ mg} = 1 \times 10^{-6} \text{ g})$$

2.5
Cell division involves the highly organized restructuring of the microtubules that form the mitotic spindle during the phases of mitosis. Microtubule function depends on dynamic instability, that is, the capacity to rapidly shorten and lengthen via polymerization/depolymermerization reactions. Cell division, which is unregulated in cancerous cells, is suppressed by taxol because this drug stabilizes microtubule structure.

Chapter 2: End-of-Chapter Questions

Review Questions

3
a. supramolecular complex—composed of molecules held together by noncovalent intermolecular forces

b. ribosome—protein-synthesizing unit composed of protein and RNA
c. channel protein—membrane protein that transports specific ions
d. carrier protein—membrane protein that transports specific molecules
e. receptor – proteins with binding sites for extracellular ligands

6

a. endotoxin—molecule that is released from membrane-bound lipids when the cell disintegrates
b. periplasmic space—the region between the outer membrane and the plasma membrane
c. biofilm—disorganized accumulations of polysaccharides that form during adherence to surfaces and growth
d. slime layer—also known as biofilm, disorganized accumulations of polysaccharides that form during adherence to surfaces and growth
e. bacterial capsule—some pathogenic bacteria possess secreted polysaccharides and proteins that allow the bacteria to avoid detection by the host immune system

9

a. glycocalyx—eukaryotic carbohydrate coat on the outside of the plasma membrane
b. extracellular matrix—secreted structural proteins and complex carbohydrates to form a gelatinous material that binds cells together
c. cell cortex—three-dimensional meshwork of proteins on the inner surface of eukaryotic plasma membranes
d. endoplasmic reticulum—a system of interconnected membranous tubules, vesicles, and sacs
e. ER lumen—space enclosed by the ER membrane

12

a. nucleoplasm—subcellular region surrounded by the nuclear envelope that harbors chromatin
b. chromatin fiber—uncondensed chromosomes
c. nuclear matrix—functions as a scaffold that organizes loops of DNA
d. nucleolus—site of rRNA synthesis and ribosome assembly
e. nuclear envelope—membrane that functions to separate DNA replication and transcription from the cytoplasm

15

a. mitochondrion—organelle that is the site of aerobic respiration
b. aerobic respiration—process in which chemical bond energy is used to synthesize ATP
c. apoptosis—programmed cell death pathway
d. outer mitochondrial membrane—porous smooth membrane of the mitochondrion
e. inner mitochondrial membrane—membrane of the mitochondrion that is impermeable to ions

18

a. cytoskeleton—supportive network of fibers, filaments and proteins inside the cell
b. microtubule—filaments formed by reversible polymerization of tubulin dimers
c. MAP—microtubule-associated proteins
d. IFT—intraflagellar transport process
e. primary cilium—nonmotile form of cilia

21

Microorganisms are required for human health with regard to improved digestion, vitamin synthesis, pathogen growth repression and robust immune system development.

24

Examples of factors that promote protein misfolding include metabolic stress (e.g., illness), oxidative stress (oxygen radical formation), inflammatory signaling processes, and genetic factors (e.g., a gene mutation that results in the synthesis of a defective protein).

27

Macromolecules of each type usually are present in low numbers. The concentration of any one macromolecule is low but adds to the overall number of molecules. This results in crowding.

30

The components of the endomembrane system are the plasma membrane, endoplasmic reticulum, Golgi apparatus, nucleus, and lysosomes. All of these control transport of ions and molecules across its membrane. Each membrane encloses an internal space that requires such control to function properly, i.e., for key biochemical reactions to take place. The compartments of the endomembrane system are connected via membranous vesicles that bud off from a donor membrane in one component in the system and fuse with the membrane of another component. For example, proteins synthesized in the RER are transferred via vesicles to the Golgi apparatus for further processing reactions.

33

Mitochondrial fission and fusion processes promote healthy mitochondrial function. Fission facilitates the production of additional energy when the cell's requirements are high and the segregation of damaged portions of mitochondria prior to their destruction. Fusion facilitates the rescue of mitochondria with minor damage by allowing the mixing of their contents with that of healthy mitochondria.

36

The nucleus is the repository of the cell's hereditary information. The nucleus also exerts a profound influence over all the cell's metabolic activities through the expression of that information.

39
The function of the Golgi apparatus is the processing, packaging, and distribution of cell products, such as glycoproteins.

42
The nuclear pore complex (NPC) is a large 120-MDa structure with a diameter of 120 nm that consists of about 100 proteins known as the nucleoporins. Filaments that extend from the cytoplasmic and nucleoplasmic side of the NPC function as docking sites for large molecules that will be transported through the pore in a GTP-dependent manner. Small substances such as ions and small proteins (40 kDa or less) readily diffuse through the NPC. Large molecules are restricted from passing through the pore.

45
Peroxisomes are small spherical membranous organelles that contain oxidative enzymes. Primary functions of peroxisomes are the generation and degradation of peroxides and the oxidation of toxic molecules. Additional functions include the synthesis of certain membrane lipids and the degradation of fatty acids and purine bases. To form peroxisomes, nuclear genes code for the enzymes and membrane proteins, which are synthesized on cytoplasmic ribosomes and then imported into preperoxisomes. The ER provides the peroxisomal membrane, and peroxins (a group of proteins) assemble the peroxisomes. [Peroxisomes are also involved in photorespiration in plants (Chapter 13 of your text.)]

48
Hepatocyte SER functions include synthesis of the lipid components of very-low-density lipoproteins (VLDL), and biotransformation reactions, which convert water-insoluble metabolites and xenobiotics into more soluble products for excretion. Striated muscle SER is called the sarcoplasmic reticulum (SR) and is a reservoir for calcium ions, the signal that triggers muscle contraction.

Fill in the Blank

51. Immune
54. Archaea
57. Physical barrier
60. Nucleus

Short Answer

63
The weak van der Waals forces provide the cell membrane with its most important property, fluidity. The function of membrane components requires that they be able to move relative to each other. These weak forces also facilitate the resealing of membrane breaks. If the membrane was covalently linked, it would be more rigid and easily break under the shearing forces of movement. In addition, incorporating other molecules into the membrane or allowing them to pass through would be impossible.

Thought Questions

66
The genes that are mutated in polycystic kidney disease encode a mechanoreceptor that monitors fluid flow in the kidney. When these genes are mutated cell division stimulation results in cyst formation.

69
The immobilization of enzymes and organelles on the cytoskeleton facilitates the highly organized set of living processes required to sustain the living state. For example, the close proximity of immobilized enzymes in a biochemical pathway allows the rapid delivery of the product of one enzyme to the active site of the next. This circumstance requires lower concentrations of reactant molecules than the time-consuming diffusion process.

72
The volume of a ribosome is calculated as follows:

$$\pi r^2 h = (3.14)(0.007\ \mu\text{m})^2(0.02\ \mu\text{m}) = 3 \times 10^{-6}\ \mu\text{m}^3$$

The volume of a bacterial cell (from question 2.42) is 1.6 μm^3. The number of ribosomes that can fit in a bacterial cell is $1.6/3 \times 10^{-6} = 5 \times 10^5$, but because they occupy only 20% of the cell's volume, divide by 5 to give 1×10^5 ribosomes per bacterial cell.

Chapter 3: In-Chapter Questions

3.1
The tetrahedral structures of the three molecules are as follows:

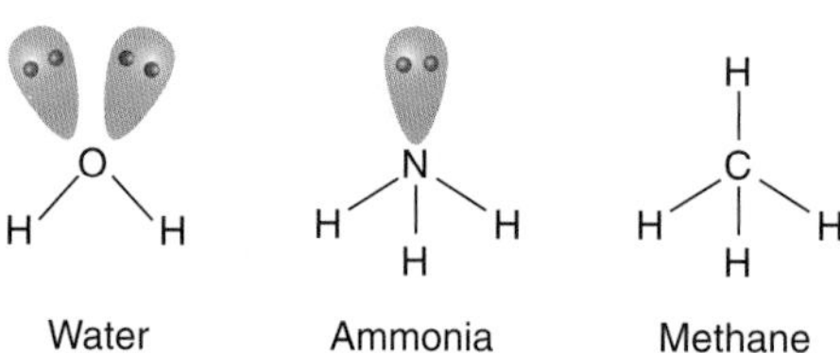

In the solid state of water, the oxygen atom has two electron pairs that form hydrogen bonds with neighboring water molecules. The nitrogen atom in ammonia has one unshared electron pair that can form a hydrogen bond with a neighboring ammonia molecule and methane has none. Note that the heats of fusion for these substances parallel the number of unshared electron pairs. Because of the ability of each ammonia molecule to form a hydrogen bond with a neighboring molecule, ammonia "ice" would be expected to be less dense than liquid ammonia.

3.2
From left to right in the illustration, the noncovalent interactions are ionic, hydrogen bonding, and van der Waals interactions.

3.3
Tendons and ligaments contain large amounts of collagen and other molecules that bind substantial amounts of structured water molecules. Water is an incompressible substance; that is, it cannot be forced to occupy a smaller space. As a result, structures containing large amounts of water can absorb relatively large amounts of force without damage.

3.4
The equilibrium shifts to the right to replace lost bicarbonate, and the acid concentration increases. The resulting condition is called acidosis.

Chapter 3: End-of-Chapter Questions

Review Questions

3
a. osmosis—spontaneous passage of solvent molecules through a semipermeable membrane that separates a solution of low solute concentration from one with high solute concentration
b. osmotic pressure—the pressure required to stop the net flow of water across a membrane
c. isotonic solution—solution that contains the same concentration of solute and water on both sides of a selectively permeable membrane
d. membrane potential—asymmetry of charge distribution on the surfaces of cell membranes that create an electrical gradient
e. hydronium ion—H_3O^+

6
Both c and d are weak acid – conjugate base pairs.

9
Osmolarity = molarity × the extent of ionization (i, the number of ions produced per ionic compound). Na_3PO_4 dissociates into four ions. Assuming 85% ionization, the osmolarity of a 1.3 M solution of Na_3PO_4 would therefore be $1.3 \times 4 \times 0.85 = 4.4$

12
One, $CH_3 - OH$ - - - - $O - H(CH_3)$

15
$\pi = iMRT$ where $\pi = 0.01$ atm
$i = 1$
$R = 0.0821$ L atm/mol K
$T = 298$ K
Solving for M:
0.01 atm = (1) (0.0821 L atm/mol K) (298 K) (M)
$M = 0.08 \times 10^{-4}$ mol/L
Solving for the molecular weight of the protein:
0.056 g/0.030 L = 1.867 g/L
1.867 g = 4.08×10^{-4} mol
1 mol of the protein = 4575.98 g = 4600 g

18
Carbon dioxide is present in the blood in sufficient quantities to make it effective as a buffer. Phosphate concentration in blood is too low for this compound to be an effective buffer. Within cells, the phosphate concentration is much higher, and it can therefore act as an effective buffer.

21
H_2CO_3 and CO_2 are both present at pH 6.4. H_2CO_3 and HCO_3^- are present at pH 8.13.

24
Hyperventilation drives the transfer of carbon dioxide from the blood. This process, which shifts the following equilibrium to the left, consumes protons, thereby making the blood more alkaline.

27
Ascorbic acid is the weak acid (HA) and sodium hydrogen ascorbate is its conjugate base (A^-). Note that a more accurate representation is H_2A and HA^-.
The number of moles of hydrogen ascorbate initially:

$$(300\text{ mL})(0.25\text{ M}) = 75\text{ mmol HA}^-$$

The number of moles of HCl added: (150 mL)(0.2 M) = 30 mmol HCl

Since HCl reacts completely with the hydrogen ascorbate, 30 mmol HCl added = 30 mmol HA^- that react with HCl to form H_2A

The number of moles of H_2A remaining after addition is:
75 mmol HA^- – 30 mmol that react with HCl =
45 mmol H_2A left over.
30 mmol of ascorbic acid were formed.
Next, use the Henderson-Hasselbalch equation ($pK_{a1} = 4.04$):
or

$$\text{pH} = pK_a + \log\frac{[\text{HAscorbate}]}{[\text{H}_2\text{Ascorbate}]} \text{ or pH} = pK_a + \log\frac{[\text{HA}^-]}{[\text{H}_2\text{A}]}$$

$$\text{pH} = 4.04 + \log\frac{[45\text{ mmol}]}{[30\text{ mmol}]}$$

$$= 4.04 + 0.18 = 4.22$$

Does a pH of 4.22 make sense? Yes, because we have more base than acid present, and 4.22 is more basic than 4.04, the pK_a. Note that this method included a shortcut: millimoles rather than molarity was used. This is valid because the total volume is the same and would cancel out. To use molarity, divide both the numerator (0.0045 mol) and the denominator (0.0030 mol) by the total volume (0.450 L).

30
In a mixture of one mole of benzoic acid ("HA") and one mole of sodium benzoate (A^-), [HA] = [A^-], and so [A^-]/[HA] = 1. Since log(1) = 0, the Henderson-Hasselbalch equation simplifies to pH = pK_a = 4.2.

$$\text{pH} = pK_a + \log\frac{[\text{A}^-]}{[\text{HA}]} \quad \text{pH} = 4.2 + \log(1) \quad \text{pH} = 4.2$$

33

When H^+ (HCl) is added to an acetic acid/acetate buffer, the H^+ will react with the acetate (A^-) to form more acetic acid (HA).

moles H^+ (HCl) added = $(1 \times 10^{-3}$ L)(1 M) = 1×10^{-3} mol H^+ added

moles of HA and A^- initially present: (1 L)(1 M HA) = 1 mol HA = 1 mol A^-

$H^+ + A^- \rightleftharpoons HA$

The number of moles of HA will increase:
1 mol HA (initial) + 0.001 mol HA (from the added HCl) = 1.001 mol HA

The number of moles of A^- will decrease:
1 mol A^- (initial) − 0.001 mol A^- (that reacted with HCl) = 0.999 mol A^-

(Note that it's not necessary to calculate the total volume and the molarity, since the volume would be in both the numerator and the denominator, and would cancel.)

$$\text{pH} = \text{p}K_a + \log\frac{[A^-]}{[HA]}$$

$$\text{pH} = 4.75 + \log[(0.999 \text{ mol } A^-)/(1.001 \text{ mol HA})]$$

$$\text{pH} = 4.75 + (-0.0009) = 4.7491 \approx 4.75$$

Compare this insignificant pH change, upon the addition of HCl to a buffer, with the addition of the same amount of HCl to water (Review Problem 3.34).

Fill in the Blank

36. Hydrogen bonding
39. The hydronium ion (H_3O^+)
42. Osmolytes
45. Buffering capacity

Short Answer

48

As water freezes, the extent of hydrogen bonding increases. However, the hydrogen bonding in the crystalline structure of ice is less efficient than in the liquid. As a result, ice is less dense than water.

Thought Questions

51

The gelatin will dehydrate as the salt becomes hydrated by the water in the gelatin.

54

The cell will undergo hemolysis. Because of the large influx of sodium ions the cell becomes hypertonic relative to blood.

57

The salts dissolved in the seawater (a hypertonic solution) pull water out of plants and will cause them to die. This is the reverse of the normal flow of water from the environment into a plant.

60

In a liquid, the molecules must be free to move over one another. In the gelatin solution, each water molecule hydrogen-bonds with two segments of the protein, locking the protein chains and the water together. Because the water molecules are no longer able to move freely, the mixture becomes semirigid.

63

The conversion of glycogen to glucose creates an increase in osmotic pressure, and water would flow into the cell. To offset this rise in osmotic pressure, ions such as sodium and potassium are pumped out of the cell. These ions would be followed by water, thus restoring cell volume.

66

Sugars contain many –OH (alcohol) groups per molecule. Alcohols are structurally similar to water (ROH vs. HOH) and have similar chemical properties. The OH groups of the alcohols "hydrate" the proteins and prevent them from aggregating.

69

The syrup is actually a mixture of the sugar and small amounts of water that remain tightly held by hydrogen bonding to the OH groups of the sugars. Such tightly held water molecules prevent the direct hydrogen bonding between –OH groups in the sugar molecules required for crystallization.

72

The equation is set up exactly the same as question 3.71.

$$69.9 \text{ J/g(1g)} + 1.03 \text{ J/g(1g)}(85.5-59.955) + 549 \text{ J/g(1g)} = 646.6 \text{ J}$$

The energy required to vaporize water is 4.7 times that of hydrogen sulfide. Most of this difference is in the heat of vaporization. In water, strong hydrogen bonds must be broken. Hydrogen sulfide does not have strong hydrogen bonds to break, hence the energy requirement is less.

75

Oxygen is a smaller atom than nitrogen. As a result, the hydrogens of the water molecule can approach the oxygen more closely to form stronger bonds. Oxygen is also more electronegative than nitrogen, resulting in an O–H bond being more polar than an N–H bond.

Chapter 4: In-Chapter Questions

4.1

$\Delta G' = \Delta G^{\circ\prime} + RT \text{ in } [\text{ADP}][\text{P}_i]/[\text{ATP}]$

where R = 8.315×10^{-3} kJ/mol•K

T = 310 K

[ADP] = 0.00135 M, [ATP] = 0.004 M,

$[\text{P}_i]$ = 0.00465 M

$\Delta G^{\circ\prime} = -30.5$ kJ/mol

$\Delta G' = -30.5$ kJ/mol + (8.315 J/mol•K)(310) ln(0.00135 M)(0.00465 M)/(0.004 M)

$\Delta G' = -30.5 + 2.577$ (ln 0.00157)
$= -30.5 - 16.64$
$= -47.14$ kJ/mol $= -47.1$ kJ/mol

4.2

Amount of ATP required to walk a mile
= (100 kcal/mi)/7.3 kcal/mol
= 13.7 mol/mi × 507 g/mol = 6945.2 g/mi = 6950 g/mi

Amount of glucose required to produce 100 kcal through ATP
= (100 kcal)/(.04)(686 kcal/mol)
= 100 kcal/274.4 kcal/mol
= 0.36 mol
= 0.36 mol × 180 g/mol = 65.6 g of glucose

Chapter 4: End-of-Chapter Questions

Review Questions

3

a. exergonic reaction—chemical reactions that release energy (i.e., have a negative free energy charge)
b. endergonic reaction—chemical reactions that absorb energy (i.e., have a positive free energy charge)
c. phosphoryl group transfer potential—the tendency of phosphorylated molecules such as ATP to undergo hydrolysis
d. dissipative system—requires that continuous work be done on the system because otherwise all natural processes will proceed toward equilibrium
e. phosphoanhydride bond—a bond formed from two molecules of phosphoric acid with the release of a water molecule.

6

For a reaction to proceed to completion the total overall $\Delta G°'$ must be negative, and there must be a common intermediate, in this case P_i. This is true of reaction b and e.

9

The principle of coupled reactions.

12

ATP + glutamate + $NH_3 \longrightarrow$ ADP + P_i + glutamine

ATP + $H_2O \longrightarrow$ ADP + P_i $\Delta G°' = -30.5$ kJ/mol

Glutamine + $H_2O \longrightarrow$ glutamate + NH_3 $\Delta G°' = -14.2$ kj/mol

Reverse the second equation and add the $\Delta G°'$ values.

ATP + $H_2O \longrightarrow$ ADP + P_i $\Delta G°' = -30.5$ kJ/mol

Glutamate + $NH_3 \longrightarrow$ glutamine + H_2O $\Delta G°' = +14.2$ kJ/mol

ATP + glutamate + $NH_3 \longrightarrow$ ADP + P_i + glutamine $\Delta G°' = -16.3$ kJ/mol

15

Under standard conditions the following statements are true: a and e.

18

Under standard conditions the following statements are true: b, d, and e.

21

The reduction of CO_2 to methane by *Methanococcus janaschii*, a hyperthermophile, is possible due to the high temperatures and pressures of the hydrothermal vent environments in which the organism thrives. This reduction is thermodynamically favorable under these conditions, but not at ambient temperature and pressure. Note that using a coupled reaction to drive CO_2 reduction forward would result in a favorable ΔG for the overall process, but would not change the ΔG for the reduction reaction alone. In other words, a coupled reaction can provide the energy to drive CO_2 reduction, but cannot convert an energy-requiring reaction into an energy-producing reaction. This is the inherent difference between an organism using ATP hydrolysis, for example, to drive CO_2 reduction forward, and an organism that obtains energy from CO_2 reduction to synthesize ATP.

24

In the absence of Mg^{2+}, increased repulsion between adjacent negative charges of ATP would cause it to have less stability than ATP with Mg^{2+} present.

27

[glucose] = $[P_i]$ = 4.8 mM = 4.8 x 10^{-3} M

[glucose-6-phosphate] = 0.25 mM

$$K_{eq} = K_{eq} = \frac{[\text{glucose}][P_i]}{[\text{glucose-6-phosphate}]}$$

$$K_{eq} = \frac{(4.8 \times 10^{-3}\text{M})(4.8 \times 10^{-3}\text{M})}{(2.5 \times 10^{-4}\text{M})}$$

$$K_{eq} = 9.2 \times 10^{-2}\ \text{M}$$

Fill in the Blank

30. Exothermic
33. Second
36. Resonance hybrid
39. Isothermic

Short Answer

42

When salt dissolves in water the ΔS term of $\Delta G = \Delta H\text{-}T\Delta S$ is strongly negative as the water molecules become ordered around the sodium and chloride ions. This term overcomes the exothermic ΔH term and the solution cools.

Thought Questions

45

A newly identified bacterial strain that could grow in the presence of arsenate opens a new debate over whether arsenate can substitute for phosphate in living organisms. Arsenic and phosphorus have similar oxidative states and its most common form 5+ (V) oxides such as arsenate ($HAsO_4^{-2}$) has similar pKa values to those of phosphate. The major argument against arsenate substituting for phosphate is arsenate-esters and arsenate-diesters are extremely unstable in aqueous solutions. This higher reactivity with water would become problematic for long-term stability of the genome of arsenate containing nucleic acids.

48

$\Delta G' = -RT \ln K_{eq}$

$-16{,}700 \text{ J/mol} = -(8.315\text{J/mol}\bullet\text{K})(298 \text{ K})\ln K_{eq}$

$-16{,}700 = -\ 2478 \ln K_{eq}$

$16{,}700/2478 = \ln K_{eq}$

$6.74 = \ln K_{eq}$

$K_{eq} = 851$ [ATP][G-6-P]/[ATP] [glucose]

51

ΔG is the most useful criterion of spontaneity because it reflects the change in entropy, which must increase for a reaction to be spontaneous.

54

Using Einstein's equation $E = mc^{2:}$

$c = 3.0 \times 10^8 \text{ m/s}^2$

$E = (0.001) \times (3.0 \times 10^8 \text{ m/s})2$

$E = 9 \times 10^{13}$ joules

Coal yields 393.3 kJ/mol = 393,300 joules/mol

$= 9 \times 10^{13}$ joules $\times$ (1/393,300 joules/mol) (1g/mol)(1kg/1000g)

= 228,832 kg

57

The balanced equation is

$C_{17}H_{35}COOH + 26O_2 \longrightarrow 18CO_2 + 18H_2O$

$\Delta H = \Delta H_{products} - \Delta H_{reactants}$

= 18 mol(–94 kcal/mol) + 18 mol(–68.4 kcal/mol) – (1 mol)(–211.4 kcal/mol)

= –1692 kcal – 1232.1 kcal + 211.4 kcal

= –2711.8 kcal = –2712 kcal

60

The sulfur of the thioester is larger than oxygen of the alcohol. So sulfur can stabilize the unshared electron pairs more easily. Consequently, there is a larger difference in energy between the reactants and products, which translates into a larger ΔG.

Chapter 5: In-Chapter Questions

1

Amino acids a and b are neutral, nonpolar, c is basic, and d is an acidic amino acid.

2

Bacteria with surface polypeptides composed of D-amino acids are resistant to degradation because the proteases, the enzymes that immune system cells use to degrade protein in foreign cells, can only catalyze the hydrolysis of peptide bonds between L-amino acids. In other words, the active sites of proteases are stereospecific; that is; they can only effectively bind peptides composed of L-amino acids.

3

The isoelectric point for the tripeptide valylcysteinyltryptophan is calculated as follows: The pKa values are (1) valyl amino group = 2.32; (2) tryptophan carboxylate group = 9.39; and (3) cysteine side chain = 8.33. The electrically neutral species of this tripeptide would contain a positively charged amino group and a negatively charged carboxylate group. The pI is calculated by adding the pKa values for the amino and carboxylate groups and dividing by two, i.e., 2.32 + 9.39/2 = 5.86. Since the pK_R for the cysteinyl side chain is more than two pH units away from the pI, it remains largely uncharged.

4

The structure of the penicillamine-cysteine disulfide is

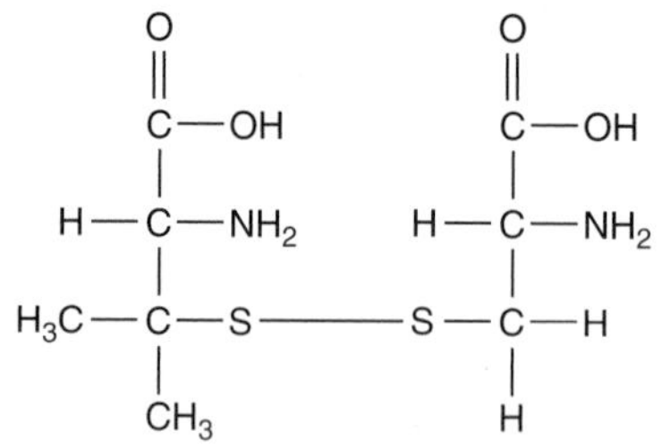

```
         O                    O
         ||                   ||
         C—OH                 C—OH
         |                    |
      H—C—NH2             H—C—NH2
         |                    |
   H3C—C—S————————————S—C—H
         |                    |
        CH3                   H
```

5

The complete structure of oxytocin is

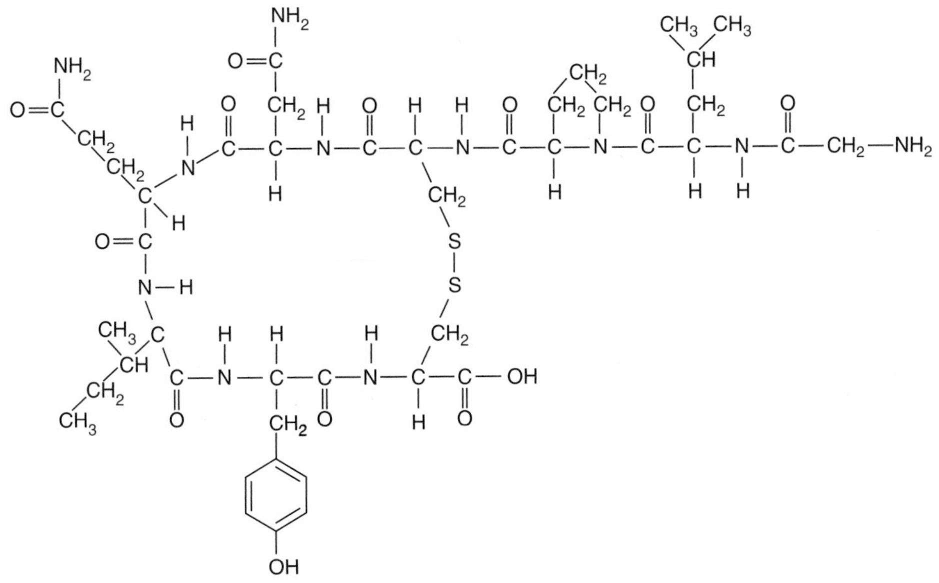

At pH 4 the terminal amino group of the glycine would be protonated to give the molecule a +1 charge. The isoelectric point of oxytocin is 5.6. Therefore at pH 9 the molecule will have a net negative charge.

6

The partial overlap in the biological properties of vasopressin and oxytocin can be explained in part by the Ile residue at position 3 of oxytocin. Presumably, given the overall similarities in size and the identical nature of six of the eight amino acid residues in the two molecules, the hydrophobic side chain of this Ile can reach and partially fit into the hydrophobic pocket of the vasopressin receptor. The degree of functional overlap between the two peptides is reduced by the leucine residue at position 8 of oxytocin because the side chain of this residue is not only neutral but smaller than the positively charged arginine residue of vasopressin. If the arginine residue of vasopressin is replaced by a lysine residue, it is expected that there will be a decrease in the molecule's binding properties because of the structural differences between the two side chains. This decrease would probably not be large, since the side chains are similar in length and are both positively charged.

7

The trait is recessive, and two copies of the aberrant gene are required for full expression of the disease. Primaquine induces the production of excess amounts of the strong oxidizing agent hydrogen peroxide. In the absence of sufficient amounts of the reducing agent NADPH, the peroxide molecules cause extensive damage to the cell. No, a higher than normal peroxide level in blood cells is damaging to the malarial parasite and is selected for in geographical regions where malaria occurs.

8

The solution to the first two illustrations is

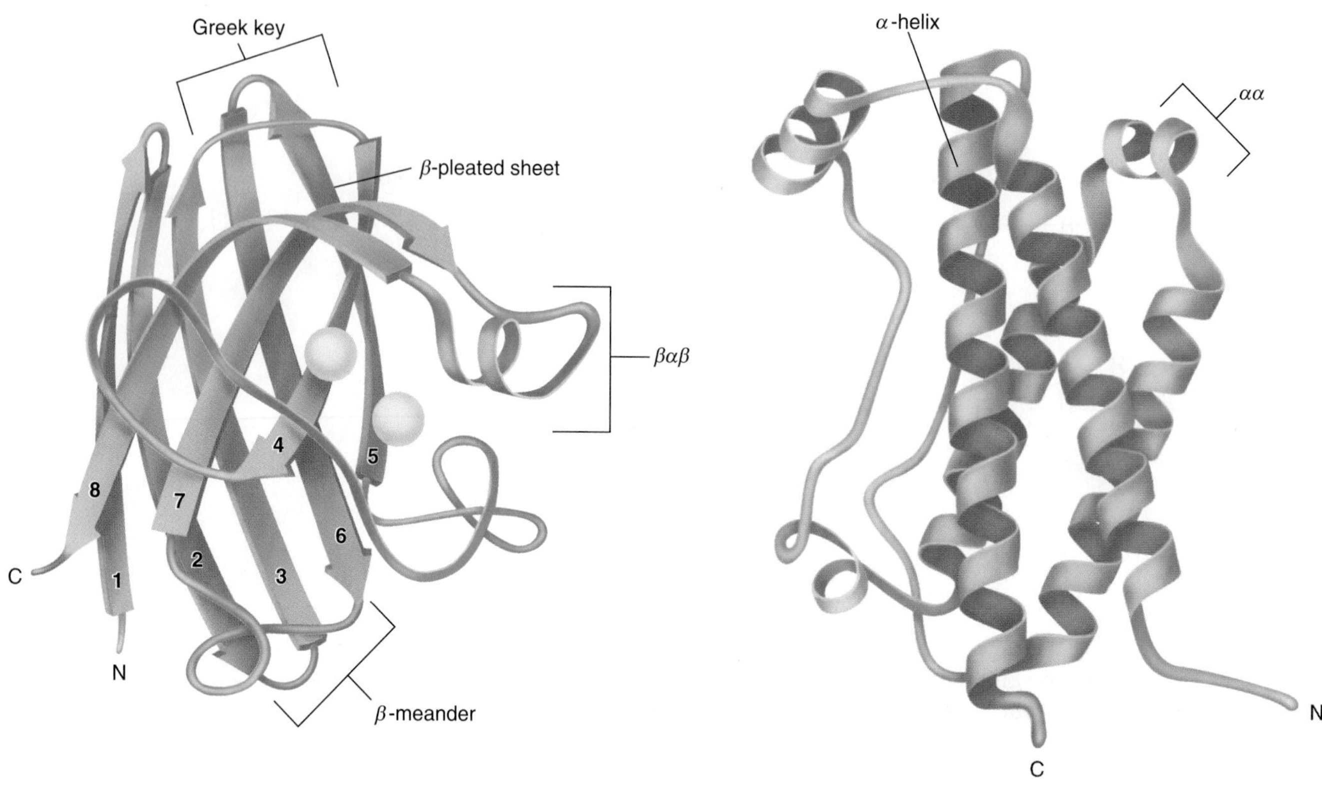

The solution to the third illustration is

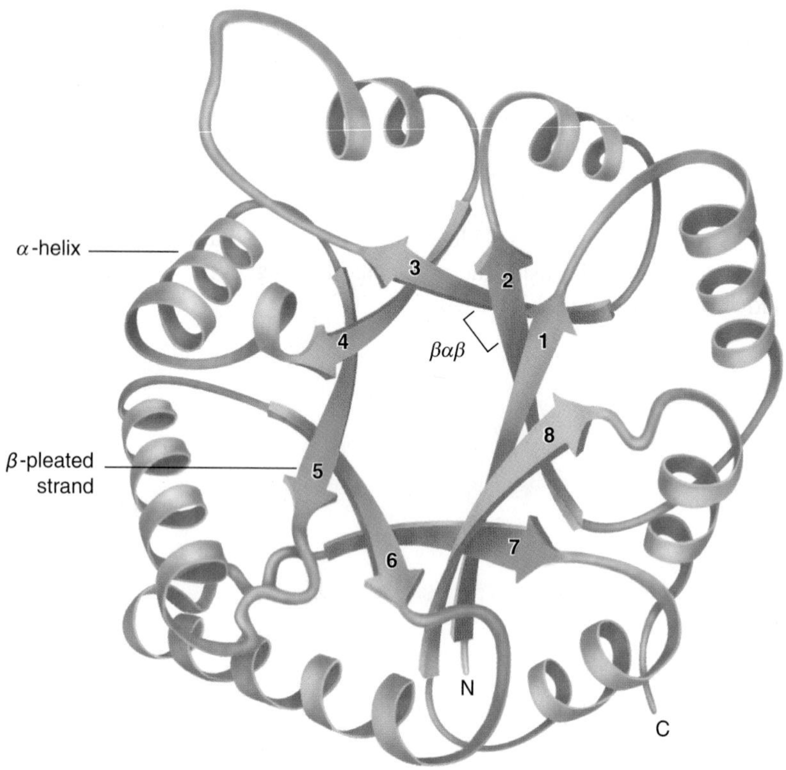

This protein is an example of an α/β-barrel structure.

9

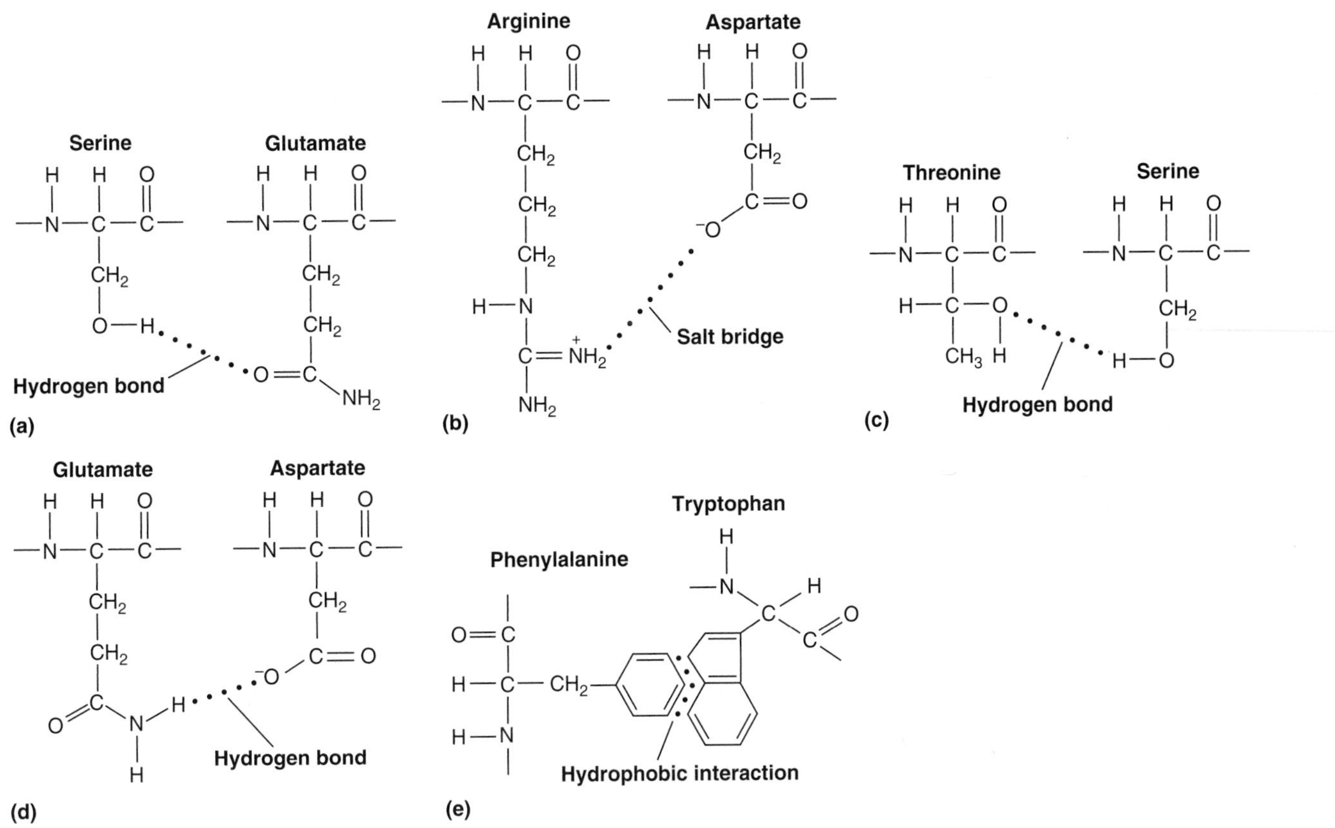

Serine Glutamate

Hydrogen bond

(a)

Arginine Aspartate

Salt bridge

(b)

Threonine Serine

Hydrogen bond

(c)

Glutamate Aspartate

Hydrogen bond

(d)

Tryptophan

Phenylalanine

Hydrophobic interaction

(e)

10

Collagen is a major structural protein found in connective tissues. Consequently, the failure of collagen molecules to form properly weakens these tissues causing diverse symptoms. Examples include cataracts, easily deformed bones, torn tendons and ligaments, and ruptured blood vessels.

11

BPG stabilizes deoxyhemoglobin. In the absence of BPG, oxyhemoglobin forms more easily. Fetal hemoglobin binds BPG poorly and, therefore, has a greater affinity for oxygen.

12

Myoglobin, composed of a single polypeptide, binds oxygen in a simple pattern—it binds the molecule tightly and releases it only when the cells' oxygen concentration is very low. The binding of oxygen by hemoglobin, a tetramer, has a more complicated sigmoidal pattern that is made possible by the noncovalent interactions among its four subunits.

Chapter 5: End-of-Chapter Questions

Review Questions

3

a. metalloprotein—proteins that contain metal ions
b. hormone—chemical signal molecules produced in one cell that regulates the function of other cells
c. holoprotein—a protein that is combined with its prosthetic group
d. intrinsically unstructured protein—a protein with partial or complete lack of ordered structure
e. kinesin—motor protein that moves vesicles and organelles along microtubules

6

a. aldimine—Schiff bases formed by an amino group reacting with an aldehyde group
b. heat shock protein—a group of conserved proteins that assist in protein folding by preventing inappropriate protein-protein interactions.
c. multifunction protein—a protein with multiple and often unrelated functions
d. protein family—composed of protein molecules that are related by amino acid sequence similarity
e. protein superfamily—proteins more distantly related

9

a. salt bridge—noncovalent bonds that occur between ionic groups of opposite charge
b. oligomer—multisubunit proteins
c. allosteric transition—ligand-induced conformation changes
d. protein denaturation—the process of structural disruption of a protein
e. amphipathic molecule—a molecule that contains both hydrophobic and hydrophilic components

12
A polypeptide is a polymer containing more than 50 amino acid residues. A protein is composed of one or more polypeptide chains. A peptide is a polymer with fewer than 50 amino acid residues.

15
The name of the molecule is cysteinylglycyltyrosine. Its abbreviated structure is:

$$^{+}H_3N\text{–Cys–Gly–Tyr–}COO^{-}$$

18
a. The amino acid sequence is a polypeptide's primary structure.
b. A β-pleated sheet is one type of secondary structure.
c. Inter- and intra-chain hydrogen bonds between N–H groups and carbonyl groups of peptide bonds are the principal feature of secondary structure. Hydrogen bonds formed between polar side chains are important in tertiary and quaternary structure.
d. Disulfide bonds are strong covalent bonds that contribute to tertiary and quaternary structure.

21
a. heat—hydrogen bonding (secondary and tertiary structure)
b. strong acid—hydrogen bonding (secondary and tertiary structure) and salt bridges (secondary and tertiary structure)
c. saturated salt solution—salt bridges (tertiary structure)
d. organic solvents—hydrophobic interactions (tertiary structure)

24
The first step in the isolation of a specific protein is the development of an assay, which allows the investigator to detect it during the purification protocol. Next, the protein, as well as other substances, are released from source tissue by cell disruption and homogenization. Preliminary purification techniques include salting out, in which large amounts of salt are used to induce protein precipitation, and dialysis, in which salts and other low molecular weight material are removed. Further purification methods, which are adapted to each research effort at the discretion of the investigator, include various types of chromatography and electrophoresis. Three chromatographic methods are: ion-exchange chromatography, gel-filtration chromatography, and affinity chromatography. Gel electrophoresis may be used to purify a protein and/or to assess the purity of a protein.

27
The primary driving force in protein folding is the requirement to achieve a low-energy state despite the decrease in entropy that occurs as the protein's three-dimensional structure becomes more ordered. Key considerations include the energy associated with different bond angles and bond rotation, the chemical properties of the amino acid side chains (e.g., whether or not the side chain will be charged at the cellular pH), and interactions between side chains. Of the noncovalent interactions that are possible, hydrophobic interactions are particularly important. (Recall that hydrophobic interactions are driven in part by the increase in entropy in the surrounding water molecules.)

30
The pKa of the amino group of Gly is 9.60 and the pKa of the carboxyl group of Alanine is 2.34. (9.60+2.34)/2 = 5.97 = pI of the dipeptide Gly-Ala. Remember the carboxyl group of Gly no longer contributes to the charge of the dipeptide because it reacted with the amino end of Ala to form a peptide bond.

Fill in the Blank

33. Amphoteric
36. Peptide bonds
39. Defense

Short Answer

42
Isoleucine and threonine have two chiral centers. The molecules are shown below.

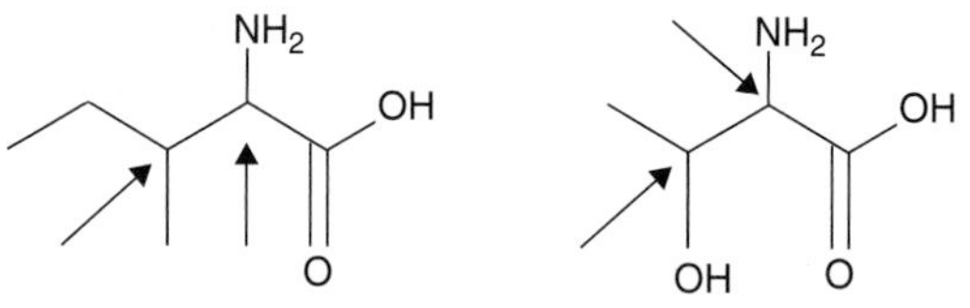

NH2 OH O
Isoleucine

NH2 OH OH O
Threonine

Arrows indicate the two chiral centers.

45
Taurocholic acid contains a strongly acidic sulfonic acid group. Its sodium salt would be ionized at the slightly alkaline pH found in the intestine. The pK_a of taurine's acidic group is significantly lower than that of the carboxylate group of glycine. As a result, taurocholic acid would be a somewhat better emulsifying agent than glycocholic acid.

Thought Questions

48
During hyperventilation, oxyhemoglobin significantly decreases.

51
The structure of each protein is different. More complex multi-domain proteins tend to require assistance. Simple single-domain proteins such as ribonuclease A spontaneous fold without the help of molecular chaperones.

54
The large size of enzymes is required to stabilize the shape and functional properties of the active site and to shield it from extraneous molecules. In addition, structural features of the protein may function in recognition processes in signaling or binding to cellular structures.

57

The hydrophobic amino acid side chains are excluded from the water and tend to cluster together. This clustering holds portions of the polypeptides in a particular conformation.

60

Glyceraldehyde-3-phosphate dehydrogenase is an example of a multifunction protein. The process of normal genetic mutation will produce changes in protein structure. Most of these changes will be neutral or detrimental. In rare instances mutations facilitate a new function.

63

The hydrophobic amino acids glycine, phenylalanine, methionine, valine, and leucine should all seek the relatively water-free interior of the decapeptide.

66

The extended α-helix probably contains amino acids with nonpolar side chains such as lysine, leucine, and phenylalanine. These need to be replaced by amino acids such as glycine and proline. Glycine's small R group permits a contiguous proline to assume a *cis* orientation, resulting in a tight turn in the peptide chain.

69

Serotonin is derived from tryptophan; dopamine comes from tyrosine.

Chapter 6: In-Chapter Questions

6.1

The amino acid residues forming the three-dimensional structure of the active site are chiral. As a result, the active site is chiral. It can bind only one isomeric form of a hexose sugar, in this case the D-isomer.

6.2

a. isomerase
b. transferase
c. lyase
d. oxidoreductase
e. ligase
f. hydrolase

6.3

The products of the degradation are the following compounds:

CH_3OH — **Methanol**

$H_2N{-}CH(CH_2C_6H_5){-}C(=O){-}OH$ — **Phenylalanine**

$H_2N{-}CH(CH_2{-}C(=O){-}OH){-}C(=O){-}OH$ — **Aspartic acid**

Cleavage of the ester bond is catalyzed by an esterase; the amide bond is cleaved by a peptidase.

6.4

$$\text{Enzyme}{-}SH + I{-}CH_2{-}C(=O){-}NH_2 \longrightarrow \text{Enzyme}{-}S{-}CH_2{-}C(=O){-}NH_2$$

6.5

Dialysis removes the formaldehyde, formic acid, and methanol that build up in the bloodstream. The bicarbonate neutralizes the acid produced and helps offset the resultant acidosis. The ethanol competitively binds with the alcohol dehydrogenase. This slows the dehydrogenation of the methanol and allows time for the kidneys to excrete it.

6.6

Menkes' syndrome—injections of copper salts into the blood would avoid intestinal malabsorption and provide the copper necessary to form adequate levels of ceruloplasmin and offset the symptoms of the disease.

6.7

Wilson's disease—zinc induces the synthesis of metallothionein, which has a high affinity for copper. Some organ damage can be averted because metallothionein sequesters copper and prevents this toxic metal from binding to and inactivating susceptible proteins and enzymes. Penicillamine forms a complex with copper in the blood. This complex is transported to the kidneys, where it is excreted.

6.8

a. cofactor
b. holoenzyme
c. apoenzyme
d. coenzyme
e. coenzyme

6.9

The patient that failed to show improvement probably had a higher level of acetylating enzymes. The patient's dosage should be based on capacity to process the drug and not on body weight.

Chapter 6: End-of-Chapter Questions

Review Questions

3

a. reaction order—the sum of the exponents on the concentration terms in the rate expression for a reaction; allows an experimenter to draw certain conclusions regarding the reactions mechanism
b. turnover number—the number of molecules of substrate converted to product in each second per mole of enzyme
c. double-displacement reaction—a reaction in which the first product is released before the second substrate binds; also referred to as a "ping pong" mechanism
d. inhibitor—a molecule that reduces an enzyme's activity
e. reaction mechanism—step-by-step description of a chemical reaction process

6

a. proenzyme—an inactive precursor of an enzyme
b. positive cooperativity—the mechanism in which the binding of one ligand to a target molecule increases the likelihood of subsequent ligand binding
c. negative cooperativity—the binding of one ligand to a target molecule, decreasing the likelihood of subsequent ligand binding
d. zymogen—an inactive form of a protoelytic enzyme
e. free radical—an atom or molecule that has an unpaired electron

9

Factors that contribute to enzyme catalysis include proximity and strain effects, electrostatic effects, acid-base catalysis, and covalent catalysis. Refer to pp. 215–218 for an explanation of each.

12

a. oxidoreductase—an enzyme that catalyzes an oxidation-reduction reaction.
b. lyase—an enzyme that catalyzes an addition or elimination reaction, adding a molecule across a double bond or removing a molecule to form a double bond.
c. ligase—an enzyme that catalyzes the joining of two molecules, often using ATP for energy.
d. transferase—an enzyme that catalyzes the transfer of a functional group from one molecule to another.
e. hydrolase—an enzyme that catalyzes a nucleophilic substitution reaction where water is the nucleophile, resulting in a cleaved bond.
f. isomerase—an enzyme that catalyzes the conversion of one isomer to another.

15

Transition metal ions are useful as enzyme cofactors because they have concentrations of positive charge, can act as Lewis acids, and can bind to two or more ligands at the same time.

18

At the start of a reaction, the concentrations of the reactants and products can be known precisely. Because equilibrium has not yet been established presumably only the forward reaction is taking place.

21

Enzymes lower the activation energy of a reaction by lowering the free energy of the transition state. The active site of the enzyme described most likely contains amino acid residues that stabilize the transition state with some or all of the following: electrostatic effects and noncovalent interactions, a shape that accepts the substrate yet eases strain in the transition state, and participation in the catalytic mechanism (e.g., by providing an acidic, basic, or nucleophilic residue to assist in acid-base or covalent catalysis). The active site should also place the substrate and reactants in close proximity to each other and in the proper orientation.

24

a. metabolon—a complex of enzymes that share intermediates of a metabolic pathway so that the product of one enzyme is in close proximity to the active site of the next enzyme in the pathway.
b. *In vivo*—in a living cell or organism.
c. *In vitro*—in "glass" (i.e., a test tube); taking place under controlled laboratory conditions as opposed to within a living cell or organism.
d. In silico—resulting from computer simulation or computer modeling.
e. metabolic flux—the rate of flow of metabolites, such as substrates, products, and intermediates, along biochemical pathways.

27

Compartmentation within eukaryotic cells is the physical separation of enzymes by a membrane (i.e., by containing certain enzymes within an organelle), or by attachment of enzymes to membranes or cytoskeletal filaments. Compartmentation (1) prevents competing reactions from occurring simultaneously and allows them to be regulated separately ("divide and control"), (2) reduces or removes diffusion barriers by locating enzymes and metabolites close to each other, (3) provides specialized reaction conditions (e.g., low pH) that would not be possible otherwise, and (4) protects other cellular components from potentially toxic reaction products ("damage control").

30

Amino acids that are capable of acting as acids or bases in enzyme catalysis are aspartic acid, glutamic acid, histidine, cysteine, tyrosine, lysine, and arginine. Of these amino acids, histidine is the most likely to be able to function as either an acid or a base, since its pK_R value, 6.0, is relatively close to physiological pH. At pH 7.6, histidine's R group would be completely ionized in aqueous solution. However, the pK_a may shift depending upon the environment of the active site. A relatively nonpolar active site would lower the pK_a to favor the neutral form of the R group (which could act as a base), while a polar active site would raise the pK_a to favor the charged form (which could act as an acid).

33
The ratio k_{cat}/K_m, called the specificity constant, is the second-order rate constant for a reaction in which $[S] << K_m$. An enzyme's specificity constant reflects the relationship between catalytic rate and substrate binding affinity. The upper limit for the enzymes k_{cat}/K_m value cannot exceed the maximal rate at which the enzyme can bind to a substrate molecule.

36
Transition metals can act as Lewis acids because they can accept electrons

39
The energy diagrams for the hypothetical reaction are as follows

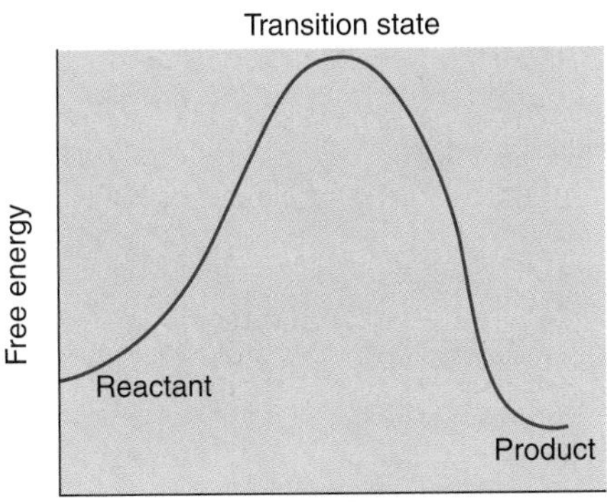

Progress of reaction
Hypothetical reaction without enzyme

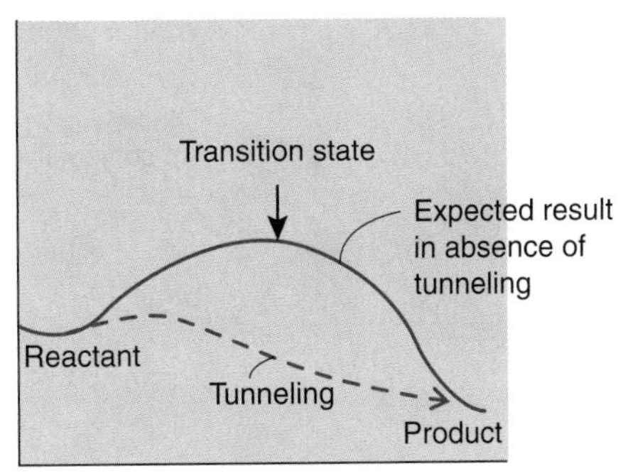

Progress of reaction
Hypothetical reaction
with enzyme and tunneling phenomenon

42
The conversion of the ketone functional group-containing dihydroxyacetone phosphate to the aldehyde functional group-containing glyceraldehyde-3-phosphate involves acid/base catalysis. The intermediate is an enediol.

H
C—OH
||
C—OH
|
CH_2—O—P(=O)(—O^-)—O^-

Enediol intermediate

Fill in the Blank

45. Oxidoreductases
48. Intermediate
51. Transferases

Short Answer

54
Activation energy is the energy that must be overcome for a chemical reaction to occur. In some reactions, when two molecules approach each other prior to a reaction there is an electronic repulsion between them. The force required to overcome this repulsion is the activation energy.

57
Assuming that the amino-acid residue-suicide substrate covalent bond is not destroyed by the hydrolysis reaction, the investigator can identify it by analyzing the enzyme hydrolysate with spectroscopic techniques.

Thought Questions

60

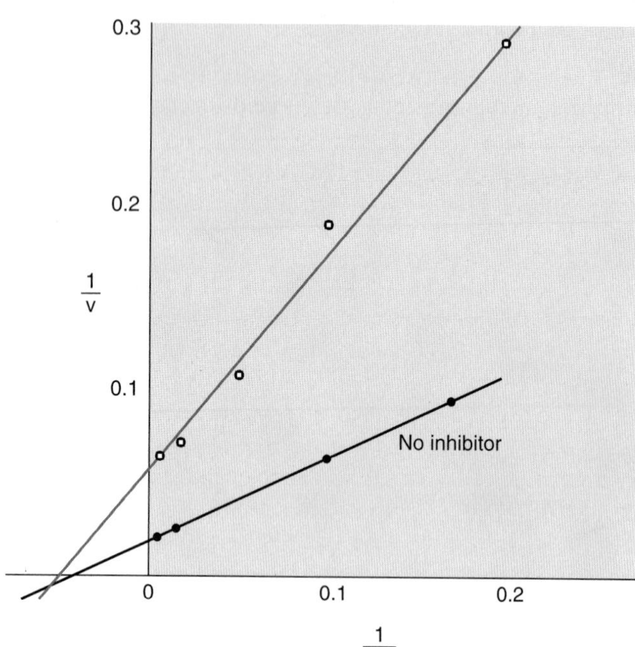

Intercept, horizontal axis $= -\frac{1}{K_m}$

Intercept, vertical axis $= \frac{1}{V_{max}}$

Slope $= \frac{K_m}{V_{max}}$

The type of inhibition being observed is noncompetitive

63

The k_{cat}/K_m: values for the alcohols are as follows:

alcohol	K_{cat}/k_m ($M^{-1}s^{-1}$)
Ethanol	0.5
1-Butanol	1.0
1-Hexanol	2.6
12-Hydroxydodecanoate	2.9
All-*trans*-retinol	3.9
Benzyl alcohol	0.2
2-Butanol	1.1×10^{-3}
Cyclohexanol	3.9×10^{-3}

Based on the k_{cat}/k_m values of these alcohols, all-*trans*-retinol is the most easilymetabolized alcohol by ADH.

66

As indicated on the graph (see answer to question 6.65), Inhibitor A is a competitive inhibitor and inhibitor B is a noncompetitive inhibitor. Competitive inhibitors have an affinity for E, and noncompetitive inhibitors have affinities for both E and ES.

69

Crowding will tend to raise activity coefficients. If the substances are not allowed to diffuse from the reaction center the concentration in terms of moles per liter can be quite low but the local concentration can be considerably higher.

72

The magnesium ion has vacant d orbitals that can act as electron pair acceptors

75

rate $= k[A]^2[B]$

[A]	[B]	Rate
0.1	0.01	1×10^6
0.1	0.02	2×10^6
0.2	0.01	4×10^6
0.2	0.02	8×10^6

78

Asp

His

The histidine protonates the OH of the alcohol making it a better leaving group.

81

a. no

b. if α was less than 1 the reaction would be faster than the uninhibited reaction

Chapter 7: In-Chapter Questions

7.1

a. Aldotetrose, b. Ketopentose, c. Ketohexose

7.2

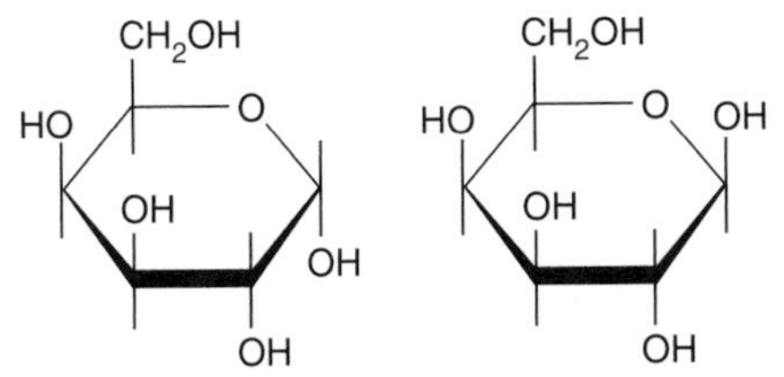

α-D-Galactose β-D-Galactose

(a)

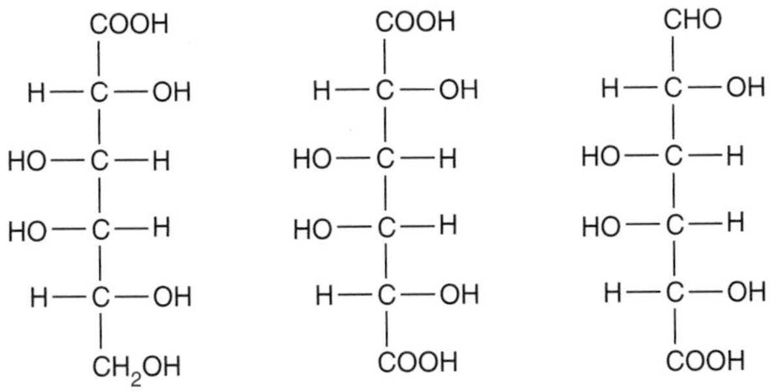

Aldonic acid Aldaric acid Uronic acid

(b)

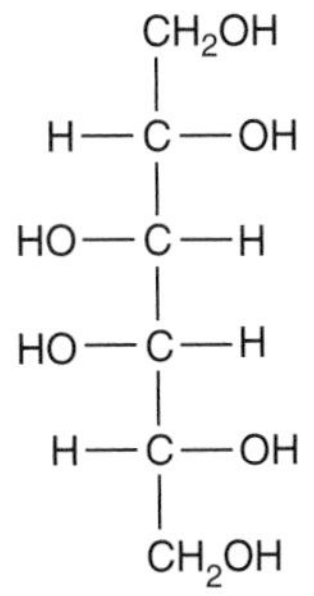

Galactitol

(c)

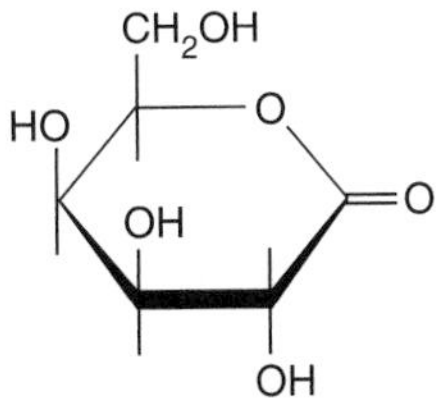

δ-Lactone of galactonic acid

(d)

7.3

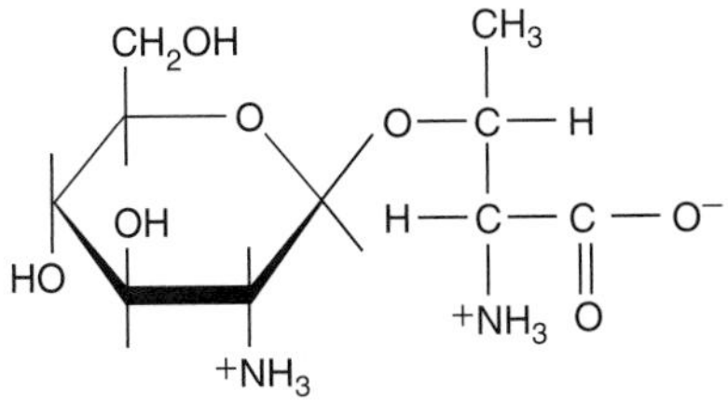

7.4

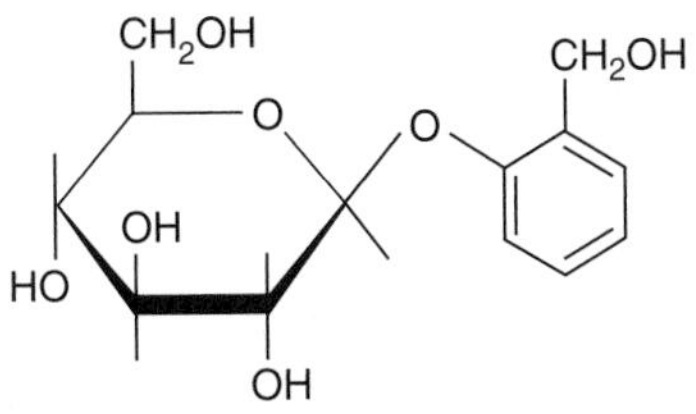

Carbohydrate Aglycone

7.5

a. glucose—reducing sugar
b. fructose—reducing sugar
c. α-methyl-D-glucoside—nonreducing
d. sucrose—nonreducing

Sugars a and b are capable of mutarotation.

7.6

The larger, insoluble glycogen molecule makes a negligible contribution to the osmotic pressure of the cell. In contrast, each molecule of an equivalent number of glucose molecules contributes to osmotic pressure. If the glucose molecules were not linked to form glycogen, the cell would burst.

7.7

In an analog system, information is encoded as a continuous signal. For example, in old-fashioned clocks the hands move continuously and not in small steps. In analog systems, small fluctuations in information processing can be meaningful. The sugar code is an analog system because of microheterogeneity, the continuous spectrum of carbohydrate structures that cells can synthesize to encode biologically relevant signaling information. In a digital system, information is represented by discrete values ("digits") of a physical quantity. Digital clocks display time as a progression of discrete numbers with no intermediate values. The DNA code is a digital system composed of four digits (bases). During protein synthesis, each DNA base triplet that ultimately codes for an amino acid has a specific meaning.

Chapter 7: End-of-Chapter Questions

Review Questions

3

a. aldonic acid—the product of the oxidation of the aldehyde group of a monosaccharide
b. uronic acid—the product formed when the terminal CH_2OH group of a monosaccharide is oxidized
c. aldaric acid—the product formed when the aldehyde and CH_2OH groups of a monosaccharide are oxidized to carboxylic acids
d. lactone—a cyclic ester
e. reducing sugar—a sugar that can be oxidized by weak oxidizing agents

6

a. Maillard reaction—the nonenzymatic glycation of protein; the nucleophilic attack of an amino nitrogen on the anomeric carbon of a reducing sugar
b. Schiff base—the imine product of a reaction between a primary amino group and a carbonyl group
c. Amadori product—a stable ketoamine formed from the rearrangement of a Schiff base; an intermediate in glycation processes that produce advanced glycation end products
d. adduct—the product of an addition reaction
e. reactive carbonyl-containing product—highly reactive molecules produced from Amadori products that cause protein cross-linkages and adduct formation; the dicarbonyl compound glyoxal (CHOCHO) is an example

9

a. glycogen—a glucose storage molecule in vertebrates; a branched polymer containing $\alpha(1{,}4)$ and $\alpha(1{,}6)$ glycosidic linkages
b. cellulose—an unbranched structural polymer produced by plants that is composed of D-glucopyranose residues linked by $\beta(1{,}4)$-glycosidic linkages
c. N-glycan—an oligosaccharide linked to a protein via a β-glycosidic bond between the core N-acetylglucosamine anomeric carbon and a side chain amide nitrogen of an asparagine residue
d. O-glycan—an oligosaccharide linked to a protein via an β-glycosidic bond to the hydroxyl oxygen of serine or threonine residues
e. glycosaminoglycan—a long unbranched heteropolysaccharide chain composed of disaccharide repeating units

12

a. Glucose and mannose are examples of epimers.
b.

c. Glucose is a reducing sugar.
d. Ribose is a monosaccharide.
e. α- and β-Glucose are anomers.
f. D-ribose and D-arabinose are diasterisomers

15

Heteroglycans are made up of more than one type of monosaccharide residue but homoglycans have only one. Examples of homoglycans and heteroglycans are starch and hyaluronic acid, respectively.

18

21

a. Carboxylate, sulfate and hydroxyl groups bind large amounts of water.
b. Hydrogen bonding is the primary type of bonding between water and glycosaminoglycans.

24

A reducing sugar reduces Cu(II) in Benedict's reagent. This reduction takes place because the hemiacetal portion of a sugar can form an aldehyde functional group, which can be oxidized to a carboxylic acid.

27

a. D-erythrose and D-threose are epimers.
b. D-glucose and D-mannose are epimers.
c. D-ribose and L-ribose are enantiomers.
d. D-allose and D-galactose are diastereomers.
e. D-glyceraldehyde and dihydroxyacetone are an aldose-ketose pair.

Fill in the Blank

30. Anomers
33. Reducing
36. Aldonic

Short Answer

39

The composition of the mixture is calculated using two simultaneous equations:

$$x + y = 1 \text{ (1) and}$$
$$112\,(x) + 19y = 53 \text{ (2)}$$

The first equation can be rearranged to yield

$$x = 1 - y \text{ (3)}$$

Substituting equation 3 into equation 2,

$$112\,(1 - y) + 19y = 53, \text{ which yields}$$
$$112 - 112y + 19y = 53$$
$$-93y = 53 - 112$$
$$-93y = -59$$

Solving for y gives

$y = -59/-93 = 0.634$ (i.e., 63% of the equilibrated mixture is in the α-D glucopyranose form).

Because $x = 1 - y$, then $1 - 0.634 = 0.375$ (i.e., 37% of the equilibrated mixture is in the β-D-glucopyranose form).

42

The sugar residues of the glycoprotein molecules in a fish's blood bind to water molecules in small ice crystals, thereby interfering with ice crystal growth.

Thought Questions

45

The thick proteoglycan coat acts to protect bacteria by preventing the binding of antibodies to their surface antigens.

48

51

a.

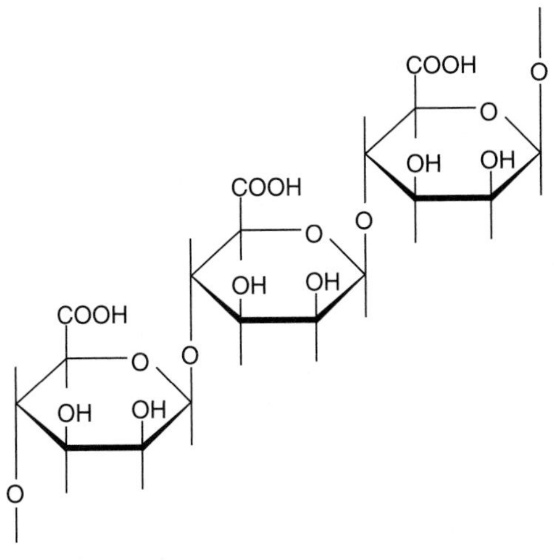

b. The polymer acts to immobilize water through extensive hydrogen bonding

54

The complete structures of the Type A and Type B antigens are as follows:

Type A

Type B

57

The conversion of glucose to fructose occurs via enol-keto tautomerism. Refer to **Figure 16**, p. 235.

60

The sulfate group has several negatively charged oxygens that are capable of hydrogen bonding. Conjugation of sulfate with the hydrophobic molecule enhances solubility because of hydrogen bonding between the sulfate oxygens and water molecules in tissue fluids.

63

Assuming a D family sugar, two structures are possible

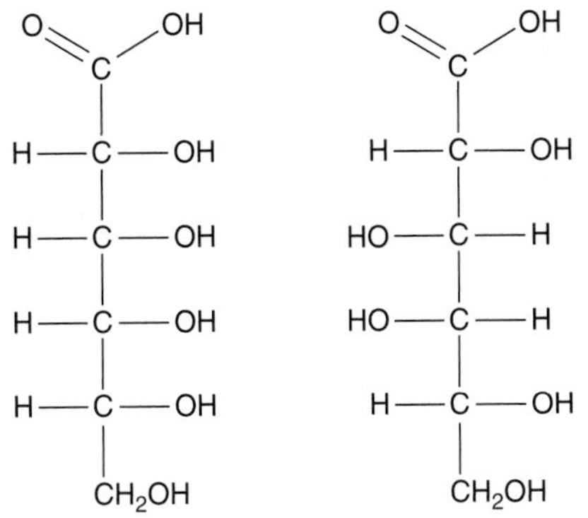

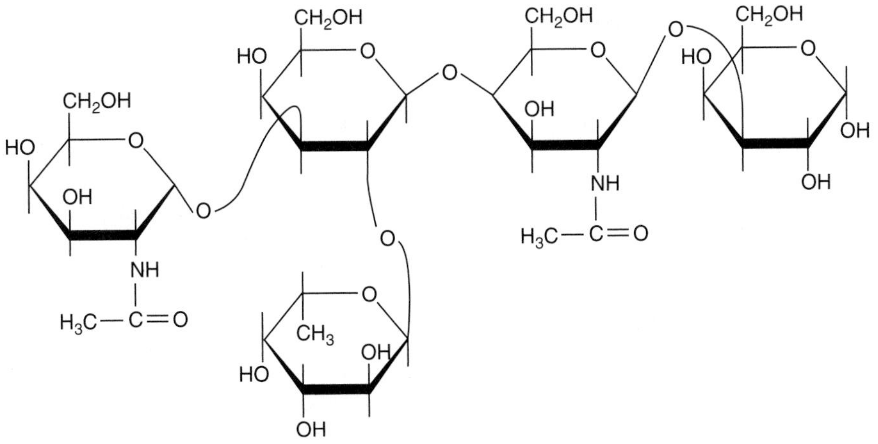

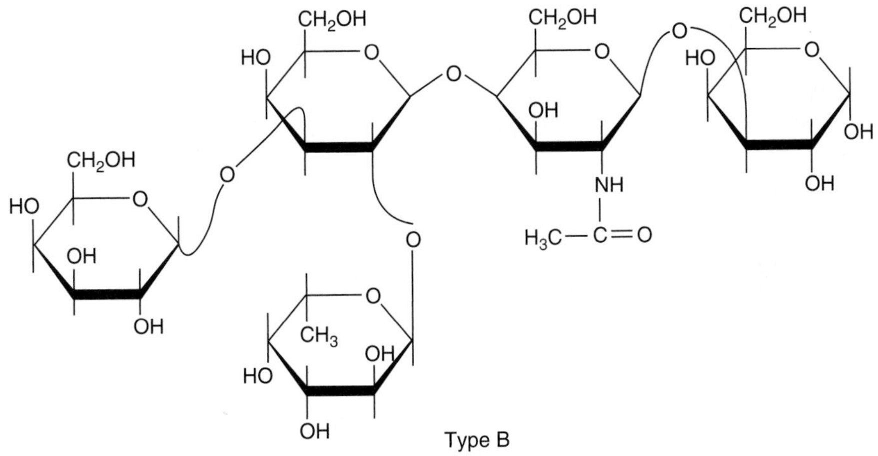

66
Three α-anomeric ring forms of 3-ketoglucose are shown. The pyranose form is the most stable because the bond angles are less strained.

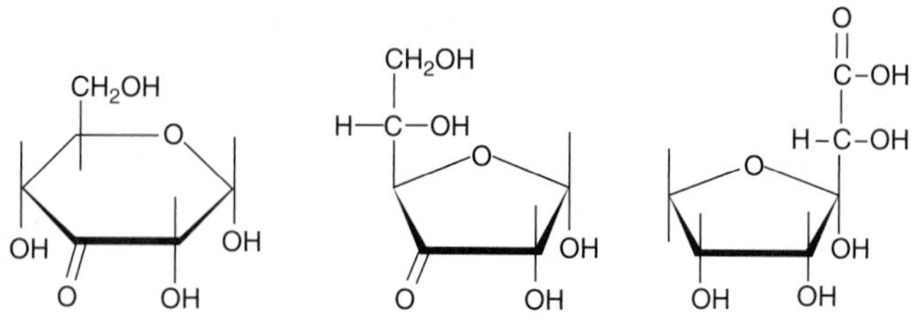

69
The structure of olestra is as follows:

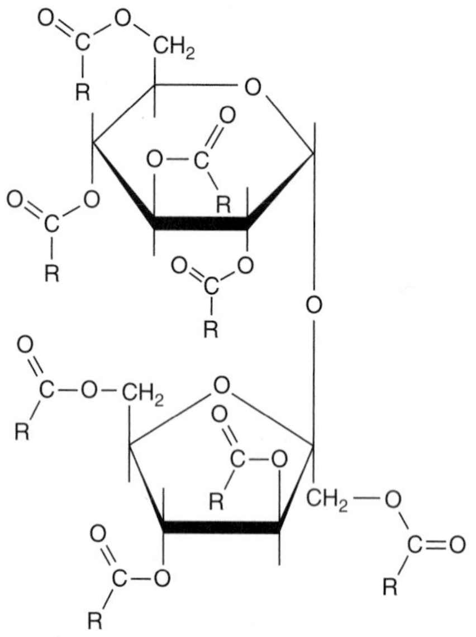

Chapter 8: In-Chapter Questions

8.1
The large excess of NADH that is produced by these reactions drives the conversion of pyruvate to lactate.

8.2
Chromium is acting as a cofactor.

8.3
In the absence of O_2, energy is produced only through glycolysis, an anaerobic process. Glycolysis produces less energy per glucose molecule than does aerobic respiration. Consequently, more glucose molecules must be metabolized to meet the energy needs of the cell. When O_2 is present, the flux of glucose through glycolysis is reduced.

8.4
At three strategic points, glycolytic and gluconeogenic reactions are catalyzed by different enzymes. For example, phosphofructokinase and fructose-1,6-diphosphatase catalyze opposing reactions. If both reactions occur simultaneously (i.e., in a futile cycle) to a significant extent, ATP hydrolysis in the reaction catalyzed by phosphofructokinase releases large amounts of heat. If the heat is not quickly dissipated, an affected individual could die of hyperthermia.

8.5
In gluconeogenesis pyruvate is converted to oxaloacetate. NADH and H^+ are required to reduce glycerate 1,3-bisphosphate to glyceraldehyde-3-phosphate. NAD^+ is the oxidized form of NADH also produced in this reaction. ATP is needed to provide the energy to carboxylate pyruvate to oxaloacetate and phosphorylate glyceraldehyde-3-phosphate to glycerate-1,3-bisphosphate. Both of these reactions also produce ADP and P_i. GTP is hydrolyzed in the conversion of oxaloacetate to phosphoenolpyruvate to yield GDP and P_i. Water is involved in the hydrolysis reactions of ATP to ADP and P_i, the conversion of phosphoenolpyruvate to 2-phosphoglycerate, and the hydrolysis of glucose-6-phosphate to glucose. Six protons are formed when four molecules of ATP and two molecules of GTP are hydrolyzed.

8.6
Without glucose-6-phosphatase activity, the individual cannot release glucose into the blood. Blood glucose levels must be maintained by frequent consumption of carbohydrate. Excess glucose-6-phosphate is converted to pyruvate, which is then reduced by NADH to form lactate.

8.7
The enzyme deficiencies prevent the breakdown of glycogen. Because the synthetic enzymes are active, some glycogen continues to be produced and causes liver enlargement. Because of the liver's strategic role in maintaining blood glucose, defective debranching enzyme causes hypoglycemia (low blood sugar).

Chapter 8: End-of-Chapter Questions

Review Questions

3

a. tautomerization—chemical reaction by which two tautomers are interconverted by the movement of a hydrogen atom and a double bond
b. tautomer—an isomer that differs from another in the location of a hydrogen atom and a double bond (e.g., keto-enol-tautomers)
c. amphibolic pathway—a metabolic pathway that functions in both anabolism and catabolism
d. electron transport system—a series of electron carrier proteins that bind reversibly to electrons at different energy levels
e. decarboxylation reaction—the removal of a carboxylic group from a carboxylic acid as carbon dioxide

6
Phosphorylation of glucose upon its entry into the cells prevents leakage of the molecule out of the cell and facilitates its binding to the active sites of enzymes.

9
The early activation of nutrient molecules in these pathways allows ATP production under varying substrate and product concentration.

12

The unique reactions in gluconeogenesis are (1) synthesis of phosphoenolpyruvate, catalyzed by pyruvate carboxylase and PEP carboxylated, (2) conversion of fructose 1,6-bisphosphate to fructose-6-phosphate, catalyzed by fructose-1-6-bisphophatase, and (3) the formation of glucose from glucose-6-phosphate, catalyzed by glucose-6-phosphatase.

15

The three principal hormones that regulate the body's levels of glucose are insulin, glucagon, and epinephrine. Insulin is an anabolic hormone that causes target cells to take up glucose from the blood and stimulates glycogenesis in liver and muscle and glycolysis in the liver. Glucagon is released when blood glucose levels are low. It stimulates liver glycogenolysis and gluconeogenesis, both of which produce glucose, which is then released into the blood. Epinephrine, released under stressful conditions, raises blood glucose levels by triggering liver glycogenolysis.

18

Substrate-level phosphorylation is the synthesis of ATP (from ADP and P_i) that is coupled to the exergonic breakdown of a high-energy organic substrate. Examples of this process in glycolysis are the conversions of glycerate-1,3-bisphosphate to glycerate-3-phosphate (by phosphoglycerate kinase) and phosphoenolpyruvate to pyruvate (by pyruvate kinase).

21

Epinephrine promotes the conversion of glycogen to glucose by activating adenylate cyclase, an enzyme whose product, cAMP, initiates a reaction cascade that activates the glycogen-degrading enzyme glycogen phosphorylase.

24

Gluconeogenesis occurs mainly in the liver. It is activated by processes that deplete blood glucose, such as fasting and exercise. Futile cycles are prevented by having the forward and reverse reactions catalyzed by different enzymes, both independently regulated.

26

Glucose can be a substrate or a product of a number of key metabolic pathways depending upon the needs of the cell, and as such, glucose plays a central role in carbohydrate metabolism. When energy is needed, glucose undergoes glycolysis to form pyruvate, which can be oxidized further either aerobically via the citric acid cycle and the electron transport chain to produce ATP, or anaerobically to form lactate (or ethanol in yeast cells and some bacteria). When energy is abundant, glucose is converted to glycogen via glycogenesis, or it may enter the pentose phosphate pathway to form pentose and other sugars. Also, pyruvate formed from glucose may react to form acetyl-CoA, a precursor for fatty acid synthesis as well as the citric acid cycle. Glucose may also be synthesized from pyruvate via gluconeogenesis, or from certain amino acids.

30

Fructose → (Fructokinase; ATP → ADP) → Fructose-1-phosphate

Fructose-1-phosphate → (Fructose-1-phosphate aldolase; ATP → ADP) → DHAP + Glyceraldehyde

Glyceraldehyde → (Glyceraldehyde kinase; ATP → ADP) → Glyceraldehyde-3-phosphate

33
In glycolysis, a dehydration occurs in the conversion of 2-phosphoglycerate to phosphoenolpyruvate, catalyzed by enolase.

36
Insulin is released in response to high blood glucose levels. In the liver, insulin leads to the inhibition of glycogenolysis and the activation of glycogenesis, resulting in a decrease in blood glucose levels. Insulin also causes an increased cellular uptake (adipocytes and muscle cells) of glucose. Glucagon is released in response to low blood glucose levels. In the liver, glucagon leads to the inhibition of glycogenesis and the activation of glycogenolysis, resulting in the release of glucose into the bloodstream.

39
The fate of pyruvate under anaerobic conditions is fermentation. In muscle cells, lactate is the only product, in yeast cells the products are ethanol and CO_2 and some microorganisms produce lactate and other acids or alcohols. All of these fermentation reactions result in the regeneration of NAD^+ so that glycolysis can continue. The fate of pyruvate under aerobic conditions is to be oxidized completely to form CO_2 and H_2O.

Fill in the Blank

42. Aldol cleavage
45. Transcription
48. Antioxidants

Short Answer

51
A possible sequence of reactions is shown below:

Two molecules of acetyl-CoA react to yield acetylacetone. The latter molecule is reduced to yield 2-buten-1-ol. Further reduction yields butanol.

54
The glucose alanine cycle has several purposes. Alanine (the product of the transamination of pyruvate), released by exercising muscle into the bloodstream, enters the liver, where it is reconverted to pyruvate, a gluconeogenesis substrate. The cycle is also used to safely transport amino nitrogen to the liver and a means of recycling α-keto acids between muscle and liver.

Thought Questions

57
Glucokinase (GK) acts as a glucose sensor because of its relatively low affinity for glucose. Since it does not usually work at high velocity GK is sensitive to small changes in blood glucose levels. In each cell type in which GK occurs it is linked to a signal transduction pathway. For example, in pancreatic β-cells, GK-catalyzed glucose phosphorylation, which occurs when blood glucose levels are high, triggers insulin release.

60
In liver, fructose is metabolized more rapidly than glucose because its metabolism bypasses two regulatory steps in the glycolytic pathway: the conversion of glucose to glucose-6-phosphate and fructose-6-phosphate to fructose-1,6-bisphosphate. Recall that fructose-1-phosphate is split into glyceraldehyde and DHAP, both of which are subsequently converted to glyceraldehyde-3-phosphate.

63
If gluconeogenesis and glycolysis were exactly the reverse of one another, futile cycles would be established and much energy would be wasted. In addition, it would be impossible for the body to store glycogen or release glucose into the blood as needed.

66
In high concentrations fructose can bypass most of the regulatory steps of the glycolytic cycle. Instead of being stored in glycogen molecules, the carbon skeletons of excess fructose molecules are converted through pyruvate and acetyl-CoA to fatty acids in triacylglycerol molecules.

69
Both glycogen and triacylglycerols are energy sources. One difference between the two molecules is in the speed with which energy can be mobilized. Glycogen can be converted to glucose and diverted to energy production very quickly (instant energy). It takes longer to mobilize the fat reserves, but once activated they provide the energy for sustained effort.

72
Refer to **Figures 8.15a** and **8.15b**, which illustrate the pentose phosphate pathway. Note that the ^{14}C label at C-2 of glucose-6-phosphate is C-1 of ribulose-5-phosphate as the result of a decarboxylation reaction. When ribulose-5-phosphate enters the nonoxidative phase of the pentose phosphate pathway the radioactive label will appear as C-1 of ribose-5-phosphate, xylulose-5-phosphate, sedoheptulose-7-phosphate and fructose-6-phosphate.

Chapter 9: In-Chapter Questions

9.1

With a $\Delta E_0'$ value of -0.345 V, the oxidation of NO_2^- is not spontaneous as written. The oxidation of ethanol is spontaneous as written because its $\Delta E_0'$ value is positive (+0.275 V).

9.2

Reactions 3, 4, and 5 are redox reactions. In reaction 3, lactate is the reducing agent and NAD^+ is the oxidizing agent. In reaction 4, cyt b (Fe^{2+}) is the reducing agent and NO_2^- the oxidizing agent. In reaction 5, NADH is the reducing agent and CH_3CHO is the oxidizing agent.

9.3

The oxidation states of the functional group carbon (indicated in bold) are:

CH_3CH_2OH	$0 - 1 - 1 + 1 = -1$
CH_3CHO	$0 - 1 + 2 = +1$
CH_3COOH	$0 + 1 + 2 = +3$

9.4

As it is incorporated into an organic molecule, the carbon atom in CO_2 is reduced.

9.5

Because of its symmetrical structure, a molecule of succinate derived from a ^{14}C-labeled acetyl-CoA is converted into two forms of oxaloacetate, one with a labeled methylene group and one with a labeled carbonyl group. ^{14}C-labeled CO_2 is not released until the third turn of the cycle when one-half of the original labeled carbon is lost (the carbonyl group derived from acetyl-CoA). The labeled carbon is further scrambled when succinyl-CoA is converted to succinate during the third and fourth turns of the cycle.

9.6

Pyruvate carboxylase converts pyruvate to oxaloacetate. If the enzyme is inactive, concentrations of pyruvate in the system rise and pyruvate is converted by NADH to lactate. Excess lactate is then excreted in the urine.

Chapter 9: End-of-Chapter Questions

Review Questions

3

a. amphibolic pathway—a metabolic pathway that functions in both anabolism and catabolism
b. anaplerotic reaction—a reaction that replenishes a substrate needed for a biochemical pathway
c. glyoxylate cycle—a modification of the citric acid cycle that occurs in plants, bacteria, and other eukaryotes; allows growth in these organisms from two carbon substrates such as ethanol, acetate, and acetyl-CoA
d. reduction potential—redox potential – a measure of the tendency of an electron donor in a redox pair to lose an electron
e. conjugate redox pair—in a half-reaction an electron donor (e.g., Cu^+) and an electron acceptor (e.g., Cu^{+2})

6

Contracting muscle converts large amounts of ATP to ADP in the muscle cells. The drop in ATP concentration stimulates two key regulatory enzymes of the citric acid cycle. (1) Citrate synthetase catalyzes the condensation of acetyl-CoA and oxaloacetate to give citrate. ATP is an allosteric inhibitor of this enzyme. As concentrations of ATP drop, this enzyme becomes more active. (2) Isocitrate dehydrogenase, which converts isocitrate to α-ketoglutarate, is inhibited by high concentrations of ATP and activated by high concentrations of ADP. A reduced ATP concentration also stimulates the conversion of pyruvate to acetyl-CoA catalyzed by pyruvate dehydrogenase.

9

The first two reactions of the glyoxylate cycle, catalyzed by citrate synthase and aconitase, also occur in the citric acid cycle. In the next two reactions isocitrate is split into succinate and glyoxylate. Glyoxylate reacts with acetyl-CoA to form malate. The cycle is completed when malate is converted to oxaloacetate by malate dehydrogenase.

12

The balanced equations for each of the reactions of the citric acid cycle are:

1. $C_2H_3O\text{-SCoA} + C_4H_2O_5^{2-} + H_2O \rightarrow \text{CoASH} + C_6H_5O_7^{3-} + H^+$
   ACETYL-COA OXALOACETATE CITRATE
2. $C_6H_5O_7^{3-} \rightarrow C_6H_5O_7^{3-}$
   CITRATE ISOCITRATE
3. $C_6H_5O_7^{3-} + NAD^+ \rightarrow C_5H_4O_5^{2-} + NADH + CO_2$
   ISOCITRATE α-KETOGLUTARATE
4. $C_5H_4O_5^{2-} + NAD^+ + \text{CoASH} \rightarrow C_4H_4O_3^{3-}\text{–SCoA} + NADH + CO_2$
   α-KETOGLUTARATE SUCCINYL-COA
5. $C_4H_4O_3^{3-}\text{–SCoA} + GDP + HPO_4^{2-} \rightarrow C_4H_4O_4^{2-} + GTP \text{ (or ATP)} + \text{CoASH}$
   SUCCINYL-COA (OR ADP) P_i SUCCINATE
6. $C_4H_4O_4^{2-} + FAD \rightarrow C_4H_2O_4^{2-} + FADH_2$
   SUCCINATE FUMARATE
7. $C_4H_2O_4^{2-} + H_2O \rightarrow C_4H_4O_5^{2-}$
   FUMARATE L-MALATE
8. $C_4H_4O_5^{2-} + NAD^+ \rightarrow C_4H_2O_5^{2-} + NADH + H^+$
   L-MALATE L-OXALOACETATE

15

The biochemical process required to obtain energy from glucose when oxygen is available are glycolysis, the conversion of pyruvate to acetyl-CoA, which then enters the citric acid cycle. The reduced coenzyme products of the cycle, NADH and $FADH_2$ then donate electrons to the electron transport system in which oxygen is the terminal electron acceptor. The energy captured by the electron transport system is then used in the synthesis of ATP.

18

Oxygen is required for the electron transport chain (ETC):

$$O_2 + NADH + H^+ \rightarrow H_2O + NAD^+$$

Without oxygen, the ETC shuts down, and NADH accumulates at the expense of NAD^+. Both high NADH and low NAD^+ levels inhibit the citric acid cycle. Low NAD^+ levels would also impact the two key regulatory enzymes that utilize NAD^+ as a substrate.

21

The roles of each enzyme, cofactor, and coenzyme of the pyruvate dehydrogenase complex are as follows. (1) Pyruvate decarboxylase (also called pyruvate dehydrogenase) decarboxylates pyruvate via the coenzyme TPP (thiamine pyrophosphate), to form HETPP (hydroxyethyl TPP) and CO_2. (2) Dihydrolipoyl transacetylase catalyzes the transfer of an acetyl group from HETPP to the coenzyme lipoic acid to regenerate TPP and form acetyl lipoic acid. A second transfer of this acetyl group from acetyl lipoic acid to the coenzyme CoASH forms acetyl CoA and dihydrolipoic acid. (3) Dihydrolipoyl dehydrogenase reoxidizes the dihydrolipoic acid to regenerate lipoic acid.

24

Unlike organisms that are capable of the glyoxylate pathway, animals cannot use two-carbon molecules as precursors in gluconeogenesis. One pathway to gluconeogenesis begins with pyruvate reacting to form oxaloacetate, then phosphoenolpyruvate. Since the decarboxylation of pyruvate to form acetyl-CoA is irreversible, and there is no biosynthetic pathway from acetyl-CoA to pyruvate, it cannot be formed from two-carbon molecules. Gluconeogenesis may use oxaloacetate as a precursor. However, oxaloacetate cannot be synthesized solely from acetyl-CoA. When acetyl-CoA enters the citric acid cycle by reacting with oxaloacetate, it adds two carbons to the cycle. However, two carbons are removed as CO_2 in subsequent reactions. Thus, there is no net addition of carbon atoms to the cycle. If oxaloacetate is used for gluconeogenesis, it must be replenished via a citric acid cycle intermediate larger than acetyl-CoA, or by its synthesis from certain amino acids. The glyoxylate cycle, not present in animals, does allow gluconeogenesis from 2-carbon molecules because it bypasses the citric acid cycle reactions that liberate CO_2.

27

Biosynthetic pathways that utilize citric acid cycle intermediates as precursors include the syntheses of: glucose from oxaloacetate; pyrimidines from oxaloacetate, purines from α-ketoglutarate, fatty acids and cholesterol from citrate; porphyrin, heme, or chlorophyll from succinyl-CoA; the amino acids Asp, Lys, Thr, Ile, and Met from oxaloacetate; and the amino acids Glu, Gln, Pro, and Arg from α-ketoglutarate.

Fill in the Blank

30. Amino acids
33. Oxaloacetate
36. Mitochondria

Short Answer

39

Calcium ions activate the regulatory enzymes of the citric acid cycle. Calcium ions are a contributing factor in the activation of the pyruvate dehydrogenase complex (PHDC) because Ca^{2+} activates the dephosphorylating enzyme pyruvate dehydrogenase phosphatase (PDP). The other two regulatory enzymes, isocitrate dehydrogenase and α-ketoglutarate, are activated when Ca^{2+} binds to the regulatory sites of both enzymes.

42

Because the enzyme pyruvate carboxylase converts pyruvate to oxaloacetate, its deficiency causes a build-up of pyruvate in the cytoplasm, where it reacts with NADH to yield lactate. When blood levels of lactate are higher than normal, the molecule is excreted in urine.

Thought Questions

45

In substrate-level phosphorylation, ADP is converted to ATP by the direct transfer of a phosphoryl group from a high energy compound. The only reaction in the citric acid cycle that involves this type of reaction is the cleavage of succinyl-CoA to form succinate, CoASH, and GTP. Another example of a substrate level phosphorylation is the glycolytic reaction that converts phosphoenolpyruvate and ADP to pyruvate and ATP.

48

Three enzymes that require thiamine are pyruvate dehydrogenase, a-ketoglutarate dehydrogenase and transketolase. Thiamine is involved in decarboxylation and acyl group transfer reactions. Absence of decarboxylation reactions would prevent pyruvate from being decarboxylated to acetyl-CoA. The body would then lack two carbon units for synthesis and energy production. Pyruvate accumulates as lactate. The overall results of thiamine deficiency are lack of energy, muscle wasting and acidosis

51

Carboxylation of pyruvate produces oxaloacetate, a citric acid cycle intermediate. Increasing the concentration of one of the intermediates stimulates the cycle and more energy is produced.

54

Inhibition of pyruvate dehydrogenase kinase, an enzyme that contributes to the regulation of pyruvate dehydrogenase, has the effect of increasing the conversion of pyruvate molecules to acetyl-CoA, thereby decreasing lactate levels.

57

$\Delta G° = -nF\,\Delta E°$

The half cells for the two reactions are: E

$S + 2H^+ + 2e^- \rightarrow H_2S$ $-0.23V$

$\frac{1}{2}O_2 + 2H^+ + 2e^- \rightarrow H_2O$ $+0.82V$

$NADH \rightarrow NAD^+ + H^+ + 2e^-$ $+0.32V$

For the formation of hydrogen sulfide the reaction would be:

$S + H^+ + NADH \rightarrow H_2S + NAD^+ = +09V$

$\Delta E° = -23\ V + 0.32\ V = +0.09\ V$

$\Delta G° = 2(0.96485\ J/\ mol\ V)(09) = 17{,}367\ J$

For the formation of water the reaction would be

$\frac{1}{2}O_2 + H^+ + NADH \rightarrow H_2O + NAD^+ = +1.14\ V$

$\Delta E° = +0.82V + 0.32V = 1.14\ V$

$\Delta G° = 2(96485\ J/\ mol\ V)(1.14\ V) = 219985\ J$

The difference in energy yield is 219985 − 17367 = 20618 kJ or 200 kJ

Chapter 10: In-Chapter Questions

10.1

a. NADH
b. $FADH_2$
c. Cyt b (reduced)
d. NADH
e. NADH

10.2

DNP is a lipophilic molecule that binds reversibly with protons. It dissipates that proton gradient in mitochondria by transferring protons across the inner membrane. The uncoupling of electron transport from oxidative phosphorylation causes the energy from food to be dissipated as heat. DNP causes liver failure because of insufficient ATP synthesis in a metabolically demanding organ.

10.3

No, for ATP synthesis to occur, the proton concentration must be higher within the inside-out mitochondrial particles. ATP synthesis requires that protons move down a concentration gradient through the base of the ATP synthetase across the membrane.

10.4

Disregarding proton leakage and assuming that the glycerol phosphate shuttle is in operation, 38 ATP would be produced from the aerobic oxidation of a glucose molecule. If the malate shuttle is in operation, only 36 ATP would be produced.

10.5

Sucrose is a disaccharide composed of glucose and fructose. As described (p. 364) the oxidation of 1 mol of glucose yields a maximum of 31 mol of ATP. Fructose, which like glucose is also partially degraded by the glycolytic pathway, also yields a maximum of 31 mol ATP. The total maximum energy yield is 62 mol of ATP per mol of sucrose.

10.6

The larger selenium atom holds its electrons less tightly than sulfur. Selenium is more easily oxidized and therefore acts as a better scavenger for oxygen than does sulfur.

10.7

The SH groups reduce hydrogen peroxide or trap hydroxyl radicals to form water. An example of a nonsulfhydryl group–containing molecule that should be capable of this activity is vitamin C or any of a number of other antioxidants (carotenoids, flavonoids, tocopherols, etc.).

10.8

The phenolic groups of both molecules are responsible for their antioxidant activity because of the ease of formation of phenoxy radicals with subsequent neutralization of electron-deficient ROS.

Chapter 10: End-of-Chapter Questions

Review Questions

3

a. chemiosmotic coupling theory—ATP synthesis is coupled to electron transport by the electrochemical proton gradient across a membrane
b. uncoupler—a molecule that uncouples ATP synthesis from electron transport; it collapses a proton gradient by transporting protons across the membrane
c. ionophore—a substance that transports cations across a membrane
d. submitochondrial particle—small membranous vesicles formed when mitochondria are subjected to sonication; used to investigate ATP synthesis
e. α,β hexmer—a component of the F_1 unit of mitochondrial ATP synthesis; catalyzes ATP synthesis

6

a. reactive oxygen species (ROS)—a reactive derivative of molecular oxygen, including superoxide radical, hydrogen peroxide, the hydroxyl radical, and singlet oxygen
b. antioxidant—a substance that prevents the oxidation of other molecules
c. oxidative stress—excessive production of reactive oxygen species
d. respiratory burst—an oxygen-consuming process in scavenger cells such as macrophages in which ROS are generated and used to kill foreign or damaged cells
e. reactive nitrogen species (RNS)—nitrogen containing radicals often classified as ROS; examples include nitric oxide, nitrogen dioxide, and peroxynitrite

9

a. radical—an atom or molecule with an unpaired electron
b. ROS—reactive oxygen species – a reactive derivative of molecular oxygen
c. RNS—reactive nitrogen species—nitrogen-containing radicals
d. GSH—glutathione—an intracellular reducing agent
e. SOD—superoxide dismutase—an enzyme that catalyzes the conversion of superoxide radical to hydrogen peroxide and oxygen

12

Principal sources of electrons for the mitochondrial electron transport system are NADH and $FADH_2$.

15

The translocation of three protons is required to drive ATP synthesis. The fourth proton drives the transport of ADP and P_i.

18

In small quantities ROS act as signal molecules. Excessive amounts of dietary antioxidants such as vitamin E compromise ROS-triggered defense mechanisms that induce increased synthesis of antioxidant enzymes.

21

The major enzyme defenses against oxidative stress are provided by superoxide dismutase, catalase, and glutathione peroxidase.

24

The advantage of dioxygen over the charged oxidizing agents, such as NO_3^-, Fe^{3+}, and SO_4^{2-}, is that O_2 readily diffuses across cell membranes. Unlike CO_2, which also diffuses across cell membranes, dioxygen is highly reactive and readily accepts electrons. Finally, O_2 is found almost everywhere on Earth's surface, and as such, is much more readily accessible than oxidizing agents such as sulfur.

27

There are four protein complexes in the mitochondrial ETC. Complex I (NADH dehydrogenase complex) catalyzes the transfer of electrons from NADH to UQ. Complex II (succinate dehydrogenase complex) mediates the transfer of electrons from succinate, an intermediate in the citric acid cycle, via FAD to UQ. Complex III (cytochrome bc_1 complex) transfers electrons from UQH_2 to cytochrome c, a mobile electron carrier protein loosely associated with the outer face of the inner mitochondrial membrane. Complex IV (cytochrome oxidase) transfers the electrons donated by cytochrome c that reduce oxygen to yield water.

30

The net reaction of the electron transport system is highly exergonic. Using a series of oxidations instead of one reaction allows a more controlled, more efficient capture of this energy, avoiding the liberation of a great deal of heat.

33

NADH dehydrogenase (Complex I) oxidizes NADH to NAD^+. When Complex I is inhibited, NADH accumulates (resulting in an increased NADH:NAD^+ ratio) and $FADH_2$, which is oxidized in Complex II, is unaffected.

Fill in the Blank

36. Respirasome
39. Uncouplers
42. Respiratory burst

Short Answer

45

Once the respiratory burst is initiated, NADPH oxidase converts O_2 and NADPH to superoxide ($O_2^{\bar{\cdot}}$), $NADP^+$, and H^+. Two molecules of superoxide react in a reaction catalyzed by superoxide dismutase to form hydrogen peroxide. Myeloperoxidase catalyzes the oxygenation of chloride ions by H_2O_2 to yield hypochlorite (^-OCl) and H_2O.

48

Methane is stable in air until an energy source such as a spark is introduced. Once this occurs, the oxygen diradical abstracts a hydrogen atom from a methane molecule to yield a hydroxyperoxide radical ($HOO^\bullet$) and a methyl radical ($CH_3^\bullet$), both of which proceed to react with other methane molecules. The chain reaction accelerates as newly formed radicals react with other methane molecules, stopping only when all fuel molecules have been consumed or O_2 is excluded from the reaction. Both heat and light are released as carbon-hydrogen bonds are broken during the chain reaction.

Thought Questions

51

Low levels of G-6-PD in combination with a high level of oxidized GSH cause high oxidative stress. Without antioxidant protection, red cell membranes become fragile, a condition that eventually causes hemolytic anemia

54

Dinitrophenol collapses the proton gradient across the mitochondrial inner membrane. The energy normally used to drive ATP synthesis is lost as heat.

57

Cyanide binds to cytochrome oxidase irreversibly. All the cytochromes in the ETC become reduced, and ATP synthesis ceases. In the case of dinitrophenol, the acidic phenol disrupts the proton gradient by shuttling protons across the membrane. None of the parts of the electron transport system are irreversibly altered by dinitrophenol. Removal of the phenol allows the system to restore itself.

60
If the cytochrome complexes were not embedded in the mitochondrial inner membrane the proton gradient could not be established and ATP synthesis would not occur.

63
If oxygen is used as the electron acceptor for the ETC, the difference in voltage between NADH and water is 1.14 V. The complete reduction of nitrate by NADH would be +1.18 V. This is almost exactly the same as for oxygen. It would be reasonable to assume that the same amount of ATP would be produced.

66
The carbon radicals generated by the attack of oxygen on an amino acid can dimerize to produce a crosslink between two amino acids.

2R′CHCONH→RCH(CONH)-CH(CONH)R

Chapter 11: In-Chapter Questions

11.1
Because steroids inhibit the release of arachidonic acid, their use shuts down the synthesis of most if not all eicosanoid molecules, hence their reputation as potent anti-inflammatory agents. Aspirin inactivates cyclooxygenase and prevents the conversion of arachidonic acid to PGG_2 the precursor of prostaglandins and thromboxanes. Aspirin is not as effective an anti-inflammatory agent as the steroids because it shuts down only a portion of eicosanoid synthetic pathways.

11.2
The product of complete hydrogenation would be hard and therefore not useful as a margarine.

11.3
When soap and grease are mixed, the hydrophobic hydrocarbon tails of the soap insert (or dissolve) into the oil droplet. The oil droplet becomes coated with soap molecules. The hydrophilic portion of the soap molecules allows the soap-oil complex to be dispersed in water.

11.4
The phospholipid of the surfactant, which possesses a polar head group and two hydrophobic acyl groups, disrupts some of the intermolecular hydrogen bonds of the water, thereby decreasing the surface tension.

11.5
Carvone and camphor are monoterpenes; abscisic acid is a sesquiterpene.

11.6
Bile salts are structurally similar to soap in that they contain a polar head group (e.g., the charged amino acid residue glycine) and a hydrophobic tail (the steroid ring system).

11.7
a. simple diffusion
b. secondary active transport or facilitated diffusion
c. primary active transport or exchanger protein
d. primary active transport or gated channel
e. fat molecules (triacylglycerols) are not directly transported across cell membranes; they must be hydrolyzed first.
f. simple diffusion.

11.8
The main stabilizing feature of biological membranes is hydrophobic interactions among the molecules in the lipid bilayer. The phospholiqids in the lipid bilayer orient themselves so that their polar head groups interact with water. Proteins in the lipid bilayer interact favorably in their hydrophobic milieu because they typically have hydrophobic amino acid residues on their outer surfaces.

11.9
The transport mechanisms discussed in the chapter fit into the following categories:
sodium channel: uniporter
glucose permease: passive uniporter
Na^+-K^+-ATPase: antiporter

Chapter 11: End-of-Chapter Questions

Review Questions

3
a. prostaglandin—an arachidonic acid derivative that contains a cylcopentane ring with hydroxyl groups at C-11 and C-15
b. thromboxane—a derivative of arachidonic acid that contains a cyclic ester
c. leukotriene—a derivative of arachidonic acid; linear molecules whose synthesis is initiated by peroxidation catalyzed by lipooxygenase
d. autocrine—refers to a hormone-like molecule that is active within the cell in which it is produced
e. anaphylaxis—an unusually severe allergic reaction triggered when an allergen binds to IgE antibodies on the surface of mast cells

6
a. phosphoglyceride—a membrane lipid composed of glycerol linked to two fatty acids, phosphate, and a polar group
b. sphingolipid—a membrane lipid that contains a long-chain amino alcohol and ceramide (a fatty acid derivative of sphingosine)
c. GPI anchor—a glycolipid (glycosylphosphatidylinositol) used to link certain proteins to membranes, preferentially in lipid rafts
d. glycolipids—a glycosphingolipid; a molecule in which a carbohydrate group is attached to ceramide through an O-glycosidic linkage
e. sphingomyelin—a type of phospholipid; the 1-hydroxyl group of ceramide is esterified to the phosphate group of phosphatidylcholine or phosphatidylethanoloamine and the amino group of sphingosine is in an amide linkage with a fatty acid

9

a. prenylation—the covalent attachment of prenyl groups (e.g., farnesyl and geranylgeranyl groups) to protein molecules
b. steroid—a derivative of triterpenes; contains four fused rings
c. digitalis—a type of cardiac glycoside; a molecule that increases the force of cardiac muscle contractions
d. lipoprotein—a conjugated protein in which lipid molecules are the prosthetic groups; a protein-lipid complex that transports water-insoluble lipids in the blood
e. apolipoprotein—the protein component of a lipoprotein

12

a. peripheral protein—a protein that is not embedded in the membrane but attached either by a covalent bond to a lipid molecule or by a noncovalent interaction with a membrane protein or lipid
b. integral protein—a protein that is embedded within a membrane
c. lipid raft—a specialized microdomain in the external leaflet of eukaryotic plasma membrane
d. passive transport—transport of a substance across membranes that requires no direct input of energy
e. active transport—the energy-requiring movement of molecules across a membrane against a concentration gradient

15

a. botulism—a form of muscle paralysis caused by a protein toxin produced by *Clostridium botulinum*
b. botulinum toxin—a protein produced by *Clostridium botulinum* that causes muscle paralysis; prevents the release of the neurotransmitter acetylcholine from the presynaptic axon of motor neurons
c. t-SNARE—target membrane specific soluble N-ethylmaleimide sensitive factor attachment protein receptors —a component of the fusion machinery required for exocytosis at nerve terminals
d. v-SNARE—vesicle specific SNARE
e. membrane fusion—the merging of two lipid bilayers

18

The protein component of plasma lipoproteins serves to solubilize the lipoproteins in the blood. It also binds to cell surface receptors that permit uptake of lipoproteins by body cells.

21

Bile salts are used to emulsify dietary fat and oils so they may be digested and then absorbed.

24

For a phospholipid to move from one side of the bilayer to the other, the polar head must move through the hydrophobic portion of the phospholipid membrane. This process requires a significant amount of energy and is therefore relatively slow.

27

Prostaglandins have major recognized roles in all the following processes: reproduction, respiration, and inflammation. (*Note*: Thromboxanes promote blood clotting, while prostaglandins inhibit platelet aggregation, thus inhibiting blood clotting. Some prostaglandins also have vasodilating or vasoconstricting activity, which may affect blood pressure.)

30

The indicated compounds are classified as follows:

a. monoterpene
b. monoterpene
c. sesquiterpene
d. polyterpene
e. diterpene
f. triterpene

33

To increase a cell's resistance to mechanical stress, increase the content of cholesterol and cardiolipin (two phospholipids linked by a glycerol) in the cell membrane.

36

The three proteins that facilitate the movement of phospholipid molecules from one side of a membrane to another are as follows. Flippase transfers phospholipids from the outer to the inner membrane leaflet. Floppase transfers phospholipids from the inner to the outer leaflet. Scramblase is a nonspecific, energy-independent redistributor of phospholipids across membranes.

39

a. In primary active transport, ATP provides energy to transport a substance *against* a concentration gradient. The Na^+-K^+ ATPase pump is a primary transporter, since it transports substances across a membrane from lower to higher concentration. Secondary active transport uses the concentration gradients that were generated by primary active transport to drive the transport of a different substance against its concentration gradient. An example of secondary active transport is the use of the gradient created by the Na^+-K^+ ATPase pump to transport glucose.
b. Passive transport is the general term for the transport of substances across a membrane by diffusion. Since passive transport occurs *down* (or *with*) a concentration gradient (i.e., from an area of high concentration to low concentration), no energy is required. Facilitated diffusion is the passive transport of substances such as ions or polar molecules that are unable to cross the membrane alone and require the presence of a protein channel or carrier. For example, nonpolar organic molecules (such as steroid hormones) and carbon dioxide cross cell membranes by passive diffusion. The red blood cell glucose transporter is an example of a carrier, and Na^+ may diffuse through a membrane only through specific Na^+ channel proteins.
c. Both carrier-mediated and channel-mediated transport are examples of facilitated diffusion. In channel-mediated transport, an integral protein forms a channel through which a specific substance may pass. In carrier-mediated transport, the substance to be transported binds to the carrier protein and causes a conformational change. This change results in the substance crossing the membrane, where it is then released by the carrier protein. See (b) for examples.

42

Detergents disrupt membranes and extract membrane proteins by solubilizing both their hydrophobic and their hydrophilic components, effectively dispersing them throughout the aqueous solvent.

Fill in the Blank

45. Integral
48. Scramblase
51. Against

Short Answer

54

HDLs (high-density lipoproteins) are a form of plasma lipoprotein that appear to be scavengers of excess cholesterol in cell membranes and cholesteryl esters from VLDLs and LDLs.

57

Toxic phospholipases are used by species such as snakes to inflict damage on prey animals. For example, PLA_2, found in snake venoms, digests cell membranes at the site of a bite and causes systemic damage such as skeletal and heart muscle necrosis.

Thought Questions

60

Most of the cholesterol in plaque results from the ingestion of LDL by the foam cells that line the arteries. High blood plasma LDL therefore promotes atherosclerosis. Because the coronary arteries are narrow, they are especially prone to occlusion by atherosclerotic plaque.

63

The hooves and lungs are subjected to much lower temperatures than the rest of the body. At these low temperatures, the membrane must be modified so that the membranes remain fluid. This can be done by increasing the unsaturation of the nonpolar tails of the membrane phospholipids.

66

Both carbohydrates and proteins contain large numbers of atoms capable of hydrogen bonding (oxygen and nitrogen). In the presence of water these materials would either dissolve or swell. Waxes on the other hand are composed of hydrophobic molecules that are resistant to the penetration of water from the leaf interior. A relatively thick wax layer prevents insect penetration.

69

Boric acid is a hard crystalline solid. One of the ways that this abrasive molecule kills insects is that when it contacts the exoskeleton the crystals cut into the wax coat. These gaps allow water to escape and the insect dies of dehydration. (Boric acid is also an internal poison in insects. When insects, which are covered with boric acid powder, groom themselves, they ingest the boric acid crystals. Boric acid interferes with the digestion of food, thereby causing starvation.)

Chapter 12: In-Chapter Questions

12.1

The triacylglycerols are emulsified in the small intestine by bile salts. They are then digested by lipases, the most important of which is pancreatic lipase. The products, fatty acids and monoacylglycerol, are transported into enterocytes and reconverted to triacylglycerol. Triacylglycerol is subsequently incorporated into chylomicrons, which are then transported into lymph via exocytosis and finally into the bloodstream for transport to the fat cells.

12.2

Although there is no cure for MCAD, symptoms can be managed primarily by ensuring that frequent feeding occurs, i.e., there are no prolonged fasts. This can be a challenge in patients who often manifest a poor appetite so intravenous feeding with glucose may be necessary. Since toxic products of fatty acid metabolites can accumulate, the diet must be low in fat. Daily consumption of carnitine supplements may also be beneficial because acyl-carnitine derivatives are excreted in urine.

12.3

a. phospholipid
b. acyl-CoA
c. carnitine

12.4

Unlike the oxidation of glucose to form pyruvate, fatty acid oxidation, which involves the citric acid cycle and the electron transport system, cannot operate in the absence of O_2.

12.5

The yield from the oxidation of stearyl-CoA is calculated as follows:

8 $FADH_2$ × 1.5 ATP/$FADH_2$ =	12 ATP
8 NADH × 2.5 ATP/NADH =	20 ATP
9 Acetyl-CoA × 10 ATP/Acetyl-CoA =	90 ATP
	122 ATP

Two ATP are required to form stearyl-CoA from stearate to give a total of 120 ATP.

12.6

Propionyl-CoA can be reversibly converted to succinyl-CoA, an intermediate in the citric acid cycle. Oxaloacetate, a downstream intermediate of this cycle can be converted to PEP. PEP is then converted to glucose via gluconeogenesis.

12.7

Adipic acid undergoes one round of β-oxidation to yield acetyl-CoA and succinyl-CoA. Succinyl-CoA is sequentially converted to oxaloacetate, PEP, and then to glucose.

12.8

Following the hydrolysis of sucrose, both monosaccharide products enter the bloodstream and travel to the liver, where fructose is converted to fructose-1-phosphate. Recall that the conversion of fructose-1-phosphate to glyceraldehyde-3-phosphate bypasses two regulatory steps. Consequently, more glycerol-phosphate and acetyl-CoA (the substrates for triacylglycerol synthesis) are produced. High blood glucose concentrations that result from this consumption of excessive amounts of sucrose trigger the release of larger than normal amounts of insulin. One of the functions of insulin is to promote fat synthesis.

12.9

a. β-Hydroxybutyrate is a product of ketone body metabolism.
b. Malonyl-CoA is the product of the reaction of acetyl-CoA and carboxybiotin that occurs during fatty acid systhesis.
c. Biotin is a carrier of CO_2 in fatty acid synthesis and several other reactions.
d. Acetyl ACP delivers acetate to the synthetic machinery of fatty acid synthesis.

12.10

The higher activity of HMG-CoA reductase in obese patients in combination with a high-calorie diet increases the synthesis of cholesterol.

Chapter 12: End-of-Chapter Questions

Review Questions

3

a. adipocytes—TG-storing cells in adipose tissue
b. MCAD—medium-chain acyl-CoA dehydrogenase – a mitochondrial enzyme that catalyzes the first reaction in the β-oxidation cycle
c. ACP—acyl carrier protein—a component of fatty acid synthase
d. α-oxidation—a mechanism for degrading branched-chain fatty acid molecules such as phytanic acid
e. odd chain oxidation—oxidation of fatty acids with an odd number of carbons; proceeds via β-oxidation to yield acetyl-CoA molecules and one propionyl-CoA

6

a. cytochrome P_{450}—one of a group of hemoproteins that when complexed with carbon monoxide absorb light at a wavelength of 450 nm; oxidizes a wide variety of hydrophobic molecules
b. mixed function oxidase—one of a group of enzymes that catalyze reactions in which one oxygen is consumed (reduced) per substrate molecule, one oxygen atom appearing in the product and the other in a molecule of water; also referred to as monooxygenases
c. flavin-containing monooxygenase—one of a group of enzymes that catalyze an NADPH and oxygen-requiring oxidation of molecules (primarily xenobiotics) bearing functional groups containing nitrogen, sulfur, or phosphorus
d. detoxication—the process by which a toxic molecule is converted to a more soluble and usually less toxic product
e. detoxification—correction of a state of toxicity; the chemical reactions that produce sobriety in an inebriated person

9

Peroxisomes have β-oxidation enzymes that are specific for long chain fatty acids, whereas mitochondria possess enzymes that are specific for short and moderate chain length fatty acids. In addition, the first reaction in the peroxisomal pathway is catalyzed by a different enzyme than the mitochondrial pathway. The $FADH_2$ produced in the first peroxisomal reaction donates its electrons to oxygen directly (forming hydrogen peroxide instead of UQ as in mitochondria). The processes are similar in that acetyl-CoA is derived from the oxidation of fatty acids.

12

Because of the presence of the methyl substituent on the β-carbon, the fatty acid first undergoes one cycle of α-oxidation. The resulting molecule, now shorter by one carbon atom, then undergoes one cycle of β-oxidation. The products of the latter process are two molecules of propionyl-CoA.

15

The yield of ATP from the oxidation of stearic acid is 122 (8 $FADH_2$ $\times$ 1.5 ATP, 8 NADH $\times$ 2.5 ATP and 9 acetyl-CoA x 10 ATP). Since the formation of stearoyl-CoA requires two ATP equivalents the net synthesis is 120 ATP. Stearic acid yields 120/18 or 6.6 ATP per carbon atom. Recall that the ratio for glucose is 5.2 ATP per carbon atom.

18

After the TG content of VLDL (very low density lipoproteins) has been depleated, the lipoprotein in referred to as an IDL (intermediate density) lipoprotein.

21

Enoyl-CoA isomerase converts the naturally occurring *cis* double bond at Δ^3 to a *trans* double bond at Δ^2, the correct position for the next round of β-oxidation.

24

The indicated bond is cleaved by glucocerebrosidase.

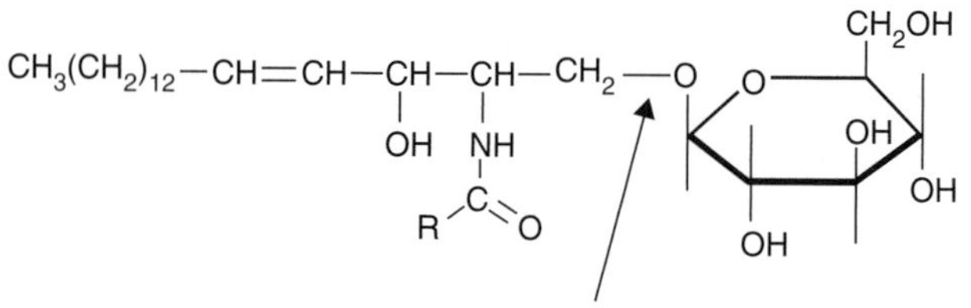

$CH_3(CH_2)_{12}$—CH=CH—CH(OH)—CH(NH—C(=O)—R)—CH_2—O—(glucose ring: CH_2OH, OH, OH, OH)

27

Review **Figure 12.31** on p 464. The bile salts are emulsifying agents that facilitate fat digestion.

30

The components of the cytochrome P_{450} electron transport system are: 1) cytochrome P_{450}, which contains heme, and 2) NADPH-cytochrome P_{450} reductase, a flavoenzyme that contains both FAD and FMN. Cyt P_{450} catalyzes oxidation reactions involving a wide variety of hydrophobic substrates. A hydroxyl group appears in each reaction. The electrons required for each Cyt P_{450}-catalyzed reaction are donated by Cyt P_{450} reductase.

33

The inability of peroxisomal thiolase to bind medium-chain acyl-CoA requires the final three-to-six β-oxidation cycles of every fatty acid to take place in one organelle, the mitochondria. This is an advantage to the cell because it allows for tighter regulation of the β-oxidation of fatty acids to form acetyl-CoA molecules. Efficiency is also increased since acetyl-CoA produced in mitochondria may enter the citric acid cycle without needing to be transported across membranes. Similarly, the NADH and $FADH_2$ may enter the ETC directly. Otherwise, there would be an energy cost associated with transporting acetyl-CoA from the peroxisomes to the mitochondria. The maximum number of ATP could not be realized from a fatty acid molecule in a peroxisome, since the NADH and $FADH_2$ produced there would not be able to enter the mitochondria to enter the ETC. A further consideration is the additional regulation by compartmentation made possible by the requirement of β-oxidation to be completed in mitochondria. For example, to prevent fatty acid synthesis from occurring at the same time as β-oxidation in the liver, the transport of fatty acids into mitochondria is inhibited when fatty acid synthesis is taking place in the cytoplasm. If β-oxidation could go to completion in peroxisomes, it would negate this control mechanism.

36

NADH donates electrons to an electron transport system composed of cytochrome b_5 reductase (a flavoprotein) and cytochrome b_5 (which contains heme). An oxygen-dependent desaturase uses these electrons (from NADH) to activate O_2. The activated O_2 oxidizes the fatty acyl-CoA to create an alkene in its hydrocarbon chain, and is reduced to form two molecules of water.

39

Since the $^{14}CO_2$ that is added to acetyl-CoA is removed in the reaction of malonyl-ACP with acetyl synthase to form acetoacetyl-ACP, no ^{14}C label appears in the eventual fatty acid products.

42

Glycerol (generated from the lipolysis of adipocyte triacylglycerols) is released into the blood and transported to the liver, where glycerol kinase converts it to glycerol-3-phosphate, which is a substrate for the synthesis of glucose, phospholipids, and triacylglycerols.

45

Of the four reactions in β-oxidation, only reactions 1 and 3 are oxidation reactions:

1) Acyl-CoA + FAD $\rightarrow$ *trans*-α,β-Enoyl-CoA + $FADH_2$
3) L-β-Hydroxyacyl-CoA + NAD^+ $\rightarrow$ β-Ketoacyl-CoA + NADH + H^+

Fill in the Blank

48. Biotin
51. Glycerol-3-phosphate
54. *S*-Adenosylmethionine

Short Answer

57

The bile acids are conjugated with glycine or taurine to yield the bile salts glycocholate or taurocholate. As with all conjugation reactions, the function of bile acid conjugation is an improvement in water solubility. These reactions serve an additional purpose in the physiology of digestion. The glycyl and tauro groups serve as a signal that prevents premature absorption of bile acids in the biliary tract and the upper region of the small intestine and promotes reabsorption in the distal ilium of the small intestine.

60

ACC1 (acetyl-CoA carboxylase 1) is a cytoplasmic enzyme in lipogenic tissues such as liver, adipose tissue, and lactating mammary gland that catalyzes the first reaction in fatty acid synthesis: the carboxylation of acetyl-CoA to yield malonyl-CoA. ACC1 is activated (i.e., ACC monomers polymerize to form high-molecular-weight polymers) when ACC1 monomers are dephosphorylated by phosphoprotein phosphatase 2A in response to insulin. ACC1 is inactivated by high glucagon or epinephrine levels. Glucagon stimulates the phosphorylation of polymerized ACC1 by AMPK and PKA to yield ACC1 monomers. Epinephrine inhibits phosphoprotein phosphatase 2A.

Thought Questions

63

a. Hydrophobic interactions are probable between the enzyme and the lipid in the micelle.
b. For phospholipases to be drawn into the micelle, they must have a hydrophobic surface.

66

Membrane phospholipids are synthesized on the cytoplasmic side of SER membrane. Because the polar head groups of phospholipid molecules make transport across the hydrophobic core of a membrane an unlikely event, a translocation mechanism is used to transfer phospholipids across the membrane to ensure balanced growth. Choline-containing phospholipids are found in high concentration on the lumenal side of ER membrane because a prominent phospholipid translocator protein called flippase preferentially transfers this class of molecule.

69

The reaction for the oxidation of one mole of palmitic acid is:

$$CH_3(CH_2)_{14}COOH + 23\ O_2 \rightarrow 16\ CO_2 + 16\ H_2O$$

One mole of palmitic acid produces a yield of 16 moles of water molecules. From this number must be subtracted 8 moles of water of which 7 moles are used in the hydration reaction in each round of the β-oxidation spiral and 1 mole is used for the hydrolysis of pyrophosphate, the reaction that drives the activation of the palmitic acid molecule to completion. The net reaction, therefore, yields a total of 8 molecules of metabolic water.

72

Fatty acids are assembled from two carbon acid synthase which ends the process at C-16. Although there is a robust capacity to elongate fatty acids, the most abundant fatty acids are C16 and C18, the product of one round of elongation process.

Chapter 13: In-Chapter Questions

13.1

Refer to **Figure 13.4** for an illustration of the relative locations of CF_oCF_1, P700 (in PSI), and P680 (in PSII) within thylakoid membrane. The Calvin cycle reactions occur in the stroma, the gel-like substance that surrounds the external surface of thylakoid membrane. CF_oCF_1 is the ATP synthase that utilizes the transmembrane proton gradient to drive ATP synthesis. P700 is the term used to indicate the special pair of chlorophyll a molecules in the reaction center of PSI that absorbs light energy and then donates the energized electrons that eventually reduce $NADP^+$. P680 is a special pair of chlorophyll a molecules in PSII that absorbs light energy and then donates energized electrons eventually to plastoquinone. The Calvin cycle enzymes are responsible for utilizing ATP and NADPH generated by the light reactions to incorporate $C0_2$ into carbohydrate molecules.

13.2

The energy of a photon is proportional to its frequency. Blue light has a higher frequency than green light and therefore has higher energy.

13.3

The presence of antenna pigments allows the light-harvesting systems of chloroplasts to collect energy from a wider range of frequencies than those absorbed by the chlorophylls. Because their absorption spectra overlap, the energy absorbed by the antenna pigments is quickly transferred to the critical chlorophylls of PSI and PSII.

13.4

Excessive light promotes the formation of ROS, which damage proteins such as D_1. β-Carotene is an antioxidant that prevents some of this damage.

13.5

a. Plastocyanin is a component of the cytochrome b_6f complex; a copper-containing protein that accepts electrons from plastoquinone.
b. β-Carotene is a carotenoid pigment that protects chlorophyll molecules from ROS.
c. Ferredoxin is a mobile, water-soluble protein that donates electrons to a flavoprotein called ferredoxin-NADP oxidoreductase.
d. Plastoquinone is a component of photosystem II that accepts electrons from pheophytin a to become plastoquinol.
e. Pheophytin a is a molecule similar in structure to chlorophyll that is a component of the electron transport pathway between PSII and PSI.
f. Lutein is a carotenoid that is a component of light-harvesting complexes.

13.6

Of the herbicides discussed, paraquat and DCMU are most hazardous to humans. Paraquat generates free radicals that can attack cell components. DCMU poisons the electron transport complex.

13.7

The hydrolysis of glycerate-l,3-bisphosphate generates 1 mol of ATP. Recall that aerobic respiration is stimulated by relatively high ADP concentrations and inhibited by relatively high ATP concentrations. Any measurable increase in ATP concentration has the effect of depressing aerobic respiration. Also recall that ATP is an inhibitor of PFK-1 and pyruvate kinase, enzymes required to channel carbon skeletons into the citric acid cycle.

Chapter 13: End-of-Chapter Questions

Review Questions

3

a. phylloquinone (A_1)—one of several electron carriers that transfers the PSI P700* energized electron to ferredoxin
b. lutein—a light absorbing and antioxidant carotenoid found in thylakoid membrane
c. Q_A—an electron carrier bound to protein in PSII; a plastoquinone
d. PQH_2—plastoquinol, the reduced form of platoquinone; oxidized to supply two electrons to plastocyanin
e. carotenoids—isoprenoid molecules that function as light-harvesting pigments and antioxidants

6

a. cytochrome b_6f complex—a multisubunit protein complex in thylakoid membrane that is similar in structure and function to cytochrome bc_1 complex in mitochondrial inner membrane; delivers electrons donated by plastoquinol to the water-soluble protein plastocyanin
b. CF_0—a membrane-spanning protein complex in the chloroplast ATP synthase that contains a proton channel
c. CF_1—the ATP synthesizing component of the chloroplast ATP synthase
d. LCHII—light-harvesting complex II – a trimer of light-harvesting proteins bound to chlorophyll a and b molecules; detachable from PSII
e. Mn_4CaO_5—a component of the oxygen-evolving complex; found on the luminal side of PSII

9

The most significant contribution of early photosynthesizing organisms to the earth's environment was the conversion of a reducing atmosphere (ammonia and methane) to an oxidizing atmosphere.

12

Excited molecules can return to its ground state in 4 ways: (1) fluorescence in which a molecule's excited state decays as it emits a photon; (2) resonance energy transfer in which excitation energy is transferred to neighboring chromophores through interaction between adjacent molecuar orbitals; (3) oxidation-reduction in which an excited electron is transferred to a neighboring molecule; and (4) radiationless decay into heat. The most important of these mechanisms in photosynthesis are resonance energy transfer, which plays an important role in the harvesting of light energy and oxidation-reduction, which is illustrated by the Z scheme. Fluorescence is used as a protective mechanism in which photons are reemitted when light absorption by a photosystem is excessive.

15

The ultimate goal of artificial photosynthesis research is to devise a cost-effective means of generating fuel such as H_2 and methane. In addition, these research efforts provide insight into the intricacies of the natural process.

18

The maximum rate of photosynthesis will occur at the λ_{max} of the photosynthesizing system. This absorption maximum should match the absorption maxima of the light absorbing pigments. See the Biochemistry in the Lab box on photosynthetic studies (located on the companion website at www.oup.com/us/mckee) for a discussion of the absorption spectrum of chlorophyll and several other pigment molecules.

21

A special pair is composed of two special chlorophyll a molecules that reside within a reaction center. In PSI the special pair is referred to as P700, because it absorbs light at 700 nm. In PSII the special pair of chlorophyll a molecules are referred to as P680.

24

Deforestation has numerous negative effects on the local environment. Among these are a disrupted water cycle (the local climate becomes drier because of the loss of transpiration by trees), significantly increased water runoff and erosion (tree roots absorb rain water and groundwater), and a degraded environment with reduced biodiversity.

27

The optimal wavelengths for photosynthesis appear to be 400–500 nm and 600–700 nm.

30

The Z scheme is a mechanism whereby electrons are transferred from water to $NADP^+$. This process produces the reducing agent NADPH required for fixing carbon dioxide in the light-independent reactions of photosynthesis. Removal of the electrons from water also results in the production of oxygen. As electrons flow from PSII to PSI, protons are pumped across the thylakoid membrane, a process that establishes the proton gradient that drives ATP synthesis.

33

If blue wavelengths are used in addition to red ones, the rate of oxygen evolution is increased. This phenomenon is known as the Emerson enhancement effect. If photosynthesis occurs in a single photosystem, then the magnitude of the enhancement should reflect the ratios of the λ_{max} in the red and blue regions. The enhancement was much greater than predicted by this ratio. The existence of a second chromophore (and by inference a second photosystem) was therefore suggested.

36

The term triose phosphate is used to describe the molecules glyceraldehyde-3-phosphate and dihydroxyacetone phosphate. Formed during the Calvin cycle, the triose phosphates are used in plants in biosynthetic processes such as the formation of sucrose, polysaccharides, fatty acids and amino acids. For a deeper understanding of triose phosphate metabolism, read the Biochemistry in Perspective box on Starch and Sucrose Metabolism on the companion website.

39

When hydrogen sulfide is the source of hydrogen atoms, the end products of photosynthesis are glucose and elemental sulfur. (The carbon and oxygen atoms in glucose are from carbon dioxide molecules.)

Fill in the Blank

42. Photosystems
45. Chromophores
48. ATP

Short Answer

51

There are two reactions in the reduction phase of the Calvin cycle. In reaction 2 of the cycle, six molecules of glycerate 3-phosphate are converted to six molecules of glycerate-1,-3-bisphosphate at the expense of 6 ATPs. In the following reaction (reaction 3 of the cycle), the six molecules of glycerate-1,3-bisphosphate are reduced by NADPH to form six molecules of glyceraldehyde-3-phosphate.

54

The ferredoxin-thioredoxin system is a series of three proteins (ferredoxin, ferredoxin-thioredoxin reductase, and thioredoxin) that transfers electrons energized by light and then donates them to enzymes with disulfide bridges. This reduction reaction either activates or inactivates certain enzymes when it converts disulfide bridges into sulfhydryl groups. For example, when light is available, the ferredoxin-thioredoxin system activates Calvin cycle enzymes such as $NADP^+$-glyceraldehyde-3-phosphate dehydrogenase. This mechanism also inactivates enzymes in opposing pathways such as glucose-6-phosphate dehydrogenase, a pentose-phosphate pathway enzyme.

Thought Questions

57
There was a significant shift from hydrogen sulfide based photosynthesis to that which oxidizes water, because water is an abundant resource, whereas hydrogen sulfide is found only in relatively small quantities.

60
Only light of a particular energy can be absorbed by photosynthetic pigments. Increasing the intensity of the light increases the number of photons present and hence can improve the rate of photosynthesis. Increasing the energy level of the light, that is, the energy of the photons, decreases the rate of photosynthesis by shifting the photons to energy levels that are not absorbed by the photosystems.

63
If sufficient carbon dioxide is already present to saturate all of the ribulose-1,5-bisphosphate carboxylase molecules, the presence of additional carbon dioxide molecules will not increase the rate of photosynthesis. In addition, photosynthesis is depressed by low light levels.

66
Under conditions of high temperature, the carbon dioxide compensation point of C3 plants rises because the oxygenase activity of rubisco increases more rapidly than the carboxylase activity.

69
Chloroplasts must be relatively large to intercept the light necessary to carry out photosynthesis

72
Because of extensive photorespiration, which competes with photosynthesis, C3 plants expend more energy per gram of carbon fixed. As a result C4 plants are more efficient.

Chapter 14: In-Chapter Questions

14.1
a. CH_3NH_2
b. NH_3
c. CH_3CH_3

14.2
As their names suggest, hemoglobin and leghemoglobin are proteins in the globin superfamily. Recall that hemoglobin is an oxygen transport protein that contains a heme group, which binds reversibly with O_2. The heme in leghemoglobin also binds to O_2. The function of leghemoglobin, the sequestration of oxygen molecules, can be deduced from the irreversible inactivation of the nitrogenase complex in root nodules by O_2.

14.3
Because of its close structural similarity to folic acid, methotrexate is a competitive inhibitor of the enzyme dihydrofolate reductase. (Recall that this enzyme converts folic acid to its biologically active form, THF.) Rapidly dividing cells require large amounts of folic acid. Methotrexate prevents the synthesis of THF, the one-carbon carrier required in nucleotide and amino acid synthesis. It is therefore toxic to rapidly dividing cells, especially those of certain tumors and normal cells that divide frequently such as hair and GI tract cells.

14.4
HO — indole — $CH_2—CH_2—\overset{+}{N}H_3$ (N–H)

Serotonin

$\xrightarrow{\textit{N}\text{-Acetyltransferase}}$

HO — indole — $CH_2CH_2NH—\overset{O}{\overset{\|}{C}}—CH_3$ (N–H)

5-Hydroxy-*N*-acetyltryptamine

$\xrightarrow[\text{SAM}]{\textit{O}\text{-Methyltransferase}}$

$CH_3—O$ — indole — $CH_2CH_2NH—\overset{O}{\overset{\|}{C}}—CH_3$ (N–H)

Melatonin

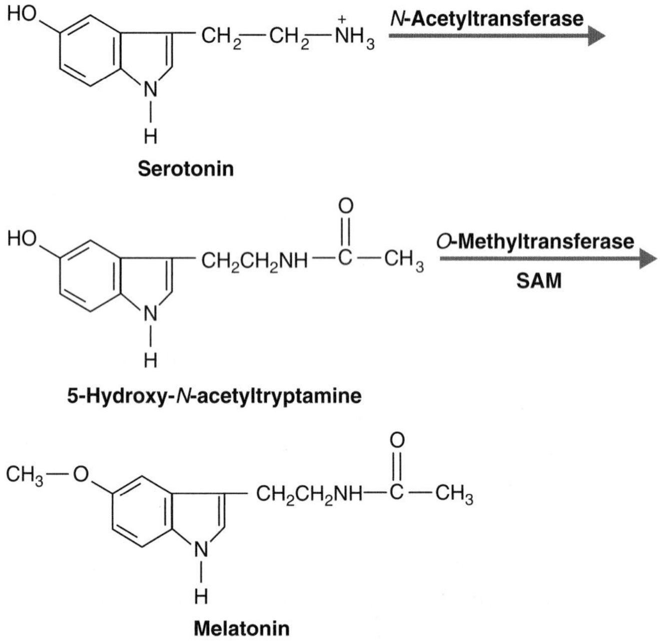

14.5

The reaction sequence is as follows:
Note that the secondary amino nitrogen alkylates more easily than the primary amino nitrogen in the guanidoacetate molecule.

Arginine + Glycine —AGAT→ Guanidoacetate + Ornithine

Guanidoacetate —GAMT (SAM → SAH)→ Creatine

Chapter 14: End-of-Chapter Questions

Review Questions

3

a. biogenic amine—an amino acid derivative that acts as a neurotransmitter (e.g., GABA and the catecholamines)
b. catecholamine—one of a class of neurotransmitters derived from tyrosine; includes dopamine, norepinephrine, and epinephrine
c. pyridoxal phosphate (PLP)—a coenzyme required in transamination reactions; derived from pyridoxine (vitamin B6)
d. urea—the major nitrogen waste molecule in mammals
e. L-Dopa—3,4-dihydroxyphenylalanine – the precursor molecule in the synthesis of the catecholamines; formed by the hydroxylation of tyrosine catalyzed by tyrosine hydroxylase

6

a. synaptic vesicle—a membrane-bound structure in neurons that contains neurotransmitter molecules
b. thioredoxin—a small protein with two thiol groups that mediates the transfer of electrons from NADPH to reduce another molecule, e.g., the reduction of ribonucleotides in the synthesis of deoxyribonucleotides
c. orotic aciduria—a rare genetic disease caused by defective UMP synthase; excessive urinary excretion of orotic acid; symptoms include anemia and growth retardation
d. *anti*-adenosine—the conformation of the nucleotide adenosine in which the base adenine is rotated outward away from the 6′-CH_2OH group of the ribose moiety
e. PRPP - 5′- phospho-α-D-ribosyl-1-pyrophosphate—the precursor molecule in the synthesis of molecules such as histidine and purine nucleotides

9

a. methotrexate—an analogue of folate; used to treat several types of cancer and autoimmune diseases
b. circadian rhythms—patterns of biological function associated with light and dark such as sleep/wake cycles
c. calmodulin (CAM)—a small calcium-binding protein that regulates a variety of enzymes
d. Lesch-Nyhan syndrome—a fatal X-linked disease caused by HGPRT (hypoxanthine guaninephosphoribosyltransferase) deficiency; excessive production of uric acid causes severe neurological symptoms
e. thioredoxin reductase—the dithiol-containing enzyme that transfers electrons from NADPH to thioredoxin

12

Nitrogen-fixing organisms solve the problem of oxygen inactivation in several ways. These are: (1) anaerobic organisms live only in anaerobic soil and are not faced with the problem of oxygen inactivation, and (2) other organisms physically separate oxygen from the nitrogenase complex. For example, many of the cyanobacteria produce specialized nitrogenase-containing cells called heterocysts. The thick cell walls of the heterocysts isolate the enzymes from atmospheric oxygen. In addition, legumes produce an oxygen binding protein called leghemoglobin, which traps oxygen before it can interact with the nitrogenase complex.

15

The concentrations of isoleucine (a) and valine (d) are typically higher in the blood leaving the liver than in the nutrient pool entering enterocytes because the liver preferentially exports these essential amino acids. The amino nitrogens of BCAA are used by other tissues for the synthesis of the nonessential amino acids. (Note that this question concerns concentrations, not total amounts. Since the liver draws amino acids from the blood to synthesize proteins, and the liver cannot replenish these EAA via synthesis, lower numbers of Ile, Leu, and Val molecules leave the liver than enter enterocytes. However, since the liver transports BCAA preferentially, their *concentration* in the blood flowing from the liver is higher.) The concentrations of serine (c) and alanine (e) are typically higher in the nutrient pool entering the enterocytes than in the blood leaving the liver because the liver uses alanine and serine to manufacture glucose for transport. Since glutamine (b) is a primary energy source for enterocytes, the concentration of glutamine is predicted to be higher in the nutrient pool entering the enterocytes. (*Note:* Recall that BCAA = branched chain amino acids and EAA are essential amino acids.)

18

a. alanine belongs to the pyruvate family
b. phenylalanine belongs to the aromatic family
c. methionine belongs to the aspartate family
d. trypthophan belongs to the aromatic family
e. histidine belongs to the histidine family
f. serine belongs to the serine family

21

The ten essential amino acids in humans are isoleucine, leucine, lysine, methionine, phenylalanine, threonine, tryptophan, and valine. In addition, histidine and arginine are essential for infants. These amino acids are essential because they cannot be synthesized in required amounts by humans and must be included in the diet.

24

The following intermediates occur during the reduction of nitrogen to yield ammonia diimine (HN=NH) and hydrazine (H_2N-NH_2).

27

Glutamate plays a central role in amino acid metabolism because it and α-ketoglutarate constitute one of the most common α-amino acid/α-keto acid pairs used in transamination reactions. Glutamate also serves as a precursor of several amino acids and as a component of polypeptides. Glutamine serves as the amino group donor in numerous biosynthetic reactions (e.g., purine, pyrimidine, and amino sugar synthesis), as a safe storage and transport form of ammonia and as a component of polypeptides.

30

The biologically active form of folic acid, referred to as tetrahydrofolate or THF, is shown below. It is formed by the reduction of folic acid with NADPH, in two reactions catalyzed by dihydrofolate reductase.

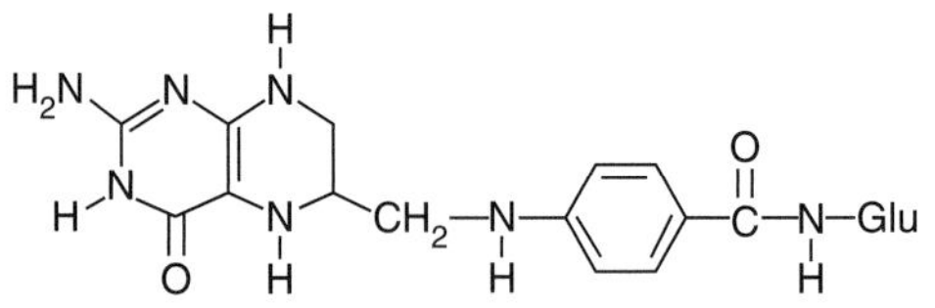

33

The carbon at position 2 of the uracil ring is derived from carbon dioxide.

36

The source of the hydrogen gas produced by the nitrogen reductase system is the same as the source of electrons: NAD(P)H.

Fill in the Blank

39. Essential
42. Tyrosine
45. Nucleoside

Short Answer

48

Orotic aciduria is so named because of the large amounts of orotate, an intermediate in the synthesis of UMP, that appear in the urine of affected individuals. Treatment consists of the administration of pyrimidine nucleotides, which inhibit orotate synthesis.

51

Nitric oxide is synthesized from arginine in a two-reaction sequence catalyzed by nitric oxide synthase. In the first reaction arginine is oxidized to L-hydroxyarginine as two electrons are transferred from NADPH to O_2 to yield the reaction product and H_2O. In the second reaction, which requires O_2 and 0.5 NADPH, L-hydroxyarginine reacts with a heme-peroxy complex (R-O-OH) to yield citrulline and NO.

Thought Questions

54

Tyrosine becomes an essential amino acid if its precursor the EAA phenylalanine is excluded from the diet or if the enzyme phenylalanine-4-monooxygenase is missing or defective or the coenzyme BH_4 is not available.

57

Glutamate is an excitatory neurotransmittor with stimulating effects on neurons that regulate bodily function such as blood pressure and body temperature. Individuals who display symptoms after consuming monsodium glutamate apparently possess efficient mechanisms for transporting glutamate across the blood-brain barrier.

60

Arginine is normally synthesized by the urea cycle. In small children, the urea cycle is not fully functional. Consequently, arginine must be obtained from external sources.

63

The conversion of cytosine to uracil occurs as follows:

Cytosine → (imino tautomer) → (H_2O) → (hydrated intermediate) → Uracil

Chapter 15: In-Chapter Questions

15.1

In newborn animals arginine will be an essential amino acid if the urea cycle is not yet fully functional.

15.2

Certain intestinal bacteria can release ammonia from urea molecules that diffuse across the membrane into the intestinal lumen. Treatment with antibiotics kills these organisms, thereby reducing blood ammonia concentration.

15.3

$$^{-}O_2S{-}CH_2{-}CH(^{+}NH_3){-}C(=O){-}O^{-} \xrightarrow{\text{Decarboxylation}} {}^{-}O_2S{-}CH_2{-}CH_2{-}\overset{+}{N}H_3 + CO_2$$

Cysteine Sulfate

$$^{-}O_2S{-}CH_2{-}CH_2{-}\overset{+}{N}H_3 \xrightarrow{\text{Oxidation}} {}^{-}O_3S{-}CH_2{-}CH_2{-}\overset{+}{N}H_3$$

Taurine

15.4

The reactions are as follows:

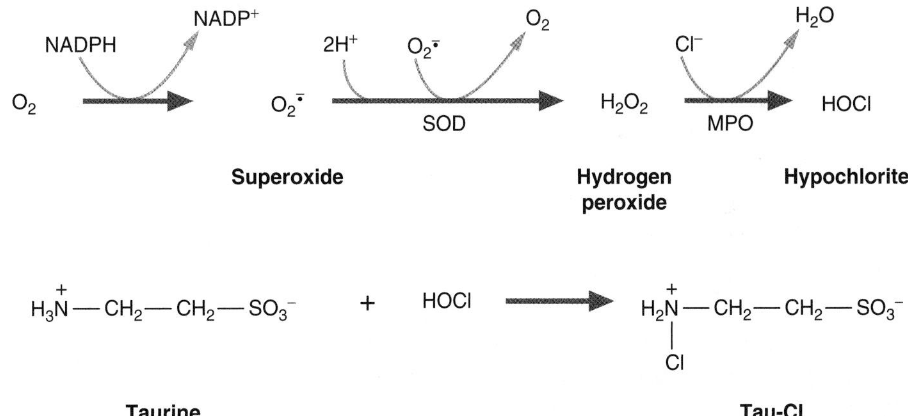

15.5

Gout is caused by high levels of uric acid. Animals that do not suffer from gout possess the enzyme urate oxidase, which converts uric acid to allantoin. Unlike uric acid, which is relatively insoluble in blood, allantoin readily dissolves and is easily excreted.

15.6

a. Urea is formed from ammonia, CO_2, and aspartate in the urea cycle.
b. Uric acid is the oxidation product of purines.
c. *β*-Alanine is produced in the degradative pathway of pyrimidines

15.7

Suggested catabolic reactions of β-alanine and β-aminoisobutyrate:

$$\underset{\beta\text{-Alanine}}{{}^{+}NH_3-CH_2-CH_2-C(=O)-O^-} \xrightarrow{\text{Deamination}} H-C(=O)-CH_2-C(=O)-O^- \xrightarrow{\text{Oxidation}} {}^{-}O-C(=O)-CH_2-C(=O)-O^-$$

$$\xrightarrow{\text{Decarboxylation}} CH_3-C(=O)-O^- \xrightarrow[\text{ATP}\ \ \text{ADP} + P_i]{\text{CoASH}} \underset{\text{Acetyl-CoA}}{CH_3-C(=O)-SCoA}$$

$$\underset{\beta\text{-Aminoisobutyrate}}{{}^{+}NH_3-CH_2-CH(CH_3)-C(=O)-O^-} \xrightarrow{\text{Deamination}} H-C(=O)-CH(CH_3)-C(=O)-O^- \xrightarrow{\text{Oxidation}} {}^{-}O-C(=O)-CH(CH_3)-C(=O)-O^-$$

$$\xrightarrow{\text{Decarboxylation}} CH_3-CH_2-C(=O)-O^- \xrightarrow{\text{Carboxylation}} \underset{\text{Succinate}}{{}^{-}O-C(=O)-CH_2-CH_2-C(=O)-O^-}$$

$$\xrightarrow[\text{ATP}\ \ \text{ADP} + P_i]{\text{CoASH}} \underset{\text{Succinyl-CoA}}{{}^{-}O-C(=O)-CH_2-CH_2-C(=O)-SCoA}$$

Chapter 15: End-of-Chapter Questions

Review Questions

3

a. L-amino acid oxidase—one of several FMN-requiring liver and kidney enzymes that convert some amino acids to α-keto acids, ammonium ion and hydrogen peroxide.
b. serine dehydratase—a hepatic pyridoxal-requiring enzyme that converts serine to pyruvate
c. bacterial urease—an enzyme produced by intestinal bacteria that hydrolyzes urea circulating in the bloodstream to form ammonia.
d. adenosine deaminase—an enzyme that releases ammonium ion from the adenine ring of AMP in a nucleotide catabolic pathway
e. glutaminase—a liver enzyme that hydrolyses glutamine to form glutamate and ammonium ion

6

a. transsulfuration pathway—the biochemical reactions that convert methionine to cysteine
b. cystathionine—an intermediate in the transsulfuration pathway; the product of the reaction of homocysteine and serine.
c. homocysteine—an intermediate in both the transsulfuration and methylation pathways.
d. PAPS—3′-phosphoadenosine-5′-phosphosulfate—a high energy sulfate donor molecule used in the synthesis of the sulfatides and proteoglycans
e. S-adenosylmethionine (SAM)—a methyl donor molecule; an intermediate in the methylation and transsulfuration pathways.

9

a. MAO—monoamine oxidase—an enzyme that catalyzes the oxidation of epinephrine, norepinephrine and dopamine, thereby inactivating them
b. PNMT—phenylethanolamine-N-methyltransferase – the SAM-requiring enzyme that catalyzes the methylation of norepinephrine to form epinephrine
c. COMT—catechol-O-methyltransferase—an enzyme that inactivates catecholamines (epinephrine, norepinephrine and dopamine) by catalyzing methylation reactions
d. NPC—Niemann-Pick disease Type C—a disease caused by the accumulation of cholesterol and glycolipids that is linked to a defective transmembrane protein; symptoms include liver and/or spleen enlargement and progressive neurological problems
e. TB—tuberculosis—an infectious disease caused by *Mycobacteriau tuberculosis*; damage to the lungs is associated with the formation of tubercles, which are aggregates of infected macrophages

12

a. T cells—a T lymphocyte—a white blood cell that bears antibody-like molecules on its surface and binds to and destroys foreign cells in cellular immunity

b. B cells—a B lymphocyte—a white blood cell that produces and secrets antibodies, the proteins that bind to foreign substances thereby initiating their destruction in the humoral immune response
c. cellular immunity—immune system processes mediated by T cells
d. humoral immunity—the immunity that results from the presence of antibodies in blood and tissue fluid; also referred to as antibody-mediated immunity
e. myoadenylate deaminase deficiency—deficiency of the skeletal muscle purine nucleotide cycle enzyme myoadenylate deaminase (AMP deaminase); affected individuals exhibit exercise–induced muscle fatigue

15

The structural features that apparently mark proteins for destruction are: (1) certain N-terminal amino acid residues (e.g., methionine or alanine), (2) peptide motif sequences (e.g., amino acid sequences with proline, glutamic acid, serine and threonine), and (3) oxidized residues (amino acid residues whose side chains have been oxidized by oxidases or ROS).

18

Purines are degraded to xanthine, which is oxidized to uric acid. Note that the structure of uric acid retains the original purine ring, which cannot be degraded by humans. A significant percentage of uric acid is excreted in the urine.

21

The first two reactions in the biochemical pathway that converts ammonium ions to urea (i.e. the formation of carbamoyl phosphate and citrulline) occur in the mitochondrial matrix. Subsequent reactions that convert citrulline to ornithine and urea occur in the cytosol. Both citrulline and ornithine are transported across the inner membrane by specific carriers.

24

Individuals with PKU lack phenylalanine hydroxylase (phenylalanine-4-monooxygenase) activity, so they cannot synthesize tyrosine from phenylalanine. Tyrosine is therefore an essential amino acid for these patients.

27

The branched chain amino acids (leu, ile, val) are metabolized primarily in muscle tissue, where they are principally used to synthesize nonessential amino acids.

30

a. uric acid— birds, reptiles and insects
b. urea—mammals
c. allantoate—bony fish
d. NH_4^+—aquatic animals
e. allantoin—some mammals

33

A vegetarian diet excludes taurine. Since domestic cats cannot synthesize taurine, they must obtain it by consuming meat. Without taurine, domestic cats become listless and die prematurely.

36

Excess amino acids present after a high protein meal are eventually converted to glutamate in the liver (see the solution for Review Question 34). Excess glutamate is converted to N-acetylglutamate, which is an allosteric activator for carbamoyl phosphate synthetase I, the most critical enzyme of the urea cycle in that NH_4^+ is one of its substrates. Since all five enzymes are controlled by substrate concentrations, higher levels of aspartate would activate argininosuccinate synthase. (A continued high protein diet results in the activation of the synthesis of all five urea cycle enzymes.)

39

Tryptophan is ketogenic because its catabolic pathway results in the formation of α-ketoadipate, which reacts further to form acetoacetyl-CoA, and then acetyl-CoA. Tryptophan is glucogenic in that one of its catabolic reactions along this pathway also produces alanine, which is converted to pyruvate (a substrate for gluconeogenesis) via a transamination reaction.

42

Humans cannot excrete waste nitrogen atoms as ammonia because of its toxicity. Urea is also toxic but much less so than ammonia. The conversion of ammonia to urea not only allows the nitrogen to be transported and excreted in much less toxic form but also prevents the large water loss that would be required by the excretion of ammonia.

Fill in the Blank

45. Protein turnover
48. Oligonucleotides
51. Cellular
54. Aspartate

Short Answer

57

β-Aminoisobutyrate is the end product of thymine catabolism. High levels of β-aminoisobutyrate occur if there is a genetic tendency for slow conversion of this molecule to succinyl-CoA or there is massive cell destruction caused by illnesses such as leukemia. Excess blood levels of β-aminoisobutyrate do not cause a gout-like illness because the molecule is water soluble.

Thought Questions

60

The energy requirements of the urea cycle are closely linked to the energy-generating citric acid cycle. Recall that fumarate, a product of the urea cycle, is easily converted to oxaloacetate, the molecule that reacts with incoming acetyl-CoA molecules. Both pathways, together referred to as the Krebs bicycle, occur within the mitochondrial matrix.

63

Tetrahydrobiopterin is a cofactor in the oxidation of phenylalanine to form tyrosine. The sustained absence of this cofactor would result in a buildup of phenylalanine and the appearance of the symptoms of PKU.

66
Because of the structural similarities to purine, caffeine is converted to a variety of derivatives by xanthine oxidase (e.g., 1-methyluric acid and 7-methylxanthine).

69
The pathway whereby uracil is converted to β-alanine is as follows

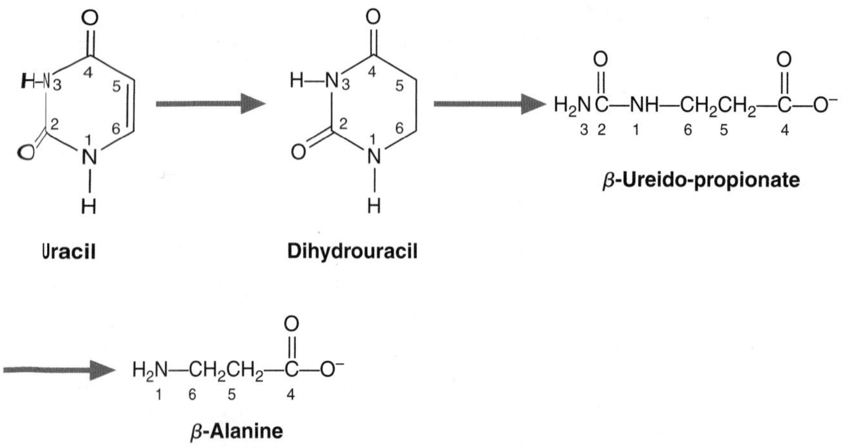

Uracil → Dihydrouracil → $H_2N\overset{O}{\overset{\|}{C}}{-}NH{-}CH_2CH_2{-}\overset{O}{\overset{\|}{C}}{-}O^-$ (β-Ureido-propionate) → $H_2N{-}CH_2CH_2{-}\overset{O}{\overset{\|}{C}}{-}O^-$ (β-Alanine)

The atoms in each molecule are numbered as indicated.

72
In the absence of insulin and/or insulin receptors, target tissues generate energy by means other than metabolizing glucose. Muscle proteins are degraded to amino acids, which are then used to generate energy for muscle contraction via the citric acid cycle.

Chapter 16: In-Chapter Questions

16.1
Several series of signal transduction components are illustrated in **Figure 8.21**. A prominent example series consist of glucagon (primary signal), glucagon receptor (receptor), adenylate cyclase (transducer), and activated protein kinase (response).

16.2
cAMP molecules would be produced by the binding of a single hormone molecule before it diffused away from the receptor site. The amplification factor for the hormone molecule is 350; that is, 350 cAMP molecules are produced for every hormone-receptor binding event.

16.3
Approximately 100,000 (or 10^5) molecules of target molecule (E_R) can be activated by a single molecule of hormone. cAMP is generated from ATP by adenylate cyclase when a hormone molecule binds to its receptor. The interaction between the receptor and adenylate cyclase is mediated by a G protein, G_s. As a consequence of hormone binding and the resulting conformational change, the receptor interacts with a nearby G_s protein. As G_s binds to the receptor, GDP dissociates. Then the binding of GTP to G_s allows one of its subunits to interact and stimulate adenylate cyclase, thus initiating cAMP synthesis. cAMP must break down quickly so the signaling mechanism can be precisely controlled.

16.4
The inhibition of GTP hydrolysis causes the subunit of G_s protein to continue activating adenylate cyclase. In intestinal cells, this enzyme activity opens chloride channels, causing loss of large amounts of chloride ions and water. The massive diarrhea caused by this process quickly leads to serious dehydration and electrolyte loss.

16.5
Both DAG and phorbol esters promote the activity of protein kinase C, which promotes cell growth and division. Phorbol esters provide initiated cells with a sustained growth advantage over normal cells. This condition is an early stage in carcinogenesis.

16.6
The high blood glucose levels in untreated diabetics result in the loss of increasingly large amounts of glucose along with water in the urine, a condition that causes dehydration. In the absence of usable glucose, the body rapidly degrades fats and proteins to generate energy, Hence Aretaeus's observation that in this disease excessive weight loss and excessive urination arc related.

16.7
Long-term fasting or low-calorie diets arc interpreted by the brain as starvation. The brain responds by lowering the body's BMR. The majority of the energy is derived from fatty acid oxidation. The glucose needed for glucose-dependent tissues is generated via gluconeogenesis at the expense of muscle protein.

16.8

As blood glucose and insulin levels in blood drop back to normal, glucagon is released from the pancreas. Glucagon acts on the liver to prevent hypoglycemia by promoting glycogenolysis and gluconeogenesis. Glucagon stimulates glycogenolysis by triggering the synthesis of cAMP, which in turn initiates a cascade of reactions that lead to the activation of glycogen phosphorylase. Increased lipolysis, hydrolysis of fat molecules, provides glycerol molecules that are substrates for gluconeogenesis.

Chapter 16: End-of-Chapter Questions

Review Questions

3

a. G Protein—one of a set of heterotrimeric GTP binding proteins; a protein that binds GTP, which activates the protein to perform a function
b. GPCR—G protein–coupled receptor—a cell surface receptor that transduces the binding of a hormone or other signal molecule into an intracellular response via the activation of a G protein
c. RTK—receptor tyrosine kinase—a transmembrane receptor that contains a cytoplasmic domain with tyrosine kinase activity that is activated when a ligand is bound to the external domain
d. growth factor—an extracellular polypeptide that stimulates cells to grow and/or undergo cell division
e. cytokine—one of a set of hormone-like polypeptides and proteins that may stimulate or inhibit cell growth or proliferation; traditionally used in reference to proteins produced by blood-forming cells and immune system cells

6

a. metabolic syndrome—a cluster of clinical disorders that includes obesity, hypertension, dyslipidemia and insulin resistance
b. hyperuricemia—high blood levels of uric acid
c. hypothalamus—an area of the brain that controls body temperature and electrolyte balance, monitors nutrient levels, and contributes to feeding behavior regulation
d. anorexigenic—appetite inhibiting
e. orexigenic—appetite stimulating

9

a. tumor promoter—a molecule that provides cells with a growth advantage over nearby cells
b. guanine nucleotide exchange factor (GEF)—a protein that mediates a conformational change in the transmembrane region of a G-protein-coupled receptor that leads to GDP/GTP exchange during G-protein activation
c. DAG—diacylglycerol, a second messenger molecule generated when PIP_2 is cleaved by phospholipase C; activates protein kinase C
d. steady state—a phase in an organism's life when the rate of anabolic processes is approximately equal to that of catabolic processes
e. IP_3—inositol-1,4,5-trisphosphate; a second messenger molecule generated when PIP_2 is cleaved by phospholipase C; binds to IP_3 receptor, a calcium channel

12

NADPH, which is formed during the pentose phosphate pathway and reactions catalyzed by isocitrate dehydrogenase and malic enzyme, is used as a reducing agent in a wide variety of synthetic reactions (e.g. amino acids, fatty acids, sphingolipids and cholesterol). The degradation of some of these molecules (e.g., fatty acids and the carbon skeletons of the amino acids) results in the synthesis of NADH, a major source of cellular energy via the mitochondrial electron transport system.

15

For several weeks after the onset of fasting, blood glucose levels are maintained via gluconeogenesis. During most of this period, amino acids derived from the breakdown of muscle proteins are the major substrates for this process, Eventually, as muscle becomes depleted, the brain switches to ketone bodies as an energy source. Consequently, the production of urea (the molecule used to dispose of the amino groups of the amino acids) declines.

18

One consequence of physical activity is the activation of the sympathetic nervous system, which in turn stimulates the adrenal gland to secrete epinephrine and norepinephrine. These hormones then activate the adipocytes enzyme hormone–sensitive lipase, which catalyzes the hydrolysis of triacylglycerol molecules to form the fatty acids used to drive muscle contraction.

21

In Alzheimer's disease there is progressive mental deterioration that is characterized by loss of memory, language skills and behavioral changes. Alzheimer's disease is a form of dementia in which brain damage is associated with aggregated β-amyloid protein outside and around neurons and aggregated tau protein inside neurons. In type 2 diabetes mellitus, pancreatic β-cell function is eventually compromised by aggregates of misfolded proteins.

24

HbA_{1c} formation is a consequence of nonenzymatic glycation of hemoglobin that occurs in the presence of high blood glucose levels. In the Maillard reaction, the aldehyde group of glucose condenses with a free amino group in a protein to form a Schiff base, the Amadori product. The Amadori product subsequently destabilizes to form a reactive carbonyl-containing product that reacts with hemoglobin molecules to form an adduct such as $HbA_{1c.}$

27

In type 1 diabetes (insulin-dependent diabetes mellitus), no insulin is produced and the action of glucagon is unopposed. Glucagon causes an increase in lipolysis in adipocytes, leading to an excess of acetyl-CoA molecules, which are converted to ketone bodies. Ketoacidosis (an excess of ketone bodies in the blood with low blood pH) is rare in type 2 diabetes (insulin-independent diabetes mellitus) in which blood levels of insulin are normal or elevated but cells are resistant to insulin. Since the action of glucagon is not completely unopposed, and there is some (although reduced) glucose uptake by cells, lipolysis is not activated as it is in type 1 diabetes, and the excess production of ketone bodies does not occur.

30

In addition to water loss, contributing factors to the relatively rapid weight loss during the first week of a prolonged diet are depletion of glycogen stores, loss of muscle protein, and lipolysis of triacylglycerol in adipocytes. Large amounts of amino acids from muscle protein are needed to provide glucose (via gluconeogenesis in the liver), the preferred energy source for the brain. Lipolysis releases fatty acids to provide an alternate energy source to glucose. Regarding the water loss, note that water bound by glycogen will be lost when glycogen is depleted. Also a molecule of water is required to hydrolyze each glycosidic, ester and peptide bond.

33

a. The intestine digests foods into nutrients that are small enough to be absorbed (e.g., sugars, fatty acids, and amino acids). The small intestine, as well as the stomach, produces ghrelin, a hormone that stimulates nutrient intake.
b. The liver has numerous roles in nutrient metabolism. Among these are monitoring blood nutrient levels, distribution of nutrients to the body's tissues (e.g., amino acids), and regulation of blood glucose via gluconeogenesis and glycogenolysis.
c. Skeletal muscle consumes glucose and fatty acids, energy sources that drive muscle contraction. During fasting skeletal muscle protein is degraded to yield alanine, which is delivered to the liver for gluconeogenesis. Cardiac muscle relies primarily on glucose and fatty acids for energy.
d. Adipose tissue cells store TGs, and releases fatty acids and glycerol into the blood for delivery to other organs.
e. The kidney uses fatty acids and glucose to meet its energy needs. It disposes of the water-soluble products of nutrient metabolism (e.g., urea).
f. The brain uses glucose as its sole fuel. During prolonged starving, the brain can adapt to use ketone bodies as an energy source.

Fill in the Blank

36. Second messengers
39. Insulin
42. Ketoacidosis

Short Answer

45

The cardiovascular system is damaged by fructose consumption by two processes, both of which stimulate atherosclerosis: (1) dyslipidemia, a consequence of hepatic fructose metabolism and (2) glycation reactions that occur about seven times as often as those of glucose and lead to AGE (advanced glycation end product) formation (see p. 249).

48

PYY is an anorexigenic (appetite-inhibiting) molecule that is produced by cells in the small intestine and colon in response to a meal. PYY exerts its action by inhibiting the appetite-stimulating NPY neurons in the hypothalamus.

Thought Questions

51

The second messenger is an effector molecule synthesized when a hormone (the first messenger) binds. It stimulates the cell to respond to the original signal. Second messengers also allow the signal to be amplified.

54

Increased mobilization of fatty acids provides an alternate energy source for muscle, thereby sparing glucose for the brain. In addition, glucagon stimulates gluconeogenesis, a pathway that utilizes amino acids derived from muscle.

57

The storage of preformed hormone molecules in secretory vesicles allows for a rapid response of the producing cells to metabolic signals. As soon as the appropriate signal is received the vesicles fuse with plasma membrane and (via exocytosis) release their contents into the bloodstream.

60

Appetite regulation in humans is a set of complex and robust mechanisms that involve several areas of the brain, such as the hypothalamus. In response to calorie restriction which is interpreted as starvation, the appetite centers of the brain respond by stimulating appetite, lowering the body's BMR (to conserve energy), lowering energy expenditures (resulting in lethargy). As a result of these and other hormone- and peptide-triggered responses, continued achievement of weight loss and maintenance of a reduced weight over time becomes very difficult, if not impossible.

63

mTORC1 is a central metabolic sensor that integrates hormonal activity, nutrient availability, energy status and stress. It stimulates anabolic processes such as protein synthesis. When levels of ATP, nutrients, and oxygen are low, AMPK inhibits mTORC1.

66

In a signal cascade the initial signal can be present in low concentrations. The message can then be amplified as the cascade progresses. In addition, an intricate multistage cascade mechanism provides opportunities for the integration of numerous cellular processes

69

Exercise promotes insulin-independent glucose uptake by muscle cells, which facilitates blood glucose control.

72
During prolonged starvation ketone body levels will eventually rise to levels that cause acidosis, a condition that causes kidney damage.

75
The unregulated conversion of fructose to triose phosphate promotes liver lipogenesis and VLDL synthesis and secretion.

Chapter 17: In-Chapter Questions

17.1
The cytosine-guanine base pair with its three hydrogen bonds is more stable than the adenine-thymine base pair. The more CG bp there are, the more stable the DNA molecule. Structure b, with the fewest CG bp, will therefore denature first.

17.2
a. Ethanol will disrupt the hydrogen bonding in the base pairs and denature the DNA.
b. Heat, which easily disrupts hydrogen bonds, will cause DNA chains to separate and denature.
c. Dimethylsulfate is an alkylating agent that can cause transversion and transition mutations.
d. Nitrous acid deaminates bases.
e. Quinaerine is an intercalating agent that can cause frame shift mutations.

17.3
The brain is especially sensitive to oxidative stress because it uses a greater proportion of oxygen than other tissues. Consequently, the chance of oxidative damage is also high. In addition, when most types of brain cells are irreversibly damaged by ROS, they cannot be replaced. In addition to hydroxyl radicals, other ROS that can contribute to oxidative stress in the brain include superoxide, hydrogen peroxide, and singlet oxygen.

17.4
In A-DNA, the dehydrated form of DNA, the base pairs are no longer at right angles to the helical axis. Instead, they tilt 20 degrees away from the horizontal as compared to B-DNA. The distance between adjacent base pairs is slightly reduced, with 11 bp per helical turn instead of the 10.4 bp that occurs in the B-form. Each turn of the double helix of A-DNA occurs in 2.5 nm instead of the 3.4 nm of B-DNA. The diameters of A-DNA and B-DNA are 2.6 and 2.4 nm, respectively. The significance of A-DNA is unclear. It has been observed that its overall appearance resembles that of RNA duplexes and the RNA-DNA hybrids that form during transcription.

With a diameter of 1.8 nm, Z-DNA is considerably slimmer than B-DNA. It is twisted into a left-handed spiral with 12 bp per turn, each of which occurs in 4.5 nm instead of the 3.4 observed in B-DNA. Segments with alternating purine and pyrimidine bases are most likely to adopt the Z-DNA configuration. In Z-DNA the bases stack in a left-handed, staggered pattern that gives this form its flattened, non-grooved surface and zigzag appearance. The significance of Z-DNA is unresolved.

H-DNA (triple helix) segments can form when a polypurine sequence is hydrogen-bonded to a polypyrimidine sequence. H-DNA, which has been observed to form under low pH conditions, is made possible by nonconventional, Hoogsteen base pairing. H-DNA may play a role in recombination.

17.5
The genome is the total set of DNA-encoded genetic information in an organism. A chromosome is a DNA molecule, usually complexed with certain proteins. Chromatin is the partially decondensed form of eukaryotic chromosomes. Nucleosomes are the repeating structural units of eukaryotic chromosomes formed by the interaction of DNA with the histones. A gene is a DNA sequence that codes for a polypeptide or an RNA molecule.

17.6
The genomes of prokaryotes are substantially smaller than those of eukaryotes. For example, the genome sizes of *E. coli* and humans are 4.6 and 30 Mb, respectively. Prokaryotic genomes are compact and continuous; that is, there are few, if any, noncoding DNA sequences. In contrast, eukaryotic DNA contains enormous amounts of noncoding sequences. Other distinguishing features of prokaryotic and eukaryotic DNA are the linkages of genes into operons in prokaryotes and intervening sequences in eukaryotic genes.

17.7
The antisense DNA sequence is 3′-CGTAAGCT TAACGTCTGAGGACGTTAAGCCGTTA-5′; the mRNA sequence is 3′-CGUAAGCUUAACGUCUGAGGACGU UAAGCCGUUA-5′, The antisense RNA sequence is 3′-GCAUUCGAAUUGCAGACUCCUGCAAUUCGGCAA U-5′.

17.8
In the original central dogma, the flow of genetic information is in one direction only, that is, from DNA to the RNA molecules, which then direct protein synthesis. The altered diagram indicates that the RNA genome of some viruses can replicate their RNA genomes (using a viral enzyme activity referred to as RNA-directed RNA polymerase) or undergo reverse transcription (i.e., synthesize DNA from an RNA sequence).

Chapter 17: End-of-Chapter Questions

Review Questions

3
a. point mutation—a change in a single nucleotide base in a DNA sequence
b. transition mutation—a DNA mutation that involves the substitution of a purine base by a different purine, or the substitution of a pyrmidine by a different pyrmidine

c. transversion mutation—a type of point mutation in which a pyrimidine is substituted for a purine and vice versa
d. silent mutatio—a point mutation that has no discernable effect on a polypeptide's function
e. missense mutation—a point mutation that results in an amino acid substitution resulting in a change in a polypeptide's function

6

a. Chargaff's rules—a set of rules describing the base composition of DNA; posits the equality of the concentration of adenine and thymine and of cytosine and guanine
b. constitutive heterochromatin—sections of DNA in eukaryotes that is permanently highly condensed and transcriptionally silent; occurs at centromeres, telomeres, transposons and repetitive sequences.
c. bacteriophage—a type of virus that infects bacteria
d. replication-dependent histones—histones that are synthesized during S phase of the cell cycle in coordination with DNA synthesis.
e. replication-independent histones—histones that are synthesized in small amounts throughout the cell cycle.

9

a. histones—a group of basic proteins found in all eukaryotes that bind to DNA to form nucleosomes
b. heterochromatin—highly condensed chromatin that is transcriptionally inactive; constitutive heterochromatin is permanently condensed; facultative herterochromatin may be decompressed in specific cell types
c. euchromatin—a less condensed form of chromatin with varying levels of transcriptional activity
d. intergenic sequences—DNA sequences in which multiple copies are arranged next to each other.
e. tandem repeats—DNA sequences in which multiple copies are arranged next to each other

12

a. DNA typing—a DNA analysis technique used to identify individuals; involves the analysis of several highly variable sequences called markers
b. short tandem repeats—DNA sequences with between 2 and 4 bp repeats; can be used to generate DNA profiles that distinguish among individuals
c. DNA profile—a unique DNA pattern of repeats of target sequences that is separated in an electrophoresis gel; used to identify individuals
d. ribozyme—atalytic RNA molecules; catalyzes self-cleavage or the cleavage of other RNA molecules
e. noncoding RNA—types of RNA other than the RNAs involved in protein synthesis (i.e. tRNAs, rRNAs and mRNAs) that act as an extensive genome regulatory network

15

Eukaryotic genomes are larger than those of prokaryotes. In contrast to prokaryotic genomes, which consist entirely of genes, the majority of eukaryotic DNA sequences do not appear to have coding functions. Unlike prokaryotic genes, most eukaryotic genes are not continuous (i.e. they usually contain introns).

18

RNA molecules differ from DNA in the following ways: (1) RNA contains ribose instead of deoxyribose, (2) the nitrogenous bases in RNA differ from those of DNA (e.g., uracil replaces thymine and several RNA bases are chemically modified), and (3) in contrast to the double helix of DNA, RNA is single-stranded.

21

There are approximately 6 million base pairs in a single human cell. Assuming that there are 10^{14} body cells, the total length of the DNA in the human body is approximately 2×10^{11} km. This estimated length is about 1000 time greater than the distance for the earth to the sun. (Note that 1 nm is 10^{-9} m.)

24

Before the publication of the Watson-Crick paper in 1953, research efforts were focused on proving that DNA is the genetic material and more recently to discover its structure. Beginning with the Watson-Crick paper, research efforts rapidly shifted to the functional properties of DNA and related cell processes. This work was eventually referred to as molecular biology.

27

The stable inheritance of DNA methylation patterns from one cell generation to the next is made possible by a class of enzymes called the maintenance methyltransferases. They methylate the cytosines in CpG-rich regions in newly synthesized DNA strands. The enzymes methylate the cytosines on the new strand at sites opposite to the methylated cytosines on the parental strand.

30

The Env trimer (a complex of three gp120/gp41 heterodimers) is a fusion machine that enables HIV to attach to and fuse with the plasma membrane of target cells. Fusion of the HIV envelope with a target cell plasma membrane is the initiating event in the infection of the target cell.

33

Water stabilizes DNA structure by binding to phosphate groups, deoxyribose 3′- and 5′-oxygen atoms, and electronegative atoms in the nucleotide bases. Also, the increased entropy of surrounding water molecules drives the hydrophobic interactions between the nucleotide bases within the helix

36

According to the histone code hypothesis, the pattern of histone modification within each DNA sequence regulates gene expression by serving as a platform for the binding of proteins that inhibit or facilitate transcription.

39

Ethyl chloride is an alkylating agent that can react with DNA bases to form ethyl derivatives.

Fill in the Blank

42. Proteome
45. Tandem repeats
48. Exons

Short Answer

51

Syncytin-1, a product of the HERV-W family of endogenous retroviral sequences, plays a critical role in the formation of the placenta. Inadequate synthesis of syncytin-1 is one of several factors that contribute to pre-eclampsia.

54

Because DNA damage has very serious consequences for mitochondria, it is expected that there are DNA repair pathways. In addition, it is also expected that mitochondria possess several antioxidant defense mechanisms. Examples of antioxidant enzymes include mitochondrial versions of superoxide dismutase, glutathione peroxidase, and peroxiredoxin (see pp. 372–373). Glutathione is an important endogenous antioxidant molecule within mitochondria. Vitamin E (α-tocopherol) obtained in the diet provides some protection to mitochondrial membranes.

Thought Questions

57

The major and minor grooves of DNA arise because the glycosidic bonds in the two hydrogen bonded strands are not exactly opposite to each other.

60

The electron withdrawing effect of the bromine increases the likelihood of enol formation of uracil. This enol mimics the hydrogen bonding pattern of cytosine. Therefore this base can be paired with guanine.

63

The histones act to shield the DNA from the action of nucleases.

66

At the crime scene, a forensic expert collects biological specimens such as blood, hair, and saliva. Once these specimens are delivered to the lab, they are analyzed and compared with the DNA of the victim. Any DNA not belonging to the victim is assumed to belong to a person, or persons, present during the time when the crime was committed. If a suspect is identified, his or her DNA profile (obtained from a swab of cheek cells or from a court-ordered blood sample) is compared with that obtained from crime scene specimens. If there is no obvious suspect, the crime scene specimens can be compared to the DNA profiles in the statewide database. This strategy has been remarkably successful in the identification of individuals later found guilty not only of recent murders, but also those from "cold cases" in which crime scene specimens had been preserved. The technology that makes this success possible includes PCR, RFLP, and STR-DNA analysis.

69

The hydrogen bonds to water molecules formed by the atoms in a phosphodiester linkage are as follows:

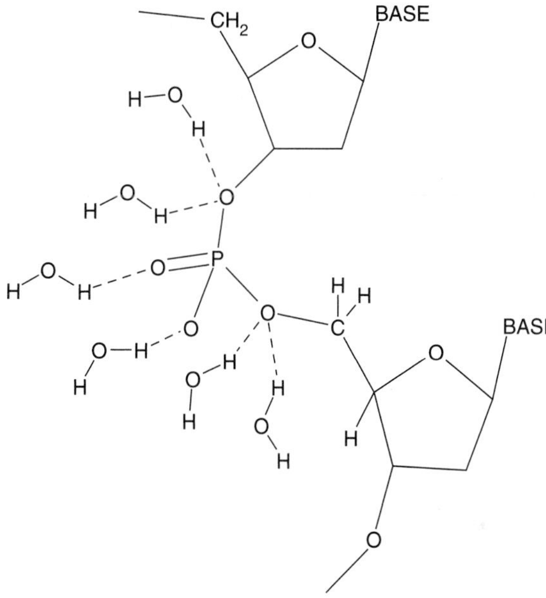

72

DNA degrades with time. Ancient fossils would have little if any intact DNA with which to reconstruct the organisms. In addition, although intact DNA is vitally important for organismal function, it is only the operating system of an organism. Without access to a living example of such an organism, it would be impossible to reconstitute the physiological structure and functional properties that are unique to a species.

Chapter 18: In-Chapter Questions

18.1

Briefly, prokaryotic DNA replication consists of DNA unwinding, RNA primer formation, DNA synthesis catalyzed by DNA polymerase and the joining of Okazaki fragments by DNA ligase. Prokaryotic DNA replication differs from the eukaryotic process in that prokaryotic replication is faster, the Okazaki fragments are longer and there is usually only one origin of replication per chromosome (eukaryotes have many per chromosome).

18.2

In excision repair short damaged sequences (e.g., thymine dimers) are excised and replaced with correct sequences. After an endonuclease deletes the damaged single-stranded sequence, a DNA polymerase activity synthesizes a replacement sequence using the undamaged strand as a template. In photoreactivation repair a photoreactivating enzyme uses light energy to repair pyrimidine dimers. In recombinational repair damaged sequences are deleted. Repair involves an exchange of an appropriate segment of the homologous DNA molecule.

18.3

When antibiotics are used in large quantities, the bacterial cells that possess resistance genes (acquired through spontaneous mutations or through intermicrobial DNA transfer mechanisms such as conjugation, transduction, and transformation) survive and even flourish. Because of antibiotic use, which acts as a selection pressure, resistant organisms (once only a minor constituent of a microbial population) become the dominant cells in their ecological niche.

18.4

Most gene duplications are apparently a consequence of accidents during genetic recombination. Examples of possible causes of gene duplication are unequal crossing over during synapses and transposition. After a gene has been duplicated, random mutations and genetic recombination may introduce variations.

18.5

Because phytochrome has been demonstrated to mediate numerous light-induced plant processes, it appears reasonable to assume that it does so in part by interacting with light-response elements (LRE) in plant cell genomes. Presumably, phytochrome influences gene expression by binding, either alone or as part of a complex, to various LREs when its chromophore is activated by light.

Chapter 18: End-of-Chapter Questions

Review Questions

3

a. replicons—a unit of the genome that contains an origin for initiating replication
b. Okazaki fragment—any of a series of deoxyribonucleotide segments that are formed during discontinuous replication of one DNA strand as the other strand is continuously replicated
c. ter region—a segment of the *E. coli* chromosome that contains DNA replication termination sequences
d. tus protein—a protein that when bound to a ter sequence facilitates DNA replication termination
e. preinitiation complex—the eukaryotic replication complex (preRC) whose formation is the first major step in DNA replication

6

a. transposition—the movement of a DNA sequence from one site in a genome to another
b. transposable element—a DNA sequence that excises itself and then inserts at another site
c. bacterial transformation—a process in which DNA fragments enter a bacterial cell and are introduced into the bacterial genome
d. transduction—the transfer of DNA segments between bacteria by bacteriophages
e. conjugation—unconventional sexual mating between bacterial cells; a donor cell transfers a DNA segment into a recipient cell through a specialized pilus.

9

a. PCR – polymerase chain reaction—a method for obtaining large numbers of DNA copies; uses Taq polymerase, a heat stable DNA polymerase
b. DNA microarray—a DNA chip used to analyze the expression of thousands of genes simultaneously
c. chromosomal jumping—a technique used to isolate clones that contain discontinuous sequences for the same chromosome
d. genome project—the process of determining the entire set of DNA base sequences of a particular organism
e. bioinformatics—the computer–based field that facilitates the analysis of biological sequence data

12

a. gene silencing—a form of posttranscriptional gene regulation in higher eukaryotes, involves short 22-nt RNAs called microRNAs
b. RNA interference—a cellular mechanism in which RNA molecules are degraded; functions in gene expression regulation and in defense against viral RNA genomes
c. tumor suppressor gene—one of a set of genes that code for proteins that actively protect cells from progressing toward cancer
d. protooncogene—a normal gene that codes for a protein involved in cell cycle regulation; promotes carcinogenesis if mutated
e. GEF—guainine nucleotide exchange factor - a protein that causes GTPases to release GDP and then bind GTP

15

The poly A tail facilitates the transport of mRNAs out of the nucleus and their subsequent translation by ribosomes. The poly A tail also serves as a timing device. As an mRNA's poly A tail becomes shorter, it eventually becomes vulnerable to degradation, a circumstance that limits its functional life.

18

The incorporation of a nucleotide into a DNA strand:

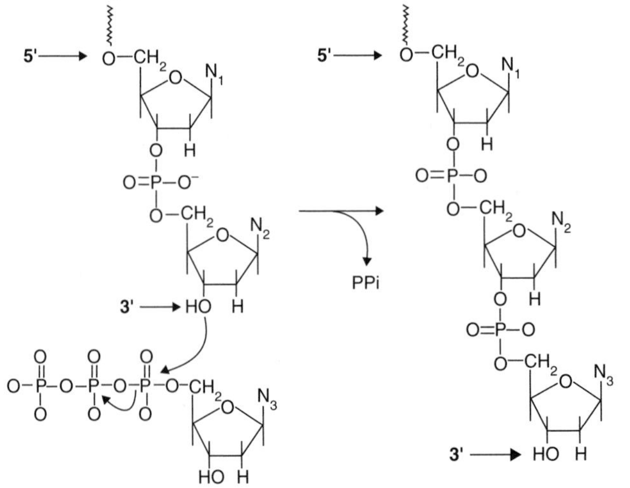

5' → O–CH2 N1
O H
O=P–O⁻
O–CH2 N2
3' → HO H
O O O
O–P–O–P–O–P–O–CH2 N3
O O O
HO H
PPi
5' → O–CH2 N1
O H
O=P–O
O–CH2 N2
O H
O=P–O
O–CH2 N3
3' → HO H

21

"Jumping genes" is the popular name for transposons. First discovered by Barbara McClintock, transposons (transposable elements) are DNA sequences that can move around the genome.

24

In bacteria general recombination is involved in transformation (DNA from one cell enters another cell and is subsequently integrated into the recipient cell's genome), transduction (a bacteriophage synthesized in a bacterial cell inadvertently carries a bacterial DNA fragment to a bacterial cell that the virus infects), and conjugation (unconventional sexual mating in which DNA from one cell enters a second cell via a sex pilus).

27

Because DNA is constantly exposed to disruptive processes, its structural integrity is highly dependent on efficient repair mechanisms. The life span of an organism is dependent on the health of its constituent cells, which is in turn dependent on the timely and accurate expression of genetic information. Consequently, the capacity of the organisms in a species to maintain the integrity of DNA molecules is an important factor in determining life span.

30

a. Transcription factors are proteins that regulate or initiate RNA synthesis by binding to specific DNA sequences called response elements.
b. RNA polymerase is an enzyme that transcribes a DNA sequence into an RNA product.
c. A promoter is a DNA sequence immediately before a gene that is recognized by RNA polymerase and signals the start point and direction of transcription.
d. A sigma factor is a bacterial protein that facilitates the binding of the core enzyme of RNA polymerase to the initiation site during transcription.
e. An enhancer is a DNA regulatory sequence that when bound to an appropriate transcription factor increases the likelihood that nearby genes will be transcribed.
f. The TATA box is the best researched example of a eukaryotic core promoter element.

33

The amplification of a single DNA molecule during 5 cycles yields 2^5 or 32 molecules.

36

Refer to **Figure 18.19**

39

In species that possess DNA photolyase, light energy captured by this enzyme's flavin and pterin chromophores is used to break the cyclobutane ring in a thymine dimer, thus converting the dimer back to two thymine monomers. The phosophodiester bond is not affected. (Humans do not possess this enzyme.)

42

The function of telomere end-binding proteins is to bind to GT-rich telomere sequences as part of the process that sequesters and stabilizes telomeres.

45

A tetragametic chimera is an individual that developed from the fusion of two nonidentical zygotes formed from four genetically distinct gametes (two eggs and two sperm).

Fill in the Blank

48. Primosome
51. Base exclusion repair
54. Transposons

Short Answer

57

Telomerase is a ribonucleoprotein that is responsible for the synthesis of short DNA segments at the ends of chromosomes after DNA replication. The added segment replaces approximately 100 to 200 nucleotides that are lost from the telomere regions during the replication process. Normal somatic cells (body cells excluding germ cells) do not usually use telomerase and, as a result, their chromosomes become gradually shorter until a critical length is reached that prevents further cell divisions. Cancer cells, in contrast, use telomerase to prevent the normal shortening process, thereby giving them a growth advantage over normal cells.

60

Because RBP2 is a demethylase, if pRB cannot bind and suppress its activity, this enzyme will demethylate histone lysines, thereby promoting transcription and possibly facilitating deregulated cell division. This is one way in which a defective RB1 contributes to tumor formation.

Thought Questions

63

DNA sequence changing processes such as genetic recombination, gene splicing, and alternate RNA splicing can allow cells to alter gene expression and expand their repertoire of proteins. Their best-best known example is antibody production in lymphocytes. The rearrangement of several possible choices for each of a number of antibody gene segments by a site-specific recombination results in the generation of an extremely large number of different antibody molecules.

66

The tanning process which occurs in response to overexposure to sunlight, is triggered by DNA damage. DNA damage in skin cells causes an accelerated aging process that is manifested as a thickened and wrinkled skin. DNA damage may also inactivate tumor suppressor genes and /or cause mutations in protooncogenes, thus increasing skin cancer risk.

69

Because the *Rb* gene codes for a tumor suppressor, retinoblastoma occurs only when both copies have been damaged or deleted. Usually a long period of time is required for random mutations to cause this event . In hereditary retinoblastoma, in which an affected individual possesses only one functional *Rb* gene, the time necessary for a random mutation to inactivate the second *Rb* gene is significantly less than that required for the inactivation of both genes that cause the nonhereditary version of the disease.

72
Once DNA has been extracted from the cells of an organism it is digested and a genomic library is created using shotgun cloning. The DNA sequence in each clone is then sequenced. Chromosome walking is used to determine overlapping sequences.

75
The RNA "copies" of the DNA can be altered to generate protein diversity without changing the DNA template sequence. In addition, RNA molecules can be used repeatedly and then disposed of without damaging the original DNA. If the DNA were used directly, the master molecule could be damaged or destroyed. In addition RNA molecules can easily leave the nucleoid or nucleus to travel to another part of the cell.

Chapter 19: In-Chapter Questions

19.1
The amino acid sequence of the beginning of the polypeptide is Met–Ser–Pro–Thr–Ala–Asp–Glu–Gly– Arg–Arg–Trp–Leu–Ile–Met–Phe. The mutation types in the altered mRNA sequences are (a) insertion of one base, (b) deletion of one base, (c) insertion of two bases, (d) deletion of three bases. The consequences of these mutations are altered amino acid sequences of the polypeptides produced from mRNA. In (a), (b), and (c) a frame shift occurs. Therefore the amino acid sequences past the mutation are different. In (d) no frame shift occurs because three bases are deleted. In this case, the only difference between the normal polypeptide and the mutated version is the deletion of a single amino acid.

19.2
Assuming that the DNA sequence given is the coding strand, the mRNA sequence is 5′-GGUUUA-3′ and the anticodons are 5′-UAA-3′. If the DNA sequence is the template strand, the mRNA sequence is 5′-UAAACC-3′ and the anticodons are 5′-GGU-3′ and 5′-UUA-3′.

19.3
The possible choices for mRNA codon base sequences for the peptide are:

Tyr—	Leu—	Thr—	Ala—
5′-UAU-3′	CUU	ACU	GCU
UAC	CUC	ACC	GCC
	CUA	ACA	GCA
	CUG	ACG	GCG
	UUA		
	UUG		

The possible choices for the DNA sequences that code for the peptide are:

Tyr—	Leu—	Thr—	Ala—
3′-ATA-5′	GAA	TGA	CGA
ATG	GAG	TGG	CGG
	GAT	TGT	CGT
	GAC	TGC	CGC
	AAT		
	AAC		

The possible choices for the tRNA anticodons that code for the peptide are:

Tyr—	Leu—	Thr—	Ala—
3′-AUA-5′	GAA	UGA	CGA
AUG	GAG	UGG	CGG
	GAU	UGU	CGU
	GAC	UGC	CGC
	AAU		
	AAG		

19.4
The formation of an ADP-ribosylated derivative of eEF-2 affects the three-dimensional structure of this protein factor. Presumably protein synthesis is arrested because the ability of eEF-2 to interact with or bind to one or more ribosomal components is altered.

19.5
After the synthesis of the plastocyanin precursor in cytoplasm, the first import signal mediates the transport of the protein into the chloroplast stroma. After this signal has been removed by a protease, a second import signal mediates the transfer of the protein into the thylakoid lumen. Plastocyanin then binds a copper atom, folds into its final three-dimensional structure, and associates with the thylakoid membrane.

Chapter 19: End-of-Chapter Questions

Review Questions

3
a. decoding center—the position in the 30S ribosomal subunit where an mRNA codon is matched to an incoming tRNA anticodon
b. peptidyl transferase center—a set of overlapping binding sites on the 50S subunit for translation factors with GTPase activity
c. GTPase associated center—a set of overlapping binding sites on the 50S subunit for translation factors with GTPase activity
d. guanine nucleotide exchange factor—a protein that removes GDP from a protein and then facilitates GTP binding
e. proton shuttle mechanism—the reaction mechanism whereby a peptide bond is formed by the nucleophilic attack of the α-amino nitrogen on the A-site aminoacyl group carbonyl carbon

6

a. proprotein—an inactive precursor protein
b. preproprotein—an inactive precursor protein with a removable signal peptide
c. disulfide exchange—a mechanism that facilitates the formation of appropriate disulfide bridges in newly synthesized proteins
d. proline hydroxylation—the ascorbic acid-requiring reaction that converts proline residues in certain connective tissue proteins to hydroxylated derivatives; required for structural integrity of connective tissue
e. proteolytic cleavage—a posttranslational removal of peptide segments by proteases; used to regulate the activity of certain polypeptides

9

a. tmRNA—a bacterial RNA molecule containing a tRNA- like domain and an mRNA-like domain that rescues ribosomes bound to damaged mRNAs
b. SECIS element—selenocysteine insertion sequence—a sequence element required in order to code for selenocysteine and located in the 3′-UTR of the mRNA for a selenocysteine-containing polypeptide
c. initiation—the beginning phase of translation
d. elongation—the polypeptide growth phase during translation of an mRNA in a ribosome.
e. termination—the phase of translation in which newly synthesized polypeptides are released from the ribosome

12

a. proteostasis network—proteins that work together as a system to control protein conformation through interactions of the proteome with molecular chaperones and degradation mediated by the ubiquitin proteasome system
b. heat shock response—the cellular response to heat and other forms of stress that involves up-regulation of genes coding for heat shock proteins as a component of a cell's repair mechanism
c. mRNP—messenger ribonucleoprotein; mRNA bound to proteins that are involved in splicing, export from the nucleus and translation
d. mTORC1—mammalian target of rapamycin complex 1; a protein complex that functions as a cellular sensor for nutrient, energy and redox and controls protein synthesis
e. NMD—nonstop-mediated mRNA decay; an mRNA surveillance system that detects and rescues stalled ribosomes

15

The major differences between prokaryotic and eukaryotic translation are speed (the prokaryotic process is significantly faster), location (the eukaryotic process is not directly coupled to transcription as prokaryotic translation is), complexity (because of their complex life styles, eukaryotes possess complex mechanisms for regulatory protein synthesis, e.g., eukaryotic translation involves a significantly larger number of protein factors than prokaryotic translation), and posttranslational modifications (eukaryotic reactions appear to be considerably more complex and varied than those observed in prokaryotes).

18

A preproprotein is the inactive precursor of a protein with a removable signal peptide. A proprotein is an inactive precursor protein. A protein is a fully functional product of translation.

21

The major differences between prokaryotic and eukaryotic translation control mechanisms are related to the complexity of eukaryotic gene expression. Features that distinguish eukaryotic expression include mRNA export (spatial separation of transcription and translation), mRNA stability (the half lives of mRNA can be modulated), negative translational control (the translation of certain mRNAs can be blocked by the binding of specific repressor proteins), initiation factor phosphorylation (mRNA translation rates are altered by certain circumstances when eIF-2 is phosphorylated), and translational frameshifting (certain mRNAs can be frameshifted so that a different polypeptide is synthesized).

24

GTP hydrolysis provides the energy that drives the movement of the peptidyl-tRNA from the A site to the P site in the ribosome. During elongation GTP hydrolysis is required for the incoming aminoacyl-tRNA to bind in the A site.

27

The large subunit contains the catalytic site for peptide bond formation. The small subunit serves as a guide for the translation factors required to regulate protein synthesis. Together the two subunits come together and form a molecular machine that polymerizes amino acids in a sequence specified by the base sequence in the mRNA molecule.

30

To ensure that proteins end up in a location appropriate to their function in a timely and predictable way, it is necessary to have a targeting mechanism. The signaling process begins with specific signal sequences, which determine where translation will be completed. Specific localization sequences and/or posttranslational modification of the product protein then ensures delivery of the protein to its target location.

33

Elongation involves three basic steps: (1) Binding of an aminoacyl-tRNA in the A site, which is effected by eEF1α-GTP; (2) transpeptidation, the nucleophilic attack of the A site α-amino group on the carbonyl carbon of the P-site in which eEF2-GTP binds to the ribosome. GTP hydrolysis drives conformational changes needed to physically move the ribosome along the mRNA.

36

Neurons of patients with Alzheimer's disease are surrounded with aggregated β-amyloid. The protein huntingtin is aggregated in Huntington's disease. Amylin is the protein that forms aggregates in the β-cells of patients with type 2 diabetes.

39

The three letter sequences listed below each amino acid in the first table are the possible mRNA sequences that code for that specific amino acid. For example, any of the four mRNA sequences listed below Ala will code for Ala. Thus, there are many correct answers to this question. Choose one three-letter sequence from each column to build an mRNA sequence that will code for the peptide. For example, one possible answer is:

mRNA 5′-GCU UCU UUU UAU UCU AAA AAA UAA GCU GAU GUU AUU-3′

cDNA 3′-CGA AGA AAA ATA AGA TTT TTT ATT CGA CTA CAA TAA-5′

Note that the sequence order must be reversed to write the sequence as "5′ ⟶ 3′":

5′ – ATT AAC CTA AGC TAA TTT TTT AGA ATA AAA AGA AGC -3′

5′ ⟶ 3′ Possible Choices for mRNA Codon Base Sequences for this Peptide

5′-Ala	Ser	Phe	Tyr	Ser	Lys	Lys	Leu	Ala	Asp	Val	Ile-3′
GCU	UCU	UUU	UAU	UCU	AAA	AAA	UUA	GCU	GAU	GUU	AUU
GCC	UCC	UUC	UAC	UCC	AAG	AAG	UUG	GCC	GAC	GUC	AUC
GCA	UCA			UAC			UCA	GCA		GUA	AUA
GCG	UCG			UCG			UCG	GCG		GUG	
	AGU			AGU			AGU				
	AGC			AGC			AGC				

3′ ⟶ 5′ Possible Choices for the DNA Sequences for this Polypeptide

Ala	Ser	Phe	Tyr	Ser	Lys	Lys	Leu	Ala	Asp	Val	Ile
3′CGA	AGA	AAA	ATA	AGA	TTT	TTT	AAT	CGA	CTA	CAA	TAA
3CGG	AGG	AAG	ATG	AGG	TTC	TTC	AAC	CGG	CTG		TAG
3CGT	AGT			AGT			GAA	CGT	CAG		TAT
3CGC	AGC			AGC			GAG	CGC	CAT		
	TCA			TCA			GAT		CAC		
	TCG			TCG			GAC				

42

Features of eukaryotic protein synthesis that help to account for the increased time required (as opposed to prokaryotic translation) include the greater quantity, variety and functioning of eukaryotic translation factors (e.g., at least 12 IF's vs 3 for prokaryotes): additional processing of mRNA (addition of a cap and a poly(A) tail; removal of introns); and the increased quantity and variety of eukaryotic posttranslational modifications, such as hydroxylation, and disulfide bond formation. Also, because eukaryotic mRNA lacks Shine-Dalgarno sequences, eukaryotic ribosomes must search for a translation start site by binding to the capped 5′ end and scan toward the 3′ end.

Fill in the Blank

45. Nonoverlapping
48. Targeting
51. Amide bond
54. Shine-Dalgarno sequence

Short Answer

57

The translation efficiency of eukaryotes, as measured by the number of polypeptides that can be synthesized per unit time, is largely made possible by the circular conformation of eukaryotic polyribosomes.

Thought Questions

60

The coding reassignment mechanisms for selenocysteine and pyrolysine are similar in many respects (e.g., the use of the reassignment sequences SECIS and PYLIS, and specific tRNAs and acyl-tRNA synthetases). A major difference between these two examples of coding reassignment is the mechanism whereby the two nonstandard amino acids are linked to their respective tRNAs. Selenocysteine-tRNA is produced from a specialized seryl-tRNA. The seryl group is converted after linkage to the tRNA to a selenocysteinyl group. In contrast pyrolysine is synthesized before it is linked to its tRNA.

63

When errors in amino acid–tRNA binding do occur, they are usually the result of similarities in amino acid structure. Several aminoacyl-tRNA synthetases possess a separate proofreading site that binds the incorrect aminoacyl–tRNA products and hydrolyzes them.

66

Posttranslational modification reactions prepare polypeptides to serve their specific functions and direct them to specific cellular or extracellular locations. Examples of these modifications include proteolytic processing (e.g., removal of signal proteins), glycosylation, methylation, phosphorylation, hydroxylation, lipophilic modifications (e.g., N-myristoylation and prenylation), and disulfide bond formation.

69

Sets of amino acids which may require proofreading include phenylalanine/tyrosine, serine/threonine, aspartate/glutamate, asparagine/glutamine, isoleucine/leucine, and glycine/alanine.

72

A two subunit ribosome is essential to ensure that all of the required elements are in place before the translational process begins. This is a physical ordering process much like an assembly line; the parts must be in place before the enzymatic activities are set in motion.

Glossary

α-tocopherol A radical scavenger belonging to a class of compounds called phenolic antioxidants.

acceptor site In RNA splicing, the upstream 3′-OH splice site.

acetal The family of organic compounds with the general formula $RCH(OR')_2$; formed from the reaction of a hemiacetal with an alcohol.

acid A molecule that can donate hydrogen ions.

acidosis A condition in which the pH of the blood is below 7.35 for a prolonged time.

activation energy The threshold energy required to produce a chemical reaction.

active site The cleft in the surface of an enzyme where a substrate binds.

active transport The energy-requiring movement of molecules across a membrane against a concentration gradient.

activity coefficient A correction factor that is used in calculating the effective concentration of a solute in solution.

acyl carrier protein A component of fatty acid synthase. Intermediates of fatty acid synthesis are linked to this molecule through a thioester linkage.

acyl group Any molecular group derived from a carboxylic acid by the removal of a hydroxyl group.

addition reaction A chemical reaction in which two molecules react to form a third and there are more groups attached to carbon atoms in the product.

adduct The product of an addition reaction.

adiponectin A peptide hormone that enhances glucose-stimulated insulin secretion and cellular responses to insulin.

A-DNA A short, compact DNA structure in which the base pairs are not at right angles to the helical axis; occurs when DNA becomes partially dehydrated.

aerobic metabolism The mechanism by which the chemical bond energy of food molecules is captured and used to drive the oxygen-dependent synthesis of adenosine triphosphate (ATP).

aerobic respiration The metabolic process in which oxygen is used to generate energy from food molecules.

aerotolerant anaerobe An organism that depends on fermentation for its energy needs and possesses protection from toxic oxygen metabolites in the form of detoxifying enzymes and antioxidant molecules.

affinity chromatography A technique in which proteins are isolated based on their capacity to bind to a specific ligand.

aldaric acid The product formed when the aldehyde and CH_2OH groups of a monosaccharide are oxidized to carboxylic acids.

aldimine An imine product of a reaction of a primary amine group with a carbonyl group; also referred to as a Schiff base.

alditol A sugar alcohol; the product of the reduction of the aldehyde or ketone group of a monosaccharide.

aldol cleavage A reverse of the aldol condensation.

aldol condensation An aldol addition reaction; the nucleophilic addition of a ketone enolate ion to an aldehyde to form a β-hydroxyketone, followed by the elimination of a water molecule.

aldonic acid The product of the oxidation of the aldehyde group of a monosaccharide.

aldose A monosaccharide with an aldehyde functional group.

aliphatic hydrocarbon A nonaromatic hydrocarbon such as methane or cyclohexane.

alkalosis A condition in which the blood pH is above 7.45 for a prolonged period of time.

alkylating agent An electrophile that reacts with a molecule that possesses an unshared pair of electrons, adding an alkyl group.

alkylation The introduction of an alkyl group into a molecule.

allosteric enzyme An enzyme whose activity is affected by the binding of effector molecules.

allosteric transition The ligand-induced conformational change in a protein.

allostery The control of protein function through ligand-binding events.

Alzheimer's disease A progressive, fatal disease that is characterized by seriously impaired intellectual functions caused by neuronal death.

amethopterin A structural analogue of folate used to treat several types of cancer; also referred to as methotrexate.

amino acid An organic molecule that contains an amino group and a carboxyl group.

amino acid pool The amino acid molecules that are immediately available in an organism for use in metabolic processes.

amino acid residue An amino acid that has been incorporated into a peptide molecule.

amphibolic pathway A metabolic pathway that functions in both anabolism and catabolism.

amphipathic molecule A molecule containing both polar and nonpolar domains.

amphoteric molecule A molecule that can act as both an acid and a base.

AMPK AMP-activated protein kinase; an important regulatory enzyme in energy metabolism.

amyloid deposits Insoluble extracellular proteinaceous debris found in the brains of patients with certain neurological diseases.

amylopectin A type of plant starch; a branched polymer containing $\alpha(1,4)$- and $\alpha(1,6)$-glycosidic linkages.

amylose A type of plant starch; an unbranched polymer of D-glucose residues linked with $\alpha(1,4)$-glycosidic linkages.

anabolic pathways A series of biochemical reactions in which large complex molecules are synthesized from smaller precursor molecules.

anaerobic organisms Organisms that do not use oxygen to generate energy.

anaerobic respiration The metabolic process in which species other than oxygen are the terminal electron acceptors in energy generation.

analogue A substance similar in structure to a naturally occurring molecule.

anaplerotic reaction A reaction that replenishes a substrate needed for a biochemical pathway.

anchor protein A molecule that facilitates the recruitment and assembly of specific sets of signal cascade proteins into complexes bound to the cytoskeleton.

anhydride The product of the condensation reaction between two carboxyl groups or two phosphate groups in which a molecule of water is eliminated.

annotation The functional identification of the genes of a genome.

anomer An isomer of a cyclic sugar that differs from another in its configuration about the hemiacetal or acetal carbon.

antenna pigment A molecule that absorbs light energy and transfers it to a reaction center during photosynthesis.

anticodon A sequence of three ribonucleotides on a tRNA molecule that is complementary to a codon on the mRNA molecule; codon-anticodon binding results in the delivery of the correct amino acid to the site of protein synthesis.

antigen Any substance able to stimulate the immune system; generally a protein or large carbohydrate.

antioxidant A substance that prevents the oxidation of other molecules.

antiparallel Aligned in opposition.

antisense strand A noncoding DNA strand that is complementary to the base sequence of an mRNA molecule transcribed from the coding DNA strand.

apoenzyme The protein portion of an enzyme that requires a cofactor to function in catalysis.

apoprotein A holoprotein without its prosthetic group.

apoptosis The genetically programmed series of events that lead to cell death.

apurinic site A nucleotide residue in a DNA molecule from which a purine base has been lost or removed.

apyrimidinic site A nucleotide residue in a DNA molecule from which a pyrimidine base has been lost or removed.

aquaporin A water channel protein.

archaea One of the three domains of living organisms: prokaryotic organisms that have the appearance of bacteria and many molecular properties that are similar to those of the eukaryotes.

aromatic hydrocarbon A molecule that contains a benzene ring or has properties similar to those exhibited by benzene.

asymmetric carbon A carbon bound to four different groups.

atherosclerosis A cardiovascular disease in which soft masses containing fatty material and cellular debris are formed in the lining of blood vessels.

attachment (att) site A short DNA sequence that facilitates site-specific recombination; also refers to an IS element.

autocrine A hormonelike molecule that is active within the cell in which it is produced.

autoimmune disease A condition in which an immune response is directed against an individual patient's own tissues.

autophagy A cellular degradation pathway in which cell components are degraded by enzymes in lysosomes.

autopoiesis A system that is autonomous, self-organizing, and self-maintaining.

autotroph An organism that transforms light energy or the energy of various chemicals into the chemical bond energy of biomolecules.

β-carotene A plant pigment molecule that acts as an absorber of light energy and as an antioxidant.

β-oxidation The catabolic pathway in which most fatty acids are degraded; acetyl-CoA is formed as the bond between the α and β carbon atoms is broken.

β_2 clamp The protein complex that promotes processivity; that is, it prevents frequent dissociation of DNA polymerase from the DNA template.

B cell A B lymphocyte; a white blood cell that produces and secretes antibodies, the proteins that bind to foreign substances, thereby initiating their destruction in the humoral immune response.

bacteria One of the three domains of life: single-celled prokaryotes with diverse capacities to exploit their environments.

bacterial artificial chromosome A derivative of a large *E. coli* plasmid used to clone DNA sequences as long as 300 kb.

base A molecule that can accept hydrogen ions.

base analogue A molecule that resembles a normal DNA nucleotide base and can substitute for it during DNA replication, leading to mutation.

base excision repair A mechanism that removes and then replaces individual nucleotides in DNA whose bases have undergone various types of damage (e.g. alkylation, deamination, or oxidation).

B-DNA The commonly found form of DNA, as the sodium salt under highly humid conditions.

bile salts Amphipathic molecules with detergent properties that are important components of bile, a yellowish green liquid that aids in the digestion of fat; a conjugated derivative of the bile acids cholic acid and deoxycholic acid.

bioenergetics The study of energy transformations in living organisms.

biogenic amine An amino acid derivative that acts as a neurotransmitter (e.g., GABA and the catecholamines).

biogeochemical cycle A pathway driven by solar and geothermal energy in which a chemical element moves throughout Earth's biotic and abiotic compartments.

bioinformatics The computer-based field that facilitates the analysis of biological sequence data.

biomolecules Molecules that make up a living organism.

bioremediation The use of biological processes to decontaminate toxic waste sites.

biotransformation A series of enzyme-catalyzed processes in which toxic and/or hydrophobic molecules are converted into (usually) less toxic and more soluble metabolites.

branched-chain amino acid One of a group of essential amino acids (leucine, isoleucine, and valine) with branched carbon skeletons.

buffer A substance that resists large pH changes when small amounts of acids or bases are added; usually a solution that contains a weak acid and its conjugate base.

C3 plants Plants that produce glycerate-3-phosphate, a three-carbon molecule, as the first stable product of photosynthesis.

C4 metabolism A photosynthetic pathway in plants such as corn and sugarcane that produces a four-carbon molecule and avoids photorespiration.

C4 plants Plants that possess mechanisms that suppress photorespiration by separating rubisco (ribulose-1,5-bisphosphate carboxylase) from atmospheric O_2.

Calvin cycle The major metabolic pathway by which carbon dioxide is incorporated into organic molecules

Cap-binding complex (CBC) A protein complex that binds to capped mRNA molecules and facilitates their translation; consists of eIF-4A (a helicase), eIF-4E (a translation initiation factor), and eIF-G (a scaffold protein); also referred to as eIF-4F.

carbanion A carbon with a negative charge.

carbocation A carbon with a positive charge.

carbon fixation The biochemical process by which inorganic carbon dioxide is incorporated into organic molecules.

carotenoid An isoprenoid molecule that functions as a light-harvesting pigment and/or protects against reactive oxygen species (ROS).

carrier protein A membrane transport protein.

catabolic pathway A series of biochemical reactions in which a large complex molecule is degraded into smaller, simpler products; in some catabolic pathways, energy is captured.

catalyst A substance that enhances the rate of a chemical reaction but is not permanently altered by the reaction.

catecholamine One of a class of neurotransmitters derived from tyrosine; includes dopamine, norepinephrine, and epinephrine.

caveolae A special type of small invagination of the plasma membrane in some cells, which contain the protein caveolin and several types of lipid molecules; involved in signal transduction and endocytosis.

caveolar endocytosis A type of clathrin-independent endocytosis; occurs most prominently in adipocytes and endothelial cells.

cDNA library A clone library of cDNA (complementary DNA) molecules produced from mRNA molecules by reverse transcription.

cell cortex The three-dimensional meshwork of proteins that reinforces the plasma membrane.

cell fractionation A technique involving homogenization and centrifugation that allows the study of cell organelles.

cellobiose A degradation product of cellulose; a disaccharide that contains two molecules of glucose linked by a $\beta(1,4)$-glycosidic bond.

cellular immunity Immune system processes mediated by T cells, a type of lymphocyte.

cellulose A polymer produced by plants that is composed of D-glucopyranose residues linked by $\beta(1,4)$-glycosidic bonds.

centromere A special region of repetitive DNA that plays a critical role in cell division; it holds the two sister chromatids together during prophase and metaphase of cell division.

chain-terminating method A technique for determining DNA base sequences that uses 2′-3′-dideoxy base analogues as chain-terminating inhibitors of DNA polymerase; also referred to as the Sanger method.

channel protein A membrane protein that contains a pore through which ions are transported.

chaperone-mediated autophagy A receptor-mediated process in which specific proteins that are bound to a chaperone complex are unfolded and then translocated into a lysosome, where they are then degraded.

chaperonins A family of molecules that control the folding and targeting of cell proteins.

Chargaff's rules A set of rules describing the base composition of DNA; posits the equality of the concentration of adenine and thymine and of cytosine and guanine.

chemiosmotic coupling theory ATP synthesis is coupled to electron transport by an electrochemical proton gradient across a membrane.

chemoautotroph An organism that transforms the energy of various chemicals into chemical bond energy.

chemoheterotroph An organism that uses preformed organic food molecules as its sole source of energy.

chemolithotroph An organism that transforms the energy in specific inorganic substances into chemical bond energy.

chemosynthesis The biochemical mechanism whereby chemical energy is extracted from certain minerals.

chiral carbon An asymmetric carbon in a molecule that has a mirror-image form.

chitin The principal structural component of the exoskeletons of arthropods and the cell walls of many fungi; a homoglycan composed of *N*-acetylglucosamine residues.

chlorophyll A magnesium-containing green pigment molecule found in plants and photosynthetic bacteria that resembles heme; absorbs light energy in photosynthesis.

chloroplast A chlorophyll-containing plastid found in the cells of algae and higher plants.

chromatin The complex of DNA and histones found in the nucleus of eukaryotic cells.

chromatin remodeling complex A multi-subunit complex that facilitates the release of the histones from nucleosomal DNA during transcription.

chromophore A molecular component that absorbs light of a specific frequency.

chromoplast A type of plastid in plants that accumulates the pigments that are responsible for the colors of leaves, flower petals, and fruits.

chromosomal jumping A technique used to isolate clones that contain discontinuous sequences from the same chromosome.

chromosome A very long DNA molecule associated with proteins that contains the genes of an organism.

chylomicron A large lipoprotein of extremely low density: transports dietary triacylglycerols and cholesteryl esters from the intestine to muscle and adipose tissue.

chylomicron remnants Chylomicrons after about 90% of the triacylglycerols have been removed by lipoprotein lipase.

***cis* isomer** An isomer in which two substituents are on the same side of the double bond.

cistron A DNA sequence that contains the coding information for a polypeptide and the signals required for ribosome function.

citric acid cycle A biochemical pathway that degrades the acetyl group of acetyl-CoA to CO_2 and H_2O as three molecules of NAD^+ and one molecule of FAD are reduced.

clamp loader The γ complex that recognizes single DNA strands with primer and transfers β_2-clamp dimer to the core polymerase.

clathrin A protein that plays a major role in the formation of coated vesicles.

clathrin-dependent endocytosis A receptor-mediated process whereby cells internalize molecules by the inward budding of plasma membrane vesicles.

coding strand The DNA strand that has the same base sequence as the RNA transcript (with thymine instead of uracil).

codon A sequence of three nucleotides in mRNA that directs the incorporation of an amino acid during protein synthesis or acts as a start or stop signal.

coenzyme A small organic molecule required in the catalytic mechanisms of certain enzymes.

coenzyme A A carrier of acetyl and acyl groups that is composed of a 3′-phosphate derivative of ADP linked to pantothenic acid (via a phosphate ester bond), which in turn is linked to β-mercaptoethylamine via an amide bond.

cofactor The nonprotein component of an enzyme (either an inorganic ion or a coenzyme) required for catalysis.

colony hybridization technique A method used to identify bacterial colonies that possess a specific recombinant DNA sequence.

competitive inhibition A reversible type of enzyme inhibition in which the inhibitor molecule competes with the substrate for occupation of the active site.

composite transposon A bacterial transposon composed of a gene and flanking IS elements.

conjugate base The anion (or molecule) that results when a weak acid loses a proton.

conjugate redox pair An electron donor and its electron acceptor form: for example, NADH and NAD^+.

conjugated protein A protein that functions only when it carries other chemical groups attached by covalent linkages or by weak interactions.

conjugation Unconventional sexual mating between bacterial cells; a donor cell transfers a DNA segment into a recipient cell through a specialized pilus.

conjugation reaction A biochemical reaction that may improve the water solubility of a molecule by converting it to a derivative that contains a water-soluble group.

consensus sequence The average of several similar DNA sequences: for example, the consensus sequence of the –10 box of *E. coli* promoter is TATAAT.

constitutive gene A routinely transcribed gene that codes for gene products required for basic cell functions.

constitutive heterochromatin Highly condensed and transcriptionally silent segments of DNA in eukaryotes, most notably repetitive sequences, telomeres, and centromeres.

contig One of a set of overlapping DNA sequences used to identify the base sequence of a region of DNA.

cooperative binding A mechanism in which binding of one ligand to a target molecule promotes the binding of other ligands.

Cori cycle A metabolic process in which lactate, produced in tissues such as muscle, is transferred to liver, where it becomes a substrate in gluconeogenesis.

cosmid Cloning vehicles that contain the γ bacteriophage cos sites incorporated into plasmid DNA sequences with one or more selectable markers.

cotranslational transfer The insertion of a polypeptide across a membrane during ongoing protein synthesis.

covalent bond The sharing of electrons between atoms.

CpG CpG dinucleotides; methylated cytosines occur predominantly in 5′-CG-3′ sequences.

CpG island Regions of the genome where CpGs constitute more than 50% of the bases.

Crabtree effect The physiological capacity of *S. cerevisiae* cells to ferment sugar to produce ethanol, which kills their competitors, and then use the ethanol as an energy source.

Crassulacean acid metabolism A photosynthetic pathway that produces a four-carbon molecule (malate) in plants that live in hot, dry regions such as deserts.

cystic fibrosis An ultimately fatal autosomal recessive disease that is caused by the missing or defective chloride channel protein CFTR.

cystic fibrosis transmembrane conductance regulator (CFTR) The plasma membrane glycoprotein that functions as a chloride channel in epithelial cells.

cytochrome P_{450} system An electron transport system that consists of two enzymes (NADPH-cytochrome P_{450} reductase and cytochrome P_{450}); involved in the oxidative metabolism of many endogenous and exogenous substances.

cytokine One of a group of hormonelike polypeptides and proteins secreted by certain immune system cells.

cytoskeleton A set of protein filaments (microtubules, microfilaments, and intermediate fibers) that maintains the cell's internal structure and allows organelles to move.

DAG Diacylglycerol; a second messenger molecule in the phosphatidylinositol pathway.

***de novo* methyltransferase** Methyltransferases that catalyze the methylation of unmodified CpGs.

decarboxylation The removal of a carboxylic group from a carboxylic acid as carbon dioxide.

decoding center The position located within the bacterial 30S ribosomal subunit where codon-anticodon base pairs form.

degeneracy The capacity of structurally different system parts to perform the same or similar functions.

denaturation A disruption of protein or nucleic acid structure caused by exposure to heat or chemicals leading to loss of biological function.

density gradient centrifugation A technique in which cell fractions are further purified by centrifugation in a density gradient.

desensitization A process in which target cells adjust to changes in stimulation by decreasing the number of cell surface receptors or by inactivating those receptors.

detoxication The process by which a toxic molecule is converted to a more soluble (and usually less toxic) product.

detoxification Correction of a state of toxicity; the chemical reactions that produce sobriety in an enebriated person.

dialysis A laboratory technique in which a semipermeable membrane is used to separate small molecules from larger ones.

diastereomers A stereoisomer that is not an enantiomer (mirror-image isomer).

dicer A nuclease that cuts pre-microRNA into mature miRNAs or initiates the gene silencing process.

differential centrifugation A cell fractionation technique in which homogenized cells are separated by centrifugal forces.

dipole A difference in charge between atoms in a molecule resulting from the unsymmetrical orientation of polar bonds.

disaccharide A glycoside composed of two monosaccharide residues.

dissipative system A system that facilitates the reduction of an energy gradient.

disulfide bridge A covalent bond formed between the sulfhydryl groups of two cysteine residues.

disulfide exchange An enzyme-catalyzed posttranslational process in which there is an interchange of disulfide bonds in a protein until the correct biologically relevant disulfide bonds are formed.

DNA fingerprinting A laboratory technique used to compare DNA banding patterns from different individuals.

DNA glycosylase A DNA repair enzyme that cleaves the N-glycosidic linkage between the damaged base and the deoxyribose component of the nucleotide.

DNA ligase An enzyme that catalyzes the formation of a covalent phosphodiester bond between the 3′-OH end of one segment and the 5′-phosphate end of another segment during DNA replication.

DNA microarray A DNA chip used to analyze the expression of thousands of genes simultaneously.

DNA profile A unique DNA pattern of repeats of target sequences that is separated in an electrophoresis gel; used to identify individuals.

DNA typing A DNA analysis technique used to identify individuals; involves the analysis of several highly variable sequences called markers.

docking protein A transmembrane protein of the rough endoplasmic reticulum that binds a signal recognition protein that is bound to a ribosome, thus triggering the resumption of protein synthesis; also referred to as signal recognition particle receptor protein.

donor site The 5′-splice site in the RNA splicing process.

double-strand break repair model A general recombination mechanism for repairing double-strand breaks in DNA utilizing homologous chromosomes; produces both crossover and noncrossover products.

downregulation The reduction in cell surface receptors in response to stimulation by specific hormone molecules.

dynein A motor protein associated with microtubules.

dyslipidemia High blood levels of total cholesterol and triacylglycerol and low HDL levels; a disorder associated with metabolic syndrome.

effector A molecule whose binding to a protein alters the protein's activity.

eicosanoid A hormonelike molecule that contains 20 carbons; most are derived from arachidonic acid; examples include prostaglandins, thromboxanes, and leukotrienes.

electron acceptor Species that accepts electrons from an electron donor during a reaction.

electron donor Species that donates electrons to an electron acceptor during a reaction.

electron transport system A series of electron carrier proteins that bind reversibly to electrons at different energy levels.

electrophile An electron-deficient species that is preferentially attracted to a region of high electron density in another species during a chemical reaction.

electrophoresis A class of techniques in which molecules are separated from each other because of differences in their net charge.

electroporation A method of introducing a cloning vector into a host cell that involves treatment with an electrical current.

electrostatic interaction Noncovalent attraction between oppositely charged atoms or groups.

elimination reaction A chemical reaction in which a double bond is formed when atoms in a molecule are removed.

elongation The polypeptide chain growth phase during translation of an mRNA in a ribosome.

emergence New and unanticipated properties in each level of organization of a system that result from interactions among the components.

emergent property A new property conferred by the complexity and dynamics of the system.

enantiomer A mirror-image stereoisomer.

endergonic process A reaction that does not spontaneously go to completion; the standard free energy change is positive and the equilibrium constant is less than 1.

endocrine hormone A hormone secreted into the bloodstream that acts on distant target cells.

endocytic cycle The continuous recycling of membrane via endocytosis and exocytosis.

endocytosis A process in which a cell takes up solutes or particles by enclosing them in vesicles pinched off from its plasma membrane.

endogenous retrovirus A decayed virus within a genome: also referred to as a LTR retrotransposon.

endomembrane system An extensive set of interconnecting internal membranes that divide the cell into functional compartments.

endoplasmic reticulum (ER) A series of membranous channels and sacs that provides a compartment separate from the cytoplasm for numerous chemical reactions.

endothermic reaction A reaction that requires energy.

enediol The intermediate formed during the isomerization reactions of monosaccharides. It contains a double bond with a hydroxyl group on each carbon of the double bond.

energy The capacity to do work.

enhancer A short DNA sequence that promotes the transcription of one or more genes when it is bound to an activator protein.

enthalpy The heat content of a system; in a biological system it is essentially equivalent to the total energy of the system.

entropy A measure of the randomness or disorder of a system; a measure of that part of the total energy in a system that is unavailable for useful work.

enzyme A biomolecule that catalyzes a biochemical reaction.

enzyme induction A process in which a signal molecule stimulates increased synthesis of a specific enzyme.

enzyme kinetics The study of the rates of enzyme-catalyzed reactions.

epidermal growth factor A protein that stimulates epithelial cells to undergo cell division.

epigenetics Heritable covalent modification-induced gene activations and repressions that do not change DNA base sequences.

epigenome The current epigenetic modifications within a cell.

epimer A molecule that differs from the configuration of another by one asymmetric carbon.

epimerization The reversible interconversion of epimers.

epimutation An alteration in the normal epigenetic pattern.

epinephrine One of several catecholamine neurotransmitters; derived from tyrosine.

epoxide An ether in which the oxygen is incorporated into a three-membered ring.

ER-associated protein degradation (ERAD) A process that targets misfolded proteins for destruction.

ER stress Stress conditions cause misfolded proteins to accumulate in the ER.

essential amino acid An amino acid that cannot be synthesized by the body and must be supplied by the diet.

essential fatty acid A fatty acid that must be supplied in the diet because it cannot be synthesized by the body; linoleic and linolenic acids in humans.

euchromatin A less condensed form of chromatin that has varying levels of transcriptional activity.

eukarya One of the three domains of life: nucleus-containing single-celled and multicellular organisms.

eukaryotic cell A living cell that possesses a true nucleus.

exergonic process A reaction that spontaneously goes to completion as written; the standard free energy change is negative, and the equilibrium constant is greater than 1.

exocytosis The secretion process in eukaryotic cells; involves the fusion of membrane-bound secretory granules with the plasma membrane.

exon The region in a split or interrupted gene that codes for RNA and ends up in the final product (e.g., mRNA).

exonuclease An enzyme that removes nucleotides from the end of the polynucleotide strand.

exothermic reaction A reaction that releases heat.

extracellular matrix (ECM) A gelatinous material, containing proteins and carbohydrates, that binds cells and tissues together.

extremophile An organism that lives under extreme conditions of temperature, pH, pressure, or ionic concentration that would easily kill most organisms.

extremozyme An enzyme that functions under extreme conditions of temperature, pressure, pH, and/or ionic concentration.

facilitated diffusion Diffusion of a substance across a membrane that is aided by a carrier.

facultative anaerobe An organism that possesses the capacity for detoxifying oxygen metabolites; energy is generated using oxygen, when available, as an electron acceptor.

facultative heterochromatin A condensed DNA sequence that can become less condensed and transcriptionally active in response to specific signaling mechanisms.

fatty acid A monocarboxylic acid usually with an even number of carbon atoms; R-COOH where R is an alkyl group.

fatty acid–binding protein An intracellular water-soluble protein whose function is to bind and transport hydrophobic fatty acids.

feedback control The control of a self-regulating system (e.g., a metabolic process or pathway) in which product influences the output of the process.

fermentation An energy-yielding process in which organic molecules serve as both donors and acceptors of electron; the anaerobic degradation of sugars.

fibrous protein A protein composed of polypeptides arranged in long sheets or fibers.

flavin adenine dinucleotide (FAD) A tightly bound prosthetic group consisting of riboflavin, D-ribitol, and adenine that functions in the class of enzymes called flavoproteins.

flavin-containing monooxygenase One of a family of NADPH- and O_2-requiring enzymes that oxidize molecules (usually xenobiotics) with nitrogen-, sulfur-, or phosphorus-containing functional groups.

flavin mononucleotide (FMN) A tightly bound prosthetic group consisting of a molecule of riboflavin and D-ribitol phosphate that functions in the class of enzymes called flavoproteins.

flavoprotein A conjugated protein in which the prosthetic group is either FMN or FAD.

fluid mosaic model The currently accepted model of cell membranes in which the membrane is a lipid bilayer with integral proteins buried in the lipid and peripheral proteins loosely attached to the membrane surface.

fluorescence A form of luminescence in which certain molecules can absorb light of one wavelength and emit light of another wavelength.

fold A core three-dimensional structure of a protein domain.

43S preinitiation complex The eukaryotic multisubunit complex composed of the 40S subunits eIF-A, eIF-2-GTP, eIF-3, and methionyl-$tRNA^{met}$ that binds to mRNA.

48S initiation complex The mature eukaryotic initiation complex that scans the mRNA in search of the start codon.

free energy The energy in a system available to do useful work.

free radical An atom or molecule that has an unpaired electron.

functional genomics The investigation of gene expression patterns.

functional group A group of atoms that undergoes characteristic reactions when attached to a carbon atom in an organic molecule or biomolecule.

G protein A heterotrimeric GTP-binding protein that acts as a molecular switch when activated by a GPCR.

gasotransmitter An endogenous gaseous molecule that acts as a signal molecule.

gel filtration chromatography A technique used to separate molecules according to their size and shape that employs a column packed with a gelatinous polymer.

gene A DNA sequence that codes for a polypeptide, rRNA, or tRNA.

gene duplication The creation of a duplicate gene or part of a gene; can result from unequal crossing over during meiosis or from retrotransposition.

gene expression The mechanism by which living organisms regulate the flow of genetic information; the control of when and if genes are transcribed.

gene silencing A form of posttranscriptional gene regulation that involves 22-nt miRNAs.

general recombination Recombination involving exchange of a pair of homologous DNA sequences; it can occur at any location on a chromosome.

gene silencing A form of posttranscriptional gene regulation that involves 22-nt miRNAs.

genetic code The set of nucleotide base triplets (codons) that code for the amino acids in proteins as well as start and stop signals.

genetics The scientific investigation of inheritance.

genome The total genetic information possessed by an organism.

genomics The investigation of entire genomes; the sequencing and characterization of genomes.

ghrelin A protein that stimulates appetite; produced by the cells of the stomach and small intestine.

globular protein A protein that adopts a globular shape.

glucagon A peptide hormone released from pancreatic α-cells; among its effects are increasing the level of glucose in blood via the breakdown of liver glycogen.

glucocorticoid A steroid hormone produced in the adrenal cortex that affects carbohydrate, protein, and lipid metabolism.

glucogenic Describing amino acids that are degraded to pyruvate or a citric acid intermediate; these amino acids are used as substrates in the synthesis of glucose in gluconeogenesis.

gluconeogenesis The synthesis of glucose from noncarbohydrate molecules.

glucose-alanine cycle A method of recycling α-keto acids between muscle and liver and for transporting ammonia to the liver.

glucosuria The presence of glucose in the urine.

glycan A polymer of monosaccharides; a polysaccharide.

glycerol phosphate shuttle A metabolic process that uses glycerol-3-phosphate to transfer electrons from NADH in the cytosol to mitochondrial FAD.

glyceroneogenesis An abbreviated version of gluconeogenesis in which glycerol-3-phosphate is synthesized from substrates other than glucose or glycerol.

glycocalyx A layer on the external surface of many eukaryotic cells that contains substantial amounts of carbohydrate-containing molecules.

glycoconjugate A molecule that possesses covalently bound carbohydrate components (e.g., glycoproteins and glycolipids).

glycoform One of several slightly different forms of a glycan component of a glycoprotein.

glycogen A glucose storage molecule in vertebrates; a branched polymer containing $\alpha(1,4)$- and $\alpha(1,6)$-glycosidic linkages.

glycogenesis A biochemical pathway that adds glucose to growing glycogen polymers when blood glucose levels are high.

glycogenolysis A biochemical pathway that removes glucose molecules from glycogen polymers when blood glucose levels are low.

glycolipid A glycosphingolipid; a molecule in which a monosaccharide, disaccharide, or oligosaccharide is attached to a ceramide through an O-glycosidic linkage.

glycolysis The enzymatic pathway that converts a glucose molecule into two molecules of pyruvate: the anaerobic process generates energy in the form of two ATP molecules and two NADH molecules.

glycome The total set of sugars and glycans that a cell or organism produces.

glycoprotein A conjugated protein in which carbohydrate molecules are covalently bound.

glycosaminoglycan A long unbranched heteropolysaccharide chain composed of disaccharide repeating units.

glycoside The acetal of a sugar.

glycosidic link An acetal linkage formed between two monosaccharides.

glyoxylate cycle A modification of the citric acid cycle that occurs in plants, bacteria, and other eukaryotes: allows growth in these organisms from two-carbon substrates such as ethanol, acetate, and acetyl-CoA.

Golgi apparatus (complex) A series of curved membranous sacs involved in packaging and distributing cell products to internal and external compartments.

GPI (glycosylphosphatidylinositol) anchor A glycolipid used to link certain proteins to membrane, preferentially in lipid rafts.

G-protein-coupled receptor (GPCR) A cell surface receptor that transduces the binding of a hormone or other signal molecule into an intracellular response via the activation of a G protein.

grana (pl) Stacks of thylakoid membrane.

granum (sing) The folded portion of the thylakoid membrane.

growth factor An extracellular polypeptide that stimulates cells to grow and/or undergo cell division.

GTPase-activating protein (GAP) A protein molecule that hydrolyzes GTP bound to a GTP-binding protein.

GTPase associated region (GAR) A set of overlapping binding sites on the 50S ribosomal subunit that activate specific translation factors with GTPase activity during protein synthesis.

guanine nucleotide exchange factor (GEF) A protein that mediates a conformational change in the transmembrane region of a G-protein-coupled receptor and leads to GDP/GTP exchange during G-protein activation.

guanylate cyclase receptor A membrane-bound or soluble receptor protein that when activated converts GTP to the second messenger molecule cyclic GMP.

heat shock protein (hsp) A protein synthesized in response to stress (e.g., high temperature).

helicase An ATP-requiring enzyme that catalyzes the unwinding of duplex DNA.

hemiacetal One of the family of organic molecules with the general formula RCH(OR)OH that is formed by the reaction of one molecule of alcohol with an aldehyde.

hemiketal One of the family of organic molecules with the general formula RRC(OR)OH that is formed by the reaction of a molecule of alcohol with a ketone.

hemoprotein A conjugated protein in which heme, an iron-containing organic group, is the prosthetic group.

Henderson-Hasselbalch equation Kinetic rate expression that defines the relationship between pH, pK_a, and the concentration of the weak acid and conjugate base components of a buffer solution.

heterochromatin Chromatin that is so highly condensed that it is transcriptionally inactive.

heteroglycan A high-molecular-weight carbohydrate polymer that contains more than one kind of monosaccharide.

heterokaryon A structure formed from the fusion of the membranes of two different cells; used to demonstrate membrane fluidity.

heterotroph An organism that obtains energy by degrading preformed food molecules usually obtained by consuming other organisms.

high-density lipoprotein A type of lipoprotein with a high protein content that is believed to scavenge excess cholesterol from cell membranes and transport it to the liver.

holoenzyme A complete enzyme consisting of an apoenzyme plus a cofactor.

holoprotein An apoprotein combined with its prosthetic group.

homeostasis The capacity of living organisms to regulate metabolic processes despite variability in their internal and external environments.

homoglycan High-molecular-weight carbohydrate polymers that contain only one type of monosaccharide.

homologous polypeptide Protein molecules whose amino acid sequences are similar; implies a common evolutionary origin.

homologous recombination Recombination involving exchange of a pair of homologous DNA sequences.

hormone A molecule produced by a specific cell that influences the function of distant target cells.

hormone response element A specific DNA sequence that binds hormone-receptor complexes; the binding of a hormone-receptor complex either enhances or diminishes the transcription of a specific gene.

hsp60 One of a family of molecular chaperones that mediate protein folding by forming a large structure composed of two stacked seven-membered rings that facilitate the ATP-dependent folding of polypeptides; also called chaperonins or Cpn 60s.

hsp70 One of a family of molecular chaperones that bind to and stabilize proteins during the early stages of the folding process.

hsp90 One of a family of molecular chaperones that finalize the folding of a limited number of client proteins; coordinates the assembly of certain protein complexes; also works with hsp70 to identify proteins damaged by oxidative or heat stress.

humoral immune response The immunity that results from the presence of antibodies in blood and tissue fluid; also referred to as an antibody-mediated immunity.

Huntington's disease An inherited, fatal neurological disease caused by an excessively long polyglutamine sequence in the protein called huntingtin.

hybridization A laboratory technique in which fragments of single-stranded DNA from different sources anneal; the rate at which a DNA hybrid forms is a measure of the similarity of the two strands.

hydration reaction A type of addition reaction in which water is added to a carbon—carbon double bond.

hydrocarbons Compounds that contain only carbon and hydrogen.

hydrogen bond The force of attraction between a hydrogen atom and a small highly electronegative atom (e.g., O or N) on another molecule or the same molecule.

hydrolase An enzyme that catalyzes reactions in which adding water cleaves bonds.

hydrolysis A chemical reaction in which molecules are cleaved by water.

hydrophilic Describing molecules or portions thereof that dissolve easily in water; hydrophilic molecules possess positive or negative charges or contain relatively large numbers of electronegative oxygen or nitrogen atoms.

hydrophobic Describing molecules that do not dissolve in water and possess few if any electronegative atoms.

hydrophobic interaction The association of nonpolar molecules when they are placed in water.

hyperammonemia A potentially fatal elevation of the concentration of ammonium ions in the blood.

hyperglycemia Blood glucose levels that are higher than normal.

hyperinsulinemia Higher than normal blood levels of insulin.

hyperosmolar hyperglycemic nonketosis Severe dehydration in non-insulin-dependent diabetics; caused by persistently high blood glucose levels.

hypertonic solution A concentrated solution with a high osmotic pressure.

hyperuricemia An abnormally high level of uric acid in the blood.

hypochromic effect The decrease in the absorption of UV light (260 nm) that occurs when purine and pyrimidine bases are incorporated into base pairs in polynucleotide sequences.

hypoglycemia Blood glucose levels that are lower than normal.

hypotonic solution A dilute solution with a low osmotic pressure.

indel An insertion or deletion mutation; occurs when from one to thousands of bases are inserted or deleted, respectively, from a DNA sequence.

inhibitor A molecule that reduces an enzyme's activity.

initiation The beginning phase of translation.

initiation complex The protein complex required to initiate the first step in the translation of ribosome-mediated mRNA.

inner mitochondrial membrane A membrane in which are embedded the respiratory complexes that are responsible for ATP synthesis.

inner nuclear membrane A membrane that encloses the nucleoplasm; contains proteins with functions such as stabilizing nuclear structure and chromatin binding.

insertional element A short DNA sequence involved in site-specific recombination; also called an IS element or att site.

insulator A DNA sequence that when bound to an insulator binding protein blocks the interaction between enhancers and promoters; also prevents the spread of heterochromatin.

insulin A peptide hormone released from pancreatic β-cells; among its many effects is the promotion of glucose uptake into the cells of certain target organs (muscle and adipose tissue).

insulin-like growth factor (IGF) A protein in humans that mediates the growth-promoting actions of growth hormone; has insulin-like properties (e.g., promotes glucose transport and fat synthesis).

insulin resistance The insensitivity of tissues to insulin; a common cause is the downregulation of insulin receptors.

integral protein A protein that is embedded within a membrane.

intercalating agents Planar molecules that insert themselves between base pairs; this action distorts the DNA chain.

interferon One of a group of glycoproteins that have nonspecific antiviral activity (e.g., stimulation of cells to produce antiviral proteins) that inhibits the synthesis of viral RNA and proteins and regulates the growth and differentiation of immune system cells.

interleukin 2 (IL-2) A member of a group of cytokines that regulate the immune system in addition to promoting cell growth and differentiation.

intermediate A species produced in the course of a reaction that exists for a finite period of time.

intermediate-density lipoprotein (IDL) A lipoprotein formed when a very-low-density lipoprotein shrinks in size and becomes more dense as a result of depletion of triacylglycerol, apolipoprotein, and phospholipid molecules.

intermediate filament A cytoskeletal component that provides cells with significant mechanical support; a flexible, strong, and relatively stable polymer (8–12 nm).

interspersed genome-wide repeats Repetitive DNA sequences that are scattered around the genome.

intrinsic termination Transcription termination that involves an RNA termination sequence that contains an inverted repeat sequence; also referred to as rho-independent termination.

intrinsically unstructured proteins (IUPs) Proteins that are partially or completely lacking in a stable three-dimensional stucture.

intron A noncoding intervening sequence in a split or interrupted gene; missing in the final RNA product.

inversion mutation A deleted DNA fragment is reinserted into its original position, but in the opposite direction.

inverted repeat A DNA sequence that is a reversed complement of another downstream sequence; defines the boundary of a transposon.

ion-exchange chromatography A technique that separates molecules on the basis of their charge.

ionophore A substance that transports cations across membranes.

IP_3 Inositol-1,4,5-triphosphate; the IP_3 receptor is a calcium channel.

irreversible inhibition A form of enzyme inhibition in which an inhibitor molecule permanently impairs an enzyme, usually through binding via a covalent bond.

isoelectric point The pH at which a protein has no net charge.

isomerase An enzyme that catalyzes the conversion of one isomer to another.

isomerization reaction A reaction that involves the intermolecular shift of atoms or groups.

isomers Molecules with the same number and types of atoms.

isoprenoid One of a class of biomolecules that contain repeating five-carbon structural units known as isoprene units: examples include terpenes and steroids.

isothermic reaction Reactions in which heat is not exchanged with the surroundings; $\Delta H = 0$.

isotonic solution A solution with exactly the same particle concentration as that inside cells; there is no net movement of water in or out of the cells.

isozyme One of two or more forms of the same enzyme activity with similar amino acid sequences.

ketal The family of organic compounds with the general formula $RRC(OR)_2$; formed from the reaction of a hemiketal with an alcohol.

ketoacidosis Acidosis caused by the excessive accumulation of ketone bodies.

ketogenesis The condition in which excess acetyl-CoA molecules are converted to acetoacetate, β-hydroxybutyrate, and acetone (referred to collectively as the ketone bodies).

ketogenic Describing amino acids degraded to form acetyl-CoA or acetoacetyl-CoA.

ketone body One of three molecules (acetone, acetoacetate, or β-hydroxybutyrate) that are produced in the liver from acetyl-CoA.

ketosis Accumulation of ketone bodies in blood and tissues.

kinesin A motor protein associated with microtubules.

Krebs bicycle A biochemical pathway in which the aspartate required in the urea cycle is generated from oxaloacetate, an intermediate in the citric acid cycle.

Krebs urea cycle The cyclic pathway that converts waste ammonia molecules along with carbon dioxide and aspartate into urea; named for its discoverer, Hans Krebs.

lactone A cyclic ester.

lactose A disaccharide found in milk; composed of one molecule of galactose linked in a β(1,4)-glycosidic bond to a molecule of glucose.

Le Chatelier's principle Law that states that when a system in equilibrium is disturbed, the equilibrium shifts to oppose the disturbance.

leaving group The group displaced during a nucleophilic substitution reaction.

lectin A carbohydrate-binding protein.

leptin A 16 kDa satiety-inducing protein secreted into the bloodstream primarily by adipose tissue.

leukotriene A biologically active molecule derived from arachidonic acid; its synthesis is initiated by a peroxidation reaction.

ligand A molecule that binds to a specific site on a larger molecule.

ligase An enzyme that catalyzes the joining of two molecules.

light harvesting The capture of light energy by pigment molecules; once absorbed, the energy is channeled to a photosynthetic reaction center.

light harvesting antenna An array of protein and chlorophyll molecules within the thylakoid membrane in chloroplasts;

transfers light energy to a chlorophyll a molecule within a photosystem reaction center.

light-independent reactions A photosynthetic pathway in which CO_2 is incorporated into carbohydrate that can occur in the absence of light; also referred to as the Calvin cycle.

light-induced repair DNA repair in which light energy is used to restore pyrimidine dimers to their original monomeric form; also referred to as photoreactivation repair.

light reactions The mechanism in photosynthesis whereby electrons are energized and subsequently used in ATP and NADPH synthesis.

limit of resolution In microscopy, the minimum distance between two separate points that allows for their discrimination.

LINE (long interspersed nuclear elements) Retrotransposons with lengths greater than 5 kb that contain a strong promoter, an integration sequence, and the coding sequences for transposition enzymes.

lipid Any of a group of biomolecules that are soluble in nonpolar solvents and insoluble in water.

lipid bilayer A biomolecular lipid layer that constitutes the structural framework of cell membranes.

lipogenesis The biosynthesis of body fat (triacylglycerol).

lipoic acid A biomolecule that contains a carboxylate group and two sulfhydryl groups that are easily oxidized or reduced; functions as an acyl group carrier in pyruvate dehydrogenase complex and α-ketoglutarate dehydrogenase complex.

lipolysis The enzyme-catalyzed hydrolysis of triacylglycerol molecules.

lipoprotein A conjugated protein in which lipid molecules are the prosthetic groups; a protein-lipid complex that transports water-insoluble lipids in the blood.

lithotroph An organism that uses specific inorganic reactions to generate energy; also known as a chemolithotroph.

London dispersion force A temporary dipole-dipole interaction.

low-density lipoprotein A type of lipoprotein that contains cholesterol, triacylglycerols, and phospholipids; transports cholesterol to peripheral tissues.

lyase An enzyme that catalyzes the cleavage of C—O, C—C, or C—N bonds, thereby producing a product that contains a double bond.

lysogeny The integration of a viral genome into a host genome.

lysosome A saclike organelle capable of degrading most biomolecules.

lytic cycle A viral life cycle in which a virus destroys its host cell.

macroautophagy A cellular pathway that uses lysosomes for bulk degradation of cytoplasmic components; also referred to as autophagy.

macromolecular crowding The dense packing of an enormous variety of macromolecules and other molecules within the interior of cells.

macromolecule A biopolymer; examples include polypeptides and DNA.

maintenance methyltransferase An enzyme that methylates newly synthesized DNA strands at sites opposite the methyl cytosine on the parental strand.

malate-aspartate shuttle A metabolic process in which the electrons from NADH in the cytoplasm are transferred to mitochondrial NAD^+; oxaloacetate is transferred by reversible conversion to malate from a mitochondrion to the cytoplasm.

maltose A degradation product of starch hydrolysis; a disaccharide composed of two glucose molecules linked by an $\alpha(1,4)$-glycosidic bond.

marker enzyme An enzyme known to be a reliable indicator of the presence of a specific organelle.

marker gene A gene whose presence can be detected, facilitating the identification of transformed cells.

mass spectrometry A technique in which molecules are vaporized and then bombarded by a high-energy electron beam, causing them to fragment as cations.

mediator A protein complex required for the transcription of almost all RNAP II promoters; a signal integration platform.

membrane potential The potential difference across the membrane of living cells; usually measured in millivolts.

Meselson-Radding model A model of general recombination that explains several phenomena not explained by the Holliday model (e.g., sometimes only one of two homologous chromosomes has a recombinant strand).

messenger RNA (mRNA) An RNA species produced by transcription that specifies the amino acid sequence of a polypeptide.

metabolic syndrome A cluster of clinical disorders that include obesity, hypertension, dyslipidemia, and insulin resistance.

metabolism The total of all chemical reactions in an organism.

metabolome The complete set of organic metabolites that are produced within a cell under the direction of the genome.

metalloprotein Conjugated proteins containing metal ions.

methotrexate A structural analogue of folate that is used in the treatment of several types of cancer; also called amethopterin.

methyl-CpG Methylated cytosine in CpG dinucleotides within CpG islands; enable the binding of CpG-binding proteins that promote heterochromatin formation.

methyl-CpG-binding protein (MeCP) Mediates chromatin-associated gene silencing by binding preferentially to the 5-MeCpG dinucleotides and recruiting histone deacetylose to the site along with histone methylases.

micelle An aggregation of molecules having a nonpolar and a polar component, leaving the polar domains facing the surrounding water.

microautophagy A process in which small amounts of cytoplasm are directly engulfed by lysosomes.

microbiota A multicellular organism's indigenous microbial flora.

microfilament A type of cytoskeletal fiber (5–7 nm) composed of polymers of globular actin (G-actin).

microheterogeneity Variations in the glycan components of each type of glycoprotein.

microribonucleoprotein (miRNP) A protein that suppresses the expression of specific genes by binding to a complementary site on an appropriate miRNA to silence translation.

microRNAs (miRNA) 22-nt ncRNAs that inhibit the translation of target mRNAs by binding to partially complementary sequences in the mRNA's 3′-UTR.

microsatellite DNA sequences of 2 to 4 bp that are tandemly repeated 10 to 20 times.

microsome A membranous vesicle derived from fragments of endoplasmic reticulum obtained by differential centrifugation.

microtubule A component of the cytoskeleton; composed of the protein tubulin.

mineralocorticoid A steroid hormone that regulates sodium and potassium metabolism.

minichromosome maintenance complex (MCM) The major DNA helicase in eukaryotes.

minisatellite Tandemly repeated sequence of about 25 bp with total lengths between 10^2 and 10^5 bp.

mismatch repair A single-strand repair mechanism that corrects helix-distorting base mispairings that are the result of replication proofreading errors or as the result of replication slippage.

missense mutation A point mutation that results in coding for a different amino acid causing a change in a polypeptide's structure and function.

mitochondrial fission The division of a mitochondrion to form two or more mitochondria within a cell.

mitochondrial fusion The merging of two or more mitochondria within a cell.

mitochondrion (pl mitochondria) An organelle possessing two membranes in which aerobic respiration occurs.

mixed anhydride An acid anhydride with two different R groups.

mixed terpenoid A biomolecule that is composed of a nonterpene component attached to an isoprenoid group.

mobile genetic element One of numerous DNA sequences that can be duplicated and move within the genome.

mobile phase The moving phase in chromatographic methods.

modular protein A protein that contains numerous duplicate or imperfect copies of one or more domains that are linked in series; also known as a mosaic protein.

modulator A ligand whose binding to an allosteric site of an enzyme alters the enzyme's activity.

module A component of a subsystem that performs a specific function.

molecular biology The science devoted to elucidating the structure and function of genomes.

molecular chaperone A molecule that assists in protein folding; many are heat shock proteins.

molecular disease A disease caused by a mutated gene.

molten globule A partially globular state of a folding polypeptide that resembles the molecule's native state.

monosaccharide A polyhydroxy aldehyde or ketone containing at least three carbon atoms.

monounsaturated Describing a fatty acid with a single double bond.

motif A unique combination of α-helix and β-pleated-sheet secondary structures that occurs in globular proteins; also known as a supersecondary structure.

motor protein Components of biological machines that bind nucleotides; nucleotide hydrolysis drives precise changes in the protein's shape.

multifunction protein A functional protein with two or more diverse and often unrelated functions.

mutarotation A spontaneous process in which the α and β forms of monosaccharides are readily interconverted.

mutation Any change in the nucleotide sequence of a gene.

myocardial infarction The interruption of the heart's blood supply leading to the death of cardiac muscle cells; a heart attack.

myosin One of a family of motor proteins that transduce ATP bond energy into unidirectional movement along actin filaments.

nascent Newly synthesized.

natively unfolded protein A functional protein with a complete lack of ordered structure.

negative cooperativity The binding of one ligand to a target molecule, decreasing the likelihood of subsequent ligand binding.

negative feedback A mechanism in a self-regulating system in which an accumulating product slows its own production.

nephrogenic diabetes insipidis An autosomal recessive disease in which the kidneys cannot produce concentrated urine.

neurotransmitter A molecule released at a nerve terminal that binds to and influences the function of other nerve cells or muscle cells.

neutral fat Triacylglycerol molecules.

***N*-glycan** An asparagine-linked oligosaccharide.

nicotinamide adenine dinucleotide (NAD) A coenzyme form of nicotinic acid containing an *N*-ribosyl derivative of nicotinamide and adenosine linked through a pyrophosphate group; occurs as the oxidized form, NAD^+, and the reduced form, NADH and is involved in electron transfer in a class of enzymes called dehydrogenases.

nicotinamide adenine dinucleotide phosphate (NADP) A coenzyme form of nicotinic acid containing an *N*-ribosyl derivative of nicotinamide and adenosine linked through a pyrophosphate group with an additional phosphate group attached at the 2′-OH group of the adenosine sugar; occurs as the oxidized form $NADP^+$ and the reduced form NADPH and is involved in electron transfer in a class of enzymes called dehydrogenases.

nitrogen fixation Conversion of molecular nitrogen (N_2) into a reduced biologically useful form (NH_3) by nitrogen-fixing microorganisms.

NMR spectroscopy A form of spectroscopy used in structural analysis of organic molecules in which the absorption of electromagnetic radiation by atomic nuclei with magnetic properties is measured.

nonalkylating agents A variety of chemicals other than the alkylating agents that can modify DNA structure.

noncoding RNAs (ncRNAs) Types of RNA other than the RNAs involved in protein synthesis (i.e., tRNAs, rRNAs, and mRNAs) that act as an extensive genome regulatory network.

noncompetitive inhibition Inhibition of an enzyme in which the inhibitor binds to both the free enzyme and the enzyme-substrate complex.

nonessential amino acid An amino acid that can be synthesized by the body.

nonessential fatty acid A fatty acid that can be synthesized by the body.

nonsense mutation A point mutation that changes the code for an amino acid into a premature stop signal.

nuclear envelope The double membrane that separates the nucleus from the cytoplasm.

nuclear lamina A dense protein meshwork attached to the inner surface of the inner nuclear membrane; involved in nuclear processes such as DNA replication, transcription, and chromatin organization.

nuclear matrix The cytoskeleton-like scaffold within the nucleus in which loops of chromatin are organized.

nuclear pore complex One of thousands of pore complexes in the nuclear envelope through which pass most of the molecules that enter or leave the nucleus of a cell.

nuclease An enzyme that hydrolyzes nucleic acid molecules to form oligonucleotides.

nucleic acid A macromolecule formed from the polymerization of nucleotides.

nucleohistone DNA complexed with histone proteins.

nucleoid In prokaryotes, an irregularly shaped region that contains a long circular DNA molecule.

nucleolus A structure revealed in the nucleus when the nucleus is stained with certain dyes: it plays a major role in the synthesis of ribosomal RNA.

nucleophile An electron-rich atom or molecule.

nucleophilic substitution A reaction in which a nucleophile substitutes for an atom or molecular group.

nucleoplasm The gelatinous substance within the nucleus that contains the cytoskeleton-like nuclear matrix and a network of chromatin fibers.

nucleoside A biomolecule composed of a pentose sugar (ribose or deoxyribose) and a nitrogenous base.

nucleosome A repeating structural element in eukaryotic chromosomes composed of a core of eight histone molecules around which about 140 base pairs of DNA are wrapped; an additional 60 base pairs connect adjacent nucleosomes.

nucleotide A biomolecule composed of a five-carbon sugar (ribose or deoxyribose), a nitrogenous base, and one or more phosphate groups.

nucleotide excision repair Bulky lesions of 2 to 30 nucleotides are removed and the resulting gap is filled; the excision enzymes appear to recognize the physical distortion rather than a specific base sequence.

nucleus A double membrane-bound organelle in eukaryotic cells that contains the cell's genome.

obligate aerobe An organism that is highly dependent on oxygen for energy production.

obligate anaerobe An organism that grows only in the absence of oxygen.

***O*-glycan** A mucin-type polysaccharide.

Okazaki fragment Any of a series of deoxyribonucleotide segments that are formed during discontinuous replication of one DNA strand as the other strand is continuously replicated.

oligomer A multisubunit protein in which some or all subunits are identical.

oligonucleotide A short nucleic acid segment that contains fewer than 50 nucleotides.

oligopeptide An amino acid polymer containing 50 or less amino acid residues.

oligosaccharide An intermediate-size carbohydrate composed of 2 to 10 monosaccharides.

omega-3-fatty acid α-Linolenic acid and its derivatives, such as eicosapentaenoic acid and docosahexaenoic acid.

omega-6-fatty acid Linoleic acid and its derivatives.

oncogene A mutated version of a protooncogene that promotes abnormal cell proliferation.

one-carbon metabolism A set of reactions in which single carbon atoms are transferred from one molecule to another.

open reading frame (ORF) A series of triplet base sequences in mRNA that does not contain a stop codon.

operator A regulatory DNA sequence to which a repressor protein binds.

operon A set of linked genes that are regulated as a unit.

optical isomer A stereoisomer that possesses one or more chiral centers.

organelle A membrane-enclosed structure within a eukaryotic cell.

origin-of-replication complex (ORC) A protein complex that binds to the DNA replication origin during the initiation phase of DNA synthesis; contains analogues of the protein DnaA.

osmosis The diffusion of solvent through a semipermeable membrane,

osmotic diuresis A process in which solutes in the urinary filtrate cause excessive loss of water and electrolytes.

osmotic pressure The pressure forcing the solvent, water, to flow across a membrane.

outer mitochondrial membrane The porous external membrane of the mitochondrion.

outer nuclear membrane A membrane surrounding the nucleus that is continuous with the endoplasmic reticulum.

oxidation The removal of electrons.

oxidation-reduction (redox) reaction A reaction involving the transfer of one or more electrons from one reactant to another.

oxidative phosphorylation The synthesis of ATP coupled to electron transport.

oxidative stress Excessive production of reactive oxygen species.

oxidize The removal of electrons from an atom or molecule.

oxidizing agent A substance that oxidizes (removes electrons from) another substance; the oxidizing agent is itself reduced in the process.

oxidoreductase An enzyme that catalyzes an oxidation-reduction reaction.

oxyanion A negatively charged oxygen atom.

palindrome A sequence that provides the same information whether it is read forward or backward; DNA palindromes contain inverted repeat sequences.

passive transport Transport of a substance across membranes that requires no direct input of energy.

Pasteur effect The observation that glucose consumption is greater under anaerobic conditions than when oxygen is available.

pentose phosphate pathway A biochemical pathway that produces NADPH, ribose, and several other sugars.

peptide An amino acid polymer with fewer than 50 amino acid residues.

peptide bond An amide linkage in an amino acid polymer.

peptidyl transfer center (PTC) The location of peptidyl transferase activity within the large subunit of bacterial ribosomes; located in a domain of 23S rRNA.

perinuclear space The space between the two membranes of the nuclear envelope.

peripheral protein A protein that is not embedded in the membrane but attached either by a covalent bond to a lipid molecule or by noncovalent interactions with a membrane protein or lipid.

pernicious anemia An illness caused by a deficiency of vitamin B_{12}; symptoms include low red blood cell count, weakness, and neurological disturbances.

peroxisome A spherical organelle in eukaryotic cells that contains oxidative enzymes; involved in such processes as fatty acid degradation, membrane lipid synthesis, and purine base degradation.

pH optimum The pH value at which an enzyme's activity is maximal.

pH scale A measure of hydrogen ion concentration; pH is the negative log of the hydrogen ion concentration in moles per liter.

phagocytosis The engulfment of foreign or damaged cells by certain white blood cells.

phase I reaction A biotransformation reaction involving oxidoreductases and hydrolases that converts hydrophobic substances into more polar molecules.

phase II reaction A biotransformation reaction in which metabolites containing appropriate functional groups are conjugated with substances such as glucuronate, glutamate, sulfate, and glutathione.

3′-phosphoadenosine-5′-phosphosulfate A high-energy sulfate donor molecule used in the biosynthesis of the sulfatides, a type of glycolipid.

phosphoglyceride A type of lipid molecule found predominantly in membrane; composed of glycerol linked to two fatty acids, phosphate, and a polar group.

phospholipid An amphipathic molecule that has a hydrophobic domain (hydrocarbon chains of fatty acid residues) and a hydrophilic (a polar head group) domain; an important structural component of membranes.

phosphoprotein A conjugated protein in which phosphate is the prosthetic group.

phosphoryl group transfer potential The tendency of a phosphorylated molecule to undergo hydrolysis.

photoautotrophs Organisms that transform light energy (usually from the sun) into chemical bond energy.

photoheterotrophs Organisms that use both light and biomolecules as energy sources.

photophosphorylation The synthesis of ATP coupled to electron transport driven by light energy.

photoreactivation repair A mechanism to repair thymine dimers using the energy of visible light.

photorespiration A light-dependent process occurring in plant cells actively engaged in photosynthesis that consumes oxygen and liberates carbon dioxide.

photosynthesis The trapping of light energy and its conversion to the chemical energy required to incorporate carbon dioxide into organic molecules.

photosystem A photosynthetic mechanism composed of light-absorbing pigments.

plasma membrane The membrane that surrounds a cell, separating it from its external environment.

plasmid A circular double-stranded DNA molecule that can exist and replicate independently of a bacterial chromosome; plasmids are stably inherited but are not required for the host cell's growth and reproduction.

plastid An organelle found in plants, algae, and some protists that contain pigments and/or storage materials such as carbohydrate.

platelet-derived growth factor A protein secreted by blood platelets during clotting; stimulates mitosis during wound healing.

point mutation A change in a single nucleotide base in a DNA sequence.

polar An unequal distribution of electrons in a bond.

polar head group A molecular group that contains phosphate or other charged or polar groups.

poly(A) binding protein (PABP) Forms a circular mRNA molecule, during the initiation phase of eukaryotic translation, by interacting with the 3′-poly(A) tail and 5′-capped end of the mRNA and eIF-G, a translation initiation factor.

polymerase chain reaction (PCR) A laboratory technique that uses a heat-stable DNA polymerase to synthesize large quantities of specific nucleotide sequences from small amounts of DNA.

polypeptide An amino acid polymer with more than 50 amino acid residues.

polysaccharide A linear or branched polymer of monosaccharides linked by glycosidic bonds.

polysome An mRNA with several ribosomes bound to it.

polyunsaturated Describing a fatty acid with two or more double bonds, usually separated by a methylene group.

positive cooperativity The mechanism in which the binding of one ligand to a target molecule increases the likelihood of subsequent ligand binding.

positive feedback The mechanism in a self-regulating system in which the product of a reaction increases its own production.

postabsorptive The phase in the feeding-fasting cycle in which nutrient levels are low.

postprandial The phase in the feeding-fasting cycle immediately after a meal: blood nutrient levels are relatively high.

posttranslational modification One of a set of reactions that alter the structure of newly synthesized polypeptides.

posttranslational translocation The transfer of previously synthesized polypeptides across the membrane of an organelle.

PPAR Peroxisome proliferator-activated receptors.

preinitiation complex A multisubunit protein complex, formed during the initiation phase of eukaryotic protein synthesis, that is able to bind to an mRNA.

preinitiation replication complex (preRC) The preliminary eukaryotic DNA replication complex; formed from the origin of replication complex and MCM complex, among other proteins.

prenylation The covalent attachment of prenyl groups (e.g., farnesyl and geranylgeranyl groups) to protein molecules.

preproprotein An inactive precursor protein with a removable signal peptide.

primary active transport A process that uses chemical energy to move molecules across a membrane.

primary cilium A nonmotile cilium that functions as a sensory organelle on the surface of most differentiated vertebrate cells.

primary structure The amino acid sequence of a polypeptide.

primase An RNA polymerase that synthesizes short RNA segments, called primers, that are required for DNA synthesis.

primer A short RNA segment required to initiate DNA synthesis.

primosome A multienzyme complex involved in the synthesis of RNA primers at various intervals along the DNA template strand during *E. coli* DNA replication.

prion Proteinaceous infectious particle: believed to be a causative agent of several acquired neurodegenerative diseases (e.g., "mad cow" disease and Creutzfeld-Jakob disease).

processivity The prevention of frequent dissociation of a polymerase from the DNA template.

proenzyme An inactive precursor of an enzyme.

prokaryotic cell A living cell that lacks a nucleus.

promoter The sequence of nucleotides immediately before a gene that is recognized by RNA polymerase and signals the start point and direction of transcription.

prophage A viral genome integrated into host cell DNA.

proprotein An inactive precursor protein.

prostaglandin An arachidonic acid derivative that contains a cyclopentane ring with hydroxyl groups at C-11 and C-15.

prosthetic group The nonprotein portion of a conjugated protein that is essential to the biological activity of the protein; often a complex organic group.

protein A macromolecule composed of one or more polypeptides.

protein family A group of protein molecules that are related by amino acid sequence similarity.

protein folding The process in which an unorganized polypeptide acquires a highly organized and relatively stable three-dimensional structure.

protein superfamily A large group of distantly related proteins; for example the globin superfamily includes the hemoglobins and myoglobins, which bind oxygen in blood and muscle cells, respectively, and the cytoglobins, which bind oxygen in the brain.

protein turnover The continuous degradation and resynthesis of protein in an organism.

proteoglycan A large molecule containing large numbers of glycosaminoglycan chains linked to a core protein molecule.

proteome The complete set of proteins produced within the cell.

proteomics The investigation of protein synthesis patterns and protein-protein interactions.

proteosome A multienzyme complex that degrades proteins linked to ubiquitin.

proteostasis A process by which cells control the folding of proteins.

proteostasis network A series of proteins organized into pathways that control the folding, trafficking, and degradation of proteins.

protocol The set of rules that specify how modules in a system interact.

protomer A component of an oligomer; may consist of one or more subunits.

protonmotive force The force arising from a gradient of protons and a membrane potential.

protooncogene A normal gene that codes for a protein involved in cell cycle regulation; promotes carcinogenesis if mutated.

pseudogene An imperfect copy of a functional gene, which is not expressed.

purine A nitrogenous base with a two-ring structure; a component of nucleotides.

pyrimidine A nitrogenous base with a single-ring structure; a component of nucleotides.

Q cycle The movement of electrons from reduced coenzyme Q, UQH_2, to cytochrome c during electron transport.

quaternary structure Association of two or more folded polypeptides to form a functional protein.

racemization The interconversion of enantiomers.

radical An atom or molecule with an unpaired electron.

reaction center The membrane-bound protein complex in a photosynthesizing cell that mediates the conversion of light energy into chemical energy.

reaction mechanism Step-by-step description of a chemical reaction process.

reactive nitrogen species (RNS) Nitrogen-containing radicals often classified as ROS; the most important are nitric oxide, nitrogen dioxide, and peroxynitrite.

reactive oxygen species (ROS) A reactive derivative of molecular oxygen, including superoxide radical, hydrogen peroxide, the hydroxyl radical, and singlet oxygen.

reading frame A set of contiguous triplet codons in an mRNA molecule.

receptor protein A protein with binding sites for extracellular ligands (signal molecules).

receptor tyrosine kinase (RTK) A transmembrane receptor that contains a cytoplasmic domain with tyrosine kinase activity that is activated when a ligand is bound to the external domain.

recombinant DNA technology A series of techniques whose essential feature is that DNA molecules obtained from various sources can be cut and spliced together.

recombination A process in which DNA molecules are cut and rejoined in new combinations.

recombinational repair A repair mechanism that can eliminate certain types of damaged DNA sequences that were not eliminated before replication; the undamaged parental strands recombine into the gap left after the removal of the damaged sequence.

redox potential A measure of the tendency of an electron donor in a redox pair to lose an electron.

reduce The addition of electrons to an atom or molecule.

reducing agent A substance that reduces the oxidation number of another reactant; the reducing agent is itself oxidized in the process.

reducing sugar A sugar that can be oxidized by weak oxidizing agents.

reduction A decrease in oxidation number of an atom or molecule.

reduction potential The tendency for a specific substance to lose or gain electrons.

reductionism A method of scientific inquiry in which a complex "whole" system is reduced to its component parts, each of which is then further broken down so that the chemical and physical characteristics of its molecules can be investigated.

redundancy The use of duplicate parts in a fail-safe mechanism in a robust system.

releasing factor A protein involved in the termination phase of translation.

replication The process in which an exact copy of parental DNA is synthesized using the polynucleotide strands of the parental DNA as templates.

replication factor C (RFC) A clamp loader protein that controls the attachment of DNA polymerase δ to each DNA strand.

replication factories Specific nuclear compartments (or nucleoids) in which DNA replication occurs.

replication fork The Y-shape region of a DNA molecule that undergoes replication; results from separation of two DNA strands.

replication licensing factor One of several proteins that bind to the origin-of-replication complex (ORC) and complete the structure of the preRC.

replication protein A (RPA) A protein that stabilizes the separated DNA strands during replication.

replicon A unit of the genome that contains an origin for initiating replication.

replisome The large complex of polypeptides, including the primosome, that replicates DNA in *E coli.*

resonance energy transfer The transfer of energy from an excited molecule to a nearby molecule, thereby exciting the second molecule.

resonance hybrid A molecule with two or more alternative structures that differ only in the position of electrons.

respirasome A functional aerobic respiration unit in the inner mitochondrial membrane; the I, III_2, $IV_{1\text{-}2}$ supercomplex has been identified in animals, plants, and fungi.

respiration A biochemical process whereby fuel molecules are oxidized and their electrons are used to generate ATP.

respiratory burst An oxygen-consuming process in scavenger cells such as macrophages in which reactive oxygen species are generated and used to kill foreign or damaged cells.

respiratory control The control of aerobic respiration by ADP concentration.

response element A DNA sequence within the promoter of genes; transcription is triggered when a specific hormone receptor complex or transcription factor binds.

restriction fragment length polymorphism (RFLP) One of a vast number of DNA sequence variations that can be used to identify individuals.

retroelement See retrotransposon.

retroposon See retrotransposon.

retrotransposon A subclass of transposons that use an RNA intermediate.

retrovirus One of a group of viruses with RNA genomes that carry the enzyme reverse transcriptase and form a DNA copy of their genome during the reproductive cycle.

reversible inhibition A form of enzyme inhibition in which the inhibitory effect of a compound can be counteracted by increasing substrate or removing the inhibitor while the enzyme remains intact.

rho-dependent termination Transcription termination in prokaryotes that requires rho factor

rho factor An ATP-dependent helicase involved in transcription termination in bacteria.

rho-independent termination A form of transcription termination in prokaryotes that does not involve rho factor; also referred to as intrinsic termination.

ribosomal RNA (rRNA) The RNA present in ribosomes. Ribosomes contain several types of single-stranded ribosomal RNA that contribute to ribosome structures and are also directly involved in protein synthesis.

ribosome A protein-RNA complex that is the site of protein biosynthesis.

ribosome recycling factor In bacteria, a tRNA-shaped protein; it binds within the A site and causes the dissociation of the ribosomal subunits after polypeptide synthesis.

riboswitch An RNA-based control mechanism made up of a specific untranslated sequence within an mRNA. Usually, the action of the riboswitch, triggered by a change in its tertiary structure induced by binding to a ligand, is to repress translation.

ribozyme A catalytic RNA molecule; catalyzes self-cleavage or the cleavage of other RNAs.

RNA Ribonucleic acid; a single-stranded, unbranched macromolecule formed from ribonucleotides; a fundamental component in protein biosynthesis.

RNA editing The alteration of the base sequence in a newly synthesized mRNA molecule; bases may be chemically modified or deleted.

RNA-induced silencing complex (RISC) The antisense strand of the short interfering RNA bound to the RISC complex targets and anneals with the viral micro RNA; nucleases within the RISC then degrade the viral miRNA sequence.

RNA interference A cellular mechanism in which RNA molecules are degraded; functions in gene expression regulation and in defense against viral RNA genomes.

RNA splicing The process in which introns are cut out and the exons are linked together to form a functional RNA product.

RNA transposon A transposable element that uses a mechanism that involves an RNA transcript; also referred to as retrotransposon.

robust Describing a system that remains stable despite perturbations of diverse types.

rough endoplasmic reticulum (RER) A type of endoplasmic reticulum that has ribosomes bound to its external surface; nascent polypeptides, containing a signal peptide, are translocated through the RER membrane.

***S*-adenosylmethionine (SAM)** The major methyl group donor in one-carbon metabolism.

salt bridge An electrostatic interaction in proteins between ionic groups of opposite charge.

salting out The decrease in protein solubility caused by an increase in the ionic strength of the solution.

satellite DNA DNA sequences that are highly repetitive; when genomic DNA is digested and centrifuged, a satellite band forms.

saturated Describes a molecule that contains no carbon–carbon double or triple bonds.

Schiff base The imine product of a reaction between a primary amino group and a carbonyl group; also referred to as an aldimine.

SDS-polyacrylamide gel electrophoresis A method for separating proteins or determining their molecular weights that employs the negatively charged detergent sodium dodecyl sulfate (SDS).

SECIS element *Se*lenocysteine *i*nsertion *s*equence element; a base sequence required at the 3′-UTR (3′ untranslated region) of the mRNAs for all selenoproteins for the insertion of selenocysteine during translation.

second messenger A molecule that mediates the action of some hormones.

secondary active transport A transport process that uses an electrochemical gradient created by the ATP-requiring pumping of ions into or out of a cell.

secondary structure The arrangement of a polypeptide chain into locally organized structures of α-helix and β-pleated sheet: secondary structure is maintained by hydrogen bonds between the amide hydrogen and the carbonyl oxygen of peptide bonds.

semiconservative replication DNA synthesis in which each polynucleotide strand serves as a template for the synthesis of a new strand.

sense strand The nontranscribed DNA strand; the DNA version of the mRNA used to synthesize the polypeptide product of a gene.

Shine-Dalgarno sequence A purine-rich sequence that occurs on an mRNA close to AUG (the initiation codon) that binds to a complementary sequence on the 30S ribosomal subunit, thereby promoting the formation of the correct preinitiation complex.

short interspersed nuclear elements (SINE) A repeating DNA sequence less than 500 bp long interspersed in mammalian genomes; SINES cannot undergo transposition without the aid of a functional LINE sequence.

short tendem repeats DNA sequences with between 2- and 4-bp repeats; can be used to generate DNA profiles that distinguish among individuals.

shotgun cloning A cloning technique in which genomic libraries are created by the random digestion of a genome.

signal hypothesis A mechanism that explains how secreted or membrane proteins are synthesized on ribosomes bound to the rough endoplasmic reticulum; a sequence of amino acid residues on the nascent polypeptide chain that mediates the insertion of the polypeptide into the RER membrane.

signal peptide A short sequence typically near the amino terminal of a polypeptide that determines its insertion into a membrane of an organelle.

signal recognition particle A large multisubunit ribonucleoprotein complex that mediates the binding of the ribosome and the emerging signal peptide to the rough endoplasmic reticulum during protein synthesis; facilitates the passage of the growing polypeptide through the RER membane.

signal transduction The mechanisms by which extracellular signals are received, amplified, and converted to a cellular response.

signaling cascade An information processing mechanism initiated by the binding of a signal molecule to a receptor and continued by a series of events involving protein conformational changes and covalent modifications that result in a response; responses include changes in enzyme activities, cytoskeletal rearrangements, cell movement, or cell cycle progression.

silencer A DNA sequence that when bound to a repressor protein prevents an RNA polymerase from binding to a nearby promoter.

silent mutation A base change in DNA that has no discernable effect.

simple diffusion A process in which each type of solute, propelled by random molecular motion, moves down a concentration gradient.

single nucleotide polymorphism A point mutation that occurs in a population to any extent.

site-directed mutagenesis A technique that introduces specific sequence changes into a cloned gene.

site-specific recombination Recombination of nonhomologous DNA sequences; a recombination mechanism between sequences with limited homology.

small interfering RNA (siRNA) A 21- to 23- nt dsRNA that plays a crucial role in RNA interference.

small nuclear ribonucleoprotein particle (snRNP) A complex of proteins and small nuclear RNA molecules that promotes RNA processing.

small nuclear RNA (snRNA) A small RNA molecule involved in the removal of introns from mRNA, rRNA and tRNA.

small nucleolar RNA (snoRNA) An RNA component of nucleolar ribonucleoprotein that facilitates chemical modifications of rRNA.

smooth endoplasmic reticulum (SER) A type of endoplasmic reticulum involved in lipid synthesis and biotransformation processes.

solvation sphere A shell of water molecules that clusters around positive and negative ions.

Southern blotting A laboratory technique in which radioactively labeled DNA or RNA probes are used to locate a complementary sequence in a DNA digest.

specificity constant In enzyme kinetics, the term k_{cat}/K_m; the second order rate constant for a reaction in which $[S] << K_m$.

sphingolipid A membrane lipid molecule that contains a long-chain amino alcohol called sphingosine; ceramide is a fatty acid derivative of sphingosine; an important component of plant and animal membranes.

sphingomyelin A type of phospholipid that contains sphingosine; the 1-hydroxyl group of ceramide is esterified to the phosphate group of phosphorylcholine or phosphorylethanolamine, and the amino group of sphingosine is in an amide linkage with a fatty acid.

spliceosome A multicomponent complex containing protein and RNA; used in the splicing phase of mRNA processing.

spontaneous chemical changes Physical or chemical processes that are accompanied by a release of energy.

SRP receptor protein A heterodimer with two GTPases on the cytoplasmic surface of the RER that binds SRP (signal recognition particle), the protein complex that facilitates the binding of a ribosome to the membrane surface during the synthesis of a signal peptide-containing polypeptide; also referred to as docking protein.

standard reduction potential The measurement of the capacity of a substance to gain or lose electrons in a galvanic cell fitted with a standard hydrogen electrode set at 0.00 V.

stationary phase The solid matrix in chromtographic techniques.

steady state A phase in an organism's life when the rate of anabolic processes is approximately equal to that of catabolic processes.

stereoisomer A molecule that has the same structural formula and bonding pattern as another but has a different arrangement of atoms in space.

steroid A derivative of triterpenes; contains four fused rings.

sterol carrier protein A cytoplasmic protein carrier for certain intermediates during cholesterol biosynthesis.

sterol regulatory element binding protein (SREBP) One of several transcription factors that are membrane proteins in the endoplasmic reticulum or the Golgi apparatus.

stroma A dense, enzyme-filled substance that surrounds the thylakoid membrane within the chloroplast.

stromal lamella A thylakoid membrane segment that interconnects two grana.

substrate The reactant in a chemical reaction that binds to an enzyme active site and is converted to a product.

substrate-level phosphorylation The synthesis of ATP from ADP by phosphorylation coupled with the exergonic breakdown of a high-energy organic substrate molecule.

subunit A polypeptide component of an oligomeric protein.

sucrose A disaccharide composed of α-glucose and β-fructose residues linked through a glycosidic bond between both anomeric carbons.

sugar A polyhydroxy aldehyde- or ketone-containing biomolecule.

superfamily The largest grouping of proteins for which common ancestry can be inferred from sequence homology.

supersecondary structure One of a set of specific combinations of α-helix and β-pleated-strand structures of protein molecules.

supraspliceosome Formed by four active spliceosomes and a pre-mRNA complex; it increases the speed and efficiency of transcript splicing and provides opportunities for intron excision proofreading.

synthesis-dependent strand annealing model A form of double-strand repair using homologous chromosomes in which there are only noncrossover products.

systems biology A field of study based on engineering principles in which the interactions between the components of living organisms are investigated; complex data sets used by systems biologists come from genomics, proteomics, and experimental sources such as protein-protein interactions and biochemical reaction fluxes.

T cell A T lymphocyte; a white blood cell that bears antibodylike molecules on its surface and binds to and destroys foreign cells in cellular immunity.

tandem repeats DNA sequences in which multiple copies are arranged next to one another; lengths of repeated sequences vary from 10 bp to more than 2000 bp.

target cell A cell that responds to the binding of a hormone or growth factor to a receptor protein.

targeting The process that directs newly synthesized proteins to their correct intracellular destination.

tautomer An isomer that differs from another in the location of a hydrogen atom and a double bond (e.g., keto-enol tautomers).

tautomerization Chemical reaction by which two tautomers are interconverted by the movement of a hydrogen atom and a double bond.

telomerase A ribonucleoprotein with an RNA component (a TG-rich repeat sequence) complementary to the telomere sequence.

telomere A structure found at both ends of a chromosome that buffers the loss of critical coding sequences after a round of DNA replication.

telomere end-binding protein (TBP) A protein that binds to and stabilizes GT-rich telomere sequences.

telomere repeat-binding factor (TRB) A protein that binds to and secures the 3′ overhang sequence of a telomere.

termination The phase of translation in which newly synthesized polypeptides are released from the ribosome.

terpene A member of a class of isoprenoids classified according to the number of isoprene residues it contains.

tertiary structure The globular three-dimensional structure of a polypeptide that results from interactions between the side chains (R groups) of the amino acid residues.

tetrahydrobiopterin (BH_4) A folic acid–like molecule that is an essential cofactor in the hydroxylation of the aromatic amino acids.

tetrahydrofolate The biologically active form of the B vitamin folic acid; a carrier of methyl, methylene, methenyl, and formyl groups in one-carbon metabolism.

thermodynamics The study of energy and its interconversion.

thermogenin See uncoupling protein.

thiamine pyrophosphate The coenzyme form of thiamine, also called vitamin B_1.

thiolytic cleavage Cleavage of a carbon-sulfur bond.

thromboxane A derivative of arachidonic acid that contains a cyclic ester.

thylakoid lumen The internal compartment created by the formation of grana.

thylakoid membrane An intricately folded internal membrane within the chloroplast.

***trans* isomer** An isomer in which two substituents are on opposite sides of the double bond.

transamination A reaction in which an amino group is transferred from one molecule to the α-carbon of an α-keto acid; the amino acid that donates the amino group is converted to the corresponding α-keto acid.

transcript An RNA molecule that is produced by the transcription of a DNA sequence.

transcription The process in which single-stranded RNA with a base sequence complementary to the template strand of DNA is synthesized.

transcription coupled repair A form of DNA repair in which repair only occurs on the DNA strand that is being transcribed.

transcription factor A protein that regulates or initiates the synthesis of specific mRNAs by binding to DNA sequences called response elements.

transcript localization The binding of mRNAs to certain cellular structures within cytoplasm that permits the creation of protein gradients within the cell.

transcriptome The complete set of RNA molecules that are produced within a cell.

transduction The transfer of DNA segments between bacteria by bacteriophages.

transfection A mechanism by which bacteriophage inadvertently transfers bacterial chromosome or plasmid sequences to a new host cell.

transfer RNA (tRNA) A small RNA molecule that binds to an amino acid and delivers it to the ribosome for incorporation into a polypeptide chain during translation.

transferase An enzyme that catalyzes the transfer of a functional group from one molecule to another.

transformation A process in which DNA fragments enter a bacterial cell and are introduced into the bacterial genome.

transgenic animal An animal that results when recombinant DNA sequences are microinjected into a fertilized ovum.

transition mutation A DNA mutation that involves the substitution of a purine base by a different purine, or the substitution of a pyrimidine by a different pyrimidine.

transition state In catalysis, the unstable intermediate formed by the enzyme that has altered the substrate so that it now shares properties of both the substrate and the product.

translation Protein synthesis; the process by which the genetic message carried by mRNAs directs the synthesis of polypeptides with the aid of ribosomes and other cell constituents.

translocation Movement of the ribosome along the mRNA during translation; also refers to a chromosomal abnormality in which a DNA fragment inserts into another site in the same chromosome or in a nonhomologous chromosome.

translocon An integral membrane protein that mediates translocation of a polypeptide.

transmembrane protein An integral membrane protein that extends completely across the membrane.

transposable DNA elements A DNA sequence that excises itself and then inserts at another site.

transposase A prokaryotic transposition enzyme, coded for by a gene within an IS element.

transposition The movement of a DNA sequence from one site in a genome to another.

transposon (transposable element) A DNA segment that carries the genes required for transposition and moves about the chromosome; sometimes the name is reserved for transposable elements that also contain genes unrelated to transposition.

transulfuration pathway A biochemical pathway that converts methionine to cysteine.

transversion mutation A type of point mutation in which a pyrimidine is substituted for a purine and vice versa.

triacylglycerol cycle A metabolic mechanism that regulates the level of fatty acids that are available to the body for energy generation and the synthesis of molecules such as phospholipids; the constant synthesis and degradation of TG.

tumor necrosis factor (TNF) A protein, toxic to tumor cells, that suppresses cell division.

tumor promoter A molecule that provides cells with a growth advantage over nearby cells.

tumor suppressor gene One of a set of genes that code for a protein that protects cells from progressing toward cancer; may inhibit a transcription factor required for cell cycle progression or may facilitate DNA repair.

turnover The rate at which all molecules in a cell are degraded and replaced with newly synthesized molecules.

turnover number The number of molecules of substrate converted to product in each second per mole of enzyme.

type 1 diabetes An autoimmune disease that destroys the insulin-producing β-cells in the pancreas; a metabolic disease in which the most obvious symptoms are hyperglycemia and dyslipidemia.

type 2 diabetes A form of diabetes mellitus in which patients are resistant to insulin.

ubiquitin A protein that is covalently attached by enzymes to proteins destined to be degraded.

ubiquitin proteosomal system An elaborate mechanism for the rapid destruction of proteins.

ubiquitination The covalent attachment of ubiquitin to proteins that are to be degraded.

uncompetitive inhibition An inhibitor binds only to the enzyme-substrate complex; a rare type of noncompetitive inhibition.

uncoupler A molecule that uncouples ATP synthesis from electron transport; it collapses a proton gradient by transporting protons across the membrane.

uncoupling protein A molecule that dissipates the proton gradient in mitochondria by translocating protons; also called thermogenin.

unfolded protein response The inhibition of new protein synthesis in the RER except for molecular chaperones; triggered by severe ER stress.

unsaturated Describes a molecule that contains one or more carbon–carbon double or triple bonds.

urea cycle A cyclic pathway in which waste ammonia molecules, carbon dioxide, and the amino nitrogen of aspartate molecules are converted to urea.

uronic acid The product formed when the terminal CH_2OH of a monosaccharide is oxidized.

van der Waals force A class of relatively weak, transient electrostatic interactions between permanent and/or induced dipoles.

vector A cloning vehicle into which a segment of foreign DNA can be spliced, ready for introduction into host cells and expression in them.

velocity The rate of a biochemical reaction; the change in the concentration of a reactant or product per unit time.

very-low-density lipoprotein (VLDL) A type of lipoprotein with a very high relative concentration of lipids; transports lipids to tissues.

vesicles Membranous sacs that bud off from a donor membrane and subsequently fuse with the membrane of another organelle or with the plasma membrane.

vesicular organelles Small spheroidal membranous sacs that contain substances that originated in the endoplasmic reticulum and Golgi apparatus or were brought into the cell by endocytosis.

vitamin An organic molecule required by organisms in minute quantities; some vitamins are coenzymes required for the function of certain cellular enzymes.

vitamin B_{12} A complex cobalt-containing molecule that is required for the N^5-methyl THF-dependent conversion of homocysteine to methionine.

wax A complex mixture of nonpolar lipids including wax esters.

wax ester One of numerous types of ester consisting of long-chain fatty acids and long-chain alcohols; a prominent constituent of most waxes.

weak acid An organic acid that does not completely dissociate in water.

weak base An organic base that has a small but measurable capacity to combine with hydrogen ions.

wobble hypothesis The hypothesis that explains why cells often have fewer tRNAs than expected; freedom in the pairing of the third base of the codon to the first base of the anticodon allows some tRNAs to pair with several codons.

work A physical change caused by a change in energy.

yeast artificial chromosome A cloning vector that can accommodate up to 100 kb of DNA; contains eukaryotic sequences that function as centromeres, telomeres, and a replication origin.

Z scheme A mechanism whereby electrons flow between photosystems II and I during photosynthesis.

Z-DNA A form of DNA that is twisted into a left-handed spiral; named for the zigzag conformation, which is slimmer than that of B-DNA.

zwitterion A neutral molecule that bears an equal number of positive and negative charges simultaneously.

zymogen The inactive form of a proteolytic enzyme.

Credits

Photos

Chapter 1

opener: monkeybusinessimages/iStock.

Chapter 2

opener: Copyright American Society for Microbiology Photo courtesy of N.J. Nyffenegger-Jann. From "Innate Recognition by Neutrophil Granulocytes Differs between Neisseria gonorrhoeae Strains Causing Local or Disseminating Infections " by Alexandra Roth, Corinna Mettheis, Petra Muenzner, Magnus Unemo, and Christof R. Hauck, in Infection and Immunity, volume 81, issue 7 (July 1 2013), pp. 2358–2370, doi:10.1128/IAI.00128-13. Reproduced with permission from American Society for Microbiology.; **2.5:** Adapted from Bruce Alberts, "Essential Cell Biology" 1998 Routledge/ Taylor & Francis; **2.6:** Adapted from Trends in Biochemical Sciences 26(10), R. John Ellis, "Macromolecular crowding: obvious but underappreciated", pp. 597–604, copyright 2001, with permission of Elsevier.; **2.8:** Adapted from Prescott, et al., Microbiology 4/e. Copyright 1999 by the McGraw-Hill Companies; **2.9a:** Adapted from David S. Goodsell, The Machinery of Life, 1998. Copyright © 1998 Spinger-Verlag. Reprinted with permission of Springer-Verlag Germany; **2.9b:** Adapted from David S. Goodsell, The Machinery of Life, 1998. Copyright © 1998 Springer-Verlag. Reprinted with permission of Springer-Verlag Germany; **2.12:** Adapted from 4/e Pearson/Benjamin Cummings, Campbell & Reece, Biology, 7th ed. 2005. Fig 7.7, p. 127; **2.13:** Courtesy of Audrey M. Glauert and G.M.W. Cook; **2.14:** Adapted from Becker, Kleinsmith & Hardin, World of the Cell, 4/e. 2000, Addison Wesley Longman. Reprinted by Permission of Pearson Education, Inc.; **2.15:** Adapted from Becker, Kleinsmith & Hardin, World of the Cell, 4/e. 2000 Addison Wesley Longman; **2.16:** Adapted from the World of the Cell, 4th ed by Wayne M. Becherk, Lewis J. Kleinsmith, and Jeff Hardin. Copyright © 2000 by Addison Wesley Longman, Inc. Reprinted with the permission of Pearson Education, Inc.; **2.18a and b:** Adapted from Hardin, J. et al., Becker's World of the Cell, 8/e. 2012 Benjamin Cummings; **2.18c:** Adapted from Cooper, G.M. and Hausman, R.E., The Cell: A Molecular Approach, 6/e. 2013 Sinauer Associates; **2.19:** P. Schulz/Biology Media/Photo Researchers Inc; **2.20 (top):** Adapted from Don Fawcett/ Photo Researchers, Inc.; **2.20 [bottom]:** Adapted from Becker, Kleinsmith & Hardin, World of the Cell, 4/e. 2000 Addison Wesley Longman; **2.21:** Adapted from Cooper, G.M. and Hausman, R.E., The Cell: A Molecular Approach, 6/e. 2013 Sinauer Associates; **2.22:** Adapted from Thomas Zeuthen, Trends in Biochemical Sciences, Vol. 26, No 2 pp. 77–79 copyright 2001 Elsevier; **2.23a:** Adapted from Youle, R.J. and van der Bliek, A.M., Mitochondrial Fission, Fusion and Stress. Science 337: 1062–1065, 31 Aug. 2012; **2.23b and c:** Adapted from Nezich, C.L. and Youle, R.J., Make or Break for Mitochondria, Life Sciences, 2013, 2: doi: 10.7554/ eLife.00804; **2.24a:** Adapted from Annual Review of Biochemistry, Vol. 52, 1983.; **2.24b:** Don W. Fawcett / Science Source; **2.25:** Becker et al, The World of the Cell, 2000 Addison Wesley Longman; **2.27a:** W. Schuler/Photo Researchers Inc; **2.27b:** © 2011 J.L. Carson/Custom Medical Stock Photo, All Rights Reserved; **2.27c:** Dr. Peter Dawson/ Science Photo Library/Photo Researches, Inc.; **2.28b and c:** Adapted from Hardin, J. et al., Becker's World of the Cell, 8/e, 2012 Benjamin Cummings; **2.29:** Donald E. Ingber, Cellular Tensegrity: Defining New Rules of Biological Design that Govern the Cytoskeleton J. Cell Science 104:613–627 1993 fig. 2, p. 615; **2A:** Adapted from Geoffrey Cooper "The Cell: A Molecular Approach" 1997 Sinauer; **2B:** Adapted from Lehringer Principles of Biochemistry by David Nelson and Michael Cox. Copyright © 2000, 1993, 1982 by Worth Publishers permissonsdept@worthpublishers.com; **2C:** United States Government http://rsb.info.nih.gov/ij/images/; **n/a:** From Microbiology: An Introduction by Gerard J. Tortora, Berdell R. Funke and Christine L. Case. Copyright 1998 The Benjamin/Cummings Publishing Company.

Chapter 3

opener: The Image Bank/Getty Images; **3.1:** rui vale sousa/Shutterstock.com; **3.2:** Adapted from Silverberg, Chemistry 2/e Copyright © The McGraw-Hill Companies; **3.9:** Adapted from R. Chang, Chemistry 7/e, McGraw Hill; **3.12:** Adapted from Pearson/Benjamin Cummings, Campbell & Reece, Biology, 7th ed. 2005. Fig 6.27b, p. 117; **3A:** Adapted from Linda Huff, American Scientist, September-October 1997 Volume 85.

Chapter 4

opener: Image Source Balck/Alamay; **4.1:** © Louise Gubb/CORBIS SABA

Chapter 5

opener: Claude Nuridsany & Marie Perennou / Science Source; **5.14a:** Adapted from "Molecules of Life," Purdue University; **5.14b:** Adapted from "Molecules of Life" Purdue University; **5.16:** Adapted from Carl Brandon & John Tooze. Introduction to Protein Structure, 2nd ed 1999. Garland 3.14, 44.; **5.18b:** Adapted from Garrett and Grisham, Biochemistry, 1996, Brooks Cole; **5.19c:** Adapted from C. Brandon & J. Tooze, Intro to Protein Structure, 2nd ed, 1999, Garland fig 5.3 p. 68; **5.19d:** Adapted from C. Brandon & Tooze, Intro to Protein Structure, 2nd ed, 1999. Garland fig 4.14c p. 58.; **5.19e:** Adapted from Brandon & Tooze, Intro to Protein Structure, 2nd ed, 1999. Garland fig 10.1 p. 176.; **5.20a:** Adapted from Annual Review of Biochemistry, vol. 45, 1976.; **5.20c:** Adapted from © 1999 from Introduction to Protein Structure by Carl Brandon and John Tooze. Reproduced by permission of Routledge, Inc., part of Taylor & Francis Group; **5.20d:** Adapted from © 1999 from Introduction to Protein Structure by Carl Brandon and John Tooze. Reproduced by permission of Routledge, Inc., part of Taylor & Francis Group; **5.20e:** Adapted from © 1999 from Introduction to Protein Structure by Carl Brandon and John Tooze. Reproduced by permission of Routledge, Inc., part of Taylor & Francis Group; **5.21:** Adapted from Lain D. Campbell and A.K. Downing, NMR of Modular Proteins Nature Structural Biology NMR Supplement p 496 fig 1; **5.23:** Adapted from a figure published in Biophysical Journal, 72, C. Reid and R.P. Rand, "Probing Protein Hydration and Conformational States in Solution," pp. 1022–1030, Copyright Elsevier (1997).; **5.24:** Adapted with permission of Harcourt College Publishers from Principles of Biochemistry with a Human Focus 1e by Reginald Garrett, 2002. Permission conveyed through Copyright Clearance Center; **5.26:** Adapted from H.J. Dyson & P.E. Wright, Intrinsically Unstructured Proteins and Their Functions, Nature reviews: molecular cell biology, p. 200 Box 2 fig a; **5.28a:** Adapted from FEBS Letters 498(2-3), Sergio T. Ferreira, Fernanda G. Ge Felices, "Protein dynamics, folding and misfolding: from basic physical chemistry to conformational diseases", copyright 2001, with permission from Elsevier.; **5.28b:** from Trends in Biochemical Sciences 25(12), Sheena E. Radford, 'Protein folding: progress made and promises ahead', pp. 611–618, copyright 2000, with permission from Elsevier.; **5.29a–c:** Adapted from Trends in Biochemical Sciences 25(12), Sheena E. Radford, "Protein folding: progress made and promises ahead", pp. 611–618, copyright 2000, with permission from Elsevier.; **5.30:** Adapted from Xu, Horwich and Stigler, "The Crystal Structure of the Asymmetric Gro-El-Gros-Es-(ADP)7 Chaperonen Complex" Nature Vol. 388 August 21, 1997 p.741–50; **5.32:** Adapted from Garrett & Grisham Biochemistry 1996 Brooks Cole; **5.35:** Adapted from K.A. Piez in B.D. Betlander. Ed., The Protein Folding Problem AAAS Selected Symposium 89 American Association for the Advancement of Science, Washington DC 1984 pp.47–61. Courtesy of the Collagen Corporation. Reproduced by permission of American Association for the Advancement of Science; **5A:** © Simon Peers and Nicholas Godley; **5C:** Adapted from Volrath, F. and Knight, D.P. "Structure and Function of the Silk Production Pathway in the Spider Nephila ebulis." Int. J. Bio. Macro. 24:243–249, 1999; **5D:** Dennis Kunkel Microscopy, Inc.; **5F:** Adapted from Christopher K. Matthews and K.E. Van Holde, "Biochemistry 2/e" 1996 Benjamin Cummings; **5H:** Adapted from Nelson & Cox Lehringer, Principles of Biochemistry, 4th ed. Freeman/Worth permissonsdept@worthpublishers .com fig 1 p. 102; **[ch 5 not numbered]:** Adapted from ©1999 from Introduction to Protein Structure by Carl Brandon and John Tooze. Reproduced by permission of Routledge, Inc. part of the Taylor & Francis Group; **online:** Adapted from Figure 16.9 p. 462 World of the Cell, 6th ed. Wayne M. Becker, Lewis J. Kleinsmith, Jeff Hardin. 2006 Pearson; **online:** Adapted from Pollard, T. D. & Earnshaw,

W. Cell Biology 2002 Saunders (Elsevier) p. 607 fig 39.3; **online:** Adapted from Pollard, T. D. & Earnshaw, W. Cell Biology 2002 Saunders (Elsevier) p. 607 fig 39.3; **online:** Adapted from a figure © Rockefeller University Press, 2004. Originally published in Journal of General Physiology. 123:642–656.

Chapter 6

6.2a: Adapted from Biochemistry, 2nd ed. By Christopher K. Matthews and K.E. Van Holde. Copyright (c) 1996 by The Benjamin/Cummings Publishing Company, Inc. Reprinted by permission of Pearson Education, Inc.; **6.2b:** Adapted from Biochemistry, 2nd ed. By Christopher K. Matthews and K.E. Van Holde. Copyright (c) 1996 by The Benjamin/Cummings Publishing Company, Inc. Reprinted by permission of Pearson Education, Inc.

Chapter 7

7.37: Adapted from S.L. Wolfe, "Cell Ultrastructure 1/e" 1985 Brooks Cole an imprint of the Wadsworth Group, a division of Thomson Learning; **7.39:** Adapted from Lehringer Principles of Biochemistry by David Nelson and Michael Cox 2000, 1993, 1982 W.H. Freeman/ Worth permissonsdept@worthpublishers.com

Chapter 8

opener: © valentinrussanov/iStock; **8.1:** Adapted from Trends in Biochemical Sciences 23(5), Bas Teusink, Michael C. Walsh, Karel van Dam, Hans V. Westerhoff, "The Danger of Metabolic Pathways with Turbo Design", pp. 162–169, copyright 1998, with permission from Elsevier; **8.9:** "Illustration of free energy changes in chemical reactions" by Eugene Hamori in Journal of Chemical Education 52(6), June 1, 1975. © American Chemical Society. Adapted with permission; **8.A:** Adapted from J.M. Thomson et al. Nature Genetics 37:630–35 1 May 2005.

Chapter 10

opener: Adapted from Hutcheon, Duncan, Ngai and Cross, Proceedings of the National Academy of the Sciences of the US, Vol 98. 2001. Reprinted with the permission of Richard L. Cross; **10.4:** Adapted from Garrett & Grisham Biochemistry 2/e, 1999, Brooks Cole; **10.6:** Adapted from Molecular Biology of the Cell by Bruce Alberts, et al. Reproduced by permission of Routledge, Inc., part of the Taylor and Francis Group. (Garland); **10.8:** Adapted from Nielson & Cox, Lehringer Priciples of Biochemistry, 4th Ed, 2005, WH Freeman, Fig 19.14, p702;
10.13: Adapted with permission of the American Association for the Advancement of Science from 'Mitochondrial Structure: Two Types of Subunits on Negatively Stained Mitochondrial Membranes' by Donald F. Parsons in Science, volume 140, issue 3570, May 31, 1963. Permission conveyed through Copyright Clearance Center; **10.14:** Adapted from Trends in Biochemical Sciences, Vol 22, Jung, Hill, Engelbrecht, pp.420–423. Reprinted with permission from Elsevier Science.; **10.15:** Adapted with permission of Harcourt College Publishers from Principles of Biochemistry with a Human Focus 1e by Reginald Garrett, 2002. Permission conveyed through Copyright Clearance Center; **10.18:** Adapted from C.R. Scriver et al, "The Metabolic and Molecular Bases of Inherited Diseases," 2001, McGraw Hill.

Chapter 11

11 opener: Adapted from Geoffrey Cooper "The Cell: A Molecular Approach" 1997 Sinauer; **11.3:** Adapted from Leonard Lessin/ Peter Arnold. Inc.; **11.3:** Adapted from Leonard Lessin/FBPA; **11.6(a),(b):** Adapted from Leonard Lessin/Peter Arnold, Inc.; **11.13:** Adapted from Leonard Lessin/ FBPA; **11.18(a)–(c):** Adapted from Leonard Lessin/ Peter Arnold, Inc.; **11.21:** Adapted from Geoffrey Cooper: The Cell: A Molecular Approach 1997 Sinauer; **11.27:** Adapted from Geoffrey Cooper "The Cell: A Molecular Approach" 1997 Sinauer;
11.29: Adapted from Lehringer Principles of Biochemistry by David Nelson and Michael M. Cox 2004 W.H. Freeman 11.20b p. 385; **11.32:** Adapted from Thomas Zeuthen, Trends in Biochemical Sciences, Vol. 26, No 2 pp. 77–79 copyright 2001 Elsevier; **n/a:** Adapted from Thomas Zeuthen, Trends in Biochemical Sciences, Vol. 26, No 2 pp. 77–79 copyright 2001 Elsevier.

Chapter 12

opener: © Martinmark / Dreamstime.com; **12.15:** Adapted from Annual Review of Biochemistry, Vol. 52, 1983; **12.19a:** with permission of Harcourt College Publishers from Principles of Biochemistry with a Human Focus 1e by Reginald Garrett, 2002. Permission conveyed through Copyright Clearance Center; **12.19b:** Adapted from International J Biochem Mol Biol 1(1): 69–89, 2010, H. Liu et al. fig. 2B, p. 70.

Chapter 13

opener: Sisse Brimberg/National Geographic Creative; **13.3a:** Adapted from Becker et al, The World of the Cell, 2000 Addison Wesley Longman; **13.3b:** Adapted from Becker et al, The World of the Cell, 2000 Addison Wesley Longman; **13.4:** Adapted from Anderson & Anderson, Trends in Biochemical Sciences Vol. 7, 1982, Elsevier Science; **13.5:** Adapted from Nelson, N. & Ben-Shem, A. The Complex Architecture of Oxygenic Photosynthesis Nature Reviews: Mol Cell Biol 5: 971–82 p. 978 fig 5a; **13.6:** Adapted from A. Melkozernov; **13.8:** Adapted from Trends in Biochemical Sciences, Vol. 22, Jung, Hill, Engelbrecht, pp. 420–423; **13.11:** Adapted from A. Melkozernov; **13.12a,b:** Adapted from A. Melkozernov; **13.16a:** Adapted from Nelson, N. & Ben-Shem, A. The Complex Architecture of Oxygenic Photosynthesis Nature Reviews: Mol Cell Biol 5: 971–82 p. 978 fig 5b; **13.16b:** Adapted from A. Melkozernov; **13.17:** Adapted from Becker et al. "The World of the Cell 4e" 2000 Addison Wesley Longman; **13.18:** Adapted from Becker et al. "The World of the Cell, 4/e" 2000. Addison Wesley Longman; **13.24:** Adapted from Becker et al, "World of the Cell, 4/e" 2000, Addison Wesley Longman; **13A:** Reprinted with permission from Macmillan Publishers Ltd: Nature reviews: microbiology, Molloy, S. Down in the Depths Nature Reviews: Microbiology 3: 582, 2005. p. 582.

Chapter 14

opener: Jon Walsh/ Photoresearchers, Inc.; **14.1a:** © NASA; **14.1b:** © NASA; **14.2:** Adapted with permission from P.C. Dossantos et.al, "Formation and Insertion of the Nitrogenase Iron-Molybdenum Cofactor" *Chemical Review* 104: 1159–1173, 2004, fig. no. 1, page no. 1161. Copyright 2005 American Chemical Society.

Chapter 15

opener: Adapted from David S. Goodsell. The Molecular Perspective Ubiquiton & the Proteasome The Oncologist 8(#):293–294, 2004. p. 2 fig. 2c, AlphaMed Press, Inc.; **15.2:** Adapted from Richsteiner, M. and Hill, C.P. Mobilizing the Proteolytic Machine: Cell Biological Roles of Proteasome Activators & Inhibitors, Trends Cell Biol 15(1):27–33 2005 p. 27,28 Fig. 1, 3.

Chapter 16

opener: © Georgia Bennett/Bennett Images. All rights reserved; **16.1:** Adapted from J. Coolman and K.H. Rahn, "Color Atlas of Biochemistry," 1999, Thieme Medical Publishers; **16.4:** Adapted from Becker et al. "The World of the Cell, 4/e" 2000, Addison Wesley Longman; **16.6:** Adapted from Cooper, G.M. and Hausman, R.E., The Cell: A Molecular Approach, 6/e. 2013 Sinauer Associates; **16.8:** Adapted from Geoffrey Cooper, "The Cell: A Molecular Approach, 1997, The Sinauer Associates; **16.10:** Adapted from T.M. Delving, "Textbook of Biochemistry with Clinical Correlations," 1999, John Wiley and Sons; **16A:** Science VU/Jackson/Visuals Unlimited, Inc.

Chapter 17

opener: wavebreakmedia/Shutterstock; **17.1:** A. Barrington Brown / Science Source; **17.9:** Reprinted by permission from Macmillan Publisher Ltd.: Nature (25 April 1953), copyright 1953 Nature Publishing Group; **17.10:** Adapted from Lehninger, Nelson, & Cox "Principles" 2000 Worth Publishers; **17.13:** Adapted from T.A. Brown, Genomes, 1999, BIOS Scientific Publishers Oxford UK; **17.15:** Adapted from Taylor and Francis; **17.16:** Thomas D. Pollard and William C. Earnshaw Cell Biology 2002 Saunders/Elsevier p.202 fig 13.6E; **17.17:** Adapted from Watson et al. Mol Biol. of the Gene, Pearson/ Benj. Cummings p. 154–155, 7.19a&b, 7.20; **17.19:** Adapted from C.K. Matthews and K.E. Van Holde Biochemistry, 2nd ed. 1996 Benjamin Cummings; **17.21:** Adapted from T.A. Brown, Genomes, 1999, BIOS Scientific Publishers Oxford UK; **17.22:** Häggström, Mikael. "Medical gallery of Mikael Häggström 2014". Wikiversity Journal of Medicine 1 (2). DOI:10.15347/wjm/2014.008. ISSN 20018762. - PANTHER Pie Chart at the PANTHER Classification System homepage. Retrieved May 25, 2011;
17B: Adapted from Watson et al. Mol Biol. Of the Gener Pearson/Benj Cummings fig 7.27 p 160;
17C: Reprinted with permission from Randy L. Jirtle, Duke University; **17D:** Adapted from Bejamin Lewin, Oxford Univ. Press & Cell Press, 2000. Reprinted from Genes VII; **17E:** Adapted from Prescott, et al., Microbiology 4/e © 1999 by the McGraw-Hill Companies; **17F:** Adapted from Recombinant DNA by J.D. Watson, M. Gilman, J. Witkowski, M. Zoller © 1992, 1983 by J.D. Watson, M. Gilman, J. Witkowski, M. Zoller. Used with the permission of W.H. Freeman and Company; **17G:** Adapted from Prescott et al. Microbiology, 4/e 1999 MGH; **17H:** Adapted from Watson et al., Recombinant DNA, 1992, WH Freeman; **17I:** Tek Image / Science Source; **17J:** Adapted from L.E. Henderson and L.O. Arthur, Scientific American, 279, Vol. 1, 1998; **17K:** Adapted from Patricia J. Wynne, "Improving HIV Therapy," Scientific American, Vol. 279.

Chapter 18

18.3: Adapted from Becker et al "The World of the Cell 4e" 2000 Addison Wesley Longman; **18.4:** Adapted from T.A. Steitz, A Mechanism for All Polymerases Nature 391:231–32 p. 231, fig. 1; **18.5:** Adapted from Oxford University Press and Cell Press 2000. Reprinted from Genes VII by Benjamin Lewin (2000) by permission of Oxford University Press; **18.6:** Adapted from Benjamin Lewin, Oxford University Press and Cell Press, 2000. Reprinted from Genes VII; **18.7:** Adapted from Erzberger, J.P., Mott, ML & Berger, JM Structural Basis for ATP dependent AnaA Assembly and Replication Origin Remodeling Nature Struc Mol Biol 12(8):676–683. p. 677, fig. 1; **18.10:** Adapted from Garrett & Grisham, Biochemistry, 3rd ed. 2005 Thomson/Brooks Cole, fig. 28.10, p. 907; **18.11:** Adapted from T. A. Brown, Genomes 3, Garland Science, New York. fig 15.22 p. 488; **18.14:** Adapted from T.A. Brown, Genomes 3, Garland Science p. 483 fig 15.15; **18.16:** Adapted from T. A. Brown Genomes 3 Garland Science p. 491 fig 15.25; **18.17:** Adapted from T. A. Brown Genomes 3 Garland Science p. 491, fig 15.25; **18.18:** Adapted from Molecular Biology of the Gene, 5th ed. by James D. Watson, et al. Copyright 2004 by Pearson Education; **18.19:** Adapted from Watson et al. Mol Biol of the Gene, Pearson/Benj Cummings p. 249, fig. 9.13; **18.25:** Adapted from A. Landy and R.A. Weisber, R.W. Hendrix, Lambda II, Coldspring Harbor Laboratory Press, 1983; **18.30:** Adapted from Lehninger, Nelson, and Cox, "Principles of Biochemistry:" 2000, Worth Publishers; **18.31:** Adapted from Lehninger, Nelson and Cox. "Principles. . ." 2000, Worth Pres; **18.33:** Adapted from Korzheva, N. et al. A Structural Model of Transcription Elongation Science 289: 619–625, fig 3 p. 622; **18.34:** Adapted from T.A. Brown, Genomes 3 2007 Garland Science p. 337, fig 12.5; **18.35:** Adapted from T.A. Brown, Genomes 3 2007, Garland Science p 337 fig 12.5; **18.36:** Adapted from G. Karp, Cell and Molecular Biology Concepts and Experiments, 4th ed. John Wiley, 2005. fig 12.47 p. 534; **18.39:** Linda A. Cicero/ Stanford News Service; **18.40:** Adapted from R.D. Kornberg The Molecular Basis of Eukaryotic Transcription PNAS USA 104:12955–61 2007 p.12957, fig. 6; **18.41:** Adapted from Chadick, J.Z. and Asturias, F.S. Structure of Eukaryotic Mediator Complexes, Trends in Biochemical Sciences 30(4):264–271, 2005. p. 267 fig 3; **18.43:** Adapted from Cooper, G.M. and Hausman, R.E., The Cell: A Molecular Approach, 6/e. 2013 Sinauer Associates; **18.46:** Adapted from the Power of Riboswitches by Jeffrey E. Barrick and Ronald R. Breaker. Copyright 2007 Scientific American; **18.47:** Adapted from Cooper, G.M. and Hausman, R.E., The Cell: A Molecular Approach, 6/e. 2013 Sinauer Associates; **18.48:** Adapted from Azubel, M. et. al. Native Spliceosomes Assembled with Pre-mRNA to Form Suprasplicesomes. J. Mol . Biol. 356: 955–66, 2006 Fig 7b p. 963; **18.51:** Adapted from Jeffrey Baereck & Ronald R. Breaker, The Power of Riboswitches, Scientific American 296(1): 50–57, 2007 p 55; **18.52:** Adapted from Terence A. Brown, Genomes, 1999. Reprinted with permission of BIOS Scientific Publishers, Oxford, UK; **18.53:** Adapted from Alberts et al., Mol Biol of the Cell, 4th ed. Garland p. 319, fig 6.27, Routledge Taylor and Francis; **18A:** Adapted from Prescott et al. Microbiology 4/e 1999 MGH; **18B:** Adapted from E.W. Nester, "Microbiology: A Human Perspective, 3/e" ©2001 McGraw Hill; **18C:** Keith V. Wood; **18E:** Adapted from R.H. Tamarin, Principles of Genetics, 6/e © 1999 The McGraw Hill Companies; **18F:** Adapted from E.W. Nester "Microbiology: A Human Perspective, 3e" © 2000 The McGraw-Hill Companies; **18G:** Adapted from Molecular Cell Biology by J. Darnell, H. Lodish, D. Baltimore, P. Matsudaira, S. Zipursky, and A. Berk. Copyright 2000, 1995, 1990, 1986 by Scientific American Books. Used with permission of W.H. Freeman and Company; **18H:** Adapted from R. H. Tamarin, Principles of Genetics, 6/e The McGraw Hill Companies; **18H:** AFFYMETRIX Inc.

Chapter 19

opener: Adapted from Harry Noller, University of California, Santa Cruz - via Phyllis Tveit Center for Molecular Biology of RNA; **19.1:** Adapted from Lodish et al. "Molecular Cell Biology" 4/e, 2000; **19.6:** Adapted from Molecular Cell Biology by J. Darnell, H. Lodish, D. Baltimore, P. Matsudaira, S. Zipursky, and A. Berk. Copyright 2000, 1995, 1990, 1986 by Scientific American Books. Used with permission of W.H. Freeman and Company; **19.13:** Adapted from L.I. Slobin "Polypeptide Chain Elongation," in Translation in Eukaryotes, edited by H. Trachsel, 1991. Copyright CRC Press, Boca Raton, Florida; **19.19:** Adapted from "The World of the Cell," Becker et al. 2000, Addison Wesley & Longman; **19.23:** Adapted from Kim, Y.E. et al., Molecular Chaperone Functions in Protein Folding and Proteostasis, Annual Review of Biochemistry, 82: 323–355 (2013); **19D:** Reprinted from *The Lancet*, 356, Rosamonde E. Banks et al, "Proteomics: New Perspectives, New Biomedical Opportunities," pp. 1749–1756, Copyright 2000, with permission from Elsevier.

Index

Page numbers followed by *f* and *t* refer to figures and tables, respectively.

B

D

N

O

Q

R

S

T

Names and Abbreviations of the Standard Amino Acids

Amino Acid	Three-Letter Abbreviations	One-Letter Abbreviations
Alanine	Ala	A
Arginine	Arg	R
Asparagine	Asn	N
Aspartic acid	Asp	D
Cysteine	Cys	C
Glutamic acid	Glu	E
Glutamine	Gln	Q
Glycine	Gly	G
Histidine	His	H
Isoleucine	Ile	I
Leucine	Leu	L
Lysine	Lys	K
Methionine	Met	M
Phenylalanine	Phe	F
Proline	Pro	P
Serine	Ser	S
Threonine	Thr	T
Tryptophan	Trp	W
Tyrosine	Tyr	Y
Valine	Val	V